Micro-
bio-
logy

To learn more about OpenStax, visit openstaxtextbooks.com
Individual print copies and bulk orders can be purchased through our website.

Books and Cover Design © 2022 Open Stax Textbooks.
Textbook content produced by OpenStax is licensed under a CCA4

Contact us:

info@openstaxtextbooks.com

Trademarks
The OpenStax name, OpenStax logo, OpenStax book covers, OpenStax CNX name, OpenStax CNX logo, OpenStax Tutor name, Openstax Tutor logo, Connexions name, Connexions logo, Rice University name, and Rice University logo are not subject to the license and may not be reproduced without the prior and express written consent

Contents

Preface   1

CHAPTER 1
An Invisible World   9
Introduction   9
1.1 What Our Ancestors Knew   10
1.2 A Systematic Approach   16
1.3 Types of Microorganisms   22
Summary   31
Review Questions   32

CHAPTER 2
How We See the Invisible World   35
Introduction   35
2.1 The Properties of Light   36
2.2 Peering Into the Invisible World   41
2.3 Instruments of Microscopy   43
2.4 Staining Microscopic Specimens   58
Summary   70
Review Questions   71

CHAPTER 3
The Cell   73
Introduction   73
3.1 Spontaneous Generation   74
3.2 Foundations of Modern Cell Theory   77
3.3 Unique Characteristics of Prokaryotic Cells   85
3.4 Unique Characteristics of Eukaryotic Cells   103
Summary   125
Review Questions   126

CHAPTER 4
Prokaryotic Diversity   131
Introduction   131
4.1 Prokaryote Habitats, Relationships, and Microbiomes   132
4.2 Proteobacteria   139
4.3 Nonproteobacteria Gram-Negative Bacteria and Phototrophic Bacteria   149
4.4 Gram-Positive Bacteria   156
4.5 Deeply Branching Bacteria   166
4.6 Archaea   167
Summary   171
Review Questions   173

CHAPTER 5
The Eukaryotes of Microbiology 177
Introduction   177
5.1 Unicellular Eukaryotic Parasites   178
5.2 Parasitic Helminths   192
5.3 Fungi   200
5.4 Algae   210
5.5 Lichens   213
Summary   216
Review Questions   217

CHAPTER 6
Acellular Pathogens 219
Introduction   219
6.1 Viruses   220
6.2 The Viral Life Cycle   229
6.3 Isolation, Culture, and Identification of Viruses   240
6.4 Viroids, Virusoids, and Prions   248
Summary   253
Review Questions   253

CHAPTER 7
Microbial Biochemistry 257
Introduction   257
7.1 Organic Molecules   258
7.2 Carbohydrates   263
7.3 Lipids   267
7.4 Proteins   272
7.5 Using Biochemistry to Identify Microorganisms   278
Summary   282
Review Questions   283

CHAPTER 8
Microbial Metabolism 287
Introduction   287
8.1 Energy, Matter, and Enzymes   288
8.2 Catabolism of Carbohydrates   295
8.3 Cellular Respiration   301
8.4 Fermentation   304
8.5 Catabolism of Lipids and Proteins   308
8.6 Photosynthesis   310
8.7 Biogeochemical Cycles   315
Summary   322
Review Questions   324

CHAPTER 9
Microbial Growth 329
Introduction   329
9.1 How Microbes Grow   330

9.2 Oxygen Requirements for Microbial Growth   345
9.3 The Effects of pH on Microbial Growth   350
9.4 Temperature and Microbial Growth   352
9.5 Other Environmental Conditions that Affect Growth   356
9.6 Media Used for Bacterial Growth   358
Summary   361
Review Questions   362

CHAPTER 10
Biochemistry of the Genome   369
Introduction   369
10.1 Using Microbiology to Discover the Secrets of Life   370
10.2 Structure and Function of DNA   381
10.3 Structure and Function of RNA   389
10.4 Structure and Function of Cellular Genomes   393
Summary   401
Review Questions   402

CHAPTER 11
Mechanisms of Microbial Genetics   407
Introduction   407
11.1 The Functions of Genetic Material   408
11.2 DNA Replication   410
11.3 RNA Transcription   419
11.4 Protein Synthesis (Translation)   422
11.5 Mutations   428
11.6 How Asexual Prokaryotes Achieve Genetic Diversity   440
11.7 Gene Regulation: Operon Theory   447
Summary   457
Review Questions   460

CHAPTER 12
Modern Applications of Microbial Genetics   467
Introduction   467
12.1 Microbes and the Tools of Genetic Engineering   468
12.2 Visualizing and Characterizing DNA, RNA, and Protein   479
12.3 Whole Genome Methods and Pharmaceutical Applications of Genetic Engineering   494
12.4 Gene Therapy   499
Summary   503
Review Questions   504

CHAPTER 13
Control of Microbial Growth   507
Introduction   507
13.1 Controlling Microbial Growth   508
13.2 Using Physical Methods to Control Microorganisms   515
13.3 Using Chemicals to Control Microorganisms   529
13.4 Testing the Effectiveness of Antiseptics and Disinfectants   546
Summary   552

Review Questions   554

CHAPTER 14
Antimicrobial Drugs   557
Introduction   557
14.1 History of Chemotherapy and Antimicrobial Discovery   558
14.2 Fundamentals of Antimicrobial Chemotherapy   562
14.3 Mechanisms of Antibacterial Drugs   566
14.4 Mechanisms of Other Antimicrobial Drugs   580
14.5 Drug Resistance   591
14.6 Testing the Effectiveness of Antimicrobials   597
14.7 Current Strategies for Antimicrobial Discovery   601
Summary   604
Review Questions   606

CHAPTER 15
Microbial Mechanisms of Pathogenicity   611
Introduction   611
15.1 Characteristics of Infectious Disease   612
15.2 How Pathogens Cause Disease   618
15.3 Virulence Factors of Bacterial and Viral Pathogens   629
15.4 Virulence Factors of Eukaryotic Pathogens   643
Summary   646
Review Questions   647

CHAPTER 16
Disease and Epidemiology   651
Introduction   651
16.1 The Language of Epidemiologists   652
16.2 Tracking Infectious Diseases   657
16.3 Modes of Disease Transmission   663
16.4 Global Public Health   673
Summary   677
Review Questions   678

CHAPTER 17
Innate Nonspecific Host Defenses   683
Introduction   683
17.1 Physical Defenses   684
17.2 Chemical Defenses   691
17.3 Cellular Defenses   700
17.4 Pathogen Recognition and Phagocytosis   708
17.5 Inflammation and Fever   713
Summary   719
Review Questions   720

CHAPTER 18
Adaptive Specific Host Defenses   725
Introduction   725

18.1 Overview of Specific Adaptive Immunity 726
18.2 Major Histocompatibility Complexes and Antigen-Presenting Cells 734
18.3 T Lymphocytes and Cellular Immunity 737
18.4 B Lymphocytes and Humoral Immunity 746
18.5 Vaccines 750
Summary 759
Review Questions 760

CHAPTER 19
Diseases of the Immune System 765
Introduction 765
19.1 Hypersensitivities 766
19.2 Autoimmune Disorders 783
19.3 Organ Transplantation and Rejection 790
19.4 Immunodeficiency 793
19.5 Cancer Immunobiology and Immunotherapy 797
Summary 799
Review Questions 800

CHAPTER 20
Laboratory Analysis of the Immune Response 803
Introduction 803
20.1 Polyclonal and Monoclonal Antibody Production 804
20.2 Detecting Antigen-Antibody Complexes 810
20.3 Agglutination Assays 822
20.4 EIAs and ELISAs 831
20.5 Fluorescent Antibody Techniques 841
Summary 848
Review Questions 849

CHAPTER 21
Skin and Eye Infections 853
Introduction 853
21.1 Anatomy and Normal Microbiota of the Skin and Eyes 854
21.2 Bacterial Infections of the Skin and Eyes 861
21.3 Viral Infections of the Skin and Eyes 877
21.4 Mycoses of the Skin 881
21.5 Protozoan and Helminthic Infections of the Skin and Eyes 886
Summary 891
Review Questions 892

CHAPTER 22
Respiratory System Infections 895
Introduction 895
22.1 Anatomy and Normal Microbiota of the Respiratory Tract 896
22.2 Bacterial Infections of the Respiratory Tract 901
22.3 Viral Infections of the Respiratory Tract 919
22.4 Respiratory Mycoses 930
Summary 938

Review Questions   939

CHAPTER 23
Urogenital System Infections   943
Introduction   943
23.1 Anatomy and Normal Microbiota of the Urogenital Tract   944
23.2 Bacterial Infections of the Urinary System   948
23.3 Bacterial Infections of the Reproductive System   954
23.4 Viral Infections of the Reproductive System   963
23.5 Fungal Infections of the Reproductive System   969
23.6 Protozoan Infections of the Urogenital System   971
Summary   975
Review Questions   976

CHAPTER 24
Digestive System Infections   979
Introduction   979
24.1 Anatomy and Normal Microbiota of the Digestive System   980
24.2 Microbial Diseases of the Mouth and Oral Cavity   986
24.3 Bacterial Infections of the Gastrointestinal Tract   993
24.4 Viral Infections of the Gastrointestinal Tract   1010
24.5 Protozoan Infections of the Gastrointestinal Tract   1017
24.6 Helminthic Infections of the Gastrointestinal Tract   1021
Summary   1033
Review Questions   1034

CHAPTER 25
Circulatory and Lymphatic System Infections   1037
Introduction   1037
25.1 Anatomy of the Circulatory and Lymphatic Systems   1038
25.2 Bacterial Infections of the Circulatory and Lymphatic Systems   1043
25.3 Viral Infections of the Circulatory and Lymphatic Systems   1065
25.4 Parasitic Infections of the Circulatory and Lymphatic Systems   1076
Summary   1088
Review Questions   1089

CHAPTER 26
Nervous System Infections   1093
Introduction   1093
26.1 Anatomy of the Nervous System   1094
26.2 Bacterial Diseases of the Nervous System   1099
26.3 Acellular Diseases of the Nervous System   1113
26.4 Fungal and Parasitic Diseases of the Nervous System   1124
Summary   1132
Review Questions   1133

Appendix A Fundamentals of Physics and Chemistry Important to Microbiology   1139

Appendix B Mathematical Basics   1149

Appendix C Metabolic Pathways   1155

Appendix D Taxonomy of Clinically Relevant Microorganisms   1163

PREFACE

Welcome to *Microbiology*, an OpenStax resource. This textbook was written to increase student access to high-quality learning materials, maintaining highest standards of academic rigor at little to no cost.

About OpenStax

OpenStax is a nonprofit based at Rice University, and it's our mission to improve student access to education. Our first openly licensed college textbook was published in 2012, and our library has since scaled to over 20 books for college and AP® Courses used by hundreds of thousands of students. Our adaptive learning technology, designed to improve learning outcomes through personalized educational paths, is being piloted in college courses throughout the country. Through our partnerships with philanthropic foundations and our alliance with other educational resource organizations, OpenStax is breaking down the most common barriers to learning and empowering students and instructors to succeed.

About OpenStax Resources

Customization

Microbiology is licensed under a Creative Commons Attribution 4.0 International (CC BY) license, which means that you can distribute, remix, and build upon the content, as long as you provide attribution to OpenStax and its content contributors.

Because our books are openly licensed, you are free to use the entire book or pick and choose the sections that are most relevant to the needs of your course. Feel free to remix the content by assigning your students certain chapters and sections in your syllabus, in the order that you prefer. You can even provide a direct link in your syllabus to the sections in the web view of your book.

Instructors also have the option of creating a customized version of their OpenStax book. The custom version can be made available to students in low-cost print or digital form through their campus bookstore. Visit your book page on openstax.org for more information.

Errata

All OpenStax textbooks undergo a rigorous review process. However, like any professional-grade textbook, errors sometimes occur. Since our books are web-based, we can make updates periodically when deemed pedagogically necessary. If you have a correction to suggest, submit it through the link on your book page on openstax.org. Subject matter experts review all errata suggestions. OpenStax is committed to remaining transparent about all updates, so you will also find a list of past errata changes on your book page on openstax.org.

Format

You can access this textbook for free in web view or PDF through openstax.org, and for a low cost in print.

About *Microbiology*

Microbiology is designed to cover the scope and sequence requirements for the single-semester Microbiology course for non-majors. The book presents the core concepts of microbiology with a focus on applications for careers in allied health. The pedagogical features of *Microbiology* make the material interesting and accessible to students while maintaining the career-application focus and scientific rigor inherent in the subject matter.

Coverage and Scope

The scope and sequence of *Microbiology* has been developed and vetted with input from numerous instructors at institutions across the US. It is designed to meet the needs of most microbiology courses for non-majors and allied health students. In addition, we have also considered the needs of institutions that offer microbiology to a mixed audience of science majors and non-majors by frequently integrating topics that may not have obvious clinical relevance, such as environmental and applied microbiology and the history of science.

With these objectives in mind, the content of this textbook has been arranged in a logical progression from fundamental to more advanced concepts. The opening chapters present an overview of the discipline, with individual chapters focusing on microscopy and cellular biology as well as each of the classifications of microorganisms. Students then explore the foundations of microbial biochemistry, metabolism, and genetics, topics that provide a basis for understanding the various means by which we can control and combat microbial growth. Beginning with Chapter 15, the focus turns to microbial pathogenicity, emphasizing how interactions between microbes and the human immune system contribute to human health and disease. The last several chapters of the text provide a survey of medical microbiology, presenting the characteristics of microbial diseases organized by body system.

A brief Table of Contents follows. While we have made every effort to align the Table of Contents with the needs of our audience, we recognize that some instructors may prefer to teach topics in a different order. A particular strength of *Microbiology* is that instructors can customize the book, adapting it to the approach that works best in their classroom.

- Chapter 1: An Invisible World
- Chapter 2: How We See the Invisible World
- Chapter 3: The Cell
- Chapter 4: Prokaryotic Diversity
- Chapter 5: The Eukaryotes of Microbiology
- Chapter 6: Acellular Pathogens
- Chapter 7: Microbial Biochemistry
- Chapter 8: Microbial Metabolism
- Chapter 9: Microbial Growth
- Chapter 10: Biochemistry of the Genome
- Chapter 11: Mechanisms of Microbial Genetics
- Chapter 12: Modern Applications of Microbial Genetics
- Chapter 13: Control of Microbial Growth
- Chapter 14: Antimicrobial Drugs
- Chapter 15: Microbial Mechanisms of Pathogenicity
- Chapter 16: Disease and Epidemiology
- Chapter 17: Innate Nonspecific Host Defenses
- Chapter 18: Adaptive Specific Host Defenses
- Chapter 19: Diseases of the Immune System
- Chapter 20: Laboratory Analysis of the Immune Response
- Chapter 21: Skin and Eye Infections
- Chapter 22: Respiratory System Infections
- Chapter 23: Urogenital System Infections
- Chapter 24: Digestive System Infections
- Chapter 25: Circulatory and Lymphatic System Infections
- Chapter 26: Nervous System Infections
- Appendix A: Fundamentals of Physics and Chemistry Important to Microbiology
- Appendix B: Mathematical Basics
- Appendix C: Metabolic Pathways
- Appendix D: Taxonomy of Clinically Relevant Microorganisms

American Society of Microbiology (ASM) Partnership

Microbiology is produced through a collaborative publishing agreement between OpenStax and the American Society for Microbiology Press. The book has been developed to align to the curriculum guidelines of the American Society for Microbiology.

About ASM

The American Society for Microbiology is the largest single life science society, composed of over 47,000 scientists and health professionals. ASM's mission is to promote and advance the microbial sciences.

ASM advances the microbial sciences through conferences, publications, certifications, and educational opportunities. It enhances laboratory capacity around the globe through training and resources and provides a network for scientists in academia, industry, and clinical settings. Additionally, ASM promotes a deeper understanding of the microbial sciences to diverse audiences and is committed to offering open-access materials through their new journals, American Academy of Microbiology reports, and textbooks.

ASM Recommended Curriculum Guidelines for Undergraduate Microbiology Education

PART 1: Concepts and Statements

Evolution

1. Cells, organelles (e.g., mitochondria and chloroplasts) and all major metabolic pathways evolved from early prokaryotic cells.
2. Mutations and horizontal gene transfer, with the immense variety of microenvironments, have selected for a huge diversity of microorganisms.
3. Human impact on the environment influences the evolution of microorganisms (e.g., emerging diseases and the selection of antibiotic resistance).
4. The traditional concept of species is not readily applicable to microbes due to asexual reproduction and the frequent occurrence of horizontal gene transfer.
5. The evolutionary relatedness of organisms is best reflected in phylogenetic trees.

Cell Structure and Function

6. The structure and function of microorganisms have been revealed by the use of microscopy (including bright field, phase contrast, fluorescent, and electron).
7. Bacteria have unique cell structures that can be targets for antibiotics, immunity and phage infection.
8. Bacteria and Archaea have specialized structures (e.g., flagella, endospores, and pili)

that often confer critical capabilities.
9. While microscopic eukaryotes (for example, fungi, protozoa and algae) carry out some of the same processes as bacteria, many of the cellular properties are fundamentally different.
10. The replication cycles of viruses (lytic and lysogenic) differ among viruses and are determined by their unique structures and genomes.

Metabolic Pathways

11. Bacteria and Archaea exhibit extensive, and often unique, metabolic diversity (e.g., nitrogen fixation, methane production, anoxygenic photosynthesis).
12. The interactions of microorganisms among themselves and with their environment are determined by their metabolic abilities (e.g., quorum sensing, oxygen consumption, nitrogen transformations).
13. The survival and growth of any microorganism in a given environment depends on its metabolic characteristics.
14. The growth of microorganisms can be controlled by physical, chemical, mechanical, or biological means.

Information Flow and Genetics

15. Genetic variations can impact microbial functions (e.g., in biofilm formation, pathogenicity and drug resistance).
16. Although the central dogma is universal in all cells, the processes of replication, transcription, and translation differ in Bacteria, Archaea, and Eukaryotes.
17. The regulation of gene expression is influenced by external and internal molecular cues and/or signals.
18. The synthesis of viral genetic material and proteins is dependent on host cells.
19. Cell genomes can be manipulated to alter cell function.

Microbial Systems

20. Microorganisms are ubiquitous and live in diverse and dynamic ecosystems.
21. Most bacteria in nature live in biofilm communities.
22. Microorganisms and their environment interact with and modify each other.
23. Microorganisms, cellular and viral, can interact with both human and nonhuman hosts in beneficial, neutral or detrimental ways.

Impact of Microorganisms

24. Microbes are essential for life as we know it and the processes that support life (e.g., in biogeochemical cycles and plant and/or animal microbiota).
25. Microorganisms provide essential models that give us fundamental knowledge about life processes.
26. Humans utilize and harness microorganisms and their products.
27. Because the true diversity of microbial life is largely unknown, its effects and potential benefits have not been fully explored.

PART 2: Competencies and Skills

Scientific Thinking

28. Ability to apply the process of science
 a. Demonstrate an ability to formulate hypotheses and design experiments based on the scientific method.
 b. Analyze and interpret results from a variety of microbiological methods and apply these methods to analogous situations.
29. Ability to use quantitative reasoning
 a. Use mathematical reasoning and graphing skills to solve problems in microbiology.
30. Ability to communicate and collaborate with other disciplines
 a. Effectively communicate fundamental concepts of microbiology in written and oral format.
 b. Identify credible scientific sources and interpret and evaluate the information therein.
31. Ability to understand the relationship between science and society
 a. Identify and discuss ethical issues in microbiology.

Microbiology Laboratory Skills

32. Properly prepare and view specimens for examination using microscopy (bright field and, if possible, phase contrast).
33. Use pure culture and selective techniques to enrich for and isolate microorganisms.
34. Use appropriate methods to identify microorganisms (media-based, molecular and serological).
35. Estimate the number of microorganisms in a sample (using, for example, direct count, viable plate count, and spectrophotometric methods).
36. Use appropriate microbiological and molecular lab equipment and methods.

37. Practice safe microbiology, using appropriate protective and emergency procedures.
38. Document and report on experimental protocols, results and conclusions.

OpenStax *Microbiology* Correlation to ASM Recommended Curriculum Guidelines for Undergraduate Microbiology Education

OpenStax *Microbiology* Correlation to ASM Curriculum Guidelines

Chapter	ASM Curriculum Guidelines
1—An Invisible World	2, 4, 5, 11, 16, 20, 23, 26, 27, 31
2—How We See the Invisible World	6, 31, 32, 33
3—The Cell	1, 2, 5, 9, 16, 21, 25, 31
4—Prokaryotic Diversity	2, 4, 8, 11, 12, 16, 20, 23, 24, 31
5—The Eukaryotes of Microbiology	2, 4, 5, 9, 12, 20, 23, 31
6—Acellular Pathogens	4, 10, 18, 23, 31
7—Microbial Biochemistry	1, 24, 33, 34
8—Microbial Metabolism	1, 11, 12, 13, 22, 24
9—Microbial Growth	12, 13, 29, 31, 33, 34, 35
10—Biochemistry of the Genome	1, 16, 25, 31
11—Mechanisms of Microbial Genetics	1, 2, 15, 16, 17, 31
12—Modern Applications of Microbial Genetics	19, 26, 31
13—Control of Microbial Growth	13, 14, 26, 31, 36, 37

OpenStax *Microbiology* Correlation to ASM Curriculum Guidelines

Chapter	ASM Curriculum Guidelines
14—Antimicrobial Drugs	3, 7, 14, 15, 23, 26, 31
15—Microbial Mechanisms of Pathogenicity	8, 9, 10, 15, 18, 23, 33
16—Disease and Epidemiology	7, 14, 23, 26, 31
17—Innate Nonspecific Host Defenses	7, 8, 23
18—Adaptive Specific Host Defenses	7, 23, 26, 31
19—Diseases of the Immune System	7, 8, 24
20—Laboratory Analysis of the Immune Response	31, 34
21—Skin and Eye Infections	8, 9, 10, 14, 18, 23, 24, 31
22—Respiratory System Infections	7, 8, 9, 14, 18, 23, 24, 31
23—Urogenital System Infections	7, 8, 9, 12, 14, 18, 22, 23, 24, 31
24—Digestive System Infections	7, 8, 9, 10, 14, 18, 23, 24, 31
25—Circulatory and Lymphatic System Infections	7, 8, 9, 14, 23, 31
26—Nervous System Infections	7, 8, 9, 14, 18, 23, 24, 31

Engaging Feature Boxes

Throughout *Microbiology*, you will find features that engage students by taking selected topics a step further. Our features include:

- **Clinical Focus.** Each chapter has a multi-part clinical case study that follows the story of a fictional patient. The case unfolds in several realistic episodes placed strategically throughout the chapter, each episode revealing new symptoms and clues about possible causes and diagnoses. The details of the case are directly related to the topics presented in the chapter, encouraging students to apply what they are learning to real-life scenarios. The final episode presents a Resolution that reveals the outcome of the case and unpacks the broader lessons to be learned.
- **Case in Point.** In addition to the Clinical Focus, many chapters also have one or more single-part case studies that serve to highlight the clinical relevance of a particular topic. These narratives are strategically placed directly after the topic of emphasis and generally conclude with a set of questions that challenge the reader to think critically about the case.
- **Micro Connections.** All chapters contain several Micro Connections feature boxes that highlight real-world applications of microbiology, drawing often-overlooked connections between microbiology and a wide range of other disciplines. While many of these connections involve medicine and healthcare, they also venture into domains such as environmental science, genetic engineering, and emerging technologies. Moreover, many Micro Connections boxes are related to current or recent events, further emphasizing the intersections between microbiology and everyday life.
- **Sigma Xi Eye on Ethics.** This unique feature, which appears in most chapters, explores an ethical issue related to chapter content. Developed in cooperation with the scientific research society Sigma Xi, each Eye on Ethics box presents students with a challenging ethical dilemma that arises at the intersection of science and healthcare. Often grounded in historical or current events, these short essays discuss multiple sides of an issue, posing questions that challenge the reader to contemplate the ethical principles that govern professionals in healthcare and the sciences.
- **Disease Profile.** This feature, which is exclusive to Chapters 21–26, highlights important connections between related diseases. Each box also includes a table cataloguing unique aspects of each disease, such as the causative agent, symptoms, portal of entry, mode of transmission, and treatment. These concise tables serve as a useful reference that students can use as a study aid.
- **Link to Learning.** This feature provides a brief introduction and a link to an online resource that students may use to further explore a topic presented in the chapter. Links typically lead to a website, interactive activity, or animation that students can investigate on their own.

Comprehensive Art Program

Our art program is designed to enhance students' understanding of concepts through clear and effective illustrations, diagrams, and photographs. Detailed drawings, comprehensive lifecycles, and clear micrographs provide visual reinforcement for concepts.

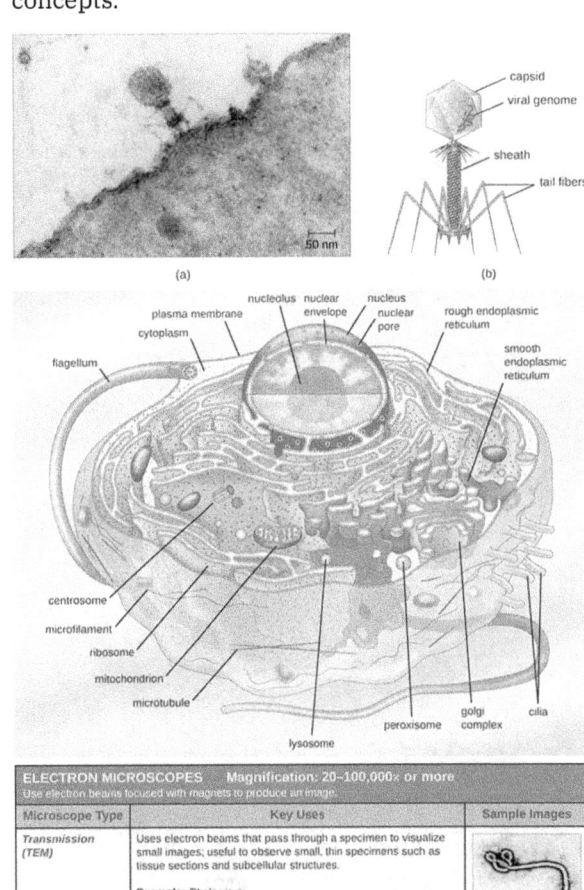

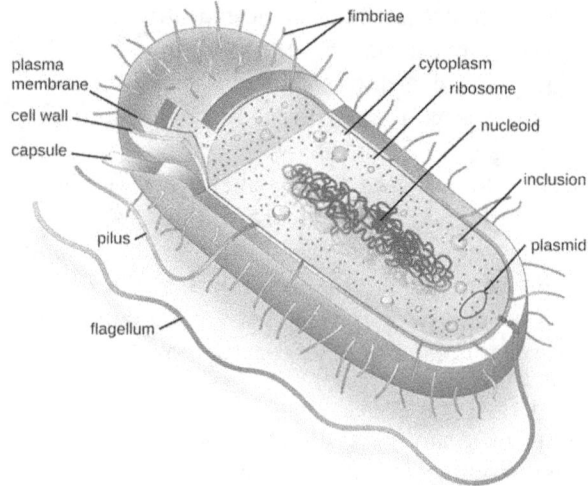

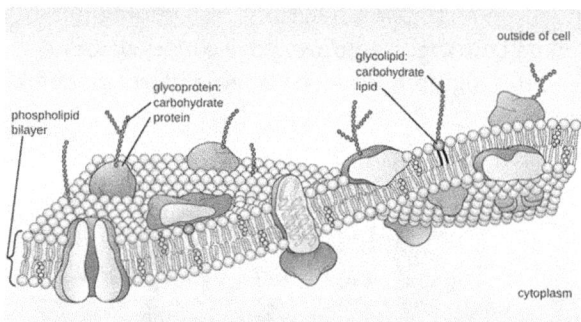

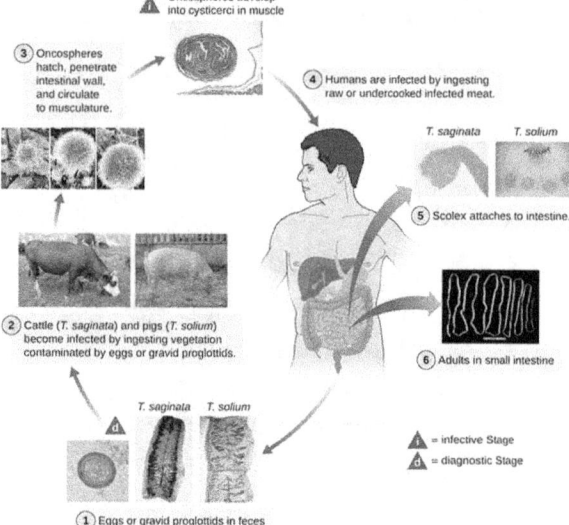

Materials That Reinforce Key Concepts

- **Learning Objectives.** Every section begins with a set of clear and concise learning objectives that are closely aligned to the content and Review Questions.
- **Summary.** The Summary distills the information in each section into a series of concise bullet points. Key Terms in the Summary are bold-faced for emphasis.
- **Key Terms.** New vocabulary is bold-faced when first introduced in the text and followed by a definition in context.
- **Check Your Understanding questions.** Each subsection of the text is punctuated by one or more comprehension-level questions. These questions encourage readers to make sure they understand what they have read before moving on to the next topic.
- **Review Questions.** Each chapter has a robust set of review questions that assesses students' mastery of the Learning Objectives. Questions are organized by format: multiple choice, matching, true/false, fill-in-the-blank, short answer, and critical thinking.

Additional Resources

Student and Instructor Resources

We've compiled additional resources for both students and instructors, including Getting Started Guides, a test bank, and an instructor answer guide. Instructor resources require a verified instructor account, which can be requested on your openstax.org log-in. Take advantage of these resources to supplement your OpenStax book.

Partner Resources

OpenStax Partners are our allies in the mission to make high-quality learning materials affordable and accessible to students and instructors everywhere. Their tools integrate seamlessly with our OpenStax titles at a low cost. To access the partner resources for your text, visit your book page on openstax.org.

About the Authors

Senior Contributing Authors

Nina Parker (Content Lead), Shenandoah University

Dr. Nina Parker received her BS and MS from the University of Michigan, and her PhD in Immunology from Ohio University. She joined Shenandoah University's Department of Biology in 1995 and serves as Associate Professor, teaching general microbiology, medical microbiology, immunology, and epidemiology to biology majors and allied health students. Prior to her academic career, Dr. Parker was trained as a Medical Technologist and received ASCP certification, experiences that drive her ongoing passion for training health professionals and those preparing for clinical laboratory work. Her areas of specialization include infectious disease, immunology, microbial pathogenesis, and medical microbiology. Dr. Parker is also deeply interested in the history of medicine and science, and pursues information about diseases often associated with regional epidemics in Virginia.

Mark Schneegurt (Lead Writer), Wichita State University

Dr. Mark A. Schneegurt is a Professor of Biological Sciences at Wichita State University and maintains joint appointments in Curriculum and Instruction and Biomedical Engineering. Dr. Schneegurt holds degrees from Rensselaer Polytechnic Institute and a Ph.D. from Brown University. He was a postdoctoral fellow at Eli Lilly and has taught and researched at Purdue University and the University of Notre Dame. His research focuses on applied and environmental microbiology, resulting in 70+ scientific publications and 150+ presentations.

Anh-Hue Thi Tu (Senior Reviewer), Georgia Southwestern State University

Dr. Anh-Hue Tu (born in Saigon, Vietnam) earned a BS in Chemistry from Baylor University and a PhD in Medical Sciences from Texas A & M Health Science Center. At the University of Alabama–Birmingham, she completed postdoctoral appointments in the areas of transcriptional regulation in *Escherichia coli* and characterization of virulence factors in *Streptococcus pneumoniae* and then became a research assistant professor working in the field of mycoplasmology. In 2004, Dr. Tu joined Georgia Southwestern State University where she currently serves as Professor, teaching various biology courses and overseeing undergraduate student research. Her areas of research interest include gene regulation, bacterial genetics, and molecular biology. Dr. Tu's teaching philosophy is to instill in her students the love of science by using critical thinking. As a teacher, she believes it is important to take technical information and express it in a way that is understandable to any student.

Brian M. Forster, Saint Joseph's University

Dr. Brian M. Forster received his BS in Biology from Binghamton University and his PhD in Microbiology from Cornell University. In 2011, he joined the faculty of Saint Joseph's University. Dr. Forster is the laboratory coordinator for the natural science laboratory-based classes designed for students who are not science majors. He teaches courses in general biology, heredity and evolution, environmental science, and microbiology for students wishing to enter nursing or allied health programs. He has publications in the *Journal of Bacteriology*, the *Journal of Microbiology & Biology Education* and *Tested Studies for Laboratory Education* (ABLE Proceedings).

Philip Lister, Central New Mexico Community College

Dr. Philip Lister earned his BS in Microbiology (1986) from Kansas State University and PhD in Medical Microbiology (1992) from Creighton University. He was a Professor of Medical Microbiology and Immunology at Creighton University (1994-2011), with appointments in the Schools of Medicine and Pharmacy. He also served as Associate Director of the Center for Research in Anti-Infectives and Biotechnology. He has published research articles, reviews, and book chapters related to antimicrobial resistance and pharmacodynamics, and has served as an Editor for the *Journal of Antimicrobial Chemotherapy*. He is currently serving as Chair of Biology and Biotechnology at Central New Mexico Community College.

Contributing Authors

Summer Allen, Brown University
Ann Auman, Pacific Lutheran University
Graciela Brelles-Mariño, Universidad Nacional de la Plata
Myriam Alhadeff Feldman, Lake Washington Institute of Technology
Paul Flowers, University of North Carolina–Pembroke
Clifton Franklund, Ferris State University
Ann Paterson, Williams Baptist University
George Pinchuk, Mississippi University for Women
Ben Rowley, University of Central Arkansas
Mark Sutherland, Hendrix College

Reviewers

Michael Angell, Eastern Michigan University
Roberto Anitori, Clark College
James Bader, Case Western Reserve University
Amy Beumer, College of William and Mary
Gilles Bolduc, Massasoit Community College
Susan Bornstein-Forst, Marian University
Nancy Boury, Iowa State University
Jennifer Brigati, Maryville College
Harold Bull, University of Saskatchewan
Evan Burkala, Oklahoma State University
Bernadette Connors, Dominican College
Richard J. Cristiano, Houston Community College–Northwest
AnnMarie DelliPizzi, Dominican College
Elisa M. LaBeau DiMenna, Central New Mexico Community College
Diane Dixon, Southeastern Oklahoma State University
Randy Durren, Longwood University

Elizabeth A. B. Emmert, Salisbury University
Karen Frederick, Marygrove College
Sharon Gusky, Northwestern Connecticut Community College
Deborah V. Harbour, College of Southern Nevada
Randall Harris, William Carey University
Diane Hartman, Baylor University
Angela Hartsock, University of Akron
Nazanin Zarabadi Hebel, Houston Community College
Heather Klenovich, Community College of Alleghany County
Kathleen Lavoie, Plattsburgh State University
Toby Mapes, Blue Ridge Community College
Barry Margulies, Towson University
Kevin M. McCabe, Columbia Gorge Community College
Karin A. Melkonian, Long Island University
Jennifer Metzler, Ball State University
Ellyn R. Mulcahy, Johnson County Community College
Jonas Okeagu, Fayetteville State University
Randall Kevin Pegg, Florida State College–Jacksonville
Judy Penn, Shoreline Community College
Lalitha Ramamoorthy, Marian University
Drew Rholl, North Park University
Hilda Rodriguez, Miami Dade College
Sean Rollins, Fitchburg State University
Sameera Sayeed, University of Pittsburgh
Pramila Sen, Houston Community College
Brian Róbert Shmaefsky, Kingwood College
Janie Sigmon, York Technical College
Denise Signorelli, College of Southern Nevada
Molly Smith, South Georgia State College–Waycross
Paula Steiert, Southwest Baptist University
Robert Sullivan, Fairfield University
Suzanne Wakim, Butte Community College
Anne Weston, Francis Crick Institute
Valencia L. Williams, West Coast University
James Wise, Chowan State University
Virginia Young, Mercer University

CHAPTER 1
An Invisible World

Figure 1.1 A veterinarian gets ready to clean a sea turtle covered in oil following the Deepwater Horizon oil spill in the Gulf of Mexico in 2010. After the spill, the population of a naturally occurring oil-eating marine bacterium called *Alcanivorax borkumensis* skyrocketed, helping to get rid of the oil. Scientists are working on ways to genetically engineer this bacterium to be more efficient in cleaning up future spills. (credit: modification of work by NOAA's National Ocean Service)

Chapter Outline

1.1 What Our Ancestors Knew

1.2 A Systematic Approach

1.3 Types of Microorganisms

INTRODUCTION From boiling thermal hot springs to deep beneath the Antarctic ice, microorganisms can be found almost everywhere on earth in great quantities. Microorganisms (or microbes, as they are also called) are small organisms. Most are so small that they cannot be seen without a microscope.

Most microorganisms are harmless to humans and, in fact, many are helpful. They play fundamental roles in ecosystems everywhere on earth, forming the backbone of many food webs. People use them to make biofuels, medicines, and even foods. Without microbes, there would be no bread, cheese, or beer. Our bodies are filled with microbes, and our skin alone is home to trillions of them.[1] Some of them we can't live without; others cause diseases that can make us sick or even kill us.

---

1 J. Hulcr et al. "A Jungle in There: Bacteria in Belly Buttons are Highly Diverse, but Predictable." *PLoS ONE* 7 no. 11 (2012): e47712. doi:10.1371/journal.pone.0047712.

Although much more is known today about microbial life than ever before, the vast majority of this invisible world remains unexplored. Microbiologists continue to identify new ways that microbes benefit and threaten humans.

1.1 What Our Ancestors Knew

Learning Objectives

By the end of this section, you will be able to:

- Describe how our ancestors improved food with the use of invisible microbes
- Describe how the causes of sickness and disease were explained in ancient times, prior to the invention of the microscope
- Describe key historical events associated with the birth of microbiology

---

Clinical Focus

Part 1

Cora, a 41-year-old lawyer and mother of two, has recently been experiencing severe headaches, a high fever, and a stiff neck. Her husband, who has accompanied Cora to see a doctor, reports that Cora also seems confused at times and unusually drowsy. Based on these symptoms, the doctor suspects that Cora may have meningitis, a potentially life-threatening infection of the tissue that surrounds the brain and spinal cord.

Meningitis has several potential causes. It can be brought on by bacteria, fungi, viruses, or even a reaction to medication or exposure to heavy metals. Although people with viral meningitis usually heal on their own, bacterial and fungal meningitis are quite serious and require treatment.

Cora's doctor orders a lumbar puncture (spinal tap) to take three samples of cerebrospinal fluid (CSF) from around the spinal cord (Figure 1.2). The samples will be sent to laboratories in three different departments for testing: clinical chemistry, microbiology, and hematology. The samples will first be visually examined to determine whether the CSF is abnormally colored or cloudy; then the CSF will be examined under a microscope to see if it contains a normal number of red and white blood cells and to check for any abnormal cell types. In the microbiology lab, the specimen will be centrifuged to concentrate any cells in a sediment; this sediment will be smeared on a slide and stained with a Gram stain. Gram staining is a procedure used to differentiate between two different types of bacteria (gram-positive and gram-negative).

About 80% of patients with bacterial meningitis will show bacteria in their CSF with a Gram stain.[2] Cora's Gram stain did not show any bacteria, but her doctor decides to prescribe her antibiotics just in case. Part of the CSF sample will be cultured—put in special dishes to see if bacteria or fungi will grow. It takes some time for most microorganisms to reproduce in sufficient quantities to be detected and analyzed.

- What types of microorganisms would be killed by antibiotic treatment?

---

2 Rebecca Buxton. "Examination of Gram Stains of Spinal Fluid—Bacterial Meningitis." *American Society for Microbiology.* 2007. http://www.microbelibrary.org/library/gram-stain/3065-examination-of-gram-stains-of-spinal-fluid-bacterial-meningitis

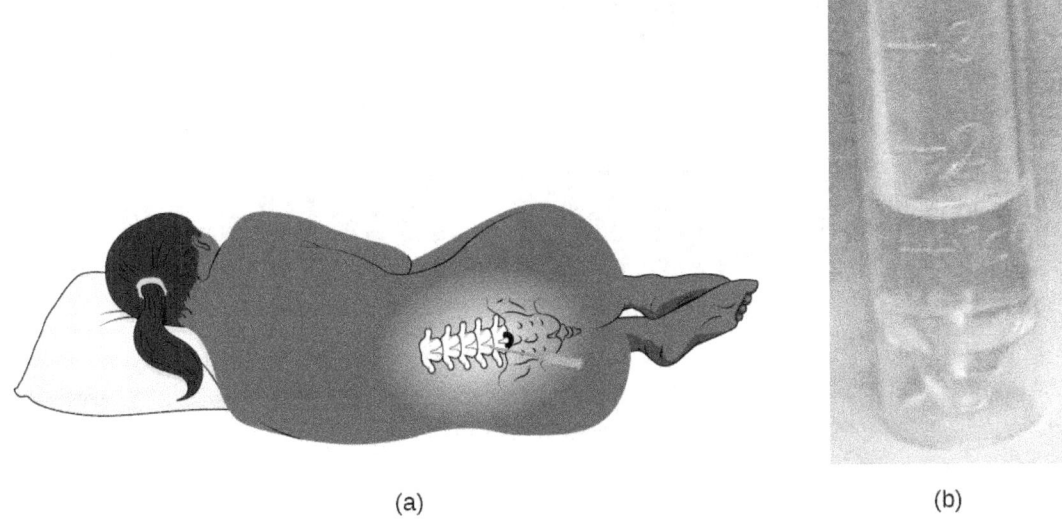

Figure 1.2 (a) A lumbar puncture is used to take a sample of a patient's cerebrospinal fluid (CSF) for testing. A needle is inserted between two vertebrae of the lower back, called the lumbar region. (b) CSF should be clear, as in this sample. Abnormally cloudy CSF may indicate an infection but must be tested further to confirm the presence of microorganisms. (credit a: modification of work by Centers for Disease Control and Prevention; credit b: modification of work by James Heilman)

Jump to the next Clinical Focus box.

Most people today, even those who know very little about microbiology, are familiar with the concept of microbes, or "germs," and their role in human health. Schoolchildren learn about bacteria, viruses, and other microorganisms, and many even view specimens under a microscope. But a few hundred years ago, before the invention of the microscope, the existence of many types of microbes was impossible to prove. By definition, **microorganisms**, or **microbes**, are very small organisms; many types of microbes are too small to see without a microscope, although some parasites and fungi are visible to the naked eye.

Humans have been living with—and using—microorganisms for much longer than they have been able to see them. Historical evidence suggests that humans have had some notion of microbial life since prehistoric times and have used that knowledge to develop foods as well as prevent and treat disease. In this section, we will explore some of the historical applications of microbiology as well as the early beginnings of microbiology as a science.

Fermented Foods and Beverages

People across the world have enjoyed fermented foods and beverages like beer, wine, bread, yogurt, cheese, and pickled vegetables for all of recorded history. Discoveries from several archeological sites suggest that even prehistoric people took advantage of fermentation to preserve and enhance the taste of food. Archaeologists studying pottery jars from a Neolithic village in China found that people were making a fermented beverage from rice, honey, and fruit as early as 7000 BC.[3]

Production of these foods and beverages requires microbial fermentation, a process that uses bacteria, mold, or yeast to convert sugars (carbohydrates) to alcohol, gases, and organic acids (Figure 1.3). While it is likely that people first learned about fermentation by accident—perhaps by drinking old milk that had curdled or old grape juice that had fermented—they later learned to harness the power of fermentation to make products like bread, cheese, and wine.

---

3 P.E. McGovern et al. "Fermented Beverages of Pre- and Proto-Historic China." *Proceedings of the National Academy of Sciences of the United States of America* 1 no. 51 (2004):17593–17598. doi:10.1073/pnas.0407921102.

Yeast fermentation yields ethanol and CO_2.

Figure 1.3 A microscopic view of *Saccharomyces cerevisiae*, the yeast responsible for making bread rise (left). Yeast is a microorganism. Its cells metabolize the carbohydrates in flour (middle) and produce carbon dioxide, which causes the bread to rise (right). (credit middle: modification of work by Janus Sandsgaard; credit right: modification of work by "MDreibelbis"/Flickr)

The Iceman Treateth

Prehistoric humans had a very limited understanding of the causes of disease, and various cultures developed different beliefs and explanations. While many believed that illness was punishment for angering the gods or was simply the result of fate, archaeological evidence suggests that prehistoric people attempted to treat illnesses and infections. One example of this is Ötzi the Iceman, a 5300-year-old mummy found frozen in the ice of the Ötzal Alps on the Austrian-Italian border in 1991. Because Ötzi was so well preserved by the ice, researchers discovered that he was infected with the eggs of the parasite *Trichuris trichiura*, which may have caused him to have abdominal pain and anemia. Researchers also found evidence of *Borrelia burgdorferi*, a bacterium that causes Lyme disease.[4] Some researchers think Ötzi may have been trying to treat his infections with the woody fruit of the *Fomitopsis betulinus* fungus, which was discovered tied to his belongings.[5] This fungus has both laxative and antibiotic properties. Ötzi was also covered in tattoos that were made by cutting incisions into his skin, filling them with herbs, and then burning the herbs.[6] There is speculation that this may have been another attempt to treat his health ailments.

Early Notions of Disease, Contagion, and Containment

Several ancient civilizations appear to have had some understanding that disease could be transmitted by things they could not see. This is especially evident in historical attempts to contain the spread of disease. For example, the Bible refers to the practice of quarantining people with leprosy and other diseases, suggesting that people understood that diseases could be communicable. Ironically, while leprosy is communicable, it is also a disease that progresses slowly. This means that people were likely quarantined after they had already spread the disease to others.

The ancient Greeks attributed disease to bad air, *mal'aria*, which they called "miasmatic odors." They developed hygiene practices that built on this idea. The Romans also believed in the miasma hypothesis and created a complex sanitation infrastructure to deal with sewage. In Rome, they built aqueducts, which brought fresh water into the city, and a giant sewer, the *Cloaca Maxima*, which carried waste away and into the river Tiber (Figure 1.4). Some researchers believe that this infrastructure helped protect the Romans from epidemics of waterborne illnesses.

---

4 A. Keller et al. "New Insights into the Tyrolean Iceman's Origin and Phenotype as Inferred by Whole-Genome Sequencing." *Nature Communications*, 3 (2012): 698. doi:10.1038/nc omms1701.

5 L. Capasso. "5300 Years Ago, the Ice Man Used Natural Laxatives and Antibiotics." *The Lancet*, 352 (1998) 9143: 1864. doi: 10.1016/s0140-6736(05)79939-6.

6 L. Capasso, L. "5300 Years Ago, the Ice Man Used Natural Laxatives and Antibiotics." *The Lancet*, 352 no. 9143 (1998): 1864. doi: 10.1016/s0140-6736(05)79939-6.

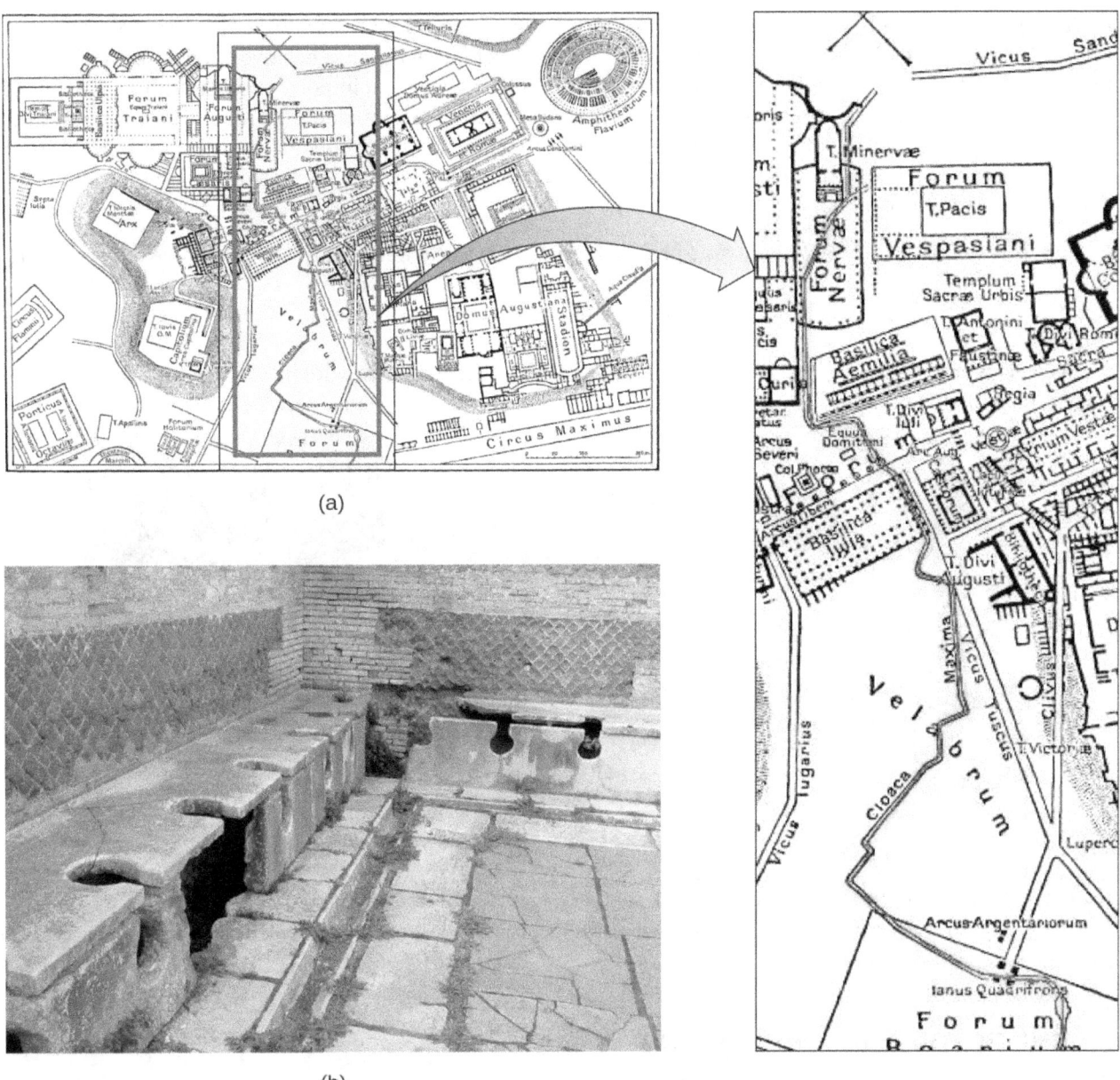

Figure 1.4 (a) The *Cloaca Maxima*, or "Greatest Sewer" (shown in red), ran through ancient Rome. It was an engineering marvel that carried waste away from the city and into the river Tiber. (b) These ancient latrines emptied into the *Cloaca Maxima*.

Even before the invention of the microscope, some doctors, philosophers, and scientists made great strides in understanding the invisible forces—what we now know as microbes—that can cause infection, disease, and death.

The Greek physician Hippocrates (460–370 BC) is considered the "father of Western medicine" (Figure 1.5). Unlike many of his ancestors and contemporaries, he dismissed the idea that disease was caused by supernatural forces. Instead, he posited that diseases had natural causes from within patients or their environments. Hippocrates and his heirs are believed to have written the *Hippocratic Corpus*, a collection of texts that make up some of the oldest surviving medical books.[7] Hippocrates is also often credited as the author of the Hippocratic Oath, taken by new physicians to pledge their dedication to diagnosing and treating patients without causing harm.

---

7 G. Pappas et al. "Insights Into Infectious Disease in the Era of Hippocrates." *International Journal of Infectious Diseases* 12 (2008) 4:347–350. doi: http://dx.doi.org/10.1016/j.ijid.2007.11.003.

While Hippocrates is considered the father of Western medicine, the Greek philosopher and historian Thucydides (460–395 BC) is considered the father of scientific history because he advocated for evidence-based analysis of cause-and-effect reasoning (Figure 1.5). Among his most important contributions are his observations regarding the Athenian plague that killed one-third of the population of Athens between 430 and 410 BC. Having survived the epidemic himself, Thucydides made the important observation that survivors did not get re-infected with the disease, even when taking care of actively sick people.[8] This observation shows an early understanding of the concept of immunity.

Marcus Terentius Varro (116–27 BC) was a prolific Roman writer who was one of the first people to propose the concept that things we cannot see (what we now call microorganisms) can cause disease (Figure 1.5). In *Res Rusticae* (*On Farming*), published in 36 BC, he said that "precautions must also be taken in neighborhood swamps . . . because certain minute creatures [*animalia minuta*] grow there which cannot be seen by the eye, which float in the air and enter the body through the mouth and nose and there cause serious diseases."[9]

(a) (b) (c)

Figure 1.5 (a) Hippocrates, the "father of Western medicine," believed that diseases had natural, not supernatural, causes. (b) The historian Thucydides observed that survivors of the Athenian plague were subsequently immune to the infection. (c) Marcus Terentius Varro proposed that disease could be caused by "certain minute creatures . . . which cannot be seen by the eye." (credit c: modification of work by Alessandro Antonelli)

✓ CHECK YOUR UNDERSTANDING

- Give two examples of foods that have historically been produced by humans with the aid of microbes.
- Explain how historical understandings of disease contributed to attempts to treat and contain disease.

The Birth of Microbiology

While the ancients may have suspected the existence of invisible "minute creatures," it wasn't until the invention of the microscope that their existence was definitively confirmed. While it is unclear who exactly invented the microscope, a Dutch cloth merchant named Antonie van Leeuwenhoek (1632–1723) was the first to develop a lens powerful enough to view microbes. In 1675, using a simple but powerful microscope, Leeuwenhoek was able to observe single-celled organisms, which he described as "animalcules" or "wee little beasties," swimming in a drop of rain water. From his drawings of these little organisms, we now know he was looking at bacteria and protists. (We will explore Leeuwenhoek's contributions to microscopy further in How We See the Invisible World.)

---

8 Thucydides. *The History of the Peloponnesian War. The Second Book.* 431 BC. Translated by Richard Crawley. http://classics.mit.edu/Thucydides/pelopwar.2.second.html.

9 Plinio Prioreschi. *A History of Medicine: Roman Medicine.* Lewiston, NY: Edwin Mellen Press, 1998: p. 215.

Nearly 200 years after van Leeuwenhoek got his first glimpse of microbes, the "Golden Age of Microbiology" spawned a host of new discoveries between 1857 and 1914. Two famous microbiologists, Louis Pasteur and Robert Koch, were especially active in advancing our understanding of the unseen world of microbes (Figure 1.6). Pasteur, a French chemist, showed that individual microbial strains had unique properties and demonstrated that fermentation is caused by microorganisms. He also invented pasteurization, a process used to kill microorganisms responsible for spoilage, and developed vaccines for the treatment of diseases, including rabies, in animals and humans. Koch, a German physician, was the first to demonstrate the connection between a single, isolated microbe and a known human disease. For example, he discovered the bacteria that cause anthrax (*Bacillus anthracis*), cholera (*Vibrio cholera*), and tuberculosis (*Mycobacterium tuberculosis*).[10] We will discuss these famous microbiologists, and others, in later chapters.

(a) (b)

Figure 1.6 (a) Louis Pasteur (1822–1895) is credited with numerous innovations that advanced the fields of microbiology and immunology. (b) Robert Koch (1843–1910) identified the specific microbes that cause anthrax, cholera, and tuberculosis.

As microbiology has developed, it has allowed the broader discipline of biology to grow and flourish in previously unimagined ways. Much of what we know about human cells comes from our understanding of microbes, and many of the tools we use today to study cells and their genetics derive from work with microbes.

⊘ CHECK YOUR UNDERSTANDING

- How did the discovery of microbes change human understanding of disease?

MICRO CONNECTIONS

Microbiology Toolbox

Because individual microbes are generally too small to be seen with the naked eye, the science of microbiology is dependent on technology that can artificially enhance the capacity of our natural senses of perception. Early microbiologists like Pasteur and Koch had fewer tools at their disposal than are found in modern laboratories, making their discoveries and innovations that much more impressive. Later chapters of this text will explore many applications of technology in depth, but for now, here is a brief overview of some of the fundamental tools of the microbiology lab.

---

10   S.M. Blevins and M.S. Bronze. "Robert Koch and the 'Golden Age' of Bacteriology." *International Journal of Infectious Diseases.* 14 no. 9 (2010): e744-e751. doi:10.1016/j.ijid.2009.12.003.

- **Microscopes** produce magnified images of microorganisms, human cells and tissues, and many other types of specimens too small to be observed with the naked eye.
- **Stains and dyes** are used to add color to microbes so they can be better observed under a microscope. Some dyes can be used on living microbes, whereas others require that the specimens be fixed with chemicals or heat before staining. Some stains only work on certain types of microbes because of differences in their cellular chemical composition.
- **Growth media** are used to grow microorganisms in a lab setting. Some media are liquids; others are more solid or gel-like. A growth medium provides nutrients, including water, various salts, a source of carbon (like glucose), and a source of nitrogen and amino acids (like yeast extract) so microorganisms can grow and reproduce. Ingredients in a growth medium can be modified to grow unique types of microorganisms.
- **A Petri dish** is a flat-lidded dish that is typically 10–11 centimeters (cm) in diameter and 1–1.5 cm high. Petri dishes made out of either plastic or glass are used to hold growth media (Figure 1.7).
- **Test tubes** are cylindrical plastic or glass tubes with rounded bottoms and open tops. They can be used to grow microbes in broth, or semisolid or solid growth media.
- A **Bunsen burner** is a metal apparatus that creates a flame that can be used to sterilize pieces of equipment. A rubber tube carries gas (fuel) to the burner. In many labs, Bunsen burners are being phased out in favor of infrared **microincinerators**, which serve a similar purpose without the safety risks of an open flame.
- An **inoculation loop** is a handheld tool that ends in a small wire loop (Figure 1.7). The loop can be used to streak microorganisms on agar in a Petri dish or to transfer them from one test tube to another. Before each use, the inoculation loop must be sterilized so cultures do not become contaminated.

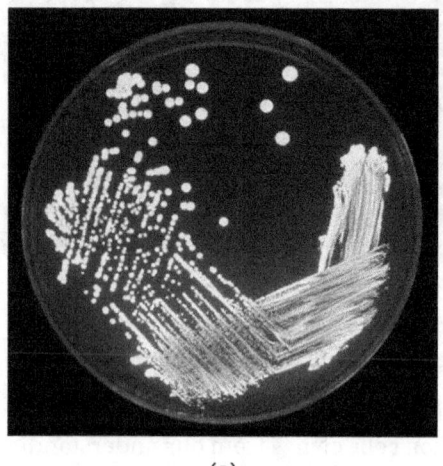
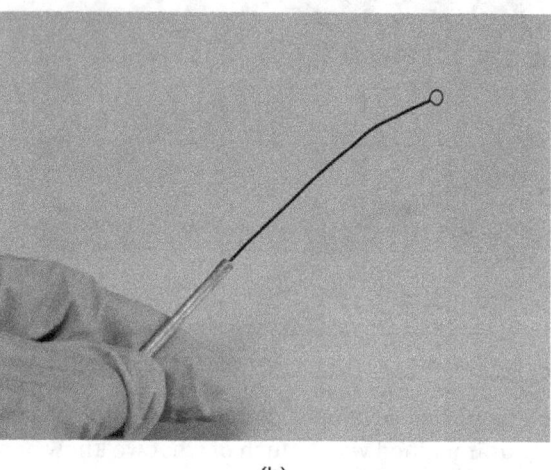

(a) (b)

Figure 1.7 (a) This Petri dish filled with agar has been streaked with *Legionella*, the bacterium responsible for causing Legionnaire's disease. (b) An inoculation loop like this one can be used to streak bacteria on agar in a Petri dish. (credit a: modification of work by Centers for Disease Control and Prevention; credit b: modification of work by Jeffrey M. Vinocur)

1.2 A Systematic Approach

Learning Objectives

By the end of this section, you will be able to:
- Describe how microorganisms are classified and distinguished as unique species
- Compare historical and current systems of taxonomy used to classify microorganisms

Once microbes became visible to humans with the help of microscopes, scientists began to realize their enormous diversity. Microorganisms vary in all sorts of ways, including their size, their appearance, and their rates of reproduction. To study this incredibly diverse new array of organisms, researchers needed a way to systematically organize them.

The Science of Taxonomy

Taxonomy is the classification, description, identification, and naming of living organisms. Classification is the practice of organizing organisms into different groups based on their shared characteristics. The most famous early taxonomist was a Swedish botanist, zoologist, and physician named Carolus Linnaeus (1701–1778). In 1735, Linnaeus published *Systema Naturae*, an 11-page booklet in which he proposed the Linnaean taxonomy, a system of categorizing and naming organisms using a standard format so scientists could discuss organisms using consistent terminology. He continued to revise and add to the book, which grew into multiple volumes (Figure 1.8).

Figure 1.8 Swedish botanist, zoologist, and physician Carolus Linnaeus developed a new system for categorizing plants and animals. In this 1853 portrait by Hendrik Hollander, Linnaeus is holding a twinflower, named *Linnaea borealis* in his honor.

In his taxonomy, Linnaeus divided the natural world into three kingdoms: animal, plant, and mineral (the mineral kingdom was later abandoned). Within the animal and plant kingdoms, he grouped organisms using a hierarchy of increasingly specific levels and sublevels based on their similarities. The names of the levels in Linnaeus's original taxonomy were kingdom, class, order, family, genus (plural: genera), and species. Species was, and continues to be, the most specific and basic taxonomic unit.

Evolving Trees of Life (Phylogenies)

With advances in technology, other scientists gradually made refinements to the Linnaean system and eventually created new systems for classifying organisms. In the 1800s, there was a growing interest in developing taxonomies that took into account the evolutionary relationships, or **phylogenies**, of all different species of organisms on earth. One way to depict these relationships is via a diagram called a phylogenetic tree (or tree of life). In these diagrams, groups of organisms are arranged by how closely related they are thought to be. In early phylogenetic trees, the relatedness of organisms was inferred by their visible similarities, such as the presence or absence of hair or the number of limbs. Now, the analysis is more complicated. Today, phylogenic analyses include genetic, biochemical, and embryological comparisons, as will be discussed later in this chapter.

Linnaeus's tree of life contained just two main branches for all living things: the animal and plant kingdoms. In 1866, Ernst Haeckel, a German biologist, philosopher, and physician, proposed another kingdom, Protista, for unicellular organisms (Figure 1.9). He later proposed a fourth kingdom, Monera, for unicellular organisms whose cells lack nuclei, like bacteria.

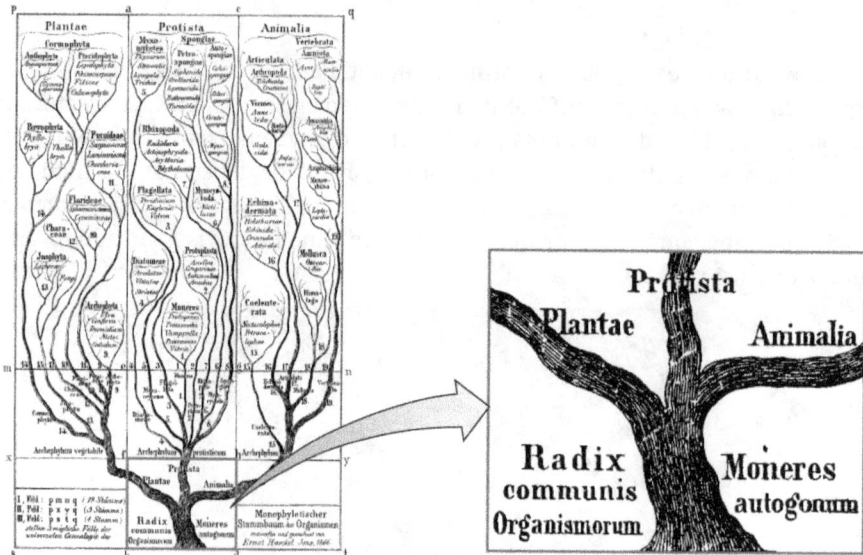

Figure 1.9 Ernst Haeckel's rendering of the tree of life, from his 1866 book *General Morphology of Organisms*, contained three kingdoms: Plantae, Protista, and Animalia. He later added a fourth kingdom, Monera, for unicellular organisms lacking a nucleus.

Nearly 100 years later, in 1969, American ecologist Robert Whittaker (1920–1980) proposed adding another kingdom—Fungi—in his tree of life. Whittaker's tree also contained a level of categorization above the kingdom level—the empire or superkingdom level—to distinguish between organisms that have membrane-bound nuclei in their cells (**eukaryotes**) and those that do not (**prokaryotes**). Empire Prokaryota contained just the Kingdom Monera. The Empire Eukaryota contained the other four kingdoms: Fungi, Protista, Plantae, and Animalia. Whittaker's five-kingdom tree was considered the standard phylogeny for many years.

Figure 1.10 shows how the tree of life has changed over time. Note that viruses are not found in any of these trees. That is because they are not made up of cells and thus it is difficult to determine where they would fit into a tree of life.

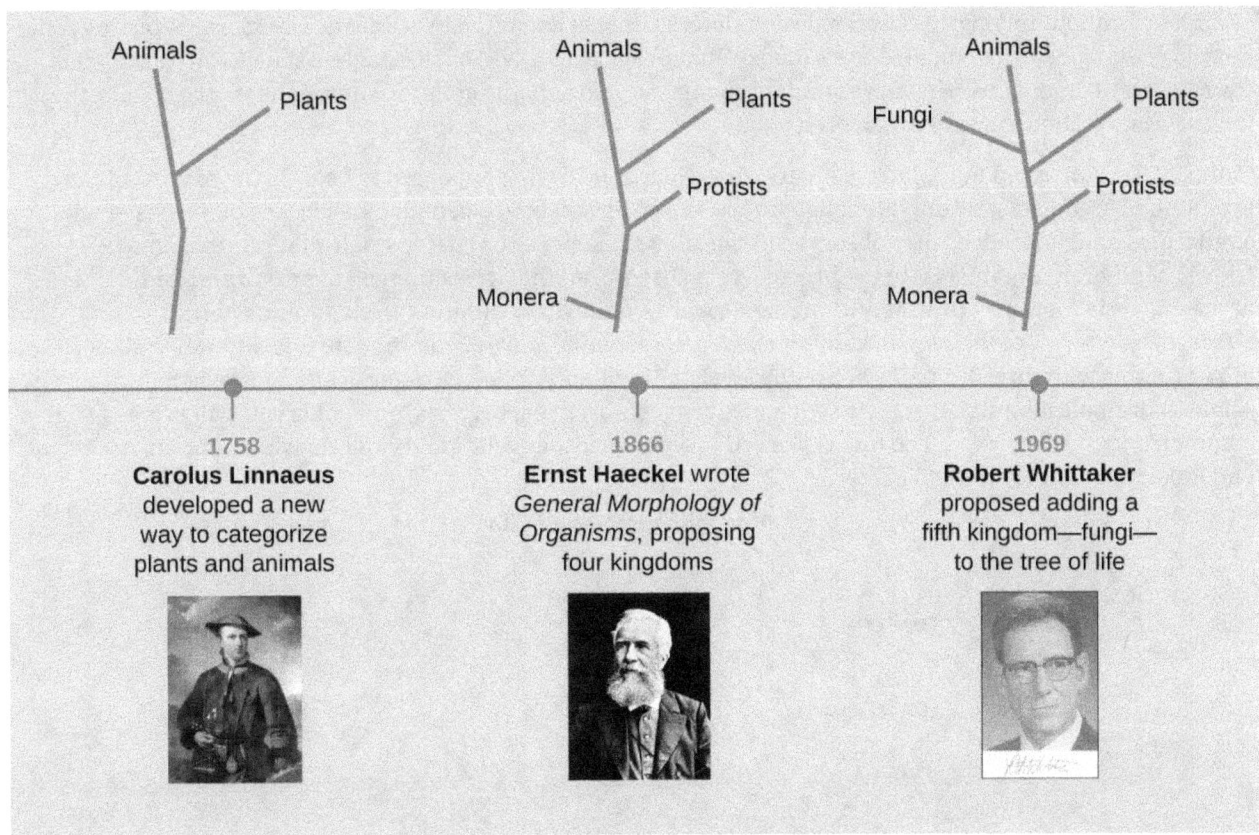

Figure 1.10 This timeline shows how the shape of the tree of life has changed over the centuries. Even today, the taxonomy of living organisms is continually being reevaluated and refined with advances in technology.

✓ CHECK YOUR UNDERSTANDING

- Briefly summarize how our evolving understanding of microorganisms has contributed to changes in the way that organisms are classified.

Clinical Focus

Part 2

Antibiotic drugs are specifically designed to kill or inhibit the growth of bacteria. But after a couple of days on antibiotics, Cora shows no signs of improvement. Also, her CSF cultures came back from the lab negative. Since bacteria or fungi were not isolated from Cora's CSF sample, her doctor rules out bacterial and fungal meningitis. Viral meningitis is still a possibility.

However, Cora now reports some troubling new symptoms. She is starting to have difficulty walking. Her muscle stiffness has spread from her neck to the rest of her body, and her limbs sometimes jerk involuntarily. In addition, Cora's cognitive symptoms are worsening. At this point, Cora's doctor becomes very concerned and orders more tests on the CSF samples.

- What types of microorganisms could be causing Cora's symptoms?

Jump to the next Clinical Focus box. Go back to the previous Clinical Focus box.

The Role of Genetics in Modern Taxonomy

Haeckel's and Whittaker's trees presented hypotheses about the phylogeny of different organisms based on readily observable characteristics. But the advent of molecular genetics in the late 20th century revealed other ways to organize phylogenetic trees. Genetic methods allow for a standardized way to compare all living organisms without relying on observable characteristics that can often be subjective. Modern taxonomy relies heavily on comparing the nucleic acids (deoxyribonucleic acid [DNA] or ribonucleic acid [RNA]) or proteins from different organisms. The more similar the nucleic acids and proteins are between two organisms, the more closely related they are considered to be.

In the 1970s, American microbiologist Carl Woese discovered what appeared to be a "living record" of the evolution of organisms. He and his collaborator George Fox created a genetics-based tree of life based on similarities and differences they observed in the gene sequences coding for small subunit ribosomal RNA (rRNA) of different organisms. In the process, they discovered that a certain type of bacteria, called archaebacteria (now known simply as archaea), were significantly different from other bacteria and eukaryotes in terms of their small subunit rRNA gene sequences. To accommodate this difference, they created a tree with three Domains above the level of Kingdom: Archaea, Bacteria, and Eukarya (Figure 1.11). Analysis of small subunit rRNA gene sequences suggests archaea, bacteria, and eukaryotes all evolved from a common ancestral cell type. The tree is skewed to show a closer evolutionary relationship between Archaea and Eukarya than they have to Bacteria.

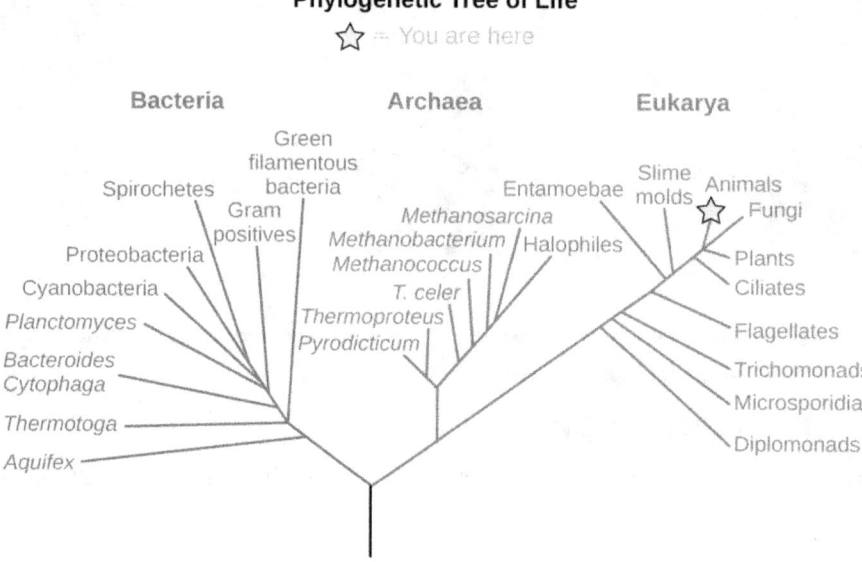

Figure 1.11 Woese and Fox's phylogenetic tree contains three domains: Bacteria, Archaea, and Eukarya. Domains Archaea and Bacteria contain all prokaryotic organisms, and Eukarya contains all eukaryotic organisms. (credit: modification of work by Eric Gaba)

Scientists continue to use analysis of RNA, DNA, and proteins to determine how organisms are related. One interesting, and complicating, discovery is that of horizontal gene transfer—when a gene of one species is absorbed into another organism's genome. Horizontal gene transfer is especially common in microorganisms and can make it difficult to determine how organisms are evolutionarily related. Consequently, some scientists now think in terms of "webs of life" rather than "trees of life."

CHECK YOUR UNDERSTANDING

- In modern taxonomy, how do scientists determine how closely two organisms are related?
- Explain why the branches on the "tree of life" all originate from a single "trunk."

Naming Microbes

In developing his taxonomy, Linnaeus used a system of **binomial nomenclature**, a two-word naming system for identifying organisms by genus and specific epithet. For example, modern humans are in the genus *Homo* and have the specific epithet name *sapiens*, so their scientific name in binomial nomenclature is *Homo sapiens*. In binomial nomenclature, the genus part of the name is always capitalized; it is followed by the specific epithet name, which is not capitalized. Both names are italicized. When referring to the species of humans, the binomial nomenclature would be *Homo sapiens*.

Taxonomic names in the 18th through 20th centuries were typically derived from Latin, since that was the common language used by scientists when taxonomic systems were first created. Today, newly discovered organisms can be given names derived from Latin, Greek, or English. Sometimes these names reflect some distinctive trait of the organism; in other cases, microorganisms are named after the scientists who discovered them. The archaeon *Haloquadratum walsbyi* is an example of both of these naming schemes. The genus, *Haloquadratum*, describes the microorganism's saltwater habitat (*halo* is derived from the Greek word for "salt") as well as the arrangement of its square cells, which are arranged in square clusters of four cells (*quadratum* is Latin for "foursquare"). The species, *walsbyi*, is named after Anthony Edward Walsby, the microbiologist who discovered *Haloquadratum walsbyi* in in 1980. While it might seem easier to give an organism a common descriptive name—like a red-headed woodpecker—we can imagine how that could become problematic. What happens when another species of woodpecker with red head coloring is discovered? The systematic nomenclature scientists use eliminates this potential problem by assigning each organism a single, unique two-word name that is recognized by scientists all over the world.

In this text, we will typically abbreviate an organism's genus and species after its first mention. The abbreviated form is simply the first initial of the genus, followed by a period and the full name of the species. For example, the bacterium *Escherichia coli* is shortened to *E. coli* in its abbreviated form. You will encounter this same convention in other scientific texts as well.

Bergey's Manuals

Whether in a tree or a web, microbes can be difficult to identify and classify. Without easily observable macroscopic features like feathers, feet, or fur, scientists must capture, grow, and devise ways to study their biochemical properties to differentiate and classify microbes. Despite these hurdles, a group of microbiologists created and updated a set of manuals for identifying and classifying microorganisms. First published in 1923 and since updated many times, *Bergey's Manual of Determinative Bacteriology* and *Bergey's Manual of Systematic Bacteriology* are the standard references for identifying and classifying different prokaryotes. (Appendix D of this textbook is partly based on Bergey's manuals; it shows how the organisms that appear in this textbook are classified.) Because so many bacteria look identical, methods based on nonvisual characteristics must be used to identify them. For example, biochemical tests can be used to identify chemicals unique to certain species. Likewise, serological tests can be used to identify specific antibodies that will react against the proteins found in certain species. Ultimately, DNA and rRNA sequencing can be used both for identifying a particular bacterial species and for classifying newly discovered species.

⊘ CHECK YOUR UNDERSTANDING

- What is binomial nomenclature and why is it a useful tool for naming organisms?
- Explain why a resource like one of Bergey's manuals would be helpful in identifying a microorganism in a sample.

 MICRO CONNECTIONS

Same Name, Different Strain

Within one species of microorganism, there can be several subtypes called strains. While different strains may be nearly identical genetically, they can have very different attributes. The bacterium *Escherichia coli* is infamous for causing food poisoning and traveler's diarrhea. However, there are actually many different strains of *E. coli*, and they vary in their ability to cause disease.

One pathogenic (disease-causing) *E. coli* strain that you may have heard of is *E. coli* O157:H7. In humans, infection from *E. coli* O157:H7 can cause abdominal cramps and diarrhea. Infection usually originates from contaminated water or food, particularly raw vegetables and undercooked meat. In the 1990s, there were several large outbreaks of *E. coli* O157:H7 thought to have originated in undercooked hamburgers.

While *E. coli* O157:H7 and some other strains have given *E. coli* a bad name, most *E. coli* strains do not cause disease. In fact, some can be helpful. Different strains of *E. coli* found naturally in our gut help us digest our food, provide us with some needed chemicals, and fight against pathogenic microbes.

LINK TO LEARNING

Learn more about phylogenetic trees by exploring the Wellcome Trust's interactive Tree of Life. The website (https://www.openstax.org/l/22wellcome) contains information, photos, and animations about many different organisms. Select two organisms to see how they are evolutionarily related.

1.3 Types of Microorganisms

Learning Objectives

By the end of this section, you will be able to:
- List the various types of microorganisms and describe their defining characteristics
- Give examples of different types of cellular and viral microorganisms and infectious agents
- Describe the similarities and differences between archaea and bacteria
- Provide an overview of the field of microbiology

Most microbes are unicellular and small enough that they require artificial magnification to be seen. However, there are some unicellular microbes that are visible to the naked eye, and some multicellular organisms that are microscopic. An object must measure about 100 micrometers (μm) to be visible without a microscope, but most microorganisms are many times smaller than that. For some perspective, consider that a typical animal cell measures roughly 10 μm across but is still microscopic. Bacterial cells are typically about 1 μm, and viruses can be 10 times smaller than bacteria (Figure 1.12). See Table 1.1 for units of length used in microbiology.

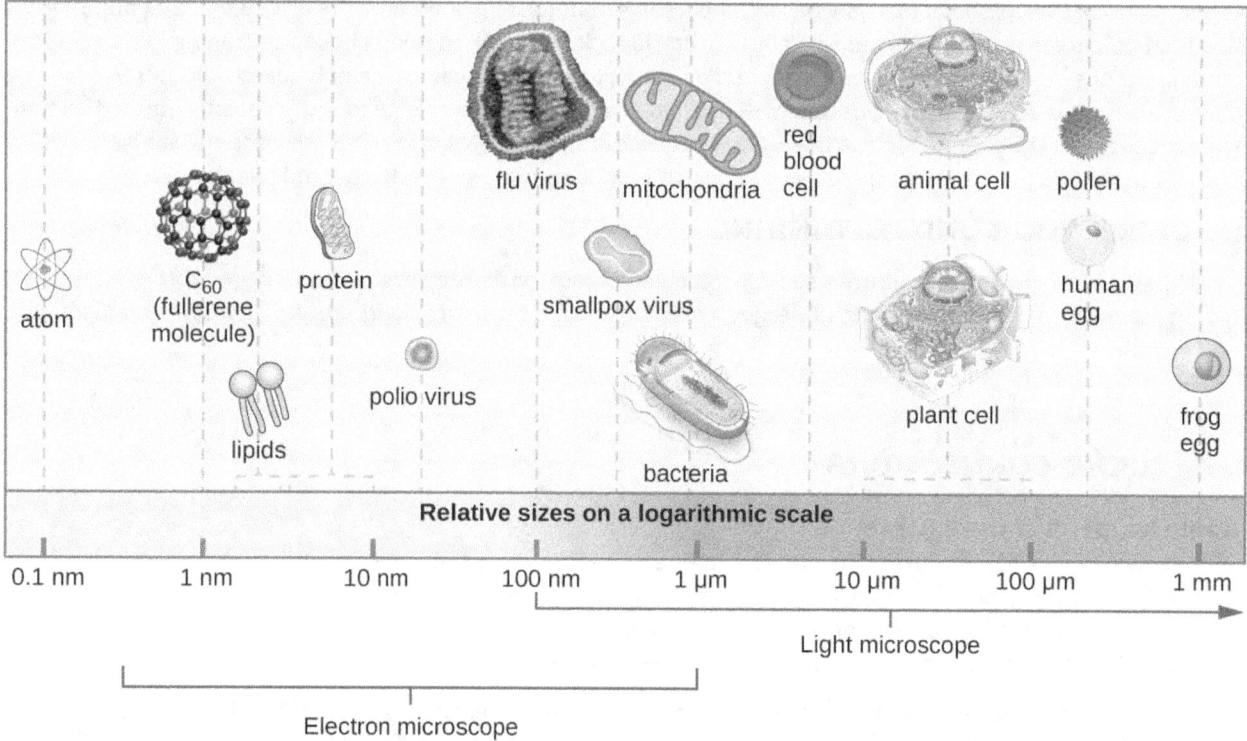

Figure 1.12 The relative sizes of various microscopic and nonmicroscopic objects. Note that a typical virus measures about 100 nm, 10 times smaller than a typical bacterium (~1 μm), which is at least 10 times smaller than a typical plant or animal cell (~10–100 μm). An object must measure about 100 μm to be visible without a microscope.

Units of Length Commonly Used in Microbiology

Metric Unit	Meaning of Prefix	Metric Equivalent
meter (m)	—	$1\ m = 10^0\ m$
decimeter (dm)	1/10	$1\ dm = 0.1\ m = 10^{-1}\ m$
centimeter (cm)	1/100	$1\ cm = 0.01\ m = 10^{-2}\ m$
millimeter (mm)	1/1000	$1\ mm = 0.001\ m = 10^{-3}\ m$
micrometer (μm)	1/1,000,000	$1\ \mu m = 0.000001\ m = 10^{-6}\ m$
nanometer (nm)	1/1,000,000,000	$1\ nm = 0.000000001\ m = 10^{-9}\ m$

Table 1.1

Microorganisms differ from each other not only in size, but also in structure, habitat, metabolism, and many other characteristics. While we typically think of microorganisms as being unicellular, there are also many multicellular organisms that are too small to be seen without a microscope. Some microbes, such as viruses, are even **acellular** (not composed of cells).

Microorganisms are found in each of the three domains of life: Archaea, Bacteria, and Eukarya. Microbes within the domains Bacteria and Archaea are all prokaryotes (their cells lack a nucleus), whereas microbes in the domain Eukarya are eukaryotes (their cells have a nucleus). Some microorganisms, such as viruses, do not fall within any of the three domains of life. In this section, we will briefly introduce each of the broad groups of microbes. Later chapters will go into greater depth about the diverse species within each group.

LINK TO LEARNING

How big is a bacterium or a virus compared to other objects? Check out this interactive website (https://www.openstax.org/l/22relsizes) to get a feel for the scale of different microorganisms.

Prokaryotic Microorganisms

Bacteria are found in nearly every habitat on earth, including within and on humans. Most bacteria are harmless or helpful, but some are **pathogens**, causing disease in humans and other animals. Bacteria are prokaryotic because their genetic material (DNA) is not housed within a true nucleus. Most bacteria have cell walls that contain peptidoglycan.

Bacteria are often described in terms of their general shape. Common shapes include spherical (coccus), rod-shaped (bacillus), or curved (spirillum, spirochete, or vibrio). Figure 1.13 shows examples of these shapes.

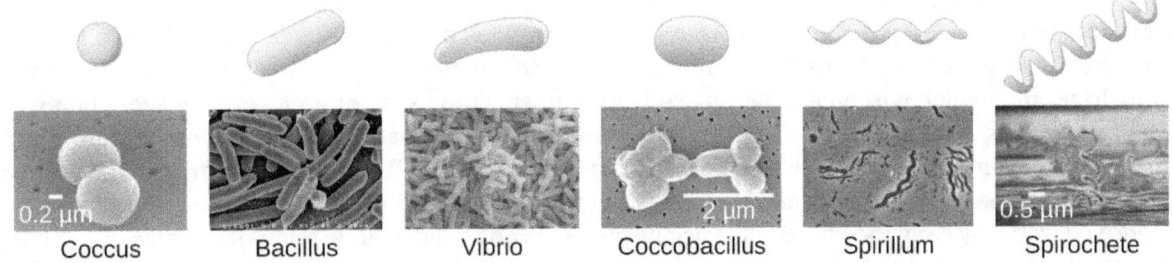

Figure 1.13 Common bacterial shapes. Note how coccobacillus is a combination of spherical (coccus) and rod-shaped (bacillus). (credit "Coccus": modification of work by Janice Haney Carr, Centers for Disease Control and Prevention; credit "Coccobacillus": modification of

work by Janice Carr, Centers for Disease Control and Prevention; credit "Spirochete": Centers for Disease Control and Prevention)

They have a wide range of metabolic capabilities and can grow in a variety of environments, using different combinations of nutrients. Some bacteria are photosynthetic, such as oxygenic cyanobacteria and anoxygenic green sulfur and green nonsulfur bacteria; these bacteria use energy derived from sunlight, and fix carbon dioxide for growth. Other types of bacteria are nonphotosynthetic, obtaining their energy from organic or inorganic compounds in their environment.

Archaea are also unicellular prokaryotic organisms. Archaea and bacteria have different evolutionary histories, as well as significant differences in genetics, metabolic pathways, and the composition of their cell walls and membranes. Unlike most bacteria, archaeal cell walls do not contain peptidoglycan, but their cell walls are often composed of a similar substance called pseudopeptidoglycan. Like bacteria, archaea are found in nearly every habitat on earth, even extreme environments that are very cold, very hot, very basic, or very acidic (Figure 1.14). Some archaea live in the human body, but none have been shown to be human pathogens.

Figure 1.14   Some archaea live in extreme environments, such as the Morning Glory pool, a hot spring in Yellowstone National Park. The color differences in the pool result from the different communities of microbes that are able to thrive at various water temperatures.

✓ CHECK YOUR UNDERSTANDING

- What are the two main types of prokaryotic organisms?
- Name some of the defining characteristics of each type.

Eukaryotic Microorganisms

The domain Eukarya contains all eukaryotes, including uni- or multicellular eukaryotes such as protists, fungi, plants, and animals. The major defining characteristic of eukaryotes is that their cells contain a nucleus.

Protists

Protists are an informal grouping of eukaryotes that are not plants, animals, or fungi. Algae and protozoa are examples of protists.

Algae (singular: alga) are protists that can be either unicellular or multicellular and vary widely in size, appearance, and habitat (Figure 1.15). Their cells are surrounded by cell walls made of cellulose, a type of carbohydrate. Algae are photosynthetic organisms that extract energy from the sun and release oxygen and carbohydrates into their environment. Because other organisms can use their waste products for energy, algae are important parts of many ecosystems. Many consumer products contain ingredients derived from algae, such as carrageenan or alginic acid, which are found in some brands of ice cream, salad dressing, beverages, lipstick, and toothpaste. A derivative of algae also plays a prominent role in the microbiology laboratory. Agar, a gel derived from algae, can be mixed with various nutrients and used to grow microorganisms in a Petri dish. Algae are also being developed as a possible source for biofuels.

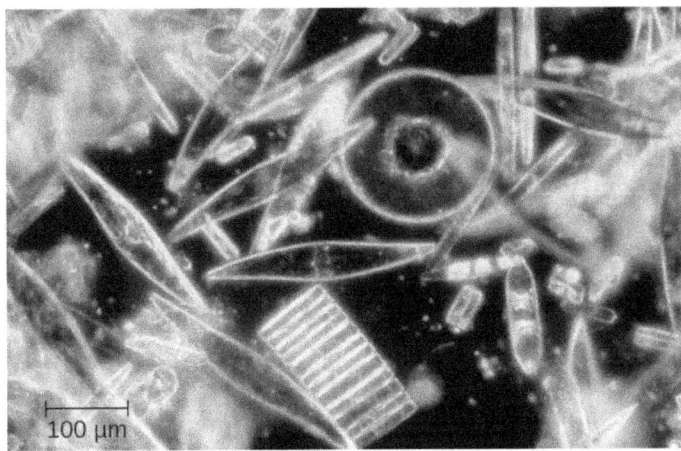

Figure 1.15 Assorted diatoms, a kind of algae, live in annual sea ice in McMurdo Sound, Antarctica. Diatoms range in size from 2 μm to 200 μm and are visualized here using light microscopy. (credit: modification of work by National Oceanic and Atmospheric Administration)

Protozoa (singular: protozoan) are protists that make up the backbone of many food webs by providing nutrients for other organisms. Protozoa are very diverse. Some protozoa move with help from hair-like structures called cilia or whip-like structures called flagella. Others extend part of their cell membrane and cytoplasm to propel themselves forward. These cytoplasmic extensions are called pseudopods ("false feet"). Some protozoa are photosynthetic; others feed on organic material. Some are free-living, whereas others are parasitic, only able to survive by extracting nutrients from a host organism. Most protozoa are harmless, but some are pathogens that can cause disease in animals or humans (Figure 1.16).

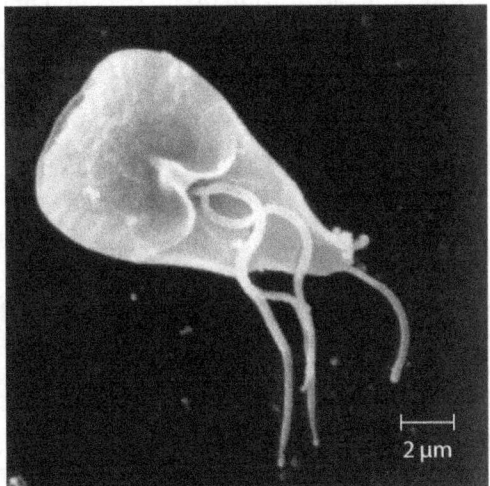

Figure 1.16 *Giardia lamblia*, an intestinal protozoan parasite that infects humans and other mammals, causing severe diarrhea. (credit: modification of work by Centers for Disease Control and Prevention)

Fungi

Fungi (singular: fungus) are also eukaryotes. Some multicellular fungi, such as mushrooms, resemble plants, but they are actually quite different. Fungi are not photosynthetic, and their cell walls are usually made out of chitin rather than cellulose.

Unicellular fungi—yeasts—are included within the study of microbiology. There are more than 1000 known species. Yeasts are found in many different environments, from the deep sea to the human navel. Some yeasts have beneficial uses, such as causing bread to rise and beverages to ferment; but yeasts can also cause food to spoil. Some even cause diseases, such as vaginal yeast infections and oral thrush (Figure 1.17).

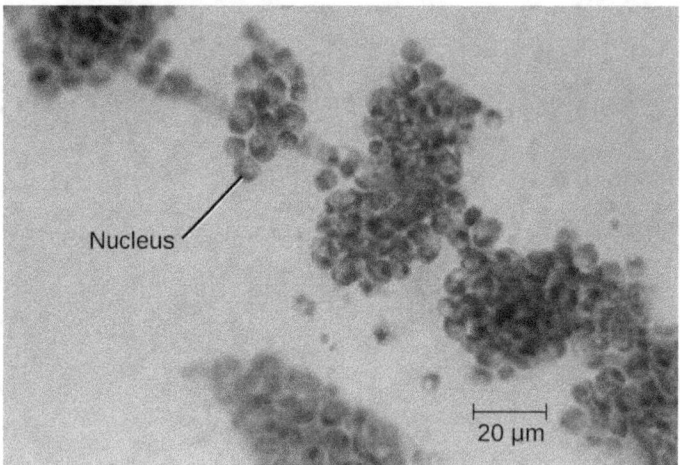

Figure 1.17 *Candida albicans* is a unicellular fungus, or yeast. It is the causative agent of vaginal yeast infections as well as oral thrush, a yeast infection of the mouth that commonly afflicts infants. *C. albicans* has a morphology similar to that of coccus bacteria; however, yeast is a eukaryotic organism (note the nuclei) and is much larger. (credit: modification of work by Centers for Disease Control and Prevention)

Other fungi of interest to microbiologists are multicellular organisms called **molds**. Molds are made up of long filaments that form visible colonies (Figure 1.18). Molds are found in many different environments, from soil to rotting food to dank bathroom corners. Molds play a critical role in the decomposition of dead plants and animals. Some molds can cause allergies, and others produce disease-causing metabolites called mycotoxins. Molds have been used to make pharmaceuticals, including penicillin, which is one of the most commonly prescribed antibiotics, and cyclosporine, used to prevent organ rejection following a transplant.

Figure 1.18 Large colonies of microscopic fungi can often be observed with the naked eye, as seen on the surface of these moldy oranges.

✓ CHECK YOUR UNDERSTANDING

- Name two types of protists and two types of fungi.
- Name some of the defining characteristics of each type.

Helminths

Multicellular parasitic worms called **helminths** are not technically microorganisms, as most are large enough to see without a microscope. However, these worms fall within the field of microbiology because diseases caused by helminths involve microscopic eggs and larvae. One example of a helminth is the guinea worm, or *Dracunculus medinensis*, which causes dizziness, vomiting, diarrhea, and painful ulcers on the legs and feet when the worm works its way out of the skin (Figure 1.19). Infection typically occurs after a person drinks water containing water fleas infected by guinea-worm larvae. In the mid-1980s, there were an estimated 3.5 million cases of guinea-worm disease, but the disease has been largely eradicated. In 2014, there were only 126 cases reported, thanks to the coordinated efforts of the World Health Organization (WHO) and other groups committed to improvements in drinking water sanitation.[11][12]

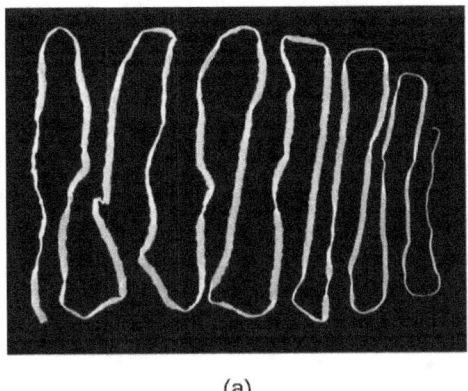 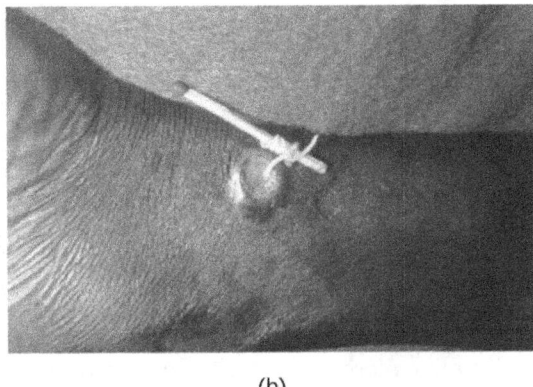

(a) (b)

Figure 1.19 (a) The beef tapeworm, *Taenia saginata*, infects both cattle and humans. *T. saginata* eggs are microscopic (around 50 μm), but adult worms like the one shown here can reach 4–10 m, taking up residence in the digestive system. (b) An adult guinea worm, *Dracunculus medinensis*, is removed through a lesion in the patient's skin by winding it around a matchstick. (credit a, b: modification of work by Centers for Disease Control and Prevention)

Viruses

Viruses are **acellular** microorganisms, which means they are not composed of cells. Essentially, a virus consists of proteins and genetic material—either DNA or RNA, but never both—that are inert outside of a host organism. However, by incorporating themselves into a host cell, viruses are able to co-opt the host's cellular mechanisms to multiply and infect other hosts.

Viruses can infect all types of cells, from human cells to the cells of other microorganisms. In humans, viruses are responsible for numerous diseases, from the common cold to deadly Ebola (Figure 1.20). However, many viruses do not cause disease.

---

[11] C. Greenaway "Dracunculiasis (Guinea Worm Disease)." *Canadian Medical Association Journal* 170 no. 4 (2004):495–500.

[12] World Health Organization. "Dracunculiasis (Guinea-Worm Disease)." *WHO.* 2015. http://www.who.int/mediacentre/factsheets/fs359/en/. Accessed October 2, 2015.

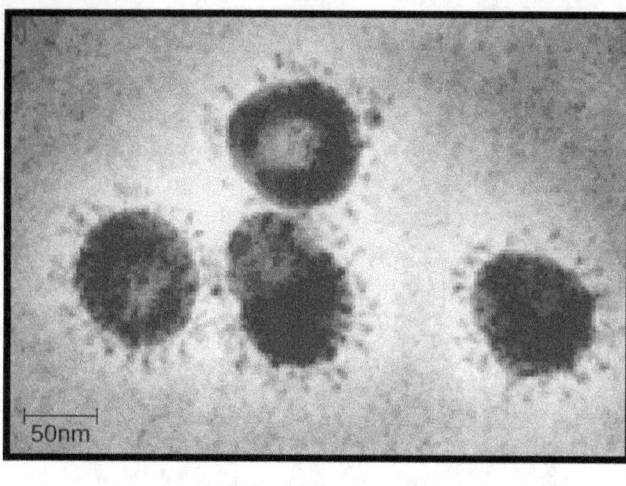 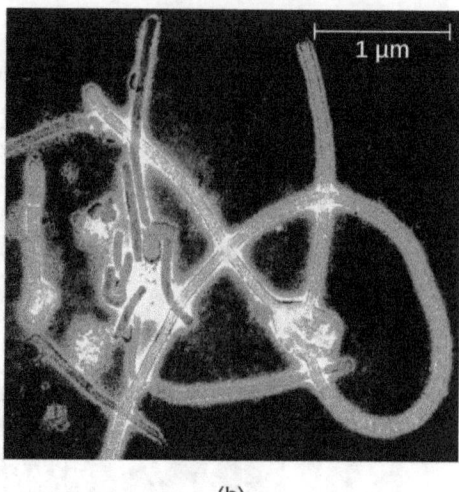

(a) (b)

Figure 1.20 (a) Members of the Coronavirus family can cause respiratory infections like the common cold, severe acute respiratory syndrome (SARS), and Middle East respiratory syndrome (MERS). Here they are viewed under a transmission electron microscope (TEM). (b) Ebolavirus, a member of the Filovirus family, as visualized using a TEM. (credit a: modification of work by Centers for Disease Control and Prevention; credit b: modification of work by Thomas W. Geisbert)

CHECK YOUR UNDERSTANDING

- Are helminths microorganisms? Explain why or why not.
- How are viruses different from other microorganisms?

Microbiology as a Field of Study

Microbiology is a broad term that encompasses the study of all different types of microorganisms. But in practice, microbiologists tend to specialize in one of several subfields. For example, **bacteriology** is the study of bacteria; **mycology** is the study of fungi; **protozoology** is the study of protozoa; **parasitology** is the study of helminths and other parasites; and **virology** is the study of viruses (Figure 1.21). **Immunology**, the study of the immune system, is often included in the study of microbiology because host–pathogen interactions are central to our understanding of infectious disease processes. Microbiologists can also specialize in certain areas of microbiology, such as clinical microbiology, environmental microbiology, applied microbiology, or food microbiology.

In this textbook, we are primarily concerned with clinical applications of microbiology, but since the various subfields of microbiology are highly interrelated, we will often discuss applications that are not strictly clinical.

 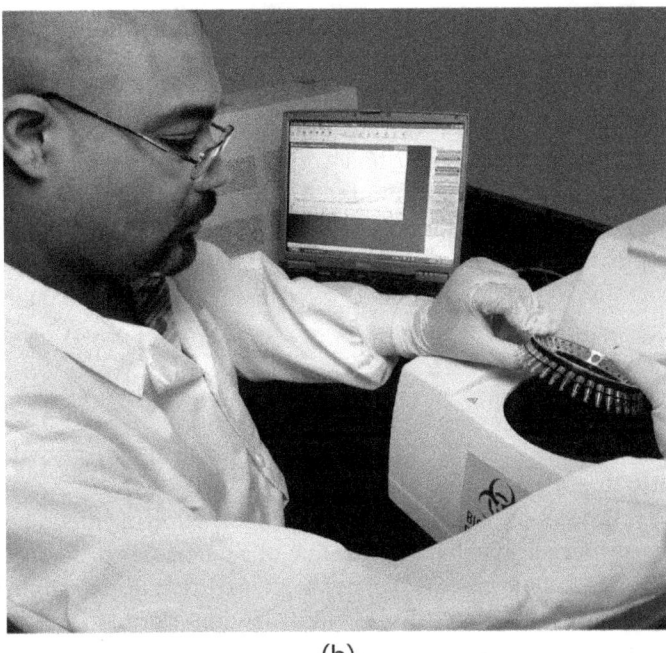

(a) (b)

Figure 1.21 (a) A virologist samples eggs from this nest to be tested for the influenza A virus, which causes avian flu in birds. (b) A biologist performs a procedure to identify an organism that causes ulcerations in humans (credit a: U.S. Fish and Wildlife Service; credit b: James Gathany / CDC; Public Domain)

Eye on Ethics

Bioethics in Microbiology

In the 1940s, the U.S. government was looking for a solution to a medical problem: the prevalence of sexually transmitted diseases (STDs) among soldiers. Several now-infamous government-funded studies used human subjects to research common STDs and treatments. In one such study, American researchers intentionally exposed more than 1300 human subjects in Guatemala to syphilis, gonorrhea, and chancroid to determine the ability of penicillin and other antibiotics to combat these diseases. Subjects of the study included Guatemalan soldiers, prisoners, prostitutes, and psychiatric patients—none of whom were informed that they were taking part in the study. Researchers exposed subjects to STDs by various methods, from facilitating intercourse with infected prostitutes to inoculating subjects with the bacteria known to cause the diseases. This latter method involved making a small wound on the subject's genitals or elsewhere on the body, and then putting bacteria directly into the wound.[13] In 2011, a U.S. government commission tasked with investigating the experiment revealed that only some of the subjects were treated with penicillin, and 83 subjects died by 1953, likely as a result of the study.[14]

Unfortunately, this is one of many horrific examples of microbiology experiments that have violated basic ethical standards. Even if this study had led to a life-saving medical breakthrough (it did not), few would argue that its methods were ethically sound or morally justifiable. But not every case is so clear cut. Professionals working in clinical settings are frequently confronted with ethical dilemmas, such as working with patients who decline a vaccine or life-saving blood transfusion. These are just two examples of life-and-death decisions that may intersect with the religious and philosophical beliefs of both the patient and the health-care professional.

---

13 Kara Rogers. "Guatemala Syphilis Experiment: American Medical Research Project". *Encylopaedia Britannica*. http://www.britannica.com/event/Guatemala-syphilis-experiment. Accessed June 24, 2015.

14 Susan Donaldson James. "Syphilis Experiments Shock, But So Do Third-World Drug Trials." *ABC World News*. August 30, 2011. http://abcnews.go.com/Health/guatemala-syphilis-experiments-shock-us-drug-trials-exploit/story?id=14414902. Accessed June 24, 2015.

No matter how noble the goal, microbiology studies and clinical practice must be guided by a certain set of ethical principles. Studies must be done with integrity. Patients and research subjects provide informed consent (not only agreeing to be treated or studied but demonstrating an understanding of the purpose of the study and any risks involved). Patients' rights must be respected. Procedures must be approved by an institutional review board. When working with patients, accurate record-keeping, honest communication, and confidentiality are paramount. Animals used for research must be treated humanely, and all protocols must be approved by an institutional animal care and use committee. These are just a few of the ethical principles explored in the *Eye on Ethics* boxes throughout this book.

Clinical Focus

Resolution

Cora's CSF samples show no signs of inflammation or infection, as would be expected with a viral infection. However, there is a high concentration of a particular protein, 14-3-3 protein, in her CSF. An electroencephalogram (EEG) of her brain function is also abnormal. The EEG resembles that of a patient with a neurodegenerative disease like Alzheimer's or Huntington's, but Cora's rapid cognitive decline is not consistent with either of these. Instead, her doctor concludes that Cora has Creutzfeldt-Jakob disease (CJD), a type of transmissible spongiform encephalopathy (TSE).

CJD is an extremely rare disease, with only about 300 cases in the United States each year. It is not caused by a bacterium, fungus, or virus, but rather by prions—which do not fit neatly into any particular category of microbe. Like viruses, prions are not found on the tree of life because they are acellular. Prions are extremely small, about one-tenth the size of a typical virus. They contain no genetic material and are composed solely of a type of abnormal protein.

CJD can have several different causes. It can be acquired through exposure to the brain or nervous-system tissue of an infected person or animal. Consuming meat from an infected animal is one way such exposure can occur. There have also been rare cases of exposure to CJD through contact with contaminated surgical equipment[15] and from cornea and growth-hormone donors who unknowingly had CJD.[16][17] In rare cases, the disease results from a specific genetic mutation that can sometimes be hereditary. However, in approximately 85% of patients with CJD, the cause of the disease is spontaneous (or sporadic) and has no identifiable cause.[18] Based on her symptoms and their rapid progression, Cora is diagnosed with sporadic CJD.

Unfortunately for Cora, CJD is a fatal disease for which there is no approved treatment. Approximately 90% of patients die within 1 year of diagnosis.[19] Her doctors focus on limiting her pain and cognitive symptoms as her disease progresses. Eight months later, Cora dies. Her CJD diagnosis is confirmed with a brain autopsy.

Go back to the previous Clinical Focus box.

---

15 Greg Botelho. "Case of Creutzfeldt-Jakob Disease Confirmed in New Hampshire." *CNN*. 2013. http://www.cnn.com/2013/09/20/health/creutzfeldt-jakob-brain-disease/.

16 P. Rudge et al. "Iatrogenic CJD Due to Pituitary-Derived Growth Hormone With Genetically Determined Incubation Times of Up to 40 Years." *Brain* 138 no. 11 (2015): 3386–3399.

17 J.G. Heckmann et al. "Transmission of Creutzfeldt-Jakob Disease via a Corneal Transplant." *Journal of Neurology, Neurosurgery & Psychiatry* 63 no. 3 (1997): 388–390.

18 National Institute of Neurological Disorders and Stroke. "Creutzfeldt-Jakob Disease Fact Sheet." *NIH*. 2015. http://www.ninds.nih.gov/disorders/cjd/detail_cjd.htm#288133058.

19 National Institute of Neurological Disorders and Stroke. "Creutzfeldt-Jakob Disease Fact Sheet." *NIH*. 2015. http://www.ninds.nih.gov/disorders/cjd/detail_cjd.htm#288133058. Accessed June 22, 2015.

SUMMARY

1.1 What Our Ancestors Knew

- **Microorganisms** (or **microbes**) are living organisms that are generally too small to be seen without a microscope.
- Throughout history, humans have used microbes to make fermented foods such as beer, bread, cheese, and wine.
- Long before the invention of the microscope, some people theorized that infection and disease were spread by living things that were too small to be seen. They also correctly intuited certain principles regarding the spread of disease and immunity.
- Antonie van Leeuwenhoek, using a microscope, was the first to actually describe observations of bacteria, in 1675.
- During the Golden Age of Microbiology (1857–1914), microbiologists, including Louis Pasteur and Robert Koch, discovered many new connections between the fields of microbiology and medicine.

1.2 A Systematic Approach

- Carolus Linnaeus developed a taxonomic system for categorizing organisms into related groups.
- **Binomial nomenclature** assigns organisms Latinized scientific names with a genus and species designation.
- A **phylogenetic tree** is a way of showing how different organisms are thought to be related to one another from an evolutionary standpoint.
- The first phylogenetic tree contained kingdoms for plants and animals; Ernst Haeckel proposed adding kingdom for protists.
- Robert Whittaker's tree contained five kingdoms: Animalia, Plantae, Protista, Fungi, and Monera.
- Carl Woese used small subunit ribosomal RNA to create a phylogenetic tree that groups organisms into three domains based on their genetic similarity.
- Bergey's manuals of determinative and systemic bacteriology are the standard references for identifying and classifying bacteria, respectively.
- Bacteria can be identified through biochemical tests, DNA/RNA analysis, and serological testing methods.

1.3 Types of Microorganisms

- Microorganisms are very diverse and are found in all three domains of life: Archaea, Bacteria, and Eukarya.
- **Archaea** and **bacteria** are classified as prokaryotes because they lack a cellular nucleus. Archaea differ from bacteria in evolutionary history, genetics, metabolic pathways, and cell wall and membrane composition.
- Archaea inhabit nearly every environment on earth, but no archaea have been identified as human pathogens.
- **Eukaryotes** studied in microbiology include algae, protozoa, fungi, and helminths.
- **Algae** are plant-like organisms that can be either unicellular or multicellular, and derive energy via photosynthesis.
- **Protozoa** are unicellular organisms with complex cell structures; most are motile.
- Microscopic **fungi** include **molds** and **yeasts**.
- **Helminths** are multicellular parasitic worms. They are included in the field of microbiology because their eggs and larvae are often microscopic.
- **Viruses** are acellular microorganisms that require a host to reproduce.
- The field of microbiology is extremely broad. Microbiologists typically specialize in one of many subfields, but all health professionals need a solid foundation in clinical microbiology.

REVIEW QUESTIONS

Multiple Choice

1. Which of the following foods is NOT made by fermentation?
   A. beer
   B. bread
   C. cheese
   D. orange juice

2. Who is considered the "father of Western medicine"?
   A. Marcus Terentius Varro
   B. Thucydides
   C. Antonie van Leeuwenhoek
   D. Hippocrates

3. Who was the first to observe "animalcules" under the microscope?
   A. Antonie van Leeuwenhoek
   B. Ötzi the Iceman
   C. Marcus Terentius Varro
   D. Robert Koch

4. Who proposed that swamps might harbor tiny, disease-causing animals too small to see?
   A. Thucydides
   B. Marcus Terentius Varro
   C. Hippocrates
   D. Louis Pasteur

5. Which of the following was NOT a kingdom in Linnaeus's taxonomy?
   A. animal
   B. mineral
   C. protist
   D. plant

6. Which of the following is a correct usage of binomial nomenclature?
   A. Homo Sapiens
   B. *homo sapiens*
   C. *Homo sapiens*
   D. *Homo Sapiens*

7. Which scientist proposed adding a kingdom for protists?
   A. Carolus Linnaeus
   B. Carl Woese
   C. Robert Whittaker
   D. Ernst Haeckel

8. Which of the following is NOT a domain in Woese and Fox's phylogenetic tree?
   A. Plantae
   B. Bacteria
   C. Archaea
   D. Eukarya

9. Which of the following is the standard resource for identifying bacteria?
   A. *Systema Naturae*
   B. Bergey's *Manual of Determinative Bacteriology*
   C. Woese and Fox's phylogenetic tree
   D. Haeckel's *General Morphology of Organisms*

10. Which of the following types of microorganisms is photosynthetic?
    A. yeast
    B. virus
    C. helminth
    D. alga

11. Which of the following is a prokaryotic microorganism?
    A. helminth
    B. protozoan
    C. cyanobacterium
    D. mold

12. Which of the following is acellular?
    A. virus
    B. bacterium
    C. fungus
    D. protozoan

13. Which of the following is a type of fungal microorganism?
    A. bacterium
    B. protozoan
    C. alga
    D. yeast

14. Which of the following is not a subfield of microbiology?
    A. bacteriology
    B. botany
    C. clinical microbiology
    D. virology

Fill in the Blank

15. Thucydides is known as the father of _____.

16. Researchers think that Ötzi the Iceman may have been infected with _____ disease.

17. The process by which microbes turn grape juice into wine is called _____.
18. In binomial nomenclature, an organism's scientific name includes its _____ and _____.
19. Whittaker proposed adding the kingdoms _____ and _____ to his phylogenetic tree.
20. _____ are organisms without membrane-bound nuclei.
21. _____ are microorganisms that are not included in phylogenetic trees because they are acellular.
22. A _____ is a disease-causing microorganism.
23. Multicellular parasitic worms studied by microbiologists are called _____.
24. The study of viruses is _____.
25. The cells of prokaryotic organisms lack a _____.

Short Answer

26. What did Thucydides learn by observing the Athenian plague?
27. Why was the invention of the microscope important for microbiology?
28. What are some ways people use microbes?
29. What is a phylogenetic tree?
30. Which of the five kingdoms in Whittaker's phylogenetic tree are prokaryotic, and which are eukaryotic?
31. What molecule did Woese and Fox use to construct their phylogenetic tree?
32. Name some techniques that can be used to identify and differentiate species of bacteria.
33. Describe the differences between bacteria and archaea.
34. Name three structures that various protozoa use for locomotion.
35. Describe the actual and relative sizes of a virus, a bacterium, and a plant or animal cell.

Critical Thinking

36. Explain how the discovery of fermented foods likely benefited our ancestors.
37. What evidence would you use to support this statement: Ancient people thought that disease was transmitted by things they could not see.
38. Why is using binomial nomenclature more useful than using common names?
39. Label the three Domains found on modern phylogenetic trees.

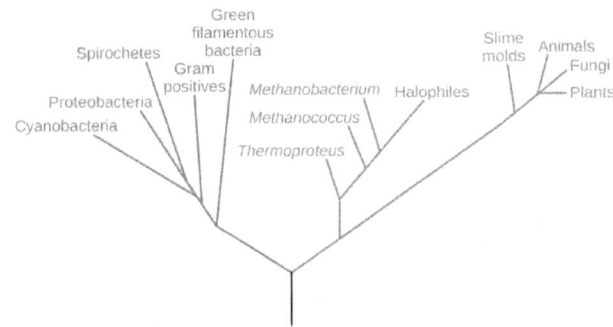

40. Contrast the behavior of a virus outside versus inside a cell.
41. Where would a virus, bacterium, animal cell, and a prion belong on this chart?

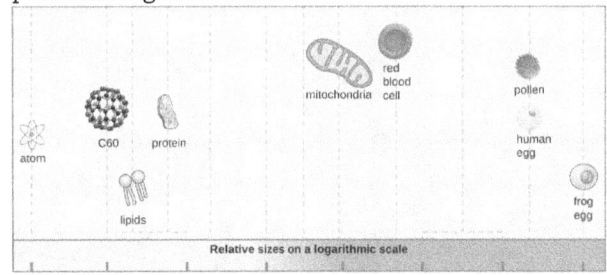

CHAPTER 2
How We See the Invisible World

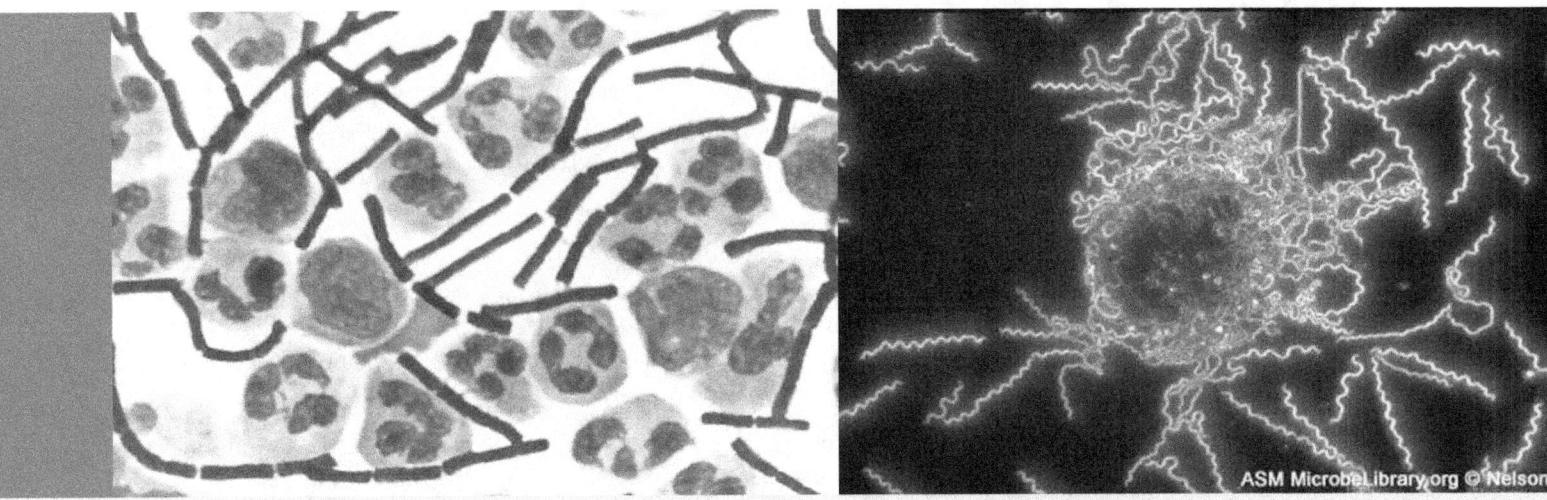

Figure 2.1 Different types of microscopy are used to visualize different structures. Brightfield microscopy (left) renders a darker image on a lighter background, producing a clear image of these *Bacillus anthracis* cells in cerebrospinal fluid (the rod-shaped bacterial cells are surrounded by larger white blood cells). Darkfield microscopy (right) increases contrast, rendering a brighter image on a darker background, as demonstrated by this image of the bacterium *Borrelia burgdorferi*, which causes Lyme disease. (credit left: modification of work by Centers for Disease Control and Prevention; credit right: modification of work by American Society for Microbiology)

Chapter Outline

2.1 The Properties of Light

2.2 Peering Into the Invisible World

2.3 Instruments of Microscopy

2.4 Staining Microscopic Specimens

INTRODUCTION When we look at a rainbow, its colors span the full spectrum of light that the human eye can detect and differentiate. Each hue represents a different frequency of visible light, processed by our eyes and brains and rendered as red, orange, yellow, green, or one of the many other familiar colors that have always been a part of the human experience. But only recently have humans developed an understanding of the properties of light that allow us to see images in color.

Over the past several centuries, we have learned to manipulate light to peer into previously invisible worlds—those too small or too far away to be seen by the naked eye. Through a microscope, we can examine microbial cells, using various techniques to manipulate color, size, and contrast in ways that help us identify species and diagnose disease.

Figure 2.1 illustrates how we can apply the properties of light to visualize and magnify images; but these stunning micrographs are just two examples of the numerous types of images we are now able to produce with different microscopic technologies. This chapter explores how various types of microscopes manipulate light in order to provide a window into the world of microorganisms. By understanding how various kinds of microscopes work, we can produce highly detailed images of microbes that can be useful for both research and clinical applications.

2.1 The Properties of Light

Learning Objectives

By the end of this section, you will be able to:
- Identify and define the characteristics of electromagnetic radiation (EMR) used in microscopy
- Explain how lenses are used in microscopy to manipulate visible and ultraviolet (UV) light

> **Clinical Focus**
>
> **Part 1**
> Cindy, a 17-year-old counselor at a summer sports camp, scraped her knee playing basketball 2 weeks ago. At the time, she thought it was only a minor abrasion that would heal, like many others before it. Instead, the wound began to look like an insect bite and has continued to become increasingly painful and swollen.
>
> The camp nurse examines the lesion and observes a large amount of pus oozing from the surface. Concerned that Cindy may have developed a potentially aggressive infection, she swabs the wound to collect a sample from the infection site. Then she cleans out the pus and dresses the wound, instructing Cindy to keep the area clean and to come back the next day. When Cindy leaves, the nurse sends the sample to the closest medical lab to be analyzed under a microscope.
>
> - What are some things we can learn about these bacteria by looking at them under a microscope?
>
> *Jump to the next Clinical Focus box.*

Visible light consists of electromagnetic waves that behave like other waves. Hence, many of the properties of light that are relevant to microscopy can be understood in terms of light's behavior as a wave. An important property of light waves is the **wavelength**, or the distance between one peak of a wave and the next peak. The height of each peak (or depth of each trough) is called the **amplitude**. In contrast, the **frequency** of the wave is the rate of vibration of the wave, or the number of wavelengths within a specified time period (Figure 2.2).

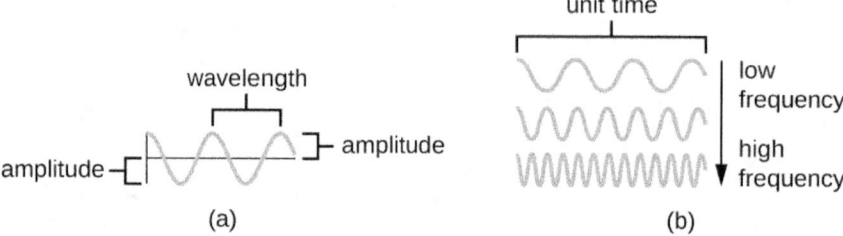

Figure 2.2 (a) The amplitude is the height of a wave, whereas the wavelength is the distance between one peak and the next. (b) These waves have different frequencies, or rates of vibration. The wave at the top has the lowest frequency, since it has the fewest peaks per unit time. The wave at the bottom has the highest frequency.

Interactions of Light

Light waves interact with materials by being reflected, absorbed, or transmitted. **Reflection** occurs when a wave bounces off of a material. For example, a red piece of cloth may reflect red light to our eyes while absorbing other colors of light. **Absorbance** occurs when a material captures the energy of a light wave. In the case of glow-in-the-dark plastics, the energy from light can be absorbed and then later re-emitted as another form of phosphorescence. Transmission occurs when a wave travels through a material, like light through glass (the process of transmission is called **transmittance**). When a material allows a large proportion of light to be transmitted, it may do so because it is thinner, or more transparent (having more **transparency** and less **opacity**). Figure 2.3 illustrates the difference between transparency and opacity.

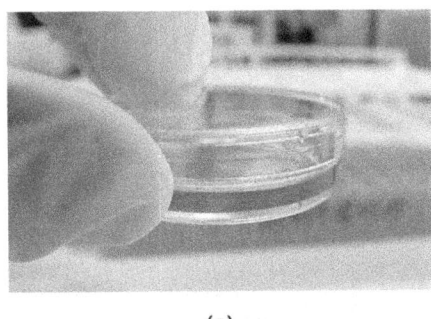 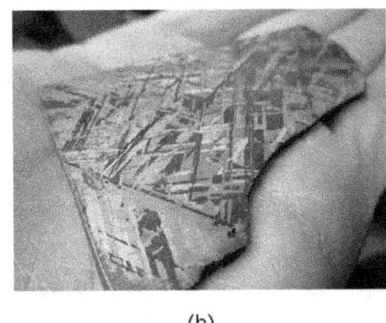

(a)   (b)

Figure 2.3 (a) A Petri dish is made of transparent plastic or glass, which allows transmission of a high proportion of light. This transparency allows us to see through the sides of the dish to view the contents. (b) This slice of an iron meteorite is opaque (i.e., it has opacity). Light is not transmitted through the material, making it impossible to see the part of the hand covered by the object. (credit a: modification of work by Umberto Salvagnin; credit b: modification of work by "Waifer X"/Flickr)

Light waves can also interact with each other by **interference**, creating complex patterns of motion. Dropping two pebbles into a puddle causes the waves on the puddle's surface to interact, creating complex interference patterns. Light waves can interact in the same way.

In addition to interfering with each other, light waves can also interact with small objects or openings by bending or scattering. This is called **diffraction**. Diffraction is larger when the object is smaller relative to the wavelength of the light (the distance between two consecutive peaks of a light wave). Often, when waves diffract in different directions around an obstacle or opening, they will interfere with each other.

✓ CHECK YOUR UNDERSTANDING

- If a light wave has a long wavelength, is it likely to have a low or high frequency?
- If an object is transparent, does it reflect, absorb, or transmit light?

Lenses and Refraction

In the context of microscopy, **refraction** is perhaps the most important behavior exhibited by light waves. Refraction occurs when light waves change direction as they enter a new medium (Figure 2.4). Different transparent materials transmit light at different speeds; thus, light can change speed when passing from one material to another. This change in speed usually also causes a change in direction (refraction), with the degree of change dependent on the angle of the incoming light.

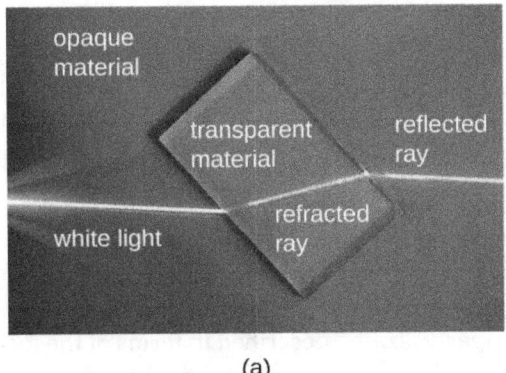 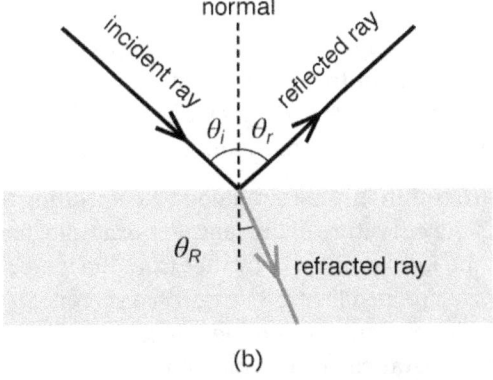

(a)   (b)

Figure 2.4 (a) Refraction occurs when light passes from one medium, such as air, to another, such as glass, changing the direction of the light rays. (b) As shown in this diagram, light rays passing from one medium to another may be either refracted or reflected.

The extent to which a material slows transmission speed relative to empty space is called the **refractive index** of that material. Large differences between the refractive indices of two materials will result in a large amount of refraction when light passes from one material to the other. For example, light moves much more slowly through water than through air, so light entering water from air can change direction greatly. We say that the water has a higher refractive index than air (Figure 2.5).

Figure 2.5   This straight pole appears to bend at an angle as it enters the water. This optical illusion is due to the large difference between the refractive indices of air and water.

When light crosses a boundary into a material with a higher refractive index, its direction turns to be closer to perpendicular to the boundary (i.e., more toward a normal to that boundary; see Figure 2.5). This is the principle behind lenses. We can think of a lens as an object with a curved boundary (or a collection of prisms) that collects all of the light that strikes it and refracts it so that it all meets at a single point called the **image point (focus)**. A convex lens can be used to magnify because it can focus at closer range than the human eye, producing a larger image. Concave lenses and mirrors can also be used in microscopes to redirect the light path. Figure 2.6 shows the **focal point** (the image point when light entering the lens is parallel) and the **focal length** (the distance to the focal point) for convex and concave lenses.

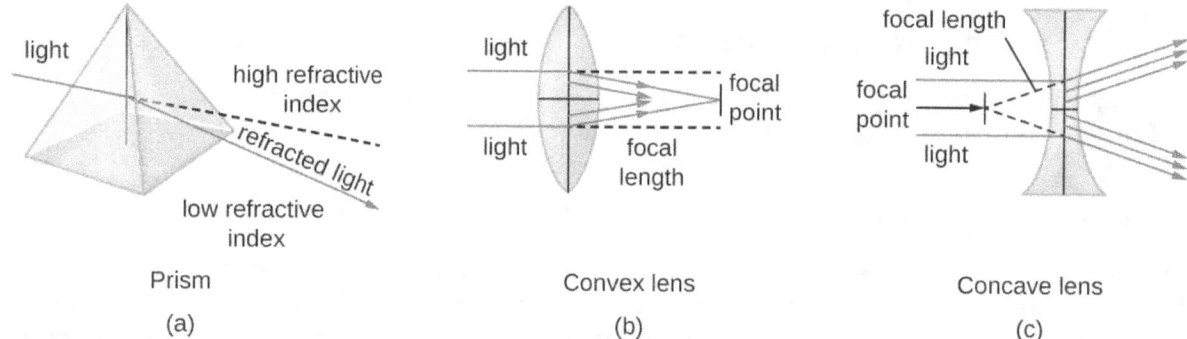

Figure 2.6   (a) A lens is like a collection of prisms, such as the one shown here. (b) When light passes through a convex lens, it is refracted toward a focal point on the other side of the lens. The focal length is the distance to the focal point. (c) Light passing through a concave lens is refracted away from a focal point in front of the lens.

The human eye contains a lens that enables us to see images. This lens focuses the light reflecting off of objects in front of the eye onto the surface of the retina, which is like a screen in the back of the eye. Artificial lenses placed in front of the eye (contact lenses, glasses, or microscopic lenses) focus light before it is focused (again) by the lens of the eye, manipulating the image that ends up on the retina (e.g., by making it appear larger).

Images are commonly manipulated by controlling the distances between the object, the lens, and the screen, as well as the curvature of the lens. For example, for a given amount of curvature, when an object is closer to the lens, the focal points are farther from the lens. As a result, it is often necessary to manipulate these distances to create a focused image on a screen. Similarly, more curvature creates image points closer to the lens and a larger image when the image is in focus. This property is often described in terms of the focal distance, or distance to the focal point.

CHECK YOUR UNDERSTANDING

- Explain how a lens focuses light at the image point.
- Name some factors that affect the focal length of a lens.

Electromagnetic Spectrum and Color

Visible light is just one form of electromagnetic radiation (EMR), a type of energy that is all around us. Other forms of EMR include microwaves, X-rays, and radio waves, among others. The different types of EMR fall on the electromagnetic spectrum, which is defined in terms of wavelength and frequency. The spectrum of visible light occupies a relatively small range of frequencies between infrared and ultraviolet light (Figure 2.7).

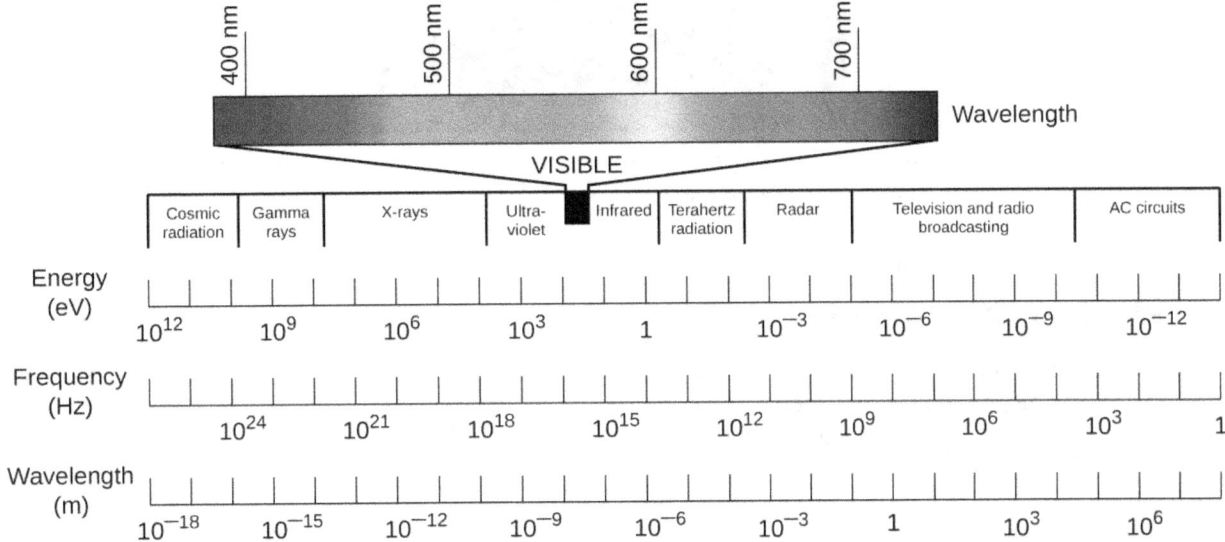

Figure 2.7 The electromagnetic spectrum ranges from high-frequency gamma rays to low-frequency radio waves. Visible light is the relatively small range of electromagnetic frequencies that can be sensed by the human eye. On the electromagnetic spectrum, visible light falls between ultraviolet and infrared light. (credit: modification of work by Johannes Ahlmann)

Whereas wavelength represents the distance between adjacent peaks of a light wave, frequency, in a simplified definition, represents the rate of oscillation. Waves with higher frequencies have shorter wavelengths and, therefore, have more oscillations per unit time than lower-frequency waves. Higher-frequency waves also contain more energy than lower-frequency waves. This energy is delivered as elementary particles called photons. Higher-frequency waves deliver more energetic photons than lower-frequency waves.

Photons with different energies interact differently with the retina. In the spectrum of visible light, each color corresponds to a particular frequency and wavelength (Figure 2.7). The lowest frequency of visible light appears as the color red, whereas the highest appears as the color violet. When the retina receives visible light of many different frequencies, we perceive this as white light. However, white light can be separated into its component colors using refraction. If we pass white light through a prism, different colors will be refracted in different directions, creating a rainbow-like spectrum on a screen behind the prism. This separation of colors is called **dispersion**, and it occurs because, for a given material, the refractive index is different for different frequencies of light.

Certain materials can refract nonvisible forms of EMR and, in effect, transform them into visible light. Certain **fluorescent** dyes, for instance, absorb ultraviolet or blue light and then use the energy to emit photons of a different color, giving off light rather than simply vibrating. This occurs because the energy absorption causes electrons to jump to higher energy states, after which they then almost immediately fall back down to their ground states, emitting specific amounts of energy as photons. Not all of the energy is emitted in a given photon, so the emitted photons will be of lower energy and, thus, of lower frequency than the absorbed ones. Thus, a dye such as Texas red may be excited by blue light, but emit red light; or a dye such as fluorescein isothiocyanate (FITC) may absorb (invisible) high-energy ultraviolet light and emit green light (Figure 2.8). In some materials, the photons may be emitted following a delay after absorption; in this case, the process is called **phosphorescence**. Glow-in-the-dark plastic works by using phosphorescent material.

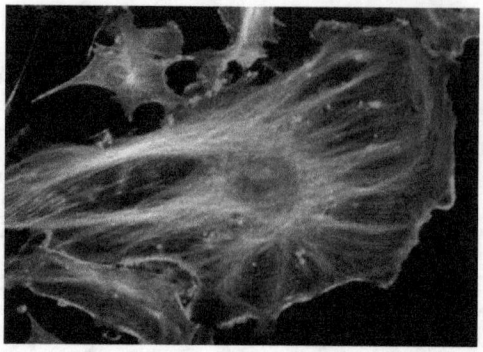

Figure 2.8 The fluorescent dyes absorbed by these bovine pulmonary artery endothelial cells emit brilliant colors when excited by ultraviolet light under a fluorescence microscope. Various cell structures absorb different dyes. The nuclei are stained blue with 4',6-diamidino-2-phenylindole (DAPI); microtubles are marked green by an antibody bound to FITC; and actin filaments are labeled red with phalloidin bound to tetramethylrhodamine (TRITC). (credit: National Institutes of Health)

CHECK YOUR UNDERSTANDING

- Which has a higher frequency: red light or green light?
- Explain why dispersion occurs when white light passes through a prism.
- Why do fluorescent dyes emit a different color of light than they absorb?

Magnification, Resolution, and Contrast

Microscopes magnify images and use the properties of light to create useful images of small objects. **Magnification** is defined as the ability of a lens to enlarge the image of an object when compared to the real object. For example, a magnification of 10× means that the image appears 10 times the size of the object as viewed with the naked eye.

Greater magnification typically improves our ability to see details of small objects, but magnification alone is not sufficient to make the most useful images. It is often useful to enhance the **resolution** of objects: the ability to tell that two separate points or objects are separate. A low-resolution image appears fuzzy, whereas a high-resolution image appears sharp. Two factors affect resolution. The first is wavelength. Shorter wavelengths are able to resolve smaller objects; thus, an electron microscope has a much higher resolution than a light microscope, since it uses an electron beam with a very short wavelength, as opposed to the long-wavelength visible light used by a light microscope. The second factor that affects resolution is **numerical aperture**, which is a measure of a lens's ability to gather light. The higher the numerical aperture, the better the resolution.

LINK TO LEARNING

Read this article (https://www.openstax.org/l/22aperture) to learn more about factors that can increase or decrease the numerical aperture of a lens.

Even when a microscope has high resolution, it can be difficult to distinguish small structures in many specimens because microorganisms are relatively transparent. It is often necessary to increase **contrast** to detect different structures in a specimen. Various types of microscopes use different features of light or electrons to increase contrast—visible differences between the parts of a specimen (see Instruments of Microscopy). Additionally, dyes that bind to some structures but not others can be used to improve the contrast between images of relatively transparent objects (see Staining Microscopic Specimens).

CHECK YOUR UNDERSTANDING

- Explain the difference between magnification and resolution.
- Explain the difference between resolution and contrast.
- Name two factors that affect resolution.

2.2 Peering Into the Invisible World

Learning Objectives

By the end of this section, you will be able to:
- Describe historical developments and individual contributions that led to the invention and development of the microscope
- Compare and contrast the features of simple and compound microscopes

Some of the fundamental characteristics and functions of microscopes can be understood in the context of the history of their use. Italian scholar Girolamo Fracastoro is regarded as the first person to formally postulate that disease was spread by tiny invisible *seminaria*, or "seeds of the contagion." In his book *De Contagione* (1546), he proposed that these seeds could attach themselves to certain objects (which he called *fomes* [cloth]) that supported their transfer from person to person. However, since the technology for seeing such tiny objects did not yet exist, the existence of the *seminaria* remained hypothetical for a little over a century—an invisible world waiting to be revealed.

Early Microscopes

Antonie van Leeuwenhoek, sometimes hailed as "the Father of Microbiology," is typically credited as the first person to have created microscopes powerful enough to view microbes (Figure 2.9). Born in the city of Delft in the Dutch Republic, van Leeuwenhoek began his career selling fabrics. However, he later became interested in lens making (perhaps to look at threads) and his innovative techniques produced microscopes that allowed him to observe microorganisms as no one had before. In 1674, he described his observations of single-celled organisms, whose existence was previously unknown, in a series of letters to the Royal Society of London. His report was initially met with skepticism, but his claims were soon verified and he became something of a celebrity in the scientific community.

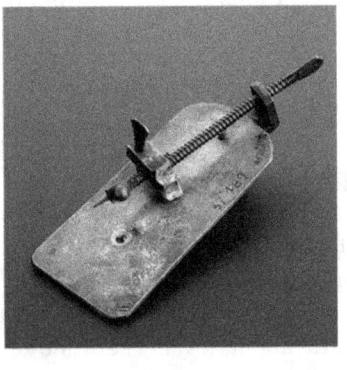

(a) (b) (c)

Figure 2.9 (a) Antonie van Leeuwenhoek (1632–1723) is credited as being the first person to observe microbes, including bacteria, which he called "animalcules" and "wee little beasties." (b) Even though van Leeuwenhoek's microscopes were simple microscopes (as seen in this replica), they were more powerful and provided better resolution than the compound microscopes of his day. (c) Though more famous for developing the telescope, Galileo Galilei (1564–1642) was also one of the pioneers of microscopy. (credit b: modification of work by "Wellcome Images"/Wikimedia Commons)

While van Leeuwenhoek is credited with the discovery of microorganisms, others before him had contributed to the development of the microscope. These included eyeglass makers in the Netherlands in the late 1500s, as well as the Italian astronomer Galileo Galilei, who used a **compound microscope** to examine insect parts (Figure 2.9). Whereas van Leeuwenhoek used a **simple microscope,** in which light is passed through just one lens, Galileo's compound microscope was more sophisticated, passing light through two sets of lenses.

Van Leeuwenhoek's contemporary, the Englishman Robert Hooke (1635–1703), also made important contributions to microscopy, publishing in his book *Micrographia* (1665) many observations using compound microscopes. Viewing a thin sample of cork through his microscope, he was the first to observe the structures that we now know as cells (Figure 2.10). Hooke described these structures as resembling "Honey-comb," and as "small Boxes or Bladders of Air," noting that each "Cavern, Bubble, or Cell" is distinct from the others (in Latin, "cell" literally means "small room"). They likely appeared to Hooke to be filled with air because the cork cells were dead, with only the rigid cell walls providing the structure.

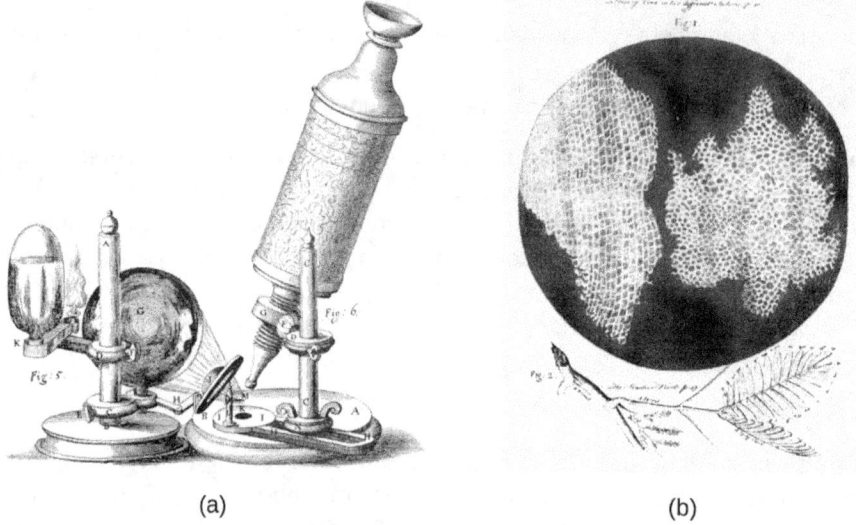

Figure 2.10 Robert Hooke used his (a) compound microscope to view (b) cork cells. Both of these engravings are from his seminal work *Micrographia*, published in 1665.

CHECK YOUR UNDERSTANDING

- Explain the difference between simple and compound microscopes.
- Compare and contrast the contributions of van Leeuwenhoek, Hooke, and Galileo to early microscopy.

MICRO CONNECTIONS

Who Invented the Microscope?

While Antonie van Leeuwenhoek and Robert Hooke generally receive much of the credit for early advances in microscopy, neither can claim to be the inventor of the microscope. Some argue that this designation should belong to Hans and Zaccharias Janssen, Dutch spectacle-makers who may have invented the telescope, the simple microscope, and the compound microscope during the late 1500s or early 1600s (Figure 2.11). Unfortunately, little is known for sure about the Janssens, not even the exact dates of their births and deaths. The Janssens were secretive about their work and never published. It is also possible that the Janssens did not invent anything at all; their neighbor, Hans Lippershey, also developed microscopes and telescopes during the same time frame, and he is often credited with inventing the telescope. The historical records from the time are as fuzzy and imprecise as the images viewed through those early lenses, and any archived records have been lost over the centuries.

By contrast, van Leeuwenhoek and Hooke can thank ample documentation of their work for their respective legacies. Like Janssen, van Leeuwenhoek began his work in obscurity, leaving behind few records. However, his friend, the prominent physician Reinier de Graaf, wrote a letter to the editor of the *Philosophical Transactions of the Royal Society of London* calling attention to van Leeuwenhoek's powerful microscopes. From 1673 onward, van Leeuwenhoek began regularly submitting letters to the Royal Society detailing his observations. In 1674, his report describing single-celled organisms produced controversy in the scientific community, but his observations were soon confirmed when the society sent a delegation to investigate his findings. He subsequently enjoyed considerable celebrity, at one point even entertaining a visit by the czar of Russia.

Similarly, Robert Hooke had his observations using microscopes published by the Royal Society in a book called *Micrographia* in 1665. The book became a bestseller and greatly increased interest in microscopy throughout much of Europe.

Figure 2.11 Zaccharias Janssen, along with his father Hans, may have invented the telescope, the simple microscope, and the compound microscope during the late 1500s or early 1600s. The historical evidence is inconclusive.

2.3 Instruments of Microscopy

Learning Objectives

By the end of this section, you will be able to:
- Identify and describe the parts of a brightfield microscope
- Calculate total magnification for a compound microscope
- Describe the distinguishing features and typical uses for various types of light microscopes, electron microscopes, and scanning probe microscopes

The early pioneers of microscopy opened a window into the invisible world of microorganisms. But microscopy continued to advance in the centuries that followed. In 1830, Joseph Jackson Lister created an essentially modern light microscope. The 20th century saw the development of microscopes that leveraged nonvisible light, such as fluorescence microscopy, which uses an ultraviolet light source, and electron microscopy, which uses short-wavelength electron beams. These advances led to major improvements in magnification, resolution, and contrast. By comparison, the relatively rudimentary microscopes of van Leeuwenhoek and his contemporaries were far less powerful than even the most basic microscopes in use today. In this section, we will survey the broad range of modern microscopic technology and common applications for each type of microscope.

Light Microscopy

Many types of microscopes fall under the category of light microscopes, which use light to visualize images. Examples of light microscopes include brightfield microscopes, darkfield microscopes, phase-contrast microscopes, differential interference contrast microscopes, fluorescence microscopes, confocal scanning laser microscopes, and two-photon microscopes. These various types of light microscopes can be used to complement each other in diagnostics and research.

Brightfield Microscopes

The **brightfield microscope**, perhaps the most commonly used type of microscope, is a compound microscope with two or more lenses that produce a dark image on a bright background. Some brightfield microscopes are **monocular** (having a single eyepiece), though most newer brightfield microscopes are **binocular** (having two eyepieces), like the one shown in Figure 2.12; in either case, each eyepiece contains a lens called an **ocular lens**. The ocular lenses typically magnify images 10 times (10×). At the other end of the body tube are a set of **objective lenses** on a rotating nosepiece. The magnification of these objective lenses typically ranges from 4× to 100×, with the magnification for each lens designated on the metal casing of the lens. The ocular and objective lenses work together to create a magnified image. The **total magnification** is the product of the ocular magnification times the objective magnification:

$$\text{ocular magnification} \times \text{objective magnification}$$

For example, if a 40× objective lens is selected and the ocular lens is 10×, the total magnification would be

$$(40\times)(10\times) = 400\times$$

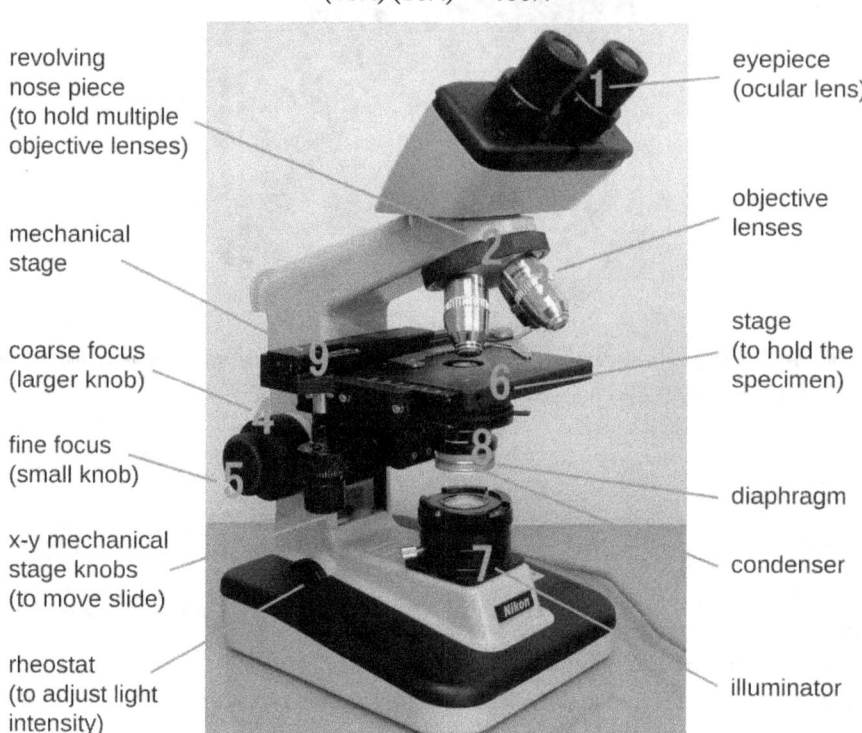

Figure 2.12 Components of a typical brightfield microscope.

The item being viewed is called a specimen. The specimen is placed on a glass slide, which is then clipped into place on the **stage** (a platform) of the microscope. Once the slide is secured, the specimen on the slide is positioned over the light using the **x-y mechanical stage knobs**. These knobs move the slide on the surface of the stage, but do not raise or lower the stage. Once the specimen is centered over the light, the stage position can be raised or lowered to focus the image. The **coarse focusing knob** is used for large-scale movements with 4× and 10× objective lenses; the **fine focusing knob** is used for small-scale movements, especially with 40× or 100× objective lenses.

When images are magnified, they become dimmer because there is less light per unit area of image. Highly magnified images produced by microscopes, therefore, require intense lighting. In a brightfield microscope, this light is provided by an **illuminator**, which is typically a high-intensity bulb below the stage. Light from the illuminator passes up through **condenser lens** (located below the stage), which focuses all of the light rays on the specimen to maximize illumination. The position of the condenser can be optimized using the attached condenser focus knob; once the optimal distance is established, the condenser should not be moved to adjust the brightness. If less-than-maximal light levels are needed, the amount of light striking the specimen can be easily adjusted by opening or closing a **diaphragm** between the condenser and the specimen. In some cases, brightness can also be adjusted using the **rheostat**, a dimmer switch that controls the intensity of the illuminator.

A brightfield microscope creates an image by directing light from the illuminator at the specimen; this light is differentially transmitted, absorbed, reflected, or refracted by different structures. Different colors can behave differently as they interact with **chromophores** (pigments that absorb and reflect particular wavelengths of light) in parts of the specimen. Often, chromophores are artificially added to the specimen using stains, which serve to increase contrast and resolution. In general, structures in the specimen will appear darker, to various extents, than the bright background, creating maximally sharp images at magnifications up to about 1000×. Further magnification would create a larger image, but without increased resolution. This allows us to see objects as small as bacteria, which are visible at about 400× or so, but not smaller objects such as viruses.

At very high magnifications, resolution may be compromised when light passes through the small amount of air between the specimen and the lens. This is due to the large difference between the refractive indices of air and glass; the air scatters the light rays before they can be focused by the lens. To solve this problem, a drop of oil can be used to fill the space between the specimen and an **oil immersion lens**, a special lens designed to be used with immersion oils. Since the oil has a refractive index very similar to that of glass, it increases the maximum angle at which light leaving the specimen can strike the lens. This increases the light collected and, thus, the resolution of the image (Figure 2.13). A variety of oils can be used for different types of light.

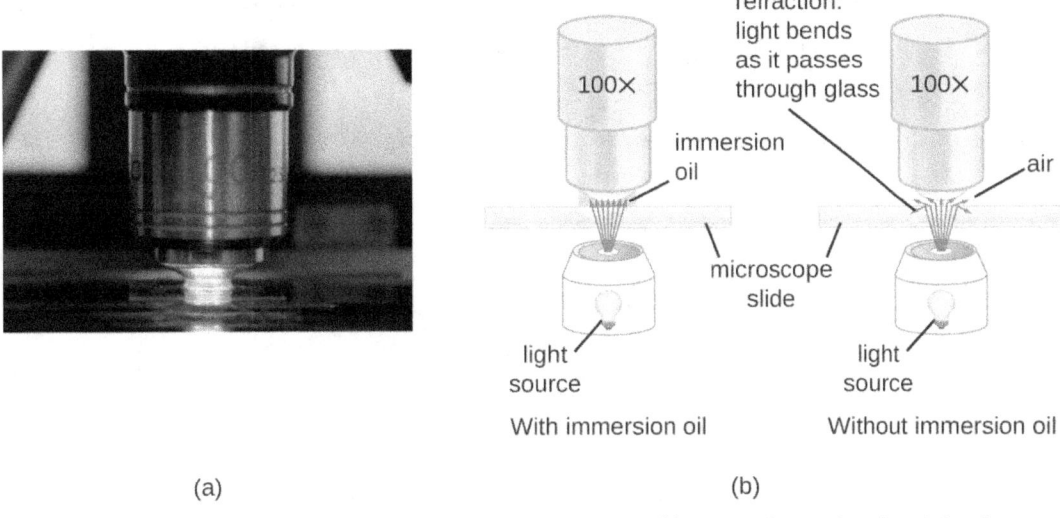

Figure 2.13 (a) Oil immersion lenses like this one are used to improve resolution. (b) Because immersion oil and glass have very similar refractive indices, there is a minimal amount of refraction before the light reaches the lens. Without immersion oil, light scatters as it passes through the air above the slide, degrading the resolution of the image.

 MICRO CONNECTIONS

Microscope Maintenance: Best Practices

Even a very powerful microscope cannot deliver high-resolution images if it is not properly cleaned and maintained. Since lenses are carefully designed and manufactured to refract light with a high degree of precision, even a slightly dirty or scratched lens will refract light in unintended ways, degrading the image of the specimen. In addition, microscopes are rather delicate instruments, and great care must be taken to avoid damaging parts and surfaces. Among other things, proper care of a microscope includes the following:

- cleaning the lenses with lens paper
- not allowing lenses to contact the slide (e.g., by rapidly changing the focus)
- protecting the bulb (if there is one) from breakage
- not pushing an objective into a slide
- not using the coarse focusing knob when using the 40× or greater objective lenses
- only using immersion oil with a specialized oil objective, usually the 100× objective
- cleaning oil from immersion lenses after using the microscope
- cleaning any oil accidentally transferred from other lenses
- covering the microscope or placing it in a cabinet when not in use

LINK TO LEARNING

Visit the online resource linked below for simulations and demonstrations involving the use of microscopes. Keep in mind that execution of specific techniques and procedures can vary depending on the specific instrument you are using. Thus, it is important to learn and practice with an actual microscope in a laboratory setting under expert supervision.

- University of Delaware's Virtual Microscope (https://www.openstax.org/l/22virtualsim)

Darkfield Microscopy

A **darkfield microscope** is a brightfield microscope that has a small but significant modification to the condenser. A small, opaque disk (about 1 cm in diameter) is placed between the illuminator and the condenser lens. This opaque light stop, as the disk is called, blocks most of the light from the illuminator as it passes through the condenser on its way to the objective lens, producing a hollow cone of light that is focused on the specimen. The only light that reaches the objective is light that has been refracted or reflected by structures in the specimen. The resulting image typically shows bright objects on a dark background (Figure 2.14).

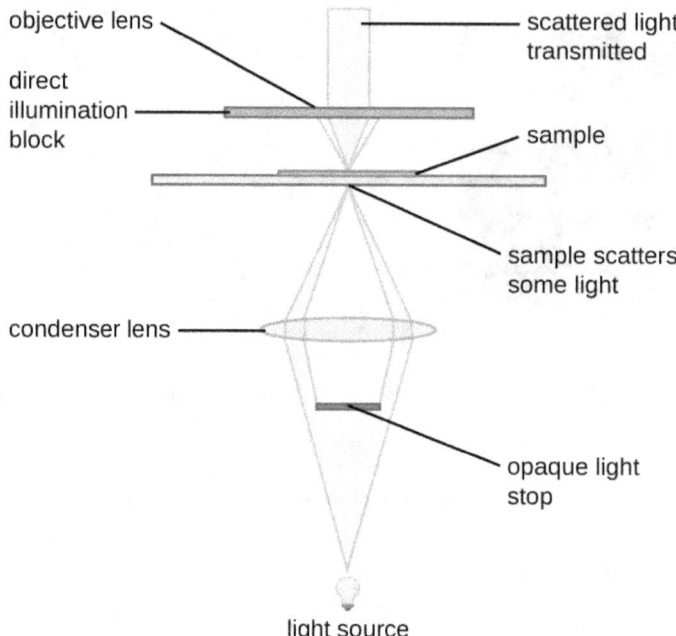

Figure 2.14 An opaque light stop inserted into a brightfield microscope is used to produce a darkfield image. The light stop blocks light traveling directly from the illuminator to the objective lens, allowing only light reflected or refracted off the specimen to reach the eye.

Darkfield microscopy can often create high-contrast, high-resolution images of specimens without the use of stains, which is particularly useful for viewing live specimens that might be killed or otherwise compromised by the stains. For example, thin spirochetes like *Treponema pallidum*, the causative agent of syphilis, can be best viewed using a darkfield microscope (Figure 2.15).

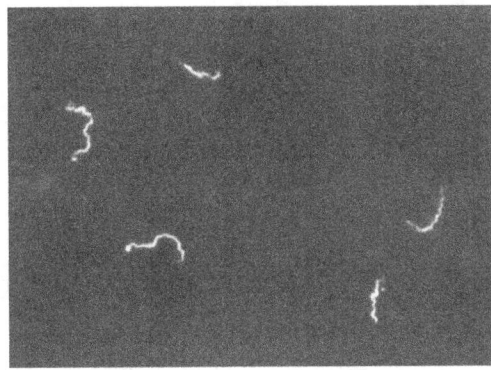

Figure 2.15 Use of a darkfield microscope allows us to view living, unstained samples of the spirochete *Treponema pallidum*. Similar to a photographic negative, the spirochetes appear bright against a dark background. (credit: Centers for Disease Control and Prevention)

CHECK YOUR UNDERSTANDING

- Identify the key differences between brightfield and darkfield microscopy.

Clinical Focus

Part 2

Wound infections like Cindy's can be caused by many different types of bacteria, some of which can spread rapidly with serious complications. Identifying the specific cause is very important to select a medication that can kill or stop the growth of the bacteria.

After calling a local doctor about Cindy's case, the camp nurse sends the sample from the wound to the closest medical laboratory. Unfortunately, since the camp is in a remote area, the nearest lab is small and poorly equipped. A more modern lab would likely use other methods to culture, grow, and identify the bacteria, but in this case, the technician decides to make a wet mount from the specimen and view it under a brightfield microscope. In a wet mount, a small drop of water is added to the slide, and a cover slip is placed over the specimen to keep it in place before it is positioned under the objective lens.

Under the brightfield microscope, the technician can barely see the bacteria cells because they are nearly transparent against the bright background. To increase contrast, the technician inserts an opaque light stop above the illuminator. The resulting darkfield image clearly shows that the bacteria cells are spherical and grouped in clusters, like grapes.

- Why is it important to identify the shape and growth patterns of cells in a specimen?
- What other types of microscopy could be used effectively to view this specimen?

Jump to the next Clinical Focus box. Go back to the previous Clinical Focus box.

Phase-Contrast Microscopes

Phase-contrast microscopes use refraction and interference caused by structures in a specimen to create high-contrast, high-resolution images without staining. It is the oldest and simplest type of microscope that creates an image by altering the wavelengths of light rays passing through the specimen. To create altered wavelength paths, an annular stop is used in the condenser. The annular stop produces a hollow cone of light that is focused on the specimen before reaching the objective lens. The objective contains a phase plate containing a phase ring. As a result, light traveling directly from the illuminator passes through the phase ring while light refracted or reflected by the specimen passes through the plate. This causes waves traveling through the ring to be about one-half of a wavelength out of phase with those passing through the plate. Because waves have peaks and troughs, they can add together (if in phase together) or cancel each other out (if out of phase). When the wavelengths are out of phase, wave troughs will cancel out wave peaks, which is called destructive interference. Structures that refract light then appear dark against a bright background of only unrefracted light. More generally, structures that differ in features such as refractive index will differ in levels of darkness (Figure 2.16).

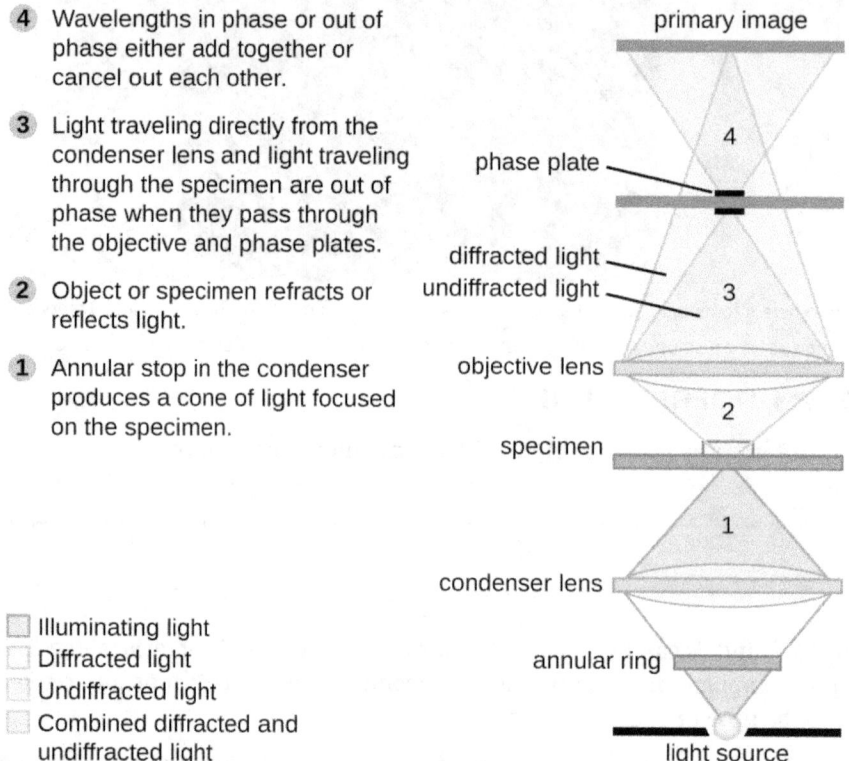

Figure 2.16 This diagram of a phase-contrast microscope illustrates phase differences between light passing through the object and background. These differences are produced by passing the rays through different parts of a phase plate. The light rays are superimposed in the image plane, producing contrast due to their interference.

Because it increases contrast without requiring stains, phase-contrast microscopy is often used to observe live specimens. Certain structures, such as organelles in eukaryotic cells and endospores in prokaryotic cells, are especially well visualized with phase-contrast microscopy (Figure 2.17).

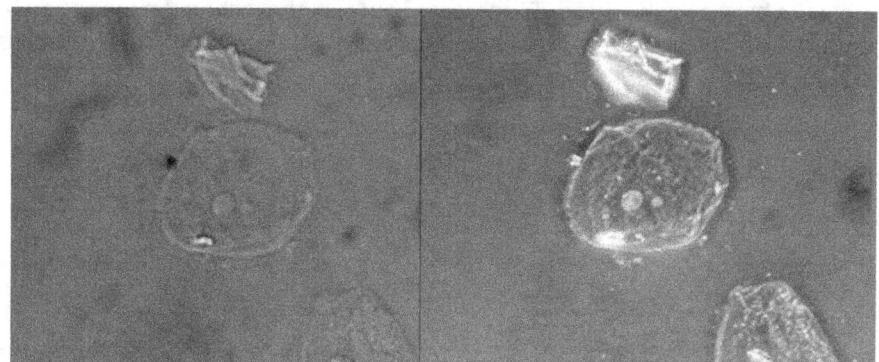

Figure 2.17 This figure compares a brightfield image (left) with a phase-contrast image (right) of the same unstained simple squamous epithelial cells. The cells are in the center and bottom right of each photograph (the irregular item above the cells is acellular debris). Notice that the unstained cells in the brightfield image are almost invisible against the background, whereas the cells in the phase-contrast image appear to glow against the background, revealing far more detail.

Differential Interference Contrast Microscopes

Differential interference contrast (DIC) microscopes (also known as Nomarski optics) are similar to phase-contrast microscopes in that they use interference patterns to enhance contrast between different features of a specimen. In a DIC microscope, two beams of light are created in which the direction of wave movement (polarization) differs. Once the beams pass through either the specimen or specimen-free space, they are recombined and effects of the specimens cause differences in the interference patterns generated by the combining of the beams. This results in high-contrast images of living organisms with a three-dimensional appearance. These microscopes are especially useful in distinguishing structures within live, unstained specimens. (Figure 2.18)

Figure 2.18 A DIC image of *Fonsecaea pedrosoi* grown on modified Leonian's agar. This fungus causes chromoblastomycosis, a chronic skin infection common in tropical and subtropical climates.

CHECK YOUR UNDERSTANDING

- What are some advantages of phase-contrast and DIC microscopy?

Fluorescence Microscopes

A **fluorescence microscope** uses fluorescent chromophores called **fluorochromes**, which are capable of absorbing energy from a light source and then emitting this energy as visible light. Fluorochromes include naturally fluorescent substances (such as chlorophylls) as well as fluorescent stains that are added to the specimen to create contrast. Dyes such as Texas red and FITC are examples of fluorochromes. Other examples include the nucleic acid dyes 4',6'-diamidino-2-phenylindole (DAPI) and acridine orange.

The microscope transmits an excitation light, generally a form of EMR with a short wavelength, such as ultraviolet or blue light, toward the specimen; the chromophores absorb the excitation light and emit visible light with longer wavelengths. The excitation light is then filtered out (in part because ultraviolet light is harmful to the eyes) so that only visible light passes through the ocular lens. This produces an image of the specimen in bright colors against a dark background.

Fluorescence microscopes are especially useful in clinical microbiology. They can be used to identify pathogens, to find particular species within an environment, or to find the locations of particular molecules and structures within a cell. Approaches have also been developed to distinguish living from dead cells using fluorescence microscopy based upon whether they take up particular fluorochromes. Sometimes, multiple fluorochromes are used on the same specimen to show different structures or features.

One of the most important applications of fluorescence microscopy is a technique called **immunofluorescence**, which is used to identify certain disease-causing microbes by observing whether antibodies bind to them. (Antibodies are protein molecules produced by the immune system that attach to specific pathogens to kill or inhibit them.) There are two approaches to this technique: direct immunofluorescence assay (DFA) and indirect immunofluorescence assay (IFA). In DFA, specific antibodies (e.g., those that the target the rabies virus) are stained with a fluorochrome. If the specimen contains the targeted pathogen, one can observe the antibodies binding to the pathogen under the fluorescent microscope. This is called a primary antibody stain because the stained antibodies attach directly to the pathogen.

In IFA, secondary antibodies are stained with a fluorochrome rather than primary antibodies. Secondary antibodies do not attach directly to the pathogen, but they do bind to primary antibodies. When the unstained primary antibodies bind to the pathogen, the fluorescent secondary antibodies can be observed binding to the primary antibodies. Thus, the secondary antibodies are attached indirectly to the pathogen. Since multiple secondary antibodies can often attach to a primary antibody, IFA increases the number of fluorescent antibodies attached to the specimen, making it easier visualize features in the specimen (Figure 2.19).

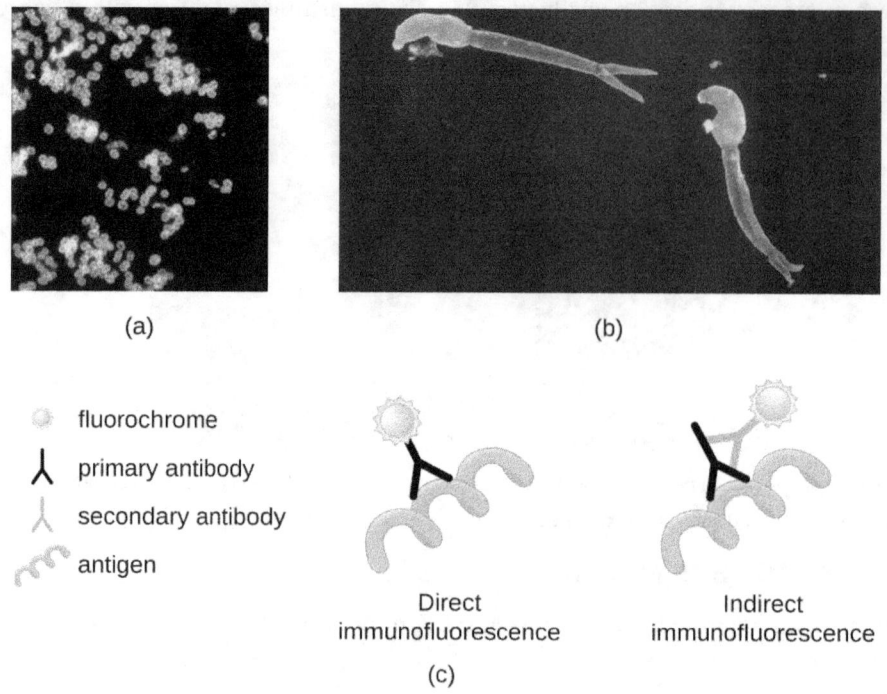

Figure 2.19 (a) A direct immunofluorescent stain is used to visualize *Neisseria gonorrhoeae*, the bacterium that causes gonorrhea. (b) An indirect immunofluorescent stain is used to visualize larvae of *Schistosoma mansoni*, a parasitic worm that causes schistosomiasis, an intestinal disease common in the tropics. (c) In direct immunofluorescence, the stain is absorbed by a primary antibody, which binds to the antigen. In indirect immunofluorescence, the stain is absorbed by a secondary antibody, which binds to a primary antibody, which, in turn, binds to the antigen. (credit a: modification of work by Centers for Disease Control and Prevention; credit b: modification of work by Centers for Disease Control and Prevention)

CHECK YOUR UNDERSTANDING

- Why must fluorochromes be used to examine a specimen under a fluorescence microscope?

Confocal Microscopes

Whereas other forms of light microscopy create an image that is maximally focused at a single distance from the observer (the depth, or z-plane), a **confocal microscope** uses a laser to scan multiple z-planes successively. This produces numerous two-dimensional, high-resolution images at various depths, which can be constructed into a three-dimensional image by a computer. As with fluorescence microscopes, fluorescent stains are generally used to increase contrast and resolution. Image clarity is further enhanced by a narrow aperture that eliminates any light that is not from the z-plane. Confocal microscopes are thus very useful for examining thick specimens such as biofilms, which can be examined alive and unfixed (Figure 2.20).

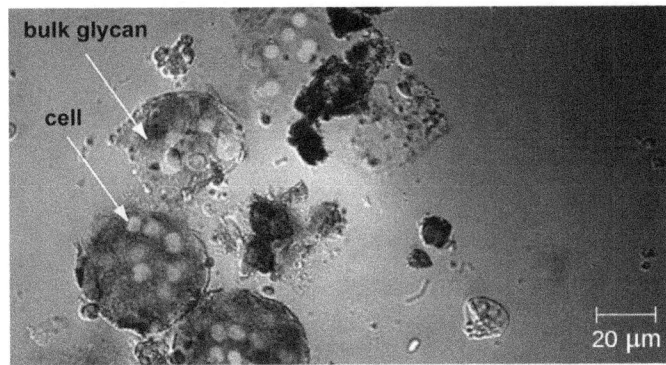

Figure 2.20 Confocal microscopy can be used to visualize structures such as this roof-dwelling cyanobacterium biofilm. (credit: modification of work by American Society for Microbiology)

LINK TO LEARNING

Explore a rotating three-dimensional view (https://www.openstax.org/l/22biofilm3d) of a biofilm as observed under a confocal microscope. After navigating to the webpage, click the "play" button to launch the video.

Two-Photon Microscopes

While the original fluorescent and confocal microscopes allowed better visualization of unique features in specimens, there were still problems that prevented optimum visualization. The effective sensitivity of fluorescence microscopy when viewing thick specimens was generally limited by out-of-focus flare, which resulted in poor resolution. This limitation was greatly reduced in the confocal microscope through the use of a confocal pinhole to reject out-of-focus background fluorescence with thin (<1 µm), unblurred optical sections. However, even the confocal microscopes lacked the resolution needed for viewing thick tissue samples. These problems were resolved with the development of the **two-photon microscope**, which uses a scanning technique, fluorochromes, and long-wavelength light (such as infrared) to visualize specimens. The low energy associated with the long-wavelength light means that two photons must strike a location at the same time to excite the fluorochrome. The low energy of the excitation light is less damaging to cells, and the long wavelength of the excitation light more easily penetrates deep into thick specimens. This makes the two-photon microscope useful for examining living cells within intact tissues—brain slices, embryos, whole organs, and even entire animals.

Currently, use of two-photon microscopes is limited to advanced clinical and research laboratories because of the high costs of the instruments. A single two-photon microscope typically costs between $300,000 and $500,000, and the lasers used to excite the dyes used on specimens are also very expensive. However, as technology improves, two-photon microscopes may become more readily available in clinical settings.

CHECK YOUR UNDERSTANDING

- What types of specimens are best examined using confocal or two-photon microscopy?

Electron Microscopy

The maximum theoretical resolution of images created by light microscopes is ultimately limited by the wavelengths of visible light. Most light microscopes can only magnify 1000×, and a few can magnify up to 1500×, but this does not begin to approach the magnifying power of an **electron microscope (EM)**, which uses short-wavelength electron beams rather than light to increase magnification and resolution.

Electrons, like electromagnetic radiation, can behave as waves, but with wavelengths of 0.005 nm, they can produce much better resolution than visible light. An EM can produce a sharp image that is magnified up to 100,000×. Thus, EMs can resolve subcellular structures as well as some molecular structures (e.g., single strands of DNA); however, electron microscopy cannot be used on living material because of the methods needed to prepare the specimens.

There are two basic types of EM: the **transmission electron microscope (TEM)** and the **scanning electron microscope (SEM)** (Figure 2.21). The TEM is somewhat analogous to the brightfield light microscope in terms of the way it functions. However, it uses an electron beam from above the specimen that is focused using a magnetic lens (rather than a glass lens) and projected through the specimen onto a detector. Electrons pass through the specimen, and then the detector captures the image (Figure 2.22).

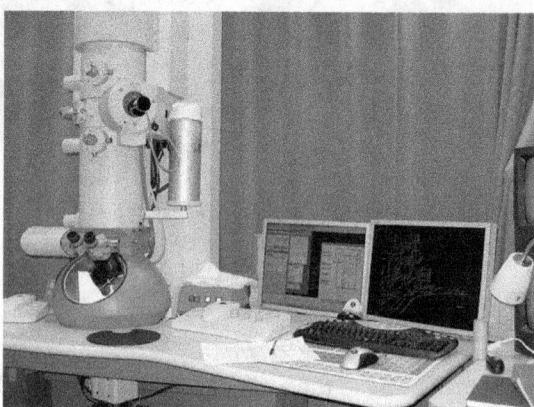

Figure 2.21 A transmission electron microscope (TEM).

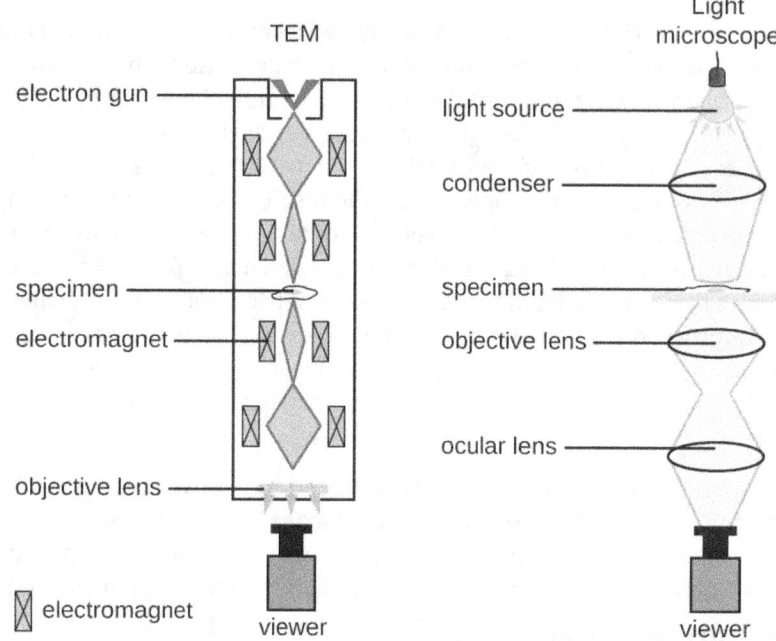

Figure 2.22 Electron microscopes use magnets to focus electron beams similarly to the way that light microscopes use lenses to focus light.

For electrons to pass through the specimen in a TEM, the specimen must be extremely thin (20–100 nm thick). The image is produced because of varying opacity in various parts of the specimen. This opacity can be enhanced by staining the specimen with materials such as heavy metals, which are electron dense. TEM requires that the beam and specimen be in a vacuum and that the specimen be very thin and dehydrated. The specific steps needed to prepare a specimen for observation under an EM are discussed in detail in the next section.

SEMs form images of surfaces of specimens, usually from electrons that are knocked off of specimens by a beam of electrons. This can create highly detailed images with a three-dimensional appearance that are displayed on a monitor (Figure 2.23). Typically, specimens are dried and prepared with fixatives that reduce artifacts, such as shriveling, that can be produced by drying, before being sputter-coated with a thin layer of metal such as gold. Whereas transmission electron microscopy requires very thin sections and allows one to see internal structures such as organelles and the interior of membranes, scanning electron microscopy can be used to view the surfaces of larger objects (such as a pollen grain) as well as the surfaces of very small samples (Figure 2.24). Some EMs can magnify an image up to 2,000,000×.[1]

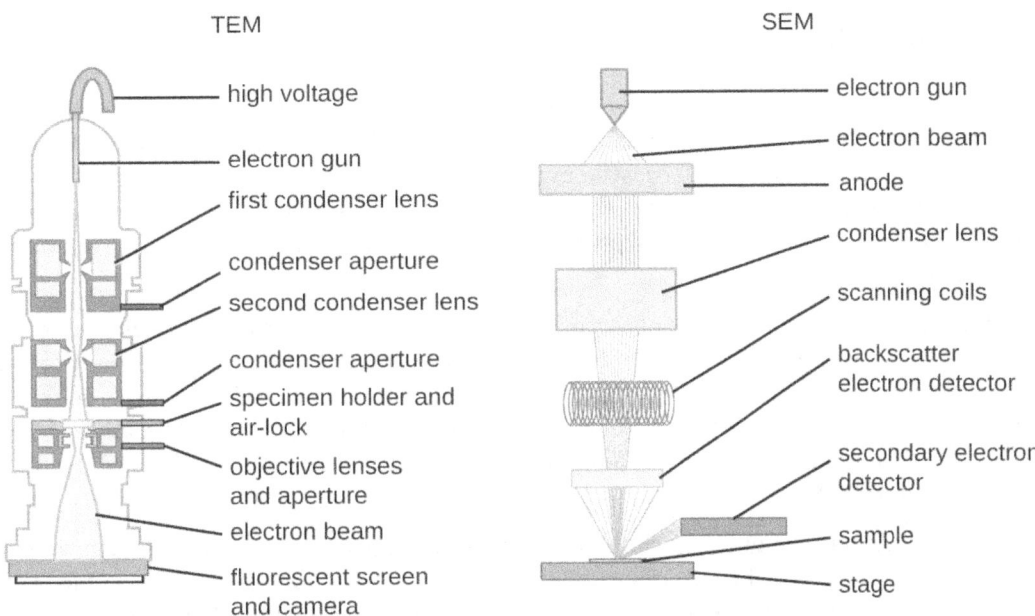

Figure 2.23 These schematic illustrations compare the components of transmission electron microscopes and scanning electron microscopes.

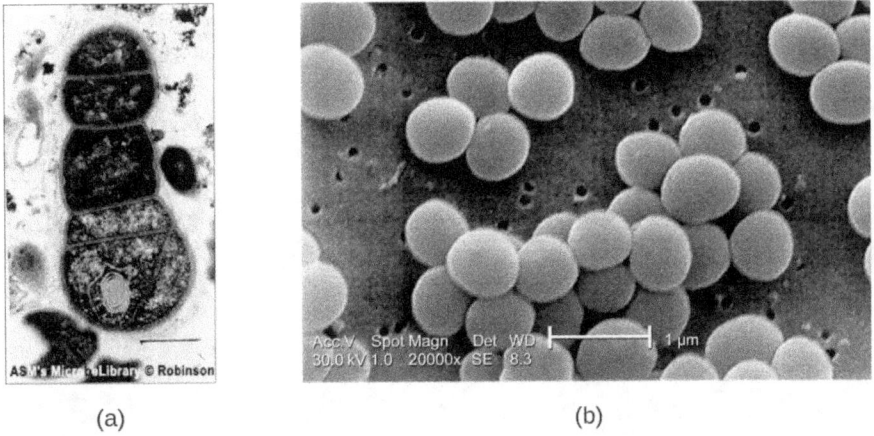

Figure 2.24 (a) This TEM image of cells in a biofilm shows well-defined internal structures of the cells because of varying levels of opacity in the specimen. (b) This color-enhanced SEM image of the bacterium *Staphylococcus aureus* illustrates the ability of scanning electron microscopy to render three-dimensional images of the surface structure of cells. (credit a: modification of work by American Society for Microbiology; credit b: modification of work by Centers for Disease Control and Prevention)

---

1 "JEM-ARM200F Transmission Electron Microscope," *JEOL USA Inc*, http://www.jeolusa.com/PRODUCTS/TransmissionElectronMicroscopes%28TEM%29/200kV/JEM-ARM200F/tabid/663/Default.aspx#195028-specifications. Accessed 8/28/2015.

CHECK YOUR UNDERSTANDING

- What are some advantages and disadvantages of electron microscopy, as opposed to light microscopy, for examining microbiological specimens?
- What kinds of specimens are best examined using TEM? SEM?

MICRO CONNECTIONS

Using Microscopy to Study Biofilms

A biofilm is a complex community of one or more microorganism species, typically forming as a slimy coating attached to a surface because of the production of an extrapolymeric substance (EPS) that attaches to a surface or at the interface between surfaces (e.g., between air and water). In nature, biofilms are abundant and frequently occupy complex niches within ecosystems (Figure 2.25). In medicine, biofilms can coat medical devices and exist within the body. Because they possess unique characteristics, such as increased resistance against the immune system and to antimicrobial drugs, biofilms are of particular interest to microbiologists and clinicians alike.

Because biofilms are thick, they cannot be observed very well using light microscopy; slicing a biofilm to create a thinner specimen might kill or disturb the microbial community. Confocal microscopy provides clearer images of biofilms because it can focus on one z-plane at a time and produce a three-dimensional image of a thick specimen. Fluorescent dyes can be helpful in identifying cells within the matrix. Additionally, techniques such as immunofluorescence and fluorescence in situ hybridization (FISH), in which fluorescent probes are used to bind to DNA, can be used.

Electron microscopy can be used to observe biofilms, but only after dehydrating the specimen, which produces undesirable artifacts and distorts the specimen. In addition to these approaches, it is possible to follow water currents through the shapes (such as cones and mushrooms) of biofilms, using video of the movement of fluorescently coated beads (Figure 2.26).

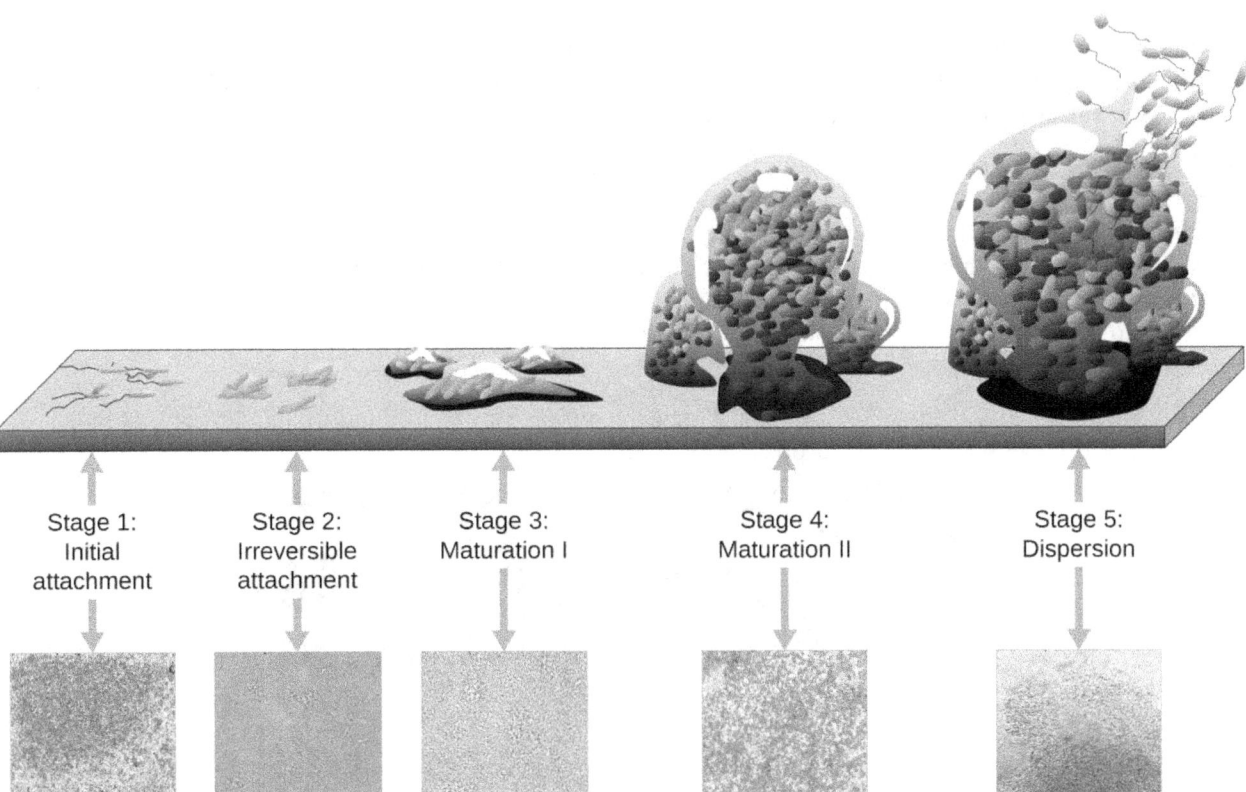

Figure 2.25 A biofilm forms when planktonic (free-floating) bacteria of one or more species adhere to a surface, produce slime, and form a colony. (credit: Public Library of Science)

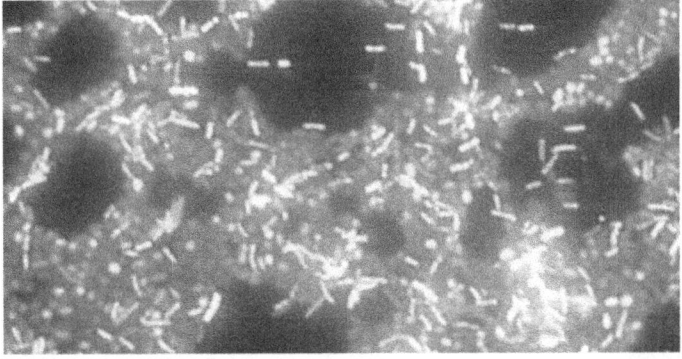

Figure 2.26 In this image, multiple species of bacteria grow in a biofilm on stainless steel (stained with DAPI for epifluorescence microscopy). (credit: Ricardo Murga, Rodney Donlan)

Scanning Probe Microscopy

A **scanning probe microscope** does not use light or electrons, but rather very sharp probes that are passed over the surface of the specimen and interact with it directly. This produces information that can be assembled into images with magnifications up to 100,000,000×. Such large magnifications can be used to observe individual atoms on surfaces. To date, these techniques have been used primarily for research rather than for diagnostics.

There are two types of scanning probe microscope: the **scanning tunneling microscope (STM)** and the **atomic force microscope (AFM)**. An STM uses a probe that is passed just above the specimen as a constant voltage bias creates the potential for an electric current between the probe and the specimen. This current occurs via quantum tunneling of electrons between the probe and the specimen, and the intensity of the current is dependent upon the distance between the probe and the specimen. The probe is moved horizontally above the surface and the intensity of the current is measured. Scanning tunneling microscopy can effectively map the structure of surfaces at a resolution at which individual atoms can be detected.

Similar to an STM, AFMs have a thin probe that is passed just above the specimen. However, rather than measuring variations in the current at a constant height above the specimen, an AFM establishes a constant current and measures variations in the height of the probe tip as it passes over the specimen. As the probe tip is passed over the specimen, forces between the atoms (van der Waals forces, capillary forces, chemical bonding, electrostatic forces, and others) cause it to move up and down. Deflection of the probe tip is determined and measured using Hooke's law of elasticity, and this information is used to construct images of the surface of the specimen with resolution at the atomic level (Figure 2.27).

Figure 2.28, Figure 2.29, and Figure 2.30 summarize the microscopy techniques for light microscopes, electron microscopes, and scanning probe microscopes, respectively.

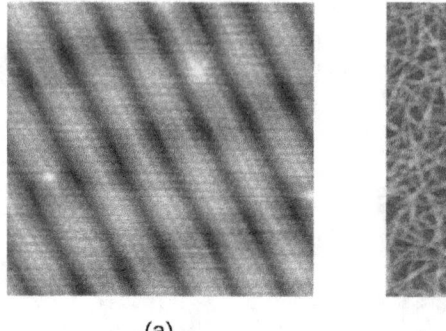
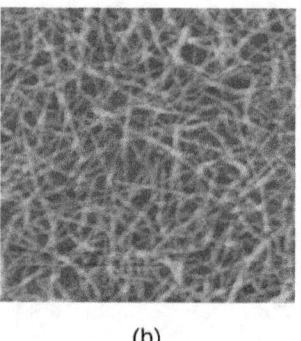

(a) (b)

Figure 2.27 STMs and AFMs allow us to view images at the atomic level. (a) This STM image of a pure gold surface shows individual atoms of gold arranged in columns. (b) This AFM image shows long, strand-like molecules of nanocellulose, a laboratory-created substance derived from plant fibers. (credit a: modification of work by "Erwinrossen"/Wikimedia Commons)

✓ CHECK YOUR UNDERSTANDING

- Which has higher magnification, a light microscope or a scanning probe microscope?
- Name one advantage and one limitation of scanning probe microscopy.

| \multicolumn{3}{l}{**LIGHT MICROSCOPES** — Magnification: up to about 1000× — Use visible or ultraviolet light to produce an image.} |

Microscope Type	Key Uses	Sample Images
Brightfield	Commonly used in a wide variety of laboratory applications as the standard microscope; produces an image on a bright background. **Example:** *Bacillus* sp. showing endospores.	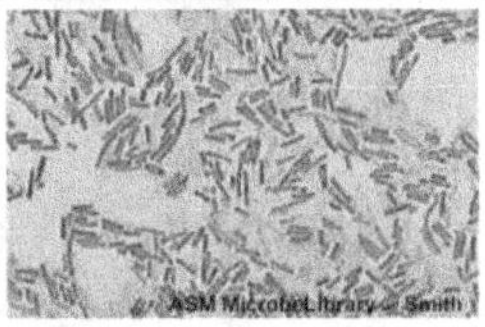
Darkfield	Increases contrast without staining by producing a bright image on a darker background; especially useful for viewing live specimens. **Example:** *Borrelia burgdorferi*	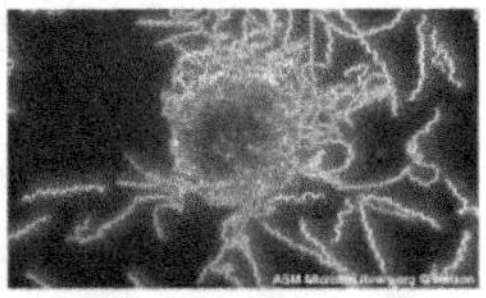
Phase contrast	Uses refraction and interference caused by structures in the specimen to create high-contrast, high-resolution images without staining, making it useful for viewing live specimens, and structures such as endospores and organelles. **Example:** *Pseudomonas* sp.	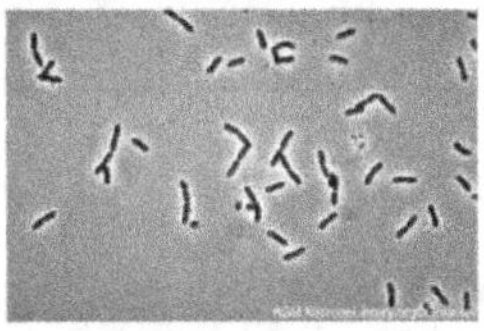
Differential interference contrast (DIC)	Uses interference patterns to enhance contrast between different features of a specimen to produce high-contrast images of living organisms with a three-dimensional appearance, making it especially useful in distinguishing structures within live, unstained specimens; images viewed reveal detailed structures within cells. **Example:** *Escherichia coli* O157:H7	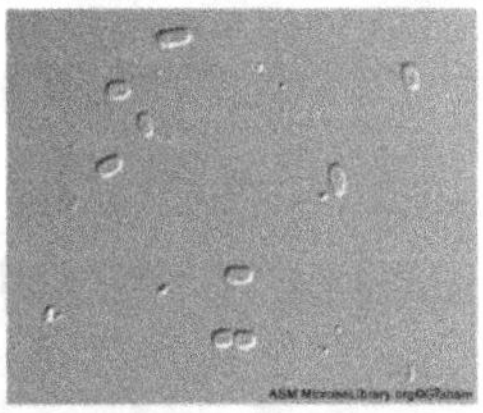
Fluorescence	Uses fluorescent stains to produce an image; can be used to identify pathogens, to find particular species, to distinguish living from dead cells, or to find locations of particular molecules within a cell; also used for immunofluorescence. **Example:** *P. putida* stained with fluorescent dyes to visualize the capsule.	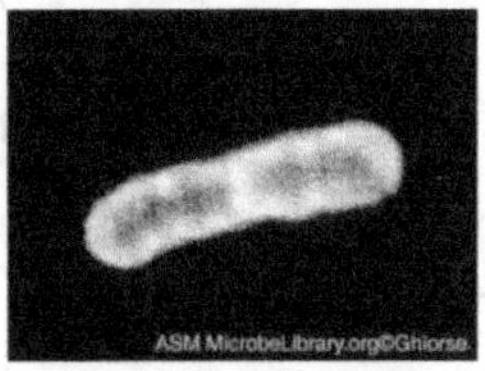
Confocal	Uses a laser to scan mutiple z-planes successively, producing numerous two-dimensional, high-resolution images at various depths that can be constructed into a three-dimensional image by a computer, making this useful for examining thick specimens such as biofilms. **Example:** *Escherichia coli* stained with acridine orange dye to show the nucleoid regions of the cells.	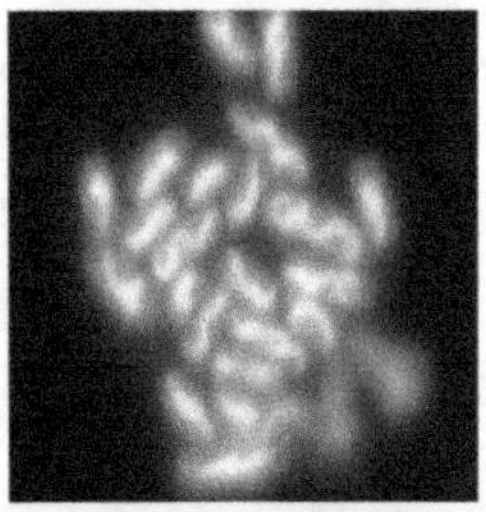
Two-photon	Uses a scanning technique, fluorochromes, and long-wavelength light (such as infrared) to penetrate deep into thick specimens such as biofilms. **Example:** Mouse intestine cells stained with fluorescent dye.	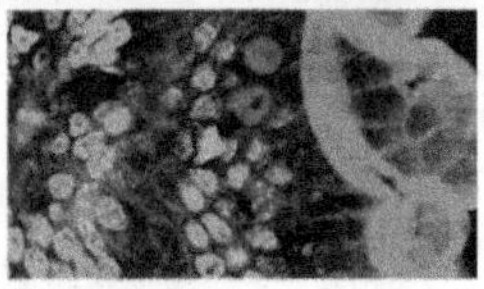

Figure 2.28 (credit "Brightfield": modification of work by American Society for Microbiology; credit "Darkfield": modification of work by American Society for Microbiology; credit "Phase contrast": modification of work by American Society for Microbiology; credit "DIC": modification of work by American Society for Microbiology; credit "Fluorescence": modification of work by American Society for Microbiology; credit "Confocal": modification of work by American Society for Microbiology; credit "Two-photon": modification of work by Alberto Diaspro, Paolo Bianchini, Giuseppe Vicidomini, Mario Faretta, Paola Ramoino, Cesare Usai)

ELECTRON MICROSCOPES	**Magnification: 20–100,000× or more**	
\multicolumn{3}{l}{Use electron beams focused with magnets to produce an image.}		
Microscope Type	**Key Uses**	**Sample Images**
Transmission (TEM)	Uses electron beams that pass through a specimen to visualize small images; useful to observe small, thin specimens such as tissue sections and subcellular structures. **Example:** *Ebola* virus	
Scanning (SEM)	Uses electron beams to visualize surfaces; useful to observe the three-dimensional surface details of specimens. **Example:** *Campylobactor jejuni*	

Figure 2.29 (credit "TEM": modification of work by American Society for Microbiology; credit "SEM": modification of work by American Society for Microbiology)

SCANNING PROBE MICROSCOPES	**Magnification: 100–100,000,000× or more**	
Use very short probes that are passed over the surface of the specimen and interact with it directly.		
Microscope Type	**Key Uses**	**Sample Images**
Scanning tunneling (STM)	Uses a probe passed horizontally at a constant distance just above the specimen while the intensity of the current is measured; can map the structure of surfaces at the atomic level; works best on conducting materials but can also be used to examine organic materials such as DNA, if fixed on a surface. **Example:** Image of surface reconstruction on a clean gold [Au(100)] surface, as visualized using scanning tunneling microscopy.	
Atomic force (AFM)	Can be used in several ways, including using a laser focused on a cantilever to measure the bending of the tip or a probe passed above the specimen while the height needed to maintain a constant current is measured; useful to observe specimens at the atomic level and can be more easily used with nonconducting samples. **Example:** AFM height image of carboxymethylated nanocellulose adsorbed on a silica surface.	

Figure 2.30

2.4 Staining Microscopic Specimens

Learning Objectives

By the end of this section, you will be able to:
- Differentiate between simple and differential stains
- Describe the unique features of commonly used stains
- Explain the procedures and name clinical applications for Gram, endospore, acid-fast, negative capsule, and flagella staining

In their natural state, most of the cells and microorganisms that we observe under the microscope lack color and contrast. This makes it difficult, if not impossible, to detect important cellular structures and their distinguishing characteristics without artificially treating specimens. We have already alluded to certain techniques involving stains and fluorescent dyes, and in this section we will discuss specific techniques for sample preparation in greater detail. Indeed, numerous methods have been developed to identify specific microbes, cellular structures, DNA sequences, or indicators of infection in tissue samples, under the microscope. Here, we will focus on the most clinically relevant techniques.

Preparing Specimens for Light Microscopy

In clinical settings, light microscopes are the most commonly used microscopes. There are two basic types of preparation used to view specimens with a light microscope: wet mounts and fixed specimens.

The simplest type of preparation is the **wet mount**, in which the specimen is placed on the slide in a drop of liquid. Some specimens, such as a drop of urine, are already in a liquid form and can be deposited on the slide using a dropper. Solid specimens, such as a skin scraping, can be placed on the slide before adding a drop of liquid to prepare the wet mount. Sometimes the liquid used is simply water, but often stains are added to enhance contrast. Once the liquid has been added to the slide, a coverslip is placed on top and the specimen is ready for examination under the microscope.

The second method of preparing specimens for light microscopy is **fixation**. The "fixing" of a sample refers to the process of attaching cells to a slide. Fixation is often achieved either by heating (heat fixing) or chemically treating the specimen. In addition to attaching the specimen to the slide, fixation also kills microorganisms in the specimen, stopping their movement and metabolism while preserving the integrity of their cellular components for observation.

To heat-fix a sample, a thin layer of the specimen is spread on the slide (called a **smear**), and the slide is then briefly heated over a heat source (Figure 2.31). Chemical fixatives are often preferable to heat for tissue specimens. Chemical agents such as acetic acid, ethanol, methanol, formaldehyde (formalin), and glutaraldehyde can denature proteins, stop biochemical reactions, and stabilize cell structures in tissue samples (Figure 2.31).

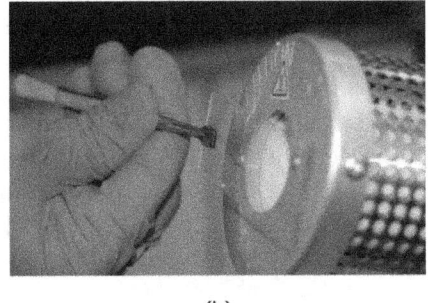

(a)   (b)   (c)

Figure 2.31 (a) A specimen can be heat-fixed by using a slide warmer like this one. (b) Another method for heat-fixing a specimen is to hold a slide with a smear over a microincinerator. (c) This tissue sample is being fixed in a solution of formalin (also known as formaldehyde). Chemical fixation kills microorganisms in the specimen, stopping degradation of the tissues and preserving their structure so that they can be examined later under the microscope. (credit a: modification of work by Nina Parker; credit b: modification of work by Nina Parker; credit c: modification of work by "University of Bristol"/YouTube)

In addition to fixation, **staining** is almost always applied to color certain features of a specimen before examining it under a light microscope. Stains, or dyes, contain salts made up of a positive ion and a negative ion. Depending on the type of dye, the positive or the negative ion may be the chromophore (the colored ion); the other, uncolored ion is called the counterion. If the chromophore is the positively charged ion, the stain is classified as a **basic dye**; if the negative ion is the chromophore, the stain is considered an **acidic dye**.

Dyes are selected for staining based on the chemical properties of the dye and the specimen being observed, which determine how the dye will interact with the specimen. In most cases, it is preferable to use a **positive stain**, a dye that will be absorbed by the cells or organisms being observed, adding color to objects of interest to make them stand out against the background. However, there are scenarios in which it is advantageous to use a **negative stain**, which is absorbed by the background but not by the cells or organisms in the specimen. Negative staining produces an outline or silhouette of the organisms against a colorful background (Figure

2.32).

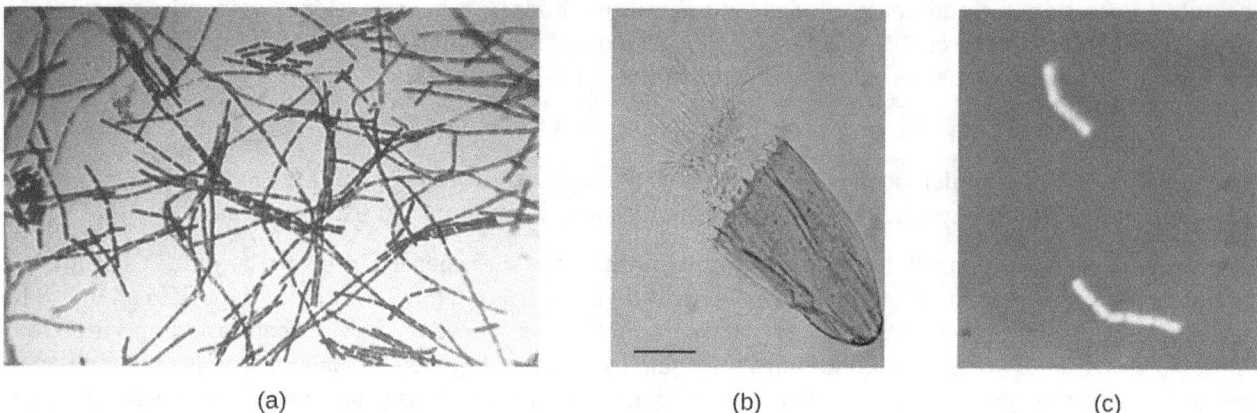

Figure 2.32 (a) These *Bacillus anthracis* cells have absorbed crystal violet, a basic positive stain. (b) This specimen of *Spinoloricus*, a microscopic marine organism, has been stained with rose bengal, a positive acidic stain. (c) These *B. megaterium* appear to be white because they have not absorbed the negative red stain applied to the slide. (credit a: modification of work by Centers for Disease Control and Prevention; credit b: modification of work by Roberto Danovaro, Antonio Pusceddu, Cristina Gambi, Iben Heiner, Reinhardt Mobjerg Kristensen; credit c: modification of work by Anh-Hue Tu)

Because cells typically have negatively charged cell walls, the positive chromophores in basic dyes tend to stick to the cell walls, making them positive stains. Thus, commonly used basic dyes such as basic fuchsin, crystal violet, malachite green, methylene blue, and safranin typically serve as positive stains. On the other hand, the negatively charged chromophores in acidic dyes are repelled by negatively charged cell walls, making them negative stains. Commonly used acidic dyes include acid fuchsin, eosin, and rose bengal. Figure 2.40 provides more detail.

Some staining techniques involve the application of only one dye to the sample; others require more than one dye. In **simple staining**, a single dye is used to emphasize particular structures in the specimen. A simple stain will generally make all of the organisms in a sample appear to be the same color, even if the sample contains more than one type of organism. In contrast, **differential staining** distinguishes organisms based on their interactions with multiple stains. In other words, two organisms in a differentially stained sample may appear to be different colors. Differential staining techniques commonly used in clinical settings include Gram staining, acid-fast staining, endospore staining, flagella staining, and capsule staining. Figure 2.41 provides more detail on these differential staining techniques.

✓ CHECK YOUR UNDERSTANDING

- Explain why it is important to fix a specimen before viewing it under a light microscope.
- What types of specimens should be chemically fixed as opposed to heat-fixed?
- Why might an acidic dye react differently with a given specimen than a basic dye?
- Explain the difference between a positive stain and a negative stain.
- Explain the difference between simple and differential staining.

Gram Staining

The **Gram stain procedure** is a differential staining procedure that involves multiple steps. It was developed by Danish microbiologist Hans Christian Gram in 1884 as an effective method to distinguish between bacteria with different types of cell walls, and even today it remains one of the most frequently used staining techniques. The steps of the Gram stain procedure are listed below and illustrated in Figure 2.33.

1. First, crystal violet, a **primary stain**, is applied to a heat-fixed smear, giving all of the cells a purple color.
2. Next, Gram's iodine, a **mordant**, is added. A mordant is a substance used to set or stabilize stains or dyes; in this case, Gram's iodine acts like a trapping agent that complexes with the crystal violet, making the crystal violet–iodine complex clump and stay contained in thick layers of peptidoglycan in the cell walls.

3. Next, a **decolorizing agent** is added, usually ethanol or an acetone/ethanol solution. Cells that have thick peptidoglycan layers in their cell walls are much less affected by the decolorizing agent; they generally retain the crystal violet dye and remain purple. However, the decolorizing agent more easily washes the dye out of cells with thinner peptidoglycan layers, making them again colorless.
4. Finally, a secondary **counterstain**, usually safranin, is added. This stains the decolorized cells pink and is less noticeable in the cells that still contain the crystal violet dye.

Gram stain process			
Gram staining steps	Cell effects	Gram-positive	Gram-negative
Step 1 **Crystal violet** *primary stain added to specimen smear.*	Stains cells purple or blue.		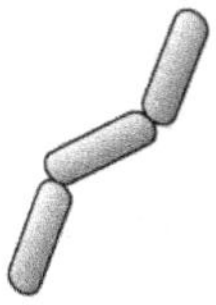
Step 2 **Iodine** *mordant makes dye less soluble so it adheres to cell walls.*	Cells remain purple or blue.		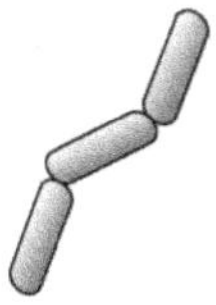
Step 3 **Alcohol** *decolorizer washes away stain from gram-negative cell walls.*	Gram-positive cells remain purple or blue. Gram-negative cells are colorless.		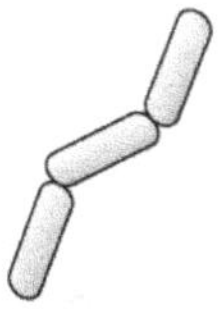
Step 4 **Safranin** *counterstain allows dye adherance to gram-negative cells.*	Gram-positive cells remain purple or blue. Gram-negative cells appear pink or red.		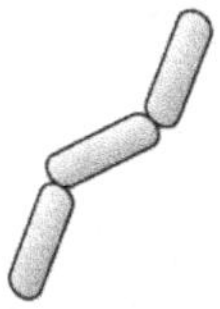

Figure 2.33 Gram-staining is a differential staining technique that uses a primary stain and a secondary counterstain to distinguish between gram-positive and gram-negative bacteria.

The purple, crystal-violet stained cells are referred to as gram-positive cells, while the red, safranin-dyed cells are gram-negative (Figure 2.34). However, there are several important considerations in interpreting the results of a Gram stain. First, older bacterial cells may have damage to their cell walls that causes them to appear gram-negative even if the species is gram-positive. Thus, it is best to use fresh bacterial cultures for Gram staining. Second, errors such as leaving on decolorizer too long can affect the results. In some cases, most cells will appear gram-positive while a few appear gram-negative (as in Figure 2.34). This suggests damage to the individual cells or that decolorizer was left on for too long; the cells should still be classified as gram-positive if they are all the same species rather than a mixed culture.

Besides their differing interactions with dyes and decolorizing agents, the chemical differences between gram-positive and gram-negative cells have other implications with clinical relevance. For example, Gram staining can help clinicians classify bacterial pathogens in a sample into categories associated with specific properties. Gram-negative bacteria tend to be more resistant to certain antibiotics than gram-positive bacteria. We will discuss this and other applications of Gram staining in more detail in later chapters.

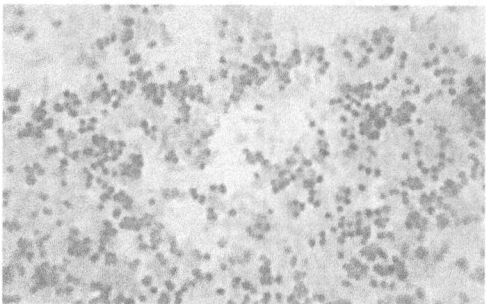

Figure 2.34 In this specimen, the gram-positive bacterium *Staphylococcus aureus* retains crystal violet dye even after the decolorizing agent is added. Gram-negative *Escherichia coli*, the most common Gram stain quality-control bacterium, is decolorized, and is only visible after the addition of the pink counterstain safranin. (credit: modification of work by Nina Parker)

CHECK YOUR UNDERSTANDING

- Explain the role of Gram's iodine in the Gram stain procedure.
- Explain the role of alcohol in the Gram stain procedure.
- What color are gram-positive and gram-negative cells, respectively, after the Gram stain procedure?

Clinical Focus

Part 3

Viewing Cindy's specimen under the darkfield microscope has provided the technician with some important clues about the identity of the microbe causing her infection. However, more information is needed to make a conclusive diagnosis. The technician decides to make a Gram stain of the specimen. This technique is commonly used as an early step in identifying pathogenic bacteria. After completing the Gram stain procedure, the technician views the slide under the brightfield microscope and sees purple, grape-like clusters of spherical cells (Figure 2.35).

- Are these bacteria gram-positive or gram-negative?
- What does this reveal about their cell walls?

Jump to the next Clinical Focus box. Go back to the previous Clinical Focus box.

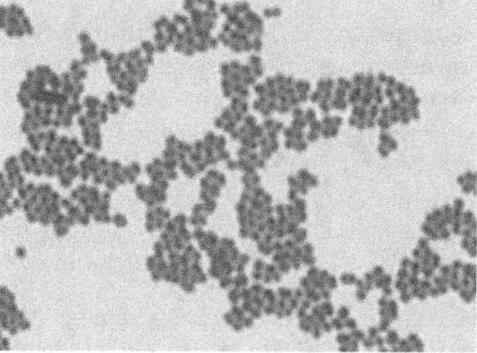

Figure 2.35 (credit: modification of work by American Society for Microbiology)

Acid-Fast Stains

Acid-fast staining is another commonly used, differential staining technique that can be an important diagnostic tool. An **acid-fast stain** is able to differentiate two types of gram-positive cells: those that have waxy mycolic acids in their cell walls, and those that do not. Two different methods for acid-fast staining are the **Ziehl-Neelsen technique** and the **Kinyoun technique**. Both use carbolfuchsin as the primary stain. The waxy, acid-fast cells retain the carbolfuchsin even after a decolorizing agent (an acid-alcohol solution) is applied. A secondary counterstain, methylene blue, is then applied, which renders non–acid-fast cells blue.

The fundamental difference between the two carbolfuchsin-based methods is whether heat is used during the primary staining process. The Ziehl-Neelsen method uses heat to infuse the carbolfuchsin into the acid-fast cells, whereas the Kinyoun method does not use heat. Both techniques are important diagnostic tools because a number of specific diseases are caused by acid-fast bacteria (AFB). If AFB are present in a tissue sample, their red or pink color can be seen clearly against the blue background of the surrounding tissue cells (Figure 2.36).

CHECK YOUR UNDERSTANDING

- Why are acid-fast stains useful?

MICRO CONNECTIONS

Using Microscopy to Diagnose Tuberculosis

Mycobacterium tuberculosis, the bacterium that causes tuberculosis, can be detected in specimens based on the presence of acid-fast bacilli. Often, a smear is prepared from a sample of the patient's sputum and then stained using the Ziehl-Neelsen technique (Figure 2.36). If acid-fast bacteria are confirmed, they are generally cultured to make a positive identification. Variations of this approach can be used as a first step in determining whether *M. tuberculosis* or other acid-fast bacteria are present, though samples from elsewhere in the body (such as urine) may contain other *Mycobacterium* species.

An alternative approach for determining the presence of *M. tuberculosis* is immunofluorescence. In this technique, fluorochrome-labeled antibodies bind to *M. tuberculosis*, if present. Antibody-specific fluorescent dyes can be used to view the mycobacteria with a fluorescence microscope.

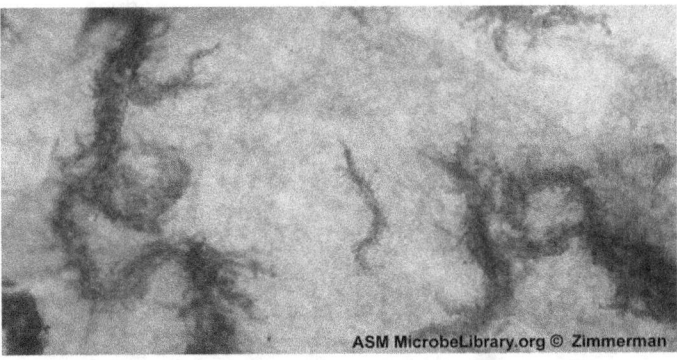

Figure 2.36 Ziehl-Neelsen staining has rendered these *Mycobacterium tuberculosis* cells red and the surrounding growth indicator medium blue. (credit: modification of work by American Society for Microbiology)

Capsule Staining

Certain bacteria and yeasts have a protective outer structure called a capsule. Since the presence of a capsule is directly related to a microbe's virulence (its ability to cause disease), the ability to determine whether cells in a sample have capsules is an important diagnostic tool. Capsules do not absorb most basic dyes; therefore, a negative staining technique (staining around the cells) is typically used for **capsule staining**. The dye stains the background but does not penetrate the capsules, which appear like halos around the borders of the cell. The specimen does not need to be heat-fixed prior to negative staining.

One common negative staining technique for identifying encapsulated yeast and bacteria is to add a few drops of India ink or nigrosin to a specimen. Other capsular stains can also be used to negatively stain encapsulated cells (Figure 2.37). Alternatively, positive and negative staining techniques can be combined to visualize capsules: The positive stain colors the body of the cell, and the negative stain colors the background but not the capsule, leaving halo around each cell.

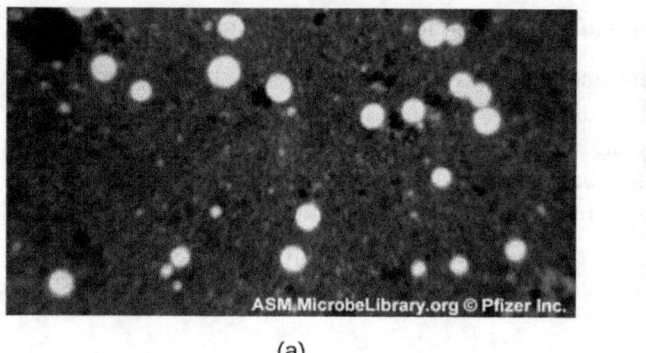 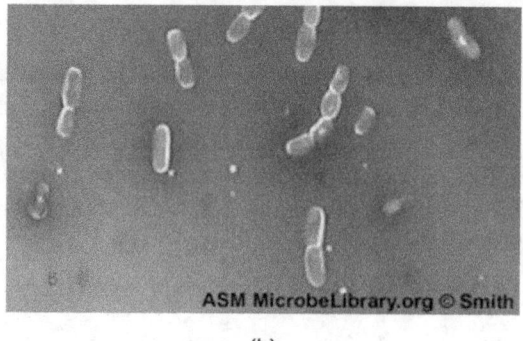

(a)   (b)

Figure 2.37   (a) India-ink was used to stain the background around these cells of the yeast *Cryptococcus neoformans*. The halos surrounding the cells are the polysaccharide capsules. (b) Crystal violet and copper sulfate dyes cannot penetrate the encapsulated *Bacillus* cells in this negatively stained sample. Encapsulated cells appear to have a light-blue halo. (credit a: modification of work by American Society for Microbiology; credit b: modification of work by American Society for Microbiology)

CHECK YOUR UNDERSTANDING

- How does negative staining help us visualize capsules?

Endospore Staining

Endospores are structures produced within certain bacterial cells that allow them to survive harsh conditions. Gram staining alone cannot be used to visualize endospores, which appear clear when Gram-stained cells are viewed. **Endospore staining** uses two stains to differentiate endospores from the rest of the cell. The Schaeffer-Fulton method (the most commonly used endospore-staining technique) uses heat to push the primary stain (malachite green) into the endospore. Washing with water decolorizes the cell, but the endospore retains the green stain. The cell is then counterstained pink with safranin. The resulting image reveals the shape and location of endospores, if they are present. The green endospores will appear either within the pink vegetative cells or as separate from the pink cells altogether. If no endospores are present, then only the pink vegetative cells will be visible (Figure 2.38).

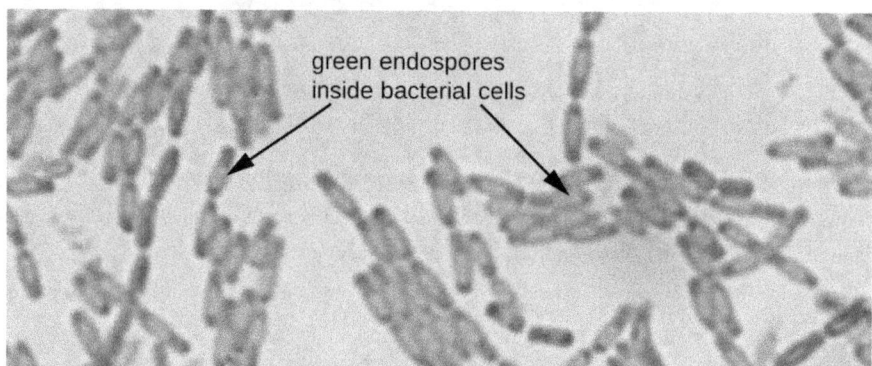

Figure 2.38   A stained preparation of *Bacillus subtilis* showing endospores as green and the vegetative cells as pink. (credit: modification of work by American Society for Microbiology)

Endospore-staining techniques are important for identifying *Bacillus* and *Clostridium*, two genera of endospore-producing bacteria that contain clinically significant species. Among others, *B. anthracis* (which causes anthrax) has been of particular interest because of concern that its spores could be used as a bioterrorism agent. *C. difficile* is a particularly important species responsible for the typically hospital-acquired infection known as "C. diff."

CHECK YOUR UNDERSTANDING

- Is endospore staining an example of positive, negative, or differential staining?

Flagella Staining

Flagella (singular: flagellum) are tail-like cellular structures used for locomotion by some bacteria, archaea, and eukaryotes. Because they are so thin, flagella typically cannot be seen under a light microscope without a specialized **flagella staining** technique. Flagella staining thickens the flagella by first applying mordant (generally tannic acid, but sometimes potassium alum), which coats the flagella; then the specimen is stained with pararosaniline (most commonly) or basic fuchsin (Figure 2.39).

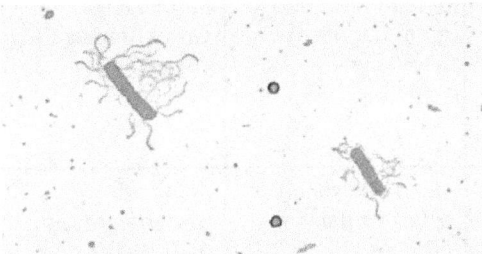

Figure 2.39 A flagella stain of *Bacillus cereus*, a common cause of foodborne illness, reveals that the cells have numerous flagella, used for locomotion. (credit: modification of work by Centers for Disease Control and Prevention)

Though flagella staining is uncommon in clinical settings, the technique is commonly used by microbiologists, since the location and number of flagella can be useful in classifying and identifying bacteria in a sample. When using this technique, it is important to handle the specimen with great care; flagella are delicate structures that can easily be damaged or pulled off, compromising attempts to accurately locate and count the number of flagella.

SIMPLE STAINS				
Stain Type	Specific Dyes	Purpose	Outcome	Sample Images
Basic stains	Methylene blue, crystal violet, malachite green, basic fuchsin, carbolfuchsin, safranin	Stain negatively charged molecules and structures, such as nucleic acids and proteins	Positive stain	
Acidic stains	Eosin, acid fuchsin, rose bengal, Congo red	Stain positively charged molecules and structures, such as proteins	Can be either a positive or negative stain, depending on the cell's chemistry.	
Negative stains	India ink, nigrosin	Stains background, not specimen	Dark background with light specimen	

Figure 2.40   (credit "basic stains": modification of work by Centers for Disease Control and Prevention; credit "Acidic stains": modification of work by Roberto Danovaro, Antonio Dell'Anno, Antonio Pusceddu, Cristina Gambi, Iben Heiner, Reinhardt Mobjerg Kristensen; credit

"Negative stains": modification of work by Anh-Hue Tu)

DIFFERENTIAL STAINS				
Stain Type	Specific Dyes	Purpose	Outcome	Sample Images
Gram stain	Uses crystal violet, Gram's iodine, ethanol (decolorizer), and safranin	Used to distinguish cells by cell-wall type (gram-positive, gram-negative)	Gram-positive cells stain purple/violet. Gram-negative cells stain pink.	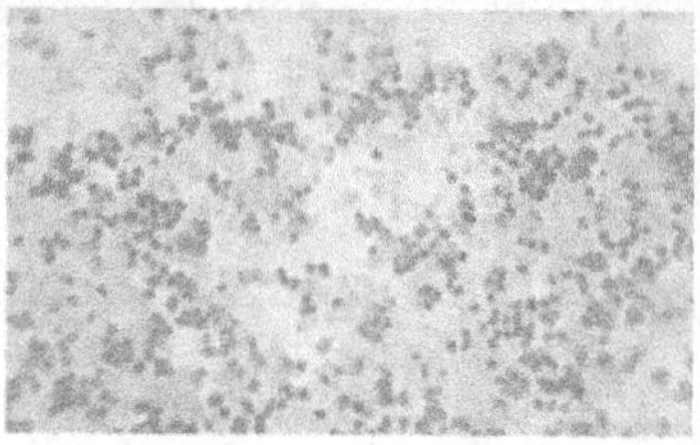
Acid-fast stain	After staining with basic fuchsin, acid-fast bacteria resist decolorization by acid-alcohol. Non acid-fast bacteria are counterstained with methylene blue.	Used to distinguish acid-fast bacteria such as *M. tuberculosis*, from non–acid-fast cells	Acid-fast bacteria are red; non–acid-fast cells are blue.	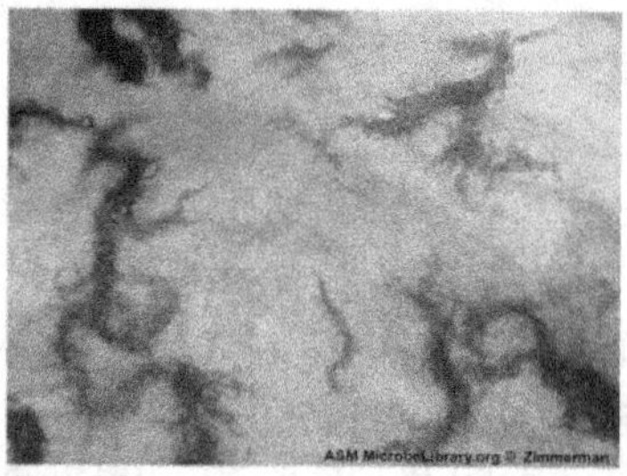
Endospore stain	Uses heat to stain endospores with malachite green (Schaeffer-Fulton procedure), then cell is washed and counterstained with safranin.	Used to distinguish organisms with endospores from those without; used to study the endospore.	Endospores appear bluish-green; other structures appear pink to red.	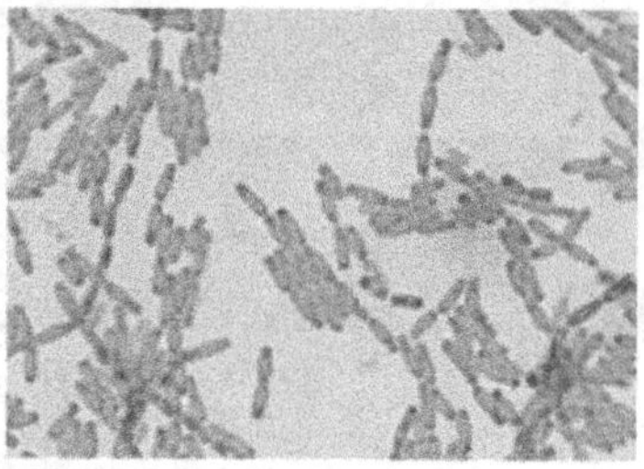
Flagella stain	Flagella are coated with a tannic acid or potassium alum mordant, then stained using either pararosaline or basic fuchsin.	Used to view and study flagella in bacteria that have them.	Flagella are visible if present.	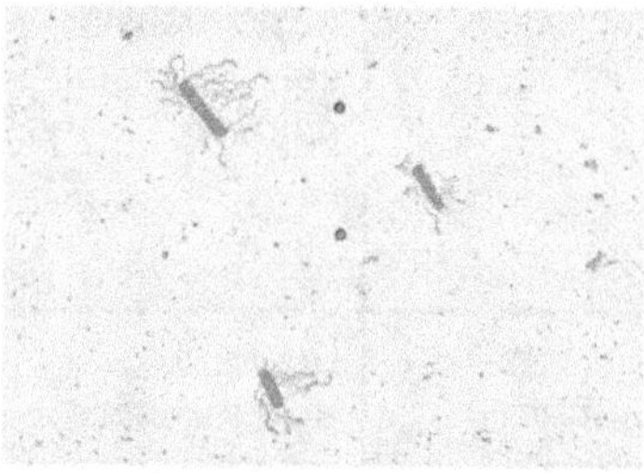
Capsule stain	Negative staining with India ink or nigrosin is used to stain the background, leaving a clear area of the cell and the capsule. Counterstaining can be used to stain the cell while leaving the capsule clear.	Used to distinguish cells with capsules from those without.	Capsules appear clear or as halos if present.	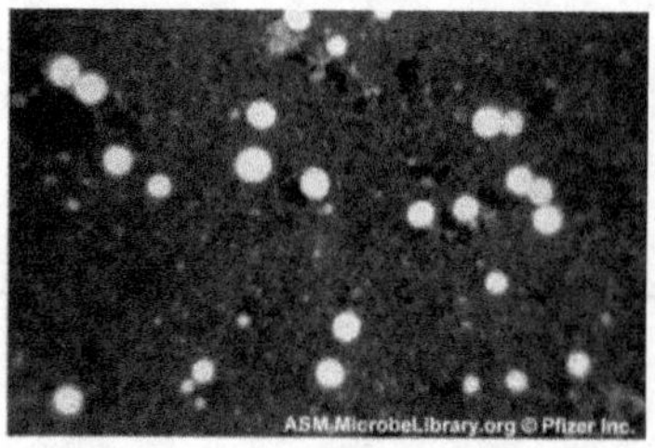

Figure 2.41 (credit "Gram stain": modification of work by Nina Parker; credit "Acid-fast stain": modification of work by American Society for Microbiology; credit "Endospore stain": modification of work by American Society for Microbiology; credit "Capsule stain" : modification of work by American Society for Microbiology; credit "Flagella stain": modification of work by Centers for Disease Control and Prevention)

Preparing Specimens for Electron Microscopy

Samples to be analyzed using a TEM must have very thin sections. But cells are too soft to cut thinly, even with diamond knives. To cut cells without damage, the cells must be embedded in plastic resin and then dehydrated through a series of soaks in ethanol solutions (50%, 60%, 70%, and so on). The ethanol replaces the water in the cells, and the resin dissolves in ethanol and enters the cell, where it solidifies. Next, **thin sections** are cut using a specialized device called an **ultramicrotome** (Figure 2.42). Finally, samples are fixed to fine copper wire or carbon-fiber grids and stained—not with colored dyes, but with substances like uranyl acetate or osmium tetroxide, which contain electron-dense heavy metal atoms.

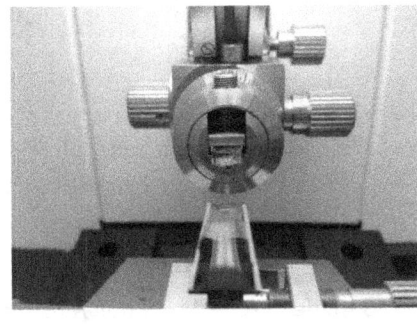
(a)

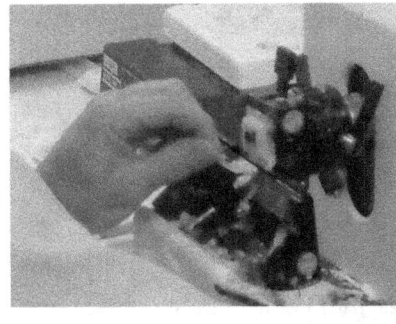
(b)

Figure 2.42 (a) An ultramicrotome used to prepare specimens for a TEM. (b) A technician uses an ultramicrotome to slice a specimen into thin sections. (credit a: modification of work by "Frost Museum"/Flickr; credit b: modification of work by U.S. Fish and Wildlife Service Northeast Region)

When samples are prepared for viewing using an SEM, they must also be dehydrated using an ethanol series. However, they must be even drier than is necessary for a TEM. Critical point drying with inert liquid carbon dioxide under pressure is used to displace the water from the specimen. After drying, the specimens are sputter-coated with metal by knocking atoms off of a palladium target, with energetic particles. Sputter-coating prevents specimens from becoming charged by the SEM's electron beam.

⊘ CHECK YOUR UNDERSTANDING

- Why is it important to dehydrate cells before examining them under an electron microscope?
- Name the device that is used to create thin sections of specimens for electron microscopy.

 MICRO CONNECTIONS

Using Microscopy to Diagnose Syphilis

The causative agent of syphilis is *Treponema pallidum*, a flexible, spiral cell (spirochete) that can be very thin (<0.15 μm) and match the refractive index of the medium, making it difficult to view using brightfield microscopy. Additionally, this species has not been successfully cultured in the laboratory on an artificial medium; therefore, diagnosis depends upon successful identification using microscopic techniques and serology (analysis of body fluids, often looking for antibodies to a pathogen). Since fixation and staining would kill the cells, darkfield microscopy is typically used for observing live specimens and viewing their movements. However, other approaches can also be used. For example, the cells can be thickened with silver particles (in tissue sections) and observed using a light microscope. It is also possible to use fluorescence or electron microscopy to view *Treponema* (Figure 2.43).

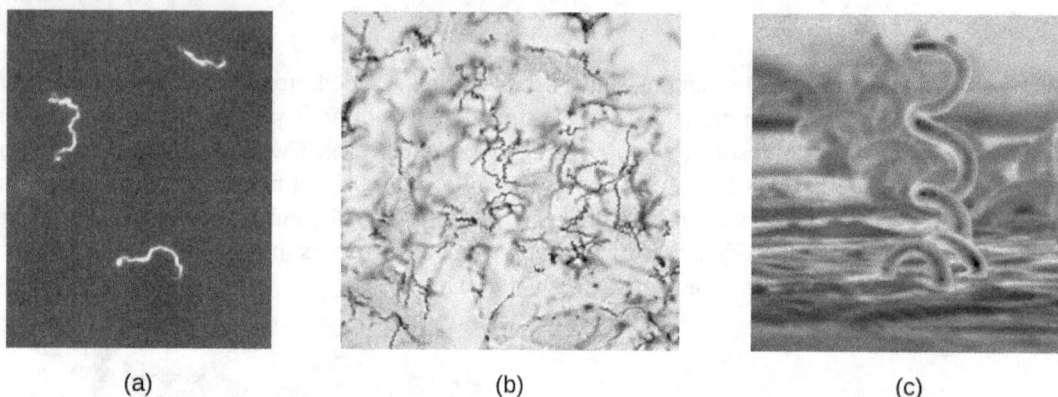

Figure 2.43 (a) Living, unstained *Treponema pallidum* spirochetes can be viewed under a darkfield microscope. (b) In this brightfield image, a modified Steiner silver stain is used to visualized *T. pallidum* spirochetes. Though the stain kills the cells, it increases the contrast to make them more visible. (c) While not used for standard diagnostic testing, *T. pallidum* can also be examined using scanning electron microscopy. (credit a: modification of work by Centers for Disease Control and Prevention; credit b: modification of work by Centers for Disease Control and Prevention; credit c: modification of work by Centers for Disease Control and Prevention)

In clinical settings, indirect immunofluorescence is often used to identify *Treponema*. A primary, unstained antibody attaches directly to the pathogen surface, and secondary antibodies "tagged" with a fluorescent stain attach to the primary antibody. Multiple secondary antibodies can attach to each primary antibody, amplifying the amount of stain attached to each *Treponema* cell, making them easier to spot (Figure 2.44).

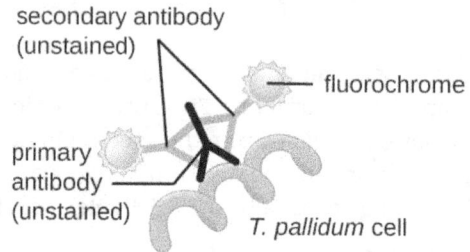

Figure 2.44 Indirect immunofluorescence can be used to identify *T. pallidum*, the causative agent of syphilis, in a specimen.

Preparation and Staining for Other Microscopes

Samples for fluorescence and confocal microscopy are prepared similarly to samples for light microscopy, except that the dyes are fluorochromes. Stains are often diluted in liquid before applying to the slide. Some dyes attach to an antibody to stain specific proteins on specific types of cells (immunofluorescence); others may attach to DNA molecules in a process called fluorescence in situ hybridization (FISH), causing cells to be stained based on whether they have a specific DNA sequence.

Sample preparation for two-photon microscopy is similar to fluorescence microscopy, except for the use of infrared dyes. Specimens for STM need to be on a very clean and atomically smooth surface. They are often mica coated with Au(111). Toluene vapor is a common fixative.

CHECK YOUR UNDERSTANDING

- What is the main difference between preparing a sample for fluorescence microscopy versus light microscopy?

LINK TO LEARNING

Cornell University's Case Studies in Microscopy (https://www.openstax.org/l/22cornellstud) offers a series of clinical problems based on real-life events. Each case study walks you through a clinical problem using appropriate techniques in microscopy at each step.

Clinical Focus

Resolution

From the results of the Gram stain, the technician now knows that Cindy's infection is caused by spherical, gram-positive bacteria that form grape-like clusters, which is typical of staphylococcal bacteria. After some additional testing, the technician determines that these bacteria are the medically important species known as *Staphylococcus aureus*, a common culprit in wound infections. Because some strains of *S. aureus* are resistant to many antibiotics, skin infections may spread to other areas of the body and become serious, sometimes even resulting in amputations or death if the correct antibiotics are not used.

After testing several antibiotics, the lab is able to identify one that is effective against this particular strain of *S. aureus*. Cindy's doctor quickly prescribes the medication and emphasizes the importance of taking the entire course of antibiotics, even if the infection appears to clear up before the last scheduled dose. This reduces the risk that any especially resistant bacteria could survive, causing a second infection or spreading to another person.

Go back to the previous Clinical Focus box.

Eye on Ethics

Microscopy and Antibiotic Resistance

As the use of antibiotics has proliferated in medicine, as well as agriculture, microbes have evolved to become more resistant. Strains of bacteria such as methicillin-resistant *S. aureus* (MRSA), which has developed a high level of resistance to many antibiotics, are an increasingly worrying problem, so much so that research is underway to develop new and more diversified antibiotics.

Fluorescence microscopy can be useful in testing the effectiveness of new antibiotics against resistant strains like MRSA. In a test of one new antibiotic derived from a marine bacterium, MC21-A (bromophene), researchers used the fluorescent dye SYTOX Green to stain samples of MRSA. SYTOX Green is often used to distinguish dead cells from living cells, with fluorescence microscopy. Live cells will not absorb the dye, but cells killed by an antibiotic will absorb the dye, since the antibiotic has damaged the bacterial cell membrane. In this particular case, MRSA bacteria that had been exposed to MC21-A did, indeed, appear green under the fluorescence microscope, leading researchers to conclude that it is an effective antibiotic against MRSA.

Of course, some argue that developing new antibiotics will only lead to even more antibiotic-resistant microbes, so-called superbugs that could spawn epidemics before new treatments can be developed. For this reason, many health professionals are beginning to exercise more discretion in prescribing antibiotics. Whereas antibiotics were once routinely prescribed for common illnesses without a definite diagnosis, doctors and hospitals are much more likely to conduct additional testing to determine whether an antibiotic is necessary and appropriate before prescribing.

A sick patient might reasonably object to this stingy approach to prescribing antibiotics. To the patient who simply wants to feel better as quickly as possible, the potential benefits of taking an antibiotic may seem to outweigh any immediate health risks that might occur if the antibiotic is ineffective. But at what point do the risks of widespread antibiotic use supersede the desire to use them in individual cases?

SUMMARY

2.1 The Properties of Light

- Light waves interacting with materials may be **reflected**, **absorbed**, or **transmitted**, depending on the properties of the material.
- Light waves can interact with each other (**interference**) or be distorted by interactions with small objects or openings (**diffraction**).
- **Refraction** occurs when light waves change speed and direction as they pass from one medium to another. Differences in the **refraction indices** of two materials determine the magnitude of directional changes when light passes from one to the other.
- A **lens** is a medium with a curved surface that refracts and focuses light to produce an image.
- Visible light is part of the **electromagnetic spectrum**; light waves of different frequencies and wavelengths are distinguished as colors by the human eye.
- A prism can separate the colors of white light (**dispersion**) because different frequencies of light have different refractive indices for a given material.
- **Fluorescent dyes** and **phosphorescent** materials can effectively transform nonvisible electromagnetic radiation into visible light.
- The power of a microscope can be described in terms of its **magnification** and **resolution**.
- Resolution can be increased by shortening wavelength, increasing the **numerical aperture** of the lens, or using stains that enhance contrast.

2.2 Peering Into the Invisible World

- **Antonie van Leeuwenhoek** is credited with the first observation of microbes, including protists and bacteria, with simple microscopes that he made.
- **Robert Hooke** was the first to describe what we now call cells.
- **Simple microscopes** have a single lens, while **compound microscopes** have multiple lenses.

2.3 Instruments of Microscopy

- Numerous types of microscopes use various technologies to generate micrographs. Most are useful for a particular type of specimen or application.
- **Light microscopy** uses lenses to focus light on a specimen to produce an image. Commonly used light microscopes include **brightfield**, **darkfield, phase-contrast, differential interference contrast, fluorescence, confocal**, and **two-photon** microscopes.
- **Electron microscopy** focuses electrons on the specimen using magnets, producing much greater magnification than light microscopy. The **transmission electron microscope (TEM)** and **scanning electron microscope (SEM)** are two common forms.
- **Scanning probe microscopy** produces images of even greater magnification by measuring feedback from sharp probes that interact with the specimen. Probe microscopes include the **scanning tunneling microscope (STM)** and the **atomic force microscope (AFM)**.

2.4 Staining Microscopic Specimens

- Samples must be properly prepared for microscopy. This may involve **staining**, **fixation**, and/or cutting **thin sections**.
- A variety of staining techniques can be used with light microscopy, including **Gram staining, acid-fast staining, capsule staining, endospore staining,** and **flagella staining**.
- Samples for TEM require very thin sections, whereas samples for SEM require sputter-coating.
- Preparation for fluorescence microscopy is similar to that for light microscopy, except that fluorochromes are used.

REVIEW QUESTIONS
Multiple Choice

1. Which of the following has the highest energy?
   A. light with a long wavelength
   B. light with an intermediate wavelength
   C. light with a short wavelength
   D. It is impossible to tell from the information given.

2. You place a specimen under the microscope and notice that parts of the specimen begin to emit light immediately. These materials can be described as _____.
   A. fluorescent
   B. phosphorescent
   C. transparent
   D. opaque

3. Who was the first to describe "cells" in dead cork tissue?
   A. Hans Janssen
   B. Zaccharias Janssen
   C. Antonie van Leeuwenhoek
   D. Robert Hooke

4. Who is the probable inventor of the compound microscope?
   A. Girolamo Fracastoro
   B. Zaccharias Janssen
   C. Antonie van Leeuwenhoek
   D. Robert Hooke

5. Which would be the best choice for viewing internal structures of a living protist such as a *Paramecium*?
   a. a brightfield microscope with a stain
   b. a brightfield microscope without a stain
   c. a darkfield microscope
   d. a transmission electron microscope

6. Which type of microscope is especially useful for viewing thick structures such as biofilms?
   a. a transmission electron microscope
   b. a scanning electron microscopes
   c. a phase-contrast microscope
   d. a confocal scanning laser microscope
   e. an atomic force microscope

7. Which type of microscope would be the best choice for viewing very small surface structures of a cell?
   a. a transmission electron microscope
   b. a scanning electron microscope
   c. a brightfield microscope
   d. a darkfield microscope
   e. a phase-contrast microscope

8. What type of microscope uses an annular stop?
   a. a transmission electron microscope
   b. a scanning electron microscope
   c. a brightfield microscope
   d. a darkfield microscope
   e. a phase-contrast microscope

9. What type of microscope uses a cone of light so that light only hits the specimen indirectly, producing a darker image on a brighter background?
   a. a transmission electron microscope
   b. a scanning electron microscope
   c. a brightfield microscope
   d. a darkfield microscope
   e. a phase-contrast microscope

10. What mordant is used in Gram staining?
    A. crystal violet
    B. safranin
    C. acid-alcohol
    D. iodine

11. What is one difference between specimen preparation for a transmission electron microscope (TEM) and preparation for a scanning electron microscope (SEM)?
    A. Only the TEM specimen requires sputter coating.
    B. Only the SEM specimen requires sputter-coating.
    C. Only the TEM specimen must be dehydrated.
    D. Only the SEM specimen must be dehydrated.

Fill in the Blank

12. When you see light bend as it moves from air into water, you are observing _____.

13. A microscope that uses multiple lenses is called a _____ microscope.

14. Chromophores that absorb and then emit light are called _____.

15. In a(n) _____ microscope, a probe located just above the specimen moves up and down in response to forces between the atoms and the tip of the probe.
16. What is the total magnification of a specimen that is being viewed with a standard ocular lens and a 40× objective lens?
17. Ziehl-Neelsen staining, a type of _____ staining, is diagnostic for *Mycobacterium tuberculosis*.
18. The _____ is used to differentiate bacterial cells based on the components of their cell walls.

Short Answer

19. Explain how a prism separates white light into different colors.
20. Why is Antonie van Leeuwenhoek's work much better known than that of Zaccharias Janssen?
21. Why did the cork cells observed by Robert Hooke appear to be empty, as opposed to being full of other structures?
22. What is the function of the condenser in a brightfield microscope?
23. **Art Connection:** Label each component of the brightfield microscope.

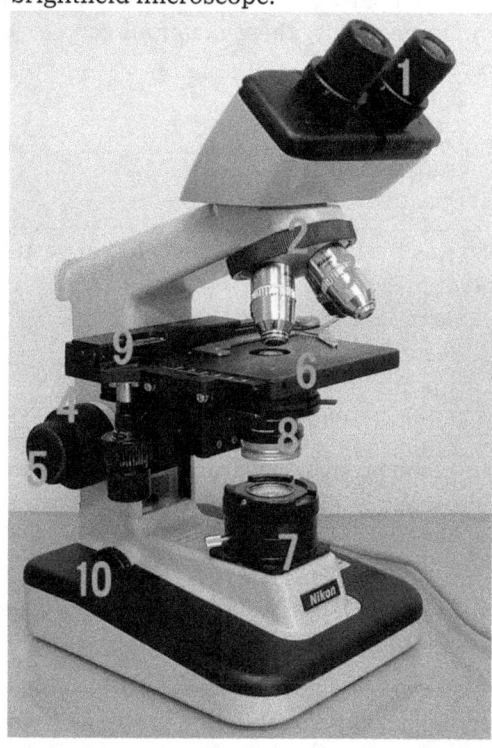

24. How could you identify whether a particular bacterial sample contained specimens with mycolic acid-rich cell walls?

Critical Thinking

25. In Figure 2.7, which of the following has the lowest energy?
    A. visible light
    B. X-rays
    C. ultraviolet rays
    D. infrared rays
26. When focusing a light microscope, why is it best to adjust the focus using the coarse focusing knob before using the fine focusing knob?
27. You need to identify structures within a cell using a microscope. However, the image appears very blurry even though you have a high magnification. What are some things that you could try to improve the resolution of the image? Describe the most basic factors that affect resolution when you first put the slide onto the stage; then consider more specific factors that could affect resolution for 40× and 100× lenses.
28. You use the Gram staining procedure to stain an L-form bacterium (a bacterium that lacks a cell wall). What color will the bacterium be after the staining procedure is finished?

CHAPTER 3
The Cell

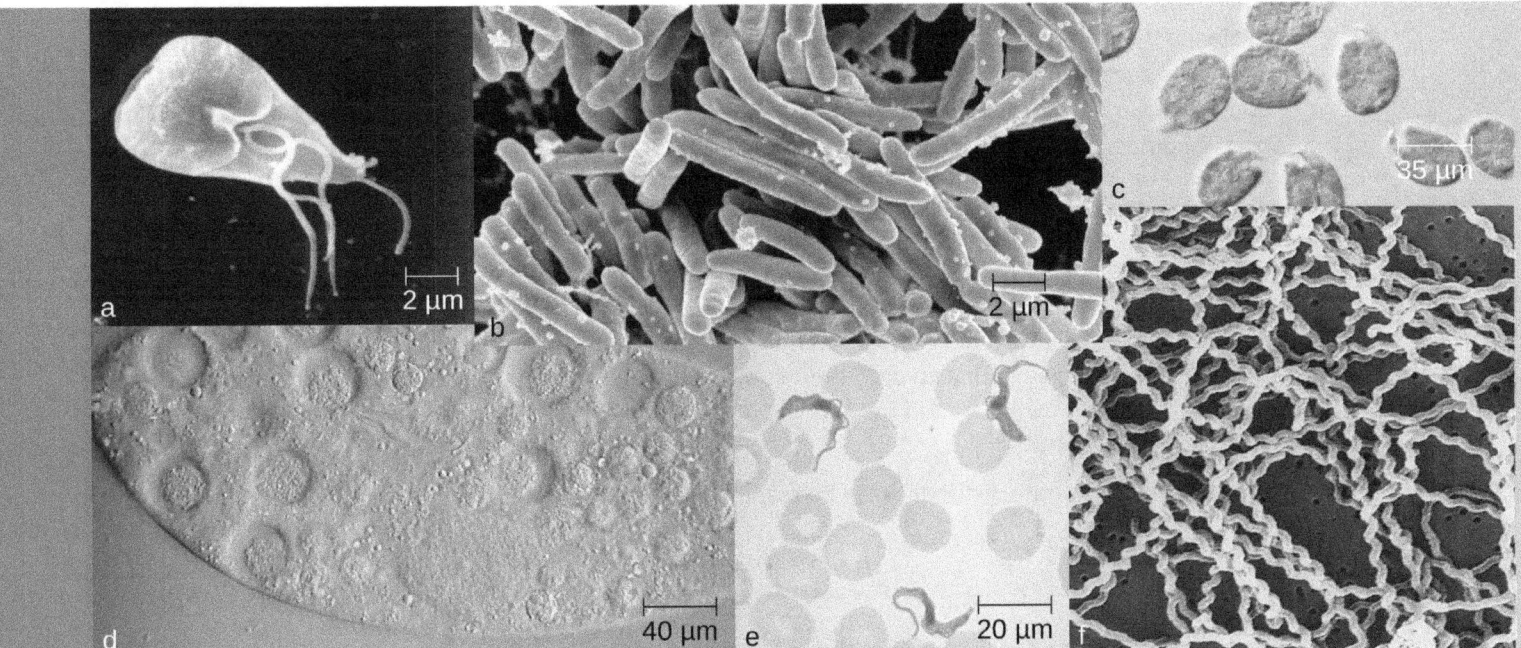

Figure 3.1 Microorganisms vary visually in their size and shape, as can be observed microscopically; but they also vary in invisible ways, such as in their metabolic capabilities. (credit a, e, f: modification of work by Centers for Disease Control and Prevention; credit b: modification of work by NIAID; credit c: modification of work by CSIRO; credit d: modification of work by "Microscopic World"/YouTube)

Chapter Outline

3.1 Spontaneous Generation

3.2 Foundations of Modern Cell Theory

3.3 Unique Characteristics of Prokaryotic Cells

3.4 Unique Characteristics of Eukaryotic Cells

INTRODUCTION Life takes many forms, from giant redwood trees towering hundreds of feet in the air to the tiniest known microbes, which measure only a few billionths of a meter. Humans have long pondered life's origins and debated the defining characteristics of life, but our understanding of these concepts has changed radically since the invention of the microscope. In the 17th century, observations of microscopic life led to the development of the cell theory: the idea that the fundamental unit of life is the cell, that all organisms contain at least one cell, and that cells only come from other cells.

Despite sharing certain characteristics, cells may vary significantly. The two main types of cells are prokaryotic cells (lacking a nucleus) and eukaryotic cells (containing a well-organized, membrane-bound nucleus). Each type of cell exhibits remarkable variety in structure, function, and metabolic activity (Figure 3.1). This chapter will focus on the historical discoveries that have shaped our current understanding of microbes, including their origins and their role in human disease. We will then explore the distinguishing structures found in prokaryotic and eukaryotic cells.

3.1 Spontaneous Generation

Learning Objectives

By the end of this section, you will be able to:
- Explain the theory of spontaneous generation and why people once accepted it as an explanation for the existence of certain types of organisms
- Explain how certain individuals (van Helmont, Redi, Needham, Spallanzani, and Pasteur) tried to prove or disprove spontaneous generation

---

Clinical Focus

Part 1

Barbara is a 19-year-old college student living in the dormitory. In January, she came down with a sore throat, headache, mild fever, chills, and a violent but unproductive (i.e., no mucus) cough. To treat these symptoms, Barbara began taking an over-the-counter cold medication, which did not seem to work. In fact, over the next few days, while some of Barbara's symptoms began to resolve, her cough and fever persisted, and she felt very tired and weak.

- What types of respiratory disease may be responsible?

Jump to the next Clinical Focus box

Humans have been asking for millennia: Where does new life come from? Religion, philosophy, and science have all wrestled with this question. One of the oldest explanations was the theory of spontaneous generation, which can be traced back to the ancient Greeks and was widely accepted through the Middle Ages.

The Theory of Spontaneous Generation

The Greek philosopher Aristotle (384–322 BC) was one of the earliest recorded scholars to articulate the theory of **spontaneous generation**, the notion that life can arise from nonliving matter. Aristotle proposed that life arose from nonliving material if the material contained *pneuma* ("vital heat"). As evidence, he noted several instances of the appearance of animals from environments previously devoid of such animals, such as the seemingly sudden appearance of fish in a new puddle of water.[1]

This theory persisted into the 17th century, when scientists undertook additional experimentation to support or disprove it. By this time, the proponents of the theory cited how frogs simply seem to appear along the muddy banks of the Nile River in Egypt during the annual flooding. Others observed that mice simply appeared among grain stored in barns with thatched roofs. When the roof leaked and the grain molded, mice appeared. Jan Baptista van Helmont, a 17th century Flemish scientist, proposed that mice could arise from rags and wheat kernels left in an open container for 3 weeks. In reality, such habitats provided ideal food sources and shelter for mouse populations to flourish.

However, one of van Helmont's contemporaries, Italian physician Francesco Redi (1626–1697), performed an experiment in 1668 that was one of the first to refute the idea that maggots (the larvae of flies) spontaneously generate on meat left out in the open air. He predicted that preventing flies from having direct contact with the meat would also prevent the appearance of maggots. Redi left meat in each of six containers (Figure 3.2). Two were open to the air, two were covered with gauze, and two were tightly sealed. His hypothesis was supported when maggots developed in the uncovered jars, but no maggots appeared in either the gauze-covered or the tightly sealed jars. He concluded that maggots could only form when flies were allowed to lay eggs in the meat, and that the maggots were the offspring of flies, not the product of spontaneous generation.

---

1 K. Zwier. "Aristotle on Spontaneous Generation." http://www.sju.edu/int/academics/cas/resources/gppc/pdf/Karen%20R.%20Zwier.pdf

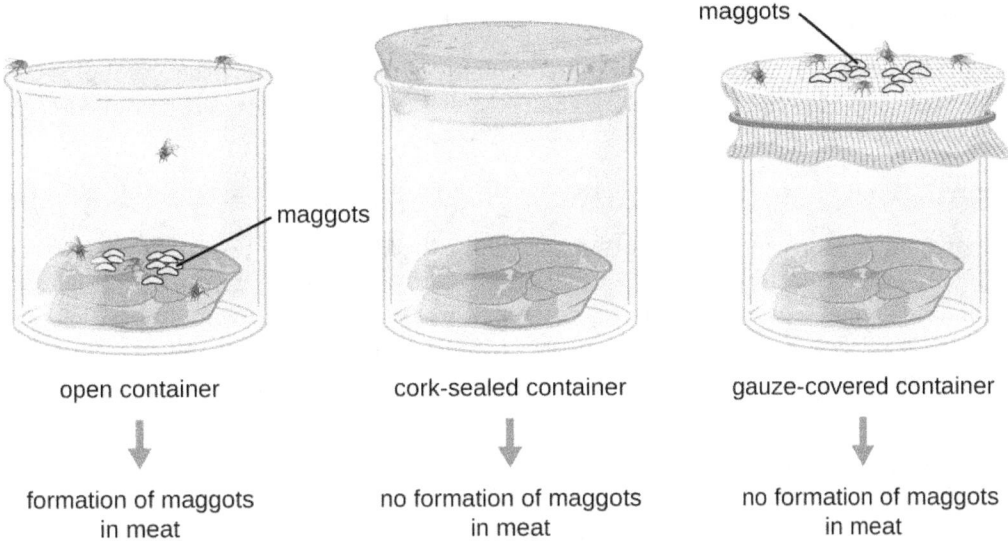

Figure 3.2 Francesco Redi's experimental setup consisted of an open container, a container sealed with a cork top, and a container covered in mesh that let in air but not flies. Maggots only appeared on the meat in the open container. However, maggots were also found on the gauze of the gauze-covered container.

In 1745, John Needham (1713–1781) published a report of his own experiments, in which he briefly boiled broth infused with plant or animal matter, hoping to kill all preexisting microbes.[2] He then sealed the flasks. After a few days, Needham observed that the broth had become cloudy and a single drop contained numerous microscopic creatures. He argued that the new microbes must have arisen spontaneously. In reality, however, he likely did not boil the broth enough to kill all preexisting microbes.

Lazzaro Spallanzani (1729–1799) did not agree with Needham's conclusions, however, and performed hundreds of carefully executed experiments using heated broth.[3] As in Needham's experiment, broth in sealed jars and unsealed jars was infused with plant and animal matter. Spallanzani's results contradicted the findings of Needham: Heated but sealed flasks remained clear, without any signs of spontaneous growth, unless the flasks were subsequently opened to the air. This suggested that microbes were introduced into these flasks from the air. In response to Spallanzani's findings, Needham argued that life originates from a "life force" that was destroyed during Spallanzani's extended boiling. Any subsequent sealing of the flasks then prevented new life force from entering and causing spontaneous generation (Figure 3.3).

Figure 3.3 (a) Francesco Redi, who demonstrated that maggots were the offspring of flies, not products of spontaneous generation. (b) John Needham, who argued that microbes arose spontaneously in broth from a "life force." (c) Lazzaro Spallanzani, whose experiments with broth aimed to disprove those of Needham.

---

2 E. Capanna. "Lazzaro Spallanzani: At the Roots of Modern Biology." *Journal of Experimental Zoology* 285 no. 3 (1999):178–196.

3 R. Mancini, M. Nigro, G. Ippolito. "Lazzaro Spallanzani and His Refutation of the Theory of Spontaneous Generation." *Le Infezioni in Medicina* 15 no. 3 (2007):199–206.

CHECK YOUR UNDERSTANDING

- Describe the theory of spontaneous generation and some of the arguments used to support it.
- Explain how the experiments of Redi and Spallanzani challenged the theory of spontaneous generation.

Disproving Spontaneous Generation

The debate over spontaneous generation continued well into the 19th century, with scientists serving as proponents of both sides. To settle the debate, the Paris Academy of Sciences offered a prize for resolution of the problem. Louis Pasteur, a prominent French chemist who had been studying microbial fermentation and the causes of wine spoilage, accepted the challenge. In 1858, Pasteur filtered air through a gun-cotton filter and, upon microscopic examination of the cotton, found it full of microorganisms, suggesting that the exposure of a broth to air was not introducing a "life force" to the broth but rather airborne microorganisms.

Later, Pasteur made a series of flasks with long, twisted necks ("swan-neck" flasks), in which he boiled broth to sterilize it (Figure 3.4). His design allowed air inside the flasks to be exchanged with air from the outside, but prevented the introduction of any airborne microorganisms, which would get caught in the twists and bends of the flasks' necks. If a life force besides the airborne microorganisms were responsible for microbial growth within the sterilized flasks, it would have access to the broth, whereas the microorganisms would not. He correctly predicted that sterilized broth in his swan-neck flasks would remain sterile as long as the swan necks remained intact. However, should the necks be broken, microorganisms would be introduced, contaminating the flasks and allowing microbial growth within the broth.

Pasteur's set of experiments irrefutably disproved the theory of spontaneous generation and earned him the prestigious Alhumbert Prize from the Paris Academy of Sciences in 1862. In a subsequent lecture in 1864, Pasteur articulated "*Omne vivum ex vivo*" ("Life only comes from life"). In this lecture, Pasteur recounted his famous swan-neck flask experiment, stating that "...life is a germ and a germ is life. Never will the doctrine of spontaneous generation recover from the mortal blow of this simple experiment."[4] To Pasteur's credit, it never has.

---

4 R. Vallery-Radot. *The Life of Pasteur*, trans. R.L. Devonshire. New York: McClure, Phillips and Co, 1902, 1:142.

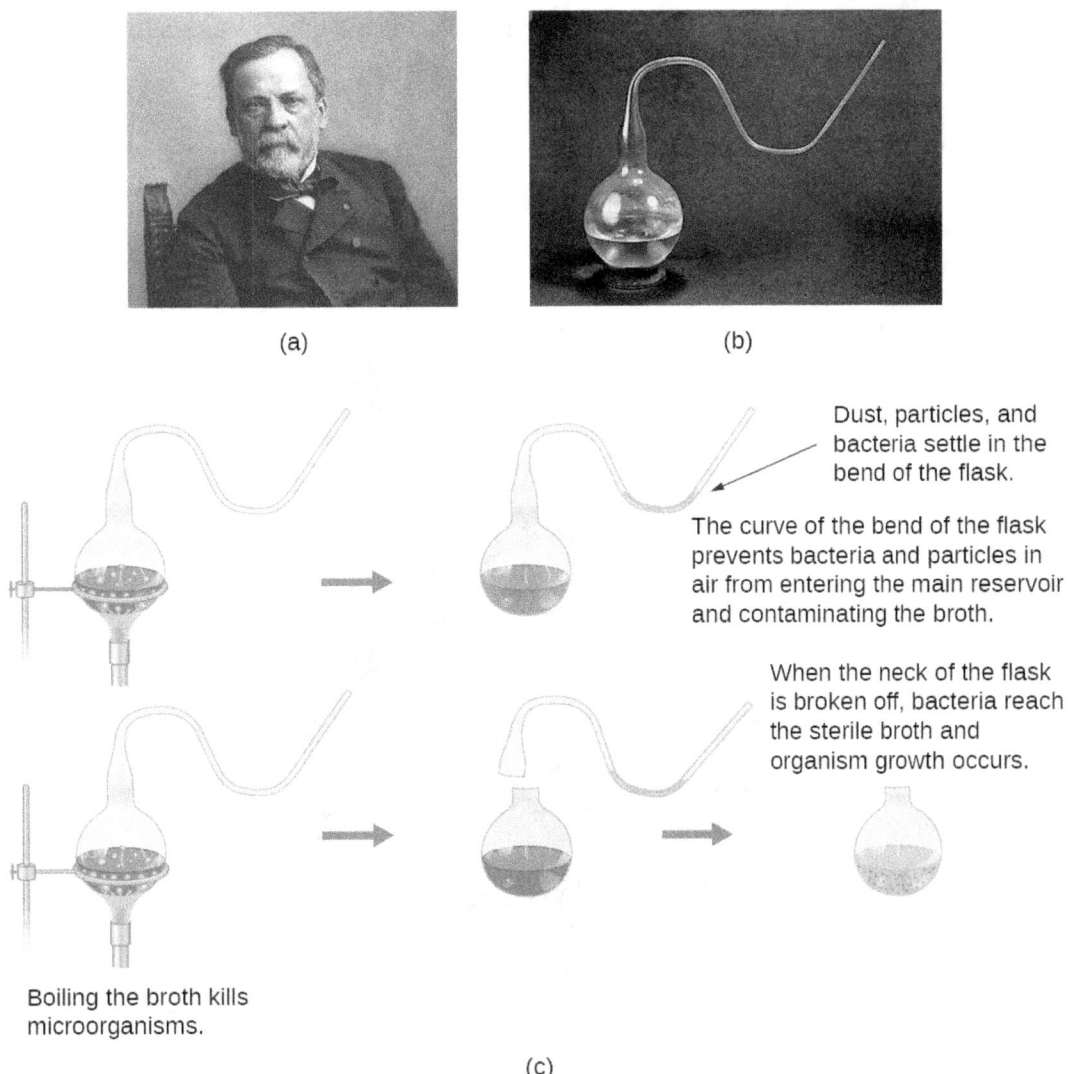

Figure 3.4 (a) French scientist Louis Pasteur, who definitively refuted the long-disputed theory of spontaneous generation. (b) The unique swan-neck feature of the flasks used in Pasteur's experiment allowed air to enter the flask but prevented the entry of bacterial and fungal spores. (c) Pasteur's experiment consisted of two parts. In the first part, the broth in the flask was boiled to sterilize it. When this broth was cooled, it remained free of contamination. In the second part of the experiment, the flask was boiled and then the neck was broken off. The broth in this flask became contaminated. (credit b: modification of work by "Wellcome Images"/Wikimedia Commons)

CHECK YOUR UNDERSTANDING
- How did Pasteur's experimental design allow air, but not microbes, to enter, and why was this important?
- What was the control group in Pasteur's experiment and what did it show?

3.2 Foundations of Modern Cell Theory

Learning Objectives

By the end of this section, you will be able to:
- Explain the key points of cell theory and the individual contributions of Hooke, Schleiden, Schwann, Remak, and Virchow
- Explain the key points of endosymbiotic theory and cite the evidence that supports this concept
- Explain the contributions of Semmelweis, Snow, Pasteur, Lister, and Koch to the development of germ theory

While some scientists were arguing over the theory of spontaneous generation, other scientists were making discoveries leading to a better understanding of what we now call the cell theory. Modern cell theory has two basic tenets:

- All cells only come from other cells (the principle of biogenesis).
- Cells are the fundamental units of organisms.

Today, these tenets are fundamental to our understanding of life on earth. However, modern cell theory grew out of the collective work of many scientists.

The Origins of Cell Theory

The English scientist Robert Hooke first used the term "cells" in 1665 to describe the small chambers within cork that he observed under a microscope of his own design. To Hooke, thin sections of cork resembled "Honey-comb," or "small Boxes or Bladders of Air." He noted that each "Cavern, Bubble, or Cell" was distinct from the others (Figure 3.5). At the time, Hooke was not aware that the cork cells were long dead and, therefore, lacked the internal structures found within living cells.

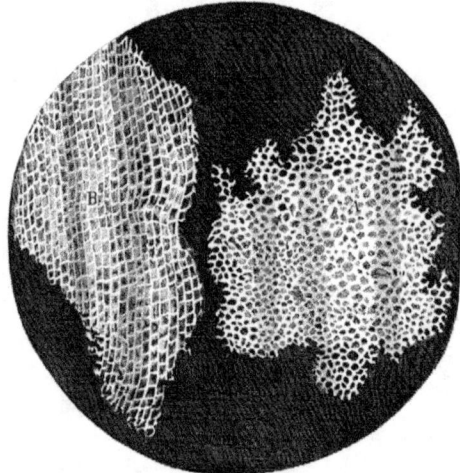

Figure 3.5 Robert Hooke (1635–1703) was the first to describe cells based upon his microscopic observations of cork. This illustration was published in his work *Micrographia*.

Despite Hooke's early description of cells, their significance as the fundamental unit of life was not yet recognized. Nearly 200 years later, in 1838, Matthias Schleiden (1804–1881), a German botanist who made extensive microscopic observations of plant tissues, described them as being composed of cells. Visualizing plant cells was relatively easy because plant cells are clearly separated by their thick cell walls. Schleiden believed that cells formed through crystallization, rather than cell division.

Theodor Schwann (1810–1882), a noted German physiologist, made similar microscopic observations of animal tissue. In 1839, after a conversation with Schleiden, Schwann realized that similarities existed between plant and animal tissues. This laid the foundation for the idea that cells are the fundamental components of plants and animals.

In the 1850s, two Polish scientists living in Germany pushed this idea further, culminating in what we recognize today as the modern cell theory. In 1852, Robert Remak (1815–1865), a prominent neurologist and embryologist, published convincing evidence that cells are derived from other cells as a result of cell division. However, this idea was questioned by many in the scientific community. Three years later, Rudolf Virchow (1821–1902), a well-respected pathologist, published an editorial essay entitled "Cellular Pathology," which popularized the concept of cell theory using the Latin phrase *omnis cellula a cellula* ("all cells arise from cells"), which is essentially the second tenet of modern cell theory.[5] Given the similarity of Virchow's work to Remak's, there is some controversy as to which scientist should receive credit for articulating cell theory. See the following Eye on Ethics feature for more about this controversy.

---

5 M. Schultz. "Rudolph Virchow." *Emerging Infectious Diseases* 14 no. 9 (2008):1480–1481.

Eye on Ethics

Science and Plagiarism

Rudolf Virchow, a prominent, Polish-born, German scientist, is often remembered as the "Father of Pathology." Well known for innovative approaches, he was one of the first to determine the causes of various diseases by examining their effects on tissues and organs. He was also among the first to use animals in his research and, as a result of his work, he was the first to name numerous diseases and created many other medical terms. Over the course of his career, he published more than 2,000 papers and headed various important medical facilities, including the Charité – Universitätsmedizin Berlin, a prominent Berlin hospital and medical school. But he is, perhaps, best remembered for his 1855 editorial essay titled "Cellular Pathology," published in *Archiv für Pathologische Anatomie und Physiologie*, a journal that Virchow himself cofounded and still exists today.

Despite his significant scientific legacy, there is some controversy regarding this essay, in which Virchow proposed the central tenet of modern cell theory—that all cells arise from other cells. Robert Remak, a former colleague who worked in the same laboratory as Virchow at the University of Berlin, had published the same idea 3 years before. Though it appears Virchow was familiar with Remak's work, he neglected to credit Remak's ideas in his essay. When Remak wrote a letter to Virchow pointing out similarities between Virchow's ideas and his own, Virchow was dismissive. In 1858, in the preface to one of his books, Virchow wrote that his 1855 publication was just an editorial piece, not a scientific paper, and thus there was no need to cite Remak's work.

By today's standards, Virchow's editorial piece would certainly be considered an act of plagiarism, since he presented Remak's ideas as his own. However, in the 19th century, standards for academic integrity were much less clear. Virchow's strong reputation, coupled with the fact that Remak was a Jew in a somewhat anti-Semitic political climate, shielded him from any significant repercussions. Today, the process of peer review and the ease of access to the scientific literature help discourage plagiarism. Although scientists are still motivated to publish original ideas that advance scientific knowledge, those who would consider plagiarizing are well aware of the serious consequences.

In academia, plagiarism represents the theft of both individual thought and research—an offense that can destroy reputations and end careers.[6][7][8][9]

---

6 B. Kisch. "Forgotten Leaders in Modern Medicine, Valentin, Gouby, Remak, Auerbach." *Transactions of the American Philosophical Society* 44 (1954):139–317.

7 H. Harris. *The Birth of the Cell*. New Haven, CT: Yale University Press, 2000:133.

8 C. Webster (ed.). *Biology, Medicine and Society 1840-1940*. Cambridge, UK; Cambridge University Press, 1981:118–119.

9 C. Zuchora-Walske. *Key Discoveries in Life Science*. Minneapolis, MN: Lerner Publishing, 2015:12–13.

(a) (b)

Figure 3.6 (a) Rudolf Virchow (1821–1902) popularized the cell theory in an 1855 essay entitled "Cellular Pathology." (b) The idea that all cells originate from other cells was first published in 1852 by his contemporary and former colleague Robert Remak (1815–1865).

CHECK YOUR UNDERSTANDING

- What are the key points of the cell theory?
- What contributions did Rudolf Virchow and Robert Remak make to the development of the cell theory?

Endosymbiotic Theory

As scientists were making progress toward understanding the role of cells in plant and animal tissues, others were examining the structures within the cells themselves. In 1831, Scottish botanist Robert Brown (1773–1858) was the first to describe observations of nuclei, which he observed in plant cells. Then, in the early 1880s, German botanist Andreas Schimper (1856–1901) was the first to describe the chloroplasts of plant cells, identifying their role in starch formation during photosynthesis and noting that they divided independent of the nucleus.

Based upon the chloroplasts' ability to reproduce independently, Russian botanist Konstantin Mereschkowski (1855–1921) suggested in 1905 that chloroplasts may have originated from ancestral photosynthetic bacteria living symbiotically inside a eukaryotic cell. He proposed a similar origin for the nucleus of plant cells. This was the first articulation of the endosymbiotic hypothesis, and would explain how eukaryotic cells evolved from ancestral bacteria.

Mereschkowski's endosymbiotic hypothesis was furthered by American anatomist Ivan Wallin (1883–1969), who began to experimentally examine the similarities between mitochondria, chloroplasts, and bacteria—in other words, to put the endosymbiotic hypothesis to the test using objective investigation. Wallin published a series of papers in the 1920s supporting the endosymbiotic hypothesis, including a 1926 publication co-authored with Mereschkowski. Wallin claimed he could culture mitochondria outside of their eukaryotic host cells. Many scientists dismissed his cultures of mitochondria as resulting from bacterial contamination. Modern genome sequencing work supports the dissenting scientists by showing that much of the genome of mitochondria had been transferred to the host cell's nucleus, preventing the mitochondria from being able to live on their own.[10][11]

---

10 T. Embley, W. Martin. "Eukaryotic Evolution, Changes, and Challenges." *Nature* Vol. 440 (2006):623–630.

11 O.G. Berg, C.G. Kurland. "Why Mitochondrial Genes Are Most Often Found in Nuclei." *Molecular Biology and Evolution* 17 no. 6

Wallin's ideas regarding the endosymbiotic hypothesis were largely ignored for the next 50 years because scientists were unaware that these organelles contained their own DNA. However, with the discovery of mitochondrial and chloroplast DNA in the 1960s, the endosymbiotic hypothesis was resurrected. Lynn Margulis (1938–2011), an American geneticist, published her ideas regarding the endosymbiotic hypothesis of the origins of mitochondria and chloroplasts in 1967.[12] In the decade leading up to her publication, advances in microscopy had allowed scientists to differentiate prokaryotic cells from eukaryotic cells. In her publication, Margulis reviewed the literature and argued that the eukaryotic organelles such as mitochondria and chloroplasts are of prokaryotic origin. She presented a growing body of microscopic, genetic, molecular biology, fossil, and geological data to support her claims.

Again, this hypothesis was not initially popular, but mounting genetic evidence due to the advent of DNA sequencing supported the **endosymbiotic theory**, which is now defined as the theory that mitochondria and chloroplasts arose as a result of prokaryotic cells establishing a symbiotic relationship within a eukaryotic host (Figure 3.7). With Margulis' initial endosymbiotic theory gaining wide acceptance, she expanded on the theory in her 1981 book *Symbiosis in Cell Evolution*. In it, she explains how endosymbiosis is a major driving factor in the evolution of organisms. More recent genetic sequencing and phylogenetic analysis show that mitochondrial DNA and chloroplast DNA are highly related to their bacterial counterparts, both in DNA sequence and chromosome structure. However, mitochondrial DNA and chloroplast DNA are reduced compared with nuclear DNA because many of the genes have moved from the organelles into the host cell's nucleus. Additionally, mitochondrial and chloroplast ribosomes are structurally similar to bacterial ribosomes, rather than to the eukaryotic ribosomes of their hosts. Last, the binary fission of these organelles strongly resembles the binary fission of bacteria, as compared with mitosis performed by eukaryotic cells. Since Margulis' original proposal, scientists have observed several examples of bacterial endosymbionts in modern-day eukaryotic cells. Examples include the endosymbiotic bacteria found within the guts of certain insects, such as cockroaches,[13] and photosynthetic bacteria-like organelles found in protists.[14]

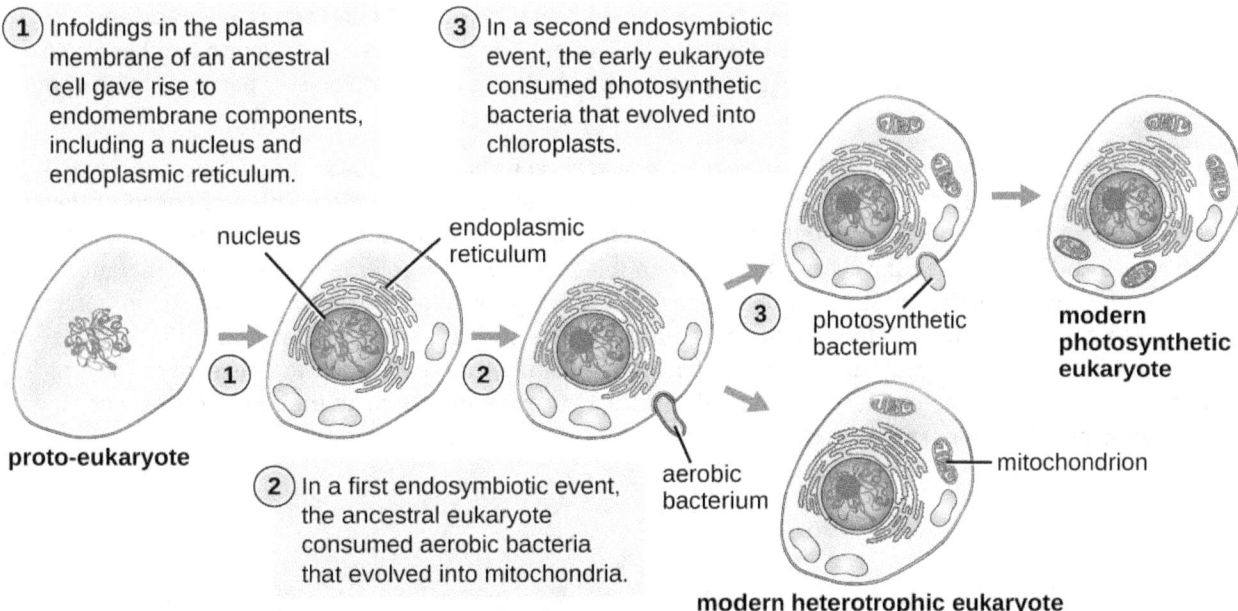

Figure 3.7 According to the endosymbiotic theory, mitochondria and chloroplasts are each derived from the uptake of bacteria. These

---

(2000):951–961.

12 L. Sagan. "On the Origin of Mitosing Cells." *Journal of Theoretical Biology* 14 no. 3 (1967):225–274.

13 A.E. Douglas. "The Microbial Dimension in Insect Nutritional Ecology." *Functional Ecology* 23 (2009):38–47.

14 J.M. Jaynes, L.P. Vernon. "The Cyanelle of *Cyanophora paradoxa*: Almost a Cyanobacterial Chloroplast." *Trends in Biochemical Sciences* 7 no. 1 (1982):22–24.

bacteria established a symbiotic relationship with their host cell that eventually led to the bacteria evolving into mitochondria and chloroplasts.

✓ CHECK YOUR UNDERSTANDING

- What does the modern endosymbiotic theory state?
- What evidence supports the endosymbiotic theory?

The Germ Theory of Disease

Prior to the discovery of microbes during the 17th century, other theories circulated about the origins of disease. For example, the ancient Greeks proposed the miasma theory, which held that disease originated from particles emanating from decomposing matter, such as that in sewage or cesspits. Such particles infected humans in close proximity to the rotting material. Diseases including the Black Death, which ravaged Europe's population during the Middle Ages, were thought to have originated in this way.

In 1546, Italian physician Girolamo Fracastoro proposed, in his essay *De Contagione et Contagiosis Morbis*, that seed-like spores may be transferred between individuals through direct contact, exposure to contaminated clothing, or through the air. We now recognize Fracastoro as an early proponent of the **germ theory of disease**, which states that diseases may result from microbial infection. However, in the 16th century, Fracastoro's ideas were not widely accepted and would be largely forgotten until the 19th century.

In 1847, Hungarian obstetrician Ignaz Semmelweis (Figure 3.8) observed that mothers who gave birth in hospital wards staffed by physicians and medical students were more likely to suffer and die from puerperal fever after childbirth (10%–20% mortality rate) than were mothers in wards staffed by midwives (1% mortality rate). Semmelweis observed medical students performing autopsies and then subsequently carrying out vaginal examinations on living patients without washing their hands in between. He suspected that the students carried disease from the autopsies to the patients they examined. His suspicions were supported by the untimely death of a friend, a physician who contracted a fatal wound infection after a postmortem examination of a woman who had died of a puerperal infection. The dead physician's wound had been caused by a scalpel used during the examination, and his subsequent illness and death closely paralleled that of the dead patient.

Although Semmelweis did not know the true cause of puerperal fever, he proposed that physicians were somehow transferring the causative agent to their patients. He suggested that the number of puerperal fever cases could be reduced if physicians and medical students simply washed their hands with chlorinated lime water before and after examining every patient. When this practice was implemented, the maternal mortality rate in mothers cared for by physicians dropped to the same 1% mortality rate observed among mothers cared for by midwives. This demonstrated that handwashing was a very effective method for preventing disease transmission. Despite this great success, many discounted Semmelweis's work at the time, and physicians were slow to adopt the simple procedure of handwashing to prevent infections in their patients because it contradicted established norms for that time period.

Figure 3.8 Ignaz Semmelweis (1818–1865) was a proponent of the importance of handwashing to prevent transfer of disease between patients by physicians.

Around the same time Semmelweis was promoting handwashing, in 1848, British physician John Snow conducted studies to track the source of cholera outbreaks in London. By tracing the outbreaks to two specific water sources, both of which were contaminated by sewage, Snow ultimately demonstrated that cholera bacteria were transmitted via drinking water. Snow's work is influential in that it represents the first known epidemiological study, and it resulted in the first known public health response to an epidemic. The work of both Semmelweis and Snow clearly refuted the prevailing miasma theory of the day, showing that disease is not only transmitted through the air but also through contaminated items.

Although the work of Semmelweis and Snow successfully showed the role of sanitation in preventing infectious disease, the cause of disease was not fully understood. The subsequent work of Louis Pasteur, Robert Koch, and Joseph Lister would further substantiate the germ theory of disease.

While studying the causes of beer and wine spoilage in 1856, Pasteur discovered properties of fermentation by microorganisms. He had demonstrated with his swan-neck flask experiments (Figure 3.4) that airborne microbes, not spontaneous generation, were the cause of food spoilage, and he suggested that if microbes were responsible for food spoilage and fermentation, they could also be responsible for causing infection. This was the foundation for the germ theory of disease.

Meanwhile, British surgeon Joseph Lister (Figure 3.9) was trying to determine the causes of postsurgical infections. Many physicians did not give credence to the idea that microbes on their hands, on their clothes, or in the air could infect patients' surgical wounds, despite the fact that 50% of surgical patients, on average, were dying of postsurgical infections.[15] Lister, however, was familiar with the work of Semmelweis and Pasteur; therefore, he insisted on handwashing and extreme cleanliness during surgery. In 1867, to further decrease the incidence of postsurgical wound infections, Lister began using carbolic acid (phenol) spray disinfectant/antiseptic during surgery. His extremely successful efforts to reduce postsurgical infection caused his techniques to become a standard medical practice.

A few years later, Robert Koch (Figure 3.9) proposed a series of postulates (Koch's postulates) based on the idea that the cause of a specific disease could be attributed to a specific microbe. Using these postulates, Koch and his colleagues were able to definitively identify the causative pathogens of specific diseases, including anthrax, tuberculosis, and cholera. Koch's "one microbe, one disease" concept was the culmination of the 19th century's paradigm shift away from miasma theory and toward the germ theory of disease. Koch's postulates are discussed more thoroughly in How Pathogens Cause Disease.

---

15 Alexander, J. Wesley. "The Contributions of Infection Control to a Century of Progress" *Annals of Surgery* 201:423-428, 1985.

(a) (b)

Figure 3.9 (a) Joseph Lister developed procedures for the proper care of surgical wounds and the sterilization of surgical equipment. (b) Robert Koch established a protocol to determine the cause of infectious disease. Both scientists contributed significantly to the acceptance of the germ theory of disease.

CHECK YOUR UNDERSTANDING

- Compare and contrast the miasma theory of disease with the germ theory of disease.
- How did Joseph Lister's work contribute to the debate between the miasma theory and germ theory and how did this increase the success of medical procedures?

Clinical Focus

Part 2

After suffering a fever, congestion, cough, and increasing aches and pains for several days, Barbara suspects that she has a case of the flu. She decides to visit the health center at her university. The PA tells Barbara that her symptoms could be due to a range of diseases, such as influenza, bronchitis, pneumonia, or tuberculosis.

During her physical examination, the PA notes that Barbara's heart rate is slightly elevated. Using a pulse oximeter, a small device that clips on her finger, he finds that Barbara has hypoxemia—a lower-than-normal level of oxygen in the blood. Using a stethoscope, the PA listens for abnormal sounds made by Barbara's heart, lungs, and digestive system. As Barbara breathes, the PA hears a crackling sound and notes a slight shortness of breath. He collects a sputum sample, noting the greenish color of the mucus, and orders a chest radiograph, which shows a "shadow" in the left lung. All of these signs are suggestive of pneumonia, a condition in which the lungs fill with mucus (Figure 3.10).

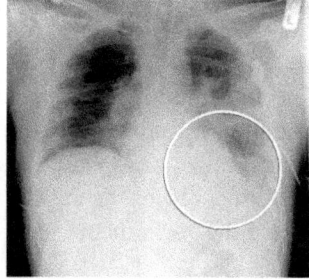 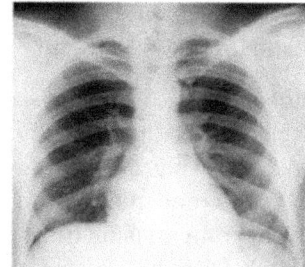

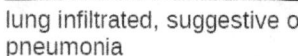

lung infiltrated, suggestive of pneumonia   normal lungs

Figure 3.10 This is a chest radiograph typical of pneumonia. Because X-ray images are negative images, a "shadow" is seen as a white area within the lung that should otherwise be black. In this case, the left lung shows a shadow as a result of pockets in the lung that have become filled with fluid. (credit left: modification of work by "Christaras A"/Wikimedia Commons)

- What kinds of infectious agents are known to cause pneumonia?

Jump to the next Clinical Focus box. Go back to the previous Clinical Focus box.

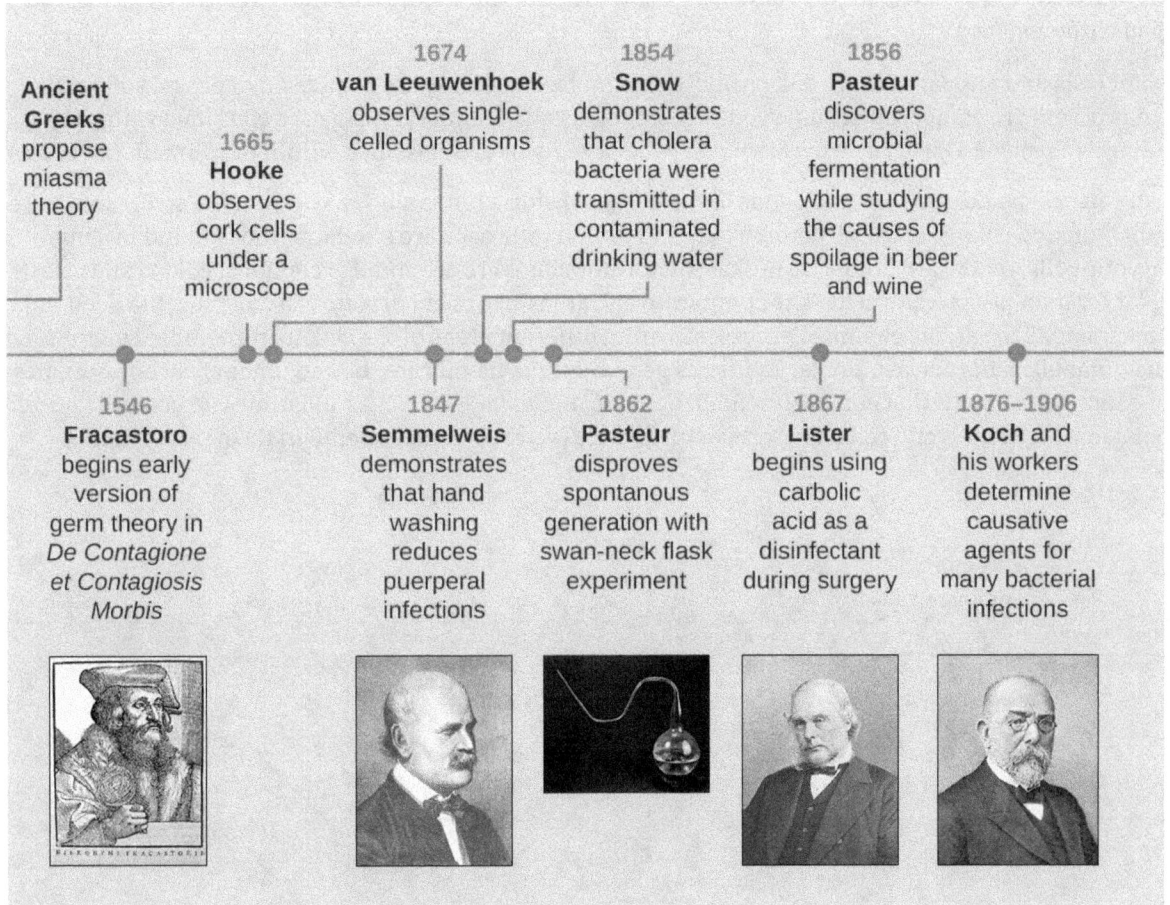

Figure 3.11 (credit "swan-neck flask": modification of work by Wellcome Images)

3.3 Unique Characteristics of Prokaryotic Cells

Learning Objectives

By the end of this section, you will be able to:
- Explain the distinguishing characteristics of prokaryotic cells
- Describe common cell morphologies and cellular arrangements typical of prokaryotic cells and explain how cells maintain their morphology
- Describe internal and external structures of prokaryotic cells in terms of their physical structure, chemical structure, and function
- Compare the distinguishing characteristics of bacterial and archaeal cells

Cell theory states that the cell is the fundamental unit of life. However, cells vary significantly in size, shape, structure, and function. At the simplest level of construction, all cells possess a few fundamental components. These include **cytoplasm** (a gel-like substance composed of water and dissolved chemicals needed for growth), which is contained within a plasma membrane (also called a cell membrane or cytoplasmic membrane); one or more chromosomes, which contain the genetic blueprints of the cell; and **ribosomes**, organelles used for the production of proteins.

Beyond these basic components, cells can vary greatly between organisms, and even within the same multicellular organism. The two largest categories of cells—**prokaryotic cells** and **eukaryotic cells**—are defined by major differences in several cell structures. Prokaryotic cells lack a nucleus surrounded by a complex nuclear membrane and generally have a single, circular chromosome located in a nucleoid. Eukaryotic cells have a nucleus surrounded by a complex nuclear membrane that contains multiple, rod-shaped chromosomes.[16]

All plant cells and animal cells are eukaryotic. Some microorganisms are composed of prokaryotic cells, whereas others are composed of eukaryotic cells. Prokaryotic microorganisms are classified within the domains Archaea and Bacteria, whereas eukaryotic organisms are classified within the domain Eukarya.

The structures inside a cell are analogous to the organs inside a human body, with unique structures suited to specific functions. Some of the structures found in prokaryotic cells are similar to those found in some eukaryotic cells; others are unique to prokaryotes. Although there are some exceptions, eukaryotic cells tend to be larger than prokaryotic cells. The comparatively larger size of eukaryotic cells dictates the need to compartmentalize various chemical processes within different areas of the cell, using complex membrane-bound organelles. In contrast, prokaryotic cells generally lack membrane-bound organelles; however, they often contain inclusions that compartmentalize their cytoplasm. Figure 3.12 illustrates structures typically associated with prokaryotic cells. These structures are described in more detail in the next section.

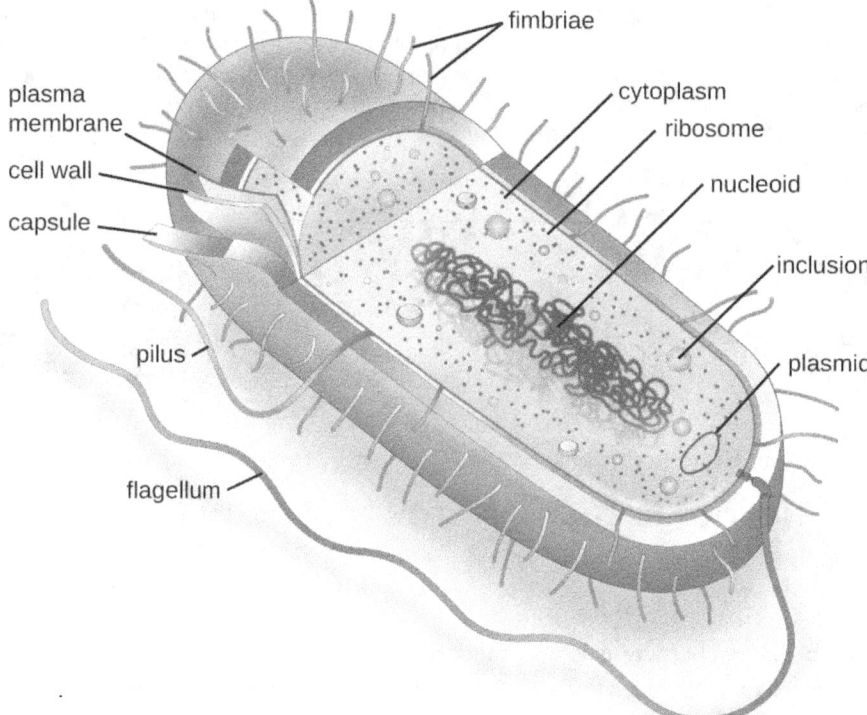

Figure 3.12 A typical prokaryotic cell contains a cell membrane, chromosomal DNA that is concentrated in a nucleoid, ribosomes, and a cell wall. Some prokaryotic cells may also possess flagella, pili, fimbriae, and capsules.

Common Cell Morphologies and Arrangements

Individual cells of a particular prokaryotic organism are typically similar in shape, or **cell morphology**. Although thousands of prokaryotic organisms have been identified, only a handful of cell morphologies are commonly seen microscopically. Figure 3.13 names and illustrates cell morphologies commonly found in prokaryotic cells. In addition to cellular shape, prokaryotic cells of the same species may group together in certain distinctive arrangements depending on the plane of cell division. Some common arrangements are shown in Figure 3.14.

---

16 Y.-H.M. Chan, W.F. Marshall. "Scaling Properties of Cell and Organelle Size." *Organogenesis* 6 no. 2 (2010):88–96.

Common Prokaryotic Cell Shapes			
Name	Description	Illustration	Image
Coccus (pl. cocci)	Round		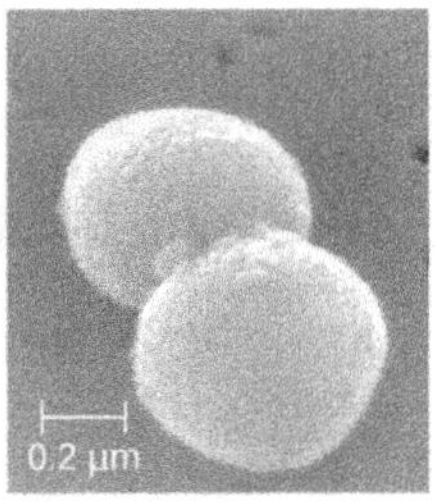
Bacillus (pl. bacilli)	Rod		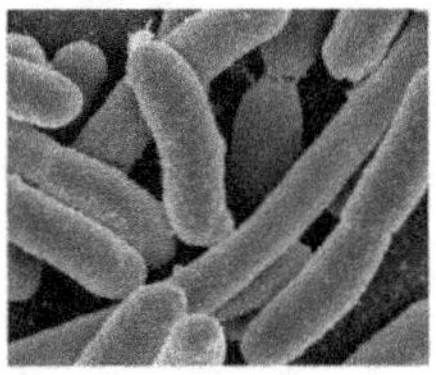
Vibrio (pl. vibrios)	Curved rod		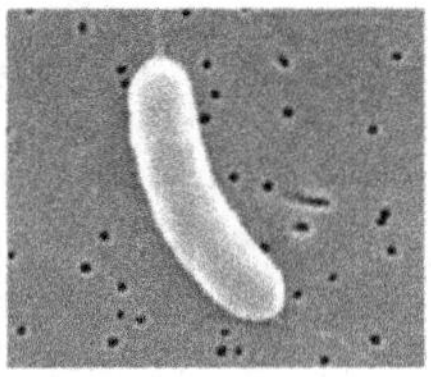
Coccobacillus (pl. coccobacilli)	Short rod		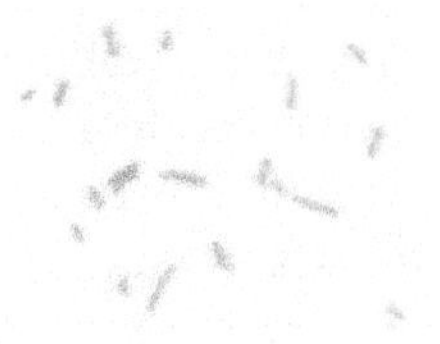
Spirillum (pl. spirilla)	Spiral		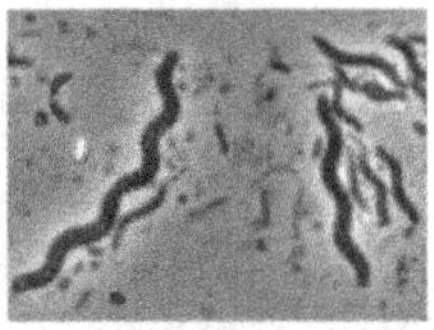
Spirochete (pl. spirochetes)	Long, loose, helical spiral		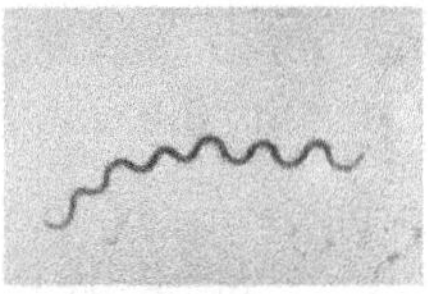

Figure 3.13 (credit "Coccus" micrograph: modification of work by Janice Haney Carr, Centers for Disease Control and Prevention; credit "Coccobacillus" micrograph: modification of work by Janice Carr, Centers for Disease Control and Prevention; credit "Spirochete" micrograph: modification of work by Centers for Disease Control and Prevention)

Common Prokaryotic Cell Arrangements		
Name	Description	Illustration
Coccus (pl. cocci)	Single coccus	
Diplococcus (pl. diplococci)	Pair of two cocci	
Tetrad (pl. tetrads)	Grouping of four cells arranged in a square	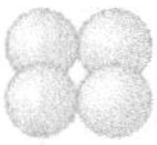
Streptococcus (pl. streptococci)	Chain of cocci	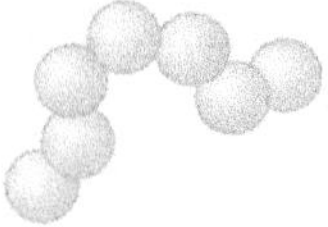
Staphylococcus (pl. staphylococci)	Cluster of cocci	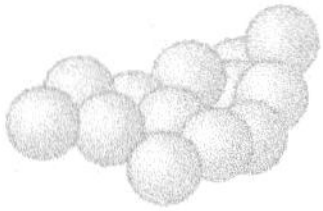
Bacillus (pl. bacilli)	Single rod	
Streptobacillus (pl. streptobacilli)	Chain of rods	

Figure 3.14

In most prokaryotic cells, morphology is maintained by the **cell wall** in combination with cytoskeletal elements. The cell wall is a structure found in most prokaryotes and some eukaryotes; it envelopes the cell membrane, protecting the cell from changes in **osmotic pressure** (Figure 3.15). Osmotic pressure occurs because of differences in the concentration of solutes on opposing sides of a semipermeable membrane. Water is able to pass through a semipermeable membrane, but solutes (dissolved molecules like salts, sugars, and other compounds) cannot. When the concentration of solutes is greater on one side of the membrane, water diffuses across the membrane from the side with the lower concentration (more water) to the side with the higher concentration (less water) until the concentrations on both sides become equal. This diffusion of water is called **osmosis**, and it can cause extreme osmotic pressure on a cell when its external environment changes.

The external environment of a cell can be described as an isotonic, hypertonic, or hypotonic medium. In an **isotonic medium**, the solute concentrations inside and outside the cell are approximately equal, so there is no net movement of water across the cell membrane. In a **hypertonic medium**, the solute concentration outside the cell exceeds that inside the cell, so water diffuses out of the cell and into the external medium. In a **hypotonic medium**, the solute concentration inside the cell exceeds that outside of the cell, so water will move by osmosis into the cell. This causes the cell to swell and potentially lyse, or burst.

The degree to which a particular cell is able to withstand changes in osmotic pressure is called tonicity. Cells that have a cell wall are better able to withstand subtle changes in osmotic pressure and maintain their shape. In hypertonic environments, cells that lack a cell wall can become dehydrated, causing **crenation**, or shriveling of the cell; the plasma membrane contracts and appears scalloped or notched (Figure 3.15). By contrast, cells that possess a cell wall undergo **plasmolysis** rather than crenation. In plasmolysis, the plasma membrane contracts and detaches from the cell wall, and there is a decrease in interior volume, but the cell wall remains intact, thus allowing the cell to maintain some shape and integrity for a period of time (Figure 3.16). Likewise, cells that lack a cell wall are more prone to lysis in hypotonic environments. The presence of a cell wall allows the cell to maintain its shape and integrity for a longer time before lysing (Figure 3.16).

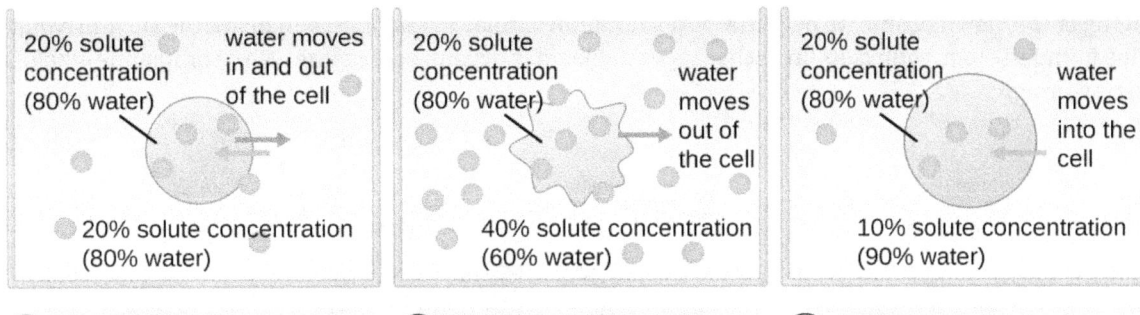

Figure 3.15 In cells that lack a cell wall, changes in osmotic pressure can lead to crenation in hypertonic environments or cell lysis in hypotonic environments.

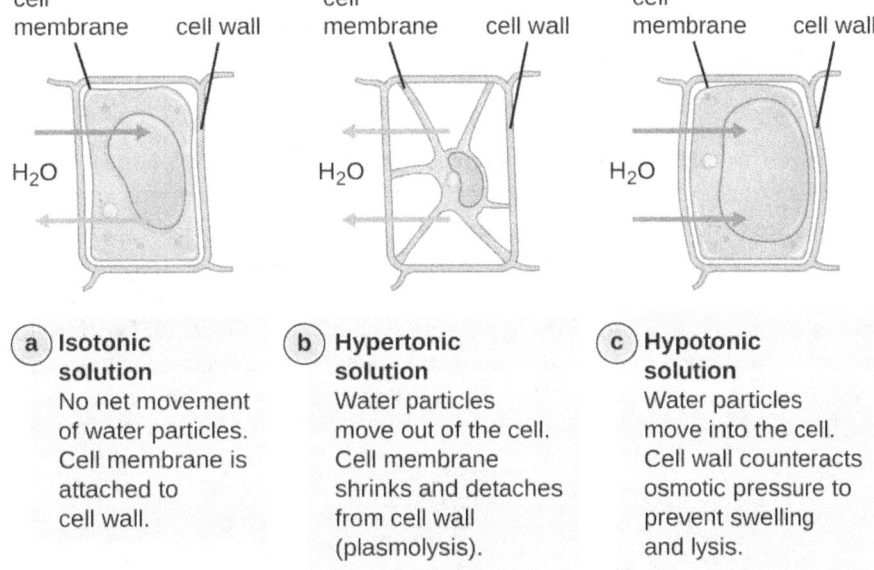

Figure 3.16 In prokaryotic cells, the cell wall provides some protection against changes in osmotic pressure, allowing it to maintain its shape longer. The cell membrane is typically attached to the cell wall in an isotonic medium (left). In a hypertonic medium, the cell membrane detaches from the cell wall and contracts (plasmolysis) as water leaves the cell. In a hypotonic medium (right), the cell wall prevents the cell membrane from expanding to the point of bursting, although lysis will eventually occur if too much water is absorbed.

✓ CHECK YOUR UNDERSTANDING

- Explain the difference between cell morphology and arrangement.
- What advantages do cell walls provide prokaryotic cells?

The Nucleoid

All cellular life has a DNA genome organized into one or more chromosomes. Prokaryotic chromosomes are typically circular, haploid (unpaired), and not bound by a complex nuclear membrane. Prokaryotic DNA and DNA-associated proteins are concentrated within the **nucleoid** region of the cell (Figure 3.17). In general, prokaryotic DNA interacts with **nucleoid-associated proteins (NAPs)** that assist in the organization and packaging of the chromosome. In bacteria, NAPs function similar to histones, which are the DNA-organizing proteins found in eukaryotic cells. In archaea, the nucleoid is organized by either NAPs or histone-like DNA organizing proteins.

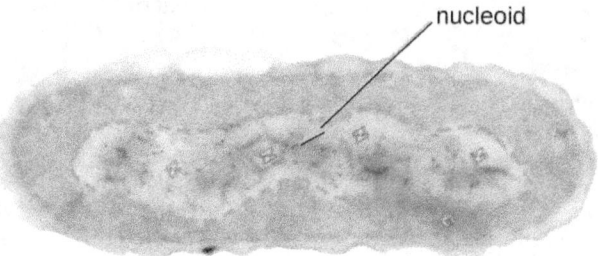

Figure 3.17 The nucleoid region (the area enclosed by the green dashed line) is a condensed area of DNA found within prokaryotic cells. Because of the density of the area, it does not readily stain and appears lighter in color when viewed with a transmission electron microscope.

Plasmids

Prokaryotic cells may also contain extrachromosomal DNA, or DNA that is not part of the chromosome. This extrachromosomal DNA is found in **plasmids**, which are small, circular, double-stranded DNA molecules. Cells that have plasmids often have hundreds of them within a single cell. Plasmids are more commonly found in bacteria; however, plasmids have been found in archaea and eukaryotic organisms. Plasmids often carry genes that confer advantageous traits such as antibiotic resistance; thus, they are important to the survival of the organism. We will discuss plasmids in more detail in Mechanisms of Microbial Genetics.

Ribosomes

All cellular life synthesizes proteins, and organisms in all three domains of life possess ribosomes, structures responsible for protein synthesis. However, ribosomes in each of the three domains are structurally different. Ribosomes, themselves, are constructed from proteins, along with ribosomal RNA (rRNA). Prokaryotic ribosomes are found in the cytoplasm. They are called **70S ribosomes** because they have a size of 70S (Figure 3.18), whereas eukaryotic cytoplasmic ribosomes have a size of 80S. (The S stands for Svedberg unit, a measure of sedimentation in an ultracentrifuge, which is based on size, shape, and surface qualities of the structure being analyzed). Although they are the same size, bacterial and archaeal ribosomes have different proteins and rRNA molecules, and the archaeal versions are more similar to their eukaryotic counterparts than to those found in bacteria.

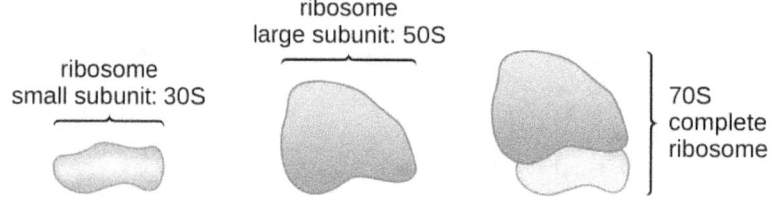

Figure 3.18 Prokaryotic ribosomes (70S) are composed of two subunits: the 30S (small subunit) and the 50S (large subunit), each of which are composed of protein and rRNA components.

Inclusions

As single-celled organisms living in unstable environments, some prokaryotic cells have the ability to store excess nutrients within cytoplasmic structures called **inclusions**. Storing nutrients in a polymerized form is advantageous because it reduces the buildup of osmotic pressure that occurs as a cell accumulates solutes. Various types of inclusions store glycogen and starches, which contain carbon that cells can access for energy. **Volutin** granules, also called **metachromatic granules** because of their staining characteristics, are inclusions that store polymerized inorganic phosphate that can be used in metabolism and assist in the formation of biofilms. Microbes known to contain volutin granules include the archaea *Methanosarcina*, the bacterium *Corynebacterium diphtheriae*, and the unicellular eukaryotic alga *Chlamydomonas*. Sulfur granules, another type of inclusion, are found in sulfur bacteria of the genus *Thiobacillus*; these granules store elemental sulfur, which the bacteria use for metabolism.

Occasionally, certain types of inclusions are surrounded by a phospholipid monolayer embedded with protein. **Polyhydroxybutyrate (PHB)**, which can be produced by species of *Bacillus* and *Pseudomonas*, is an example of an inclusion that displays this type of monolayer structure. Industrially, PHB has also been used as a source of biodegradable polymers for bioplastics. Several different types of inclusions are shown in Figure 3.19.

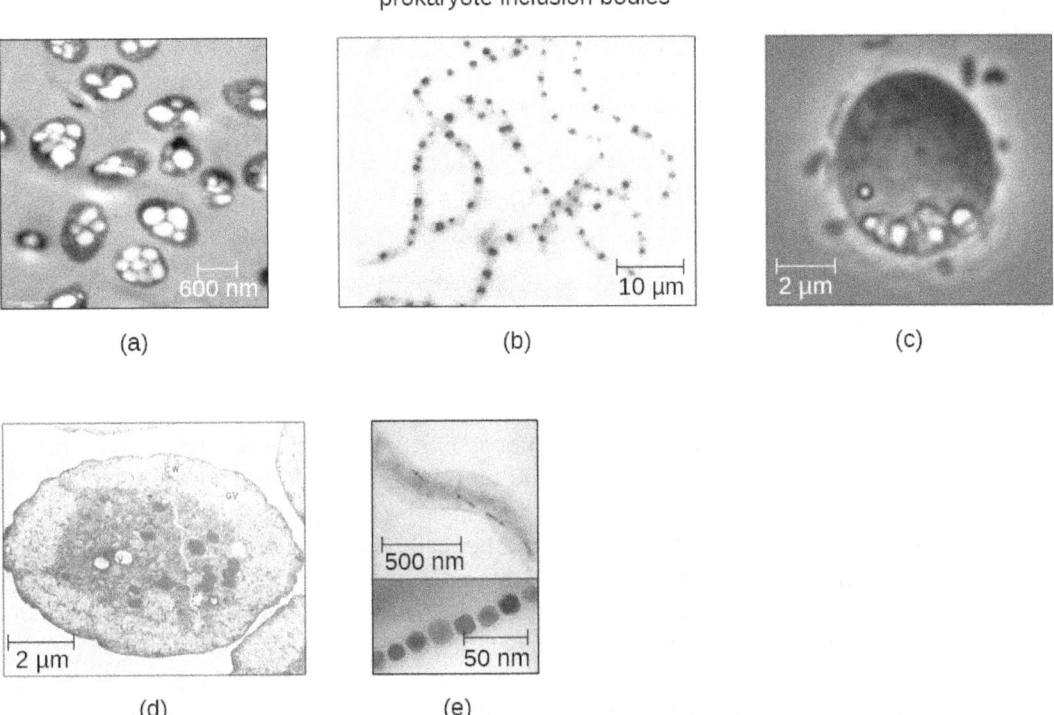

Figure 3.19 Prokaryotic cells may have various types of inclusions. (a) A transmission electron micrograph of polyhydroxybutryrate lipid droplets. (b) A light micrograph of volutin granules. (c) A phase-contrast micrograph of sulfur granules. (d) A transmission electron micrograph of gas vacuoles. (e) A transmission electron micrograph of magnetosomes. (credit b, c, d: modification of work by American Society for Microbiology)

Some prokaryotic cells have other types of inclusions that serve purposes other than nutrient storage. For example, some prokaryotic cells produce gas vacuoles, accumulations of small, protein-lined vesicles of gas. These gas vacuoles allow the prokaryotic cells that synthesize them to alter their buoyancy so that they can adjust their location in the water column. Magnetotactic bacteria, such as *Magnetospirillum magnetotacticum*, contain **magnetosomes**, which are inclusions of magnetic iron oxide or iron sulfide surrounded by a lipid layer. These allow cells to align along a magnetic field, aiding their movement (Figure 3.19). Cyanobacteria such as *Anabaena cylindrica* and bacteria such as *Halothiobacillus neapolitanus* produce **carboxysome** inclusions. Carboxysomes are composed of outer shells of thousands of protein subunits. Their interior is filled with ribulose-1,5-bisphosphate carboxylase/oxygenase (RuBisCO) and carbonic anhydrase. Both of these compounds are used for carbon metabolism. Some prokaryotic cells also possess carboxysomes that sequester functionally related enzymes in one location. These structures are considered proto-organelles because they compartmentalize important compounds or chemical reactions, much like many eukaryotic organelles.

Endospores

Bacterial cells are generally observed as **vegetative cells**, but some genera of bacteria have the ability to form **endospores**, structures that essentially protect the bacterial genome in a dormant state when environmental conditions are unfavorable. Endospores (not to be confused with the reproductive spores formed by fungi) allow some bacterial cells to survive long periods without food or water, as well as exposure to chemicals, extreme temperatures, and even radiation. Table 3.1 compares the characteristics of vegetative cells and endospores.

Characteristics of Vegetative Cells versus Endospores

Vegetative Cells	Endospores
Sensitive to extreme temperatures and radiation	Resistant to extreme temperatures and radiation
Gram-positive	Do not absorb Gram stain, only special endospore stains (see Staining Microscopic Specimens)
Normal water content and enzymatic activity	Dehydrated; no metabolic activity
Capable of active growth and metabolism	Dormant; no growth or metabolic activity

Table 3.1

The process by which vegetative cells transform into endospores is called **sporulation**, and it generally begins when nutrients become depleted or environmental conditions become otherwise unfavorable (Figure 3.20). The process begins with the formation of a septum in the vegetative bacterial cell. The septum divides the cell asymmetrically, separating a DNA forespore from the mother cell. The forespore, which will form the core of the endospore, is essentially a copy of the cell's chromosomes, and is separated from the mother cell by a second membrane. A cortex gradually forms around the forespore by laying down layers of calcium and dipicolinic acid between membranes. A protein spore coat then forms around the cortex while the DNA of the mother cell disintegrates. Further maturation of the endospore occurs with the formation of an outermost exosporium. The endospore is released upon disintegration of the mother cell, completing sporulation.

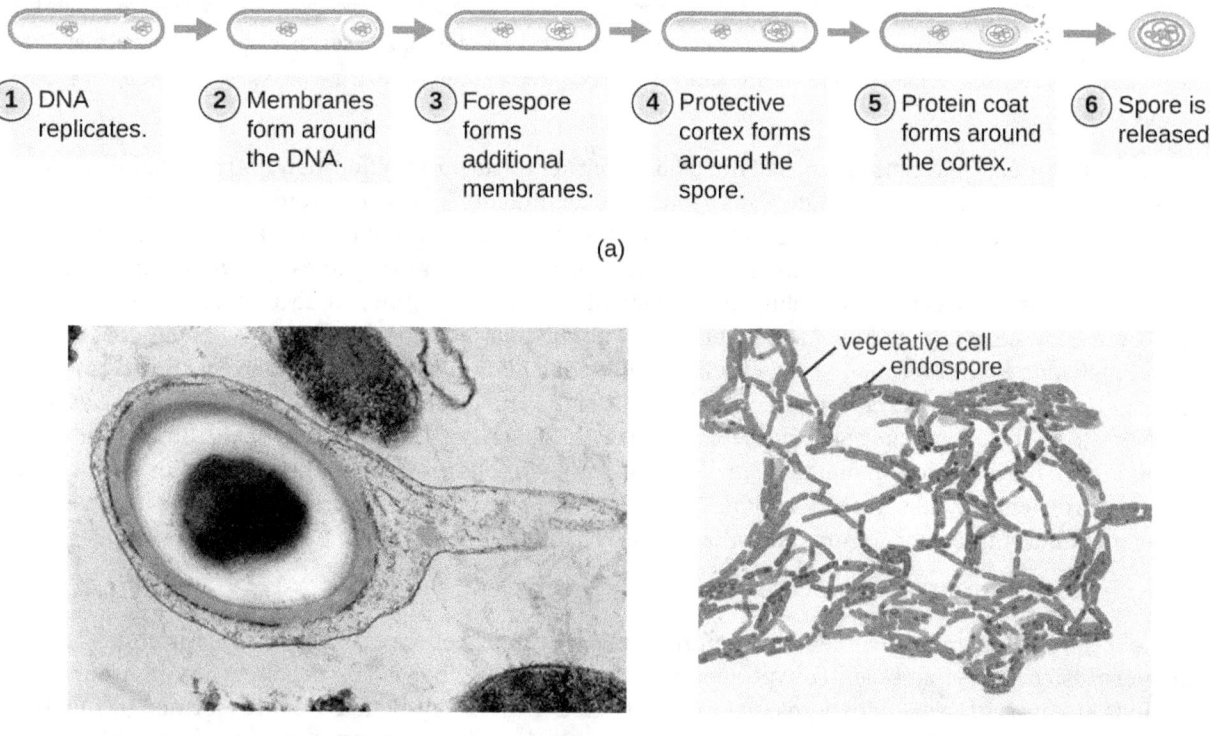

Figure 3.20 (a) Sporulation begins following asymmetric cell division. The forespore becomes surrounded by a double layer of membrane,

a cortex, and a protein spore coat, before being released as a mature endospore upon disintegration of the mother cell. (b) An electron micrograph of a *Carboxydothermus hydrogenoformans* endospore. (c) These *Bacillus spp.* cells are undergoing sporulation. The endospores have been visualized using Malachite Green spore stain. (credit b: modification of work by Jonathan Eisen)

Endospores of certain species have been shown to persist in a dormant state for extended periods of time, up to thousands of years.[17] However, when living conditions improve, endospores undergo **germination**, reentering a vegetative state. After germination, the cell becomes metabolically active again and is able to carry out all of its normal functions, including growth and cell division.

Not all bacteria have the ability to form endospores; however, there are a number of clinically significant endospore-forming gram-positive bacteria of the genera *Bacillus* and *Clostridium*. These include *B. anthracis*, the causative agent of anthrax, which produces endospores capable of survive for many decades[18]; *C. tetani* (causes tetanus); *C. difficile* (causes pseudomembranous colitis); *C. perfringens* (causes gas gangrene); and *C. botulinum* (causes botulism). Pathogens such as these are particularly difficult to combat because their endospores are so hard to kill. Special sterilization methods for endospore-forming bacteria are discussed in Control of Microbial Growth.

✓ CHECK YOUR UNDERSTANDING

- What is an inclusion?
- What is the function of an endospore?

Plasma Membrane

Structures that enclose the cytoplasm and internal structures of the cell are known collectively as the **cell envelope**. In prokaryotic cells, the structures of the cell envelope vary depending on the type of cell and organism. Most (but not all) prokaryotic cells have a cell wall, but the makeup of this cell wall varies. All cells (prokaryotic and eukaryotic) have a **plasma membrane** (also called **cytoplasmic membrane** or **cell membrane**) that exhibits selective permeability, allowing some molecules to enter or leave the cell while restricting the passage of others.

The structure of the plasma membrane is often described in terms of the **fluid mosaic model**, which refers to the ability of membrane components to move fluidly within the plane of the membrane, as well as the mosaic-like composition of the components, which include a diverse array of lipid and protein components (Figure 3.21). The plasma membrane structure of most bacterial and eukaryotic cell types is a bilayer composed mainly of phospholipids formed with ester linkages and proteins. These phospholipids and proteins have the ability to move laterally within the plane of the membranes as well as between the two phospholipid layers.

---

17 F. Rothfuss, M Bender, R Conrad. "Survival and Activity of Bacteria in a Deep, Aged Lake Sediment (Lake Constance)." *Microbial Ecology* 33 no. 1 (1997):69–77.

18 R. Sinclair et al. "Persistence of Category A Select Agents in the Environment." *Applied and Environmental Microbiology* 74 no. 3 (2008):555–563.

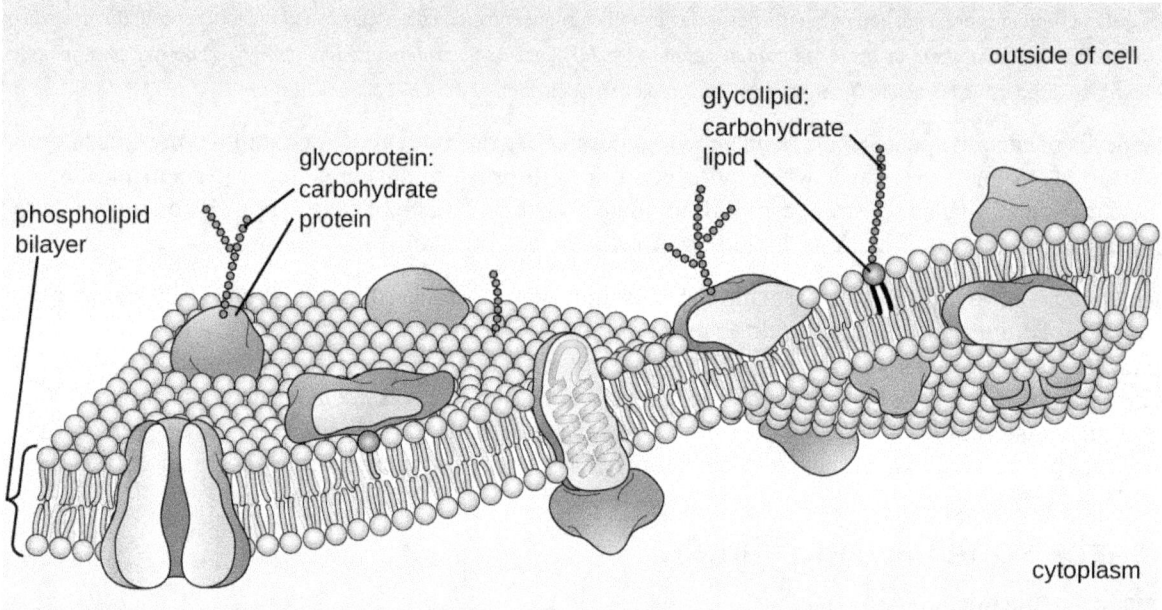

Figure 3.21 The bacterial plasma membrane is a phospholipid bilayer with a variety of embedded proteins that perform various functions for the cell. Note the presence of glycoproteins and glycolipids, whose carbohydrate components extend out from the surface of the cell. The abundance and arrangement of these proteins and lipids can vary greatly between species.

Archaeal membranes are fundamentally different from bacterial and eukaryotic membranes in a few significant ways. First, archaeal membrane phospholipids are formed with ether linkages, in contrast to the ester linkages found in bacterial or eukaryotic cell membranes. Second, archaeal phospholipids have branched chains, whereas those of bacterial and eukaryotic cells are straight chained. Finally, although some archaeal membranes can be formed of bilayers like those found in bacteria and eukaryotes, other archaeal plasma membranes are lipid monolayers.

Proteins on the cell's surface are important for a variety of functions, including cell-to-cell communication, and sensing environmental conditions and pathogenic virulence factors. Membrane proteins and phospholipids may have carbohydrates (sugars) associated with them and are called glycoproteins or glycolipids, respectively. These glycoprotein and glycolipid complexes extend out from the surface of the cell, allowing the cell to interact with the external environment (Figure 3.21). Glycoproteins and glycolipids in the plasma membrane can vary considerably in chemical composition among archaea, bacteria, and eukaryotes, allowing scientists to use them to characterize unique species.

Plasma membranes from different cells types also contain unique phospholipids, which contain fatty acids. As described in Using Biochemistry to Identify Microorganisms, phospholipid-derived fatty acid analysis (PLFA) profiles can be used to identify unique types of cells based on differences in fatty acids. Archaea, bacteria, and eukaryotes each have a unique PFLA profile.

Membrane Transport Mechanisms

One of the most important functions of the plasma membrane is to control the transport of molecules into and out of the cell. Internal conditions must be maintained within a certain range despite any changes in the external environment. The transport of substances across the plasma membrane allows cells to do so.

Cells use various modes of transport across the plasma membrane. For example, molecules moving from a higher concentration to a lower concentration with the concentration gradient are transported by simple diffusion, also known as passive transport (Figure 3.22). Some small molecules, like carbon dioxide, may cross the membrane bilayer directly by simple diffusion. However, charged molecules, as well as large molecules, need the help of carriers or channels in the membrane. These structures ferry molecules across the membrane, a process known as facilitated diffusion (Figure 3.23).

Active transport occurs when cells move molecules across their membrane *against* concentration gradients

([Figure 3.24](#)). A major difference between passive and active transport is that active transport requires adenosine triphosphate (ATP) or other forms of energy to move molecules "uphill." Therefore, active transport structures are often called "pumps."

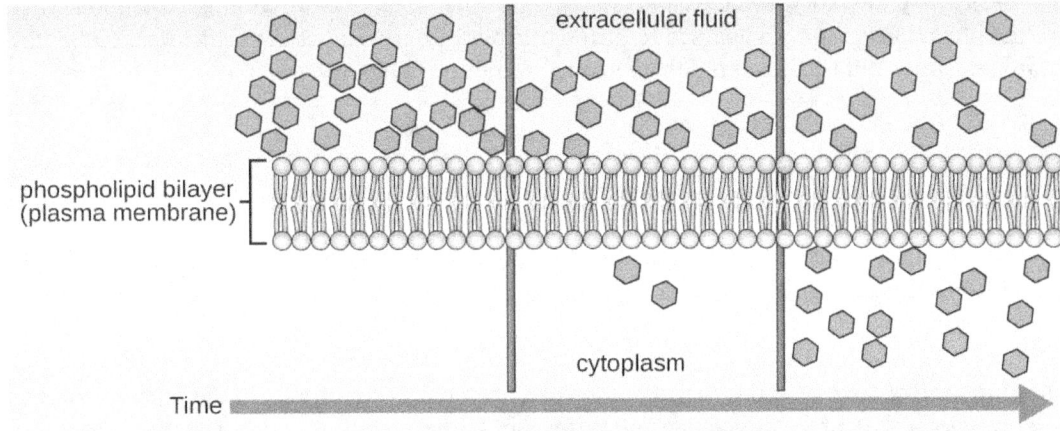

simple diffusion

Figure 3.22 Simple diffusion down a concentration gradient directly across the phospholipid bilayer. (credit: modification of work by Mariana Ruiz Villareal)

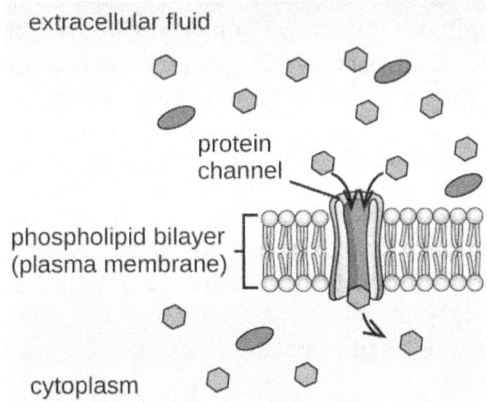

facilitated diffusion

Figure 3.23 Facilitated diffusion down a concentration gradient through a membrane protein. (credit: modification of work by Mariana Ruiz Villareal)

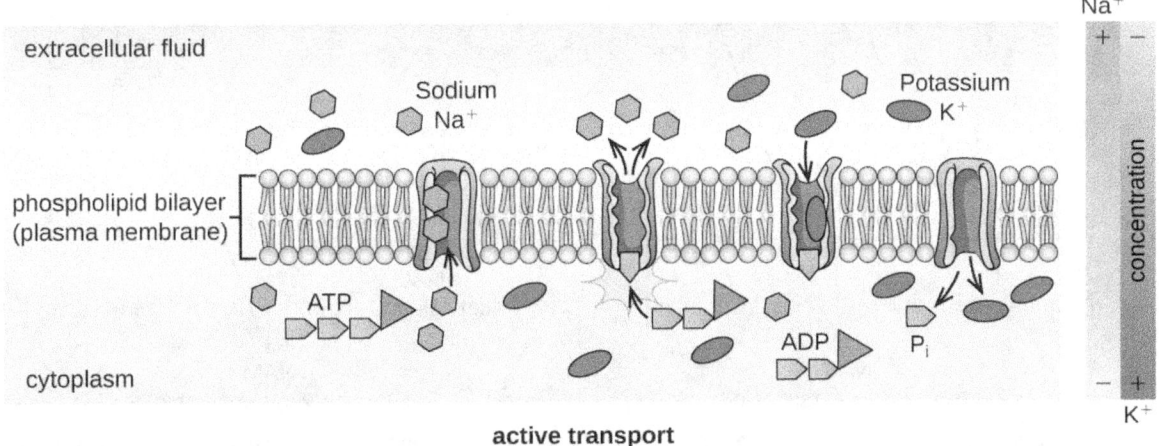

active transport

Figure 3.24 Active transport against a concentration gradient via a membrane pump that requires energy. (credit: modification of work by Mariana Ruiz Villareal)

Group translocation also transports substances into bacterial cells. In this case, as a molecule moves into a cell against its concentration gradient, it is chemically modified so that it does not require transport against an unfavorable concentration gradient. A common example of this is the bacterial phosphotransferase system, a series of carriers that phosphorylates (i.e., adds phosphate ions to) glucose or other sugars upon entry into cells. Since the phosphorylation of sugars is required during the early stages of sugar metabolism, the phosphotransferase system is considered to be an energy neutral system.

Photosynthetic Membrane Structures

Some prokaryotic cells, namely cyanobacteria and photosynthetic bacteria, have membrane structures that enable them to perform photosynthesis. These structures consist of an infolding of the plasma membrane that encloses photosynthetic pigments such as green **chlorophylls** and bacteriochlorophylls. In cyanobacteria, these membrane structures are called thylakoids; in photosynthetic bacteria, they are called chromatophores, lamellae, or chlorosomes.

Cell Wall

The primary function of the cell wall is to protect the cell from harsh conditions in the outside environment. When present, there are notable similarities and differences among the cell walls of archaea, bacteria, and eukaryotes.

The major component of bacterial cell walls is called **peptidoglycan** (or murein); it is only found in bacteria. Structurally, peptidoglycan resembles a layer of meshwork or fabric (Figure 3.25). Each layer is composed of long chains of alternating molecules of N-acetylglucosamine (NAG) and N-acetylmuramic acid (NAM). The structure of the long chains has significant two-dimensional tensile strength due to the formation of peptide bridges that connect NAG and NAM within each peptidoglycan layer. In gram-negative bacteria, tetrapeptide chains extending from each NAM unit are directly cross-linked, whereas in gram-positive bacteria, these tetrapeptide chains are linked by pentaglycine cross-bridges. Peptidoglycan subunits are made inside of the bacterial cell and then exported and assembled in layers, giving the cell its shape.

Since peptidoglycan is unique to bacteria, many antibiotic drugs are designed to interfere with peptidoglycan synthesis, weakening the cell wall and making bacterial cells more susceptible to the effects of osmotic pressure (see Mechanisms of Antibacterial Drugs). In addition, certain cells of the human immune system are able "recognize" bacterial pathogens by detecting peptidoglycan on the surface of a bacterial cell; these cells then engulf and destroy the bacterial cell, using enzymes such as lysozyme, which breaks down and digests the peptidoglycan in their cell walls (see Pathogen Recognition and Phagocytosis).

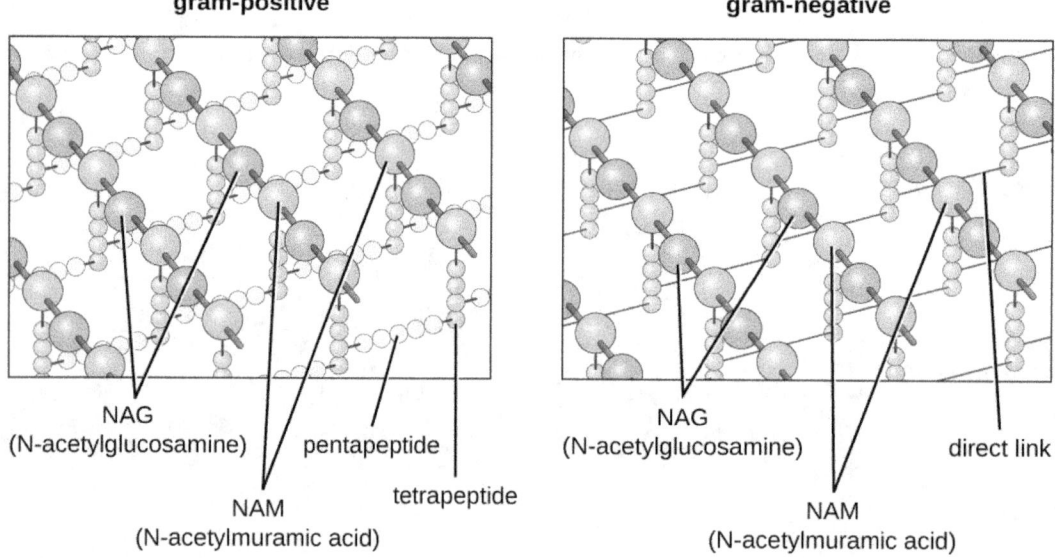

Figure 3.25 Peptidoglycan is composed of polymers of alternating NAM and NAG subunits, which are cross-linked by peptide bridges linking NAM subunits from various glycan chains. This provides the cell wall with tensile strength in two dimensions.

The Gram staining protocol (see Staining Microscopic Specimens) is used to differentiate two common types of cell wall structures (Figure 3.26). Gram-positive cells have a cell wall consisting of many layers of peptidoglycan totaling 30–100 nm in thickness. These peptidoglycan layers are commonly embedded with teichoic acids (TAs), carbohydrate chains that extend through and beyond the peptidoglycan layer.[19] TA is thought to stabilize peptidoglycan by increasing its rigidity. TA also plays a role in the ability of pathogenic gram-positive bacteria such as *Streptococcus* to bind to certain proteins on the surface of host cells, enhancing their ability to cause infection. In addition to peptidoglycan and TAs, bacteria of the family Mycobacteriaceae have an external layer of waxy **mycolic acids** in their cell wall; as described in Staining Microscopic Specimens, these bacteria are referred to as acid-fast, since acid-fast stains must be used to penetrate the mycolic acid layer for purposes of microscopy (Figure 3.27).

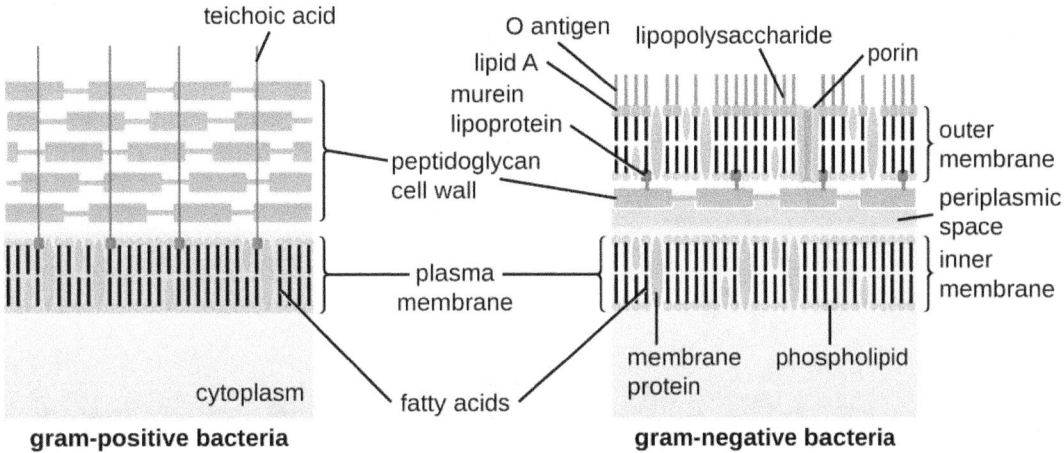

Figure 3.26 Bacteria contain two common cell wall structural types. Gram-positive cell walls are structurally simple, containing a thick layer of peptidoglycan with embedded teichoic acid external to the plasma membrane.[20] Gram-negative cell walls are structurally more complex, containing three layers: the inner membrane, a thin layer of peptidoglycan, and an outer membrane containing lipopolysaccharide. (credit: modification of work by "Franciscosp2"/Wikimedia Commons)

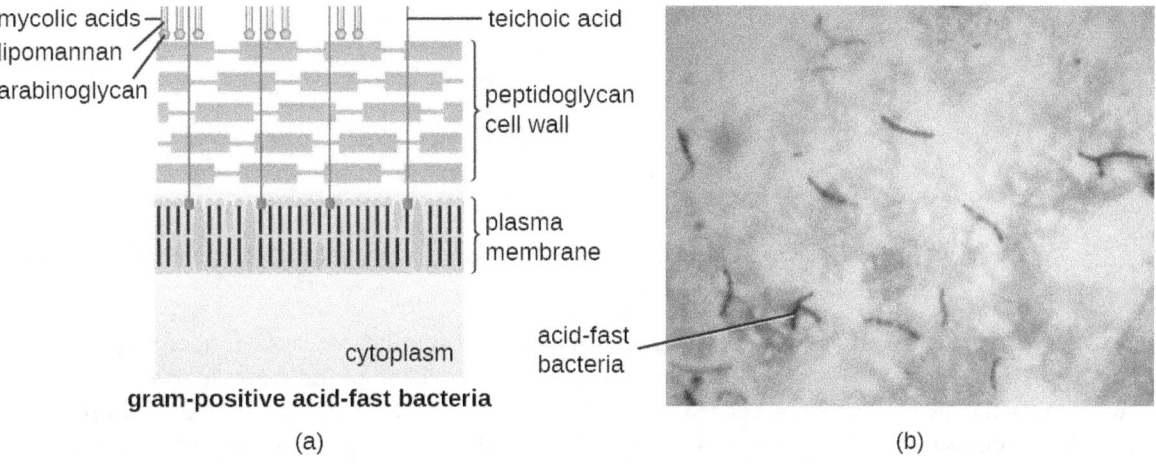

Figure 3.27 (a) Some gram-positive bacteria, including members of the Mycobacteriaceae, produce waxy mycolic acids found exterior to their structurally-distinct peptidoglycan. (b) The acid-fast staining protocol detects the presence of cell walls that are rich in mycolic acid. Acid-fast cells are stained red by carbolfuschin. (credit a: modification of work by "Franciscosp2"/Wikimedia Commons; credit b: modification of work by Centers for Disease Control and Prevention)

---

19 T.J. Silhavy, D. Kahne, S. Walker. "The Bacterial Cell Envelope." *Cold Spring Harbor Perspectives in Biology* 2 no. 5 (2010):a000414.

20 B. Zuber et al. "Granular Layer in the Periplasmic Space of Gram-Positive Bacteria and Fine Structures of *Enterococcus gallinarum* and *Streptococcus gordonii* Septa Revealed by Cryo-Electron Microscopy of Vitreous Sections." *Journal of Bacteriology* 188 no. 18 (2006):6652–6660

Gram-negative cells have a much thinner layer of peptidoglycan (no more than about 4 nm thick[21]) than gram-positive cells, and the overall structure of their cell envelope is more complex. In gram-negative cells, a gel-like matrix occupies the **periplasmic space** between the cell wall and the plasma membrane, and there is a second lipid bilayer called the **outer membrane**, which is external to the peptidoglycan layer (Figure 3.26). This outer membrane is attached to the peptidoglycan by murein lipoprotein. The outer leaflet of the outer membrane contains the molecule **lipopolysaccharide (LPS)**, which functions as an endotoxin in infections involving gram-negative bacteria, contributing to symptoms such as fever, hemorrhaging, and septic shock. Each LPS molecule is composed of Lipid A, a core polysaccharide, and an O side chain that is composed of sugar-like molecules that comprise the external face of the LPS (Figure 3.28). The composition of the O side chain varies between different species and strains of bacteria. Parts of the O side chain called antigens can be detected using serological or immunological tests to identify specific pathogenic strains like *Escherichia coli* O157:H7, a deadly strain of bacteria that causes bloody diarrhea and kidney failure.

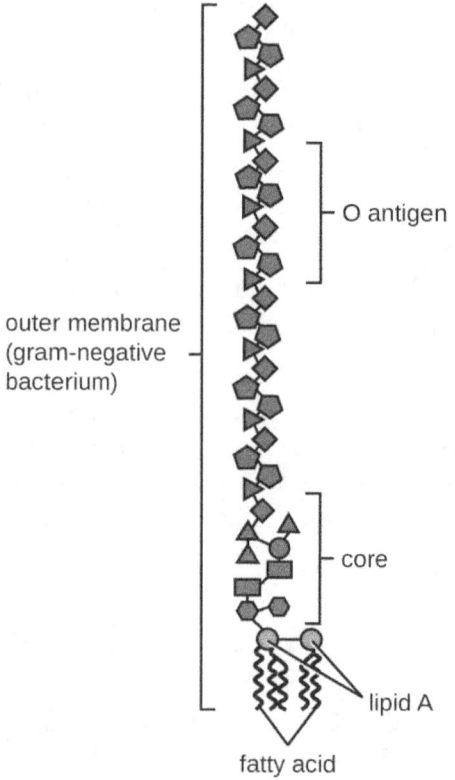

Figure 3.28 The outer membrane of a gram-negative bacterial cell contains lipopolysaccharide (LPS), a toxin composed of Lipid A embedded in the outer membrane, a core polysaccharide, and the O side chain.

Archaeal cell wall structure differs from that of bacteria in several significant ways. First, archaeal cell walls do not contain peptidoglycan; instead, they contain a similar polymer called pseudopeptidoglycan (pseudomurein) in which NAM is replaced with a different subunit. Other archaea may have a layer of glycoproteins or polysaccharides that serves as the cell wall instead of pseudopeptidoglycan. Last, as is the case with some bacterial species, there are a few archaea that appear to lack cell walls entirely.

Glycocalyces and S-Layers

Although most prokaryotic cells have cell walls, some may have additional cell envelope structures exterior to the cell wall, such as glycocalyces and S-layers. A **glycocalyx** is a sugar coat, of which there are two important types: capsules and slime layers. A **capsule** is an organized layer located outside of the cell wall and usually composed of polysaccharides or proteins (Figure 3.29). A **slime layer** is a less tightly organized layer that is only loosely attached to the cell wall and can be more easily washed off. Slime layers may be composed of

---

21 L. Gana, S. Chena, G.J. Jensena. "Molecular Organization of Gram-Negative Peptidoglycan." *Proceedings of the National Academy of Sciences of the United States of America* 105 no. 48 (2008):18953–18957.

polysaccharides, glycoproteins, or glycolipids.

Glycocalyces allows cells to adhere to surfaces, aiding in the formation of biofilms (colonies of microbes that form in layers on surfaces). In nature, most microbes live in mixed communities within biofilms, partly because the biofilm affords them some level of protection. Biofilms generally hold water like a sponge, preventing desiccation. They also protect cells from predation and hinder the action of antibiotics and disinfectants. All of these properties are advantageous to the microbes living in a biofilm, but they present challenges in a clinical setting, where the goal is often to eliminate microbes.

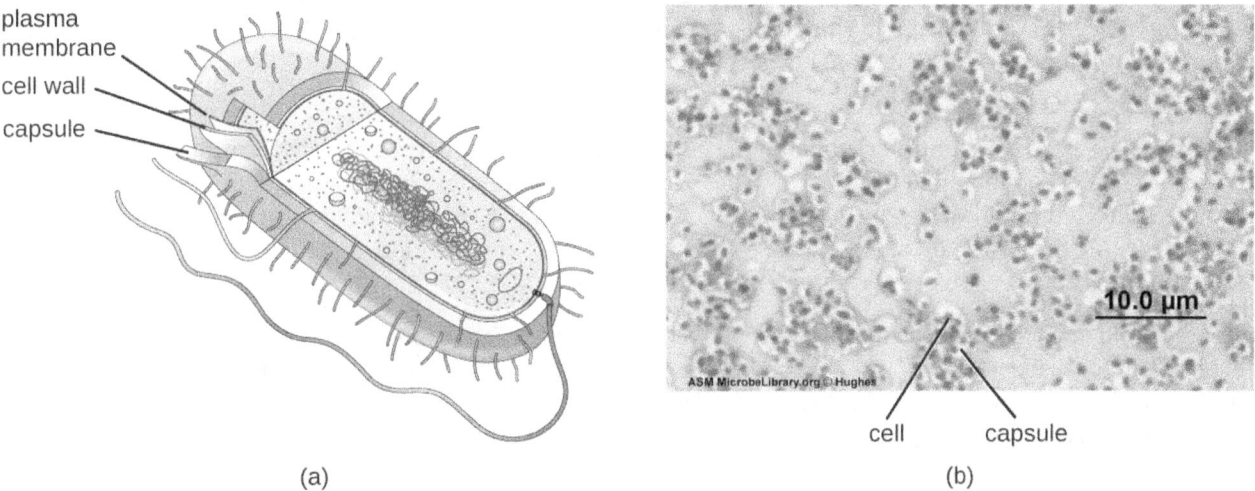

Figure 3.29 (a) Capsules are a type of glycocalyx composed of an organized layer of polysaccharides. (b) A capsule stain of *Pseudomonas aeruginosa*, a bacterial pathogen capable of causing many different types of infections in humans. (credit b: modification of work by American Society for Microbiology)

The ability to produce a capsule can contribute to a microbe's pathogenicity (ability to cause disease) because the capsule can make it more difficult for phagocytic cells (such as white blood cells) to engulf and kill the microorganism. *Streptococcus pneumoniae*, for example, produces a capsule that is well known to aid in this bacterium's pathogenicity. As explained in Staining Microscopic specimens, capsules are difficult to stain for microscopy; negative staining techniques are typically used.

An **S-layer** is another type of cell envelope structure; it is composed of a mixture of structural proteins and glycoproteins. In bacteria, S-layers are found outside the cell wall, but in some archaea, the S-layer serves *as* the cell wall. The exact function of S-layers is not entirely understood, and they are difficult to study; but available evidence suggests that they may play a variety of functions in different prokaryotic cells, such as helping the cell withstand osmotic pressure and, for certain pathogens, interacting with the host immune system.

> Clinical Focus
>
> **Part 3**
> After diagnosing Barbara with pneumonia, the PA writes her a prescription for amoxicillin, a commonly-prescribed type of penicillin derivative. More than a week later, despite taking the full course as directed, Barbara still feels weak and is not fully recovered, although she is still able to get through her daily activities. She returns to the health center for a follow-up visit.
>
> Many types of bacteria, fungi, and viruses can cause pneumonia. Amoxicillin targets the peptidoglycan of bacterial cell walls. Since the amoxicillin has not resolved Barbara's symptoms, the PA concludes that the causative agent probably lacks peptidoglycan, meaning that the pathogen could be a virus, a fungus, or a bacterium that lacks peptidoglycan. Another possibility is that the pathogen is a bacterium containing peptidoglycan but has developed resistance to amoxicillin.

- How can the PA definitively identify the cause of Barbara's pneumonia?
- What form of treatment should the PA prescribe, given that the amoxicillin was ineffective?

Jump to the next Clinical Focus box. Go back to the previous Clinical Focus box.

Filamentous Appendages

Many bacterial cells have protein appendages embedded within their cell envelopes that extend outward, allowing interaction with the environment. These appendages can attach to other surfaces, transfer DNA, or provide movement. Filamentous appendages include fimbriae, pili, and flagella.

Fimbriae and Pili

Fimbriae and pili are structurally similar and, because differentiation between the two is problematic, these terms are often used interchangeably.[22][23] The term **fimbriae** commonly refers to short bristle-like proteins projecting from the cell surface by the hundreds. Fimbriae enable a cell to attach to surfaces and to other cells. For pathogenic bacteria, adherence to host cells is important for colonization, infectivity, and virulence. Adherence to surfaces is also important in biofilm formation.

The term **pili** (singular: pilus) commonly refers to longer, less numerous protein appendages that aid in attachment to surfaces (Figure 3.30). A specific type of pilus, called the **F pilus** or **sex pilus**, is important in the transfer of DNA between bacterial cells, which occurs between members of the same generation when two cells physically transfer or exchange parts of their respective genomes (see How Asexual Prokaryotes Achieve Genetic Diversity).

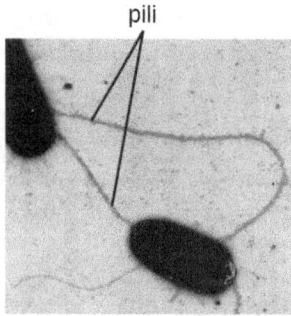

Figure 3.30 Bacteria may produce two different types of protein appendages that aid in surface attachment. Fimbriae typically are more numerous and shorter, whereas pili (shown here) are longer and less numerous per cell. (credit: modification of work by American Society for Microbiology)

 MICRO CONNECTIONS

Group A Strep

Before the structure and function of the various components of the bacterial cell envelope were well understood, scientists were already using cell envelope characteristics to classify bacteria. In 1933, Rebecca Lancefield proposed a method for serotyping various β-hemolytic strains of *Streptococcus* species using an agglutination assay, a technique using the clumping of bacteria to detect specific cell-surface antigens. In doing so, Lancefield discovered that one group of *S. pyogenes*, found in Group A, was associated with a variety of human diseases. She determined that various strains of Group A strep could be distinguished from each other based on variations in specific cell surface proteins that she named M proteins.

---

22   J.A. Garnetta et al. "Structural Insights Into the Biogenesis and Biofilm Formation by the *Escherichia coli* Common Pilus." *Proceedings of the National Academy of Sciences of the United States of America* 109 no. 10 (2012):3950–3955.

23   T. Proft, E.N. Baker. "Pili in Gram-Negative and Gram-Positive Bacteria—Structure, Assembly and Their Role in Disease." *Cellular and Molecular Life Sciences* 66 (2009):613.

Today, more than 80 different strains of Group A strep have been identified based on M proteins. Various strains of Group A strep are associated with a wide variety of human infections, including streptococcal pharyngitis (strep throat), impetigo, toxic shock syndrome, scarlet fever, rheumatic fever, and necrotizing fasciitis. The M protein is an important virulence factor for Group A strep, helping these strains evade the immune system. Changes in M proteins appear to alter the infectivity of a particular strain of Group A strep.

Flagella

Flagella are structures used by cells to move in aqueous environments. Bacterial flagella act like propellers. They are stiff spiral filaments composed of flagellin protein subunits that extend outward from the cell and spin in solution. The **basal body** is the motor for the flagellum and is embedded in the plasma membrane (Figure 3.31). A hook region connects the basal body to the filament. Gram-positive and gram-negative bacteria have different basal body configurations due to differences in cell wall structure.

Different types of motile bacteria exhibit different arrangements of flagella (Figure 3.32). A bacterium with a singular flagellum, typically located at one end of the cell (polar), is said to have a **monotrichous** flagellum. An example of a monotrichously flagellated bacterial pathogen is *Vibrio cholerae*, the gram-negative bacterium that causes cholera. Cells with **amphitrichous** flagella have a flagellum or tufts of flagella at each end. An example is *Spirillum minor*, the cause of spirillary (Asian) rat-bite fever or sodoku. Cells with **lophotrichous** flagella have a tuft at one end of the cell. The gram-negative bacillus *Pseudomonas aeruginosa*, an opportunistic pathogen known for causing many infections, including "swimmer's ear" and burn wound infections, has lophotrichous flagella. Flagella that cover the entire surface of a bacterial cell are called **peritrichous** flagella. The gram-negative bacterium *E. coli* shows a peritrichous arrangement of flagella.

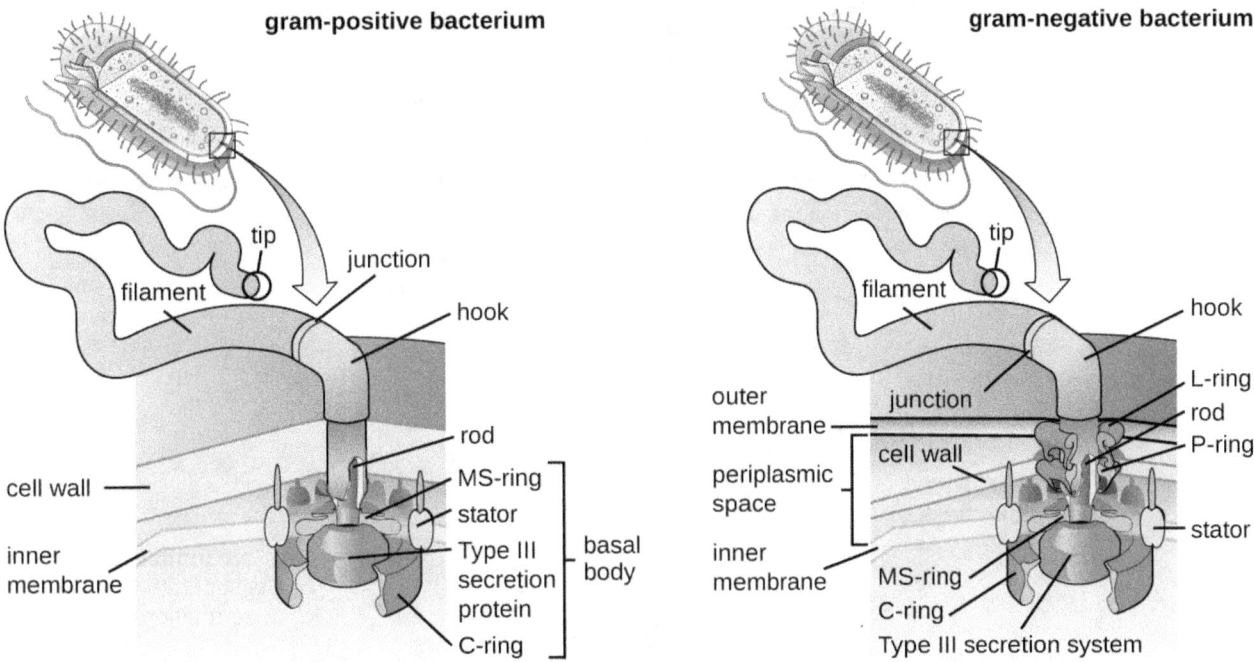

Figure 3.31 The basic structure of a bacterial flagellum consists of a basal body, hook, and filament. The basal body composition and arrangement differ between gram-positive and gram-negative bacteria. (credit: modification of work by "LadyofHats"/Mariana Ruiz Villareal)

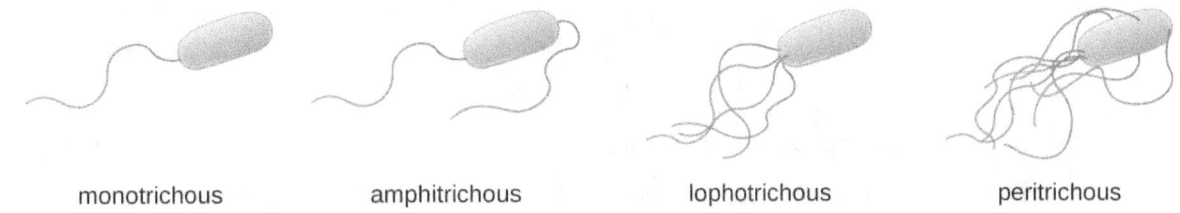

Figure 3.32 Flagellated bacteria may exhibit multiple arrangements of their flagella. Common arrangements include monotrichous,

amphitrichous, lophotrichous, or peritrichous.

Directional movement depends on the configuration of the flagella. Bacteria can move in response to a variety of environmental signals, including light (**phototaxis**), magnetic fields (**magnetotaxis**) using magnetosomes, and, most commonly, chemical gradients (**chemotaxis**). Purposeful movement toward a chemical attractant, like a food source, or away from a repellent, like a poisonous chemical, is achieved by increasing the length of **runs** and decreasing the length of **tumbles**. When running, flagella rotate in a counterclockwise direction, allowing the bacterial cell to move forward. In a peritrichous bacterium, the flagella are all bundled together in a very streamlined way (Figure 3.33), allowing for efficient movement. When tumbling, flagella are splayed out while rotating in a clockwise direction, creating a looping motion and preventing meaningful forward movement but reorienting the cell toward the direction of the attractant. When an attractant exists, runs and tumbles still occur; however, the length of runs is longer, while the length of the tumbles is reduced, allowing overall movement toward the higher concentration of the attractant. When no chemical gradient exists, the lengths of runs and tumbles are more equal, and overall movement is more random (Figure 3.34).

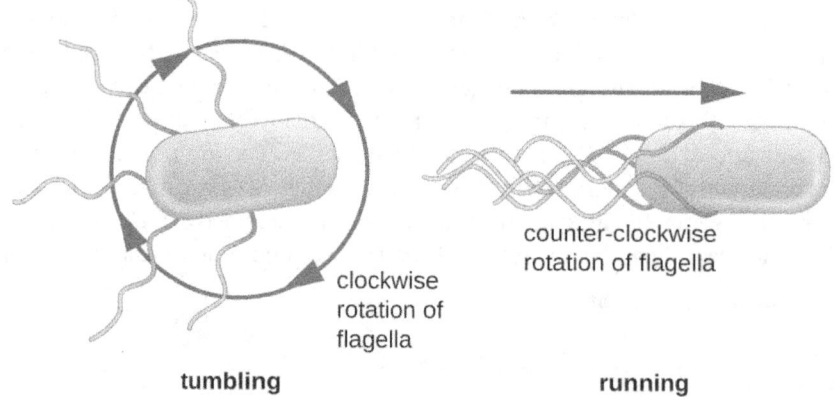

Figure 3.33 Bacteria achieve directional movement by changing the rotation of their flagella. In a cell with peritrichous flagella, the flagella bundle when they rotate in a counterclockwise direction, resulting in a run. However, when the flagella rotate in a clockwise direction, the flagella are no longer bundled, resulting in tumbles.

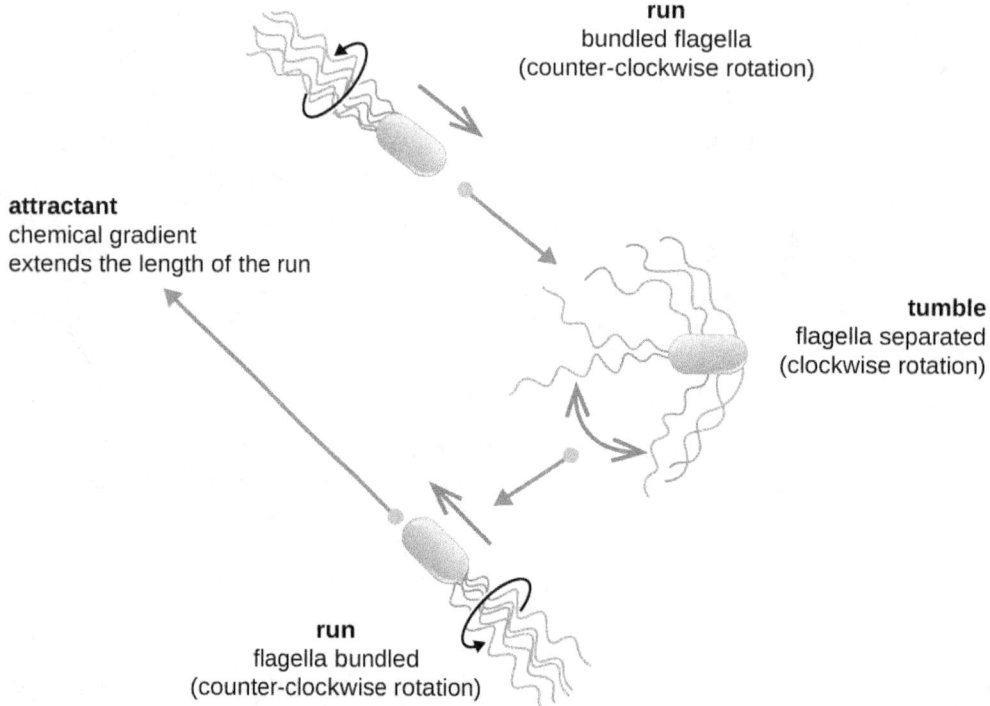

Figure 3.34 Without a chemical gradient, flagellar rotation cycles between counterclockwise (run) and clockwise (tumble) with no overall directional movement. However, when a chemical gradient of an attractant exists, the length of runs is extended, while the length of

tumbles is decreased. This leads to chemotaxis: an overall directional movement toward the higher concentration of the attractant.

CHECK YOUR UNDERSTANDING

- What is the peptidoglycan layer and how does it differ between gram-positive and gram-negative bacteria?
- Compare and contrast monotrichous, amphitrichous, lophotrichous, and peritrichous flagella.

3.4 Unique Characteristics of Eukaryotic Cells

Learning Objectives

By the end of this section, you will be able to:
- Explain the distinguishing characteristics of eukaryotic cells
- Describe internal and external structures of eukaryotic cells in terms of their physical structure, chemical structure, and function
- Identify and describe structures and organelles unique to eukaryotic cells
- Compare and contrast similar structures found in prokaryotic and eukaryotic cells
- Describe the processes of eukaryotic mitosis and meiosis, and compare to prokaryotic binary fission

Eukaryotic organisms include protozoans, algae, fungi, plants, and animals. Some eukaryotic cells are independent, single-celled microorganisms, whereas others are part of multicellular organisms. The cells of eukaryotic organisms have several distinguishing characteristics. Above all, eukaryotic cells are defined by the presence of a nucleus surrounded by a complex nuclear membrane. Also, eukaryotic cells are characterized by the presence of membrane-bound organelles in the cytoplasm. Organelles such as mitochondria, the endoplasmic reticulum (ER), Golgi apparatus, lysosomes, and peroxisomes are held in place by the **cytoskeleton**, an internal network that supports transport of intracellular components and helps maintain cell shape (Figure 3.35). The genome of eukaryotic cells is packaged in multiple, rod-shaped chromosomes as opposed to the single, circular-shaped chromosome that characterizes most prokaryotic cells. Table 3.2 compares the characteristics of eukaryotic cell structures with those of bacteria and archaea.

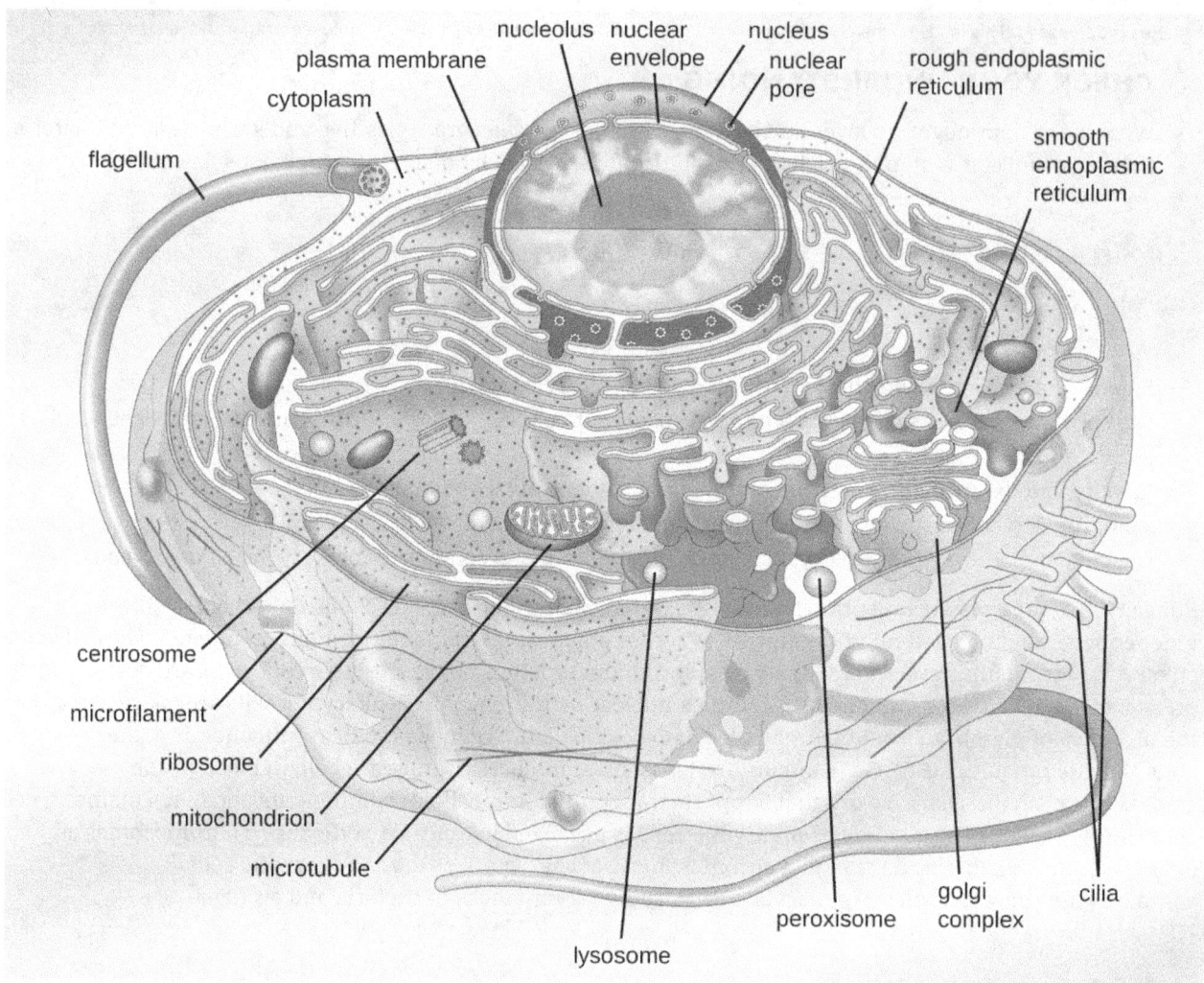

Figure 3.35 An illustration of a generalized, single-celled eukaryotic organism. Note that cells of eukaryotic organisms vary greatly in terms of structure and function, and a particular cell may not have all of the structures shown here.

Summary of Cell Structures

Cell Structure	Prokaryotes		Eukaryotes
	Bacteria	Archaea	
Size	~0.5–1 μm	~0.5–1 μm	~5–20 μm
Surface area-to-volume ratio	High	High	Low
Nucleus	No	No	Yes
Genome characteristics	• Single chromosome • Circular • Haploid • Lacks histones	• Single chromosome • Circular • Haploid • Contains histones	• Multiple chromosomes • Linear • Haploid or diploid • Contains histones

Summary of Cell Structures

Cell Structure	Prokaryotes		Eukaryotes
	Bacteria	Archaea	
Cell division	Binary fission	Binary fission	Mitosis, meiosis
Membrane lipid composition	• Ester-linked • Straight-chain fatty acids • Bilayer	• Ether-linked • Branched isoprenoids • Bilayer or monolayer	• Ester-linked • Straight-chain fatty acids • Sterols • Bilayer
Cell wall composition	• Peptidoglycan, or • None	• Pseudopeptidoglycan, or • Glycopeptide, or • Polysaccharide, or • Protein (S-layer), or • None	• Cellulose (plants, some algae) • Chitin (molluscs, insects, crustaceans, and fungi) • Silica (some algae) • Most others lack cell walls
Motility structures	Rigid spiral flagella composed of flagellin	Rigid spiral flagella composed of archaeal flagellins	Flexible flagella and cilia composed of microtubules
Membrane-bound organelles	No	No	Yes
Endomembrane system	No	No	Yes (ER, Golgi, lysosomes)
Ribosomes	70S	70S	• 80S in cytoplasm and rough ER • 70S in mitochondria, chloroplasts

Table 3.2

Cell Morphologies

Eukaryotic cells display a wide variety of different cell morphologies. Possible shapes include spheroid, ovoid, cuboidal, cylindrical, flat, lenticular, fusiform, discoidal, crescent, ring stellate, and polygonal (Figure 3.36). Some eukaryotic cells are irregular in shape, and some are capable of changing shape. The shape of a particular type of eukaryotic cell may be influenced by factors such as its primary function, the organization of its cytoskeleton, the viscosity of its cytoplasm, the rigidity of its cell membrane or cell wall (if it has one), and the physical pressure exerted on it by the surrounding environment and/or adjoining cells.

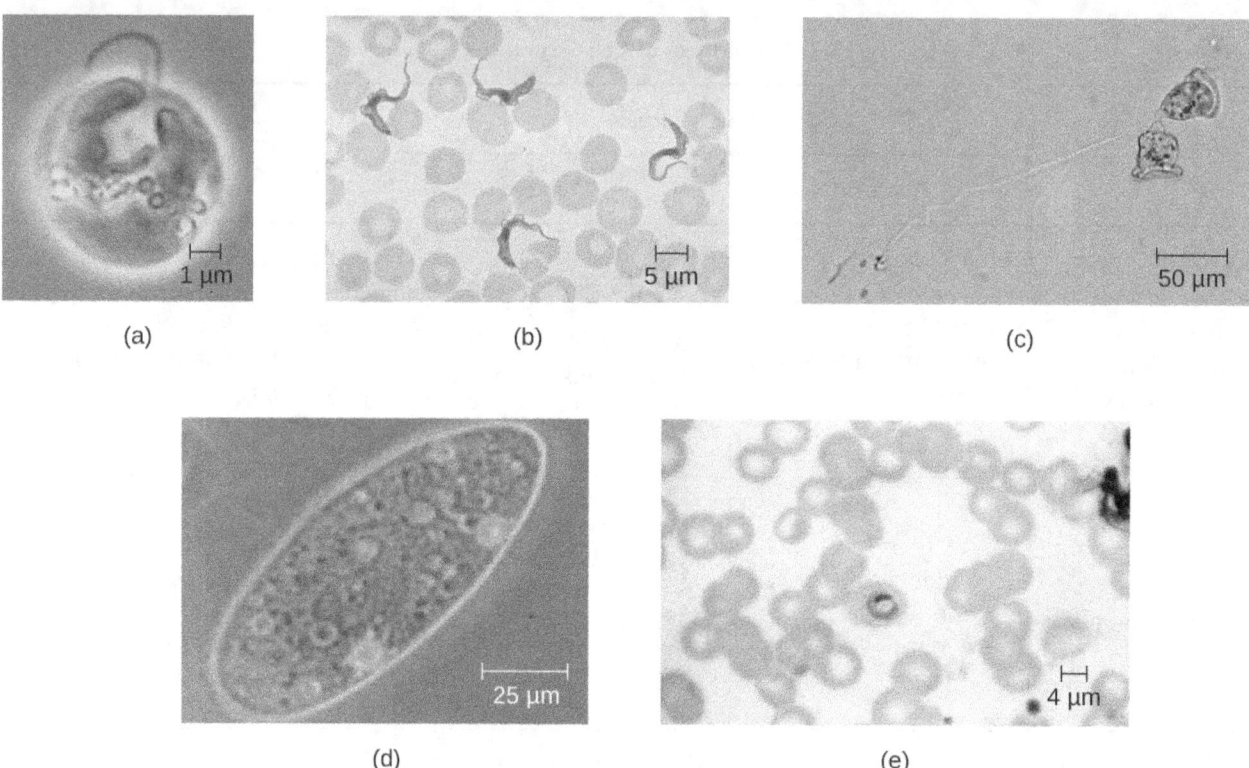

Figure 3.36 Eukaryotic cells come in a variety of cell shapes. (a) Spheroid *Chromulina* alga. (b) Fusiform shaped *Trypanosoma*. (c) Bell-shaped *Vorticella*. (d) Ovoid *Paramecium*. (e) Ring-shaped *Plasmodium ovale*. (credit a: modification of work by NOAA; credit b, e: modification of work by Centers for Disease Control and Prevention)

✓ CHECK YOUR UNDERSTANDING

- Identify two differences between eukaryotic and prokaryotic cells.

Nucleus

Unlike prokaryotic cells, in which DNA is loosely contained in the nucleoid region, eukaryotic cells possess a **nucleus**, which is surrounded by a complex nuclear membrane that houses the DNA genome (Figure 3.37). By containing the cell's DNA, the nucleus ultimately controls all activities of the cell and also serves an essential role in reproduction and heredity. Eukaryotic cells typically have their DNA organized into multiple linear chromosomes. The DNA within the nucleus is highly organized and condensed to fit inside the nucleus, which is accomplished by wrapping the DNA around proteins called histones.

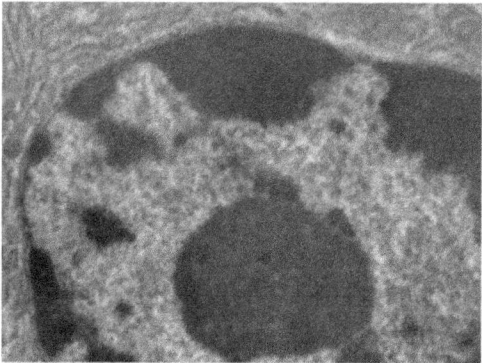

Figure 3.37 Eukaryotic cells have a well-defined nucleus surrounded by a nuclear membrane. The nucleus of this mammalian lung cell is located in the bottom right corner of the image. The large, dark, oval-shaped structure within the nucleus is the nucleolus.

Although most eukaryotic cells have only one nucleus, exceptions exist. For example, protozoans of the genus

Paramecium typically have two complete nuclei: a small nucleus that is used for reproduction (micronucleus) and a large nucleus that directs cellular metabolism (macronucleus). Additionally, some fungi transiently form cells with two nuclei, called heterokaryotic cells, during sexual reproduction. Cells whose nuclei divide, but whose cytoplasm does not, are called **coenocytes**.

The nucleus is bound by a complex **nuclear membrane**, often called the **nuclear envelope**, that consists of two distinct lipid bilayers that are contiguous with each other (Figure 3.38). Despite these connections between the inner and outer membranes, each membrane contains unique lipids and proteins on its inner and outer surfaces. The nuclear envelope contains nuclear pores, which are large, rosette-shaped protein complexes that control the movement of materials into and out of the nucleus. The overall shape of the nucleus is determined by the **nuclear lamina**, a meshwork of intermediate filaments found just inside the nuclear envelope membranes. Outside the nucleus, additional intermediate filaments form a looser mesh and serve to anchor the nucleus in position within the cell.

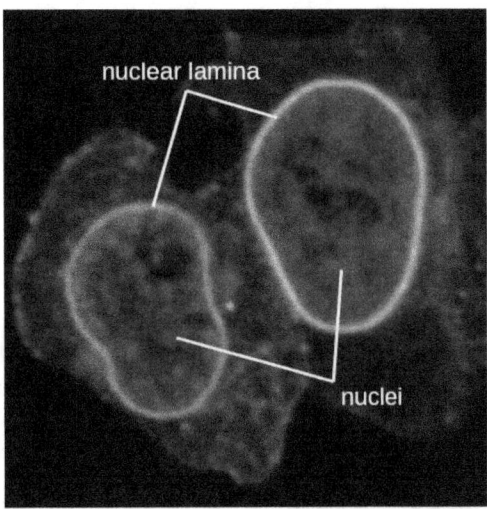

Figure 3.38   In this fluorescent microscope image, all the intermediate filaments have been stained with a bright green fluorescent stain. The nuclear lamina is the intense bright green ring around the faint red nuclei.

Eukaryotes are able to multiply through asexual reproduction, during which a single parent cell becomes two identical daughter cells. This process of clonal reproduction is called **mitosis**. Although mitosis may sound similar to asexual binary fission in prokaryotes, the processes are very different. In contrast to the single chromosome in most prokaryotes, eukaryotic cells possess multiple chromosomes that must be replicated and strategically divided between daughter cells. Therefore, mitosis is a much more complex cellular process than binary fission.

The eukaryotic cell cycle is an ordered and carefully regulated series of events involving cell growth, DNA replication, and cell division to produce two clonal daughter cells. One "turn" or cycle of the cell cycle consist of two general phases: interphase and the mitotic phase. (Figure 3.39). During **interphase**, the cell is not dividing, but rather is undergoing normal growth processes and DNA is replicated preparing for cell division. The three stages of interphase are called G1, S, and G2.

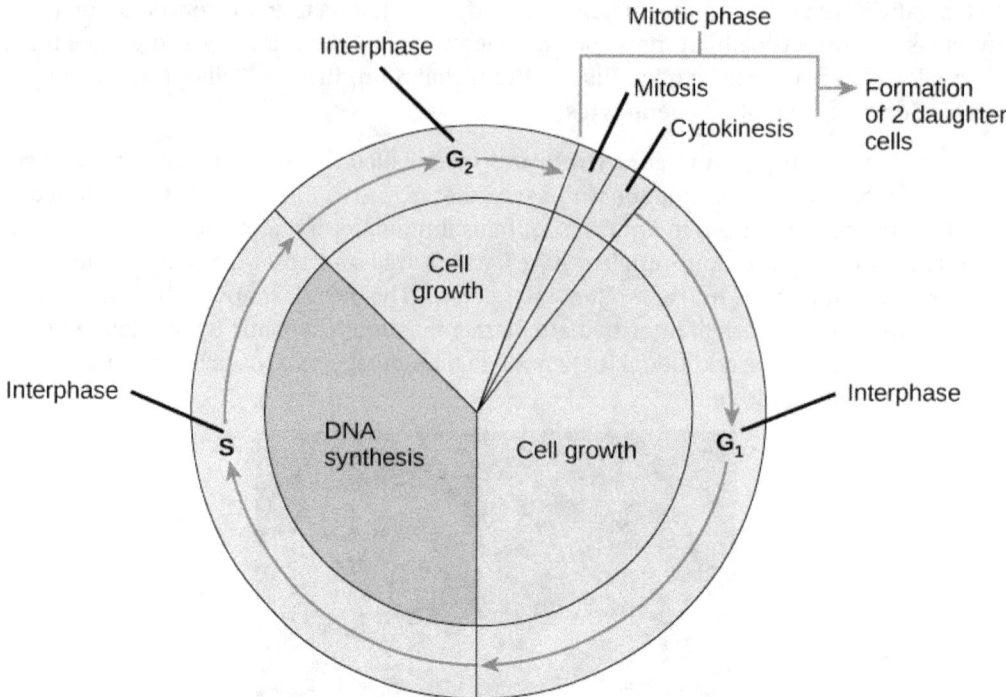

Figure 3.39 The cell cycle consists of interphase and the mitotic phase. During interphase, the cell grows and the nuclear DNA is duplicated. Interphase is followed by the mitotic phase. During the mitotic phase, the duplicated chromosomes are segregated and distributed into daughter nuclei. The cytoplasm is usually divided as well, resulting in two daughter cells.

The **mitotic phase** is a multistep process during which the duplicated chromosomes are aligned, separated, move to opposite poles of the cell, and then are divided into two identical daughter cells. The first portion of the mitotic phase is called **karyokinesis**, or nuclear division. Karyokinesis is divided into a series of phases—prophase, prometaphase, metaphase, anaphase, and telophase—that result in the division of the cell nucleus (Figure 3.40). The second portion of the mitotic phase, called **cytokinesis**, is the physical separation of the cytoplasmic components into the two daughter cells.

LINK TO LEARNING

Go to this University of Arizona website about the stages of mitosis to learn more.

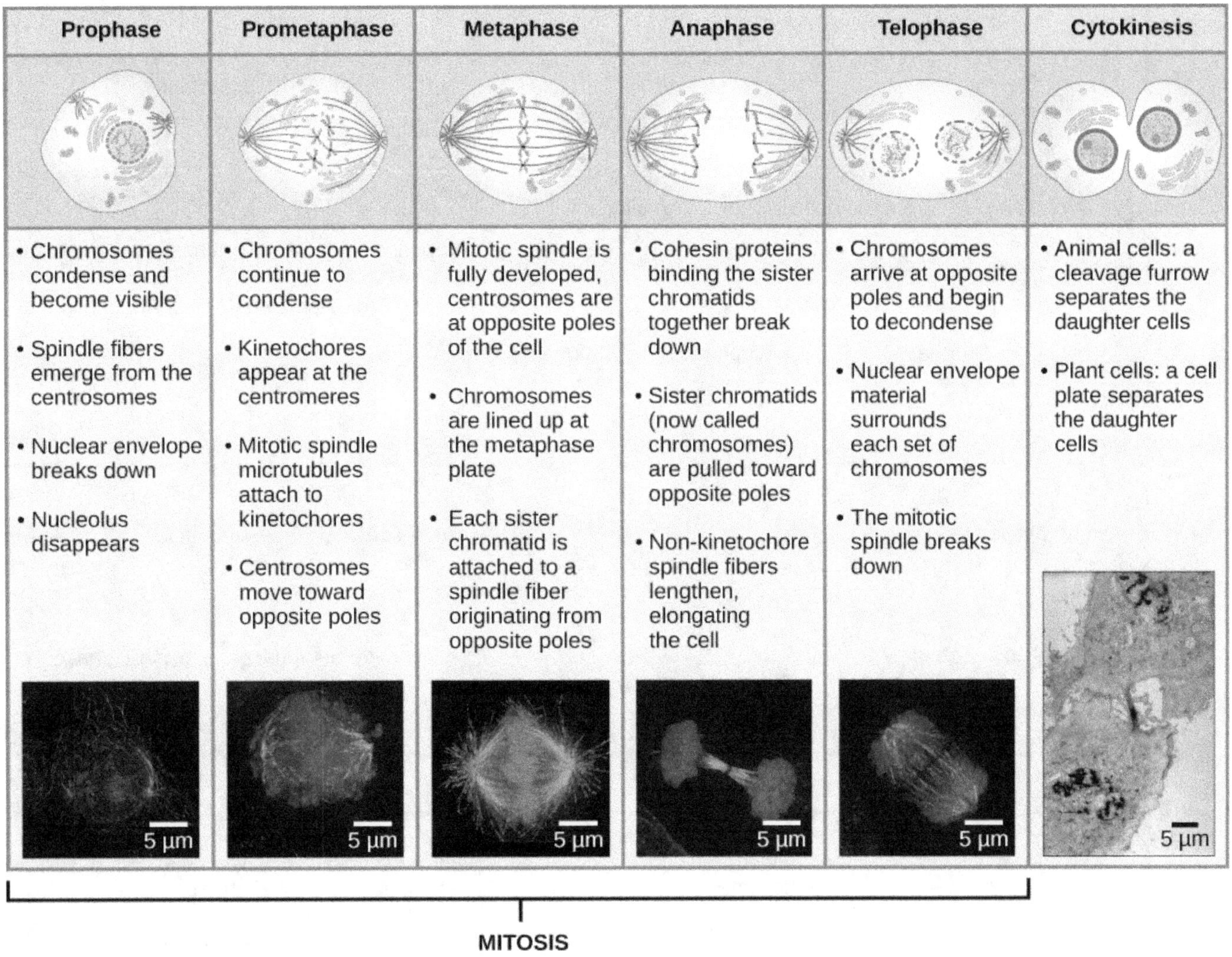

Prophase	Prometaphase	Metaphase	Anaphase	Telophase	Cytokinesis
• Chromosomes condense and become visible • Spindle fibers emerge from the centrosomes • Nuclear envelope breaks down • Nucleolus disappears	• Chromosomes continue to condense • Kinetochores appear at the centromeres • Mitotic spindle microtubules attach to kinetochores • Centrosomes move toward opposite poles	• Mitotic spindle is fully developed, centrosomes are at opposite poles of the cell • Chromosomes are lined up at the metaphase plate • Each sister chromatid is attached to a spindle fiber originating from opposite poles	• Cohesin proteins binding the sister chromatids together break down • Sister chromatids (now called chromosomes) are pulled toward opposite poles • Non-kinetochore spindle fibers lengthen, elongating the cell	• Chromosomes arrive at opposite poles and begin to decondense • Nuclear envelope material surrounds each set of chromosomes • The mitotic spindle breaks down	• Animal cells: a cleavage furrow separates the daughter cells • Plant cells: a cell plate separates the daughter cells

MITOSIS

Figure 3.40 Karyokinesis (or mitosis) is divided into five stages—prophase, prometaphase, metaphase, anaphase, and telophase. The pictures at the bottom were taken by fluorescence microscopy (hence, the black background) of cells artificially stained by fluorescent dyes: blue fluorescence indicates DNA (chromosomes) and green fluorescence indicates microtubules (spindle apparatus). (credit "mitosis drawings": modification of work by Mariana Ruiz Villareal; credit "micrographs": modification of work by Roy van Heesbeen; credit "cytokinesis micrograph": Wadsworth Center/New York State Department of Health; scale-bar data from Matt Russell)

In addition to the mitotic asexual reproduction described above, most eukaryotic microorganisms also have the option of sexual reproduction involving **meiosis**. Although mitosis and meiosis both require DNA replication, nuclear division, and share procedural similarities, there are important differences between the process and outcomes (Figure 3.41).

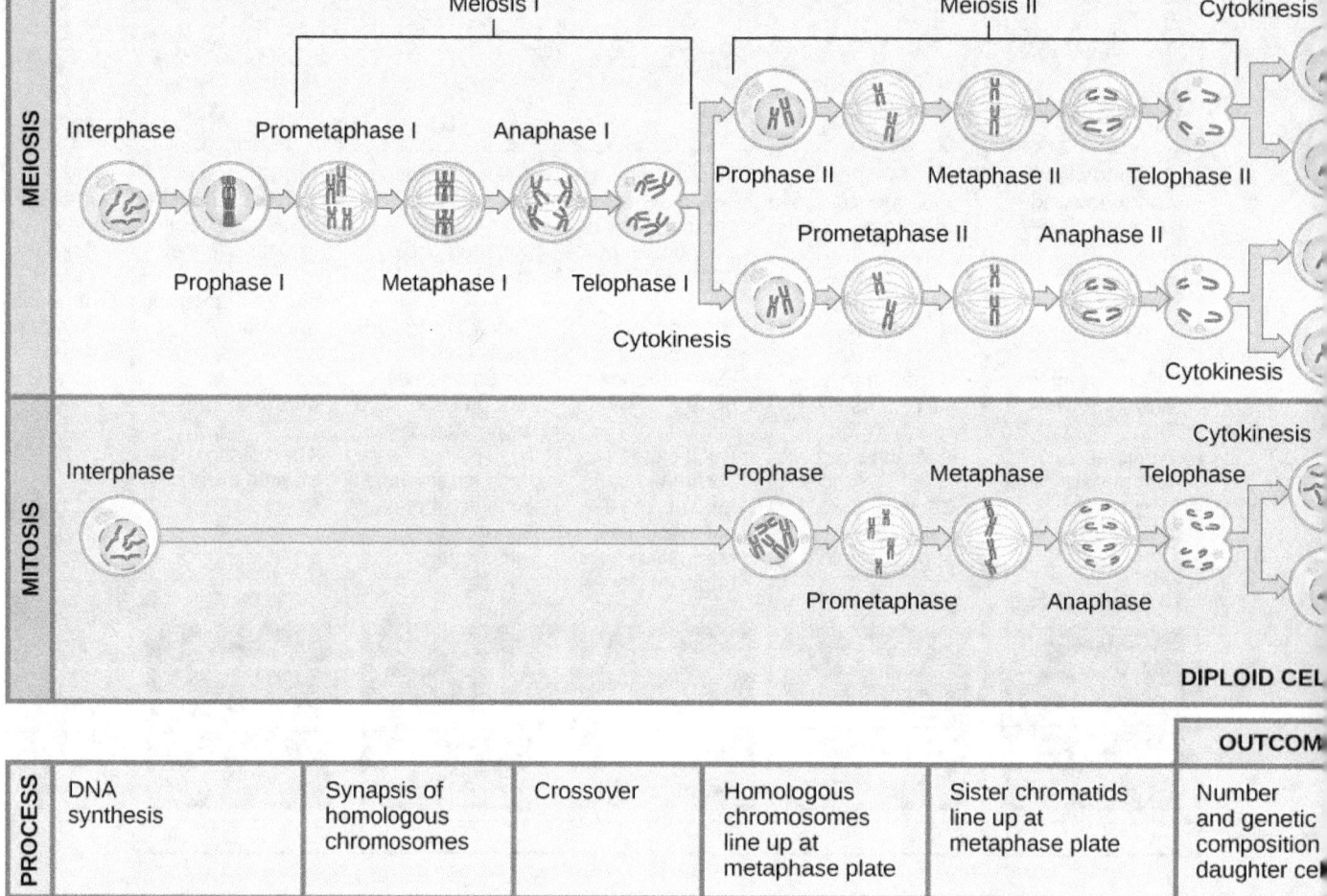

Figure 3.41 Meiosis and mitosis are both preceded by one cycle of DNA replication; however, meiosis includes two nuclear divisions. The four daughter cells resulting from meiosis are haploid and genetically distinct. The daughter cells resulting from mitosis are diploid and identical to the parent cell.

In contrast to the single nuclear division that completes mitosis, meiosis involves two separate nuclear divisions. Rather than creating two clonal daughter cells, the goal of meiosis is to create four genetically-distinct gametes, with each gamete possessing half the number of chromosomes found in the original cell. This strategic chromosome reduction is essential for the fertilization that occurs during sexual reproduction to produce in a zygote with a full complement of chromosomes.

Nucleolus

The **nucleolus** is a dense region within the nucleus where ribosomal RNA (rRNA) biosynthesis occurs. In addition, the nucleolus is also the site where assembly of ribosomes begins. Preribosomal complexes are assembled from rRNA and proteins in the nucleolus; they are then transported out to the cytoplasm, where ribosome assembly is completed (Figure 3.42).

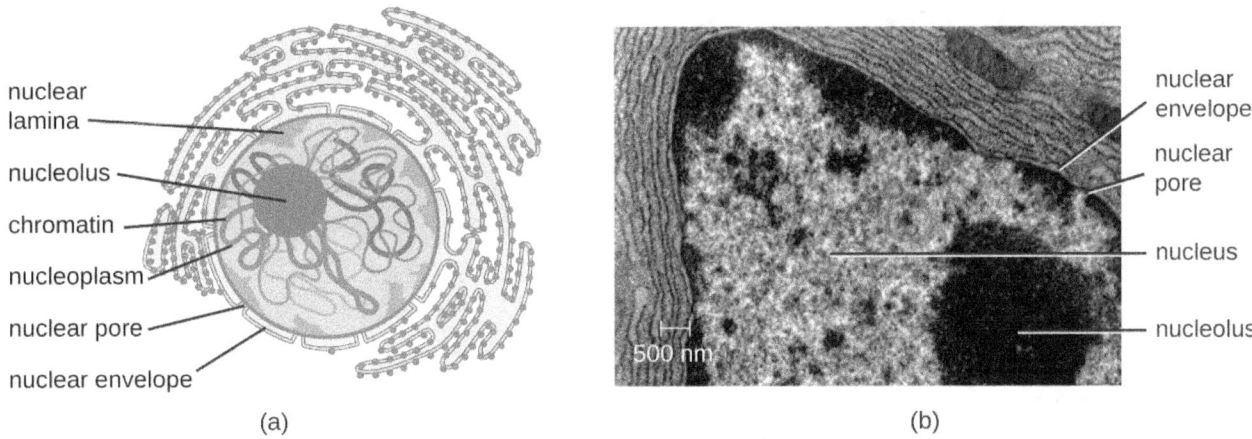

Figure 3.42 (a) The nucleolus is the dark, dense area within the nucleus. It is the site of rRNA synthesis and preribosomal assembly. (b) Electron micrograph showing the nucleolus.

Ribosomes

Ribosomes found in eukaryotic organelles such as mitochondria or chloroplasts have 70S ribosomes—the same size as prokaryotic ribosomes. However, nonorganelle-associated ribosomes in eukaryotic cells are **80S ribosomes**, composed of a 40S small subunit and a 60S large subunit. In terms of size and composition, this makes them distinct from the ribosomes of prokaryotic cells.

The two types of nonorganelle-associated eukaryotic ribosomes are defined by their location in the cell: **free ribosomes** and **membrane-bound ribosomes**. Free ribosomes are found in the cytoplasm and serve to synthesize water-soluble proteins; membrane-bound ribosomes are found attached to the rough endoplasmic reticulum and make proteins for insertion into the cell membrane or proteins destined for export from the cell.

The differences between eukaryotic and prokaryotic ribosomes are clinically relevant because certain antibiotic drugs are designed to target one or the other. For example, cycloheximide targets eukaryotic action, whereas chloramphenicol targets prokaryotic ribosomes.[24] Since human cells are eukaryotic, they generally are not harmed by antibiotics that destroy the prokaryotic ribosomes in bacteria. However, sometimes negative side effects may occur because mitochondria in human cells contain prokaryotic ribosomes.

Endomembrane System

The **endomembrane system**, unique to eukaryotic cells, is a series of membranous tubules, sacs, and flattened disks that synthesize many cell components and move materials around within the cell (Figure 3.43). Because of their larger cell size, eukaryotic cells require this system to transport materials that cannot be dispersed by diffusion alone. The endomembrane system comprises several organelles and connections between them, including the endoplasmic reticulum, Golgi apparatus, lysosomes, and vesicles.

---

24 A.E. Barnhill, M.T. Brewer, S.A. Carlson. "Adverse Effects of Antimicrobials via Predictable or Idiosyncratic Inhibition of Host Mitochondrial Components." *Antimicrobial Agents and Chemotherapy* 56 no. 8 (2012):4046–4051.

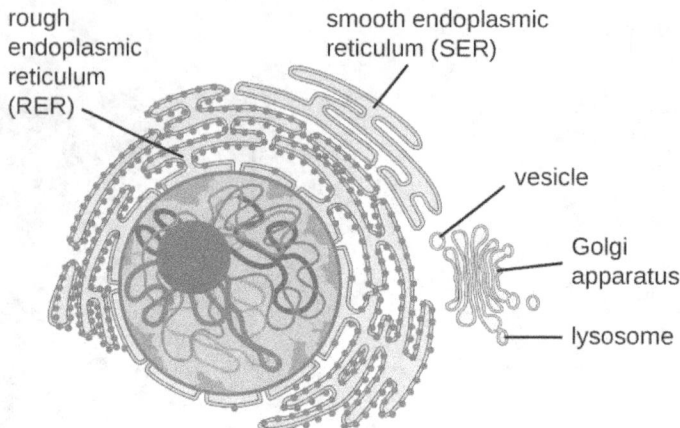

Figure 3.43 The endomembrane system is composed of a series of membranous intracellular structures that facilitate movement of materials throughout the cell and to the cell membrane.

Endoplasmic Reticulum

The **endoplasmic reticulum (ER)** is an interconnected array of tubules and **cisternae** (flattened sacs) with a single lipid bilayer (Figure 3.44). The spaces inside of the cisternae are called **lumen** of the ER. There are two types of ER, **rough endoplasmic reticulum (RER)** and **smooth endoplasmic reticulum (SER)**. These two different types of ER are sites for the synthesis of distinctly different types of molecules. RER is studded with ribosomes bound on the cytoplasmic side of the membrane. These ribosomes make proteins destined for the plasma membrane (Figure 3.44). Following synthesis, these proteins are inserted into the membrane of the RER. Small sacs of the RER containing these newly synthesized proteins then bud off as **transport vesicles** and move either to the Golgi apparatus for further processing, directly to the plasma membrane, to the membrane of another organelle, or out of the cell. Transport vesicles are single-lipid, bilayer, membranous spheres with hollow interiors that carry molecules. SER does not have ribosomes and, therefore, appears "smooth." It is involved in biosynthesis of lipids, carbohydrate metabolism, and detoxification of toxic compounds within the cell.

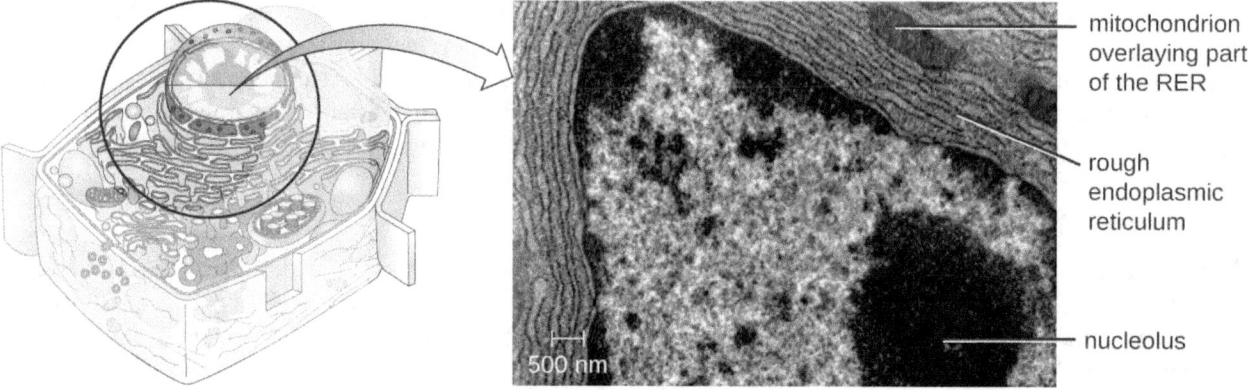

Figure 3.44 The rough endoplasmic reticulum is studded with ribosomes for the synthesis of membrane proteins (which give it its rough appearance).

Golgi Apparatus

The Golgi apparatus was discovered within the endomembrane system in 1898 by Italian scientist Camillo Golgi (1843–1926), who developed a novel staining technique that showed stacked membrane structures within the cells of *Plasmodium*, the causative agent of malaria. The **Golgi apparatus** is composed of a series of membranous disks called dictyosomes, each having a single lipid bilayer, that are stacked together (Figure 3.45).

Enzymes in the Golgi apparatus modify lipids and proteins transported from the ER to the Golgi, often adding carbohydrate components to them, producing glycolipids, glycoproteins, or proteoglycans. Glycolipids and glycoproteins are often inserted into the plasma membrane and are important for signal recognition by other cells or infectious particles. Different types of cells can be distinguished from one another by the structure and arrangement of the glycolipids and glycoproteins contained in their plasma membranes. These glycolipids and glycoproteins commonly also serve as cell surface receptors.

Transport vesicles leaving the ER fuse with a Golgi apparatus on its receiving, or *cis*, face. The proteins are processed within the Golgi apparatus, and then additional transport vesicles containing the modified proteins and lipids pinch off from the Golgi apparatus on its outgoing, or *trans*, face. These outgoing vesicles move to and fuse with the plasma membrane or the membrane of other organelles.

Exocytosis is the process by which **secretory vesicles** (spherical membranous sacs) release their contents to the cell's exterior (Figure 3.45). All cells have constitutive secretory pathways in which secretory vesicles transport soluble proteins that are released from the cell continually (constitutively). Certain specialized cells also have regulated secretory pathways, which are used to store soluble proteins in secretory vesicles. Regulated secretion involves substances that are only released in response to certain events or signals. For example, certain cells of the human immune system (e.g., mast cells) secrete histamine in response to the presence of foreign objects or pathogens in the body. Histamine is a compound that triggers various mechanisms used by the immune system to eliminate pathogens.

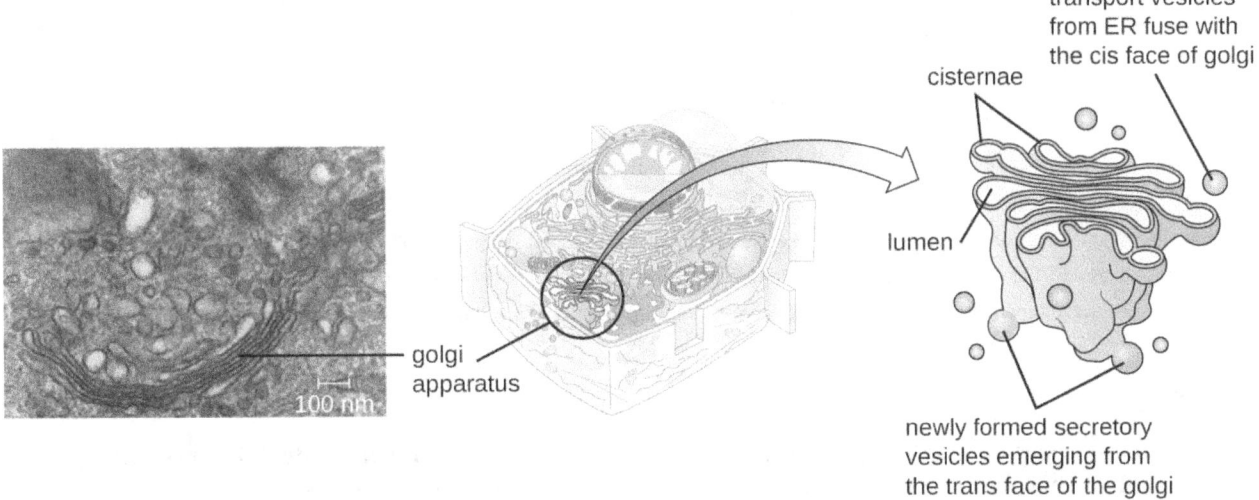

Figure 3.45 A transmission electron micrograph (left) of a Golgi apparatus in a white blood cell. The illustration (right) shows the cup-shaped, stacked disks and several transport vesicles. The Golgi apparatus modifies lipids and proteins, producing glycolipids and glycoproteins, respectively, which are commonly inserted into the plasma membrane.

Lysosomes

In the 1960s, Belgian scientist Christian de Duve (1917–2013) discovered **lysosomes**, membrane-bound organelles of the endomembrane system that contain digestive enzymes. Certain types of eukaryotic cells use lysosomes to break down various particles, such as food, damaged organelles or cellular debris, microorganisms, or immune complexes. Compartmentalization of the digestive enzymes within the lysosome allows the cell to efficiently digest matter without harming the cytoplasmic components of the cell.

⊘ CHECK YOUR UNDERSTANDING

- Name the components of the endomembrane system and describe the function of each component.

Peroxisomes

Christian de Duve is also credited with the discovery of **peroxisomes**, membrane-bound organelles that are not part of the endomembrane system (Figure 3.46). Peroxisomes form independently in the cytoplasm from the synthesis of peroxin proteins by free ribosomes and the incorporation of these peroxin proteins into existing peroxisomes. Growing peroxisomes then divide by a process similar to binary fission.

Peroxisomes were first named for their ability to produce hydrogen peroxide, a highly reactive molecule that helps to break down molecules such as uric acid, amino acids, and fatty acids. Peroxisomes also possess the enzyme catalase, which can degrade hydrogen peroxide. Along with the SER, peroxisomes also play a role in lipid biosynthesis. Like lysosomes, the compartmentalization of these degradative molecules within an organelle helps protect the cytoplasmic contents from unwanted damage.

The peroxisomes of certain organisms are specialized to meet their particular functional needs. For example, glyoxysomes are modified peroxisomes of yeasts and plant cells that perform several metabolic functions, including the production of sugar molecules. Similarly, glycosomes are modified peroxisomes made by certain trypanosomes, the pathogenic protozoans that cause Chagas disease and African sleeping sickness.

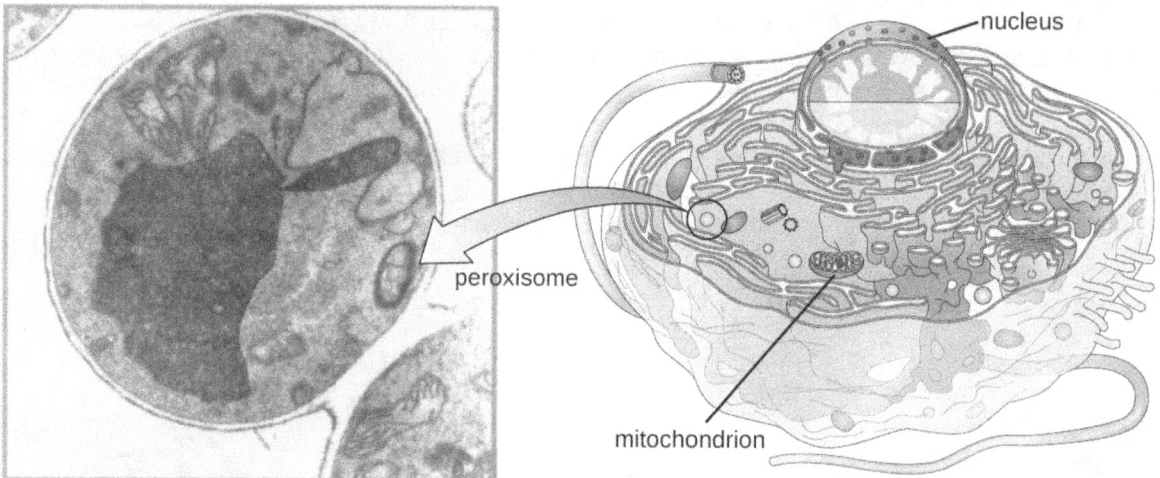

Figure 3.46 A transmission electron micrograph (left) of a cell containing a peroxisome. The illustration (right) shows the location of peroxisomes in a cell. These eukaryotic structures play a role in lipid biosynthesis and breaking down various molecules. They may also have other specialized functions depending on the cell type. (credit "micrograph": modification of work by American Society for Microbiology)

Cytoskeleton

Eukaryotic cells have an internal cytoskeleton made of **microfilaments**, **intermediate filaments**, and **microtubules**. This matrix of fibers and tubes provides structural support as well as a network over which materials can be transported within the cell and on which organelles can be anchored (Figure 3.47). For example, the process of exocytosis involves the movement of a vesicle via the cytoskeletal network to the plasma membrane, where it can release its contents.

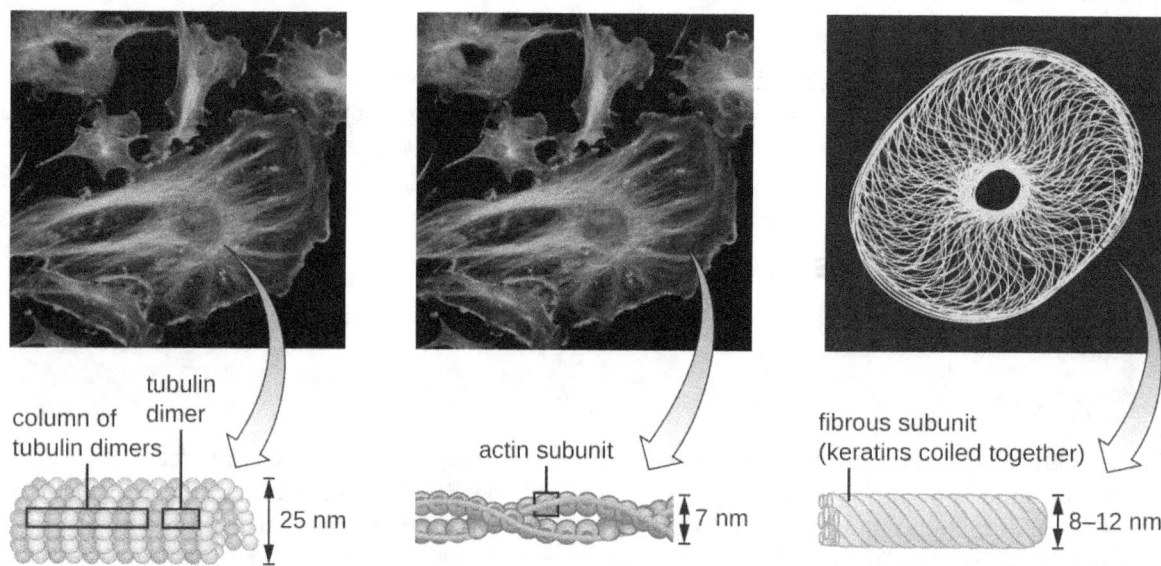

Figure 3.47 The cytoskeleton is a network of microfilaments, intermediate filaments, and microtubules found throughout the cytoplasm

of a eukaryotic cell. In these fluorescently labeled animal cells, the microtubules are green, the actin microfilaments are red, the nucleus is blue, and keratin (a type of intermediate filament) is yellow.

Microfilaments are composed of two intertwined strands of actin, each composed of **actin** monomers forming filamentous cables 6 nm in diameter[25] (Figure 3.48). The actin filaments work together with motor proteins, like myosin, to effect muscle contraction in animals or the amoeboid movement of some eukaryotic microbes. In ameboid organisms, actin can be found in two forms: a stiffer, polymerized, gel form and a more fluid, unpolymerized soluble form. Actin in the gel form creates stability in the ectoplasm, the gel-like area of cytoplasm just inside the plasma membrane of ameboid protozoans.

Temporary extensions of the cytoplasmic membrane called **pseudopodia** (meaning "false feet") are produced through the forward flow of soluble actin filaments into the pseudopodia, followed by the gel-sol cycling of the actin filaments, resulting in cell motility. Once the cytoplasm extends outward, forming a pseudopodium, the remaining cytoplasm flows up to join the leading edge, thereby creating forward locomotion. Beyond amoeboid movement, microfilaments are also involved in a variety of other processes in eukaryotic cells, including cytoplasmic streaming (the movement or circulation of cytoplasm within the cell), cleavage furrow formation during cell division, and muscle movement in animals (Figure 3.48). These functions are the result of the dynamic nature of microfilaments, which can polymerize and depolymerize relatively easily in response to cellular signals, and their interactions with molecular motors in different types of eukaryotic cells.

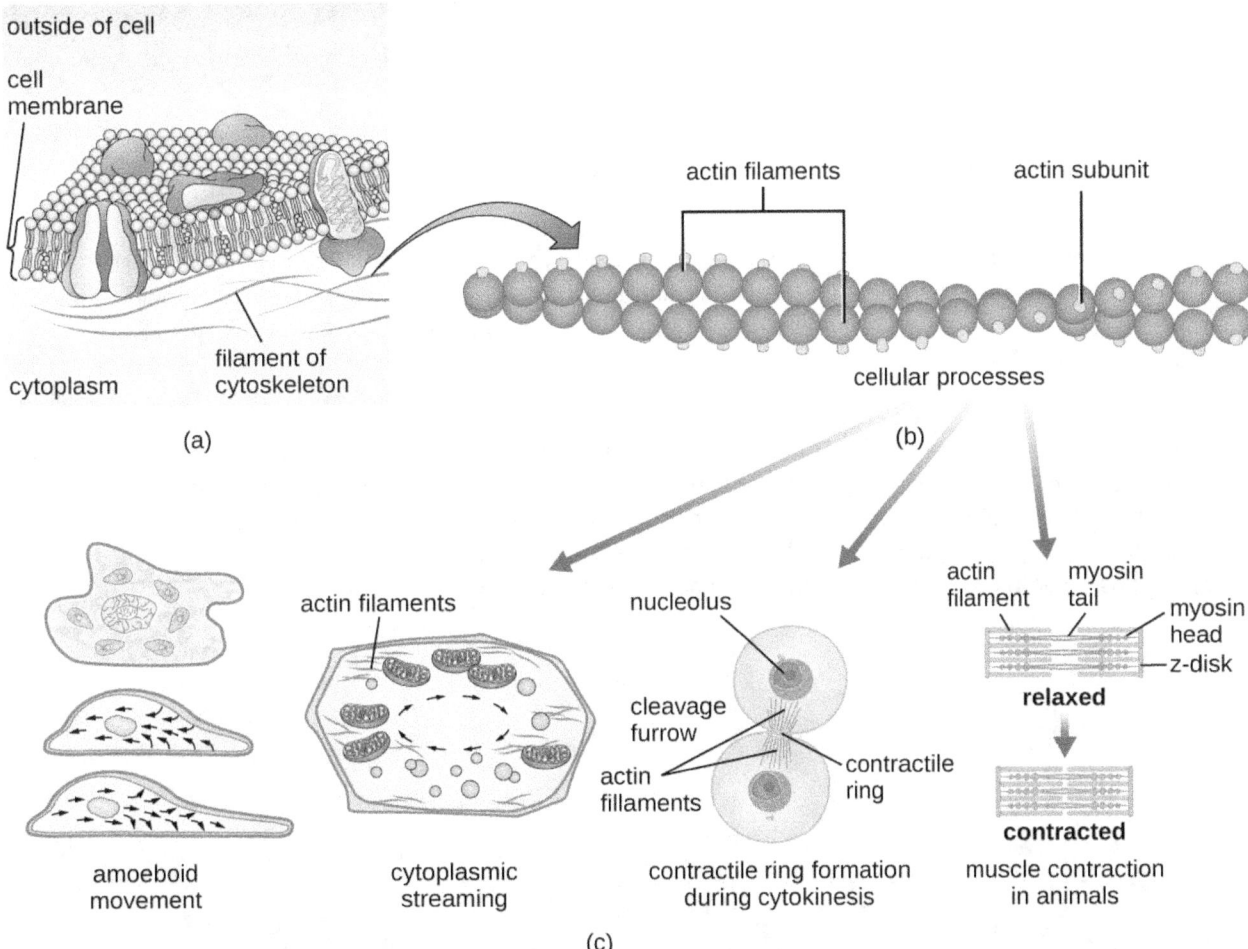

Figure 3.48 (a) A microfilament is composed of a pair of actin filaments. (b) Each actin filament is a string of polymerized actin monomers. (c) The dynamic nature of actin, due to its polymerization and depolymerization and its association with myosin, allows microfilaments to be involved in a variety of cellular processes, including ameboid movement, cytoplasmic streaming, contractile ring formation during cell

---

25 Fuchs E, Cleveland DW. "A Structural Scaffolding of Intermediate Filaments in Health and Disease." *Science* 279 no. 5350 (1998):514–519.

division, and muscle contraction in animals.

Intermediate filaments (Figure 3.49) are a diverse group of cytoskeletal filaments that act as cables within the cell. They are termed "intermediate" because their 10-nm diameter is thicker than that of actin but thinner than that of microtubules.[26] They are composed of several strands of polymerized subunits that, in turn, are made up of a wide variety of monomers. Intermediate filaments tend to be more permanent in the cell and maintain the position of the nucleus. They also form the nuclear lamina (lining or layer) just inside the nuclear envelope. Additionally, intermediate filaments play a role in anchoring cells together in animal tissues. The intermediate filament protein desmin is found in desmosomes, the protein structures that join muscle cells together and help them resist external physical forces. The intermediate filament protein keratin is a structural protein found in hair, skin, and nails.

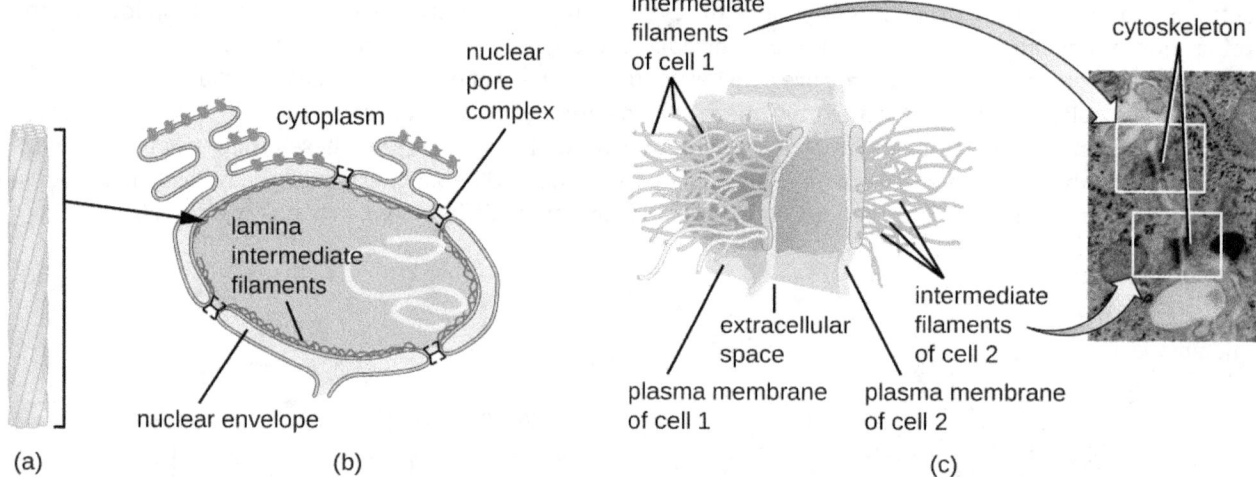

Figure 3.49 (a) Intermediate filaments are composed of multiple strands of polymerized subunits. They are more permanent than other cytoskeletal structures and serve a variety of functions. (b) Intermediate filaments form much of the nuclear lamina. (c) Intermediate filaments form the desmosomes between cells in some animal tissues. (credit c "illustration": modification of work by Mariana Ruiz Villareal)

Microtubules (Figure 3.50) are a third type of cytoskeletal fiber composed of tubulin dimers (α tubulin and β tubulin). These form hollow tubes 23 nm in diameter that are used as girders within the cytoskeleton.[27] Like microfilaments, microtubules are dynamic and have the ability to rapidly assemble and disassemble. Microtubules also work with motor proteins (such as dynein and kinesin) to move organelles and vesicles around within the cytoplasm. Additionally, microtubules are the main components of eukaryotic flagella and cilia, composing both the filament and the basal body components (Figure 3.57).

---

26 E. Fuchs, D.W. Cleveland. "A Structural Scaffolding of Intermediate Filaments in Health and Disease." *Science* 279 no. 5350 (1998):514–519.

27 E. Fuchs, D.W. Cleveland. "A Structural Scaffolding of Intermediate Filaments in Health and Disease." *Science* 279 no. 5350 (1998):514–519.

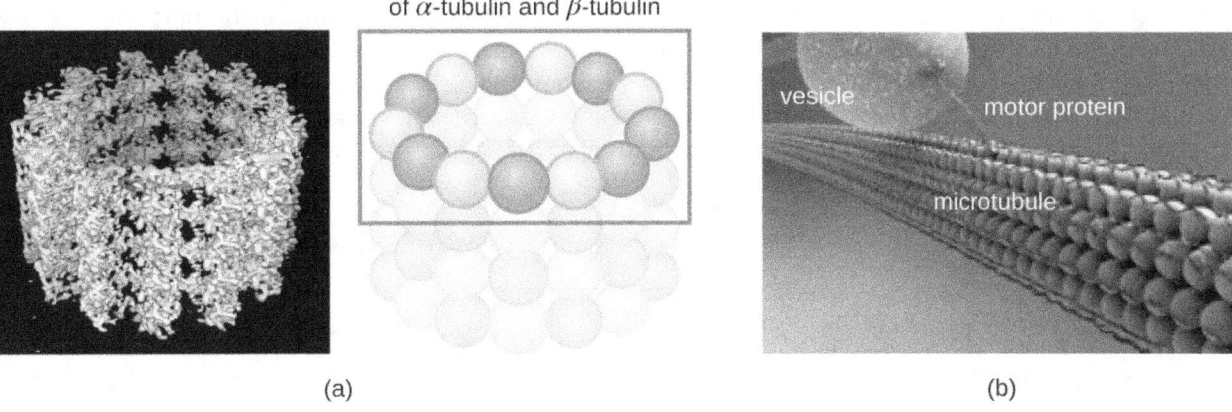

Figure 3.50 (a) Microtubules are hollow structures composed of polymerized tubulin dimers. (b) They are involved in several cellular processes, including the movement of organelles throughout the cytoplasm. Motor proteins carry organelles along microtubule tracks that crisscross the entire cell. (credit b: modification of work by National Institute on Aging)

In addition, microtubules are involved in cell division, forming the mitotic spindle that serves to separate chromosomes during mitosis and meiosis. The mitotic spindle is produced by two **centrosomes**, which are essentially microtubule-organizing centers, at opposite ends of the cell. Each centrosome is composed of a pair of **centrioles** positioned at right angles to each other, and each centriole is an array of nine parallel microtubules arranged in triplets (Figure 3.51).

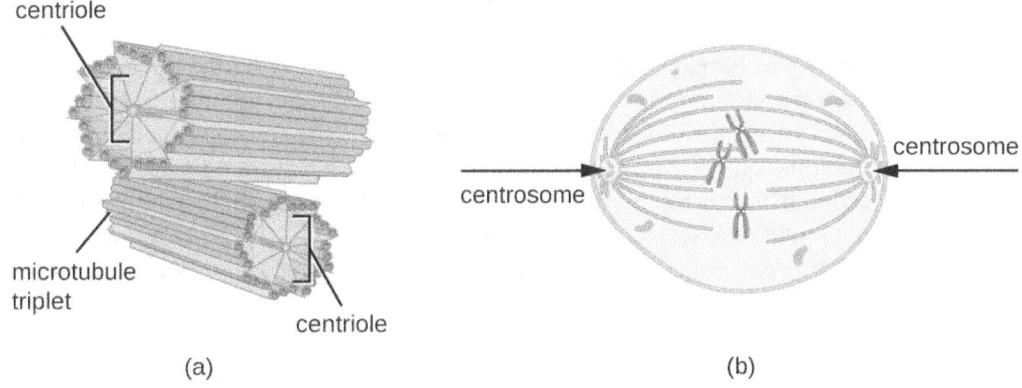

Figure 3.51 (a) A centrosome is composed of two centrioles positioned at right angles to each other. Each centriole is composed of nine triplets of microtubules held together by accessory proteins. (b) In animal cells, the centrosomes (arrows) serve as microtubule-organizing centers of the mitotic spindle during mitosis.

✓ CHECK YOUR UNDERSTANDING

- Compare and contrast the three types of cytoskeletal structures described in this section.

Mitochondria

The large, complex organelles in which aerobic cellular respiration occurs in eukaryotic cells are called **mitochondria** (Figure 3.52). The term "mitochondrion" was first coined by German microbiologist Carl Benda in 1898 and was later connected with the process of respiration by Otto Warburg in 1913. Scientists during the 1960s discovered that mitochondria have their own genome and 70S ribosomes. The mitochondrial genome was found to be bacterial, when it was sequenced in 1976. These findings ultimately supported the endosymbiotic theory proposed by Lynn Margulis, which states that mitochondria originally arose through an endosymbiotic event in which a bacterium capable of aerobic cellular respiration was taken up by phagocytosis into a host cell and remained as a viable intracellular component.

Each mitochondrion has two lipid membranes. The outer membrane is a remnant of the original host cell's membrane structures. The inner membrane was derived from the bacterial plasma membrane. The electron transport chain for aerobic respiration uses integral proteins embedded in the inner membrane. The **mitochondrial matrix**, corresponding to the location of the original bacterium's cytoplasm, is the current location of many metabolic enzymes. It also contains mitochondrial DNA and 70S ribosomes. Invaginations of the inner membrane, called cristae, evolved to increase surface area for the location of biochemical reactions. The folding patterns of the cristae differ among various types of eukaryotic cells and are used to distinguish different eukaryotic organisms from each other.

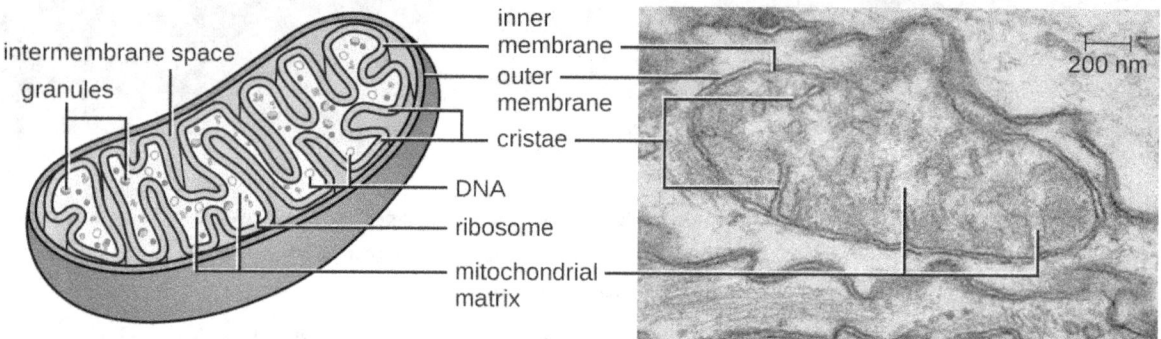

Figure 3.52 Each mitochondrion is surrounded by two membranes, the inner of which is extensively folded into cristae and is the site of the intermembrane space. The mitochondrial matrix contains the mitochondrial DNA, ribosomes, and metabolic enzymes. The transmission electron micrograph of a mitochondrion, on the right, shows both membranes, including cristae and the mitochondrial matrix. (credit "micrograph": modification of work by Matthew Britton; scale-bar data from Matt Russell)

Chloroplasts

Plant cells and algal cells contain **chloroplasts**, the organelles in which photosynthesis occurs (Figure 3.53). All chloroplasts have at least three membrane systems: the outer membrane, the inner membrane, and the thylakoid membrane system. Inside the outer and inner membranes is the chloroplast **stroma**, a gel-like fluid that makes up much of a chloroplast's volume, and in which the **thylakoid** system floats. The thylakoid system is a highly dynamic collection of folded membrane sacs. It is where the green photosynthetic pigment chlorophyll is found and the light reactions of photosynthesis occur. In most plant chloroplasts, the thylakoids are arranged in stacks called grana (singular: granum), whereas in some algal chloroplasts, the thylakoids are free floating.

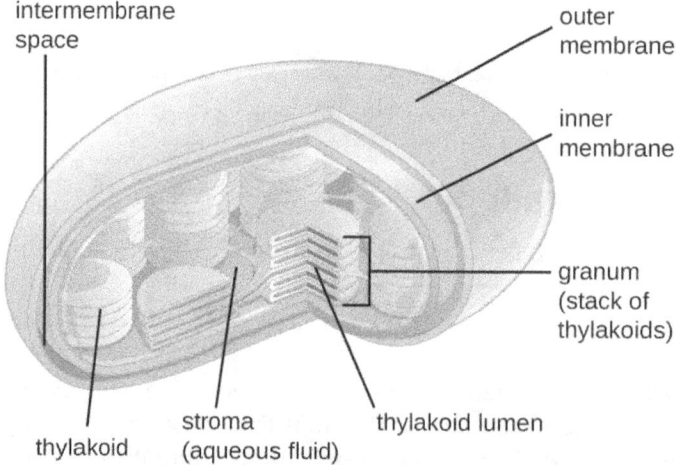

Figure 3.53 Photosynthesis takes place in chloroplasts, which have an outer membrane and an inner membrane. Stacks of thylakoids called grana form a third membrane layer.

Other organelles similar to mitochondria have arisen in other types of eukaryotes, but their roles differ. Hydrogenosomes are found in some anaerobic eukaryotes and serve as the location of anaerobic hydrogen production. Hydrogenosomes typically lack their own DNA and ribosomes. Kinetoplasts are a variation of the mitochondria found in some eukaryotic pathogens. In these organisms, each cell has a single, long, branched mitochondrion in which kinetoplast DNA, organized as multiple circular pieces of DNA, is found concentrated at one pole of the cell.

 MICRO CONNECTIONS

Mitochondria-Related Organelles in Protozoan Parasites

Many protozoans, including several protozoan parasites that cause infections in humans, can be identified by their unusual appearance. Distinguishing features may include complex cell morphologies, the presence of unique organelles, or the absence of common organelles. The protozoan parasites *Giardia lamblia* and *Trichomonas vaginalis* are two examples.

G. lamblia, a frequent cause of diarrhea in humans and many other animals, is an anaerobic parasite that possesses two nuclei and several flagella. Its Golgi apparatus and endoplasmic reticulum are greatly reduced, and it lacks mitochondria completely. However, it does have organelles known as mitosomes, double-membrane-bound organelles that appear to be severely reduced mitochondria. This has led scientists to believe that *G. lamblia*'s ancestors once possessed mitochondria that evolved to become mitosomes. *T. vaginalis*, which causes the sexually transmitted infection vaginitis, is another protozoan parasite that lacks conventional mitochondria. Instead, it possesses hydrogenosomes, mitochondrial-related, double-membrane-bound organelles that produce molecular hydrogen used in cellular metabolism. Scientists believe that hydrogenosomes, like mitosomes, also evolved from mitochondria.[28]

Plasma Membrane

The plasma membrane of eukaryotic cells is similar in structure to the prokaryotic plasma membrane in that it is composed mainly of phospholipids forming a bilayer with embedded peripheral and integral proteins (Figure 3.54). These membrane components move within the plane of the membrane according to the fluid mosaic model. However, unlike the prokaryotic membrane, eukaryotic membranes contain sterols, including cholesterol, that alter membrane fluidity. Additionally, many eukaryotic cells contain some specialized lipids, including sphingolipids, which are thought to play a role in maintaining membrane stability as well as being involved in signal transduction pathways and cell-to-cell communication.

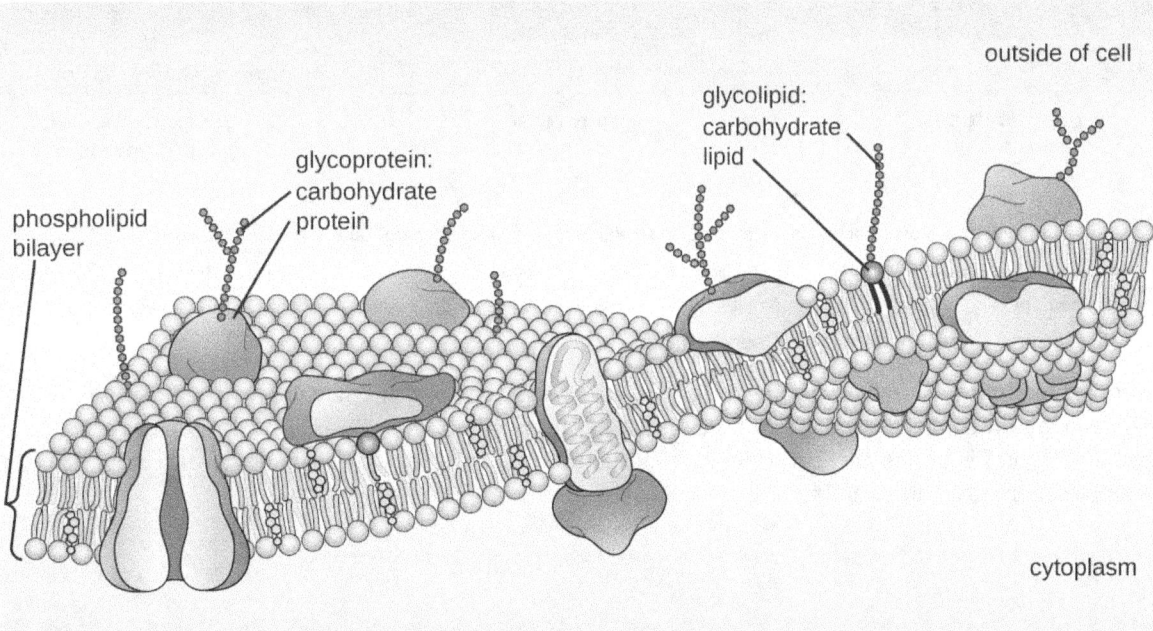

Figure 3.54 The eukaryotic plasma membrane is composed of a lipid bilayer with many embedded or associated proteins. It contains cholesterol for the maintenance of membrane, as well as glycoproteins and glycolipids that are important in the recognition other cells or

---

28 N. Yarlett, J.H.P. Hackstein. "Hydrogenosomes: One Organelle, Multiple Origins." *BioScience* 55 no. 8 (2005):657–658.

pathogens.

Membrane Transport Mechanisms

The processes of simple diffusion, facilitated diffusion, and active transport are used in both eukaryotic and prokaryotic cells. However, eukaryotic cells also have the unique ability to perform various types of **endocytosis**, the uptake of matter through plasma membrane invagination and vacuole/vesicle formation (Figure 3.55). A type of endocytosis involving the engulfment of large particles through membrane invagination is called **phagocytosis**, which means "cell eating." In phagocytosis, particles (or other cells) are enclosed in a pocket within the membrane, which then pinches off from the membrane to form a vacuole that completely surrounds the particle. Another type of endocytosis is called **pinocytosis**, which means "cell drinking." In pinocytosis, small, dissolved materials and liquids are taken into the cell through small vesicles. Saprophytic fungi, for example, obtain their nutrients from dead and decaying matter largely through pinocytosis.

Receptor-mediated endocytosis is a type of endocytosis that is initiated by specific molecules called ligands when they bind to cell surface receptors on the membrane. Receptor-mediated endocytosis is the mechanism that peptide and amine-derived hormones use to enter cells and is also used by various viruses and bacteria for entry into host cells.

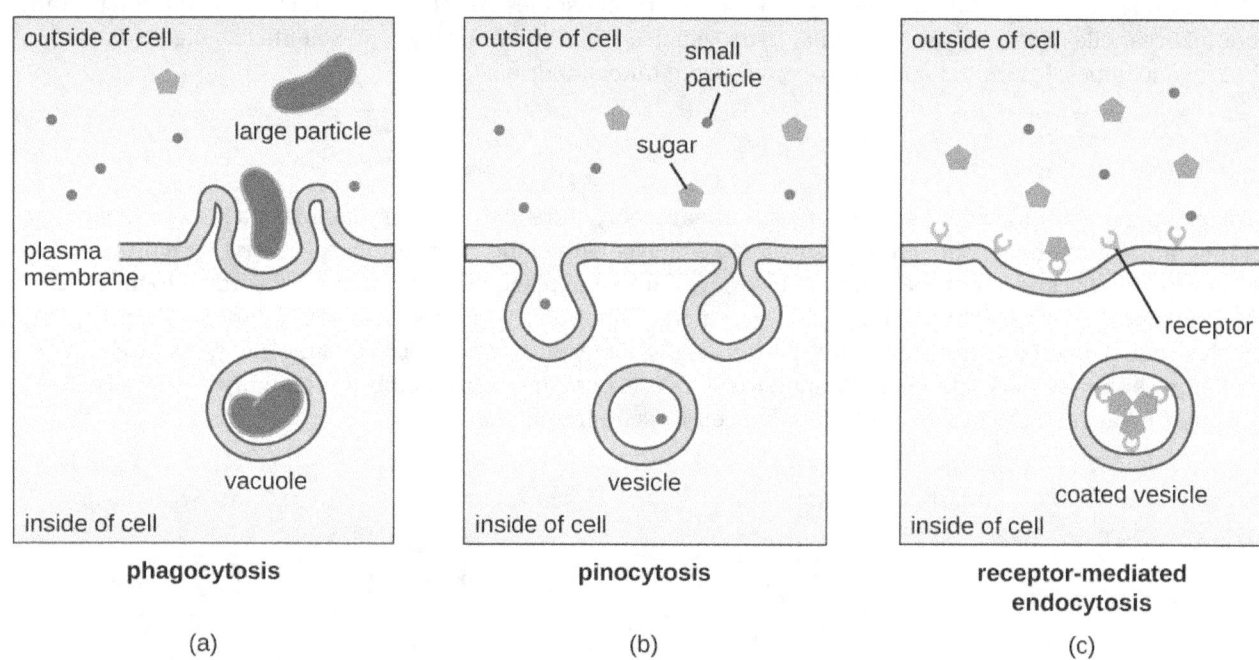

Figure 3.55 Three variations of endocytosis are shown. (a) In phagocytosis, the cell membrane surrounds the particle and pinches off to form an intracellular vacuole. (b) In pinocytosis, the cell membrane surrounds a small volume of fluid and pinches off, forming a vesicle. (c) In receptor-mediated endocytosis, the uptake of substances is targeted to a specific substance (a ligand) that binds at the receptor on the external cell membrane. (credit: modification of work by Mariana Ruiz Villarreal)

The process by which secretory vesicles release their contents to the cell's exterior is called **exocytosis**. Vesicles move toward the plasma membrane and then meld with the membrane, ejecting their contents out of the cell. Exocytosis is used by cells to remove waste products and may also be used to release chemical signals that can be taken up by other cells.

Cell Wall

In addition to a plasma membrane, some eukaryotic cells have a cell wall. Cells of fungi, algae, plants, and even some protists have cell walls. Depending upon the type of eukaryotic cell, cell walls can be made of a wide range of materials, including cellulose (fungi and plants); biogenic silica, calcium carbonate, agar, and carrageenan (protists and algae); or chitin (fungi). In general, all cell walls provide structural stability for the cell and protection from environmental stresses such as desiccation, changes in osmotic pressure, and traumatic injury.[29]

Extracellular Matrix

Cells of animals and some protozoans do not have cell walls to help maintain shape and provide structural stability. Instead, these types of eukaryotic cells produce an **extracellular matrix** for this purpose. They secrete a sticky mass of carbohydrates and proteins into the spaces between adjacent cells (Figure 3.56). Some protein components assemble into a basement membrane to which the remaining extracellular matrix components adhere. Proteoglycans typically form the bulky mass of the extracellular matrix while fibrous proteins, like collagen, provide strength. Both proteoglycans and collagen are attached to fibronectin proteins, which, in turn, are attached to integrin proteins. These integrin proteins interact with transmembrane proteins in the plasma membranes of eukaryotic cells that lack cell walls.

In animal cells, the extracellular matrix allows cells within tissues to withstand external stresses and transmits signals from the outside of the cell to the inside. The amount of extracellular matrix is quite extensive in various types of connective tissues, and variations in the extracellular matrix can give different types of tissues their distinct properties. In addition, a host cell's extracellular matrix is often the site where microbial pathogens attach themselves to establish infection. For example, *Streptococcus pyogenes*, the bacterium that causes strep throat and various other infections, binds to fibronectin in the extracellular matrix of the cells lining the oropharynx (upper region of the throat).

---

[29] M. Dudzick. "Protists." OpenStax CNX. November 27, 2013. http://cnx.org/contents/f7048bb6-e462-459b-805c-ef291cf7049c@1

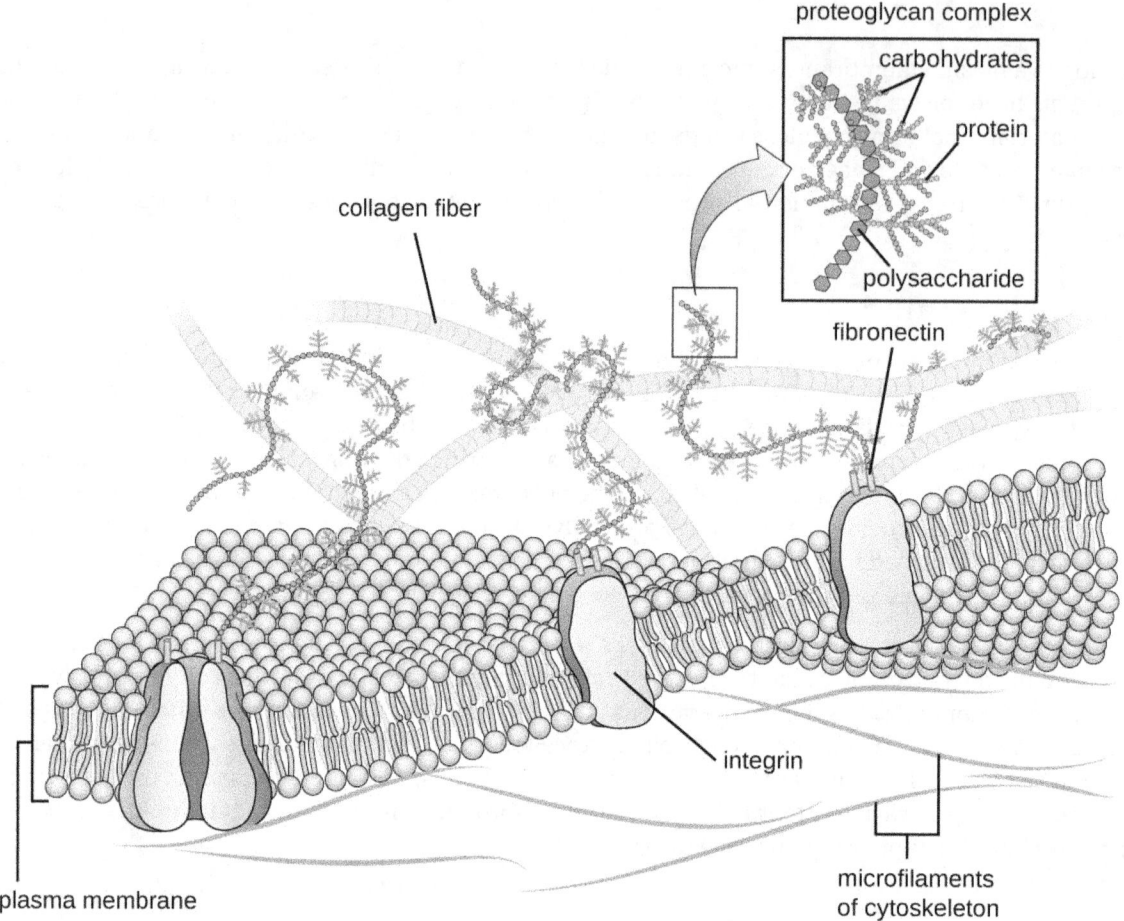

Figure 3.56 The extracellular matrix is composed of protein and carbohydrate components. It protects cells from physical stresses and transmits signals arriving at the outside edges of the tissue to cells deeper within the tissue.

Flagella and Cilia

Some eukaryotic cells use **flagella** for locomotion; however, eukaryotic flagella are structurally distinct from those found in prokaryotic cells. Whereas the prokaryotic flagellum is a stiff, rotating structure, a eukaryotic flagellum is more like a flexible whip composed of nine parallel pairs of microtubules surrounding a central pair of microtubules. This arrangement is referred to as a 9+2 array (Figure 3.57). The parallel microtubules use **dynein** motor proteins to move relative to each other, causing the flagellum to bend.

Cilia (singular: **cilium**) are a similar external structure found in some eukaryotic cells. Unique to eukaryotes, cilia are shorter than flagella and often cover the entire surface of a cell; however, they are structurally similar to flagella (a 9+2 array of microtubules) and use the same mechanism for movement. A structure called a **basal body** is found at the base of each cilium and flagellum. The basal body, which attaches the cilium or flagellum to the cell, is composed of an array of triplet microtubules similar to that of a centriole but embedded in the plasma membrane. Because of their shorter length, cilia use a rapid, flexible, waving motion. In addition to motility, cilia may have other functions such as sweeping particles past or into cells. For example, ciliated protozoans use the sweeping of cilia to move food particles into their mouthparts, and ciliated cells in the mammalian respiratory tract beat in synchrony to sweep mucus and debris up and out of the lungs (Figure 3.57).

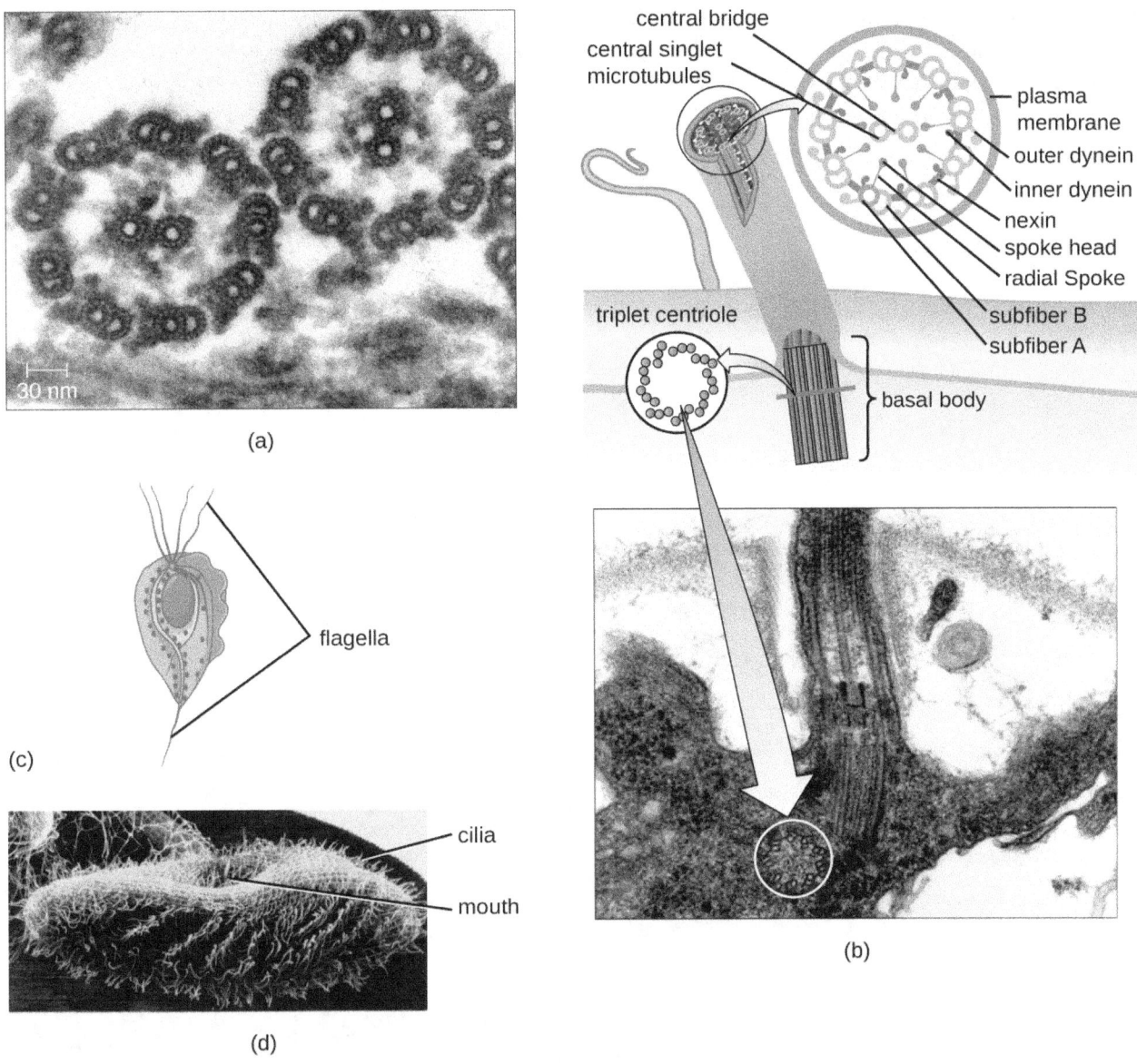

Figure 3.57 (a) Eukaryotic flagella and cilia are composed of a 9+2 array of microtubules, as seen in this transmission electron micrograph cross-section. (b) The sliding of these microtubules relative to each other causes a flagellum to bend. (c) An illustration of *Trichomonas vaginalis*, a flagellated protozoan parasite that causes vaginitis. (d) Many protozoans, like this *Paramecium*, have numerous cilia that aid in locomotion as well as in feeding. Note the mouth opening shown here. (credit d: modification of work by University of Vermont/National Institutes of Health)

CHECK YOUR UNDERSTANDING

- Explain how the cellular envelope of eukaryotic cells compares to that of prokaryotic cells.
- Explain the difference between eukaryotic and prokaryotic flagella.

Clinical Focus

Resolution

Since amoxicillin has not resolved Barbara's case of pneumonia, the PA prescribes another antibiotic, azithromycin, which targets bacterial ribosomes rather than peptidoglycan. After taking the azithromycin as directed, Barbara's symptoms resolve and she finally begins to feel like herself again. Presuming no drug resistance to amoxicillin was involved, and given the effectiveness of azithromycin, the causative agent of Barbara's pneumonia is most likely *Mycoplasma pneumoniae*. Even though this bacterium is a prokaryotic cell, it is not inhibited by amoxicillin because it does not have a cell wall and, therefore, does not make peptidoglycan.

Go back to the previous Clinical Focus box.

SUMMARY

3.1 Spontaneous Generation

- The theory of **spontaneous generation** states that life arose from nonliving matter. It was a long-held belief dating back to Aristotle and the ancient Greeks.
- Experimentation by Francesco Redi in the 17th century presented the first significant evidence refuting spontaneous generation by showing that flies must have access to meat for maggots to develop on the meat. Prominent scientists designed experiments and argued both in support of (John Needham) and against (Lazzaro Spallanzani) spontaneous generation.
- Louis Pasteur is credited with conclusively disproving the theory of spontaneous generation with his famous swan-neck flask experiment. He subsequently proposed that "life only comes from life."

3.2 Foundations of Modern Cell Theory

- Although cells were first observed in the 1660s by Robert Hooke, **cell theory** was not well accepted for another 200 years. The work of scientists such as Schleiden, Schwann, Remak, and Virchow contributed to its acceptance.
- **Endosymbiotic theory** states that mitochondria and chloroplasts, organelles found in many types of organisms, have their origins in bacteria. Significant structural and genetic information support this theory.
- The **miasma theory of disease** was widely accepted until the 19th century, when it was replaced by the **germ theory of disease** thanks to the work of Semmelweis, Snow, Pasteur, Lister, and Koch, and others.

3.3 Unique Characteristics of Prokaryotic Cells

- Prokaryotic cells differ from eukaryotic cells in that their genetic material is contained in a **nucleoid** rather than a membrane-bound nucleus. In addition, prokaryotic cells generally lack membrane-bound organelles.
- Prokaryotic cells of the same species typically share a similar **cell morphology** and **cellular arrangement**.
- Most prokaryotic cells have a **cell wall** that helps the organism maintain cellular morphology and protects it against changes in osmotic pressure.
- Outside of the nucleoid, prokaryotic cells may contain extrachromosomal DNA in **plasmids**.
- Prokaryotic **ribosomes** that are found in the cytoplasm have a size of 70S.
- Some prokaryotic cells have **inclusions** that store nutrients or chemicals for other uses.
- Some prokaryotic cells are able to form **endospores** through **sporulation** to survive in a dormant state when conditions are unfavorable. Endospores can **germinate**, transforming back into **vegetative cells** when conditions improve.
- In prokaryotic cells, the **cell envelope** includes a **plasma membrane** and usually a cell wall.
- Bacterial membranes are composed of phospholipids with integral or peripheral proteins. The fatty acid components of these phospholipids are ester-linked and are often used to identify specific types of bacteria. The proteins serve a variety of functions, including transport, cell-to-cell communication, and sensing environmental conditions. Archaeal membranes are distinct in that they are composed of fatty acids that are ether-linked to phospholipids.
- Some molecules can move across the bacterial membrane by simple diffusion, but most large molecules must be actively transported through membrane structures using cellular energy.
- Prokaryotic cell walls may be composed of **peptidoglycan** (bacteria) or **pseudopeptidoglycan** (archaea).
- Gram-positive bacterial cells are characterized by a thick **peptidoglycan** layer, whereas gram-negative bacterial cells are characterized by a thin peptidoglycan layer surrounded by an outer membrane.
- Some prokaryotic cells produce **glycocalyx** coatings, such as **capsules** and **slime layers**, that aid in attachment to surfaces and/or evasion of the host immune system.
- Some prokaryotic cells have **fimbriae** or **pili**, filamentous appendages that aid in attachment to surfaces. Pili are also used in the transfer of genetic material between cells.
- Some prokaryotic cells use one or more **flagella** to move through water. **Peritrichous** bacteria, which have numerous flagella, use **runs** and **tumbles** to move purposefully in the direction of a chemical attractant.

3.4 Unique Characteristics of Eukaryotic Cells

- Eukaryotic cells are defined by the presence of a

- **nucleus** containing the DNA genome and bound by a **nuclear membrane** (or **nuclear envelope**) composed of two lipid bilayers that regulate transport of materials into and out of the nucleus through nuclear pores.
- Eukaryotic cell morphologies vary greatly and may be maintained by various structures, including the cytoskeleton, the cell membrane, and/or the cell wall
- The **nucleolus**, located in the nucleus of eukaryotic cells, is the site of ribosomal synthesis and the first stages of ribosome assembly.
- Eukaryotic cells contain **80S ribosomes** in the rough endoplasmic reticulum (**membrane bound-ribosomes**) and cytoplasm (**free ribosomes**). They contain 70s ribosomes in mitochondria and chloroplasts.
- Eukaryotic cells have evolved an **endomembrane** system, containing membrane-bound organelles involved in transport. These include vesicles, the endoplasmic reticulum, and the Golgi apparatus.
- The **smooth endoplasmic reticulum** plays a role in lipid biosynthesis, carbohydrate metabolism, and detoxification of toxic compounds. The **rough endoplasmic reticulum** contains membrane-bound 80S ribosomes that synthesize proteins destined for the cell membrane
- The **Golgi apparatus** processes proteins and lipids, typically through the addition of sugar molecules, producing glycoproteins or glycolipids, components of the plasma membrane that are used in cell-to-cell communication.
- **Lysosomes** contain digestive enzymes that break down small particles ingested by **endocytosis**, large particles or cells ingested by **phagocytosis**, and damaged intracellular components.
- The **cytoskeleton**, composed of **microfilaments**, **intermediate filaments**, and **microtubules**, provides structural support in eukaryotic cells and serves as a network for transport of intracellular materials.
- **Centrosomes** are microtubule-organizing centers important in the formation of the mitotic spindle in mitosis.
- **Mitochondria** are the site of cellular respiration. They have two membranes: an outer membrane and an inner membrane with cristae. The mitochondrial matrix, within the inner membrane, contains the mitochondrial DNA, 70S ribosomes, and metabolic enzymes.
- The plasma membrane of eukaryotic cells is structurally similar to that found in prokaryotic cells, and membrane components move according to the fluid mosaic model. However, eukaryotic membranes contain sterols, which alter membrane fluidity, as well as glycoproteins and glycolipids, which help the cell recognize other cells and infectious particles.
- In addition to active transport and passive transport, eukaryotic cell membranes can take material into the cell via **endocytosis**, or expel matter from the cell via **exocytosis.**
- Cells of fungi, algae, plants, and some protists have a **cell wall,** whereas cells of animals and some protozoans have a sticky **extracellular matrix** that provides structural support and mediates cellular signaling.
- Eukaryotic flagella are structurally distinct from prokaryotic flagella but serve a similar purpose (locomotion). **Cilia** are structurally similar to eukaryotic flagella, but shorter; they may be used for locomotion, feeding, or movement of extracellular particles.

REVIEW QUESTIONS
Multiple Choice

1. Which of the following individuals argued in favor of the theory of spontaneous generation?
   A. Francesco Redi
   B. Louis Pasteur
   C. John Needham
   D. Lazzaro Spallanzani

2. Which of the following individuals is credited for definitively refuting the theory of spontaneous generation using broth in swan-neck flask?
   A. Aristotle
   B. Jan Baptista van Helmont
   C. John Needham
   D. Louis Pasteur

3. Which of the following scientists experimented with raw meat, maggots, and flies in an attempt to disprove the theory of spontaneous generation?
   A. Aristotle
   B. Lazzaro Spallanzani
   C. Antonie van Leeuwenhoek
   D. Francesco Redi

4. Which of the following individuals did not contribute to the establishment of cell theory?
   A. Girolamo Fracastoro
   B. Matthias Schleiden
   C. Robert Remak
   D. Robert Hooke

5. Whose proposal of the endosymbiotic theory of mitochondrial and chloroplast origin was ultimately accepted by the greater scientific community?
   A. Rudolf Virchow
   B. Ignaz Semmelweis
   C. Lynn Margulis
   D. Theodor Schwann

6. Which of the following developed a set of postulates for determining whether a particular disease is caused by a particular pathogen?
   A. John Snow
   B. Robert Koch
   C. Joseph Lister
   D. Louis Pasteur

7. Which of the following terms refers to a prokaryotic cell that is comma shaped?
   A. coccus
   B. coccobacilli
   C. vibrio
   D. spirillum

8. Which bacterial structures are important for adherence to surfaces? (Select all that apply.)
   A. endospores
   B. cell walls
   C. fimbriae
   D. capsules
   E. flagella

9. Which of the following cell wall components is unique to gram-negative cells?
   A. lipopolysaccharide
   B. teichoic acid
   C. mycolic acid
   D. peptidoglycan

10. Which of the following terms refers to a bacterial cell having a single tuft of flagella at one end?
    A. monotrichous
    B. amphitrichous
    C. peritrichous
    D. lophotrichous

11. Bacterial cell walls are primarily composed of which of the following?
    A. phospholipid
    B. protein
    C. carbohydrate
    D. peptidoglycan

12. Which of the following organelles is not part of the endomembrane system?
    A. endoplasmic reticulum
    B. Golgi apparatus
    C. lysosome
    D. peroxisome

13. Which type of cytoskeletal fiber is important in the formation of the nuclear lamina?
    A. microfilaments
    B. intermediate filaments
    C. microtubules
    D. fibronectin

14. Sugar groups may be added to proteins in which of the following?
    A. smooth endoplasmic reticulum
    B. rough endoplasmic reticulum
    C. Golgi apparatus
    D. lysosome

15. Which of the following structures of a eukaryotic cell is not likely derived from endosymbiotic bacterium?
    A. mitochondrial DNA
    B. mitochondrial ribosomes
    C. inner membrane
    D. outer membrane

16. Which type of nutrient uptake involves the engulfment of small dissolved molecules into vesicles?
    A. active transport
    B. pinocytosis
    C. receptor-mediated endocytosis
    D. facilitated diffusion

17. Which of the following is not composed of microtubules?
    A. desmosomes
    B. centrioles
    C. eukaryotic flagella
    D. eukaryotic cilia

True/False

18. Exposure to air is necessary for microbial growth.
19. Bacteria have 80S ribosomes each composed of a 60S large subunit and a 40S small subunit.
20. Mitochondria in eukaryotic cells contain ribosomes that are structurally similar to those found in prokaryotic cells.

Fill in the Blank

21. The assertion that "life only comes from life" was stated by Louis Pasteur in regard to his experiments that definitively refuted the theory of _____.
22. John Snow is known as the Father of _____.
23. The _____ theory states that disease may originate from proximity to decomposing matter and is not due to person-to-person contact.
24. The scientist who first described cells was _____.
25. Prokaryotic cells that are rod-shaped are called _____.
26. The type of inclusion containing polymerized inorganic phosphate is called _____.
27. Peroxisomes typically produce _____, a harsh chemical that helps break down molecules.
28. Microfilaments are composed of _____ monomers.

Short Answer

29. Explain in your own words Pasteur's swan-neck flask experiment.
30. Explain why the experiments of Needham and Spallanzani yielded in different results even though they used similar methodologies.
31. How did the explanation of Virchow and Remak for the origin of cells differ from that of Schleiden and Schwann?
32. What evidence exists that supports the endosymbiotic theory?
33. What were the differences in mortality rates due to puerperal fever that Ignaz Semmelweis observed? How did he propose to reduce the occurrence of puerperal fever? Did it work?
34. What is the direction of water flow for a bacterial cell living in a hypotonic environment? How do cell walls help bacteria living in such environments?
35. How do bacterial flagella respond to a chemical gradient of an attractant to move toward a higher concentration of the chemical?

36. Label the parts of the prokaryotic cell.

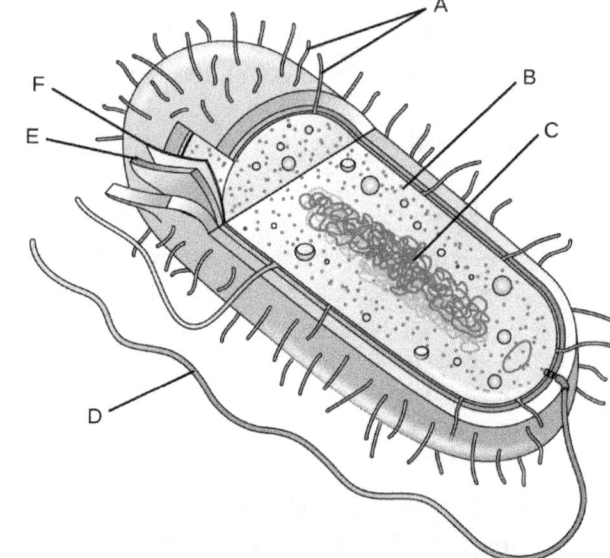

37. What existing evidence supports the theory that mitochondria are of prokaryotic origin?
38. Why do eukaryotic cells require an endomembrane system?
39. Name at least two ways that prokaryotic flagella are different from eukaryotic flagella.

Critical Thinking

40. What would the results of Pasteur's swan-neck flask experiment have looked like if they supported the theory of spontaneous generation?
41. Why are mitochondria and chloroplasts unable to multiply outside of a host cell?
42. Why was the work of Snow so important in supporting the germ theory?
43. Which of the following slides is a good example of staphylococci?

  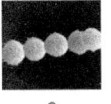 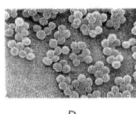

A B C D

Figure 3.58 (credit a: modification of work by U.S. Department of Agriculture; credit b: modification of work by Centers for Disease Control and Prevention; credit c: modification of work by NIAID)

44. Provide some examples of bacterial structures that might be used as antibiotic targets and explain why.
45. The causative agent of botulism, a deadly form of food poisoning, is an endospore-forming bacterium called *Clostridium botulinim*. Why might it be difficult to kill this bacterium in contaminated food?
46. Label the lettered parts of this eukaryotic cell.

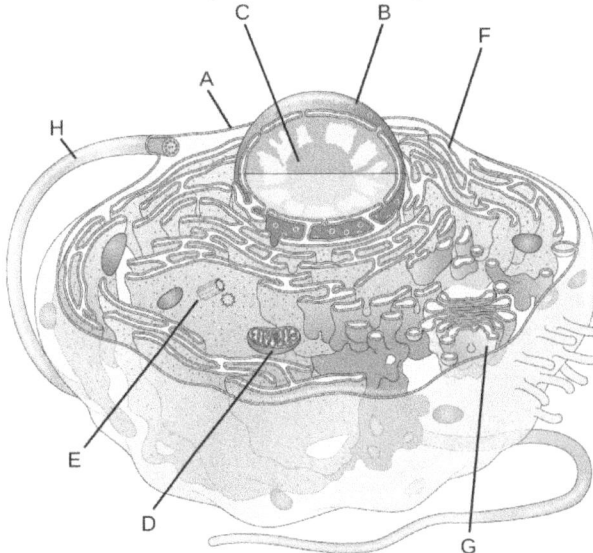

47. How are peroxisomes more like mitochondria than like the membrane-bound organelles of the endomembrane system? How do they differ from mitochondria?
48. Why must the functions of both lysosomes and peroxisomes be compartmentalized?

CHAPTER 4
Prokaryotic Diversity

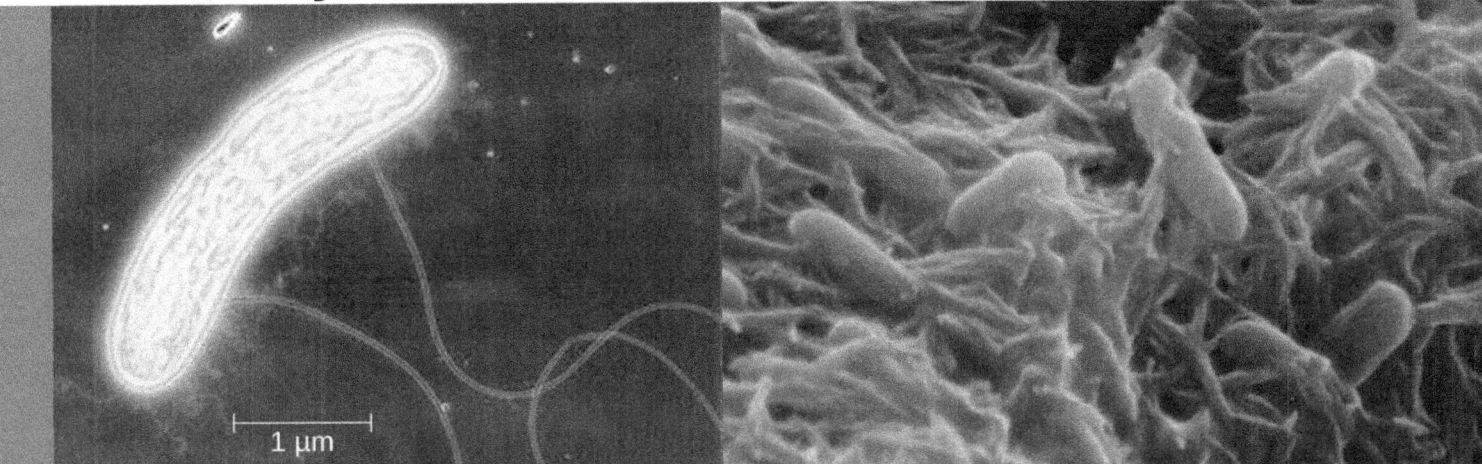

Figure 4.1 The bacterium *Shewanella* lives in the deep sea, where there is little oxygen diffused in the water. It is able to survive in this harsh environment by attaching to the sea floor and using long appendages, called "nanocables," to sense oxygen. (credit a: modification of work by NASA; credit b: modification of work by Liza Gross)

Chapter Outline

4.1 Prokaryote Habitats, Relationships, and Microbiomes

4.2 Proteobacteria

4.3 Nonproteobacteria Gram-Negative Bacteria and Phototrophic Bacteria

4.4 Gram-Positive Bacteria

4.5 Deeply Branching Bacteria

4.6 Archaea

INTRODUCTION Scientists have studied prokaryotes for centuries, but it wasn't until 1966 that scientist Thomas Brock (1926–) discovered that certain bacteria can live in boiling water. This led many to wonder whether prokaryotes may also live in other extreme environments, such as at the bottom of the ocean, at high altitudes, or inside volcanoes, or even on other planets.

Prokaryotes have an important role in changing, shaping, and sustaining the entire biosphere. They can produce proteins and other substances used by molecular biologists in basic research and in medicine and industry. For example, the bacterium *Shewanella* lives in the deep sea, where oxygen is scarce. It grows long appendages, which have special sensors used to seek the limited oxygen in its environment. It can also digest toxic waste and generate electricity. Other species of prokaryotes can produce more oxygen than the entire Amazon rainforest, while still others supply plants, animals, and humans with usable forms of nitrogen; and inhabit our body, protecting us from harmful microorganisms and producing some vitally important substances. This chapter will examine the diversity, structure, and function of prokaryotes.

4.1 Prokaryote Habitats, Relationships, and Microbiomes

Learning Objectives

By the end of this section, you will be able to:
- Identify and describe unique examples of prokaryotes in various habitats on earth
- Identify and describe symbiotic relationships
- Compare normal/commensal/resident microbiota to transient microbiota
- Explain how prokaryotes are classified

Clinical Focus

Part 1

Marsha, a 20-year-old university student, recently returned to the United States from a trip to Nigeria, where she had interned as a medical assistant for an organization working to improve access to laboratory services for tuberculosis testing. When she returned, Marsha began to feel fatigue, which she initially attributed to jet lag. However, the fatigue persisted, and Marsha soon began to experience other bothersome symptoms, such as occasional coughing, night sweats, loss of appetite, and a low-grade fever of 37.4 °C (99.3 °F).

Marsha expected her symptoms would subside in a few days, but instead, they gradually became more severe. About two weeks after returning home, she coughed up some sputum and noticed that it contained blood and small whitish clumps resembling cottage cheese. Her fever spiked to 38.2 °C (100.8 °F), and she began feeling sharp pains in her chest when breathing deeply. Concerned that she seemed to be getting worse, Marsha scheduled an appointment with her physician.

- Could Marsha's symptoms be related to her overseas travel, even several weeks after returning home?

Jump to the next Clinical Focus box.

All living organisms are classified into three domains of life: Archaea, Bacteria, and Eukarya. In this chapter, we will focus on the domains Archaea and Bacteria. Archaea and bacteria are unicellular prokaryotic organisms. Unlike eukaryotes, they have no nuclei or any other membrane-bound organelles.

Prokaryote Habitats and Functions

Prokaryotes are ubiquitous. They can be found everywhere on our planet, even in hot springs, in the Antarctic ice shield, and under extreme pressure two miles under water. One bacterium, *Paracoccus denitrificans*, has even been shown to survive when scientists removed it from its native environment (soil) and used a centrifuge to subject it to forces of gravity as strong as those found on the surface of Jupiter.

Prokaryotes also are abundant on and within the human body. According to a report by National Institutes of Health, prokaryotes, especially bacteria, outnumber human cells 10:1.[1] More recent studies suggest the ratio could be closer to 1:1, but even that ratio means that there are a great number of bacteria within the human body.[2] Bacteria thrive in the human mouth, nasal cavity, throat, ears, gastrointestinal tract, and vagina. Large colonies of bacteria can be found on healthy human skin, especially in moist areas (armpits, navel, and areas behind ears). However, even drier areas of the skin are not free from bacteria.

The existence of prokaryotes is very important for the stability and thriving of ecosystems. For example, they are a necessary part of soil formation and stabilization processes through the breakdown of organic matter

---

1 Medical Press. "Mouth Bacteria Can Change Their Diet, Supercomputers Reveal." August 12, 2014. http://medicalxpress.com/news/2014-08-mouth-bacteria-diet-supercomputers-reveal.html. Accessed February 24, 2015.

2 A. Abbott. "Scientists Bust Myth That Our Bodies Have More Bacteria Than Human Cells: Decades-Old Assumption about Microbiota Revisited." *Nature*. http://www.nature.com/news/scientists-bust-myth-that-our-bodies-have-more-bacteria-than-human-cells-1.19136. Accessed June 3, 2016.

and development of biofilms. One gram of soil contains up to 10 billion microorganisms (most of them prokaryotic) belonging to about 1,000 species. Many species of bacteria use substances released from plant roots, such as acids and carbohydrates, as nutrients. The bacteria metabolize these plant substances and release the products of bacterial metabolism back to the soil, forming humus and thus increasing the soil's fertility. In salty lakes such as the Dead Sea (Figure 4.2), salt-loving halobacteria decompose dead brine shrimp and nourish young brine shrimp and flies with the products of bacterial metabolism.

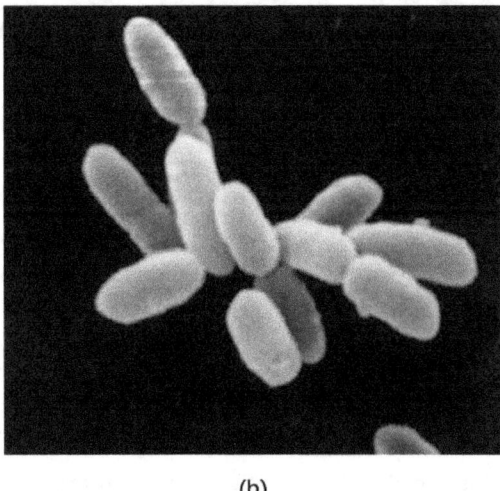

(a)　　　　　　　　　　　　　　　　　　　　　　　　　　(b)

Figure 4.2 (a) Some prokaryotes, called halophiles, can thrive in extremely salty environments such as the Dead Sea, pictured here. (b) The archaeon *Halobacterium salinarum*, shown here in an electron micrograph, is a halophile that lives in the Dead Sea. (credit a: modification of work by Jullen Menichini; credit b: modification of work by NASA)

In addition to living in the ground and the water, prokaryotic microorganisms are abundant in the air, even high in the atmosphere. There may be up to 2,000 different kinds of bacteria in the air, similar to their diversity in the soil.

Prokaryotes can be found everywhere on earth because they are extremely resilient and adaptable. They are often metabolically flexible, which means that they might easily switch from one energy source to another, depending on the availability of the sources, or from one metabolic pathway to another. For example, certain prokaryotic cyanobacteria can switch from a conventional type of lipid metabolism, which includes production of fatty aldehydes, to a different type of lipid metabolism that generates biofuel, such as fatty acids and wax esters. Groundwater bacteria store complex high-energy carbohydrates when grown in pure groundwater, but they metabolize these molecules when the groundwater is enriched with phosphates. Some bacteria get their energy by reducing sulfates into sulfides, but can switch to a different metabolic pathway when necessary, producing acids and free hydrogen ions.

Prokaryotes perform functions vital to life on earth by capturing (or "fixing") and recycling elements like carbon and nitrogen. Organisms such as animals require organic carbon to grow, but, unlike prokaryotes, they are unable to use inorganic carbon sources like carbon dioxide. Thus, animals rely on prokaryotes to convert carbon dioxide into organic carbon products that they can use. This process of converting carbon dioxide to organic carbon products is called carbon fixation.

Plants and animals also rely heavily on prokaryotes for nitrogen fixation, the conversion of atmospheric nitrogen into ammonia, a compound that some plants can use to form many different biomolecules necessary to their survival. Bacteria in the genus *Rhizobium*, for example, are nitrogen-fixing bacteria; they live in the roots of legume plants such as clover, alfalfa, and peas (Figure 4.3). Ammonia produced by *Rhizobium* helps these plants to survive by enabling them to make building blocks of nucleic acids. In turn, these plants may be eaten by animals—sustaining their growth and survival—or they may die, in which case the products of nitrogen fixation will enrich the soil and be used by other plants.

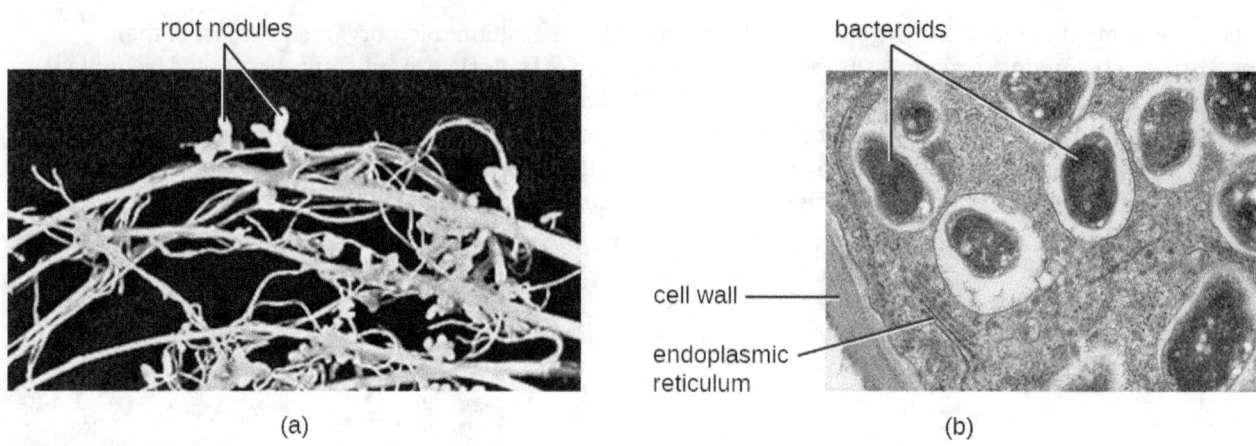

Figure 4.3 (a) Nitrogen-fixing bacteria such as *Rhizobium* live in the root nodules of legumes such as clover. (b) This micrograph of the root nodule shows bacteroids (bacterium-like cells or modified bacterial cells) within the plant cells. The bacteroids are visible as darker ovals within the larger plant cell. (credit a: modification of work by USDA)

Another positive function of prokaryotes is in cleaning up the environment. Recently, some researchers focused on the diversity and functions of prokaryotes in manmade environments. They found that some bacteria play a unique role in degrading toxic chemicals that pollute water and soil.[3]

Despite all of the positive and helpful roles prokaryotes play, some are human pathogens that may cause illness or infection when they enter the body. In addition, some bacteria can contaminate food, causing spoilage or foodborne illness, which makes them subjects of concern in food preparation and safety. Less than 1% of prokaryotes (all of them bacteria) are thought to be human pathogens, but collectively these species are responsible for a large number of the diseases that afflict humans.

Besides pathogens, which have a direct impact on human health, prokaryotes also affect humans in many indirect ways. For example, prokaryotes are now thought to be key players in the processes of climate change. In recent years, as temperatures in the earth's polar regions have risen, soil that was formerly frozen year-round (permafrost) has begun to thaw. Carbon trapped in the permafrost is gradually released and metabolized by prokaryotes. This produces massive amounts of carbon dioxide and methane, greenhouse gases that escape into the atmosphere and contribute to the greenhouse effect.

✓ CHECK YOUR UNDERSTANDING

- In what types of environments can prokaryotes be found?
- Name some ways that plants and animals rely on prokaryotes.

Symbiotic Relationships

As we have learned, prokaryotic microorganisms can associate with plants and animals. Often, this association results in unique relationships between organisms. For example, bacteria living on the roots or leaves of a plant get nutrients from the plant and, in return, produce substances that protect the plant from pathogens. On the other hand, some bacteria are plant pathogens that use mechanisms of infection similar to bacterial pathogens of animals and humans.

Prokaryotes live in a **community**, or a group of interacting populations of organisms. A population is a group of individual organisms belonging to the same biological species and limited to a certain geographic area. Populations can have **cooperative interactions**, which benefit the populations, or **competitive interactions**, in which one population competes with another for resources. The study of these interactions between microbial populations and their environment is called **microbial ecology**.

Any interaction between different species that are associated with each other within a community is called

---

3 A.M. Kravetz "Unique Bacteria Fights Man-Made Chemical Waste." 2012. http://www.livescience.com/25181-bacteria-strain-cleans-up-toxins-nsf-bts.html. Accessed March 9, 2015.

symbiosis. Such interactions fall along a continuum between opposition and cooperation. Interactions in a symbiotic relationship may be beneficial or harmful, or have no effect on one or both of the species involved. Table 4.1 summarizes the main types of symbiotic interactions among prokaryotes.

Types of Symbiotic Relationships

Type	Population A	Population B
Mutualism	Benefitted	Benefitted
Amensalism	Harmed	Unaffected
Commensalism	Benefitted	Unaffected
Neutralism	Unaffected	Unaffected
Parasitism	Benefitted	Harmed

Table 4.1

When two species benefit from each other, the symbiosis is called **mutualism** (or syntropy, or crossfeeding). For example, humans have a mutualistic relationship with the bacterium *Bacteroides thetaiotaomicron*, which lives in the intestinal tract. *Bacteroides thetaiotaomicron* digests complex polysaccharide plant materials that human digestive enzymes cannot break down, converting them into monosaccharides that can be absorbed by human cells. Humans also have a mutualistic relationship with certain strains of *Escherichia coli*, another bacterium found in the gut. *E. coli* relies on intestinal contents for nutrients, and humans derive certain vitamins from *E. coli*, particularly vitamin K, which is required for the formation of blood clotting factors. (This is only true for some strains of *E. coli*, however. Other strains are pathogenic and do not have a mutualistic relationship with humans.)

A type of symbiosis in which one population harms another but remains unaffected itself is called **amensalism**. In the case of bacteria, some amensalist species produce bactericidal substances that kill other species of bacteria. The microbiota of the skin is composed of a variety of bacterial species, including *Staphylococcus epidermidis* and *Propionibacterium acnes*. Although both species have the potential to cause infectious diseases when protective barriers are breached, they both produce a variety of antibacterial bacteriocins and bacteriocin-like compounds. *S. epidermidis* and *P. acnes* are unaffected by the bacteriocins and bacteriocin-like compounds they produce, but these compounds can target and kill other potential pathogens.

In another type of symbiosis, called **commensalism**, one organism benefits while the other is unaffected. This occurs when the bacterium *Staphylococcus epidermidis* uses the dead cells of the human skin as nutrients. Billions of these bacteria live on our skin, but in most cases (especially when our immune system is healthy), we do not react to them in any way. *S. epidermidis* provides an excellent example of how the classifications of symbiotic relationships are not always distinct. One could also consider the symbiotic relationship of *S. epidermidis* with humans as mutualism. Humans provide a food source of dead skin cells to the bacterium, and in turn the production of bacteriocin can provide an defense against potential pathogens.

If neither of the symbiotic organisms is affected in any way, we call this type of symbiosis **neutralism**. An example of neutralism is the coexistence of metabolically active (vegetating) bacteria and endospores (dormant, metabolically passive bacteria). For example, the bacterium *Bacillus anthracis* typically forms endospores in soil when conditions are unfavorable. If the soil is warmed and enriched with nutrients, some *B. anthracis* endospores germinate and remain in symbiosis with other species of endospores that have not germinated.

A type of symbiosis in which one organism benefits while harming the other is called **parasitism**. The relationship between humans and many pathogenic prokaryotes can be characterized as parasitic because these organisms invade the body, producing toxic substances or infectious diseases that cause harm. Diseases such as tetanus, diphtheria, pertussis, tuberculosis, and leprosy all arise from interactions between bacteria and humans.

Scientists have coined the term **microbiome** to refer to all prokaryotic and eukaryotic microorganisms and their genetic material that are associated with a certain organism or environment. Within the human microbiome, there are **resident microbiota** and **transient microbiota**. The resident microbiota consists of microorganisms that constantly live in or on our bodies. The term transient microbiota refers to microorganisms that are only temporarily found in the human body, and these may include pathogenic microorganisms. Hygiene and diet can alter both the resident and transient microbiota.

The resident microbiota is amazingly diverse, not only in terms of the variety of species but also in terms of the preference of different microorganisms for different areas of the human body. For example, in the human mouth, there are thousands of commensal or mutualistic species of bacteria. Some of these bacteria prefer to inhabit the surface of the tongue, whereas others prefer the internal surface of the cheeks, and yet others prefer the front or back teeth or gums. The inner surface of the cheek has the least diverse microbiota because of its exposure to oxygen. By contrast, the crypts of the tongue and the spaces between teeth are two sites with limited oxygen exposure, so these sites have more diverse microbiota, including bacteria living in the absence of oxygen (e.g., *Bacteroides*, *Fusobacterium*). Differences in the oral microbiota between randomly chosen human individuals are also significant. Studies have shown, for example, that the prevalence of such bacteria as *Streptococcus*, *Haemophilus*, *Neisseria*, and others was dramatically different when compared between individuals.[4]

There are also significant differences between the microbiota of different sites of the same human body. The inner surface of the cheek has a predominance of *Streptococcus*, whereas in the throat, the palatine tonsil, and saliva, there are two to three times fewer *Streptococcus*, and several times more *Fusobacterium*. In the plaque removed from gums, the predominant bacteria belong to the genus *Fusobacterium*. However, in the intestine, both *Streptococcus* and *Fusobacterium* disappear, and the genus *Bacteroides* becomes predominant.

Not only can the microbiota vary from one body site to another, the microbiome can also change over time within the same individual. Humans acquire their first inoculations of normal flora during natural birth and shortly after birth. Before birth, there is a rapid increase in the population of *Lactobacillus* spp. in the vagina, and this population serves as the first colonization of microbiota during natural birth. After birth, additional microbes are acquired from health-care providers, parents, other relatives, and individuals who come in contact with the baby. This process establishes a microbiome that will continue to evolve over the course of the individual's life as new microbes colonize and are eliminated from the body. For example, it is estimated that within a 9-hour period, the microbiota of the small intestine can change so that half of the microbial inhabitants will be different.[5] The importance of the initial *Lactobacillus* colonization during vaginal child birth is highlighted by studies demonstrating a higher incidence of diseases in individuals born by cesarean section, compared to those born vaginally. Studies have shown that babies born vaginally are predominantly colonized by vaginal lactobacillus, whereas babies born by cesarean section are more frequently colonized by microbes of the normal skin microbiota, including common hospital-acquired pathogens.

Throughout the body, resident microbiotas are important for human health because they occupy niches that might be otherwise taken by pathogenic microorganisms. For instance, *Lactobacillus* spp. are the dominant bacterial species of the normal vaginal microbiota for most women. lactobacillus produce lactic acid, contributing to the acidity of the vagina and inhibiting the growth of pathogenic yeasts. However, when the population of the resident microbiota is decreased for some reason (e.g., because of taking antibiotics), the pH of the vagina increases, making it a more favorable environment for the growth of yeasts such as *Candida albicans*. Antibiotic therapy can also disrupt the microbiota of the intestinal tract and respiratory tract, increasing the risk for secondary infections and/or promoting the long-term carriage and shedding of

---

4 E.M. Bik et al. "Bacterial Diversity in the Oral Cavity of 10 Healthy Individuals." *The ISME Journal* 4 no. 8 (2010):962–974.

5 C.C. Booijink et al. "High Temporal and Intra-Individual Variation Detected in the Human Ileal Microbiota." *Environmental Microbiology* 12 no. 12 (2010):3213–3227.

pathogens.

✓ CHECK YOUR UNDERSTANDING
- Explain the difference between cooperative and competitive interactions in microbial communities.
- List the types of symbiosis and explain how each population is affected.

Taxonomy and Systematics

Assigning prokaryotes to a certain species is challenging. They do not reproduce sexually, so it is not possible to classify them according to the presence or absence of interbreeding. Also, they do not have many morphological features. Traditionally, the classification of prokaryotes was based on their shape, staining patterns, and biochemical or physiological differences. More recently, as technology has improved, the nucleotide sequences in genes have become an important criterion of microbial classification.

In 1923, American microbiologist David Hendricks Bergey (1860–1937) published *A Manual in Determinative Bacteriology*. With this manual, he attempted to summarize the information about the kinds of bacteria known at that time, using Latin binomial classification. Bergey also included the morphological, physiological, and biochemical properties of these organisms. His manual has been updated multiple times to include newer bacteria and their properties. It is a great aid in bacterial taxonomy and methods of characterization of bacteria. A more recent sister publication, the five-volume *Bergey's Manual of Systematic Bacteriology*, expands on Bergey's original manual. It includes a large number of additional species, along with up-to-date descriptions of the taxonomy and biological properties of all named prokaryotic taxa. This publication incorporates the approved names of bacteria as determined by the List of Prokaryotic Names with Standing in Nomenclature (LPSN).

🔗 LINK TO LEARNING

The 7th edition (published in 1957) of *Bergey's Manual of Determinative Bacteriology* is now available (https://openstax.org/l/22mandeterbact) . More recent, updated editions are available in print. You can also access a searchable database (https://openstax.org/l/22databmicrefst) of microbial reference strains, published by the American Type Culture Collection (ATCC).

Classification by Staining Patterns

According to their staining patterns, which depend on the properties of their cell walls, bacteria have traditionally been classified into gram-positive, gram-negative, and "atypical," meaning neither gram-positive nor gram-negative. As explained in Staining Microscopic Specimens, gram-positive bacteria possess a thick peptidoglycan cell wall that retains the primary stain (crystal violet) during the decolorizing step; they remain purple after the gram-stain procedure because the crystal violet dominates the light red/pink color of the secondary counterstain, safranin. In contrast, gram-negative bacteria possess a thin peptidoglycan cell wall that does not prevent the crystal violet from washing away during the decolorizing step; therefore, they appear light red/pink after staining with the safranin. Bacteria that cannot be stained by the standard Gram stain procedure are called atypical bacteria. Included in the atypical category are species of *Mycoplasma* and *Chlamydia*. *Rickettsia* are also considered atypical because they are too small to be evaluated by the Gram stain.

More recently, scientists have begun to further classify gram-negative and gram-positive bacteria. They have added a special group of deeply branching bacteria based on a combination of physiological, biochemical, and genetic features. They also now further classify gram-negative bacteria into Proteobacteria, *Cytophaga-Flavobacterium-Bacteroides* (CFB), and spirochetes.

The deeply branching bacteria are thought to be a very early evolutionary form of bacteria (see Deeply Branching Bacteria). They live in hot, acidic, ultraviolet-light-exposed, and anaerobic (deprived of oxygen) conditions. Proteobacteria is a phylum of very diverse groups of gram-negative bacteria; it includes some important human pathogens (e.g., *E. coli* and *Bordetella pertussis*). The CFB group of bacteria includes components of the normal human gut microbiota, like *Bacteroides*. The spirochetes are spiral-shaped bacteria and include the pathogen *Treponema pallidum*, which causes syphilis. We will characterize these groups of bacteria in more detail later in the chapter.

Based on their prevalence of guanine and cytosine nucleotides, gram-positive bacteria are also classified into low G+C and high G+C gram-positive bacteria. The low G+C gram-positive bacteria have less than 50% of guanine and cytosine nucleotides in their DNA. They include human pathogens, such as those that cause anthrax (*Bacillus anthracis*), tetanus (*Clostridium tetani*), and listeriosis (*Listeria monocytogenes*). High G+C gram-positive bacteria, which have more than 50% guanine and cytosine nucleotides in their DNA, include the bacteria that cause diphtheria (*Corynebacterium diphtheriae*), tuberculosis (*Mycobacterium tuberculosis*), and other diseases.

The classifications of prokaryotes are constantly changing as new species are being discovered. We will describe them in more detail, along with the diseases they cause, in later sections and chapters.

 CHECK YOUR UNDERSTANDING

- How do scientists classify prokaryotes?

 MICRO CONNECTIONS

Human Microbiome Project

The Human Microbiome Project was launched by the National Institutes of Health (NIH) in 2008. One main goal of the project is to create a large repository of the gene sequences of important microbes found in humans, helping biologists and clinicians understand the dynamics of the human microbiome and the relationship between the human microbiota and diseases. A network of labs working together has been compiling the data from swabs of several areas of the skin, gut, and mouth from hundreds of individuals.

One of the challenges in understanding the human microbiome has been the difficulty of culturing many of the microbes that inhabit the human body. It has been estimated that we are only able to culture 1% of the bacteria in nature and that we are unable to grow the remaining 99%. To address this challenge, researchers have used metagenomic analysis, which studies genetic material harvested directly from microbial communities, as opposed to that of individual species grown in a culture. This allows researchers to study the genetic material of all microbes in the microbiome, rather than just those that can be cultured.[6]

One important achievement of the Human Microbiome Project is establishing the first reference database on microorganisms living in and on the human body. Many of the microbes in the microbiome are beneficial, but some are not. It was found, somewhat unexpectedly, that all of us have some serious microbial pathogens in our microbiota. For example, the conjunctiva of the human eye contains 24 genera of bacteria and numerous pathogenic species.[7] A healthy human mouth contains a number of species of the genus *Streptococcus*, including pathogenic species *S. pyogenes* and *S. pneumoniae*.[8] This raises the question of why certain prokaryotic organisms exist commensally in certain individuals but act as deadly pathogens in others. Also unexpected was the number of organisms that had never been cultured. For example, in one metagenomic study of the human gut microbiota, 174 new species of bacteria were identified.[9]

Another goal for the near future is to characterize the human microbiota in patients with different diseases and to find out whether there are any relationships between the contents of an individual's microbiota and risk

---

6 National Institutes of Health. "Human Microbiome Project. Overview." http://commonfund.nih.gov/hmp/overview. Accessed June 7, 2016.

7 Q. Dong et al. "Diversity of Bacteria at Healthy Human Conjunctiva." *Investigative Ophthalmology & Visual Science* 52 no. 8 (2011):5408–5413.

8 F.E. Dewhirst et al. "The Human Oral Microbiome." *Journal of Bacteriology* 192 no. 19 (2010):5002–5017.

9 J.C. Lagier et al. "Microbial Culturomics: Paradigm Shift in the Human Gut Microbiome Study." *Clinical Microbiology and Infection* 18 no. 12 (2012):1185–1193.

for or susceptibility to specific diseases. Analyzing the microbiome in a person with a specific disease may reveal new ways to fight diseases.

4.2 Proteobacteria

Learning Objectives

By the end of this section, you will be able to:
- Describe the unique features of each class within the phylum Proteobacteria: Alphaproteobacteria, Betaproteobacteria, Gammaproteobacteria, Deltaproteobacteria, and Epsilonproteobacteria
- Give an example of a bacterium in each class of Proteobacteria

In 1987, the American microbiologist Carl Woese (1928–2012) suggested that a large and diverse group of bacteria that he called "purple bacteria and their relatives" should be defined as a separate phylum within the domain Bacteria based on the similarity of the nucleotide sequences in their genome.[10] This phylum of gram-negative bacteria subsequently received the name **Proteobacteria**. It includes many bacteria that are part of the normal human microbiota as well as many pathogens. The Proteobacteria are further divided into five classes: Alphaproteobacteria, Betaproteobacteria, Gammaproteobacteria, Deltaproteobacteria, and Epsilonproteobacteria (Appendix D).

Alphaproteobacteria

The first class of Proteobacteria is the **Alphaproteobacteria**, many of which are obligate or facultative intracellular bacteria. Some species are characterized as **oligotroph**s, organisms capable of living in low-nutrient environments such as deep oceanic sediments, glacial ice, or deep undersurface soil.

Among the Alphaproteobacteria are rickettsias, **obligate intracellular pathogen**s, that require part of their life cycle to occur inside other cells called host cells. When not growing inside a host cell, *Rickettsia* are metabolically inactive outside the host cell. They cannot synthesize their own adenosine triphosphate (ATP), and, therefore, rely on cells for their energy needs.

Rickettsia spp. include a number of serious human pathogens. For example, *R. rickettsii* causes Rocky Mountain spotted fever, a life-threatening form of meningoencephalitis (inflammation of the membranes that wrap the brain). *R. rickettsii* infects ticks and can be transmitted to humans via a bite from an infected tick (Figure 4.4).

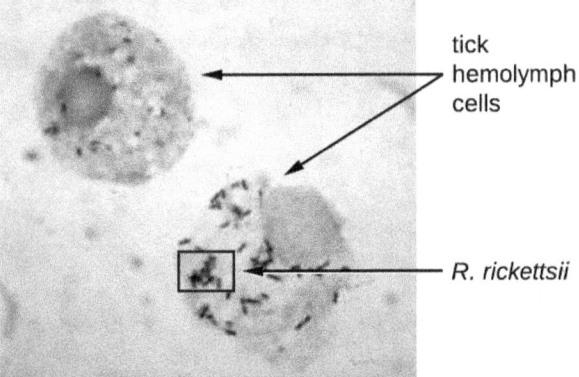

Figure 4.4 Rickettsias require special staining methods to see them under a microscope. Here, *R. rickettsii*, which causes Rocky Mountain spotted fever, is shown infecting the cells of a tick. (credit: modification of work by Centers for Disease Control and Prevention)

Another species of *Rickettsia*, *R. prowazekii*, is spread by lice. It causes epidemic typhus, a severe infectious disease common during warfare and mass migrations of people. *R. prowazekii* infects human endothelium cells, causing inflammation of the inner lining of blood vessels, high fever, abdominal pain, and sometimes delirium. A relative, *R. typhi*, causes a less severe disease known as murine or endemic typhus, which is still

---

10   C.R. Woese. "Bacterial Evolution." *Microbiological Review* 51 no. 2 (1987):221–271.

observed in the southwestern United States during warm seasons.

Table 4.2 summarizes the characteristics of important genera of Alphaproteobacteria.

C. trachomatis is a human pathogen that causes trachoma, a disease of the eyes, often leading to blindness. *C. trachomatis* also causes the sexually transmitted disease lymphogranuloma venereum (LGV). This disease is often mildly symptomatic, manifesting as regional lymph node swelling, or it may be asymptomatic, but it is extremely contagious and is common on college campuses.

Class Alphaproteobacteria		
Genus	Microscopic Morphology	Unique Characteristics
Agrobacterium	Gram-negative bacillus	Plant pathogen; one species, *A. tumefaciens*, causes tumors in plants
Bartonella	Gram-negative, pleomorphic, flagellated coccobacillus	Facultative intracellular bacteria, transmitted by lice and fleas, cause trench fever and cat scratch disease in humans
Brucella	Gram-negative, small, flagellated coccobacillus	Facultative intracellular bacteria, transmitted by contaminated milk from infected cows, cause brucellosis in cattle and humans
Caulobacter	Gram-negative bacillus	Used in studies on cellular adaptation and differentiation because of its peculiar life cycle (during cell division, forms "swarm" cells and "stalked" cells)
Coxiella	Small, gram-negative bacillus	Obligatory intracellular bacteria; cause Q fever; potential for use as biological weapon
Ehrlichia	Very small, gram-negative, coccoid or ovoid bacteria	Obligatory intracellular bacteria; can be transported from cell to cell; transmitted by ticks; cause ehrlichiosis (destruction of white blood cells and inflammation) in humans and dogs
Hyphomicrobium	Gram-negative bacilli; grows from a stalk	Similar to *Caulobacter*
Methylocystis	Gram-negative, coccoid or short bacilli	Nitrogen-fixing aerobic bacteria
Rhizobium	Gram-negative, rectangular bacilli with rounded ends forming clusters	Nitrogen-fixing bacteria that live in soil and form symbiotic relationship with roots of legumes (e.g., clover, alfalfa, and beans)
Rickettsia	Gram-negative, highly pleomorphic bacteria (may be cocci, rods, or threads)	Obligate intracellular bacteria; transmitted by ticks; may cause Rocky Mountain spotted fever and typhus

Table 4.2

⊘ CHECK YOUR UNDERSTANDING
- What is a common characteristic among the human pathogenic Alphaproteobacteria?

Betaproteobacteria

Betaproteobacteria are a diverse group of bacteria. The different bacterial species within this group utilize a wide range of metabolic strategies and can survive in a range of environments. Some genera include species that are human pathogens, able to cause severe, sometimes life-threatening disease. The genus *Neisseria*, for example, includes the bacteria *N. gonorrhoeae*, the causative agent of the STI gonorrhea, and *N. meningitides*, the causative agent of bacterial meningitis.

Neisseria are cocci that live on mucosal surfaces of the human body. They are fastidious, or difficult to culture, and they require high levels of moisture, nutrient supplements, and carbon dioxide. Also, *Neisseria* are microaerophilic, meaning that they require low levels of oxygen. For optimal growth and for the purposes of identification, *Neisseria* spp. are grown on chocolate agar (i.e., agar supplemented by partially hemolyzed red blood cells). Their characteristic pattern of growth in culture is diplococcal: pairs of cells resembling coffee beans (Figure 4.5).

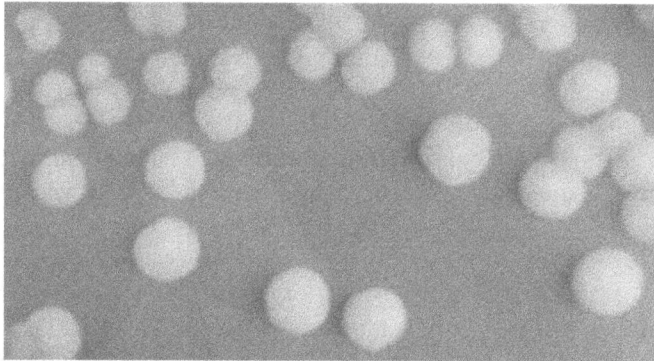

Figure 4.5 *Neisseria meningitidis* growing in colonies on a chocolate agar plate. (credit: Centers for Disease Control and Prevention)

The pathogen responsible for pertussis (whooping cough) is also a member of Betaproteobacteria. The bacterium *Bordetella pertussis*, from the order Burkholderiales, produces several toxins that paralyze the movement of cilia in the human respiratory tract and directly damage cells of the respiratory tract, causing a severe cough.

Table 4.3 summarizes the characteristics of important genera of Betaproteobacteria.

Class Betaproteobacteria

Example Genus	Microscopic Morphology	Unique Characteristics
Bordetella	A small, gram-negative coccobacillus	Aerobic, very fastidious; *B. pertussis* causes pertussis (whooping cough)
Burkholderia	Gram-negative bacillus	Aerobic, aquatic, cause diseases in horses and humans (especially patients with cystic fibrosis); agents of nosocomial infections
Leptothrix	Gram-negative, sheathed, filamentous bacillus	Aquatic; oxidize iron and manganese; can live in wastewater treatment plants and clog pipes
Neisseria	Gram-negative, coffee bean-shaped coccus forming pairs	Require moisture and high concentration of carbon dioxide; oxidase positive, grow on chocolate agar; pathogenic species cause gonorrhea and meningitis

| Class Betaproteobacteria |||
Example Genus	Microscopic Morphology	Unique Characteristics
Thiobacillus	Gram-negative bacillus	Thermophilic, acidophilic, strictly aerobic bacteria; oxidize iron and sulfur

Table 4.3

⊘ CHECK YOUR UNDERSTANDING

- Why are *Neisseria* classified as microaerophilic?

Clinical Focus

Part 2

When Marsha finally went to the doctor's office, the physician listened to her breathing through a stethoscope. He heard some crepitation (a crackling sound) in her lungs, so he ordered a chest radiograph and asked the nurse to collect a sputum sample for microbiological evaluation and cytology. The radiologic evaluation found cavities, opacities, and a particular pattern of distribution of abnormal material (Figure 4.6).

- What are some possible diseases that could be responsible for Marsha's radiograph results?

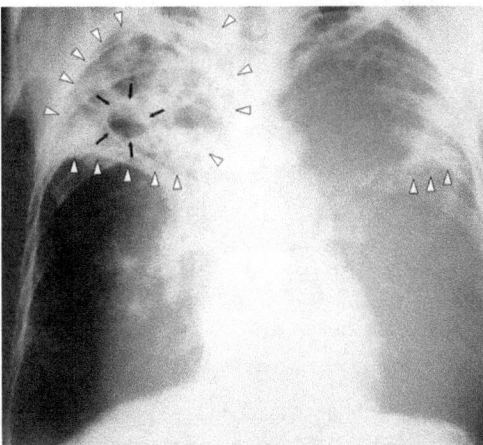

Figure 4.6 This anteroposterior radiograph shows the presence of bilateral pulmonary infiltrate (white triangles) and "caving formation" (black arrows) present in the right apical region. (credit: Centers for Disease Control and Prevention)

Jump to the next Clinical Focus box. Go back to the previous Clinical Focus box.

Gammaproteobacteria

The most diverse class of gram-negative bacteria is **Gammaproteobacteria**, and it includes a number of human pathogens. For example, a large and diverse family, *Pseudomonaceae*, includes the genus *Pseudomonas*. Within this genus is the species *P. aeruginosa*, a pathogen responsible for diverse infections in various regions of the body. *P. aeruginosa* is a strictly aerobic, nonfermenting, highly motile bacterium. It often infects wounds and burns, can be the cause of chronic urinary tract infections, and can be an important cause of respiratory infections in patients with cystic fibrosis or patients on mechanical ventilators. Infections by *P. aeruginosa* are often difficult to treat because the bacterium is resistant to many antibiotics and has a remarkable ability to form biofilms. Other representatives of *Pseudomonas* include the fluorescent (glowing) bacterium *P. fluorescens* and the soil bacteria *P. putida*, which is known for its ability to degrade xenobiotics (substances not naturally produced or found in living organisms).

The *Pasteurellaceae* also includes several clinically relevant genera and species. This family includes several bacteria that are human and/or animal pathogens. For example, *Pasteurella haemolytica* causes severe pneumonia in sheep and goats. *P. multocida* is a species that can be transmitted from animals to humans through bites, causing infections of the skin and deeper tissues. The genus *Haemophilus* contains two human pathogens, *H. influenzae* and *H. ducreyi*. Despite its name, *H. influenzae* does not cause influenza (which is a viral disease). *H. influenzae* can cause both upper and lower respiratory tract infections, including sinusitis, bronchitis, ear infections, and pneumonia. Before the development of effective vaccination, strains of *H. influenzae* were a leading cause of more invasive diseases, like meningitis in children. *H. ducreyi* causes the STI known as chancroid.

The order Vibrionales includes the human pathogen *Vibrio cholerae*. This comma-shaped aquatic bacterium thrives in highly alkaline environments like shallow lagoons and sea ports. A toxin produced by *V. cholerae* causes hypersecretion of electrolytes and water in the large intestine, leading to profuse watery diarrhea and dehydration. *V. parahaemolyticus* is also a cause of gastrointestinal disease in humans, whereas *V. vulnificus* causes serious and potentially life-threatening cellulitis (infection of the skin and deeper tissues) and blood-borne infections. Another representative of Vibrionales, *Aliivibrio fischeri*, engages in a symbiotic relationship with squid. The squid provides nutrients for the bacteria to grow and the bacteria produce bioluminescence that protects the squid from predators (Figure 4.7).

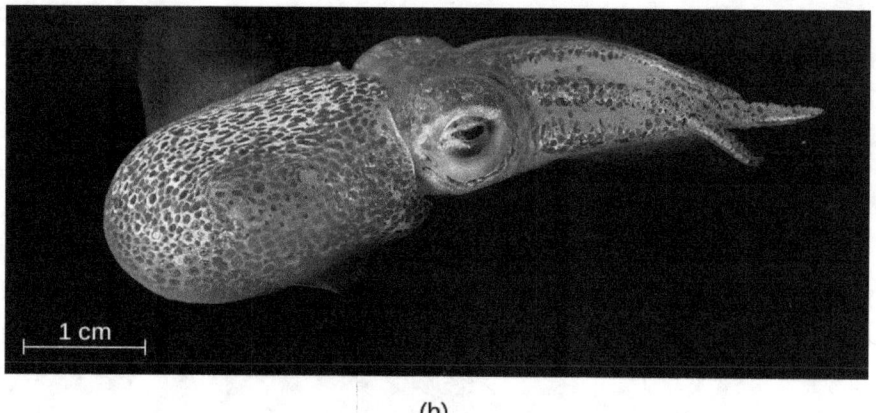

(a) (b)

Figure 4.7 (a) *Aliivibrio fischeri* is a bioluminescent bacterium. (b) *A. fischeri* colonizes and lives in a mutualistic relationship with the Hawaiian bobtail squid (*Euprymna scolopes*). (credit a: modification of work by American Society for Microbiology; credit b: modification of work by Margaret McFall-Ngai)

The genus *Legionella* also belongs to the Gammaproteobacteria. *L. pneumophila*, the pathogen responsible for Legionnaires disease, is an aquatic bacterium that tends to inhabit pools of warm water, such as those found in the tanks of air conditioning units in large buildings (Figure 4.8). Because the bacteria can spread in aerosols, outbreaks of Legionnaires disease often affect residents of a building in which the water has become contaminated with *Legionella*. In fact, these bacteria derive their name from the first known outbreak of Legionnaires disease, which occurred in a hotel hosting an American Legion veterans' association convention in Philadelphia in 1976.

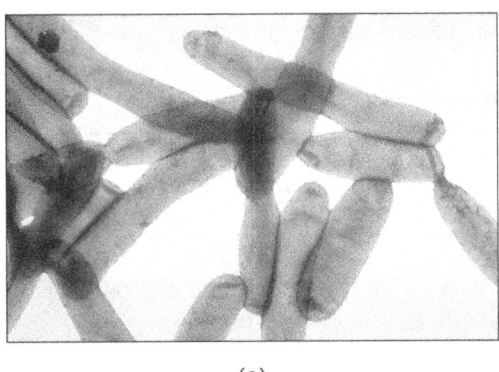 

(a) (b)

Figure 4.8 (a) *Legionella pneumophila*, the causative agent of Legionnaires disease, thrives in warm water. (b) Outbreaks of Legionnaires disease often originate in the air conditioning units of large buildings when water in or near the system becomes contaminated with *L. pneumophila*. (credit a: modification of work by Centers for Disease Control and Prevention)

Enterobacteriaceae is a large family of **enteric** (intestinal) bacteria belonging to the Gammaproteobacteria. They are facultative anaerobes and are able to ferment carbohydrates. Within this family, microbiologists recognize two distinct categories. The first category is called the coliforms, after its prototypical bacterium species, *Escherichia coli*. Coliforms are able to ferment lactose completely (i.e., with the production of acid and gas). The second category, noncoliforms, either cannot ferment lactose or can only ferment it incompletely (producing either acid or gas, but not both). The noncoliforms include some notable human pathogens, such as *Salmonella* spp., *Shigella* spp., and *Yersinia pestis*.

E. coli has been perhaps the most studied bacterium since it was first described in 1886 by Theodor Escherich (1857–1911). Many strains of E. coli are in mutualistic relationships with humans. However, some strains produce a potentially deadly toxin called Shiga toxin. Shiga toxin is one of the most potent bacterial toxins identified. Upon entering target cells, Shiga toxin interacts with ribosomes, stopping protein synthesis. Lack of protein synthesis leads to cellular death and hemorrhagic colitis, characterized by inflammation of intestinal tract and bloody diarrhea. In the most severe cases, patients can develop a deadly hemolytic uremic syndrome. Other E. coli strains may cause traveler's diarrhea, a less severe but very widespread disease.

The genus *Salmonella*, which belongs to the noncoliform group of *Enterobacteriaceae*, is interesting in that there is still no consensus about how many species it includes. Scientists have reclassified many of the groups they once thought to be species as **serotypes** (also called serovars), which are strains or variations of the same species of bacteria. Their classification is based on patterns of reactivity by animal antisera against molecules on the surface of the bacterial cells. A number of serotypes of *Salmonella* can cause salmonellosis, characterized by inflammation of the small and the large intestine, accompanied by fever, vomiting, and diarrhea. The species *S. enterobacterica* (serovar *typhi*) causes typhoid fever, with symptoms including fever, abdominal pain, and skin rashes (Figure 4.9).

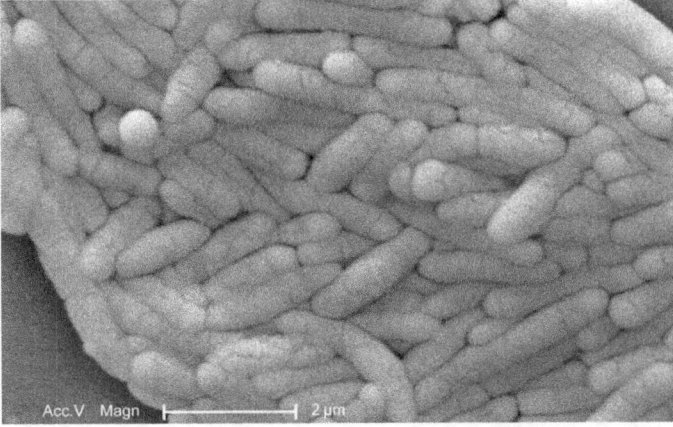

Figure 4.9 *Salmonella typhi* is the causative agent of typhoid fever. (credit: Centers for Disease Control and Prevention)

Table 4.4 summarizes the characteristics of important genera of Gammaproteobacteria.

Class Gammaproteobacteria

Example Genus	Microscopic Morphology	Unique Characteristics
Beggiatoa	Gram-negative bacteria; disc-shaped or cylindrical	Aquatic, live in water with high content of hydrogen disulfide; can cause problems for sewage treatment
Enterobacter	Gram-negative bacillus	Facultative anaerobe; cause urinary and respiratory tract infections in hospitalized patients; implicated in the pathogenesis of obesity
Erwinia	Gram-negative bacillus	Plant pathogen causing leaf spots and discoloration; may digest cellulose; prefer relatively low temperatures (25–30 °C)
Escherichia	Gram-negative bacillus	Facultative anaerobe; inhabit the gastrointestinal tract of warm-blooded animals; some strains are mutualists, producing vitamin K; others, like serotype *E. coli* O157:H7, are pathogens; *E. coli* has been a model organism for many studies in genetics and molecular biology
Hemophilus	Gram-negative bacillus	Pleomorphic, may appear as coccobacillus, aerobe, or facultative anaerobe; grow on blood agar; pathogenic species can cause respiratory infections, chancroid, and other diseases
Klebsiella	Gram-negative bacillus; appears rounder and thicker than other members of *Enterobacteriaceae*	Facultative anaerobe, encapsulated, nonmotile; pathogenic species may cause pneumonia, especially in people with alcoholism
Legionella	Gram-negative bacillus	Fastidious, grow on charcoal-buffered yeast extract; *L. pneumophila* causes Legionnaires disease
Methylomonas	Gram-negative bacillus	Use methane as source of carbon and energy
Proteus	Gram-negative bacillus (pleomorphic)	Common inhabitants of the human gastrointestinal tract; motile; produce urease; opportunistic pathogens; may cause urinary tract infections and sepsis
Pseudomonas	Gram-negative bacillus	Aerobic; versatile; produce yellow and blue pigments, making them appear green in culture; opportunistic, antibiotic-resistant pathogens may cause wound infections, hospital-acquired infections, and secondary infections in patients with cystic fibrosis

Class Gammaproteobacteria

Example Genus	Microscopic Morphology	Unique Characteristics
Serratia	Gram-negative bacillus	Motile; may produce red pigment; opportunistic pathogens responsible for a large number of hospital-acquired infections
Shigella	Gram-negative bacillus	Nonmotile; dangerously pathogenic; produce Shiga toxin, which can destroy cells of the gastrointestinal tract; can cause dysentery
Vibrio	Gram-negative, comma- or curved rod-shaped bacteria	Inhabit seawater; flagellated, motile; may produce toxin that causes hypersecretion of water and electrolytes in the gastrointestinal tract; some species may cause serious wound infections
Yersinia	Gram-negative bacillus	Carried by rodents; human pathogens; *Y. pestis* causes bubonic plague and pneumonic plague; *Y. enterocolitica* can be a pathogen causing diarrhea in humans

Table 4.4

CHECK YOUR UNDERSTANDING

- List two families of Gammaproteobacteria.

Deltaproteobacteria

The **Deltaproteobacteria** is a small class of gram-negative Proteobacteria that includes sulfate-reducing bacteria (SRBs), so named because they use sulfate as the final electron acceptor in the electron transport chain. Few SRBs are pathogenic. However, the SRB *Desulfovibrio orale* is associated with periodontal disease (disease of the gums).

Deltaproteobacteria also includes the genus *Bdellovibrio*, species of which are parasites of other gram-negative bacteria. *Bdellovibrio* invades the cells of the host bacterium, positioning itself in the periplasm, the space between the plasma membrane and the cell wall, feeding on the host's proteins and polysaccharides. The infection is lethal for the host cells.

Another type of Deltaproteobacteria, myxobacteria, lives in the soil, scavenging inorganic compounds. Motile and highly social, they interact with other bacteria within and outside their own group. They can form multicellular, macroscopic "fruiting bodies" (Figure 4.10), structures that are still being studied by biologists and bacterial ecologists.[11] These bacteria can also form metabolically inactive myxospores.

---

11 H. Reichenbach. "Myxobacteria, Producers of Novel Bioactive Substances." *Journal of Industrial Microbiology & Biotechnology* 27 no. 3 (2001):149–156.

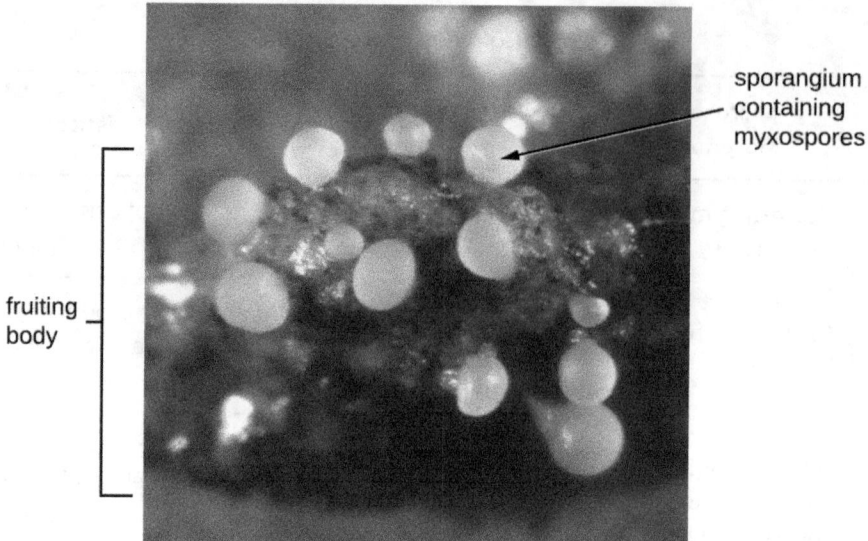

Figure 4.10 Myxobacteria form fruiting bodies. (credit: modification of work by Michiel Vos)

Table 4.5 summarizes the characteristics of several important genera of Deltaproteobacteria.

Class Deltaproteobacteria

Genus	Microscopic Morphology	Unique characteristics
Bdellovibrio	Gram-negative, comma-shaped rod	Obligate aerobes; motile; parasitic (infecting other bacteria)
Desulfovibrio (formerly *Desufuromonas*)	Gram-negative, comma-shaped rod	Reduce sulfur; can be used for removal of toxic and radioactive waste
Myxobacterium	Gram-negative, coccoid bacteria forming colonies (swarms)	Live in soil; can move by gliding; used as a model organism for studies of intercellular communication (signaling)

Table 4.5

CHECK YOUR UNDERSTANDING

- What type of Deltaproteobacteria forms fruiting bodies?

Epsilonproteobacteria

The smallest class of Proteobacteria is **Epsilonproteobacteria**, which are gram-negative microaerophilic bacteria (meaning they only require small amounts of oxygen in their environment). Two clinically relevant genera of Epsilonproteobacteria are *Campylobacter* and *Helicobacter*, both of which include human pathogens. *Campylobacter* can cause food poisoning that manifests as severe enteritis (inflammation in the small intestine). This condition, caused by the species *C. jejuni*, is rather common in developed countries, usually because of eating contaminated poultry products. Chickens often harbor *C. jejuni* in their gastrointestinal tract and feces, and their meat can become contaminated during processing.

Within the genus *Helicobacter*, the helical, flagellated bacterium *H. pylori* has been identified as a beneficial member of the stomach microbiota, but it is also the most common cause of chronic gastritis and ulcers of the stomach and duodenum (Figure 4.11). Studies have also shown that *H. pylori* is linked to stomach cancer.[12] *H. pylori* is somewhat unusual in its ability to survive in the highly acidic environment of the stomach. It produces urease and other enzymes that modify its environment to make it less acidic.

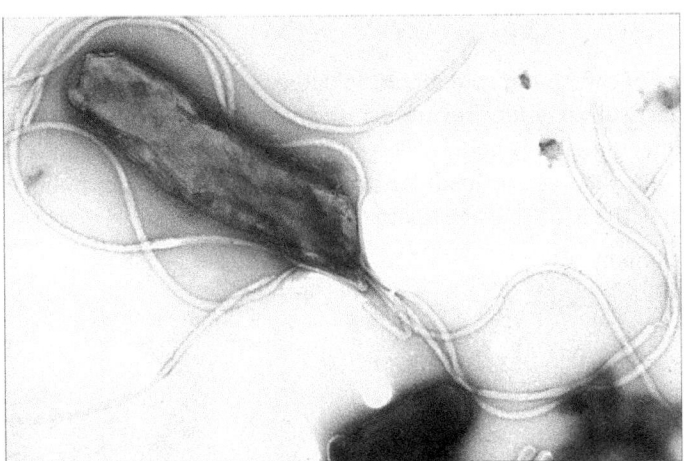

Figure 4.11 *Helicobacter pylori* can cause chronic gastritis, which can lead to ulcers and stomach cancer.

Table 4.6 summarizes the characteristics of the most clinically relevant genera of Epsilonproteobacteria.

Class Epsilonproteobacteria

Example Genus	Microscopic Morphology	Unique Characteristics
Campylobacter	Gram-negative, spiral-shaped rod	Aerobic (microaerophilic); often infects chickens; may infect humans via undercooked meat, causing severe enteritis
Helicobacter	Gram-negative, spiral-shaped rod	Aerobic (microaerophilic) bacterium; can damage the inner lining of the stomach, causing chronic gastritis, peptic ulcers, and stomach cancer

Table 4.6

✓ CHECK YOUR UNDERSTANDING
- Name two Epsilonproteobacteria that cause gastrointestinal disorders.

4.3 Nonproteobacteria Gram-Negative Bacteria and Phototrophic Bacteria

Learning Objectives

By the end of this section, you will be able to:
- Describe the unique features of nonproteobacteria gram-negative bacteria
- Give an example of a nonproteobacteria bacterium in each category
- Describe the unique features of phototrophic bacteria
- Identify phototrophic bacteria

---

12 S. Suerbaum, P. Michetti. "*Helicobacter pylori* infection." *New England Journal of Medicine* 347 no. 15 (2002):1175–1186.

The majority of the gram-negative bacteria belong to the phylum Proteobacteria, discussed in the previous section. Those that do not are called the nonproteobacteria. In this section, we will describe four classes of gram-negative nonproteobacteria: *Chlamydia*, the spirochetes, the CFB group, and the Planctomycetes. A diverse group of phototrophic bacteria that includes Proteobacteria and nonproteobacteria will be discussed at the end of this section.

Chlamydia

Chlamydia is another taxon of the Alphaproteobacteria. Members of this genus are gram-negative, obligate intracellular pathogens that are extremely resistant to the cellular defenses, giving them the ability to spread from host to host rapidly via elementary bodies. The metabolically and reproductively inactive **elementary bodies** are the endospore-like form of intracellular bacteria that enter an epithelial cell, where they become active. Figure 4.12 illustrates the life cycle of *Chlamydia*.

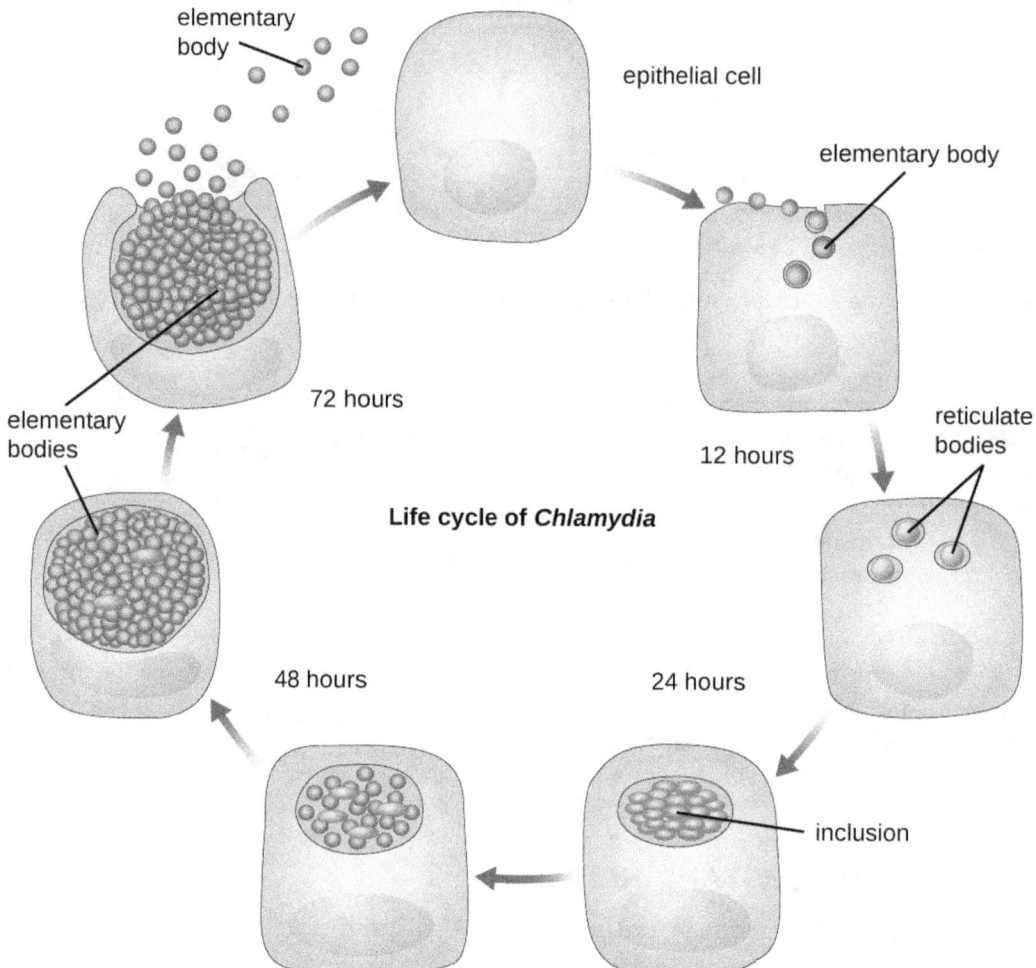

Figure 4.12 *Chlamydia* begins infection of a host when the metabolically inactive elementary bodies enter an epithelial cell. Once inside the host cell, the elementary bodies turn into active reticulate bodies. The reticulate bodies multiply and release more elementary bodies when the cell dies after the *Chlamydia* uses all of the host cell's ATP. (credit: modification of work by Centers for Disease Control and Prevention)

Spirochetes

Spirochetes are characterized by their long (up to 250 μm), spiral-shaped bodies. Most **spirochetes** are also very thin, which makes it difficult to examine gram-stained preparations under a conventional brightfield microscope. Darkfield fluorescent microscopy is typically used instead. Spirochetes are also difficult or even impossible to culture. They are highly motile, using their axial filament to propel themselves. The axial filament is similar to a flagellum, but it wraps around the cell and runs inside the cell body of a spirochete in the periplasmic space between the outer membrane and the plasma membrane (Figure 4.13).

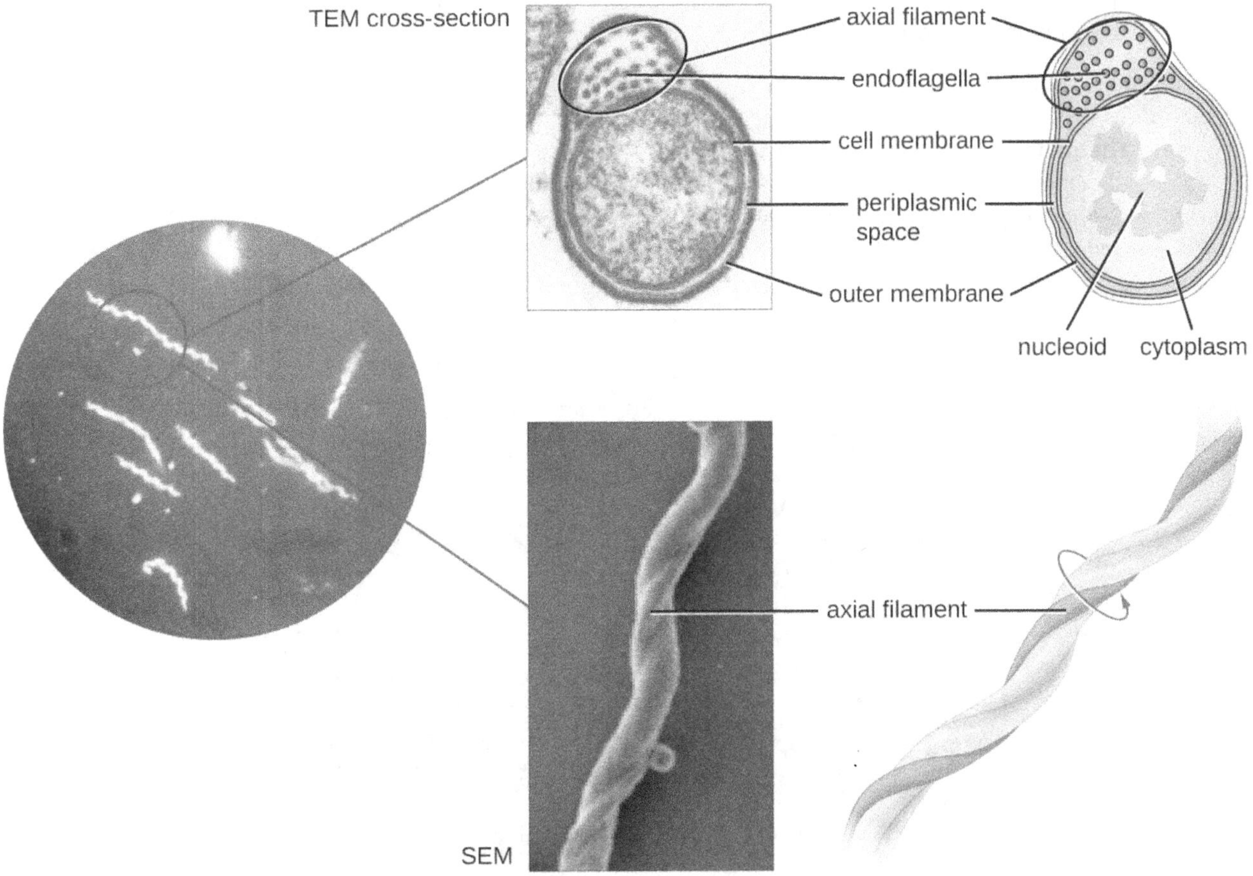

Figure 4.13 Spirochetes are typically observed using darkfield microscopy (left). However, electron microscopy (top center, bottom center) provides a more detailed view of their cellular morphology. The flagella found between the inner and outer membranes of spirochetes wrap around the bacterium, causing a twisting motion used for locomotion. (credit "spirochetes" micrograph: modification of work by Centers for Disease Control and Prevention; credit "SEM/TEM": modification of work by Guyard C, Raffel SJ, Schrumpf ME, Dahlstrom E, Sturdevant D, Ricklefs SM, Martens C, Hayes SF, Fischer ER, Hansen BT, Porcella SF, Schwan TG)

Several genera of spirochetes include human pathogens. For example, the genus *Treponema* includes a species *T. pallidum*, which is further classified into four subspecies: *T. pallidum pallidum*, *T. pallidum pertenue*, *T. pallidum carateum*, and *T. pallidum endemicum*. The subspecies *T. pallidum pallidum* causes the sexually transmitted infection known as syphilis, the third most prevalent sexually transmitted bacterial infection in the United States, after chlamydia and gonorrhea. The other subspecies of *T. pallidum* cause tropical infectious diseases of the skin, bones, and joints.

Another genus of spirochete, *Borrelia*, contains a number of pathogenic species. *B. burgdorferi* causes Lyme disease, which is transmitted by several genera of ticks (notably *Ixodes* and *Amblyomma*) and often produces a "bull's eye" rash, fever, fatigue, and, sometimes, debilitating arthritis. *B. recurrens* causes a condition known as relapsing fever. Appendix D lists the genera, species, and related diseases for spirochetes.

✓ CHECK YOUR UNDERSTANDING

- Why do scientists typically use darkfield fluorescent microscopy to visualize spirochetes?

Cytophaga, Fusobacterium, and *Bacteroides*

The gram-negative nonproteobacteria of the genera *Cytophaga*, *Fusobacterium*, and *Bacteroides* are classified together as a phylum and called the **CFB group**. Although they are phylogenetically diverse, bacteria of the CFB group share some similarities in the sequence of nucleotides in their DNA. They are rod-shaped bacteria adapted to anaerobic environments, such as the tissue of the gums, gut, and rumen of ruminating animals. CFB bacteria are avid fermenters, able to process cellulose in rumen, thus enabling ruminant animals to obtain carbon and energy from grazing.

Cytophaga are motile aquatic bacteria that glide. *Fusobacteria* inhabit the human mouth and may cause severe infectious diseases. The largest genus of the CFB group is *Bacteroides*, which includes dozens of species that are prevalent inhabitants of the human large intestine, making up about 30% of the entire gut microbiome (Figure 4.14). One gram of human feces contains up to 100 billion *Bacteroides* cells. Most *Bacteroides* are mutualistic. They benefit from nutrients they find in the gut, and humans benefit from their ability to prevent pathogens from colonizing the large intestine. Indeed, when populations of *Bacteroides* are reduced in the gut—as often occurs when a patient takes antibiotics—the gut becomes a more favorable environment for pathogenic bacteria and fungi, which can cause secondary infections.

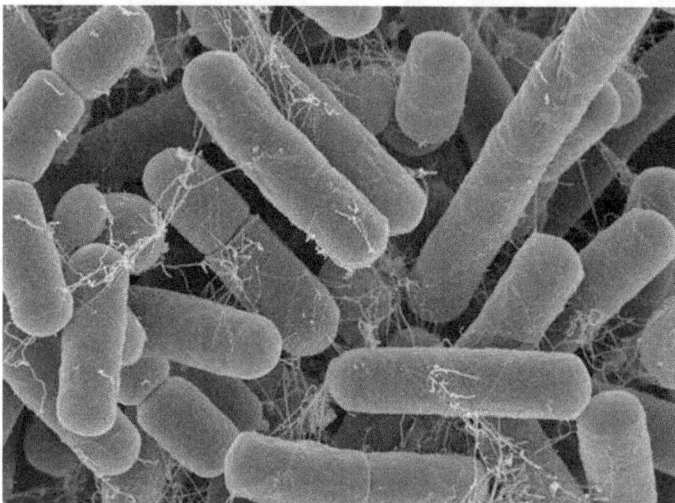

Figure 4.14 *Bacteroides* comprise up to 30% of the normal microbiota in the human gut. (credit: NOAA)

Only a few species of *Bacteroides* are pathogenic. *B. melaninogenicus*, for example, can cause wound infections in patients with weakened immune systems.

✓ CHECK YOUR UNDERSTANDING

- Why are *Cytophaga*, *Fusobacterium*, and *Bacteroides* classified together as the CFB group?

Planctomycetes

The Planctomycetes are found in aquatic environments, inhabiting freshwater, saltwater, and brackish water. Planctomycetes are unusual in that they reproduce by budding, meaning that instead of one maternal cell splitting into two equal daughter cells in the process of binary fission, the mother cell forms a bud that detaches from the mother cell and lives as an independent cell. These so-called swarmer cells are motile and not attached to a surface. However, they will soon differentiate into sessile (immobile) cells with an appendage called a holdfast that allows them to attach to surfaces in the water (Figure 4.15). Only the sessile cells are able to reproduce.

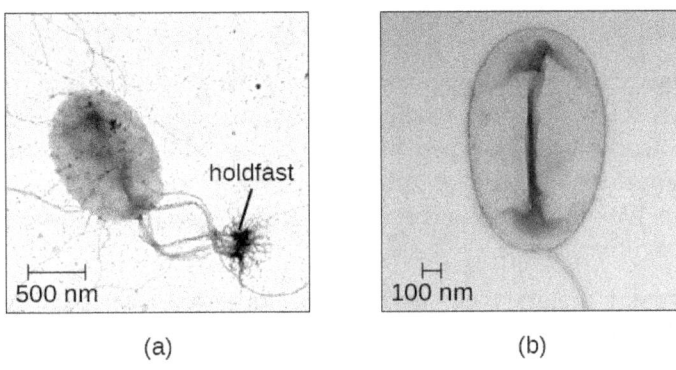

Figure 4.15 (a) Sessile Planctomycetes have a holdfast that allows them to adhere to surfaces in aquatic environments. (b) Swarmers are motile and lack a holdfast. (credit: modification of work by American Society for Microbiology)

Table 4.7 summarizes the characteristics of some of the most clinically relevant genera of nonproteobacteria.

Nonproteobacteria

Example Genus	Microscopic Morphology	Unique Characteristics
Chlamydia	Gram-negative, coccoid or ovoid bacterium	Obligatory intracellular bacteria; some cause chlamydia, trachoma, and pneumonia
Bacteroides	Gram-negative bacillus	Obligate anaerobic bacteria; abundant in the human gastrointestinal tract; usually mutualistic, although some species are opportunistic pathogens
Cytophaga	Gram-negative bacillus	Motile by gliding; live in soil or water; decompose cellulose; may cause disease in fish
Fusobacterium	Gram-negative bacillus with pointed ends	Anaerobic; form; biofilms; some species cause disease in humans (periodontitis, ulcers)
Leptospira	Spiral-shaped bacterium (spirochetes); gram negative-like (better viewed by darkfield microscopy); very thin	Aerobic, abundant in shallow water reservoirs; infect rodents and domestic animals; can be transmitted to humans by infected animals' urine; may cause severe disease
Borrelia	Gram-negative-like spirochete; very thin; better viewed by darkfield microscopy	*B. burgdorferi* causes Lyme disease and *B. recurrens* causes relapsing fever
Treponema	Gram-negative-like spirochete; very thin; better viewed by darkfield microscopy	Motile; do not grow in culture; *T. pallidum* (subspecies *T. pallidum pallidum*) causes syphilis

Table 4.7

CHECK YOUR UNDERSTANDING

- How do Planctomycetes reproduce?

Phototrophic Bacteria

The **phototrophic bacteria** are a large and diverse category of bacteria that do not represent a taxon but, rather, a group of bacteria that use sunlight as their primary source of energy. This group contains both Proteobacteria and nonproteobacteria. They use solar energy to synthesize ATP through photosynthesis. When they produce oxygen, they perform oxygenic photosynthesis. When they do not produce oxygen, they perform anoxygenic photosynthesis. With the exception of some cyanobacteria, the majority of phototrophic bacteria perform anoxygenic photosynthesis.

One large group of phototrophic bacteria includes the purple or green bacteria that perform photosynthesis with the help of **bacteriochlorophylls**, which are green, purple, or blue pigments similar to chlorophyll in plants. Some of these bacteria have a varying amount of red or orange pigments called carotenoids. Their color varies from orange to red to purple to green (Figure 4.16), and they are able to absorb light of various wavelengths. Traditionally, these bacteria are classified into sulfur and nonsulfur bacteria; they are further differentiated by color.

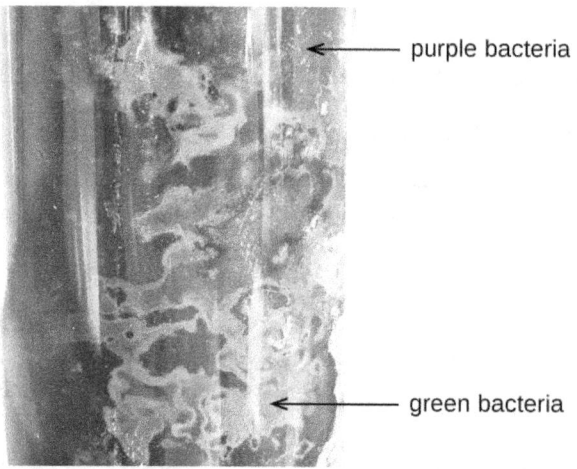

Figure 4.16 Purple and green sulfur bacteria use bacteriochlorophylls to perform photosynthesis.

The sulfur bacteria perform anoxygenic photosynthesis, using sulfites as electron donors and releasing free elemental sulfur. Nonsulfur bacteria use organic substrates, such as succinate and malate, as donors of electrons.

The **purple sulfur bacteria** oxidize hydrogen sulfide into elemental sulfur and sulfuric acid and get their purple color from the pigments bacteriochlorophylls and carotenoids. Bacteria of the genus *Chromatium* are purple sulfur Gammaproteobacteria. These microorganisms are strict anaerobes and live in water. They use carbon dioxide as their only source of carbon, but their survival and growth are possible only in the presence of sulfites, which they use as electron donors. *Chromatium* has been used as a model for studies of bacterial photosynthesis since the 1950s.[13]

The **green sulfur bacteria** use sulfide for oxidation and produce large amounts of green bacteriochlorophyll. The genus *Chlorobium* is a green sulfur bacterium that is implicated in climate change because it produces methane, a greenhouse gas. These bacteria use at least four types of chlorophyll for photosynthesis. The most prevalent of these, bacteriochlorophyll, is stored in special vesicle-like organelles called chlorosomes.

Purple nonsulfur bacteria are similar to purple sulfur bacteria, except that they use hydrogen rather than hydrogen sulfide for oxidation. Among the **purple nonsulfur bacteria** is the genus *Rhodospirillum*. These microorganisms are facultative anaerobes, which are actually pink rather than purple, and can metabolize ("fix") nitrogen. They may be valuable in the field of biotechnology because of their potential ability to produce biological plastic and hydrogen fuel.[14]

---

13 R.C. Fuller et al. "Carbon Metabolism in *Chromatium*." *Journal of Biological Chemistry* 236 (1961):2140–2149.

14 T.T. Selao et al. "Comparative Proteomic Studies in *Rhodospirillum rubrum* Grown Under Different Nitrogen Conditions." *Journal*

The **green nonsulfur bacteria** are similar to green sulfur bacteria but they use substrates other than sulfides for oxidation. *Chloroflexus* is an example of a green nonsulfur bacterium. It often has an orange color when it grows in the dark, but it becomes green when it grows in sunlight. It stores bacteriochlorophyll in chlorosomes, similar to *Chlorobium*, and performs anoxygenic photosynthesis, using organic sulfites (low concentrations) or molecular hydrogen as electron donors, so it can survive in the dark if oxygen is available. *Chloroflexus* does not have flagella but can glide, like *Cytophaga*. It grows at a wide range of temperatures, from 35 °C to 70 °C, thus can be thermophilic.

Another large, diverse group of phototrophic bacteria compose the phylum **Cyanobacteria**; they get their blue-green color from the chlorophyll contained in their cells (Figure 4.17). Species of this group perform oxygenic photosynthesis, producing megatons of gaseous oxygen. Scientists hypothesize that cyanobacteria played a critical role in the change of our planet's anoxic atmosphere 1–2 billion years ago to the oxygen-rich environment we have today.[15]

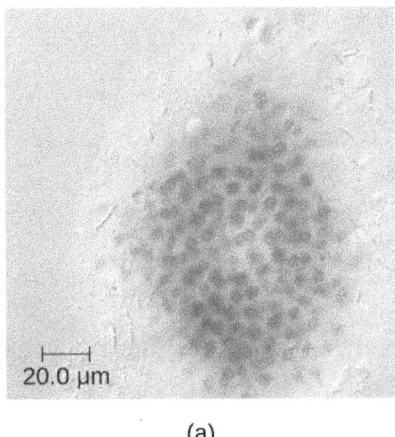
(a)

(b)

Figure 4.17 (a) *Microcystis aeruginosa* is a type of cyanobacteria commonly found in freshwater environments. (b) In warm temperatures, *M. aeruginosa* and other cyanobacteria can multiply rapidly and produce neurotoxins, resulting in blooms that are harmful to fish and other aquatic animals. (credit a: modification of work by Dr. Barry H. Rosen/U.S. Geological Survey; credit b: modification of work by NOAA)

Cyanobacteria have other remarkable properties. Amazingly adaptable, they thrive in many habitats, including marine and freshwater environments, soil, and even rocks. They can live at a wide range of temperatures, even in the extreme temperatures of the Antarctic. They can live as unicellular organisms or in colonies, and they can be filamentous, forming sheaths or biofilms. Many of them fix nitrogen, converting molecular nitrogen into nitrites and nitrates that other bacteria, plants, and animals can use. The reactions of nitrogen fixation occur in specialized cells called heterocysts.

Photosynthesis in Cyanobacteria is oxygenic, using the same type of chlorophyll a found in plants and algae as the primary photosynthetic pigment. Cyanobacteria also use phycocyanin and cyanophycin, two secondary photosynthetic pigments that give them their characteristic blue color. They are located in special organelles called phycobilisomes and in folds of the cellular membrane called thylakoids, which are remarkably similar to the photosynthetic apparatus of plants. Scientists hypothesize that plants originated from endosymbiosis of ancestral eukaryotic cells and ancestral photosynthetic bacteria.[16] Cyanobacteria are also an interesting object of research in biochemistry,[17] with studies investigating their potential as biosorbents[18] and products of human nutrition.[19]

---

of Proteome Research 7 no. 8 (2008):3267–3275.

15 A. De los Rios et al. "Ultrastructural and Genetic Characteristics of Endolithic Cyanobacterial Biofilms Colonizing Antarctic Granite Rocks." *FEMS Microbiology Ecology* 59 no. 2 (2007):386–395.

16 T. Cavalier-Smith. "Membrane Heredity and Early Chloroplast Evolution." *Trends in Plant Science* 5 no. 4 (2000):174–182.

17 S. Zhang, D.A. Bryant. "The Tricarboxylic Acid Cycle in Cyanobacteria." *Science* 334 no. 6062 (2011):1551–1553.

18 A. Cain et al. "Cyanobacteria as a Biosorbent for Mercuric Ion." *Bioresource Technology* 99 no. 14 (2008):6578–6586.

19 C.S. Ku et al. "Edible Blue-Green Algae Reduce the Production of Pro-Inflammatory Cytokines by Inhibiting NF-κB Pathway in

Unfortunately, cyanobacteria can sometimes have a negative impact on human health. Genera such as *Microcystis* can form harmful cyanobacterial blooms, forming dense mats on bodies of water and producing large quantities of toxins that can harm wildlife and humans. These toxins have been implicated in tumors of the liver and diseases of the nervous system in animals and humans.[20]

Table 4.8 summarizes the characteristics of important phototrophic bacteria.

Phototrophic Bacteria

Phylum	Class	Example Genus or Species	Common Name	Oxygenic or Anoxygenic	Sulfur Deposition
Cyanobacteria	Cyanophyceae	*Microcystis aeruginosa*	Blue-green bacteria	Oxygenic	None
Chlorobi	Chlorobia	*Chlorobium*	Green sulfur bacteria	Anoxygenic	Outside the cell
Chloroflexi (Division)	Chloroflexi	*Chloroflexus*	Green nonsulfur bacteria	Anoxygenic	None
Proteobacteria	Alphaproteobacteria	*Rhodospirillum*	Purple nonsulfur bacteria	Anoxygenic	None
Proteobacteria	Betaproteobacteria	*Rhodocyclus*	Purple nonsulfur bacteria	Anoxygenic	None
Proteobacteria	Gammaproteobacteria	*Chromatium*	Purple sulfur bacteria	Anoxygenic	Inside the cell

Table 4.8

✓ CHECK YOUR UNDERSTANDING

- What characteristic makes phototrophic bacteria different from other prokaryotes?

4.4 Gram-Positive Bacteria

Learning Objectives

By the end of this section, you will be able to:
- Describe the unique features of each category of high G+C and low G+C gram-positive bacteria
- Identify similarities and differences between high G+C and low G+C bacterial groups
- Give an example of a bacterium of high G+C and low G+C group commonly associated with each category

Prokaryotes are identified as gram-positive if they have a multiple layer matrix of peptidoglycan forming the cell wall. Crystal violet, the primary stain of the Gram stain procedure, is readily retained and stabilized within

---

Macrophages and Splenocytes." *Biochimica et Biophysica Acta* 1830 no. 4 (2013):2981–2988.

20 I. Stewart et al. Cyanobacterial Poisoning in Livestock, Wild Mammals and Birds – an Overview. *Advances in Experimental Medicine and Biology* 619 (2008):613–637.

this matrix, causing gram-positive prokaryotes to appear purple under a brightfield microscope after Gram staining. For many years, the retention of Gram stain was one of the main criteria used to classify prokaryotes, even though some prokaryotes did not readily stain with either the primary or secondary stains used in the Gram stain procedure.

Advances in nucleic acid biochemistry have revealed additional characteristics that can be used to classify gram-positive prokaryotes, namely the guanine to cytosine ratios (G+C) in DNA and the composition of 16S rRNA subunits. Microbiologists currently recognize two distinct groups of gram-positive, or weakly staining gram-positive, prokaryotes. The class Actinobacteria comprises the **high G+C gram-positive bacteria**, which have more than 50% guanine and cytosine nucleotides in their DNA. The class Bacilli comprises **low G+C gram-positive bacteria**, which have less than 50% of guanine and cytosine nucleotides in their DNA.

Actinobacteria: High G+C Gram-Positive Bacteria

The name Actinobacteria comes from the Greek words for *rays* and *small rod*, but Actinobacteria are very diverse. Their microscopic appearance can range from thin filamentous branching rods to coccobacilli. Some Actinobacteria are very large and complex, whereas others are among the smallest independently living organisms. Most Actinobacteria live in the soil, but some are aquatic. The vast majority are aerobic. One distinctive feature of this group is the presence of several different peptidoglycans in the cell wall.

The genus *Actinomyces* is a much studied representative of Actinobacteria. *Actinomyces* spp. play an important role in soil ecology, and some species are human pathogens. A number of *Actinomyces* spp. inhabit the human mouth and are opportunistic pathogens, causing infectious diseases like periodontitis (inflammation of the gums) and oral abscesses. The species *A. israelii* is an anaerobe notorious for causing endocarditis (inflammation of the inner lining of the heart) (Figure 4.18).

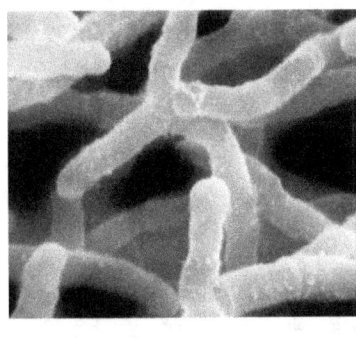
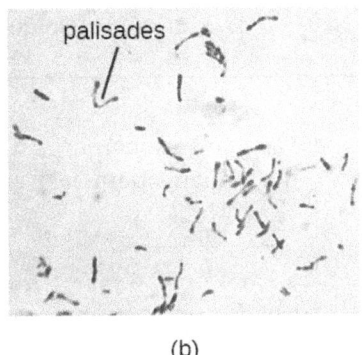
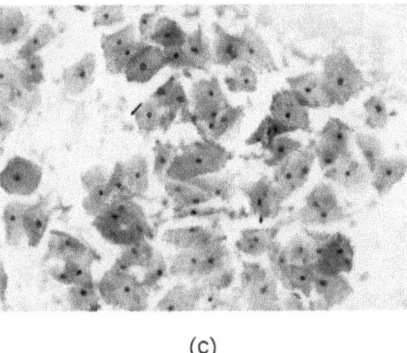

(a) (b) (c)

Figure 4.18 (a) *Actinomyces israelii* (false-color scanning electron micrograph [SEM]) has a branched structure. (b) *Corynebacterium diphtheria* causes the deadly disease diphtheria. Note the distinctive palisades. (c) The gram-variable bacterium *Gardnerella vaginalis* causes bacterial vaginosis in women. This micrograph shows a Pap smear from a woman with vaginosis. (credit a: modification of work by "GrahamColm"/Wikimedia Commons; credit b: modification of work by Centers for Disease Control and Prevention; credit c: modification of work by Mwakigonja AR, Torres LM, Mwakyoma HA, Kaaya EE)

The genus *Mycobacterium* is represented by bacilli covered with a mycolic acid coat. This waxy coat protects the bacteria from some antibiotics, prevents them from drying out, and blocks penetration by Gram stain reagents (see Staining Microscopic Specimens). Because of this, a special acid-fast staining procedure is used to visualize these bacteria. The genus *Mycobacterium* is an important cause of a diverse group of infectious diseases. *M. tuberculosis* is the causative agent of tuberculosis, a disease that primarily impacts the lungs but can infect other parts of the body as well. It has been estimated that one-third of the world's population has been infected with *M. tuberculosis* and millions of new infections occur each year. Treatment of *M. tuberculosis* is challenging and requires patients to take a combination of drugs for an extended time. Complicating treatment even further is the development and spread of multidrug-resistant strains of this pathogen.

Another pathogenic species, *M. leprae*, is the cause of Hansen's disease (leprosy), a chronic disease that impacts peripheral nerves and the integrity of the skin and mucosal surface of the respiratory tract. Loss of pain sensation and the presence of skin lesions increase susceptibility to secondary injuries and infections with other pathogens.

Bacteria in the genus *Corynebacterium* contain diaminopimelic acid in their cell walls, and microscopically often form *palisades*, or pairs of rod-shaped cells resembling the letter V. Cells may contain metachromatic granules, intracellular storage of inorganic phosphates that are useful for identification of *Corynebacterium*. The vast majority of *Corynebacterium* spp. are nonpathogenic; however, *C. diphtheria* is the causative agent of diphtheria, a disease that can be fatal, especially in children (Figure 4.18). *C. diphtheria* produces a toxin that forms a pseudomembrane in the patient's throat, causing swelling, difficulty breathing, and other symptoms that can become serious if untreated.

The genus *Bifidobacterium* consists of filamentous anaerobes, many of which are commonly found in the gastrointestinal tract, vagina, and mouth. In fact, *Bifidobacterium* spp. constitute a substantial part of the human gut microbiota and are frequently used as probiotics and in yogurt production.

The genus *Gardnerella*, contains only one species, *G. vaginalis*. This species is defined as "gram-variable" because its small coccobacilli do not show consistent results when Gram stained (Figure 4.18). Based on its genome, it is placed into the high G+C gram-positive group. *G. vaginalis* can cause bacterial vaginosis in women; symptoms are typically mild or even undetectable, but can lead to complications during pregnancy.

Table 4.9 summarizes the characteristics of some important genera of Actinobacteria. Additional information on Actinobacteria appears in Appendix D.

Actinobacteria: High G+C Gram-Positive

Example Genus	Microscopic Morphology	Unique Characteristics
Actinomyces	Gram-positive bacillus; in colonies, shows fungus-like threads (hyphae)	Facultative anaerobes; in soil, decompose organic matter; in the human mouth, may cause gum disease
Arthrobacter	Gram-positive bacillus (at the exponential stage of growth) or coccus (in stationary phase)	Obligate aerobes; divide by "snapping," forming V-like pairs of daughter cells; degrade phenol, can be used in bioremediation
Bifidobacterium	Gram-positive, filamentous actinobacterium	Anaerobes commonly found in human gut microbiota
Corynebacterium	Gram-positive bacillus	Aerobes or facultative anaerobes; form palisades; grow slowly; require enriched media in culture; *C. diphtheriae* causes diphtheria
Frankia	Gram-positive, fungus-like (filamentous) bacillus	Nitrogen-fixing bacteria; live in symbiosis with legumes
Gardnerella	Gram-variable coccobacillus	Colonize the human vagina, may alter the microbial ecology, thus leading to vaginosis

Actinobacteria: High G+C Gram-Positive		
Example Genus	Microscopic Morphology	Unique Characteristics
Micrococcus	Gram-positive coccus, form microscopic clusters	Ubiquitous in the environment and on the human skin; oxidase-positive (as opposed to morphologically similar *S. aureus*); some are opportunistic pathogens
Mycobacterium	Gram-positive, acid-fast bacillus	Slow growing, aerobic, resistant to drying and phagocytosis; covered with a waxy coat made of mycolic acid; *M. tuberculosis* causes tuberculosis; *M. leprae* causes leprosy
Nocardia	Weakly gram-positive bacillus; forms acid-fast branches	May colonize the human gingiva; may cause severe pneumonia and inflammation of the skin
Propionibacterium	Gram-positive bacillus	Aerotolerant anaerobe; slow-growing; *P. acnes* reproduces in the human sebaceous glands and may cause or contribute to acne
Rhodococcus	Gram-positive bacillus	Strict aerobe; used in industry for biodegradation of pollutants; *R. fascians* is a plant pathogen, and *R. equi* causes pneumonia in foals
Streptomyces	Gram-positive, fungus-like (filamentous) bacillus	Very diverse genus (>500 species); aerobic, spore-forming bacteria; scavengers, decomposers found in soil (give the soil its "earthy" odor); used in pharmaceutical industry as antibiotic producers (more than two-thirds of clinically useful antibiotics)

Table 4.9

✓ CHECK YOUR UNDERSTANDING

- What is one distinctive feature of Actinobacteria?

Low G+C Gram-positive Bacteria

The low G+C gram-positive bacteria have less than 50% guanine and cytosine in their DNA, and this group of bacteria includes a number of genera of bacteria that are pathogenic.

> ### Clinical Focus
>
> **Part 3**
> Based on her symptoms, Marsha's doctor suspected that she had a case of tuberculosis. Although less common in the United States, tuberculosis is still extremely common in many parts of the world, including Nigeria. Marsha's work there in a medical lab likely exposed her to *Mycobacterium tuberculosis*, the bacterium that causes tuberculosis.
>
> Marsha's doctor ordered her to stay at home, wear a respiratory mask, and confine herself to one room as

much as possible. He also said that Marsha had to take one semester off school. He prescribed isoniazid and rifampin, antibiotics used in a drug cocktail to treat tuberculosis, which Marsha was to take three times a day for at least three months.

- Why did the doctor order Marsha to stay home for three months?

Jump to the next Clinical Focus box. Go back to the previous Clinical Focus box.

Clostridia

One large and diverse class of low G+C gram-positive bacteria is Clostridia. The best studied genus of this class is *Clostridium*. These rod-shaped bacteria are generally obligate anaerobes that produce endospores and can be found in anaerobic habitats like soil and aquatic sediments rich in organic nutrients. The endospores may survive for many years.

Clostridium spp. produce more kinds of protein toxins than any other bacterial genus, and several species are human pathogens. *C. perfringens* is the third most common cause of food poisoning in the United States and is the causative agent of an even more serious disease called gas gangrene. Gas gangrene occurs when *C. perfringens* endospores enter a wound and germinate, becoming viable bacterial cells and producing a toxin that can cause the necrosis (death) of tissue. *C. tetani*, which causes tetanus, produces a neurotoxin that is able to enter neurons, travel to regions of the central nervous system where it blocks the inhibition of nerve impulses involved in muscle contractions, and cause a life-threatening spastic paralysis. *C. botulinum* produces botulinum neurotoxin, the most lethal biological toxin known. Botulinum toxin is responsible for rare but frequently fatal cases of botulism. The toxin blocks the release of acetylcholine in neuromuscular junctions, causing flaccid paralysis. In very small concentrations, botulinum toxin has been used to treat muscle pathologies in humans and in a cosmetic procedure to eliminate wrinkles. *C. difficile* is a common source of hospital-acquired infections (Figure 4.19) that can result in serious and even fatal cases of colitis (inflammation of the large intestine). Infections often occur in patients who are immunosuppressed or undergoing antibiotic therapy that alters the normal microbiota of the gastrointestinal tract. Appendix D lists the genera, species, and related diseases for Clostridia.

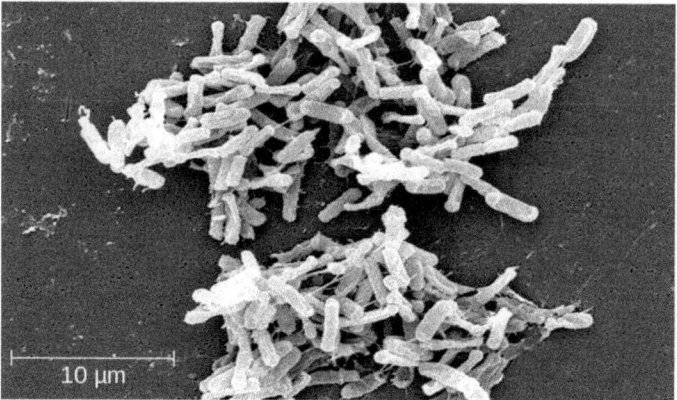

Figure 4.19 *Clostridium difficile*, a gram-positive, rod-shaped bacterium, causes severe colitis and diarrhea, often after the normal gut microbiota is eradicated by antibiotics. (credit: modification of work by Centers for Disease Control and Prevention)

Lactobacillales

The order Lactobacillales comprises low G+C gram-positive bacteria that include both bacilli and cocci in the genera *Lactobacillus*, *Leuconostoc*, *Enterococcus*, and *Streptococcus*. Bacteria of the latter three genera typically are spherical or ovoid and often form chains.

Streptococcus, the name of which comes from the Greek word for *twisted chain*, is responsible for many types of infectious diseases in humans. Species from this genus, often referred to as streptococci, are usually classified by serotypes called Lancefield groups, and by their ability to lyse red blood cells when grown on blood agar.

S. pyogenes belongs to the Lancefield group A, β-hemolytic *Streptococcus*. This species is considered a pyogenic pathogen because of the associated pus production observed with infections it causes (Figure 4.20). *S. pyogenes* is the most common cause of bacterial pharyngitis (strep throat); it is also an important cause of various skin infections that can be relatively mild (e.g., impetigo) or life threatening (e.g., necrotizing fasciitis, also known as flesh eating disease), life threatening.

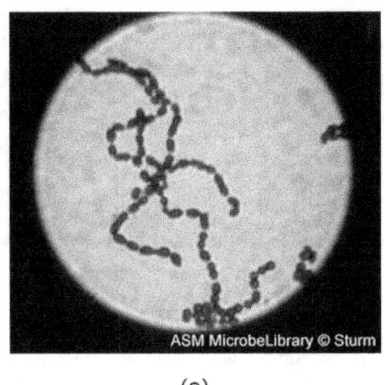
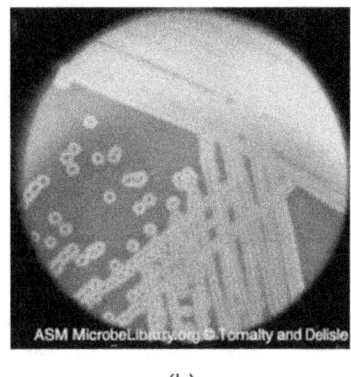

(a) (b)

Figure 4.20 (a) A gram-stained specimen of *Streptococcus pyogenes* shows the chains of cocci characteristic of this organism's morphology. (b) *S. pyogenes* on blood agar shows characteristic lysis of red blood cells, indicated by the halo of clearing around colonies. (credit a, b: modification of work by American Society for Microbiology)

The nonpyogenic (i.e., not associated with pus production) streptococci are a group of streptococcal species that are not a taxon but are grouped together because they inhabit the human mouth. The nonpyogenic streptococci do not belong to any of the Lancefield groups. Most are commensals, but a few, such as *S. mutans*, are implicated in the development of dental caries.

S. pneumoniae (commonly referred to as pneumococcus), is a *Streptococcus* species that also does not belong to any Lancefield group. *S. pneumoniae* cells appear microscopically as diplococci, pairs of cells, rather than the long chains typical of most streptococci. Scientists have known since the 19th century that *S. pneumoniae* causes pneumonia and other respiratory infections. However, this bacterium can also cause a wide range of other diseases, including meningitis, septicemia, osteomyelitis, and endocarditis, especially in newborns, the elderly, and patients with immunodeficiency.

Bacilli

The name of the class Bacilli suggests that it is made up of bacteria that are bacillus in shape, but it is a morphologically diverse class that includes bacillus-shaped and cocccus-shaped genera. Among the many genera in this class are two that are very important clinically: *Bacillus* and *Staphylococcus*.

Bacteria in the genus *Bacillus* are bacillus in shape and can produce endospores. They include aerobes or facultative anaerobes. A number of *Bacillus* spp. are used in various industries, including the production of antibiotics (e.g., barnase), enzymes (e.g., alpha-amylase, BamH1 restriction endonuclease), and detergents (e.g., subtilisin).

Two notable pathogens belong to the genus *Bacillus*. *B. anthracis* is the pathogen that causes anthrax, a severe disease that affects wild and domesticated animals and can spread from infected animals to humans. Anthrax manifests in humans as charcoal-black ulcers on the skin, severe enterocolitis, pneumonia, and brain damage due to swelling. If untreated, anthrax is lethal. *B. cereus*, a closely related species, is a pathogen that may cause food poisoning. It is a rod-shaped species that forms chains. Colonies appear milky white with irregular shapes when cultured on blood agar (Figure 4.21). One other important species is *B. thuringiensis*. This bacterium produces a number of substances used as insecticides because they are toxic for insects.

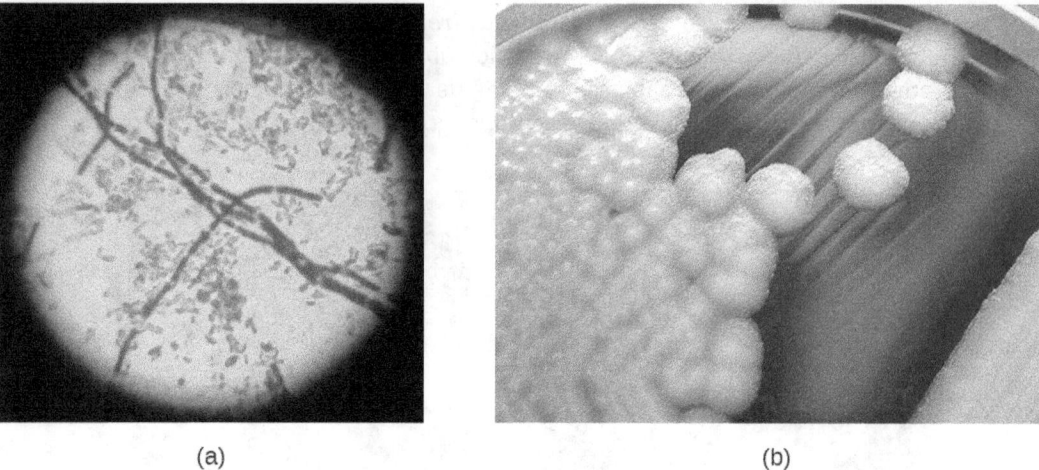

(a) (b)

Figure 4.21 (a) In this gram-stained specimen, the violet rod-shaped cells forming chains are the gram-positive bacteria *Bacillus cereus*. The small, pink cells are the gram-negative bacteria *Escherichia coli*. (b) In this culture, white colonies of *B. cereus* have been grown on sheep blood agar. (credit a: modification of work by "Bibliomaniac 15"/Wikimedia Commons; credit b: modification of work by Centers for Disease Control and Prevention)

The genus *Staphylococcus* also belongs to the class Bacilli, even though its shape is coccus rather than a bacillus. The name *Staphylococcus* comes from a Greek word for *bunches of grapes*, which describes their microscopic appearance in culture (Figure 4.22). *Staphylococcus* spp. are facultative anaerobic, halophilic, and nonmotile. The two best-studied species of this genus are *S. epidermidis* and *S. aureus*.

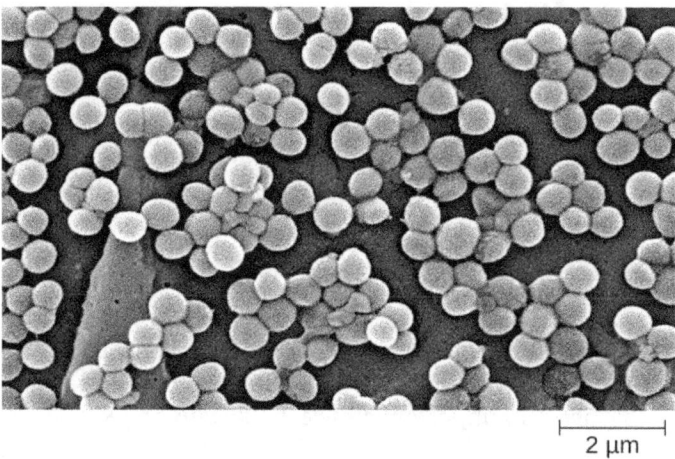

Figure 4.22 This SEM of *Staphylococcus aureus* illustrates the typical "grape-like" clustering of cells. (credit: modification of work by Centers for Disease Control and Prevention)

S. epidermidis, whose main habitat is the human skin, is thought to be nonpathogenic for humans with healthy immune systems, but in patients with immunodeficiency, it may cause infections in skin wounds and prostheses (e.g., artificial joints, heart valves). *S. epidermidis* is also an important cause of infections associated with intravenous catheters. This makes it a dangerous pathogen in hospital settings, where many patients may be immunocompromised.

Strains of *S. aureus* cause a wide variety of infections in humans, including skin infections that produce boils, carbuncles, cellulitis, or impetigo. Certain strains of *S. aureus* produce a substance called enterotoxin, which can cause severe enteritis, often called staph food poisoning. Some strains of *S. aureus* produce the toxin responsible for toxic shock syndrome, which can result in cardiovascular collapse and death.

Many strains of *S. aureus* have developed resistance to antibiotics. Some antibiotic-resistant strains are designated as methicillin-resistant *S. aureus* (MRSA) and vancomycin-resistant *S. aureus* (VRSA). These strains are some of the most difficult to treat because they exhibit resistance to nearly all available antibiotics, not just methicillin and vancomycin. Because they are difficult to treat with antibiotics, infections can be lethal. MRSA and VRSA are also contagious, posing a serious threat in hospitals, nursing homes, dialysis facilities, and other places where there are large populations of elderly, bedridden, and/or immunocompromised patients. Appendix D lists the genera, species, and related diseases for bacilli.

Mycoplasmas

Although *Mycoplasma* spp. do not possess a cell wall and, therefore, are not stained by Gram-stain reagents, this genus is still included with the low G+C gram-positive bacteria. The genus *Mycoplasma* includes more than 100 species, which share several unique characteristics. They are very small cells, some with a diameter of about 0.2 µm, which is smaller than some large viruses. They have no cell walls and, therefore, are **pleomorphic**, meaning that they may take on a variety of shapes and can even resemble very small animal cells. Because they lack a characteristic shape, they can be difficult to identify. One species, *M. pneumoniae*, causes the mild form of pneumonia known as "walking pneumonia" or "atypical pneumonia." This form of pneumonia is typically less severe than forms caused by other bacteria or viruses.

Table 4.10 summarizes the characteristics of notable genera low G+C Gram-positive bacteria.

Bacilli: Low G+C Gram-Positive Bacteria

Example Genus	Microscopic Morphology	Unique Characteristics
Bacillus	Large, gram-positive bacillus	Aerobes or facultative anaerobes; form endospores; *B. anthracis* causes anthrax in cattle and humans, *B. cereus* may cause food poisoning
Clostridium	Gram-positive bacillus	Strict anaerobes; form endospores; all known species are pathogenic, causing tetanus, gas gangrene, botulism, and colitis
Enterococcus	Gram-positive coccus; forms microscopic pairs in culture (resembling *Streptococcus pneumoniae*)	Anaerobic aerotolerant bacteria, abundant in the human gut, may cause urinary tract and other infections in the nosocomial environment
Lactobacillus	Gram-positive bacillus	Facultative anaerobes; ferment sugars into lactic acid; part of the vaginal microbiota; used as probiotics
Leuconostoc	Gram-positive coccus; may form microscopic chains in culture	Fermenter, used in food industry to produce sauerkraut and kefir
Mycoplasma	The smallest bacteria; appear pleomorphic under electron microscope	Have no cell wall; classified as low G+C Gram-positive bacteria because of their genome; *M. pneumoniae* causes "walking" pneumonia
Staphylococcus	Gram-positive coccus; forms microscopic clusters in culture that resemble bunches of grapes	Tolerate high salt concentration; facultative anaerobes; produce catalase; *S. aureus* can also produce coagulase and toxins responsible for local (skin) and generalized infections

Bacilli: Low G+C Gram-Positive Bacteria

Example Genus	Microscopic Morphology	Unique Characteristics
Streptococcus	Gram-positive coccus; forms chains or pairs in culture	Diverse genus; classified into groups based on sharing certain antigens; some species cause hemolysis and may produce toxins responsible for human local (throat) and generalized disease
Ureaplasma	Similar to Mycoplasma	Part of the human vaginal and lower urinary tract microbiota; may cause inflammation, sometimes leading to internal scarring and infertility

Table 4.10

✓ CHECK YOUR UNDERSTANDING

- Name some ways in which streptococci are classified.
- Name one pathogenic low G+C gram-positive bacterium and a disease it causes.

Clinical Focus

Resolution

Marsha's sputum sample was sent to the microbiology lab to confirm the identity of the microorganism causing her infection. The lab also performed antimicrobial susceptibility testing (AST) on the sample to confirm that the physician has prescribed the correct antimicrobial drugs.

Direct microscopic examination of the sputum revealed acid-fast bacteria (AFB) present in Marsha's sputum. When placed in culture, there were no signs of growth for the first 8 days, suggesting that microorganism was either dead or growing very slowly. Slow growth is a distinctive characteristic of *M. tuberculosis*.

After four weeks, the lab microbiologist observed distinctive colorless granulated colonies (Figure 4.23). The colonies contained AFB showing the same microscopic characteristics as those revealed during the direct microscopic examination of Marsha's sputum. To confirm the identification of the AFB, samples of the colonies were analyzed using nucleic acid hybridization, or direct nucleic acid amplification (NAA) testing. When a bacterium is acid-fast, it is classified in the family *Mycobacteriaceae*. DNA sequencing of variable genomic regions of the DNA extracted from these bacteria revealed that it was high G+C. This fact served to finalize Marsha's diagnosis as infection with *M. tuberculosis*. After nine months of treatment with the drugs prescribed by her doctor, Marsha made a full recovery.

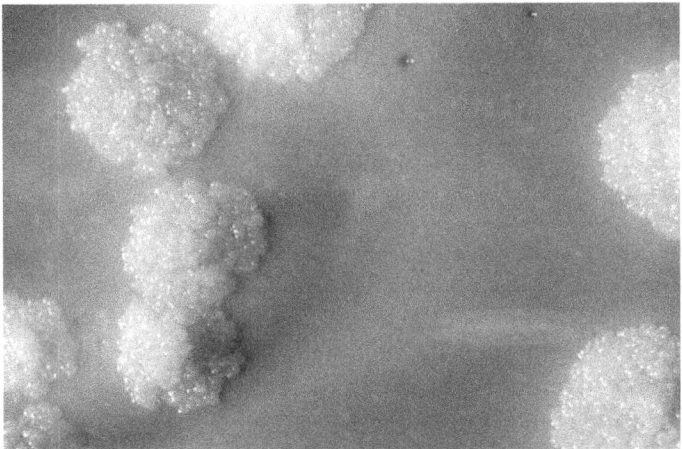

Figure 4.23 *M. tuberculosis* grows on Löwenstein-Jensen (LJ) agar in distinct colonies. (credit: Centers for Disease Control and Prevention)

Go back to the previous Clinical Focus box.

Eye on Ethics

Biopiracy and Bioprospecting

In 1969, an employee of a Swiss pharmaceutical company was vacationing in Norway and decided to collect some soil samples. He took them back to his lab, and the Swiss company subsequently used the fungus *Tolypocladium inflatum* in those samples to develop cyclosporine A, a drug widely used in patients who undergo tissue or organ transplantation. The Swiss company earns more than $1 billion a year for production of cyclosporine A, yet Norway receives nothing in return—no payment to the government or benefit for the Norwegian people. Despite the fact the cyclosporine A saves numerous lives, many consider the means by which the soil samples were obtained to be an act of "biopiracy," essentially a form of theft. Do the ends justify the means in a case like this?

Nature is full of as-yet-undiscovered bacteria and other microorganisms that could one day be used to develop new life-saving drugs or treatments.[21] Pharmaceutical and biotechnology companies stand to reap huge profits from such discoveries, but ethical questions remain. To whom do biological resources belong? Should companies who invest (and risk) millions of dollars in research and development be required to share revenue or royalties for the right to access biological resources?

Compensation is not the only issue when it comes to bioprospecting. Some communities and cultures are philosophically opposed to bioprospecting, fearing unforeseen consequences of collecting genetic or biological material. Native Hawaiians, for example, are very protective of their unique biological resources.

For many years, it was unclear what rights government agencies, private corporations, and citizens had when it came to collecting samples of microorganisms from public land. Then, in 1993, the Convention on Biological Diversity granted each nation the rights to any genetic and biological material found on their own land. Scientists can no longer collect samples without a prior arrangement with the land owner for compensation. This convention now ensures that companies act ethically in obtaining the samples they use to create their products.

---

21 J. Andre. *Bioethics as Practice*. Chapel Hill, NC: University of North Carolina Press, 2002.

4.5 Deeply Branching Bacteria

Learning Objectives

By the end of this section, you will be able to:
- Describe the unique features of deeply branching bacteria
- Give examples of significant deeply branching bacteria

On a phylogenetic tree (see A Systematic Approach), the trunk or root of the tree represents a common ancient evolutionary ancestor, often called the last universal common ancestor (LUCA), and the branches are its evolutionary descendants. Scientists consider the **deeply branching bacteria**, such as the genus *Acetothermus*, to be the first of these non-LUCA forms of life produced by evolution some 3.5 billion years ago. When placed on the phylogenetic tree, they stem from the common root of life, deep and close to the LUCA root—hence the name "deeply branching" (Figure 4.24).

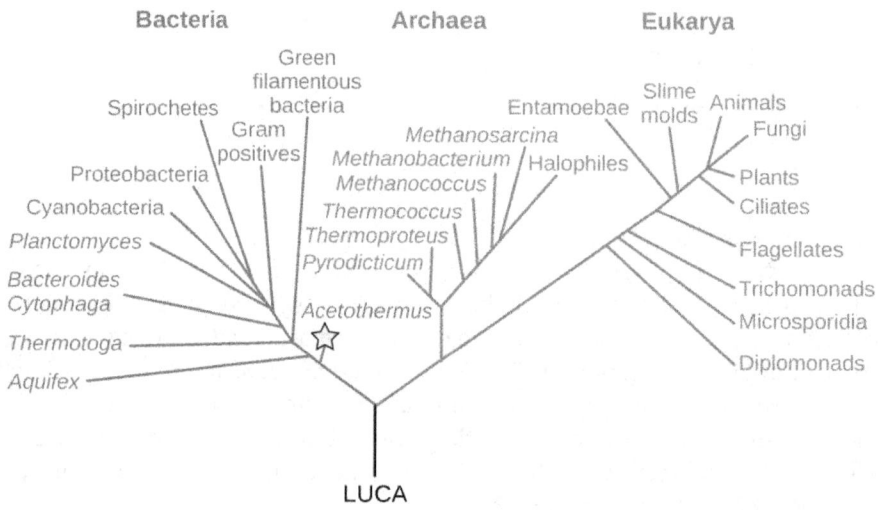

Figure 4.24 The star on this phylogenetic tree of life shows the position of the deeply branching bacteria *Acetothermus*. (credit: modification of work by Eric Gaba)

The deeply branching bacteria may provide clues regarding the structure and function of ancient and now extinct forms of life. We can hypothesize that ancient bacteria, like the deeply branching bacteria that still exist, were thermophiles or hyperthermophiles, meaning that they thrived at very high temperatures. *Acetothermus paucivorans*, a gram-negative anaerobic bacterium discovered in 1988 in sewage sludge, is a thermophile growing at an optimal temperature of 58 °C.[22] Scientists have determined it to be the deepest branching bacterium, or the closest evolutionary relative of the LUCA (Figure 4.24).

The class Aquificae includes deeply branching bacteria that are adapted to the harshest conditions on our planet, resembling the conditions thought to dominate the earth when life first appeared. Bacteria from the genus *Aquifex* are hyperthermophiles, living in hot springs at a temperature higher than 90 °C. The species *A. pyrophilus* thrives near underwater volcanoes and thermal ocean vents, where the temperature of water (under high pressure) can reach 138 °C. *Aquifex* bacteria use inorganic substances as nutrients. For example, *A. pyrophilus* can reduce oxygen, and it is able to reduce nitrogen in anaerobic conditions. They also show a remarkable resistance to ultraviolet light and ionizing radiation. Taken together, these observations support the hypothesis that the ancient ancestors of deeply branching bacteria began evolving more than 3 billion years ago, when the earth was hot and lacked an atmosphere, exposing the bacteria to nonionizing and ionizing radiation.

---

22 G. Dietrich et al. "*Acetothermus paucivorans*, gen. nov., sp. Nov., a Strictly Anaerobic, Thermophilic Bacterium From Sewage Sludge, Fermenting Hexoses to Acetate, CO_2, and H_2." *Systematic and Applied Microbiology* 10 no. 2 (1988):174–179.

The class Thermotogae is represented mostly by hyperthermophilic, as well as some mesophilic (preferring moderate temperatures), anaerobic gram-negative bacteria whose cells are wrapped in a peculiar sheath-like outer membrane called a toga. The thin layer of peptidoglycan in their cell wall has an unusual structure; it contains diaminopimelic acid and D-lysine. These bacteria are able to use a variety of organic substrates and produce molecular hydrogen, which can be used in industry. The class contains several genera, of which the best known is the genus *Thermotoga*. One species of this genus, *T. maritima*, lives near the thermal ocean vents and thrives in temperatures of 90 °C; another species, *T. subterranea*, lives in underground oil reservoirs.

Finally, the deeply branching bacterium *Deinococcus radiodurans* belongs to a genus whose name is derived from a Greek word meaning *terrible berry*. Nicknamed "Conan the Bacterium," *D. radiodurans* is considered a polyextremophile because of its ability to survive under the many different kinds of extreme conditions—extreme heat, drought, vacuum, acidity, and radiation. It owes its name to its ability to withstand doses of ionizing radiation that kill all other known bacteria; this special ability is attributed to some unique mechanisms of DNA repair.

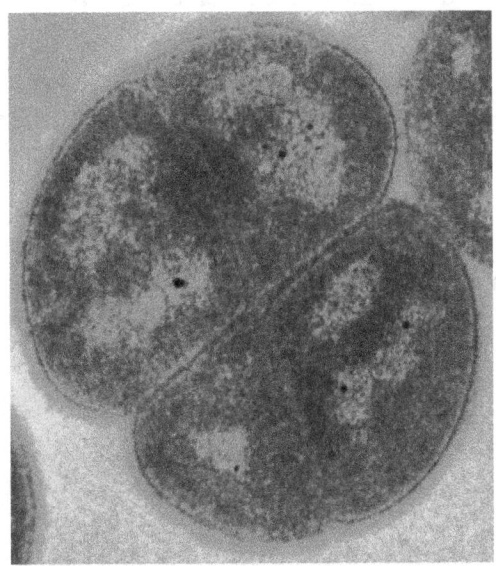

Figure 4.25 *Deinococcus radiodurans*, or "Conan the Bacterium," survives in the harshest conditions on earth.

4.6 Archaea

Learning Objectives

By the end of this section, you will be able to:
- Describe the unique features of each category of Archaea
- Explain why archaea might not be associated with human microbiomes or pathology
- Give common examples of archaea commonly associated with unique environmental habitats

Like organisms in the domain Bacteria, organisms of the domain **Archaea** are all unicellular organisms. However, archaea differ structurally from bacteria in several significant ways, as discussed in Unique Characteristics of Prokaryotic Cells. To summarize:

- The archaeal cell membrane is composed of ether linkages with branched isoprene chains (as opposed to the bacterial cell membrane, which has ester linkages with unbranched fatty acids).
- Archaeal cell walls lack peptidoglycan, but some contain a structurally similar substance called pseudopeptidoglycan or pseudomurein.
- The genomes of Archaea are larger and more complex than those of bacteria.

Domain Archaea is as diverse as domain Bacteria, and its representatives can be found in any habitat. Some archaea are mesophiles, and many are extremophiles, preferring extreme hot or cold, extreme salinity, or other conditions that are hostile to most other forms of life on earth. Their metabolism is adapted to the harsh environments, and they can perform methanogenesis, for example, which bacteria and eukaryotes cannot.

The size and complexity of the archaeal genome makes it difficult to classify. Most taxonomists agree that within the Archaea, there are currently five major phyla: Crenarchaeota, Euryarchaeota, Korarchaeota, Nanoarchaeota, and Thaumarchaeota. There are likely many other archaeal groups that have not yet been systematically studied and classified.

With few exceptions, archaea are not present in the human microbiota, and none are currently known to be associated with infectious diseases in humans, animals, plants, or microorganisms. However, many play important roles in the environment and may thus have an indirect impact on human health.

Crenarchaeota

Crenarchaeota is a class of Archaea that is extremely diverse, containing genera and species that differ vastly in their morphology and requirements for growth. All Crenarchaeota are aquatic organisms, and they are thought to be the most abundant microorganisms in the oceans. Most, but not all, Crenarchaeota are hyperthermophiles; some of them (notably, the genus *Pyrolobus*) are able to grow at temperatures up to 113 °C.[23]

Archaea of the genus *Sulfolobus* (Figure 4.26) are thermophiles that prefer temperatures around 70–80°C and acidophiles that prefer a pH of 2–3.[24] *Sulfolobus* can live in aerobic or anaerobic environments. In the presence of oxygen, *Sulfolobus* spp. use metabolic processes similar to those of heterotrophs. In anaerobic environments, they oxidize sulfur to produce sulfuric acid, which is stored in granules. *Sulfolobus* spp. are used in biotechnology for the production of thermostable and acid-resistant proteins called affitins.[25] Affitins can bind and neutralize various antigens (molecules found in toxins or infectious agents that provoke an immune response from the body).

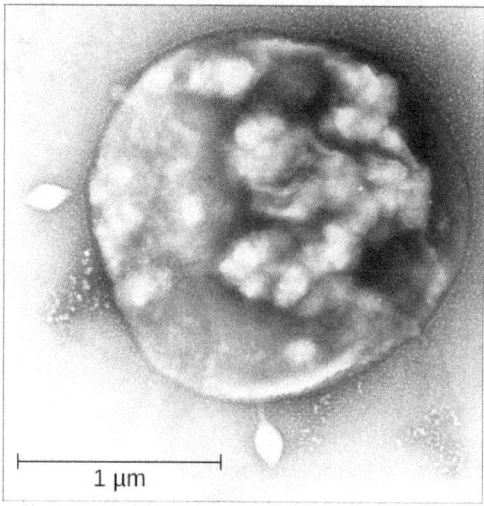

Figure 4.26 *Sulfolobus*, an archaeon of the class Crenarchaeota, oxidizes sulfur and stores sulfuric acid in its granules.

Another genus, *Thermoproteus*, is represented by strictly anaerobic organisms with an optimal growth temperature of 85 °C. They have flagella and, therefore, are motile. *Thermoproteus* has a cellular membrane in which lipids form a monolayer rather than a bilayer, which is typical for archaea. Its metabolism is autotrophic. To synthesize ATP, *Thermoproteus* spp. reduce sulfur or molecular hydrogen and use carbon dioxide or carbon monoxide as a source of carbon. *Thermoproteus* is thought to be the deepest-branching genus of Archaea, and

---

23 E. Blochl et al."*Pyrolobus fumani*, gen. and sp. nov., represents a novel group of Archaea, extending the upper temperature limit for life to 113°C." *Extremophiles* 1 (1997):14–21.

24 T.D. Brock et al. "*Sulfolobus*: A New Genus of Sulfur-Oxidizing Bacteria Living at Low pH and High Temperature." *Archiv für Mikrobiologie* 84 no. 1 (1972):54–68.

25 S. Pacheco et al. "Affinity Transfer to the Archaeal Extremophilic Sac7d Protein by Insertion of a CDR." *Protein Engineering Design and Selection* 27 no. 10 (2014):431-438.

thus is a living example of some of our planet's earliest forms of life.

✓ CHECK YOUR UNDERSTANDING

- What types of environments do Crenarchaeota prefer?

Euryarchaeota

The phylum Euryarchaeota includes several distinct classes. Species in the classes Methanobacteria, Methanococci, and Methanomicrobia represent Archaea that can be generally described as methanogens. Methanogens are unique in that they can reduce carbon dioxide in the presence of hydrogen, producing methane. They can live in the most extreme environments and can reproduce at temperatures varying from below freezing to boiling. Methanogens have been found in hot springs as well as deep under ice in Greenland. Methanogens also produce gases in ruminants and humans. Some scientists have even hypothesized that **methanogens** may inhabit the planet Mars because the mixture of gases produced by methanogens resembles the makeup of the Martian atmosphere.[26]

The class Halobacteria (which was named before scientists recognized the distinction between Archaea and Bacteria) includes halophilic ("salt-loving") archaea. Halobacteria require a very high concentrations of sodium chloride in their aquatic environment. The required concentration is close to saturation, at 36%; such environments include the Dead Sea as well as some salty lakes in Antarctica and south-central Asia. One remarkable feature of these organisms is that they perform photosynthesis using the protein bacteriorhodopsin, which gives them, and the bodies of water they inhabit, a beautiful purple color (Figure 4.27).

Figure 4.27 Halobacteria growing in these salt ponds gives them a distinct purple color. (credit: modification of work by Tony Hisgett)

Notable species of Halobacteria include *Halobacterium salinarum*, which may be the oldest living organism on earth; scientists have isolated its DNA from fossils that are 250 million years old.[27] Another species, *Haloferax volcanii*, shows a very sophisticated system of ion exchange, which enables it to balance the concentration of salts at high temperatures.

✓ CHECK YOUR UNDERSTANDING

- Where do Halobacteria live?

---

26 R.R. Britt "Crater Critters: Where Mars Microbes Might Lurk." http://www.space.com/1880-crater-critters-mars-microbes-lurk.html. Accessed April 7, 2015.

27 H. Vreeland et al. "Fatty acid and DA Analyses of Permian Bacterium Isolated From Ancient Salt Crystals Reveal Differences With Their Modern Relatives." *Extremophiles* 10 (2006):71–78.

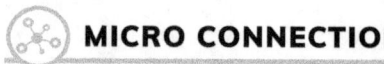 **MICRO CONNECTIONS**

Finding a Link Between Archaea and Disease

Archaea are not known to cause any disease in humans, animals, plants, bacteria, or in other archaea. Although this makes sense for the extremophiles, not all archaea live in extreme environments. Many genera and species of Archaea are mesophiles, so they can live in human and animal microbiomes, although they rarely do. As we have learned, some methanogens exist in the human gastrointestinal tract. Yet we have no reliable evidence pointing to any archaean as the causative agent of any human disease.

Still, scientists have attempted to find links between human disease and archaea. For example, in 2004, Lepp et al. presented evidence that an archaean called *Methanobrevibacter oralis* inhabits the gums of patients with periodontal disease. The authors suggested that the activity of these methanogens causes the disease.[28] However, it was subsequently shown that there was no causal relationship between *M. oralis* and periodontitis. It seems more likely that periodontal disease causes an enlargement of anaerobic regions in the mouth that are subsequently populated by *M. oralis*.[29]

There remains no good answer as to why archaea do not seem to be pathogenic, but scientists continue to speculate and hope to find the answer.

---

[28] P.W. Lepp et al. "Methanogenic Archaea and Human Gum Disease." *Proceedings of the National Academies of Science of the United States of America* 101 no. 16 (2004):6176–6181.

[29] R.I. Aminov. "Role of Archaea in Human Disease." *Frontiers in Cellular and Infection Microbiology* 3 (2013):42.

SUMMARY

4.1 Prokaryote Habitats, Relationships, and Microbiomes

- Prokaryotes are unicellular microorganisms whose cells have no nucleus.
- Prokaryotes can be found everywhere on our planet, even in the most extreme environments.
- Prokaryotes are very flexible metabolically, so they are able to adjust their feeding to the available natural resources.
- Prokaryotes live in **communities** that interact among themselves and with large organisms that they use as hosts (including humans).
- The totality of forms of prokaryotes (particularly bacteria) living on the human body is called the human microbiome, which varies between regions of the body and individuals, and changes over time.
- The totality of forms of prokaryotes (particularly bacteria) living in a certain region of the human body (e.g., mouth, throat, gut, eye, vagina) is called the **microbiota** of this region.
- Prokaryotes are classified into domains Archaea and Bacteria.
- In recent years, the traditional approaches to classification of prokaryotes have been supplemented by approaches based on molecular genetics.

4.2 Proteobacteria

- **Proteobacteria** is a phylum of gram-negative bacteria discovered by Carl Woese in the 1980s based on nucleotide sequence homology.
- Proteobacteria are further classified into the classes alpha-, beta-, gamma-, delta- and epsilonproteobacteria, each class having separate orders, families, genera, and species.
- **Alphaproteobacteria** include several obligate and facultative intracellular pathogens, including the rickettsias. Some Alphaproteobacteria can convert atmospheric nitrogen to nitrites, making nitrogen usable for other forms of life.
- **Betaproteobacteria** are a diverse group of bacteria that include human pathogens of the genus *Neisseria* and the species *Bordetella pertussis*.
- **Gammaproteobacteria** are the largest and the most diverse group of Proteobacteria. Many are human pathogens that are aerobes or facultative anaerobes. Some Gammaproteobacteria are **enteric** bacteria that may be coliform or noncoliform. *Escherichia coli*, a member of Gammaproteobacteria, is perhaps the most studied bacterium.
- **Deltaproteobacteria** make up a small group able to reduce sulfate or elemental sulfur. Some are scavengers and form myxospores, with multicellular fruiting bodies.
- **Epsilonproteobacteria** make up the smallest group of Proteobacteria. The genera *Campylobacter* and *Helicobacter* are human pathogens.

4.3 Nonproteobacteria Gram-Negative Bacteria and Phototrophic Bacteria

- Gram-negative nonproteobacteria include the taxa **spirochetes**; the *Chlamydia*, *Cytophaga*, *Fusobacterium*, *Bacteroides* group; Planctomycetes; and many representatives of **phototrophic bacteria.**
- Spirochetes are motile, spiral bacteria with a long, narrow body; they are difficult or impossible to culture.
- Several genera of spirochetes contain human pathogens that cause such diseases as syphilis and Lyme disease.
- *Cytophaga*, *Fusobacterium*, and *Bacteroides* are classified together as a phylum called the **CFB group**. They are rod-shaped anaerobic organoheterotrophs and avid fermenters. *Cytophaga* are aquatic bacteria with the gliding motility. *Fusobacteria* inhabit the human mouth and may cause severe infectious diseases. *Bacteroides* are present in vast numbers in the human gut, most of them being mutualistic but some are pathogenic.
- Planctomycetes are aquatic bacteria that reproduce by budding; they may form large colonies, and develop a holdfast.
- Phototrophic bacteria are not a taxon but, rather, a group categorized by their ability to use the energy of sunlight. They include Proteobacteria and nonproteobacteria, as well as sulfur and nonsulfur bacteria colored purple or green.
- Sulfur bacteria perform anoxygenic photosynthesis, using sulfur compounds as donors of electrons, whereas nonsulfur bacteria use organic compounds (succinate, malate) as donors of electrons.
- Some phototrophic bacteria are able to fix

nitrogen, providing the usable forms of nitrogen to other organisms.
- **Cyanobacteria** are oxygen-producing bacteria thought to have played a critical role in the forming of the earth's atmosphere.

4.4 Gram-Positive Bacteria

- Gram-positive bacteria are a very large and diverse group of microorganisms. Understanding their taxonomy and knowing their unique features is important for diagnostics and treatment of infectious diseases.
- Gram-positive bacteria are classified into **high G+C gram-positive** and **low G+C gram-positive** bacteria, based on the prevalence of guanine and cytosine nucleotides in their genome
- Actinobacteria is the taxonomic name of the class of high G+C gram-positive bacteria. This class includes the genera *Actinomyces, Arthrobacter, Corynebacterium, Frankia, Gardnerella, Micrococcus, Mycobacterium, Nocardia, Propionibacterium, Rhodococcus,* and *Streptomyces*. Some representatives of these genera are used in industry; others are human or animal pathogens.
- Examples of high G+C gram-positive bacteria that are human pathogens include *Mycobacterium tuberculosis*, which causes tuberculosis; *M. leprae*, which causes leprosy (Hansen's disease); and *Corynebacterium diphtheriae*, which causes diphtheria.
- *Clostridia* spp. are low G+C gram-positive bacteria that are generally obligate anaerobes and can form endospores. Pathogens in this genus include *C. perfringens* (gas gangrene), *C. tetani* (tetanus), and *C. botulinum* (botulism).
- Lactobacillales include the genera *Enterococcus, Lactobacillus, Leuconostoc,* and *Streptococcus*. *Streptococcus* is responsible for many human diseases, including pharyngitis (strep throat), scarlet fever, rheumatic fever, glomerulonephritis, pneumonia, and other respiratory infections.
- Bacilli is a taxonomic class of low G+C gram-positive bacteria that include rod-shaped and coccus-shaped species, including the genera *Bacillus* and *Staphylococcus*. *B. anthracis* causes anthrax, *B. cereus* may cause opportunistic infections of the gastrointestinal tract, and *S. aureus* strains can cause a wide range of infections and diseases, many of which are highly resistant to antibiotics.
- *Mycoplasma* spp. are very small, **pleomorphic** low G+C gram-positive bacteria that lack cell walls. *M. pneumoniae* causes atypical pneumonia.

4.5 Deeply Branching Bacteria

- **Deeply branching bacteria** are phylogenetically the most ancient forms of life, being the closest to the last universal common ancestor.
- Deeply branching bacteria include many species that thrive in extreme environments that are thought to resemble conditions on earth billions of years ago
- Deeply branching bacteria are important for our understanding of evolution; some of them are used in industry

4.6 Archaea

- **Archaea** are unicellular, prokaryotic microorganisms that differ from bacteria in their genetics, biochemistry, and ecology.
- Some archaea are extremophiles, living in environments with extremely high or low temperatures, or extreme salinity.
- Only archaea are known to produce methane. Methane-producing archaea are called **methanogens**.
- Halophilic archaea prefer a concentration of salt close to saturation and perform photosynthesis using bacteriorhodopsin.
- Some archaea, based on fossil evidence, are among the oldest organisms on earth.
- Archaea do not live in great numbers in human microbiomes and are not known to cause disease.

REVIEW QUESTIONS
Multiple Choice

1. The term prokaryotes refers to which of the following?
   A. very small organisms
   B. unicellular organisms that have no nucleus
   C. multicellular organisms
   D. cells that resemble animal cells more than plant cells

2. The term microbiota refers to which of the following?
   A. all microorganisms of the same species
   B. all of the microorganisms involved in a symbiotic relationship
   C. all microorganisms in a certain region of the human body
   D. all microorganisms in a certain geographic region

3. Which of the following refers to the type of interaction between two prokaryotic populations in which one population benefits and the other is not affected?
   A. mutualism
   B. commensalism
   C. parasitism
   D. neutralism

4. Which of the following describes Proteobacteria in domain Bacteria?
   A. phylum
   B. class
   C. species
   D. genus

5. Which of the following Alphaproteobacteria is the cause of Rocky Mountain spotted fever and typhus?
   A. Bartonella
   B. Coxiella
   C. Rickettsia
   D. Ehrlichia
   E. Brucella

6. Class Betaproteobacteria includes all but which of the following genera?
   A. *Neisseria.*
   B. *Bordetella.*
   C. *Leptothrix.*
   D. *Campylobacter.*

7. *Haemophilus influenzae* is a common cause of which of the following?
   A. influenza
   B. dysentery
   C. upper respiratory tract infections
   D. hemophilia

8. Which of the following is the organelle that spirochetes use to propel themselves?
   A. plasma membrane
   B. axial filament
   C. pilum
   D. fimbria

9. Which of the following bacteria are the most prevalent in the human gut?
   A. cyanobacteria
   B. staphylococci
   C. *Borrelia*
   D. *Bacteroides*

10. Which of the following refers to photosynthesis performed by bacteria with the use of water as the donor of electrons?
    A. oxygenic
    B. anoxygenic
    C. heterotrophic
    D. phototrophic

11. Which of the following bacterial species is classified as high G+C gram-positive?
    A. *Corynebacterium diphtheriae*
    B. *Staphylococcus aureus*
    C. *Bacillus anthracis*
    D. *Streptococcus pneumonia*

12. The term "deeply branching" refers to which of the following?
    A. the cellular shape of deeply branching bacteria
    B. the position in the evolutionary tree of deeply branching bacteria
    C. the ability of deeply branching bacteria to live in deep ocean waters
    D. the pattern of growth in culture of deeply branching bacteria

13. Which of these deeply branching bacteria is considered a polyextremophile?
    A. *Aquifex pyrophilus*
    B. *Deinococcus radiodurans*
    C. *Staphylococcus aureus*
    D. *Mycobacterium tuberculosis*

14. Archaea and Bacteria are most similar in terms of their _____.
    A. genetics
    B. cell wall structure
    C. ecology
    D. unicellular structure

15. Which of the following is true of archaea that produce methane?
    A. They reduce carbon dioxide in the presence of nitrogen.
    B. They live in the most extreme environments.
    C. They are always anaerobes.
    D. They have been discovered on Mars.

True/False

16. Among prokaryotes, there are some that can live in every environment on earth.

Fill in the Blank

17. When prokaryotes live as interacting communities in which one population benefits to the harm of the other, the type of symbiosis is called _____.
18. The domain _____ does not include prokaryotes.
19. Pathogenic bacteria that are part of the transient microbiota can sometimes be eliminated by _____ therapy.
20. Nitrogen-fixing bacteria provide other organisms with usable nitrogen in the form of _____.
21. Rickettsias are _____ intracellular bacteria.
22. The species _____, which belongs to Epsilonproteobacteria, causes peptic ulcers of the stomach and duodenum.
23. The genus *Salmonella* belongs to the class _____ and includes pathogens that cause salmonellosis and typhoid fever.
24. The bacterium that causes syphilis is called _____.
25. Bacteria in the genus *Rhodospirillum* that use hydrogen for oxidation and fix nitrogen are _____ bacteria.
26. *Streptococcus* is the _____ of bacteria that is responsible for many human diseases.
27. One species of *Streptococcus*, *S. pyogenes*, is a classified as a _____ pathogen due to the characteristic production of pus in infections it causes.
28. *Propionibacterium* belongs to _____ G+C gram-positive bacteria. One of its species is used in the food industry and another causes acne.
29. The length of the branches of the evolutionary tree characterizes the evolutionary _____ between organisms.
30. The deeply branching bacteria are thought to be the form of life closest to the last universal _____ _____.
31. Many of the deeply branching bacteria are aquatic and hyperthermophilic, found near underwater volcanoes and thermal ocean _____.
32. The deeply branching bacterium *Deinococcus radiodurans* is able to survive exposure to high doses of _____.
33. _____ is a genus of Archaea. Its optimal environmental temperature ranges from 70 °C to 80 °C, and its optimal pH is 2–3. It oxidizes sulfur and produces sulfuric acid.
34. _____ was once thought to be the cause of periodontal disease, but, more recently, the causal relationship between this archaean and the disease was not confirmed.

Short Answer

35. Compare commensalism and amensalism.
36. Give an example of the changes of human microbiota that result from medical intervention.
37. What is the metabolic difference between coliforms and noncoliforms? Which category contains several species of intestinal pathogens?
38. Why are *Mycoplasma* and *Chlamydia* classified as obligate intracellular pathogens?

39. Explain the term CFB group and name the genera that this group includes.
40. Name and briefly describe the bacterium that causes Lyme disease.
41. Characterize the phylum Cyanobacteria.
42. Name and describe two types of *S. aureus* that show multiple antibiotic resistance.
43. Briefly describe the significance of deeply branching bacteria for basic science and for industry.
44. What is thought to account for the unique radiation resistance of *D. radiodurans*?
45. What accounts for the purple color in salt ponds inhabited by halophilic archaea?
46. What evidence supports the hypothesis that some archaea live on Mars?

Critical Thinking

47. The cell shown is found in the human stomach and is now known to cause peptic ulcers. What is the name of this bacterium?

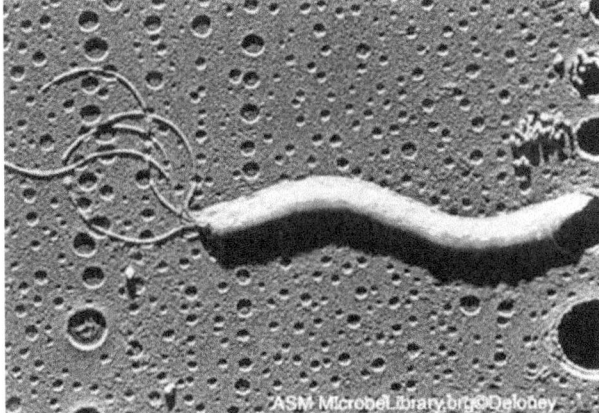

Figure 4.28 (credit: American Society for Microbiology)

48. The microscopic growth pattern shown is characteristic of which genus of bacteria?

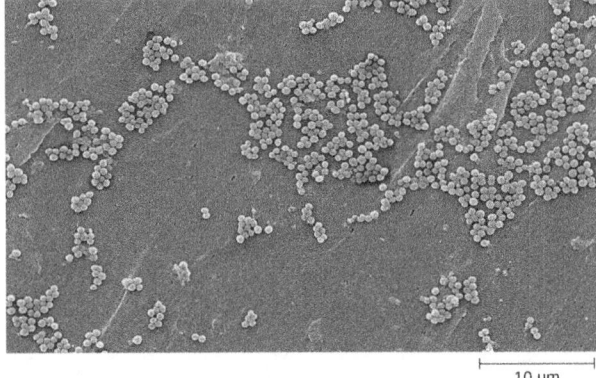

Figure 4.29 (credit: modification of work by Janice Haney Carr/Centers for Disease Control and Prevention)

49. What is the connection between this methane bog and archaea?

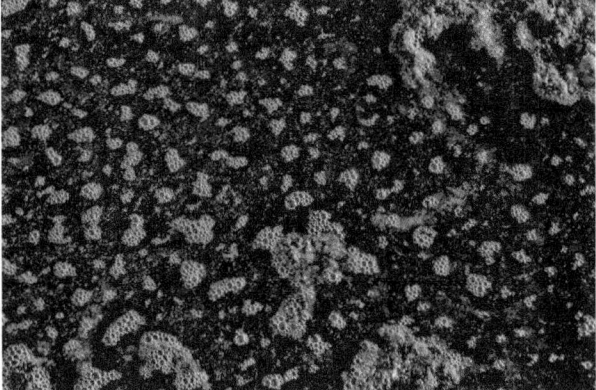

Figure 4.30 (credit: Chad Skeers)

CHAPTER 5
The Eukaryotes of Microbiology

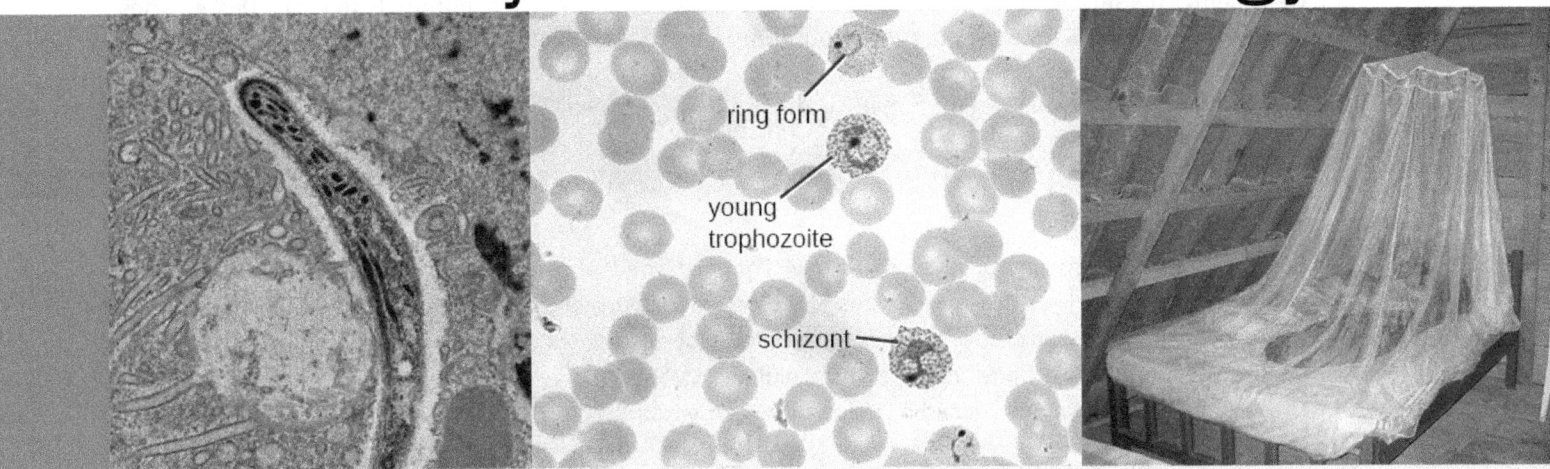

Figure 5.1 Malaria is a disease caused by a eukaryotic parasite transmitted to humans by mosquitos. Micrographs (left and center) show a sporozoite life stage, trophozoites, and a schizont in a blood smear. On the right is depicted a primary defense against mosquito-borne illnesses like malaria—mosquito netting. (credit left: modification of work by Ute Frevert; credit middle: modification of work by Centers for Disease Control and Prevention; credit right: modification of work by Tjeerd Wiersma)

Chapter Outline

5.1 Unicellular Eukaryotic Parasites

5.2 Parasitic Helminths

5.3 Fungi

5.4 Algae

5.5 Lichens

INTRODUCTION Although bacteria and viruses account for a large number of the infectious diseases that afflict humans, many serious illnesses are caused by eukaryotic organisms. One example is malaria, which is caused by *Plasmodium*, a eukaryotic organism transmitted through mosquito bites. Malaria is a major cause of morbidity (illness) and mortality (death) that threatens 3.4 billion people worldwide.[1] In severe cases, organ failure and blood or metabolic abnormalities contribute to medical emergencies and sometimes death. Even after initial recovery, relapses may occur years later. In countries where malaria is endemic, the disease represents a major public health challenge that can place a tremendous strain on developing economies.

---

1 Centers for Disease Control and Prevention. "Impact of Malaria." September 22, 2015. http://www.cdc.gov/malaria/malaria_worldwide/impact.html. Accessed January 18, 2016.

Worldwide, major efforts are underway to reduce malaria infections. Efforts include the distribution of insecticide-treated bed nets and the spraying of pesticides. Researchers are also making progress in their efforts to develop effective vaccines.[2] The President's Malaria Initiative, started in 2005, supports prevention and treatment. The Bill and Melinda Gates Foundation has a large initiative to eliminate malaria. Despite these efforts, malaria continues to cause long-term morbidity (such as intellectual disabilities in children) and mortality (especially in children younger than 5 years), so we still have far to go.

5.1 Unicellular Eukaryotic Parasites

Learning Objectives

By the end of this section, you will be able to:
- Summarize the general characteristics of unicellular eukaryotic parasites
- Describe the general life cycles and modes of reproduction in unicellular eukaryotic parasites
- Identify challenges associated with classifying unicellular eukaryotes
- Explain the taxonomic scheme used for unicellular eukaryotes
- Give examples of infections caused by unicellular eukaryotes

---

Clinical Focus

Part 1

Upon arriving home from school, 7-year-old Sarah complains that a large spot on her arm will not stop itching. She keeps scratching at it, drawing the attention of her parents. Looking more closely, they see that it is a red circular spot with a raised red edge (Figure 5.2). The next day, Sarah's parents take her to their doctor, who examines the spot using a Wood's lamp. A Wood's lamp produces ultraviolet light that causes the spot on Sarah's arm to fluoresce, which confirms what the doctor already suspected: Sarah has a case of ringworm.

Sarah's mother is mortified to hear that her daughter has a "worm." How could this happen?

- What are some likely ways that Sarah might have contracted ringworm?

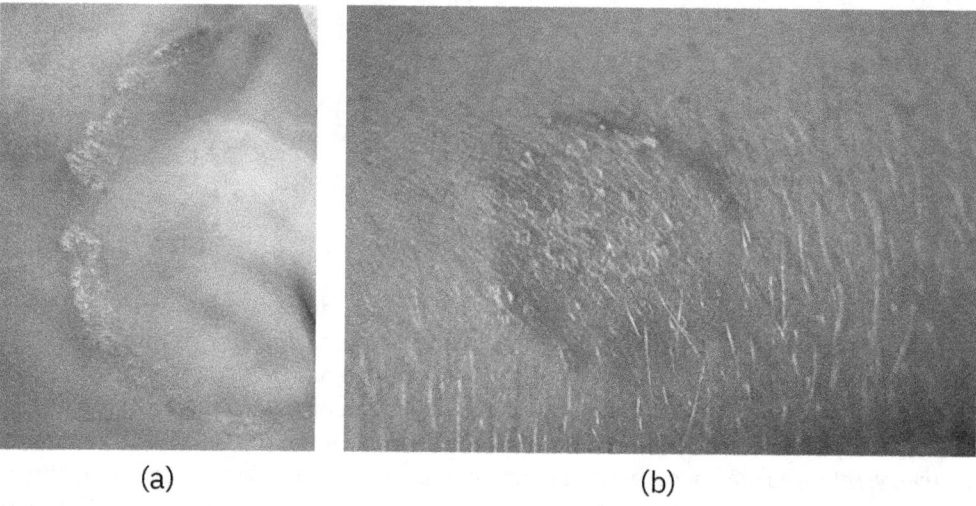

Figure 5.2 Ringworm presents as a raised ring, which is gray or brown on brown or black skin (a), and red on lighter skin (b). (Credit: Centers for Disease Control and Prevention)

Jump to the next Clinical Focus box.

---

2 RTS, S Clinical Trials Partnership. "Efficacy and safety of RTS,S/AS01 malaria vaccine with or without a booster dose in infants and children in Africa: final results of a phase 3, individually randomised, controlled trial." The Lancet 23 April 2015. DOI: http://dx.doi.org/10.1016/S0140-6736(15)60721-8.

Eukaryotic microbes are an extraordinarily diverse group, including species with a wide range of life cycles, morphological specializations, and nutritional needs. Although more diseases are caused by viruses and bacteria than by microscopic eukaryotes, these eukaryotes are responsible for some diseases of great public health importance. For example, the protozoal disease malaria was responsible for 584,000 deaths worldwide (primarily children in Africa) in 2013, according to the World Health Organization (WHO). The protist parasite *Giardia* causes a diarrheal illness (giardiasis) that is easily transmitted through contaminated water supplies. In the United States, *Giardia* is the most common human intestinal parasite (Figure 5.3). Although it may seem surprising, parasitic worms are included within the study of microbiology because identification depends on observation of microscopic adult worms or eggs. Even in developed countries, these worms are important parasites of humans and of domestic animals. There are fewer fungal pathogens, but these are important causes of illness, as well. On the other hand, fungi have been important in producing antimicrobial substances such as penicillin. In this chapter, we will examine characteristics of protists, worms, and fungi while considering their roles in causing disease.

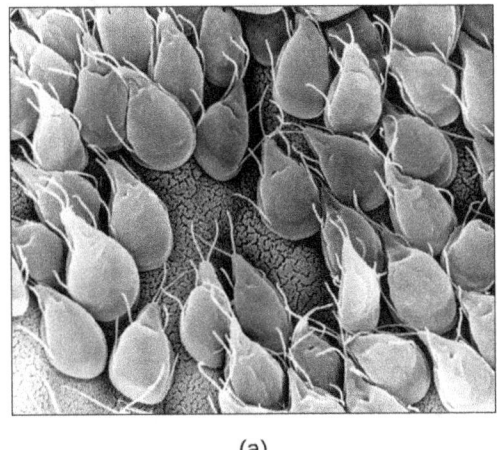 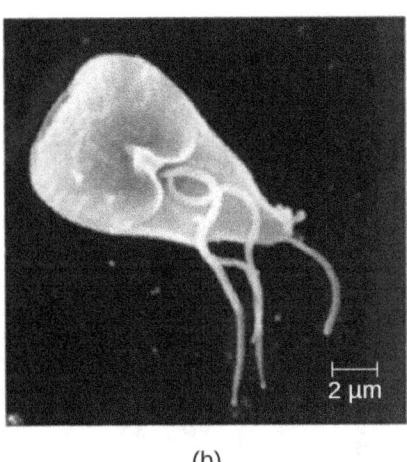

(a) (b)

Figure 5.3 (a) A scanning electron micrograph shows many *Giardia* parasites in the trophozoite, or feeding stage, in a gerbil intestine. (b) An individual trophozoite of *G. lamblia*, visualized here in a scanning electron micrograph. This waterborne protist causes severe diarrhea when ingested. (credit a, b: modification of work by Centers for Disease Control and Prevention)

Characteristics of Protists

The word *protist* is a historical term that is now used informally to refer to a diverse group of microscopic eukaryotic organisms. It is not considered a formal taxonomic term because the organisms it describes do not have a shared evolutionary origin. Historically, the protists were informally grouped into the "animal-like" protozoans, the "plant-like" algae, and the "fungus-like" protists such as water molds. These three groups of protists differ greatly in terms of their basic characteristics. For example, algae are photosynthetic organisms that can be unicellular or multicellular. Protozoa, on the other hand, are nonphotosynthetic, motile organisms that are always unicellular. Other informal terms may also be used to describe various groups of protists. For example, microorganisms that drift or float in water, moved by currents, are referred to as **plankton**. Types of plankton include **zooplankton**, which are motile and nonphotosynthetic, and **phytoplankton**, which are photosynthetic.

Protozoans inhabit a wide variety of habitats, both aquatic and terrestrial. Many are free-living, while others are parasitic, carrying out a life cycle within a host or hosts and potentially causing illness. There are also beneficial symbionts that provide metabolic services to their hosts. During the feeding and growth part of their life cycle, they are called **trophozoites**; these feed on small particulate food sources such as bacteria. While some types of protozoa exist exclusively in the trophozoite form, others can develop from trophozoite to an encapsulated cyst stage when environmental conditions are too harsh for the trophozoite. A **cyst** is a cell with a protective wall, and the process by which a trophozoite becomes a cyst is called **encystment**. When conditions become more favorable, these cysts are triggered by environmental cues to become active again through **excystment**.

One protozoan genus capable of encystment is *Eimeria*, which includes some human and animal pathogens.

Figure 5.4 illustrates the life cycle of *Eimeria*.

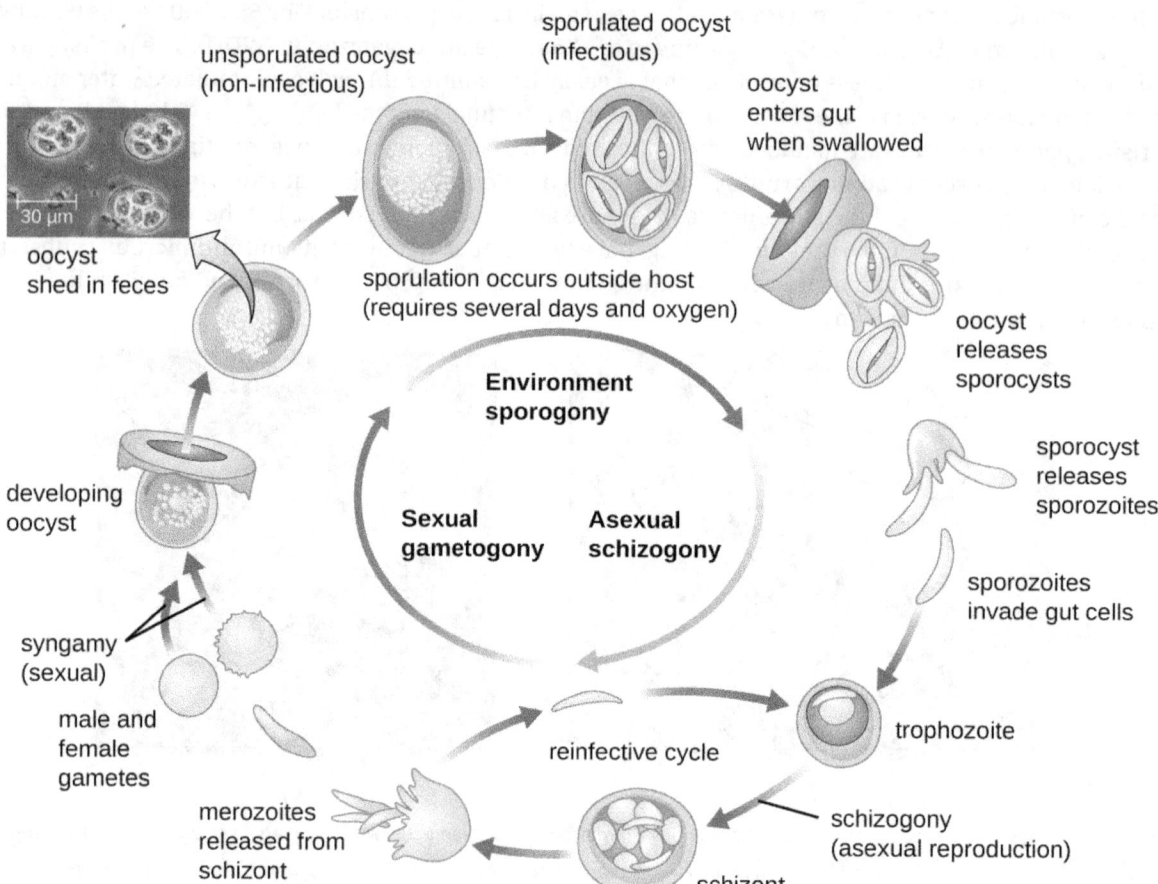

Figure 5.4 In the sexual/asexual life cycle of *Eimeria*, oocysts (inset) are shed in feces and may cause disease when ingested by a new host. (credit "life cycle," "micrograph": modification of work by USDA)

Protozoans have a variety of reproductive mechanisms. Some protozoans reproduce asexually and others reproduce sexually; still others are capable of both sexual and asexual reproduction. In protozoans, asexual reproduction occurs by binary fission, budding, or schizogony. In **schizogony**, the nucleus of a cell divides multiple times before the cell divides into many smaller cells. The products of schizogony are called merozoites and they are stored in structures known as schizonts. Protozoans may also reproduce sexually, which increases genetic diversity and can lead to complex life cycles. Protozoans can produce haploid gametes that fuse through **syngamy**. However, they can also exchange genetic material by joining to exchange DNA in a process called conjugation. This is a different process than the conjugation that occurs in bacteria. The term protist conjugation refers to a true form of eukaryotic sexual reproduction between two cells of different mating types. It is found in **ciliates**, a group of protozoans, and is described later in this subsection.

All protozoans have a plasma membrane, or **plasmalemma**, and some have bands of protein just inside the membrane that add rigidity, forming a structure called the **pellicle**. Some protists, including protozoans, have distinct layers of cytoplasm under the membrane. In these protists, the outer gel layer (with microfilaments of actin) is called the **ectoplasm**. Inside this layer is a sol (fluid) region of cytoplasm called the **endoplasm**. These structures contribute to complex cell shapes in some protozoans, whereas others (such as amoebas) have more flexible shapes (Figure 5.5).

Different groups of protozoans have specialized feeding structures. They may have a specialized structure for taking in food through phagocytosis, called a **cytostome**, and a specialized structure for the exocytosis of wastes called a **cytoproct**. Oral grooves leading to cytostomes are lined with hair-like cilia to sweep in food particles. Protozoans are heterotrophic. Protozoans that are **holozoic** ingest whole food particles through phagocytosis. Forms that are **saprozoic** ingest small, soluble food molecules.

Many protists have whip-like flagella or hair-like cilia made of microtubules that can be used for locomotion (Figure 5.5). Other protists use cytoplasmic extensions known as pseudopodia ("false feet") to attach the cell to a surface; they then allow cytoplasm to flow into the extension, thus moving themselves forward.

Protozoans have a variety of unique organelles and sometimes lack organelles found in other cells. Some have **contractile vacuoles**, organelles that can be used to move water out of the cell for osmotic regulation (salt and water balance) (Figure 5.5). Mitochondria may be absent in parasites or altered to kinetoplastids (modified mitochondria) or hydrogenosomes (see Unique Characteristics of Prokaryotic Cells for more discussion of these structures).

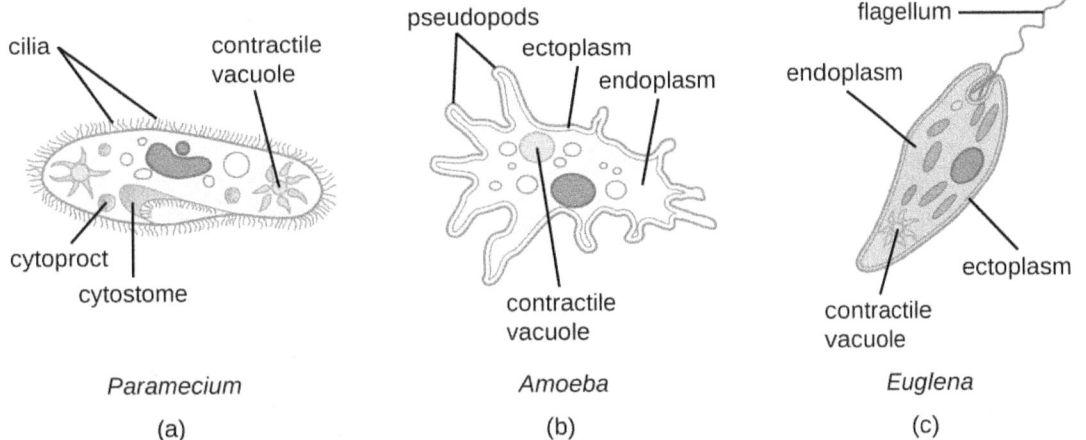

Figure 5.5 (a) *Paramecium* spp. have hair-like appendages called cilia for locomotion. (b) *Amoeba* spp. use lobe-like pseudopodia to anchor the cell to a solid surface and pull forward. (c) *Euglena* spp. use a whip-like structure called a flagellum to propel the cell.

CHECK YOUR UNDERSTANDING

- What is the sequence of events in reproduction by schizogony and what are the cells produced called?

Taxonomy of Protists

The protists are a **polyphyletic** group, meaning they lack a shared evolutionary origin. Since the current taxonomy is based on evolutionary history (as determined by biochemistry, morphology, and genetics), protists are scattered across many different taxonomic groups within the domain Eukarya. Eukarya is currently divided into six supergroups that are further divided into subgroups, as illustrated in (Figure 5.6). In this section, we will primarily be concerned with the supergroups Amoebozoa, Excavata, and Chromalveolata; these supergroups include many protozoans of clinical significance. The supergroups Opisthokonta and Rhizaria also include some protozoans, but few of clinical significance. In addition to protozoans, Opisthokonta also includes animals and fungi, some of which we will discuss in Parasitic Helminths and Fungi. Some examples of the Archaeplastida will be discussed in Algae. Figure 5.7 and Figure 5.8 summarize the characteristics of each supergroup and subgroup and list representatives of each.

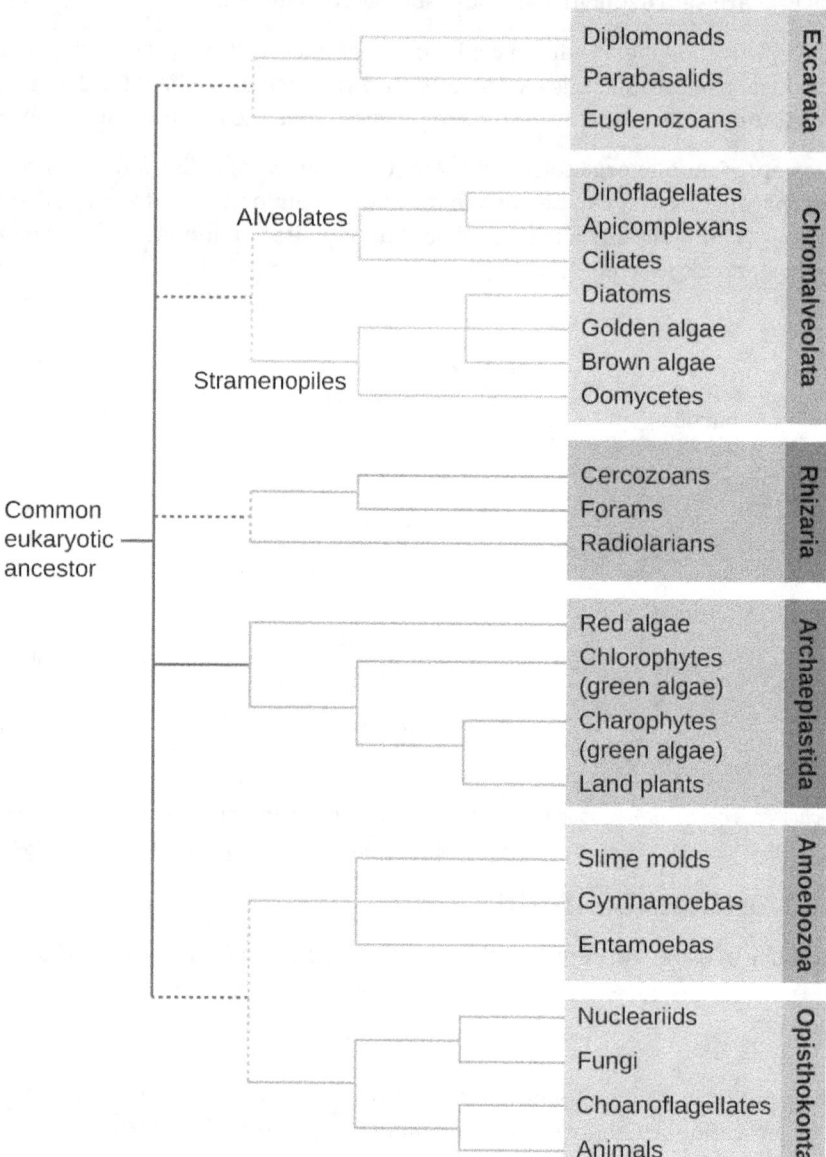

Figure 5.6 This tree shows a proposed classification of the domain Eukarya based on evolutionary relationships. Currently, the domain Eukarya is divided into six supergroups. Within each supergroup are multiple kingdoms. Dotted lines indicate suggested evolutionary relationships that remain under debate.

| The Eukaryote Supergroups and Some Examples |||||
Supergroup	Subgroups	Distinguishing Features	Examples	Clinical Notes
Excavata	Fornicata	Form cysts Pair of equal nuclei No mitochondria Often parasitic Four free flagella	*Giardia lamblia*	Giardiasis
	Parabasalids	No mitochondria Four free flagella One attached flagellum No cysts Parasitic or symbiotic Basal bodies Kinetoplastids	*Trichomonas*	Trichomoniasis
	Euglenozoans	Photosynthetic or heterotrophic Flagella	*Euglena*	N/a
			Trypanosoma	African sleeping sickness, Chagas disease
			Leishmania	Leishmaniasis
Chromalveolata	Dinoflagellates	Cellulose theca Two dissimilar flagella	*Gonyaulax*	Red tides
			Alexandrium	Paralytic shellfish poisoning
			Pfiesteria	Harmful algal blooms
	Apicomplexans	Intracellular parasite Apical organelles	*Plasmodium*	Malaria
			Cryptosporidium	Cryptosporidiosis
			Theileria (Babesia)	Babesiosis
			Toxoplasma	Toxoplasmosis
	Ciliates	Cilia	*Balantidium*	Balantidiasis
			Paramecium	N/a
			Stentor	N/a
	Öomycetes/peronosporomycetes	"Water molds" Generally diploid Cellulose cell walls	*Phytophthora*	Diseases in crops

Figure 5.7

The Eukaryote Supergroups and Some Examples (continued)				
Supergroup	Subgroups	Distinguishing Features	Examples	Clinical Notes
Rhizaria	Foraminifera	Amoeboid Threadlike pseudopodia Calcium carbonate shells	*Astrolonche*	N/a
	Radiolaria	Amoeboid Threadlike pseudopodia Silica shells	*Actinomma*	N/a
	Cercozoa	Amoeboid Threadlike pseudopodia Complex shells Parasitic forms	*Spongospora subterranea*	Powdery scab (potato disease)
			Plasmodiophora brassicae	Cabbage clubroot
Archaeplastida	Red algae	Chlorophyll *a* Phycoerythrin Phycocyanin Floridean starch Agar in cell walls	*Gelidium*	Source of agar
			Gracilaria	Source of agar
	Chlorophytes	Chlorophyll *a* Chlorophyll *b* Cellulose cell walls Starch storage	*Acetabularia*	N/a
			Ulva	N/a
Amoebozoa	Slime molds	Plasmodial and cellular forms	*Dictyostelium*	N/a
	Entamoebas	Trophozoites Form cysts	*Entamoeba*	Amoebiasis
			Naegleria	Primary amoebic meningoencephalitis
			Acanthamoeba	Keratitis, granulomatous amoebic encephalitis
Opisthokonta	Fungi	Chitin cell walls Unicellular or multicellular Often hyphae	Zygomycetes	Zygomycosis
			Ascomycetes	Candidiasis
			Basidiomycetes	Cryptococcosis
			Microsporidia	Microsporidiosis
	Animals	Multicellular heterotrophs No cell walls	Nematoda	Trichinosis; hookworm and pinworm infections
			Trematoda	Schistosomiasis
			Cestoda	Tapeworm infections

Figure 5.8

✓ CHECK YOUR UNDERSTANDING

- Which supergroups contain the clinically significant protists?

Amoebozoa

The supergroup Amoebozoa includes protozoans that use amoeboid movement. Actin microfilaments produce pseudopodia, into which the remainder of the protoplasm flows, thereby moving the organism. The genus *Entamoeba* includes commensal or parasitic species, including the medically important *E. histolytica*, which is transmitted by cysts in feces and is the primary cause of amoebic dysentery. Another member of this group that is pathogenic to humans is *Acanthamoeba*, which can cause keratitis (corneal inflammation) and blindness. The notorious "brain eating amoeba," *Naegleria fowleri*, is a considered a distant relative of the

Amoebozoa and is classified in the phylum Percolozoa.

The Eumycetozoa are an unusual group of organisms called slime molds, which have previously been classified as animals, fungi, and plants (Figure 5.9). Slime molds can be divided into two types: cellular slime molds and plasmodial slime molds. The cellular slime molds exist as individual amoeboid cells that periodically aggregate into a mobile slug. The aggregate then forms a fruiting body that produces haploid spores. Plasmodial slime molds exist as large, multinucleate amoeboid cells that form reproductive stalks to produce spores that divide into gametes. One cellular slime mold, *Dictyostelium discoideum*, has been an important study organism for understanding cell differentiation, because it has both single-celled and multicelled life stages, with the cells showing some degree of differentiation in the multicelled form. Figure 5.10 and Figure 5.11 illustrate the life cycles of cellular and plasmodial slime molds, respectively.

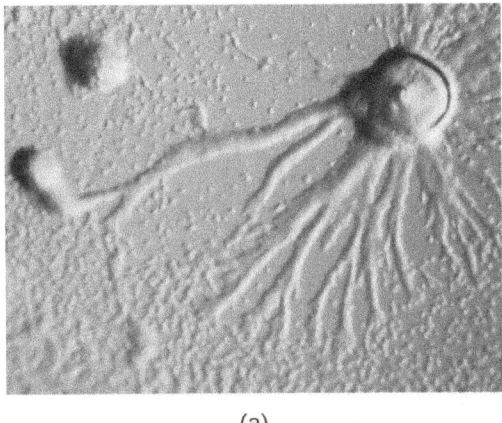
(a)

(b)

Figure 5.9 (a) The cellular slime mold *Dictyostelium discoideum* can be grown on agar in a Petri dish. In this image, individual amoeboid cells (visible as small spheres) are streaming together to form an aggregation that is beginning to rise in the upper right corner of the image. The primitively multicellular aggregation consists of individual cells that each have their own nucleus. (b) *Fuligo septica* is a plasmodial slime mold. This brightly colored organism consists of a large cell with many nuclei.

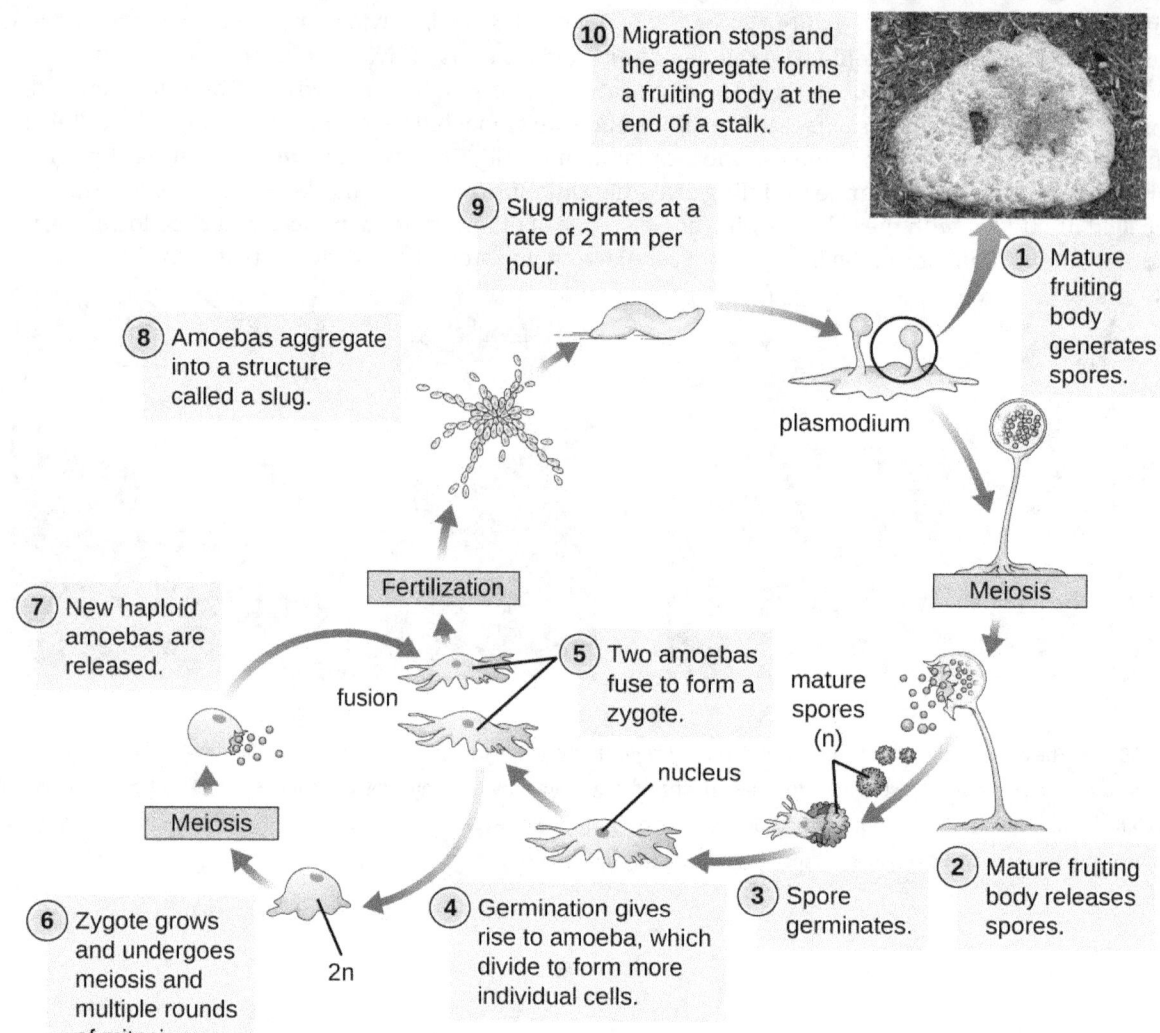

Figure 5.10 The life cycle of the cellular slime mold *Dictyostelium discoideum* primarily involves individual amoebas but includes the formation of a multinucleate plasmodium formed from a uninucleate zygote (the result of the fusion of two individual amoeboid cells). The plasmodium is able to move and forms a fruiting body that generates haploid spores. (credit "photo": modification of work by "thatredhead4"/Flickr)

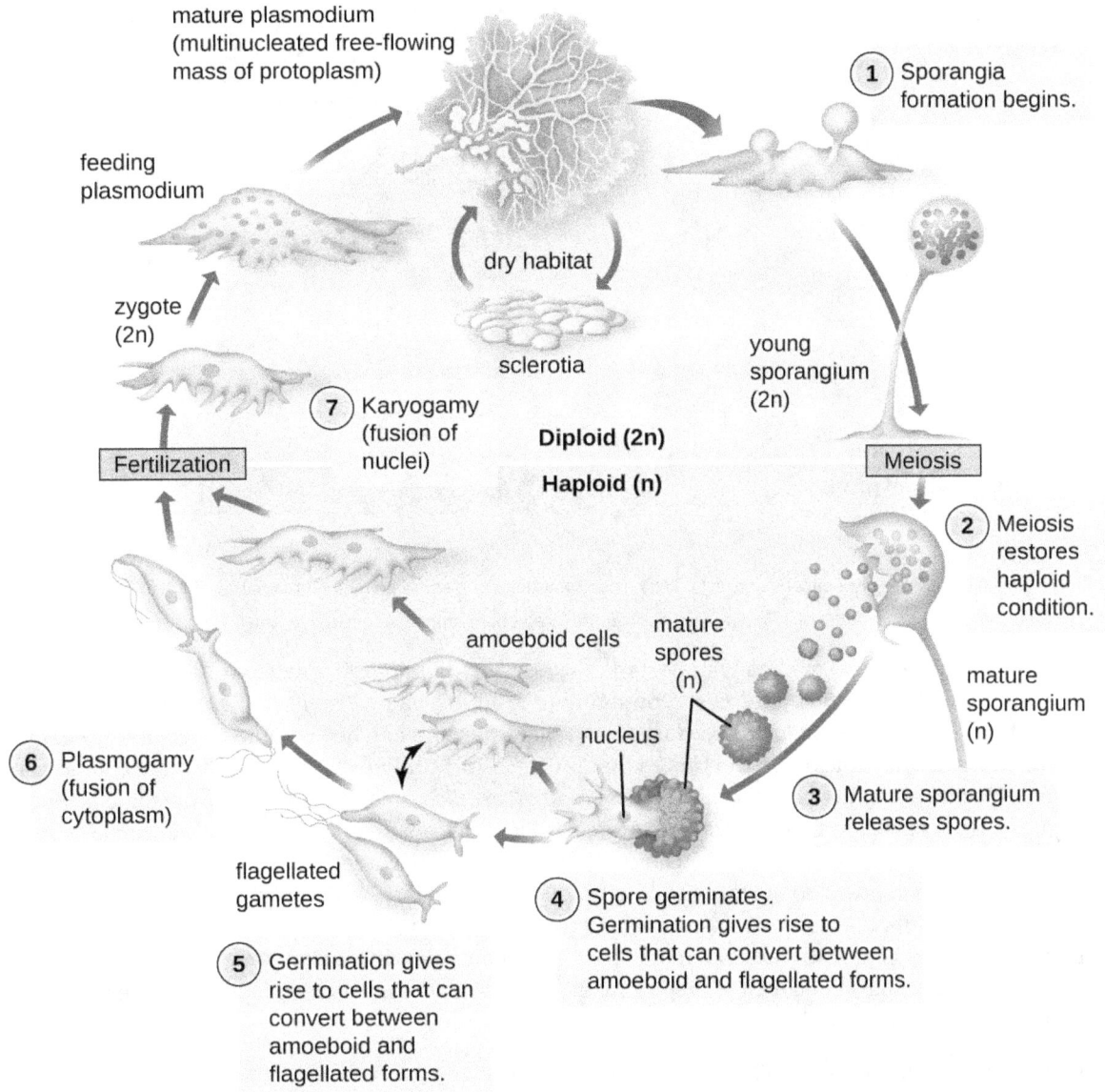

Figure 5.11 Plasmodial slime molds exist as large multinucleate amoeboid cells that form reproductive stalks to produce spores that divide into gametes.

Chromalveolata

The supergroup Chromalveolata is united by similar origins of its members' plastids and includes the apicomplexans, ciliates, diatoms, and dinoflagellates, among other groups (we will cover the diatoms and dinoflagellates in Algae). The apicomplexans are intra- or extracellular parasites that have an apical complex at one end of the cell. The apical complex is a concentration of organelles, vacuoles, and microtubules that allows the parasite to enter host cells (Figure 5.12). Apicomplexans have complex life cycles that include an infective sporozoite that undergoes schizogony to make many merozoites (see the example in Figure 5.4). Many are capable of infecting a variety of animal cells, from insects to livestock to humans, and their life cycles often depend on transmission between multiple hosts. The genus *Plasmodium* is an example of this group.

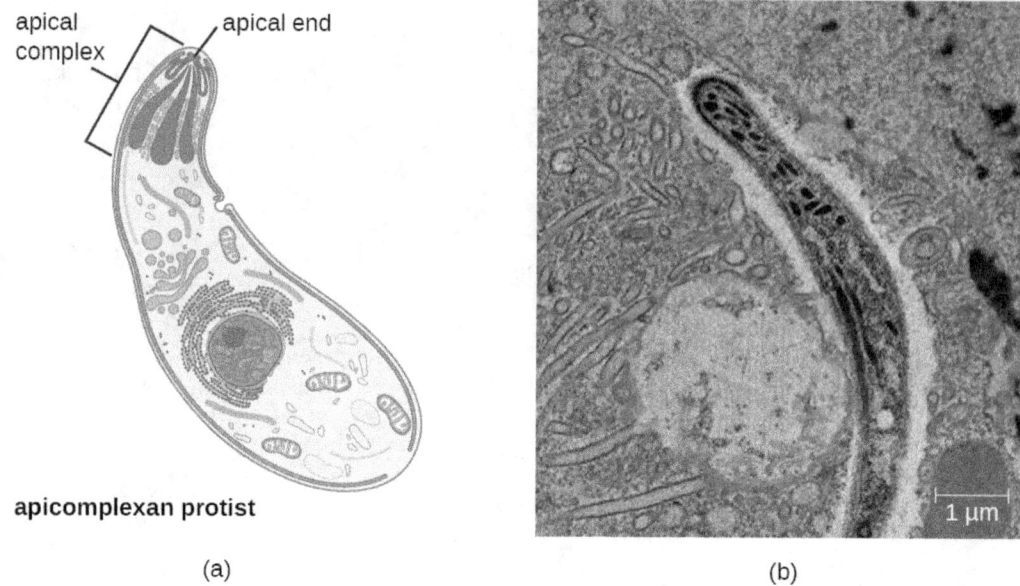

Figure 5.12 (a) Apicomplexans are parasitic protists. They have a characteristic apical complex that enables them to infect host cells. (b) A colorized electron microscope image of a *Plasmodium* sporozoite. (credit b: modification of work by Ute Frevert)

Other apicomplexans are also medically important. *Cryptosporidium parvum* causes intestinal symptoms and can cause epidemic diarrhea when the cysts contaminate drinking water. *Theileria (Babesia) microti*, transmitted by the tick *Ixodes scapularis*, causes recurring fever that can be fatal and is becoming a common transfusion-transmitted pathogen in the United States (*Theileria* and *Babesia* are closely related genera and there is some debate about the best classification). Finally, *Toxoplasma gondii* causes toxoplasmosis and can be transmitted from cat feces, unwashed fruit and vegetables, or from undercooked meat. Because toxoplasmosis can be associated with serious birth defects, pregnant women need to be aware of this risk and use caution if they are exposed to the feces of potentially infected cats. A national survey found the frequency of individuals with antibodies for toxoplasmosis (and thus who presumably have a current latent infection) in the United States to be 11%. Rates are much higher in other countries, including some developed countries.[3] There is also evidence and a good deal of theorizing that the parasite may be responsible for altering infected humans' behavior and personality traits.[4]

The ciliates (Ciliaphora), also within the Chromalveolata, are a large, very diverse group characterized by the presence of cilia on their cell surface. Although the cilia may be used for locomotion, they are often used for feeding, as well, and some forms are nonmotile. *Balantidium coli* (Figure 5.13) is the only parasitic ciliate that affects humans by causing intestinal illness, although it rarely causes serious medical issues except in the immunocompromised (those having a weakened immune system). Perhaps the most familiar ciliate is *Paramecium*, a motile organism with a clearly visible cytostome and cytoproct that is often studied in biology laboratories (Figure 5.14). Another ciliate, *Stentor*, is sessile and uses its cilia for feeding (Figure 5.15). Generally, these organisms have a **micronucleus** that is diploid, somatic, and used for sexual reproduction by conjugation. They also have a **macronucleus** that is derived from the micronucleus; the macronucleus becomes polyploid (multiple sets of duplicate chromosomes), and has a reduced set of metabolic genes.

Ciliates are able to reproduce through conjugation, in which two cells attach to each other. In each cell, the diploid micronuclei undergo meiosis, producing eight haploid nuclei each. Then, all but one of the haploid micronuclei and the macronucleus disintegrate; the remaining (haploid) micronucleus undergoes mitosis. The two cells then exchange one micronucleus each, which fuses with the remaining micronucleus present to form a new, genetically different, diploid micronucleus. The diploid micronucleus undergoes two mitotic divisions, so each cell has four micronuclei, and two of the four combine to form a new macronucleus. The chromosomes

---

3 J. Flegr et al. "Toxoplasmosis—A Global Threat. Correlation of Latent Toxoplasmosis With Specific Disease Burden in a Set of 88 Countries." *PloS ONE* 9 no. 3 (2014):e90203.

4 J. Flegr. "Effects of Toxoplasma on Human Behavior." *Schizophrenia Bull* 33, no. 3 (2007):757–760.

in the macronucleus then replicate repeatedly, the macronucleus reaches its polyploid state, and the two cells separate. The two cells are now genetically different from each other and from their previous versions.

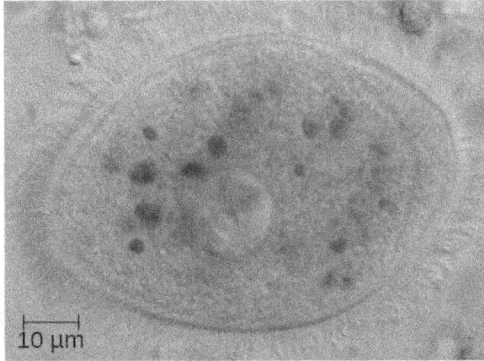

Figure 5.13 This specimen of the ciliate *Balantidium coli* is a trophozoite form isolated from the gut of a primate. *B. coli* is the only ciliate capable of parasitizing humans. (credit: modification of work by Kouassi RYW, McGraw SW, Yao PK, Abou-Bacar A, Brunet J, Pesson B, Bonfoh B, N'goran EK & Candolfi E)

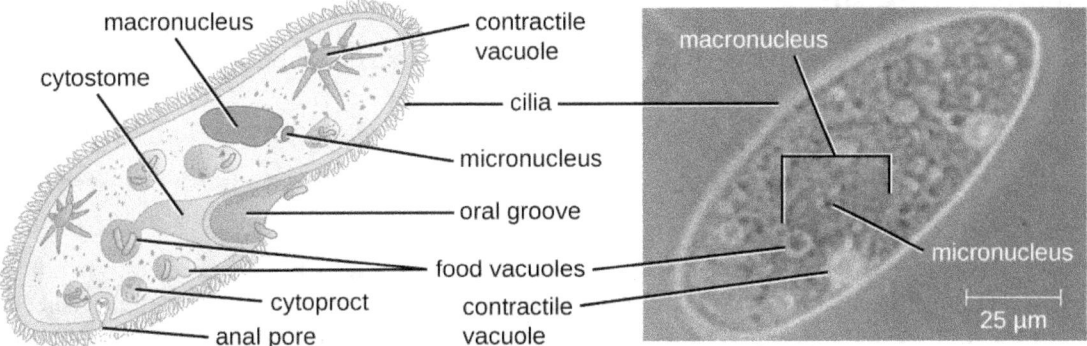

Figure 5.14 *Paramecium* has a primitive mouth (called an oral groove) to ingest food, and an anal pore to excrete it. Contractile vacuoles allow the organism to excrete excess water. Cilia enable the organism to move.

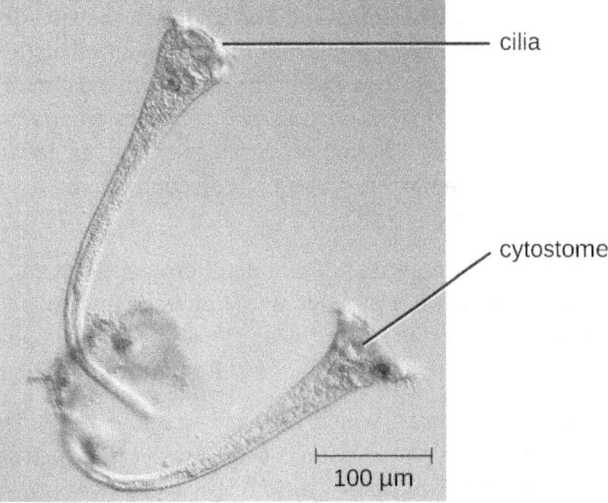

Figure 5.15 This differential interference contrast micrograph (magnification: ×65) of *Stentor roeselie* shows cilia present on the margins of the structure surrounding the cytostome; the cilia move food particles. (credit: modification of work by "picturepest"/Flickr)

Öomycetes have similarities to fungi and were once classified with them. They are also called water molds. However, they differ from fungi in several important ways. Öomycetes have cell walls of cellulose (unlike the chitinous cell walls of fungi) and they are generally diploid, whereas the dominant life forms of fungi are typically haploid. *Phytophthora*, the plant pathogen found in the soil that caused the Irish potato famine, is classified within this group (Figure 5.16).

Figure 5.16 A saprobic oomycete, or water mold, engulfs a dead insect. (credit: modification of work by Thomas Bresson)

LINK TO LEARNING

Explore the procedures for detecting the presence of an apicomplexan in a public water supply, at this (https://openstax.org/l/22detpreapicom) website.

This video (https://openstax.org/l/22feedstentor) shows the feeding of *Stentor*.

Excavata

The third and final supergroup to be considered in this section is the Excavata, which includes primitive eukaryotes and many parasites with limited metabolic abilities. These organisms have complex cell shapes and structures, often including a depression on the surface of the cell called an excavate. The group Excavata includes the subgroups Fornicata, Parabasalia, and Euglenozoa. The Fornicata lack mitochondria but have flagella. This group includes *Giardia lamblia* (also known as *G. intestinalis* or *G. duodenalis*), a widespread pathogen that causes diarrheal illness and can be spread through cysts from feces that contaminate water supplies (Figure 5.3). Parabasalia are frequent animal endosymbionts; they live in the guts of animals like termites and cockroaches. They have basal bodies and modified mitochondria (kinetoplastids). They also have a large, complex cell structure with an undulating membrane and often have many flagella. The trichomonads (a subgroup of the Parabasalia) include pathogens such as *Trichomonas vaginalis*, which causes the human sexually transmitted disease trichomoniasis. Trichomoniasis often does not cause symptoms in men, but men are able to transmit the infection. In women, it causes vaginal discomfort and discharge and may cause complications in pregnancy if left untreated.

The Euglenozoa are common in the environment and include photosynthetic and nonphotosynthetic species. Members of the genus *Euglena* are typically not pathogenic. Their cells have two flagella, a pellicle, a **stigma** (eyespot) to sense light, and chloroplasts for photosynthesis (Figure 5.17). The pellicle of *Euglena* is made of a series of protein bands surrounding the cell; it supports the cell membrane and gives the cell shape.

The Euglenozoa also include the trypanosomes, which are parasitic pathogens. The genus *Trypanosoma* includes *T. brucei*, which causes African trypanosomiasis (African sleeping sickness and *T. cruzi*, which causes American trypanosomiasis (Chagas disease). These tropical diseases are spread by insect bites. In African sleeping sickness, *T. brucei* colonizes the blood and the brain after being transmitted via the bite of a tsetse fly (*Glossina* spp.) (Figure 5.18). The early symptoms include confusion, difficulty sleeping, and lack of coordination. Left untreated, it is fatal.

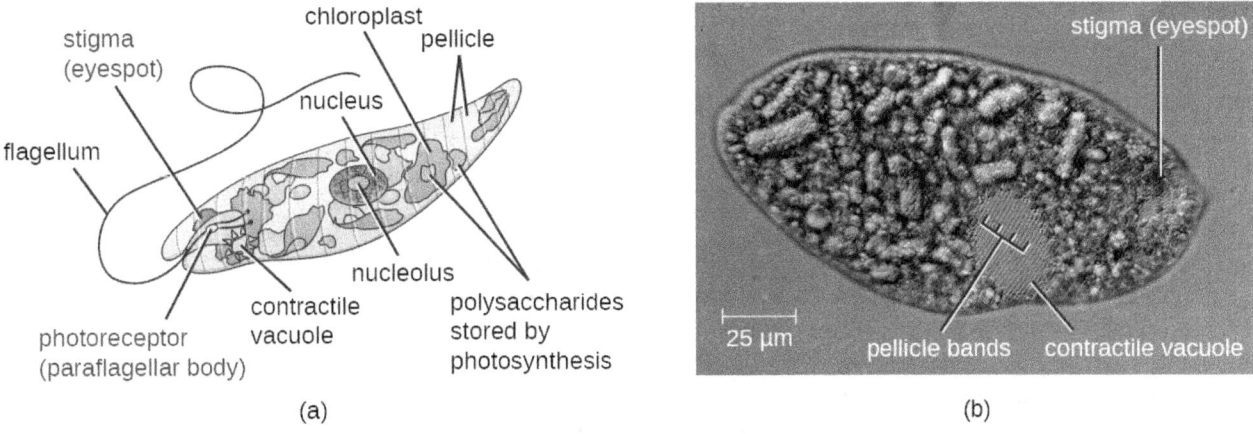

Figure 5.17 (a) This illustration of a *Euglena* shows the characteristic structures, such as the stigma and flagellum. (b) The pellicle, under the cell membrane, gives the cell its distinctive shape and is visible in this image as delicate parallel striations over the surface of the entire cell (especially visible over the grey contractile vacuole). (credit a: modification of work by Claudio Miklos; credit b: modification of work by David Shykind)

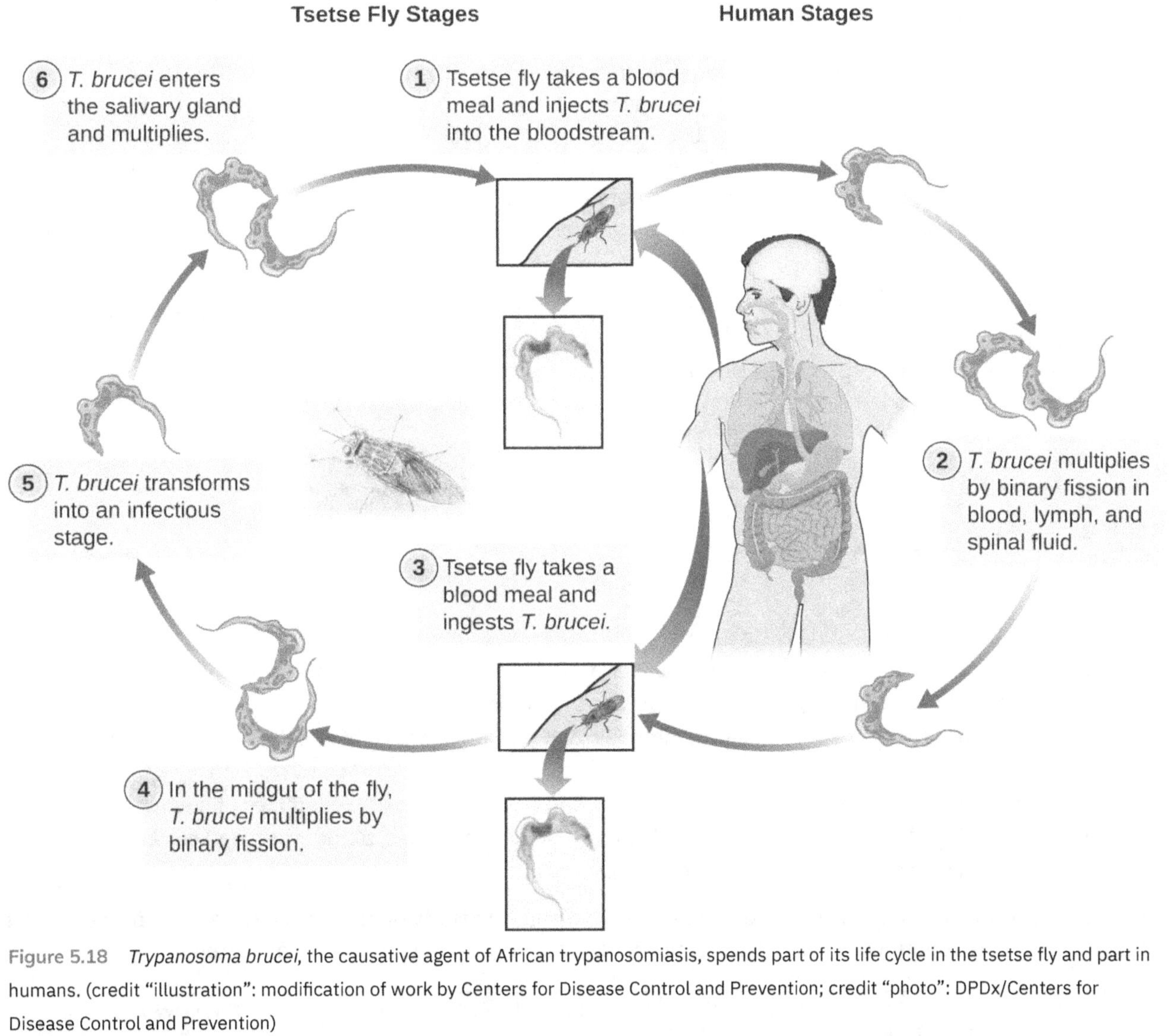

Figure 5.18 *Trypanosoma brucei*, the causative agent of African trypanosomiasis, spends part of its life cycle in the tsetse fly and part in humans. (credit "illustration": modification of work by Centers for Disease Control and Prevention; credit "photo": DPDx/Centers for Disease Control and Prevention)

Chagas' disease originated and is most common in Latin America. The disease is transmitted by *Triatoma* spp., insects often called "kissing bugs," and affects either the heart tissue or tissues of the digestive system.

Untreated cases can eventually lead to heart failure or significant digestive or neurological disorders.

The genus *Leishmania* includes trypanosomes that cause disfiguring skin disease and sometimes systemic illness as well.

> ### Eye on Ethics
>
> **Neglected Parasites**
>
> The Centers for Disease Control and Prevention (CDC) is responsible for identifying public health priorities in the United States and developing strategies to address areas of concern. As part of this mandate, the CDC has officially identified five parasitic diseases it considers to have been neglected (i.e., not adequately studied). These neglected parasitic infections (NPIs) include toxoplasmosis, Chagas disease, toxocariasis (a nematode infection transmitted primarily by infected dogs), cysticercosis (a disease caused by a tissue infection of the tapeworm *Taenia solium*), and trichomoniasis (a sexually transmitted disease caused by the parabasalid *Trichomonas vaginalis*).
>
> The decision to name these specific diseases as NPIs means that the CDC will devote resources toward improving awareness and developing better diagnostic testing and treatment through studies of available data. The CDC may also advise on treatment of these diseases and assist in the distribution of medications that might otherwise be difficult to obtain.[5]
>
> Of course, the CDC does not have unlimited resources, so by prioritizing these five diseases, it is effectively deprioritizing others. Given that many Americans have never heard of many of these NPIs, it is fair to ask what criteria the CDC used in prioritizing diseases. According to the CDC, the factors considered were the number of people infected, the severity of the illness, and whether the illness can be treated or prevented. Although several of these NPIs may seem to be more common outside the United States, the CDC argues that many cases in the United States likely go undiagnosed and untreated because so little is known about these diseases.[6]
>
> What criteria should be considered when prioritizing diseases for purposes of funding or research? Are those identified by the CDC reasonable? What other factors could be considered? Should government agencies like the CDC have the same criteria as private pharmaceutical research labs? What are the ethical implications of deprioritizing other potentially neglected parasitic diseases such as leishmaniasis?
>
>

5.2 Parasitic Helminths

Learning Objectives

By the end of this section, you will be able to:
- Explain why we include the study of parasitic worms within the discipline of microbiology
- Compare the basic morphology of the major groups of parasitic helminthes
- Describe the characteristics of parasitic nematodes, and give an example of infective eggs and infective larvae
- Describe the characteristics of parasitic trematodes and cestodes, and give examples of each
- Identify examples of the primary causes of infections due to nematodes, trematodes, and cestodes
- Classify parasitic worms according to major groups

Parasitic helminths are animals that are often included within the study of microbiology because many species of these worms are identified by their microscopic eggs and larvae. There are two major groups of parasitic

---

5 Centers for Disease Control and Prevention. "Neglected Parasitic Infections (NPIs) in the United States." http://www.cdc.gov/parasites/npi/. Last updated July 10, 2014.

6 Centers for Disease Control and Prevention. "Fact Sheet: Neglected Parasitic Infections in the United States." http://www.cdc.gov/parasites/resources/pdf/npi_factsheet.pdf

helminths: the roundworms (Nematoda) and flatworms (Platyhelminthes). Of the many species that exist in these groups, about half are parasitic and some are important human pathogens. As animals, they are multicellular and have organ systems. However, the parasitic species often have limited digestive tracts, nervous systems, and locomotor abilities. Parasitic forms may have complex reproductive cycles with several different life stages and more than one type of host. Some are **monoecious**, having both male and female reproductive organs in a single individual, while others are **dioecious**, each having either male or female reproductive organs.

Nematoda (Roundworms)

Phylum **Nematoda** (the roundworms) is a diverse group containing more than 15,000 species, of which several are important human parasites (Figure 5.19). These unsegmented worms have a full digestive system even when parasitic. Some are common intestinal parasites, and their eggs can sometimes be identified in feces or around the anus of infected individuals. *Ascaris lumbricoides* is the largest nematode intestinal parasite found in humans; females may reach lengths greater than 1 meter. *A. lumbricoides* is also very widespread, even in developed nations, although it is now a relatively uncommon problem in the United States. It may cause symptoms ranging from relatively mild (such as a cough and mild abdominal pain) to severe (such as intestinal blockage and impaired growth).

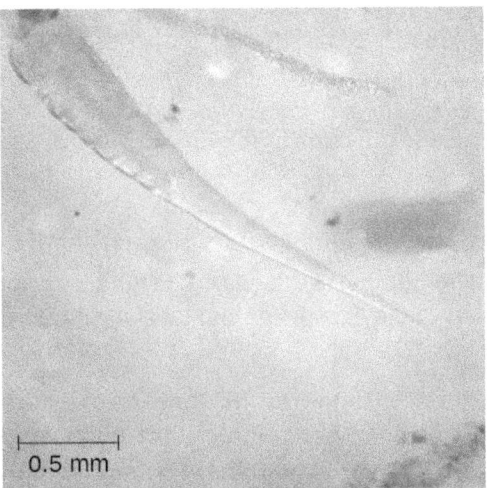

Figure 5.19   A micrograph of the nematode *Enterobius vermicularis*, also known as the pinworm. (credit: modification of work by Centers for Disease Control and Prevention)

Of all nematode infections in the United States, pinworm (caused by *Enterobius vermicularis*) is the most common. Pinworm causes sleeplessness and itching around the anus, where the female worms lay their eggs during the night. *Toxocara canis* and *T. cati* are nematodes found in dogs and cats, respectively, that can be transmitted to humans, causing toxocariasis. Antibodies to these parasites have been found in approximately 13.9% of the U.S. population, suggesting that exposure is common.[7] Infection can cause larval migrans, which can result in vision loss and eye inflammation, or fever, fatigue, coughing, and abdominal pain, depending on whether the organism infects the eye or the viscera. Another common nematode infection is hookworm, which is caused by *Necator americanus* (the New World or North American hookworm) and *Ancylostoma duodenale* (the Old World hookworm). Symptoms of hookworm infection can include abdominal pain, diarrhea, loss of appetite, weight loss, fatigue, and anemia.

Trichinellosis, also called trichinosis, caused by *Trichinella spiralis*, is contracted by consuming undercooked meat, which releases the larvae and allows them to encyst in muscles. Infection can cause fever, muscle pains, and digestive system problems; severe infections can lead to lack of coordination, breathing and heart problems, and even death. Finally, heartworm in dogs and other animals is caused by the nematode *Dirofilaria immitis*, which is transmitted by mosquitoes. Symptoms include fatigue and cough; when left untreated, death

---

7   Won K, Kruszon-Moran D, Schantz P, Jones J. "National seroprevalence and risk factors for zoonotic Toxocara spp. infection." In: Abstracts of the 56th American Society of Tropical Medicine and Hygiene; Philadelphia, Pennsylvania; 2007 Nov 4-8.

may result.

> **Clinical Focus**
>
> **Part 2**
>
> The physician explains to Sarah's mother that ringworm can be transferred between people through touch. "It's common in school children, because they often come in close contact with each other, but anyone can become infected," he adds. "Because you can transfer it through objects, locker rooms and public pools are also a potential source of infection. It's very common among wrestlers and athletes in other contact sports."
>
> Looking very uncomfortable, Sarah says to her mother "I want this worm out of me."
>
> The doctor laughs and says, "Sarah, you're in luck because ringworm is just a name; it is not an actual worm. You have nothing wriggling around under your skin."
>
> "Then what is it?" asks Sarah.
>
> - What type of pathogen causes ringworm?
>
> *Jump to the next Clinical Focus box. Go back to the previous Clinical Focus box.*

✓ CHECK YOUR UNDERSTANDING

- What is the most common nematode infection in the United States?

Platyhelminths (Flatworms)

Phylum **Platyhelminthes** (the platyhelminths) are flatworms. This group includes the flukes, tapeworms, and the turbellarians, which include planarians. The flukes and tapeworms are medically important parasites (Figure 5.20).

The **flukes** (trematodes) are nonsegmented flatworms that have an oral sucker (Figure 5.21) (and sometimes a second ventral sucker) and attach to the inner walls of intestines, lungs, large blood vessels, or the liver. Trematodes have complex life cycles, often with multiple hosts. Several important examples are the liver flukes (*Clonorchis* and *Opisthorchis*), the intestinal fluke (*Fasciolopsis buski*), and the oriental lung fluke (*Paragonimus westermani*). Schistosomiasis is a serious parasitic disease, considered second in the scale of its impact on human populations only to malaria. The parasites *Schistosoma mansoni*, *S. haematobium*, and *S. japonicum*, which are found in freshwater snails, are responsible for schistosomiasis (Figure 5.22). Immature forms burrow through the skin into the blood. They migrate to the lungs, then to the liver and, later, other organs. Symptoms include anemia, malnutrition, fever, abdominal pain, fluid buildup, and sometimes death.

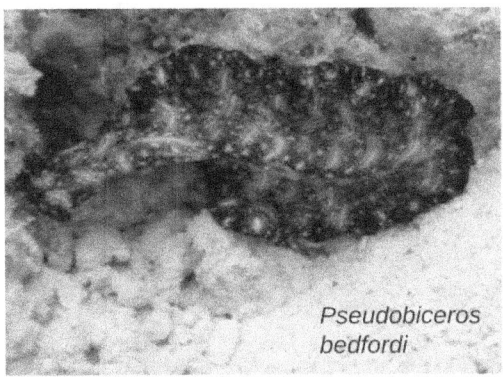
(a) Class Turbellaria

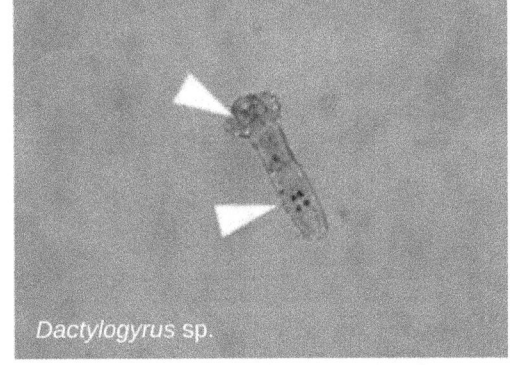
(b) Class Monogenea

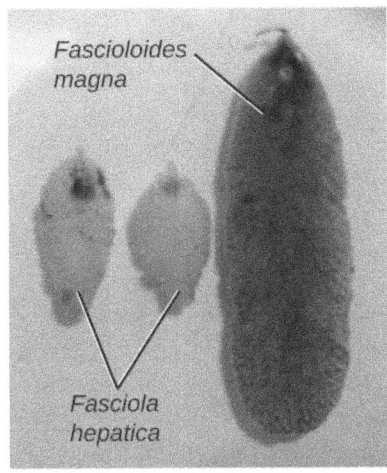
(c) Class Trematoda

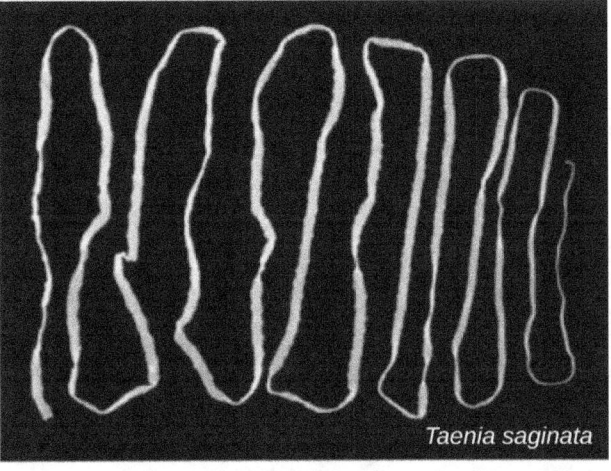
(d) Class Cestoda

Figure 5.20 Phylum Platyhelminthes is divided into four classes. (a) Class Turbellaria includes the Bedford's flatworm (*Pseudobiceros bedfordi*), which is about 8–10 cm long. (b) The parasitic class Monogenea includes *Dactylogyrus* spp. Worms in this genus are commonly called gill flukes. The specimen pictured here is about 0.2 mm long and has two anchors, indicated by arrows, that it uses to latch onto the gills of host fish. (c) The Trematoda class includes the common liver fluke *Fasciola hepatica* and the giant liver fluke *Fascioloides magna* (right). The *F. magna* specimen shown here is about 7 cm long. (d) Class Cestoda includes tapeworms such as *Taenia saginata*, which infects both cattle and humans and can reach lengths of 4–10 meters; the specimen shown here is about 4 meters long. (credit c: modification of work by "Flukeman"/Wikimedia Commons)

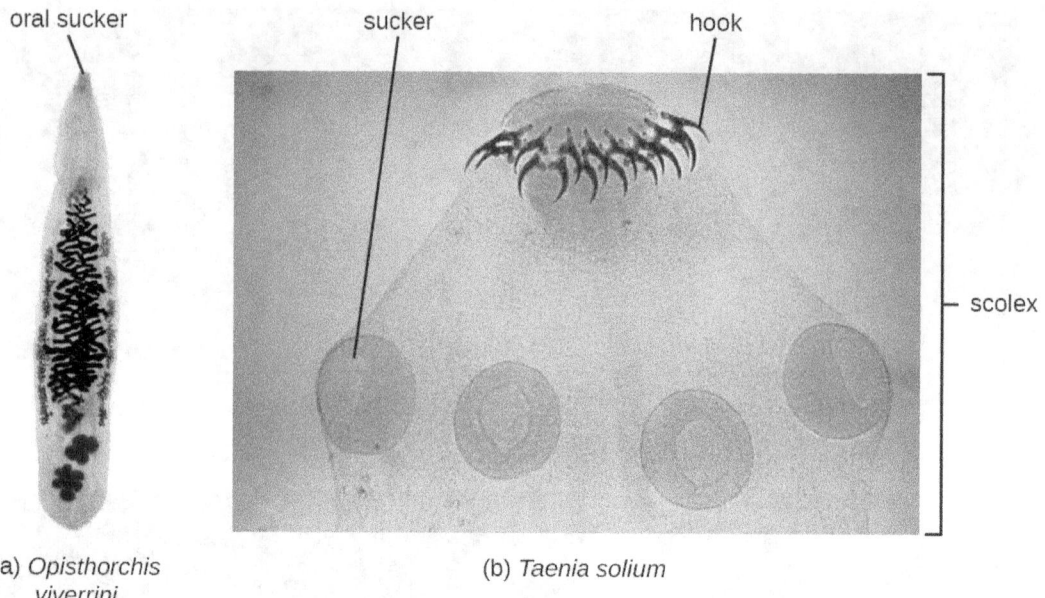

(a) *Opisthorchis viverrini* (b) *Taenia solium*

Figure 5.21 (a) The oral sucker is visible on the anterior end of this liver fluke, *Opisthorchis viverrini*. (b) This micrograph shows the scolex of the cestode *Taenia solium*, also known as the pork tapeworm. The visible suckers and hooks allow the worm to attach itself to the inner wall of the intestine. (credit a: modification of work by Sripa B, Kaewkes S, Sithithaworn P, Mairiang E, Laha T, and Smout M; credit b: modification of work by Centers for Disease Control and Prevention)

The other medically important group of platyhelminths are commonly known as **tapeworms** (cestodes) and are segmented flatworms that may have suckers or hooks at the **scolex** (head region) (Figure 5.21). Tapeworms use these suckers or hooks to attach to the wall of the small intestine. The body of the worm is made up of segments called **proglottids** that contain reproductive structures; these detach when the gametes are fertilized, releasing gravid proglottids with eggs. Tapeworms often have an intermediate host that consumes the eggs, which then hatch into a larval form called an oncosphere. The oncosphere migrates to a particular tissue or organ in the intermediate host, where it forms cysticerci. After being eaten by the definitive host, the cysticerci develop into adult tapeworms in the host's digestive system (Figure 5.23). *Taenia saginata* (the beef tapeworm) and *T. solium* (the pork tapeworm) enter humans through ingestion of undercooked, contaminated meat. The adult worms develop and reside in the intestine, but the larval stage may migrate and be found in other body locations such as skeletal and smooth muscle. The beef tapeworm is relatively benign, although it can cause digestive problems and, occasionally, allergic reactions. The pork tapeworm can cause more serious problems when the larvae leave the intestine and colonize other tissues, including those of the central nervous system. *Diphylobothrium latum* is the largest human tapeworm and can be ingested in undercooked fish. It can grow to a length of 15 meters. *Echinococcus granulosus,* the dog tapeworm, can parasitize humans and uses dogs as an important host.

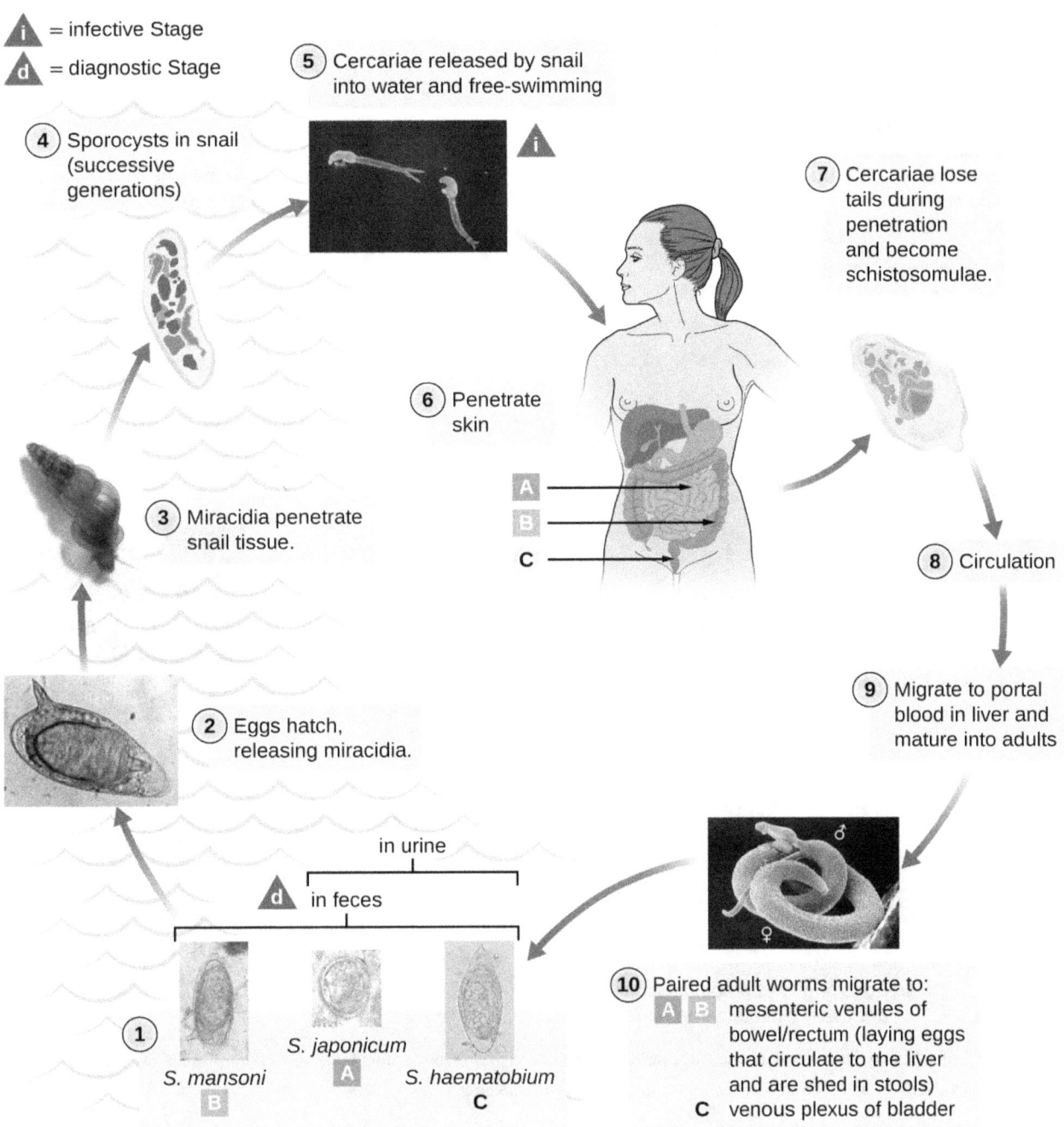

Figure 5.22 The life cycle of *Schistosoma* spp. includes several species of water snails, which serve as secondary hosts. The parasite is transmitted to humans through contact with contaminated water and takes up residence in the veins of the digestive system. Eggs escape the host in the urine or feces and infect a snail to complete the life cycle. (credit "illustration": modification of work by Centers for Disease Control and Prevention; credit "step 3 photo": modification of work by Fred A. Lewis, Yung-san Liang, Nithya Raghavan & Matty Knight)

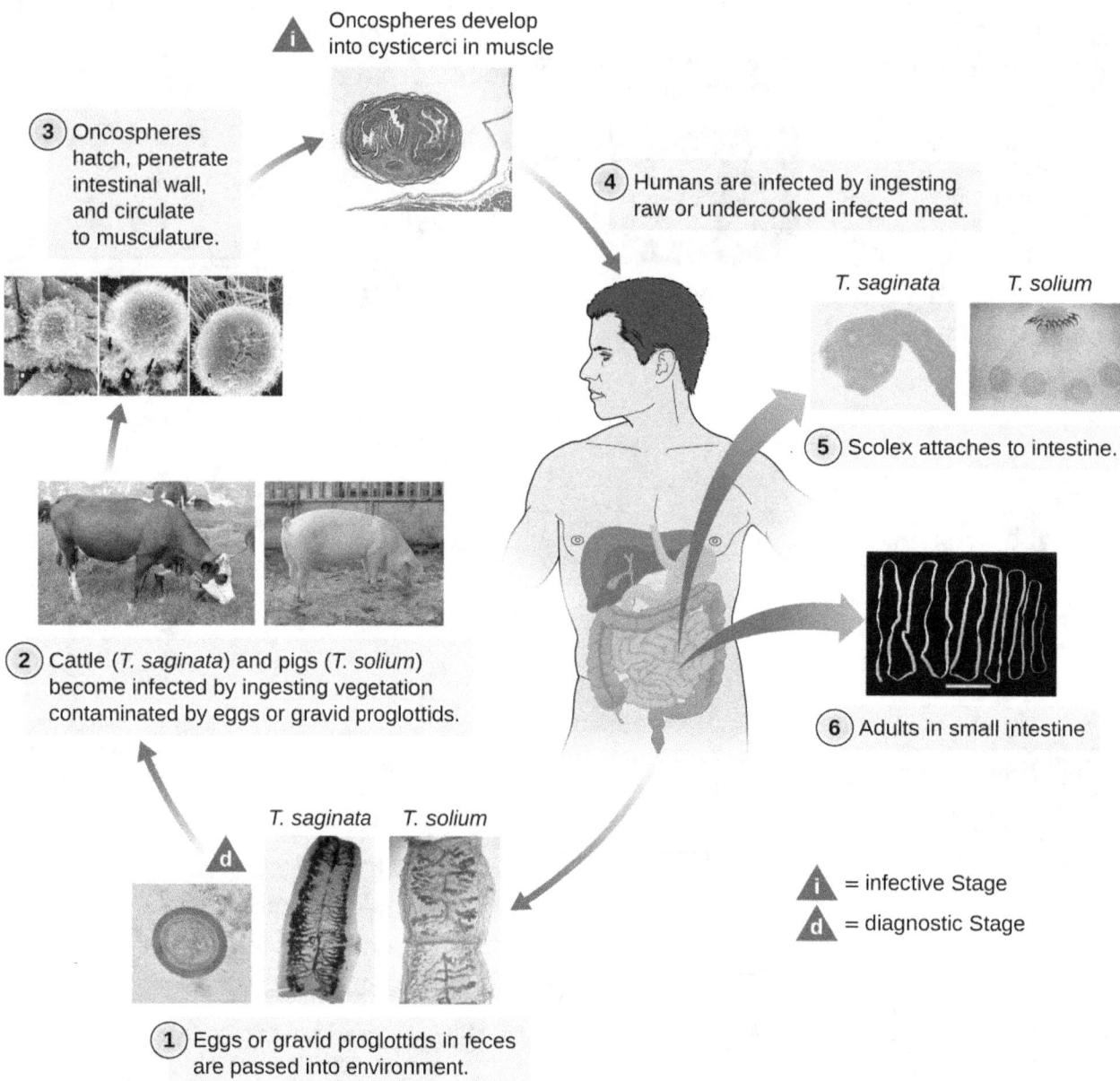

Figure 5.23 Life cycle of a tapeworm. (credit "illustration": modification of work by Centers for Disease Control and Prevention; credit "step 3 micrographs": modification of work by American Society for Microbiology)

CHECK YOUR UNDERSTANDING

- What group of medically important flatworms is segmented and what group is unsegmented?

MICRO CONNECTIONS

Food for Worms?

For residents of temperate, developed countries, it may be difficult to imagine just how common helminth infections are in the human population. In fact, they are quite common and even occur frequently in the United States. Worldwide, approximately 807–1,221 million people are infected with *Ascaris lumbricoides* (perhaps one-sixth of the human population) and far more are infected if all nematode species are considered.[8] Rates of infection are relatively high even in industrialized nations. Approximately 604–795 million people are infected with whipworm (*Trichuris*) worldwide (*Trichuris* can also infect dogs), and 576–740 million people are infected with hookworm (*Necator americanus* and *Ancylostoma duodenale*).[9] *Toxocara*, a nematode parasite of dogs and cats, is also able to infect humans. It is widespread in the United States, with about 10,000 symptomatic cases annually. However, one study found 14% of the population (more than 40 million Americans) was seropositive, meaning they had been exposed to the parasite at one time. More than 200 million people have schistosomiasis worldwide. Most of the World Health Organization (WHO) neglected tropical diseases are helminths. In some cases, helminths may cause subclinical illnesses, meaning the symptoms are so mild that that they go unnoticed. In other cases, the effects may be more severe or chronic, leading to fluid accumulation and organ damage. With so many people affected, these parasites constitute a major global public health concern.

MICRO CONNECTIONS

Eradicating the Guinea Worm

Dracunculiasis, or Guinea worm disease, is caused by a nematode called *Dracunculus medinensis*. When people consume contaminated water, water fleas (small crustaceans) containing the nematode larvae may be ingested. These larvae migrate out of the intestine, mate, and move through the body until females eventually emerge (generally through the feet). While Guinea worm disease is rarely fatal, it is extremely painful and can be accompanied by secondary infections and edema (Figure 5.24).

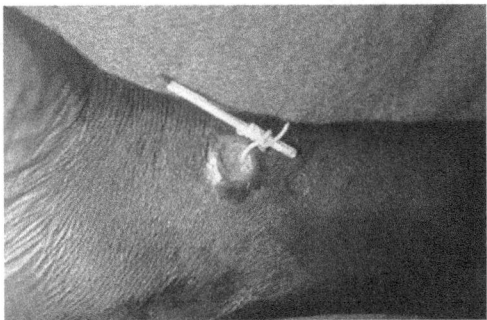

Figure 5.24 The Guinea worm can be removed from a leg vein of an infected person by gradually winding it around a stick, like this matchstick. (credit: Centers for Disease Control and Prevention)

---

8 Fenwick, A. "The global burden of neglected tropical diseases." *Public health* 126 no.3 (Mar 2012): 233–6.

9 de Silva, N., et. al. (2003). "Soil-transmitted helminth infections: updating the global picture". *Trends in Parasitology* 19 (December 2003): 547–51.

An eradication campaign led by WHO, the CDC, the United Nations Children's Fund (UNICEF), and the Carter Center (founded by former U.S. president Jimmy Carter) has been extremely successful in reducing cases of dracunculiasis. This has been possible because diagnosis is straightforward, there is an inexpensive method of control, there is no animal reservoir, the water fleas are not airborne (they are restricted to still water), the disease is geographically limited, and there has been a commitment from the governments involved. Additionally, no vaccines or medication are required for treatment and prevention. In 1986, 3.5 million people were estimated to be affected. After the eradication campaign, which included helping people in affected areas learn to filter water with cloth, only four countries continue to report the disease (Chad, Mali, South Sudan, and Ethiopia) with a total of 126 cases reported to WHO in 2014.[10]

5.3 Fungi

Learning Objectives

By the end of this section, you will be able to:

- Explain why the study of fungi such as yeast and molds is within the discipline of microbiology
- Describe the unique characteristics of fungi
- Describe examples of asexual and sexual reproduction of fungi
- Compare the major groups of fungi in this chapter, and give examples of each
- Identify examples of the primary causes of infections due to yeasts and molds
- Identify examples of toxin-producing fungi
- Classify fungal organisms according to major groups

The fungi comprise a diverse group of organisms that are heterotrophic and typically saprozoic. In addition to the well-known macroscopic fungi (such as mushrooms and molds), many unicellular yeasts and spores of macroscopic fungi are microscopic. For this reason, fungi are included within the field of microbiology.

Fungi are important to humans in a variety of ways. Both microscopic and macroscopic fungi have medical relevance, with some pathogenic species that can cause **mycoses** (illnesses caused by fungi). Some pathogenic fungi are opportunistic, meaning that they mainly cause infections when the host's immune defenses are compromised and do not normally cause illness in healthy individuals. Fungi are important in other ways. They act as decomposers in the environment, and they are critical for the production of certain foods such as cheeses. Fungi are also major sources of antibiotics, such as penicillin from the fungus *Penicillium*.

Characteristics of Fungi

Fungi have well-defined characteristics that set them apart from other organisms. Most multicellular fungal bodies, commonly called molds, are made up of filaments called **hyphae**. Hyphae can form a tangled network called a **mycelium** and form the **thallus** (body) of fleshy fungi. Hyphae that have walls between the cells are called **septate hyphae**; hyphae that lack walls and cell membranes between the cells are called nonseptate or **coenocytic hyphae**). (Figure 5.25).

---

10 World Health Organization. "South Sudan Reports Zero Cases of Guinea-Worm Disease for Seventh Consecutive Month." 2016. http://www.who.int/dracunculiasis/no_new_case_for_seventh_consecutive_months/en/. Accessed May 2, 2016.

septate hyphae coenocytic (nonseptate) hyphae pseudohyphae

molds   yeast cells

Figure 5.25 Multicellular fungi (molds) form hyphae, which may be septate or nonseptate. Unicellular fungi (yeasts) cells form pseudohyphae from individual yeast cells.

In contrast to molds, yeasts are unicellular fungi. The **budding yeasts** reproduce asexually by budding off a smaller daughter cell; the resulting cells may sometimes stick together as a short chain or **pseudohypha** (Figure 5.25).

Some fungi are dimorphic, having more than one appearance during their life cycle. These **dimorphic fungi** may be able to appear as yeasts or molds, which can be important for infectivity. They are capable of changing their appearance in response to environmental changes such as nutrient availability or fluctuations in temperature, growing as a mold, for example, at 25 °C (77 °F), and as yeast cells at 37 °C (98.6 °F). This ability helps dimorphic fungi to survive in diverse environments. Two examples of dimorphic yeasts are the human pathogens Histoplasma capsulatum and Candida albicans. H. capsulatum causes the lung disease histoplasmosis, and C. albicans is associated with vaginal yeast infections, oral thrush, and candidiasis of the skin (Figure 5.26).

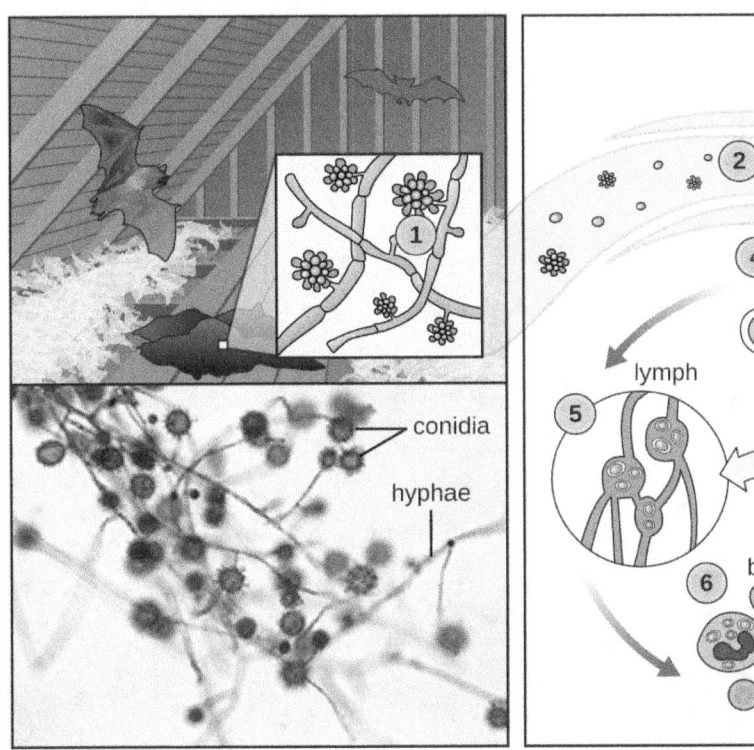
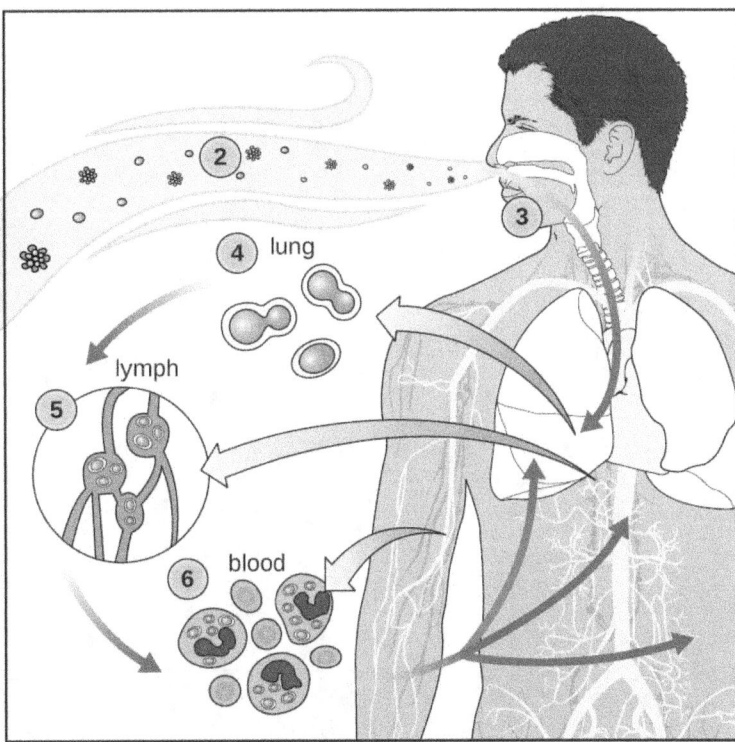

Figure 5.26 *Histoplasma capsulatum* is a dimorphic fungus that grows in soil exposed to bird feces or bat feces (guano) (top left). It can change forms to survive at different temperatures. In the outdoors, it typically grows as a mycelium (as shown in the micrograph, bottom left), but when the spores are inhaled (right), it responds to the high internal temperature of the body (37 °C [98.6 °F]) by turning into a yeast that can multiply in the lungs, causing the chronic lung disease histoplasmosis. (credit: modification of work by Centers for Disease

Control and Prevention)

There are notable unique features in fungal cell walls and membranes. Fungal cell walls contain **chitin**, as opposed to the cellulose found in the cell walls of plants and many protists. Additionally, whereas animals have cholesterol in their cell membranes, fungal cell membranes have different sterols called ergosterols. Ergosterols are often exploited as targets for antifungal drugs.

Fungal life cycles are unique and complex. Fungi reproduce sexually either through cross- or self-fertilization. Haploid fungi form hyphae that have gametes at the tips. Two different mating types (represented as "+ type" and "– type") are involved. The cytoplasms of the + and – type gametes fuse (in an event called plasmogamy), producing a cell with two distinct nuclei (a **dikaryotic** cell). Later, the nuclei fuse (in an event called karyogamy) to create a diploid zygote. The zygote undergoes meiosis to form **spores** that germinate to start the haploid stage, which eventually creates more haploid mycelia (Figure 5.27). Depending on the taxonomic group, these sexually produced spores are known as zygospores (in Zygomycota), ascospores (in Ascomycota), or basidiospores (in Basidiomycota) (Figure 5.28).

Fungi may also exhibit asexual reproduction by mitosis, mitosis with budding, fragmentation of hyphae, and formation of asexual spores by mitosis. These spores are specialized cells that, depending on the organism, may have unique characteristics for survival, reproduction, and dispersal. Fungi exhibit several types of asexual spores and these can be important in classification.

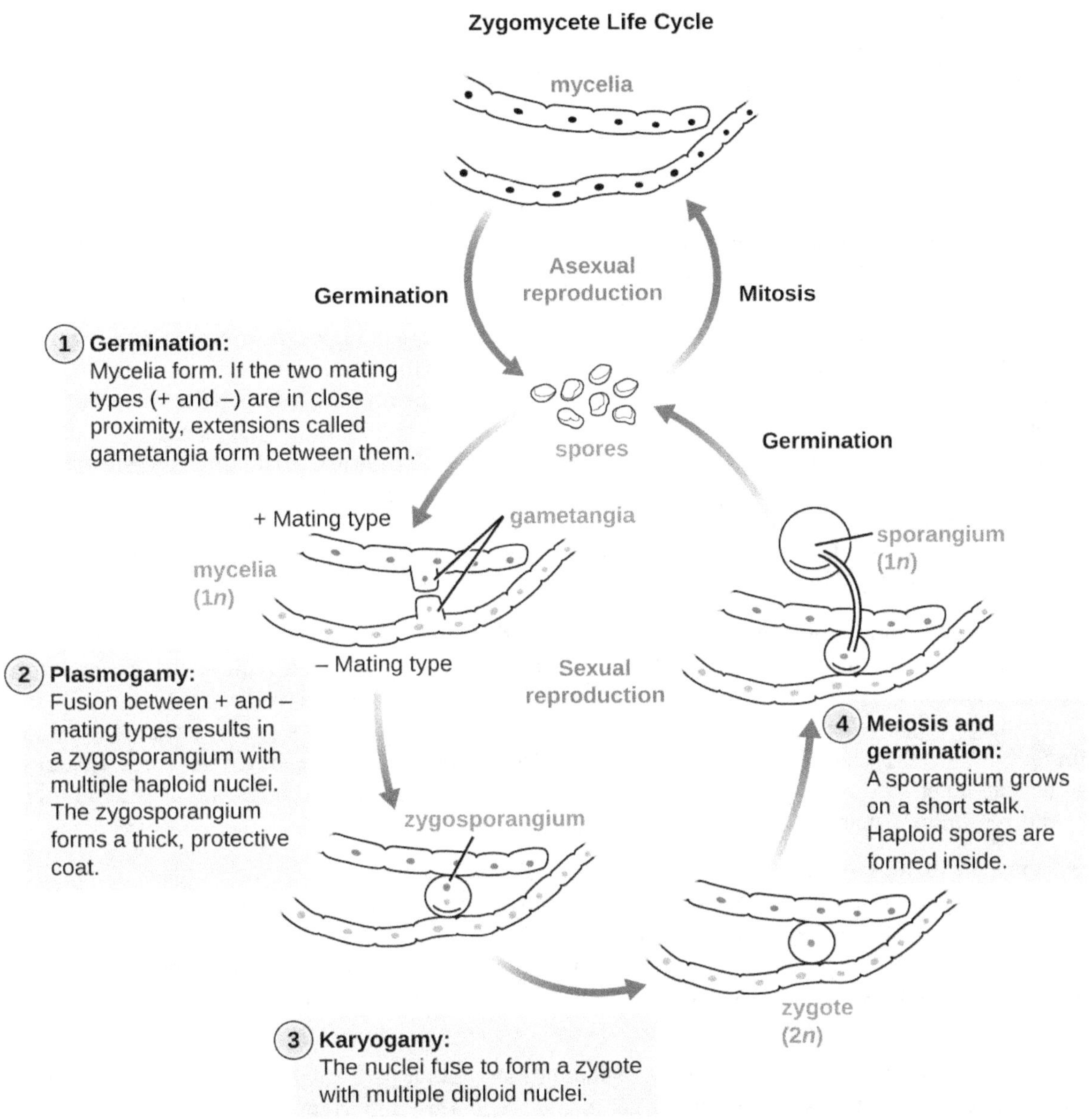

Zygomycete Life Cycle
mycelia
Germination
Asexual reproduction
Mitosis
1 Germination:
Mycelia form. If the two mating types (+ and –) are in close proximity, extensions called gametangia form between them.
spores
Germination
+ Mating type
gametangia
mycelia (1n)
sporangium (1n)
2 Plasmogamy:
Fusion between + and – mating types results in a zygosporangium with multiple haploid nuclei. The zygosporangium forms a thick, protective coat.
– Mating type
Sexual reproduction
4 Meiosis and germination:
A sporangium grows on a short stalk. Haploid spores are formed inside.
zygosporangium
zygote (2n)
3 Karyogamy:
The nuclei fuse to form a zygote with multiple diploid nuclei.

Figure 5.27 Zygomycetes have sexual and asexual life cycles. In the sexual life cycle, + and – mating types conjugate to form a zygosporangium.

sporangium
hyphae
sporangium
25 µm
(a)
(b)

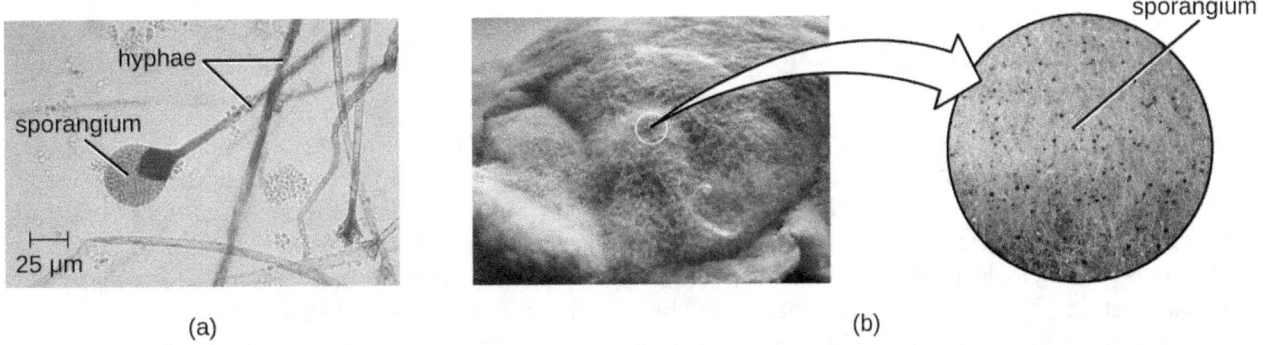

Figure 5.28 These images show asexually produced spores. (a) This brightfield micrograph shows the release of spores from a sporangium at the end of a hypha called a sporangiophore. The organism is a *Mucor* sp. fungus, a mold often found indoors. (b) Sporangia grow at the ends of stalks, which appear as the white fuzz seen on this bread mold, *Rhizopus stolonifer*. The tips of bread mold are the dark, spore-containing sporangia. (credit a: modification of work by Centers for Disease Control and Prevention; credit b right: modification of work by “Andrew”/Flickr)

CHECK YOUR UNDERSTANDING

- Is a dimorphic fungus a yeast or a mold? Explain.

Fungal Diversity

The fungi are very diverse, comprising seven major groups. Not all of the seven groups contain pathogens. Some of these groups are generally associated with plants and include plant pathogens. For example, Urediniomycetes and Ustilagomycetes include the plant rusts and smuts, respectively. These form reddish or dark masses, respectively, on plants as rusts (red) or smuts (dark). Some species have substantial economic impact because of their ability to reduce crop yields. Glomeromycota includes the mycorrhizal fungi, important symbionts with plant roots that can promote plant growth by acting like an extended root system. The Glomeromycota are obligate symbionts, meaning that they can only survive when associated with plant roots; the fungi receive carbohydrates from the plant and the plant benefits from the increased ability to take up nutrients and minerals from the soil. The Chytridiomycetes (chytrids) are small fungi, but are extremely ecologically important. Chytrids are generally aquatic and have flagellated, motile gametes; specific types are implicated in amphibian declines around the world. Because of their medical importance, we will focus on Zygomycota, Ascomycota, Basidiomycota, and Microsporidia. Figure 5.33 summarizes the characteristics of these medically important groups of fungi.

The Zygomycota (zygomycetes) are mainly saprophytes with coenocytic hyphae and haploid nuclei. They use sporangiospores for asexual reproduction. The group name comes from the **zygospores** that they use for sexual reproduction (Figure 5.27), which have hard walls formed from the fusion of reproductive cells from two individuals. Zygomycetes are important for food science and as crop pathogens. One example is *Rhizopus stolonifer* (Figure 5.28), an important bread mold that also causes rice seedling blight. *Mucor* is a genus of fungi that can potentially cause necrotizing infections in humans, although most species are intolerant of temperatures found in mammalian bodies (Figure 5.28).

The Ascomycota include fungi that are used as food (edible mushrooms, morels, and truffles), others that are common causes of food spoilage (bread molds and plant pathogens), and still others that are human pathogens. Ascomycota may have septate hyphae and cup-shaped fruiting bodies called **ascocarps**. Some genera of Ascomycota use sexually produced **ascospores** as well as asexual spores called **conidia**, but sexual phases have not been discovered or described for others. Some produce an **ascus** containing ascospores within an ascocarp (Figure 5.29).

Examples of the Ascomycota include several bread molds and minor pathogens, as well as species capable of causing more serious mycoses. Species in the genus *Aspergillus* are important causes of allergy and infection, and are useful in research and in the production of certain fermented alcoholic beverages such as Japanese sake. The fungus *Aspergillus flavus*, a contaminant of nuts and stored grains, produces an **aflatoxin** that is both a toxin and the most potent known natural carcinogen. *Neurospora crassa* is of particular use in genetics research because the spores produced by meiosis are kept inside the ascus in a row that reflects the cell divisions that produced them, giving a direct view of segregation and assortment of genes (Figure 5.30). *Penicillium* produces the antibiotic penicillin (Figure 5.29).

Many species of ascomycetes are medically important. A large number of species in the genera *Trichophyton*, *Microsporum*, and *Epidermophyton* are dermatophytes, pathogenic fungi capable of causing skin infections such as athlete's foot, jock itch, and ringworm. *Blastomyces dermatitidis* is a dimorphic fungus that can cause blastomycosis, a respiratory infection that, if left untreated, can become disseminated to other body sites, sometimes leading to death. Another important respiratory pathogen is the dimorphic fungus *Histoplasma capsulatum* (Figure 5.26), which is associated with birds and bats in the Ohio and Mississippi river valleys. *Coccidioides immitis* causes the serious lung disease Valley fever. *Candida albicans*, the most common cause of vaginal and other yeast infections, is also an ascomycete fungus; it is a part of the normal microbiota of the skin, intestine, genital tract, and ear (Figure 5.29). Ascomycetes also cause plant diseases, including ergot infections, Dutch elm disease, and powdery mildews.

Saccharomyces yeasts, including the baker's yeast *S. cerevisiae*, are unicellular ascomycetes with haploid and diploid stages (Figure 5.31). This and other *Saccharomyces* species are used for brewing beer.

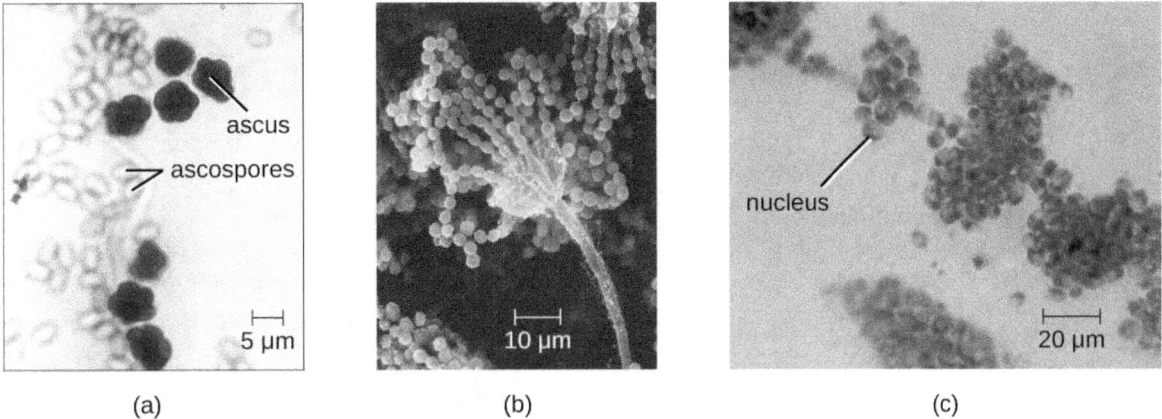

Figure 5.29 (a) This brightfield micrograph shows ascospores being released from asci in the fungus *Talaromyces flavus* var. *flavus*. (b) This electron micrograph shows the conidia (spores) borne on the conidiophore of *Aspergillus*, a type of toxic fungus found mostly in soil and plants. (c) This brightfield micrograph shows the yeast *Candida albicans*, the causative agent of candidiasis and thrush. (credit a, b, c: modification of work by Centers for Disease Control and Prevention)

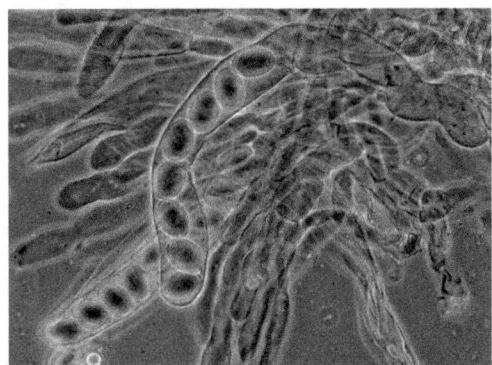

Figure 5.30 These ascospores, lined up within an ascus, are produced sexually. (credit: Peter G. Werner)

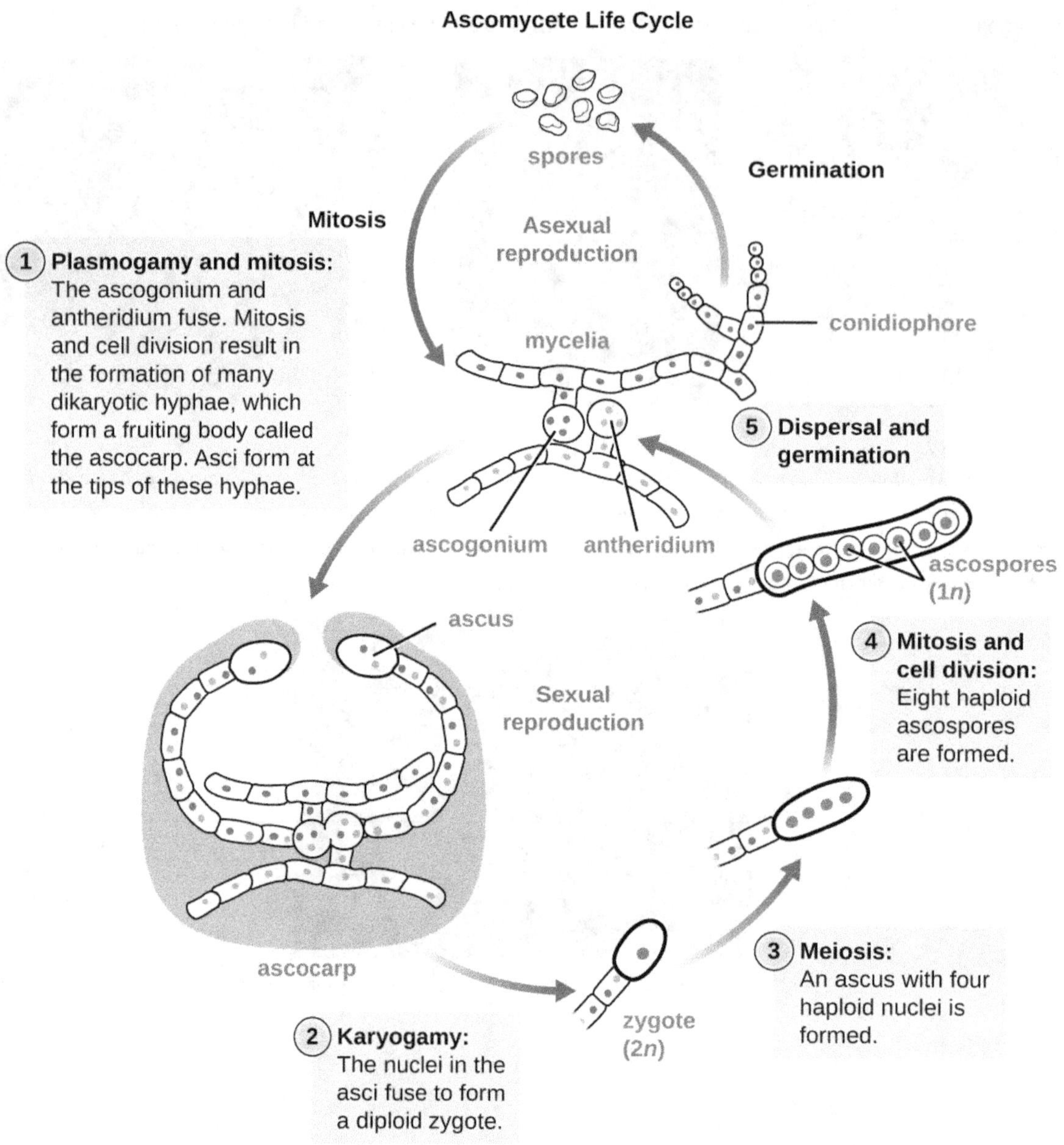
Ascomycete Life Cycle
spores
Germination
Mitosis
Asexual reproduction
1 Plasmogamy and mitosis: The ascogonium and antheridium fuse. Mitosis and cell division result in the formation of many dikaryotic hyphae, which form a fruiting body called the ascocarp. Asci form at the tips of these hyphae.
mycelia
conidiophore
5 Dispersal and germination
ascogonium
antheridium
ascospores (1n)
ascus
4 Mitosis and cell division: Eight haploid ascospores are formed.
Sexual reproduction
ascocarp
zygote (2n)
3 Meiosis: An ascus with four haploid nuclei is formed.
2 Karyogamy: The nuclei in the asci fuse to form a diploid zygote.

Figure 5.31 The life cycle of an ascomycete is characterized by the production of asci during the sexual phase. The haploid phase is the predominant phase of the life cycle. Whether spores are produced through sexual or asexual processes, they can germinate into haploid hyphae.

The Basidiomycota (basidiomycetes) are fungi that have **basidia** (club-shaped structures) that produce **basidiospores** (spores produced through budding) within fruiting bodies called **basidiocarps** (Figure 5.32). They are important as decomposers and as food. This group includes rusts, stinkhorns, puffballs, and mushrooms. Several species are of particular importance. *Cryptococcus neoformans*, a fungus commonly found as a yeast in the environment, can cause serious lung infections when inhaled by individuals with weakened immune systems. The edible meadow mushroom, *Agricus campestris*, is a basidiomycete, as is the poisonous mushroom *Amanita phalloides*, known as the death cap. The deadly toxins produced by *A. phalloides* have been used to study transcription.

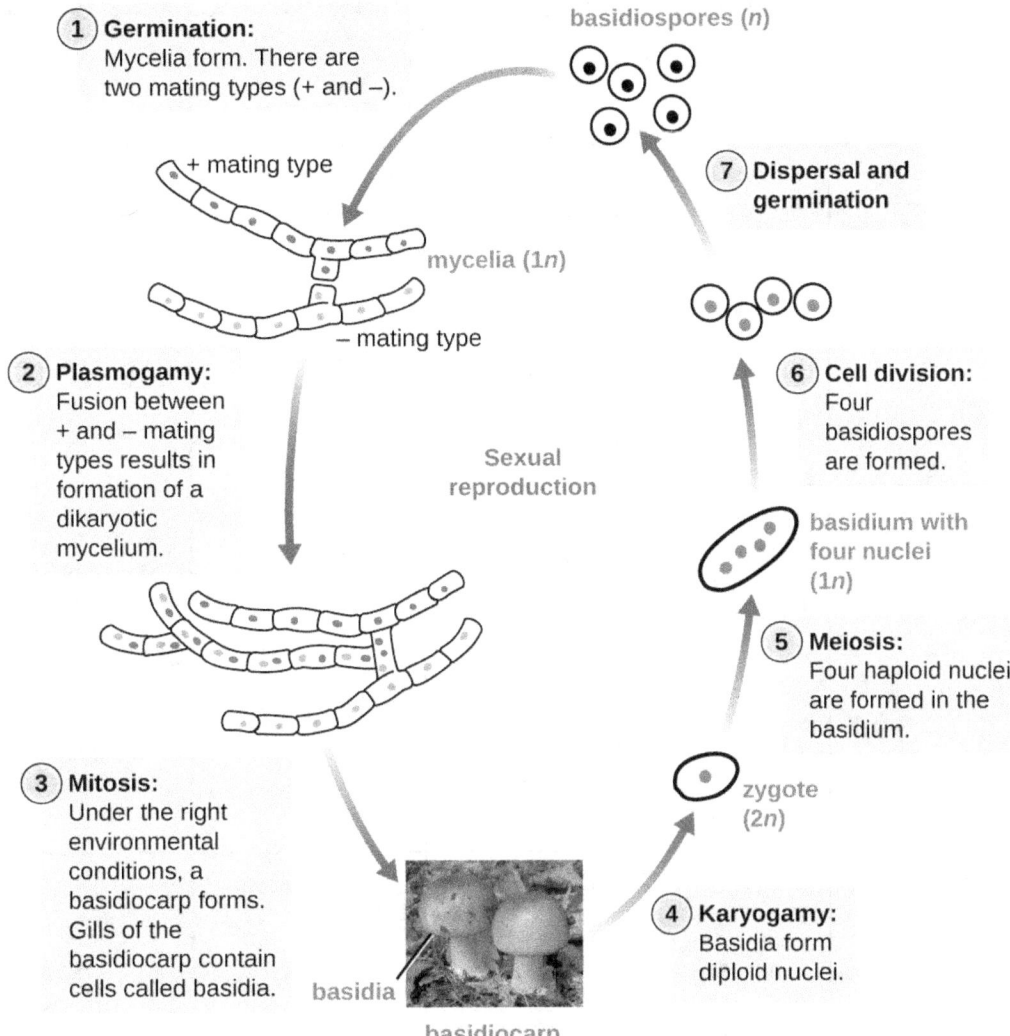

Figure 5.32 The life cycle of a basidiomycete alternates a haploid generation with a prolonged stage in which two nuclei (dikaryon) are present in the hyphae.

Finally, the **Microsporidia** are unicellular fungi that are obligate intracellular parasites. They lack mitochondria, peroxisomes, and centrioles, but their spores release a unique **polar tubule** that pierces the host cell membrane to allow the fungus to gain entry into the cell. A number of microsporidia are human pathogens, and infections with microsporidia are called microsporidiosis. One pathogenic species is *Enterocystozoan bieneusi*, which can cause symptoms such as diarrhea, cholecystitis (inflammation of the gall bladder), and in rare cases, respiratory illness.

Select Groups of Fungi				
Group	Characteristics	Examples	Medically Important Species	Image
Ascomycota	Septate hyphae Ascus with ascospores in ascocarp Conidiospores	Cup fungi Edible mushrooms Morels Truffles *Neurospora* *Penicillium*	*Aspergillus* spp. *Trichophyton* spp. *Microsporum* spp. *Epidermophyton* spp. *Blastomyces dermititidis* *Histoplasma capsulatum*	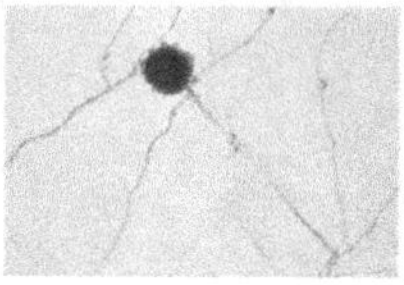 *Aspergillus niger*
Basidiomycota	Basidia produce basidiospores in a basidiocarp	Club fungi Rusts Stinkhorns Puffballs Mushrooms *Cryptococcus neoformans* *Amanita phalloides*	*Crytococcus neoformans*	 *Amanita phalloides*
Microsporidia	Lack mitochondria, perioxisomes, centrioles Spores produce a polar tube	*Enterocystozoan bieneusi*	*Enterocystozoan bieneusi*	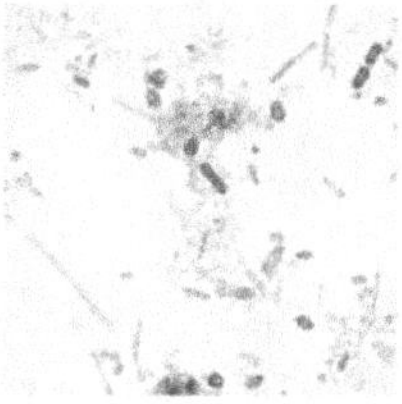 Microsporidia (unidentified)
Zygomycota	Mainly saprophytes Coenocytic hyphae Haploid nuclei Zygospores	*Rhizopus stolonifera*	*Mucor* spp.	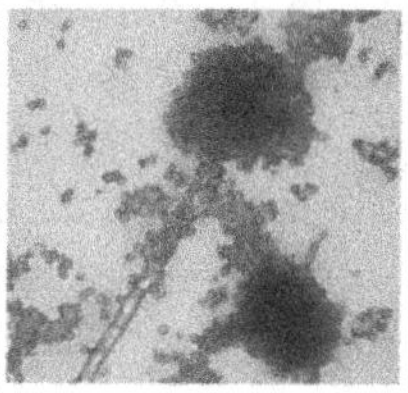 *Rhizopus* sp.

Figure 5.33 (credit "Ascomycota": modification of work by Dr. Lucille Georg, Centers for Disease Control and Prevention; credit "Microsporidia": modification of work by Centers for Disease Control and Prevention)

CHECK YOUR UNDERSTANDING

- Which group of fungi appears to be associated with the greatest number of human diseases?

 **MICRO CONNECTIONS**

Eukaryotic Pathogens in Eukaryotic Hosts

When we think about antimicrobial medications, antibiotics such as penicillin often come to mind. Penicillin and related antibiotics interfere with the synthesis of peptidoglycan cell walls, which effectively targets bacterial cells. These antibiotics are useful because humans (like all eukaryotes) do not have peptidoglycan cell walls.

Developing medications that are effective against eukaryotic cells but not harmful to human cells is more difficult. Despite huge morphological differences, the cells of humans, fungi, and protists are similar in terms of their ribosomes, cytoskeletons, and cell membranes. As a result, it is more challenging to develop medications that target protozoans and fungi in the same way that antibiotics target prokaryotes.

Fungicides have relatively limited modes of action. Because fungi have ergosterols (instead of cholesterol) in their cell membranes, the different enzymes involved in sterol production can be a target of some medications. The azole and morpholine fungicides interfere with the synthesis of membrane sterols. These are used widely in agriculture (fenpropimorph) and clinically (e.g., miconazole). Some antifungal medications target the chitin cell walls of fungi. Despite the success of these compounds in targeting fungi, antifungal medications for systemic infections still tend to have more toxic side effects than antibiotics for bacteria.

Clinical Focus

Part 3

Sarah is relieved the ringworm is not an actual worm, but wants to know what it really is. The physician explains that ringworm is a fungus. He tells her that she will not see mushrooms popping out of her skin, because this fungus is more like the invisible part of a mushroom that hides in the soil. He reassures her that they are going to get the fungus out of her too.

The doctor cleans and then carefully scrapes the lesion to place a specimen on a slide. By looking at it under a microscope, the physician is able to confirm that a fungal infection is responsible for Sarah's lesion. In Figure 5.34, it is possible to see macro- and microconidia in *Trichophyton rubrum*. Cell walls are also visible. Even if the pathogen resembled a helminth under the microscope, the presence of cell walls would rule out the possibility because animal cells lack cell walls.

The doctor prescribes an antifungal cream for Sarah's mother to apply to the ringworm. Sarah's mother asks, "What should we do if it doesn't go away?"

- Can all forms of ringworm be treated with the same antifungal medication?

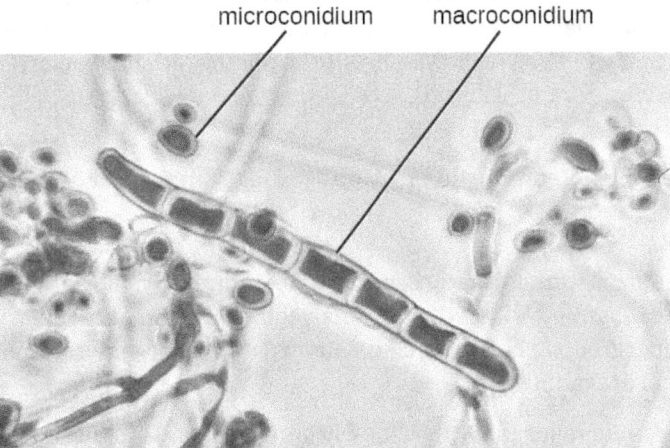

Figure 5.34 This micrograph shows hyphae (macroconidium) and microconidia of *Trichophyton rubrum*, a dermatophyte responsible

> for fungal infections of the skin. (credit: modification of work by Centers for Disease Control and Prevention)
>
> *Jump to the next Clinical Focus box. Go back to the previous Clinical Focus box.*

5.4 Algae

Learning Objectives

By the end of this section, you will be able to:
- Explain why algae are included within the discipline of microbiology
- Describe the unique characteristics of algae
- Identify examples of toxin-producing algae
- Compare the major groups of algae in this chapter, and give examples of each
- Classify algal organisms according to major groups

The **algae** are autotrophic protists that can be unicellular or multicellular. These organisms are found in the supergroups Chromalveolata (dinoflagellates, diatoms, golden algae, and brown algae) and Archaeplastida (red algae and green algae). They are important ecologically and environmentally because they are responsible for the production of approximately 70% of the oxygen and organic matter in aquatic environments. Some types of algae, even those that are microscopic, are regularly eaten by humans and other animals. Additionally, algae are the source for **agar**, agarose, and **carrageenan**, solidifying agents used in laboratories and in food production. Although algae are typically not pathogenic, some produce toxins. Harmful **algal blooms**, which occur when algae grow quickly and produce dense populations, can produce high concentrations of toxins that impair liver and nervous-system function in aquatic animals and humans.

Like protozoans, algae often have complex cell structures. For instance, algal cells can have one or more chloroplasts that contain structures called **pyrenoids** to synthesize and store starch. The chloroplasts themselves differ in their number of membranes, indicative of secondary or rare tertiary endosymbiotic events. Primary chloroplasts have two membranes—one from the original cyanobacteria that the ancestral eukaryotic cell engulfed, and one from the plasma membrane of the engulfing cell. Chloroplasts in some lineages appear to have resulted from secondary endosymbiosis, in which another cell engulfed a green or red algal cell that already had a primary chloroplast within it. The engulfing cell destroyed everything except the chloroplast and possibly the cell membrane of its original cell, leaving three or four membranes around the chloroplast. Different algal groups have different pigments, which are reflected in common names such as red algae, brown algae, and green algae.

Some algae, the seaweeds, are macroscopic and may be confused with plants. Seaweeds can be red, brown, or green, depending on their photosynthetic pigments. Green algae, in particular, share some important similarities with land plants; however, there are also important distinctions. For example, seaweeds do not have true tissues or organs like plants do. Additionally, seaweeds do not have a waxy cuticle to prevent desiccation. Algae can also be confused with cyanobacteria, photosynthetic bacteria that bear a resemblance to algae; however, cyanobacteria are prokaryotes (see Nonproteobacteria Gram-negative Bacteria and Phototrophic Bacteria).

Algae have a variety of life cycles. Reproduction may be asexual by mitosis or sexual using gametes.

Algal Diversity

Although the algae and protozoa were formerly separated taxonomically, they are now mixed into supergroups. The algae are classified within the Chromalveolata and the Archaeplastida. Although the Euglenozoa (within the supergroup Excavata) include photosynthetic organisms, these are not considered algae because they feed and are motile.

The dinoflagellates and stramenopiles fall within the Chromalveolata. The **dinoflagellates** are mostly marine organisms and are an important component of plankton. They have a variety of nutritional types and may be phototrophic, heterotrophic, or mixotrophic. Those that are photosynthetic use chlorophyll a, chlorophyll c_2, and other photosynthetic pigments (Figure 5.35). They generally have two flagella, causing them to whirl (in fact, the name dinoflagellate comes from the Greek word for "whirl": *dini*). Some have cellulose plates forming a hard outer covering, or **theca**, as armor. Additionally, some dinoflagellates produce neurotoxins that can cause paralysis in humans or fish. Exposure can occur through contact with water containing the dinoflagellate toxins or by feeding on organisms that have eaten dinoflagellates.

When a population of dinoflagellates becomes particularly dense, a **red tide** (a type of harmful algal bloom) can occur. Red tides cause harm to marine life and to humans who consume contaminated marine life. Major toxin producers include *Gonyaulax* and *Alexandrium,* both of which cause paralytic shellfish poisoning. Another species, *Pfiesteria piscicida*, is known as a fish killer because, at certain parts of its life cycle, it can produce toxins harmful to fish and it appears to be responsible for a suite of symptoms, including memory loss and confusion, in humans exposed to water containing the species.

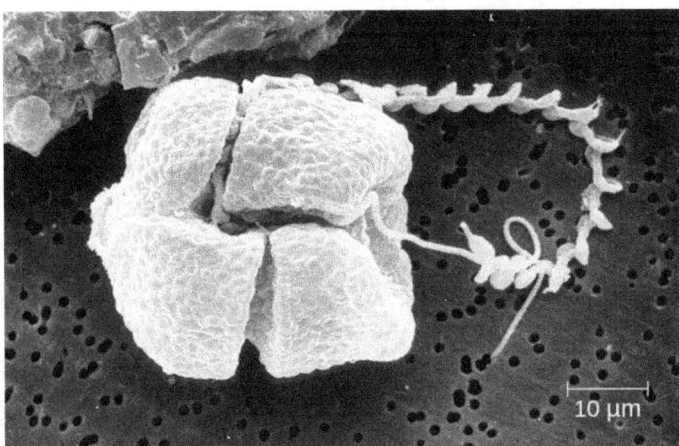

Figure 5.35   The dinoflagellates exhibit great diversity in shape. Many are encased in cellulose armor and have two flagella that fit in grooves between the plates. Movement of these two perpendicular flagella causes a spinning motion. (credit: modification of work by CSIRO)

The **stramenopiles** include the golden algae (Chrysophyta), the brown algae (Phaeophyta), and the **diatoms** (Bacillariophyta). Stramenopiles have chlorophyll *a*, chlorophyll c_1/c_2, and fucoxanthin as photosynthetic pigments. Their storage carbohydrate is chrysolaminarin. While some lack cell walls, others have scales. Diatoms have **frustules**, which are outer cell walls of crystallized silica; their fossilized remains are used to produce diatomaceous earth, which has a range of uses such as filtration and insulation. Additionally, diatoms can reproduce sexually and asexually, and the male gametes of centric diatoms have flagella providing directed movement to seek female gametes for sexual reproduction.

Brown algae (Phaeophyta) are multicellular marine seaweeds. Some can be extremely large, such as the giant kelp (*Laminaria*). They have leaf-like blades, stalks, and structures called holdfasts that are used to attach to substrate. However, these are not true leaves, stems, or roots (Figure 5.36). Their photosynthetic pigments are chlorophyll *a*, chlorophyll *c*, β-carotene, and fucoxanthine. They use laminarin as a storage carbohydrate.

The Archaeplastids include the green algae (Chlorophyta), the red algae (Rhodophyta), another group of green algae (Charophyta), and the land plants. The Charaphyta are the most similar to land plants because they share a mechanism of cell division and an important biochemical pathway, among other traits that the other groups do not have. Like land plants, the Charophyta and Chlorophyta have chlorophyll *a* and chlorophyll *b* as photosynthetic pigments, cellulose cell walls, and starch as a carbohydrate storage molecule. *Chlamydomonas* is a green alga that has a single large chloroplast, two flagella, and a stigma (eyespot); it is important in molecular biology research (Figure 5.37).

Chlorella is a nonmotile, large, unicellular alga, and *Acetabularia* is an even larger unicellular green alga. The size of these organisms challenges the idea that all cells are small, and they have been used in genetics research since Joachim Hämmerling (1901–1980) began to work with them in 1943. *Volvox* is a colonial, unicellular alga (Figure 5.37). A larger, multicellular green alga is *Ulva*, also known as the sea lettuce because of its large, edible, green blades. The range of life forms within the Chlorophyta—from unicellular to various levels of coloniality to multicellular forms—has been a useful research model for understanding the evolution of multicellularity. The red algae are mainly multicellular but include some unicellular forms. They have rigid cell walls containing agar or carrageenan, which are useful as food solidifying agents and as a solidifier added to growth media for microbes.

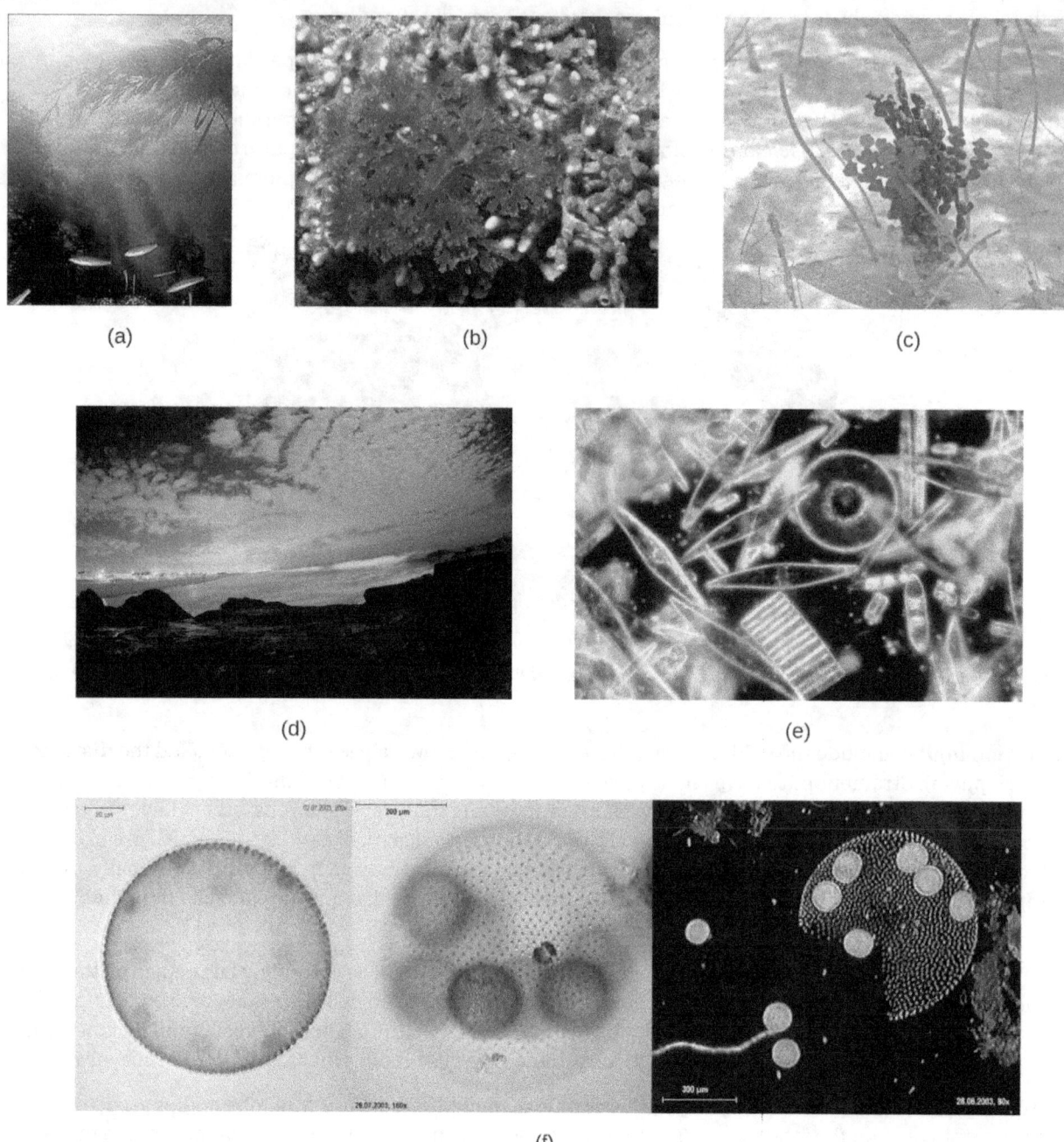

Figure 5.36 (a) These large multicellular kelps are members of the brown algae. Note the "leaves" and "stems" that make them appear similar to green plants. (b) This is a species of red algae that is also multicellular. (c) The green alga *Halimeda incrassata*, shown here growing on the sea floor in shallow water, appears to have plant-like structures, but is not a true plant. (d) Bioluminesence, visible in the cresting wave in this picture, is a phenomenon of certain dinoflagellates. (e) Diatoms (pictured in this micrograph) produce silicaceous tests (skeletons) that form diatomaceous earths. (f) Colonial green algae, like volvox in these three micrographs, exhibit simple cooperative associations of cells. (credit a, e: modification of work by NOAA; credit b: modification of work by Ed Bierman; credit c: modification of work by James St. John; credit d: modification of work by "catalano82"/Flickr; credit f: modification of work by Dr. Ralf Wagner)

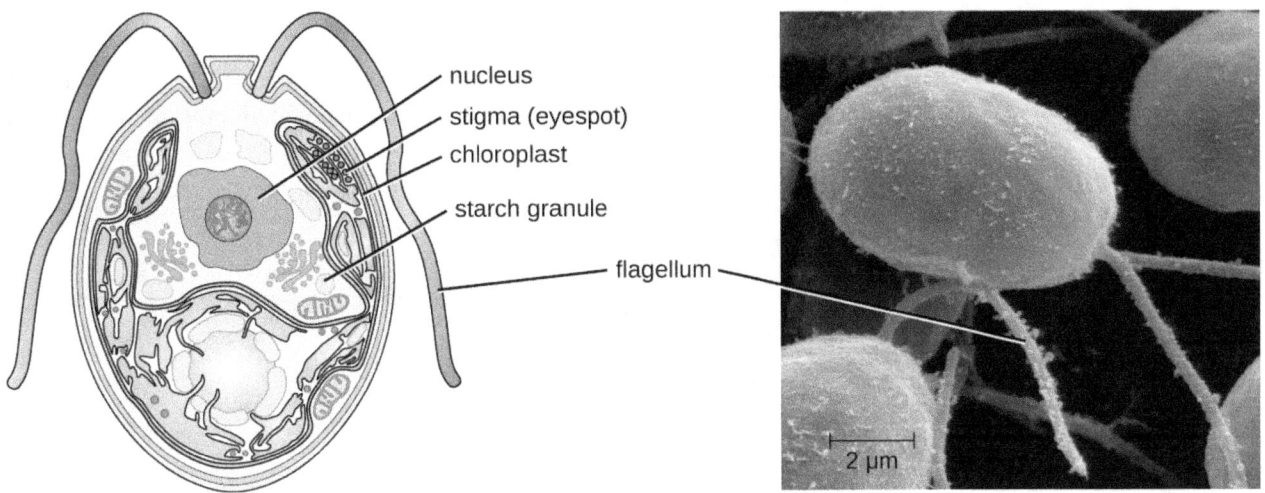

Figure 5.37 *Chlamydomonas* is a unicellular green alga.

CHECK YOUR UNDERSTANDING

- Which groups of algae are associated with harmful algal blooms?

5.5 Lichens

Learning Objectives

By the end of this section, you will be able to:
- Explain why lichens are included in the study of microbiology
- Describe the unique characteristics of a lichen and the role of each partner in the symbiotic relationship of a lichen
- Describe ways in which lichens are beneficial to the environment

No one has to worry about getting sick from a lichen infection, but lichens are interesting from a microbiological perspective and they are an important component of most terrestrial ecosystems. Lichens provide opportunities for study of close relationships between unrelated microorganisms. Lichens contribute to soil production by breaking down rock, and they are early colonizers in soilless environments such as lava flows. The cyanobacteria in some lichens can fix nitrogen and act as a nitrogen source in some environments. Lichens are also important soil stabilizers in some desert environments and they are an important winter food source for caribou. Finally, lichens produce compounds that have antibacterial effects, and further research may discover compounds that are medically useful to humans.

Characteristics

A **lichen** is a combination of two organisms, a green alga or cyanobacterium and fungus, living in a symbiotic relationship. Whereas algae normally grow only in aquatic or extremely moist environments, lichens can potentially be found on almost any surface (especially rocks) or as **epiphytes** (meaning that they grow on other plants).

In some ways, the symbiotic relationship between lichens and algae seems like a mutualism (a relationship in which both organisms benefit). The fungus can obtain photosynthates from the algae or cyanobacterium and the algae or cyanobacterium can grow in a drier environment than it could otherwise tolerate. However, most scientists consider this symbiotic relationship to be a controlled parasitism (a relationship in which one organism benefits and the other is harmed) because the photosynthetic organism grows less well than it would without the fungus. It is important to note that such symbiotic interactions fall along a continuum between conflict and cooperation.

Lichens are slow growing and can live for centuries. They have been used in foods and to extract chemicals as dyes or antimicrobial substances. Some are very sensitive to pollution and have been used as environmental indicators.

Lichens have a body called a thallus, an outer, tightly packed fungal layer called a **cortex**, and an inner, loosely packed fungal layer called a **medulla** (Figure 5.38). Lichens use hyphal bundles called **rhizines** to attach to the substrate.

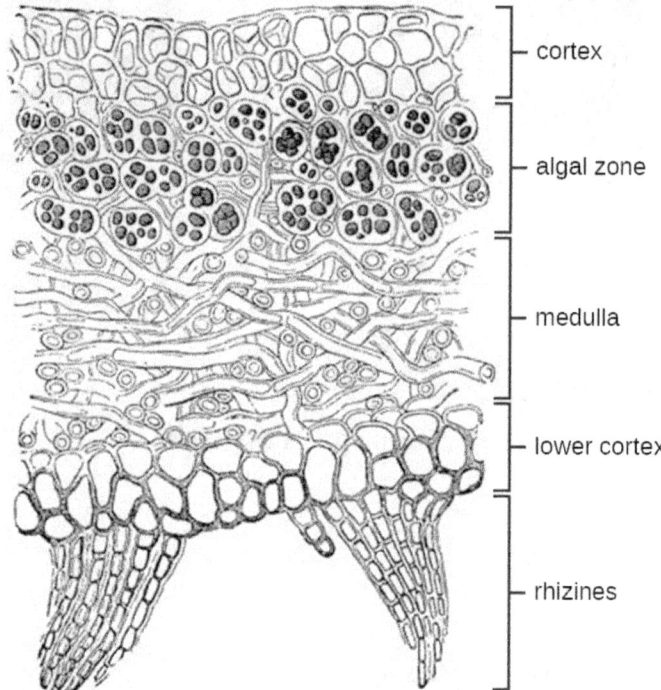

Figure 5.38 This cross-section of a lichen thallus shows its various components. The upper cortex of fungal hyphae provides protection. Photosynthesis occurs in the algal zone. The medulla consists of fungal hyphae. The lower cortex also provides protection. The rhizines anchor the thallus to the substrate.

Lichen Diversity

Lichens are classified as fungi and the fungal partners belong to the Ascomycota and Basidiomycota. Lichens can also be grouped into types based on their morphology. There are three major types of lichens, although other types exist as well. Lichens that are tightly attached to the substrate, giving them a crusty appearance, are called **crustose lichens**. Those that have leaf-like lobes are **foliose lichens**; they may only be attached at one point in the growth form, and they also have a second cortex below the medulla. Finally, **fruticose lichens** have rounded structures and an overall branched appearance. Figure 5.39 shows an example of each of the forms of lichens.

Figure 5.39 Examples of the three types of lichens are shown here. (a) This is a crustose lichen found mostly on marine rocks, *Caloplaca*

marina. (b) This is a foliose lichen, *Flavoparmelia caperata*. (c) This is a fruticose lichen, *Letharia vulpina*, which is sufficiently poisonous that it was once used to make arrowheads. (credit b, c: modification of work by Jason Hollinger)

✓ CHECK YOUR UNDERSTANDING

- What types of organisms are found in lichens?
- What are the three growth forms of lichens?

> Clinical Focus
>
> **Resolution**
>
> Sarah's mother asks the doctor what she should do if the cream prescribed for Sarah's ringworm does not work. The doctor explains that ringworm is a general term for a condition caused by multiple species. The first step is to take a scraping for examination under the microscope, which the doctor has already done. He explains that he has identified the infection as a fungus, and that the antifungal cream works against the most common fungi associated with ringworm. However, the cream may not work against some species of fungus. If the cream is not working after a couple of weeks, Sarah should come in for another visit, at which time the doctor will take steps to identify the species of the fungus.
>
> Positive identification of dermatophytes requires culturing. For this purpose, Sabouraud's agar may be used. In the case of Sarah's infection, which cleared up within 2 weeks of treatment, the culture would have a granular texture and would appear pale pink on top and red underneath. These features suggest that the fungus is *Trichophyton rubrum*, a common cause of ringworm.
>
> Go back to the previous Clinical Focus box.

SUMMARY

5.1 Unicellular Eukaryotic Parasites

- **Protists** are a diverse, **polyphyletic** group of eukaryotic organisms.
- Protists may be unicellular or multicellular. They vary in how they get their nutrition, morphology, method of locomotion, and mode of reproduction.
- Important structures of protists include **contractile vacuoles**, cilia, flagella, **pellicles**, and pseudopodia; some lack organelles such as mitochondria.
- Taxonomy of protists is changing rapidly as relationships are reassessed using newer techniques.
- The protists include important pathogens and parasites.

5.2 Parasitic Helminths

- Helminth parasites are included within the study of microbiology because they are often identified by looking for microscopic eggs and larvae.
- The two major groups of helminth parasites are the roundworms (Nematoda) and the flatworms (Platyhelminthes).
- Nematodes are common intestinal parasites often transmitted through undercooked foods, although they are also found in other environments.
- Platyhelminths include **tapeworms** and **flukes**, which are often transmitted through undercooked meat.

5.3 Fungi

- The fungi include diverse saprotrophic eukaryotic organisms with chitin cell walls
- Fungi can be unicellular or multicellular; some (like yeast) and fungal spores are microscopic, whereas some are large and conspicuous
- Reproductive types are important in distinguishing fungal groups
- Medically important species exist in the four fungal groups Zygomycota, Ascomycota, Basidiomycota, and Microsporidia
- Members of Zygomycota, Ascomycota, and Basidiomycota produce deadly toxins
- Important differences in fungal cells, such as ergosterols in fungal membranes, can be targets for antifungal medications, but similarities between human and fungal cells make it difficult to find targets for medications and these medications often have toxic adverse effects

5.4 Algae

- Algae are a diverse group of photosynthetic eukaryotic protists
- Algae may be unicellular or multicellular
- Large, multicellular algae are called seaweeds but are not plants and lack plant-like tissues and organs
- Although algae have little pathogenicity, they may be associated with toxic **algal blooms** that can and aquatic wildlife and contaminate seafood with toxins that cause paralysis
- Algae are important for producing **agar**, which is used as a solidifying agent in microbiological media, and **carrageenan**, which is used as a solidifying agent

5.5 Lichens

- **Lichens** are a symbiotic association between a fungus and an algae or a cyanobacterium
- The symbiotic association found in lichens is currently considered to be a controlled **parasitism**, in which the fungus benefits and the algae or cyanobacterium is harmed
- Lichens are slow growing and can live for centuries in a variety of habitats
- Lichens are environmentally important, helping to create soil, providing food, and acting as indicators of air pollution

REVIEW QUESTIONS
Multiple Choice

1. Which genus includes the causative agent for malaria?
   A. *Euglena*
   B. *Paramecium*
   C. *Plasmodium*
   D. *Trypanosoma*

2. Which protist is a concern because of its ability to contaminate water supplies and cause diarrheal illness?
   A. *Plasmodium vivax*
   B. *Toxoplasma gondii*
   C. *Giardia lamblia*
   D. *Trichomonas vaginalis*

3. A fluke is classified within which of the following?
   A. Nematoda
   B. Rotifera
   C. Platyhelminthes
   D. Annelida

4. A nonsegmented worm is found during a routine colonoscopy of an individual who reported having abdominal cramps, nausea, and vomiting. This worm is likely which of the following?
   A. nematode
   B. fluke
   C. trematode
   D. annelid

5. A segmented worm has male and female reproductive organs in each segment. Some use hooks to attach to the intestinal wall. Which type of worm is this?
   A. fluke
   B. nematode
   C. cestode
   D. annelid

6. Mushrooms are a type of which of the following?
   A. conidia
   B. ascus
   C. polar tubule
   D. basidiocarp

7. Which of the following is the most common cause of human yeast infections?
   A. *Candida albicans*
   B. *Blastomyces dermatitidis*
   C. *Cryptococcus neoformans*
   D. *Aspergillus fumigatus*

8. Which of the following is an ascomycete fungus associated with bat droppings that can cause a respiratory infection if inhaled?
   A. *Candida albicans*
   B. *Histoplasma capsulatum*
   C. *Rhizopus stolonifera*
   D. *Trichophyton rubrum*

9. Which polysaccharide found in red algal cell walls is a useful solidifying agent?
   A. chitin
   B. cellulose
   C. phycoerythrin
   D. agar

10. Which is the term for the hard outer covering of some dinoflagellates?
    A. theca
    B. thallus
    C. mycelium
    D. shell

11. Which protists are associated with red tides?
    A. red algae
    B. brown algae
    C. dinoflagellates
    D. green algae

12. You encounter a lichen with leafy structures. Which term describes this lichen?
    A. crustose
    B. foliose
    C. fruticose
    D. agarose

13. Which of the following is the term for the outer layer of a lichen?
    A. the cortex
    B. the medulla
    C. the thallus
    D. the theca

14. The fungus in a lichen is which of the following?
    A. a basidiomycete
    B. an ascomycete
    C. a zygomycete
    D. an apicomplexan

Fill in the Blank

15. The plasma membrane of a protist is called the _____.
16. Animals belong to the same supergroup as the kingdom _____.
17. Flukes are in class _____.
18. A species of worm in which there are distinct male and female individuals is described as _____.
19. Nonseptate hyphae are also called _____.
20. Unicellular fungi are called _____.
21. Some fungi have proven medically useful because they can be used to produce _____.
22. Structures in chloroplasts used to synthesize and store starch are called _____.
23. Algae with chloroplasts with three or four membranes are a result of _____ _____.

Short Answer

24. What are kinetoplastids?
25. Aside from a risk of birth defects, what other effect might a toxoplasmosis infection have?
26. What is the function of the ciliate macronucleus?
27. What is the best defense against tapeworm infection?
28. Which genera of fungi are common dermatophytes (fungi that cause skin infections)?
29. What is a dikaryotic cell?
30. What is a distinctive feature of diatoms?
31. Why are algae not considered parasitic?
32. Which groups contain the multicellular algae?
33. What are three ways that lichens are environmentally valuable?

Critical Thinking

34. The protist shown has which of the following?
    A. pseudopodia
    B. flagella
    C. a shell
    D. cilia

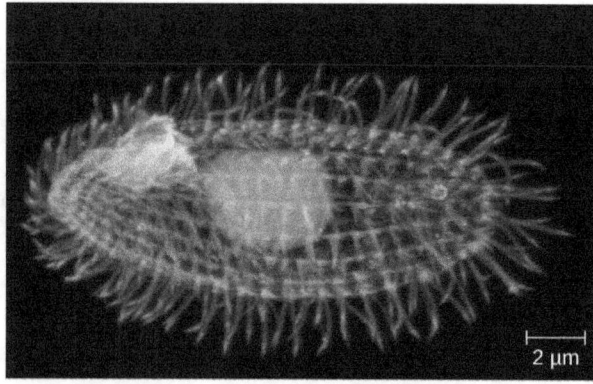

Figure 5.40 (credit: modification of work by Richard Robinson)

35. Protist taxonomy has changed greatly in recent years as relationships have been re-examined using newer approaches. How do newer approaches differ from older approaches?
36. What characteristics might make you think a protist could be pathogenic? Are certain nutritional characteristics, methods of locomotion, or morphological differences likely to be associated with the ability to cause disease?
37. Given the life cycle of the *Schistosoma* parasite, suggest a method of prevention of the disease.
38. Which of the drawings shows septate hyphae?

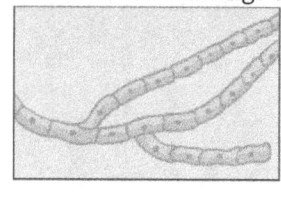 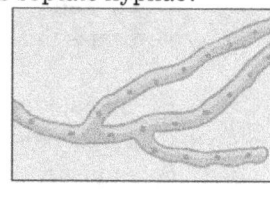

A   B

39. Explain the benefit of research into the pathways involved in the synthesis of chitin in fungi.

CHAPTER 6
Acellular Pathogens

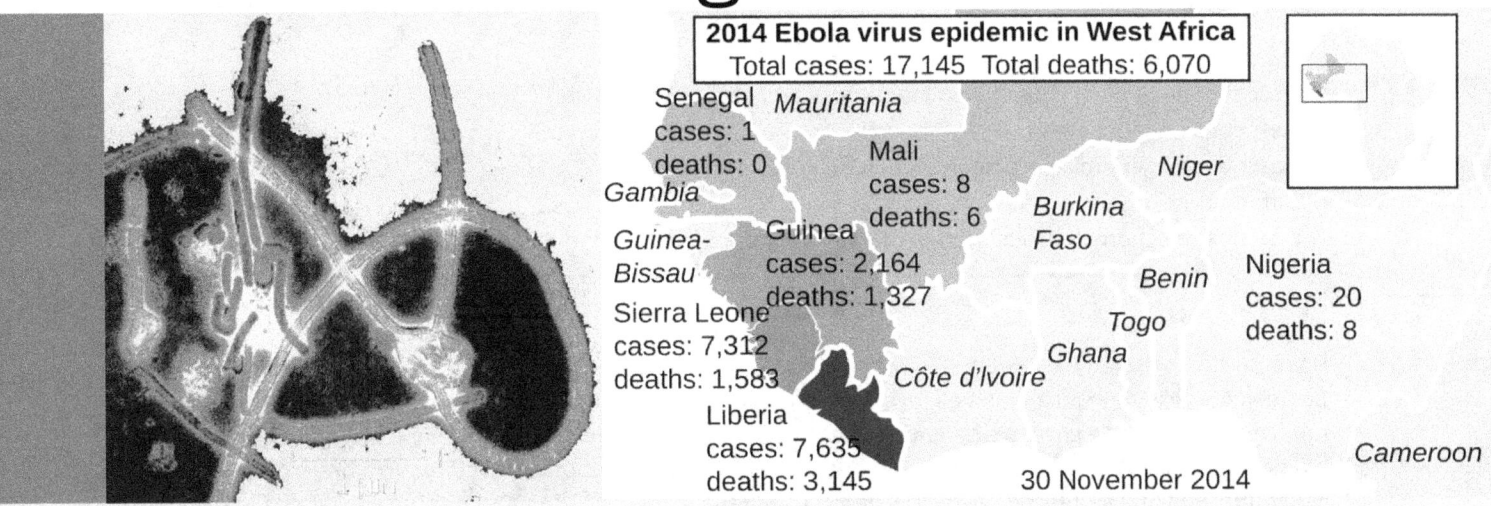

Figure 6.1 The year 2014 saw the first large-scale outbreak of Ebola virus (electron micrograph, left) in human populations in West Africa (right). Such epidemics are now widely reported and documented, but viral epidemics are sure to have plagued human populations since the origin of our species. (credit left: modification of work by Thomas W. Geisbert)

Chapter Outline

6.1 Viruses

6.2 The Viral Life Cycle

6.3 Isolation, Culture, and Identification of Viruses

6.4 Viroids, Virusoids, and Prions

INTRODUCTION Public health measures in the developed world have dramatically reduced mortality from viral epidemics. But when epidemics do occur, they can spread quickly with global air travel. In 2009, an outbreak of H1N1 influenza spread across various continents. In early 2014, cases of Ebola in Guinea led to a massive epidemic in western Africa. This included the case of an infected man who traveled to the United States, sparking fears the epidemic might spread beyond Africa.

Until the late 1930s and the advent of the electron microscope, no one had seen a virus. Yet treatments for preventing or curing viral infections were used and developed long before that. Historical records suggest that by the 17th century, and perhaps earlier, inoculation (also known as variolation) was being used to prevent the viral disease smallpox in various parts of the world. By the late 18th century, Englishman Edward Jenner was inoculating patients with cowpox to prevent smallpox, a technique he coined *vaccination*.[1]

Today, the structure and genetics of viruses are well defined, yet new discoveries continue to reveal their complexities. In this chapter, we will learn about the structure, classification, and cultivation of viruses, and how they impact their hosts. In addition, we will learn about other infective particles such as viroids and

---

1 S. Riedel "Edward Jenner and the History of Smallpox and Vaccination." *Baylor University Medical Center Proceedings* 18, no. 1 (January 2005): 21–25.

prions.

6.1 Viruses

Learning Objectives

By the end of this section, you will be able to:
- Describe the general characteristics of viruses as pathogens
- Describe viral genomes
- Describe the general characteristics of viral life cycles
- Differentiate among bacteriophages, plant viruses, and animal viruses
- Describe the characteristics used to identify viruses as obligate intracellular parasites

---

Clinical Focus

Part 1

David, a 45-year-old journalist, has just returned to the U.S. from travels in Russia, China, and Africa. He is not feeling well, so he goes to his general practitioner complaining of weakness in his arms and legs, fever, headache, noticeable agitation, and minor discomfort. He thinks it may be related to a dog bite he suffered while interviewing a Chinese farmer. He is experiencing some prickling and itching sensations at the site of the bite wound, but he tells the doctor that the dog seemed healthy and that he had not been concerned until now. The doctor ordered a culture and sensitivity test to rule out bacterial infection of the wound, and the results came back negative for any possible pathogenic bacteria.

- Based on this information, what additional tests should be performed on the patient?
- What type of treatment should the doctor recommend?

Jump to the next Clinical Focus box.

Despite their small size, which prevented them from being seen with light microscopes, the discovery of a filterable component smaller than a bacterium that causes tobacco mosaic disease (TMD) dates back to 1892.[2] At that time, Dmitri Ivanovski, a Russian botanist, discovered the source of TMD by using a porcelain filtering device first invented by Charles Chamberland and Louis Pasteur in Paris in 1884. Porcelain Chamberland filters have a pore size of 0.1 μm, which is small enough to remove all bacteria ≥0.2 μm from any liquids passed through the device. An extract obtained from TMD-infected tobacco plants was made to determine the cause of the disease. Initially, the source of the disease was thought to be bacterial. It was surprising to everyone when Ivanovski, using a Chamberland filter, found that the cause of TMD was not removed after passing the extract through the porcelain filter. So if a bacterium was not the cause of TMD, what could be causing the disease? Ivanovski concluded the cause of TMD must be an extremely small bacterium or bacterial spore. Other scientists, including Martinus Beijerinck, continued investigating the cause of TMD. It was Beijerinck, in 1899, who eventually concluded the causative agent was not a bacterium but, instead, possibly a chemical, like a biological poison we would describe today as a toxin. As a result, the word *virus*, Latin for poison, was used to describe the cause of TMD a few years after Ivanovski's initial discovery. Even though he was not able to see the virus that caused TMD, and did not realize the cause was not a bacterium, Ivanovski is credited as the original discoverer of viruses and a founder of the field of virology.

Today, we can see viruses using electron microscopes (Figure 6.2) and we know much more about them. Viruses are distinct biological entities; however, their evolutionary origin is still a matter of speculation. In terms of taxonomy, they are not included in the tree of life because they are **acellular** (not consisting of cells). In order to survive and reproduce, viruses must infect a cellular host, making them obligate intracellular parasites. The genome of a virus enters a host cell and directs the production of the viral components, proteins and nucleic acids, needed to form new virus particles called **virions**. New virions are made in the host cell by

---

2 H. Lecoq. "[Discovery of the First Virus, the Tobacco Mosaic Virus: 1892 or 1898?]." *Comptes Rendus de l'Academie des Sciences – Serie III – Sciences de la Vie* 324, no. 10 (2001): 929–933.

assembly of viral components. The new virions transport the viral genome to another host cell to carry out another round of infection. Table 6.1 summarizes the properties of viruses.

Characteristics of Viruses
Infectious, acellular pathogens
Obligate intracellular parasites with host and cell-type specificity
DNA or RNA genome (never both)
Genome is surrounded by a protein capsid and, in some cases, a phospholipid membrane studded with viral glycoproteins
Lack genes for many products needed for successful reproduction, requiring exploitation of host-cell genomes to reproduce

Table 6.1

(a)

(b)

Figure 6.2 (a) Tobacco mosaic virus (TMV) viewed with transmission electron microscope. (b) Plants infected with tobacco mosaic disease (TMD), caused by TMV. (credit a: modification of work by USDA Agricultural Research Service—scale-bar data from Matt Russell; credit b: modification of work by USDA Forest Service, Department of Plant Pathology Archive North Carolina State University)

CHECK YOUR UNDERSTANDING

- Why was the first virus investigated mistaken for a toxin?

Hosts and Viral Transmission

Viruses can infect every type of host cell, including those of plants, animals, fungi, protists, bacteria, and archaea. Most viruses will only be able to infect the cells of one or a few species of organism. This is called the **host range**. However, having a wide host range is not common and viruses will typically only infect specific hosts and only specific cell types within those hosts. The viruses that infect bacteria are called **bacteriophages**, or simply phages. The word *phage* comes from the Greek word for devour. Other viruses are just identified by their host group, such as animal or plant viruses. Once a cell is infected, the effects of the virus can vary depending on the type of virus. Viruses may cause abnormal growth of the cell or cell death, alter the cell's genome, or cause little noticeable effect in the cell.

Viruses can be transmitted through direct contact, indirect contact with fomites, or through a **vector**: an animal that transmits a pathogen from one host to another. Arthropods such as mosquitoes, ticks, and flies, are typical vectors for viral diseases, and they may act as **mechanical vectors** or **biological vectors**. Mechanical transmission occurs when the arthropod carries a viral pathogen on the outside of its body and transmits it to a new host by physical contact. Biological transmission occurs when the arthropod carries the viral pathogen inside its body and transmits it to the new host through biting.

In humans, a wide variety of viruses are capable of causing various infections and diseases. Some of the deadliest emerging pathogens in humans are viruses, yet we have few treatments or drugs to deal with viral infections, making them difficult to eradicate.

Viruses that can be transmitted from an animal host to a human host can cause zoonoses. For example, the avian influenza virus originates in birds, but can cause disease in humans. Reverse zoonoses are caused by infection of an animal by a virus that originated in a human.

MICRO CONNECTIONS

Fighting Bacteria with Viruses

The emergence of superbugs, or multidrug resistant bacteria, has become a major challenge for pharmaceutical companies and a serious health-care problem. According to a 2013 report by the US Centers for Disease Control and Prevention (CDC), more than 2 million people are infected with drug-resistant bacteria in the US annually, resulting in at least 23,000 deaths.[3] The continued use and overuse of antibiotics will likely lead to the evolution of even more drug-resistant strains.

One potential solution is the use of phage therapy, a procedure that uses bacteria-killing viruses (bacteriophages) to treat bacterial infections. Phage therapy is not a new idea. The discovery of bacteriophages dates back to the early 20th century, and phage therapy was first used in Europe in 1915 by the English bacteriologist Frederick Twort.[4] However, the subsequent discovery of penicillin and other antibiotics led to the near abandonment of this form of therapy, except in the former Soviet Union and a few countries in Eastern Europe. Interest in phage therapy outside of the countries of the former Soviet Union is only recently re-emerging because of the rise in antibiotic-resistant bacteria.[5]

Phage therapy has some advantages over antibiotics in that phages kill only one specific bacterium, whereas antibiotics kill not only the pathogen but also beneficial bacteria of the normal microbiota. Development of new antibiotics is also expensive for drug companies and for patients, especially for those who live in countries with high poverty rates.

Phages have also been used to prevent food spoilage. In 2006, the US Food and Drug Administration approved the use of a solution containing six bacteriophages that can be sprayed on lunch meats such as bologna, ham, and turkey to kill *Listeria monocytogenes*, a bacterium responsible for listeriosis, a form of food poisoning. Some consumers have concerns about the use of phages on foods, however, especially given the rising popularity of organic products. Foods that have been treated with phages must declare "bacteriophage preparation" in the list of ingredients or include a label declaring that the meat has been "treated with antimicrobial solution to reduce microorganisms."[6]

CHECK YOUR UNDERSTANDING

- Why do humans not have to be concerned about the presence of bacteriophages in their food?
- What are three ways that viruses can be transmitted between hosts?

---

3 US Department of Health and Human Services, Centers for Disease Control and Prevention. "Antibiotic Resistance Threats in the United States, 2013." http://www.cdc.gov/drugresistance/pdf/ar-threats-2013-508.pdf (accessed September 22, 2015).

4 M. Clokie et al. "Phages in Nature." *Bacteriophage* 1, no. 1 (2011): 31–45.

5 A. Sulakvelidze et al. "Bacteriophage Therapy." *Antimicrobial Agents and Chemotherapy* 45, no. 3 (2001): 649–659.

6 US Food and Drug Administration. "FDA Approval of *Listeria*-specific Bacteriophage Preparation on Ready-to-Eat (RTE) Meat and Poultry Products." http://www.fda.gov/food/ingredientspackaginglabeling/ucm083572.htm (accessed September 22, 2015).

Viral Structures

In general, virions (viral particles) are small and cannot be observed using a regular light microscope. They are much smaller than prokaryotic and eukaryotic cells; this is an adaptation allowing viruses to infect these larger cells (see Figure 6.3). The size of a virion can range from 20 nm for small viruses up to 900 nm for typical, large viruses (see Figure 6.4). Recent discoveries, however, have identified new giant viral species, such as *Pandoravirus salinus* and *Pithovirus sibericum*, with sizes approaching that of a bacterial cell.[7]

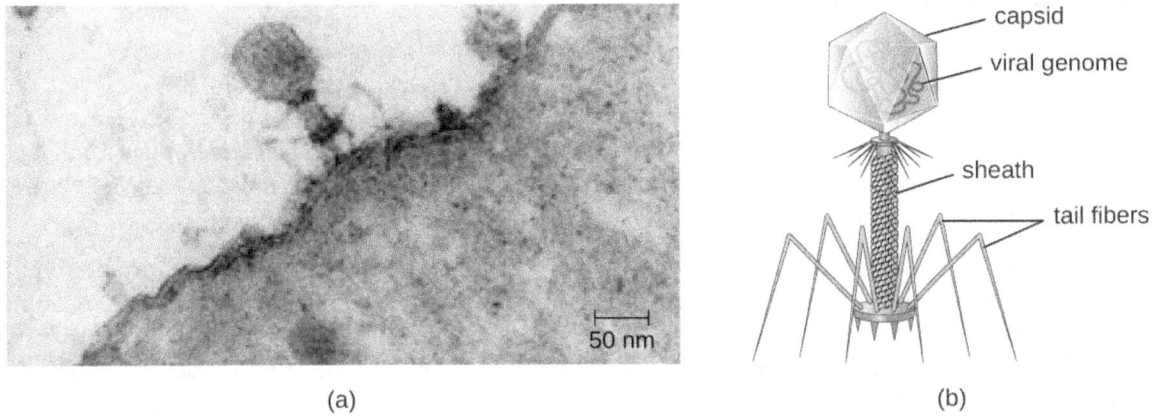

Figure 6.3 (a) In this transmission electron micrograph, a bacteriophage (a virus that infects bacteria) is dwarfed by the bacterial cell it infects. (b) An illustration of the bacteriophage in the micrograph. (credit a: modification of work by U.S. Department of Energy, Office of Science, LBL, PBD)

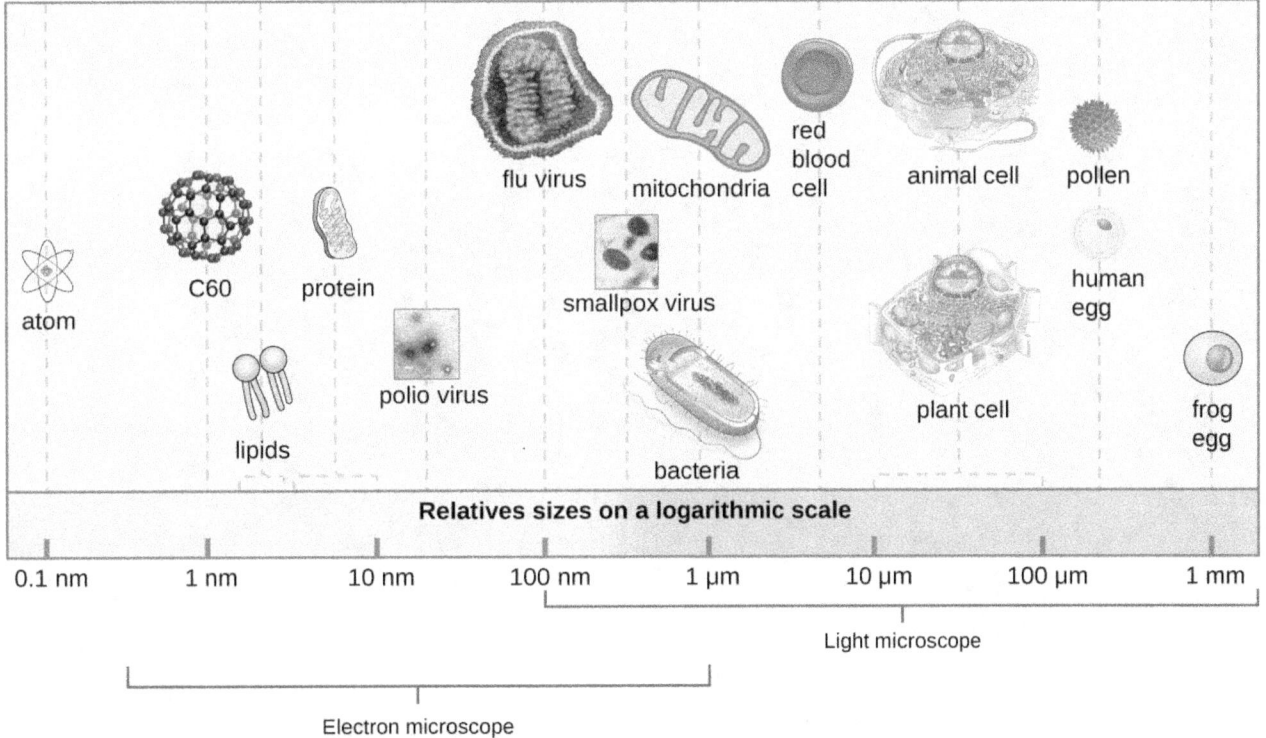

Figure 6.4 The size of a virus is small relative to the size of most bacterial and eukaryotic cells and their organelles.

In 1935, after the development of the electron microscope, Wendell Stanley was the first scientist to crystallize the structure of the tobacco mosaic virus and discovered that it is composed of RNA and protein. In 1943, he isolated *Influenza B virus*, which contributed to the development of an influenza (flu) vaccine. Stanley's

---

7 N. Philippe et al. "Pandoraviruses: Amoeba Viruses with Genomes up to 2.5 Mb Reaching that of Parasitic Eukaryotes." *Science* 341, no. 6143 (2013): 281–286.

discoveries unlocked the mystery of the nature of viruses that had been puzzling scientists for over 40 years and his contributions to the field of virology led to him being awarded the Nobel Prize in 1946.

As a result of continuing research into the nature of viruses, we now know they consist of a nucleic acid (either RNA or DNA, but never both) surrounded by a protein coat called a **capsid** (see Figure 6.5). The interior of the capsid is not filled with cytosol, as in a cell, but instead it contains the bare necessities in terms of genome and enzymes needed to direct the synthesis of new virions. Each capsid is composed of protein subunits called **capsomeres** made of one or more different types of capsomere proteins that interlock to form the closely packed capsid.

There are two categories of viruses based on general composition. Viruses formed from only a nucleic acid and capsid are called **naked viruses** or **nonenveloped viruses**. Viruses formed with a nucleic-acid packed capsid surrounded by a lipid layer are called **enveloped viruses** (see Figure 6.5). The **viral envelope** is a small portion of phospholipid membrane obtained as the virion buds from a host cell. The viral envelope may either be intracellular or cytoplasmic in origin.

Extending outward and away from the capsid on some naked viruses and enveloped viruses are protein structures called **spikes**. At the tips of these spikes are structures that allow the virus to attach and enter a cell, like the influenza virus hemagglutinin spikes (H) or enzymes like the neuraminidase (N) influenza virus spikes that allow the virus to detach from the cell surface during release of new virions. Influenza viruses are often identified by their H and N spikes. For example, H1N1 influenza viruses were responsible for the pandemics in 1918 and 2009,[8] H2N2 for the pandemic in 1957, and H3N2 for the pandemic in 1968.

---

8 J. Cohen. "What's Old Is New: 1918 Virus Matches 2009 H1N1 Strain. *Science* 327, no. 5973 (2010): 1563–1564.

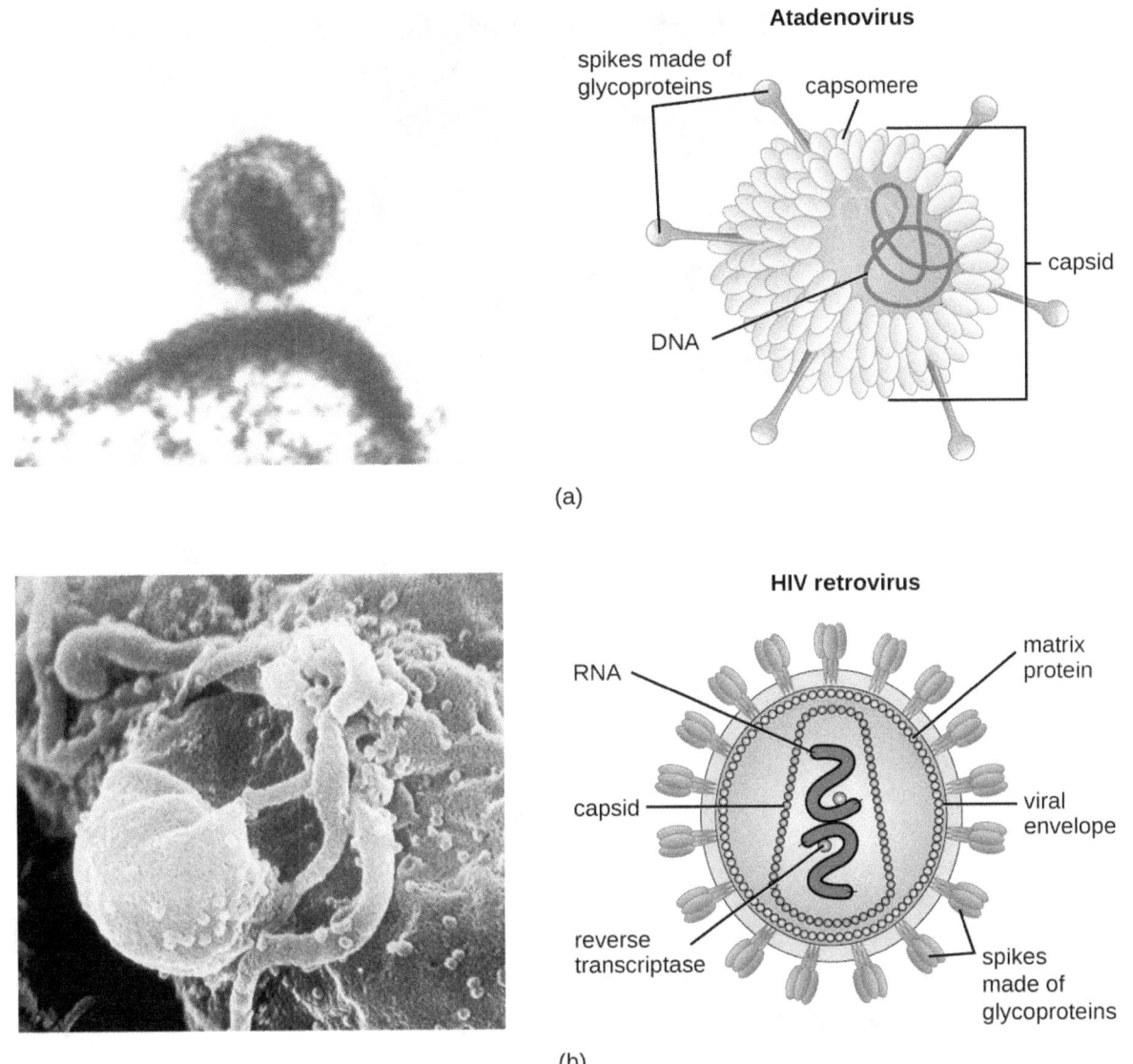

Figure 6.5 (a) The naked atadenovirus uses spikes made of glycoproteins from its capsid to bind to host cells. (b) The enveloped human immunodeficiency virus uses spikes made of glycoproteins embedded in its envelope to bind to host cells (credit a "micrograph": modification of work by NIAID; credit b "micrograph": modification of work by Centers for Disease Control and Prevention)

Viruses vary in the shape of their capsids, which can be either **helical**, **polyhedral**, or **complex**. A helical capsid forms the shape of tobacco mosaic virus (TMV), a naked helical virus, and Ebola virus, an enveloped helical virus. The capsid is cylindrical or rod shaped, with the genome fitting just inside the length of the capsid. Polyhedral capsids form the shapes of poliovirus and rhinovirus, and consist of a nucleic acid surrounded by a polyhedral (many-sided) capsid in the form of an icosahedron. An **icosahedral** capsid is a three-dimensional, 20-sided structure with 12 vertices. These capsids somewhat resemble a soccer ball. Both helical and polyhedral viruses can have envelopes. Viral shapes seen in certain types of bacteriophages, such as T4 phage, and poxviruses, like vaccinia virus, may have features of both polyhedral and helical viruses so they are described as a complex viral shape (see Figure 6.6). In the bacteriophage complex form, the genome is located within the polyhedral head and the **sheath** connects the head to the **tail fibers** and **tail pins** that help the virus attach to receptors on the host cell's surface. Poxviruses that have complex shapes are often brick shaped, with intricate surface characteristics not seen in the other categories of capsid.

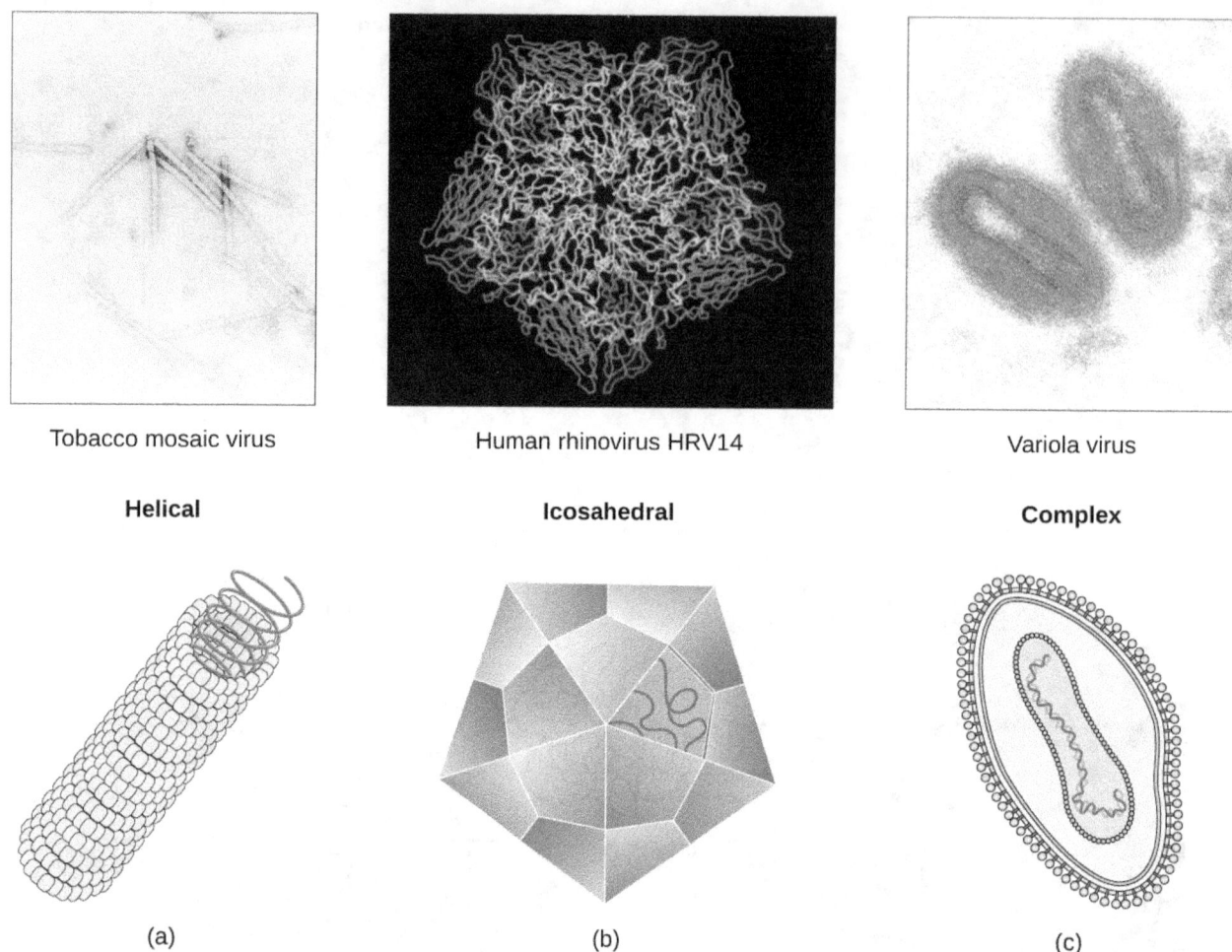

Figure 6.6 Viral capsids can be (a) helical, (b) polyhedral, or (c) have a complex shape. (credit a "micrograph": modification of work by USDA ARS; credit b "micrograph": modification of work by U.S. Department of Energy)

CHECK YOUR UNDERSTANDING

- Which types of viruses have spikes?

Classification and Taxonomy of Viruses

Although viruses are not classified in the three domains of life, their numbers are great enough to require classification. Since 1971, the International Union of Microbiological Societies Virology Division has given the task of developing, refining, and maintaining a universal virus taxonomy to the International Committee on Taxonomy of Viruses (ICTV). Since viruses can mutate so quickly, it can be difficult to classify them into a genus and a species epithet using the binomial nomenclature system. Thus, the ICTV's viral nomenclature system classifies viruses into families and genera based on viral genetics, chemistry, morphology, and mechanism of multiplication. To date, the ICTV has classified known viruses in seven orders, 96 families, and 350 genera. Viral family names end in -*viridae* (e.g, *Parvoviridae*) and genus names end in –*virus* (e.g., *Parvovirus*). The names of viral orders, families, and genera are all italicized. When referring to a viral species, we often use a genus and species epithet such as *Pandoravirus dulcis* or *Pandoravirus salinus*.

The Baltimore classification system is an alternative to ICTV nomenclature. The Baltimore system classifies viruses according to their genomes (DNA or RNA, single versus double stranded, and mode of replication). This system thus creates seven groups of viruses that have common genetics and biology.

LINK TO LEARNING

Explore the latest virus taxonomy (https://www.openstax.org/l/22virustaxon) at the ICTV website.

Aside from formal systems of nomenclature, viruses are often informally grouped into categories based on chemistry, morphology, or other characteristics they share in common. Categories may include naked or enveloped structure, single-stranded (ss) or double-stranded (ds) DNA or ss or ds RNA genomes, segmented or nonsegmented genomes, and positive-strand (+) or negative-strand (−) RNA. For example, herpes viruses can be classified as a dsDNA enveloped virus; human immunodeficiency virus (HIV) is a +ssRNA enveloped virus, and tobacco mosaic virus is a +ssRNA virus. Other characteristics such as host specificity, tissue specificity, capsid shape, and special genes or enzymes may also be used to describe groups of similar viruses. Table 6.2 lists some of the most common viruses that are human pathogens by genome type.

Common Pathogenic Viruses

Genome	Family	Example Virus	Clinical Features
dsDNA, enveloped	Poxviridae	Orthopoxvirus	Skin papules, pustules, lesions
	Poxviridae	Parapoxvirus	Skin lesions
	Herpesviridae	Simplexvirus	Cold sores, genital herpes, sexually transmitted disease
dsDNA, naked	Adenoviridae	Atadenovirus	Respiratory infection (common cold)
	Papillomaviridae	Papillomavirus	Genital warts, cervical, vulvar, or vaginal cancer
	Reoviridae	Reovirus	Gastroenteritis severe diarrhea (stomach flu)
ssDNA, naked	Parvoviridae	Adeno-associated dependoparvovirus A	Respiratory tract infection
	Parvoviridae	Adeno-associated dependoparvovirus B	Respiratory tract infection
dsRNA, naked	Reoviridae	Rotavirus	Gastroenteritis
+ssRNA, naked	Picornaviridae	Enterovirus C	Poliomyelitis
	Picornaviridae	Rhinovirus	Upper respiratory tract infection (common cold)
	Picornaviridae	Hepatovirus	Hepatitis
+ssRNA, enveloped	Togaviridae	Alphavirus	Encephalitis, hemorrhagic fever
	Togaviridae	Rubivirus	Rubella

Common Pathogenic Viruses

Genome	Family	Example Virus	Clinical Features
−ssRNA, enveloped	Retroviridae	Lentivirus	Acquired immune deficiency syndrome (AIDS)
	Filoviridae	Zaire Ebolavirus	Hemorrhagic fever
	Orthomyxoviridae	Influenzavirus A, B, C	Flu
	Rhabdoviridae	Lyssavirus	Rabies

Table 6.2

CHECK YOUR UNDERSTANDING

- What are the types of virus genomes?

Classification of Viral Diseases

While the ICTV has been tasked with the biological classification of viruses, it has also played an important role in the classification of diseases caused by viruses. To facilitate the tracking of virus-related human diseases, the ICTV has created classifications that link to the International Classification of Diseases (ICD), the standard taxonomy of disease that is maintained and updated by the World Health Organization (WHO). The ICD assigns an alphanumeric code of up to six characters to every type of viral infection, as well as all other types of diseases, medical conditions, and causes of death. This ICD code is used in conjunction with two other coding systems (the Current Procedural Terminology, and the Healthcare Common Procedure Coding System) to categorize patient conditions for treatment and insurance reimbursement.

For example, when a patient seeks treatment for a viral infection, ICD codes are routinely used by clinicians to order laboratory tests and prescribe treatments specific to the virus suspected of causing the illness. This ICD code is then used by medical laboratories to identify tests that must be performed to confirm the diagnosis. The ICD code is used by the health-care management system to verify that all treatments and laboratory work performed are appropriate for the given virus. Medical coders use ICD codes to assign the proper code for procedures performed, and medical billers, in turn, use this information to process claims for reimbursement by insurance companies. Vital-records keepers use ICD codes to record cause of death on death certificates, and epidemiologists used ICD codes to calculate morbidity and mortality statistics.

CHECK YOUR UNDERSTANDING

- Identify two locations where you would likely find an ICD code.

Clinical Focus

Part 2

David's doctor was concerned that his symptoms included prickling and itching at the site of the dog bite; these sensations could be early symptoms of rabies. Several tests are available to diagnose rabies in live patients, but no single antemortem test is adequate. The doctor decided to take samples of David's blood, saliva, and skin for testing. The skin sample was taken from the nape of the neck (posterior side of the neck near the hairline). It was about 6-mm long and contained at least 10 hair follicles, including the superficial cutaneous nerve. An immunofluorescent staining technique was used on the skin biopsy specimen to detect rabies antibodies in the cutaneous nerves at the base of the hair follicles. A test was also performed on a serum sample from David's blood to determine whether any antibodies for the rabies virus had been produced.

Meanwhile, the saliva sample was used for reverse transcriptase-polymerase chain reaction (RT-PCR) analysis, a test that can detect the presence of viral nucleic acid (RNA). The blood tests came back positive for the presence of rabies virus antigen, prompting David's doctor to prescribe prophylactic treatment. David is given a series of intramuscular injections of human rabies immunoglobulin along with a series of rabies vaccines.

- Why does the immunofluorescent technique look for rabies antibodies rather than the rabies virus itself?
- If David has contracted rabies, what is his prognosis?

Jump to the next Clinical Focus box. Go back to the previous Clinical Focus box.

6.2 The Viral Life Cycle

Learning Objectives

By the end of this section, you will be able to:
- Describe the lytic and lysogenic life cycles
- Describe the replication process of animal viruses
- Describe unique characteristics of retroviruses and latent viruses
- Discuss human viruses and their virus-host cell interactions
- Explain the process of transduction
- Describe the replication process of plant viruses

All viruses depend on cells for reproduction and metabolic processes. By themselves, viruses do not encode for all of the enzymes necessary for viral replication. But within a host cell, a virus can commandeer cellular machinery to produce more viral particles. Bacteriophages replicate only in the cytoplasm, since prokaryotic cells do not have a nucleus or organelles. In eukaryotic cells, most DNA viruses can replicate inside the nucleus, with an exception observed in the large DNA viruses, such as the poxviruses, that can replicate in the cytoplasm. With a few exceptions, RNA viruses that infect animal cells replicate in the cytoplasm. An important exception that will be highlighted later is Influenza virus.

The Life Cycle of Viruses with Prokaryote Hosts

The life cycle of bacteriophages has been a good model for understanding how viruses affect the cells they infect, since similar processes have been observed for eukaryotic viruses, which can cause immediate death of the cell or establish a latent or chronic infection. **Virulent phages** typically lead to the death of the cell through cell lysis. **Temperate phages**, on the other hand, can become part of a host chromosome and are replicated with the cell genome until such time as they are induced to make newly assembled viruses, or **progeny viruses**.

The Lytic Cycle

During the **lytic cycle** of virulent phage, the bacteriophage takes over the cell, reproduces new phages, and destroys the cell. T-even phage is a good example of a well-characterized class of virulent phages. There are five stages in the bacteriophage lytic cycle (see Figure 6.7). **Attachment** is the first stage in the infection process in which the phage interacts with specific bacterial surface receptors (e.g., lipopolysaccharides and OmpC protein on host surfaces). Most phages have a narrow host range and may infect one species of bacteria or one strain within a species. This unique recognition can be exploited for targeted treatment of bacterial infection by phage therapy or for phage typing to identify unique bacterial subspecies or strains. The second stage of infection is entry or **penetration**. This occurs through contraction of the tail sheath, which acts like a hypodermic needle to inject the viral genome through the cell wall and membrane. The phage head and remaining components remain outside the bacteria.

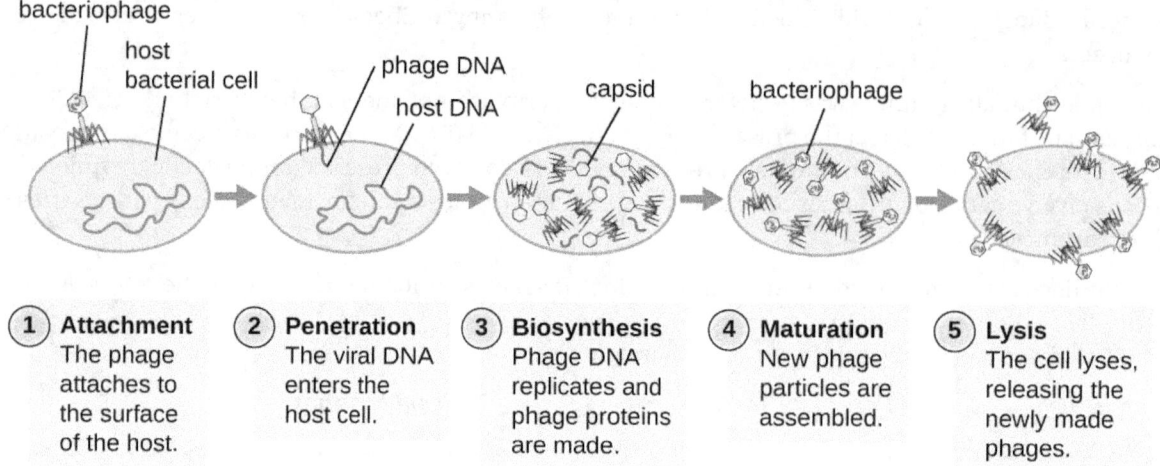

Figure 6.7 A virulent phage shows only the lytic cycle pictured here. In the lytic cycle, the phage replicates and lyses the host cell.

The third stage of infection is **biosynthesis** of new viral components. After entering the host cell, the virus synthesizes virus-encoded endonucleases to degrade the bacterial chromosome. It then hijacks the host cell to replicate, transcribe, and translate the necessary viral components (capsomeres, sheath, base plates, tail fibers, and viral enzymes) for the assembly of new viruses. Polymerase genes are usually expressed early in the cycle, while capsid and tail proteins are expressed later. During the **maturation** phase, new virions are created. To liberate free phages, the bacterial cell wall is disrupted by phage proteins such as holin or lysozyme. The final stage is release. Mature viruses burst out of the host cell in a process called **lysis** and the progeny viruses are liberated into the environment to infect new cells.

The Lysogenic Cycle

In a **lysogenic cycle**, the phage genome also enters the cell through attachment and penetration. A prime example of a phage with this type of life cycle is the lambda phage. During the lysogenic cycle, instead of killing the host, the phage genome integrates into the bacterial chromosome and becomes part of the host. The integrated phage genome is called a **prophage**. A bacterial host with a prophage is called a **lysogen**. The process in which a bacterium is infected by a temperate phage is called **lysogeny**. It is typical of temperate phages to be latent or inactive within the cell. As the bacterium replicates its chromosome, it also replicates the phage's DNA and passes it on to new daughter cells during reproduction. The presence of the phage may alter the phenotype of the bacterium, since it can bring in extra genes (e.g., toxin genes that can increase bacterial virulence). This change in the host phenotype is called **lysogenic conversion** or **phage conversion**. Some bacteria, such as *Vibrio cholerae* and *Clostridium botulinum*, are less virulent in the absence of the prophage. The phages infecting these bacteria carry the toxin genes in their genome and enhance the virulence of the host when the toxin genes are expressed. In the case of *V. cholera*, phage encoded toxin can cause severe diarrhea; in *C. botulinum*, the toxin can cause paralysis. During lysogeny, the prophage will persist in the host chromosome until **induction**, which results in the excision of the viral genome from the host chromosome. After induction has occurred the temperate phage can proceed through a lytic cycle and then undergo lysogeny in a newly infected cell (see Figure 6.8).

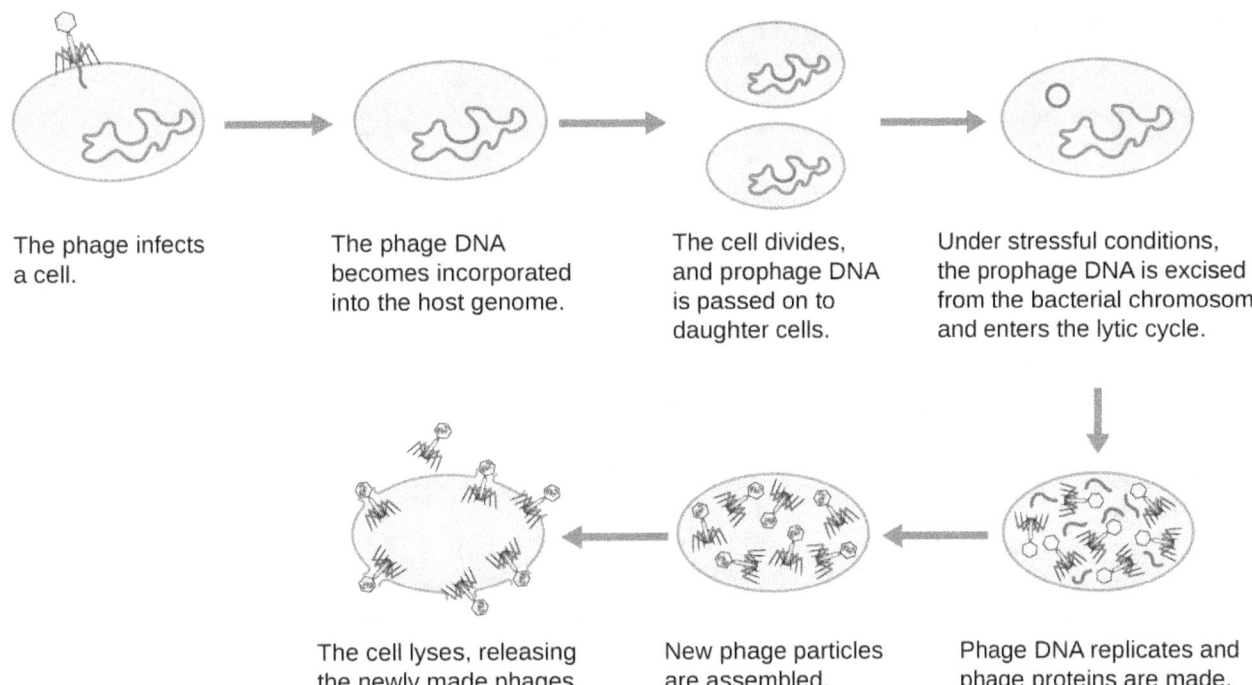

Figure 6.8 A temperate bacteriophage has both lytic and lysogenic cycles. In the lysogenic cycle, phage DNA is incorporated into the host genome, forming a prophage, which is passed on to subsequent generations of cells. Environmental stressors such as starvation or exposure to toxic chemicals may cause the prophage to be excised and enter the lytic cycle.

LINK TO LEARNING

This video (https://www.openstax.org/l/22lyticcycle) illustrates the lytic cycle.

CHECK YOUR UNDERSTANDING

- Is a latent phage undetectable in a bacterium?

Transduction

Transduction occurs when a bacteriophage transfers bacterial DNA from one bacterium to another during sequential infections. There are two types of transduction: generalized and specialized transduction. During the lytic cycle of viral replication, the virus hijacks the host cell, degrades the host chromosome, and makes more viral genomes. As it assembles and packages DNA into the phage head, packaging occasionally makes a mistake. Instead of packaging viral DNA, it takes a random piece of host DNA and inserts it into the capsid. Once released, this virion will then inject the former host's DNA into a newly infected host. The asexual transfer of genetic information can allow for DNA recombination to occur, thus providing the new host with new genes (e.g., an antibiotic-resistance gene, or a sugar-metabolizing gene). **Generalized transduction** occurs when a random piece of bacterial chromosomal DNA is transferred by the phage during the lytic cycle. **Specialized transduction** occurs at the end of the lysogenic cycle, when the prophage is excised and the bacteriophage enters the lytic cycle. Since the phage is integrated into the host genome, the prophage can replicate as part of the host. However, some conditions (e.g., ultraviolet light exposure or chemical exposure) stimulate the prophage to undergo induction, causing the phage to excise from the genome, enter the lytic cycle, and produce new phages to leave host cells. During the process of excision from the host chromosome, a phage may occasionally remove some bacterial DNA near the site of viral integration. The phage and host DNA from one end or both ends of the integration site are packaged within the capsid and are transferred to the new, infected host. Since the DNA transferred by the phage is not randomly packaged but is instead a specific piece of DNA near the site of integration, this mechanism of gene transfer is referred to as specialized transduction (see Figure 6.9). The DNA can then recombine with host chromosome, giving the latter new characteristics. Transduction seems to play an important role in the evolutionary process of bacteria, giving them a mechanism for asexual exchange of genetic information.

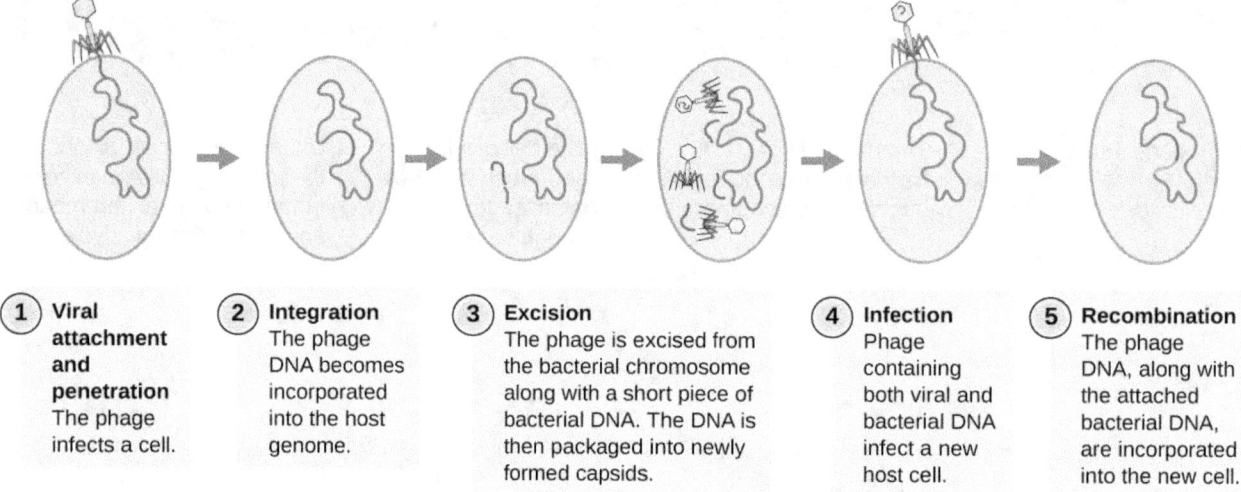

Figure 6.9 This flowchart illustrates the mechanism of specialized transduction. An integrated phage excises, bringing with it a piece of the DNA adjacent to its insertion point. On reinfection of a new bacterium, the phage DNA integrates along with the genetic material acquired from the previous host.

CHECK YOUR UNDERSTANDING

- Which phage life cycle is associated with which forms of transduction?

Life Cycle of Viruses with Animal Hosts

Lytic animal viruses follow similar infection stages to bacteriophages: attachment, penetration, biosynthesis, maturation, and release (see Figure 6.10). However, the mechanisms of penetration, nucleic-acid biosynthesis, and release differ between bacterial and animal viruses. After binding to host receptors, animal viruses enter through endocytosis (engulfment by the host cell) or through membrane fusion (viral envelope with the host cell membrane). Many viruses are host specific, meaning they only infect a certain type of host; and most viruses only infect certain types of cells within tissues. This specificity is called a **tissue tropism**. Examples of this are demonstrated by the poliovirus, which exhibits tropism for the tissues of the brain and spinal cord, or the influenza virus, which has a primary tropism for the respiratory tract.

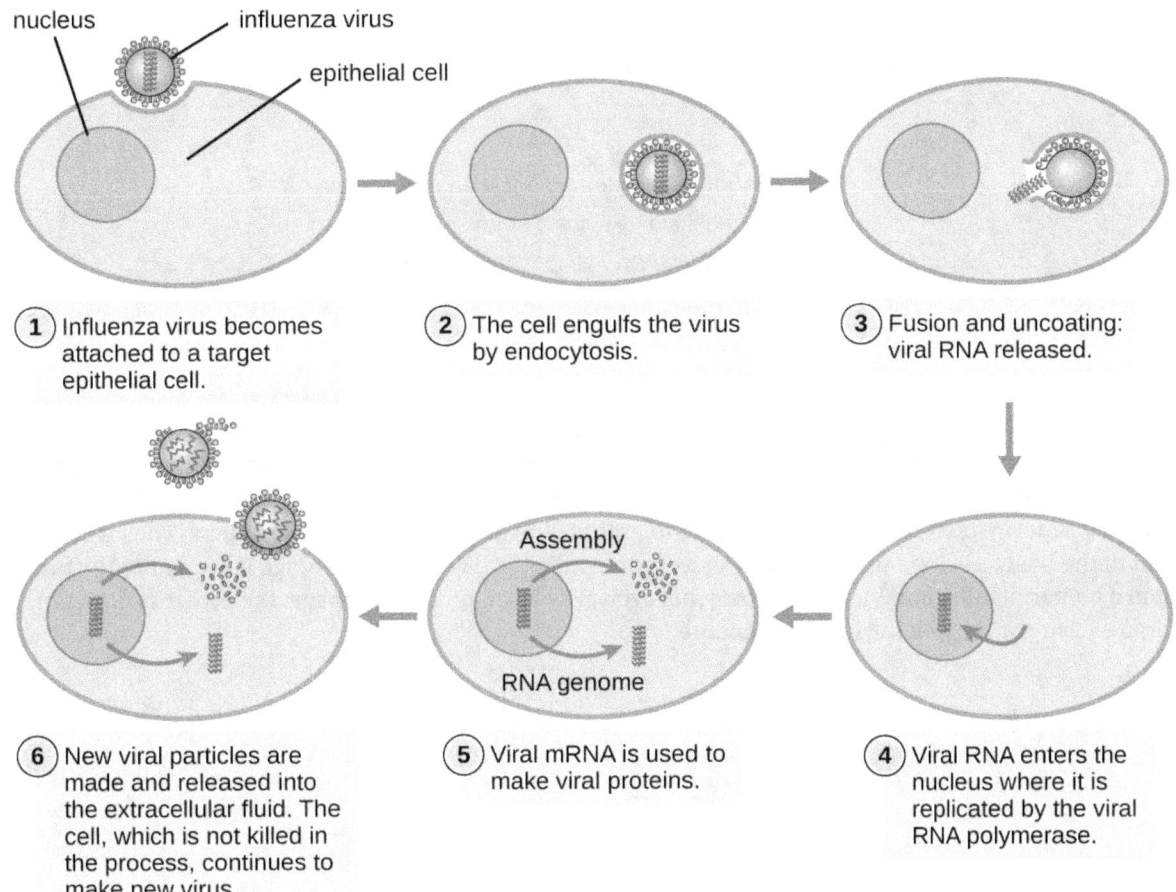

Figure 6.10 Influenza virus is one of the few RNA viruses that replicates in the nucleus of cells. In influenza virus infection, viral glycoproteins attach the virus to a host epithelial cell. As a result, the virus is engulfed. Viral RNA and viral proteins are made and assembled into new virions that are released by budding.

Animal viruses do not always express their genes using the normal flow of genetic information—from DNA to RNA to protein. Some viruses have a dsDNA genome like cellular organisms and can follow the normal flow. However, others may have ssDNA, dsRNA, or ssRNA genomes. The nature of the genome determines how the genome is replicated and expressed as viral proteins. If a genome is ssDNA, host enzymes will be used to synthesize a second strand that is complementary to the genome strand, thus producing dsDNA. The dsDNA can now be replicated, transcribed, and translated similar to host DNA.

If the viral genome is RNA, a different mechanism must be used. There are three types of RNA genome: dsRNA, **positive (+) single-strand (+ssRNA)** or **negative (−) single-strand RNA (−ssRNA)**. If a virus has a +ssRNA genome, it can be translated directly to make viral proteins. Viral genomic +ssRNA acts like cellular mRNA. However, if a virus contains a −ssRNA genome, the host ribosomes cannot translate it until the −ssRNA is replicated into +ssRNA by viral RNA-dependent RNA polymerase (RdRP) (see Figure 6.11). The RdRP is brought in by the virus and can be used to make +ssRNA from the original −ssRNA genome. The RdRP is also an important enzyme for the replication of dsRNA viruses, because it uses the negative strand of the double-stranded genome as a template to create +ssRNA. The newly synthesized +ssRNA copies can then be translated by cellular ribosomes.

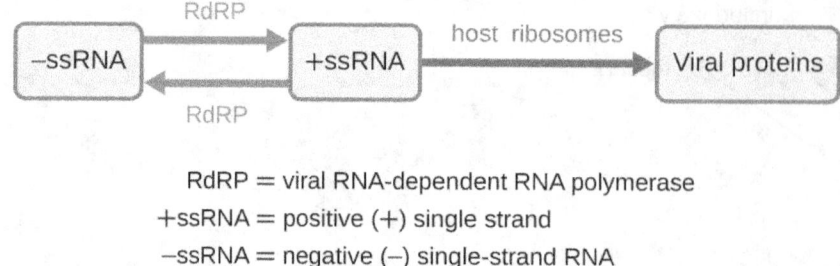

Figure 6.11 RNA viruses can contain +ssRNA that can be directly read by the ribosomes to synthesize viral proteins. Viruses containing −ssRNA must first use the −ssRNA as a template for the synthesis of +ssRNA before viral proteins can be synthesized.

An alternative mechanism for viral nucleic acid synthesis is observed in the **retroviruses**, which are +ssRNA viruses (see Figure 6.12). Single-stranded RNA viruses such as HIV carry a special enzyme called **reverse transcriptase** within the capsid that synthesizes a complementary ssDNA (cDNA) copy using the +ssRNA genome as a template. The ssDNA is then made into dsDNA, which can integrate into the host chromosome and become a permanent part of the host. The integrated viral genome is called a **provirus**. The virus now can remain in the host for a long time to establish a chronic infection. The provirus stage is similar to the prophage stage in a bacterial infection during the lysogenic cycle. However, unlike prophage, the provirus does not undergo excision after splicing into the genome.

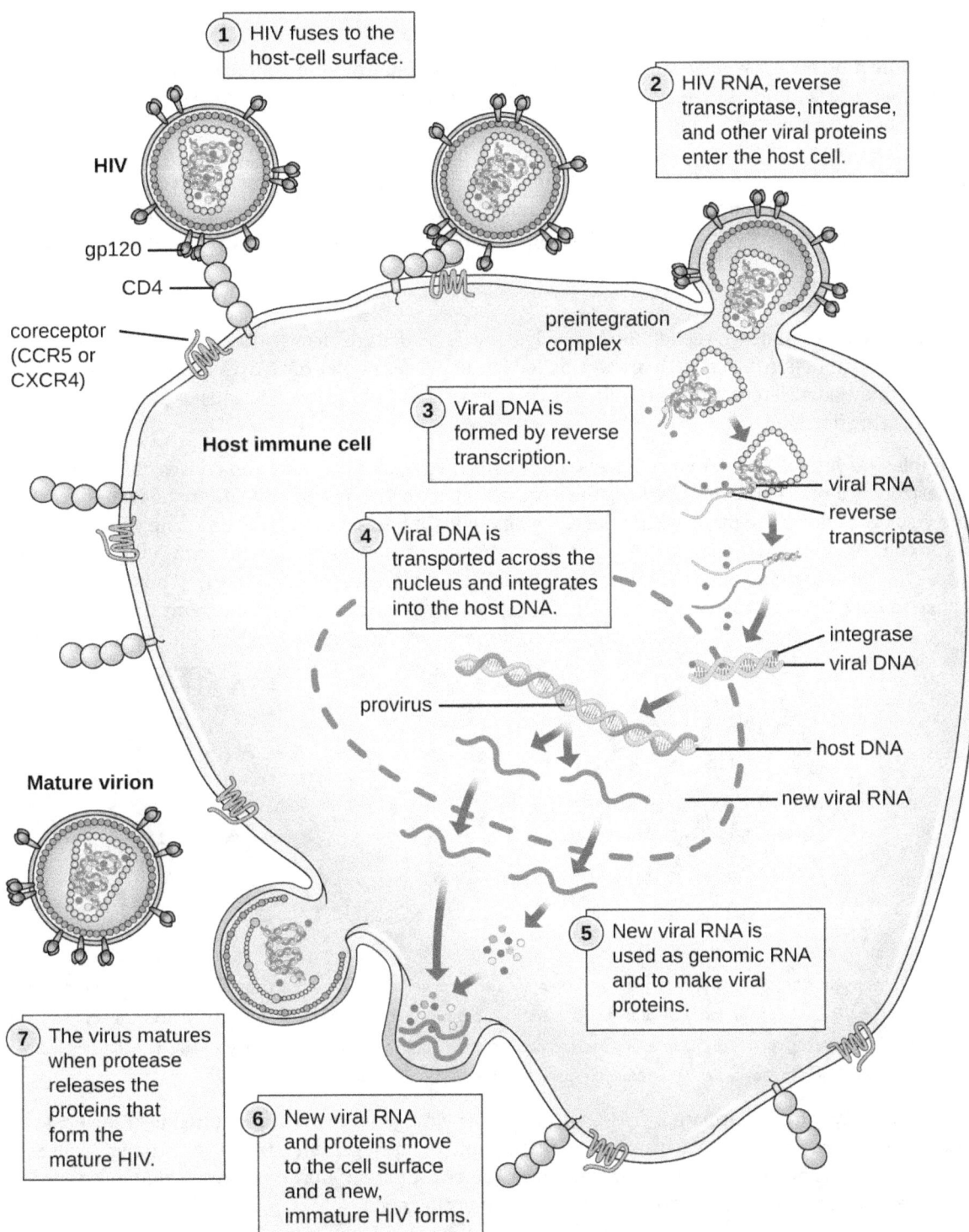

Figure 6.12 HIV, an enveloped, icosahedral retrovirus, attaches to a cell surface receptor of an immune cell and fuses with the cell membrane. Viral contents are released into the cell, where viral enzymes convert the single-stranded RNA genome into DNA and incorporate it into the host genome. (credit: modification of work by NIAID, NIH)

CHECK YOUR UNDERSTANDING

- Is RNA-dependent RNA polymerase made from a viral gene or a host gene?

Persistent Infections

Persistent infection occurs when a virus is not completely cleared from the system of the host but stays in certain tissues or organs of the infected person. The virus may remain silent or undergo productive infection without seriously harming or killing the host. Mechanisms of persistent infection may involve the regulation of the viral or host gene expressions or the alteration of the host immune response. The two primary categories of persistent infections are latent infection and chronic infection. Examples of viruses that cause latent infections include herpes simplex virus (oral and genital herpes), varicella-zoster virus (chickenpox and shingles), and Epstein-Barr virus (mononucleosis). Hepatitis C virus and HIV are two examples of viruses that cause long-term chronic infections.

Latent Infection

Not all animal viruses undergo replication by the lytic cycle. There are viruses that are capable of remaining hidden or dormant inside the cell in a process called latency. These types of viruses are known as **latent viruses** and may cause latent infections. Viruses capable of latency may initially cause an acute infection before becoming dormant.

For example, the varicella-zoster virus infects many cells throughout the body and causes chickenpox, characterized by a rash of blisters covering the skin. About 10 to 12 days postinfection, the disease resolves and the virus goes dormant, living within nerve-cell ganglia for years. During this time, the virus does not kill the nerve cells or continue replicating. It is not clear why the virus stops replicating within the nerve cells and expresses few viral proteins but, in some cases, typically after many years of dormancy, the virus is reactivated and causes a new disease called shingles (Figure 6.13). Whereas chickenpox affects many areas throughout the body, shingles is a nerve cell-specific disease emerging from the ganglia in which the virus was dormant.

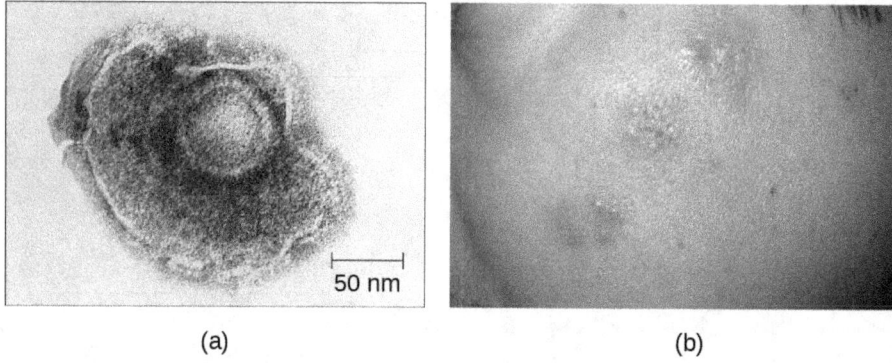

(a) (b)

Figure 6.13 (a) Varicella-zoster, the virus that causes chickenpox, has an enveloped icosahedral capsid visible in this transmission electron micrograph. Its double-stranded DNA genome becomes incorporated in the host DNA. (b) After a period of latency, the virus can reactivate in the form of shingles, usually manifesting as a painful, localized rash on one side of the body. (credit a: modification of work by Erskine Palmer and B.G. Partin—scale-bar data from Matt Russell; credit b: Paulo O / Flickr (CC-BY)

Latent viruses may remain dormant by existing as circular viral genome molecules outside of the host chromosome. Others become proviruses by integrating into the host genome. During dormancy, viruses do not cause any symptoms of disease and may be difficult to detect. A patient may be unaware that he or she is carrying the virus unless a viral diagnostic test has been performed.

Chronic Infection

A chronic infection is a disease with symptoms that are recurrent or persistent over a long time. Some viral infections can be chronic if the body is unable to eliminate the virus. HIV is an example of a virus that produces a chronic infection, often after a long period of latency. Once a person becomes infected with HIV, the virus can be detected in tissues continuously thereafter, but untreated patients often experience no symptoms for years. However, the virus maintains chronic persistence through several mechanisms that interfere with immune function, including preventing expression of viral antigens on the surface of infected cells, altering immune cells themselves, restricting expression of viral genes, and rapidly changing viral antigens through mutation. Eventually, the damage to the immune system results in progression of the disease leading to acquired immunodeficiency syndrome (AIDS). The various mechanisms that HIV uses to avoid being cleared by the immune system are also used by other chronically infecting viruses, including the hepatitis C virus.

✓ CHECK YOUR UNDERSTANDING

- In what two ways can a virus manage to maintain a persistent infection?

Life Cycle of Viruses with Plant Hosts

Plant viruses are more similar to animal viruses than they are to bacteriophages. Plant viruses may be enveloped or non-enveloped. Like many animal viruses, plant viruses can have either a DNA or RNA genome and be single stranded or double stranded. However, most plant viruses do not have a DNA genome; the majority have a +ssRNA genome, which acts like messenger RNA (mRNA). Only a minority of plant viruses have other types of genomes.

Plant viruses may have a narrow or broad host range. For example, the citrus tristeza virus infects only a few plants of the *Citrus* genus, whereas the cucumber mosaic virus infects thousands of plants of various plant families. Most plant viruses are transmitted by contact between plants, or by fungi, nematodes, insects, or other arthropods that act as mechanical vectors. However, some viruses can only be transferred by a specific type of insect vector; for example, a particular virus might be transmitted by aphids but not whiteflies. In some cases, viruses may also enter healthy plants through wounds, as might occur due to pruning or weather damage.

Viruses that infect plants are considered biotrophic parasites, which means that they can establish an infection without killing the host, similar to what is observed in the lysogenic life cycles of bacteriophages. Viral infection can be asymptomatic (latent) or can lead to cell death (lytic infection). The life cycle begins with the penetration of the virus into the host cell. Next, the virus is uncoated within the cytoplasm of the cell when the capsid is removed. Depending on the type of nucleic acid, cellular components are used to replicate the viral genome and synthesize viral proteins for assembly of new virions. To establish a systemic infection, the virus must enter a part of the vascular system of the plant, such as the phloem. The time required for systemic infection may vary from a few days to a few weeks depending on the virus, the plant species, and the environmental conditions. The virus life cycle is complete when it is transmitted from an infected plant to a healthy plant.

✓ CHECK YOUR UNDERSTANDING

- What is the structure and genome of a typical plant virus?

Viral Growth Curve

Unlike the growth curve for a bacterial population, the growth curve for a virus population over its life cycle does not follow a sigmoidal curve. During the initial stage, an inoculum of virus causes infection. In the **eclipse phase**, viruses bind and penetrate the cells with no virions detected in the medium. The chief difference that next appears in the viral growth curve compared to a bacterial growth curve occurs when virions are released from the lysed host cell at the same time. Such an occurrence is called a **burst**, and the number of virions per bacterium released is described as the **burst size**. In a one-step multiplication curve for bacteriophage, the host cells lyse, releasing many viral particles to the medium, which leads to a very steep rise in **viral titer** (the number of virions per unit volume). If no viable host cells remain, the viral particles begin to degrade during the decline of the culture (see Figure 6.14).

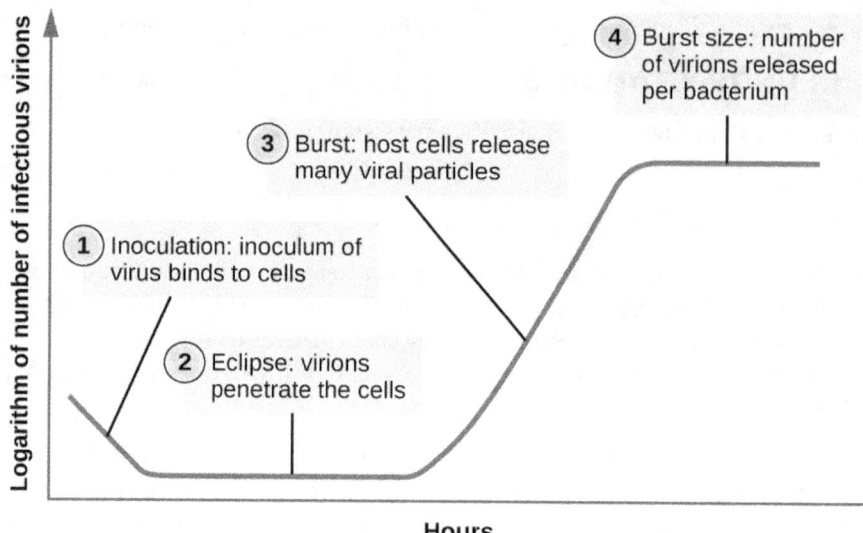

Figure 6.14 The one-step multiplication curve for a bacteriophage population follows three steps: 1) inoculation, during which the virions attach to host cells; 2) eclipse, during which entry of the viral genome occurs; and 3) burst, when sufficient numbers of new virions are produced and emerge from the host cell. The burst size is the maximum number of virions produced per bacterium.

CHECK YOUR UNDERSTANDING

- What aspect of the life cycle of a virus leads to the sudden increase in the growth curve?

Eye on Ethics

Unregistered Treatments

Ebola is incurable and deadly. The outbreak in West Africa in 2014 was unprecedented, dwarfing other human Ebola epidemics in the level of mortality. Of 24,666 suspected or confirmed cases reported, 10,179 people died.[9]

No approved treatments or vaccines for Ebola are available. While some drugs have shown potential in laboratory studies and animal models, they have not been tested in humans for safety and effectiveness. Not only are these drugs untested or unregistered but they are also in short supply.

Given the great suffering and high mortality rates, it is fair to ask whether unregistered and untested medications are better than none at all. Should such drugs be dispensed and, if so, who should receive them, in light of their extremely limited supplies? Is it ethical to treat untested drugs on patients with Ebola? On the other hand, is it ethical to withhold potentially life-saving drugs from dying patients? Or should the drugs perhaps be reserved for health-care providers working to contain the disease?

In August 2014, two infected US aid workers and a Spanish priest were treated with ZMapp, an unregistered drug that had been tested in monkeys but not in humans. The two American aid workers recovered, but the priest died. Later that month, the WHO released a report on the ethics of treating patients with the drug. Since Ebola is often fatal, the panel reasoned that it is ethical to give the unregistered drugs and unethical to withhold them for safety concerns. This situation is an example of "compassionate use" outside the well-established system of regulation and governance of therapies.

---

9 World Health Organization. "WHO Ebola Data and Statistics." March 18, 2005. http://apps.who.int/gho/data/view.ebola-sitrep.ebola-summary-20150318?lang=en

Case in Point

Ebola in the US

On September 24, 2014, Thomas Eric Duncan arrived at the Texas Health Presbyterian Hospital in Dallas complaining of a fever, headache, vomiting, and diarrhea—symptoms commonly observed in patients with the cold or the flu. After examination, an emergency department doctor diagnosed him with sinusitis, prescribed some antibiotics, and sent him home. Two days later, Duncan returned to the hospital by ambulance. His condition had deteriorated and additional blood tests confirmed that he has been infected with the Ebola virus.

Further investigations revealed that Duncan had just returned from Liberia, one of the countries in the midst of a severe Ebola epidemic. On September 15, nine days before he showed up at the hospital in Dallas, Duncan had helped transport an Ebola-stricken neighbor to a hospital in Liberia. The hospital continued to treat Duncan, but he died several days after being admitted.

The timeline of the Duncan case is indicative of the life cycle of the Ebola virus. The incubation time for Ebola ranges from 2 days to 21 days. Nine days passed between Duncan's exposure to the virus infection and the appearance of his symptoms. This corresponds, in part, to the eclipse period in the growth of the virus population. During the eclipse phase, Duncan would have been unable to transmit the disease to others. However, once an infected individual begins exhibiting symptoms, the disease becomes very contagious. Ebola virus is transmitted through direct contact with droplets of bodily fluids such as saliva, blood, and vomit. Duncan could conceivably have transmitted the disease to others at any time after he began having symptoms, presumably some time before his arrival at the hospital in Dallas. Once a hospital realizes a patient like Duncan is infected with Ebola virus, the patient is immediately quarantined, and public health officials initiate a back trace to identify everyone with whom a patient like Duncan might have interacted during the period in which he was showing symptoms.

Public health officials were able to track down 10 high-risk individuals (family members of Duncan) and 50 low-risk individuals to monitor them for signs of infection. None contracted the disease. However, one of the nurses charged with Duncan's care did become infected. This, along with Duncan's initial misdiagnosis, made it clear that US hospitals needed to provide additional training to medical personnel to prevent a possible Ebola outbreak in the US.

- What types of training can prepare health professionals to contain emerging epidemics like the Ebola outbreak of 2014?
- What is the difference between a contagious pathogen and an infectious pathogen?

Figure 6.15 Researchers working with Ebola virus use layers of defenses against accidental infection, including protective clothing, breathing systems, and negative air-pressure cabinets for bench work. (credit: modification of work by Randal J. Schoepp)

LINK TO LEARNING

For additional information about Ebola, please visit the CDC (https://www.openstax.org/l/22ebolacdc) website.

6.3 Isolation, Culture, and Identification of Viruses

Learning Objectives

By the end of this section, you will be able to:
- Discuss why viruses were originally described as filterable agents
- Describe the cultivation of viruses and specimen collection and handling
- Compare in vivo and in vitro techniques used to cultivate viruses

At the beginning of this chapter, we described how porcelain Chamberland filters with pores small enough to allow viruses to pass through were used to discover TMV. Today, porcelain filters have been replaced with membrane filters and other devices used to isolate and identify viruses.

Isolation of Viruses

Unlike bacteria, many of which can be grown on an artificial nutrient medium, viruses require a living host cell for replication. Infected host cells (eukaryotic or prokaryotic) can be cultured and grown, and then the growth medium can be harvested as a source of virus. Virions in the liquid medium can be separated from the host cells by either centrifugation or filtration. Filters can physically remove anything present in the solution that is larger than the virions; the viruses can then be collected in the filtrate (see Figure 6.16).

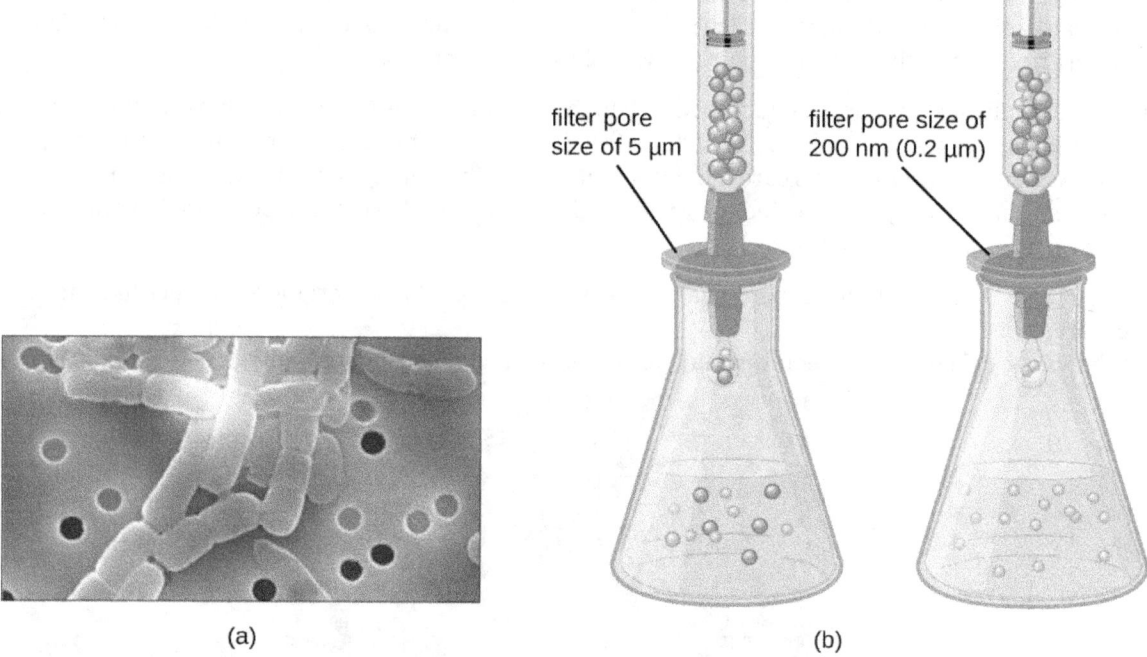

Figure 6.16 Membrane filters can be used to remove cells or viruses from a solution. (a) This scanning electron micrograph shows rod-shaped bacterial cells captured on the surface of a membrane filter. Note differences in the comparative size of the membrane pores and bacteria. Viruses will pass through this filter. (b) The size of the pores in the filter determines what is captured on the surface of the filter (animal [red] and bacteria [blue]) and removed from liquid passing through. Note the viruses (green) pass through the finer filter. (credit a: modification of work by U.S. Department of Energy)

✓ CHECK YOUR UNDERSTANDING

- What size filter pore is needed to collect a virus?

Cultivation of Viruses

Viruses can be grown **in vivo** (within a whole living organism, plant, or animal) or **in vitro** (outside a living organism in cells in an artificial environment. Flat horizontal cell culture flasks (see Figure 6.17(a)) are a common vessel used for in vitro work. Bacteriophages can be grown in the presence of a dense layer of bacteria (also called a **bacterial lawn**) grown in a 0.7 % soft agar in a Petri dish or flat (horizontal) flask (see Figure 6.17(b)). As the phage kills the bacteria, many plaques are observed among the cloudy bacterial lawn.

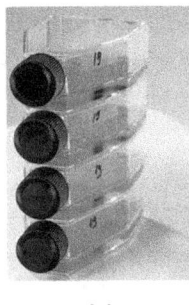
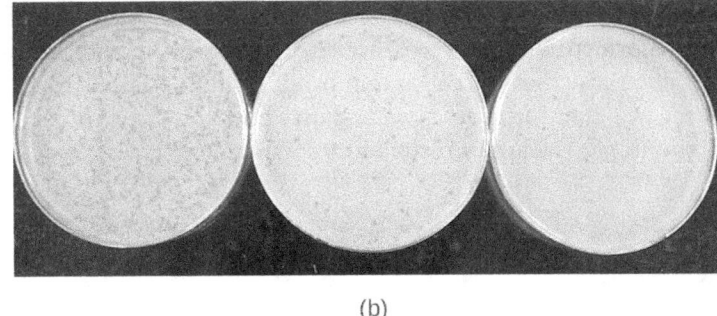

(a) (b)

Figure 6.17 (a) Flasks like this may be used to culture human or animal cells for viral culturing. (b) These plates contain bacteriophage T4 grown on an *Escherichia coli* lawn. Clear plaques are visible where host bacterial cells have been lysed. Viral titers increase on the plates to the left. (credit a: modification of work by National Institutes of Health; credit b: modification of work by American Society for Microbiology)

Animal viruses require cells within a host animal or tissue-culture cells derived from an animal. Animal virus cultivation is important for 1) identification and diagnosis of pathogenic viruses in clinical specimens, 2) production of vaccines, and 3) basic research studies. In vivo host sources can be a developing embryo in an embryonated bird's egg (e.g., chicken, turkey) or a whole animal. For example, most of the influenza vaccine manufactured for annual flu vaccination programs is cultured in hens' eggs.

The embryo or host animal serves as an incubator for viral replication (see Figure 6.18). Location within the embryo or host animal is important. Many viruses have a tissue tropism, and must therefore be introduced into a specific site for growth. Within an embryo, target sites include the amniotic cavity, the chorioallantoic membrane, or the yolk sac. Viral infection may damage tissue membranes, producing lesions called pox; disrupt embryonic development; or cause the death of the embryo.

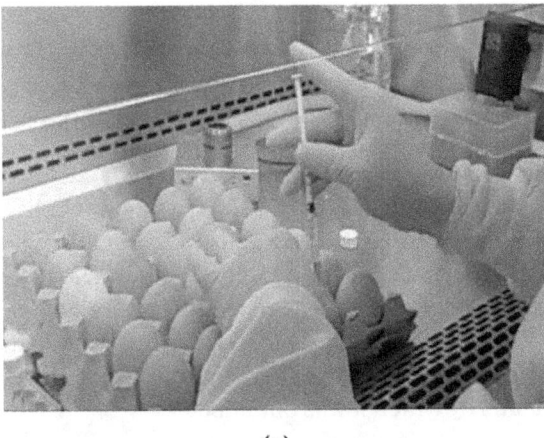
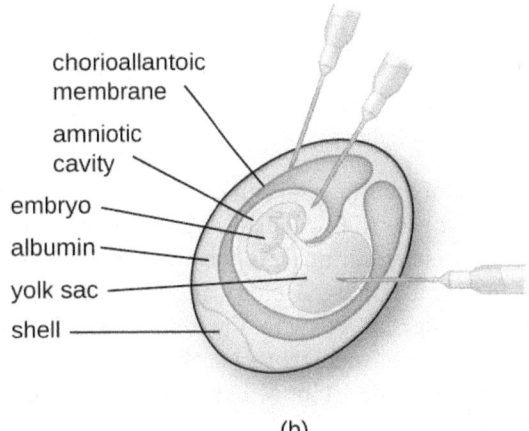

(a) (b)

Figure 6.18 (a) The cells within chicken eggs are used to culture different types of viruses. (b) Viruses can be replicated in various locations within the egg, including the chorioallantoic membrane, the amniotic cavity, and the yolk sac. (credit a: modification of work by "Chung Hoang"/YouTube)

For in vitro studies, various types of cells can be used to support the growth of viruses. A primary cell culture is freshly prepared from animal organs or tissues. Cells are extracted from tissues by mechanical scraping or mincing to release cells or by an enzymatic method using trypsin or collagenase to break up tissue and release single cells into suspension. Because of anchorage-dependence requirements, primary cell cultures require a liquid culture medium in a Petri dish or tissue-culture flask so cells have a solid surface such as glass or plastic for attachment and growth. Primary cultures usually have a limited life span. When cells in a primary culture undergo mitosis and a sufficient density of cells is produced, cells come in contact with other cells. When this cell-to-cell-contact occurs, mitosis is triggered to stop. This is called contact inhibition and it prevents the density of the cells from becoming too high. To prevent contact inhibition, cells from the primary cell culture must be transferred to another vessel with fresh growth medium. This is called a secondary cell culture. Periodically, cell density must be reduced by pouring off some cells and adding fresh medium to provide space and nutrients to maintain cell growth. In contrast to primary cell cultures, continuous cell lines, usually derived from transformed cells or tumors, are often able to be subcultured many times or even grown indefinitely (in which case they are called immortal). Continuous cell lines may not exhibit anchorage dependency (they will grow in suspension) and may have lost their contact inhibition. As a result, continuous cell lines can grow in piles or lumps resembling small tumor growths (see Figure 6.19).

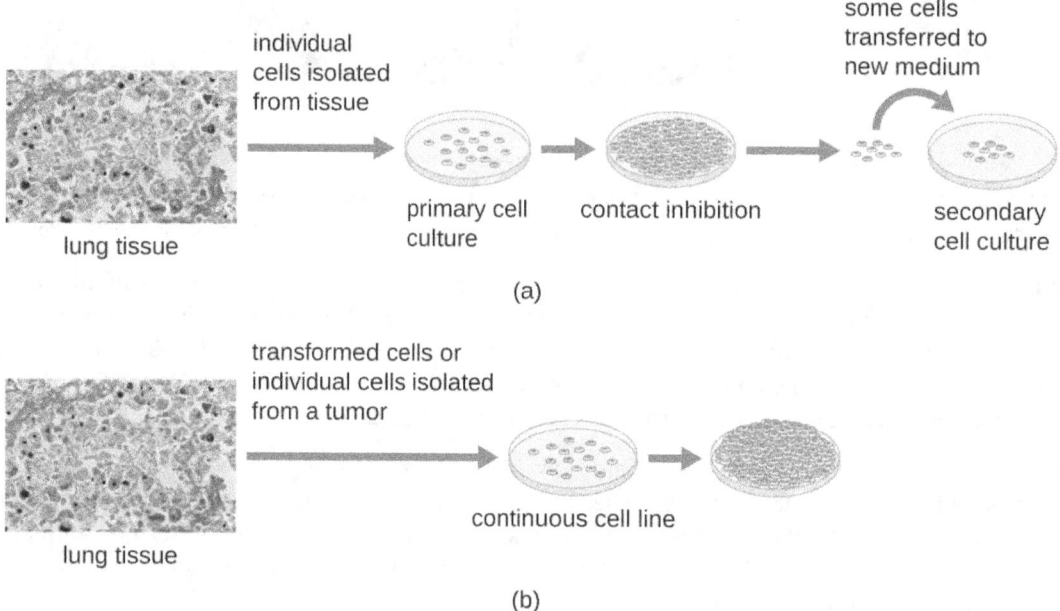

Figure 6.19 Cells for culture are prepared by separating them from their tissue matrix. (a) Primary cell cultures grow attached to the surface of the culture container. Contact inhibition slows the growth of the cells once they become too dense and begin touching each other. At this point, growth can only be sustained by making a secondary culture. (b) Continuous cell cultures are not affected by contact inhibition. They continue to grow regardless of cell density. (credit "micrographs": modification of work by Centers for Disease Control and Prevention)

An example of an immortal cell line is the HeLa cell line, which was originally cultivated from tumor cells obtained from Henrietta Lacks, a patient who died of cervical cancer in 1951. HeLa cells were the first continuous tissue-culture cell line and were used to establish tissue culture as an important technology for research in cell biology, virology, and medicine. Prior to the discovery of HeLa cells, scientists were not able to establish tissue cultures with any reliability or stability. More than six decades later, this cell line is still alive and being used for medical research. See Eye on Ethics: The Immortal Cell Line of Henrietta Lacks to read more about this important cell line and the controversial means by which it was obtained.

✓ CHECK YOUR UNDERSTANDING

- What property of cells makes periodic dilutions of primary cell cultures necessary?

Eye on Ethics

The Immortal Cell Line of Henrietta Lacks

In January 1951, Henrietta Lacks, a 30-year-old African American woman from Baltimore, was diagnosed with cervical cancer at Johns Hopkins Hospital. We now know her cancer was caused by the human papillomavirus (HPV). Cytopathic effects of the virus altered the characteristics of her cells in a process called transformation, which gives the cells the ability to divide continuously. This ability, of course, resulted in a cancerous tumor that eventually killed Mrs. Lacks in October at age 31. Before her death, samples of her cancerous cells were taken without her knowledge or permission. The samples eventually ended up in the possession of Dr. George Gey, a biomedical researcher at Johns Hopkins University. Gey was able to grow some of the cells from Lacks's sample, creating what is known today as the immortal HeLa cell line. These cells have the ability to live and grow indefinitely and, even today, are still widely used in many areas of research.

According to Lacks's husband, neither Henrietta nor the family gave the hospital permission to collect her tissue specimen. Indeed, the family was not aware until 20 years after Lacks's death that her cells were still alive and actively being used for commercial and research purposes. Yet HeLa cells have been pivotal in numerous research discoveries related to polio, cancer, and AIDS, among other diseases. The cells have also been commercialized, although they have never themselves been patented. Despite this, Henrietta Lacks's estate has never benefited from the use of the cells, although, in 2013, the Lacks family was given control over the publication of the genetic sequence of her cells.

This case raises several bioethical issues surrounding patients' informed consent and the right to know. At the time Lacks's tissues were taken, there were no laws or guidelines about informed consent. Does that mean she was treated fairly at the time? Certainly by today's standards, the answer would be no. Harvesting tissue or organs from a dying patient without consent is not only considered unethical but illegal, regardless of whether such an act could save other patients' lives. Is it ethical, then, for scientists to continue to use Lacks's tissues for research, even though they were obtained illegally by today's standards?

Ethical or not, Lacks's cells are widely used today for so many applications that it is impossible to list them all. Is this a case in which the ends justify the means? Would Lacks be pleased to know about her contribution to science and the millions of people who have benefited? Would she want her family to be compensated for the commercial products that have been developed using her cells? Or would she feel violated and exploited by the researchers who took part of her body without her consent? Because she was never asked, we will never know.

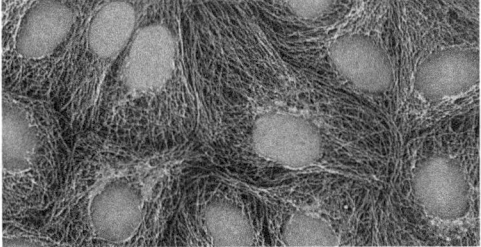

Figure 6.20 A multiphoton fluorescence image of HeLa cells in culture. Various fluorescent stains have been used to show the DNA (cyan), microtubules (green), and Golgi apparatus (orange). (credit: modification of work by National Institutes of Health)

Detection of a Virus

Regardless of the method of cultivation, once a virus has been introduced into a whole host organism, embryo, or tissue-culture cell, a sample can be prepared from the infected host, embryo, or cell line for further analysis under a brightfield, electron, or fluorescent microscope. **Cytopathic effects (CPEs)** are distinct observable cell abnormalities due to viral infection. CPEs can include loss of adherence to the surface of the container, changes in cell shape from flat to round, shrinkage of the nucleus, vacuoles in the cytoplasm, fusion of cytoplasmic membranes and the formation of multinucleated syncytia, inclusion bodies in the nucleus or cytoplasm, and complete cell lysis (see Figure 6.21).

Further pathological changes include viral disruption of the host genome and altering normal cells into transformed cells, which are the types of cells associated with carcinomas and sarcomas. The type or severity of the CPE depends on the type of virus involved. Figure 6.21 lists CPEs for specific viruses.

Cytopathic Effects of Specific Viruses		
Virus	**Cytopathic Effect**	**Example**
Paramyxovirus	Syncytium and faint basophilic cytoplasmic inclusion bodies	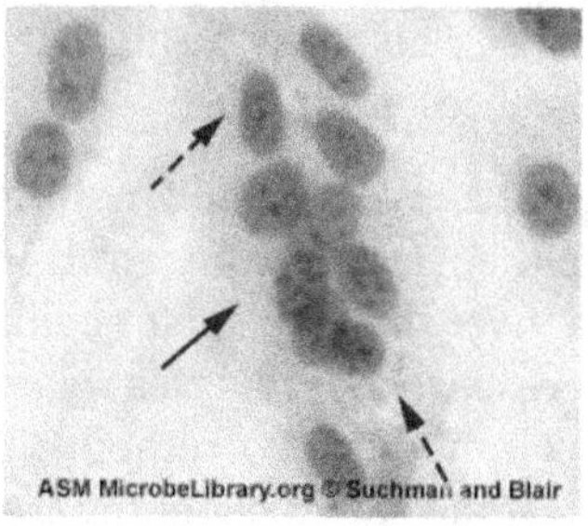
Poxvirus	Pink eosinophilic cytoplasmic inclusion bodies (arrows) and cell swelling	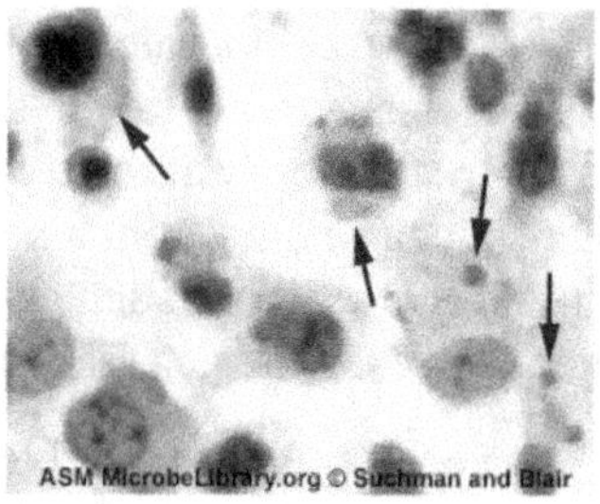
Herpesvirus	Cytoplasmic stranding (arrow) and nuclear inclusion bodies (dashed arrow)	
Adenovirus	Cell enlargement, rounding, and distinctive "grape-like" clusters	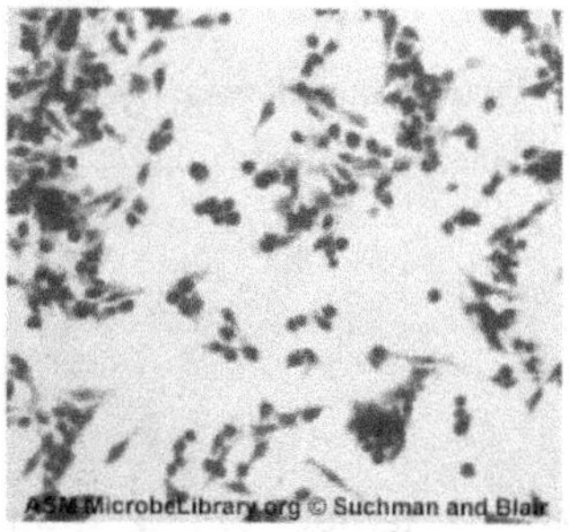

Figure 6.21 (credit "micrographs": modification of work by American Society for Microbiology)

LINK TO LEARNING

Watch this video (https://www.openstax.org/l/22virusesoncell) to learn about the effects of viruses on cells.

Hemagglutination Assay

A serological assay is used to detect the presence of certain types of viruses in patient serum. Serum is the straw-colored liquid fraction of blood plasma from which clotting factors have been removed. Serum can be used in a direct assay called a hemagglutination assay to detect specific types of viruses in the patient's sample. Hemagglutination is the agglutination (clumping) together of erythrocytes (red blood cells). Many viruses produce surface proteins or spikes called hemagglutinins that can bind to receptors on the membranes of erythrocytes and cause the cells to agglutinate. Hemagglutination is observable without using the microscope, but this method does not always differentiate between infectious and noninfectious viral particles, since both can agglutinate erythrocytes.

To identify a specific pathogenic virus using hemagglutination, we must use an indirect approach. Proteins called antibodies, generated by the patient's immune system to fight a specific virus, can be used to bind to components such as hemagglutinins that are uniquely associated with specific types of viruses. The binding of the antibodies with the hemagglutinins found on the virus subsequently prevent erythrocytes from directly interacting with the virus. So when erythrocytes are added to the antibody-coated viruses, there is no appearance of agglutination; agglutination has been inhibited. We call these types of indirect assays for virus-specific antibodies hemagglutination inhibition (HAI) assays. HAI can be used to detect the presence of antibodies specific to many types of viruses that may be causing or have caused an infection in a patient even months or years after infection (see Figure 6.22). This assay is described in greater detail in Agglutination Assays.

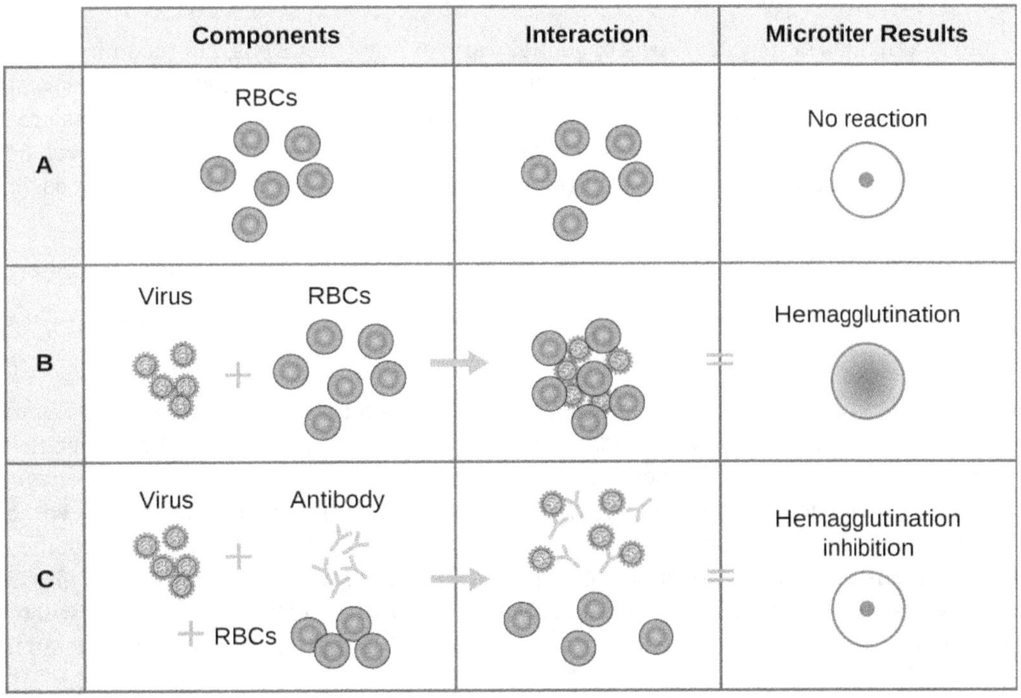

Figure 6.22 This chart shows the possible outcomes of a hemagglutination test. Row A: Erythrocytes do not bind together and will sink to the bottom of the well plate; this becomes visible as a red dot in the center of the well. Row B: Many viruses have hemagglutinins that causes agglutination of erythrocytes; the resulting hemagglutination forms a lattice structure that results in red color throughout the well. Row C: Virus-specific antibody, the viruses, and the erythrocytes are added to the well plate. The virus-specific antibodies inhibit agglutination, as can be seen as a red dot in the bottom of the well. (credit: modification of work by Centers for Disease Control and Prevention)

CHECK YOUR UNDERSTANDING

- What is the outcome of a positive HIA test?

Nucleic Acid Amplification Test

Nucleic acid amplification tests (NAAT) are used in molecular biology to detect unique nucleic acid sequences of viruses in patient samples. Polymerase chain reaction (PCR) is an NAAT used to detect the presence of viral DNA in a patient's tissue or body fluid sample. PCR is a technique that amplifies (i.e., synthesizes many copies) of a viral DNA segment of interest. Using PCR, short nucleotide sequences called primers bind to specific sequences of viral DNA, enabling identification of the virus.

Reverse transcriptase-PCR (RT-PCR) is an NAAT used to detect the presence of RNA viruses. RT-PCR differs from PCR in that the enzyme reverse transcriptase (RT) is used to make a cDNA from the small amount of viral RNA in the specimen. The cDNA can then be amplified by PCR. Both PCR and RT-PCR are used to detect and confirm the presence of the viral nucleic acid in patient specimens.

Case in Point

HPV Scare

Michelle, a 21-year-old nursing student, came to the university clinic worried that she might have been exposed to a sexually transmitted disease (STD). Her sexual partner had recently developed several bumps on the base of his penis. He had put off going to the doctor, but Michelle suspects they are genital warts caused by HPV. She is especially concerned because she knows that HPV not only causes warts but is a prominent cause of cervical cancer. She and her partner always use condoms for contraception, but she is not confident that this precaution will protect her from HPV.

Michelle's physician finds no physical signs of genital warts or any other STDs, but recommends that Michelle get a Pap smear along with an HPV test. The Pap smear will screen for abnormal cervical cells and the CPEs associated with HPV; the HPV test will test for the presence of the virus. If both tests are negative, Michelle can be more assured that she most likely has not become infected with HPV. However, her doctor suggests it might be wise for Michelle to get vaccinated against HPV to protect herself from possible future exposure.

- Why does Michelle's physician order two different tests instead of relying on one or the other?

Enzyme Immunoassay

Enzyme immunoassays (EIAs) rely on the ability of antibodies to detect and attach to specific biomolecules called antigens. The detecting antibody attaches to the target antigen with a high degree of specificity in what might be a complex mixture of biomolecules. Also included in this type of assay is a colorless enzyme attached to the detecting antibody. The enzyme acts as a tag on the detecting antibody and can interact with a colorless substrate, leading to the production of a colored end product. EIAs often rely on layers of antibodies to capture and react with antigens, all of which are attached to a membrane filter (see Figure 6.23). EIAs for viral antigens are often used as preliminary screening tests. If the results are positive, further confirmation will require tests with even greater sensitivity, such as a western blot or an NAAT. EIAs are discussed in more detail in EIAs and ELISAs.

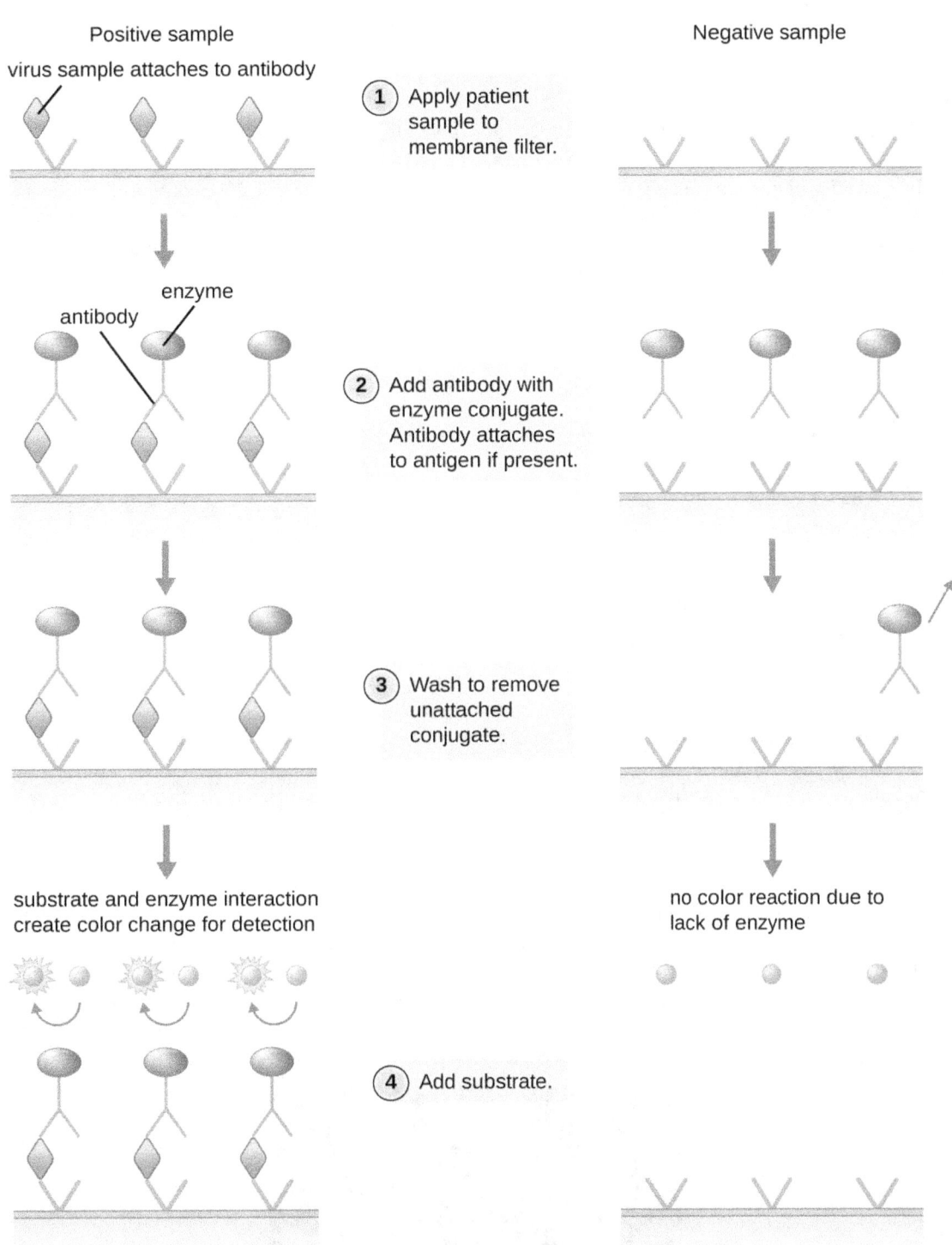

Figure 6.23 Similar to rapid, over-the-counter pregnancy tests, EIAs for viral antigens require a few drops of diluted patient serum or plasma applied to a membrane filter. The membrane filter has been previously modified and embedded with antibody to viral antigen and internal controls. Antibody conjugate is added to the filter, with the targeted antibody attached to the antigen (in the case of a positive test). Excess conjugate is washed off the filter. Substrate is added to activate the enzyme-mediated reaction to reveal the color change of a positive test. (credit: modification of work by "Cavitri"/Wikimedia Commons)

CHECK YOUR UNDERSTANDING

- What typically indicates a positive EIA test?

> **Clinical Focus**
>
> **Part 3**
>
> Along with the RT/PCR analysis, David's saliva was also collected for viral cultivation. In general, no single diagnostic test is sufficient for antemortem diagnosis, since the results will depend on the sensitivity of the assay, the quantity of virions present at the time of testing, and the timing of the assay, since release of virions in the saliva can vary. As it turns out, the result was negative for viral cultivation from the saliva. This is not surprising to David's doctor, because one negative result is not an absolute indication of the absence of infection. It may be that the number of virions in the saliva is low at the time of sampling. It is not unusual to repeat the test at intervals to enhance the chance of detecting higher virus loads.
>
> - Should David's doctor modify his course of treatment based on these test results?
>
> *Jump to the* next *Clinical Focus box. Go back to the* previous *Clinical Focus box.*

6.4 Viroids, Virusoids, and Prions

Learning Objectives

By the end of this section, you will be able to:
- Describe viroids and their unique characteristics
- Describe virusoids and their unique characteristics
- Describe prions and their unique characteristics

Research attempts to discover the causative agents of previously uninvestigated diseases have led to the discovery of nonliving disease agents quite different from viruses. These include particles consisting only of RNA or only of protein that, nonetheless, are able to self-propagate at the expense of a host—a key similarity to viruses that allows them to cause disease conditions. To date, these discoveries include viroids, virusoids, and the proteinaceous prions.

Viroids

In 1971, Theodor Diener, a pathologist working at the Agriculture Research Service, discovered an acellular particle that he named a viroid, meaning "virus-like." **Viroids** consist only of a short strand of circular RNA capable of self-replication. The first viroid discovered was found to cause potato tuber spindle disease, which causes slower sprouting and various deformities in potato plants (see Figure 6.24). Like viruses, potato spindle tuber viroids (PSTVs) take control of the host machinery to replicate their RNA genome. Unlike viruses, viroids do not have a protein coat to protect their genetic information.

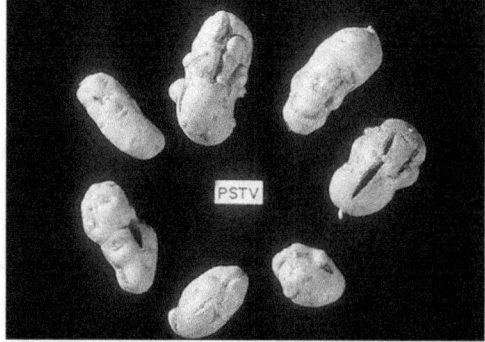

Figure 6.24 These potatoes have been infected by the potato spindle tuber viroid (PSTV), which is typically spread when infected knives are used to cut healthy potatoes, which are then planted. (credit: Pamela Roberts, University of Florida Institute of Food and Agricultural Sciences, USDA ARS)

Viroids can result in devastating losses of commercially important agricultural food crops grown in fields and orchards. Since the discovery of PSTV, other viroids have been discovered that cause diseases in plants.

Tomato planta macho viroid (TPMVd) infects tomato plants, which causes loss of chlorophyll, disfigured and brittle leaves, and very small tomatoes, resulting in loss of productivity in this field crop. Avocado sunblotch viroid (ASBVd) results in lower yields and poorer-quality fruit. ASBVd is the smallest viroid discovered thus far that infects plants. Peach latent mosaic viroid (PLMVd) can cause necrosis of flower buds and branches, and wounding of ripened fruit, which leads to fungal and bacterial growth in the fruit. PLMVd can also cause similar pathological changes in plums, nectarines, apricots, and cherries, resulting in decreased productivity in these orchards, as well. Viroids, in general, can be dispersed mechanically during crop maintenance or harvesting, vegetative reproduction, and possibly via seeds and insects, resulting in a severe drop in food availability and devastating economic consequences.

CHECK YOUR UNDERSTANDING

- What is the genome of a viroid made of?

Virusoids

A second type of pathogenic RNA that can infect commercially important agricultural crops are the **virusoids**, which are subviral particles best described as non–self-replicating ssRNAs. RNA replication of virusoids is similar to that of viroids but, unlike viroids, virusoids require that the cell also be infected with a specific "helper" virus. There are currently only five described types of virusoids and their associated helper viruses. The helper viruses are all from the family of Sobemoviruses. An example of a helper virus is the subterranean clover mottle virus, which has an associated virusoid packaged inside the viral capsid. Once the helper virus enters the host cell, the virusoids are released and can be found free in plant cell cytoplasm, where they possess ribozyme activity. The helper virus undergoes typical viral replication independent of the activity of the virusoid. The virusoid genomes are small, only 220 to 388 nucleotides long. A virusoid genome does not code for any proteins, but instead serves only to replicate virusoid RNA.

Virusoids belong to a larger group of infectious agents called satellite RNAs, which are similar pathogenic RNAs found in animals. Unlike the plant virusoids, satellite RNAs may encode for proteins; however, like plant virusoids, satellite RNAs must coinfect with a helper virus to replicate. One satellite RNA that infects humans and that has been described by some scientists as a virusoid is the hepatitis delta virus (HDV), which, by some reports, is also called hepatitis delta virusoid. Much larger than a plant virusoid, HDV has a circular, ssRNA genome of 1,700 nucleotides and can direct the biosynthesis of HDV-associated proteins. The HDV helper virus is the hepatitis B virus (HBV). Coinfection with HBV and HDV results in more severe pathological changes in the liver during infection, which is how HDV was first discovered.

CHECK YOUR UNDERSTANDING

- What is the main difference between a viroid and a virusoid?

Prions

At one time, scientists believed that any infectious particle must contain DNA or RNA. Then, in 1982, Stanley Prusiner, a medical doctor studying scrapie (a fatal, degenerative disease in sheep) discovered that the disease was caused by proteinaceous infectious particles, or **prions**. Because proteins are acellular and do not contain DNA or RNA, Prusiner's findings were originally met with resistance and skepticism; however, his research was eventually validated, and he received the Nobel Prize in Physiology or Medicine in 1997.

A prion is a misfolded rogue form of a normal protein (PrPc) found in the cell. This rogue prion protein (PrPsc), which may be caused by a genetic mutation or occur spontaneously, can be infectious, stimulating other endogenous normal proteins to become misfolded, forming plaques (see Figure 6.25). Today, prions are known to cause various forms of **transmissible spongiform encephalopathy** (TSE) in human and animals. TSE is a rare degenerative disorder that affects the brain and nervous system. The accumulation of rogue proteins causes the brain tissue to become sponge-like, killing brain cells and forming holes in the tissue, leading to brain damage, loss of motor coordination, and dementia (see Figure 6.26). Infected individuals are mentally impaired and become unable to move or speak. There is no cure, and the disease progresses rapidly, eventually leading to death within a few months or years.

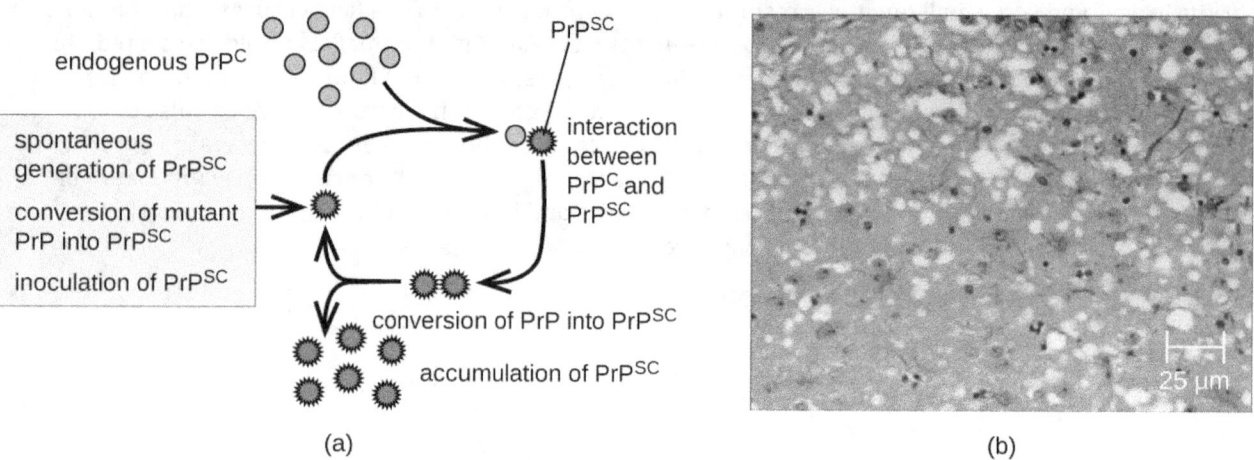

Figure 6.25 Endogenous normal prion protein (PrPc) is converted into the disease-causing form (PrPsc) when it encounters this variant form of the protein. PrPsc may arise spontaneously in brain tissue, especially if a mutant form of the protein is present, or it may originate from misfolded prions consumed in food that eventually find their way into brain tissue. (credit b: modification of work by USDA)

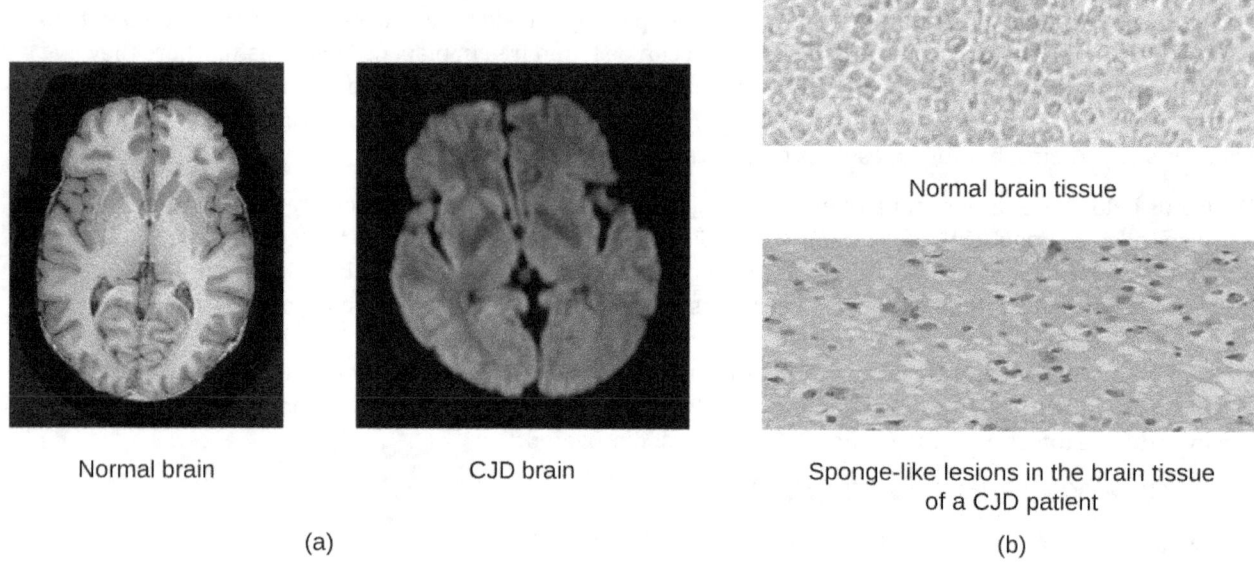

Figure 6.26 Creutzfeldt-Jakob disease (CJD) is a fatal disease that causes degeneration of neural tissue. (a) These brain scans compare a normal brain to one with CJD. (b) Compared to a normal brain, the brain tissue of a CJD patient is full of sponge-like lesions, which result from abnormal formations of prion protein. (credit a (right): modification of work by Dr. Laughlin Dawes; credit b (top): modification of work by Suzanne Wakim; credit b (bottom): modification of work by Centers for Disease Control and Prevention)

TSEs in humans include kuru, fatal familial insomnia, Gerstmann-Straussler-Scheinker disease, and Creutzfeldt-Jakob disease (see Figure 6.26). TSEs in animals include mad cow disease, scrapie (in sheep and goats), and chronic wasting disease (in elk and deer). TSEs can be transmitted between animals and from animals to humans by eating contaminated meat or animal feed. Transmission between humans can occur through heredity (as is often the case with GSS and CJD) or by contact with contaminated tissue, as might occur during a blood transfusion or organ transplant. There is no evidence for transmission via casual contact with an infected person. Table 6.3 lists TSEs that affect humans and their modes of transmission.

Transmissible Spongiform Encephalopathies (TSEs) in Humans

Disease	Mechanism(s) of Transmission[10]
Sporadic CJD (sCJD)	Not known; possibly by alteration of normal prior protein (PrP) to rogue form due to somatic mutation
Variant CJD (vCJD)	Eating contaminated cattle products and by secondary bloodborne transmission
Familial CJD (fCJD)	Mutation in germline PrP gene
Iatrogenic CJD (iCJD)	Contaminated neurosurgical instruments, corneal graft, gonadotrophic hormone, and, secondarily, by blood transfusion
Kuru	Eating infected meat through ritualistic cannibalism
Gerstmann-Straussler-Scheinker disease (GSS)	Mutation in germline PrP gene
Fatal familial insomnia (FFI)	Mutation in germline PrP gene

Table 6.3

Prions are extremely difficult to destroy because they are resistant to heat, chemicals, and radiation. Even standard sterilization procedures do not ensure the destruction of these particles. Currently, there is no treatment or cure for TSE disease, and contaminated meats or infected animals must be handled according to federal guidelines to prevent transmission.

CHECK YOUR UNDERSTANDING

- Does a prion have a genome?

LINK TO LEARNING

For more information on the handling of animals and prion-contaminated materials, visit the guidelines published on the CDC (https://www.openstax.org/l/22cdccontaminat) and WHO (https://www.openstax.org/l/22whocontaminat) websites.

Clinical Focus

Resolution

A few days later, David's doctor receives the results of the immunofluorescence test on his skin sample. The test is negative for rabies antigen. A second viral antigen test on his saliva sample also comes back negative. Despite these results, the doctor decides to continue David's current course of treatment. Given the positive RT-PCR test, it is best not to rule out a possible rabies infection.

Near the site of the bite, David receives an injection of rabies immunoglobulin, which attaches to and inactivates any rabies virus that may be present in his tissues. Over the next 14 days, he receives a series of

---

10 National Institute of Neurological Disorders and Stroke. "Creutzfeldt-Jakob Disease Fact Sheet." http://www.ninds.nih.gov/disorders/cjd/detail_cjd.htm (accessed December 31, 2015).

four rabies-specific vaccinations in the arm. These vaccines activate David's immune response and help his body recognize and fight the virus. Thankfully, with treatment, David symptoms improve and he makes a full recovery.

Not all rabies cases have such a fortunate outcome. In fact, rabies is usually fatal once the patient starts to exhibit symptoms, and postbite treatments are mainly palliative (i.e., sedation and pain management).

Go back to the previous Clinical Focus box.

SUMMARY

6.1 Viruses

- Viruses are generally ultramicroscopic, typically from 20 nm to 900 nm in length. Some large viruses have been found.
- **Virions** are acellular and consist of a nucleic acid, DNA or RNA, but not both, surrounded by a protein **capsid**. There may also be a phospholipid membrane surrounding the capsid.
- Viruses are obligate intracellular parasites.
- Viruses are known to infect various types of cells found in plants, animals, fungi, protists, bacteria, and archaea. Viruses typically have limited **host ranges** and infect specific cell types.
- Viruses may have **helical**, **polyhedral**, or **complex** shapes.
- Classification of viruses is based on morphology, type of nucleic acid, host range, cell specificity, and enzymes carried within the virion.
- Like other diseases, viral diseases are classified using ICD codes.

6.2 The Viral Life Cycle

- Many viruses target specific hosts or tissues. Some may have more than one host.
- Many viruses follow several stages to infect host cells. These stages include **attachment, penetration, uncoating, biosynthesis, maturation,** and **release**.
- Bacteriophages have a **lytic** or **lysogenic cycle**. The lytic cycle leads to the death of the host, whereas the lysogenic cycle leads to integration of phage into the host genome.
- Bacteriophages inject DNA into the host cell, whereas animal viruses enter by endocytosis or membrane fusion.
- Animal viruses can undergo **latency**, similar to lysogeny for a bacteriophage.
- The majority of plant viruses are positive-strand ssRNA and can undergo latency, chronic, or lytic infection, as observed for animal viruses.
- The growth curve of bacteriophage populations is a **one-step multiplication curve** and not a sigmoidal curve, as compared to the bacterial growth curve.
- Bacteriophages transfer genetic information between hosts using either **generalized** or **specialized transduction**.

6.3 Isolation, Culture, and Identification of Viruses

- Viral cultivation requires the presence of some form of host cell (whole organism, embryo, or cell culture).
- Viruses can be isolated from samples by filtration.
- Viral filtrate is a rich source of released virions.
- Bacteriophages are detected by presence of clear **plaques** on bacterial lawn.
- Animal and plant viruses are detected by **cytopathic effects**, molecular techniques (PCR, RT-PCR), enzyme immunoassays, and serological assays (hemagglutination assay, hemagglutination inhibition assay).

6.4 Viroids, Virusoids, and Prions

- Other acellular agents such as **viroids**, **virusoids**, and **prions** also cause diseases. Viroids consist of small, naked ssRNAs that cause diseases in plants. Virusoids are ssRNAs that require other helper viruses to establish an infection. Prions are proteinaceous infectious particles that cause **transmissible spongiform encephalopathies**.
- Prions are extremely resistant to chemicals, heat, and radiation.
- There are no treatments for prion infection.

REVIEW QUESTIONS

Multiple Choice

1. The component(s) of a virus that is/are extended from the envelope for attachment is/are the:
   A. capsomeres
   B. spikes
   C. nucleic acid
   D. viral whiskers

2. Which of the following does a virus lack? Select all that apply.
   A. ribosomes
   B. metabolic processes
   C. nucleic acid
   D. glycoprotein

3. The envelope of a virus is derived from the host's
   A. nucleic acids
   B. membrane structures
   C. cytoplasm
   D. genome

4. In naming viruses, the family name ends with _____ and genus name ends with _____.
   A. -virus; -viridae
   B. -viridae; -virus
   C. -virion; virus
   D. -virus; virion

5. What is another name for a nonenveloped virus?
   A. enveloped virus
   B. provirus
   C. naked virus
   D. latent virus

6. Which of the following leads to the destruction of the host cells?
   A. lysogenic cycle
   B. lytic cycle
   C. prophage
   D. temperate phage

7. A virus obtains its envelope during which of the following phases?
   A. attachment
   B. penetration
   C. assembly
   D. release

8. Which of the following components is brought into a cell by HIV?
   A. a DNA-dependent DNA polymerase
   B. RNA polymerase
   C. ribosome
   D. reverse transcriptase

9. A positive-strand RNA virus:
   A. must first be converted to a mRNA before it can be translated.
   B. can be used directly to translate viral proteins.
   C. will be degraded by host enzymes.
   D. is not recognized by host ribosomes.

10. What is the name for the transfer of genetic information from one bacterium to another bacterium by a phage?
    A. transduction
    B. penetration
    C. excision
    D. translation

11. Which of the followings cannot be used to culture viruses?
    A. tissue culture
    B. liquid medium only
    C. embryo
    D. animal host

12. Which of the following tests can be used to detect the presence of a specific virus?
    A. EIA
    B. RT-PCR
    C. PCR
    D. all of the above

13. Which of the following is NOT a cytopathic effect?
    A. transformation
    B. cell fusion
    C. mononucleated cell
    D. inclusion bodies

14. Which of these infectious agents do not have nucleic acid?
    A. viroids
    B. viruses
    C. bacteria
    D. prions

15. Which of the following is true of prions?
    A. They can be inactivated by boiling at 100 °C.
    B. They contain a capsid.
    C. They are a rogue form of protein, PrP.
    D. They can be reliably inactivated by an autoclave.

True/False

16. True or False: Scientists have identified viruses that are able to infect fungal cells.

Fill in the Blank

17. A virus that infects a bacterium is called a/an _____.

18. A/an _____ virus possesses characteristics of both a polyhedral and helical virus.
19. A virus containing only nucleic acid and a capsid is called a/an _____ virus or _____ virus.
20. The _____ _____ on the bacteriophage allow for binding to the bacterial cell.
21. An enzyme from HIV that can make a copy of DNA from RNA is called _____.
22. For lytic viruses, _____ is a phase during a viral growth curve when the virus is not detected.
23. Viruses can be diagnosed and observed using a(n) _____ microscope.
24. Cell abnormalities resulting from a viral infection are called _____ _____.
25. Both viroids and virusoids have a(n) _____ genome, but virusoids require a(n) _____ to reproduce.

Short Answer

26. Discuss the geometric differences among helical, polyhedral, and complex viruses.
27. What was the meaning of the word "virus" in the 1880s and why was it used to describe the cause of tobacco mosaic disease?
28. Briefly explain the difference between the mechanism of entry of a T-even bacteriophage and an animal virus.
29. Discuss the difference between generalized and specialized transduction.
30. Differentiate between lytic and lysogenic cycles.
31. Briefly explain the various methods of culturing viruses.
32. Describe the disease symptoms observed in animals infected with prions.

Critical Thinking

33. Name each labeled part of the illustrated bacteriophage.

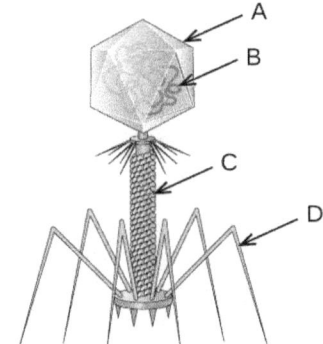

34. In terms of evolution, which do you think arises first? The virus or the host? Explain your answer.
35. Do you think it is possible to create a virus in the lab? Imagine that you are a mad scientist. Describe how you would go about creating a new virus.
36. Label the five stages of a bacteriophage infection in the figure:

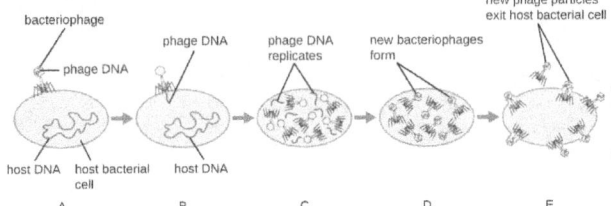

37. Bacteriophages have lytic and lysogenic cycles. Discuss the advantages and disadvantages for the phage.
38. How does reverse transcriptase aid a retrovirus in establishing a chronic infection?
39. Discuss some methods by which plant viruses are transmitted from a diseased plant to a healthy one.
40. Label the components indicated by arrows.

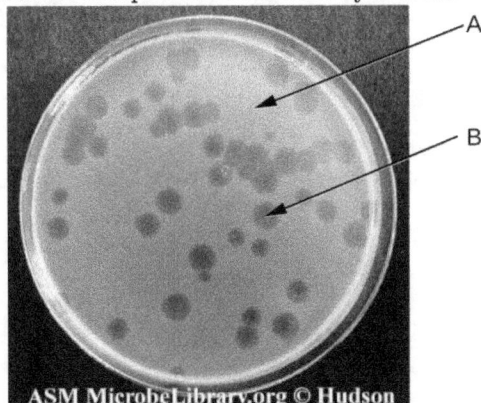

Figure 6.27 (credit: modification of work by American Society for Microbiology)

41. What are some characteristics of the viruses that are similar to a computer virus?
42. Does a prion replicate? Explain.

CHAPTER 7
Microbial Biochemistry

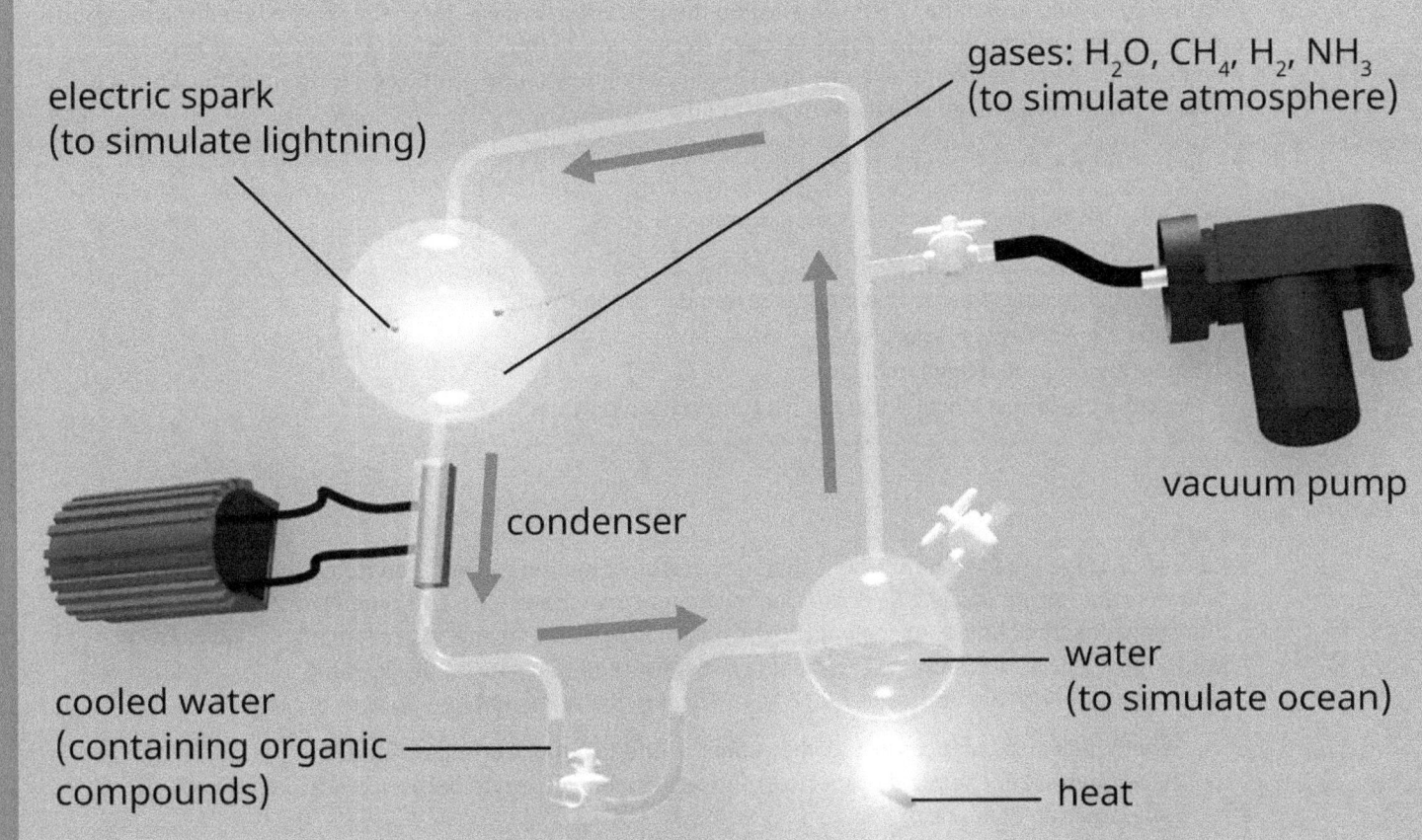

Figure 7.1 Scientist Stanley Miller and Harold Urey demonstrated that organic compounds may have originated naturally from inorganic matter. The Miller-Urey experiment illustrated here simulated the effects of lightning on chemical compounds found in the earth's early atmosphere. The resulting reactions yielded amino acids, the chemical building blocks of proteins. (credit: modification of work by Courtney Harrington)

Chapter Outline

7.1 Organic Molecules

7.2 Carbohydrates

7.3 Lipids

7.4 Proteins

7.5 Using Biochemistry to Identify Microorganisms

INTRODUCTION The earth is estimated to be 4.6 billion years old, but for the first 2 billion years, the atmosphere lacked oxygen, without which the earth could not support life as we know it. One hypothesis about how life emerged on earth involves the concept of a "primordial soup." This idea proposes that life began in a body of water when metals and gases from the atmosphere combined with a source of energy, such as lightning or ultraviolet light, to form the carbon compounds that are the chemical building blocks of life. In 1952, Stanley Miller (1930–2007), a graduate student at the University of Chicago, and his professor Harold Urey (1893–1981), set out to confirm this hypothesis in a now-famous experiment. Miller and Urey combined what they believed to be the major components of the earth's early atmosphere—water (H_2O), methane (CH_4), hydrogen (H_2), and ammonia (NH_3)—and sealed them in a sterile flask. Next, they heated the flask to produce water vapor and passed electric sparks through the mixture to mimic lightning in the atmosphere (Figure 7.1). When they analyzed the contents of the flask a week later, they found amino acids, the structural units of proteins—molecules essential to the function of all organisms.

7.1 Organic Molecules

Learning Objectives

By the end of this section, you will be able to:
- Identify common elements and structures found in organic molecules
- Explain the concept of isomerism
- Identify examples of functional groups
- Describe the role of functional groups in synthesizing polymers

---

Clinical Focus

Part 1

Penny is a 16-year-old student who visited her doctor, complaining about an itchy skin rash. She had a history of allergic episodes. The doctor looked at her sun-tanned skin and asked her if she switched to a different sunscreen. She said she had, so the doctor diagnosed an allergic eczema. The symptoms were mild so the doctor told Penny to avoid using the sunscreen that caused the reaction and prescribed an over-the-counter moisturizing cream to keep her skin hydrated and to help with itching.

- What kinds of substances would you expect to find in a moisturizing cream?
- What physical or chemical properties of these substances would help alleviate itching and inflammation of the skin?

Jump to the next Clinical Focus box.

Biochemistry is the discipline that studies the chemistry of life, and its objective is to explain form and function based on chemical principles. Organic chemistry is the discipline devoted to the study of carbon-based chemistry, which is the foundation for the study of biomolecules and the discipline of biochemistry. Both biochemistry and organic chemistry are based on the concepts of general chemistry, some of which are presented in Appendix A.

Elements in Living Cells

The most abundant element in cells is hydrogen (H), followed by carbon (C), oxygen (O), nitrogen (N), phosphorous (P), and sulfur (S). We call these elements **macronutrients**, and they account for about 99% of the dry weight of cells. Some elements, such as sodium (Na), potassium (K), magnesium (Mg), zinc (Zn), iron (Fe), calcium (Ca), molybdenum (Mo), copper (Cu), cobalt (Co), manganese (Mn), or vanadium (V), are required by some cells in very small amounts and are called **micronutrients** or **trace elements**. All of these elements are essential to the function of many biochemical reactions, and, therefore, are essential to life.

The four most abundant elements in living matter (C, N, O, and H) have low atomic numbers and are thus light elements capable of forming strong bonds with other atoms to produce molecules (Figure 7.2). Carbon forms four chemical bonds, whereas nitrogen forms three, oxygen forms two, and hydrogen forms one. When bonded together within molecules, oxygen, sulfur, and nitrogen often have one or more "lone pairs" of electrons that play important roles in determining many of the molecules' physical and chemical properties (see Appendix A). These traits in combination permit the formation of a vast number of diverse molecular species necessary to form the structures and enable the functions of living organisms.

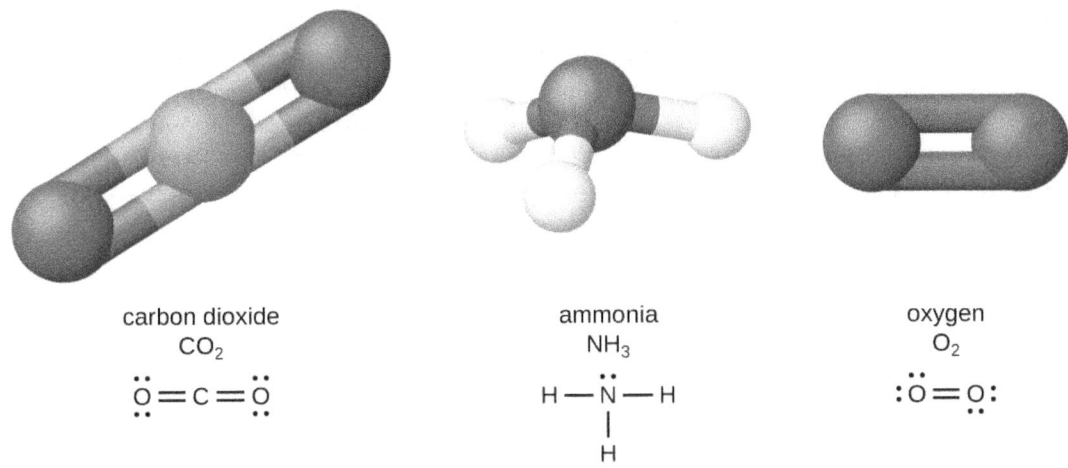

Figure 7.2 Some common molecules include carbon dioxide, ammonia, and oxygen, which consist of combinations of oxygen atoms (red spheres), carbon atoms (gray spheres), hydrogen atoms (white spheres), or nitrogen atoms (blue spheres).

Living organisms contain inorganic compounds (mainly water and salts; see Appendix A) and organic molecules. Organic molecules contain carbon; inorganic compounds do not. Carbon oxides and carbonates are exceptions; they contain carbon but are considered inorganic because they do not contain hydrogen. The atoms of an **organic molecule** are typically organized around chains of carbon atoms.

Inorganic compounds make up 1%–1.5% of the dry weight of living cells. They are small, simple compounds that play important roles in the cell, although they do not form cell structures. Most of the carbon found in organic molecules originates from inorganic carbon sources such as carbon dioxide captured via carbon fixation by microorganisms.

✓ CHECK YOUR UNDERSTANDING

- Describe the most abundant elements in nature.
- What are the differences between organic and inorganic molecules?

Organic Molecules and Isomerism

Organic molecules in organisms are generally larger and more complex than inorganic molecules. Their carbon skeletons are held together by covalent bonds. They form the cells of an organism and perform the chemical reactions that facilitate life. All of these molecules, called **biomolecules** because they are part of living matter, contain carbon, which is the building block of life. Carbon is a very unique element in that it has four valence electrons in its outer orbitals and can form four single covalent bonds with up to four other atoms at the same time (see Appendix A). These atoms are usually oxygen, hydrogen, nitrogen, sulfur, phosphorous, and carbon itself; the simplest organic compound is methane, in which carbon binds only to hydrogen (Figure 7.3).

As a result of carbon's unique combination of size and bonding properties, carbon atoms can bind together in large numbers, thus producing a chain or **carbon skeleton**. The carbon skeleton of organic molecules can be straight, branched, or ring shaped (cyclic). Organic molecules are built on chains of carbon atoms of varying lengths; most are typically very long, which allows for a huge number and variety of compounds. No other element has the ability to form so many different molecules of so many different sizes and shapes.

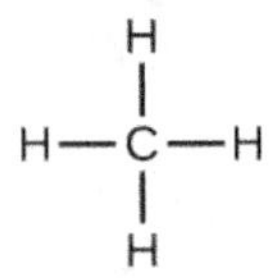

Figure 7.3 A carbon atom can bond with up to four other atoms. The simplest organic molecule is methane (CH_4), depicted here.

Molecules with the same atomic makeup but different structural arrangement of atoms are called **isomers**.

The concept of isomerism is very important in chemistry because the structure of a molecule is always directly related to its function. Slight changes in the structural arrangements of atoms in a molecule may lead to very different properties. Chemists represent molecules by their **structural formula**, which is a graphic representation of the molecular structure, showing how the atoms are arranged. Compounds that have identical molecular formulas but differ in the bonding sequence of the atoms are called **structural isomers**. The monosaccharides glucose, galactose, and fructose all have the same molecular formula, $C_6H_{12}O_6$, but we can see from Figure 7.4 that the atoms are bonded together differently.

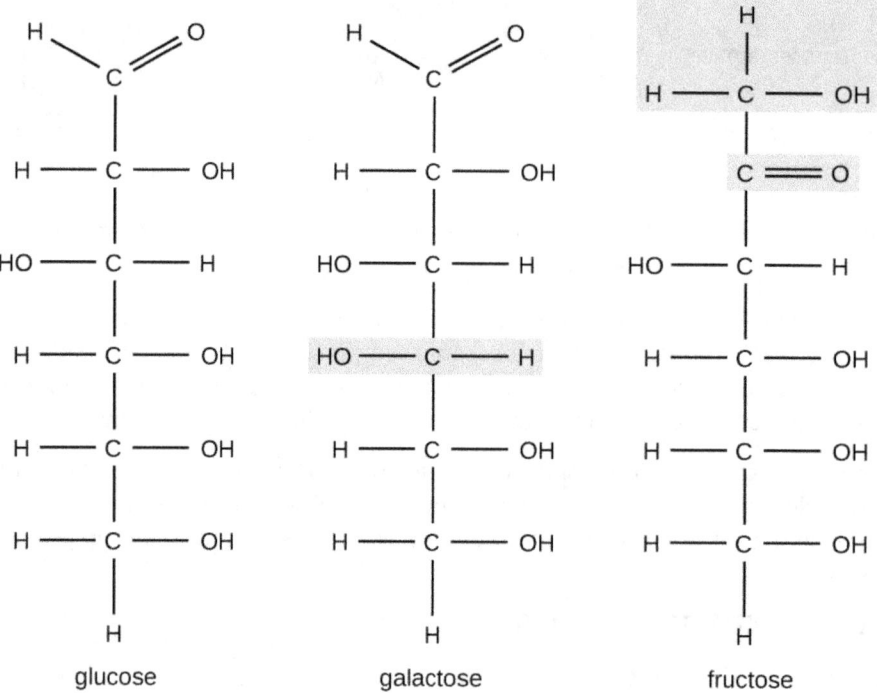

Figure 7.4 Glucose, galactose, and fructose have the same chemical formula ($C_6H_{12}O_6$), but these structural isomers differ in their physical and chemical properties.

Isomers that differ in the spatial arrangements of atoms are called **stereoisomers**; one unique type is **enantiomers**. The properties of enantiomers were originally discovered by Louis Pasteur in 1848 while using a microscope to analyze crystallized fermentation products of wine. Enantiomers are molecules that have the characteristic of **chirality**, in which their structures are nonsuperimposable mirror images of each other. Chirality is an important characteristic in many biologically important molecules, as illustrated by the examples of structural differences in the enantiomeric forms of the monosaccharide glucose or the amino acid alanine (Figure 7.5).

Many organisms are only able to use one enantiomeric form of certain types of molecules as nutrients and as building blocks to make structures within a cell. Some enantiomeric forms of amino acids have distinctly different tastes and smells when consumed as food. For example, L-aspartame, commonly called aspartame, tastes sweet, whereas D-aspartame is tasteless. Drug enantiomers can have very different pharmacologic affects. For example, the compound methorphan exists as two enantiomers, one of which acts as an antitussive (*dextro*methorphan, a cough suppressant), whereas the other acts as an analgesic (*levo*methorphan, a drug similar in effect to codeine).

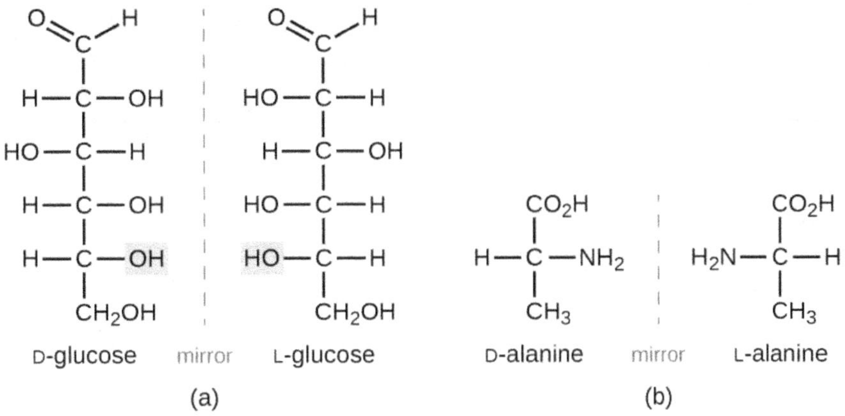

Figure 7.5 Enantiomers are stereoisomers that exhibit chirality. Their chemical structures are nonsuperimposable mirror images of each other. (a) D-glucose and L-glucose are monosaccharides that are enantiomers. (b) The enantiomers D-alanine and L-alanine are enantiomers found in bacterial cell walls and human cells, respectively.

Enantiomers are also called optical isomers because they can rotate the plane of polarized light. Some of the crystals Pasteur observed from wine fermentation rotated light clockwise whereas others rotated the light counterclockwise. Today, we denote enantiomers that rotate polarized light clockwise (+) as *d* forms, and the mirror image of the same molecule that rotates polarized light counterclockwise (−) as the *l* form. The *d* and *l* labels are derived from the Latin words *dexter* (on the right) and *laevus* (on the left), respectively. These two different optical isomers often have very different biological properties and activities. Certain species of molds, yeast, and bacteria, such as *Rhizopus*, *Yarrowia*, and *Lactobacillus* spp., respectively, can only metabolize one type of optical isomer; the opposite isomer is not suitable as a source of nutrients. Another important reason to be aware of optical isomers is the therapeutic use of these types of chemicals for drug treatment, because some microorganisms can only be affected by one specific optical isomer.

✓ CHECK YOUR UNDERSTANDING

- We say that life is carbon based. What makes carbon so suitable to be part of all the macromolecules of living organisms?

Biologically Significant Functional Groups

In addition to containing carbon atoms, biomolecules also contain **functional groups**—groups of atoms within molecules that are categorized by their specific chemical composition and the chemical reactions they perform, regardless of the molecule in which the group is found. Some of the most common functional groups are listed in Figure 7.6. In the formulas, the symbol R stands for "residue" and represents the remainder of the molecule. R might symbolize just a single hydrogen atom or it may represent a group of many atoms. Notice that some functional groups are relatively simple, consisting of just one or two atoms, while some comprise two of these simpler functional groups. For example, a carbonyl group is a functional group composed of a carbon atom double bonded to an oxygen atom: C=O. It is present in several classes of organic compounds as part of larger functional groups such as ketones, aldehydes, carboxylic acids, and amides. In ketones, the carbonyl is present as an internal group, whereas in aldehydes it is a terminal group.

Common Functional Groups Found in Biomolecules		
Name	Functional Group	Compounds
Aldehyde	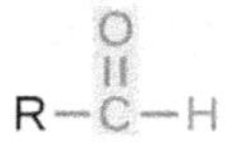 R–C(=O)–H	Carbohydrates
Amide	 R–C(=O)–N(H)–R'	Proteins
Amino	R–NH$_2$	Amino acids, proteins
Carbonyl	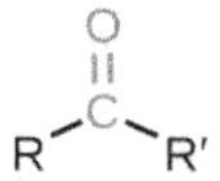 R–C(=O)–R'	Ketones, aldehydes, carboxylic acids, amides
Carboxylic acid	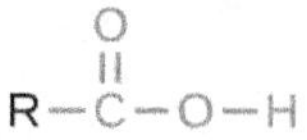 R–C(=O)–O–H	Amino acids, proteins, fatty acids
Ester	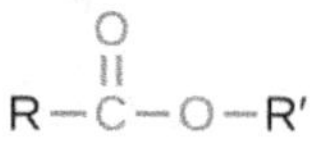 R–C(=O)–O–R'	Lipids, nucleic acids
Ether	R–O–R'	Disaccharides, polysaccharides, lipids
Hydroxyl	R–O–H	Alcohols, monosaccharides, amino acids, nucleic acids
Ketone	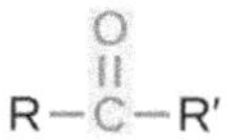 R–C(=O)–R'	Carbohydrates
Methyl	R–CH$_3$	Methylated compounds such as methyl alcohols and methyl esters
Phosphate	R–PO$_3$H$_2$	Nucleic acids, phospholipids, ATP
Sulfhydryl	R–S–H	Amino acids, proteins

*Functional groups are represented in pink. Ketone and aldehyde both contain a carbonyl group, highlighted in blue.

Figure 7.6

Macromolecules

Carbon chains form the skeletons of most organic molecules. Functional groups combine with the chain to form biomolecules. Because these biomolecules are typically large, we call them **macromolecules**. Many biologically relevant macromolecules are formed by linking together a great number of identical, or very similar, smaller organic molecules. The smaller molecules act as building blocks and are called **monomers**, and the macromolecules that result from their linkage are called **polymers**. Cells and cell structures include four main groups of carbon-containing macromolecules: polysaccharides, proteins, lipids, and nucleic acids. The first three groups of molecules will be studied throughout this chapter. The biochemistry of nucleic acids will be discussed in Biochemistry of the Genome.

Of the many possible ways that monomers may be combined to yield polymers, one common approach encountered in the formation of biological macromolecules is **dehydration synthesis**. In this chemical reaction, monomer molecules bind end to end in a process that results in the formation of water molecules as a byproduct:

$$H\text{—monomer—}OH + H\text{—monomer—}OH \longrightarrow H\text{—monomer—monomer—}OH + H_2O$$

Figure 7.7 shows dehydration synthesis of glucose binding together to form maltose and a water molecule. Table 7.1 summarizes macromolecules and some of their functions.

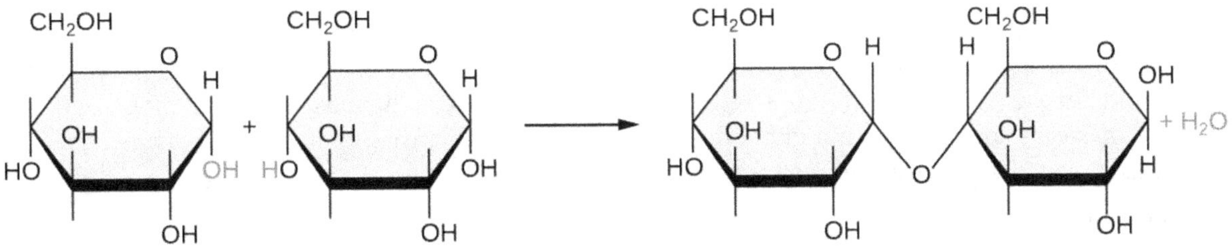

Figure 7.7 In this dehydration synthesis reaction, two molecules of glucose are linked together to form maltose. In the process, a water molecule is formed.

Some Functions of Macromolecules

Macromolecule	Functions
Carbohydrates	Energy storage, receptors, food, structural role in plants, fungal cell walls, exoskeletons of insects
Lipids	Energy storage, membrane structure, insulation, hormones, pigments
Nucleic acids	Storage and transfer of genetic information
Proteins	Enzymes, structure, receptors, transport, structural role in the cytoskeleton of a cell and the extracellular matrix

Table 7.1

CHECK YOUR UNDERSTANDING

- What is the byproduct of a dehydration synthesis reaction?

7.2 Carbohydrates

Learning Objectives

By the end of this section, you will be able to:
- Give examples of monosaccharides and polysaccharides
- Describe the function of monosaccharides and polysaccharides within a cell

The most abundant biomolecules on earth are **carbohydrates**. From a chemical viewpoint, carbohydrates are primarily a combination of carbon and water, and many of them have the empirical formula $(CH_2O)_n$, where n is the number of repeated units. This view represents these molecules simply as "hydrated" carbon atom chains in which water molecules attach to each carbon atom, leading to the term "carbohydrates." Although all carbohydrates contain carbon, hydrogen, and oxygen, there are some that also contain nitrogen, phosphorus, and/or sulfur. Carbohydrates have myriad different functions. They are abundant in terrestrial ecosystems, many forms of which we use as food sources. These molecules are also vital parts of macromolecular structures that store and transmit genetic information (i.e., DNA and RNA). They are the basis of biological polymers that impart strength to various structural components of organisms (e.g., cellulose and chitin), and they are the primary source of energy storage in the form of starch and glycogen.

Monosaccharides: The Sweet Ones

In biochemistry, carbohydrates are often called **saccharides**, from the Greek *sakcharon*, meaning sugar, although not all the saccharides are sweet. The simplest carbohydrates are called **monosaccharides**, or simple sugars. They are the building blocks (monomers) for the synthesis of polymers or complex carbohydrates, as will be discussed further in this section. Monosaccharides are classified based on the number of carbons in the molecule. General categories are identified using a prefix that indicates the number of carbons and the suffix –*ose*, which indicates a saccharide; for example, triose (three carbons), tetrose (four carbons), pentose (five carbons), and hexose (six carbons) (Figure 7.8). The hexose D-glucose is the most abundant monosaccharide in nature. Other very common and abundant hexose monosaccharides are galactose, used to make the disaccharide milk sugar lactose, and the fruit sugar fructose.

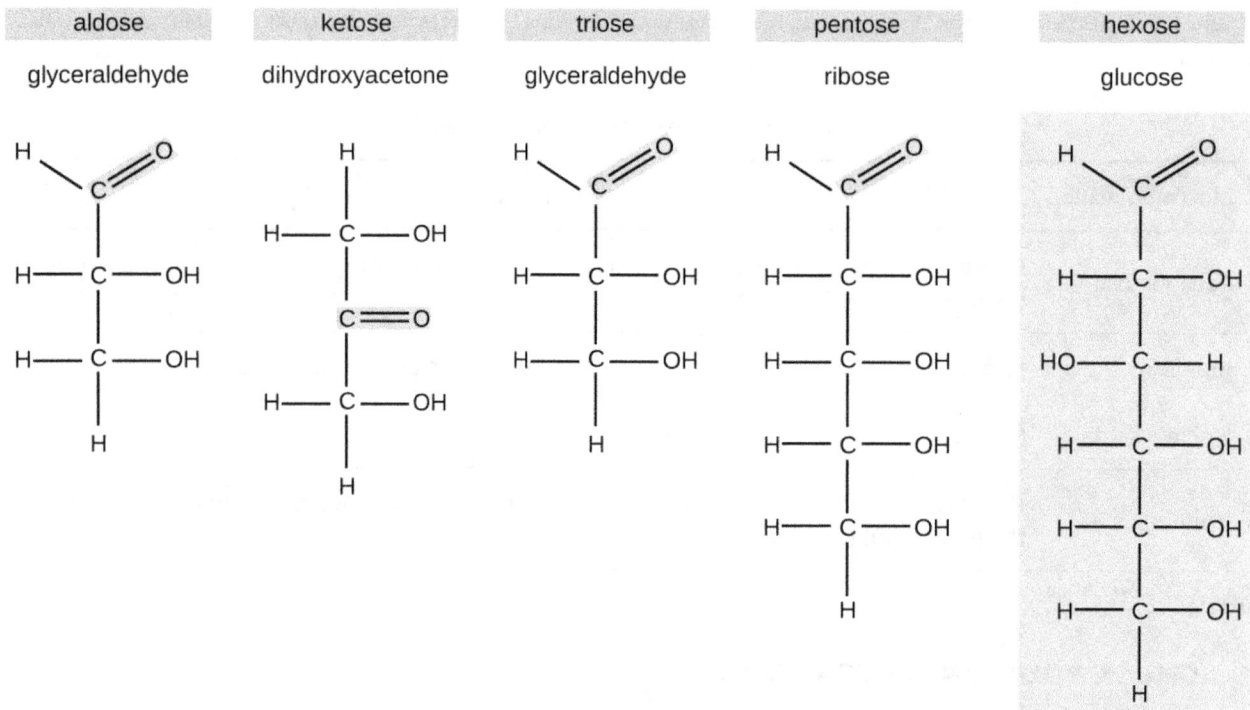

Figure 7.8 Monosaccharides are classified based on the position of the carbonyl group and the number of carbons in the backbone.

Monosaccharides of four or more carbon atoms are typically more stable when they adopt cyclic, or ring, structures. These ring structures result from a chemical reaction between functional groups on opposite ends of the sugar's flexible carbon chain, namely the carbonyl group and a relatively distant hydroxyl group. Glucose, for example, forms a six-membered ring (Figure 7.9).

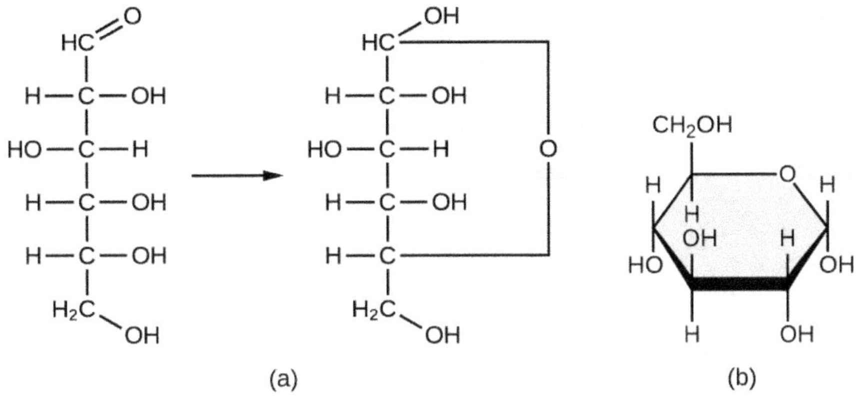

Figure 7.9 (a) A linear monosaccharide (glucose in this case) forms a cyclic structure. (b) This illustration shows a more realistic depiction of the cyclic monosaccharide structure. Note in these cyclic structural diagrams, the carbon atoms composing the ring are not explicitly shown.

✓ CHECK YOUR UNDERSTANDING

- Why do monosaccharides form ring structures?

Disaccharides

Two monosaccharide molecules may chemically bond to form a **disaccharide**. The name given to the covalent bond between the two monosaccharides is a **glycosidic bond**. Glycosidic bonds form between hydroxyl groups of the two saccharide molecules, an example of the dehydration synthesis described in the previous section of this chapter:

monosaccharide—OH + HO—monosaccharide ⟶ monosaccharide—O—monosaccharide
                                                                    disaccharide

Common disaccharides are the grain sugar maltose, made of two glucose molecules; the milk sugar lactose, made of a galactose and a glucose molecule; and the table sugar sucrose, made of a glucose and a fructose molecule (Figure 7.10).

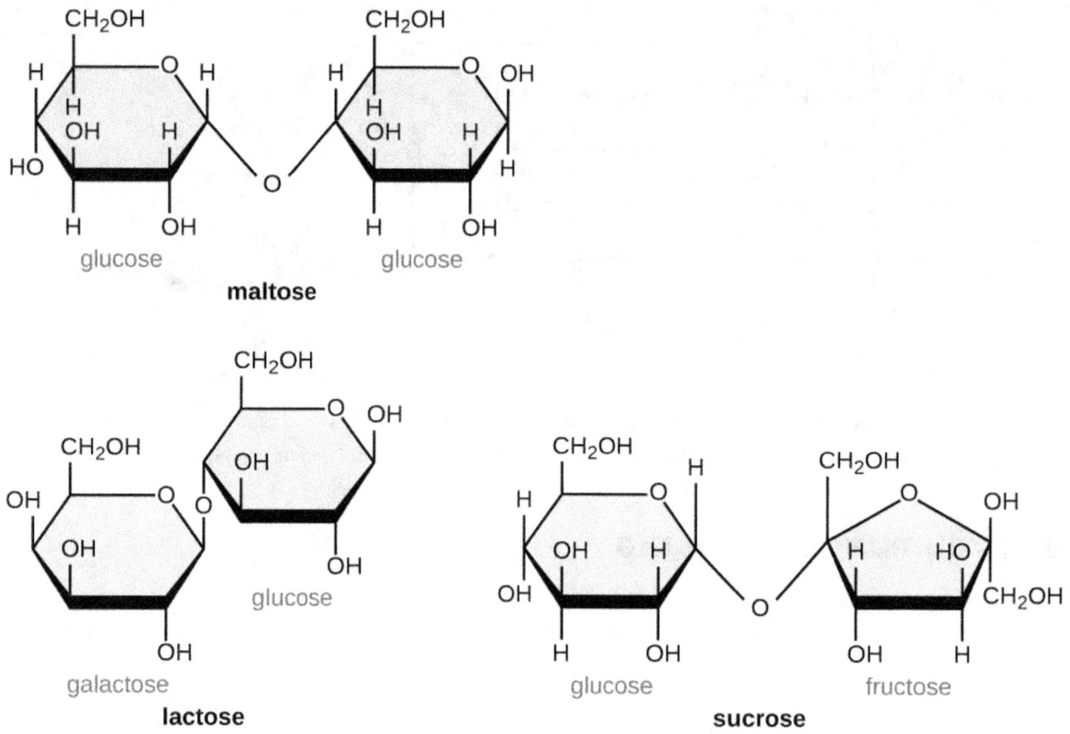

Figure 7.10 Common disaccharides include maltose, lactose, and sucrose.

Polysaccharides

Polysaccharides, also called glycans, are large polymers composed of hundreds of monosaccharide monomers. Unlike mono- and disaccharides, **polysaccharides** are not sweet and, in general, they are not soluble in water. Like disaccharides, the monomeric units of polysaccharides are linked together by glycosidic bonds.

Polysaccharides are very diverse in their structure. Three of the most biologically important polysaccharides—**starch**, **glycogen**, and **cellulose**—are all composed of repetitive glucose units, although they differ in their structure (Figure 7.11). Cellulose consists of a linear chain of glucose molecules and is a common structural component of cell walls in plants and other organisms. Glycogen and starch are branched polymers; glycogen is the primary energy-storage molecule in animals and bacteria, whereas plants primarily store energy in starch. The orientation of the glycosidic linkages in these three polymers is different as well and, as a consequence, linear and branched macromolecules have different properties.

Modified glucose molecules can be fundamental components of other structural polysaccharides. Examples of these types of structural polysaccharides are N-acetyl glucosamine (NAG) and N-acetyl muramic acid (NAM) found in bacterial cell wall peptidoglycan. Polymers of NAG form chitin, which is found in fungal cell walls and in the exoskeleton of insects.

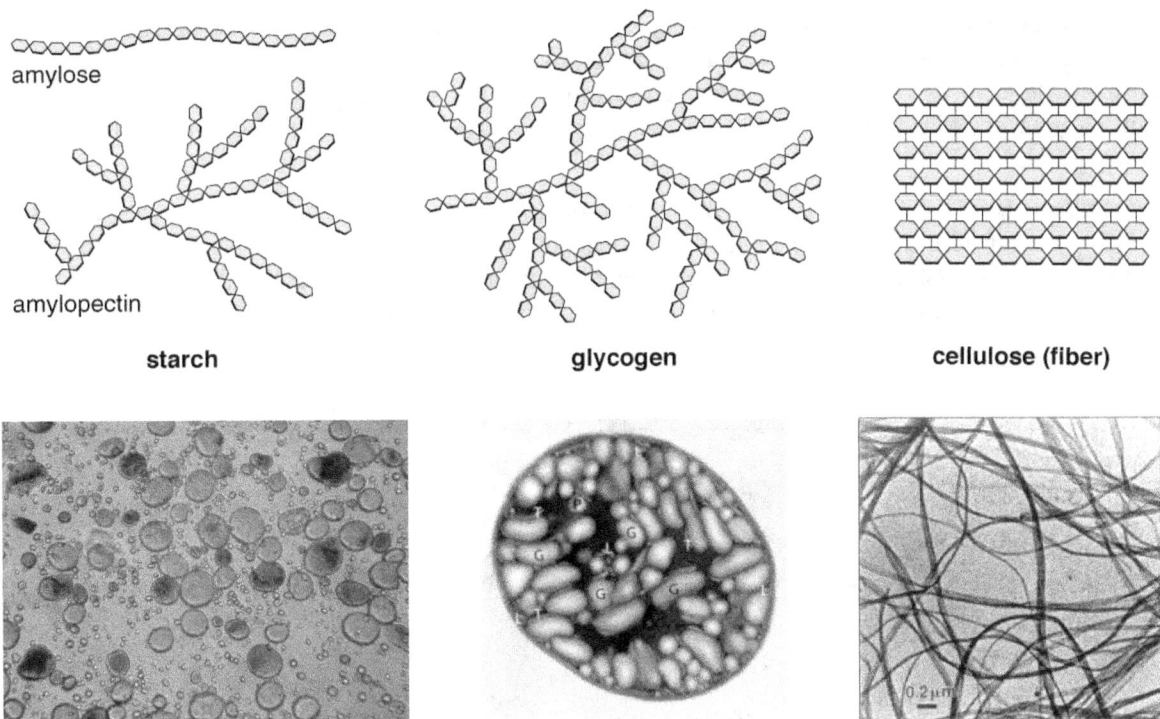

Figure 7.11 Starch, glycogen, and cellulose are three of the most important polysaccharides. In the top row, hexagons represent individual glucose molecules. Micrographs (bottom row) show wheat starch granules stained with iodine (left), glycogen granules (G) inside the cell of a cyanobacterium (middle), and bacterial cellulose fibers (right). (credit "iodine granules": modification of work by Kiselov Yuri; credit "glycogen granules": modification of work by Stöckel J, Elvitigala TR, Liberton M, Pakrasi HB; credit "cellulose": modification of work by American Society for Microbiology)

CHECK YOUR UNDERSTANDING

- What are the most biologically important polysaccharides and why are they important?

7.3 Lipids

Learning Objectives

By the end of this section, you will be able to:
- Describe the chemical composition of lipids
- Describe the unique characteristics and diverse structures of lipids
- Compare and contrast triacylglycerides (triglycerides) and phospholipids.
- Describe how phospholipids are used to construct biological membranes.

Although they are composed primarily of carbon and hydrogen, **lipid** molecules may also contain oxygen, nitrogen, sulfur, and phosphorous. Lipids serve numerous and diverse purposes in the structure and functions of organisms. They can be a source of nutrients, a storage form for carbon, energy-storage molecules, or structural components of membranes and hormones. Lipids comprise a broad class of many chemically distinct compounds, the most common of which are discussed in this section.

Fatty Acids and Triacylglycerides

The **fatty acid**s are lipids that contain long-chain hydrocarbons terminated with a carboxylic acid functional group. Because the long hydrocarbon chain, fatty acids are **hydrophobic** ("water fearing") or nonpolar. Fatty acids with hydrocarbon chains that contain only single bonds are called **saturated fatty acids** because they have the greatest number of hydrogen atoms possible and are, therefore, "saturated" with hydrogen. Fatty acids with hydrocarbon chains containing at least one double bond are called **unsaturated fatty acid**s because they have fewer hydrogen atoms. Saturated fatty acids have a straight, flexible carbon backbone, whereas unsaturated fatty acids have "kinks" in their carbon skeleton because each double bond causes a rigid bend of the carbon skeleton. These differences in saturated versus unsaturated fatty acid structure result in different properties for the corresponding lipids in which the fatty acids are incorporated. For example, lipids containing saturated fatty acids are solids at room temperature, whereas lipids containing unsaturated fatty acids are liquids.

A **triacylglycerol**, or **triglyceride**, is formed when three fatty acids are chemically linked to a glycerol molecule (Figure 7.12). Triglycerides are the primary components of adipose tissue (body fat), and are major constituents of sebum (skin oils). They play an important metabolic role, serving as efficient energy-storage molecules that can provide more than double the caloric content of both carbohydrates and proteins.

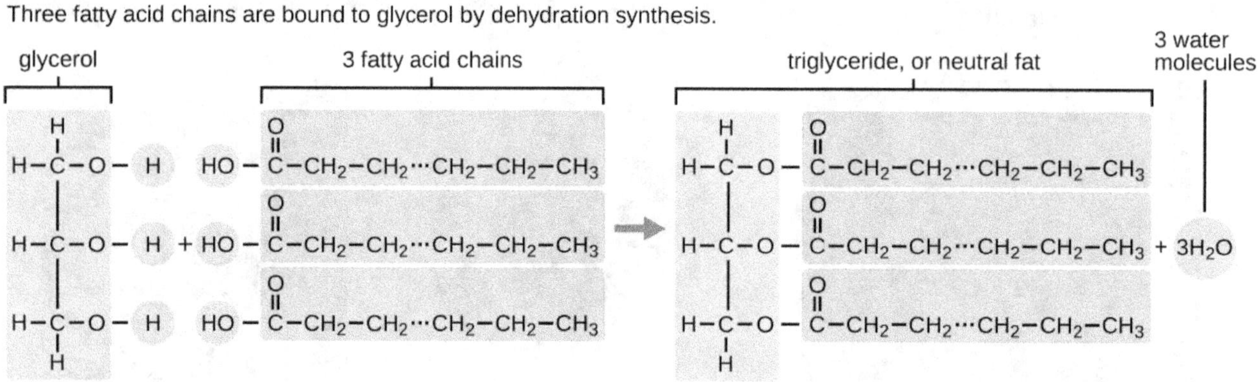

Three fatty acid chains are bound to glycerol by dehydration synthesis.

Figure 7.12 Triglycerides are composed of a glycerol molecule attached to three fatty acids by a dehydration synthesis reaction.

CHECK YOUR UNDERSTANDING

- Explain why fatty acids with hydrocarbon chains that contain only single bonds are called saturated fatty acids.

Phospholipids and Biological Membranes

Triglycerides are classified as simple lipids because they are formed from just two types of compounds: glycerol and fatty acids. In contrast, complex lipids contain at least one additional component, for example, a phosphate group (**phospholipids**) or a carbohydrate moiety (**glycolipids**). Figure 7.13 depicts a typical phospholipid composed of two fatty acids linked to glycerol (a diglyceride). The two fatty acid carbon chains may be both saturated, both unsaturated, or one of each. Instead of another fatty acid molecule (as for triglycerides), the third binding position on the glycerol molecule is occupied by a modified phosphate group.

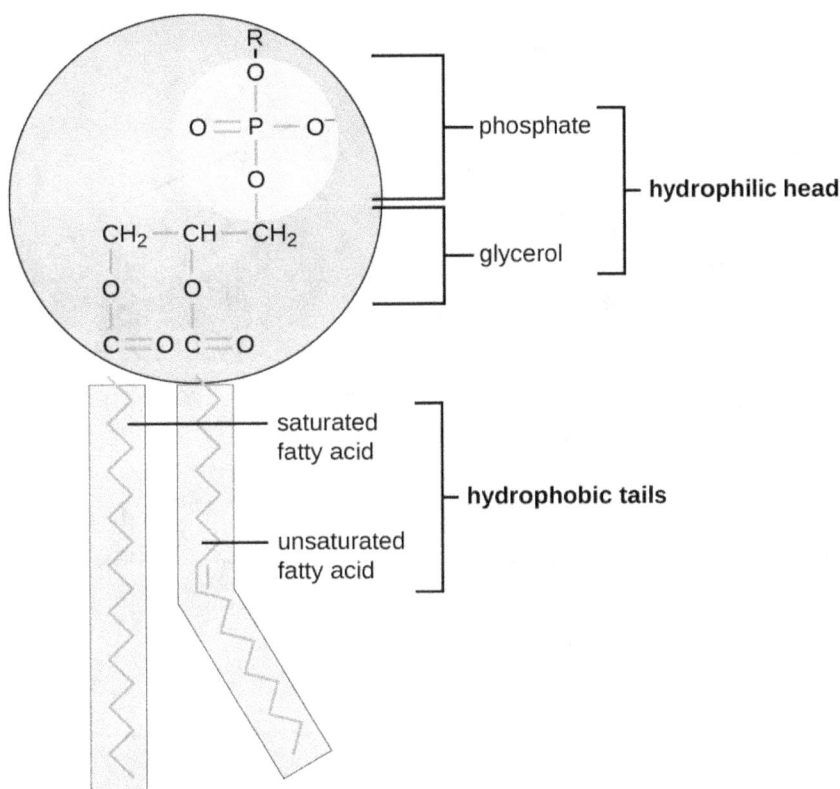

Figure 7.13 This illustration shows a phospholipid with two different fatty acids, one saturated and one unsaturated, bonded to the glycerol molecule. The unsaturated fatty acid has a slight kink in its structure due to the double bond.

The molecular structure of lipids results in unique behavior in aqueous environments. Figure 7.12 depicts the structure of a triglyceride. Because all three substituents on the glycerol backbone are long hydrocarbon chains, these compounds are nonpolar and not significantly attracted to polar water molecules—they are hydrophobic. Conversely, phospholipids such as the one shown in Figure 7.13 have a negatively charged phosphate group. Because the phosphate is charged, it is capable of strong attraction to water molecules and thus is **hydrophilic**, or "water loving." The hydrophilic portion of the phospholipid is often referred to as a polar "head," and the long hydrocarbon chains as nonpolar "tails." A molecule presenting a hydrophobic portion and a hydrophilic moiety is said to be **amphipathic**. Notice the "R" designation within the hydrophilic head depicted in Figure 7.13, indicating that a polar head group can be more complex than a simple phosphate moiety. Glycolipids are examples in which carbohydrates are bonded to the lipids' head groups.

The amphipathic nature of phospholipids enables them to form uniquely functional structures in aqueous environments. As mentioned, the polar heads of these molecules are strongly attracted to water molecules, and the nonpolar tails are not. Because of their considerable lengths, these tails are, in fact, strongly attracted to one another. As a result, energetically stable, large-scale assemblies of phospholipid molecules are formed in which the hydrophobic tails congregate within enclosed regions, shielded from contact with water by the polar heads (Figure 7.14). The simplest of these structures are **micelles**, spherical assemblies containing a hydrophobic interior of phospholipid tails and an outer surface of polar head groups. Larger and more complex structures are created from **lipid-bilayer** sheets, or **unit membranes**, which are large, two-dimensional assemblies of phospholipids congregated tail to tail. The cell membranes of nearly all organisms are made from lipid-bilayer sheets, as are the membranes of many intracellular components. These sheets may also form lipid-bilayer spheres that are the structural basis of vesicles and liposomes, subcellular components that play a role in numerous physiological functions.

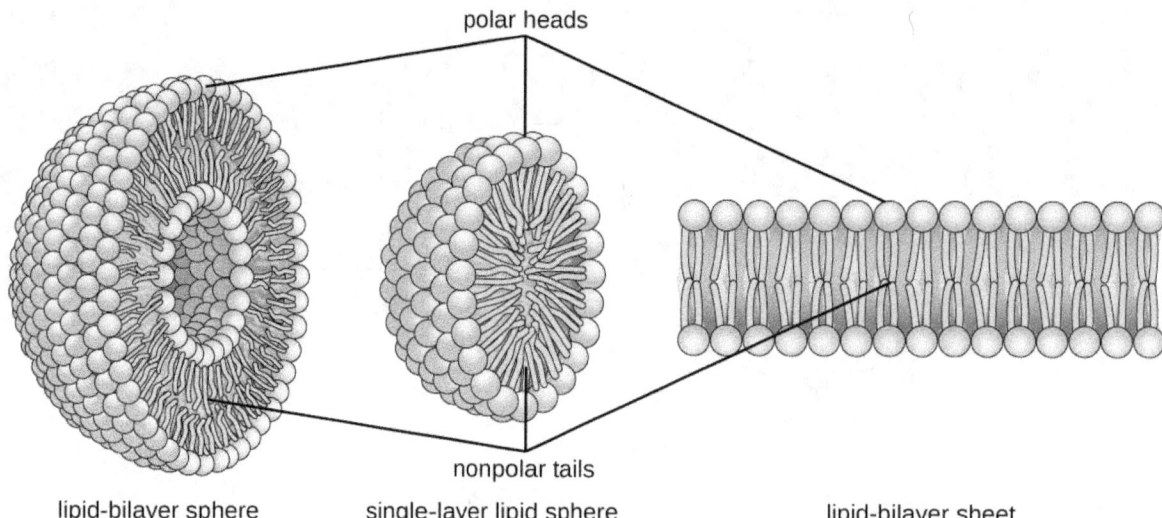

Figure 7.14 Phospholipids tend to arrange themselves in aqueous solution forming liposomes, micelles, or lipid bilayer sheets. (credit: modification of work by Mariana Ruiz Villarreal)

✓ CHECK YOUR UNDERSTANDING

- How is the amphipathic nature of phospholipids significant?

Isoprenoids and Sterols

The **isoprenoids** are branched lipids, also referred to as terpenoids, that are formed by chemical modifications of the isoprene molecule (Figure 7.15). These lipids play a wide variety of physiological roles in plants and animals, with many technological uses as pharmaceuticals (capsaicin), pigments (e.g., orange beta carotene, xanthophylls), and fragrances (e.g., menthol, camphor, limonene [lemon fragrance], and pinene [pine fragrance]). Long-chain isoprenoids are also found in hydrophobic oils and waxes. Waxes are typically water resistant and hard at room temperature, but they soften when heated and liquefy if warmed adequately. In humans, the main wax production occurs within the sebaceous glands of hair follicles in the skin, resulting in a secreted material called sebum, which consists mainly of triacylglycerol, wax esters, and the hydrocarbon squalene. There are many bacteria in the microbiota on the skin that feed on these lipids. One of the most prominent bacteria that feed on lipids is *Propionibacterium acnes*, which uses the skin's lipids to generate short-chain fatty acids and is involved in the production of acne.

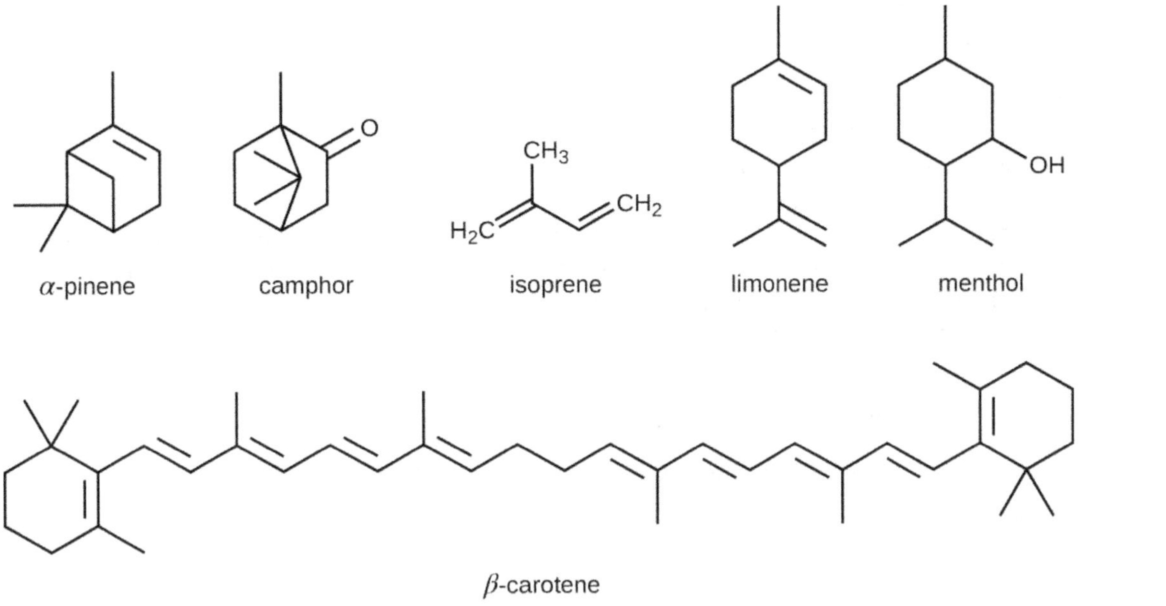

Figure 7.15 Five-carbon isoprene molecules are chemically modified in various ways to yield isoprenoids.

Another type of lipids are **steroids**, complex, ringed structures that are found in cell membranes; some function as hormones. The most common types of steroids are **sterols**, which are steroids containing an OH group. These are mainly hydrophobic molecules, but also have hydrophilic hydroxyl groups. The most common sterol found in animal tissues is cholesterol. Its structure consists of four rings with a double bond in one of the rings, and a hydroxyl group at the sterol-defining position. The function of cholesterol is to strengthen cell membranes in eukaryotes and in bacteria without cell walls, such as *Mycoplasma*. Prokaryotes generally do not produce cholesterol, although bacteria produce similar compounds called hopanoids, which are also multiringed structures that strengthen bacterial membranes (Figure 7.16). Fungi and some protozoa produce a similar compound called ergosterol, which strengthens the cell membranes of these organisms.

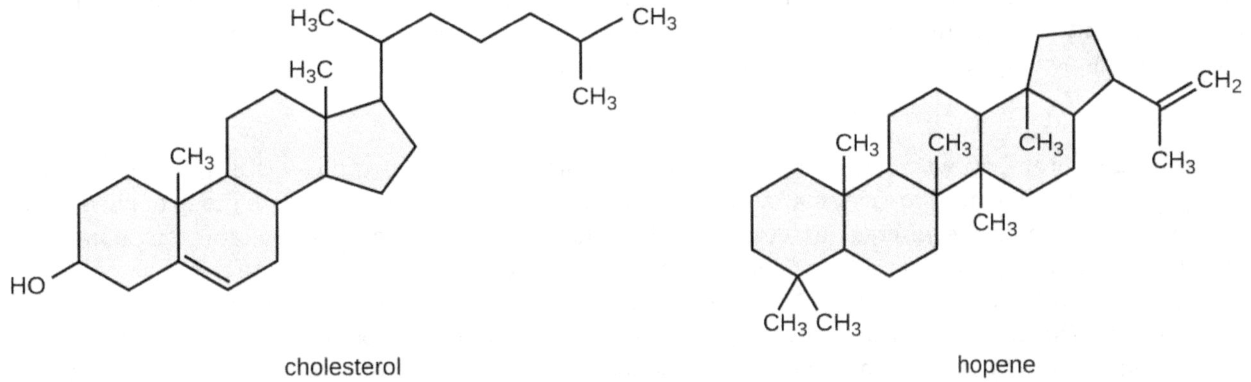

Figure 7.16 Cholesterol and hopene (a hopanoid compound) are molecules that reinforce the structure of the cell membranes in eukaryotes and prokaryotes, respectively.

🔗 LINK TO LEARNING

Liposomes
This video (https://openstax.org/l/22liposomes) provides additional information about phospholipids and liposomes.

✓ CHECK YOUR UNDERSTANDING

- How are isoprenoids used in technology?

> **Clinical Focus**
>
> **Part 2**
>
> The moisturizing cream prescribed by Penny's doctor was a topical corticosteroid cream containing hydrocortisone. Hydrocortisone is a synthetic form of cortisol, a corticosteroid hormone produced in the adrenal glands, from cholesterol. When applied directly to the skin, it can reduce inflammation and temporarily relieve minor skin irritations, itching, and rashes by reducing the secretion of histamine, a compound produced by cells of the immune system in response to the presence of pathogens or other foreign substances. Because histamine triggers the body's inflammatory response, the ability of hydrocortisone to reduce the local production of histamine in the skin effectively suppresses the immune system and helps limit inflammation and accompanying symptoms such as pruritus (itching) and rashes.
>
> - Does the corticosteroid cream treat the cause of Penny's rash, or just the symptoms?
>
> *Jump to the next Clinical Focus box. Go back to the previous Clinical Focus box.*

7.4 Proteins

Learning Objectives

By the end of this section, you will be able to:
- Describe the fundamental structure of an amino acid
- Describe the chemical structures of proteins
- Summarize the unique characteristics of proteins

At the beginning of this chapter, a famous experiment was described in which scientists synthesized amino acids under conditions simulating those present on earth long before the evolution of life as we know it. These compounds are capable of bonding together in essentially any number, yielding molecules of essentially any size that possess a wide array of physical and chemical properties and perform numerous functions vital to all organisms. The molecules derived from amino acids can function as structural components of cells and subcellular entities, as sources of nutrients, as atom- and energy-storage reservoirs, and as functional species such as hormones, enzymes, receptors, and transport molecules.

Amino Acids and Peptide Bonds

An **amino acid** is an organic molecule in which a hydrogen atom, a carboxyl group (–COOH), and an amino group (–NH$_2$) are all bonded to the same carbon atom, the so-called α carbon. The fourth group bonded to the α carbon varies among the different amino acids and is called a residue or a **side chain**, represented in structural formulas by the letter *R*. A residue is a monomer that results when two or more amino acids combine and remove water molecules. The primary structure of a protein, a peptide chain, is made of amino acid residues. The unique characteristics of the functional groups and *R* groups allow these components of the amino acids to form hydrogen, ionic, and disulfide bonds, along with polar/nonpolar interactions needed to form secondary, tertiary, and quaternary protein structures. These groups are composed primarily of carbon, hydrogen, oxygen, nitrogen, and sulfur, in the form of hydrocarbons, acids, amides, alcohols, and amines. A few examples illustrating these possibilities are provided in Figure 7.17.

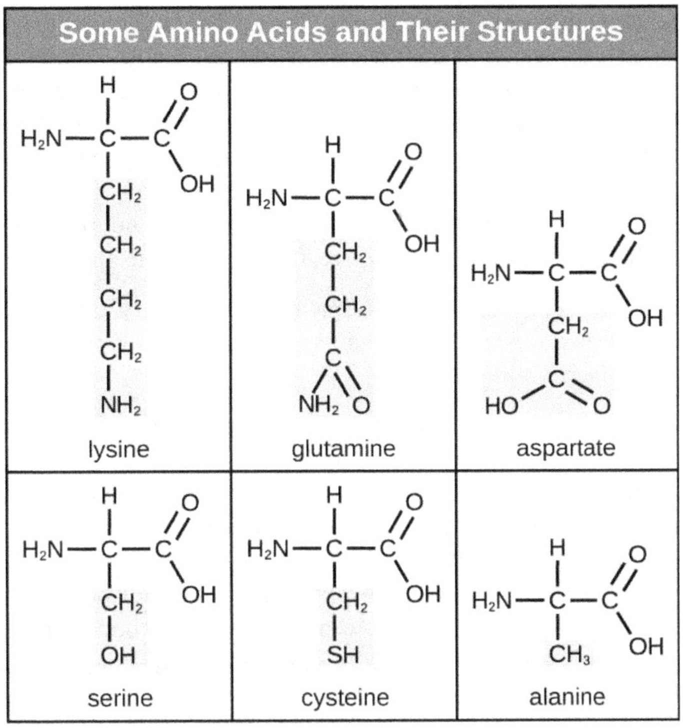

Figure 7.17 Some Amino Acids and Their Structures (lysine, glutamine, aspartate, serine, cysteine, alanine). *Blue shading indicates R group.

Amino acids may chemically bond together by reaction of the carboxylic acid group of one molecule with the amine group of another. This reaction forms a **peptide bond** and a water molecule and is another example of dehydration synthesis (Figure 7.18). Molecules formed by chemically linking relatively modest numbers of amino acids (approximately 50 or fewer) are called peptides, and prefixes are often used to specify these numbers: dipeptides (two amino acids), tripeptides (three amino acids), and so forth. More generally, the approximate number of amino acids is designated: **oligopeptides** are formed by joining up to approximately 20 amino acids, whereas **polypeptides** are synthesized from up to approximately 50 amino acids. When the number of amino acids linked together becomes very large, or when multiple polypeptides are used as building subunits, the macromolecules that result are called **proteins**. The continuously variable length (the number of monomers) of these biopolymers, along with the variety of possible R groups on each amino acid, allows for a nearly unlimited diversity in the types of proteins that may be formed.

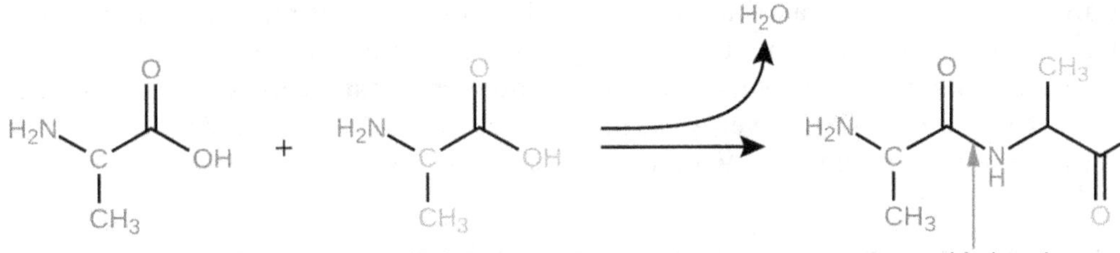

Figure 7.18 Peptide bond formation is a dehydration synthesis reaction. The carboxyl group of the first amino acid (alanine) is linked to the amino group of the incoming second amino acid (alanine). In the process, a molecule of water is released.

CHECK YOUR UNDERSTANDING

- How many amino acids are in polypeptides?

Protein Structure

The size (length) and specific amino acid sequence of a protein are major determinants of its shape, and the shape of a protein is critical to its function. For example, in the process of biological nitrogen fixation (see Biogeochemical Cycles), soil microorganisms collectively known as rhizobia symbiotically interact with roots of legume plants such as soybeans, peanuts, or beans to form a novel structure called a nodule on the plant roots. The plant then produces a carrier protein called leghemoglobin, a protein that carries nitrogen or oxygen. Leghemoglobin binds with a very high affinity to its substrate oxygen at a specific region of the protein where the shape and amino acid sequence are appropriate (the active site). If the shape or chemical environment of the active site is altered, even slightly, the substrate may not be able to bind as strongly, or it may not bind at all. Thus, for the protein to be fully active, it must have the appropriate shape for its function.

Protein structure is categorized in terms of four levels: primary, secondary, tertiary, and quaternary. The **primary structure** is simply the sequence of amino acids that make up the polypeptide chain. Figure 7.19 depicts the primary structure of a protein.

The chain of amino acids that defines a protein's primary structure is not rigid, but instead is flexible because of the nature of the bonds that hold the amino acids together. When the chain is sufficiently long, hydrogen bonding may occur between amine and carbonyl functional groups within the peptide backbone (excluding the R side group), resulting in localized folding of the polypeptide chain into helices and sheets. These shapes constitute a protein's **secondary structure**. The most common secondary structures are the α-helix and β-pleated sheet. In the **α-helix** structure, the helix is held by hydrogen bonds between the oxygen atom in a carbonyl group of one amino acid and the hydrogen atom of the amino group that is just four amino acid units farther along the chain. In the **β-pleated sheet**, the pleats are formed by similar hydrogen bonds between continuous sequences of carbonyl and amino groups that are further separated on the backbone of the polypeptide chain (Figure 7.20).

The next level of protein organization is the **tertiary structure**, which is the large-scale three-dimensional shape of a single polypeptide chain. Tertiary structure is determined by interactions between amino acid residues that are far apart in the chain. A variety of interactions give rise to protein tertiary structure, such as **disulfide bridge**s, which are bonds between the sulfhydryl (–SH) functional groups on amino acid side groups; hydrogen bonds; ionic bonds; and hydrophobic interactions between nonpolar side chains. All these interactions, weak and strong, combine to determine the final three-dimensional shape of the protein and its function (Figure 7.21).

The process by which a polypeptide chain assumes a large-scale, three-dimensional shape is called protein folding. Folded proteins that are fully functional in their normal biological role are said to possess a **native structure**. When a protein loses its three-dimensional shape, it may no longer be functional. These unfolded proteins are **denatured**. Denaturation implies the loss of the secondary structure and tertiary structure (and, if present, the quaternary structure) without the loss of the primary structure.

Some proteins are assemblies of several separate polypeptides, also known as protein subunits. These proteins function adequately only when all subunits are present and appropriately configured. The interactions that hold these subunits together constitute the **quaternary structure** of the protein. The overall quaternary structure is stabilized by relatively weak interactions. Hemoglobin, for example, has a quaternary structure of four globular protein subunits: two α and two β polypeptides, each one containing an iron-based heme (Figure 7.22).

Another important class of proteins is the **conjugated proteins** that have a nonprotein portion. If the conjugated protein has a carbohydrate attached, it is called a **glycoprotein**. If it has a lipid attached, it is called a **lipoprotein**. These proteins are important components of membranes. Figure 7.23 summarizes the four levels of protein structure.

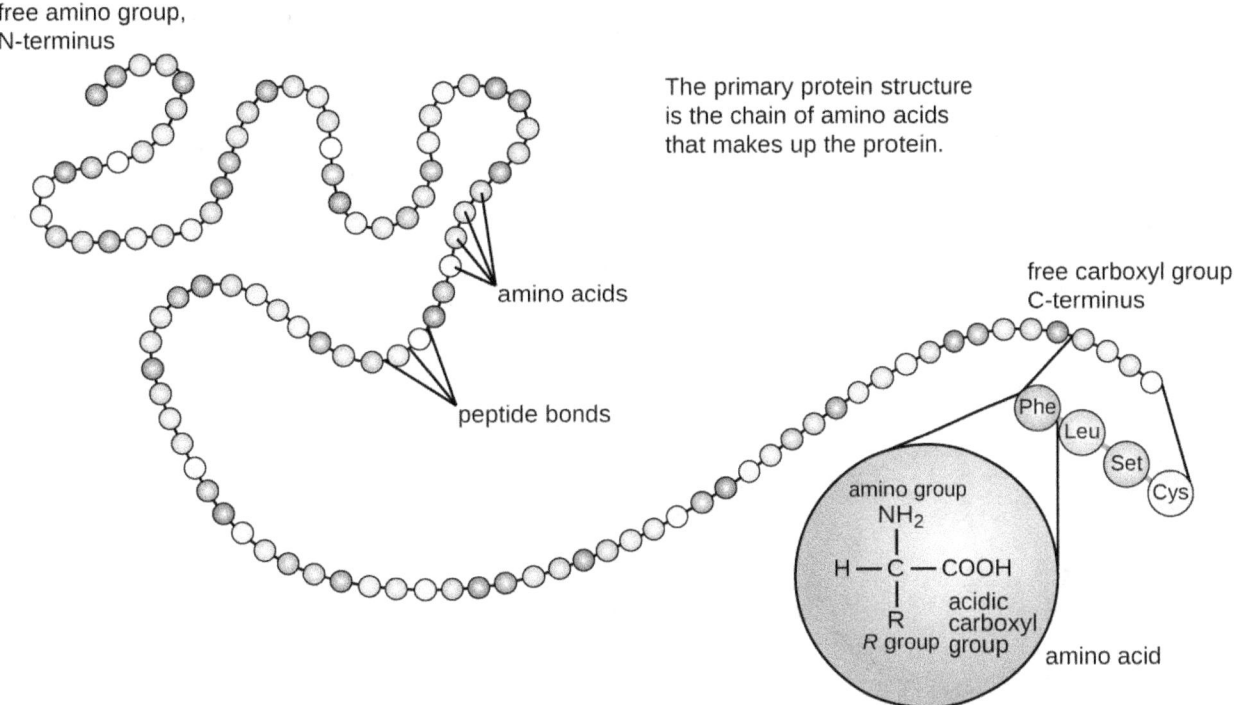

Figure 7.19 The primary structure of a protein is the sequence of amino acids. (credit: modification of work by National Human Genome Research Institute)

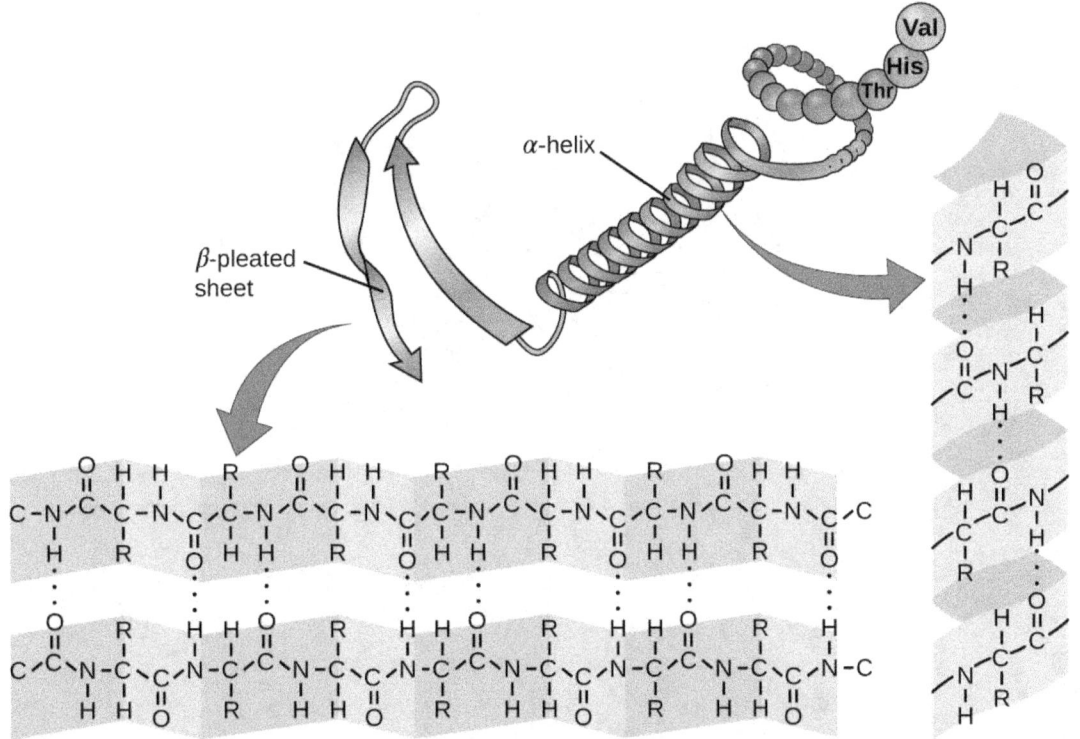

Figure 7.20 The secondary structure of a protein may be an α-helix or a β-pleated sheet, or both.

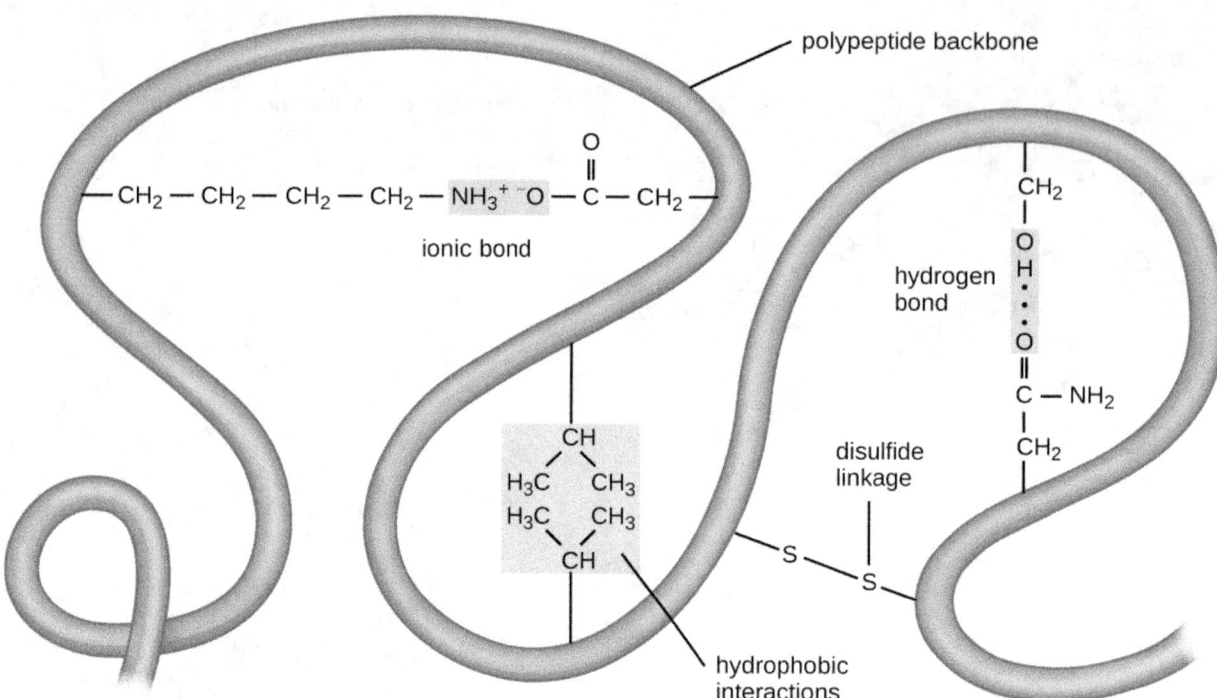

Figure 7.21 The tertiary structure of proteins is determined by a variety of attractive forces, including hydrophobic interactions, ionic bonding, hydrogen bonding, and disulfide linkages.

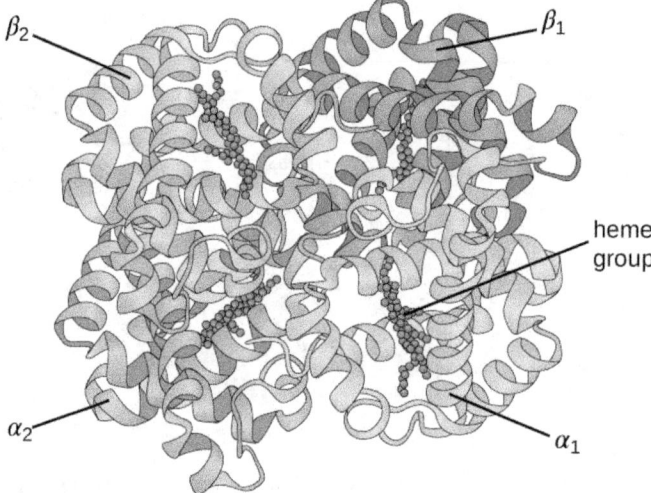

Figure 7.22 A hemoglobin molecule has two α and two β polypeptides together with four heme groups.

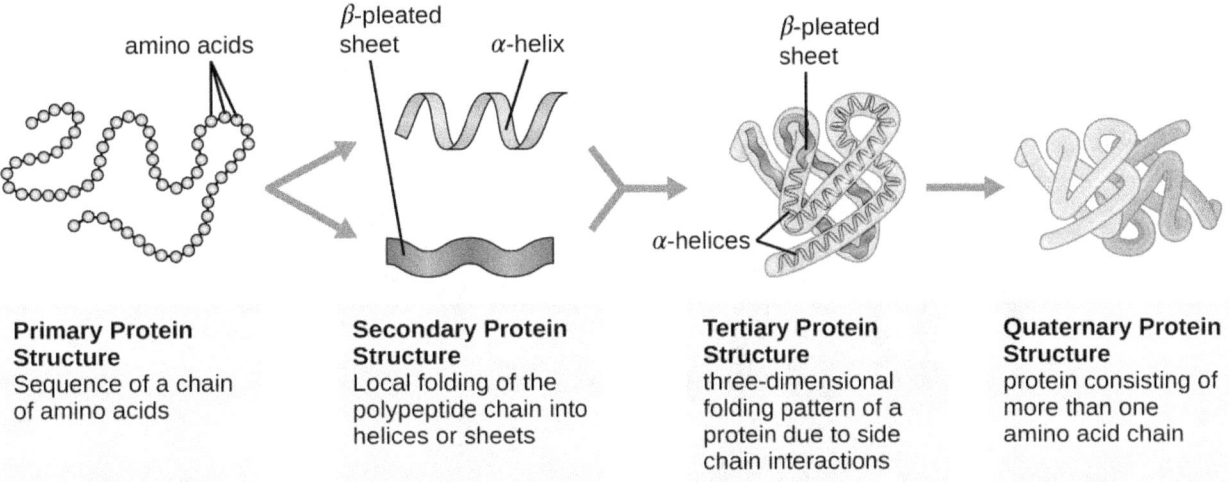

Figure 7.23 Protein structure has four levels of organization. (credit: modification of work by National Human Genome Research Institute)

CHECK YOUR UNDERSTANDING

- What can happen if a protein's primary, secondary, tertiary, or quaternary structure is changed?

 MICRO CONNECTIONS

Primary Structure, Dysfunctional Proteins, and Cystic Fibrosis

Proteins associated with biological membranes are classified as extrinsic or intrinsic. Extrinsic proteins, also called peripheral proteins, are loosely associated with one side of the membrane. Intrinsic proteins, or integral proteins, are embedded in the membrane and often function as part of transport systems as transmembrane proteins. Cystic fibrosis (CF) is a human genetic disorder caused by a change in the transmembrane protein. It affects mostly the lungs but may also affect the pancreas, liver, kidneys, and intestine. CF is caused by a loss of the amino acid phenylalanine in a cystic fibrosis transmembrane protein (CFTR). The loss of one amino acid changes the primary structure of a protein that normally helps transport salt and water in and out of cells (Figure 7.24).

The change in the primary structure prevents the protein from functioning properly, which causes the body to produce unusually thick mucus that clogs the lungs and leads to the accumulation of sticky mucus. The mucus obstructs the pancreas and stops natural enzymes from helping the body break down food and absorb vital nutrients.

In the lungs of individuals with cystic fibrosis, the altered mucus provides an environment where bacteria can thrive. This colonization leads to the formation of biofilms in the small airways of the lungs. The most common pathogens found in the lungs of patients with cystic fibrosis are *Pseudomonas aeruginosa* (Figure 7.25) and *Burkholderia cepacia*. *Pseudomonas* differentiates within the biofilm in the lung and forms large colonies, called "mucoid" *Pseudomonas*. The colonies have a unique pigmentation that shows up in laboratory tests (Figure 7.25) and provides physicians with the first clue that the patient has CF (such colonies are rare in healthy individuals).

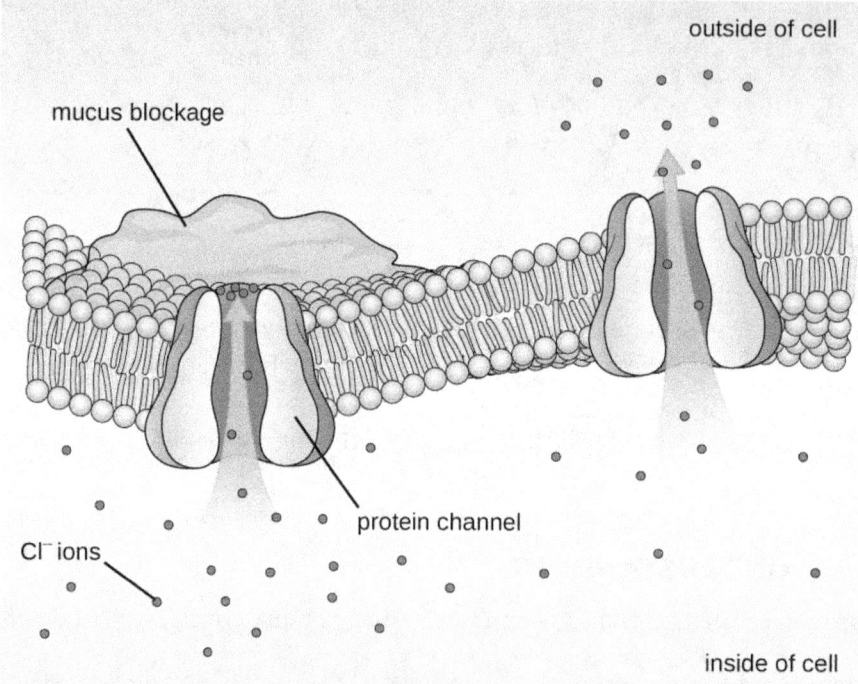

Figure 7.24 The normal CFTR protein is a channel protein that helps salt (sodium chloride) move in and out of cells.

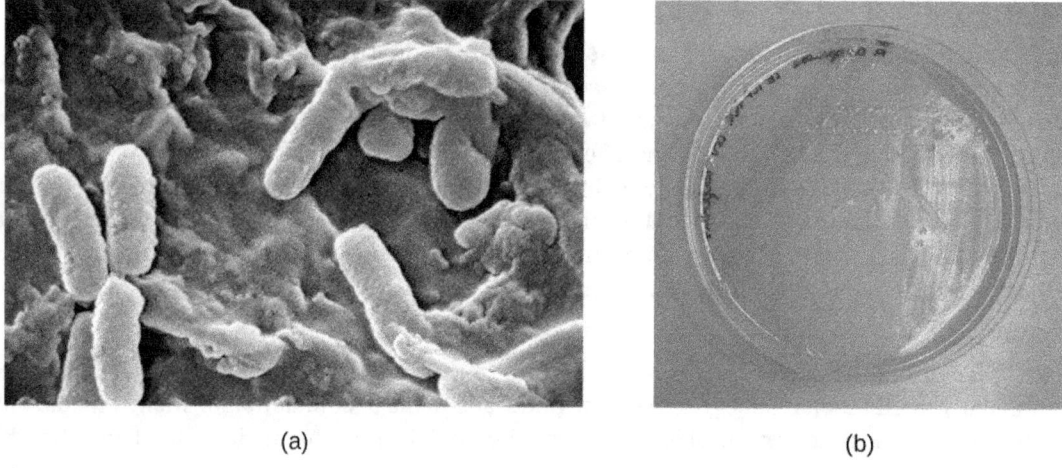

Figure 7.25 (a) A scanning electron micrograph shows the opportunistic bacterium *Pseudomonas aeruginosa*. (b) Pigment-producing *P. aeruginosa* on cetrimide agar shows the green pigment called pyocyanin. (credit a: modification of work by the Centers for Disease Control and Prevention)

LINK TO LEARNING

For more information about cystic fibrosis, visit the Cystic Fibrosis Foundation (https://openstax.org/l/22cystfibrofoun) website.

7.5 Using Biochemistry to Identify Microorganisms

Learning Objectives

By the end of this section, you will be able to:
- Describe examples of biosynthesis products within a cell that can be detected to identify bacteria

Accurate identification of bacterial isolates is essential in a clinical microbiology laboratory because the results often inform decisions about treatment that directly affect patient outcomes. For example, cases of food poisoning require accurate identification of the causative agent so that physicians can prescribe appropriate treatment. Likewise, it is important to accurately identify the causative pathogen during an outbreak of disease so that appropriate strategies can be employed to contain the epidemic.

There are many ways to detect, characterize, and identify microorganisms. Some methods rely on phenotypic biochemical characteristics, while others use genotypic identification. The biochemical characteristics of a bacterium provide many traits that are useful for classification and identification. Analyzing the nutritional and metabolic capabilities of the bacterial isolate is a common approach for determining the genus and the species of the bacterium. Some of the most important metabolic pathways that bacteria use to survive will be discussed in Microbial Metabolism. In this section, we will discuss a few methods that use biochemical characteristics to identify microorganisms.

Some microorganisms store certain compounds as granules within their cytoplasm, and the contents of these granules can be used for identification purposes. For example, poly-β-hydroxybutyrate (PHB) is a carbon- and energy-storage compound found in some nonfluorescent bacteria of the genus *Pseudomonas*. Different species within this genus can be classified by the presence or the absence of PHB and fluorescent pigments. The human pathogen *P. aeruginosa* and the plant pathogen *P. syringae* are two examples of fluorescent *Pseudomonas* species that do not accumulate PHB granules.

Other systems rely on biochemical characteristics to identify microorganisms by their biochemical reactions, such as carbon utilization and other metabolic tests. In small laboratory settings or in teaching laboratories, those assays are carried out using a limited number of test tubes. However, more modern systems, such as the one developed by Biolog, Inc., are based on panels of biochemical reactions performed simultaneously and analyzed by software. Biolog's system identifies cells based on their ability to metabolize certain biochemicals and on their physiological properties, including pH and chemical sensitivity. It uses all major classes of biochemicals in its analysis. Identifications can be performed manually or with the semi- or fully automated instruments.

Another automated system identifies microorganisms by determining the specimen's mass spectrum and then comparing it to a database that contains known mass spectra for thousands of microorganisms. This method is based on matrix-assisted laser desorption/ionization time-of-flight mass spectrometry (MALDI-TOF) and uses disposable MALDI plates on which the microorganism is mixed with a specialized matrix reagent (Figure 7.26). The sample/reagent mixture is irradiated with a high-intensity pulsed ultraviolet laser, resulting in the ejection of gaseous ions generated from the various chemical constituents of the microorganism. These gaseous ions are collected and accelerated through the mass spectrometer, with ions traveling at a velocity determined by their mass-to-charge ratio (m/z), thus, reaching the detector at different times. A plot of detector signal versus m/z yields a mass spectrum for the organism that is uniquely related to its biochemical composition. Comparison of the mass spectrum to a library of reference spectra obtained from identical analyses of known microorganisms permits identification of the unknown microbe.

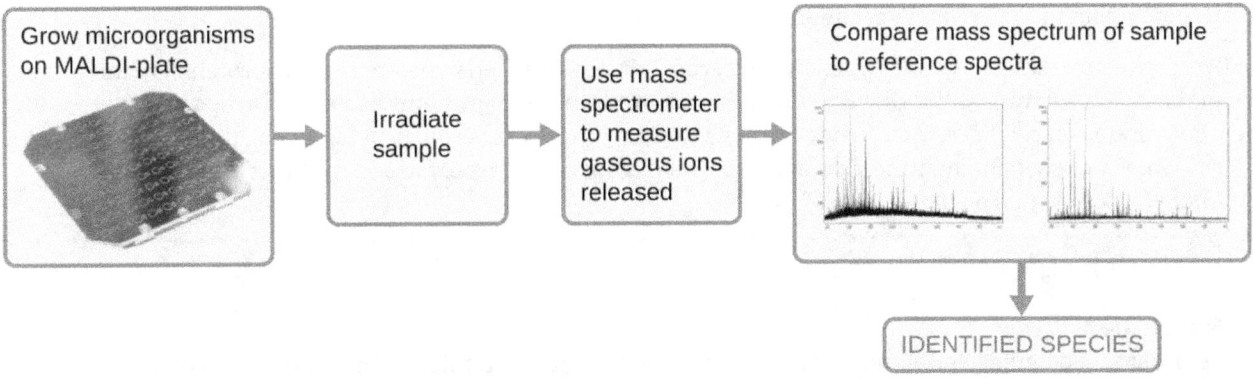

Figure 7.26 MALDI-TOF methods are now routinely used for diagnostic procedures in clinical microbiology laboratories. This technology is able to rapidly identify some microorganisms that cannot be readily identified by more traditional methods. (credit "MALDI plate photo": modification of work by Chen Q, Liu T, Chen G; credit "graphs": modification of work by Bailes J, Vidal L, Ivanov DA, Soloviev M)

Microbes can also be identified by measuring their unique lipid profiles. As we have learned, fatty acids of lipids can vary in chain length, presence or absence of double bonds, and number of double bonds, hydroxyl groups, branches, and rings. To identify a microbe by its lipid composition, the fatty acids present in their membranes are analyzed. A common biochemical analysis used for this purpose is a technique used in clinical, public health, and food laboratories. It relies on detecting unique differences in fatty acids and is called **fatty acid methyl ester (FAME) analysis**. In a FAME analysis, fatty acids are extracted from the membranes of microorganisms, chemically altered to form volatile methyl esters, and analyzed by gas chromatography (GC). The resulting GC chromatogram is compared with reference chromatograms in a database containing data for thousands of bacterial isolates to identify the unknown microorganism (Figure 7.27).

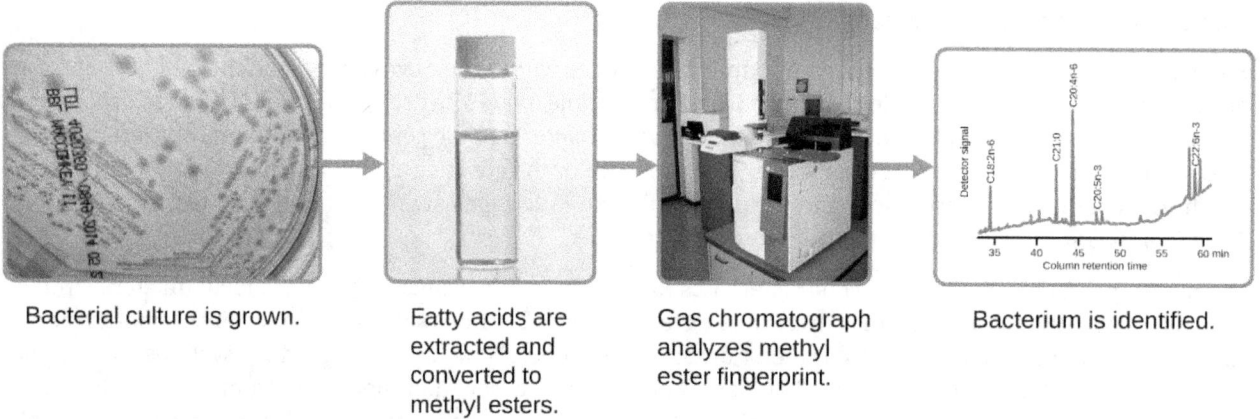

Figure 7.27 Fatty acid methyl ester (FAME) analysis in bacterial identification results in a chromatogram unique to each bacterium. Each peak in the gas chromatogram corresponds to a particular fatty acid methyl ester and its height is proportional to the amount present in the cell. (credit "culture": modification of work by the Centers for Disease Control and Prevention; credit "graph": modification of work by Zhang P. and Liu P.)

A related method for microorganism identification is called **phospholipid-derived fatty acids (PLFA) analysis**. Membranes are mostly composed of phospholipids, which can be saponified (hydrolyzed with alkali) to release the fatty acids. The resulting fatty acid mixture is then subjected to FAME analysis, and the measured lipid profiles can be compared with those of known microorganisms to identify the unknown microorganism.

Bacterial identification can also be based on the proteins produced under specific growth conditions within the human body. These types of identification procedures are called **proteomic analysis**. To perform proteomic analysis, proteins from the pathogen are first separated by high-pressure liquid chromatography (HPLC), and the collected fractions are then digested to yield smaller peptide fragments. These peptides are identified by mass spectrometry and compared with those of known microorganisms to identify the unknown microorganism in the original specimen.

Microorganisms can also be identified by the carbohydrates attached to proteins (glycoproteins) in the plasma membrane or cell wall. Antibodies and other carbohydrate-binding proteins can attach to specific carbohydrates on cell surfaces, causing the cells to clump together. Serological tests (e.g., the Lancefield groups tests, which are used for identification of *Streptococcus* species) are performed to detect the unique carbohydrates located on the surface of the cell.

Clinical Focus

Resolution
Penny stopped using her new sunscreen and applied the corticosteroid cream to her rash as directed. However, after several days, her rash had not improved and actually seemed to be getting worse. She made a follow-up appointment with her doctor, who observed a bumpy red rash and pus-filled blisters around

hair follicles (Figure 7.28). The rash was especially concentrated in areas that would have been covered by a swimsuit. After some questioning, Penny told the physician that she had recently attended a pool party and spent some time in a hot tub. In light of this new information, the doctor suspected a case of hot tub rash, an infection frequently caused by the bacterium *Pseudomonas aeruginosa*, an opportunistic pathogen that can thrive in hot tubs and swimming pools, especially when the water is not sufficiently chlorinated. *P. aeruginosa* is the same bacterium that is associated with infections in the lungs of patients with cystic fibrosis.

The doctor collected a specimen from Penny's rash to be sent to the clinical microbiology lab. Confirmatory tests were carried out to distinguish *P. aeruginosa* from enteric pathogens that can also be present in pool and hot-tub water. The test included the production of the blue-green pigment pyocyanin on cetrimide agar and growth at 42 °C. Cetrimide is a selective agent that inhibits the growth of other species of microbial flora and also enhances the production of *P. aeruginosa* pigments pyocyanin and fluorescein, which are a characteristic blue-green and yellow-green, respectively.

Tests confirmed the presence of *P. aeruginosa* in Penny's skin sample, but the doctor decided not to prescribe an antibiotic. Even though *P. aeruginosa* is a bacterium, *Pseudomonas* species are generally resistant to many antibiotics. Luckily, skin infections like Penny's are usually self-limiting; the rash typically lasts about 2 weeks and resolves on its own, with or without medical treatment. The doctor advised Penny to wait it out and keep using the corticosteroid cream. The cream will not kill the *P. aeruginosa* on Penny's skin, but it should calm her rash and minimize the itching by suppressing her body's inflammatory response to the bacteria.

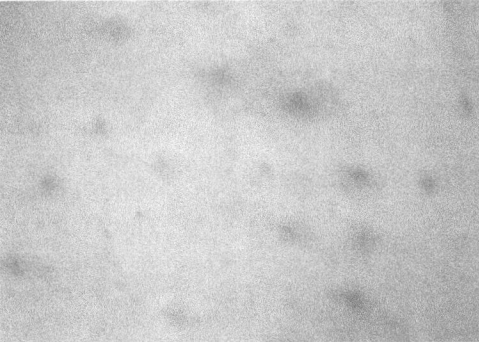

Figure 7.28 Exposure to *Pseudomonas aeruginosa* in the water of a pool or hot tub can sometimes cause a skin infection that manifests as "hot tub rash." (credit: modification of work by "Lsupellmel"/Wikimedia Commons)

Go back to the previous Clinical Focus box.

SUMMARY

7.1 Organic Molecules

- The most abundant elements in cells are hydrogen, carbon, oxygen, nitrogen, phosphorus, and sulfur.
- Life is carbon based. Each carbon atom can bind to another one producing a **carbon skeleton** that can be straight, branched, or ring shaped.
- The same numbers and types of atoms may bond together in different ways to yield different molecules called **isomers**. Isomers may differ in the bonding sequence of their atoms (**structural isomers**) or in the spatial arrangement of atoms whose bonding sequences are the same (**stereoisomers**), and their physical and chemical properties may vary slightly or drastically.
- **Functional groups** confer specific chemical properties to molecules bearing them. Common functional groups in biomolecules are hydroxyl, methyl, carbonyl, carboxyl, amino, phosphate, and sulfhydryl.
- **Macromolecules** are **polymers** assembled from individual units, the **monomers**, which bind together like building blocks. Many biologically significant macromolecules are formed by **dehydration synthesis**, a process in which monomers bind together by combining their functional groups and generating water molecules as byproducts.

7.2 Carbohydrates

- **Carbohydrates**, the most abundant biomolecules on earth, are widely used by organisms for structural and energy-storage purposes.
- Carbohydrates include individual sugar molecules (**monosaccharides**) as well as two or more molecules chemically linked by **glycosidic bonds**. **Monosaccharides** are classified based on the number of carbons the molecule as trioses (3 C), tetroses (4 C), pentoses (5 C), and hexoses (6 C). They are the building blocks for the synthesis of polymers or complex carbohydrates.
- **Disaccharides** such as sucrose, lactose, and maltose are molecules composed of two monosaccharides linked together by a glycosidic bond.
- **Polysaccharides**, or **glycans**, are polymers composed of hundreds of monosaccharide monomers linked together by glycosidic bonds. The energy-storage polymers **starch** and **glycogen** are examples of polysaccharides and are all composed of branched chains of glucose molecules.
- The polysaccharide **cellulose** is a common structural component of the cell walls of organisms. Other structural polysaccharides, such as N-acetyl glucosamine (NAG) and N-acetyl muramic acid (NAM), incorporate modified glucose molecules and are used in the construction of peptidoglycan or chitin.

7.3 Lipids

- **Lipids** are composed mainly of carbon and hydrogen, but they can also contain oxygen, nitrogen, sulfur, and phosphorous. They provide nutrients for organisms, store carbon and energy, play structural roles in membranes, and function as hormones, pharmaceuticals, fragrances, and pigments.
- Fatty acids are long-chain hydrocarbons with a carboxylic acid functional group. Their relatively long nonpolar hydrocarbon chains make them **hydrophobic**. Fatty acids with no double bonds are **saturated**; those with double bonds are **unsaturated**.
- Fatty acids chemically bond to glycerol to form structurally essential lipids such as **triglycerides** and **phospholipids.** Triglycerides comprise three fatty acids bonded to glycerol, yielding a hydrophobic molecule. Phospholipids contain both hydrophobic hydrocarbon chains and polar head groups, making them **amphipathic** and capable of forming uniquely functional large scale structures.
- Biological membranes are large-scale structures based on phospholipid bilayers that provide hydrophilic exterior and interior surfaces suitable for aqueous environments, separated by an intervening hydrophobic layer. These bilayers are the structural basis for cell membranes in most organisms, as well as subcellular components such as vesicles.
- **Isoprenoids** are lipids derived from isoprene molecules that have many physiological roles and a variety of commercial applications.
- A wax is a long-chain isoprenoid that is typically water resistant; an example of a wax-containing substance is sebum, produced by sebaceous glands in the skin. **Steroids** are lipids with complex, ringed structures that function as structural components of cell membranes and

as hormones. **Sterols** are a subclass of steroids containing a hydroxyl group at a specific location on one of the molecule's rings; one example is cholesterol.
- Bacteria produce hopanoids, structurally similar to cholesterol, to strengthen bacterial membranes. Fungi and protozoa produce a strengthening agent called ergosterol.

7.4 Proteins

- Amino acids are small molecules essential to all life. Each has an α carbon to which a hydrogen atom, carboxyl group, and amine group are bonded. The fourth bonded group, represented by R, varies in chemical composition, size, polarity, and charge among different amino acids, providing variation in properties.
- **Peptides** are polymers formed by the linkage of amino acids via dehydration synthesis. The bonds between the linked amino acids are called **peptide bonds.** The number of amino acids linked together may vary from a few to many.
- **Proteins** are polymers formed by the linkage of a very large number of amino acids. They perform many important functions in a cell, serving as nutrients and enzymes; storage molecules for carbon, nitrogen, and energy; and structural components.
- The structure of a protein is a critical determinant of its function and is described by a graduated classification: **primary, secondary, tertiary,** and **quaternary**. The **native structure** of a protein may be disrupted by **denaturation,** resulting in loss of its higher-order structure and its biological function.
- Some proteins are formed by several separate protein subunits, the interaction of these subunits composing the **quaternary structure** of the protein complex.
- **Conjugated proteins** have a nonpolypeptide portion that can be a carbohydrate (forming a **glycoprotein**) or a lipid fraction (forming a **lipoprotein**). These proteins are important components of membranes.

7.5 Using Biochemistry to Identify Microorganisms

- Accurate identification of bacteria is essential in a clinical laboratory for diagnostic and management of epidemics, pandemics, and food poisoning caused by bacterial outbreaks.
- The phenotypic identification of microorganisms involves using observable traits, including profiles of structural components such as lipids, biosynthetic products such as sugars or amino acids, or storage compounds such as poly-β-hydroxybutyrate.
- An unknown microbe may be identified from the unique mass spectrum produced when it is analyzed by **matrix assisted laser desorption/ionization time of flight mass spectrometry (MALDI-TOF)**.
- Microbes can be identified by determining their lipid compositions, using **fatty acid methyl esters (FAME)** or **phospholipid-derived fatty acids (PLFA) analysis**.
- **Proteomic analysis**, the study of all accumulated proteins of an organism; can also be used for bacterial identification.
- Glycoproteins in the plasma membrane or cell wall structures can bind to lectins or antibodies and can be used for identification.

REVIEW QUESTIONS
Multiple Choice

1. Which of these elements is *not* a micronutrient?
   A. C
   B. Ca
   C. Co
   D. Cu

2. Which of the following is the name for molecules whose structures are nonsuperimposable mirror images?
   A. structural isomers
   B. monomers
   C. polymers
   D. enantiomers

3. By definition, carbohydrates contain which elements?
   A. carbon and hydrogen
   B. carbon, hydrogen, and nitrogen
   C. carbon, hydrogen, and oxygen
   D. carbon and oxygen

4. Monosaccharides may link together to form polysaccharides by forming which type of bond?
   A. hydrogen
   B. peptide
   C. ionic
   D. glycosidic

5. Which of the following describes lipids?
   A. a source of nutrients for organisms
   B. energy-storage molecules
   C. molecules having structural role in membranes
   D. molecules that are part of hormones and pigments
   E. all of the above

6. Molecules bearing both polar and nonpolar groups are said to be which of the following?
   A. hydrophilic
   B. amphipathic
   C. hydrophobic
   D. polyfunctional

7. Which of the following groups varies among different amino acids?
   A. hydrogen atom
   B. carboxyl group
   C. R group
   D. amino group

8. The amino acids present in proteins differ in which of the following?
   A. size
   B. shape
   C. side groups
   D. all of the above

9. Which of the following bonds are not involved in tertiary structure?
   A. peptide bonds
   B. ionic bonds
   C. hydrophobic interactions
   D. hydrogen bonds

10. Which of the following characteristics/compounds is not considered to be a phenotypic biochemical characteristic used of microbial identification?
    A. poly-β-hydroxybutyrate
    B. small-subunit (16S) rRNA gene
    C. carbon utilization
    D. lipid composition

11. Proteomic analysis is a methodology that deals with which of the following?
    A. the analysis of proteins functioning as enzymes within the cell
    B. analysis of transport proteins in the cell
    C. the analysis of integral proteins of the cell membrane
    D. the study of all accumulated proteins of an organism

12. Which method involves the generation of gas phase ions from intact microorganisms?
    A. FAME
    B. PLFA
    C. MALDI-TOF
    D. Lancefield group testing

13. Which method involves the analysis of membrane-bound carbohydrates?
    A. FAME
    B. PLFA
    C. MALDI-TOF
    D. Lancefield group testing

14. Which method involves conversion of a microbe's lipids to volatile compounds for analysis by gas chromatography?
    A. FAME
    B. proteomic analysis
    C. MALDI-TOF
    D. Lancefield group testing

True/False

15. Aldehydes, amides, carboxylic acids, esters, and ketones all contain carbonyl groups.
16. Two molecules containing the same types and numbers of atoms but different bonding sequences are called enantiomers.
17. Lipids are a naturally occurring group of substances that are not soluble in water but are freely soluble in organic solvents.
18. Fatty acids having no double bonds are called "unsaturated."

19. A triglyceride is formed by joining three glycerol molecules to a fatty acid backbone in a dehydration reaction.
20. A change in one amino acid in a protein sequence always results in a loss of function.
21. MALDI-TOF relies on obtaining a unique mass spectrum for the bacteria tested and then checking the acquired mass spectrum against the spectrum databases registered in the analysis software to identify the microorganism.
22. Lancefield group tests can identify microbes using antibodies that specifically bind cell-surface proteins.

Matching

23. Match each polysaccharide with its description.

    ___chitin   A. energy storage polymer in plants

    ___glycogen   B. structural polymer found in plants

    ___starch   C. structural polymer found in cell walls of fungi and exoskeletons of some animals

    ___cellulose   D. energy storage polymer found in animal cells and bacteria

Fill in the Blank

24. Waxes contain esters formed from long-chain _____ and saturated _____, and they may also contain substituted hydrocarbons.
25. Cholesterol is the most common member of the _____ group, found in animal tissues; it has a tetracyclic carbon ring system with a _____ bond in one of the rings and one free _____ group.
26. The sequence of amino acids in a protein is called its _____.
27. Denaturation implies the loss of the _____ and _____ structures without the loss of the _____ structure.
28. A FAME analysis involves the conversion of _____ to more volatile _____ for analysis using _____.

Short Answer

29. Why are carbon, nitrogen, oxygen, and hydrogen the most abundant elements in living matter and, therefore, considered macronutrients?
30. Identify the functional group in each of the depicted structural formulas.

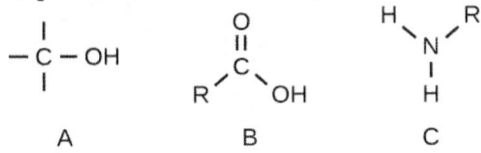

31. What are monosaccharides, disaccharides, and polysaccharides?
32. Describe the structure of a typical phospholipid. Are these molecules polar or nonpolar?
33. Compare MALDI-TOF, FAME, and PLFA, and explain how each technique would be used to identify pathogens.

Critical Thinking

34. The structural formula shown corresponds to penicillin G, a narrow-spectrum antibiotic that is given intravenously or intramuscularly as a treatment for several bacterial diseases. The antibiotic is produced by fungi of the genus *Penicillium*. (a) Identify three major functional groups in this molecule that each comprise two simpler functional groups. (b) Name the two simpler functional groups composing each of the major functional groups identified in (a).

35. The figure depicts the structural formulas of glucose, galactose, and fructose. (a) Circle the functional groups that classify the sugars either an aldose or a ketose, and identify each sugar as one or the other. (b) The chemical formula of these compounds is the same, although the structural formula is different. What are such compounds called?

36. Structural diagrams for the linear and cyclic forms of a monosaccharide are shown. (a) What is the molecular formula for this monosaccharide? (Count the C, H and O atoms in each to confirm that these two molecules have the same formula, and report this formula.) (b) Identify which hydroxyl group in the linear structure undergoes the ring-forming reaction with the carbonyl group.

37. The term "dextrose" is commonly used in medical settings when referring to the biologically relevant isomer of the monosaccharide glucose. Explain the logic of this alternative name.

38. Microorganisms can thrive under many different conditions, including high-temperature environments such as hot springs. To function properly, cell membranes have to be in a fluid state. How do you expect the fatty acid content (saturated versus unsaturated) of bacteria living in high-temperature environments might compare with that of bacteria living in more moderate temperatures?

39. Heating a protein sufficiently may cause it to denature. Considering the definition of denaturation, what does this statement say about the strengths of peptide bonds in comparison to hydrogen bonds?

40. The image shown represents a tetrapeptide. (a) How many peptide bonds are in this molecule? (b) Identify the side groups of the four amino acids composing this peptide.

CHAPTER 8
Microbial Metabolism

Figure 8.1 Prokaryotes have great metabolic diversity with important consequences to other forms of life. Acidic mine drainage (left) is a serious environmental problem resulting from the introduction of water and oxygen to sulfide-oxidizing bacteria during mining processes. These bacteria produce large amounts of sulfuric acid as a byproduct of their metabolism, resulting in a low-pH environment that can kill many aquatic plants and animals. On the other hand, some prokaryotes are essential to other life forms. Root nodules of many plants (right) house nitrogen-fixing bacteria that convert atmospheric nitrogen into ammonia, providing a usable nitrogen source for these plants. (credit left: modification of work by D. Hardesty, USGS Columbia Environment Research Center; credit right: modification of work by Celmow SR, Clairmont L, Madsen LH, and Guinel FC)

Chapter Outline

8.1 Energy, Matter, and Enzymes

8.2 Catabolism of Carbohydrates

8.3 Cellular Respiration

8.4 Fermentation

8.5 Catabolism of Lipids and Proteins

8.6 Photosynthesis

8.7 Biogeochemical Cycles

INTRODUCTION Throughout earth's history, microbial metabolism has been a driving force behind the development and maintenance of the planet's biosphere. Eukaryotic organisms such as plants and animals typically depend on organic molecules for energy, growth, and reproduction. Prokaryotes, on the other hand, can metabolize a wide range of organic as well as inorganic matter, from complex organic molecules like cellulose to inorganic molecules and ions such as atmospheric nitrogen (N_2), molecular hydrogen (H_2), sulfide (S^{2-}), manganese (II) ions (Mn^{2+}), ferrous iron (Fe^{2+}), and ferric iron (Fe^{3+}), to name a few. By metabolizing such substances, microbes chemically convert them to other forms. In some cases, microbial metabolism produces chemicals that can be harmful to other organisms; in others, it produces substances that are essential to the metabolism and survival of other life forms (Figure 8.1).

8.1 Energy, Matter, and Enzymes

Learning Objectives

By the end of this section, you will be able to:
- Define and describe metabolism
- Compare and contrast autotrophs and heterotrophs
- Describe the importance of oxidation-reduction reactions in metabolism
- Describe why ATP, FAD, NAD^+, and $NADP^+$ are important in a cell
- Identify the structure and structural components of an enzyme
- Describe the differences between competitive and noncompetitive enzyme inhibitors

---

Clinical Focus

Part 1

Hannah is a 15-month-old girl from Washington state. She is spending the summer in Gambia, where her parents are working for a nongovernmental organization. About 3 weeks after her arrival in Gambia, Hannah's appetite began to diminish and her parents noticed that she seemed unusually sluggish, fatigued, and confused. She also seemed very irritable when she was outdoors, especially during the day. When she began vomiting, her parents figured she had caught a 24-hour virus, but when her symptoms persisted, they took her to a clinic. The local physician noticed that Hannah's reflexes seemed abnormally slow, and when he examined her eyes with a light, she seemed unusually light sensitive. She also seemed to be experiencing a stiff neck.

- What are some possible causes of Hannah's symptoms?

Jump to the next Clinical Focus box.

The term used to describe all of the chemical reactions inside a cell is **metabolism** (Figure 8.2). Cellular processes such as the building or breaking down of complex molecules occur through series of stepwise, interconnected chemical reactions called metabolic pathways. Reactions that are spontaneous and release energy are **exergonic reaction**s, whereas **endergonic reaction**s require energy to proceed. The term **anabolism** refers to those endergonic metabolic pathways involved in biosynthesis, converting simple molecular building blocks into more complex molecules, and fueled by the use of cellular energy. Conversely, the term **catabolism** refers to exergonic pathways that break down complex molecules into simpler ones. Molecular energy stored in the bonds of complex molecules is released in catabolic pathways and harvested in such a way that it can be used to produce high-energy molecules, which are used to drive anabolic pathways. Thus, in terms of energy and molecules, cells are continually balancing catabolism with anabolism.

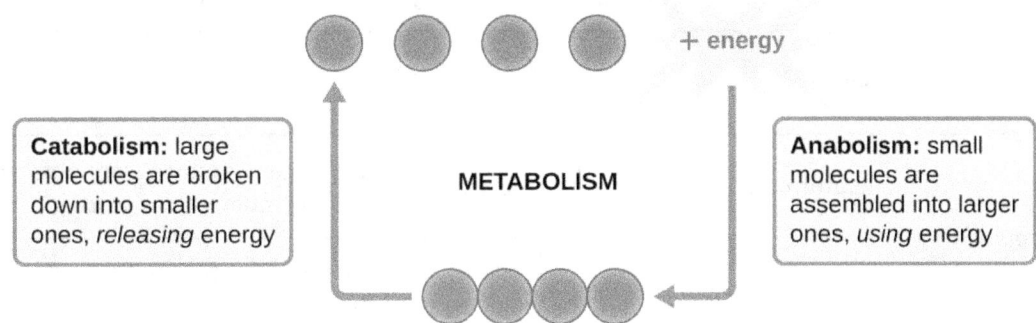

Figure 8.2 Metabolism includes catabolism and anabolism. Anabolic pathways require energy to synthesize larger molecules. Catabolic pathways generate energy by breaking down larger molecules. Both types of pathways are required for maintaining the cell's energy balance.

Classification by Carbon and Energy Source

Organisms can be identified according to the source of carbon they use for metabolism as well as their energy source. The prefixes auto- ("self") and hetero- ("other") refer to the origins of the carbon sources various organisms can use. Organisms that convert inorganic carbon dioxide (CO_2) into organic carbon compounds are **autotrophs**. Plants and cyanobacteria are well-known examples of autotrophs. Conversely, **heterotrophs** rely on more complex organic carbon compounds as nutrients; these are provided to them initially by autotrophs. Many organisms, ranging from humans to many prokaryotes, including the well-studied *Escherichia coli*, are heterotrophic.

Organisms can also be identified by the energy source they use. All energy is derived from the transfer of electrons, but the source of electrons differs between various types of organisms. The prefixes photo- ("light") and chemo- ("chemical") refer to the energy sources that various organisms use. Those that get their energy for electron transfer from light are **phototrophs**, whereas **chemotrophs** obtain energy for electron transfer by breaking chemical bonds. There are two types of chemotrophs: **organotrophs** and **lithotrophs**. Organotrophs, including humans, fungi, and many prokaryotes, are chemotrophs that obtain energy from organic compounds. Lithotrophs ("litho" means "rock") are chemotrophs that get energy from inorganic compounds, including hydrogen sulfide (H_2S) and reduced iron. Lithotrophy is unique to the microbial world.

The strategies used to obtain both carbon and energy can be combined for the classification of organisms according to nutritional type. Most organisms are chemoheterotrophs because they use organic molecules as both their electron and carbon sources. Table 8.1 summarizes this and the other classifications.

Classifications of Organisms by Energy and Carbon Source

Classifications		Energy Source	Carbon Source	Examples
Chemotrophs	Chemoautotrophs	Chemical	Inorganic	Hydrogen-, sulfur-, iron-, nitrogen-, and carbon monoxide-oxidizing bacteria
	Chemoheterotrophs	Chemical	Organic compounds	All animals, most fungi, protozoa, and bacteria
Phototrophs	Photoautotrophs	Light	Inorganic	All plants, algae, cyanobacteria, and green and purple sulfur bacteria
	Photoheterotrophs	Light	Organic compounds	Green and purple nonsulfur bacteria, heliobacteria

Table 8.1

✓ CHECK YOUR UNDERSTANDING

- Explain the difference between catabolism and anabolism.
- Explain the difference between autotrophs and heterotrophs.

Oxidation and Reduction in Metabolism

The transfer of electrons between molecules is important because most of the energy stored in atoms and used to fuel cell functions is in the form of high-energy electrons. The transfer of energy in the form of electrons allows the cell to transfer and use energy incrementally; that is, in small packages rather than a single, destructive burst. Reactions that remove electrons from donor molecules, leaving them oxidized, are **oxidation reactions**; those that add electrons to acceptor molecules, leaving them reduced, are **reduction reactions**. Because electrons can move from one molecule to another, oxidation and reduction occur in tandem. These pairs of reactions are called oxidation-reduction reactions, or **redox reactions**.

Energy Carriers: NAD^+, $NADP^+$, FAD, and ATP

The energy released from the breakdown of the chemical bonds within nutrients can be stored either through the reduction of electron carriers or in the bonds of adenosine triphosphate (ATP). In living systems, a small class of compounds functions as mobile **electron carriers**, molecules that bind to and shuttle high-energy electrons between compounds in pathways. The principal electron carriers we will consider originate from the B vitamin group and are derivatives of nucleotides; they are **nicotinamide adenine dinucleotide**, **nicotine adenine dinucleotide phosphate**, and **flavin adenine dinucleotide**. These compounds can be easily reduced or oxidized. Nicotinamide adenine dinucleotide (NAD^+/NADH) is the most common mobile electron carrier used in catabolism. NAD^+ is the oxidized form of the molecule; NADH is the reduced form of the molecule. Nicotine adenine dinucleotide phosphate ($NADP^+$), the oxidized form of an NAD^+ variant that contains an extra phosphate group, is another important electron carrier; it forms **NADPH** when reduced. The oxidized form of flavin adenine dinucleotide is **FAD**, and its reduced form is **$FADH_2$**. Both NAD^+/NADH and FAD/$FADH_2$ are extensively used in energy extraction from sugars during catabolism in chemoheterotrophs, whereas $NADP^+$/NADPH plays an important role in anabolic reactions and photosynthesis. Collectively, $FADH_2$, NADH, and NADPH are often referred to as having reducing power due to their ability to donate electrons to various chemical reactions.

A living cell must be able to handle the energy released during catabolism in a way that enables the cell to store energy safely and release it for use only as needed. Living cells accomplish this by using the compound

adenosine triphosphate (ATP). ATP is often called the "energy currency" of the cell, and, like currency, this versatile compound can be used to fill any energy need of the cell. At the heart of ATP is a molecule of **adenosine monophosphate (AMP)**, which is composed of an adenine molecule bonded to a ribose molecule and a single phosphate group. Ribose is a five-carbon sugar found in RNA, and AMP is one of the nucleotides in RNA. The addition of a second phosphate group to this core molecule results in the formation of **adenosine diphosphate (ADP)**; the addition of a third phosphate group forms ATP (Figure 8.3). Adding a phosphate group to a molecule, a process called phosphorylation, requires energy. Phosphate groups are negatively charged and thus repel one another when they are arranged in series, as they are in ADP and ATP. This repulsion makes the ADP and ATP molecules inherently unstable. Thus, the bonds between phosphate groups (one in ADP and two in ATP) are called **high-energy phosphate bonds**. When these high-energy bonds are broken to release one phosphate (called **inorganic phosphate [P$_i$]**) or two connected phosphate groups (called **pyrophosphate [PP$_i$]**) from ATP through a process called dephosphorylation, energy is released to drive endergonic reactions (Figure 8.4).

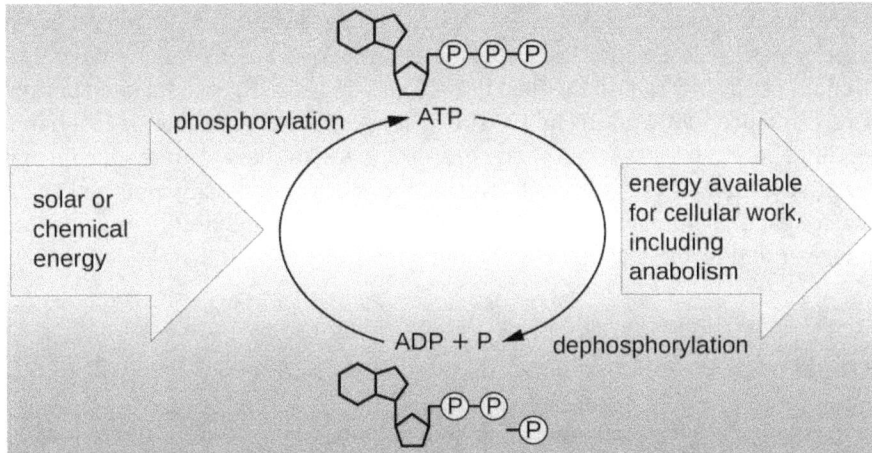

Figure 8.3 The energy released from dephosphorylation of ATP is used to drive cellular work, including anabolic pathways. ATP is regenerated through phosphorylation, harnessing the energy found in chemicals or from sunlight. (credit: modification of work by Robert Bear, David Rintoul)

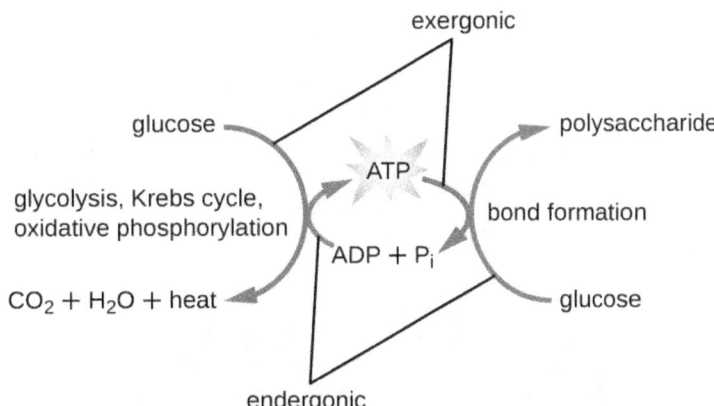

Figure 8.4 Exergonic reactions are coupled to endergonic ones, making the combination favorable. Here, the endergonic reaction of ATP phosphorylation is coupled to the exergonic reactions of catabolism. Similarly, the exergonic reaction of ATP dephosphorylation is coupled to the endergonic reaction of polysaccharide formation, an example of anabolism.

✓ CHECK YOUR UNDERSTANDING

- What is the function of an electron carrier?

Enzyme Structure and Function

A substance that helps speed up a chemical reaction is a **catalyst**. Catalysts are not used or changed during chemical reactions and, therefore, are reusable. Whereas inorganic molecules may serve as catalysts for a wide range of chemical reactions, proteins called **enzymes** serve as catalysts for biochemical reactions inside cells. Enzymes thus play an important role in controlling cellular metabolism.

An enzyme functions by lowering the **activation energy** of a chemical reaction inside the cell. Activation energy is the energy needed to form or break chemical bonds and convert reactants to products (Figure 8.5). Enzymes lower the activation energy by binding to the reactant molecules and holding them in such a way as to speed up the reaction.

The chemical reactants to which an enzyme binds are called **substrates**, and the location within the enzyme where the substrate binds is called the enzyme's **active site**. The characteristics of the amino acids near the active site create a very specific chemical environment within the active site that induces suitability to binding, albeit briefly, to a specific substrate (or substrates). Due to this jigsaw puzzle-like match between an enzyme and its substrates, enzymes are known for their specificity. In fact, as an enzyme binds to its substrate(s), the enzyme structure changes slightly to find the best fit between the transition state (a structural intermediate between the substrate and product) and the active site, just as a rubber glove molds to a hand inserted into it. This active-site modification in the presence of substrate, along with the simultaneous formation of the transition state, is called induced fit (Figure 8.6). Overall, there is a specifically matched enzyme for each substrate and, thus, for each chemical reaction; however, there is some flexibility as well. Some enzymes have the ability to act on several different structurally related substrates.

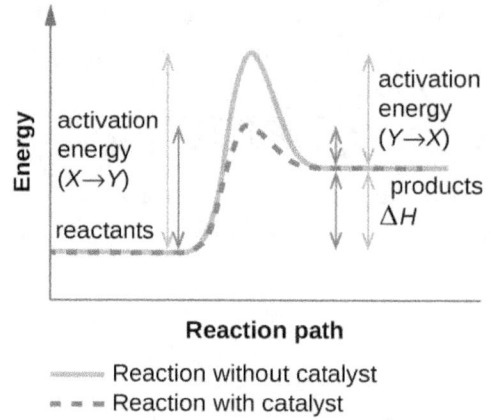

Figure 8.5 Enzymes lower the activation energy of a chemical reaction.

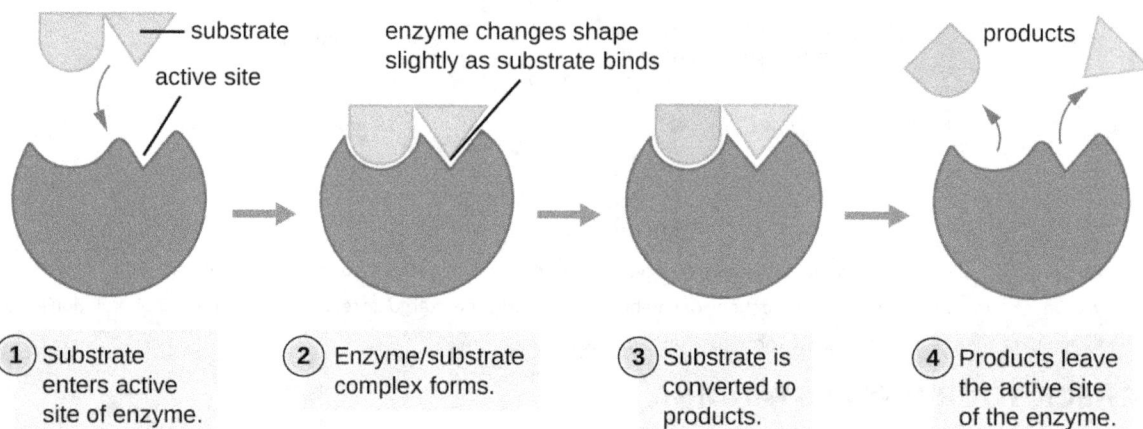

Figure 8.6 According to the induced-fit model, the active site of the enzyme undergoes conformational changes upon binding with the substrate.

Enzymes are subject to influences by local environmental conditions such as pH, substrate concentration, and temperature. Although increasing the environmental temperature generally increases reaction rates, enzyme catalyzed or otherwise, increasing or decreasing the temperature outside of an optimal range can affect chemical bonds within the active site, making them less well suited to bind substrates. High temperatures will eventually cause enzymes, like other biological molecules, to denature, losing their three-dimensional structure and function. Enzymes are also suited to function best within a certain pH range, and, as with temperature, extreme environmental pH values (acidic or basic) can cause enzymes to denature. Active-site amino-acid side chains have their own acidic or basic properties that are optimal for catalysis and, therefore, are sensitive to changes in pH.

Another factor that influences enzyme activity is substrate concentration: Enzyme activity is increased at higher concentrations of substrate until it reaches a saturation point at which the enzyme can bind no additional substrate. Overall, enzymes are optimized to work best under the environmental conditions in which the organisms that produce them live. For example, while microbes that inhabit hot springs have enzymes that work best at high temperatures, human pathogens have enzymes that work best at 37°C. Similarly, while enzymes produced by most organisms work best at a neutral pH, microbes growing in acidic environments make enzymes optimized to low pH conditions, allowing for their growth at those conditions.

Many enzymes do not work optimally, or even at all, unless bound to other specific nonprotein helper molecules, either temporarily through ionic or hydrogen bonds or permanently through stronger covalent bonds. Binding to these molecules promotes optimal conformation and function for their respective enzymes. Two types of helper molecules are **cofactor**s and **coenzyme**s. Cofactors are inorganic ions such as iron (Fe^{2+}) and magnesium (Mg^{2+}) that help stabilize enzyme conformation and function. One example of an enzyme that requires a metal ion as a cofactor is the enzyme that builds DNA molecules, DNA polymerase, which requires a bound zinc ion (Zn^{2+}) to function.

Coenzymes are organic helper molecules that are required for enzyme action. Like enzymes, they are not consumed and, hence, are reusable. The most common sources of coenzymes are dietary vitamins. Some vitamins are precursors to coenzymes and others act directly as coenzymes.

Some cofactors and coenzymes, like coenzyme A (CoA), often bind to the enzyme's active site, aiding in the chemistry of the transition of a substrate to a product (Figure 8.7). In such cases, an enzyme lacking a necessary cofactor or coenzyme is called an **apoenzyme** and is inactive. Conversely, an enzyme with the necessary associated cofactor or coenzyme is called a **holoenzyme** and is active. NADH and ATP are also both examples of commonly used coenzymes that provide high-energy electrons or phosphate groups, respectively, which bind to enzymes, thereby activating them.

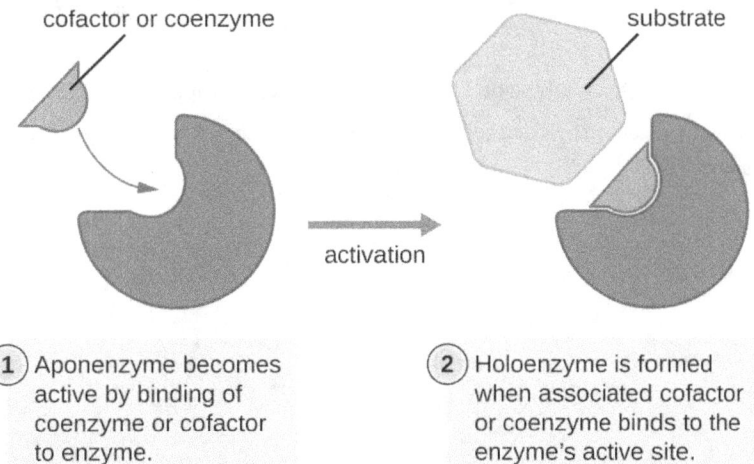

Figure 8.7 The binding of a coenzyme or cofactor to an apoenzyme is often required to form an active holoenzyme.

CHECK YOUR UNDERSTANDING

- What role do enzymes play in a chemical reaction?

Enzyme Inhibitors

Enzymes can be regulated in ways that either promote or reduce their activity. There are many different kinds of molecules that inhibit or promote enzyme function, and various mechanisms exist for doing so (Figure 8.8). A **competitive inhibitor** is a molecule similar enough to a substrate that it can compete with the substrate for binding to the active site by simply blocking the substrate from binding. For a competitive inhibitor to be effective, the inhibitor concentration needs to be approximately equal to the substrate concentration. Sulfa drugs provide a good example of competitive competition. They are used to treat bacterial infections because they bind to the active site of an enzyme within the bacterial folic acid synthesis pathway. When present in a sufficient dose, a sulfa drug prevents folic acid synthesis, and bacteria are unable to grow because they cannot synthesize DNA, RNA, and proteins. Humans are unaffected because we obtain folic acid from our diets.

On the other hand, a **noncompetitive (allosteric) inhibitor** binds to the enzyme at an **allosteric site**, a location other than the active site, and still manages to block substrate binding to the active site by inducing a conformational change that reduces the affinity of the enzyme for its substrate (Figure 8.9). Because only one inhibitor molecule is needed per enzyme for effective inhibition, the concentration of inhibitors needed for noncompetitive inhibition is typically much lower than the substrate concentration.

In addition to allosteric inhibitors, there are **allosteric activators** that bind to locations on an enzyme away from the active site, inducing a conformational change that increases the affinity of the enzyme's active site(s) for its substrate(s).

Allosteric control is an important mechanism of regulation of metabolic pathways involved in both catabolism and anabolism. In a most efficient and elegant way, cells have evolved also to use the products of their own metabolic reactions for **feedback inhibition** of enzyme activity. Feedback inhibition involves the use of a pathway product to regulate its own further production. The cell responds to the abundance of specific products by slowing production during anabolic or catabolic reactions (Figure 8.9).

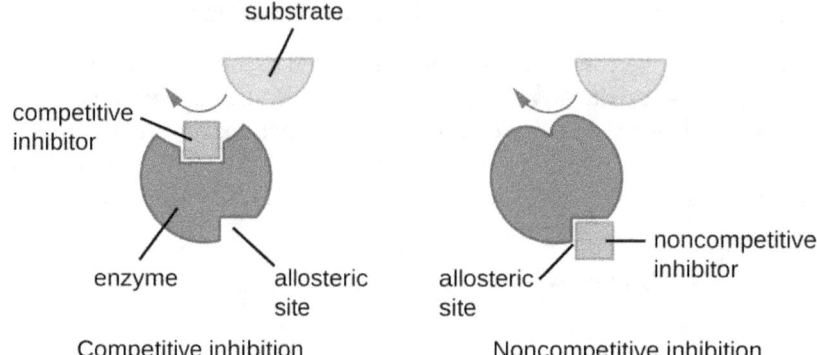

Figure 8.8 Enzyme activity can be regulated by either competitive inhibitors, which bind to the active site, or noncompetitive inhibitors, which bind to an allosteric site.

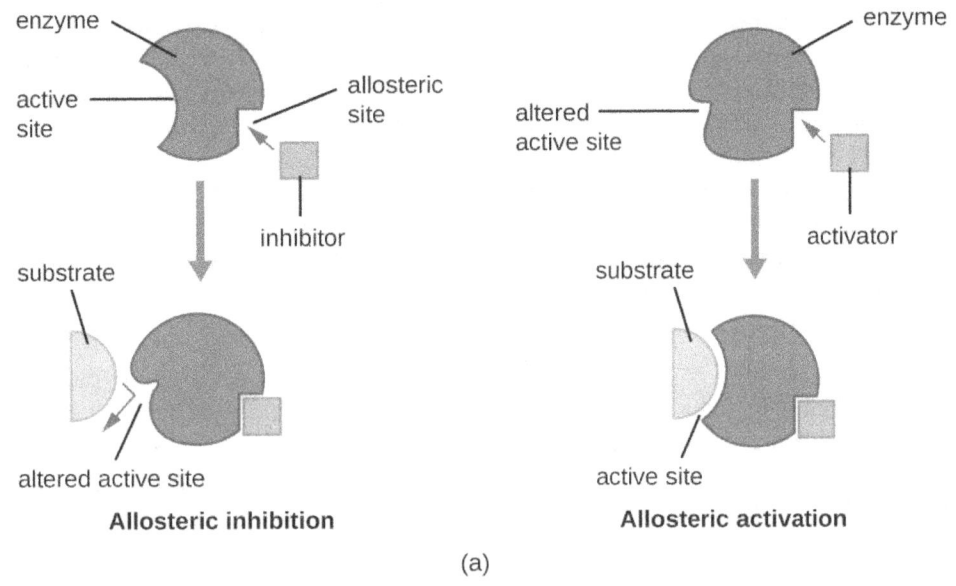
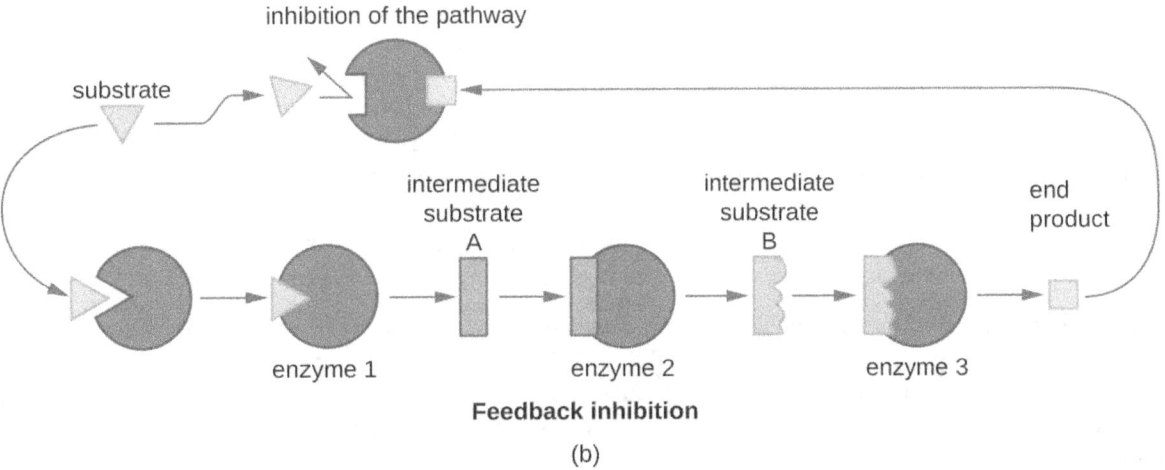

Figure 8.9 (a) Binding of an allosteric inhibitor reduces enzyme activity, but binding of an allosteric activator increases enzyme activity. (b) Feedback inhibition, where the end product of the pathway serves as a noncompetitive inhibitor to an enzyme early in the pathway, is an important mechanism of allosteric regulation in cells.

CHECK YOUR UNDERSTANDING

- Explain the difference between a competitive inhibitor and a noncompetitive inhibitor.

8.2 Catabolism of Carbohydrates

Learning Objectives

By the end of this section, you will be able to:
- Describe why glycolysis is not oxygen dependent
- Define and describe the net yield of three-carbon molecules, ATP, and NADH from glycolysis
- Explain how three-carbon pyruvate molecules are converted into two-carbon acetyl groups that can be funneled into the Krebs cycle.
- Define and describe the net yield of CO_2, GTP/ATP, $FADH_2$, and NADH from the Krebs cycle
- Explain how intermediate carbon molecules of the Krebs cycle can be used in a cell

Extensive enzyme pathways exist for breaking down carbohydrates to capture energy in ATP bonds. In addition, many catabolic pathways produce intermediate molecules that are also used as building blocks for anabolism. Understanding these processes is important for several reasons. First, because the main metabolic processes involved are common to a wide range of chemoheterotrophic organisms, we can learn a great deal about human metabolism by studying metabolism in more easily manipulated bacteria like *E. coli*. Second, because animal and human pathogens are also chemoheterotrophs, learning about the details of metabolism in these bacteria, including possible differences between bacterial and human pathways, is useful for the diagnosis of pathogens as well as for the discovery of antimicrobial therapies targeting specific pathogens. Last, learning specifically about the pathways involved in chemoheterotrophic metabolism also serves as a basis for comparing other more unusual metabolic strategies used by microbes. Although the chemical source of electrons initiating electron transfer is different between chemoheterorophs and chemoautotrophs, many similar processes are used in both types of organisms.

The typical example used to introduce concepts of metabolism to students is carbohydrate catabolism. For chemoheterotrophs, our examples of metabolism start with the catabolism of polysaccharides such as glycogen, starch, or cellulose. Enzymes such as amylase, which breaks down glycogen or starch, and cellulases, which break down cellulose, can cause the hydrolysis of glycosidic bonds between the glucose monomers in these polymers, releasing glucose for further catabolism.

Glycolysis

For bacteria, eukaryotes, and most archaea, **glycolysis** is the most common pathway for the catabolism of glucose; it produces energy, reduced electron carriers, and precursor molecules for cellular metabolism. Every living organism carries out some form of glycolysis, suggesting this mechanism is an ancient universal metabolic process. The process itself does not use oxygen; however, glycolysis can be coupled with additional metabolic processes that are either aerobic or anaerobic. Glycolysis takes place in the cytoplasm of prokaryotic and eukaryotic cells. It begins with a single six-carbon glucose molecule and ends with two molecules of a three-carbon sugar called pyruvate. Pyruvate may be broken down further after glycolysis to harness more energy through aerobic or anaerobic respiration, but many organisms, including many microbes, may be unable to respire; for these organisms, glycolysis may be their only source of generating ATP.

The type of glycolysis found in animals and that is most common in microbes is the **Embden-Meyerhof-Parnas (EMP) pathway**, named after Gustav Embden (1874–1933), Otto Meyerhof (1884–1951), and Jakub Parnas (1884–1949). Glycolysis using the EMP pathway consists of two distinct phases (Figure 8.10). The first part of the pathway, called the energy investment phase, uses energy from two ATP molecules to modify a glucose molecule so that the six-carbon sugar molecule can be split evenly into two phosphorylated three-carbon molecules called glyceraldehyde 3-phosphate (G3P). The second part of the pathway, called the energy payoff phase, extracts energy by oxidizing G3P to pyruvate, producing four ATP molecules and reducing two molecules of NAD^+ to two molecules of NADH, using electrons that originated from glucose. (A discussion and illustration of the full EMP pathway with chemical structures and enzyme names appear in Appendix C.)

The ATP molecules produced during the energy payoff phase of glycolysis are formed by **substrate-level phosphorylation** (Figure 8.11), one of two mechanisms for producing ATP. In substrate-level phosphorylation, a phosphate group is removed from an organic molecule and is directly transferred to an available ADP molecule, producing ATP. During glycolysis, high-energy phosphate groups from the intermediate molecules are added to ADP to make ATP.

Overall, in this process of glycolysis, the net gain from the breakdown of a single glucose molecule is:

- two ATP molecules
- two NADH molecule, and
- two pyruvate molecules.

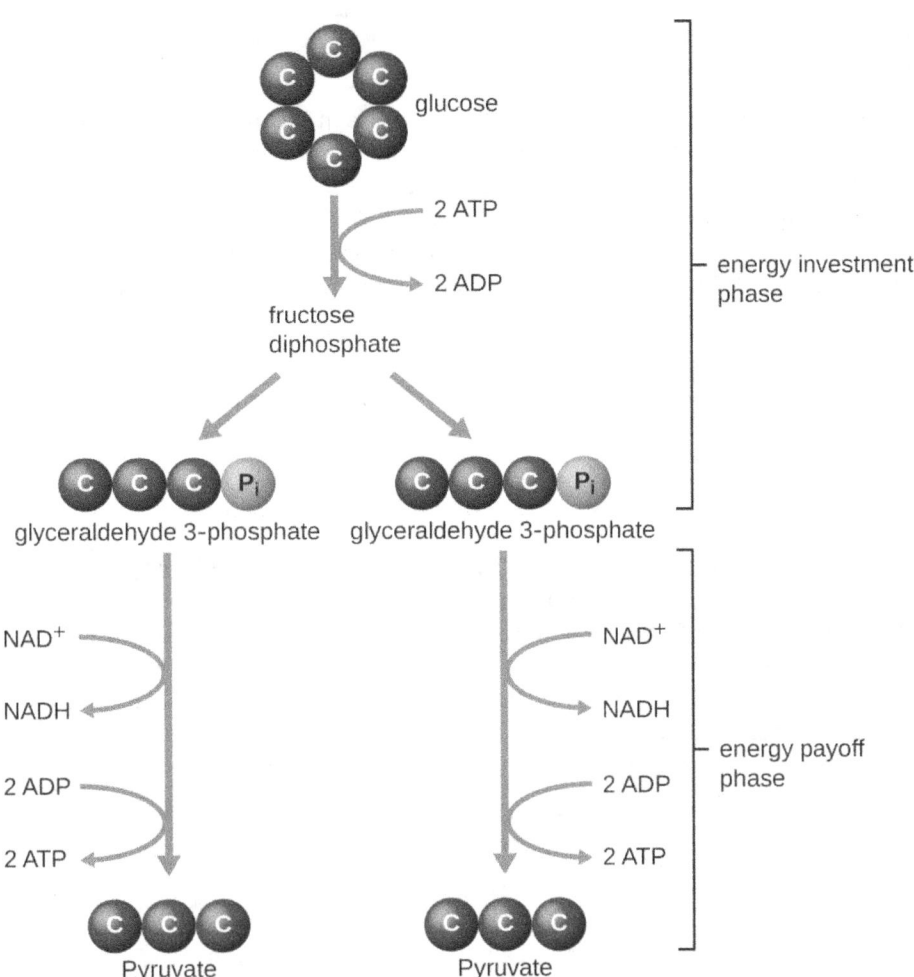

Figure 8.10 The energy investment phase of the Embden-Meyerhof-Parnas glycolysis pathway uses two ATP molecules to phosphorylate glucose, forming two glyceraldehyde 3-phosphate (G3P) molecules. The energy payoff phase harnesses the energy in the G3P molecules, producing four ATP molecules, two NADH molecules, and two pyruvates.

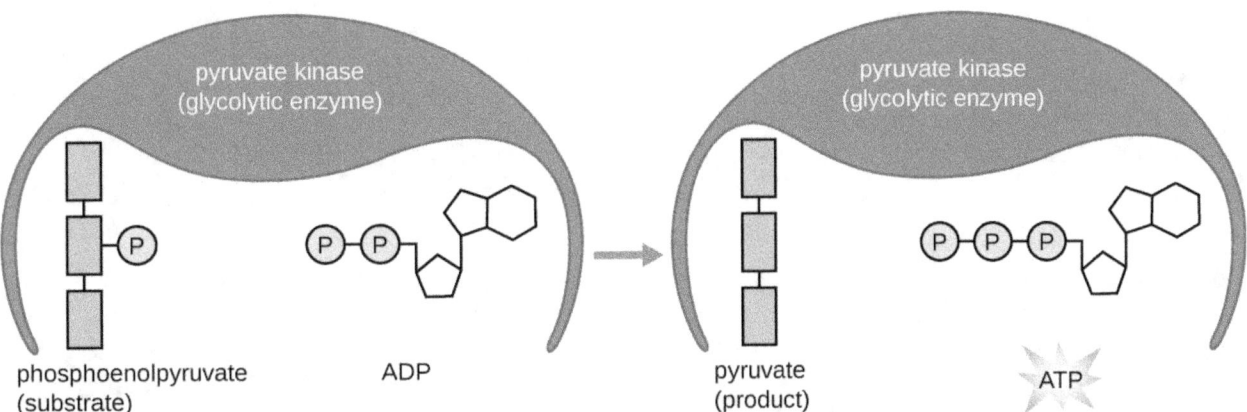

Figure 8.11 The ATP made during glycolysis is a result of substrate-level phosphorylation. One of the two enzymatic reactions in the energy payoff phase of Embden Meyerhof-Parnas glycolysis that produce ATP in this way is shown here.

Other Glycolytic Pathways

When we refer to glycolysis, unless otherwise indicated, we are referring to the EMP pathway used by animals and many bacteria. However, some prokaryotes use alternative glycolytic pathways. One important alternative is the **Entner-Doudoroff (ED) pathway**, named after its discoverers Nathan Entner and Michael Doudoroff (1911–1975). Although some bacteria, including the opportunistic gram-negative pathogen *Pseudomonas* *aeruginosa*, contain only the ED pathway for glycolysis, other bacteria, like *E. coli*, have the ability to use either the ED pathway or the EMP pathway.

A third type of glycolytic pathway that occurs in all cells, which is quite different from the previous two pathways, is the **pentose phosphate pathway** (PPP) also called the **phosphogluconate pathway** or the **hexose monophosphate shunt**. Evidence suggests that the PPP may be the most ancient universal glycolytic pathway. The intermediates from the PPP are used for the biosynthesis of nucleotides and amino acids. Therefore, this glycolytic pathway may be favored when the cell has need for nucleic acid and/or protein synthesis, respectively. A discussion and illustration of the complete ED pathway and PPP with chemical structures and enzyme names appear in Appendix C.

✓ CHECK YOUR UNDERSTANDING

- When might an organism use the ED pathway or the PPP for glycolysis?

Transition Reaction, Coenzyme A, and the Krebs Cycle

Glycolysis produces pyruvate, which can be further oxidized to capture more energy. For pyruvate to enter the next oxidative pathway, it must first be decarboxylated by the enzyme complex pyruvate dehydrogenase to a two-carbon acetyl group in the **transition reaction**, also called the **bridge reaction** (see Appendix C and Figure 8.12). In the transition reaction, electrons are also transferred to NAD^+ to form NADH. To proceed to the next phase of this metabolic process, the comparatively tiny two-carbon acetyl must be attached to a very large carrier compound called coenzyme A (CoA). The transition reaction occurs in the mitochondrial matrix of eukaryotes; in prokaryotes, it occurs in the cytoplasm because prokaryotes lack membrane-enclosed organelles.

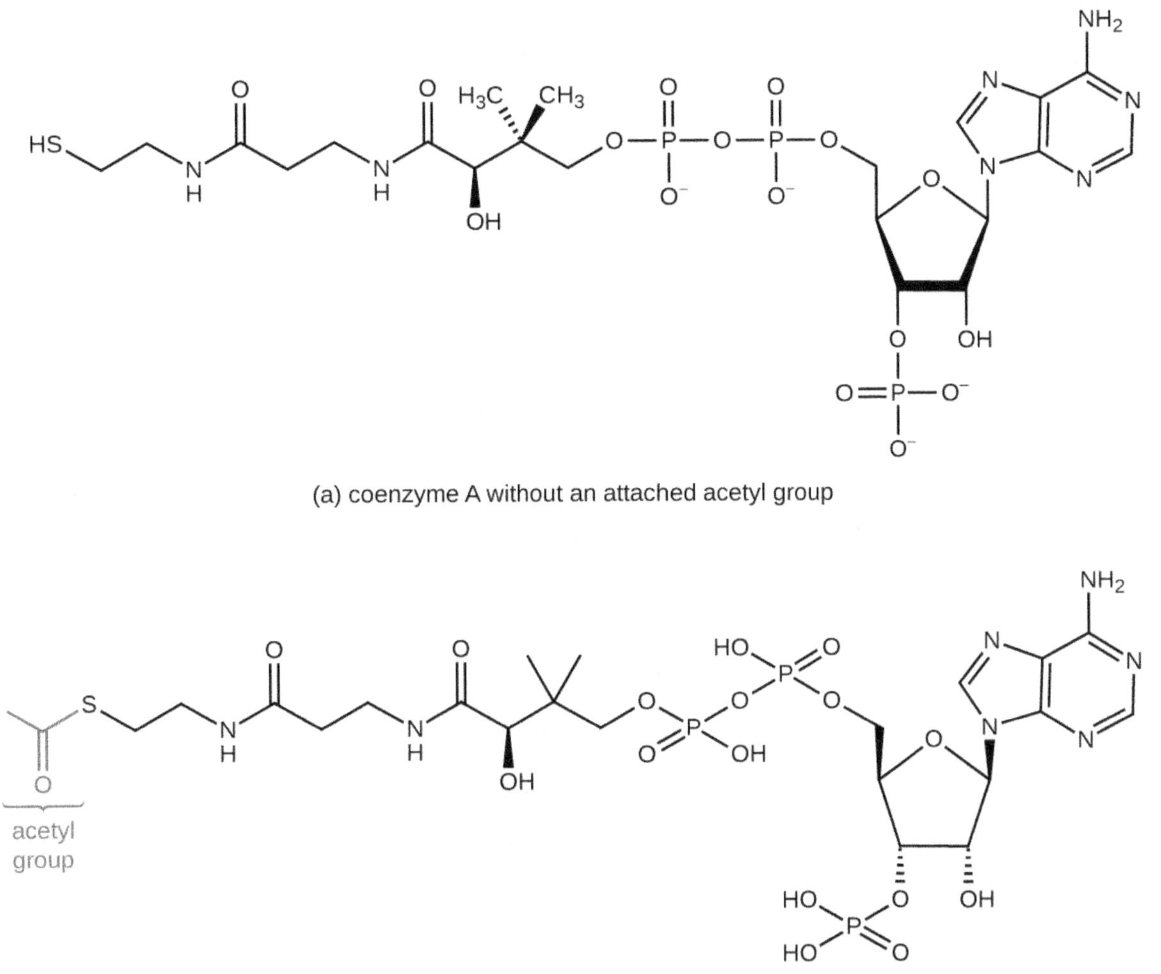

(a) coenzyme A without an attached acetyl group

(b) coenzyme A with an attached acetyl group

Figure 8.12 (a) Coenzyme A is shown here without an attached acetyl group. (b) Coenzyme A is shown here with an attached acetyl group.

The **Krebs cycle** transfers remaining electrons from the acetyl group produced during the transition reaction to electron carrier molecules, thus reducing them. The Krebs cycle also occurs in the cytoplasm of prokaryotes along with glycolysis and the transition reaction, but it takes place in the mitochondrial matrix of eukaryotic cells where the transition reaction also occurs. The Krebs cycle is named after its discoverer, British scientist Hans Adolf Krebs (1900–1981) and is also called the **citric acid cycle**, or the **tricarboxylic acid cycle (TCA)** because citric acid has three carboxyl groups in its structure. Unlike glycolysis, the Krebs cycle is a closed loop: The last part of the pathway regenerates the compound used in the first step (Figure 8.13). The eight steps of the cycle are a series of chemical reactions that capture the two-carbon acetyl group (the CoA carrier does not enter the Krebs cycle) from the transition reaction, which is added to a four-carbon intermediate in the Krebs cycle, producing the six-carbon intermediate citric acid (giving the alternate name for this cycle). As one turn of the cycle returns to the starting point of the four-carbon intermediate, the cycle produces two CO_2 molecules, one ATP molecule (or an equivalent, such as guanosine triphosphate [GTP]) produced by substrate-level phosphorylation, and three molecules of NADH and one of $FADH_2$. (A discussion and detailed illustration of the full Krebs cycle appear in Appendix C.)

Although many organisms use the Krebs cycle as described as part of glucose metabolism, several of the intermediate compounds in the Krebs cycle can be used in synthesizing a wide variety of important cellular molecules, including amino acids, chlorophylls, fatty acids, and nucleotides; therefore, the cycle is both anabolic and catabolic (Figure 8.14).

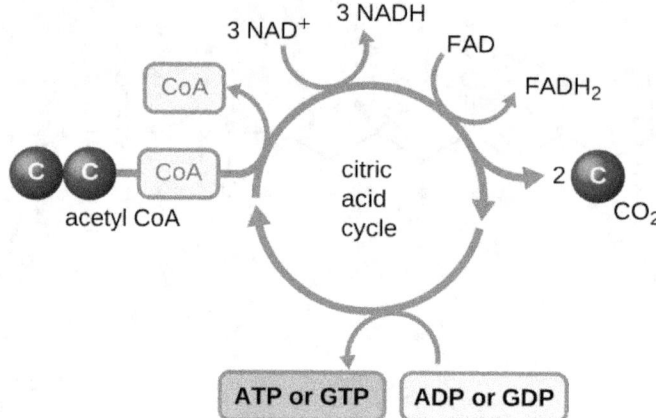

Figure 8.13 The Krebs cycle, also known as the citric acid cycle, is summarized here. Note incoming two-carbon acetyl results in the main outputs per turn of two CO_2, three NADH, one $FADH_2$, and one ATP (or GTP) molecules made by substrate-level phosphorylation. Two turns of the Krebs cycle are required to process all of the carbon from one glucose molecule.

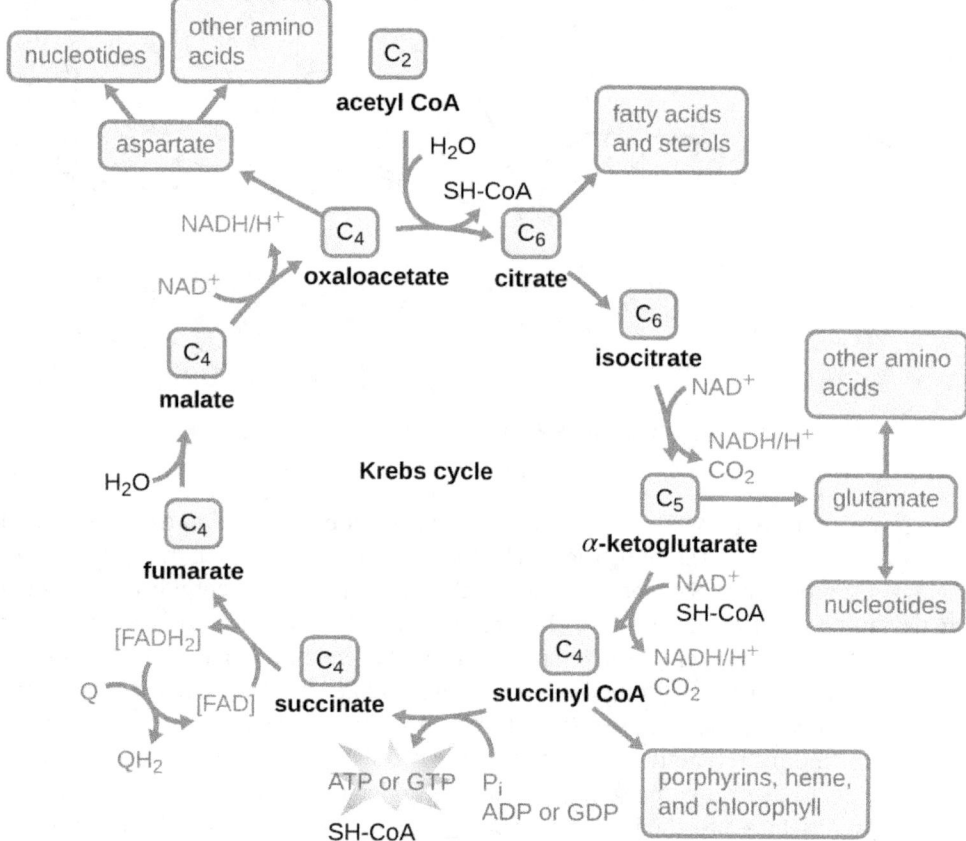

Figure 8.14 Many organisms use intermediates from the Krebs cycle, such as amino acids, fatty acids, and nucleotides, as building blocks for biosynthesis.

8.3 Cellular Respiration

Learning Objectives

By the end of this section, you will be able to:
- Compare and contrast the electron transport system location and function in a prokaryotic cell and a eukaryotic cell
- Compare and contrast the differences between substrate-level and oxidative phosphorylation
- Explain the relationship between chemiosmosis and proton motive force
- Describe the function and location of ATP synthase in a prokaryotic versus eukaryotic cell
- Compare and contrast aerobic and anaerobic respiration

We have just discussed two pathways in glucose catabolism—glycolysis and the Krebs cycle—that generate ATP by substrate-level phosphorylation. Most ATP, however, is generated during a separate process called **oxidative phosphorylation**, which occurs during cellular respiration. Cellular respiration begins when electrons are transferred from NADH and $FADH_2$—made in glycolysis, the transition reaction, and the Krebs cycle—through a series of chemical reactions to a final inorganic electron acceptor (either oxygen in aerobic respiration or non-oxygen inorganic molecules in anaerobic respiration). These electron transfers take place on the inner part of the cell membrane of prokaryotic cells or in specialized protein complexes in the inner membrane of the mitochondria of eukaryotic cells. The energy of the electrons is harvested to generate an electrochemical gradient across the membrane, which is used to make ATP by oxidative phosphorylation.

Electron Transport System

The **electron transport system (ETS)** is the last component involved in the process of cellular respiration; it comprises a series of membrane-associated protein complexes and associated mobile accessory electron carriers (Figure 8.15). Electron transport is a series of chemical reactions that resembles a bucket brigade in that electrons from NADH and $FADH_2$ are passed rapidly from one ETS electron carrier to the next. These carriers can pass electrons along in the ETS because of their **redox potential**. For a protein or chemical to accept electrons, it must have a more positive redox potential than the electron donor. Therefore, electrons move from electron carriers with more negative redox potential to those with more positive redox potential. The four major classes of electron carriers involved in both eukaryotic and prokaryotic electron transport systems are the cytochromes, flavoproteins, iron-sulfur proteins, and the quinones.

In **aerobic respiration**, the final electron acceptor (i.e., the one having the most positive redox potential) at the end of the ETS is an oxygen molecule (O_2) that becomes reduced to water (H_2O) by the final ETS carrier. This electron carrier, **cytochrome oxidase**, differs between bacterial types and can be used to differentiate closely related bacteria for diagnoses. For example, the gram-negative opportunist *Pseudomonas aeruginosa* and the gram-negative cholera-causing *Vibrio cholerae* use cytochrome c oxidase, which can be detected by the oxidase test, whereas other gram-negative Enterobacteriaceae, like *E. coli*, are negative for this test because they produce different cytochrome oxidase types.

There are many circumstances under which aerobic respiration is not possible, including any one or more of the following:

- The cell lacks genes encoding an appropriate cytochrome oxidase for transferring electrons to oxygen at the end of the electron transport system.
- The cell lacks genes encoding enzymes to minimize the severely damaging effects of dangerous oxygen radicals produced during aerobic respiration, such as hydrogen peroxide (H_2O_2) or superoxide (O_2^-).
- The cell lacks a sufficient amount of oxygen to carry out aerobic respiration.

One possible alternative to aerobic respiration is **anaerobic respiration**, using an inorganic molecule other than oxygen as a final electron acceptor. There are many types of anaerobic respiration found in bacteria and archaea. Denitrifiers are important soil bacteria that use nitrate (NO_3^-) and nitrite (NO_2^-) as final electron acceptors, producing nitrogen gas (N_2). Many aerobically respiring bacteria, including *E. coli*, switch to using nitrate as a final electron acceptor and producing nitrite when oxygen levels have been depleted.

Microbes using anaerobic respiration commonly have an intact Krebs cycle, so these organisms can access the energy of the NADH and FADH$_2$ molecules formed. However, anaerobic respirers use altered ETS carriers encoded by their genomes, including distinct complexes for electron transfer to their final electron acceptors. Smaller electrochemical gradients are generated from these electron transfer systems, so less ATP is formed through anaerobic respiration.

✓ CHECK YOUR UNDERSTANDING

- Do both aerobic respiration and anaerobic respiration use an electron transport chain?

Chemiosmosis, Proton Motive Force, and Oxidative Phosphorylation

In each transfer of an electron through the ETS, the electron loses energy, but with some transfers, the energy is stored as potential energy by using it to pump hydrogen ions (H$^+$) across a membrane. In prokaryotic cells, H$^+$ is pumped to the outside of the cytoplasmic membrane (called the periplasmic space in gram-negative and gram-positive bacteria), and in eukaryotic cells, they are pumped from the mitochondrial matrix across the inner mitochondrial membrane into the intermembrane space. There is an uneven distribution of H$^+$ across the membrane that establishes an electrochemical gradient because H$^+$ ions are positively charged (electrical) and there is a higher concentration (chemical) on one side of the membrane. This electrochemical gradient formed by the accumulation of H$^+$ (also known as a proton) on one side of the membrane compared with the other is referred to as the **proton motive force** (PMF). Because the ions involved are H$^+$, a pH gradient is also established, with the side of the membrane having the higher concentration of H$^+$ being more acidic. Beyond the use of the PMF to make ATP, as discussed in this chapter, the PMF can also be used to drive other energetically unfavorable processes, including nutrient transport and flagella rotation for motility.

The potential energy of this electrochemical gradient generated by the ETS causes the H$^+$ to diffuse across a membrane (the plasma membrane in prokaryotic cells and the inner membrane in mitochondria in eukaryotic cells). This flow of hydrogen ions across the membrane, called **chemiosmosis**, must occur through a channel in the membrane via a membrane-bound enzyme complex called **ATP synthase** (Figure 8.15). The tendency for movement in this way is much like water accumulated on one side of a dam, moving through the dam when opened. ATP synthase (like a combination of the intake and generator of a hydroelectric dam) is a complex protein that acts as a tiny generator, turning by the force of the H$^+$ diffusing through the enzyme, down their electrochemical gradient from where there are many mutually repelling H$^+$ to where there are fewer H$^+$. In prokaryotic cells, H$^+$ flows from the outside of the cytoplasmic membrane into the cytoplasm, whereas in eukaryotic mitochondria, H$^+$ flows from the intermembrane space to the mitochondrial matrix. The turning of the parts of this molecular machine regenerates ATP from ADP and inorganic phosphate (P$_i$) by oxidative phosphorylation, a second mechanism for making ATP that harvests the potential energy stored within an electrochemical gradient.

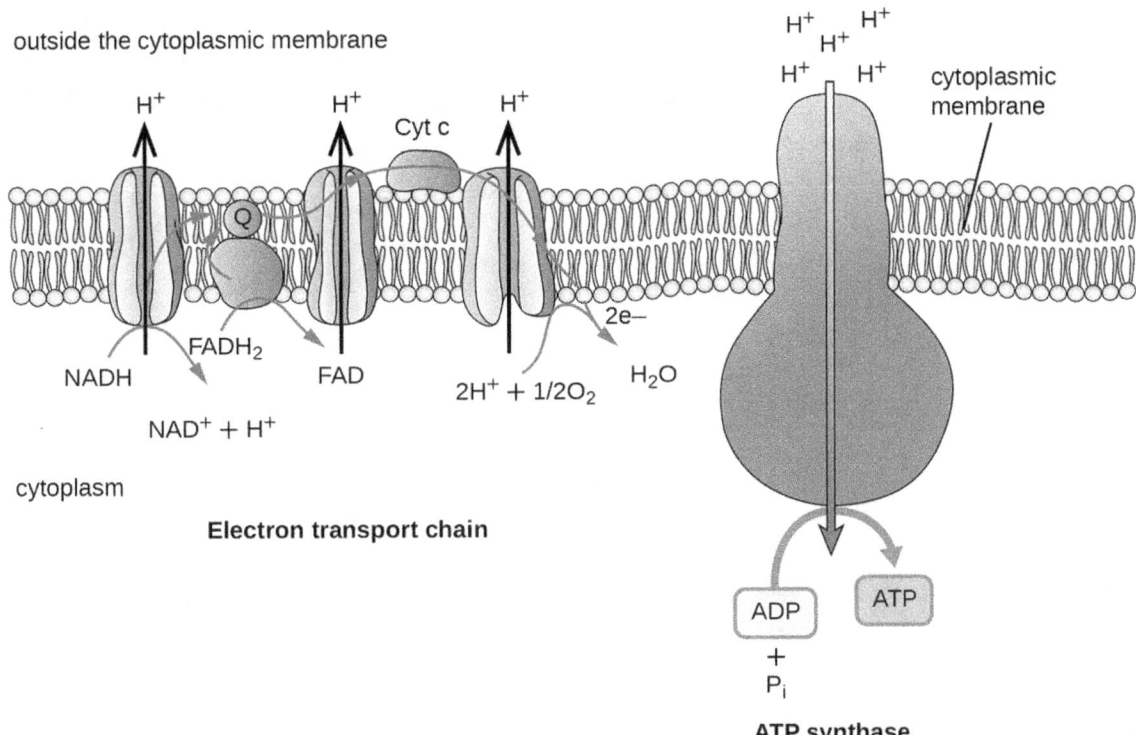

Figure 8.15 The bacterial electron transport chain is a series of protein complexes, electron carriers, and ion pumps that is used to pump H⁺ out of the bacterial cytoplasm into the extracellular space. H⁺ flows back down the electrochemical gradient into the bacterial cytoplasm through ATP synthase, providing the energy for ATP production by oxidative phosphorylation.(credit: modification of work by Klaus Hoffmeier)

The number of ATP molecules generated from the catabolism of glucose varies. For example, the number of hydrogen ions that the electron transport system complexes can pump through the membrane varies between different species of organisms. In aerobic respiration in mitochondria, the passage of electrons from one molecule of NADH generates enough proton motive force to make three ATP molecules by oxidative phosphorylation, whereas the passage of electrons from one molecule of FADH$_2$ generates enough proton motive force to make only two ATP molecules. Thus, the 10 NADH molecules made per glucose during glycolysis, the transition reaction, and the Krebs cycle carry enough energy to make 30 ATP molecules, whereas the two FADH$_2$ molecules made per glucose during these processes provide enough energy to make four ATP molecules. Overall, the theoretical maximum yield of ATP made during the complete aerobic respiration of glucose is 38 molecules, with four being made by substrate-level phosphorylation and 34 being made by oxidative phosphorylation (Figure 8.16). In reality, the total ATP yield is usually less, ranging from one to 34 ATP molecules, depending on whether the cell is using aerobic respiration or anaerobic respiration; in eukaryotic cells, some energy is expended to transport intermediates from the cytoplasm into the mitochondria, affecting ATP yield.

Figure 8.16 summarizes the theoretical maximum yields of ATP from various processes during the complete aerobic respiration of one glucose molecule.

Source	Carbon Flow	Molecules of Reduced Coenzymes Produced	Net ATP Molecules Made by Substrate-Level Phosphorylation	Net ATP Molecules Made by Oxidative Phosphorylation	Theoretical Maximum Yield of ATP Molecules
Glycolysis (EMP)	Glucose (6C) ⟶ 2 pyruvates (3C)	2 NADH	2 ATP	6 ATP from 2 NADH	8
Transition reaction	2 pyruvates (3C) ⟶ 2 acetyl (2C) + 2 CO_2	2 NADH		6 ATP from 2 NADH	6
Krebs cycle	2 acetyl (2C) ⟶ 4 CO_2	6 NADH 2 $FADH_2$	2 ATP	18 ATP from 6 NADH 4 ATP from 2 $FADH_2$	24
Total:	glucose (6C) ⟶ 6 CO_2	10 NADH 2 $FADH_2$	4 ATP	34 ATP	38 ATP

Figure 8.16

✓ CHECK YOUR UNDERSTANDING

- What are the functions of the proton motive force?

8.4 Fermentation

Learning Objectives

By the end of this section, you will be able to:
- Define fermentation and explain why it does not require oxygen
- Describe the fermentation pathways and their end products and give examples of microorganisms that use these pathways
- Compare and contrast fermentation and anaerobic respiration

Many cells are unable to carry out respiration because of one or more of the following circumstances:

1. The cell lacks a sufficient amount of any appropriate, inorganic, final electron acceptor to carry out cellular respiration.
2. The cell lacks genes to make appropriate complexes and electron carriers in the electron transport system.
3. The cell lacks genes to make one or more enzymes in the Krebs cycle.

Whereas lack of an appropriate inorganic final electron acceptor is environmentally dependent, the other two conditions are genetically determined. Thus, many prokaryotes, including members of the clinically important genus *Streptococcus*, are permanently incapable of respiration, even in the presence of oxygen. Conversely, many prokaryotes are facultative, meaning that, should the environmental conditions change to provide an appropriate inorganic final electron acceptor for respiration, organisms containing all the genes required to do so will switch to cellular respiration for glucose metabolism because respiration allows for much greater ATP production per glucose molecule.

If respiration does not occur, NADH must be reoxidized to NAD$^+$ for reuse as an electron carrier for glycolysis, the cell's only mechanism for producing any ATP, to continue. Some living systems use an organic molecule (commonly pyruvate) as a final electron acceptor through a process called **fermentation**. Fermentation does not involve an electron transport system and does not directly produce any additional ATP beyond that produced during glycolysis by substrate-level phosphorylation. Organisms carrying out fermentation, called fermenters, produce a maximum of two ATP molecules per glucose during glycolysis. Table 8.2 compares the final electron acceptors and methods of ATP synthesis in aerobic respiration, anaerobic respiration, and fermentation. Note that the number of ATP molecules shown for glycolysis assumes the Embden-Meyerhof-Parnas pathway. The number of ATP molecules made by substrate-level phosphorylation (SLP) versus oxidative phosphorylation (OP) are indicated.

Comparison of Respiration Versus Fermentation

Type of Metabolism	Example	Final Electron Acceptor	Pathways Involved in ATP Synthesis (Type of Phosphorylation)	Maximum Yield of ATP Molecules
Aerobic respiration	*Staphylococcus aureus*	O_2	EMP glycolysis (SLP) Krebs cycle (SLP) Electron transport and chemiosmosis (OP):	2 2 34
			Total	**38**
Anaerobic respiration	*Paracoccus denitrificans*	NO_3^-, SO_4^{-2}, Fe^{+3}, CO_2, other inorganics	EMP glycolysis (SLP) Krebs cycle (SLP) Electron transport and chemiosmosis (OP):	2 2 1–32
			Total	**5–36**
Fermentation	*Candida albicans*	Organics (usually pyruvate)	EMP glycolysis (SLP) Fermentation	2 0
			Total	**2**

Table 8.2

Microbial fermentation processes have been manipulated by humans and are used extensively in the production of various foods and other commercial products, including pharmaceuticals. Microbial fermentation can also be useful for identifying microbes for diagnostic purposes.

Fermentation by some bacteria, like those in yogurt and other soured food products, and by animals in muscles during oxygen depletion, is lactic acid fermentation. The chemical reaction of lactic acid fermentation is as follows:

$$\text{Pyruvate} + \text{NADH} \leftrightarrow \text{lactic acid} + \text{NAD}^+$$

Bacteria of several gram-positive genera, including *Lactobacillus*, *Leuconostoc*, and *Streptococcus*, are collectively known as the lactic acid bacteria (LAB), and various strains are important in food production. During yogurt and cheese production, the highly acidic environment generated by lactic acid fermentation denatures proteins contained in milk, causing it to solidify. When lactic acid is the only fermentation product, the process is said to be **homolactic fermentation**; such is the case for *Lactobacillus delbrueckii* and *S. thermophiles* used in yogurt production. However, many bacteria perform **heterolactic fermentation**, producing a mixture of lactic acid, ethanol and/or acetic acid, and CO_2 as a result, because of their use of the branched pentose phosphate pathway instead of the EMP pathway for glycolysis. One important heterolactic fermenter is *Leuconostoc mesenteroides*, which is used for souring vegetables like cucumbers and cabbage, producing pickles and sauerkraut, respectively.

Lactic acid bacteria are also important medically. The production of low pH environments within the body inhibits the establishment and growth of pathogens in these areas. For example, the vaginal microbiota is composed largely of lactic acid bacteria, but when these bacteria are reduced, yeast can proliferate, causing a yeast infection. Additionally, lactic acid bacteria are important in maintaining the health of the gastrointestinal tract and, as such, are the primary component of probiotics.

Another familiar fermentation process is alcohol fermentation, which produces ethanol. The ethanol fermentation reaction is shown in Figure 8.17. In the first reaction, the enzyme pyruvate decarboxylase removes a carboxyl group from pyruvate, releasing CO_2 gas while producing the two-carbon molecule acetaldehyde. The second reaction, catalyzed by the enzyme alcohol dehydrogenase, transfers an electron from NADH to acetaldehyde, producing ethanol and NAD^+. The ethanol fermentation of pyruvate by the yeast *Saccharomyces cerevisiae* is used in the production of alcoholic beverages and also makes bread products rise due to CO_2 production. Outside of the food industry, ethanol fermentation of plant products is important in biofuel production.

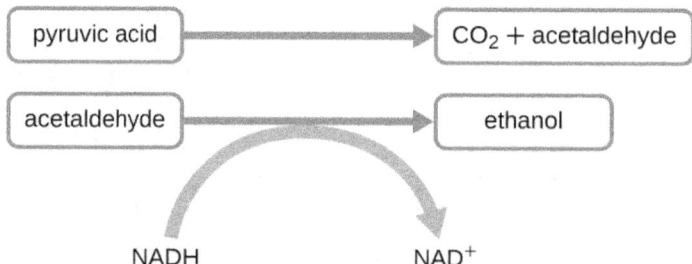

Figure 8.17 The chemical reactions of alcohol fermentation are shown here. Ethanol fermentation is important in the production of alcoholic beverages and bread.

Beyond lactic acid fermentation and alcohol fermentation, many other fermentation methods occur in prokaryotes, all for the purpose of ensuring an adequate supply of NAD^+ for glycolysis (Table 8.3). Without these pathways, glycolysis would not occur and no ATP would be harvested from the breakdown of glucose. It should be noted that most forms of fermentation besides homolactic fermentation produce gas, commonly CO_2 and/or hydrogen gas. Many of these different types of fermentation pathways are also used in food production and each results in the production of different organic acids, contributing to the unique flavor of a particular fermented food product. The propionic acid produced during propionic acid fermentation contributes to the distinctive flavor of Swiss cheese, for example.

Several fermentation products are important commercially outside of the food industry. For example, chemical solvents such as acetone and butanol are produced during acetone-butanol-ethanol fermentation. Complex organic pharmaceutical compounds used in antibiotics (e.g., penicillin), vaccines, and vitamins are produced through mixed acid fermentation. Fermentation products are used in the laboratory to differentiate various bacteria for diagnostic purposes. For example, enteric bacteria are known for their ability to perform mixed acid fermentation, reducing the pH, which can be detected using a pH indicator. Similarly, the bacterial production of acetoin during butanediol fermentation can also be detected. Gas production from fermentation can also be seen in an inverted Durham tube that traps produced gas in a broth culture.

Microbes can also be differentiated according to the substrates they can ferment. For example, *E. coli* can ferment lactose, forming gas, whereas some of its close gram-negative relatives cannot. The ability to ferment the sugar alcohol sorbitol is used to identify the pathogenic enterohemorrhagic O157:H7 strain of *E. coli* because, unlike other *E. coli* strains, it is unable to ferment sorbitol. Last, mannitol fermentation differentiates the mannitol-fermenting *Staphylococcus aureus* from other non–mannitol-fermenting staphylococci.

Common Fermentation Pathways

Pathway	End Products	Example Microbes	Commercial Products
Acetone-butanol-ethanol	Acetone, butanol, ethanol, CO_2	*Clostridium acetobutylicum*	Commercial solvents, gasoline alternative
Alcohol	Ethanol, CO_2	*Candida, Saccharomyces*	Beer, bread
Butanediol	Formic and lactic acid; ethanol; acetoin; 2,3 butanediol; CO_2; hydrogen gas	*Klebsiella, Enterobacter*	Chardonnay wine
Butyric acid	Butyric acid, CO_2, hydrogen gas	*Clostridium butyricum*	Butter
Lactic acid	Lactic acid	*Streptococcus, Lactobacillus*	Sauerkraut, yogurt, cheese
Mixed acid	Acetic, formic, lactic, and succinic acids; ethanol, CO_2, hydrogen gas	*Escherichia, Shigella*	Vinegar, cosmetics, pharmaceuticals
Propionic acid	Acetic acid, propionic acid, CO_2	*Propionibacterium, Bifidobacterium*	Swiss cheese

Table 8.3

✓ CHECK YOUR UNDERSTANDING

- When would a metabolically versatile microbe perform fermentation rather than cellular respiration?

 MICRO CONNECTIONS

Identifying Bacteria by Using API Test Panels

Identification of a microbial isolate is essential for the proper diagnosis and appropriate treatment of patients. Scientists have developed techniques that identify bacteria according to their biochemical characteristics. Typically, they either examine the use of specific carbon sources as substrates for fermentation or other metabolic reactions, or they identify fermentation products or specific enzymes present in reactions. In the past, microbiologists have used individual test tubes and plates to conduct biochemical testing. However, scientists, especially those in clinical laboratories, now more frequently use plastic, disposable, multitest panels that contain a number of miniature reaction tubes, each typically including a specific substrate and pH indicator. After inoculation of the test panel with a small sample of the microbe in question and incubation, scientists can compare the results to a database that includes the expected results for specific biochemical reactions for known microbes, thus enabling rapid identification of a sample microbe. These test panels have allowed scientists to reduce costs while improving efficiency and reproducibility by performing a larger number of tests simultaneously.

Many commercial, miniaturized biochemical test panels cover a number of clinically important groups of bacteria and yeasts. One of the earliest and most popular test panels is the Analytical Profile Index (API) panel invented in the 1970s. Once some basic laboratory characterization of a given strain has been performed, such as determining the strain's Gram morphology, an appropriate test strip that contains 10 to 20 different biochemical tests for differentiating strains within that microbial group can be used. Currently, the various API strips can be used to quickly and easily identify more than 600 species of bacteria, both aerobic and anaerobic, and approximately 100 different types of yeasts. Based on the colors of the reactions when metabolic end products are present, due to the presence of pH indicators, a metabolic profile is created from the results (Figure 8.18). Microbiologists can then compare the sample's profile to the database to identify the specific microbe.

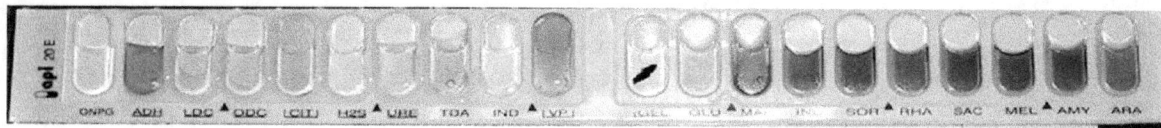

Figure 8.18 The API 20NE test strip is used to identify specific strains of gram-negative bacteria outside the Enterobacteriaceae. Here is an API 20NE test strip result for *Photobacterium damselae* ssp. *piscicida*.

Clinical Focus

Part 2

Many of Hannah's symptoms are consistent with several different infections, including influenza and pneumonia. However, her sluggish reflexes along with her light sensitivity and stiff neck suggest some possible involvement of the central nervous system, perhaps indicating meningitis. Meningitis is an infection of the cerebrospinal fluid (CSF) around the brain and spinal cord that causes inflammation of the meninges, the protective layers covering the brain. Meningitis can be caused by viruses, bacteria, or fungi. Although all forms of meningitis are serious, bacterial meningitis is particularly serious. Bacterial meningitis may be caused by several different bacteria, but the bacterium *Neisseria meningitidis*, a gram-negative, bean-shaped diplococcus, is a common cause and leads to death within 1 to 2 days in 5% to 10% of patients.

Given the potential seriousness of Hannah's conditions, her physician advised her parents to take her to the hospital in the Gambian capital of Banjul and there have her tested and treated for possible meningitis. After a 3-hour drive to the hospital, Hannah was immediately admitted. Physicians took a blood sample and performed a lumbar puncture to test her CSF. They also immediately started her on a course of the antibiotic ceftriaxone, the drug of choice for treatment of meningitis caused by *N. meningitidis*, without waiting for laboratory test results.

- How might biochemical testing be used to confirm the identity of *N. meningitidis*?
- Why did Hannah's doctors decide to administer antibiotics without waiting for the test results?

Jump to the next Clinical Focus box. Go back to the previous Clinical Focus box.

8.5 Catabolism of Lipids and Proteins

Learning Objectives

By the end of this section, you will be able to:
- Describe how lipids are catabolized
- Describe how lipid catabolism can be used to identify microbes
- Describe how proteins are catabolized
- Describe how protein catabolism can be used to identify bacteria

Previous sections have discussed the catabolism of glucose, which provides energy to living cells, as well as how polysaccharides like glycogen, starch, and cellulose are degraded to glucose monomers. But microbes consume more than just carbohydrates for food. In fact, the microbial world is known for its ability to degrade a wide range of molecules, both naturally occurring and those made by human processes, for use as carbon sources. In this section, we will see that the pathways for both lipid and protein catabolism connect to those used for carbohydrate catabolism, eventually leading into glycolysis, the transition reaction, and the Krebs cycle pathways. Metabolic pathways should be considered to be porous—that is, substances enter from other pathways, and intermediates leave for other pathways. These pathways are not closed systems. Many of the substrates, intermediates, and products in a particular pathway are reactants in other pathways.

Lipid Catabolism

Triglycerides are a form of long-term energy storage in animals. They are made of glycerol and three fatty acids (see Figure 7.12). Phospholipids compose the cell and organelle membranes of all organisms except the archaea. Phospholipid structure is similar to triglycerides except that one of the fatty acids is replaced by a phosphorylated head group (see Figure 7.13). Triglycerides and phospholipids are broken down first by releasing fatty acid chains (and/or the phosphorylated head group, in the case of phospholipids) from the three-carbon glycerol backbone. The reactions breaking down triglycerides are catalyzed by **lipases** and those involving phospholipids are catalyzed by **phospholipases**. These enzymes contribute to the virulence of certain microbes, such as the bacterium *Staphylococcus aureus* and the fungus *Cryptococcus neoformans*. These microbes use phospholipases to destroy lipids and phospholipids in host cells and then use the catabolic products for energy (see Virulence Factors of Bacterial and Viral Pathogens).

The resulting products of lipid catabolism, glycerol and fatty acids, can be further degraded. Glycerol can be phosphorylated to glycerol-3-phosphate and easily converted to glyceraldehyde 3-phosphate, which continues through glycolysis. The released fatty acids are catabolized in a process called **β-oxidation**, which sequentially removes two-carbon acetyl groups from the ends of fatty acid chains, reducing NAD^+ and FAD to produce NADH and $FADH_2$, respectively, whose electrons can be used to make ATP by oxidative phosphorylation. The acetyl groups produced during β-oxidation are carried by coenzyme A to the Krebs cycle, and their movement through this cycle results in their degradation to CO_2, producing ATP by substrate-level phosphorylation and additional NADH and $FADH_2$ molecules (see Appendix C for a detailed illustration of β-oxidation).

Other types of lipids can also be degraded by certain microbes. For example, the ability of certain pathogens, like *Mycobacterium tuberculosis*, to degrade cholesterol contributes to their virulence. The side chains of cholesterol can be easily removed enzymatically, but degradation of the remaining fused rings is more problematic. The four fused rings are sequentially broken in a multistep process facilitated by specific enzymes, and the resulting products, including pyruvate, can be further catabolized in the Krebs cycle.

✓ CHECK YOUR UNDERSTANDING

- How can lipases and phospholipases contribute to virulence in microbes?

Protein Catabolism

Proteins are degraded through the concerted action of a variety of microbial **protease** enzymes. Extracellular proteases cut proteins internally at specific amino acid sequences, breaking them down into smaller peptides that can then be taken up by cells. Some clinically important pathogens can be identified by their ability to produce a specific type of extracellular protease. For example, the production of the extracellular protease gelatinase by members of the genera *Proteus* and *Serratia* can be used to distinguish them from other gram-negative enteric bacteria. Following inoculation and growth of microbes in gelatin broth, degradation of the gelatin protein due to gelatinase production prevents solidification of gelatin when refrigerated. Other pathogens can be distinguished by their ability to degrade casein, the main protein found in milk. When grown on skim milk agar, production of the extracellular protease caseinase causes degradation of casein, which appears as a zone of clearing around the microbial growth. Caseinase production by the opportunist pathogen *Pseudomonas aeruginosa* can be used to distinguish it from other related gram-negative bacteria.

After extracellular protease degradation and uptake of peptides in the cell, the peptides can then be broken down further into individual amino acids by additional intracellular proteases, and each amino acid can be enzymatically deaminated to remove the amino group. The remaining molecules can then enter the transition reaction or the Krebs cycle.

CHECK YOUR UNDERSTANDING

- How can protein catabolism help identify microbes?

Clinical Focus

Part 3

Because bacterial meningitis progresses so rapidly, Hannah's doctors had decided to treat her aggressively with antibiotics, based on empirical observation of her symptoms. However, laboratory testing to confirm the cause of Hannah's meningitis was still important for several reasons. *N. meningitidis* is an infectious pathogen that can be spread from person to person through close contact; therefore, if tests confirm *N. meningitidis* as the cause of Hannah's symptoms, Hannah's parents and others who came into close contact with her might need to be vaccinated or receive prophylactic antibiotics to lower their risk of contracting the disease. On the other hand, if it turns out that *N. meningitidis* is not the cause, Hannah's doctors might need to change her treatment.

The clinical laboratory performed a Gram stain on Hannah's blood and CSF samples. The Gram stain showed the presence of a bean-shaped gram-negative diplococcus. The technician in the hospital lab cultured Hannah's blood sample on both blood agar and chocolate agar, and the bacterium that grew on both media formed gray, nonhemolytic colonies. Next, he performed an oxidase test on this bacterium and determined that it was oxidase positive. Last, he examined the repertoire of sugars that the bacterium could use as a carbon source and found that the bacterium was positive for glucose and maltose use but negative for lactose and sucrose use. All of these test results are consistent with characteristics of *N. meningitidis*.

- What do these test results tell us about the metabolic pathways of *N. meningitidis*?
- Why do you think that the hospital used these biochemical tests for identification in lieu of molecular analysis by DNA testing?

Jump to the next Clinical Focus box. Go back to the previous Clinical Focus box.

8.6 Photosynthesis

Learning Objectives

By the end of this section, you will be able to:
- Describe the function and locations of photosynthetic pigments in eukaryotes and prokaryotes
- Describe the major products of the light-dependent and light-independent reactions
- Describe the reactions that produce glucose in a photosynthetic cell
- Compare and contrast cyclic and noncyclic photophosphorylation

Heterotrophic organisms ranging from *E. coli* to humans rely on the chemical energy found mainly in carbohydrate molecules. Many of these carbohydrates are produced by **photosynthesis**, the biochemical process by which phototrophic organisms convert solar energy (sunlight) into chemical energy. Although photosynthesis is most commonly associated with plants, microbial photosynthesis is also a significant supplier of chemical energy, fueling many diverse ecosystems. In this section, we will focus on microbial photosynthesis.

Photosynthesis takes place in two sequential stages: the light-dependent reactions and the light-independent reactions (Figure 8.19). In the **light-dependent reactions**, energy from sunlight is absorbed by pigment molecules in photosynthetic membranes and converted into stored chemical energy. In the **light-independent reactions**, the chemical energy produced by the light-dependent reactions is used to drive the assembly of sugar molecules using CO_2; however, these reactions are still light dependent because the products of the light-dependent reactions necessary for driving them are short-lived. The light-dependent reactions produce ATP and either NADPH or NADH to temporarily store energy. These energy carriers are used in the light-independent reactions to drive the energetically unfavorable process of "fixing" inorganic CO_2 in an organic form, sugar.

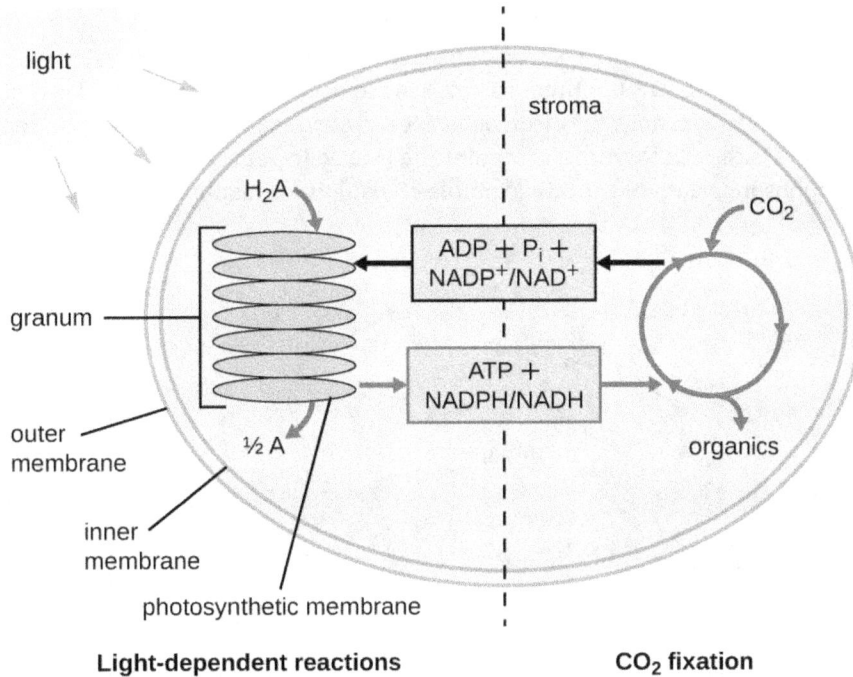

Figure 8.19 The light-dependent reactions of photosynthesis (left) convert light energy into chemical energy, forming ATP and NADPH. These products are used by the light-independent reactions to fix CO_2, producing organic carbon molecules.

Photosynthetic Structures in Eukaryotes and Prokaryotes

In all phototrophic eukaryotes, photosynthesis takes place inside a **chloroplast**, an organelle that arose in eukaryotes by endosymbiosis of a photosynthetic bacterium (see Unique Characteristics of Eukaryotic Cells). These chloroplasts are enclosed by a double membrane with inner and outer layers. Within the chloroplast is a third membrane that forms stacked, disc-shaped photosynthetic structures called thylakoids (Figure 8.20). A stack of thylakoids is called a granum, and the space surrounding the granum within the chloroplast is called stroma.

Photosynthetic membranes in prokaryotes, by contrast, are not organized into distinct membrane-enclosed organelles; rather, they are infolded regions of the plasma membrane. In cyanobacteria, for example, these infolded regions are also referred to as thylakoids. In either case, embedded within the thylakoid membranes or other photosynthetic bacterial membranes are **photosynthetic pigment** molecules organized into one or more photosystems, where light energy is actually converted into chemical energy.

Photosynthetic pigments within the photosynthetic membranes are organized into **photosystems**, each of which is composed of a light-harvesting (antennae) complex and a reaction center. The **light-harvesting complex** consists of multiple proteins and associated pigments that each may absorb light energy and, thus, become excited. This energy is transferred from one pigment molecule to another until eventually (after about a millionth of a second) it is delivered to the reaction center. Up to this point, only energy—not electrons—has been transferred between molecules. The **reaction center** contains a pigment molecule that can undergo oxidation upon excitation, actually giving up an electron. It is at this step in photosynthesis that light energy is converted into an excited electron.

Different kinds of light-harvesting pigments absorb unique patterns of wavelengths (colors) of visible light. Pigments reflect or transmit the wavelengths they cannot absorb, making them appear the corresponding color. Examples of photosynthetic pigments (molecules used to absorb solar energy) are bacteriochlorophylls (green, purple, or red), carotenoids (orange, red, or yellow), chlorophylls (green), phycocyanins (blue), and phycoerythrins (red). By having mixtures of pigments, an organism can absorb energy from more wavelengths. Because photosynthetic bacteria commonly grow in competition for sunlight, each type of photosynthetic bacteria is optimized for harvesting the wavelengths of light to which it is commonly exposed, leading to stratification of microbial communities in aquatic and soil ecosystems by light quality and penetration.

Once the light harvesting complex transfers the energy to the reaction center, the reaction center delivers its high-energy electrons, one by one, to an electron carrier in an electron transport system, and electron transfer through the ETS is initiated. The ETS is similar to that used in cellular respiration and is embedded within the photosynthetic membrane. Ultimately, the electron is used to produce NADH or NADPH. The electrochemical gradient that forms across the photosynthetic membrane is used to generate ATP by chemiosmosis through the process of photophosphorylation, another example of oxidative phosphorylation (Figure 8.21).

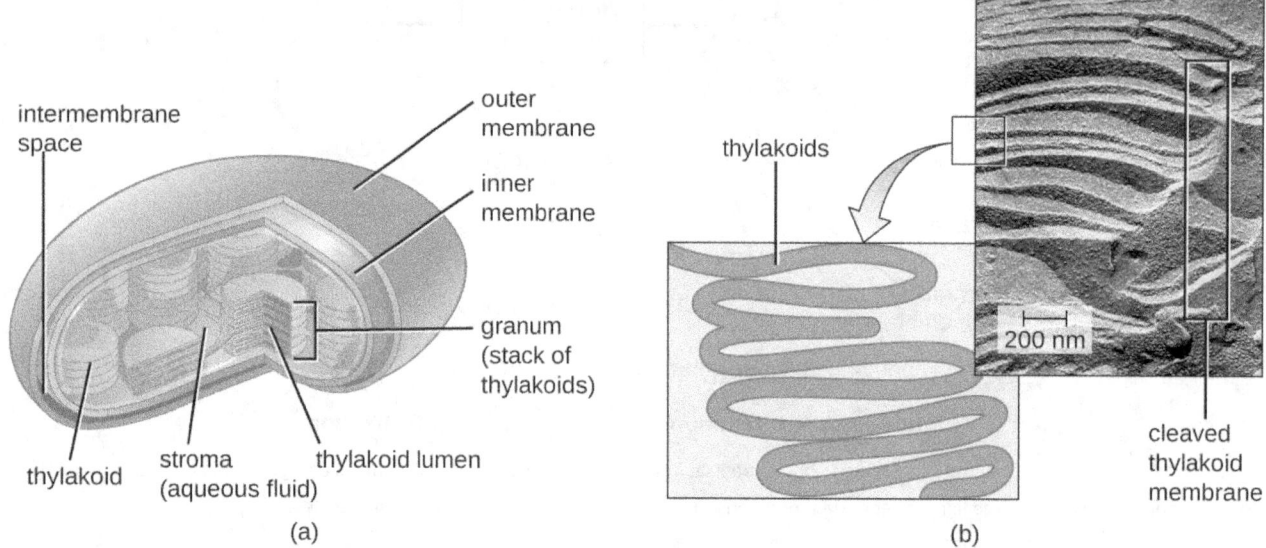

Figure 8.20 (a) Photosynthesis in eukaryotes takes place in chloroplasts, which contain thylakoids stacked into grana. (b) A photosynthetic prokaryote has infolded regions of the plasma membrane that function like thylakoids. (credit: scale bar data from Matt Russell.)

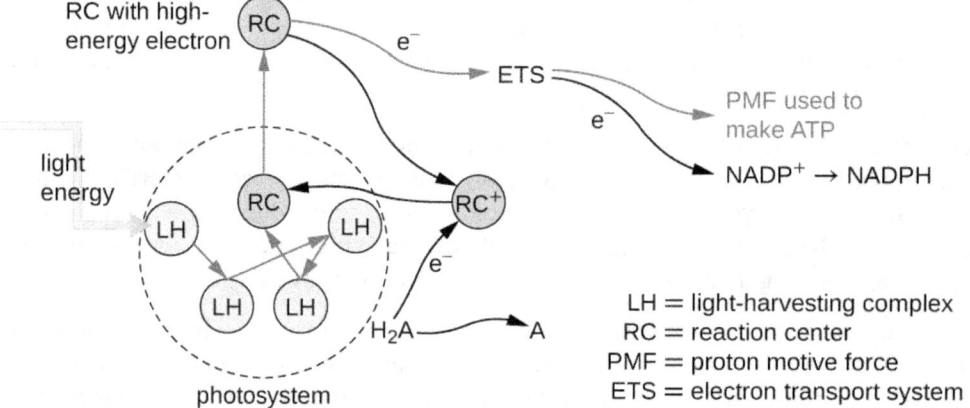

Figure 8.21 This figure summarizes how a photosystem works. Light harvesting (LH) pigments absorb light energy, converting it to chemical energy. The energy is passed from one LH pigment to another until it reaches a reaction center (RC) pigment, exciting an electron. This high-energy electron is lost from the RC pigment and passed through an electron transport system (ETS), ultimately producing NADH or NADPH and ATP. A reduced molecule (H_2A) donates an electron, replacing electrons to the electron-deficient RC pigment.

CHECK YOUR UNDERSTANDING

- In a phototrophic eukaryote, where does photosynthesis take place?

Oxygenic and Anoxygenic Photosynthesis

For photosynthesis to continue, the electron lost from the reaction center pigment must be replaced. The source of this electron (H_2A) differentiates the **oxygenic photosynthesis** of plants and cyanobacteria from **anoxygenic photosynthesis** carried out by other types of bacterial phototrophs (Figure 8.22). In oxygenic photosynthesis, H_2O is split and supplies the electron to the reaction center. Because oxygen is generated as a byproduct and is released, this type of photosynthesis is referred to as oxygenic photosynthesis. However, when other reduced compounds serve as the electron donor, oxygen is not generated; these types of photosynthesis are called anoxygenic photosynthesis. Hydrogen sulfide (H_2S) or thiosulfate ($S_2O_3^{2-}$) can serve as the electron donor, generating elemental sulfur and sulfate (SO_4^{2-}) ions, respectively, as a result.

Photosystems have been classified into two types: photosystem I (PSI) and photosystem II (PSII) (Figure 8.23). Cyanobacteria and plant chloroplasts have both photosystems, whereas anoxygenic photosynthetic bacteria use only one of the photosystems. Both photosystems are excited by light energy simultaneously. If the cell requires both ATP and NADPH for biosynthesis, then it will carry out **noncyclic photophosphorylation**. Upon passing of the PSII reaction center electron to the ETS that connects PSII and PSI, the lost electron from the PSII reaction center is replaced by the splitting of water. The excited PSI reaction center electron is used to reduce $NADP^+$ to NADPH and is replaced by the electron exiting the ETS. The flow of electrons in this way is called the **Z-scheme**.

If a cell's need for ATP is significantly greater than its need for NADPH, it may bypass the production of reducing power through **cyclic photophosphorylation**. Only PSI is used during cyclic photophosphorylation; the high-energy electron of the PSI reaction center is passed to an ETS carrier and then ultimately returns to the oxidized PSI reaction center pigment, thereby reducing it.

Oxygenic photosynthesis

$$6CO_2 + 12H_2O + \text{light energy} \longrightarrow C_6H_{12}O_6 + 6O_2 + 6H_2O$$

carbon dioxide, water → glucose, oxygen, water

Anoxygenic photosynthesis

$$CO_2 + 2H_2A + \text{light energy} \longrightarrow [CH_2O] + 2A + H_2O$$

carbon dioxide, electron donor* → carbohydrate, water

*$H_2A = H_2O, H_2S, H_2$, or other electron donor

Figure 8.22 Eukaryotes and cyanobacteria carry out oxygenic photosynthesis, producing oxygen, whereas other bacteria carry out anoxygenic photosynthesis, which does not produce oxygen.

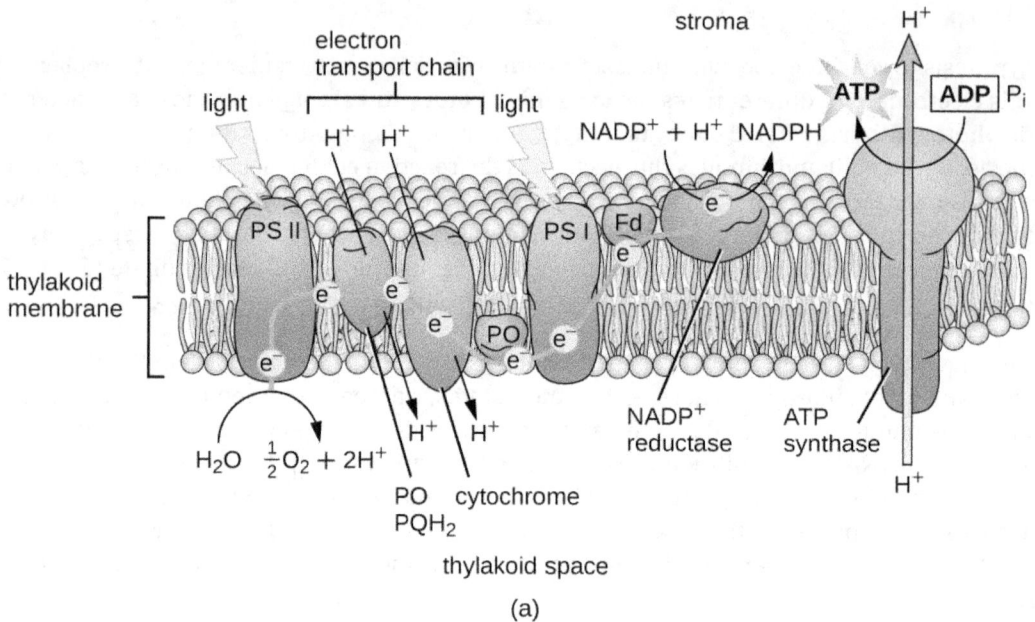

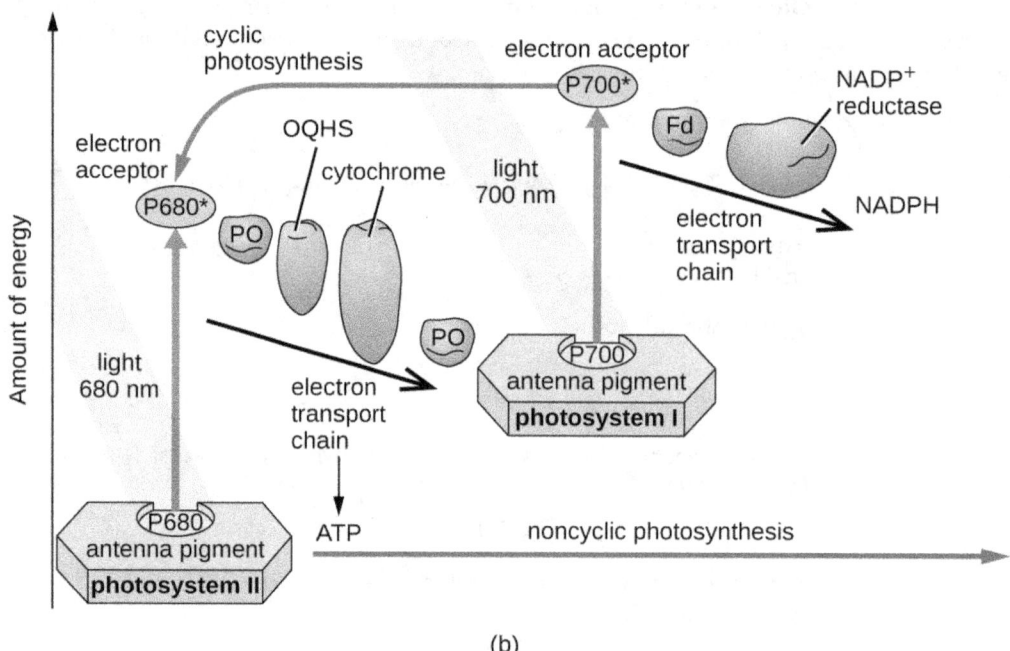

Figure 8.23 (a) PSI and PSII are found on the thylakoid membrane. The high-energy electron from PSII is passed to an ETS, which generates a proton motive force for ATP synthesis by chemiosmosis, and ultimately replaces the electron lost by the PSI reaction center. The PSI reaction center electron is used to make NADPH. (b) When both ATP and NADPH are required, noncyclic photophosphorylation (in cyanobacteria and plants) provides both. The electron flow described here is referred to as the Z-scheme (shown in yellow in [a]). When the cell's ATP needs outweigh those for NADPH, cyanobacteria and plants will use only PSI, and its reaction center electron is passed to the ETS to generate a proton motive force used for ATP synthesis.

CHECK YOUR UNDERSTANDING

- Why would a photosynthetic bacterium have different pigments?

Light-Independent Reactions

After the energy from the sun is converted into chemical energy and temporarily stored in ATP and NADPH molecules (having lifespans of millionths of a second), photoautotrophs have the fuel needed to build multicarbon carbohydrate molecules, which can survive for hundreds of millions of years, for long-term energy storage. The carbon comes from CO_2, the gas that is a waste product of cellular respiration.

The **Calvin-Benson cycle** (named for Melvin Calvin [1911–1997] and Andrew Benson [1917–2015]), the biochemical pathway used for fixation of CO_2, is located within the cytoplasm of photosynthetic bacteria and in the stroma of eukaryotic chloroplasts. The light-independent reactions of the Calvin cycle can be organized into three basic stages: fixation, reduction, and regeneration (see Appendix C for a detailed illustration of the Calvin cycle).

- **Fixation**: The enzyme **ribulose bisphosphate carboxylase (RuBisCO)** catalyzes the addition of a CO_2 to ribulose bisphosphate (RuBP). This results in the production of 3-phosphoglycerate (3-PGA).
- **Reduction**: Six molecules of both ATP and NADPH (from the light-dependent reactions) are used to convert 3-PGA into glyceraldehyde 3-phosphate (G3P). Some G3P is then used to build glucose.
- **Regeneration**: The remaining G3P not used to synthesize glucose is used to regenerate RuBP, enabling the system to continue CO_2 fixation. Three more molecules of ATP are used in these regeneration reactions.

The Calvin cycle is used extensively by plants and photoautotrophic bacteria, and the enzyme RuBisCO is said to be the most plentiful enzyme on earth, composing 30%–50% of the total soluble protein in plant chloroplasts.[1] However, besides its prevalent use in photoautotrophs, the Calvin cycle is also used by many nonphotosynthetic chemoautotrophs to fix CO_2. Additionally, other bacteria and archaea use alternative systems for CO_2 fixation. Although most bacteria using Calvin cycle alternatives are chemoautotrophic, certain green sulfur photoautotrophic bacteria have been also shown to use an alternative CO_2 fixation pathway.

CHECK YOUR UNDERSTANDING

- Describe the three stages of the Calvin cycle.

8.7 Biogeochemical Cycles

Learning Objectives

By the end of this section, you will be able to:
- Define and describe the importance of microorganisms in the biogeochemical cycles of carbon, nitrogen, and sulfur
- Define and give an example of bioremediation

Energy flows directionally through ecosystems, entering as sunlight for phototrophs or as inorganic molecules for chemoautotrophs. The six most common elements associated with organic molecules—carbon, hydrogen, nitrogen, oxygen, phosphorus, and sulfur—take a variety of chemical forms and may exist for long periods in the atmosphere, on land, in water, or beneath earth's surface. Geologic processes, such as erosion, water drainage, the movement of the continental plates, and weathering, all are involved in the cycling of elements on earth. Because geology and chemistry have major roles in the study of this process, the recycling of inorganic matter between living organisms and their nonliving environment is called a **biogeochemical cycle**. Here, we will focus on the function of microorganisms in these cycles, which play roles at each step, most frequently interconverting oxidized versions of molecules with reduced ones.

Carbon Cycle

Carbon is one of the most important elements to living organisms, as shown by its abundance and presence in all organic molecules. The carbon cycle exemplifies the connection between organisms in various ecosystems.

---

1 A. Dhingra et al. "Enhanced Translation of a Chloroplast-Expressed *RbcS* Gene Restores Small Subunit Levels and Photosynthesis in Nuclear *RbcS* Antisense Plants." *Proceedings of the National Academy of Sciences of the United States of America* 101 no. 16 (2004):6315–6320.

Carbon is exchanged between heterotrophs and autotrophs within and between ecosystems primarily by way of atmospheric CO_2, a fully oxidized version of carbon that serves as the basic building block that autotrophs use to build multicarbon, high-energy organic molecules such as glucose. Photoautotrophs and chemoautotrophs harness energy from the sun and from inorganic chemical compounds, respectively, to covalently bond carbon atoms together into reduced organic compounds whose energy can be later accessed through the processes of respiration and fermentation (Figure 8.24).

Overall, there is a constant exchange of CO_2 between the heterotrophs (which produce CO_2 as a result of respiration or fermentation) and the autotrophs (which use the CO_2 for fixation). Autotrophs also respire or ferment, consuming the organic molecules they form; they do not fix carbon for heterotrophs, but rather use it for their own metabolic needs.

Bacteria and archaea that use methane as their carbon source are called methanotrophs. Reduced one-carbon compounds like methane accumulate in certain anaerobic environments when CO_2 is used as a terminal electron acceptor in anaerobic respiration by archaea called methanogens. Some methanogens also ferment acetate (two carbons) to produce methane and CO_2. Methane accumulation due to methanogenesis occurs in both natural anaerobic soil and aquatic environments; methane accumulation also occurs as a result of animal husbandry because methanogens are members of the normal microbiota of ruminants. Environmental methane accumulation due to methanogenesis is of consequence because it is a strong greenhouse gas, and methanotrophs help to reduce atmospheric methane levels.

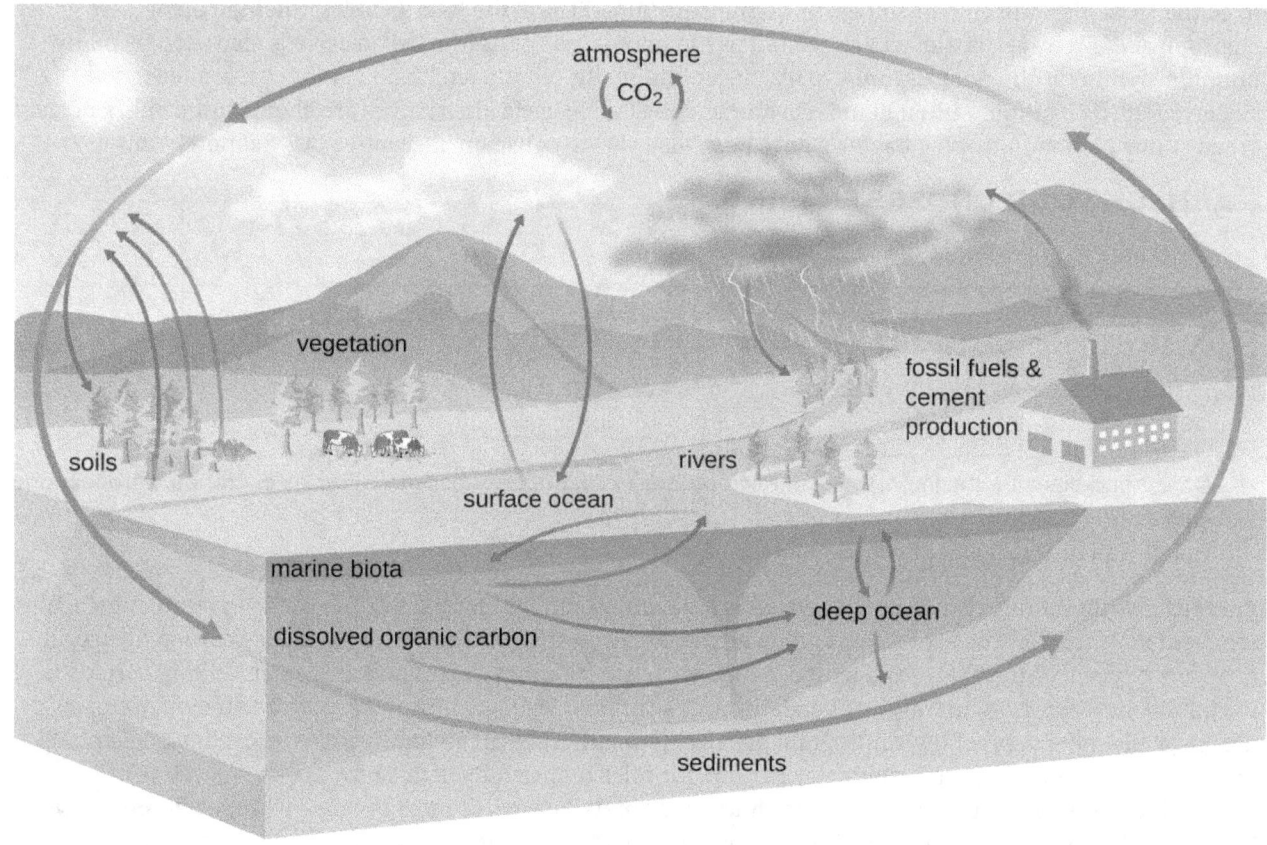

Carbon cycle

Figure 8.24 This figure summarizes the carbon cycle. Eukaryotes participate in aerobic respiration, fermentation, and oxygenic photosynthesis. Prokaryotes participate in all the steps shown. (credit: modification of work by NOAA)

CHECK YOUR UNDERSTANDING

- Describe the interaction between heterotrophs and autotrophs in the carbon cycle.

Nitrogen Cycle

Many biological macromolecules, including proteins and nucleic acids, contain nitrogen; however, getting nitrogen into living organisms is difficult. Prokaryotes play essential roles in the nitrogen cycle (Figure 8.25), transforming nitrogen between various forms for their own needs, benefiting other organisms indirectly. Plants and phytoplankton cannot incorporate nitrogen from the atmosphere (where it exists as tightly bonded, triple covalent N_2), even though this molecule composes approximately 78% of the atmosphere. Nitrogen enters the living world through free-living and symbiotic bacteria, which incorporate nitrogen into their macromolecules through specialized biochemical pathways called **nitrogen fixation**. Cyanobacteria in aquatic ecosystems fix inorganic nitrogen (from nitrogen gas) into ammonia (NH_3) that can be easily incorporated into biological macromolecules. *Rhizobium* bacteria (Figure 8.1) also fix nitrogen and live symbiotically in the root nodules of legumes (such as beans, peanuts, and peas), providing them with needed organic nitrogen while receiving fixed carbon as sugar in exchange. Free-living bacteria, such as members of the genus *Azotobacter*, are also able to fix nitrogen.

The nitrogen that enters living systems by nitrogen fixation is eventually converted from organic nitrogen back into nitrogen gas by microbes through three steps: ammonification, nitrification, and denitrification. In terrestrial systems, the first step is the ammonification process, in which certain bacteria and fungi convert nitrogenous waste from living animals or from the remains of dead organisms into ammonia (NH_3). This ammonia is then oxidized to nitrite $\left(NO_2^-\right)$, then to nitrate $\left(NO_3^-\right)$, by nitrifying soil bacteria such as members of the genus *Nitrosomonas*, through the process of nitrification. Last, the process of denitrification occurs, whereby soil bacteria, such as members of the genera *Pseudomonas* and *Clostridium*, use nitrate as a terminal electron acceptor in anaerobic respiration, converting it into nitrogen gas that reenters the atmosphere. A similar process occurs in the marine nitrogen cycle, where these three processes are performed by marine bacteria and archaea.

Human activity releases nitrogen into the environment by the use of artificial fertilizers that contain nitrogen and phosphorus compounds, which are then washed into lakes, rivers, and streams by surface runoff. A major effect from fertilizer runoff is saltwater and freshwater eutrophication, in which nutrient runoff causes the overgrowth and subsequent death of aquatic algae, making water sources anaerobic and inhospitable for the survival of aquatic organisms.

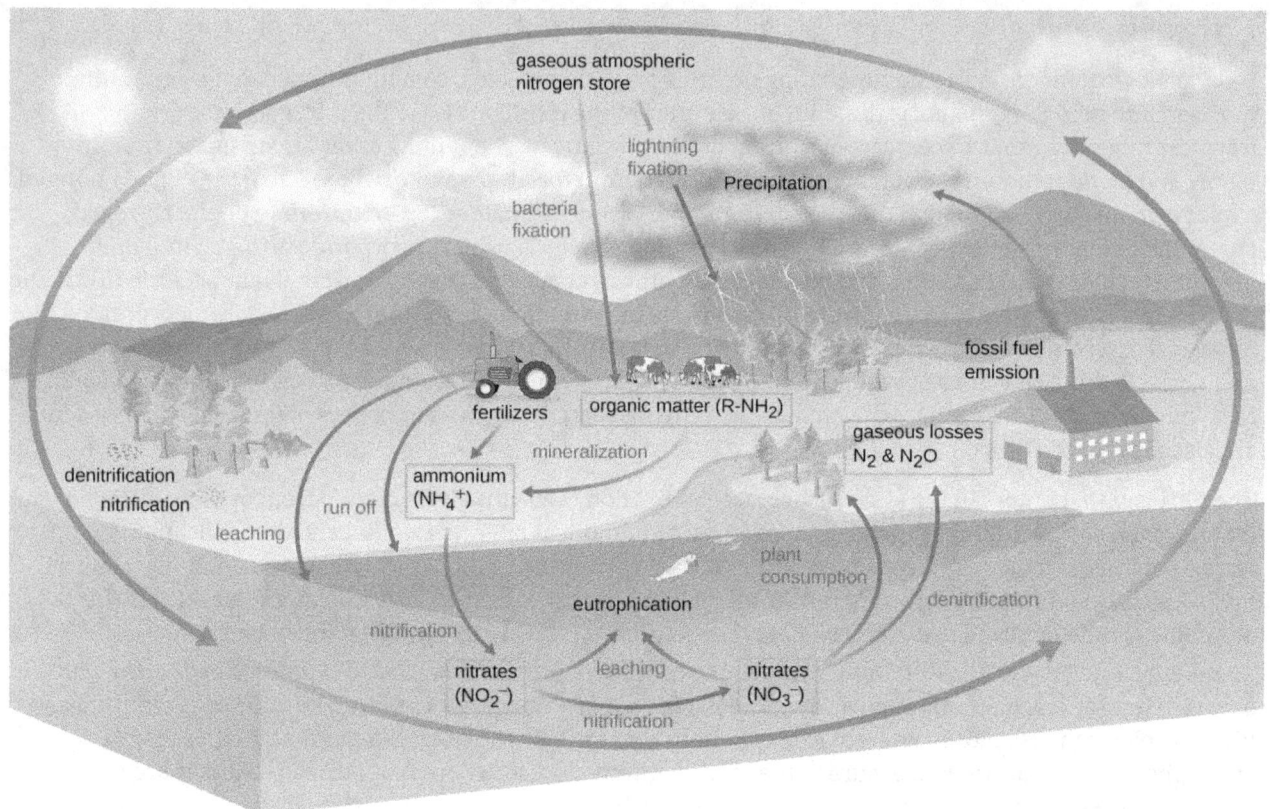

Nitrogen cycle

Figure 8.25 This figure summarizes the nitrogen cycle. Note that specific groups of prokaryotes each participate in every step in the cycle. (credit: modification of work by NOAA)

CHECK YOUR UNDERSTANDING

- What are the four steps of the nitrogen cycle?

LINK TO LEARNING

To learn more about the nitrogen cycle, visit the PBS (https://openstax.org/l/22nitrogencycle) website.

Sulfur Cycle

Sulfur is an essential element for the macromolecules of living organisms. As part of the amino acids cysteine and methionine, it is involved in the formation of proteins. It is also found in several vitamins necessary for the synthesis of important biological molecules like coenzyme A. Several groups of microbes are responsible for carrying out processes involved in the sulfur cycle (Figure 8.26). Anoxygenic photosynthetic bacteria as well as chemoautotrophic archaea and bacteria use hydrogen sulfide as an electron donor, oxidizing it first to elemental sulfur (S^0), then to sulfate (SO_4^{2-}). This leads to stratification of hydrogen sulfide in soil, with levels increasing at deeper, more anaerobic depths.

Many bacteria and plants can use sulfate as a sulfur source. Decomposition dead organisms by fungi and bacteria remove sulfur groups from amino acids, producing hydrogen sulfide, returning inorganic sulfur to the environment.

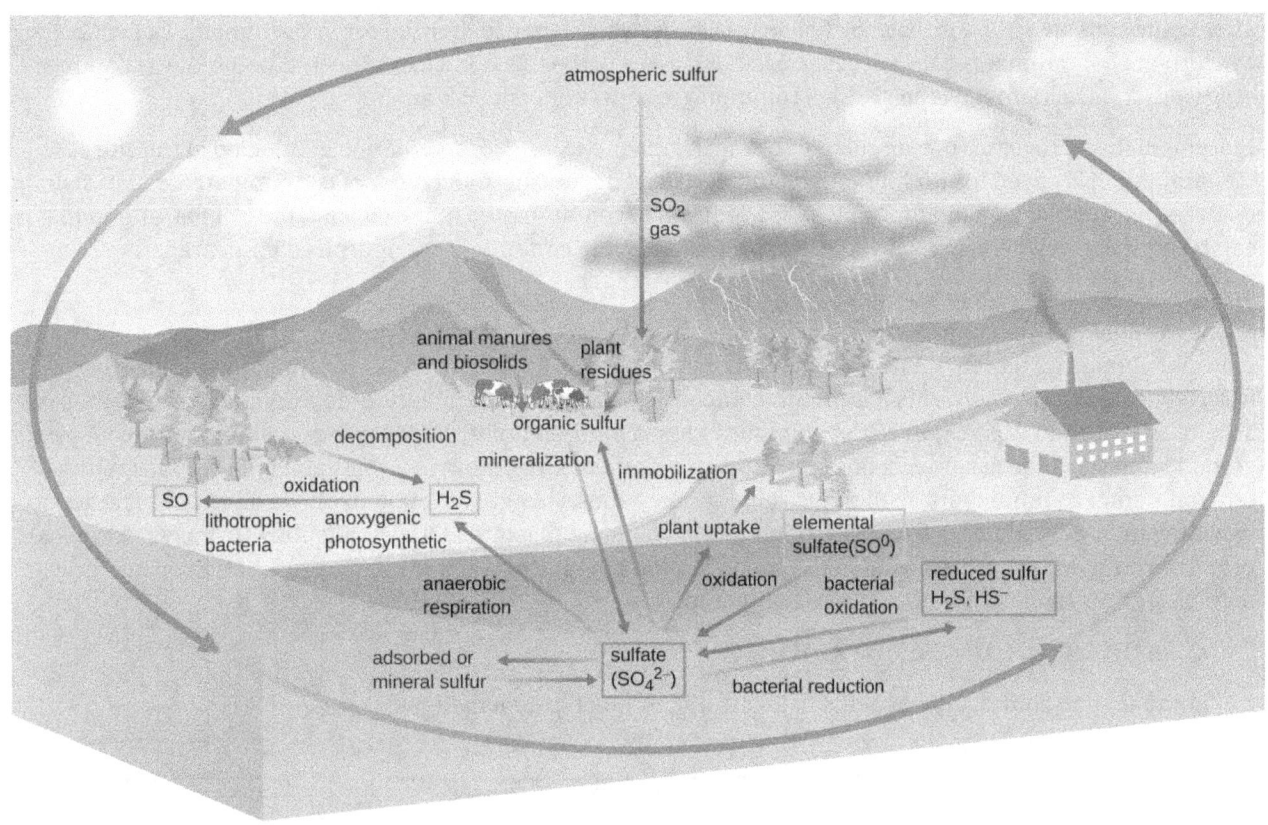

Sulfur cycle

Figure 8.26 This figure summarizes the sulfur cycle. Note that specific groups of prokaryotes each may participate in every step in the cycle. (credit: modification of work by NOAA)

✓ CHECK YOUR UNDERSTANDING

- Which groups of microbes carry out the sulfur cycle?

Other Biogeochemical Cycles

Beyond their involvement in the carbon, nitrogen, and sulfur cycles, prokaryotes are involved in other biogeochemical cycles as well. Like the carbon, nitrogen, and sulfur cycles, several of these additional biogeochemical cycles, such as the iron (Fe), manganese (Mn), and chromium (Cr) cycles, also involve redox chemistry, with prokaryotes playing roles in both oxidation and reduction. Several other elements undergo chemical cycles that do not involve redox chemistry. Examples of these are phosphorus (P), calcium (Ca), and silica (Si) cycles. The cycling of these elements is particularly important in oceans because large quantities of these elements are incorporated into the exoskeletons of marine organisms. These biogeochemical cycles do not involve redox chemistry but instead involve fluctuations in the solubility of compounds containing calcium, phosphorous, and silica. The overgrowth of naturally occurring microbial communities is typically limited by the availability of nitrogen (as previously mentioned), phosphorus, and iron. Human activities introducing excessive amounts of iron, nitrogen, or phosphorus (typically from detergents) may lead to eutrophication.

Bioremediation

Microbial **bioremediation** leverages microbial metabolism to remove **xenobiotics** or other pollutants. Xenobiotics are compounds synthesized by humans and introduced into the environment in much higher concentrations than would naturally occur. Such environmental contamination may involve adhesives, dyes, flame retardants, lubricants, oil and petroleum products, organic solvents, pesticides, and products of the combustion of gasoline and oil. Many xenobiotics resist breakdown, and some accumulate in the food chain after being consumed or absorbed by fish and wildlife, which, in turn, may be eaten by humans. Of particular concern are contaminants like polycyclic aromatic hydrocarbon (PAH), a carcinogenic xenobiotic found in crude oil, and trichloroethylene (TCE), a common groundwater contaminant.

Bioremediation processes can be categorized as in situ or ex situ. Bioremediation conducted at the site of contamination is called in situ bioremediation and does not involve movement of contaminated material. In contrast, ex situ bioremediation involves the removal of contaminated material from the original site so that it can be treated elsewhere, typically in a large, lined pit where conditions are optimized for degradation of the contaminant.

Some bioremediation processes rely on microorganisms that are indigenous to the contaminated site or material. Enhanced bioremediation techniques, which may be applied to either in situ or ex situ processing, involve the addition of nutrients and/or air to encourage the growth of pollution-degrading microbes; they may also involve the addition of non-native microbes known for their ability to degrade contaminants. For example, certain bacteria of the genera *Rhodococcus* and *Pseudomonas* are known for their ability to degrade many environmental contaminants, including aromatic compounds like those found in oil, down to CO_2. The genes encoding their degradatory enzymes are commonly found on plasmids. Others, like *Alcanivorax borkumensis*, produce surfactants that are useful in the solubilization of the hydrophobic molecules found in oil, making them more accessible to other microbes for degradation.

✓ CHECK YOUR UNDERSTANDING

- Compare and contrast the benefits of in situ and ex situ bioremediation.

Clinical Focus

Resolution

Although there is a DNA test specific for *Neisseria meningitidis*, it is not practical for use in some developing countries because it requires expensive equipment and a high level of expertise to perform. The hospital in Banjul was not equipped to perform DNA testing. Biochemical testing, however, is much less expensive and is still effective for microbial identification.

Fortunately for Hannah, her symptoms began to resolve with antibiotic therapy. Patients who survive bacterial meningitis often suffer from long-term complications such as brain damage, hearing loss, and seizures, but after several weeks of recovery, Hannah did not seem to be exhibiting any long-term effects and her behavior returned to normal. Because of her age, her parents were advised to monitor her closely for any signs of developmental issues and have her regularly evaluated by her pediatrician.

N. meningitidis is found in the normal respiratory microbiota in 10%–20% of the human population.[2] In most cases, it does not cause disease, but for reasons not fully understood, the bacterium can sometimes invade the bloodstream and cause infections in other areas of the body, including the brain. The disease is more common in infants and children, like Hannah.

The prevalence of meningitis caused by *N. meningitidis* is particularly high in the so-called meningitis belt, a region of sub-Saharan African that includes 26 countries stretching from Senegal to Ethiopia (Figure 8.27). The reasons for this high prevalence are not clear, but several factors may contribute to higher rates of transmission, such as the dry, dusty climate; overcrowding and low standards of living; and the relatively low immunocompetence and nutritional status of the population.[3] A vaccine against four bacterial strains of *N. meningitidis* is available. Vaccination is recommended for 11- and 12-year-old children, with a booster at age 16 years. Vaccination is also recommended for young people who live in close quarters with others (e.g., college dormitories, military barracks), where the disease is more easily transmitted. Travelers visiting the "meningitis belt" should also be vaccinated, especially during the dry season (December through June) when the prevalence is highest.[4,5]

---

2 Centers for Disease Control and Prevention. "Meningococcal Disease: Causes and Transmission." http://www.cdc.gov/meningococcal/about/causes-transmission.html. Accessed September 12, 2016.

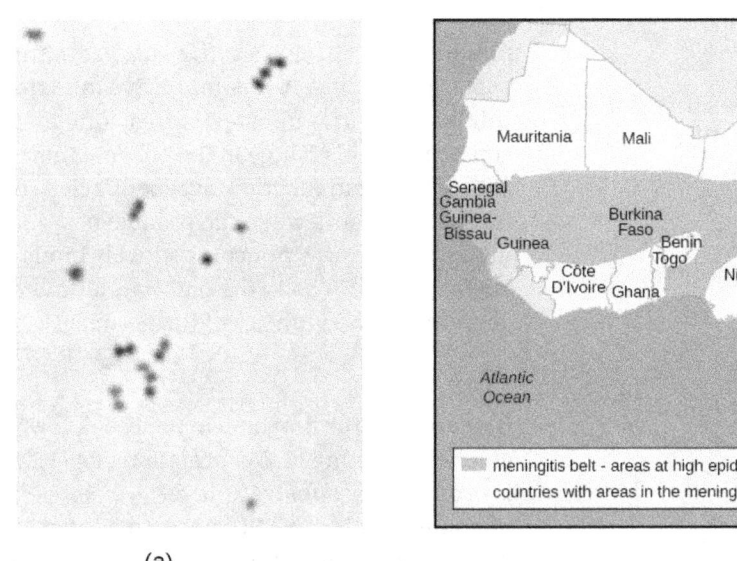

Figure 8.27 (a) *Neisseria meningitidis* is a gram-negative diplococcus, as shown in this gram-stained sample. (b) The "meningitis belt" is the area of sub-Saharan Africa with high prevalence of meningitis caused by *N. meningitidis*. (credit a, b: modification of work by Centers for Disease Control and Prevention)

Go back to the previous Clinical Focus box.

---

3 Centers for Disease Control and Prevention. "Meningococcal Disease in Other Countries." http://www.cdc.gov/meningococcal/global.html. Accessed September 12, 2016.

4 Centers for Disease Control and Prevention. "Health Information for Travelers to the Gambia: Traveler View." http://wwwnc.cdc.gov/travel/destinations/traveler/none/the-gambia. Accessed September 12, 2016.

5 Centers for Disease Control and Prevention. "Meningococcal: Who Needs to Be Vaccinated?" http://www.cdc.gov/vaccines/vpd-vac/mening/who-vaccinate.htm. Accessed September 12, 2016.

SUMMARY

8.1 Energy, Matter, and Enzymes

- **Metabolism** includes chemical reactions that break down complex molecules (**catabolism**) and those that build complex molecules (**anabolism**).
- Organisms may be classified according to their source of carbon. **Autotrophs** convert inorganic carbon dioxide into organic carbon; **heterotrophs** use fixed organic carbon compounds.
- Organisms may also be classified according to their energy source. **Phototrophs** obtain their energy from light. **Chemotrophs** get their energy from chemical compounds. **Organotrophs** use organic molecules, and **lithotrophs** use inorganic chemicals.
- Cellular **electron carriers** accept high-energy electrons from foods and later serve as electron donors in subsequent **redox reactions**. FAD/$FADH_2$, NAD^+/NADH, and $NADP^+$/NADPH are important electron carriers.
- **Adenosine triphosphate (ATP)** serves as the energy currency of the cell, safely storing chemical energy in its two **high-energy phosphate bonds** for later use to drive processes requiring energy.
- **Enzymes** are biological **catalysts** that increase the rate of chemical reactions inside cells by lowering the activation energy required for the reaction to proceed.
- In nature, **exergonic reactions** do not require energy beyond activation energy to proceed, and they release energy. They may proceed without enzymes, but at a slow rate. Conversely, **endergonic reactions** require energy beyond activation energy to occur. In cells, endergonic reactions are coupled to exergonic reactions, making the combination energetically favorable.
- **Substrates** bind to the enzyme's **active site**. This process typically alters the structures of both the active site and the substrate, favoring transition-state formation; this is known as **induced fit**.
- **Cofactors** are inorganic ions that stabilize enzyme conformation and function. **Coenzymes** are organic molecules required for proper enzyme function and are often derived from vitamins. An enzyme lacking a cofactor or coenzyme is an **apoenzyme**; an enzyme with a bound cofactor or coenzyme is a **holoenzyme**.
- **Competitive inhibitors** regulate enzymes by binding to an enzyme's active site, preventing substrate binding. **Noncompetitive (allosteric) inhibitors** bind to **allosteric sites**, inducing a conformational change in the enzyme that prevents it from functioning. **Feedback inhibition** occurs when the product of a metabolic pathway noncompetitively binds to an enzyme early on in the pathway, ultimately preventing the synthesis of the product.

8.2 Catabolism of Carbohydrates

- **Glycolysis** is the first step in the breakdown of glucose, resulting in the formation of ATP, which is produced by **substrate-level phosphorylation**; NADH; and two pyruvate molecules. Glycolysis does not use oxygen and is not oxygen dependent.
- After glycolysis, a three-carbon pyruvate is decarboxylated to form a two-carbon acetyl group, coupled with the formation of NADH. The acetyl group is attached to a large carrier compound called coenzyme A.
- After the transition step, coenzyme A transports the two-carbon acetyl to the **Krebs cycle**, where the two carbons enter the cycle. Per turn of the cycle, one acetyl group derived from glycolysis is further oxidized, producing three NADH molecules, one $FADH_2$, and one ATP by **substrate-level phosphorylation**, and releasing two CO_2 molecules.
- The Krebs cycle may be used for other purposes. Many of the intermediates are used to synthesize important cellular molecules, including amino acids, chlorophylls, fatty acids, and nucleotides.

8.3 Cellular Respiration

- Most ATP generated during the cellular respiration of glucose is made by **oxidative phosphorylation**.
- An **electron transport system (ETS)** is composed of a series of membrane-associated protein complexes and associated mobile accessory electron carriers. The ETS is embedded in the cytoplasmic membrane of prokaryotes and the inner mitochondrial membrane of eukaryotes.
- Each ETS complex has a different redox potential, and electrons move from electron carriers with more negative redox potential to those with more positive redox potential.
- To carry out **aerobic respiration**, a cell requires

oxygen as the final electron acceptor. A cell also needs a complete Krebs cycle, an appropriate cytochrome oxidase, and oxygen detoxification enzymes to prevent the harmful effects of oxygen radicals produced during aerobic respiration.
- Organisms performing **anaerobic respiration** use alternative electron transport system carriers for the ultimate transfer of electrons to the final non-oxygen electron acceptors.
- Microbes show great variation in the composition of their electron transport systems, which can be used for diagnostic purposes to help identify certain pathogens.
- As electrons are passed from NADH and $FADH_2$ through an ETS, the electron loses energy. This energy is stored through the pumping of H^+ across the membrane, generating a **proton motive force**.
- The energy of this proton motive force can be harnessed by allowing hydrogen ions to diffuse back through the membrane by **chemiosmosis** using **ATP synthase**. As hydrogen ions diffuse through down their electrochemical gradient, components of ATP synthase spin, making ATP from ADP and P_i by oxidative phosphorylation.
- Aerobic respiration forms more ATP (a maximum of 34 ATP molecules) during oxidative phosphorylation than does anaerobic respiration (between one and 32 ATP molecules).

8.4 Fermentation

- Fermentation uses an organic molecule as a final electron acceptor to regenerate NAD^+ from NADH so that glycolysis can continue.
- Fermentation does not involve an electron transport system, and no ATP is made by the fermentation process directly. Fermenters make very little ATP—only two ATP molecules per glucose molecule during glycolysis.
- Microbial fermentation processes have been used for the production of foods and pharmaceuticals, and for the identification of microbes.
- During lactic acid fermentation, pyruvate accepts electrons from NADH and is reduced to lactic acid. Microbes performing **homolactic fermentation** produce only lactic acid as the fermentation product; microbes performing **heterolactic fermentation** produce a mixture of lactic acid, ethanol and/or acetic acid, and CO_2.
- Lactic acid production by the normal microbiota prevents growth of pathogens in certain body regions and is important for the health of the gastrointestinal tract.
- During ethanol fermentation, pyruvate is first decarboxylated (releasing CO_2) to acetaldehyde, which then accepts electrons from NADH, reducing acetaldehyde to ethanol. Ethanol fermentation is used for the production of alcoholic beverages, for making bread products rise, and for biofuel production.
- Fermentation products of pathways (e.g., propionic acid fermentation) provide distinctive flavors to food products. Fermentation is used to produce chemical solvents (acetone-butanol-ethanol fermentation) and pharmaceuticals (mixed acid fermentation).
- Specific types of microbes may be distinguished by their fermentation pathways and products. Microbes may also be differentiated according to the substrates they are able to ferment.

8.5 Catabolism of Lipids and Proteins

- Collectively, microbes have the ability to degrade a wide variety of carbon sources besides carbohydrates, including lipids and proteins. The catabolic pathways for all of these molecules eventually connect into glycolysis and the Krebs cycle.
- Several types of lipids can be microbially degraded. Triglycerides are degraded by extracellular **lipases**, releasing fatty acids from the glycerol backbone. Phospholipids are degraded by **phospholipases**, releasing fatty acids and the phosphorylated head group from the glycerol backbone. Lipases and phospholipases act as virulence factors for certain pathogenic microbes.
- Fatty acids can be further degraded inside the cell through **β-oxidation**, which sequentially removes two-carbon acetyl groups from the ends of fatty acid chains.
- Protein degradation involves extracellular **proteases** that degrade large proteins into smaller peptides. Detection of the extracellular proteases gelatinase and caseinase can be used to differentiate clinically relevant bacteria.

8.6 Photosynthesis

- Heterotrophs depend on the carbohydrates produced by autotrophs, many of which are photosynthetic, converting solar energy into chemical energy.

- Different photosynthetic organisms use different mixtures of **photosynthetic pigments**, which increase the range of the wavelengths of light an organism can absorb.
- **Photosystems** (PSI and PSII) each contain a **light-harvesting complex**, composed of multiple proteins and associated pigments that absorb light energy. The **light-dependent reactions** of photosynthesis convert solar energy into chemical energy, producing ATP and NADPH or NADH to temporarily store this energy.
- In **oxygenic photosynthesis**, H_2O serves as the electron donor to replace the reaction center electron, and oxygen is formed as a byproduct. In **anoxygenic photosynthesis**, other reduced molecules like H_2S or thiosulfate may be used as the electron donor; as such, oxygen is not formed as a byproduct.
- **Noncyclic photophosphorylation** is used in oxygenic photosynthesis when there is a need for both ATP and NADPH production. If a cell's needs for ATP outweigh its needs for NADPH, then it may carry out **cyclic photophosphorylation** instead, producing only ATP.
- The **light-independent reactions** of photosynthesis use the ATP and NADPH from the light-dependent reactions to fix CO_2 into organic sugar molecules.

8.7 Biogeochemical Cycles

- The recycling of inorganic matter between living organisms and their nonliving environment is called a **biogeochemical cycle**. Microbes play significant roles in these cycles.

- In the **carbon cycle**, heterotrophs degrade reduced organic molecule to produce carbon dioxide, whereas autotrophs fix carbon dioxide to produce organics. **Methanogens** typically form methane by using CO_2 as a final electron acceptor during anaerobic respiration; methanotrophs oxidize the methane, using it as their carbon source.
- In the **nitrogen cycle**, nitrogen-fixing bacteria convert atmospheric nitrogen into ammonia (ammonification). The ammonia can then be oxidized to nitrite and nitrate (nitrification). Nitrates can then be assimilated by plants. Soil bacteria convert nitrate back to nitrogen gas (denitrification).
- In **sulfur cycling**, many anoxygenic photosynthesizers and chemoautotrophs use hydrogen sulfide as an electron donor, producing elemental sulfur and then sulfate; sulfate-reducing bacteria and archaea then use sulfate as a final electron acceptor in anaerobic respiration, converting it back to hydrogen sulfide.
- Human activities that introduce excessive amounts of naturally limited nutrients (like iron, nitrogen, or phosphorus) to aquatic systems may lead to eutrophication.
- Microbial **bioremediation** is the use of microbial metabolism to remove or degrade **xenobiotics** and other environmental contaminants and pollutants. Enhanced bioremediation techniques may involve the introduction of non-native microbes specifically chosen or engineered for their ability to degrade contaminants.

REVIEW QUESTIONS

Multiple Choice

1. Which of the following is an organism that obtains its energy from the transfer of electrons originating from chemical compounds and its carbon from an inorganic source?
   A. chemoautotroph
   B. chemoheterotroph
   C. photoheterotroph
   D. photoautotroph

2. Which of the following molecules is reduced?
   A. NAD^+
   B. FAD
   C. O_2
   D. NADPH

3. Enzymes work by which of the following?
   A. increasing the activation energy
   B. reducing the activation energy
   C. making exergonic reactions endergonic
   D. making endergonic reactions exergonic

4. To which of the following does a competitive inhibitor most structurally resemble?
   A. the active site
   B. the allosteric site
   C. the substrate
   D. a coenzyme

5. Which of the following are organic molecules that help enzymes work correctly?
   A. cofactors
   B. coenzymes
   C. holoenzymes
   D. apoenzymes

6. During which of the following is ATP not made by substrate-level phosphorylation?
   A. Embden-Meyerhof pathway
   B. Transition reaction
   C. Krebs cycle
   D. Entner-Doudoroff pathway

7. Which of the following products is made during Embden-Meyerhof glycolysis?
   A. NAD+
   B. pyruvate
   C. CO_2
   D. two-carbon acetyl

8. During the catabolism of glucose, which of the following is produced only in the Krebs cycle?
   A. ATP
   B. NADH
   C. NADPH
   D. $FADH_2$

9. Which of the following is not a name for the cycle resulting in the conversion of a two-carbon acetyl to one ATP, two CO_2, one $FADH_2$, and three NADH molecules?
   A. Krebs cycle
   B. tricarboxylic acid cycle
   C. Calvin cycle
   D. citric acid cycle

10. Which is the location of electron transports systems in prokaryotes?
    A. the outer mitochondrial membrane
    B. the cytoplasm
    C. the inner mitochondrial membrane
    D. the cytoplasmic membrane

11. Which is the source of the energy used to make ATP by oxidative phosphorylation?
    A. oxygen
    B. high-energy phosphate bonds
    C. the proton motive force
    D. P_i

12. A cell might perform anaerobic respiration for which of the following reasons?
    A. It lacks glucose for degradation.
    B. It lacks the transition reaction to convert pyruvate to acetyl-CoA.
    C. It lacks Krebs cycle enzymes for processing acetyl-CoA to CO_2.
    D. It lacks a cytochrome oxidase for passing electrons to oxygen.

13. In prokaryotes, which of the following is true?
    A. As electrons are transferred through an ETS, H^+ is pumped out of the cell.
    B. As electrons are transferred through an ETS, H^+ is pumped into the cell.
    C. As protons are transferred through an ETS, electrons are pumped out of the cell.
    D. As protons are transferred through an ETS, electrons are pumped into the cell.

14. Which of the following is not an electron carrier within an electron transport system?
    A. flavoprotein
    B. ATP synthase
    C. ubiquinone
    D. cytochrome oxidase

15. Which of the following is the purpose of fermentation?
    A. to make ATP
    B. to make carbon molecule intermediates for anabolism
    C. to make NADH
    D. to make NAD^+

16. Which molecule typically serves as the final electron acceptor during fermentation?
    A. oxygen
    B. NAD^+
    C. pyruvate
    D. CO_2

17. Which fermentation product is important for making bread rise?
    A. ethanol
    B. CO_2
    C. lactic acid
    D. hydrogen gas

18. Which of the following is not a commercially important fermentation product?
    A. ethanol
    B. pyruvate
    C. butanol
    D. penicillin

19. Which of the following molecules is not produced during the breakdown of phospholipids?
    A. glucose
    B. glycerol
    C. acetyl groups
    D. fatty acids

20. Caseinase is which type of enzyme?
    A. phospholipase
    B. lipase
    C. extracellular protease
    D. intracellular protease

21. Which of the following is the first step in triglyceride degradation?
    A. removal of fatty acids
    B. β-oxidation
    C. breakage of fused rings
    D. formation of smaller peptides

22. During the light-dependent reactions, which molecule loses an electron?
    A. a light-harvesting pigment molecule
    B. a reaction center pigment molecule
    C. NADPH
    D. 3-phosphoglycerate

23. In prokaryotes, in which direction are hydrogen ions pumped by the electron transport system of photosynthetic membranes?
    A. to the outside of the plasma membrane
    B. to the inside (cytoplasm) of the cell
    C. to the stroma
    D. to the intermembrane space of the chloroplast

24. Which of the following does not occur during cyclic photophosphorylation in cyanobacteria?
    A. electron transport through an ETS
    B. photosystem I use
    C. ATP synthesis
    D. NADPH formation

25. Which of the following are two products of the light-dependent reactions?
    A. glucose and NADPH
    B. NADPH and ATP
    C. glyceraldehyde 3-phosphate and CO_2
    D. glucose and oxygen

26. Which of the following is the group of archaea that can use CO_2 as their final electron acceptor during anaerobic respiration, producing CH_4?
    A. methylotrophs
    B. methanotrophs
    C. methanogens
    D. anoxygenic photosynthesizers

27. Which of the following processes is not involved in the conversion of organic nitrogen to nitrogen gas?
    A. nitrogen fixation
    B. ammonification
    C. nitrification
    D. denitrification

28. Which of the following processes produces hydrogen sulfide?
    A. anoxygenic photosynthesis
    B. oxygenic photosynthesis
    C. anaerobic respiration
    D. chemoautrophy

29. The biogeochemical cycle of which of the following elements is based on changes in solubility rather than redox chemistry?
    A. carbon
    B. sulfur
    C. nitrogen
    D. phosphorus

True/False

30. Competitive inhibitors bind to allosteric sites.
31. Glycolysis requires oxygen or another inorganic final electron acceptor to proceed.
32. All organisms that use aerobic cellular respiration have cytochrome oxidase.
33. Photosynthesis always results in the formation of oxygen.
34. There are many naturally occurring microbes that have the ability to degrade several of the compounds found in oil.

Matching

35. Match the fermentation pathway with the correct commercial product it is used to produce:

___ acetone-butanol-ethanol fermentation

___ alcohol fermentation

___ lactic acid fermentation

___ mixed acid fermentation

___ propionic acid fermentation

a. bread

b. pharmaceuticals

c. Swiss cheese

d. yogurt

e. industrial solvents

Fill in the Blank

36. Processes in which cellular energy is used to make complex molecules from simpler ones are described as _____.
37. The loss of an electron from a molecule is called _____.
38. The part of an enzyme to which a substrate binds is called the _____.
39. Per turn of the Krebs cycle, one acetyl is oxidized, forming ____ CO_2, ____ ATP, ____ NADH, and ____ $FADH_2$ molecules.
40. Most commonly, glycolysis occurs by the _____ pathway.
41. The final ETS complex used in aerobic respiration that transfers energy-depleted electrons to oxygen to form H_2O is called _____.
42. The passage of hydrogen ions through _____ down their electrochemical gradient harnesses the energy needed for ATP synthesis by oxidative phosphorylation.
43. The microbe responsible for ethanol fermentation for the purpose of producing alcoholic beverages is _____.
44. _____ results in the production of a mixture of fermentation products, including lactic acid, ethanol and/or acetic acid, and CO_2.
45. Fermenting organisms make ATP through the process of _____.
46. The process by which two-carbon units are sequentially removed from fatty acids, producing acetyl-CoA, $FADH_2$, and NADH is called _____.
47. The NADH and $FADH_2$ produced during β-oxidation are used to make _____.
48. _____ is a type of medium used to detect the production of an extracellular protease called caseinase.
49. The enzyme responsible for CO_2 fixation during the Calvin cycle is called _____.
50. The types of pigment molecules found in plants, algae, and cyanobacteria are _____ and _____.
51. The molecule central to the carbon cycle that is exchanged within and between ecosystems, being produced by heterotrophs and used by autotrophs, is _____.
52. The use of microbes to remove pollutants from a contaminated system is called _____.

Short Answer

53. In cells, can an oxidation reaction happen in the absence of a reduction reaction? Explain.
54. What is the function of molecules like NAD^+/NADH and FAD/$FADH_2$ in cells?

55. What is substrate-level phosphorylation? When does it occur during the breakdown of glucose to CO_2?
56. Why is the Krebs cycle important in both catabolism and anabolism?
57. What is the relationship between chemiosmosis and the proton motive force?
58. How does oxidative phosphorylation differ from substrate-level phosphorylation?
59. How does the location of ATP synthase differ between prokaryotes and eukaryotes? Where do protons accumulate as a result of the ETS in each cell type?
60. Why are some microbes, including *Streptococcus* spp., unable to perform aerobic respiration, even in the presence of oxygen?
61. How can fermentation be used to differentiate various types of microbes?
62. How are the products of lipid and protein degradation connected to glucose metabolism pathways?
63. What is the general strategy used by microbes for the degradation of macromolecules?
64. Why would an organism perform cyclic phosphorylation instead of noncyclic phosphorylation?
65. What is the function of photosynthetic pigments in the light-harvesting complex?
66. Why must autotrophic organisms also respire or ferment in addition to fixing CO_2?
67. How can human activity lead to eutrophication?

Critical Thinking

68. What would be the consequences to a cell of having a mutation that knocks out coenzyme A synthesis?
69. The bacterium *E. coli* is capable of performing aerobic respiration, anaerobic respiration, and fermentation. When would it perform each process and why? How is ATP made in each case?
70. Do you think that β-oxidation can occur in an organism incapable of cellular respiration? Why or why not?
71. Is life dependent on the carbon fixation that occurs during the light-independent reactions of photosynthesis? Explain.
72. In considering the symbiotic relationship between *Rhizobium* species and their plant hosts, what metabolic activity does each organism perform that benefits the other member of the pair?

CHAPTER 9
Microbial Growth

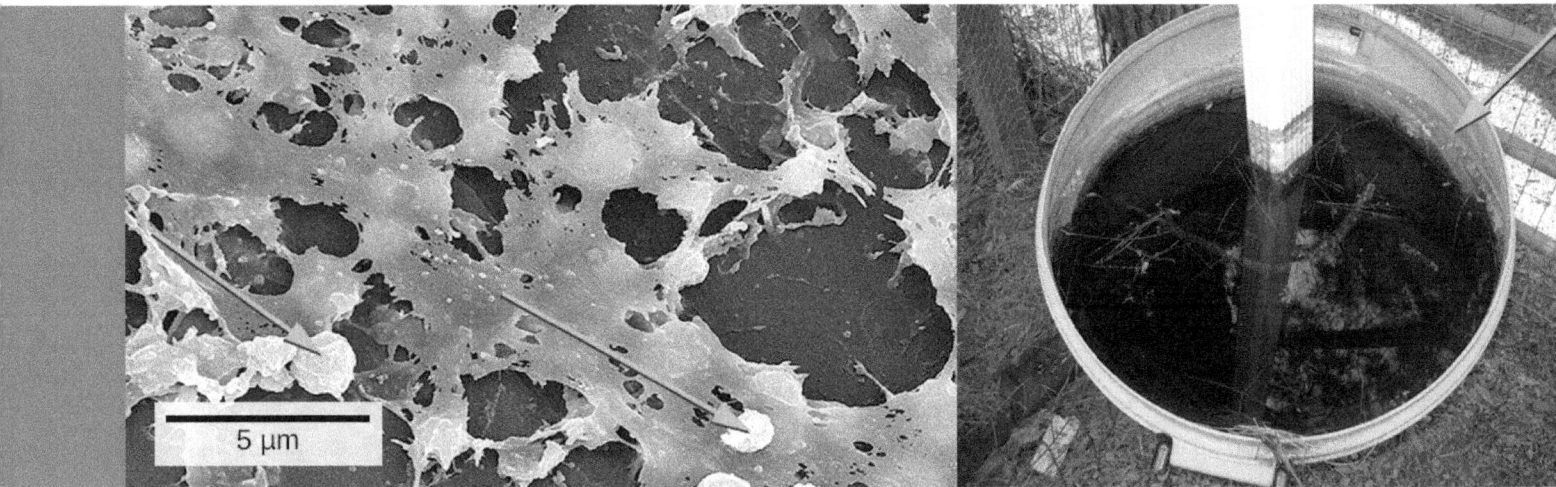

Figure 9.1 Medical devices that are inserted into a patient's body often become contaminated with a thin biofilm of microorganisms enmeshed in the sticky material they secrete. The electron micrograph (left) shows the inside walls of an in-dwelling catheter. Arrows point to the round cells of *Staphylococcus aureus* bacteria attached to the layers of extracellular substrate. The garbage can (right) served as a rain collector. The arrow points to a green biofilm on the sides of the container. (credit left: modification of work by Centers for Disease Control and Prevention; credit right: modification of work by NASA)

Chapter Outline

9.1 How Microbes Grow

9.2 Oxygen Requirements for Microbial Growth

9.3 The Effects of pH on Microbial Growth

9.4 Temperature and Microbial Growth

9.5 Other Environmental Conditions that Affect Growth

9.6 Media Used for Bacterial Growth

INTRODUCTION We are all familiar with the slimy layer on a pond surface or that makes rocks slippery. These are examples of biofilms—microorganisms embedded in thin layers of matrix material (Figure 9.1). Biofilms were long considered random assemblages of cells and had little attention from researchers. Recently, progress in visualization and biochemical methods has revealed that biofilms are an organized ecosystem within which many cells, usually of different species of bacteria, fungi, and algae, interact through cell signaling and coordinated responses. The biofilm provides a protected environment in harsh conditions and aids colonization by microorganisms. Biofilms also have clinical importance. They form on medical devices, resist routine cleaning and sterilization, and cause health-acquired infections. Within the body, biofilms form on the teeth as plaque, in the lungs of patients with cystic fibrosis, and on the cardiac tissue of patients with endocarditis. The slime layer helps protect the cells from host immune defenses and antibiotic treatments.

Studying biofilms requires new approaches. Because of the cells' adhesion properties, many of the methods for culturing and counting cells that are explored in this chapter are not easily applied to biofilms. This is the beginning of a new era of challenges and rewarding insight into the ways that microorganisms grow and thrive in nature.

9.1 How Microbes Grow

Learning Objectives

By the end of this section, you will be able to:
- Define the generation time for growth based on binary fission
- Identify and describe the activities of microorganisms undergoing typical phases of binary fission (simple cell division) in a growth curve
- Explain several laboratory methods used to determine viable and total cell counts in populations undergoing exponential growth
- Describe examples of cell division not involving binary fission, such as budding or fragmentation
- Describe the formation and characteristics of biofilms
- Identify health risks associated with biofilms and how they are addressed
- Describe quorum sensing and its role in cell-to-cell communication and coordination of cellular activities

---

Clinical Focus

Part 1

Jeni, a 24-year-old pregnant woman in her second trimester, visits a clinic with complaints of high fever, 38.9 °C (102 °F), fatigue, and muscle aches—typical flu-like signs and symptoms. Jeni exercises regularly and follows a nutritious diet with emphasis on organic foods, including raw milk that she purchases from a local farmer's market. All of her immunizations are up to date. However, the health-care provider who sees Jeni is concerned and orders a blood sample to be sent for testing by the microbiology laboratory.

- Why is the health-care provider concerned about Jeni's signs and symptoms?

Jump to the next Clinical Focus box

The bacterial cell cycle involves the formation of new cells through the replication of DNA and partitioning of cellular components into two daughter cells. In prokaryotes, reproduction is always asexual, although extensive genetic recombination in the form of horizontal gene transfer takes place, as will be explored in a different chapter. Most bacteria have a single circular chromosome; however, some exceptions exist. For example, *Borrelia burgdorferi*, the causative agent of Lyme disease, has a linear chromosome.

Binary Fission

The most common mechanism of cell replication in bacteria is a process called **binary fission**, which is depicted in Figure 9.2. Before dividing, the cell grows and increases its number of cellular components. Next, the replication of DNA starts at a location on the circular chromosome called the origin of replication, where the chromosome is attached to the inner cell membrane. Replication continues in opposite directions along the chromosome until the terminus is reached.

The center of the enlarged cell constricts until two daughter cells are formed, each offspring receiving a complete copy of the parental genome and a division of the cytoplasm (cytokinesis). This process of cytokinesis and cell division is directed by a protein called FtsZ. FtsZ assembles into a Z ring on the cytoplasmic membrane (Figure 9.3). The Z ring is anchored by FtsZ-binding proteins and defines the division plane between the two daughter cells. Additional proteins required for cell division are added to the Z ring to form a structure called the divisome. The divisome activates to produce a peptidoglycan cell wall and build a **septum** that divides the two daughter cells. The daughter cells are separated by the division septum, where all of the cells' outer layers (the cell wall and outer membranes, if present) must be remodeled to complete division. For example, we know that specific enzymes break bonds between the monomers in peptidoglycans and allow addition of new subunits along the division septum.

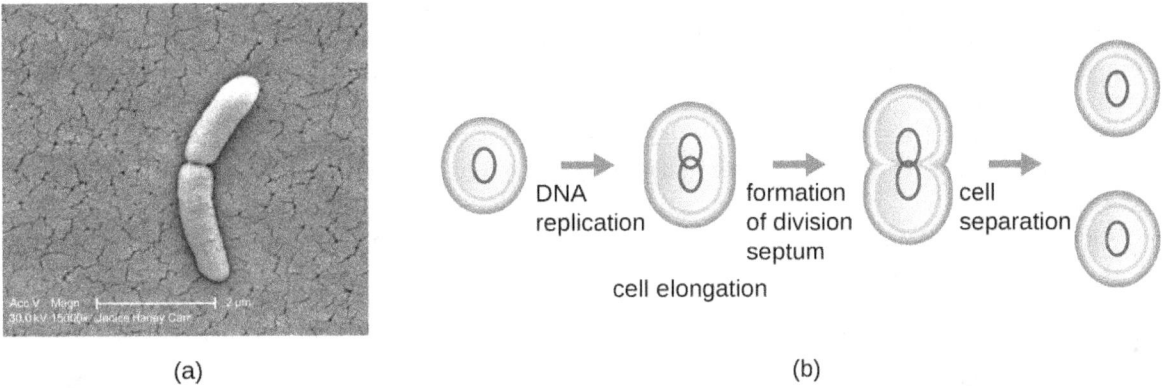

(a) (b)

Figure 9.2 (a) The electron micrograph depicts two cells of *Salmonella typhimurium* after a binary fission event. (b) Binary fission in bacteria starts with the replication of DNA as the cell elongates. A division septum forms in the center of the cell. Two daughter cells of similar size form and separate, each receiving a copy of the original chromosome. (credit a: modification of work by Centers for Disease Control and Prevention)

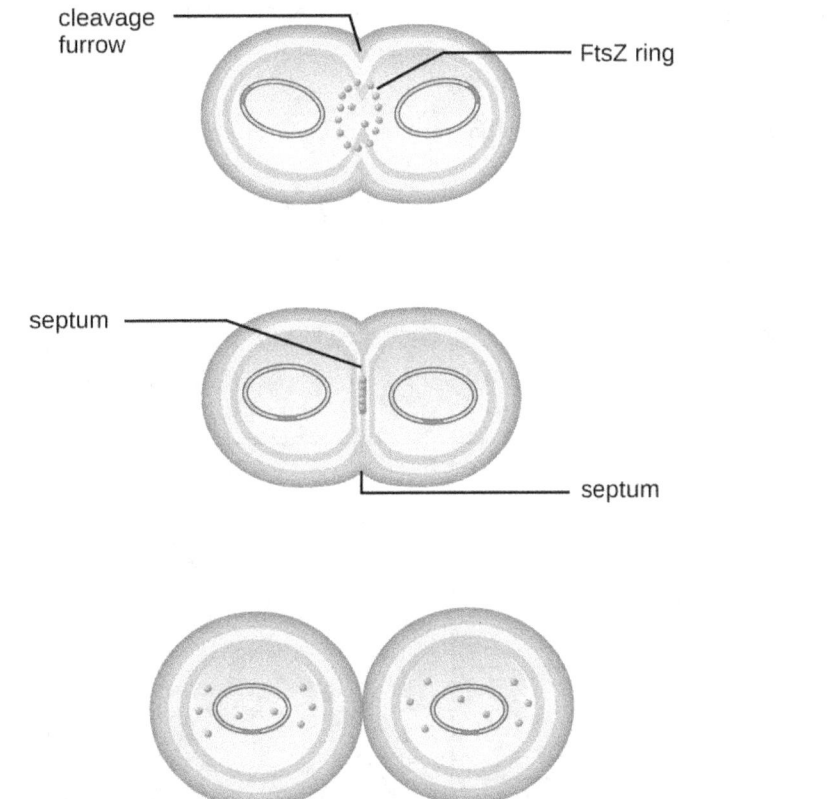

Figure 9.3 FtsZ proteins assemble to form a Z ring that is anchored to the plasma membrane. The Z ring pinches the cell envelope to separate the cytoplasm of the new cells.

CHECK YOUR UNDERSTANDING

- What is the name of the protein that assembles into a Z ring to initiate cytokinesis and cell division?

Generation Time

In eukaryotic organisms, the generation time is the time between the same points of the life cycle in two successive generations. For example, the typical generation time for the human population is 25 years. This definition is not practical for bacteria, which may reproduce rapidly or remain dormant for thousands of years.

In prokaryotes (Bacteria and Archaea), the **generation time** is also called the **doubling time** and is defined as the time it takes for the population to double through one round of binary fission. Bacterial doubling times vary enormously. Whereas *Escherichia coli* can double in as little as 20 minutes under optimal growth conditions in the laboratory, bacteria of the same species may need several days to double in especially harsh environments. Most pathogens grow rapidly, like *E. coli*, but there are exceptions. For example, *Mycobacterium tuberculosis*, the causative agent of tuberculosis, has a generation time of between 15 and 20 hours. On the other hand, *M. leprae*, which causes Hansen's disease (leprosy), grows much more slowly, with a doubling time of 14 days.

 MICRO CONNECTIONS

Calculating Number of Cells

It is possible to predict the number of cells in a population when they divide by binary fission at a constant rate. As an example, consider what happens if a single cell divides every 30 minutes for 24 hours. The diagram in Figure 9.4 shows the increase in cell numbers for the first three generations.

The number of cells increases exponentially and can be expressed as 2^n, where n is the number of generations. If cells divide every 30 minutes, after 24 hours, 48 divisions would have taken place. If we apply the formula 2^n, where n is equal to 48, the single cell would give rise to 2^{48} or 281,474,976,710,656 cells at 48 generations (24 hours). When dealing with such huge numbers, it is more practical to use scientific notation. Therefore, we express the number of cells as 2.8×10^{14} cells.

In our example, we used one cell as the initial number of cells. For any number of starting cells, the formula is adapted as follows:

$$N_n = N_0 2^n$$

N_n is the number of cells at any generation n, N_0 is the initial number of cells, and n is the number of generations.

Number of generations (n)	Number of cells	Each division adds two new cells
0	1	
1	2	
2	4	
3	8	

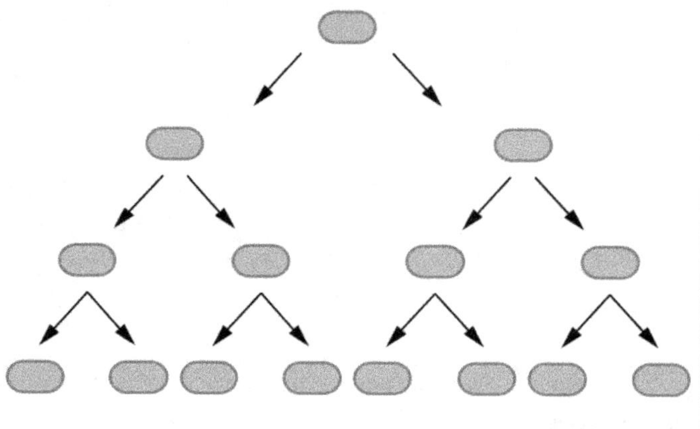

Figure 9.4 The parental cell divides and gives rise to two daughter cells. Each of the daughter cells, in turn, divides, giving a total of four cells in the second generation and eight cells in the third generation. Each division doubles the number of cells.

CHECK YOUR UNDERSTANDING

- With a doubling time of 30 minutes and a starting population size of 1×10^5 cells, how many cells will be present after 2 hours, assuming no cell death?

The Growth Curve

Microorganisms grown in closed culture (also known as a batch culture), in which no nutrients are added and most waste is not removed, follow a reproducible growth pattern referred to as the **growth curve**. An example of a batch culture in nature is a pond in which a small number of cells grow in a closed environment. The **culture density** is defined as the number of cells per unit volume. In a closed environment, the culture density is also a measure of the number of cells in the population. Infections of the body do not always follow the growth curve, but correlations can exist depending upon the site and type of infection. When the number of live cells is plotted against time, distinct phases can be observed in the curve (Figure 9.5).

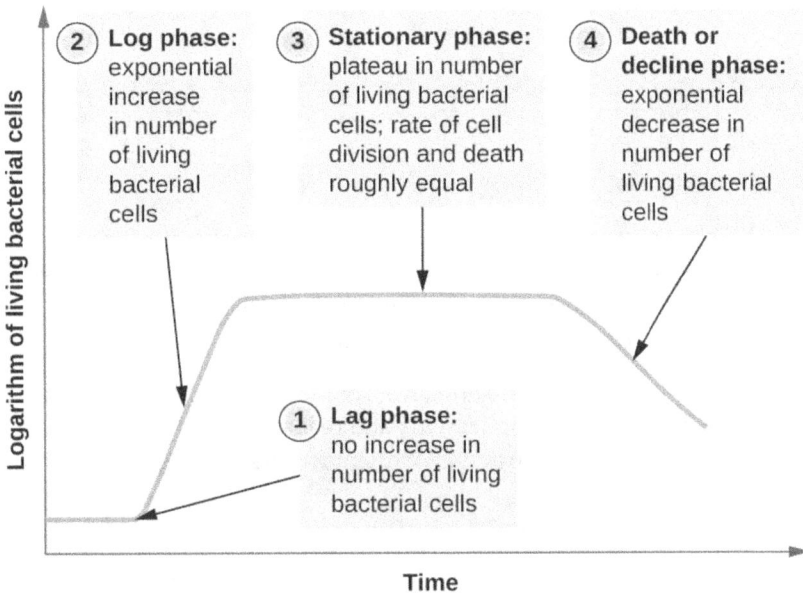

Figure 9.5 The growth curve of a bacterial culture is represented by the logarithm of the number of live cells plotted as a function of time. The graph can be divided into four phases according to the slope, each of which matches events in the cell. The four phases are lag, log, stationary, and death.

The Lag Phase

The beginning of the growth curve represents a small number of cells, referred to as an **inoculum**, that are added to a fresh **culture medium**, a nutritional broth that supports growth. The initial phase of the growth curve is called the **lag phase**, during which cells are gearing up for the next phase of growth. The number of cells does not change during the lag phase; however, cells grow larger and are metabolically active, synthesizing proteins needed to grow within the medium. If any cells were damaged or shocked during the transfer to the new medium, repair takes place during the lag phase. The duration of the lag phase is determined by many factors, including the species and genetic make-up of the cells, the composition of the medium, and the size of the original inoculum.

The Log Phase

In the **logarithmic (log) growth phase**, sometimes called exponential growth phase, the cells are actively dividing by binary fission and their number increases exponentially. For any given bacterial species, the generation time under specific growth conditions (nutrients, temperature, pH, and so forth) is genetically determined, and this generation time is called the **intrinsic growth rate**. During the log phase, the relationship between time and number of cells is not linear but exponential; however, the growth curve is often plotted on a semilogarithmic graph, as shown in Figure 9.6, which gives the appearance of a linear relationship.

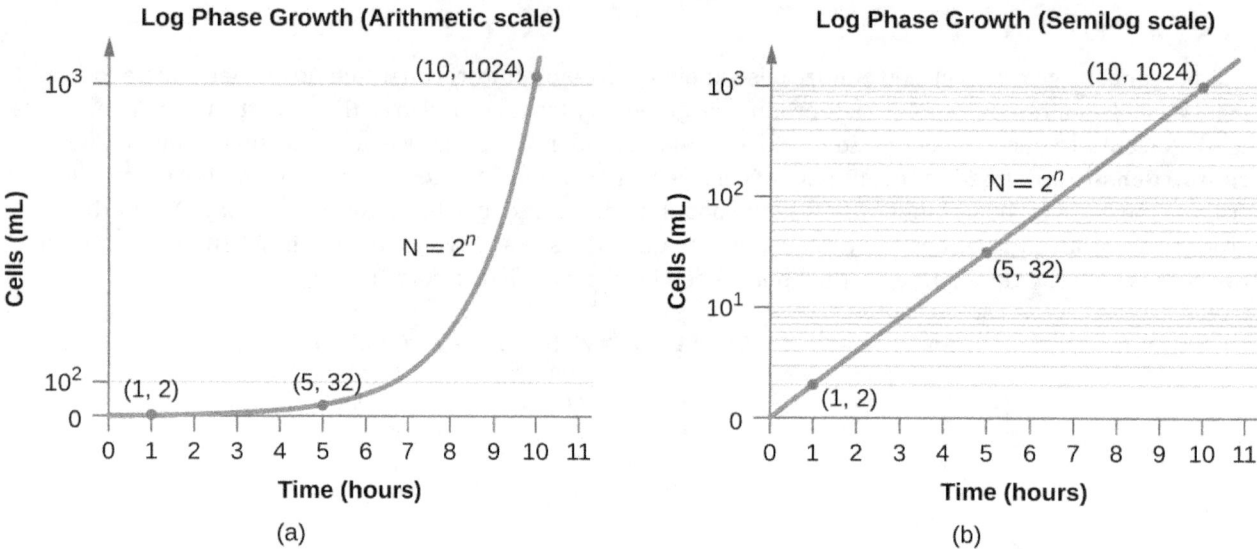

Figure 9.6 Both graphs illustrate population growth during the log phase for a bacterial sample with an initial population of one cell and a doubling time of 1 hour. (a) When plotted on an arithmetic scale, the growth rate resembles a curve. (b) When plotted on a semilogarithmic scale (meaning the values on the y-axis are logarithmic), the growth rate appears linear.

Cells in the log phase show constant growth rate and uniform metabolic activity. For this reason, cells in the log phase are preferentially used for industrial applications and research work. The log phase is also the stage where bacteria are the most susceptible to the action of disinfectants and common antibiotics that affect protein, DNA, and cell-wall synthesis.

Stationary Phase

As the number of cells increases through the log phase, several factors contribute to a slowing of the growth rate. Waste products accumulate and nutrients are gradually used up. In addition, gradual depletion of oxygen begins to limit aerobic cell growth. This combination of unfavorable conditions slows and finally stalls population growth. The total number of live cells reaches a plateau referred to as the **stationary phase** (Figure 9.5). In this phase, the number of new cells created by cell division is now equivalent to the number of cells dying; thus, the total population of living cells is relatively stagnant. The culture density in a stationary culture is constant. The culture's carrying capacity, or maximum culture density, depends on the types of microorganisms in the culture and the specific conditions of the culture; however, carrying capacity is constant for a given organism grown under the same conditions.

During the stationary phase, cells switch to a survival mode of metabolism. As growth slows, so too does the synthesis of peptidoglycans, proteins, and nucleic-acids; thus, stationary cultures are less susceptible to antibiotics that disrupt these processes. In bacteria capable of producing endospores, many cells undergo sporulation during the stationary phase. Secondary metabolites, including antibiotics, are synthesized in the stationary phase. In certain pathogenic bacteria, the stationary phase is also associated with the expression of virulence factors, products that contribute to a microbe's ability to survive, reproduce, and cause disease in a host organism. For example, quorum sensing in *Staphylococcus aureus* initiates the production of enzymes that can break down human tissue and cellular debris, clearing the way for bacteria to spread to new tissue where nutrients are more plentiful.

The Death Phase

As a culture medium accumulates toxic waste and nutrients are exhausted, cells die in greater and greater numbers. Soon, the number of dying cells exceeds the number of dividing cells, leading to an exponential decrease in the number of cells (Figure 9.5). This is the aptly named **death phase**, sometimes called the decline phase. Many cells lyse and release nutrients into the medium, allowing surviving cells to maintain viability and form endospores. A few cells, the so-called **persisters**, are characterized by a slow metabolic rate. Persister cells are medically important because they are associated with certain chronic infections, such as tuberculosis, that do not respond to antibiotic treatment.

Sustaining Microbial Growth

The growth pattern shown in Figure 9.5 takes place in a closed environment; nutrients are not added and waste and dead cells are not removed. In many cases, though, it is advantageous to maintain cells in the logarithmic phase of growth. One example is in industries that harvest microbial products. A chemostat (Figure 9.7) is used to maintain a continuous culture in which nutrients are supplied at a steady rate. A controlled amount of air is mixed in for aerobic processes. Bacterial suspension is removed at the same rate as nutrients flow in to maintain an optimal growth environment.

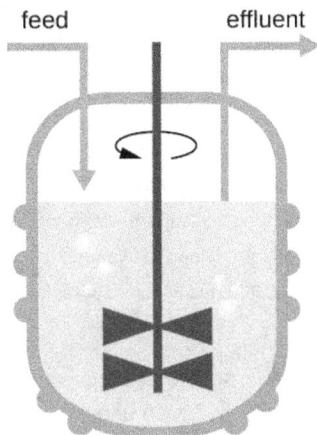

Figure 9.7 A chemostat is a culture vessel fitted with an opening to add nutrients (feed) and an outlet to remove contents (effluent), effectively diluting toxic wastes and dead cells. The addition and removal of fluids is adjusted to maintain the culture in the logarithmic phase of growth. If aerobic bacteria are grown, suitable oxygen levels are maintained.

✓ CHECK YOUR UNDERSTANDING

- During which phase does growth occur at the fastest rate?
- Name two factors that limit microbial growth.

Measurement of Bacterial Growth

Estimating the number of bacterial cells in a sample, known as a bacterial count, is a common task performed by microbiologists. The number of bacteria in a clinical sample serves as an indication of the extent of an infection. Quality control of drinking water, food, medication, and even cosmetics relies on estimates of bacterial counts to detect contamination and prevent the spread of disease. Two major approaches are used to measure cell number. The direct methods involve counting cells, whereas the indirect methods depend on the measurement of cell presence or activity without actually counting individual cells. Both direct and indirect methods have advantages and disadvantages for specific applications.

Direct Cell Count

Direct cell count refers to counting the cells in a liquid culture or colonies on a plate. It is a direct way of estimating how many organisms are present in a sample. Let's look first at a simple and fast method that requires only a specialized slide and a compound microscope.

The simplest way to count bacteria is called the **direct microscopic cell count**, which involves transferring a known volume of a culture to a calibrated slide and counting the cells under a light microscope. The calibrated slide is called a **Petroff-Hausser chamber** (Figure 9.8) and is similar to a hemocytometer used to count red blood cells. The central area of the counting chamber is etched into squares of various sizes. A sample of the culture suspension is added to the chamber under a coverslip that is placed at a specific height from the surface of the grid. It is possible to estimate the concentration of cells in the original sample by counting individual cells in a number of squares and determining the volume of the sample observed. The area of the squares and the height at which the coverslip is positioned are specified for the chamber. The concentration must be corrected for dilution if the sample was diluted before enumeration.

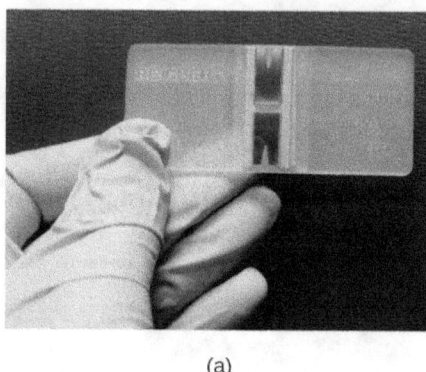 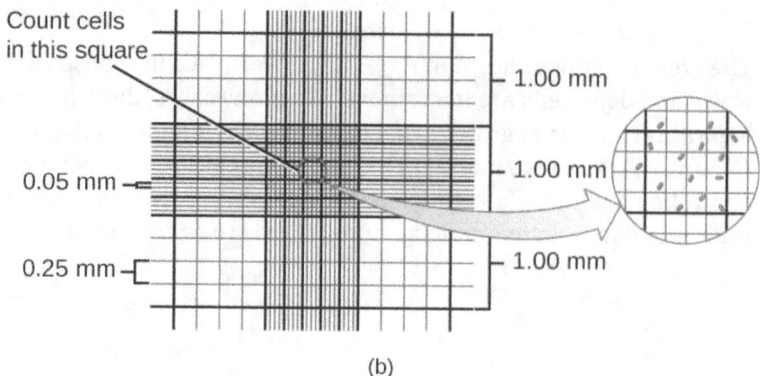

Figure 9.8 (a) A Petroff-Hausser chamber is a special slide designed for counting the bacterial cells in a measured volume of a sample. A grid is etched on the slide to facilitate precision in counting. (b) This diagram illustrates the grid of a Petroff-Hausser chamber, which is made up of squares of known areas. The enlarged view shows the square within which bacteria (red cells) are counted. If the coverslip is 0.2 mm above the grid and the square has an area of 0.04 mm², then the volume is 0.008 mm³, or 0.000008 mL. Since there are 10 cells inside the square, the density of bacteria is 10 cells/0.000008 mL, which equates to 1,250,000 cells/mL. (credit a: modification of work by Jeffrey M. Vinocur)

Cells in several small squares must be counted and the average taken to obtain a reliable measurement. The advantages of the chamber are that the method is easy to use, relatively fast, and inexpensive. On the downside, the counting chamber does not work well with dilute cultures because there may not be enough cells to count.

Using a counting chamber does not necessarily yield an accurate count of the number of live cells because it is not always possible to distinguish between live cells, dead cells, and debris of the same size under the microscope. However, newly developed fluorescence staining techniques make it possible to distinguish viable and dead bacteria. These viability stains (or live stains) bind to nucleic acids, but the primary and secondary stains differ in their ability to cross the cytoplasmic membrane. The primary stain, which fluoresces green, can penetrate intact cytoplasmic membranes, staining both live and dead cells. The secondary stain, which fluoresces red, can stain a cell only if the cytoplasmic membrane is considerably damaged. Thus, live cells fluoresce green because they only absorb the green stain, whereas dead cells appear red because the red stain displaces the green stain on their nucleic acids (Figure 9.9).

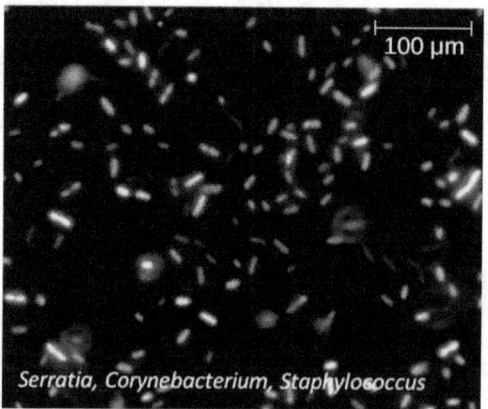

Figure 9.9 Fluorescence staining can be used to differentiate between viable and dead bacterial cells in a sample for purposes of counting. Viable cells are stained green, whereas dead cells are stained red. (credit: modification of work by Emerson J, Adams R, Bentancourt Román C, Brooks B, Coil D, Dahlhousen K, Ganz H, et al.)

Another technique uses an electronic cell counting device (Coulter counter) to detect and count the changes in electrical resistance in a saline solution. A glass tube with a small opening is immersed in an electrolyte solution. A first electrode is suspended in the glass tube. A second electrode is located outside of the tube. As cells are drawn through the small aperture in the glass tube, they briefly change the resistance measured between the two electrodes and the change is recorded by an electronic sensor (Figure 9.10); each resistance change represents a cell. The method is rapid and accurate within a range of concentrations; however, if the culture is too concentrated, more than one cell may pass through the aperture at any given time and skew the results. This method also does not differentiate between live and dead cells.

Direct counts provide an estimate of the total number of cells in a sample. However, in many situations, it is important to know the number of live, or **viable**, cells. Counts of live cells are needed when assessing the extent of an infection, the effectiveness of antimicrobial compounds and medication, or contamination of food and water.

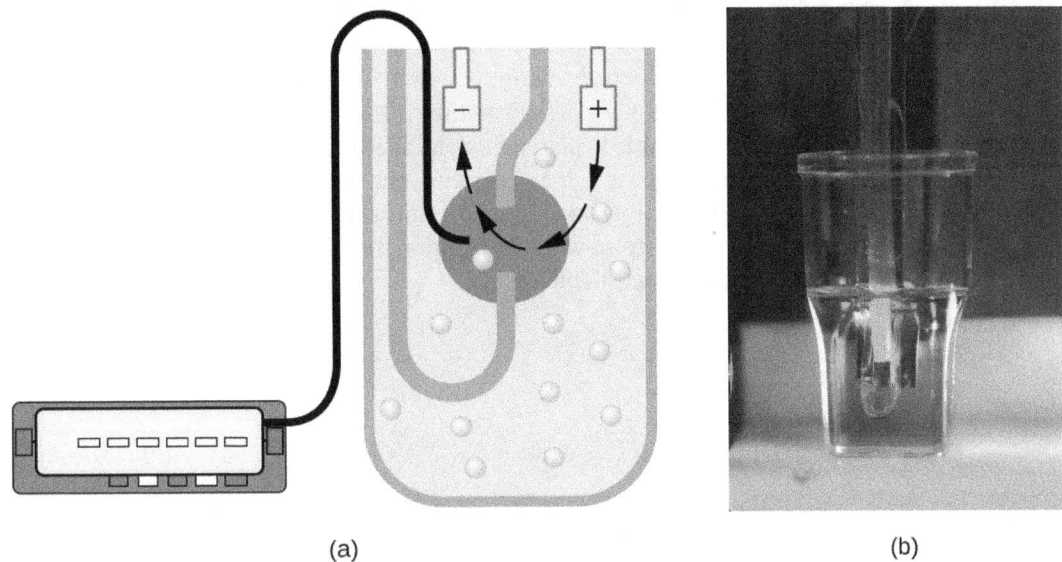

Figure 9.10 A Coulter counter is an electronic device that counts cells. It measures the change in resistance in an electrolyte solution that takes place when a cell passes through a small opening in the inside container wall. A detector automatically counts the number of cells passing through the opening. (credit b: modification of work by National Institutes of Health)

✓ CHECK YOUR UNDERSTANDING

- Why would you count the number of cells in more than one square in the Petroff-Hausser chamber to estimate cell numbers?
- In the viability staining method, why do dead cells appear red?

Plate Count

The **viable plate count**, or simply plate count, is a count of viable or live cells. It is based on the principle that viable cells replicate and give rise to visible colonies when incubated under suitable conditions for the specimen. The results are usually expressed as **colony-forming unit**s per milliliter (CFU/mL) rather than cells per milliliter because more than one cell may have landed on the same spot to give rise to a single colony. Furthermore, samples of bacteria that grow in clusters or chains are difficult to disperse and a single colony may represent several cells. Some cells are described as viable but nonculturable and will not form colonies on solid media. For all these reasons, the viable plate count is considered a low estimate of the actual number of live cells. These limitations do not detract from the usefulness of the method, which provides estimates of live bacterial numbers.

Microbiologists typically count plates with 30–300 colonies. Samples with too few colonies (<30) do not give statistically reliable numbers, and overcrowded plates (>300 colonies) make it difficult to accurately count individual colonies. Also, counts in this range minimize occurrences of more than one bacterial cell forming a single colony. Thus, the calculated CFU is closer to the true number of live bacteria in the population.

There are two common approaches to inoculating plates for viable counts: the pour plate and the spread plate methods. Although the final inoculation procedure differs between these two methods, they both start with a serial dilution of the culture.

Serial Dilution

The **serial dilution** of a culture is an important first step before proceeding to either the pour plate or spread plate method. The goal of the serial dilution process is to obtain plates with CFUs in the range of 30–300, and the process usually involves several dilutions in multiples of 10 to simplify calculation. The number of serial dilutions is chosen according to a preliminary estimate of the culture density. Figure 9.11 illustrates the serial dilution method.

A fixed volume of the original culture, 1.0 mL, is added to and thoroughly mixed with the first dilution tube solution, which contains 9.0 mL of sterile broth. This step represents a dilution factor of 10, or 1:10, compared with the original culture. From this first dilution, the same volume, 1.0 mL, is withdrawn and mixed with a fresh tube of 9.0 mL of dilution solution. The dilution factor is now 1:100 compared with the original culture. This process continues until a series of dilutions is produced that will bracket the desired cell concentration for accurate counting. From each tube, a sample is plated on solid medium using either the **pour plate method** (Figure 9.12) or the **spread plate method** (Figure 9.13). The plates are incubated until colonies appear. Two to three plates are usually prepared from each dilution and the numbers of colonies counted on each plate are averaged. In all cases, thorough mixing of samples with the dilution medium (to ensure the cell distribution in the tube is random) is paramount to obtaining reliable results.

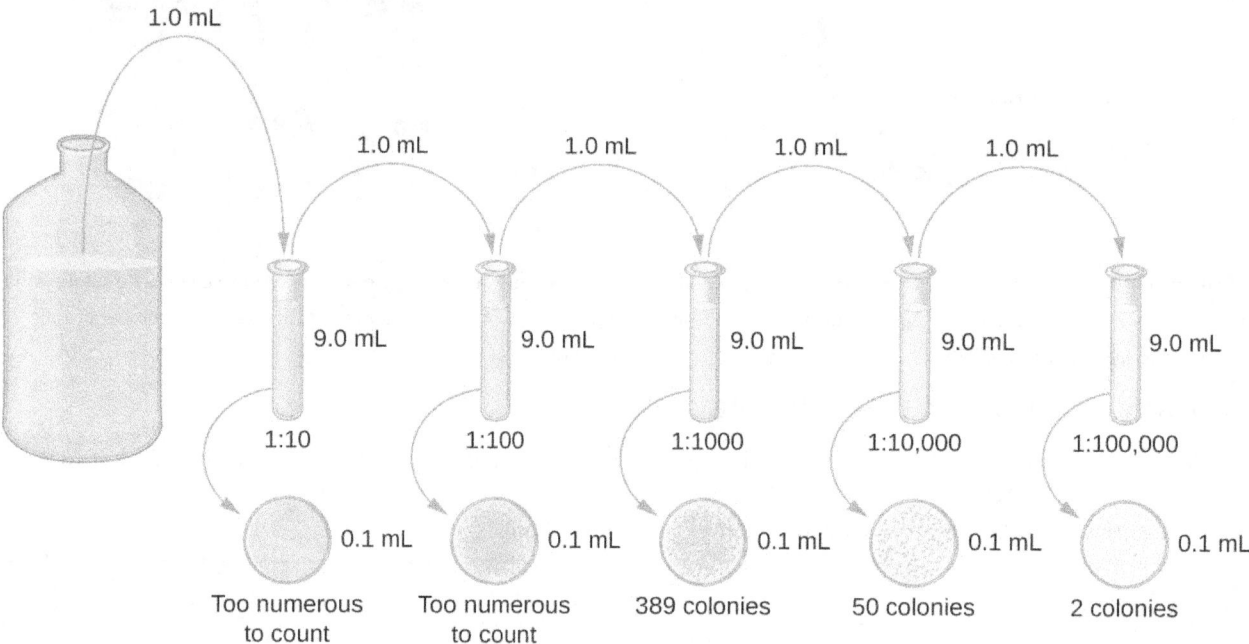

Figure 9.11 Serial dilution involves diluting a fixed volume of cells mixed with dilution solution using the previous dilution as an inoculum. The result is dilution of the original culture by an exponentially growing factor. (credit: modification of work by "Leberechtc"/Wikimedia Commons)

The dilution factor is used to calculate the number of cells in the original cell culture. In our example, an average of 50 colonies was counted on the plates obtained from the 1:10,000 dilution. Because only 0.1 mL of suspension was pipetted on the plate, the multiplier required to reconstitute the original concentration is 10 × 10,000. The number of CFU per mL is equal to 50 × 10 × 10,000 = 5,000,000. The number of bacteria in the culture is estimated as 5 million cells/mL. The colony count obtained from the 1:1000 dilution was 389, well below the expected 500 for a 10-fold difference in dilutions. This highlights the issue of inaccuracy when colony counts are greater than 300 and more than one bacterial cell grows into a single colony.

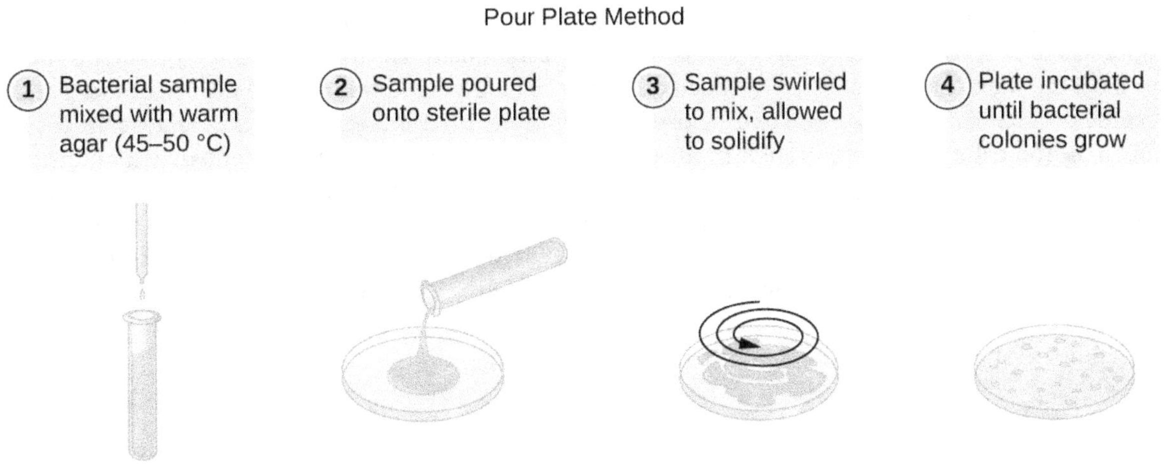

Pour Plate Method

1. Bacterial sample mixed with warm agar (45–50 °C)
2. Sample poured onto sterile plate
3. Sample swirled to mix, allowed to solidify
4. Plate incubated until bacterial colonies grow

Figure 9.12   In the pour plate method of cell counting, the sample is mixed in liquid warm agar (45–50 °C) poured into a sterile Petri dish and further mixed by swirling. This process is repeated for each serial dilution prepared. The resulting colonies are counted and provide an estimate of the number of cells in the original volume sampled.

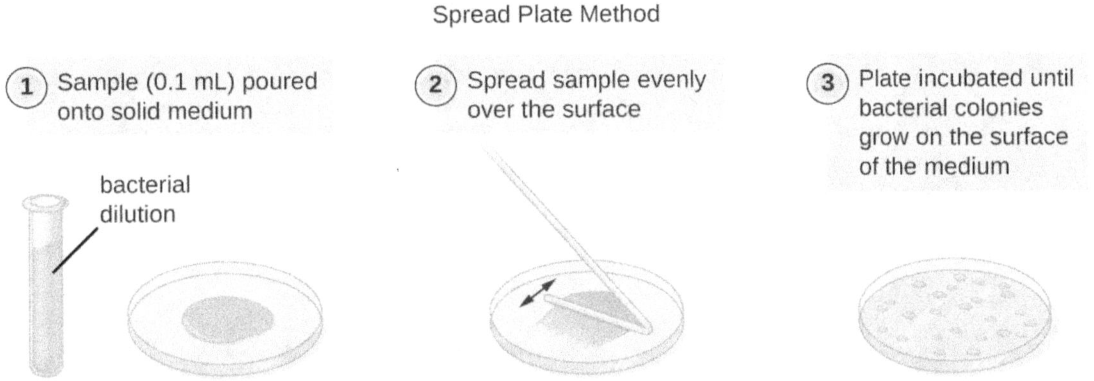

Spread Plate Method

1. Sample (0.1 mL) poured onto solid medium
2. Spread sample evenly over the surface
3. Plate incubated until bacterial colonies grow on the surface of the medium

bacterial dilution

Figure 9.13   In the spread plate method of cell counting, the sample is poured onto solid agar and then spread using a sterile spreader. This process is repeated for each serial dilution prepared. The resulting colonies are counted and provide an estimate of the number of cells in the original volume samples.

A very dilute sample—drinking water, for example—may not contain enough organisms to use either of the plate count methods described. In such cases, the original sample must be concentrated rather than diluted before plating. This can be accomplished using a modification of the plate count technique called the **membrane filtration technique**. Known volumes are vacuum-filtered aseptically through a membrane with a pore size small enough to trap microorganisms. The membrane is transferred to a Petri plate containing an appropriate growth medium. Colonies are counted after incubation. Calculation of the cell density is made by dividing the cell count by the volume of filtered liquid.

LINK TO LEARNING

Watch this video (https://openstax.org/l/22serdilpltcvid) for demonstrations of serial dilutions and spread plate techniques.

The Most Probable Number

The number of microorganisms in dilute samples is usually too low to be detected by the plate count methods described thus far. For these specimens, microbiologists routinely use the **most probable number (MPN) method**, a statistical procedure for estimating of the number of viable microorganisms in a sample. Often used for water and food samples, the MPN method evaluates detectable growth by observing changes in turbidity or color due to metabolic activity.

A typical application of MPN method is the estimation of the number of coliforms in a sample of pond water.

Coliforms are gram-negative rod bacteria that ferment lactose. The presence of coliforms in water is considered a sign of contamination by fecal matter. For the method illustrated in Figure 9.14, a series of three dilutions of the water sample is tested by inoculating five lactose broth tubes with 10 mL of sample, five lactose broth tubes with 1 mL of sample, and five lactose broth tubes with 0.1 mL of sample. The lactose broth tubes contain a pH indicator that changes color from red to yellow when the lactose is fermented. After inoculation and incubation, the tubes are examined for an indication of coliform growth by a color change in media from red to yellow. The first set of tubes (10-mL sample) showed growth in all the tubes; the second set of tubes (1 mL) showed growth in two tubes out of five; in the third set of tubes, no growth is observed in any of the tubes (0.1-mL dilution). The numbers 5, 2, and 0 are compared with Figure B1 in Appendix B, which has been constructed using a probability model of the sampling procedure. From our reading of the table, we conclude that 49 is the most probable number of bacteria per 100 mL of pond water.no lo

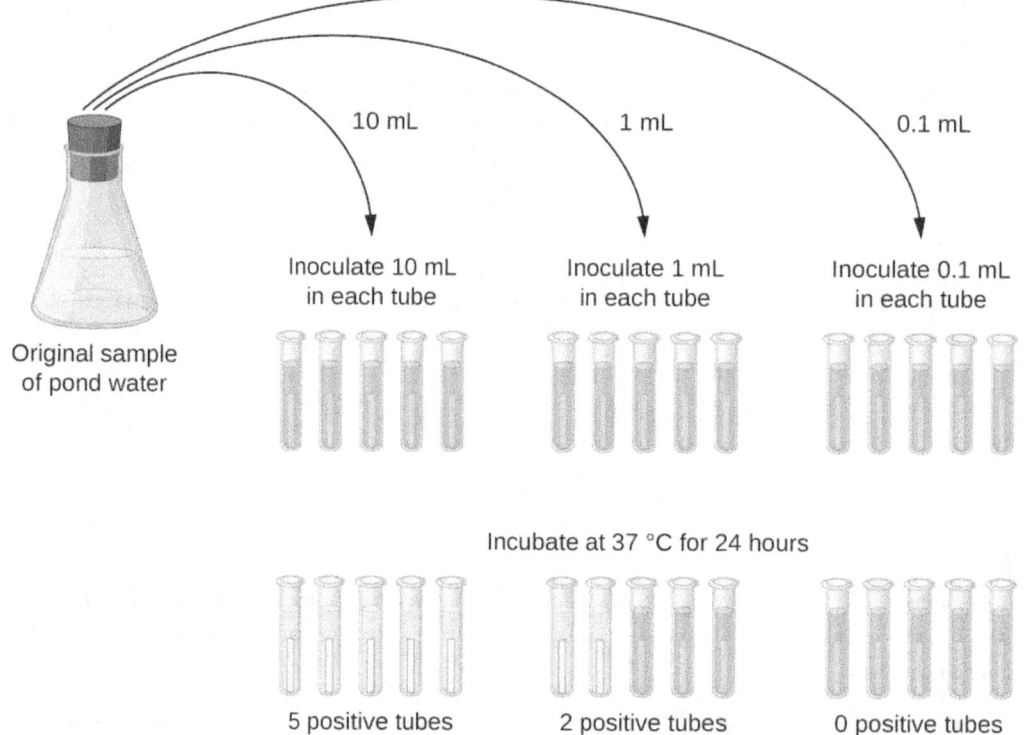

Figure 9.14 In the most probable number method, sets of five lactose broth tubes are inoculated with three different volumes of pond water: 10 mL, 1 mL, and 0.1mL. Bacterial growth is assessed through a change in the color of the broth from red to yellow as lactose is fermented.

✓ CHECK YOUR UNDERSTANDING

- What is a colony-forming unit?
- What two methods are frequently used to estimate bacterial numbers in water samples?

Indirect Cell Counts

Besides direct methods of counting cells, other methods, based on an indirect detection of cell density, are commonly used to estimate and compare cell densities in a culture. The foremost approach is to measure the **turbidity** (cloudiness) of a sample of bacteria in a liquid suspension. The laboratory instrument used to measure turbidity is called a spectrophotometer (Figure 9.15). In a spectrophotometer, a light beam is transmitted through a bacterial suspension, the light passing through the suspension is measured by a detector, and the amount of light passing through the sample and reaching the detector is converted to either percent transmission or a logarithmic value called absorbance (optical density). As the numbers of bacteria in a suspension increase, the turbidity also increases and causes less light to reach the detector. The decrease in light passing through the sample and reaching the detector is associated with a decrease in percent transmission and increase in absorbance measured by the spectrophotometer.

Measuring turbidity is a fast method to estimate cell density as long as there are enough cells in a sample to produce turbidity. It is possible to correlate turbidity readings to the actual number of cells by performing a viable plate count of samples taken from cultures having a range of absorbance values. Using these values, a calibration curve is generated by plotting turbidity as a function of cell density. Once the calibration curve has been produced, it can be used to estimate cell counts for all samples obtained or cultured under similar conditions and with densities within the range of values used to construct the curve.

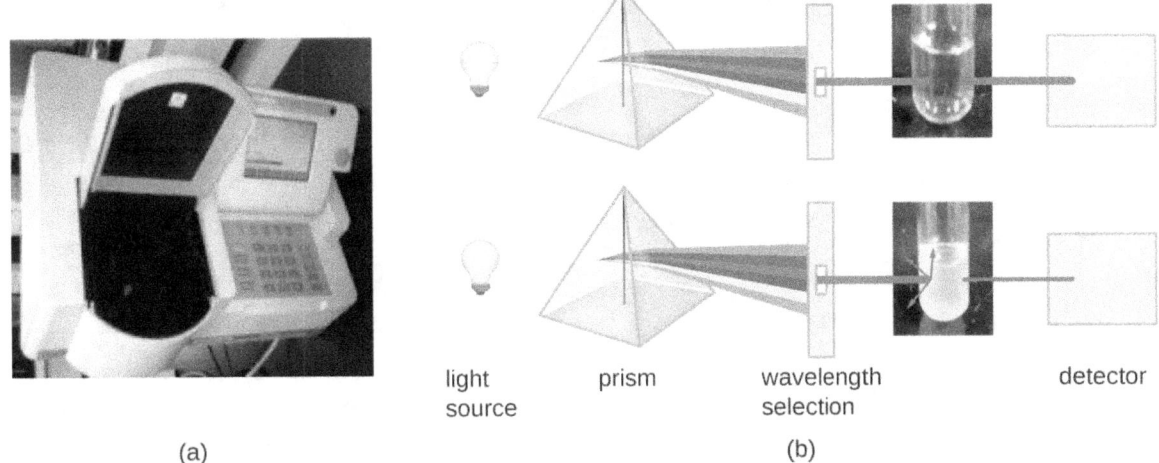

Figure 9.15 (a) A spectrophotometer is commonly used to measure the turbidity of a bacterial cell suspension as an indirect measure of cell density. (b) A spectrophotometer works by splitting white light from a source into a spectrum. The spectrophotometer allows choice of the wavelength of light to use for the measurement. The optical density (turbidity) of the sample will depend on the wavelength, so once one wavelength is chosen, it must be used consistently. The filtered light passes through the sample (or a control with only medium) and the light intensity is measured by a detector. The light passing into a suspension of bacteria is scattered by the cells in such a way that some fraction of it never reaches the detector. This scattering happens to a far lesser degree in the control tube with only the medium. (credit a: modification of work by Hwang HS, Kim MS; credit b "test tube photos": modification of work by Suzanne Wakim)

Measuring dry weight of a culture sample is another indirect method of evaluating culture density without directly measuring cell counts. The cell suspension used for weighing must be concentrated by filtration or centrifugation, washed, and then dried before the measurements are taken. The degree of drying must be standardized to account for residual water content. This method is especially useful for filamentous microorganisms, which are difficult to enumerate by direct or viable plate count.

As we have seen, methods to estimate viable cell numbers can be labor intensive and take time because cells must be grown. Recently, indirect ways of measuring live cells have been developed that are both fast and easy to implement. These methods measure cell activity by following the production of metabolic products or disappearance of reactants. Adenosine triphosphate (ATP) formation, biosynthesis of proteins and nucleic acids, and consumption of oxygen can all be monitored to estimate the number of cells.

CHECK YOUR UNDERSTANDING

- What is the purpose of a calibration curve when estimating cell count from turbidity measurements?
- What are the newer indirect methods of counting live cells?

Alternative Patterns of Cell Division

Binary fission is the most common pattern of cell division in prokaryotes, but it is not the only one. Other mechanisms usually involve asymmetrical division (as in budding) or production of spores in aerial filaments.

In some cyanobacteria, many nucleoids may accumulate in an enlarged round cell or along a filament, leading to the generation of many new cells at once. The new cells often split from the parent filament and float away in a process called **fragmentation** (Figure 9.16). Fragmentation is commonly observed in the Actinomycetes, a group of gram-positive, anaerobic bacteria commonly found in soil. Another curious example of cell division in prokaryotes, reminiscent of live birth in animals, is exhibited by the giant bacterium *Epulopiscium*. Several daughter cells grow fully in the parent cell, which eventually disintegrates, releasing the new cells to the environment. Other species may form a long narrow extension at one pole in a process called **budding**. The tip of the extension swells and forms a smaller cell, the bud that eventually detaches from the parent cell. Budding is most common in yeast (Figure 9.16), but it is also observed in prosthecate bacteria and some cyanobacteria.

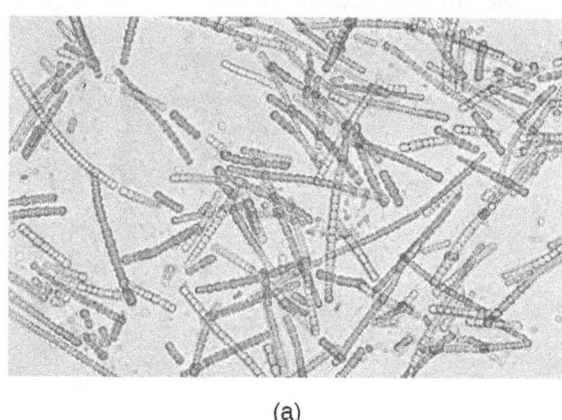
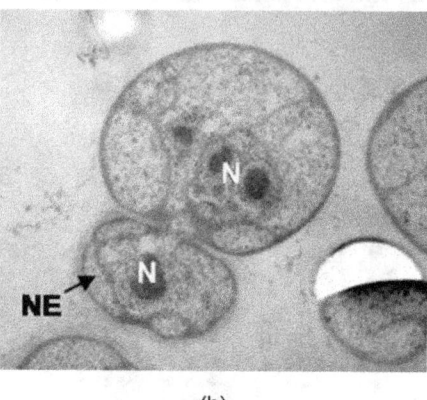

(a)   (b)

Figure 9.16   (a) Filamentous cyanobacteria, like those pictured here, replicate by fragmentation. (b) In this electron micrograph, cells of the bacterium *Gemmata obscuriglobus* are budding. The larger cell is the mother cell. Labels indicate the nucleoids (N) and the still-forming nuclear envelope (NE) of the daughter cell. (credit a: modification of work by CSIRO; credit b: modification of work by Kuo-Chang Lee, Rick I Webb and John A Fuerst)

The soil bacteria *Actinomyces* grow in long filaments divided by septa, similar to the mycelia seen in fungi, resulting in long cells with multiple nucleoids. Environmental signals, probably related to low nutrient availability, lead to the formation of aerial filaments. Within these aerial filaments, elongated cells divide simultaneously. The new cells, which contain a single nucleoid, develop into spores that give rise to new colonies.

CHECK YOUR UNDERSTANDING

- Identify at least one difference between fragmentation and budding.

Biofilms

In nature, microorganisms grow mainly in **biofilms**, complex and dynamic ecosystems that form on a variety of environmental surfaces, from industrial conduits and water treatment pipelines to rocks in river beds. Biofilms are not restricted to solid surface substrates, however. Almost any surface in a liquid environment containing some minimal nutrients will eventually develop a biofilm. Microbial mats that float on water, for example, are biofilms that contain large populations of photosynthetic microorganisms. Biofilms found in the human mouth may contain hundreds of bacterial species. Regardless of the environment where they occur, biofilms are not random collections of microorganisms; rather, they are highly structured communities that provide a selective advantage to their constituent microorganisms.

Biofilm Structure

Observations using confocal microscopy have shown that environmental conditions influence the overall structure of biofilms. Filamentous biofilms called streamers form in rapidly flowing water, such as freshwater streams, eddies, and specially designed laboratory flow cells that replicate growth conditions in fast-moving fluids. The streamers are anchored to the substrate by a "head" and the "tail" floats downstream in the current. In still or slow-moving water, biofilms mainly assume a mushroom-like shape. The structure of biofilms may also change with other environmental conditions such as nutrient availability.

Detailed observations of biofilms under confocal laser and scanning electron microscopes reveal clusters of microorganisms embedded in a matrix interspersed with open water channels. The extracellular matrix consists of **extracellular polymeric substances (EPS)** secreted by the organisms in the biofilm. The extracellular matrix represents a large fraction of the biofilm, accounting for 50%–90% of the total dry mass. The properties of the EPS vary according to the resident organisms and environmental conditions.

EPS is a hydrated gel composed primarily of polysaccharides and containing other macromolecules such as proteins, nucleic acids, and lipids. It plays a key role in maintaining the integrity and function of the biofilm. Channels in the EPS allow movement of nutrients, waste, and gases throughout the biofilm. This keeps the cells hydrated, preventing desiccation. EPS also shelters organisms in the biofilm from predation by other microbes or cells (e.g., protozoans, white blood cells in the human body).

Biofilm Formation

Free-floating microbial cells that live in an aquatic environment are called **planktonic** cells. The formation of a biofilm essentially involves the attachment of planktonic cells to a substrate, where they become **sessile** (attached to a surface). This occurs in stages, as depicted in Figure 9.17. The first stage involves the attachment of planktonic cells to a surface coated with a conditioning film of organic material. At this point, attachment to the substrate is reversible, but as cells express new phenotypes that facilitate the formation of EPS, they transition from a planktonic to a sessile lifestyle. The biofilm develops characteristic structures, including an extensive matrix and water channels. Appendages such as fimbriae, pili, and flagella interact with the EPS, and microscopy and genetic analysis suggest that such structures are required for the establishment of a mature biofilm. In the last stage of the biofilm life cycle, cells on the periphery of the biofilm revert to a planktonic lifestyle, sloughing off the mature biofilm to colonize new sites. This stage is referred to as dispersal.

Figure 9.17 Stages in the formation and life cycle of a biofilm. (credit: modification of work by Public Library of Science and American Society for Microbiology)

Within a biofilm, different species of microorganisms establish metabolic collaborations in which the waste product of one organism becomes the nutrient for another. For example, aerobic microorganisms consume oxygen, creating anaerobic regions that promote the growth of anaerobes. This occurs in many polymicrobial infections that involve both aerobic and anaerobic pathogens.

The mechanism by which cells in a biofilm coordinate their activities in response to environmental stimuli is called **quorum sensing**. Quorum sensing—which can occur between cells of different species within a biofilm—enables microorganisms to detect their cell density through the release and binding of small, diffusible molecules called **autoinducers**. When the cell population reaches a critical threshold (a quorum), these autoinducers initiate a cascade of reactions that activate genes associated with cellular functions that are beneficial only when the population reaches a critical density. For example, in some pathogens, synthesis of virulence factors only begins when enough cells are present to overwhelm the immune defenses of the host. Although mostly studied in bacterial populations, quorum sensing takes place between bacteria and eukaryotes and between eukaryotic cells such as the fungus *Candida albicans*, a common member of the human microbiota that can cause infections in immunocompromised individuals.

The signaling molecules in quorum sensing belong to two major classes. Gram-negative bacteria communicate mainly using N-acylated homoserine lactones, whereas gram-positive bacteria mostly use small peptides (Figure 9.18). In all cases, the first step in quorum sensing consists of the binding of the autoinducer to its specific receptor only when a threshold concentration of signaling molecules is reached. Once binding to the receptor takes place, a cascade of signaling events leads to changes in gene expression. The result is the activation of biological responses linked to quorum sensing, notably an increase in the production of signaling molecules themselves, hence the term autoinducer.

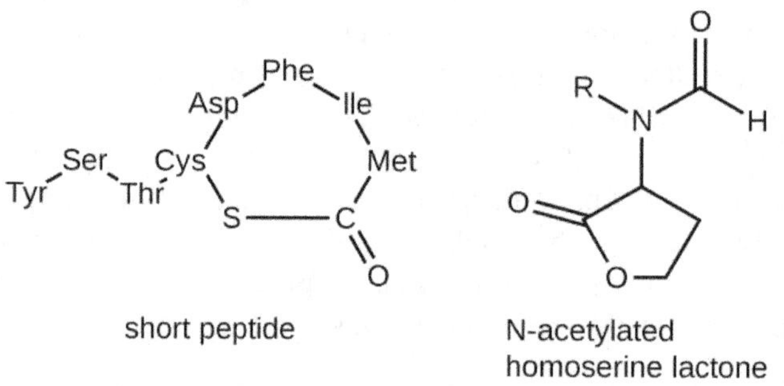

Figure 9.18 Short peptides in gram-positive bacteria and N-acetylated homoserine lactones in gram-negative bacteria act as autoinducers in quorum sensing and mediate the coordinated response of bacterial cells. The R side chain of the N-acetylated homoserine lactone is specific for the species of gram-negative bacteria. Some secreted homoserine lactones are recognized by more than one species.

Biofilms and Human Health

The human body harbors many types of biofilms, some beneficial and some harmful. For example, the layers of normal microbiota lining the intestinal and respiratory mucosa play a role in warding off infections by pathogens. However, other biofilms in the body can have a detrimental effect on health. For example, the plaque that forms on teeth is a biofilm that can contribute to dental and periodontal disease. Biofilms can also form in wounds, sometimes causing serious infections that can spread. The bacterium *Pseudomonas aeruginosa* often colonizes biofilms in the airways of patients with cystic fibrosis, causing chronic and sometimes fatal infections of the lungs. Biofilms can also form on medical devices used in or on the body, causing infections in patients with in-dwelling catheters, artificial joints, or contact lenses.

Pathogens embedded within biofilms exhibit a higher resistance to antibiotics than their free-floating counterparts. Several hypotheses have been proposed to explain why. Cells in the deep layers of a biofilm are metabolically inactive and may be less susceptible to the action of antibiotics that disrupt metabolic activities. The EPS may also slow the diffusion of antibiotics and antiseptics, preventing them from reaching cells in the deeper layers of the biofilm. Phenotypic changes may also contribute to the increased resistance exhibited by bacterial cells in biofilms. For example, the increased production of efflux pumps, membrane-embedded proteins that actively extrude antibiotics out of bacterial cells, have been shown to be an important mechanism of antibiotic resistance among biofilm-associated bacteria. Finally, biofilms provide an ideal environment for the exchange of extrachromosomal DNA, which often includes genes that confer antibiotic resistance.

✓ CHECK YOUR UNDERSTANDING

- What is the matrix of a biofilm composed of?
- What is the role of quorum sensing in a biofilm?

9.2 Oxygen Requirements for Microbial Growth

Learning Objectives

By the end of this section, you will be able to:
- Interpret visual data demonstrating minimum, optimum, and maximum oxygen or carbon dioxide requirements for growth
- Identify and describe different categories of microbes with requirements for growth with or without oxygen: obligate aerobe, obligate anaerobe, facultative anaerobe, aerotolerant anaerobe, microaerophile, and capnophile
- Give examples of microorganisms for each category of growth requirements

Ask most people "What are the major requirements for life?" and the answers are likely to include water and oxygen. Few would argue about the need for water, but what about oxygen? Can there be life without oxygen?

The answer is that molecular oxygen (O_2) is not always needed. The earliest signs of life are dated to a period when conditions on earth were highly reducing and free oxygen gas was essentially nonexistent. Only after cyanobacteria started releasing oxygen as a byproduct of photosynthesis and the capacity of iron in the oceans for taking up oxygen was exhausted did oxygen levels increase in the atmosphere. This event, often referred to as the Great Oxygenation Event or the Oxygen Revolution, caused a massive extinction. Most organisms could not survive the powerful oxidative properties of **reactive oxygen species** (ROS), highly unstable ions and molecules derived from partial reduction of oxygen that can damage virtually any macromolecule or structure with which they come in contact. Singlet oxygen ($O_2\cdot$), superoxide $\left(O_2^-\right)$, peroxides (H_2O_2), hydroxyl radical (OH·), and hypochlorite ion (OCl^-), the active ingredient of household bleach, are all examples of ROS. The organisms that were able to detoxify reactive oxygen species harnessed the high electronegativity of oxygen to produce free energy for their metabolism and thrived in the new environment.

Oxygen Requirements of Microorganisms

Many ecosystems are still free of molecular oxygen. Some are found in extreme locations, such as deep in the ocean or in earth's crust; others are part of our everyday landscape, such as marshes, bogs, and sewers. Within the bodies of humans and other animals, regions with little or no oxygen provide an anaerobic environment for microorganisms. (Figure 9.19).

(a)

(b)

Figure 9.19 Anaerobic environments are still common on earth. They include environments like (a) a bog where undisturbed dense sediments are virtually devoid of oxygen, and (b) the rumen (the first compartment of a cow's stomach), which provides an oxygen-free incubator for methanogens and other obligate anaerobic bacteria. (credit a: modification of work by National Park Service; credit b: modification of work by US Department of Agriculture)

We can easily observe different requirements for molecular oxygen by growing bacteria in **thioglycolate tube cultures**. A test-tube culture starts with autoclaved **thioglycolate medium** containing a low percentage of agar to allow motile bacteria to move throughout the medium. Thioglycolate has strong reducing properties and autoclaving flushes out most of the oxygen. The tubes are inoculated with the bacterial cultures to be tested and incubated at an appropriate temperature. Over time, oxygen slowly diffuses throughout the thioglycolate tube culture from the top. Bacterial density increases in the area where oxygen concentration is best suited for the growth of that particular organism.

The growth of bacteria with varying oxygen requirements in thioglycolate tubes is illustrated in Figure 9.20. In tube A, all the growth is seen at the top of the tube. The bacteria are **obligate (strict) aerobe**s that cannot grow without an abundant supply of oxygen. Tube B looks like the opposite of tube A. Bacteria grow at the bottom of tube B. Those are **obligate anaerobe**s, which are killed by oxygen. Tube C shows heavy growth at the top of the tube and growth throughout the tube, a typical result with **facultative anaerobe**s. Facultative anaerobes are organisms that thrive in the presence of oxygen but also grow in its absence by relying on fermentation or anaerobic respiration, if there is a suitable electron acceptor other than oxygen and the organism is able to perform anaerobic respiration. The **aerotolerant anaerobe**s in tube D are indifferent to the presence of oxygen. They do not use oxygen because they usually have a fermentative metabolism, but they are not harmed by the presence of oxygen as obligate anaerobes are. Tube E on the right shows a "Goldilocks" culture. The oxygen level has to be just right for growth, not too much and not too little. These **microaerophile**s are bacteria that require a minimum level of oxygen for growth, about 1%–10%, well below the 21% found in the atmosphere.

Examples of obligate aerobes are *Mycobacterium tuberculosis*, the causative agent of tuberculosis and *Micrococcus luteus*, a gram-positive bacterium that colonizes the skin. *Neisseria meningitidis*, the causative agent of severe bacterial meningitis, and *N. gonorrhoeae*, the causative agent of sexually transmitted gonorrhea, are also obligate aerobes.

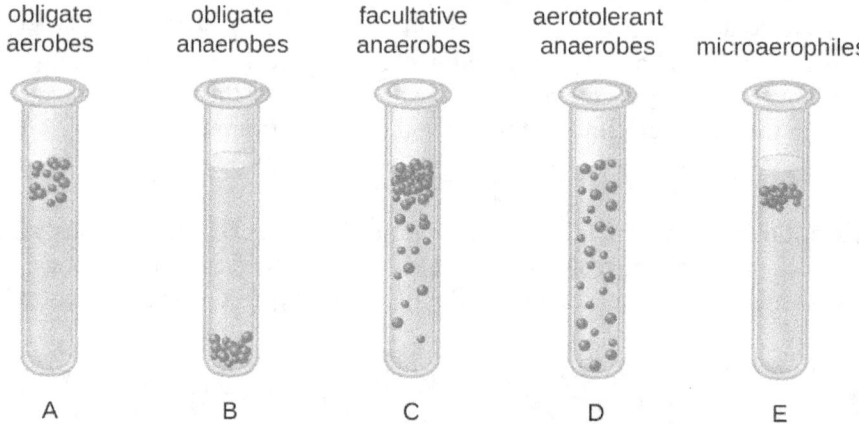

Figure 9.20 Diagram of bacterial cell distribution in thioglycolate tubes.

Many obligate anaerobes are found in the environment where anaerobic conditions exist, such as in deep sediments of soil, still waters, and at the bottom of the deep ocean where there is no photosynthetic life. Anaerobic conditions also exist naturally in the intestinal tract of animals. Obligate anaerobes, mainly *Bacteroidetes*, represent a large fraction of the microbes in the human gut. Transient anaerobic conditions exist when tissues are not supplied with blood circulation; they die and become an ideal breeding ground for obligate anaerobes. Another type of obligate anaerobe encountered in the human body is the gram-positive, rod-shaped *Clostridium* spp. Their ability to form endospores allows them to survive in the presence of oxygen. One of the major causes of health-acquired infections is *C. difficile*, known as C. diff. Prolonged use of antibiotics for other infections increases the probability of a patient developing a secondary *C. difficile* infection. Antibiotic treatment disrupts the balance of microorganisms in the intestine and allows the colonization of the gut by *C. difficile*, causing a significant inflammation of the colon.

Other clostridia responsible for serious infections include *C. tetani*, the agent of tetanus, and *C. perfringens*, which causes gas gangrene. In both cases, the infection starts in necrotic tissue (dead tissue that is not supplied with oxygen by blood circulation). This is the reason that deep puncture wounds are associated with tetanus. When tissue death is accompanied by lack of circulation, gangrene is always a danger.

The study of obligate anaerobes requires special equipment. Obligate anaerobic bacteria must be grown under conditions devoid of oxygen. The most common approach is culture in an **anaerobic jar** (Figure 9.21).

Anaerobic jars include chemical packs that remove oxygen and release carbon dioxide (CO_2). An **anaerobic chamber** is an enclosed box from which all oxygen is removed. Gloves sealed to openings in the box allow handling of the cultures without exposing the culture to air (Figure 9.21).

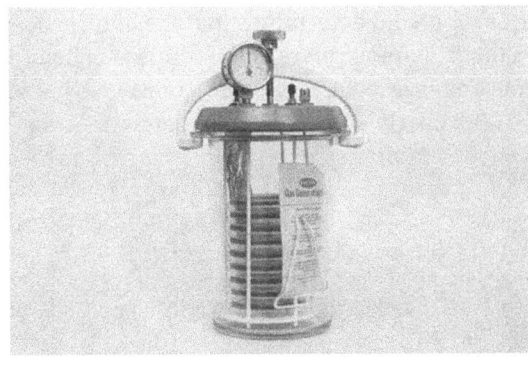
(a)

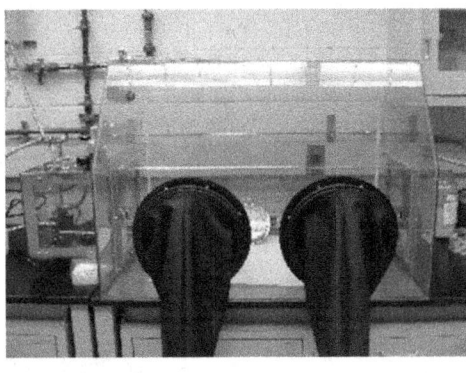
(b)

Figure 9.21 (a) An anaerobic jar is pictured that is holding nine Petri plates supporting cultures. (b) Openings in the side of an anaerobic box are sealed by glove-like sleeves that allow for the handling of cultures inside the box. (credit a: modification of work by Centers for Disease Control and Prevention; credit b: modification of work by NIST)

Staphylococci and Enterobacteriaceae are examples of facultative anaerobes. Staphylococci are found on the skin and upper respiratory tract. Enterobacteriaceae are found primarily in the gut and upper respiratory tract but can sometimes spread to the urinary tract, where they are capable of causing infections. It is not unusual to see mixed bacterial infections in which the facultative anaerobes use up the oxygen, creating an environment for the obligate anaerobes to flourish.

Examples of aerotolerant anaerobes include lactobacilli and streptococci, both found in the oral microbiota. *Campylobacter jejuni*, which causes gastrointestinal infections, is an example of a microaerophile and is grown under low-oxygen conditions.

The **optimum oxygen concentration**, as the name implies, is the ideal concentration of oxygen for a particular microorganism. The lowest concentration of oxygen that allows growth is called the **minimum permissive oxygen concentration**. The highest tolerated concentration of oxygen is the **maximum permissive oxygen concentration**. The organism will not grow outside the range of oxygen levels found between the minimum and maximum permissive oxygen concentrations.

⊘ CHECK YOUR UNDERSTANDING

- Would you expect the oldest bacterial lineages to be aerobic or anaerobic?
- Which bacteria grow at the top of a thioglycolate tube, and which grow at the bottom of the tube?

Case in Point

An Unwelcome Anaerobe

Charles is a retired bus driver who developed type 2 diabetes over 10 years ago. Since his retirement, his lifestyle has become very sedentary and he has put on a substantial amount of weight. Although he has felt tingling and numbness in his left foot for a while, he has not been worried because he thought his foot was simply "falling asleep." Recently, a scratch on his foot does not seem to be healing and is becoming increasingly ugly. Because the sore did not bother him much, Charles figured it could not be serious until his daughter noticed a purplish discoloration spreading on the skin and oozing (Figure 9.22). When he was finally seen by his physician, Charles was rushed to the operating room. His open sore, or ulcer, is the result of a diabetic foot.

The concern here is that gas gangrene may have taken hold in the dead tissue. The most likely agent of gas gangrene is *Clostridium perfringens*, an endospore-forming, gram-positive bacterium. It is an obligate anaerobe that grows in tissue devoid of oxygen. Since dead tissue is no longer supplied with oxygen by the circulatory system, the dead tissue provides pockets of ideal environment for the growth of *C. perfringens*.

A surgeon examines the ulcer and radiographs of Charles's foot and determines that the bone is not yet infected. The wound will have to be surgically debrided (debridement refers to the removal of dead and infected tissue) and a sample sent for microbiological lab analysis, but Charles will not have to have his foot amputated. Many diabetic patients are not so lucky. In 2008, nearly 70,000 diabetic patients in the United States lost a foot or limb to amputation, according to statistics from the Centers for Disease Control and Prevention.[1]

- Which growth conditions would you recommend for the detection of *C. perfringens*?

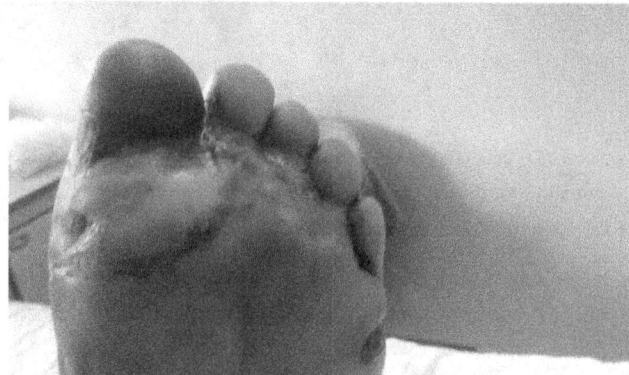

Figure 9.22 This clinical photo depicts ulcers on the foot of a diabetic patient. Dead tissue accumulating in ulcers can provide an ideal growth environment for the anaerobe *C. perfringens*, a causative agent of gas gangrene. (Credit: Phalinn Ooi / Wikimedia Commons (CC-BY))

Detoxification of Reactive Oxygen Species

Aerobic respiration constantly generates reactive oxygen species (ROS), byproducts that must be detoxified. Even organisms that do not use aerobic respiration need some way to break down some of the ROS that may form from atmospheric oxygen. Three main enzymes break down those toxic byproducts: superoxide dismutase, peroxidase, and catalase. Each one catalyzes a different reaction. Reactions of type seen in Reaction 1 are catalyzed by **peroxidase**s.

$$(1) \quad X - (2H^+) + H_2O_2 \rightarrow \text{oxidized-}X + 2H_2O$$

In these reactions, an electron donor (reduced compound; e.g., reduced nicotinamide adenine dinucleotide [NADH]) oxidizes hydrogen peroxide, or other peroxides, to water. The enzymes play an important role by limiting the damage caused by peroxidation of membrane lipids. Reaction 2 is mediated by the enzyme **superoxide dismutase** (SOD) and breaks down the powerful superoxide anions generated by aerobic metabolism:

$$(2) \quad 2O_2^- + 2H^+ \rightarrow H_2O_2 + O_2$$

The enzyme **catalase** converts hydrogen peroxide to water and oxygen as shown in Reaction 3.

$$(3) \quad 2H_2O_2 \rightarrow 2H_2O + O_2$$

Obligate anaerobes usually lack all three enzymes. Aerotolerant anaerobes do have SOD but no catalase. Reaction 3, shown occurring in Figure 9.23, is the basis of a useful and rapid test to distinguish streptococci, which are aerotolerant and do not possess catalase, from staphylococci, which are facultative anaerobes. A sample of culture rapidly mixed in a drop of 3% hydrogen peroxide will release bubbles if the culture is

---

1 Centers for Disease Control and Prevention. "Living With Diabetes: Keep Your Feet Healthy." http://www.cdc.gov/Features/DiabetesFootHealth/

catalase positive.

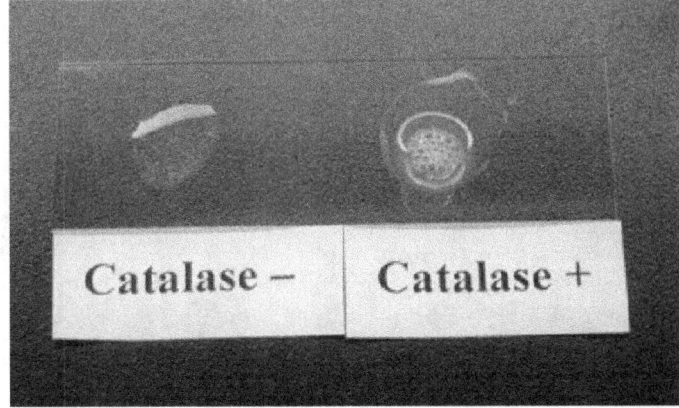

Figure 9.23 The catalase test detects the presence of the enzyme catalase by noting whether bubbles are released when hydrogen peroxide is added to a culture sample. Compare the positive result (right) with the negative result (left). (credit: Centers for Disease Control and Prevention)

Bacteria that grow best in a higher concentration of CO_2 and a lower concentration of oxygen than present in the atmosphere are called **capnophiles**. One common approach to grow capnophiles is to use a **candle jar**. A candle jar consists of a jar with a tight-fitting lid that can accommodate the cultures and a candle. After the cultures are added to the jar, the candle is lit and the lid closed. As the candle burns, it consumes most of the oxygen present and releases CO_2.

CHECK YOUR UNDERSTANDING

- What substance is added to a sample to detect catalase?
- What is the function of the candle in a candle jar?

Clinical Focus

Part 2

The health-care provider who saw Jeni was concerned primarily because of her pregnancy. Her condition enhances the risk for infections and makes her more vulnerable to those infections. The immune system is downregulated during pregnancy, and pathogens that cross the placenta can be very dangerous for the fetus. A note on the provider's order to the microbiology lab mentions a suspicion of infection by *Listeria monocytogenes*, based on the signs and symptoms exhibited by the patient.

Jeni's blood samples are streaked directly on sheep blood agar, a medium containing tryptic soy agar enriched with 5% sheep blood. (Blood is considered sterile; therefore, competing microorganisms are not expected in the medium.) The inoculated plates are incubated at 37 °C for 24 to 48 hours. Small grayish colonies surrounded by a clear zone emerge. Such colonies are typical of *Listeria* and other pathogens such as streptococci; the clear zone surrounding the colonies indicates complete lysis of blood in the medium, referred to as beta-hemolysis (Figure 9.24). When tested for the presence of catalase, the colonies give a positive response, eliminating *Streptococcus* as a possible cause. Furthermore, a Gram stain shows short gram-positive bacilli. Cells from a broth culture grown at room temperature displayed the tumbling motility characteristic of *Listeria* (Figure 9.24). All of these clues lead the lab to positively confirm the presence of *Listeria* in Jeni's blood samples.

- How serious is Jeni's condition and what is the appropriate treatment?

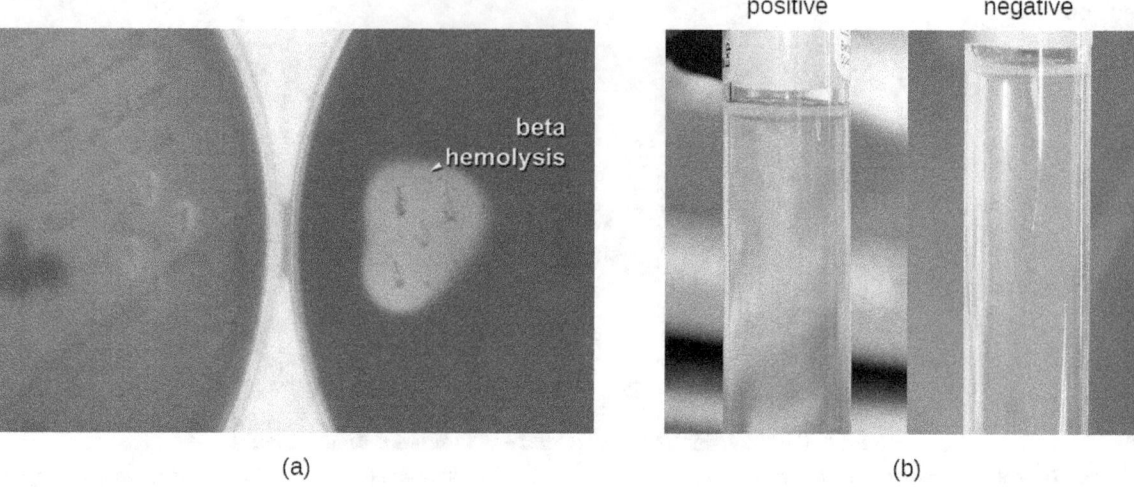

Figure 9.24 (a) A sample blood agar test showing beta-hemolysis. (b) A sample motility test showing both positive and negative results. (credit a: modification of work by Centers for Disease Control and Prevention; credit b: modification of work by "VeeDunn"/Flickr)

Jump to the next Clinical Focus box. Go back to the previous Clinical Focus box.

9.3 The Effects of pH on Microbial Growth

Learning Objectives

By the end of this section, you will be able to:
- Illustrate and briefly describe minimum, optimum, and maximum pH requirements for growth
- Identify and describe the different categories of microbes with pH requirements for growth: acidophiles, neutrophiles, and alkaliphiles
- Give examples of microorganisms for each category of pH requirement

Yogurt, pickles, sauerkraut, and lime-seasoned dishes all owe their tangy taste to a high acid content (Figure 9.25). Recall that acidity is a function of the concentration of hydrogen ions [H^+] and is measured as pH. Environments with pH values below 7.0 are considered acidic, whereas those with pH values above 7.0 are considered basic. Extreme pH affects the structure of all macromolecules. The hydrogen bonds holding together strands of DNA break up at high pH. Lipids are hydrolyzed by an extremely basic pH. The proton motive force responsible for production of ATP in cellular respiration depends on the concentration gradient of H^+ across the plasma membrane (see Cellular Respiration). If H^+ ions are neutralized by hydroxide ions, the concentration gradient collapses and impairs energy production. But the component most sensitive to pH in the cell is its workhorse, the protein. Moderate changes in pH modify the ionization of amino-acid functional groups and disrupt hydrogen bonding, which, in turn, promotes changes in the folding of the molecule, promoting denaturation and destroying activity.

Figure 9.25 Lactic acid bacteria that ferment milk into yogurt or transform vegetables in pickles thrive at a pH close to 4.0. Sauerkraut and dishes such as pico de gallo owe their tangy flavor to their acidity. Acidic foods have been a mainstay of the human diet for centuries, partly because most microbes that cause food spoilage grow best at a near neutral pH and do not tolerate acidity well. (credit "yogurt": modification of work by "nina.jsc"/Flickr; credit "pickles": modification of work by Noah Sussman; credit "sauerkraut": modification of work

by Jesse LaBuff; credit "pico de gallo": modification of work by "regan76"/Flickr)

The **optimum growth pH** is the most favorable pH for the growth of an organism. The lowest pH value that an organism can tolerate is called the **minimum growth pH** and the highest pH is the **maximum growth pH**. These values can cover a wide range, which is important for the preservation of food and to microorganisms' survival in the stomach. For example, the optimum growth pH of *Salmonella* spp. is 7.0–7.5, but the minimum growth pH is closer to 4.2.

Most bacteria are **neutrophile**s, meaning they grow optimally at a pH within one or two pH units of the neutral pH of 7 (see Figure 9.26). Most familiar bacteria, like *Escherichia coli*, staphylococci, and *Salmonella* spp. are neutrophiles and do not fare well in the acidic pH of the stomach. However, there are pathogenic strains of *E. coli*, *S. typhi*, and other species of intestinal pathogens that are much more resistant to stomach acid. In comparison, fungi thrive at slightly acidic pH values of 5.0–6.0.

Microorganisms that grow optimally at pH less than 5.55 are called **acidophiles**. For example, the sulfur-oxidizing *Sulfolobus* spp. isolated from sulfur mud fields and hot springs in Yellowstone National Park are extreme acidophiles. These archaea survive at pH values of 2.5–3.5. Species of the archaean genus *Ferroplasma* live in acid mine drainage at pH values of 0–2.9. *Lactobacillus* bacteria, which are an important part of the normal microbiota of the vagina, can tolerate acidic environments at pH values 3.5–6.8 and also contribute to the acidity of the vagina (pH of 4, except at the onset of menstruation) through their metabolic production of lactic acid. The vagina's acidity plays an important role in inhibiting other microbes that are less tolerant of acidity. Acidophilic microorganisms display a number of adaptations to survive in strong acidic environments. For example, proteins show increased negative surface charge that stabilizes them at low pH. Pumps actively eject H^+ ions out of the cells. The changes in the composition of membrane phospholipids probably reflect the need to maintain membrane fluidity at low pH.

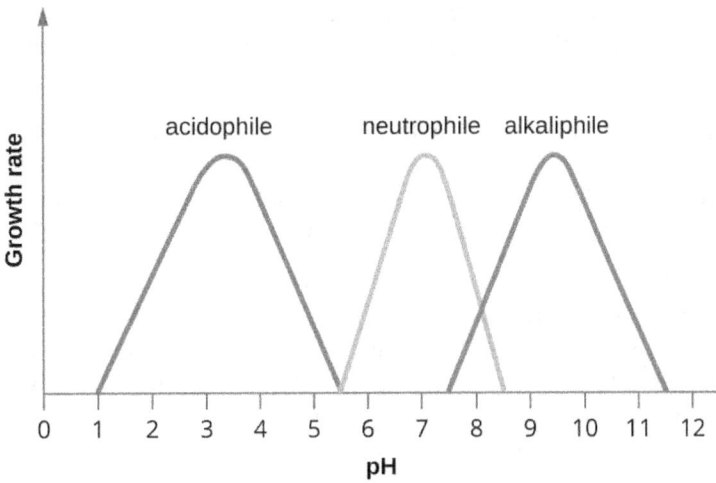

Figure 9.26 The curves show the approximate pH ranges for the growth of the different classes of pH-specific prokaryotes. Each curve has an optimal pH and extreme pH values at which growth is much reduced. Most bacteria are neutrophiles and grow best at near-neutral pH (center curve). Acidophiles have optimal growth at pH values near 3 and alkaliphiles have optimal growth at pH values above 9.

At the other end of the spectrum are **alkaliphile**s, microorganisms that grow best at pH between 8.0 and 10.5. *Vibrio cholerae*, the pathogenic agent of cholera, grows best at the slightly basic pH of 8.0; it can survive pH values of 11.0 but is inactivated by the acid of the stomach. When it comes to survival at high pH, the bright pink archaean *Natronobacterium*, found in the soda lakes of the African Rift Valley, may hold the record at a pH of 10.5 (Figure 9.27). Extreme alkaliphiles have adapted to their harsh environment through evolutionary modification of lipid and protein structure and compensatory mechanisms to maintain the proton motive force in an alkaline environment. For example, the alkaliphile *Bacillus firmus* derives the energy for transport reactions and motility from a Na^+ ion gradient rather than a proton motive force. Many enzymes from alkaliphiles have a higher isoelectric point, due to an increase in the number of basic amino acids, than homologous enzymes from neutrophiles.

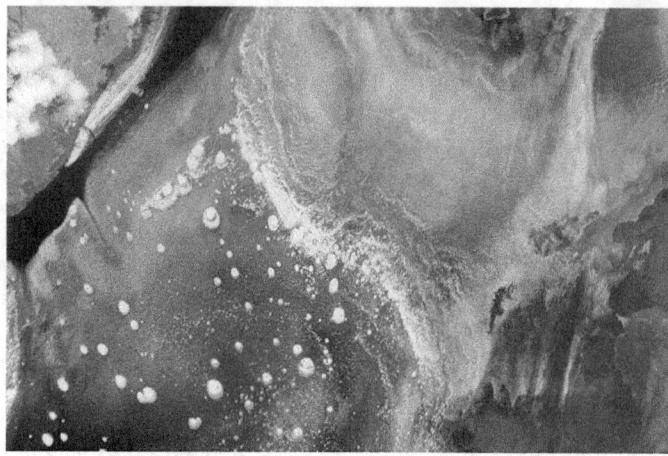

Figure 9.27 View from space of Lake Natron in Tanzania. The pink color is due to the pigmentation of the extreme alkaliphilic and halophilic microbes that colonize the lake. (credit: NASA)

MICRO CONNECTIONS

Survival at the Low pH of the Stomach

Peptic ulcers (or stomach ulcers) are painful sores on the stomach lining. Until the 1980s, they were believed to be caused by spicy foods, stress, or a combination of both. Patients were typically advised to eat bland foods, take anti-acid medications, and avoid stress. These remedies were not particularly effective, and the condition often recurred. This all changed dramatically when the real cause of most peptic ulcers was discovered to be a slim, corkscrew-shaped bacterium, *Helicobacter pylori*. This organism was identified and isolated by Barry Marshall and Robin Warren, whose discovery earned them the Nobel Prize in Medicine in 2005.

The ability of *H. pylori* to survive the low pH of the stomach would seem to suggest that it is an extreme acidophile. As it turns out, this is not the case. In fact, *H. pylori* is a neutrophile. So, how does it survive in the stomach? Remarkably, *H. pylori* creates a microenvironment in which the pH is nearly neutral. It achieves this by producing large amounts of the enzyme urease, which breaks down urea to form NH_4^+ and CO_2. The ammonium ion raises the pH of the immediate environment.

This metabolic capability of *H. pylori* is the basis of an accurate, noninvasive test for infection. The patient is given a solution of urea containing radioactively labeled carbon atoms. If *H. pylori* is present in the stomach, it will rapidly break down the urea, producing radioactive CO_2 that can be detected in the patient's breath. Because peptic ulcers may lead to gastric cancer, patients who are determined to have *H. pylori* infections are treated with antibiotics.

CHECK YOUR UNDERSTANDING

- What effect do extremes of pH have on proteins?
- What pH-adaptive type of bacteria would most human pathogens be?

9.4 Temperature and Microbial Growth

Learning Objectives

By the end of this section, you will be able to:
- Illustrate and briefly describe minimum, optimum, and maximum temperature requirements for growth
- Identify and describe different categories of microbes with temperature requirements for growth: psychrophile, psychrotrophs, mesophile, thermophile, hyperthermophile
- Give examples of microorganisms in each category of temperature tolerance

When the exploration of Lake Whillans started in Antarctica, researchers did not expect to find much life.

Constant subzero temperatures and lack of obvious sources of nutrients did not seem to be conditions that would support a thriving ecosystem. To their surprise, the samples retrieved from the lake showed abundant microbial life. In a different but equally harsh setting, bacteria grow at the bottom of the ocean in sea vents (Figure 9.28), where temperatures can reach 340 °C (700 °F).

Microbes can be roughly classified according to the range of temperature at which they can grow. The growth rates are the highest at the **optimum growth temperature** for the organism. The lowest temperature at which the organism can survive and replicate is its **minimum growth temperature**. The highest temperature at which growth can occur is its **maximum growth temperature**. The following ranges of permissive growth temperatures are approximate only and can vary according to other environmental factors.

Organisms categorized as **mesophile**s ("middle loving") are adapted to moderate temperatures, with optimal growth temperatures ranging from room temperature (about 20 °C) to about 45 °C. As would be expected from the core temperature of the human body, 37 °C (98.6 °F), normal human microbiota and pathogens (e.g., *E. coli*, *Salmonella* spp., and *Lactobacillus* spp.) are mesophiles.

Organisms called **psychrotrophs**, also known as psychrotolerant, prefer cooler environments, from a high temperature of 25 °C to refrigeration temperature about 4 °C. They are found in many natural environments in temperate climates. They are also responsible for the spoilage of refrigerated food.

Clinical Focus

Resolution

The presence of *Listeria* in Jeni's blood suggests that her symptoms are due to listeriosis, an infection caused by *L. monocytogenes*. Listeriosis is a serious infection with a 20% mortality rate and is a particular risk to Jeni's fetus. A sample from the amniotic fluid cultured for the presence of *Listeria* gave negative results. Because the absence of organisms does not rule out the possibility of infection, a molecular test based on the nucleic acid amplification of the 16S ribosomal RNA of *Listeria* was performed to confirm that no bacteria crossed the placenta. Fortunately, the results from the molecular test were also negative.

Jeni was admitted to the hospital for treatment and recovery. She received a high dose of two antibiotics intravenously for 2 weeks. The preferred drugs for the treatment of listeriosis are ampicillin or penicillin G with an aminoglycoside antibiotic. Resistance to common antibiotics is still rare in *Listeria* and antibiotic treatment is usually successful. She was released to home care after a week and fully recovered from her infection.

L. monocytogenes is a gram-positive short rod found in soil, water, and food. It is classified as a psychrophile and is halotolerant. Its ability to multiply at refrigeration temperatures (4–10 °C) and its tolerance for high concentrations of salt (up to 10% sodium chloride [NaCl]) make it a frequent source of food poisoning. Because *Listeria* can infect animals, it often contaminates food such as meat, fish, or dairy products. Contamination of commercial foods can often be traced to persistent biofilms that form on manufacturing equipment that is not sufficiently cleaned.

Listeria infection is relatively common among pregnant women because the elevated levels of progesterone downregulate the immune system, making them more vulnerable to infection. The pathogen can cross the placenta and infect the fetus, often resulting in miscarriage, stillbirth, or fatal neonatal infection. Pregnant women are thus advised to avoid consumption of soft cheeses, refrigerated cold cuts, smoked seafood, and unpasteurized dairy products. Because *Listeria* bacteria can easily be confused with diphtheroids, another common group of gram-positive rods, it is important to alert the laboratory when listeriosis is suspected.

Go back to the previous Clinical Focus box.

The organisms retrieved from arctic lakes such as Lake Whillans are considered extreme **psychrophiles** (cold loving). Psychrophiles are microorganisms that can grow at 0 °C and below, have an optimum growth temperature close to

15 °C, and usually do not survive at temperatures above 20 °C. They are found in permanently cold environments such as the deep waters of the oceans. Because they are active at low temperature, psychrophiles and psychrotrophs are important decomposers in cold climates.

Organisms that grow at optimum temperatures of 50 °C to a maximum of 80 °C are called **thermophiles** ("heat loving"). They do not multiply at room temperature. Thermophiles are widely distributed in hot springs, geothermal soils, and manmade environments such as garden compost piles where the microbes break down kitchen scraps and vegetal material. Examples of thermophiles include *Thermus aquaticus* and *Geobacillus* spp. Higher up on the extreme temperature scale we find the **hyperthermophiles**, which are characterized by growth ranges from 80 °C to a maximum of 110 °C, with some extreme examples that survive temperatures above 121 °C, the average temperature of an autoclave. The hydrothermal vents at the bottom of the ocean are a prime example of extreme environments, with temperatures reaching an estimated 340 °C (Figure 9.28). Microbes isolated from the vents achieve optimal growth at temperatures higher than 100 °C. Noteworthy examples are *Pyrobolus* and *Pyrodictium*, archaea that grow at 105 °C and survive autoclaving. Figure 9.29 shows the typical skewed curves of temperature-dependent growth for the categories of microorganisms we have discussed.

Figure 9.28 A black smoker at the bottom of the ocean belches hot, chemical-rich water, and heats the surrounding waters. Sea vents provide an extreme environment that is nonetheless teeming with macroscopic life (the red tubeworms) supported by an abundant microbial ecosystem. (credit: NOAA)

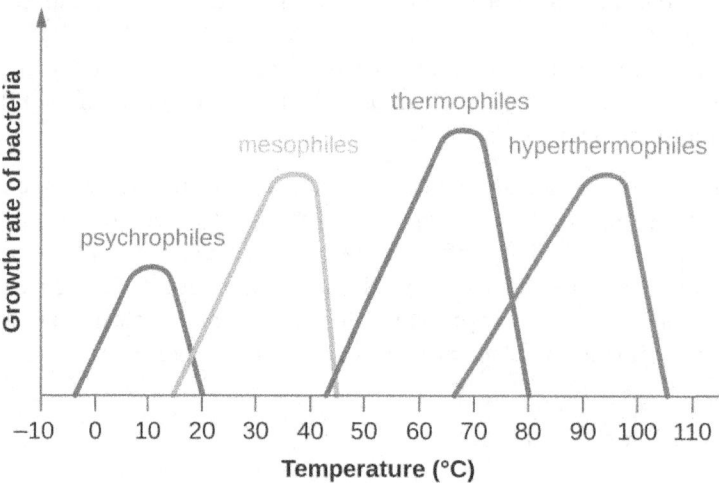

Figure 9.29 The graph shows growth rate of bacteria as a function of temperature. Notice that the curves are skewed toward the optimum temperature. The skewing of the growth curve is thought to reflect the rapid denaturation of proteins as the temperature rises past the optimum for growth of the microorganism.

Life in extreme environments raises fascinating questions about the adaptation of macromolecules and metabolic processes. Very low temperatures affect cells in many ways. Membranes lose their fluidity and are damaged by ice crystal formation. Chemical reactions and diffusion slow considerably. Proteins become too rigid to catalyze reactions and may undergo denaturation. At the opposite end of the temperature spectrum, heat denatures proteins and nucleic acids. Increased fluidity impairs metabolic processes in membranes. Some of the practical applications of the destructive effects of heat on microbes are sterilization by steam, pasteurization, and incineration of inoculating loops. Proteins in psychrophiles are, in general, rich in hydrophobic residues, display an increase in flexibility, and have a lower number of secondary stabilizing bonds when compared with homologous proteins from mesophiles. Antifreeze proteins and solutes that decrease the freezing temperature of the cytoplasm are common. The lipids in the membranes tend to be unsaturated to increase fluidity. Growth rates are much slower than those encountered at moderate temperatures. Under appropriate conditions, mesophiles and even thermophiles can survive freezing. Liquid cultures of bacteria are mixed with sterile glycerol solutions and frozen to –80 °C for long-term storage as stocks. Cultures can withstand freeze drying (lyophilization) and then be stored as powders in sealed ampules to be reconstituted with broth when needed.

Macromolecules in thermophiles and hyperthermophiles show some notable structural differences from what is observed in the mesophiles. The ratio of saturated to polyunsaturated lipids increases to limit the fluidity of the cell membranes. Their DNA sequences show a higher proportion of guanine–cytosine nitrogenous bases, which are held together by three hydrogen bonds in contrast to adenine and thymine, which are connected in the double helix by two hydrogen bonds. Additional secondary structures, ionic and covalent bonds, as well as the replacement of key amino acids to stabilize folding, contribute to the resistance of proteins to denaturation. The so-called thermoenzymes purified from thermophiles have important practical applications. For example, amplification of nucleic acids in the polymerase chain reaction (PCR) depends on the thermal stability of *Taq* polymerase, an enzyme isolated from *T. aquaticus*. Degradation enzymes from thermophiles are added as ingredients in hot-water detergents, increasing their effectiveness.

✓ CHECK YOUR UNDERSTANDING

- What temperature requirements do most bacterial human pathogens have?
- What DNA adaptation do thermophiles exhibit?

Eye on Ethics

Feeding the World...and the World's Algae

Artificial fertilizers have become an important tool in food production around the world. They are responsible for many of the gains of the so-called green revolution of the 20th century, which has allowed the planet to feed many of its more than 7 billion people. Artificial fertilizers provide nitrogen and phosphorus, key limiting nutrients, to crop plants, removing the normal barriers that would otherwise limit the rate of growth. Thus, fertilized crops grow much faster, and farms that use fertilizer produce higher crop yields.

However, careless use and overuse of artificial fertilizers have been demonstrated to have significant negative impacts on aquatic ecosystems, both freshwater and marine. Fertilizers that are applied at inappropriate times or in too-large quantities allow nitrogen and phosphorus compounds to escape use by crop plants and enter drainage systems. Inappropriate use of fertilizers in residential settings can also contribute to nutrient loads, which find their way to lakes and coastal marine ecosystems. As water warms and nutrients are plentiful, microscopic algae bloom, often changing the color of the water because of the high cell density.

Most algal blooms are not directly harmful to humans or wildlife; however, they can cause harm indirectly. As the algal population expands and then dies, it provides a large increase in organic matter to the bacteria that live in deep water. With this large supply of nutrients, the population of nonphotosynthetic microorganisms explodes, consuming available oxygen and creating "dead zones" where animal life has virtually disappeared.

Depletion of oxygen in the water is not the only damaging consequence of some algal blooms. The algae that produce red tides in the Gulf of Mexico, *Karenia brevis*, secrete potent toxins that can kill fish and other organisms and also accumulate in shellfish. Consumption of contaminated shellfish can cause severe neurological and gastrointestinal symptoms in humans. Shellfish beds must be regularly monitored for the presence of the toxins, and harvests are often shut down when it is present, incurring economic costs to the fishery. Cyanobacteria, which can form blooms in marine and freshwater ecosystems, produce toxins called microcystins, which can cause allergic reactions and liver damage when ingested in drinking water or during swimming. Recurring cyanobacterial algal blooms in Lake Erie (Figure 9.30) have forced municipalities to issue drinking water bans for days at a time because of unacceptable toxin levels.

This is just a small sampling of the negative consequences of algal blooms, red tides, and dead zones. Yet the benefits of crop fertilizer—the main cause of such blooms—are difficult to dispute. There is no easy solution to this dilemma, as a ban on fertilizers is not politically or economically feasible. In lieu of this, we must advocate for responsible use and regulation in agricultural and residential contexts, as well as the restoration of wetlands, which can absorb excess fertilizers before they reach lakes and oceans.

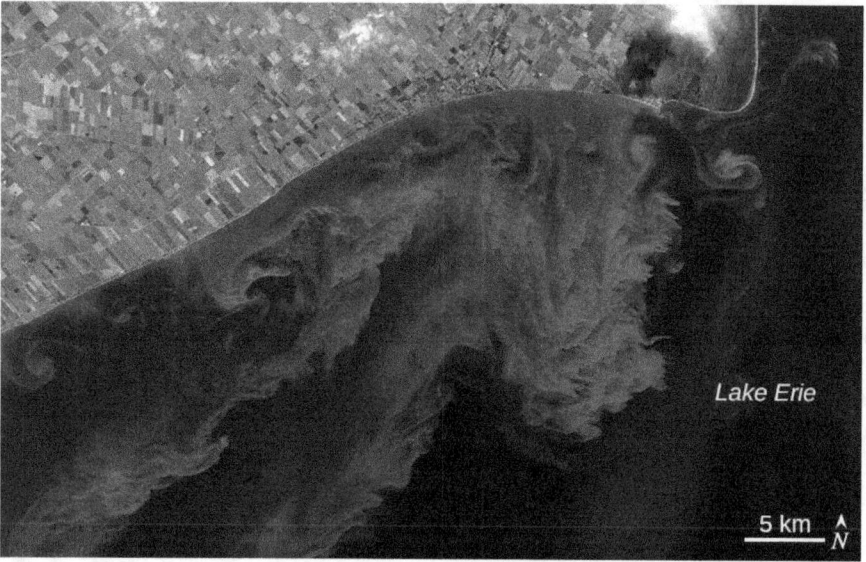

Figure 9.30 Heavy rains cause runoff of fertilizers into Lake Erie, triggering extensive algal blooms, which can be observed along the shoreline. Notice the brown unplanted and green planted agricultural land on the shore. (credit: NASA)

LINK TO LEARNING

This video (https://openstax.org/l/22algaebloomvid) discusses algal blooms and dead zones in more depth.

9.5 Other Environmental Conditions that Affect Growth

Learning Objectives

By the end of this section, you will be able to:
- Identify and describe different categories of microbes with specific growth requirements other than oxygen, pH, and temperature, such as altered barometric pressure, osmotic pressure, humidity, and light
- Give at least one example microorganism for each category of growth requirement

Microorganisms interact with their environment along more dimensions than pH, temperature, and free oxygen levels, although these factors require significant adaptations. We also find microorganisms adapted to varying levels of salinity, barometric pressure, humidity, and light.

Osmotic and Barometric Pressure

Most natural environments tend to have lower solute concentrations than the cytoplasm of most microorganisms. Rigid cell walls protect the cells from bursting in a dilute environment. Not much protection is available against high osmotic pressure. In this case, water, following its concentration gradient, flows out of the cell. This results in plasmolysis (the shrinking of the protoplasm away from the intact cell wall) and cell death. This fact explains why brines and layering meat and fish in salt are time-honored methods of preserving food. Microorganisms called **halophiles** ("salt loving") actually require high salt concentrations for growth. These organisms are found in marine environments where salt concentrations hover at 3.5%. Extreme halophilic microorganisms, such as the red alga *Dunaliella salina* and the archaeal species *Halobacterium* in Figure 9.31, grow in hypersaline lakes such as the Great Salt Lake, which is 3.5–8 times saltier than the ocean, and the Dead Sea, which is 10 times saltier than the ocean.

Figure 9.31 Photograph taken from space of the Great Salt Lake in Utah. The purple color is caused by high density of the alga *Dunaliella* and the archaean *Halobacterium* spp. (credit: NASA)

Dunaliella spp. counters the tremendous osmotic pressure of the environment with a high cytoplasmic concentration of glycerol and by actively pumping out salt ions. *Halobacterium* spp. accumulates large concentrations of K^+ and other ions in its cytoplasm. Its proteins are designed for high salt concentrations and lose activity at salt concentrations below 1–2 M. Although most **halotolerant** organisms, for example *Halomonas* spp. in salt marshes, do not need high concentrations of salt for growth, they will survive and divide in the presence of high salt. Not surprisingly, the staphylococci, micrococci, and corynebacteria that colonize our skin tolerate salt in their environment. Halotolerant pathogens are an important cause of food-borne illnesses because they survive and multiply in salty food. For example, the halotolerant bacteria *S. aureus*, *Bacillus cereus*, and *V. cholerae* produce dangerous enterotoxins and are major causes of food poisoning.

Microorganisms depend on available water to grow. Available moisture is measured as water activity (a_w), which is the ratio of the vapor pressure of the medium of interest to the vapor pressure of pure distilled water; therefore, the a_w of water is equal to 1.0. Bacteria require high a_w (0.97–0.99), whereas fungi can tolerate drier environments; for example, the range of a_w for growth of *Aspergillus* spp. is 0.8–0.75. Decreasing the water content of foods by drying, as in jerky, or through freeze-drying or by increasing osmotic pressure, as in brine and jams, are common methods of preventing spoilage.

Microorganisms that require high atmospheric pressure for growth are called **barophiles**. The bacteria that live at the bottom of the ocean must be able to withstand great pressures. Because it is difficult to retrieve intact specimens and reproduce such growth conditions in the laboratory, the characteristics of these microorganisms are largely unknown.

Light

Photoautotrophs, such as cyanobacteria or green sulfur bacteria, and photoheterotrophs, such as purple nonsulfur bacteria, depend on sufficient light intensity at the wavelengths absorbed by their pigments to grow and multiply. Energy from light is captured by pigments and converted into chemical energy that drives carbon fixation and other metabolic processes. The portion of the electromagnetic spectrum that is absorbed by these organisms is defined as photosynthetically active radiation (PAR). It lies within the visible light spectrum ranging from 400 to 700 nanometers (nm) and extends in the near infrared for some photosynthetic bacteria. A number of accessory pigments, such as fucoxanthin in brown algae and phycobilins in cyanobacteria, widen the useful range of wavelengths for photosynthesis and compensate for the low light levels available at greater depths of water. Other microorganisms, such as the archaea of the class Halobacteria, use light energy to drive their proton and sodium pumps. The light is absorbed by a pigment protein complex called bacteriorhodopsin, which is similar to the eye pigment rhodopsin. Photosynthetic bacteria are present not only in aquatic environments but also in soil and in symbiosis with fungi in lichens. The peculiar watermelon snow is caused by a microalga *Chlamydomonas nivalis*, a green alga rich in a secondary red carotenoid pigment (astaxanthin) which gives the pink hue to the snow where the alga grows.

✓ CHECK YOUR UNDERSTANDING

- Which photosynthetic pigments were described in this section?
- What is the fundamental stress of a hypersaline environment for a cell?

9.6 Media Used for Bacterial Growth

Learning Objectives

By the end of this section, you will be able to:
- Identify and describe culture media for the growth of bacteria, including examples of all-purpose media, enriched, selective, differential, defined, and enrichment media

The study of microorganisms is greatly facilitated if we are able to culture them, that is, to keep reproducing populations alive under laboratory conditions. Culturing many microorganisms is challenging because of highly specific nutritional and environmental requirements and the diversity of these requirements among different species.

Nutritional Requirements

The number of available media to grow bacteria is considerable. Some media are considered general all-purpose media and support growth of a large variety of organisms. A prime example of an all-purpose medium is tryptic soy broth (TSB). Specialized media are used in the identification of bacteria and are supplemented with dyes, pH indicators, or antibiotics. One type, **enriched media**, contains growth factors, vitamins, and other essential nutrients to promote the growth of **fastidious organisms**, organisms that cannot make certain nutrients and require them to be added to the medium. When the complete chemical composition of a medium is known, it is called a **chemically defined medium**. For example, in EZ medium, all individual chemical components are identified and the exact amounts of each is known. In **complex media**, which contain extracts and digests of yeasts, meat, or plants, the precise chemical composition of the medium is not known. Amounts of individual components are undetermined and variable. Nutrient broth, tryptic soy broth, and brain heart infusion, are all examples of complex media.

Media that inhibit the growth of unwanted microorganisms and support the growth of the organism of interest by supplying nutrients and reducing competition are called **selective media**. An example of a selective medium is MacConkey agar. It contains bile salts and crystal violet, which interfere with the growth of many gram-positive bacteria and favor the growth of gram-negative bacteria, particularly the Enterobacteriaceae. These species are commonly named enterics, reside in the intestine, and are adapted to the presence of bile salts. The **enrichment culture**s foster the preferential growth of a desired microorganism that represents a fraction of the organisms present in an inoculum. For example, if we want to isolate bacteria that break down crude oil, hydrocarbonoclastic bacteria, sequential subculturing in a medium that supplies carbon only in the form of crude oil will enrich the cultures with oil-eating bacteria. The **differential media** make it easy to distinguish colonies of different bacteria by a change in the color of the colonies or the color of the medium. Color changes are the result of end products created by interaction of bacterial enzymes with differential substrates in the medium or, in the case of hemolytic reactions, the lysis of red blood cells in the medium. In Figure 9.32, the differential fermentation of lactose can be observed on MacConkey agar. The lactose fermenters produce acid, which turns the medium and the colonies of strong fermenters hot pink. The medium is supplemented with the pH indicator neutral red, which turns to hot pink at low pH. Selective and differential media can be combined and play an important role in the identification of bacteria by biochemical methods.

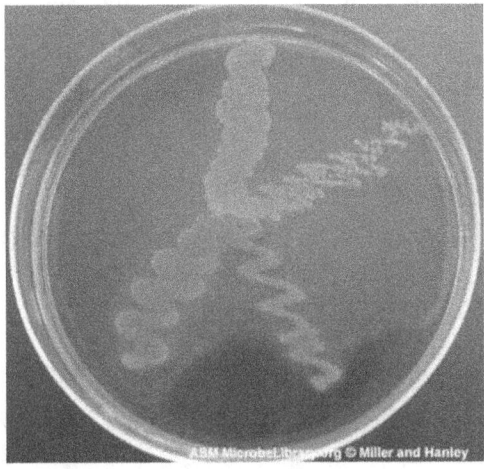

Figure 9.32 On this MacConkey agar plate, the lactose-fermenter *E. coli* colonies are bright pink. *Serratia marcescens*, which does not ferment lactose, forms a cream-colored streak on the tan medium. (credit: American Society for Microbiology)

CHECK YOUR UNDERSTANDING

- Distinguish complex and chemically defined media.
- Distinguish selective and enrichment media.

LINK TO LEARNING

Compare the compositions of EZ medium (https://openstax.org/l/22EZMedium) and sheep blood (https://openstax.org/l/22bloodagar) agar.

Case in Point

The End-of-Year Picnic

The microbiology department is celebrating the end of the school year in May by holding its traditional picnic on the green. The speeches drag on for a couple of hours, but finally all the faculty and students can dig into the food: chicken salad, tomatoes, onions, salad, and custard pie. By evening, the whole department, except for two vegetarian students who did not eat the chicken salad, is stricken with nausea, vomiting, retching, and abdominal cramping. Several individuals complain of diarrhea. One patient shows signs of shock (low blood pressure). Blood and stool samples are collected from patients, and an analysis of all foods served at the meal is conducted.

Bacteria can cause gastroenteritis (inflammation of the stomach and intestinal tract) either by colonizing and replicating in the host, which is considered an infection, or by secreting toxins, which is considered intoxication. Signs and symptoms of infections are typically delayed, whereas intoxication manifests within hours, as happened after the picnic.

Blood samples from the patients showed no signs of bacterial infection, which further suggests that this was a case of intoxication. Since intoxication is due to secreted toxins, bacteria are not usually detected in blood or stool samples. MacConkey agar and sorbitol-MacConkey agar plates and xylose-lysine-deoxycholate (XLD) plates were inoculated with stool samples and did not reveal any unusually colored colonies, and no black colonies or white colonies were observed on XLD. All lactose fermenters on MacConkey agar also ferment sorbitol. These results ruled out common agents of food-borne illnesses: *E. coli*, *Salmonella* spp., and *Shigella* spp.

Analysis of the chicken salad revealed an abnormal number of gram-positive cocci arranged in clusters (Figure 9.33). A culture of the gram-positive cocci releases bubbles when mixed with hydrogen peroxide. The culture turned mannitol salt agar yellow after a 24-hour incubation.

All the tests point to *Staphylococcus aureus* as the organism that secreted the toxin. Samples from the salad showed the presence of gram-positive cocci bacteria in clusters. The colonies were positive for catalase. The bacteria grew on mannitol salt agar fermenting mannitol, as shown by the change to yellow of the medium. The pH indicator in mannitol salt agar is phenol red, which turns to yellow when the medium is acidified by the products of fermentation.

The toxin secreted by *S. aureus* is known to cause severe gastroenteritis. The organism was probably introduced into the salad during preparation by the food handler and multiplied while the salad was kept in the warm ambient temperature during the speeches.

- What are some other factors that might have contributed to rapid growth of *S. aureus* in the chicken salad?
- Why would *S. aureus* not be inhibited by the presence of salt in the chicken salad?

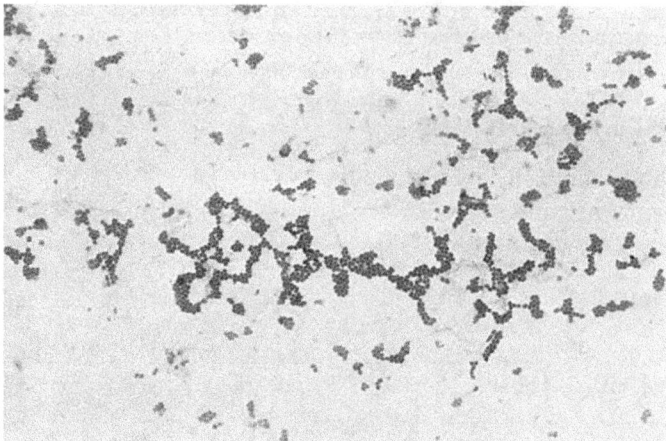

Figure 9.33 Gram-positive cocci in clusters. (credit: Centers for Disease Control and Prevention)

SUMMARY

9.1 How Microbes Grow

- Most bacterial cells divide by **binary fission**. **Generation time** in bacterial growth is defined as the **doubling time** of the population.
- Cells in a closed system follow a pattern of growth with four phases: **lag**, **logarithmic (exponential)**, **stationary**, and **death**.
- Cells can be counted by **direct viable cell count**. The **pour plate** and **spread plate** methods are used to plate **serial dilutions** into or onto, respectively, agar to allow counting of viable cells that give rise to **colony-forming units**. **Membrane filtration** is used to count live cells in dilute solutions. The **most probable cell number (MPN)** method allows estimation of cell numbers in cultures without using solid media.
- Indirect methods can be used to estimate **culture density** by measuring **turbidity** of a culture or live cell density by measuring metabolic activity.
- Other patterns of cell division include multiple nucleoid formation in cells; asymmetric division, as in **budding**; and the formation of hyphae and terminal spores.
- **Biofilms** are communities of microorganisms enmeshed in a matrix of **extracellular polymeric substance**. The formation of a biofilm occurs when **planktonic** cells attach to a substrate and become **sessile**. Cells in biofilms coordinate their activity by communicating through **quorum sensing**.
- Biofilms are commonly found on surfaces in nature and in the human body, where they may be beneficial or cause severe infections. Pathogens associated with biofilms are often more resistant to antibiotics and disinfectants.

9.2 Oxygen Requirements for Microbial Growth

- Aerobic and anaerobic environments can be found in diverse niches throughout nature, including different sites within and on the human body.
- Microorganisms vary in their requirements for molecular oxygen. **Obligate aerobes** depend on aerobic respiration and use oxygen as a terminal electron acceptor. They cannot grow without oxygen.
- **Obligate anaerobes** cannot grow in the presence of oxygen. They depend on fermentation and anaerobic respiration using a final electron acceptor other than oxygen.
- **Facultative anaerobes** show better growth in the presence of oxygen but will also grow without it.
- Although **aerotolerant anaerobes** do not perform aerobic respiration, they can grow in the presence of oxygen. Most aerotolerant anaerobes test negative for the enzyme **catalase**.
- **Microaerophiles** need oxygen to grow, albeit at a lower concentration than 21% oxygen in air.
- **Optimum oxygen concentration** for an organism is the oxygen level that promotes the fastest growth rate. The **minimum permissive oxygen concentration** and the **maximum permissive oxygen concentration** are, respectively, the lowest and the highest oxygen levels that the organism will tolerate.
- **Peroxidase**, **superoxide dismutase**, and **catalase** are the main enzymes involved in the detoxification of the **reactive oxygen species**. Superoxide dismutase is usually present in a cell that can tolerate oxygen. All three enzymes are usually detectable in cells that perform aerobic respiration and produce more ROS.
- A **capnophile** is an organism that requires a higher than atmospheric concentration of CO_2 to grow.

9.3 The Effects of pH on Microbial Growth

- Bacteria are generally **neutrophiles**. They grow best at neutral pH close to 7.0.
- **Acidophiles** grow optimally at a pH near 3.0. **Alkaliphiles** are organisms that grow optimally between a pH of 8 and 10.5. Extreme acidophiles and alkaliphiles grow slowly or not at all near neutral pH.
- Microorganisms grow best at their **optimum growth pH**. Growth occurs slowly or not at all below the **minimum growth pH** and above the **maximum growth pH**.

9.4 Temperature and Microbial Growth

- Microorganisms thrive at a wide range of temperatures; they have colonized different natural environments and have adapted to extreme temperatures. Both extreme cold and hot temperatures require evolutionary adjustments to macromolecules and biological

processes.
- **Psychrophiles** grow best in the temperature range of 0–15 °C whereas **psychrotrophs** thrive between 4°C and 25 °C.
- **Mesophiles** grow best at moderate temperatures in the range of 20 °C to about 45 °C. Pathogens are usually mesophiles.
- **Thermophiles** and **hyperthemophiles** are adapted to life at temperatures above 50 °C.
- Adaptations to cold and hot temperatures require changes in the composition of membrane lipids and proteins.

9.5 Other Environmental Conditions that Affect Growth

- **Halophiles** require high salt concentration in the medium, whereas **halotolerant** organisms can grow and multiply in the presence of high salt but do not require it for growth.
- Halotolerant pathogens are an important source of foodborne illnesses because they contaminate foods preserved in salt.
- Photosynthetic bacteria depend on visible light for energy.
- Most bacteria, with few exceptions, require high moisture to grow.

9.6 Media Used for Bacterial Growth

- **Chemically defined media** contain only chemically known components.
- **Selective media** favor the growth of some microorganisms while inhibiting others.
- **Enriched media** contain added essential nutrients a specific organism needs to grow
- **Differential media** help distinguish bacteria by the color of the colonies or the change in the medium.

REVIEW QUESTIONS
Multiple Choice

1. Which of the following methods would be used to measure the concentration of bacterial contamination in processed peanut butter?
   A. turbidity measurement
   B. total plate count
   C. dry weight measurement
   D. direct counting of bacteria on a calibrated slide under the microscope

2. In which phase would you expect to observe the most endospores in a *Bacillus* cell culture?
   A. death phase
   B. lag phase
   C. log phase
   D. log, lag, and death phases would all have roughly the same number of endospores.

3. During which phase would penicillin, an antibiotic that inhibits cell-wall synthesis, be most effective?
   A. death phase
   B. lag phase
   C. log phase
   D. stationary phase

4. Which of the following is the best definition of generation time in a bacterium?
   A. the length of time it takes to reach the log phase
   B. the length of time it takes for a population of cells to double
   C. the time it takes to reach stationary phase
   D. the length of time of the exponential phase

5. What is the function of the Z ring in binary fission?
   A. It controls the replication of DNA.
   B. It forms a contractile ring at the septum.
   C. It separates the newly synthesized DNA molecules.
   D. It mediates the addition of new peptidoglycan subunits.

6. If a culture starts with 50 cells, how many cells will be present after five generations with no cell death?
   A. 200
   B. 400
   C. 1600
   D. 3200

7. Filamentous cyanobacteria often divide by which of the following?
   A. budding
   B. mitosis
   C. fragmentation
   D. formation of endospores

8. Which is a reason for antimicrobial resistance being higher in a biofilm than in free-floating bacterial cells?
   A. The EPS allows faster diffusion of chemicals in the biofilm.
   B. Cells are more metabolically active at the base of a biofilm.
   C. Cells are metabolically inactive at the base of a biofilm.
   D. The structure of a biofilm favors the survival of antibiotic resistant cells.

9. Quorum sensing is used by bacterial cells to determine which of the following?
   A. the size of the population
   B. the availability of nutrients
   C. the speed of water flow
   D. the density of the population

10. Which of the following statements about autoinducers is incorrect?
    A. They bind directly to DNA to activate transcription.
    B. They can activate the cell that secreted them.
    C. N-acylated homoserine lactones are autoinducers in gram-negative cells.
    D. Autoinducers may stimulate the production of virulence factors.

11. An inoculated thioglycolate medium culture tube shows dense growth at the surface and turbidity throughout the rest of the tube. What is your conclusion?
    A. The organisms die in the presence of oxygen
    B. The organisms are facultative anaerobes.
    C. The organisms should be grown in an anaerobic chamber.
    D. The organisms are obligate aerobes.

12. An inoculated thioglycolate medium culture tube is clear throughout the tube except for dense growth at the bottom of the tube. What is your conclusion?
    A. The organisms are obligate anaerobes.
    B. The organisms are facultative anaerobes.
    C. The organisms are aerotolerant.
    D. The organisms are obligate aerobes.

13. *Pseudomonas aeruginosa* is a common pathogen that infects the airways of patients with cystic fibrosis. It does not grow in the absence of oxygen. The bacterium is probably which of the following?
    A. an aerotolerant anaerobe
    B. an obligate aerobe
    C. an obligate anaerobe
    D. a facultative anaerobe

14. *Streptococcus mutans* is a major cause of cavities. It resides in the gum pockets, does not have catalase activity, and can be grown outside of an anaerobic chamber. The bacterium is probably which of the following?
    A. a facultative anaerobe
    B. an obligate aerobe
    C. an obligate anaerobe
    D. an aerotolerant anaerobe

15. Why do the instructions for the growth of *Neisseria gonorrhoeae* recommend a CO_2-enriched atmosphere?
    A. It uses CO_2 as a final electron acceptor in respiration.
    B. It is an obligate anaerobe.
    C. It is a capnophile.
    D. It fixes CO_2 through photosynthesis.

16. Bacteria that grow in mine drainage at pH 1–2 are probably which of the following?
    A. alkaliphiles
    B. acidophiles
    C. neutrophiles
    D. obligate anaerobes

17. Bacteria isolated from Lake Natron, where the water pH is close to 10, are which of the following?
    A. alkaliphiles
    B. facultative anaerobes
    C. neutrophiles
    D. obligate anaerobes

18. In which environment are you most likely to encounter an acidophile?
    A. human blood at pH 7.2
    B. a hot vent at pH 1.5
    C. human intestine at pH 8.5
    D. milk at pH 6.5

19. A soup container was forgotten in the refrigerator and shows contamination. The contaminants are probably which of the following?
    A. thermophiles
    B. acidophiles
    C. mesophiles
    D. psychrotrophs

20. Bacteria isolated from a hot tub at 39 °C are probably which of the following?
    A. thermophiles
    B. psychrotrophs
    C. mesophiles
    D. hyperthermophiles

21. In which environment are you most likely to encounter a hyperthermophile?
    A. hot tub
    B. warm ocean water in Florida
    C. hydrothermal vent at the bottom of the ocean
    D. human body

22. Which of the following environments would harbor psychrophiles?
    A. mountain lake with a water temperature of 12 °C
    B. contaminated plates left in a 35 °C incubator
    C. yogurt cultured at room temperature
    D. salt pond in the desert with a daytime temperature of 34 °C

23. Which of the following is the reason jams and dried meats often do not require refrigeration to prevent spoilage?
    A. low pH
    B. toxic alkaline chemicals
    C. naturally occurring antibiotics
    D. low water activity

24. Bacteria living in salt marshes are most likely which of the following?
    A. acidophiles
    B. barophiles
    C. halotolerant
    D. thermophiles

25. EMB agar is a medium used in the identification and isolation of pathogenic bacteria. It contains digested meat proteins as a source of organic nutrients. Two indicator dyes, eosin and methylene blue, inhibit the growth of gram-positive bacteria and distinguish between lactose fermenting and nonlactose fermenting organisms. Lactose fermenters form metallic green or deep purple colonies, whereas the nonlactose fermenters form completely colorless colonies. EMB agar is an example of which of the following?
    A. a selective medium only
    B. a differential medium only
    C. a selective medium and a chemically defined medium
    D. a selective medium, a differential medium, and a complex medium

26. *Haemophilus influenzae* must be grown on chocolate agar, which is blood agar treated with heat to release growth factors in the medium. *H. influenzae* is described as _____.
    A. an acidophile
    B. a thermophile
    C. an obligate anaerobe
    D. fastidious

Matching

27. Match the definition with the name of the growth phase in the growth curve.

    ___Number of dying cells is higher than the number of cells dividing   A. Lag phase

    ___Number of new cells equal to number of dying cells   B. Log phase

    ___New enzymes to use available nutrients are induced   C. Stationary phase

    ___Binary fission is occurring at maximum rate   D. Death phase

29. Match the type of bacterium with its environment. Each choice may be used once, more than once, or not at all. Put the appropriate letter beside the environment.

    ___psychotroph   A. food spoiling in refrigerator

    ___mesophile   B. hydrothermal vent

    ___thermophile   C. deep ocean waters

    ___hyperthermophile   D. human pathogen

    ___psychrophile   E. garden compost

28. Four tubes are illustrated with cultures grown in a medium that slows oxygen diffusion. Match the culture tube with the correct type of bacteria from the following list: facultative anaerobe, obligate anaerobe, microaerophile, aerotolerant anaerobe, obligate aerobe.

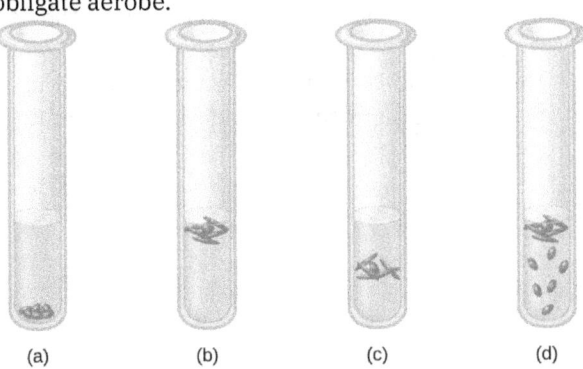

(a)   (b)   (c)   (d)

Fill in the Blank

30. Direct count of total cells can be performed using a _____ or a _____.
31. The _____ method allows direct count of total cells growing on solid medium.
32. A statistical estimate of the number of live cells in a liquid is usually done by _____.
33. For this indirect method of estimating the growth of a culture, you measure _____ using a spectrophotometer.
34. Active growth of a culture may be estimated indirectly by measuring the following products of cell metabolism: _____ or _____.
35. A bacterium that thrives in a soda lake where the average pH is 10.5 can be classified as a(n) _____.
36. *Lactobacillus acidophilus* grows best at pH 4.5. It is considered a(n) _____.
37. A bacterium that thrives in the Great Salt Lake but not in fresh water is probably a _____.
38. Bacteria isolated from the bottom of the ocean need high atmospheric pressures to survive. They are _____.
39. *Staphylococcus aureus* can be grown on multipurpose growth medium or on mannitol salt agar that contains 7.5% NaCl. The bacterium is _____.

40. Blood agar contains many unspecified nutrients, supports the growth of a large number of bacteria, and allows differentiation of bacteria according to hemolysis (breakdown of blood). The medium is _____ and _____.

41. Rogosa agar contains yeast extract. The pH is adjusted to 5.2 and discourages the growth of many microorganisms; however, all the colonies look similar. The medium is _____ and _____.

Short Answer

42. Why is it important to measure the transmission of light through a control tube with only broth in it when making turbidity measures of bacterial cultures?
43. In terms of counting cells, what does a plating method accomplish that an electronic cell counting method does not?
44. Order the following stages of the development of a biofilm from the earliest to the last step.
    1. secretion of EPS
    2. reversible attachment
    3. dispersal
    4. formation of water channels
    5. irreversible attachment
45. Infections among hospitalized patients are often related to the presence of a medical device in the patient. Which conditions favor the formation of biofilms on in-dwelling catheters and prostheses?
46. Why are some obligate anaerobes able to grow in tissues (e.g., gum pockets) that are not completely free of oxygen?
47. Why should *Haemophilus influenzae* be grown in a candle jar?
48. In terms of oxygen requirements, what type of organism would most likely be responsible for a foodborne illness associated with canned foods?
49. Which macromolecule in the cell is most sensitive to changes in pH?
50. Which metabolic process in the bacterial cell is particularly challenging at high pH?
51. How are hyperthermophile's proteins adapted to the high temperatures of their environment?
52. Why would NASA be funding microbiology research in Antarctica?
53. Fish sauce is a salty condiment produced using fermentation. What type of organism is likely responsible for the fermentation of the fish sauce?
54. What is the major difference between an enrichment culture and a selective culture?

Critical Thinking

55. A patient in the hospital has an intravenous catheter inserted to allow for the delivery of medications, fluids, and electrolytes. Four days after the catheter is inserted, the patient develops a fever and an infection in the skin around the catheter. Blood cultures reveal that the patient has a blood-borne infection. Tests in the clinical laboratory identify the blood-borne pathogen as *Staphylococcus epidermidis*, and antibiotic susceptibility tests are performed to provide doctors with essential information for selecting the best drug for treatment of the infection. Antibacterial chemotherapy is initiated and delivered through the intravenous catheter that was originally inserted into the patient. Within 7 days, the skin infection is gone, blood cultures are negative for *S. epidermidis*, and the antibacterial chemotherapy is discontinued. However, 2 days after discontinuing the antibacterial chemotherapy, the patient develops another fever and skin infection and the blood cultures are positive for the same strain of *S. epidermidis* that had been isolated the previous week. This time, doctors remove the intravenous catheter and administer oral antibiotics, which successfully treat both the skin and blood-borne infection caused by *S. epidermidis*. Furthermore, the infection does not return after discontinuing the oral antibacterial chemotherapy. What are some possible reasons why intravenous chemotherapy failed to completely cure the patient despite laboratory tests showing the bacterial strain was susceptible to the prescribed antibiotic? Why might the second round of antibiotic therapy have been more successful? Justify your answers.
56. Why are autoinducers small molecules?
57. Refer to Figure B1 in Appendix B. If the results from a pond water sample were recorded as 3, 2, 1, what would be the MPN of bacteria in 100 mL of pond water?
58. Refer to Figure 9.15. Why does turbidity lose reliability at high cell concentrations when the culture reaches the stationary phase?
59. A microbiology instructor prepares cultures for a gram-staining practical laboratory by inoculating growth medium with a gram-positive coccus (nonmotile) and a gram-negative rod (motile). The goal is to demonstrate staining of a mixed culture. The flask is incubated at 35 °C for 24 hours without aeration. A sample is stained and reveals only gram-negative rods. Both cultures are known facultative anaerobes. Give a likely reason for success of the gram-negative rod. Assume that the cultures have comparable intrinsic growth rates.
60. People who use proton pumps inhibitors or antacids are more prone to infections of the gastrointestinal tract. Can you explain the observation in light of what you have learned?
61. The bacterium that causes Hansen's disease (leprosy), *Mycobacterium leprae*, infects mostly the extremities of the body: hands, feet, and nose. Can you make an educated guess as to its optimum temperature of growth?
62. Refer to Figure 9.29. Some hyperthermophiles can survive autoclaving temperatures. Are they a concern in health care?
63. *Haemophilus, influenzae* grows best at 35–37 °C with ~5% CO_2 (or in a candle-jar) and requires hemin (X factor) and nicotinamide-adenine-dinucleotide (NAD, also known as V factor) for growth. (Centers for Disease Control and Prevention, World Health Organization. "*CDC Laboratory Methods for the Diagnosis of Meningitis Caused by* Neisseria meningitidis, Streptococcus pneumoniae, *and* Haemophilus influenza. WHO Manual, 2nd edition." 2011. http://www.cdc.gov/meningitis/lab-manual/full-manual.pdf) Using the vocabulary learned in this chapter, describe *H. influenzae*.

CHAPTER 10
Biochemistry of the Genome

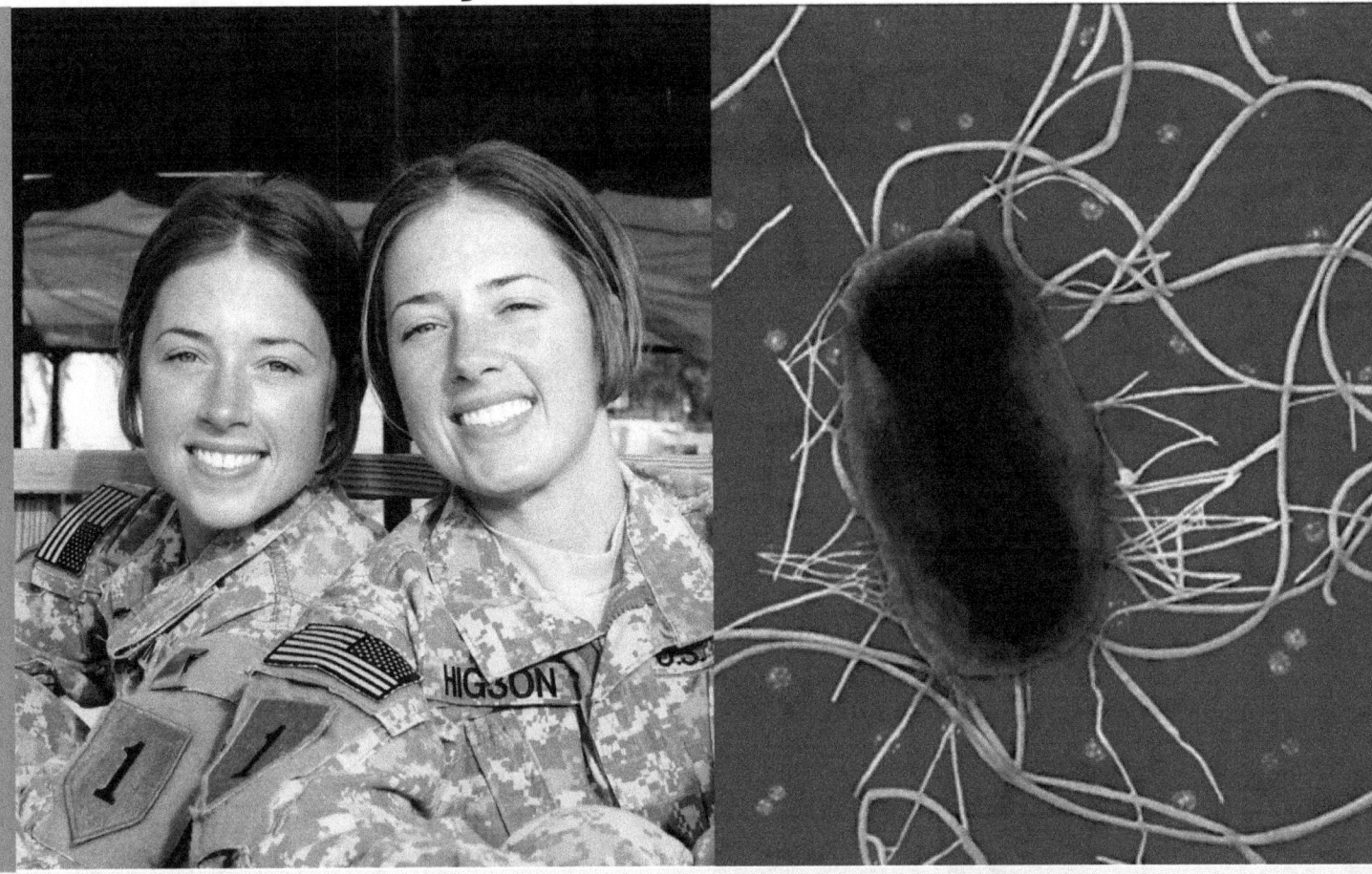

Figure 10.1 Siblings within a family share some genes with each other and with each parent. Identical twins, however, are genetically identical. Bacteria like *Escherichia coli* may acquire genes encoding virulence factors, converting them into pathogenic strains, like this uropathogenic *E. coli*. (Credit left: Army twins, Myrtle Beach, S.C. / U.S. Army; Public Domain; credit right: modification of work by American Society for Microbiology)

Chapter Outline

10.1 Using Microbiology to Discover the Secrets of Life

10.2 Structure and Function of DNA

10.3 Structure and Function of RNA

10.4 Structure and Function of Cellular Genomes

INTRODUCTION Children inherit some characteristics from each parent. Siblings typically look similar to each other, but not exactly the same—except in the case of identical twins. How can we explain these phenomena? The answers lie in heredity (the transmission of traits from one generation to the next) and genetics (the science of heredity). Because humans reproduce sexually, 50% of a child's genes come from the mother's egg cell and the remaining 50% from the father's sperm cell. Sperm and egg are formed through the process of meiosis, where DNA recombination occurs. Thus, there is no predictable pattern as to which 50% comes from which parent. Thus, siblings have only some genes, and their associated characteristics, in common. Identical twins are the exception, because they are genetically identical.

Genetic differences among related microbes also dictate many observed biochemical and virulence differences. For example, some strains of the bacterium *Escherichia coli* are harmless members of the normal microbiota in the human gastrointestinal tract. Other strains of the same species have genes that give them the ability to cause disease. In bacteria, such genes are not inherited via sexual reproduction, as in humans. Often, they are transferred via plasmids, small circular pieces of double-stranded DNA that can be exchanged between prokaryotes.

10.1 Using Microbiology to Discover the Secrets of Life

Learning Objectives

By the end of this section, you will be able to:
- Describe the discovery of nucleic acid and nucleotides
- Explain the historical experiments that led to the characterization of DNA
- Describe how microbiology and microorganisms have been used to discover the biochemistry of genes
- Explain how scientists established the link between DNA and heredity

---

Clinical Focus

Part 1

Alex is a 22-year-old college student who vacationed in Puerta Vallarta, Mexico, for spring break. Unfortunately, two days after flying home to Ohio, he began to experience abdominal cramping and extensive watery diarrhea. Because of his discomfort, he sought medical attention at a large Cincinnati hospital nearby.

- What types of infections or other conditions may be responsible?

Jump to the next Clinical Focus box.

Through the early 20th century, DNA was not yet recognized as the genetic material responsible for heredity, the passage of traits from one generation to the next. In fact, much of the research was dismissed until the mid-20th century. The scientific community believed, incorrectly, that the process of inheritance involved a blending of parental traits that produced an intermediate physical appearance in offspring; this hypothetical process appeared to be correct because of what we know now as continuous variation, which results from the action of many genes to determine a particular characteristic, like human height. Offspring appear to be a "blend" of their parents' traits when we look at characteristics that exhibit continuous variation. The blending theory of inheritance asserted that the original parental traits were lost or absorbed by the blending in the offspring, but we now know that this is not the case.

Two separate lines of research, begun in the mid to late 1800s, ultimately led to the discovery and characterization of DNA and the foundations of genetics, the science of heredity. These lines of research began to converge in the 1920s, and research using microbial systems ultimately resulted in significant contributions to elucidating the molecular basis of genetics.

Discovery and Characterization of DNA

Modern understanding of DNA has evolved from the discovery of nucleic acid to the development of the double-helix model. In the 1860s, Friedrich Miescher (1844–1895), a physician by profession, was the first person to isolate phosphorus-rich chemicals from leukocytes (white blood cells) from the pus on used bandages from a local surgical clinic. He named these chemicals (which would eventually be known as RNA and DNA) "nuclein" because they were isolated from the nuclei of the cells. His student Richard Altmann (1852–1900) subsequently termed it "nucleic acid" 20 years later when he discovered the acidic nature of nuclein. In the last two decades of the 19th century, German biochemist Albrecht Kossel (1853–1927) isolated and characterized the five different nucleotide bases composing nucleic acid. These are adenine, guanine, cytosine, thymine (in DNA), and uracil (in RNA). Kossell received the Nobel Prize in Physiology or Medicine in 1910 for his work on nucleic acids and for his considerable work on proteins, including the discovery of histidine.

Foundations of Genetics

Despite the discovery of DNA in the late 1800s, scientists did not make the association with heredity for many more decades. To make this connection, scientists, including a number of microbiologists, performed many experiments on plants, animals, and bacteria.

Mendel's Pea Plants

While Miescher was isolating and discovering DNA in the 1860s, Austrian monk and botanist Johann Gregor Mendel (1822–1884) was experimenting with garden peas, demonstrating and documenting basic patterns of inheritance, now known as Mendel's laws.

In 1856, Mendel began his decade-long research into inheritance patterns. He used the diploid garden pea, *Pisum sativum*, as his primary model system because it naturally self-fertilizes and is highly inbred, producing "true-breeding" pea plant lines—plants that always produce offspring that look like the parent. By experimenting with true-breeding pea plants, Mendel avoided the appearance of unexpected traits in offspring that might occur if he used plants that were not true-breeding. Mendel performed hybridizations, which involve mating two true-breeding individuals (P generation) that have different traits, and examined the characteristics of their offspring (first filial generation, F_1) as well as the offspring of self-fertilization of the F_1 generation (second filial generation, F_2) (Figure 10.2).

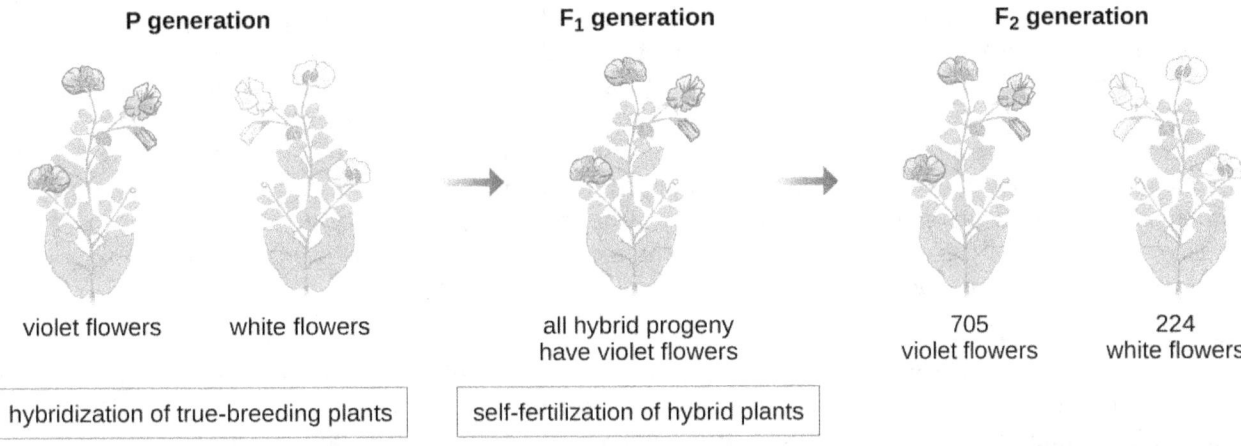

Figure 10.2 In one of his experiments on inheritance patterns, Mendel crossed plants that were true-breeding for violet flower color with plants true-breeding for white flower color (the P generation). The resulting hybrids in the F_1 generation all had violet flowers. In the F_2 generation, approximately three-quarters of the plants had violet flowers, and one-quarter had white flowers.

In 1865, Mendel presented the results of his experiments with nearly 30,000 pea plants to the local natural history society. He demonstrated that traits are transmitted faithfully from parents to offspring independently of other traits. In 1866, he published his work, "Experiments in Plant Hybridization,"[1] in the *Proceedings of the Natural History Society of Brünn*. Mendel's work went virtually unnoticed by the scientific community, which believed, incorrectly, in the theory of blending of traits in continuous variation.

He was not recognized for his extraordinary scientific contributions during his lifetime. In fact, it was not until 1900 that his work was rediscovered, reproduced, and revitalized by scientists on the brink of discovering the chromosomal basis of heredity.

---

1 J.G. Mendel. "Versuche über Pflanzenhybriden." *Verhandlungen des naturforschenden Vereines in Brünn, Bd. Abhandlungen* 4 (1865):3–7. (For English translation, see http://www.mendelweb.org/Mendel.plain.html)

The Chromosomal Theory of Inheritance

Mendel carried out his experiments long before chromosomes were visualized under a microscope. However, with the improvement of microscopic techniques during the late 1800s, cell biologists could stain and visualize subcellular structures with dyes and observe their actions during meiosis. They were able to observe chromosomes replicating, condensing from an amorphous nuclear mass into distinct X-shaped bodies and migrating to separate cellular poles. The speculation that chromosomes might be the key to understanding heredity led several scientists to examine Mendel's publications and re-evaluate his model in terms of the behavior of chromosomes during mitosis and meiosis.

In 1902, Theodor Boveri (1862–1915) observed that in sea urchins, nuclear components (chromosomes) determined proper embryonic development. That same year, Walter Sutton (1877–1916) observed the separation of chromosomes into daughter cells during meiosis. Together, these observations led to the development of the Chromosomal Theory of Inheritance, which identified chromosomes as the genetic material responsible for Mendelian inheritance.

Despite compelling correlations between the behavior of chromosomes during meiosis and Mendel's observations, the Chromosomal Theory of Inheritance was proposed long before there was any direct evidence that traits were carried on chromosomes. Thomas Hunt Morgan (1866–1945) and his colleagues spent several years carrying out crosses with the fruit fly, *Drosophila melanogaster*. They performed meticulous microscopic observations of fly chromosomes and correlated these observations with resulting fly characteristics. Their work provided the first experimental evidence to support the Chromosomal Theory of Inheritance in the early 1900s. In 1915, Morgan and his "Fly Room" colleagues published *The Mechanism of Mendelian Heredity*, which identified chromosomes as the cellular structures responsible for heredity. For his many significant contributions to genetics, Morgan received the Nobel Prize in Physiology or Medicine in 1933.

In the late 1920s, Barbara McClintock (1902–1992) developed chromosomal staining techniques to visualize and differentiate between the different chromosomes of maize (corn). In the 1940s and 1950s, she identified a breakage event on chromosome 9, which she named the dissociation locus (*Ds*). *Ds* could change position within the chromosome. She also identified an activator locus (*Ac*). *Ds* chromosome breakage could be activated by an *Ac* element (transposase enzyme). At first, McClintock's finding of these jumping genes, which we now call transposons, was not accepted by the scientific community. It wasn't until the 1960s and later that transposons were discovered in bacteriophages, bacteria, and *Drosophila*. Today, we know that transposons are mobile segments of DNA that can move within the genome of an organism. They can regulate gene expression, protein expression, and virulence (ability to cause disease).

Microbes and Viruses in Genetic Research

Microbiologists have also played a crucial part in our understanding of genetics. Experimental organisms such as Mendel's garden peas, Morgan's fruit flies, and McClintock's corn had already been used successfully to pave the way for an understanding of genetics. However, microbes and viruses were (and still are) excellent model systems for the study of genetics because, unlike peas, fruit flies, and corn, they are propagated more easily in the laboratory, growing to high population densities in a small amount of space and in a short time. In addition, because of their structural simplicity, microbes and viruses are more readily manipulated genetically.

Fortunately, despite significant differences in size, structure, reproduction strategies, and other biological characteristics, there is biochemical unity among all organisms; they have in common the same underlying molecules responsible for heredity and the use of genetic material to give cells their varying characteristics. In the words of French scientist Jacques Monod, "What is true for *E. coli* is also true for the elephant," meaning that the biochemistry of life has been maintained throughout evolution and is shared in all forms of life, from simple unicellular organisms to large, complex organisms. This biochemical continuity makes microbes excellent models to use for genetic studies.

In a clever set of experiments in the 1930s and 1940s, German scientist Joachim Hämmerling (1901–1980), using the single-celled alga *Acetabularia* as a microbial model, established that the genetic information in a eukaryotic cell is housed within the nucleus. *Acetabularia* spp. are unusually large algal cells that grow asymmetrically, forming a "foot" containing the nucleus, which is used for substrate attachment; a stalk; and an umbrella-like cap—structures that can all be easily seen with the naked eye. In an early set of experiments, Hämmerling removed either the cap or the foot of the cells and observed whether new caps or feet were regenerated (Figure 10.3). He found that when the foot of these cells was removed, new feet did not grow; however, when caps were removed from the cells, new caps were regenerated. This suggested that the hereditary information was located in the nucleus-containing foot of each cell.

Acetabularia

(a)

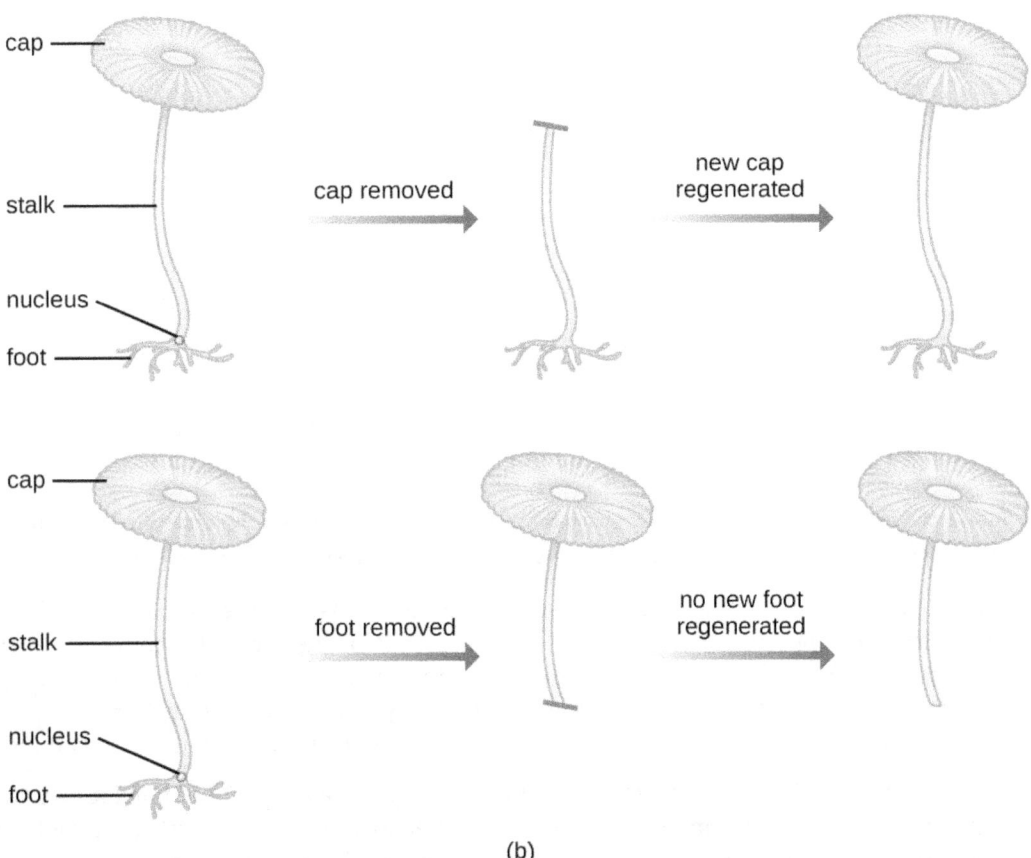

(b)

Figure 10.3 (a) The cells of the single-celled alga *Acetabularia* measure 2–6 cm and have a cell morphology that can be observed with the naked eye. Each cell has a cap, a stalk, and a foot, which contains the nucleus. (b) Hämmerling found that if he removed the cap, a new cap would regenerate; but if he removed the foot, a new foot would not regenerate. He concluded that the genetic information needed for regeneration was found in the nucleus. (credit a: modification of work by James St. John)

In another set of experiments, Hämmerling used two species of *Acetabularia* that have different cap morphologies, *A. crenulata* and *A. mediterranea* (Figure 10.4). He cut the caps from both types of cells and then grafted the stalk from an *A. crenulata* onto an *A. mediterranea* foot, and vice versa. Over time, he observed that the grafted cell with the *A. crenulata* foot and *A. mediterranea* stalk developed a cap with the *A. crenulata* morphology. Conversely, the grafted cell with the *A. mediterranea* foot and *A. crenulata* stalk developed a cap with the *A. mediterranea* morphology. He microscopically confirmed the presence of nuclei in the feet of these cells and attributed the development of these cap morphologies to the nucleus of each grafted cell. Thus, he showed experimentally that the nucleus was the location of genetic material that dictated a cell's properties.

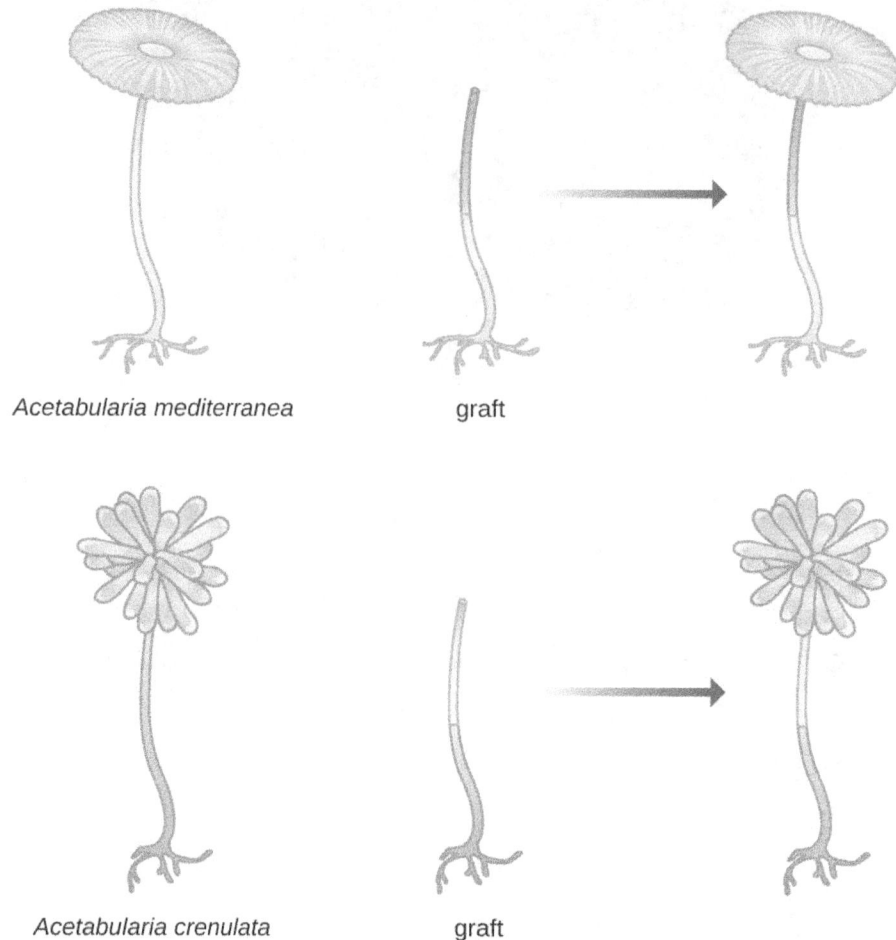

Figure 10.4 In a second set of experiments, Hämmerling used two morphologically different species and grafted stalks from each species to the feet of the other. He found that the properties of the regenerated caps were dictated by the species of the nucleus-containing foot.

Another microbial model, the red bread mold *Neurospora crassa*, was used by George Beadle and Edward Tatum to demonstrate the relationship between genes and the proteins they encode. Beadle had worked with fruit flies in Morgan's laboratory but found them too complex to perform certain types of experiments. *N. crassa*, on the other hand, is a simpler organism and has the ability to grow on a minimal medium because it contains enzymatic pathways that allow it to use the medium to produce its own vitamins and amino acids.

Beadle and Tatum irradiated the mold with X-rays to induce changes to a sequence of nucleic acids, called mutations. They mated the irradiated mold spores and attempted to grow them on both a complete medium and a minimal medium. They looked for mutants that grew on a complete medium, supplemented with vitamins and amino acids, but did not grow on the minimal medium lacking these supplements. Such molds theoretically contained mutations in the genes that encoded biosynthetic pathways. Upon finding such mutants, they systematically tested each to determine which vitamin or amino acid it was unable to produce (Figure 10.5) and published this work in 1941.[2]

---

2 G.W. Beadle, E.L. Tatum. "Genetic Control of Biochemical Reactions in *Neurospora*." Proceedings of the National Academy of

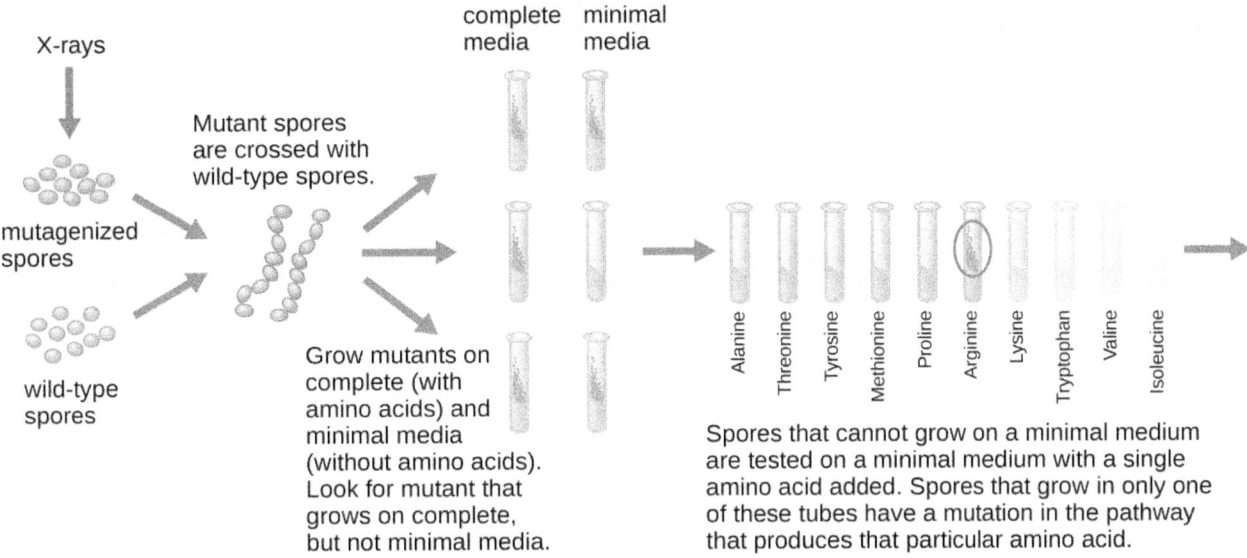

Figure 10.5 Beadle and Tatum's experiment involved the mating of irradiated and nonirradiated mold spores. These spores were grown on both complete medium and a minimal medium to determine which amino acid or vitamin the mutant was unable to produce on its own.

Subsequent work by Beadle, Tatum, and colleagues showed that they could isolate different classes of mutants that required a particular supplement, like the amino acid arginine (Figure 10.6). With some knowledge of the arginine biosynthesis pathway, they identified three classes of arginine mutants by supplementing the minimal medium with intermediates (citrulline or ornithine) in the pathway. The three mutants differed in their abilities to grow in each of the media, which led the group of scientists to propose, in 1945, that each type of mutant had a defect in a different gene in the arginine biosynthesis pathway. This led to the so-called one gene–one enzyme hypothesis, which suggested that each gene encodes one enzyme.

Subsequent knowledge about the processes of transcription and translation led scientists to revise this to the "one gene–one polypeptide" hypothesis. Although there are some genes that do not encode polypeptides (but rather encode for transfer RNAs [tRNAs] or ribosomal RNAs [rRNAs], which we will discuss later), the one gene–one enzyme hypothesis is true in many cases, especially in microbes. Beadle and Tatum's discovery of the link between genes and corresponding characteristics earned them the 1958 Nobel Prize in Physiology and Medicine and has since become the basis for modern molecular genetics.

Sciences 27 no. 11 (1941):499–506.

Beadle and Tatum Experiments				
Bread Mold	Minimal Medium (MM)	MM + Ornithine	MM + Citrulline	MM + Arginine
Wild type	grew	grew	grew	grew
Mutant 1	did not grow	grew	grew	grew
Mutant 2	did not grow	did not grow	grew	grew
Mutant 3	did not grow	did not grow	did not grow	grew

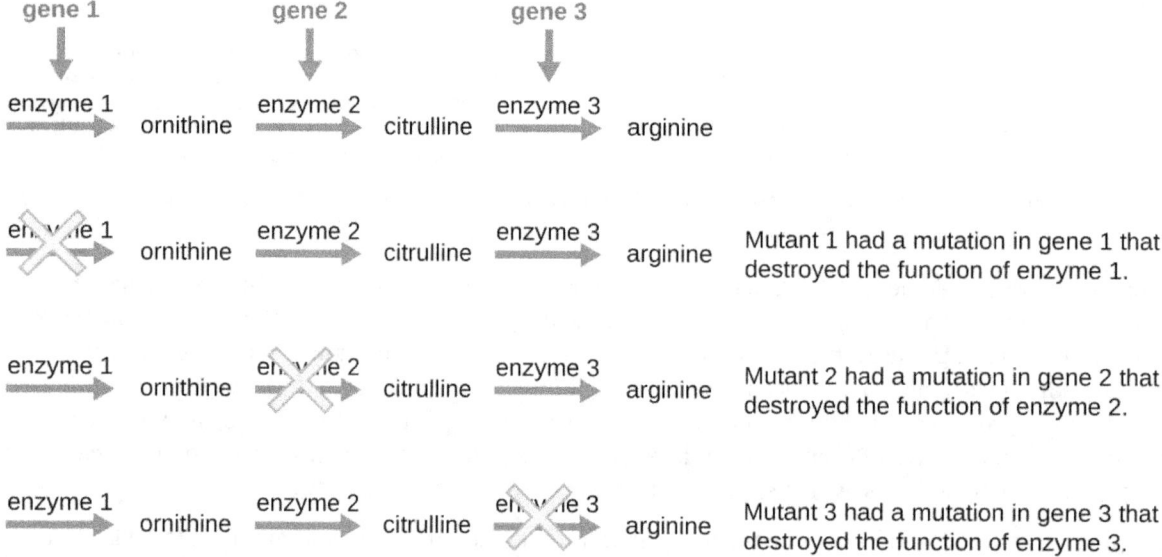

Figure 10.6 Three classes of arginine mutants were identified, each differing in their ability to grow in the presence of intermediates in the arginine biosynthesis pathway. From this, Beadle and Tatum concluded that each mutant was defective in a different gene encoding a different enzyme in the arginine biosynthesis pathway, leading to them to their one gene–one enzyme hypothesis.

LINK TO LEARNING

To learn more about the experiments of Beadle and Tatum, visit this website (https://www.openstax.org/l/22expbeatatum) from the DNA Learning Center.

CHECK YOUR UNDERSTANDING

- What organism did Morgan and his colleagues use to develop the Chromosomal Theory of Inheritance? What traits did they track?
- What did Hämmerling prove with his experiments on *Acetabularia*?

DNA as the Molecule Responsible for Heredity

By the beginning of the 20th century, a great deal of work had already been done on characterizing DNA and establishing the foundations of genetics, including attributing heredity to chromosomes found within the nucleus. Despite all of this research, it was not until well into the 20th century that these lines of research converged and scientists began to consider that DNA could be the genetic material that offspring inherited from their parents. DNA, containing only four different nucleotides, was thought to be structurally too simple to encode such complex genetic information. Instead, protein was thought to have the complexity required to serve as cellular genetic information because it is composed of 20 different amino acids that could be combined in a huge variety of combinations. Microbiologists played a pivotal role in the research that determined that DNA is the molecule responsible for heredity.

Griffith's Transformation Experiments

British bacteriologist Frederick Griffith (1879–1941) was perhaps the first person to show that hereditary information could be transferred from one cell to another "horizontally" (between members of the same generation), rather than "vertically" (from parent to offspring). In 1928, he reported the first demonstration of bacterial transformation, a process in which external DNA is taken up by a cell, thereby changing its characteristics.[3] He was working with two strains of *Streptococcus pneumoniae*, a bacterium that causes pneumonia: a rough (R) strain and a smooth (S) strain. The R strain is nonpathogenic and lacks a capsule on its outer surface; as a result, colonies from the R strain appear rough when grown on plates. The S strain is pathogenic and has a capsule outside its cell wall, allowing it to escape phagocytosis by the host immune system. The capsules cause colonies from the S strain to appear smooth when grown on plates.

In a series of experiments, Griffith analyzed the effects of live R, live S, and heat-killed S strains of *S. pneumoniae* on live mice (Figure 10.7). When mice were injected with the live S strain, the mice died. When he injected the mice with the live R strain or the heat-killed S strain, the mice survived. But when he injected the mice with a mixture of live R strain and heat-killed S strain, the mice died. Upon isolating the live bacteria from the dead mouse, he only recovered the S strain of bacteria. When he then injected this isolated S strain into fresh mice, the mice died. Griffith concluded that something had passed from the heat-killed S strain into the live R strain and "transformed" it into the pathogenic S strain; he called this the "transforming principle." These experiments are now famously known as Griffith's transformation experiments.

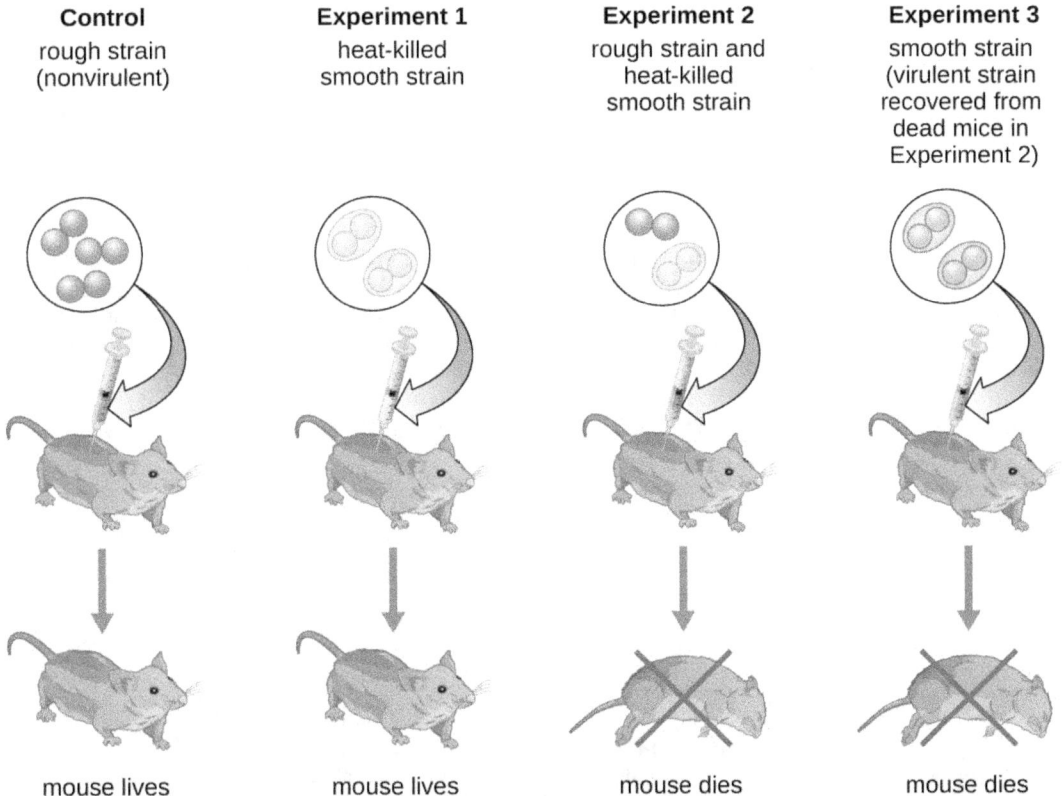

Figure 10.7 In his famous series of experiments, Griffith used two strains of *S. pneumoniae*. The S strain is pathogenic and causes death. Mice injected with the nonpathogenic R strain or the heat-killed S strain survive. However, a combination of the heat-killed S strain and the live R strain causes the mice to die. The S strain recovered from the dead mouse showed that something had passed from the heat-killed S strain to the R strain, transforming the R strain into an S strain in the process.

In 1944, Oswald Avery, Colin MacLeod, and Maclyn McCarty were interested in exploring Griffith's transforming principle further. They isolated the S strain from infected dead mice, heat-killed it, and

---

3 F. Griffith. "The Significance of Pneumococcal Types." *Journal of Hygiene* 27 no. 2 (1928):8–159.

inactivated various components of the S extract, conducting a systematic elimination study (Figure 10.8). They used enzymes that specifically degraded proteins, RNA, and DNA and mixed the S extract with each of these individual enzymes. Then, they tested each extract/enzyme combination's resulting ability to transform the R strain, as observed by the diffuse growth of the S strain in culture media and confirmed visually by growth on plates. They found that when DNA was degraded, the resulting mixture was no longer able to transform the R strain bacteria, whereas no other enzymatic treatment was able to prevent transformation. This led them to conclude that DNA was the transforming principle. Despite their results, many scientists did not accept their conclusion, instead believing that there were protein contaminants within their extracts.

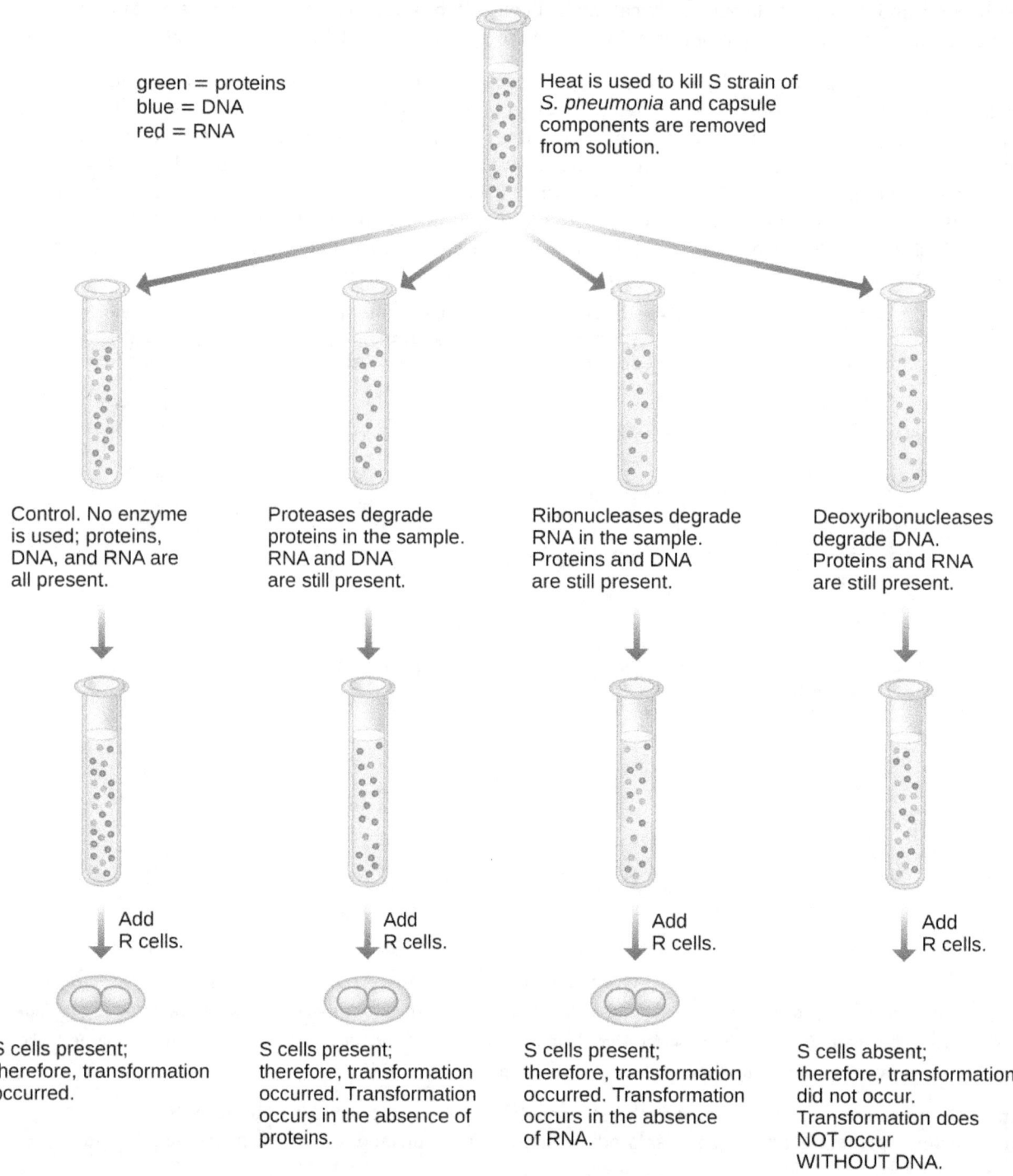

Figure 10.8 Oswald Avery, Colin MacLeod, and Maclyn McCarty followed up on Griffith's experiment and experimentally determined that

the transforming principle was DNA.

✓ CHECK YOUR UNDERSTANDING

- How did Avery, MacLeod, and McCarty's experiments show that DNA was the transforming principle first described by Griffith?

Hershey and Chase's Proof of DNA as Genetic Material

Alfred Hershey and Martha Chase performed their own experiments in 1952 and were able to provide confirmatory evidence that DNA, not protein, was the genetic material (Figure 10.9).[4] Hershey and Chase were studying a bacteriophage, a virus that infects bacteria. Viruses typically have a simple structure: a protein coat, called the capsid, and a nucleic acid core that contains the genetic material, either DNA or RNA (see Viruses). The particular bacteriophage they were studying was the T2 bacteriophage, which infects *E. coli* cells. As we now know today, T2 attaches to the surface of the bacterial cell and then it injects its nucleic acids inside the cell. The phage DNA makes multiple copies of itself using the host machinery, and eventually the host cell bursts, releasing a large number of bacteriophages.

Hershey and Chase labeled the protein coat in one batch of phage using radioactive sulfur, ^{35}S, because sulfur is found in the amino acids methionine and cysteine but not in nucleic acids. They labeled the DNA in another batch using radioactive phosphorus, ^{32}P, because phosphorus is found in DNA and RNA but not typically in protein.

Each batch of phage was allowed to infect the cells separately. After infection, Hershey and Chase put each phage bacterial suspension in a blender, which detached the phage coats from the host cell, and spun down the resulting suspension in a centrifuge. The heavier bacterial cells settled down and formed a pellet, whereas the lighter phage particles stayed in the supernatant. In the tube with the protein labeled, the radioactivity remained only in the supernatant. In the tube with the DNA labeled, the radioactivity was detected only in the bacterial cells. Hershey and Chase concluded that it was the phage DNA that was injected into the cell that carried the information to produce more phage particles, thus proving that DNA, not proteins, was the source of the genetic material. As a result of their work, the scientific community more broadly accepted DNA as the molecule responsible for heredity.

---

[4] A.D. Hershey, M. Chase. "Independent Functions of Viral Protein and Nucleic Acid in Growth of Bacteriophage." *Journal of General Physiology* 36 no. 1 (1952):39–56.

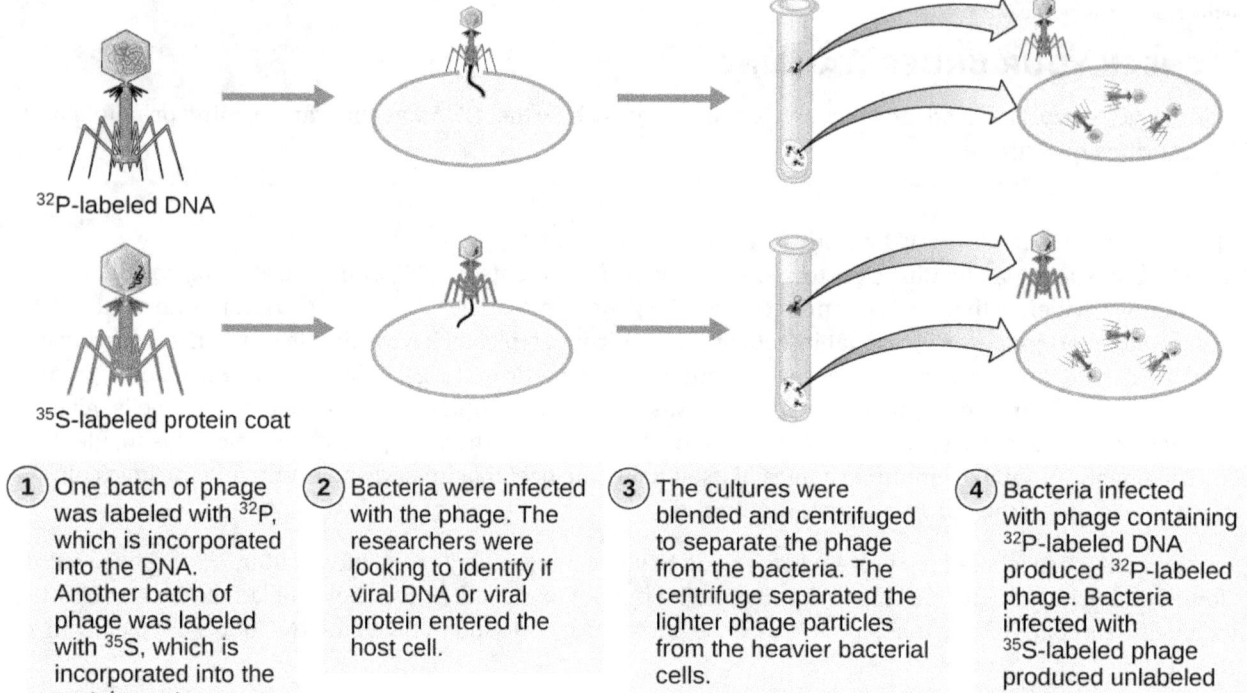

Figure 10.9 Martha Chase and Alfred Hershey conducted an experiment separately labeling the DNA and proteins of the T2 bacteriophage to determine which component was the genetic material responsible for the production of new phage particles.

By the time Hershey and Chase published their experiment in the early 1950s, microbiologists and other scientists had been researching heredity for over 80 years. Building on one another's research during that time culminated in the general agreement that DNA was the genetic material responsible for heredity (Figure 10.10). This knowledge set the stage for the age of molecular biology to come and the significant advancements in biotechnology and systems biology that we are experiencing today.

LINK TO LEARNING

To learn more about the experiments involved in the history of genetics and the discovery of DNA as the genetic material of cells, visit this website (https://www.openstax.org/l/22dnalearncen) from the DNA Learning Center.

CHECK YOUR UNDERSTANDING

- How did Hershey and Chase use microbes to prove that DNA is genetic material?

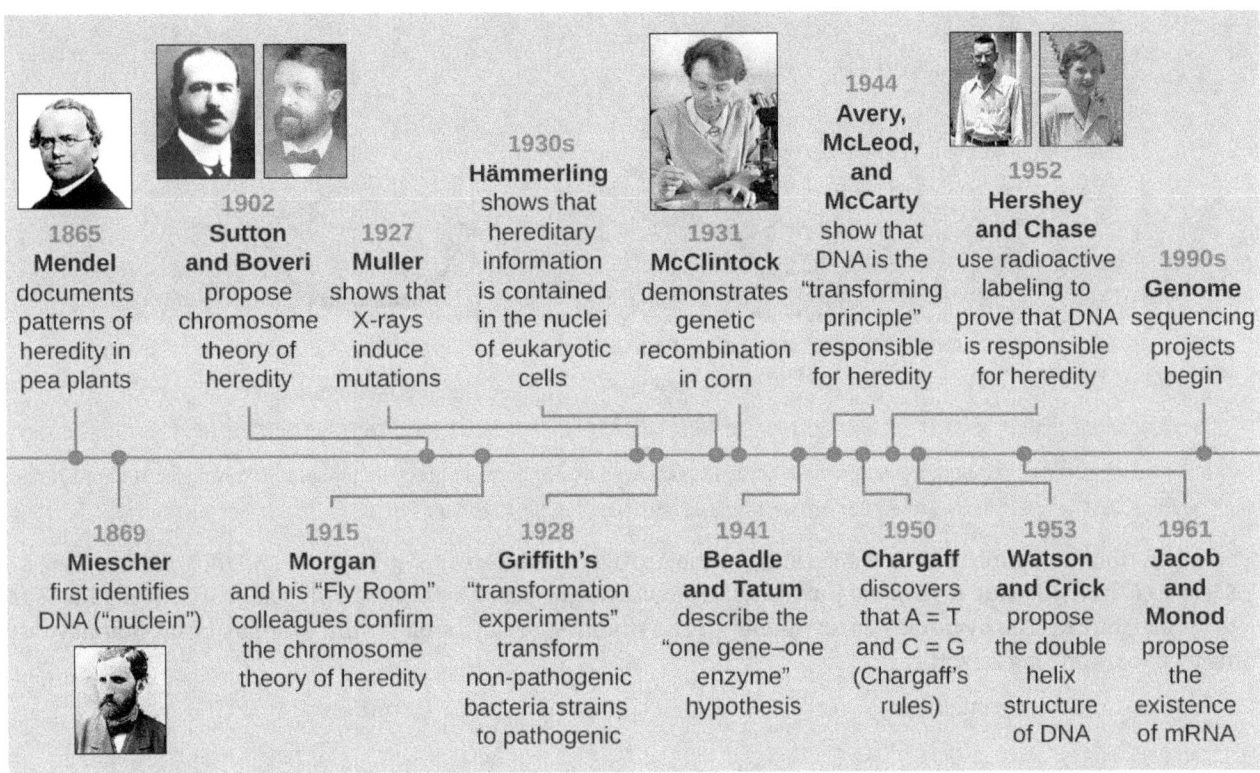

Figure 10.10 A timeline of key events leading up to the identification of DNA as the molecule responsible for heredity

10.2 Structure and Function of DNA

Learning Objectives

By the end of this section, you will be able to:
- Describe the biochemical structure of deoxyribonucleotides
- Identify the base pairs used in the synthesis of deoxyribonucleotides
- Explain why the double helix of DNA is described as antiparallel

In Microbial Metabolism, we discussed the microbial catabolism of three classes of macromolecules: proteins, lipids and carbohydrates. In this chapter, we will discuss the genetic role of a fourth class of molecules: nucleic acids. Like other macromolecules, **nucleic acid**s are composed of monomers, called **nucleotide**s, which are polymerized to form large strands. Each nucleic acid strand contains certain nucleotides that appear in a certain order within the strand, called its **base sequence**. The base sequence of **deoxyribonucleic acid (DNA)** is responsible for carrying and retaining the hereditary information in a cell. In Mechanisms of Microbial Genetics, we will discuss in detail the ways in which DNA uses its own base sequence to direct its own synthesis, as well as the synthesis of RNA and proteins, which, in turn, gives rise to products with diverse structure and function. In this section, we will discuss the basic structure and function of DNA.

DNA Nucleotides

The building blocks of nucleic acids are nucleotides. Nucleotides that compose DNA are called **deoxyribonucleotides**. The three components of a deoxyribonucleotide are a five-carbon sugar called deoxyribose, a phosphate group, and a **nitrogenous base**, a nitrogen-containing ring structure that is responsible for complementary base pairing between nucleic acid strands (Figure 10.11). The carbon atoms of the five-carbon deoxyribose are numbered 1′, 2′, 3′, 4′, and 5′ (1′ is read as "one prime"). A nucleoside comprises the five-carbon sugar and nitrogenous base.

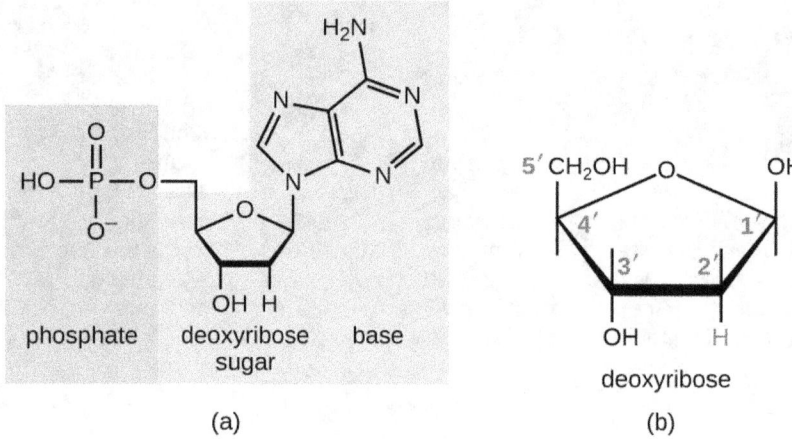

Figure 10.11 (a) Each deoxyribonucleotide is made up of a sugar called deoxyribose, a phosphate group, and a nitrogenous base—in this case, adenine. (b) The five carbons within deoxyribose are designated as 1′, 2′, 3′, 4′, and 5′.

The deoxyribonucleotide is named according to the nitrogenous bases (Figure 10.12). The nitrogenous bases **adenine** (A) and **guanine** (G) are the **purines**; they have a double-ring structure with a six-carbon ring fused to a five-carbon ring. The **pyrimidines**, **cytosine** (C) and **thymine** (T), are smaller nitrogenous bases that have only a six-carbon ring structure.

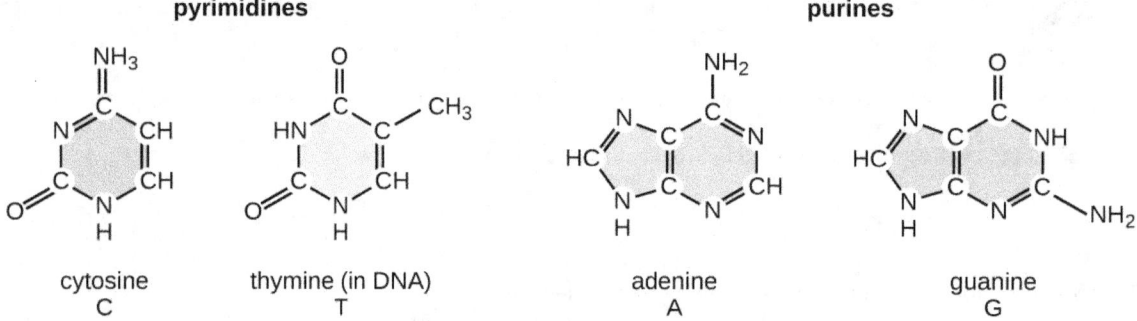

Figure 10.12 Nitrogenous bases within DNA are categorized into the two-ringed purines adenine and guanine and the single-ringed pyrimidines cytosine and thymine. Thymine is unique to DNA.

Individual nucleoside triphosphates combine with each other by covalent bonds known as 5′-3′ **phosphodiester bonds**, or linkages whereby the phosphate group attached to the 5′ carbon of the sugar of one nucleotide bonds to the hydroxyl group of the 3′ carbon of the sugar of the next nucleotide. Phosphodiester bonding between nucleotides forms the **sugar-phosphate backbone**, the alternating sugar-phosphate structure composing the framework of a nucleic acid strand (Figure 10.13). During the polymerization process, deoxynucleotide triphosphates (dNTP) are used. To construct the sugar-phosphate backbone, the two terminal phosphates are released from the dNTP as a pyrophosphate. The resulting strand of nucleic acid has a free phosphate group at the 5′ carbon end and a free hydroxyl group at the 3′ carbon end. The two unused phosphate groups from the nucleotide triphosphate are released as pyrophosphate during phosphodiester bond formation. Pyrophosphate is subsequently hydrolyzed, releasing the energy used to drive nucleotide polymerization.

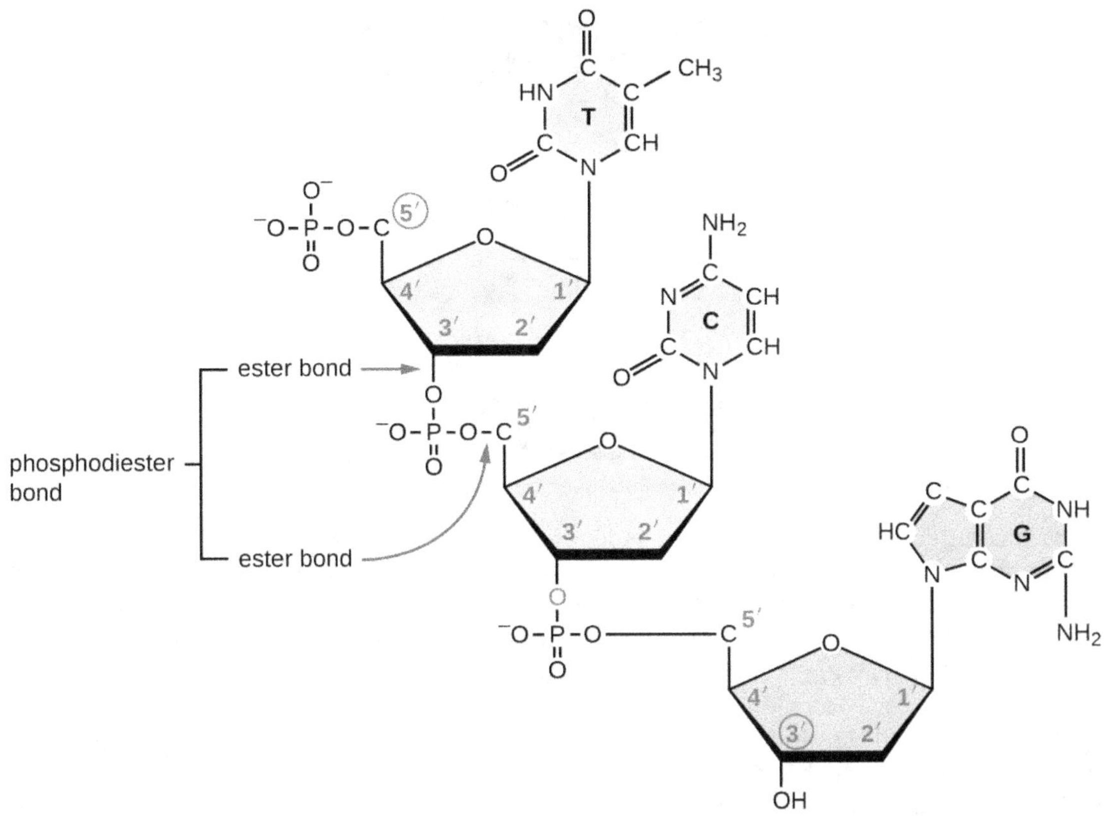

Figure 10.13 Phosphodiester bonds form between the phosphate group attached to the 5' carbon of one nucleotide and the hydroxyl group of the 3' carbon in the next nucleotide, bringing about polymerization of nucleotides in to nucleic acid strands. Note the 5' and 3' ends of this nucleic acid strand.

CHECK YOUR UNDERSTANDING

- What is meant by the 5' and 3' ends of a nucleic acid strand?

Discovering the Double Helix

By the early 1950s, considerable evidence had accumulated indicating that DNA was the genetic material of cells, and now the race was on to discover its three-dimensional structure. Around this time, Austrian biochemist Erwin Chargaff[5] (1905–2002) examined the content of DNA in different species and discovered that adenine, thymine, guanine, and cytosine were not found in equal quantities, and that it varied from species to species, but not between individuals of the same species. He found that the amount of adenine was very close to equaling the amount of thymine, and the amount of cytosine was very close to equaling the amount of guanine, or A = T and G = C. These relationships are also known as Chargaff's rules.

Other scientists were also actively exploring this field during the mid-20th century. In 1952, American scientist Linus Pauling (1901–1994) was the world's leading structural chemist and odds-on favorite to solve the structure of DNA. Pauling had earlier discovered the structure of protein α helices, using X-ray diffraction, and, based upon X-ray diffraction images of DNA made in his laboratory, he proposed a triple-stranded model of DNA.[6] At the same time, British researchers Rosalind Franklin (1920–1958) and her graduate student R.G. Gosling were also using X-ray diffraction to understand the structure of DNA (Figure 10.14). It was Franklin's scientific expertise that resulted in the production of more well-defined X-ray diffraction images of DNA that would clearly show the overall double-helix structure of DNA.

---

5 N. Kresge et al. "Chargaff's Rules: The Work of Erwin Chargaff." *Journal of Biological Chemistry* 280 (2005):e21.
6 L. Pauling, "A Proposed Structure for the Nucleic Acids." *Proceedings of the National Academy of Science of the United States of*

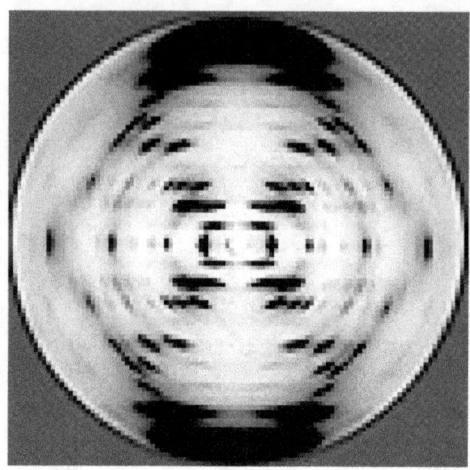

Figure 10.14 The X-ray diffraction pattern of DNA shows its helical nature. (credit: National Institutes of Health)

James Watson (1928–), an American scientist, and Francis Crick (1916–2004), a British scientist, were working together in the 1950s to discover DNA's structure. They used Chargaff's rules and Franklin and Wilkins' X-ray diffraction images of DNA fibers to piece together the purine-pyrimidine pairing of the double helical DNA molecule (Figure 10.15). In April 1953, Watson and Crick published their model of the DNA double helix in *Nature*.[7] The same issue additionally included papers by Wilkins and colleagues,[8] as well as by Franklin and Gosling,[9] each describing different aspects of the molecular structure of DNA. In 1962, James Watson, Francis Crick, and Maurice Wilkins were awarded the Nobel Prize in Physiology and Medicine. Unfortunately, by then Franklin had died, and Nobel prizes at the time were not awarded posthumously. Work continued, however, on learning about the structure of DNA. In 1973, Alexander Rich (1924–2015) and colleagues were able to analyze DNA crystals to confirm and further elucidate DNA structure.[10]

Figure 10.15 In 1953, James Watson and Francis Crick built this model of the structure of DNA, shown here on display at the Science Museum in London.

America 39 no. 2 (1953):84–97.

7 J.D. Watson, F.H.C. Crick. "A Structure for Deoxyribose Nucleic Acid." *Nature* 171 no. 4356 (1953):737–738.

8 M.H.F. Wilkins et al. "Molecular Structure of Deoxypentose Nucleic Acids." *Nature* 171 no. 4356 (1953):738–740.

9 R. Franklin, R.G. Gosling. "Molecular Configuration in Sodium Thymonucleate." *Nature* 171 no. 4356 (1953):740–741.

10 R.O. Day et al. "A Crystalline Fragment of the Double Helix: The Structure of the Dinucleoside Phosphate Guanylyl-3',5'-Cytidine." *Proceedings of the National Academy of Sciences of the United States of America* 70 no. 3 (1973):849–853.

CHECK YOUR UNDERSTANDING

- Which scientists are given most of the credit for describing the molecular structure of DNA?

DNA Structure

Watson and Crick proposed that DNA is made up of two strands that are twisted around each other to form a right-handed helix. The two DNA strands are **antiparallel**, such that the 3' end of one strand faces the 5' end of the other (Figure 10.16). The 3' end of each strand has a free hydroxyl group, while the 5' end of each strand has a free phosphate group. The sugar and phosphate of the polymerized nucleotides form the backbone of the structure, whereas the nitrogenous bases are stacked inside. These nitrogenous bases on the interior of the molecule interact with each other, base pairing.

Analysis of the diffraction patterns of DNA has determined that there are approximately 10 bases per turn in DNA. The asymmetrical spacing of the sugar-phosphate backbones generates major grooves (where the backbone is far apart) and minor grooves (where the backbone is close together) (Figure 10.16). These grooves are locations where proteins can bind to DNA. The binding of these proteins can alter the structure of DNA, regulate replication, or regulate transcription of DNA into RNA.

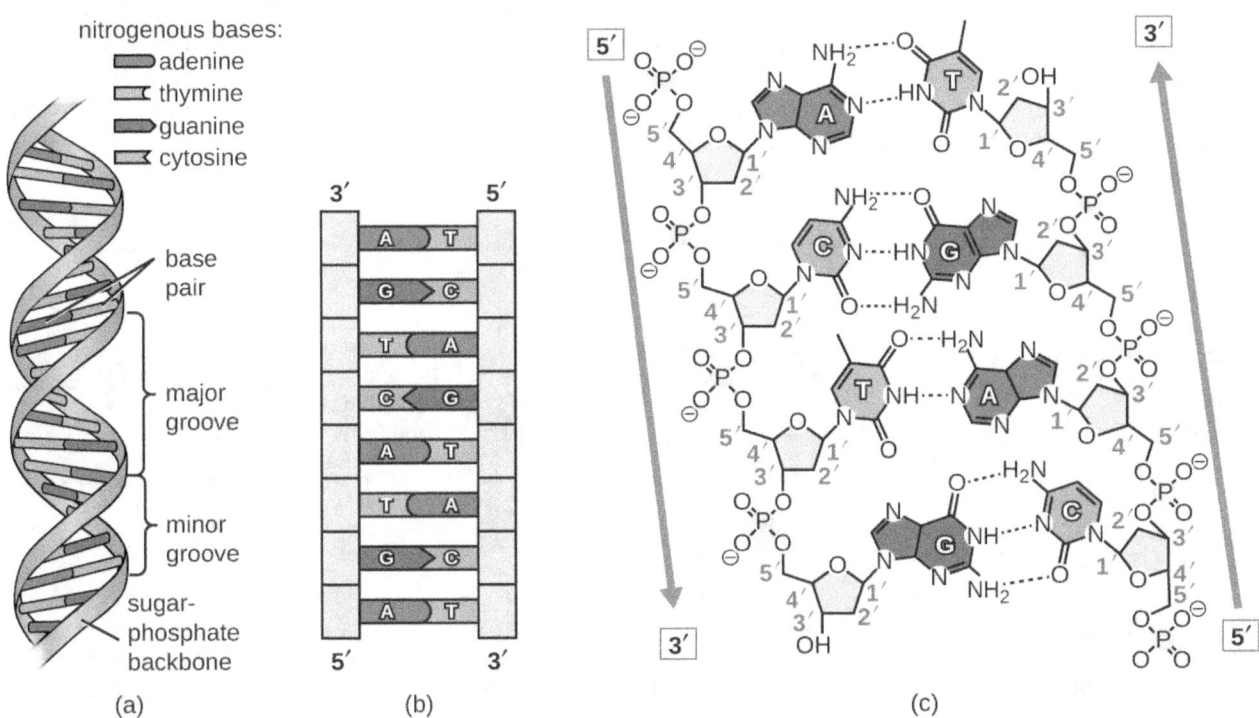

Figure 10.16 Watson and Crick proposed the double helix model for DNA. (a) The sugar-phosphate backbones are on the outside of the double helix and purines and pyrimidines form the "rungs" of the DNA helix ladder. (b) The two DNA strands are antiparallel to each other. (c) The direction of each strand is identified by numbering the carbons (1 through 5) in each sugar molecule. The 5' end is the one where carbon #5 is not bound to another nucleotide; the 3' end is the one where carbon #3 is not bound to another nucleotide.

Base pairing takes place between a purine and pyrimidine. In DNA, adenine (A) and thymine (T) are **complementary base pairs**, and cytosine (C) and guanine (G) are also complementary base pairs, explaining Chargaff's rules (Figure 10.17). The base pairs are stabilized by hydrogen bonds; adenine and thymine form two hydrogen bonds between them, whereas cytosine and guanine form three hydrogen bonds between them.

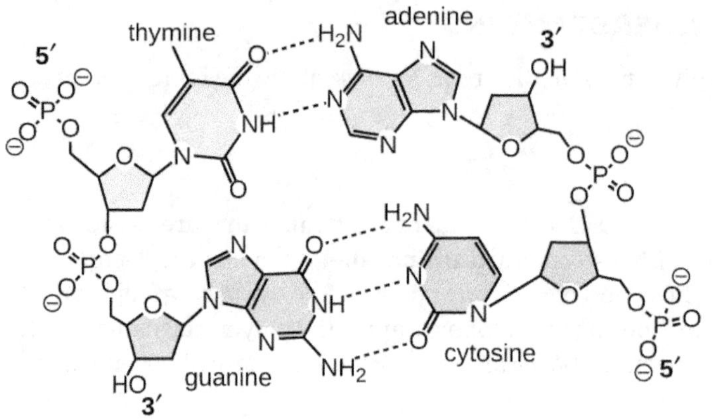

Figure 10.17 Hydrogen bonds form between complementary nitrogenous bases on the interior of DNA.

In the laboratory, exposing the two DNA strands of the double helix to high temperatures or to certain chemicals can break the hydrogen bonds between complementary bases, thus separating the strands into two separate single strands of DNA (single-stranded DNA [ssDNA]). This process is called DNA denaturation and is analogous to protein denaturation, as described in Proteins. The ssDNA strands can also be put back together as double-stranded DNA (dsDNA), through reannealing or renaturing by cooling or removing the chemical denaturants, allowing these hydrogen bonds to reform. The ability to artificially manipulate DNA in this way is the basis for several important techniques in biotechnology (Figure 10.18). Because of the additional hydrogen bonding between the C = G base pair, DNA with a high GC content is more difficult to denature than DNA with a lower GC content.

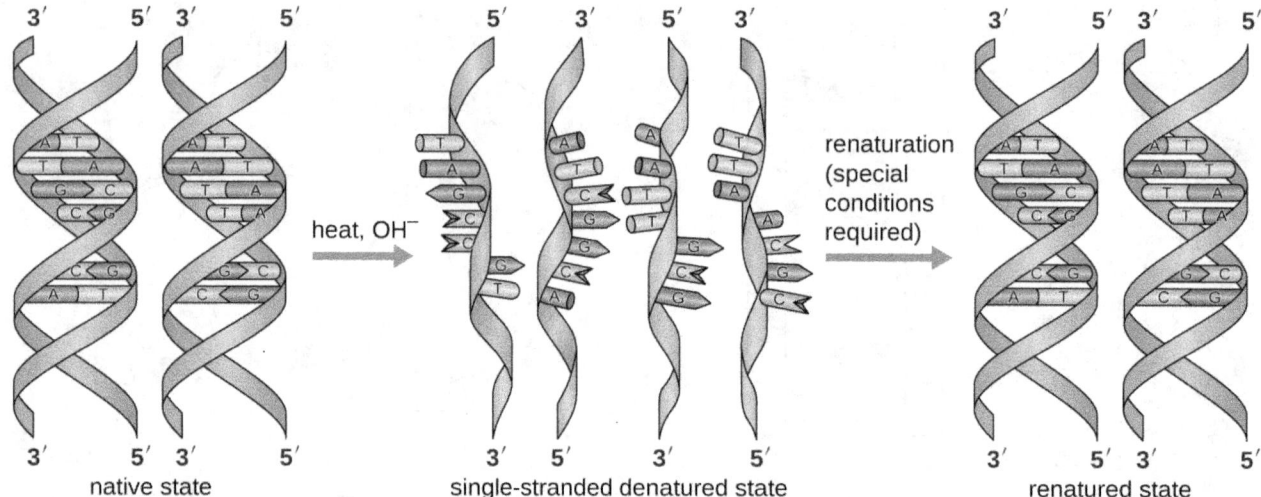

Figure 10.18 In the laboratory, the double helix can be denatured to single-stranded DNA through exposure to heat or chemicals, and then renatured through cooling or removal of chemical denaturants to allow the DNA strands to reanneal. (credit: modification of work by Hernández-Lemus E, Nicasio-Collazo LA, Castañeda-Priego R)

LINK TO LEARNING

View an animation (https://www.openstax.org/l/22dnastruanim) on DNA structure from the DNA Learning Center to learn more.

CHECK YOUR UNDERSTANDING

- What are the two complementary base pairs of DNA and how are they bonded together?

DNA Function

DNA stores the information needed to build and control the cell. The transmission of this information from mother to daughter cells is called **vertical gene transfer** and it occurs through the process of DNA replication. DNA is replicated when a cell makes a duplicate copy of its DNA, then the cell divides, resulting in the correct distribution of one DNA copy to each resulting cell. DNA can also be enzymatically degraded and used as a source of nucleosides and nucleotides for the cell. Unlike other macromolecules, DNA does not serve a structural role in cells.

CHECK YOUR UNDERSTANDING

- How does DNA transmit genetic information to offspring?

Eye on Ethics

Paving the Way for Women in Science and Health Professions

Historically, women have been underrepresented in the sciences and in medicine, and often their pioneering contributions have gone relatively unnoticed. For example, although Rosalind Franklin performed the X-ray diffraction studies demonstrating the double helical structure of DNA, it is Watson and Crick who became famous for this discovery, building on her data. There still remains great controversy over whether their acquisition of her data was appropriate and whether personality conflicts and gender bias contributed to the delayed recognition of her significant contributions. Similarly, Barbara McClintock did pioneering work in maize (corn) genetics from the 1930s through 1950s, discovering transposons (jumping genes), but she was not recognized until much later, receiving a Nobel Prize in Physiology or Medicine in 1983 (Figure 10.19).

Today, women still remain underrepresented in many fields of science and medicine. While more than half of the undergraduate degrees in science are awarded to women, only 46% of doctoral degrees in science are awarded to women. In academia, the number of women at each level of career advancement continues to decrease, with women holding less than one-third of the positions of Ph.D.-level scientists in tenure-track positions, and less than one-quarter of the full professorships at 4-year colleges and universities.[11] Even in the health professions, like nearly all other fields, women are often underrepresented in many medical careers and earn significantly less than their male counterparts, as shown in a 2013 study published by the *Journal of the American Medical Association*.[12]

Why do such disparities continue to exist and how do we break these cycles? The situation is complex and likely results from the combination of various factors, including how society conditions the behaviors of girls from a young age and supports their interests, both professionally and personally. Some have suggested that women do not belong in the laboratory, including Nobel Prize winner Tim Hunt, whose 2015 public comments suggesting that women are too emotional for science[13] were met with widespread condemnation.

---

11 N.H. Wolfinger "For Female Scientists, There's No Good Time to Have Children." *The Atlantic* July 29, 2013. http://www.theatlantic.com/sexes/archive/2013/07/for-female-scientists-theres-no-good-time-to-have-children/278165/.

12 S.A. Seabury et al. "Trends in the Earnings of Male and Female Health Care Professionals in the United States, 1987 to 2010." *Journal of the American Medical Association Internal Medicine* 173 no. 18 (2013):1748–1750.

13 E. Chung. "Tim Hunt, Sexism and Science: The Real 'Trouble With Girls' in Labs." *CBC News Technology and Science*, June 12, 2015. http://www.cbc.ca/news/technology/tim-hunt-sexism-and-science-the-real-trouble-with-girls-in-labs-1.3110133. Accessed 8/4/2016.

Perhaps girls should be supported more from a young age in the areas of science and math (Figure 10.19). Science, technology, engineering, and mathematics (STEM) programs sponsored by the American Association of University Women (AAUW)[14] and National Aeronautics and Space Administration (NASA)[15] are excellent examples of programs that offer such support. Contributions by women in science should be made known more widely to the public, and marketing targeted to young girls should include more images of historically and professionally successful female scientists and medical professionals, encouraging all bright young minds, including girls and women, to pursue careers in science and medicine.

(a)

(b)

Figure 10.19 (a) Barbara McClintock's work on maize genetics in the 1930s through 1950s resulted in the discovery of transposons, but its significance was not recognized at the time. (b) Efforts to appropriately mentor and to provide continued societal support for women in science and medicine may someday help alleviate some of the issues preventing gender equality at all levels in science and medicine. (credit a: modification of work by Smithsonian Institution; credit b: modification of work by Haynie SL, Hinkle AS, Jones NL, Martin CA, Olsiewski PJ, Roberts MF)

Clinical Focus

Part 2

Based upon his symptoms, Alex's physician suspects that he is suffering from a foodborne illness that he acquired during his travels. Possibilities include bacterial infection (e.g., enterotoxigenic *E. coli*, *Vibrio cholerae*, *Campylobacter jejuni*, *Salmonella*), viral infection (rotavirus or norovirus), or protozoan infection (*Giardia lamblia*, *Cryptosporidium parvum*, or *Entamoeba histolytica*).

His physician orders a stool sample to identify possible causative agents (e.g., bacteria, cysts) and to look for the presence of blood because certain types of infectious agents (like *C. jejuni*, *Salmonella*, and *E. histolytica*) are associated with the production of bloody stools.

Alex's stool sample showed neither blood nor cysts. Following analysis of his stool sample and based upon his recent travel history, the hospital physician suspected that Alex was suffering from traveler's diarrhea caused by enterotoxigenic *E. coli* (ETEC), the causative agent of most traveler's diarrhea. To verify the diagnosis and rule out other possibilities, Alex's physician ordered a diagnostic lab test of his stool sample to look for DNA sequences encoding specific virulence factors of ETEC. The physician instructed Alex to

---

14 American Association of University Women. "Building a STEM Pipeline for Girls and Women." http://www.aauw.org/what-we-do/stem-education/. Accessed June 10, 2016.

15 National Aeronautics and Space Administration. "Outreach Programs: Women and Girls Initiative." http://women.nasa.gov/outreach-programs/. Accessed June 10, 2016.

drink lots of fluids to replace what he was losing and discharged him from the hospital.

ETEC produces several plasmid-encoded virulence factors that make it pathogenic compared with typical *E. coli*. These include the secreted toxins heat-labile enterotoxin (LT) and heat-stabile enterotoxin (ST), as well as colonization factor (CF). Both LT and ST cause the excretion of chloride ions from intestinal cells to the intestinal lumen, causing a consequent loss of water from intestinal cells, resulting in diarrhea. CF encodes a bacterial protein that aids in allowing the bacterium to adhere to the lining of the small intestine.

- Why did Alex's physician use genetic analysis instead of either isolation of bacteria from the stool sample or direct Gram stain of the stool sample alone?

Jump to the next Clinical Focus box. Go back to the previous Clinical Focus box.

10.3 Structure and Function of RNA

Learning Objectives

By the end of this section, you will be able to:
- Describe the biochemical structure of ribonucleotides
- Describe the similarities and differences between RNA and DNA
- Describe the functions of the three main types of RNA used in protein synthesis
- Explain how RNA can serve as hereditary information

Structurally speaking, **ribonucleic acid (RNA)**, is quite similar to DNA. However, whereas DNA molecules are typically long and double stranded, RNA molecules are much shorter and are typically single stranded. RNA molecules perform a variety of roles in the cell but are mainly involved in the process of protein synthesis (translation) and its regulation.

RNA Structure

RNA is typically single stranded and is made of **ribonucleotides** that are linked by phosphodiester bonds. A ribonucleotide in the RNA chain contains ribose (the pentose sugar), one of the four nitrogenous bases (A, U, G, and C), and a phosphate group. The subtle structural difference between the sugars gives DNA added stability, making DNA more suitable for storage of genetic information, whereas the relative instability of RNA makes it more suitable for its more short-term functions. The RNA-specific pyrimidine **uracil** forms a complementary base pair with adenine and is used instead of the thymine used in DNA. Even though RNA is single stranded, most types of RNA molecules show extensive intramolecular base pairing between complementary sequences within the RNA strand, creating a predictable three-dimensional structure essential for their function (Figure 10.20 and Figure 10.21).

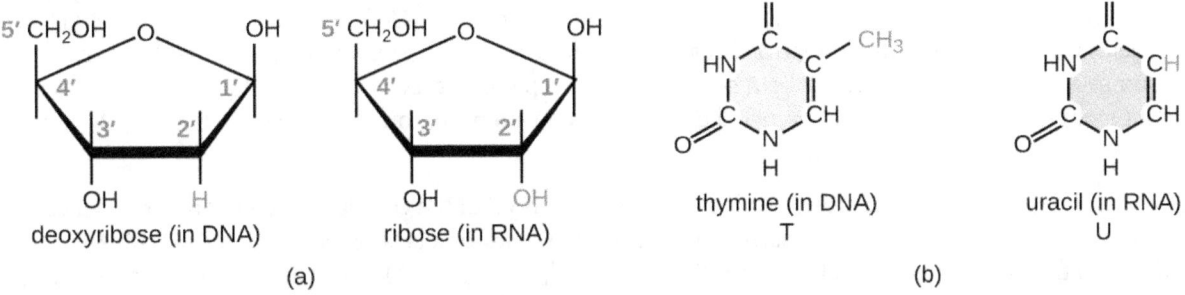

Figure 10.20 (a) Ribonucleotides contain the pentose sugar ribose instead of the deoxyribose found in deoxyribonucleotides. (b) RNA contains the pyrimidine uracil in place of thymine found in DNA.

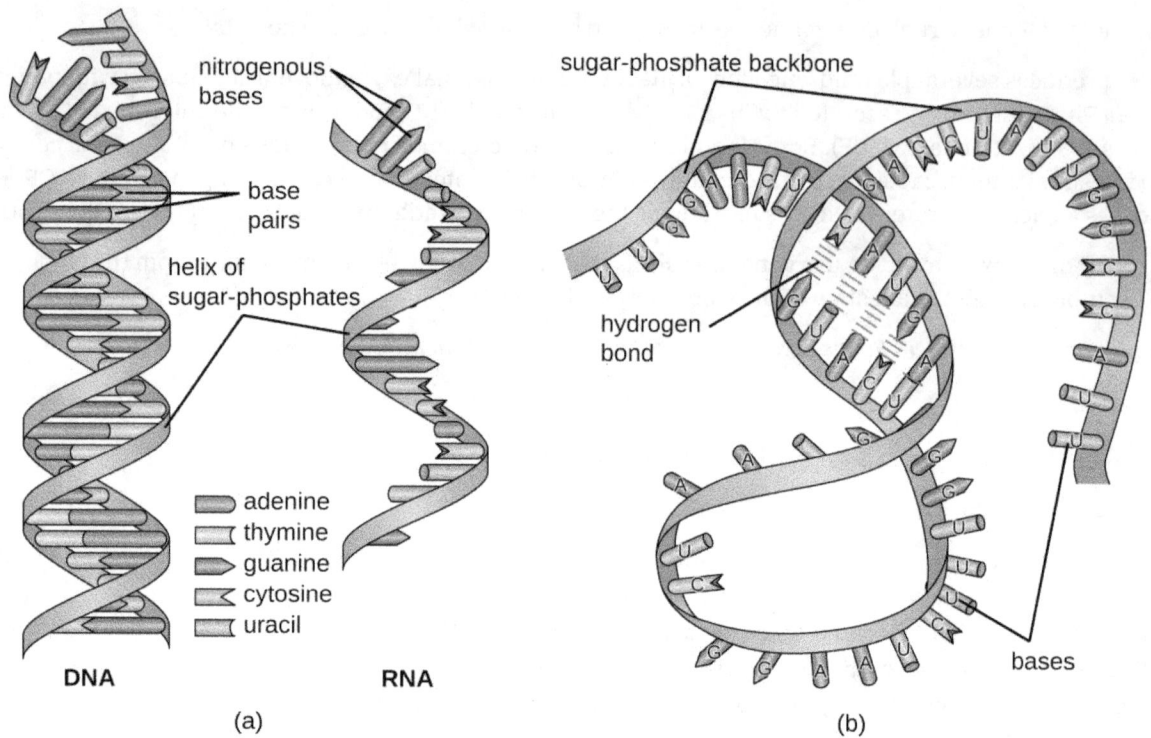

Figure 10.21 (a) DNA is typically double stranded, whereas RNA is typically single stranded. (b) Although it is single stranded, RNA can fold upon itself, with the folds stabilized by short areas of complementary base pairing within the molecule, forming a three-dimensional structure.

CHECK YOUR UNDERSTANDING

- How does the structure of RNA differ from the structure of DNA?

Functions of RNA in Protein Synthesis

Cells access the information stored in DNA by creating RNA to direct the synthesis of proteins through the process of translation. Proteins within a cell have many functions, including building cellular structures and serving as enzyme catalysts for cellular chemical reactions that give cells their specific characteristics. The three main types of RNA directly involved in protein synthesis are **messenger RNA (mRNA), ribosomal RNA (rRNA),** and **transfer RNA (tRNA).**

In 1961, French scientists François Jacob and Jacques Monod hypothesized the existence of an intermediary between DNA and its protein products, which they called messenger RNA.[16] Evidence supporting their hypothesis was gathered soon afterwards showing that information from DNA is transmitted to the ribosome for protein synthesis using mRNA. If DNA serves as the complete library of cellular information, mRNA serves as a photocopy of specific information needed at a particular point in time that serves as the instructions to make a protein.

The mRNA carries the message from the DNA, which controls all of the cellular activities in a cell. If a cell requires a certain protein to be synthesized, the gene for this product is "turned on" and the mRNA is synthesized through the process of transcription (see RNA Transcription). The mRNA then interacts with ribosomes and other cellular machinery (Figure 10.22) to direct the synthesis of the protein it encodes during the process of translation (see Protein Synthesis). mRNA is relatively unstable and short-lived in the cell, especially in prokaryotic cells, ensuring that proteins are only made when needed.

---

16 A. Rich. "The Era of RNA Awakening: Structural Biology of RNA in the Early Years." *Quarterly Reviews of Biophysics* 42 no. 2 (2009):117–137.

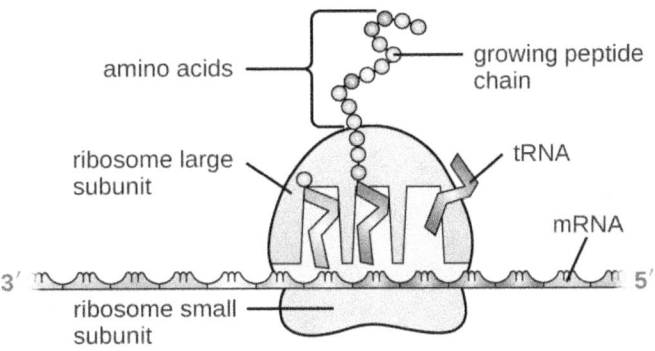

Figure 10.22 A generalized illustration of how mRNA and tRNA are used in protein synthesis within a cell.

rRNA and tRNA are stable types of RNA. In prokaryotes and eukaryotes, tRNA and rRNA are encoded in the DNA, then copied into long RNA molecules that are cut to release smaller fragments containing the individual mature RNA species. In eukaryotes, synthesis, cutting, and assembly of rRNA into ribosomes takes place in the nucleolus region of the nucleus, but these activities occur in the cytoplasm of prokaryotes. Neither of these types of RNA carries instructions to direct the synthesis of a polypeptide, but they play other important roles in protein synthesis.

Ribosomes are composed of rRNA and protein. As its name suggests, rRNA is a major constituent of ribosomes, composing up to about 60% of the ribosome by mass and providing the location where the mRNA binds. The rRNA ensures the proper alignment of the mRNA, tRNA, and the ribosomes; the rRNA of the ribosome also has an enzymatic activity (peptidyl transferase) and catalyzes the formation of the peptide bonds between two aligned amino acids during protein synthesis. Although rRNA had long been thought to serve primarily a structural role, its catalytic role within the ribosome was proven in 2000.[17] Scientists in the laboratories of Thomas Steitz (1940–) and Peter Moore (1939–) at Yale University were able to crystallize the ribosome structure from *Haloarcula marismortui*, a halophilic archaeon isolated from the Dead Sea. Because of the importance of this work, Steitz shared the 2009 Nobel Prize in Chemistry with other scientists who made significant contributions to the understanding of ribosome structure.

Transfer RNA is the third main type of RNA and one of the smallest, usually only 70–90 nucleotides long. It carries the correct amino acid to the site of protein synthesis in the ribosome. It is the base pairing between the tRNA and mRNA that allows for the correct amino acid to be inserted in the polypeptide chain being synthesized (Figure 10.23). Any mutations in the tRNA or rRNA can result in global problems for the cell because both are necessary for proper protein synthesis (Table 10.1).

---

17 P. Nissen et al. "The Structural Basis of Ribosome Activity in Peptide Bond Synthesis." *Science* 289 no. 5481 (2000):920–930.

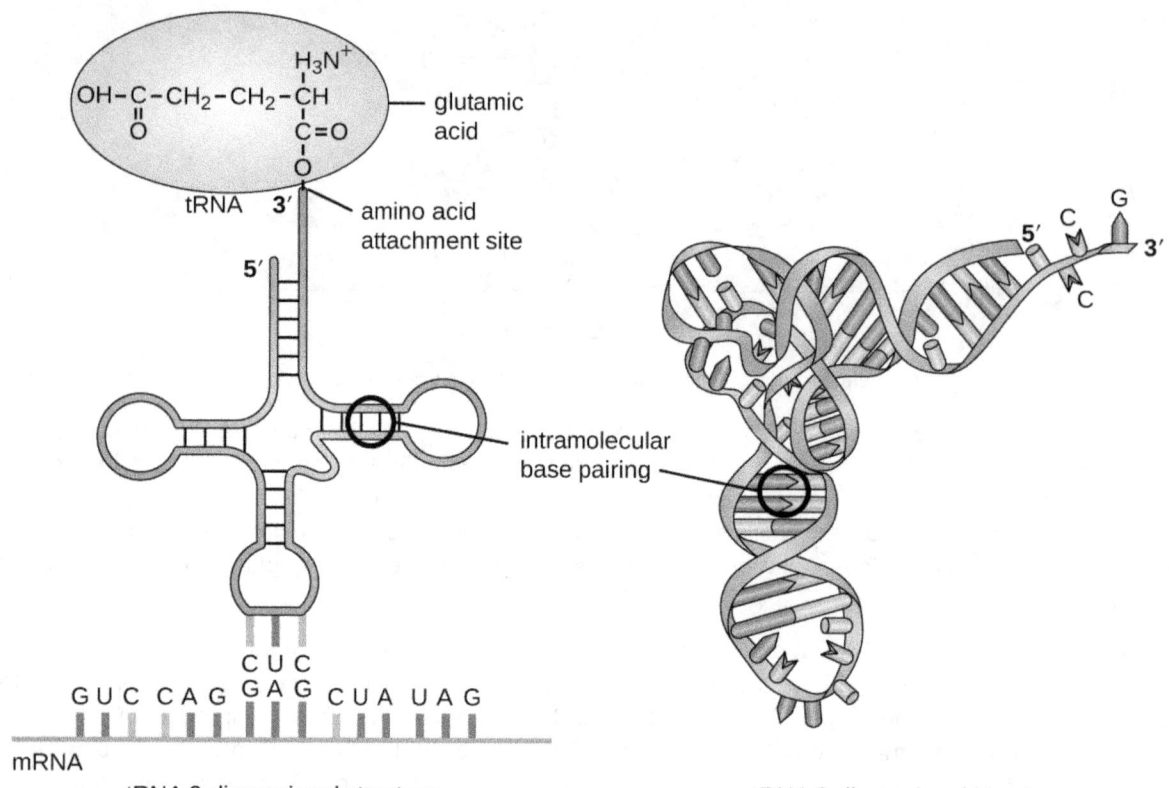

Figure 10.23 A tRNA molecule is a single-stranded molecule that exhibits significant intracellular base pairing, giving it its characteristic three-dimensional shape.

Structure and Function of RNA

	mRNA	rRNA	tRNA
Structure	Short, unstable, single-stranded RNA corresponding to a gene encoded within DNA	Longer, stable RNA molecules composing 60% of ribosome's mass	Short (70-90 nucleotides), stable RNA with extensive intramolecular base pairing; contains an amino acid binding site and an mRNA binding site
Function	Serves as intermediary between DNA and protein; used by ribosome to direct synthesis of protein it encodes	Ensures the proper alignment of mRNA, tRNA, and ribosome during protein synthesis; catalyzes peptide bond formation between amino acids	Carries the correct amino acid to the site of protein synthesis in the ribosome

Table 10.1

CHECK YOUR UNDERSTANDING

- What are the functions of the three major types of RNA molecules involved in protein synthesis?

RNA as Hereditary Information

Although RNA does not serve as the hereditary information in most cells, RNA does hold this function for many viruses that do not contain DNA. Thus, RNA clearly does have the additional capacity to serve as genetic information. Although RNA is typically single stranded within cells, there is significant diversity in viruses. Rhinoviruses, which cause the common cold; influenza viruses; and the Ebola virus are single-stranded RNA viruses. Rotaviruses, which cause severe gastroenteritis in children and other immunocompromised individuals, are examples of double-stranded RNA viruses. Because double-stranded RNA is uncommon in eukaryotic cells, its presence serves as an indicator of viral infection. The implications for a virus having an RNA genome instead of a DNA genome are discussed in more detail in Viruses.

10.4 Structure and Function of Cellular Genomes

Learning Objectives

By the end of this section, you will be able to:
- Define gene and genotype and differentiate genotype from phenotype
- Describe chromosome structure and packaging
- Compare prokaryotic and eukaryotic chromosomes
- Explain why extrachromosomal DNA is important in a cell

Thus far, we have discussed the structure and function of individual pieces of DNA and RNA. In this section, we will discuss how all of an organism's genetic material—collectively referred to as its **genome**—is organized inside of the cell. Since an organism's genetics to a large extent dictate its characteristics, it should not be surprising that organisms differ in the arrangement of their DNA and RNA.

Genotype versus Phenotype

All cellular activities are encoded within a cell's DNA. The sequence of bases within a DNA molecule represents the genetic information of the cell. Segments of DNA molecules are called **gene**s, and individual genes contain the instructional code necessary for synthesizing various proteins, enzymes, or stable RNA molecules.

The full collection of genes that a cell contains within its genome is called its **genotype**. However, a cell does not express all of its genes simultaneously. Instead, it turns on (expresses) or turns off certain genes when necessary. The set of genes being expressed at any given point in time determines the cell's activities and its observable characteristics, referred to as its **phenotype**. Genes that are always expressed are known as constitutive genes; some constitutive genes are known as housekeeping genes because they are necessary for the basic functions of the cell.

While the genotype of a cell remains constant, the phenotype may change in response to environmental signals (e.g., changes in temperature or nutrient availability) that affect which nonconstitutive genes are expressed. For example, the oral bacterium *Streptococcus mutans* produces a sticky slime layer that allows it to adhere to teeth, forming dental plaque; however, the genes that control the production of the slime layer are only expressed in the presence of sucrose (table sugar). Thus, while the genotype of *S. mutans* is constant, its phenotype changes depending on the presence and absence of sugar in its environment. Temperature can also regulate gene expression. For example, the gram-negative bacterium *Serratia marcescens*, a pathogen frequently associated with hospital-acquired infections, produces a red pigment at 28 °C but not at 37 °C, the normal internal temperature of the human body (Figure 10.24).

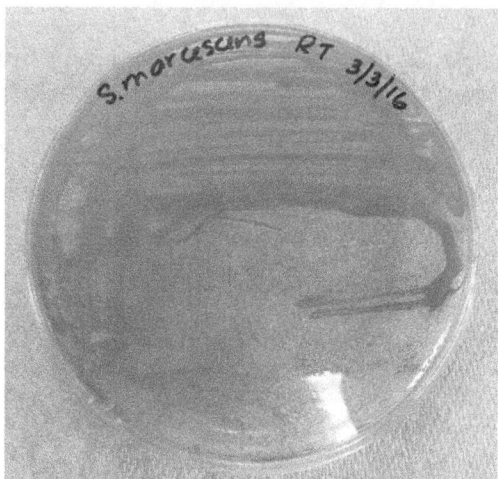 

Figure 10.24 Both plates contain strains of *Serratia marcescens* that have the gene for red pigment. However, this gene is expressed at 28 °C (left) but not at 37 °C (right). (credit: modification of work by Ann Auman)

Organization of Genetic Material

The vast majority of an organism's genome is organized into the cell's **chromosomes**, which are discrete DNA structures within cells that control cellular activity. Recall that while eukaryotic chromosomes are housed in the membrane-bound nucleus, most prokaryotes contain a single, circular chromosome that is found in an area of the cytoplasm called the nucleoid (see Unique Characteristics of Prokaryotic Cells). A chromosome may contain several thousand genes.

Organization of Eukaryotic Chromosome

Chromosome structure differs somewhat between eukaryotic and prokaryotic cells. Eukaryotic chromosomes are typically linear, and eukaryotic cells contain multiple distinct chromosomes. Many eukaryotic cells contain two copies of each chromosome and, therefore, are **diploid**.

The length of a chromosome greatly exceeds the length of the cell, so a chromosome needs to be packaged into a very small space to fit within the cell. For example, the combined length of all of the 3 billion base pairs[18] of DNA of the human genome would measure approximately 2 meters if completely stretched out, and some eukaryotic genomes are many times larger than the human genome. DNA **supercoiling** refers to the process by which DNA is twisted to fit inside the cell. Supercoiling may result in DNA that is either underwound (less than one turn of the helix per 10 base pairs) or overwound (more than one turn per 10 base pairs) from its normal relaxed state. Proteins known to be involved in supercoiling include **topoisomerases**; these enzymes help maintain the structure of supercoiled chromosomes, preventing overwinding of DNA during certain cellular processes like DNA replication.

During **DNA packaging**, DNA-binding proteins called **histones** perform various levels of DNA wrapping and attachment to scaffolding proteins. The combination of DNA with these attached proteins is referred to as **chromatin**. In eukaryotes, the packaging of DNA by histones may be influenced by environmental factors that affect the presence of methyl groups on certain cytosine nucleotides of DNA. The influence of environmental factors on DNA packaging is called epigenetics. Epigenetics is another mechanism for regulating gene expression without altering the sequence of nucleotides. Epigenetic changes can be maintained through multiple rounds of cell division and, therefore, can be heritable.

LINK TO LEARNING

View this animation (https://www.openstax.org/l/22dnapackanim) from the DNA Learning Center to learn more about on DNA packaging in eukaryotes.

---

18 National Human Genome Research Institute. "The Human Genome Project Completion: Frequently Asked Questions." https://www.genome.gov/11006943. Accessed June 10, 2016

Organization of Prokaryotic Chromosomes

Chromosomes in bacteria and archaea are usually circular, and a prokaryotic cell typically contains only a single chromosome within the nucleoid. Because the chromosome contains only one copy of each gene, prokaryotes are **haploid**. As in eukaryotic cells, DNA supercoiling is necessary for the genome to fit within the prokaryotic cell. The DNA in the bacterial chromosome is arranged in several supercoiled domains. As with eukaryotes, topoisomerases are involved in supercoiling DNA. DNA gyrase is a type of topoisomerase, found in bacteria and some archaea, that helps prevent the overwinding of DNA. (Some antibiotics kill bacteria by targeting DNA gyrase.) In addition, histone-like proteins bind DNA and aid in DNA packaging. Other proteins bind to the origin of replication, the location in the chromosome where DNA replication initiates. Because different regions of DNA are packaged differently, some regions of chromosomal DNA are more accessible to enzymes and thus may be used more readily as templates for gene expression. Interestingly, several bacteria, including *Helicobacter pylori* and *Shigella flexneri*, have been shown to induce epigenetic changes in their hosts upon infection, leading to chromatin remodeling that may cause long-term effects on host immunity.[19]

✓ CHECK YOUR UNDERSTANDING

- What is the difference between a cell's genotype and its phenotype?
- How does DNA fit inside cells?

Noncoding DNA

In addition to genes, a genome also contains many regions of **noncoding DNA** that do not encode proteins or stable RNA products. Noncoding DNA is commonly found in areas prior to the start of coding sequences of genes as well as in intergenic regions (i.e., DNA sequences located between genes) (Figure 10.25).

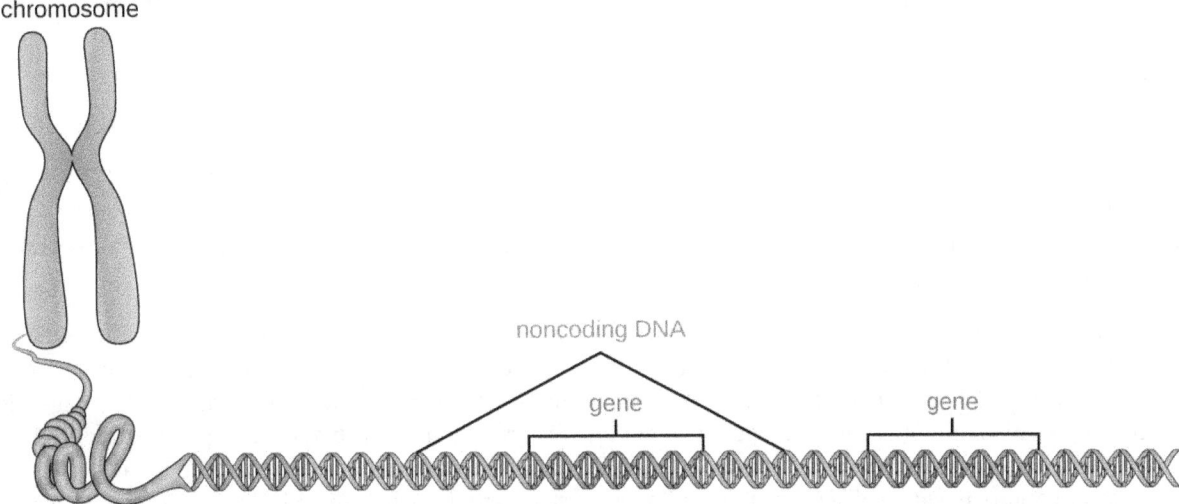

Figure 10.25 Chromosomes typically have a significant amount of noncoding DNA, often found in intergenic regions.

Prokaryotes appear to use their genomes very efficiently, with only an average of 12% of the genome being taken up by noncoding sequences. In contrast, noncoding DNA can represent about 98% of the genome in eukaryotes, as seen in humans, but the percentage of noncoding DNA varies between species.[20] These noncoding DNA regions were once referred to as "junk DNA"; however, this terminology is no longer widely accepted because scientists have since found roles for some of these regions, many of which contribute to the regulation of transcription or translation through the production of small noncoding RNA molecules, DNA packaging, and chromosomal stability. Although scientists may not fully understand the roles of all noncoding regions of DNA, it is generally believed that they do have purposes within the cell.

---

19 H. Bierne et al. "Epigenetics and Bacterial Infections." *Cold Spring Harbor Perspectives in Medicine* 2 no. 12 (2012):a010272.
20 R.J. Taft et al. "The Relationship between Non-Protein-Coding DNA and Eukaryotic Complexity." *Bioessays* 29 no. 3 (2007):288–299.

CHECK YOUR UNDERSTANDING

- What is the role of noncoding DNA?

Extrachromosomal DNA

Although most DNA is contained within a cell's chromosomes, many cells have additional molecules of DNA outside the chromosomes, called **extrachromosomal DNA**, that are also part of its genome. The genomes of eukaryotic cells would also include the chromosomes from any organelles such as mitochondria and/or chloroplasts that these cells maintain (Figure 10.26). The maintenance of circular chromosomes in these organelles is a vestige of their prokaryotic origins and supports the endosymbiotic theory (see Foundations of Modern Cell Theory). In some cases, genomes of certain DNA viruses can also be maintained independently in host cells during latent viral infection. In these cases, these viruses are another form of extrachromosomal DNA. For example, the human papillomavirus (HPV) may be maintained in infected cells in this way.

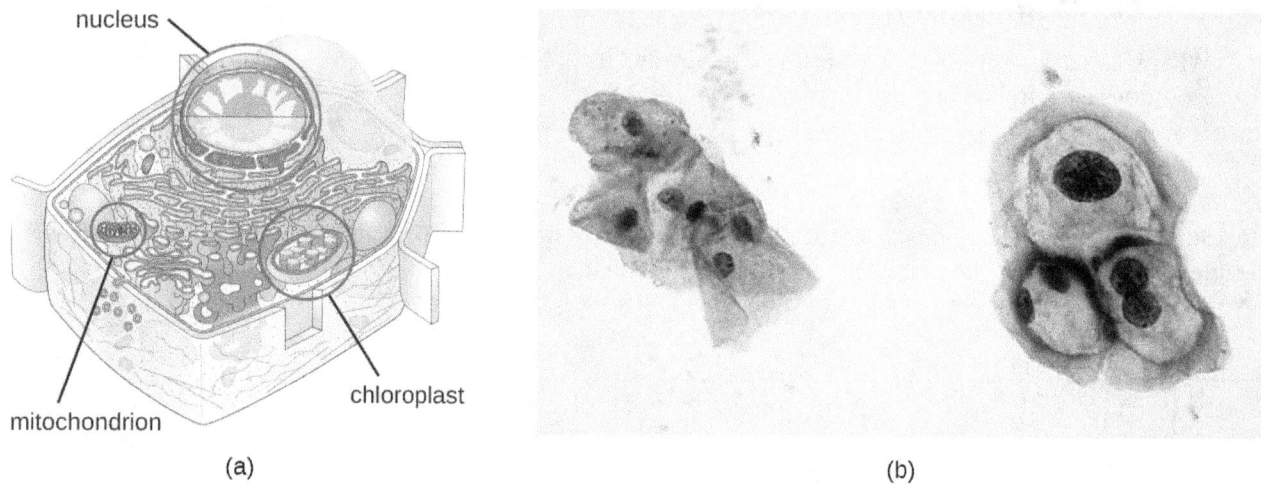

Figure 10.26 The genome of a eukaryotic cell consists of the chromosome housed in the nucleus, and extrachromosomal DNA found in the mitochondria (all cells) and chloroplasts (plants and algae). The cells shown in (b) represent cells obtained from a pap smear. The cells on the left are normal squamous cells whereas the cells on the right are infected with human papillomavirus and show enlarged nuclei with increased staining (hyperchromasia).

Besides chromosomes, some prokaryotes also have smaller loops of DNA called plasmids that may contain one or a few genes not essential for normal growth (Figure 3.12). Bacteria can exchange these plasmids with other bacteria in a process known as horizontal gene transfer (HGT). The exchange of genetic material on plasmids sometimes provides microbes with new genes beneficial for growth and survival under special conditions. In some cases, genes obtained from plasmids may have clinical implications, encoding virulence factors that give a microbe the ability to cause disease or make a microbe resistant to certain antibiotics. Plasmids are also used heavily in genetic engineering and biotechnology as a way to move genes from one cell to another. The role of plasmids in horizontal gene transfer and biotechnology will be discussed further in Mechanisms of Microbial Genetics and Modern Applications of Microbial Genetics.

CHECK YOUR UNDERSTANDING

- How are plasmids involved in antibiotic resistance?

Case in Point

Lethal Plasmids

Maria, a 20-year-old anthropology student from Texas, recently became ill in the African nation of Botswana, where she was conducting research as part of a study-abroad program. Maria's research was focused on traditional African methods of tanning hides for the production of leather. Over a period of three weeks, she visited a tannery daily for several hours to observe and participate in the tanning process. One day, after returning from the tannery, Maria developed a fever, chills, and a headache, along with chest pain, muscle aches, nausea, and other flu-like symptoms. Initially, she was not concerned, but when her fever spiked and she began to cough up blood, her African host family became alarmed and rushed her to the hospital, where her condition continued to worsen.

After learning about her recent work at the tannery, the physician suspected that Maria had been exposed to anthrax. He ordered a chest X-ray, a blood sample, and a spinal tap, and immediately started her on a course of intravenous penicillin. Unfortunately, lab tests confirmed the physician's presumptive diagnosis. Maria's chest X-ray exhibited pleural effusion, the accumulation of fluid in the space between the pleural membranes, and a Gram stain of her blood revealed the presence of gram-positive, rod-shaped bacteria in short chains, consistent with *Bacillus anthracis*. Blood and bacteria were also shown to be present in her cerebrospinal fluid, indicating that the infection had progressed to meningitis. Despite supportive treatment and aggressive antibiotic therapy, Maria slipped into an unresponsive state and died three days later.

Anthrax is a disease caused by the introduction of endospores from the gram-positive bacterium *B. anthracis* into the body. Once infected, patients typically develop meningitis, often with fatal results. In Maria's case, she inhaled the endospores while handling the hides of animals that had been infected.

The genome of *B. anthracis* illustrates how small structural differences can lead to major differences in virulence. In 2003, the genomes of *B. anthracis* and *Bacillus cereus*, a similar but less pathogenic bacterium of the same genus, were sequenced and compared.[21] Researchers discovered that the 16S rRNA gene sequences of these bacteria are more than 99% identical, meaning that they are actually members of the same species despite their traditional classification as separate species. Although their chromosomal sequences also revealed a great deal of similarity, several virulence factors of *B. anthracis* were found to be encoded on two large plasmids not found in *B. cereus*. The plasmid pX01 encodes a three-part toxin that suppresses the host immune system, whereas the plasmid pX02 encodes a capsular polysaccharide that further protects the bacterium from the host immune system (Figure 10.27). Since *B. cereus* lacks these plasmids, it does not produce these virulence factors, and although it is still pathogenic, it is typically associated with mild cases of diarrhea from which the body can quickly recover. Unfortunately for Maria, the presence of these toxin-encoding plasmids in *B. anthracis* gives it its lethal virulence.

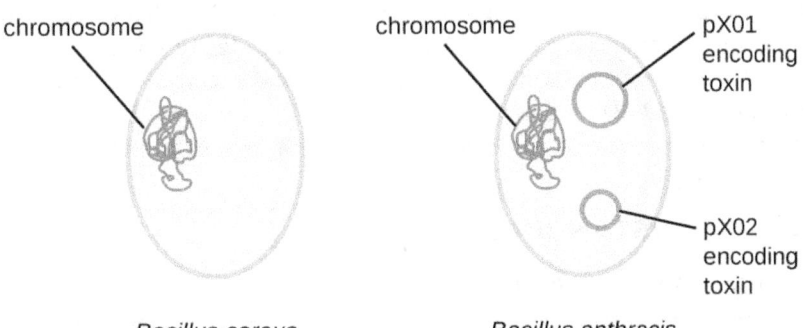

Figure 10.27 Genome sequencing of *Bacillus anthracis* and its close relative *B. cereus* reveals that the pathogenicity of *B. anthracis* is due to the maintenance of two plasmids, pX01 and pX02, which encode virulence factors.

- What do you think would happen to the pathogenicity of *B. anthracis* if it lost one or both of its plasmids?

---

[21] N. Ivanova et al. "Genome Sequence of *Bacillus cereus* and Comparative Analysis with *Bacillus anthracis*." *Nature* 423 no. 6935 (2003):87–91.

> **Clinical Focus**
>
> **Resolution**
>
> Within 24 hours, the results of the diagnostic test analysis of Alex's stool sample revealed that it was positive for heat-labile enterotoxin (LT), heat-stabile enterotoxin (ST), and colonization factor (CF), confirming the hospital physician's suspicion of ETEC. During a follow-up with Alex's family physician, this physician noted that Alex's symptoms were not resolving quickly and he was experiencing discomfort that was preventing him from returning to classes. The family physician prescribed Alex a course of ciprofloxacin to resolve his symptoms. Fortunately, the ciprofloxacin resolved Alex's symptoms within a few days.
>
> Alex likely got his infection from ingesting contaminated food or water. Emerging industrialized countries like Mexico are still developing sanitation practices that prevent the contamination of water with fecal material. Travelers in such countries should avoid the ingestion of undercooked foods, especially meats, seafood, vegetables, and unpasteurized dairy products. They should also avoid use of water that has not been treated; this includes drinking water, ice cubes, and even water used for brushing teeth. Using bottled water for these purposes is a good alternative. Good hygiene (handwashing) can also aid the prevention of an ETEC infection. Alex had not been careful about his food or water consumption, which led to his illness.
>
> Alex's symptoms were very similar to those of cholera, caused by the gram-negative bacterium *Vibrio cholerae*, which also produces a toxin similar to ST and LT. At some point in the evolutionary history of ETEC, a nonpathogenic strain of *E. coli* similar to those typically found in the gut may have acquired the genes encoding the ST and LT toxins from *V. cholerae*. The fact that the genes encoding those toxins are encoded on extrachromosomal plasmids in ETEC supports the idea that these genes were acquired by *E. coli* and are likely maintained in bacterial populations through horizontal gene transfer.
>
> Go back to the *previous* Clinical Focus box.

Viral Genomes

Viral genomes exhibit significant diversity in structure. Some viruses have genomes that consist of DNA as their genetic material. This DNA may be single stranded, as exemplified by human parvoviruses, or double stranded, as seen in the herpesviruses and poxviruses. Additionally, although all cellular life uses DNA as its genetic material, some viral genomes are made of either single-stranded or double-stranded RNA molecules, as we have discussed. Viral genomes are typically smaller than most bacterial genomes, encoding only a few genes, because they rely on their hosts to carry out many of the functions required for their replication. The diversity of viral genome structures and their implications for viral replication life cycles are discussed in more detail in The Viral Life Cycle.

✓ CHECK YOUR UNDERSTANDING

- Why do viral genomes vary widely among viruses?

 MICRO CONNECTIONS

Genome Size Matters

There is great variation in size of genomes among different organisms. Most eukaryotes maintain multiple chromosomes; humans, for example have 23 pairs, giving them 46 chromosomes. Despite being large at 3 billion base pairs, the human genome is far from the largest genome. Plants often maintain very large genomes, up to 150 billion base pairs, and commonly are polyploid, having multiple copies of each chromosome.

The size of bacterial genomes also varies considerably, although they tend to be smaller than eukaryotic genomes (Figure 10.28). Some bacterial genomes may be as small as only 112,000 base pairs. Often, the size of a bacterium's genome directly relates to how much the bacterium depends on its host for survival. When a bacterium relies on the host cell to carry out certain functions, it loses the genes encoding the abilities to carry out those functions itself. These types of bacterial endosymbionts are reminiscent of the prokaryotic origins of mitochondria and chloroplasts.

From a clinical perspective, obligate and facultative intracellular pathogens also tend to have small genomes (some around 1 million base pairs). Because host cells can supply most of their nutrients, they tend to have a reduced number of genes encoding metabolic functions, making their cultivation in the laboratory difficult if not impossible Due to their small sizes, the genomes of organisms like *Mycoplasma genitalium* (580,000 base pairs), *Chlamydia trachomatis* (1.0 million), *Rickettsia prowazekii* (1.1 million), and *Treponema pallidum* (1.1 million) were some of the earlier bacterial genomes sequenced. Respectively, these pathogens cause urethritis and pelvic inflammation, chlamydia, typhus, and syphilis.

Whereas obligate intracellular pathogens have unusually small genomes, other bacteria with a great variety of metabolic and enzymatic capabilities have unusually large bacterial genomes. *Pseudomonas aeruginosa*, for example, is a bacterium commonly found in the environment and is able to grow on a wide range of substrates. Its genome contains 6.3 million base pairs, giving it a high metabolic ability and the ability to produce virulence factors that cause several types of opportunistic infections.

Interestingly, there has been significant variability in genome size in viruses as well, ranging from 3,500 base pairs to 2.5 million base pairs, significantly exceeding the size of many bacterial genomes. The great variation observed in viral genome sizes further contributes to the great diversity of viral genome characteristics already discussed.

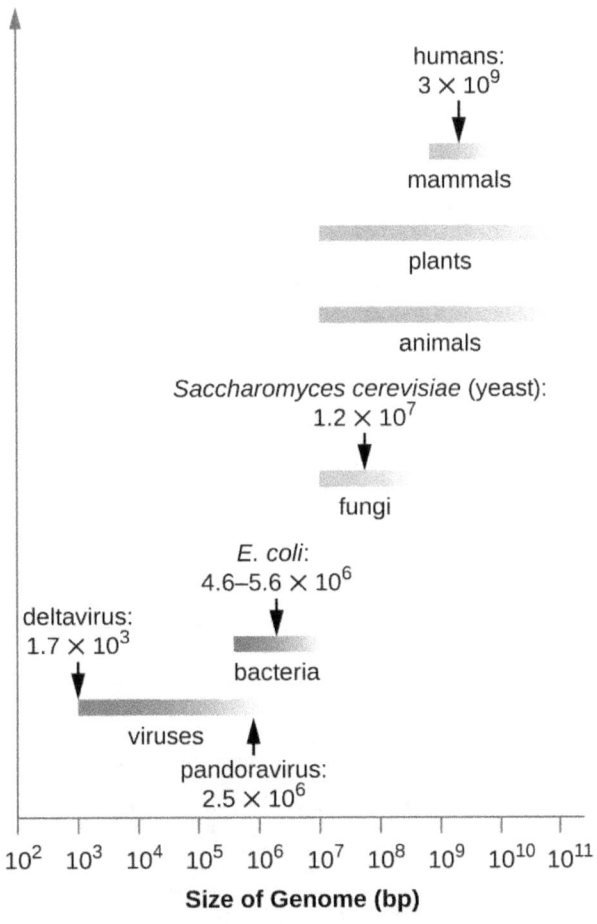

Figure 10.28 There is great variability as well as overlap among the genome sizes of various groups of organisms and viruses.

LINK TO LEARNING

Visit the genome database (https://www.openstax.org/l/22NCBIgendata) of the National Center for Biotechnology Information (NCBI) to see the genomes that have been sequenced and their sizes.

SUMMARY

10.1 Using Microbiology to Discover the Secrets of Life

- **DNA** was discovered and characterized long before its role in heredity was understood. Microbiologists played significant roles in demonstrating that DNA is the hereditary information found within cells.
- In the 1850s and 1860s, Gregor Mendel experimented with true-breeding garden peas to demonstrate the **heritability** of specific observable traits.
- In 1869, Friedrich Miescher isolated and purified a compound rich in phosphorus from the nuclei of white blood cells; he named the compound nuclein. Miescher's student Richard Altmann discovered its acidic nature, renaming it **nucleic acid**. Albrecht Kossell characterized the **nucleotide bases** found within nucleic acids.
- Although Walter Sutton and Theodor Boveri proposed the **Chromosomal Theory of Inheritance** in 1902, it was not scientifically demonstrated until the 1915 publication of the work of Thomas Hunt Morgan and his colleagues.
- Using *Acetabularia*, a large algal cell, as his model system, Joachim Hämmerling demonstrated in the 1930s and 1940s that the nucleus was the location of hereditary information in these cells.
- In the 1940s, George Beadle and Edward Tatum used the mold *Neurospora crassa* to show that each protein's production was under the control of a single gene, demonstrating the **"one gene–one enzyme" hypothesis**.
- In 1928, Frederick Griffith showed that dead encapsulated bacteria could pass genetic information to live nonencapsulated bacteria and transform them into harmful strains. In 1944, Oswald Avery, Colin McLeod, and Maclyn McCarty identified the compound as DNA.
- The nature of DNA as the molecule that stores genetic information was unequivocally demonstrated in the experiment of Alfred Hershey and Martha Chase published in 1952. Labeled DNA from bacterial viruses entered and infected bacterial cells, giving rise to more viral particles. The labeled protein coats did not participate in the transmission of genetic information.

10.2 Structure and Function of DNA

- **Nucleic acids** are composed of **nucleotides**, each of which contains a pentose sugar, a phosphate group, and a **nitrogenous base**. **Deoxyribonucleotides** within DNA contain **deoxyribose** as the pentose sugar.
- DNA contains the **pyrimidines cytosine** and **thymine**, and the **purines adenine** and **guanine**.
- **Nucleotides** are linked together by phosphodiester bonds between the 5′ phosphate group of one nucleotide and the 3′ hydroxyl group of another. A **nucleic acid strand** has a free phosphate group at the 5′ end and a free hydroxyl group at the 3′ end.
- Chargaff discovered that the amount of **adenine** is approximately equal to the amount of **thymine** in DNA, and that the amount of the **guanine** is approximately equal to **cytosine**. These relationships were later determined to be due to complementary base pairing.
- Watson and Crick, building on the work of Chargaff, Franklin and Gosling, and Wilkins, proposed the double helix model and base pairing for DNA structure.
- DNA is composed of two complementary strands oriented **antiparallel** to each other with the **phosphodiester backbones** on the exterior of the molecule. The nitrogenous bases of each strand face each other and complementary bases hydrogen bond to each other, stabilizing the double helix.
- Heat or chemicals can break the hydrogen bonds between complementary bases, denaturing DNA. Cooling or removing chemicals can lead to renaturation or reannealing of DNA by allowing hydrogen bonds to reform between complementary bases.
- DNA stores the instructions needed to build and control the cell. This information is transmitted from parent to offspring through **vertical gene transfer**.

10.3 Structure and Function of RNA

- **Ribonucleic acid (RNA)** is typically single stranded and contains ribose as its pentose sugar and the pyrimidine uracil instead of thymine. An RNA strand can undergo significant intramolecular base pairing to take on a three-dimensional structure.
- There are three main types of RNA, all involved

in protein synthesis.
- Messenger RNA (**mRNA**) serves as the intermediary between DNA and the synthesis of protein products during translation.
- Ribosomal RNA (**rRNA**) is a type of stable RNA that is a major constituent of ribosomes. It ensures the proper alignment of the mRNA and the ribosomes during protein synthesis and catalyzes the formation of the peptide bonds between two aligned amino acids during protein synthesis.
- Transfer RNA (**tRNA**) is a small type of stable RNA that carries an amino acid to the corresponding site of protein synthesis in the ribosome. It is the base pairing between the tRNA and mRNA that allows for the correct amino acid to be inserted in the polypeptide chain being synthesized.
- Although RNA is not used for long-term genetic information in cells, many viruses do use RNA as their genetic material.

10.4 Structure and Function of Cellular Genomes

- The entire genetic content of a cell is its **genome**.
- **Genes** code for proteins, or stable RNA molecules, each of which carries out a specific function in the cell.
- Although the **genotype** that a cell possesses remains constant, expression of genes is dependent on environmental conditions.
- A **phenotype** is the observable characteristics of a cell (or organism) at a given point in time and results from the complement of genes currently being used.

- The majority of genetic material is organized into **chromosomes** that contain the DNA that controls cellular activities.
- Prokaryotes are typically haploid, usually having a single circular chromosome found in the nucleoid. Eukaryotes are diploid; DNA is organized into multiple linear chromosomes found in the nucleus.
- Supercoiling and DNA packaging using DNA binding proteins allows lengthy molecules to fit inside a cell. Eukaryotes and archaea use histone proteins, and bacteria use different proteins with similar function.
- Prokaryotic and eukaryotic genomes both contain **noncoding DNA**, the function of which is not well understood. Some noncoding DNA appears to participate in the formation of small noncoding RNA molecules that influence gene expression; some appears to play a role in maintaining chromosomal structure and in DNA packaging.
- **Extrachromosomal DNA** in eukaryotes includes the chromosomes found within organelles of prokaryotic origin (mitochondria and chloroplasts) that evolved by endosymbiosis. Some viruses may also maintain themselves extrachromosomally.
- Extrachromosomal DNA in prokaryotes is commonly maintained as **plasmids** that encode a few nonessential genes that may be helpful under specific conditions. Plasmids can be spread through a bacterial community by horizontal gene transfer.
- Viral genomes show extensive variation and may be composed of either RNA or DNA, and may be either double or single stranded.

REVIEW QUESTIONS

Multiple Choice

1. Frederick Griffith infected mice with a combination of dead R and live S bacterial strains. What was the outcome, and why did it occur?
   A. The mice will live. Transformation was not required.
   B. The mice will die. Transformation of genetic material from R to S was required.
   C. The mice will live. Transformation of genetic material from S to R was required.
   D. The mice will die. Transformation was not required.

2. Why was the alga *Acetabularia* a good model organism for Joachim Hämmerling to use to identify the location of genetic material?
   A. It lacks a nuclear membrane.
   B. It self-fertilizes.
   C. It is a large, asymmetrical, single cell easy to see with the naked eye.
   D. It makes a protein capsid.

3. Which of the following best describes the results from Hershey and Chase's experiment using bacterial viruses with ^{35}S-labeled proteins or ^{32}P-labeled DNA that are consistent with protein being the molecule responsible for hereditary?
   A. After infection with the ^{35}S-labeled viruses and centrifugation, only the pellet would be radioactive.
   B. After infection with the ^{35}S-labeled viruses and centrifugation, both the pellet and the supernatant would be radioactive.
   C. After infection with the ^{32}P-labeled viruses and centrifugation, only the pellet would be radioactive.
   D. After infection with the ^{32}P-labeled viruses and centrifugation, both the pellet and the supernatant would be radioactive.

4. Which method did Morgan and colleagues use to show that hereditary information was carried on chromosomes?
   A. statistical predictions of the outcomes of crosses using true-breeding parents
   B. correlations between microscopic observations of chromosomal movement and the characteristics of offspring
   C. transformation of nonpathogenic bacteria to pathogenic bacteria
   D. mutations resulting in distinct defects in metabolic enzymatic pathways

5. According to Beadle and Tatum's "one gene–one enzyme" hypothesis, which of the following enzymes will eliminate the transformation of hereditary material from pathogenic bacteria to nonpathogenic bacteria?
   A. carbohydrate-degrading enzymes
   B. proteinases
   C. ribonucleases
   D. deoxyribonucleases

6. Which of the following is not found within DNA?
   A. thymine
   B. phosphodiester bonds
   C. complementary base pairing
   D. amino acids

7. If 30% of the bases within a DNA molecule are adenine, what is the percentage of thymine?
   A. 20%
   B. 25%
   C. 30%
   D. 35%

8. Which of the following statements about base pairing in DNA is incorrect?
   A. Purines always base pairs with pyrimidines.
   B. Adenine binds to guanine.
   C. Base pairs are stabilized by hydrogen bonds.
   D. Base pairing occurs at the interior of the double helix.

9. If a DNA strand contains the sequence 5'-ATTCCGGATCGA-3', which of the following is the sequence of the complementary strand of DNA?
   A. 5'-TAAGGCCTAGCT-3'
   B. 5'-ATTCCGGATCGA-3'
   C. 3'-TAACCGGTACGT-5'
   D. 5'-TCGATCCGGAAT-3'

10. During denaturation of DNA, which of the following happens?
    A. Hydrogen bonds between complementary bases break.
    B. Phosphodiester bonds break within the sugar-phosphate backbone.
    C. Hydrogen bonds within the sugar-phosphate backbone break.
    D. Phosphodiester bonds between complementary bases break.

11. Which of the following types of RNA codes for a protein?
    a. dsRNA
    b. mRNA
    c. rRNA
    d. tRNA

12. A nucleic acid is purified from a mixture. The molecules are relatively small, contain uracil, and most are covalently bound to an amino acid. Which of the following was purified?
    A. DNA
    B. mRNA
    C. rRNA
    D. tRNA

13. Which of the following types of RNA is known for its catalytic abilities?
    A. dsRNA
    B. mRNA
    C. rRNA
    D. tRNA

14. Ribosomes are composed of rRNA and what other component?
    A. protein
    B. carbohydrates
    C. DNA
    D. mRNA

15. Which of the following may use RNA as its genome?
    A. a bacterium
    B. an archaeon
    C. a virus
    D. a eukaryote

16. Which of the following correctly describes the structure of the typical eukaryotic genome?
    A. diploid
    B. linear
    C. singular
    D. double stranded

17. Which of the following is typically found as part of the prokaryotic genome?
    A. chloroplast DNA
    B. linear chromosomes
    C. plasmids
    D. mitochondrial DNA

18. *Serratia marcescens* cells produce a red pigment at room temperature. The red color of the colonies is an example of which of the following?
    A. genotype
    B. phenotype
    C. change in DNA base composition
    D. adaptation to the environment

19. Which of the following genes would not likely be encoded on a plasmid?
    A. genes encoding toxins that damage host tissue
    B. genes encoding antibacterial resistance
    C. gene encoding enzymes for glycolysis
    D. genes encoding enzymes for the degradation of an unusual substrate

20. Histones are DNA binding proteins that are important for DNA packaging in which of the following?
    A. double-stranded and single-stranded DNA viruses
    B. archaea and bacteria
    C. bacteria and eukaryotes
    D. eukaryotes and archaea

True/False

21. The work of Rosalind Franklin and R.G. Gosling was important in demonstrating the helical nature of DNA.
22. The A-T base pair has more hydrogen bonding than the C-G base pair.
23. Ribosomes are composed mostly of RNA.
24. Double-stranded RNA is commonly found inside cells.
25. Within an organism, phenotypes may change while genotypes remain constant.
26. Noncoding DNA has no biological purpose.

Matching

27. Match the correct molecule with its description:

    ___tRNA   A. is a major component of ribosome
    ___rRNA   B. is a copy of the information in a gene
    ___mRNA   C. carries an amino acid to the ribosome

Fill in the Blank

28. The element _____ is unique to nucleic acids compared with other macromolecules.
29. In the late 1800s and early 1900s, the macromolecule thought to be responsible for heredity was _____.
30. The end of a nucleic acid strand with a free phosphate group is called the _____.
31. Plasmids are typically transferred among members of a bacterial community by _____ gene transfer.

Short Answer

32. Why do bacteria and viruses make good model systems for various genetic studies?
33. Why was nucleic acid disregarded for so long as the molecule responsible for the transmission of hereditary information?
34. Bacteriophages inject their genetic material into host cells, whereas animal viruses enter host cells completely. Why was it important to use a bacteriophage in the Hershey–Chase experiment rather than an animal virus?
35. What is the role of phosphodiester bonds within the sugar-phosphate backbone of DNA?
36. What is meant by the term "antiparallel?"
37. Why is DNA with a high GC content more difficult to denature than that with a low GC content?
38. What are the differences between DNA nucleotides and RNA nucleotides?
39. How is the information stored within the base sequence of DNA used to determine a cell's properties?
40. How do complementary base pairs contribute to intramolecular base pairing within an RNA molecule?
41. If an antisense RNA has the sequence 5'AUUCGAAUGC3', what is the sequence of the mRNA to which it will bind? Be sure to label the 5' and 3' ends of the molecule you draw.
42. Why does double-stranded RNA (dsRNA) stimulate RNA interference?
43. What are some differences in chromosomal structures between prokaryotes and eukaryotes?
44. How do prokaryotes and eukaryotes manage to fit their lengthy DNA inside of cells? Why is this necessary?
45. What are some functions of noncoding DNA?
46. In the chromatin of eukaryotic cells, which regions of the chromosome would you expect to be more compact: the regions that contain genes being actively copied into RNA or those that contain inactive genes?

Critical Thinking

47. In the figure shown, if the nuclei were contained within the stalks of *Acetabularia*, what types of caps would you expect from the pictured grafts?

Acetabularia mediterranea

Acetabularia crenulata

graft A   graft B

48. Why are Hershey and Chase credited with identifying DNA as the carrier of heredity even though DNA had been discovered many years before?
49. A certain DNA sample is found to have a makeup consisting of 22% thymine. Use Chargaff's rules to fill in the percentages for the other three nitrogenous bases.

adenine	guanine	thymine	cytosine
___%	___%	22%	___%

50. In considering the structure of the DNA double helix, how would you expect the structure to differ if there was base pairing between two purines? Between two pyrimidines?
51. Identify the location of mRNA, rRNA, and tRNA in the figure.

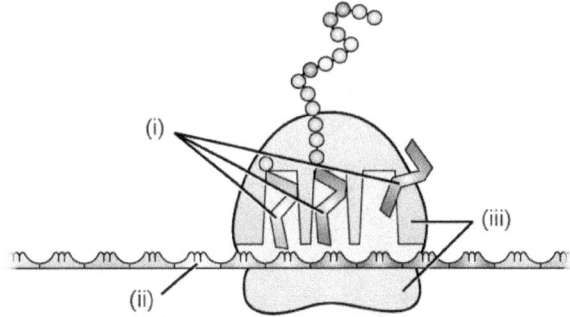

52. Why does it make sense that tRNA and rRNA molecules are more stable than mRNA molecules?
53. A new type of bacteriophage has been isolated and you are in charge of characterizing its genome. The base composition of the bacteriophage is A (15%), C (20%), T (35%), and G (30%). What can you conclude about the genome of the virus?

Chapter 11
Mechanisms of Microbial Genetics

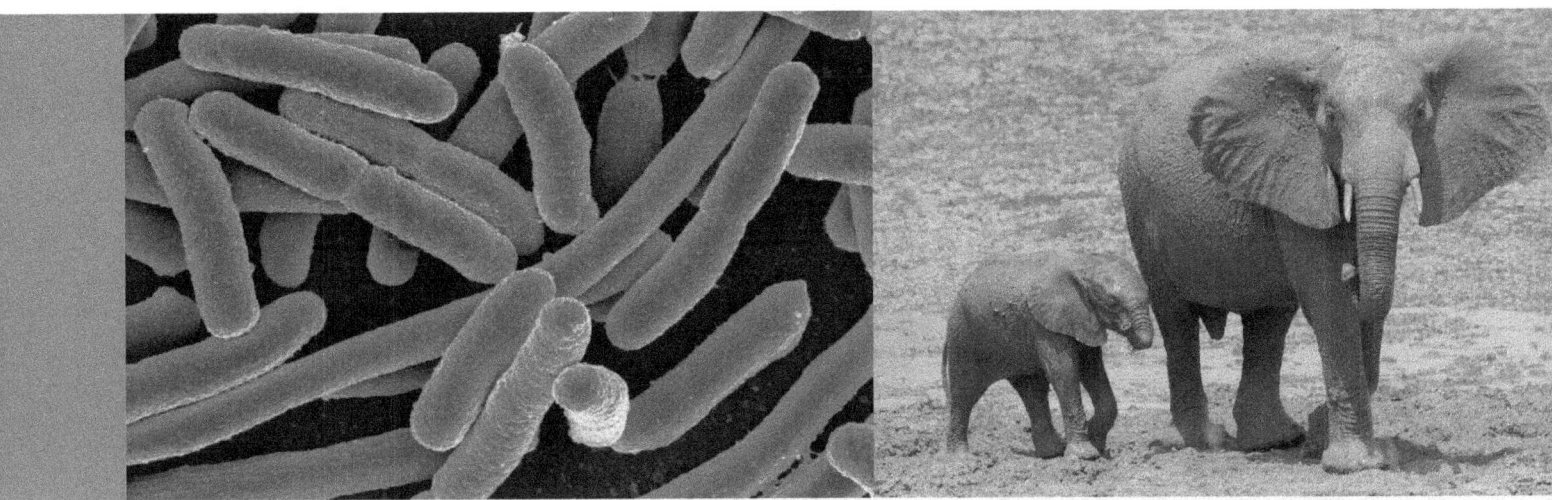

Figure 11.1 *Escherichia coli* (left) may not appear to have much in common with an elephant (right), but the genetic blueprints for these vastly different organisms are both encoded in DNA. (credit left: modification of work by NIAID; credit right: modification of work by Tom Lubbock)

Chapter Outline

11.1 The Functions of Genetic Material

11.2 DNA Replication

11.3 RNA Transcription

11.4 Protein Synthesis (Translation)

11.5 Mutations

11.6 How Asexual Prokaryotes Achieve Genetic Diversity

11.7 Gene Regulation: Operon Theory

INTRODUCTION In 1954, French scientist and future Nobel laureate Jacques Monod (1910–1976) famously said, "What is true in *E. coli* is true in the elephant," suggesting that the biochemistry of life was maintained throughout evolution and is shared in all forms of known life. Since Monod's famous statement, we have learned a great deal about the mechanisms of gene regulation, expression, and replication in living cells. All cells use DNA for information storage, share the same genetic code, and use similar mechanisms to replicate and express it. Although many aspects of genetics are universally shared, variations do exist among contemporary genetic systems. We now know that within the shared overall theme of the genetic mechanism, there are significant differences among the three domains of life: Eukarya, Archaea, and Bacteria. Additionally, viruses, cellular parasites but not themselves living cells, show dramatic variation in their genetic material and the replication and gene expression processes. Some of these differences have allowed us to engineer clinical tools such as antibiotics and antiviral drugs that specifically inhibit the reproduction of pathogens yet are harmless to their hosts.

11.1 The Functions of Genetic Material

Learning Objectives

By the end of this section, you will be able to:

- Explain the two functions of the genome
- Explain the meaning of the central dogma of molecular biology
- Differentiate between genotype and phenotype and explain how environmental factors influence phenotype

---

Clinical Focus

Part 1

Mark is 60-year-old software engineer who suffers from type II diabetes, which he monitors and keeps under control largely through diet and exercise. One spring morning, while doing some gardening, he scraped his lower leg while walking through blackberry brambles. He continued working all day in the yard and did not bother to clean the wound and treat it with antibiotic ointment until later that evening. For the next 2 days, his leg became increasingly red, swollen, and warm to the touch. It was sore not only on the surface, but deep in the muscle. After 24 hours, Mark developed a fever and stiffness in the affected leg. Feeling increasingly weak, he called a neighbor, who drove him to the emergency department.

- Did Mark wait too long to seek medical attention? At what point do his signs and symptoms warrant seeking medical attention?
- What types of infections or other conditions might be responsible for Mark's symptoms?

Jump to the next Clinical Focus box.

DNA serves two essential functions that deal with cellular information. First, DNA is the genetic material responsible for inheritance and is passed from parent to offspring for all life on earth. To preserve the integrity of this genetic information, DNA must be replicated with great accuracy, with minimal errors that introduce changes to the DNA sequence. A genome contains the full complement of DNA within a cell and is organized into smaller, discrete units called genes that are arranged on chromosomes and plasmids. The second function of DNA is to direct and regulate the construction of the proteins necessary to a cell for growth and reproduction in a particular cellular environment.

A gene is composed of DNA that is "read" or transcribed to produce an RNA molecule during the process of transcription. One major type of RNA molecule, called messenger RNA (mRNA), provides the information for the ribosome to catalyze protein synthesis in a process called translation. The processes of transcription and translation are collectively referred to as **gene expression**. Gene expression is the synthesis of a specific protein with a sequence of amino acids that is encoded in the gene. The flow of genetic information from DNA to RNA to protein is described by the **central dogma** (Figure 11.2). This central dogma of molecular biology further elucidates the mechanism behind Beadle and Tatum's "one gene-one enzyme" hypothesis (see Using Microorganisms to Discover the Secrets of Life). Each of the processes of replication, transcription, and translation includes the stages of 1) initiation, 2) elongation (polymerization), and 3) termination. These stages will be described in more detail in this chapter.

Figure 11.2 The central dogma states that DNA encodes messenger RNA, which, in turn, encodes protein.

A cell's genotype is the full collection of genes it contains, whereas its phenotype is the set of observable characteristics that result from those genes. The phenotype is the product of the array of proteins being produced by the cell at a given time, which is influenced by the cell's genotype as well as interactions with the cell's environment. Genes code for proteins that have functions in the cell. Production of a specific protein encoded by an individual gene often results in a distinct phenotype for the cell compared with the phenotype without that protein. For this reason, it is also common to refer to the genotype of an individual gene and its phenotype. Although a cell's genotype remains constant, not all genes are used to direct the production of their proteins simultaneously. Cells carefully regulate expression of their genes, only using genes to make specific proteins when those proteins are needed (Figure 11.3).

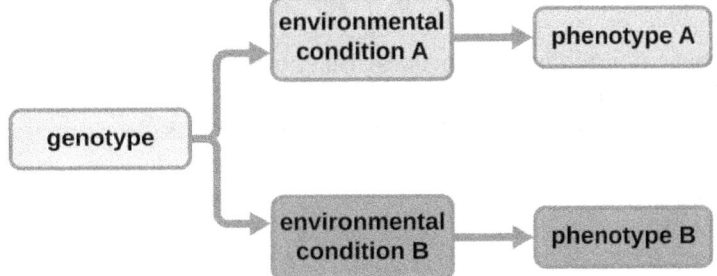

Figure 11.3 Phenotype is determined by the specific genes within a genotype that are expressed under specific conditions. Although multiple cells may have the same genotype, they may exhibit a wide range of phenotypes resulting from differences in patterns of gene expression in response to different environmental conditions.

✓ CHECK YOUR UNDERSTANDING

- What are the two functions of DNA?
- Distinguish between the genotype and phenotype of a cell.
- How can cells have the same genotype but differ in their phenotype?

Eye on Ethics

Use and Abuse of Genome Data

Why can some humans harbor opportunistic pathogens like *Haemophilus influenzae*, *Staphylococcus aureus*, or *Streptococcus pyogenes*, in their upper respiratory tracts but remain asymptomatic carriers, while other individuals become seriously ill when infected? There is evidence suggesting that differences in susceptibility to infection between patients may be a result, at least in part, of genetic differences between human hosts. For example, genetic differences in human leukocyte antigens (HLAs) and red blood cell antigens among hosts have been implicated in different immune responses and resulting disease progression from infection with *H. influenzae*.

Because the genetic interplay between pathogen and host may contribute to disease outcomes, understanding differences in genetic makeup between individuals may be an important clinical tool. Ecological genomics is a relatively new field that seeks to understand how the genotypes of different organisms interact with each other in nature. The field answers questions about how gene expression of one organism affects gene expression of another. Medical applications of ecological genomics will focus on how pathogens interact with specific individuals, as opposed to humans in general. Such analyses would allow medical professionals to use knowledge of an individual's genotype to apply more individualized plans for treatment and prevention of disease.

With the advent of next-generation sequencing, it is relatively easy to obtain the entire genomic sequences of pathogens; a bacterial genome can be sequenced in as little as a day.[1] The speed and cost of sequencing the human genome has also been greatly reduced and, already, individuals can submit samples to receive extensive reports on their personal genetic traits, including ancestry and carrier status for various genetic diseases. As sequencing technologies progress further, such services will continue to become less expensive, more extensive, and quicker.

---

1 D.J. Edwards, K.E. Holt. "Beginner's Guide to Comparative Bacterial Genome Analysis Using Next-Generation Sequence Data." *Microbial Informatics and Experimentation* 3 no. 1 (2013):2.

However, as this day quickly approaches, there are many ethical concerns with which society must grapple. For example, should genome sequencing be a standard practice for everybody? Should it be required by law or by employers if it will lower health-care costs? If one refuses genome sequencing, does he or she forfeit his or her right to health insurance coverage? For what purposes should the data be used? Who should oversee proper use of these data? If genome sequencing reveals predisposition to a particular disease, do insurance companies have the right to increase rates? Will employers treat an employee differently? Knowing that environmental influences also affect disease development, how should the data on the presence of a particular disease-causing allele in an individual be used ethically? The Genetic Information Nondiscrimination Act of 2008 (GINA) currently prohibits discriminatory practices based on genetic information by both health insurance companies and employers. However, GINA does not cover life, disability, or long-term care insurance policies. Clearly, all members of society must continue to engage in conversations about these issues so that such genomic data can be used to improve health care while simultaneously protecting an individual's rights.

11.2 DNA Replication

Learning Objectives

By the end of this section, you will be able to:
- Explain the meaning of semiconservative DNA replication
- Explain why DNA replication is bidirectional and includes both a leading and lagging strand
- Explain why Okazaki fragments are formed
- Describe the process of DNA replication and the functions of the enzymes involved
- Identify the differences between DNA replication in bacteria and eukaryotes
- Explain the process of rolling circle replication

The elucidation of the structure of the double helix by James Watson and Francis Crick in 1953 provided a hint as to how DNA is copied during the process of **replication**. Separating the strands of the double helix would provide two templates for the synthesis of new complementary strands, but exactly how new DNA molecules were constructed was still unclear. In one model, **semiconservative replication**, the two strands of the double helix separate during DNA replication, and each strand serves as a template from which the new complementary strand is copied; after replication, each double-stranded DNA includes one parental or "old" strand and one "new" strand. There were two competing models also suggested: conservative and dispersive, which are shown in Figure 11.4.

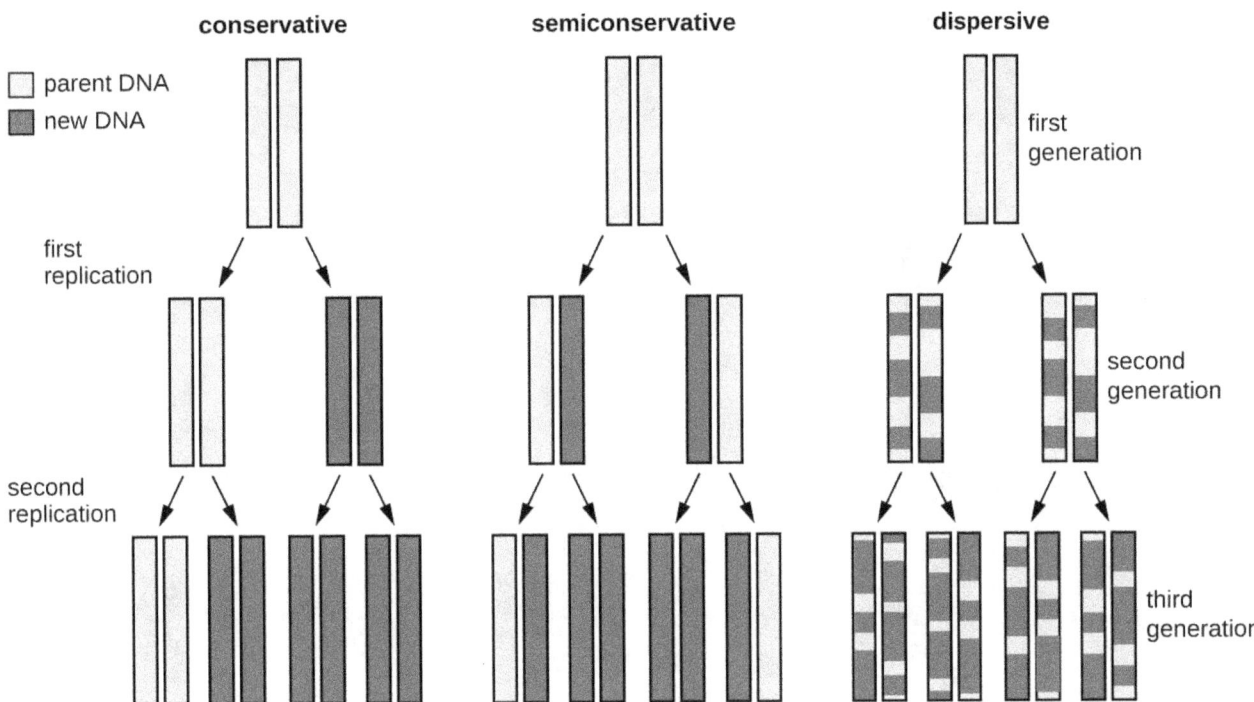

Figure 11.4 There were three models suggested for DNA replication. In the conservative model, parental DNA strands (blue) remained associated in one DNA molecule while new daughter strands (red) remained associated in newly formed DNA molecules. In the semiconservative model, parental strands separated and directed the synthesis of a daughter strand, with each resulting DNA molecule being a hybrid of a parental strand and a daughter strand. In the dispersive model, all resulting DNA strands have regions of double-stranded parental DNA and regions of double-stranded daughter DNA.

Matthew Meselson (1930–) and Franklin Stahl (1929–) devised an experiment in 1958 to test which of these models correctly represents DNA replication (Figure 11.5). They grew *E. coli* for several generations in a medium containing a "heavy" isotope of nitrogen (^{15}N) that was incorporated into nitrogenous bases and, eventually, into the DNA. This labeled the parental DNA. The *E. coli* culture was then shifted into a medium containing ^{14}N and allowed to grow for one generation. The cells were harvested and the DNA was isolated. The DNA was separated by ultracentrifugation, during which the DNA formed bands according to its density. DNA grown in ^{15}N would be expected to form a band at a higher density position than that grown in ^{14}N. Meselson and Stahl noted that after one generation of growth in ^{14}N, the single band observed was intermediate in position in between DNA of cells grown exclusively in ^{15}N or ^{14}N. This suggested either a semiconservative or dispersive mode of replication. Some cells were allowed to grow for one more generation in ^{14}N and spun again. The DNA harvested from cells grown for two generations in ^{14}N formed two bands: one DNA band was at the intermediate position between ^{15}N and ^{14}N, and the other corresponded to the band of ^{14}N DNA. These results could only be explained if DNA replicates in a semiconservative manner. Therefore, the other two models were ruled out. As a result of this experiment, we now know that during DNA replication, each of the two strands that make up the double helix serves as a template from which new strands are copied. The new strand will be complementary to the parental or "old" strand. The resulting DNA molecules have the same sequence and are divided equally into the two daughter cells.

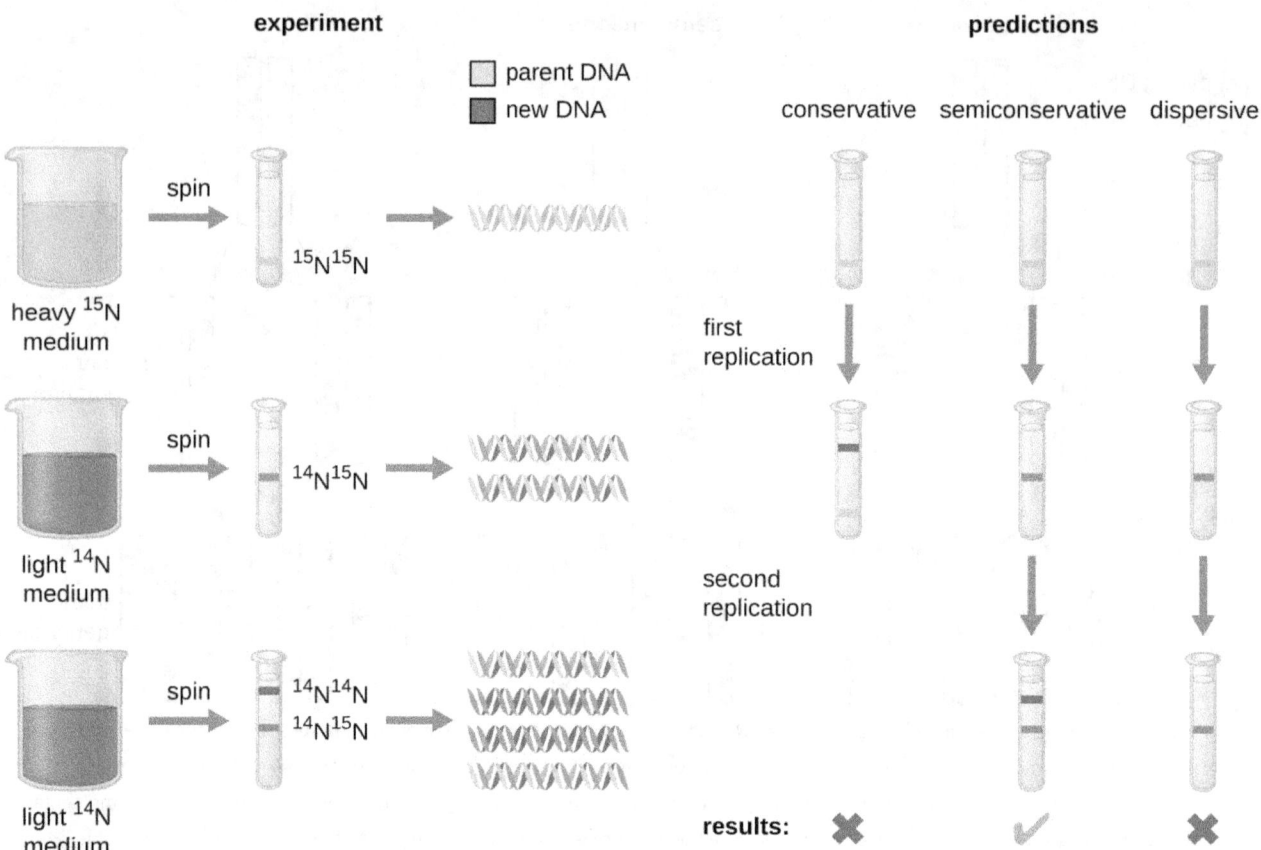

Figure 11.5 Meselson and Stahl experimented with *E. coli* grown first in heavy nitrogen (^{15}N) then in ^{14}N. DNA grown in ^{15}N (blue band) was heavier than DNA grown in ^{14}N (red band), and sedimented to a lower level on ultracentrifugation. After one round of replication, the DNA sedimented halfway between the ^{15}N and ^{14}N levels (purple band), ruling out the conservative model of replication. After a second round of replication, the dispersive model of replication was ruled out. These data supported the semiconservative replication model.

CHECK YOUR UNDERSTANDING

- What would have been the conclusion of Meselson and Stahl's experiment if, after the first generation, they had found two bands of DNA?

DNA Replication in Bacteria

DNA replication has been well studied in bacteria primarily because of the small size of the genome and the mutants that are available. *E. coli* has 4.6 million base pairs (Mbp) in a single circular chromosome and all of it is replicated in approximately 42 minutes, starting from a single origin of replication and proceeding around the circle bidirectionally (i.e., in both directions). This means that approximately 1000 nucleotides are added per second. The process is quite rapid and occurs with few errors.

DNA replication uses a large number of proteins and enzymes (Table 11.1). One of the key players is the enzyme **DNA polymerase**, also known as DNA pol. In bacteria, three main types of DNA polymerases are known: DNA pol I, DNA pol II, and DNA pol III. It is now known that DNA pol III is the enzyme required for DNA synthesis; DNA pol I and DNA pol II are primarily required for repair. DNA pol III adds deoxyribonucleotides each complementary to a nucleotide on the template strand, one by one to the 3'-OH group of the growing DNA chain. The addition of these nucleotides requires energy. This energy is present in the bonds of three phosphate groups attached to each nucleotide (a triphosphate nucleotide), similar to how energy is stored in the phosphate bonds of adenosine triphosphate (ATP) (Figure 11.6). When the bond between the phosphates is broken and diphosphate is released, the energy released allows for the formation of a covalent phosphodiester bond by dehydration synthesis between the incoming nucleotide and the free 3'-OH group on the growing DNA strand.

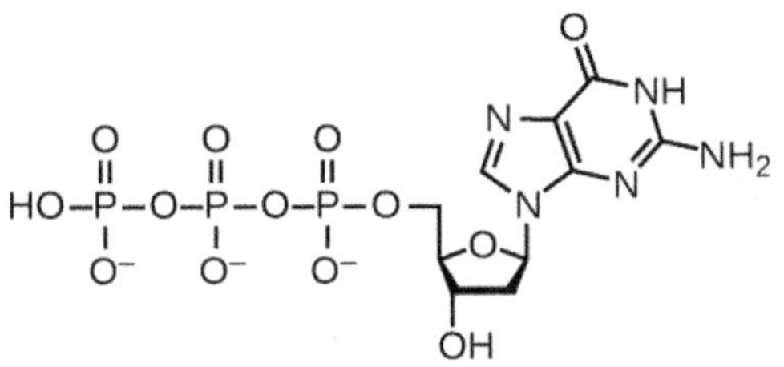

Figure 11.6 This structure shows the guanosine triphosphate deoxyribonucleotide that is incorporated into a growing DNA strand by cleaving the two end phosphate groups from the molecule and transferring the energy to the sugar phosphate bond. The other three nucleotides form analogous structures.

Initiation

The **initiation of replication** occurs at specific nucleotide sequence called the **origin of replication**, where various proteins bind to begin the replication process. *E. coli* has a single origin of replication (as do most prokaryotes), called *oriC*, on its one chromosome. The origin of replication is approximately 245 base pairs long and is rich in adenine-thymine (AT) sequences.

Some of the proteins that bind to the origin of replication are important in making single-stranded regions of DNA accessible for replication. Chromosomal DNA is typically wrapped around histones (in eukaryotes and archaea) or histone-like proteins (in bacteria), and is **supercoiled**, or extensively wrapped and twisted on itself. This packaging makes the information in the DNA molecule inaccessible. However, enzymes called topoisomerases change the shape and supercoiling of the chromosome. For bacterial DNA replication to begin, the supercoiled chromosome is relaxed by **topoisomerase II**, also called **DNA gyrase**. An enzyme called **helicase** then separates the DNA strands by breaking the hydrogen bonds between the nitrogenous base pairs. Recall that AT sequences have fewer hydrogen bonds and, hence, have weaker interactions than guanine-cytosine (GC) sequences. These enzymes require ATP hydrolysis. As the DNA opens up, Y-shaped structures called **replication forks** are formed. Two replication forks are formed at the origin of replication, allowing for bidirectional replication and formation of a structure that looks like a bubble when viewed with a transmission electron microscope; as a result, this structure is called a **replication bubble**. The DNA near each replication fork is coated with **single-stranded binding proteins** to prevent the single-stranded DNA from rewinding into a double helix.

Once single-stranded DNA is accessible at the origin of replication, DNA replication can begin. However, DNA pol III is able to add nucleotides only in the 5' to 3' direction (a new DNA strand can be only extended in this direction). This is because DNA polymerase requires a free 3'-OH group to which it can add nucleotides by forming a covalent phosphodiester bond between the 3'-OH end and the 5' phosphate of the next nucleotide. This also means that it cannot add nucleotides if a free 3'-OH group is not available, which is the case for a single strand of DNA. The problem is solved with the help of an RNA sequence that provides the free 3'-OH end. Because this sequence allows the start of DNA synthesis, it is appropriately called the **primer**. The primer is five to 10 nucleotides long and complementary to the parental or template DNA. It is synthesized by RNA **primase**, which is an RNA polymerase. Unlike DNA polymerases, RNA polymerases do not need a free 3'-OH group to synthesize an RNA molecule. Now that the primer provides the free 3'-OH group, DNA polymerase III can now extend this RNA primer, adding DNA nucleotides one by one that are complementary to the template strand (Figure 11.4).

Elongation

During **elongation in DNA replication**, the addition of nucleotides occurs at its maximal rate of about 1000 nucleotides per second. DNA polymerase III can only extend in the 5' to 3' direction, which poses a problem at the replication fork. The DNA double helix is antiparallel; that is, one strand is oriented in the 5' to 3' direction and the other is oriented in the 3' to 5' direction (see Structure and Function of DNA). During replication, one strand, which is complementary to the 3' to 5' parental DNA strand, is synthesized continuously toward the replication fork because polymerase can add nucleotides in this direction. This continuously synthesized strand is known as the **leading strand**. The other strand, complementary to the 5' to 3' parental DNA, grows away from the replication fork, so the polymerase must move back toward the replication fork to begin adding bases to a new primer, again in the direction away from the replication fork. It does so until it bumps into the previously synthesized strand and then it moves back again (Figure 11.7). These steps produce small DNA sequence fragments known as **Okazaki fragments**, each separated by RNA primer. Okazaki fragments are named after the Japanese research team and married couple Reiji and Tsuneko Okazaki, who first discovered them in 1966. The strand with the Okazaki fragments is known as the **lagging strand**, and its synthesis is said to be discontinuous.

The leading strand can be extended from one primer alone, whereas the lagging strand needs a new primer for each of the short Okazaki fragments. The overall direction of the lagging strand will be 3' to 5', and that of the leading strand 5' to 3'. A protein called the sliding clamp holds the DNA polymerase in place as it continues to add nucleotides. The sliding clamp is a ring-shaped protein that binds to the DNA and holds the polymerase in place. Beyond its role in initiation, topoisomerase also prevents the overwinding of the DNA double helix ahead of the replication fork as the DNA is opening up; it does so by causing temporary nicks in the DNA helix and then resealing it. As synthesis proceeds, the RNA primers are replaced by DNA. The primers are removed by the **exonuclease** activity of DNA polymerase I, and the gaps are filled in. The nicks that remain between the newly synthesized DNA (that replaced the RNA primer) and the previously synthesized DNA are sealed by the enzyme **DNA ligase** that catalyzes the formation of covalent phosphodiester linkage between the 3'-OH end of one DNA fragment and the 5' phosphate end of the other fragment, stabilizing the sugar-phosphate backbone of the DNA molecule.

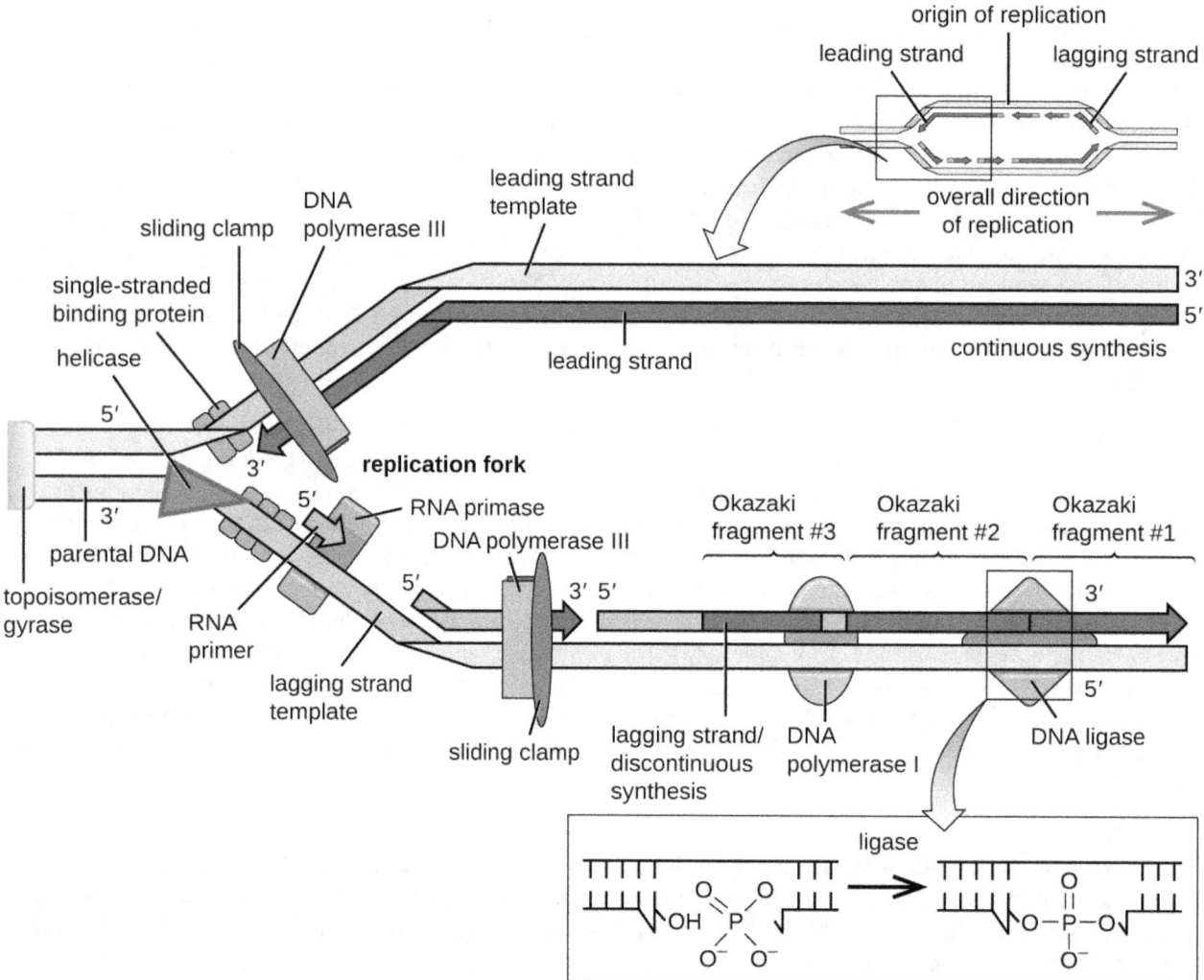

Figure 11.7 At the origin of replication, topoisomerase II relaxes the supercoiled chromosome. Two replication forks are formed by the opening of the double-stranded DNA at the origin, and helicase separates the DNA strands, which are coated by single-stranded binding proteins to keep the strands separated. DNA replication occurs in both directions. An RNA primer complementary to the parental strand is synthesized by RNA primase and is elongated by DNA polymerase III through the addition of nucleotides to the 3'-OH end. On the leading

strand, DNA is synthesized continuously, whereas on the lagging strand, DNA is synthesized in short stretches called Okazaki fragments. RNA primers within the lagging strand are removed by the exonuclease activity of DNA polymerase I, and the Okazaki fragments are joined by DNA ligase.

Termination

Once the complete chromosome has been replicated, **termination of DNA replication** must occur. Although much is known about initiation of replication, less is known about the termination process. Following replication, the resulting complete circular genomes of prokaryotes are concatenated, meaning that the circular DNA chromosomes are interlocked and must be separated from each other. This is accomplished through the activity of bacterial topoisomerase IV, which introduces double-stranded breaks into DNA molecules, allowing them to separate from each other; the enzyme then reseals the circular chromosomes. The resolution of concatemers is an issue unique to prokaryotic DNA replication because of their circular chromosomes. Because both bacterial DNA gyrase and topoisomerase IV are distinct from their eukaryotic counterparts, these enzymes serve as targets for a class of antimicrobial drugs called quinolones.

The Molecular Machinery Involved in Bacterial DNA Replication

Enzyme or Factor	Function
DNA pol I	Exonuclease activity removes RNA primer and replaces it with newly synthesized DNA
DNA pol III	Main enzyme that adds nucleotides in the 5' to 3' direction
Helicase	Opens the DNA helix by breaking hydrogen bonds between the nitrogenous bases
Ligase	Seals the gaps between the Okazaki fragments on the lagging strand to create one continuous DNA strand
Primase	Synthesizes RNA primers needed to start replication
Single-stranded binding proteins	Bind to single-stranded DNA to prevent hydrogen bonding between DNA strands, reforming double-stranded DNA
Sliding clamp	Helps hold DNA pol III in place when nucleotides are being added
Topoisomerase II (DNA gyrase)	Relaxes supercoiled chromosome to make DNA more accessible for the initiation of replication; helps relieve the stress on DNA when unwinding, by causing breaks and then resealing the DNA
Topoisomerase IV	Introduces single-stranded break into concatenated chromosomes to release them from each other, and then reseals the DNA

Table 11.1

✓ CHECK YOUR UNDERSTANDING

- Which enzyme breaks the hydrogen bonds holding the two strands of DNA together so that replication can occur?
- Is it the lagging strand or the leading strand that is synthesized in the direction toward the opening of the replication fork?

- Which enzyme is responsible for removing the RNA primers in newly replicated bacterial DNA?

DNA Replication in Eukaryotes

Eukaryotic genomes are much more complex and larger than prokaryotic genomes and are typically composed of multiple linear chromosomes (Table 11.2). The human genome, for example, has 3 billion base pairs per haploid set of chromosomes, and 6 billion base pairs are inserted during replication. There are multiple origins of replication on each eukaryotic chromosome (Figure 11.8); the human genome has 30,000 to 50,000 origins of replication. The rate of replication is approximately 100 nucleotides per second—10 times slower than prokaryotic replication.

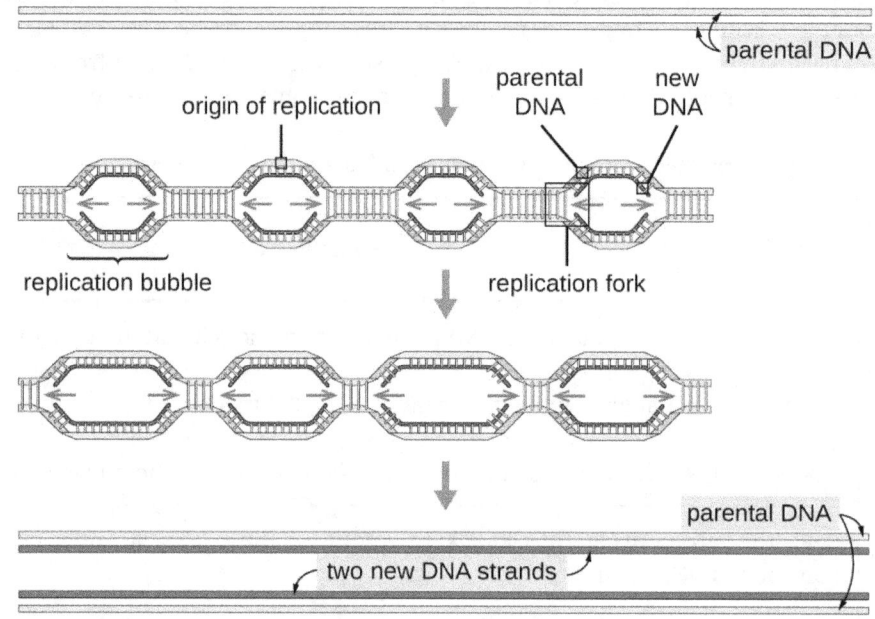

Figure 11.8 Eukaryotic chromosomes are typically linear, and each contains multiple origins of replication.

The essential steps of replication in eukaryotes are the same as in prokaryotes. Before replication can start, the DNA has to be made available as a template. Eukaryotic DNA is highly supercoiled and packaged, which is facilitated by many proteins, including histones (see Structure and Function of Cellular Genomes). At the origin of replication, a prereplication complex composed of several proteins, including helicase, forms and recruits other enzymes involved in the initiation of replication, including topoisomerase to relax supercoiling, single-stranded binding protein, RNA primase, and DNA polymerase. Following initiation of replication, in a process similar to that found in prokaryotes, elongation is facilitated by eukaryotic DNA polymerases. The leading strand is continuously synthesized by the eukaryotic polymerase enzyme pol δ, while the lagging strand is synthesized by pol ε. A sliding clamp protein holds the DNA polymerase in place so that it does not fall off the DNA. The enzyme ribonuclease H (RNase H), instead of a DNA polymerase as in bacteria, removes the RNA primer, which is then replaced with DNA nucleotides. The gaps that remain are sealed by DNA ligase.

Because eukaryotic chromosomes are linear, one might expect that their replication would be more straightforward. As in prokaryotes, the eukaryotic DNA polymerase can add nucleotides only in the 5' to 3' direction. In the leading strand, synthesis continues until it reaches either the end of the chromosome or another replication fork progressing in the opposite direction. On the lagging strand, DNA is synthesized in short stretches, each of which is initiated by a separate primer. When the replication fork reaches the end of the linear chromosome, there is no place to make a primer for the DNA fragment to be copied at the end of the chromosome. These ends thus remain unpaired and, over time, they may get progressively shorter as cells continue to divide.

The ends of the linear chromosomes are known as **telomere**s and consist of noncoding repetitive sequences. The telomeres protect coding sequences from being lost as cells continue to divide. In humans, a six base-pair sequence, TTAGGG, is repeated 100 to 1000 times to form the telomere. The discovery of the enzyme

telomerase (Figure 11.9) clarified our understanding of how chromosome ends are maintained. Telomerase contains a catalytic part and a built-in RNA template. It attaches to the end of the chromosome, and complementary bases to the RNA template are added on the 3' end of the DNA strand. Once the 3' end of the lagging strand template is sufficiently elongated, DNA polymerase can add the nucleotides complementary to the ends of the chromosomes. In this way, the ends of the chromosomes are replicated. In humans, telomerase is typically active in germ cells and adult stem cells; it is not active in adult somatic cells and may be associated with the aging of these cells. Eukaryotic microbes including fungi and protozoans also produce telomerase to maintain chromosomal integrity. For her discovery of telomerase and its action, Elizabeth Blackburn (1948–) received the Nobel Prize for Medicine or Physiology in 2009.

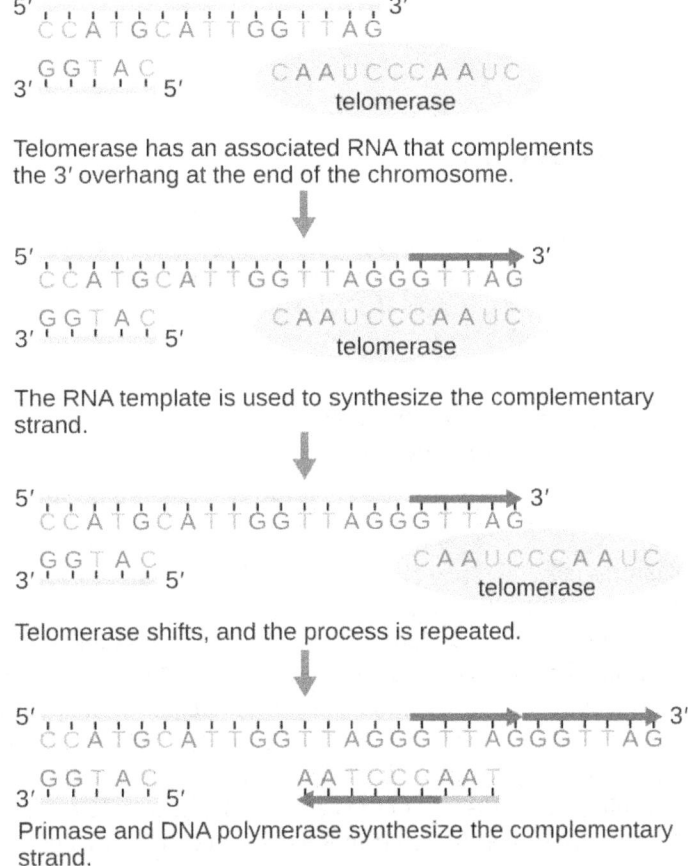

Figure 11.9 In eukaryotes, the ends of the linear chromosomes are maintained by the action of the telomerase enzyme.

Comparison of Bacterial and Eukaryotic Replication

Property	Bacteria	Eukaryotes
Genome structure	Single circular chromosome	Multiple linear chromosomes
Number of origins per chromosome	Single	Multiple
Rate of replication	1000 nucleotides per second	100 nucleotides per second
Telomerase	Not present	Present
RNA primer removal	DNA pol I	RNase H

Comparison of Bacterial and Eukaryotic Replication

Property	Bacteria	Eukaryotes
Strand elongation	DNA pol III	pol δ, pol ε

Table 11.2

🔗 LINK TO LEARNING

This animation (https://openstax.org/l/22DNAreplicani) compares the process of prokaryotic and eukaryotic DNA replication.

✓ CHECK YOUR UNDERSTANDING

- How does the origin of replication differ between eukaryotes and prokaryotes?
- What polymerase enzymes are responsible for DNA synthesis during eukaryotic replication?
- What is found at the ends of the chromosomes in eukaryotes and why?

DNA Replication of Extrachromosomal Elements: Plasmids and Viruses

To copy their nucleic acids, plasmids and viruses frequently use variations on the pattern of DNA replication described for prokaryote genomes. For more information on the wide range of viral replication strategies, see The Viral Life Cycle.

Rolling Circle Replication

Whereas many bacterial plasmids (see Unique Characteristics of Prokaryotic Cells) replicate by a process similar to that used to copy the bacterial chromosome, other plasmids, several bacteriophages, and some viruses of eukaryotes use **rolling circle replication** (Figure 11.10). The circular nature of plasmids and the circularization of some viral genomes on infection make this possible. Rolling circle replication begins with the enzymatic nicking of one strand of the double-stranded circular molecule at the double-stranded origin (dso) site. In bacteria, DNA polymerase III binds to the 3'-OH group of the nicked strand and begins to unidirectionally replicate the DNA using the un-nicked strand as a template, displacing the nicked strand as it does so. Completion of DNA replication at the site of the original nick results in full displacement of the nicked strand, which may then recircularize into a single-stranded DNA molecule. RNA primase then synthesizes a primer to initiate DNA replication at the single-stranded origin (sso) site of the single-stranded DNA (ssDNA) molecule, resulting in a double-stranded DNA (dsDNA) molecule identical to the other circular DNA molecule.

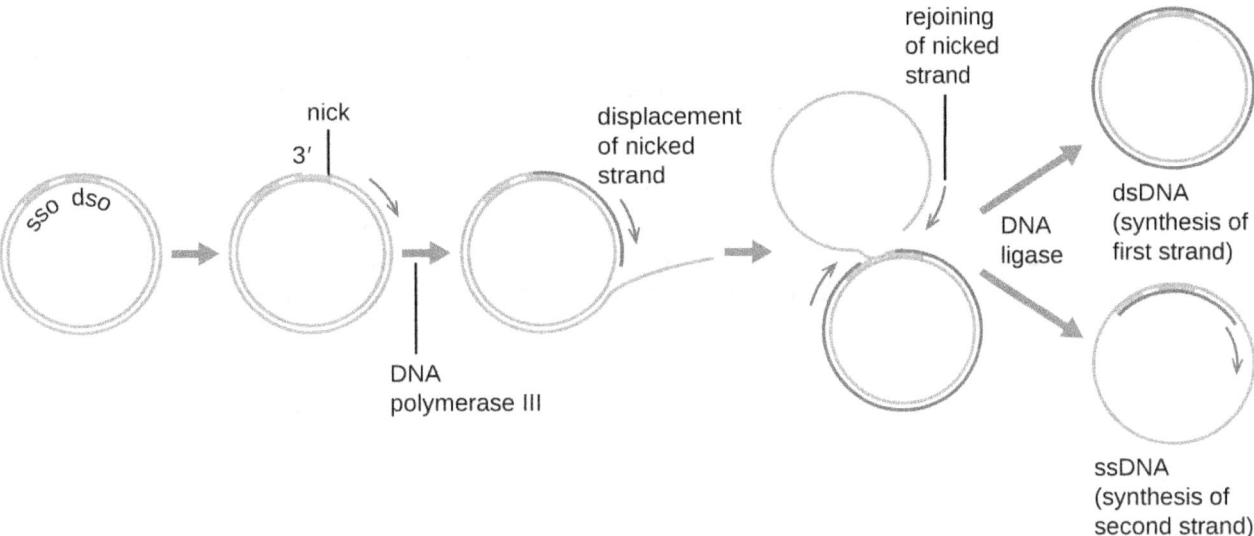

Figure 11.10 The process of rolling circle replication results in the synthesis of a single new copy of the circular DNA molecule, as shown here.

CHECK YOUR UNDERSTANDING

- Is there a lagging strand in rolling circle replication? Why or why not?

11.3 RNA Transcription

Learning Objectives

By the end of this section, you will be able to:
- Explain how RNA is synthesized using DNA as a template
- Distinguish between transcription in prokaryotes and eukaryotes

During the process of **transcription**, the information encoded within the DNA sequence of one or more genes is transcribed into a strand of RNA, also called an **RNA transcript**. The resulting single-stranded RNA molecule, composed of ribonucleotides containing the bases adenine (A), cytosine (C), guanine (G), and uracil (U), acts as a mobile molecular copy of the original DNA sequence. Transcription in prokaryotes and in eukaryotes requires the DNA double helix to partially unwind in the region of RNA synthesis. The unwound region is called a **transcription bubble**. Transcription of a particular gene always proceeds from one of the two DNA strands that acts as a template, the so-called **antisense strand**. The RNA product is complementary to the template strand of DNA and is almost identical to the nontemplate DNA strand, or the **sense strand**. The only difference is that in RNA, all of the T nucleotides are replaced with U nucleotides; during RNA synthesis, U is incorporated when there is an A in the complementary antisense strand.

Transcription in Bacteria

Bacteria use the same RNA polymerase to transcribe all of their genes. Like DNA polymerase, **RNA polymerase** adds nucleotides one by one to the 3'-OH group of the growing nucleotide chain. One critical difference in activity between DNA polymerase and RNA polymerase is the requirement for a 3'-OH onto which to add nucleotides: DNA polymerase requires such a 3'-OH group, thus necessitating a primer, whereas RNA polymerase does not. During transcription, a ribonucleotide complementary to the DNA template strand is added to the growing RNA strand and a covalent phosphodiester bond is formed by dehydration synthesis between the new nucleotide and the last one added. In *E. coli*, RNA polymerase comprises six polypeptide subunits, five of which compose the polymerase core enzyme responsible for adding RNA nucleotides to a growing strand. The sixth subunit is known as sigma (σ). The σ factor enables RNA polymerase to bind to a specific promoter, thus allowing for the transcription of various genes. There are various σ factors that allow for transcription of various genes.

Initiation

The **initiation of transcription** begins at a **promoter**, a DNA sequence onto which the transcription machinery binds and initiates transcription. The nucleotide pair in the DNA double helix that corresponds to the site from which the first 5' RNA nucleotide is transcribed is the initiation site. Nucleotides preceding the initiation site are designated "upstream," whereas nucleotides following the initiation site are called "downstream" nucleotides. In most cases, promoters are located just upstream of the genes they regulate. Although promoter sequences vary among bacterial genomes, a few elements are conserved. At the −10 and −35 positions within the DNA prior to the initiation site (designated +1), there are two promoter consensus sequences, or regions that are similar across all promoters and across various bacterial species. The −10 consensus sequence, called the TATA box, is TATAAT. The −35 sequence is recognized and bound by σ.

Elongation

The **elongation in transcription** phase begins when the σ subunit dissociates from the polymerase, allowing the core enzyme to synthesize RNA complementary to the DNA template in a 5' to 3' direction at a rate of approximately 40 nucleotides per second. As elongation proceeds, the DNA is continuously unwound ahead of the core enzyme and rewound behind it (Figure 11.11).

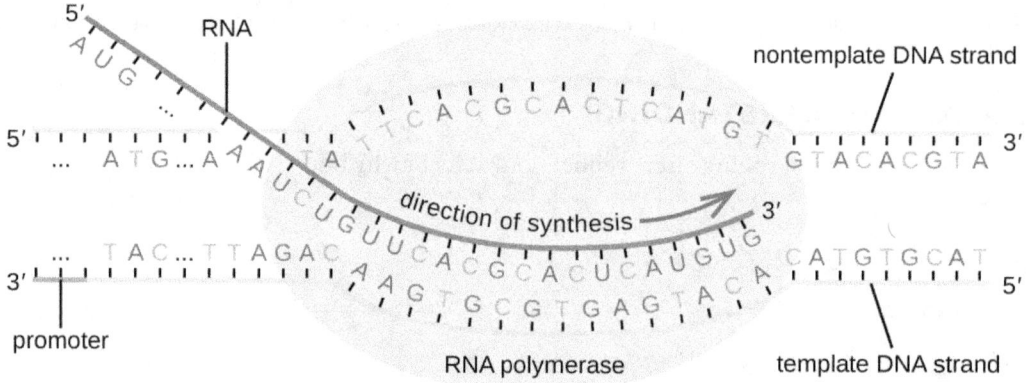

Figure 11.11 During elongation, the bacterial RNA polymerase tracks along the DNA template, synthesizes mRNA in the 5' to 3' direction, and unwinds and rewinds the DNA as it is read.

Termination

Once a gene is transcribed, the bacterial polymerase must dissociate from the DNA template and liberate the newly made RNA. This is referred to as **termination of transcription**. The DNA template includes repeated nucleotide sequences that act as termination signals, causing RNA polymerase to stall and release from the DNA template, freeing the RNA transcript.

> **CHECK YOUR UNDERSTANDING**
> - Where does σ factor of RNA polymerase bind DNA to start transcription?
> - What occurs to initiate the polymerization activity of RNA polymerase?
> - Where does the signal to end transcription come from?

Transcription in Eukaryotes

Prokaryotes and eukaryotes perform fundamentally the same process of transcription, with a few significant differences (see Table 11.3). Eukaryotes use three different polymerases, RNA polymerases I, II, and III, all structurally distinct from the bacterial RNA polymerase. Each transcribes a different subset of genes. Interestingly, archaea contain a single RNA polymerase that is more closely related to eukaryotic RNA polymerase II than to its bacterial counterpart. Eukaryotic mRNAs are also usually monocistronic, meaning that they each encode only a single polypeptide, whereas prokaryotic mRNAs of bacteria and archaea are commonly **polycistronic**, meaning that they encode multiple polypeptides.

The most important difference between prokaryotes and eukaryotes is the latter's membrane-bound nucleus, which influences the ease of use of RNA molecules for protein synthesis. With the genes bound in a nucleus, the eukaryotic cell must transport protein-encoding RNA molecules to the cytoplasm to be translated. Protein-encoding **primary transcript**s, the RNA molecules directly synthesized by RNA polymerase, must undergo several processing steps to protect these RNA molecules from degradation during the time they are transferred from the nucleus to the cytoplasm and translated into a protein. For example, eukaryotic mRNAs may last for several hours, whereas the typical prokaryotic mRNA lasts no more than 5 seconds.

The primary transcript (also called pre-mRNA) is first coated with RNA-stabilizing proteins to protect it from degradation while it is processed and exported out of the nucleus. The first type of processing begins while the primary transcript is still being synthesized; a special 7-methylguanosine nucleotide, called the **5' cap**, is added to the 5' end of the growing transcript. In addition to preventing degradation, factors involved in subsequent protein synthesis recognize the cap, which helps initiate translation by ribosomes. Once elongation is complete, another processing enzyme then adds a string of approximately 200 adenine nucleotides to the 3' end, called the **poly-A tail**. This modification further protects the pre-mRNA from degradation and signals to cellular factors that the transcript needs to be exported to the cytoplasm.

Eukaryotic genes that encode polypeptides are composed of coding sequences called **exon**s (*ex*-on signifies that they are *ex*pressed) and intervening sequences called **intron**s (*int*-ron denotes their *int*ervening role). Transcribed RNA sequences corresponding to introns do not encode regions of the functional polypeptide and are removed from the pre-mRNA during processing. It is essential that all of the intron-encoded RNA sequences are completely and precisely removed from a pre-mRNA before protein synthesis so that the exon-encoded RNA sequences are properly joined together to code for a functional polypeptide. If the process errs by even a single nucleotide, the sequences of the rejoined exons would be shifted, and the resulting polypeptide would be nonfunctional. The process of removing intron-encoded RNA sequences and reconnecting those encoded by exons is called **RNA splicing** and is facilitated by the action of a **spliceosome** containing small nuclear ribonucleo proteins (snRNPs). Intron-encoded RNA sequences are removed from the pre-mRNA while it is still in the nucleus. Although they are not translated, introns appear to have various functions, including gene regulation and mRNA transport. On completion of these modifications, the mature transcript, the mRNA that encodes a polypeptide, is transported out of the nucleus, destined for the cytoplasm for translation. Introns can be spliced out differently, resulting in various exons being included or excluded from the final mRNA product. This process is known as alternative splicing. The advantage of alternative splicing is that different types of mRNA transcripts can be generated, all derived from the same DNA sequence. In recent years, it has been shown that some archaea also have the ability to splice their pre-mRNA.

Comparison of Transcription in Bacteria Versus Eukaryotes

Property	Bacteria	Eukaryotes
Number of polypeptides encoded per mRNA	Monocistronic or polycistronic	Exclusively monocistronic
Strand elongation	core + σ = holoenzyme	RNA polymerases I, II, or III
Addition of 5' cap	No	Yes
Addition of 3' poly-A tail	No	Yes
Splicing of pre-mRNA	No	Yes

Table 11.3

 LINK TO LEARNING

Visualize how mRNA splicing (https://openstax.org/l/22mrnasplice) happens by watching the process in action in this video. See how introns are removed during RNA splicing (https://openstax.org/l/22rnasplice) here.

✓ CHECK YOUR UNDERSTANDING

- In eukaryotic cells, how is the RNA transcript from a gene for a protein modified after it is transcribed?
- Do exons or introns contain information for protein sequences?

Clinical Focus

Part 2

In the emergency department, a nurse told Mark that he had made a good decision to come to the hospital because his symptoms indicated an infection that had gotten out of control. Mark's symptoms had progressed, with the area of skin affected and the amount of swelling increasing. Within the affected area, a rash had begun, blistering and small gas pockets underneath the outermost layer of skin had formed, and some of the skin was becoming gray. Based on the putrid smell of the pus draining from one of the blisters, the rapid progression of the infection, and the visual appearance of the affected skin, the physician immediately began treatment for necrotizing fasciitis. Mark's physician ordered a culture of the fluid draining from the blister and also ordered blood work, including a white blood cell count.

Mark was admitted to the intensive care unit and began intravenous administration of a broad-spectrum antibiotic to try to minimize further spread of the infection. Despite antibiotic therapy, Mark's condition deteriorated quickly. Mark became confused and dizzy. Within a few hours of his hospital admission, his blood pressure dropped significantly and his breathing became shallower and more rapid. Additionally, blistering increased, with the blisters intensifying in color to purplish black, and the wound itself seemed to be progressing rapidly up Mark's leg.

- What are possible causative agents of Mark's necrotizing fasciitis?
- What are some possible explanations for why the antibiotic treatment does not seem to be working?

Jump to the next Clinical Focus box. Go back to the previous Clinical Focus box.

11.4 Protein Synthesis (Translation)

Learning Objectives

By the end of this section, you will be able to:
- Describe the genetic code and explain why it is considered almost universal
- Explain the process of translation and the functions of the molecular machinery of translation
- Compare translation in eukaryotes and prokaryotes

The synthesis of proteins consumes more of a cell's energy than any other metabolic process. In turn, proteins account for more mass than any other macromolecule of living organisms. They perform virtually every function of a cell, serving as both functional (e.g., enzymes) and structural elements. The process of **translation**, or **protein synthesis**, the second part of gene expression, involves the decoding by a ribosome of an mRNA message into a polypeptide product.

The Genetic Code

Translation of the mRNA template converts nucleotide-based genetic information into the "language" of amino acids to create a protein product. A protein sequence consists of 20 commonly occurring amino acids. Each amino acid is defined within the mRNA by a triplet of nucleotides called a **codon**. The relationship between an mRNA codon and its corresponding amino acid is called the **genetic code**.

The three-nucleotide code means that there is a total of 64 possible combinations (4^3, with four different nucleotides possible at each of the three different positions within the codon). This number is greater than the number of amino acids and a given amino acid is encoded by more than one codon (Figure 11.12). This redundancy in the genetic code is called **degeneracy**. Typically, whereas the first two positions in a codon are important for determining which amino acid will be incorporated into a growing polypeptide, the third position, called the **wobble position**, is less critical. In some cases, if the nucleotide in the third position is changed, the same amino acid is still incorporated.

Whereas 61 of the 64 possible triplets code for amino acids, three of the 64 codons do not code for an amino acid; they terminate protein synthesis, releasing the polypeptide from the translation machinery. These are called **stop codons** or **nonsense codons**. Another codon, AUG, also has a special function. In addition to specifying the amino acid methionine, it also typically serves as the **start codon** to initiate translation. The **reading frame**, the way nucleotides in mRNA are grouped into codons, for translation is set by the AUG start codon near the 5' end of the mRNA. Each set of three nucleotides following this start codon is a codon in the mRNA message.

The genetic code is nearly universal. With a few exceptions, virtually all species use the same genetic code for protein synthesis, which is powerful evidence that all extant life on earth shares a common origin. However, unusual amino acids such as selenocysteine and pyrrolysine have been observed in archaea and bacteria. In the case of selenocysteine, the codon used is UGA (normally a stop codon). However, UGA can encode for selenocysteine using a stem-loop structure (known as the selenocysteine insertion sequence, or SECIS element), which is found at the 3' untranslated region of the mRNA. Pyrrolysine uses a different stop codon, UAG. The incorporation of pyrrolysine requires the *pylS* gene and a unique transfer RNA (tRNA) with a CUA anticodon.

second letter

first letter	U	C	A	G	third letter
U	UUU Phe UUC Phe UUA Leu UUG Leu	UCU Ser UCC Ser UCA Ser UCG Ser	UAU Tyr UAC Tyr UAA stop UAG stop	UGU Cys UGC Cys UGA stop UGG Trp	U C A G
C	CUU Leu CUC Leu CUA Leu CUG Leu	CCU Pro CCC Pro CCA Pro CCG Pro	CAU His CAC His CAA Gln CAG Gln	CGU Arg CGC Arg CGA Arg CGG Arg	U C A G
A	AUU Ile AUC Ile AUA Ile AUG Met	ACU Thr ACC Thr ACA Thr ACG Thr	AAU Asn AAC Asn AAA Lys AAG Lys	AGU Ser AGC Ser AGA Arg AGG Arg	U C A G
G	GUU Val GUC Val GUA Val GUG Val	GCU Ala GCC Ala GCA Ala GCG Ala	GAU Asp GAC Asp GAA Glu GAG Glu	GGU Gly GGC Gly GGA Gly GGG Gly	U C A G

Figure 11.12 This figure shows the genetic code for translating each nucleotide triplet in mRNA into an amino acid or a termination signal in a nascent protein. The first letter of a codon is shown vertically on the left, the second letter of a codon is shown horizontally across the top, and the third letter of a codon is shown vertically on the right. (credit: modification of work by National Institutes of Health)

CHECK YOUR UNDERSTANDING

- How many bases are in each codon?
- What amino acid is coded for by the codon AAU?
- What happens when a stop codon is reached?

The Protein Synthesis Machinery

In addition to the mRNA template, many molecules and macromolecules contribute to the process of translation. The composition of each component varies across taxa; for instance, ribosomes may consist of different numbers of ribosomal RNAs (rRNAs) and polypeptides depending on the organism. However, the general structures and functions of the protein synthesis machinery are comparable from bacteria to human cells. Translation requires the input of an mRNA template, ribosomes, tRNAs, and various enzymatic factors.

Ribosomes

A ribosome is a complex macromolecule composed of catalytic rRNAs (called ribozymes) and structural rRNAs, as well as many distinct polypeptides. Mature rRNAs make up approximately 50% of each ribosome. Prokaryotes have 70S ribosomes, whereas eukaryotes have 80S ribosomes in the cytoplasm and rough endoplasmic reticulum, and 70S ribosomes in mitochondria and chloroplasts. Ribosomes dissociate into large and small subunits when they are not synthesizing proteins and reassociate during the **initiation of translation**. In *E. coli*, the small subunit is described as 30S (which contains the 16S rRNA subunit), and the large subunit is 50S (which contains the 5S and 23S rRNA subunits), for a total of 70S (Svedberg units are not additive). Eukaryote ribosomes have a small 40S subunit (which contains the 18S rRNA subunit) and a large 60S subunit (which contains the 5S, 5.8S and 28S rRNA subunits), for a total of 80S. The small subunit is responsible for binding the mRNA template, whereas the large subunit binds tRNAs (discussed in the next subsection).

Each mRNA molecule is simultaneously translated by many ribosomes, all synthesizing protein in the same direction: reading the mRNA from 5' to 3' and synthesizing the polypeptide from the N terminus to the C terminus. The complete structure containing an mRNA with multiple associated ribosomes is called a **polyribosome** (or **polysome**). In both bacteria and archaea, before transcriptional termination occurs, each protein-encoding transcript is already being used to begin synthesis of numerous copies of the encoded polypeptide(s) because the processes of transcription and translation can occur concurrently, forming polyribosomes (Figure 11.13). The reason why transcription and translation can occur simultaneously is because both of these processes occur in the same 5' to 3' direction, they both occur in the cytoplasm of the cell, and because the RNA transcript is not processed once it is transcribed. This allows a prokaryotic cell to respond to an environmental signal requiring new proteins very quickly. In contrast, in eukaryotic cells, simultaneous transcription and translation is not possible. Although polyribosomes also form in eukaryotes, they cannot do so until RNA synthesis is complete and the RNA molecule has been modified and transported out of the nucleus.

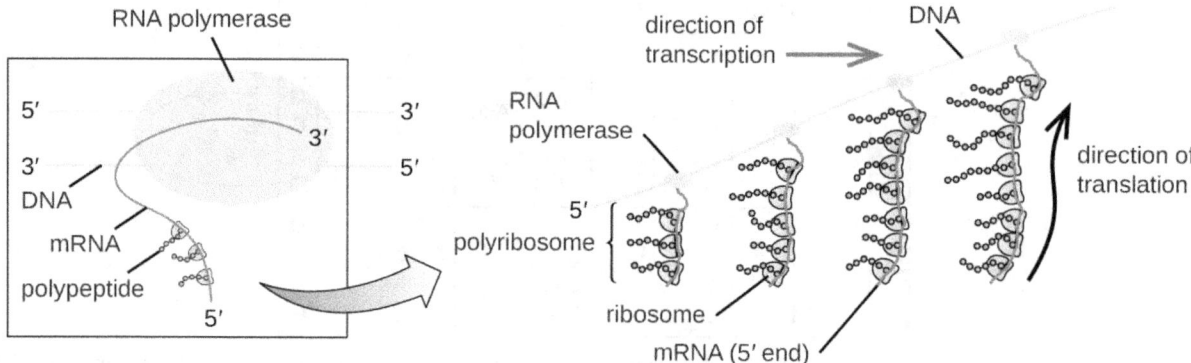

Figure 11.13 In prokaryotes, multiple RNA polymerases can transcribe a single bacterial gene while numerous ribosomes concurrently translate the mRNA transcripts into polypeptides. In this way, a specific protein can rapidly reach a high concentration in the bacterial cell.

Transfer RNAs

Transfer RNAs (tRNAs) are structural RNA molecules and, depending on the species, many different types of tRNAs exist in the cytoplasm. Bacterial species typically have between 60 and 90 types. Serving as adaptors, each tRNA type binds to a specific codon on the mRNA template and adds the corresponding amino acid to the polypeptide chain. Therefore, tRNAs are the molecules that actually "translate" the language of RNA into the language of proteins. As the adaptor molecules of translation, it is surprising that tRNAs can fit so much specificity into such a small package. The tRNA molecule interacts with three factors: aminoacyl tRNA synthetases, ribosomes, and mRNA.

Mature tRNAs take on a three-dimensional structure when complementary bases exposed in the single-stranded RNA molecule hydrogen bond with each other (Figure 11.14). This shape positions the amino-acid binding site, called the **CCA amino acid binding end**, which is a cytosine-cytosine-adenine sequence at the 3' end of the tRNA, and the **anticodon** at the other end. The anticodon is a three-nucleotide sequence that bonds with an mRNA codon through complementary base pairing.

An amino acid is added to the end of a tRNA molecule through the process of tRNA "charging," during which each tRNA molecule is linked to its correct or **cognate amino acid** by a group of enzymes called **aminoacyl tRNA synthetase**s. At least one type of aminoacyl tRNA synthetase exists for each of the 20 amino acids. During this process, the amino acid is first activated by the addition of adenosine monophosphate (AMP) and then transferred to the tRNA, making it a **charged tRNA**, and AMP is released.

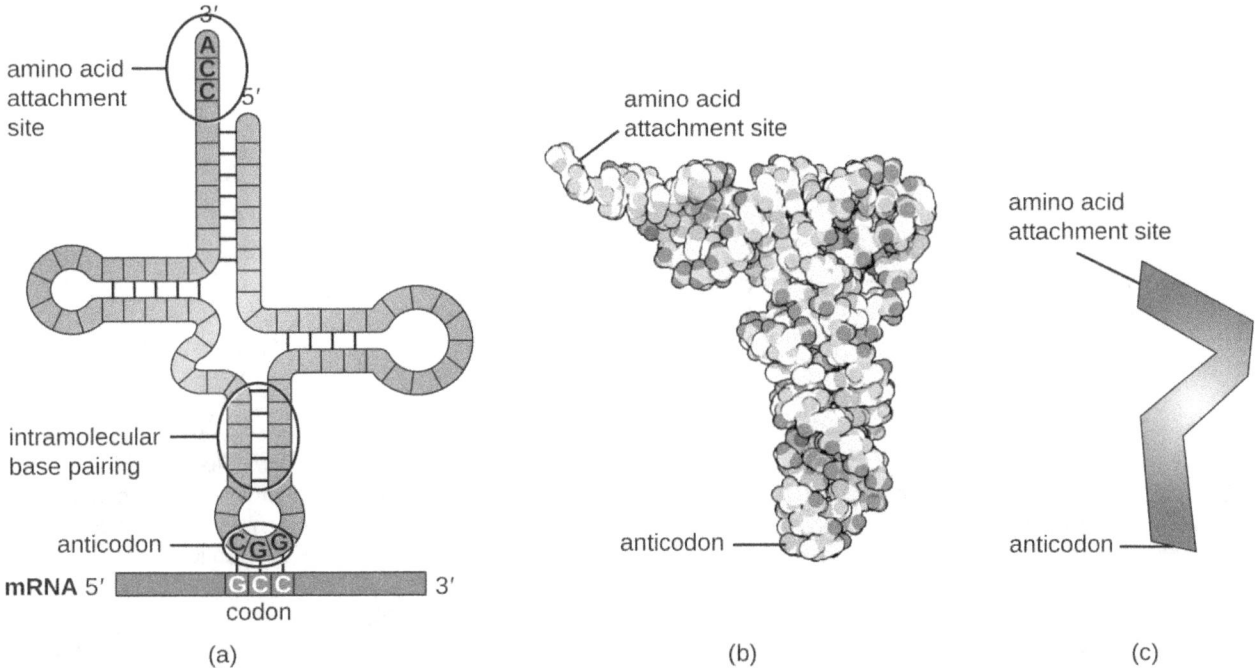

Figure 11.14 (a) After folding caused by intramolecular base pairing, a tRNA molecule has one end that contains the anticodon, which interacts with the mRNA codon, and the CCA amino acid binding end. (b) A space-filling model is helpful for visualizing the three-dimensional shape of tRNA. (c) Simplified models are useful when drawing complex processes such as protein synthesis.

✓ CHECK YOUR UNDERSTANDING

- Describe the structure and composition of the prokaryotic ribosome.
- In what direction is the mRNA template read?
- Describe the structure and function of a tRNA.

The Mechanism of Protein Synthesis

Translation is similar in prokaryotes and eukaryotes. Here we will explore how translation occurs in *E. coli*, a representative prokaryote, and specify any differences between bacterial and eukaryotic translation.

Initiation

The **initiation of protein synthesis** begins with the formation of an initiation complex. In *E. coli*, this complex involves the small 30S ribosome, the mRNA template, three **initiation factors** that help the ribosome assemble correctly, guanosine triphosphate (GTP) that acts as an energy source, and a special initiator tRNA carrying *N*-formyl-methionine (fMet-tRNA^{fMet}) (Figure 11.15). The initiator tRNA interacts with the start codon AUG of the mRNA and carries a formylated methionine (fMet). Because of its involvement in initiation, fMet is inserted at the beginning (N terminus) of every polypeptide chain synthesized by *E. coli*. In *E. coli* mRNA, a leader sequence upstream of the first AUG codon, called the Shine-Dalgarno sequence (also known as the ribosomal binding site AGGAGG), interacts through complementary base pairing with the rRNA molecules that compose the ribosome. This interaction anchors the 30S ribosomal subunit at the correct location on the mRNA template. At this point, the 50S ribosomal subunit then binds to the initiation complex, forming an intact ribosome.

In eukaryotes, initiation complex formation is similar, with the following differences:

- The initiator tRNA is a different specialized tRNA carrying methionine, called Met-tRNAi
- Instead of binding to the mRNA at the Shine-Dalgarno sequence, the eukaryotic initiation complex recognizes the 5' cap of the eukaryotic mRNA, then tracks along the mRNA in the 5' to 3' direction until the AUG start codon is recognized. At this point, the 60S subunit binds to the complex of Met-tRNAi, mRNA, and the 40S subunit.

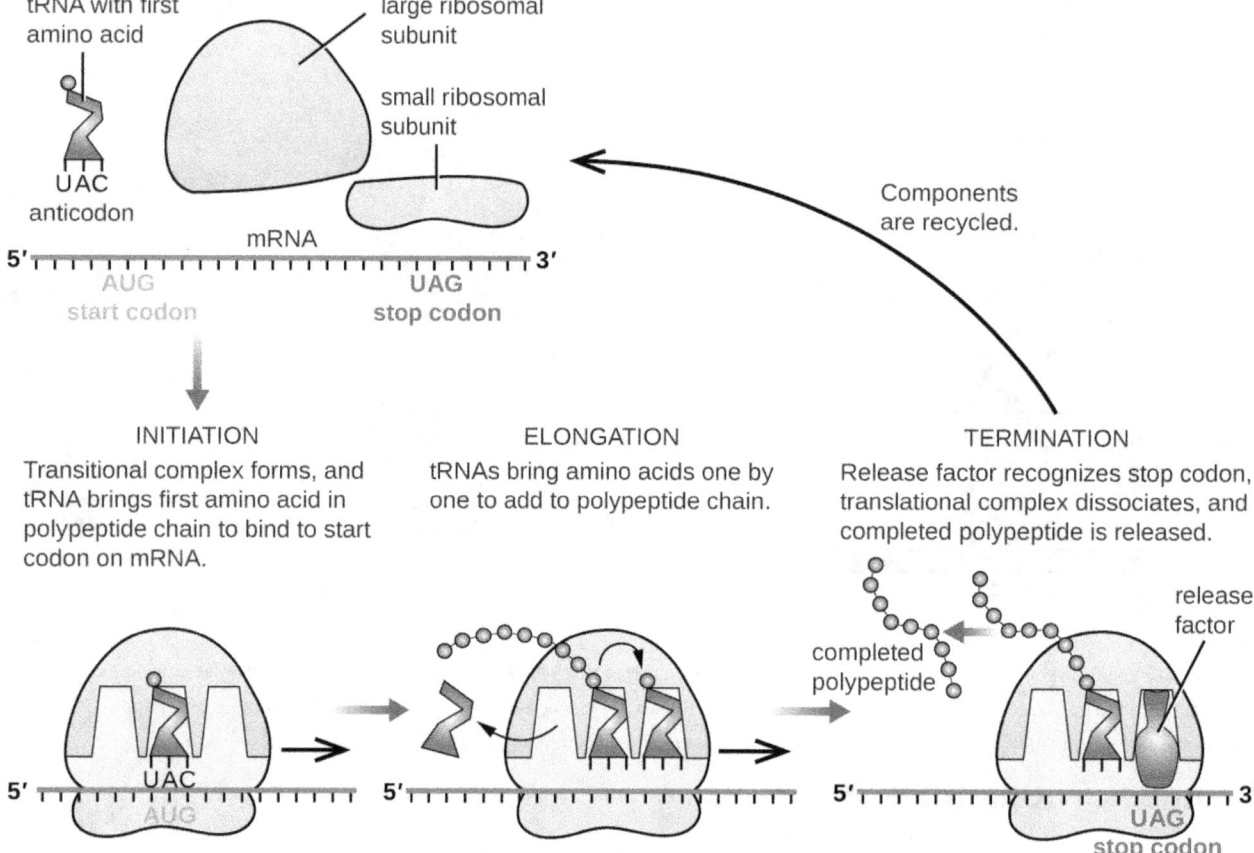

Figure 11.15 Translation in bacteria begins with the formation of the initiation complex, which includes the small ribosomal subunit, the mRNA, the initiator tRNA carrying N-formyl-methionine, and initiation factors. Then the 50S subunit binds, forming an intact ribosome.

Elongation

In prokaryotes and eukaryotes, the basics of **elongation of translation** are the same. In *E. coli*, the binding of the 50S ribosomal subunit to produce the intact ribosome forms three functionally important ribosomal sites: The **A (aminoacyl) site** binds incoming charged aminoacyl tRNAs. The **P (peptidyl) site** binds charged tRNAs carrying amino acids that have formed peptide bonds with the growing polypeptide chain but have not yet dissociated from their corresponding tRNA. The **E (exit) site** releases dissociated tRNAs so that they can be recharged with free amino acids. There is one notable exception to this assembly line of tRNAs: During initiation complex formation, bacterial fMet–tRNA^{fMet} or eukaryotic Met-tRNAi enters the P site directly without first entering the A site, providing a free A site ready to accept the tRNA corresponding to the first codon after the AUG.

Elongation proceeds with single-codon movements of the ribosome each called a translocation event. During each translocation event, the charged tRNAs enter at the A site, then shift to the P site, and then finally to the E site for removal. Ribosomal movements, or steps, are induced by conformational changes that advance the ribosome by three bases in the 3' direction. Peptide bonds form between the amino group of the amino acid attached to the A-site tRNA and the carboxyl group of the amino acid attached to the P-site tRNA. The formation of each peptide bond is catalyzed by **peptidyl transferase**, an RNA-based ribozyme that is integrated into the 50S ribosomal subunit. The amino acid bound to the P-site tRNA is also linked to the growing polypeptide chain. As the ribosome steps across the mRNA, the former P-site tRNA enters the E site, detaches from the amino acid, and is expelled. Several of the steps during elongation, including binding of a charged aminoacyl tRNA to the A site and translocation, require energy derived from GTP hydrolysis, which is catalyzed by specific elongation factors. Amazingly, the *E. coli* translation apparatus takes only 0.05 seconds to add each amino acid, meaning that a 200 amino-acid protein can be translated in just 10 seconds.

Termination

The **termination of translation** occurs when a nonsense codon (UAA, UAG, or UGA) is encountered for which there is no complementary tRNA. On aligning with the A site, these nonsense codons are recognized by release factors in prokaryotes and eukaryotes that result in the P-site amino acid detaching from its tRNA, releasing the newly made polypeptide. The small and large ribosomal subunits dissociate from the mRNA and from each other; they are recruited almost immediately into another translation initiation complex.

In summary, there are several key features that distinguish prokaryotic gene expression from that seen in eukaryotes. These are illustrated in Figure 11.16 and listed in Figure 11.17.

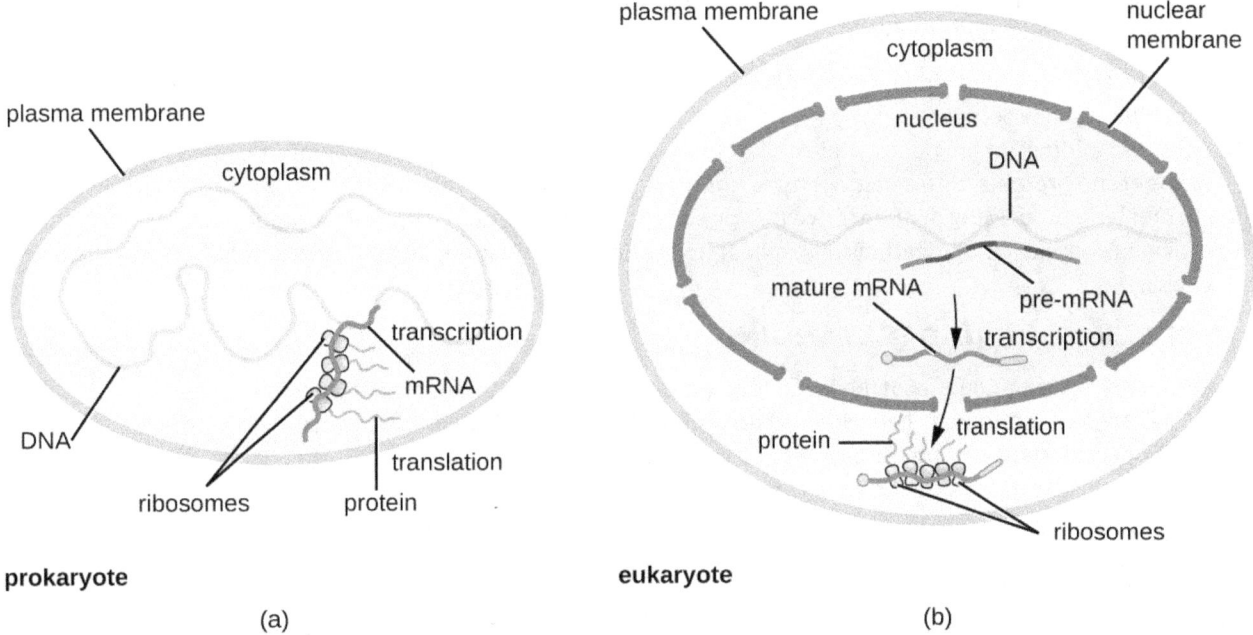

Figure 11.16 (a) In prokaryotes, the processes of transcription and translation occur simultaneously in the cytoplasm, allowing for a rapid cellular response to an environmental cue. (b) In eukaryotes, transcription is localized to the nucleus and translation is localized to the cytoplasm, separating these processes and necessitating RNA processing for stability.

Comparison of Translation in Bacteria Versus Eukaryotes		
Property	Bacteria	Eukaryotes
Ribosomes	70S • 30S (small subunit) with 16S rRNA subunit • 50S (large subunit) with 5S and 23S rRNA subunits	80S • 40S (small subunit) with 18S rRNA subunit • 60S (large subunit) with 5S, 5.8S, and 28S rRNA subunits
Amino acid carried by initiator tRNA	fMet	Met
Shine-Dalgarno sequence in mRNA	Present	Absent
Simultaneous transcription and translation	Yes	No

Figure 11.17

Protein Targeting, Folding, and Modification

During and after translation, polypeptides may need to be modified before they are biologically active. Post-translational modifications include:

1. removal of translated signal sequences—short tails of amino acids that aid in directing a protein to a specific cellular compartment
2. proper "folding" of the polypeptide and association of multiple polypeptide subunits, often facilitated by chaperone proteins, into a distinct three-dimensional structure
3. proteolytic processing of an inactive polypeptide to release an active protein component, and
4. various chemical modifications (e.g., phosphorylation, methylation, or glycosylation) of individual amino acids.

✓ CHECK YOUR UNDERSTANDING

- What are the components of the initiation complex for translation in prokaryotes?
- What are two differences between initiation of prokaryotic and eukaryotic translation?
- What occurs at each of the three active sites of the ribosome?
- What causes termination of translation?

11.5 Mutations

Learning Objectives

By the end of this section, you will be able to:
- Compare point mutations and frameshift mutations
- Describe the differences between missense, nonsense, and silent mutations
- Describe the differences between light and dark repair
- Explain how different mutagens act
- Explain why the Ames test can be used to detect carcinogens
- Analyze sequences of DNA and identify examples of types of mutations

A **mutation** is a heritable change in the DNA sequence of an organism. The resulting organism, called a **mutant**, may have a recognizable change in phenotype compared to the **wild type**, which is the phenotype most commonly observed in nature. A change in the DNA sequence is conferred to mRNA through transcription, and may lead to an altered amino acid sequence in a protein on translation. Because proteins carry out the vast majority of cellular functions, a change in amino acid sequence in a protein may lead to an altered phenotype for the cell and organism.

Effects of Mutations on DNA Sequence

There are several types of mutations that are classified according to how the DNA molecule is altered. One type, called a **point mutation**, affects a single base and most commonly occurs when one base is substituted or replaced by another. Mutations also result from the addition of one or more bases, known as an **insertion**, or the removal of one or more bases, known as a **deletion**.

✓ CHECK YOUR UNDERSTANDING

- What type of a mutation occurs when a gene has two fewer nucleotides in its sequence?

Effects of Mutations on Protein Structure and Function

Point mutations may have a wide range of effects on protein function (Figure 11.18). As a consequence of the degeneracy of the genetic code, a point mutation will commonly result in the same amino acid being incorporated into the resulting polypeptide despite the sequence change. This change would have no effect on the protein's structure, and is thus called a **silent mutation**. A **missense mutation** results in a different amino acid being incorporated into the resulting polypeptide. The effect of a missense mutation depends on how chemically different the new amino acid is from the wild-type amino acid. The location of the changed amino acid within the protein also is important. For example, if the changed amino acid is part of the enzyme's active site, then the effect of the missense mutation may be significant. Many missense mutations result in proteins that are still functional, at least to some degree. Sometimes the effects of missense mutations may be only apparent under certain environmental conditions; such missense mutations are called **conditional mutation**s. Rarely, a missense mutation may be beneficial. Under the right environmental conditions, this type of mutation may give the organism that harbors it a selective advantage. Yet another type of point mutation, called a **nonsense mutation**, converts a codon encoding an amino acid (a sense codon) into a stop codon (a nonsense codon). Nonsense mutations result in the synthesis of proteins that are shorter than the wild type and typically not functional.

Deletions and insertions also cause various effects. Because codons are triplets of nucleotides, insertions or deletions in groups of three nucleotides may lead to the insertion or deletion of one or more amino acids and may not cause significant effects on the resulting protein's functionality. However, **frameshift mutation**s, caused by insertions or deletions of a number of nucleotides that are not a multiple of three are extremely problematic because a shift in the reading frame results (Figure 11.18). Because ribosomes read the mRNA in triplet codons, frameshift mutations can change every amino acid after the point of the mutation. The new reading frame may also include a stop codon before the end of the coding sequence. Consequently, proteins made from genes containing frameshift mutations are nearly always nonfunctional.

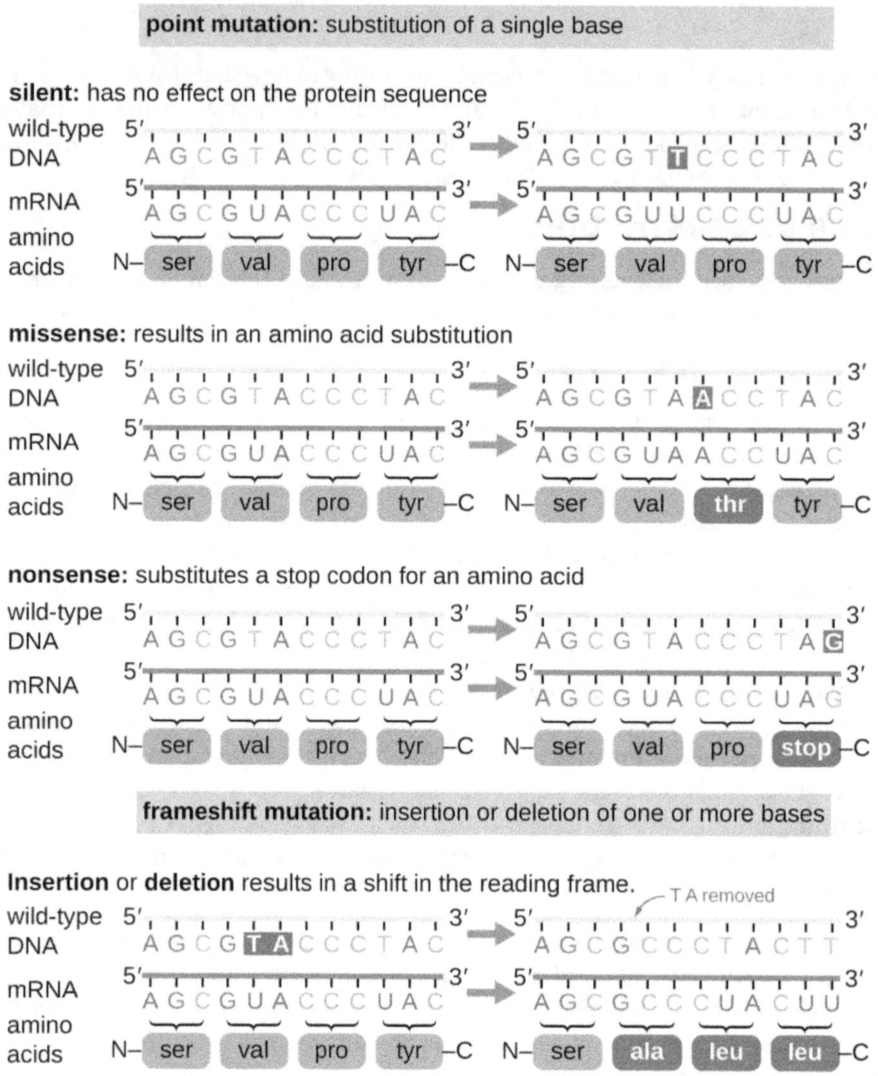

Figure 11.18 Mutations can lead to changes in the protein sequence encoded by the DNA.

CHECK YOUR UNDERSTANDING

- What are the reasons a nucleotide change in a gene for a protein might not have any effect on the phenotype of that gene?
- Is it possible for an insertion of three nucleotides together after the fifth nucleotide in a protein-coding gene to produce a protein that is shorter than normal? How or how not?

MICRO CONNECTIONS

A Beneficial Mutation

Since the first case of infection with human immunodeficiency virus (HIV) was reported in 1981, nearly 40 million people have died from HIV infection,[2] the virus that causes acquired immune deficiency syndrome (AIDS). The virus targets helper T cells that play a key role in bridging the innate and adaptive immune response, infecting and killing cells normally involved in the body's response to infection. There is no cure for HIV infection, but many drugs have been developed to slow or block the progression of the virus. Although individuals around the world may be infected, the highest prevalence among people 15–49 years old is in sub-Saharan Africa, where nearly one person in 20 is infected, accounting for greater than 70% of the infections worldwide[3] (Figure 11.19). Unfortunately, this is also a part of the world where prevention strategies and drugs to treat the infection are the most lacking.

In recent years, scientific interest has been piqued by the discovery of a few individuals from northern Europe who are resistant to HIV infection. In 1998, American geneticist Stephen J. O'Brien at the National Institutes of Health (NIH) and colleagues published the results of their genetic analysis of more than 4,000 individuals. These indicated that many individuals of Eurasian descent (up to 14% in some ethnic groups) have a deletion mutation, called CCR5-delta 32, in the gene encoding CCR5. CCR5 is a coreceptor found on the surface of T cells that is necessary for many strains of the virus to enter the host cell. The mutation leads to the production of a receptor to which HIV cannot effectively bind and thus blocks viral entry. People homozygous for this mutation have greatly reduced susceptibility to HIV infection, and those who are heterozygous have some protection from infection as well.

It is not clear why people of northern European descent, specifically, carry this mutation, but its prevalence seems to be highest in northern Europe and steadily decreases in populations as one moves south. Research indicates that the mutation has been present since before HIV appeared and may have been selected for in European populations as a result of exposure to the plague or smallpox. This mutation may protect individuals from plague (caused by the bacterium *Yersinia pestis*) and smallpox (caused by the variola virus) because this receptor may also be involved in these diseases. The age of this mutation is a matter of debate, but estimates suggest it appeared between 1875 years to 225 years ago, and may have been spread from Northern Europe through Viking invasions.

This exciting finding has led to new avenues in HIV research, including looking for drugs to block CCR5 binding to HIV in individuals who lack the mutation. Although DNA testing to determine which individuals carry the CCR5-delta 32 mutation is possible, there are documented cases of individuals homozygous for the mutation contracting HIV. For this reason, DNA testing for the mutation is not widely recommended by public health officials so as not to encourage risky behavior in those who carry the mutation. Nevertheless, inhibiting the binding of HIV to CCR5 continues to be a valid strategy for the development of drug therapies for those infected with HIV.

---

2 World Health Organization. " Global Health Observatory (GHO) Data, HIV/AIDS." http://www.who.int/gho/hiv/en/. Accessed August 5, 2016.

3 World Health Organization. " Global Health Observatory (GHO) Data, HIV/AIDS." http://www.who.int/gho/hiv/en/. Accessed August 5, 2016.

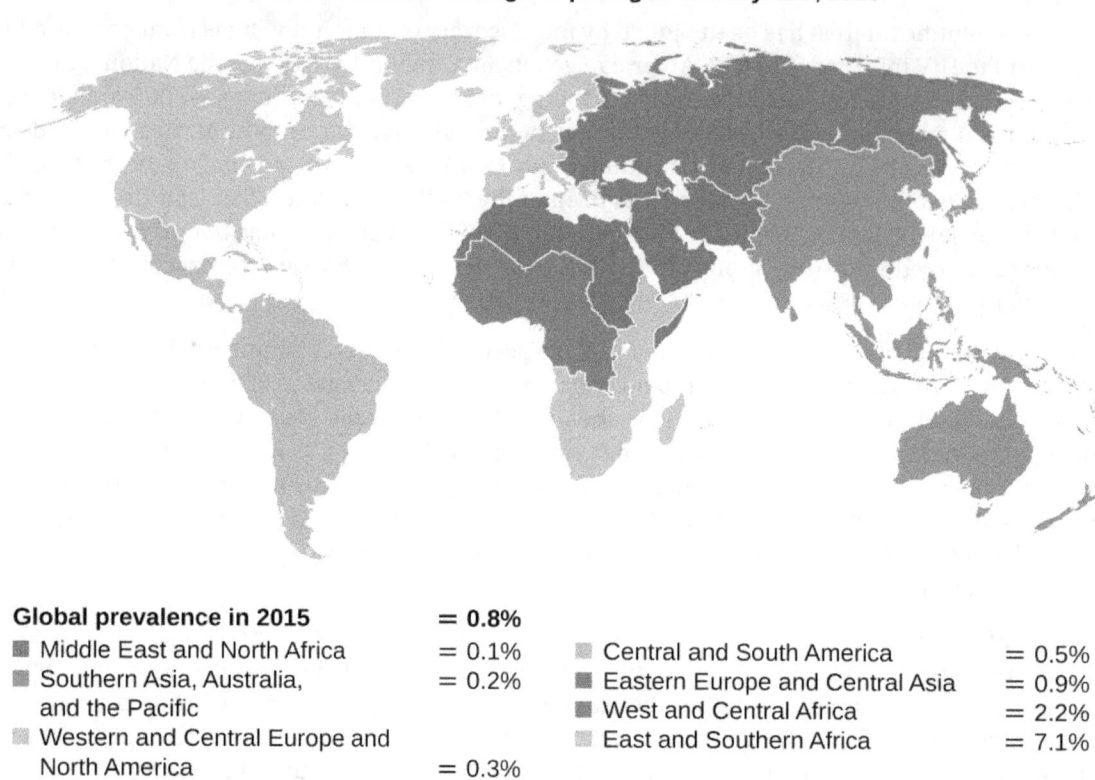

Figure 11.19 HIV is highly prevalent in sub-Saharan Africa, but its prevalence is quite low in some other parts of the world.

Causes of Mutations

Mistakes in the process of DNA replication can cause **spontaneous mutations** to occur. The error rate of DNA polymerase is one incorrect base per billion base pairs replicated. Exposure to **mutagens** can cause **induced mutations**, which are various types of chemical agents or radiation (Table 11.4). Exposure to a mutagen can increase the rate of mutation more than 1000-fold. Mutagens are often also **carcinogens**, agents that cause cancer. However, whereas nearly all carcinogens are mutagenic, not all mutagens are necessarily carcinogens.

Chemical Mutagens

Various types of chemical mutagens interact directly with DNA either by acting as nucleoside analogs or by modifying nucleotide bases. Chemicals called **nucleoside analogs** are structurally similar to normal nucleotide bases and can be incorporated into DNA during replication (Figure 11.20). These base analogs induce mutations because they often have different base-pairing rules than the bases they replace. Other chemical mutagens can modify normal DNA bases, resulting in different base-pairing rules. For example, nitrous acid deaminates cytosine, converting it to uracil. Uracil then pairs with adenine in a subsequent round of replication, resulting in the conversion of a GC base pair to an AT base pair. Nitrous acid also deaminates adenine to hypoxanthine, which base pairs with cytosine instead of thymine, resulting in the conversion of a TA base pair to a CG base pair.

Chemical mutagens known as **intercalating agents** work differently. These molecules slide between the stacked nitrogenous bases of the DNA double helix, distorting the molecule and creating atypical spacing between nucleotide base pairs (Figure 11.21). As a result, during DNA replication, DNA polymerase may either skip replicating several nucleotides (creating a deletion) or insert extra nucleotides (creating an insertion). Either outcome may lead to a frameshift mutation. Combustion products like polycyclic aromatic hydrocarbons are particularly dangerous intercalating agents that can lead to mutation-caused cancers. The intercalating agents ethidium bromide and acridine orange are commonly used in the laboratory to stain DNA for visualization and are potential mutagens.

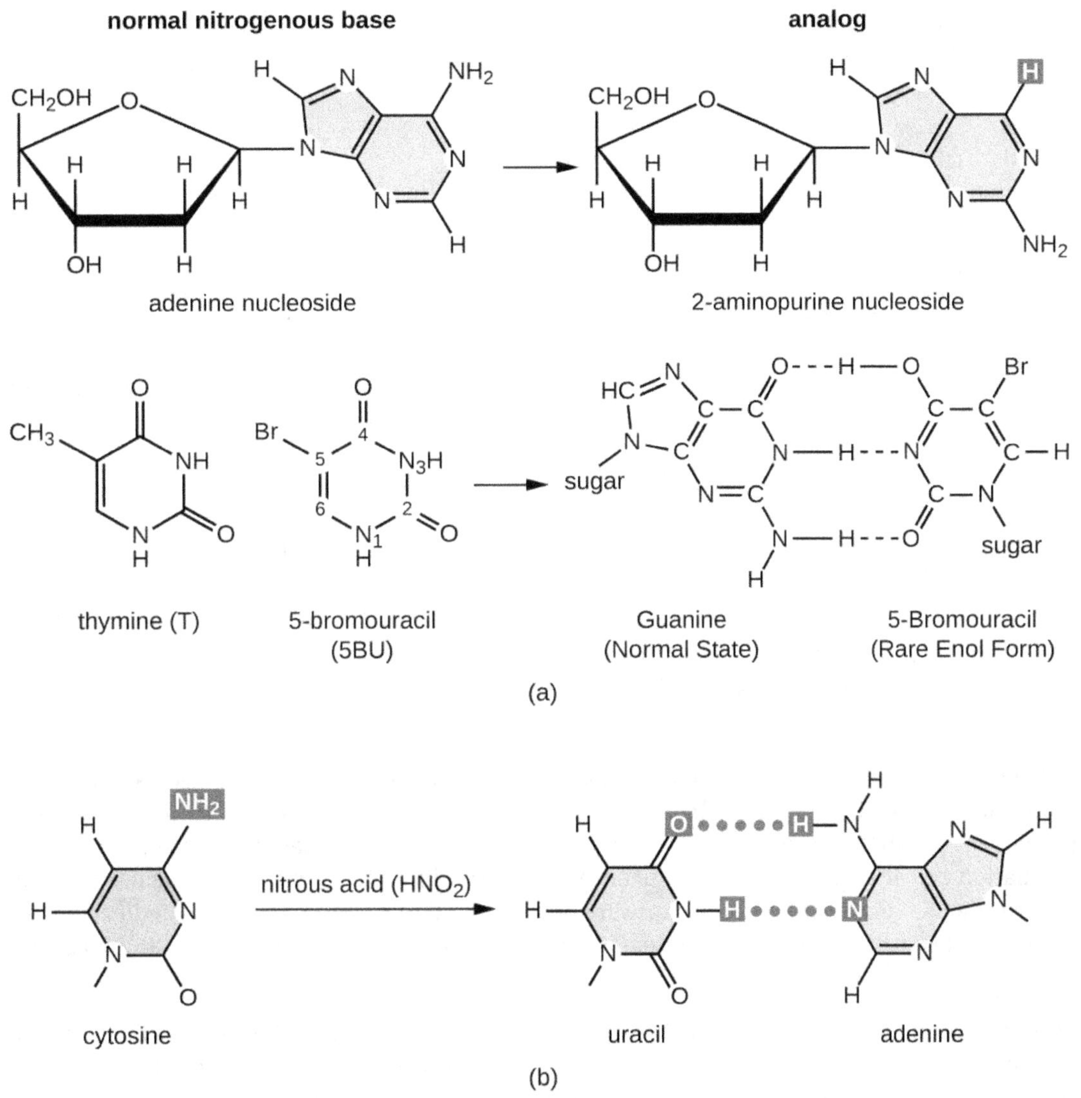

Figure 11.20 (a) 2-aminopurine nucleoside (2AP) structurally is a nucleoside analog to adenine nucleoside, whereas 5-bromouracil (5BU) is a nucleoside analog to thymine nucleoside. 2AP base pairs with C, converting an AT base pair to a GC base pair after several rounds of replication. 5BU pairs with G, converting an AT base pair to a GC base pair after several rounds of replication. (b) Nitrous acid is a different type of chemical mutagen that modifies already existing nucleoside bases like C to produce U, which base pairs with A. This chemical modification, as shown here, results in converting a CG base pair to a TA base pair.

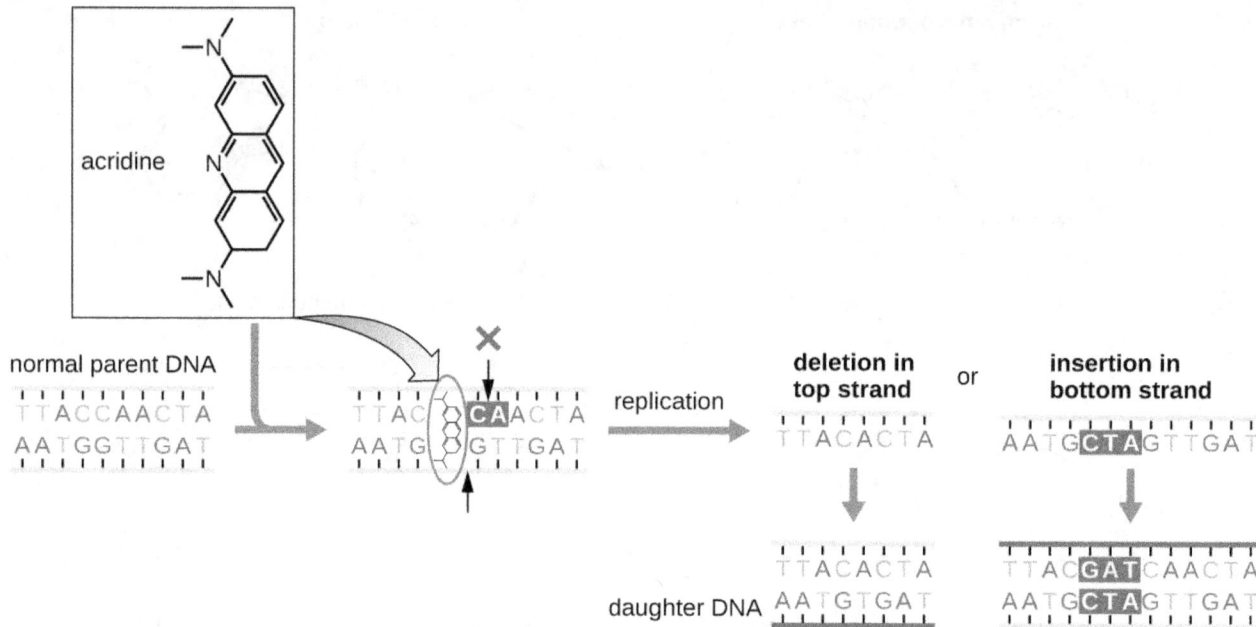

Figure 11.21 Intercalating agents, such as acridine, introduce atypical spacing between base pairs, resulting in DNA polymerase introducing either a deletion or an insertion, leading to a potential frameshift mutation.

Radiation

Exposure to either ionizing or nonionizing radiation can each induce mutations in DNA, although by different mechanisms. Strong **ionizing radiation** like X-rays and gamma rays can cause single- and double-stranded breaks in the DNA backbone through the formation of hydroxyl radicals on radiation exposure (Figure 11.22). Ionizing radiation can also modify bases; for example, the deamination of cytosine to uracil, analogous to the action of nitrous acid.[4] Ionizing radiation exposure is used to kill microbes to sterilize medical devices and foods, because of its dramatic nonspecific effect in damaging DNA, proteins, and other cellular components (see Using Physical Methods to Control Microorganisms).

Nonionizing radiation, like ultraviolet light, is not energetic enough to initiate these types of chemical changes. However, **nonionizing radiation** can induce dimer formation between two adjacent pyrimidine bases, commonly two thymines, within a nucleotide strand. During **thymine dimer** formation, the two adjacent thymines become covalently linked and, if left unrepaired, both DNA replication and transcription are stalled at this point. DNA polymerase may proceed and replicate the dimer incorrectly, potentially leading to frameshift or point mutations.

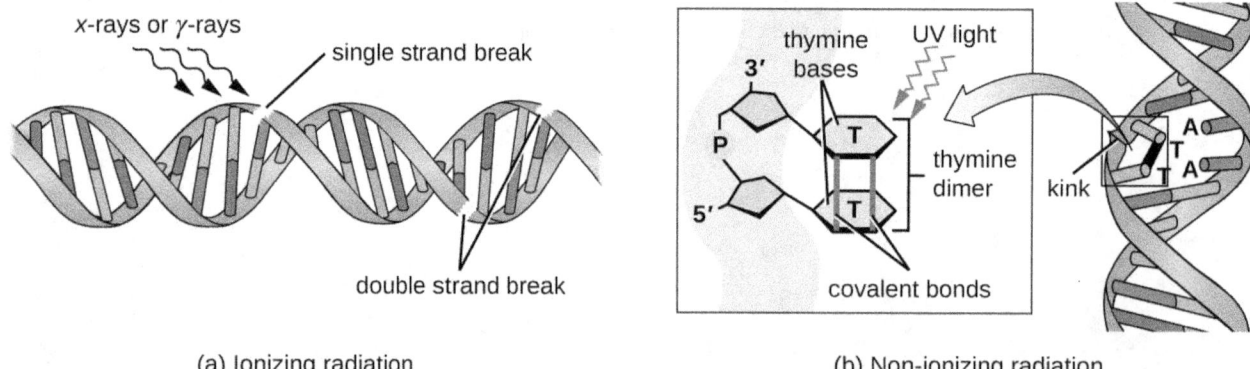

Figure 11.22 (a) Ionizing radiation may lead to the formation of single-stranded and double-stranded breaks in the sugar-phosphate backbone of DNA, as well as to the modification of bases (not shown). (b) Nonionizing radiation like ultraviolet light can lead to the

---

4 K.R. Tindall et al. "Changes in DNA Base Sequence Induced by Gamma-Ray Mutagenesis of Lambda Phage and Prophage." *Genetics* 118 no. 4 (1988):551–560.

formation of thymine dimers, which can stall replication and transcription and introduce frameshift or point mutations.

A Summary of Mutagenic Agents

Mutagenic Agents	Mode of Action	Effect on DNA	Resulting Type of Mutation
Nucleoside analogs			
2-aminopurine	Is inserted in place of A but base pairs with C	Converts AT to GC base pair	Point
5-bromouracil	Is inserted in place of T but base pairs with G	Converts AT to GC base pair	Point
Nucleotide-modifying agent			
Nitrous oxide	Deaminates C to U	Converts GC to AT base pair	Point
Intercalating agents			
Acridine orange, ethidium bromide, polycyclic aromatic hydrocarbons	Distorts double helix, creates unusual spacing between nucleotides	Introduces small deletions and insertions	Frameshift
Ionizing radiation			
X-rays, γ-rays	Forms hydroxyl radicals	Causes single- and double-strand DNA breaks	Repair mechanisms may introduce mutations
X-rays, γ-rays	Modifies bases (e.g., deaminating C to U)	Converts GC to AT base pair	Point
Nonionizing radiation			
Ultraviolet	Forms pyrimidine (usually thymine) dimers	Causes DNA replication errors	Frameshift or point

Table 11.4

✓ CHECK YOUR UNDERSTANDING

- How does a base analog introduce a mutation?
- How does an intercalating agent introduce a mutation?
- What type of mutagen causes thymine dimers?

DNA Repair

The process of DNA replication is highly accurate, but mistakes can occur spontaneously or be induced by mutagens. Uncorrected mistakes can lead to serious consequences for the phenotype. Cells have developed several repair mechanisms to minimize the number of mutations that persist.

Proofreading

Most of the mistakes introduced during DNA replication are promptly corrected by most DNA polymerases through a function called proofreading. In proofreading, the DNA polymerase reads the newly added base, ensuring that it is complementary to the corresponding base in the template strand before adding the next one. If an incorrect base has been added, the enzyme makes a cut to release the wrong nucleotide and a new base is added.

Mismatch Repair

Some errors introduced during replication are corrected shortly after the replication machinery has moved. This mechanism is called mismatch repair. The enzymes involved in this mechanism recognize the incorrectly added nucleotide, excise it, and replace it with the correct base. One example is the methyl-directed mismatch repair in *E. coli*. The DNA is hemimethylated. This means that the parental strand is methylated while the newly synthesized daughter strand is not. It takes several minutes before the new strand is methylated. Proteins MutS, MutL, and MutH bind to the hemimethylated site where the incorrect nucleotide is found. MutH cuts the nonmethylated strand (the new strand). An exonuclease removes a portion of the strand (including the incorrect nucleotide). The gap formed is then filled in by DNA pol III and ligase.

Repair of Thymine Dimers

Because the production of thymine dimers is common (many organisms cannot avoid ultraviolet light), mechanisms have evolved to repair these lesions. In **nucleotide excision repair** (also called dark repair), enzymes remove the pyrimidine dimer and replace it with the correct nucleotides (Figure 11.23). In *E. coli*, the DNA is scanned by an enzyme complex. If a distortion in the double helix is found that was introduced by the pyrimidine dimer, the enzyme complex cuts the sugar-phosphate backbone several bases upstream and downstream of the dimer, and the segment of DNA between these two cuts is then enzymatically removed. DNA pol I replaces the missing nucleotides with the correct ones and DNA ligase seals the gap in the sugar-phosphate backbone.

The **direct repair** (also called light repair) of thymine dimers occurs through the process of **photoreactivation** in the presence of visible light. An enzyme called photolyase recognizes the distortion in the DNA helix caused by the thymine dimer and binds to the dimer. Then, in the presence of visible light, the photolyase enzyme changes conformation and breaks apart the thymine dimer, allowing the thymines to again correctly base pair with the adenines on the complementary strand. Photoreactivation appears to be present in all organisms, with the exception of placental mammals, including humans. Photoreactivation is particularly important for organisms chronically exposed to ultraviolet radiation, like plants, photosynthetic bacteria, algae, and corals, to prevent the accumulation of mutations caused by thymine dimer formation.

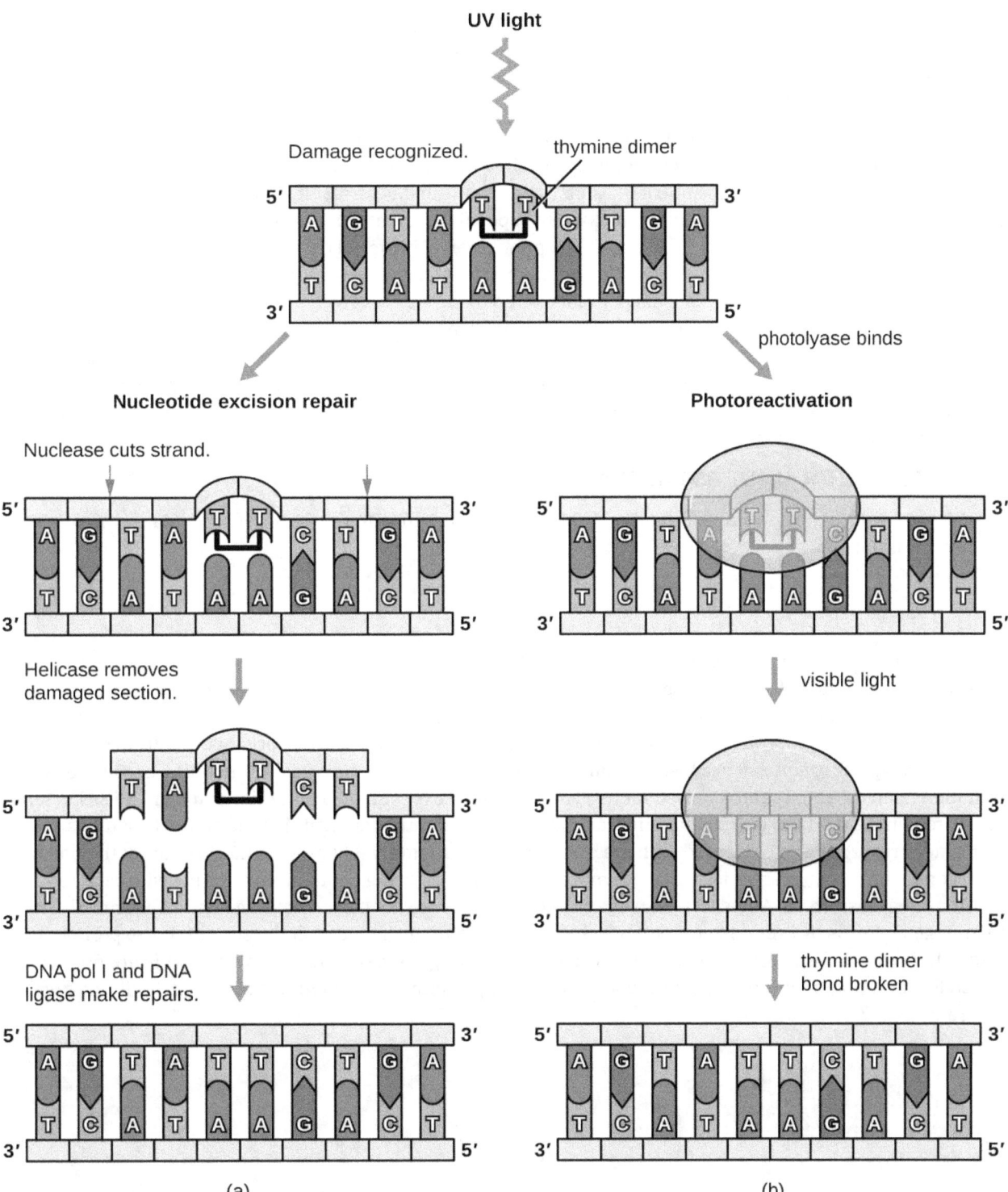

Figure 11.23 Bacteria have two mechanisms for repairing thymine dimers. (a) In nucleotide excision repair, an enzyme complex recognizes the distortion in the DNA complex around the thymine dimer and cuts and removes the damaged DNA strand. The correct nucleotides are replaced by DNA pol I and the nucleotide strand is sealed by DNA ligase. (b) In photoreactivation, the enzyme photolyase binds to the thymine dimer and, in the presence of visible light, breaks apart the dimer, restoring the base pairing of the thymines with complementary adenines on the opposite DNA strand.

CHECK YOUR UNDERSTANDING

- During mismatch repair, how does the enzyme recognize which is the new and which is the old strand?

- What type of mutation does photolyase repair?

Identifying Bacterial Mutants

One common technique used to identify bacterial mutants is called **replica plating**. This technique is used to detect nutritional mutants, called **auxotroph**s, which have a mutation in a gene encoding an enzyme in the biosynthesis pathway of a specific nutrient, such as an amino acid. As a result, whereas wild-type cells retain the ability to grow normally on a medium lacking the specific nutrient, auxotrophs are unable to grow on such a medium. During replica plating (Figure 11.24), a population of bacterial cells is mutagenized and then plated as individual cells on a complex nutritionally complete plate and allowed to grow into colonies. Cells from these colonies are removed from this master plate, often using sterile velvet. This velvet, containing cells, is then pressed in the same orientation onto plates of various media. At least one plate should also be nutritionally complete to ensure that cells are being properly transferred between the plates. The other plates lack specific nutrients, allowing the researcher to discover various auxotrophic mutants unable to produce specific nutrients. Cells from the corresponding colony on the nutritionally complete plate can be used to recover the mutant for further study.

✓ CHECK YOUR UNDERSTANDING

- Why are cells plated on a nutritionally complete plate in addition to nutrient-deficient plates when looking for a mutant?

The Ames Test

The **Ames test**, developed by Bruce Ames (1928–) in the 1970s, is a method that uses bacteria for rapid, inexpensive screening of the carcinogenic potential of new chemical compounds. The test measures the mutation rate associated with exposure to the compound, which, if elevated, may indicate that exposure to this compound is associated with greater cancer risk. The Ames test uses as the test organism a strain of *Salmonella typhimurium* that is a histidine auxotroph, unable to synthesize its own histidine because of a mutation in an essential gene required for its synthesis. After exposure to a potential mutagen, these bacteria are plated onto a medium lacking histidine, and the number of mutants regaining the ability to synthesize histidine is recorded and compared with the number of such mutants that arise in the absence of the potential mutagen (Figure 11.25). Chemicals that are more mutagenic will bring about more mutants with restored histidine synthesis in the Ames test. Because many chemicals are not directly mutagenic but are metabolized to mutagenic forms by liver enzymes, rat liver extract is commonly included at the start of this experiment to mimic liver metabolism. After the Ames test is conducted, compounds identified as mutagenic are further tested for their potential carcinogenic properties by using other models, including animal models like mice and rats.

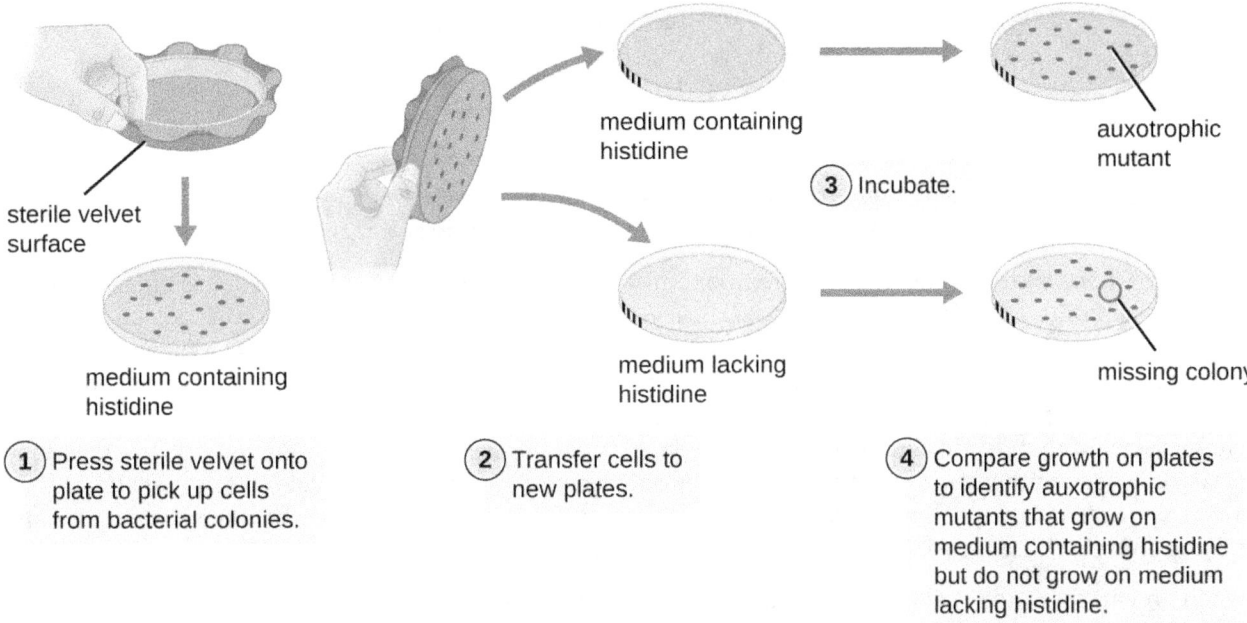

Figure 11.24 Identification of auxotrophic mutants, like histidine auxotrophs, is done using replica plating. After mutagenesis, colonies that grow on nutritionally complete medium but not on medium lacking histidine are identified as histidine auxotrophs.

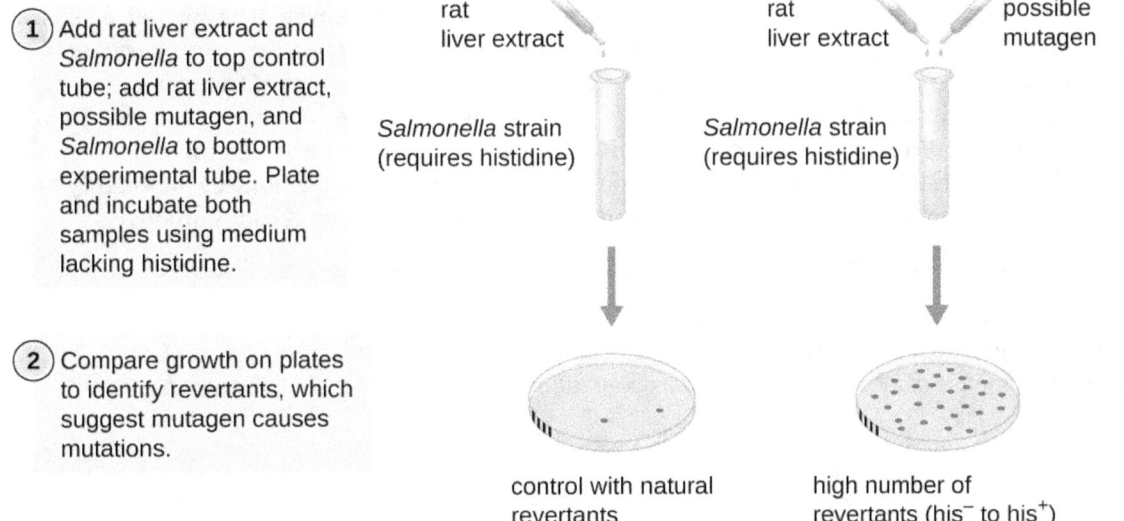

Figure 11.25 The Ames test is used to identify mutagenic, potentially carcinogenic chemicals. A *Salmonella* histidine auxotroph is used as the test strain, exposed to a potential mutagen/carcinogen. The number of reversion mutants capable of growing in the absence of supplied histidine is counted and compared with the number of natural reversion mutants that arise in the absence of the potential mutagen.

✓ CHECK YOUR UNDERSTANDING

- What mutation is used as an indicator of mutation rate in the Ames test?
- Why can the Ames test work as a test for carcinogenicity?

11.6 How Asexual Prokaryotes Achieve Genetic Diversity

Learning Objectives

By the end of this section, you will be able to:
- Compare the processes of transformation, transduction, and conjugation
- Explain how asexual gene transfer results in prokaryotic genetic diversity
- Explain the structure and consequences for bacterial genetic diversity of transposons

Typically, when we consider genetic transfer, we think of **vertical gene transfer**, the transmission of genetic information from generation to generation. Vertical gene transfer is by far the main mode of transmission of genetic information in all cells. In sexually reproducing organisms, crossing-over events and independent assortment of individual chromosomes during meiosis contribute to genetic diversity in the population. Genetic diversity is also introduced during sexual reproduction, when the genetic information from two parents, each with different complements of genetic information, are combined, producing new combinations of parental genotypes in the diploid offspring. The occurrence of mutations also contributes to genetic diversity in a population. Genetic diversity of offspring is useful in changing or inconsistent environments and may be one reason for the evolutionary success of sexual reproduction.

When prokaryotes and eukaryotes reproduce asexually, they transfer a nearly identical copy of their genetic material to their offspring through vertical gene transfer. Although asexual reproduction produces more offspring more quickly, any benefits of diversity among those offspring are lost. How then do organisms whose dominant reproductive mode is asexual create genetic diversity? In prokaryotes, **horizontal gene transfer (HGT)**, the introduction of genetic material from one organism to another organism within the same generation, is an important way to introduce genetic diversity. HGT allows even distantly related species to share genes, influencing their phenotypes. It is thought that HGT is more prevalent in prokaryotes but that only a small fraction of the prokaryotic genome may be transferred by this type of transfer at any one time. As the phenomenon is investigated more thoroughly, it may be revealed to be even more common. Many scientists believe that HGT and mutation are significant sources of genetic variation, the raw material for the process of natural selection, in prokaryotes. Although HGT is more common among evolutionarily related organisms, it may occur between any two species that live together in a natural community.

HGT in prokaryotes is known to occur by the three primary mechanisms that are illustrated in Figure 11.26:

1. Transformation: naked DNA is taken up from the environment
2. Transduction: genes are transferred between cells in a virus (see The Viral Life Cycle)
3. Conjugation: use of a hollow tube called a conjugation pilus to transfer genes between cells

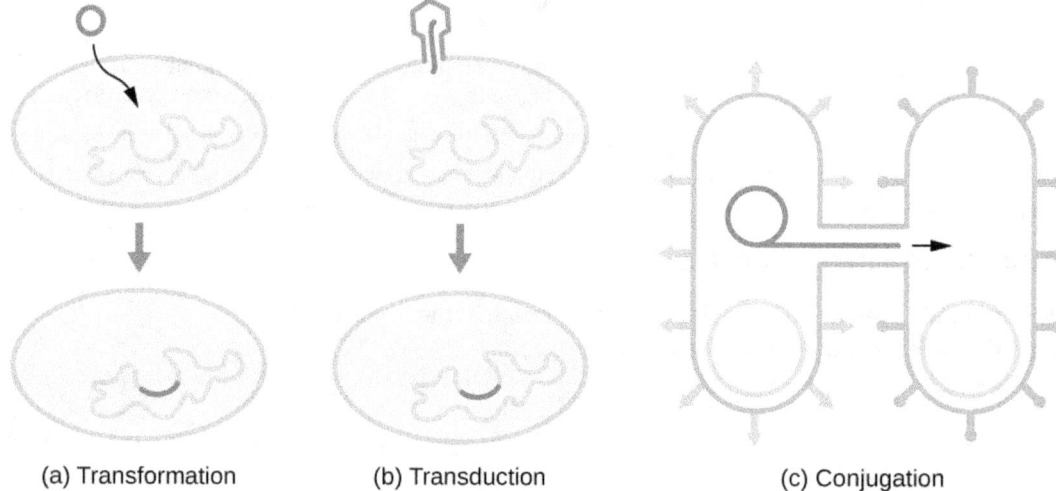

(a) Transformation   (b) Transduction   (c) Conjugation

Figure 11.26 There are three prokaryote-specific mechanisms leading to horizontal gene transfer in prokaryotes. a) In transformation, the cell takes up DNA directly from the environment. The DNA may remain separate as a plasmid or be incorporated into the host genome. b) In transduction, a bacteriophage injects DNA that is a hybrid of viral DNA and DNA from a previously infected bacterial cell. c) In conjugation, DNA is transferred between cells through a cytoplasmic bridge after a conjugation pilus draws the two cells close enough to form the bridge.

CHECK YOUR UNDERSTANDING

- What are three ways sexual reproduction introduces genetic variation into offspring?
- What is a benefit of asexual reproduction?
- What are the three mechanisms of horizontal gene transfer in prokaryotes?

Transformation

Frederick Griffith was the first to demonstrate the process of transformation. In 1928, he showed that live, nonpathogenic *Streptococcus pneumoniae* bacteria could be transformed into pathogenic bacteria through exposure to a heat-killed pathogenic strain. He concluded that some sort of agent, which he called the "transforming principle," had been passed from the dead pathogenic bacteria to the live, nonpathogenic bacteria. In 1944, Oswald Avery (1877–1955), Colin MacLeod (1909–1972), and Maclyn McCarty (1911–2005) demonstrated that the transforming principle was DNA (see Using Microorganisms to Discover the Secrets of Life).

In **transformation**, the prokaryote takes up naked DNA found in its environment and that is derived from other cells that have lysed on death and released their contents, including their genome, into the environment. Many bacteria are naturally competent, meaning that they actively bind to environmental DNA, transport it across their cell envelopes into their cytoplasm, and make it single stranded. Typically, double-stranded foreign DNA within cells is destroyed by nucleases as a defense against viral infection. However, these nucleases are usually ineffective against single-stranded DNA, so this single-stranded DNA within the cell has the opportunity to recombine into the bacterial genome. A molecule of DNA that contains fragments of DNA from different organisms is called recombinant DNA. (Recombinant DNA will be discussed in more detail in Microbes and the Tools of Genetic Engineering.) If the bacterium incorporates the new DNA into its own genome through recombination, the bacterial cell may gain new phenotypic properties. For example, if a nonpathogenic bacterium takes up DNA for a toxin gene from a pathogen and then incorporates it into its chromosome, it, too, may become pathogenic. Plasmid DNA may also be taken up by competent bacteria and confer new properties to the cell. Overall, transformation in nature is a relatively inefficient process because environmental DNA levels are low because of the activity of nucleases that are also released during cellular lysis. Additionally, genetic recombination is inefficient at incorporating new DNA sequences into the genome.

In nature, bacterial transformation is an important mechanism for the acquisition of genetic elements encoding virulence factors and antibiotic resistance. Genes encoding resistance to antimicrobial compounds have been shown to be widespread in nature, even in environments not influenced by humans. These genes, which allow microbes living in mixed communities to compete for limited resources, can be transferred within a population by transformation, as well as by the other processes of HGT. In the laboratory, we can exploit the natural process of bacterial transformation for genetic engineering to make a wide variety of medicinal products, as discussed in Microbes and the Tools of Genetic Engineering.

CHECK YOUR UNDERSTANDING

- Why does a bacterial cell make environmental DNA brought into the cell into a single-stranded form?

Transduction

Viruses that infect bacteria (bacteriophages) may also move short pieces of chromosomal DNA from one bacterium to another in a process called **transduction** (see Figure 6.9). Recall that in generalized transduction, any piece of chromosomal DNA may be transferred to a new host cell by accidental packaging of chromosomal DNA into a phage head during phage assembly. By contrast, specialized transduction results from the imprecise excision of a lysogenic prophage from the bacterial chromosome such that it carries with it a piece of the bacterial chromosome from either side of the phage's integration site to a new host cell. As a result, the host may acquire new properties. This process is called lysogenic conversion. Of medical significance, a lysogenic phage may carry with it a virulence gene to its new host. Once inserted into the new host's chromosome, the new host may gain pathogenicity. Several pathogenic bacteria, including *Corynebacterium diphtheriae* (the causative agent of diphtheria) and *Clostridium botulinum* (the causative agent of botulism), are virulent because of the introduction of toxin-encoding genes by lysogenic bacteriophages, affirming the clinical relevance of transduction in the exchange of genes involved in infectious disease. Archaea have their own viruses that translocate genetic material from one individual to another.

✓ CHECK YOUR UNDERSTANDING

- What is the agent of transduction of prokaryotic cells?
- In specialized transduction, where does the transducing piece of DNA come from?

Case in Point

The Clinical Consequences of Transduction

Paul, a 23-year-old relief worker from Atlanta, traveled to Haiti in 2011 to provide aid following the 2010 earthquake. After working there for several weeks, he suddenly began experiencing abdominal distress, including severe cramping, nausea, vomiting, and watery diarrhea. He also began to experience intense muscle cramping. At a local clinic, the physician suspected that Paul's symptoms were caused by cholera because there had been a cholera outbreak after the earthquake. Because cholera is transmitted by the fecal-oral route, breaches in sanitation infrastructure, such as often occur following natural disasters, may precipitate outbreaks. The physician confirmed the presumptive diagnosis using a cholera dipstick test. He then prescribed Paul a single dose of doxycycline, as well as oral rehydration salts, instructing him to drink significant amounts of clean water.

Cholera is caused by the gram-negative curved rod *Vibrio cholerae* (Figure 11.27). Its symptoms largely result from the production of the cholera toxin (CT), which ultimately activates a chloride transporter to pump chloride ions out of the epithelial cells into the gut lumen. Water then follows the chloride ions, causing the prolific watery diarrhea characteristic of cholera. The gene encoding the cholera toxin is incorporated into the bacterial chromosome of *V. cholerae* through infection of the bacterium with the lysogenic filamentous CTX phage, which carries the CT gene and introduces it into the chromosome on integration of the prophage. Thus, pathogenic strains of *V. cholerae* result from horizontal gene transfer by specialized transduction.

- Why are outbreaks of cholera more common as a result of a natural disaster?
- Why is muscle cramping a common symptom of cholera? Why is treatment with oral rehydration salts so important for the treatment of cholera?
- In areas stricken by cholera, what are some strategies that people could use to prevent disease transmission?

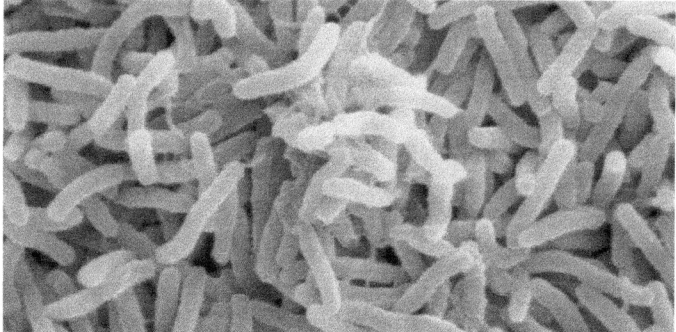

Figure 11.27 A scanning electron micrograph of *Vibrio cholerae* shows its characteristic curved rod shape.

Conjugation

In **conjugation**, DNA is directly transferred from one prokaryote to another by means of a **conjugation pilus**, which brings the organisms into contact with one another. In *E. coli*, the genes encoding the ability to conjugate are located on a bacterial plasmid called the **F plasmid**, also known as the **fertility factor**, and the conjugation pilus is called the **F pilus**. The F-plasmid genes encode both the proteins composing the F pilus and those involved in rolling circle replication of the plasmid. Cells containing the F plasmid, capable of forming an F pilus, are called **F⁺ cells** or **donor cells**, and those lacking an F plasmid are called **F⁻ cells** or **recipient cells**.

Conjugation of the F Plasmid

During typical conjugation in *E. coli*, the F pilus of an F⁺ cell comes into contact with an F⁻ cell and retracts, bringing the two cell envelopes into contact (Figure 11.28). Then a cytoplasmic bridge forms between the two cells at the site of the conjugation pilus. As rolling circle replication of the F plasmid occurs in the F⁺ cell, a single-stranded copy of the F plasmid is transferred through the cytoplasmic bridge to the F⁻ cell, which then synthesizes the complementary strand, making it double stranded. The F⁻ cell now becomes an F⁺ cell capable of making its own conjugation pilus. Eventually, in a mixed bacterial population containing both F⁺ and F⁻ cells, all cells will become F⁺ cells. Genes on the *E. coli* F plasmid also encode proteins preventing conjugation between F⁺ cells.

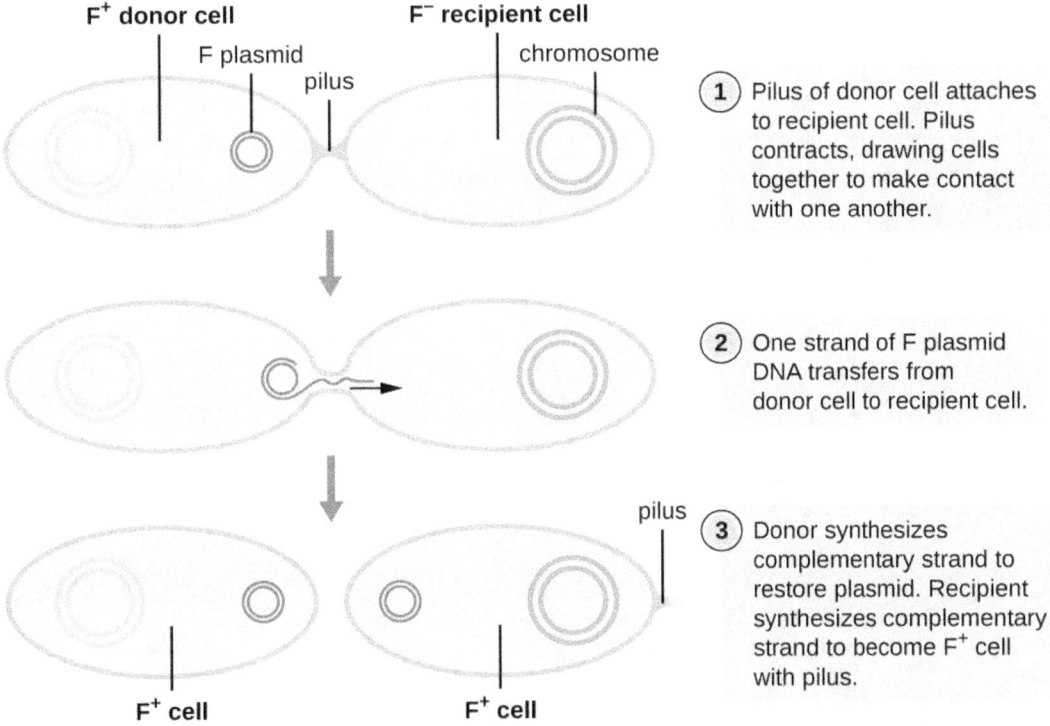

Figure 11.28 Typical conjugation of the F plasmid from an F⁺ cell to an F⁻ cell is brought about by the conjugation pilus bringing the two cells into contact. A single strand of the F plasmid is transferred to the F⁻ cell, which is then made double stranded.

Conjugation of F' and Hfr Cells

Although typical conjugation in *E. coli* results in the transfer of the F-plasmid DNA only, conjugation may also transfer chromosomal DNA. This is because the F plasmid occasionally integrates into the bacterial chromosome through recombination between the plasmid and the chromosome, forming an **Hfr cell** (Figure 11.29). "Hfr" refers to the high frequency of recombination seen when recipient F⁻ cells receive genetic information from Hfr cells through conjugation. Similar to the imprecise excision of a prophage during specialized transduction, the integrated F plasmid may also be imprecisely excised from the chromosome, producing an **F' plasmid** that carries with it some chromosomal DNA adjacent to the integration site. On conjugation, this DNA is introduced to the recipient cell and may be either maintained as part of the F' plasmid or be recombined into the recipient cell's bacterial chromosome.

Hfr cells may also treat the bacterial chromosome like an enormous F plasmid and attempt to transfer a copy of it to a recipient F⁻ cell. Because the bacterial chromosome is so large, transfer of the entire chromosome takes a long time (Figure 11.30). However, contact between bacterial cells during conjugation is transient, so it is unusual for the entire chromosome to be transferred. Host chromosomal DNA near the integration site of the F plasmid, displaced by the unidirectional process of rolling circle replication, is more likely to be transferred and recombined into a recipient cell's chromosome than host genes farther away. Thus, the relative location of bacterial genes on the Hfr cell's genome can be mapped based on when they are transferred through conjugation. As a result, prior to the age of widespread bacterial genome sequencing, distances on prokaryotic genome maps were often measured in minutes.

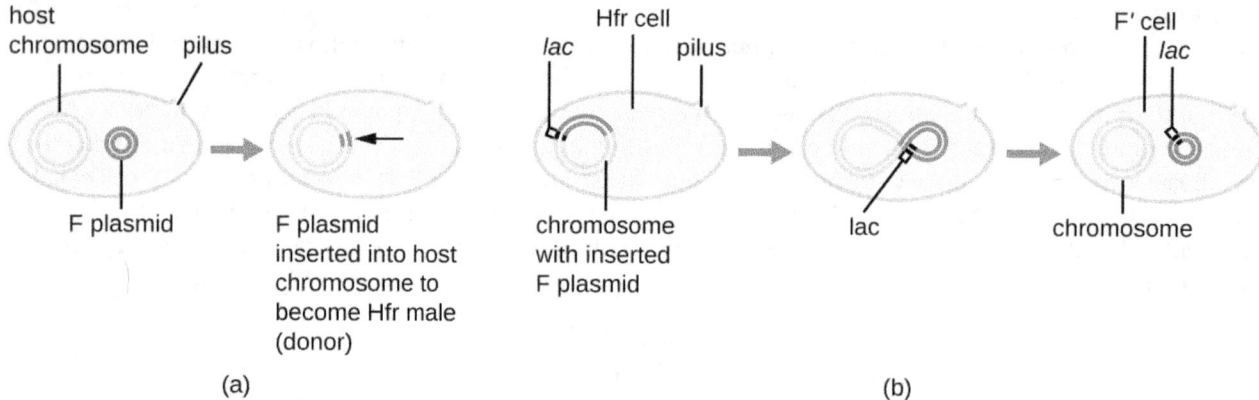

Figure 11.29 (a) The F plasmid can occasionally integrate into the bacterial chromosome, producing an Hfr cell. (b) Imprecise excision of the F plasmid from the chromosome of an Hfr cell may lead to the production of an F' plasmid that carries chromosomal DNA adjacent to the integration site. This F' plasmid can be transferred to an F⁻ cell by conjugation.

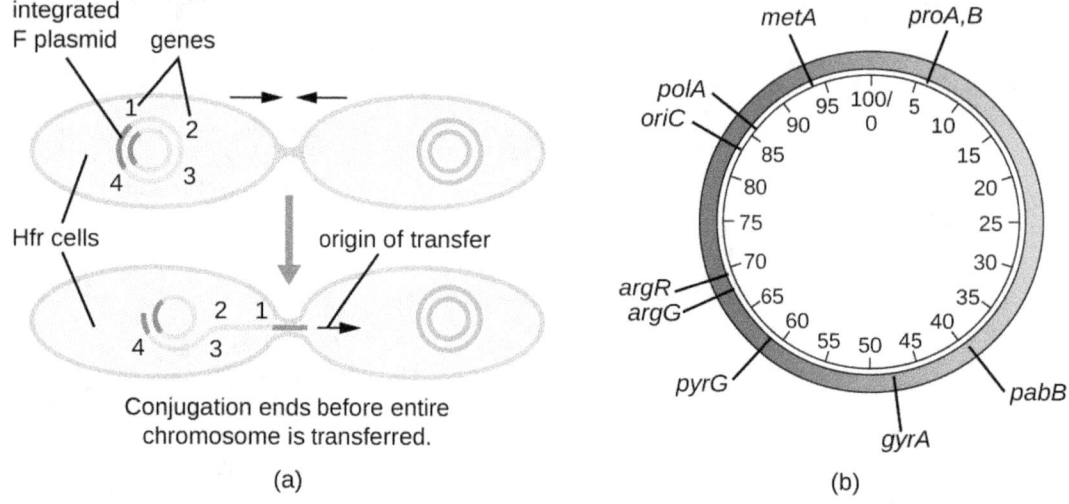

Figure 11.30 (a) An Hfr cell may attempt to transfer the entire bacterial chromosome to an F⁻ cell, treating the chromosome like an extremely large F plasmid. However, contact between cells during conjugation is temporary. Chromosomal genes closest to the integration site (gene 1) that are first displaced during rolling circle replication will be transferred more quickly than genes far away from the integration site (gene 4). Hence, they are more likely to be recombined into the recipient F⁻ cell's chromosome. (b) The time it takes for a gene to be transferred, as detected by recombination into the F⁻ cell's chromosome, can be used to generate a map of the bacterial genome, such as this genomic map of *E. coli*. Note that it takes approximately 100 minutes for the entire genome (4.6 Mbp) of an Hfr strain of *E. coli* to be transferred by conjugation.

Consequences and Applications of Conjugation

Plasmids are an important type of extrachromosomal DNA element in bacteria and, in those cells that harbor them, are considered to be part of the bacterial genome. From a clinical perspective, plasmids often code for genes involved in virulence. For example, genes encoding proteins that make a bacterial cell resistant to a particular antibiotic are encoded on **R plasmids**. R plasmids, in addition to their genes for antimicrobial resistance, contain genes that control conjugation and transfer of the plasmid. R plasmids are able to transfer between cells of the same species and between cells of different species. Single R plasmids commonly contain multiple genes conferring resistance to multiple antibiotics.

Genes required for the production of various toxins and molecules important for colonization during infection may also be found encoded on plasmids. For example, verotoxin-producing strains of *E. coli* (VTEC) appear to have acquired the genes encoding the Shiga toxin from its gram-negative relative *Shigella dysenteriae* through the acquisition of a large plasmid encoding this toxin. VTEC causes severe diarrheal disease that may result in hemolytic uremic syndrome (HUS), which may be lead to kidney failure and death.

In nonclinical settings, bacterial genes that encode metabolic enzymes needed to degrade specialized atypical compounds like polycyclic aromatic hydrocarbons (PAHs) are also frequently encoded on plasmids. Additionally, certain plasmids have the ability to move from bacterial cells to other cell types, like those of plants and animals, through mechanisms distinct from conjugation. Such mechanisms and their use in genetic engineering are covered in Modern Applications of Microbial Genetics.

LINK TO LEARNING

Click through this animation (https://openstax.org/l/22conjuganim) to learn more about the process of conjugation.

CHECK YOUR UNDERSTANDING

- What type of replication occurs during conjugation?
- What occurs to produce an Hfr *E. coli* cell?
- What types of traits are encoded on plasmids?

Transposition

Genetic elements called **transposons** (transposable elements), or "jumping genes," are molecules of DNA that include special inverted repeat sequences at their ends and a gene encoding the enzyme transposase (Figure 11.31). Transposons allow the entire sequence to independently excise from one location in a DNA molecule and integrate into the DNA elsewhere through a process called **transposition**. Transposons were originally discovered in maize (corn) by American geneticist Barbara McClintock (1902–1992) in the 1940s. Transposons have since been found in all types of organisms, both prokaryotes and eukaryotes. Thus, unlike the three previous mechanisms discussed, transposition is not prokaryote-specific. Most transposons are nonreplicative, meaning they move in a "cut-and-paste" fashion. Some may be replicative, however, retaining their location in the DNA while making a copy to be inserted elsewhere ("copy and paste"). Because transposons can move within a DNA molecule, from one DNA molecule to another, or even from one cell to another, they have the ability to introduce genetic diversity. Movement within the same DNA molecule can alter phenotype by inactivating or activating a gene.

Transposons may carry with them additional genes, moving these genes from one location to another with them. For example, bacterial transposons can relocate antibiotic resistance genes, moving them from chromosomes to plasmids. This mechanism has been shown to be responsible for the colocalization of multiple antibiotic resistance genes on a single R plasmid in *Shigella* strains causing bacterial dysentery. Such an R plasmid can then be easily transferred among a bacterial population through the process of conjugation.

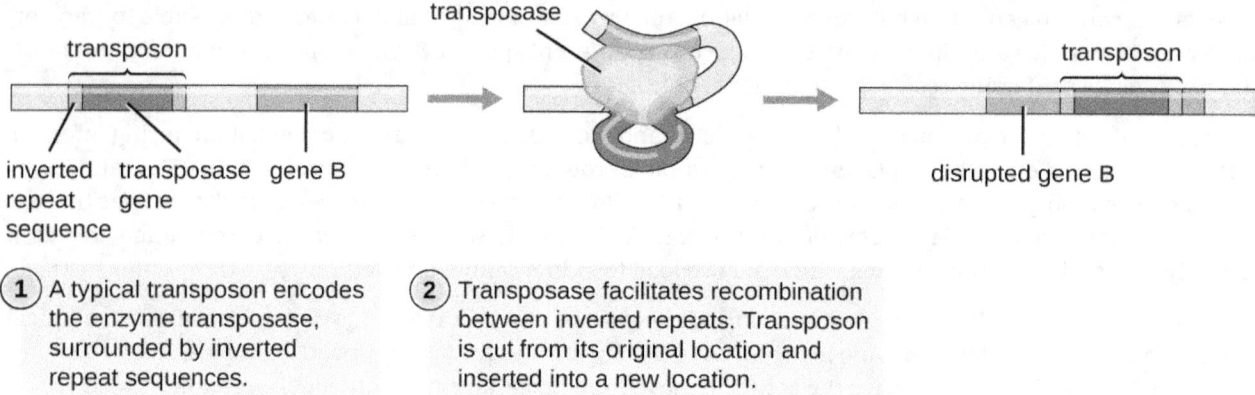

1. A typical transposon encodes the enzyme transposase, surrounded by inverted repeat sequences.

2. Transposase facilitates recombination between inverted repeats. Transposon is cut from its original location and inserted into a new location.

Figure 11.31 Transposons are segments of DNA that have the ability to move from one location to another because they code for the enzyme transposase. In this example, a nonreplicative transposon has disrupted gene B. The consequence of that the transcription of gene B may now have been interrupted.

CHECK YOUR UNDERSTANDING

- What are two ways a transposon can affect the phenotype of a cell it moves to?

Table 11.5 summarizes the processes discussed in this section.

Summary of Mechanisms of Genetic Diversity in Prokaryotes

Term	Definition
Conjugation	Transfer of DNA through direct contact using a conjugation pilus
Transduction	Mechanism of horizontal gene transfer in bacteria in which genes are transferred through viral infection
Transformation	Mechanism of horizontal gene transfer in which naked environmental DNA is taken up by a bacterial cell
Transposition	Process whereby DNA independently excises from one location in a DNA molecule and integrates elsewhere

Table 11.5

Clinical Focus

Part 3

Despite continued antibiotic treatment, Mark's infection continued to progress rapidly. The infected region continued to expand, and he had to be put on a ventilator to help him breathe. Mark's physician ordered surgical removal of the infected tissue. Following an initial surgery, Mark's wound was monitored daily to ensure that the infection did not return, but it continued to spread.

After two additional rounds of surgery, the infection finally seemed to be contained. A few days later, Mark was removed from the ventilator and was able to breathe on his own. However, he had lost a great deal of skin and soft tissue on his lower leg.

- Why does the removal of infected tissue stem the infection?

- What are some likely complications of this method of treatment?

Jump to the next Clinical Focus box. Go back to the previous Clinical Focus box.

11.7 Gene Regulation: Operon Theory

Learning Objectives

By the end of this section, you will be able to:
- Compare inducible operons and repressible operons
- Describe why regulation of operons is important

Each nucleated cell in a multicellular organism contains copies of the same DNA. Similarly, all cells in two pure bacterial cultures inoculated from the same starting colony contain the same DNA, with the exception of changes that arise from spontaneous mutations. If each cell in a multicellular organism has the same DNA, then how is it that cells in different parts of the organism's body exhibit different characteristics? Similarly, how is it that the same bacterial cells within two pure cultures exposed to different environmental conditions can exhibit different phenotypes? In both cases, each genetically identical cell does not turn on, or express, the same set of genes. Only a subset of proteins in a cell at a given time is expressed.

Genomic DNA contains both structural genes, which encode products that serve as cellular structures or enzymes, and regulatory genes, which encode products that regulate gene expression. The expression of a gene is a highly regulated process. Whereas regulating gene expression in multicellular organisms allows for cellular differentiation, in single-celled organisms like prokaryotes, it primarily ensures that a cell's resources are not wasted making proteins that the cell does not need at that time.

Elucidating the mechanisms controlling gene expression is important to the understanding of human health. Malfunctions in this process in humans lead to the development of cancer and other diseases. Understanding the interaction between the gene expression of a pathogen and that of its human host is important for the understanding of a particular infectious disease. Gene regulation involves a complex web of interactions within a given cell among signals from the cell's environment, signaling molecules within the cell, and the cell's DNA. These interactions lead to the expression of some genes and the suppression of others, depending on circumstances.

Prokaryotes and eukaryotes share some similarities in their mechanisms to regulate gene expression; however, gene expression in eukaryotes is more complicated because of the temporal and spatial separation between the processes of transcription and translation. Thus, although most regulation of gene expression occurs through transcriptional control in prokaryotes, regulation of gene expression in eukaryotes occurs at the transcriptional level and post-transcriptionally (after the primary transcript has been made).

Prokaryotic Gene Regulation

In bacteria and archaea, structural proteins with related functions are usually encoded together within the genome in a block called an **operon** and are transcribed together under the control of a single promoter, resulting in the formation of a polycistronic transcript (Figure 11.32). In this way, regulation of the transcription of all of the structural genes encoding the enzymes that catalyze the many steps in a single biochemical pathway can be controlled simultaneously, because they will either all be needed at the same time, or none will be needed. For example, in *E. coli*, all of the structural genes that encode enzymes needed to use lactose as an energy source lie next to each other in the lactose (or *lac*) operon under the control of a single promoter, the *lac* promoter. French scientists François Jacob (1920–2013) and Jacques Monod at the Pasteur Institute were the first to show the organization of bacterial genes into operons, through their studies on the *lac* operon of *E. coli*. For this work, they won the Nobel Prize in Physiology or Medicine in 1965. Although eukaryotic genes are not organized into operons, prokaryotic operons are excellent models for learning about gene regulation generally. There are some gene clusters in eukaryotes that function similar to operons. Many of the principles can be applied to eukaryotic systems and contribute to our understanding of changes in gene expression in eukaryotes that can result pathological changes such as cancer.

Each operon includes DNA sequences that influence its own transcription; these are located in a region called the regulatory region. The regulatory region includes the promoter and the region surrounding the promoter, to which **transcription factors**, proteins encoded by regulatory genes, can bind. Transcription factors influence the binding of RNA polymerase to the promoter and allow its progression to transcribe structural genes. A **repressor** is a transcription factor that suppresses transcription of a gene in response to an external stimulus by binding to a DNA sequence within the regulatory region called the **operator**, which is located between the RNA polymerase binding site of the promoter and the transcriptional start site of the first structural gene. Repressor binding physically blocks RNA polymerase from transcribing structural genes. Conversely, an **activator** is a transcription factor that increases the transcription of a gene in response to an external stimulus by facilitating RNA polymerase binding to the promoter. An **inducer**, a third type of regulatory molecule, is a small molecule that either activates or represses transcription by interacting with a repressor or an activator.

In prokaryotes, there are examples of operons whose gene products are required rather consistently and whose expression, therefore, is unregulated. Such operons are **constitutively expressed**, meaning they are transcribed and translated continuously to provide the cell with constant intermediate levels of the protein products. Such genes encode enzymes involved in housekeeping functions required for cellular maintenance, including DNA replication, repair, and expression, as well as enzymes involved in core metabolism. In contrast, there are other prokaryotic operons that are expressed only when needed and are regulated by repressors, activators, and inducers.

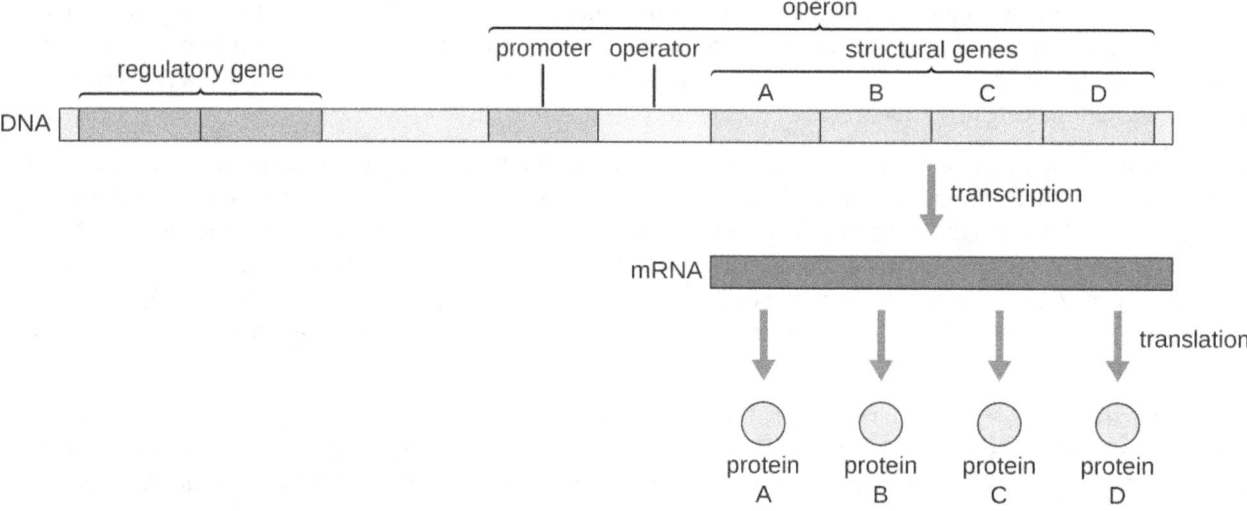

Figure 11.32 In prokaryotes, structural genes of related function are often organized together on the genome and transcribed together under the control of a single promoter. The operon's regulatory region includes both the promoter and the operator. If a repressor binds to the operator, then the structural genes will not be transcribed. Alternatively, activators may bind to the regulatory region, enhancing transcription.

CHECK YOUR UNDERSTANDING

- What are the parts in the DNA sequence of an operon?
- What types of regulatory molecules are there?

Regulation by Repression

Prokaryotic operons are commonly controlled by the binding of repressors to operator regions, thereby preventing the transcription of the structural genes. Such operons are classified as either **repressible operons** or inducible operons. Repressible operons, like the tryptophan (*trp*) operon, typically contain genes encoding enzymes required for a biosynthetic pathway. As long as the product of the pathway, like tryptophan, continues to be required by the cell, a repressible operon will continue to be expressed. However, when the product of the biosynthetic pathway begins to accumulate in the cell, removing the need for the cell to continue to make more, the expression of the operon is repressed. Conversely, **inducible operons**, like the *lac* operon of *E. coli*, often contain genes encoding enzymes in a pathway involved in the metabolism of a specific substrate like lactose. These enzymes are only required when that substrate is available, thus expression of the operons is typically induced only in the presence of the substrate.

The *trp* Operon: A Repressible Operon

E. coli can synthesize tryptophan using enzymes that are encoded by five structural genes located next to each other in the *trp* operon (Figure 11.33). When environmental tryptophan is low, the operon is turned on. This means that transcription is initiated, the genes are expressed, and tryptophan is synthesized. However, if tryptophan is present in the environment, the *trp* operon is turned off. Transcription does not occur and tryptophan is not synthesized.

When tryptophan is not present in the cell, the repressor by itself does not bind to the operator; therefore, the operon is active and tryptophan is synthesized. However, when tryptophan accumulates in the cell, two tryptophan molecules bind to the *trp* repressor molecule, which changes its shape, allowing it to bind to the *trp* operator. This binding of the active form of the *trp* repressor to the operator blocks RNA polymerase from transcribing the structural genes, stopping expression of the operon. Thus, the actual product of the biosynthetic pathway controlled by the operon regulates the expression of the operon.

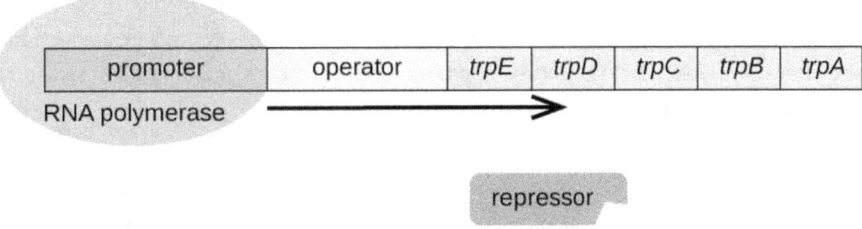

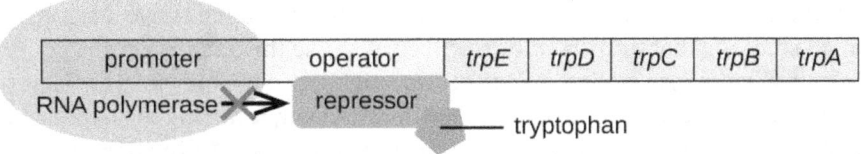

Figure 11.33 The five structural genes needed to synthesize tryptophan in *E. coli* are located next to each other in the *trp* operon. When tryptophan is absent, the repressor protein does not bind to the operator, and the genes are transcribed. When tryptophan is plentiful, tryptophan binds the repressor protein at the operator sequence. This physically blocks the RNA polymerase from transcribing the tryptophan biosynthesis genes.

🔗 LINK TO LEARNING

Watch this video (https://openstax.org/l/22trpoperon) to learn more about the *trp* operon.

The *lac* Operon: An Inducible Operon

The *lac* operon is an example of an inducible operon that is also subject to activation in the absence of glucose (Figure 11.34). The *lac* operon encodes three structural genes necessary to acquire and process the disaccharide lactose from the environment, breaking it down into the simple sugars glucose and galactose. For the *lac* operon to be expressed, lactose must be present. This makes sense for the cell because it would be energetically wasteful to create the enzymes to process lactose if lactose was not available.

In the absence of lactose, the *lac* repressor is bound to the operator region of the *lac* operon, physically preventing RNA polymerase from transcribing the structural genes. However, when lactose is present, the lactose inside the cell is converted to allolactose. Allolactose serves as an inducer molecule, binding to the repressor and changing its shape so that it is no longer able to bind to the operator DNA. Removal of the repressor in the presence of lactose allows RNA polymerase to move through the operator region and begin transcription of the *lac* structural genes.

In the absence of lactose, the *lac* repressor binds the operator, and transcription is blocked.

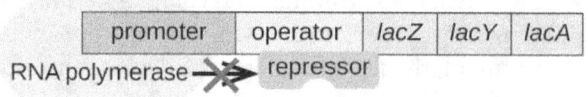

In the presence of lactose, the *lac* repressor is released from the operator, and transcription proceeds at a slow rate.

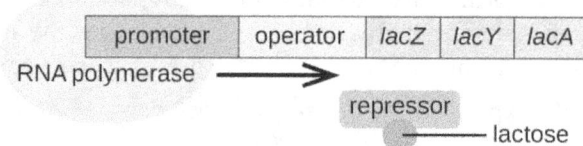

Figure 11.34 The three structural genes that are needed to degrade lactose in *E. coli* are located next to each other in the *lac* operon. When lactose is absent, the repressor protein binds to the operator, physically blocking the RNA polymerase from transcribing the *lac* structural genes. When lactose is available, a lactose molecule binds the repressor protein, preventing the repressor from binding to the operator sequence, and the genes are transcribed.

The *lac* Operon: Activation by Catabolite Activator Protein

Bacteria typically have the ability to use a variety of substrates as carbon sources. However, because glucose is usually preferable to other substrates, bacteria have mechanisms to ensure that alternative substrates are only used when glucose has been depleted. Additionally, bacteria have mechanisms to ensure that the genes encoding enzymes for using alternative substrates are expressed only when the alternative substrate is available. In the 1940s, Jacques Monod was the first to demonstrate the preference for certain substrates over others through his studies of *E. coli*'s growth when cultured in the presence of two different substrates simultaneously. Such studies generated diauxic growth curves, like the one shown in Figure 11.35. Although the preferred substrate glucose is used first, *E. coli* grows quickly and the enzymes for lactose metabolism are absent. However, once glucose levels are depleted, growth rates slow, inducing the expression of the enzymes needed for the metabolism of the second substrate, lactose. Notice how the growth rate in lactose is slower, as indicated by the lower steepness of the growth curve.

The ability to switch from glucose use to another substrate like lactose is a consequence of the activity of an enzyme called Enzyme IIA (EIIA). When glucose levels drop, cells produce less ATP from catabolism (see Catabolism of Carbohydrates), and EIIA becomes phosphorylated. Phosphorylated EIIA activates adenylyl cyclase, an enzyme that converts some of the remaining ATP to **cyclic AMP (cAMP)**, a cyclic derivative of AMP and important signaling molecule involved in glucose and energy metabolism in *E. coli*. As a result, cAMP levels begin to rise in the cell (Figure 11.36).

The *lac* operon also plays a role in this switch from using glucose to using lactose. When glucose is scarce, the accumulating cAMP caused by increased adenylyl cyclase activity binds to **catabolite activator protein (CAP)**, also known as cAMP receptor protein (CRP). The complex binds to the promoter region of the *lac* operon (Figure 11.37). In the regulatory regions of these operons, a CAP binding site is located upstream of the RNA polymerase binding site in the promoter. Binding of the CAP-cAMP complex to this site increases the binding ability of RNA polymerase to the promoter region to initiate the transcription of the structural genes. Thus, in the case of the *lac* operon, for transcription to occur, lactose must be present (removing the lac repressor protein) and glucose levels must be depleted (allowing binding of an activating protein). When glucose levels are high, there is catabolite repression of operons encoding enzymes for the metabolism of alternative substrates. Because of low cAMP levels under these conditions, there is an insufficient amount of the CAP-cAMP complex to activate transcription of these operons. See Table 11.6 for a summary of the regulation of the lac operon.

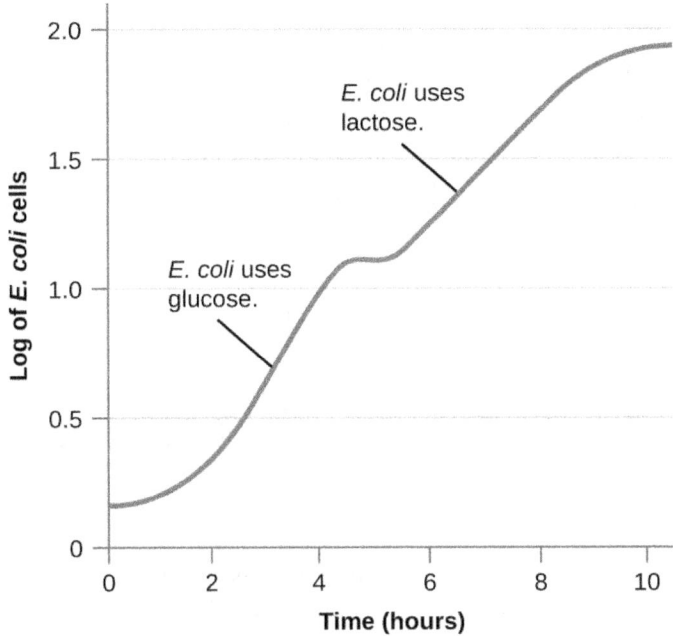

Figure 11.35 When grown in the presence of two substrates, *E. coli* uses the preferred substrate (in this case glucose) until it is depleted. Then, enzymes needed for the metabolism of the second substrate are expressed and growth resumes, although at a slower rate.

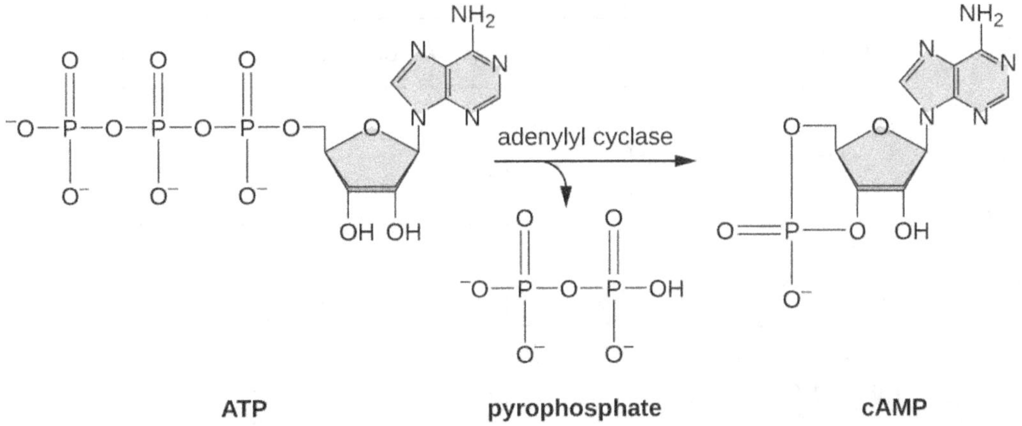

Figure 11.36 When ATP levels decrease due to depletion of glucose, some remaining ATP is converted to cAMP by adenylyl cyclase. Thus, increased cAMP levels signal glucose depletion.

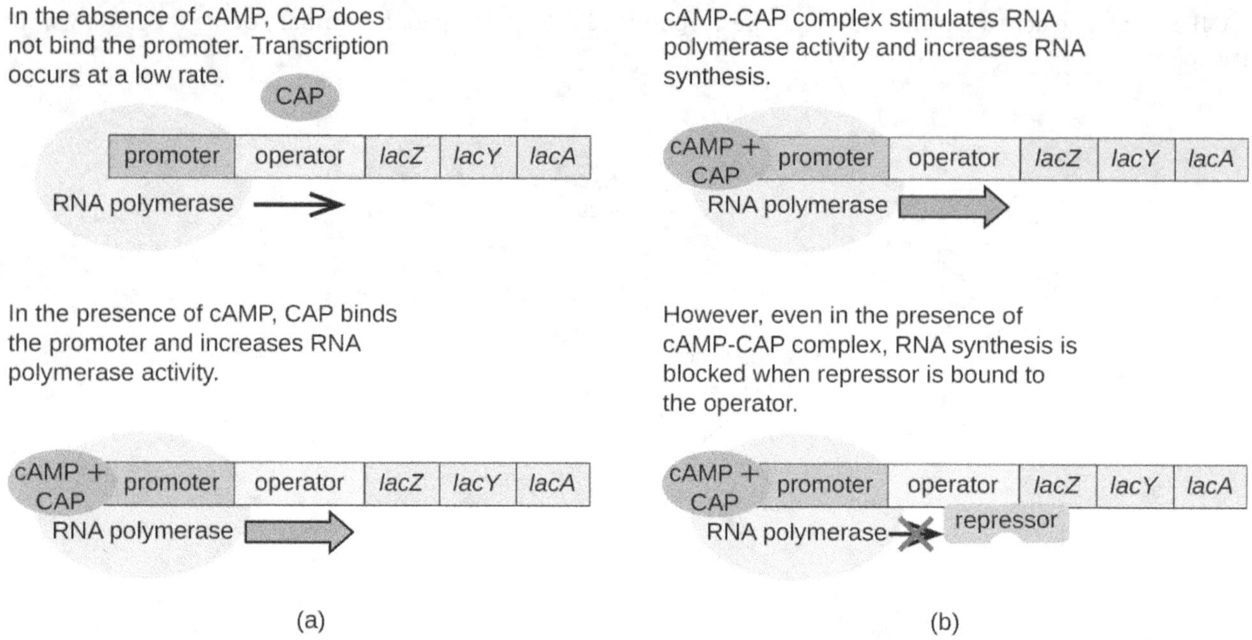
In the absence of cAMP, CAP does not bind the promoter. Transcription occurs at a low rate.
CAP
promoter
operator
lacZ
lacY
lacA
RNA polymerase
cAMP-CAP complex stimulates RNA polymerase activity and increases RNA synthesis.
cAMP + CAP
In the presence of cAMP, CAP binds the promoter and increases RNA polymerase activity.
However, even in the presence of cAMP-CAP complex, RNA synthesis is blocked when repressor is bound to the operator.
repressor
(a)
(b)

Figure 11.37 (a) In the presence of cAMP, CAP binds to the promoters of operons, like the *lac* operon, that encode genes for enzymes for the use of alternate substrates. (b) For the *lac* operon to be expressed, there must be activation by cAMP-CAP as well as removal of the lac repressor from the operator.

Conditions Affecting Transcription of the *lac* Operon

Glucose	CAP binds	Lactose	Repressor binds	Transcription
+	–	–	+	No
+	–	+	–	Some
–	+	–	+	No
–	+	+	–	Yes

Table 11.6

LINK TO LEARNING

Watch an animated tutorial (https://openstax.org/l/22lacoperon) about the workings of lac operon here.

CHECK YOUR UNDERSTANDING

- What affects the binding of the *trp* operon repressor to the operator?
- How and when is the behavior of the *lac* repressor protein altered?
- In addition to being repressible, how else is the *lac* operon regulated?

Global Responses of Prokaryotes

In prokaryotes, there are also several higher levels of gene regulation that have the ability to control the transcription of many related operons simultaneously in response to an environmental signal. A group of operons all controlled simultaneously is called a regulon.

Alarmones

When sensing impending stress, prokaryotes alter the expression of a wide variety of operons to respond in coordination. They do this through the production of **alarmones**, which are small intracellular nucleotide derivatives. Alarmones change which genes are expressed and stimulate the expression of specific stress-response genes. The use of alarmones to alter gene expression in response to stress appears to be important in pathogenic bacteria. On encountering host defense mechanisms and other harsh conditions during infection, many operons encoding virulence genes are upregulated in response to alarmone signaling. Knowledge of these responses is key to being able to fully understand the infection process of many pathogens and to the development of therapies to counter this process.

Alternate σ Factors

Since the σ subunit of bacterial RNA polymerase confers specificity as to which promoters should be transcribed, altering the **σ factor** used is another way for bacteria to quickly and globally change what regulons are transcribed at a given time. The σ factor recognizes sequences within a bacterial promoter, so different σ factors will each recognize slightly different promoter sequences. In this way, when the cell senses specific environmental conditions, it may respond by changing which σ factor it expresses, degrading the old one and producing a new one to transcribe the operons encoding genes whose products will be useful under the new environmental condition. For example, in sporulating bacteria of the genera *Bacillus* and *Clostridium* (which include many pathogens), a group of σ factors controls the expression of the many genes needed for sporulation in response to sporulation-stimulating signals.

✓ CHECK YOUR UNDERSTANDING

- What is the name given to a collection of operons that can be regulated as a group?
- What type of stimulus would trigger the transcription of a different σ factor?

Additional Methods of Regulation in Bacteria: Attenuation and Riboswitches

Although most gene expression is regulated at the level of transcription initiation in prokaryotes, there are also mechanisms to control both the completion of transcription as well as translation concurrently. Since their discovery, these mechanisms have been shown to control the completion of transcription and translation of many prokaryotic operons. Because these mechanisms link the regulation of transcription and translation directly, they are specific to prokaryotes, because these processes are physically separated in eukaryotes.

One such regulatory system is **attenuation**, whereby secondary stem-loop structures formed within the 5' end of an mRNA being transcribed determine if transcription to complete the synthesis of this mRNA will occur and if this mRNA will be used for translation. Beyond the transcriptional repression mechanism already discussed, attenuation also controls expression of the *trp* operon in *E. coli* (Figure 11.38). The *trp* operon regulatory region contains a leader sequence called *trpL* between the operator and the first structural gene, which has four stretches of RNA that can base pair with each other in different combinations. When a terminator stem-loop forms, transcription terminates, releasing RNA polymerase from the mRNA. However, when an antiterminator stem-loop forms, this prevents the formation of the terminator stem-loop, so RNA polymerase can transcribe the structural genes.

A related mechanism of concurrent regulation of transcription and translation in prokaryotes is the use of a **riboswitch**, a small region of noncoding RNA found within the 5' end of some prokaryotic mRNA molecules (Figure 11.39. A riboswitch may bind to a small intracellular molecule to stabilize certain secondary structures of the mRNA molecule. The binding of the small molecule determines which stem-loop structure forms, thus influencing the completion of mRNA synthesis and protein synthesis.

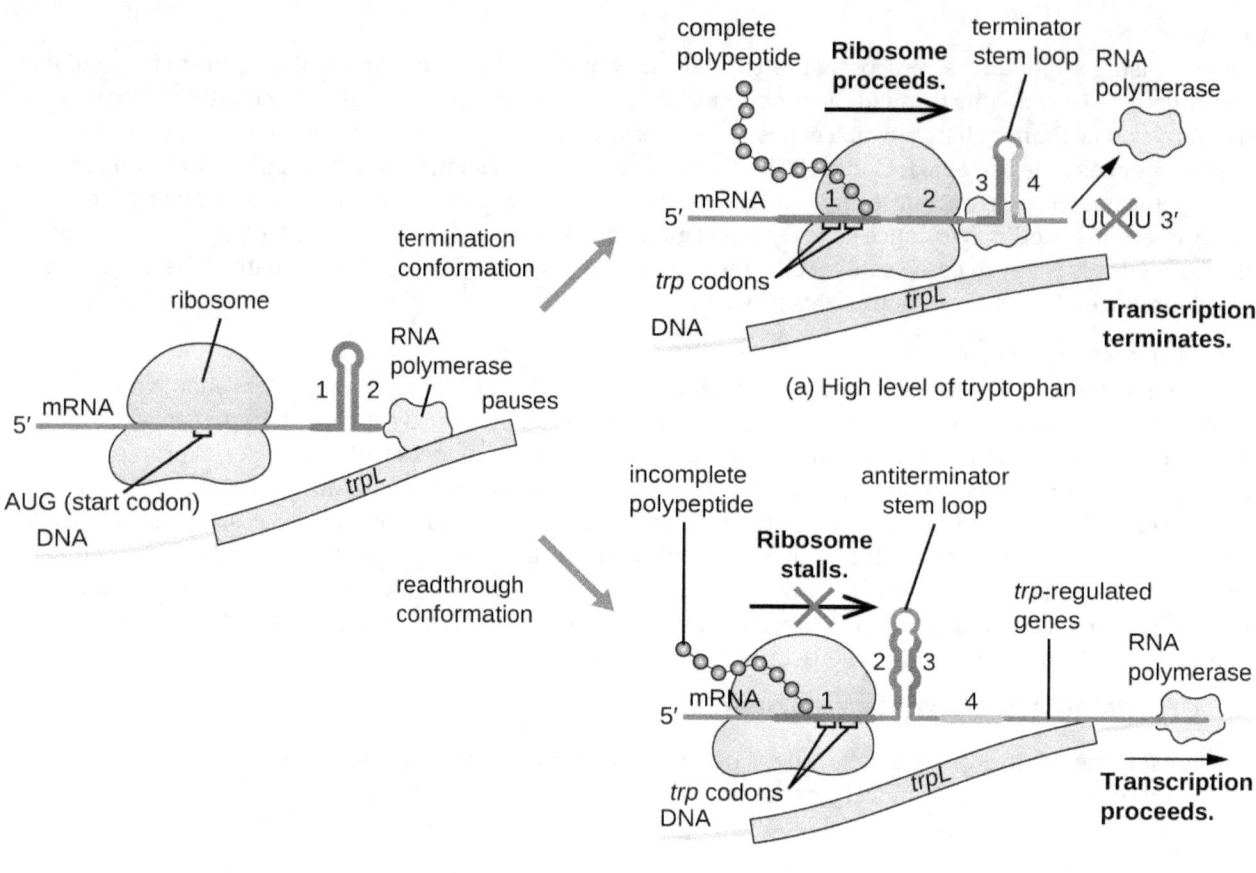

Figure 11.38 When tryptophan is plentiful, translation of the short leader peptide encoded by *trpL* proceeds, the terminator loop between regions 3 and 4 forms, and transcription terminates. When tryptophan levels are depleted, translation of the short leader peptide stalls at region 1, allowing regions 2 and 3 to form an antiterminator loop, and RNA polymerase can transcribe the structural genes of the *trp* operon.

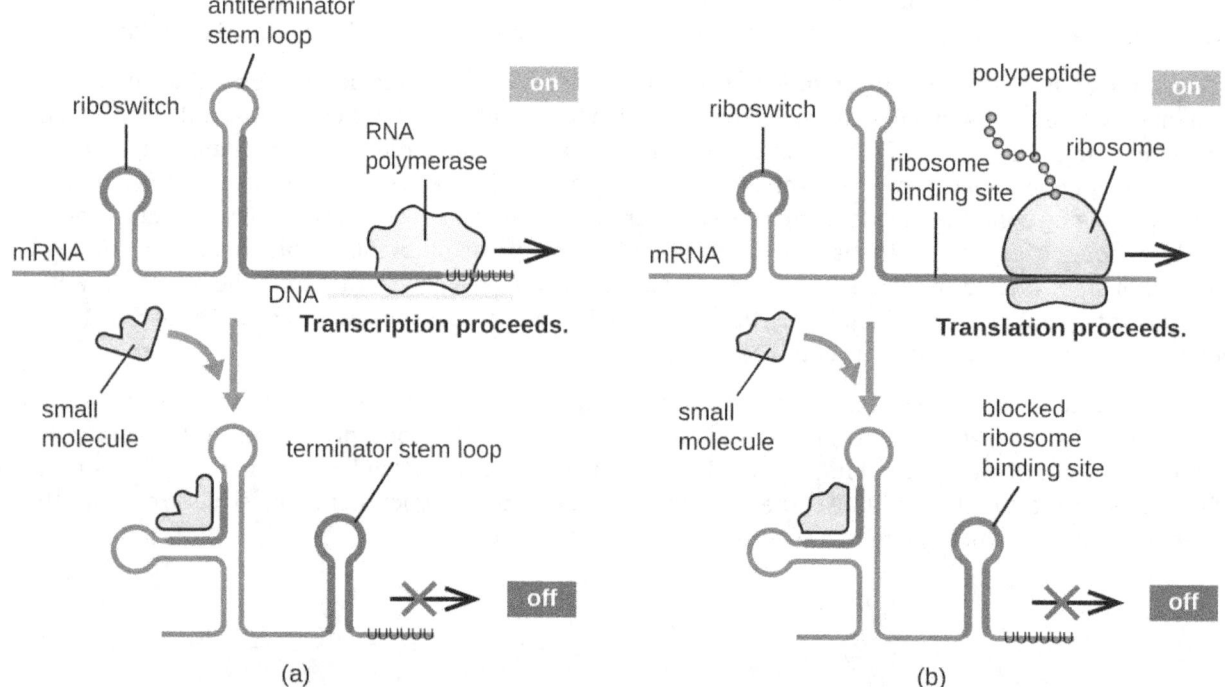

Figure 11.39 Riboswitches found within prokaryotic mRNA molecules can bind to small intracellular molecules, stabilizing certain RNA

structures, influencing either the completion of the synthesis of the mRNA molecule itself (left) or the protein made using that mRNA (right).

Other Factors Affecting Gene Expression in Prokaryotes and Eukaryotes

Although the focus on our discussion of transcriptional control used prokaryotic operons as examples, eukaryotic transcriptional control is similar in many ways. As in prokaryotes, eukaryotic transcription can be controlled through the binding of transcription factors including repressors and activators. Interestingly, eukaryotic transcription can be influenced by the binding of proteins to regions of DNA, called enhancers, rather far away from the gene, through DNA looping facilitated between the enhancer and the promoter (Figure 11.40). Overall, regulating transcription is a highly effective way to control gene expression in both prokaryotes and eukaryotes. However, the control of gene expression in eukaryotes in response to environmental and cellular stresses can be accomplished in additional ways without the binding of transcription factors to regulatory regions.

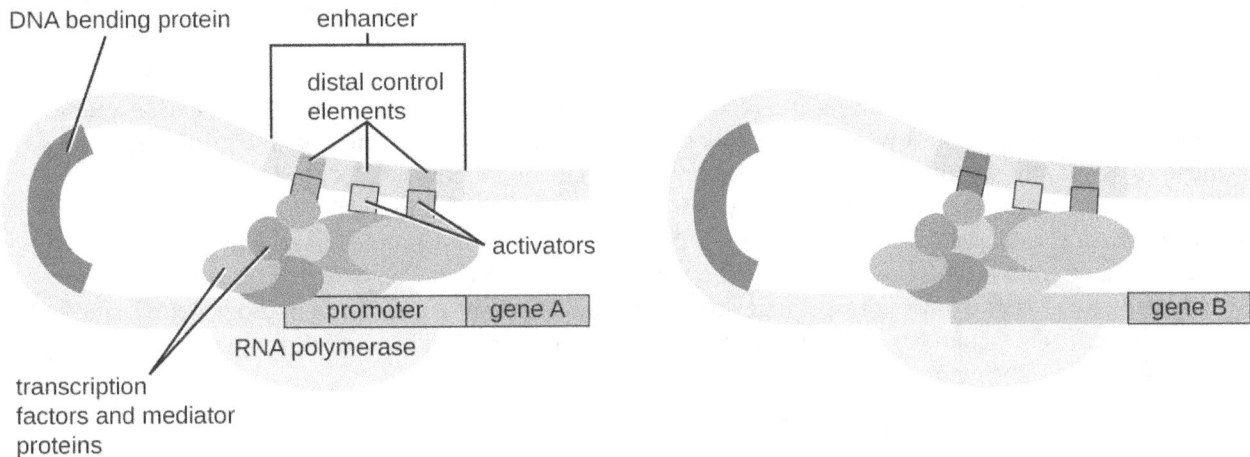

Figure 11.40   In eukaryotes, an enhancer is a DNA sequence that promotes transcription. Each enhancer is made up of short DNA sequences called distal control elements. Activators bound to the distal control elements interact with mediator proteins and transcription factors. Two different genes may have the same promoter but different distal control elements, enabling differential gene expression.

DNA-Level Control

In eukaryotes, the DNA molecules or associated histones can be chemically modified in such a way as to influence transcription; this is called **epigenetic regulation**. Methylation of certain cytosine nucleotides in DNA in response to environmental factors has been shown to influence use of such DNA for transcription, with DNA methylation commonly correlating to lowered levels of gene expression. Additionally, in response to environmental factors, histone proteins for packaging DNA can also be chemically modified in multiple ways, including acetylation and deacetylation, influencing the packaging state of DNA and thus affecting the availability of loosely wound DNA for transcription. These chemical modifications can sometimes be maintained through multiple rounds of cell division, making at least some of these epigenetic changes heritable.

LINK TO LEARNING

This video (https://openstax.org/l/22epigreg) describes how epigenetic regulation controls gene expression.

CHECK YOUR UNDERSTANDING

- What stops or allows transcription to proceed when attenuation is operating?
- What determines the state of a riboswitch?
- Describe the function of an enhancer.
- Describe two mechanisms of epigenetic regulation in eukaryotes.

Clinical Focus

Resolution

Although Mark survived his bout with necrotizing fasciitis, he would now have to undergo a skin-grafting surgery, followed by long-term physical therapy. Based on the amount of muscle mass he lost, it is unlikely that his leg will return to full strength, but his physical therapist is optimistic that he will regain some use of his leg.

Laboratory testing revealed the causative agent of Mark's infection was a strain of *Staphylococcus aureus*. As required by law, Mark's case was reported to the state health department and ultimately to the Centers for Disease Control and Prevention (CDC). At the CDC, the strain of *Staphylococcus aureus* strep isolated from Mark was analyzed more thoroughly for methicillin resistance.

Methicillin resistance is genetically coded and is increasing among strains of *S. aureus* through horizontal gene transfer. Strains of *S. aureus* that are resistant to *methicillin* are typically resistant to virtually all *beta-lactam* antibiotics and other classes of antibiotics as well. In necrotizing fasciitis, blood flow to the infected area is typically limited because of the action of various genetically encoded bacterial toxins. This is why there is typically little to no bleeding as a result of the incision test. Unfortunately, these bacterial toxins limit the effectiveness of intravenous antibiotics in clearing infection from the skin and underlying tissue, meaning that antibiotic resistance alone does not explain the ineffectiveness of Mark's treatment. Nevertheless, intravenous antibiotic therapy was warranted to help minimize the possible outcome of sepsis, which is a common outcome of necrotizing fasciitis. Through genomic analysis by the CDC of the strain isolated from Mark, several of the important virulence genes were shown to be encoded within pathogenicity islands that were associated with prophages. Horizontal transfer of pathogenicity island-encoded virulence factors between strains of S. aureus has been shown to occur through induction of prophage and can be induced by treatment with antibiotics.

Go back to the previous Clinical Focus box.

SUMMARY

11.1 The Functions of Genetic Material

- DNA serves two important cellular functions: It is the genetic material passed from parent to offspring and it serves as the information to direct and regulate the construction of the proteins necessary for the cell to perform all of its functions.
- The **central dogma** states that DNA organized into genes specifies the sequences of messenger RNA (mRNA), which, in turn, specifies the amino acid sequence of proteins.
- The genotype of a cell is the full collection of genes a cell contains. Not all genes are used to make proteins simultaneously. The phenotype is a cell's observable characteristics resulting from the proteins it is producing at a given time under specific environmental conditions.

11.2 DNA Replication

- The DNA replication process is **semiconservative**, which results in two DNA molecules, each having one parental strand of DNA and one newly synthesized strand.
- In bacteria, the **initiation of replication** occurs at the **origin of replication**, where **supercoiled** DNA is unwound by **DNA gyrase**, made single-stranded by **helicase**, and bound by **single-stranded binding protein** to maintain its single-stranded state. **Primase** synthesizes a short RNA **primer**, providing a free 3'-OH group to which **DNA polymerase III** can add DNA nucleotides.
- During **elongation**, the **leading strand** of DNA is synthesized continuously from a single primer. The **lagging strand** is synthesized discontinuously in short **Okazaki fragments**, each requiring its own primer. The RNA primers are removed and replaced with DNA nucleotides by bacterial **DNA polymerase I**, and **DNA ligase** seals the gaps between these fragments.
- **Termination** of replication in bacteria involves the resolution of circular DNA concatemers by topoisomerase IV to release the two copies of the circular chromosome.
- Eukaryotes typically have multiple linear chromosomes, each with multiple origins of replication. Overall, replication in eukaryotes is similar to that in prokaryotes.
- The linear nature of eukaryotic chromosomes necessitates **telomeres** to protect genes near the end of the chromosomes. **Telomerase** extends telomeres, preventing their degradation, in some cell types.
- **Rolling circle replication** is a type of rapid unidirectional DNA synthesis of a circular DNA molecule used for the replication of some plasmids.

11.3 RNA Transcription

- During **transcription**, the information encoded in DNA is used to make RNA.
- **RNA polymerase** synthesizes RNA, using the antisense strand of the DNA as template by adding complementary RNA nucleotides to the 3' end of the growing strand.
- RNA polymerase binds to DNA at a sequence called a **promoter** during the **initiation of transcription**.
- Genes encoding proteins of related functions are frequently transcribed under the control of a single promoter in prokaryotes, resulting in the formation of a **polycistronic mRNA** molecule that encodes multiple polypeptides.
- Unlike DNA polymerase, RNA polymerase does not require a 3'-OH group to add nucleotides, so a **primer** is not needed during initiation.
- **Termination of transcription** in bacteria occurs when the RNA polymerase encounters specific DNA sequences that lead to stalling of the polymerase. This results in release of RNA polymerase from the DNA template strand, freeing the **RNA transcript**.
- Eukaryotes have three different RNA polymerases. Eukaryotes also have monocistronic mRNA, each encoding only a single polypeptide.
- Eukaryotic primary transcripts are processed in several ways, including the addition of a **5' cap** and a 3'-**poly-A tail**, as well as **splicing**, to generate a mature mRNA molecule that can be transported out of the nucleus and that is protected from degradation.

11.4 Protein Synthesis (Translation)

- In **translation**, polypeptides are synthesized using mRNA sequences and cellular machinery, including tRNAs that match mRNA **codons** to specific amino acids and ribosomes composed of RNA and proteins that catalyze the reaction.
- The **genetic code** is **degenerate** in that several mRNA codons code for the same amino acids. The genetic code is almost universal among

living organisms.
- Prokaryotic (70S) and cytoplasmic eukaryotic (80S) ribosomes are each composed of a large subunit and a small subunit of differing sizes between the two groups. Each subunit is composed of rRNA and protein. Organelle ribosomes in eukaryotic cells resemble prokaryotic ribosomes.
- Some 60 to 90 species of tRNA exist in bacteria. Each tRNA has a three-nucleotide **anticodon** as well as a binding site for a **cognate amino acid**. All tRNAs with a specific anticodon will carry the same amino acid.
- **Initiation** of translation occurs when the small ribosomal subunit binds with **initiation factors** and an initiator tRNA at the **start codon** of an mRNA, followed by the binding to the initiation complex of the large ribosomal subunit.
- In prokaryotic cells, the start codon codes for N-formyl-methionine carried by a special initiator tRNA. In eukaryotic cells, the start codon codes for methionine carried by a special initiator tRNA. In addition, whereas ribosomal binding of the mRNA in prokaryotes is facilitated by the Shine-Dalgarno sequence within the mRNA, eukaryotic ribosomes bind to the 5' cap of the mRNA.
- During the **elongation** stage of translation, a **charged tRNA** binds to mRNA in the **A site** of the ribosome; a peptide bond is catalyzed between the two adjacent amino acids, breaking the bond between the first amino acid and its tRNA; the ribosome moves one codon along the mRNA; and the first tRNA is moved from the **P site** of the ribosome to the **E site** and leaves the ribosomal complex.
- **Termination** of translation occurs when the ribosome encounters a **stop codon**, which does not code for a tRNA. Release factors cause the polypeptide to be released, and the ribosomal complex dissociates.
- In prokaryotes, transcription and translation may be coupled, with translation of an mRNA molecule beginning as soon as transcription allows enough mRNA exposure for the binding of a ribosome, prior to transcription termination. Transcription and translation are not coupled in eukaryotes because transcription occurs in the nucleus, whereas translation occurs in the cytoplasm or in association with the rough endoplasmic reticulum.
- Polypeptides often require one or more **post-translational modifications** to become biologically active.

11.5 Mutations

- A **mutation** is a heritable change in DNA. A mutation may lead to a change in the amino-acid sequence of a protein, possibly affecting its function.
- A **point mutation** affects a single base pair. A point mutation may cause a **silent mutation** if the mRNA codon codes for the same amino acid, a **missense mutation** if the mRNA codon codes for a different amino acid, or a **nonsense mutation** if the mRNA codon becomes a stop codon.
- Missense mutations may retain function, depending on the chemistry of the new amino acid and its location in the protein. Nonsense mutations produce truncated and frequently nonfunctional proteins.
- A **frameshift mutation** results from an insertion or deletion of a number of nucleotides that is not a multiple of three. The change in reading frame alters every amino acid after the point of the mutation and results in a nonfunctional protein.
- **Spontaneous mutations** occur through DNA replication errors, whereas **induced mutations** occur through exposure to a **mutagen**.
- Mutagenic agents are frequently carcinogenic but not always. However, nearly all carcinogens are mutagenic.
- Chemical mutagens include base analogs and chemicals that modify existing bases. In both cases, mutations are introduced after several rounds of DNA replication.
- **Ionizing radiation,** such as X-rays and γ-rays, leads to breakage of the phosphodiester backbone of DNA and can also chemically modify bases to alter their base-pairing rules.
- **Nonionizing radiation** like ultraviolet light may introduce pyrimidine (thymine) dimers, which, during DNA replication and transcription, may introduce frameshift or point mutations.
- Cells have mechanisms to repair naturally occurring mutations. DNA polymerase has proofreading activity. Mismatch repair is a process to repair incorrectly incorporated bases after DNA replication has been completed.
- Pyrimidine dimers can also be repaired. In **nucleotide excision repair (dark repair)**, enzymes recognize the distortion introduced by the pyrimidine dimer and replace the damaged strand with the correct bases, using the

undamaged DNA strand as a template. Bacteria and other organisms may also use **direct repair**, in which the photolyase enzyme, in the presence of visible light, breaks apart the pyrimidines.
- Through comparison of growth on the complete plate and lack of growth on media lacking specific nutrients, specific loss-of-function mutants called **auxotrophs** can be identified.
- The **Ames test** is an inexpensive method that uses auxotrophic bacteria to measure mutagenicity of a chemical compound. Mutagenicity is an indicator of carcinogenic potential.

11.6 How Asexual Prokaryotes Achieve Genetic Diversity

- **Horizontal gene transfer** is an important way for asexually reproducing organisms like prokaryotes to acquire new traits.
- There are three mechanisms of horizontal gene transfer typically used by bacteria: **transformation**, **transduction**, and **conjugation**.
- Transformation allows for competent cells to take up naked DNA, released from other cells on their death, into their cytoplasm, where it may recombine with the host genome.
- In **generalized transduction**, any piece of chromosomal DNA may be transferred by accidental packaging of the degraded host chromosome into a phage head. In **specialized transduction**, only chromosomal DNA adjacent to the integration site of a lysogenic phage may be transferred as a result of imprecise excision of the prophage.
- Conjugation is mediated by the **F plasmid,** which encodes a **conjugation pilus** that brings an F plasmid-containing **F⁺ cell** into contact with an **F⁻ cell**.
- The rare integration of the F plasmid into the bacterial chromosome, generating an **Hfr cell**, allows for transfer of chromosomal DNA from the donor to the recipient. Additionally, imprecise excision of the F plasmid from the chromosome may generate an F' plasmid that may be transferred to a recipient by conjugation.
- Conjugation transfer of **R plasmids** is an important mechanism for the spread of antibiotic resistance in bacterial communities.
- **Transposons** are molecules of DNA with inverted repeats at their ends that also encode the enzyme transposase, allowing for their movement from one location in DNA to another. Although found in both prokaryotes and eukaryotes, transposons are clinically relevant in bacterial pathogens for the movement of virulence factors, including antibiotic resistance genes.

11.7 Gene Regulation: Operon Theory

- **Gene expression** is a tightly regulated process.
- Gene expression in prokaryotes is largely regulated at the point of transcription. Gene expression in eukaryotes is additionally regulated post-transcriptionally.
- Prokaryotic structural genes of related function are often organized into **operons**, all controlled by transcription from a single promoter. The regulatory region of an operon includes the promoter itself and the region surrounding the promoter to which transcription factors can bind to influence transcription.
- Although some operons are **constitutively expressed**, most are subject to regulation through the use of **transcription factors** (repressors and activators). A **repressor** binds to an **operator**, a DNA sequence within the regulatory region between the RNA polymerase binding site in the promoter and first structural gene, thereby physically blocking transcription of these operons. An **activator** binds within the regulatory region of an operon, helping RNA polymerase bind to the promoter, thereby enhancing the transcription of this operon. An **inducer** influences transcription through interacting with a repressor or activator.
- The *trp* operon is a classic example of a **repressible operon**. When tryptophan accumulates, tryptophan binds to a repressor, which then binds to the operator, preventing further transcription.
- The *lac* operon is a classic example an **inducible operon**. When lactose is present in the cell, it is converted to allolactose. Allolactose acts as an inducer, binding to the repressor and preventing the repressor from binding to the operator. This allows transcription of the structural genes.
- The *lac* operon is also subject to activation. When glucose levels are depleted, some cellular ATP is converted into cAMP, which binds to the **catabolite activator protein (CAP)**. The cAMP-CAP complex activates transcription of the *lac* operon. When glucose levels are high, its presence prevents transcription of the *lac*

operon and other operons by **catabolite repression**.
- Small intracellular molecules called **alarmones** are made in response to various environmental stresses, allowing bacteria to control the transcription of a group of operons, called a regulon.
- Bacteria have the ability to change which **σ factor** of RNA polymerase they use in response to environmental conditions to quickly and globally change which regulons are transcribed.
- Prokaryotes have regulatory mechanisms, including **attenuation** and the use of **riboswitches**, to simultaneously control the completion of transcription and translation from that transcript. These mechanisms work through the formation of stem loops in the 5' end of an mRNA molecule currently being synthesized.
- There are additional points of regulation of gene expression in prokaryotes and eukaryotes. In eukaryotes, **epigenetic regulation** by chemical modification of DNA or histones, and regulation of RNA processing are two methods.

REVIEW QUESTIONS
Multiple Choice

1. DNA does all but which of the following?
   A. serves as the genetic material passed from parent to offspring
   B. remains constant despite changes in environmental conditions
   C. provides the instructions for the synthesis of messenger RNA
   D. is read by ribosomes during the process of translation

2. According to the central dogma, which of the following represents the flow of genetic information in cells?
   A. protein to DNA to RNA
   B. DNA to RNA to protein
   C. RNA to DNA to protein
   D. DNA to protein to RNA

3. Which of the following is the enzyme that replaces the RNA nucleotides in a primer with DNA nucleotides?
   A. DNA polymerase III
   B. DNA polymerase I
   C. primase
   D. helicase

4. Which of the following is not involved in the initiation of replication?
   A. ligase
   B. DNA gyrase
   C. single-stranded binding protein
   D. primase

5. Which of the following enzymes involved in DNA replication is unique to eukaryotes?
   A. helicase
   B. DNA polymerase
   C. ligase
   D. telomerase

6. Which of the following would be synthesized using 5'-CAGTTCGGA-3' as a template?
   A. 3'-AGGCTTGAC-4'
   B. 3'-TCCGAACTG-5'
   C. 3'-GTCAAGCCT-5'
   D. 3'-CAGTTCGGA-5'

7. During which stage of bacterial transcription is the σ subunit of the RNA polymerase involved?
   A. initiation
   B. elongation
   C. termination
   D. splicing

8. Which of the following components is involved in the initiation of transcription?
   A. primer
   B. origin
   C. promoter
   D. start codon

9. Which of the following is not a function of the 5' cap and 3' poly-A tail of a mature eukaryotic mRNA molecule?
   A. to facilitate splicing
   B. to prevent mRNA degradation
   C. to aid export of the mature transcript to the cytoplasm
   D. to aid ribosome binding to the transcript

10. Mature mRNA from a eukaryote would contain each of these features except which of the following?
    A. exon-encoded RNA
    B. intron-encoded RNA
    C. 5' cap
    D. 3' poly-A tail

11. Which of the following is the name of the three-base sequence in the mRNA that binds to a tRNA molecule?
    A. P site
    B. codon
    C. anticodon
    D. CCA binding site

12. Which component is the last to join the initiation complex during the initiation of translation?
    A. the mRNA molecule
    B. the small ribosomal subunit
    C. the large ribosomal subunit
    D. the initiator tRNA

13. During elongation in translation, to which ribosomal site does an incoming charged tRNA molecule bind?
    A. A site
    B. P site
    C. E site
    D. B site

14. Which of the following is the amino acid that appears at the N-terminus of all newly translated prokaryotic and eukaryotic polypeptides?
    A. tryptophan
    B. methionine
    C. selenocysteine
    D. glycine

15. When the ribosome reaches a nonsense codon, which of the following occurs?
    A. a methionine is incorporated
    B. the polypeptide is released
    C. a peptide bond forms
    D. the A site binds to a charged tRNA

16. Which of the following is a change in the sequence that leads to formation of a stop codon?
    A. missense mutation
    B. nonsense mutation
    C. silent mutation
    D. deletion mutation

17. The formation of pyrimidine dimers results from which of the following?
    A. spontaneous errors by DNA polymerase
    B. exposure to gamma radiation
    C. exposure to ultraviolet radiation
    D. exposure to intercalating agents

18. Which of the following is an example of a frameshift mutation?
    A. a deletion of a codon
    B. missense mutation
    C. silent mutation
    D. deletion of one nucleotide

19. Which of the following is the type of DNA repair in which thymine dimers are directly broken down by the enzyme photolyase?
    A. direct repair
    B. nucleotide excision repair
    C. mismatch repair
    D. proofreading

20. Which of the following regarding the Ames test is true?
    A. It is used to identify newly formed auxotrophic mutants.
    B. It is used to identify mutants with restored biosynthetic activity.
    C. It is used to identify spontaneous mutants.
    D. It is used to identify mutants lacking photoreactivation activity.

21. Which is the mechanism by which improper excision of a prophage from a bacterial chromosome results in packaging of bacterial genes near the integration site into a phage head?
    A. conjugation
    B. generalized transduction
    C. specialized transduction
    D. transformation

22. Which of the following refers to the uptake of naked DNA from the surrounding environment?
    A. conjugation
    B. generalized transduction
    C. specialized transduction
    D. transformation

23. The F plasmid is involved in which of the following processes?
    A. conjugation
    B. transduction
    C. transposition
    D. transformation

24. Which of the following refers to the mechanism of horizontal gene transfer naturally responsible for the spread of antibiotic resistance genes within a bacterial population?
    A. conjugation
    B. generalized transduction
    C. specialized transduction
    D. transformation

25. An operon of genes encoding enzymes in a biosynthetic pathway is likely to be which of the following?
    A. inducible
    B. repressible
    C. constitutive
    D. monocistronic

26. An operon encoding genes that are transcribed and translated continuously to provide the cell with constant intermediate levels of the protein products is said to be which of the following?
    A. repressible
    B. inducible
    C. constitutive
    D. activated

27. Which of the following conditions leads to maximal expression of the *lac* operon?
    A. lactose present, glucose absent
    B. lactose present, glucose present
    C. lactose absent, glucose absent
    D. lactose absent, glucose present

28. Which of the following is a type of regulation of gene expression unique to eukaryotes?
    A. attenuation
    B. use of alternate σ factor
    C. chemical modification of histones
    D. alarmones

True/False

29. Cells are always producing proteins from every gene they possess.
30. More primers are used in lagging strand synthesis than in leading strand synthesis.
31. Each codon within the genetic code encodes a different amino acid.
32. Carcinogens are typically mutagenic.
33. Asexually reproducing organisms lack mechanisms for generating genetic diversity within a population.

Fill in the Blank

34. The process of making an RNA copy of a gene is called _____.
35. A cell's _____ remains constant whereas its phenotype changes in response to environmental influences.
36. The enzyme responsible for relaxing supercoiled DNA to allow for the initiation of replication is called _____.
37. Unidirectional replication of a circular DNA molecule like a plasmid that involves nicking one DNA strand and displacing it while synthesizing a new strand is called _____.
38. A _____ mRNA is one that codes for multiple polypeptides.
39. The protein complex responsible for removing intron-encoded RNA sequences from primary transcripts in eukaryotes is called the _____.
40. The third position within a codon, in which changes often result in the incorporation of the same amino acid into the growing polypeptide, is called the _____.
41. The enzyme that adds an amino acid to a tRNA molecule is called _____.
42. A chemical mutagen that is structurally similar to a nucleotide but has different base-pairing rules is called a _____.
43. The enzyme used in light repair to split thymine dimers is called _____.
44. The phenotype of an organism that is most commonly observed in nature is called the _____.
45. A small DNA molecule that has the ability to independently excise from one location in a larger DNA molecule and integrate into the DNA elsewhere is called a _____.

46. _____ is a group of mechanisms that allow for the introduction of genetic material from one organism to another organism within the same generation.
47. The DNA sequence, to which repressors may bind, that lies between the promoter and the first structural gene is called the _____.
48. The prevention of expression of operons encoding substrate use pathways for substrates other than glucose when glucose is present is called _____.

Short Answer

49. Can two observably different cells have the same genotype? Explain.
50. Why is primase required for DNA replication?
51. What is the role of single-stranded binding protein in DNA replication?
52. Below is a DNA sequence. Envision that this is a section of a DNA molecule that has separated in preparation for replication, so you are only seeing one DNA strand. Construct the complementary DNA sequence (indicating 5' and 3' ends).
    DNA sequence: 3'-T A C T G A C T G A C G A T C-5'
53. What is the purpose of RNA processing in eukaryotes? Why don't prokaryotes require similar processing?
54. Below is a DNA sequence. Envision that this is a section of a DNA molecule that has separated in preparation for transcription, so you are only seeing the antisense strand. Construct the mRNA sequence transcribed from this template.
    Antisense DNA strand: 3'-T A C T G A C T G A C G A T C-5'
55. Why does translation terminate when the ribosome reaches a stop codon? What happens?
56. How does the process of translation differ between prokaryotes and eukaryotes?
57. What is meant by the genetic code being nearly universal?
58. Below is an antisense DNA sequence. Translate the mRNA molecule synthesized using the genetic code, recording the resulting amino acid sequence, indicating the N and C termini.
    Antisense DNA strand: 3'-T A C T G A C T G A C G A T C-5'
59. Why is it more likely that insertions or deletions will be more detrimental to a cell than point mutations?
60. Briefly describe two ways in which chromosomal DNA from a donor cell may be transferred to a recipient cell during the process of conjugation.
61. Describe what happens when a nonsense mutation is introduced into the gene encoding transposase within a transposon.
62. What are two ways that bacteria can influence the transcription of multiple different operons simultaneously in response to a particular environmental condition?

Critical Thinking

63. A pure culture of an unknown bacterium was streaked onto plates of a variety of media. You notice that the colony morphology is strikingly different on plates of minimal media with glucose compared to that seen on trypticase soy agar plates. How can you explain these differences in colony morphology?
64. Review Figure 11.4 and Figure 11.5. Why was it important that Meselson and Stahl continue their experiment to at least two rounds of replication after isotopic labeling of the starting DNA with ^{15}N, instead of stopping the experiment after only one round of replication?
65. If deoxyribonucleotides that lack the 3'-OH groups are added during the replication process, what do you expect will occur?
66. Predict the effect of an alteration in the sequence of nucleotides in the −35 region of a bacterial promoter.

67. Label the following in the figure: ribosomal E, P, and A sites; mRNA; codons; anticodons; growing polypeptide; incoming amino acid; direction of translocation; small ribosomal unit; large ribosomal unit.

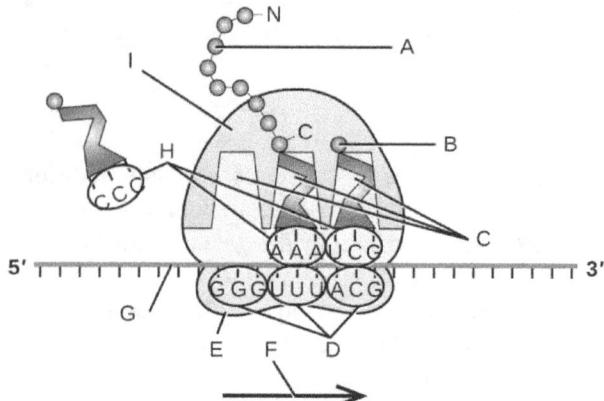

68. Prior to the elucidation of the genetic code, prominent scientists, including Francis Crick, had predicted that each mRNA codon, coding for one of the 20 amino acids, needed to be at least three nucleotides long. Why is it not possible for codons to be any shorter?

69. Below are several DNA sequences that are mutated compared with the wild-type sequence: 3'-T A C T G A C T G A C G A T C-5'. Envision that each is a section of a DNA molecule that has separated in preparation for transcription, so you are only seeing the template strand. Construct the complementary DNA sequences (indicating 5' and 3' ends) for each mutated DNA sequence, then transcribe (indicating 5' and 3' ends) the template strands, and translate the mRNA molecules using the genetic code, recording the resulting amino acid sequence (indicating the N and C termini). What type of mutation is each?

Mutated DNA Template Strand #1: 3'-T A C T G T C T G A C G A T C-5'
Complementary DNA sequence:
mRNA sequence transcribed from template:
Amino acid sequence of peptide:
Type of mutation:

Mutated DNA Template Strand #2: 3'-T A C G G A C T G A C G A T C-5'
Complementary DNA sequence:
mRNA sequence transcribed from template:
Amino acid sequence of peptide:
Type of mutation:

Mutated DNA Template Strand #3: 3'-T A C T G A C T G A C T A T C-5'
Complementary DNA sequence:
mRNA sequence transcribed from template:
Amino acid sequence of peptide:
Type of mutation:

Mutated DNA Template Strand #4: 3'-T A C G A C T G A C T A T C-5'
Complementary DNA sequence:
mRNA sequence transcribed from template:
Amino acid sequence of peptide:
Type of mutation:

70. Why do you think the Ames test is preferable to the use of animal models to screen chemical compounds for mutagenicity?

71. The following figure is from Monod's original work on diauxic growth showing the growth of E. coli in the simultaneous presence of xylose and glucose as the only carbon sources. Explain what is happening at points A–D with respect to the carbon source being used for growth, and explain whether the xylose-use operon is being expressed (and why). Note that expression of the enzymes required for xylose use is regulated in a manner similar to the expression of the enzymes required for lactose use.

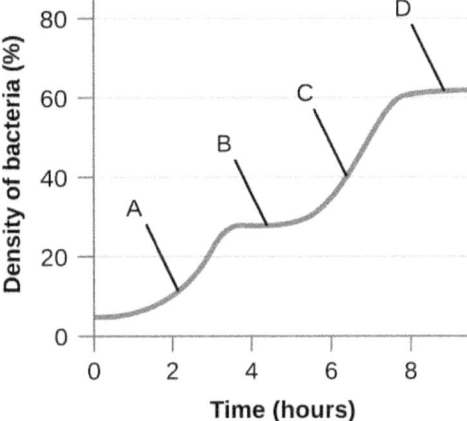

CHAPTER 12
Modern Applications of Microbial Genetics

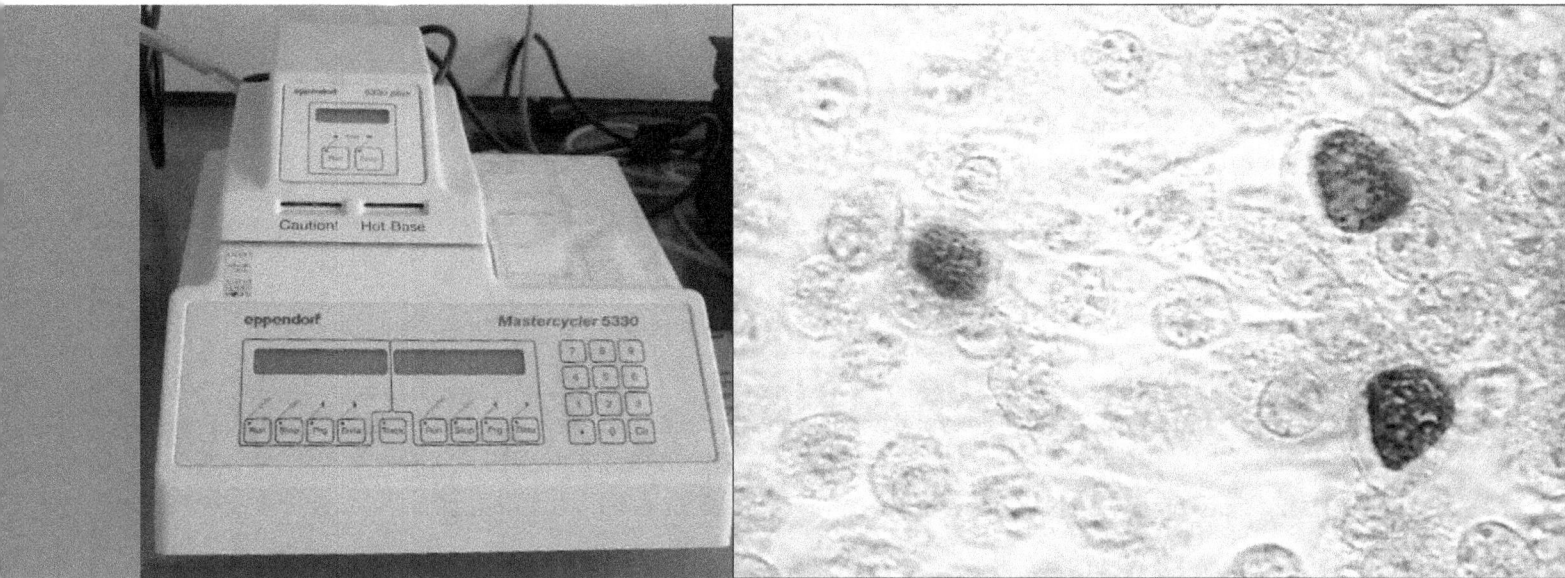

Figure 12.1 A thermal cycler (left) is used during a polymerase chain reaction (PCR). PCR amplifies the number of copies of DNA and can assist in diagnosis of infections caused by microbes that are difficult to culture, such as *Chlamydia trachomatis* (right). *C. trachomatis* causes chlamydia, the most common sexually transmitted disease in the United States, and trachoma, the world's leading cause of preventable blindness. (credit right: modification of work by Centers for Disease Control and Prevention)

Chapter Outline

12.1 Microbes and the Tools of Genetic Engineering

12.2 Visualizing and Characterizing DNA, RNA, and Protein

12.3 Whole Genome Methods and Pharmaceutical Applications of Genetic Engineering

12.4 Gene Therapy

INTRODUCTION Watson and Crick's identification of the structure of DNA in 1953 was the seminal event in the field of genetic engineering. Since the 1970s, there has been a veritable explosion in scientists' ability to manipulate DNA in ways that have revolutionized the fields of biology, medicine, diagnostics, forensics, and industrial manufacturing. Many of the molecular tools discovered in recent decades have been produced using prokaryotic microbes. In this chapter, we will explore some of those tools, especially as they relate to applications in medicine and health care.

As an example, the thermal cycler in Figure 12.1 is used to perform a diagnostic technique called the polymerase chain reaction (PCR), which relies on DNA polymerase enzymes from thermophilic bacteria. Other molecular tools, such as restriction enzymes and plasmids obtained from microorganisms, allow scientists to insert genes from humans or other organisms into microorganisms. The microorganisms are then grown on an industrial scale to synthesize products such as insulin, vaccines, and biodegradable polymers. These are just a few of the numerous applications of microbial genetics that we will explore in this chapter.

12.1 Microbes and the Tools of Genetic Engineering

Learning Objectives

By the end of this section, you will be able to:
- Identify tools of molecular genetics that are derived from microorganisms
- Describe the methods used to create recombinant DNA molecules
- Describe methods used to introduce DNA into prokaryotic cells
- List the types of genomic libraries and describe their uses
- Describe the methods used to introduce DNA into eukaryotic cells

---

Clinical Focus

Part 1

Kayla, a 24-year-old electrical engineer and running enthusiast, just moved from Arizona to New Hampshire to take a new job. On her weekends off, she loves to explore her new surroundings, going for long runs in the pine forests. In July she spent a week hiking through the mountains. In early August, Kayla developed a low fever, headache, and mild muscle aches, and she felt a bit fatigued. Not thinking much of it, she took some ibuprofen to combat her symptoms and vowed to get more rest.

- What types of medical conditions might be responsible for Kayla's symptoms?

Jump to the next Clinical Focus box.

The science of using living systems to benefit humankind is called **biotechnology**. Technically speaking, the domestication of plants and animals through farming and breeding practices is a type of biotechnology. However, in a contemporary sense, we associate biotechnology with the direct alteration of an organism's genetics to achieve desirable traits through the process of **genetic engineering**. Genetic engineering involves the use of **recombinant DNA technology**, the process by which a DNA sequence is manipulated *in vitro*, thus creating **recombinant DNA molecules** that have new combinations of genetic material. The recombinant DNA is then introduced into a host organism. If the DNA that is introduced comes from a different species, the host organism is now considered to be **transgenic**.

One example of a transgenic microorganism is the bacterial strain that produces human insulin (Figure 12.2). The insulin gene from humans was inserted into a plasmid. This recombinant DNA plasmid was then inserted into bacteria. As a result, these transgenic microbes are able to produce and secrete human insulin. Many prokaryotes are able to acquire foreign DNA and incorporate functional genes into their own genome through "mating" with other cells (conjugation), viral infection (transduction), and taking up DNA from the environment (transformation). Recall that these mechanisms are examples of horizontal gene transfer—the transfer of genetic material between cells of the same generation.

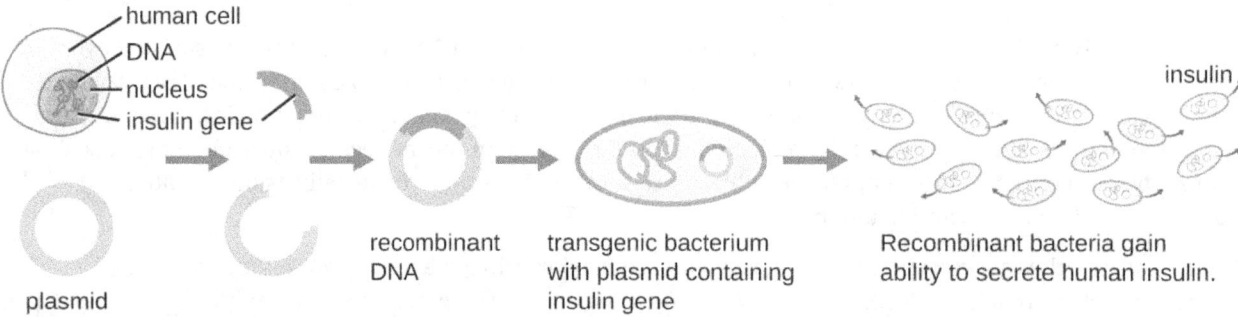

Figure 12.2 Recombinant DNA technology is the artificial recombination of DNA from two organisms. In this example, the human insulin gene is inserted into a bacterial plasmid. This recombinant plasmid can then be used to transform bacteria, which gain the ability to

produce the insulin protein.

Molecular Cloning

Herbert Boyer and Stanley Cohen first demonstrated the complete **molecular cloning** process in 1973 when they successfully cloned genes from the African clawed frog (*Xenopus laevis*) into a bacterial plasmid that was then introduced into the bacterial host *Escherichia coli*. Molecular cloning is a set of methods used to construct recombinant DNA and incorporate it into a host organism; it makes use of a number of molecular tools that are derived from microorganisms.

Restriction Enzymes and Ligases

In recombinant DNA technology, DNA molecules are manipulated using naturally occurring enzymes derived mainly from bacteria and viruses. The creation of recombinant DNA molecules is possible due to the use of naturally occurring **restriction endonucleases (restriction enzymes)**, bacterial enzymes produced as a protection mechanism to cut and destroy foreign cytoplasmic DNA that is most commonly a result of bacteriophage infection. Stewart Linn and Werner Arber discovered restriction enzymes in their 1960s studies of how *E. coli* limits bacteriophage replication on infection. Today, we use restriction enzymes extensively for cutting DNA fragments that can then be spliced into another DNA molecule to form recombinant molecules. Each restriction enzyme cuts DNA at a characteristic **recognition site**, a specific, usually palindromic, DNA sequence typically between four to six base pairs in length. A palindrome is a sequence of letters that reads the same forward as backward. (The word "level" is an example of a palindrome.) Palindromic DNA sequences contain the same base sequences in the 5′ to 3′ direction on one strand as in the 5′ to 3′ direction on the complementary strand. A restriction enzyme recognizes the DNA palindrome and cuts each backbone at identical positions in the palindrome. Some restriction enzymes cut to produce molecules that have complementary overhangs (**sticky ends**) while others cut without generating such overhangs, instead producing **blunt ends** (Figure 12.3).

Molecules with complementary sticky ends can easily **anneal**, or form hydrogen bonds between complementary bases, at their sticky ends. The annealing step allows **hybridization** of the single-stranded overhangs. Hybridization refers to the joining together of two complementary single strands of DNA. Blunt ends can also attach together, but less efficiently than sticky ends due to the lack of complementary overhangs facilitating the process. In either case, **ligation** by DNA ligase can then rejoin the two sugar-phosphate backbones of the DNA through covalent bonding, making the molecule a continuous double strand. In 1972, Paul Berg, a Stanford biochemist, was the first to produce a recombinant DNA molecule using this technique, combining the SV40 monkey virus with *E. coli* bacteriophage lambda to create a hybrid.

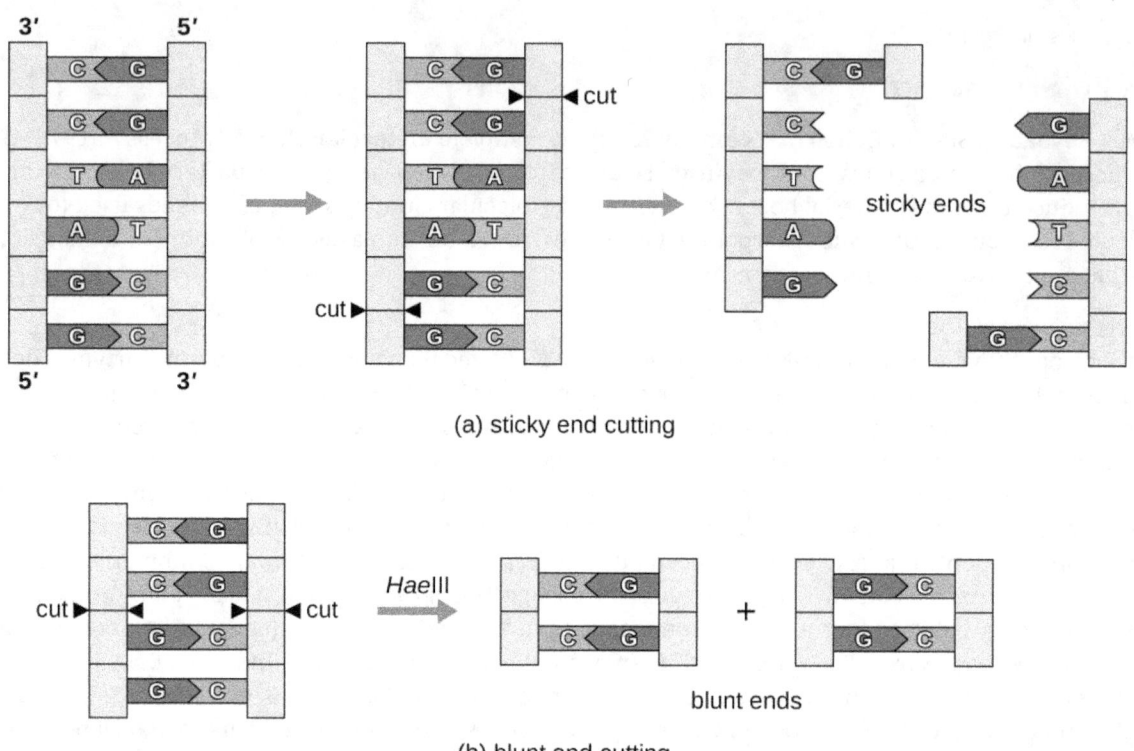

Figure 12.3 (a) In this six-nucleotide restriction enzyme site, recognized by the enzyme *Bam*HI, notice that the sequence reads the same in the 5′ to 3′ direction on both strands. This is known as a palindrome. The cutting of the DNA by the restriction enzyme at the sites (indicated by the black arrows) produces DNA fragments with sticky ends. Another piece of DNA cut with the same restriction enzyme could attach to one of these sticky ends, forming a recombinant DNA molecule. (b) This four-nucleotide recognition site also exhibits a palindromic sequence. The cutting of the DNA by the restriction enzyme *Hae*III at the indicated sites produces DNA fragments with blunt ends. Any other piece of blunt DNA could attach to one of the blunt ends produced, forming a recombinant DNA molecule.

Plasmids

After restriction digestion, genes of interest are commonly inserted into plasmids, small pieces of typically circular, double-stranded DNA that replicate independently of the bacterial chromosome (see Unique Characteristics of Prokaryotic Cells). In recombinant DNA technology, plasmids are often used as **vectors**, DNA molecules that carry DNA fragments from one organism to another. Plasmids used as vectors can be genetically engineered by researchers and scientific supply companies to have specialized properties, as illustrated by the commonly used plasmid vector pUC19 (Figure 12.4). Some plasmid vectors contain genes that confer antibiotic resistance; these resistance genes allow researchers to easily find plasmid-containing colonies by plating them on media containing the corresponding antibiotic. The antibiotic kills all host cells that do not harbor the desired plasmid vector, but those that contain the vector are able to survive and grow.

Plasmid vectors used for cloning typically have a **polylinker site**, or **multiple cloning site (MCS)**. A polylinker site is a short sequence containing multiple unique restriction enzyme recognition sites that are used for inserting DNA into the plasmid after restriction digestion of both the DNA and the plasmid. Having these multiple restriction enzyme recognition sites within the polylinker site makes the plasmid vector versatile, so it can be used for many different cloning experiments involving different restriction enzymes.

This polylinker site is often found within a **reporter gene**, another gene sequence artificially engineered into the plasmid that encodes a protein that allows for visualization of DNA insertion. The reporter gene allows a researcher to distinguish host cells that contain recombinant plasmids with cloned DNA fragments from host cells that only contain the non-recombinant plasmid vector. The most common reporter gene used in plasmid vectors is the bacterial *lacZ* gene encoding beta-galactosidase, an enzyme that naturally degrades lactose but can also degrade a colorless synthetic analog X-gal, thereby producing blue colonies on X-gal–containing media. The *lacZ* reporter gene is disabled when the recombinant DNA is spliced into the plasmid. Because the LacZ protein is not produced when the gene is disabled, X-gal is not degraded and white colonies are produced, which can then be isolated. This **blue-white screening** method is described later and shown in Figure 12.5. In addition to these features, some plasmids come pre-digested and with an enzyme linked to the linearized plasmid to aid in ligation after the insertion of foreign DNA fragments.

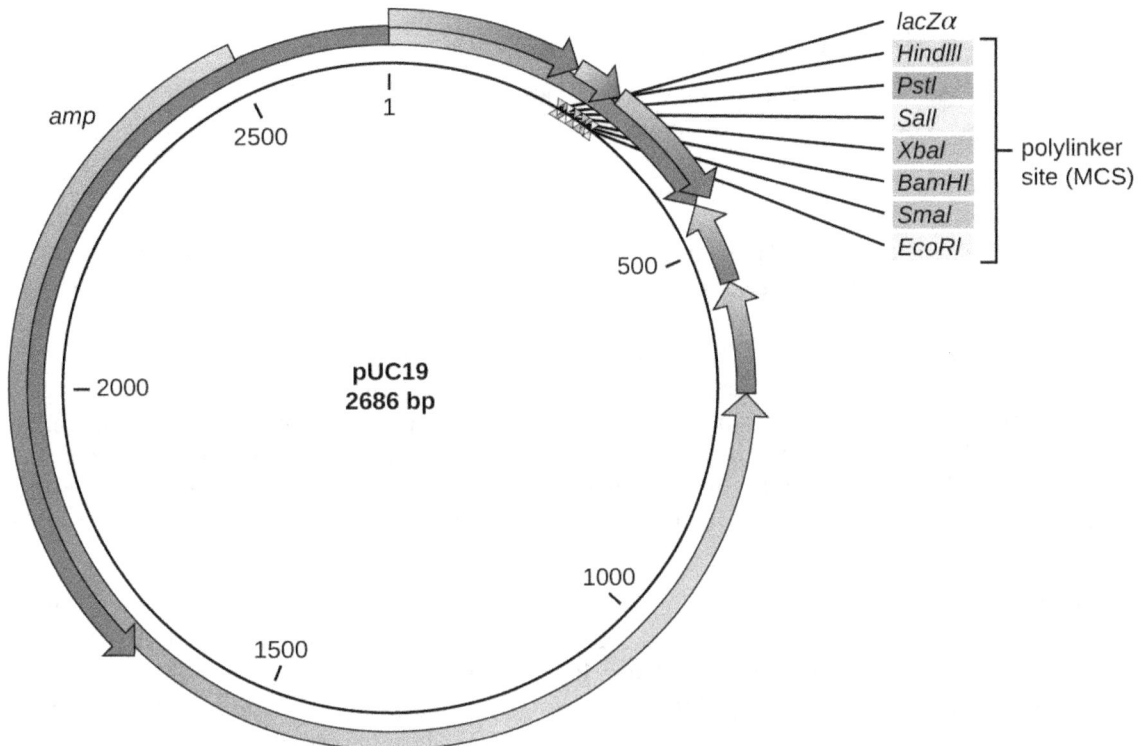

Figure 12.4 The artificially constructed plasmid vector pUC19 is commonly used for cloning foreign DNA. Arrows indicate the directions in which the genes are transcribed. Note the polylinker site, containing multiple unique restriction enzyme recognition sites, found within the *lacZ* reporter gene. Also note the ampicillin (*amp*) resistance gene encoded on the plasmid.

Molecular Cloning using Transformation

The most commonly used mechanism for introducing engineered plasmids into a bacterial cell is transformation, a process in which bacteria take up free DNA from their surroundings. In nature, free DNA typically comes from other lysed bacterial cells; in the laboratory, free DNA in the form of recombinant plasmids is introduced to the cell's surroundings.

Some bacteria, such as *Bacillus* spp., are naturally competent, meaning they are able to take up foreign DNA. However, not all bacteria are naturally competent. In most cases, bacteria must be made artificially competent in the laboratory by increasing the permeability of the cell membrane. This can be achieved through chemical treatments that neutralize charges on the cell membrane or by exposing the bacteria to an electric field that creates microscopic pores in the cell membrane. These methods yield chemically competent or electrocompetent bacteria, respectively.

Following the transformation protocol, bacterial cells are plated onto an antibiotic-containing medium to inhibit the growth of the many host cells that were not transformed by the plasmid conferring antibiotic resistance. A technique called **blue-white screening** is then used for *lacZ*-encoding plasmid vectors such as pUC19. Blue colonies have a functional beta-galactosidase enzyme because the *lacZ* gene is uninterrupted, with no foreign DNA inserted into the polylinker site. These colonies typically result from the digested, linearized plasmid religating to itself. However, white colonies lack a functional beta-galactosidase enzyme, indicating the insertion of foreign DNA within the polylinker site of the plasmid vector, thus disrupting the *lacZ* gene. Thus, white colonies resulting from this blue-white screening contain plasmids with an insert and can be further screened to characterize the foreign DNA. To be sure the correct DNA was incorporated into the plasmid, the DNA insert can then be sequenced.

LINK TO LEARNING

View an animation of molecular cloning (https://openstax.org/l/22moleclonani) from the DNA Learning Center.

CHECK YOUR UNDERSTANDING

- In blue-white screening, what does a blue colony mean and why is it blue?

Molecular Cloning Using Conjugation or Transduction

The bacterial process of conjugation (see How Asexual Prokaryotes Achieve Genetic Diversity) can also be manipulated for molecular cloning. F plasmids, or fertility plasmids, are transferred between bacterial cells through the process of conjugation. Recombinant DNA can be transferred by conjugation when bacterial cells containing a recombinant F plasmid are mixed with compatible bacterial cells lacking the plasmid. F plasmids encode a surface structure called an F pilus that facilitates contact between a cell containing an F plasmid and one without an F plasmid. On contact, a cytoplasmic bridge forms between the two cells and the F-plasmid-containing cell replicates its plasmid, transferring a copy of the recombinant F plasmid to the recipient cell. Once it has received the recombinant F plasmid, the recipient cell can produce its own F pilus and facilitate transfer of the recombinant F plasmid to an additional cell. The use of conjugation to transfer recombinant F plasmids to recipient cells is another effective way to introduce recombinant DNA molecules into host cells.

Alternatively, bacteriophages can be used to introduce recombinant DNA into host bacterial cells through a manipulation of the transduction process (see How Asexual Prokaryotes Achieve Genetic Diversity). In the laboratory, DNA fragments of interest can be engineered into **phagemids**, which are plasmids that have phage sequences that allow them to be packaged into bacteriophages. Bacterial cells can then be infected with these bacteriophages so that the recombinant phagemids can be introduced into the bacterial cells. Depending on the type of phage, the recombinant DNA may be integrated into the host bacterial genome (lysogeny), or it may exist as a plasmid in the host's cytoplasm.

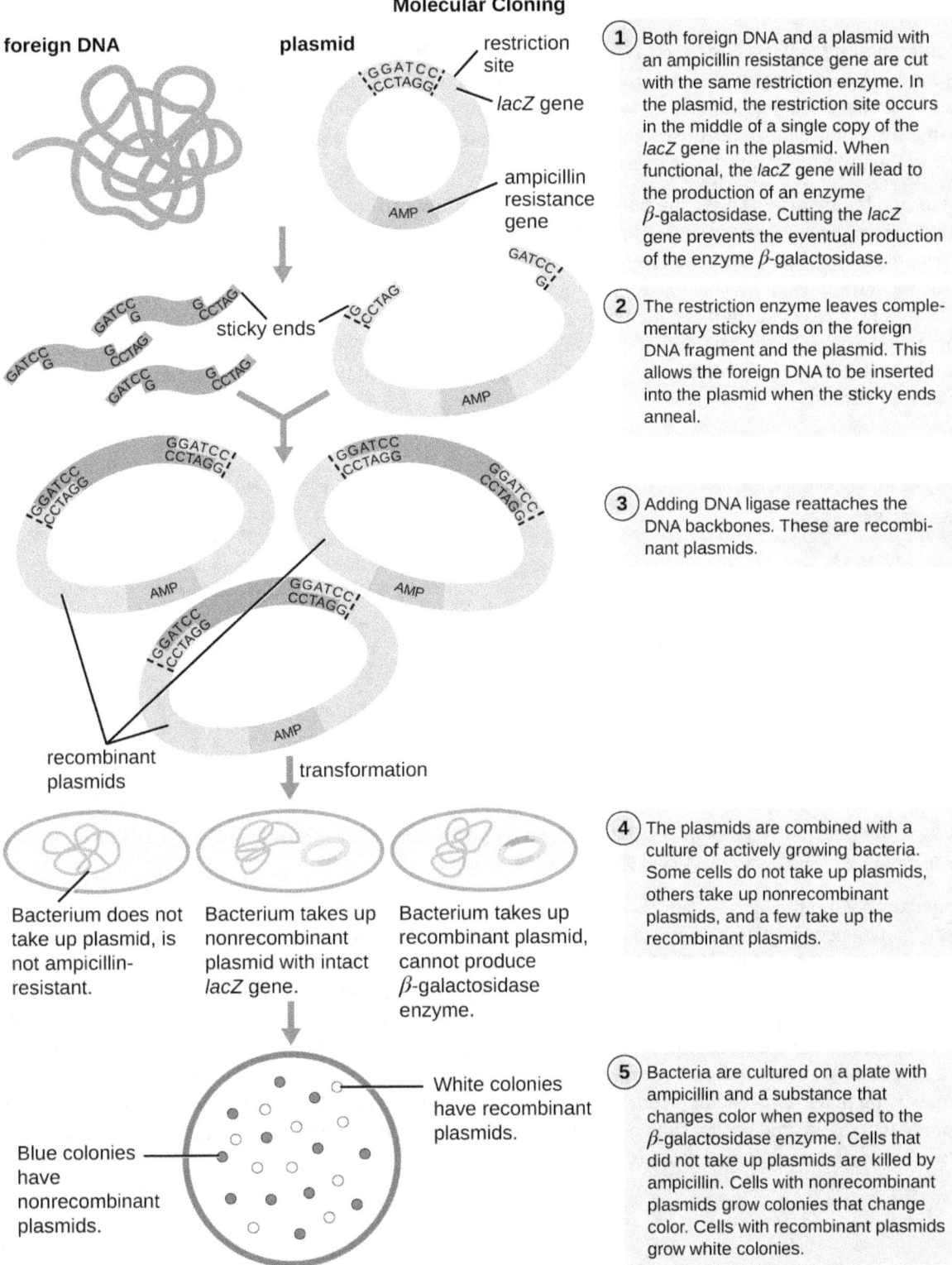

Figure 12.5 The steps involved in molecular cloning using bacterial transformation are outlined in this graphic flowchart.

CHECK YOUR UNDERSTANDING

- What is the original function of a restriction enzyme?
- What two processes are exploited to get recombinant DNA into a bacterial host cell?
- Distinguish the uses of an antibiotic resistance gene and a reporter gene in a plasmid vector.

Creating a Genomic Library

Molecular cloning may also be used to generate a **genomic library**. The library is a complete (or nearly complete) copy of an organism's genome contained as recombinant DNA plasmids engineered into unique clones of bacteria. Having such a library allows a researcher to create large quantities of each fragment by growing the bacterial host for that fragment. These fragments can be used to determine the sequence of the DNA and the function of any genes present.

One method for generating a genomic library is to ligate individual restriction enzyme-digested genomic fragments into plasmid vectors cut with the same restriction enzyme (Figure 12.6). After transformation into a bacterial host, each transformed bacterial cell takes up a single recombinant plasmid and grows into a colony of cells. All of the cells in this colony are identical **clones** and carry the same recombinant plasmid. The resulting library is a collection of colonies, each of which contains a fragment of the original organism's genome, that are each separate and distinct and can each be used for further study. This makes it possible for researchers to screen these different clones to discover the one containing a gene of interest from the original organism's genome.

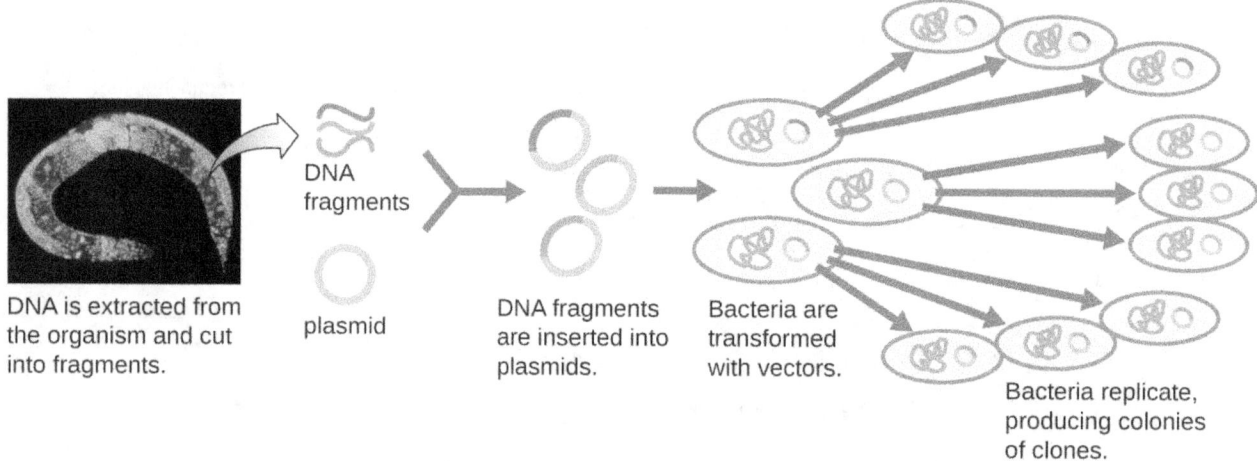

Figure 12.6 The generation of a genomic library facilitates the discovery of the genomic DNA fragment that contains a gene of interest. (credit "micrograph": modification of work by National Institutes of Health)

To construct a genomic library using larger fragments of genomic DNA, an *E. coli* bacteriophage, such as lambda, can be used as a host (Figure 12.7). Genomic DNA can be sheared or enzymatically digested and ligated into a pre-digested bacteriophage lambda DNA vector. Then, these recombinant phage DNA molecules can be packaged into phage particles and used to infect *E. coli* host cells on a plate. During infection within each cell, each recombinant phage will make many copies of itself and lyse the *E. coli* lawn, forming a plaque. Thus, each plaque from a phage library represents a unique recombinant phage containing a distinct genomic DNA fragment. Plaques can then be screened further to look for genes of interest. One advantage to producing a library using phages instead of plasmids is that a phage particle holds a much larger insert of foreign DNA compared with a plasmid vector, thus requiring a much smaller number of cultures to fully represent the entire genome of the original organism.

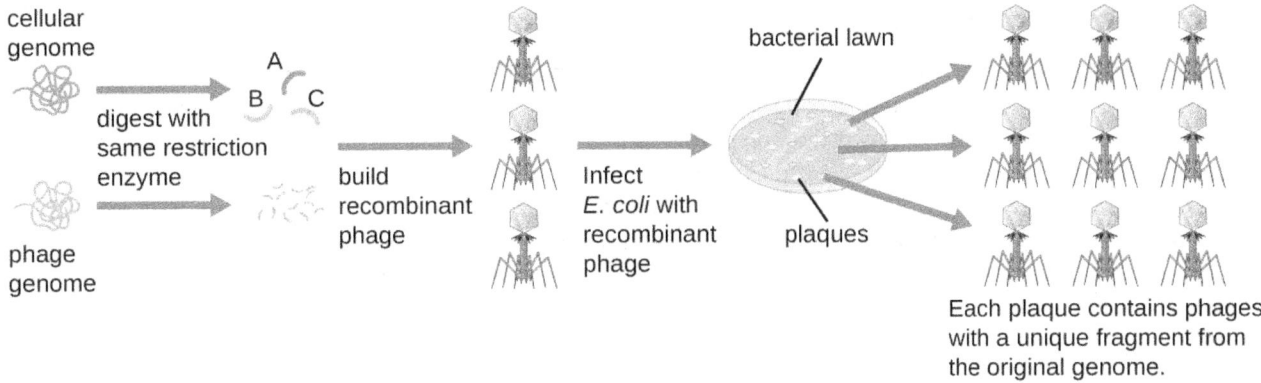

Figure 12.7 Recombinant phage DNA molecules are made by ligating digested phage particles with fragmented genomic DNA molecules. These recombinant phage DNA molecules are packaged into phage particles and allowed to infect a bacterial lawn. Each plaque represents a unique recombinant DNA molecule that can be further screened for genes of interest.

To focus on the expressed genes in an organism or even a tissue, researchers construct libraries using the organism's messenger RNA (mRNA) rather than its genomic DNA. Whereas all cells in a single organism will have the same genomic DNA, different tissues express different genes, producing different complements of mRNA. For example, all human cells' genomic DNA contains the gene for insulin, but only cells in the pancreas express mRNA directing the production of insulin. Because mRNA cannot be cloned directly, in the laboratory mRNA must be used as a template by the retroviral enzyme reverse transcriptase to make **complementary DNA (cDNA)**. A cell's full complement of mRNA can be reverse-transcribed into cDNA molecules, which can be used as a template for DNA polymerase to make double-stranded DNA copies; these fragments can subsequently be ligated into either plasmid vectors or bacteriophage to produce a cDNA library. The benefit of a cDNA library is that it contains DNA from only the expressed genes in the cell. This means that the introns, control sequences such as promoters, and DNA not destined to be translated into proteins are not represented in the library. The focus on translated sequences means that the library cannot be used to study the sequence and structure of the genome in its entirety. The construction of a cDNA genomic library is shown in Figure 12.8.

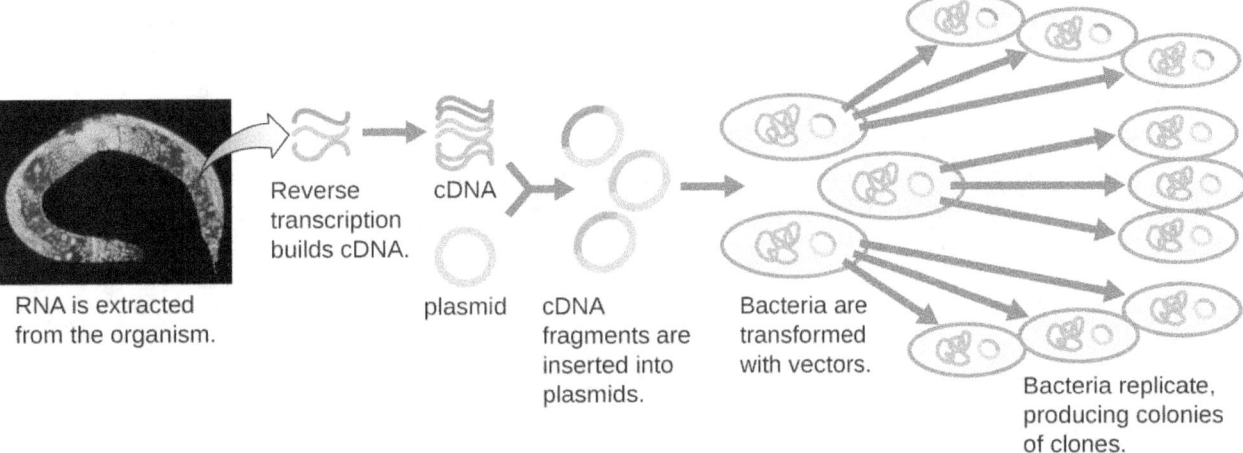

Figure 12.8 Complementary DNA (cDNA) is made from mRNA by the retroviral enzyme reverse transcriptase, converted into double-stranded copies, and inserted into either plasmid vectors or bacteriophage, producing a cDNA library. (credit "micrograph": modification of work by National Institutes of Health)

✓ CHECK YOUR UNDERSTANDING

- What are the hosts for the genomic libraries described?
- What is cDNA?

Introducing Recombinant Molecules into Eukaryotic Hosts

The use of bacterial hosts for genetic engineering laid the foundation for recombinant DNA technology; however, researchers have also had great interest in genetically engineering eukaryotic cells, particularly those of plants and animals. The introduction of recombinant DNA molecules into eukaryotic hosts is called **transfection**. Genetically engineered plants, called transgenic plants, are of significant interest for agricultural and pharmaceutical purposes. The first transgenic plant sold commercially was the Flavr Savr delayed-ripening tomato, which came to market in 1994. Genetically engineered livestock have also been successfully produced, resulting, for example, in pigs with increased nutritional value[1] and goats that secrete pharmaceutical products in their milk.[2]

Electroporation

Compared to bacterial cells, eukaryotic cells tend to be less amenable as hosts for recombinant DNA molecules. Because eukaryotes are typically neither competent to take up foreign DNA nor able to maintain plasmids, transfection of eukaryotic hosts is far more challenging and requires more intrusive techniques for success. One method used for transfecting cells in cell culture is called **electroporation**. A brief electric pulse induces the formation of transient pores in the phospholipid bilayers of cells through which the gene can be introduced. At the same time, the electric pulse generates a short-lived positive charge on one side of the cell's interior and a negative charge on the opposite side; the charge difference draws negatively charged DNA molecules into the cell (Figure 12.9).

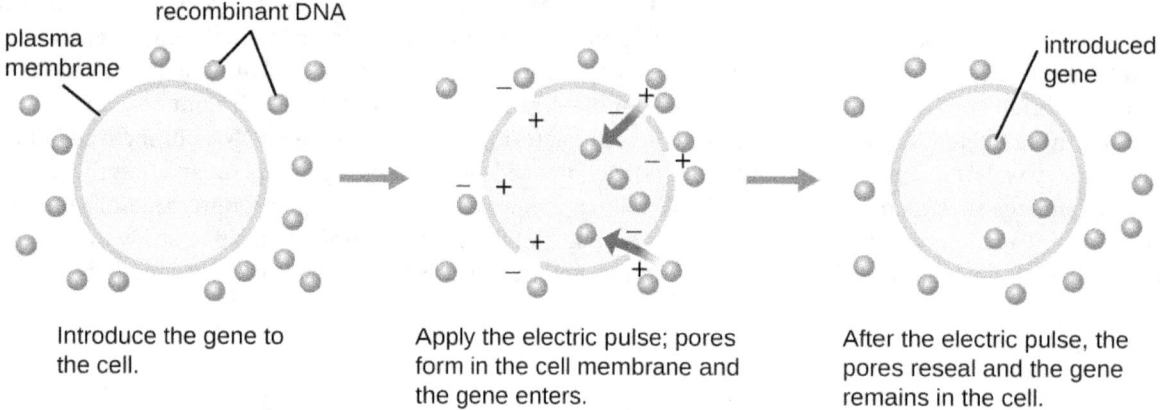

Figure 12.9 Electroporation is one laboratory technique used to introduce DNA into eukaryotic cells.

Microinjection

An alternative method of transfection is called **microinjection**. Because eukaryotic cells are typically larger than those of prokaryotes, DNA fragments can sometimes be directly injected into the cytoplasm using a glass micropipette, as shown in Figure 12.10.

---

1 Liangxue Lai, Jing X. Kang, Rongfeng Li, Jingdong Wang, William T. Witt, Hwan Yul Yong, Yanhong Hao et al. "Generation of Cloned Transgenic Pigs Rich in Omega-3 Fatty Acids." *Nature Biotechnology* 24 no. 4 (2006): 435–436.

2 Raylene Ramos Moura, Luciana Magalhães Melo, and Vicente José de Figueirêdo Freitas. "Production of Recombinant Proteins in Milk of Transgenic and Non-Transgenic Goats." *Brazilian Archives of Biology and Technology* 54 no. 5 (2011): 927–938.

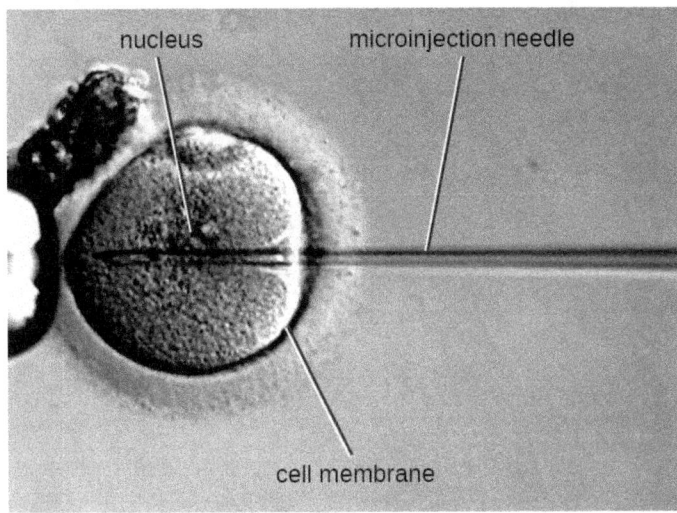

Figure 12.10 Microinjection is another technique for introducing DNA into eukaryotic cells. A microinjection needle containing recombinant DNA is able to penetrate both the cell membrane and nuclear envelope.

Gene Guns

Transfecting plant cells can be even more difficult than animal cells because of their thick cell walls. One approach involves treating plant cells with enzymes to remove their cell walls, producing protoplasts. Then, a **gene gun** is used to shoot gold or tungsten particles coated with recombinant DNA molecules into the plant protoplasts at high speeds. Recipient protoplast cells can then recover and be used to generate new transgenic plants (Figure 12.11).

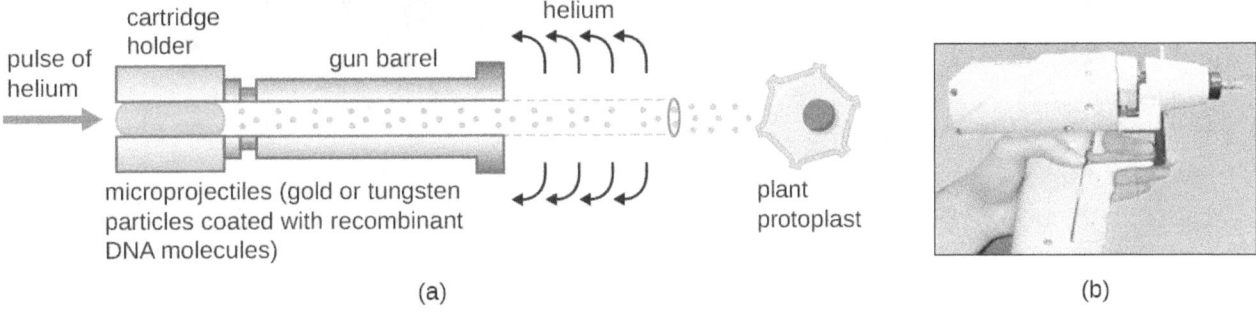

Figure 12.11 Heavy-metal particles coated with recombinant DNA are shot into plant protoplasts using a gene gun. The resulting transformed cells are allowed to recover and can be used to generate recombinant plants. (a) A schematic of a gene gun. (b) A photograph of a gene gun. (credit a, b: modification of work by JA O'Brien, SC Lummis)

Shuttle Vectors

Another method of transfecting plants involves **shuttle vectors**, plasmids that can move between bacterial and eukaryotic cells. The **tumor-inducing (T_i) plasmids** originating from the bacterium *Agrobacterium tumefaciens* are commonly used as shuttle vectors for incorporating genes into plants (Figure 12.12). In nature, the T_i plasmids of *A. tumefaciens* cause plants to develop tumors when they are transferred from bacterial cells to plant cells. Researchers have been able to manipulate these naturally occurring plasmids to remove their tumor-causing genes and insert desirable DNA fragments. The resulting recombinant T_i plasmids can be transferred into the plant genome through the natural transfer of T_i plasmids from the bacterium to the plant host. Once inside the plant host cell, the gene of interest recombines into the plant cell's genome.

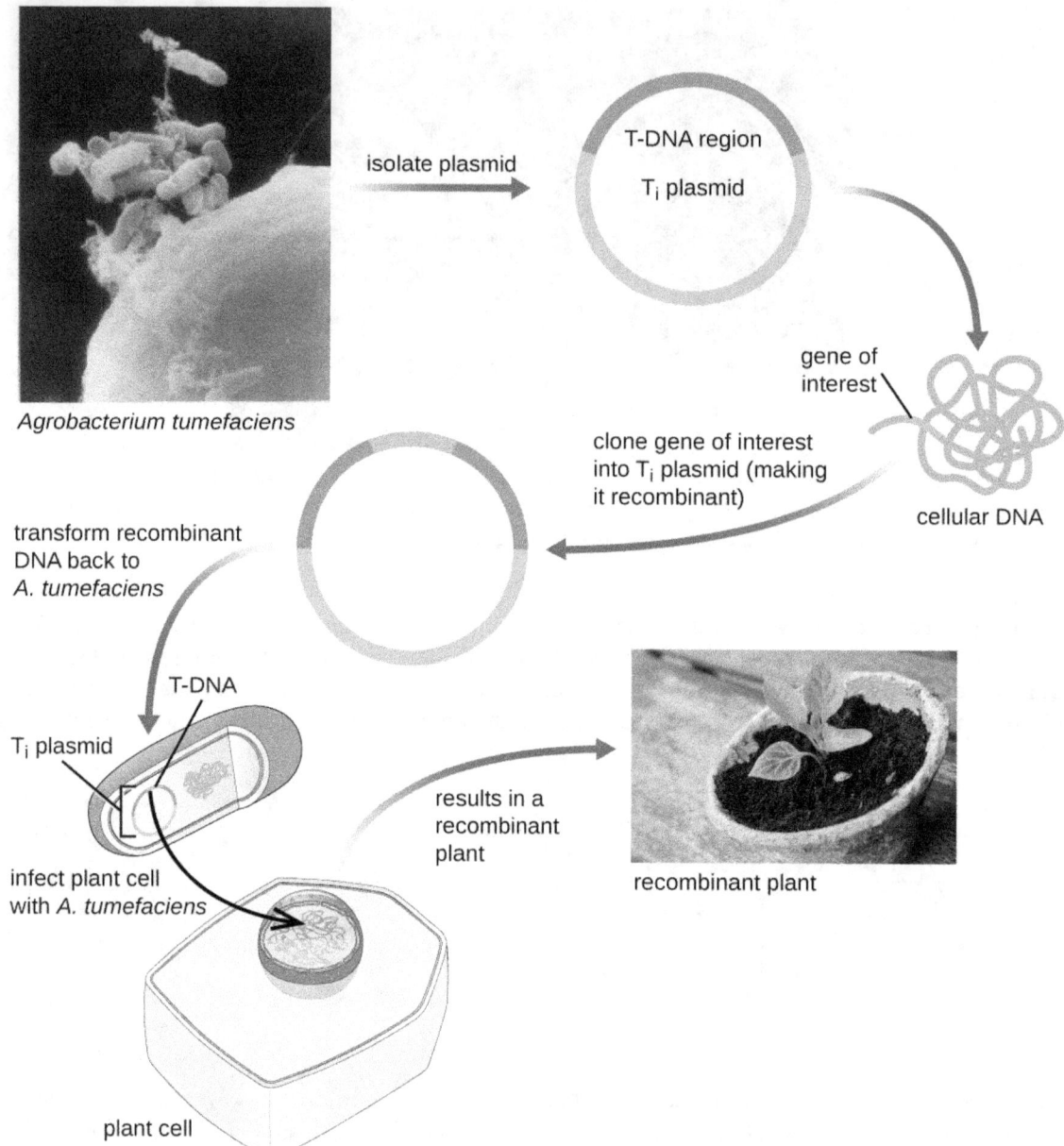

Figure 12.12 The T$_i$ plasmid of *Agrobacterium tumefaciens* is a useful shuttle vector for the uptake of genes of interest into plant cells. The gene of interest is cloned into the T$_i$ plasmid, which is then introduced into plant cells. The gene of interest then recombines into the plant cell's genome, allowing for the production of transgenic plants.

Viral Vectors

Viral vectors can also be used to transfect eukaryotic cells. In fact, this method is often used in gene therapy (see Gene Therapy) to introduce healthy genes into human patients suffering from diseases that result from genetic mutations. Viral genes can be deleted and replaced with the gene to be delivered to the patient;[3] the virus then infects the host cell and delivers the foreign DNA into the genome of the targeted cell. Adenoviruses are often used for this purpose because they can be grown to high titer and can infect both nondividing and dividing host cells. However, use of viral vectors for gene therapy can pose some risks for patients, as discussed in Gene Therapy.

---

3 William S.M. Wold and Karoly Toth. "Adenovirus Vectors for Gene Therapy, Vaccination and Cancer Gene Therapy." *Current Gene Therapy 13* no. 6 (2013): 421.

CHECK YOUR UNDERSTANDING
- What are the methods used to introduce recombinant DNA vectors into animal cells?
- Compare and contrast shuttle vectors and viral vectors.

12.2 Visualizing and Characterizing DNA, RNA, and Protein

Learning Objectives

By the end of this section, you will be able to:
- Explain the use of nucleic acid probes to visualize specific DNA sequences
- Explain the use of gel electrophoresis to separate DNA fragments
- Explain the principle of restriction fragment length polymorphism analysis and its uses
- Compare and contrast Southern and northern blots
- Explain the principles and uses of microarray analysis
- Describe the methods uses to separate and visualize protein variants
- Explain the method and uses of polymerase chain reaction and DNA sequencing

The sequence of a DNA molecule can help us identify an organism when compared to known sequences housed in a database. The sequence can also tell us something about the function of a particular part of the DNA, such as whether it encodes a particular protein. Comparing **protein signatures**—the expression levels of specific arrays of proteins—between samples is an important method for evaluating cellular responses to a multitude of environmental factors and stresses. Analysis of protein signatures can reveal the identity of an organism or how a cell is responding during disease.

The DNA and proteins of interest are microscopic and typically mixed in with many other molecules including DNA or proteins irrelevant to our interests. Many techniques have been developed to isolate and characterize molecules of interest. These methods were originally developed for research purposes, but in many cases they have been simplified to the point that routine clinical use is possible. For example, many pathogens, such as the bacterium *Helicobacter pylori*, which causes stomach ulcers, can be detected using protein-based tests. In addition, an increasing number of highly specific and accurate DNA amplification-based identification assays can now detect pathogens such as antibiotic-resistant enteric bacteria, herpes simplex virus, varicella-zoster virus, and many others.

Molecular Analysis of DNA

In this subsection, we will outline some of the basic methods used for separating and visualizing specific fragments of DNA that are of interest to a scientist. Some of these methods do not require knowledge of the complete sequence of the DNA molecule. Before the advent of rapid DNA sequencing, these methods were the only ones available to work with DNA, but they still form the basic arsenal of tools used by molecular geneticists to study the body's responses to microbial and other diseases.

Nucleic Acid Probing

DNA molecules are small, and the information contained in their sequence is invisible. How does a researcher isolate a particular stretch of DNA, or having isolated it, determine what organism it is from, what its sequence is, or what its function is? One method to identify the presence of a certain DNA sequence uses artificially constructed pieces of DNA called probes. Probes can be used to identify different bacterial species in the environment and many DNA probes are now available to detect pathogens clinically. For example, DNA probes are used to detect the vaginal pathogens *Candida albicans*, *Gardnerella vaginalis*, and *Trichomonas vaginalis*.

To screen a genomic library for a particular gene or sequence of interest, researchers must know something about that gene. If researchers have a portion of the sequence of DNA for the gene of interest, they can design a **DNA probe**, a single-stranded DNA fragment that is complementary to part of the gene of interest and different from other DNA sequences in the sample. The DNA probe may be synthesized chemically by commercial laboratories, or it may be created by cloning, isolating, and denaturing a DNA fragment from a living organism. In either case, the DNA probe must be labeled with a molecular tag or beacon, such as a radioactive phosphorus atom (as is used for **autoradiography**) or a fluorescent dye (as is used in fluorescent *in situ* hybridization, or FISH), so that the probe and the DNA it binds to can be seen (Figure 12.13). The DNA sample being probed must also be denatured to make it single-stranded so that the single-stranded DNA probe can anneal to the single-stranded DNA sample at locations where their sequences are complementary. While these techniques are valuable for diagnosis, their direct use on sputum and other bodily samples may be problematic due to the complex nature of these samples. DNA often must first be isolated from bodily samples through chemical extraction methods before a DNA probe can be used to identify pathogens.

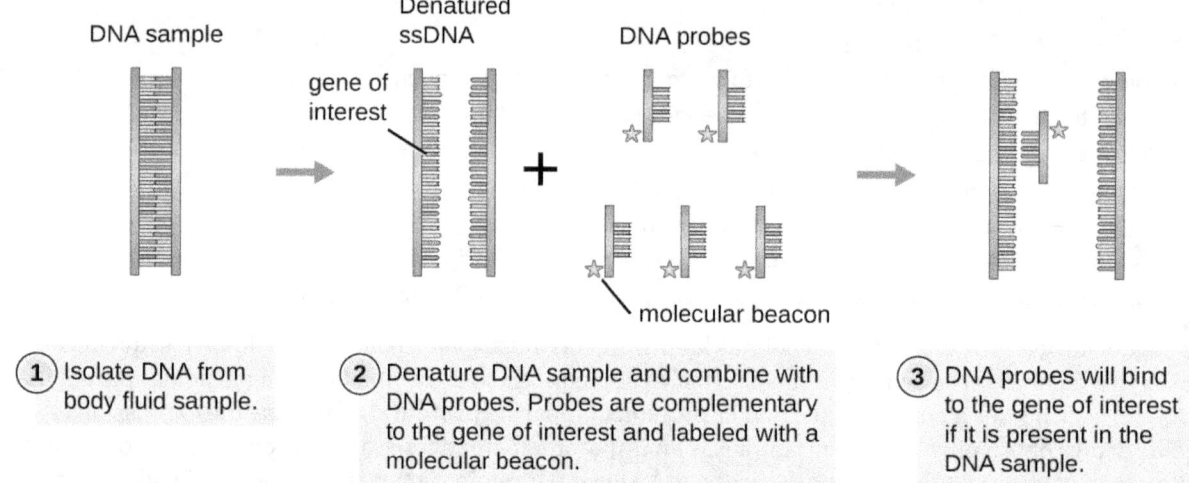

Figure 12.13 DNA probes can be used to confirm the presence of a suspected pathogen in patient samples. This diagram illustrates how a DNA probe can be used to search for a gene of interest associated with the suspected pathogen.

Clinical Focus

Part 2

The mild, flu-like symptoms that Kayla is experiencing could be caused by any number of infectious agents. In addition, several non-infectious autoimmune conditions, such as multiple sclerosis, systemic lupus erythematosus (SLE), and amyotrophic lateral sclerosis (ALS), also have symptoms that are consistent with Kayla's early symptoms. However, over the course of several weeks, Kayla's symptoms worsened. She began to experience joint pain in her knees, heart palpitations, and a strange limpness in her facial muscles. In addition, she suffered from a stiff neck and painful headaches. Reluctantly, she decided it was time to seek medical attention.

- Do Kayla's new symptoms provide any clues as to what type of infection or other medical condition she may have?
- What tests or tools might a health-care provider use to pinpoint the pathogen causing Kayla's symptoms?

Jump to the next Clinical Focus box. Go back to the previous Clinical Focus box.

Agarose Gel Electrophoresis

There are a number of situations in which a researcher might want to physically separate a collection of DNA fragments of different sizes. A researcher may also digest a DNA sample with a restriction enzyme to form fragments. The resulting size and fragment distribution pattern can often yield useful information about the sequence of DNA bases that can be used, much like a bar-code scan, to identify the individual or species to which the DNA belongs.

Gel electrophoresis is a technique commonly used to separate biological molecules based on size and biochemical characteristics, such as charge and polarity. **Agarose gel electrophoresis** is widely used to separate DNA (or RNA) of varying sizes that may be generated by restriction enzyme digestion or by other means, such as the PCR (Figure 12.14).

Due to its negatively charged backbone, DNA is strongly attracted to a positive electrode. In agarose gel electrophoresis, the gel is oriented horizontally in a buffer solution. Samples are loaded into sample wells on the side of the gel closest to the negative electrode, then drawn through the molecular sieve of the agarose matrix toward the positive electrode. The agarose matrix impedes the movement of larger molecules through the gel, whereas smaller molecules pass through more readily. Thus, the distance of migration is inversely correlated to the size of the DNA fragment, with smaller fragments traveling a longer distance through the gel. Sizes of DNA fragments within a sample can be estimated by comparison to fragments of known size in a DNA ladder also run on the same gel. To separate very large DNA fragments, such as chromosomes or viral genomes, agarose gel electrophoresis can be modified by periodically alternating the orientation of the electric field during pulsed-field gel electrophoresis (PFGE). In PFGE, smaller fragments can reorient themselves and migrate slightly faster than larger fragments and this technique can thus serve to separate very large fragments that would otherwise travel together during standard agarose gel electrophoresis. In any of these electrophoresis techniques, the locations of the DNA or RNA fragments in the gel can be detected by various methods. One common method is adding ethidium bromide, a stain that inserts into the nucleic acids at non-specific locations and can be visualized when exposed to ultraviolet light. Other stains that are safer than ethidium bromide, a potential carcinogen, are now available.

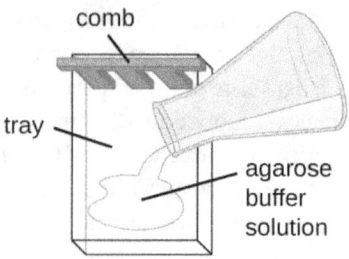
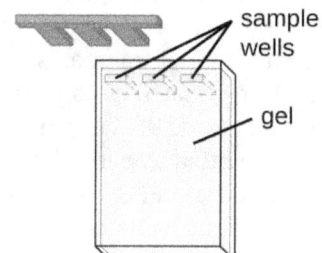
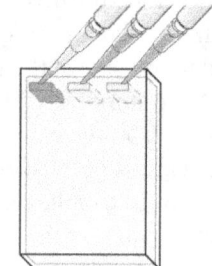
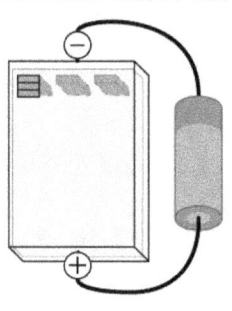
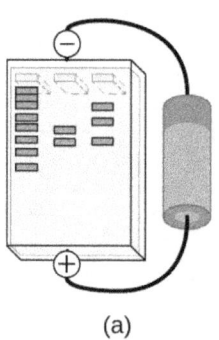
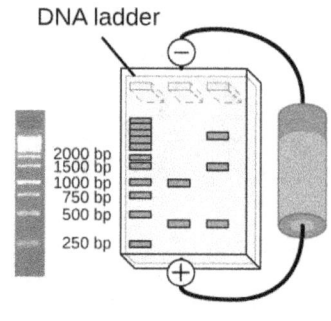

(a)

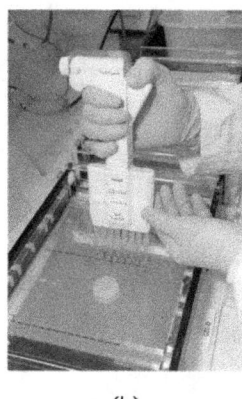
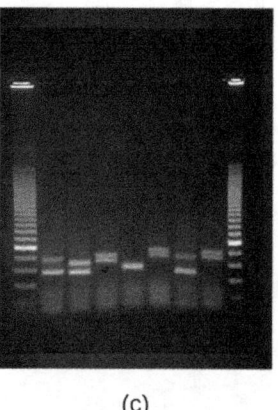

(b)   (c)

Figure 12.14 (a) The process of agarose gel electrophoresis. (b) A researcher loading samples into a gel. (c) This photograph shows a completed electrophoresis run on an agarose gel. The DNA ladder is located in lanes 1 and 9. Seven samples are located in lanes 2 through 8. The gel was stained with ethidium bromide and photographed under ultraviolet light. (credit a: modification of work by Magnus Manske; credit b: modification of work by U.S. Department of Agriculture; credit c: modification of work by James Jacob)

Restriction Fragment Length Polymorphism (RFLP) Analysis

Restriction enzyme recognition sites are short (only a few nucleotides long), sequence-specific palindromes, and may be found throughout the genome. Thus, differences in DNA sequences in the genomes of individuals will lead to differences in distribution of restriction-enzyme recognition sites that can be visualized as distinct banding patterns on a gel after agarose gel electrophoresis. **Restriction fragment length polymorphism (RFLP)** analysis compares DNA banding patterns of different DNA samples after restriction digestion (Figure 12.15).

RFLP analysis has many practical applications in both medicine and forensic science. For example, epidemiologists use RFLP analysis to track and identify the source of specific microorganisms implicated in outbreaks of food poisoning or certain infectious diseases. RFLP analysis can also be used on human DNA to determine inheritance patterns of chromosomes with variant genes, including those associated with heritable diseases or to establish paternity.

Forensic scientists use RFLP analysis as a form of DNA fingerprinting, which is useful for analyzing DNA obtained from crime scenes, suspects, and victims. DNA samples are collected, the numbers of copies of the sample DNA molecules are increased using PCR, and then subjected to restriction enzyme digestion and agarose gel electrophoresis to generate specific banding patterns. By comparing the banding patterns of samples collected from the crime scene against those collected from suspects or victims, investigators can definitively determine whether DNA evidence collected at the scene was left behind by suspects or victims.

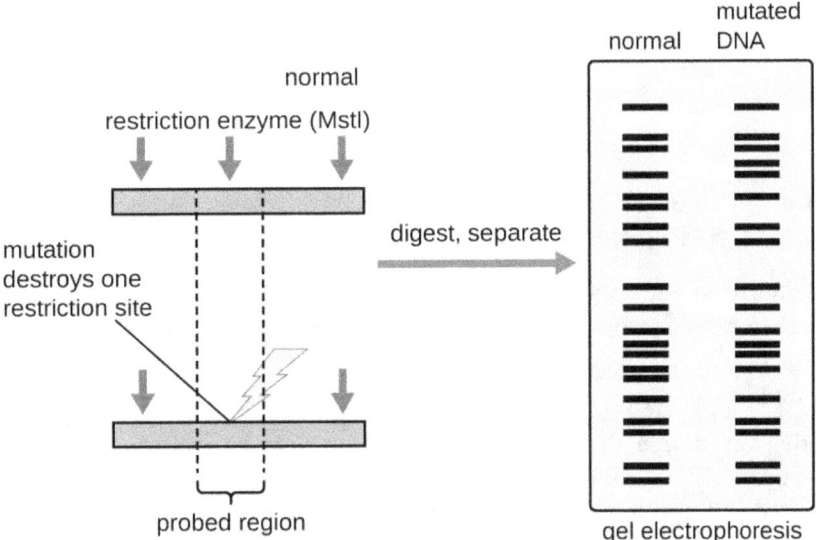

Figure 12.15 RFLP analysis can be used to differentiate DNA sequences. In this example, a normal chromosome is digested into two fragments, whereas digestion of a mutated chromosome produces only one fragment. The small red arrows pointing to the two different chromosome segments show the locations of the restriction enzyme recognition sites. After digestion and agarose gel electrophoresis, the banding patterns reflect the change by showing the loss of two shorter bands and the gain of a longer band. (credit: modification of work by National Center for Biotechnology Information)

Southern Blots and Modifications

Several molecular techniques capitalize on sequence complementarity and hybridization between nucleic acids of a sample and DNA probes. Typically, probing nucleic-acid samples within a gel is unsuccessful because as the DNA probe soaks into a gel, the sample nucleic acids within the gel diffuse out. Thus, blotting techniques are commonly used to transfer nucleic acids to a thin, positively charged membrane made of nitrocellulose or nylon. In the **Southern blot** technique, developed by Sir Edwin Southern in 1975, DNA fragments within a sample are first separated by agarose gel electrophoresis and then transferred to a membrane through capillary action (Figure 12.16). The DNA fragments that bind to the surface of the membrane are then exposed to a specific single-stranded DNA probe labeled with a radioactive or fluorescent molecular beacon to aid in detection. Southern blots may be used to detect the presence of certain DNA sequences in a given DNA sample. Once the target DNA within the membrane is visualized, researchers can cut out the portion of the membrane containing the fragment to recover the DNA fragment of interest.

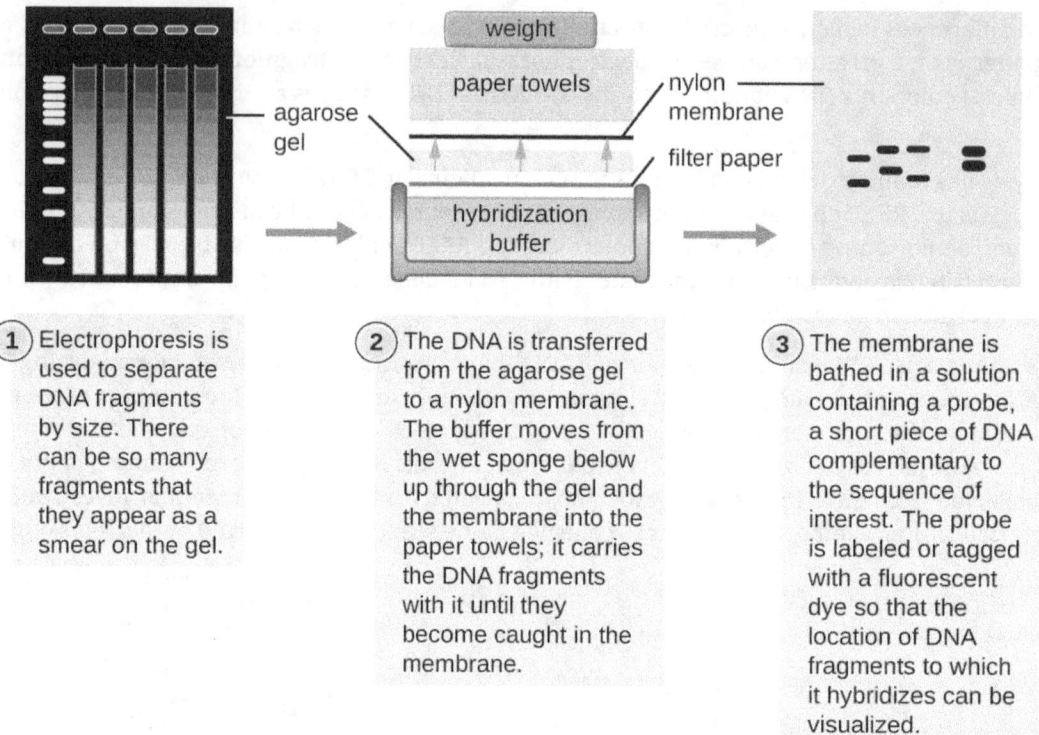

Figure 12.16 In the Southern blot technique, DNA fragments are first separated by agarose gel electrophoresis, then transferred by capillary action to a nylon membrane, which is then soaked with a DNA probe tagged with a molecular beacon for easy visualization.

Variations of the Southern blot—the dot blot, slot blot, and the spot blot—do not involve electrophoresis, but instead concentrate DNA from a sample into a small location on a membrane. After hybridization with a DNA probe, the signal intensity detected is measured, allowing the researcher to estimate the amount of target DNA present within the sample.

A colony blot is another variation of the Southern blot in which colonies representing different clones in a genomic library are transferred to a membrane by pressing the membrane onto the culture plate. The cells on the membrane are lysed and the membrane can then be probed to determine which colonies within a genomic library harbor the target gene. Because the colonies on the plate are still growing, the cells of interest can be isolated from the plate.

In the **northern blot**, another variation of the Southern blot, RNA (not DNA) is immobilized on the membrane and probed. Northern blots are typically used to detect the amount of mRNA made through gene expression within a tissue or organism sample.

Microarray Analysis

Another technique that capitalizes on the hybridization between complementary nucleic acid sequences is called **microarray analysis**. Microarray analysis is useful for the comparison of gene-expression patterns between different cell types—for example, cells infected with a virus versus uninfected cells, or cancerous cells versus healthy cells (Figure 12.17).

Typically, DNA or cDNA from an experimental sample is deposited on a glass slide alongside known DNA sequences. Each slide can hold more than 30,000 different DNA fragment types. Distinct DNA fragments (encompassing an organism's entire genomic library) or cDNA fragments (corresponding to an organism's full complement of expressed genes) can be individually spotted on a glass slide.

Once deposited on the slide, genomic DNA or mRNA can be isolated from the two samples for comparison. If mRNA is isolated, it is reverse-transcribed to cDNA using reverse transcriptase. Then the two samples of genomic DNA or cDNA are labeled with different fluorescent dyes (typically red and green). The labeled genomic DNA samples are then combined in equal amounts, added to the microarray chip, and allowed to hybridize to complementary spots on the microarray.

Hybridization of sample genomic DNA molecules can be monitored by measuring the intensity of fluorescence at particular spots on the microarray. Differences in the amount of hybridization between the samples can be readily observed. If only one sample's nucleic acids hybridize to a particular spot on the microarray, then that spot will appear either green or red. However, if both samples' nucleic acids hybridize, then the spot will appear yellow due to the combination of the red and green dyes.

Although microarray technology allows for a holistic comparison between two samples in a short time, it requires sophisticated (and expensive) detection equipment and analysis software. Because of the expense, this technology is typically limited to research settings. Researchers have used microarray analysis to study how gene expression is affected in organisms that are infected by bacteria or viruses or subjected to certain chemical treatments.

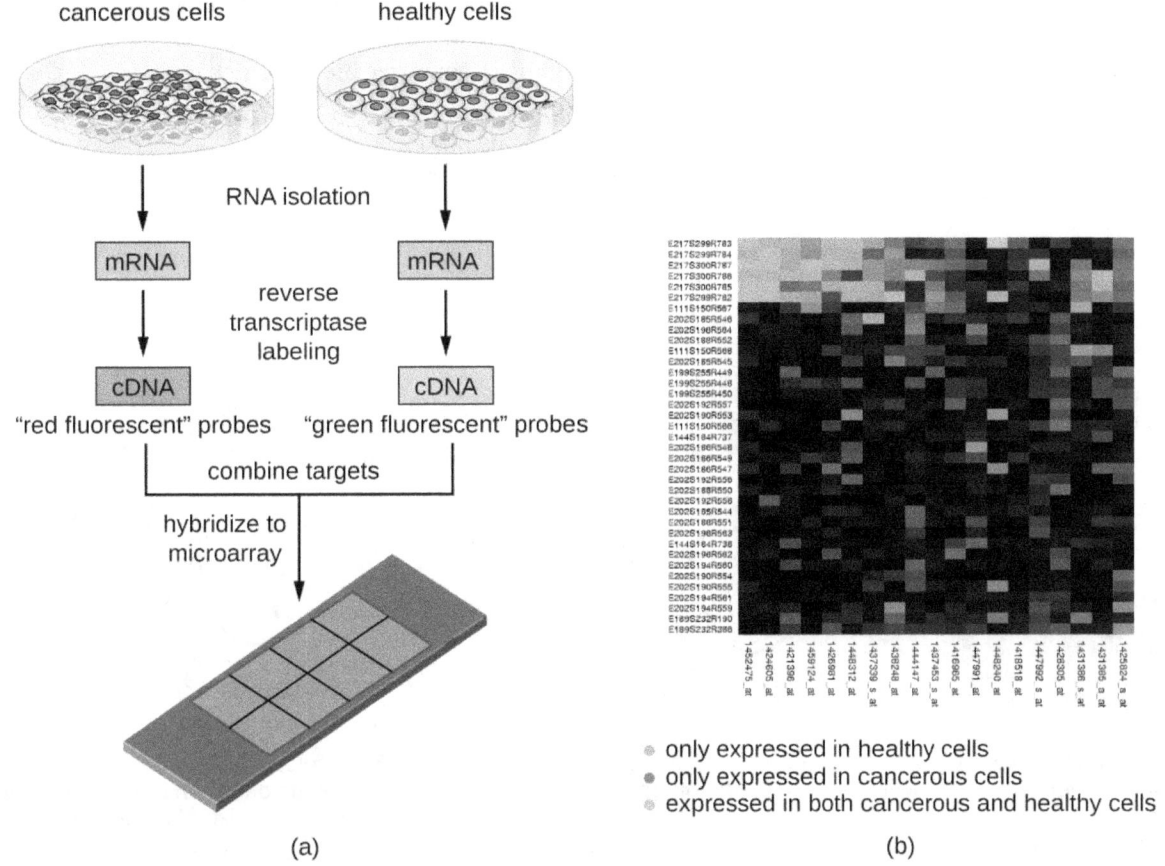

Figure 12.17 (a) The steps in microarray analysis are illustrated. Here, gene expression patterns are compared between cancerous cells and healthy cells. (b) Microarray information can be expressed as a heat map. Genes are shown on the left side; different samples are shown across the bottom. Genes expressed only in cancer cells are shown in varying shades of red; genes expressed only in normal cells are shown in varying shades of green. Genes that are expressed in both cancerous and normal cells are shown in yellow.

LINK TO LEARNING

Explore microchip technology (https://openstax.org/l/22intwebmictec) at this interactive website.

CHECK YOUR UNDERSTANDING

- What does a DNA probe consist of?
- Why is a Southern blot used after gel electrophoresis of a DNA digest?

Molecular Analysis of Proteins

In many cases it may not be desirable or possible to study DNA or RNA directly. Proteins can provide species-specific information for identification as well as important information about how and whether a cell or tissue is responding to the presence of a pathogenic microorganism. Various proteins require different methods for isolation and characterization.

Polyacrylamide Gel Electrophoresis

A variation of gel electrophoresis, called **polyacrylamide gel electrophoresis (PAGE)**, is commonly used for separating proteins. In PAGE, the gel matrix is finer and composed of polyacrylamide instead of agarose. Additionally, PAGE is typically performed using a vertical gel apparatus (Figure 12.18). Because of the varying charges associated with amino acid side chains, PAGE can be used to separate intact proteins based on their net charges. Alternatively, proteins can be denatured and coated with a negatively charged detergent called sodium dodecyl sulfate (SDS), masking the native charges and allowing separation based on size only. PAGE can be further modified to separate proteins based on two characteristics, such as their charges at various pHs as well as their size, through the use of two-dimensional PAGE. In any of these cases, following electrophoresis, proteins are visualized through staining, commonly with either Coomassie blue or a silver stain.

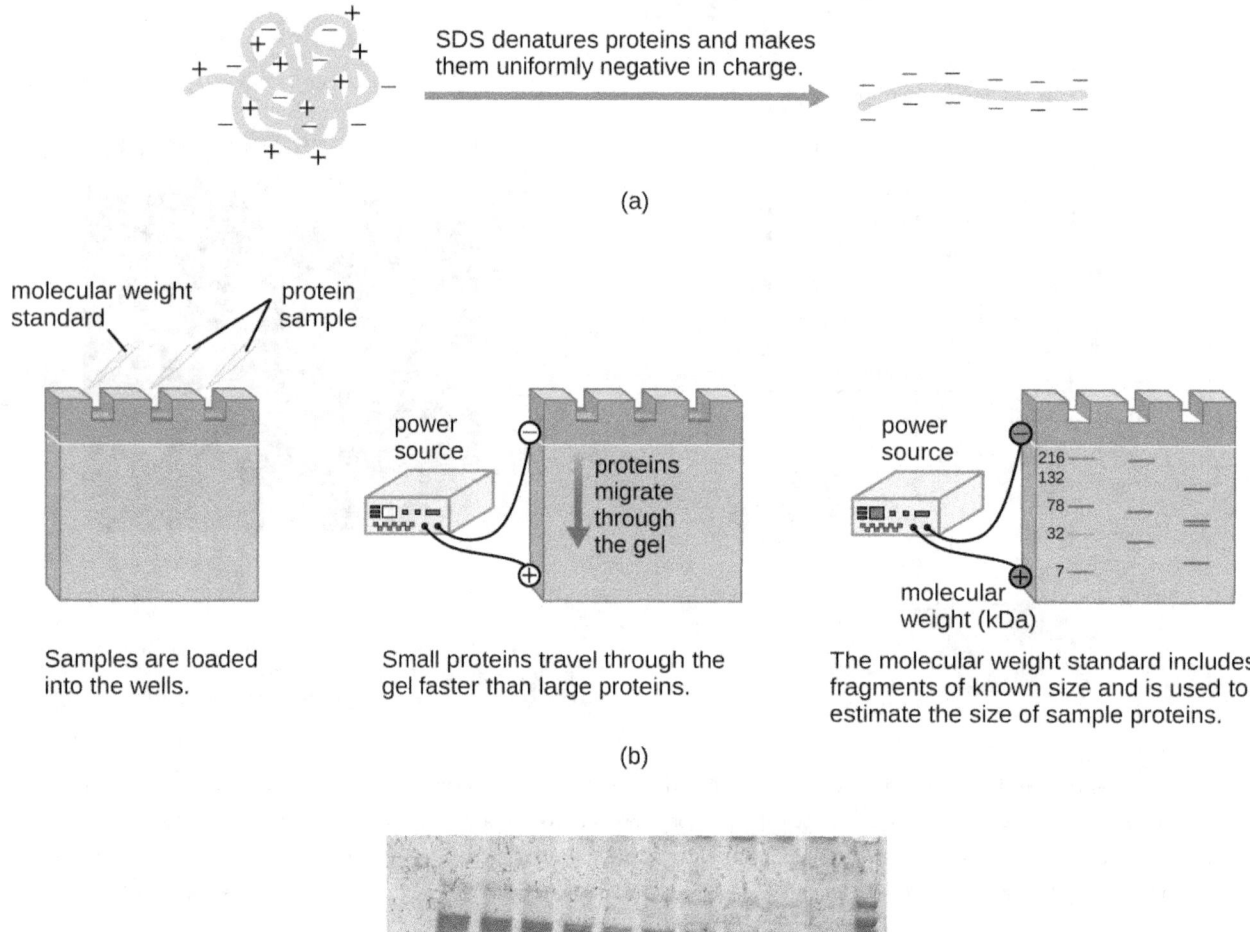

Figure 12.18 (a) SDS is a detergent that denatures proteins and masks their native charges, making them uniformly negatively charged. (b) The process of SDS-PAGE is illustrated in these steps. (c) A photograph of an SDS-PAGE gel shows Coomassie stained bands where proteins of different size have migrated along the gel in response to the applied voltage. A size standard lane is visible on the right side of

the gel. (credit b: modification of work by "GeneEd"/YouTube)

CHECK YOUR UNDERSTANDING

- On what basis are proteins separated in SDS-PAGE?

Clinical Focus

Part 3

When Kayla described her symptoms, her physician at first suspected bacterial meningitis, which is consistent with her headaches and stiff neck. However, she soon ruled this out as a possibility because meningitis typically progresses more quickly than what Kayla was experiencing. Many of her symptoms still paralleled those of amyotrophic lateral sclerosis (ALS) and systemic lupus erythematosus (SLE), and the physician also considered Lyme disease a possibility given how much time Kayla spends in the woods. Kayla did not recall any recent tick bites (the typical means by which Lyme disease is transmitted) and she did not have the typical bull's-eye rash associated with Lyme disease (Figure 12.19). However, 20–30% of patients with Lyme disease never develop this rash, so the physician did not want to rule it out.

Kayla's doctor ordered an MRI of her brain, a complete blood count to test for anemia, blood tests assessing liver and kidney function, and additional tests to confirm or rule out SLE or Lyme disease. Her test results were inconsistent with both SLE and ALS, and the result of the test looking for Lyme disease antibodies was "equivocal," meaning inconclusive. Having ruled out ALS and SLE, Kayla's doctor decided to run additional tests for Lyme disease.

- Why would Kayla's doctor still suspect Lyme disease even if the test results did not detect Lyme antibodies in the blood?
- What type of molecular test might be used for the detection of blood antibodies to Lyme disease?

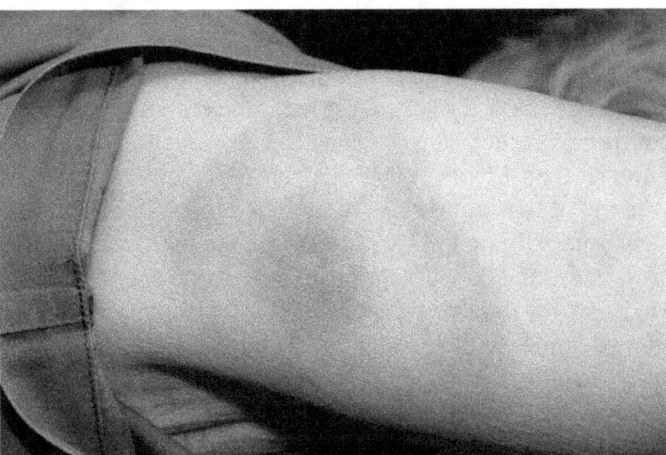

Figure 12.19 A bulls-eye rash is one of the common symptoms of Lyme diseases, but up to 30% of infected individuals never develop a rash. (credit: Centers for Disease Control and Prevention)

Jump to the next Clinical Focus box. Go back to the previous Clinical Focus box.

Amplification-Based DNA Analysis Methods

Various methods can be used for obtaining sequences of DNA, which are useful for studying disease-causing organisms. With the advent of rapid sequencing technology, our knowledge base of the entire genomes of pathogenic organisms has grown phenomenally. We start with a description of the polymerase chain reaction, which is not a sequencing method but has allowed researchers and clinicians to obtain the large quantities of DNA needed for sequencing and other studies. The polymerase chain reaction eliminates the dependence we once had on cells to make multiple copies of DNA, achieving the same result through relatively simple reactions outside the cell.

Polymerase Chain Reaction (PCR)

Most methods of DNA analysis, such as restriction enzyme digestion and agarose gel electrophoresis, or DNA sequencing require large amounts of a specific DNA fragment. In the past, large amounts of DNA were produced by growing the host cells of a genomic library. However, libraries take time and effort to prepare and DNA samples of interest often come in minute quantities. The **polymerase chain reaction (PCR)** permits rapid amplification in the number of copies of specific DNA sequences for further analysis (Figure 12.20). One of the most powerful techniques in molecular biology, PCR was developed in 1983 by Kary Mullis while at Cetus Corporation. PCR has specific applications in research, forensic, and clinical laboratories, including:

- determining the sequence of nucleotides in a specific region of DNA
- amplifying a target region of DNA for cloning into a plasmid vector
- identifying the source of a DNA sample left at a crime scene
- analyzing samples to determine paternity
- comparing samples of ancient DNA with modern organisms
- determining the presence of difficult to culture, or unculturable, microorganisms in humans or environmental samples

PCR is an *in vitro* laboratory technique that takes advantage of the natural process of DNA replication. The heat-stable DNA polymerase enzymes used in PCR are derived from hyperthermophilic prokaryotes. *Taq DNA polymerase*, commonly used in PCR, is derived from the *Thermus aquaticus* bacterium isolated from a hot spring in Yellowstone National Park. DNA replication requires the use of primers for the initiation of replication to have free 3′-hydroxyl groups available for the addition of nucleotides by DNA polymerase. However, while primers composed of RNA are normally used in cells, DNA primers are used for PCR. **DNA primers** are preferable due to their stability, and DNA primers with known sequences targeting a specific DNA region can be chemically synthesized commercially. These DNA primers are functionally similar to the DNA probes used for the various hybridization techniques described earlier, binding to specific targets due to complementarity between the target DNA sequence and the primer.

PCR occurs over multiple cycles, each containing three steps: denaturation, annealing, and extension. Machines called thermal cyclers are used for PCR; these machines can be programmed to automatically cycle through the temperatures required at each step (Figure 12.1). First, double-stranded template DNA containing the target sequence is denatured at approximately 95 °C. The high temperature required to physically (rather than enzymatically) separate the DNA strands is the reason the heat-stable DNA polymerase is required. Next, the temperature is lowered to approximately 50 °C. This allows the DNA primers complementary to the ends of the target sequence to anneal (stick) to the template strands, with one primer annealing to each strand. Finally, the temperature is raised to 72 °C, the optimal temperature for the activity of the heat-stable DNA polymerase, allowing for the addition of nucleotides to the primer using the single-stranded target as a template. Each cycle doubles the number of double-stranded target DNA copies. Typically, PCR protocols include 25–40 cycles, allowing for the amplification of a single target sequence by tens of millions to over a trillion.

Natural DNA replication is designed to copy the entire genome, and initiates at one or more origin sites. Primers are constructed during replication, not before, and do not consist of a few specific sequences. PCR targets specific regions of a DNA sample using sequence-specific primers. In recent years, a variety of isothermal PCR amplification methods that circumvent the need for thermal cycling have been developed, taking advantage of accessory proteins that aid in the DNA replication process. As the development of these methods continues and their use becomes more widespread in research, forensic, and clinical labs, thermal cyclers may become obsolete.

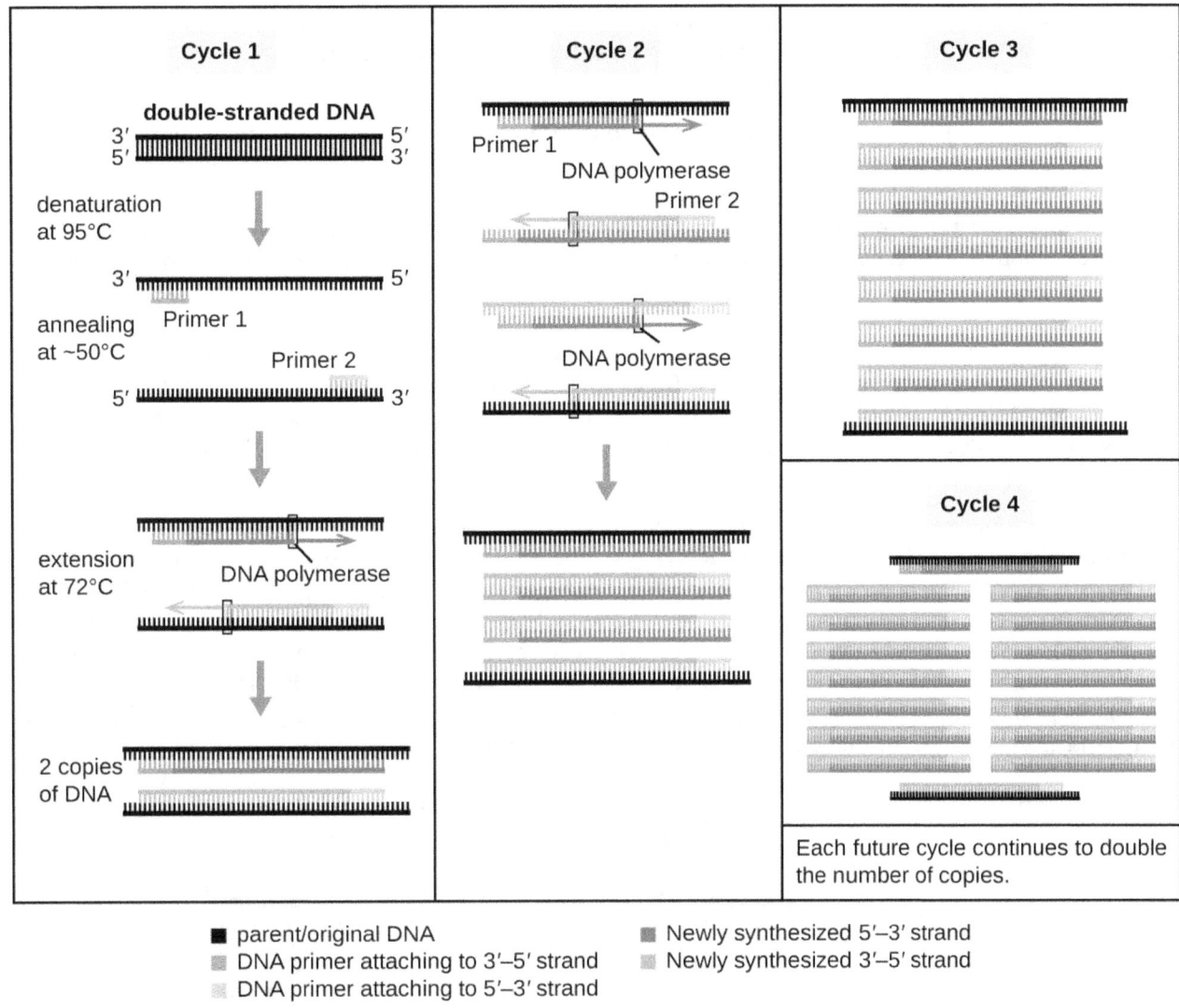

Figure 12.20 The polymerase chain reaction (PCR) is used to produce many copies of a specific sequence of DNA.

LINK TO LEARNING

Deepen your understanding of the polymerase chain reaction by viewing this animation (https://openstax.org/l/22polychareami) and working through an interactive (https://openstax.org/l/22intexerpolchr) exercise.

PCR Variations

Several later modifications to PCR further increase the utility of this technique. **Reverse transcriptase PCR (RT-PCR)** is used for obtaining DNA copies of a specific mRNA molecule. RT-PCR begins with the use of the reverse transcriptase enzyme to convert mRNA molecules into cDNA. That cDNA is then used as a template for traditional PCR amplification. RT-PCR can detect whether a specific gene has been expressed in a sample. Another recent application of PCR is **real-time PCR**, also known as **quantitative PCR (qPCR)**. Standard PCR and RT-PCR protocols are not quantitative because any one of the reagents may become limiting before all of the cycles within the protocol are complete, and samples are only analyzed at the end. Because it is not possible to determine when in the PCR or RT-PCR protocol a given reagent has become limiting, it is not possible to know how many cycles were completed prior to this point, and thus it is not possible to determine how many original template molecules were present in the sample at the start of PCR. In qPCR, however, the use of fluorescence allows one to monitor the increase in a double-stranded template during a PCR reaction as it occurs. These kinetics data can then be used to quantify the amount of the original target sequence. The use of qPCR in recent years has further expanded the capabilities of PCR, allowing researchers to determine the number of DNA copies, and sometimes organisms, present in a sample. In clinical settings, qRT-PCR is used to determine viral load in HIV-positive patients to evaluate the effectiveness of their therapy.

DNA Sequencing

A basic sequencing technique is the **chain termination method**, also known as the **dideoxy method** or the **Sanger DNA sequencing method**, developed by Frederick Sanger in 1972. The chain termination method involves DNA replication of a single-stranded template with the use of a DNA primer to initiate synthesis of a complementary strand, DNA polymerase, a mix of the four regular deoxynucleotide (dNTP) monomers, and a small proportion of dideoxynucleotides (ddNTPs), each labeled with a molecular beacon. The ddNTPs are monomers missing a hydroxyl group (–OH) at the site at which another nucleotide usually attaches to form a chain (Figure 12.21). Every time a ddNTP is randomly incorporated into the growing complementary strand, it terminates the process of DNA replication for that particular strand. This results in multiple short strands of replicated DNA that are each terminated at a different point during replication. When the reaction mixture is subjected to gel electrophoresis, the multiple newly replicated DNA strands form a ladder of differing sizes. Because the ddNTPs are labeled, each band on the gel reflects the size of the DNA strand when the ddNTP terminated the reaction.

In Sanger's day, four reactions were set up for each DNA molecule being sequenced, each reaction containing only one of the four possible ddNTPs. Each ddNTP was labeled with a radioactive phosphorus molecule. The products of the four reactions were then run in separate lanes side by side on long, narrow PAGE gels, and the bands of varying lengths were detected by autoradiography. Today, this process has been simplified with the use of ddNTPs, each labeled with a different colored fluorescent dye or fluorochrome (Figure 12.22), in one sequencing reaction containing all four possible ddNTPs for each DNA molecule being sequenced (Figure 12.23). These fluorochromes are detected by fluorescence spectroscopy. Determining the fluorescence color of each band as it passes by the detector produces the nucleotide sequence of the template strand.

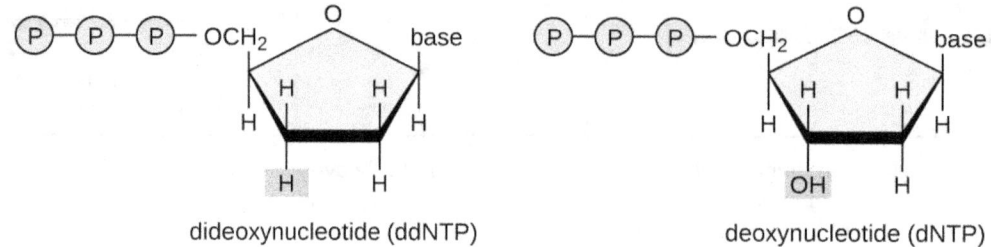

Figure 12.21 A dideoxynucleotide is similar in structure to a deoxynucleotide, but is missing the 3' hydroxyl group (indicated by the shaded box). When a dideoxynucleotide is incorporated into a DNA strand, DNA synthesis stops.

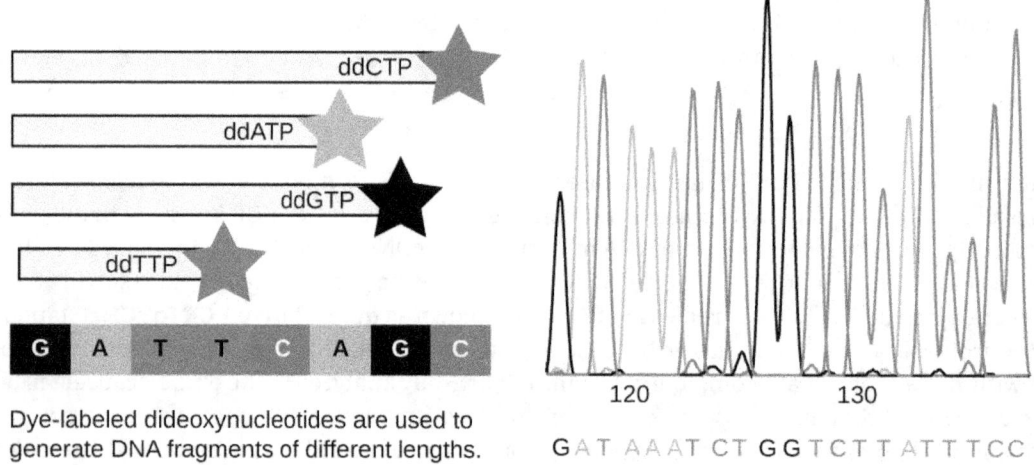

Figure 12.22 Frederick Sanger's dideoxy chain termination method is illustrated, using ddNTPs tagged with fluorochromes. Using ddNTPs, a mixture of DNA fragments of every possible size, varying in length by only one nucleotide, can be generated. The DNA is separated on the basis of size and each band can be detected with a fluorescence detector.

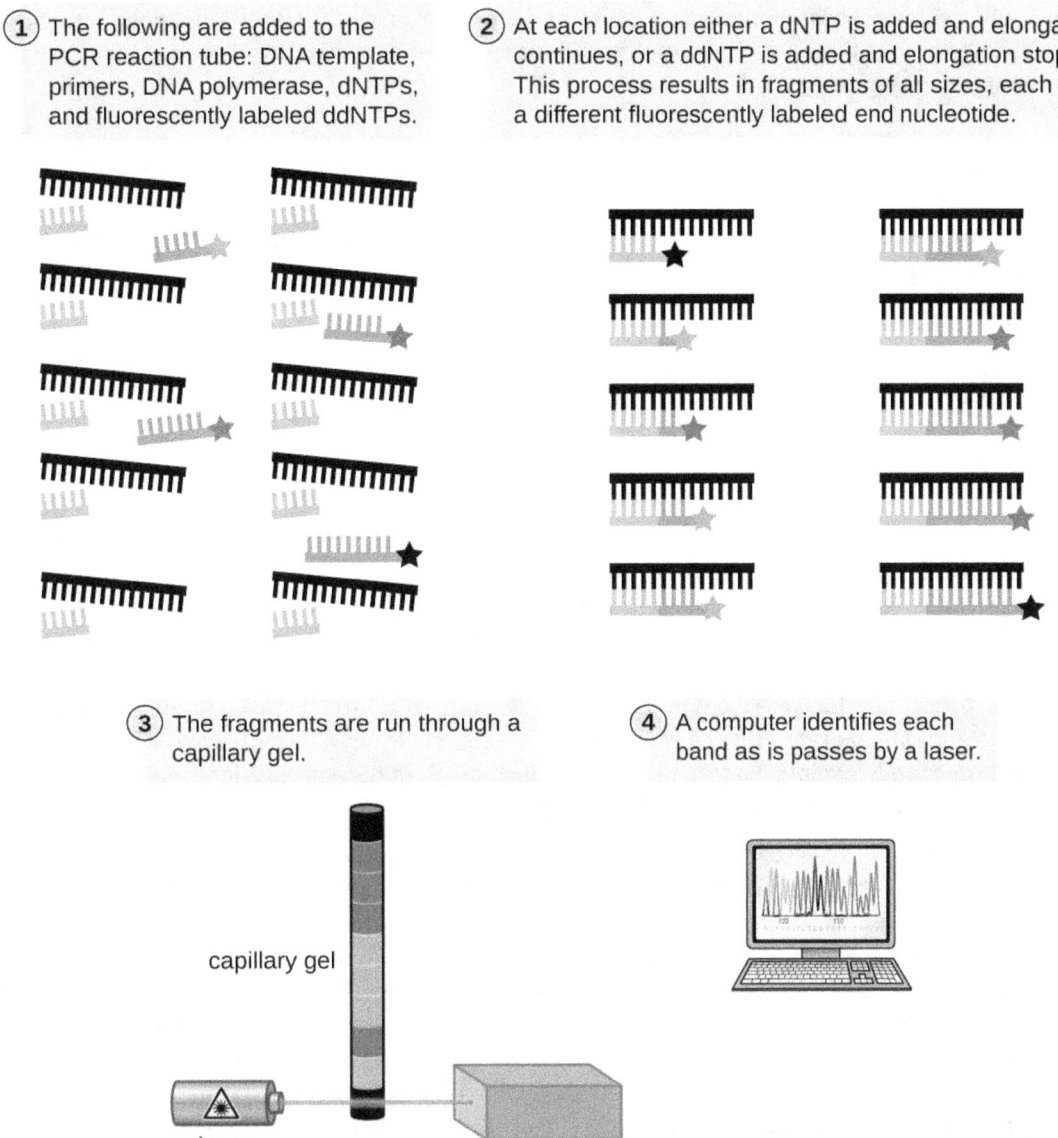

Figure 12.23 This diagram summarizes the Sanger sequencing method using fluorochrome-labeled ddNTPs and capillary gel electrophoresis.

Since 2005, automated sequencing techniques used by laboratories fall under the umbrella of **next generation sequencing**, which is a group of automated techniques used for rapid DNA sequencing. These methods have revolutionized the field of molecular genetics because the low-cost sequencers can generate sequences of hundreds of thousands or millions of short fragments (25 to 600 base pairs) just in one day. Although several variants of next generation sequencing technologies are made by different companies (for example, 454 Life Sciences' pyrosequencing and Illumina's Solexa technology), they all allow millions of bases to be sequenced quickly, making the sequencing of entire genomes relatively easy, inexpensive, and commonplace. In **454 sequencing (pyrosequencing)**, for example, a DNA sample is fragmented into 400–600-bp single-strand fragments, modified with the addition of DNA adapters to both ends of each fragment. Each DNA fragment is then immobilized on a bead and amplified by PCR, using primers designed to anneal to the adapters, creating a bead containing many copies of that DNA fragment. Each bead is then put into a separate well containing sequencing enzymes. To the well, each of the four nucleotides is added one after the other; when each one is incorporated, pyrophosphate is released as a byproduct of polymerization, emitting a small flash of light that is recorded by a detector. This provides the order of nucleotides incorporated as a new strand of DNA is made and is an example of synthesis sequencing. Next generation sequencers use sophisticated software to get through the cumbersome process of putting all the fragments in order. Overall, these technologies continue to advance rapidly, decreasing the cost of sequencing and increasing the availability of sequence data from a wide variety of organisms quickly.

The National Center for Biotechnology Information houses a widely used genetic sequence database called GenBank where researchers deposit genetic information for public use. Upon publication of sequence data, researchers upload it to GenBank, giving other researchers access to the information. The collaboration allows researchers to compare newly discovered or unknown sample sequence information with the vast array of sequence data that already exists.

LINK TO LEARNING

View an animation (https://openstax.org/l/22454seqanim) about 454 sequencing to deepen your understanding of this method.

Case in Point

Using a NAAT to Diagnose a *C. difficile* Infection

Javier, an 80-year-old patient with a history of heart disease, recently returned home from the hospital after undergoing an angioplasty procedure to insert a stent into a cardiac artery. To minimize the possibility of infection, Javier was administered intravenous broad-spectrum antibiotics during and shortly after his procedure. He was released four days after the procedure, but a week later, he began to experience mild abdominal cramping and watery diarrhea several times a day. He lost his appetite, became severely dehydrated, and developed a fever. He also noticed blood in his stool. Javier's wife called the physician, who instructed her to take him to the emergency room immediately.

The hospital staff ran several tests and found that Javier's kidney creatinine levels were elevated compared with the levels in his blood, indicating that his kidneys were not functioning well. Javier's symptoms suggested a possible infection with *Clostridium difficile*, a bacterium that is resistant to many antibiotics. The hospital collected and cultured a stool sample to look for the production of toxins A and B by *C. difficile*, but the results came back negative. However, the negative results were not enough to rule out a *C. difficile* infection because culturing of *C. difficile* and detection of its characteristic toxins can be difficult, particularly in some types of samples. To be safe, they proceeded with a diagnostic nucleic acid amplification test (NAAT). Currently NAATs are the clinical diagnostician's gold standard for detecting the genetic material of a pathogen. In Javier's case, qPCR was used to look for the gene encoding *C. difficile* toxin B (*tcdB*). When the qPCR analysis came back positive, the attending physician concluded that Javier was indeed suffering from a *C. difficile* infection and immediately prescribed the antibiotic vancomycin, to be administered intravenously. The antibiotic cleared the infection and Javier made a full recovery.

Because infections with *C. difficile* were becoming widespread in Javier's community, his sample was further analyzed to see whether the specific strain of *C. difficile* could be identified. Javier's stool sample was subjected to ribotyping and repetitive sequence-based PCR (rep-PCR) analysis. In ribotyping, a short sequence of DNA between the 16S rRNA and 23S rRNA genes is amplified and subjected to restriction digestion (Figure 12.24). This sequence varies between strains of *C. difficile*, so restriction enzymes will cut in different places. In rep-PCR, DNA primers designed to bind to short sequences commonly found repeated within the *C. difficile* genome were used for PCR. Following restriction digestion, agarose gel electrophoresis was performed in both types of analysis to examine the banding patterns that resulted from each procedure (Figure 12.25). Rep-PCR can be used to further subtype various ribotypes, increasing resolution for detecting differences between strains. The ribotype of the strain infecting Javier was found to be ribotype 27, a strain known for its increased virulence, resistance to antibiotics, and increased prevalence in the United States, Canada, Japan, and Europe.[4]

---

4 Patrizia Spigaglia, Fabrizio Barbanti, Anna Maria Dionisi, and Paola Mastrantonio. "*Clostridium difficile* Isolates Resistant to Fluoroquinolones in Italy: Emergence of PCR Ribotype 018." *Journal of Clinical Microbiology* 48 no. 8 (2010): 2892–2896.

- How do banding patterns differ between strains of *C. difficile*?
- Why do you think laboratory tests were unable to detect toxin production directly?

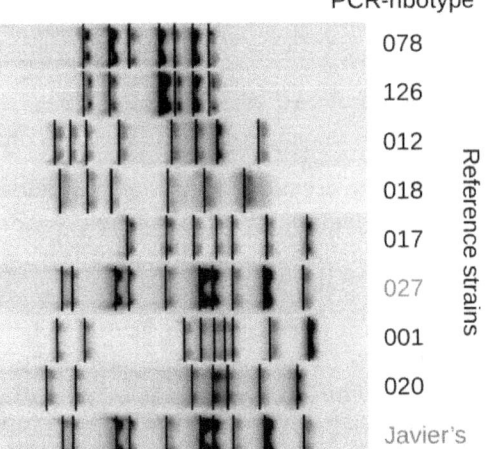

Figure 12.24 A gel showing PCR products of various *Clostridium difficile* strains. Javier's sample is shown at the bottom; note that it matches ribotype 27 in the reference set. (credit: modification of work by American Society for Microbiology)

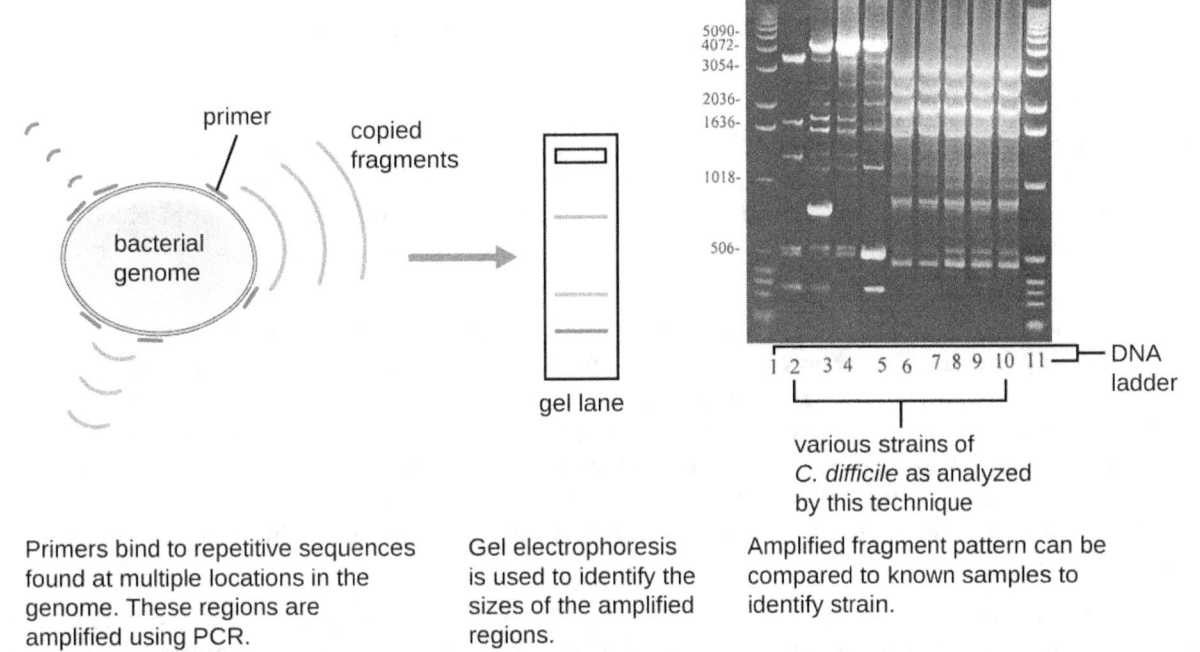

Primers bind to repetitive sequences found at multiple locations in the genome. These regions are amplified using PCR.

Gel electrophoresis is used to identify the sizes of the amplified regions.

Amplified fragment pattern can be compared to known samples to identify strain.

Figure 12.25 Strains of infectious bacteria, such as *C. difficile,* can be identified by molecular analysis. PCR ribotyping is commonly used to identify particular *C. difficile* strains. Rep-PCR is an alternate molecular technique that is also used to identify particular *C. difficile* strains. (credit b: modification of work by American Society for Microbiology)

CHECK YOUR UNDERSTANDING

- How is PCR similar to the natural DNA replication process in cells? How is it different?
- Compare RT-PCR and qPCR in terms of their respective purposes.
- In chain-termination sequencing, how is the identity of each nucleotide in a sequence determined?

12.3 Whole Genome Methods and Pharmaceutical Applications of Genetic Engineering

Learning Objectives

By the end of this section, you will be able to:
- Explain the uses of genome-wide comparative analyses
- Summarize the advantages of genetically engineered pharmaceutical products

Advances in molecular biology have led to the creation of entirely new fields of science. Among these are fields that study aspects of whole genomes, collectively referred to as whole-genome methods. In this section, we'll provide a brief overview of the whole-genome fields of genomics, transcriptomics, and proteomics.

Genomics, Transcriptomics, and Proteomics

The study and comparison of entire genomes, including the complete set of genes and their nucleotide sequence and organization, is called **genomics**. This field has great potential for future medical advances through the study of the human genome as well as the genomes of infectious organisms. Analysis of microbial genomes has contributed to the development of new antibiotics, diagnostic tools, vaccines, medical treatments, and environmental cleanup techniques.

The field of **transcriptomics** is the science of the entire collection of mRNA molecules produced by cells. Scientists compare gene expression patterns between infected and uninfected host cells, gaining important information about the cellular responses to infectious disease. Additionally, transcriptomics can be used to monitor the gene expression of virulence factors in microorganisms, aiding scientists in better understanding pathogenic processes from this viewpoint.

When genomics and transcriptomics are applied to entire microbial communities, we use the terms **metagenomics** and **metatranscriptomics**, respectively. Metagenomics and metatranscriptomics allow researchers to study genes and gene expression from a collection of multiple species, many of which may not be easily cultured or cultured at all in the laboratory. A DNA microarray (discussed in the previous section) can be used in metagenomics studies.

Another up-and-coming clinical application of genomics and transcriptomics is **pharmacogenomics**, also called **toxicogenomics**, which involves evaluating the effectiveness and safety of drugs on the basis of information from an individual's genomic sequence. Genomic responses to drugs can be studied using experimental animals (such as laboratory rats or mice) or live cells in the laboratory before embarking on studies with humans. Changes in gene expression in the presence of a drug can sometimes be an early indicator of the potential for toxic effects. Personal genome sequence information may someday be used to prescribe medications that will be most effective and least toxic on the basis of the individual patient's genotype.

The study of **proteomics** is an extension of genomics that allows scientists to study the entire complement of proteins in an organism, called the proteome. Even though all cells of a multicellular organism have the same set of genes, cells in various tissues produce different sets of proteins. Thus, the genome is constant, but the proteome varies and is dynamic within an organism. Proteomics may be used to study which proteins are expressed under various conditions within a single cell type or to compare protein expression patterns between different organisms.

The most prominent disease being studied with proteomic approaches is cancer, but this area of study is also being applied to infectious diseases. Research is currently underway to examine the feasibility of using proteomic approaches to diagnose various types of hepatitis, tuberculosis, and HIV infection, which are rather difficult to diagnose using currently available techniques.[5]

---

5 E.O. List, D.E. Berryman, B. Bower, L. Sackmann-Sala, E. Gosney, J. Ding, S. Okada, and J.J. Kopchick. "The Use of Proteomics to Study Infectious Diseases." *Infectious Disorders-Drug Targets* (Formerly *Current Drug Targets-Infectious Disorders*) 8 no. 1 (2008): 31–45.

A recent and developing proteomic analysis relies on identifying proteins called **biomarkers**, whose expression is affected by the disease process. Biomarkers are currently being used to detect various forms of cancer as well as infections caused by pathogens such as *Yersinia pestis* and *Vaccinia virus*.[6]

Other "-omic" sciences related to genomics and proteomics include metabolomics, glycomics, and lipidomics, which focus on the complete set of small-molecule metabolites, sugars, and lipids, respectively, found within a cell. Through these various global approaches, scientists continue to collect, compile, and analyze large amounts of genetic information. This emerging field of **bioinformatics** can be used, among many other applications, for clues to treating diseases and understanding the workings of cells.

Additionally, researchers can use reverse genetics, a technique related to classic mutational analysis, to determine the function of specific genes. Classic methods of studying gene function involved searching for the genes responsible for a given phenotype. Reverse genetics uses the opposite approach, starting with a specific DNA sequence and attempting to determine what phenotype it produces. Alternatively, scientists can attach known genes (called reporter genes) that encode easily observable characteristics to genes of interest, and the location of expression of such genes of interest can be easily monitored. This gives the researcher important information about what the gene product might be doing or where it is located in the organism. Common reporter genes include bacterial *lacZ*, which encodes beta-galactosidase and whose activity can be monitored by changes in colony color in the presence of X-gal as previously described, and the gene encoding the jellyfish protein green fluorescent protein (GFP) whose activity can be visualized in colonies under ultraviolet light exposure (Figure 12.26).

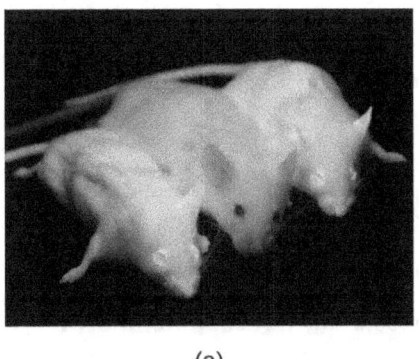
(a)
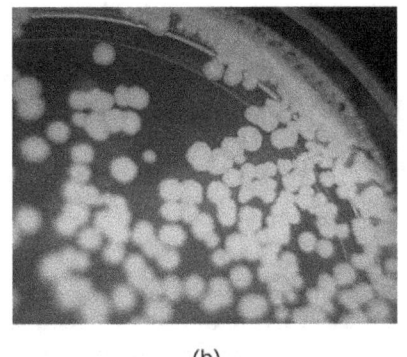
(b)
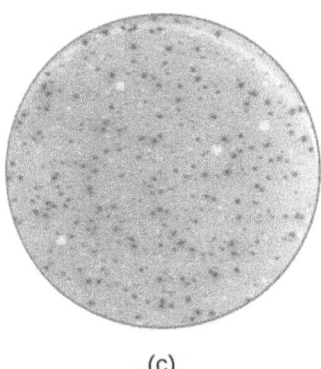
(c)

Figure 12.26 (a) The gene encoding green fluorescence protein is a commonly used reporter gene for monitoring gene expression patterns in organisms. Under ultraviolet light, GFP fluoresces. Here, two mice are expressing GFP, while the middle mouse is not. (b) GFP can be used as a reporter gene in bacteria as well. Here, a plate containing bacterial colonies expressing GFP is shown. (c) Blue-white screening in bacteria is accomplished through the use of the *lacZ* reporter gene, followed by plating of bacteria onto medium containing X-gal. Cleavage of X-gal by the LacZ enzyme results in the formation of blue colonies. (credit a: modification of work by Ingrid Moen, Charlotte Jevne, Jian Wang, Karl-Henning Kalland, Martha Chekenya, Lars A Akslen, Linda Sleire, Per Ø Enger, Rolf K Reed, Anne M Øyan, Linda EB Stuhr; credit b: modification of work by "2.5JIGEN.com"/Flickr; credit c: modification of work by American Society for Microbiology)

CHECK YOUR UNDERSTANDING

- How is genomics different from traditional genetics?
- If you wanted to study how two different cells in the body respond to an infection, what –omics field would you apply?
- What are the biomarkers uncovered in proteomics used for?

---

6 Mohan Natesan, and Robert G. Ulrich. "Protein Microarrays and Biomarkers of Infectious Disease." *International Journal of Molecular Sciences 11* no. 12 (2010): 5165–5183.

> **Clinical Focus**
>
> **Resolution**
>
> Because Kayla's symptoms were persistent and serious enough to interfere with daily activities, Kayla's physician decided to order some laboratory tests. The physician collected samples of Kayla's blood, cerebrospinal fluid (CSF), and synovial fluid (from one of her swollen knees) and requested PCR analysis on all three samples. The PCR tests on the CSF and synovial fluid came back positive for the presence of *Borrelia burgdorferi*, the bacterium that causes Lyme disease.
>
> Kayla's physician immediately prescribed a full course of the antibiotic doxycycline. Fortunately, Kayla recovered fully within a few weeks and did not suffer from the long-term symptoms of post-treatment Lyme disease syndrome (PTLDS), which affects 10–20% of Lyme disease patients. To prevent future infections, Kayla's physician advised her to use insect repellant and wear protective clothing during her outdoor adventures. These measures can limit exposure to Lyme-bearing ticks, which are common in many regions of the United States during the warmer months of the year. Kayla was also advised to make a habit of examining herself for ticks after returning from outdoor activities, as prompt removal of a tick greatly reduces the chances of infection.
>
> Lyme disease is often difficult to diagnose. *B. burgdorferi* is not easily cultured in the laboratory, and the initial symptoms can be very mild and resemble those of many other diseases. But left untreated, the symptoms can become quite severe and debilitating. In addition to two antibody tests, which were inconclusive in Kayla's case, and the PCR test, a Southern blot could be used with *B. burgdorferi*-specific DNA probes to identify DNA from the pathogen. Sequencing of surface protein genes of *Borrelia* species is also being used to identify strains within the species that may be more readily transmitted to humans or cause more severe disease.
>
> *Go back to the previous Clinical Focus box.*

Recombinant DNA Technology and Pharmaceutical Production

Genetic engineering has provided a way to create new pharmaceutical products called **recombinant DNA pharmaceuticals**. Such products include antibiotic drugs, vaccines, and hormones used to treat various diseases. Table 12.1 lists examples of recombinant DNA products and their uses.

For example, the naturally occurring antibiotic synthesis pathways of various *Streptomyces* spp., long known for their antibiotic production capabilities, can be modified to improve yields or to create new antibiotics through the introduction of genes encoding additional enzymes. More than 200 new antibiotics have been generated through the targeted inactivation of genes and the novel combination of antibiotic synthesis genes in antibiotic-producing *Streptomyces* hosts.[7]

Genetic engineering is also used to manufacture subunit vaccines, which are safer than other vaccines because they contain only a single antigenic molecule and lack any part of the genome of the pathogen (see Vaccines). For example, a vaccine for hepatitis B is created by inserting a gene encoding a hepatitis B surface protein into a yeast; the yeast then produces this protein, which the human immune system recognizes as an antigen. The hepatitis B antigen is purified from yeast cultures and administered to patients as a vaccine. Even though the vaccine does not contain the hepatitis B virus, the presence of the antigenic protein stimulates the immune system to produce antibodies that will protect the patient against the virus in the event of exposure.[8][9]

---

[7] Jose-Luis Adrio and Arnold L. Demain. "Recombinant Organisms for Production of Industrial Products." *Bioengineered Bugs 1* no. 2 (2010): 116–131.

[8] U.S. Department of Health and Human Services. "Types of Vaccines." 2013. http://www.vaccines.gov/more_info/types/#subunit. Accessed May 27, 2016.

[9] The Internet Drug List. *Recombivax*. 2015. http://www.rxlist.com/recombivax-drug.htm. Accessed May 27, 2016.

Genetic engineering has also been important in the production of other therapeutic proteins, such as insulin, interferons, and human growth hormone, to treat a variety of human medical conditions. For example, at one time, it was possible to treat diabetes only by giving patients pig insulin, which caused allergic reactions due to small differences between the proteins expressed in human and pig insulin. However, since 1978, recombinant DNA technology has been used to produce large-scale quantities of human insulin using *E. coli* in a relatively inexpensive process that yields a more consistently effective pharmaceutical product. Scientists have also genetically engineered *E. coli* capable of producing human growth hormone (HGH), which is used to treat growth disorders in children and certain other disorders in adults. The HGH gene was cloned from a cDNA library and inserted into *E. coli* cells by cloning it into a bacterial vector. Eventually, genetic engineering will be used to produce DNA vaccines and various gene therapies, as well as customized medicines for fighting cancer and other diseases.

Some Genetically Engineered Pharmaceutical Products and Applications

Recombinant DNA Product	Application
Atrial natriuretic peptide	Treatment of heart disease (e.g., congestive heart failure), kidney disease, high blood pressure
DNase	Treatment of viscous lung secretions in cystic fibrosis
Erythropoietin	Treatment of severe anemia with kidney damage
Factor VIII	Treatment of hemophilia
Hepatitis B vaccine	Prevention of hepatitis B infection
Human growth hormone	Treatment of growth hormone deficiency, Turner's syndrome, burns
Human insulin	Treatment of diabetes
Interferons	Treatment of multiple sclerosis, various cancers (e.g., melanoma), viral infections (e.g., Hepatitis B and C)
Tetracenomycins	Used as antibiotics
Tissue plasminogen activator	Treatment of pulmonary embolism in ischemic stroke, myocardial infarction

Table 12.1

✓ CHECK YOUR UNDERSTANDING

- What bacterium has been genetically engineered to produce human insulin for the treatment of diabetes?
- Explain how microorganisms can be engineered to produce vaccines.

RNA Interference Technology

In Structure and Function of RNA, we described the function of mRNA, rRNA, and tRNA. In addition to these types of RNA, cells also produce several types of small noncoding RNA molecules that are involved in the regulation of gene expression. These include **antisense RNA** molecules, which are complementary to regions of specific mRNA molecules found in both prokaryotes and eukaryotic cells. Non-coding RNA molecules play a major role in **RNA interference (RNAi)**, a natural regulatory mechanism by which mRNA molecules are prevented from guiding the synthesis of proteins. RNA interference of specific genes results from the base pairing of short, single-stranded antisense RNA molecules to regions within complementary mRNA molecules, preventing protein synthesis. Cells use RNA interference to protect themselves from viral invasion, which may introduce double-stranded RNA molecules as part of the viral replication process (Figure 12.27).

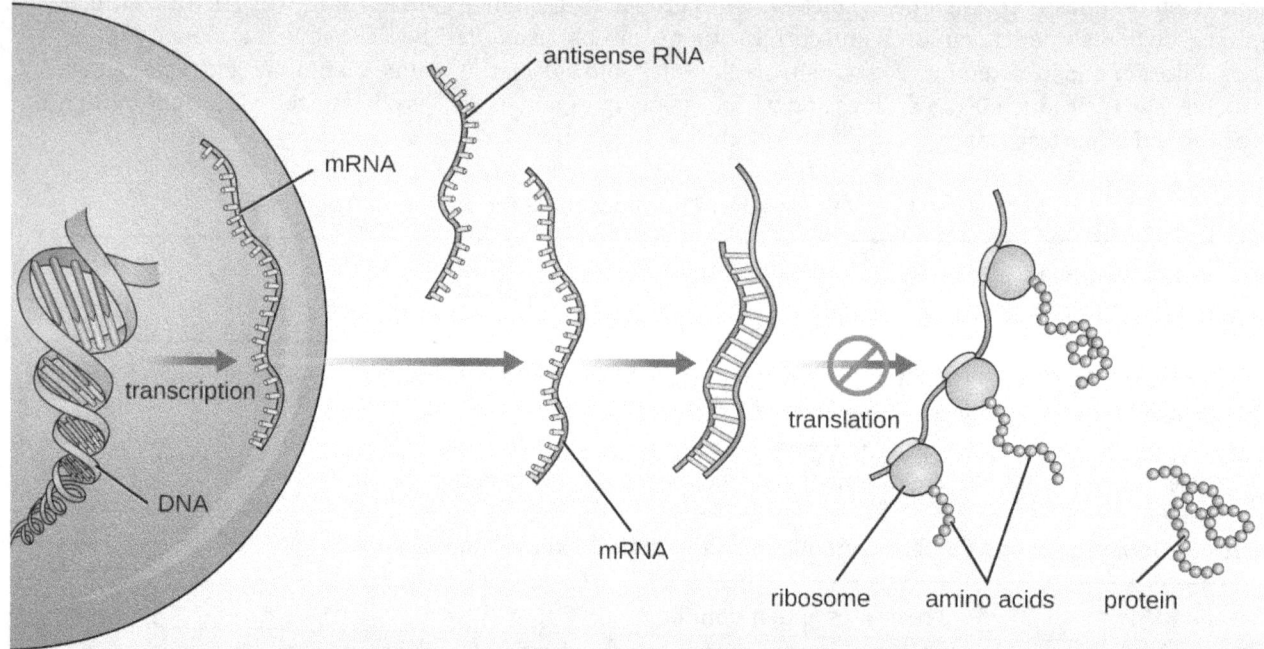

Figure 12.27 Cells like the eukaryotic cell shown in this diagram commonly make small antisense RNA molecules with sequences complementary to specific mRNA molecules. When an antisense RNA molecule is bound to an mRNA molecule, the mRNA can no longer be used to direct protein synthesis. (credit: modification of work by Robinson R)

Researchers are currently developing techniques to mimic the natural process of RNA interference as a way to treat viral infections in eukaryotic cells. RNA interference technology involves using small interfering RNAs (siRNAs) or microRNAs (miRNAs) (Figure 12.28). siRNAs are completely complementary to the mRNA transcript of a specific gene of interest while miRNAs are mostly complementary. These double-stranded RNAs are bound to DICER, an endonuclease that cleaves the RNA into short molecules (approximately 20 nucleotides long). The RNAs are then bound to RNA-induced silencing complex (RISC), a ribonucleoprotein. The siRNA-RISC complex binds to mRNA and cleaves it. For miRNA, only one of the two strands binds to RISC. The miRNA-RISC complex then binds to mRNA, inhibiting translation. If the miRNA is completely complementary to the target gene, then the mRNA can be cleaved. Taken together, these mechanisms are known as **gene silencing**.

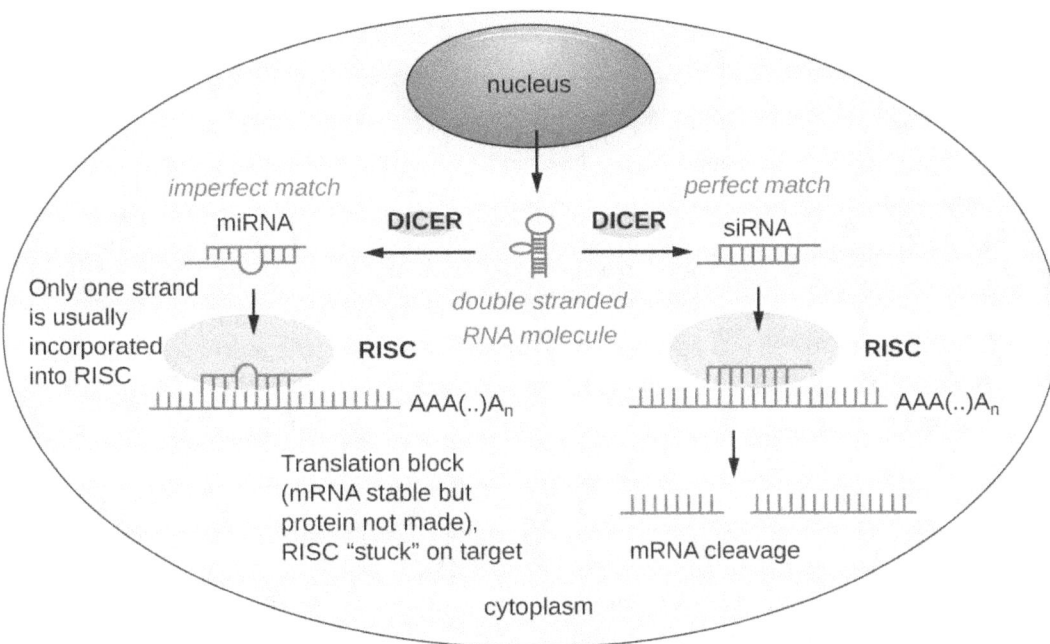

Figure 12.28 This diagram illustrates the process of using siRNA or miRNA in a eukaryotic cell to silence genes involved in the pathogenesis of various diseases. (credit: modification of work by National Center for Biotechnology Information)

12.4 Gene Therapy

Learning Objectives

By the end of this section, you will be able to:
- Summarize the mechanisms, risks, and potential benefits of gene therapy
- Identify ethical issues involving gene therapy and the regulatory agencies that provide oversight for clinical trials
- Compare somatic-cell and germ-line gene therapy

Many types of genetic engineering have yielded clear benefits with few apparent risks. Few would question, for example, the value of our now abundant supply of human insulin produced by genetically engineered bacteria. However, many emerging applications of genetic engineering are much more controversial, often because their potential benefits are pitted against significant risks, real or perceived. This is certainly the case for **gene therapy**, a clinical application of genetic engineering that may one day provide a cure for many diseases but is still largely an experimental approach to treatment.

Mechanisms and Risks of Gene Therapy

Human diseases that result from genetic mutations are often difficult to treat with drugs or other traditional forms of therapy because the signs and symptoms of disease result from abnormalities in a patient's genome. For example, a patient may have a genetic mutation that prevents the expression of a specific protein required for the normal function of a particular cell type. This is the case in patients with Severe Combined Immunodeficiency (SCID), a genetic disease that impairs the function of certain white blood cells essential to the immune system.

Gene therapy attempts to correct genetic abnormalities by introducing a nonmutated, functional gene into the patient's genome. The nonmutated gene encodes a functional protein that the patient would otherwise be unable to produce. Viral vectors such as adenovirus are sometimes used to introduce the functional gene; part of the viral genome is removed and replaced with the desired gene (Figure 12.29). More advanced forms of gene therapy attempt to correct the mutation at the original site in the genome, such as is the case with treatment of SCID.

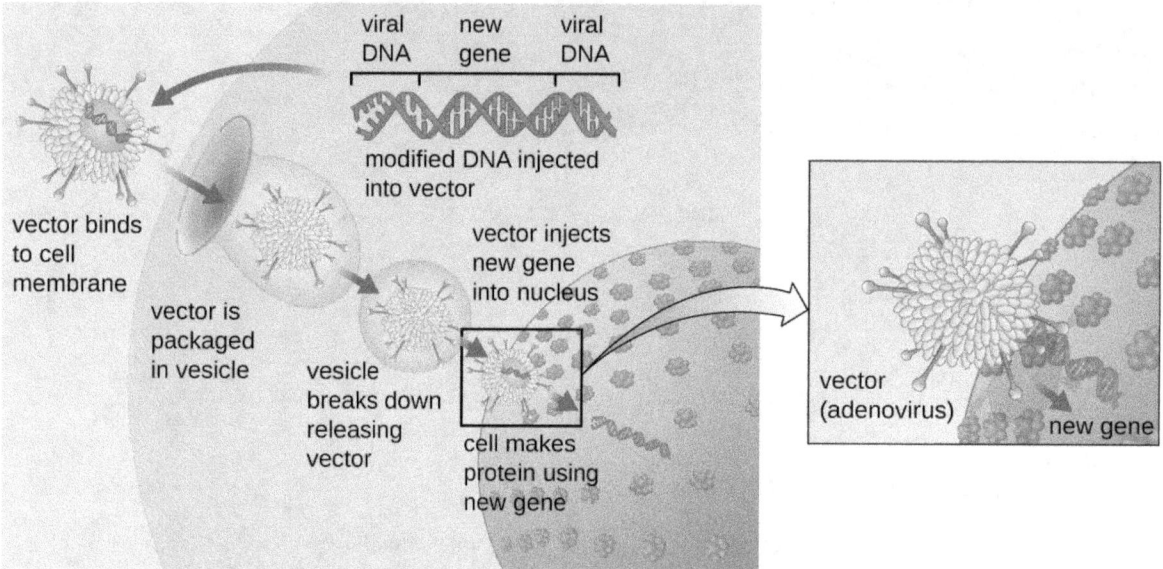

Figure 12.29 Gene therapy using an adenovirus vector can be used to treat or cure certain genetic diseases in which a patient has a defective gene. (credit: modification of work by National Institutes of Health)

So far, gene therapies have proven relatively ineffective, with the possible exceptions of treatments for cystic fibrosis and adenosine deaminase deficiency, a type of SCID. Other trials have shown the clear hazards of attempting genetic manipulation in complex multicellular organisms like humans. In some patients, the use of an adenovirus vector can trigger an unanticipated inflammatory response from the immune system, which may lead to organ failure. Moreover, because viruses can often target multiple cell types, the virus vector may infect cells not targeted for the therapy, damaging these other cells and possibly leading to illnesses such as cancer. Another potential risk is that the modified virus could revert to being infectious and cause disease in the patient. Lastly, there is a risk that the inserted gene could unintentionally inactivate another important gene in the patient's genome, disrupting normal cell cycling and possibly leading to tumor formation and cancer. Because gene therapy involves so many risks, candidates for gene therapy need to be fully informed of these risks before providing informed consent to undergo the therapy.

> Case in Point
>
> **Gene Therapy Gone Wrong**
> The risks of gene therapy were realized in the 1999 case of Jesse Gelsinger, an 18-year-old patient who received gene therapy as part of a clinical trial at the University of Pennsylvania. Jesse received gene therapy for a condition called ornithine transcarbamylase (OTC) deficiency, which leads to ammonia accumulation in the blood due to deficient ammonia processing. Four days after the treatment, Jesse died after a massive immune response to the adenovirus vector.[10]
>
> Until that point, researchers had not really considered an immune response to the vector to be a legitimate risk, but on investigation, it appears that the researchers had some evidence suggesting that this was a possible outcome. Prior to Jesse's treatment, several other human patients had suffered side effects of the treatment, and three monkeys used in a trial had died as a result of inflammation and clotting disorders. Despite this information, it appears that neither Jesse nor his family were made aware of these outcomes when they consented to the therapy. Jesse's death was the first patient death due to a gene therapy treatment and resulted in the immediate halting of the clinical trial in which he was involved, the subsequent halting of all other gene therapy trials at the University of Pennsylvania, and the investigation of all other gene therapy trials in the United States. As a result, the regulation and oversight of gene therapy

---

10 Barbara Sibbald. "Death but One Unintended Consequence of Gene-Therapy Trial." *Canadian Medical Association Journal 164* no. 11 (2001): 1612–1612.

overall was reexamined, resulting in new regulatory protocols that are still in place today.

CHECK YOUR UNDERSTANDING
- Explain how gene therapy works in theory.
- Identify some risks of gene therapy.

Oversight of Gene Therapy

Presently, there is significant oversight of gene therapy clinical trials. At the federal level, three agencies regulate gene therapy in parallel: the Food and Drug Administration (FDA), the Office of Human Research Protection (OHRP), and the Recombinant DNA Advisory Committee (RAC) at the National Institutes of Health (NIH). Along with several local agencies, these federal agencies interact with the institutional review board to ensure that protocols are in place to protect patient safety during clinical trials. Compliance with these protocols is enforced mostly on the local level in cooperation with the federal agencies. Gene therapies are currently under the most extensive federal and local review compared to other types of therapies, which are more typically only under the review of the FDA. Some researchers believe that these extensive regulations actually inhibit progress in gene therapy research. In 2013, the Institute of Medicine (now the National Academy of Medicine) called upon the NIH to relax its review of gene therapy trials in most cases.[11] However, ensuring patient safety continues to be of utmost concern.

Ethical Concerns

Beyond the health risks of gene therapy, the ability to genetically modify humans poses a number of ethical issues related to the limits of such "therapy." While current research is focused on gene therapy for genetic diseases, scientists might one day apply these methods to manipulate other genetic traits not perceived as desirable. This raises questions such as:

- Which genetic traits are worthy of being "corrected"?
- Should gene therapy be used for cosmetic reasons or to enhance human abilities?
- Should genetic manipulation be used to impart desirable traits to the unborn?
- Is everyone entitled to gene therapy, or could the cost of gene therapy create new forms of social inequality?
- Who should be responsible for regulating and policing inappropriate use of gene therapies?

The ability to alter reproductive cells using gene therapy could also generate new ethical dilemmas. To date, the various types of gene therapies have been targeted to somatic cells, the non-reproductive cells within the body. Because somatic cell traits are not inherited, any genetic changes accomplished by somatic-cell gene therapy would not be passed on to offspring. However, should scientists successfully introduce new genes to germ cells (eggs or sperm), the resulting traits could be passed on to offspring. This approach, called germ-line gene therapy, could potentially be used to combat heritable diseases, but it could also lead to unintended consequences for future generations. Moreover, there is the question of informed consent, because those impacted by germ-line gene therapy are unborn and therefore unable to choose whether they receive the therapy. For these reasons, the U.S. government does not currently fund research projects investigating germ-line gene therapies in humans.

---

11 Kerry Grens. "Report: Ease Gene Therapy Reviews." *The Scientist*, December 9, 2013. http://www.the-scientist.com/?articles.view/articleNo/38577/title/Report--Ease-Gene-Therapy-Reviews/. Accessed May 27, 2016.

> Eye on Ethics
>
> **Risky Gene Therapies**
>
> While there are currently no gene therapies on the market in the United States, many are in the pipeline and it is likely that some will eventually be approved. With recent advances in gene therapies targeting p53, a gene whose somatic cell mutations have been implicated in over 50% of human cancers,[12] cancer treatments through gene therapies could become much more widespread once they reach the commercial market.
>
> Bringing any new therapy to market poses ethical questions that pit the expected benefits against the risks. How quickly should new therapies be brought to the market? How can we ensure that new therapies have been sufficiently tested for safety and effectiveness before they are marketed to the public? The process by which new therapies are developed and approved complicates such questions, as those involved in the approval process are often under significant pressure to get a new therapy approved even in the face of significant risks.
>
> To receive FDA approval for a new therapy, researchers must collect significant laboratory data from animal trials and submit an Investigational New Drug (IND) application to the FDA's Center for Drug Evaluation and Research (CDER). Following a 30-day waiting period during which the FDA reviews the IND, clinical trials involving human subjects may begin. If the FDA perceives a problem prior to or during the clinical trial, the FDA can order a "clinical hold" until any problems are addressed. During clinical trials, researchers collect and analyze data on the therapy's effectiveness and safety, including any side effects observed. Once the therapy meets FDA standards for effectiveness and safety, the developers can submit a New Drug Application (NDA) that details how the therapy will be manufactured, packaged, monitored, and administered.
>
> Because new gene therapies are frequently the result of many years (even decades) of laboratory and clinical research, they require a significant financial investment. By the time a therapy has reached the clinical trials stage, the financial stakes are high for pharmaceutical companies and their shareholders. This creates potential conflicts of interest that can sometimes affect the objective judgment of researchers, their funders, and even trial participants. The Jesse Gelsinger case (see Case in Point: Gene Therapy Gone Wrong) is a classic example. Faced with a life-threatening disease and no reasonable treatments available, it is easy to see why a patient might be eager to participate in a clinical trial no matter the risks. It is also easy to see how a researcher might view the short-term risks for a small group of study participants as a small price to pay for the potential benefits of a game-changing new treatment.
>
> Gelsinger's death led to increased scrutiny of gene therapy, and subsequent negative outcomes of gene therapy have resulted in the temporary halting of clinical trials pending further investigation. For example, when children in France treated with gene therapy for SCID began to develop leukemia several years after treatment, the FDA temporarily stopped clinical trials of similar types of gene therapy occurring in the United States.[13] Cases like these highlight the need for researchers and health professionals not only to value human well-being and patients' rights over profitability, but also to maintain scientific objectivity when evaluating the risks and benefits of new therapies.

CHECK YOUR UNDERSTANDING

- Why is gene therapy research so tightly regulated?
- What is the main ethical concern associated with germ-line gene therapy?

---

12 Zhen Wang and Yi Sun. "Targeting p53 for Novel Anticancer Therapy." *Translational Oncology 3*, no. 1 (2010): 1–12.
13 Erika Check. "Gene Therapy: A Tragic Setback." *Nature 420* no. 6912 (2002): 116–118.

SUMMARY

12.1 Microbes and the Tools of Genetic Engineering

- **Biotechology** is the science of utilizing living systems to benefit humankind. In recent years, the ability to directly alter an organism's genome through **genetic engineering** has been made possible due to advances in **recombinant DNA technology,** which allows researchers to create **recombinant DNA molecules** with new combinations of genetic material.
- **Molecular cloning** involves methods used to construct recombinant DNA and facilitate their replication in host organisms. These methods include the use of **restriction enzymes** (to cut both foreign DNA and **plasmid vectors),** **ligation** (to paste fragments of DNA together), and the introduction of recombinant DNA into a host organism (often bacteria).
- **Blue-white screening** allows selection of bacterial transformants that contain recombinant plasmids using the phenotype of a **reporter gene** that is disabled by insertion of the DNA fragment.
- **Genomic libraries** can be made by cloning genomic fragments from one organism into plasmid vectors or into bacteriophage.
- **cDNA libraries** can be generated to represent the mRNA molecules expressed in a cell at a given point.
- **Transfection** of eukaryotic hosts can be achieved through various methods using **electroporation, gene guns, microinjection, shuttle vectors**, and **viral vectors**.

12.2 Visualizing and Characterizing DNA, RNA, and Protein

- Finding a gene of interest within a sample requires the use of a single-stranded **DNA probe** labeled with a molecular beacon (typically radioactivity or fluorescence) that can hybridize with a complementary single-stranded nucleic acid in the sample.
- **Agarose gel electrophoresis** allows for the separation of DNA molecules based on size.
- **Restriction fragment length polymorphism (RFLP)** analysis allows for the visualization by agarose gel electrophoresis of distinct variants of a DNA sequence caused by differences in restriction sites.
- **Southern blot** analysis allows researchers to find a particular DNA sequence within a sample whereas **northern blot** analysis allows researchers to detect a particular mRNA sequence expressed in a sample.
- **Microarray technology** is a nucleic acid hybridization technique that allows for the examination of many thousands of genes at once to find differences in genes or gene expression patterns between two samples of genomic DNA or cDNA,
- **Polyacrylamide gel electrophoresis (PAGE)** allows for the separation of proteins by size, especially if native protein charges are masked through pretreatment with SDS.
- **Polymerase chain reaction** allows for the rapid amplification of a specific DNA sequence. Variations of PCR can be used to detect mRNA expression (**reverse transcriptase PCR**) or to quantify a particular sequence in the original sample (**real-time PCR**).
- Although the development of **Sanger DNA sequencing** was revolutionary, advances in **next generation sequencing** allow for the rapid and inexpensive sequencing of the genomes of many organisms, accelerating the volume of new sequence data.

12.3 Whole Genome Methods and Pharmaceutical Applications of Genetic Engineering

- The science of **genomics** allows researchers to study organisms on a holistic level and has many applications of medical relevance.
- **Transcriptomics** and **proteomics** allow researchers to compare gene expression patterns between different cells and shows great promise in better understanding global responses to various conditions.
- The various –omics technologies complement each other and together provide a more complete picture of an organism's or microbial community's (**metagenomics**) state.
- The analysis required for large data sets produced through genomics, transcriptomics, and **proteomics** has led to the emergence of **bioinformatics**.
- **Reporter genes** encoding easily observable characteristics are commonly used to track gene expression patterns of genes of unknown function.
- The use of recombinant DNA technology has revolutionized the pharmaceutical industry,

allowing for the rapid production of high-quality **recombinant DNA pharmaceuticals** used to treat a wide variety of human conditions.
- **RNA interference** technology has great promise as a method of treating viral infections by silencing the expression of specific genes

12.4 Gene Therapy

- While gene therapy shows great promise for the treatment of genetic diseases, there are also significant risks involved.
- There is considerable federal and local regulation of the development of gene therapies by pharmaceutical companies for use in humans.
- Before gene therapy use can increase dramatically, there are many ethical issues that need to be addressed by the medical and research communities, politicians, and society at large.

REVIEW QUESTIONS

Multiple Choice

1. Which of the following is required for repairing the phosphodiester backbone of DNA during molecular cloning?
   a. cDNA
   b. reverse transcriptase
   c. restriction enzymes
   d. DNA ligase

2. All of the following are processes used to introduce DNA molecules into bacterial cells *except*:
   a. transformation
   b. transduction
   c. transcription
   d. conjugation

3. The enzyme that uses RNA as a template to produce a DNA copy is called:
   a. a restriction enzyme
   b. DNA ligase
   c. reverse transcriptase
   d. DNA polymerase

4. In blue-white screening, what do blue colonies represent?
   a. cells that have not taken up the plasmid vector
   b. cells with recombinant plasmids containing a new insert
   c. cells containing empty plasmid vectors
   d. cells with a non-functional *lacZ* gene

5. The T_i plasmid is used for introducing genes into:
   a. animal cells
   b. plant cells
   c. bacteriophages
   d. *E. coli* cells

6. Which technique is used to separate protein fragments based on size?
   a. polyacrylamide gel electrophoresis
   b. Southern blot
   c. agarose gel electrophoresis
   d. polymerase chain reaction

7. Which technique uses restriction enzyme digestion followed by agarose gel electrophoresis to generate a banding pattern for comparison to another sample processed in the same way?
   a. qPCR
   b. RT-PCR
   c. RFLP
   d. 454 sequencing

8. All of the following techniques involve hybridization between single-stranded nucleic acid molecules *except*:
   a. Southern blot analysis
   b. RFLP analysis
   c. northern blot analysis
   d. microarray analysis

9. The science of studying the entire collection of mRNA molecules produced by cells, allowing scientists to monitor differences in gene expression patterns between cells, is called:
   a. genomics
   b. transcriptomics
   c. proteomics
   d. pharmacogenomics

10. The science of studying genomic fragments from microbial communities, allowing researchers to study genes from a collection of multiple species, is called:
    a. pharmacogenomics
    b. transcriptomics
    c. metagenomics
    d. proteomics

11. The insulin produced by recombinant DNA technology is
    a. a combination of *E. coli* and human insulin.
    b. identical to human insulin produced in the pancreas.
    c. cheaper but less effective than pig insulin for treating diabetes.
    d. engineered to be more effective than human insulin.

12. At what point can the FDA halt the development or use of gene therapy?
    a. on submission of an IND application
    b. during clinical trials
    c. after manufacturing and marketing of the approved therapy
    d. all of the answers are correct

True/False

13. Recombination is a process not usually observed in nature.
14. It is generally easier to introduce recombinant DNA into prokaryotic cells than into eukaryotic cells.
15. In agarose gel electrophoresis, DNA will be attracted to the negative electrode.
16. RNA interference does not influence the sequence of genomic DNA.

Fill in the Blank

17. The process of introducing DNA molecules into eukaryotic cells is called _____.
18. The _____ blot technique is used to find an RNA fragment within a sample that is complementary to a DNA probe.
19. The PCR step during which the double-stranded template molecule becomes single-stranded is called _____.
20. The sequencing method involving the incorporation of ddNTPs is called _____.
21. The application of genomics to evaluate the effectiveness and safety of drugs on the basis of information from an individual's genomic sequence is called _____.
22. A gene whose expression can be easily visualized and monitored is called a _____.
23. _____ is a common viral vector used in gene therapy for introducing a new gene into a specifically targeted cell type.

Short Answer

24. Name three elements incorporated into a plasmid vector for efficient cloning.
25. When would a scientist want to generate a cDNA library instead of a genomic library?
26. What is one advantage of generating a genomic library using phages instead of plasmids?
27. Why is it important that a DNA probe be labeled with a molecular beacon?
28. When separating proteins strictly by size, why is exposure to SDS first required?
29. Why must the DNA polymerase used during PCR be heat-stable?
30. If all cellular proteins are encoded by the cell's genes, what information does proteomics provide that genomics cannot?
31. Briefly describe the risks associated with somatic cell gene therapy.

Critical Thinking

32. Is biotechnology always associated with genetic engineering? Explain your answer.
33. Which is more efficient: blunt-end cloning or sticky-end cloning? Why?
34. Suppose you are working in a molecular biology laboratory and are having difficulty performing the PCR successfully. You decide to double-check the PCR protocol programmed into the thermal cycler and discover that the annealing temperature was programmed to be 65 °C instead of 50 °C, as you had intended. What effects would this mistake have on the PCR reaction? Refer to Figure 12.20.

35. What is the advantage of microarray analysis over northern blot analysis in monitoring changes in gene expression?
36. What is the difference between reverse transcriptase PCR (RT-PCR) and real-time quantitative PCR (qPCR)?
37. What are some advantages of cloning human genes into bacteria to treat human diseases caused by specific protein deficiencies?
38. Compare the ethical issues involved in the use of somatic cell gene therapy and germ-line gene therapy.

CHAPTER 13
Control of Microbial Growth

Location	Average number CFUs per 6.5 × 6.5 cm area
Door latch	256
Door lock	14
Door lock control	182
Door handle	29
Window control	4
Cruise control button	69
Steering wheel	239
Interior steering wheel	390
Radio volume knob	99
Gear shifter	115
Center console	506

Figure 13.1 Most environments, including cars, are not sterile. A study[1] analyzed 11 locations within 18 different cars to determine the number of microbial colony-forming units (CFUs) present. The center console harbored by far the most microbes (506 CFUs), possibly because that is where drinks are placed (and often spilled). Frequently touched sites also had high concentrations. (credit "photo": modification of work by Jeff Wilcox)

Chapter Outline

13.1 Controlling Microbial Growth

13.2 Using Physical Methods to Control Microorganisms

13.3 Using Chemicals to Control Microorganisms

13.4 Testing the Effectiveness of Antiseptics and Disinfectants

INTRODUCTION How clean is clean? People wash their cars and vacuum the carpets, but most would not want to eat from these surfaces. Similarly, we might eat with silverware cleaned in a dishwasher, but we could not use the same dishwasher to clean surgical instruments. As these examples illustrate, "clean" is a relative term. Car washing, vacuuming, and dishwashing all reduce the microbial load on the items treated, thus making them "cleaner." But whether they are "clean enough" depends on their intended use. Because people do not normally eat from cars or carpets, these items do not require the same level of cleanliness that silverware does. Likewise, because silverware is not used for invasive surgery, these utensils do not require the same level of cleanliness as surgical equipment, which requires sterilization to prevent infection.

Why not play it safe and sterilize everything? Sterilizing everything we come in contact with is impractical, as well as potentially dangerous. As this chapter will demonstrate, sterilization protocols often require time- and labor-intensive treatments that may degrade the quality of the item being treated or have toxic effects on users. Therefore, the user must consider the item's intended application when choosing a cleaning method to ensure that it is "clean enough."

---

1 R.E. Stephenson et al. "Elucidation of Bacteria Found in Car Interiors and Strategies to Reduce the Presence of Potential Pathogens." *Biofouling* 30 no. 3 (2014):337–346.

13.1 Controlling Microbial Growth

Learning Objectives

By the end of this section, you will be able to:
- Compare disinfectants, antiseptics, and sterilants
- Describe the principles of controlling the presence of microorganisms through sterilization and disinfection
- Differentiate between microorganisms of various biological safety levels and explain methods used for handling microbes at each level

---

Clinical Focus

Part 1

Roberta is a 46-year-old real estate agent who recently underwent a cholecystectomy (surgery to remove painful gallstones). The surgery was performed laparoscopically with the aid of a duodenoscope, a specialized endoscope that allows surgeons to see inside the body with the aid of a tiny camera. On returning home from the hospital, Roberta developed abdominal pain and a high fever. She also experienced a burning sensation during urination and noticed blood in her urine. She notified her surgeon of these symptoms, per her postoperative instructions.

- What are some possible causes of Roberta's symptoms?

Jump to the next Clinical Focus box.

To prevent the spread of human disease, it is necessary to control the growth and abundance of microbes in or on various items frequently used by humans. Inanimate items, such as doorknobs, toys, or towels, which may harbor microbes and aid in disease transmission, are called **fomites**. Two factors heavily influence the level of cleanliness required for a particular fomite and, hence, the protocol chosen to achieve this level. The first factor is the application for which the item will be used. For example, invasive applications that require insertion into the human body require a much higher level of cleanliness than applications that do not. The second factor is the level of resistance to antimicrobial treatment by potential pathogens. For example, foods preserved by canning often become contaminated with the bacterium *Clostridium botulinum*, which produces the neurotoxin that causes botulism. Because *C. botulinum* can produce endospores that can survive harsh conditions, extreme temperatures and pressures must be used to eliminate the endospores. Other organisms may not require such extreme measures and can be controlled by a procedure such as washing clothes in a laundry machine.

Laboratory Biological Safety Levels

For researchers or laboratory personnel working with pathogens, the risks associated with specific pathogens determine the levels of cleanliness and control required. The Centers for Disease Control and Prevention (CDC) and the National Institutes of Health (NIH) have established four classification levels, called "biological safety levels" (BSLs). Various organizations around the world, including the World Health Organization (WHO) and the European Union (EU), use a similar classification scheme. According to the CDC, the BSL is determined by the agent's infectivity, ease of transmission, and potential disease severity, as well as the type of work being done with the agent.[2]

Each BSL requires a different level of biocontainment to prevent contamination and spread of infectious agents to laboratory personnel and, ultimately, the community. For example, the lowest BSL, BSL-1, requires the fewest precautions because it applies to situations with the lowest risk for microbial infection.

BSL-1 agents are those that generally do not cause infection in healthy human adults. These include noninfectious bacteria, such as nonpathogenic strains of *Escherichia coli* and *Bacillus subtilis*, and viruses

---

[2] US Centers for Disease Control and Prevention. "Recognizing the Biosafety Levels." http://www.cdc.gov/training/quicklearns/biosafety/. Accessed June 7, 2016.

known to infect animals other than humans, such as baculoviruses (insect viruses). Because working with BSL-1 agents poses very little risk, few precautions are necessary. Laboratory workers use standard aseptic technique and may work with these agents at an open laboratory bench or table, wearing personal protective equipment (PPE) such as a laboratory coat, goggles, and gloves, as needed. Other than a sink for handwashing and doors to separate the laboratory from the rest of the building, no additional modifications are needed.

Agents classified as BSL-2 include those that pose moderate risk to laboratory workers and the community, and are typically "indigenous," meaning that they are commonly found in that geographical area. These include bacteria such as *Staphylococcus aureus* and *Salmonella* spp., and viruses like hepatitis, mumps, and measles viruses. BSL-2 laboratories require additional precautions beyond those of BSL-1, including restricted access; required PPE, including a face shield in some circumstances; and the use of biological safety cabinets for procedures that may disperse agents through the air (called "aerosolization"). BSL-2 laboratories are equipped with self-closing doors, an eyewash station, and an **autoclave**, which is a specialized device for sterilizing materials with pressurized steam before use or disposal. BSL-1 laboratories may also have an autoclave.

BSL-3 agents have the potential to cause lethal infections by inhalation. These may be either indigenous or "exotic," meaning that they are derived from a foreign location, and include pathogens such as *Mycobacterium tuberculosis*, *Bacillus anthracis*, West Nile virus, and human immunodeficiency virus (HIV). Because of the serious nature of the infections caused by BSL-3 agents, laboratories working with them require restricted access. Laboratory workers are under medical surveillance, possibly receiving vaccinations for the microbes with which they work. In addition to the standard PPE already mentioned, laboratory personnel in BSL-3 laboratories must also wear a respirator and work with microbes and infectious agents in a biological safety cabinet at all times. BSL-3 laboratories require a hands-free sink, an eyewash station near the exit, and two sets of self-closing and locking doors at the entrance. These laboratories are equipped with directional airflow, meaning that clean air is pulled through the laboratory from clean areas to potentially contaminated areas. This air cannot be recirculated, so a constant supply of clean air is required.

BSL-4 agents are the most dangerous and often fatal. These microbes are typically exotic, are easily transmitted by inhalation, and cause infections for which there are no treatments or vaccinations. Examples include Ebola virus and Marburg virus, both of which cause hemorrhagic fevers, and smallpox virus. There are only a small number of laboratories in the United States and around the world appropriately equipped to work with these agents. In addition to BSL-3 precautions, laboratory workers in BSL-4 facilities must also change their clothing on entering the laboratory, shower on exiting, and decontaminate all material on exiting. While working in the laboratory, they must either wear a full-body protective suit with a designated air supply or conduct all work within a biological safety cabinet with a high-efficiency particulate air (HEPA)-filtered air supply and a doubly HEPA-filtered exhaust. If wearing a suit, the air pressure within the suit must be higher than that outside the suit, so that if a leak in the suit occurs, laboratory air that may be contaminated cannot be drawn into the suit (Figure 13.2). The laboratory itself must be located either in a separate building or in an isolated portion of a building and have its own air supply and exhaust system, as well as its own decontamination system. The BSLs are summarized in Figure 13.3.

Figure 13.2 A protective suit like this one is an additional precaution for those who work in BSL-4 laboratories. This suit has its own air supply and maintains a positive pressure relative to the outside, so that if a leak occurs, air will flow out of the suit, not into it from the laboratory. (Credit: James Gathany, CDC, public domain)

Biosafety Levels			
Biological Safety Levels	Description	Examples	CDC Classification
BSL-4	Microbes are dangerous and exotic, posing a high risk of aerosol-transmitted infections, which are frequently fatal without treatment or vaccines. Few labs are at this level.	Ebola and Marburg viruses	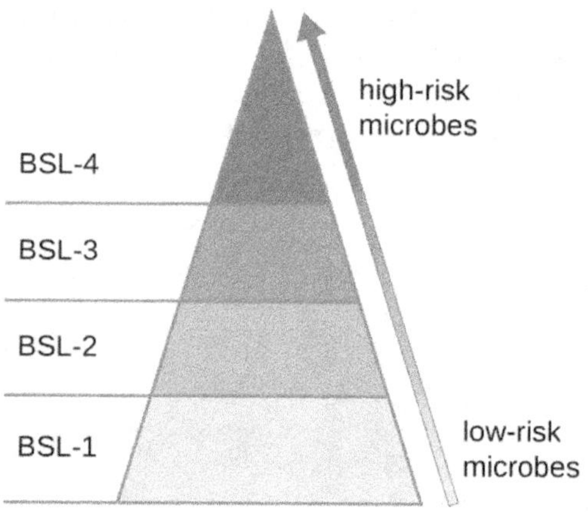 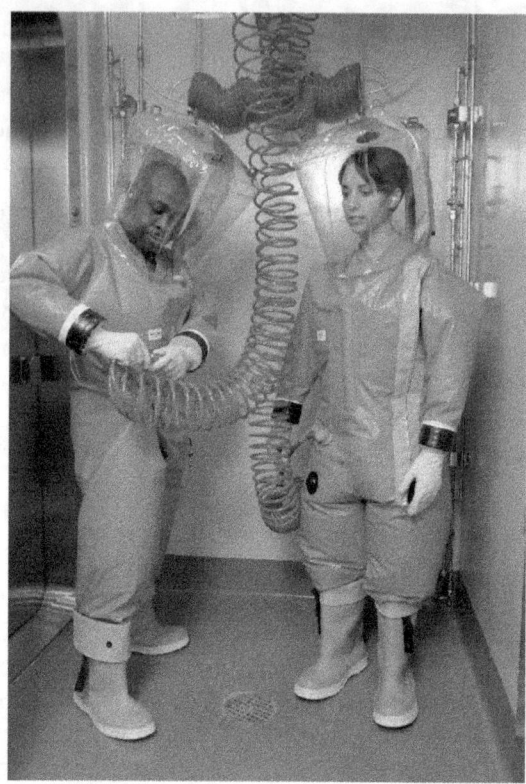
BSL-3	Microbes are indigenous or exotic and cause serious or potentially lethal diseases through respiratory transmission.	*Mycobacterium tuberculosis*	
BSL-2	Microbes are typically indigenous and are associated with diseases of varying severity. They pose moderate risk to workers and the environment.	*Staphylococcus aureus*	
BSL-1	Microbes are not known to cause disease in healthy hosts and pose minimal risk to workers and the environment.	Nonpathogenic strains of *Escherichia coli*	

Figure 13.3 The CDC classifies infectious agents into four biosafety levels based on potential risk to laboratory personnel and the

community. Each level requires a progressively greater level of precaution. (credit "pyramid": modification of work by Centers for Disease Control and Prevention)

LINK TO LEARNING

To learn more (https://openstax.org/l/22cdcfourbsls) about the four BSLs, visit the CDC's website.

CHECK YOUR UNDERSTANDING

- What are some factors used to determine the BSL necessary for working with a specific pathogen?

Sterilization

The most extreme protocols for microbial control aim to achieve **sterilization**: the complete removal or killing of all vegetative cells, endospores, and viruses from the targeted item or environment. Sterilization protocols are generally reserved for laboratory, medical, manufacturing, and food industry settings, where it may be imperative for certain items to be completely free of potentially infectious agents. Sterilization can be accomplished through either physical means, such as exposure to high heat, pressure, or filtration through an appropriate filter, or by chemical means. Chemicals that can be used to achieve sterilization are called **sterilants**. Sterilants effectively kill all microbes and viruses, and, with appropriate exposure time, can also kill endospores.

For many clinical purposes, **aseptic technique** is necessary to prevent contamination of sterile surfaces. Aseptic technique involves a combination of protocols that collectively maintain sterility, or **asepsis**, thus preventing contamination of the patient with microbes and infectious agents. Failure to practice aseptic technique during many types of clinical procedures may introduce microbes to the patient's body and put the patient at risk for **sepsis**, a systemic inflammatory response to an infection that results in high fever, increased heart and respiratory rates, shock, and, possibly, death. Medical procedures that carry risk of contamination must be performed in a **sterile field**, a designated area that is kept free of all vegetative microbes, endospores, and viruses. Sterile fields are created according to protocols requiring the use of sterilized materials, such as packaging and drapings, and strict procedures for washing and application of sterilants. Other protocols are followed to maintain the sterile field while the medical procedure is being performed.

One food sterilization protocol, **commercial sterilization**, uses heat at a temperature low enough to preserve food quality but high enough to destroy common pathogens responsible for food poisoning, such as *C. botulinum*. Because *C. botulinum* and its endospores are commonly found in soil, they may easily contaminate crops during harvesting, and these endospores can later germinate within the anaerobic environment once foods are canned. Metal cans of food contaminated with *C. botulinum* will bulge due to the microbe's production of gases; contaminated jars of food typically bulge at the metal lid. To eliminate the risk for *C. botulinum* contamination, commercial food-canning protocols are designed with a large margin of error. They assume an impossibly large population of endospores (10^{12} per can) and aim to reduce this population to 1 endospore per can to ensure the safety of canned foods. For example, low- and medium-acid foods are heated to 121 °C for a minimum of 2.52 minutes, which is the time it would take to reduce a population of 10^{12} endospores per can down to 1 endospore at this temperature. Even so, commercial sterilization does not eliminate the presence of all microbes; rather, it targets those pathogens that cause spoilage and foodborne diseases, while allowing many nonpathogenic organisms to survive. Therefore, "sterilization" is somewhat of a misnomer in this context, and commercial sterilization may be more accurately described as "quasi-sterilization."

CHECK YOUR UNDERSTANDING

- What is the difference between sterilization and aseptic technique?

LINK TO LEARNING

The Association of Surgical Technologists publishes standards (https://openstax.org/l/22ASTstanasepte) for aseptic technique, including creating and maintaining a sterile field.

Other Methods of Control

Sterilization protocols require procedures that are not practical, or necessary, in many settings. Various other methods are used in clinical and nonclinical settings to reduce the microbial load on items. Although the terms for these methods are often used interchangeably, there are important distinctions (Figure 13.4).

The process of **disinfection** inactivates most microbes on the surface of a fomite by using antimicrobial chemicals or heat. Because some microbes remain, the disinfected item is not considered sterile. Ideally, **disinfectant**s should be fast acting, stable, easy to prepare, inexpensive, and easy to use. An example of a natural disinfectant is vinegar; its acidity kills most microbes. Chemical disinfectants, such as chlorine bleach or products containing chlorine, are used to clean nonliving surfaces such as laboratory benches, clinical surfaces, and bathroom sinks. Typical disinfection does not lead to sterilization because endospores tend to survive even when all vegetative cells have been killed.

Unlike disinfectants, **antiseptic**s are antimicrobial chemicals safe for use on living skin or tissues. Examples of antiseptics include hydrogen peroxide and isopropyl alcohol. The process of applying an antiseptic is called **antisepsis**. In addition to the characteristics of a good disinfectant, antiseptics must also be selectively effective against microorganisms and able to penetrate tissue deeply without causing tissue damage.

The type of protocol required to achieve the desired level of cleanliness depends on the particular item to be cleaned. For example, those used clinically are categorized as critical, semicritical, and noncritical. Critical items must be sterile because they will be used inside the body, often penetrating sterile tissues or the bloodstream; examples of **critical item**s include surgical instruments, catheters, and intravenous fluids. Gastrointestinal endoscopes and various types of equipment for respiratory therapies are examples of **semicritical item**s; they may contact mucous membranes or nonintact skin but do not penetrate tissues. Semicritical items do not typically need to be sterilized but do require a high level of disinfection. Items that may contact but not penetrate intact skin are **noncritical item**s; examples are bed linens, furniture, crutches, stethoscopes, and blood pressure cuffs. These articles need to be clean but not highly disinfected.

The act of handwashing is an example of **degerming**, in which microbial numbers are significantly reduced by gently scrubbing living tissue, most commonly skin, with a mild chemical (e.g., soap) to avoid the transmission of pathogenic microbes. Wiping the skin with an alcohol swab at an injection site is another example of degerming. These degerming methods remove most (but not all) microbes from the skin's surface.

The term **sanitization** refers to the cleansing of fomites to remove enough microbes to achieve levels deemed safe for public health. For example, commercial dishwashers used in the food service industry typically use very hot water and air for washing and drying; the high temperatures kill most microbes, sanitizing the dishes. Surfaces in hospital rooms are commonly sanitized using a chemical disinfectant to prevent disease transmission between patients. Figure 13.4 summarizes common protocols, definitions, applications, and agents used to control microbial growth.

\multicolumn{4}{c}{**Common Protocols for Control of Microbial Growth**}
Protocol	**Definition**	**Common Application**	**Common Agents**
For Use on Fomites			
Disinfection	Reduces or destroys microbial load of an inanimate item through application of heat or antimicrobial chemicals	Cleaning surfaces like laboratory benches, clinical surfaces, and bathrooms	Chlorine bleach, phenols (e.g., Lysol), glutaraldehyde
Sanitization	Reduces microbial load of an inanimate item to safe public health levels through application of heat or antimicrobial chemicals	Commercial dishwashing of eating utensils, cleaning public restrooms	Detergents containing phosphates (e.g., Finish), industrial-strength cleaners containing quaternary ammonium compounds
Sterilization	Completely eliminates all vegetative cells, endospores, and viruses from an inanimate item	Preparation of surgical equipment and of needles used for injection	Pressurized steam (autoclave), chemicals, radiation
For Use on Living Tissue			
Antisepsis	Reduces microbial load on skin or tissue through application of an antimicrobial chemical	Cleaning skin broken due to injury; cleaning skin before surgery	Boric acid, isopropyl alcohol, hydrogen peroxide, iodine (betadine)
Degerming	Reduces microbial load on skin or tissue through gentle to firm scrubbing and the use of mild chemicals	Handwashing	Soap, alcohol swab

Figure 13.4

CHECK YOUR UNDERSTANDING

- What is the difference between a disinfectant and an antiseptic?
- Which is most effective at removing microbes from a product: sanitization, degerming, or sterilization? Explain.

Clinical Focus

Part 2

Roberta's physician suspected that a bacterial infection was responsible for her sudden-onset high fever, abdominal pain, and bloody urine. Based on these symptoms, the physician diagnosed a urinary tract infection (UTI). A wide variety of bacteria may cause UTIs, which typically occur when bacteria from the lower gastrointestinal tract are introduced to the urinary tract. However, Roberta's recent gallstone surgery caused the physician to suspect that she had contracted a nosocomial (hospital-acquired) infection during her surgery. The physician took a urine sample and ordered a urine culture to check for the presence of white blood cells, red blood cells, and bacteria. The results of this test would help determine the cause of the infection. The physician also prescribed a course of the antibiotic ciprofloxacin, confident that it would clear Roberta's infection.

- What are some possible ways that bacteria could have been introduced to Roberta's urinary tract during her surgery?

Jump to the next Clinical Focus box. Go back to the previous Clinical Focus box.

Measuring Microbial Control

Physical and chemical methods of microbial control that kill the targeted microorganism are identified by the suffix *-cide* (or *-cidal*). The prefix indicates the type of microbe or infectious agent killed by the treatment method: **bactericides** kill bacteria, **viricides** kill or inactivate viruses, and **fungicides** kill fungi. Other methods do not kill organisms but, instead, stop their growth, making their population static; such methods are identified by the suffix *-stat* (or *-static*). For example, **bacteriostatic** treatments inhibit the growth of bacteria, whereas **fungistatic** treatments inhibit the growth of fungi. Factors that determine whether a particular treatment is *-cidal* or *-static* include the types of microorganisms targeted, the concentration of the chemical used, and the nature of the treatment applied.

Although *-static* treatments do not actually kill infectious agents, they are often less toxic to humans and other animals, and may also better preserve the integrity of the item treated. Such treatments are typically sufficient to keep the microbial population of an item in check. The reduced toxicity of some of these *-static* chemicals also allows them to be impregnated safely into plastics to prevent the growth of microbes on these surfaces. Such plastics are used in products such as toys for children and cutting boards for food preparation. When used to treat an infection, *-static* treatments are typically sufficient in an otherwise healthy individual, preventing the pathogen from multiplying, thus allowing the individual's immune system to clear the infection.

The degree of microbial control can be evaluated using a **microbial death curve** to describe the progress and effectiveness of a particular protocol. When exposed to a particular microbial control protocol, a fixed percentage of the microbes within the population will die. Because the rate of killing remains constant even when the population size varies, the percentage killed is more useful information than the absolute number of microbes killed. Death curves are often plotted as semilog plots just like microbial growth curves because the reduction in microorganisms is typically logarithmic (Figure 13.5). The amount of time it takes for a specific protocol to produce a one order-of-magnitude decrease in the number of organisms, or the death of 90% of the population, is called the **decimal reduction time (DRT)** or **D-value**.

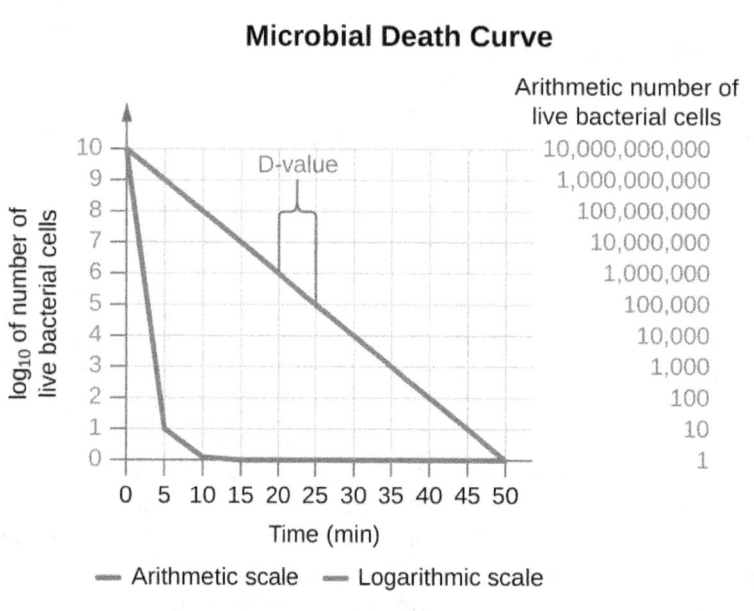

Time (min)	Number of cells
0	10^{10}
5	10^{9}
10	10^{8}
15	10^{7}
20	10^{6}
25	10^{5}
30	10^{4}
35	10^{3}
40	10^{2}
45	10^{1}
50	10^{0}

Figure 13.5 Microbial death is logarithmic and easily observed using a semilog plot instead of an arithmetic one. The decimal reduction time (D-value) is the time it takes to kill 90% of the population (a 1-log decrease in the total population) when exposed to a specific microbial control protocol, as indicated by the purple bracket.

Several factors contribute to the effectiveness of a disinfecting agent or microbial control protocol. First, as demonstrated in Figure 13.5, the length of time of exposure is important. Longer exposure times kill more microbes. Because microbial death of a population exposed to a specific protocol is logarithmic, it takes longer to kill a high-population load than a low-population load exposed to the same protocol. A shorter treatment time (measured in multiples of the D-value) is needed when starting with a smaller number of organisms. Effectiveness also depends on the susceptibility of the agent to that disinfecting agent or protocol. The concentration of disinfecting agent or intensity of exposure is also important. For example, higher temperatures and higher concentrations of disinfectants kill microbes more quickly and effectively. Conditions that limit contact between the agent and the targeted cells cells—for example, the presence of bodily fluids, tissue, organic debris (e.g., mud or feces), or biofilms on surfaces—increase the cleaning time or intensity of the microbial control protocol required to reach the desired level of cleanliness. All these factors must be considered when choosing the appropriate protocol to control microbial growth in a given situation.

✓ CHECK YOUR UNDERSTANDING

- What are two possible reasons for choosing a bacteriostatic treatment over a bactericidal one?
- Name at least two factors that can compromise the effectiveness of a disinfecting agent.

13.2 Using Physical Methods to Control Microorganisms

Learning Objectives

By the end of this section, you will be able to:

- Understand and compare various physical methods of controlling microbial growth, including heating, refrigeration, freezing, high-pressure treatment, desiccation, lyophilization, irradiation, and filtration

For thousands of years, humans have used various physical methods of microbial control for food preservation. Common control methods include the application of high temperatures, radiation, filtration, and desiccation (drying), among others. Many of these methods nonspecifically kill cells by disrupting membranes, changing membrane permeability, or damaging proteins and nucleic acids by denaturation, degradation, or chemical modification. Various physical methods used for microbial control are described in this section.

Heat

Heating is one of the most common—and oldest—forms of microbial control. It is used in simple techniques like cooking and canning. Heat can kill microbes by altering their membranes and denaturing proteins. The **thermal death point (TDP)** of a microorganism is the lowest temperature at which all microbes are killed in a 10-minute exposure. Different microorganisms will respond differently to high temperatures, with some (e.g., endospore-formers such as *C. botulinum*) being more heat tolerant. A similar parameter, the **thermal death time (TDT)**, is the length of time needed to kill all microorganisms in a sample at a given temperature. These parameters are often used to describe sterilization procedures that use high heat, such as autoclaving. Boiling is one of the oldest methods of moist-heat control of microbes, and it is typically quite effective at killing vegetative cells and some viruses. However, boiling is less effective at killing endospores; some endospores are able to survive up to 20 hours of boiling. Additionally, boiling may be less effective at higher altitudes, where the boiling point of water is lower and the boiling time needed to kill microbes is therefore longer. For these reasons, boiling is not considered a useful sterilization technique in the laboratory or clinical setting.

Many different heating protocols can be used for sterilization in the laboratory or clinic, and these protocols can be broken down into two main categories: **dry-heat sterilization** and **moist-heat sterilization**. Aseptic technique in the laboratory typically involves some dry-heat sterilization protocols using direct application of high heat, such as sterilizing inoculating loops (Figure 13.6). Incineration at very high temperatures destroys all microorganisms. Dry heat can also be applied for relatively long periods of time (at least 2 hours) at temperatures up to 170 °C by using a dry-heat sterilizer, such as an oven. However, moist-heat sterilization is typically the more effective protocol because it penetrates cells better than dry heat does.

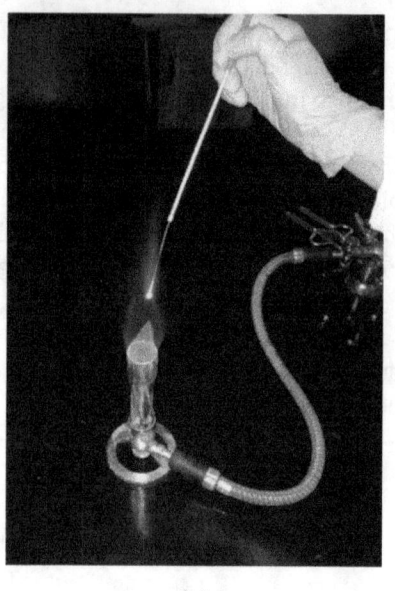 

Figure 13.6 (a) Sterilizing a loop, often referred to as "flaming a loop," is a common component of aseptic technique in the microbiology laboratory and is used to incinerate any microorganisms on the loop. (b) Alternatively, a bactericinerator may be used to reduce aerosolization of microbes and remove the presence of an open flame in the laboratory. These are examples of dry-heat sterilization by the direct application of high heat capable of incineration. (credit a: modification of work by Anh-Hue Tu; credit b: modification of work by Brian Forster)

Autoclaves

Autoclaves rely on moist-heat sterilization. They are used to raise temperatures above the boiling point of water to sterilize items such as surgical equipment from vegetative cells, viruses, and especially endospores, which are known to survive boiling temperatures, without damaging the items. Charles Chamberland (1851–1908) designed the modern autoclave in 1879 while working in the laboratory of Louis Pasteur. The autoclave is still considered the most effective method of sterilization (Figure 13.7). Outside laboratory and clinical settings, large industrial autoclaves called **retorts** allow for moist-heat sterilization on a large scale.

In general, the air in the chamber of an autoclave is removed and replaced with increasing amounts of steam trapped within the enclosed chamber, resulting in increased interior pressure and temperatures above the boiling point of water. The two main types of autoclaves differ in the way that air is removed from the chamber. In gravity displacement autoclaves, steam is introduced into the chamber from the top or sides. Air, which is heavier than steam, sinks to the bottom of the chamber, where it is forced out through a vent. Complete displacement of air is difficult, especially in larger loads, so longer cycles may be required for such loads. In prevacuum sterilizers, air is removed completely using a high-speed vacuum before introducing steam into the chamber. Because air is more completely eliminated, the steam can more easily penetrate wrapped items. Many autoclaves are capable of both gravity and prevacuum cycles, using the former for the decontamination of waste and sterilization of media and unwrapped glassware, and the latter for sterilization of packaged instruments.

Figure 13.7 A technician sterilizes a sample using an autoclave. (Credit: Martha Cooper / Picryl; Public Domain.

Standard operating temperatures for autoclaves are 121 °C or, in some cases, 132 °C, typically at a pressure of 15 to 20 pounds per square inch (psi). The length of exposure depends on the volume and nature of material being sterilized, but it is typically 20 minutes or more, with larger volumes requiring longer exposure times to ensure sufficient heat transfer to the materials being sterilized. The steam must directly contact the liquids or dry materials being sterilized, so containers are left loosely closed and instruments are loosely wrapped in paper or foil. The key to autoclaving is that the temperature must be high enough to kill endospores to achieve complete sterilization.

Because sterilization is so important to safe medical and laboratory protocols, quality control is essential. Autoclaves may be equipped with recorders to document the pressures and temperatures achieved during each run. Additionally, internal indicators of various types should be autoclaved along with the materials to be sterilized to ensure that the proper sterilization temperature has been reached (Figure 13.8). One common type of indicator is the use of heat-sensitive autoclave tape, which has white stripes that turn black when the appropriate temperature is achieved during a successful autoclave run. This type of indicator is relatively inexpensive and can be used during every run. However, autoclave tape provides no indication of length of exposure, so it cannot be used as an indicator of sterility. Another type of indicator, a biological indicator spore test, uses either a strip of paper or a liquid suspension of the endospores of *Geobacillus stearothermophilus* to determine whether the endospores are killed by the process. The endospores of the obligate thermophilic bacterium *G. stearothermophilus* are the gold standard used for this purpose because of their extreme heat resistance. Biological spore indicators can also be used to test the effectiveness of other sterilization protocols, including ethylene oxide, dry heat, formaldehyde, gamma radiation, and hydrogen peroxide plasma sterilization using either *G. stearothermophilus, Bacillus atrophaeus, B. subtilis,* or *B. pumilus* spores. In the case of validating autoclave function, the endospores are incubated after autoclaving to ensure no viable endospores remain. Bacterial growth subsequent to endospore germination can be monitored by biological indicator spore tests that detect acid metabolites or fluorescence produced by enzymes derived from viable *G. stearothermophilus*. A third type of autoclave indicator is the Diack tube, a glass ampule containing a temperature-sensitive pellet that melts at the proper sterilization temperature. Spore strips or Diack tubes are used periodically to ensure the autoclave is functioning properly.

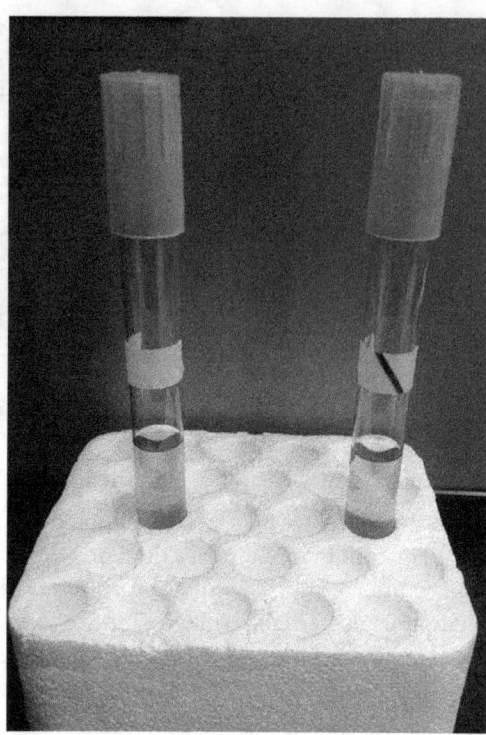

Figure 13.8 The white strips on autoclave tape (left tube) turn dark during a successful autoclave run (right tube). (credit: modification of work by Brian Forster)

Pasteurization

Although complete sterilization is ideal for many medical applications, it is not always practical for other applications and may also alter the quality of the product. Boiling and autoclaving are not ideal ways to control microbial growth in many foods because these methods may ruin the consistency and other organoleptic (sensory) qualities of the food. Pasteurization is a form of microbial control for food that uses heat but does not render the food sterile. Traditional **pasteurization** kills pathogens and reduces the number of spoilage-causing microbes while maintaining food quality. The process of pasteurization was first developed by Louis Pasteur in the 1860s as a method for preventing the spoilage of beer and wine. Today, pasteurization is most commonly used to kill heat-sensitive pathogens in milk and other food products (e.g., apple juice and honey) (Figure 13.9). However, because pasteurized food products are not sterile, they will eventually spoil.

The methods used for milk pasteurization balance the temperature and the length of time of treatment. One method, **high-temperature short-time (HTST) pasteurization**, exposes milk to a temperature of 72 °C for 15 seconds, which lowers bacterial numbers while preserving the quality of the milk. An alternative is **ultra-high-temperature (UHT) pasteurization**, in which the milk is exposed to a temperature of 138 °C for 2 or more seconds. UHT pasteurized milk can be stored for a long time in sealed containers without being refrigerated; however, the very high temperatures alter the proteins in the milk, causing slight changes in the taste and smell. Still, this method of pasteurization is advantageous in regions where access to refrigeration is limited.

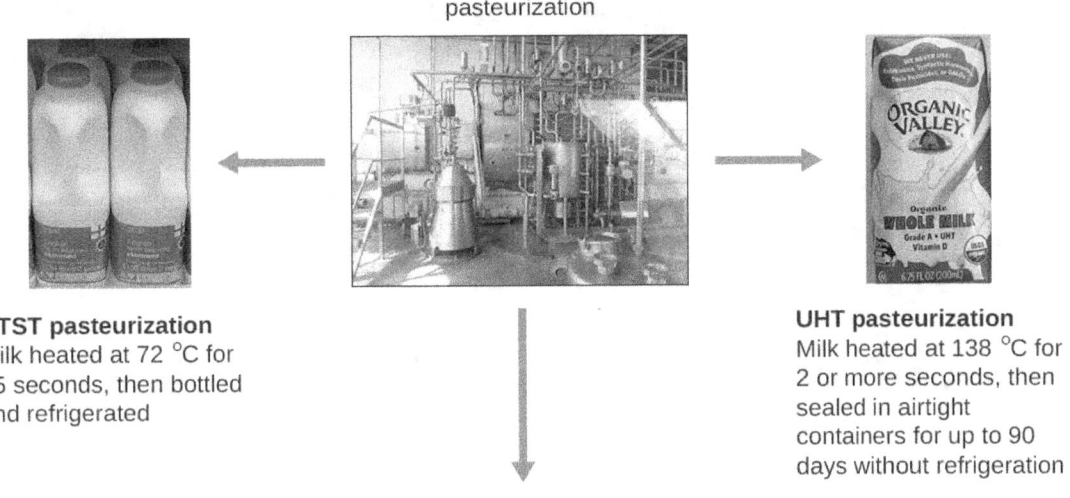

Figure 13.9 Two different methods of pasteurization, HTST and UHT, are commonly used to kill pathogens associated with milk spoilage. (credit left: modification of work by Mark Hillary; credit right: modification of work by Kerry Ceszyk)

✓ CHECK YOUR UNDERSTANDING

- In an autoclave, how are temperatures above boiling achieved?
- How would the onset of spoilage compare between HTST-pasteurized and UHT-pasteurized milk?
- Why is boiling not used as a sterilization method in a clinical setting?

Refrigeration and Freezing

Just as high temperatures are effective for controlling microbial growth, exposing microbes to low temperatures can also be an easy and effective method of microbial control, with the exception of psychrophiles, which prefer cold temperatures (see Temperature and Microbial Growth). Refrigerators used in home kitchens or in the laboratory maintain temperatures between 0 °C and 7 °C. This temperature range inhibits microbial metabolism, slowing the growth of microorganisms significantly and helping preserve refrigerated products such as foods or medical supplies. Certain types of laboratory cultures can be preserved by refrigeration for later use.

Freezing below −2 °C may stop microbial growth and even kill susceptible organisms. According to the US Department of Agriculture (USDA), the only safe ways that frozen foods can be thawed are in the refrigerator, immersed in cold water changed every 30 minutes, or in the microwave, keeping the food at temperatures not conducive for bacterial growth.[3] In addition, halted bacterial growth can restart in thawed foods, so thawed foods should be treated like fresh perishables.

Bacterial cultures and medical specimens requiring long-term storage or transport are often frozen at ultra-low temperatures of −70 °C or lower. These ultra-low temperatures can be achieved by storing specimens on dry ice in an ultra-low freezer or in special liquid nitrogen tanks, which maintain temperatures lower than −196 °C (Figure 13.10).

---

3 US Department of Agriculture. "Freezing and Food Safety." 2013. http://www.fsis.usda.gov/wps/portal/fsis/topics/food-safety-education/get-answers/food-safety-fact-sheets/safe-food-handling/freezing-and-food-safety/CT_Index. Accessed June 8, 2016.

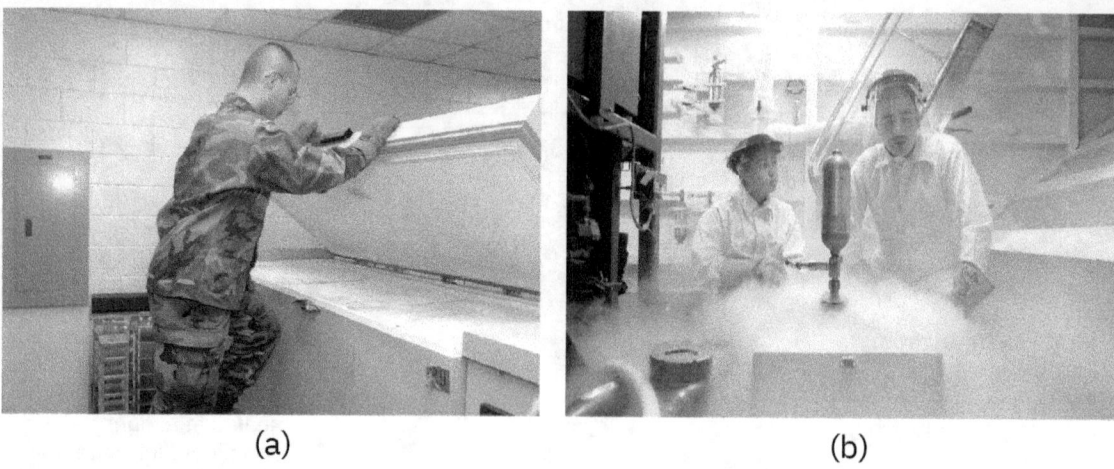

Figure 13.10 Cultures and other medical specimens can be stored for long periods at ultra-low temperatures. (a) An ultra-low freezer maintains temperatures at or below −70 °C. (b) Two people stand in a room with a large nitrogen storage unit emitting a large amount of fog. (credit a: modification of work by "Expert Infantry"/Flickr; credit b: Credit: US Navy; Public Domain.)

✓ CHECK YOUR UNDERSTANDING

- Does placing food in a refrigerator kill bacteria on the food?

Pressure

Exposure to high pressure kills many microbes. In the food industry, high-pressure processing (also called pascalization) is used to kill bacteria, yeast, molds, parasites, and viruses in foods while maintaining food quality and extending shelf life. The application of high pressure between 100 and 800 MPa (sea level atmospheric pressure is about 0.1 MPa) is sufficient to kill vegetative cells by protein denaturation, but endospores may survive these pressures.[4][5]

In clinical settings, hyperbaric oxygen therapy is sometimes used to treat infections. In this form of therapy, a patient breathes pure oxygen at a pressure higher than normal atmospheric pressure, typically between 1 and 3 atmospheres (atm). This is achieved by placing the patient in a hyperbaric chamber or by supplying the pressurized oxygen through a breathing tube. Hyperbaric oxygen therapy helps increase oxygen saturation in tissues that become hypoxic due to infection and inflammation. This increased oxygen concentration enhances the body's immune response by increasing the activities of neutrophils and macrophages, white blood cells that fight infections. Increased oxygen levels also contribute to the formation of toxic free radicals that inhibit the growth of oxygen-sensitive or anaerobic bacteria like as *Clostridium perfringens*, a common cause of gas gangrene. In *C. perfringens* infections, hyperbaric oxygen therapy can also reduce secretion of a bacterial toxin that causes tissue destruction. Hyperbaric oxygen therapy also seems to enhance the effectiveness of antibiotic treatments. Unfortunately, some rare risks include oxygen toxicity and effects on delicate tissues, such as the eyes, middle ear, and lungs, which may be damaged by the increased air pressure.

High pressure processing is not commonly used for disinfection or sterilization of fomites. Although the application of pressure and steam in an autoclave is effective for killing endospores, it is the high temperature achieved, and not the pressure directly, that results in endospore death.

---

4 C. Ferstl. "High Pressure Processing: Insights on Technology and Regulatory Requirements." Food for Thought/White Paper. Series Volume 10. Livermore, CA: The National Food Lab; July 2013.

5 US Food and Drug Administration. "Kinetics of Microbial Inactivation for Alternative Food Processing Technologies: High Pressure Processing." 2000. http://www.fda.gov/Food/FoodScienceResearch/SafePracticesforFoodProcesses/ucm101456.htm. Accessed July 19, 2106.

Case in Point

A Streak of Bad Potluck

One Monday in spring 2015, an Ohio woman began to experience blurred, double vision; difficulty swallowing; and drooping eyelids. She was rushed to the emergency department of her local hospital. During the examination, she began to experience abdominal cramping, nausea, paralysis, dry mouth, weakness of facial muscles, and difficulty speaking and breathing. Based on these symptoms, the hospital's incident command center was activated, and Ohio public health officials were notified of a possible case of botulism. Meanwhile, other patients with similar symptoms began showing up at other local hospitals. Because of the suspicion of botulism, antitoxin was shipped overnight from the CDC to these medical facilities, to be administered to the affected patients. The first patient died of respiratory failure as a result of paralysis, and about half of the remaining victims required additional hospitalization following antitoxin administration, with at least two requiring ventilators for breathing.

Public health officials investigated each of the cases and determined that all of the patients had attended the same church potluck the day before. Moreover, they traced the source of the outbreak to a potato salad made with home-canned potatoes. More than likely, the potatoes were canned using boiling water, a method that allows endospores of *Clostridium botulinum* to survive. *C. botulinum* produces botulinum toxin, a neurotoxin that is often deadly once ingested. According to the CDC, the Ohio case was the largest botulism outbreak in the United States in nearly 40 years.[6]

Killing *C. botulinum* endospores requires a minimum temperature of 116 °C (240 °F), well above the boiling point of water. This temperature can only be reached in a pressure canner, which is recommended for home canning of low-acid foods such as meat, fish, poultry, and vegetables (Figure 13.11). Additionally, the CDC recommends boiling home-canned foods for about 10 minutes before consumption. Since the botulinum toxin is heat labile (meaning that it is denatured by heat), 10 minutes of boiling will render nonfunctional any botulinum toxin that the food may contain.

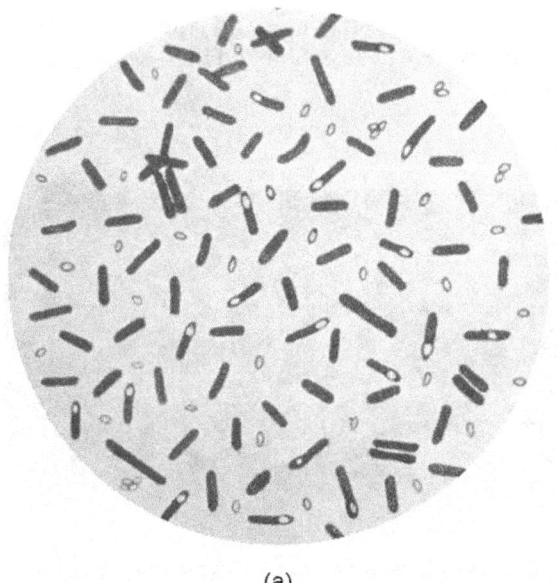

(a) (b)

Figure 13.11 (a) *Clostridium botulinum* is the causative agent of botulism. (b) A pressure canner is recommended for home canning because endospores of *C. botulinum* can survive temperatures above the boiling point of water. (credit a: modification of work by Centers for Disease Control and Prevention; credit b: modification of work by National Center for Home Food Preservation)

---

6 CL McCarty et al. "Large Outbreak of Botulism Associated with a Church Potluck Meal-Ohio, 2015." *Morbidity and Mortality Weekly Report* 64, no. 29 (2015):802–803.

LINK TO LEARNING

To learn more (https://openstax.org/l/22cdccanathome) about proper home-canning techniques, visit the CDC's website.

Desiccation

Drying, also known as **desiccation** or dehydration, is a method that has been used for millennia to preserve foods such as raisins, prunes, and jerky. It works because all cells, including microbes, require water for their metabolism and survival. Although drying controls microbial growth, it might not kill all microbes or their endospores, which may start to regrow when conditions are more favorable and water content is restored.

In some cases, foods are dried in the sun, relying on evaporation to achieve desiccation. Freeze-drying, or **lyophilization**, is another method of dessication in which an item is rapidly frozen ("snap-frozen") and placed under vacuum so that water is lost by sublimation. Lyophilization combines both exposure to cold temperatures and desiccation, making it quite effective for controlling microbial growth. In addition, lyophilization causes less damage to an item than conventional desiccation and better preserves the item's original qualities. Lyophilized items may be stored at room temperature if packaged appropriately to prevent moisture acquisition. Lyophilization is used for preservation in the food industry and is also used in the laboratory for the long-term storage and transportation of microbial cultures.

The water content of foods and materials, called the **water activity**, can be lowered without physical drying by the addition of solutes such as salts or sugars. At very high concentrations of salts or sugars, the amount of available water in microbial cells is reduced dramatically because water will be drawn from an area of low solute concentration (inside the cell) to an area of high solute concentration (outside the cell) (Figure 13.12). Many microorganisms do not survive these conditions of high osmotic pressure. Honey, for example, is 80% sucrose, an environment in which very few microorganisms are capable of growing, thereby eliminating the need for refrigeration. Salted meats and fish, like ham and cod, respectively, were critically important foods before the age of refrigeration. Fruits were preserved by adding sugar, making jams and jellies. However, certain microbes, such as molds and yeasts, tend to be more tolerant of desiccation and high osmotic pressures, and, thus, may still contaminate these types of foods.

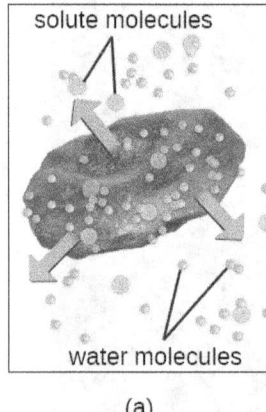
(a)

(b)

Figure 13.12 (a) The addition of a solute creates a hypertonic environment, drawing water out of cells. (b) Some foods can be dried directly, like raisins and jerky. Other foods are dried with the addition of salt, as in the case of salted fish, or sugar, as in the case of jam. (credit a: modification of work by "Bruce Blaus"/Wikimedia Commons; credit raisins: modification of work by Christian Schnettelker; credit jerky: modification of work by Larry Jacobsen; credit salted fish: modification of work by "The Photographer"/Wikimedia Commons; credit jam: modification of work by Kim Becker)

CHECK YOUR UNDERSTANDING

- How does the addition of salt or sugar to food affect its water activity?

Radiation

Radiation in various forms, from high-energy radiation to sunlight, can be used to kill microbes or inhibit their growth. **Ionizing radiation** includes X-rays, gamma rays, and high-energy electron beams. Ionizing radiation is strong enough to pass into the cell, where it alters molecular structures and damages cell components. For example, ionizing radiation introduces double-strand breaks in DNA molecules. This may directly cause DNA mutations to occur, or mutations may be introduced when the cell attempts to repair the DNA damage. As these mutations accumulate, they eventually lead to cell death.

Both X-rays and gamma rays easily penetrate paper and plastic and can therefore be used to sterilize many packaged materials. In the laboratory, ionizing radiation is commonly used to sterilize materials that cannot be autoclaved, such as plastic Petri dishes and disposable plastic inoculating loops. For clinical use, ionizing radiation is used to sterilize gloves, intravenous tubing, and other latex and plastic items used for patient care. Ionizing radiation is also used for the sterilization of other types of delicate, heat-sensitive materials used clinically, including tissues for transplantation, pharmaceutical drugs, and medical equipment.

In Europe, gamma irradiation for food preservation is widely used, although it has been slow to catch on in the United States (see the Micro Connections box on this topic). Packaged dried spices are also often gamma-irradiated. Because of their ability to penetrate paper, plastic, thin sheets of wood and metal, and tissue, great care must be taken when using X-rays and gamma irradiation. These types of ionizing irradiation cannot penetrate thick layers of iron or lead, so these metals are commonly used to protect humans who may be potentially exposed.

Another type of radiation, **nonionizing radiation**, is commonly used for disinfection and uses less energy than ionizing radiation. It does not penetrate cells or packaging. Ultraviolet (UV) light is one example; it causes thymine dimers to form between adjacent thymines within a single strand of DNA (Figure 13.13). When DNA polymerase encounters the thymine dimer, it does not always incorporate the appropriate complementary nucleotides (two adenines), and this leads to formation of mutations that can ultimately kill microorganisms.

UV light can be used effectively by both consumers and laboratory personnel to control microbial growth. UV lamps are now commonly incorporated into water purification systems for use in homes. In addition, small portable UV lights are commonly used by campers to purify water from natural environments before drinking. Germicidal lamps are also used in surgical suites, biological safety cabinets, and transfer hoods, typically emitting UV light at a wavelength of 260 nm. Because UV light does not penetrate surfaces and will not pass through plastics or glass, cells must be exposed directly to the light source.

Sunlight has a very broad spectrum that includes UV and visible light. In some cases, sunlight can be effective against certain bacteria because of both the formation of thymine dimers by UV light and by the production of reactive oxygen products induced in low amounts by exposure to visible light.

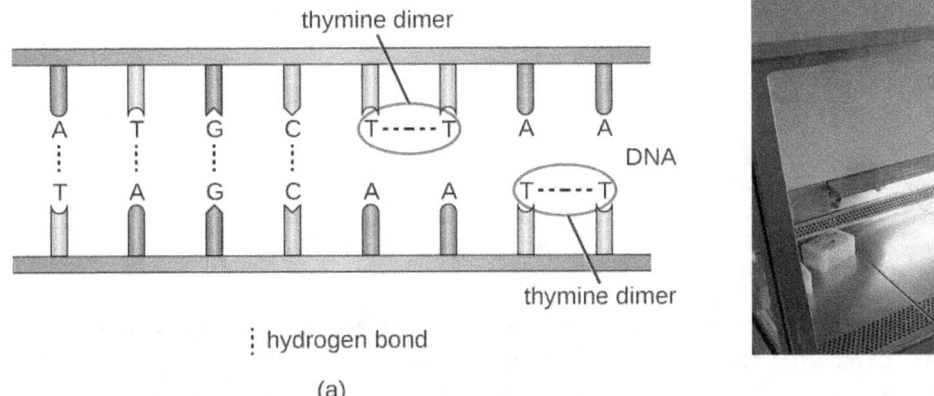
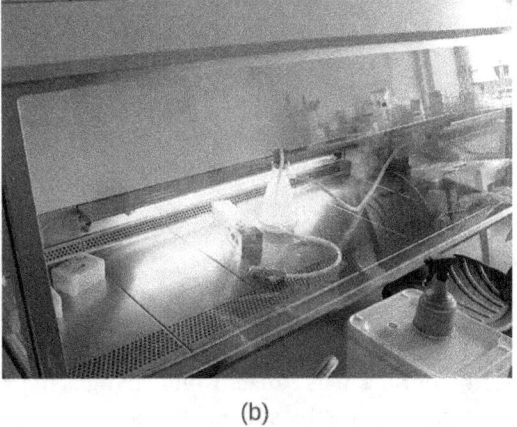

Figure 13.13 (a) UV radiation causes the formation of thymine dimers in DNA, leading to lethal mutations in the exposed microbes. (b) Germicidal lamps that emit UV light are commonly used in the laboratory to disinfect equipment.

CHECK YOUR UNDERSTANDING

- What are two advantages of ionizing radiation as a sterilization method?
- How does the effectiveness of ionizing radiation compare with that of nonionizing radiation?

MICRO CONNECTIONS

Irradiated Food: Would You Eat That?

Of all the ways to prevent food spoilage and foodborne illness, gamma irradiation may be the most unappetizing. Although gamma irradiation is a proven method of eliminating potentially harmful microbes from food, the public has yet to buy in. Most of their concerns, however, stem from misinformation and a poor understanding of the basic principles of radiation.

The most common method of irradiation is to expose food to cobalt-60 or cesium-137 by passing it through a radiation chamber on a conveyor belt. The food does not directly contact the radioactive material and does not become radioactive itself. Thus, there is no risk for exposure to radioactive material through eating gamma-irradiated foods. Additionally, irradiated foods are not significantly altered in terms of nutritional quality, aside from the loss of certain vitamins, which is also exacerbated by extended storage. Alterations in taste or smell may occur in irradiated foods with high fat content, such as fatty meats and dairy products, but this effect can be minimized by using lower doses of radiation at colder temperatures.

In the United States, the CDC, Environmental Protection Agency (EPA), and the Food and Drug Administration (FDA) have deemed irradiation safe and effective for various types of meats, poultry, shellfish, fresh fruits and vegetables, eggs with shells, and spices and seasonings. Gamma irradiation of foods has also been approved for use in many other countries, including France, the Netherlands, Portugal, Israel, Russia, China, Thailand, Belgium, Australia, and South Africa. To help ameliorate consumer concern and assist with education efforts, irradiated foods are now clearly labeled and marked with the international irradiation symbol, called the "radura" (Figure 13.14). Consumer acceptance seems to be rising, as indicated by several recent studies.[7]

Figure 13.14 (a) Foods are exposed to gamma radiation by passage on a conveyor belt through a radiation chamber. (b) Gamma-irradiated foods must be clearly labeled and display the irradiation symbol, known as the "radura." (credit a, b: modification of work by U.S. Department of Agriculture)

Sonication

The use of high-frequency ultrasound waves to disrupt cell structures is called **sonication**. Application of ultrasound waves causes rapid changes in pressure within the intracellular liquid; this leads to cavitation, the formation of bubbles inside the cell, which can disrupt cell structures and eventually cause the cell to lyse or collapse. Sonication is useful in the laboratory for efficiently lysing cells to release their contents for further

---

[7] AM Johnson et al. "Consumer Acceptance of Electron-Beam Irradiated Ready-to-Eat Poultry Meats." *Food Processing Preservation*, 28 no. 4 (2004):302–319.

research; outside the laboratory, sonication is used for cleaning surgical instruments, lenses, and a variety of other objects such as coins, tools, and musical instruments.

Filtration

Filtration is a method of physically separating microbes from samples. Air is commonly filtered through **high-efficiency particulate air (HEPA) filter**s (Figure 13.15). HEPA filters have effective pore sizes of 0.3 µm, small enough to capture bacterial cells, endospores, and many viruses, as air passes through these filters, nearly sterilizing the air on the other side of the filter. HEPA filters have a variety of applications and are used widely in clinical settings, in cars and airplanes, and even in the home. For example, they may be found in vacuum cleaners, heating and air-conditioning systems, and air purifiers.

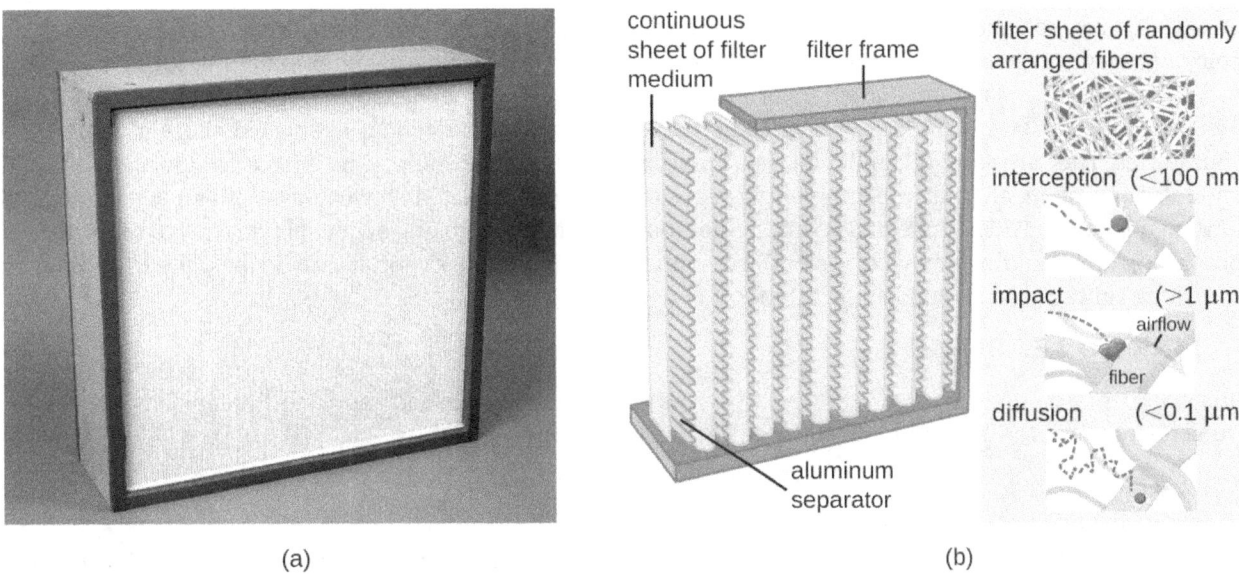

Figure 13.15 (a) HEPA filters like this one remove microbes, endospores, and viruses as air flows through them. (b) A schematic of a HEPA filter. (credit a: modification of work by CSIRO; credit b: modification of work by "LadyofHats"/Mariana Ruiz Villareal)

Biological Safety Cabinets

Biological safety cabinets are a good example of the use of HEPA filters. HEPA filters in biological safety cabinets (BSCs) are used to remove particulates in the air either entering the cabinet (air intake), leaving the cabinet (air exhaust), or treating both the intake and exhaust. Use of an air-intake HEPA filter prevents environmental contaminants from entering the BSC, creating a clean area for handling biological materials. Use of an air-exhaust HEPA filter prevents laboratory pathogens from contaminating the laboratory, thus maintaining a safe work area for laboratory personnel.

There are three classes of BSCs: I, II, and III. Each class is designed to provide a different level of protection for laboratory personnel and the environment; BSC II and III are also designed to protect the materials or devices in the cabinet. Table 13.1 summarizes the level of safety provided by each class of BSC for each BSL.

Biological Risks and BSCs

Biological Risk Assessed	BSC Class	Protection of Personnel	Protection of Environment	Protection of Product
BSL-1, BSL-2, BSL-3	I	Yes	Yes	No
BSL-1, BSL-2, BSL-3	II	Yes	Yes	Yes

Biological Risks and BSCs

Biological Risk Assessed	BSC Class	Protection of Personnel	Protection of Environment	Protection of Product
BSL-4	III; II when used in suit room with suit	Yes	Yes	Yes

Table 13.1

Class I BSCs protect laboratory workers and the environment from a low to moderate risk for exposure to biological agents used in the laboratory. Air is drawn into the cabinet and then filtered before exiting through the building's exhaust system. Class II BSCs use directional air flow and partial barrier systems to contain infectious agents. Class III BSCs are designed for working with highly infectious agents like those used in BSL-4 laboratories. They are gas tight, and materials entering or exiting the cabinet must be passed through a double-door system, allowing the intervening space to be decontaminated between uses. All air is passed through one or two HEPA filters and an air incineration system before being exhausted directly to the outdoors (not through the building's exhaust system). Personnel can manipulate materials inside the Class III cabinet by using long rubber gloves sealed to the cabinet.

LINK TO LEARNING

This video (https://openstax.org/l/22BSCsdesvideo) shows how BSCs are designed and explains how they protect personnel, the environment, and the product.

Filtration in Hospitals

HEPA filters are also commonly used in hospitals and surgical suites to prevent contamination and the spread of airborne microbes through ventilation systems. HEPA filtration systems may be designed for entire buildings or for individual rooms. For example, burn units, operating rooms, or isolation units may require special HEPA-filtration systems to remove opportunistic pathogens from the environment because patients in these rooms are particularly vulnerable to infection.

Membrane Filters

Filtration can also be used to remove microbes from liquid samples using **membrane filtration**. Membrane filters for liquids function similarly to HEPA filters for air. Typically, membrane filters that are used to remove bacteria have an effective pore size of 0.2 µm, smaller than the average size of a bacterium (1 µm), but filters with smaller pore sizes are available for more specific needs. Membrane filtration is useful for removing bacteria from various types of heat-sensitive solutions used in the laboratory, such as antibiotic solutions and vitamin solutions. Large volumes of culture media may also be filter sterilized rather than autoclaved to protect heat-sensitive components. Often when filtering small volumes, syringe filters are used, but vacuum filters are typically used for filtering larger volumes (Figure 13.16).

 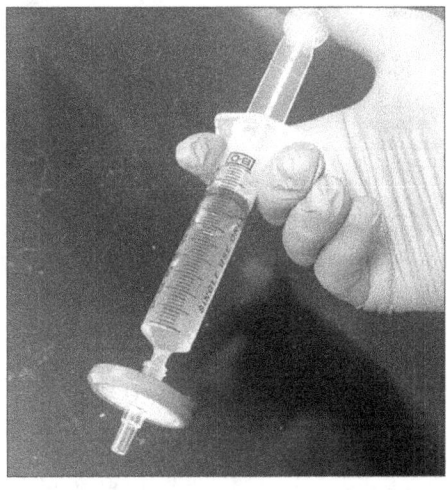

(a)   (b)

Figure 13.16 Membrane filters come in a variety of sizes, depending on the volume of solution being filtered. (a) Larger volumes are filtered in units like these. The solution is drawn through the filter by connecting the unit to a vacuum. (b) Smaller volumes are often filtered using syringe filters, which are units that fit on the end of a syringe. In this case, the solution is pushed through by depressing the syringe's plunger. (credit a, b: modification of work by Brian Forster)

CHECK YOUR UNDERSTANDING

- Would membrane filtration with a 0.2-µm filter likely remove viruses from a solution? Explain.
- Name at least two common uses of HEPA filtration in clinical or laboratory settings.

Figure 13.17 and Figure 13.18 summarize the physical methods of control discussed in this section.

Physical Methods of Control

Method	Conditions	Mode of Action	Example Uses
Heat			
Boiling	100 °C at sea level	Denatures proteins and alters membranes	Cooking, personal use, preparing certain laboratory media
Dry-heat oven	170 °C for 2 hours	Denatures proteins and alters membranes, dehydration, desiccation	Sterilization of heat-stable medical and laboratory equipment and glassware
Incineration	Exposure to flame	Destroy by burning	Flaming loop, microincinerator
Autoclave	Typical settings: 121 °C for 15 minutes at 15 pounds per square inch (psi)	Denatures proteins and alters membranes	Sterilization of microbiological media, heat-stable medical and laboratory equipment, and other heat-stable items
Pasteurization	Can vary. One type is 72 °C for 15 seconds (HTST)	Denatures proteins and alters membranes	Prevents spoilage of milk, apple juice, honey, and other ingestible liquids
Cold			
Refrigeration	0 °C to 7 °C	Inhibits metabolism (slows or arrests cell division)	Preservation of food or laboratory materials (solutions, cultures)
Freezing	Below −2 °C	Stops metabolism, may kill microbes	Long-term storage of food, laboratory cultures, or medical specimens
Pressure			
High-pressure processing	100–800 MPa	Denatures proteins and can cause cell lysis	Preservation of food
Hyberbaric oxygen therapy	Air pressure three times higher than normal	Inhibits metabolism and growth of anaerobic microbes	Treatment of certain infections (e.g., gas gangrene)
Desiccation			
Simple desiccation	Drying	Inhibits metabolism	Dried fruits, jerky
Reduce water activity	Addition of salt or water	Inhibits metabolism and can cause lysis	Salted meats and fish, honey, jams and jellies
Lyophilization	Rapid freezing under vacuum	Inhibits metabolism	Preservation of food, laboratory cultures, or reagents
Radiation			
Ionizing radiation	Exposure to X-rays or gamma rays	Alters molecular structures, introduces double-strand breaks into DNA	Sterilization of spices and heat-sensitive laboratory and medical items; used for food sterilization in Europe but not widely accepted in US
Nonionizing radiation	Exposure to ultraviolet light	Introduces thymine dimers, leading to mutations	Disinfection of surfaces in laboratories and rooms in health-care environment, and disinfection of water and air

Figure 13.17

Physical Methods of Control (continued)			
Method	Conditions	Mode of Action	Example Uses
Sonication			
Sonication	Exposure to ultrasonic waves	Cavitation (formation of empty space) disrupts cells, lysing them	Laboratory research to lyse cells; cleaning jewelry, lenses, and equipment
Filtration			
HEPA filtration	Use of high-efficiency particulate air (HEPA) filter with 0.3 μm pore size	Physically removes microbes from air	Laboratory biological safety cabinets, operating rooms, isolation units, heating and air conditioning systems, vacuum cleaners
Membrane filtration	Use of membrane filter with 0.2-μm or smaller pore size	Physically removes microbes from liquid solutions	Removal of bacteria from heat-sensitive solutions like vitamins, antibiotics, and media with heat-sensitive components

Figure 13.18

13.3 Using Chemicals to Control Microorganisms

Learning Objectives

By the end of this section, you will be able to:
- Understand and compare various chemicals used to control microbial growth, including their uses, advantages and disadvantages, chemical structure, and mode of action

In addition to physical methods of microbial control, chemicals are also used to control microbial growth. A wide variety of chemicals can be used as disinfectants or antiseptics. When choosing which to use, it is important to consider the type of microbe targeted; how clean the item needs to be; the disinfectant's effect on the item's integrity; its safety to animals, humans, and the environment; its expense; and its ease of use. This section describes the variety of chemicals used as disinfectants and antiseptics, including their mechanisms of action and common uses.

Phenolics

In the 1800s, scientists began experimenting with a variety of chemicals for disinfection. In the 1860s, British surgeon Joseph Lister (1827–1912) began using carbolic acid, known as phenol, as a disinfectant for the treatment of surgical wounds (see Foundations of Modern Cell Theory). In 1879, Lister's work inspired the American chemist Joseph Lawrence (1836–1909) to develop Listerine, an alcohol-based mixture of several related compounds that is still used today as an oral antiseptic. Today, carbolic acid is no longer used as a surgical disinfectant because it is a skin irritant, but the chemical compounds found in antiseptic mouthwashes and throat lozenges are called **phenolics**.

Chemically, phenol consists of a benzene ring with an –OH group, and phenolics are compounds that have this group as part of their chemical structure (Figure 13.19). Phenolics such as thymol and eucalyptol occur naturally in plants. Other phenolics can be derived from creosote, a component of coal tar. Phenolics tend to be stable, persistent on surfaces, and less toxic than phenol. They inhibit microbial growth by denaturing proteins and disrupting membranes.

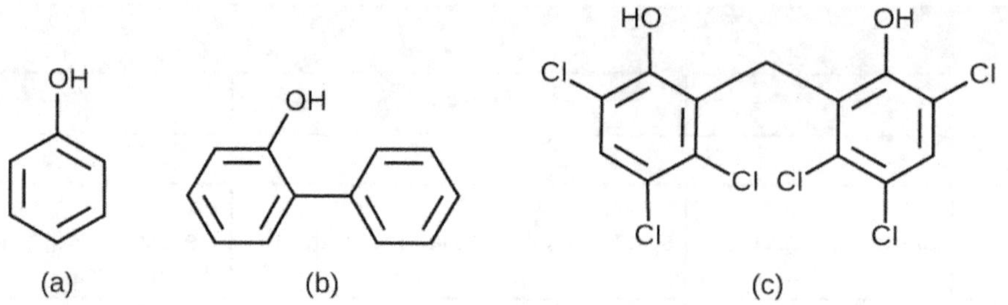

Figure 13.19 Phenol and phenolic compounds have been used to control microbial growth. (a) Chemical structure of phenol, also known as carbolic acid. (b) o-Phenylphenol, a type of phenolic, has been used as a disinfectant as well as to control bacterial and fungal growth on harvested citrus fruits. (c) Hexachlorophene, another phenol, known as a bisphenol (two rings), is the active ingredient in pHisoHex.

Since Lister's time, several phenolic compounds have been used to control microbial growth. Phenolics like cresols (methylated phenols) and o-phenylphenol were active ingredients in various formulations of Lysol since its invention in 1889. o-Phenylphenol was also commonly used in agriculture to control bacterial and fungal growth on harvested crops, especially citrus fruits, but its use in the United States is now far more limited. The bisphenol hexachlorophene, a disinfectant, is the active ingredient in pHisoHex, a topical cleansing detergent widely used for handwashing in hospital settings. pHisoHex is particularly effective against gram-positive bacteria, including those causing staphylococcal and streptococcal skin infections. pHisoHex was formerly used for bathing infants, but this practice has been discontinued because it has been shown that exposure to hexachlorophene can lead to neurological problems.

Triclosan is another bisphenol compound that has seen widespread application in antibacterial products over the last several decades. Initially used in toothpastes, triclosan has also been used in hand soaps and impregnated into a wide variety of other products, including cutting boards, knives, shower curtains, clothing, and concrete, to make them antimicrobial. However, in 2016 the FDA banned the marketing of over-the-counter antiseptic products containing triclosan and 18 other chemicals. This ruling was based on the lack of evidence of safety or efficacy, as well as concerns about the health risks of long-term exposure (See Micro Connections below). In 2019 the FDA issued an updated ban ruling to included 28 chemicals. Rulings on benzalkonium chloride, ethyl alcohol, and isopropyl alcohol have been deferred to allow for the submission of additional safety and efficacy data.[8]

 MICRO CONNECTIONS

Triclosan: Antibacterial Overkill?

Hand soaps and other cleaning products are often marketed as "antibacterial," suggesting that they provide a level of cleanliness superior to that of conventional soaps and cleansers. But are the antibacterial ingredients in these products really safe and effective?

About 75% of antibacterial liquid hand soaps and 30% of bar soaps contain the chemical triclosan, a phenolic, (Figure 13.20).[9] Triclosan blocks an enzyme in the bacterial fatty acid-biosynthesis pathway that is not found in the comparable human pathway. Although the use of triclosan in the home increased dramatically during the 1990s, more than 40 years of research by the FDA have turned up no conclusive evidence that washing with triclosan-containing products provides increased health benefits compared with washing with traditional soap. Although some studies indicate that fewer bacteria may remain on a person's hands after washing with triclosan-based soap, compared with traditional soap, no evidence points to any reduction in the transmission of bacteria that cause respiratory and gastrointestinal illness. In short, soaps with triclosan may remove or kill a few more germs but not enough to reduce the spread of disease.

---

8   US Food and Drug Administration. "FDA Issues Final Rule on Safety and Effectiveness of Antibacterial Soaps." 2016. https://www.fda.gov/news-events/press-announcements/fda-issues-final-rule-safety-and-effectiveness-antibacterial-soaps. Accessed October 29, 2020.

9   J. Stromberg. "Five Reasons Why You Should Probably Stop Using Antibacterial Soap." *Smithsonian.com* January 3, 2014. http://www.smithsonianmag.com/science-nature/five-reasons-why-you-should-probably-stop-using-antibacterial-

Perhaps more disturbing, some clear risks associated with triclosan-based soaps have come to light. The widespread use of triclosan has led to an increase in triclosan-resistant bacterial strains, including those of clinical importance, such as *Salmonella enterica*; this resistance may render triclosan useless as an antibacterial in the long run.[10][11] Bacteria can easily gain resistance to triclosan through a change to a single gene encoding the targeted enzyme in the bacterial fatty acid-synthesis pathway. Other disinfectants with a less specific mode of action are much less prone to engendering resistance because it would take much more than a single genetic change.

Use of triclosan over the last several decades has also led to a buildup of the chemical in the environment. Triclosan in hand soap is directly introduced into wastewater and sewage systems as a result of the handwashing process. There, its antibacterial properties can inhibit or kill bacteria responsible for the decomposition of sewage, causing septic systems to clog and back up. Eventually, triclosan in wastewater finds its way into surface waters, streams, lakes, sediments, and soils, disrupting natural populations of bacteria that carry out important environmental functions, such as inhibiting algae. Triclosan also finds its way into the bodies of amphibians and fish, where it can act as an endocrine disruptor. Detectable levels of triclosan have also been found in various human bodily fluids, including breast milk, plasma, and urine.[12] In fact, a study conducted by the CDC found detectable levels of triclosan in the urine of 75% of 2,517 people tested in 2003–2004.[13] This finding is even more troubling given the evidence that triclosan may affect immune function in humans.[14]

In December 2013, the FDA gave soap manufacturers until 2016 to prove that antibacterial soaps provide a significant benefit over traditional soaps; if unable to do so, manufacturers will be forced to remove these products from the market.

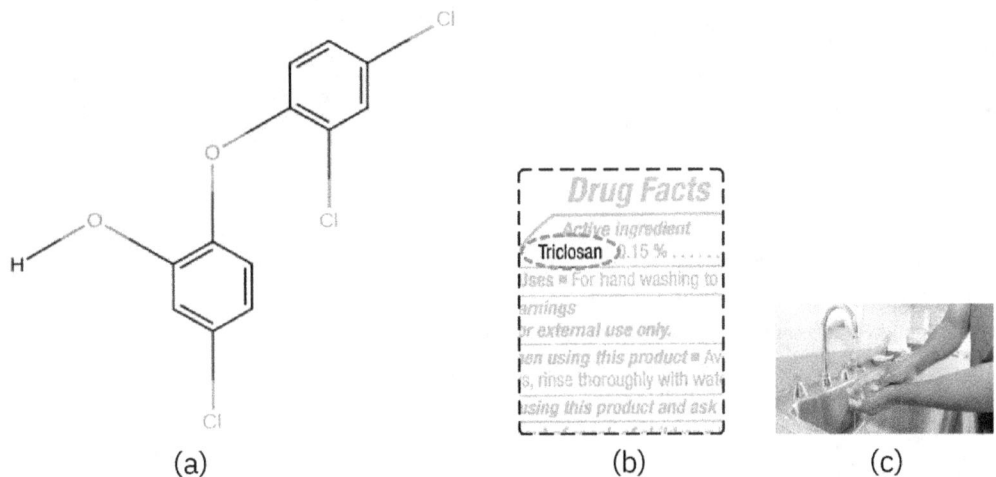

Figure 13.20 Triclosan is a common ingredient in antibacterial soaps despite evidence that it poses environmental and health risks and offers no significant health benefit compared to conventional soaps. (credit b: modification of work by FDA; c: Michelle Gigante/US Air Force; Public Domain.)

---

soap-180948078/?no-ist. Accessed June 9, 2016.

10 SP Yazdankhah et al. "Triclosan and Antimicrobial Resistance in Bacteria: An Overview." *Microbial Drug Resistance* 12 no. 2 (2006):83–90.

11 L. Birošová, M. Mikulášová. "Development of Triclosan and Antibiotic Resistance in *Salmonella enterica* serovar Typhimurium." *Journal of Medical Microbiology* 58 no. 4 (2009):436–441.

12 AB Dann, A. Hontela. "Triclosan: Environmental Exposure, Toxicity and Mechanisms of Action." *Journal of Applied Toxicology* 31 no. 4 (2011):285–311.

13 US Centers for Disease Control and Prevention. "Triclosan Fact Sheet." 2013. http://www.cdc.gov/biomonitoring/Triclosan_FactSheet.html. Accessed June 9, 2016.

14 EM Clayton et al. "The Impact of Bisphenol A and Triclosan on Immune Parameters in the US Population, NHANES 2003-2006." *Environmental Health Perspectives* 119 no. 3 (2011):390.

CHECK YOUR UNDERSTANDING

- Why is triclosan more like an antibiotic than a traditional disinfectant?

Heavy Metals

Some of the first chemical disinfectants and antiseptics to be used were heavy metals. Heavy metals kill microbes by binding to proteins, thus inhibiting enzymatic activity (Figure 13.21). Heavy metals are oligodynamic, meaning that very small concentrations show significant antimicrobial activity. Ions of heavy metals bind to sulfur-containing amino acids strongly and bioaccumulate within cells, allowing these metals to reach high localized concentrations. This causes proteins to denature.

Heavy metals are not selectively toxic to microbial cells. They may bioaccumulate in human or animal cells, as well, and excessive concentrations can have toxic effects on humans. If too much silver accumulates in the body, for example, it can result in a condition called argyria, in which the skin turns irreversibly blue-gray. One way to reduce the potential toxicity of heavy metals is by carefully controlling the duration of exposure and concentration of the heavy metal.

(a)
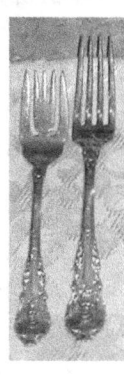
(b)
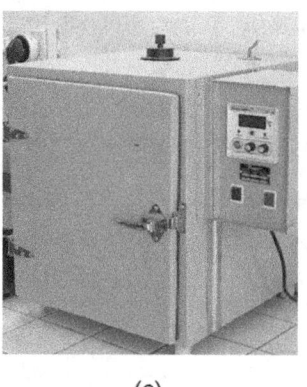
(c)
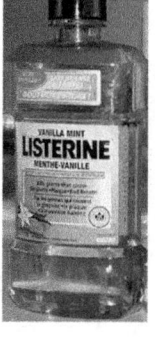
(d)
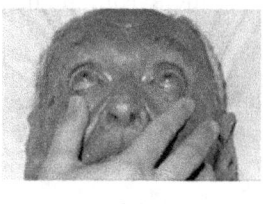
(e)

Figure 13.21 Heavy metals denature proteins, impairing cell function and, thus, giving them strong antimicrobial properties. (a) Copper in fixtures like this door handle kills microbes that otherwise might accumulate on frequently touched surfaces. (b) Eating utensils contain small amounts of silver to inhibit microbial growth. (c) Copper commonly lines incubators to minimize contamination of cell cultures stored inside. (d) Antiseptic mouthwashes commonly contain zinc chloride. (e) This patient is suffering from argyria, an irreversible condition caused by bioaccumulation of silver in the body. (credit b: modification of work by "Shoshanah"/Flickr; credit e: modification of work by Herbert L. Fred and Hendrik A. van Dijk)

Mercury

Mercury is an example of a heavy metal that has been used for many years to control microbial growth. It was used for many centuries to treat syphilis. Mercury compounds like mercuric chloride are mainly bacteriostatic and have a very broad spectrum of activity. Various forms of mercury bind to sulfur-containing amino acids within proteins, inhibiting their functions.

In recent decades, the use of such compounds has diminished because of mercury's toxicity. It is toxic to the central nervous, digestive, and renal systems at high concentrations, and has negative environmental effects, including bioaccumulation in fish. Topical antiseptics such as mercurochrome, which contains mercury in low concentrations, and merthiolate, a **tincture** (a solution of mercury dissolved in alcohol) were once commonly used. However, because of concerns about using mercury compounds, these antiseptics are no longer sold in the United States.

Silver

Silver has long been used as an antiseptic. In ancient times, drinking water was stored in silver jugs.[15] Silvadene cream is commonly used to treat topical wounds and is particularly helpful in preventing infection in burn wounds. Silver nitrate drops were once routinely applied to the eyes of newborns to protect against ophthalmia neonatorum, eye infections that can occur due to exposure to pathogens in the birth canal, but antibiotic creams are more now commonly used. Silver is often combined with antibiotics, making the antibiotics thousands of times more effective.[16] Silver is also commonly incorporated into catheters and bandages, rendering them antimicrobial; however, there is evidence that heavy metals may also enhance selection for antibiotic resistance.[17]

Copper, Nickel, and Zinc

Several other heavy metals also exhibit antimicrobial activity. Copper sulfate is a common algicide used to control algal growth in swimming pools and fish tanks. The use of metallic copper to minimize microbial growth is also becoming more widespread. Copper linings in incubators help reduce contamination of cell cultures. The use of copper pots for water storage in underdeveloped countries is being investigated as a way to combat diarrheal diseases. Copper coatings are also becoming popular for frequently handled objects such as doorknobs, cabinet hardware, and other fixtures in health-care facilities in an attempt to reduce the spread of microbes.

Nickel and zinc coatings are now being used in a similar way. Other forms of zinc, including zinc chloride and zinc oxide, are also used commercially. Zinc chloride is quite safe for humans and is commonly found in mouthwashes, substantially increasing their length of effectiveness. Zinc oxide is found in a variety of products, including topical antiseptic creams such as calamine lotion, diaper ointments, baby powder, and dandruff shampoos.

✓ CHECK YOUR UNDERSTANDING

- Why are many heavy metals both antimicrobial and toxic to humans?

Halogens

Other chemicals commonly used for disinfection are the halogens iodine, chlorine, and fluorine. Iodine works by oxidizing cellular components, including sulfur-containing amino acids, nucleotides, and fatty acids, and destabilizing the macromolecules that contain these molecules. It is often used as a topical tincture, but it may cause staining or skin irritation. An **iodophor** is a compound of iodine complexed with an organic molecule, thereby increasing iodine's stability and, in turn, its efficacy. One common iodophor is povidone-iodine, which includes a wetting agent that releases iodine relatively slowly. Betadine is a brand of povidone-iodine commonly used as a hand scrub by medical personnel before surgery and for topical antisepsis of a patient's skin before incision (Figure 13.22).

---

15 N. Silvestry-Rodriguez et al. "Silver as a Disinfectant." In *Reviews of Environmental Contamination and Toxicology*, pp. 23-45. Edited by GW Ware and DM Whitacre. New York: Springer, 2007.

16 B. Owens. "Silver Makes Antibiotics Thousands of Times More Effective." *Nature* June 19 2013. http://www.nature.com/news/silver-makes-antibiotics-thousands-of-times-more-effective-1.13232

17 C. Seiler, TU Berendonk. "Heavy Metal Driven Co-Selection of Antibiotic Resistance in Soil and Water Bodies Impacted by Agriculture and Aquaculture." *Frontiers in Microbiology* 3 (2012):399.

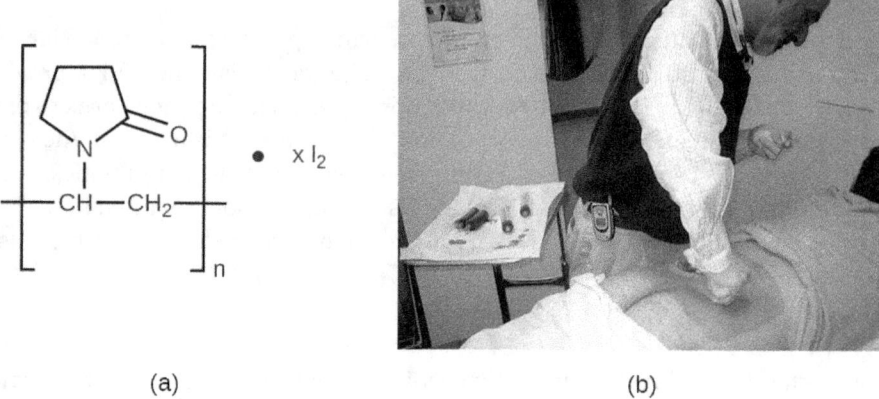

Figure 13.22 (a) Betadine is a solution of the iodophor povidone-iodine. (b) It is commonly used as a topical antiseptic on a patient's skin before incision during surgery. (credit b: modification of work by Andrew Ratto)

Chlorine is another halogen commonly used for disinfection. When chlorine gas is mixed with water, it produces a strong oxidant called hypochlorous acid, which is uncharged and enters cells easily. Chlorine gas is commonly used in municipal drinking water and wastewater treatment plants, with the resulting hypochlorous acid producing the actual antimicrobial effect. Those working at water treatment facilities need to take great care to minimize personal exposure to chlorine gas. Sodium hypochlorite is the chemical component of common household bleach, and it is also used for a wide variety of disinfecting purposes. Hypochlorite salts, including sodium and calcium hypochlorites, are used to disinfect swimming pools. Chlorine gas, sodium hypochlorite, and calcium hypochlorite are also commonly used disinfectants in the food processing and restaurant industries to reduce the spread of foodborne diseases. Workers in these industries also need to take care to use these products correctly to ensure their own safety as well as the safety of consumers. A recent joint statement published by the Food and Agriculture Organization (FAO) of the United Nations and WHO indicated that none of the many beneficial uses of chlorine products in food processing to reduce the spread of foodborne illness posed risks to consumers.[18]

Another class of chlorinated compounds called chloramines are widely used as disinfectants. Chloramines are relatively stable, releasing chlorine over long periods time. Chloramines are derivatives of ammonia by substitution of one, two, or all three hydrogen atoms with chlorine atoms (Figure 13.23).

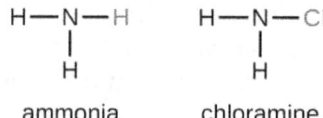

Figure 13.23 Monochloroamine, one of the chloramines, is derived from ammonia by the replacement of one hydrogen atom with a chlorine atom.

Chloramines and other cholorine compounds may be used for disinfection of drinking water, and chloramine tablets are frequently used by the military for this purpose. After a natural disaster or other event that compromises the public water supply, the CDC recommends disinfecting tap water by adding small amounts of regular household bleach. Recent research suggests that sodium dichloroisocyanurate (NaDCC) may also be a good alternative for drinking water disinfection. Currently, NaDCC tablets are available for general use and for use by the military, campers, or those with emergency needs; for these uses, NaDCC is preferable to chloramine tablets. Chlorine dioxide, a gaseous agent used for fumigation and sterilization of enclosed areas, is also commonly used for the disinfection of water.

Although chlorinated compounds are relatively effective disinfectants, they have their disadvantages. Some may irritate the skin, nose, or eyes of some individuals, and they may not completely eliminate certain hardy organisms from contaminated drinking water. The protozoan parasite *Cryptosporidium*, for example, has a

---

18 World Health Organization. "Benefits and Risks of the Use of Chlorine-Containing Disinfectants in Food Production and Food Processing: Report of a Joint FAO/WHO Expert Meeting." Geneva, Switzerland: World Health Organization, 2009.

protective outer shell that makes it resistant to chlorinated disinfectants. Thus, boiling of drinking water in emergency situations is recommended when possible.

The halogen fluorine is also known to have antimicrobial properties that contribute to the prevention of dental caries (cavities).[19] Fluoride is the main active ingredient of toothpaste and is also commonly added to tap water to help communities maintain oral health. Chemically, fluoride can become incorporated into the hydroxyapatite of tooth enamel, making it more resistant to corrosive acids produced by the fermentation of oral microbes. Fluoride also enhances the uptake of calcium and phosphate ions in tooth enamel, promoting remineralization. In addition to strengthening enamel, fluoride also seems to be bacteriostatic. It accumulates in plaque-forming bacteria, interfering with their metabolism and reducing their production of the acids that contribute to tooth decay.

✓ CHECK YOUR UNDERSTANDING

- What is a benefit of a chloramine over hypochlorite for disinfecting?

Alcohols

Alcohols make up another group of chemicals commonly used as disinfectants and antiseptics. They work by rapidly denaturing proteins, which inhibits cell metabolism, and by disrupting membranes, which leads to cell lysis. Once denatured, the proteins may potentially refold if enough water is present in the solution. Alcohols are typically used at concentrations of about 70% aqueous solution and, in fact, work better in aqueous solutions than 100% alcohol solutions. This is because alcohols coagulate proteins. In higher alcohol concentrations, rapid coagulation of surface proteins prevents effective penetration of cells. The most commonly used alcohols for disinfection are ethyl alcohol (ethanol) and isopropyl alcohol (isopropanol, rubbing alcohol) (Figure 13.24).

Alcohols tend to be bactericidal and fungicidal, but may also be viricidal for enveloped viruses only. Although alcohols are not sporicidal, they do inhibit the processes of sporulation and germination. Alcohols are volatile and dry quickly, but they may also cause skin irritation because they dehydrate the skin at the site of application. One common clinical use of alcohols is swabbing the skin for degerming before needle injection. Alcohols also are the active ingredients in instant hand sanitizers, which have gained popularity in recent years. The alcohol in these hand sanitizers works both by denaturing proteins and by disrupting the microbial cell membrane, but will not work effectively in the presence of visible dirt.

Last, alcohols are used to make tinctures with other antiseptics, such as the iodine tinctures discussed previously in this chapter. All in all, alcohols are inexpensive and quite effective for the disinfection of a broad range of vegetative microbes. However, one disadvantage of alcohols is their high volatility, limiting their effectiveness to immediately after application.

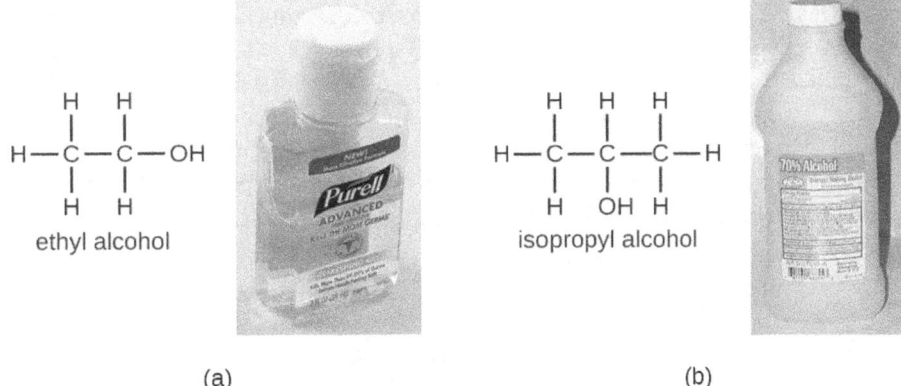

Figure 13.24 (a) Ethyl alcohol, the intoxicating ingredient found in alcoholic drinks, is also used commonly as a disinfectant. (b) Isopropyl alcohol, also called rubbing alcohol, has a related molecular structure and is another commonly used disinfectant. (credit a photo:

---

19   RE Marquis. "Antimicrobial Actions of Fluoride for Oral Bacteria." *Canadian Journal of Microbiology* 41 no. 11 (1995):955–964.

modification of work by D Coetzee; credit b photo: modification of work by Craig Spurrier)

✓ CHECK YOUR UNDERSTANDING

- Name at least three advantages of alcohols as disinfectants.
- Describe several specific applications of alcohols used in disinfectant products.

Surfactants

Surface-active agents, or **surfactants**, are a group of chemical compounds that lower the surface tension of water. Surfactants are the major ingredients in soaps and detergents. Soaps are salts of long-chain fatty acids and have both polar and nonpolar regions, allowing them to interact with polar and nonpolar regions in other molecules (Figure 13.25). They can interact with nonpolar oils and grease to create emulsions in water, loosening and lifting away dirt and microbes from surfaces and skin. Soaps do not kill or inhibit microbial growth and so are not considered antiseptics or disinfectants. However, proper use of soaps mechanically carries away microorganisms, effectively degerming a surface. Some soaps contain added bacteriostatic agents such as triclocarban or cloflucarban, compounds structurally related to triclosan, that introduce antiseptic or disinfectant properties to the soaps.

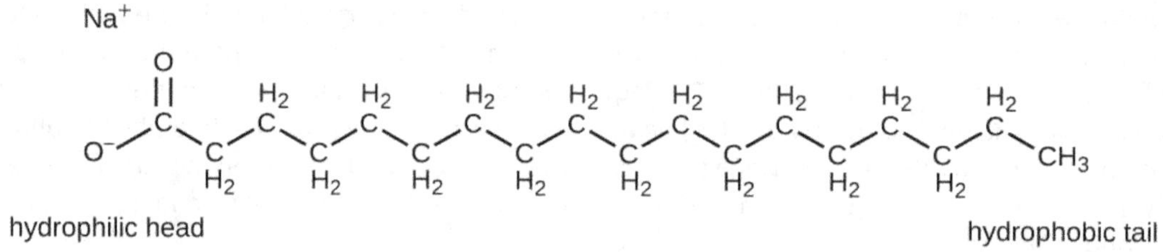

Figure 13.25 Soaps are the salts (sodium salt in the illustration) of fatty acids and have the ability to emulsify lipids, fats, and oils by interacting with water through their hydrophilic heads and with the lipid at their hydrophobic tails.

Soaps, however, often form films that are difficult to rinse away, especially in hard water, which contains high concentrations of calcium and magnesium mineral salts. Detergents contain synthetic surfactant molecules with both polar and nonpolar regions that have strong cleansing activity but are more soluble, even in hard water, and, therefore, leave behind no soapy deposits. Anionic detergents, such as those used for laundry, have a negatively charged anion at one end attached to a long hydrophobic chain, whereas cationic detergents have a positively charged cation instead. Cationic detergents include an important class of disinfectants and antiseptics called the **quaternary ammonium salts (quats),** named for the characteristic quaternary nitrogen atom that confers the positive charge (Figure 13.26). Overall, quats have properties similar to phospholipids, having hydrophilic and hydrophobic ends. As such, quats have the ability to insert into the bacterial phospholipid bilayer and disrupt membrane integrity. The cationic charge of quats appears to confer their antimicrobial properties, which are diminished when neutralized. Quats have several useful properties. They are stable, nontoxic, inexpensive, colorless, odorless, and tasteless. They tend to be bactericidal by disrupting membranes. They are also active against fungi, protozoans, and enveloped viruses, but endospores are unaffected. In clinical settings, they may be used as antiseptics or to disinfect surfaces. Mixtures of quats are also commonly found in household cleaners and disinfectants, including many current formulations of Lysol brand products, which contain benzalkonium chlorides as the active ingredients. Benzalkonium chlorides, along with the quat cetylpyrimidine chloride, are also found in products such as skin antiseptics, oral rinses, and mouthwashes.

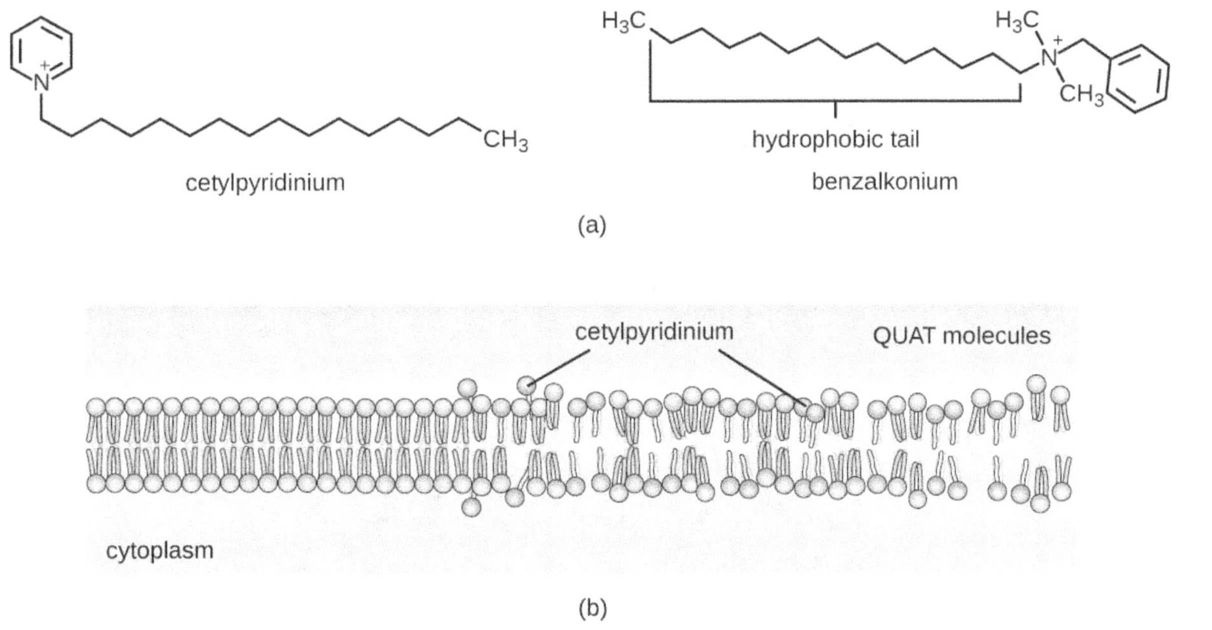

Figure 13.26 (a) Two common quats are benzylalkonium chloride and cetylpyrimidine chloride. Note the hydrophobic nonpolar carbon chain at one end and the nitrogen-containing cationic component at the other end. (b) Quats are able to infiltrate the phospholipid plasma membranes of bacterial cells and disrupt their integrity, leading to death of the cell.

CHECK YOUR UNDERSTANDING

- Why are soaps not considered disinfectants?

MICRO CONNECTIONS

Handwashing the Right Way

Handwashing is critical for public health and should be emphasized in a clinical setting. For the general public, the CDC recommends handwashing before, during, and after food handling; before eating; before and after interacting with someone who is ill; before and after treating a wound; after using the toilet or changing diapers; after coughing, sneezing, or blowing the nose; after handling garbage; and after interacting with an animal, its feed, or its waste. Figure 13.27 illustrates the five steps of proper handwashing recommended by the CDC.

Handwashing is even more important for health-care workers, who should wash their hands thoroughly between every patient contact, after the removal of gloves, after contact with bodily fluids and potentially infectious fomites, and before and after assisting a surgeon with invasive procedures. Even with the use of proper surgical attire, including gloves, scrubbing for surgery is more involved than routine handwashing. The goal of surgical scrubbing is to reduce the normal microbiota on the skin's surface to prevent the introduction of these microbes into a patient's surgical wounds.

There is no single widely accepted protocol for surgical scrubbing. Protocols for length of time spent scrubbing may depend on the antimicrobial used; health-care workers should always check the manufacturer's recommendations. According to the Association of Surgical Technologists (AST), surgical scrubs may be performed with or without the use of brushes (Figure 13.27).

CDC handwashing recommendations for the general public

(1) **Wet** your hands with clean, running water (warm or cold), turn off the tap, and apply soap.

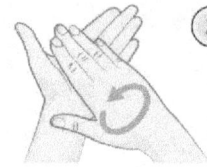
(2) **Lather** your hands by rubbing them together with the soap. Be sure to lather the backs of your hands, between your fingers, and under your nails.

(3) **Scrub** your hands for at least 20 seconds. Need a timer? Hum the "Happy Birthday" song from beginning to end twice.

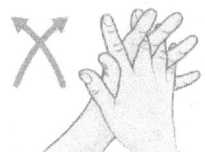
(4) **Rinse** your hands well under clean, running water.

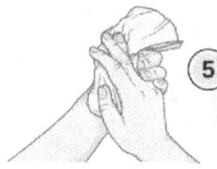
(5) **Dry** your hands using a clean towel or air-dry them.

(a)

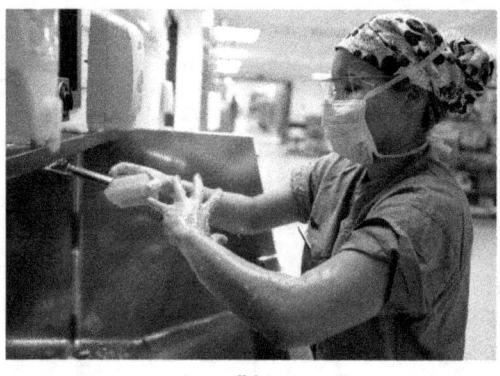

(b)

Figure 13.27 (a) The CDC recommends five steps as part of typical handwashing for the general public. (b) Surgical scrubbing is more extensive, requiring scrubbing starting from the fingertips, extending to the hands and forearms, and then up beyond the elbows, as shown here. (credit a: modification of work by World Health Organization; credit b: b: Staff Sgt. Kevin Iinuma / US Air Force; Public Domain)

LINK TO LEARNING

To learn more (https://openstax.org/l/22CDChandwash) about proper handwashing, visit the CDC's website.

Bisbiguanides

Bisbiguanides were first synthesized in the 20th century and are cationic (positively charged) molecules known for their antiseptic properties (Figure 13.28). One important **bisbiguanide** antiseptic is chlorhexidine. It has broad-spectrum activity against yeasts, gram-positive bacteria, and gram-negative bacteria, with the exception of *Pseudomonas aeruginosa*, which may develop resistance on repeated exposure.[20] Chlorhexidine disrupts cell membranes and is bacteriostatic at lower concentrations or bactericidal at higher concentrations, in which it actually causes the cells' cytoplasmic contents to congeal. It also has activity against enveloped viruses. However, chlorhexidine is poorly effective against *Mycobacterium tuberculosis* and nonenveloped viruses, and it is not sporicidal. Chlorhexidine is typically used in the clinical setting as a surgical scrub and for other handwashing needs for medical personnel, as well as for topical antisepsis for patients before surgery or needle injection. It is more persistent than iodophors, providing long-lasting antimicrobial activity. Chlorhexidine solutions may also be used as oral rinses after oral procedures or to treat gingivitis. Another bisbiguanide, alexidine, is gaining popularity as a surgical scrub and an oral rinse because it acts faster than chlorhexidine.

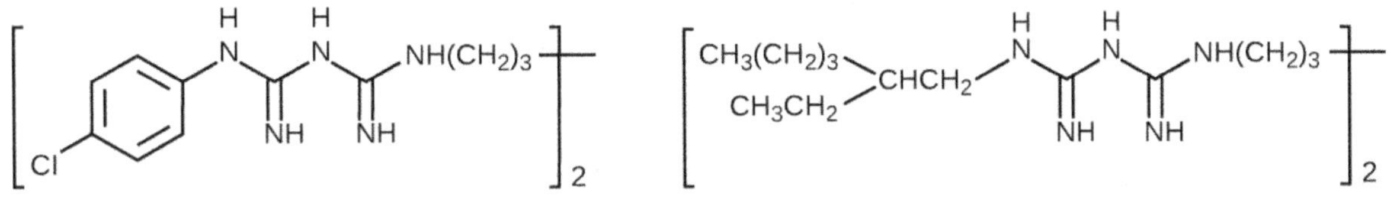

Figure 13.28 The bisbiguanides chlorhexadine and alexidine are cationic antiseptic compounds commonly used as surgical scrubs.

✓ CHECK YOUR UNDERSTANDING

- What two effects does chlorhexidine have on bacterial cells?

Alkylating Agents

The **alkylating agent**s are a group of strong disinfecting chemicals that act by replacing a hydrogen atom within a molecule with an alkyl group (C_nH_{2n+1}), thereby inactivating enzymes and nucleic acids (Figure 13.29). The alkylating agent formaldehyde (CH_2OH) is commonly used in solution at a concentration of 37% (known as formalin) or as a gaseous disinfectant and biocide. It is a strong, broad-spectrum disinfectant and biocide that has the ability to kill bacteria, viruses, fungi, and endospores, leading to sterilization at low temperatures, which is sometimes a convenient alternative to the more labor-intensive heat sterilization methods. It also cross-links proteins and has been widely used as a chemical fixative. Because of this, it is used for the storage of tissue specimens and as an embalming fluid. It also has been used to inactivate infectious agents in vaccine preparation. Formaldehyde is very irritating to living tissues and is also carcinogenic; therefore, it is not used as an antiseptic.

Glutaraldehyde is structurally similar to formaldehyde but has two reactive aldehyde groups, allowing it to act more quickly than formaldehyde. It is commonly used as a 2% solution for sterilization and is marketed under the brand name Cidex. It is used to disinfect a variety of surfaces and surgical and medical equipment. However, similar to formaldehyde, glutaraldehyde irritates the skin and is not used as an antiseptic.

A new type of disinfectant gaining popularity for the disinfection of medical equipment is o-phthalaldehyde (OPA), which is found in some newer formulations of Cidex and similar products, replacing glutaraldehyde. o-Phthalaldehyde also has two reactive aldehyde groups, but they are linked by an aromatic bridge. o-Phthalaldehyde is thought to work similarly to glutaraldehyde and formaldehyde, but is much less irritating to skin and nasal passages, produces a minimal odor, does not require processing before use, and is more effective against mycobacteria.

---

20 L. Thomas et al. "Development of Resistance to Chlorhexidine Diacetate in *Pseudomonas aeruginosa* and the Effect of a 'Residual' Concentration." *Journal of Hospital Infection* 46 no. 4 (2000):297–303.

Ethylene oxide is a type of alkylating agent that is used for gaseous sterilization. It is highly penetrating and can sterilize items within plastic bags such as catheters, disposable items in laboratories and clinical settings (like packaged Petri dishes), and other pieces of equipment. Ethylene oxide exposure is a form of cold sterilization, making it useful for the sterilization of heat-sensitive items. Great care needs to be taken with the use of ethylene oxide, however; it is carcinogenic, like the other alkylating agents, and is also highly explosive. With careful use and proper aeration of the products after treatment, ethylene oxide is highly effective, and ethylene oxide sterilizers are commonly found in medical settings for sterilizing packaged materials.

β-Propionolactone is an alkylating agent with a different chemical structure than the others already discussed. Like other alkylating agents, β-propionolactone binds to DNA, thereby inactivating it (Figure 13.29). It is a clear liquid with a strong odor and has the ability to kill endospores. As such, it has been used in either liquid form or as a vapor for the sterilization of medical instruments and tissue grafts, and it is a common component of vaccines, used to maintain their sterility. It has also been used for the sterilization of nutrient broth, as well as blood plasma, milk, and water. It is quickly metabolized by animals and humans to lactic acid. It is also an irritant, however, and may lead to permanent damage of the eyes, kidneys, or liver. Additionally, it has been shown to be carcinogenic in animals; thus, precautions are necessary to minimize human exposure to β-propionolactone.[21]

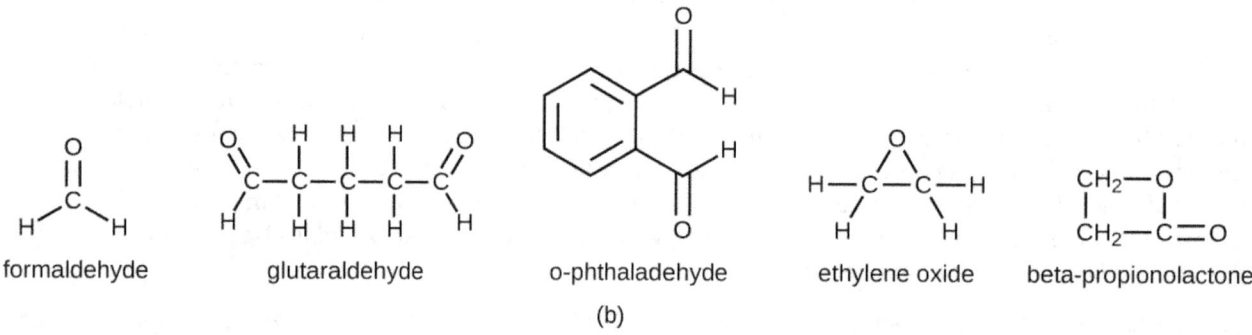

Figure 13.29 (a) Alkylating agents replace hydrogen atoms with alkyl groups. Here, guanine is alkylated, resulting in its hydrogen bonding with thymine, instead of cytosine. (b) The chemical structures of several alkylating agents.

CHECK YOUR UNDERSTANDING

- What chemical reaction do alkylating agents participate in?
- Why are alkylating agents not used as antiseptics?

---

21 Institute of Medicine. "Long-Term Health Effects of Participation in Project SHAD (Shipboard Hazard and Defense)." Washington, DC: The National Academies Press, 2007.

MICRO CONNECTIONS

Diehard Prions

Prions, the acellular, misfolded proteins responsible for incurable and fatal diseases such as kuru and Creutzfeldt-Jakob disease (see Viroids, Virusoids, and Prions), are notoriously difficult to destroy. Prions are extremely resistant to heat, chemicals, and radiation. They are also extremely infectious and deadly; thus, handling and disposing of prion-infected items requires extensive training and extreme caution.

Typical methods of disinfection can reduce but not eliminate the infectivity of prions. Autoclaving is not completely effective, nor are chemicals such as phenol, alcohols, formalin, and β-propiolactone. Even when fixed in formalin, affected brain and spinal cord tissues remain infectious.

Personnel who handle contaminated specimens or equipment or work with infected patients must wear a protective coat, face protection, and cut-resistant gloves. Any contact with skin must be immediately washed with detergent and warm water without scrubbing. The skin should then be washed with 1 N NaOH or a 1:10 dilution of bleach for 1 minute. Contaminated waste must be incinerated or autoclaved in a strong basic solution, and instruments must be cleaned and soaked in a strong basic solution.

LINK TO LEARNING

For more information on the handling of animals and prion-contaminated materials, visit the guidelines published on the WHO (https://openstax.org/l/22WHOhandanipri) website.

Peroxygens

Peroxygens are strong oxidizing agents that can be used as disinfectants or antiseptics. The most widely used **peroxygen** is hydrogen peroxide (H_2O_2), which is often used in solution to disinfect surfaces and may also be used as a gaseous agent. Hydrogen peroxide solutions are inexpensive skin antiseptics that break down into water and oxygen gas, both of which are environmentally safe. This decomposition is accelerated in the presence of light, so hydrogen peroxide solutions typically are sold in brown or opaque bottles. One disadvantage of using hydrogen peroxide as an antiseptic is that it also causes damage to skin that may delay healing or lead to scarring. Contact lens cleaners often include hydrogen peroxide as a disinfectant.

Hydrogen peroxide works by producing free radicals that damage cellular macromolecules. Hydrogen peroxide has broad-spectrum activity, working against gram-positive and gram-negative bacteria (with slightly greater efficacy against gram-positive bacteria), fungi, viruses, and endospores. However, bacteria that produce the oxygen-detoxifying enzymes catalase or peroxidase may have inherent tolerance to low hydrogen peroxide concentrations (Figure 13.30). To kill endospores, the length of exposure or concentration of solutions of hydrogen peroxide must be increased. Gaseous hydrogen peroxide has greater efficacy and can be used as a sterilant for rooms or equipment.

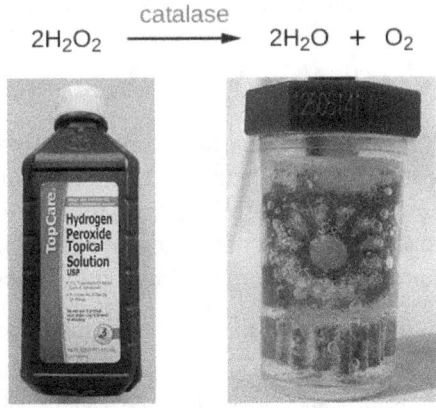

Figure 13.30 Catalase enzymatically converts highly reactive hydrogen peroxide (H_2O_2) into water and oxygen. Hydrogen peroxide can be

used to clean wounds. Hydrogen peroxide is used to sterilize items such as contact lenses. (credit photos: modification of work by Kerry Ceszyk)

Plasma, a hot, ionized gas, described as the fourth state of matter, is useful for sterilizing equipment because it penetrates surfaces and kills vegetative cells and endospores. Hydrogen peroxide and peracetic acid, another commonly used peroxygen, each may be introduced as a plasma. Peracetic acid can be used as a liquid or plasma sterilant insofar as it readily kills endospores, is more effective than hydrogen peroxide even at rather low concentrations, and is immune to inactivation by catalases and peroxidases. It also breaks down to environmentally innocuous compounds; in this case, acetic acid and oxygen.

Other examples of peroxygens include benzoyl peroxide and carbamide peroxide. Benzoyl peroxide is a peroxygen that used in acne medication solutions. It kills the bacterium *Propionibacterium acnes*, which is associated with acne. Carbamide peroxide, an ingredient used in toothpaste, is a peroxygen that combats oral biofilms that cause tooth discoloration and halitosis (bad breath).[22] Last, ozone gas is a peroxygen with disinfectant qualities and is used to clean air or water supplies. Overall, peroxygens are highly effective and commonly used, with no associated environmental hazard.

CHECK YOUR UNDERSTANDING

- How do peroxides kill cells?

Supercritical Fluids

Within the last 15 years, the use of **supercritical fluids**, especially supercritical carbon dioxide ($scCO_2$), has gained popularity for certain sterilizing applications. When carbon dioxide is brought to approximately 10 times atmospheric pressure, it reaches a supercritical state that has physical properties between those of liquids and gases. Materials put into a chamber in which carbon dioxide is pressurized in this way can be sterilized because of the ability of $scCO_2$ to penetrate surfaces.

Supercritical carbon dioxide works by penetrating cells and forming carbonic acid, thereby lowering the cell pH considerably. This technique is effective against vegetative cells and is also used in combination with peracetic acid to kill endospores. Its efficacy can also be augmented with increased temperature or by rapid cycles of pressurization and depressurization, which more likely produce cell lysis.

Benefits of $scCO_2$ include the nonreactive, nontoxic, and nonflammable properties of carbon dioxide, and this protocol is effective at low temperatures. Unlike other methods, such as heat and irradiation, that can degrade the object being sterilized, the use of $scCO_2$ preserves the object's integrity and is commonly used for treating foods (including spices and juices) and medical devices such as endoscopes. It is also gaining popularity for disinfecting tissues such as skin, bones, tendons, and ligaments prior to transplantation. $scCO_2$ can also be used for pest control because it can kill insect eggs and larvae within products.

CHECK YOUR UNDERSTANDING

- Why is the use of supercritical carbon dioxide gaining popularity for commercial and medical uses?

Chemical Food Preservatives

Chemical preservatives are used to inhibit microbial growth and minimize spoilage in some foods. Commonly used chemical preservatives include sorbic acid, benzoic acid, and propionic acid, and their more soluble salts potassium sorbate, sodium benzoate, and calcium propionate, all of which are used to control the growth of molds in acidic foods. Each of these preservatives is nontoxic and readily metabolized by humans. They are also flavorless, so they do not compromise the flavor of the foods they preserve.

Sorbic and benzoic acids exhibit increased efficacy as the pH decreases. Sorbic acid is thought to work by

---

22 Yao, C.S. et al. "In vitro antibacterial effect of carbamide peroxide on oral biofilm." *Journal of Oral Microbiology* Jun 12, 2013. http://www.ncbi.nlm.nih.gov/pmc/articles/PMC3682087/. doi: 10.3402/jom.v5i0.20392.

inhibiting various cellular enzymes, including those in the citric acid cycle, as well as catalases and peroxidases. It is added as a preservative in a wide variety of foods, including dairy, bread, fruit, and vegetable products. Benzoic acid is found naturally in many types of fruits and berries, spices, and fermented products. It is thought to work by decreasing intracellular pH, interfering with mechanisms such as oxidative phosphorylation and the uptake of molecules such as amino acids into cells. Foods preserved with benzoic acid or sodium benzoate include fruit juices, jams, ice creams, pastries, soft drinks, chewing gum, and pickles.

Propionic acid is thought to both inhibit enzymes and decrease intracellular pH, working similarly to benzoic acid. However, propionic acid is a more effective preservative at a higher pH than either sorbic acid or benzoic acid. Propionic acid is naturally produced by some cheeses during their ripening and is added to other types of cheese and baked goods to prevent mold contamination. It is also added to raw dough to prevent contamination by the bacterium *Bacillus mesentericus*, which causes bread to become ropy.

Other commonly used chemical preservatives include sulfur dioxide and nitrites. Sulfur dioxide prevents browning of foods and is used for the preservation of dried fruits; it has been used in winemaking since ancient times. Sulfur dioxide gas dissolves in water readily, forming sulfites. Although sulfites can be metabolized by the body, some people have sulfite allergies, including asthmatic reactions. Additionally, sulfites degrade thiamine, an important nutrient in some foods. The mode of action of sulfites is not entirely clear, but they may interfere with the disulfide bond (see Figure 7.21) formation in proteins, inhibiting enzymatic activity. Alternatively, they may reduce the intracellular pH of the cell, interfering with proton motive force-driven mechanisms.

Nitrites are added to processed meats to maintain color and stop the germination of *Clostridium botulinum* endospores. Nitrites are reduced to nitric oxide, which reacts with heme groups and iron-sulfur groups. When nitric oxide reacts with the heme group within the myoglobin of meats, a red product forms, giving meat its red color. Alternatively, it is thought that when nitric acid reacts with the iron-sulfur enzyme ferredoxin within bacteria, this electron transport-chain carrier is destroyed, preventing ATP synthesis. Nitrosamines, however, are carcinogenic and can be produced through exposure of nitrite-preserved meats (e.g., hot dogs, lunch meat, breakfast sausage, bacon, meat in canned soups) to heat during cooking.

Natural Chemical Food Preservatives

The discovery of natural antimicrobial substances produced by other microbes has added to the arsenal of preservatives used in food. Nisin is an antimicrobial peptide produced by the bacterium *Lactococcus lactis* and is particularly effective against gram-positive organisms. Nisin works by disrupting cell wall production, leaving cells more prone to lysis. It is used to preserve cheeses, meats, and beverages.

Natamycin is an antifungal macrolide antibiotic produced by the bacterium *Streptomyces natalensis*. It was approved by the FDA in 1982 and is used to prevent fungal growth in various types of dairy products, including cottage cheese, sliced cheese, and shredded cheese. Natamycin is also used for meat preservation in countries outside the United States.

✓ CHECK YOUR UNDERSTANDING

- What are the advantages and drawbacks of using sulfites and nitrites as food preservatives?

Chemical Disinfectants

Chemical	Mode of Action	Example Uses
Phenolics		

Chemical Disinfectants

Chemical	Mode of Action	Example Uses
Cresols o-Phenylphenol Hexachlorophene Triclosan	Denature proteins and disrupt membranes	Disinfectant in Lysol Prevent contamination of crops (citrus) Antibacterial soap pHisoHex for handwashing in hospitals

Metals

Mercury Silver Copper Nickel Zinc	Bind to proteins and inhibit enzyme activity	Topical antiseptic Treatment of wounds and burns Prevention of eye infections in newborns Antibacterial in catheters and bandages Mouthwash Algicide for pools and fish tanks Containers for long-term water storage

Halogens

Iodine Chlorine Fluorine	Oxidation and destabilization of cellular macromolecules	Topical antiseptic Hand scrub for medical personnel Water disinfectant Water treatment plants Household bleach Food processing Prevention of dental carries

Alcohols

Ethanol Isopropanol	Denature proteins and disrupt membranes	Disinfectant Antiseptic

Surfactants

Quaternary ammonium salts	Lowers surface tension of water to help with washing away of microbes, and disruption of cell membranes	Soaps and detergent Disinfectant Antiseptic Mouthwash

Bisbiguanides

Chemical Disinfectants

Chemical	Mode of Action	Example Uses
Chlorhexidine Alexidine	Disruption of cell membranes	Oral rinse Hand scrub for medical personnel

Alkylating Agents

Chemical	Mode of Action	Example Uses
Formaldehyde Glutaraldehyde o-Phthalaldehyde Ethylene oxide β-Propionolactone	Inactivation of enzymes and nucleic acid	Disinfectant Tissue specimen storage Embalming Sterilization of medical equipment Vaccine component for sterility

Peroxygens

Chemical	Mode of Action	Example Uses
Hydrogen peroxide Peracetic acid Benzoyl peroxide Carbamide peroxide Ozone gas	Oxidation and destabilization of cellular macromolecules	Antiseptic Disinfectant Acne medication Toothpaste ingredient

Supercritical Gases

Chemical	Mode of Action	Example Uses
Carbon dioxide	Penetrates cells, forms carbonic acid, lowers intracellular pH	Food preservation Disinfection of medical devices Disinfection of transplant tissues

Chemical Food Preservatives

Chemical	Mode of Action	Example Uses
Sorbic acid Benzoic acid Propionic acid Potassium sorbate Sodium benzoate Calcium propionate Sulfur dioxide Nitrites	Decrease pH and inhibit enzymatic function	Preservation of food products

Natural Food Preservatives

Chemical Disinfectants

Chemical	Mode of Action	Example Uses
Nisin Natamycin	Inhibition of cell wall synthesis (Nisin)	Preservation of dairy products, meats, and beverages

13.4 Testing the Effectiveness of Antiseptics and Disinfectants

Learning Objectives

By the end of this section, you will be able to:
- Describe why the phenol coefficient is used
- Compare and contrast the disk-diffusion, use-dilution, and in-use methods for testing the effectiveness of antiseptics, disinfectants, and sterilants

The effectiveness of various chemical disinfectants is reflected in the terms used to describe them. Chemical disinfectants are grouped by the power of their activity, with each category reflecting the types of microbes and viruses its component disinfectants are effective against. High-level germicides have the ability to kill vegetative cells, fungi, viruses, and endospores, leading to sterilization, with extended use. Intermediate-level germicides, as their name suggests, are less effective against endospores and certain viruses, and low-level germicides kill only vegetative cells and certain enveloped viruses, and are ineffective against endospores.

However, several environmental conditions influence the potency of an antimicrobial agent and its effectiveness. For example, length of exposure is particularly important, with longer exposure increasing efficacy. Similarly, the concentration of the chemical agent is also important, with higher concentrations being more effective than lower ones. Temperature, pH, and other factors can also affect the potency of a disinfecting agent.

One method to determine the effectiveness of a chemical agent includes swabbing surfaces before and after use to confirm whether a sterile field was maintained during use. Additional tests are described in the sections that follow. These tests allow for the maintenance of appropriate disinfection protocols in clinical settings, controlling microbial growth to protect patients, health-care workers, and the community.

Phenol Coefficient

The effectiveness of a disinfectant or antiseptic can be determined in a number of ways. Historically, a chemical agent's effectiveness was often compared with that of phenol, the first chemical agent used by Joseph Lister. In 1903, British chemists Samuel Rideal (1863–1929) and J. T. Ainslie Walker (1868–1930) established a protocol to compare the effectiveness of a variety of chemicals with that of phenol, using as their test organisms *Staphylococcus aureus* (a gram-positive bacterium) and *Salmonella enterica* serovar Typhi (a gram-negative bacterium). They exposed the test bacteria to the antimicrobial chemical solutions diluted in water for 7.5 minutes. They then calculated a phenol coefficient for each chemical for each of the two bacteria tested. A **phenol coefficient** of 1.0 means that the chemical agent has about the same level of effectiveness as phenol. A chemical agent with a phenol coefficient of less than 1.0 is less effective than phenol. An example is formalin, with phenol coefficients of 0.3 (*S. aureus*) and 0.7 (*S. enterica* serovar Typhi). A chemical agent with a phenol coefficient greater than 1.0 is more effective than phenol, such as chloramine, with phenol coefficients of 133 and 100, respectively. Although the phenol coefficient was once a useful measure of effectiveness, it is no longer commonly used because the conditions and organisms used were arbitrarily chosen.

⊘ CHECK YOUR UNDERSTANDING

- What are the differences between the three levels of disinfectant effectiveness?

Disk-Diffusion Method

The **disk-diffusion method** involves applying different chemicals to separate, sterile filter paper disks (Figure 13.31). The disks are then placed on an agar plate that has been inoculated with the targeted bacterium and the chemicals diffuse out of the disks into the agar where the bacteria have been inoculated. As the "lawn" of bacteria grows, zones of inhibition of microbial growth are observed as clear areas around the disks. Although there are other factors that contribute to the sizes of zones of inhibition (e.g., whether the agent is water soluble and able to diffuse in the agar), larger zones typically correlate to increased inhibition effectiveness of the chemical agent. The diameter across each zone is measured in millimeters.

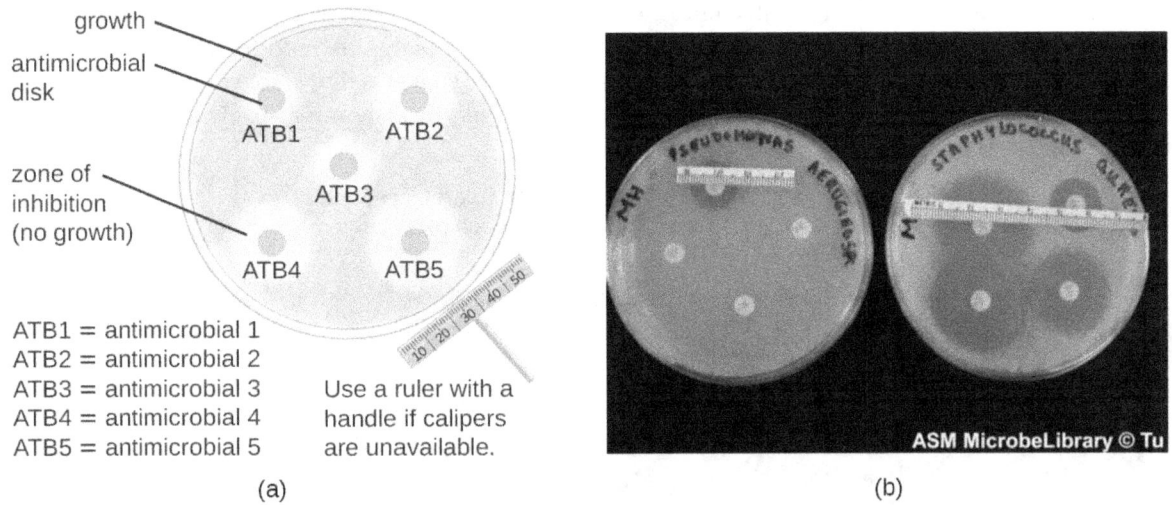

Figure 13.31 A disk-diffusion assay is used to determine the effectiveness of chemical agents against a particular microbe. (a) A plate is inoculated with various antimicrobial discs. The zone of inhibition around each disc indicates how effective that antimicrobial is against the particular species being tested. (b) On these plates, four antimicrobial agents are tested for efficacy in killing *Pseudomonas aeruginosa* (left) and *Staphylococcus aureus* (right). These antimicrobials are much more effective at killing *S. aureus*, as indicated by the size of the zones of inhibition. (credit b: modification of work by American Society for Microbiology)

✓ CHECK YOUR UNDERSTANDING

- When comparing the activities of two disinfectants against the same microbe, using the disk-diffusion assay, and assuming both are water soluble and can easily diffuse in the agar, would a more effective disinfectant have a larger zone of inhibition or a smaller one?

Use-Dilution Test

Other methods are also used for measuring the effectiveness of a chemical agent in clinical settings. The **use-dilution test** is commonly used to determine a chemical's disinfection effectiveness on an inanimate surface. For this test, a cylinder of stainless steel is dipped in a culture of the targeted microorganism and then dried. The cylinder is then dipped in solutions of disinfectant at various concentrations for a specified amount of time. Finally, the cylinder is transferred to a new test tube containing fresh sterile medium that does not contain disinfectant, and this test tube is incubated. Bacterial survival is demonstrated by the presence of turbidity in the medium, whereas killing of the target organism on the cylinder by the disinfectant will produce no turbidity.

The Association of Official Agricultural Chemists International (AOAC), a nonprofit group that establishes many protocol standards, has determined that a minimum of 59 of 60 replicates must show no growth in such a test to achieve a passing result, and the results must be repeatable from different batches of disinfectant and when performed on different days. Disinfectant manufacturers perform use-dilution tests to validate the efficacy claims for their products, as designated by the EPA.

CHECK YOUR UNDERSTANDING

- Is the use-dilution test performed in a clinical setting? Why?

In-Use Test

An **in-use test** can determine whether an actively used solution of disinfectant in a clinical setting is microbially contaminated (Figure 13.32). A 1-mL sample of the used disinfectant is diluted into 9 mL of sterile broth medium that also contains a compound to inactivate the disinfectant. Ten drops, totaling approximately 0.2 mL of this mixture, are then inoculated onto each of two agar plates. One plate is incubated at 37 °C for 3 days and the other is incubated at room temperature for 7 days. The plates are monitored for growth of microbial colonies. Growth of five or more colonies on either plate suggests that viable microbial cells existed in the disinfectant solution and that it is contaminated. Such in-use tests monitor the effectiveness of disinfectants in the clinical setting.

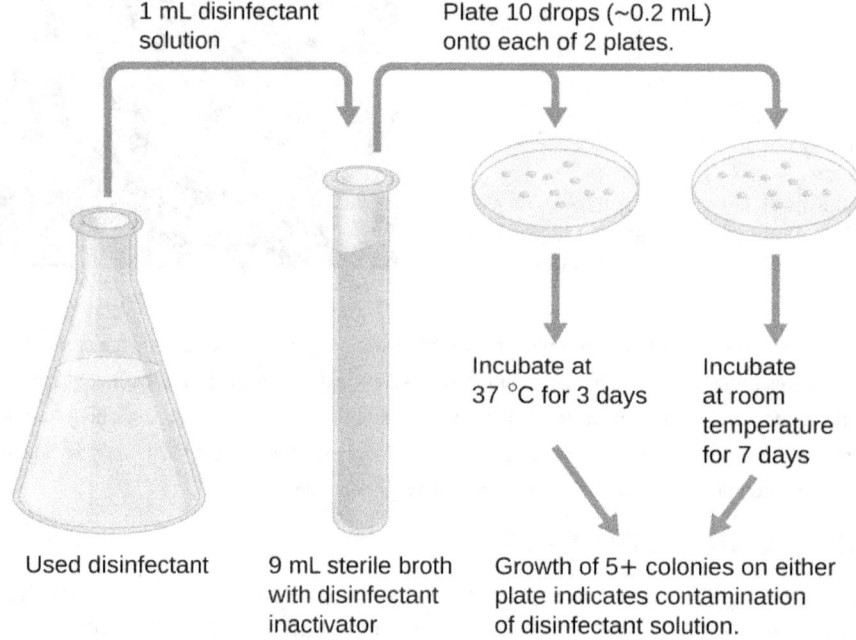

Figure 13.32 Used disinfectant solutions in a clinical setting can be checked with the in-use test for contamination with microbes.

CHECK YOUR UNDERSTANDING

- What does a positive in-use test indicate?

Clinical Focus

Resolution

Despite antibiotic treatment, Roberta's symptoms worsened. She developed pyelonephritis, a severe kidney infection, and was rehospitalized in the intensive care unit (ICU). Her condition continued to deteriorate, and she developed symptoms of septic shock. At this point, her physician ordered a culture from her urine to determine the exact cause of her infection, as well as a drug sensitivity test to determine what antibiotics would be effective against the causative bacterium. The results of this test indicated resistance to a wide range of antibiotics, including the carbapenems, a class of antibiotics that are used as the last resort for many types of bacterial infections. This was an alarming outcome, suggesting that Roberta's infection was caused by a so-called superbug: a bacterial strain that has developed resistance to the majority of commonly used antibiotics. In this case, the causative agent belonged to the carbapenem-resistant Enterobacteriaceae (CRE), a drug-resistant family of bacteria normally found in the digestive system (Figure 13.33). When CRE is introduced to other body systems, as might occur through improperly cleaned surgical instruments, catheters, or endoscopes, aggressive infections can occur.

CRE infections are notoriously difficult to treat, with a 40%–50% fatality rate. To treat her kidney infection and septic shock, Roberta was treated with dialysis, intravenous fluids, and medications to maintain blood pressure and prevent blood clotting. She was also started on aggressive treatment with intravenous administration of a new drug called tigecycline, which has been successful in treating infections caused by drug-resistant bacteria.

After several weeks in the ICU, Roberta recovered from her CRE infection. However, public health officials soon noticed that Roberta's case was not isolated. Several patients who underwent similar procedures at the same hospital also developed CRE infections, some dying as a result. Ultimately, the source of the infection was traced to the duodenoscopes used in the procedures. Despite the hospital staff meticulously following manufacturer protocols for disinfection, bacteria, including CRE, remained within the instruments and were introduced to patients during procedures.

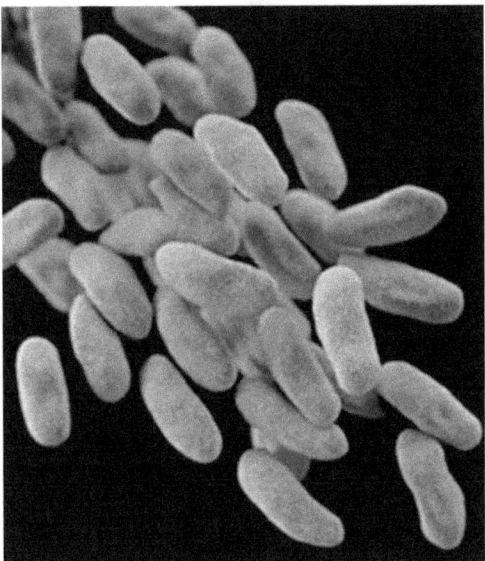

Figure 13.33 CRE is an extremely drug-resistant strain of bacteria that is typically associated with nosocomial infections. (credit: Centers for Disease Control and Prevention)

Go back to the previous Clinical Focus box.

Eye on Ethics

Who Is Responsible?

Carbapenem-resistant Enterobacteriaceae infections due to contaminated endoscopes have become a high-profile problem in recent years. Several CRE outbreaks have been traced to endoscopes, including a case at Ronald Reagan UCLA Medical Center in early 2015 in which 179 patients may have been exposed to a contaminated endoscope. Seven of the patients developed infections, and two later died. Several lawsuits have been filed against Olympus, the manufacturer of the endoscopes. Some claim that Olympus did not obtain FDA approval for design changes that may have led to contamination, and others claim that the manufacturer knowingly withheld information from hospitals concerning defects in the endoscopes.

Lawsuits like these raise difficult-to-answer questions about liability. Invasive procedures are inherently risky, but negative outcomes can be minimized by strict adherence to established protocols. Who is responsible, however, when negative outcomes occur due to flawed protocols or faulty equipment? Can hospitals or health-care workers be held liable if they have strictly followed a flawed procedure? Should manufacturers be held liable—and perhaps be driven out of business—if their lifesaving equipment fails or is found defective? What is the government's role in ensuring that use and maintenance of medical equipment and protocols are fail-safe?

Protocols for cleaning or sterilizing medical equipment are often developed by government agencies like the FDA, and other groups, like the AOAC, a nonprofit scientific organization that establishes many protocols for standard use globally. These procedures and protocols are then adopted by medical device and equipment manufacturers. Ultimately, the end-users (hospitals and their staff) are responsible for following these procedures and can be held liable if a breach occurs and patients become ill from improperly cleaned equipment.

Unfortunately, protocols are not infallible, and sometimes it takes negative outcomes to reveal their flaws. In 2008, the FDA had approved a disinfection protocol for endoscopes, using glutaraldehyde (at a lower concentration when mixed with phenol), o-phthalaldehyde, hydrogen peroxide, peracetic acid, and a mix of hydrogen peroxide with peracetic acid. However, subsequent CRE outbreaks from endoscope use showed that this protocol alone was inadequate.

As a result of CRE outbreaks, hospitals, manufacturers, and the FDA are investigating solutions. Many hospitals are instituting more rigorous cleaning procedures than those mandated by the FDA. Manufacturers are looking for ways to redesign duodenoscopes to minimize hard-to-reach crevices where bacteria can escape disinfectants, and the FDA is updating its protocols. In February 2015, the FDA added new recommendations for careful hand cleaning of the duodenoscope elevator mechanism (the location where microbes are most likely to escape disinfection), and issued more careful documentation about quality control of disinfection protocols (Figure 13.34).

There is no guarantee that new procedures, protocols, or equipment will completely eliminate the risk for infection associated with endoscopes. Yet these devices are used successfully in 500,000–650,000 procedures annually in the United States, many of them lifesaving. At what point do the risks outweigh the benefits of these devices, and who should be held responsible when negative outcomes occur?

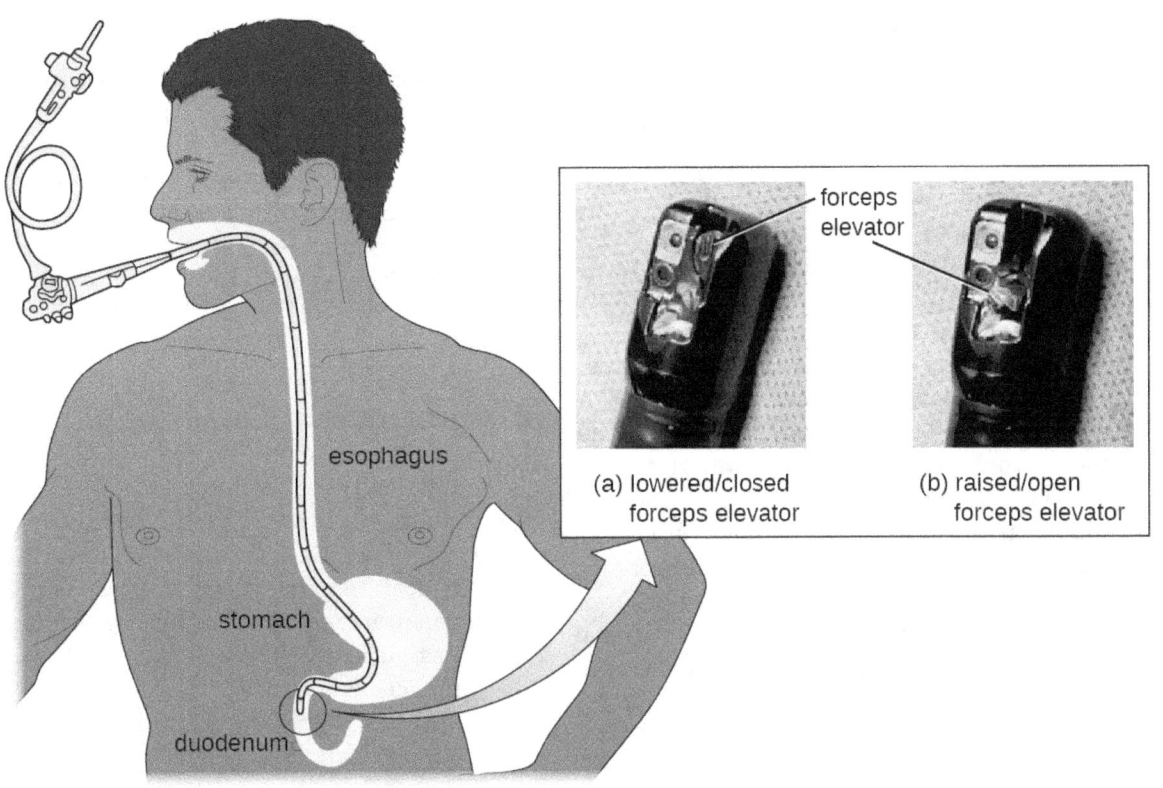

Figure 13.34 The elevator mechanism in a duodenoscope contains crevices that are difficult to disinfect. Pathogens that survive disinfection protocols can be passed from one patient to another, causing serious infections. (credit "photos": modification of work by Centers for Disease Control and Prevention)

SUMMARY

13.1 Controlling Microbial Growth

- Inanimate items that may harbor microbes and aid in their transmission are called **fomites**. The level of cleanliness required for a fomite depends both on the item's use and the infectious agent with which the item may be contaminated.
- The CDC and the NIH have established four **biological safety levels (BSLs)** for laboratories performing research on infectious agents. Each level is designed to protect laboratory personnel and the community. These BSLs are determined by the agent's infectivity, ease of transmission, and potential disease severity, as well as the type of work being performed with the agent.
- **Disinfection** removes potential pathogens from a fomite, whereas **antisepsis** uses antimicrobial chemicals safe enough for tissues; in both cases, microbial load is reduced, but microbes may remain unless the chemical used is strong enough to be a **sterilant**.
- The amount of cleanliness (**sterilization** versus high-level disinfection versus general cleanliness) required for items used clinically depends on whether the item will come into contact with sterile tissues (**critical item**), mucous membranes (**semicritical item**), or intact skin (**noncritical item**).
- Medical procedures with a risk for contamination should be carried out in a **sterile field** maintained by proper **aseptic technique** to prevent **sepsis**.
- Sterilization is necessary for some medical applications as well as in the food industry, where endospores of *Clostridium botulinum* are killed through **commercial sterilization** protocols.
- Physical or chemical methods to control microbial growth that result in death of the microbe are indicated by the suffixes *-cide* or *-cidal* (e.g., as with **bactericides**, **viricides**, and **fungicides**), whereas those that inhibit microbial growth are indicated by the suffixes *-stat* or *-static* (e.g., **bacteriostatic**, **fungistatic**).
- **Microbial death curves** display the logarithmic decline of living microbes exposed to a method of microbial control. The time it takes for a protocol to yield a 1-log (90%) reduction in the microbial population is the **decimal reduction time**, or **D-value**.
- When choosing a microbial control protocol, factors to consider include the length of exposure time, the type of microbe targeted, its susceptibility to the protocol, the intensity of the treatment, the presence of organics that may interfere with the protocol, and the environmental conditions that may alter the effectiveness of the protocol.

13.2 Using Physical Methods to Control Microorganisms

- Heat is a widely used and highly effective method for controlling microbial growth.
- **Dry-heat sterilization** protocols are used commonly in aseptic techniques in the laboratory. However, **moist-heat sterilization** is typically the more effective protocol because it penetrates cells better than dry heat does.
- **Pasteurization** is used to kill pathogens and reduce the number of microbes that cause food spoilage. **High-temperature, short-time pasteurization** is commonly used to pasteurize milk that will be refrigerated; **ultra-high temperature pasteurization** can be used to pasteurize milk for long-term storage without refrigeration.
- Refrigeration slows microbial growth; freezing stops growth, killing some organisms. Laboratory and medical specimens may be frozen on dry ice or at ultra-low temperatures for storage and transport.
- High-pressure processing can be used to kill microbes in food. Hyperbaric oxygen therapy to increase oxygen saturation has also been used to treat certain infections.
- **Desiccation** has long been used to preserve foods and is accelerated through the addition of salt or sugar, which decrease water activity in foods.
- **Lyophilization** combines cold exposure and desiccation for the long-term storage of foods and laboratory materials, but microbes remain and can be rehydrated.
- **Ionizing radiation**, including gamma irradiation, is an effective way to sterilize heat-sensitive and packaged materials. **Nonionizing radiation**, like ultraviolet light, is unable to penetrate surfaces but is useful for surface sterilization.
- **HEPA** filtration is commonly used in hospital ventilation systems and biological safety cabinets in laboratories to prevent transmission of airborne microbes. **Membrane filtration** is

commonly used to remove bacteria from heat-sensitive solutions.

13.3 Using Chemicals to Control Microorganisms

- **Heavy metals**, including mercury, silver, copper, and zinc, have long been used for disinfection and preservation, although some have toxicity and environmental risks associated with them.
- **Halogens**, including chlorine, fluorine, and iodine, are also commonly used for disinfection. Chlorine compounds, including **sodium hypochlorite**, **chloramines**, and **chlorine dioxide**, are commonly used for water disinfection. Iodine, in both **tincture** and **iodophor** forms, is an effective antiseptic.
- **Alcohols**, including ethyl alcohol and isopropyl alcohol, are commonly used antiseptics that act by denaturing proteins and disrupting membranes.
- **Phenolics** are stable, long-acting disinfectants that denature proteins and disrupt membranes. They are commonly found in household cleaners, mouthwashes, and hospital disinfectants, and are also used to preserve harvested crops.
- The phenolic compound **triclosan**, found in antibacterial soaps, plastics, and textiles is technically an antibiotic because of its specific mode of action of inhibiting bacterial fatty-acid synthesis..
- **Surfactants**, including soaps and detergents, lower the surface tension of water to create emulsions that mechanically carry away microbes. Soaps are long-chain fatty acids, whereas detergents are synthetic surfactants.
- **Quaternary ammonium compounds** (**quats**) are cationic detergents that disrupt membranes. They are used in household cleaners, skin disinfectants, oral rinses, and mouthwashes.
- **Bisbiguanides** disrupt cell membranes, causing cell contents to gel. **Chlorhexidine** and **alexidine** are commonly used for surgical scrubs, for handwashing in clinical settings, and in prescription oral rinses.
- **Alkylating agents** effectively sterilize materials at low temperatures but are carcinogenic and may also irritate tissue. **Glutaraldehyde** and **o-phthalaldehyde** are used as hospital disinfectants but not as antiseptics. **Formaldehyde** is used for the storage of tissue specimens, as an embalming fluid, and in vaccine preparation to inactivate infectious agents. **Ethylene oxide** is a gas sterilant that can permeate heat-sensitive packaged materials, but it is also explosive and carcinogenic.
- **Peroxygens**, including **hydrogen peroxide**, **peracetic acid**, **benzoyl peroxide**, and ozone gas, are strong oxidizing agents that produce free radicals in cells, damaging their macromolecules. They are environmentally safe and are highly effective disinfectants and antiseptics.
- Pressurized carbon dioxide in the form of a **supercritical fluid** easily permeates packaged materials and cells, forming carbonic acid and lowering intracellular pH. Supercritical carbon dioxide is nonreactive, nontoxic, nonflammable, and effective at low temperatures for sterilization of medical devices, implants, and transplanted tissues.
- Chemical preservatives are added to a variety of foods. **Sorbic acid**, **benzoic acid**, **propionic acid**, and their more soluble salts inhibit enzymes or reduce intracellular pH.
- **Sulfites** are used in winemaking and food processing to prevent browning of foods.
- **Nitrites** are used to preserve meats and maintain color, but cooking nitrite-preserved meats may produce carcinogenic nitrosamines.
- **Nisin** and **natamycin** are naturally produced preservatives used in cheeses and meats. Nisin is effective against gram-positive bacteria and natamycin against fungi.

13.4 Testing the Effectiveness of Antiseptics and Disinfectants

- Chemical disinfectants are grouped by the types of microbes and infectious agents they are effective against. **High-level germicides** kill vegetative cells, fungi, viruses, and endospores, and can ultimately lead to sterilization. **Intermediate-level germicides** cannot kill all viruses and are less effective against endospores. **Low-level germicides** kill vegetative cells and some enveloped viruses, but are ineffective against endospores.
- The effectiveness of a disinfectant is influenced by several factors, including length of exposure, concentration of disinfectant, temperature, and pH.
- Historically, the effectiveness of a chemical disinfectant was compared with that of phenol at killing *Staphylococcus aureus* and *Salmonella enterica* serovar Typhi, and a **phenol coefficient**

was calculated.
- The **disk-diffusion method** is used to test the effectiveness of a chemical disinfectant against a particular microbe.
- The **use-dilution test** determines the effectiveness of a disinfectant on a surface. **In-use tests** can determine whether disinfectant solutions are being used correctly in clinical settings.

REVIEW QUESTIONS
Multiple Choice

1. Which of the following types of medical items requires sterilization?
   A. needles
   B. bed linens
   C. respiratory masks
   D. blood pressure cuffs

2. Which of the following is suitable for use on tissues for microbial control to prevent infection?
   A. disinfectant
   B. antiseptic
   C. sterilant
   D. water

3. Which biosafety level is appropriate for research with microbes or infectious agents that pose moderate risk to laboratory workers and the community, and are typically indigenous?
   A. BSL-1
   B. BSL-2
   C. BSL-3
   D. BSL-4

4. Which of the following best describes a microbial control protocol that inhibits the growth of molds and yeast?
   A. bacteriostatic
   B. fungicidal
   C. bactericidal
   D. fungistatic

5. The decimal reduction time refers to the amount of time it takes to which of the following?
   A. reduce a microbial population by 10%
   B. reduce a microbial population by 0.1%
   C. reduce a microbial population by 90%
   D. completely eliminate a microbial population

6. Which of the following methods brings about cell lysis due to cavitation induced by rapid localized pressure changes?
   A. microwaving
   B. gamma irradiation
   C. ultraviolet radiation
   D. sonication

7. Which of the following terms is used to describe the time required to kill all of the microbes within a sample at a given temperature?
   A. D-value
   B. thermal death point
   C. thermal death time
   D. decimal reduction time

8. Which of the following microbial control methods does not actually kill microbes or inhibit their growth but instead removes them physically from samples?
   A. filtration
   B. desiccation
   C. lyophilization
   D. nonionizing radiation

9. Which of the following refers to a disinfecting chemical dissolved in alcohol?
   A. iodophor
   B. tincture
   C. phenolic
   D. peroxygen

10. Which of the following peroxygens is widely used as a household disinfectant, is inexpensive, and breaks down into water and oxygen gas?
    A. hydrogen peroxide
    B. peracetic acid
    C. benzoyl peroxide
    D. ozone

11. Which of the following chemical food preservatives is used in the wine industry but may cause asthmatic reactions in some individuals?
    A. nitrites
    B. sulfites
    C. propionic acid
    D. benzoic acid

12. Bleach is an example of which group of chemicals used for disinfection?
    A. heavy metals
    B. halogens
    C. quats
    D. bisbiguanides

13. Which chemical disinfectant works by methylating enzymes and nucleic acids and is known for being toxic and carcinogenic?
    A. sorbic acid
    B. triclosan
    C. formaldehyde
    D. hexaclorophene

14. Which type of test is used to determine whether disinfectant solutions actively used in a clinical setting are being used correctly?
    A. disk-diffusion assay
    B. phenol coefficient test
    C. in-use test
    D. use-dilution test

15. The effectiveness of chemical disinfectants has historically been compared to that of which of the following?
    A. phenol
    B. ethyl alcohol
    C. bleach
    D. formaldehyde

16. Which of the following refers to a germicide that can kill vegetative cells and certain enveloped viruses but not endospores?
    A. high-level germicide
    B. intermediate-level germicide
    C. low-level germicide
    D. sterilant

True/False

17. Sanitization leaves an object free of microbes.
18. Ionizing radiation can penetrate surfaces, but nonionizing radiation cannot.
19. Moist-heat sterilization protocols require the use of higher temperatures for longer periods of time than do dry-heat sterilization protocols do.
20. Soaps are classified as disinfectants.
21. Mercury-based compounds have fallen out of favor for use as preservatives and antiseptics.

Fill in the Blank

22. A medical item that comes into contact with intact skin and does not penetrate sterile tissues or come into contact with mucous membranes is called a(n) _____ item.
23. The goal of _____ _____ protocols is to rid canned produce of *Clostridium botulinum* endospores.
24. In an autoclave, the application of pressure to _____ is increased to allow the steam to achieve temperatures above the boiling point of water.
25. Doorknobs and other surfaces in clinical settings are often coated with _____, _____, or _____ to prevent the transmission of microbes.
26. If a chemical disinfectant is more effective than phenol, then its phenol coefficient would be _____ than 1.0.
27. If used for extended periods of time, _____ germicides may lead to sterility.
28. In the disk-diffusion assay, a large zone of inhibition around a disk to which a chemical disinfectant has been applied indicates _____ of the test microbe to the chemical disinfectant.

Short Answer

29. What are some characteristics of microbes and infectious agents that would require handling in a BSL-3 laboratory?
30. What is the purpose of degerming? Does it completely eliminate microbes?
31. What are some factors that alter the effectiveness of a disinfectant?

32. What is the advantage of HTST pasteurization compared with sterilization? What is an advantage of UHT treatment?
33. How does the addition of salt or sugar help preserve food?
34. Which is more effective at killing microbes: autoclaving or freezing? Explain.
35. Which solution of ethyl alcohol is more effective at inhibiting microbial growth: a 70% solution or a 100% solution? Why?
36. When might a gas treatment be used to control microbial growth instead of autoclaving? What are some examples?
37. What is the advantage of using an iodophor rather than iodine or an iodine tincture?
38. Why were chemical disinfectants once commonly compared with phenol?
39. Why is length of exposure to a chemical disinfectant important for its activity?

Critical Thinking

40. When plotting microbial death curves, how might they look different for bactericidal versus bacteriostatic treatments?
41. What are the benefits of cleaning something to a level of cleanliness beyond what is required? What are some possible disadvantages of doing so?
42. In 2001, endospores of *Bacillus anthracis*, the causative agent of anthrax, were sent to government officials and news agencies via the mail. In response, the US Postal Service began to irradiate mail with UV light. Was this an effective strategy? Why or why not?
43. Looking at Figure 13.29 and reviewing the functional groups in Figure 7.6, which alkylating agent shown lacks an aldehyde group?
44. Do you think naturally produced antimicrobial products like nisin and natamycin should replace sorbic acid for food preservation? Why or why not?
45. Why is the use of skin disinfecting compounds required for surgical scrubbing and not for everyday handwashing?
46. What are some advantages of use-dilution and in-use tests compared with the disk-diffusion assay?

CHAPTER 14
Antimicrobial Drugs

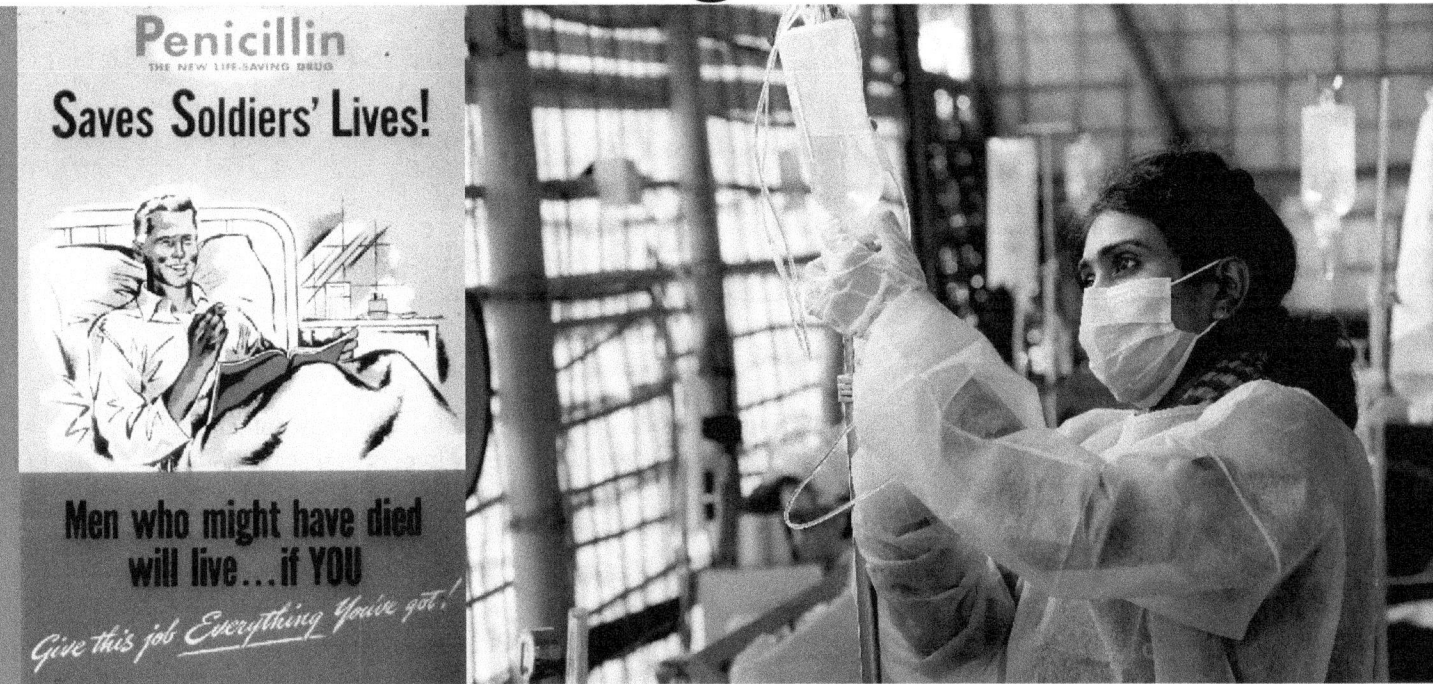

Figure 14.1 First mass produced in the 1940s, penicillin was instrumental in saving millions of lives during World War II and was considered a wonder drug.[1] Today, overprescription of antibiotics (especially for childhood illnesses) has contributed to the evolution of drug-resistant pathogens. (credit left: modification of work by Chemical Heritage Foundation; Credit right: DFID / Flickr; CC-BY)

Chapter Outline

14.1 History of Chemotherapy and Antimicrobial Discovery

14.2 Fundamentals of Antimicrobial Chemotherapy

14.3 Mechanisms of Antibacterial Drugs

14.4 Mechanisms of Other Antimicrobial Drugs

14.5 Drug Resistance

14.6 Testing the Effectiveness of Antimicrobials

14.7 Current Strategies for Antimicrobial Discovery

INTRODUCTION In nature, some microbes produce substances that inhibit or kill other microbes that might otherwise compete for the same resources. Humans have successfully exploited these abilities, using microbes to mass-produce substances that can be used as antimicrobial drugs. Since their discovery, antimicrobial drugs have saved countless lives, and they remain an essential tool for treating and controlling infectious

---

1 "Treatment of War Wounds: A Historical Review." *Clinical Orthopaedics and Related Research* 467 no. 8 (2009):2168–2191.

disease. But their widespread and often unnecessary use has had an unintended side effect: the rise of multidrug-resistant microbial strains. In this chapter, we will discuss how antimicrobial drugs work, why microbes develop resistance, and what health professionals can do to encourage responsible use of antimicrobials.

14.1 History of Chemotherapy and Antimicrobial Discovery

Learning Objectives

By the end of this section, you will be able to:
- Compare and contrast natural, semisynthetic, and synthetic antimicrobial drugs
- Describe the chemotherapeutic approaches of ancient societies
- Describe the historically important individuals and events that led to the development of antimicrobial drugs

---

Clinical Focus

Part 1

Marisa, a 52-year-old woman, was suffering from severe abdominal pain, swollen lymph nodes, fatigue, and a fever. She had just returned home from visiting extended family in her native country of Cambodia. While abroad, she received medical care in neighboring Vietnam for a compressed spinal cord. She still had discomfort when leaving Cambodia, but the pain increased as her trip home continued and her husband drove her straight from the airport to the emergency room.

Her doctor considers whether Marisa could be suffering from appendicitis, a urinary tract infection (UTI), or pelvic inflammatory disease (PID). However, each of those conditions is typically preceded or accompanied by additional symptoms. He considers the treatment she received in Vietnam for her compressed spinal cord, but abdominal pain is not usually associated with spinal cord compression. He examines her health history further.

- What type of infection or other condition may be responsible?
- What type of lab tests might the doctor order?

Jump to the next Clinical Focus box.

Most people associate the term chemotherapy with treatments for cancer. However, chemotherapy is actually a broader term that refers to any use of chemicals or drugs to treat disease. Chemotherapy may involve drugs that target cancerous cells or tissues, or it may involve **antimicrobial drugs** that target infectious microorganisms. Antimicrobial drugs typically work by destroying or interfering with microbial structures and enzymes, either killing microbial cells or inhibiting their growth. But before we examine how these drugs work, we will briefly explore the history of humans' use of antimicrobials for the purpose of chemotherapy.

Use of Antimicrobials in Ancient Societies

Although the discovery of antimicrobials and their subsequent widespread use is commonly associated with modern medicine, there is evidence that humans have been exposed to antimicrobial compounds for millennia. Chemical analyses of the skeletal remains of people from Nubia[2] (now found in present-day Sudan) dating from between 350 and 550 AD have shown residue of the antimicrobial agent tetracycline in high enough quantities to suggest the purposeful fermentation of tetracycline-producing *Streptomyces* during the beer-making process. The resulting beer, which was thick and gruel-like, was used to treat a variety of ailments in both adults and children, including gum disease and wounds. The antimicrobial properties of certain plants may also have been recognized by various cultures around the world, including Indian and Chinese herbalists (Figure 14.2) who have long used plants for a wide variety of medical purposes. Healers of many cultures understood the antimicrobial properties of fungi and their use of moldy bread or other mold-containing products to treat wounds has been well documented for centuries.[3] Today, while about 80% of the world's population still relies on plant-derived medicines,[4] scientists are now discovering the active compounds conferring the medicinal benefits contained in many of these traditionally used plants.

Figure 14.2 For millennia, Chinese herbalists have used many different species of plants for the treatment of a wide variety of human ailments.

CHECK YOUR UNDERSTANDING

- Give examples of how antimicrobials were used in ancient societies.

The First Antimicrobial Drugs

Societies relied on traditional medicine for thousands of years; however, the first half of the 20th century brought an era of strategic drug discovery. In the early 1900s, the German physician and scientist Paul Ehrlich (1854–1915) set out to discover or synthesize chemical compounds capable of killing infectious microbes without harming the patient. In 1909, after screening more than 600 arsenic-containing compounds, Ehrlich's assistant Sahachiro Hata (1873–1938) found one such "magic bullet." Compound 606 targeted the bacterium *Treponema pallidum*, the causative agent of syphilis. Compound 606 was found to successfully cure syphilis in rabbits and soon after was marketed under the name Salvarsan as a remedy for the disease in humans (Figure 14.3). Ehrlich's innovative approach of systematically screening a wide variety of compounds remains a common strategy for the discovery of new antimicrobial agents even today.

---

2 M.L. Nelson et al. "Brief Communication: Mass Spectroscopic Characterization of Tetracycline in the Skeletal Remains of an Ancient Population from Sudanese Nubia 350–550 CE." *American Journal of Physical Anthropology* 143 no. 1 (2010):151–154.

3 M. Wainwright. "Moulds in Ancient and More Recent Medicine." *Mycologist* 3 no. 1 (1989):21–23.

4 S. Verma, S.P. Singh. "Current and Future Status of Herbal Medicines." *Veterinary World* 1 no. 11 (2008):347–350.

Figure 14.3 Paul Ehrlich was influential in the discovery of Compound 606, an antimicrobial agent that proved to be an effective treatment for syphilis.

A few decades later, German scientists Josef Klarer, Fritz Mietzsch, and Gerhard Domagk discovered the antibacterial activity of a synthetic dye, prontosil, that could treat streptococcal and staphylococcal infections in mice. Domagk's own daughter was one of the first human recipients of the drug, which completely cured her of a severe streptococcal infection that had resulted from a poke with an embroidery needle. Gerhard Domagk (1895–1964) was awarded the Nobel Prize in Medicine in 1939 for his work with prontosil and sulfanilamide, the active breakdown product of prontosil in the body. Sulfanilamide, the first synthetic antimicrobial created, served as the foundation for the chemical development of a family of sulfa drugs. A **synthetic antimicrobial** is a drug that is developed from a chemical not found in nature. The success of the sulfa drugs led to the discovery and production of additional important classes of synthetic antimicrobials, including the quinolines and oxazolidinones.

A few years before the discovery of prontosil, scientist Alexander Fleming (1881–1955) made his own accidental discovery that turned out to be monumental. In 1928, Fleming returned from holiday and examined some old plates of staphylococci in his research laboratory at St. Mary's Hospital in London. He observed that contaminating mold growth (subsequently identified as a strain of *Penicillium notatum*) inhibited staphylococcal growth on one plate. Fleming, therefore, is credited with the discovery of **penicillin**, the first **natural antibiotic**, (Figure 14.4). Further experimentation showed that penicillin from the mold was antibacterial against streptococci, meningococci, and *Corynebacterium diphtheriae*, the causative agent of diphtheria.

Fleming and his colleagues were credited with discovering and identifying penicillin, but its isolation and mass production were accomplished by a team of researchers at Oxford University under the direction of Howard Florey (1898–1968) and Ernst Chain (1906–1979) (Figure 14.4). In 1940, the research team purified penicillin and reported its success as an antimicrobial agent against streptococcal infections in mice. Their subsequent work with human subjects also showed penicillin to be very effective. Because of their important work, Fleming, Florey, and Chain were awarded the Nobel Prize in Physiology and Medicine in 1945.

In the early 1940s, scientist Dorothy Hodgkin (1910–1994), who studied crystallography at Oxford University, used X-rays to analyze the structure of a variety of natural products. In 1946, she determined the structure of penicillin, for which she was awarded the Nobel Prize in Chemistry in 1964. Once the structure was understood, scientists could modify it to produce a variety of semisynthetic penicillins. A **semisynthetic antimicrobial** is a chemically modified derivative of a natural antibiotic. The chemical modifications are generally designed to increase the range of bacteria targeted, increase stability, decrease toxicity, or confer other properties beneficial for treating infections.

Penicillin is only one example of a natural antibiotic. Also in the 1940s, Selman Waksman (1888–1973) (Figure 14.5), a prominent soil microbiologist at Rutgers University, led a research team that discovered several antimicrobials, including actinomycin, streptomycin, and neomycin. The discoveries of these antimicrobials stemmed from Waksman's study of fungi and the Actinobacteria, including soil bacteria in the genus *Streptomyces*, known for their natural production of a wide variety of antimicrobials. His work earned him the Nobel Prize in Physiology and Medicine in 1952. The actinomycetes are the source of more than half of all natural antibiotics[5] and continue to serve as an excellent reservoir for the discovery of novel antimicrobial agents. Some researchers argue that we have not yet come close to tapping the full antimicrobial potential of this group.[6]

(a)   (b)

Figure 14.4 (a) Alexander Fleming was the first to discover a naturally produced antimicrobial, penicillin, in 1928. (b) Howard Florey and Ernst Chain discovered how to scale up penicillin production. Then they figured out how to purify it and showed its efficacy as an antimicrobial in animal and human trials in the early 1940s.

Figure 14.5 Selman Waksman was the first to show the vast antimicrobial production capabilities of a group of soil bacteria, the actinomycetes.

---

5 J. Berdy. "Bioactive Microbial Metabolites." *The Journal of Antibiotics* 58 no. 1 (2005):1–26.
6 M. Baltz. "Antimicrobials from Actinomycetes: Back to the Future." *Microbe* 2 no. 3 (2007):125–131.

CHECK YOUR UNDERSTANDING

- Why is the soil a reservoir for antimicrobial resistance genes?

14.2 Fundamentals of Antimicrobial Chemotherapy

Learning Objectives

By the end of this section, you will be able to:
- Contrast bacteriostatic versus bactericidal antibacterial activities
- Contrast broad-spectrum drugs versus narrow-spectrum drugs
- Explain the significance of superinfections
- Discuss the significance of dosage and the route of administration of a drug
- Identify factors and variables that can influence the side effects of a drug
- Describe the significance of positive and negative interactions between drugs

Several factors are important in choosing the most appropriate antimicrobial drug therapy, including bacteriostatic versus bactericidal mechanisms, spectrum of activity, dosage and route of administration, the potential for side effects, and the potential interactions between drugs. The following discussion will focus primarily on antibacterial drugs, but the concepts translate to other antimicrobial classes.

Bacteriostatic Versus Bactericidal

Antibacterial drugs can be either **bacteriostatic** or bactericidal in their interactions with target bacteria. Bacteriostatic drugs cause a reversible inhibition of growth, with bacterial growth restarting after elimination of the drug. By contrast, **bactericidal** drugs kill their target bacteria. The decision of whether to use a bacteriostatic or bactericidal drugs depends on the type of infection and the immune status of the patient. In a patient with strong immune defenses, bacteriostatic and bactericidal drugs can be effective in achieving clinical cure. However, when a patient is immunocompromised, a bactericidal drug is essential for the successful treatment of infections. Regardless of the immune status of the patient, life-threatening infections such as acute endocarditis require the use of a bactericidal drug.

Spectrum of Activity

The spectrum of activity of an antibacterial drug relates to diversity of targeted bacteria. A **narrow-spectrum antimicrobial** targets only specific subsets of bacterial pathogens. For example, some narrow-spectrum drugs only target gram-positive bacteria, whereas others target only gram-negative bacteria. If the pathogen causing an infection has been identified, it is best to use a narrow-spectrum antimicrobial and minimize collateral damage to the normal microbiota. A **broad-spectrum antimicrobial** targets a wide variety of bacterial pathogens, including both gram-positive and gram-negative species, and is frequently used as empiric therapy to cover a wide range of potential pathogens while waiting on the laboratory identification of the infecting pathogen. Broad-spectrum antimicrobials are also used for polymicrobic infections (mixed infection with multiple bacterial species), or as prophylactic prevention of infections with surgery/invasive procedures. Finally, broad-spectrum antimicrobials may be selected to treat an infection when a narrow-spectrum drug fails because of development of drug resistance by the target pathogen.

The risk associated with using broad-spectrum antimicrobials is that they will also target a broad spectrum of the normal microbiota, increasing the risk of a **superinfection**, a secondary infection in a patient having a preexisting infection. A superinfection develops when the antibacterial intended for the preexisting infection kills the protective microbiota, allowing another pathogen resistant to the antibacterial to proliferate and cause a secondary infection (Figure 14.6). Common examples of superinfections that develop as a result of antimicrobial usage include yeast infections (candidiasis) and pseudomembranous colitis caused by *Clostridium difficile*, which can be fatal.

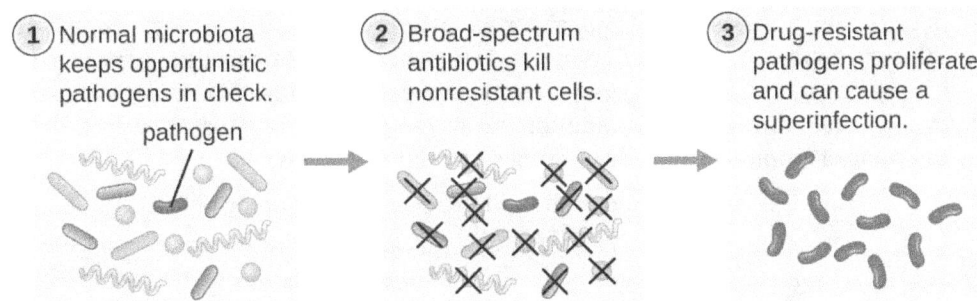

Figure 14.6 Broad-spectrum antimicrobial use may lead to the development of a superinfection. (credit: modification of work by Centers for Disease Control and Prevention)

CHECK YOUR UNDERSTANDING

- What is a superinfection and how does one arise?

Dosage and Route of Administration

The amount of medication given during a certain time interval is the **dosage**, and it must be determined carefully to ensure that optimum therapeutic drug levels are achieved at the site of infection without causing significant toxicity (side effects) to the patient. Each drug class is associated with a variety of potential side effects, and some of these are described for specific drugs later in this chapter. Despite best efforts to optimize dosing, allergic reactions and other potentially serious side effects do occur. Therefore, the goal is to select the optimum dosage that will minimize the risk of side effects while still achieving clinical cure, and there are important factors to consider when selecting the best dose and dosage interval. For example, in children, dose is based upon the patient's mass. However, the same is not true for adults and children 12 years of age and older, for which there is typically a single standard dose regardless of the patient's mass. With the great variability in adult body mass, some experts have argued that mass should be considered for all patients when determining appropriate dosage.[7] An additional consideration is how drugs are metabolized and eliminated from the body. In general, patients with a history of liver or kidney dysfunction may experience reduced drug metabolism or clearance from the body, resulting in increased drug levels that may lead to toxicity and make them more prone to side effects.

There are also some factors specific to the drugs themselves that influence appropriate dose and time interval between doses. For example, the half-life, or rate at which 50% of a drug is eliminated from the plasma, can vary significantly between drugs. Some drugs have a short half-life of only 1 hour and must be given multiple times a day, whereas other drugs have half-lives exceeding 12 hours and can be given as a single dose every 24 hours. Although a longer half-life can be considered an advantage for an antibacterial when it comes to convenient dosing intervals, the longer half-life can also be a concern for a drug that has serious side effects because drug levels may remain toxic for a longer time. Last, some drugs are dose dependent, meaning they are more effective when administered in large doses to provide high levels for a short time at the site of infection. Others are time dependent, meaning they are more effective when lower optimum levels are maintained over a longer period of time.

The **route of administration**, the method used to introduce a drug into the body, is also an important consideration for drug therapy. Drugs that can be administered orally are generally preferred because patients can more conveniently take these drugs at home. However, some drugs are not absorbed easily from the gastrointestinal (GI) tract into the bloodstream. These drugs are often useful for treating diseases of the intestinal tract, such as tapeworms treated with niclosamide, or for decontaminating the bowel, as with colistin. Some drugs that are not absorbed easily, such as bacitracin, polymyxin, and several antifungals, are available as topical preparations for treatment of superficial skin infections. Sometimes, patients may not initially be able to take oral medications because of their illness (e.g., vomiting, intubation for respirator).

---

7 M.E. Falagas, D.E. Karageorgopoulos. "Adjustment of Dosing of Antimicrobial Agents for Bodyweight in Adults." *The Lancet* 375 no. 9710 (2010):248–251.

When this occurs, and when a chosen drug is not absorbed in the GI tract, administration of the drug by a parenteral route (intravenous or intramuscular injection) is preferred and typically is performed in healthcare settings. For most drugs, the plasma levels achieved by intravenous administration is substantially higher than levels achieved by oral or intramuscular administration, and this can also be an important consideration when choosing the route of administration for treating an infection (Figure 14.7).

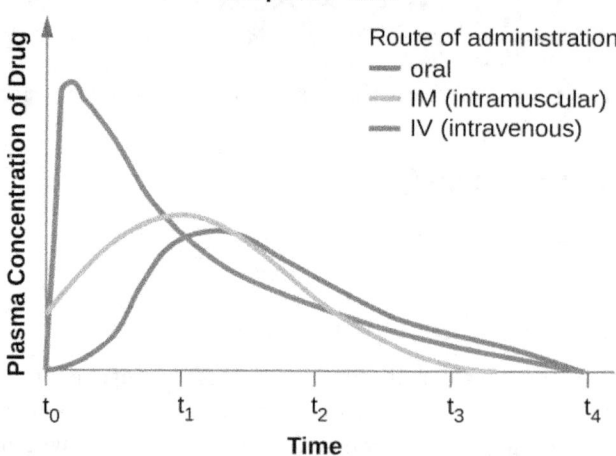

Figure 14.7 On this graph, t_0 represents the time at which a drug dose is administered. The curves illustrate how plasma concentration of the drug changes over specific intervals of time (t_1 through t_4). As the graph shows, when a drug is administered intravenously, the concentration peaks very quickly and then gradually decreases. When drugs are administered orally or intramuscularly, it takes longer for the concentration to reach its peak.

✓ CHECK YOUR UNDERSTANDING

- List five factors to consider when determining the dosage of a drug.
- Name some typical side effects associated with drugs and identify some factors that might contribute to these side effects.

Drug Interactions

For the optimum treatment of some infections, two antibacterial drugs may be administered together to provide a synergistic interaction that is better than the efficacy of either drug alone. A classic example of synergistic combinations is trimethoprim and sulfamethoxazole (Bactrim). Individually, these two drugs provide only bacteriostatic inhibition of bacterial growth, but combined, the drugs are bactericidal.

Whereas synergistic drug interactions provide a benefit to the patient, antagonistic interactions produce harmful effects. Antagonism can occur between two antimicrobials or between antimicrobials and nonantimicrobials being used to treat other conditions. The effects vary depending on the drugs involved, but antagonistic interactions may cause loss of drug activity, decreased therapeutic levels due to increased metabolism and elimination, or increased potential for toxicity due to decreased metabolism and elimination. As an example, some antibacterials are absorbed most effectively from the acidic environment of the stomach. If a patient takes antacids, however, this increases the pH of the stomach and negatively impacts the absorption of these antimicrobials, decreasing their effectiveness in treating an infection. Studies have also shown an association between use of some antimicrobials and failure of oral contraceptives.[8]

✓ CHECK YOUR UNDERSTANDING

- Explain the difference between synergistic and antagonistic drug interactions.

---

8 B.D. Dickinson et al. "Drug Interactions between Oral Contraceptives and Antibiotics." *Obstetrics & Gynecology* 98, no. 5 (2001):853–860.

Eye on Ethics

Resistance Police

In the United States and many other countries, most antimicrobial drugs are self-administered by patients at home. Unfortunately, many patients stop taking antimicrobials once their symptoms dissipate and they feel better. If a 10-day course of treatment is prescribed, many patients only take the drug for 5 or 6 days, unaware of the negative consequences of not completing the full course of treatment. A shorter course of treatment not only fails to kill the target organisms to expected levels, it also selects for drug-resistant variants within the target population and within the patient's microbiota.

Patients' nonadherence especially amplifies drug resistance when the recommended course of treatment is long. Treatment for tuberculosis (TB) is a case in point, with the recommended treatment lasting from 6 months to a year. The CDC estimates that about one-third of the world's population is infected with TB, most living in underdeveloped or underserved regions where antimicrobial drugs are available over the counter. In such countries, there may be even lower rates of adherence than in developed areas. Nonadherence leads to antibiotic resistance and more difficulty in controlling pathogens. As a direct result, the emergence of multidrug-resistant and extensively drug-resistant strains of TB is becoming a huge problem.

Overprescription of antimicrobials also contributes to antibiotic resistance. Patients often demand antibiotics for diseases that do not require them, like viral colds and ear infections. Pharmaceutical companies aggressively market drugs to physicians and clinics, making it easy for them to give free samples to patients, and some pharmacies even offer certain antibiotics free to low-income patients with a prescription.

In recent years, various initiatives have aimed to educate parents and clinicians about the judicious use of antibiotics. However, a recent study showed that, between 2000 and 2013, the parental expectation for antimicrobial prescriptions for children actually increased (Figure 14.8).

One possible solution is a regimen called directly observed therapy (DOT), which involves the supervised administration of medications to patients. Patients are either required to visit a health-care facility to receive their medications, or health-care providers must administer medication in patients' homes or another designated location. DOT has been implemented in many cases for the treatment of TB and has been shown to be effective; indeed, DOT is an integral part of WHO's global strategy for eradicating TB.[9],[10] But is this a practical strategy for all antibiotics? Would patients taking penicillin, for example, be more or less likely to adhere to the full course of treatment if they had to travel to a health-care facility for each dose? And who would pay for the increased cost associated with DOT? When it comes to overprescription, should someone be policing physicians or drug companies to enforce best practices? What group should assume this responsibility, and what penalties would be effective in discouraging overprescription?

---

[9] Centers for Disease Control and Prevention. "Tuberculosis (TB)." http://www.cdc.gov/tb/education/ssmodules/module9/ss9reading2.htm. Accessed June 2, 2016.

[10] World Health Organization. "Tuberculosis (TB): The Five Elements of DOTS." http://www.who.int/tb/dots/whatisdots/en/. Accessed June 2, 2016.

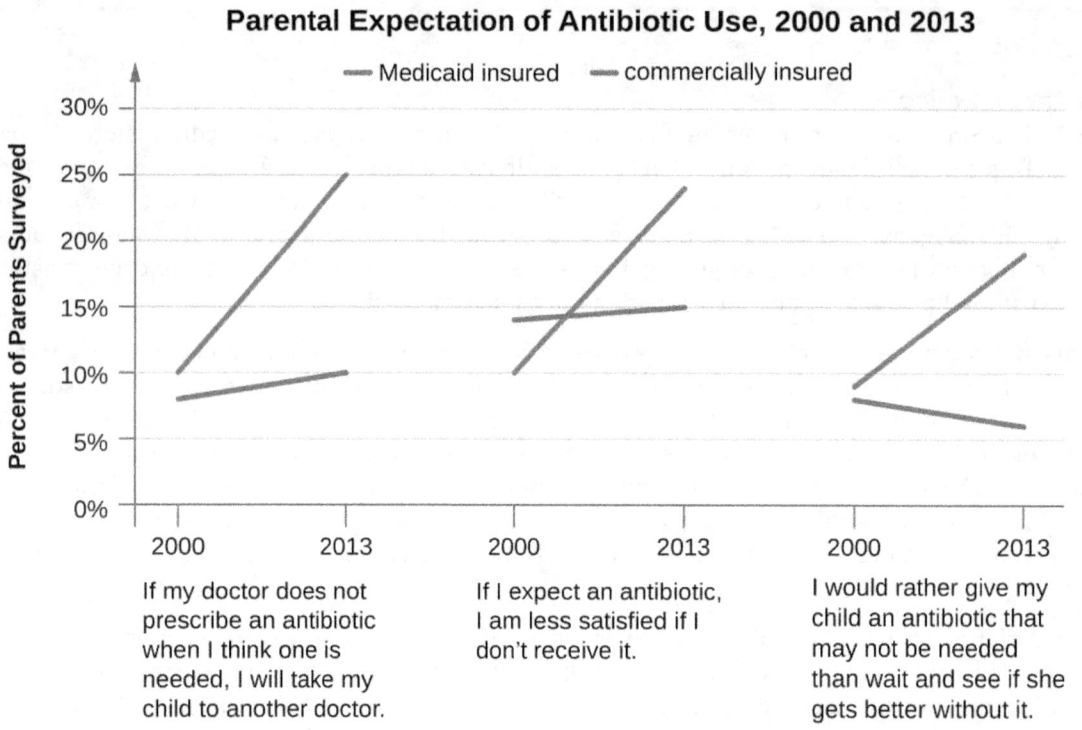

Figure 14.8 This graph indicates trends in parental expectations related to prescription of antibiotics based on a recent study.[11] Among parents of Medicaid-insured children, there was a clear upward trend in parental expectations for prescription antibiotics. Expectations were relatively stable (and lesser) among parents whose children were commercially insured, suggesting that these parents were somewhat better informed than those with Medicaid-insured children.

14.3 Mechanisms of Antibacterial Drugs

Learning Objective

- Describe the mechanisms of action associated with drugs that inhibit cell wall biosynthesis, protein synthesis, membrane function, nucleic acid synthesis, and metabolic pathways

An important quality for an antimicrobial drug is **selective toxicity**, meaning that it selectively kills or inhibits the growth of microbial targets while causing minimal or no harm to the host. Most antimicrobial drugs currently in clinical use are antibacterial because the prokaryotic cell provides a greater variety of unique targets for selective toxicity, in comparison to fungi, parasites, and viruses. Each class of antibacterial drugs has a unique **mode of action** (the way in which a drug affects microbes at the cellular level), and these are summarized in Figure 14.9 and Table 14.1.

---

11 Vaz, L.E., et al. "Prevalence of Parental Misconceptions About Antibiotic Use." *Pediatrics* 136 no.2 (August 2015). DOI: 10.1542/peds.2015-0883.

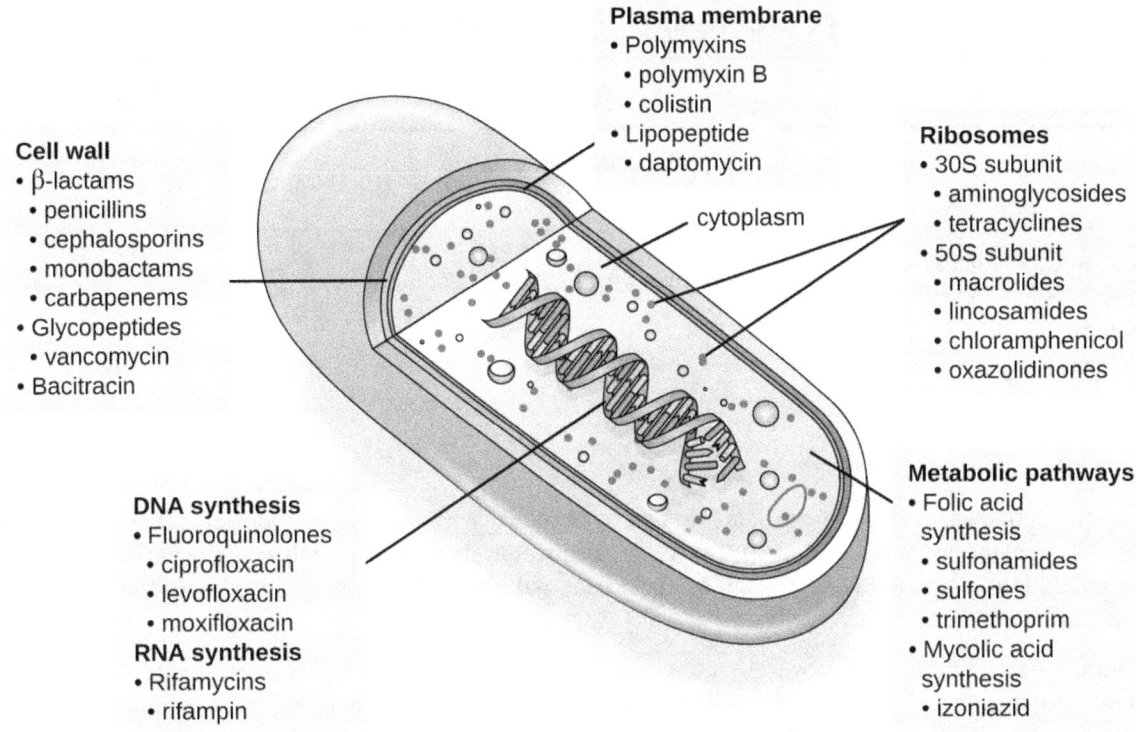

Figure 14.9 There are several classes of antibacterial compounds that are typically classified based on their bacterial target.

Common Antibacterial Drugs by Mode of Action

Mode of Action	Target	Drug Class
Inhibit cell wall biosynthesis	Penicillin-binding proteins	β-lactams: penicillins, cephalosporins, monobactams, carbapenems
	Peptidoglycan subunits	Glycopeptides
	Peptidoglycan subunit transport	Bacitracin
Inhibit biosynthesis of proteins	30S ribosomal subunit	Aminoglycosides, tetracyclines
	50S ribosomal subunit	Macrolides, lincosamides, chloramphenicol, oxazolidinones
Disrupt membranes	Lipopolysaccharide, inner and outer membranes	Polymyxin B, colistin, daptomycin
Inhibit nucleic acid synthesis	RNA	Rifamycin
	DNA	Fluoroquinolones
Antimetabolites	Folic acid synthesis enzyme	Sulfonamides, trimethoprim

Common Antibacterial Drugs by Mode of Action

Mode of Action	Target	Drug Class
	Mycolic acid synthesis enzyme	Isonicotinic acid hydrazide
Mycobacterial adenosine triphosphate (ATP) synthase inhibitor	Mycobacterial ATP synthase	Diarylquinoline

Table 14.1

Inhibitors of Cell Wall Biosynthesis

Several different classes of antibacterials block steps in the biosynthesis of peptidoglycan, making cells more susceptible to osmotic lysis (Table 14.2). Therefore, antibacterials that target cell wall biosynthesis are bactericidal in their action. Because human cells do not make peptidoglycan, this mode of action is an excellent example of selective toxicity.

Penicillin, the first antibiotic discovered, is one of several antibacterials within a class called **β-lactams**. This group of compounds includes the penicillins, cephalosporins, monobactams, and carbapenems, and is characterized by the presence of a β-lactam ring found within the central structure of the drug molecule (Figure 14.10). The β-lactam antibacterials block the crosslinking of peptide chains during the biosynthesis of new peptidoglycan in the bacterial cell wall. They are able to block this process because the β-lactam structure is similar to the structure of the peptidoglycan subunit component that is recognized by the crosslinking transpeptidase enzyme, also known as a penicillin-binding protein (PBP). Although the β-lactam ring must remain unchanged for these drugs to retain their antibacterial activity, strategic chemical changes to the R groups have allowed for development of a wide variety of semisynthetic β-lactam drugs with increased potency, expanded spectrum of activity, and longer half-lives for better dosing, among other characteristics.

Penicillin G and penicillin V are natural antibiotics from fungi and are primarily active against gram-positive bacterial pathogens, and a few gram-negative bacterial pathogens such as *Pasteurella multocida*. Figure 14.10 summarizes the semisynthetic development of some of the penicillins. Adding an amino group ($-NH_2$) to penicillin G created the aminopenicillins (i.e., ampicillin and amoxicillin) that have increased spectrum of activity against more gram-negative pathogens. Furthermore, the addition of a hydroxyl group (-OH) to amoxicillin increased acid stability, which allows for improved oral absorption. Methicillin is a semisynthetic penicillin that was developed to address the spread of enzymes (penicillinases) that were inactivating the other penicillins. Changing the R group of penicillin G to the more bulky dimethoxyphenyl group provided protection of the β-lactam ring from enzymatic destruction by penicillinases, giving us the first penicillinase-resistant penicillin.

Similar to the penicillins, **cephalosporins** contain a β-lactam ring (Figure 14.10) and block the transpeptidase activity of penicillin-binding proteins. However, the β-lactam ring of cephalosporins is fused to a six-member ring, rather than the five-member ring found in penicillins. This chemical difference provides cephalosporins with an increased resistance to enzymatic inactivation by **β-lactamases**. The drug cephalosporin C was originally isolated from the fungus *Cephalosporium acremonium* in the 1950s and has a similar spectrum of activity to that of penicillin against gram-positive bacteria but is active against more gram-negative bacteria than penicillin. Another important structural difference is that cephalosporin C possesses two R groups, compared with just one R group for penicillin, and this provides for greater diversity in chemical alterations and development of semisynthetic cephalosporins. The family of semisynthetic cephalosporins is much larger than the penicillins, and these drugs have been classified into generations based primarily on their spectrum of activity, increasing in spectrum from the narrow-spectrum, first-generation cephalosporins to the broad-spectrum, fourth-generation cephalosporins. A new fifth-generation cephalosporin has been developed that is active against methicillin-resistant *Staphylococcus aureus* (MRSA).

The carbapenems and monobactams also have a β-lactam ring as part of their core structure, and they inhibit the transpeptidase activity of penicillin-binding proteins. The only monobactam used clinically is aztreonam. It is a narrow-spectrum antibacterial with activity only against gram-negative bacteria. In contrast, the carbapenem family includes a variety of semisynthetic drugs (imipenem, meropenem, and doripenem) that provide very broad-spectrum activity against gram-positive and gram-negative bacterial pathogens.

The drug **vancomycin**, a member of a class of compounds called the **glycopeptides**, was discovered in the 1950s as a natural antibiotic from the actinomycete *Amycolatopsis orientalis*. Similar to the β-lactams, vancomycin inhibits cell wall biosynthesis and is bactericidal. However, in contrast to the β-lactams, the structure of vancomycin is not similar to that of cell-wall peptidoglycan subunits and does not directly inactivate penicillin-binding proteins. Rather, vancomycin is a very large, complex molecule that binds to the end of the peptide chain of cell wall precursors, creating a structural blockage that prevents the cell wall subunits from being incorporated into the growing N-acetylglucosamine and N-acetylmuramic acid (NAM-NAG) backbone of the peptidoglycan structure (transglycosylation). Vancomycin also structurally blocks transpeptidation. Vancomycin is bactericidal against gram-positive bacterial pathogens, but it is not active against gram-negative bacteria because of its inability to penetrate the protective outer membrane.

The drug **bacitracin** consists of a group of structurally similar peptide antibiotics originally isolated from *Bacillus subtilis*. Bacitracin blocks the activity of a specific cell-membrane molecule that is responsible for the movement of peptidoglycan precursors from the cytoplasm to the exterior of the cell, ultimately preventing their incorporation into the cell wall. Bacitracin is effective against a wide range of bacteria, including gram-positive organisms found on the skin, such as *Staphylococcus* and *Streptococcus*. Although it may be administered orally or intramuscularly in some circumstances, bacitracin has been shown to be nephrotoxic (damaging to the kidneys). Therefore, it is more commonly combined with neomycin and polymyxin in topical ointments such as Neosporin.

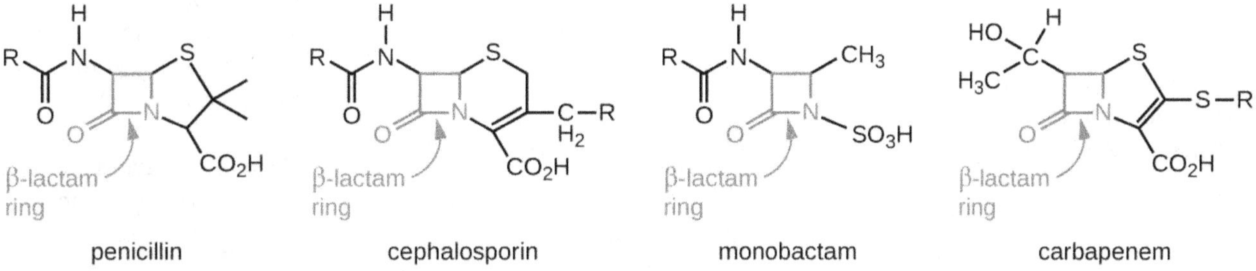

	penicillin	cephalosporin	monobactam	carbapenem
R group	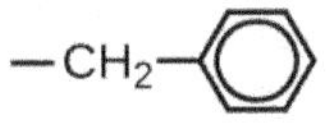 $-CH_2-\bigcirc$	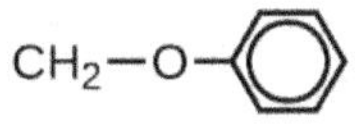 $CH_2-O-\bigcirc$	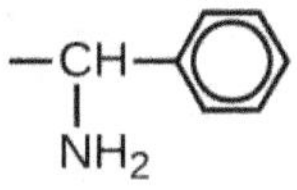 $-CH(NH_2)-\bigcirc$	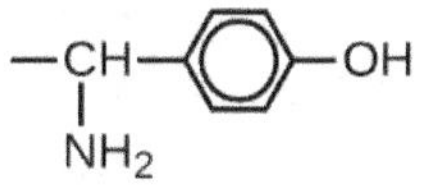 $-CH(NH_2)-\bigcirc-OH$	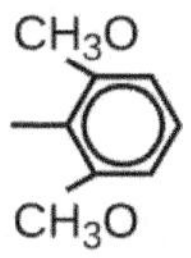 $CH_3O-\bigcirc-CH_3O$
Drug name	penicillin G	penicillin V	ampicillin	amoxicillin	methicillin
Spectrum of activity	G+ and a few G−	similar to penicillin G	G+ and more G− than penicillin	similar to ampicillin	G+ only, including β-lactamase producers
Route of administration	parenteral	oral	parenteral and oral	oral (better than ampicillin)	parenteral

Figure 14.10 Penicillins, cephalosporins, monobactams, and carbapenems all contain a β-lactam ring, the site of attack by inactivating β-

lactamase enzymes. Although they all share the same nucleus, various penicillins differ from each other in the structure of their R groups. Chemical changes to the R groups provided increased spectrum of activity, acid stability, and resistance to β-lactamase degradation.

Drugs that Inhibit Bacterial Cell Wall Synthesis

Mechanism of Action	Drug Class	Specific Drugs	Natural or Semisynthetic	Spectrum of Activity
Interact directly with PBPs and inhibit transpeptidase activity	Penicillins	Penicillin G, penicillin V	Natural	Narrow-spectrum against gram-positive and a few gram-negative bacteria
		Ampicillin, amoxicillin	Semisynthetic	Narrow-spectrum against gram-positive bacteria but with increased gram-negative spectrum
		Methicillin	Semisynthetic	Narrow-spectrum against gram-positive bacteria only, including strains producing penicillinase
	Cephalosporins	Cephalosporin C	Natural	Narrow-spectrum similar to penicillin but with increased gram-negative spectrum
		First-generation cephalosporins	Semisynthetic	Narrow-spectrum similar to cephalosporin C
		Second-generation cephalosporins	Semisynthetic	Narrow-spectrum but with increased gram-negative spectrum compared with first generation
		Third- and fourth-generation cephalosporins	Semisynthetic	Broad-spectrum against gram-positive and gram-negative bacteria, including some β-lactamase producers
		Fifth-generation cephalosporins	Semisynthetic	Broad-spectrum against gram-positive and gram-negative bacteria, including MRSA

Drugs that Inhibit Bacterial Cell Wall Synthesis

Mechanism of Action	Drug Class	Specific Drugs	Natural or Semisynthetic	Spectrum of Activity
	Monobactams	Aztreonam	Semisynthetic	Narrow-spectrum against gram-negative bacteria, including some β-lactamase producers
	Carbapenems	Imipenem, meropenem, doripenem	Semisynthetic	Broadest spectrum of the β-lactams against gram-positive and gram-negative bacteria, including many β-lactamase producers
Large molecules that bind to the peptide chain of peptidoglycan subunits, blocking transglycosylation and transpeptidation	Glycopeptides	Vancomycin	Natural	Narrow spectrum against gram-positive bacteria only, including multidrug-resistant strains
Block transport of peptidoglycan subunits across cytoplasmic membrane	Bacitracin	Bacitracin	Natural	Broad-spectrum against gram-positive and gram-negative bacteria

Table 14.2

✓ CHECK YOUR UNDERSTANDING

- Describe the mode of action of β-lactams.

Inhibitors of Protein Biosynthesis

The cytoplasmic ribosomes found in animal cells (80S) are structurally distinct from those found in bacterial cells (70S), making protein biosynthesis a good selective target for antibacterial drugs. Several types of protein biosynthesis inhibitors are discussed in this section and are summarized in Figure 14.11.

Protein Synthesis Inhibitors That Bind the 30S Subunit

Aminoglycosides are large, highly polar antibacterial drugs that bind to the 30S subunit of bacterial ribosomes, impairing the proofreading ability of the ribosomal complex. This impairment causes mismatches between codons and anticodons, resulting in the production of proteins with incorrect amino acids and shortened proteins that insert into the cytoplasmic membrane. Disruption of the cytoplasmic membrane by the faulty proteins kills the bacterial cells. The **aminoglycosides**, which include drugs such as streptomycin, gentamicin, neomycin, and kanamycin, are potent broad-spectrum antibacterials. However, aminoglycosides have been shown to be nephrotoxic (damaging to kidney), neurotoxic (damaging to the nervous system), and ototoxic (damaging to the ear).

Another class of antibacterial compounds that bind to the 30S subunit is the **tetracyclines**. In contrast to aminoglycosides, these drugs are bacteriostatic and inhibit protein synthesis by blocking the association of tRNAs with the ribosome during translation. Naturally occurring tetracyclines produced by various strains of

Streptomyces were first discovered in the 1940s, and several semisynthetic tetracyclines, including doxycycline and tigecycline have also been produced. Although the tetracyclines are broad spectrum in their coverage of bacterial pathogens, side effects that can limit their use include phototoxicity, permanent discoloration of developing teeth, and liver toxicity with high doses or in patients with kidney impairment.

Protein Synthesis Inhibitors That Bind the 50S Subunit

There are several classes of antibacterial drugs that work through binding to the 50S subunit of bacterial ribosomes. The macrolide antibacterial drugs have a large, complex ring structure and are part of a larger class of naturally produced secondary metabolites called polyketides, complex compounds produced in a stepwise fashion through the repeated addition of two-carbon units by a mechanism similar to that used for fatty acid synthesis. Macrolides are broad-spectrum, bacteriostatic drugs that block elongation of proteins by inhibiting peptide bond formation between specific combinations of amino acids. The first macrolide was **erythromycin**. It was isolated in 1952 from *Streptomyces erythreus* and prevents translocation. Semisynthetic macrolides include azithromycin and telithromycin. Compared with erythromycin, **azithromycin** has a broader spectrum of activity, fewer side effects, and a significantly longer half-life (1.5 hours for erythromycin versus 68 hours for azithromycin) that allows for once-daily dosing and a short 3-day course of therapy (i.e., Zpac formulation) for most infections. Telithromycin is the first semisynthetic within the class known as ketolides. Although telithromycin shows increased potency and activity against macrolide-resistant pathogens, the US Food and Drug Administration (FDA) has limited its use to treatment of community-acquired pneumonia and requires the strongest "black box warning" label for the drug because of serious hepatotoxicity.

The **lincosamides** include the naturally produced **lincomycin** and semisynthetic **clindamycin**. Although structurally distinct from macrolides, lincosamides are similar in their mode of action to the macrolides through binding to the 50S ribosomal subunit and preventing peptide bond formation. Lincosamides are particularly active against streptococcal and staphylococcal infections.

The drug **chloramphenicol** represents yet another structurally distinct class of antibacterials that also bind to the 50S ribosome, inhibiting peptide bond formation. Chloramphenicol, produced by *Streptomyces venezuelae*, was discovered in 1947; in 1949, it became the first broad-spectrum antibiotic that was approved by the FDA. Although it is a natural antibiotic, it is also easily synthesized and was the first antibacterial drug synthetically mass produced. As a result of its mass production, broad-spectrum coverage, and ability to penetrate into tissues efficiently, chloramphenicol was historically used to treat a wide range of infections, from meningitis to typhoid fever to conjunctivitis. Unfortunately, serious side effects, such as lethal gray baby syndrome, and suppression of bone marrow production, have limited its clinical role. Chloramphenicol also causes anemia in two different ways. One mechanism involves the targeting of mitochondrial ribosomes within hematopoietic stem cells, causing a reversible, dose-dependent suppression of blood cell production. Once chloramphenicol dosing is discontinued, blood cell production returns to normal. This mechanism highlights the similarity between 70S ribosomes of bacteria and the 70S ribosomes within our mitochondria. The second mechanism of anemia is idiosyncratic (i.e., the mechanism is not understood), and involves an irreversible lethal loss of blood cell production known as aplastic anemia. This mechanism of aplastic anemia is not dose dependent and can develop after therapy has stopped. Because of toxicity concerns, chloramphenicol usage in humans is now rare in the United States and is limited to severe infections unable to be treated by less toxic antibiotics. Because its side effects are much less severe in animals, it is used in veterinary medicine.

The **oxazolidinones**, including linezolid, are a new broad-spectrum class of synthetic protein synthesis inhibitors that bind to the 50S ribosomal subunit of both gram-positive and gram-negative bacteria. However, their mechanism of action seems somewhat different from that of the other 50S subunit-binding protein synthesis inhibitors already discussed. Instead, they seem to interfere with formation of the initiation complex (association of the 50S subunit, 30S subunit, and other factors) for translation, and they prevent translocation of the growing protein from the ribosomal A site to the P site. Table 14.3 summarizes the protein synthesis inhibitors.

Major classes of protein synthesis–inhibiting antibacterials

Chloramphenicol, macrolides, and lincosamides
- Bind to the 50S ribosomal subunit
- Prevent peptide bond formation
- Stop protein synthesis

Aminoglycosides
- Bind to the 30S ribosomal subunit
- Impair proofreading, resulting in production of faulty proteins

Tetracyclines
- Bind to the 30S ribosomal subunit
- Block the binding of tRNAs, thereby inhibiting protein synthesis

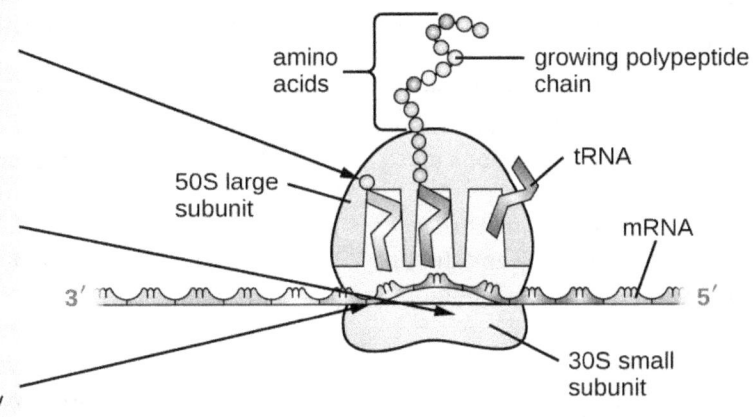

Figure 14.11 The major classes of protein synthesis inhibitors target the 30S or 50S subunits of cytoplasmic ribosomes.

Drugs That Inhibit Bacterial Protein Synthesis

Molecular Target	Mechanism of Action	Drug Class	Specific Drugs	Bacteriostatic or Bactericidal	Spectrum of Activity
30S subunit	Causes mismatches between codons and anticodons, leading to faulty proteins that insert into and disrupt cytoplasmic membrane	Aminoglycosides	Streptomycin, gentamicin, neomycin, kanamycin	Bactericidal	Broad spectrum
	Blocks association of tRNAs with ribosome	Tetracyclines	Tetracycline, doxycycline, tigecycline	Bacteriostatic	Broad spectrum
50S subunit	Blocks peptide bond formation between amino acids	Macrolides	Erythromycin, azithromycin, telithromycin	Bacteriostatic	Broad spectrum
		Lincosamides	Lincomycin, clindamycin	Bacteriostatic	Narrow spectrum
		Not applicable	Chloramphenicol	Bacteriostatic	Broad spectrum
	Interferes with the formation of the initiation complex between 50S and 30S subunits and other factors.	Oxazolidinones	Linezolid	Bacteriostatic	Broad spectrum

Table 14.3

CHECK YOUR UNDERSTANDING

- Compare and contrast the different types of protein synthesis inhibitors.

Inhibitors of Membrane Function

A small group of antibacterials target the bacterial membrane as their mode of action (Table 14.4). The **polymyxins** are natural polypeptide antibiotics that were first discovered in 1947 as products of *Bacillus polymyxa*; only polymyxin B and polymyxin E (**colistin**) have been used clinically. They are lipophilic with detergent-like properties and interact with the lipopolysaccharide component of the outer membrane of gram-negative bacteria, ultimately disrupting both their outer and inner membranes and killing the bacterial cells. Unfortunately, the membrane-targeting mechanism is not a selective toxicity, and these drugs also target and damage the membrane of cells in the kidney and nervous system when administered systemically. Because of these serious side effects and their poor absorption from the digestive tract, polymyxin B is used in over-the-counter topical antibiotic ointments (e.g., Neosporin), and oral colistin was historically used only for bowel decontamination to prevent infections originating from bowel microbes in immunocompromised patients or for those undergoing certain abdominal surgeries. However, the emergence and spread of multidrug-resistant pathogens has led to increased use of intravenous colistin in hospitals, often as a drug of last resort to treat serious infections. The antibacterial **daptomycin** is a cyclic lipopeptide produced by *Streptomyces roseosporus* that seems to work like the polymyxins, inserting in the bacterial cell membrane and disrupting it. However, in contrast to polymyxin B and colistin, which target only gram-negative bacteria, daptomycin specifically targets gram-positive bacteria. It is typically administered intravenously and seems to be well tolerated, showing reversible toxicity in skeletal muscles.

Drugs That Inhibit Bacterial Membrane Function

Mechanism of Action	Drug Class	Specific Drugs	Spectrum of Activity	Clinical Use
Interacts with lipopolysaccharide in the outer membrane of gram-negative bacteria, killing the cell through the eventual disruption of the outer membrane and cytoplasmic membrane	Polymyxins	Polymyxin B	Narrow spectrum against gram-negative bacteria, including multidrug-resistant strains	Topical preparations to prevent infections in wounds
		Polymyxin E (colistin)	Narrow spectrum against gram-negative bacteria, including multidrug-resistant strains	Oral dosing to decontaminate bowels to prevent infections in immunocompromised patients or patients undergoing invasive surgery/procedures.
				Intravenous dosing to treat serious systemic infections caused by multidrug-resistant pathogens
Inserts into the cytoplasmic membrane of gram-positive bacteria, disrupting the membrane and killing the cell	Lipopeptide	Daptomycin	Narrow spectrum against gram-positive bacteria, including multidrug-resistant strains	Complicated skin and skin-structure infections and bacteremia caused by gram-positive pathogens, including MRSA

Table 14.4

✓ CHECK YOUR UNDERSTANDING

- How do polymyxins inhibit membrane function?

Inhibitors of Nucleic Acid Synthesis

Some antibacterial drugs work by inhibiting nucleic acid synthesis (Table 14.5). For example, **metronidazole** is a semisynthetic member of the nitroimidazole family that is also an antiprotozoan. It interferes with DNA replication in target cells. The drug **rifampin** is a semisynthetic member of the rifamycin family and functions by blocking RNA polymerase activity in bacteria. The RNA polymerase enzymes in bacteria are structurally different from those in eukaryotes, providing for selective toxicity against bacterial cells. It is used for the treatment of a variety of infections, but its primary use, often in a cocktail with other antibacterial drugs, is against mycobacteria that cause tuberculosis. Despite the selectivity of its mechanism, rifampin can induce liver enzymes to increase metabolism of other drugs being administered (antagonism), leading to hepatotoxicity (liver toxicity) and negatively influencing the bioavailability and therapeutic effect of the companion drugs.

One member of the quinolone family, a group of synthetic antimicrobials, is **nalidixic acid**. It was discovered in 1962 as a byproduct during the synthesis of chloroquine, an antimalarial drug. Nalidixic acid selectively inhibits the activity of bacterial DNA gyrase, blocking DNA replication. Chemical modifications to the original quinolone backbone have resulted in the production of **fluoroquinolones**, like ciprofloxacin and levofloxacin, which also inhibit the activity of DNA gyrase. Ciprofloxacin and levofloxacin are effective against a broad spectrum of gram-positive or gram-negative bacteria, and are among the most commonly prescribed antibiotics used to treat a wide range of infections, including urinary tract infections, respiratory infections, abdominal infections, and skin infections. However, despite their selective toxicity against DNA gyrase, side effects associated with different fluoroquinolones include phototoxicity, neurotoxicity, cardiotoxicity, glucose metabolism dysfunction, and increased risk for tendon rupture.

Drugs That Inhibit Bacterial Nucleic Acid Synthesis

Mechanisms of Action	Drug Class	Specific Drugs	Spectrum of activity	Clinical Use
Inhibits bacterial RNA polymerase activity and blocks transcription, killing the cell	Rifamycin	Rifampin	Narrow spectrum with activity against gram-positive and limited numbers of gram-negative bacteria. Also active against *Mycobacterium tuberculosis*.	Combination therapy for treatment of tuberculosis
Inhibits the activity of DNA gyrase and blocks DNA replication, killing the cell	Fluoroquinolones	Ciprofloxacin, ofloxacin, moxifloxacin	Broad spectrum against gram-positive and gram-negative bacteria	Wide variety of skin and systemic infections

Table 14.5

✓ CHECK YOUR UNDERSTANDING

- Why do inhibitors of bacterial nucleic acid synthesis not target host cells?

Inhibitors of Metabolic Pathways

Some synthetic drugs control bacterial infections by functioning as **antimetabolites**, competitive inhibitors for bacterial metabolic enzymes (Table 14.6). The **sulfonamides (sulfa drugs)** are the oldest synthetic antibacterial agents and are structural analogues of *para*-aminobenzoic acid (PABA), an early intermediate in folic acid synthesis (Figure 14.12). By inhibiting the enzyme involved in the production of dihydrofolic acid, sulfonamides block bacterial biosynthesis of folic acid and, subsequently, pyrimidines and purines required for nucleic acid synthesis. This mechanism of action provides bacteriostatic inhibition of growth against a wide spectrum of gram-positive and gram-negative pathogens. Because humans obtain folic acid from food instead of synthesizing it intracellularly, sulfonamides are selectively toxic for bacteria. However, allergic reactions to sulfa drugs are common. The sulfones are structurally similar to sulfonamides but are not commonly used today except for the treatment of Hansen's disease (leprosy).

Trimethoprim is a synthetic antimicrobial compound that serves as an antimetabolite within the same folic acid synthesis pathway as sulfonamides. However, **trimethoprim** is a structural analogue of dihydrofolic acid and inhibits a later step in the metabolic pathway (Figure 14.12). Trimethoprim is used in combination with the sulfa drug sulfamethoxazole to treat urinary tract infections, ear infections, and bronchitis. As discussed, the combination of trimethoprim and sulfamethoxazole is an example of antibacterial synergy. When used alone, each antimetabolite only decreases production of folic acid to a level where bacteriostatic inhibition of growth occurs. However, when used in combination, inhibition of both steps in the metabolic pathway decreases folic acid synthesis to a level that is lethal to the bacterial cell. Because of the importance of folic acid during fetal development, sulfa drugs and trimethoprim use should be carefully considered during early pregnancy.

The drug **isoniazid** is an antimetabolite with specific toxicity for mycobacteria and has long been used in combination with rifampin or streptomycin in the treatment of tuberculosis. It is administered as a prodrug, requiring activation through the action of an intracellular bacterial peroxidase enzyme, forming isoniazid-nicotinamide adenine dinucleotide (NAD) and isoniazid-nicotinamide adenine dinucleotide phosphate (NADP), ultimately preventing the synthesis of mycolic acid, which is essential for mycobacterial cell walls. Possible side effects of isoniazid use include hepatotoxicity, neurotoxicity, and hematologic toxicity (anemia).

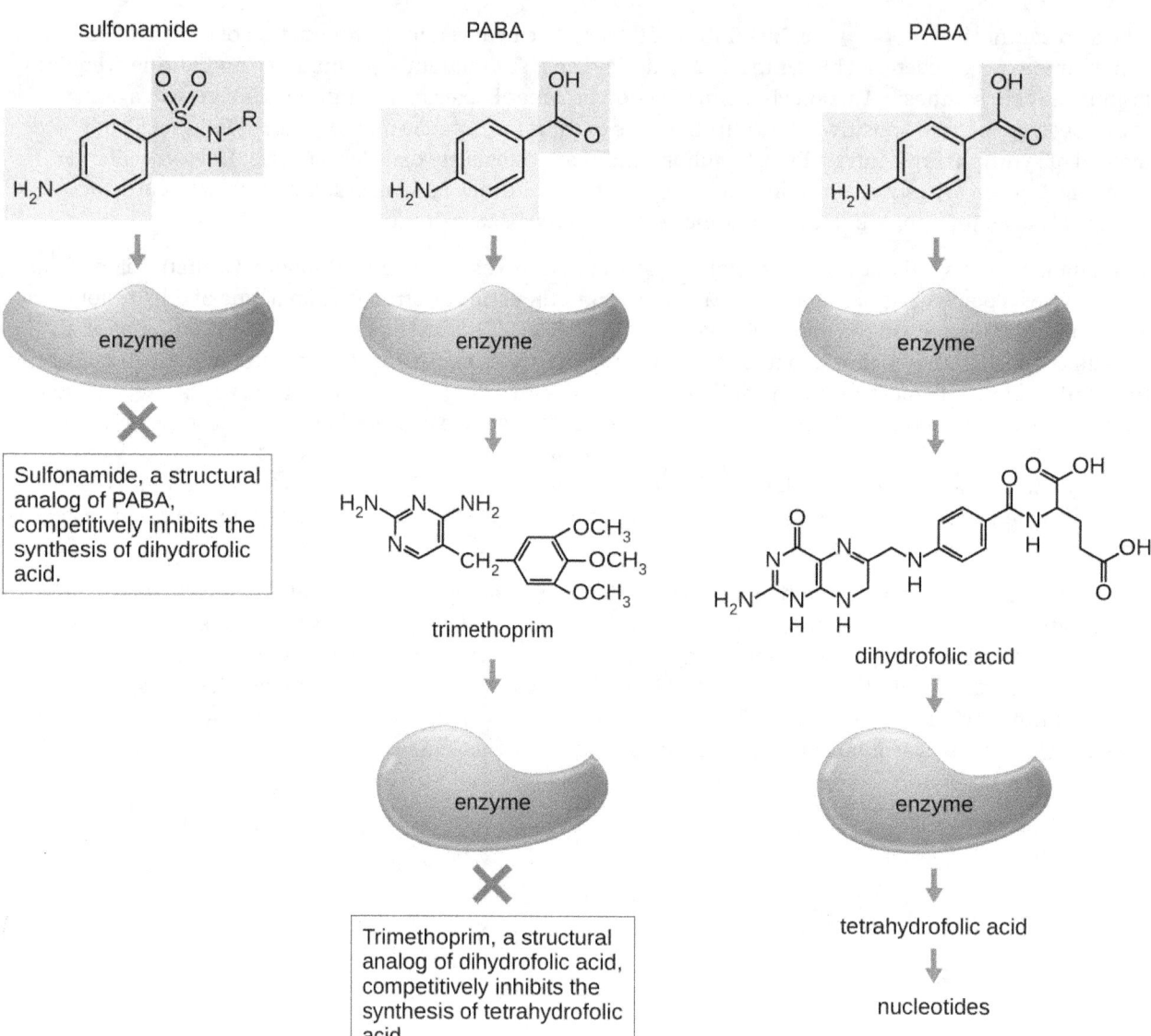

Figure 14.12 Sulfonamides and trimethoprim are examples of antimetabolites that interfere in the bacterial synthesis of folic acid by blocking purine and pyrimidine biosynthesis, thus inhibiting bacterial growth.

Antimetabolite Drugs

Metabolic Pathway Target	Mechanism of Action	Drug Class	Specific Drugs	Spectrum of Activity
Folic acid synthesis	Inhibits the enzyme involved in production of dihydrofolic acid	Sulfonamides	Sulfamethoxazole	Broad spectrum against gram-positive and gram-negative bacteria
		Sulfones	Dapsone	
	Inhibits the enzyme involved in the production of tetrahydrofolic acid	Not applicable	Trimethoprim	Broad spectrum against gram-positive and gram-negative bacteria

Antimetabolite Drugs

Metabolic Pathway Target	Mechanism of Action	Drug Class	Specific Drugs	Spectrum of Activity
Mycolic acid synthesis	Interferes with the synthesis of mycolic acid	Not applicable	Isoniazid	Narrow spectrum against *Mycobacterium* spp., including *M. tuberculosis*

Table 14.6

CHECK YOUR UNDERSTANDING

- How do sulfonamides and trimethoprim selectively target bacteria?

Inhibitor of ATP Synthase

Bedaquiline, representing the synthetic antibacterial class of compounds called the diarylquinolines, uses a novel mode of action that specifically inhibits mycobacterial growth. Although the specific mechanism has yet to be elucidated, this compound appears to interfere with the function of ATP synthases, perhaps by interfering with the use of the hydrogen ion gradient for ATP synthesis by oxidative phosphorylation, leading to reduced ATP production. Due to its side effects, including hepatotoxicity and potentially lethal heart arrhythmia, its use is reserved for serious, otherwise untreatable cases of tuberculosis.

LINK TO LEARNING

To learn more about the general principles of antimicrobial therapy and bacterial modes of action, read this article (https://openstax.org/l/22AntiMicrob) for a general overview of antimicrobial therapy strategies and read this article (https://openstax.org/l/22AntiMicRes) on the mechanism of action of antimicrobial classes.

Clinical Focus

Part 2

Reading thorough Marisa's health history, the doctor noticed that during her hospitalization in Vietnam, she was catheterized and received the antimicrobial drugs ceftazidime and metronidazole. Upon learning this, the doctor ordered a CT scan of Marisa's abdomen to rule out appendicitis; the doctor also requested blood work to see if she had an elevated white blood cell count, and ordered a urine analysis test and urine culture to look for the presence of white blood cells, red blood cells, and bacteria.

Marisa's urine sample came back positive for the presence of bacteria, indicating a urinary tract infection (UTI). The doctor prescribed ciprofloxacin. In the meantime, her urine was cultured to grow the bacterium for further testing.

- What types of antimicrobials are typically prescribed for UTIs?
- Based upon the antimicrobial drugs she was given in Vietnam, which of the antimicrobials for treatment of a UTI would you predict to be ineffective?

Jump to the next Clinical Focus box. Go back to the previous Clinical Focus box.

14.4 Mechanisms of Other Antimicrobial Drugs

Learning Objective

- Explain the differences between modes of action of drugs that target fungi, protozoa, helminths, and viruses

Because fungi, protozoa, and helminths are eukaryotic, their cells are very similar to human cells, making it more difficult to develop drugs with selective toxicity. Additionally, viruses replicate within human host cells, making it difficult to develop drugs that are selectively toxic to viruses or virus-infected cells. Despite these challenges, there are antimicrobial drugs that target fungi, protozoa, helminths, and viruses, and some even target more than one type of microbe. Table 14.7, Table 14.8, Table 14.9, and Table 14.10 provide examples for antimicrobial drugs in these various classes.

Antifungal Drugs

The most common mode of action for antifungal drugs is the disruption of the cell membrane. Antifungals take advantage of small differences between fungi and humans in the biochemical pathways that synthesize sterols. The sterols are important in maintaining proper membrane fluidity and, hence, proper function of the cell membrane. For most fungi, the predominant membrane sterol is ergosterol. Because human cell membranes use cholesterol, instead of ergosterol, antifungal drugs that target ergosterol synthesis are selectively toxic (Figure 14.13).

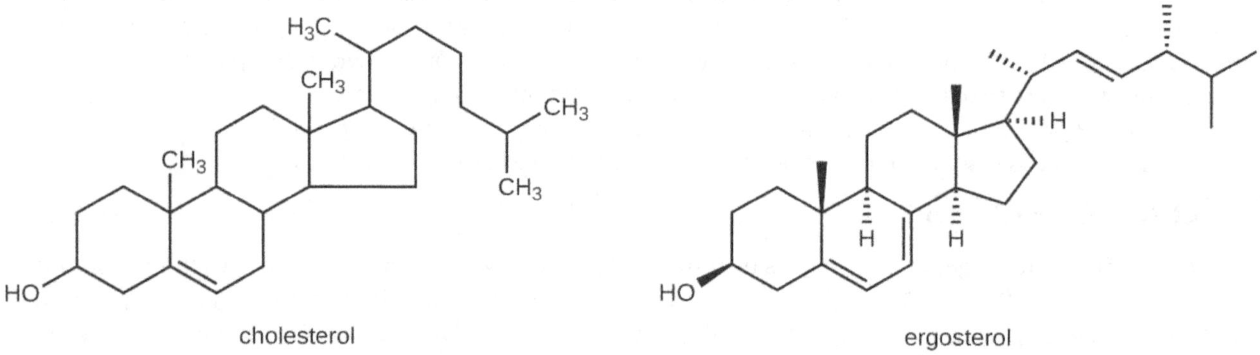

Figure 14.13 The predominant sterol found in human cells is cholesterol, whereas the predominant sterol found in fungi is ergosterol, making ergosterol a good target for antifungal drug development.

The **imidazoles** are synthetic fungicides that disrupt ergosterol biosynthesis; they are commonly used in medical applications and also in agriculture to keep seeds and harvested crops from molding. Examples include miconazole, ketoconazole, and clotrimazole, which are used to treat fungal skin infections such as ringworm, specifically tinea pedis (athlete's foot), tinea cruris (jock itch), and tinea corporis. These infections are commonly caused by dermatophytes of the genera *Trichophyton*, *Epidermophyton*, and *Microsporum*. Miconazole is also used predominantly for the treatment of vaginal yeast infections caused by the fungus *Candida*, and ketoconazole is used for the treatment of tinea versicolor and dandruff, which both can be caused by the fungus *Malassezia*.

The **triazole** drugs, including **fluconazole**, also inhibit ergosterol biosynthesis. However, they can be administered orally or intravenously for the treatment of several types of systemic yeast infections, including oral thrush and cryptococcal meningitis, both of which are prevalent in patients with AIDS. The triazoles also exhibit more selective toxicity, compared with the imidazoles, and are associated with fewer side effects.

The **allylamines**, a structurally different class of synthetic antifungal drugs, inhibit an earlier step in ergosterol biosynthesis. The most commonly used allylamine is **terbinafine** (marketed under the brand name Lamisil), which is used topically for the treatment of dermatophytic skin infections like athlete's foot, ringworm, and jock itch. Oral treatment with terbinafine is also used for the treatment of fingernail and toenail fungus, but it can be associated with the rare side effect of hepatotoxicity.

The **polyenes** are a class of antifungal agents naturally produced by certain actinomycete soil bacteria and are structurally related to macrolides. These large, lipophilic molecules bind to ergosterol in fungal cytoplasmic membranes, thus creating pores. Common examples include nystatin and amphotericin B. Nystatin is typically used as a topical treatment for yeast infections of the skin, mouth, and vagina, but may also be used for intestinal fungal infections. The drug **amphotericin B** is used for systemic fungal infections like aspergillosis, cryptococcal meningitis, histoplasmosis, blastomycosis, and candidiasis. Amphotericin B was the only antifungal drug available for several decades, but its use is associated with some serious side effects, including nephrotoxicity (kidney toxicity).

Amphotericin B is often used in combination with flucytosine, a fluorinated pyrimidine analog that is converted by a fungal-specific enzyme into a toxic product that interferes with both DNA replication and protein synthesis in fungi. Flucytosine is also associated with hepatotoxicity (liver toxicity) and bone marrow depression.

Beyond targeting ergosterol in fungal cell membranes, there are a few antifungal drugs that target other fungal structures (Figure 14.14). The echinocandins, including caspofungin, are a group of naturally produced antifungal compounds that block the synthesis of β(1→3) glucan found in fungal cell walls but not found in human cells. This drug class has the nickname "penicillin for fungi." Caspofungin is used for the treatment of aspergillosis as well as systemic yeast infections.

Although chitin is only a minor constituent of fungal cell walls, it is also absent in human cells, making it a selective target. The polyoxins and nikkomycins are naturally produced antifungals that target chitin synthesis. Polyoxins are used to control fungi for agricultural purposes, and nikkomycin Z is currently under development for use in humans to treat yeast infections and Valley fever (coccidioidomycosis), a fungal disease prevalent in the southwestern US.[12]

The naturally produced antifungal griseofulvin is thought to specifically disrupt fungal cell division by interfering with microtubules involved in spindle formation during mitosis. It was one of the first antifungals, but its use is associated with hepatotoxicity. It is typically administered orally to treat various types of dermatophytic skin infections when other topical antifungal treatments are ineffective.

There are a few drugs that act as antimetabolites against fungal processes. For example, atovaquone, a representative of the naphthoquinone drug class, is a semisynthetic antimetabolite for fungal and protozoal versions of a mitochondrial cytochrome important in electron transport. Structurally, it is an analog of coenzyme Q, with which it competes for electron binding. It is particularly useful for the treatment of *Pneumocystis* pneumonia caused by *Pneumocystis jirovecii*. The antibacterial sulfamethoxazole-trimethoprim combination also acts as an antimetabolite against *P. jirovecii*.

Table 14.7 shows the various therapeutic classes of antifungal drugs, categorized by mode of action, with examples of each.

---

[12] Centers for Disease Control and Prevention. "Valley Fever: Awareness Is Key." http://www.cdc.gov/features/valleyfever/. Accessed June 1, 2016.

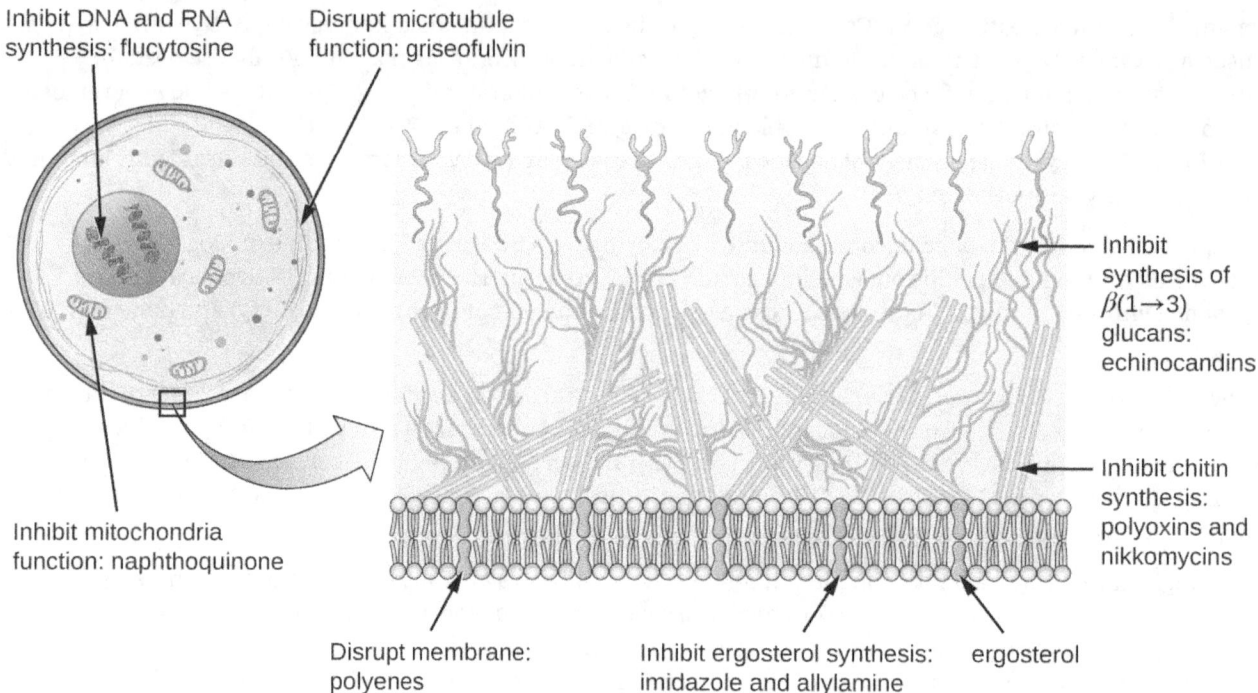

Figure 14.14 Antifungal drugs target several different cell structures. (credit right: modification of work by "Maya and Rike"/Wikimedia Commons)

Common Antifungal Drugs

Mechanism of Action	Drug Class	Specific Drugs	Clinical Uses
Inhibit ergosterol synthesis	Imidazoles	Miconazole, ketoconazole, clotrimazole	Fungal skin infections and vaginal yeast infections
	Triazoles	Fluconazole	Systemic yeast infections, oral thrush, and cryptococcal meningitis
	Allylamines	Terbinafine	Dermatophytic skin infections (athlete's foot, ring worm, jock itch), and infections of fingernails and toenails
Bind ergosterol in the cell membrane and create pores that disrupt the membrane	Polyenes	Nystatin	Used topically for yeast infections of skin, mouth, and vagina; also used for fungal infections of the intestine
		Amphotericin B	Variety systemic fungal infections
Inhibit cell wall synthesis	Echinocandins	Caspofungin	Aspergillosis and systemic yeast infections
	Not applicable	Nikkomycin Z	Coccidioidomycosis (Valley fever) and yeast infections

Common Antifungal Drugs

Mechanism of Action	Drug Class	Specific Drugs	Clinical Uses
Inhibit microtubules and cell division	Not applicable	Griseofulvin	Dermatophytic skin infections

Table 14.7

✓ CHECK YOUR UNDERSTANDING

- How is disruption of ergosterol biosynthesis an effective mode of action for antifungals?

Case in Point

Treating a Fungal Infection of the Lungs

Jack, a 48-year-old engineer, is HIV positive but generally healthy thanks to antiretroviral therapy (ART). However, after a particularly intense week at work, he developed a fever and a dry cough. He assumed that he just had a cold or mild flu due to overexertion and didn't think much of it. However, after about a week, he began to experience fatigue, weight loss, and shortness of breath. He decided to visit his physician, who found that Jack had a low level of blood oxygenation. The physician ordered blood testing, a chest X-ray, and the collection of an induced sputum sample for analysis. His X-ray showed a fine cloudiness and several pneumatoceles (thin-walled pockets of air), which indicated *Pneumocystis* pneumonia (PCP), a type of pneumonia caused by the fungus *Pneumocystis jirovecii*. Jack's physician admitted him to the hospital and prescribed Bactrim, a combination of sulfamethoxazole and trimethoprim, to be administered intravenously.

P. jirovecii is a yeast-like fungus with a life cycle similar to that of protozoans. As such, it was classified as a protozoan until the 1980s. It lives only in the lung tissue of infected persons and is transmitted from person to person, with many people exposed as children. Typically, *P. jirovecii* only causes pneumonia in immunocompromised individuals. Healthy people may carry the fungus in their lungs with no symptoms of disease. PCP is particularly problematic among HIV patients with compromised immune systems.

PCP is usually treated with oral or intravenous Bactrim, but atovaquone or pentamidine (another antiparasitic drug) are alternatives. If not treated, PCP can progress, leading to a collapsed lung and nearly 100% mortality. Even with antimicrobial drug therapy, PCP still is responsible for 10% of HIV-related deaths.

The cytological examination, using direct immunofluorescence assay (DFA), of a smear from Jack's sputum sample confirmed the presence of *P. jirovecii* (Figure 14.15). Additionally, the results of Jack's blood tests revealed that his white blood cell count had dipped, making him more susceptible to the fungus. His physician reviewed his ART regimen and made adjustments. After a few days of hospitalization, Jack was released to continue his antimicrobial therapy at home. With the adjustments to his ART therapy, Jack's CD4 counts began to increase and he was able to go back to work.

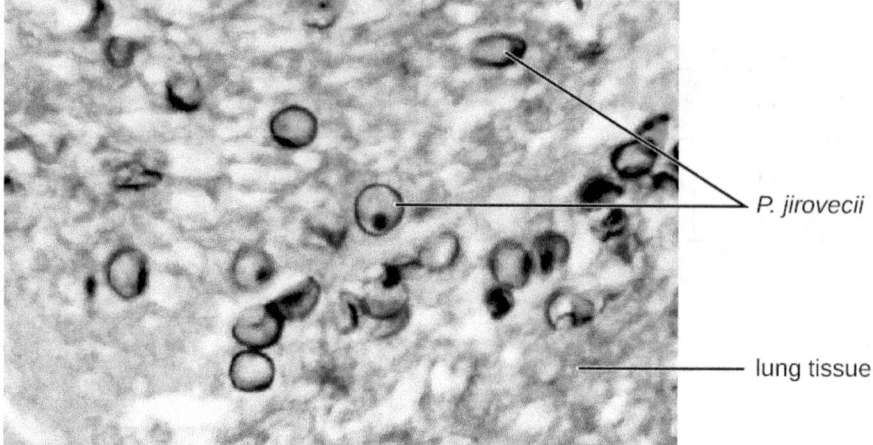

Figure 14.15 Microscopic examination of an induced sputum sample or bronchoaveolar lavage sample typically reveals the organism, as shown here. (credit: modification of work by the Centers for Disease Control and Prevention)

Antiprotozoan Drugs

There are a few mechanisms by which antiprotozoan drugs target infectious protozoans (Table 14.9). Some are antimetabolites, such as atovaquone, proguanil, and artemisinins. Atovaquone, in addition to being antifungal, blocks electron transport in protozoans and is used for the treatment of protozoan infections including malaria, babesiosis, and toxoplasmosis. Proguanil is another synthetic antimetabolite that is processed in parasitic cells into its active form, which inhibits protozoan folic acid synthesis. It is often used in combination with atovaquone, and the combination is marketed as Malarone for both malaria treatment and prevention.

Artemisinin, a plant-derived antifungal first discovered by Chinese scientists in the 1970s, is quite effective against malaria. Semisynthetic derivatives of **artemisinin** are more water soluble than the natural version, which makes them more bioavailable. Although the exact mechanism of action is unclear, artemisinins appear to act as prodrugs that are metabolized by target cells to produce reactive oxygen species (ROS) that damage target cells. Due to the rise in resistance to antimalarial drugs, artemisinins are also commonly used in combination with other antimalarial compounds in artemisinin-based combination therapy (ACT).

Several antimetabolites are used for the treatment of toxoplasmosis caused by the parasite *Toxoplasma gondii*. The synthetic sulfa drug sulfadiazine competitively inhibits an enzyme in folic acid production in parasites and can be used to treat malaria and toxoplasmosis. Pyrimethamine is a synthetic drug that inhibits a different enzyme in the folic acid production pathway and is often used in combination with sulfadoxine (another sulfa drug) for the treatment of malaria or in combination with sulfadiazine for the treatment of toxoplasmosis. Side effects of pyrimethamine include decreased bone marrow activity that may cause increased bruising and low red blood cell counts. When toxicity is a concern, spiramycin, a macrolide protein synthesis inhibitor, is typically administered for the treatment of toxoplasmosis.

Two classes of antiprotozoan drugs interfere with nucleic acid synthesis: nitroimidazoles and quinolines. Nitroimidazoles, including semisynthetic metronidazole, which was discussed previously as an antibacterial drug, and synthetic tinidazole, are useful in combating a wide variety of protozoan pathogens, such as *Giardia lamblia*, *Entamoeba histolytica*, and *Trichomonas vaginalis*. Upon introduction into these cells in low-oxygen environments, nitroimidazoles become activated and introduce DNA strand breakage, interfering with DNA replication in target cells. Unfortunately, metronidazole is associated with carcinogenesis (the development of cancer) in humans.

Another type of synthetic antiprotozoan drug that has long been thought to specifically interfere with DNA replication in certain pathogens is **pentamidine**. It has historically been used for the treatment of African sleeping sickness (caused by the protozoan *Trypanosoma brucei*) and leishmaniasis (caused by protozoa of the genus *Leishmania*), but it is also an alternative treatment for the fungus *Pneumocystis*. Some studies indicate that it specifically binds to the DNA found within kinetoplasts (kDNA; long mitochondrion-like structures unique to trypanosomes), leading to the cleavage of kDNA. However, nuclear DNA of both the parasite and host remain unaffected. It also appears to bind to tRNA, inhibiting the addition of amino acids to tRNA, thus preventing protein synthesis. Possible side effects of pentamidine use include pancreatic dysfunction and liver damage.

The **quinolines** are a class of synthetic compounds related to quinine, which has a long history of use against malaria. Quinolines are thought to interfere with heme detoxification, which is necessary for the parasite's effective breakdown of hemoglobin into amino acids inside red blood cells. The synthetic derivatives chloroquine, quinacrine (also called mepacrine), and mefloquine are commonly used as antimalarials, and chloroquine is also used to treat amebiasis typically caused by *Entamoeba histolytica*. Long-term prophylactic use of chloroquine or mefloquine may result in serious side effects, including hallucinations or cardiac issues. Patients with glucose-6-phosphate dehydrogenase deficiency experience severe anemia when treated with chloroquine.

Common Antiprotozoan Drugs

Mechanism of Action	Drug Class	Specific Drugs	Clinical Uses
Inhibit electron transport in mitochondria	Naphthoquinone	Atovaquone	Malaria, babesiosis, and toxoplasmosis
Inhibit folic acid synthesis	Not applicable	Proquanil	Combination therapy with atovaquone for malaria treatment and prevention
Inhibit folic acid synthesis	Sulfonamide	Sulfadiazine	Malaria and toxoplasmosis
Inhibit folic acid synthesis	Not applicable	Pyrimethamine	Combination therapy with sulfadoxine (sulfa drug) for malaria
Produces damaging reactive oxygen species	Not applicable	Artemisinin	Combination therapy to treat malaria
Inhibit DNA synthesis	Nitroimidazoles	Metronidazole, tinidazole	Infections caused by *Giardia lamblia*, *Entamoeba histolytica*, and *Trichomonas vaginalis*
Inhibit DNA synthesis	Not applicable	Pentamidine	African sleeping sickness and leishmaniasis
Inhibit heme detoxification	Quinolines	Chloroquine	Malaria and infections with *E. histolytica*
Inhibit heme detoxification	Quinolines	Mepacrine, mefloquine	Malaria

Table 14.8

✓ CHECK YOUR UNDERSTANDING

- List two modes of action for antiprotozoan drugs.

Antihelminthic Drugs

Because helminths are multicellular eukaryotes like humans, developing drugs with selective toxicity against them is extremely challenging. Despite this, several effective classes have been developed (Table 14.9). Synthetic **benzimidazoles**, like **mebendazole** and **albendazole**, bind to helminthic β-tubulin, preventing microtubule formation. Microtubules in the intestinal cells of the worms seem to be particularly affected, leading to a reduction in glucose uptake. Besides their activity against a broad range of helminths, benzimidazoles are also active against many protozoans, fungi, and viruses, and their use for inhibiting mitosis and cell cycle progression in cancer cells is under study.[13] Possible side effects of their use include liver damage and bone marrow suppression.

The avermectins are members of the macrolide family that were first discovered from a Japanese soil isolate, *Streptomyces avermectinius*. A more potent semisynthetic derivative of avermectin is **ivermectin**, which binds to glutamate-gated chloride channels specific to invertebrates including helminths, blocking neuronal transmission and causing starvation, paralysis, and death of the worms. Ivermectin is used to treat roundworm diseases, including onchocerciasis (also called river blindness, caused by the worm *Onchocerca volvulus*) and strongyloidiasis (caused by the worm *Strongyloides stercoralis* or *S. fuelleborni*). Ivermectin also can also treat parasitic insects like mites, lice, and bed bugs, and is nontoxic to humans.

Niclosamide is a synthetic drug that has been used for over 50 years to treat tapeworm infections. Although its mode of action is not entirely clear, niclosamide appears to inhibit ATP formation under anaerobic conditions and inhibit oxidative phosphorylation in the mitochondria of its target pathogens. Niclosamide is not absorbed from the gastrointestinal tract, thus it can achieve high localized intestinal concentrations in patients. Recently, it has been shown to also have antibacterial, antiviral, and antitumor activities.[14,15,16]

Another synthetic antihelminthic drug is **praziquantel**, which used for the treatment of parasitic tapeworms and liver flukes, and is particularly useful for the treatment of schistosomiasis (caused by blood flukes from three genera of *Schistosoma*). Its mode of action remains unclear, but it appears to cause the influx of calcium into the worm, resulting in intense spasm and paralysis of the worm. It is often used as a preferred alternative to niclosamide in the treatment of tapeworms when gastrointestinal discomfort limits niclosamide use.

The thioxanthenones, another class of synthetic drugs structurally related to quinine, exhibit antischistosomal activity by inhibiting RNA synthesis. The thioxanthenone lucanthone and its metabolite hycanthone were the first used clinically, but serious neurological, gastrointestinal, cardiovascular, and hepatic side effects led to their discontinuation. Oxamniquine, a less toxic derivative of hycanthone, is only effective against *S. mansoni*, one of the three species known to cause schistosomiasis in humans. Praziquantel was developed to target the other two schistosome species, but concerns about increasing resistance have renewed interest in developing additional derivatives of oxamniquine to target all three clinically important schistosome species.

---

[13] B. Chu et al. "A Benzimidazole Derivative Exhibiting Antitumor Activity Blocks EGFR and HER2 Activity and Upregulates DR5 in Breast Cancer Cells." *Cell Death and Disease* 6 (2015):e1686

[14] J.-X. Pan et al. "Niclosamide, An Old Antihelminthic Agent, Demonstrates Antitumor Activity by Blocking Multiple Signaling Pathways of Cancer Stem Cells." *Chinese Journal of Cancer* 31 no. 4 (2012):178–184.

[15] F. Imperi et al. "New Life for an Old Drug: The Anthelmintic Drug Niclosamide Inhibits *Pseudomonas aeruginosa* Quorum Sensing." *Antimicrobial Agents and Chemotherapy* 57 no. 2 (2013):996-1005.

[16] A. Jurgeit et al. "Niclosamide Is a Proton Carrier and Targets Acidic Endosomes with Broad Antiviral Effects." *PLoS Pathogens* 8 no. 10 (2012):e1002976.

Common Antihelminthic Drugs

Mechanism of Action	Drug Class	Specific Drugs	Clinical Uses
Inhibit microtubule formation, reducing glucose uptake	Benzimidazoles	Mebendazole, albendazole	Variety of helminth infections
Block neuronal transmission, causing paralysis and starvation	Avermectins	Ivermectin	Roundworm diseases, including river blindness and strongyloidiasis, and treatment of parasitic insects
Inhibit ATP production	Not applicable	Niclosamide	Intestinal tapeworm infections
Induce calcium influx	Not applicable	Praziquantel	Schistosomiasis (blood flukes)
Inhibit RNA synthesis	Thioxanthenones	Lucanthone, hycanthone, oxamniquine	Schistosomiasis (blood flukes)

Table 14.9

CHECK YOUR UNDERSTANDING

- Why are antihelminthic drugs difficult to develop?

Antiviral Drugs

Unlike the complex structure of fungi, protozoa, and helminths, viral structure is simple, consisting of nucleic acid, a protein coat, viral enzymes, and, sometimes, a lipid envelope. Furthermore, viruses are obligate intracellular pathogens that use the host's cellular machinery to replicate. These characteristics make it difficult to develop drugs with selective toxicity against viruses.

Many antiviral drugs are nucleoside analogs and function by inhibiting nucleic acid biosynthesis. For example, **acyclovir** (marketed as Zovirax) is a synthetic analog of the nucleoside guanosine (Figure 14.16). It is activated by the herpes simplex viral enzyme thymidine kinase and, when added to a growing DNA strand during replication, causes chain termination. Its specificity for virus-infected cells comes from both the need for a viral enzyme to activate it and the increased affinity of the activated form for viral DNA polymerase compared to host cell DNA polymerase. Acyclovir and its derivatives are frequently used for the treatment of herpes virus infections, including genital herpes, chickenpox, shingles, Epstein-Barr virus infections, and cytomegalovirus infections. Acyclovir can be administered either topically or systemically, depending on the infection. One possible side effect of its use includes nephrotoxicity. The drug adenine-arabinoside, marketed as vidarabine, is a synthetic analog to deoxyadenosine that has a mechanism of action similar to that of acyclovir. It is also effective for the treatment of various human herpes viruses. However, because of possible side effects involving low white blood cell counts and neurotoxicity, treatment with acyclovir is now preferred.

Ribavirin, another synthetic guanosine analog, works by a mechanism of action that is not entirely clear. It appears to interfere with both DNA and RNA synthesis, perhaps by reducing intracellular pools of guanosine triphosphate (GTP). Ribavarin also appears to inhibit the RNA polymerase of hepatitis C virus. It is primarily used for the treatment of the RNA viruses like hepatitis C (in combination therapy with interferon) and respiratory syncytial virus. Possible side effects of ribavirin use include anemia and developmental effects on unborn children in pregnant patients. In recent years, another nucleotide analog, sofosbuvir (Solvaldi), has also been developed for the treatment of hepatitis C. Sofosbuvir is a uridine analog that interferes with viral polymerase activity. It is commonly coadministered with ribavirin, with and without interferon.

Inhibition of nucleic acid synthesis is not the only target of synthetic antivirals. Although the mode of action of **amantadine** and its relative **rimantadine** are not entirely clear, these drugs appear to bind to a transmembrane protein that is involved in the escape of the influenza virus from endosomes. Blocking escape of the virus also prevents viral RNA release into host cells and subsequent viral replication. Increasing resistance has limited the use of amantadine and rimantadine in the treatment of influenza A. Use of amantadine can result in neurological side effects, but the side effects of rimantadine seem less severe. Interestingly, because of their effects on brain chemicals such as dopamine and NMDA (N-methyl D-aspartate), amantadine and rimantadine are also used for the treatment of Parkinson's disease.

Neuraminidase inhibitors, including olsetamivir (Tamiflu), zanamivir (Relenza), and peramivir (Rapivab), specifically target influenza viruses by blocking the activity of influenza virus neuraminidase, preventing the release of the virus from infected cells. These three antivirals can decrease flu symptoms and shorten the duration of illness, but they differ in their modes of administration: olsetamivir is administered orally, zanamivir is inhaled, and peramivir is administered intravenously. Resistance to these neuraminidase inhibitors still seems to be minimal.

Pleconaril is a synthetic antiviral under development that showed promise for the treatment of picornaviruses. Use of **pleconaril** for the treatment of the common cold caused by rhinoviruses was not approved by the FDA in 2002 because of lack of proven effectiveness, lack of stability, and association with irregular menstruation. Its further development for this purpose was halted in 2007. However, pleconaril is still being investigated for use in the treatment of life-threatening complications of enteroviruses, such as meningitis and sepsis. It is also being investigated for use in the global eradication of a specific enterovirus, polio.[17] Pleconaril seems to work by binding to the viral capsid and preventing the uncoating of viral particles inside host cells during viral infection.

Viruses with complex life cycles, such as HIV, can be more difficult to treat. First, HIV targets CD4-positive white blood cells, which are necessary for a normal immune response to infection. Second, HIV is a retrovirus, meaning that it converts its RNA genome into a DNA copy that integrates into the host cell's genome, thus hiding within host cell DNA. Third, the HIV reverse transcriptase lacks proofreading activity and introduces mutations that allow for rapid development of antiviral drug resistance. To help prevent the emergence of resistance, a combination of specific synthetic antiviral drugs is typically used in ART for HIV (Figure 14.17).

The **reverse transcriptase inhibitors** block the early step of converting viral RNA genome into DNA, and can include competitive nucleoside analog inhibitors (e.g., azidothymidine/zidovudine, or AZT) and non-nucleoside noncompetitive inhibitors (e.g., etravirine) that bind reverse transcriptase and cause an inactivating conformational change. Drugs called **protease inhibitors** (e.g., ritonavir) block the processing of viral proteins and prevent viral maturation. Protease inhibitors are also being developed for the treatment of other viral types.[18] For example, simeprevir (Olysio) has been approved for the treatment of hepatitis C and is administered with ribavirin and interferon in combination therapy. The **integrase inhibitors** (e.g., raltegravir), block the activity of the HIV integrase responsible for the recombination of a DNA copy of the viral genome into the host cell chromosome. Additional drug classes for HIV treatment include the CCR5 antagonists and the **fusion inhibitors** (e.g., enfuviritide), which prevent the binding of HIV to the host cell coreceptor (chemokine receptor type 5 [CCR5]) and the merging of the viral envelope with the host cell membrane, respectively. Table 14.10 shows the various therapeutic classes of antiviral drugs, categorized by mode of action, with examples of each.

---

17 M.J. Abzug. "The Enteroviruses: Problems in Need of Treatments." *Journal of Infection* 68 no. S1 (2014):108–14.
18 B.L. Pearlman. "Protease Inhibitors for the Treatment of Chronic Hepatitis C Genotype-1 Infection: The New Standard of Care." *Lancet Infectious Diseases* 12 no. 9 (2012):717–728.

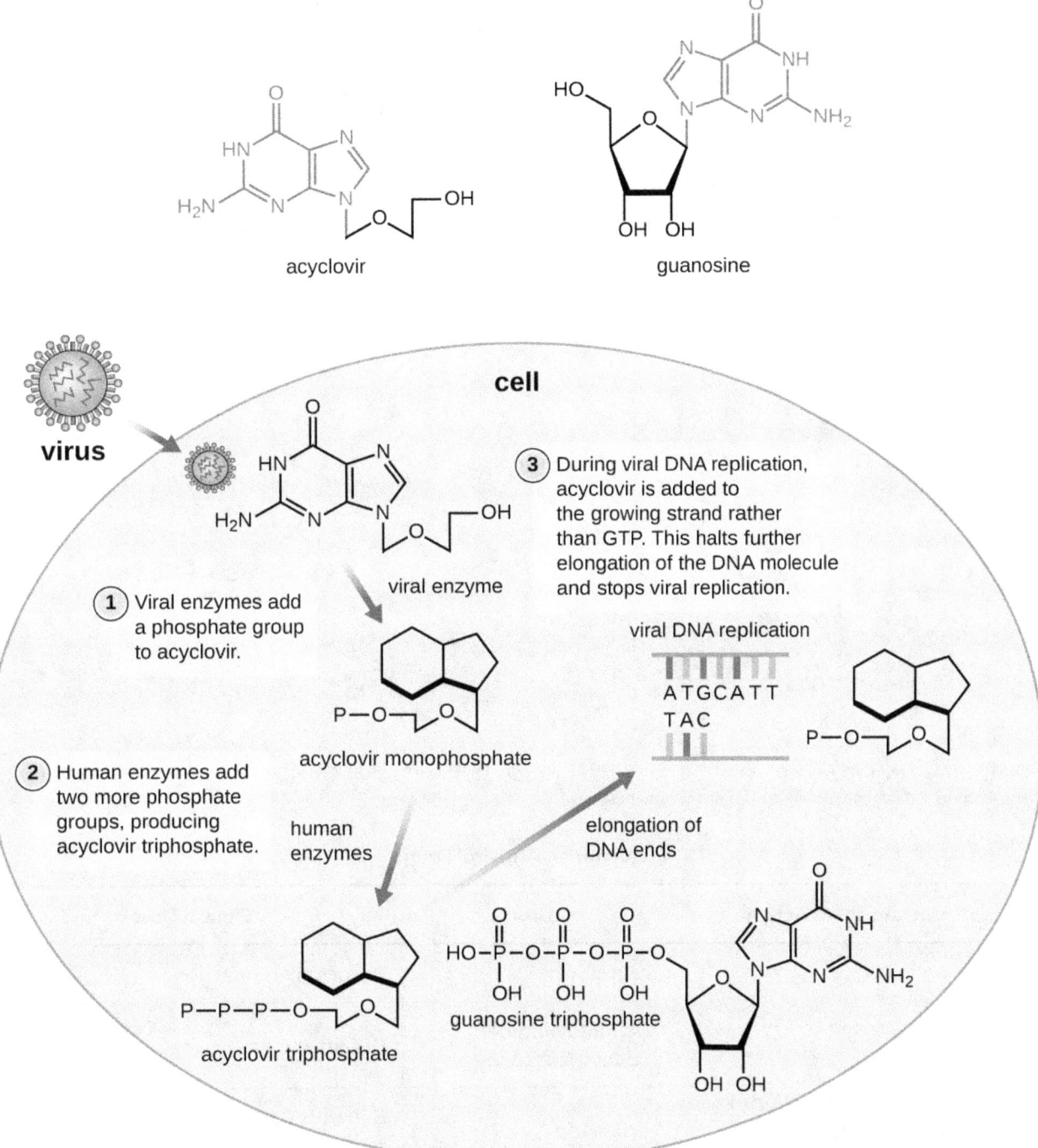

Figure 14.16 Acyclovir is a structural analog of guanosine. It is specifically activated by the viral enzyme thymidine kinase and then preferentially binds to viral DNA polymerase, leading to chain termination during DNA replication.

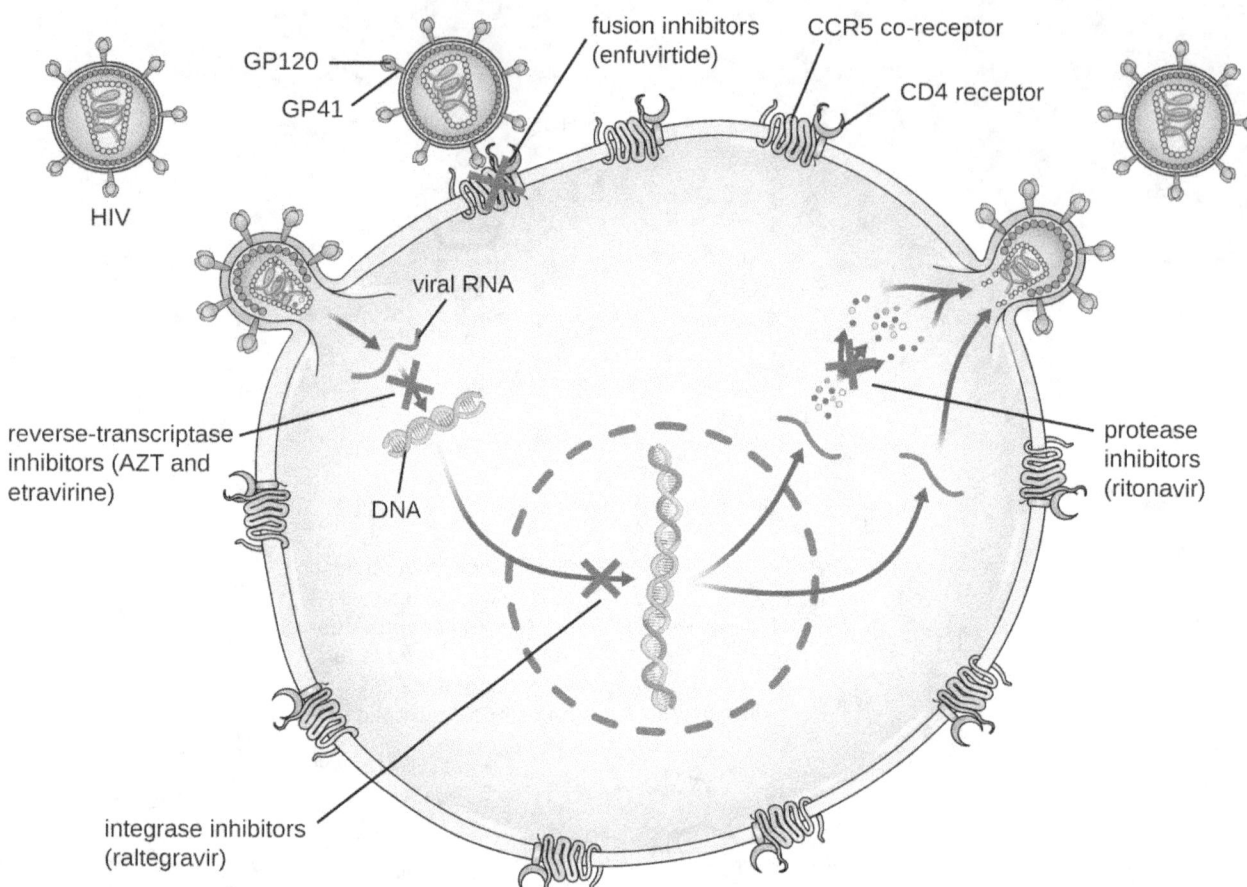

Figure 14.17 Antiretroviral therapy (ART) is typically used for the treatment of HIV. The targets of drug classes currently in use are shown here. (credit: modification of work by Thomas Splettstoesser)

Common Antiviral Drugs

Mechanism of Action	Drug	Clinical Uses
Nucleoside analog inhibition of nucleic acid synthesis	Acyclovir	Herpes virus infections
	Azidothymidine/ zidovudine (AZT)	HIV infections
	Ribavirin	Hepatitis C virus and respiratory syncytial virus infections
	Vidarabine	Herpes virus infections
	Sofosbuvir	Hepatitis C virus infections
Non-nucleoside noncompetitive inhibition	Etravirine	HIV infections
Inhibit escape of virus from endosomes	Amantadine, rimantadine	Infections with influenza virus

Common Antiviral Drugs

Mechanism of Action	Drug	Clinical Uses
Inhibit neuraminadase	Olsetamivir, zanamivir, peramivir	Infections with influenza virus
Inhibit viral uncoating	Pleconaril	Serious enterovirus infections
Inhibition of protease	Ritonavir	HIV infections
Inhibition of protease	Simeprevir	Hepatitis C virus infections
Inhibition of integrase	Raltegravir	HIV infections
Inhibition of membrane fusion	Enfuviritide	HIV infections

Table 14.10

CHECK YOUR UNDERSTANDING

- Why is HIV difficult to treat with antivirals?

LINK TO LEARNING

To learn more about the various classes of antiretroviral drugs used in the ART of HIV infection, explore each of the drugs in the HIV drug classes provided by US Department of Health and Human Services at this (https://openstax.org/l/22HIVUSDepthea) website.

14.5 Drug Resistance

Learning Objectives

By the end of this section, you will be able to:
- Explain the concept of drug resistance
- Describe how microorganisms develop or acquire drug resistance
- Describe the different mechanisms of antimicrobial drug resistance

Antimicrobial resistance is not a new phenomenon. In nature, microbes are constantly evolving in order to overcome the antimicrobial compounds produced by other microorganisms. Human development of antimicrobial drugs and their widespread clinical use has simply provided another selective pressure that promotes further evolution. Several important factors can accelerate the evolution of **drug resistance**. These include the overuse and misuse of antimicrobials, inappropriate use of antimicrobials, subtherapeutic dosing, and patient noncompliance with the recommended course of treatment.

Exposure of a pathogen to an antimicrobial compound can select for chromosomal mutations conferring resistance, which can be transferred vertically to subsequent microbial generations and eventually become predominant in a microbial population that is repeatedly exposed to the antimicrobial. Alternatively, many genes responsible for drug resistance are found on plasmids or in transposons that can be transferred easily between microbes through horizontal gene transfer (see How Asexual Prokaryotes Achieve Genetic Diversity). Transposons also have the ability to move resistance genes between plasmids and chromosomes to further promote the spread of resistance.

Mechanisms for Drug Resistance

There are several common mechanisms for drug resistance, which are summarized in Figure 14.18. These mechanisms include enzymatic modification of the drug, modification of the antimicrobial target, and prevention of drug penetration or accumulation.

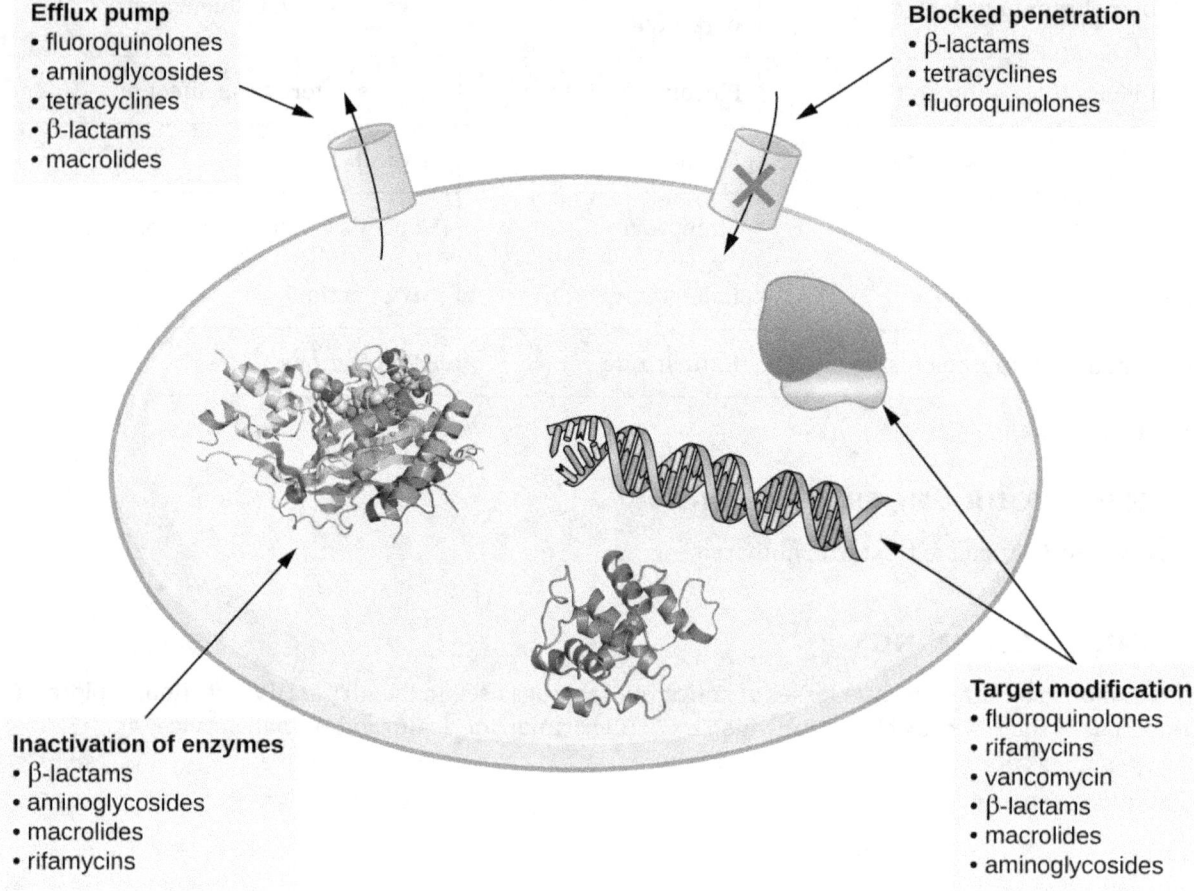

Figure 14.18 There are multiple strategies that microbes use to develop resistance to antimicrobial drugs. (Not shown: target overproduction, target mimicry, and enzymatic bypass). (credit: modification of work by Gerard D Wright)

Drug Modification or Inactivation

Resistance genes may code for enzymes that chemically modify an antimicrobial, thereby inactivating it, or destroy an antimicrobial through hydrolysis. Resistance to many types of antimicrobials occurs through this mechanism. For example, aminoglycoside resistance can occur through enzymatic transfer of chemical groups to the drug molecule, impairing the binding of the drug to its bacterial target. For β-lactams, bacterial resistance can involve the enzymatic hydrolysis of the β-lactam bond within the β-lactam ring of the drug molecule. Once the β-lactam bond is broken, the drug loses its antibacterial activity. This mechanism of resistance is mediated by β-lactamases, which are the most common mechanism of β-lactam resistance. Inactivation of rifampin commonly occurs through glycosylation, phosphorylation, or adenosine diphosphate (ADP) ribosylation, and resistance to macrolides and lincosamides can also occur due to enzymatic inactivation of the drug or modification.

Prevention of Cellular Uptake or Efflux

Microbes may develop resistance mechanisms that involve inhibiting the accumulation of an antimicrobial drug, which then prevents the drug from reaching its cellular target. This strategy is common among gram-negative pathogens and can involve changes in outer membrane lipid composition, porin channel selectivity, and/or porin channel concentrations. For example, a common mechanism of carbapenem resistance among *Pseudomonas aeruginosa* is to decrease the amount of its OprD porin, which is the primary portal of entry for carbapenems through the outer membrane of this pathogen. Additionally, many gram-positive and gram-negative pathogenic bacteria produce efflux pumps that actively transport an antimicrobial drug out of the cell and prevent the accumulation of drug to a level that would be antibacterial. For example, resistance to β-lactams, tetracyclines, and fluoroquinolones commonly occurs through active efflux out of the cell, and it is rather common for a single efflux pump to have the ability to translocate multiple types of antimicrobials.

Target Modification

Because antimicrobial drugs have very specific targets, structural changes to those targets can prevent drug binding, rendering the drug ineffective. Through spontaneous mutations in the genes encoding antibacterial drug targets, bacteria have an evolutionary advantage that allows them to develop resistance to drugs. This mechanism of resistance development is quite common. Genetic changes impacting the active site of penicillin-binding proteins (PBPs) can inhibit the binding of β-lactam drugs and provide resistance to multiple drugs within this class. This mechanism is very common among strains of *Streptococcus pneumoniae*, which alter their own PBPs through genetic mechanisms. In contrast, strains of *Staphylococcus aureus* develop resistance to methicillin (MRSA) through the acquisition of a new low-affinity PBP, rather than structurally alter their existing PBPs. Not only does this new low-affinity PBP provide resistance to methicillin but it provides resistance to virtually all β-lactam drugs, with the exception of the newer fifth-generation cephalosporins designed specifically to kill MRSA. Other examples of this resistance strategy include alterations in

- ribosome subunits, providing resistance to macrolides, tetracyclines, and aminoglycosides;
- lipopolysaccharide (LPS) structure, providing resistance to polymyxins;
- RNA polymerase, providing resistance to rifampin;
- DNA gyrase, providing resistance to fluoroquinolones;
- metabolic enzymes, providing resistance to sulfa drugs, sulfones, and trimethoprim; and
- peptidoglycan subunit peptide chains, providing resistance to glycopeptides.

Target Overproduction or Enzymatic Bypass

When an antimicrobial drug functions as an antimetabolite, targeting a specific enzyme to inhibit its activity, there are additional ways that microbial resistance may occur. First, the microbe may overproduce the target enzyme such that there is a sufficient amount of antimicrobial-free enzyme to carry out the proper enzymatic reaction. Second, the bacterial cell may develop a bypass that circumvents the need for the functional target enzyme. Both of these strategies have been found as mechanisms of sulfonamide resistance. Vancomycin resistance among *S. aureus* has been shown to involve the decreased cross-linkage of peptide chains in the bacterial cell wall, which provides an increase in targets for vancomycin to bind to in the outer cell wall. Increased binding of vancomycin in the outer cell wall provides a blockage that prevents free drug molecules from penetrating to where they can block new cell wall synthesis.

Target Mimicry

A recently discovered mechanism of resistance called target mimicry involves the production of proteins that prevent drugs from binding to their bacterial cellular targets. For example, *fluoroquinolone* resistance by *Mycobacterium tuberculosis* can involve the production of a protein that resembles DNA. This protein is called MfpA (Mycobacterium fluoroquinolone resistance protein A). The mimicry of DNA by MfpA results in DNA gyrase binding to MfpA, preventing the binding of fluoroquinolones to DNA gyrase.

✓ CHECK YOUR UNDERSTANDING

- List several mechanisms for drug resistance.

Multidrug-Resistant Microbes and Cross Resistance

From a clinical perspective, our greatest concerns are **multidrug-resistant microbes (MDRs)** and cross resistance. MDRs are colloquially known as "superbugs" and carry one or more resistance mechanism(s), making them resistant to multiple antimicrobials. In **cross-resistance**, a single resistance mechanism confers resistance to multiple antimicrobial drugs. For example, having an efflux pump that can export multiple antimicrobial drugs is a common way for microbes to be resistant to multiple drugs by using a single resistance mechanism. In recent years, several clinically important superbugs have emerged, and the CDC reports that superbugs are responsible for more than 2 million infections in the US annually, resulting in at least 23,000 fatalities.[19] Several of the superbugs discussed in the following sections have been dubbed the ESKAPE pathogens. This acronym refers to the names of the pathogens (*Enterococcus faecium*, *Staphylococcus aureus*, *Klebsiella pneumoniae*, *Acinetobacter baumannii*, *Pseudomonas aeruginosa* and *Enterobacter* spp.) but it is also fitting in that these pathogens are able to "escape" many conventional forms of antimicrobial therapy. As such, infections by ESKAPE pathogens can be difficult to treat and they cause a large number of nosocomial infections.

Methicillin-Resistant *Staphylococcus aureus* (MRSA)

Methicillin, a semisynthetic penicillin, was designed to resist inactivation by β-lactamases. Unfortunately, soon after the introduction of methicillin to clinical practice, methicillin-resistant strains of *S. aureus* appeared and started to spread. The mechanism of resistance, acquisition of a new low-affinity PBP, provided *S. aureus* with resistance to all available β-lactams. Strains of **methicillin-resistant *S. aureus* (MRSA)** are widespread opportunistic pathogens and a particular concern for skin and other wound infections, but may also cause pneumonia and septicemia. Although originally a problem in health-care settings (hospital-acquired MRSA [HA-MRSA]), MRSA infections are now also acquired through contact with contaminated members of the general public, called community-associated MRSA (CA-MRSA). Approximately one-third of the population carries *S. aureus* as a member of their normal nasal microbiota without illness, and about 6% of these strains are methicillin resistant.[20][21]

 MICRO CONNECTIONS

Clavulanic Acid: Penicillin's Little Helper

With the introduction of penicillin in the early 1940s, and its subsequent mass production, society began to think of antibiotics as miracle cures for a wide range of infectious diseases. Unfortunately, as early as 1945, penicillin resistance was first documented and started to spread. Greater than 90% of current *S. aureus* clinical isolates are resistant to penicillin.[22]

Although developing new antimicrobial drugs is one solution to this problem, scientists have explored new approaches, including the development of compounds that inactivate resistance mechanisms. The development of clavulanic acid represents an early example of this strategy. Clavulanic acid is a molecule produced by the bacterium *Streptococcus clavuligerus*. It contains a β-lactam ring, making it structurally similar to penicillin and other β-lactams, but shows no clinical effectiveness when administered on its own. Instead, clavulanic acid binds irreversibly within the active site of β-lactamases and prevents them from inactivating a coadministered penicillin.

---

19 Centers for Disease Control and Prevention. "Antibiotic/Antimicrobial Resistance." http://www.cdc.gov/drugresistance/index.html. Accessed June 2, 2016.

20 A.S. Kalokhe et al. "Multidrug-Resistant Tuberculosis Drug Susceptibility and Molecular Diagnostic Testing: A Review of the Literature. *American Journal of the Medical Sciences* 345 no. 2 (2013):143–148.

21 Centers for Disease Control and Prevention. "Methicillin-Resistant *Staphylococcus aureus* (MRSA): General Information About MRSA in the Community." http://www.cdc.gov/mrsa/community/index.html. Accessed June 2, 2016

22 F.D. Lowy. "Antimicrobial Resistance: The Example of *Staphylococcus aureus*." *Journal of Clinical Investigation* 111 no. 9 (2003):1265–1273.

Clavulanic acid was first developed in the 1970s and was mass marketed in combination with amoxicillin beginning in the 1980s under the brand name Augmentin. As is typically the case, resistance to the amoxicillin-clavulanic acid combination soon appeared. Resistance most commonly results from bacteria increasing production of their β-lactamase and overwhelming the inhibitory effects of clavulanic acid, mutating their β-lactamase so it is no longer inhibited by clavulanic acid, or from acquiring a new β-lactamase that is not inhibited by clavulanic acid. Despite increasing resistance concerns, clavulanic acid and related β-lactamase inhibitors (sulbactam and tazobactam) represent an important new strategy: the development of compounds that directly inhibit antimicrobial resistance-conferring enzymes.

Vancomycin-Resistant Enterococci and *Staphylococcus aureus*

Vancomycin is only effective against gram-positive organisms, and it is used to treat wound infections, septic infections, endocarditis, and meningitis that are caused by pathogens resistant to other antibiotics. It is considered one of the last lines of defense against such resistant infections, including MRSA. With the rise of antibiotic resistance in the 1970s and 1980s, vancomycin use increased, and it is not surprising that we saw the emergence and spread of **vancomycin-resistant enterococci (VRE)**, **vancomycin-resistant *S. aureus* (VRSA)**, and **vancomycin-intermediate *S. aureus* (VISA)**. The mechanism of vancomycin resistance among enterococci is target modification involving a structural change to the peptide component of the peptidoglycan subunits, preventing vancomycin from binding. These strains are typically spread among patients in clinical settings by contact with health-care workers and contaminated surfaces and medical equipment.

VISA and VRSA strains differ from each other in the mechanism of resistance and the degree of resistance each mechanism confers. VISA strains exhibit intermediate resistance, with a minimum inhibitory concentration (MIC) of 4–8 µg/mL, and the mechanism involves an increase in vancomycin targets. VISA strains decrease the crosslinking of peptide chains in the cell wall, providing an increase in vancomycin targets that trap vancomycin in the outer cell wall. In contrast, VRSA strains acquire vancomycin resistance through horizontal transfer of resistance genes from VRE, an opportunity provided in individuals coinfected with both VRE and MRSA. VRSA exhibit a higher level of resistance, with MICs of 16 µg/mL or higher.[23] In the case of all three types of vancomycin-resistant bacteria, rapid clinical identification is necessary so proper procedures to limit spread can be implemented. The oxazolidinones like linezolid are useful for the treatment of these vancomycin-resistant, opportunistic pathogens, as well as MRSA.

Extended-Spectrum β-Lactamase–Producing Gram-Negative Pathogens

Gram-negative pathogens that produce **extended-spectrum β-lactamases (ESBLs)** show resistance well beyond just penicillins. The spectrum of β-lactams inactivated by ESBLs provides for resistance to all penicillins, cephalosporins, monobactams, and the β-lactamase-inhibitor combinations, but not the carbapenems. An even greater concern is that the genes encoding for ESBLs are usually found on mobile plasmids that also contain genes for resistance to other drug classes (e.g., fluoroquinolones, aminoglycosides, tetracyclines), and may be readily spread to other bacteria by horizontal gene transfer. These multidrug-resistant bacteria are members of the intestinal microbiota of some individuals, but they are also important causes of opportunistic infections in hospitalized patients, from whom they can be spread to other people.

Carbapenem-Resistant Gram-Negative Bacteria

The occurrence of **carbapenem-resistant Enterobacteriaceae (CRE)** and carbapenem resistance among other gram-negative bacteria (e.g., *P. aeruginosa*, *Acinetobacter baumannii*, *Stenotrophomonas maltophila*) is a growing health-care concern. These pathogens develop resistance to carbapenems through a variety of mechanisms, including production of carbapenemases (broad-spectrum β-lactamases that inactivate all β-lactams, including carbapenems), active efflux of carbapenems out of the cell, and/or prevention of carbapenem entry through porin channels. Similar to concerns with ESBLs, carbapenem-resistant, gram-negative pathogens are usually resistant to multiple classes of antibacterials, and some have even developed

---

23 Centers for Disease Control and Prevention. "Healthcare-Associated Infections (HIA): General Information about VISA/VRSA." http://www.cdc.gov/HAI/organisms/visa_vrsa/visa_vrsa.html. Accessed June 2, 2016.

pan-resistance (resistance to all available antibacterials). Infections with carbapenem-resistant, gram-negative pathogens commonly occur in health-care settings through interaction with contaminated individuals or medical devices, or as a result of surgery.

Multidrug-Resistant *Mycobacterium tuberculosis*

The emergence of **multidrug-resistant *Mycobacterium tuberculosis* (MDR-TB)** and **extensively drug-resistant *Mycobacterium tuberculosis* (XDR-TB)** is also of significant global concern. MDR-TB strains are resistant to both rifampin and isoniazid, the drug combination typically prescribed for treatment of tuberculosis. XDR-TB strains are additionally resistant to any fluoroquinolone and at least one of three other drugs (amikacin, kanamycin, or capreomycin) used as a second line of treatment, leaving these patients very few treatment options. Both types of pathogens are particularly problematic in immunocompromised persons, including those suffering from HIV infection. The development of resistance in these strains often results from the incorrect use of antimicrobials for tuberculosis treatment, selecting for resistance.

CHECK YOUR UNDERSTANDING

- How does drug resistance lead to superbugs?

LINK TO LEARNING

To learn more about the top 18 drug-resistant threats (https://openstax.org/l/22CDC18drugres) to the US, visit the CDC's website.

 MICRO CONNECTIONS

Factory Farming and Drug Resistance

Although animal husbandry has long been a major part of agriculture in America, the rise of concentrated animal feeding operations (CAFOs) since the 1950s has brought about some new environmental issues, including the contamination of water and air with biological waste, and ethical issues regarding animal rights also are associated with growing animals in this way. Additionally, the increase in CAFOs involves the extensive use of antimicrobial drugs in raising livestock. Antimicrobials are used to prevent the development of infectious disease in the close quarters of CAFOs; however, the majority of antimicrobials used in factory farming are for the promotion of growth—in other words, to grow larger animals.

The mechanism underlying this enhanced growth remains unclear. These antibiotics may not necessarily be the same as those used clinically for humans, but they are structurally related to drugs used for humans. As a result, use of antimicrobial drugs in animals can select for antimicrobial resistance, with these resistant bacteria becoming cross-resistant to drugs typically used in humans. For example, tylosin use in animals appears to select for bacteria also cross-resistant to other macrolides, including erythromycin, commonly used in humans.

Concentrations of the drug-resistant bacterial strains generated by CAFOs become increased in water and soil surrounding these farms. If not directly pathogenic in humans, these resistant bacteria may serve as a reservoir of mobile genetic elements that can then pass resistance genes to human pathogens. Fortunately, the cooking process typically inactivates any antimicrobials remaining in meat, so humans typically are not directly ingesting these drugs. Nevertheless, many people are calling for more judicious use of these drugs, perhaps charging farmers user fees to reduce indiscriminate use. In fact, in 2012, the FDA published guidelines for farmers who voluntarily phase out the use of antimicrobial drugs except under veterinary supervision and when necessary to ensure animal health. Although following the guidelines is voluntary at this time, the FDA does recommend what it calls "judicious" use of antimicrobial drugs in food-producing animals in an effort to decrease antimicrobial resistance.

Clinical Focus

Part 3

Unfortunately, Marisa's urinary tract infection did not resolve with ciprofloxacin treatment. Laboratory testing showed that her infection was caused by a strain of *Klebsiella pneumoniae* with significant antimicrobial resistance. The resistance profile of this *K. pneumoniae* included resistance to the carbapenem class of antibacterials, a group of β-lactams that is typically reserved for the treatment of highly resistant bacteria. *K. pneumoniae* is an opportunistic, capsulated, gram-negative rod that may be a member of the normal microbiota of the intestinal tract, but may also cause a number of diseases, including pneumonia and UTIs.

Specific laboratory tests looking for carbapenemase production were performed on Marisa's samples and came back positive. Based upon this result, in combination with her health history, production of a carbapenemase known as the New Delhi Metallo-β-lactamase (NDM) was suspected. Although the origin of the NDM carbapenemase is not completely known, many patients infected with NDM-containing strains have travel histories involving hospitalizations in India or surrounding countries.

- How would doctors determine which types of antimicrobial drugs should be administered?

Jump to the next Clinical Focus box. Go back to the previous Clinical Focus box.

14.6 Testing the Effectiveness of Antimicrobials

Learning Objectives

By the end of this section, you will be able to:
- Describe how the Kirby-Bauer disk diffusion test determines the susceptibility of a microbe to an antibacterial drug.
- Explain the significance of the minimal inhibitory concentration and the minimal bactericidal concentration relative to the effectiveness of an antimicrobial drug.

Testing the effectiveness of antimicrobial drugs against specific organisms is important in identifying their spectrum of activity and the therapeutic dosage. This type of test, generally described as antimicrobial susceptibility testing (AST), is commonly performed in a clinical laboratory. In this section, we will discuss common methods of testing the effectiveness of antimicrobials.

The Kirby-Bauer Disk Diffusion Test

The **Kirby-Bauer disk diffusion test** has long been used as a starting point for determining the susceptibility of specific microbes to various antimicrobial drugs. The Kirby-Bauer assay starts with a Mueller-Hinton agar plate on which a confluent lawn is inoculated with a patient's isolated bacterial pathogen. Filter paper disks impregnated with known amounts of antibacterial drugs to be tested are then placed on the agar plate. As the bacterial inoculum grows, antibiotic diffuses from the circular disk into the agar and interacts with the growing bacteria. Antibacterial activity is observed as a clear circular **zone of inhibition** around the drug-impregnated disk, similar to the disk-diffusion assay depicted in Figure 13.31. The diameter of the zone of inhibition, measured in millimeters and compared to a standardized chart, determines the susceptibility or resistance of the bacterial pathogen to the drug.

There are multiple factors that determine the size of a zone of inhibition in this assay, including drug solubility, rate of drug diffusion through agar, the thickness of the agar medium, and the drug concentration impregnated into the disk. Due to a lack of standardization of these factors, interpretation of the Kirby-Bauer disk diffusion assay provides only limited information on susceptibility and resistance to the drugs tested. The assay cannot distinguish between bacteriostatic and bactericidal activities, and differences in zone sizes cannot be used to compare drug potencies or efficacies. Comparison of zone sizes to a standardized chart will only provide information on the antibacterials to which a bacterial pathogen is susceptible or resistant.

CHECK YOUR UNDERSTANDING

- How does one use the information from a Kirby-Bauer assay to predict the therapeutic effectiveness of an antimicrobial drug in a patient?

MICRO CONNECTIONS

Antibiograms: Taking Some of the Guesswork Out of Prescriptions

Unfortunately, infectious diseases don't take a time-out for lab work. As a result, physicians rarely have the luxury of conducting susceptibility testing before they write a prescription. Instead, they rely primarily on the empirical evidence (i.e., the signs and symptoms of disease) and their professional experience to make an educated guess as to the diagnosis, causative agent(s), and drug most likely to be effective. This approach allows treatment to begin sooner so the patient does not have to wait for lab test results. In many cases, the prescription is effective; however, in an age of increased antimicrobial resistance, it is becoming increasingly more difficult to select the most appropriate empiric therapy. Selecting an inappropriate empiric therapy not only puts the patient at risk but may promote greater resistance to the drug prescribed.

Recently, studies have shown that antibiograms are useful tools in the decision-making process of selecting appropriate empiric therapy. An **antibiogram** is a compilation of local antibiotic susceptibility data broken down by bacterial pathogen. In a November 2014 study published in the journal *Infection Control and Hospital Epidemiology*, researchers determined that 85% of the prescriptions ordered in skilled nursing facilities were decided upon empirically, but only 35% of those prescriptions were deemed appropriate when compared with the eventual pathogen identification and susceptibility profile obtained from the clinical laboratory. However, in one nursing facility where use of antibiograms was implemented to direct selection of empiric therapy, appropriateness of empiric therapy increased from 32% before antibiogram implementation to 45% after implementation of antibiograms.[24] Although these data are preliminary, they do suggest that health-care facilities can reduce the number of inappropriate prescriptions by using antibiograms to select empiric therapy, thus benefiting patients and minimizing opportunities for antimicrobial resistance to develop.

LINK TO LEARNING

Visit this website to view an interactive antibiogram (https://openstax.org/l/22StanUnintanti) provided by Stanford University.

Dilution Tests

As discussed, the limitations of the Kirby-Bauer disk diffusion test do not allow for a direct comparison of antibacterial potencies to guide selection of the best therapeutic choice. However, antibacterial dilution tests can be used to determine a particular drug's **minimal inhibitory concentration (MIC)**, the lowest concentration of drug that inhibits visible bacterial growth, and **minimal bactericidal concentration (MBC)**, the lowest drug concentration that kills ≥99.9% of the starting inoculum. Determining these concentrations helps identify the correct drug for a particular pathogen. For the macrobroth dilution assay, a dilution series of the drug in broth is made in test tubes and the same number of cells of a test bacterial strain is added to each tube (Figure 14.19). The MIC is determined by examining the tubes to find the lowest drug concentration that inhibits visible growth; this is observed as turbidity (cloudiness) in the broth. Tubes with no visible growth are then inoculated onto agar media without antibiotic to determine the MBC. Generally, serum levels of an antibacterial should be at least three to five times above the MIC for treatment of an infection.

The MIC assay can also be performed using 96-well microdilution trays, which allow for the use of small volumes and automated dispensing devices, as well as the testing of multiple antimicrobials and/or

---

24 J.P. Furuno et al. "Using Antibiograms to Improve Antibiotic Prescribing in Skilled Nursing Facilities." *Infection Control and Hospital Epidemiology* 35 no. Suppl S3 (2014):S56–61.

microorganisms in one tray (Figure 14.20). MICs are interpreted as the lowest concentration that inhibits visible growth, the same as for the macrobroth dilution in test tubes. Growth may also be interpreted visually or by using a spectrophotometer or similar device to detect turbidity or a color change if an appropriate biochemical substrate that changes color in the presence of bacterial growth is also included in each well.

The **Etest** is an alternative method used to determine MIC, and is a combination of the Kirby-Bauer disk diffusion test and dilution methods. Similar to the Kirby-Bauer assay, a confluent lawn of a bacterial isolate is inoculated onto the surface of an agar plate. Rather than using circular disks impregnated with one concentration of drug, however, commercially available plastic strips that contain a gradient of an antibacterial are placed on the surface of the inoculated agar plate (Figure 14.21). As the bacterial inoculum grows, antibiotic diffuses from the plastic strips into the agar and interacts with the bacterial cells. Because the rate of drug diffusion is directly related to concentration, an elliptical zone of inhibition is observed with the Etest drug gradient, rather than a circular zone of inhibition observed with the Kirby-Bauer assay. To interpret the results, the intersection of the elliptical zone with the gradient on the drug-containing strip indicates the MIC. Because multiple strips containing different antimicrobials can be placed on the same plate, the MIC of multiple antimicrobials can be determined concurrently and directly compared. However, unlike the macrobroth and microbroth dilution methods, the MBC cannot be determined with the Etest.

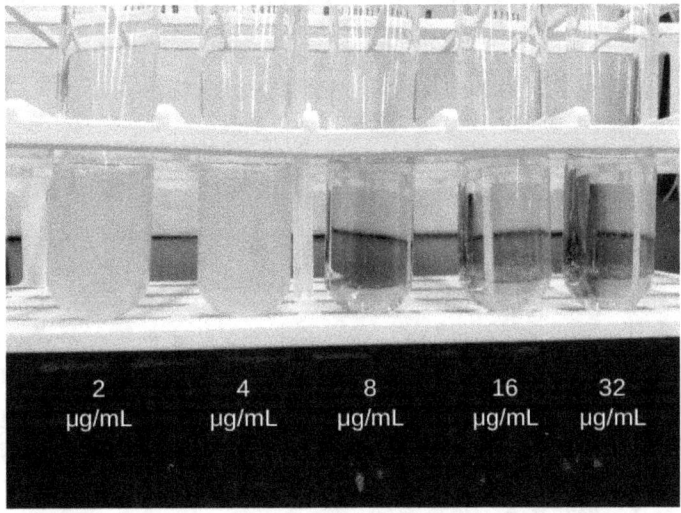

Figure 14.19 In a dilution test, the lowest dilution that inhibits turbidity (cloudiness) is the MIC. In this example, the MIC is 8 µg/mL. Broth from samples without turbidity can be inoculated onto plates lacking the antimicrobial drug. The lowest dilution that kills ≥99.9% of the starting inoculum is observed on the plates is the MBC. (credit: modification of work by Suzanne Wakim)

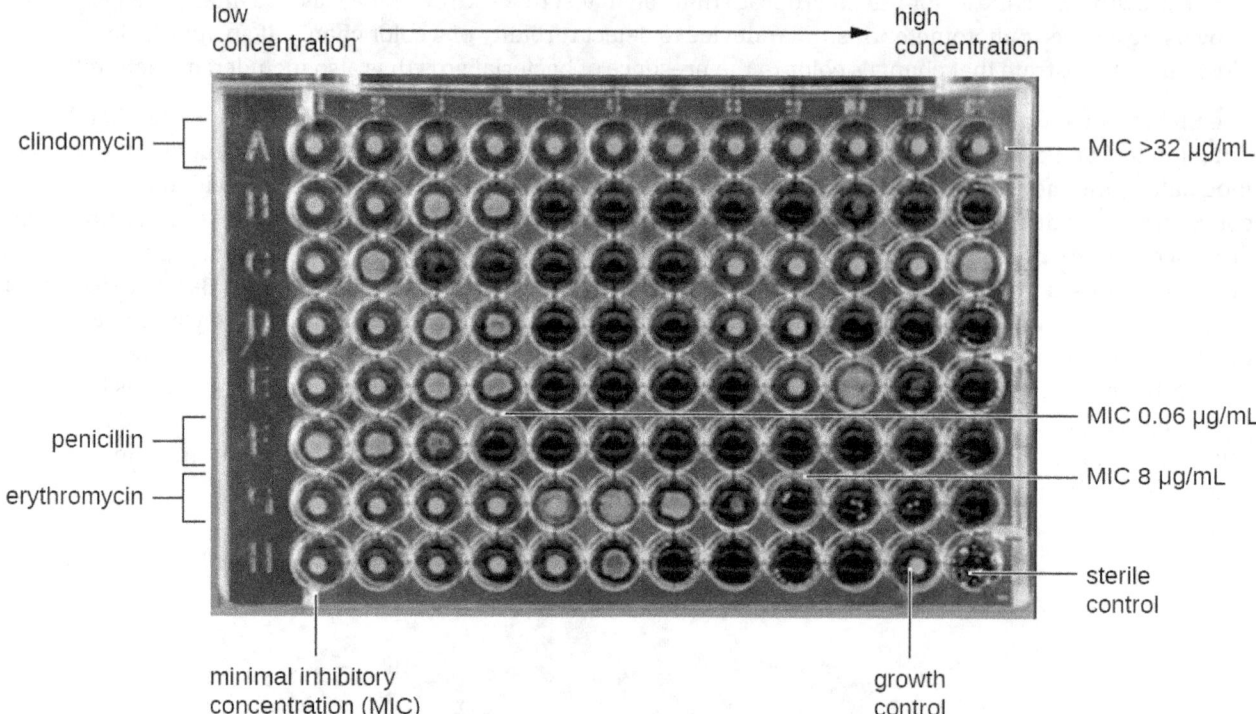

Figure 14.20 A microdilution tray can also be used to determine MICs of multiple antimicrobial drugs in a single assay. In this example, the drug concentrations increase from left to right and the rows with clindamycin, penicillin, and erythromycin have been indicated to the left of the plate. For penicillin and erythromycin, the lowest concentrations that inhibited visible growth are indicated by red circles and were 0.06 µg/mL for penicillin and 8 µg/mL for erythromycin. For clindamycin, visible bacterial growth was observed at every concentration up to 32 µg/mL and the MIC is interpreted as >32 µg/mL. (credit: modification of work by Centers for Disease Control and Prevention)

Figure 14.21 The Etest can be used to determine the MIC of an antibiotic. In this Etest, vancomycin is shown to have a MIC of 1.5 µg/mL against *Staphylococcus aureus*.

CHECK YOUR UNDERSTANDING

- Compare and contrast MIC and MBC.

Clinical Focus

Resolution

Marisa's UTI was likely caused by the catheterizations she had in Vietnam. Most bacteria that cause UTIs are members of the normal gut microbiota, but they can cause infections when introduced to the urinary tract, as might have occurred when the catheter was inserted. Alternatively, if the catheter itself was not sterile, bacteria on its surface could have been introduced into Marisa's body. The antimicrobial therapy Marisa received in Cambodia may also have been a complicating factor because it may have selected for antimicrobial-resistant strains already present in her body. These bacteria would have already contained genes for antimicrobial resistance, either acquired by spontaneous mutation or through horizontal gene transfer, and, therefore, had the best evolutionary advantage for adaptation and growth in the presence of the antimicrobial therapy. As a result, one of these resistant strains may have been subsequently introduced into her urinary tract.

Laboratory testing at the CDC confirmed that the strain of *Klebsiella pneumoniae* from Marisa's urine sample was positive for the presence of NDM, a very active carbapenemase that is beginning to emerge as a new problem in antimicrobial resistance. While NDM-positive strains are resistant to a wide range of antimicrobials, they have shown susceptibility to tigecycline (structurally related to tetracycline) and the polymyxins B and E (colistin).

To prevent her infection from spreading, Marisa was isolated from the other patients in a separate room. All hospital staff interacting with her were advised to follow strict protocols to prevent surface and equipment contamination. This would include especially stringent hand hygiene practices and careful disinfection of all items coming into contact with her.

Marisa's infection finally responded to tigecycline and eventually cleared. She was discharged a few weeks after admission, and a follow-up stool sample showed her stool to be free of NDM-containing *K. pneumoniae*, meaning that she was no longer harboring the highly resistant bacterium.

Go back to the previous Clinical Focus box.

14.7 Current Strategies for Antimicrobial Discovery

Learning Objectives

By the end of this section, you will be able to:
- Describe the methods and strategies used for discovery of new antimicrobial agents.

With the continued evolution and spread of antimicrobial resistance, and now the identification of pan-resistant bacterial pathogens, the search for new antimicrobials is essential for preventing the postantibiotic era. Although development of more effective semisynthetic derivatives is one strategy, resistance to them develops rapidly because bacterial pathogens are already resistant to earlier-generation drugs in the family and can easily mutate and develop resistance to the new semisynthetic drugs. Today, scientists continue to hunt for new antimicrobial compounds and explore new avenues of antimicrobial discovery and synthesis. They check large numbers of soils and microbial products for antimicrobial activity by using high-throughput screening methods, which use automation to test large numbers of samples simultaneously. The recent development of the iChip[25] allows researchers to investigate the antimicrobial-producing capabilities of soil microbes that are difficult to grow by standard cultivation techniques in the laboratory. Rather than grow the microbes in the laboratory, they are grown in situ—right in the soil. Use of the iChip has resulted in the discovery of teixobactin, a novel antimicrobial from Mount Ararat, Turkey. Teixobactin targets two distinct steps in gram-positive cell wall synthesis and for which antimicrobial resistance appears not yet to have evolved.

Although soils have been widely examined, other environmental niches have not been tested as fully. Since 70% of the earth is covered with water, marine environments could be mined more fully for the presence of antimicrobial-producing microbes. In addition, researchers are using combinatorial chemistry, a method for making a very large number of related compounds from simple precursors, and testing them for antimicrobial activity. An additional strategy that needs to be explored further is the development of compounds that inhibit resistance mechanisms and restore the activity of older drugs, such as the strategy described earlier for β-lactamase inhibitors like clavulanic acid. Finally, developing inhibitors of virulence factor production and

---

25 L. Losee et al. "A New Antibiotic Kills Pathogens Without Detectable Resistance." *Nature* 517 no. 7535 (2015):455–459.

function could be a very important avenue. Although this strategy would not be directly antibacterial, drugs that slow the progression of an infection could provide an advantage for the immune system and could be used successfully in combination with antimicrobial drugs.

CHECK YOUR UNDERSTANDING

- What are new sources and strategies for developing drugs to fight infectious diseases?

Eye on Ethics

The (Free?) Market for New Antimicrobials

There used to be plenty of antimicrobial drugs on the market to treat infectious diseases. However, the spread of antimicrobial resistance has created a need for new antibiotics to replace those that are no longer as effective as they once were. Unfortunately, pharmaceutical companies are not particularly motivated to fill this need. As of 2009, all but five pharmaceutical companies had moved away from antimicrobial drug development.[26] As a result, the number of FDA approvals of new antimicrobials has fallen drastically in recent decades (Figure 14.22).

Given that demand usually encourages supply, one might expect pharmaceutical companies to be rushing to get back in the business of developing new antibiotics. But developing new drugs is a lengthy process and requires large investments in research and development. Pharmaceutical companies can typically get a higher return on their investment by developing products for chronic, nonmicrobial diseases like diabetes; such drugs must be taken for life, and therefore generate more long-term revenue than an antibiotic that does its job in a week or two. But what will happen when drugs like vancomycin, a superantimicrobial reserved for use as a last resort, begin to lose their effectiveness against ever more drug-resistant superbugs? Will drug companies wait until all antibiotics have become useless before beginning to look for new ones?

Recently, it has been suggested that large pharmaceutical companies should be given financial incentives to pursue such research. In September 2014, the White House released an executive order entitled "Combating Antibiotic Resistant Bacteria," calling upon various government agencies and the private sector to work together to "accelerate basic and applied research and development for new antimicrobials, other therapeutics, and vaccines."[27] As a result, as of March 2015, President Obama's proposed fiscal year 2016 budget doubled the amount of federal funding to $1.2 billion for "combating and preventing antibiotic resistance," which includes money for antimicrobial research and development.[28] Similar suggestions have also been made on a global scale. In December 2014, a report chaired by former Goldman Sachs economist Jim O'Neill was published in *The Review on Antimicrobial Resistance*.[29]

These developments reflect the growing belief that for-profit pharmaceutical companies must be subsidized to encourage development of new antimicrobials. But some ask whether pharmaceutical development should be motivated by profit at all. Given that millions of lives may hang in the balance, some might argue that drug companies have an ethical obligation to devote their research and development efforts to high-utility drugs, as opposed to highly profitable ones. Yet this obligation conflicts with the fundamental goals of a for-profit company. Are government subsidies enough to ensure that drug companies make the public interest a priority, or should government agencies assume responsibility for

---

26 H.W. Boucher et al. "Bad Bugs, No Drugs: No ESKAPE! An Update from the Infectious Diseases Society of America." *Clinical Infectious Diseases* 48 no. 1 (2009):1–12.

27 The White House. *National Action Plan for Combating Antibiotic-Resistant Bacteria.* Washington, DC: The White House, 2015.

28 White House Office of the Press Secretary. "Fact Sheet: Obama Administration Releases National Action Plan to Combat Antibiotic-Resistant Bacteria." March 27, 2015. https://www.whitehouse.gov/the-press-office/2015/03/27/fact-sheet-obama-administration-releases-national-action-plan-combat-ant

29 Review on Antimicrobial Resistance. http://amr-review.org. Accessed June 1, 2016.

developing critical drugs that may have little or no return on investment?

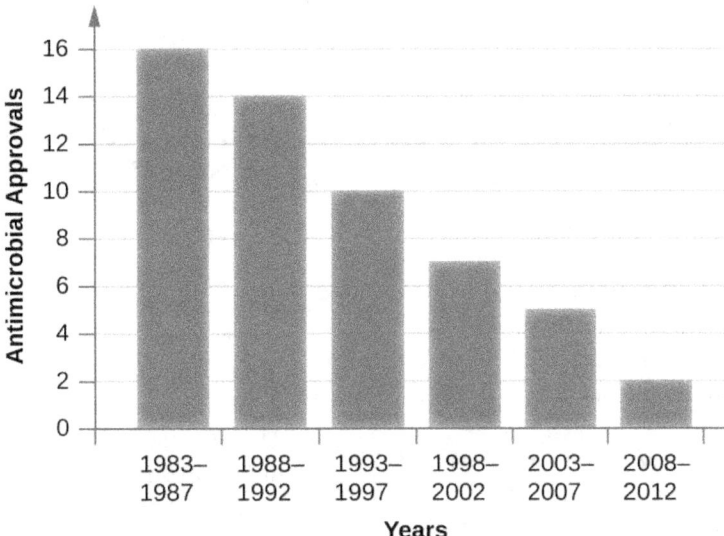

Figure 14.22 In recent decades, approvals of new antimicrobials by the FDA have steadily fallen. In the five-year period from 1983–1987, 16 new antimicrobial drugs were approved, compared to just two from 2008–2012.

LINK TO LEARNING

To further examine the scope of the problem, view this (https://openstax.org/l/22PBSDecAntimic) video.

To learn more (https://openstax.org/l/22MSUAntResLeaH) about the history of antimicrobial drug discovery, visit Michigan State University's Antimicrobial Resistance Learning Site.

SUMMARY

14.1 History of Chemotherapy and Antimicrobial Discovery

- **Antimicrobial drugs** produced by purposeful fermentation and/or contained in plants have been used as traditional medicines in many cultures for millennia.
- The purposeful and systematic search for a chemical "magic bullet" that specifically target infectious microbes was initiated by Paul Ehrlich in the early 20th century.
- The discovery of the **natural antibiotic**, penicillin, by Alexander Fleming in 1928 started the modern age of antimicrobial discovery and research.
- Sulfanilamide, the first **synthetic antimicrobial**, was discovered by Gerhard Domagk and colleagues and is a breakdown product of the synthetic dye, prontosil.

14.2 Fundamentals of Antimicrobial Chemotherapy

- Antimicrobial drugs can be **bacteriostatic** or **bactericidal**, and these characteristics are important considerations when selecting the most appropriate drug.
- The use of **narrow-spectrum** antimicrobial drugs is preferred in many cases to avoid **superinfection** and the development of antimicrobial resistance.
- **Broad-spectrum** antimicrobial use is warranted for serious systemic infections when there is no time to determine the causative agent, when narrow-spectrum antimicrobials fail, or for the treatment or prevention of infections with multiple types of microbes.
- The **dosage** and **route of administration** are important considerations when selecting an antimicrobial to treat and infection. Other considerations include the patient's age, mass, ability to take oral medications, liver and kidney function, and possible interactions with other drugs the patient may be taking.

14.3 Mechanisms of Antibacterial Drugs

- Antibacterial compounds exhibit **selective toxicity**, largely due to differences between prokaryotic and eukaryotic cell structure.
- Cell wall synthesis inhibitors, including the **β-lactams**, the **glycopeptides**, and **bacitracin**, interfere with peptidoglycan synthesis, making bacterial cells more prone to osmotic lysis.
- There are a variety of broad-spectrum, bacterial protein synthesis inhibitors that selectively target the prokaryotic 70S ribosome, including those that bind to the 30S subunit (**aminoglycosides** and **tetracyclines**) and others that bind to the 50S subunit (**macrolides**, **lincosamides**, **chloramphenicol**, and **oxazolidinones**).
- **Polymyxins** are lipophilic polypeptide antibiotics that target the lipopolysaccharide component of gram-negative bacteria and ultimately disrupt the integrity of the outer and inner membranes of these bacteria.
- The nucleic acid synthesis inhibitors rifamycins and **fluoroquinolones** target bacterial RNA transcription and DNA replication, respectively.
- Some antibacterial drugs are **antimetabolites**, acting as competitive inhibitors for bacterial metabolic enzymes. **Sulfonamides** and **trimethoprim** are antimetabolites that interfere with bacterial folic acid synthesis. **Isoniazid** is an antimetabolite that interferes with mycolic acid synthesis in mycobacteria.

14.4 Mechanisms of Other Antimicrobial Drugs

- Because fungi, protozoans, and helminths are eukaryotic organisms like human cells, it is more challenging to develop antimicrobial drugs that specifically target them. Similarly, it is hard to target viruses because human viruses replicate inside of human cells.
- **Antifungal drugs** interfere with ergosterol synthesis, bind to ergosterol to disrupt fungal cell membrane integrity, or target cell wall-specific components or other cellular proteins.
- **Antiprotozoan drugs** increase cellular levels of reactive oxygen species, interfere with protozoal DNA replication (nuclear versus kDNA, respectively), and disrupt heme detoxification.
- **Antihelminthic drugs** disrupt helminthic and protozoan microtubule formation; block neuronal transmissions; inhibit anaerobic ATP formation and/or oxidative phosphorylation; induce a calcium influx in tapeworms, leading to spasms and paralysis; and interfere with RNA synthesis in schistosomes.
- **Antiviral drugs** inhibit viral entry, inhibit viral uncoating, inhibit nucleic acid biosynthesis, prevent viral escape from endosomes in host cells, and prevent viral release from infected

cells.
- Because it can easily mutate to become drug resistant, HIV is typically treated with a combination of several **antiretroviral drugs**, which may include **reverse transcriptase inhibitors, protease inhibitors, integrase inhibitors**, and drugs that interfere with viral binding and fusion to initiate infection.

14.5 Drug Resistance

- **Antimicrobial resistance** is on the rise and is the result of selection of drug-resistant strains in clinical environments, the overuse and misuse of antibacterials, the use of subtherapeutic doses of antibacterial drugs, and poor patient compliance with antibacterial drug therapies.
- Drug resistance genes are often carried on plasmids or in transposons that can undergo vertical transfer easily and between microbes through horizontal gene transfer.
- Common modes of antimicrobial drug resistance include drug modification or inactivation, prevention of cellular uptake or efflux, target modification, target overproduction or enzymatic bypass, and target mimicry.
- Problematic microbial strains showing extensive antimicrobial resistance are emerging; many of these strains can reside as members of the normal microbiota in individuals but also can cause opportunistic infection. The transmission of many of these highly resistant microbial strains often occurs in clinical settings, but can also be community-acquired.

14.6 Testing the Effectiveness of Antimicrobials

- The **Kirby-Bauer disk diffusion** test helps determine the susceptibility of a microorganism to various antimicrobial drugs. However, the **zones of inhibition** measured must be correlated to known standards to determine susceptibility and resistance, and do not provide information on bactericidal versus bacteriostatic activity, or allow for direct comparison of drug potencies.
- Antibiograms are useful for monitoring local trends in antimicrobial resistance/susceptibility and for directing appropriate selection of empiric antibacterial therapy.
- There are several laboratory methods available for determining the **minimum inhibitory concentration (MIC)** of an antimicrobial drug against a specific microbe. The **minimal bactericidal concentration (MBC)** can also be determined, typically as a follow-up experiment to MIC determination using the tube dilution method.

14.7 Current Strategies for Antimicrobial Discovery

- Current research into the development of antimicrobial drugs involves the use of high-throughput screening and combinatorial chemistry technologies.
- New technologies are being developed to discover novel antibiotics from soil microorganisms that cannot be cultured by standard laboratory methods.
- Additional strategies include searching for antibiotics from sources other than soil, identifying new antibacterial targets, using combinatorial chemistry to develop novel drugs, developing drugs that inhibit resistance mechanisms, and developing drugs that target virulence factors and hold infections in check.

REVIEW QUESTIONS

Multiple Choice

1. A scientist discovers that a soil bacterium he has been studying produces an antimicrobial that kills gram-negative bacteria. She isolates and purifies the antimicrobial compound, then chemically converts a chemical side chain to a hydroxyl group. When she tests the antimicrobial properties of this new version, she finds that this antimicrobial drug can now also kill gram-positive bacteria. The new antimicrobial drug with broad-spectrum activity is considered to be which of the following?
   A. resistant
   B. semisynthetic
   C. synthetic
   D. natural

2. Which of the following antimicrobial drugs is synthetic?
   A. sulfanilamide
   B. penicillin
   C. actinomycin
   D. neomycin

3. Which of the following combinations would most likely contribute to the development of a superinfection?
   A. long-term use of narrow-spectrum antimicrobials
   B. long-term use of broad-spectrum antimicrobials
   C. short-term use of narrow-spectrum antimicrobials
   D. short-term use of broad-spectrum antimicrobials

4. Which of the following routes of administration would be appropriate and convenient for home administration of an antimicrobial to treat a systemic infection?
   A. oral
   B. intravenous
   C. topical
   D. parenteral

5. Which clinical situation would be appropriate for treatment with a narrow-spectrum antimicrobial drug?
   A. treatment of a polymicrobic mixed infection in the intestine
   B. prophylaxis against infection after a surgical procedure
   C. treatment of strep throat caused by culture identified *Streptococcus pyogenes*
   D. empiric therapy of pneumonia while waiting for culture results

6. Which of the following terms refers to the ability of an antimicrobial drug to harm the target microbe without harming the host?
   A. mode of action
   B. therapeutic level
   C. spectrum of activity
   D. selective toxicity

7. Which of the following is not a type of β-lactam antimicrobial?
   A. penicillins
   B. glycopeptides
   C. cephalosporins
   D. monobactams

8. Which of the following does not bind to the 50S ribosomal subunit?
   A. tetracyclines
   B. lincosamides
   C. macrolides
   D. chloramphenicol

9. Which of the following antimicrobials inhibits the activity of DNA gyrase?
   A. polymyxin B
   B. clindamycin
   C. nalidixic acid
   D. rifampin

10. Which of the following is not an appropriate target for antifungal drugs?
    A. ergosterol
    B. chitin
    C. cholesterol
    D. β(1→3) glucan

11. Which of the following drug classes specifically inhibits neuronal transmission in helminths?
    A. quinolines
    B. avermectins
    C. amantadines
    D. imidazoles

12. Which of the following is a nucleoside analog commonly used as a reverse transcriptase inhibitor in the treatment of HIV?
    A. acyclovir
    B. ribavirin
    C. adenine-arabinoside
    D. azidothymidine

13. Which of the following is an antimalarial drug that is thought to increase ROS levels in target cells?
    A. artemisinin
    B. amphotericin b
    C. praziquantel
    D. pleconaril

14. Which of the following resistance mechanisms describes the function of β-lactamase?
    A. efflux pump
    B. target mimicry
    C. drug inactivation
    D. target overproduction

15. Which of the following resistance mechanisms is commonly effective against a wide range of antimicrobials in multiple classes?
    A. efflux pump
    B. target mimicry
    C. target modification
    D. target overproduction

16. Which of the following resistance mechanisms is the most nonspecific to a particular class of antimicrobials?
    A. drug modification
    B. target mimicry
    C. target modification
    D. efflux pump

17. Which of the following types of drug-resistant bacteria do not typically persist in individuals as a member of their intestinal microbiota?
    A. MRSA
    B. VRE
    C. CRE
    D. ESBL-producing bacteria

18. In the Kirby-Bauer disk diffusion test, the _____ of the zone of inhibition is measured and used for interpretation.
    A. diameter
    B. microbial population
    C. circumference
    D. depth

19. Which of the following techniques cannot be used to determine the minimum inhibitory concentration of an antimicrobial drug against a particular microbe?
    A. Etest
    B. microbroth dilution test
    C. Kirby-Bauer disk diffusion test
    D. macrobroth dilution test

20. The utility of an antibiogram is that it shows antimicrobial susceptibility trends
    A. over a large geographic area.
    B. for an individual patient.
    C. in research laboratory strains.
    D. in a localized population.

21. Which of the following has yielded compounds with the most antimicrobial activity?
    A. water
    B. air
    C. volcanoes
    D. soil

True/False

22. Narrow-spectrum antimicrobials are commonly used for prophylaxis following surgery.
23. β-lactamases can degrade vancomycin.
24. Echinocandins, known as "penicillin for fungi," target β(1→3) glucan in fungal cell walls.
25. If drug A produces a larger zone of inhibition than drug B on the Kirby-Bauer disk diffusion test, drug A should always be prescribed.
26. The rate of discovery of antimicrobial drugs has decreased significantly in recent decades.

Fill in the Blank

27. The group of soil bacteria known for their ability to produce a wide variety of antimicrobials is called the _____.
28. The bacterium known for causing pseudomembranous colitis, a potentially deadly superinfection, is _____.
29. Selective toxicity antimicrobials are easier to develop against bacteria because they are _____ cells, whereas human cells are eukaryotic.
30. Antiviral drugs, like Tamiflu and Relenza, that are effective against the influenza virus by preventing viral escape from host cells are called _____.
31. *Staphylococcus aureus*, including MRSA strains, may commonly be carried as a normal member of the _____ microbiota in some people.
32. The method that can determine the MICs of multiple antimicrobial drugs against a microbial strain using a single agar plate is called the _____.

Short Answer

33. Where do antimicrobials come from naturally? Why?
34. Why was Salvarsan considered to be a "magic bullet" for the treatment of syphilis?
35. When prescribing antibiotics, what aspects of the patient's health history should the clinician ask about and why?
36. When is using a broad-spectrum antimicrobial drug warranted?
37. If human cells and bacterial cells perform transcription, how are the rifamycins specific for bacterial infections?
38. What bacterial structural target would make an antibacterial drug selective for gram-negative bacteria? Provide one example of an antimicrobial compound that targets this structure.
39. How does the biology of HIV necessitate the need to treat HIV infections with multiple drugs?
40. Niclosamide is insoluble and thus is not readily absorbed from the stomach into the bloodstream. How does the insolubility of niclosamide aid its effectiveness as a treatment for tapeworm infection?
41. Why does the length of time of antimicrobial treatment for tuberculosis contribute to the rise of resistant strains?
42. What is the difference between multidrug resistance and cross-resistance?
43. How is the information from a Kirby-Bauer disk diffusion test used for the recommendation of the clinical use of an antimicrobial drug?
44. What is the difference between MIC and MBC?

Critical Thinking

45. In nature, why do antimicrobial-producing microbes commonly also have antimicrobial resistance genes?
46. Why are yeast infections a common type of superinfection that results from long-term use of broad-spectrum antimicrobials?
47. Too often patients will stop taking antimicrobial drugs before the prescription is finished. What are factors that cause a patient to stop too soon, and what negative impacts could this have?
48. In considering the cell structure of prokaryotes compared with that of eukaryotes, propose one possible reason for side effects in humans due to treatment of bacterial infections with protein synthesis inhibitors.
49. Which of the following molecules is an example of a nucleoside analog?

50. Why can't drugs used to treat influenza, like amantadines and neuraminidase inhibitors, be used to treat a wider variety of viral infections?
51. Can an Etest be used to find the MBC of a drug? Explain.
52. Who should be responsible for discovering and developing new antibiotics? Support your answer with reasoning.

CHAPTER 15
Microbial Mechanisms of Pathogenicity

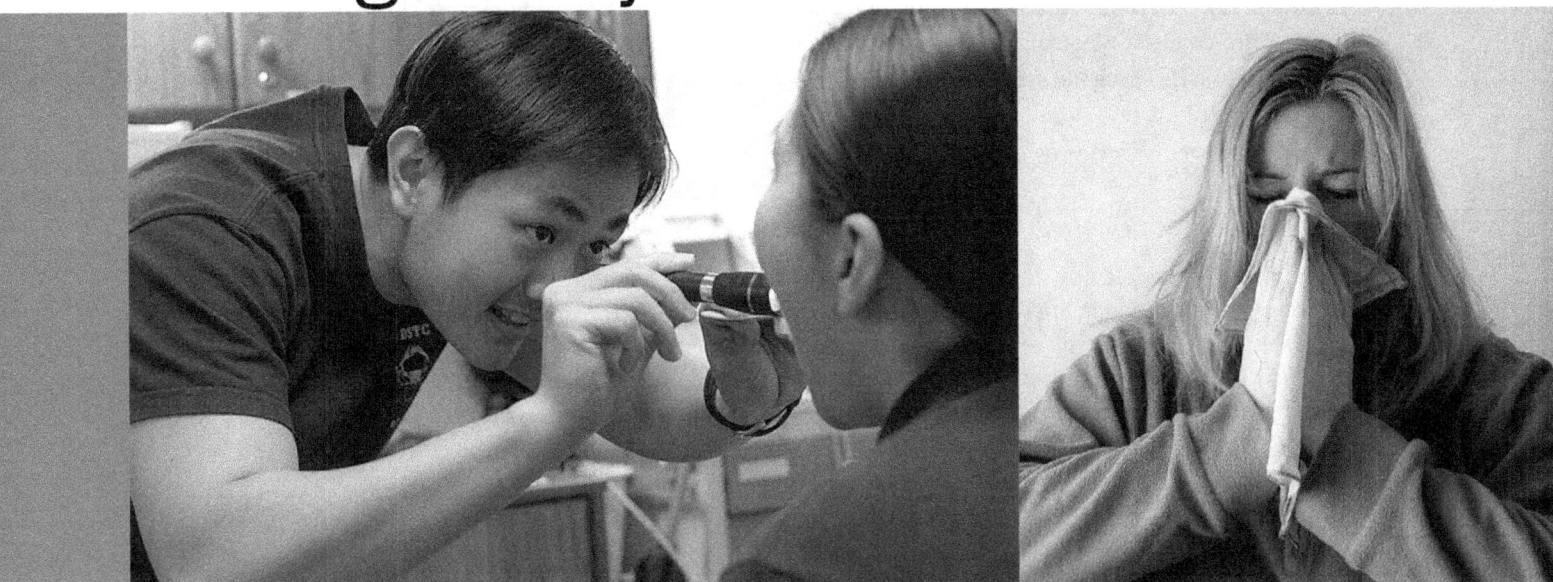

Figure 15.1 Although medical professionals rely heavily on signs and symptoms to diagnose disease and prescribe treatment, many diseases can produce similar signs and symptoms. (credit left: modification of work by U.S. Navy)

Chapter Outline

15.1 Characteristics of Infectious Disease

15.2 How Pathogens Cause Disease

15.3 Virulence Factors of Bacterial and Viral Pathogens

15.4 Virulence Factors of Eukaryotic Pathogens

INTRODUCTION Jane woke up one spring morning feeling not quite herself. Her throat felt a bit dry and she was sniffling. She wondered why she felt so lousy. Was it because of a change in the weather? The pollen count? Was she coming down with something? Did she catch a bug from her coworker who sneezed on her in the elevator yesterday?

The signs and symptoms we associate with illness can have many different causes. Sometimes they are the direct result of a pathogenic infection, but in other cases they result from a response by our immune system to a pathogen or another perceived threat. For example, in response to certain pathogens, the immune system may release pyrogens, chemicals that cause the body temperature to rise, resulting in a fever. This response creates a less-than-favorable environment for the pathogen, but it also makes us feel sick.

Medical professionals rely heavily on analysis of signs and symptoms to determine the cause of an ailment and prescribe treatment. In some cases, signs and symptoms alone are enough to correctly identify the causative agent of a disease, but since few diseases produce truly unique symptoms, it is often necessary to confirm the identity of the infectious agent by other direct and indirect diagnostic methods.

15.1 Characteristics of Infectious Disease

Learning Objectives

By the end of this section, you will be able to:
- Distinguish between signs and symptoms of disease
- Explain the difference between a communicable disease and a noncommunicable disease
- Compare different types of infectious diseases, including iatrogenic, nosocomial, and zoonotic diseases
- Identify and describe the stages of an acute infectious disease in terms of number of pathogens present and severity of signs and symptoms

---

Clinical Focus

Part 1

Michael, a 10-year-old boy in generally good health, went to a birthday party on Sunday with his family. He ate many different foods but was the only one in the family to eat the undercooked hot dogs served by the hosts. Monday morning, he woke up feeling achy and nauseous, and he was running a fever of 38 °C (100.4 °F). His parents, assuming Michael had caught the flu, made him stay home from school and limited his activities. But after 4 days, Michael began to experience severe headaches, and his fever spiked to 40 °C (104 °F). Growing worried, his parents finally decide to take Michael to a nearby clinic.

- What signs and symptoms is Michael experiencing?
- What do these signs and symptoms tell us about the stage of Michael's disease?

Jump to the next Clinical Focus box.

A **disease** is any condition in which the normal structure or functions of the body are damaged or impaired. Physical injuries or disabilities are not classified as disease, but there can be several causes for disease, including infection by a pathogen, genetics (as in many cancers or deficiencies), noninfectious environmental causes, or inappropriate immune responses. Our focus in this chapter will be on infectious diseases, although when diagnosing infectious diseases, it is always important to consider possible noninfectious causes.

Signs and Symptoms of Disease

An **infection** is the successful colonization of a host by a microorganism. Infections can lead to disease, which causes signs and symptoms resulting in a deviation from the normal structure or functioning of the host. Microorganisms that can cause disease are known as pathogens.

The **signs** of disease are objective and measurable, and can be directly observed by a clinician. Vital signs, which are used to measure the body's basic functions, include body temperature (normally 37 °C [98.6 °F]), heart rate (normally 60–100 beats per minute), breathing rate (normally 12–18 breaths per minute), and blood pressure (normally between 90/60 and 120/80 mm Hg). Changes in any of the body's vital signs may be indicative of disease. For example, having a fever (a body temperature significantly higher than 37 °C or 98.6 °F) is a sign of disease because it can be measured.

In addition to changes in vital signs, other observable conditions may be considered signs of disease. For example, the presence of antibodies in a patient's serum (the liquid portion of blood that lacks clotting factors) can be observed and measured through blood tests and, therefore, can be considered a sign. However, it is important to note that the presence of antibodies is not always a sign of an active disease. Antibodies can remain in the body long after an infection has resolved; also, they may develop in response to a pathogen that is in the body but not currently causing disease.

Unlike signs, **symptoms** of disease are subjective. Symptoms are felt or experienced by the patient, but they cannot be clinically confirmed or objectively measured. Examples of symptoms include nausea, loss of appetite, and pain. Such symptoms are important to consider when diagnosing disease, but they are subject to memory bias and are difficult to measure precisely. Some clinicians attempt to quantify symptoms by asking patients to assign a numerical value to their symptoms. For example, the Wong-Baker Faces pain-rating scale asks patients to rate their pain on a scale of 0–10. An alternative method of quantifying pain is measuring skin conductance fluctuations. These fluctuations reflect sweating due to skin sympathetic nerve activity resulting from the stressor of pain.[1]

A specific group of signs and symptoms characteristic of a particular disease is called a **syndrome**. Many syndromes are named using a nomenclature based on signs and symptoms or the location of the disease. Table 15.1 lists some of the prefixes and suffixes commonly used in naming syndromes.

Nomenclature of Symptoms

Affix	Meaning	Example
cyto-	cell	cytopenia: reduction in the number of blood cells
hepat-	of the liver	hepatitis: inflammation of the liver
-pathy	disease	neuropathy: a disease affecting nerves
-emia	of the blood	bacteremia: presence of bacteria in blood
-itis	inflammation	colitis: inflammation of the colon
-lysis	destruction	hemolysis: destruction of red blood cells
-oma	tumor	lymphoma: cancer of the lymphatic system
-osis	diseased or abnormal condition	leukocytosis: abnormally high number of white blood cells
-derma	of the skin	keratoderma: a thickening of the skin

Table 15.1

Clinicians must rely on signs and on asking questions about symptoms, medical history, and the patient's recent activities to identify a particular disease and the potential causative agent. Diagnosis is complicated by the fact that different microorganisms can cause similar signs and symptoms in a patient. For example, an individual presenting with symptoms of diarrhea may have been infected by one of a wide variety of pathogenic microorganisms. Bacterial pathogens associated with diarrheal disease include *Vibrio cholerae*, *Listeria monocytogenes*, *Campylobacter jejuni*, and enteropathogenic *Escherichia coli* (EPEC). Viral pathogens associated with diarrheal disease include norovirus and rotavirus. Parasitic pathogens associated with diarrhea include *Giardia lamblia* and *Cryptosporidium parvum*. Likewise, fever is indicative of many types of infection, from the common cold to the deadly Ebola hemorrhagic fever.

Finally, some diseases may be **asymptomatic** or **subclinical**, meaning they do not present any noticeable signs or symptoms. For example, most individual infected with herpes simplex virus remain asymptomatic and are unaware that they have been infected.

---

[1] F. Savino et al. "Pain Assessment in Children Undergoing Venipuncture: The Wong–Baker Faces Scale Versus Skin Conductance Fluctuations." *PeerJ* 1 (2013):e37; https://peerj.com/articles/37/

CHECK YOUR UNDERSTANDING
- Explain the difference between signs and symptoms.

Classifications of Disease

The World Health Organization's (WHO) International Classification of Diseases (ICD) is used in clinical fields to classify diseases and monitor morbidity (the number of cases of a disease) and mortality (the number of deaths due to a disease). In this section, we will introduce terminology used by the ICD (and in health-care professions in general) to describe and categorize various types of disease.

An **infectious disease** is any disease caused by the direct effect of a pathogen. A pathogen may be cellular (bacteria, parasites, and fungi) or acellular (viruses, viroids, and prions). Some infectious diseases are also **communicable**, meaning they are capable of being spread from person to person through either direct or indirect mechanisms. Some infectious communicable diseases are also considered **contagious** diseases, meaning they are easily spread from person to person. Not all contagious diseases are equally so; the degree to which a disease is contagious usually depends on how the pathogen is transmitted. For example, measles is a highly contagious viral disease that can be transmitted when an infected person coughs or sneezes and an uninfected person breathes in droplets containing the virus. Gonorrhea is not as contagious as measles because transmission of the pathogen (*Neisseria gonorrhoeae*) requires close intimate contact (usually sexual) between an infected person and an uninfected person.

Diseases that are contracted as the result of a medical procedure are known as **iatrogenic diseases**. Iatrogenic diseases can occur after procedures involving wound treatments, catheterization, or surgery if the wound or surgical site becomes contaminated. For example, an individual treated for a skin wound might acquire necrotizing fasciitis (an aggressive, "flesh-eating" disease) if bandages or other dressings became contaminated by *Clostridium perfringens* or one of several other bacteria that can cause this condition.

Diseases acquired in hospital settings are known as **nosocomial diseases**. Several factors contribute to the prevalence and severity of nosocomial diseases. First, sick patients bring numerous pathogens into hospitals, and some of these pathogens can be transmitted easily via improperly sterilized medical equipment, bed sheets, call buttons, door handles, or by clinicians, nurses, or therapists who do not wash their hands before touching a patient. Second, many hospital patients have weakened immune systems, making them more susceptible to infections. Compounding this, the prevalence of antibiotics in hospital settings can select for drug-resistant bacteria that can cause very serious infections that are difficult to treat.

Certain infectious diseases are not transmitted between humans directly but can be transmitted from animals to humans. Such a disease is called **zoonotic disease** (or **zoonosis**). According to WHO, a zoonosis is a disease that occurs when a pathogen is transferred from a vertebrate animal to a human; however, sometimes the term is defined more broadly to include diseases transmitted by all animals (including invertebrates). For example, rabies is a viral zoonotic disease spread from animals to humans through bites and contact with infected saliva. Many other zoonotic diseases rely on insects or other arthropods for transmission. Examples include yellow fever (transmitted through the bite of mosquitoes infected with yellow fever virus) and Rocky Mountain spotted fever (transmitted through the bite of ticks infected with *Rickettsia rickettsii*).

In contrast to communicable infectious diseases, a **noncommunicable** infectious disease is not spread from one person to another. One example is tetanus, caused by *Clostridium tetani*, a bacterium that produces endospores that can survive in the soil for many years. This disease is typically only transmitted through contact with a skin wound; it cannot be passed from an infected person to another person. Similarly, Legionnaires disease is caused by *Legionella pneumophila*, a bacterium that lives within amoebae in moist locations like water-cooling towers. An individual may contract Legionnaires disease via contact with the contaminated water, but once infected, the individual cannot pass the pathogen to other individuals.

In addition to the wide variety of noncommunicable infectious diseases, **noninfectious diseases** (those not caused by pathogens) are an important cause of morbidity and mortality worldwide. Noninfectious diseases can be caused by a wide variety factors, including genetics, the environment, or immune system dysfunction, to name a few. For example, sickle cell anemia is an inherited disease caused by a genetic mutation that can be passed from parent to offspring (Figure 15.2). Other types of noninfectious diseases are listed in Table 15.2.

Types of Noninfectious Diseases

Type	Definition	Example
Inherited	A genetic disease	Sickle cell anemia
Congenital	Disease that is present at or before birth	Down syndrome
Degenerative	Progressive, irreversible loss of function	Parkinson disease (affecting central nervous system)
Nutritional deficiency	Impaired body function due to lack of nutrients	Scurvy (vitamin C deficiency)
Endocrine	Disease involving malfunction of glands that release hormones to regulate body functions	Hypothyroidism – thyroid does not produce enough thyroid hormone, which is important for metabolism
Neoplastic	Abnormal growth (benign or malignant)	Some forms of cancer
Idiopathic	Disease for which the cause is unknown	Idiopathic juxtafoveal retinal telangiectasia (dilated, twisted blood vessels in the retina of the eye)

Table 15.2

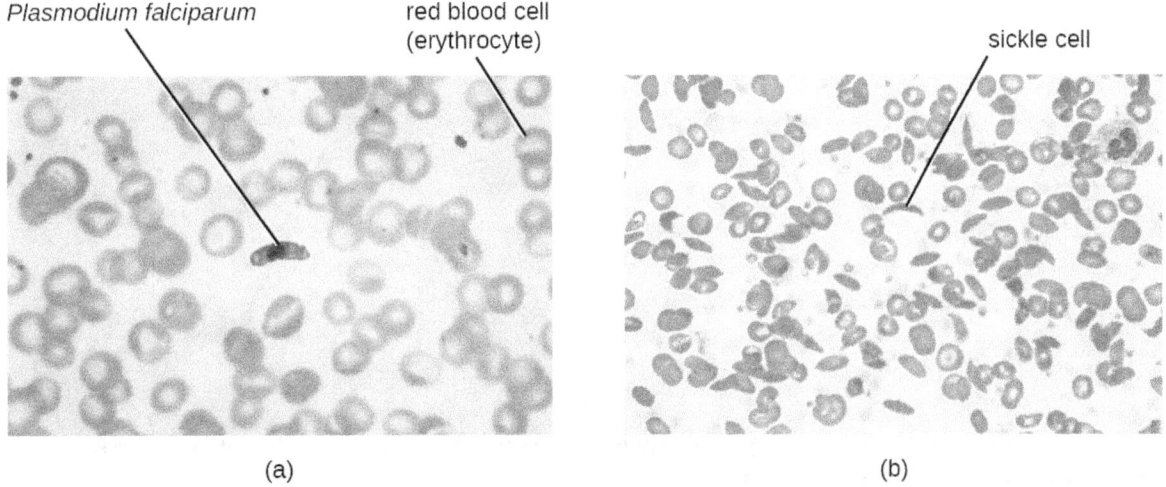

Figure 15.2 Blood smears showing two diseases of the blood. (a) Malaria is an infectious, zoonotic disease caused by the protozoan pathogen *Plasmodium falciparum* (shown here) and several other species of the genus *Plasmodium*. It is transmitted by mosquitoes to humans. (b) Sickle cell disease is a noninfectious genetic disorder that results in abnormally shaped red blood cells, which can stick together and obstruct the flow of blood through the circulatory system. It is not caused by a pathogen, but rather a genetic mutation. (credit a: modification of work by Centers for Disease Control and Prevention; credit b: modification of work by Ed Uthman)

LINK TO LEARNING

Lists of common infectious diseases can be found at the following Centers for Disease Control and Prevention (https://openstax.org/l/22CDCdis) (CDC), World Health Organization (https://openstax.org/l/22WHOdis) (WHO), and International Classification of Diseases (https://openstax.org/l/22WHOclass) websites.

CHECK YOUR UNDERSTANDING
- Describe how a disease can be infectious but not contagious.
- Explain the difference between iatrogenic disease and nosocomial disease.

Periods of Disease

The five periods of disease (sometimes referred to as stages or phases) include the incubation, prodromal, illness, decline, and convalescence periods (Figure 15.3). The **incubation period** occurs in an acute disease after the initial entry of the pathogen into the host (patient). It is during this time the pathogen begins multiplying in the host. However, there are insufficient numbers of pathogen particles (cells or viruses) present to cause signs and symptoms of disease. Incubation periods can vary from a day or two in acute disease to months or years in chronic disease, depending upon the pathogen. Factors involved in determining the length of the incubation period are diverse, and can include strength of the pathogen, strength of the host immune defenses, site of infection, type of infection, and the size infectious dose received. During this incubation period, the patient is unaware that a disease is beginning to develop.

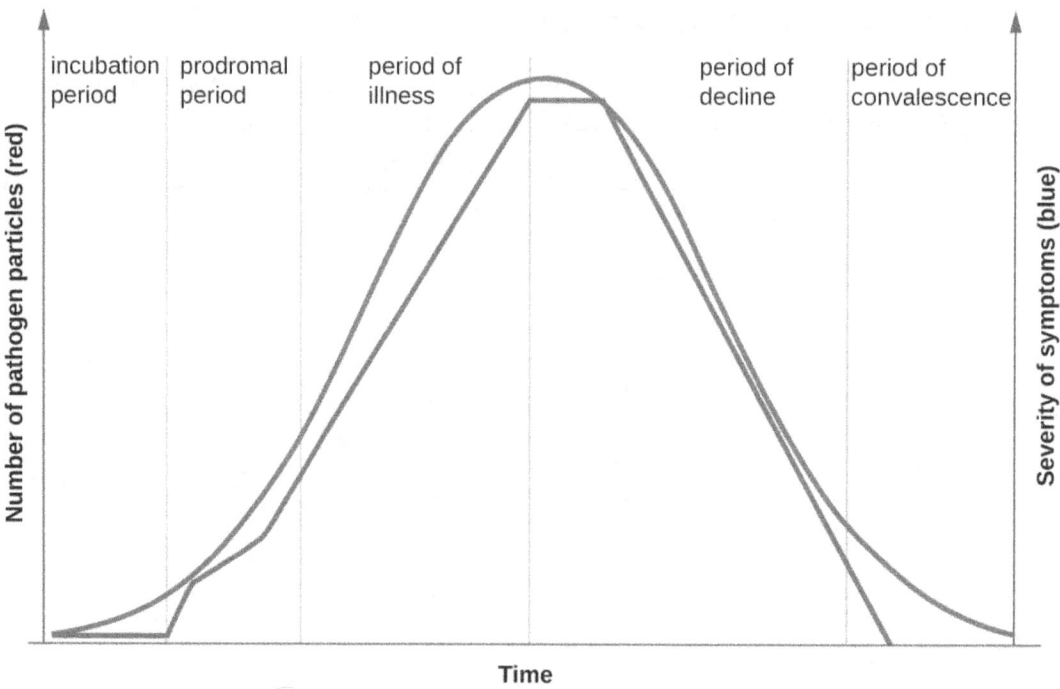

Figure 15.3 The progression of an infectious disease can be divided into five periods, which are related to the number of pathogen particles (red) and the severity of signs and symptoms (blue).

The **prodromal period** occurs after the incubation period. During this phase, the pathogen continues to multiply and the host begins to experience general signs and symptoms of illness, which typically result from activation of the immune system, such as fever, pain, soreness, swelling, or inflammation. Usually, such signs and symptoms are too general to indicate a particular disease. Following the prodromal period is the **period of illness**, during which the signs and symptoms of disease are most obvious and severe.

The period of illness is followed by the **period of decline**, during which the number of pathogen particles begins to decrease, and the signs and symptoms of illness begin to decline. However, during the decline period, patients may become susceptible to developing secondary infections because their immune systems have been weakened by the primary infection. The final period is known as the **period of convalescence**. During this stage, the patient generally returns to normal functions, although some diseases may inflict permanent damage that the body cannot fully repair.

Infectious diseases can be contagious during all five of the periods of disease. Which periods of disease are more likely to associated with transmissibility of an infection depends upon the disease, the pathogen, and the mechanisms by which the disease develops and progresses. For example, with meningitis (infection of the lining of brain), the periods of infectivity depend on the type of pathogen causing the infection. Patients with bacterial meningitis are contagious during the incubation period for up to a week before the onset of the prodromal period, whereas patients with viral meningitis become contagious when the first signs and symptoms of the prodromal period appear. With many viral diseases associated with rashes (e.g., chickenpox, measles, rubella, roseola), patients are contagious during the incubation period up to a week before the rash develops. In contrast, with many respiratory infections (e.g., colds, influenza, diphtheria, strep throat, and pertussis) the patient becomes contagious with the onset of the prodromal period. Depending upon the pathogen, the disease, and the individual infected, transmission can still occur during the periods of decline, convalescence, and even long after signs and symptoms of the disease disappear. For example, an individual recovering from a diarrheal disease may continue to carry and shed the pathogen in feces for some time, posing a risk of transmission to others through direct contact or indirect contact (e.g., through contaminated objects or food).

✓ CHECK YOUR UNDERSTANDING

- Name some of the factors that can affect the length of the incubation period of a particular disease.

Acute and Chronic Diseases

The duration of the period of illness can vary greatly, depending on the pathogen, effectiveness of the immune response in the host, and any medical treatment received. For an **acute disease**, pathologic changes occur over a relatively short time (e.g., hours, days, or a few weeks) and involve a rapid onset of disease conditions. For example, influenza (caused by Influenzavirus) is considered an acute disease because the incubation period is approximately 1–2 days. Infected individuals can spread influenza to others for approximately 5 days after becoming ill. After approximately 1 week, individuals enter the period of decline.

For a **chronic disease**, pathologic changes can occur over longer time spans (e.g., months, years, or a lifetime). For example, chronic gastritis (inflammation of the lining of the stomach) is caused by the gram-negative bacterium *Helicobacter pylori*. *H. pylori* is able to colonize the stomach and persist in its highly acidic environment by producing the enzyme urease, which modifies the local acidity, allowing the bacteria to survive indefinitely.[2] Consequently, *H. pylori* infections can recur indefinitely unless the infection is cleared using antibiotics.[3] Hepatitis B virus can cause a chronic infection in some patients who do not eliminate the virus after the acute illness. A chronic infection with hepatitis B virus is characterized by the continued production of infectious virus for 6 months or longer after the acute infection, as measured by the presence of viral antigen in blood samples.

In **latent diseases**, as opposed to chronic infections, the causal pathogen goes dormant for extended periods of time with no active replication. Examples of diseases that go into a latent state after the acute infection include herpes (herpes simplex viruses [HSV-1 and HSV-2]), chickenpox (varicella-zoster virus [VZV]), and mononucleosis (Epstein-Barr virus [EBV]). HSV-1, HSV-2, and VZV evade the host immune system by residing in a latent form within cells of the nervous system for long periods of time, but they can reactivate to become active infections during times of stress and immunosuppression. For example, an initial infection by VZV may result in a case of childhood chickenpox, followed by a long period of latency. The virus may reactivate decades later, causing episodes of shingles in adulthood. EBV goes into latency in B cells of the immune system and possibly epithelial cells; it can reactivate years later to produce B-cell lymphoma.

---

2 J.G. Kusters et al. Pathogenesis of *Helicobacter pylori* Infection. *Clinical Microbiology Reviews* 19 no. 3 (2006):449–490.

3 N.R. Salama et al. "Life in the Human Stomach: Persistence Strategies of the Bacterial Pathogen *Helicobacter pylori.*" *Nature Reviews Microbiology* 11 (2013):385–399.

CHECK YOUR UNDERSTANDING
- Explain the difference between latent disease and chronic disease.

15.2 How Pathogens Cause Disease

Learning Objectives

By the end of this section, you will be able to:
- Summarize Koch's postulates and molecular Koch's postulates, respectively, and explain their significance and limitations
- Explain the concept of pathogenicity (virulence) in terms of infectious and lethal dose
- Distinguish between primary and opportunistic pathogens and identify specific examples of each
- Summarize the stages of pathogenesis
- Explain the roles of portals of entry and exit in the transmission of disease and identify specific examples of these portals

For most infectious diseases, the ability to accurately identify the causative pathogen is a critical step in finding or prescribing effective treatments. Today's physicians, patients, and researchers owe a sizable debt to the physician Robert Koch (1843–1910), who devised a systematic approach for confirming causative relationships between diseases and specific pathogens.

Koch's Postulates

In 1884, Koch published four postulates (Table 15.3) that summarized his method for determining whether a particular microorganism was the cause of a particular disease. Each of Koch's postulates represents a criterion that must be met before a disease can be positively linked with a pathogen. In order to determine whether the criteria are met, tests are performed on laboratory animals and cultures from healthy and diseased animals are compared (Figure 15.4).

Koch's Postulates
(1) The suspected pathogen must be found in every case of disease and not be found in healthy individuals.
(2) The suspected pathogen can be isolated and grown in pure culture.
(3) A healthy test subject infected with the suspected pathogen must develop the same signs and symptoms of disease as seen in postulate 1.
(4) The pathogen must be re-isolated from the new host and must be identical to the pathogen from postulate 2.

Table 15.3

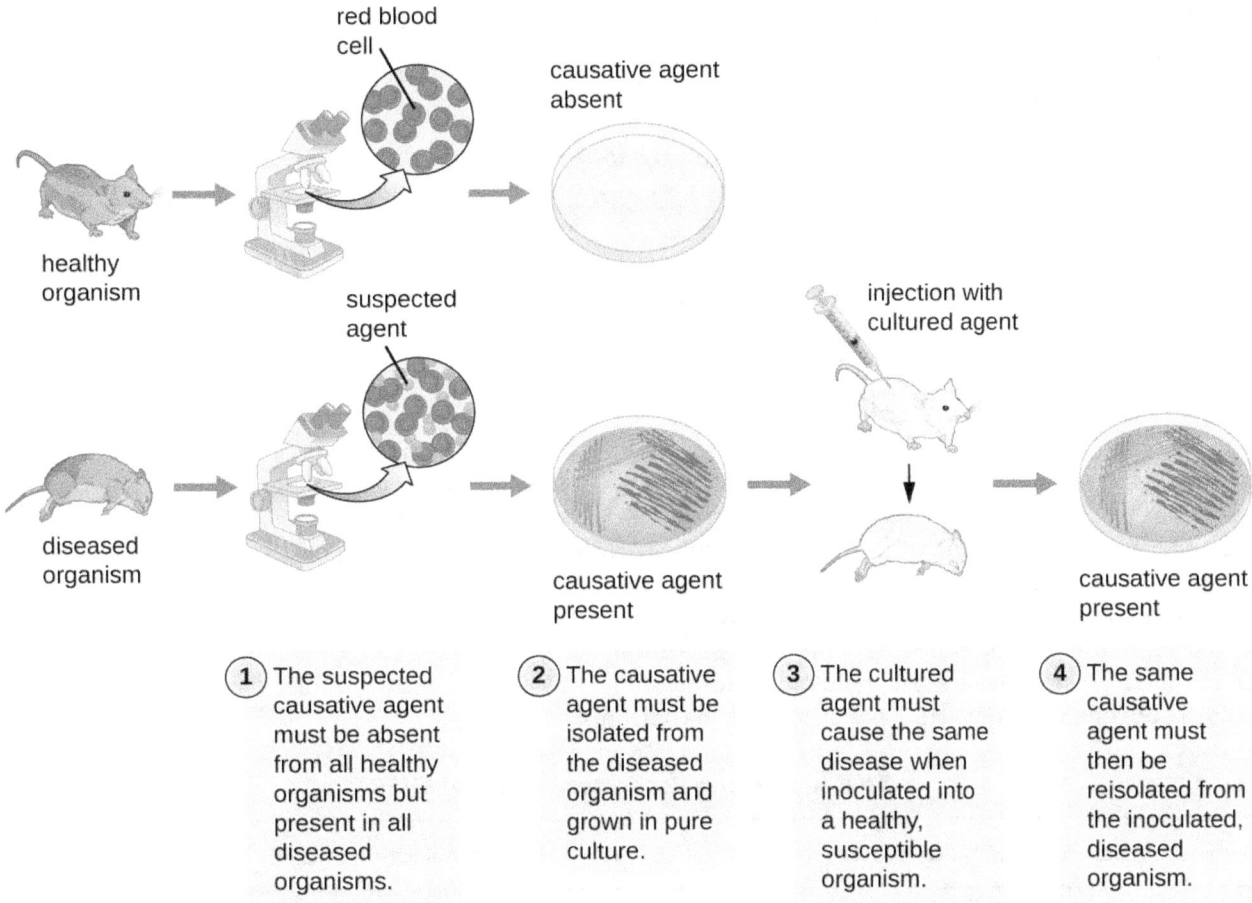

Figure 15.4 The steps for confirming that a pathogen is the cause of a particular disease using Koch's postulates.

In many ways, Koch's postulates are still central to our current understanding of the causes of disease. However, advances in microbiology have revealed some important limitations in Koch's criteria. Koch made several assumptions that we now know are untrue in many cases. The first relates to postulate 1, which assumes that pathogens are only found in diseased, not healthy, individuals. This is not true for many pathogens. For example, *H. pylori*, described earlier in this chapter as a pathogen causing chronic gastritis, is also part of the normal microbiota of the stomach in many healthy humans who never develop gastritis. It is estimated that upwards of 50% of the human population acquires *H. pylori* early in life, with most maintaining it as part of the normal microbiota for the rest of their life without ever developing disease.

Koch's second faulty assumption was that all healthy test subjects are equally susceptible to disease. We now know that individuals are not equally susceptible to disease. Individuals are unique in terms of their microbiota and the state of their immune system at any given time. The makeup of the resident microbiota can influence an individual's susceptibility to an infection. Members of the normal microbiota play an important role in immunity by inhibiting the growth of transient pathogens. In some cases, the microbiota may prevent a pathogen from establishing an infection; in others, it may not prevent an infection altogether but may influence the severity or type of signs and symptoms. As a result, two individuals with the same disease may not always present with the same signs and symptoms. In addition, some individuals have stronger immune systems than others. Individuals with immune systems weakened by age or an unrelated illness are much more susceptible to certain infections than individuals with strong immune systems.

Koch also assumed that all pathogens are microorganisms that can be grown in pure culture (postulate 2) and that animals could serve as reliable models for human disease. However, we now know that not all pathogens can be grown in pure culture, and many human diseases cannot be reliably replicated in animal hosts. Viruses and certain bacteria, including *Rickettsia* and *Chlamydia*, are obligate intracellular pathogens that can grow only when inside a host cell. If a microbe cannot be cultured, a researcher cannot move past postulate 2. Likewise, without a suitable nonhuman host, a researcher cannot evaluate postulate 2 without deliberately infecting humans, which presents obvious ethical concerns. AIDS is an example of such a disease because the human immunodeficiency virus (HIV) only causes disease in humans.

CHECK YOUR UNDERSTANDING

- Briefly summarize the limitations of Koch's postulates.

Molecular Koch's Postulates

In 1988, Stanley Falkow (1934–) proposed a revised form of Koch's postulates known as molecular Koch's postulates. These are listed in the left column of Table 15.4. The premise for molecular Koch's postulates is not in the ability to isolate a particular pathogen but rather to identify a gene that may cause the organism to be pathogenic.

Falkow's modifications to Koch's original postulates explain not only infections caused by intracellular pathogens but also the existence of pathogenic strains of organisms that are usually nonpathogenic. For example, the predominant form of the bacterium *Escherichia coli* is a member of the normal microbiota of the human intestine and is generally considered harmless. However, there are pathogenic strains of *E. coli* such as enterotoxigenic *E. coli* (ETEC) and enterohemorrhagic *E. coli* (O157:H7) (EHEC). We now know ETEC and EHEC exist because of the acquisition of new genes by the once-harmless *E. coli*, which, in the form of these pathogenic strains, is now capable of producing toxins and causing illness. The pathogenic forms resulted from minor genetic changes. The right-side column of Table 15.4 illustrates how molecular Koch's postulates can be applied to identify EHEC as a pathogenic bacterium.

Molecular Koch's Postulates Applied to EHEC

Molecular Koch's Postulates	Application to EHEC
(1) The phenotype (sign or symptom of disease) should be associated only with pathogenic strains of a species.	EHEC causes intestinal inflammation and diarrhea, whereas nonpathogenic strains of *E. coli* do not.
(2) Inactivation of the suspected gene(s) associated with pathogenicity should result in a measurable loss of pathogenicity.	One of the genes in EHEC encodes for Shiga toxin, a bacterial toxin (poison) that inhibits protein synthesis. Inactivating this gene reduces the bacteria's ability to cause disease.
(3) Reversion of the inactive gene should restore the disease phenotype.	By adding the gene that encodes the toxin back into the genome (e.g., with a phage or plasmid), EHEC's ability to cause disease is restored.

Table 15.4

As with Koch's original postulates, the molecular Koch's postulates have limitations. For example, genetic manipulation of some pathogens is not possible using current methods of molecular genetics. In a similar vein, some diseases do not have suitable animal models, which limits the utility of both the original and molecular postulates.

CHECK YOUR UNDERSTANDING

- Explain the differences between Koch's original postulates and the molecular Koch's postulates.

Pathogenicity and Virulence

The ability of a microbial agent to cause disease is called **pathogenicity**, and the degree to which an organism is pathogenic is called **virulence**. Virulence is a continuum. On one end of the spectrum are organisms that are avirulent (not harmful) and on the other are organisms that are highly virulent. Highly virulent pathogens will almost always lead to a disease state when introduced to the body, and some may even cause multi-organ and body system failure in healthy individuals. Less virulent pathogens may cause an initial infection, but may not always cause severe illness. Pathogens with low virulence would more likely result in mild signs and symptoms of disease, such as low-grade fever, headache, or muscle aches. Some individuals might even be asymptomatic.

An example of a highly virulent microorganism is *Bacillus anthracis*, the pathogen responsible for anthrax. *B. anthracis* can produce different forms of disease, depending on the route of transmission (e.g., cutaneous injection, inhalation, ingestion). The most serious form of anthrax is inhalation anthrax. After *B. anthracis* spores are inhaled, they germinate. An active infection develops and the bacteria release potent toxins that cause edema (fluid buildup in tissues), hypoxia (a condition preventing oxygen from reaching tissues), and necrosis (cell death and inflammation). Signs and symptoms of inhalation anthrax include high fever, difficulty breathing, vomiting and coughing up blood, and severe chest pains suggestive of a heart attack. With inhalation anthrax, the toxins and bacteria enter the bloodstream, which can lead to multi-organ failure and death of the patient. If a gene (or genes) involved in pathogenesis is inactivated, the bacteria become less virulent or nonpathogenic.

Virulence of a pathogen can be quantified using controlled experiments with laboratory animals. Two important indicators of virulence are the **median infectious dose (ID_{50})** and the **median lethal dose (LD_{50})**, both of which are typically determined experimentally using animal models. The ID_{50} is the number of pathogen cells or virions required to cause active infection in 50% of inoculated animals. The LD_{50} is the number of pathogenic cells, virions, or amount of toxin required to kill 50% of infected animals. To calculate these values, each group of animals is inoculated with one of a range of known numbers of pathogen cells or virions. In graphs like the one shown in Figure 15.5, the percentage of animals that have been infected (for ID_{50}) or killed (for LD_{50}) is plotted against the concentration of pathogen inoculated. Figure 15.5 represents data graphed from a hypothetical experiment measuring the LD_{50} of a pathogen. Interpretation of the data from this graph indicates that the LD_{50} of the pathogen for the test animals is 10^4 pathogen cells or virions (depending upon the pathogen studied).

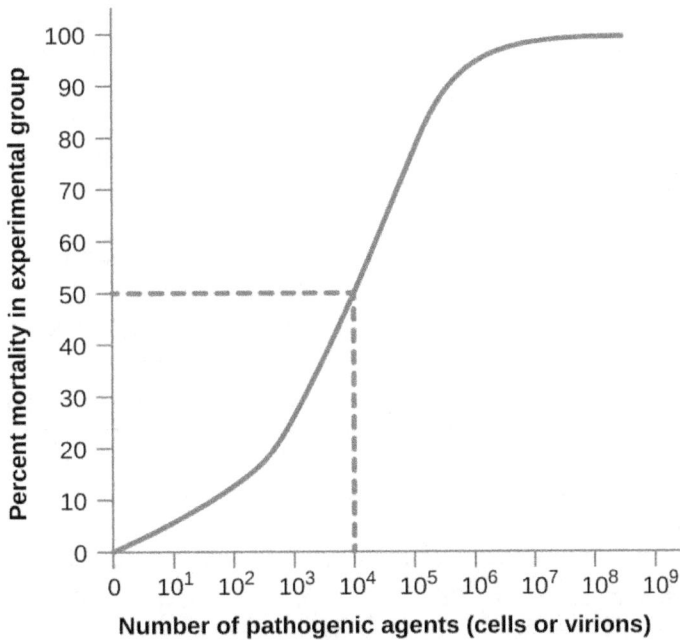

Figure 15.5 A graph like this is used to determine LD_{50} by plotting pathogen concentration against the percent of infected test animals that have died. In this example, the LD_{50} = 10^4 pathogenic particles.

Table 15.5 lists selected foodborne pathogens and their ID_{50} values in humans (as determined from epidemiologic data and studies on human volunteers). Keep in mind that these are *median* values. The actual infective dose for an individual can vary widely, depending on factors such as route of entry; the age, health, and immune status of the host; and environmental and pathogen-specific factors such as susceptibility to the acidic pH of the stomach. It is also important to note that a pathogen's infective dose does not necessarily correlate with disease severity. For example, just a single cell of *Salmonella enterica* serotype Typhimurium can result in an active infection. The resultant disease, *Salmonella* gastroenteritis or salmonellosis, can cause nausea, vomiting, and diarrhea, but has a mortality rate of less than 1% in healthy adults. In contrast, *S. enterica* serotype Typhi has a much higher ID_{50}, typically requiring as many as 1,000 cells to produce infection. However, this serotype causes typhoid fever, a much more systemic and severe disease that has a mortality rate as high as 10% in untreated individuals.

ID_{50} for Selected Foodborne Diseases[4]

Pathogen	ID_{50}
Viruses	
Hepatitis A virus	10–100
Norovirus	1–10
Rotavirus	10–100
Bacteria	
Escherichia coli, enterohemorrhagic (EHEC, serotype O157)	10–100
E. coli, enteroinvasive (EIEC)	200–5,000
E. coli, enteropathogenic (EPEC)	10,000,000–10,000,000,000
E. coli, enterotoxigenic (ETEC)	10,000,000–10,000,000,000
Salmonella enterica serovar Typhi	<1,000
S. enterica serovar Typhimurium	≥1
Shigella dysenteriae	10–200
Vibrio cholerae (serotypes O139, O1)	1,000,000
V. parahemolyticus	100,000,000
Protozoa	
Giardia lamblia	1
Cryptosporidium parvum	10–100

Table 15.5

---

[4] Food and Drug Administration. "Bad Bug Book, Foodborne Pathogenic Microorganisms and Natural Toxins." 2nd ed. Silver Spring, MD: US Food and Drug Administration; 2012.

CHECK YOUR UNDERSTANDING

- What is the difference between a pathogen's infective dose and lethal dose?
- Which is more closely related to the severity of a disease?

Primary Pathogens versus Opportunistic Pathogens

Pathogens can be classified as either primary pathogens or opportunistic pathogens. A **primary pathogen** can cause disease in a host regardless of the host's resident microbiota or immune system. An **opportunistic pathogen**, by contrast, can only cause disease in situations that compromise the host's defenses, such as the body's protective barriers, immune system, or normal microbiota. Individuals susceptible to opportunistic infections include the very young, the elderly, women who are pregnant, patients undergoing chemotherapy, people with immunodeficiencies (such as acquired immunodeficiency syndrome [AIDS]), patients who are recovering from surgery, and those who have had a breach of protective barriers (such as a severe wound or burn).

An example of a primary pathogen is enterohemorrhagic *E. coli* (EHEC), which produces a virulence factor known as Shiga toxin. This toxin inhibits protein synthesis, leading to severe and bloody diarrhea, inflammation, and renal failure, even in patients with healthy immune systems. *Staphylococcus epidermidis*, on the other hand, is an opportunistic pathogen that is among the most frequent causes of nosocomial disease.[5] *S. epidermidis* is a member of the normal microbiota of the skin, where it is generally avirulent. However, in hospitals, it can also grow in biofilms that form on catheters, implants, or other devices that are inserted into the body during surgical procedures. Once inside the body, *S. epidermidis* can cause serious infections such as endocarditis, and it produces virulence factors that promote the persistence of such infections.

Other members of the normal microbiota can also cause opportunistic infections under certain conditions. This often occurs when microbes that reside harmlessly in one body location end up in a different body system, where they cause disease. For example, *E. coli* normally found in the large intestine can cause a urinary tract infection if it enters the bladder. This is the leading cause of urinary tract infections among women.

Members of the normal microbiota may also cause disease when a shift in the environment of the body leads to overgrowth of a particular microorganism. For example, the yeast *Candida* is part of the normal microbiota of the skin, mouth, intestine, and vagina, but its population is kept in check by other organisms of the microbiota. If an individual is taking antibacterial medications, however, bacteria that would normally inhibit the growth of *Candida* can be killed off, leading to a sudden growth in the population of *Candida*, which is not affected by antibacterial medications because it is a fungus. An overgrowth of *Candida* can manifest as oral thrush (growth on mouth, throat, and tongue), a vaginal yeast infection, or cutaneous candidiasis. Other scenarios can also provide opportunities for *Candida* infections. Untreated diabetes can result in a high concentration of glucose in the saliva, which provides an optimal environment for the growth of *Candida*, resulting in thrush. Immunodeficiencies such as those seen in patients with HIV, AIDS, and cancer also lead to higher incidence of thrush. Vaginal yeast infections can result from decreases in estrogen levels during the menstruation or menopause. The amount of glycogen available to lactobacilli in the vagina is controlled by levels of estrogen; when estrogen levels are low, lactobacilli produce less lactic acid. The resultant increase in vaginal pH allows overgrowth of *Candida* in the vagina.

CHECK YOUR UNDERSTANDING

- Explain the difference between a primary pathogen and an opportunistic pathogen.
- Describe some conditions under which an opportunistic infection can occur.

---

5 M. Otto. "*Staphylococcus epidermidis*--The 'Accidental' Pathogen." *Nature Reviews Microbiology* 7 no. 8 (2009):555–567.

Stages of Pathogenesis

To cause disease, a pathogen must successfully achieve four steps or stages of pathogenesis: exposure (contact), adhesion (colonization), invasion, and infection. The pathogen must be able to gain entry to the host, travel to the location where it can establish an infection, evade or overcome the host's immune response, and cause damage (i.e., disease) to the host. In many cases, the cycle is completed when the pathogen exits the host and is transmitted to a new host.

Exposure

An encounter with a potential pathogen is known as **exposure** or **contact**. The food we eat and the objects we handle are all ways that we can come into contact with potential pathogens. Yet, not all contacts result in infection and disease. For a pathogen to cause disease, it needs to be able to gain access into host tissue. An anatomic site through which pathogens can pass into host tissue is called a **portal of entry**. These are locations where the host cells are in direct contact with the external environment. Major portals of entry are identified in Figure 15.6 and include the skin, mucous membranes, and parenteral routes.

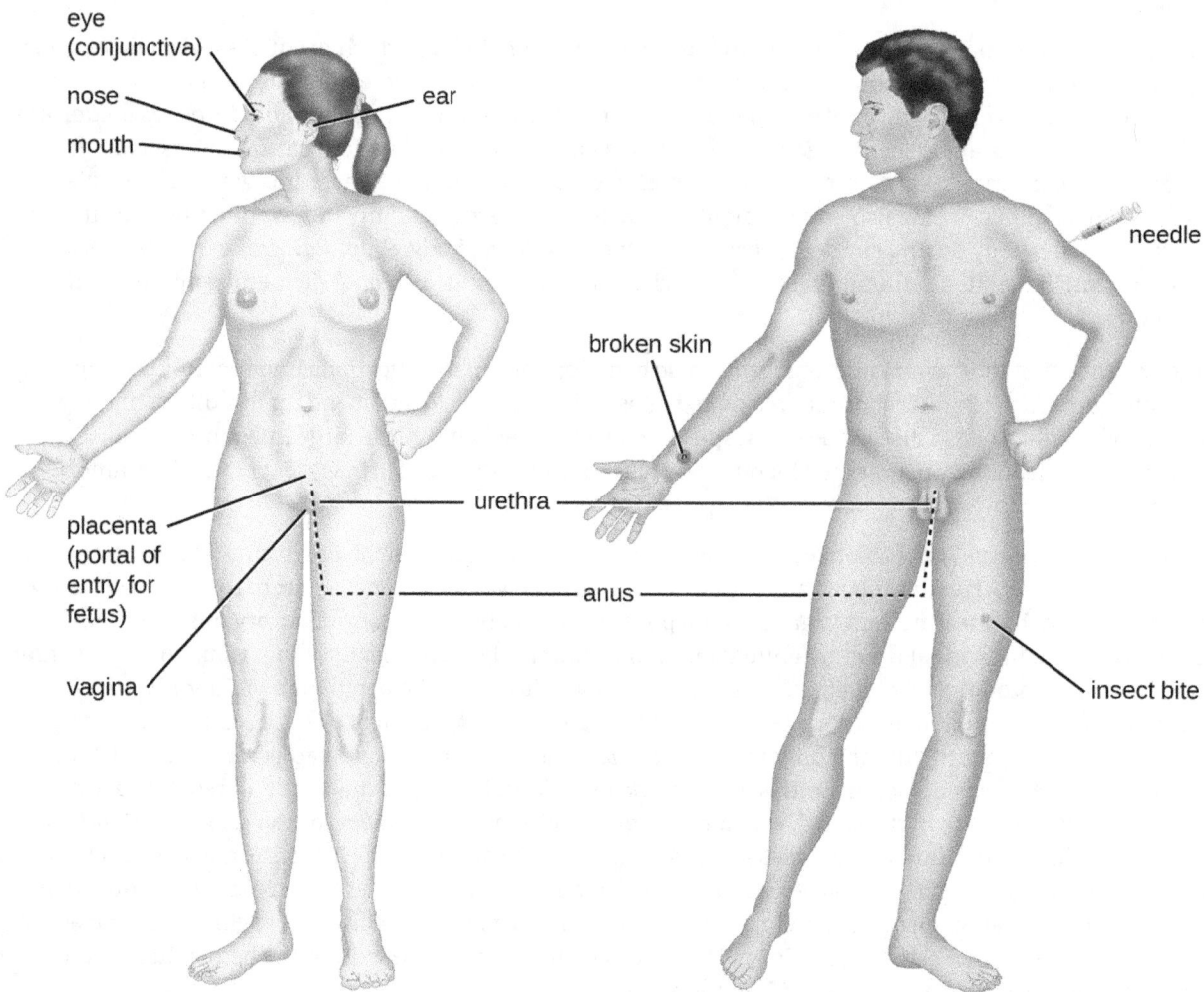

Figure 15.6 Shown are different portals of entry where pathogens can gain access into the body. With the exception of the placenta, many of these locations are directly exposed to the external environment.

Mucosal surfaces are the most important portals of entry for microbes; these include the mucous membranes of the respiratory tract, the gastrointestinal tract, and the genitourinary tract. Although most mucosal surfaces are in the interior of the body, some are contiguous with the external skin at various body openings, including the eyes, nose, mouth, urethra, and anus.

Most pathogens are suited to a particular portal of entry. A pathogen's portal specificity is determined by the organism's environmental adaptions and by the enzymes and toxins they secrete. The respiratory and gastrointestinal tracts are particularly vulnerable portals of entry because particles that include microorganisms are constantly inhaled or ingested, respectively.

Pathogens can also enter through a breach in the protective barriers of the skin and mucous membranes. Pathogens that enter the body in this way are said to enter by the **parenteral route**. For example, the skin is a good natural barrier to pathogens, but breaks in the skin (e.g., wounds, insect bites, animal bites, needle pricks) can provide a parenteral portal of entry for microorganisms.

In pregnant women, the placenta normally prevents microorganisms from passing from the mother to the fetus. However, a few pathogens are capable of crossing the blood-placental barrier. The gram-positive bacterium *Listeria monocytogenes*, which causes the foodborne disease listeriosis, is one example that poses a serious risk to the fetus and can sometimes lead to spontaneous abortion. Other pathogens that can pass the placental barrier to infect the fetus are known collectively by the acronym TORCH (Table 15.6).

Transmission of infectious diseases from mother to baby is also a concern at the time of birth when the baby passes through the birth canal. Babies whose mothers have active chlamydia or gonorrhea infections may be exposed to the causative pathogens in the vagina, which can result in eye infections that lead to blindness. To prevent this, it is standard practice to administer antibiotic drops to infants' eyes shortly after birth.

Pathogens Capable of Crossing the Placental Barrier (TORCH Infections)

	Disease	Pathogen
T	Toxoplasmosis	*Toxoplasma gondii* (protozoan)
O[6]	Syphilis Chickenpox Hepatitis B HIV Fifth disease (erythema infectiosum)	*Treponema pallidum* (bacterium) Varicella-zoster virus (human herpesvirus 3) Hepatitis B virus (hepadnavirus) Retrovirus Parvovirus B19
R	Rubella (German measles)	Togavirus
C	Cytomegalovirus	Human herpesvirus 5
H	Herpes	Herpes simplex viruses (HSV) 1 and 2

Table 15.6

Clinical Focus

Part 2

At the clinic, a physician takes down Michael's medical history and asks about his activities and diet over the past week. Upon learning that Michael became sick the day after the party, the physician orders a blood test to check for pathogens associated with foodborne diseases. After tests confirm that presence of a gram-positive rod in Michael's blood, he is given an injection of a broad-spectrum antibiotic and sent to a nearby hospital, where he is admitted as a patient. There he is to receive additional intravenous antibiotic therapy and fluids.

- Is this bacterium in Michael's blood part of normal microbiota?
- What portal of entry did the bacteria use to cause this infection?

---

[6] The O in TORCH stands for "other."

Jump to the next Clinical Focus box. Go back to the previous Clinical Focus box.

Adhesion

Following the initial exposure, the pathogen adheres at the portal of entry. The term **adhesion** refers to the capability of pathogenic microbes to attach to the cells of the body using adhesion factors, and different pathogens use various mechanisms to adhere to the cells of host tissues.

Molecules (either proteins or carbohydrates) called **adhesins** are found on the surface of certain pathogens and bind to specific receptors (glycoproteins) on host cells. Adhesins are present on the fimbriae and flagella of bacteria, the cilia of protozoa, and the capsids or membranes of viruses. Protozoans can also use hooks and barbs for adhesion; spike proteins on viruses also enhance viral adhesion. The production of glycocalyces (slime layers and capsules) (Figure 15.7), with their high sugar and protein content, can also allow certain bacterial pathogens to attach to cells.

Biofilm growth can also act as an adhesion factor. A biofilm is a community of bacteria that produce a glycocalyx, known as extrapolymeric substance (EPS), that allows the biofilm to attach to a surface. Persistent *Pseudomonas aeruginosa* infections are common in patients suffering from cystic fibrosis, burn wounds, and middle-ear infections (otitis media) because *P. aeruginosa* produces a biofilm. The EPS allows the bacteria to adhere to the host cells and makes it harder for the host to physically remove the pathogen. The EPS not only allows for attachment but provides protection against the immune system and antibiotic treatments, preventing antibiotics from reaching the bacterial cells within the biofilm. In addition, not all bacteria in a biofilm are rapidly growing; some are in stationary phase. Since antibiotics are most effective against rapidly growing bacteria, portions of bacteria in a biofilm are protected against antibiotics.[7]

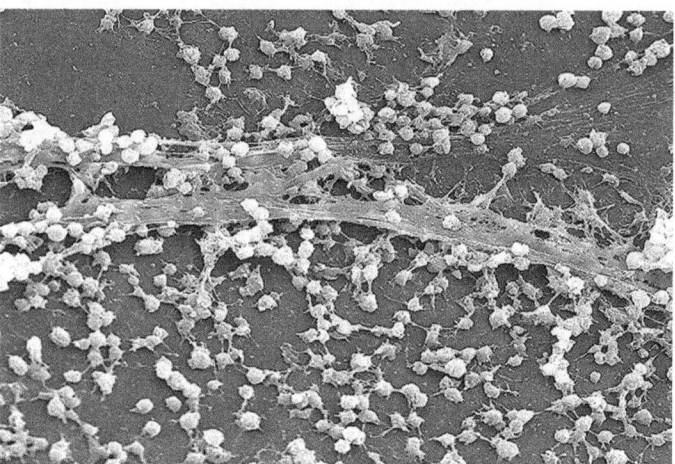

Figure 15.7 Glycocalyx produced by bacteria in a biofilm allows the cells to adhere to host tissues and to medical devices such as the catheter surface shown here. (credit: modification of work by Centers for Disease Control and Prevention)

Invasion

Once adhesion is successful, **invasion** can proceed. Invasion involves the dissemination of a pathogen throughout local tissues or the body. Pathogens may produce exoenzymes or toxins, which serve as virulence factors that allow them to colonize and damage host tissues as they spread deeper into the body. Pathogens may also produce virulence factors that protect them against immune system defenses. A pathogen's specific virulence factors determine the degree of tissue damage that occurs. Figure 15.8 shows the invasion of *H. pylori* into the tissues of the stomach, causing damage as it progresses.

---

[7] D. Davies. "Understanding Biofilm Resistance to Antibacterial Agents." *Nature Reviews Drug Discovery* 2 (2003):114–122.

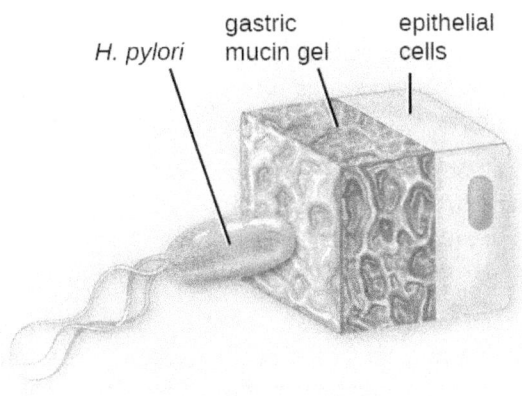
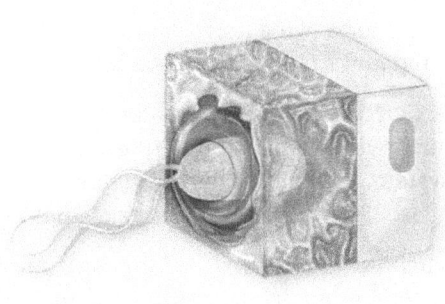

Contact with stomach acid keeps the mucin lining the epithelial cell layer in a spongy gel-like state. This consistency is impermeable to the bacterium *H. pylori*.

The bacterium releases urease, which neutralizes the stomach acid. This causes the mucin to liquefy, and the bacterium can swim right through it.

Figure 15.8 *H. pylori* is able to invade the lining of the stomach by producing virulence factors that enable it pass through the mucin layer covering epithelial cells. (credit: modification of work by Zina Deretsky, National Science Foundation)

Intracellular pathogens achieve invasion by entering the host's cells and reproducing. Some are obligate intracellular pathogens (meaning they can only reproduce inside of host cells) and others are facultative intracellular pathogens (meaning they can reproduce either inside or outside of host cells). By entering the host cells, intracellular pathogens are able to evade some mechanisms of the immune system while also exploiting the nutrients in the host cell.

Entry to a cell can occur by endocytosis. For most kinds of host cells, pathogens use one of two different mechanisms for endocytosis and entry. One mechanism relies on effector proteins secreted by the pathogen; these effector proteins trigger entry into the host cell. This is the method that *Salmonella* and *Shigella* use when invading intestinal epithelial cells. When these pathogens come in contact with epithelial cells in the intestine, they secrete effector molecules that cause protrusions of membrane ruffles that bring the bacterial cell in. This process is called membrane ruffling. The second mechanism relies on surface proteins expressed on the pathogen that bind to receptors on the host cell, resulting in entry. For example, *Yersinia pseudotuberculosis* produces a surface protein known as invasin that binds to beta-1 integrins expressed on the surface of host cells.

Some host cells, such as white blood cells and other phagocytes of the immune system, actively endocytose pathogens in a process called phagocytosis. Although phagocytosis allows the pathogen to gain entry to the host cell, in most cases, the host cell kills and degrades the pathogen by using digestive enzymes. Normally, when a pathogen is ingested by a phagocyte, it is enclosed within a phagosome in the cytoplasm; the phagosome fuses with a lysosome to form a phagolysosome, where digestive enzymes kill the pathogen (see Pathogen Recognition and Phagocytosis). However, some intracellular pathogens have the ability to survive and multiply within phagocytes. Examples include *Listeria monocytogenes* and *Shigella*; these bacteria produce proteins that lyse the phagosome before it fuses with the lysosome, allowing the bacteria to escape into the phagocyte's cytoplasm where they can multiply. Bacteria such as *Mycobacterium tuberculosis*, *Legionella pneumophila*, and *Salmonella* species use a slightly different mechanism to evade being digested by the phagocyte. These bacteria prevent the fusion of the phagosome with the lysosome, thus remaining alive and dividing within the phagosome.

Infection

Following invasion, successful multiplication of the pathogen leads to infection. Infections can be described as local, focal, or systemic, depending on the extent of the infection. A **local infection** is confined to a small area of the body, typically near the portal of entry. For example, a hair follicle infected by *Staphylococcus aureus* infection may result in a boil around the site of infection, but the bacterium is largely contained to this small location. Other examples of local infections that involve more extensive tissue involvement include urinary tract infections confined to the bladder or pneumonia confined to the lungs.

In a **focal infection**, a localized pathogen, or the toxins it produces, can spread to a secondary location. For example, a dental hygienist nicking the gum with a sharp tool can lead to a local infection in the gum by *Streptococcus* bacteria of the normal oral microbiota. These *Streptococcus* spp. may then gain access to the bloodstream and make their way to other locations in the body, resulting in a secondary infection.

When an infection becomes disseminated throughout the body, we call it a **systemic infection**. For example, infection by the varicella-zoster virus typically gains entry through a mucous membrane of the upper respiratory system. It then spreads throughout the body, resulting in the classic red skin lesions associated with chickenpox. Since these lesions are not sites of initial infection, they are signs of a systemic infection.

Sometimes a **primary infection**, the initial infection caused by one pathogen, can lead to a **secondary infection** by another pathogen. For example, the immune system of a patient with a primary infection by HIV becomes compromised, making the patient more susceptible to secondary diseases like oral thrush and others caused by opportunistic pathogens. Similarly, a primary infection by Influenzavirus damages and decreases the defense mechanisms of the lungs, making patients more susceptible to a secondary pneumonia by a bacterial pathogen like *Haemophilus influenzae* or *Streptococcus pneumoniae*. Some secondary infections can even develop as a result of treatment for a primary infection. Antibiotic therapy targeting the primary pathogen can cause collateral damage to the normal microbiota, creating an opening for opportunistic pathogens (see Case in Point: A Secondary Yeast Infection).

> ### Case in Point
>
> **A Secondary Yeast Infection**
>
> Anita, a 36-year-old mother of three, goes to an urgent care center complaining of pelvic pressure, frequent and painful urination, abdominal cramps, and occasional blood-tinged urine. Suspecting a urinary tract infection (UTI), the physician requests a urine sample and sends it to the lab for a urinalysis. Since it will take approximately 24 hours to get the results of the culturing, the physician immediately starts Anita on the antibiotic ciprofloxacin. The next day, the microbiology lab confirms the presence of *E. coli* in Anita's urine, which is consistent with the presumptive diagnosis. However, the antimicrobial susceptibility test indicates that ciprofloxacin would not effectively treat Anita's UTI, so the physician prescribes a different antibiotic.
>
> After taking her antibiotics for 1 week, Anita returns to the clinic complaining that the prescription is not working. Although the painful urination has subsided, she is now experiencing vaginal itching, burning, and discharge. After a brief examination, the physician explains to Anita that the antibiotics were likely successful in killing the *E. coli* responsible for her UTI; however, in the process, they also wiped out many of the "good" bacteria in Anita's normal microbiota. The new symptoms that Anita has reported are consistent with a secondary yeast infection by *Candida albicans*, an opportunistic fungus that normally resides in the vagina but is inhibited by the bacteria that normally reside in the same environment.
>
> To confirm this diagnosis, a microscope slide of a direct vaginal smear is prepared from the discharge to check for the presence of yeast. A sample of the discharge accompanies this slide to the microbiology lab to determine if there has been an increase in the population of yeast causing vaginitis. After the microbiology lab confirms the diagnosis, the physician prescribes an antifungal drug for Anita to use to eliminate her secondary yeast infection.
>
> - Why was *Candida* not killed by the antibiotics prescribed for the UTI?

CHECK YOUR UNDERSTANDING

- List three conditions that could lead to a secondary infection.

Transmission of Disease

For a pathogen to persist, it must put itself in a position to be transmitted to a new host, leaving the infected host through a **portal of exit** (Figure 15.9). As with portals of entry, many pathogens are adapted to use a particular portal of exit. Similar to portals of entry, the most common portals of exit include the skin and the respiratory, urogenital, and gastrointestinal tracts. Coughing and sneezing can expel pathogens from the respiratory tract. A single sneeze can send thousands of virus particles into the air. Secretions and excretions can transport pathogens out of other portals of exit. Feces, urine, semen, vaginal secretions, tears, sweat, and shed skin cells can all serve as vehicles for a pathogen to leave the body. Pathogens that rely on insect vectors for transmission exit the body in the blood extracted by a biting insect. Similarly, some pathogens exit the body in blood extracted by needles.

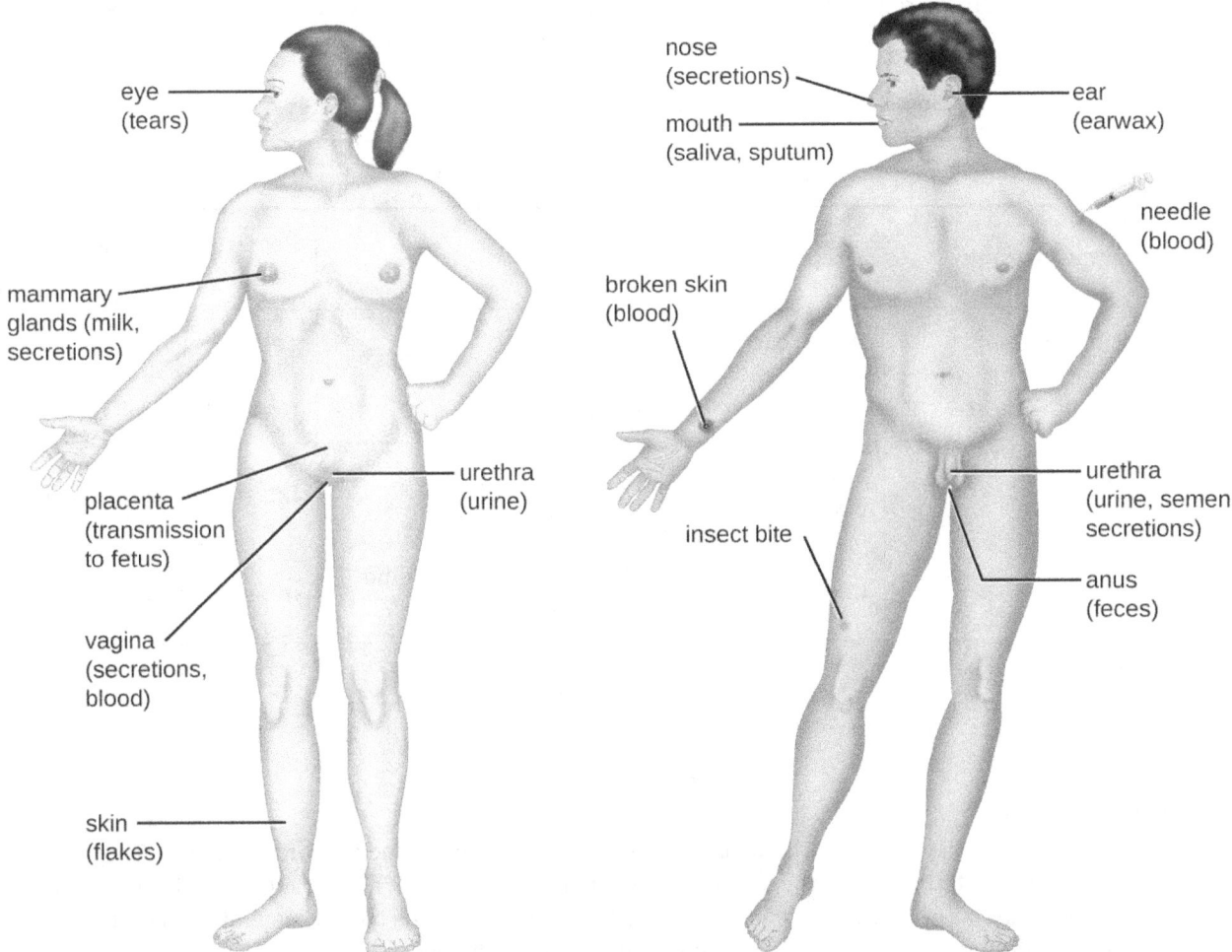

Figure 15.9 Pathogens leave the body of an infected host through various portals of exit to infect new hosts.

15.3 Virulence Factors of Bacterial and Viral Pathogens

Learning Objectives

By the end of this section, you will be able to:
- Explain how virulence factors contribute to signs and symptoms of infectious disease
- Differentiate between endotoxins and exotoxins
- Describe and differentiate between various types of exotoxins
- Describe the mechanisms viruses use for adhesion and antigenic variation

In the previous section, we explained that some pathogens are more virulent than others. This is due to the unique **virulence factors** produced by individual pathogens, which determine the extent and severity of disease they may cause. A pathogen's virulence factors are encoded by genes that can be identified using molecular Koch's postulates. When genes encoding virulence factors are inactivated, virulence in the pathogen is diminished. In this section, we examine various types and specific examples of virulence factors and how they contribute to each step of pathogenesis.

Virulence Factors for Adhesion

As discussed in the previous section, the first two steps in pathogenesis are exposure and adhesion. Recall that an adhesin is a protein or glycoprotein found on the surface of a pathogen that attaches to receptors on the host cell. Adhesins are found on bacterial, viral, fungal, and protozoan pathogens. One example of a bacterial adhesin is type 1 fimbrial adhesin, a molecule found on the tips of fimbriae of enterotoxigenic *E. coli* (ETEC). Recall that fimbriae are hairlike protein bristles on the cell surface. Type 1 fimbrial adhesin allows the fimbriae of ETEC cells to attach to the mannose glycans expressed on intestinal epithelial cells. Table 15.7 lists common adhesins found in some of the pathogens we have discussed or will be seeing later in this chapter.

Some Bacterial Adhesins and Their Host Attachment Sites

Pathogen	Disease	Adhesin	Attachment Site
Streptococcus pyogenes	Strep throat	Protein F	Respiratory epithelial cells
Streptococcus mutans	Dental caries	Adhesin P1	Teeth
Neisseria gonorrhoeae	Gonorrhea	Type IV pili	Urethral epithelial cells
Enterotoxigenic *E. coli* (ETEC)	Traveler's diarrhea	Type 1 fimbriae	Intestinal epithelial cells
Vibrio cholerae	Cholera	N-methylphenylalanine pili	Intestinal epithelial cells

Table 15.7

> ### Clinical Focus
>
> ### Part 3
>
> The presence of bacteria in Michael's blood is a sign of infection, since blood is normally sterile. There is no indication that the bacteria entered the blood through an injury. Instead, it appears the portal of entry was the gastrointestinal route. Based on Michael's symptoms, the results of his blood test, and the fact that Michael was the only one in the family to partake of the hot dogs, the physician suspects that Michael is suffering from a case of listeriosis.
>
> *Listeria monocytogenes*, the facultative intracellular pathogen that causes listeriosis, is a common contaminant in ready-to-eat foods such as lunch meats and dairy products. Once ingested, these bacteria invade intestinal epithelial cells and translocate to the liver, where they grow inside hepatic cells. Listeriosis is fatal in about one in five normal healthy people, and mortality rates are slightly higher in patients with pre-existing conditions that weaken the immune response. A cluster of virulence genes encoded on a pathogenicity island is responsible for the pathogenicity of *L. monocytogenes*. These genes are regulated by a transcriptional factor known as peptide chain release factor 1 (PrfA). One of the genes regulated by PrfA is *hyl*, which encodes a toxin known as listeriolysin O (LLO), which allows the bacterium to escape vacuoles upon entry into a host cell. A second gene regulated by PrfA is *actA*, which encodes for a surface protein known as actin assembly-inducing protein (ActA). ActA is expressed on the surface of *Listeria* and polymerizes host actin. This enables the bacterium to produce actin tails, move around the cell's cytoplasm, and spread from cell to cell without exiting into the extracellular compartment.

> Michael's condition has begun to worsen. He is now experiencing a stiff neck and hemiparesis (weakness of one side of the body). Concerned that the infection is spreading, the physician decides to conduct additional tests to determine what is causing these new symptoms.
>
> - What kind of pathogen causes listeriosis, and what virulence factors contribute to the signs and symptoms Michael is experiencing?
> - Is it likely that the infection will spread from Michael's blood? If so, how might this explain his new symptoms?
>
> Jump to the next Clinical Focus box. Go back to the previous Clinical Focus box.

Bacterial Exoenzymes and Toxins as Virulence Factors

After exposure and adhesion, the next step in pathogenesis is invasion, which can involve enzymes and toxins. Many pathogens achieve invasion by entering the bloodstream, an effective means of dissemination because blood vessels pass close to every cell in the body. The downside of this mechanism of dispersal is that the blood also includes numerous elements of the immune system. Various terms ending in –emia are used to describe the presence of pathogens in the bloodstream. The presence of bacteria in blood is called **bacteremia**. Bacteremia involving pyogens (pus-forming bacteria) is called pyemia. When viruses are found in the blood, it is called **viremia**. The term **toxemia** describes the condition when toxins are found in the blood. If bacteria are both present and multiplying in the blood, this condition is called **septicemia**.

Patients with septicemia are described as **septic**, which can lead to **shock**, a life-threatening decrease in blood pressure (systolic pressure <90 mm Hg) that prevents cells and organs from receiving enough oxygen and nutrients. Some bacteria can cause shock through the release of toxins (virulence factors that can cause tissue damage) and lead to low blood pressure. Gram-negative bacteria are engulfed by immune system phagocytes, which then release tumor necrosis factor, a molecule involved in inflammation and fever. Tumor necrosis factor binds to blood capillaries to increase their permeability, allowing fluids to pass out of blood vessels and into tissues, causing swelling, or edema (Figure 15.10). With high concentrations of tumor necrosis factor, the inflammatory reaction is severe and enough fluid is lost from the circulatory system that blood pressure decreases to dangerously low levels. This can have dire consequences because the heart, lungs, and kidneys rely on normal blood pressure for proper function; thus, multi-organ failure, shock, and death can occur.

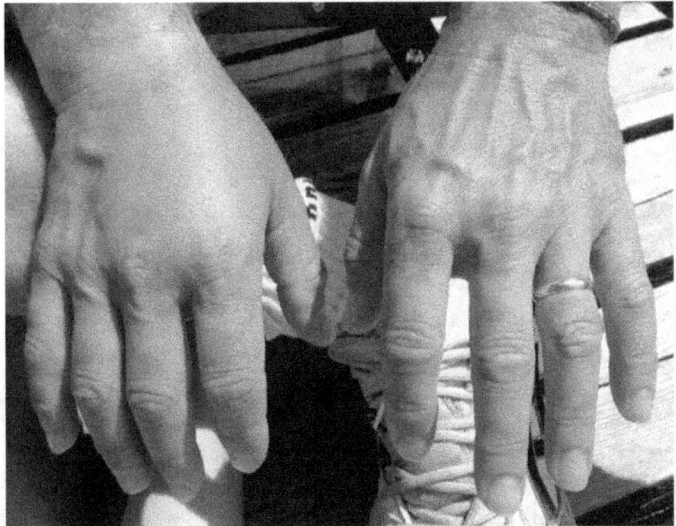

Figure 15.10 This patient has edema in the tissue of the right hand. Such swelling can occur when bacteria cause the release of pro-inflammatory molecules from immune cells and these molecules cause an increased permeability of blood vessels, allowing fluid to escape the bloodstream and enter tissue.

Exoenzymes

Some pathogens produce extracellular enzymes, or **exoenzymes**, that enable them to invade host cells and deeper tissues. Exoenzymes have a wide variety of targets. Some general classes of exoenzymes and associated pathogens are listed in Table 15.8. Each of these exoenzymes functions in the context of a particular tissue structure to facilitate invasion or support its own growth and defend against the immune system. For example, **hyaluronidase** S, an enzyme produced by pathogens like *Staphylococcus aureus*, *Streptococcus pyogenes*, and *Clostridium perfringens*, degrades the glycoside hyaluronan (hyaluronic acid), which acts as an intercellular cement between adjacent cells in connective tissue (Figure 15.11). This allows the pathogen to pass through the tissue layers at the portal of entry and disseminate elsewhere in the body (Figure 15.11).

Some Classes of Exoenzymes and Their Targets

Class	Example	Function
Glycohydrolases	Hyaluronidase S in *Staphylococcus aureus*	Degrades hyaluronic acid that cements cells together to promote spreading through tissues
Nucleases	DNAse produced by *S. aureus*	Degrades DNA released by dying cells (bacteria and host cells) that can trap the bacteria, thus promoting spread
Phospholipases	Phospholipase C of *Bacillus anthracis*	Degrades phospholipid bilayer of host cells, causing cellular lysis, and degrade membrane of phagosomes to enable escape into the cytoplasm
Proteases	Collagenase in *Clostridium perfringens*	Degrades collagen in connective tissue to promote spread

Table 15.8

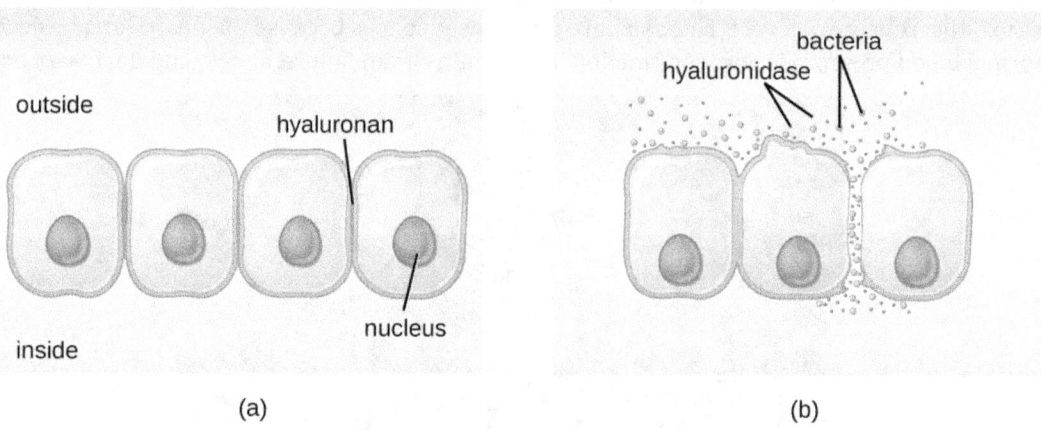

Figure 15.11 (a) Hyaluronan is a polymer found in the layers of epidermis that connect adjacent cells. (b) Hyaluronidase produced by bacteria degrades this adhesive polymer in the extracellular matrix, allowing passage between cells that would otherwise be blocked.

Pathogen-produced nucleases, such as **DNAse** produced by *S. aureus*, degrade extracellular DNA as a means of escape and spreading through tissue. As bacterial and host cells die at the site of infection, they lyse and release their intracellular contents. The DNA chromosome is the largest of the intracellular molecules, and masses of extracellular DNA can trap bacteria and prevent their spread. *S. aureus* produces a DNAse to degrade the mesh of extracellular DNA so it can escape and spread to adjacent tissues. This strategy is also used by *S. aureus* and other pathogens to degrade and escape webs of extracellular DNA produced by immune system phagocytes to trap the bacteria.

Enzymes that degrade the phospholipids of cell membranes are called phospholipases. Their actions are specific in regard to the type of phospholipids they act upon and where they enzymatically cleave the molecules. The pathogen responsible for anthrax, *B. anthracis,* produces phospholipase C. When *B. anthracis* is ingested by phagocytic cells of the immune system, phospholipase C degrades the membrane of the phagosome before it can fuse with the lysosome, allowing the pathogen to escape into the cytoplasm and multiply. Phospholipases can also target the membrane that encloses the phagosome within phagocytic cells. As described earlier in this chapter, this is the mechanism used by intracellular pathogens such as *L. monocytogenes* and *Rickettsia* to escape the phagosome and multiply within the cytoplasm of phagocytic cells. The role of phospholipases in bacterial virulence is not restricted to phagosomal escape. Many pathogens produce phospholipases that act to degrade cell membranes and cause lysis of target cells. These phospholipases are involved in lysis of red blood cells, white blood cells, and tissue cells.

Bacterial pathogens also produce various protein-digesting enzymes, or proteases. Proteases can be classified according to their substrate target (e.g., serine proteases target proteins with the amino acid serine) or if they contain metals in their active site (e.g., zinc metalloproteases contain a zinc ion, which is necessary for enzymatic activity).

One example of a protease that contains a metal ion is the exoenzyme **collagenase**. Collagenase digests collagen, the dominant protein in connective tissue. Collagen can be found in the extracellular matrix, especially near mucosal membranes, blood vessels, nerves, and in the layers of the skin. Similar to hyaluronidase, collagenase allows the pathogen to penetrate and spread through the host tissue by digesting this connective tissue protein. The collagenase produced by the gram-positive bacterium *Clostridium perfringens*, for example, allows the bacterium to make its way through the tissue layers and subsequently enter and multiply in the blood (septicemia). *C. perfringens* then uses toxins and a phospholipase to cause cellular lysis and necrosis. Once the host cells have died, the bacterium produces gas by fermenting the muscle carbohydrates. The widespread necrosis of tissue and accompanying gas are characteristic of the condition known as gas gangrene (Figure 15.12).

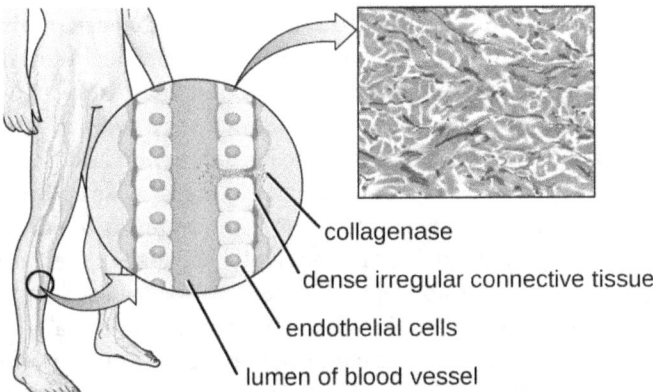

Figure 15.12 The illustration depicts a blood vessel with a single layer of endothelial cells surrounding the lumen and dense connective tissue (shown in red) surrounding the endothelial cell layer. Collagenase produced by *C. perfringens* degrades the collagen between the endothelial cells, allowing the bacteria to enter the bloodstream. (credit illustration: modification of work by Bruce Blaus; credit micrograph: Micrograph provided by the Regents of University of Michigan Medical School © 2012)

LINK TO LEARNING

Two types of cell death are apoptosis and necrosis. Visit this website (https://openstax.org/l/22CellDeath) to learn more about the differences between these mechanisms of cell death and their causes.

Toxins

In addition to exoenzymes, certain pathogens are able to produce **toxins**, biological poisons that assist in their ability to invade and cause damage to tissues. The ability of a pathogen to produce toxins to cause damage to host cells is called **toxigenicity**.

Toxins can be categorized as endotoxins or exotoxins. The lipopolysaccharide (LPS) found on the outer membrane of gram-negative bacteria is called **endotoxin** (Figure 15.13). During infection and disease, gram-negative bacterial pathogens release endotoxin either when the cell dies, resulting in the disintegration of the membrane, or when the bacterium undergoes binary fission. The lipid component of endotoxin, lipid A, is responsible for the toxic properties of the LPS molecule. Lipid A is relatively conserved across different genera of gram-negative bacteria; therefore, the toxic properties of lipid A are similar regardless of the gram-negative pathogen. In a manner similar to that of tumor necrosis factor, lipid A triggers the immune system's inflammatory response (see Inflammation and Fever). If the concentration of endotoxin in the body is low, the inflammatory response may provide the host an effective defense against infection; on the other hand, high concentrations of endotoxin in the blood can cause an excessive inflammatory response, leading to a severe drop in blood pressure, multi-organ failure, and death.

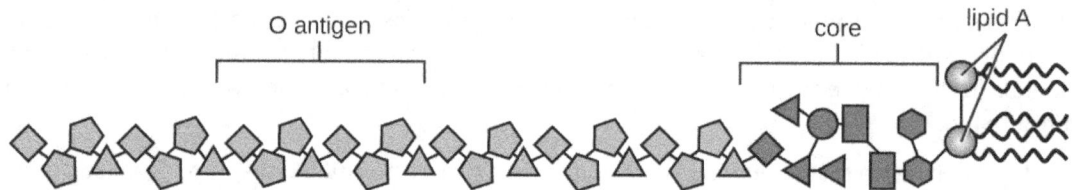

Figure 15.13 Lipopolysaccharide is composed of lipid A, a core glycolipid, and an O-specific polysaccharide side chain. Lipid A is the toxic component that promotes inflammation and fever.

A classic method of detecting endotoxin is by using the *Limulus* amebocyte lysate (LAL) test. In this procedure, the blood cells (amebocytes) of the horseshoe crab (*Limulus polyphemus*) is mixed with a patient's serum. The amebocytes will react to the presence of any endotoxin. This reaction can be observed either chromogenically (color) or by looking for coagulation (clotting reaction) to occur within the serum. An alternative method that has been used is an enzyme-linked immunosorbent assay (ELISA) that uses antibodies to detect the presence of endotoxin.

Unlike the toxic lipid A of endotoxin, **exotoxins** are protein molecules that are produced by a wide variety of living pathogenic bacteria. Although some gram-negative pathogens produce exotoxins, the majority are produced by gram-positive pathogens. Exotoxins differ from endotoxin in several other key characteristics, summarized in Table 15.9. In contrast to endotoxin, which stimulates a general systemic inflammatory response when released, exotoxins are much more specific in their action and the cells they interact with. Each exotoxin targets specific receptors on specific cells and damages those cells through unique molecular mechanisms. Endotoxin remains stable at high temperatures, and requires heating at 121 °C (250 °F) for 45 minutes to inactivate. By contrast, most exotoxins are heat labile because of their protein structure, and many are denatured (inactivated) at temperatures above 41 °C (106 °F). As discussed earlier, endotoxin can stimulate a lethal inflammatory response at very high concentrations and has a measured LD_{50} of 0.24 mg/kg. By contrast, very small concentrations of exotoxins can be lethal. For example, botulinum toxin, which causes botulism, has an LD_{50} of 0.000001 mg/kg (240,000 times more lethal than endotoxin).

Comparison of Endotoxin and Exotoxins Produced by Bacteria

Characteristic	Endotoxin	Exotoxin
Source	Gram-negative bacteria	Gram-positive (primarily) and gram-negative bacteria
Composition	Lipid A component of lipopolysaccharide	Protein
Effect on host	General systemic symptoms of inflammation and fever	Specific damage to cells dependent upon receptor-mediated targeting of cells and specific mechanisms of action
Heat stability	Heat stable	Most are heat labile, but some are heat stable

Comparison of Endotoxin and Exotoxins Produced by Bacteria

Characteristic	Endotoxin	Exotoxin
LD_{50}	High	Low

Table 15.9

The exotoxins can be grouped into three categories based on their target: intracellular targeting, membrane disrupting, and superantigens. Table 15.10 provides examples of well-characterized toxins within each of these three categories.

Some Common Exotoxins and Associated Bacterial Pathogens

Category	Example	Pathogen	Mechanism and Disease
Intracellular-targeting toxins	Cholera toxin	*Vibrio cholerae*	Activation of adenylate cyclase in intestinal cells, causing increased levels of cyclic adenosine monophosphate (cAMP) and secretion of fluids and electrolytes out of cell, causing diarrhea
	Tetanus toxin	*Clostridium tetani*	Inhibits the release of inhibitory neurotransmitters in the central nervous system, causing spastic paralysis
	Botulinum toxin	*Clostridium botulinum*	Inhibits release of the neurotransmitter acetylcholine from neurons, resulting in flaccid paralysis
	Diphtheria toxin	*Corynebacterium diphtheriae*	Inhibition of protein synthesis, causing cellular death
Membrane-disrupting toxins	Streptolysin	*Streptococcus pyogenes*	Proteins that assemble into pores in cell membranes, disrupting their function and killing the cell
	Pneumolysin	*Streptococcus pneumoniae*	
	Alpha-toxin	*Staphylococcus aureus*	
	Alpha-toxin	*Clostridium perfringens*	Phospholipases that degrade cell membrane phospholipids, disrupting membrane function and killing the cell
	Phospholipase C	*Pseudomonas aeruginosa*	
	Beta-toxin	*Staphylococcus aureus*	

Some Common Exotoxins and Associated Bacterial Pathogens

Category	Example	Pathogen	Mechanism and Disease
Superantigens	Toxic shock syndrome toxin	*Staphylococcus aureus*	Stimulates excessive activation of immune system cells and release of cytokines (chemical mediators) from immune system cells. Life-threatening fever, inflammation, and shock are the result.
	Streptococcal mitogenic exotoxin	*Streptococcus pyogenes*	
	Streptococcal pyrogenic toxins	*Streptococcus pyogenes*	

Table 15.10

The **intracellular targeting toxins** comprise two components: A for activity and B for binding. Thus, these types of toxins are known as **A-B exotoxins** (Figure 15.14). The B component is responsible for the cellular specificity of the toxin and mediates the initial attachment of the toxin to specific cell surface receptors. Once the A-B toxin binds to the host cell, it is brought into the cell by endocytosis and entrapped in a vacuole. The A and B subunits separate as the vacuole acidifies. The A subunit then enters the cell cytoplasm and interferes with the specific internal cellular function that it targets.

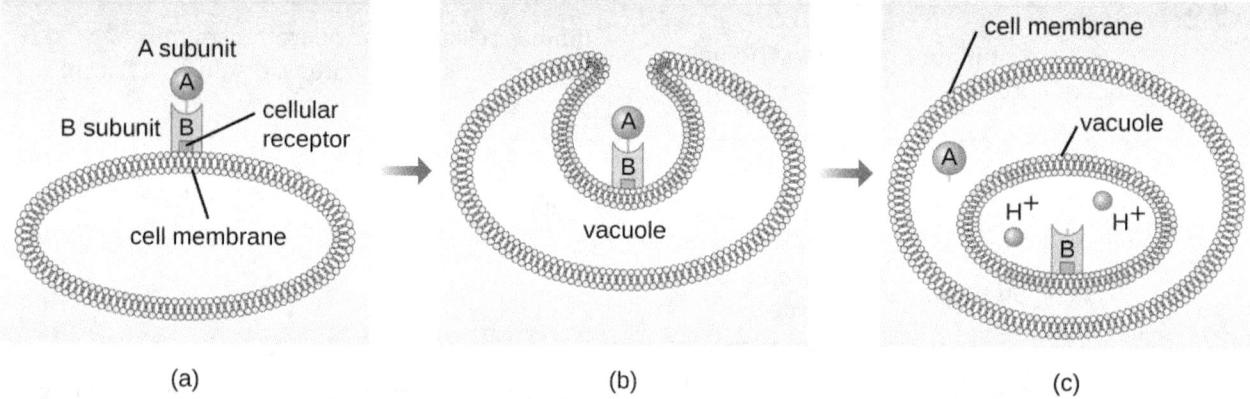

Figure 15.14   (a) In A-B toxins, the B component binds to the host cell through its interaction with specific cell surface receptors. (b) The toxin is brought in through endocytosis. (c) Once inside the vacuole, the A component (active component) separates from the B component and the A component gains access to the cytoplasm. (credit: modification of work by "Biology Discussion Forum"/YouTube)

Four unique examples of A-B toxins are the diphtheria, cholera, botulinum, and tetanus toxins. The diphtheria toxin is produced by the gram-positive bacterium *Corynebacterium diphtheriae*, the causative agent of nasopharyngeal and cutaneous diphtheria. After the A subunit of the diphtheria toxin separates and gains access to the cytoplasm, it facilitates the transfer of adenosine diphosphate (ADP)-ribose onto an elongation-factor protein (EF-2) that is needed for protein synthesis. Hence, diphtheria toxin inhibits protein synthesis in the host cell, ultimately killing the cell (Figure 15.15).

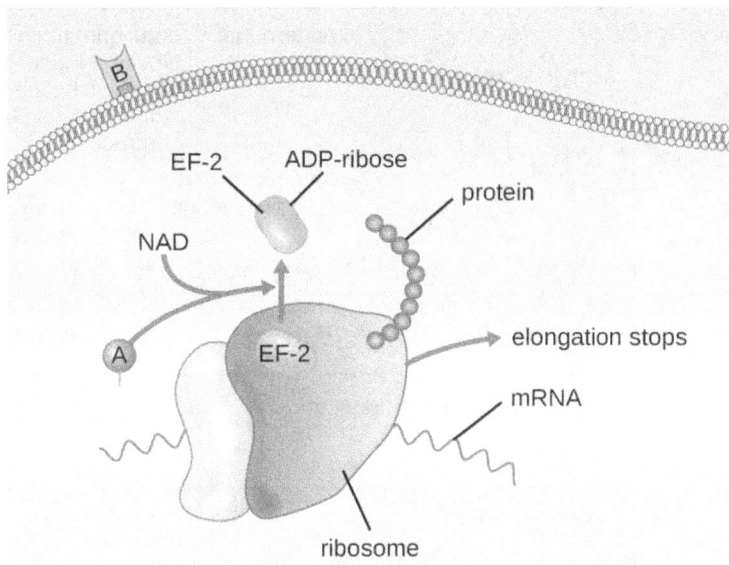

Figure 15.15 The mechanism of the diphtheria toxin inhibiting protein synthesis. The A subunit inactivates elongation factor 2 by transferring an ADP-ribose. This stops protein elongation, inhibiting protein synthesis and killing the cell.

Cholera toxin is an **enterotoxin** produced by the gram-negative bacterium *Vibrio cholerae* and is composed of one A subunit and five B subunits. The mechanism of action of the cholera toxin is complex. The B subunits bind to receptors on the intestinal epithelial cell of the small intestine. After gaining entry into the cytoplasm of the epithelial cell, the A subunit activates an intracellular G protein. The activated G protein, in turn, leads to the activation of the enzyme adenyl cyclase, which begins to produce an increase in the concentration of cyclic AMP (a secondary messenger molecule). The increased cAMP disrupts the normal physiology of the intestinal epithelial cells and causes them to secrete excessive amounts of fluid and electrolytes into the lumen of the intestinal tract, resulting in severe "rice-water stool" diarrhea characteristic of cholera.

Botulinum toxin (also known as botox) is a neurotoxin produced by the gram-positive bacterium *Clostridium botulinum*. It is the most acutely toxic substance known to date. The toxin is composed of a light A subunit and heavy protein chain B subunit. The B subunit binds to neurons to allow botulinum toxin to enter the neurons at the neuromuscular junction. The A subunit acts as a protease, cleaving proteins involved in the neuron's release of acetylcholine, a neurotransmitter molecule. Normally, neurons release acetylcholine to induce muscle fiber contractions. The toxin's ability to block acetylcholine release results in the inhibition of muscle contractions, leading to muscle relaxation. This has the potential to stop breathing and cause death. Because of its action, low concentrations of botox are used for cosmetic and medical procedures, including the removal of wrinkles and treatment of overactive bladder.

LINK TO LEARNING

Click this link (https://openstax.org/l/22pathochol) to see an animation of how the cholera toxin functions.

Click this link (https://openstax.org/l/22Botulin) to see an animation of how the botulinum toxin functions.

Another neurotoxin is tetanus toxin, which is produced by the gram-positive bacterium *Clostridium tetani*. This toxin also has a light A subunit and heavy protein chain B subunit. Unlike botulinum toxin, tetanus toxin binds to inhibitory interneurons, which are responsible for release of the inhibitory neurotransmitters glycine and gamma-aminobutyric acid (GABA). Normally, these neurotransmitters bind to neurons at the neuromuscular junction, resulting in the inhibition of acetylcholine release. Tetanus toxin inhibits the release of glycine and GABA from the interneuron, resulting in permanent muscle contraction. The first symptom is typically stiffness of the jaw (lockjaw). Violent muscle spasms in other parts of the body follow, typically culminating with respiratory failure and death. Figure 15.16 shows the actions of both botulinum and tetanus toxins.

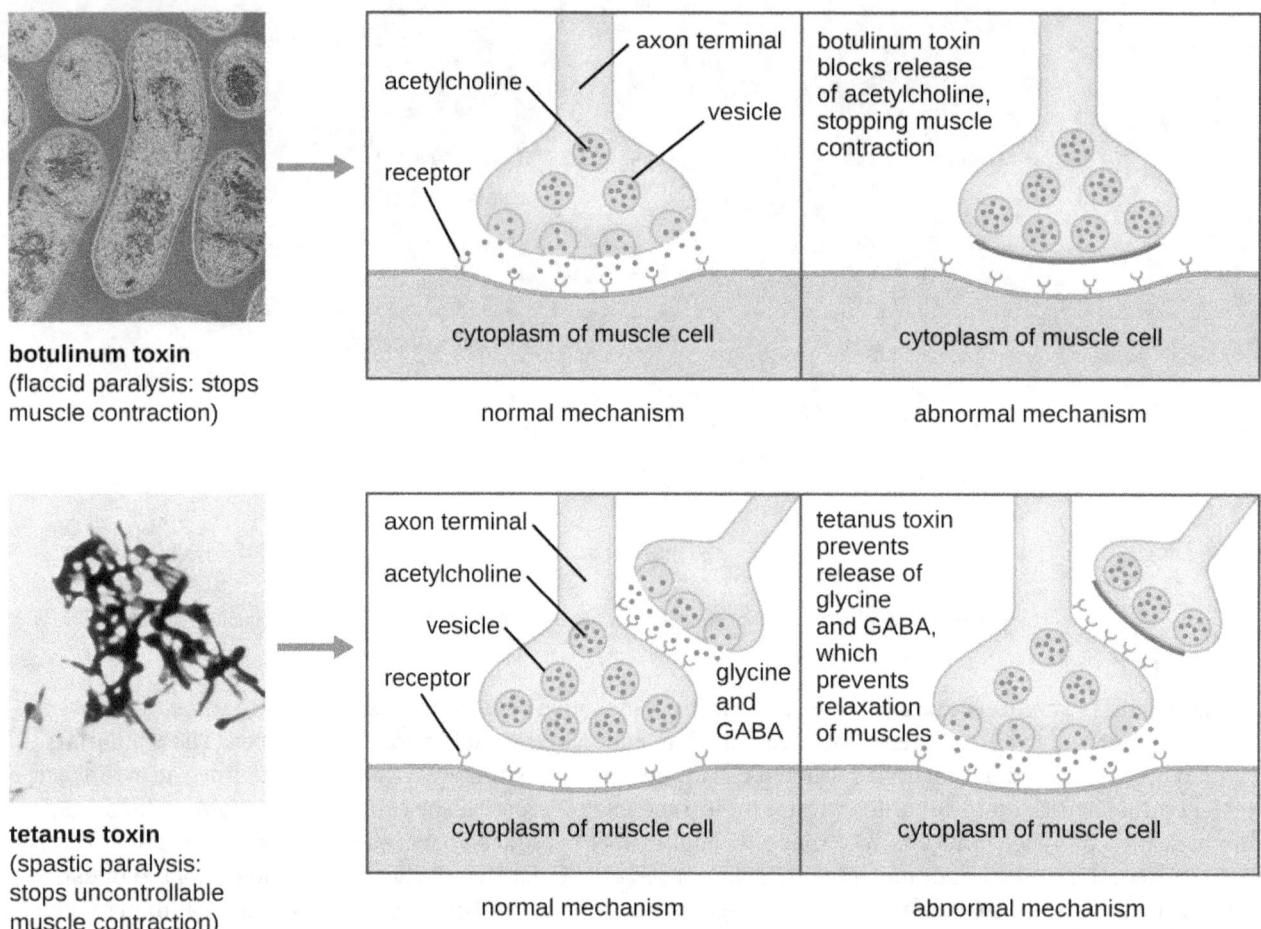

Figure 15.16 Mechanisms of botulinum and tetanus toxins. (credit micrographs: modification of work by Centers for Disease Control and Prevention)

Membrane-disrupting toxins affect cell membrane function either by forming pores or by disrupting the phospholipid bilayer in host cell membranes. Two types of membrane-disrupting exotoxins are **hemolysins** and leukocidins, which form pores in cell membranes, causing leakage of the cytoplasmic contents and cell lysis. These toxins were originally thought to target red blood cells (erythrocytes) and white blood cells (leukocytes), respectively, but we now know they can affect other cells as well. The gram-positive bacterium *Streptococcus pyogenes* produces streptolysins, water-soluble hemolysins that bind to the cholesterol moieties in the host cell membrane to form a pore. The two types of streptolysins, O and S, are categorized by their ability to cause hemolysis in erythrocytes in the absence or presence of oxygen. Streptolysin O is not active in the presence of oxygen, whereas streptolysin S is active in the presence of oxygen. Other important pore-forming membrane-disrupting toxins include alpha toxin of *Staphylococcus aureus* and pneumolysin of *Streptococcus pneumoniae*.

Bacterial phospholipases are **membrane-disrupting toxins** that degrade the phospholipid bilayer of cell membranes rather than forming pores. We have already discussed the phospholipases associated with *B. anthracis*, *L. pneumophila*, and *Rickettsia* species that enable these bacteria to effect the lysis of phagosomes. These same phospholipases are also hemolysins. Other phospholipases that function as hemolysins include the alpha toxin of *Clostridium perfringens*, phospholipase C of *P. aeruginosa*, and beta toxin of *Staphylococcus aureus*.

Some strains of *S. aureus* also produce a leukocidin called Panton-Valentine leukocidin (PVL). PVL consists of two subunits, S and F. The S component acts like the B subunit of an A-B exotoxin in that it binds to glycolipids on the outer plasma membrane of animal cells. The F-component acts like the A subunit of an A-B exotoxin and carries the enzymatic activity. The toxin inserts and assembles into a pore in the membrane. Genes that encode PVL are more frequently present in *S. aureus* strains that cause skin infections and pneumonia.[8] PVL promotes skin infections by causing edema, erythema (reddening of the skin due to blood vessel dilation), and skin necrosis. PVL has also been shown to cause necrotizing pneumonia. PVL promotes pro-inflammatory and cytotoxic effects on alveolar leukocytes. This results in the release of enzymes from the leukocytes, which, in turn, cause damage to lung tissue.

The third class of exotoxins is the **superantigens**. These are exotoxins that trigger an excessive, nonspecific stimulation of immune cells to secrete cytokines (chemical messengers). The excessive production of cytokines, often called a cytokine storm, elicits a strong immune and inflammatory response that can cause life-threatening high fevers, low blood pressure, multi-organ failure, shock, and death. The prototype superantigen is the toxic shock syndrome toxin of *S. aureus*. Most toxic shock syndrome cases are associated with vaginal colonization by toxin-producing *S. aureus* in menstruating women; however, colonization of other body sites can also occur. Some strains of *Streptococcus pyogenes* also produce superantigens; they are referred to as the streptococcal mitogenic exotoxins and the streptococcal pyrogenic toxins.

CHECK YOUR UNDERSTANDING

- Describe how exoenzymes contribute to bacterial invasion.
- Explain the difference between exotoxins and endotoxin.
- Name the three classes of exotoxins.

Virulence Factors for Survival in the Host and Immune Evasion

Evading the immune system is also important to invasiveness. Bacteria use a variety of virulence factors to evade phagocytosis by cells of the immune system. For example, many bacteria produce capsules, which are used in adhesion but also aid in immune evasion by preventing ingestion by phagocytes. The composition of the capsule prevents immune cells from being able to adhere and then phagocytose the cell. In addition, the capsule makes the bacterial cell much larger, making it harder for immune cells to engulf the pathogen (Figure 15.17). A notable capsule-producing bacterium is the gram-positive pathogen *Streptococcus pneumoniae*, which causes pneumococcal pneumonia, meningitis, septicemia, and other respiratory tract infections. Encapsulated strains of *S. pneumoniae* are more virulent than nonencapsulated strains and are more likely to invade the bloodstream and cause septicemia and meningitis.

Some pathogens can also produce proteases to protect themselves against phagocytosis. As described in Adaptive Specific Host Defenses, the human immune system produces antibodies that bind to surface molecules found on specific bacteria (e.g., capsules, fimbriae, flagella, LPS). This binding initiates phagocytosis and other mechanisms of antibacterial killing and clearance. Proteases combat antibody-mediated killing and clearance by attacking and digesting the antibody molecules (Figure 15.17).

In addition to capsules and proteases, some bacterial pathogens produce other virulence factors that allow them to evade the immune system. The fimbriae of certain species of *Streptococcus* contain M protein, which alters the surface of *Streptococcus* and inhibits phagocytosis by blocking the binding of the complement molecules that assist phagocytes in ingesting bacterial pathogens. The acid-fast bacterium *Mycobacterium tuberculosis* (the causative agent of tuberculosis) produces a waxy substance known as mycolic acid in its cell envelope. When it is engulfed by phagocytes in the lung, the protective mycolic acid coat enables the bacterium to resist some of the killing mechanisms within the phagolysosome.

---

8 V. Meka. "Panton-Valentine Leukocidin." http://www.antimicrobe.org/h04c.files/history/PVL-S-aureus.asp

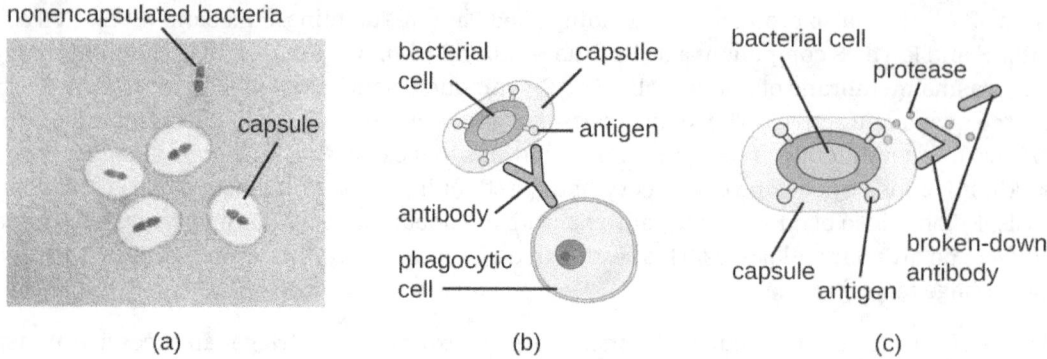

Figure 15.17 (a) A micrograph of capsules around bacterial cells. (b) Antibodies normally function by binding to antigens, molecules on the surface of pathogenic bacteria. Phagocytes then bind to the antibody, initiating phagocytosis. (c) Some bacteria also produce proteases, virulence factors that break down host antibodies to evade phagocytosis. (credit a: modification of work by Centers for Disease Control and Prevention)

Some bacteria produce virulence factors that promote infection by exploiting molecules naturally produced by the host. For example, most strains of *Staphylococcus aureus* produce the exoenzyme **coagulase**, which exploits the natural mechanism of blood clotting to evade the immune system. Normally, blood clotting is triggered in response to blood vessel damage; platelets begin to plug the clot, and a cascade of reactions occurs in which fibrinogen, a soluble protein made by the liver, is cleaved into fibrin. Fibrin is an insoluble, thread-like protein that binds to blood platelets, cross-links, and contracts to form a mesh of clumped platelets and red blood cells. The resulting clot prevents further loss of blood from the damaged blood vessels. However, if bacteria release coagulase into the bloodstream, the fibrinogen-to-fibrin cascade is triggered in the absence of blood vessel damage. The resulting clot coats the bacteria in fibrin, protecting the bacteria from exposure to phagocytic immune cells circulating in the bloodstream.

Whereas coagulase causes blood to clot, kinases have the opposite effect by triggering the conversion of plasminogen to plasmin, which is involved in the digestion of fibrin clots. By digesting a clot, kinases allow pathogens trapped in the clot to escape and spread, similar to the way that collagenase, hyaluronidase, and DNAse facilitate the spread of infection. Examples of kinases include staphylokinases and streptokinases, produced by *Staphylococcus aureus* and *Streptococcus pyogenes*, respectively. It is intriguing that *S. aureus* can produce both coagulase to promote clotting and staphylokinase to stimulate the digestion of clots. The action of the coagulase provides an important protective barrier from the immune system, but when nutrient supplies are diminished or other conditions signal a need for the pathogen to escape and spread, the production of staphylokinase can initiate this process.

A final mechanism that pathogens can use to protect themselves against the immune system is called **antigenic variation**, which is the alteration of surface proteins so that a pathogen is no longer recognized by the host's immune system. For example, the bacterium *Borrelia burgdorferi*, the causative agent of Lyme disease, contains a surface lipoprotein known as VlsE. Because of genetic recombination during DNA replication and repair, this bacterial protein undergoes antigenic variation. Each time fever occurs, the VlsE protein in *B. burgdorferi* can differ so much that antibodies against previous VlsE sequences are not effective. It is believed that this variation in the VlsE contributes to the ability *B. burgdorferi* to cause chronic disease. Another important human bacterial pathogen that uses antigenic variation to avoid the immune system is *Neisseria gonorrhoeae*, which causes the sexually transmitted disease gonorrhea. This bacterium is well known for its ability to undergo antigenic variation of its type IV pili to avoid immune defenses.

⊘ CHECK YOUR UNDERSTANDING

- Name at least two ways that a capsule provides protection from the immune system.
- Besides capsules, name two other virulence factors used by bacteria to evade the immune system.

Clinical Focus

Resolution

Based on Michael's reported symptoms of stiff neck and hemiparesis, the physician suspects that the infection may have spread to his nervous system. The physician decides to order a spinal tap to look for any bacteria that may have invaded the meninges and cerebrospinal fluid (CSF), which would normally be sterile. To perform the spinal tap, Michael's lower back is swabbed with an iodine antiseptic and then covered with a sterile sheet. The needle is aseptically removed from the manufacturer's sealed plastic packaging by the clinician's gloved hands. The needle is inserted and a small volume of fluid is drawn into an attached sample tube. The tube is removed, capped and a prepared label with Michael's data is affixed to it. This STAT (urgent or immediate analysis required) specimen is divided into three separate sterile tubes, each with 1 mL of CSF. These tubes are immediately taken to the hospital's lab, where they are analyzed in the clinical chemistry, hematology, and microbiology departments. The preliminary results from all three departments indicate there is a cerebrospinal infection occurring, with the microbiology department reporting the presence of a gram-positive rod in Michael's CSF.

These results confirm what his physician had suspected: Michael's new symptoms are the result of meningitis, acute inflammation of the membranes that protect the brain and spinal cord. Because meningitis can be life threatening and because the first antibiotic therapy was not effective in preventing the spread of infection, Michael is prescribed an aggressive course of two antibiotics, ampicillin and gentamicin, to be delivered intravenously. Michael remains in the hospital for several days for supportive care and for observation. After a week, he is allowed to return home for bed rest and oral antibiotics. After 3 weeks of this treatment, he makes a full recovery.

Go back to the previous Clinical Focus box.

Viral Virulence

Although viral pathogens are not similar to bacterial pathogens in terms of structure, some of the properties that contribute to their virulence are similar. Viruses use adhesins to facilitate adhesion to host cells, and certain enveloped viruses rely on antigenic variation to avoid the host immune defenses. These virulence factors are discussed in more detail in the following sections.

Viral Adhesins

One of the first steps in any viral infection is adhesion of the virus to specific receptors on the surface of cells. This process is mediated by adhesins that are part of the viral capsid or membrane envelope. The interaction of viral adhesins with specific cell receptors defines the tropism (preferential targeting) of viruses for specific cells, tissues, and organs in the body. The spike protein hemagglutinin found on Influenzavirus is an example of a viral adhesin; it allows the virus to bind to the sialic acid on the membrane of host respiratory and intestinal cells. Another viral adhesin is the glycoprotein gp20, found on HIV. For HIV to infect cells of the immune system, it must interact with two receptors on the surface of cells. The first interaction involves binding between gp120 and the CD4 cellular marker that is found on some essential immune system cells. However, before viral entry into the cell can occur, a second interaction between gp120 and one of two chemokine receptors (CCR5 and CXCR4) must occur. Table 15.11 lists the adhesins for some common viral pathogens and the specific sites to which these adhesins allow viruses to attach.

Some Viral Adhesins and Their Host Attachment Sites

Pathogen	Disease	Adhesin	Attachment Site
Influenzavirus	Influenza	Hemagglutinin	Sialic acid of respiratory and intestinal cells

Some Viral Adhesins and Their Host Attachment Sites

Pathogen	Disease	Adhesin	Attachment Site
Herpes simplex virus I or II	Oral herpes, genital herpes	Glycoproteins gB, gC, gD	Heparan sulfate on mucosal surfaces of the mouth and genitals
Human immunodeficiency virus	HIV/AIDS	Glycoprotein gp120	CD4 and CCR5 or CXCR4 of immune system cells

Table 15.11

Antigenic Variation in Viruses

Antigenic variation also occurs in certain types of enveloped viruses, including influenza viruses, which exhibit two forms of antigenic variation: **antigenic drift** and **antigenic shift** (Figure 15.18). Antigenic drift is the result of point mutations causing slight changes in the spike proteins hemagglutinin (H) and neuraminidase (N). On the other hand, antigenic shift is a major change in spike proteins due to gene reassortment. This reassortment for antigenic shift occurs typically when two different influenza viruses infect the same host.

The rate of antigenic variation in influenza viruses is very high, making it difficult for the immune system to recognize the many different strains of Influenzavirus. Although the body may develop immunity to one strain through natural exposure or vaccination, antigenic variation results in the continual emergence of new strains that the immune system will not recognize. This is the main reason that vaccines against Influenzavirus must be given annually. Each year's influenza vaccine provides protection against the most prevalent strains for that year, but new or different strains may be more prevalent the following year.

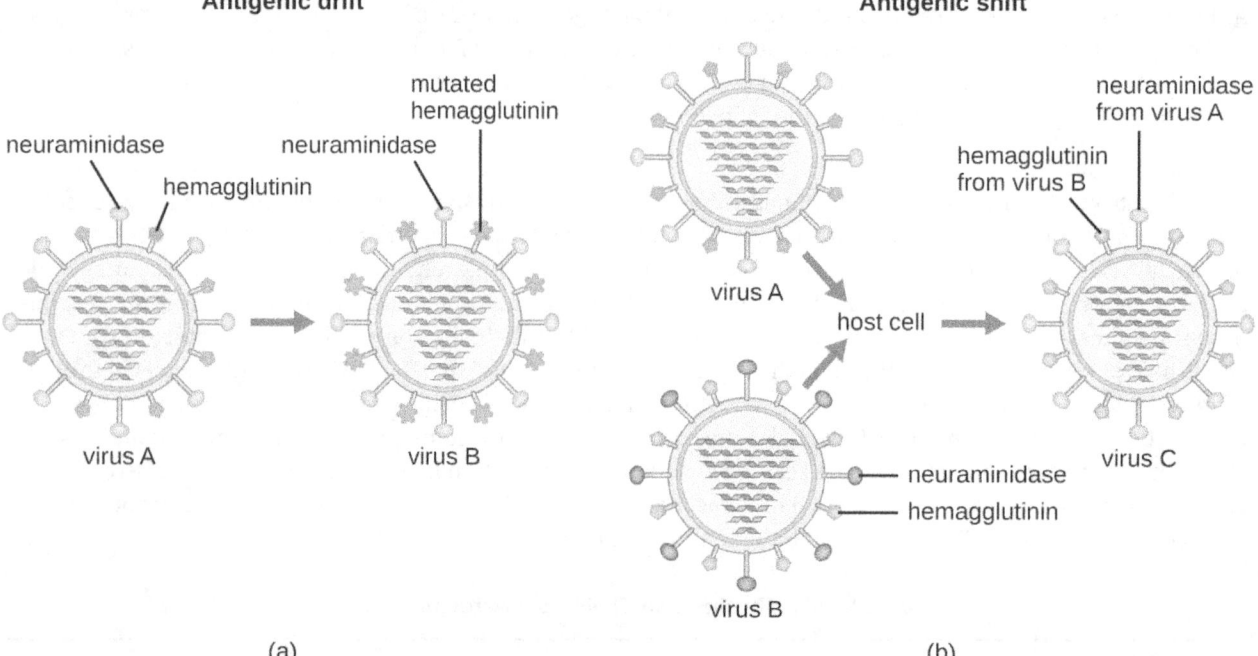

Figure 15.18 Antigenic drift and antigenic shift in influenza viruses. (a) In antigenic drift, mutations in the genes for the surface proteins neuraminidase and/or hemagglutinin result in small antigenic changes over time. (b) In antigenic shift, simultaneous infection of a cell with two different influenza viruses results in mixing of the genes. The resultant virus possesses a mixture of the proteins of the original viruses. Influenza pandemics can often be traced to antigenic shifts.

LINK TO LEARNING

For another explanation of how antigenic shift and drift (https://openstax.org/l/22Antigenic) occur, watch this video.

CHECK YOUR UNDERSTANDING

- Describe the role of adhesins in viral tropism.
- Explain the difference between antigenic drift and antigenic shift.

15.4 Virulence Factors of Eukaryotic Pathogens

Learning Objectives

By the end of this section, you will be able to:
- Describe virulence factors unique to fungi and parasites
- Compare virulence factors of fungi and bacteria
- Explain the difference between protozoan parasites and helminths
- Describe how helminths evade the host immune system

Although fungi and parasites are important pathogens causing infectious diseases, their pathogenic mechanisms and virulence factors are not as well characterized as those of bacteria. Despite the relative lack of detailed mechanisms, the stages of pathogenesis and general mechanisms of virulence involved in disease production by these pathogens are similar to those of bacteria.

Fungal Virulence

Pathogenic fungi can produce virulence factors that are similar to the bacterial virulence factors that have been discussed earlier in this chapter. In this section, we will look at the virulence factors associated with species of *Candida, Cryptococcus, Claviceps,* and *Aspergillus*.

Candida albicans is an opportunistic fungal pathogen and causative agent of oral thrush, vaginal yeast infections, and cutaneous candidiasis. *Candida* produces adhesins (surface glycoproteins) that bind to the phospholipids of epithelial and endothelial cells. To assist in spread and tissue invasion, *Candida* produces proteases and phospholipases (i.e., exoenzymes). One of these proteases degrades keratin, a structural protein found on epithelial cells, enhancing the ability of the fungus to invade host tissue. In animal studies, it has been shown that the addition of a protease inhibitor led to attenuation of *Candida* infection.[9] Similarly, the phospholipases can affect the integrity of host cell membranes to facilitate invasion.

The main virulence factor for *Cryptococcus*, a fungus that causes pneumonia and meningitis, is capsule production. The polysaccharide glucuronoxylomannan is the principal constituent of the *Cryptococcus* capsule. Similar to encapsulated bacterial cells, encapsulated *Cryptococcus* cells are more resistant to phagocytosis than nonencapsulated *Cryptococcus*, which are effectively phagocytosed and, therefore, less virulent.

Like some bacteria, many fungi produce exotoxins. Fungal toxins are called **mycotoxins**. *Claviceps purpurea*, a fungus that grows on rye and related grains, produces a mycotoxin called ergot toxin, an alkaloid responsible for the disease known as ergotism. There are two forms of ergotism: gangrenous and convulsive. In gangrenous ergotism, the ergot toxin causes vasoconstriction, resulting in improper blood flow to the extremities, eventually leading to gangrene. A famous outbreak of gangrenous ergotism occurred in Eastern Europe during the 5th century AD due to the consumption of rye contaminated with *C. purpurea*. In convulsive ergotism, the toxin targets the central nervous system, causing mania and hallucinations.

---

9 K. Fallon et al. "Role of Aspartic Proteases in Disseminated *Candida albicans* Infection in Mice." *Infection and Immunity* 65 no. 2 (1997):551–556.

The mycotoxin aflatoxin is a virulence factor produced by the fungus *Aspergillus*, an opportunistic pathogen that can enter the body via contaminated food or by inhalation. Inhalation of the fungus can lead to the chronic pulmonary disease aspergillosis, characterized by fever, bloody sputum, and/or asthma. Aflatoxin acts in the host as both a mutagen (a substance that causes mutations in DNA) and a **carcinogen** (a substance involved in causing cancer), and has been associated with the development of liver cancer. Aflatoxin has also been shown to cross the blood-placental barrier.[10] A second mycotoxin produced by *Aspergillus* is gliotoxin. This toxin promotes virulence by inducing host cells to self-destruct and by evading the host's immune response by inhibiting the function of phagocytic cells as well as the pro-inflammatory response. Like *Candida*, *Aspergillus* also produces several proteases. One is elastase, which breaks down the protein elastin found in the connective tissue of the lung, leading to the development of lung disease. Another is catalase, an enzyme that protects the fungus from hydrogen peroxide produced by the immune system to destroy pathogens.

CHECK YOUR UNDERSTANDING

- List virulence factors common to bacteria and fungi.
- What functions do mycotoxins perform to help fungi survive in the host?

Protozoan Virulence

Protozoan pathogens are unicellular eukaryotic parasites that have virulence factors and pathogenic mechanisms analogous to prokaryotic and viral pathogens, including adhesins, toxins, antigenic variation, and the ability to survive inside phagocytic vesicles.

Protozoans often have unique features for attaching to host cells. The protozoan *Giardia lamblia*, which causes the intestinal disease giardiasis, uses a large adhesive disc composed of microtubules to attach to the intestinal mucosa. During adhesion, the flagella of *G. lamblia* move in a manner that draws fluid out from under the disc, resulting in an area of lower pressure that facilitates adhesion to epithelial cells. *Giardia* does not invade the intestinal cells but rather causes inflammation (possibly through the release of cytopathic substances that cause damage to the cells) and shortens the intestinal villi, inhibiting absorption of nutrients.

Some protozoans are capable of antigenic variation. The obligate intracellular pathogen *Plasmodium falciparum* (one of the causative agents of malaria) resides inside red blood cells, where it produces an adhesin membrane protein known as PfEMP1. This protein is expressed on the surface of the infected erythrocytes, causing blood cells to stick to each other and to the walls of blood vessels. This process impedes blood flow, sometimes leading to organ failure, anemia, jaundice (yellowing of skin and sclera of the eyes due to buildup of bilirubin from lysed red blood cells), and, subsequently, death. Although PfEMP1 can be recognized by the host's immune system, antigenic variations in the structure of the protein over time prevent it from being easily recognized and eliminated. This allows malaria to persist as a chronic infection in many individuals.

The virulence factors of *Trypanosoma brucei*, the causative agent of African sleeping sickness, include the abilities to form capsules and undergo antigenic variation. *T. brucei* evades phagocytosis by producing a dense glycoprotein coat that resembles a bacterial capsule. Over time, host antibodies are produced that recognize this coat, but *T. brucei* is able to alter the structure of the glycoprotein to evade recognition.

CHECK YOUR UNDERSTANDING

- Explain how antigenic variation by protozoan pathogens helps them survive in the host.

Helminth Virulence

Helminths, or parasitic worms, are multicellular eukaryotic parasites that depend heavily on virulence factors that allow them to gain entry to host tissues. For example, the aquatic larval form of *Schistosoma mansoni*, which causes schistosomiasis, penetrates intact skin with the aid of proteases that degrade skin proteins, including elastin.

---

10 C.P. Wild et al. "In-utero exposure to aflatoxin in west Africa." *Lancet* 337 no. 8757 (1991):1602.

To survive within the host long enough to perpetuate their often-complex life cycles, helminths need to evade the immune system. Some helminths are so large that the immune system is ineffective against them. Others, such as adult roundworms (which cause trichinosis, ascariasis, and other diseases), are protected by a tough outer cuticle.

Over the course of their life cycles, the surface characteristics of the parasites vary, which may help prevent an effective immune response. Some helminths express polysaccharides called glycans on their external surface; because these glycans resemble molecules produced by host cells, the immune system fails to recognize and attack the helminth as a foreign body. This "glycan gimmickry," as it has been called, serves as a protective cloak that allows the helminth to escape detection by the immune system.[11]

In addition to evading host defenses, helminths can actively suppress the immune system. *S. mansoni*, for example, degrades host antibodies with proteases. Helminths produce many other substances that suppress elements of both innate nonspecific and adaptive specific host defenses. They also release large amounts of material into the host that may locally overwhelm the immune system or cause it to respond inappropriately.

CHECK YOUR UNDERSTANDING

- Describe how helminths avoid being destroyed by the host immune system.

---

[11] I. van Die, R.D. Cummings. "Glycan Gimmickry by Parasitic Helminths: A Strategy for Modulating the Host Immune Response?" *Glycobiology* 20 no. 1 (2010):2–12.

SUMMARY

15.1 Characteristics of Infectious Disease

- In an **infection**, a microorganism enters a host and begins to multiply. Some infections cause **disease**, which is any deviation from the normal function or structure of the host.
- **Signs** of a disease are objective and are measured. **Symptoms** of a disease are subjective and are reported by the patient.
- Diseases can either be **noninfectious** (due to genetics and environment) or **infectious** (due to pathogens). Some infectious diseases are **communicable** (transmissible between individuals) or **contagious** (easily transmissible between individuals); others are **noncommunicable**, but may be contracted via contact with environmental reservoirs or animals (**zoonoses**).
- **Nosocomial diseases** are contracted in hospital settings, whereas **iatrogenic disease** are the direct result of a medical procedure.
- An **acute disease** is short in duration, whereas a **chronic disease** lasts for months or years. **Latent diseases** last for years, but are distinguished from chronic diseases by the lack of active replication during extended dormant periods.
- The periods of disease include the **incubation period**, the **prodromal period**, the **period of illness**, the **period of decline**, and the **period of convalescence**. These periods are marked by changes in the number of infectious agents and the severity of signs and symptoms.

15.2 How Pathogens Cause Disease

- **Koch's postulates** are used to determine whether a particular microorganism is a pathogen. **Molecular Koch's postulates** are used to determine what genes contribute to a pathogen's ability to cause disease.
- **Virulence**, the degree to which a pathogen can cause disease, can be quantified by calculating either the ID_{50} or LD_{50} of a pathogen on a given population.
- **Primary pathogens** are capable of causing pathological changes associated with disease in a healthy individual, whereas **opportunistic pathogens** can only cause disease when the individual is compromised by a break in protective barriers or immunosuppression.
- Infections and disease can be caused by pathogens in the environment or microbes in an individual's **resident microbiota**.
- Infections can be classified as **local**, **focal**, or **systemic** depending on the extent to which the pathogen spreads in the body.
- A **secondary infection** can sometimes occur after the host's defenses or normal microbiota are compromised by a **primary infection** or antibiotic treatment.
- Pathogens enter the body through **portals of entry** and leave through **portals of exit**. The stages of pathogenesis include **exposure**, **adhesion**, **invasion**, **infection**, and **transmission**.

15.3 Virulence Factors of Bacterial and Viral Pathogens

- **Virulence factors** contribute to a pathogen's ability to cause disease.
- **Exoenzymes** and **toxins** allow pathogens to invade host tissue and cause tissue damage. Exoenzymes are classified according to the macromolecule they target and exotoxins are classified based on their mechanism of action.
- Bacterial toxins include **endotoxin** and **exotoxins**. Endotoxin is the lipid A component of the LPS of the gram-negative cell envelope. Exotoxins are proteins secreted mainly by gram-positive bacteria, but also are secreted by gram-negative bacteria.
- Bacterial pathogens may evade the host immune response by producing **capsules** to avoid phagocytosis, surviving the intracellular environment of phagocytes, degrading antibodies, or through **antigenic variation**.
- Viral pathogens use adhesins for initiating infections and antigenic variation to avoid immune defenses.
- Influenza viruses use both **antigenic drift** and **antigenic shift** to avoid being recognized by the immune system.

15.4 Virulence Factors of Eukaryotic Pathogens

- Fungal and parasitic pathogens use pathogenic mechanisms and virulence factors that are similar to those of bacterial pathogens
- Fungi initiate infections through the interaction of adhesins with receptors on host cells. Some fungi produce toxins and exoenzymes involved in disease production and capsules that provide

protection of phagocytosis.
- Protozoa adhere to target cells through complex mechanisms and can cause cellular damage through release of cytopathic substances. Some protozoa avoid the immune system through antigenic variation and production of capsules.
- Helminthic worms are able to avoid the immune system by coating their exteriors with glycan molecules that make them look like host cells or by suppressing the immune system.

REVIEW QUESTIONS
Multiple Choice

1. Which of the following would be a sign of an infection?
   A. muscle aches
   B. headache
   C. fever
   D. nausea

2. Which of the following is an example of a noncommunicable infectious disease?
   A. infection with a respiratory virus
   B. food poisoning due to a preformed bacterial toxin in food
   C. skin infection acquired from a dog bite
   D. infection acquired from the stick of a contaminated needle

3. During an oral surgery, the surgeon nicked the patient's gum with a sharp instrument. This allowed *Streptococcus*, a bacterium normally present in the mouth, to gain access to the blood. As a result, the patient developed bacterial endocarditis (an infection of the heart). Which type of disease is this?
   A. iatrogenic
   B. nosocomial
   C. vectors
   D. zoonotic

4. Which period is the stage of disease during which the patient begins to present general signs and symptoms?
   A. convalescence
   B. incubation
   C. illness
   D. prodromal

5. A communicable disease that can be easily transmitted from person to person is which type of disease?
   A. contagious
   B. iatrogenic
   C. acute
   D. nosocomial

6. Which of the following is a pathogen that could not be identified by the original Koch's postulates?
   A. *Staphylococcus aureus*
   B. *Pseudomonas aeruginosa*
   C. Human immunodeficiency virus
   D. *Salmonella enterica* serovar Typhimurium

7. Pathogen A has an ID_{50} of 50 particles, pathogen B has an ID_{50} of 1,000 particles, and pathogen C has an ID_{50} of 1×10^6 particles. Which pathogen is most virulent?
   A. pathogen A
   B. pathogen B
   C. pathogen C

8. Which of the following choices lists the steps of pathogenesis in the correct order?
   A. invasion, infection, adhesion, exposure
   B. adhesion, exposure, infection, invasion
   C. exposure, adhesion, invasion, infection
   D. disease, infection, exposure, invasion

9. Which of the following would be a virulence factor of a pathogen?
   A. a surface protein allowing the pathogen to bind to host cells
   B. a secondary host the pathogen can infect
   C. a surface protein the host immune system recognizes
   D. the ability to form a provirus

10. You have recently identified a new toxin. It is produced by a gram-negative bacterium. It is composed mostly of protein, has high toxicity, and is not heat stable. You also discover that it targets liver cells. Based on these characteristics, how would you classify this toxin?
    A. superantigen
    B. endotoxin
    C. exotoxin
    D. leukocidin

11. Which of the following applies to hyaluronidase?
    A. It acts as a spreading factor.
    B. It promotes blood clotting.
    C. It is an example of an adhesin.
    D. It is produced by immune cells to target pathogens.

12. Phospholipases are enzymes that do which of the following?
    A. degrade antibodies
    B. promote pathogen spread through connective tissue.
    C. degrade nucleic acid to promote spread of pathogen
    D. degrade cell membranes to allow pathogens to escape phagosomes

13. Which of the following is a major virulence factor for the fungal pathogen *Cryptococcus*?
    A. hemolysin
    B. capsule
    C. collagenase
    D. fimbriae

14. Which of the following pathogens undergoes antigenic variation to avoid immune defenses?
    A. *Candida*
    B. *Cryptococcus*
    C. *Plasmodium*
    D. *Giardia*

Fill in the Blank

15. A difference between an acute disease and chronic disease is that chronic diseases have an extended period of _____.

16. A person steps on a rusty nail and develops tetanus. In this case, the person has acquired a(n) _____ disease.

17. A(n) _____ pathogen causes disease only when conditions are favorable for the microorganism because of transfer to an inappropriate body site or weakened immunity in an individual.

18. The concentration of pathogen needed to kill 50% of an infected group of test animals is the _____.

19. A(n) _____ infection is a small region of infection from which a pathogen may move to another part of the body to establish a second infection.

20. Cilia, fimbriae, and pili are all examples of structures used by microbes for _____.

21. The glycoprotein adhesion gp120 on HIV must interact with _____ on some immune cells as the first step in the process of infecting the cell.

22. Adhesins are usually located on _____ of the pathogen and are composed mainly of _____ and _____.

23. The Shiga and diphtheria toxins target _____ in host cells.

24. Antigenic _____ is the result of reassortment of genes responsible for the production of influenza virus spike proteins between different virus particles while in the same host, whereas antigenic _____ is the result of point mutations in the spike proteins.

25. *Candida* can invade tissue by producing the exoenzymes _____ and _____.

26. The larval form of *Schistosoma mansoni* uses a _____ to help it gain entry through intact skin.

Short Answer

27. Brian goes to the hospital after not feeling well for a week. He has a fever of 38 °C (100.4 °F) and complains of nausea and a constant migraine. Distinguish between the signs and symptoms of disease in Brian's case.

28. Describe the virulence factors associated with the fungal pathogen *Aspergillus*.

29. Explain how helminths evade the immune system.

Critical Thinking

30. Two periods of acute disease are the periods of illness and period of decline. (a) In what way are both of these periods similar? (b) In terms of quantity of pathogen, in what way are these periods different? (c) What initiates the period of decline?

31. In July 2015, a report (C. Owens. "*P. aeruginosa* survives in sinks 10 years after hospital outbreak." 2015. http://www.healio.com/infectious-disease/nosocomial-infections/news/online/%7B5afba909-56d9-48cc-a9b0-ffe4568161e8%7D/p-aeruginosa-survives-in-sinks-10-years-after-hospital-outbreak) was released indicating the gram-negative bacterium *Pseudomonas aeruginosa* was found on hospital sinks 10 years after the initial outbreak in a neonatal intensive care unit. *P. aeruginosa* usually causes localized ear and eye infections but can cause pneumonia or septicemia in vulnerable individuals like newborn babies. Explain how the current discovery of the presence of this reported *P. aeruginosa* could lead to a recurrence of nosocomial disease.

32. Diseases that involve biofilm-producing bacteria are of serious concern. They are not as easily treated compared with those involving free-floating (or planktonic) bacteria. Explain three reasons why biofilm formers are more pathogenic.

33. A microbiologist has identified a new gram-negative pathogen that causes liver disease in rats. She suspects that the bacterium's fimbriae are a virulence factor. Describe how molecular Koch's postulates could be used to test this hypothesis.

34. Acupuncture is a form of alternative medicine that is used for pain relief. Explain how acupuncture could facilitate exposure to pathogens.

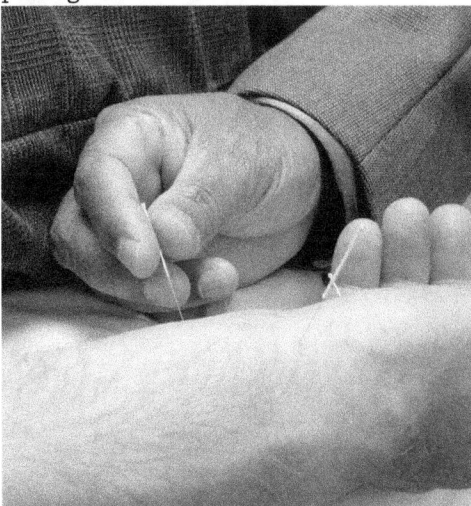

35. Two types of toxins are hemolysins and leukocidins. (a) How are these toxins similar? (b) How do they differ?

36. Imagine that a mutation in the gene encoding the cholera toxin was made. This mutation affects the A-subunit, preventing it from interacting with any host protein. (a) Would the toxin be able to enter into the intestinal epithelial cell? (b) Would the toxin be able to cause diarrhea?

CHAPTER 16
Disease and Epidemiology

Figure 16.1 Signs like this may seem self-explanatory today, but a few short centuries ago, people lacked a basic understanding of how diseases spread. Microbiology has greatly contributed to the field of epidemiology, which focuses on containing the spread of disease. (credit: modification of work by Tony Webster)

Chapter Outline

16.1 The Language of Epidemiologists

16.2 Tracking Infectious Diseases

16.3 Modes of Disease Transmission

16.4 Global Public Health

INTRODUCTION In the United States and other developed nations, public health is a key function of government. A healthy citizenry is more productive, content, and prosperous; high rates of death and disease, on the other hand, can severely hamper economic productivity and foster social and political instability. The burden of disease makes it difficult for citizens to work consistently, maintain employment, and accumulate wealth to better their lives and support a growing economy.

In this chapter, we will explore the intersections between microbiology and epidemiology, the science that underlies public health. Epidemiology studies how disease originates and spreads throughout a population, with the goal of preventing outbreaks and containing them when they do occur. Over the past two centuries, discoveries in epidemiology have led to public health policies that have transformed life in developed nations, leading to the eradication (or near eradication) of many diseases that were once causes of great human suffering and premature death. However, the work of epidemiologists is far from finished. Numerous diseases continue to plague humanity, and new diseases are always emerging. Moreover, in the developing world, lack of infrastructure continues to pose many challenges to efforts to contain disease.

16.1 The Language of Epidemiologists

Learning Objectives

By the end of this section, you will be able to:
- Explain the difference between prevalence and incidence of disease
- Distinguish the characteristics of sporadic, endemic, epidemic, and pandemic diseases
- Explain the use of Koch's postulates and their modifications to determine the etiology of disease
- Explain the relationship between epidemiology and public health

---

Clinical Focus

Part 1

In late November and early December, a hospital in western Florida started to see a spike in the number of cases of acute gastroenteritis-like symptoms. Patients began arriving at the emergency department complaining of excessive bouts of emesis (vomiting) and diarrhea (with no blood in the stool). They also complained of abdominal pain and cramping, and most were severely dehydrated. Alarmed by the number of cases, hospital staff made some calls and learned that other regional hospitals were also seeing 10 to 20 similar cases per day.

- What are some possible causes of this outbreak?
- In what ways could these cases be linked, and how could any suspected links be confirmed?

Jump to the next Clinical Focus box.

The field of **epidemiology** concerns the geographical distribution and timing of infectious disease occurrences and how they are transmitted and maintained in nature, with the goal of recognizing and controlling outbreaks. The science of epidemiology includes **etiology** (the study of the causes of disease) and investigation of disease transmission (mechanisms by which a disease is spread).

Analyzing Disease in a Population

Epidemiological analyses are always carried out with reference to a population, which is the group of individuals that are at risk for the disease or condition. The population can be defined geographically, but if only a portion of the individuals in that area are susceptible, additional criteria may be required. Susceptible individuals may be defined by particular behaviors, such as intravenous drug use, owning particular pets, or membership in an institution, such as a college. Being able to define the population is important because most measures of interest in epidemiology are made with reference to the size of the population.

The state of being diseased is called **morbidity**. Morbidity in a population can be expressed in a few different ways. Morbidity or total morbidity is expressed in numbers of individuals without reference to the size of the population. The **morbidity rate** can be expressed as the number of diseased individuals out of a standard number of individuals in the population, such as 100,000, or as a percent of the population.

There are two aspects of morbidity that are relevant to an epidemiologist: a disease's **prevalence** and its **incidence**. Prevalence is the number, or proportion, of individuals with a particular illness in a given population at a point in time. For example, the Centers for Disease Control and Prevention (CDC) estimated that in 2012, there were about 1.2 million people 13 years and older with an active human immunodeficiency virus (HIV) infection. Expressed as a proportion, or rate, this is a prevalence of 467 infected persons per 100,000 in the population.[1] On the other hand, incidence is the number or proportion of *new* cases in a period of time. For the same year and population, the CDC estimates that there were 43,165 newly diagnosed cases of HIV infection, which is an incidence of 13.7 new cases per 100,000 in the population.[2] The relationship between incidence and prevalence can be seen in Figure 16.2. For a chronic disease like HIV infection, prevalence will generally be higher than incidence because it represents the cumulative number of new cases over many years minus the number of cases that are no longer active (e.g., because the patient died or was cured).

In addition to morbidity rates, the incidence and prevalence of **mortality** (death) may also be reported. A mortality rate can be expressed as the percentage of the population that has died from a disease or as the number of deaths per 100,000 persons (or other suitable standard number).

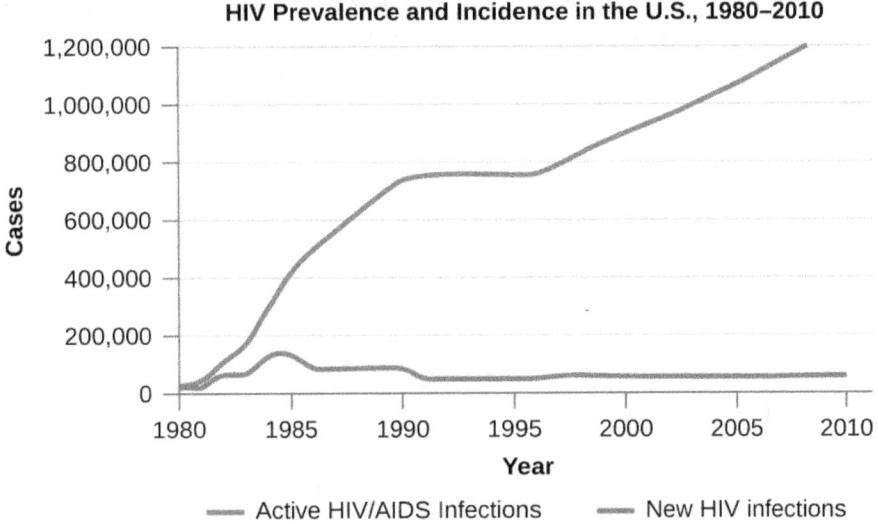

Figure 16.2 This graph compares the incidence of HIV (the number of new cases reported each year) with the prevalence (the total number of cases each year). Prevalence and incidence can also be expressed as a rate or proportion for a given population.

CHECK YOUR UNDERSTANDING

- Explain the difference between incidence and prevalence.
- Describe how morbidity and mortality rates are expressed.

Patterns of Incidence

Diseases that are seen only occasionally, and usually without geographic concentration, are called **sporadic diseases**. Examples of sporadic diseases include tetanus, rabies, and plague. In the United States, *Clostridium tetani*, the bacterium that causes tetanus, is ubiquitous in the soil environment, but incidences of infection occur only rarely and in scattered locations because most individuals are vaccinated, clean wounds appropriately, or are only rarely in a situation that would cause infection.[3] Likewise in the United States there are a few scattered cases of plague each year, usually contracted from rodents in rural areas in the western states.[4]

---

1 H. Irene Hall, Qian An, Tian Tang, Ruiguang Song, Mi Chen, Timothy Green, and Jian Kang. "Prevalence of Diagnosed and Undiagnosed HIV Infection—United States, 2008–2012." *Morbidity and Mortality Weekly Report* 64, no. 24 (2015): 657–662.

2 Centers for Disease Control and Prevention. "Diagnoses of HIV Infection in the United States and Dependent Areas, 2014." *HIV Surveillance Report* 26 (2015).

Diseases that are constantly present (often at a low level) in a population within a particular geographic region are called **endemic diseases**. For example, malaria is endemic to some regions of Brazil, but is not endemic to the United States.

Diseases for which a larger than expected number of cases occurs in a short time within a geographic region are called **epidemic diseases**. Influenza is a good example of a commonly epidemic disease. Incidence patterns of influenza tend to rise each winter in the northern hemisphere. These seasonal increases are expected, so it would not be accurate to say that influenza is epidemic every winter; however, some winters have an usually large number of seasonal influenza cases in particular regions, and such situations would qualify as epidemics (Figure 16.3 and Figure 16.4).

An epidemic disease signals the breakdown of an equilibrium in disease frequency, often resulting from some change in environmental conditions or in the population. In the case of influenza, the disruption can be due to antigenic shift or drift (see Virulence Factors of Bacterial and Viral Pathogens), which allows influenza virus strains to circumvent the acquired immunity of their human hosts.

An epidemic that occurs on a worldwide scale is called a **pandemic disease**. For example, HIV/AIDS is a pandemic disease and novel influenza virus strains often become pandemic.

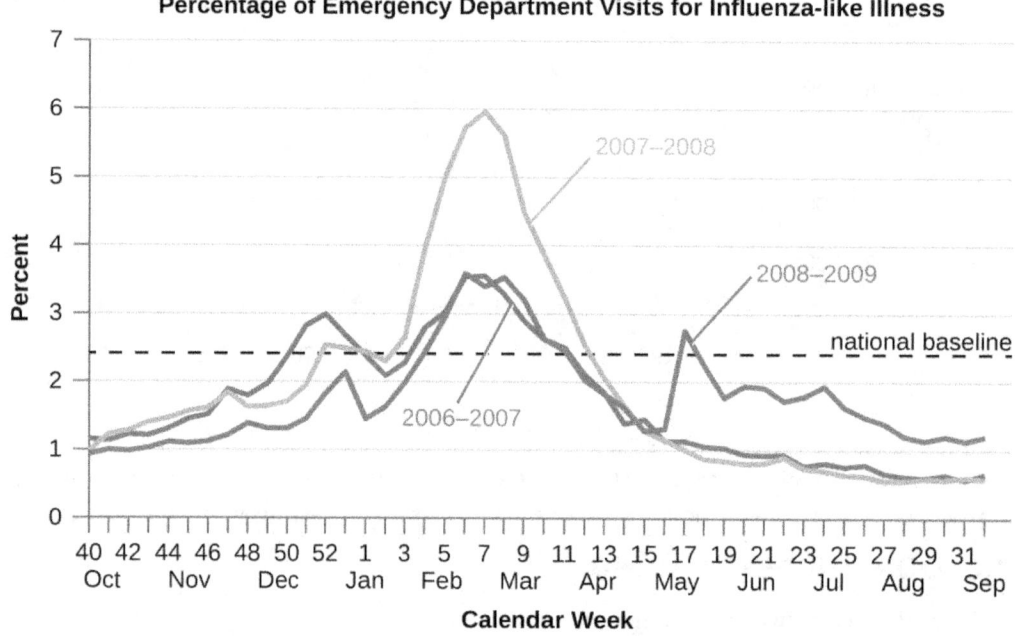

Figure 16.3 The 2007–2008 influenza season in the United States saw much higher than normal numbers of visits to emergency departments for influenza-like symptoms as compared to the previous and the following years. (credit: modification of work by Centers for Disease Control and Prevention)

---

3 Centers for Disease Control and Prevention. "Tetanus Surveillance—United States, 2001–2008." *Morbidity and Mortality Weekly Report* 60, no. 12 (2011): 365.

4 Centers for Disease Control and Prevention. "Plague in the United States." 2015. http://www.cdc.gov/plague/maps. Accessed June 1, 2016.

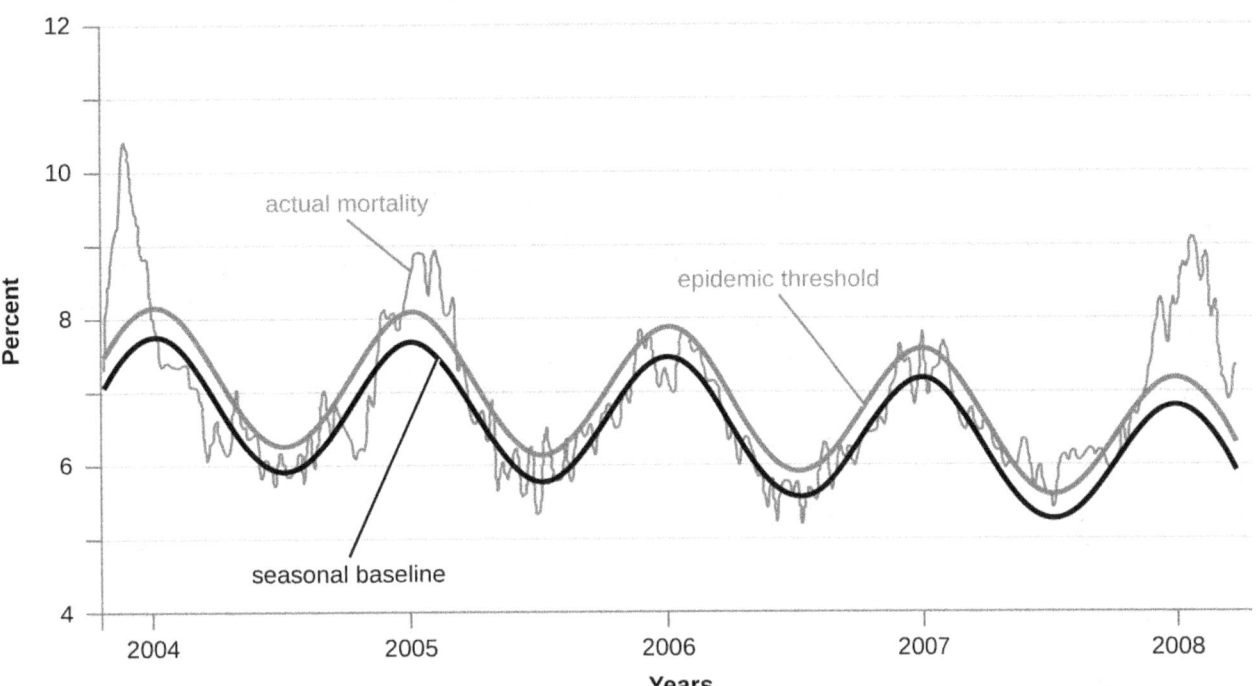

Figure 16.4 The seasonal epidemic threshold (blue curve) is set by the CDC-based data from the previous five years. When actual mortality rates exceed this threshold, a disease is considered to be epidemic. As this graph shows, pneumonia- and influenza-related mortality saw pronounced epidemics during the winters of 2003–2004, 2005, and 2008. (credit: modification of work by Centers for Disease Control and Prevention)

CHECK YOUR UNDERSTANDING

- Explain the difference between sporadic and endemic disease.
- Explain the difference between endemic and epidemic disease.

Clinical Focus

Part 2

Hospital physicians suspected that some type of food poisoning was to blame for the sudden post-Thanksgiving outbreak of gastroenteritis in western Florida. Over a two-week period, 254 cases were observed, but by the end of the first week of December, the epidemic ceased just as quickly as it had started. Suspecting a link between the cases based on the localized nature of the outbreak, hospitals handed over their medical records to the regional public health office for study.

Laboratory testing of stool samples had indicated that the infections were caused by *Salmonella* bacteria. Patients ranged from children as young as three to seniors in their late eighties. Cases were nearly evenly split between males and females. Across the region, there had been three confirmed deaths in the outbreak, all due to severe dehydration. In each of the fatal cases, the patients had not sought medical care until their symptoms were severe; also, all of the deceased had preexisting medical conditions such as congestive heart failure, diabetes, or high blood pressure.

After reviewing the medical records, epidemiologists with the public health office decided to conduct interviews with a randomly selected sample of patients.

- What conclusions, if any, can be drawn from the medical records?
- What would epidemiologists hope to learn by interviewing patients? What kinds of questions might

they ask?

Jump to the next Clinical Focus box. Go back to the previous Clinical Focus box.

Etiology

When studying an epidemic, an epidemiologist's first task is to determine the cause of the disease, called the **etiologic agent** or **causative agent**. Connecting a disease to a specific pathogen can be challenging because of the extra effort typically required to demonstrate direct causation as opposed to a simple association. It is not enough to observe an association between a disease and a suspected pathogen; controlled experiments are needed to eliminate other possible causes. In addition, pathogens are typically difficult to detect when there is no immediate clue as to what is causing the outbreak. Signs and symptoms of disease are also commonly nonspecific, meaning that many different agents can give rise to the same set of signs and symptoms. This complicates diagnosis even when a causative agent is familiar to scientists.

Robert Koch was the first scientist to specifically demonstrate the causative agent of a disease (anthrax) in the late 1800s. Koch developed four criteria, now known as Koch's postulates, which had to be met in order to positively link a disease with a pathogenic microbe. Without Koch's postulates, the Golden Age of Microbiology would not have occurred. Between 1876 and 1905, many common diseases were linked with their etiologic agents, including cholera, diphtheria, gonorrhea, meningitis, plague, syphilis, tetanus, and tuberculosis. Today, we use the molecular Koch's postulates, a variation of Koch's original postulates that can be used to establish a link between the disease state and virulence traits unique to a pathogenic strain of a microbe. Koch's original postulates and molecular Koch's postulates were described in more detail in How Pathogens Cause Disease.

✓ CHECK YOUR UNDERSTANDING

- List some challenges to determining the causative agent of a disease outbreak.

The Role of Public Health Organizations

The main national public health agency in the United States is the **Centers for Disease Control and Prevention (CDC)**, an agency of the Department of Health and Human Services. The CDC is charged with protecting the public from disease and injury. One way that the CDC carries out this mission is by overseeing the National Notifiable Disease Surveillance System (NNDSS) in cooperation with regional, state, and territorial public health departments. The NNDSS monitors diseases considered to be of public health importance on a national scale. Such diseases are called **notifiable diseases** or **reportable diseases** because all cases must be reported to the CDC. A physician treating a patient with a notifiable disease is legally required to submit a report on the case. Notifiable diseases include HIV infection, measles, West Nile virus infections, and many others. Some states have their own lists of notifiable diseases that include diseases beyond those on the CDC's list.

Notifiable diseases are tracked by epidemiological studies and the data is used to inform health-care providers and the public about possible risks. The CDC publishes the **Morbidity and Mortality Weekly Report (MMWR)**, which provides physicians and health-care workers with updates on public health issues and the latest data pertaining to notifiable diseases. Table 16.1 is an example of the kind of data contained in the *MMWR*.

Incidence of Four Notifiable Diseases in the United States, Week Ending January 2, 2016

Disease	Current Week (Jan 2, 2016)	Median of Previous 52 Weeks	Maximum of Previous 52 Weeks	Cumulative Cases 2015
Campylobacteriosis	406	869	1,385	46,618

Incidence of Four Notifiable Diseases in the United States, Week Ending January 2, 2016

Disease	Current Week (Jan 2, 2016)	Median of Previous 52 Weeks	Maximum of Previous 52 Weeks	Cumulative Cases 2015
Chlamydia trachomatis infection	11,024	28,562	31,089	1,425,303
Giardiasis	115	230	335	11,870
Gonorrhea	3,207	7,155	8,283	369,926

Table 16.1

LINK TO LEARNING

The current *Morbidity and Mortality Weekly Report* (https://openstax.org/l/22mortweekrep) is available online.

CHECK YOUR UNDERSTANDING

- Describe how health agencies obtain data about the incidence of diseases of public health importance.

16.2 Tracking Infectious Diseases

Learning Objectives

By the end of this section, you will be able to:
- Explain the research approaches used by the pioneers of epidemiology
- Explain how descriptive, analytical, and experimental epidemiological studies go about determining the cause of morbidity and mortality

Epidemiology has its roots in the work of physicians who looked for patterns in disease occurrence as a way to understand how to prevent it. The idea that disease could be transmitted was an important precursor to making sense of some of the patterns. In 1546, Girolamo Fracastoro first proposed the germ theory of disease in his essay *De Contagione et Contagiosis Morbis*, but this theory remained in competition with other theories, such as the miasma hypothesis, for many years (see What Our Ancestors Knew). Uncertainty about the cause of disease was not an absolute barrier to obtaining useful knowledge from patterns of disease. Some important researchers, such as Florence Nightingale, subscribed to the miasma hypothesis. The transition to acceptance of the germ theory during the 19th century provided a solid mechanistic grounding to the study of disease patterns. The studies of 19th century physicians and researchers such as John Snow, Florence Nightingale, Ignaz Semmelweis, Joseph Lister, Robert Koch, Louis Pasteur, and others sowed the seeds of modern epidemiology.

Pioneers of Epidemiology

John Snow (Figure 16.5) was a British physician known as the father of epidemiology for determining the source of the 1854 Broad Street cholera epidemic in London. Based on observations he had made during an earlier cholera outbreak (1848–1849), Snow proposed that cholera was spread through a fecal-oral route of transmission and that a microbe was the infectious agent. He investigated the 1854 cholera epidemic in two ways. First, suspecting that contaminated water was the source of the epidemic, Snow identified the source of water for those infected. He found a high frequency of cholera cases among individuals who obtained their water from the River Thames downstream from London. This water contained the refuse and sewage from London and settlements upstream. He also noted that brewery workers did not contract cholera and on investigation found the owners provided the workers with beer to drink and stated that they likely did not drink water.[5] Second, he also painstakingly mapped the incidence of cholera and found a high frequency among those individuals using a particular water pump located on Broad Street. In response to Snow's advice, local officials removed the pump's handle,[6] resulting in the containment of the Broad Street cholera epidemic.

Snow's work represents an early epidemiological study and it resulted in the first known public health response to an epidemic. Snow's meticulous case-tracking methods are now common practice in studying disease outbreaks and in associating new diseases with their causes. His work further shed light on unsanitary sewage practices and the effects of waste dumping in the Thames. Additionally, his work supported the germ theory of disease, which argued disease could be transmitted through contaminated items, including water contaminated with fecal matter.

Snow's work illustrated what is referred to today as a **common source spread** of infectious disease, in which there is a single source for all of the individuals infected. In this case, the single source was the contaminated well below the Broad Street pump. Types of common source spread include point source spread, continuous common source spread, and intermittent common source spread. In **point source spread** of infectious disease, the common source operates for a short time period—less than the incubation period of the pathogen. An example of point source spread is a single contaminated potato salad at a group picnic. In **continuous common source spread**, the infection occurs for an extended period of time, longer than the incubation period. An example of continuous common source spread would be the source of London water taken downstream of the city, which was continuously contaminated with sewage from upstream. Finally, with **intermittent common source spread**, infections occur for a period, stop, and then begin again. This might be seen in infections from a well that was contaminated only after large rainfalls and that cleared itself of contamination after a short period.

In contrast to common source spread, **propagated spread** occurs through direct or indirect person-to-person contact. With propagated spread, there is no single source for infection; each infected individual becomes a source for one or more subsequent infections. With propagated spread, unless the spread is stopped immediately, infections occur for longer than the incubation period. Although point sources often lead to large-scale but localized outbreaks of short duration, propagated spread typically results in longer duration outbreaks that can vary from small to large, depending on the population and the disease (Figure 16.6). In addition, because of person-to-person transmission, propagated spread cannot be easily stopped at a single source like point source spread.

---

5 John Snow. *On the Mode of Communication of Cholera. Second edition, Much Enlarged.* John Churchill, 1855.

6 John Snow. "The Cholera near Golden-Wquare, and at Deptford." *Medical Times and Gazette* 9 (1854): 321–322. http://www.ph.ucla.edu/epi/snow/choleragoldensquare.html.

Figure 16.5 (a) John Snow (1813–1858), British physician and father of epidemiology. (b) Snow's detailed mapping of cholera incidence led to the discovery of the contaminated water pump on Broad street (red square) responsible for the 1854 cholera epidemic. (credit a: modification of work by "Rsabbatini"/Wikimedia Commons)

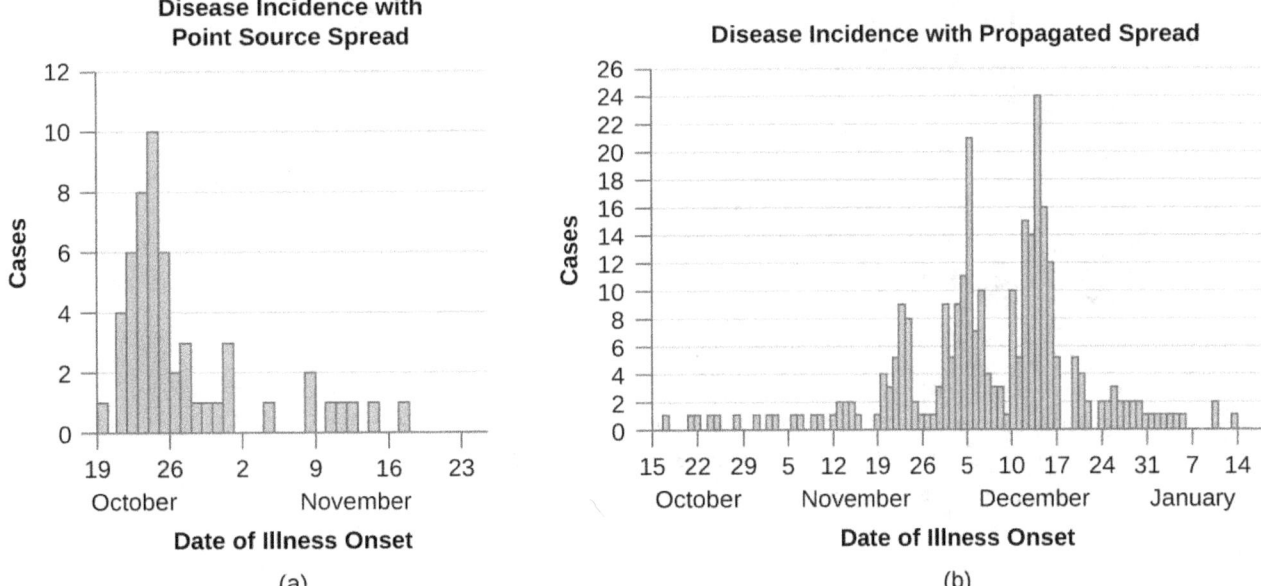

Figure 16.6 (a) Outbreaks that can be attributed to point source spread often have a short duration. (b) Outbreaks attributed to propagated spread can have a more extended duration. (credit a, b: modification of work by Centers for Disease Control and Prevention)

Florence Nightingale's work is another example of an early epidemiological study. In 1854, Nightingale was part of a contingent of nurses dispatched by the British military to care for wounded soldiers during the Crimean War. Nightingale kept meticulous records regarding the causes of illness and death during the war. Her recordkeeping was a fundamental task of what would later become the science of epidemiology. Her analysis of the data she collected was published in 1858. In this book, she presented monthly frequency data on causes of death in a wedge chart histogram (Figure 16.7). This graphical presentation of data, unusual at the time, powerfully illustrated that the vast majority of casualties during the war occurred not due to wounds sustained in action but to what Nightingale deemed preventable infectious diseases. Often these diseases occurred because of poor sanitation and lack of access to hospital facilities. Nightingale's findings led to many reforms in the British military's system of medical care.

Joseph Lister provided early epidemiological evidence leading to good public health practices in clinics and hospitals. These settings were notorious in the mid-1800s for fatal infections of surgical wounds at a time when the germ theory of disease was not yet widely accepted (see Foundations of Modern Cell Theory). Most physicians did not wash their hands between patient visits or clean and sterilize their surgical tools. Lister, however, discovered the disinfecting properties of carbolic acid, also known as phenol (see Using Chemicals to Control Microorganisms). He introduced several disinfection protocols that dramatically lowered post-surgical infection rates.[7] He demanded that surgeons who worked for him use a 5% carbolic acid solution to clean their surgical tools between patients, and even went so far as to spray the solution onto bandages and over the surgical site during operations (Figure 16.8). He also took precautions not to introduce sources of infection from his skin or clothing by removing his coat, rolling up his sleeves, and washing his hands in a dilute solution of carbolic acid before and during the surgery.

 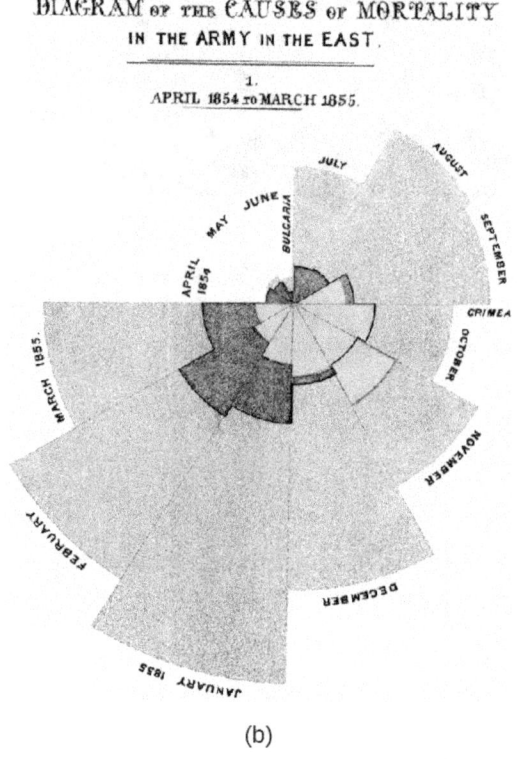

(a) (b)

Figure 16.7 (a) Florence Nightingale reported on the data she collected as a nurse in the Crimean War. (b) Nightingale's diagram shows the number of fatalities in soldiers by month of the conflict from various causes. The total number dead in a particular month is equal to the area of the wedge for that month. The colored sections of the wedge represent different causes of death: wounds (pink), preventable infectious diseases (gray), and all other causes (brown).

---

7 O.M. Lidwell. "Joseph Lister and Infection from the Air." *Epidemiology and Infection* 99 (1987): 569–578. http://www.ncbi.nlm.nih.gov/pmc/articles/PMC2249236/pdf/epidinfect00006-0004.pdf.

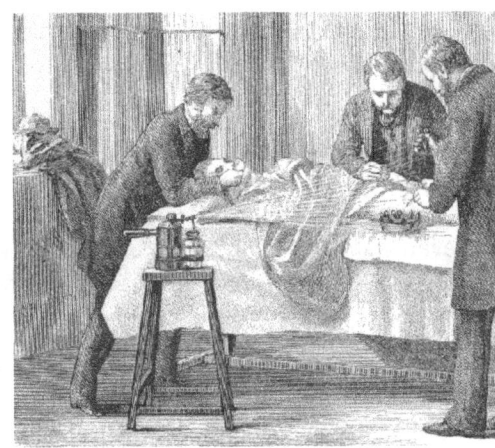

Figure 16.8 Joseph Lister initiated the use of a carbolic acid (phenol) during surgeries. This illustration of a surgery shows a pressurized canister of carbolic acid being sprayed over the surgical site.

LINK TO LEARNING

John Snow's own account of his work (https://openstax.org/l/22JohnSnowacco) has additional links and information.

This CDC resource (https://openstax.org/l/22CDCpointsourc) further breaks down the pattern expected from a point-source outbreak.

Learn more about Nightingale's wedge chart (https://openstax.org/l/22nightwedgecha) here.

CHECK YOUR UNDERSTANDING

- Explain the difference between common source spread and propagated spread of disease.
- Describe how the observations of John Snow, Florence Nightingale, and Joseph Lister led to improvements in public health.

Types of Epidemiological Studies

Today, epidemiologists make use of study designs, the manner in which data are gathered to test a hypothesis, similar to those of researchers studying other phenomena that occur in populations. These approaches can be divided into observational studies (in which subjects are not manipulated) and experimental studies (in which subjects are manipulated). Collectively, these studies give modern-day epidemiologists multiple tools for exploring the connections between infectious diseases and the populations of susceptible individuals they might infect.

Observational Studies

In an **observational study**, data are gathered from study participants through measurements (such as physiological variables like white blood cell count), or answers to questions in interviews (such as recent travel or exercise frequency). The subjects in an observational study are typically chosen at random from a population of affected or unaffected individuals. However, the subjects in an observational study are in no way manipulated by the researcher. Observational studies are typically easier to carry out than experimental studies, and in certain situations they may be the only studies possible for ethical reasons.

Observational studies are only able to measure associations between disease occurrence and possible causative agents; they do not necessarily prove a causal relationship. For example, suppose a study finds an association between heavy coffee drinking and lower incidence of skin cancer. This might suggest that coffee prevents skin cancer, but there may be another unmeasured factor involved, such as the amount of sun exposure the participants receive. If it turns out that coffee drinkers work more in offices and spend less time outside in the sun than those who drink less coffee, then it may be possible that the lower rate of skin cancer is due to less sun exposure, not to coffee consumption. The observational study cannot distinguish between these two potential causes.

There are several useful approaches in observational studies. These include methods classified as descriptive epidemiology and analytical epidemiology. **Descriptive epidemiology** gathers information about a disease outbreak, the affected individuals, and how the disease has spread over time in an exploratory stage of study. This type of study will involve interviews with patients, their contacts, and their family members; examination of samples and medical records; and even histories of food and beverages consumed. Such a study might be conducted while the outbreak is still occurring. Descriptive studies might form the basis for developing a hypothesis of causation that could be tested by more rigorous observational and experimental studies.

Analytical epidemiology employs carefully selected groups of individuals in an attempt to more convincingly evaluate hypotheses about potential causes for a disease outbreak. The selection of cases is generally made at random, so the results are not biased because of some common characteristic of the study participants. Analytical studies may gather their data by going back in time (retrospective studies), or as events unfold forward in time (prospective studies).

Retrospective studies gather data from the past on present-day cases. Data can include things like the medical history, age, gender, or occupational history of the affected individuals. This type of study examines associations between factors chosen or available to the researcher and disease occurrence.

Prospective studies follow individuals and monitor their disease state during the course of the study. Data on the characteristics of the study subjects and their environments are gathered at the beginning and during the study so that subjects who become ill may be compared with those who do not. Again, the researchers can look for associations between the disease state and variables that were measured during the study to shed light on possible causes.

Analytical studies incorporate groups into their designs to assist in teasing out associations with disease. Approaches to group-based analytical studies include cohort studies, case-control studies, and cross-sectional studies. The **cohort method** examines groups of individuals (called cohorts) who share a particular characteristic. For example, a cohort might consist of individuals born in the same year and the same place; or it might consist of people who practice or avoid a particular behavior, e.g., smokers or nonsmokers. In a cohort study, cohorts can be followed prospectively or studied retrospectively. If only a single cohort is followed, then the affected individuals are compared with the unaffected individuals in the same group. Disease outcomes are recorded and analyzed to try to identify correlations between characteristics of individuals in the cohort and disease incidence. Cohort studies are a useful way to determine the causes of a condition without violating the ethical prohibition of exposing subjects to a risk factor. Cohorts are typically identified and defined based on suspected risk factors to which individuals have already been exposed through their own choices or circumstances.

Case-control studies are typically retrospective and compare a group of individuals with a disease to a similar group of individuals without the disease. Case-control studies are far more efficient than cohort studies because researchers can deliberately select subjects who are already affected with the disease as opposed to waiting to see which subjects from a random sample will develop a disease.

A **cross-sectional study** analyzes randomly selected individuals in a population and compares individuals affected by a disease or condition to those unaffected at a single point in time. Subjects are compared to look for associations between certain measurable variables and the disease or condition. Cross-sectional studies are also used to determine the prevalence of a condition.

Experimental Studies

Experimental epidemiology uses laboratory or clinical studies in which the investigator manipulates the study subjects to study the connections between diseases and potential causative agents or to assess treatments. Examples of treatments might be the administration of a drug, the inclusion or exclusion of different dietary items, physical exercise, or a particular surgical procedure. Animals or humans are used as test subjects. Because **experimental studies** involve manipulation of subjects, they are typically more difficult and sometimes impossible for ethical reasons.

Koch's postulates require experimental interventions to determine the causative agent for a disease. Unlike observational studies, experimental studies can provide strong evidence supporting cause because other factors are typically held constant when the researcher manipulates the subject. The outcomes for one group receiving the treatment are compared to outcomes for a group that does not receive the treatment but is treated the same in every other way. For example, one group might receive a regimen of a drug administered as a pill, while the untreated group receives a placebo (a pill that looks the same but has no active ingredient). Both groups are treated as similarly as possible except for the administration of the drug. Because other variables are held constant in both the treated and the untreated groups, the researcher is more certain that any change in the treated group is a result of the specific manipulation.

Experimental studies provide the strongest evidence for the etiology of disease, but they must also be designed carefully to eliminate subtle effects of bias. Typically, experimental studies with humans are conducted as double-blind studies, meaning neither the subjects nor the researchers know who is a treatment case and who is not. This design removes a well-known cause of bias in research called the placebo effect, in which knowledge of the treatment by either the subject or the researcher can influence the outcomes.

CHECK YOUR UNDERSTANDING

- Describe the advantages and disadvantages of observational studies and experimental studies.
- Explain the ways that groups of subjects can be selected for analytical studies.

Clinical Focus

Part 3

Since laboratory tests had confirmed *Salmonella*, a common foodborne pathogen, as the etiologic agent, epidemiologists suspected that the outbreak was caused by contamination at a food processing facility serving the region. Interviews with patients focused on food consumption during and after the Thanksgiving holiday, corresponding with the timing of the outbreak. During the interviews, patients were asked to list items consumed at holiday gatherings and describe how widely each item was consumed among family members and relatives. They were also asked about the sources of food items (e.g., brand, location of purchase, date of purchase). By asking such questions, health officials hoped to identify patterns that would lead back to the source of the outbreak.

Analysis of the interview responses eventually linked almost all of the cases to consumption of a holiday dish known as the turducken—a chicken stuffed inside a duck stuffed inside a turkey. Turducken is a dish not generally consumed year-round, which would explain the spike in cases just after the Thanksgiving holiday. Additional analysis revealed that the turduckens consumed by the affected patients were purchased already stuffed and ready to be cooked. Moreover, the pre-stuffed turduckens were all sold at the same regional grocery chain under two different brand names. Upon further investigation, officials traced both brands to a single processing plant that supplied stores throughout the Florida panhandle.

- Is this an example of common source spread or propagated spread?
- What next steps would the public health office likely take after identifying the source of the outbreak?

Jump to the next Clinical Focus box. Go back to the previous Clinical Focus box.

16.3 Modes of Disease Transmission

Learning Objectives

By the end of this section, you will be able to:
- Describe the different types of disease reservoirs
- Compare contact, vector, and vehicle modes of transmission
- Identify important disease vectors
- Explain the prevalence of nosocomial infections

Understanding how infectious pathogens spread is critical to preventing infectious disease. Many pathogens require a living host to survive, while others may be able to persist in a dormant state outside of a living host. But having infected one host, all pathogens must also have a mechanism of transfer from one host to another or they will die when their host dies. Pathogens often have elaborate adaptations to exploit host biology, behavior, and ecology to live in and move between hosts. Hosts have evolved defenses against pathogens, but because their rates of evolution are typically slower than their pathogens (because their generation times are longer), hosts are usually at an evolutionary disadvantage. This section will explore where pathogens survive—both inside and outside hosts—and some of the many ways they move from one host to another.

Reservoirs and Carriers

For pathogens to persist over long periods of time they require **reservoirs** where they normally reside. Reservoirs can be living organisms or nonliving sites. Nonliving reservoirs can include soil and water in the environment. These may naturally harbor the organism because it may grow in that environment. These environments may also become contaminated with pathogens in human feces, pathogens shed by intermediate hosts, or pathogens contained in the remains of intermediate hosts.

Pathogens may have mechanisms of dormancy or resilience that allow them to survive (but typically not to reproduce) for varying periods of time in nonliving environments. For example, *Clostridium tetani* survives in the soil and in the presence of oxygen as a resistant endospore. Although many viruses are soon destroyed once in contact with air, water, or other non-physiological conditions, certain types are capable of persisting outside of a living cell for varying amounts of time. For example, a study that looked at the ability of influenza viruses to infect a cell culture after varying amounts of time on a banknote showed survival times from 48 hours to 17 days, depending on how they were deposited on the banknote.[8] On the other hand, cold-causing rhinoviruses are somewhat fragile, typically surviving less than a day outside of physiological fluids.

A human acting as a reservoir of a pathogen may or may not be capable of transmitting the pathogen, depending on the stage of infection and the pathogen. To help prevent the spread of disease among school children, the CDC has developed guidelines based on the risk of transmission during the course of the disease. For example, children with chickenpox are considered contagious for five days from the start of the rash, whereas children with most gastrointestinal illnesses should be kept home for 24 hours after the symptoms disappear.

An individual capable of transmitting a pathogen without displaying symptoms is referred to as a carrier. A **passive carrier** is contaminated with the pathogen and can mechanically transmit it to another host; however, a passive carrier is not infected. For example, a health-care professional who fails to wash his hands after seeing a patient harboring an infectious agent could become a passive carrier, transmitting the pathogen to another patient who becomes infected.

By contrast, an **active carrier** is an infected individual who can transmit the disease to others. An active carrier may or may not exhibit signs or symptoms of infection. For example, active carriers may transmit the disease during the incubation period (before they show signs and symptoms) or the period of convalescence (after symptoms have subsided). Active carriers who do not present signs or symptoms of disease despite infection are called **asymptomatic carriers**. Pathogens such as hepatitis B virus, herpes simplex virus, and HIV are frequently transmitted by asymptomatic carriers. Mary Mallon, better known as Typhoid Mary, is a famous historical example of an asymptomatic carrier. An Irish immigrant, Mallon worked as a cook for households in and around New York City between 1900 and 1915. In each household, the residents developed typhoid fever (caused by *Salmonella typhi*) a few weeks after Mallon started working. Later investigations determined that Mallon was responsible for at least 122 cases of typhoid fever, five of which were fatal.[9] See Eye on Ethics: Typhoid Mary for more about the Mallon case.

---

8 Yves Thomas, Guido Vogel, Werner Wunderli, Patricia Suter, Mark Witschi, Daniel Koch, Caroline Tapparel, and Laurent Kaiser. "Survival of Influenza Virus on Banknotes." *Applied and Environmental Microbiology* 74, no. 10 (2008): 3002–3007.

9 Filio Marineli, Gregory Tsoucalas, Marianna Karamanou, and George Androutsos. "Mary Mallon (1869–1938) and the History of Typhoid Fever." *Annals of Gastroenterology* 26 (2013): 132–134. http://www.ncbi.nlm.nih.gov/pmc/articles/PMC3959940/pdf/AnnGastroenterol-26-132.pdf.

A pathogen may have more than one living reservoir. In zoonotic diseases, animals act as reservoirs of human disease and transmit the infectious agent to humans through direct or indirect contact. In some cases, the disease also affects the animal, but in other cases the animal is asymptomatic.

In parasitic infections, the parasite's preferred host is called the **definitive host**. In parasites with complex life cycles, the definitive host is the host in which the parasite reaches sexual maturity. Some parasites may also infect one or more **intermediate hosts** in which the parasite goes through several immature life cycle stages or reproduces asexually.

LINK TO LEARNING

George Soper, the sanitary engineer who traced the typhoid outbreak to Mary Mallon, gives an account (https://openstax.org/l/22geosopcurtyp) of his investigation, an example of descriptive epidemiology, in "The Curious Career of Typhoid Mary."

CHECK YOUR UNDERSTANDING

- List some nonliving reservoirs for pathogens.
- Explain the difference between a passive carrier and an active carrier.

Transmission

Regardless of the reservoir, transmission must occur for an infection to spread. First, transmission from the reservoir to the individual must occur. Then, the individual must transmit the infectious agent to other susceptible individuals, either directly or indirectly. Pathogenic microorganisms employ diverse transmission mechanisms.

Contact Transmission

Contact transmission includes direct contact or indirect contact. Person-to-person transmission is a form of **direct contact transmission**. Here the agent is transmitted by physical contact between two individuals (Figure 16.9) through actions such as touching, kissing, sexual intercourse, or droplet sprays. Direct contact can be categorized as vertical, horizontal, or droplet transmission. **Vertical direct contact transmission** occurs when pathogens are transmitted from mother to child during pregnancy, birth, or breastfeeding. Other kinds of direct contact transmission are called **horizontal direct contact transmission**. Often, contact between mucous membranes is required for entry of the pathogen into the new host, although skin-to-skin contact can lead to mucous membrane contact if the new host subsequently touches a mucous membrane. Contact transmission may also be site-specific; for example, some diseases can be transmitted by sexual contact but not by other forms of contact.

When an individual coughs or sneezes, small droplets of mucus that may contain pathogens are ejected. This leads to direct **droplet transmission**, which refers to droplet transmission of a pathogen to a new host over distances of one meter or less. A wide variety of diseases are transmitted by droplets, including influenza and many forms of pneumonia. Transmission over distances greater than one meter is called airborne transmission.

Indirect contact transmission involves inanimate objects called fomites that become contaminated by pathogens from an infected individual or reservoir (Figure 16.10). For example, an individual with the common cold may sneeze, causing droplets to land on a fomite such as a tablecloth or carpet, or the individual may wipe her nose and then transfer mucus to a fomite such as a doorknob or towel. Transmission occurs indirectly when a new susceptible host later touches the fomite and transfers the contaminated material to a susceptible portal of entry. Fomites can also include objects used in clinical settings that are not properly sterilized, such as syringes, needles, catheters, and surgical equipment. Pathogens transmitted indirectly via such fomites are a major cause of healthcare-associated infections (see Controlling Microbial Growth).

Figure 16.9 Direct contact transmission of pathogens can occur through physical contact. Many pathogens require contact with a mucous membrane to enter the body, but the host may transfer the pathogen from another point of contact (e.g., hand) to a mucous membrane (e.g., mouth or eye). (credit left: modification of work by Lisa Doehnert)

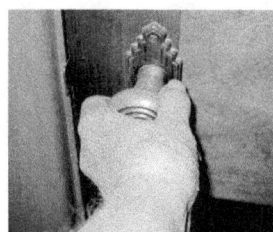

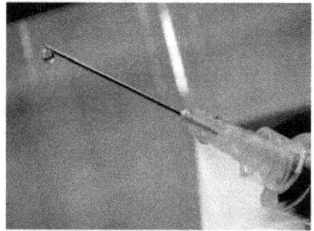

Figure 16.10 Fomites are nonliving objects that facilitate the indirect transmission of pathogens. Contaminated doorknobs, towels, and syringes are all common examples of fomites. (credit left: modification of work by Kate Ter Haar; credit middle: modification of work by Vernon Swanepoel; credit right: modification of work by "Zaldylmg"/Flickr)

Vehicle Transmission

The term **vehicle transmission** refers to the transmission of pathogens through vehicles such as water, food, and air. Water contamination through poor sanitation methods leads to waterborne transmission of disease. Waterborne disease remains a serious problem in many regions throughout the world. The World Health Organization (WHO) estimates that contaminated drinking water is responsible for more than 500,000 deaths each year.[10] Similarly, food contaminated through poor handling or storage can lead to foodborne transmission of disease (Figure 16.11).

Dust and fine particles known as aerosols, which can float in the air, can carry pathogens and facilitate the airborne transmission of disease. For example, dust particles are the dominant mode of transmission of hantavirus to humans. Hantavirus is found in mouse feces, urine, and saliva, but when these substances dry, they can disintegrate into fine particles that can become airborne when disturbed; inhalation of these particles can lead to a serious and sometimes fatal respiratory infection.

Although droplet transmission over short distances is considered contact transmission as discussed above, longer distance transmission of droplets through the air is considered vehicle transmission. Unlike larger particles that drop quickly out of the air column, fine mucus droplets produced by coughs or sneezes can remain suspended for long periods of time, traveling considerable distances. In certain conditions, droplets desiccate quickly to produce a droplet nucleus that is capable of transmitting pathogens; air temperature and humidity can have an impact on effectiveness of airborne transmission.

Tuberculosis is often transmitted via airborne transmission when the causative agent, *Mycobacterium tuberculosis*, is released in small particles with coughs. Because tuberculosis requires as few as 10 microbes to initiate a new infection, patients with tuberculosis must be treated in rooms equipped with special ventilation, and anyone entering the room should wear a mask.

---

10 World Health Organization. Fact sheet No. 391—*Drinking Water*. June 2005. http://www.who.int/mediacentre/factsheets/fs391/en.

Figure 16.11 Food is an important vehicle of transmission for pathogens, especially of the gastrointestinal and upper respiratory systems. Notice the glass shield above the food trays, designed to prevent pathogens ejected in coughs and sneezes from entering the food. (credit: Fort George G. Meade Public Affairs Office)

Clinical Focus

Resolution

After identifying the source of the contaminated turduckens, the Florida public health office notified the CDC, which requested an expedited inspection of the facility by state inspectors. Inspectors found that a machine used to process the chicken was contaminated with *Salmonella* as a result of substandard cleaning protocols. Inspectors also found that the process of stuffing and packaging the turduckens prior to refrigeration allowed the meat to remain at temperatures conducive to bacterial growth for too long. The contamination and the delayed refrigeration led to vehicle (food) transmission of the bacteria in turduckens.

Based on these findings, the plant was shut down for a full and thorough decontamination. All turduckens produced in the plant were recalled and pulled from store shelves ahead of the December holiday season, preventing further outbreaks.

Go back to the previous Clinical Focus Box.

Vector Transmission

Diseases can also be transmitted by a mechanical or biological vector, an animal (typically an arthropod) that carries the disease from one host to another. **Mechanical transmission** is facilitated by a **mechanical vector**, an animal that carries a pathogen from one host to another without being infected itself. For example, a fly may land on fecal matter and later transmit bacteria from the feces to food that it lands on; a human eating the food may then become infected by the bacteria, resulting in a case of diarrhea or dysentery (Figure 16.12).

Biological transmission occurs when the pathogen reproduces within a **biological vector** that transmits the pathogen from one host to another (Figure 16.12). Arthropods are the main vectors responsible for biological transmission (Figure 16.13). Most arthropod vectors transmit the pathogen by biting the host, creating a wound that serves as a portal of entry. The pathogen may go through part of its reproductive cycle in the gut or salivary glands of the arthropod to facilitate its transmission through the bite. For example, hemipterans (called "kissing bugs" or "assassin bugs") transmit Chagas disease to humans by defecating when they bite, after which the human scratches or rubs the infected feces into a mucous membrane or break in the skin.

Biological insect vectors include mosquitoes, which transmit malaria and other diseases, and lice, which transmit typhus. Other arthropod vectors can include arachnids, primarily ticks, which transmit Lyme disease and other diseases, and mites, which transmit scrub typhus and rickettsial pox. Biological transmission, because it involves survival and reproduction within a parasitized vector, complicates the biology of the pathogen and its transmission. There are also important non-arthropod vectors of disease, including mammals and birds. Various species of mammals can transmit rabies to humans, usually by means of a bite that transmits the rabies virus. Chickens and other domestic poultry can transmit avian influenza to humans through direct or indirect contact with avian influenza virus A shed in the birds' saliva, mucous, and feces.

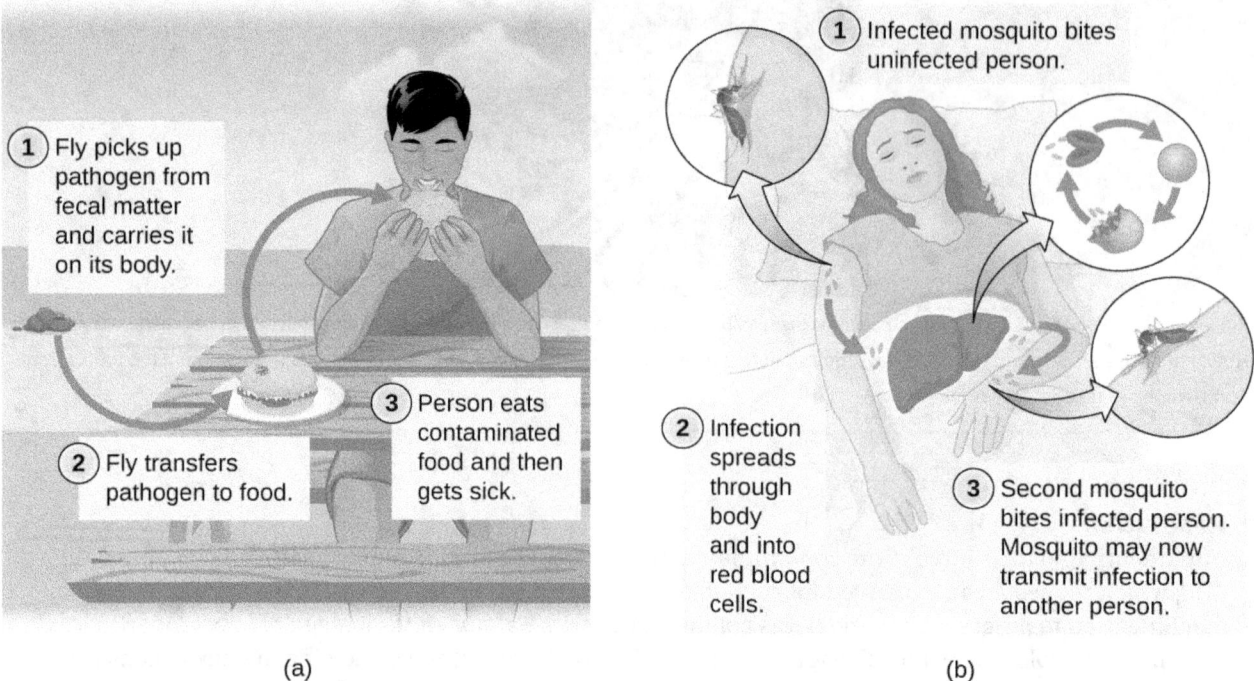

Figure 16.12 (a) A mechanical vector carries a pathogen on its body from one host to another, not as an infection. (b) A biological vector carries a pathogen from one host to another after becoming infected itself.

Common Arthropod Vectors and Select Pathogens			
Vector	Species	Pathogen	Disease
Black fly 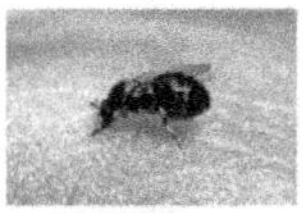	*Simulium* spp.	*Onchocerca volvulus*	Onchocerciasis (river blindness)
Flea 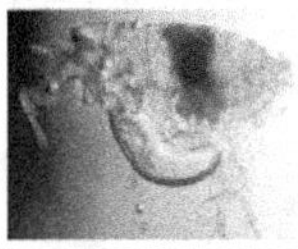	*Xenopsylla cheopis*	*Rickettsia typhi*	Murine typhus
		Yersinia pestis	Plague
Kissing bug 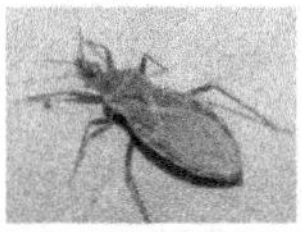	*Triatoma* spp.	*Trypanosoma cruzi*	Chagas disease
Louse 	*Pediculus humanus humanus*	*Bartonella quintana*	Trench fever
		Borrelia recurrentis	Relapsing fever
		Rickettsia prowazekii	Typhus
Mite (chigger) 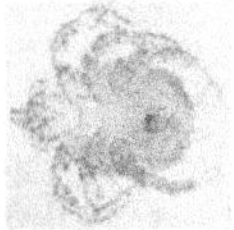	*Leptotrombidium* spp.	*Orientia tsutsugamushi*	Scrub typhus
	Liponyssoides sanguineus	*Rickettsia akari*	Rickettsialpox
Mosquito 	*Aedes* spp., *Haemagogus* spp.	Yellow fever virus	Yellow fever
	Anopheles spp.	*Plasmodium falciparum*	Malaria
	Culex pipiens	West Nile virus	West Nile disease
Sand fly 	*Phlebotomus* spp.	*Leishmania* spp.	Leishmaniasis
Tick 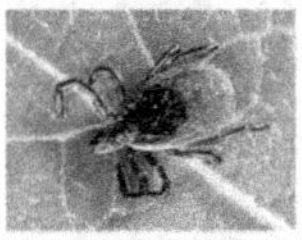	*Ixodes* spp.	*Borrelia* spp.	Lyme disease
	Dermacentor spp. and others	*Rickettsia rickettsii*	Rocky Mountain spotted fever
Tsetse fly 	*Glossina* spp.	*Trypanosoma brucei*	African trypanosomiasis (sleeping sickness)

Figure 16.13 (credit "Black fly", "Tick", "Tsetse fly": modification of work by USDA; credit: "Flea": modification of work by Centers for Disease Control and Prevention; credit: "Louse", "Mosquito", "Sand fly": modification of work by James Gathany, Centers for Disease Control and Prevention; credit "Kissing bug": modification of work by Glenn Seplak; credit "Mite": modification of work by Michael Wunderli)

CHECK YOUR UNDERSTANDING

- Describe how diseases can be transmitted through the air.
- Explain the difference between a mechanical vector and a biological vector.

Eye on Ethics

Using GMOs to Stop the Spread of Zika

In 2016, an epidemic of the Zika virus was linked to a high incidence of birth defects in South America and Central America. As winter turned to spring in the northern hemisphere, health officials correctly predicted the virus would spread to North America, coinciding with the breeding season of its major vector, the *Aedes aegypti* mosquito.

The range of the *A. aegypti* mosquito extends well into the southern United States (Figure 16.14). Because these same mosquitoes serve as vectors for other problematic diseases (dengue fever, yellow fever, and others), various methods of mosquito control have been proposed as solutions. Chemical pesticides have been used effectively in the past, and are likely to be used again; but because chemical pesticides can have negative impacts on the environment, some scientists have proposed an alternative that involves genetically engineering *A. aegypti* so that it cannot reproduce. This method, however, has been the subject of some controversy.

One method that has worked in the past to control pests, with little apparent downside, has been sterile male introductions. This method controlled the screw-worm fly pest in the southwest United States and fruit fly pests of fruit crops. In this method, males of the target species are reared in the lab, sterilized with radiation, and released into the environment where they mate with wild females, who subsequently bear no live offspring. Repeated releases shrink the pest population.

A similar method, taking advantage of recombinant DNA technology,[11] introduces a dominant lethal allele into male mosquitoes that is suppressed in the presence of tetracycline (an antibiotic) during laboratory rearing. The males are released into the environment and mate with female mosquitoes. Unlike the sterile male method, these matings produce offspring, but they die as larvae from the lethal gene in the absence of tetracycline in the environment. As of 2016, this method has yet to be implemented in the United States, but a UK company tested the method in Piracicaba, Brazil, and found an 82% reduction in wild *A. aegypti* larvae and a 91% reduction in dengue cases in the treated area.[12] In August 2016, amid news of Zika infections in several Florida communities, the FDA gave the UK company permission to test this same mosquito control method in Key West, Florida, pending compliance with local and state regulations and a referendum in the affected communities.

The use of genetically modified organisms (GMOs) to control a disease vector has its advocates as well as its opponents. In theory, the system could be used to drive the *A. aegypti* mosquito extinct—a noble goal according to some, given the damage they do to human populations.[13] But opponents of the idea are concerned that the gene could escape the species boundary of *A. aegypti* and cause problems in other species, leading to unforeseen ecological consequences. Opponents are also wary of the program because it is being administered by a for-profit corporation, creating the potential for conflicts of interest that would have to be tightly regulated; and it is not clear how any unintended consequences of the program could be reversed.

---

11 Blandine Massonnet-Bruneel, Nicole Corre-Catelin, Renaud Lacroix, Rosemary S. Lees, Kim Phuc Hoang, Derric Nimmo, Luke Alphey, and Paul Reiter. "Fitness of Transgenic Mosquito *Aedes aegypti* Males Carrying a Dominant Lethal Genetic System." *PLOS ONE* 8, no. 5 (2013): e62711.

12 Richard Levine. "Cases of Dengue Drop 91 Percent Due to Genetically Modified Mosquitoes." *Entomology Today*. https://entomologytoday.org/2016/07/14/cases-of-dengue-drop-91-due-to-genetically-modified-mosquitoes.

13 Olivia Judson. "A Bug's Death." *The New York Times*, September 25, 2003. http://www.nytimes.com/2003/09/25/opinion/a-bug-

There are other epidemiological considerations as well. *Aedes aegypti* is apparently not the only vector for the Zika virus. *Aedes albopictus*, the Asian tiger mosquito, is also a vector for the Zika virus.[14] *A. albopictus* is now widespread around the planet including much of the United States (Figure 16.14). Many other mosquitoes have been found to harbor Zika virus, though their capacity to act as vectors is unknown.[15] Genetically modified strains of *A. aegypti* will not control the other species of vectors. Finally, the Zika virus can apparently be transmitted sexually between human hosts, from mother to child, and possibly through blood transfusion. All of these factors must be considered in any approach to controlling the spread of the virus.

Clearly there are risks and unknowns involved in conducting an open-environment experiment of an as-yet poorly understood technology. But allowing the Zika virus to spread unchecked is also risky. Does the threat of a Zika epidemic justify the ecological risk of genetically engineering mosquitos? Are current methods of mosquito control sufficiently ineffective or harmful that we need to try untested alternatives? These are the questions being put to public health officials now.

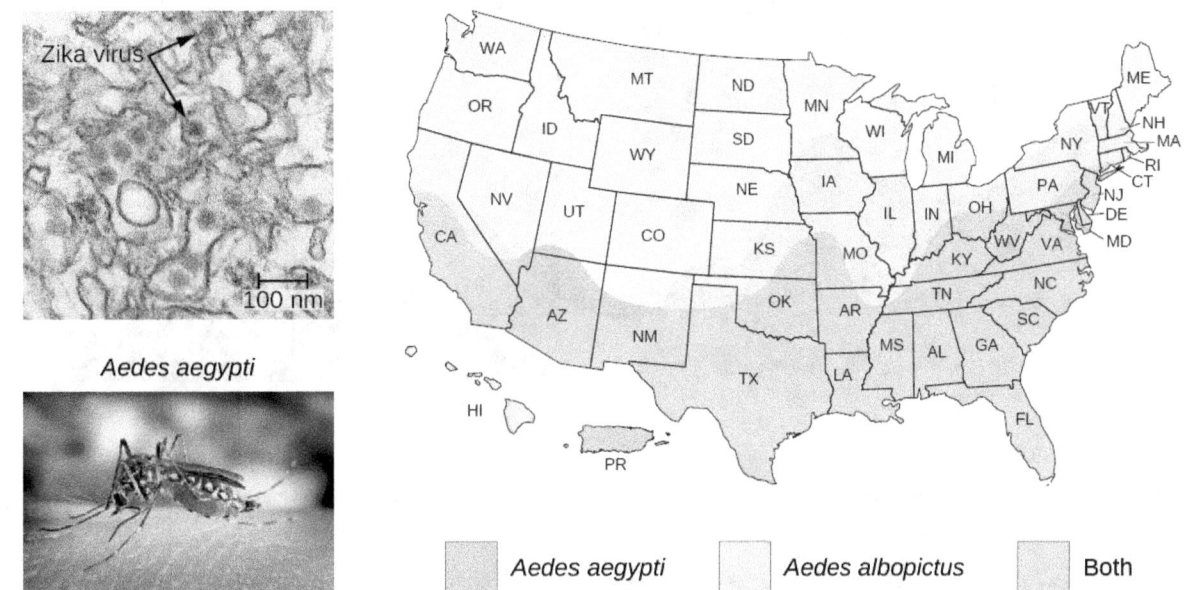

Figure 16.14 The Zika virus is an enveloped virus transmitted by mosquitoes, especially *Aedes aegypti*. The range of this mosquito includes much of the United States, from the Southwest and Southeast to as far north as the Mid-Atlantic. The range of *A. albopictus*, another vector, extends even farther north to New England and parts of the Midwest. (credit micrograph: modification of work by Cynthia Goldsmith, Centers for Disease Control and Prevention; credit photo: modification of work by James Gathany, Centers for Disease Control and Prevention; credit map: modification of work by Centers for Disease Control and Prevention)

Quarantining

Individuals suspected or known to have been exposed to certain contagious pathogens may be **quarantined**, or isolated to prevent transmission of the disease to others. Hospitals and other health-care facilities generally set up special wards to isolate patients with particularly hazardous diseases such as tuberculosis or Ebola (Figure 16.15). Depending on the setting, these wards may be equipped with special air-handling methods, and

---

s-death.html.

14 Gilda Grard, Mélanie Caron, Illich Manfred Mombo, Dieudonné Nkoghe, Statiana Mboui Ondo, Davy Jiolle, Didier Fontenille, Christophe Paupy, and Eric Maurice Leroy. "Zika Virus in Gabon (Central Africa)–2007: A New Threat from *Aedes albopictus*?" *PLOS Neglected Tropical Diseases* 8, no. 2 (2014): e2681.

15 Constância F.J. Ayres. "Identification of Zika Virus Vectors and Implications for Control." *The Lancet Infectious Diseases* 16, no. 3 (2016): 278–279.

personnel may implement special protocols to limit the risk of transmission, such as personal protective equipment or the use of chemical disinfectant sprays upon entry and exit of medical personnel.

The duration of the quarantine depends on factors such as the incubation period of the disease and the evidence suggestive of an infection. The patient may be released if signs and symptoms fail to materialize when expected or if preventive treatment can be administered in order to limit the risk of transmission. If the infection is confirmed, the patient may be compelled to remain in isolation until the disease is no longer considered contagious.

In the United States, public health authorities may only quarantine patients for certain diseases, such as cholera, diphtheria, infectious tuberculosis, and strains of influenza capable of causing a pandemic. Individuals entering the United States or moving between states may be quarantined by the CDC if they are suspected of having been exposed to one of these diseases. Although the CDC routinely monitors entry points to the United States for crew or passengers displaying illness, quarantine is rarely implemented.

(a)

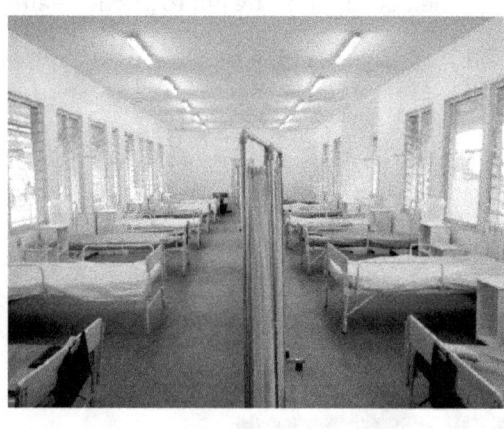
(b)

Figure 16.15 (a) The Aeromedical Biological Containment System (ABCS) is a module designed by the CDC and Department of Defense specifically for transporting highly contagious patients by air. (b) An isolation ward for Ebola patients in Lagos, Nigeria. (credit a: modification of work by Centers for Disease Control and Prevention; credit b: modification of work by CDC Global)

Healthcare-Associated (Nosocomial) Infections

Hospitals, retirement homes, and prisons attract the attention of epidemiologists because these settings are associated with increased incidence of certain diseases. Higher rates of transmission may be caused by characteristics of the environment itself, characteristics of the population, or both. Consequently, special efforts must be taken to limit the risks of infection in these settings.

Infections acquired in health-care facilities, including hospitals, are called **nosocomial infections** or **healthcare-associated infections (HAI)**. HAIs are often connected with surgery or other invasive procedures that provide the pathogen with access to the portal of infection. For an infection to be classified as an HAI, the patient must have been admitted to the health-care facility for a reason other than the infection. In these settings, patients suffering from primary disease are often afflicted with compromised immunity and are more susceptible to secondary infection and opportunistic pathogens.

In 2011, more than 720,000 HAIs occurred in hospitals in the United States, according to the CDC. About 22% of these HAIs occurred at a surgical site, and cases of pneumonia accounted for another 22%; urinary tract infections accounted for an additional 13%, and primary bloodstream infections 10%.[16] Such HAIs often occur when pathogens are introduced to patients' bodies through contaminated surgical or medical equipment, such as catheters and respiratory ventilators. Health-care facilities seek to limit nosocomial infections through training and hygiene protocols such as those described in Control of Microbial Growth.

---

16 Centers for Disease Control and Prevention. "HAI Data and Statistics." 2016. http://www.cdc.gov/hai/surveillance. Accessed Jan 2, 2016.

CHECK YOUR UNDERSTANDING
- Give some reasons why HAIs occur.

16.4 Global Public Health

Learning Objectives

By the end of this section, you will be able to:
- Describe the entities involved in international public health and their activities
- Identify and differentiate between emerging and reemerging infectious diseases

A large number of international programs and agencies are involved in efforts to promote global public health. Among their goals are developing infrastructure in health care, public sanitation, and public health capacity; monitoring infectious disease occurrences around the world; coordinating communications between national public health agencies in various countries; and coordinating international responses to major health crises. In large part, these international efforts are necessary because disease-causing microorganisms know no national boundaries.

The World Health Organization (WHO)

International public health issues are coordinated by the **World Health Organization (WHO)**, an agency of the United Nations. Of its roughly $4 billion budget for 2015–16[17], about $1 billion was funded by member states and the remaining $3 billion by voluntary contributions. In addition to monitoring and reporting on infectious disease, WHO also develops and implements strategies for their control and prevention. WHO has had a number of successful international public health campaigns. For example, its vaccination program against smallpox, begun in the mid-1960s, resulted in the global eradication of the disease by 1980. WHO continues to be involved in infectious disease control, primarily in the developing world, with programs targeting malaria, HIV/AIDS, and tuberculosis, among others. It also runs programs to reduce illness and mortality that occur as a result of violence, accidents, lifestyle-associated illnesses such as diabetes, and poor health-care infrastructure.

WHO maintains a global alert and response system that coordinates information from member nations. In the event of a public health emergency or epidemic, it provides logistical support and coordinates international response to the emergency. The United States contributes to this effort through the CDC. The CDC carries out international monitoring and public health efforts, mainly in the service of protecting US public health in an increasingly connected world. Similarly, the European Union maintains a Health Security Committee that monitors disease outbreaks within its member countries and internationally, coordinating with WHO.

CHECK YOUR UNDERSTANDING
- Name the organizations that participate in international public health monitoring.

Emerging and Reemerging Infectious Diseases

Both WHO and some national public health agencies such as the CDC monitor and prepare for **emerging infectious diseases**. An emerging infectious disease is either new to the human population or has shown an increase in prevalence in the previous twenty years. Whether the disease is new or conditions have changed to cause an increase in frequency, its status as emerging implies the need to apply resources to understand and control its growing impact.

Emerging diseases may change their frequency gradually over time, or they may experience sudden epidemic growth. The importance of vigilance was made clear during the Ebola hemorrhagic fever epidemic in western Africa through 2014–2015. Although health experts had been aware of the Ebola virus since the 1970s, an outbreak on such a large scale had never happened before (Figure 16.16). Previous human epidemics had

---

17 World Health Organization. "Programme Budget 2014–2015." http://www.who.int/about/finances-accountability/budget/en.

been small, isolated, and contained. Indeed, the gorilla and chimpanzee populations of western Africa had suffered far worse from Ebola than the human population. The pattern of small isolated human epidemics changed in 2014. Its high transmission rate, coupled with cultural practices for treatment of the dead and perhaps its emergence in an urban setting, caused the disease to spread rapidly, and thousands of people died. The international public health community responded with a large emergency effort to treat patients and contain the epidemic.

Emerging diseases are found in all countries, both developed and developing (Table 16.2). Some nations are better equipped to deal with them. National and international public health agencies watch for epidemics like the Ebola outbreak in developing countries because those countries rarely have the health-care infrastructure and expertise to deal with large outbreaks effectively. Even with the support of international agencies, the systems in western Africa struggled to identify and care for the sick and control spread. In addition to the altruistic goal of saving lives and assisting nations lacking in resources, the global nature of transportation means that an outbreak anywhere can spread quickly to every corner of the planet. Managing an epidemic in one location—its source—is far easier than fighting it on many fronts.

Ebola is not the only disease that needs to be monitored in the global environment. In 2015, WHO set priorities on several emerging diseases that had a high probability of causing epidemics and that were poorly understood (and thus urgently required research and development efforts).

A **reemerging infectious disease** is a disease that is increasing in frequency after a previous period of decline. Its reemergence may be a result of changing conditions or old prevention regimes that are no longer working. Examples of such diseases are drug-resistant forms of tuberculosis, bacterial pneumonia, and malaria. Drug-resistant strains of the bacteria causing gonorrhea and syphilis are also becoming more widespread, raising concerns of untreatable infections.

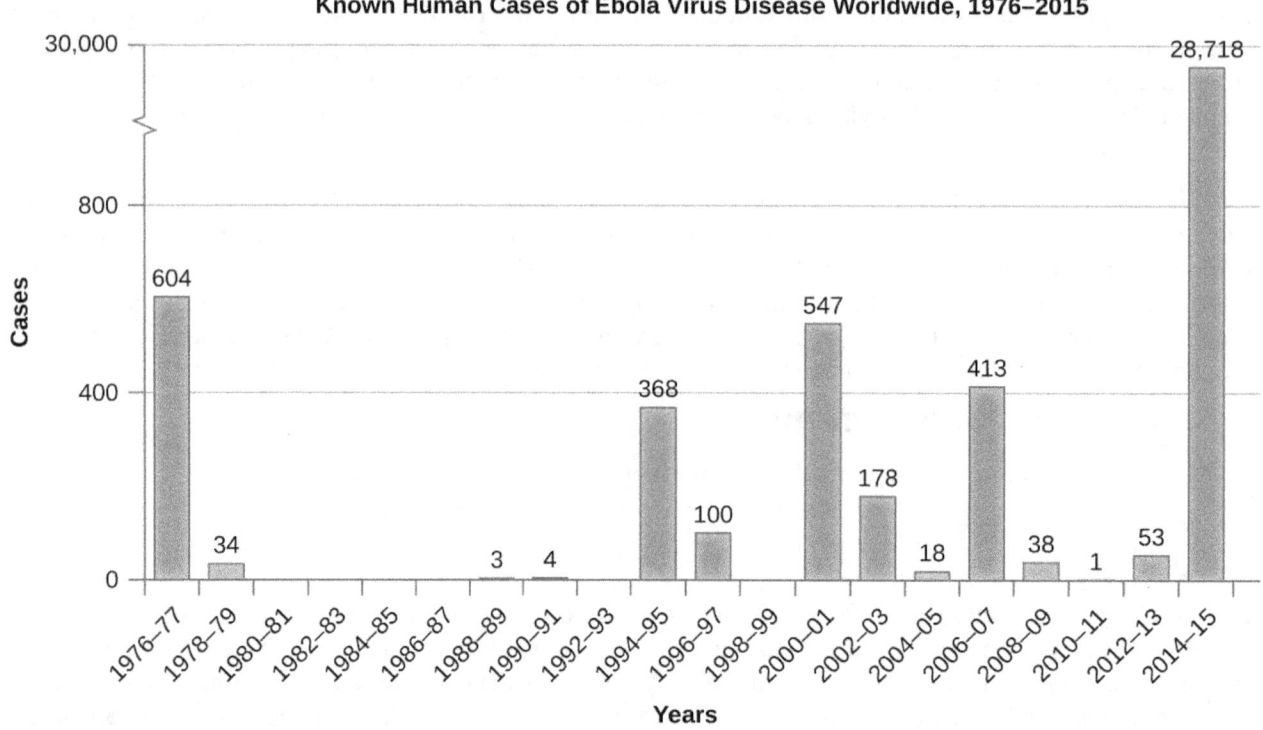

Figure 16.16   Even before the Ebola epidemic of 2014–15, Ebola was considered an emerging disease because of several smaller outbreaks between the mid-1990s and 2000s.

Some Emerging and Reemerging Infectious Diseases

Disease	Pathogen	Year Discovered	Affected Regions	Transmission
AIDS	HIV	1981	Worldwide	Contact with infected body fluids
Chikungunya fever	Chikungunya virus	1952	Africa, Asia, India; spreading to Europe and the Americas	Mosquito-borne
Ebola virus disease	Ebola virus	1976	Central and Western Africa	Contact with infected body fluids
H1N1 Influenza (swine flu)	H1N1 virus	2009	Worldwide	Droplet transmission
Lyme disease	*Borrelia burgdorferi* bacterium	1981	Northern hemisphere	From mammal reservoirs to humans by tick vectors
West Nile virus disease	West Nile virus	1937	Africa, Australia, Canada to Venezuela, Europe, Middle East, Western Asia	Mosquito-borne

Table 16.2

CHECK YOUR UNDERSTANDING

- Explain why it is important to monitor emerging infectious diseases.
- Explain how a bacterial disease could reemerge, even if it had previously been successfully treated and controlled.

MICRO CONNECTIONS

SARS Outbreak and Identification

On November 16, 2002, the first case of a SARS outbreak was reported in Guangdong Province, China. The patient exhibited influenza-like symptoms such as fever, cough, myalgia, sore throat, and shortness of breath. As the number of cases grew, the Chinese government was reluctant to openly communicate information about the epidemic with the World Health Organization (WHO) and the international community. The slow reaction of Chinese public health officials to this new disease contributed to the spread of the epidemic within and later outside China. In April 2003, the Chinese government finally responded with a huge public health effort involving quarantines, medical checkpoints, and massive cleaning projects. Over 18,000 people were quarantined in Beijing alone. Large funding initiatives were created to improve health-care facilities, and dedicated outbreak teams were created to coordinate the response. By August 16, 2003, the last SARS patients were released from a hospital in Beijing nine months after the first case was reported in China.

In the meantime, SARS spread to other countries on its way to becoming a global pandemic. Though the infectious agent had yet to be identified, it was thought to be an influenza virus. The disease was named SARS, an acronym for severe acute respiratory syndrome, until the etiologic agent could be identified. Travel restrictions to Southeast Asia were enforced by many countries. By the end of the outbreak, there were 8,098 cases and 774 deaths worldwide. China and Hong Kong were hit hardest by the epidemic, but Taiwan, Singapore, and Toronto, Canada, also saw significant numbers of cases.

Fortunately, timely public health responses in many countries effectively suppressed the outbreak and led to its eventual containment. For example, the disease was introduced to Canada in February 2003 by an infected traveler from Hong Kong, who died shortly after being hospitalized. By the end of March, hospital isolation and home quarantine procedures were in place in the Toronto area, stringent anti-infection protocols were introduced in hospitals, and the media were actively reporting on the disease. Public health officials tracked down contacts of infected individuals and quarantined them. A total of 25,000 individuals were quarantined in the city. Thanks to the vigorous response of the Canadian public health community, SARS was brought under control in Toronto by June, a mere four months after it was introduced.

In 2003, WHO established a collaborative effort to identify the causative agent of SARS, which has now been identified as a coronavirus that was associated with horseshoe bats. The genome of the SARS virus was sequenced and published by researchers at the CDC and in Canada in May 2003, and in the same month researchers in the Netherlands confirmed the etiology of the disease by fulfilling Koch's postulates for the SARS coronavirus. The last known case of SARS worldwide was reported in 2004.

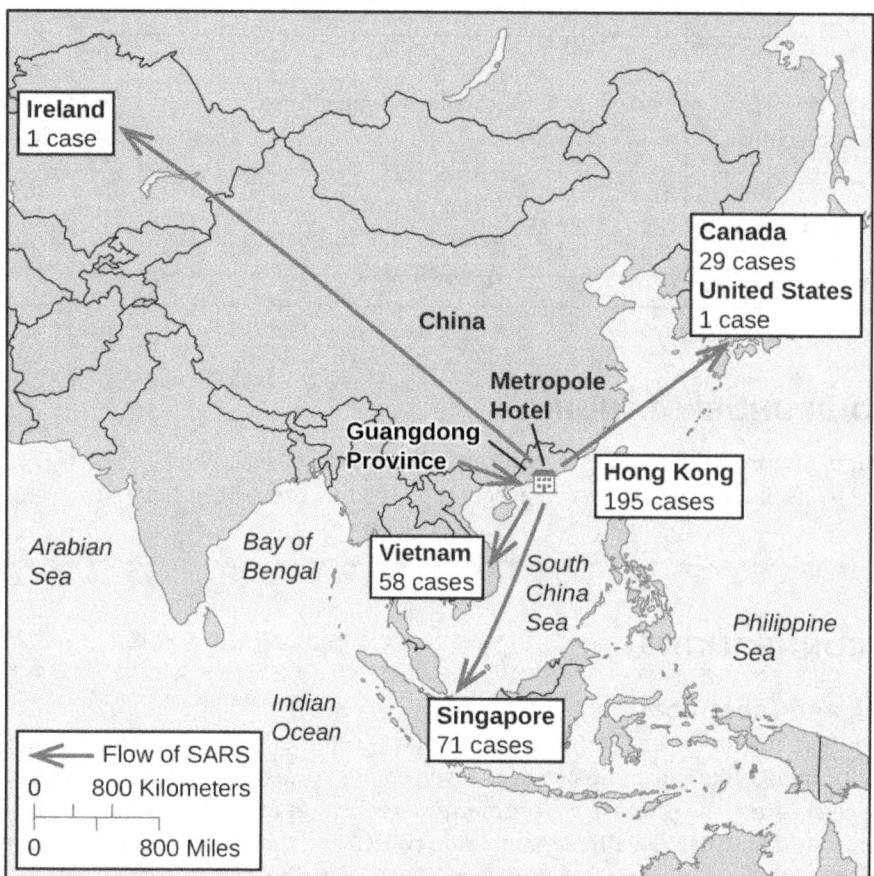

Figure 16.17 This map shows the spread of SARS as of March 28, 2003. (credit: modification of work by Central Intelligence Agency)

LINK TO LEARNING

This database (https://openstax.org/l/22dataoutinfdis) of reports chronicles outbreaks of infectious disease around the world. It was on this system that the first information about the SARS outbreak in China emerged.

The CDC publishes Emerging Infectious Diseases (https://openstax.org/l/22CDCEmerinfdis), a monthly journal available online.

SUMMARY

16.1 The Language of Epidemiologists

- **Epidemiology** is the science underlying public health.
- **Morbidity** means being in a state of illness, whereas **mortality** refers to death; both **morbidity rates** and **mortality rates** are of interest to epidemiologists.
- **Incidence** is the number of new cases (morbidity or mortality), usually expressed as a proportion, during a specified time period; **prevalence** is the total number affected in the population, again usually expressed as a proportion.
- **Sporadic diseases** only occur rarely and largely without a geographic focus. **Endemic diseases** occur at a constant (and often low) level within a population. **Epidemic diseases** and **pandemic diseases** occur when an outbreak occurs on a significantly larger than expected level, either locally or globally, respectively.
- **Koch's postulates** specify the procedure for confirming a particular pathogen as the etiologic agent of a particular disease. Koch's postulates have limitations in application if the microbe cannot be isolated and cultured or if there is no animal host for the microbe. In this case, molecular Koch's postulates would be utilized.
- In the United States, the **Centers for Disease Control and Prevention** monitors **notifiable diseases** and publishes weekly updates in the *Morbidity and Mortality Weekly Report*.

16.2 Tracking Infectious Diseases

- Early pioneers of epidemiology such as John Snow, Florence Nightingale, and Joseph Lister, studied disease at the population level and used data to disrupt disease transmission.
- **Descriptive epidemiology** studies rely on case analysis and patient histories to gain information about outbreaks, frequently while they are still occurring.
- **Retrospective epidemiology** studies use historical data to identify associations with the disease state of present cases. **Prospective epidemiology** studies gather data and follow cases to find associations with future disease states.
- **Analytical epidemiology** studies are observational studies that are carefully designed to compare groups and uncover associations between environmental or genetic factors and disease.
- **Experimental epidemiology** studies generate strong evidence of causation in disease or treatment by manipulating subjects and comparing them with control subjects.

16.3 Modes of Disease Transmission

- **Reservoirs** of human disease can include the human and animal populations, soil, water, and inanimate objects or materials.
- **Contact transmission** can be **direct** or **indirect** through physical contact with either an infected host (direct) or contact with a fomite that an infected host has made contact with previously (indirect).
- Vector transmission occurs when a living organism carries an infectious agent on its body (**mechanical**) or as an infection host itself (**biological**), to a new host.
- **Vehicle transmission** occurs when a substance, such as soil, water, or air, carries an infectious agent to a new host.
- **Healthcare-associated infections (HAI)**, or **nosocomial infections**, are acquired in a clinical setting. Transmission is facilitated by medical interventions and the high concentration of susceptible, immunocompromised individuals in clinical settings.

16.4 Global Public Health

- The **World Health Organization (WHO)** is an agency of the United Nations that collects and analyzes data on disease occurrence from member nations. WHO also coordinates public health programs and responses to international health emergencies.
- **Emerging diseases** are those that are new to human populations or that have been increasing in the past two decades. **Reemerging diseases** are those that are making a resurgence in susceptible populations after previously having been controlled in some geographic areas.

REVIEW QUESTIONS

Multiple Choice

1. Which is the most common type of biological vector of human disease?
   a. viruses
   b. bacteria
   c. mammals
   d. arthropods

2. A mosquito bites a person who subsequently develops a fever and abdominal rash. What type of transmission would this be?
   a. mechanical vector transmission
   b. biological vector transmission
   c. direct contact transmission
   d. vehicle transmission

3. Cattle are allowed to pasture in a field that contains the farmhouse well, and the farmer's family becomes ill with a gastrointestinal pathogen after drinking the water. What type of transmission of infectious agents would this be?
   a. biological vector transmission
   b. direct contact transmission
   c. indirect contact transmission
   d. vehicle transmission

4. A blanket from a child with chickenpox is likely to be contaminated with the virus that causes chickenpox (Varicella-zoster virus). What is the blanket called?
   a. fomite
   b. host
   c. pathogen
   d. vector

5. Which of the following would NOT be considered an emerging disease?
   a. Ebola hemorrhagic fever
   b. West Nile virus fever/encephalitis
   c. Zika virus disease
   d. Tuberculosis

6. Which of the following would NOT be considered a reemerging disease?
   a. Drug-resistant tuberculosis
   b. Drug-resistant gonorrhea
   c. Malaria
   d. West Nile virus fever/encephalitis

7. Which of the following factors can lead to reemergence of a disease?
   a. A mutation that allows it to infect humans
   b. A period of decline in vaccination rates
   c. A change in disease reporting procedures
   d. Better education on the signs and symptoms of the disease

8. Why are emerging diseases with very few cases the focus of intense scrutiny?
   a. They tend to be more deadly
   b. They are increasing and therefore not controlled
   c. They naturally have higher transmission rates
   d. They occur more in developed countries

Matching

9. Match each term with its description.

___ sporadic disease

___ endemic disease

___ pandemic disease

___ morbidity rate

___ mortality rate

A. the number of disease cases per 100,000 individuals

B. a disease in higher than expected numbers around the world

C. the number of deaths from a disease for every 10,000 individuals

D. a disease found occasionally in a region with cases occurring mainly in isolation from each other

E. a disease found regularly in a region

10. Match each type of epidemiology study with its description.

___ experimental

___ analytical

___ prospective

___ descriptive

___ retrospective

A. examination of past case histories and medical test results conducted on patients in an outbreak

B. examination of current case histories, interviews with patients and their contacts, interpretation of medical test results; frequently conducted while outbreak is still in progress

C. use of a set of test subjects (human or animal) and control subjects that are treated the same as the test subjects except for the specific treatment being studied

D. observing groups of individuals to look for associations with disease

E. a comparison of a cohort of individuals through the course of the study

11. Match each pioneer of epidemiology with his or her contribution.

___Florence Nightingale — A. determined the source of a cholera outbreak in London

___Robert Koch — B. showed that surgical wound infection rates could be dramatically reduced by using carbolic acid to disinfect surgical tools, bandages, and surgical sites

___Joseph Lister — C. compiled data on causes of mortality in soldiers, leading to innovations in military medical care

___John Snow — D. developed a methodology for conclusively determining the etiology of disease

Fill in the Blank

12. The _____ collects data and conducts epidemiologic studies in the United States.

13. _____ occurs when an infected individual passes the infection on to other individuals, who pass it on to still others, increasing the penetration of the infection into the susceptible population.

14. A batch of food contaminated with botulism exotoxin, consumed at a family reunion by most of the members of a family, would be an example of a _____ outbreak.

15. A patient in the hospital with a urinary catheter develops a bladder infection. This is an example of a(n) _____ infection.

16. A _____ is an animal that can transfer infectious pathogens from one host to another.

17. The _____ collects data and conducts epidemiologic studies at the global level.

Short Answer

18. During an epidemic, why might the prevalence of a disease at a particular time not be equal to the sum of the incidences of the disease?

19. In what publication would you find data on emerging/reemerging diseases in the United States?

20. What activity did John Snow conduct, other than mapping, that contemporary epidemiologists also use when trying to understand how to control a disease?

21. Differentiate between droplet vehicle transmission and airborne transmission.

Critical Thinking

22. Why might an epidemiological population in a state not be the same size as the number of people in a state? Use an example.

23. Many people find that they become ill with a cold after traveling by airplane. The air circulation systems of commercial aircraft use HEPA filters that should remove any infectious agents that pass through them. What are the possible reasons for increased incidence of colds after flights?

24. An Atlantic crossing by boat from England to New England took 60–80 days in the 18th century. In the late 19th century the voyage took less than a week. How do you think these time differences for travel might have impacted the spread of infectious diseases from Europe to the Americas, or vice versa?

CHAPTER 17
Innate Nonspecific Host Defenses

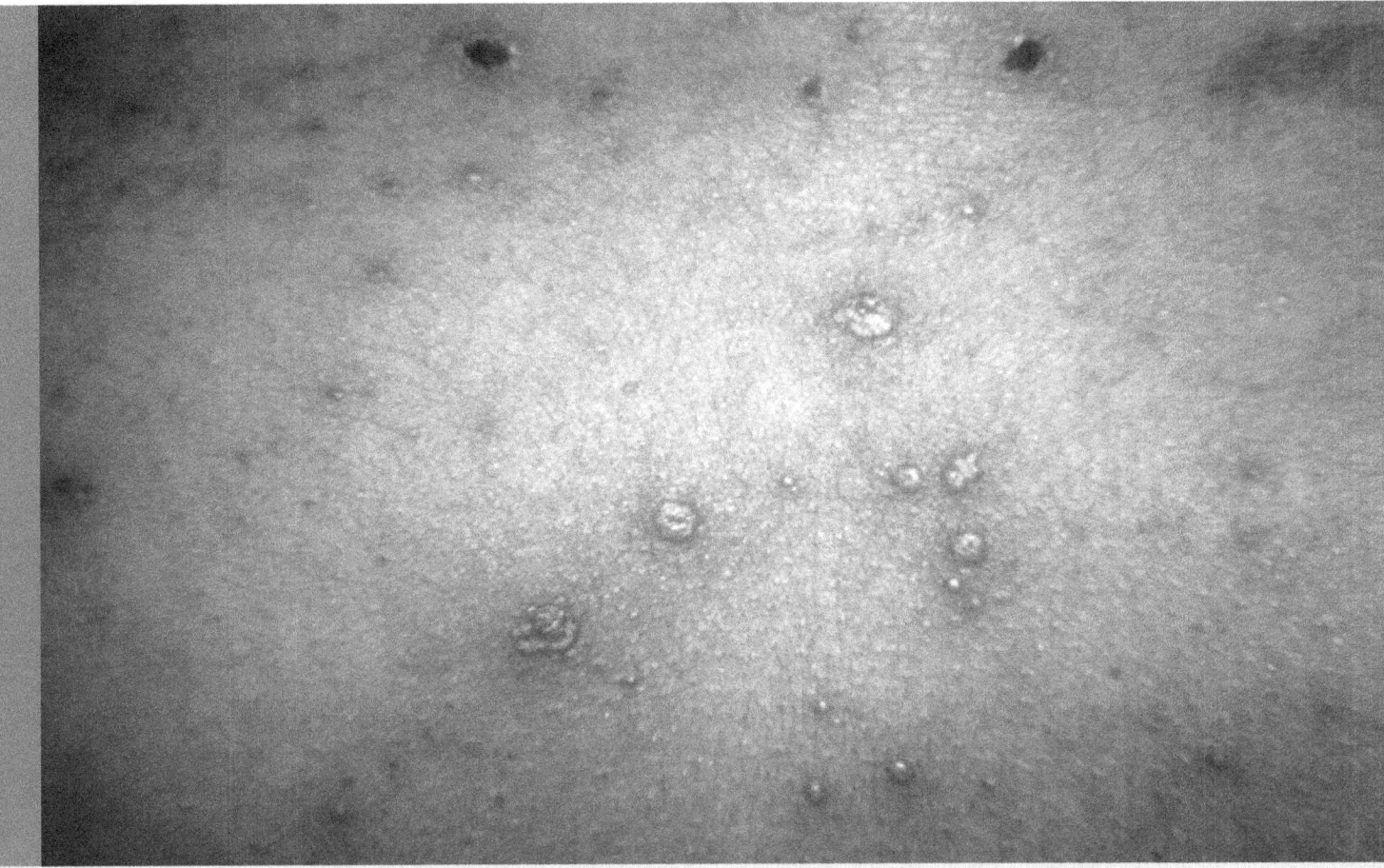

Figure 17.1 Varicella, or chickenpox, is caused by the highly contagious varicella-zoster virus. The characteristic rash seen here is partly a result of inflammation associated with the body's immune response to the virus. Inflammation is a response mechanism of innate immunity that helps the body fight off a wide range of infections. (credit: John Noble / CDC; Public Domain)

Chapter Outline

17.1 Physical Defenses

17.2 Chemical Defenses

17.3 Cellular Defenses

17.4 Pathogen Recognition and Phagocytosis

17.5 Inflammation and Fever

INTRODUCTION Despite relatively constant exposure to pathogenic microbes in the environment, humans do not generally suffer from constant infection or disease. Under most circumstances, the body is able to defend itself from the threat of infection thanks to a complex immune system designed to repel, kill, and expel disease-causing invaders. Immunity as a whole can be described as two interrelated parts: nonspecific innate immunity, which is the subject of this chapter, and specific adaptive host defenses, which are discussed in the next chapter.

The nonspecific innate immune response provides a first line of defense that can often prevent infections from gaining a solid foothold in the body. These defenses are described as *nonspecific* because they do not target any specific pathogen; rather, they defend against a wide range of potential pathogens. They are called *innate* because they are built-in mechanisms of the human organism. Unlike the specific adaptive defenses, they are not acquired over time and they have no "memory" (they do not improve after repeated exposures to specific pathogens).

Broadly speaking, nonspecific innate defenses provide an immediate (or very rapid) response against potential pathogens. However, these responses are neither perfect nor impenetrable. They can be circumvented by pathogens on occasion, and sometimes they can even cause damage to the body, contributing to the signs and symptoms of infection (Figure 17.1).

17.1 Physical Defenses

Learning Objectives

By the end of this section, you will be able to:
- Describe the various physical barriers and mechanical defenses that protect the human body against infection and disease
- Describe the role of microbiota as a first-line defense against infection and disease

---

Clinical Focus

Part 1

Angela, a 25-year-old female patient in the emergency department, is having some trouble communicating verbally because of shortness of breath. A nurse observes constriction and swelling of the airway and labored breathing. The nurse asks Angela if she has a history of asthma or allergies. Angela shakes her head no, but there is fear in her eyes. With some difficulty, she explains that her father died suddenly at age 27, when she was just a little girl, of a similar respiratory attack. The underlying cause had never been identified.

- What are some possible causes of constriction and swelling of the airway?
- What causes swelling of body tissues in general?

Jump to the next Clinical Focus box.

Nonspecific innate immunity can be characterized as a multifaceted system of defenses that targets invading pathogens in a nonspecific manner. In this chapter, we have divided the numerous defenses that make up this system into three categories: physical defenses, chemical defenses, and cellular defenses. However, it is important to keep in mind that these defenses do not function independently, and the categories often overlap. Table 17.1 provides an overview of the nonspecific defenses discussed in this chapter.

Overview of Nonspecific Innate Immune Defenses

Physical defenses	Physical barriers
	Mechanical defenses

Overview of Nonspecific Innate Immune Defenses

	Microbiome
	Chemicals and enzymes in body fluids
	Antimicrobial peptides
Chemical defenses	Plasma protein mediators
	Cytokines
	Inflammation-eliciting mediators
Cellular defenses	Granulocytes
	Agranulocytes

Table 17.1

Physical defenses provide the body's most basic form of nonspecific defense. They include physical barriers to microbes, such as the skin and mucous membranes, as well as mechanical defenses that physically remove microbes and debris from areas of the body where they might cause harm or infection. In addition, the microbiome provides a measure of physical protection against disease, as microbes of the normal microbiota compete with pathogens for nutrients and cellular binding sites necessary to cause infection.

Physical Barriers

Physical barriers play an important role in preventing microbes from reaching tissues that are susceptible to infection. At the cellular level, barriers consist of cells that are tightly joined to prevent invaders from crossing through to deeper tissue. For example, the endothelial cells that line blood vessels have very tight cell-to-cell junctions, blocking microbes from gaining access to the bloodstream. Cell junctions are generally composed of cell membrane proteins that may connect with the extracellular matrix or with complementary proteins from neighboring cells. Tissues in various parts of the body have different types of cell junctions. These include tight junctions, desmosomes, and gap junctions, as illustrated in Figure 17.2. Invading microorganisms may attempt to break down these substances chemically, using enzymes such as proteases that can cause structural damage to create a point of entry for pathogens.

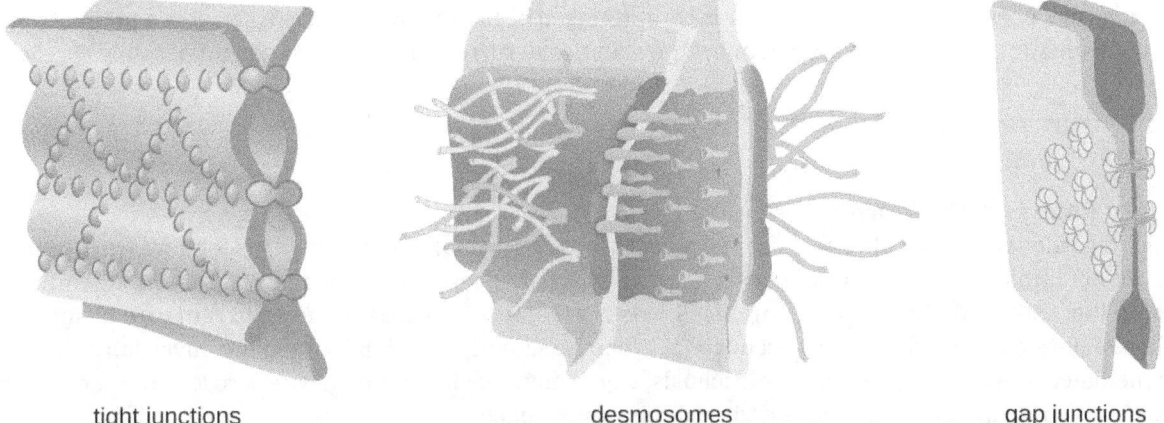

Figure 17.2 There are multiple types of cell junctions in human tissue, three of which are shown here. Tight junctions rivet two adjacent cells together, preventing or limiting material exchange through the spaces between them. Desmosomes have intermediate fibers that act like shoelaces, tying two cells together, allowing small materials to pass through the resulting spaces. Gap junctions are channels between

two cells that permit their communication via signals. (credit: modification of work by Mariana Ruiz Villareal)

The Skin Barrier

One of the body's most important physical barriers is the skin barrier, which is composed of three layers of closely packed cells. The thin upper layer is called the epidermis. A second, thicker layer, called the dermis, contains hair follicles, sweat glands, nerves, and blood vessels. A layer of fatty tissue called the hypodermis lies beneath the dermis and contains blood and lymph vessels (Figure 17.3).

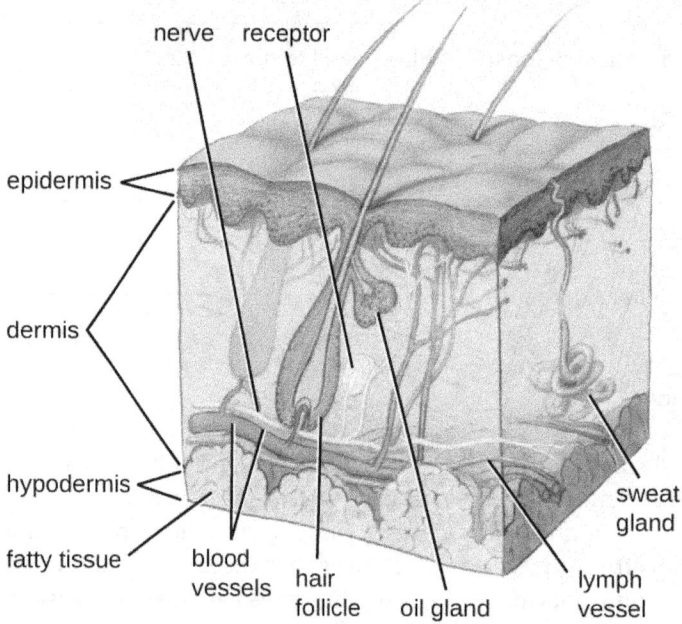

Figure 17.3 Human skin has three layers, the epidermis, the dermis, and the hypodermis, which provide a thick barrier between microbes outside the body and deeper tissues. Dead skin cells on the surface of the epidermis are continually shed, taking with them microbes on the skin's surface. (credit: modification of work by National Institutes of Health)

The topmost layer of skin, the epidermis, consists of cells that are packed with keratin. These dead cells remain as a tightly connected, dense layer of protein-filled cell husks on the surface of the skin. The keratin makes the skin's surface mechanically tough and resistant to degradation by bacterial enzymes. Fatty acids on the skin's surface create a dry, salty, and acidic environment that inhibits the growth of some microbes and is highly resistant to breakdown by bacterial enzymes. In addition, the dead cells of the epidermis are frequently shed, along with any microbes that may be clinging to them. Shed skin cells are continually replaced with new cells from below, providing a new barrier that will soon be shed in the same way.

Infections can occur when the skin barrier is compromised or broken. A wound can serve as a point of entry for opportunistic pathogens, which can infect the skin tissue surrounding the wound and possibly spread to deeper tissues.

Case in Point

Every Rose Has its Thorn

Mike, a gardener from southern California, recently noticed a small red bump on his left forearm. Initially, he did not think much of it, but soon it grew larger and then ulcerated (opened up), becoming a painful lesion that extended across a large part of his forearm (Figure 17.4). He went to an urgent care facility, where a physician asked about his occupation. When he said he was a landscaper, the physician immediately suspected a case of sporotrichosis, a type of fungal infection known as rose gardener's disease because it often afflicts landscapers and gardening enthusiasts.

Under most conditions, fungi cannot produce skin infections in healthy individuals. Fungi grow filaments known as hyphae, which are not particularly invasive and can be easily kept at bay by the physical barriers of the skin and mucous membranes. However, small wounds in the skin, such as those caused by thorns, can provide an opening for opportunistic pathogens like *Sporothrix schenkii,* a soil-dwelling fungus and the causative agent of rose gardener's disease. Once it breaches the skin barrier, *S. schenkii* can infect the skin and underlying tissues, producing ulcerated lesions like Mike's. Compounding matters, other pathogens may enter the infected tissue, causing secondary bacterial infections.

Luckily, rose gardener's disease is treatable. Mike's physician wrote him a prescription for some antifungal drugs as well as a course of antibiotics to combat secondary bacterial infections. His lesions eventually healed, and Mike returned to work with a new appreciation for gloves and protective clothing.

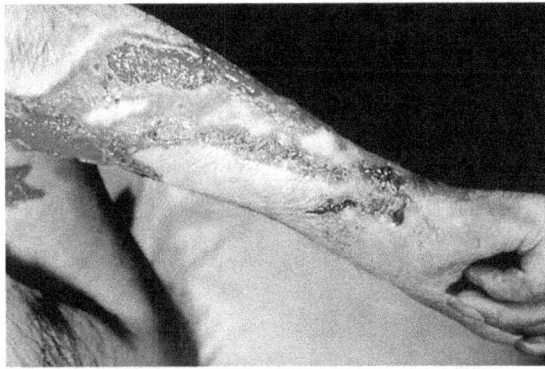 

Figure 17.4 Rose gardener's disease can occur when the fungus *Sporothrix schenkii* breaches the skin through small cuts, such as might be inflicted by thorns. (credit left: modification of work by Elisa Self; credit right: modification of work by Centers for Disease Control and Prevention)

Mucous Membranes

The **mucous membranes** lining the nose, mouth, lungs, and urinary and digestive tracts provide another nonspecific barrier against potential pathogens. Mucous membranes consist of a layer of epithelial cells bound by tight junctions. The epithelial cells secrete a moist, sticky substance called **mucus**, which covers and protects the more fragile cell layers beneath it and traps debris and particulate matter, including microbes. Mucus secretions also contain antimicrobial peptides.

In many regions of the body, mechanical actions serve to flush mucus (along with trapped or dead microbes) out of the body or away from potential sites of infection. For example, in the respiratory system, inhalation can bring microbes, dust, mold spores, and other small airborne debris into the body. This debris becomes trapped in the mucus lining the respiratory tract, a layer known as the mucociliary blanket. The epithelial cells lining the upper parts of the respiratory tract are called **ciliated epithelial cells** because they have hair-like appendages known as cilia. Movement of the cilia propels debris-laden mucus out and away from the lungs. The expelled mucus is then swallowed and destroyed in the stomach, or coughed up, or sneezed out (Figure 17.5). This system of removal is often called the **mucociliary escalator**.

Figure 17.5 This scanning electron micrograph shows ciliated and nonciliated epithelial cells from the human trachea. The mucociliary escalator pushes mucus away from the lungs, along with any debris or microorganisms that may be trapped in the sticky mucus, and the mucus moves up to the esophagus where it can be removed by swallowing.

The mucociliary escalator is such an effective barrier to microbes that the lungs, the lowermost (and most sensitive) portion of the respiratory tract, were long considered to be a sterile environment in healthy individuals. Only recently has research suggested that healthy lungs may have a small normal microbiota. Disruption of the mucociliary escalator by the damaging effects of smoking or diseases such as cystic fibrosis can lead to increased colonization of bacteria in the lower respiratory tract and frequent infections, which highlights the importance of this physical barrier to host defenses.

Like the respiratory tract, the digestive tract is a portal of entry through which microbes enter the body, and the mucous membranes lining the digestive tract provide a nonspecific physical barrier against ingested microbes. The intestinal tract is lined with epithelial cells, interspersed with mucus-secreting goblet cells (Figure 17.6). This mucus mixes with material received from the stomach, trapping foodborne microbes and debris. The mechanical action of **peristalsis**, a series of muscular contractions in the digestive tract, moves the sloughed mucus and other material through the intestines, rectum, and anus, excreting the material in feces.

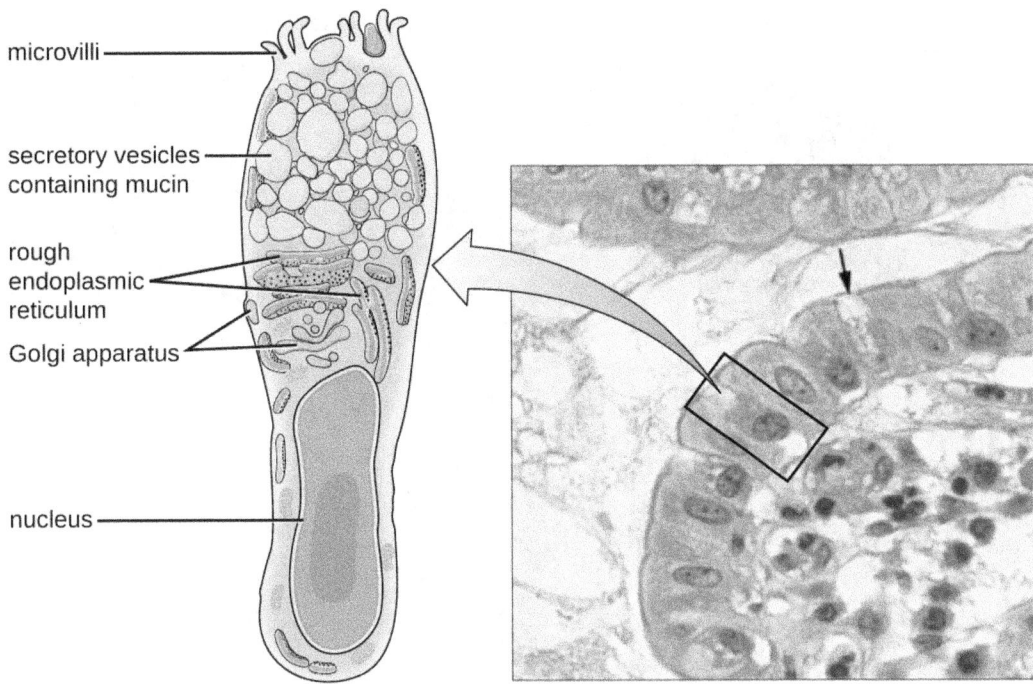

Figure 17.6 Goblet cells produce and secrete mucus. The arrows in this micrograph point to the mucus-secreting goblet cells (magnification 1600×) in the intestinal epithelium. (credit micrograph: Micrograph provided by the Regents of University of Michigan Medical School © 2012)

Endothelia

The epithelial cells lining the urogenital tract, blood vessels, lymphatic vessels, and certain other tissues are known as **endothelia**. These tightly packed cells provide a particularly effective frontline barrier against invaders. The endothelia of the **blood-brain barrier**, for example, protect the central nervous system (CNS), which consists of the brain and the spinal cord. The CNS is one of the most sensitive and important areas of the body, as microbial infection of the CNS can quickly lead to serious and often fatal inflammation. The cell junctions in the blood vessels traveling through the CNS are some of the tightest and toughest in the body, preventing any transient microbes in the bloodstream from entering the CNS. This keeps the cerebrospinal fluid that surrounds and bathes the brain and spinal cord sterile under normal conditions.

⊘ CHECK YOUR UNDERSTANDING

- Describe how the mucociliary escalator functions.
- Name two places you would find endothelia.

Mechanical Defenses

In addition to physical barriers that keep microbes out, the body has a number of mechanical defenses that physically remove pathogens from the body, preventing them from taking up residence. We have already discussed several examples of mechanical defenses, including the shedding of skin cells, the expulsion of mucus via the mucociliary escalator, and the excretion of feces through intestinal peristalsis. Other important examples of mechanical defenses include the flushing action of urine and tears, which both serve to carry microbes away from the body. The flushing action of urine is largely responsible for the normally sterile environment of the urinary tract, which includes the kidneys, ureters, and urinary bladder. Urine passing out of the body washes out transient microorganisms, preventing them from taking up residence. The eyes also have physical barriers and mechanical mechanisms for preventing infections. The eyelashes and eyelids prevent dust and airborne microorganisms from reaching the surface of the eye. Any microbes or debris that make it past these physical barriers may be flushed out by the mechanical action of blinking, which bathes the eye in tears, washing debris away (Figure 17.7).

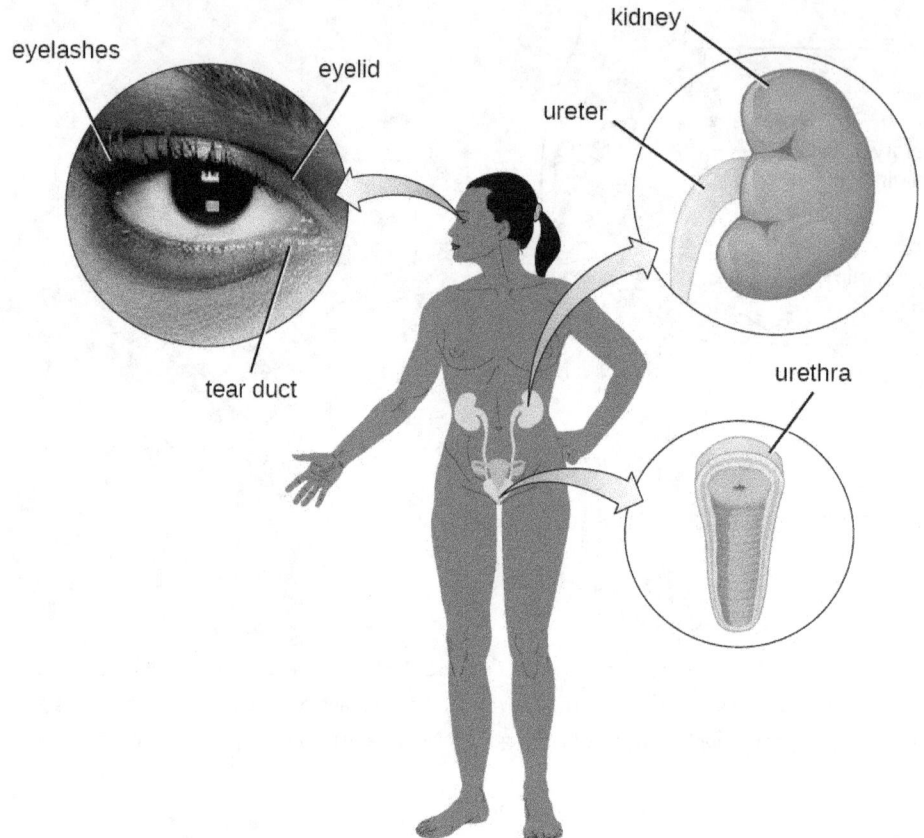

Figure 17.7 Tears flush microbes away from the surface of the eye. Urine washes microbes out of the urinary tract as it passes through; as a result, the urinary system is normally sterile.

✓ CHECK YOUR UNDERSTANDING

- Name two mechanical defenses that protect the eyes.

Microbiome

In various regions of the body, resident microbiota serve as an important first-line defense against invading pathogens. Through their occupation of cellular binding sites and competition for available nutrients, the resident microbiota prevent the critical early steps of pathogen attachment and proliferation required for the establishment of an infection. For example, in the vagina, members of the resident microbiota compete with opportunistic pathogens like the yeast *Candida*. This competition prevents infections by limiting the availability of nutrients, thus inhibiting the growth of *Candida*, keeping its population in check. Similar competitions occur between the microbiota and potential pathogens on the skin, in the upper respiratory tract, and in the gastrointestinal tract. As will be discussed later in this chapter, the resident microbiota also contribute to the chemical defenses of the innate nonspecific host defenses.

The importance of the normal microbiota in host defenses is highlighted by the increased susceptibility to infectious diseases when the microbiota is disrupted or eliminated. Treatment with antibiotics can significantly deplete the normal microbiota of the gastrointestinal tract, providing an advantage for pathogenic bacteria to colonize and cause diarrheal infection. In the case of diarrhea caused by *Clostridium difficile*, the infection can be severe and potentially lethal. One strategy for treating *C. difficile* infections is fecal transplantation, which involves the transfer of fecal material from a donor (screened for potential pathogens) into the intestines of the recipient patient as a method of restoring the normal microbiota and combating *C. difficile* infections.

Table 17.2 provides a summary of the physical defenses discussed in this section.

Physical Defenses of Nonspecific Innate Immunity

Defense	Examples	Function
Cellular barriers	Skin, mucous membranes, endothelial cells	Deny entry to pathogens
Mechanical defenses	Shedding of skin cells, mucociliary sweeping, peristalsis, flushing action of urine and tears	Remove pathogens from potential sites of infection
Microbiome	Resident bacteria of the skin, upper respiratory tract, gastrointestinal tract, and genitourinary tract	Compete with pathogens for cellular binding sites and nutrients

Table 17.2

CHECK YOUR UNDERSTANDING

- List two ways resident microbiota defend against pathogens.

17.2 Chemical Defenses

Learning Objectives

By the end of this section, you will be able to:
- Describe how enzymes in body fluids provide protection against infection or disease
- List and describe the function of antimicrobial peptides, complement components, cytokines, and acute-phase proteins
- Describe similarities and differences among classic, alternate, and lectin complement pathways

In addition to physical defenses, the innate nonspecific immune system uses a number of **chemical mediators** that inhibit microbial invaders. The term "chemical mediators" encompasses a wide array of substances found in various body fluids and tissues throughout the body. Chemical mediators may work alone or in conjunction with each other to inhibit microbial colonization and infection.

Some chemical mediators are endogenously produced, meaning they are produced by human body cells; others are produced exogenously, meaning that they are produced by certain microbes that are part of the microbiome. Some mediators are produced continually, bathing the area in the antimicrobial substance; others are produced or activated primarily in response to some stimulus, such as the presence of microbes.

Chemical and Enzymatic Mediators Found in Body Fluids

Fluids produced by the skin include examples of both endogenous and exogenous mediators. Sebaceous glands in the dermis secrete an oil called sebum that is released onto the skin surface through hair follicles. This sebum is an endogenous mediator, providing an additional layer of defense by helping seal off the pore of the hair follicle, preventing bacteria on the skin's surface from invading sweat glands and surrounding tissue (Figure 17.8). Certain members of the microbiome, such as the bacterium *Propionibacterium acnes* and the fungus *Malassezia*, among others, can use lipase enzymes to degrade sebum, using it as a food source. This produces oleic acid, which creates a mildly acidic environment on the surface of the skin that is inhospitable to many pathogenic microbes. Oleic acid is an example of an exogenously produced mediator because it is produced by resident microbes and not directly by body cells.

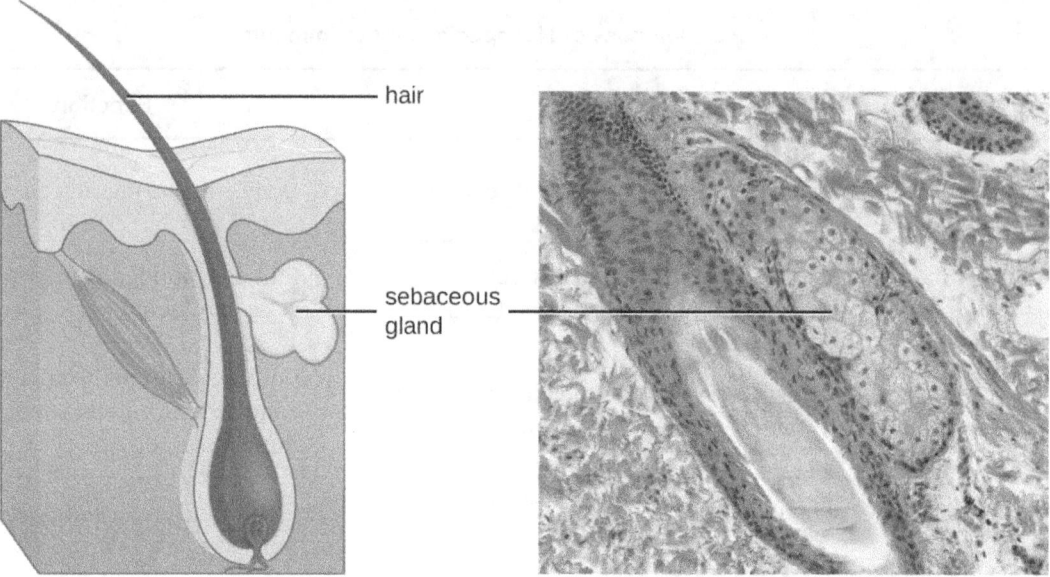

Figure 17.8 Sebaceous glands secrete sebum, a chemical mediator that lubricates and protect the skin from invading microbes. Sebum is also a food source for resident microbes that produce oleic acid, an exogenously produced mediator. (credit micrograph: Micrograph provided by the Regents of University of Michigan Medical School © 2012)

Environmental factors that affect the microbiota of the skin can have a direct impact on the production of chemical mediators. Low humidity or decreased sebum production, for example, could make the skin less habitable for microbes that produce oleic acid, thus making the skin more susceptible to pathogens normally inhibited by the skin's low pH. Many skin moisturizers are formulated to counter such effects by restoring moisture and essential oils to the skin.

The digestive tract also produces a large number of chemical mediators that inhibit or kill microbes. In the oral cavity, saliva contains mediators such as lactoperoxidase enzymes, and mucus secreted by the esophagus contains the antibacterial enzyme lysozyme. In the stomach, highly acidic gastric fluid kills most microbes. In the lower digestive tract, the intestines have pancreatic and intestinal enzymes, antibacterial peptides (cryptins), bile produced from the liver, and specialized Paneth cells that produce lysozyme. Together, these mediators are able to eliminate most pathogens that manage to survive the acidic environment of the stomach.

In the urinary tract, urine flushes microbes out of the body during urination. Furthermore, the slight acidity of urine (the average pH is about 6) inhibits the growth of many microbes and potential pathogens in the urinary tract.

The female reproductive system employs lactate, an exogenously produced chemical mediator, to inhibit microbial growth. The cells and tissue layers composing the vagina produce glycogen, a branched and more complex polymer of glucose. Lactobacilli in the area ferment glycogen to produce lactate, lowering the pH in the vagina and inhibiting transient microbiota, opportunistic pathogens like *Candida* (a yeast associated with vaginal infections), and other pathogens responsible for sexually transmitted diseases.

In the eyes, tears contain the chemical mediators lysozyme and lactoferrin, both of which are capable of eliminating microbes that have found their way to the surface of the eyes. Lysozyme cleaves the bond between NAG and NAM in peptidoglycan, a component of the cell wall in bacteria. It is more effective against gram-positive bacteria, which lack the protective outer membrane associated with gram-negative bacteria. Lactoferrin inhibits microbial growth by chemically binding and sequestering iron. This effectually starves many microbes that require iron for growth.

In the ears, cerumen (earwax) exhibits antimicrobial properties due to the presence of fatty acids, which lower the pH to between 3 and 5.

The respiratory tract uses various chemical mediators in the nasal passages, trachea, and lungs. The mucus produced in the nasal passages contains a mix of antimicrobial molecules similar to those found in tears and saliva (e.g., lysozyme, lactoferrin, lactoperoxidase). Secretions in the trachea and lungs also contain lysozyme and lactoferrin, as well as a diverse group of additional chemical mediators, such as the lipoprotein complex called surfactant, which has antibacterial properties.

✓ CHECK YOUR UNDERSTANDING

- Explain the difference between endogenous and exogenous mediators
- Describe how pH affects antimicrobial defenses

Antimicrobial Peptides

The **antimicrobial peptides (AMPs)** are a special class of nonspecific cell-derived mediators with broad-spectrum antimicrobial properties. Some AMPs are produced routinely by the body, whereas others are primarily produced (or produced in greater quantities) in response to the presence of an invading pathogen. Research has begun exploring how AMPs can be used in the diagnosis and treatment of disease.

AMPs may induce cell damage in microorganisms in a variety of ways, including by inflicting damage to membranes, destroying DNA and RNA, or interfering with cell-wall synthesis. Depending on the specific antimicrobial mechanism, a particular AMP may inhibit only certain groups of microbes (e.g., gram-positive or gram-negative bacteria) or it may be more broadly effective against bacteria, fungi, protozoa, and viruses. Many AMPs are found on the skin, but they can also be found in other regions of the body.

A family of AMPs called defensins can be produced by epithelial cells throughout the body as well as by cellular defenses such as macrophages and neutrophils (see Cellular Defenses). Defensins may be secreted or act inside host cells; they combat microorganisms by damaging their plasma membranes. AMPs called bacteriocins are produced exogenously by certain members of the resident microbiota within the gastrointestinal tract. The genes coding for these types of AMPs are often carried on plasmids and can be passed between different species within the resident microbiota through lateral or horizontal gene transfer.

There are numerous other AMPs throughout the body. The characteristics of a few of the more significant AMPs are summarized in Table 17.3.

Characteristics of Selected Antimicrobial Peptides (AMPs)

AMP	Secreted by	Body site	Pathogens inhibited	Mode of action
Bacteriocins	Resident microbiota	Gastrointestinal tract	Bacteria	Disrupt membrane
Cathelicidin	Epithelial cells, macrophages, and other cell types	Skin	Bacteria and fungi	Disrupts membrane
Defensins	Epithelial cells, macrophages, neutrophils	Throughout the body	Fungi, bacteria, and many viruses	Disrupt membrane
Dermicidin	Sweat glands	Skin	Bacteria and fungi	Disrupts membrane integrity and ion channels
Histatins	Salivary glands	Oral cavity	Fungi	Disrupt intracellular function

Table 17.3

✓ CHECK YOUR UNDERSTANDING

- Why are antimicrobial peptides (AMPs) considered nonspecific defenses?

Plasma Protein Mediators

Many nonspecific innate immune factors are found in **plasma**, the fluid portion of blood. Plasma contains electrolytes, sugars, lipids, and proteins, each of which helps to maintain homeostasis (i.e., stable internal body functioning), and contains the proteins involved in the clotting of blood. Additional proteins found in blood plasma, such as acute-phase proteins, complement proteins, and cytokines, are involved in the nonspecific innate immune response.

 MICRO CONNECTIONS

Plasma versus Serum

There are two terms for the fluid portion of blood: plasma and serum. How do they differ if they are both fluid and lack cells? The fluid portion of blood left over after coagulation (blood cell clotting) has taken place is serum. Although molecules such as many vitamins, electrolytes, certain sugars, complement proteins, and antibodies are still present in serum, clotting factors are largely depleted. Plasma, conversely, still contains all the clotting elements. To obtain plasma from blood, an anticoagulant must be used to prevent clotting. Examples of anticoagulants include heparin and ethylene diamine tetraacetic acid (EDTA). Because clotting is inhibited, once obtained, the sample must be gently spun down in a centrifuge. The heavier, denser blood cells form a pellet at the bottom of a centrifuge tube, while the fluid plasma portion, which is lighter and less dense, remains above the cell pellet.

Acute-Phase Proteins

The **acute-phase proteins** are another class of antimicrobial mediators. Acute-phase proteins are primarily produced in the liver and secreted into the blood in response to inflammatory molecules from the immune system. Examples of acute-phase proteins include C-reactive protein, serum amyloid A, ferritin, transferrin, fibrinogen, and mannose-binding lectin. Each of these proteins has a different chemical structure and inhibits or destroys microbes in some way (Table 17.4).

Some Acute-Phase Proteins and Their Functions

C-reactive protein	Coats bacteria (opsonization), preparing them for ingestion by phagocytes
Serum amyloid A	
Ferritin	Bind and sequester iron, thereby inhibiting the growth of pathogens
Transferrin	
Fibrinogen	Involved in formation of blood clots that trap bacterial pathogens
Mannose-binding lectin	Activates complement cascade

Table 17.4

The Complement System

The **complement system** is a group of plasma protein mediators that can act as an innate nonspecific defense while also serving to connect innate and adaptive immunity (discussed in the next chapter). The complement system is composed of more than 30 proteins (including C1 through C9) that normally circulate as precursor proteins in blood. These precursor proteins become activated when stimulated or triggered by a variety of factors, including the presence of microorganisms. Complement proteins are considered part of innate nonspecific immunity because they are always present in the blood and tissue fluids, allowing them to be activated quickly. Also, when activated through the alternative pathway (described later in this section), complement proteins target pathogens in a nonspecific manner.

The process by which circulating complement precursors become functional is called **complement activation**. This process is a cascade that can be triggered by one of three different mechanisms, known as the alternative, classical, and lectin pathways.

The alternative pathway is initiated by the spontaneous activation of the complement protein C3. The hydrolysis of C3 produces two products, C3a and C3b. When no invader microbes are present, C3b is very quickly degraded in a hydrolysis reaction using the water in the blood. However, if invading microbes are present, C3b attaches to the surface of these microbes. Once attached, C3b will recruit other complement proteins in a cascade (Figure 17.9).

The classical pathway provides a more efficient mechanism of activating the complement cascade, but it depends upon the production of antibodies by the specific adaptive immune defenses. To initiate the classical pathway, a specific antibody must first bind to the pathogen to form an antibody-antigen complex. This activates the first protein in the complement cascade, the C1 complex. The C1 complex is a multipart protein complex, and each component participates in the full activation of the overall complex. Following recruitment and activation of the C1 complex, the remaining classical pathway complement proteins are recruited and activated in a cascading sequence (Figure 17.9).

The lectin activation pathway is similar to the classical pathway, but it is triggered by the binding of mannose-binding lectin, an acute-phase protein, to carbohydrates on the microbial surface. Like other acute-phase proteins, lectins are produced by liver cells and are commonly upregulated in response to inflammatory signals received by the body during an infection (Figure 17.9).

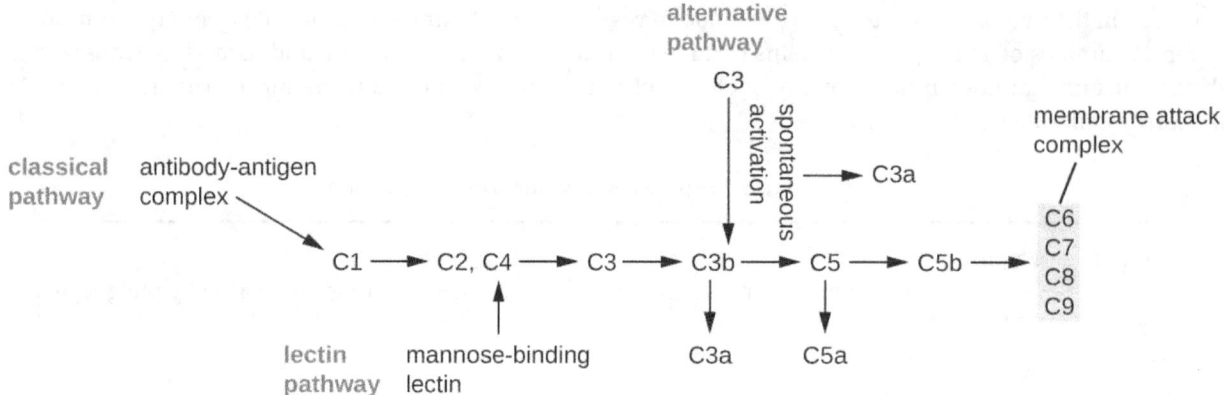

Figure 17.9 The three complement activation pathways have different triggers, as shown here, but all three result in the activation of the complement protein C3, which produces C3a and C3b. The latter binds to the surface of the target cell and then works with other complement proteins to cleave C5 into C5a and C5b. C5b also binds to the cell surface and then recruits C6 through C9; these molecules form a ring structure called the membrane attack complex (MAC), which punches through the cell membrane of the invading pathogen, causing it to swell and burst.

Although each complement activation pathway is initiated in a different way, they all provide the same protective outcomes: opsonization, inflammation, chemotaxis, and cytolysis. The term **opsonization** refers to the coating of a pathogen by a chemical substance (called an **opsonin**) that allows phagocytic cells to recognize, engulf, and destroy it more easily. Opsonins from the complement cascade include C1q, C3b, and C4b. Additional important opsonins include mannose-binding proteins and antibodies. The complement fragments C3a and C5a are well-characterized anaphylatoxins with potent proinflammatory functions. Anaphylatoxins activate mast cells, causing degranulation and the release of inflammatory chemical signals, including mediators that cause vasodilation and increased vascular permeability. C5a is also one of the most potent chemoattractants for neutrophils and other white blood cells, cellular defenses that will be discussed in the next section.

The complement proteins C6, C7, C8, and C9 assemble into a **membrane attack complex (MAC)**, which allows C9 to polymerize into pores in the membranes of gram-negative bacteria. These pores allow water, ions, and other molecules to move freely in and out of the targeted cells, eventually leading to cell lysis and death of the pathogen (Figure 17.9). However, the MAC is only effective against gram-negative bacteria; it cannot penetrate the thick layer of peptidoglycan associated with cell walls of gram-positive bacteria. Since the MAC does not pose a lethal threat to gram-positive bacterial pathogens, complement-mediated opsonization is more important for their clearance.

Cytokines

Cytokines are soluble proteins that act as communication signals between cells. In a nonspecific innate immune response, various cytokines may be released to stimulate production of chemical mediators or other cell functions, such as cell proliferation, cell differentiation, inhibition of cell division, apoptosis, and chemotaxis.

When a cytokine binds to its target receptor, the effect can vary widely depending on the type of cytokine and the type of cell or receptor to which it has bound. The function of a particular cytokine can be described as autocrine, paracrine, or endocrine (Figure 17.10). In **autocrine function**, the same cell that releases the cytokine is the recipient of the signal; in other words, autocrine function is a form of self-stimulation by a cell. In contrast, **paracrine function** involves the release of cytokines from one cell to other nearby cells, stimulating some response from the recipient cells. Last, **endocrine function** occurs when cells release cytokines into the bloodstream to be carried to target cells much farther away.

CYTOKINES: Molecular Messengers		
Autocrine	**Paracrine**	**Endocrine**
Same cell secretes and receives cytokine signal.	Cytokine signal secreted to a nearby cell.	Cytokine signal secreted to circulatory system; travels to distant cells.

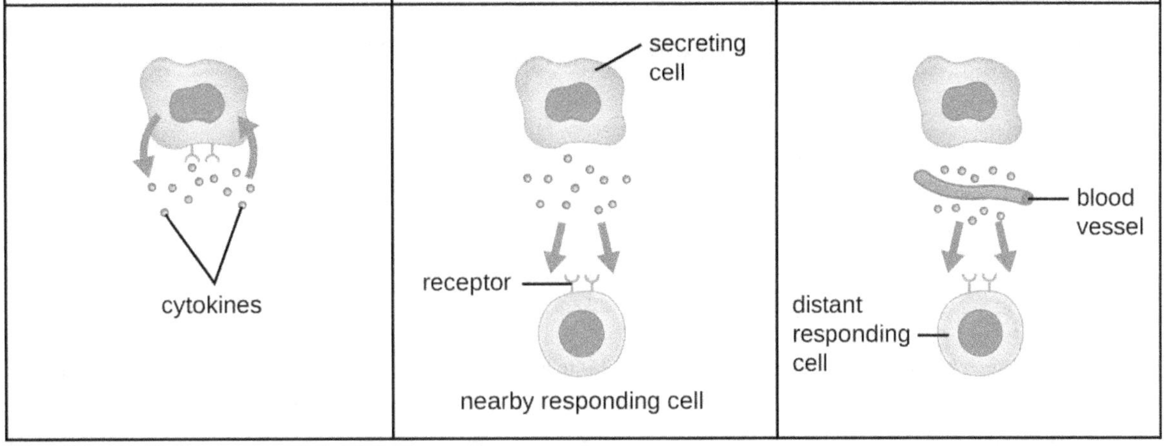

Figure 17.10 Autocrine, paracrine, and endocrine actions describe which cells are targeted by cytokines and how far the cytokines must travel to bind to their intended target cells' receptors.

Three important classes of cytokines are the interleukins, chemokines, and interferons. The **interleukins** were originally thought to be produced only by leukocytes (white blood cells) and to only stimulate leukocytes, thus the reasons for their name. Although interleukins are involved in modulating almost every function of the immune system, their role in the body is not restricted to immunity. Interleukins are also produced by and stimulate a variety of cells unrelated to immune defenses.

The **chemokines** are chemotactic factors that recruit leukocytes to sites of infection, tissue damage, and inflammation. In contrast to more general chemotactic factors, like complement factor C5a, chemokines are very specific in the subsets of leukocytes they recruit.

Interferons are a diverse group of immune signaling molecules and are especially important in our defense against viruses. Type I **interferons** (interferon-α and interferon-β) are produced and released by cells infected with virus. These interferons stimulate nearby cells to stop production of mRNA, destroy RNA already produced, and reduce protein synthesis. These cellular changes inhibit viral replication and production of mature virus, slowing the spread of the virus. Type I interferons also stimulate various immune cells involved in viral clearance to more aggressively attack virus-infected cells. Type II interferon (interferon-γ) is an important activator of immune cells (Figure 17.11).

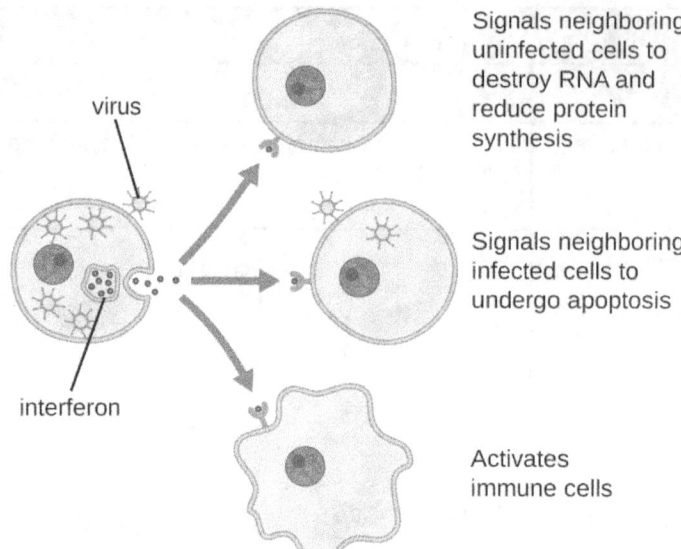

Figure 17.11 Interferons are cytokines released by a cell infected with a virus. Interferon-α and interferon-β signal uninfected neighboring cells to inhibit mRNA synthesis, destroy RNA, and reduce protein synthesis (top arrow). Interferon-α and interferon-β also promote apoptosis in cells infected with the virus (middle arrow). Interferon-γ alerts neighboring immune cells to an attack (bottom arrow). Although interferons do not cure the cell releasing them or other infected cells, which will soon die, their release may prevent additional cells from becoming infected, thus stemming the infection.

Inflammation-Eliciting Mediators

Many of the chemical mediators discussed in this section contribute in some way to inflammation and fever, which are nonspecific immune responses discussed in more detail in Inflammation and Fever. Cytokines stimulate the production of acute-phase proteins such as C-reactive protein and mannose-binding lectin in the liver. These acute-phase proteins act as opsonins, activating complement cascades through the lectin pathway.

Some cytokines also bind mast cells and basophils, inducing them to release **histamine**, a proinflammatory compound. Histamine receptors are found on a variety of cells and mediate proinflammatory events, such as bronchoconstriction (tightening of the airways) and smooth muscle contraction.

In addition to histamine, mast cells may release other chemical mediators, such as **leukotrienes**. Leukotrienes are lipid-based proinflammatory mediators that are produced from the metabolism of arachidonic acid in the cell membrane of leukocytes and tissue cells. Compared with the proinflammatory effects of histamine, those of leukotrienes are more potent and longer lasting. Together, these chemical mediators can induce coughing, vomiting, and diarrhea, which serve to expel pathogens from the body.

Certain cytokines also stimulate the production of prostaglandins, chemical mediators that promote the inflammatory effects of kinins and histamines. Prostaglandins can also help to set the body temperature higher, leading to fever, which promotes the activities of white blood cells and slightly inhibits the growth of pathogenic microbes (see Inflammation and Fever).

Another inflammatory mediator, **bradykinin**, contributes to edema, which occurs when fluids and leukocytes leak out of the bloodstream and into tissues. It binds to receptors on cells in the capillary walls, causing the capillaries to dilate and become more permeable to fluids.

✓ CHECK YOUR UNDERSTANDING

- What do the three complement activation pathways have in common?
- Explain autocrine, paracrine, and endocrine signals.
- Name two important inflammation-eliciting mediators.

Clinical Focus

Part 2

To relieve the constriction of her airways, Angela is immediately treated with antihistamines and administered corticosteroids through an inhaler, and then monitored for a period of time. Though her condition does not worsen, the drugs do not seem to be alleviating her condition. She is admitted to the hospital for further observation, testing, and treatment.

Following admission, a clinician conducts allergy testing to try to determine if something in her environment might be triggering an allergic inflammatory response. A doctor orders blood analysis to check for levels of particular cytokines. A sputum sample is also taken and sent to the lab for microbial staining, culturing, and identification of pathogens that could be causing an infection.

- Which aspects of the innate immune system could be contributing to Angela's airway constriction?
- Why was Angela treated with antihistamines?
- Why would the doctor be interested in levels of cytokines in Angela's blood?

Jump to the next Clinical Focus box. Go back to the previous Clinical Focus box.

Table 17.5 provides a summary of the chemical defenses discussed in this section.

Chemical Defenses of Nonspecific Innate Immunity

Defense	Examples	Function
Chemicals and enzymes in body fluids	Sebum from sebaceous glands	Provides oil barrier protecting hair follicle pores from pathogens
	Oleic acid from sebum and skin microbiota	Lowers pH to inhibit pathogens
	Lysozyme in secretions	Kills bacteria by attacking cell wall
	Acid in stomach, urine, and vagina	Inhibits or kills bacteria
	Digestive enzymes and bile	Kill bacteria
	Lactoferrin and transferrin	Bind and sequester iron, inhibiting bacterial growth
	Surfactant in lungs	Kills bacteria
Antimicrobial peptides	Defensins, bacteriocins, dermicidin, cathelicidin, histatins,	Kill bacteria by attacking membranes or interfering with cell functions
Plasma protein mediators	Acute-phase proteins (C-reactive protein, serum amyloid A, ferritin, fibrinogen, transferrin, and mannose-binding lectin)	Inhibit the growth of bacteria and assist in the trapping and killing of bacteria
	Complements C3b and C4b	Opsonization of pathogens to aid phagocytosis

Chemical Defenses of Nonspecific Innate Immunity		
Defense	Examples	Function
	Complement C5a	Chemoattractant for phagocytes
	Complements C3a and C5a	Proinflammatory anaphylatoxins
Cytokines	Interleukins	Stimulate and modulate most functions of immune system
	Chemokines	Recruit white blood cells to infected area
	Interferons	Alert cells to viral infection, induce apoptosis of virus-infected cells, induce antiviral defenses in infected and nearby uninfected cells, stimulate immune cells to attack virus-infected cells
Inflammation-eliciting mediators	Histamine	Promotes vasodilation, bronchoconstriction, smooth muscle contraction, increased secretion and mucus production
	Leukotrienes	Promote inflammation; stronger and longer lasting than histamine
	Prostaglandins	Promote inflammation and fever
	Bradykinin	Increases vasodilation and vascular permeability, leading to edema

Table 17.5

17.3 Cellular Defenses

Learning Objectives

By the end of this section, you will be able to:
- Identify and describe the components of blood
- Explain the process by which the formed elements of blood are formed (hematopoiesis)
- Describe the characteristics of formed elements found in peripheral blood, as well as their respective functions within the innate immune system

In the previous section, we discussed some of the chemical mediators found in plasma, the fluid portion of blood. The nonfluid portion of blood consists of various types of formed elements, so called because they are all formed from the same stem cells found in bone marrow. The three major categories of formed elements are: red blood cells (RBCs), also called **erythrocytes**; **platelets**, also called **thrombocytes**; and white blood cells (WBCs), also called **leukocytes**.

Red blood cells are primarily responsible for carrying oxygen to tissues. Platelets are cellular fragments that participate in blood clot formation and tissue repair. Several different types of WBCs participate in various nonspecific mechanisms of innate and adaptive immunity. In this section, we will focus primarily on the innate mechanisms of various types of WBCs.

Hematopoiesis

All of the formed elements of blood are derived from pluripotent hematopoietic stem cells (HSCs) in the bone marrow. As the HSCs make copies of themselves in the bone marrow, individual cells receive different cues from the body that control how they develop and mature. As a result, the HSCs differentiate into different types of blood cells that, once mature, circulate in peripheral blood. This process of differentiation, called **hematopoiesis**, is shown in more detail in Figure 17.12.

In terms of sheer numbers, the vast majority of HSCs become erythrocytes. Much smaller numbers become leukocytes and platelets. Leukocytes can be further subdivided into **granulocytes**, which are characterized by numerous granules visible in the cytoplasm, and agranulocytes, which lack granules. Figure 17.13 provides an overview of the various types of formed elements, including their relative numbers, primary function, and lifespans.

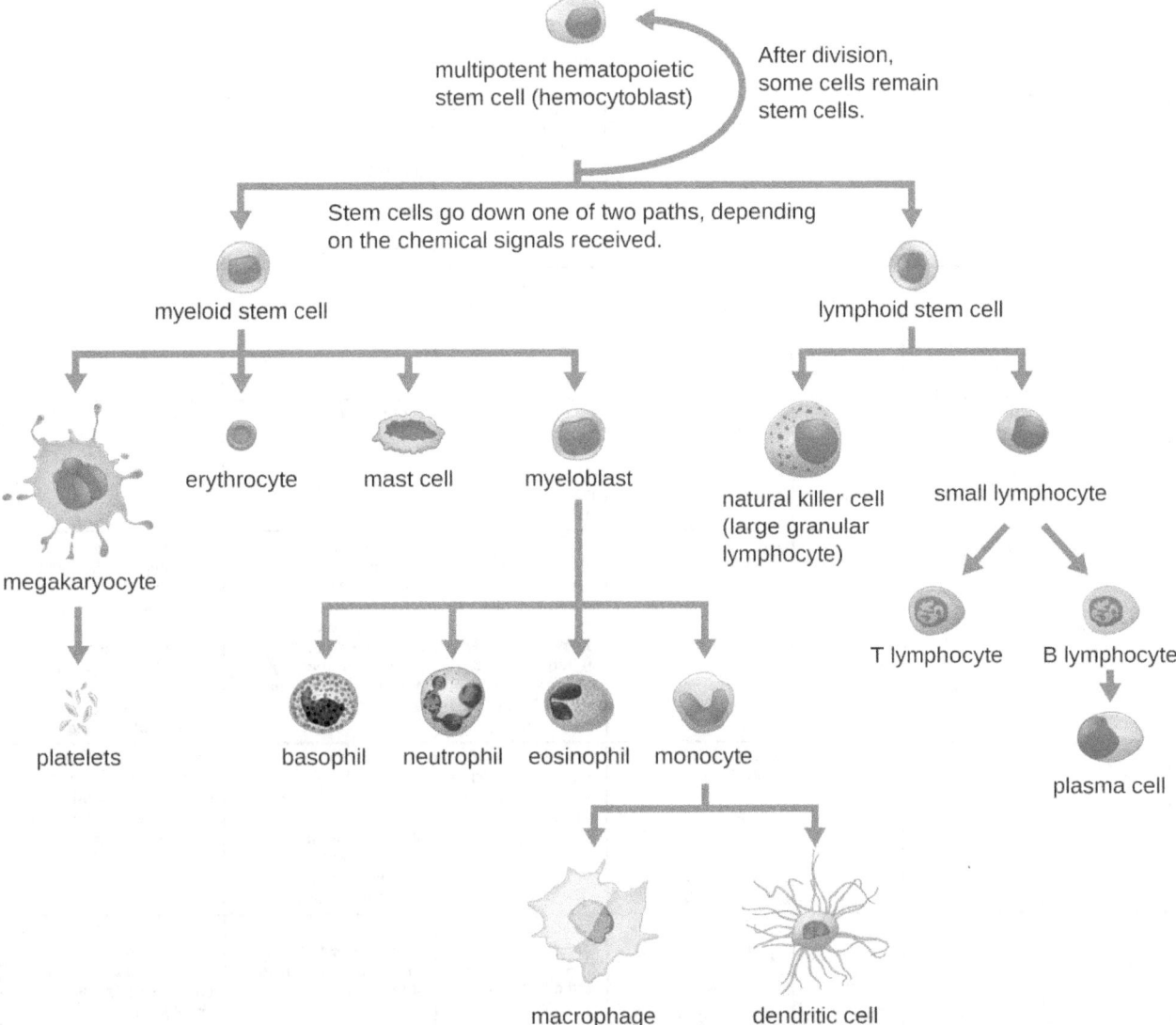

Figure 17.12   All the formed elements of the blood arise by differentiation of hematopoietic stem cells in the bone marrow.

Formed Element	Major Subtypes	Numbers Present per Microliter (μL) and Mean (Range)	Appearance in a Standard Blood Smear	Summary of Functions	Comments
Erythrocytes (red blood cells)		5.2 million (4.4–6.0 million)	Flattened biconcave disk; no nucleus; pale red	Transport oxygen and some carbon dioxide between tissue and lungs	Lifespan of approximately 120 days
Leukocytes (white blood cells)		7000 (5000–10,000)	Obvious dark-staining nucleus	All function in body defenses	Exit capillaries and move into tissues; lifespan of usually a few hours or days
	Granulocytes, including neutrophils, eosinophils, and basophils	Total leukocytes (%) — 4360 (1800–9950)	Abundant granules in cytoplasm; nucleus normally lobed	Nonspecific (innate) resistance to disease	Classified according to membrane-bound granules in cytoplasm
	Neutrophils	50–70 — 4150 (1800–7300)	Nucleus lobes increase with age; pale lilac granules	Phagocytic; particularly effective against bacteria; release cytotoxic chemicals from granules	Most common leukocyte; lifespan of minutes to days
	Eosinophils	1–3 — 165 (0–700)	Nucleus generally two-lobed; bright red-orange granules	Phagocytic cells; particularly active with antigen-antibody complexes; release degradative enzymes, toxic proteins and antihistamines; combat parasitic infections	Lifespan of minutes to days
	Basophils	<1 — 44 (0–150)	Nucleus generally two-lobed but difficult to see due to presence of heavy, dense, dark purple granules	Pro-inflammatory	Least common leukocyte; lifespan unknown
	Agranulocytes, including lymphocytes and monocytes	2640 (1700–4950)	Lack abundant granules in cytoplasm; have a simple-shaped nucleus that may be indented	Body defenses	Group consists of two major cell types from different lineages
	Lymphocytes	20–40 — 2185 (1500–4000)	Spherical cells with a single, often large, nucleus occupying much of the cell's volume; stains purple; seen in large (natural killer cells) and small (B and T cells) variants	Primarily specific (adaptive) immunity: T cells directly attack other cells (cellular immunity); B cells release antibodies (humoral immunity); natural killer cells are similar to T cells but nonspecific	Initial cells originate in bone marrow, but secondary production occurs in lymphatic tissue; several distinct subtypes; memory cells form after exposure to a pathogen and rapidly increase responses to subsequent exposure; lifespan of many years
	Monocytes	1–6 — 455 (200–950)	Largest leukocyte; has an indented or horseshoe-shaped nucleus	Very effective phagocytic cells engulfing pathogens or worn-out cells; also serve as antigen-presenting cells (APCs) or other components of the immune system	Produced in red bone marrow; referred to as macrophages and dendritic cells after leaving the circulation
Platelets		350,000 (150,000–500,000)	Cellular fragments surrounded by a plasma membrane and containing granules; stains purple	Hemostasis; release growth factors for repair and healing of tissue	Formed from megakaryocytes that remain in the red bone marrow and shed platelets into circulation

Figure 17.13 Formed elements of blood include erythrocytes (red blood cells), leukocytes (white blood cells), and platelets.

Granulocytes

The various types of granulocytes can be distinguished from one another in a blood smear by the appearance of their nuclei and the contents of their granules, which confer different traits, functions, and staining properties. The **neutrophils**, also called **polymorphonuclear neutrophils (PMNs)**, have a nucleus with three to five lobes and small, numerous, lilac-colored granules. Each lobe of the nucleus is connected by a thin strand of material to the other lobes. The **eosinophils** have fewer lobes in the nucleus (typically 2–3) and larger granules that stain reddish-orange. The **basophils** have a two-lobed nucleus and large granules that stain dark blue or purple (Figure 17.14).

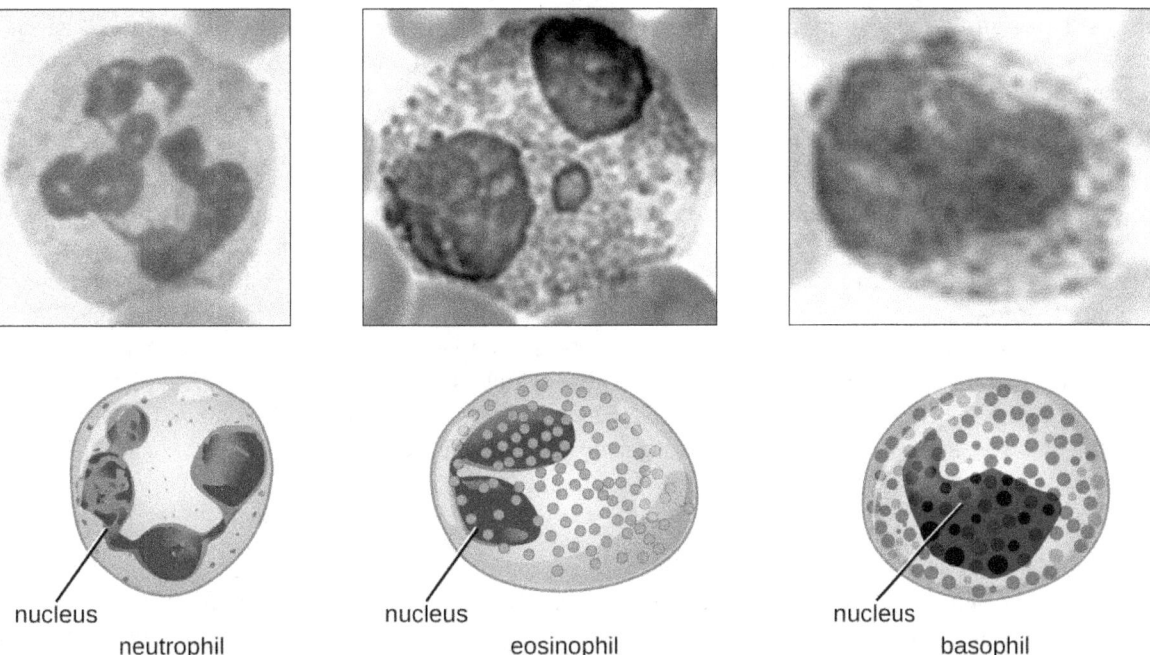

Figure 17.14 Granulocytes can be distinguished by the number of lobes in their nuclei and the staining properties of their granules. (credit "neutrophil" micrograph: modification of work by Ed Uthman)

Neutrophils (PMNs)

Neutrophils (PMNs) are frequently involved in the elimination and destruction of extracellular bacteria. They are capable of migrating through the walls of blood vessels to areas of bacterial infection and tissue damage, where they seek out and kill infectious bacteria. PMN granules contain a variety of defensins and hydrolytic enzymes that help them destroy bacteria through phagocytosis (described in more detail in Pathogen Recognition and Phagocytosis) In addition, when many neutrophils are brought into an infected area, they can be stimulated to release toxic molecules into the surrounding tissue to better clear infectious agents. This is called degranulation.

Another mechanism used by neutrophils is neutrophil extracellular traps (NETs), which are extruded meshes of chromatin that are closely associated with antimicrobial granule proteins and components. Chromatin is DNA with associated proteins (usually histone proteins, around which DNA wraps for organization and packing within a cell). By creating and releasing a mesh or lattice-like structure of chromatin that is coupled with antimicrobial proteins, the neutrophils can mount a highly concentrated and efficient attack against nearby pathogens. Proteins frequently associated with NETs include lactoferrin, gelatinase, cathepsin G, and myeloperoxidase. Each has a different means of promoting antimicrobial activity, helping neutrophils eliminate pathogens. The toxic proteins in NETs may kill some of the body's own cells along with invading pathogens. However, this collateral damage can be repaired after the danger of the infection has been eliminated.

As neutrophils fight an infection, a visible accumulation of leukocytes, cellular debris, and bacteria at the site of infection can be observed. This buildup is what we call **pus** (also known as purulent or suppurative discharge or drainage). The presence of pus is a sign that the immune defenses have been activated against an infection; historically, some physicians believed that inducing pus formation could actually promote the healing of wounds. The practice of promoting "laudable pus" (by, for instance, wrapping a wound in greasy wool soaked in wine) dates back to the ancient physician Galen in the 2nd century AD, and was practiced in variant forms until the 17th century (though it was not universally accepted). Today, this method is no longer practiced because we now know that it is not effective. Although a small amount of pus formation can indicate a strong immune response, artificially inducing pus formation does not promote recovery.

Eosinophils

Eosinophils are granulocytes that protect against protozoa and helminths; they also play a role in allergic reactions. The granules of eosinophils, which readily absorb the acidic reddish dye eosin, contain histamine, degradative enzymes, and a compound known as major basic protein (MBP) (Figure 17.14). MBP binds to the surface carbohydrates of parasites, and this binding is associated with disruption of the cell membrane and membrane permeability.

Basophils

Basophils have cytoplasmic granules of varied size and are named for their granules' ability to absorb the basic dye methylene blue (Figure 17.14). Their stimulation and degranulation can result from multiple triggering events. Activated complement fragments C3a and C5a, produced in the activation cascades of complement proteins, act as anaphylatoxins by inducing degranulation of basophils and inflammatory responses. This cell type is important in allergic reactions and other responses that involve inflammation. One of the most abundant components of basophil granules is histamine, which is released along with other chemical factors when the basophil is stimulated. These chemicals can be chemotactic and can help to open the gaps between cells in the blood vessels. Other mechanisms for basophil triggering require the assistance of antibodies, as discussed in B Lymphocytes and Humoral Immunity.

Mast Cells

Hematopoiesis also gives rise to **mast cells**, which appear to be derived from the same common myeloid progenitor cell as neutrophils, eosinophils, and basophils. Functionally, mast cells are very similar to basophils, containing many of the same components in their granules (e.g., histamine) and playing a similar role in allergic responses and other inflammatory reactions. However, unlike basophils, mast cells leave the circulating blood and are most frequently found residing in tissues. They are often associated with blood vessels and nerves or found close to surfaces that interface with the external environment, such as the skin and mucous membranes in various regions of the body (Figure 17.15).

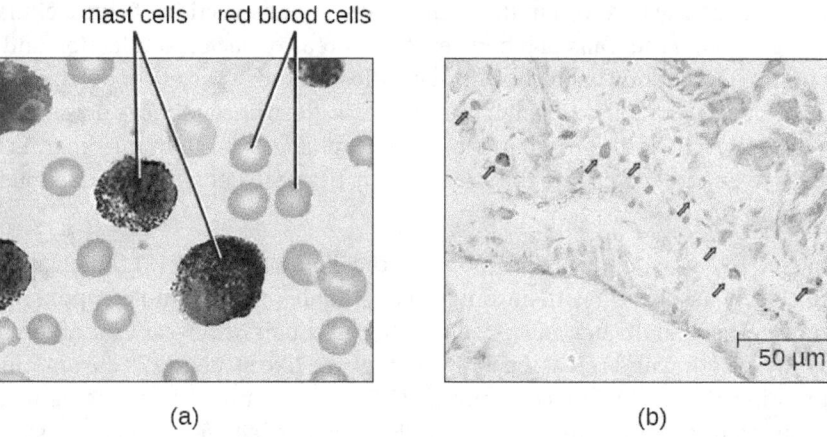

Figure 17.15 Mast cells function similarly to basophils by inducing and promoting inflammatory responses. (a) This figure shows mast cells in blood. In a blood smear, they are difficult to differentiate from basophils (b). Unlike basophils, mast cells migrate from the blood into various tissues. (credit right: modification of work by Greenland JR, Xu X, Sayah DM, Liu FC, Jones KD, Looney MR, Caughey GH)

✓ CHECK YOUR UNDERSTANDING

- Describe the granules and nuclei of neutrophils, eosinophils, basophils, and mast cells.
- Name three antimicrobial mechanisms of neutrophils

> **Clinical Focus**
>
> **Part 3**
>
> Angela's tests come back negative for all common allergens, and her sputum samples contain no abnormal presence of pathogenic microbes or elevated levels of members of the normal respiratory microbiota. She does, however, have elevated levels of inflammatory cytokines in her blood.
>
> The swelling of her airway has still not responded to treatment with antihistamines or corticosteroids. Additional blood work shows that Angela has a mildly elevated white blood cell count but normal antibody levels. Also, she has a lower-than-normal level of the complement protein C4.
>
> - What does this new information reveal about the cause of Angela's constricted airways?
> - What are some possible conditions that could lead to low levels of complement proteins?
>
> *Jump to the next Clinical Focus box. Go back to the previous Clinical Focus box.*

Agranulocytes

As their name suggests, **agranulocytes** lack visible granules in the cytoplasm. Agranulocytes can be categorized as lymphocytes or monocytes (Figure 17.13). Among the lymphocytes are natural killer cells, which play an important role in nonspecific innate immune defenses. Lymphocytes also include the B cells and T cells, which are discussed in the next chapter because they are central players in the specific adaptive immune defenses. The monocytes differentiate into macrophages and dendritic cells, which are collectively referred to as the mononuclear phagocyte system.

Natural Killer Cells

Most lymphocytes are primarily involved in the specific adaptive immune response, and thus will be discussed in the following chapter. An exception is the **natural killer cells (NK cells)**; these mononuclear lymphocytes use nonspecific mechanisms to recognize and destroy cells that are abnormal in some way. Cancer cells and cells infected with viruses are two examples of cellular abnormalities that are targeted by NK cells. Recognition of such cells involves a complex process of identifying inhibitory and activating molecular markers on the surface of the target cell. Molecular markers that make up the major histocompatibility complex (MHC) are expressed by healthy cells as an indication of "self." This will be covered in more detail in next chapter. NK cells are able to recognize normal MHC markers on the surface of healthy cells, and these MHC markers serve as an inhibitory signal preventing NK cell activation. However, cancer cells and virus-infected cells actively diminish or eliminate expression of MHC markers on their surface. When these MHC markers are diminished or absent, the NK cell interprets this as an abnormality and a cell in distress. This is one part of the NK cell activation process (Figure 17.16). NK cells are also activated by binding to activating molecular molecules on the target cell. These activating molecular molecules include "altered self" or "nonself" molecules. When a NK cell recognizes a decrease in inhibitory normal MHC molecules and an increase in activating molecules on the surface of a cell, the NK cell will be activated to eliminate the cell in distress.

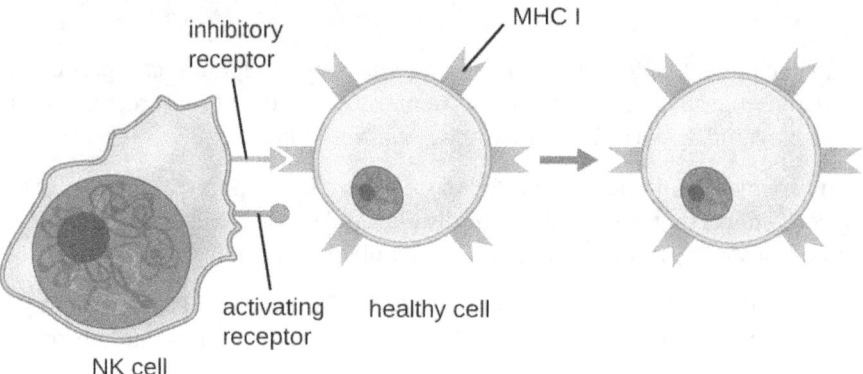
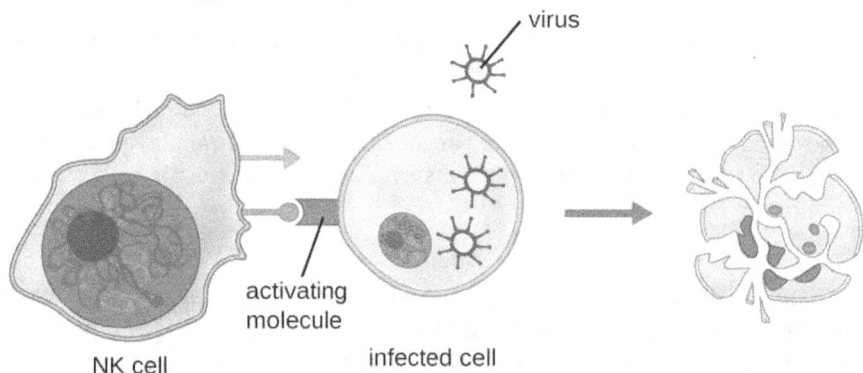

Figure 17.16 Natural killer (NK) cells are inhibited by the presence of the major histocompatibility cell (MHC) receptor on healthy cells. Cancer cells and virus-infected cells have reduced expression of MHC and increased expression of activating molecules. When a NK cell recognizes decreased MHC and increased activating molecules, it will kill the abnormal cell.

Once a cell has been recognized as a target, the NK cell can use several different mechanisms to kill its target. For example, it may express cytotoxic membrane proteins and cytokines that stimulate the target cell to undergo apoptosis, or controlled cell suicide. NK cells may also use perforin-mediated cytotoxicity to induce apoptosis in target cells. This mechanism relies on two toxins released from granules in the cytoplasm of the NK cell: **perforin**, a protein that creates pores in the target cell, and **granzymes**, proteases that enter through the pores into the target cell's cytoplasm, where they trigger a cascade of protein activation that leads to apoptosis. The NK cell binds to the abnormal target cell, releases its destructive payload, and detaches from the target cell. While the target cell undergoes apoptosis, the NK cell synthesizes more perforin and proteases to use on its next target.

NK cells contain these toxic compounds in granules in their cytoplasm. When stained, the granules are azurophilic and can be visualized under a light microscope (Figure 17.17). Even though they have granules, NK cells are not considered granulocytes because their granules are far less numerous than those found in true granulocytes. Furthermore, NK cells have a different lineage than granulocytes, arising from lymphoid rather than myeloid stem cells (Figure 17.12).

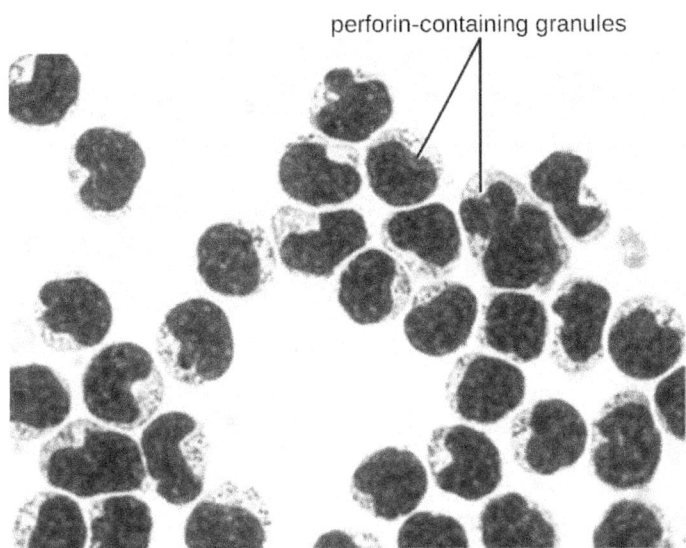

Figure 17.17 Natural killer cell with perforin-containing granules. (credit: modification of work by Rolstad B)

Monocytes

The largest of the white blood cells, **monocytes** have a nucleus that lacks lobes, and they also lack granules in the cytoplasm (Figure 17.18). Nevertheless, they are effective phagocytes, engulfing pathogens and apoptotic cells to help fight infection.

When monocytes leave the bloodstream and enter a specific body tissue, they differentiate into tissue-specific phagocytes called **macrophages** and **dendritic cells**. They are particularly important residents of lymphoid tissue, as well as nonlymphoid sites and organs. Macrophages and dendritic cells can reside in body tissues for significant lengths of time. Macrophages in specific body tissues develop characteristics suited to the particular tissue. Not only do they provide immune protection for the tissue in which they reside but they also support normal function of their neighboring tissue cells through the production of cytokines. Macrophages are given tissue-specific names, and a few examples of tissue-specific macrophages are listed in Table 17.6. Dendritic cells are important sentinels residing in the skin and mucous membranes, which are portals of entry for many pathogens. Monocytes, macrophages, and dendritic cells are all highly phagocytic and important promoters of the immune response through their production and release of cytokines. These cells provide an essential bridge between innate and adaptive immune responses, as discussed in the next section as well as the next chapter.

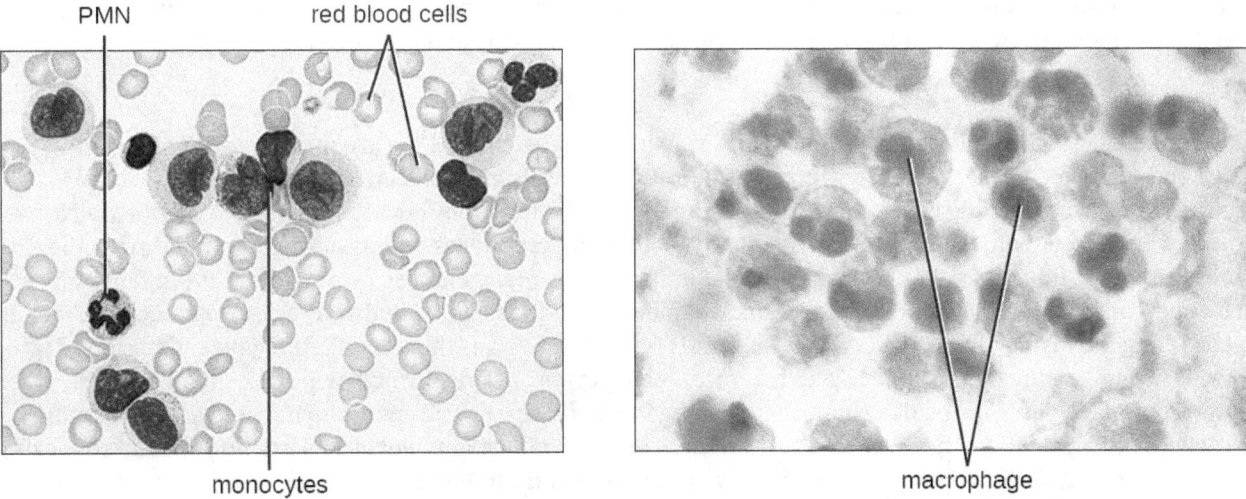

Figure 17.18 Monocytes are large, agranular white blood cells with a nucleus that lacks lobes. When monocytes leave the bloodstream, they differentiate and become macrophages with tissue-specific properties. (credit left: modification of work by Armed Forces Institute of Pathology; credit right: modification of work by Centers for Disease Control and Prevention)

Macrophages Found in Various Body Tissues

Tissue	Macrophage
Brain and central nervous system	Microglial cells
Liver	Kupffer cells
Lungs	Alveolar macrophages (dust cells)
Peritoneal cavity	Peritoneal macrophages

Table 17.6

CHECK YOUR UNDERSTANDING

- Describe the signals that activate natural killer cells.
- What is the difference between monocytes and macrophages?

17.4 Pathogen Recognition and Phagocytosis

Learning Objectives

By the end of this section, you will be able to:
- Explain how leukocytes migrate from peripheral blood into infected tissues
- Explain the mechanisms by which leukocytes recognize pathogens
- Explain the process of phagocytosis and the mechanisms by which phagocytes destroy and degrade pathogens

Several of the cell types discussed in the previous section can be described as phagocytes—cells whose main function is to seek, ingest, and kill pathogens. This process, called phagocytosis, was first observed in starfish in the 1880s by Nobel Prize-winning zoologist Ilya Metchnikoff (1845–1916), who made the connection to white blood cells (WBCs) in humans and other animals. At the time, Pasteur and other scientists believed that WBCs were spreading pathogens rather than killing them (which is true for some diseases, such as tuberculosis). But in most cases, phagocytes provide a strong, swift, and effective defense against a broad range of microbes, making them a critical component of innate nonspecific immunity. This section will focus on the mechanisms by which phagocytes are able to seek, recognize, and destroy pathogens.

Extravasation (Diapedesis) of Leukocytes

Some phagocytes are leukocytes (WBCs) that normally circulate in the bloodstream. To reach pathogens located in infected tissue, leukocytes must pass through the walls of small capillary blood vessels within tissues. This process, called **extravasation**, or **diapedesis**, is initiated by complement factor C5a, as well as cytokines released into the immediate vicinity by resident macrophages and tissue cells responding to the presence of the infectious agent (Figure 17.19). Similar to C5a, many of these cytokines are proinflammatory and chemotactic, and they bind to cells of small capillary blood vessels, initiating a response in the endothelial cells lining the inside of the blood vessel walls. This response involves the upregulation and expression of various cellular adhesion molecules and receptors. Leukocytes passing through will stick slightly to the adhesion molecules, slowing down and rolling along the blood vessel walls near the infected area. When they reach a cellular junction, they will bind to even more of these adhesion molecules, flattening out and squeezing through the cellular junction in a process known as **transendothelial migration**. This mechanism of "rolling adhesion" allows leukocytes to exit the bloodstream and enter the infected areas, where they can begin phagocytosing the invading pathogens.

Note that extravasation does not occur in arteries or veins. These blood vessels are surrounded by thicker, multilayer protective walls, in contrast to the thin single-cell-layer walls of capillaries. Furthermore, the blood flow in arteries is too turbulent to allow for rolling adhesion. Also, some leukocytes tend to respond to an infection more quickly than others. The first to arrive typically are neutrophils, often within hours of a bacterial infection. By contract, monocytes may take several days to leave the bloodstream and differentiate into macrophages.

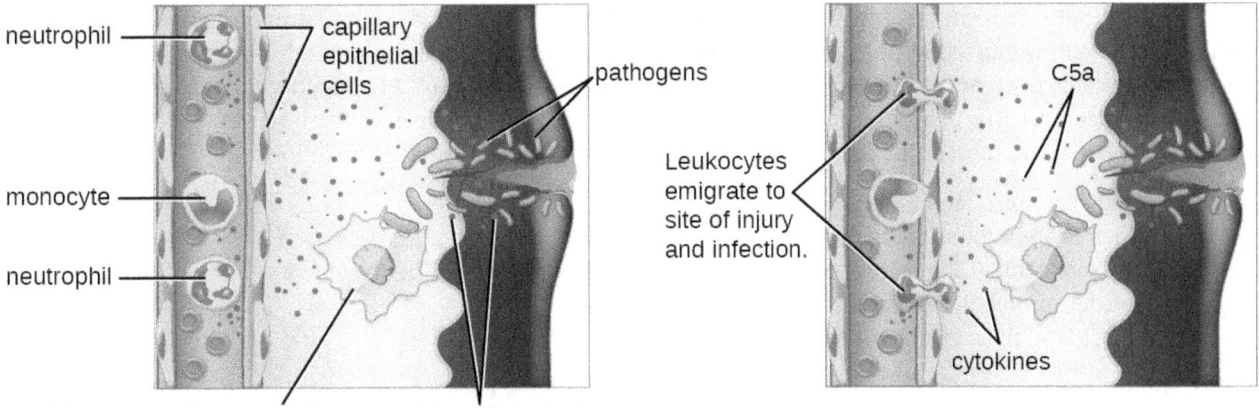

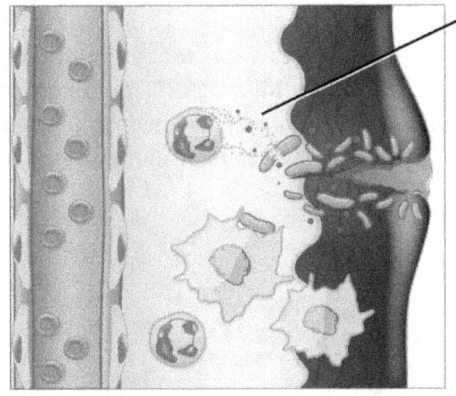

Figure 17.19 Damaged cells and macrophages that have ingested pathogens release cytokines that are proinflammatory and chemotactic for leukocytes. In addition, activation of complement at the site of infection results in production of the chemotactic and proinflammatory C5a. Leukocytes exit the blood vessel and follow the chemoattractant signal of cytokines and C5a to the site of infection. Granulocytes such as neutrophils release chemicals that destroy pathogens. They are also capable of phagocytosis and intracellular killing of bacterial pathogens.

LINK TO LEARNING

Watch the following videos on leukocyte extravasation (https://openstax.org/l/22leukextrvid) and leukocyte rolling (https://openstax.org/l/22leukrollvid) to learn more.

CHECK YOUR UNDERSTANDING

- Explain the role of adhesion molecules in the process of extravasation.

Pathogen Recognition

As described in the previous section, opsonization of pathogens by antibody; complement factors C1q, C3b, and C4b; and lectins can assist phagocytic cells in recognition of pathogens and attachment to initiate phagocytosis. However, not all pathogen recognition is opsonin dependent. Phagocytes can also recognize molecular structures that are common to many groups of pathogenic microbes. Such structures are called **pathogen-associated molecular patterns (PAMPs)**. Common PAMPs include the following:

- peptidoglycan, found in bacterial cell walls;
- flagellin, a protein found in bacterial flagella;
- lipopolysaccharide (LPS) from the outer membrane of gram-negative bacteria;
- lipopeptides, molecules expressed by most bacteria; and
- nucleic acids such as viral DNA or RNA.

Like numerous other PAMPs, these substances are integral to the structure of broad classes of microbes.

The structures that allow phagocytic cells to detect PAMPs are called **pattern recognition receptors (PRRs)**. One group of PRRs is the **toll-like receptors (TLRs)**, which bind to various PAMPs and communicate with the nucleus of the phagocyte to elicit a response. Many TLRs (and other PRRs) are located on the surface of a phagocyte, but some can also be found embedded in the membranes of interior compartments and organelles (Figure 17.20). These interior PRRs can be useful for the binding and recognition of intracellular pathogens that may have gained access to the inside of the cell before phagocytosis could take place. Viral nucleic acids, for example, might encounter an interior PRR, triggering production of the antiviral cytokine interferon.

In addition to providing the first step of pathogen recognition, the interaction between PAMPs and PRRs on macrophages provides an intracellular signal that activates the phagocyte, causing it to transition from a dormant state of readiness and slow proliferation to a state of hyperactivity, proliferation, production/secretion of cytokines, and enhanced intracellular killing. PRRs on macrophages also respond to chemical distress signals from damaged or stressed cells. This allows macrophages to extend their responses beyond protection from infectious diseases to a broader role in the inflammatory response initiated from injuries or other diseases.

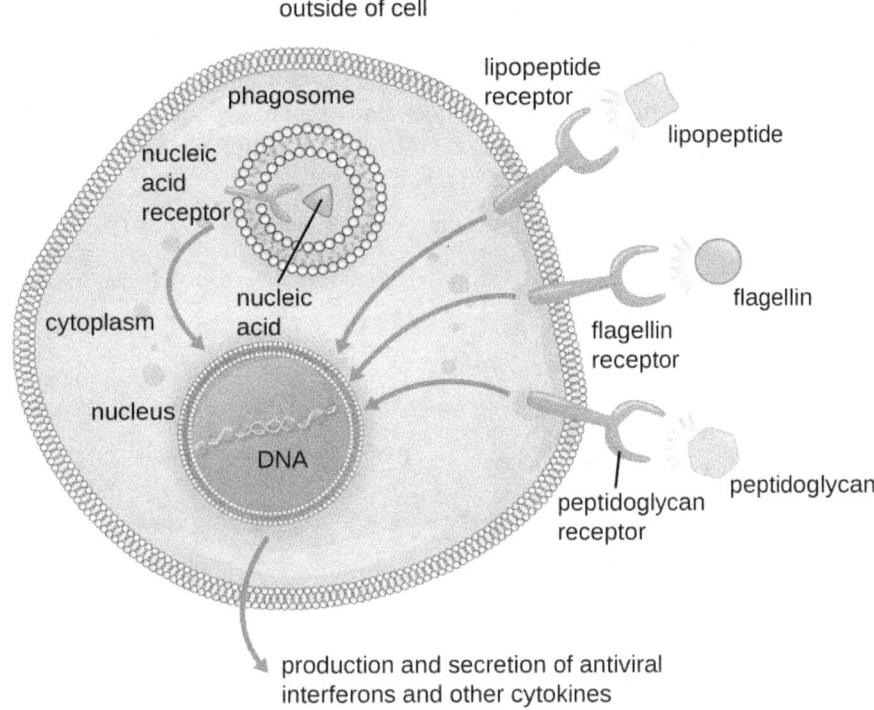

Figure 17.20 Phagocytic cells contain pattern recognition receptors (PRRs) capable of recognizing various pathogen-associated molecular patterns (PAMPs). These PRRs can be found on the plasma membrane or in internal phagosomes. When a PRR recognizes a PAMP, it sends a signal to the nucleus that activates genes involved in phagocytosis, cellular proliferation, production and secretion of antiviral interferons and proinflammatory cytokines, and enhanced intracellular killing.

CHECK YOUR UNDERSTANDING

- Name four pathogen-associated molecular patterns (PAMPs).
- Describe the process of phagocyte activation.

Pathogen Degradation

Once pathogen recognition and attachment occurs, the pathogen is engulfed in a vesicle and brought into the internal compartment of the phagocyte in a process called **phagocytosis** (Figure 17.21). PRRs can aid in phagocytosis by first binding to the pathogen's surface, but phagocytes are also capable of engulfing nearby items even if they are not bound to specific receptors. To engulf the pathogen, the phagocyte forms a pseudopod that wraps around the pathogen and then pinches it off into a membrane vesicle called a **phagosome**. Acidification of the phagosome (pH decreases to the range of 4–5) provides an important early antibacterial mechanism. The phagosome containing the pathogen fuses with one or more lysosomes, forming a **phagolysosome**. Formation of the phagolysosome enhances the acidification, which is essential for activation of pH-dependent digestive lysosomal enzymes and production of hydrogen peroxide and toxic reactive oxygen species. Lysosomal enzymes such as lysozyme, phospholipase, and proteases digest the pathogen. Other enzymes are involved in a respiratory burst. During the respiratory burst, phagocytes will increase their uptake and consumption of oxygen, but not for energy production. The increased oxygen consumption is focused on the production of superoxide anion, hydrogen peroxide, hydroxyl radicals, and other reactive oxygen species that are antibacterial.

In addition to the reactive oxygen species produced by the respiratory burst, reactive nitrogen compounds with cytotoxic (cell-killing) potential can also form. For example, nitric oxide can react with superoxide to form peroxynitrite, a highly reactive nitrogen compound with degrading capabilities similar to those of the reactive oxygen species. Some phagocytes even contain an internal storehouse of microbicidal defensin proteins (e.g., neutrophil granules). These destructive forces can be released into the area around the cell to degrade microbes externally. Neutrophils, especially, can be quite efficient at this secondary antimicrobial mechanism.

Once degradation is complete, leftover waste products are excreted from the cell in an exocytic vesicle. However, it is important to note that not all remains of the pathogen are excreted as waste. Macrophages and dendritic cells are also antigen-presenting cells involved in the specific adaptive immune response. These cells further process the remains of the degraded pathogen and present key antigens (specific pathogen proteins) on their cellular surface. This is an important step for stimulation of some adaptive immune responses, as will be discussed in more detail in the next chapter.

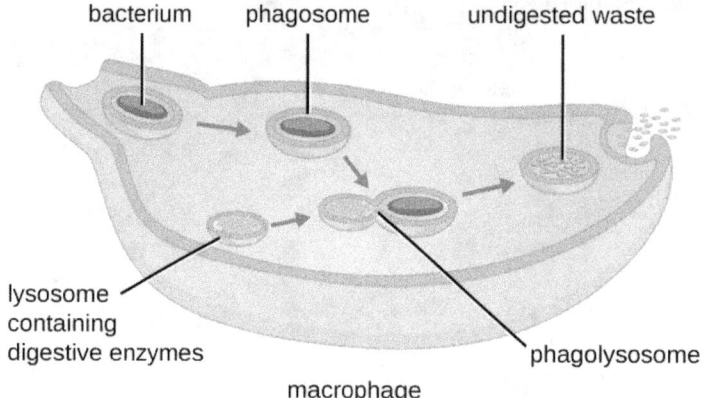

Figure 17.21 The stages of phagocytosis include the engulfment of a pathogen, the formation of a phagosome, the digestion of the pathogenic particle in the phagolysosome, and the expulsion of undigested materials from the cell.

LINK TO LEARNING

Visit this link (https://openstax.org/l/22phagpathvid) to view a phagocyte chasing and engulfing a pathogen.

CHECK YOUR UNDERSTANDING

- What is the difference between a phagosome and a lysosome?

MICRO CONNECTIONS

When Phagocytosis Fails

Although phagocytosis successfully destroys many pathogens, some are able to survive and even exploit this defense mechanism to multiply in the body and cause widespread infection. Protozoans of the genus *Leishmania* are one example. These obligate intracellular parasites are flagellates transmitted to humans by the bite of a sand fly. Infections cause serious and sometimes disfiguring sores and ulcers in the skin and other tissues (Figure 17.22). Worldwide, an estimated 1.3 million people are newly infected with leishmaniasis annually.[1]

Salivary peptides from the sand fly activate host macrophages at the site of their bite. The classic or alternate pathway for complement activation ensues with C3b opsonization of the parasite. *Leishmania* cells are phagocytosed, lose their flagella, and multiply in a form known as an amastigote (Leishman-Donovan body) within the phagolysosome. Although many other pathogens are destroyed in the phagolysosome, survival of the *Leishmania* amastigotes is maintained by the presence of surface lipophosphoglycan and acid phosphatase. These substances inhibit the macrophage respiratory burst and lysosomal enzymes. The parasite then multiplies inside the cell and lyses the infected macrophage, releasing the amastigotes to infect other macrophages within the same host. Should another sand fly bite an infected person, it might ingest amastigotes and then transmit them to another individual through another bite.

There are several different forms of leishmaniasis. The most common is a localized cutaneous form of the

---

1 World Health Organization. "Leishmaniasis." 2016. http://www.who.int/mediacentre/factsheets/fs375/en/.

illness caused by *L. tropica*, which typically resolves spontaneously over time but with some significant lymphocyte infiltration and permanent scarring. A mucocutaneous form of the disease, caused by *L. viannia brasilienfsis*, produces lesions in the tissue of the nose and mouth and can be life threatening. A visceral form of the illness can be caused by several of the different *Leishmania* species. It affects various organ systems and causes abnormal enlargement of the liver and spleen. Irregular fevers, anemia, liver dysfunction, and weight loss are all signs and symptoms of visceral leishmaniasis. If left untreated, it is typically fatal.

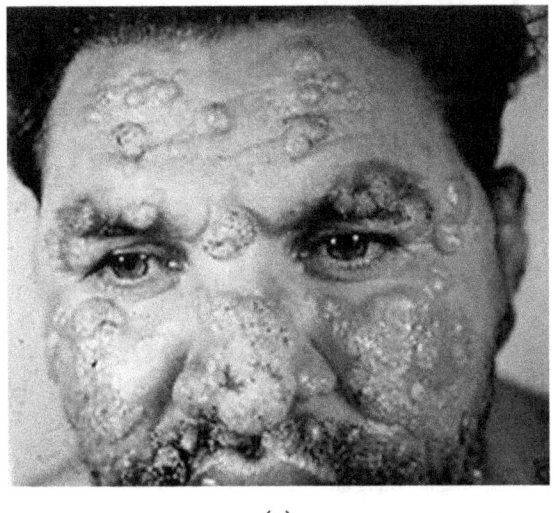
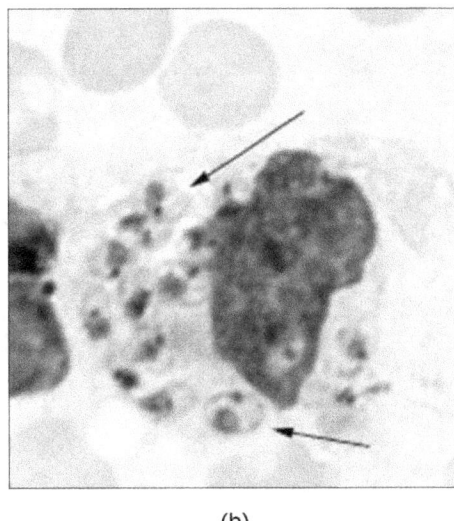

(a) (b)

Figure 17.22 (a) Cutaneous leishmaniasis is a disfiguring disease caused by the intracellular flagellate *Leishmania tropica*, transmitted by the bite of a sand fly. (b) This light micrograph of a sample taken from a skin lesion shows a large cell, which is a macrophage infected with *L. tropica* amastigotes (arrows). The amastigotes have lost their flagella but their nuclei are visible. Soon the amastigotes will lyse the macrophage and be engulfed by other phagocytes, spreading the infection. (credit a: modification of work by Otis Historical Archives of "National Museum of Health & Medicine"; credit b: modification of work by Centers for Disease Control and Prevention)

17.5 Inflammation and Fever

Learning Objectives

By the end of this section, you will be able to:
- Identify the signs of inflammation and fever and explain why they occur
- Explain the advantages and risks posed by inflammatory responses

The inflammatory response, or **inflammation**, is triggered by a cascade of chemical mediators and cellular responses that may occur when cells are damaged and stressed or when pathogens successfully breach the physical barriers of the innate immune system. Although inflammation is typically associated with negative consequences of injury or disease, it is a necessary process insofar as it allows for recruitment of the cellular defenses needed to eliminate pathogens, remove damaged and dead cells, and initiate repair mechanisms. Excessive inflammation, however, can result in local tissue damage and, in severe cases, may even become deadly.

Acute Inflammation

An early, if not immediate, response to tissue injury is acute inflammation. Immediately following an injury, vasoconstriction of blood vessels will occur to minimize blood loss. The amount of vasoconstriction is related to the amount of vascular injury, but it is usually brief. Vasoconstriction is followed by vasodilation and increased vascular permeability, as a direct result of the release of histamine from resident mast cells. Increased blood flow and vascular permeability can dilute toxins and bacterial products at the site of injury or infection. They also contribute to the five observable signs associated with the inflammatory response: **erythema** (redness), **edema** (swelling), heat, pain, and altered function. Vasodilation and increased vascular permeability are also associated with an influx of phagocytes at the site of injury and/or infection. This can enhance the inflammatory response because phagocytes may release proinflammatory chemicals when they are activated by cellular distress signals released from damaged cells, by PAMPs, or by opsonins on the surface of pathogens. Activation of the complement system can further enhance the inflammatory response through the production of the anaphylatoxin C5a. Figure 17.23 illustrates a typical case of acute inflammation at the site of a skin wound.

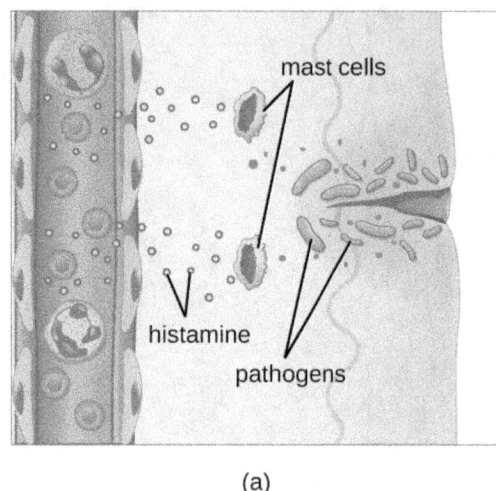
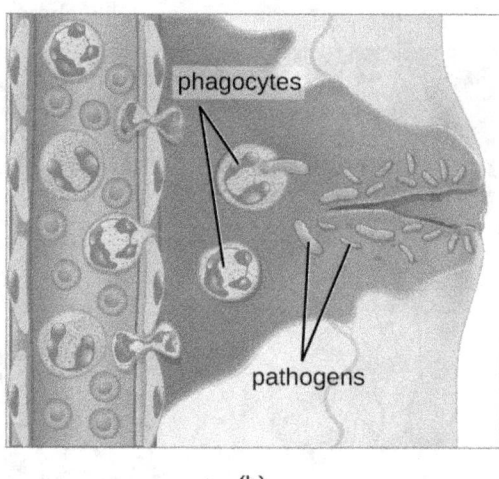

(a)   (b)

Figure 17.23 (a) Mast cells detect injury to nearby cells and release histamine, initiating an inflammatory response. (b) Histamine increases blood flow to the wound site, and increased vascular permeability allows fluid, proteins, phagocytes, and other immune cells to enter infected tissue. These events result in the swelling and reddening of the injured site, and the increased blood flow to the injured site causes it to feel warm. Inflammation is also associated with pain due to these events stimulating nerve pain receptors in the tissue. The interaction of phagocyte PRRs with cellular distress signals and PAMPs and opsonins on the surface of pathogens leads to the release of more proinflammatory chemicals, enhancing the inflammatory response.

During the period of inflammation, the release of bradykinin causes capillaries to remain dilated, flooding tissues with fluids and leading to edema. Increasing numbers of neutrophils are recruited to the area to fight pathogens. As the fight rages on, pus forms from the accumulation of neutrophils, dead cells, tissue fluids, and lymph. Typically, after a few days, macrophages will help to clear out this pus. Eventually, tissue repair can begin in the wounded area.

Chronic Inflammation

When acute inflammation is unable to clear an infectious pathogen, chronic inflammation may occur. This often results in an ongoing (and sometimes futile) lower-level battle between the host organism and the pathogen. The wounded area may heal at a superficial level, but pathogens may still be present in deeper tissues, stimulating ongoing inflammation. Additionally, chronic inflammation may be involved in the progression of degenerative neurological diseases such as Alzheimer's and Parkinson's, heart disease, and metastatic cancer.

Chronic inflammation may lead to the formation of **granulomas**, pockets of infected tissue walled off and surrounded by WBCs. Macrophages and other phagocytes wage an unsuccessful battle to eliminate the pathogens and dead cellular materials within a granuloma. One example of a disease that produces chronic inflammation is tuberculosis, which results in the formation of granulomas in lung tissues. A tubercular granuloma is called a tubercle (Figure 17.24). Tuberculosis will be covered in more detail in Bacterial Infections of the Respiratory Tract.

Chronic inflammation is not just associated with bacterial infections. Chronic inflammation can be an important cause of tissue damage from viral infections. The extensive scarring observed with hepatitis C infections and liver cirrhosis is the result of chronic inflammation.

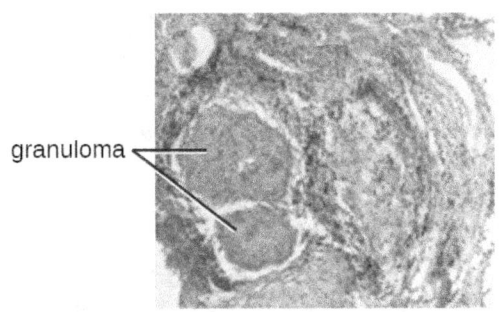

Figure 17.24 A tubercle is a granuloma in the lung tissue of a patient with tuberculosis. In this micrograph, white blood cells (stained purple) have walled off a pocket of tissue infected with *Mycobacterium tuberculosis*. Granulomas also occur in many other forms of disease. (credit: modification of work by Piotrowski WJ, Górski P, Duda-Szymańska J, Kwiatkowska S)

CHECK YOUR UNDERSTANDING

- Name the five signs of inflammation.
- Is a granuloma an acute or chronic form of inflammation? Explain.

 MICRO CONNECTIONS

Chronic Edema

In addition to granulomas, chronic inflammation can also result in long-term edema. A condition known as lymphatic filariasis (also known as elephantiasis) provides an extreme example. Lymphatic filariasis is caused by microscopic nematodes (parasitic worms) whose larvae are transmitted between human hosts by mosquitoes. Adult worms live in the lymphatic vessels, where their presence stimulates infiltration by lymphocytes, plasma cells, eosinophils, and thrombocytes (a condition known as lymphangitis). Because of the chronic nature of the illness, granulomas, fibrosis, and blocking of the lymphatic system may eventually occur. Over time, these blockages may worsen with repeated infections over decades, leading to skin thickened with edema and fibrosis. Lymph (extracellular tissue fluid) may spill out of the lymphatic areas and back into tissues, causing extreme swelling (Figure 17.25). Secondary bacterial infections commonly follow. Because it is a disease caused by a parasite, eosinophilia (a dramatic rise in the number of eosinophils in the blood) is characteristic of acute infection. However, this increase in antiparasite granulocytes is not sufficient to clear the infection in many cases.

Lymphatic filariasis affects an estimated 120 million people worldwide, mostly concentrated in Africa and Asia.[2] Improved sanitation and mosquito control can reduce transmission rates.

---

2 Centers for Disease Control and Prevention. "Parasites–Lymphatic Filiariasis." 2016. http://www.cdc.gov/parasites/lymphaticfilariasis/gen_info/faqs.html.

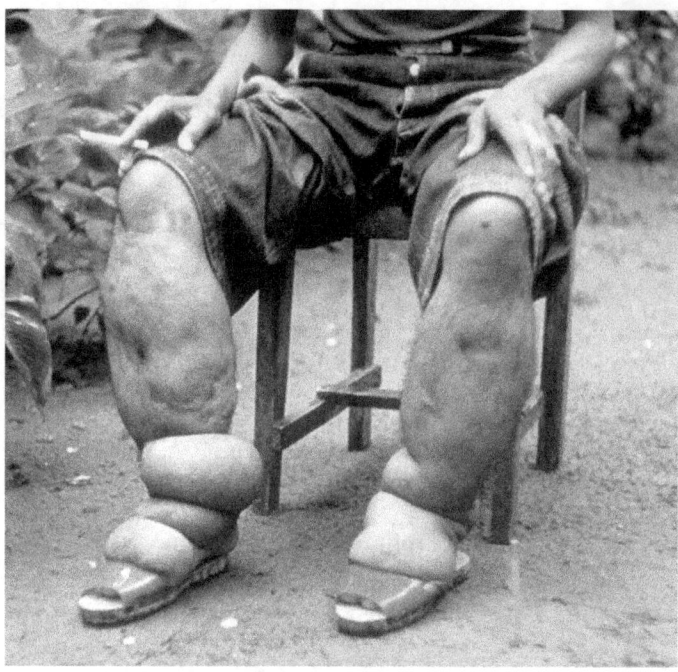

Figure 17.25 Elephantiasis (chronic edema) of the legs due to filariasis. (credit: modification of work by Centers for Disease Control and Prevention)

Fever

A **fever** is an inflammatory response that extends beyond the site of infection and affects the entire body, resulting in an overall increase in body temperature. Body temperature is normally regulated and maintained by the hypothalamus, an anatomical section of the brain that functions to maintain homeostasis in the body. However, certain bacterial or viral infections can result in the production of **pyrogens**, chemicals that effectively alter the "thermostat setting" of the hypothalamus to elevate body temperature and cause fever. Pyrogens may be exogenous or endogenous. For example, the endotoxin lipopolysaccharide (LPS), produced by gram-negative bacteria, is an exogenous pyrogen that may induce the leukocytes to release endogenous pyrogens such as interleukin-1 (IL-1), IL-6, interferon-γ (IFN-γ), and tumor necrosis factor (TNF). In a cascading effect, these molecules can then lead to the release of prostaglandin E2 (PGE_2) from other cells, resetting the hypothalamus to initiate fever (Figure 17.26).

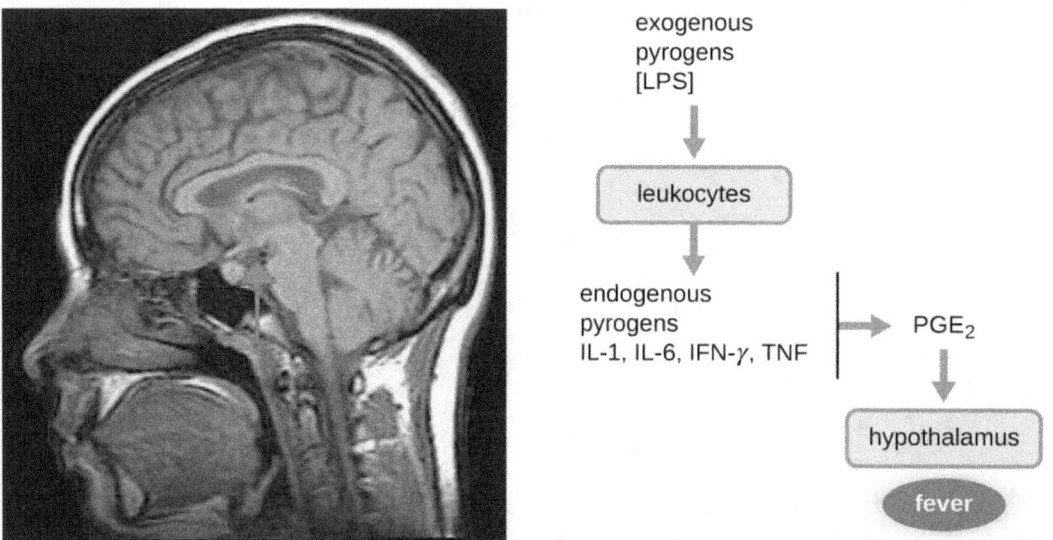

Figure 17.26 The role of the hypothalamus in the inflammatory response. Macrophages recognize pathogens in an area and release

cytokines that trigger inflammation. The cytokines also send a signal up the vagus nerve to the hypothalamus.

Like other forms of inflammation, a fever enhances the innate immune defenses by stimulating leukocytes to kill pathogens. The rise in body temperature also may inhibit the growth of many pathogens since human pathogens are mesophiles with optimum growth occurring around 35 °C (95 °F). In addition, some studies suggest that fever may also stimulate release of iron-sequestering compounds from the liver, thereby starving out microbes that rely on iron for growth.[3]

During fever, the skin may appear pale due to vasoconstriction of the blood vessels in the skin, which is mediated by the hypothalamus to divert blood flow away from extremities, minimizing the loss of heat and raising the core temperature. The hypothalamus will also stimulate shivering of muscles, another effective mechanism of generating heat and raising the core temperature.

The **crisis phase** occurs when the fever breaks. The hypothalamus stimulates vasodilation, resulting in a return of blood flow to the skin and a subsequent release of heat from the body. The hypothalamus also stimulates sweating, which cools the skin as the sweat evaporates.

Although a low-level fever may help an individual overcome an illness, in some instances, this immune response can be too strong, causing tissue and organ damage and, in severe cases, even death. The inflammatory response to bacterial superantigens is one scenario in which a life-threatening fever may develop. Superantigens are bacterial or viral proteins that can cause an excessive activation of T cells from the specific adaptive immune defense, as well as an excessive release of cytokines that overstimulates the inflammatory response. For example, *Staphylococcus aureus* and *Streptococcus pyogenes* are capable of producing superantigens that cause toxic shock syndrome and scarlet fever, respectively. Both of these conditions can be associated with very high, life-threatening fevers in excess of 42 °C (108 °F).

CHECK YOUR UNDERSTANDING

- Explain the difference between exogenous and endogenous pyrogens.
- How does a fever inhibit pathogens?

Clinical Focus

Resolution

Given her father's premature death, Angela's doctor suspects that she has hereditary angioedema, a genetic disorder that compromises the function of C1 inhibitor protein. Patients with this genetic abnormality may have occasional episodes of swelling in various parts of the body. In Angela's case, the swelling has occurred in the respiratory tract, leading to difficulty breathing. Swelling may also occur in the gastrointestinal tract, causing abdominal cramping, diarrhea, and vomiting, or in the muscles of the face or limbs. This swelling may be nonresponsive to steroid treatment and is often misdiagnosed as an allergy.

Because there are three types of hereditary angioedema, the doctor orders a more specific blood test to look for levels of C1-INH, as well as a functional assay of Angela's C1 inhibitors. The results suggest that Angela has type I hereditary angioedema, which accounts for 80%–85% of all cases. This form of the disorder is caused by a deficiency in C1 esterase inhibitors, the proteins that normally help suppress activation of the complement system. When these proteins are deficient or nonfunctional, overstimulation of the system can lead to production of inflammatory anaphylatoxins, which results in swelling and fluid buildup in tissues.

There is no cure for hereditary angioedema, but timely treatment with purified and concentrated C1-INH from blood donors can be effective, preventing tragic outcomes like the one suffered by Angela's father. A number of therapeutic drugs, either currently approved or in late-stage human trials, may also be

---

3 N. Parrow et al. "Sequestration and Scavenging of Iron in Infection." *Infection and Immunity* 81 no. 10 (2013):3503–3514

considered as options for treatment in the near future. These drugs work by inhibiting inflammatory molecules or the receptors for inflammatory molecules.

Thankfully, Angela's condition was quickly diagnosed and treated. Although she may experience additional episodes in the future, her prognosis is good and she can expect to live a relatively normal life provided she seeks treatment at the onset of symptoms.

Go back to the previous Clinical Focus box.

SUMMARY

17.1 Physical Defenses

- **Nonspecific innate immunity** provides a first line of defense against infection by nonspecifically blocking entry of microbes and targeting them for destruction or removal from the body.
- The physical defenses of innate immunity include physical barriers, mechanical actions that remove microbes and debris, and the microbiome, which competes with and inhibits the growth of pathogens.
- The skin, mucous membranes, and endothelia throughout the body serve as physical barriers that prevent microbes from reaching potential sites of infection. Tight cell junctions in these tissues prevent microbes from passing through.
- Microbes trapped in dead skin cells or **mucus** are removed from the body by mechanical actions such as shedding of skin cells, mucociliary sweeping, coughing, **peristalsis**, and flushing of bodily fluids (e.g., urination, tears)
- The resident microbiota provide a physical defense by occupying available cellular binding sites and competing with pathogens for available nutrients.

17.2 Chemical Defenses

- Numerous **chemical mediators** produced endogenously and exogenously exhibit nonspecific antimicrobial functions.
- Many chemical mediators are found in body fluids such as sebum, saliva, mucus, gastric and intestinal fluids, urine, tears, cerumen, and vaginal secretions.
- **Antimicrobial peptides (AMPs)** found on the skin and in other areas of the body are largely produced in response to the presence of pathogens. These include dermcidin, cathelicidin, defensins, histatins, and bacteriocins.
- **Plasma** contains various proteins that serve as chemical mediators, including **acute-phase proteins, complement proteins**, and **cytokines**.
- The **complement system** involves numerous precursor proteins that circulate in plasma. These proteins become activated in a cascading sequence in the presence of microbes, resulting in the **opsonization** of pathogens, chemoattraction of leukocytes, induction of inflammation, and cytolysis through the formation of a **membrane attack complex (MAC)**.
- **Cytokines** are proteins that facilitate various nonspecific responses by innate immune cells, including production of other chemical mediators, cell proliferation, cell death, and differentiation.
- Cytokines play a key role in the inflammatory response, triggering production of inflammation-eliciting mediators such as acute-phase proteins, **histamine**, leukotrienes, **prostaglandins**, and **bradykinin**.

17.3 Cellular Defenses

- The **formed elements** of the blood include red blood cells (**erythrocytes**), white blood cells (**leukocytes**), and **platelets (thrombocytes)**. Of these, leukocytes are primarily involved in the immune response.
- All formed elements originate in the bone marrow as stem cells (HSCs) that differentiate through **hematopoiesis**.
- **Granulocytes** are leukocytes characterized by a lobed nucleus and granules in the cytoplasm. These include **neutrophils (PMNs)**, **eosinophils**, and **basophils**.
- Neutrophils are the leukocytes found in the largest numbers in the bloodstream and they primarily fight bacterial infections.
- Eosinophils target parasitic infections. Eosinophils and basophils are involved in allergic reactions. Both release histamine and other proinflammatory compounds from their granules upon stimulation.
- **Mast cells** function similarly to basophils but can be found in tissues outside the bloodstream.
- **Natural killer (NK)** cells are lymphocytes that recognize and kill abnormal or infected cells by releasing proteins that trigger apoptosis.
- **Monocytes** are large, mononuclear leukocytes that circulate in the bloodstream. They may leave the bloodstream and take up residence in body tissues, where they differentiate and become tissue-specific **macrophages** and **dendritic cells**.

17.4 Pathogen Recognition and Phagocytosis

- Phagocytes are cells that recognize pathogens and destroy them through phagocytosis.
- Recognition often takes place by the use of

- phagocyte receptors that bind molecules commonly found on pathogens, known as **pathogen-associated molecular patterns (PAMPs)**.
- The receptors that bind PAMPs are called **pattern recognition receptors**, or **PRRs**. **Toll-like receptors (TLRs)** are one type of PRR found on phagocytes.
- **Extravasation** of white blood cells from the bloodstream into infected tissue occurs through the process of **transendothelial migration**.
- Phagocytes degrade pathogens through **phagocytosis**, which involves engulfing the pathogen, killing and digesting it within a **phagolysosome**, and then excreting undigested matter.

17.5 Inflammation and Fever

- **Inflammation** results from the collective response of chemical mediators and cellular defenses to an injury or infection.
- **Acute inflammation** is short lived and localized to the site of injury or infection. **Chronic inflammation** occurs when the inflammatory response is unsuccessful, and may result in the formation of **granulomas** (e.g., with tuberculosis) and scarring (e.g., with hepatitis C viral infections and liver cirrhosis).
- The five cardinal signs of inflammation are **erythema**, **edema**, heat, pain, and altered function. These largely result from innate responses that draw increased blood flow to the injured or infected tissue.
- **Fever** is a system-wide sign of inflammation that raises the body temperature and stimulates the immune response.
- Both inflammation and fever can be harmful if the inflammatory response is too severe.

REVIEW QUESTIONS

Multiple Choice

1. Which of the following best describes the innate nonspecific immune system?
   A. a targeted and highly specific response to a single pathogen or molecule
   B. a generalized and nonspecific set of defenses against a class or group of pathogens
   C. a set of barrier mechanisms that adapts to specific pathogens after repeated exposure
   D. the production of antibody molecules against pathogens

2. Which of the following constantly sheds dead cells along with any microbes that may be attached to those cells?
   A. epidermis
   B. dermis
   C. hypodermis
   D. mucous membrane

3. Which of the following uses a particularly dense suite of tight junctions to prevent microbes from entering the underlying tissue?
   A. the mucociliary escalator
   B. the epidermis
   C. the blood-brain barrier
   D. the urethra

4. Which of the following serve as chemical signals between cells and stimulate a wide range of nonspecific defenses?
   A. cytokines
   B. antimicrobial peptides
   C. complement proteins
   D. antibodies

5. Bacteriocins and defensins are types of which of the following?
   A. leukotrienes
   B. cytokines
   C. inflammation-eliciting mediators
   D. antimicrobial peptides

6. Which of the following chemical mediators is secreted onto the surface of the skin?
   A. cerumen
   B. sebum
   C. gastric acid
   D. prostaglandin

7. Identify the complement activation pathway that is triggered by the binding of an acute-phase protein to a pathogen.
   A. classical
   B. alternate
   C. lectin
   D. cathelicidin

8. Histamine, leukotrienes, prostaglandins, and bradykinin are examples of which of the following?
   A. chemical mediators primarily found in the digestive system
   B. chemical mediators that promote inflammation
   C. antimicrobial peptides found on the skin
   D. complement proteins that form MACs

9. White blood cells are also referred to as which of the following?
   A. platelets
   B. erythrocytes
   C. leukocytes
   D. megakaryocytes

10. Hematopoiesis occurs in which of the following?
    A. liver
    B. bone marrow
    C. kidneys
    D. central nervous system

11. Granulocytes are which type of cell?
    A. lymphocyte
    B. erythrocyte
    C. megakaryocyte
    D. leukocyte

12. PAMPs would be found on the surface of which of the following?
    A. pathogen
    B. phagocyte
    C. skin cell
    D. blood vessel wall

13. _____ on phagocytes bind to PAMPs on bacteria, which triggers the uptake and destruction of the bacterial pathogens?
    A. PRRs
    B. AMPs
    C. PAMPs
    D. PMNs

14. Which of the following best characterizes the mode of pathogen recognition for opsonin-dependent phagocytosis?
    A. Opsonins produced by a pathogen attract phagocytes through chemotaxis.
    B. A PAMP on the pathogen's surface is recognized by a phagocyte's toll-like receptors.
    C. A pathogen is first coated with a molecule such as a complement protein, which allows it to be recognized by phagocytes.
    D. A pathogen is coated with a molecule such as a complement protein that immediately lyses the cell.

15. Which refers to swelling as a result of inflammation?
    A. erythema
    B. edema
    C. granuloma
    D. vasodilation

16. Which type of inflammation occurs at the site of an injury or infection?
    A. acute
    B. chronic
    C. endogenous
    D. exogenous

Matching

17. Match each cell type with its description.

___natural killer cell

___basophil

___macrophage

___eosinophil

A. stains with basic dye methylene blue, has large amounts of histamine in granules, and facilitates allergic responses and inflammation

B. stains with acidic dye eosin, has histamine and major basic protein in granules, and facilitates responses to protozoa and helminths

C. recognizes abnormal cells, binds to them, and releases perforin and granzyme molecules, which induce apoptosis

D. large agranular phagocyte that resides in tissues such as the brain and lungs

18. Match each cellular defense with the infection it would most likely target.

___natural killer cell

___neutrophil

___eosinophil

A. virus-infected cell

B. tapeworm in the intestines

C. bacteria in a skin lesion

Fill in the Blank

19. The muscular contraction of the intestines that results in movement of material through the digestive tract is called _____.

20. _____ are the hair-like appendages of cells lining parts of the respiratory tract that sweep debris away from the lungs.

21. Secretions that bathe and moisten the interior of the intestines are produced by _____ cells.

22. _____ are antimicrobial peptides produced by members of the normal microbiota.

23. _____ is the fluid portion of a blood sample that has been drawn in the presence of an anticoagulant compound.

24. The process by which cells are drawn or attracted to an area by a microbe invader is known as _____.

25. Platelets are also called _____.

26. The cell in the bone marrow that gives rise to all other blood cell types is the _____.

27. PMNs are another name for _____.

28. Kupffer cells residing in the liver are a type of _____.

29. _____ are similar to basophils, but reside in tissues rather than circulating in the blood.

30. _____, also known as diapedesis, refers to the exit from the bloodstream of neutrophils and other circulating leukocytes.

31. Toll-like receptors are examples of _____.

32. A(n) _____ is a walled-off area of infected tissue that exhibits chronic inflammation.

33. The _____ is the part of the body responsible for regulating body temperature.

34. Heat and redness, or _____, occur when the small blood vessels in an inflamed area dilate (open up), bringing more blood much closer to the surface of the skin.

Short Answer

35. Differentiate a physical barrier from a mechanical removal mechanism and give an example of each.
36. Identify some ways that pathogens can breach the physical barriers of the innate immune system.
37. Differentiate the main activation methods of the classic, alternative, and lectin complement cascades.
38. What are the four protective outcomes of complement activation?
39. Explain the difference between plasma and the formed elements of the blood.
40. List three ways that a neutrophil can destroy an infectious bacterium.
41. Briefly summarize the events leading up to and including the process of transendothelial migration.
42. Differentiate exogenous and endogenous pyrogens, and provide an example of each.

Critical Thinking

43. Neutrophils can sometimes kill human cells along with pathogens when they release the toxic contents of their granules into the surrounding tissue. Likewise, natural killer cells target human cells for destruction. Explain why it is advantageous for the immune system to have cells that can kill human cells as well as pathogens.
44. Refer to Figure 17.13. In a blood smear taken from a healthy patient, which type of leukocyte would you expect to observe in the highest numbers?
45. If a gram-negative bacterial infection reaches the bloodstream, large quantities of LPS can be released into the blood, resulting in a syndrome called septic shock. Death due to septic shock is a real danger. The overwhelming immune and inflammatory responses that occur with septic shock can cause a perilous drop in blood pressure; intravascular blood clotting; development of thrombi and emboli that block blood vessels, leading to tissue death; failure of multiple organs; and death of the patient. Identify and characterize two to three therapies that might be useful in stopping the dangerous events and outcomes of septic shock once it has begun, given what you have learned about inflammation and innate immunity in this chapter.
46. In Lubeck, Germany, in 1930, a group of 251 infants was accidentally administered a tainted vaccine for tuberculosis that contained live *Mycobacterium tuberculosis*. This vaccine was administered orally, directly exposing the infants to the deadly bacterium. Many of these infants contracted tuberculosis, and some died. However, 44 of the infants never contracted tuberculosis. Based on your knowledge of the innate immune system, what innate defenses might have inhibited *M. tuberculosis* enough to prevent these infants from contracting the disease?

CHAPTER 18
Adaptive Specific Host Defenses

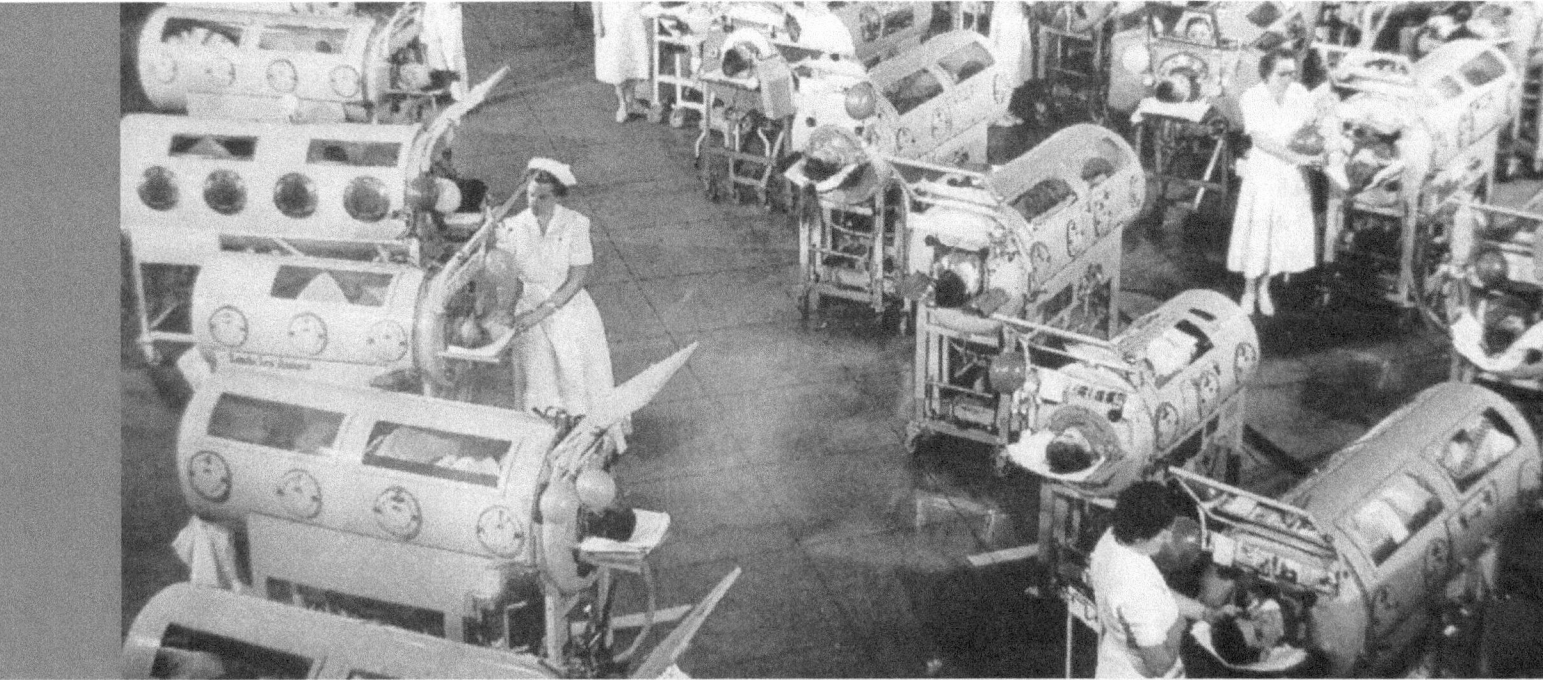

Figure 18.1 Polio was once a common disease with potentially serious consequences, including paralysis. Vaccination has all but eliminated the disease from most countries around the world. An iron-lung ward, such as the one shown in this 1953 photograph, housed patients paralyzed from polio and unable to breathe for themselves.

Chapter Outline

18.1 Overview of Specific Adaptive Immunity

18.2 Major Histocompatibility Complexes and Antigen-Presenting Cells

18.3 T Lymphocytes and Cellular Immunity

18.4 B Lymphocytes and Humoral Immunity

18.5 Vaccines

INTRODUCTION People living in developed nations and born in the 1960s or later may have difficulty understanding the once heavy burden of devastating infectious diseases. For example, smallpox, a deadly viral disease, once destroyed entire civilizations but has since been eradicated. Thanks to the vaccination efforts by multiple groups, including the World Health Organization, Rotary International, and the United Nations Children's Fund (UNICEF), smallpox has not been diagnosed in a patient since 1977. Polio is another excellent example. This crippling viral disease paralyzed patients, who were often kept alive in "iron lung wards" as recently as the 1950s (Figure 18.1). Today, vaccination against polio has nearly eradicated the disease. Vaccines have also reduced the prevalence of once-common infectious diseases such as chickenpox, German measles, measles, mumps, and whooping cough. The success of these and other vaccines is due to the very specific and adaptive host defenses that are the focus of this chapter.

Innate Nonspecific Host Defenses described innate immunity against microbial pathogens. Higher animals, such as humans, also possess an adaptive immune defense, which is highly specific for individual microbial pathogens. This specific adaptive immunity is acquired through active infection or vaccination and serves as an important defense against pathogens that evade the defenses of innate immunity.

18.1 Overview of Specific Adaptive Immunity

Learning Objectives

By the end of this section, you will be able to:

- Define memory, primary response, secondary response, and specificity
- Distinguish between humoral and cellular immunity
- Differentiate between antigens, epitopes, and haptens
- Describe the structure and function of antibodies and distinguish between the different classes of antibodies

---

Clinical Focus

Part 1

Olivia, a one-year old infant, is brought to the emergency room by her parents, who report her symptoms: excessive crying, irritability, sensitivity to light, unusual lethargy, and vomiting. A physician feels swollen lymph nodes in Olivia's throat and armpits. In addition, the area of the abdomen over the spleen is swollen and tender.

- What do these symptoms suggest?
- What tests might be ordered to try to diagnose the problem?

Jump to the next Clinical Focus box.

Adaptive immunity is defined by two important characteristics: **specificity** and **memory**. Specificity refers to the adaptive immune system's ability to target specific pathogens, and memory refers to its ability to quickly respond to pathogens to which it has previously been exposed. For example, when an individual recovers from chickenpox, the body develops a *memory* of the infection that will *specifically* protect it from the causative agent, the varicella-zoster virus, if it is exposed to the virus again later.

Specificity and memory are achieved by essentially programming certain cells involved in the immune response to respond rapidly to subsequent exposures of the pathogen. This programming occurs as a result of the first exposure to a pathogen or vaccine, which triggers a **primary response**. Subsequent exposures result in a **secondary response** that is faster and stronger as a result of the body's memory of the first exposure (Figure 18.2). This secondary response, however, is specific to the pathogen in question. For example, exposure to one virus (e.g., varicella-zoster virus) will not provide protection against other viral diseases (e.g., measles, mumps, or polio).

Adaptive specific immunity involves the actions of two distinct cell types: **B lymphocytes (B cells)** and **T lymphocytes (T cells)**. Although B cells and T cells arise from a common hematopoietic stem cell differentiation pathway (see Figure 17.12), their sites of maturation and their roles in adaptive immunity are very different.

B cells mature in the bone marrow and are responsible for the production of glycoproteins called **antibodies**, or **immunoglobulins**. Antibodies are involved in the body's defense against pathogens and toxins in the extracellular environment. Mechanisms of adaptive specific immunity that involve B cells and antibody production are referred to as **humoral immunity**. The maturation of T cells occurs in the thymus. T cells function as the central orchestrator of both innate and adaptive immune responses. They are also responsible for destruction of cells infected with intracellular pathogens. The targeting and destruction of intracellular pathogens by T cells is called cell-mediated immunity, or **cellular immunity**.

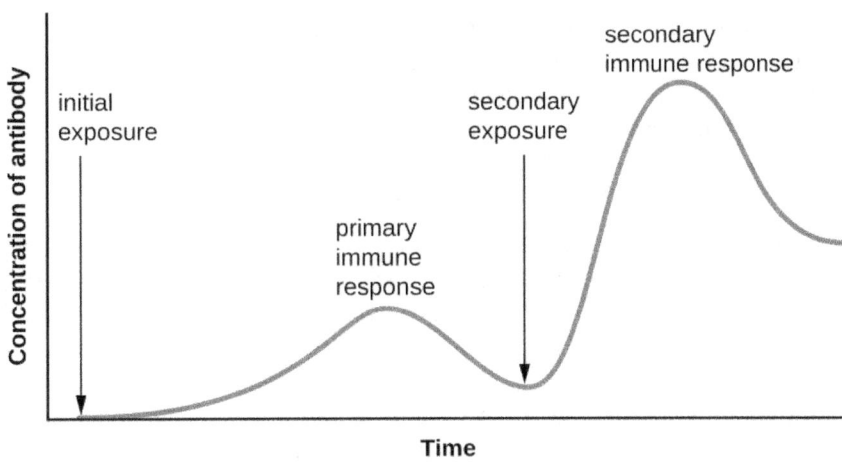

Figure 18.2 This graph illustrates the primary and secondary immune responses related to antibody production after an initial and secondary exposure to an antigen. Notice that the secondary response is faster and provides a much higher concentration of antibody.

CHECK YOUR UNDERSTANDING

- List the two defining characteristics of adaptive immunity.
- Explain the difference between a primary and secondary immune response.
- How do humoral and cellular immunity differ?

Antigens

Activation of the adaptive immune defenses is triggered by pathogen-specific molecular structures called **antigens**. Antigens are similar to the pathogen-associated molecular patterns (PAMPs) discussed in Pathogen Recognition and Phagocytosis; however, whereas PAMPs are molecular structures found on numerous pathogens, antigens are unique to a specific pathogen. The antigens that stimulate adaptive immunity to chickenpox, for example, are unique to the varicella-zoster virus but significantly different from the antigens associated with other viral pathogens.

The term *antigen* was initially used to describe molecules that stimulate the production of antibodies; in fact, the term comes from a combination of the words antibody and generator, and a molecule that stimulates antibody production is said to be **antigenic**. However, the role of antigens is not limited to humoral immunity and the production of antibodies; antigens also play an essential role in stimulating cellular immunity, and for this reason antigens are sometimes more accurately referred to as **immunogens**. In this text, however, we will typically refer to them as antigens.

Pathogens possess a variety of structures that may contain antigens. For example, antigens from bacterial cells may be associated with their capsules, cell walls, fimbriae, flagella, or pili. Bacterial antigens may also be associated with extracellular toxins and enzymes that they secrete. Viruses possess a variety of antigens associated with their capsids, envelopes, and the spike structures they use for attachment to cells.

Antigens may belong to any number of molecular classes, including carbohydrates, lipids, nucleic acids, proteins, and combinations of these molecules. Antigens of different classes vary in their ability to stimulate adaptive immune defenses as well as in the type of response they stimulate (humoral or cellular). The structural complexity of an antigenic molecule is an important factor in its antigenic potential. In general, more complex molecules are more effective as antigens. For example, the three-dimensional complex structure of proteins make them the most effective and potent antigens, capable of stimulating both humoral and cellular immunity. In comparison, carbohydrates are less complex in structure and therefore less effective as antigens; they can only stimulate humoral immune defenses. Lipids and nucleic acids are the least antigenic molecules, and in some cases may only become antigenic when combined with proteins or carbohydrates to form glycolipids, lipoproteins, or nucleoproteins.

One reason the three-dimensional complexity of antigens is so important is that antibodies and T cells do not recognize and interact with an entire antigen but with smaller exposed regions on the surface of antigens called **epitopes**. A single antigen may possess several different epitopes (Figure 18.3), and different antibodies may bind to different epitopes on the same antigen (Figure 18.4). For example, the bacterial flagellum is a large, complex protein structure that can possess hundreds or even thousands of epitopes with unique three-dimensional structures. Moreover, flagella from different bacterial species (or even strains of the same species) contain unique epitopes that can only be bound by specific antibodies.

An antigen's size is another important factor in its antigenic potential. Whereas large antigenic structures like flagella possess multiple epitopes, some molecules are too small to be antigenic by themselves. Such molecules, called **haptens**, are essentially free epitopes that are not part of the complex three-dimensional structure of a larger antigen. For a hapten to become antigenic, it must first attach to a larger carrier molecule (usually a protein) to produce a conjugate antigen. The hapten-specific antibodies produced in response to the conjugate antigen are then able to interact with unconjugated free hapten molecules. Haptens are not known to be associated with any specific pathogens, but they are responsible for some allergic responses. For example, the hapten urushiol, a molecule found in the oil of plants that cause poison ivy, causes an immune response that can result in a severe rash (called contact dermatitis). Similarly, the hapten penicillin can cause allergic reactions to drugs in the penicillin class.

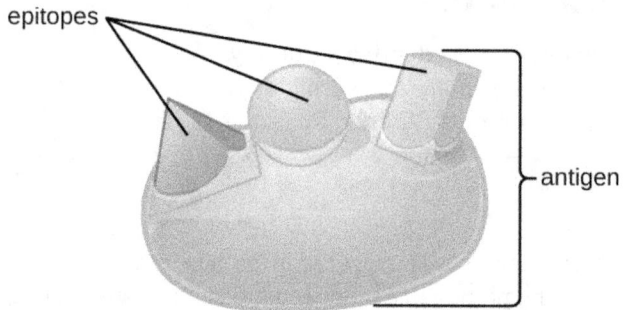

Figure 18.3 An antigen is a macromolecule that reacts with components of the immune system. A given antigen may contain several motifs that are recognized by immune cells.

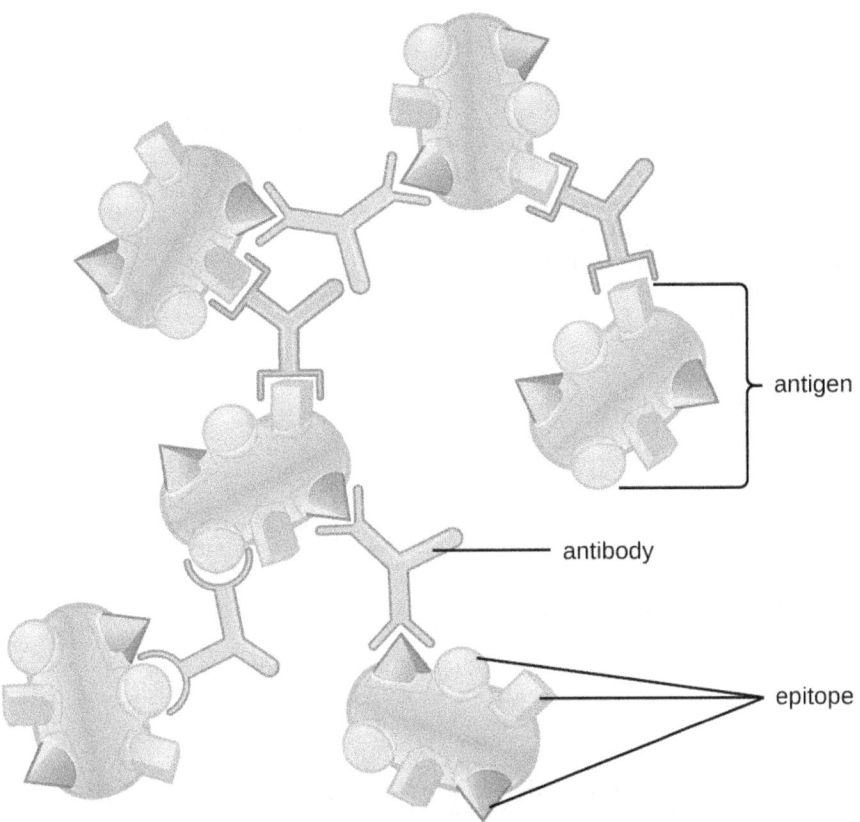

Figure 18.4 A typical protein antigen has multiple epitopes, shown by the ability of three different antibodies to bind to different epitopes of the same antigen.

✓ CHECK YOUR UNDERSTANDING

- What is the difference between an antigen and an epitope?
- What factors affect an antigen's antigenic potential?
- Why are haptens typically not antigenic, and how do they become antigenic?

Antibodies

Antibodies (also called immunoglobulins) are glycoproteins that are present in both the blood and tissue fluids. The basic structure of an antibody monomer consists of four protein chains held together by disulfide bonds (Figure 18.5). A disulfide bond is a covalent bond between the sulfhydryl *R* groups found on two cysteine amino acids. The two largest chains are identical to each other and are called the **heavy chains**. The two smaller chains are also identical to each other and are called the **light chains**. Joined together, the heavy and light chains form a basic Y-shaped structure.

The two 'arms' of the Y-shaped antibody molecule are known as the **Fab region**, for "fragment of antigen binding." The far end of the Fab region is the variable region, which serves as the site of antigen binding. The amino acid sequence in the variable region dictates the three-dimensional structure, and thus the specific three-dimensional epitope to which the Fab region is capable of binding. Although the epitope specificity of the Fab regions is identical for each arm of a single antibody molecule, this region displays a high degree of variability between antibodies with different epitope specificities. Binding to the Fab region is necessary for neutralization of pathogens, agglutination or aggregation of pathogens, and antibody-dependent cell-mediated cytotoxicity.

The constant region of the antibody molecule includes the trunk of the Y and lower portion of each arm of the Y. The trunk of the Y is also called the **Fc region**, for "fragment of crystallization," and is the site of complement factor binding and binding to phagocytic cells during antibody-mediated opsonization.

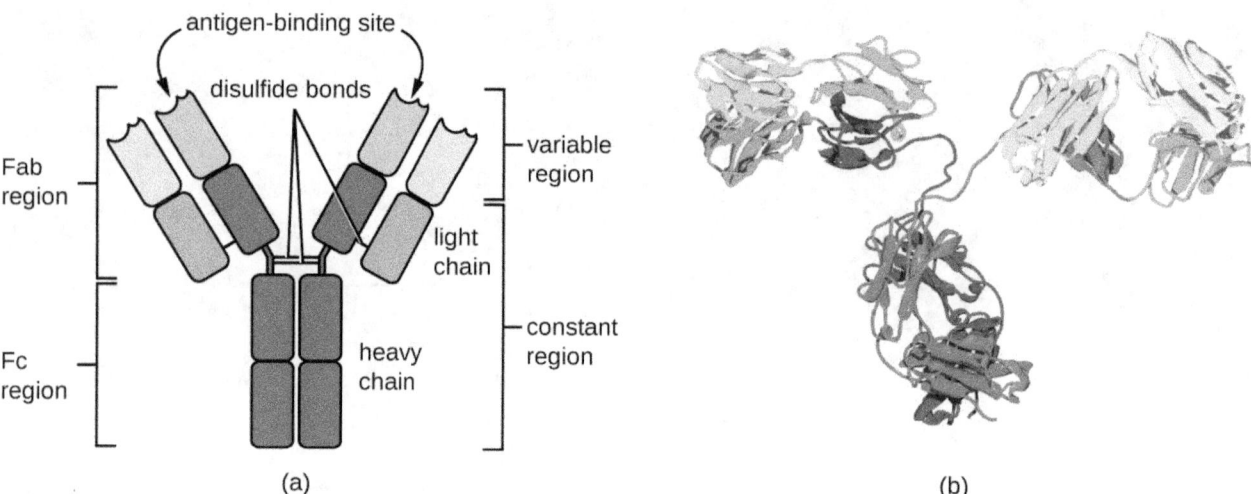

Figure 18.5 (a) The typical four-chain structure of a generic antibody monomer. (b) The corresponding three-dimensional structure of the antibody IgG. (credit b: modification of work by Tim Vickers)

CHECK YOUR UNDERSTANDING

- Describe the different functions of the Fab region and the Fc region.

Antibody Classes

The constant region of an antibody molecule determines its class, or isotype. The five classes of antibodies are IgG, IgM, IgA, IgD, and IgE. Each class possesses unique heavy chains designated by Greek letters γ, μ, α, δ, and ε, respectively. Antibody classes also exhibit important differences in abundance in serum, arrangement, body sites of action, functional roles, and size (Figure 18.6).

IgG is a monomer that is by far the most abundant antibody in human blood, accounting for about 80% of total serum antibody. IgG penetrates efficiently into tissue spaces, and is the only antibody class with the ability to cross the placental barrier, providing passive immunity to the developing fetus during pregnancy. IgG is also the most versatile antibody class in terms of its role in the body's defense against pathogens.

IgM is initially produced in a monomeric membrane-bound form that serves as an antigen-binding receptor on B cells. The secreted form of IgM assembles into a pentamer with five monomers of IgM bound together by a protein structure called the J chain. Although the location of the J chain relative to the Fc regions of the five monomers prevents IgM from performing some of the functions of IgG, the ten available Fab sites associated with a pentameric IgM make it an important antibody in the body's arsenal of defenses. IgM is the first antibody produced and secreted by B cells during the primary and secondary immune responses, making pathogen-specific IgM a valuable diagnostic marker during active or recent infections.

IgA accounts for about 13% of total serum antibody, and secretory IgA is the most common and abundant antibody class found in the mucus secretions that protect the mucous membranes. IgA can also be found in other secretions such as breast milk, tears, and saliva. Secretory IgA is assembled into a dimeric form with two monomers joined by a protein structure called the secretory component. One of the important functions of secretory IgA is to trap pathogens in mucus so that they can later be eliminated from the body.

Similar to IgM, **IgD** is a membrane-bound monomer found on the surface of B cells, where it serves as an antigen-binding receptor. However, IgD is not secreted by B cells, and only trace amounts are detected in serum. These trace amounts most likely come from the degradation of old B cells and the release of IgD molecules from their cytoplasmic membranes.

IgE is the least abundant antibody class in serum. Like IgG, it is secreted as a monomer, but its role in adaptive immunity is restricted to anti-parasitic defenses. The Fc region of IgE binds to basophils and mast cells. The Fab region of the bound IgE then interacts with specific antigen epitopes, causing the cells to release potent pro-inflammatory mediators. The inflammatory reaction resulting from the activation of mast cells and basophils aids in the defense against parasites, but this reaction is also central to allergic reactions (see Diseases of the Immune System).

Properties	IgG monomer	IgM pentamer	Secretory IgA dimer	IgD monomer	IgE monomer
Structure	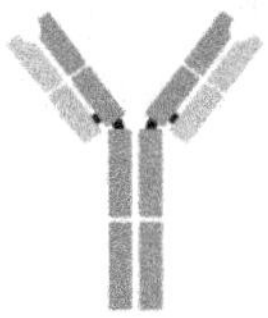	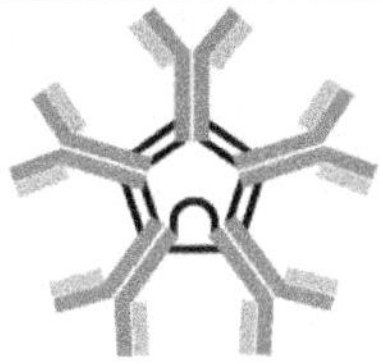	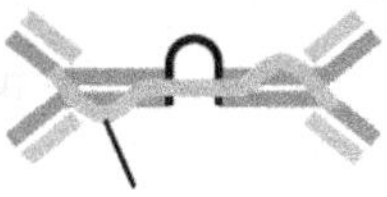 Secretory component	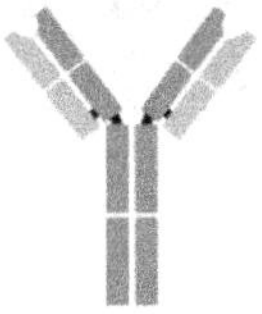	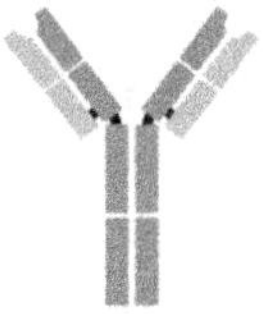
Heavy chains	γ	μ	α	δ	ε
Number of antigen-binding sites	2	10	4	2	2
Molecular weight (Daltons)	150,000	900,000	385,000	180,000	200,000
Percentage of total antibody in serum	80%	6%	13% (monomer)	<1%	<1%
Crosses placenta	yes	no	no	no	no
Fixes complement	yes	yes	no	no	no
Fc binds to	phagocytes				mast cells and basophils
Function	Neutralization, agglutination, complement activation, opsonization, and antibody-dependent cell-mediated cyotoxicity.	Neutralization, agglutination, and complement activation. The monomer form serves as the B-cell receptor.	Neutralization and trapping of pathogens in mucus.	B-cell receptor.	Activation of basophils and mast cells against parasites and allergens.

Figure 18.6

✓ CHECK YOUR UNDERSTANDING

- What part of an antibody molecule determines its class?
- What class of antibody is involved in protection against parasites?
- Describe the difference in structure between IgM and IgG.

Antigen-Antibody Interactions

Different classes of antibody play important roles in the body's defense against pathogens. These functions include neutralization of pathogens, opsonization for phagocytosis, agglutination, complement activation, and antibody-dependent cell-mediated cytotoxicity. For most of these functions, antibodies also provide an important link between adaptive specific immunity and innate nonspecific immunity.

Neutralization involves the binding of certain antibodies (IgG, IgM, or IgA) to epitopes on the surface of pathogens or toxins, preventing their attachment to cells. For example, Secretory IgA can bind to specific pathogens and block initial attachment to intestinal mucosal cells. Similarly, specific antibodies can bind to certain toxins, blocking them from attaching to target cells and thus neutralizing their toxic effects. Viruses can be neutralized and prevented from infecting a cell by the same mechanism (Figure 18.7).

As described in Chemical Defenses, opsonization is the coating of a pathogen with molecules, such as complement factors, C-reactive protein, and serum amyloid A, to assist in phagocyte binding to facilitate phagocytosis. IgG antibodies also serve as excellent opsonins, binding their Fab sites to specific epitopes on the surface of pathogens. Phagocytic cells such as macrophages, dendritic cells, and neutrophils have receptors on their surfaces that recognize and bind to the Fc portion of the IgG molecules; thus, IgG helps such phagocytes attach to and engulf the pathogens they have bound (Figure 18.8).

Agglutination or aggregation involves the cross-linking of pathogens by antibodies to create large aggregates (Figure 18.9). IgG has two Fab antigen-binding sites, which can bind to two separate pathogen cells, clumping them together. When multiple IgG antibodies are involved, large aggregates can develop; these aggregates are easier for the kidneys and spleen to filter from the blood and easier for phagocytes to ingest for destruction. The pentameric structure of IgM provides ten Fab binding sites per molecule, making it the most efficient antibody for agglutination.

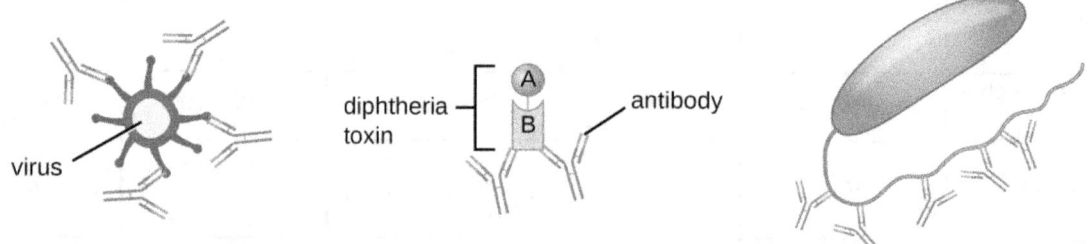

Figure 18.7 Neutralization involves the binding of specific antibodies to antigens found on bacteria, viruses, and toxins, preventing them from attaching to target cells.

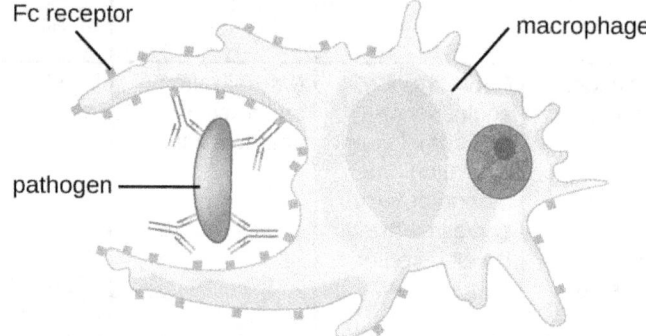

Figure 18.8 Antibodies serve as opsonins and inhibit infection by tagging pathogens for destruction by macrophages, dendritic cells, and neutrophils. These phagocytic cells use Fc receptors to bind to IgG-opsonized pathogens and initiate the first step of attachment before phagocytosis.

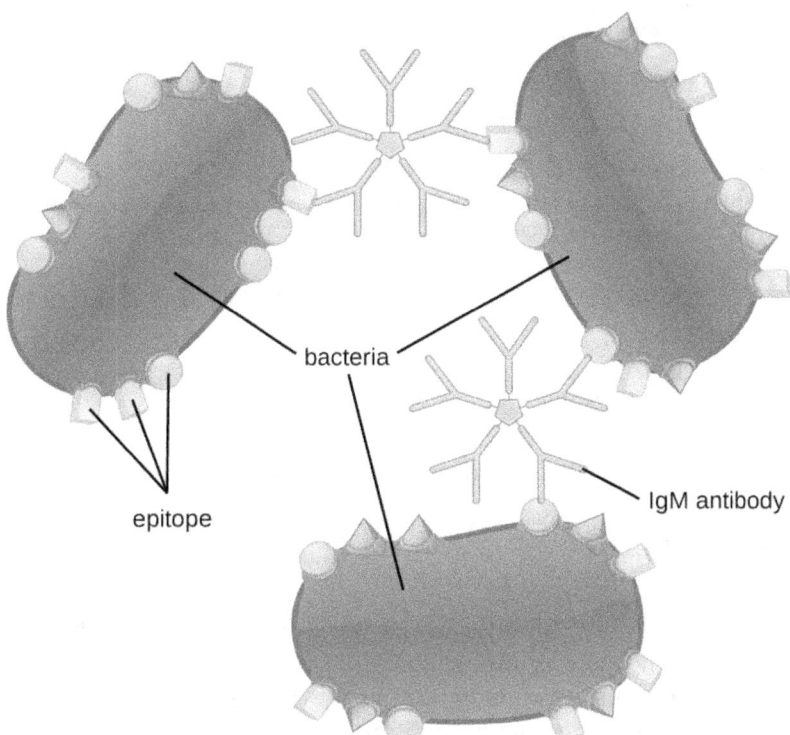

Figure 18.9 Antibodies, especially IgM antibodies, agglutinate bacteria by binding to epitopes on two or more bacteria simultaneously. When multiple pathogens and antibodies are present, aggregates form when the binding sites of antibodies bind with separate pathogens.

Another important function of antibodies is activation of the complement cascade. As discussed in the previous chapter, the complement system is an important component of the innate defenses, promoting the inflammatory response, recruiting phagocytes to site of infection, enhancing phagocytosis by opsonization, and killing gram-negative bacterial pathogens with the membrane attack complex (MAC). Complement activation can occur through three different pathways (see Figure 17.9), but the most efficient is the classical pathway, which requires the initial binding of IgG or IgM antibodies to the surface of a pathogen cell, allowing for recruitment and activation of the C1 complex.

Yet another important function of antibodies is **antibody-dependent cell-mediated cytotoxicity (ADCC)**, which enhances killing of pathogens that are too large to be phagocytosed. This process is best characterized for natural killer cells (NK cells), as shown in Figure 18.10, but it can also involve macrophages and eosinophils. ADCC occurs when the Fab region of an IgG antibody binds to a large pathogen; Fc receptors on effector cells (e.g., NK cells) then bind to the Fc region of the antibody, bringing them into close proximity with the target pathogen. The effector cell then secretes powerful cytotoxins (e.g., perforin and granzymes) that kill the pathogen.

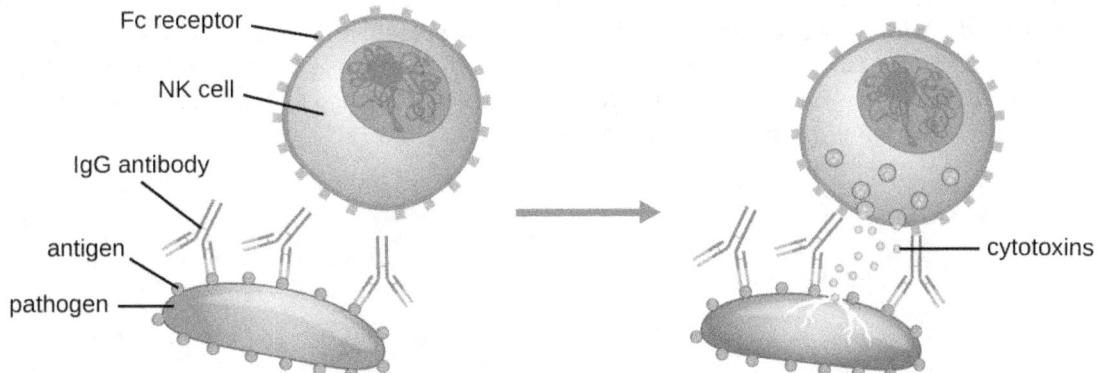

Figure 18.10 In this example of ADCC, antibodies bind to a large pathogenic cell that is too big for phagocytosis and then bind to Fc receptors on the membrane of a natural killer cell. This interaction brings the NK cell into close proximity, where it can kill the pathogen

through release of lethal extracellular cytotoxins.

✓ CHECK YOUR UNDERSTANDING

- Where is IgA normally found?
- Which class of antibody crosses the placenta, providing protection to the fetus?
- Compare the mechanisms of opsonization and antibody-dependent cell-mediated cytotoxicity.

18.2 Major Histocompatibility Complexes and Antigen-Presenting Cells

Learning Objectives

By the end of this section, you will be able to:
- Identify cells that express MHC I and/or MHC II molecules and describe the structures and cellular location of MHC I and MHC II molecules
- Identify the cells that are antigen-presenting cells
- Describe the process of antigen processing and presentation with MHC I and MHC II

As discussed in Cellular Defenses, major histocompatibility complex (MHC) molecules are expressed on the surface of healthy cells, identifying them as normal and "self" to natural killer (NK) cells. MHC molecules also play an important role in the presentation of foreign antigens, which is a critical step in the activation of T cells and thus an important mechanism of the adaptive immune system.

Major Histocompatibility Complex Molecules

The **major histocompatibility complex** (**MHC**) is a collection of genes coding for MHC molecules found on the surface of all nucleated cells of the body. In humans, the MHC genes are also referred to as human leukocyte antigen (HLA) genes. Mature red blood cells, which lack a nucleus, are the only cells that do not express MHC molecules on their surface.

There are two classes of MHC molecules involved in adaptive immunity, MHC I and MHC II (Figure 18.11). **MHC I** molecules are found on all nucleated cells; they present normal self-antigens as well as abnormal or nonself pathogens to the effector T cells involved in cellular immunity. In contrast, **MHC II** molecules are only found on macrophages, dendritic cells, and B cells; they present abnormal or nonself pathogen antigens for the initial activation of T cells.

Both types of MHC molecules are transmembrane glycoproteins that assemble as dimers in the cytoplasmic membrane of cells, but their structures are quite different. MHC I molecules are composed of a longer α protein chain coupled with a smaller $β_2$ microglobulin protein, and only the α chain spans the cytoplasmic membrane. The α chain of the MHC I molecule folds into three separate domains: $α_1$, $α_2$ and $α_3$. MHC II molecules are composed of two protein chains (an α and a β chain) that are approximately similar in length. Both chains of the MHC II molecule possess portions that span the plasma membrane, and each chain folds into two separate domains: $α_1$ and $α_2$, and $β_1$, and $β_2$. In order to present abnormal or non-self-antigens to T cells, MHC molecules have a cleft that serves as the antigen-binding site near the "top" (or outermost) portion of the MHC-I or MHC-II dimer. For MHC I, the antigen-binding cleft is formed by the $α_1$ and $α_2$ domains, whereas for MHC II, the cleft is formed by the $α_1$ and $β_1$ domains (Figure 18.11).

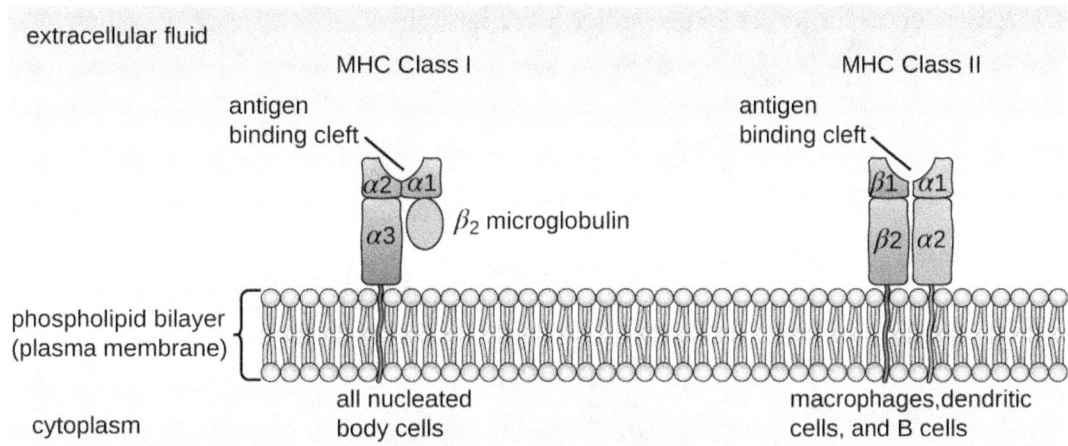

Figure 18.11 MHC I are found on all nucleated body cells, and MHC II are found on macrophages, dendritic cells, and B cells (along with MHC I). The antigen-binding cleft of MHC I is formed by domains α_1 and α_2. The antigen-binding cleft of MHC II is formed by domains α_1 and β_1.

CHECK YOUR UNDERSTANDING

- Compare the structures of the MHC I and MHC II molecules.

Antigen-Presenting Cells (APCs)

All nucleated cells in the body have mechanisms for processing and presenting antigens in association with MHC molecules. This signals the immune system, indicating whether the cell is normal and healthy or infected with an intracellular pathogen. However, only macrophages, dendritic cells, and B cells have the ability to present antigens specifically for the purpose of activating T cells; for this reason, these types of cells are sometimes referred to as **antigen-presenting cells (APCs)**.

While all APCs play a similar role in adaptive immunity, there are some important differences to consider. Macrophages and dendritic cells are phagocytes that ingest and kill pathogens that penetrate the first-line barriers (i.e., skin and mucous membranes). B cells, on the other hand, do not function as phagocytes but play a primary role in the production and secretion of antibodies. In addition, whereas macrophages and dendritic cells recognize pathogens through nonspecific receptor interactions (e.g., PAMPs, toll-like receptors, and receptors for opsonizing complement or antibody), B cells interact with foreign pathogens or their free antigens using antigen-specific immunoglobulin as receptors (monomeric IgD and IgM). When the immunoglobulin receptors bind to an antigen, the B cell internalizes the antigen by endocytosis before processing and presentting the antigen to T cells.

Antigen Presentation with MHC II Molecules

MHC II molecules are only found on the surface of APCs. Macrophages and dendritic cells use similar mechanisms for processing and presentation of antigens and their epitopes in association with MHC II; B cells use somewhat different mechanisms that will be described further in B Lymphocytes and Humoral Immunity. For now, we will focus on the steps of the process as they pertain to dendritic cells.

After a dendritic cell recognizes and attaches to a pathogen cell, the pathogen is internalized by phagocytosis and is initially contained within a phagosome. Lysosomes containing antimicrobial enzymes and chemicals fuse with the phagosome to create a phagolysosome, where degradation of the pathogen for antigen processing begins. Proteases (protein-degrading) are especially important in antigen processing because only protein antigen epitopes are presented to T cells by MHC II (Figure 18.12).

APCs do not present all possible epitopes to T cells; only a selection of the most antigenic or immunodominant epitopes are presented. The mechanism by which epitopes are selected for processing and presentation by an APC is complicated and not well understood; however, once the most antigenic, immunodominant epitopes have been processed, they associate within the antigen-binding cleft of MHC II molecules and are translocated to the cell surface of the dendritic cell for presentation to T cells.

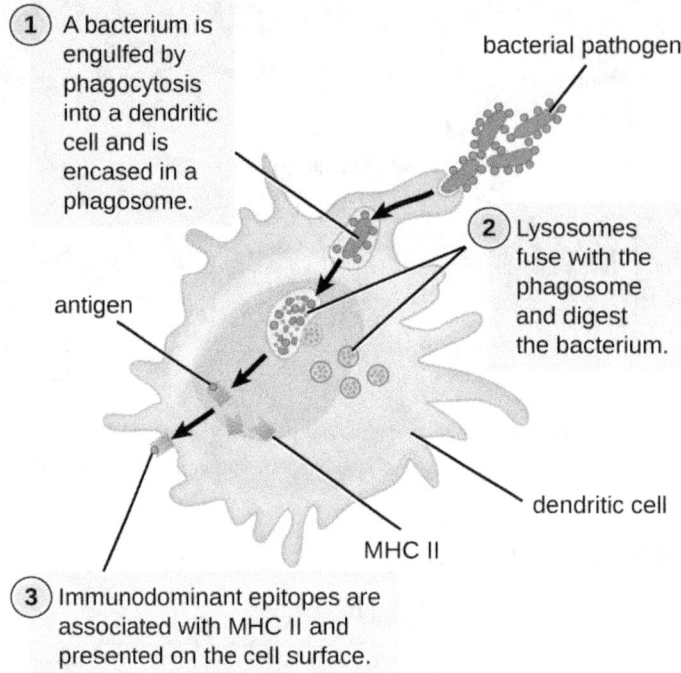

Figure 18.12 A dendritic cell phagocytoses a bacterial cell and brings it into a phagosome. Lysosomes fuse with the phagosome to create a phagolysosome, where antimicrobial chemicals and enzymes degrade the bacterial cell. Proteases process bacterial antigens, and the most antigenic epitopes are selected and presented on the cell's surface in conjunction with MHC II molecules. T cells recognize the presented antigens and are thus activated.

✓ CHECK YOUR UNDERSTANDING

- What are the three kinds of APCs?
- What role to MHC II molecules play in antigen presentation?
- What is the role of antigen presentation in adaptive immunity?

Antigen Presentation with MHC I Molecules

MHC I molecules, found on all normal, healthy, nucleated cells, signal to the immune system that the cell is a normal "self" cell. In a healthy cell, proteins normally found in the cytoplasm are degraded by proteasomes (enzyme complexes responsible for degradation and processing of proteins) and processed into self-antigen epitopes; these self-antigen epitopes bind within the MHC I antigen-binding cleft and are then presented on the cell surface. Immune cells, such as NK cells, recognize these self-antigens and do not target the cell for destruction. However, if a cell becomes infected with an intracellular pathogen (e.g., a virus), protein antigens specific to the pathogen are processed in the proteasomes and bind with MHC I molecules for presentation on the cell surface. This presentation of pathogen-specific antigens with MHC I signals that the infected cell must be targeted for destruction along with the pathogen.

Before elimination of infected cells can begin, APCs must first activate the T cells involved in cellular immunity. If an intracellular pathogen directly infects the cytoplasm of an APC, then the processing and presentation of antigens can occur as described (in proteasomes and on the cell surface with MHC I). However, if the intracellular pathogen does not directly infect APCs, an alternative strategy called **cross-presentation** is utilized. In cross-presentation, antigens are brought into the APC by mechanisms normally leading to presentation with MHC II (i.e., through phagocytosis), but the antigen is presented on an MHC I molecule for CD8 T cells. The exact mechanisms by which cross-presentation occur are not yet well understood, but it appears that cross-presentation is primarily a function of dendritic cells and not macrophages or B cells.

CHECK YOUR UNDERSTANDING

- Compare and contrast antigen processing and presentation associated with MHC I and MHC II molecules.
- What is cross-presentation, and when is it likely to occur?

18.3 T Lymphocytes and Cellular Immunity

Learning Objectives

By the end of this section, you will be able to:
- Describe the process of T-cell maturation and thymic selection
- Explain the genetic events that lead to diversity of T-cell receptors
- Compare and contrast the various classes and subtypes of T cells in terms of activation and function
- Explain the mechanism by which superantigens effect unregulated T-cell activation

As explained in Overview of Specific Adaptive Immunity, the antibodies involved in humoral immunity often bind pathogens and toxins before they can attach to and invade host cells. Thus, humoral immunity is primarily concerned with fighting pathogens in extracellular spaces. However, pathogens that have already gained entry to host cells are largely protected from the humoral antibody-mediated defenses. Cellular immunity, on the other hand, targets and eliminates intracellular pathogens through the actions of T lymphocytes, or T cells (Figure 18.13). T cells also play a more central role in orchestrating the overall adaptive immune response (humoral as well as cellular) along with the cellular defenses of innate immunity.

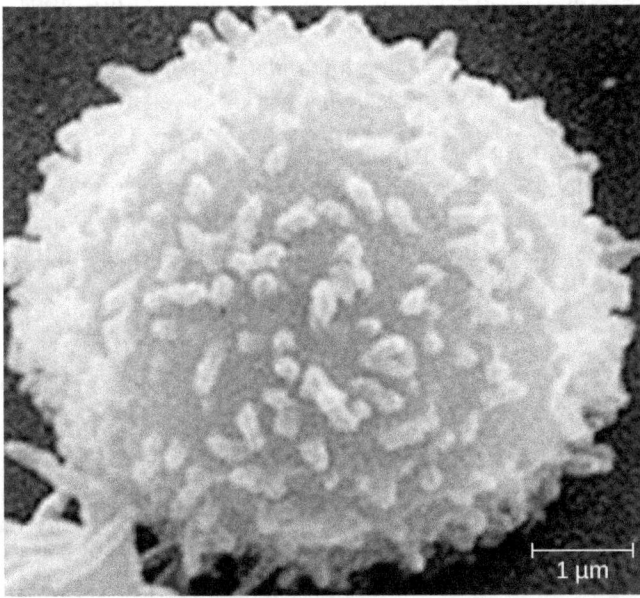

Figure 18.13 This scanning electron micrograph shows a T lymphocyte, which is responsible for the cell-mediated immune response. The spike-like membrane structures increase surface area, allowing for greater interaction with other cell types and their signals. (credit: modification of work by NCI)

T Cell Production and Maturation

T cells, like all other white blood cells involved in innate and adaptive immunity, are formed from multipotent hematopoietic stem cells (HSCs) in the bone marrow (see Figure 17.12). However, unlike the white blood cells of innate immunity, eventual T cells differentiate first into lymphoid stem cells that then become small, immature lymphocytes, sometimes called lymphoblasts. The first steps of differentiation occur in the red marrow of bones (Figure 18.14), after which immature T lymphocytes enter the bloodstream and travel to the thymus for the final steps of maturation (Figure 18.15). Once in the thymus, the immature T lymphocytes are referred to as thymocytes.

The maturation of thymocytes within the thymus can be divided into three critical steps of positive and negative selection, collectively referred to as **thymic selection**. The first step of thymic selection occurs in the cortex of the thymus and involves the development of a functional T-cell receptor (TCR) that is required for activation by APCs. Thymocytes with defective TCRs are removed by negative selection through the induction of **apoptosis** (programmed controlled cell death). The second step of thymic selection also occurs in the cortex and involves the positive selection of thymocytes that will interact appropriately with MHC molecules. Thymocytes that can interact appropriately with MHC molecules receive a positive stimulation that moves them further through the process of maturation, whereas thymocytes that do not interact appropriately are not stimulated and are eliminated by apoptosis. The third and final step of thymic selection occurs in both the cortex and medulla and involves negative selection to remove self-reacting thymocytes, those that react to self-antigens, by apoptosis. This final step is sometimes referred to as **central tolerance** because it prevents self-reacting T cells from reaching the bloodstream and potentially causing autoimmune disease, which occurs when the immune system attacks healthy "self" cells.

Despite central tolerance, some self-reactive T cells generally escape the thymus and enter the peripheral bloodstream. Therefore, a second line of defense called **peripheral tolerance** is needed to protect against autoimmune disease. Peripheral tolerance involves mechanisms of **anergy** and inhibition of self-reactive T cells by **regulatory T cells**. Anergy refers to a state of nonresponsiveness to antigen stimulation. In the case of self-reactive T cells that escape the thymus, lack of an essential co-stimulatory signal required for activation causes anergy and prevents autoimmune activation. Regulatory T cells participate in peripheral tolerance by inhibiting the activation and function of self-reactive T cells and by secreting anti-inflammatory cytokines.

It is not completely understood what events specifically direct maturation of thymocytes into regulatory T cells. Current theories suggest the critical events may occur during the third step of thymic selection, when most self-reactive T cells are eliminated. Regulatory T cells may receive a unique signal that is below the threshold required to target them for negative selection and apoptosis. Consequently, these cells continue to mature and then exit the thymus, armed to inhibit the activation of self-reactive T cells.

It has been estimated that the three steps of thymic selection eliminate 98% of thymocytes. The remaining 2% that exit the thymus migrate through the bloodstream and lymphatic system to sites of secondary lymphoid organs/tissues, such as the lymph nodes, spleen, and tonsils (Figure 18.15), where they await activation through the presentation of specific antigens by APCs. Until they are activated, they are known as **mature naïve T cells**.

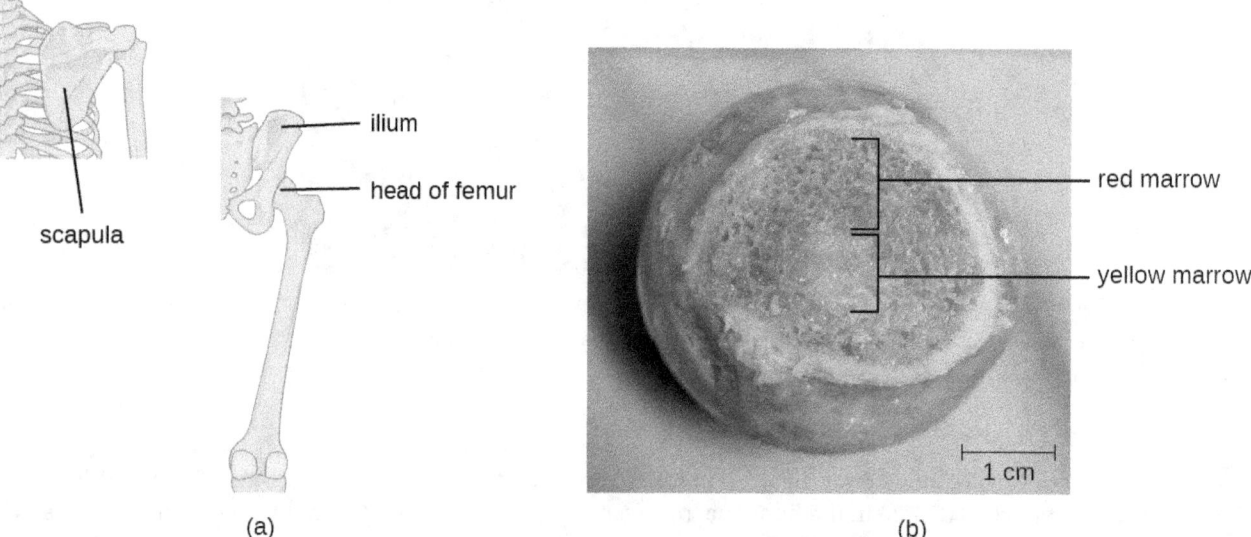

Figure 18.14 (a) Red bone marrow can be found in the head of the femur (thighbone) and is also present in the flat bones of the body, such as the ilium and the scapula. (b) Red bone marrow is the site of production and differentiation of many formed elements of blood, including erythrocytes, leukocytes, and platelets. The yellow bone marrow is populated primarily with adipose cells.

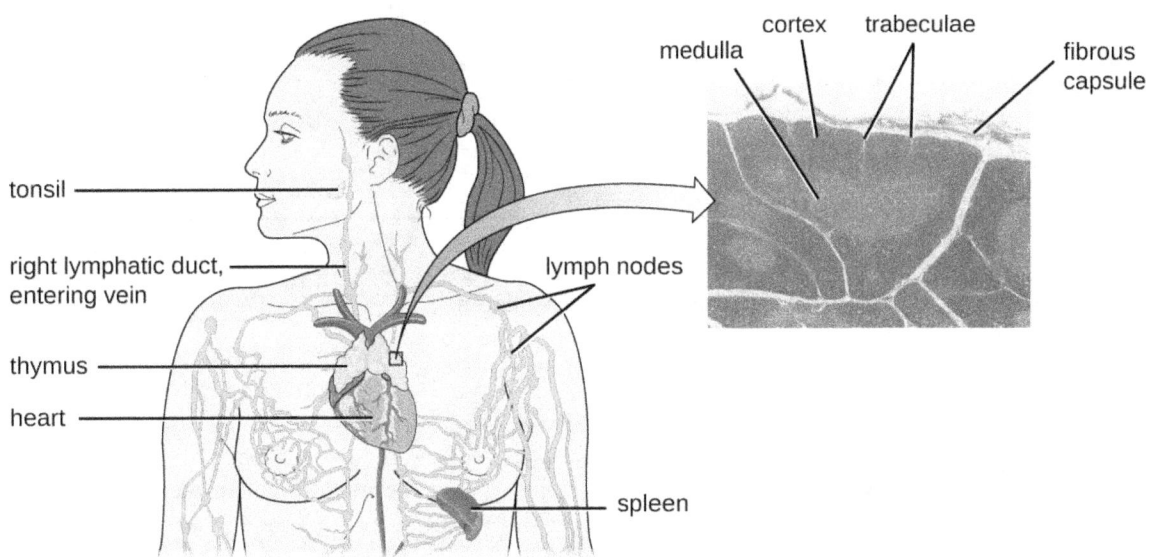

Figure 18.15 The thymus is a bi-lobed, H-shaped glandular organ that is located just above the heart. It is surrounded by a fibrous capsule of connective tissue. The darkly staining cortex and the lighter staining medulla of individual lobules are clearly visible in the light micrograph of the thymus of a newborn (top right, LM × 100). (credit micrograph: modification of micrograph provided by the Regents of University of Michigan Medical School © 2012)

✓ CHECK YOUR UNDERSTANDING

- What anatomical sites are involved in T cell production and maturation?
- What are the three steps involved in thymic selection?
- Why are central tolerance and peripheral tolerance important? What do they prevent?

Classes of T Cells

T cells can be categorized into three distinct classes: helper T cells, regulatory T cells, and cytotoxic T cells. These classes are differentiated based on their expression of certain surface molecules, their mode of activation, and their functional roles in adaptive immunity (Table 18.1).

All T cells produce **cluster of differentiation (CD) molecules**, cell surface glycoproteins that can be used to identify and distinguish between the various types of white blood cells. Although T cells can produce a variety of CD molecules, CD4 and CD8 are the two most important used for differentiation of the classes. Helper T cells and regulatory T cells are characterized by the expression of CD4 on their surface, whereas cytotoxic T cells are characterized by the expression of CD8.

Classes of T cells can also be distinguished by the specific MHC molecules and APCs with which they interact for activation. Helper T cells and regulatory T cells can only be activated by APCs presenting antigens associated with MHC II. In contrast, cytotoxic T cells recognize antigens presented in association with MHC I, either by APCs or by nucleated cells infected with an intracellular pathogen.

The different classes of T cells also play different functional roles in the immune system. **Helper T cells** serve as the central orchestrators that help activate and direct functions of humoral and cellular immunity. In addition, helper T cells enhance the pathogen-killing functions of macrophages and NK cells of innate immunity. In contrast, the primary role of regulatory T cells is to prevent undesirable and potentially damaging immune responses. Their role in peripheral tolerance, for example, protects against autoimmune disorders, as discussed earlier. Finally, **cytotoxic T cells** are the primary effector cells for cellular immunity. They recognize and target cells that have been infected by intracellular pathogens, destroying infected cells along with the pathogens inside.

Classes of T Cells

Class	Surface CD Molecules	Activation	Functions
Helper T cells	CD4	APCs presenting antigens associated with MHC II	Orchestrate humoral and cellular immunity
			Involved in the activation of macrophages and NK cells
Regulatory T cells	CD4	APCs presenting antigens associated with MHC II	Involved in peripheral tolerance and prevention of autoimmune responses
Cytotoxic T cells	CD8	APCs or infected nucleated cells presenting antigens associated with MHC I	Destroy cells infected with intracellular pathogens

Table 18.1

CHECK YOUR UNDERSTANDING

- What are the unique functions of the three classes of T cells?
- Which T cells can be activated by antigens presented by cells other than APCs?

T-Cell Receptors

For both helper T cells and cytotoxic T cells, activation is a complex process that requires the interactions of multiple molecules and exposure to cytokines. The **T-cell receptor (TCR)** is involved in the first step of pathogen epitope recognition during the activation process.

The TCR comes from the same receptor family as the antibodies IgD and IgM, the antigen receptors on the B cell membrane surface, and thus shares common structural elements. Similar to antibodies, the TCR has a variable region and a constant region, and the variable region provides the antigen-binding site (Figure 18.16). However, the structure of TCR is smaller and less complex than the immunoglobulin molecules (Figure 18.5). Whereas immunoglobulins have four peptide chains and Y-shaped structures, the TCR consists of just two peptide chains (α and β chains), both of which span the cytoplasmic membrane of the T cell.

TCRs are epitope-specific, and it has been estimated that 25 million T cells with unique epitope-binding TCRs are required to protect an individual against a wide range of microbial pathogens. Because the human genome only contains about 25,000 genes, we know that each specific TCR cannot be encoded by its own set of genes. This raises the question of how such a vast population of T cells with millions of specific TCRs can be achieved. The answer is a process called genetic rearrangement, which occurs in the thymus during the first step of thymic selection.

The genes that code for the variable regions of the TCR are divided into distinct gene segments called variable (V), diversity (D), and joining (J) segments. The genes segments associated with the α chain of the TCR consist 70 or more different V_α segments and 61 different J_α segments. The gene segments associated with the β chain of the TCR consist of 52 different V_β segments, two different D_β segments, and 13 different J_β segments. During the development of the functional TCR in the thymus, genetic rearrangement in a T cell brings together one V_α segment and one J_α segment to code for the variable region of the α chain. Similarly, genetic rearrangement brings one of the V_β segments together with one of the D_β segments and one of thetJ_β segments to code for the variable region of the β chain. All the possible combinations of rearrangements between different segments of

V, D, and J provide the genetic diversity required to produce millions of TCRs with unique epitope-specific variable regions.

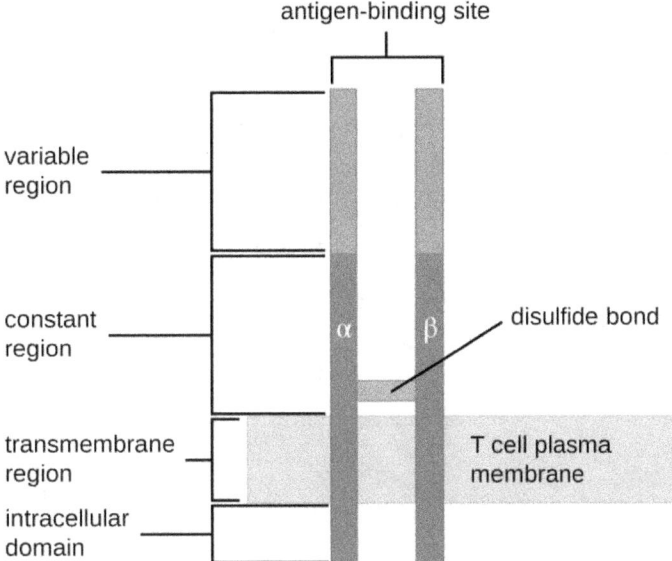

Figure 18.16 A T-cell receptor spans the cytoplasmic membrane and projects variable binding regions into the extracellular space to bind processed antigens associated with MHC I or MHC II molecules.

✓ CHECK YOUR UNDERSTANDING

- What are the similarities and differences between TCRs and immunoglobulins?
- What process is used to provide millions of unique TCR binding sites?

Activation and Differentiation of Helper T Cells

Helper T cells can only be activated by APCs presenting processed foreign epitopes in association with MHC II. The first step in the activation process is TCR recognition of the specific foreign epitope presented within the MHC II antigen-binding cleft. The second step involves the interaction of CD4 on the helper T cell with a region of the MHC II molecule separate from the antigen-binding cleft. This second interaction anchors the MHC II-TCR complex and ensures that the helper T cell is recognizing both the foreign ("nonself") epitope and "self" antigen of the APC; both recognitions are required for activation of the cell. In the third step, the APC and T cell secrete cytokines that activate the helper T cell. The activated helper T cell then proliferates, dividing by mitosis to produce clonal naïve helper T cells that differentiate into subtypes with different functions (Figure 18.17).

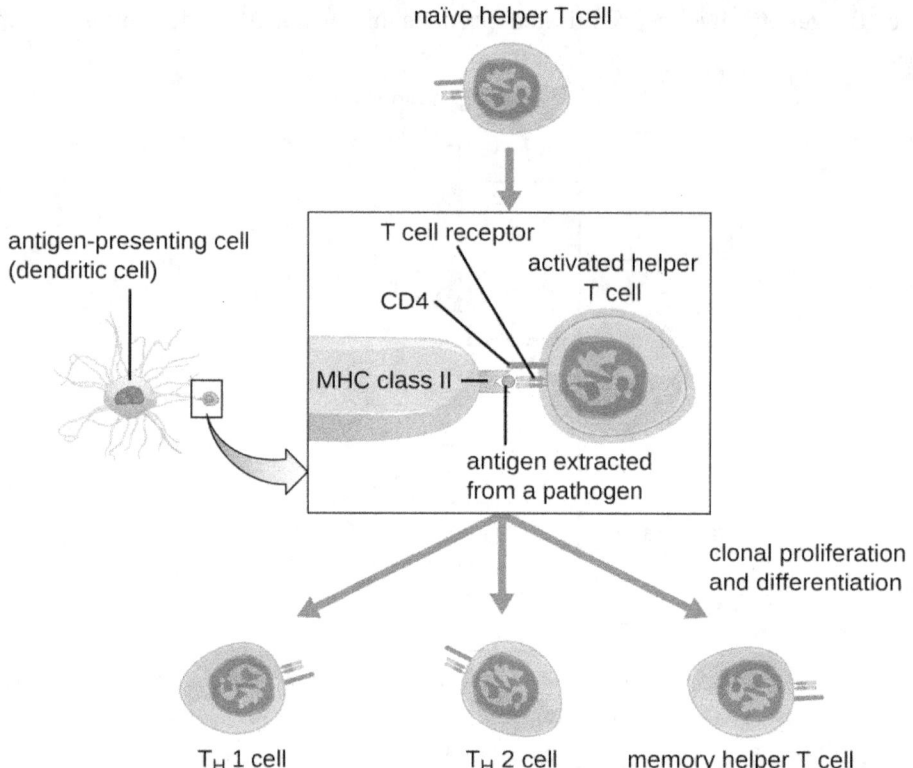

Figure 18.17 This illustration depicts the activation of a naïve (unactivated) helper T cell by an antigen-presenting cell and the subsequent proliferation and differentiation of the activated T cell into different subtypes.

Activated helper T cells can differentiate into one of four distinct subtypes, summarized in Table 18.2. The differentiation process is directed by APC-secreted cytokines. Depending on which APC-secreted cytokines interact with an activated helper T cell, the cell may differentiate into a T helper 1 (T_H1) cell, a T helper 2 (T_H2) cell, or a memory helper T cell. The two types of helper T cells are relatively short-lived **effector cells**, meaning that they perform various functions of the immediate immune response. In contrast, **memory helper T cells** are relatively long lived; they are programmed to "remember" a specific antigen or epitope in order to mount a rapid, strong, secondary response to subsequent exposures.

T_H1 cells secrete their own cytokines that are involved in stimulating and orchestrating other cells involved in adaptive and innate immunity. For example, they stimulate cytotoxic T cells, enhancing their killing of infected cells and promoting differentiation into memory cytotoxic T cells. T_H1 cells also stimulate macrophages and neutrophils to become more effective in their killing of intracellular bacteria. They can also stimulate NK cells to become more effective at killing target cells.

T_H2 cells play an important role in orchestrating the humoral immune response through their secretion of cytokines that activate B cells and direct B cell differentiation and antibody production. Various cytokines produced by T_H2 cells orchestrate antibody class switching, which allows B cells to switch between the production of IgM, IgG, IgA, and IgE as needed to carry out specific antibody functions and to provide pathogen-specific humoral immune responses.

A third subtype of helper T cells called **T_H17 cells** was discovered through observations that immunity to some infections is not associated with T_H1 or T_H2 cells. T_H17 cells and the cytokines they produce appear to be specifically responsible for the body's defense against chronic mucocutaneous infections. Patients who lack sufficient T_H17 cells in the mucosa (e.g., HIV patients) may be more susceptible to bacteremia and gastrointestinal infections.[1]

---

1 Blaschitz C., Raffatellu M. "Th17 cytokines and the gut mucosal barrier." J Clin Immunol. 2010 Mar; 30(2):196-203. doi: 10.1007/s10875-010-9368-7.

Subtypes of Helper T Cells

Subtype	Functions
T_H1 cells	Stimulate cytotoxic T cells and produce memory cytotoxic T cells
	Stimulate macrophages and neutrophils (PMNs) for more effective intracellular killing of pathogens
	Stimulate NK cells to kill more effectively
T_H2 cells	Stimulate B cell activation and differentiation into plasma cells and memory B cells
	Direct antibody class switching in B cells
T_H17 cells	Stimulate immunity to specific infections such as chronic mucocutaneous infections
Memory helper T cells	"Remember" a specific pathogen and mount a strong, rapid secondary response upon re-exposure

Table 18.2

Activation and Differentiation of Cytotoxic T Cells

Cytotoxic T cells (also referred to as cytotoxic T lymphocytes, or CTLs) are activated by APCs in a three-step process similar to that of helper T cells. The key difference is that the activation of cytotoxic T cells involves recognition of an antigen presented with MHC I (as opposed to MHC II) and interaction of CD8 (as opposed to CD4) with the receptor complex. After the successful co-recognition of foreign epitope and self-antigen, the production of cytokines by the APC and the cytotoxic T cell activate clonal proliferation and differentiation. Activated cytotoxic T cells can differentiate into effector cytotoxic T cells that target pathogens for destruction or memory cells that are ready to respond to subsequent exposures.

As noted, proliferation and differentiation of cytotoxic T cells is also stimulated by cytokines secreted from T_H1 cells activated by the same foreign epitope. The co-stimulation that comes from these T_H1 cells is provided by secreted cytokines. Although it is possible for activation of cytotoxic T cells to occur without stimulation from T_H1 cells, the activation is not as effective or long-lasting.

Once activated, cytotoxic T cells serve as the effector cells of cellular immunity, recognizing and kill cells infected with intracellular pathogens through a mechanism very similar to that of NK cells. However, whereas NK cells recognize nonspecific signals of cell stress or abnormality, cytotoxic T cells recognize infected cells through antigen presentation of pathogen-specific epitopes associated with MHC I. Once an infected cell is recognized, the TCR of the cytotoxic T cell binds to the epitope and releases perforin and granzymes that destroy the infected cell (Figure 18.18). Perforin is a protein that creates pores in the target cell, and **granzymes** are proteases that enter the pores and induce apoptosis. This mechanism of programmed cell death is a controlled and efficient means of destroying and removing infected cells without releasing the pathogens inside to infect neighboring cells, as might occur if the infected cells were simply lysed.

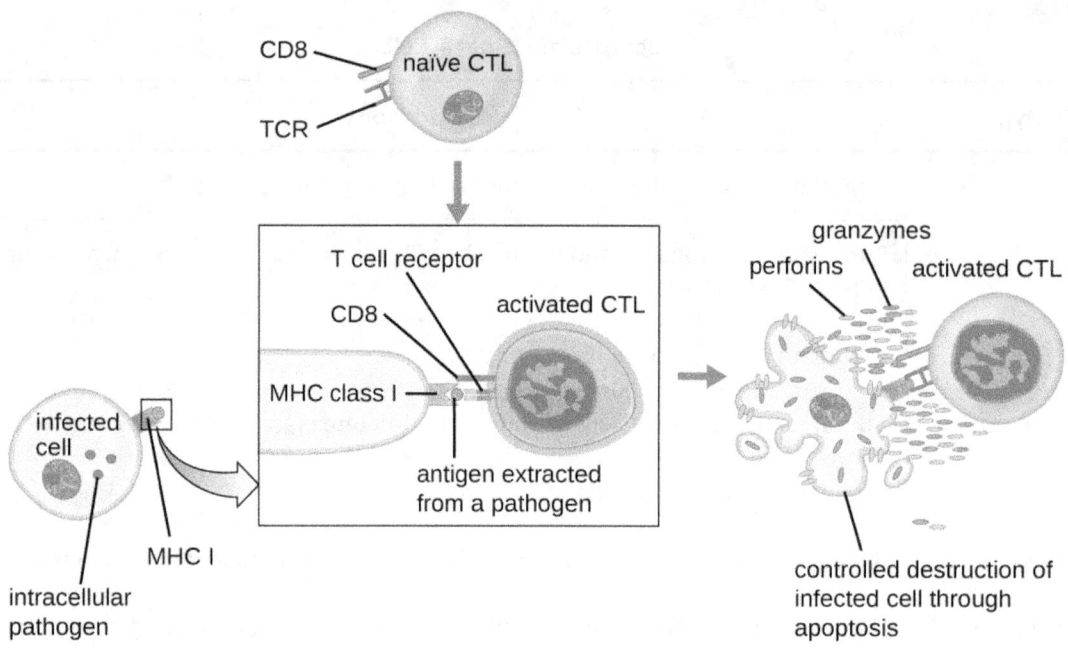

Figure 18.18 This figure illustrates the activation of a naïve (unactivated) cytotoxic T cell (CTL) by an antigen-presenting MHC I molecule on an infected body cell. Once activated, the CTL releases perforin and granzymes that invade the infected cell and induce controlled cell death, or apoptosis.

LINK TO LEARNING

In this video (https://www.openstax.org/l/22cytoTcellapop) , you can see a cytotoxic T cell inducing apoptosis in a target cell.

CHECK YOUR UNDERSTANDING

- Compare and contrast the activation of helper T cells and cytotoxic T cells.
- What are the different functions of helper T cell subtypes?
- What is the mechanism of CTL-mediated destruction of infected cells?

Superantigens and Unregulated Activation of T Cells

When T cell activation is controlled and regulated, the result is a protective response that is effective in combating infections. However, if T cell activation is unregulated and excessive, the result can be life-threatening. Certain bacterial and viral pathogens produce toxins known as superantigens (see Virulence Factors of Bacterial and Viral Pathogens) that can trigger such an unregulated response. Known bacterial superantigens include toxic shock syndrome toxin (TSST), staphylococcal enterotoxins, streptococcal pyrogenic toxins, streptococcal superantigen, and the streptococcal mitogenic exotoxin. Viruses known to produce superantigens include Epstein-Barr virus (human herpesvirus 4), cytomegalovirus (human herpesvirus 5), and others.

The mechanism of T cell activation by superantigens involves their simultaneous binding to MHC II molecules of APCs and the variable region of the TCR β chain. This binding occurs outside of the antigen-binding cleft of MHC II, so the superantigen will bridge together and activate MHC II and TCR without specific foreign epitope recognition (Figure 18.19). The result is an excessive, uncontrolled release of cytokines, often called a **cytokine storm**, which stimulates an excessive inflammatory response. This can lead to a dangerous decrease in blood pressure, shock, multi-organ failure, and potentially, death.

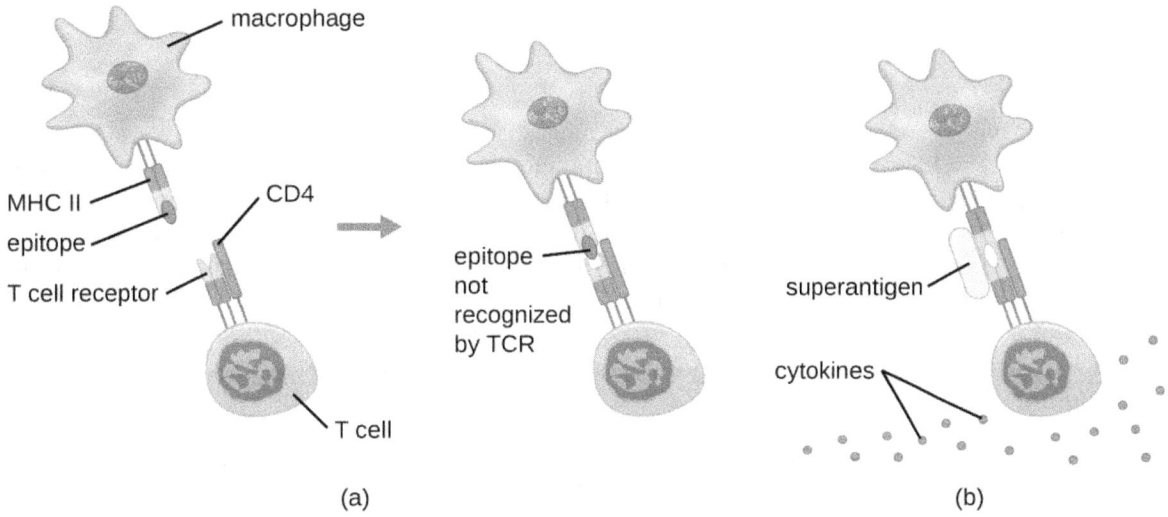

Figure 18.19 (a) The macrophage in this figure is presenting a foreign epitope that does not match the TCR of the T cell. Because the T cell does not recognize the epitope, it is not activated. (b) The macrophage in this figure is presenting a superantigen that is not recognized by the TCR of the T cell, yet the superantigen still is able to bridge and bind the MHC II and TCR molecules. This nonspecific, uncontrolled activation of the T cell results in an excessive release of cytokines that activate other T cells and cause excessive inflammation. (credit: modification of work by "Microbiotic"/YouTube)

CHECK YOUR UNDERSTANDING

- What are examples of superantigens?
- How does a superantigen activate a helper T cell?
- What effect does a superantigen have on a T cell?

Case in Point

Superantigens

Melissa, an otherwise healthy 22-year-old woman, is brought to the emergency room by her concerned boyfriend. She complains of a sudden onset of high fever, vomiting, diarrhea, and muscle aches. In her initial interview, she tells the attending physician that she is on hormonal birth control and also is two days into the menstruation portion of her cycle. She is on no other medications and is not abusing any drugs or alcohol. She is not a smoker. She is not diabetic and does not currently have an infection of any kind to her knowledge.

While waiting in the emergency room, Melissa's blood pressure begins to drop dramatically and her mental state deteriorates to general confusion. The physician believes she is likely suffering from toxic shock syndrome (TSS). TSS is caused by the toxin TSST-1, a superantigen associated with *Staphylococcus aureus*, and improper tampon use is a common cause of infections leading to TSS. The superantigen inappropriately stimulates widespread T cell activation and excessive cytokine release, resulting in a massive and systemic inflammatory response that can be fatal.

Vaginal or cervical swabs may be taken to confirm the presence of the microbe, but these tests are not critical to perform based on Melissa's symptoms and medical history. The physician prescribes rehydration, supportive therapy, and antibiotics to stem the bacterial infection. She also prescribes drugs to increase Melissa's blood pressure. Melissa spends three days in the hospital undergoing treatment; in addition, her kidney function is monitored because of the high risk of kidney failure associated with TSS. After 72 hours, Melissa is well enough to be discharged to continue her recovery at home.

- In what way would antibiotic therapy help to combat a superantigen?

> **Clinical Focus**
>
> **Part 2**
>
> Olivia's swollen lymph nodes, abdomen, and spleen suggest a strong immune response to a systemic infection in progress. In addition, little Olivia is reluctant to turn her head and appears to be experiencing severe neck pain. The physician orders a complete blood count, blood culture, and lumbar puncture. The cerebrospinal fluid (CSF) obtained appears cloudy and is further evaluated by Gram stain assessment and culturing for potential bacterial pathogens. The complete blood count indicates elevated numbers of white blood cells in Olivia's bloodstream. The white blood cell increases are recorded at 28.5 K/µL (normal range: 6.0–17.5 K/µL). The neutrophil percentage was recorded as 60% (normal range: 23–45%). Glucose levels in the CSF were registered at 30 mg/100 mL (normal range: 50–80 mg/100 mL). The WBC count in the CSF was 1,163/mm^3 (normal range: 5–20/mm^3).
>
> - Based on these results, do you have a preliminary diagnosis?
> - What is a recommended treatment based on this preliminary diagnosis?
>
> *Jump to the* next *Clinical Focus box. Go back to the* previous *Clinical Focus box.*

18.4 B Lymphocytes and Humoral Immunity

Learning Objectives

By the end of this section, you will be able to:

- Describe the production and maturation of B cells
- Compare the structure of B-cell receptors and T-cell receptors
- Compare T-dependent and T-independent activation of B cells
- Compare the primary and secondary antibody responses

Humoral immunity refers to mechanisms of the adaptive immune defenses that are mediated by antibodies secreted by B lymphocytes, or B cells. This section will focus on B cells and discuss their production and maturation, receptors, and mechanisms of activation.

B Cell Production and Maturation

Like T cells, B cells are formed from multipotent hematopoietic stem cells (HSCs) in the bone marrow and follow a pathway through lymphoid stem cell and lymphoblast (see Figure 17.12). Unlike T cells, however, lymphoblasts destined to become B cells do not leave the bone marrow and travel to the thymus for maturation. Rather, eventual B cells continue to mature in the bone marrow.

The first step of B cell maturation is an assessment of the functionality of their antigen-binding receptors. This occurs through positive selection for B cells with normal functional receptors. A mechanism of negative selection is then used to eliminate self-reacting B cells and minimize the risk of autoimmunity. Negative selection of self-reacting B cells can involve elimination by apoptosis, editing or modification of the receptors so they are no longer self-reactive, or induction of anergy in the B cell. Immature B cells that pass the selection in the bone marrow then travel to the spleen for their final stages of maturation. There they become **naïve mature B cells**, i.e., mature B cells that have not yet been activated.

✓ CHECK YOUR UNDERSTANDING

- Compare the maturation of B cells with the maturation of T cells.

B-Cell Receptors

Like T cells, B cells possess antigen-specific receptors with diverse specificities. Although they rely on T cells for optimum function, B cells can be activated without help from T cells. **B-cell receptors (BCRs)** for naïve mature B cells are membrane-bound monomeric forms of IgD and IgM. They have two identical heavy chains and two identical light chains connected by disulfide bonds into a basic "Y" shape (Figure 18.20). The trunk of the Y-shaped molecule, the constant region of the two heavy chains, spans the B cell membrane. The two antigen-binding sites exposed to the exterior of the B cell are involved in the binding of specific pathogen epitopes to initiate the activation process. It is estimated that each naïve mature B cell has upwards of 100,000 BCRs on its membrane, and each of these BCRs has an identical epitope-binding specificity.

In order to be prepared to react to a wide range of microbial epitopes, B cells, like T cells, use genetic rearrangement of hundreds of gene segments to provide the necessary diversity of receptor specificities. The variable region of the BCR heavy chain is made up of V, D, and J segments, similar to the β chain of the TCR. The variable region of the BCR light chain is made up of V and J segments, similar to the α chain of the TCR. Genetic rearrangement of all possible combinations of V-J-D (heavy chain) and V-J (light chain) provides for millions of unique antigen-binding sites for the BCR and for the antibodies secreted after activation.

One important difference between BCRs and TCRs is the way they can interact with antigenic epitopes. Whereas TCRs can only interact with antigenic epitopes that are presented within the antigen-binding cleft of MHC I or MHC II, BCRs do not require antigen presentation with MHC; they can interact with epitopes on free antigens or with epitopes displayed on the surface of intact pathogens. Another important difference is that TCRs only recognize protein epitopes, whereas BCRs can recognize epitopes associated with different molecular classes (e.g., proteins, polysaccharides, lipopolysaccharides).

Activation of B cells occurs through different mechanisms depending on the molecular class of the antigen. Activation of a B cell by a protein antigen requires the B cell to function as an APC, presenting the protein epitopes with MHC II to helper T cells. Because of their dependence on T cells for activation of B cells, protein antigens are classified as **T-dependent antigens**. In contrast, polysaccharides, lipopolysaccharides, and other nonprotein antigens are considered **T-independent antigens** because they can activate B cells without antigen processing and presentation to T cells.

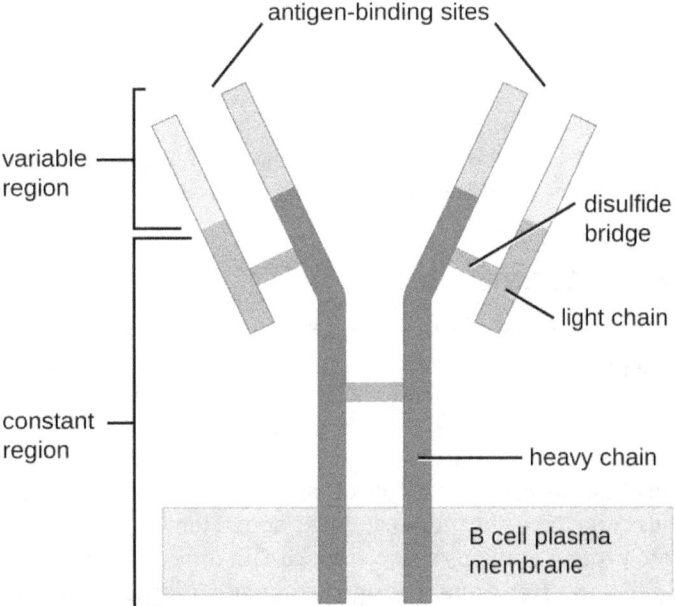

Figure 18.20 B-cell receptors are embedded in the membranes of B cells. The variable regions of all of the receptors on a single cell bind the same specific antigen.

⊘ CHECK YOUR UNDERSTANDING

- What types of molecules serve as the BCR?
- What are the differences between TCRs and BCRs with respect to antigen recognition?
- Which molecule classes are T-dependent antigens and which are T-independent antigens?

T Cell-Independent Activation of B cells

Activation of B cells without the cooperation of helper T cells is referred to as T cell-independent activation and occurs when BCRs interact with T-independent antigens. T-independent antigens (e.g., polysaccharide capsules, lipopolysaccharide) have repetitive epitope units within their structure, and this repetition allows for the cross-linkage of multiple BCRs, providing the first signal for activation (Figure 18.21). Because T cells are not involved, the second signal has to come from other sources, such as interactions of toll-like receptors with PAMPs or interactions with factors from the complement system.

Once a B cell is activated, it undergoes clonal proliferation and daughter cells differentiate into plasma cells. **Plasma cells** are antibody factories that secrete large quantities of antibodies. After differentiation, the surface BCRs disappear and the plasma cell secretes pentameric IgM molecules that have the same antigen specificity as the BCRs (Figure 18.21).

The T cell-independent response is short-lived and does not result in the production of memory B cells. Thus it will not result in a secondary response to subsequent exposures to T-independent antigens.

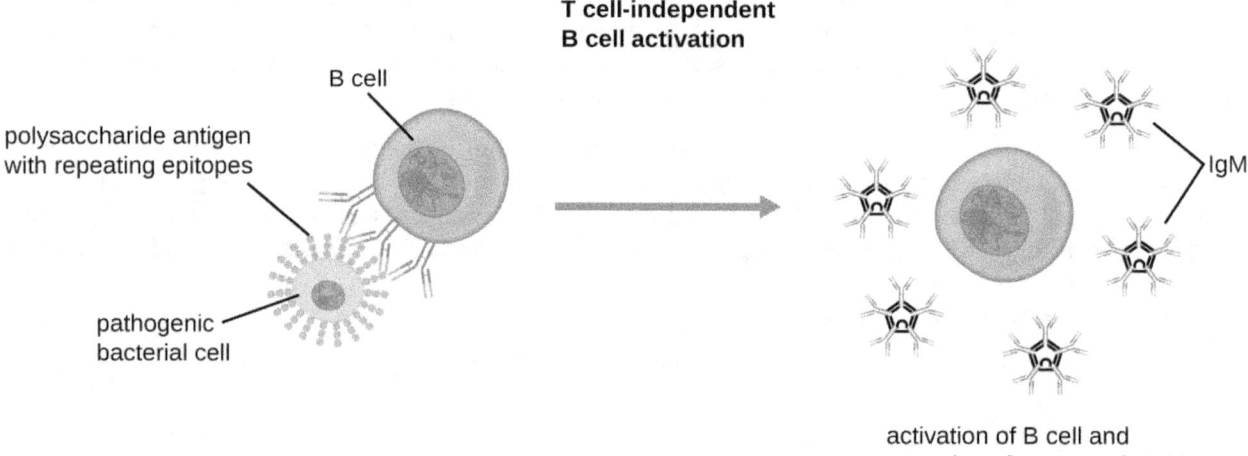

Figure 18.21 T-independent antigens have repeating epitopes that can induce B cell recognition and activation without involvement from T cells. A second signal, such as interaction of TLRs with PAMPs (not shown), is also required for activation of the B cell. Once activated, the B cell proliferates and differentiates into antibody-secreting plasma cells.

✓ CHECK YOUR UNDERSTANDING

- What are the two signals required for T cell-independent activation of B cells?
- What is the function of a plasma cell?

T Cell-Dependent Activation of B cells

T cell-dependent activation of B cells is more complex than T cell-independent activation, but the resulting immune response is stronger and develops memory. T cell-dependent activation can occur either in response to free protein antigens or to protein antigens associated with an intact pathogen. Interaction between the BCRs on a naïve mature B cell and a free protein antigen stimulate internalization of the antigen, whereas interaction with antigens associated with an intact pathogen initiates the extraction of the antigen from the pathogen before internalization. Once internalized inside the B cell, the protein antigen is processed and presented with MHC II. The presented antigen is then recognized by helper T cells specific to the same antigen. The TCR of the helper T cell recognizes the foreign antigen, and the T cell's CD4 molecule interacts with MHC II on the B cell. The coordination between B cells and helper T cells that are specific to the same antigen is referred to as **linked recognition**.

Once activated by linked recognition, T_H2 cells produce and secrete cytokines that activate the B cell and cause proliferation into clonal daughter cells. After several rounds of proliferation, additional cytokines provided by the T_H2 cells stimulate the differentiation of activated B cell clones into **memory B cells**, which will quickly respond to subsequent exposures to the same protein epitope, and plasma cells that lose their membrane

BCRs and initially secrete pentameric IgM (Figure 18.22).

After initial secretion of IgM, cytokines secreted by T_H2 cells stimulate the plasma cells to switch from IgM production to production of IgG, IgA, or IgE. This process, called **class switching** or isotype switching, allows plasma cells cloned from the same activated B cell to produce a variety of antibody classes with the same epitope specificity. Class switching is accomplished by genetic rearrangement of gene segments encoding the constant region, which determines an antibody's class. The variable region is not changed, so the new class of antibody retains the original epitope specificity.

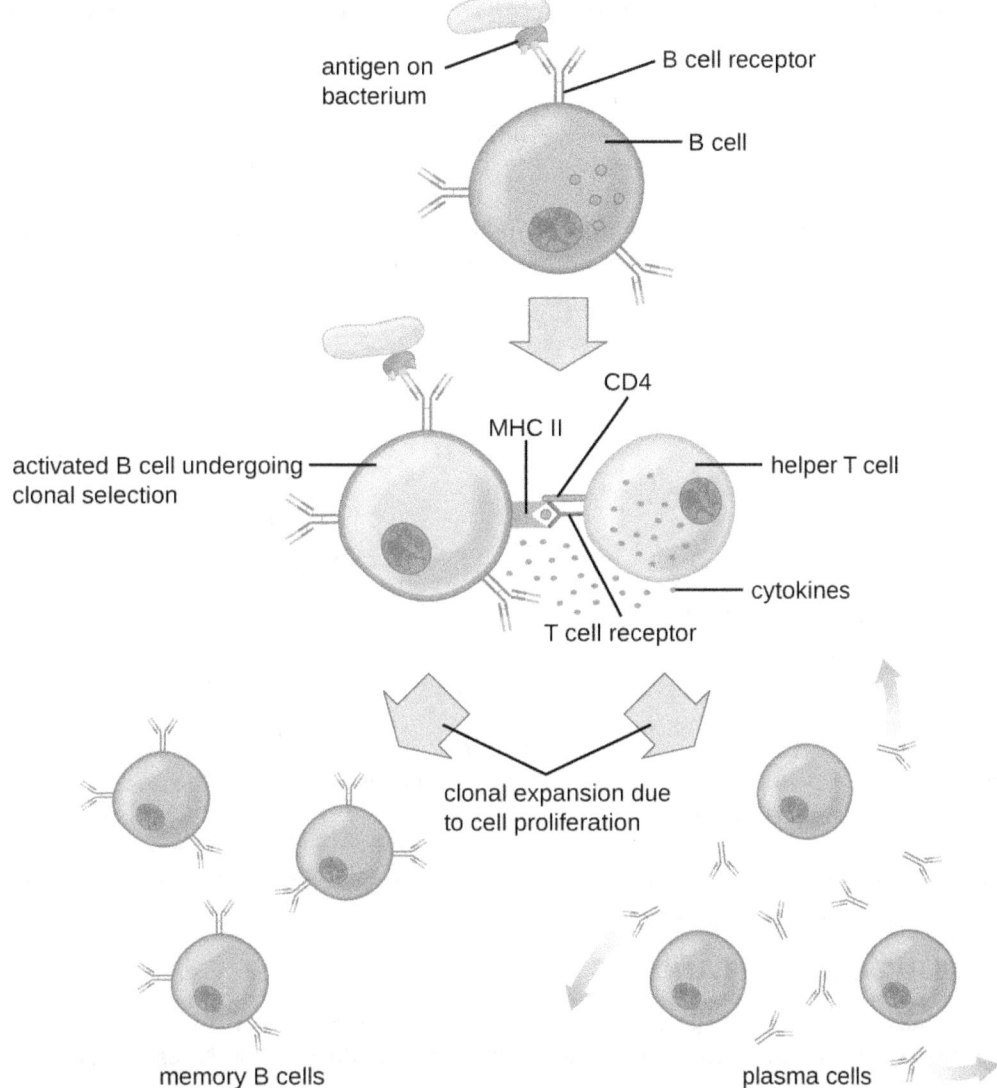

Figure 18.22 In T cell-dependent activation of B cells, the B cell recognizes and internalizes an antigen and presents it to a helper T cell that is specific to the same antigen. The helper T cell interacts with the antigen presented by the B cell, which activates the T cell and stimulates the release of cytokines that then activate the B cell. Activation of the B cell triggers proliferation and differentiation into B cells and plasma cells.

✓ CHECK YOUR UNDERSTANDING

- What steps are required for T cell-dependent activation of B cells?
- What is antibody class switching and why is it important?

Primary and Secondary Responses

T cell-dependent activation of B cells plays an important role in both the primary and secondary responses associated with adaptive immunity. With the first exposure to a protein antigen, a T cell-dependent primary antibody response occurs. The initial stage of the primary response is a **lag period**, or latent period, of approximately 10 days, during which no antibody can be detected in serum. This lag period is the time required for all of the steps of the primary response, including naïve mature B cell binding of antigen with BCRs, antigen processing and presentation, helper T cell activation, B cell activation, and clonal proliferation. The end of the lag period is characterized by a rise in IgM levels in the serum, as T_H2 cells stimulate B cell differentiation into plasma cells. IgM levels reach their peak around 14 days after primary antigen exposure; at about this same time, T_H2 stimulates antibody class switching, and IgM levels in serum begin to decline. Meanwhile, levels of IgG increase until they reach a peak about three weeks into the primary response (Figure 18.23).

During the primary response, some of the cloned B cells are differentiated into memory B cells programmed to respond to subsequent exposures. This secondary response occurs more quickly and forcefully than the primary response. The lag period is decreased to only a few days and the production of IgG is significantly higher than observed for the primary response (Figure 18.23). In addition, the antibodies produced during the secondary response are more effective and bind with higher affinity to the targeted epitopes. Plasma cells produced during secondary responses live longer than those produced during the primary response, so levels of specific antibody remain elevated for a longer period of time.

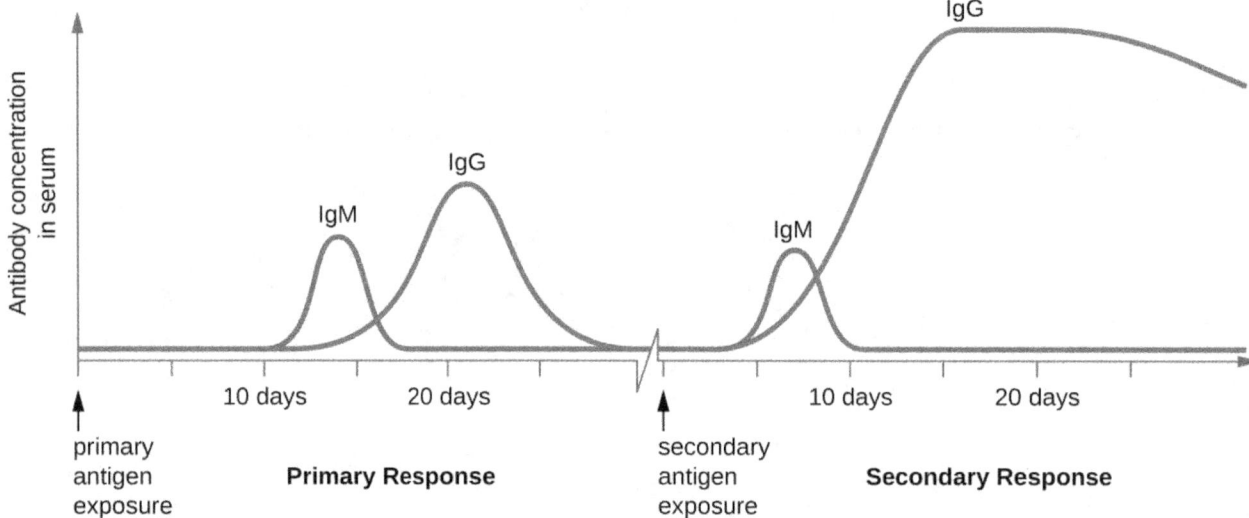

Figure 18.23 Compared to the primary response, the secondary antibody response occurs more quickly and produces antibody levels that are higher and more sustained. The secondary response mostly involves IgG.

✓ CHECK YOUR UNDERSTANDING

- What events occur during the lag period of the primary antibody response?
- Why do antibody levels remain elevated longer during the secondary antibody response?

18.5 Vaccines

Learning Objectives

By the end of this section, you will be able to:
- Compare the various kinds of artificial immunity
- Differentiate between variolation and vaccination
- Describe different types of vaccines and explain their respective advantages and disadvantages

For many diseases, prevention is the best form of treatment, and few strategies for disease prevention are as effective as vaccination. Vaccination is a form of artificial immunity. By artificially stimulating the adaptive immune defenses, a vaccine triggers memory cell production similar to that which would occur during a primary response. In so doing, the patient is able to mount a strong secondary response upon exposure to the pathogen—but without having to first suffer through an initial infection. In this section, we will explore several different kinds of artificial immunity along with various types of vaccines and the mechanisms by which they induce artificial immunity.

Classifications of Adaptive Immunity

All forms of adaptive immunity can be described as either active or passive. **Active immunity** refers to the activation of an individual's own adaptive immune defenses, whereas **passive immunity** refers to the transfer of adaptive immune defenses from another individual or animal. Active and passive immunity can be further subdivided based on whether the protection is acquired naturally or artificially.

Natural active immunity is adaptive immunity that develops after natural exposure to a pathogen (Figure 18.24). Examples would include the lifelong immunity that develops after recovery from a chickenpox or measles infection (although an acute infection is not always necessary to activate adaptive immunity). The length of time that an individual is protected can vary substantially depending upon the pathogen and antigens involved. For example, activation of adaptive immunity by protein spike structures during an intracellular viral infection can activate lifelong immunity, whereas activation by carbohydrate capsule antigens during an extracellular bacterial infection may activate shorter-term immunity.

Natural passive immunity involves the natural passage of antibodies from a mother to her child before and after birth. IgG is the only antibody class that can cross the placenta from mother's blood to the fetal blood supply. Placental transfer of IgG is an important passive immune defense for the infant, lasting up to six months after birth. Secretory IgA can also be transferred from mother to infant through breast milk.

Artificial passive immunity refers to the transfer of antibodies produced by a donor (human or animal) to another individual. This transfer of antibodies may be done as a prophylactic measure (i.e., to prevent disease after exposure to a pathogen) or as a strategy for treating an active infection. For example, artificial passive immunity is commonly used for post-exposure prophylaxis against rabies, hepatitis A, hepatitis B, and chickenpox (in high risk individuals). Active infections treated by artificial passive immunity include cytomegalovirus infections in immunocompromised patients and Ebola virus infections. In 1995, eight patients in the Democratic Republic of the Congo with active Ebola infections were treated with blood transfusions from patients who were recovering from Ebola. Only one of the eight patients died (a 12.5% mortality rate), which was much lower than the expected 80% mortality rate for Ebola in untreated patients.[2] Artificial passive immunity is also used for the treatment of diseases caused by bacterial toxins, including tetanus, botulism, and diphtheria.

Artificial active immunity is the foundation for vaccination. It involves the activation of adaptive immunity through the deliberate exposure of an individual to weakened or inactivated pathogens, or preparations consisting of key pathogen antigens.

---

[2] K. Mupapa, M. Massamba, K. Kibadi, K. Kivula, A. Bwaka, M. Kipasa, R. Colebunders, J. J. Muyembe-Tamfum. "Treatment of Ebola Hemorrhagic Fever with Blood Transfusions from Convalescent Patients." *Journal of Infectious Diseases* 179 Suppl. (1999): S18–S23.

Mechanisms of Acquisition of Immunity		
	Natural acquired	**Artificial acquired**
Passive	Immunity acquired from antibodies passed in breast milk or through placenta 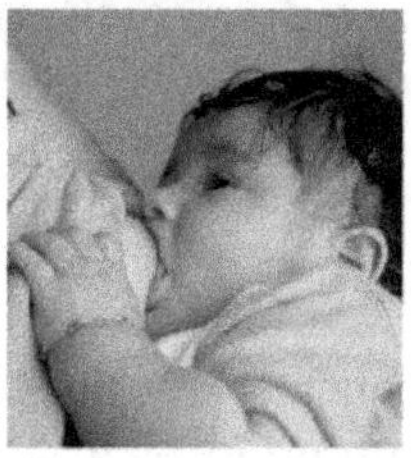	Immunity gained through antibodies harvested from another person or an animal 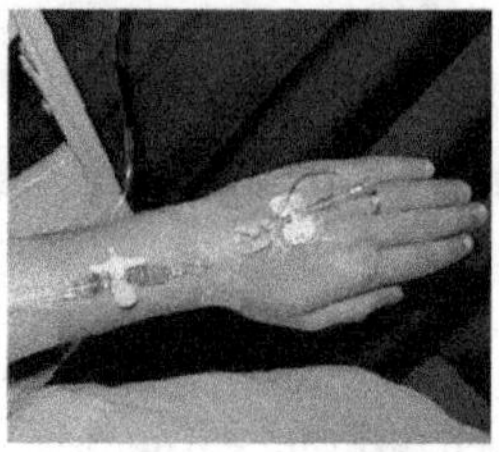
Active	Immunity gained through illness and recovery 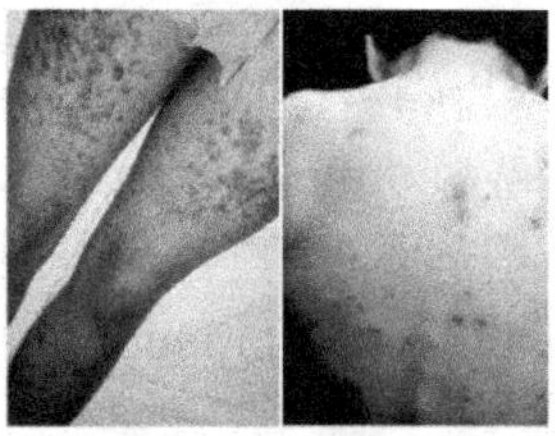	Immunity acquired through a vaccine 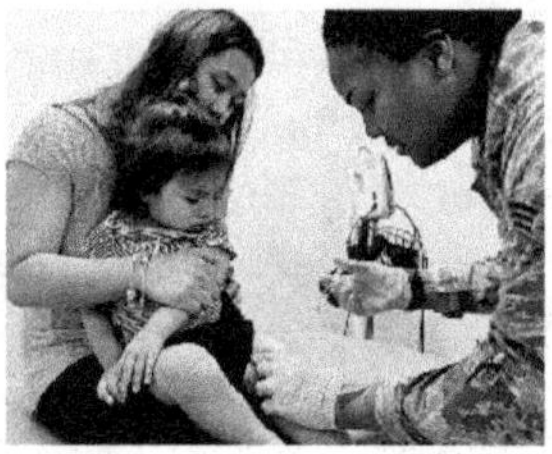

Figure 18.24 The four classifications of immunity. (credit top left photo: modification of work by USDA; credit top right photo: modification of work by "Michaelberry"/Wikimedia; credit bottom left photo: modification of work by Centers for Disease Control and Prevention; credit bottom right photo: Airman 1st Class Destinee Doughert / U.S. Air Force; Public Domain)

CHECK YOUR UNDERSTANDING

- What is the difference between active and passive immunity?
- What kind of immunity is conferred by a vaccine?

Herd Immunity

The four kinds of immunity just described result from an individual's adaptive immune system. For any given disease, an individual may be considered immune or susceptible depending on his or her ability to mount an effective immune response upon exposure. Thus, any given population is likely to have some individuals who are immune and other individuals who are susceptible. If a population has very few susceptible individuals, even those susceptible individuals will be protected by a phenomenon called **herd immunity**. Herd immunity has nothing to do with an individual's ability to mount an effective immune response; rather, it occurs because there are too few susceptible individuals in a population for the disease to spread effectively.

Vaccination programs create herd immunity by greatly reducing the number of susceptible individuals in a population. Even if some individuals in the population are not vaccinated, as long as a certain percentage is immune (either naturally or artificially), the few susceptible individuals are unlikely to be exposed to the pathogen. However, because new individuals are constantly entering populations (for example, through birth or relocation), vaccination programs are necessary to maintain herd immunity.

Eye on Ethics

Vaccination: Obligation or Choice

A growing number of parents are choosing not to vaccinate their children. They are dubbed "antivaxxers," and the majority of them believe that vaccines are a cause of autism (or other disease conditions), a link that has now been thoroughly disproven. Others object to vaccines on religious or moral grounds (e.g., the argument that Gardasil vaccination against HPV may promote sexual promiscuity), on personal ethical grounds (e.g., a conscientious objection to any medical intervention), or on political grounds (e.g., the notion that mandatory vaccinations are a violation of individual liberties).[3]

It is believed that this growing number of unvaccinated individuals has led to new outbreaks of whooping cough and measles. We would expect that herd immunity would protect those unvaccinated in our population, but herd immunity can only be maintained if enough individuals are being vaccinated.

Vaccination is clearly beneficial for public health. But from the individual parent's perspective the view can be murkier. Vaccines, like all medical interventions, have associated risks, and while the risks of vaccination may be extremely low compared to the risks of infection, parents may not always understand or accept the consensus of the medical community. Do such parents have a right to withhold vaccination from their children? Should they be allowed to put their children—and society at large—at risk?

Many governments insist on childhood vaccinations as a condition for entering public school, but it has become easy in most states to opt out of the requirement or to keep children out of the public system. Since the 1970s, West Virginia and Mississippi have had in place a stringent requirement for childhood vaccination, without exceptions, and neither state has had a case of measles since the early 1990s. California lawmakers recently passed a similar law in response to a measles outbreak in 2015, making it much more difficult for parents to opt out of vaccines if their children are attending public schools. Given this track record and renewed legislative efforts, should other states adopt similarly strict requirements?

What role should health-care providers play in promoting or enforcing universal vaccination? Studies have shown that many parents' minds can be changed in response to information delivered by health-care workers, but is it the place of health-care workers to try to persuade parents to have their children vaccinated? Some health-care providers are understandably reluctant to treat unvaccinated patients. Do they have the right to refuse service to patients who decline vaccines? Do insurance companies have the right to deny coverage to antivaxxers? These are all ethical questions that policymakers may be forced to address as more parents skirt vaccination norms.

Variolation and Vaccination

Thousands of years ago, it was first recognized that individuals who survived a smallpox infection were immune to subsequent infections. The practice of inoculating individuals to actively protect them from smallpox appears to have originated in the 10th century in China, when the practice of **variolation** was described (Figure 18.25). Variolation refers to the deliberate inoculation of individuals with infectious material from scabs or pustules of smallpox victims. Infectious materials were either injected into the skin or introduced through the nasal route. The infection that developed was usually milder than naturally acquired smallpox, and recovery from the milder infection provided protection against the more serious disease.

Although the majority of individuals treated by variolation developed only mild infections, the practice was not without risks. More serious and sometimes fatal infections did occur, and because smallpox was contagious, infections resulting from variolation could lead to epidemics. Even so, the practice of variolation for smallpox prevention spread to other regions, including India, Africa, and Europe.

---

3 Elizabeth Yale. "Why Anti-Vaccination Movements Can Never Be Tamed." *Religion & Politics*, July 22, 2014. http://religionandpolitics.org/2014/07/22/why-anti-vaccination-movements-can-never-be-tamed.

Figure 18.25 Variolation for smallpox originated in the Far East and the practice later spread to Europe and Africa. This Japanese relief depicts a patient receiving a smallpox variolation from the physician Ogata Shunsaku (1748–1810).

Although variolation had been practiced for centuries, the English physician Edward Jenner (1749–1823) is generally credited with developing the modern process of vaccination. Jenner observed that milkmaids who developed cowpox, a disease similar to smallpox but milder, were immune to the more serious smallpox. This led Jenner to hypothesize that exposure to a less virulent pathogen could provide immune protection against a more virulent pathogen, providing a safer alternative to variolation. In 1796, Jenner tested his hypothesis by obtaining infectious samples from a milkmaid's active cowpox lesion and injecting the materials into a young boy (Figure 18.26). The boy developed a mild infection that included a low-grade fever, discomfort in his axillae (armpit) and loss of appetite. When the boy was later infected with infectious samples from smallpox lesions, he did not contract smallpox.[4] This new approach was termed **vaccination**, a name deriving from the use of cowpox (Latin *vacca* meaning "cow") to protect against smallpox. Today, we know that Jenner's vaccine worked because the cowpox virus is genetically and antigenically related to the *Variola* viruses that caused smallpox. Exposure to cowpox antigens resulted in a primary response and the production of memory cells that identical or related epitopes of Variola virus upon a later exposure to smallpox.

The success of Jenner's smallpox vaccination led other scientists to develop vaccines for other diseases. Perhaps the most notable was Louis Pasteur, who developed vaccines for rabies, cholera, and anthrax. During the 20th and 21st centuries, effective vaccines were developed to prevent a wide range of diseases caused by viruses (e.g., chickenpox and shingles, hepatitis, measles, mumps, polio, and yellow fever) and bacteria (e.g., diphtheria, pneumococcal pneumonia, tetanus, and whooping cough,).

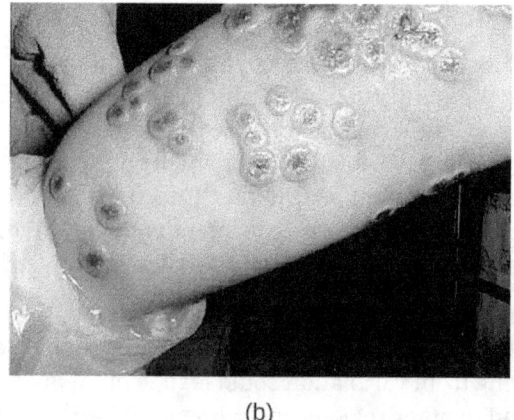

(a) (b)

Figure 18.26 (a) A painting of Edward Jenner depicts a cow and a milkmaid in the background. (b) Lesions on a patient infected with cowpox, a zoonotic disease caused by a virus closely related to the one that causes smallpox. (credit b: modification of work by the Centers

---

4 N. J. Willis. "Edward Jenner and the Eradication of Smallpox." *Scottish Medical Journal* 42 (1997): 118–121.

for Disease Control and Prevention)

✓ CHECK YOUR UNDERSTANDING

- What is the difference between variolation and vaccination for smallpox?
- Explain why vaccination is less risky than variolation.

Classes of Vaccines

For a vaccine to provide protection against a disease, it must expose an individual to pathogen-specific antigens that will stimulate a protective adaptive immune response. By its very nature, this entails some risk. As with any pharmaceutical drug, vaccines have the potential to cause adverse effects. However, the ideal vaccine causes no severe adverse effects and poses no risk of contracting the disease that it is intended to prevent. Various types of vaccines have been developed with these goals in mind. These different classes of vaccines are described in the next section and summarized in Table 18.3.

Live Attenuated Vaccines

Live attenuated vaccines expose an individual to a weakened strain of a pathogen with the goal of establishing a subclinical infection that will activate the adaptive immune defenses. Pathogens are attenuated to decrease their virulence using methods such as genetic manipulation (to eliminate key virulence factors) or long-term culturing in an unnatural host or environment (to promote mutations and decrease virulence).

By establishing an active infection, live attenuated vaccines stimulate a more comprehensive immune response than some other types of vaccines. Live attenuated vaccines activate both cellular and humoral immunity and stimulate the development of memory for long-lasting immunity. In some cases, vaccination of one individual with a live attenuated pathogen can even lead to natural transmission of the attenuated pathogen to other individuals. This can cause the other individuals to also develop an active, subclinical infection that activates their adaptive immune defenses.

Disadvantages associated with live attenuated vaccines include the challenges associated with long-term storage and transport as well as the potential for a patient to develop signs and symptoms of disease during the active infection (particularly in immunocompromised patients). There is also a risk of the attenuated pathogen reverting back to full virulence. Table 18.3 lists examples live attenuated vaccines.

Inactivated Vaccines

Inactivated vaccines contain whole pathogens that have been killed or inactivated with heat, chemicals, or radiation. For inactivated vaccines to be effective, the inactivation process must not affect the structure of key antigens on the pathogen.

Because the pathogen is killed or inactive, inactivated vaccines do not produce an active infection, and the resulting immune response is weaker and less comprehensive than that provoked by a live attenuated vaccine. Typically the response involves only humoral immunity, and the pathogen cannot be transmitted to other individuals. In addition, inactivated vaccines usually require higher doses and multiple boosters, possibly causing inflammatory reactions at the site of injection.

Despite these disadvantages, inactivated vaccines do have the advantages of long-term storage stability and ease of transport. Also, there is no risk of causing severe active infections. However, inactivated vaccines are not without their side effects. Table 18.3 lists examples of inactivated vaccines.

Subunit Vaccines

Whereas live attenuated and inactive vaccines expose an individual to a weakened or dead pathogen, **subunit vaccines** only expose the patient to the key antigens of a pathogen—not whole cells or viruses. Subunit vaccines can be produced either by chemically degrading a pathogen and isolating its key antigens or by producing the antigens through genetic engineering. Because these vaccines contain only the essential antigens of a pathogen, the risk of side effects is relatively low. Table 18.3 lists examples of subunit vaccines.

Toxoid Vaccines

Like subunit vaccines, **toxoid vaccines** do not introduce a whole pathogen to the patient; they contain inactivated bacterial toxins, called toxoids. Toxoid vaccines are used to prevent diseases in which bacterial toxins play an important role in pathogenesis. These vaccines activate humoral immunity that neutralizes the toxins. Table 18.3 lists examples of toxoid vaccines.

Conjugate Vaccines

A **conjugate vaccine** is a type of subunit vaccine that consists of a protein conjugated to a capsule polysaccharide. Conjugate vaccines have been developed to enhance the efficacy of subunit vaccines against pathogens that have protective polysaccharide capsules that help them evade phagocytosis, causing invasive infections that can lead to meningitis and other serious conditions. The subunit vaccines against these pathogens introduce T-independent capsular polysaccharide antigens that result in the production of antibodies that can opsonize the capsule and thus combat the infection; however, children under the age of two years do not respond effectively to these vaccines. Children do respond effectively when vaccinated with the conjugate vaccine, in which a protein with T-dependent antigens is conjugated to the capsule polysaccharide. The conjugated protein-polysaccharide antigen stimulates production of antibodies against both the protein and the capsule polysaccharide. Table 18.3 lists examples of conjugate vaccines.

Classes of Vaccines

Class	Description	Advantages	Disadvantages	Examples
Live attenuated	Weakened strain of whole pathogen	Cellular and humoral immunity	Difficult to store and transport	Chickenpox, German measles, measles, mumps, tuberculosis, typhoid fever, yellow fever
		Long-lasting immunity	Risk of infection in immunocompromised patients	
		Transmission to contacts	Risk of reversion	
Inactivated	Whole pathogen killed or inactivated with heat, chemicals, or radiation	Ease of storage and transport	Weaker immunity (humoral only)	Cholera, hepatitis A, influenza, plague, rabies
		No risk of severe active infection	Higher doses and more boosters required	
Subunit	Immunogenic antigens	Lower risk of side effects	Limited longevity	Anthrax, hepatitis B, influenza, meningitis, papillomavirus, pneumococcal pneumonia, whooping cough
			Multiple doses required	
			No protection against antigenic variation	
Toxoid	Inactivated bacterial toxin	Humoral immunity to neutralize toxin	Does not prevent infection	Botulism, diphtheria, pertussis, tetanus

Classes of Vaccines

Class	Description	Advantages	Disadvantages	Examples
Conjugate	Capsule polysaccharide conjugated to protein	T-dependent response to capsule	Costly to produce	Meningitis (*Haemophilus influenzae*, *Streptococcus pneumoniae*, *Neisseria meningitides*)
			No protection against antigenic variation	
		Better response in young children	May interfere with other vaccines	

Table 18.3

CHECK YOUR UNDERSTANDING

- What is the risk associated with a live attenuated vaccine?
- Why is a conjugated vaccine necessary in some cases?

MICRO CONNECTIONS

DNA Vaccines

DNA vaccines represent a relatively new and promising approach to vaccination. A DNA vaccine is produced by incorporating genes for antigens into a recombinant plasmid vaccine. Introduction of the DNA vaccine into a patient leads to uptake of the recombinant plasmid by some of the patient's cells, followed by transcription and translation of antigens and presentation of these antigens with MHC I to activate adaptive immunity. This results in the stimulation of both humoral and cellular immunity without the risk of active disease associated with live attenuated vaccines.

Although most DNA vaccines for humans are still in development, it is likely that they will become more prevalent in the near future as researchers are working on engineering DNA vaccines that will activate adaptive immunity against several different pathogens at once. First-generation DNA vaccines tested in the 1990s looked promising in animal models but were disappointing when tested in human subjects. Poor cellular uptake of the DNA plasmids was one of the major problems impacting their efficacy. Trials of second-generation DNA vaccines have been more promising thanks to new techniques for enhancing cellular uptake and optimizing antigens. DNA vaccines for various cancers and viral pathogens such as HIV, HPV, and hepatitis B and C are currently in development.

Some DNA vaccines are already in use. In 2005, a DNA vaccine against West Nile virus was approved for use in horses in the United States. Canada has also approved a DNA vaccine to protect fish from infectious hematopoietic necrosis virus.[5] A DNA vaccine against Japanese encephalitis virus was approved for use in humans in 2010 in Australia.[6]

---

5 M. Alonso and J. C. Leong. "Licensed DNA Vaccines Against Infectious Hematopoietic Necrosis Virus (IHNV)." *Recent Patents on DNA & Gene Sequences (Discontinued)* 7 no. 1 (2013): 62–65, issn 1872-2156/2212-3431. doi 10.2174/1872215611307010009.

6 S.B. Halstead and S. J. Thomas. "New Japanese Encephalitis Vaccines: Alternatives to Production in Mouse Brain." *Expert Review of Vaccines* 10 no. 3 (2011): 355–64.

> **Clinical Focus**
>
> **Resolution**
>
> Based on Olivia's symptoms, her physician made a preliminary diagnosis of bacterial meningitis without waiting for positive identification from the blood and CSF samples sent to the lab. Olivia was admitted to the hospital and treated with intravenous broad-spectrum antibiotics and rehydration therapy. Over the next several days, her condition began to improve, and new blood samples and lumbar puncture samples showed an absence of microbes in the blood and CSF with levels of white blood cells returning to normal. During this time, the lab produced a positive identification of *Neisseria meningitidis*, the causative agent of meningococcal meningitis, in her original CSF sample.
>
> *N. meningitidis* produces a polysaccharide capsule that serves as a virulence factor. *N. meningitidis* tends to affect infants after they begin to lose the natural passive immunity provided by maternal antibodies. At one year of age, Olivia's maternal IgG antibodies would have disappeared, and she would not have developed memory cells capable of recognizing antigens associated with the polysaccharide capsule of the *N. meningitidis*. As a result, her adaptive immune system was unable to produce protective antibodies to combat the infection, and without antibiotics she may not have survived. Olivia's infection likely would have been avoided altogether had she been vaccinated. A conjugate vaccine to prevent meningococcal meningitis is available and approved for infants as young as two months of age. However, current vaccination schedules in the United States recommend that the vaccine be administered at age 11–12 with a booster at age 16.
>
> Go back to the *previous* Clinical Focus box.

LINK TO LEARNING

In countries with developed public health systems, many vaccines are routinely administered to children and adults. Vaccine schedules are changed periodically, based on new information and research results gathered by public health agencies. In the United States, the CDC publishes schedules and other updated information (https://www.openstax.org/l/22CDCVacSched) about vaccines.

SUMMARY

18.1 Overview of Specific Adaptive Immunity

- **Adaptive immunity** is an acquired defense against foreign pathogens that is characterized by **specificity** and **memory**. The first exposure to an antigen stimulates a **primary response**, and subsequent exposures stimulate a faster and strong **secondary response.**
- Adaptive immunity is a dual system involving **humoral immunity** (antibodies produced by B cells) and **cellular immunity** (T cells directed against intracellular pathogens).
- **Antigens**, also called **immunogens**, are molecules that activate adaptive immunity. A single antigen possesses smaller **epitopes**, each capable of inducing a specific adaptive immune response.
- An antigen's ability to stimulate an immune response depends on several factors, including its molecular class, molecular complexity, and size.
- **Antibodies** (**immunoglobulins**) are Y-shaped glycoproteins with two Fab sites for binding antigens and an Fc portion involved in complement activation and opsonization.
- The five classes of antibody are **IgM, IgG, IgA, IgE**, and **IgD**, each differing in size, arrangement, location within the body, and function. The five primary functions of antibodies are neutralization, opsonization, agglutination, complement activation, and antibody-dependent cell-mediated cytotoxicity (ADCC).

18.2 Major Histocompatibility Complexes and Antigen-Presenting Cells

- **Major histocompatibility complex (MHC)** is a collection of genes coding for glycoprotein molecules expressed on the surface of all nucleated cells.
- **MHC I** molecules are expressed on all nucleated cells and are essential for presentation of normal "self" antigens. Cells that become infected by intracellular pathogens can present foreign antigens on MHC I as well, marking the infected cell for destruction.
- **MHC II** molecules are expressed only on the surface of **antigen-presenting cells** (macrophages, dendritic cells, and B cells). Antigen presentation with MHC II is essential for the activation of T cells.
- **Antigen-presenting cells (APCs)** primarily ingest pathogens by phagocytosis, destroy them in the phagolysosomes, process the protein antigens, and select the most antigenic/immunodominant epitopes with MHC II for presentation to T cells.
- **Cross-presentation** is a mechanism of antigen presentation and T-cell activation used by dendritic cells not directly infected by the pathogen; it involves phagocytosis of the pathogen but presentation on MHC I rather than MHC II.

18.3 T Lymphocytes and Cellular Immunity

- Immature T lymphocytes are produced in the red bone marrow and travel to the thymus for maturation.
- **Thymic selection** is a three-step process of negative and positive selection that determines which T cells will mature and exit the thymus into the peripheral bloodstream.
- **Central tolerance** involves negative selection of self-reactive T cells in the thymus, and **peripheral tolerance** involves **anergy** and **regulatory T cells** that prevent self-reactive immune responses and autoimmunity.
- The **TCR** is similar in structure to immunoglobulins, but less complex. Millions of unique epitope-binding TCRs are encoded through a process of genetic rearrangement of V, D, and J gene segments.
- T cells can be divided into three classes—**helper T cells, cytotoxic T cells,** and **regulatory T cells**—based on their expression of CD4 or CD8, the MHC molecules with which they interact for activation, and their respective functions.
- Activated helper T cells differentiate into T_H1, T_H2, T_H17, or **memory T cell subtypes**. Differentiation is directed by the specific cytokines to which they are exposed. T_H1, T_H2, and T_H17 perform different functions related to stimulation of adaptive and innate immune defenses. Memory T cells are long-lived cells that can respond quickly to secondary exposures.
- Once activated, cytotoxic T cells target and kill cells infected with intracellular pathogens. Killing requires recognition of specific pathogen epitopes presented on the cell surface using MHC I molecules. Killing is mediated by

perforin and **granzymes** that induce apoptosis.
- **Superantigens** are bacterial or viral proteins that cause a nonspecific activation of helper T cells, leading to an excessive release of cytokines (**cytokine storm**) and a systemic, potentially fatal inflammatory response.

18.4 B Lymphocytes and Humoral Immunity

- **B lymphocytes** or **B cells** produce antibodies involved in humoral immunity. B cells are produced in the bone marrow, where the initial stages of maturation occur, and travel to the spleen for final steps of maturation into naïve mature B cells.
- **B-cell receptors (BCRs)** are membrane-bound monomeric forms of IgD and IgM that bind specific antigen epitopes with their Fab antigen-binding regions. Diversity of antigen binding specificity is created by genetic rearrangement of V, D, and J segments similar to the mechanism used for TCR diversity.
- Protein antigens are called **T-dependent antigens** because they can only activate B cells with the cooperation of helper T cells. Other molecule classes do not require T cell cooperation and are called **T-independent antigens**.
- **T cell-independent activation** of B cells involves cross-linkage of BCRs by repetitive nonprotein antigen epitopes. It is characterized by the production of IgM by **plasma cells** and does not produce memory B cells.
- **T cell-dependent activation** of B cells involves processing and presentation of protein antigens to helper T cells, activation of the B cells by cytokines secreted from activated T_H2 cells, and plasma cells that produce different classes of antibodies as a result of **class switching**. **Memory B cells** are also produced.
- Secondary exposures to T-dependent antigens result in a secondary antibody response initiated by memory B cells. The secondary response develops more quickly and produces higher and more sustained levels of antibody with higher affinity for the specific antigen.

18.5 Vaccines

- Adaptive immunity can be divided into four distinct classifications: **natural active immunity, natural passive immunity, artificial passive immunity,** and **artificial active immunity.**
- Artificial active immunity is the foundation for **vaccination** and vaccine development. Vaccination programs not only confer artificial immunity on individuals, but also foster **herd immunity** in populations.
- **Variolation** against smallpox originated in the 10th century in China, but the procedure was risky because it could cause the disease it was intended to prevent. Modern vaccination was developed by Edward Jenner, who developed the practice of inoculating patients with infectious materials from cowpox lesions to prevent smallpox.
- **Live attenuated vaccines** and **inactivated vaccines** contain whole pathogens that are weak, killed, or inactivated. **Subunit vaccines, toxoid vaccines,** and **conjugate vaccines** contain acellular components with antigens that stimulate an immune response.

REVIEW QUESTIONS
Multiple Choice

1. Antibodies are produced by _____.
    a. plasma cells
    b. T cells
    c. bone marrow
    d. Macrophages

2. Cellular adaptive immunity is carried out by _____.
    a. B cells
    b. T cells
    c. bone marrow
    d. neutrophils

3. A single antigen molecule may be composed of many individual _____.
    a. T-cell receptors
    b. B-cell receptors
    c. MHC II
    d. epitopes

4. Which class of molecules is the most antigenic?
    a. polysaccharides
    b. lipids
    c. proteins
    d. carbohydrates

5. MHC I molecules present
   a. processed foreign antigens from proteasomes.
   b. processed self-antigens from phagolysosome.
   c. antibodies.
   d. T cell antigens.

6. MHC II molecules present
   a. processed self-antigens from proteasomes.
   b. processed foreign antigens from phagolysosomes.
   c. antibodies.
   d. T cell receptors.

7. Which type of antigen-presenting molecule is found on all nucleated cells?
   a. MHC II
   b. MHC I
   c. antibodies
   d. B-cell receptors

8. Which type of antigen-presenting molecule is found only on macrophages, dendritic cells, and B cells?
   a. MHC I
   b. MHC II
   c. T-cell receptors
   d. B-cell receptors

9. What is a superantigen?
   a. a protein that is highly efficient at stimulating a single type of productive and specific T cell response
   b. a protein produced by antigen-presenting cells to enhance their presentation capabilities
   c. a protein produced by T cells as a way of increasing the antigen activation they receive from antigen-presenting cells
   d. a protein that activates T cells in a nonspecific and uncontrolled manner

10. To what does the TCR of a helper T cell bind?
    a. antigens presented with MHC I molecules
    b. antigens presented with MHC II molecules
    c. free antigen in a soluble form
    d. haptens only

11. Cytotoxic T cells will bind with their TCR to which of the following?
    a. antigens presented with MHC I molecules
    b. antigens presented with MHC II molecules
    c. free antigen in a soluble form
    d. haptens only

12. A _____ molecule is a glycoprotein used to identify and distinguish white blood cells.
    a. T-cell receptor
    b. B-cell receptor
    c. MHC I
    d. cluster of differentiation

13. Name the T helper cell subset involved in antibody production.
    a. T_H1
    b. T_H2
    c. T_H17
    d. CTL

14. Which of the following would be a T-dependent antigen?
    a. lipopolysaccharide
    b. glycolipid
    c. protein
    d. carbohydrate

15. Which of the following would be a BCR?
    a. CD4
    b. MHC II
    c. MHC I
    d. IgD

16. Which of the following does not occur during the lag period of the primary antibody response?
    a. activation of helper T cells
    b. class switching to IgG
    c. presentation of antigen with MHC II
    d. binding of antigen to BCRs

17. A patient is bitten by a dog with confirmed rabies infection. After treating the bite wound, the physician injects the patient with antibodies that are specific for the rabies virus to prevent the development of an active infection. This is an example of:
    a. Natural active immunity
    b. Artificial active immunity
    c. Natural passive immunity
    d. Artificial passive immunity

18. A patient gets a cold, and recovers a few days later. The patient's classmates come down with the same cold roughly a week later, but the original patient does not get the same cold again. This is an example of:
    a. Natural active immunity
    b. Artificial active immunity
    c. Natural passive immunity
    d. Artificial passive immunity

Matching

19. Match the antibody class with its description.

___IgA   A. This class of antibody is the only one that can cross the placenta.

___IgD   B. This class of antibody is the first to appear after activation of B cells.

___IgE   C. This class of antibody is involved in the defense against parasitic infections and involved in allergic responses.

___IgG   D. This class of antibody is found in very large amounts in mucus secretions.

___IgM   E. This class of antibody is not secreted by B cells but is expressed on the surface of naïve B cells.

20. Match each type of vaccine with the corresponding example.

___ inactivated vaccine

___ live attenuated vaccine

___ toxoid vaccine

___ subunit vaccine

A. Weakened influenza virions that can only replicate in the slightly lower temperatures of the nasal passages are sprayed into the nose. They do not cause serious flu symptoms, but still produce an active infection that induces a protective adaptive immune response.

B. Tetanus toxin molecules are harvested and chemically treated to render them harmless. They are then injected into a patient's arm.

C. Influenza virus particles grown in chicken eggs are harvested and chemically treated to render them noninfectious. These immunogenic particles are then purified and packaged and administered as an injection.

D. The gene for hepatitis B virus surface antigen is inserted into a yeast genome. The modified yeast is grown and the virus protein is produced, harvested, purified, and used in a vaccine.

Fill in the Blank

21. There are two critically important aspects of adaptive immunity. The first is specificity, while the second is _____.
22. _____ immunity involves the production of antibody molecules that bind to specific antigens.
23. The heavy chains of an antibody molecule contain _____ region segments, which help to determine its class or isotype.
24. The variable regions of the heavy and light chains form the _____ sites of an antibody.
25. MHC molecules are used for antigen _____ to T cells.
26. MHC II molecules are made up of two subunits (α and β) of approximately equal size, whereas MHC I molecules consist of a larger α subunit and a smaller subunit called _____.
27. A _____ T cell will become activated by presentation of foreign antigen associated with an MHC I molecule.
28. A _____ T cell will become activated by presentation of foreign antigen in association with an MHC II molecule.
29. A TCR is a protein dimer embedded in the plasma membrane of a T cell. The _____ region of each of the two protein chains is what gives it the capability to bind to a presented antigen.
30. Peripheral tolerance mechanisms function on T cells after they mature and exit the _____.
31. Both _____ and effector T cells are produced during differentiation of activated T cells.
32. _____ antigens can stimulate B cells to become activated but require cytokine assistance delivered by helper T cells.
33. T-independent antigens can stimulate B cells to become activated and secrete antibodies without assistance from helper T cells. These antigens possess _____ antigenic epitopes that cross-link BCRs.
34. A(n) _____ pathogen is in a weakened state; it is still capable of stimulating an immune response but does not cause a disease.
35. _____ immunity occurs when antibodies from one individual are harvested and given to another to protect against disease or treat active disease.
36. In the practice of _____, scabs from smallpox victims were used to immunize susceptible individuals against smallpox.

Short Answer

37. What is the difference between humoral and cellular adaptive immunity?
38. What is the difference between an antigen and a hapten?
39. Describe the mechanism of antibody-dependent cell-mediated cytotoxicity.
40. What is the basic difference in effector function between helper and cytotoxic T cells?
41. What necessary interactions are required for activation of helper T cells and activation/effector function of cytotoxic T cells?
42. Briefly compare the pros and cons of inactivated versus live attenuated vaccines.

Critical Thinking

43. Which mechanism of antigen presentation would be used to present antigens from a cell infected with a virus?
44. Which pathway of antigen presentation would be used to present antigens from an extracellular bacterial infection?
45. A patient lacks the ability to make functioning T cells because of a genetic disorder. Would this patient's B cells be able to produce antibodies in response to an infection? Explain your answer.

CHAPTER 19
Diseases of the Immune System

Figure 19.1 Bee stings and other allergens can cause life-threatening, systemic allergic reactions. Sensitive individuals may need to carry an epinephrine auto-injector (e.g., EpiPen) in case of a sting. A bee-sting allergy is an example of an immune response that is harmful to the host rather than protective; epinephrine counteracts the severe drop in blood pressure that can result from the immune response. (credit right: modification of work by Carol Bleistine)

Chapter Outline

19.1 Hypersensitivities

19.2 Autoimmune Disorders

19.3 Organ Transplantation and Rejection

19.4 Immunodeficiency

19.5 Cancer Immunobiology and Immunotherapy

INTRODUCTION An allergic reaction is an immune response to a type of antigen called an allergen. Allergens can be found in many different items, from peanuts and insect stings to latex and some drugs. Unlike other kinds of antigens, allergens are not necessarily associated with pathogenic microbes, and many allergens provoke no immune response at all in most people.

Allergic responses vary in severity. Some are mild and localized, like hay fever or hives, but others can result in systemic, life-threatening reactions. Anaphylaxis, for example, is a rapidly developing allergic reaction that can cause a dangerous drop in blood pressure and severe swelling of the throat that may close off the airway.

Allergies are just one example of how the immune system—the system normally responsible for preventing disease—can actually cause or mediate disease symptoms. In this chapter, we will further explore allergies and other disorders of the immune system, including hypersensitivity reactions, autoimmune diseases, transplant rejection, and diseases associated with immunodeficiency.

19.1 Hypersensitivities

Learning Objectives

By the end of this section, you will be able to:
- Identify and compare the distinguishing characteristics, mechanisms, and major examples of type I, II, III, and IV hypersensitivities

---

Clinical Focus

Part 1

Kerry, a 40-year-old airline pilot, has made an appointment with her primary care physician to discuss a rash that develops whenever she spends time in the sun. As she explains to her physician, it does not seem like sunburn. She is careful not to spend too much time in the sun and she uses sunscreen. Despite these precautions, the rash still appears, manifesting as red, raised patches that get slightly scaly. The rash persists for 7 to 10 days each time, and it seems to largely go away on its own. Lately, the rashes have also begun to appear on her cheeks and above her eyes on either side of her forehead.

- Is Kerry right to be concerned, or should she simply be more careful about sun exposure?
- Are there conditions that might be brought on by sun exposure that Kerry's physician should be considering?

Jump to the next Clinical Focus box.

In Adaptive Specific Host Defenses, we discussed the mechanisms by which adaptive immune defenses, both humoral and cellular, protect us from infectious diseases. However, these same protective immune defenses can also be responsible for undesirable reactions called **hypersensitivity** reactions. Hypersensitivity reactions are classified by their immune mechanism.

- Type I hypersensitivity reactions involve immunoglobulin E (IgE) antibody against soluble antigen, triggering mast cell degranulation.
- Type II hypersensitivity reactions involve IgG and IgM antibodies directed against cellular antigens, leading to cell damage mediated by other immune system effectors.
- Type III hypersensitivity reactions involve the interactions of IgG, IgM, and, occasionally, IgA[1] antibodies with antigen to form immune complexes. Accumulation of immune complexes in tissue leads to tissue damage mediated by other immune system effectors.
- Type IV hypersensitivity reactions are T-cell–mediated reactions that can involve tissue damage mediated by activated macrophages and cytotoxic T cells.

Type I Hypersensitivities

When a presensitized individual is exposed to an **allergen**, it can lead to a rapid immune response that occurs almost immediately. Such a response is called an **allergy** and is classified as a **type I hypersensitivity**. Allergens may be seemingly harmless substances such as animal dander, molds, or pollen. Allergens may also be substances considered innately more hazardous, such as insect venom or therapeutic drugs. Food intolerances can also yield allergic reactions as individuals become sensitized to foods such as peanuts or shellfish (Figure 19.2). Regardless of the allergen, the first exposure activates a primary IgE antibody response that sensitizes an individual to type I hypersensitivity reaction upon subsequent exposure.

---

1 D.S. Strayer et al (eds). *Rubin's Pathology: Clinicopathologic Foundations of Medicine.* 7th ed. 2Philadelphia, PA: Lippincott, Williams & Wilkins, 2014.

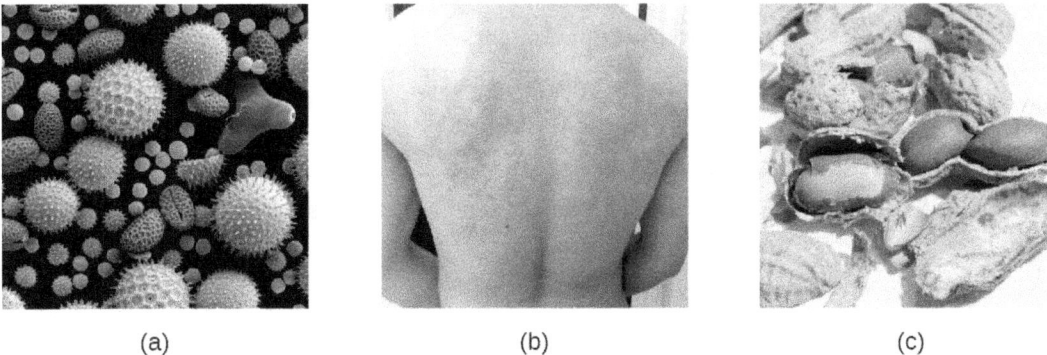

Figure 19.2 (a) Allergens in plant pollen, shown here in a colorized electron micrograph, may trigger allergic rhinitis or hay fever in sensitive individuals. (b) Skin rashes are often associated with allergic reactions. (c) Peanuts can be eaten safely by most people but can provoke severe allergic reactions in sensitive individuals.

For susceptible individuals, a first exposure to an allergen activates a strong T_H2 cell response (Figure 19.3). Cytokines interleukin (IL)-4 and IL-13 from the T_H2 cells activate B cells specific to the same allergen, resulting in clonal proliferation, differentiation into plasma cells, and antibody-class switch from production of IgM to production of IgE. The fragment crystallizable (Fc) regions of the IgE antibodies bind to specific receptors on the surface of mast cells throughout the body. It is estimated that each mast cell can bind up to 500,000 IgE molecules, with each IgE molecule having two allergen-specific fragment antigen-binding (Fab) sites available for binding allergen on subsequent exposures. By the time this occurs, the allergen is often no longer present and there is no allergic reaction, but the mast cells are primed for a subsequent exposure and the individual is sensitized to the allergen.

On subsequent exposure, allergens bind to multiple IgE molecules on mast cells, cross-linking the IgE molecules. Within minutes, this cross-linking of IgE activates the mast cells and triggers **degranulation**, a reaction in which the contents of the granules in the mast cell are released into the extracellular environment. Preformed components that are released from granules include histamine, serotonin, and bradykinin (Table 19.1). The activated mast cells also release newly formed lipid mediators (leukotrienes and prostaglandins from membrane arachadonic acid metabolism) and cytokines such as tumor necrosis factor (Table 19.2).

The chemical mediators released by mast cells collectively cause the inflammation and signs and symptoms associated with type I hypersensitivity reactions. Histamine stimulates mucus secretion in nasal passages and tear formation from lacrimal glands, promoting the runny nose and watery eyes of allergies. Interaction of histamine with nerve endings causes itching and sneezing. The vasodilation caused by several of the mediators can result in hives, headaches, angioedema (swelling that often affects the lips, throat, and tongue), and hypotension (low blood pressure). Bronchiole constriction caused by some of the chemical mediators leads to wheezing, dyspnea (difficulty breathing), coughing, and, in more severe cases, cyanosis (bluish color to the skin or mucous membranes). Vomiting can result from stimulation of the vomiting center in the cerebellum by histamine and serotonin. Histamine can also cause relaxation of intestinal smooth muscles and diarrhea.

Selected Preformed Components of Mast Cell Granules

Granule Component	Activity
Heparin	Stimulates the generation of bradykinin, which causes increased vascular permeability, vasodilation, bronchiole constriction, and increased mucus secretion
Histamine	Causes smooth-muscle contraction, increases vascular permeability, increases mucus and tear formation

Selected Preformed Components of Mast Cell Granules

Granule Component	Activity
Serotonin	Increases vascular permeability, causes vasodilation and smooth-muscle contraction

Table 19.1

Selected Newly Formed Chemical Mediators of Inflammation and Allergic Response

Chemical Mediator	Activity
Leukotriene	Causes smooth-muscle contraction and mucus secretion, increases vascular permeability
Prostaglandin	Causes smooth-muscle contraction and vasodilation
TNF-α (cytokine)	Causes inflammation and stimulates cytokine production by other cell types

Table 19.2

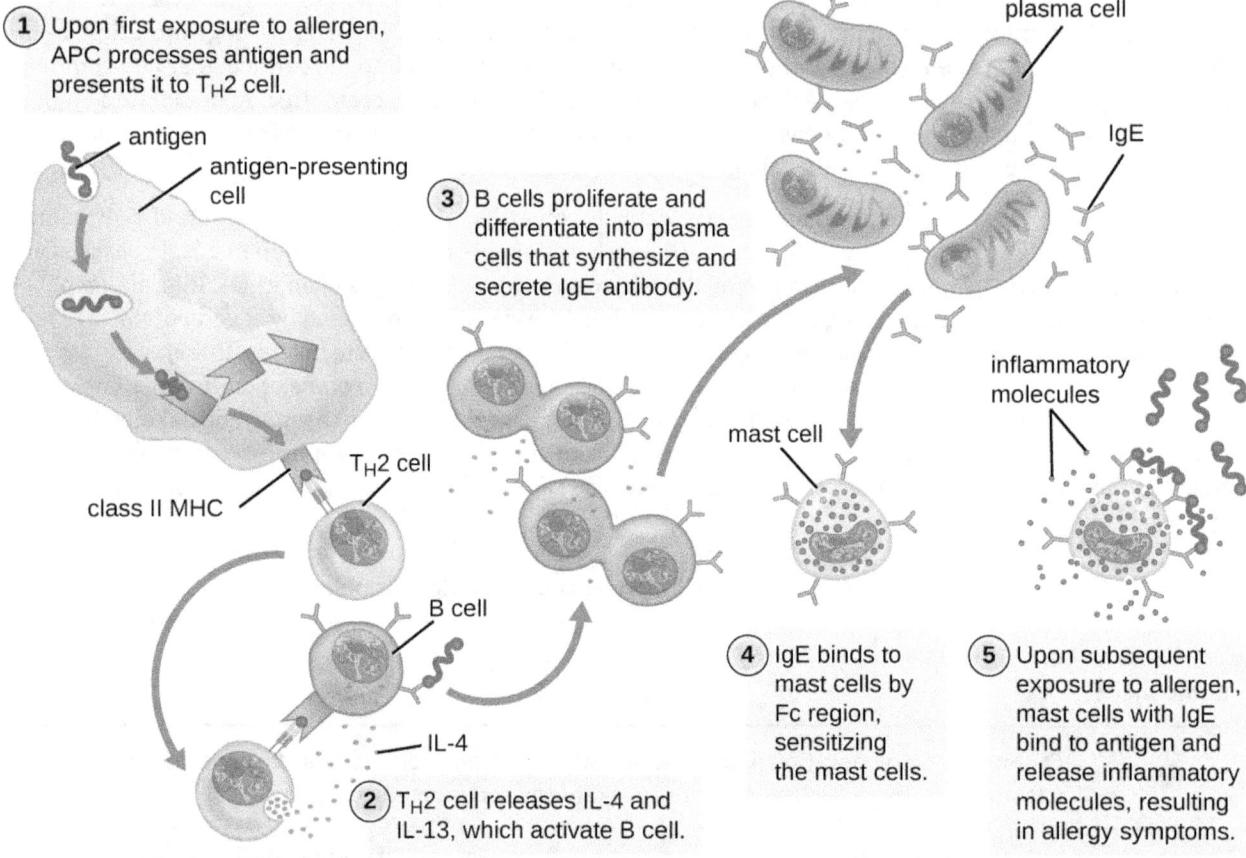

Figure 19.3 On first exposure to an allergen in a susceptible individual, antigen-presenting cells process and present allergen epitopes with major histocompatibility complex (MHC) II to T helper cells. B cells also process and present the same allergen epitope to T_H2 cells, which release cytokines IL-4 and IL-13 to stimulate proliferation and differentiation into IgE-secreting plasma cells. The IgE molecules

bind to mast cells with their Fc region, sensitizing the mast cells for activation with subsequent exposure to the allergen. With each subsequent exposure, the allergen cross-links IgE molecules on the mast cells, activating the mast cells and causing the release of preformed chemical mediators from granules (degranulation), as well as newly formed chemical mediators that collectively cause the signs and symptoms of type I hypersensitivity reactions.

Type I hypersensitivity reactions can be either localized or systemic. Localized type I hypersensitivity reactions include hay fever rhinitis, hives, and asthma (Table 19.3). Systemic type I hypersensitivity reactions are referred to as **anaphylaxis** or **anaphylactic shock**. Although anaphylaxis shares many symptoms common with the localized type I hypersensitivity reactions, the swelling of the tongue and trachea, blockage of airways, dangerous drop in blood pressure, and development of shock can make anaphylaxis especially severe and life-threatening. In fact, death can occur within minutes of onset of signs and symptoms.

Late-phase reactions in type I hypersensitivities may develop 4–12 hours after the early phase and are mediated by eosinophils, neutrophils, and lymphocytes that have been recruited by chemotactic factors released from mast cells. Activation of these recruited cells leads to the release of more chemical mediators that cause tissue damage and late-phase symptoms of swelling and redness of the skin, coughing, wheezing, and nasal discharge.

Individuals who possess genes for maladaptive traits, such as intense type I hypersensitivity reactions to otherwise harmless components of the environment, would be expected to suffer reduced reproductive success. With this kind of evolutionary selective pressure, such traits would not be expected to persist in a population. This suggests that type I hypersensitivities may have an adaptive function. There is evidence that the IgE produced during type I hypersensitivity reactions is actually meant to counter helminth infections[2]. Helminths are one of few organisms that possess proteins that are targeted by IgE. In addition, there is evidence that helminth infections at a young age reduce the likelihood of type I hypersensitivities to innocuous substances later in life. Thus it may be that allergies are an unfortunate consequence of strong selection in the mammalian lineage or earlier for a defense against parasitic worms.

Type I Hypersensitivities

Common Name	Cause	Signs and Symptoms
Allergy-induced asthma	Inhalation of allergens	Constriction of bronchi, labored breathing, coughing, chills, body aches
Anaphylaxis	Systemic reaction to allergens	Hives, itching, swelling of tongue and throat, nausea, vomiting, low blood pressure, shock
Hay fever	Inhalation of mold or pollen	Runny nose, watery eyes, sneezing
Hives (urticaria)	Food or drug allergens, insect stings	Raised, bumpy skin rash with itching; bumps may converge into large raised areas

Table 19.3

✓ CHECK YOUR UNDERSTANDING

- What are the cells that cause a type I hypersensitivity reaction?
- Describe the differences between immediate and late-phase type I hypersensitivity reactions.

---

2 C.M. Fitzsimmons et al. "Helminth Allergens, Parasite-Specific IgE, and Its Protective Role in Human Immunity." *Frontier in Immunology* 5 (2015):47.

- List the signs and symptoms of anaphylaxis.

 MICRO CONNECTIONS

The Hygiene Hypothesis

In most modern societies, good hygiene is associated with regular bathing, and good health with cleanliness. But some recent studies suggest that the association between health and clean living may be a faulty one. Some go so far as to suggest that children should be encouraged to play in the dirt—or even eat dirt[3]—for the benefit of their health. This recommendation is based on the so-called hygiene hypothesis, which proposes that childhood exposure to antigens from a diverse range of microbes leads to a better-functioning immune system later in life.

The hygiene hypothesis was first suggested in 1989 by David Strachan[4], who observed an inverse relationship between the number of older children in a family and the incidence of hay fever. Although hay fever in children had increased dramatically during the mid-20th century, incidence was significantly lower in families with more children. Strachan proposed that the lower incidence of allergies in large families could be linked to infections acquired from older siblings, suggesting that these infections made children less susceptible to allergies. Strachan also argued that trends toward smaller families and a greater emphasis on cleanliness in the 20th century had decreased exposure to pathogens and thus led to higher overall rates of allergies, asthma, and other immune disorders.

Other researchers have observed an inverse relationship between the incidence of immune disorders and infectious diseases that are now rare in industrialized countries but still common in less industrialized countries.[5] In developed nations, children under the age of 5 years are not exposed to many of the microbes, molecules, and antigens they almost certainly would have encountered a century ago. The lack of early challenges to the immune system by organisms with which humans and their ancestors evolved may result in failures in immune system functioning later in life.

Type II (Cytotoxic) Hypersensitivities

Immune reactions categorized as **type II hypersensitivities**, or cytotoxic hypersensitivities, are mediated by IgG and IgM antibodies binding to cell-surface antigens or matrix-associated antigens on basement membranes. These antibodies can either activate complement, resulting in an inflammatory response and lysis of the targeted cells, or they can be involved in antibody-dependent cell-mediated cytotoxicity (ADCC) with cytotoxic T cells.

In some cases, the antigen may be a self-antigen, in which case the reaction would also be described as an autoimmune disease. (Autoimmune diseases are described in Autoimmune Disorders). In other cases, antibodies may bind to naturally occurring, but exogenous, cell-surface molecules such as antigens associated with blood typing found on red blood cells (RBCs). This leads to the coating of the RBCs by antibodies, activation of the complement cascade, and complement-mediated lysis of RBCs, as well as opsonization of RBCs for phagocytosis. Two examples of type II hypersensitivity reactions involving RBCs are hemolytic transfusion reaction (HTR) and hemolytic disease of the newborn (HDN). These type II hypersensitivity reactions, which will be discussed in greater detail, are summarized in Table 19.4.

Immunohematology is the study of blood and blood-forming tissue in relation to the immune response. Antibody-initiated responses against blood cells are type II hypersensitivities, thus falling into the field of immunohematology. For students first learning about immunohematology, understanding the immunological

---

3 S.T. Weiss. "Eat Dirt—The Hygiene Hypothesis and Allergic Diseases." *New England Journal of Medicine* 347 no. 12 (2002):930–931.

4 D.P. Strachan "Hay Fever, Hygiene, and Household Size." *British Medical Journal* 299 no. 6710 (1989):1259.

5 H. Okada et al. "The 'Hygiene Hypothesis' for Autoimmune and Allergic Diseases: An Update." *Clinical & Experimental Immunology* 160 no. 1 (2010):1–9.

mechanisms involved is made even more challenging by the complex nomenclature system used to identify different blood-group antigens, often called blood types. The first blood-group antigens either used alphabetical names or were named for the first person known to produce antibodies to the red blood cell antigen (e.g., Kell, Duffy, or Diego). However, in 1980, the International Society of Blood Transfusion (ISBT) Working Party on Terminology created a standard for blood-group terminology in an attempt to more consistently identify newly discovered blood group antigens. New antigens are now given a number and assigned to a blood-group system, collection, or series. However, even with this effort, blood-group nomenclature is still inconsistent.

Common Type II Hypersensitivities

Common Name	Cause	Signs and Symptoms
Hemolytic disease of the newborn (HDN)	IgG from mother crosses the placenta, targeting the fetus' RBCs for destruction	Anemia, edema, enlarged liver or spleen, hydrops (fluid in body cavity), leading to death of newborn in severe cases
Hemolytic transfusion reactions (HTR)	IgG and IgM bind to antigens on transfused RBCs, targeting donor RBCs for destruction	Fever, jaundice, hypotension, disseminated intravascular coagulation, possibly leading to kidney failure and death

Table 19.4

ABO Blood Group Incompatibility

The recognition that individuals have different blood types was first described by Karl Landsteiner (1868–1943) in the early 1900s, based on his observation that serum from one person could cause a clumping of RBCs from another. These studies led Landsteiner to the identification of four distinct blood types. Subsequent research by other scientists determined that the four blood types were based on the presence or absence of surface carbohydrates "A" and "B," and this provided the foundation for the **ABO blood group system** that is still in use today (Figure 19.4). The functions of these antigens are unknown, but some have been associated with normal biochemical functions of the cell. Furthermore, ABO blood types are inherited as alleles (one from each parent), and they display patterns of dominant and codominant inheritance. The alleles for A and B blood types are codominant to each other, and both are dominant over blood type O. Therefore, individuals with genotypes of AA or AO have type A blood and express the A carbohydrate antigen on the surface of their RBCs. People with genotypes of BB or BO have type B blood and express the B carbohydrate antigen on the surface of their RBCs. Those with a genotype of AB have type AB blood and express both A and B carbohydrate antigens on the surface of their RBCs. Finally, individuals with a genotype of OO have type O blood and lack A and B carbohydrate on the surface of their RBCs.

It is important to note that the RBCs of all four ABO blood types share a common protein receptor molecule, and it is the addition of specific carbohydrates to the protein receptors that determines A, B, and AB blood types. The genes that are inherited for the A, B, and AB blood types encode enzymes that add the carbohydrate component to the protein receptor. Individuals with O blood type still have the protein receptor but lack the enzymes that would add carbohydrates that would make their red blood cell type A, B, or AB.

IgM antibodies in plasma that cross-react with blood group antigens not present on an individual's own RBCs are called **isohemagglutinin**s (Figure 19.4). Isohemagglutinins are produced within the first few weeks after birth and persist throughout life. These antibodies are produced in response to exposure to environmental antigens from food and microorganisms. A person with type A blood has A antigens on the surface of their RBCs and will produce anti-B antibodies to environmental antigens that resemble the carbohydrate component of B antigens. A person with type B blood has B antigens on the surface of their RBCs and will produce anti-A antibodies to environmental antigens that are similar to the carbohydrate component of A antigens. People with blood type O lack both A and B antigens on their RBCs and, therefore, produce both anti-A and anti-B antibodies. Conversely, people with AB blood type have both A and B antigens on their RBCs and, therefore, lack anti-A and anti-B antibodies.

	Blood Type			
	A	B	AB	O
Red blood cell type	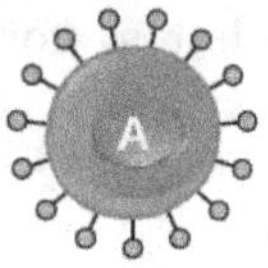 A	 B	 AB	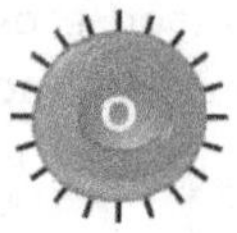 O
Isohemagglutinins	 Anti-B	 Anti-A	None	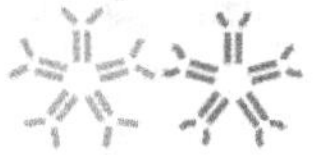 Anti-A and Anti-B
Antigens on red blood cell	A antigen	B antigen	A and B antigens	None

Figure 19.4

A patient may require a blood transfusion because they lack sufficient RBCs (anemia) or because they have experienced significant loss of blood volume through trauma or disease. Although the blood transfusion is given to help the patient, it is essential that the patient receive a transfusion with matching ABO blood type. A transfusion with an incompatible ABO blood type may lead to a strong, potentially lethal type II hypersensitivity cytotoxic response called **hemolytic transfusion reaction (HTR)** (Figure 19.5).

For instance, if a person with type B blood receives a transfusion of type A blood, their anti-A antibodies will bind to and agglutinate the transfused RBCs. In addition, activation of the classical complement cascade will lead to a strong inflammatory response, and the complement membrane attack complex (MAC) will mediate massive hemolysis of the transfused RBCs. The debris from damaged and destroyed RBCs can occlude blood vessels in the alveoli of the lungs and the glomeruli of the kidneys. Within 1 to 24 hours of an incompatible transfusion, the patient experiences fever, chills, pruritus (itching), urticaria (hives), dyspnea, hemoglobinuria (hemoglobin in the urine), and hypotension (low blood pressure). In the most serious reactions, dangerously low blood pressure can lead to shock, multi-organ failure, and death of the patient.

Hospitals, medical centers, and associated clinical laboratories typically use hemovigilance systems to minimize the risk of HTRs due to clerical error. Hemovigilance systems are procedures that track transfusion information from the donor source and blood products obtained to the follow-up of recipient patients. Hemovigilance systems used in many countries identify HTRs and their outcomes through mandatory reporting (e.g., to the Food and Drug Administration in the United States), and this information is valuable to help prevent such occurrences in the future. For example, if an HTR is found to be the result of laboratory or clerical error, additional blood products collected from the donor at that time can be located and labeled correctly to avoid additional HTRs. As a result of these measures, HTR-associated deaths in the United States occur in about one per 2 million transfused units.[6]

---

6 E.C. Vamvakas, M.A. Blajchman. "Transfusion-Related Mortality: The Ongoing Risks of Allogeneic Blood Transfusion and the Available Strategies for Their Prevention." *Blood* 113 no. 15 (2009):3406–3417.

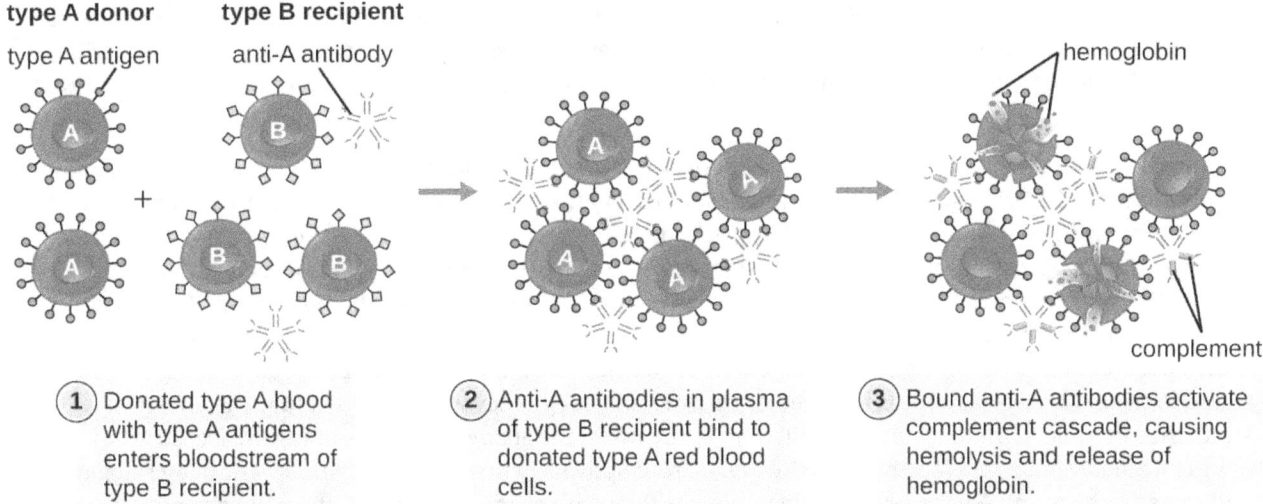

Figure 19.5 A type II hypersensitivity hemolytic transfusion reaction (HTR) leading to hemolytic anemia. Blood from a type A donor is administered to a patient with type B blood. The anti-A isohemagglutinin IgM antibodies in the recipient bind to and agglutinate the incoming donor type A red blood cells. The bound anti-A antibodies activate the classical complement cascade, resulting in destruction of the donor red blood cells.

Rh Factors

Many different types of erythrocyte antigens have been discovered since the description of the ABO red cell antigens. The second most frequently described RBC antigens are **Rh factor**s, named after the rhesus macaque (*Macaca mulatta*) factors identified by Karl Landsteiner and Alexander Weiner in 1940. The Rh system of RBC antigens is the most complex and immunogenic blood group system, with more than 50 specificities identified to date. Of all the Rh antigens, the one designated Rho (Weiner) or D (Fisher-Race) is the most immunogenic. Cells are classified as Rh positive (Rh+) if the Rho/D antigen is present or as Rh negative (Rh−) if the Rho/D antigen is absent. In contrast to the carbohydrate molecules that distinguish the ABO blood groups and are the targets of IgM isohemagglutinins in HTRs, the Rh factor antigens are proteins. As discussed in B Lymphocytes and Humoral Immunity, protein antigens activate B cells and antibody production through a T-cell–dependent mechanism, and the T_H2 cells stimulate class switching from IgM to other antibody classes. In the case of Rh factor antigens, T_H2 cells stimulate class switching to IgG, and this has important implications for the mechanism of HDN.

Like ABO incompatibilities, blood transfusions from a donor with the wrong Rh factor antigens can cause a type II hypersensitivity HTR. However, in contrast to the IgM isohemagglutinins produced early in life through exposure to environmental antigens, production of anti-Rh factor antibodies requires the exposure of an individual with Rh− blood to Rh+ positive RBCs and activation of a primary antibody response. Although this primary antibody response can cause an HTR in the transfusion patient, the hemolytic reaction would be delayed up to 2 weeks during the extended lag period of a primary antibody response (B Lymphocytes and Humoral Immunity). However, if the patient receives a subsequent transfusion with Rh+ RBCs, a more rapid HTR would occur with anti-Rh factor antibody already present in the blood. Furthermore, the rapid secondary antibody response would provide even more anti-Rh factor antibodies for the HTR.

Rh factor incompatibility between mother and fetus can also cause a type II hypersensitivity hemolytic reaction, referred to as **hemolytic disease of the newborn (HDN)** (Figure 19.6). If an Rh− woman carries an Rh+ baby to term, the mother's immune system can be exposed to Rh+ fetal red blood cells. This exposure will usually occur during the last trimester of pregnancy and during the delivery process. If this exposure occurs, the Rh+ fetal RBCs will activate a primary adaptive immune response in the mother, and anti-Rh factor IgG antibodies will be produced. IgG antibodies are the only class of antibody that can cross the placenta from mother to fetus; however, in most cases, the first Rh+ baby is unaffected by these antibodies because the first exposure typically occurs late enough in the pregnancy that the mother does not have time to mount a sufficient primary antibody response before the baby is born.

If a subsequent pregnancy with an Rh+ fetus occurs, however, the mother's second exposure to the Rh factor antigens causes a strong secondary antibody response that produces larger quantities of anti-Rh factor IgG. These antibodies can cross the placenta from mother to fetus and cause HDN, a potentially lethal condition for the baby (Figure 19.6).

Prior to the development of techniques for diagnosis and prevention, Rh factor incompatibility was the most common cause of HDN, resulting in thousands of infant deaths each year worldwide.[7] For this reason, the Rh factors of prospective parents are regularly screened, and treatments have been developed to prevent HDN caused by Rh incompatibility. To prevent Rh factor-mediated HDN, human Rho(D) immune globulin (e.g., RhoGAM) is injected intravenously or intramuscularly into the mother during the 28th week of pregnancy and within 72 hours after delivery. Additional doses may be administered after events that may result in transplacental hemorrhage (e.g., umbilical blood sampling, chorionic villus sampling, abdominal trauma, amniocentesis). This treatment is initiated during the first pregnancy with an Rh+ fetus. The anti-Rh antibodies in Rho(D) immune globulin will bind to the Rh factor of any fetal RBCs that gain access to the mother's bloodstream, preventing these Rh+ cells from activating the mother's primary antibody response. Without a primary anti-Rh factor antibody response, the next pregnancy with an Rh+ will have minimal risk of HDN. However, the mother will need to be retreated with Rho(D) immune globulin during that pregnancy to prevent a primary anti-Rh antibody response that could threaten subsequent pregnancies.

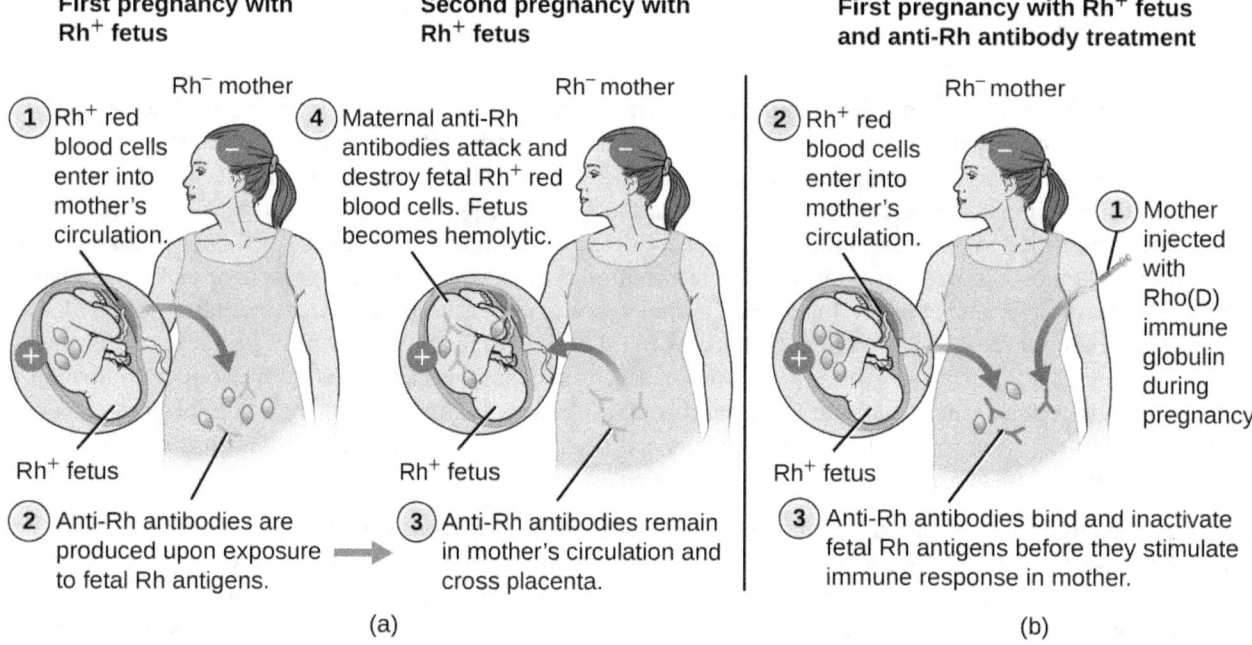

Figure 19.6 (a) When an Rh− mother has an Rh+ fetus, fetal erythrocytes are introduced into the mother's circulatory system before or during birth, leading to production of anti-Rh IgG antibodies. These antibodies remain in the mother and, if she becomes pregnant with a second Rh+ baby, they can cross the placenta and attach to fetal Rh+ erythrocytes. Complement-mediated hemolysis of fetal erythrocytes results in a lack of sufficient cells for proper oxygenation of the fetus. (b) HDN can be prevented by administering Rho(D) immune globulin during and after each pregnancy with an Rh+ fetus. The immune globulin binds fetal Rh+ RBCs that gain access to the mother's bloodstream, preventing activation of her primary immune response.

LINK TO LEARNING

Use this interactive Blood Typing Game (https://openstax.org/l/22actbloodtyping) to reinforce your knowledge of blood typing.

---

[7] G. Reali. "Forty Years of Anti-D Immunoprophylaxis." *Blood Transfusion* 5 no. 1 (2007):3–6.

CHECK YOUR UNDERSTANDING

- What happens to cells that possess incompatible antigens in a type II hypersensitivity reaction?
- Describe hemolytic disease of the newborn and explain how it can be prevented.

> ### Clinical Focus
>
> **Part 2**
>
> Kerry's primary care physician is not sure why Kerry seems to develop rashes after spending time in the sun, so she orders a urinalysis and basic blood tests. The results reveal that Kerry has proteinuria (abnormal protein levels in the urine), hemoglobinuria (excess hemoglobin in the urine), and a low hematocrit (RBC count). These tests suggest that Kerry is suffering from a mild bout of hemolytic anemia. The physician suspects that the problem might be autoimmune, so she refers Kerry to a rheumatologist for additional testing and diagnosis.
>
> - Rheumatologists specialize in musculoskeletal diseases such as arthritis, osteoporosis, and joint pain. Why might Kerry's physician refer her to this particular type of specialist even though she is exhibiting none of these symptoms?
>
> *Jump to the next Clinical Focus box. Go back to the previous Clinical Focus box.*

Type III Hypersensitivities

Type III hypersensitivities are immune-complex reactions that were first characterized by Nicolas Maurice Arthus (1862–1945) in 1903. To produce antibodies for experimental procedures, Arthus immunized rabbits by injecting them with serum from horses. However, while immunizing rabbits repeatedly with horse serum, Arthus noticed a previously unreported and unexpected localized subcutaneous hemorrhage with edema at the site of injection. This reaction developed within 3 to 10 hours after injection. This localized reaction to non-self serum proteins was called an **Arthus reaction**. An Arthus reaction occurs when soluble antigens bind with IgG in a ratio that results in the accumulation of antigen-antibody aggregates called **immune complex**es.

A unique characteristic of **type III hypersensitivity** is antibody excess (primarily IgG), coupled with a relatively low concentration of antigen, resulting in the formation of small immune complexes that deposit on the surface of the epithelial cells lining the inner lumen of small blood vessels or on the surfaces of tissues (Figure 19.7). This immune complex accumulation leads to a cascade of inflammatory events that include the following:

1. IgG binding to antibody receptors on localized mast cells, resulting in mast-cell degranulation
2. Complement activation with production of pro-inflammatory C3a and C5a (see Chemical Defenses)
3. Increased blood-vessel permeability with chemotactic recruitment of neutrophils and macrophages

Because these immune complexes are not an optimal size and are deposited on cell surfaces, they cannot be phagocytosed in the usual way by neutrophils and macrophages, which, in turn, are often described as "frustrated." Although phagocytosis does not occur, neutrophil degranulation results in the release of lysosomal enzymes that cause extracellular destruction of the immune complex, damaging localized cells in the process. Activation of coagulation pathways also occurs, resulting in thrombi (blood clots) that occlude blood vessels and cause ischemia that can lead to vascular necrosis and localized hemorrhage.

Systemic type III hypersensitivity (**serum sickness**) occurs when immune complexes deposit in various body sites, resulting in a more generalized systemic inflammatory response. These immune complexes involve non-self proteins such as antibodies produced in animals for artificial passive immunity (see Vaccines), certain drugs, or microbial antigens that are continuously released over time during chronic infections (e.g., subacute bacterial endocarditis, chronic viral hepatitis). The mechanisms of serum sickness are similar to those described in localized type III hypersensitivity but involve widespread activation of mast cells, complement, neutrophils, and macrophages, which causes tissue destruction in areas such as the kidneys, joints, and blood vessels. As a result of tissue destruction, symptoms of serum sickness include chills, fever, rash, vasculitis, and arthritis. Development of glomerulonephritis or hepatitis is also possible.

Autoimmune diseases such as systemic lupus erythematosus (SLE) and rheumatoid arthritis can also involve damaging type III hypersensitivity reactions when auto-antibodies form immune complexes with self antigens. These conditions are discussed in Autoimmune Disorders.

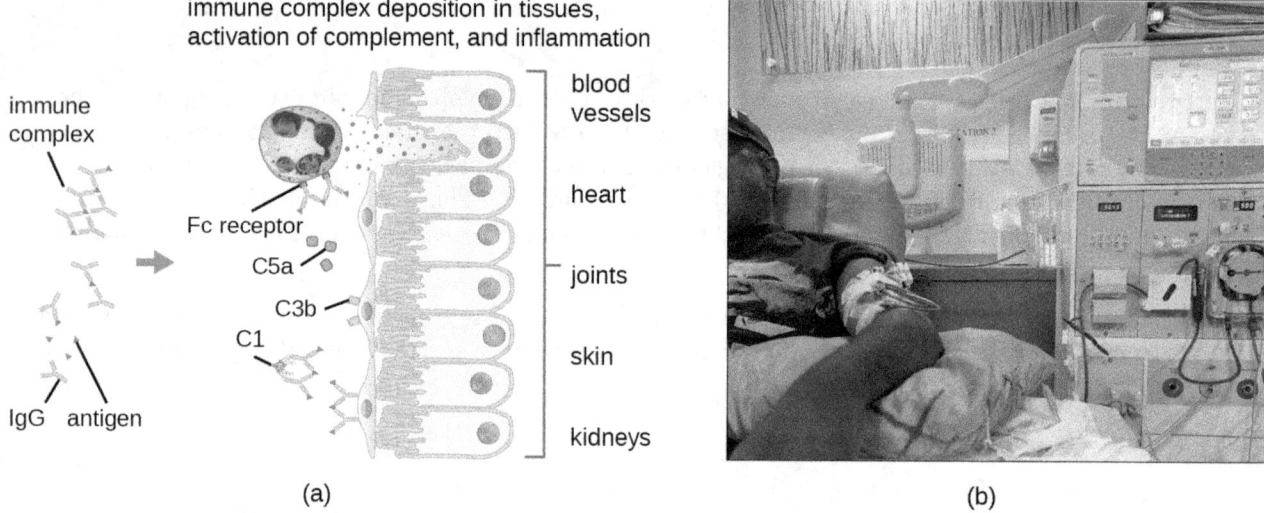

Figure 19.7 Type III hypersensitivities and the systems they affect. (a) Immune complexes form and deposit in tissue. Complement activation, stimulation of an inflammatory response, and recruitment and activation of neutrophils result in damage to blood vessels, heart tissue, joints, skin, and/or kidneys. (b) If the kidneys are damaged by a type III hypersensitivity reaction, dialysis may be required. (Credit b: Tech. Sgt. Bennie J. Davis III / US Air Force; Public Domain)

✓ CHECK YOUR UNDERSTANDING

- Why is antibody excess important in type III hypersensitivity?
- Describe the differences between the Arthus reaction and serum sickness.

 MICRO CONNECTIONS

Diphtheria Antitoxin

Antibacterial sera are much less commonly used now than in the past, having been replaced by toxoid vaccines. However, a diphtheria antitoxin produced in horses is one example of such a treatment that is still used in some parts of the world. Although it is not licensed by the FDA for use in the United States, diphtheria antitoxin can be used to treat cases of diphtheria, which are caused by the bacterium *Corynebacterium diphtheriae*.[8] The treatment is not without risks, however. Serum sickness can occur when the patient develops an immune response to non-self horse proteins. Immune complexes are formed between the horse proteins and circulating antibodies when the two exist in certain proportions. These immune complexes can deposit in organs, causing damage such as arthritis, nephritis, rash, and fever. Serum sickness is usually transient with no permanent damage unless the patient is chronically exposed to the antigen, which can then result in irreversible damage to body sites such as joints and kidneys. Over time, phagocytic cells such as macrophages are able to clear the horse serum antigens, which results in improvement of the patient's condition and a decrease in symptoms as the immune response dissipates.

---

8 Centers for Disease Control and Prevention. "Diphtheria Antitoxin." http://www.cdc.gov/diphtheria/dat.html. Accessed March 25, 2016.

Clinical Focus

Part 3

Kerry does not make it to the rheumatologist. She has a seizure as she is leaving her primary care physician's office. She is quickly rushed to the emergency department, where her primary care physician relates her medical history and recent test results. The emergency department physician calls in the rheumatologist on staff at the hospital for consultation. Based on the symptoms and test results, the rheumatologist suspects that Kerry has lupus and orders a pair of blood tests: an antinuclear antibody test (ANA) to look for antibodies that bind to DNA and another test that looks for antibodies that bind to a self-antigen called the Smith antigen (Sm).

- Based on the blood tests ordered, what type of reaction does the rheumatologist suspect is causing Kerry's seizure?

Jump to the next Clinical Focus box. Go back to the previous Clinical Focus box.

Type IV Hypersensitivities

Type IV hypersensitivities are not mediated by antibodies like the other three types of hypersensitivities. Rather, **type IV hypersensitivities** are regulated by T cells and involve the action of effector cells. These types of hypersensitivities can be organized into three subcategories based on T-cell subtype, type of antigen, and the resulting effector mechanism (Table 19.5).

In the first type IV subcategory, CD4 T_H1-mediated reactions are described as delayed-type hypersensitivities (DTH). The sensitization step involves the introduction of antigen into the skin and phagocytosis by local antigen presenting cells (APCs). The APCs activate helper T cells, stimulating clonal proliferation and differentiation into memory T_H1 cells. Upon subsequent exposure to the antigen, these sensitized memory T_H1 cells release cytokines that activate macrophages, and activated macrophages are responsible for much of the tissue damage. Examples of this T_H1-mediated hypersensitivity are observed in tuberculin the Mantoux skin test and **contact dermatitis**, such as occurs in latex allergy reactions.

In the second type IV subcategory, CD4 T_H2-mediated reactions result in chronic asthma or chronic allergic rhinitis. In these cases, the soluble antigen is first inhaled, resulting in eosinophil recruitment and activation with the release of cytokines and inflammatory mediators.

In the third type IV subcategory, CD8 cytotoxic T lymphocyte (CTL)-mediated reactions are associated with tissue transplant rejection and contact dermatitis (Figure 19.8). For this form of cell-mediated hypersensitivity, APCs process and present the antigen with MHC I to naïve CD8 T cells. When these naïve CD8 T cells are activated, they proliferate and differentiate into CTLs. Activated T_H1 cells can also enhance the activation of the CTLs. The activated CTLs then target and induce granzyme-mediated apoptosis in cells presenting the same antigen with MHC I. These target cells could be "self" cells that have absorbed the foreign antigen (such as with contact dermatitis due to poison ivy), or they could be transplanted tissue cells displaying foreign antigen from the donor.

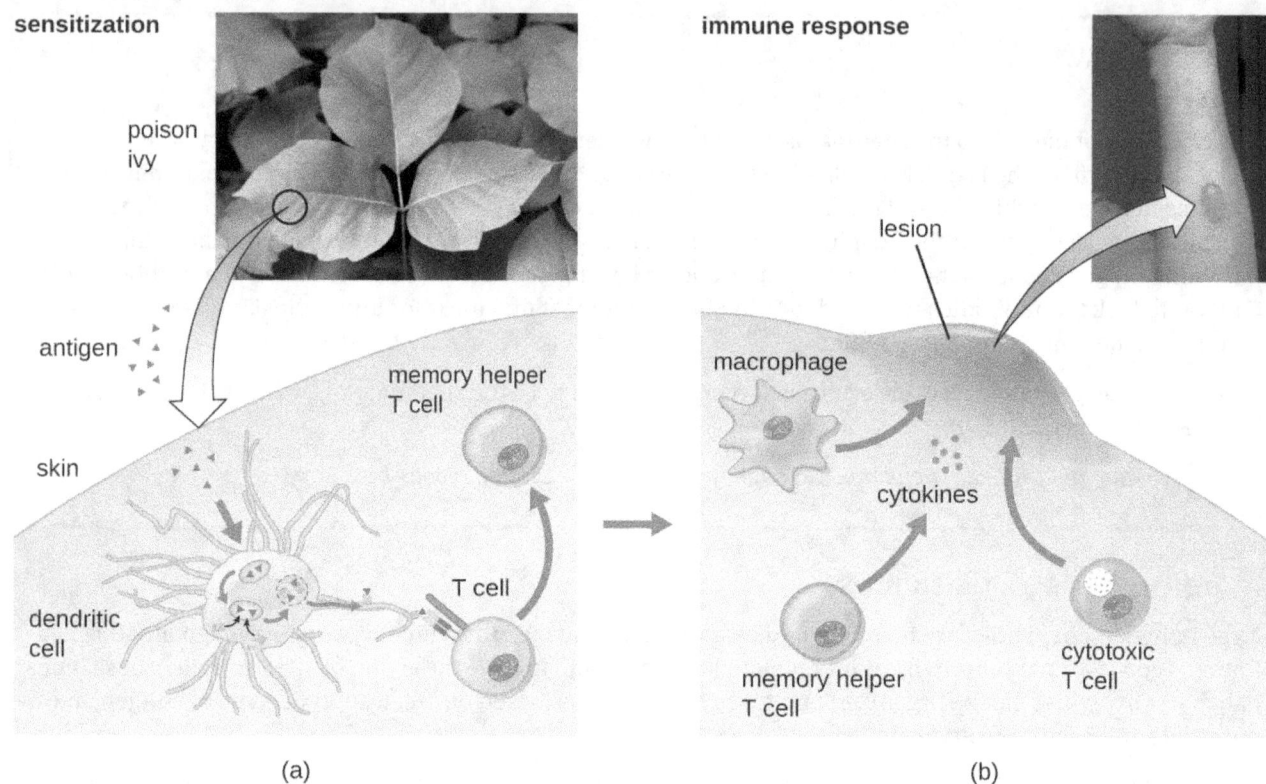

Figure 19.8 Exposure to hapten antigens in poison ivy can cause contact dermatitis, a type IV hypersensitivity. (a) The first exposure to poison ivy does not result in a reaction. However, sensitization stimulates helper T cells, leading to production of memory helper T cells that can become reactivated on future exposures. (b) Upon secondary exposure, the memory helper T cells become reactivated, producing inflammatory cytokines that stimulate macrophages and cytotoxic T cells to induce an inflammatory lesion at the exposed site. This lesion, which will persist until the allergen is removed, can inflict significant tissue damage if it continues long enough.

Type IV Hypersensitivities

Subcategory	Antigen	Effector Mechanism	Examples
1	Soluble antigen	Activated macrophages damage tissue and promote inflammatory response	Contact dermatitis (e.g., exposure to latex) and delayed-type hypersensitivity (e.g., tuberculin reaction)
2	Soluble antigen	Eosinophil recruitment and activation release cytokines and pro-inflammatory chemicals	Chronic asthma and chronic allergic rhinitis
3	Cell-associated antigen	CTL-mediated cytotoxicity	Contact dermatitis (e.g., contact with poison ivy) and tissue-transplant rejection

Table 19.5

✓ CHECK YOUR UNDERSTANDING

- Describe the three subtypes of type IV hypersensitivity.
- Explain how T cells contribute to tissue damage in type IV hypersensitivity.

MICRO CONNECTIONS

Using Delayed Hypersensitivity to Test for TB

Austrian pediatrician Clemans von Pirquet (1874–1929) first described allergy mechanisms, including type III serum sickness.[9] His interest led to the development of a test for tuberculosis (TB), using the tuberculin antigen, based on earlier work identifying the TB pathogen performed by Robert Koch. Pirquet's method involved scarification, which results in simultaneous multiple punctures, using a device with an array of needles to break the skin numerous times in a small area. The device Pirquet used was similar to the tine test device with four needles seen in Figure 19.9.

The tips of all the needles in the array are coated with tuberculin, a protein extract of TB bacteria, effectively introducing the tuberculin into the skin. One to 3 days later, the area can be examined for a delayed hypersensitivity reaction, signs of which include swelling and redness.

As you can imagine, scarification was not a pleasant experience,[10] and the numerous skin punctures put the patient at risk of developing bacterial infection of the skin. Mantoux modified Pirquet's test to use a single subcutaneous injection of purified tuberculin material. A positive test, which is indicated by a delayed localized swelling at the injection site, does not necessarily mean that the patient is currently infected with active TB. Because type IV (delayed-type) hypersensitivity is mediated by reactivation of memory T cells, such cells may have been created recently (due to an active current infection) or years prior (if a patient had TB and had spontaneously cleared it, or if it had gone into latency). However, the test can be used to confirm infection in cases in which symptoms in the patient or findings on a radiograph suggest its presence.

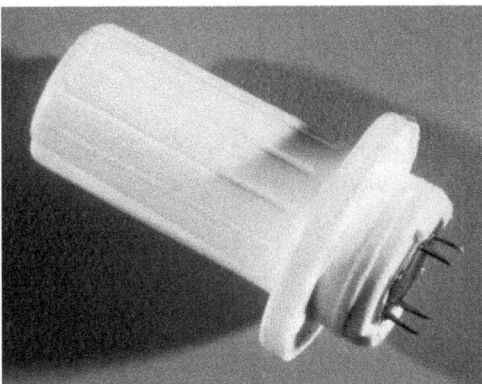

Figure 19.9 The modern version of Pirquet's scarification is the tine test, which uses devices like this to administer tuberculin antigen into the skin, usually on the inside of the forearm. The tine test is considered less reliable than the Mantoux test. (credit: modification of work by the Centers for Disease Control and Prevention)

Hypersensitivity Pneumonitis

Some disease caused by hypersensitivities are not caused exclusively by one type. For example, **hypersensitivity pneumonitis (HP)**, which is often an occupational or environmental disease, occurs when the lungs become inflamed due to an allergic reaction to inhaled dust, endospores, bird feathers, bird droppings, molds, or chemicals. HP goes by many different names associated with various forms of exposure (Figure 19.10). HP associated with bird droppings is sometimes called pigeon fancier's lung or poultry worker's lung—both common in bird breeders and handlers. Cheese handler's disease, farmer's lung, sauna takers' disease, and hot-tub lung are other names for HP associated with exposure to molds in various environments.

---

9 B. Huber "100 Jahre Allergie: Clemens von Pirquet–sein Allergiebegriff und das ihm zugrunde liegende Krankheitsverständnis." *Wiener Klinische Wochenschrift* 118 no. 19–20 (2006):573–579.

10 C.A. Stewart. "The Pirquet Test: Comparison of the Scarification and the Puncture Methods of Application." *Archives of Pediatrics & Adolescent Medicine* 35 no. 3 (1928):388–391.

Pathology associated with HP can be due to both type III (mediated by immune complexes) and type IV (mediated by T_H1 cells and macrophages) hypersensitivities. Repeated exposure to allergens can cause alveolitis due to the formation of immune complexes in the alveolar wall of the lung accompanied by fluid accumulation, and the formation of granulomas and other lesions in the lung as a result of T_H1-mediated macrophage activation. Alveolitis with fluid and granuloma formation results in poor oxygen perfusion in the alveoli, which, in turn, can cause symptoms such as coughing, dyspnea, chills, fever, sweating, myalgias, headache, and nausea. Symptoms may occur as quickly as 2 hours after exposure and can persist for weeks if left untreated.

(a)

(b)

Figure 19.10 Occupational exposure to dust, mold, and other allergens can result in hypersensitivity pneumonitis. (a) People exposed daily to large numbers of birds may be susceptible to poultry worker's lung. (b) Workers in a cheese factory may become sensitized to different types of molds and develop cheese handler's disease. (credit a: modification of work by The Global Orphan Project)

CHECK YOUR UNDERSTANDING

- Explain why hypersensitivity pneumonitis is considered an occupational disease.

Figure 19.11 summarizes the mechanisms and effects of each type of hypersensitivity discussed in this section.

	Hypersensitivity Types and Their Mechanisms			
	Type I	Type II	Type III	Type IV
Immune reactant	IgE	IgG or IgM	IgG and IgM	T cells
Antigen form	Soluble antigen	Cell-bound antigen	Soluble antigen	Soluble or cell-bound antigen
Mechanism of activation	Allergen-specific IgE antibodies bind to mast cells via their Fc receptor. When the specific allergen binds to the IgE, cross-linking of IgE induces degranulation of mast cells.	IgG or IgM antibody binds to cellular antigen, leading to complement activation and cell lysis. IgG can also mediate ADCC with cytotoxic T cells, natural killer cells, macrophages, and neutrophils.	Antigen-antibody complexes are deposited in tissues. Complement activation provides inflammatory mediators and recruits neutrophils. Enzymes released from neutrophils damage tissue.	$T_H 1$ cells secrete cytokines, which activate macrophages and cytotoxic T cells.
Examples of hypersensitivity reactions	Local and systemic anaphylaxis, seasonal hay fever, food allergies, and drug allergies	Red blood cell destruction after transfusion with mismatched blood types or during hemolytic disease of the newborn.	Post-streptococcal glomerulonephritis, rheumatoid arthritis, and systemic lupus erythematosus	Contact dermatitis, type I diabetes mellitus, and multiple sclerosis

Figure 19.11 Components of the immune system cause four types of hypersensitivities. Notice that types I–III are B-cell/antibody-mediated hypersensitivities, whereas type IV hypersensitivity is exclusively a T-cell phenomenon.

Diagnosis of Hypersensitivities

Diagnosis of type I hypersensitivities is a complex process requiring several diagnostic tests in addition to a well-documented patient history. Serum IgE levels can be measured, but elevated IgE alone does not confirm allergic disease. As part of the process to identify the antigens responsible for a type I reaction allergy, testing through a prick puncture skin test (PPST) or an intradermal test can be performed. PPST is carried out with the introduction of allergens in a series of superficial skin pricks on the patient's back or arms (Figure 19.12). PPSTs are considered to be the most convenient and least expensive way to diagnose allergies, according to the US Joint Council of Allergy and the European Academy of Allergy and Immunology. The second type of testing, the intradermal test, requires injection into the dermis with a small needle. This needle, also known as a tuberculin needle, is attached to a syringe containing a small amount of allergen. Both the PPST and the intradermal tests are observed for 15–20 minutes for a **wheal-flare reaction** to the allergens. Measurement of any wheal (a raised, itchy bump) and flare (redness) within minutes indicates a type I hypersensitivity, and the larger the wheal-flare reaction, the greater the patient's sensitivity to the allergen.

Type III hypersensitivities can often be misdiagnosed because of their nonspecific inflammatory nature. The symptoms are easily visible, but they may be associated with any of a number of other diseases. A strong, comprehensive patient history is crucial to proper and accurate diagnosis. Tests used to establish the diagnosis of hypersensitivity pneumonitis (resulting from type III hypersensitivity) include bronchoalveolar lavage (BAL), pulmonary function tests, and high-resolution computed tomography (HRCT).

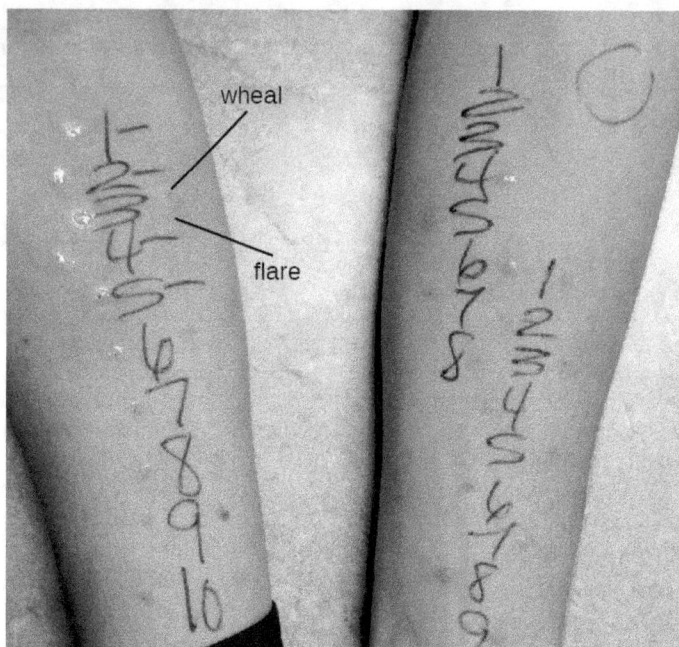

Figure 19.12 Results of an allergy skin-prick test to test for type I hypersensitivity to a group of potential allergens. A positive result is indicated by a raised area (wheal) and surrounding redness (flare). (credit: modification of work by "OakleyOriginals"/Flickr)

CHECK YOUR UNDERSTANDING

- Describe the prick puncture skin test.
- Explain why type III hypersensitivities can be difficult to diagnose.

Treatments of Hypersensitivities

Allergic reactions can be treated in various ways. Prevention of allergic reactions can be achieved by **desensitization** (hyposensitization) therapy, which can be used to reduce the hypersensitivity reaction through repeated injections of allergens. Extremely dilute concentrations of known allergens (determined from the allergen tests) are injected into the patient at prescribed intervals (e.g., weekly). The quantity of allergen delivered by the shots is slowly increased over a buildup period until an effective dose is determined and that dose is maintained for the duration of treatment, which can last years. Patients are usually encouraged to remain in the doctor's office for 30 minutes after receiving the injection in case the allergens administered cause a severe systemic reaction. Doctors' offices that administer desensitization therapy must be prepared to provide resuscitation and drug treatment in the case of such an event.

Desensitization therapy is used for insect sting allergies and environmental allergies. The allergy shots elicit the production of different interleukins and IgG antibody responses instead of IgE. When excess allergen-specific IgG antibodies are produced and bind to the allergen, they can act as **blocking antibodies** to neutralize the allergen before it can bind IgE on mast cells. There are early studies using oral therapy for desensitization of food allergies that are promising.[11,12] These studies involve feeding children who have allergies tiny amounts of the allergen (e.g., peanut flour) or related proteins over time. Many of the subjects show reduced severity of reaction to the food allergen after the therapy.

There are also therapies designed to treat severe allergic reactions. Emergency systemic anaphylaxis is treated initially with an epinephrine injection, which can counteract the drop in blood pressure. Individuals with known severe allergies often carry a self-administering auto-injector that can be used in case of exposure to

---

11 C.L. Schneider et al. "A Pilot Study of Omalizumab to Facilitate Rapid Oral Desensitization in High-Risk Peanut-Allergic Patients." *Journal of Allergy and Clinical Immunology* 132 no. 6 (2013):1368–1374.

12 P. Varshney et al. "A Randomized Controlled Study of Peanut Oral Immunotherapy: Clinical Desensitization and Modulation of the Allergic Response." *Journal of Allergy and Clinical Immunology* 127 no. 3 (2011):654–660.

the allergen (e.g., an insect sting or accidental ingestion of a food that causes a severe reaction). By self-administering an epinephrine shot (or sometimes two), the patient can stem the reaction long enough to seek medical attention. Follow-up treatment generally involves giving the patient antihistamines and slow-acting corticosteroids for several days after the reaction to prevent potential late-phase reactions. However, the effects of antihistamine and corticosteroid treatment are not well studied and are used based on theoretical considerations.

Treatment of milder allergic reactions typically involves antihistamines and other anti-inflammatory drugs. A variety of antihistamine drugs are available, in both prescription and over-the-counter strengths. There are also antileukotriene and antiprostaglandin drugs that can be used in tandem with antihistamine drugs in a combined (and more effective) therapy regime.

Treatments of type III hypersensitivities include preventing further exposure to the antigen and the use of anti-inflammatory drugs. Some conditions can be resolved when exposure to the antigen is prevented. Anti-inflammatory corticosteroid inhalers can also be used to diminish inflammation to allow lung lesions to heal. Systemic corticosteroid treatment, oral or intravenous, is also common for type III hypersensitivities affecting body systems. Treatment of hypersensitivity pneumonitis includes avoiding the allergen, along with the possible addition of prescription steroids such as prednisone to reduce inflammation.

Treatment of type IV hypersensitivities includes antihistamines, anti-inflammatory drugs, analgesics, and, if possible, eliminating further exposure to the antigen.

⊘ CHECK YOUR UNDERSTANDING

- Describe desensitization therapy.
- Explain the role of epinephrine in treatment of hypersensitivity reactions.

19.2 Autoimmune Disorders

Learning Objectives

By the end of this section, you will be able to:
- Explain why autoimmune disorders develop
- Provide a few examples of organ-specific and systemic autoimmune diseases

In 1970, artist Walt Kelly developed a poster promoting Earth Day, featuring a character from *Pogo*, his daily newspaper comic strip. In the poster, Pogo looks out across a litter-strewn forest and says wryly, "We have met the enemy and he is us." Pogo was not talking about the human immune system, but he very well could have been. Although the immune system protects the body by attacking invading "enemies" (pathogens), in some cases, the immune system can mistakenly identify the body's own cells as the enemy, resulting in **autoimmune disease**.

Autoimmune diseases are those in which the body is attacked by its own specific adaptive immune response. In normal, healthy states, the immune system induces **tolerance**, which is a lack of an anti-self immune response. However, with autoimmunity, there is a loss of immune tolerance, and the mechanisms responsible for autoimmune diseases include type II, III, and IV hypersensitivity reactions. Autoimmune diseases can have a variety of mixed symptoms that flare up and disappear, making diagnosis difficult.

The causes of autoimmune disease are a combination of the individual's genetic makeup and the effect of environmental influences, such as sunlight, infections, medications, and environmental chemicals. However, the vagueness of this list reflects our poor understanding of the etiology of these diseases. Except in a very few specific diseases, the initiation event(s) of most autoimmune states has not been fully characterized.

There are several possible causes for the origin of autoimmune diseases and autoimmunity is likely due to several factors. Evidence now suggests that regulatory T and B cells play an essential role in the maintenance of tolerance and prevention of autoimmune responses. The regulatory T cells are especially important for inhibiting autoreactive T cells that are not eliminated during thymic selection and escape the thymus (see T Lymphocytes and Cellular Immunity). In addition, antigen mimicry between pathogen antigens and our own self antigens can lead to cross-reactivity and autoimmunity. Hidden self-antigens may become exposed because of trauma, drug interactions, or disease states, and trigger an autoimmune response. All of these factors could contribute to autoimmunity. Ultimately, damage to tissues and organs in the autoimmune disease state comes as a result of inflammatory responses that are inappropriate; therefore, treatment often includes immunosuppressive drugs and corticosteroids.

Organ-Specific Autoimmune Diseases

Some autoimmune diseases are considered organ specific, meaning that the immune system targets specific organs or tissues. Examples of organ-specific autoimmune diseases include celiac disease, Graves disease, Hashimoto thyroiditis, type I diabetes mellitus, and Addison disease.

Celiac Disease

Celiac disease is largely a disease of the small intestine, although other organs may be affected. People in their 30s and 40s, and children are most commonly affected, but **celiac disease** can begin at any age. It results from a reaction to proteins, commonly called gluten, found mainly in wheat, barley, rye, and some other grains. The disease has several genetic causes (predispositions) and poorly understood environmental influences. On exposure to gluten, the body produces various autoantibodies and an inflammatory response. The inflammatory response in the small intestine leads to a reduction in the depth of the microvilli of the mucosa, which hinders absorption and can lead to weight loss and anemia. The disease is also characterized by diarrhea and abdominal pain, symptoms that are often misdiagnosed as irritable bowel syndrome.

Diagnosis of celiac disease is accomplished from serological tests for the presence of primarily IgA antibodies to components of gluten, the transglutinaminase enzyme, and autoantibodies to endomysium, a connective tissue surrounding muscle fibers. Serological tests are typically followed up with endoscopy and biopsy of the duodenal mucosa. Serological screening surveys have found about 1% of individuals in the United Kingdom are positive even though they do not all display symptoms.[13] This early recognition allows for more careful monitoring and prevention of severe disease.

Celiac disease is treated with complete removal of gluten-containing foods from the diet, which results in improved symptoms and reduced risk of complications. Other theoretical approaches include breeding grains that do not contain the immunologically reactive components or developing dietary supplements that contain enzymes that break down the protein components that cause the immune response.[14]

Disorders of the Thyroid

Graves disease is the most common cause of hyperthyroidism in the United States. Symptoms of Graves disease result from the production of thyroid-stimulating immunoglobulin (TSI) also called TSH-receptor antibody. TSI targets and binds to the receptor for thyroid stimulating hormone (TSH), which is naturally produced by the pituitary gland. TSI may cause conflicting symptoms because it may stimulate the thyroid to make too much thyroid hormone or block thyroid hormone production entirely, making diagnosis more difficult. Signs and symptoms of Graves disease include heat intolerance, rapid and irregular heartbeat, weight loss, goiter (a swollen thyroid gland, protruding under the skin of the throat [Figure 19.13]) and exophthalmia (bulging eyes) often referred to as Graves ophthalmopathy (Figure 19.14).

The most common cause of hypothyroidism in the United States is **Hashimoto thyroiditis**, also called chronic lymphocytic thyroiditis. Patients with Hashimoto thyroiditis often develop a spectrum of different diseases because they are more likely to develop additional autoimmune diseases such as Addison disease (discussed later in this section), type 1 diabetes, rheumatoid arthritis, and celiac disease. Hashimoto thyroiditis is a T_H1 cell-mediated disease that occurs when the thyroid gland is attacked by cytotoxic lymphocytes, macrophages, and autoantibodies. This autoimmune response leads to numerous symptoms that include goiter (Figure 19.13), cold intolerance, muscle weakness, painful and stiff joints, depression, and memory loss.

---

13 D.A. Van Heel, J. West. "Recent Advances in Coeliac Disease." *Gut* 55 no. 7 (2006):1037–1046.

14 ibid.

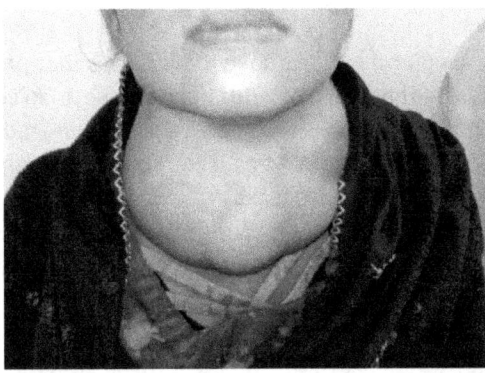

Figure 19.13　Goiter, a hypertrophy of the thyroid, is a symptom of Graves disease and Hashimoto thyroiditis.

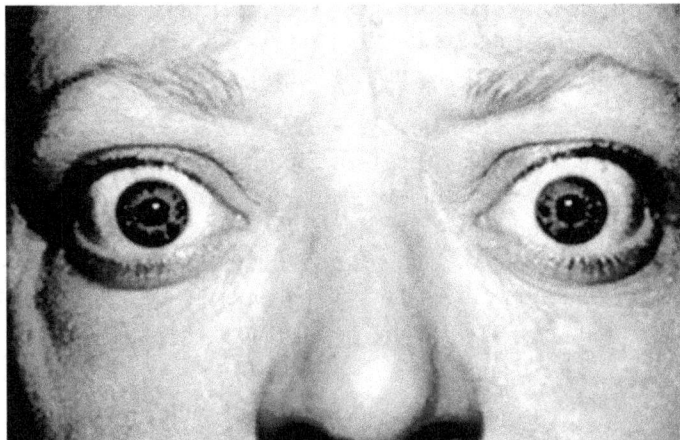

Figure 19.14　Exophthalmia, or Graves ophthalmopathy, is a sign of Graves disease. (credit: modification of work by Jonathan Trobe, University of Michigan Kellogg Eye Center)

Type 1 Diabetes

Juvenile diabetes, or **type 1 diabetes mellitus**, is usually diagnosed in children and young adults. It is a T-cell-dependent autoimmune disease characterized by the selective destruction of the β cells of the islets of Langerhans in the pancreas by CD4 T_H1-mediated CD8 T cells, anti-β-cell antibodies, and macrophage activity. There is also evidence that viral infections can have either a potentiating or inhibitory role in the development of type 1 diabetes (T1D) mellitus. The destruction of the β cells causes a lack of insulin production by the pancreas. In T1D, β-cell destruction may take place over several years, but symptoms of hyperglycemia, extreme increase in thirst and urination, weight loss, and extreme fatigue usually have a sudden onset, and diagnosis usually does not occur until most β cells have already been destroyed.

Autoimmune Addison Disease

Destruction of the adrenal glands (the glands lying above the kidneys that produce glucocorticoids, mineralocorticoids, and sex steroids) is the cause of **Addison disease**, also called primary adrenal insufficiency (PAI). Today, up to 80% of Addison disease cases are diagnosed as autoimmune Addison disease (AAD), which is caused by an autoimmune response to adrenal tissues disrupting adrenal function. Disruption of adrenal function causes impaired metabolic processes that require normal steroid hormone levels, causing signs and symptoms throughout the body. There is evidence that both humoral and CD4 T_H1-driven CD8 T-cell–mediated immune mechanisms are directed at the adrenal cortex in AAD. There is also evidence that the autoimmune response is associated with autoimmune destruction of other endocrine glands as well, such as the pancreas and thyroid, conditions collectively referred to as autoimmune polyendocrine syndromes (APS). In up to 80% of patients with AAD, antibodies are produced to three enzymes involved in steroid synthesis: 21-hydroxylase (21-OH), 17α-hydroxylase, and cholesterol side-chain–cleaving enzyme.[15] The most common autoantibody found in AAD is to 21-OH, and antibodies to any of the key enzymes for steroid production are diagnostic for AAD. The adrenal cortex cells are targeted, destroyed, and replaced with fibrous tissue by immune-mediated inflammation. In some patients, at least 90% of the adrenal cortex is destroyed before symptoms become diagnostic.

Symptoms of AAD include weakness, nausea, decreased appetite, weight loss, hyperpigmentation (Figure 19.15), hyperkalemia (elevated blood potassium levels), hyponatremia (decreased blood sodium levels), hypoglycemia (decreased levels of blood sugar), hypotension (decreased blood pressure), anemia, lymphocytosis (decreased levels of white blood cells), and fatigue. Under extreme stress, such as surgery, accidental trauma, or infection, patients with AAD may experience an adrenal crisis that causes the patient to vomit, experience abdominal pain, back or leg cramps, and even severe hypotension leading to shock.

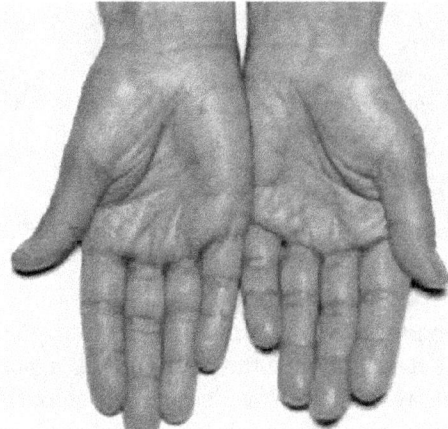

Figure 19.15   Hyperpigmentation is a sign of Addison disease. (credit: modification of work by Petros Perros)

✓ CHECK YOUR UNDERSTANDING

- What are the names of autoimmune diseases that interfere with hormone gland function?
- Describe how the mechanisms of Graves disease and Hashimoto thyroiditis differ.
- Name the cells that are destroyed in type 1 diabetes mellitus and describe the result.

Systemic Autoimmune Diseases

Whereas organ-specific autoimmune diseases target specific organs or tissues, **systemic autoimmune disease**s are more generalized, targeting multiple organs or tissues throughout the body. Examples of systemic autoimmune diseases include multiple sclerosis, myasthenia gravis, psoriasis, rheumatoid arthritis, and systemic lupus erythematosus.

---

15 P. Martorell et al. "Autoimmunity in Addison's Disease." *Netherlands Journal of Medicine* 60 no. 7 (2002):269–275.

Multiple Sclerosis

Multiple sclerosis (MS) is an autoimmune central nervous system disease that affects the brain and spinal cord. Lesions in multiple locations within the central nervous system are a hallmark of **multiple sclerosis** and are caused by infiltration of immune cells across the blood-brain barrier. The immune cells include T cells that promote inflammation, demyelination, and neuron degeneration, all of which disrupt neuronal signaling. Symptoms of MS include visual disturbances; muscle weakness; difficulty with coordination and balance; sensations such as numbness, prickling, or "pins and needles"; and cognitive and memory problems.

Myasthenia Gravis

Autoantibodies directed against acetylcholine receptors (AChRs) in the synaptic cleft of neuromuscular junctions lead to **myasthenia gravis** (Figure 19.16). Anti-AChR antibodies are high-affinity IgGs and their synthesis requires activated CD4 T cells to interact with and stimulate B cells. Once produced, the anti-AChR antibodies affect neuromuscular transmission by at least three mechanisms:

- Complement binding and activation at the neuromuscular junction
- Accelerated AChR endocytosis of molecules cross-linked by antibodies
- Functional AChR blocking, which prevents normal acetylcholine attachment to, and activation of, AChR

Regardless of the mechanism, the effect of anti-AChR is extreme muscle weakness and potentially death through respiratory arrest in severe cases.

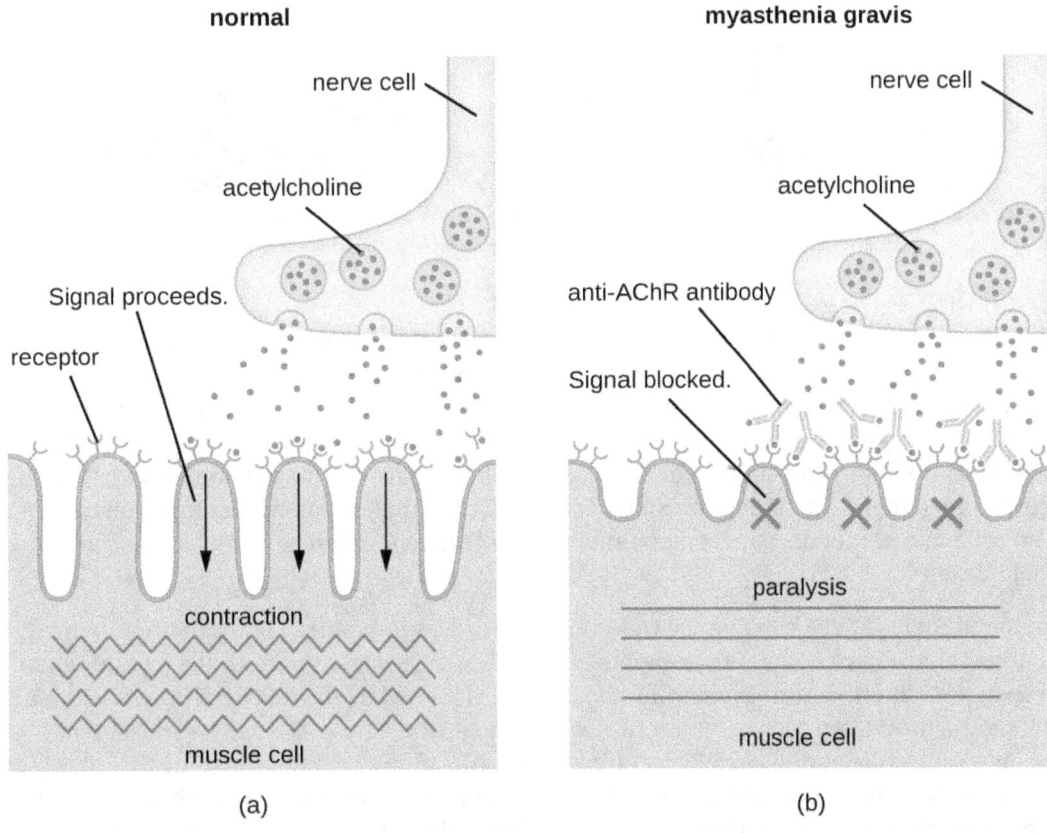

Figure 19.16 Myasthenia gravis and impaired muscle contraction. (a) Normal release of the neurotransmitter acetylcholine stimulates muscle contraction. (b) In myasthenia gravis, autoantibodies block the receptors for acetylcholine (AChr) on muscle cells, resulting in paralysis.

Psoriasis

Psoriasis is a skin disease that causes itchy or sore patches of thick, red skin with silvery scales on elbows, knees, scalp, back, face, palms, feet, and sometimes other areas. Some individuals with **psoriasis** also get a form of arthritis called psoriatic arthritis, in which the joints can become inflamed. Psoriasis results from the complex interplay between keratinocytes, dendritic cells, and T cells, and the cytokines produced by these various cells. In a process called cell turnover, skin cells that grow deep in the skin rise to the surface.

Normally, this process takes a month. In psoriasis, as a result of cytokine activation, cell turnover happens in just a few days. The thick inflamed patches of skin that are characteristic of psoriasis develop because the skin cells rise too fast.

Rheumatoid Arthritis

The most common chronic inflammatory joint disease is **rheumatoid arthritis** (RA) (Figure 19.17) and it is still a major medical challenge because of unsolved questions related to the environmental and genetic causes of the disease. RA involves type III hypersensitivity reactions and the activation of CD4 T cells, resulting in chronic release of the inflammatory cytokines IL-1, IL-6, and tumor necrosis factor-α (TNF-α). The activated CD4 T cells also stimulate the production of rheumatoid factor (RF) antibodies and anticyclic citrullinated peptide antibodies (anti-CCP) that form immune complexes. Increased levels of acute-phase proteins, such as C-reactive protein (CRP), are also produced as part of the inflammatory process and participate in complement fixation with the antibodies on the immune complexes. The formation of immune complexes and reaction to the immune factors cause an inflammatory process in joints, particularly in the hands, feet, and legs. Diagnosis of RA is based on elevated levels of RF, anti-CCP, quantitative CRP, and the erythrocyte sedimentation rate (ESR) (modified Westergren). In addition, radiographs, ultrasound, or magnetic resonance imaging scans can identify joint damage, such as erosions, a loss of bone within the joint, and narrowing of joint space.

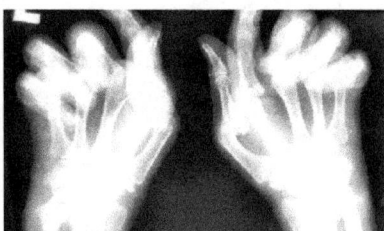 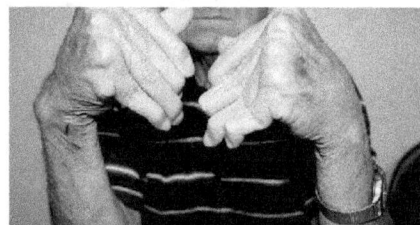

Figure 19.17 The radiograph (left) and photograph (right) show damage to the hands typical of rheumatoid arthritis. (credit right: modification of work by "handarmdoc"/Flickr)

Systemic Lupus Erythematosus

The damage and pathology of **systemic lupus erythematosus (SLE)** is caused by type III hypersensitivity reactions. Autoantibodies produced in SLE are directed against nuclear and cytoplasmic proteins. Antinuclear antibodies (ANAs) are present in more than 95% of patients with SLE,[16] with additional autoantibodies including anti-double–stranded DNA (ds-DNA) and anti-Sm antibodies (antibodies to small nuclear ribonucleoprotein). Anti-ds-DNA and anti-Sm antibodies are unique to patients with SLE; thus, their presence is included in the classification criteria of SLE. Cellular interaction with autoantibodies leads to nuclear and cellular destruction, with components released after cell death leading to the formation of immune complexes.

Because autoantibodies in SLE can target a wide variety of cells, symptoms of SLE can occur in many body locations. However, the most common symptoms include fatigue, fever with no other cause, hair loss, and a sunlight-sensitive "butterfly" or wolf-mask (lupus) rash that is found in about 50% of people with SLE (Figure 19.18). The rash is most often seen over the cheeks and bridge of the nose, but can be widespread. Other symptoms may appear depending on affected areas. The joints may be affected, leading to arthritis of the fingers, hands, wrists, and knees. Effects on the brain and nervous system can lead to headaches, numbness, tingling, seizures, vision problems, and personality changes. There may also be abdominal pain, nausea, vomiting, arrhythmias, shortness of breath, and blood in the sputum. Effects on the skin can lead to additional areas of skin lesions, and vasoconstriction can cause color changes in the fingers when they are cold (Raynaud phenomenon). Effects on the kidneys can lead to edema in the legs and weight gain. A diagnosis of SLE depends on identification of four of 11 of the most common symptoms and confirmed production of an array of autoantibodies unique to SLE. A positive test for ANAs alone is not diagnostic.

---

16 C.C. Mok, C.S. Lau. "Pathogenesis of Systemic Lupus Erythematosus." *Journal of Clinical Pathology* 56 no. 7 (2003):481–490.

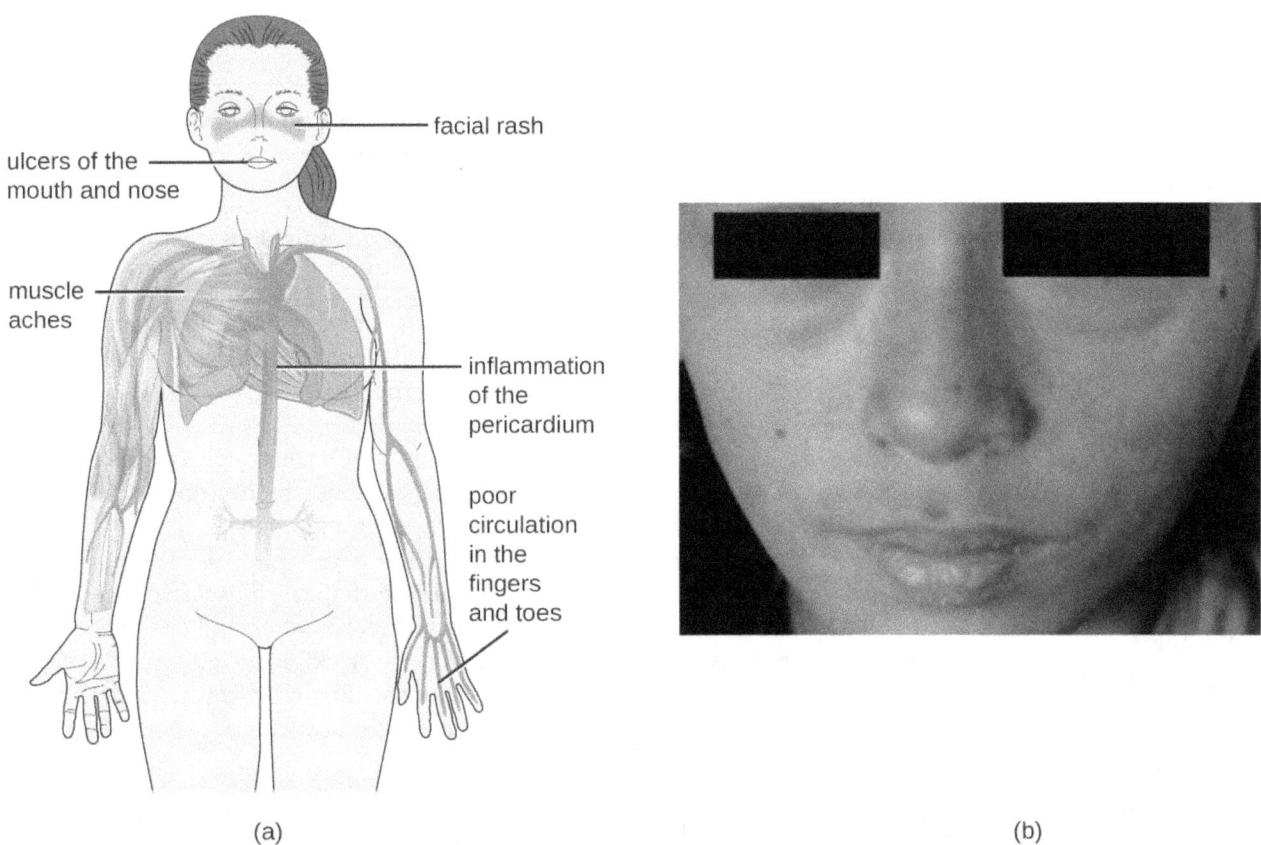

Figure 19.18 (a) Systemic lupus erythematosus is characterized by autoimmunity to the individual's own DNA and/or proteins. (b) This patient is presenting with a butterfly rash, one of the characteristic signs of lupus. (credit a: modification of work by Mikael Häggström; credit b: modification of work by Shrestha D, Dhakal AK, Shiva RK, Shakya A, Shah SC, Shakya H)

CHECK YOUR UNDERSTANDING

- List the ways antibodies contribute to the pathogenesis of myasthenia gravis.
- Explain why rheumatoid arthritis is considered a type III hypersensitivity.
- Describe the symptoms of systemic lupus erythematosus and explain why they affect so many different parts of the body.
- What is recognized as an antigen in myasthenia gravis?

Table 19.6 summarizes the causes, signs, and symptoms of select autoimmune diseases.

Select Autoimmune Diseases

Disease	Cause	Signs and Symptoms
Addison disease	Destruction of adrenal gland cells by cytotoxic T cells	Weakness, nausea, hypotension, fatigue; adrenal crisis with severe pain in abdomen, lower back, and legs; circulatory system collapse, kidney failure
Celiac disease	Antibodies to gluten become autoantibodies that target cells of the small intestine	Severe diarrhea, abdominal pain, anemia, malnutrition

Select Autoimmune Diseases

Disease	Cause	Signs and Symptoms
Diabetes mellitus (type I)	Cytotoxic T-cell destruction of the insulin-producing β cells of the pancreas	Hyperglycemia, extreme increase in thirst and urination, weight loss, extreme fatigue
Graves disease	Autoantibodies target thyroid-stimulating hormone receptors, resulting in overstimulation of the thyroid	Hyperthyroidism with rapid and irregular heartbeat, heat intolerance, weight loss, goiter, exophthalmia
Hashimoto thyroiditis	Thyroid gland is attacked by cytotoxic T cells, lymphocytes, macrophages, and autoantibodies	Thyroiditis with goiter, cold intolerance, muscle weakness, painful and stiff joints, depression, memory loss
Multiple sclerosis (MS)	Cytotoxic T-cell destruction of the myelin sheath surrounding nerve axons in the central nervous system	Visual disturbances, muscle weakness, impaired coordination and balance, numbness, prickling or "pins and needles" sensations, impaired cognitive function and memory
Myasthenia gravis	Autoantibodies directed against acetylcholine receptors within the neuromuscular junction	Extreme muscle weakness eventually leading to fatal respiratory arrest
Psoriasis	Cytokine activation of keratinocytes causes rapid and excessive epidermal cell turnover	Itchy or sore patches of thick, red skin with silvery scales; commonly affects elbows, knees, scalp, back, face, palms, feet
Rheumatoid arthritis	Autoantibodies, immune complexes, complement activation, phagocytes, and T cells damage membranes and bone in joints	Joint inflammation, pain and disfigurement, chronic systemic inflammation
Systemic lupus erythematosus (SLE)	Autoantibodies directed against nuclear and cytoplasmic molecules form immune complexes that deposit in tissues. Phagocytic cells and complement activation cause tissue damage and inflammation	Fatigue, fever, joint pain and swelling, hair loss, anemia, clotting, a sunlight-sensitive "butterfly" rash, skin lesions, photosensitivity, decreased kidney function, memory loss, confusion, depression

Table 19.6

19.3 Organ Transplantation and Rejection

Learning Objectives

By the end of this section, you will be able to:
- Explain why human leukocyte antigens (HLAs) are important in tissue transplantation
- Explain the types of grafts possible and their potential for interaction with the immune system
- Describe what occurs during graft-versus-host disease (GVHD)

A graft is the transplantation of an organ or tissue to a different location, with the goal of replacing a missing or damaged organ or tissue. Grafts are typically moved without their attachments to the circulatory system and must reestablish these, in addition to the other connections and interactions with their new surrounding tissues. There are different types of grafts depending on the source of the new tissue or organ. Tissues that are transplanted from one genetically distinct individual to another within the same species are called **allografts**. An interesting variant of the allograft is an **isograft**, in which tissue from one twin is transplanted to another. As long as the twins are monozygotic (therefore, essentially genetically identical), the transplanted tissue is virtually never rejected. If tissues are transplanted from one area on an individual to another area on the same individual (e.g., a skin graft on a burn patient), it is known as an **autograft**. If tissues from an animal are transplanted into a human, this is called a **xenograft**.

Transplant Rejection

The different types of grafts described above have varying risks for rejection (Table 19.7). Rejection occurs when the recipient's immune system recognizes the donor tissue as foreign (non-self), triggering an immune response. The major histocompatibility complex markers MHC I and MHC II, more specifically identified as human leukocyte antigens (HLAs), play a role in transplant rejection. The HLAs expressed in tissue transplanted from a genetically different individual or species may be recognized as non-self molecules by the host's dendritic cells. If this occurs, the dendritic cells will process and present the foreign HLAs to the host's helper T cells and cytotoxic T cells, thereby activating them. Cytotoxic T cells then target and kill the grafted cells through the same mechanism they use to kill virus-infected cells; helper T cells may also release cytokines that activate macrophages to kill graft cells.

Types of Tissue and Organ Grafts and Their Complications

Graft	Procedure	Complications
Autograft	From self to self	No rejection concerns
Isograft	From identical twin to twin	Little concern of rejection
Allograft	From relative or nonrelative to individual	Rejection possible
Xenograft	From animal to human	Rejection possible

Table 19.7

With the three highly polymorphic MHC I genes in humans (*HLA-A*, *HLA-B*, and *HLA-C*) determining compatibility, each with many alleles segregating in a population, odds are extremely low that a randomly chosen donor will match a recipient's six-allele genotype (the two alleles at each locus are expressed codominantly). This is why a parent or a sibling may be the best donor in many situations—a genetic match between the MHC genes is much more likely and the organ is much less likely to be rejected.

Although matching all of the MHC genes can lower the risk for rejection, there are a number of additional gene products that also play a role in stimulating responses against grafted tissue. Because of this, no non-self grafted tissue is likely to completely avoid rejection. However, the more similar the MHC gene match, the more likely the graft is to be tolerated for a longer time. Most transplant recipients, even those with tissues well matched to their MHC genes, require treatment with immunosuppressant drugs for the rest of their lives. This can make them more vulnerable than the general population to complications from infectious diseases. It can also result in transplant-related malignancies because the body's normal defenses against cancer cells are being suppressed.

CHECK YOUR UNDERSTANDING

- What part of the immune response is responsible for graft rejection?

- Explain why blood relatives are preferred as organ donors.
- Describe the role of immunosuppression in transplantation.

Graft-versus-Host Disease

A form of rejection called **graft-versus-host disease (GVHD)** primarily occurs in recipients of bone marrow transplants and peripheral blood stem cells. GHVD presents a unique situation because the transplanted tissue is capable of producing immune cells; APCs in the donated bone marrow may recognize the host cells as non-self, leading to activation of the donor cytotoxic T cells. Once activated, the donor's T cells attack the recipient cells, causing acute GVHD.

Acute GVHD typically develops within weeks after a bone marrow transplant, causing tissue damage affecting the skin, gastrointestinal tract, liver, and eyes. In addition, acute GVHD may also lead to a cytokine storm, an unregulated secretion of cytokines that may be fatal. In addition to acute GVHD, there is also the risk for chronic GVHD developing months after the bone marrow transplant. The mechanisms responsible for chronic GVHD are not well understood.

To minimize the risk of GVHD, it is critically important to match the HLAs of the host and donor as closely as possible in bone marrow transplants. In addition, the donated bone marrow is processed before grafting to remove as many donor APCs and T cells as possible, leaving mostly hematopoietic stem cells.

CHECK YOUR UNDERSTANDING

- Why does GVHD occur in specifically in bone marrow transplants?
- What cells are responsible for GVHD?

The Future of Transplantation

Historically speaking, the practice of transplanting tissues—and the complications that can accompany such procedures—is a relatively recent development. It was not until 1954 that the first successful organ transplantation between two humans was achieved. Yet the field of organ transplantation has progressed rapidly since that time.

Nonetheless, the practice of transplanting non-self tissues may soon become obsolete. Scientists are now attempting to develop methods by which new organs may be grown *in vitro* from an individual's own harvested cells to replace damaged or abnormal ones. Because organs produced in this way would contain the individual's own cells, they could be transplanted into the individual without risk for rejection.

An alternative approach that is gaining renewed research interest is genetic modification of donor animals, such as pigs, to provide transplantable organs that do not elicit an immune response in the recipient. The approach involves excising the genes in the pig (in the embryo) that are most responsible for the rejection reaction after transplantation. Finding these genes and effectively removing them is a challenge, however. So too is identifying and neutralizing risks from viral sequences that might be embedded in the pig genome, posing a risk for infection in the human recipient.

LINK TO LEARNING

There are currently more than a dozen different tissues and organs used in human transplantations. Learn more about them at this (https://openstax.org/l/22organstransp) website.

Clinical Focus

Resolution

Kerry's tests come back positive, confirming a diagnosis of lupus, a disease that occurs 10 times more frequently in women than men. SLE cannot be cured, but there are various therapies available for reducing and managing its symptoms. Specific therapies are prescribed based on the particular symptoms presenting in the patient. Kerry's rheumatologist starts her therapy with a low dose of corticosteroids to reduce her rashes. She also prescribes a low dose of hydroxychloroquine, an anti-inflammatory drug that is used to treat inflammation in patients with RA, childhood arthritis, SLE, and other autoimmune diseases. Although the mechanism of action of hydroxychloroquine is not well defined, it appears that this drug interferes with the processes of antigen processing and activation of autoimmunity. Because of its mechanism, the effects of hydroxychloroquine are not as immediate as that of other anti-inflammatory drugs, but it is still considered a good companion therapy for SLE. Kerry's doctor also advises her to limit her exposure to sunlight, because photosensitivity to sunlight may precipitate rashes.

Over the next 6 months, Kerry follows her treatment plan and her symptoms do not return. However, future flare-ups are likely to occur. She will need to continue her treatment for the rest of her life and seek medical attention whenever new symptoms develop.

Go back to the *previous* Clinical Focus box.

19.4 Immunodeficiency

Learning Objectives

By the end of this section, you will be able to:
- Compare the causes of primary and secondary immunodeficiencies
- Describe treatments for primary and secondary immunodeficiencies

Immunodeficiencies are inherited (primary) or acquired (secondary) disorders in which elements of host immune defenses are either absent or functionally defective. In developed countries, most immunodeficiencies are inherited, and they are usually first seen in the clinic as recurrent or overwhelming infections in infants. However, on a global scale, malnutrition is the most common cause of immunodeficiency and would be categorized as an acquired immunodeficiency. Acquired immunodeficiencies are more likely to develop later in life, and the pathogenic mechanisms of many remain obscure.

Primary Immunodeficiency

Primary immunodeficiencies, which number more than 250, are caused by inherited defects of either nonspecific innate or specific adaptive immune defenses. In general, patients born with primary immunodeficiency (PI) commonly have an increased susceptibility to infection. This susceptibility can become apparent shortly after birth or in early childhood for some individuals, whereas other patients develop symptoms later in life. Some primary immunodeficiencies are due to a defect of a single cellular or humoral component of the immune system; others may result from defects of more than one component. Examples of primary immunodeficiencies include chronic granulomatous disease, X-linked agammaglobulinemia, selective IgA deficiency, and severe combined immunodeficiency disease.

Chronic Granulomatous Disease

The causes of **chronic granulomatous disease** (CGD) are defects in the NADPH oxidase system of phagocytic cells, including neutrophils and macrophages, that prevent the production of superoxide radicals in phagolysosomes. The inability to produce superoxide radicals impairs the antibacterial activity of phagocytes. As a result, infections in patients with CGD persist longer, leading to a chronic local inflammation called a granuloma. Microorganisms that are the most common causes of infections in patients with CGD include *Aspergillus* spp., *Staphylococcus aureus*, *Chromobacterium violaceum*, *Serratia marcescens*, and *Salmonella typhimurium*.

X-Linked Agammaglobulinemia

Deficiencies in B cells due to defective differentiation lead to a lack of specific antibody production known as **X-linked agammaglobulinemia**. In 1952, Ogden C. Bruton (1908–2003) described the first immunodeficiency in a boy whose immune system failed to produce antibodies. This defect is inherited on the X chromosome and is characterized by the absence of immunoglobulin in the serum; it is called Bruton X-linked agammaglobulinemia (XLA). The defective gene, *BTK*, in XLA is now known to encode a tyrosine kinase called Bruton tyrosine kinase (Btk). In patients whose B cells are unable to produce sufficient amounts of Btk, the B-cell maturation and differentiation halts at the pre-B-cell stage of growth. B-cell maturation and differentiation beyond the pre-B-cell stage of growth is required for immunoglobulin production. Patients who lack antibody production suffer from recurrent infections almost exclusively due to extracellular pathogens that cause pyogenic infections: *Haemophilus influenzae*, *Streptococcus pneumoniae*, *S. pyogenes*, and *S. aureus*. Because cell-mediated immunity is not impaired, these patients are not particularly vulnerable to infections caused by viruses or intracellular pathogens.

Selective IgA Deficiency

The most common inherited form of immunoglobulin deficiency is **selective IgA deficiency**, affecting about one in 800 people. Individuals with selective IgA deficiency produce normal levels of IgG and IgM, but are not able to produce secretory IgA. IgA deficiency predisposes these individuals to lung and gastrointestinal infections for which secretory IgA is normally an important defense mechanism. Infections in the lungs and gastrointestinal tract can involve a variety of pathogens, including *H. influenzae*, *S. pneumoniae*, *Moraxella catarrhalis*, *S. aureus*, *Giardia lamblia*, or pathogenic strains of *Escherichia coli*.

Severe Combined Immunodeficiency

Patients who suffer from **severe combined immunodeficiency (SCID)** have B-cell and T-cell defects that impair T-cell dependent antibody responses as well as cell-mediated immune responses. Patients with SCID also cannot develop immunological memory, so vaccines provide them no protection, and live attenuated vaccines (e.g., for varicella-zoster, measles virus, rotavirus, poliovirus) can actually cause the infection they are intended to prevent. The most common form is X-linked SCID, which accounts for nearly 50% of all cases and occurs primarily in males. Patients with SCID are typically diagnosed within the first few months of life after developing severe, often life-threatening, opportunistic infection by *Candida* spp., *Pneumocystis jirovecii*, or pathogenic strains of *E. coli*.

Without treatment, babies with SCID do not typically survive infancy. In some cases, a bone marrow transplant may successfully correct the defects in lymphocyte development that lead to the SCID phenotype, by replacing the defective component. However, this treatment approach is not without risks, as demonstrated by the famous case of David Vetter (1971–1984), better known as "Bubble Boy" (Figure 19.19). Vetter, a patient with SCID who lived in a protective plastic bubble to prevent exposure to opportunistic microbes, received a bone marrow transplant from his sister. Because of a latent Epstein-Barr virus infection in her bone marrow, however, he developed mononucleosis and died of Burkitt lymphoma at the age of 12 years.

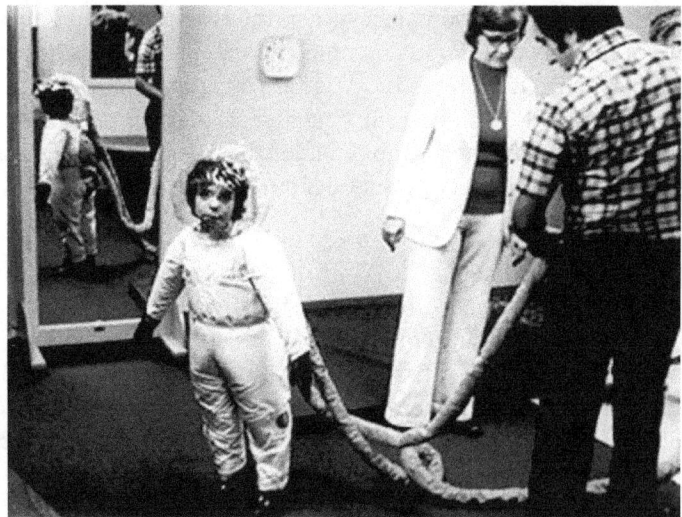

Figure 19.19 David Vetter, popularly known as "The Bubble Boy," was born with SCID and lived most of his life isolated inside a plastic bubble. Here he is shown outside the bubble in a suit specially built for him by NASA. (credit: NASA Johnson Space Center)

CHECK YOUR UNDERSTANDING

- What is the fundamental cause of a primary immunodeficiency?
- Explain why patients with chronic granulomatous disease are especially susceptible to bacterial infections.
- Explain why individuals with selective IgA deficiency are susceptible to respiratory and gastrointestinal infections.

Secondary Immunodeficiency

A **secondary immunodeficiency** occurs as a result an acquired impairment of function of B cells, T cells, or both. Secondary immunodeficiencies can be caused by:

- Systemic disorders such as diabetes mellitus, malnutrition, hepatitis, or HIV infection
- Immunosuppressive treatments such as cytotoxic chemotherapy, bone marrow ablation before transplantation, or radiation therapy
- Prolonged critical illness due to infection, surgery, or trauma in the very young, elderly, or hospitalized patients

Unlike primary immunodeficiencies, which have a genetic basis, secondary immunodeficiencies are often reversible if the underlying cause is resolved. Patients with secondary immunodeficiencies develop an increased susceptibility to an otherwise benign infection by opportunistic pathogens such as *Candida* spp., *P. jirovecii*, and *Cryptosporidium*.

HIV infection and the associated **acquired immunodeficiency syndrome (AIDS)** are the best-known secondary immunodeficiencies. AIDS is characterized by profound CD4 T-cell lymphopenia (decrease in lymphocytes). The decrease in CD4 T cells is the result of various mechanisms, including HIV-induced pyroptosis (a type of apoptosis that stimulates an inflammatory response), viral cytopathic effect, and cytotoxicity to HIV-infected cells.

The most common cause of secondary immunodeficiency worldwide is severe malnutrition, which affects both innate and adaptive immunity. More research and information are needed for the more common causes of secondary immunodeficiency; however, the number of new discoveries in AIDS research far exceeds that of any other single cause of secondary immunodeficiency. AIDS research has paid off extremely well in terms of discoveries and treatments; increased research into the most common cause of immunodeficiency, malnutrition, would likely be as beneficial.

CHECK YOUR UNDERSTANDING

- What is the most common cause of secondary immunodeficiencies?
- Explain why secondary immunodeficiencies can sometimes be reversed.

Case in Point

An Immunocompromised Host

Benjamin, a 50-year-old male patient who has been receiving chemotherapy to treat his chronic myelogenous leukemia (CML), a disease characterized by massive overproduction of nonfunctional, malignant myelocytic leukocytes that crowd out other, healthy leukocytes, is seen in the emergency department. He is complaining of a productive, wet cough, dyspnea, and fatigue. On examination, his pulse is 120 beats per minute (bpm) (normal range is 60–100 bpm) and weak, and his blood pressure is 90/60 mm Hg (normal is 120/80 mm Hg). During auscultation, a distinct crackling can be heard in his lungs as he breathes, and his pulse-oximeter level (a measurement of blood-oxygen saturation) is 80% (normal is 95%–100%). He has a fever; his temperature is 38.9 °C (102 °F). Sputum cultures and blood samples are obtained and sent to the lab, but Benjamin goes into respiratory distress and dies before the results can be obtained.

Benjamin's death was a result of a combination of his immune system being compromised by his leukemia and his chemotherapy treatment further weakening his ability to mount an immune response. CML (and leukemia in general) and corresponding chemotherapy cause a decrease in the number of leukocytes capable of normal function, leading to secondary immunodeficiency. This increases the risk for opportunistic bacterial, viral, protozoal, and fungal infections that could include *Staphylococcus*, enteroviruses, *Pneumocystis*, *Giardia*, or *Candida*. Benjamin's symptoms were suggestive of bacterial pneumonia, but his leukemia and chemotherapy likely complicated and contributed to the severity of the pneumonia, resulting in his death. Because his leukemia was overproducing certain white blood cells, and those overproduced white blood cells were largely nonfunctional or abnormal in their function, he did not have the proper immune system blood cells to help him fight off the infection.

Table 19.8 summarizes primary and secondary immunodeficiencies, their effects on immune function, and typical outcomes.

Primary and Secondary Immunodeficiencies

	Disease	Effect on Immune Function	Outcomes
Primary immunodeficiencies	Chronic granulomatous disease	Impaired killing of bacteria within the phagolysosome of neutrophils and macrophages	Chronic infections and granulomas
	Selective IgA deficiency	Inability to produce secretory IgA	Predisposition to lung and gastrointestinal infections
	Severe combined immunodeficiency disease (SCID)	Deficient humoral and cell-mediated immune responses	Early development of severe and life-threatening opportunistic infections
	X-linked agammaglobulinemia	Flawed differentiation of B cells and absence of specific antibodies	Recurrent infections almost exclusively due to pathogens that cause pyogenic infections
Secondary immunodeficiencies	Immunosuppressive therapies (e.g., chemotherapy, radiotherapy)	Impaired humoral and/or cell-mediated immune responses	Opportunistic infections, rare cancers
	Malnutrition	Impaired humoral and/or cell-mediated immune responses	Opportunistic infections, rare cancers
	Viral infection (e.g., HIV)	Impaired cell-mediated immune responses due to CD4 T-cell lymphopenia	Opportunistic infections, rare cancers

Table 19.8

19.5 Cancer Immunobiology and Immunotherapy

Learning Objectives

By the end of this section, you will be able to:
- Explain how the adaptive specific immune response responds to tumors
- Discuss the risks and benefits of tumor vaccines

Cancer involves a loss of the ability of cells to control their cell cycle, the stages each eukaryotic cell goes through as it grows and then divides. When this control is lost, the affected cells rapidly divide and often lose the ability to differentiate into the cell type appropriate for their location in the body. In addition, they lose contact inhibition and can start to grow on top of each other. This can result in formation of a **tumor**. It is important to make a distinction here: The term "cancer" is used to describe the diseases resulting from loss of cell-cycle regulation and subsequent cell proliferation. But the term "tumor" is more general. A "tumor" is an abnormal mass of cells, and a tumor can be benign (not cancerous) or malignant (cancerous).

Traditional cancer treatment uses radiation and/or chemotherapy to destroy cancer cells; however, these treatments can have unwanted side effects because they harm normal cells as well as cancer cells. Newer, promising therapies attempt to enlist the patient's immune system to target cancer cells specifically. It is known that the immune system can recognize and destroy cancerous cells, and some researchers and immunologists also believe, based on the results of their experiments, that many cancers are eliminated by the body's own defenses before they can become a health problem. This idea is not universally accepted by researchers, however, and needs further investigation for verification.

Cell-Mediated Response to Tumors

Cell-mediated immune responses can be directed against cancer cells, many of which do not have the normal complement of self-proteins, making them a target for elimination. Abnormal cancer cells may also present tumor antigens. These tumor antigens are not a part of the screening process used to eliminate lymphocytes during development; thus, even though they are self-antigens, they can stimulate and drive adaptive immune responses against abnormal cells.

Presentation of tumor antigens can stimulate naïve helper T cells to become activated by cytokines such as IL-12 and differentiate into T_H1 cells. T_H1 cells release cytokines that can activate natural killer (NK) cells and enhance the killing of activated cytotoxic T cells. Both NK cells and cytotoxic T cells can recognize and target cancer cells, and induce apoptosis through the action of perforins and granzymes. In addition, activated cytotoxic T cells can bind to cell-surface proteins on abnormal cells and induce apoptosis by a second killing mechanism called the CD95 (Fas) cytotoxic pathway.

Despite these mechanisms for removing cancerous cells from the body, cancer remains a common cause of death. Unfortunately, malignant tumors tend to actively suppress the immune response in various ways. In some cancers, the immune cells themselves are cancerous. In leukemia, lymphocytes that would normally facilitate the immune response become abnormal. In other cancers, the cancerous cells can become resistant to induction of apoptosis. This may occur through the expression of membrane proteins that shut off cytotoxic T cells or that induce regulatory T cells that can shut down immune responses.

The mechanisms by which cancer cells alter immune responses are still not yet fully understood, and this is a very active area of research. As scientists' understanding of adaptive immunity improves, cancer therapies that harness the body's immune defenses may someday be more successful in treating and eliminating cancer.

✓ CHECK YOUR UNDERSTANDING

- How do cancer cells suppress the immune system?
- Describe how the immune system recognizes and destroys cancer cells.

Cancer Vaccines

There are two types of cancer vaccines: preventive and therapeutic. Preventive vaccines are used to prevent cancer from occurring, whereas therapeutic vaccines are used to treat patients with cancer. Most preventive cancer vaccines target viral infections that are known to lead to cancer. These include vaccines against human papillomavirus (HPV) and hepatitis B, which help prevent cervical and liver cancer, respectively.

Most therapeutic cancer vaccines are in the experimental stage. They exploit tumor-specific antigens to stimulate the immune system to selectively attack cancer cells. Specifically, they aim to enhance T_H1 function and interaction with cytotoxic T cells, which, in turn, results in more effective attack on abnormal tumor cells. In some cases, researchers have used genetic engineering to develop antitumor vaccines in an approach similar to that used for DNA vaccines (see Micro Connections: DNA vaccines). The vaccine contains a recombinant plasmid with genes for tumor antigens; theoretically, the tumor gene would not induce new cancer because it is not functional, but it could trick the immune system into targeting the tumor gene product as a foreign invader.

The first FDA-approved therapeutic cancer vaccine was sipuleucel-T (Provenge), approved in 2010 to treat certain cases of prostate cancer.[17] This unconventional vaccine is custom designed using the patient's own cells. APCs are removed from the patient and cultured with a tumor-specific molecule; the cells are then returned to the patient. This approach appears to enhance the patient's immune response against the cancer cells. Another therapeutic cancer vaccine (talimogene laherparepvec, also called T-VEC or Imlygic) was approved by the FDA in 2015 for treatment of melanoma, a form of skin cancer. This vaccine contains a virus that is injected into tumors, where it infects and lyses the tumor cells. The virus also induces a response in lesions or tumors besides those into which the vaccine is injected, indicating that it is stimulating a more general (as opposed to local) antitumor immune response in the patient.

CHECK YOUR UNDERSTANDING

- Explain the difference between preventative and therapeutic cancer vaccines.
- Describe at least two different approaches to developing therapeutic anti-cancer vaccines.

MICRO CONNECTIONS

Using Viruses to Cure Cancer

Viruses typically destroy the cells they infect—a fact responsible for any number of human diseases. But the cell-killing powers of viruses may yet prove to be the cure for some types of cancer, which is generally treated by attempting to rid the body of cancerous cells. Several clinical trials are studying the effects of viruses targeted at cancer cells. Reolysin, a drug currently in testing phases, uses reoviruses (respiratory enteric orphan viruses) that can infect and kill cells that have an activated Ras-signaling pathway, a common mutation in cancerous cells. Viruses such as rubeola (the measles virus) can also be genetically engineered to aggressively attack tumor cells. These modified viruses not only bind more specifically to receptors overexpressed on cancer cells, they also carry genes driven by promoters that are only turned on within cancer cells. Herpesvirus and others have also been modified in this way.

---

17 National Institutes of Health, National Cancer Institute. "Cancer Vaccines." http://www.cancer.gov/about-cancer/causes-prevention/vaccines-fact-sheet#q8. Accessed on May 20, 2016.

SUMMARY

19.1 Hypersensitivities

- An **allergy** is an adaptive immune response, sometimes life-threatening, to an **allergen**.
- **Type I hypersensitivity** requires sensitization of mast cells with IgE, involving an initial IgE antibody response and IgE attachment to mast cells. On second exposure to an allergen, cross-linking of IgE molecules on mast cells triggers degranulation and release of preformed and newly formed chemical mediators of inflammation. Type I hypersensitivity may be localized and relatively minor (hives and hay fever) or system-wide and dangerous (systemic **anaphylaxis**).
- **Type II hypersensitivities** result from antibodies binding to antigens on cells and initiating cytotoxic responses. Examples include **hemolytic transfusion reaction** and **hemolytic disease of the newborn**.
- **Type III hypersensitivities** result from formation and accumulation of **immune complexes** in tissues, stimulating damaging inflammatory responses.
- **Type IV hypersensitivities** are not mediated by antibodies, but by helper T-cell activation of macrophages, eosinophils, and cytotoxic T cells.

19.2 Autoimmune Disorders

- **Autoimmune diseases** result from a breakdown in immunological tolerance. The actual induction event(s) for autoimmune states are largely unknown.
- Some autoimmune diseases attack specific organs, whereas others are more systemic.
- Organ-specific autoimmune diseases include **celiac disease, Graves disease, Hashimoto thyroiditis, type I diabetes mellitus,** and **Addison disease**.
- Systemic autoimmune diseases include **multiple sclerosis, myasthenia gravis, psoriasis, rheumatoid arthritis,** and **systemic lupus erythematosus**.
- Treatments for autoimmune diseases generally involve anti-inflammatory and immunosuppressive drugs.

19.3 Organ Transplantation and Rejection

- Grafts and transplants can be classified as **autografts, isografts, allografts,** or **xenografts** based on the genetic differences between the donor's and recipient's tissues.
- Genetic differences, especially among the MHC (HLA) genes, will dictate the likelihood that **rejection** of the transplanted tissue will occur.
- Transplant recipients usually require immunosuppressive therapy to avoid rejection, even with good genetic matching. This can create additional problems when immune responses are needed to fight off infectious agents and prevent cancer.
- **Graft-versus-host disease** can occur in bone marrow transplants, as the mature T cells in the transplant itself recognize the recipient's tissues as foreign.
- Transplantation methods and technology have improved greatly in recent decades and may move into new areas with the use of stem cell technology to avoid the need for genetic matching of MHC molecules.

19.4 Immunodeficiency

- **Primary immunodeficiencies** are caused by genetic abnormalities; **secondary immunodeficiencies** are acquired through disease, diet, or environmental exposures
- Primary immunodeficiencies may result from flaws in phagocyte killing of innate immunity, or impairment of T cells and B cells.
- Primary immunodeficiencies include chronic granulomatous disease, X-linked agammaglobulinemia, selective IgA deficiency, and severe combined immunodeficiency disease.
- Secondary immunodeficiencies result from environmentally induced defects in B cells and/or T cells.
- Causes for secondary immunodeficiencies include malnutrition, viral infection, diabetes, prolonged infections, and chemical or radiation exposure.

19.5 Cancer Immunobiology and Immunotherapy

- Cancer results from a loss of control of the cell cycle, resulting in uncontrolled cell proliferation and a loss of the ability to differentiate.
- Adaptive and innate immune responses are engaged by **tumor** antigens, self-molecules only found on abnormal cells. These adaptive responses stimulate helper T cells to activate cytotoxic T cells and NK cells of innate

- immunity that will seek and destroy cancer cells.
- New anticancer therapies are in development that will exploit natural adaptive immunity anticancer responses. These include external stimulation of cytotoxic T cells and therapeutic vaccines that assist or enhance the immune response.

REVIEW QUESTIONS

Multiple Choice

1. Which of the following is the type of cell largely responsible for type I hypersensitivity responses?
   A. erythrocyte
   B. mast cell
   C. T lymphocyte
   D. antibody

2. Type I hypersensitivities require which of the following initial priming events to occur?
   A. sensitization
   B. secondary immune response
   C. cellular trauma
   D. degranulation

3. Which of the following are the main mediators/initiators of type II hypersensitivity reactions?
   A. antibodies
   B. mast cells
   C. erythrocytes
   D. histamines

4. Inflammatory molecules are released by mast cells in type I hypersensitivities; type II hypersensitivities, however, are characterized by which of the following?
   A. cell lysis (cytotoxicity)
   B. strong antibody reactions against antigens
   C. leukotriene release upon stimulation
   D. localized tissue reactions, such as hives

5. An immune complex is an aggregate of which of the following?
   A. antibody molecules
   B. antigen molecules
   C. antibody and antigen molecules
   D. histamine molecules

6. Which of the following is a common treatment for type III hypersensitivity reactions?
   A. anti-inflammatory steroid treatments
   B. antihistamine treatments
   C. hyposensitization injections of allergens
   D. RhoGAM injections

7. Which of the following induces a type III hypersensitivity?
   A. release of inflammatory molecules from mast cells
   B. accumulation of immune complexes in tissues and small blood vessels
   C. destruction of cells bound by antigens
   D. destruction of cells bound by antibodies

8. Which one of the following is not an example of a type IV hypersensitivity?
   A. latex allergy
   B. Contact dermatitis (e.g., contact with poison ivy)
   C. a positive tuberculin skin test
   D. hemolytic disease of the newborn

9. Which of the following is an example of an organ-specific autoimmune disease?
   A. rheumatoid arthritis
   B. psoriasis
   C. Addison disease
   D. myasthenia gravis

10. Which of the following is an example of a systemic autoimmune disease?
    A. Hashimoto thyroiditis
    B. type I diabetes mellitus
    C. Graves disease
    D. myasthenia gravis

11. Which of the following is a genetic disease that results in lack of production of antibodies?
    A. agammaglobulinemia
    B. myasthenia gravis
    C. HIV/AIDS
    D. chronic granulomatous disease

12. Which of the following is a genetic disease that results in almost no adaptive immunity due to lack of B and/or T cells?
    A. agammaglobulinemia
    B. severe combined immunodeficiency
    C. HIV/AIDS
    D. chronic granulomatous disease

13. All but which one of the following are examples of secondary immunodeficiencies?
    A. HIV/AIDS
    B. malnutrition
    C. chronic granulomatous disease
    D. immunosuppression due to measles infection

14. Cancer results when a mutation leads to which of the following?
    A. cell death
    B. apoptosis
    C. loss of cell-cycle control
    D. shutdown of the cell cycle

15. Tumor antigens are _____ that are inappropriately expressed and found on abnormal cells.
    A. self antigens
    B. foreign antigens
    C. antibodies
    D. T-cell receptors

Matching

16. Match the graft with its description.

    ___autograft   A. donor is a different species than the recipient

    ___allograft   B. donor and recipient are the same individual

    ___xenograft   C. donor is an identical twin of the recipient

    ___isograft   D. donor is the same species as the recipient, but genetically different

Fill in the Blank

17. Antibodies involved in type I hypersensitivities are of the _____ class.
18. Allergy shots work by shifting antibody responses to produce _____ antibodies.
19. A person who is blood type A would have IgM hemagglutinin antibodies against type _____ red blood cells in their plasma.
20. The itchy and blistering rash that develops with contact to poison ivy is caused by a type _____ hypersensitivity reaction.
21. The thyroid-stimulating immunoglobulin that acts like thyroid-stimulating hormone and causes Graves disease is an antibody to the _____.
22. For a transplant to have the best chances of avoiding rejection, the genes coding for the _____ molecules should be closely matched between donor and recipient.
23. Because it is a "transplant" that can include APCs and T cells from the donor, a bone marrow transplant may induce a very specific type of rejection known as _____ disease.
24. Diseases due to _____ abnormalities are termed primary immunodeficiencies.
25. A secondary immunodeficiency is _____, rather than genetic.
26. A _____ cancer vaccine is one that stops the disease from occurring in the first place.
27. A _____ cancer vaccine is one that will help to treat the disease after it has occurred.

Short Answer

28. Although both type I and type II hypersensitivities involve antibodies as immune effectors, different mechanisms are involved with these different hypersensitivities. Differentiate the two.
29. What types of antibodies are most common in type III hypersensitivities, and why?
30. Why is a parent usually a better match for transplanted tissue to a donor than a random individual of the same species?
31. Compare the treatments for primary and secondary immunodeficiencies.
32. How can tumor antigens be effectively targeted without inducing an autoimmune (anti-self) response?

Critical Thinking

33. Patients are frequently given instructions to avoid allergy medications for a period of time prior to allergy testing. Why would this be important?
34. In some areas of the world, a tuberculosis vaccine known as bacillus Calmette-Guérin (BCG) is used. It is not used in the United States. Every person who has received this vaccine and mounted a protective response will have a positive reaction in a tuberculin skin test. Why? What does this mean for the usefulness of this skin test in those countries where this vaccine is used?

CHAPTER 20
Laboratory Analysis of the Immune Response

Figure 20.1 Lab-on-a-chip technology allows immunological assays to be miniaturized so tests can be done rapidly with minimum quantities of expensive reagents. The chips contain tiny flow tubes to allow movement of fluids by capillary action, reactions sites with embedded reagents, and data output through electronic sensors. (credit: modification of work by Maggie Bartlett, NHGRI)

Chapter Outline

20.1 Polyclonal and Monoclonal Antibody Production

20.2 Detecting Antigen-Antibody Complexes

20.3 Agglutination Assays

20.4 EIAs and ELISAs

20.5 Fluorescent Antibody Techniques

INTRODUCTION Many laboratory tests are designed to confirm a presumptive diagnosis by detecting antibodies specific to a suspected pathogen. Unfortunately, many such tests are time-consuming and expensive. That is now changing, however, with the development of new, miniaturized technologies that are fast and inexpensive. For example, researchers at Columbia University are developing a "lab-on-a-chip" technology that will test a single drop of blood for 15 different infectious diseases, including HIV and syphilis, in a matter of minutes.[1] The blood is pulled through tiny capillaries into reaction chambers where the patient's antibodies mix with reagents. A chip reader that attaches to a cell phone analyzes the results and sends them to the patient's healthcare provider. Currently the device is being field tested in Rwanda to check pregnant women for chronic diseases. Researchers estimate that the chip readers will sell for about $100 and individual chips for $1.[2]

20.1 Polyclonal and Monoclonal Antibody Production

Learning Objectives

By the end of this section, you will be able to:
- Compare the method of development, use, and characteristics of monoclonal and polyclonal antibodies
- Explain the nature of antibody cross-reactivity and why this is less of a problem with monoclonal antibodies

---

Clinical Focus

Part 1

In an unfortunate incident, a healthcare worker struggling with addiction was caught stealing syringes of painkillers and replacing them with syringes filled with unknown substances. The hospital immediately fired the employee and had him arrested; however, two patients that he had worked with later tested positive for HIV.

While there was no proof that the infections originated from the tainted syringes, the hospital's public health physician took immediate steps to determine whether any other patients had been put at risk. Although the worker had only been employed for a short time, it was determined that he had come into contact with more than 1300 patients. The hospital decided to contact all of these patients and have them tested for HIV.

- Why does the hospital feel it is necessary to test every patient for HIV?
- What types of tests can be used to determine if a patient has HIV?

Jump to the next Clinical Focus box.

In addition to being crucial for our normal immune response, antibodies provide powerful tools for research and diagnostic purposes. The high specificity of antibodies makes them an excellent tool for detecting and quantifying a broad array of targets, from drugs to serum proteins to microorganisms. With *in vitro* assays, antibodies can be used to precipitate soluble antigens, agglutinate (clump) cells, opsonize and kill bacteria with the assistance of complement, and neutralize drugs, toxins, and viruses.

An antibody's **specificity** results from the antigen-binding site formed within the variable regions—regions of the antibody that have unique patterns of amino acids that can only bind to target antigens with a molecular sequence that provides complementary charges and noncovalent bonds. There are limitations to antibody specificity, however. Some antigens are so chemically similar that cross-reactivity occurs; in other words, antibodies raised against one antigen bind to a chemically similar but different antigen. Consider an antigen that consists of a single protein with multiple epitopes (Figure 20.2). This single protein may stimulate the production of many different antibodies, some of which may bind to chemically identical epitopes on other

---

1 Chin, Curtis D. et al., "Mobile Device for Disease Diagnosis and Data Tracking in Resource-Limited Settings," *Clinical Chemistry* 59, no. 4 (2013): 629-40.

2 Evarts, H., "Fast, Low-Cost Device Uses the Cloud to Speed Up Testing for HIV and More," January 24, 2013. Accessed July 14, 2016. http://engineering.columbia.edu/fast-low-cost-device-uses-cloud-speed-diagnostic-testing-hiv-and-more.

proteins.

Cross-reactivity is more likely to occur between antibodies and antigens that have low **affinity** or **avidity**. Affinity, which can be determined experimentally, is a measure of the binding strength between an antibody's binding site and an epitope, whereas avidity is the total strength of all the interactions in an antibody-antigen complex (which may have more than one bonding site). Avidity is influenced by affinity as well as the structural arrangements of the epitope and the variable regions of the antibody. If an antibody has a high affinity/avidity for a specific antigen, it is less likely to cross-react with an antigen for which it has a lower affinity/avidity.

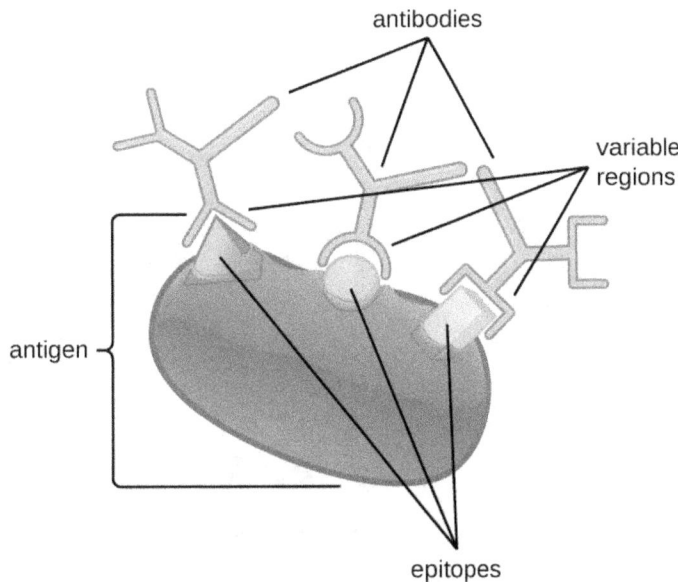

Figure 20.2 An antibody binds to a specific region on an antigen called an epitope. A single antigen can have multiple epitopes for different, specific antibodies.

✓ CHECK YOUR UNDERSTANDING

- What property makes antibodies useful for research and clinical diagnosis?
- What is cross-reactivity and why does it occur?

Producing Polyclonal Antibodies

Antibodies used for research and diagnostic purposes are often obtained by injecting a lab animal such as a rabbit or a goat with a specific antigen. Within a few weeks, the animal's immune system will produce high levels of antibodies specific for the antigen. These antibodies can be harvested in an **antiserum**, which is whole serum collected from an animal following exposure to an antigen. Because most antigens are complex structures with multiple epitopes, they result in the production of multiple antibodies in the lab animal. This so-called **polyclonal antibody** response is also typical of the response to infection by the human immune system. Antiserum drawn from an animal will thus contain antibodies from multiple clones of B cells, with each B cell responding to a specific epitope on the antigen (Figure 20.3).

Lab animals are usually injected at least twice with antigen when being used to produce antiserum. The second injection will activate memory cells that make class IgG antibodies against the antigen. The memory cells also undergo **affinity maturation**, resulting in a pool of antibodies with higher average affinity. Affinity maturation occurs because of mutations in the immunoglobulin gene variable regions, resulting in B cells with slightly altered antigen-binding sites. On re-exposure to the antigen, those B cells capable of producing antibody with higher affinity antigen-binding sites will be stimulated to proliferate and produce more antibody than their lower-affinity peers. An adjuvant, which is a chemical that provokes a generalized activation of the immune system that stimulates greater antibody production, is often mixed with the antigen prior to injection.

Antiserum obtained from animals will not only contain antibodies against the antigen artificially introduced in the laboratory, but it will also contain antibodies to any other antigens to which the animal has been exposed during its lifetime. For this reason, antisera must first be "purified" to remove other antibodies before using the antibodies for research or diagnostic assays.

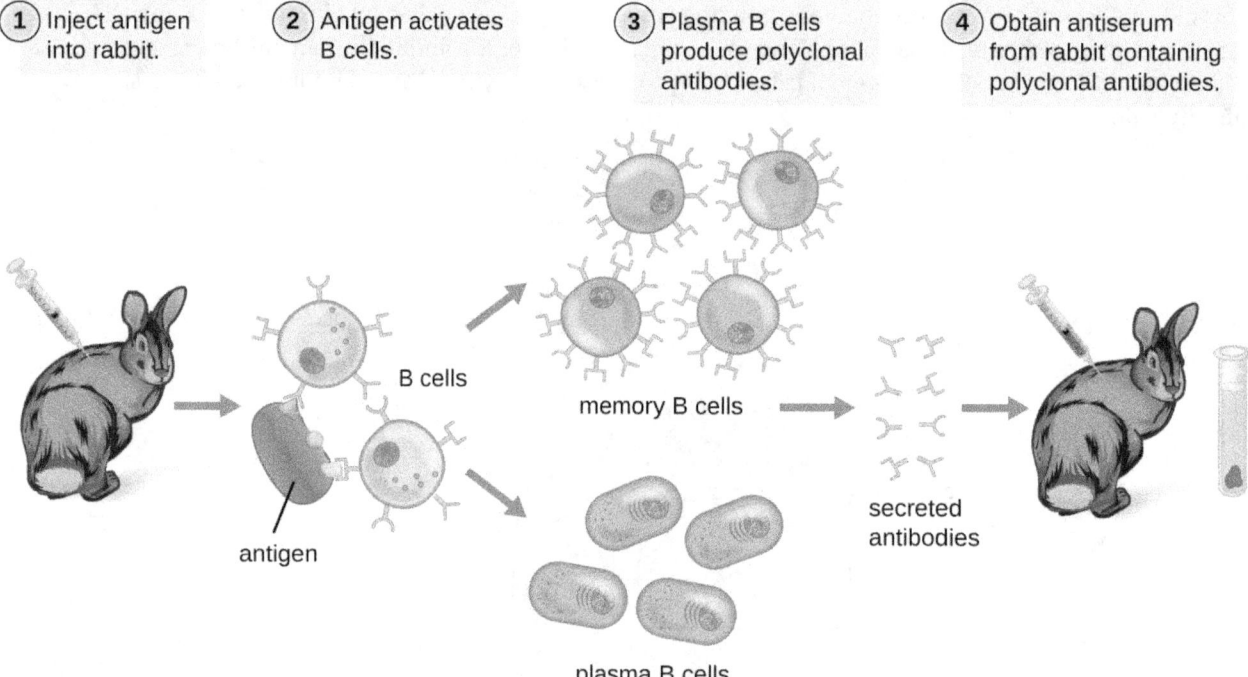

Figure 20.3 This diagram illustrates the process for harvesting polyclonal antibodies produced in response to an antigen.

Clinical Uses of Polyclonal Antisera

Polyclonal antisera are used in many clinical tests that are designed to determine whether a patient is producing antibodies in response to a particular pathogen. While these tests are certainly powerful diagnostic tools, they have their limitations, because they are an indirect means of determining whether a particular pathogen is present. Tests based on a polyclonal response can sometimes lead to a **false-positive** result—in other words, a test that confirms the presence of an antigen that is, in fact, not present. Antibody-based tests can also result in a **false-negative** result, which occurs when the test fails to detect an antibody that is, in fact, present.

The accuracy of antibody tests can be described in terms of **test sensitivity** and **test specificity**. Test sensitivity is the probability of getting a positive test result when the patient is indeed infected. If a test has high sensitivity, the probability of a false negative is low. Test specificity, on the other hand, is the probability of getting a negative test result when the patient is not infected. If a test has high specificity, the probability of a false positive is low.

False positives often occur due to cross-reactivity, which can occur when epitopes from a different pathogen are similar to those found on the pathogen being tested for. For this reason, antibody-based tests are often used only as screening tests; if the results are positive, other confirmatory tests are used to make sure that the results were not a false positive.

For example, a blood sample from a patient suspected of having hepatitis C can be screened for the virus using antibodies that bind to antigens on hepatitis C virus. If the patient is indeed infected with hepatitis C virus, the antibodies will bind to the antigens, yielding a positive test result. If the patient is not infected with hepatitic C virus, the antibodies will generally not bind to anything and the test should be negative; however, a false positive may occur if the patient has been previously infected by any of a variety of pathogens that elicit antibodies that cross-react with the hepatitis C virus antigens. Antibody tests for hepatitis C have high sensitivity (a low probability of a false negative) but low specificity (a high probability of a false positive). Thus, patients who test positive must have a second, confirmatory test to rule out the possibility of a false positive. The confirmatory test is a more expensive and time-consuming test that directly tests for the presence of hepatitis C viral RNA in the blood. Only after the confirmatory test comes back positive can the patient be definitively diagnosed with a hepatitis C infection. Antibody-based tests can result in a false negative if, for any reason, the patient's immune system has not produced detectable levels of antibodies. For some diseases, it may take several weeks following infection before the immune system produces enough antibodies to cross the detection threshold of the assay. In immunocompromised patients, the immune system may not be capable of producing a detectable level of antibodies.

Another limitation of using antibody production as an indicator of disease is that antibodies in the blood will persist long after the infection has been cleared. Depending on the type of infection, antibodies will be present for many months; sometimes, they may be present for the remainder of the patient's life. Thus, a positive antibody-based test only means that the patient was infected at some point in time; it does not prove that the infection is active.

In addition to their role in diagnosis, polyclonal antisera can activate complement, detect the presence of bacteria in clinical and food industry settings, and perform a wide array of precipitation reactions that can detect and quantify serum proteins, viruses, or other antigens. However, with the many specificities of antibody present in a polyclonal antiserum, there is a significant likelihood that the antiserum will cross-react with antigens to which the individual was never exposed. Therefore, we must always account for the possibility of false-positive results when working with a polyclonal antiserum.

✓ CHECK YOUR UNDERSTANDING

- What is a false positive and what are some reasons that false positives occur?
- What is a false negative and what are some reasons that false positives occur?
- If a patient tests negative on a highly sensitive test, what is the likelihood that the person is infected with the pathogen?

Producing Monoclonal Antibodies

Some types of assays require better antibody specificity and affinity than can be obtained using a polyclonal antiserum. To attain this high specificity, all of the antibodies must bind with high affinity to a single epitope. This high specificity can be provided by **monoclonal antibodies (mAbs)**. Table 20.1 compares some of the important characteristics of monoclonal and polyclonal antibodies.

Unlike polyclonal antibodies, which are produced in live animals, monoclonal antibodies are produced *in vitro* using tissue-culture techniques. mAbs are produced by immunizing an animal, often a mouse, multiple times with a specific antigen. B cells from the spleen of the immunized animal are then removed. Since normal B cells are unable to proliferate forever, they are fused with immortal, cancerous B cells called myeloma cells, to yield **hybridoma** cells. All of the cells are then placed in a selective medium that allows only the hybridomas to grow; unfused myeloma cells cannot grow, and any unfused B cells die off. The hybridomas, which are capable of growing continuously in culture while producing antibodies, are then screened for the desired mAb. Those producing the desired mAb are grown in tissue culture; the culture medium is harvested periodically and mAbs are purified from the medium. This is a very expensive and time-consuming process. It may take weeks of culturing and many liters of media to provide enough mAbs for an experiment or to treat a single patient. mAbs are expensive (Figure 20.4).

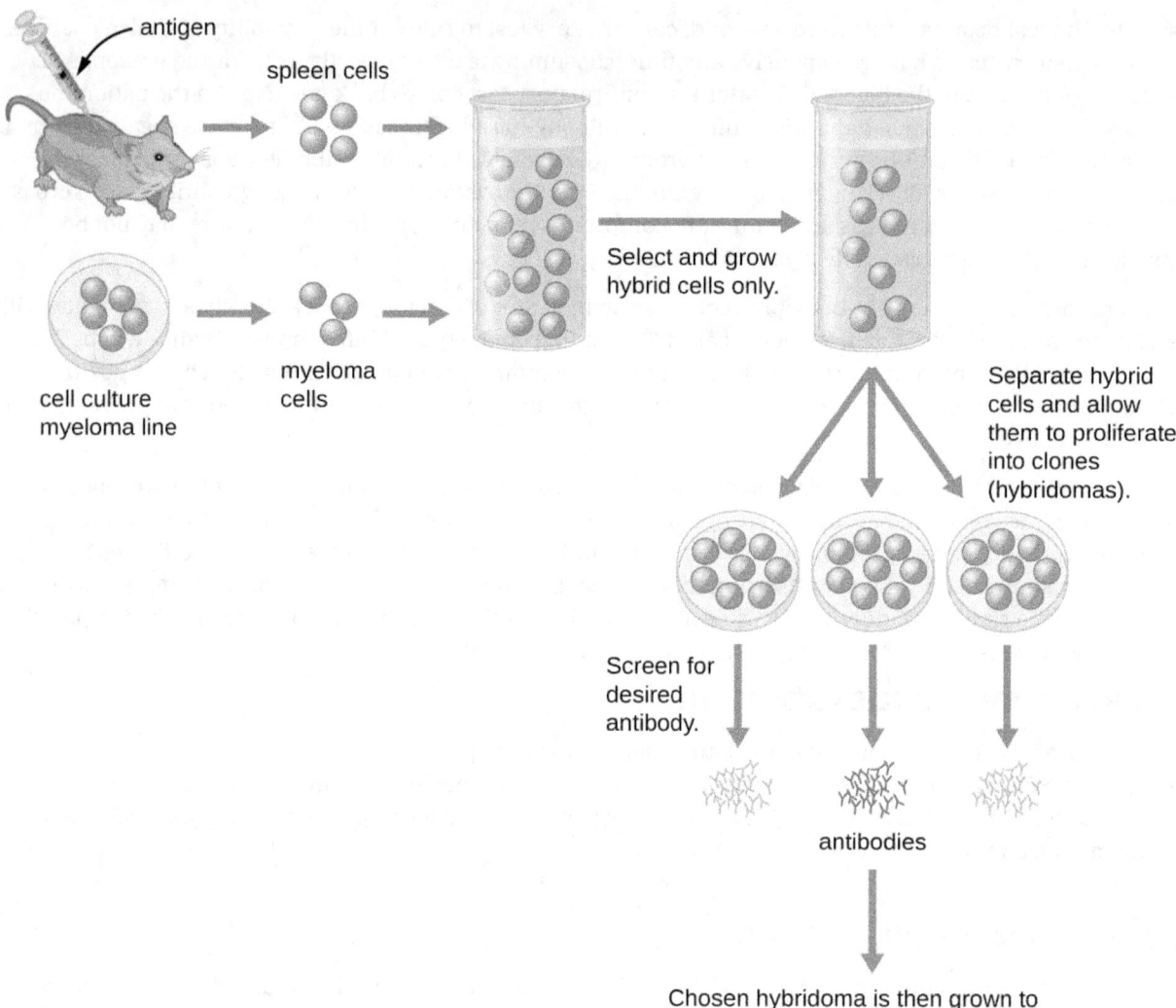

Figure 20.4 Monoclonal antibodies (mAbs) are produced by introducing an antigen to a mouse and then fusing polyclonal B cells from the mouse's spleen to myeloma cells. The resulting hybridoma cells are cultured and continue to produce antibodies to the antigen. Hybridomas producing the desired mAb are then grown in large numbers on a selective medium that is periodically harvested to obtain the desired mAbs.

Characteristics of Polyclonal and Monoclonal Antibodies

Monoclonal Antibodies	Polyclonal Antibodies
Expensive production	Inexpensive production
Long production time	Rapid production
Large quantities of specific antibodies	Large quantities of nonspecific antibodies
Recognize a single epitope on an antigen	Recognize multiple epitopes on an antigen

Characteristics of Polyclonal and Monoclonal Antibodies	
Monoclonal Antibodies	**Polyclonal Antibodies**
Production is continuous and uniform once the hybridoma is made	Different batches vary in composition

Table 20.1

Clinical Uses of Monoclonal Antibodies

Since the most common methods for producing monoclonal antibodies use mouse cells, it is necessary to create **humanized monoclonal antibodies** for human clinical use. Mouse antibodies cannot be injected repeatedly into humans, because the immune system will recognize them as being foreign and will respond to them with neutralizing antibodies. This problem can be minimized by genetically engineering the antibody in the mouse B cell. The variable regions of the mouse light and heavy chain genes are ligated to human constant regions, and the chimeric gene is then transferred into a host cell. This allows production of a mAb that is mostly "human" with only the antigen-binding site being of mouse origin.

Humanized mAbs have been successfully used to treat cancer with minimal side effects. For example, the humanized monoclonal antibody drug Herceptin has been helpful for the treatment of some types of breast cancer. There have also been a few preliminary trials of humanized mAb for the treatment of infectious diseases, but none of these treatments are currently in use. In some cases, mAbs have proven too specific to treat infectious diseases, because they recognize some serovars of a pathogen but not others. Using a cocktail of multiple mAbs that target different strains of the pathogen can address this problem. However, the great cost associated with mAb production is another challenge that has prevented mAbs from becoming practical for use in treating microbial infections.[3]

One promising technology for inexpensive mAbs is the use of genetically engineered plants to produce antibodies (or **plantibodies**). This technology transforms plant cells into antibody factories rather than relying on tissue culture cells, which are expensive and technically demanding. In some cases, it may even be possible to deliver these antibodies by having patients eat the plants rather than by extracting and injecting the antibodies. For example, in 2013, a research group cloned antibody genes into plants that had the ability to neutralize an important toxin from bacteria that can cause severe gastrointestinal disease.[4] Eating the plants could potentially deliver the antibodies directly to the toxin.

✓ CHECK YOUR UNDERSTANDING

- How are humanized monoclonal antibodies produced?
- What does the "monoclonal" of monoclonal antibodies mean?

---

3 Saylor, Carolyn, Ekaterina Dadachova and Arturo Casadevall, "Monoclonal Antibody-Based Therapies for Microbial Diseases," *Vaccine* 27 (2009): G38-G46.

4 Nakanishi, Katsuhiro et al., "Production of Hybrid-IgG/IgA Plantibodies with Neutralizing Activity against Shiga Toxin 1," *PloS One* 8, no. 11 (2013): e80712.

 **MICRO CONNECTIONS**

Using Monoclonal Antibodies to Combat Ebola

During the 2014–2015 Ebola outbreak in West Africa, a few Ebola-infected patients were treated with ZMapp, a drug that had been shown to be effective in trials done in rhesus macaques only a few months before.[5] ZMapp is a combination of three mAbs produced by incorporating the antibody genes into tobacco plants using a viral vector. By using three mAbs, the drug is effective across multiple strains of the virus. Unfortunately, there was only enough ZMapp to treat a tiny number of patients.

While the current technology is not adequate for producing large quantities of ZMapp, it does show that plantibodies—plant-produced mAbs—are feasible for clinical use, potentially cost effective, and worth further development. The last several years have seen an explosion in the number of new mAb-based drugs for the treatment of cancer and infectious diseases; however, the widespread use of such drugs is currently inhibited by their exorbitant cost, especially in underdeveloped parts of the world, where a single dose might cost more than the patient's lifetime income. Developing methods for cloning antibody genes into plants could reduce costs dramatically.

20.2 Detecting Antigen-Antibody Complexes

Learning Objectives

By the end of this section, you will be able to:
- Describe various types of assays used to find antigen-antibody complexes
- Describe the circumstances under which antigen-antibody complexes precipitate out of solution
- Explain how antibodies in patient serum can be used to diagnose disease

Laboratory tests to detect antibodies and antigens outside of the body (e.g., in a test tube) are called *in vitro* assays. When both antibodies and their corresponding antigens are present in a solution, we can often observe a precipitation reaction in which large complexes (lattices) form and settle out of solution. In the next several sections, we will discuss several common *in vitro* assays.

Precipitin Reactions

A visible antigen-antibody complex is called a **precipitin**, and *in vitro* assays that produce a precipitin are called precipitin reactions. A precipitin reaction typically involves adding soluble antigens to a test tube containing a solution of antibodies. Each antibody has two arms, each of which can bind to an epitope. When an antibody binds to two antigens, the two antigens become bound together by the antibody. A lattice can form as antibodies bind more and more antigens together, resulting in a precipitin (Figure 20.5). Most precipitin tests use a polyclonal antiserum rather than monoclonal antibodies because polyclonal antibodies can bind to multiple epitopes, making lattice formation more likely. Although mAbs may bind some antigens, the binding will occur less often, making it much less likely that a visible precipitin will form.

---

5 Qiu, Xiangguo et al., "Reversion of Advanced Ebola Virus Disease in Nonhuman Primates with ZMapp," *Nature* 514 (2014): 47–53.

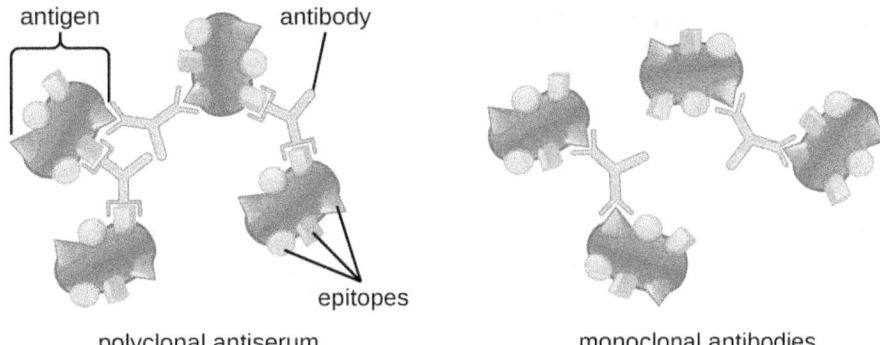

Figure 20.5 Polyclonal antiserum binds to multiple epitopes on an antigen, leading to lattice formation that results in a visible precipitin. Monoclonal antibodies can only bind to a single epitope; therefore, less binding occurs and lattice formation generally does not occur.

The amount of precipitation also depends on several other factors. For example, precipitation is enhanced when the antibodies have a high affinity for the antigen. While most antibodies bind antigen with high affinity, even high-affinity binding uses relatively weak noncovalent bonds, so that individual interactions will often break and new interactions will occur.

In addition, for precipitin formation to be visible, there must be an optimal ratio of antibody to antigen. The optimal ratio is not likely to be a 1:1 antigen-to-antibody ratio; it can vary dramatically, depending on the number of epitopes on the antigen and the class of antibody. Some antigens may have only one or two epitopes recognized by the antiserum, whereas other antigens may have many different epitopes and/or multiple instances of the same epitope on a single antigen molecule.

Figure 20.6 illustrates how the ratio of antigen and antibody affects the amount of precipitation. To achieve the optimal ratio, antigen is slowly added to a solution containing antibodies, and the amount of precipitin is determined qualitatively. Initially, there is not enough antigen to produce visible lattice formation; this is called the zone of antibody excess. As more antigen is added, the reaction enters the **equivalence zone** (or zone of equivalence), where both the optimal antigen-antibody interaction and maximal precipitation occur. If even more antigen were added, the amount of antigen would become excessive and actually cause the amount of precipitation to decline.

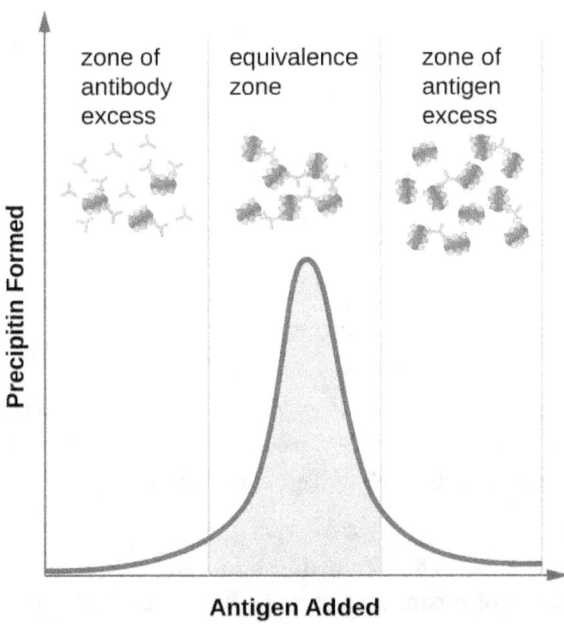

Figure 20.6 As antigen is slowly added to a solution containing a constant amount antibody, the amount of precipitin increases as the antibody-to-antigen ratio approaches the equivalence zone and decreases once the proportion of antigen exceeds the optimal ratio.

CHECK YOUR UNDERSTANDING
- What is a precipitin?
- Why do polyclonal antisera produce a better precipitin reaction?

Precipitin Ring Test

A variety of techniques allow us to use precipitin formation to quantify either antigen concentration or the amount of antibody present in an antiserum. One such technique is the **precipitin ring test** (Figure 20.7), which is used to determine the relative amount of antigen-specific antibody in a sample of serum. To perform this test, a set of test tubes is prepared by adding an antigen solution to the bottom of each tube. Each tube receives the same volume of solution, and the concentration of antigens is constant (e.g., 1 mg/mL). Next, glycerol is added to the antigen solution in each test tube, followed by a serial dilution of the antiserum. The glycerol prevents mixing of the antiserum with the antigen solution, allowing antigen-antibody binding to take place only at the interface of the two solutions. The result is a visible ring of precipitin in the tubes that have an antigen-antibody ratio within the equivalence zone. This highest dilution with a visible ring is used to determine the **titer** of the antibodies. The titer is the reciprocal of the highest dilution showing a positive result, expressed as a whole number. In Figure 20.7, the titer is 16.

While a measurement of titer does not tell us in absolute terms how much antibody is present, it does give a measure of biological activity, which is often more important than absolute amount. In this example, it would not be useful to know what mass of IgG were present in the antiserum, because there are many different specificities of antibody present; but it is important for us to know how much of the antibody activity in a patient's serum is directed against the antigen of interest (e.g., a particular pathogen or allergen).

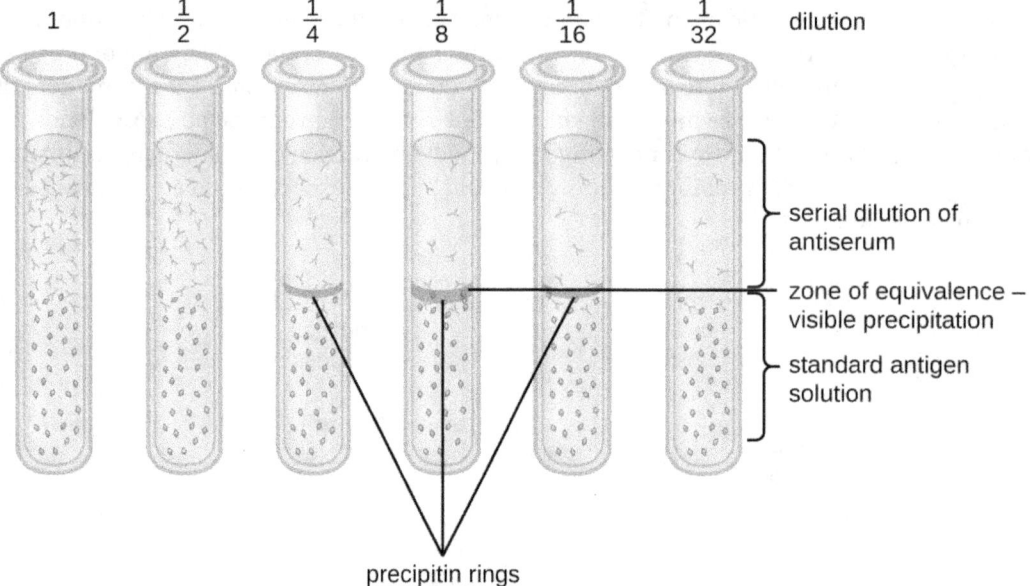

Figure 20.7 A precipitin ring test is performed using a standard antigen solution in the bottom of the tube and a serial dilution of antiserum in the top of the tube. Glycerol prevents the two solutions from mixing so that precipitation only occurs at the interface. A visible ring of precipitation is seen in the 1/4, 1/8, and 1/16 dilutions, indicating that these concentrations are within the equivalence zone. Since 1/16 is the highest dilution in which a precipitin is observed, the titer is the reciprocal, or 16.

Ouchterlony Assay

While the precipitin ring test provides insights into antibody-antigen interactions, it also has some drawbacks. It requires the use of large amounts of serum, and great care must be taken to avoid mixing the solutions and disrupting the ring. Performing a similar test in an agar gel matrix can minimize these problems. This type of assay is variously called **double immunodiffusion** or the **Ouchterlony assay** for Orjan Ouchterlony,[6] who first described the technique in 1948.

When agar is highly purified, it produces a clear, colorless gel. Holes are punched in the gel to form wells, and antigen and antisera are added to neighboring wells. Proteins are able to diffuse through the gel, and precipitin arcs form between the wells at the zone of equivalence. Because the precipitin lattice is too large to diffuse through the gel, the arcs are firmly locked in place and easy to see (Figure 20.8).

Although there are now more sensitive and quantitative methods of detecting antibody-antigen interactions, the Ouchterlony test provides a rapid and qualitative way of determining whether an antiserum has antibodies against a particular antigen. The Ouchterlony test is particularly useful when looking for cross-reactivity. We can check an antiserum against a group of closely related antigens and see which combinations form precipitin arcs.

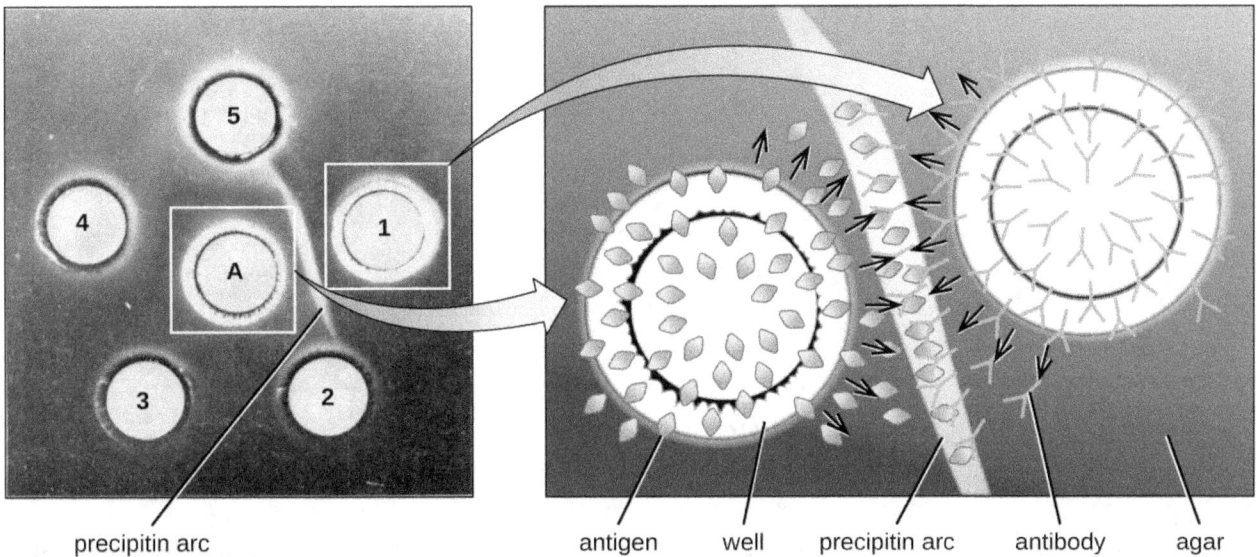

Figure 20.8 The Ouchterlony test places antigen (well A) and antisera (wells 1 through 5) in a gel. The antibodies and antigen diffuse through the gel, causing a precipitin arc to form at the zone of equivalence. In this example, only the antiserum in well 1 contains antibodies to the antigen. The resulting precipitin arc is stable because the lattice is too large to diffuse through the gel. (credit left: modification of work by Higgins PJ, Tong C, Borenfreund E, Okin RS, Bendich A)

Radial Immunodiffusion Assay

The **radial immunodiffusion** (RID) assay is similar to the Ouchterlony assay but is used to precisely quantify antigen concentration rather than to compare different antigens. In this assay, the antiserum is added to tempered agar (liquid agar at slightly above 45 °C), which is poured into a small petri dish or onto a glass slide and allowed to cool. Wells are cut in the cooled agar, and antigen is then added to the wells and allowed to diffuse. As the antigen and antibody interact, they form a zone of precipitation. The square of the diameter of the zone of precipitation is directly proportional to the concentration of antigen. By measuring the zones of precipitation produced by samples of known concentration (see the outer ring of samples in Figure 20.9), we can prepare a standard curve for determining the concentration of an unknown solution. The RID assay is a also useful test for determining the concentration of many serum proteins such as the C3 and C4 complement proteins, among others.

---

6 Ouchterlony, Örjan, "In Vitro Method for Testing the Toxin-Producing Capacity of Diphtheria Bacteria," *Acta Pathologica Microbiologica Scandinavica* 26, no. 4 (1949): 516-24.

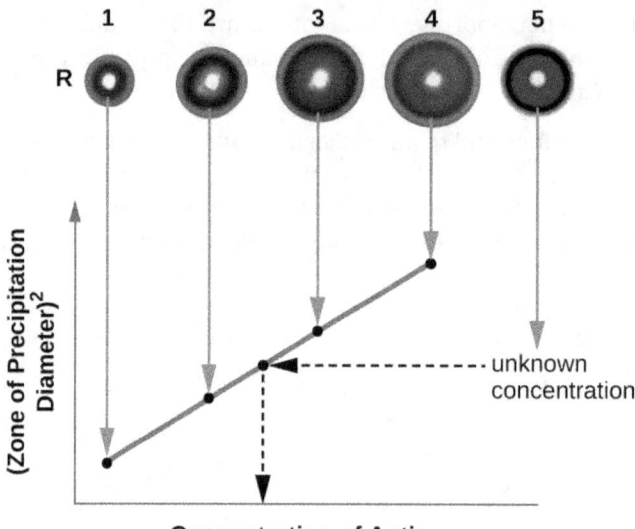

Figure 20.9 In this radial immunodiffusion (RID) assay, an antiserum is mixed with the agar before it is cooled, and solutions containing antigen are added to each well in increasing concentrations (wells 1–4). An antigen solution of an unknown concentration is added to well 5. The zones of precipitation are measured and plotted against a standard curve to determine the antigen concentration of the unknown sample. (credit circles: modification of work by Kangwa M, Yelemane V, Polat AN, Gorrepati KD, Grasselli M, Fernández-Lahore M)

CHECK YOUR UNDERSTANDING

- Why does a precipitin ring form in a precipitin ring test, and what are some reasons why a ring might not form?
- Compare and contrast the techniques used in an Ouchterlony assay and a radial immunodiffusion assay.

Flocculation Assays

A flocculation assay is similar to a precipitin reaction except that it involves insoluble antigens such as lipids. A **flocculant** is similar to a precipitin in that there is a visible lattice of antigen and antibody, but because lipids are insoluble in aqueous solution, they cannot precipitate. Instead of precipitation, flocculation (foaming) is observed in the test tube fluid.

 MICRO CONNECTIONS

Using Flocculation to Test for Syphilis

Syphilis is a sexually transmitted infection that can cause severe, chronic disease in adults. In addition, it is readily passed from infected mothers to their newborns during pregnancy and childbirth, often resulting in stillbirth or serious long-term health problems for the infant. Unfortunately, syphilis can also be difficult to diagnose in expectant mothers, because it is often asymptomatic, especially in women. In addition, the causative agent, the bacterium *Treponema pallidum*, is both difficult to grow on conventional lab media and too small to see using routine microcopy. For these reasons, presumptive diagnoses of syphilis are generally confirmed indirectly in the laboratory using tests that detect antibodies to treponemal antigens.

In 1906, German scientist August von Wassermann (1866–1925) introduced the first test for syphilis that relied on detecting anti-treponemal antibodies in the patient's blood. The antibodies detected in the Wassermann test were antiphospholipid antibodies that are nonspecific to *T. pallidum*. Their presence can assist in the diagnosis of syphilis, but because they are nonspecific, they can also lead to false-positive results in patients with other diseases and autoimmune conditions. The original Wasserman test has been modified over the years to minimize false-positives and is now known as the Venereal Disease Research Lab test, better known by its acronym, the VDRL test.

To perform the VDRL test, patient serum or cerebral spinal fluid is placed on a slide with a mixture of cardiolipin (an antigenic phospholipid found in the mitochondrial membrane of various pathogens), lecithin, and cholesterol. The lecithin and cholesterol stabilize the reaction and diminish false positives. Antitreponemal antibodies from an infected patient's serum will bind cardiolipin and form a flocculant. Although the VDRL test is more specific than the original Wassermann assay, false positives may still occur in patients with autoimmune diseases that cause extensive cell damage (e.g., systemic lupus erythematosus).

Neutralization Assay

To cause infection, viruses must bind to receptors on host cells. Antiviral antibodies can neutralize viral infections by coating the virions, blocking the binding (Figure 18.7). This activity neutralizes virions and can result in the formation of large antibody-virus complexes (which are readily removed by phagocytosis) or by antibody binding to the virus and blocking its binding to host cell receptors. This neutralization activity is the basis of neutralization assays, sensitive assays used for diagnoses of viral infections.

When viruses infect cells, they often cause damage (cytopathic effects) that may include lysis of the host cells. Cytopathic effects can be visualized by growing host cells in a petri dish, covering the cells with a thin layer of agar, and then adding virus (see Isolation, Culture, and Identification of Viruses). The virus will diffuse very slowly through the agar. A virus will enter a host cell, proliferate (causing cell damage), be released from the dead host cell, and then move to neighboring cells. As more and more cells die, plaques of dead cells will form (Figure 20.10).

During the course of a viral infection, the patient will mount an antibody response to the virus, and we can quantify those antibodies using a plaque reduction assay. To perform the assay, a serial dilution is carried out on a serum sample. Each dilution is then mixed with a standardized amount of the suspect virus. Any virus-specific antibodies in the serum will neutralize some of the virus. The suspensions are then added to host cells in culture to allow any nonneutralized virus to infect the cells and form plaques after several days. The titer is defined as the reciprocal of the highest dilution showing a 50% reduction in plaques. Titer is always expressed as a whole number. For example, if a 1/64 dilution was the highest dilution to show 50% plaque reduction, then the titer is 64.

The presence of antibodies in the patient's serum does not tell us whether the patient is currently infected or was infected in the past. Current infections can be identified by waiting two weeks and testing another serum sample. A four-fold increase in neutralizing titer in this second sample indicates a new infection.

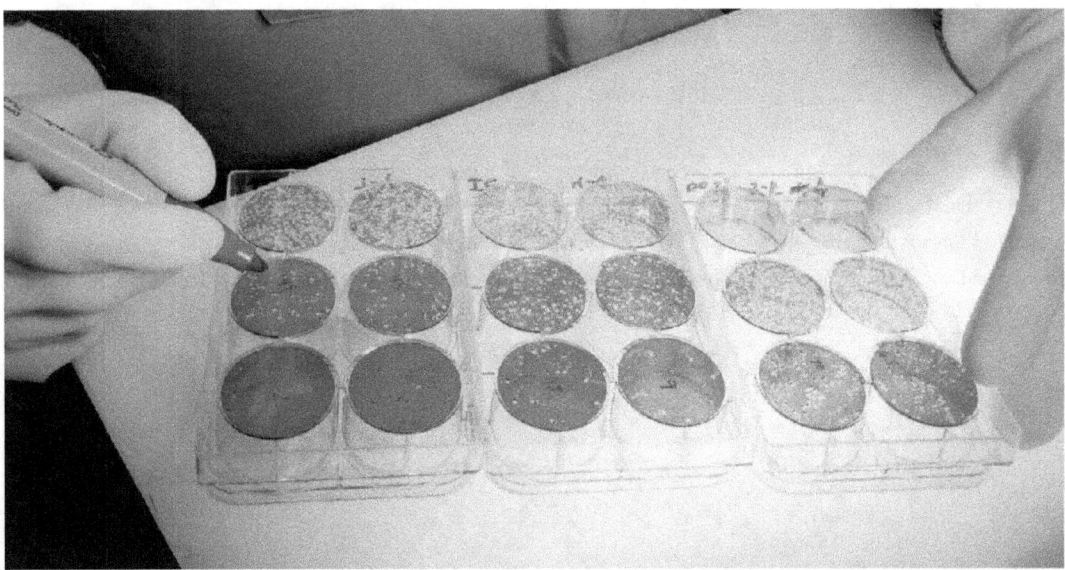

Figure 20.10 In a neutralization assay, antibodies in patient serum neutralize viruses added to the wells, preventing the formation of plaques. In the assay pictured, the wells with numerous plaques (white patches) contain a low concentration of antibodies. The wells with relatively few plaques have a high concentration of antibodies. (credit: modification of work by Centers for Disease Control and Prevention)

CHECK YOUR UNDERSTANDING

- In a neutralization assay, if a patient's serum has high numbers of antiviral antibodies, would you expect to see more or fewer plaques?

Immunoelectrophoresis

When a patient has elevated protein levels in the blood or is losing protein in the urine, a clinician will often order a polyacrylamide gel electrophoresis (PAGE) assay (see Visualizing and Characterizing DNA, RNA, and Protein). This assay compares the relative abundance of the various types of serum proteins. Abnormal protein electrophoresis patterns can be further studied using **immunoelectrophoresis (IEP)**. The IEP begins by running a PAGE. Antisera against selected serum proteins are added to troughs running parallel to the electrophoresis track, forming precipitin arcs similar to those seen in an Ouchterlony assay (Figure 20.11). This allows the identification of abnormal immunoglobulin proteins in the sample.

IEP is particularly useful in the diagnosis of multiple myeloma, a cancer of antibody-secreting cells. Patients with multiple myeloma cannot produce healthy antibodies; instead they produce abnormal antibodies that are monoclonal proteins (M proteins). Thus, patients with multiple myeloma will present with elevated serum protein levels that show a distinct band in the gamma globulin region of a protein electrophoresis gel and a sharp spike (in M protein) on the densitometer scan rather than the normal broad smear (Figure 20.12). When antibodies against the various types of antibody heavy and light chains are used to form precipitin arcs, the M protein will cause distinctly skewed arcs against one class of heavy chain and one class of light chain as seen in Figure 20.11.

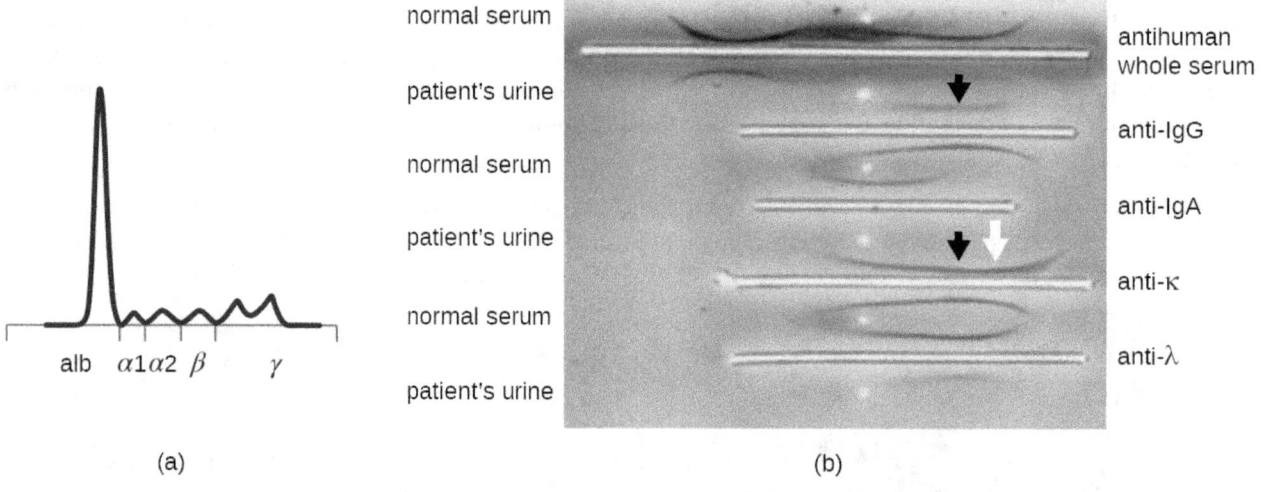

Figure 20.11 (a) This graph shows normal measurements of serum proteins. (b) This photograph shows an immunoelectrophoresis of urine. After electrophoresis, antisera were added to the troughs and the precipitin arcs formed, illustrating the distribution of specific proteins. The skewed arcs (arrows) help to diagnose multiple myeloma. (credit a, b: modification of work by Izawa S, Akimoto T, Ikeuchi H, Kusano E, Nagata D)

MICRO CONNECTIONS

Protein Electrophoresis and the Characterization of Immunoglobulin Structure

The advent of electrophoresis ultimately led to researching and understanding the structure of antibodies. When Swedish biochemist Arne Tiselius (1902–1971) published the first protein electrophoresis results in 1937,[7] he could identify the protein albumin (the smallest and most abundant serum protein) by the sharp band it produced in the gel. The other serum proteins could not be resolved in a simple protein electrophoresis, so he named the three broad bands, with many proteins in each band, alpha, beta, and gamma globulins. Two years later, American immunologist Elvin Kabat (1914–2000) traveled to Sweden to work with Tiselius using this new technique and showed that antibodies migrated as gamma globulins.[8] With this new understanding in hand, researchers soon learned that multiple myeloma, because it is a cancer of antibody-secreting cells, could be tentatively diagnosed by the presence of a large M spike in the gamma-globulin region by protein electrophoresis. Prior to this discovery, studies on immunoglobulin structure had been minimal, because of the difficulty of obtaining pure samples to study. Sera from multiple myeloma patients proved to be an excellent source of highly enriched monoclonal immunoglobulin, providing the raw material for studies over the next 20-plus years that resulted in the elucidation of the structure of immunoglobulin.

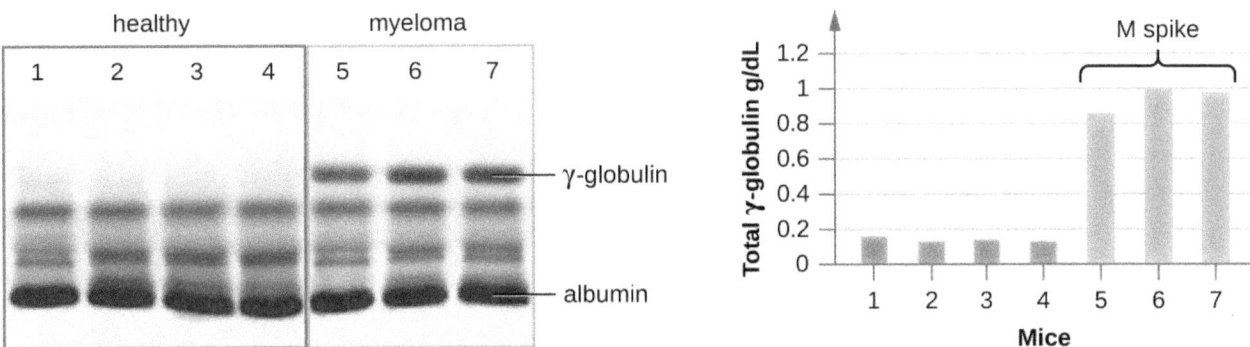

Figure 20.12 Electrophoresis patterns of myeloma (right) and normal sera (left). The proteins have been stained; when the density of each band is quantified by densitometry, the data produce the bar graph on the right. Both gels show the expected dense band of albumin at the bottom and an abnormal spike in the gamma-globulin region. (credit: modification of work by Soodgupta D, Hurchla MA, Jiang M, Zheleznyak A, Weilbaecher KN, Anderson CJ, Tomasson MH, Shokeen M)

CHECK YOUR UNDERSTANDING

- In general, what does an immunoelectrophoresis assay accomplish?

Immunoblot Assay: The Western Blot

After performing protein gel electrophoresis, specific proteins can be identified in the gel using antibodies. This technique is known as the **western blot**. Following separation of proteins by PAGE, the protein antigens in the gel are transferred to and immobilized on a nitrocellulose membrane. This membrane can then be exposed to a primary antibody produced to specifically bind to the protein of interest. A second antibody equipped with a molecular beacon will then bind to the first. These secondary antibodies are coupled to another molecule such as an enzyme or a **fluorophore** (a molecule that fluoresces when excited by light). When using antibodies coupled to enzymes, a **chromogenic substrate** for the enzyme is added. This substrate is usually colorless but will develop color in the presence of the antibody. The fluorescence or substrate

---

[7] Tiselius, Arne, "Electrophoresis of Serum Globulin: Electrophoretic Analysis of Normal and Immune Sera," *Biochemical Journal* 31, no. 9 (1937): 1464.

[8] Tiselius, Arne and Elvin A. Kabat. "An Electrophoretic Study of Immune Sera and Purified Antibody Preparations," *The Journal of Experimental Medicine* 69, no. 1 (1939): 119-31.

coloring identifies the location of the specific protein in the membrane to which the antibodies are bound (Figure 20.13).

Typically, polyclonal antibodies are used for western blot assays. They are more sensitive than mAbs because of their ability to bind to various epitopes of the primary antigen, and the signal from polyclonal antibodies is typically stronger than that from mAbs. Monoclonal antibodies can also be used; however, they are much more expensive to produce and are less sensitive, since they are only able to recognize one specific epitope.

Several variations of the western blot are useful in research. In a southwestern blot, proteins are separated by SDS-PAGE, blotted onto a nitrocellulose membrane, allowed to renature, and then probed with a fluorescently or radioactively labeled DNA probe; the purpose of the southwestern is to identify specific DNA-protein interactions. Far-western blots are carried out to determine protein-protein interactions between immobilized proteins (separated by SDS-PAGE, blotted onto a nitrocellulose membrane, and allowed to renature) and non-antibody protein probes. The bound non-antibody proteins that interact with the immobilized proteins in a far-western blot may be detected by radiolabeling, fluorescence, or the use of an antibody with an enzymatic molecular beacon.

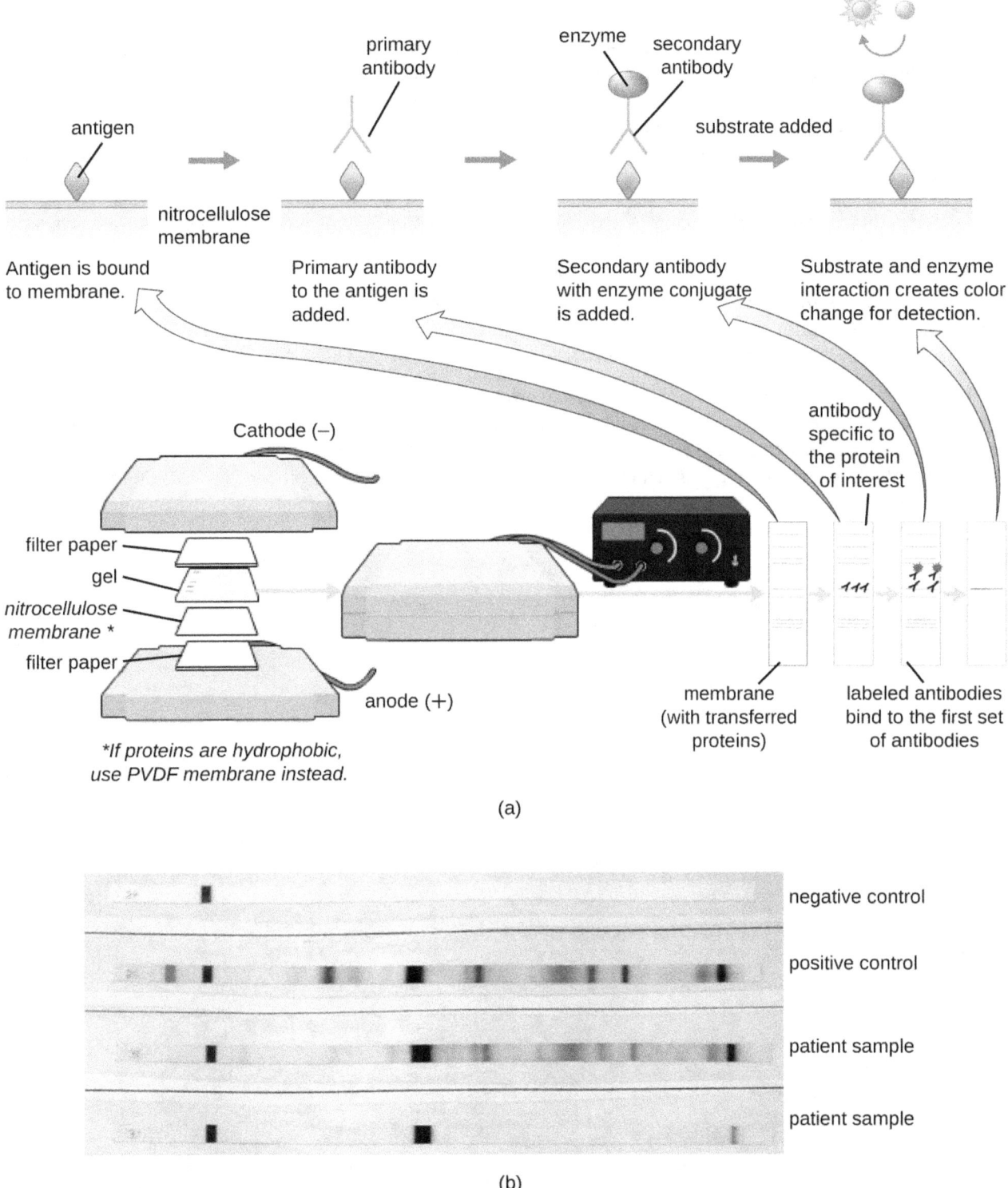

Figure 20.13 (a) This diagram summarizes the process of western blotting. Antibodies are used to identify specific bands on the protein gel. (b) A western blot test for antibodies against HIV. The top strip is the negative control; the next strip is the positive control. The bottom two strips are patient serum samples containing antibodies. (credit a: modification of work by "Bensaccount"/Wikimedia Commons)

CHECK YOUR UNDERSTANDING

- What is the function of the enzyme in the immunoblot assay?

Complement-Mediated Immunoassay

One of the key functions of antibodies is the activation (fixation) of complement. When antibody binds to bacteria, for example, certain complement proteins recognize the bound antibody and activate the complement cascade. In response, other complement proteins bind to the bacteria where some serve as opsonins to increase the efficiency of phagocytosis and others create holes in gram-negative bacterial cell membranes, causing lysis. This lytic activity can be used to detect the presence of antibodies against specific antigens in the serum.

Red blood cells are good indicator cells to use when evaluating complement-mediated cytolysis. Hemolysis of red blood cells releases hemoglobin, which is a brightly colored pigment, and hemolysis of even a small number of red cells will cause the solution to become noticeably pink (Figure 20.14). This characteristic plays a role in the **complement fixation test**, which allows the detection of antibodies against specific pathogens. The complement fixation test can be used to check for antibodies against pathogens that are difficult to culture in the lab such as fungi, viruses, or the bacteria *Chlamydia*.

To perform the complement fixation test, antigen from a pathogen is added to patient serum. If antibodies to the antigen are present, the antibody will bind the antigen and fix all the available complement. When red blood cells and antibodies against red blood cells are subsequently added to the mix, there will be no complement left to lyse the red cells. Thus, if the solution remains clear, the test is positive. If there are no antipathogen antibodies in the patient's serum, the added antibodies will activate the complement to lyse the red cells, yielding a negative test (Figure 20.14).

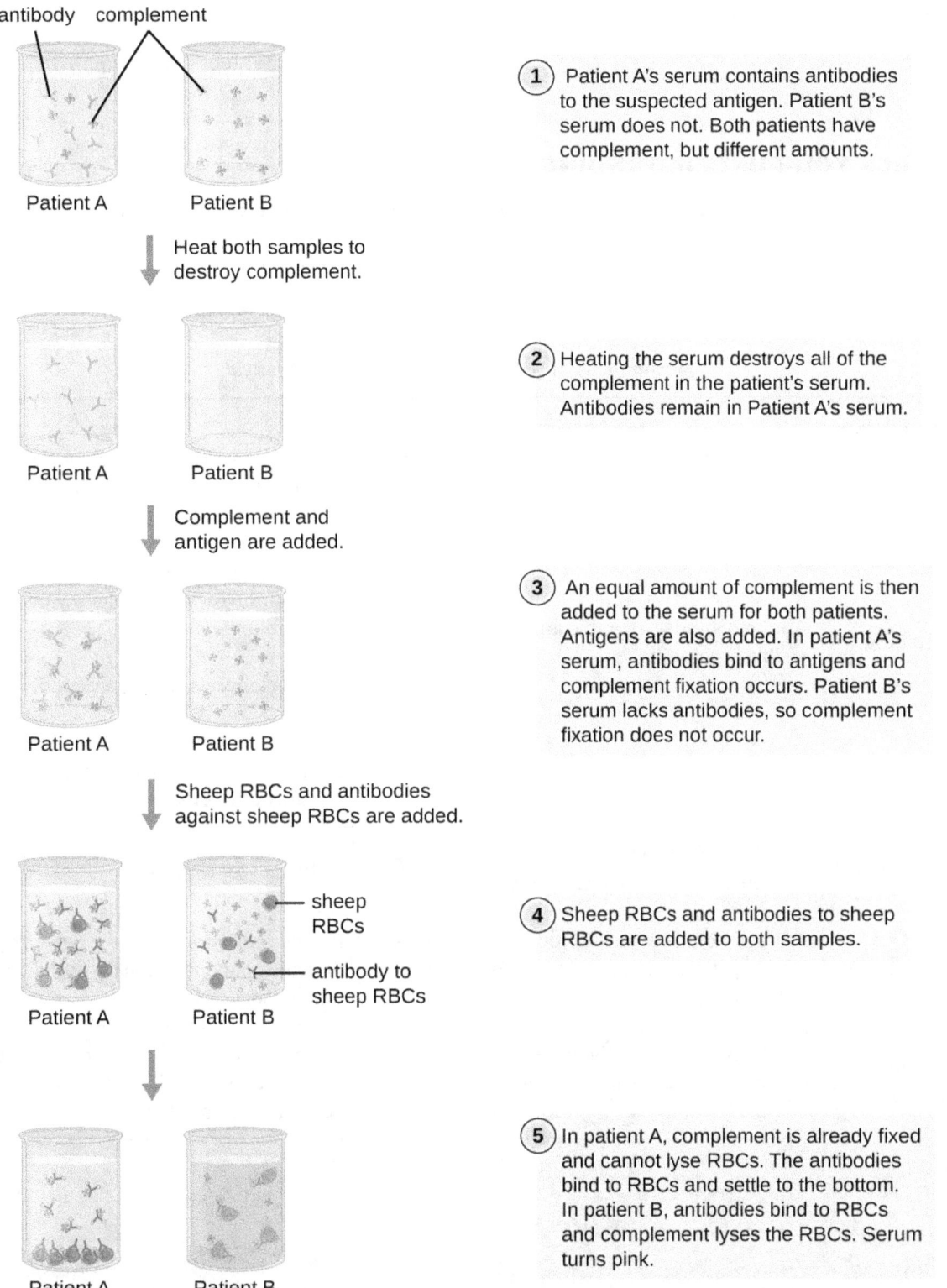

Figure 20.14 The complement fixation test is used to determine whether a patient's serum contains antibodies to a specific antigen. If it does, complement fixation will occur, and there will be no complement available to lyse the antibody-bound sheep red blood cells that are added to the solution in the next step. If the sample does not contain antibodies to the antigen, hemolysis of the sheep blood cells will be observed.

LINK TO LEARNING

View this video (https://openstax.org/l/22complfixatst) to see an outline of the steps of the complement fixation test.

CHECK YOUR UNDERSTANDING

- In a complement fixation test, if the serum turns pink, does the patient have antibodies to the antigen or not? Explain.

Table 20.2 summarizes the various types of antibody-antigen assays discussed in this section.

Mechanisms of Select Antibody-Antigen Assays

Type of Assay	Mechanism	Examples
Precipitation	Antibody binds to soluble antigen, forming a visible precipitin	Precipitin ring test to visualize lattice formation in solution
		Immunoelectrophoresis to examine distribution of antigens following electrophoresis
		Ouchterlony assay to compare diverse antigens
		Radial immunodiffusion assay to quantify antigens
Flocculation	Antibody binds to insoluble molecules in suspension, forming visible aggregates	VDRL test for syphilis
Neutralization	Antibody binds to virus, blocking viral entry into target cells and preventing formation of plaques	Plaque reduction assay for detecting presence of neutralizing antibodies in patient sera
Complement activation	Antibody binds to antigen, inducing complement activation and leaving no complement to lyse red blood cells	Complement fixation test for patient antibodies against hard-to-culture bacteria such as *Chlamydia*

Table 20.2

20.3 Agglutination Assays

Learning Objectives

By the end of this section, you will be able to:
- Compare direct and indirect agglutination
- Identify various uses of hemagglutination in the diagnosis of disease
- Explain how blood types are determined
- Explain the steps used to cross-match blood to be used in a transfusion

In addition to causing precipitation of soluble molecules and flocculation of molecules in suspension, antibodies can also clump together cells or particles (e.g., antigen-coated latex beads) in a process called

agglutination (Figure 18.9). Agglutination can be used as an indicator of the presence of antibodies against bacteria or red blood cells. Agglutination assays are usually quick and easy to perform on a glass slide or **microtiter plate** (Figure 20.15). Microtiter plates have an array of wells to hold small volumes of reagents and to observe reactions (e.g., agglutination) either visually or using a specially designed spectrophotometer. The wells come in many different sizes for assays involving different volumes of reagents.

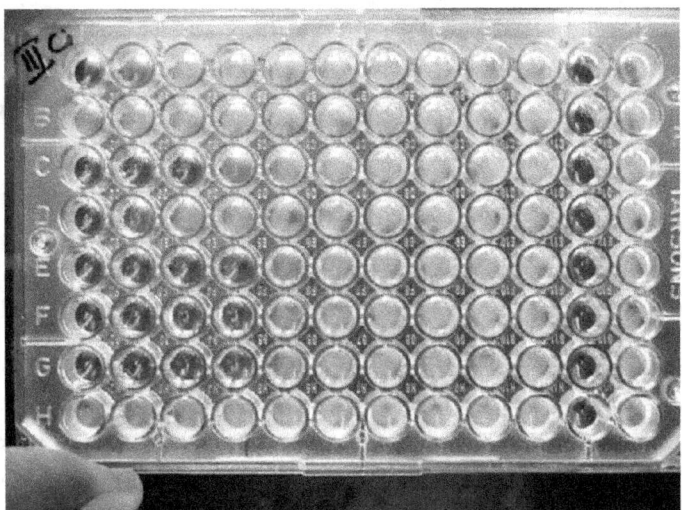

Figure 20.15 Microtiter plates are used for conducting numerous reactions simultaneously in an array of wells. (credit: modification of work by "Microrao"/Wikimedia)

Agglutination of Bacteria and Viruses

The use of agglutination tests to identify streptococcal bacteria was developed in the 1920s by Rebecca Lancefield working with her colleagues A.R. Dochez and Oswald Avery.[9] She used antibodies to identify M protein, a virulence factor on streptococci that is necessary for the bacteria's ability to cause strep throat. Production of antibodies against M protein is crucial in mounting a protective response against the bacteria.

Lancefield used antisera to show that different strains of the same species of streptococci express different versions of M protein, which explains why children can come down with strep throat repeatedly. Lancefield classified beta-hemolytic streptococci into many groups based on antigenic differences in group-specific polysaccharides located in the bacterial cell wall. The strains are called **serovars** because they are differentiated using antisera. Identifying the serovars present in a disease outbreak is important because some serovars may cause more severe disease than others.

The method developed by Lancefield is a **direct agglutination assay**, since the bacterial cells themselves agglutinate. A similar strategy is more commonly used today when identifying serovars of bacteria and viruses; however, to improve visualization of the agglutination, the antibodies may be attached to inert latex beads. This technique is called an **indirect agglutination assay** (or latex fixation assay), because the agglutination of the beads is a marker for antibody binding to some other antigen (Figure 20.16). Indirect assays can be used to detect the presence of either antibodies or specific antigens.

---

9 Lancefield, Rebecca C., "The Antigenic Complex of *Streptococcus haemoliticus*. I. Demonstration of a Type-Specific Substance in Extracts of *Streptococcus haemolyticus*," *The Journal of Experimental Medicine* 47, no. 1 (1928): 91-103.

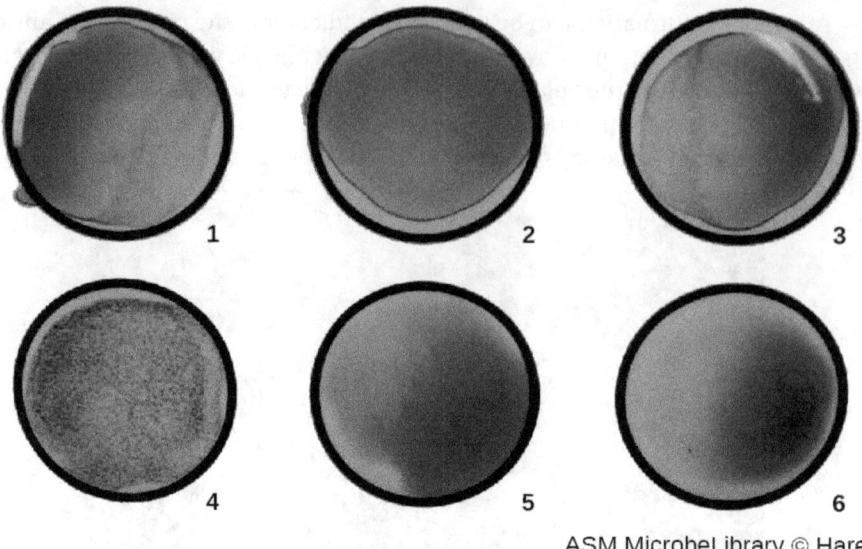

ASM MicrobeLibrary © Hare

Figure 20.16 Antibodies against six different serovars of Group A strep were attached to latex beads. Each of the six antibody preparations was mixed with bacteria isolated from a patient. The tiny clumps seen in well 4 are indicative of agglutination, which is absent from all other wells. This indicates that the serovar associated with well 4 is present in the patient sample. (credit: modification of work by American Society for Microbiology)

To identify antibodies in a patient's serum, the antigen of interest is attached to latex beads. When mixed with patient serum, the antibodies will bind the antigen, cross-linking the latex beads and causing the beads to agglutinate indirectly; this indicates the presence of the antibody (Figure 20.17). This technique is most often used when looking for IgM antibodies, because their structure provides maximum cross-linking. One widely used example of this assay is a test for rheumatoid factor (RF) to confirm a diagnosis of rheumatoid arthritis. RF is, in fact, the presence of IgM antibodies that bind to the patient's own IgG. RF will agglutinate IgG-coated latex beads.

In the reverse test, soluble antigens can be detected in a patient's serum by attaching specific antibodies (commonly mAbs) to the latex beads and mixing this complex with the serum (Figure 20.17).

Agglutination tests are widely used in underdeveloped countries that may lack appropriate facilities for culturing bacteria. For example, the Widal test, used for the diagnosis of typhoid fever, looks for agglutination of *Salmonella enterica* subspecies *typhi* in patient sera. The Widal test is rapid, inexpensive, and useful for monitoring the extent of an outbreak; however, it is not as accurate as tests that involve culturing of the bacteria. The Widal test frequently produces false positives in patients with previous infections with other subspecies of *Salmonella*, as well as false negatives in patients with hyperproteinemia or immune deficiencies.

In addition, agglutination tests are limited by the fact that patients generally do not produce detectable levels of antibody during the first week (or longer) of an infection. A patient is said to have undergone **seroconversion** when antibody levels reach the threshold for detection. Typically, seroconversion coincides with the onset of signs and symptoms of disease. However, in an HIV infection, for example, it generally takes 3 weeks for seroconversion to take place, and in some instances, it may take much longer.

Similar to techniques for the precipitin ring test and plaque assays, it is routine to prepare serial two-fold dilutions of the patient's serum and determine the titer of agglutinating antibody present. Since antibody levels change over time in both primary and secondary immune responses, by checking samples over time, changes in antibody titer can be detected. For example, a comparison of the titer during the acute phase of an infection versus the titer from the convalescent phase will distinguish whether an infection is current or has occurred in the past. It is also possible to monitor how well the patient's immune system is responding to the pathogen.

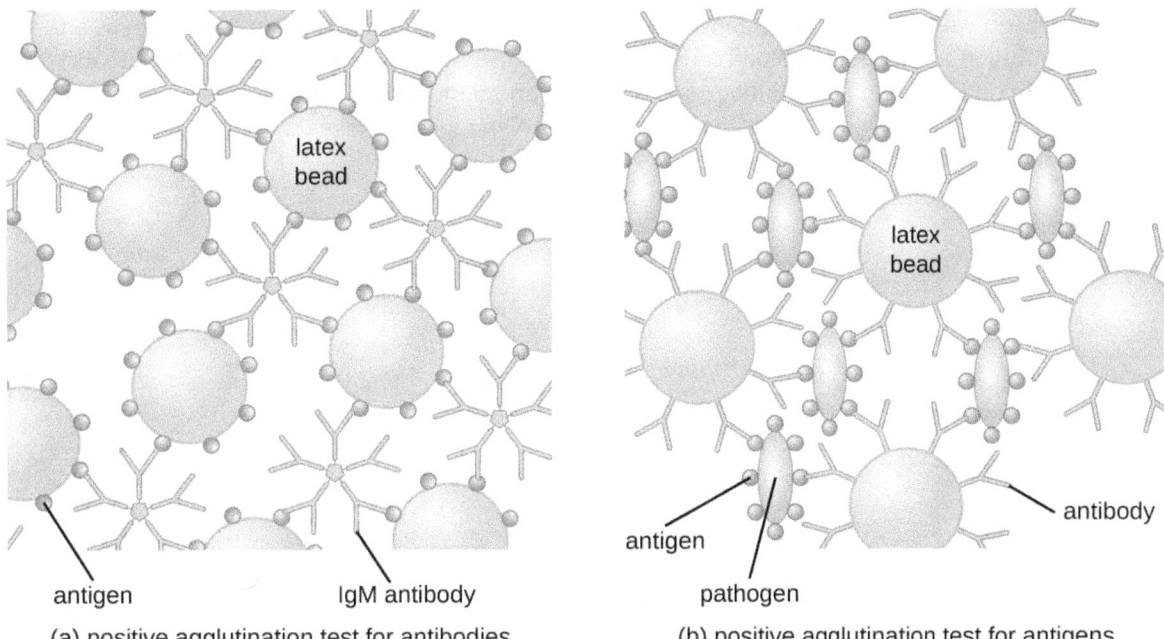

Figure 20.17 (a) Latex beads coated with an antigen will agglutinate when mixed with patient serum if the serum contains IgM antibodies against the antigen. (b) Latex beads coated with antibodies will agglutinate when mixed with patient serum if the serum contains antigens specific to the antibodies.

LINK TO LEARNING

Watch this video (https://openstax.org/l/22agglrealatbe) that demonstrates agglutination reactions with latex beads.

CHECK YOUR UNDERSTANDING

- How is agglutination used to distinguish serovars from each other?
- In a latex bead assay to test for antibodies in a patient's serum, with what are the beads coated?
- What has happened when a patient has undergone seroconversion?

Hemagglutination

Agglutination of red blood cells is called **hemagglutination**. One common assay that uses hemagglutination is the **direct Coombs' test**, also called the **direct antihuman globulin test (DAT)**, which generally looks for nonagglutinating antibodies. The test can also detect complement attached to red blood cells.

The Coombs' test is often employed when a newborn has jaundice, yellowing of the skin caused by high blood concentrations of bilirubin, a product of the breakdown of hemoglobin in the blood. The Coombs' test is used to determine whether the child's red blood cells have been bound by the mother's antibodies. These antibodies would activate complement, leading to red blood cell lysis and the subsequent jaundice. Other conditions that can cause positive direct Coombs' tests include hemolytic transfusion reactions, autoimmune hemolytic anemia, infectious mononucleosis (caused by Epstein-Barr virus), syphilis, and *Mycoplasma* pneumonia. A positive direct Coombs' test may also be seen in some cancers and as an allergic reaction to some drugs (e.g., penicillin).

The antibodies bound to red blood cells in these conditions are most often IgG, and because of the orientation of the antigen-binding sites on IgG and the comparatively large size of a red blood cell, it is unlikely that any visible agglutination will occur. However, the presence of IgG bound to red blood cells can be detected by adding **Coombs' reagent**, an antiserum containing antihuman IgG antibodies (that may be combined with anti-complement) (Figure 20.18). The Coombs' reagent links the IgG attached to neighboring red blood cells and thus promotes agglutination.

There is also an **indirect Coombs' test** known as the **indirect antiglobulin test (IAT)**. This screens an individual for antibodies against red blood cell antigens (other than the A and B antigens) that are unbound in a patient's serum (Figure 20.18). IAT can be used to screen pregnant women for antibodies that may cause hemolytic disease of the newborn. It can also be used prior to giving blood transfusions. More detail on how the IAT is performed is discussed below.

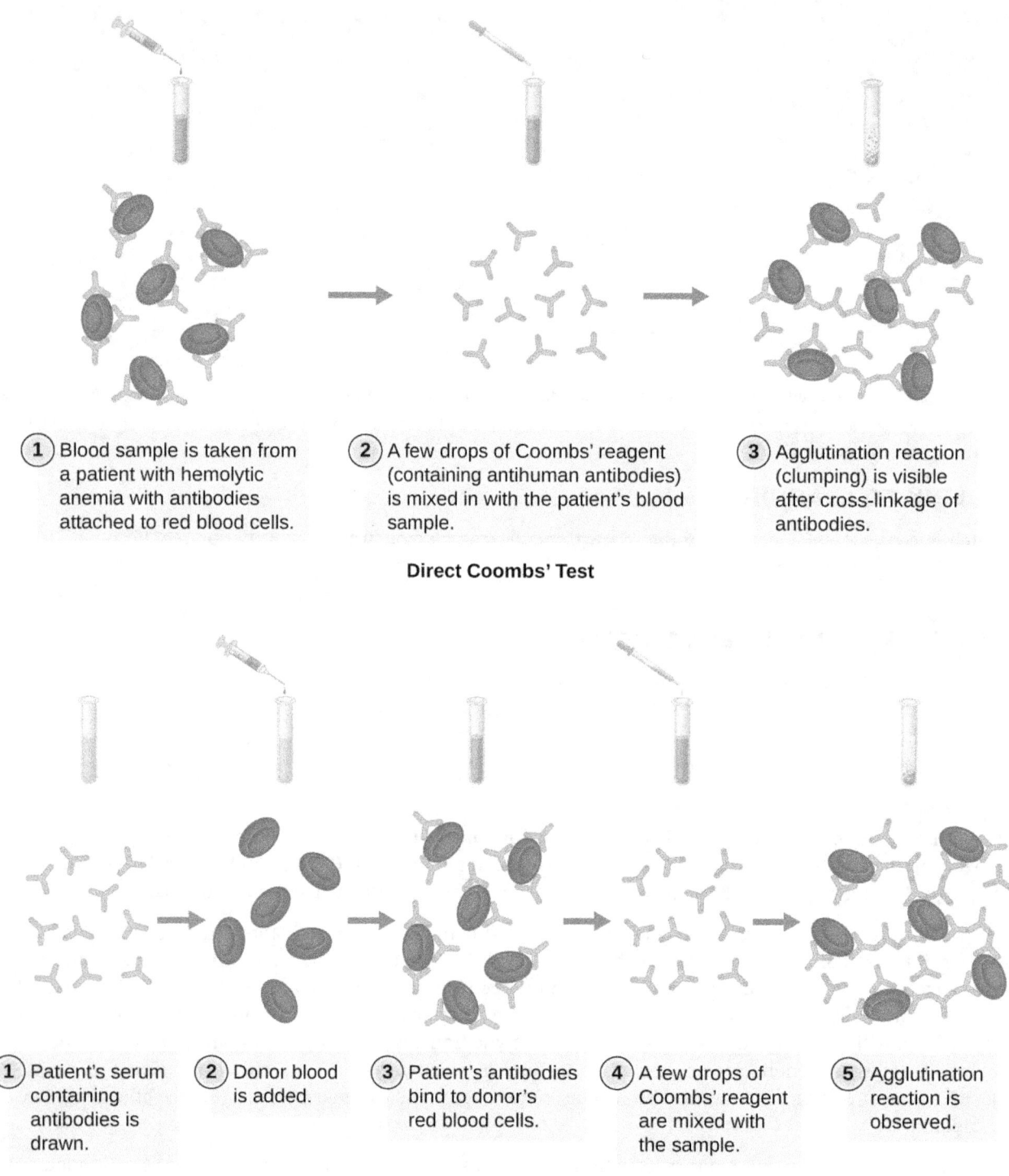

Figure 20.18 The steps in direct and indirect Coombs' tests are shown in the illustration.

Antibodies that bind to red blood cells are not the only cause of hemagglutination. Some viruses also bind to red blood cells, and this binding can cause agglutination when the viruses cross-link the red blood cells. For example, influenza viruses have two different types of viral spikes called neuraminidase (N) and hemagglutinin (H), the latter named for its ability to agglutinate red blood cells (see Viruses). Thus, we can use red blood cells to detect the presence of influenza virus by **direct hemagglutination assays** (HA), in which the virus causes visible agglutination of red blood cells. The mumps and rubella viruses can also be detected using HA.

Most frequently, a serial dilution viral agglutination assay is used to measure the titer or estimate the amount of virus produced in cell culture or for vaccine production. A viral titer can be determined using a direct HA by making a serial dilution of the sample containing the virus, starting with a high concentration of sample that is then diluted in a series of wells. The highest dilution producing visible agglutination is the titer. The assay is carried out in a microtiter plate with V- or round-bottomed wells. In the presence of agglutinating viruses, the red blood cells and virus clump together and produce a diffuse mat over the bottom of the well. In the absence of virus, the red blood cells roll or sediment to the bottom of the well and form a dense pellet, which is why flat-bottomed wells cannot be used (Figure 20.19).

A modification of the HA assay can be used to determine the titer of antiviral antibodies. The presence of these antibodies in a patient's serum or in a lab-produced antiserum will neutralize the virus and block it from agglutinating the red cells, making this a **viral hemagglutination inhibition assay** (HIA). In this assay, patient serum is mixed with a standardized amount of virus. After a short incubation, a standardized amount of red blood cells is added and hemagglutination is observed. The titer of the patient's serum is the highest dilution that blocks agglutination (Figure 20.20).

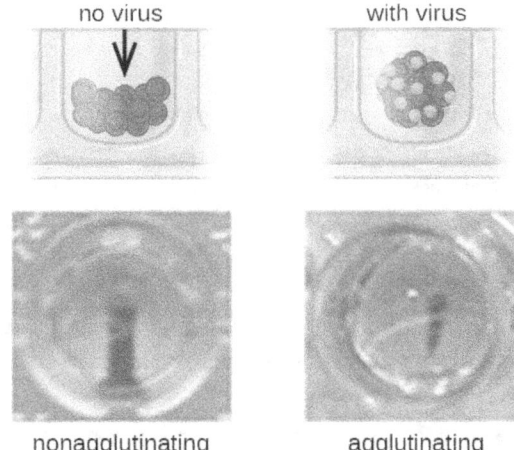

Figure 20.19 A viral suspension is mixed with a standardized amount of red blood cells. No agglutination of red blood cells is visible when the virus is absent, and the cells form a compact pellet at the bottom of the well. In the presence of virus, a diffuse pink precipitate forms in the well. (credit bottom: modification of work by American Society for Microbiology)

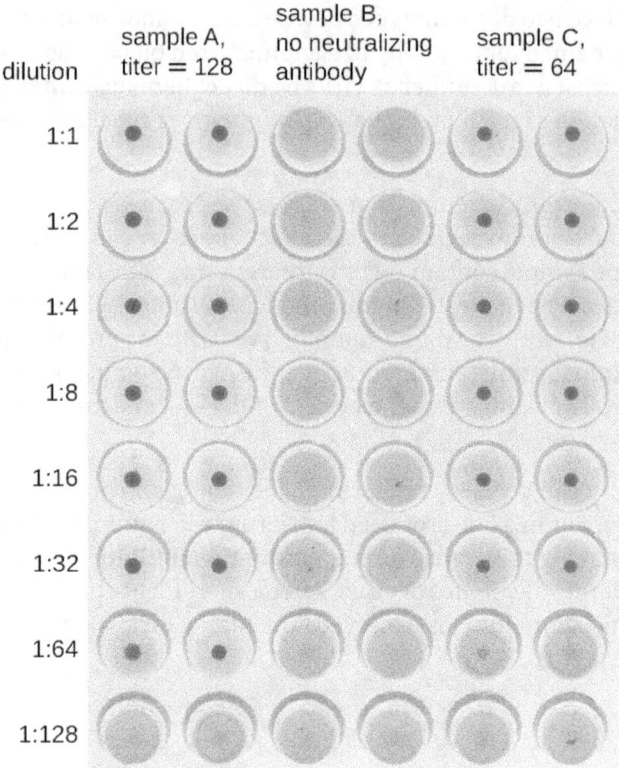

Figure 20.20 In this HIA, serum containing antibodies to influenzavirus underwent serial two-fold dilutions in a microtiter plate. Red blood cells were then added to the wells. Agglutination only occurred in those wells where the antibodies were too dilute to neutralize the virus. The highest dilution of patient serum that blocks agglutination is the titer of antibody in the patient's serum. In the case of this test, Sample A shows a titer of 64, and Sample C shows a titer of 32. (credit: modification of work by Evan Burkala)

CHECK YOUR UNDERSTANDING
- What is the mechanism by which viruses are detected in a hemagglutination assay?
- Which hemagglutination result tells us the titer of virus in a sample?

Eye on Ethics

Animals in the Laboratory

Much of what we know today about the human immune system has been learned through research conducted using animals—primarily, mammals—as models. Besides research, mammals are also used for the production of most of the antibodies and other immune system components needed for immunodiagnostics. Vaccines, diagnostics, therapies, and translational medicine in general have all been developed through research with animal models.

Consider some of the common uses of laboratory animals for producing immune system components. Guinea pigs are used as a source of complement, and mice are the primary source of cells for making mAbs. These mAbs can be used in research and for therapeutic purposes. Antisera are raised in a variety of species, including horses, sheep, goats, and rabbits. When producing an antiserum, the animal will usually be injected at least twice, and adjuvants may be used to boost the antibody response. The larger animals used for making antisera will have blood harvested repeatedly over long periods of time, with little harm to the animals, but that is not usually the case for rabbits. Although we can obtain a few milliliters of blood from the ear veins of rabbits, we usually need larger volumes, which results in the deaths of the animals.

We also use animals for the study of disease. The only way to grow *Treponema pallidum* for the study of syphilis is in living animals. Many viruses can be grown in cell culture, but growth in cell culture tells us very little about how the immune system will respond to the virus. When working on a newly discovered disease, we still employ Koch's postulates, which require causing disease in lab animals using pathogens from pure culture as a crucial step in proving that a particular microorganism is the cause of a disease. Studying the proliferation of bacteria and viruses in animal hosts, and how the host immune system responds, has been central to microbiological research for well over 100 years.

While the practice of using laboratory animals is essential to scientific research and medical diagnostics, many people strongly object to the exploitation of animals for human benefit. This ethical argument is not a new one—indeed, one of Charles Darwin's daughters was an active antivivisectionist (vivisection is the practice of cutting or dissecting a live animal to study it). Most scientists acknowledge that there should be limits on the extent to which animals can be exploited for research purposes. Ethical considerations have led the National Institutes of Health (NIH) to develop strict regulations on the types of research that may be performed. These regulations also include guidelines for the humane treatment of lab animals, setting standards for their housing, care, and euthanization. The NIH document "Guide for the Care and Use of Laboratory Animals" makes it clear that the use of animals in research is a privilege granted by society to researchers.

The NIH guidelines are based on the principle of the three R's: replace, refine, and reduce. Researchers should strive to *replace* animal models with nonliving models, *replace* vertebrates with invertebrates whenever possible, or use computer-models when applicable. They should *refine* husbandry and experimental procedures to reduce pain and suffering, and use experimental designs and procedures that *reduce* the number of animals needed to obtain the desired information. To obtain funding, researchers must satisfy NIH reviewers that the research justifies the use of animals and that their use is in accordance with the guidelines.

At the local level, any facility that uses animals and receives federal funding must have an Institutional Animal Care and Use Committee (IACUC) that ensures that the NIH guidelines are being followed. The IACUC must include researchers, administrators, a veterinarian, and at least one person with no ties to the institution, that is, a concerned citizen. This committee also performs inspections of laboratories and protocols. For research involving human subjects, an Institutional Review Board (IRB) ensures that proper guidelines are followed.

LINK TO LEARNING

Visit this site (https://openstax.org/l/22NIHcareuseani) to view the NIH Guide for the Care and Use of Laboratory Animals.

Blood Typing and Cross-Matching

In addition to antibodies against bacteria and viruses to which they have previously been exposed, most individuals also carry antibodies against blood types other than their own. There are presently 33 immunologically important blood-type systems, many of which are restricted within various ethnic groups or rarely result in the production of antibodies. The most important and perhaps best known are the ABO and Rh blood groups (see Figure 19.4).

When units of blood are being considered for transfusion, pretransfusion blood testing must be performed. For the blood unit, commercially prepared antibodies against the A, B, and Rh antigens are mixed with red blood cells from the units to initially confirm that the blood type on the unit is accurate. Once a unit of blood has been requested for transfusion, it is vitally important to make sure the donor (unit of blood) and recipient (patient) are compatible for these crucial antigens. In addition to confirming the blood type of the unit, the patient's blood type is also confirmed using the same commercially prepared antibodies to A, B, and Rh. For example, as shown in Figure 20.21, if the donor blood is A-positive, it will agglutinate with the anti-A antiserum and with the anti-Rh antiserum. If no agglutination is observed with any of the sera, then the blood type would be O-negative.

Following determination of the blood type, immediately prior to releasing the blood for transfusion, a **cross-match** is performed in which a small aliquot of the donor red blood cells are mixed with serum from the patient awaiting transfusion. If the patient does have antibodies against the donor red blood cells, hemagglutination will occur. To confirm any negative test results and check for sensitized red blood cells, Coombs' reagent may be added to the mix to facilitate visualization of the antibody-red blood cell interaction.

Under some circumstances, a minor cross-match may be performed as well. In this assay, a small aliquot of donor serum is mixed with patient red blood cells. This allows the detection of agglutinizing antibodies in the donor serum. This test is rarely necessary because transfusions generally use packed red blood cells with most of the plasma removed by centrifugation.

Red blood cells have many other antigens in addition to ABO and Rh. While most people are unlikely to have antibodies against these antigens, women who have had multiple pregnancies or patients who have had multiple transfusions may have them because of repeated exposure. For this reason, an **antibody screen** test is used to determine if such antibodies are present. Patient serum is checked against commercially prepared, pooled, type O red blood cells that express these antigens. If agglutination occurs, the antigen to which the patient is responding must be identified and determined not to be present in the donor unit.

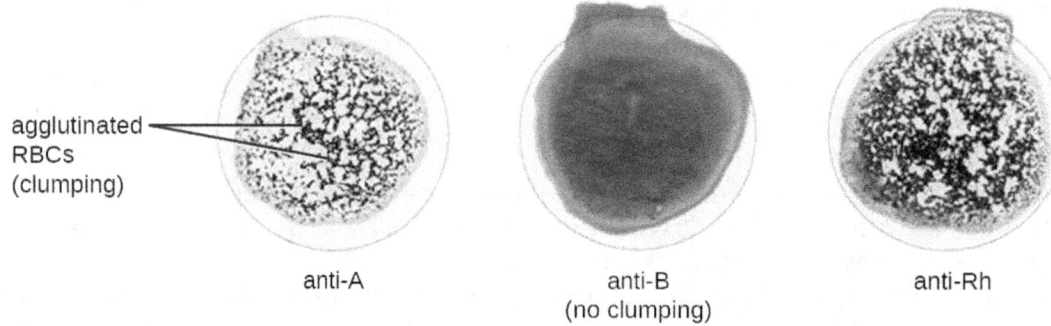

Figure 20.21 This sample of a commercially produced "bedside" card enables quick typing of both a recipient's and donor's blood before transfusion. The card contains three reaction sites or wells. One is coated with an anti-A antibody, one with an anti-B antibody, and one with an anti-Rh antibody. Agglutination of red blood cells in a given site indicates a positive identification of the blood antigens: in this case, A and Rh antigens for blood type A-positive.

✓ CHECK YOUR UNDERSTANDING

- If a patient's blood agglutinates with anti-B serum, what is the patient's blood type?
- What is a cross-match assay, and why is it performed?

Table 20.3 summarizes the various kinds of agglutination assays discussed in this section.

Mechanisms of Select Antibody-Antigen Assays

Type of Assay	Mechanism	Example
Agglutination	Direct: Antibody is used to clump bacterial cells or other large structures	Serotyping bacteria
	Indirect: Latex beads are coupled with antigen or antibody to look for antibody or antigen, respectively, in patient serum	Confirming the presence of rheumatoid factor (IgM-binding Ig) in patient serum

Mechanisms of Select Antibody-Antigen Assays

Type of Assay	Mechanism	Example
Hemagglutination	Direct: Some bacteria and viruses cross-link red blood cells and clump them together	Diagnosing influenza, mumps, and measles
	Direct Coombs' test (DAT): Detects nonagglutinating antibodies or complement proteins on red blood cells *in vivo*	Checking for maternal antibodies binding to neonatal red blood cells
	Indirect Coombs' test (IAT): Screens an individual for antibodies against red blood cell antigens (other than the A and B antigens) that are unbound in a patient's serum *in vitro*	Performing pretransfusion blood testing
	Viral hemagglutination inhibition: Uses antibodies from a patient to inhibit viral agglutination	Diagnosing various viral diseases by the presence of patient antibodies against the virus
	Blood typing and cross-matching: Detects ABO, Rh, and minor antigens in the blood	Matches donor blood to recipient immune requirements

Table 20.3

20.4 EIAs and ELISAs

Learning Objectives

By the end of this section, you will be able to:
- Explain the differences and similarities between EIA, FEIA, and ELISA
- Describe the difference and similarities between immunohistochemistry and immunocytochemistry
- Describe the different purposes of direct and indirect ELISA

Similar to the western blot, **enzyme immunoassays (EIAs)** use antibodies to detect the presence of antigens. However, EIAs differ from western blots in that the assays are conducted in microtiter plates or *in vivo* rather than on an absorbent membrane. There are many different types of EIAs, but they all involve an antibody molecule whose constant region binds an enzyme, leaving the variable region free to bind its specific antigen. The addition of a substrate for the enzyme allows the antigen to be visualized or quantified (Figure 20.22).

In EIAs, the substrate for the enzyme is most often a chromogen, a colorless molecule that is converted into a colored end product. The most widely used enzymes are alkaline phosphatase and horseradish peroxidase for which appropriate substrates are readily available. In some EIAs, the substrate is a **fluorogen**, a nonfluorescent molecule that the enzyme converts into a fluorescent form. EIAs that utilize a fluorogen are called **fluorescent enzyme immunoassays (FEIAs)**. Fluorescence can be detected by either a fluorescence microscope or a spectrophotometer.

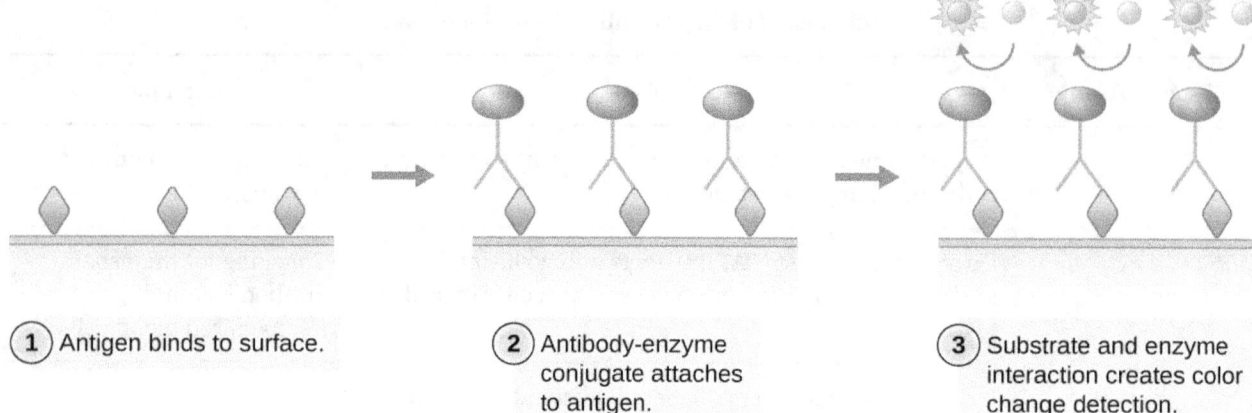

Figure 20.22 Enzyme immunoassays, such as the direct ELISA shown here, use an enzyme-antibody conjugate to deliver a detectable substrate to the site of an antigen. The substrate may be a colorless molecule that is converted into a colored end product or an inactive fluorescent molecule that fluoresces after enzyme activation. (credit: modification of work by "Cavitri"/Wikimedia Commons)

MICRO CONNECTIONS

The MMR Titer

The MMR vaccine is a combination vaccine that provides protection against measles, mumps, and rubella (German measles). Most people receive the MMR vaccine as children and thus have antibodies against these diseases. However, for various reasons, even vaccinated individuals may become susceptible to these diseases again later in life. For example, some children may receive only one round of the MMR vaccine instead of the recommended two. In addition, the titer of protective antibodies in an individual's body may begin to decline with age or as the result of some medical conditions.

To determine whether the titer of antibody in an individual's bloodstream is sufficient to provide protection, an MMR titer test can be performed. The test is a simple immunoassay that can be done quickly with a blood sample. The results of the test will indicate whether the individual still has immunity or needs another dose of the MMR vaccine.

Submitting to an MMR titer is often a pre-employment requirement for healthcare workers, especially those who will frequently be in contact with young children or immunocompromised patients. Were a healthcare worker to become infected with measles, mumps, or rubella, the individual could easily pass these diseases on to susceptible patients, leading to an outbreak. Depending on the results of the MMR titer, healthcare workers might need to be revaccinated prior to beginning work.

Immunostaining

One powerful use of EIA is **immunostaining**, in which antibody-enzyme conjugates enhance microscopy. **Immunohistochemistry (IHC)** is used for examining whole tissues. As seen in Figure 20.23, a section of tissue can be stained to visualize the various cell types. In this example, a mAb against CD8 was used to stain CD8 cells in a section of tonsil tissue. It is now possible to count the number of CD8 cells, determine their relative numbers versus the other cell types present, and determine the location of these cells within this tissue. Such data would be useful for studying diseases such as AIDS, in which the normal function of CD8 cells is crucial for slowing disease progression.

Immunocytochemistry (ICC) is another valuable form of immunostaining. While similar to IHC, in ICC, extracellular matrix material is stripped away, and the cell membrane is etched with alcohol to make it permeable to antibodies. This allows antibodies to pass through the cell membrane and bind to specific targets inside the cell. Organelles, cytoskeletal components, and other intracellular structures can be visualized in this way. While some ICC techniques use EIA, the enzyme can be replaced with a fluorescent molecule, making it a fluorescent immunoassay.

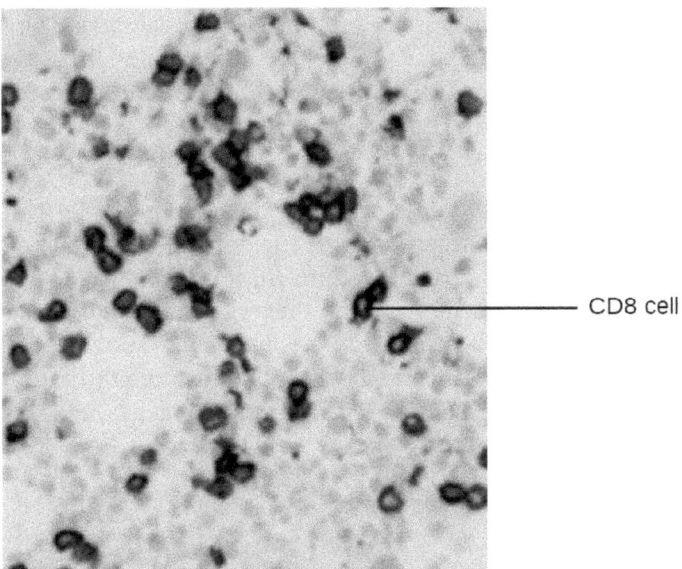

Figure 20.23 Enzyme-linked antibodies against CD8 were used to stain the CD8 cells in this preparation of bone marrow using a chromogen. (credit: modification of work by Yamashita M, Fujii Y, Ozaki K, Urano Y, Iwasa M, Nakamura S, Fujii S, Abe M, Sato Y, Yoshino T)

⊘ CHECK YOUR UNDERSTANDING

- What is the difference between immunohistochemistry and immunocytochemistry?
- What must be true of the product of the enzymatic reaction used in immunohistochemistry?

Enzyme-linked Immunosorbent Assays (ELISAs)

The **enzyme-linked immunosorbent assays (ELISAs)** are widely used EIAs. In the **direct ELISA**, antigens are immobilized in the well of a microtiter plate. An antibody that is specific for a particular antigen and is conjugated to an enzyme is added to each well. If the antigen is present, then the antibody will bind. After washing to remove any unbound antibodies, a colorless substrate (chromogen) is added. The presence of the enzyme converts the substrate into a colored end product (Figure 20.22). While this technique is faster because it only requires the use of one antibody, it has the disadvantage that the signal from a direct ELISA is lower (lower sensitivity).

In a **sandwich ELISA**, the goal is to use antibodies to precisely quantify specific antigen present in a solution, such as antigen from a pathogen, a serum protein, or a hormone from the blood or urine to list just a few examples. The first step of a sandwich ELISA is to add the **primary antibody** to all the wells of a microtiter plate (Figure 20.24). The antibody sticks to the plastic by hydrophobic interactions. After an appropriate incubation time, any unbound antibody is washed away. Comparable washes are used between each of the subsequent steps to ensure that only specifically bound molecules remain attached to the plate. A blocking protein is then added (e.g., albumin or the milk protein casein) to bind the remaining nonspecific protein-binding sites in the well. Some of the wells will receive known amounts of antigen to allow the construction of a standard curve, and unknown antigen solutions are added to the other wells. The primary antibody captures the antigen and, following a wash, the **secondary antibody** is added, which is a polyclonal antibody that is conjugated to an enzyme. After a final wash, a colorless substrate (chromogen) is added, and the enzyme converts it into a colored end product. The color intensity of the sample caused by the end product is measured with a spectrophotometer. The amount of color produced (measured as absorbance) is directly proportional to the amount of enzyme, which in turn is directly proportional to the captured antigen. ELISAs are extremely sensitive, allowing antigen to be quantified in the nanogram (10^{-9} g) per mL range.

In an **indirect ELISA**, we quantify antigen-specific antibody rather than antigen. We can use indirect ELISA to detect antibodies against many types of pathogens, including *Borrelia burgdorferi* (Lyme disease) and HIV. There are three important differences between indirect and direct ELISAs as shown in Figure 20.25. Rather than using antibody to capture antigen, the indirect ELISA starts with attaching known antigen (e.g., peptides from HIV) to the bottom of the microtiter plate wells. After blocking the unbound sites on the plate, patient serum is added; if antibodies are present (primary antibody), they will bind the antigen. After washing away any unbound proteins, the secondary antibody with its conjugated enzyme is directed against the primary antibody (e.g., antihuman immunoglobulin). The secondary antibody allows us to quantify how much antigen-specific antibody is present in the patient's serum by the intensity of the color produced from the conjugated enzyme-chromogen reaction.

As with several other tests for antibodies discussed in this chapter, there is always concern about cross-reactivity with antibodies directed against some other antigen, which can lead to false-positive results. Thus, we cannot definitively diagnose an HIV infection (or any other type of infection) based on a single indirect ELISA assay. We must confirm any suspected positive test, which is most often done using either an immunoblot that actually identifies the presence of specific peptides from the pathogen or a test to identify the nucleic acids associated with the pathogen, such as reverse transcriptase PCR (RT-PCR) or a nucleic acid antigen test.

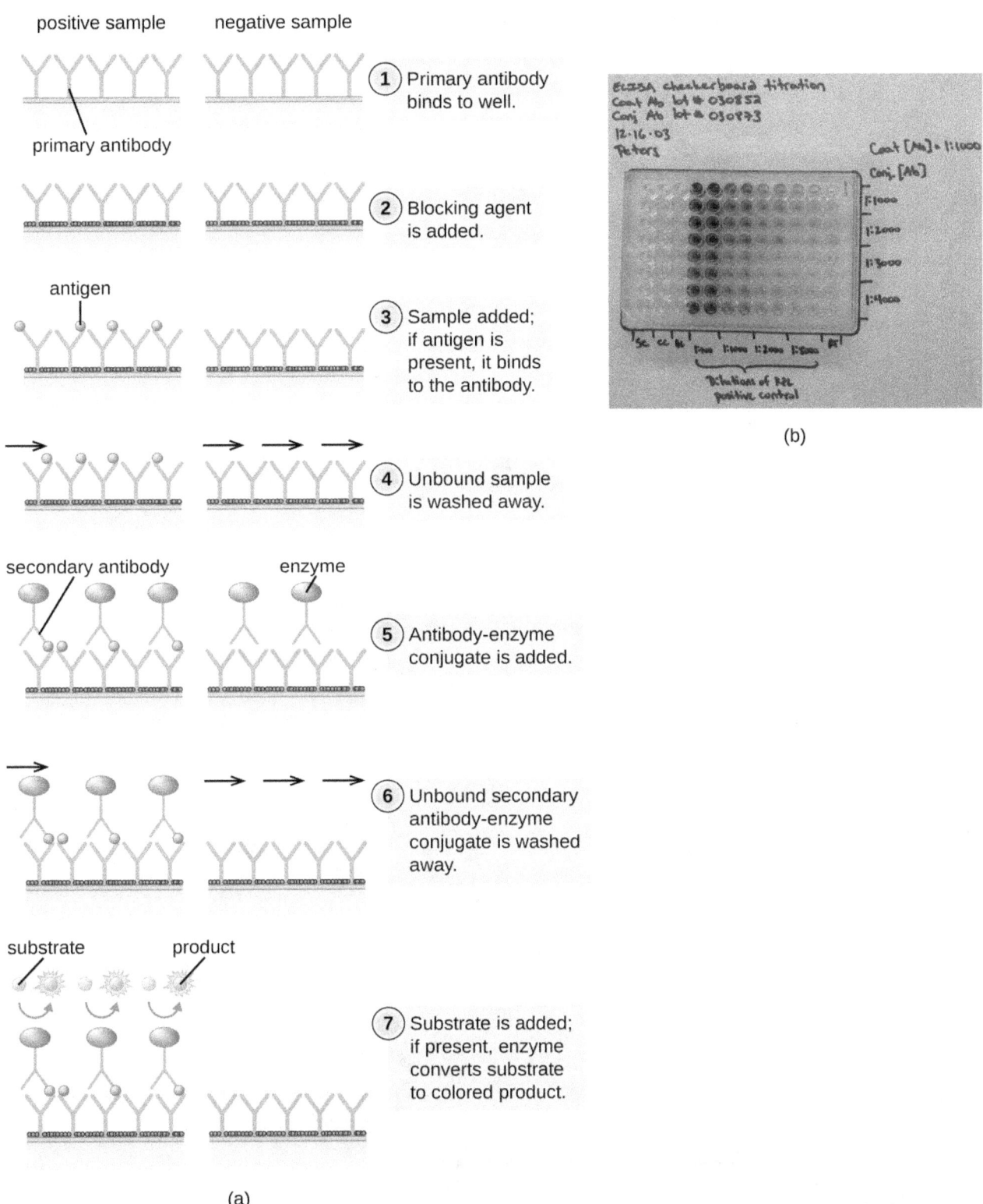

Figure 20.24 (a) In a sandwich ELISA, a primary antibody is used to first capture an antigen with the primary antibody. A secondary antibody conjugated to an enzyme that also recognizes epitopes on the antigen is added. After the addition of the chromogen, a spectrophotometer measures the absorbance of end product, which is directly proportional to the amount of captured antigen. (b) An ELISA plate shows dilutions of antibodies (left) and antigens (bottom). Higher concentrations result in a darker final color. (credit b: modification of work by U.S. Fish and Wildlife Service Pacific Region)

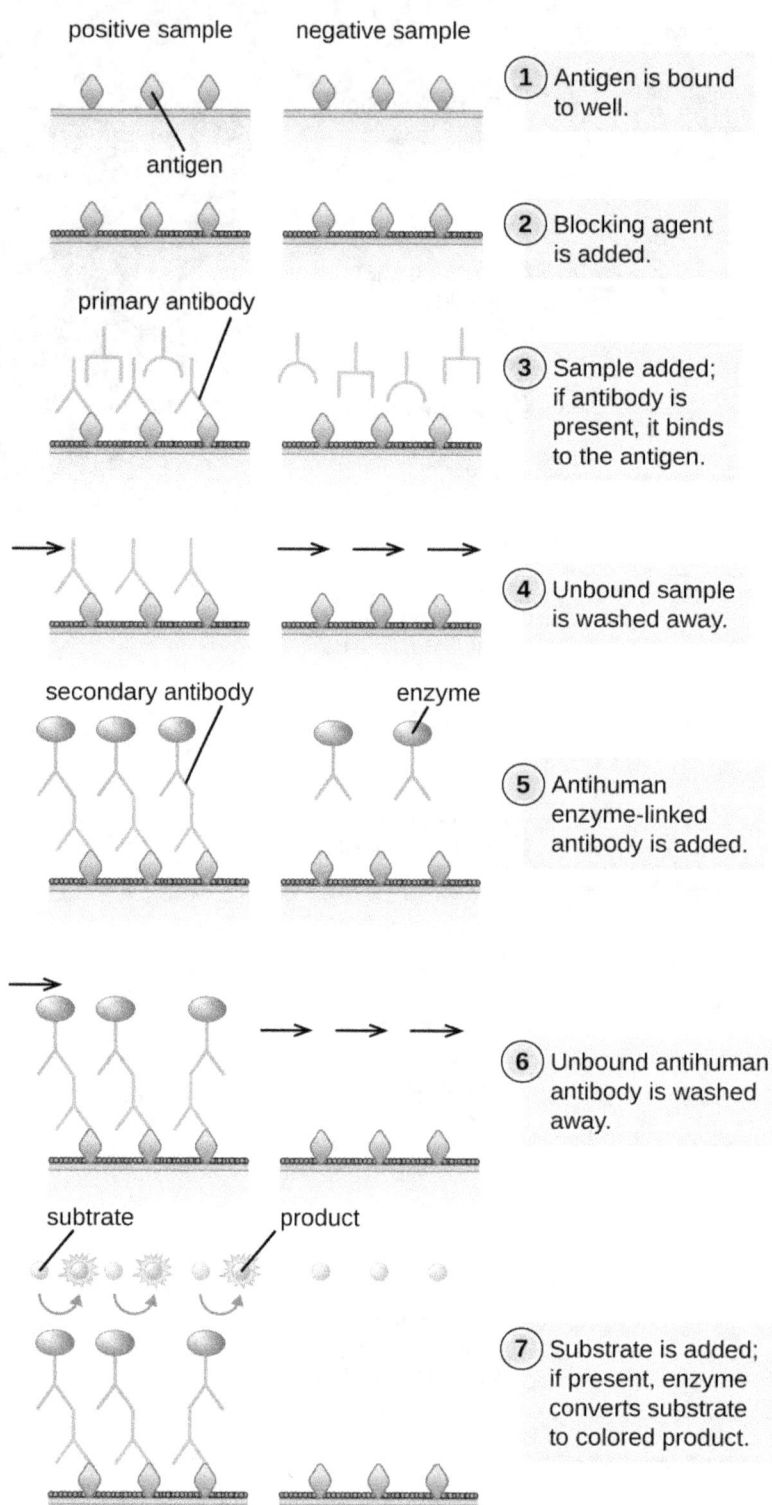

Figure 20.25 The indirect ELISA is used to quantify antigen-specific antibodies in patient serum for disease diagnosis. Antigen from the suspected disease agent is attached to microtiter plates. The primary antibody comes from the patient's serum, which is subsequently bound by the enzyme-conjugated secondary antibody. Measuring the production of end product allows us to detect or quantify the amount of antigen-specific antibody present in the patient's serum.

CHECK YOUR UNDERSTANDING

- What is the purpose of the secondary antibody in a direct ELISA?

- What do the direct and indirect ELISAs quantify?

Clinical Focus

Part 2

Although contacting and testing the 1300 patients for HIV would be time consuming and expensive, administrators hoped to minimize the hospital's liability by proactively seeking out and treating potential victims of the rogue employee's crime. Early detection of HIV is important, and prompt treatment can slow the progression of the disease.

There are a variety of screening tests for HIV, but the most widely used is the indirect ELISA. As with other indirect ELISAs, the test works by attaching antigen (in this case, HIV peptides) to a well in a 96-well plate. If the patient is HIV positive, anti-HIV antibodies will bind to the antigen and be identified by the second antibody-enzyme conjugate.

- How accurate is an indirect ELISA test for HIV, and what factors could impact the test's accuracy?
- Should the hospital use any other tests to confirm the results of the indirect ELISA?

Jump to the previous Clinical Focus box. Jump to the next Clinical Focus box.

Immunofiltration and Immunochromatographic Assays

For some situations, it may be necessary to detect or quantify antigens or antibodies that are present at very low concentration in solution. Immunofiltration techniques have been developed to make this possible. In **immunofiltration**, a large volume of fluid is passed through a porous membrane into an absorbent pad. An antigen attached to the porous membrane will capture antibody as it passes; alternatively, we can also attach an antibody to the membrane to capture antigen.

The method of immunofiltration has been adapted in the development of **immunochromatographic assays**, commonly known as **lateral flow tests** or strip tests. These tests are quick and easy to perform, making them popular for point-of-care use (i.e., in the doctor's office) or in-home use. One example is the TORCH test that allows doctors to screen pregnant women or newborns for infection by an array of viruses and other pathogens (*Toxoplasma*, other viruses, rubella, cytomegalovirus, herpes simplex). In-home pregnancy tests are another widely used example of a lateral flow test (Figure 20.26). Immunofiltration tests are also popular in developing countries, because they are inexpensive and do not require constant refrigeration of the dried reagents. However, the technology is also built into some sophisticated laboratory equipment.

In lateral flow tests (Figure 20.27), fluids such as urine are applied to an absorbent pad on the test strip. The fluid flows by capillary action and moves through a stripe of beads with antibodies attached to their surfaces. The fluid in the sample actually hydrates the reagents, which are present in a dried state in the stripe. Antibody-coated beads made of latex or tiny gold particles will bind antigens in the test fluid. The antibody-antigen complexes then flow over a second stripe that has immobilized antibody against the antigen; this stripe will retain the beads that have bound antigen. A third control stripe binds any beads. A red color (from gold particles) or blue (from latex beads) developing at the test line indicates a positive test. If the color only develops at the control line, the test is negative.

Like ELISA techniques, lateral flow tests take advantage of antibody sandwiches, providing sensitivity and specificity. While not as quantitative as ELISA, these tests have the advantage of being fast, inexpensive, and not dependent on special equipment. Thus, they can be performed anywhere by anyone. There are some concerns about putting such powerful diagnostic tests into the hands of people who may not understand the tests' limitations, such as the possibility of false-positive results. While home pregnancy tests have become widely accepted, at-home antibody-detection tests for diseases like HIV have raised some concerns in the medical community. Some have questioned whether self-administration of such tests should be allowed in the absence of medical personnel who can explain the test results and order appropriate confirmatory tests. However, with growing numbers of lateral flow tests becoming available, and the rapid development of lab-on-a-chip technology (Figure 20.1), home medical tests are likely to become even more commonplace in the future.

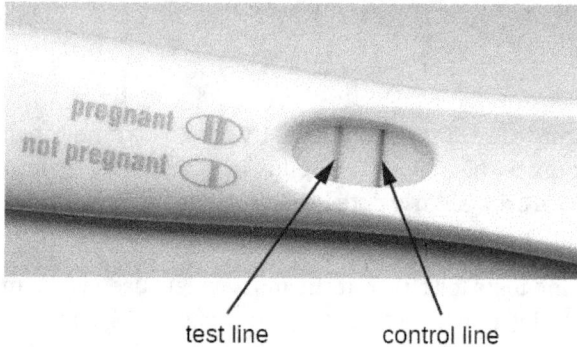

Figure 20.26 A lateral flow test detecting pregnancy-related hormones in urine. The control stripe verifies the validity of the test and the test line determines the presence of pregnancy-related hormones in the urine. (credit: modification of work by Klaus Hoffmeier)

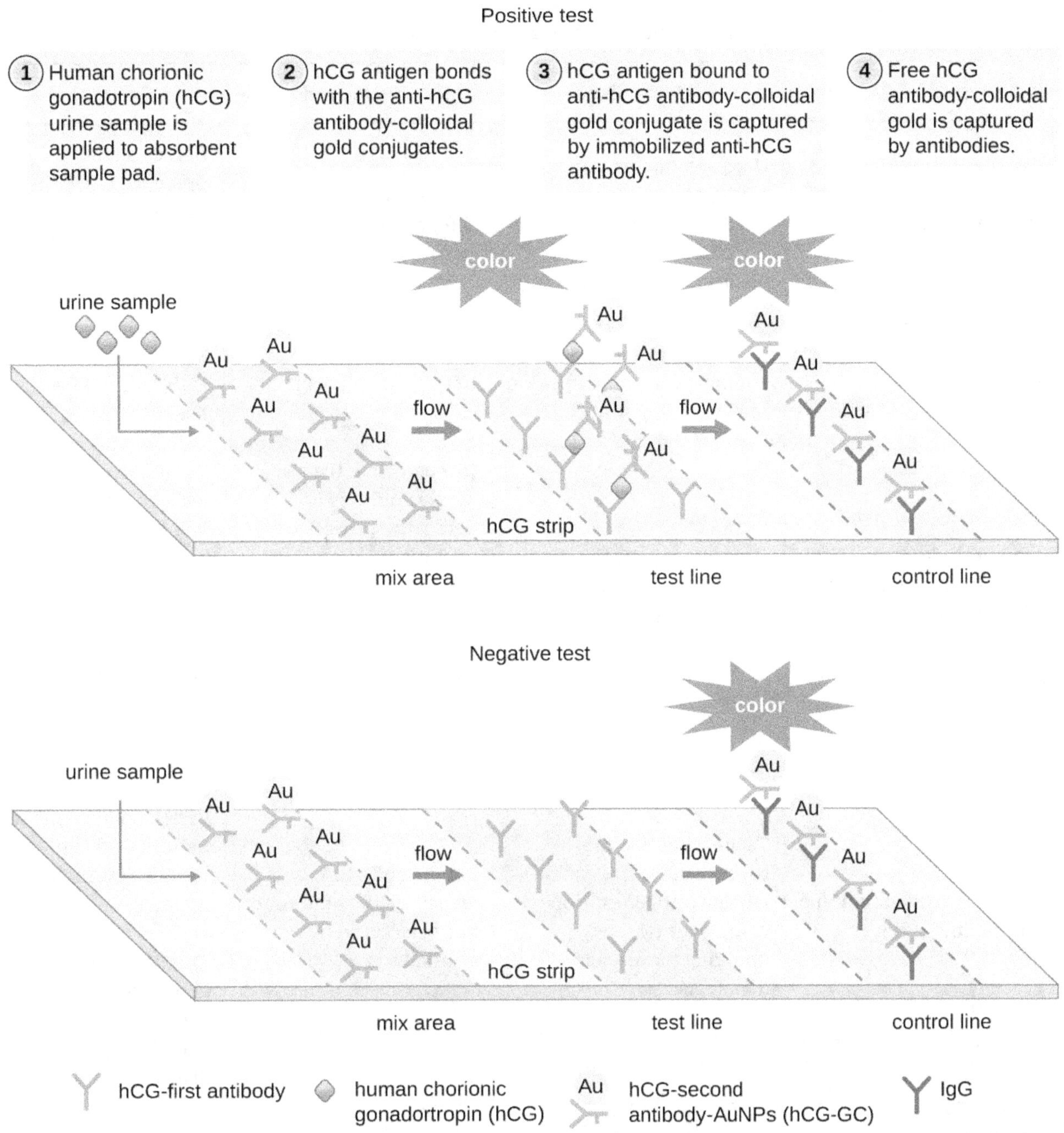
Positive test
1 Human chorionic gonadotropin (hCG) urine sample is applied to absorbent sample pad.
2 hCG antigen bonds with the anti-hCG antibody-colloidal gold conjugates.
3 hCG antigen bound to anti-hCG antibody-colloidal gold conjugate is captured by immobilized anti-hCG antibody.
4 Free hCG antibody-colloidal gold is captured by antibodies.
color
color
urine sample
Au
flow
hCG strip
mix area
test line
control line
Negative test
color
urine sample
Au
flow
hCG strip
mix area
test line
control line
hCG-first antibody
human chorionic gonadortropin (hCG)
Au
hCG-second antibody-AuNPs (hCG-GC)
IgG

Figure 20.27 Immunochromatographic assays, or lateral flow tests, allow the testing of antigen in a dilute solution. As the fluid flows through the test strip, it rehydrates the reagents. Antibodies conjugated to small particles bind the antigen in the first stripe and then flow onto the second stripe where they are bound by a second, fixed antibody. This produces a line of color, depending on the color of the beads. The third, control stripe binds beads as well to indicate that the test is working properly. (credit: modification of work by Yeh CH, Zhao ZQ, Shen PL, Lin YC)

CHECK YOUR UNDERSTANDING

- What physical process does the lateral flow method require to function?
- Explain the purpose of the third strip in a lateral flow assay.

Table 20.4 compares some of the key mechanisms and examples of some of the EIAs discussed in this section as well as immunoblots, which were discussed in Detecting Antigen-Antibody Complexes.

Immunoblots & Enzyme Immunoassays

Type of Assay	Mechanism	Specific Procedures	Examples
Immunoblots	Uses enzyme-antibody conjugates to identify specific proteins that have been transferred to an absorbent membrane	Western blot: Detects the presence of a particular protein	Detecting the presence of HIV peptides (or peptides from other infectious agents) in patient sera
Immunostaining	Uses enzyme-antibody conjugates to stain specific molecules on or in cells	Immunohistochemistry: Used to stain specific cells in a tissue	Stain for presence of CD8 cells in host tissue
Enzyme-linked immunosorbent assay (ELISA)	Uses enzyme-antibody conjugates to quantify target molecules	Direct ELISA: Uses a single antibody to detect the presence of an antigen	Detection of HIV antigen p24 up to one month after being infected
		Indirect ELISA: Measures the amount of antibody produced against an antigen	Detection of HIV antibodies in serum
Immunochromatographic (lateral flow) assays	Techniques use the capture of flowing, color-labeled antigen-antibody complexes by fixed antibody for disease diagnosis	Sandwich ELISA: Measures the amount of antigen bound by the antibody	Detection of antibodies for various pathogens in patient sera (e.g., rapid strep, malaria dipstick)
			Pregnancy test detecting human chorionic gonadotrophin in urine

Table 20.4

Clinical Focus

Part 3

Although the indirect ELISA for HIV is a sensitive assay, there are several complicating considerations. First, if an infected person is tested too soon after becoming infected, the test can yield false-negative results. The seroconversion window is generally about three weeks, but in some cases, it can be more than two months.

In addition to false negatives, false positives can also occur, usually due to previous infections with other viruses that induce cross-reacting antibodies. The false-positive rate depends on the particular brand of test used, but 0.5% is not unusual.[10] Because of the possibility of a false positive, all positive tests are followed up with a confirmatory test. This confirmatory test is often an immunoblot (western blot) in which HIV peptides from the patient's blood are identified using an HIV-specific mAb-enzyme conjugate. A positive western blot would confirm an HIV infection and a negative blot would confirm the absence of HIV despite the positive ELISA.

Unfortunately, western blots for HIV antigens often yield indeterminant results, in which case, they neither confirm nor invalidate the results of the indirect ELISA. In fact, the rate of indeterminants can be 10–49% (which is why, combined with their cost, western blots are not used for screening). Similar to the indirect ELISA, an indeterminant western blot can occur because of cross-reactivity or previous viral infections, vaccinations, or autoimmune diseases.

- Of the 1300 patients being tested, how many false-positive ELISA tests would be expected?
- Of the false positives, how many indeterminant western blots could be expected?
- How would the hospital address any cases in which a patient's western blot was indeterminant?

Jump to the previous Clinical Focus box. Jump to the next Clinical Focus box.

20.5 Fluorescent Antibody Techniques

Learning Objectives

By the end of this section, you will be able to:
- Describe the benefits of immunofluorescent antibody assays in comparison to nonfluorescent assays
- Compare direct and indirect fluorescent antibody assays
- Explain how a flow cytometer can be used to quantify specific subsets of cells present in a complex mixture of cell types
- Explain how a fluorescence-activated cell sorter can be used to separate unique types of cells

Rapid visualization of bacteria from a clinical sample such as a throat swab or sputum can be achieved through **fluorescent antibody (FA) techniques** that attach a fluorescent marker (fluorogen) to the constant region of an antibody, resulting in a reporter molecule that is quick to use, easy to see or measure, and able to bind to target markers with high specificity. We can also label cells, allowing us to precisely quantify particular subsets of cells or even purify these subsets for further research.

As with the enzyme assays, FA methods may be direct, in which a labeled mAb binds an antigen, or indirect, in which secondary polyclonal antibodies bind patient antibodies that react to a prepared antigen. Applications of these two methods were demonstrated in Figure 2.19. FA methods are also used in automated cell counting and sorting systems to enumerate or segregate labeled subpopulations of cells in a sample.

Direct Fluorescent Antibody Techniques

Direct fluorescent antibody (DFA) tests use a fluorescently labeled mAb to bind and illuminate a target antigen. DFA tests are particularly useful for the rapid diagnosis of bacterial diseases. For example, fluorescence-labeled antibodies against *Streptococcus pyogenes* (group A strep) can be used to obtain a diagnosis of strep throat from a throat swab. The diagnosis is ready in a matter of minutes, and the patient can be started on antibiotics before even leaving the clinic. DFA techniques may also be used to diagnose pneumonia caused by *Mycoplasma pneumoniae* or *Legionella pneumophila* from sputum samples (Figure 20.28). The fluorescent antibodies bind to the bacteria on a microscope slide, allowing ready detection of the bacteria using a fluorescence microscope. Thus, the DFA technique is valuable for visualizing certain bacteria that are difficult to isolate or culture from patient samples.

---

10 Thomas, Justin G., Victor Jaffe, Judith Shaffer, and Jose Abreu, "HIV Testing: US Recommendations 2014," *Osteopathic Family Physician* 6, no. 6 (2014).

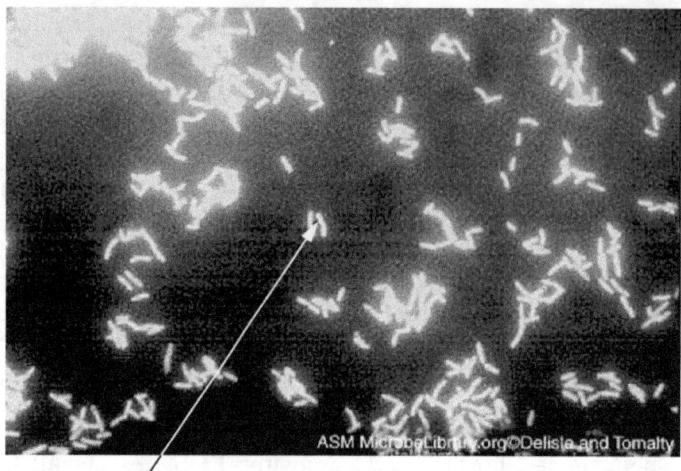
Fluorecein-labeled antibody attached to *Legionella* bacilli

Figure 20.28 A green fluorescent mAb against *L. pneumophila* is used here to visualize and identify bacteria from a smear of a sample from the respiratory tract of a pneumonia patient. (credit: modification of work by American Society for Microbiology)

LINK TO LEARNING

Watch the animation (https://openstax.org/l/22dirfluorant) on this page to review the procedures of the direct fluorescent antibody test.

CHECK YOUR UNDERSTANDING

- In a direct fluorescent antibody test, what does the fluorescent antibody bind to?

Indirect Fluorescent Antibody Techniques

Indirect fluorescent antibody (IFA) tests (Figure 20.29) are used to look for antibodies in patient serum. For example, an IFA test for the diagnosis of syphilis uses *T. pallidum* cells isolated from a lab animal (the bacteria cannot be grown on lab media) and a smear prepared on a glass slide. Patient serum is spread over the smear and anti-treponemal antibodies, if present, are allowed to bind. The serum is washed off and a secondary antibody added. The secondary antibody is an antihuman immunoglobulin conjugated to a fluorogen. On examination, the *T. pallidum* bacteria will only be visible if they have been bound by the antibodies from the patient's serum.

The IFA test for syphilis provides an important complement to the VDRL test discussed in Detecting Antigen-Antibody Complexes. The VDRL is more likely to generate false-positive reactions than the IFA test; however, the VDRL is a better test for determining whether an infection is currently active.

IFA tests are also useful for the diagnosis of autoimmune diseases. For example, systemic lupus erythematosus (SLE) (see Autoimmune Disorders) is characterized by elevated expression levels of antinuclear antibodies (ANA). These autoantibodies can be expressed against a variety of DNA-binding proteins and even against DNA itself. Because autoimmunity is often difficult to diagnose, especially early in disease progression, testing for ANA can be a valuable clue in making a diagnosis and starting appropriate treatment.

The IFA for ANA begins by fixing cells grown in culture to a glass slide and making them permeable to antibody. The slides are then incubated with serial dilutions of serum from the patient. After incubation, the slide is washed to remove unbound proteins, and the fluorescent antibody (antihuman IgG conjugated to a fluorogen) added. After an incubation and wash, the cells can be examined for fluorescence evident around the nucleus (Figure 20.30). The titer of ANA in the serum is determined by the highest dilution showing fluorescence. Because many healthy people express ANA, the American College of Rheumatology recommends that the titer must be at least 1:40 in the presence of symptoms involving two or more organ systems to be considered indicative of SLE.[11]

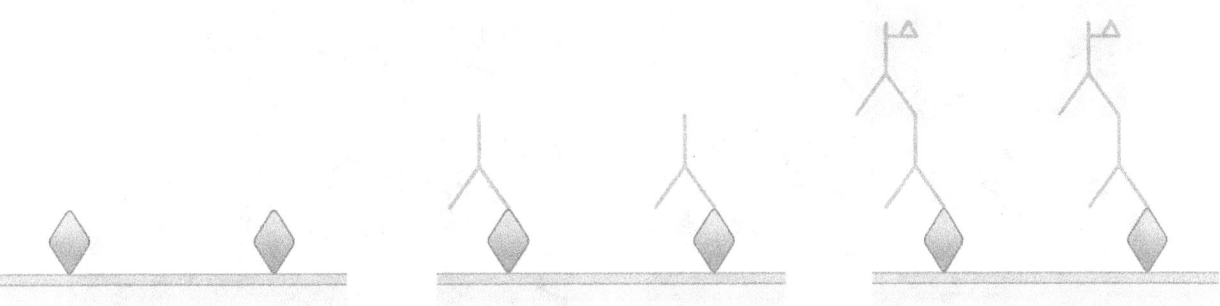

① Antigen is fixed to a surface.

② Patient serum is added; if antibodies are present, they bind to the antigen.

③ Secondary antibody (with florescent label) is added; if patient antibodies are present, the secondary antibody binds to the patient antibodies.

(a)

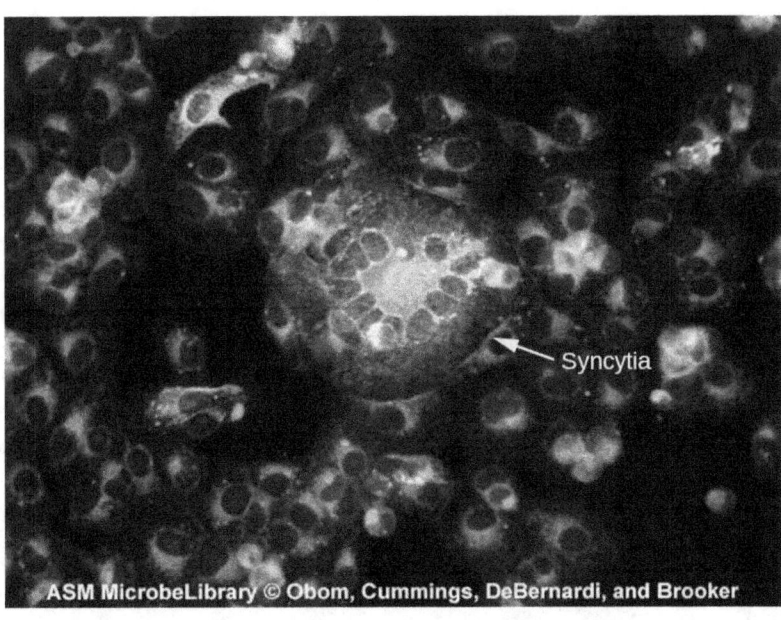

(b)

Figure 20.29 (a) The IFA test is used to detect antigen-specific antibodies by allowing them to bind to antigen fixed to a surface and then illuminating these complexes with a secondary antibody-fluorogen conjugate. (b) In this example of a micrograph of an indirect fluorescent antibody test, a patient's antibodies to the measles virus bind to viral antigens present on inactivated measles-infected cells affixed to a slide. Secondary antibodies bind the patient's antibodies and carry a fluorescent molecule. (credit b: modification of work by American Society for Microbiology)

---

11 Gill, James M., ANNA M. Quisel, PETER V. Rocca, and DENE T. Walters. "Diagnosis of systemic lupus erythematosus." *American family physician* 68, no. 11 (2003): 2179-2186.

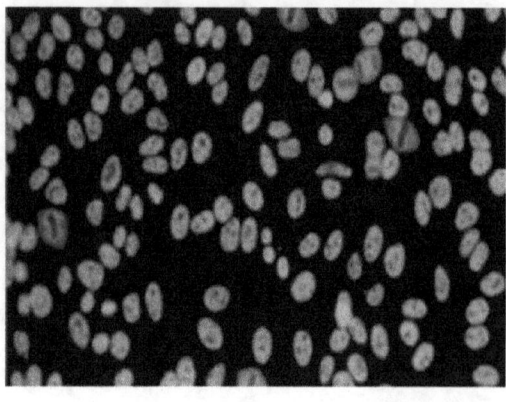

diseased   healthy control

Figure 20.30 In this test for antinuclear antibodies (ANA), cells are exposed to serum from a patient suspected of making ANA and then to a fluorescent mAb specific for human immunoglobulin. As a control, serum from a healthy patient is also used. Visible fluorescence around the nucleus demonstrates the presence of ANA in the patient's serum. In the healthy control where lower levels of ANA are produced, very faint green is detected. (credit left, right: modification of work by Al-Hussaini AA, Alzahrani MD, Alenizi AS, Suliman NM, Khan MA, Alharbi SA, Chentoufi AA)

✓ CHECK YOUR UNDERSTANDING

- In an indirect fluorescent antibody test, what does the fluorescent antibody bind to?
- What is the ANA test looking for?

Flow Cytometry

Fluorescently labeled antibodies can be used to quantify cells of a specific type in a complex mixture using **flow cytometry** (Figure 20.31), an automated, cell-counting system that detects fluorescing cells as they pass through a narrow tube one cell at a time. For example, in HIV infections, it is important to know the level of CD4 T cells in the patient's blood; if the numbers fall below 500 per µL of blood, the patient becomes more likely to acquire opportunistic infections; below 200 per µL, the patient can no longer mount a useful adaptive immune response at all. The analysis begins by incubating a mixed-cell population (e.g., white blood cells from a donor) with a fluorescently labeled mAb specific for a subpopulation of cells (e.g., anti-CD4). Some experiments look at two cell markers simultaneously by adding a different fluorogen to the appropriate mAb. The cells are then introduced to the flow cytometer through a narrow capillary that forces the cells to pass in single file. A laser is used to activate the fluorogen. The fluorescent light radiates out in all directions, so the fluorescence detector can be positioned at an angle from the incident laser light.

Figure 20.31 shows the obscuration bar in front of the forward-scatter detector that prevents laser light from hitting the detector. As a cell passes through the laser bar, the forward-scatter detector detects light scattered around the obscuration bar. The scattered light is transformed into a voltage pulse, and the cytometer counts a cell. The fluorescence from a labeled cell is detected by the side-scatter detectors. The light passes through various dichroic mirrors such that the light emitted from the fluorophore is received by the correct detector.

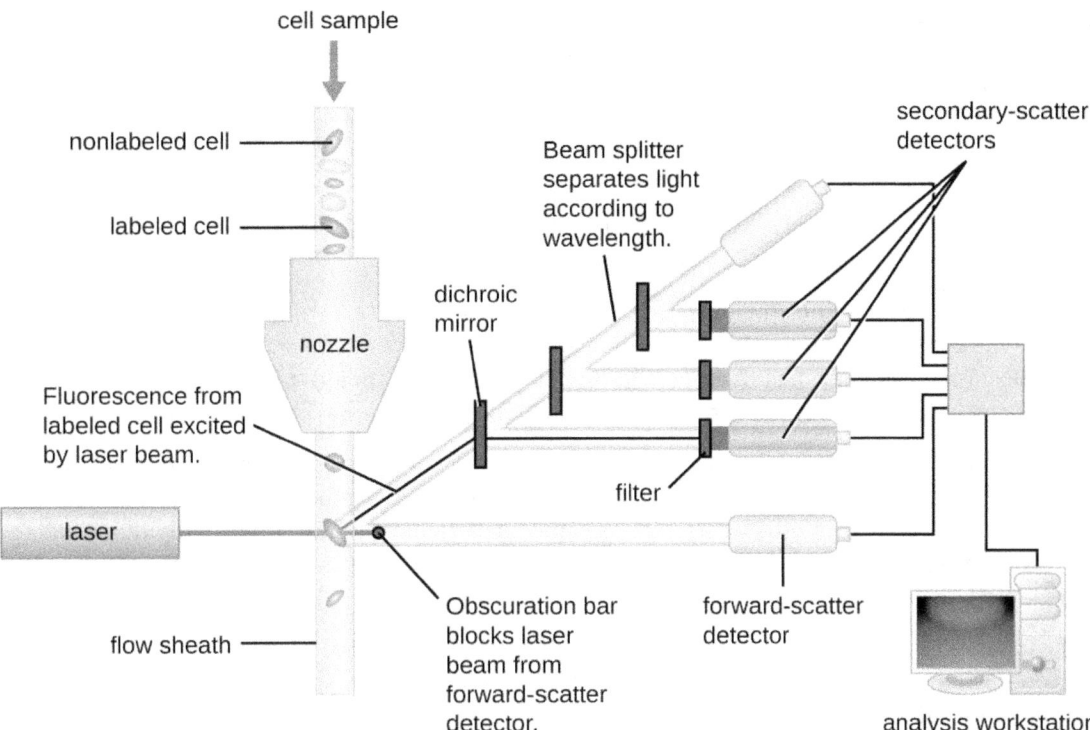

Figure 20.31 In flow cytometry, a mixture of fluorescently labeled and unlabeled cells passes through a narrow capillary. A laser excites the fluorogen, and the fluorescence intensity of each cell is measured by a detector. (credit: modification of work by "Kierano"/Wikimedia Commons)

Data are collected from both the forward- and side-scatter detectors. One way these data can be presented is in the form of a histogram. The forward scatter is placed on the y-axis (to represent the number of cells), and the side scatter is placed on the x-axis (to represent the fluoresence of each cell). The scaling for the x-axis is logarithmic, so fluorescence intensity increases by a factor of 10 with each unit increase along the axis. Figure 20.32 depicts an example in which a culture of cells is combined with an antibody attached to a fluorophore to detect CD8 cells and then analyzed by flow cytometry. The histogram has two peaks. The peak on the left has lower fluorescence readings, representing the subset of the cell population (approximately 30 cells) that does not fluoresce; hence, they are not bound by antibody and therefore do not express CD8. The peak on the right has higher fluorescence readings, representing the subset of the cell population (approximately 100 cells) that show fluorescence; hence, they are bound by the antibody and therefore do express CD8.

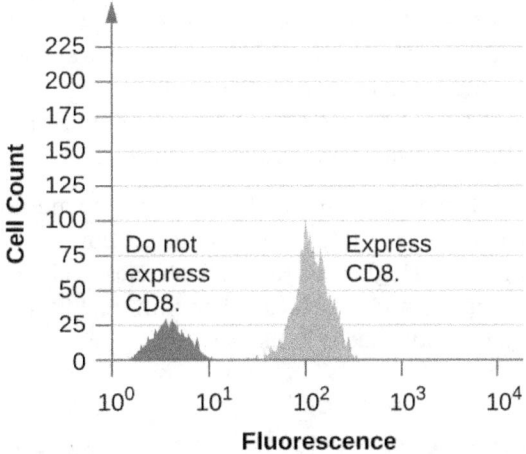

Figure 20.32 Flow cytometry data are often compiled as a histogram. In the histogram, the area under each peak is proportional to the number of cells in each population. The x-axis is the relative fluorescence expressed by the cells (on a log scale), and the y-axis represents the number of cells at a particular level of fluorescence.

CHECK YOUR UNDERSTANDING

- What is the purpose of the laser in a flow cytometer?
- In the output from a flow cytometer, the area under the histogram is equivalent to what?

Clinical Focus

Resolution

After notifying all 1300 patients, the hospital begins scheduling HIV screening. Appointments were scheduled a minimum of 3 weeks after the patient's last hospital visit to minimize the risk of false negatives. Because some false positives were anticipated, the public health physician set up a counseling protocol for any patient whose indirect ELISA came back positive.

Of the 1300 patients, eight tested positive using the ELISA. Five of these tests were invalidated by negative western blot tests, but one western blot came back positive, confirming that the patient had indeed contracted HIV. The two remaining western blots came back indeterminate. These individuals had to submit to a third test, a PCR, to confirm the presence or absence of HIV sequences. Luckily, both patients tested negative.

As for the lone patient confirmed to have HIV, the tests cannot prove or disprove any connection to the syringes compromised by the former hospital employee. Even so, the hospital's insurance will fully cover the patient's treatment, which began immediately.

Although we now have drugs that are typically effective at controlling the progression of HIV and AIDS, there is still no cure. If left untreated, or if the drug regimen fails, the patient will experience a gradual decline in the number of CD4 helper T cells, resulting in severe impairment of all adaptive immune functions. Even moderate declines of helper T cell numbers can result in immunodeficiency, leaving the patient susceptible to opportunistic infections. To monitor the status of the patient's helper T cells, the hospital will use flow cytometry. This sensitive test allows physicians to precisely determine the number of helper T cells so they can adjust treatment if the number falls below 500 cells/µL.

Jump to the previous Clinical Focus box.

Cell Sorting Using Immunofluorescence

The flow cytometer and immunofluorescence can also be modified to sort cells from a single sample into purified subpopulations of cells for research purposes. This modification of the flow cytometer is called a **fluorescence-activated cell sorter (FACS)**. In a FACS, fluorescence by a cell induces the device to put a charge on a droplet of the transporting fluid containing that cell. The charge is specific to the wavelength of the fluorescent light, which allows for differential sorting by those different charges. The sorting is accomplished by an electrostatic deflector that moves the charged droplet containing the cell into one collecting vessel or another. The process results in highly purified subpopulations of cells.

One limitation of a FACS is that it only works on isolated cells. Thus, the method would work in sorting white blood cells, since they exist as isolated cells. But for cells in a tissue, flow cytometry can only be applied if we can excise the tissue and separate it into single cells (using proteases to cleave cell-cell adhesion molecules) without disrupting cell integrity. This method may be used on tumors, but more often, immunohistochemistry and immunocytochemistry are used to study cells in tissues.

LINK TO LEARNING

Watch videos to learn more about how flow cytometry (https://openstax.org/l/22flowcytometry) and a FACS (https://openstax.org/l/22FACSwork) work.

CHECK YOUR UNDERSTANDING

- In fluorescence activated cell sorting, what characteristic of the target cells allows them to be separated?

Table 20.5 compares the mechanisms of the fluorescent antibody techniques discussed in this section.

Fluorescent Antibody Techniques

Type of Assay	Mechanism	Examples
Direct fluorescent antibody (DFA)	Uses fluorogen-antibody conjugates to label bacteria from patient samples	Visualizing *Legionella pneumophila* from a throat swab
Indirect fluorescent antibody (IFA)	Detects disease-specific antibodies in patent serum	Diagnosing syphilis; detecting antinuclear antibodies (ANA) for lupus and other autoimmune diseases
Flow cytometry	Labels cell membranes with fluorogen-antibody conjugate markers excited by a laser; machine counts the cell and records the relative fluorescence	Counting the number of fluorescently labeled CD4 or CD8 cells in a sample
Fluorescence activated cell sorter (FACS)	Form of flow cytometry that both counts cells and physically separates them into pools of high and low fluorescence cells	Sorting cancer cells

Table 20.5

SUMMARY

20.1 Polyclonal and Monoclonal Antibody Production

- Antibodies bind with high **specificity** to antigens used to challenge the immune system, but they may also show **cross-reactivity** by binding to other antigens that share chemical properties with the original antigen.
- Injection of an antigen into an animal will result in a **polyclonal antibody** response in which different antibodies are produced that react with the various epitopes on the antigen.
- **Polyclonal antisera** are useful for some types of laboratory assays, but other assays require more specificity. Diagnostic tests that use polyclonal antisera are typically only used for screening because of the possibility of **false-positive** and **false-negative** results.
- **Monoclonal antibodies** provide higher specificity than polyclonal antisera because they bind to a single epitope and usually have high **affinity**.
- Monoclonal antibodies are typically produced by culturing antibody-secreting **hybridomas** derived from mice. mAbs are currently used to treat cancer, but their exorbitant cost has prevented them from being used more widely to treat infectious diseases. Still, their potential for laboratory and clinical use is driving the development of new, cost-effective solutions such as **plantibodies**.

20.2 Detecting Antigen-Antibody Complexes

- When present in the correct ratio, antibody and antigen will form a **precipitin**, or lattice that precipitates out of solution.
- A **precipitin ring test** can be used to visualize lattice formation in solution. The **Ouchterlony assay** demonstrates lattice formation in a gel. The **radial immunodiffusion** assay is used to quantify antigen by measuring the size of a precipitation zone in a gel infused with antibodies.
- Insoluble antigens in suspension will form **flocculants** when bound by antibodies. This is the basis of the VDRL test for syphilis in which anti-treponemal antibodies bind to cardiolipin in suspension.
- Viral infections can be detected by quantifying virus-neutralizing antibodies in a patient's serum.
- Different antibody classes in plasma or serum are identified by using **immunoelectrophoresis**.
- The presence of specific antigens (e.g., bacterial or viral proteins) in serum can be demonstrated by **western blot** assays, in which the proteins are transferred to a nitrocellulose membrane and identified using labeled antibodies.
- In the complement fixation test, complement is used to detect antibodies against various pathogens.

20.3 Agglutination Assays

- Antibodies can agglutinate cells or large particles into a visible matrix. **Agglutination** tests are often done on cards or in **microtiter plates** that allow multiple reactions to take place side by side using small volumes of reagents.
- Using antisera against certain proteins allows identification of **serovars** within species of bacteria.
- Detecting antibodies against a pathogen can be a powerful tool for diagnosing disease, but there is a period of time before patients go through **seroconversion** and the level of antibodies becomes detectable.
- Agglutination of latex beads in **indirect agglutination assays** can be used to detect the presence of specific antigens or specific antibodies in patient serum.
- The presence of some antibacterial and antiviral antibodies can be confirmed by the use of the direct **Coombs' test**, which uses Coombs' reagent to cross-link antibodies bound to red blood cells and facilitate **hemagglutination**.
- Some viruses and bacteria will bind and agglutinate red blood cells; this interaction is the basis of the **direct hemagglutination assay**, most often used to determine the titer of virus in solution.
- **Neutralization assays** quantify the level of virus-specific antibody by measuring the decrease in hemagglutination observed after mixing patient serum with a standardized amount of virus.
- Hemagglutination assays are also used to screen and **cross-match** donor and recipient blood to ensure that the transfusion recipient does not have antibodies to antigens in the donated blood.

20.4 EIAs and ELISAs

- **Enzyme immunoassays (EIA)** are used to visualize and quantify antigens. They use an antibody conjugated to an enzyme to bind the antigen, and the enzyme converts a substrate into an observable end product. The substrate may be either a chromogen or a fluorogen.
- **Immunostaining** is an EIA technique for visualizing cells in a tissue (**immunohistochemistry**) or examining intracellular structures (**immunocytochemistry**).
- **Direct ELISA** is used to quantify an antigen in solution. The primary antibody captures the antigen, and the secondary antibody delivers an enzyme. Production of end product from the chromogenic substrate is directly proportional to the amount of captured antigen.
- **Indirect ELISA** is used to detect antibodies in patient serum by attaching antigen to the well of a microtiter plate, allowing the patient (primary) antibody to bind the antigen and an enzyme-conjugated secondary antibody to detect the primary antibody.
- **Immunofiltration and immunochromatographic assays** are used in **lateral flow tests**, which can be used to diagnose pregnancy and various diseases by detecting color-labeled antigen-antibody complexes in urine or other fluid samples

20.5 Fluorescent Antibody Techniques

- **Immunofluorescence** assays use antibody-fluorogen conjugates to illuminate antigens for easy, rapid detection.
- **Direct immunofluorescence** can be used to detect the presence of bacteria in clinical samples such as sputum.
- **Indirect immunofluorescence** detects the presence of antigen-specific antibodies in patient sera. The fluorescent antibody binds to the antigen-specific antibody rather than the antigen.
- The use of indirect immunofluorescence assays to detect **antinuclear antibodies** is an important tool in the diagnosis of several autoimmune diseases.
- **Flow cytometry** uses fluorescent mAbs against cell-membrane proteins to quantify specific subsets of cells in complex mixtures.
- **Fluorescence-activated cell sorters** are an extension of flow cytometry in which fluorescence intensity is used to physically separate cells into high and low fluorescence populations.

REVIEW QUESTIONS

Multiple Choice

1. For many uses in the laboratory, polyclonal antibodies work well, but for some types of assays, they lack sufficient _____ because they cross-react with inappropriate antigens.
   a. specificity
   b. sensitivity
   c. accuracy
   d. reactivity

2. How are monoclonal antibodies produced?
   a. Antibody-producing B cells from a mouse are fused with myeloma cells and then the cells are grown in tissue culture.
   b. A mouse is injected with an antigen and then antibodies are harvested from its serum.
   c. They are produced by the human immune system as a natural response to an infection.
   d. They are produced by a mouse's immune system as a natural response to an infection.

3. The formation of _____ is a positive result in the VDRL test.
   a. flocculant
   b. precipitin
   c. coagulation
   d. a bright pink color

4. The titer of a virus neutralization test is the highest dilution of patient serum
   a. in which there is no detectable viral DNA.
   b. in which there is no detectable viral protein.
   c. that completely blocks plaque formation.
   d. that reduces plaque formation by at least 50%.

5. In the Ouchterlony assay, we see a sharp precipitin arc form between antigen and antiserum. Why does this arc remain visible for a long time?
   a. The antibody molecules are too large to diffuse through the agar.
   b. The precipitin lattice is too large to diffuse through the agar.
   c. Methanol, added once the arc forms, denatures the protein and blocks diffusion.
   d. The antigen molecules are chemically coupled to the gel matrix.

6. We use antisera to distinguish between various _____ within a species of bacteria.
   a. isotypes
   b. serovars
   c. subspecies
   d. lines

7. When using antisera to characterize bacteria, we will often link the antibodies to _____ to better visualize the agglutination.
   a. latex beads
   b. red blood cells
   c. other bacteria
   d. white blood cells

8. The antibody screening test that is done along with pretransfusion blood typing is used to ensure that the recipient
   a. does not have a previously undetected bacterial or viral infection.
   b. is not immunocompromised.
   c. actually does have the blood type stated in the online chart.
   d. is not making antibodies against antigens outside the ABO or Rh systems.

9. The direct Coombs' test is designed to detect when people have a disease that causes them to
   a. have an excessively high fever.
   b. quit making antibodies.
   c. make too many red blood cells.
   d. produce antibodies that bind to their own red blood cells.

10. Viral hemagglutination assays only work with certain types of viruses because
    a. the virus must be able to cross-link red blood cells directly.
    b. the virus must be able to lyse red blood cells.
    c. the virus must not be able to lyse red blood cells.
    d. other viruses are too dangerous to work with in a clinical lab setting.

11. In an enzyme immunoassay, the enzyme
    a. is bound by the antibody's antigen-binding site.
    b. is attached to the well of a microtiter plate.
    c. is conjugated to the suspect antigen.
    d. is bound to the constant region of the secondary antibody.

12. When using an EIA to study microtubules or other structures inside a cell, we first chemically fix the cell and then treat the cells with alcohol. What is the purpose of this alcohol treatment?
    a. It makes holes in the cell membrane large enough for antibodies to pass.
    b. It makes the membrane sticky so antibodies will bind and be taken up by receptor-mediated endocytosis.
    c. It removes negative charges from the membrane, which would otherwise repulse the antibodies.
    d. It prevents nonspecific binding of the antibodies to the cell membrane.

13. In a lateral-flow pregnancy test, you see a blue band form on the control line and no band form on the test line. This is probably a _____ test for pregnancy.
    a. positive
    b. false-positive
    c. false-negative
    d. negative

14. When performing an FEIA, the fluorogen replaces the _____ that is used in an EIA.
    a. antigen
    b. chromogenic substrate
    c. enzyme
    d. secondary antibody

15. Suppose you need to quantify the level of CD8 T cells in the blood of a patient recovering from influenza. You treat a sample of the patient's white blood cells using a fluorescent mAb against CD8, pass the cells through a flow cytometer, and produce the histogram shown below. The area under the peak to the left (blue) is three times greater than the area of the peak on the right (red). What can you determine from these data?

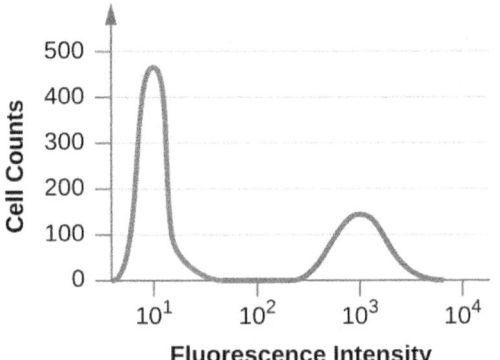

a. There are no detectable CD8 cells.
b. There are three times as many CD4 cells than CD8 cells.
c. There are three times as many CD8 cells than CD4 cells.
d. CD8 cells make up about one-fourth of the total number of cells.

16. In the data described in the previous question, the average fluorescence intensity of cells in the second (red) peak is about _____ that in the first (blue) peak.
a. three times
b. 100 times
c. one-third
d. 1000 times

17. In a direct fluorescent antibody test, which of the following would we most likely be looking for using a fluorescently-labeled mAb?
a. bacteria in a patient sample
b. bacteria isolated from a patient and grown on agar plates
c. antiserum from a patient smeared onto a glass slide
d. antiserum from a patient that had bound to antigen-coated beads

Fill in the Blank

18. When we inject an animal with the same antigen a second time a few weeks after the first, _____ takes place, which means the antibodies produced after the second injection will on average bind the antigen more tightly.

19. When using mAbs to treat disease in humans, the mAbs must first be _____ by replacing the mouse constant region DNA with human constant region DNA.

20. If we used normal mouse mAbs to treat human disease, multiple doses would cause the patient to respond with _____ against the mouse antibodies.

21. A polyclonal response to an infection occurs because most antigens have multiple _____.

22. When slowly adding antigen to an antiserum, the amount of precipitin would gradually increase until reaching the _____; addition of more antigen after this point would actually decrease the amount of precipitin.

23. The radial immunodiffusion test quantifies antigen by mixing _____ into a gel and then allowing antigen to diffuse out from a well cut in the gel.

24. In the major cross-match, we mix _____ with the donor red blood cells and look for agglutination.

25. Coombs' reagent is an antiserum with antibodies that bind to human _____.

26. To detect antibodies against bacteria in the bloodstream using an EIA, we would run a(n) _____, which we would start by attaching antigen from the bacteria to the wells of a microtiter plate.

27. In flow cytometry, cell subsets are labeled using a fluorescent antibody to a membrane protein. The fluorogen is activated by a(n) _____ as the cells pass by the detectors.

28. Fluorescence in a flow cytometer is measured by a detector set at an angle to the light source. There is also an in-line detector that can detect cell clumps or _____.

Short Answer

29. Describe two reasons why polyclonal antibodies are more likely to exhibit cross-reactivity than monoclonal antibodies.
30. Explain why hemolysis in the complement fixation test is a negative test for infection.
31. What is meant by the term "neutralizing antibodies," and how can we quantify this effect using the viral neutralization assay?
32. Explain why the titer of a direct hemagglutination assay is the highest dilution that still causes hemagglutination, whereas in the viral hemagglutination inhibition assay, the titer is the highest dilution at which hemagglutination is not observed.
33. Why would a doctor order a direct Coombs' test when a baby is born with jaundice?
34. Why is it important in a sandwich ELISA that the antigen has multiple epitopes? And why might it be advantageous to use polyclonal antisera rather than mAb in this assay?
35. The pregnancy test strip detects the presence of human chorionic gonadotrophin in urine. This hormone is initially produced by the fetus and later by the placenta. Why is the test strip preferred for this test rather than using either a direct or indirect ELISA with their more quantifiable results?

Critical Thinking

36. Suppose you were screening produce in a grocery store for the presence of *E. coli* contamination. Would it be better to use a polyclonal anti-*E. coli* antiserum or a mAb against an *E. coli* membrane protein? Explain.
37. Both IgM and IgG antibodies can be used in precipitation reactions. However, one of these immunoglobulin classes will form precipitates at much lower concentrations than the other. Which class is this, and why is it so much more efficient in this regard?
38. When shortages of donated blood occur, O-negative blood may be given to patients, even if they have a different blood type. Why is this the case? If O-negative blood supplies were depleted, what would be the next-best choice for a patient with a different blood type in critical need of a transfusion? Explain your answers.
39. Label the primary and secondary antibodies, and discuss why the production of end product will be proportional to the amount of antigen.

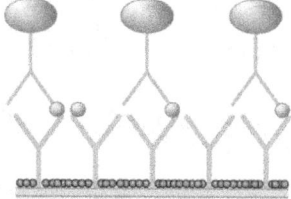

40. A patient suspected of having syphilis is tested using both the VDRL test and IFA. The IFA test comes back positive, but the VDRL test is negative. What is the most likely reason for these results?
41. A clinician suspects that a patient with pneumonia may be infected by *Legionella pneumophila*. Briefly describe two reasons why a DFA test might be better for detecting this pathogen than standard bacteriology techniques.

CHAPTER 21
Skin and Eye Infections

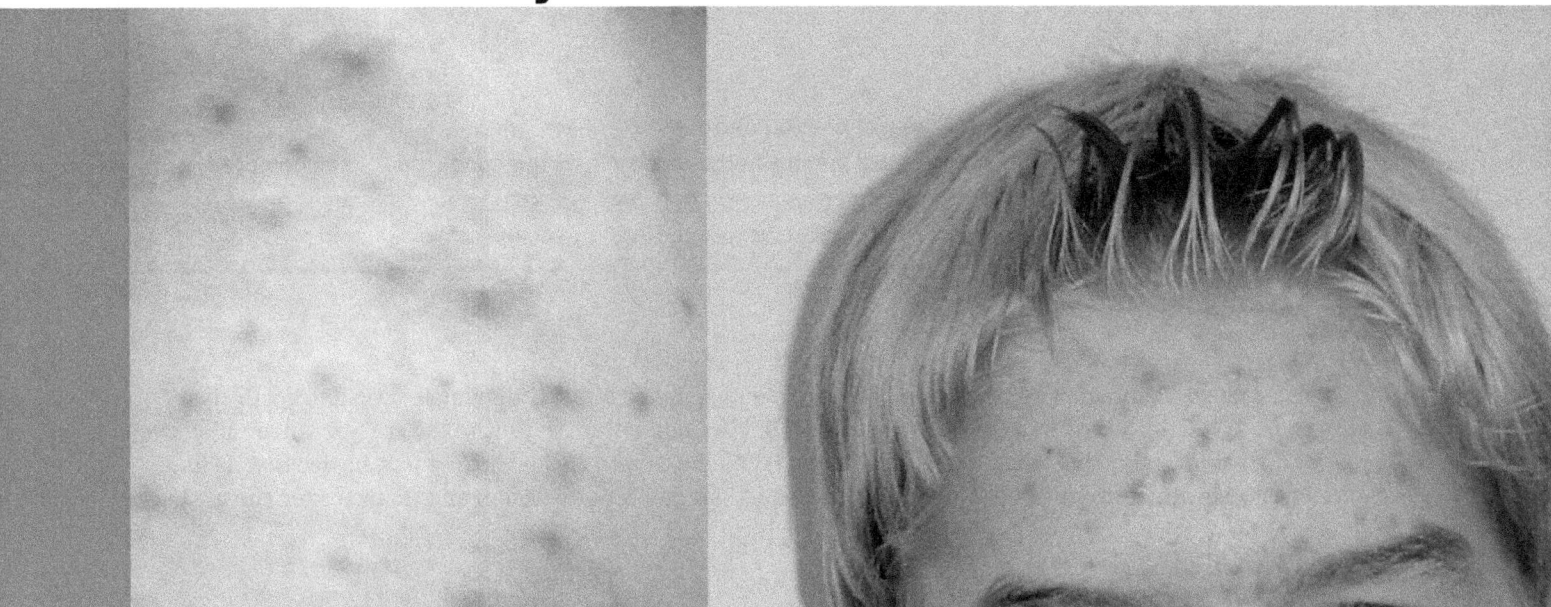

Figure 21.1 The skin is an important barrier to pathogens, but it can also develop infections. These raised lesions (left) are typical of folliculitis, a condition that results from the inflammation of hair follicles. Acne lesions (right) also result from inflammation of hair follicles. In this case, the inflammation results when hair follicles become clogged with complex lipids, fatty acids, and dead skin cells, producing a favorable environment for bacteria.

Chapter Outline

21.1 Anatomy and Normal Microbiota of the Skin and Eyes

21.2 Bacterial Infections of the Skin and Eyes

21.3 Viral Infections of the Skin and Eyes

21.4 Mycoses of the Skin

21.5 Protozoan and Helminthic Infections of the Skin and Eyes

INTRODUCTION The human body is covered in skin, and like most coverings, skin is designed to protect what is underneath. One of its primary purposes is to prevent microbes in the surrounding environment from invading underlying tissues and organs. But in spite of its role as a protective covering, skin is not itself immune from infection. Certain pathogens and toxins can cause severe infections or reactions when they come in contact with the skin. Other pathogens are opportunistic, breaching the skin's natural defenses through cuts, wounds, or a disruption of normal microbiota resulting in an infection in the surrounding skin and tissue. Still other pathogens enter the body via different routes—through the respiratory or digestive systems, for example—but cause reactions that manifest as skin rashes or lesions.

Nearly all humans experience skin infections to some degree. Many of these conditions are, as the name suggests, "skin deep," with symptoms that are local and non-life-threatening. At some point, almost everyone must endure conditions like acne, athlete's foot, and minor infections of cuts and abrasions, all of which result from infections of the skin. But not all skin infections are quite so innocuous. Some can become invasive, leading to systemic infection or spreading over large areas of skin, potentially becoming life-threatening.

21.1 Anatomy and Normal Microbiota of the Skin and Eyes

Learning Objectives

By the end of this section, you will be able to:
- Describe the major anatomical features of the skin and eyes
- Compare and contrast the microbiomes of various body sites, such as the hands, back, feet, and eyes
- Explain how microorganisms overcome defenses of skin and eyes in order to cause infection
- Describe general signs and symptoms of disease associated with infections of the skin and eyes

---

Clinical Focus

Part 1

Sam, a college freshman with a bad habit of oversleeping, nicked himself shaving in a rush to get to class on time. At the time, he didn't think twice about it. But two days later, he noticed the cut was surrounded by a reddish area of skin that was warm to the touch. When the wound started oozing pus, he decided he had better stop by the university's clinic. The doctor took a sample from the lesion and then cleaned the area.

- What type of microbe could be responsible for Sam's infection?

Jump to the next Clinical Focus box.

Human skin is an important part of the innate immune system. In addition to serving a wide range of other functions, the skin serves as an important barrier to microbial invasion. Not only is it a physical barrier to penetration of deeper tissues by potential pathogens, but it also provides an inhospitable environment for the growth of many pathogens. In this section, we will provide a brief overview of the anatomy and normal microbiota of the skin and eyes, along with general symptoms associated with skin and eye infections.

Layers of the Skin

Human skin is made up of several layers and sublayers. The two main layers are the **epidermis** and the **dermis**. These layers cover a third layer of tissue called the **hypodermis**, which consists of fibrous and adipose connective tissue (Figure 21.2).

The epidermis is the outermost layer of the skin, and it is relatively thin. The exterior surface of the epidermis, called the **stratum corneum**, primarily consists of dead skin cells. This layer of dead cells limits direct contact between the outside world and live cells. The stratum corneum is rich in **keratin**, a tough, fibrous protein that is also found in hair and nails. Keratin helps make the outer surface of the skin relatively tough and waterproof. It also helps to keep the surface of the skin dry, which reduces microbial growth. However, some microbes are still able to live on the surface of the skin, and some of these can be shed with dead skin cells in the process of **desquamation**, which is the shedding and peeling of skin that occurs as a normal process but that may be accelerated when infection is present.

Beneath the epidermis lies a thicker skin layer called the dermis. The dermis contains connective tissue and embedded structures such as blood vessels, nerves, and muscles. Structures called **hair follicles** (from which hair grows) are located within the dermis, even though much of their structure consists of epidermal tissue. The dermis also contains the two major types of glands found in human skin: **sweat glands** (tubular glands that produce sweat) and **sebaceous glands** (which are associated with hair follicles and produce **sebum**, a lipid-rich substance containing proteins and minerals).

Perspiration (sweat) provides some moisture to the epidermis, which can increase the potential for microbial growth. For this reason, more microbes are found on the regions of the skin that produce the most sweat, such as the skin of the underarms and groin. However, in addition to water, sweat also contains substances that inhibit microbial growth, such as salts, lysozyme, and antimicrobial peptides. Sebum also serves to protect the skin and reduce water loss. Although some of the lipids and fatty acids in sebum inhibit microbial growth, sebum contains compounds that provide nutrition for certain microbes.

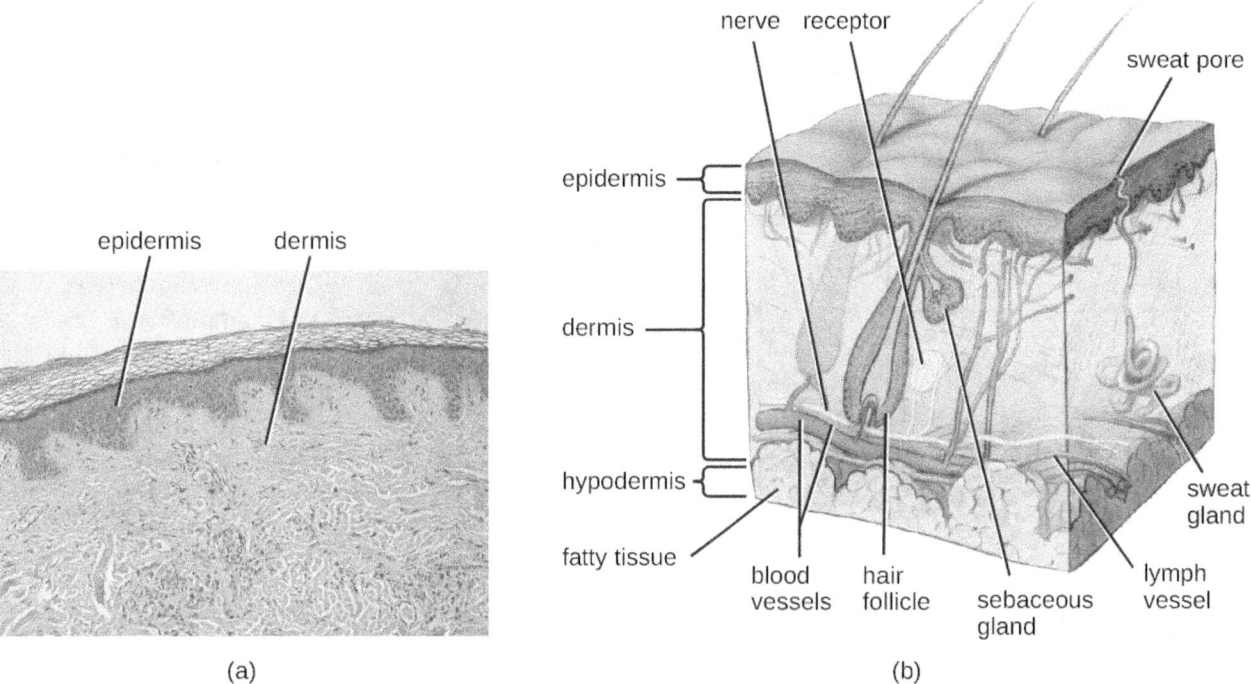

Figure 21.2 (a) A micrograph of a section through human skin shows the epidermis and dermis. (b) The major layers of human skin are the epidermis, dermis, and hypodermis. (credit b: modification of work by National Cancer Institute)

✓ CHECK YOUR UNDERSTANDING

- How does desquamation help with preventing infections?

Normal Microbiota of the Skin

The skin is home to a wide variety of normal microbiota, consisting of commensal organisms that derive nutrition from skin cells and secretions such as sweat and sebum. The normal microbiota of skin tends to inhibit transient-microbe colonization by producing antimicrobial substances and outcompeting other microbes that land on the surface of the skin. This helps to protect the skin from pathogenic infection.

The skin's properties differ from one region of the body to another, as does the composition of the skin's microbiota. The availability of nutrients and moisture partly dictates which microorganisms will thrive in a particular region of the skin. Relatively moist skin, such as that of the nares (nostrils) and underarms, has a much different microbiota than the dryer skin on the arms, legs, hands, and top of the feet. Some areas of the skin have higher densities of sebaceous glands. These sebum-rich areas, which include the back, the folds at the side of the nose, and the back of the neck, harbor distinct microbial communities that are less diverse than those found on other parts of the body.

Different types of bacteria dominate the dry, moist, and sebum-rich regions of the skin. The most abundant microbes typically found in the dry and sebaceous regions are Betaproteobacteria and Propionibacteria, respectively. In the moist regions, *Corynebacterium* and *Staphylococcus* are most commonly found (Figure 21.3). Viruses and fungi are also found on the skin, with *Malassezia* being the most common type of fungus found as part of the normal microbiota. The role and populations of viruses in the microbiota, known as viromes, are still not well understood, and there are limitations to the techniques used to identify them. However, Circoviridae, Papillomaviridae, and Polyomaviridae appear to be the most common residents in the healthy skin virome.[1][2][3]

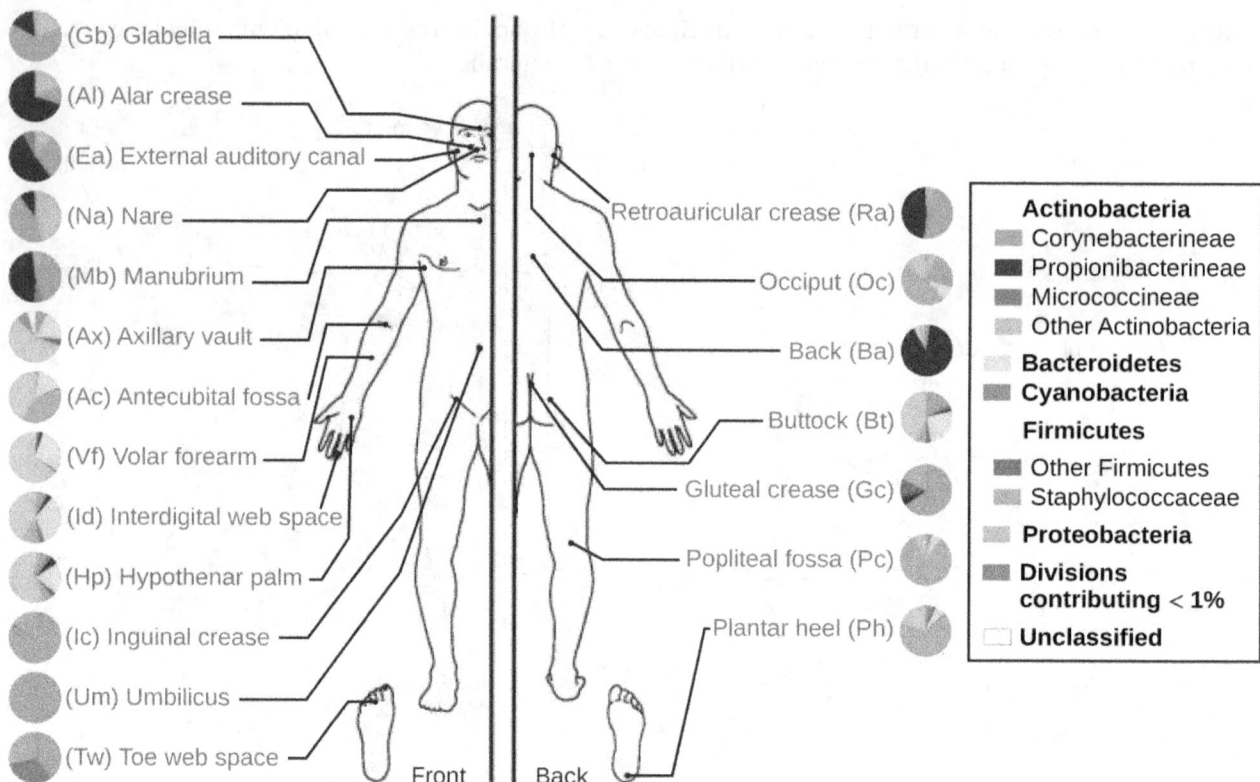

Figure 21.3 The normal microbiota varies on different regions of the skin, especially in dry versus moist areas. The figure shows the major organisms commonly found in different locations of a healthy individual's skin and external mucosa. Note that there is significant variation among individuals. (credit: modification of work by National Human Genome Research Institute)

CHECK YOUR UNDERSTANDING

- What are the four most common bacteria that are part of the normal skin microbiota?

Infections of the Skin

While the microbiota of the skin can play a protective role, it can also cause harm in certain cases. Often, an opportunistic pathogen residing in the skin microbiota of one individual may be transmitted to another individual more susceptible to an infection. For example, methicillin-resistant *Staphylococcus aureus* (MRSA) can often take up residence in the nares of health care workers and hospital patients; though harmless on intact, healthy skin, MRSA can cause infections if introduced into other parts of the body, as might occur during surgery or via a post-surgical incision or wound. This is one reason why clean surgical sites are so important.

Injury or damage to the skin can allow microbes to enter deeper tissues, where nutrients are more abundant and the environment is more conducive to bacterial growth. Wound infections are common after a puncture or laceration that damages the physical barrier of the skin. Microbes may infect structures in the dermis, such as hair follicles and glands, causing a localized infection, or they may reach the bloodstream, which can lead to a systemic infection.

In some cases, infectious microbes can cause a variety of rashes or lesions that differ in their physical

---

1 Belkaid, Y., and J.A. Segre. "Dialogue Between Skin Microbiota and Immunity," *Science* 346 (2014) 6212:954–959.

2 Foulongne, Vincent, et al. "Human Skin Microbiota: High Diversity of DNA Viruses Identified on the Human Skin by High Throughput Sequencing." *PLoS ONE* (2012) 7(6): e38499. doi: 10.1371/journal.pone.0038499.

3 Robinson, C.M., and J.K. Pfeiffer. "Viruses and the Microbiota." *Annual Review of Virology* (2014) 1:55–59. doi: 10.1146/annurev-virology-031413-085550.

characteristics. These rashes can be the result of inflammation reactions or direct responses to toxins produced by the microbes. Table 21.1 lists some of the medical terminology used to describe skin lesions and rashes based on their characteristics; Figure 21.4 and Figure 21.5 illustrate some of the various types of skin lesions. It is important to note that many different diseases can lead to skin conditions of very similar appearance; thus the terms used in the table are generally not exclusive to a particular type of infection or disease.

Some Medical Terms Associated with Skin Lesions and Rashes

Term	Definition
abscess	localized collection of pus
bulla (pl., bullae)	fluid-filled blister no more than 5 mm in diameter
carbuncle	deep, pus-filled abscess generally formed from multiple furuncles
crust	dried fluids from a lesion on the surface of the skin
cyst	encapsulated sac filled with fluid, semi-solid matter, or gas, typically located just below the upper layers of skin
folliculitis	a localized rash due to inflammation of hair follicles
furuncle (boil)	pus-filled abscess due to infection of a hair follicle
macules	smooth spots of discoloration on the skin
papules	small raised bumps on the skin
pseudocyst	lesion that resembles a cyst but with a less defined boundary
purulent	pus-producing; suppurative
pustules	fluid- or pus-filled bumps on the skin
pyoderma	any suppurative (pus-producing) infection of the skin
suppurative	producing pus; purulent
ulcer	break in the skin; open sore
vesicle	small, fluid-filled lesion
wheal	swollen, inflamed skin that itches or burns, such as from an insect bite

Table 21.1

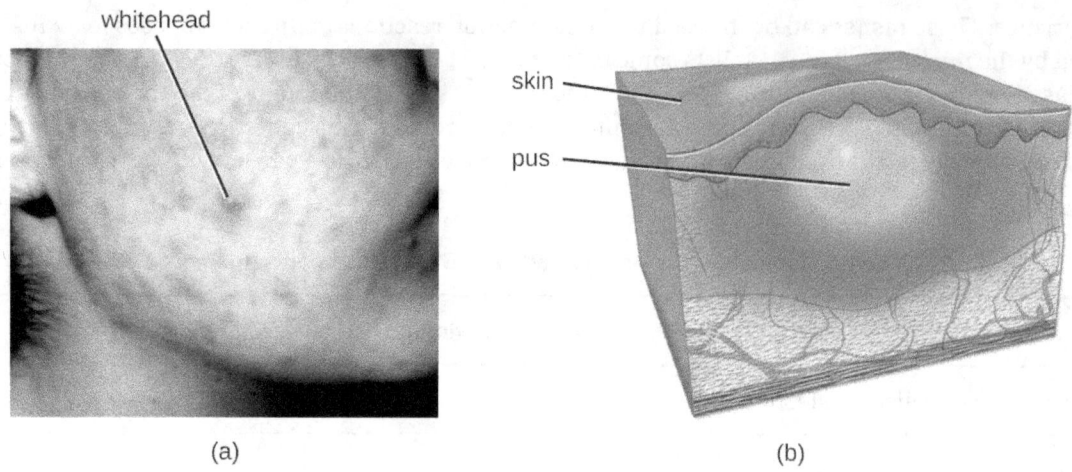

Figure 21.4 (a) Acne is a bacterial infection of the skin that manifests as a rash of inflamed hair follicles (folliculitis). The large whitehead near the center of the cheek is an infected hair follicle that has become purulent (or suppurative), leading to the formation of a furuncle. (b) An abscess is a pus-filled lesion. (credit b: modification of work by Bruce Blaus)

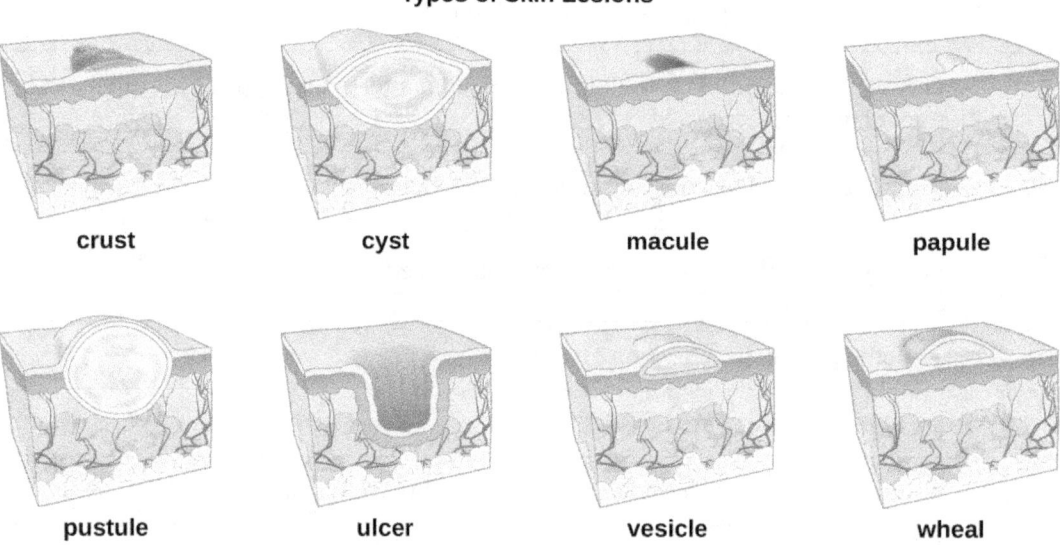

Figure 21.5 Numerous causes can lead to skin lesions of various types, some of which are very similar in appearance. (credit: modification of work by Bruce Blaus)

CHECK YOUR UNDERSTANDING

- How can asymptomatic health care workers transmit bacteria such as MRSA to patients?

Anatomy and Microbiota of the Eye

Although the eye and skin have distinct anatomy, they are both in direct contact with the external environment. An important component of the eye is the nasolacrimal drainage system, which serves as a conduit for the fluid of the eye, called tears. Tears flow from the external eye to the nasal cavity by the lacrimal apparatus, which is composed of the structures involved in tear production (Figure 21.6). The **lacrimal gland**, above the eye, secretes tears to keep the eye moist. There are two small openings, one on the inside edge of the upper eyelid and one on the inside edge of the lower eyelid, near the nose. Each of these openings is called a **lacrimal punctum**. Together, these lacrimal puncta collect tears from the eye that are then conveyed through **lacrimal ducts** to a reservoir for tears called the **lacrimal sac,** also known as the dacrocyst or tear sac.

From the sac, tear fluid flows via a **nasolacrimal duct** to the inner nose. Each nasolacrimal duct is located underneath the skin and passes through the bones of the face into the nose. Chemicals in tears, such as defensins, lactoferrin, and lysozyme, help to prevent colonization by pathogens. In addition, mucins facilitate removal of microbes from the surface of the eye.

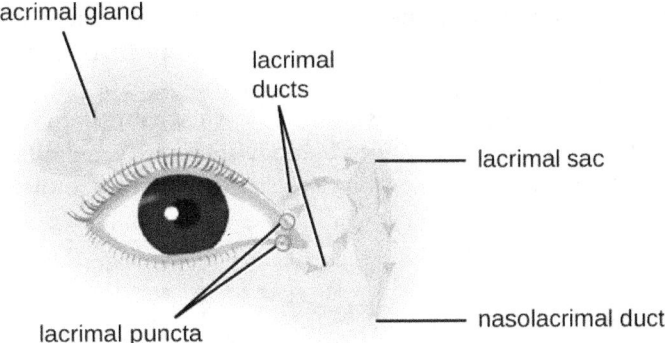

Figure 21.6 The lacrimal apparatus includes the structures of the eye associated with tear production and drainage. (credit: modification of work by "Evidence Based Medical Educator Inc."/YouTube)

The surfaces of the eyeball and inner eyelid are mucous membranes called **conjunctiva**. The normal conjunctival microbiota has not been well characterized, but does exist. One small study (part of the Ocular Microbiome project) found twelve genera that were consistently present in the conjunctiva.[4] These microbes are thought to help defend the membranes against pathogens. However, it is still unclear which microbes may be transient and which may form a stable microbiota.[5]

Use of contact lenses can cause changes in the normal microbiota of the conjunctiva by introducing another surface into the natural anatomy of the eye. Research is currently underway to better understand how contact lenses may impact the normal microbiota and contribute to eye disease.

The watery material inside of the eyeball is called the vitreous humor. Unlike the conjunctiva, it is protected from contact with the environment and is almost always sterile, with no normal microbiota (Figure 21.7).

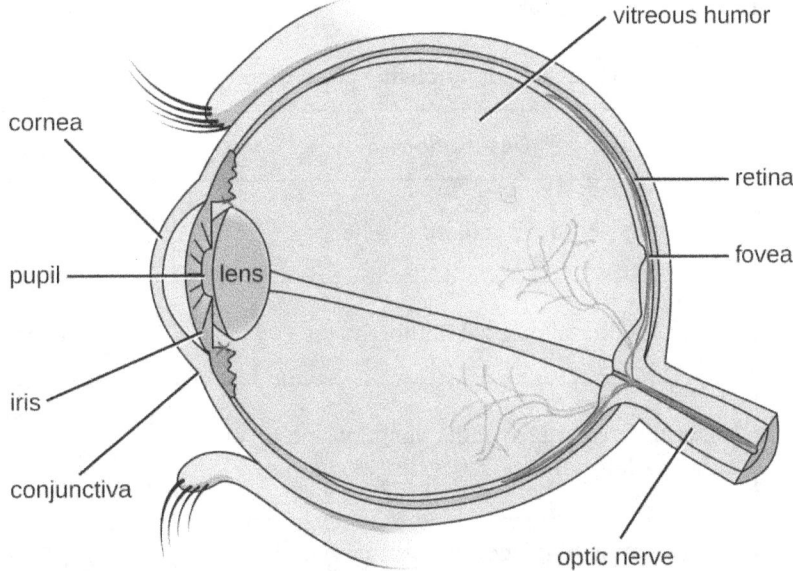

Figure 21.7 Some microbes live on the conjunctiva of the human eye, but the vitreous humor is sterile.

---

4 Abelson, M.B., Lane, K., and Slocum, C.. "The Secrets of Ocular Microbiomes." *Review of Ophthalmology* June 8, 2015. http://www.reviewofophthalmology.com/content/t/ocular_disease/c/55178. Accessed Sept 14, 2016.

5 Shaikh-Lesko, R. "Visualizing the Ocular Microbiome." *The Scientist* May 12, 2014. http://www.the-scientist.com/?articles.view/articleNo/39945/title/Visualizing-the-Ocular-Microbiome. Accessed Sept 14, 2016.

Infections of the Eye

The conjunctiva is a frequent site of infection of the eye; like other mucous membranes, it is also a common portal of entry for pathogens. Inflammation of the conjunctiva is called **conjunctivitis**, although it is commonly known as pinkeye because of the pink appearance in the eye. Infections of deeper structures, beneath the cornea, are less common (Figure 21.8). Conjunctivitis occurs in multiple forms. It may be acute or chronic. Acute purulent conjunctivitis is associated with pus formation, while acute hemorrhagic conjunctivitis is associated with bleeding in the conjunctiva. The term **blepharitis** refers to an inflammation of the eyelids, while **keratitis** refers to an inflammation of the cornea (Figure 21.8); **keratoconjunctivitis** is an inflammation of both the cornea and the conjunctiva, and **dacryocystitis** is an inflammation of the lacrimal sac that can often occur when a nasolacrimal duct is blocked.

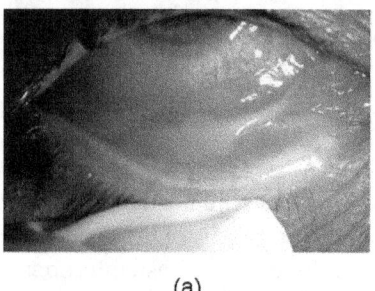
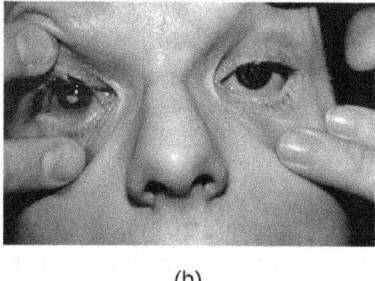
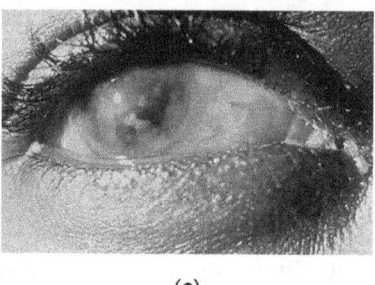

(a) (b) (c)

Figure 21.8 (a) Conjunctivitis is inflammation of the conjunctiva. (b) Blepharitis is inflammation of the eyelids. (c) Keratitis is inflammation of the cornea. (credit a: modification of work by Lopez-Prats MJ, Sanz Marco E, Hidalgo-Mora JJ, Garcia-Delpech S, Diaz-Llopis M; credit b, c: modification of work by Centers for Disease Control and Prevention)

Infections leading to conjunctivitis, blepharitis, keratoconjunctivitis, or dacryocystitis may be caused by bacteria or viruses, but allergens, pollutants, or chemicals can also irritate the eye and cause inflammation of various structures. Viral infection is a more likely cause of conjunctivitis in cases with symptoms such as fever and watery discharge that occurs with upper respiratory infection and itchy eyes. Table 21.2 summarizes some common forms of conjunctivitis and blepharitis.

Types of Conjunctivities and Blepharitis

Condition	Description	Causative Agent(s)
Acute purulent conjunctivitis	Conjunctivitis with purulent discharge	Bacterial (*Haemophilus*, *Staphylococcus*)
Acute hemorrhagic conjunctivitis	Involves subconjunctival hemorrhages	Viral (Picornaviradae)
Acute ulcerative blepharitis	Infection involving eyelids; pustules and ulcers may develop	Bacterial (*Staphylococcal*) or viral (herpes simplex, varicella-zoster, etc.)
Follicular conjunctivitis	Inflammation of the conjunctiva with nodules (dome-shaped structures that are red at the base and pale on top)	Viral (adenovirus and others); environmental irritants
Dacryocystitis	Inflammation of the lacrimal sac often associated with a plugged nasolacrimal duct	Bacterial (*Haemophilus*, *Staphylococcus*, *Streptococcus*)

Types of Conjunctivities and Blepharitis

Condition	Description	Causative Agent(s)
Keratitis	Inflammation of cornea	Bacterial, viral, or protozoal; environmental irritants
Keratoconjunctivitis	Inflammation of cornea and conjunctiva	Bacterial, viral (adenoviruses), or other causes (including dryness of the eye)
Nonulcerative blepharitis	Inflammation, irritation, redness of the eyelids without ulceration	Environmental irritants; allergens
Papillary conjunctivitis	Inflammation of the conjunctiva; nodules and papillae with red tops develop	Environmental irritants; allergens

Table 21.2

CHECK YOUR UNDERSTANDING
- How does the lacrimal apparatus help to prevent eye infections?

21.2 Bacterial Infections of the Skin and Eyes

Learning Objectives

By the end of this section, you will be able to:
- Identify the most common bacterial pathogens that cause infections of the skin and eyes
- Compare the major characteristics of specific bacterial diseases affecting the skin and eyes

Despite the skin's protective functions, infections are common. Gram-positive *Staphylococcus* spp. and *Streptococcus* spp. are responsible for many of the most common skin infections. However, many skin conditions are not strictly associated with a single pathogen. Opportunistic pathogens of many types may infect skin wounds, and individual cases with identical symptoms may result from different pathogens or combinations of pathogens.

In this section, we will examine some of the most important bacterial infections of the skin and eyes and discuss how biofilms can contribute to and exacerbate such infections. Key features of bacterial skin and eye infections are also summarized in the Disease Profile boxes throughout this section.

Staphylococcal Infections of the Skin

Staphylococcus species are commonly found on the skin, with *S. epidermidis* and *S. hominis* being prevalent in the normal microbiota. *S. aureus* is also commonly found in the nasal passages and on healthy skin, but pathogenic strains are often the cause of a broad range of infections of the skin and other body systems.

S. aureus is quite contagious. It is spread easily through skin-to-skin contact, and because many people are chronic nasal carriers (asymptomatic individuals who carry *S. aureus* in their nares), the bacteria can easily be transferred from the nose to the hands and then to fomites or other individuals. Because it is so contagious, *S. aureus* is prevalent in most community settings. This prevalence is particularly problematic in hospitals, where antibiotic-resistant strains of the bacteria may be present, and where immunocompromised patients may be more susceptible to infection. Resistant strains include methicillin-resistant *S. aureus* (MRSA), which can be acquired through health-care settings (hospital-acquired MRSA, or HA-MRSA) or in the community (community-acquired MRSA, or CA-MRSA). Hospital patients often arrive at health-care facilities already colonized with antibiotic-resistant strains of *S. aureus* that can be transferred to health-care providers and other patients. Some hospitals have attempted to detect these individuals in order to institute prophylactic measures, but they have had mixed success (see Eye on Ethics: Screening Patients for MRSA).

When a staphylococcal infection develops, choice of medication is important. As discussed above, many staphylococci (such as MRSA) are resistant to some or many antibiotics. Thus, antibiotic sensitivity is measured to identify the most suitable antibiotic. However, even before receiving the results of sensitivity analysis, suspected *S. aureus* infections are often initially treated with drugs known to be effective against MRSA, such as trimethoprim-sulfamethoxazole (TMP/SMZ), clindamycin, a tetracycline (doxycycline or minocycline), or linezolid.

The pathogenicity of staphylococcal infections is often enhanced by characteristic chemicals secreted by some strains. Staphylococcal virulence factors include hemolysins called **staphylolysins**, which are cytotoxic for many types of cells, including skin cells and white blood cells. Virulent strains of *S. aureus* are also coagulase-positive, meaning they produce coagulase, a plasma-clotting protein that is involved in abscess formation. They may also produce leukocidins, which kill white blood cells and can contribute to the production of pus and Protein A, which inhibits phagocytosis by binding to the constant region of antibodies. Some virulent strains of *S. aureus* also produce other toxins, such as toxic shock syndrome toxin-1 (see Virulence Factors of Bacterial and Viral Pathogens).

To confirm the causative agent of a suspected staphylococcal skin infection, samples from the wound are cultured. Under the microscope, gram-positive *Staphylococcus* species have cellular arrangements that form grapelike clusters; when grown on blood agar, colonies have a unique pigmentation ranging from opaque white to cream. A catalase test is used to distinguish *Staphylococcus* from *Streptococcus*, which is also a genus of gram-positive cocci and a common cause of skin infections. *Staphylococcus* species are catalase-positive while *Streptococcus* species are catalase-negative.

Other tests are performed on samples from the wound in order to distinguish coagulase-positive species of *Staphylococcus* (CoPS) such as *S. aureus* from common coagulase-negative species (CoNS) such as *S. epidermidis*. Although CoNS are less likely than CoPS to cause human disease, they can cause infections when they enter the body, as can sometimes occur via catheters, indwelling medical devices, and wounds. Passive agglutination testing can be used to distinguish CoPS from CoNS. If the sample is coagulase-positive, the sample is generally presumed to contain *S. aureus*. Additional genetic testing would be necessary to identify the particular strain of *S. aureus*.

Another way to distinguish CoPS from CoNS is by culturing the sample on mannitol salt agar (MSA). *Staphylococcus* species readily grow on this medium because they are tolerant of the high concentration of sodium chloride (7.5% NaCl). However, CoPS such as *S. aureus* ferment mannitol (which will be evident on a MSA plate), whereas CoNS such as *S. epidermidis* do not ferment mannitol but can be distinguished by the fermentation of other sugars such as lactose, malonate, and raffinose (Figure 21.9).

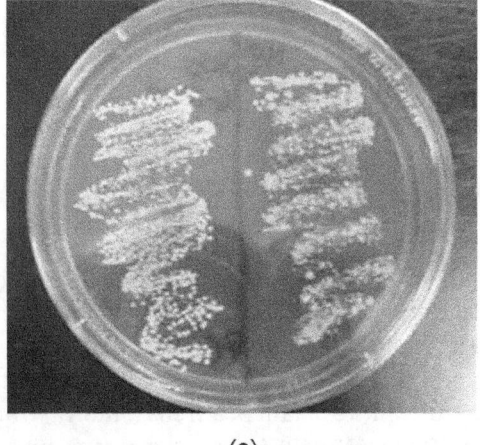
(a)

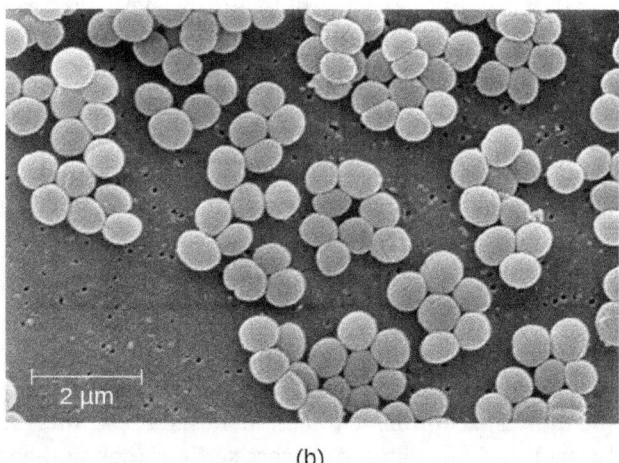
(b)

Figure 21.9 (a) A mannitol salt agar plate is used to distinguish different species of staphylococci. In this plate, *S. aureus* is on the left and

S. epidermidis is in the right. Because *S. aureus* is capable of fermenting mannitol, it produces acids that cause the color to change to yellow. (b) This scanning electron micrograph shows the characteristic grapelike clusters of *S. aureus*. (credit a: modification of work by "ScienceProfOnline"/YouTube; credit b: modification of work by Centers for Disease Control and Prevention)

Eye on Ethics

Screening Patients for MRSA

According to the CDC, 86% of invasive MRSA infections are associated in some way with healthcare, as opposed to being community-acquired. In hospitals and clinics, asymptomatic patients who harbor MRSA may spread the bacteria to individuals who are more susceptible to serious illness.

In an attempt to control the spread of MRSA, hospitals have tried screening patients for MRSA. If patients test positive following a nasal swab test, they can undergo decolonization using chlorhexidine washes or intranasal mupirocin. Some studies have reported substantial reductions in MRSA disease following implementation of these protocols, while others have not. This is partly because there is no standard protocol for these procedures. Several different MRSA identification tests may be used, some involving slower culturing techniques and others rapid testing. Other factors, such as the effectiveness of general hand-washing protocols, may also play a role in helping to prevent MRSA transmission. There are still other questions that need to be addressed: How frequently should patients be screened? Which individuals should be tested? From where on the body should samples be collected? Will increased resistance develop from the decolonization procedures?

Even if identification and decolonization procedures are perfected, ethical questions will remain. Should patients have the right to decline testing? Should a patient who tests positive for MRSA have the right to decline the decolonization procedure, and if so, should hospitals have the right to refuse treatment to the patient? How do we balance the individual's right to receive care with the rights of other patients who could be exposed to disease as a result?

Superficial Staphylococcal Infections

S. aureus is often associated with **pyoderma**, skin infections that are **purulent**. Pus formation occurs because many strains of *S. aureus* produce leukocidins, which kill white blood cells. These purulent skin infections may initially manifest as **folliculitis**, but can lead to **furuncles** or deeper abscesses called **carbuncles**.

Folliculitis generally presents as bumps and pimples that may be itchy, red, and/or pus-filled. In some cases, folliculitis is self-limiting, but if it continues for more than a few days, worsens, or returns repeatedly, it may require medical treatment. Sweat, skin injuries, ingrown hairs, tight clothing, irritation from shaving, and skin conditions can all contribute to folliculitis. Avoidance of tight clothing and skin irritation can help to prevent infection, but topical antibiotics (and sometimes other treatments) may also help. Folliculitis can be identified by skin inspection; treatment is generally started without first culturing and identifying the causative agent.

In contrast, furuncles (boils) are deeper infections (Figure 21.10). They are most common in those individuals (especially young adults and teenagers) who play contact sports, share athletic equipment, have poor nutrition, live in close quarters, or have weakened immune systems. Good hygiene and skin care can often help to prevent furuncles from becoming more infective, and they generally resolve on their own. However, if furuncles spread, increase in number or size, or lead to systemic symptoms such as fever and chills, then medical care is needed. They may sometimes need to be drained (at which time the pathogens can be cultured) and treated with antibiotics.

When multiple boils develop into a deeper lesion, it is called a carbuncle (Figure 21.10). Because carbuncles are deeper, they are more commonly associated with systemic symptoms and a general feeling of illness. Larger, recurrent, or worsening carbuncles require medical treatment, as do those associated with signs of illness such as fever. Carbuncles generally need to be drained and treated with antibiotics. While carbuncles are relatively easy to identify visually, culturing and laboratory analysis of the wound may be recommended for some infections because antibiotic resistance is relatively common.

Proper hygiene is important to prevent these types of skin infections or to prevent the progression of existing infections.

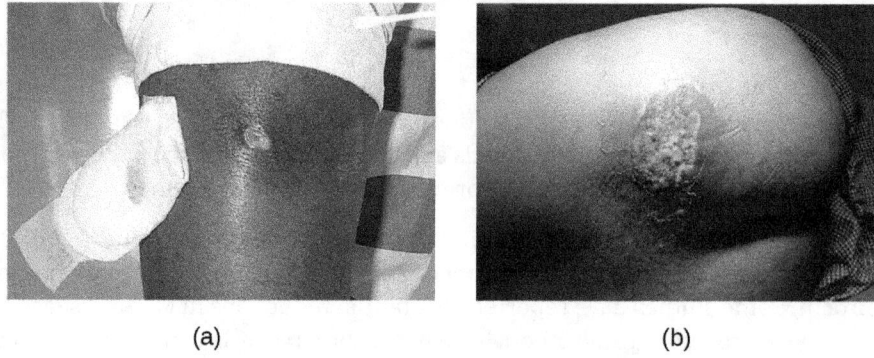

Figure 21.10 Furuncles (boils) and carbuncles are infections of the skin often caused by *Staphylococcus* bacteria. (a) A furuncle contains pus and exhibits swelling. (b) A carbuncle is a pus-filled lesion that is typically deeper than the furuncle. It often forms from multiple furuncles. (Credit a: Public Health Image Library / CDC; Public Domain; credit b: modification of work by "Drvgaikwad"/Wikimedia Commons)

Staphylococcal scalded skin syndrome (SSSS) is another superficial infection caused by *S. aureus* that is most commonly seen in young children, especially infants. Bacterial exotoxins first produce **erythema** (redness of the skin) and then severe peeling of the skin, as might occur after scalding (Figure 21.11). SSSS is diagnosed by examining characteristics of the skin (which may rub off easily), using blood tests to check for elevated white blood cell counts, culturing, and other methods. Intravenous antibiotics and fluid therapy are used as treatment.

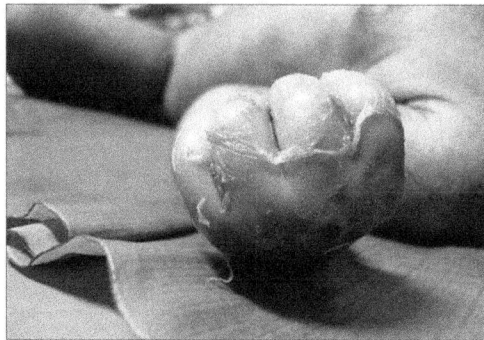

Figure 21.11 A newborn with staphylococcal scalded skin syndrome (SSSS), which results in large regions of peeling, dead skin. (credit: modification of work by D Jeyakumari, R Gopal, M Eswaran, and C MaheshKumar)

Impetigo

The skin infection **impetigo** causes the formation of vesicles, pustules, and possibly bullae, often around the nose and mouth. Bullae are large, fluid-filled blisters that measure at least 5 mm in diameter. Impetigo can be diagnosed as either nonbullous or bullous. In nonbullous impetigo, vesicles and pustules rupture and become encrusted sores. Typically the crust is yellowish, often with exudate draining from the base of the lesion. In bullous impetigo, the bullae fill and rupture, resulting in larger, draining, encrusted lesions (Figure 21.12).

Especially common in children, impetigo is particularly concerning because it is highly contagious. Impetigo can be caused by *S. aureus* alone, by *Streptococcus pyogenes* alone, or by coinfection of *S. aureus* and *S. pyogenes*. Impetigo is often diagnosed through observation of its characteristic appearance, although culture and susceptibility testing may also be used.

Topical or oral antibiotic treatment is typically effective in treating most cases of impetigo. However, cases caused by *S. pyogenes* can lead to serious sequelae (pathological conditions resulting from infection, disease, injury, therapy, or other trauma) such as acute glomerulonephritis (AGN), which is severe inflammation in the kidneys.

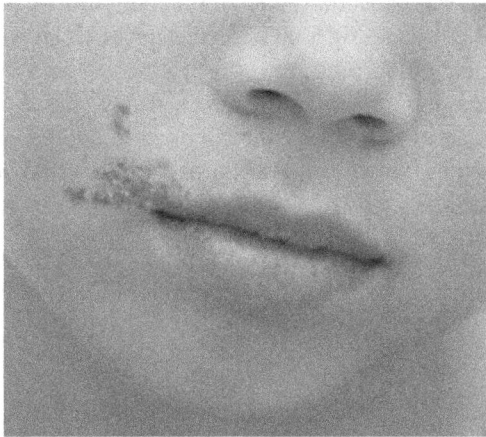

Figure 21.12 Impetigo is characterized by vesicles, pustules, or bullae that rupture, producing encrusted sores. (credit: modification of work by FDA)

Nosocomial *S. epidermidis* Infections

Though not as virulent as *S. aureus*, the staphylococcus *S. epidermidis* can cause serious opportunistic infections. Such infections usually occur only in hospital settings. *S. epidermidis* is usually a harmless resident of the normal skin microbiota. However, health-care workers can inadvertently transfer *S. epidermidis* to medical devices that are inserted into the body, such as catheters, prostheses, and indwelling medical devices. Once it has bypassed the skin barrier, *S. epidermidis* can cause infections inside the body that can be difficult to treat. Like *S. aureus*, *S. epidermidis* is resistant to many antibiotics, and localized infections can become systemic if not treated quickly. To reduce the risk of nosocomial (hospital-acquired) *S. epidermidis*, health-care workers must follow strict procedures for handling and sterilizing medical devices before and during surgical procedures.

⊘ CHECK YOUR UNDERSTANDING

- Why are *Staphylococcus aureus* infections often purulent?

Streptococcal Infections of the Skin

Streptococcus are gram-positive cocci with a microscopic morphology that resembles chains of bacteria. Colonies are typically small (1–2 mm in diameter), translucent, entire edge, with a slightly raised elevation that can be either nonhemolytic, alpha-hemolytic, or beta-hemolytic when grown on blood agar (Figure 21.13). Additionally, they are facultative anaerobes that are catalase-negative.

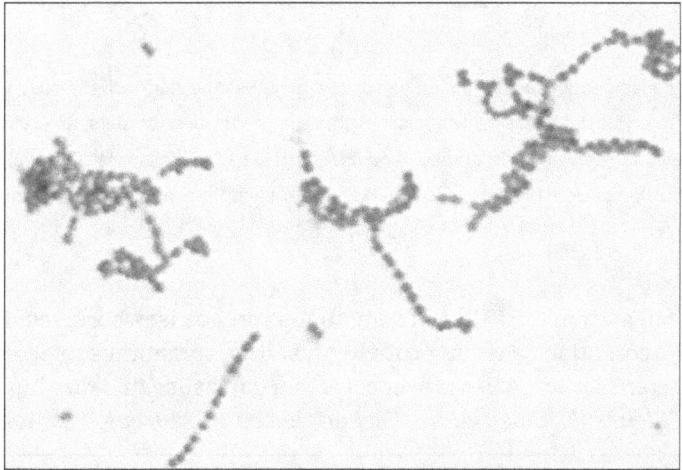

Figure 21.13 *Streptococcus pyogenes* forms chains of cocci. (credit: modification of work by Centers for Disease Control and Prevention)

The genus *Streptococcus* includes important pathogens that are categorized in serological Lancefield groups based on the distinguishing characteristics of their surface carbohydrates. The most clinically important streptococcal species in humans is *S. pyogenes*, also known as group A streptococcus (GAS). *S. pyogenes* produces a variety of extracellular enzymes, including streptolysins O and S, hyaluronidase, and streptokinase. These enzymes can aid in transmission and contribute to the inflammatory response.[6] *S. pyogenes* also produces a capsule and **M protein**, a streptococcal cell wall protein. These virulence factors help the bacteria to avoid phagocytosis while provoking a substantial immune response that contributes to symptoms associated with streptococcal infections.

S. pyogenes causes a wide variety of diseases not only in the skin, but in other organ systems as well. Examples of diseases elsewhere in the body include pharyngitis and scarlet fever, which will be covered in later chapters.

Cellulitis, Erysipelas, and Erythema Nosodum

Common streptococcal conditions of the skin include cellulitis, erysipelas, and erythema nodosum. An infection that develops in the dermis or hypodermis can cause **cellulitis**, which presents as a reddened area of the skin that is warm to the touch and painful. The causative agent is often *S. pyogenes*, which may breach the epidermis through a cut or abrasion, although cellulitis may also be caused by staphylococci. *S. pyogenes* can also cause **erysipelas**, a condition that presents as a large, intensely inflamed patch of skin involving the dermis (often on the legs or face). These infections can be **suppurative**, which results in a bullous form of erysipelas. Streptococcal and other pathogens may also cause a condition called **erythema nodosum**, characterized by inflammation in the subcutaneous fat cells of the hypodermis. It sometimes results from a streptococcal infection, though other pathogens can also cause the condition. It is not suppurative, but leads to red nodules on the skin, most frequently on the shins (Figure 21.14).

In general, streptococcal infections are best treated through identification of the specific pathogen followed by treatment based upon that particular pathogen's susceptibility to different antibiotics. Many immunological tests, including agglutination reactions and ELISAs, can be used to detect streptococci. Penicillin is commonly prescribed for treatment of cellulitis and erysipelas because resistance is not widespread in streptococci at this time. In most patients, erythema nodosum is self-limiting and is not treated with antimicrobial drugs. Recommended treatments may include nonsteroidal anti-inflammatory drugs (NSAIDs), cool wet compresses, elevation, and bed rest.

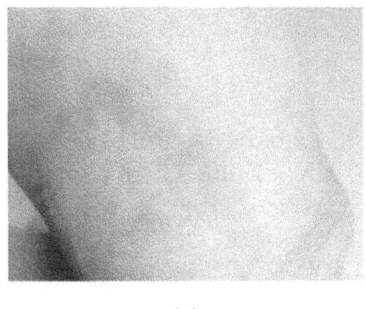 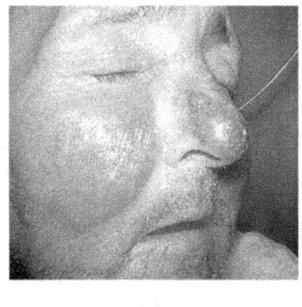 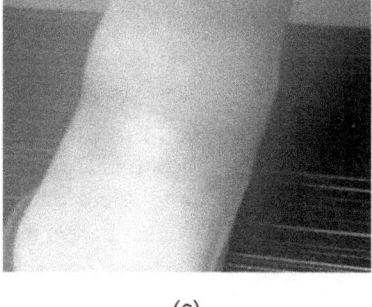

(a)   (b)   (c)

Figure 21.14 *S. pyogenes* can cause a variety of skin conditions once it breaches the skin barrier through a cut or wound. (a) Cellulitis presents as a painful, red rash. (b) Erysipelas presents as a raised rash, usually with clear borders. (c) Erythema nodosum is characterized by red lumps or nodules, typically on the lower legs. (credit a: modification of work by "Bassukas ID, Gaitanis G, Zioga A, Boboyianni C, Stergiopoulou C; credit b: modification of work by Centers for Disease Control and Prevention; credit c: modification of work by Dean C, Crow WT)

Necrotizing Fasciitis

Streptococcal infections that start in the skin can sometimes spread elsewhere, resulting in a rare but potentially life-threatening condition called **necrotizing fasciitis**, sometimes referred to as flesh-eating bacterial syndrome. *S. pyogenes* is one of several species that can cause this rare but potentially-fatal condition; others include *Klebsiella*, *Clostridium*, *Escherichia coli*, *S. aureus*, and *Aeromonas hydrophila*.

---

6   Starr, C.R. and Engelberg N.C. "Role of Hyaluronidase in Subcutaneous Spread and Growth of Group A Streptococcus." *Infection and Immunity* 2006(7:1): 40–48. doi: 10.1128/IAI.74.1.40-48.2006.

Necrotizing fasciitis occurs when the fascia, a thin layer of connective tissue between the skin and muscle, becomes infected. Severe invasive necrotizing fasciitis due to *Streptococcus pyogenes* occurs when virulence factors that are responsible for adhesion and invasion overcome host defenses. *S. pyogenes* invasins allow bacterial cells to adhere to tissues and establish infection. Bacterial proteases unique to *S. pyogenes* aggressively infiltrate and destroy host tissues, inactivate complement, and prevent neutrophil migration to the site of infection. The infection and resulting tissue death can spread very rapidly, as large areas of skin become detached and die. Treatment generally requires debridement (surgical removal of dead or infected tissue) or amputation of infected limbs to stop the spread of the infection; surgical treatment is supplemented with intravenous antibiotics and other therapies (Figure 21.15).

Necrotizing fasciitis does not always originate from a skin infection; in some cases there is no known portal of entry. Some studies have suggested that experiencing a blunt force trauma can increase the risk of developing streptococcal necrotizing fasciitis.[7]

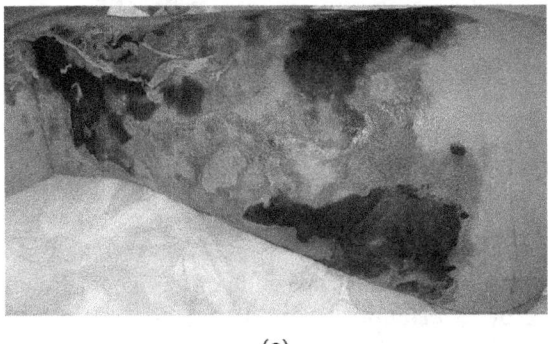
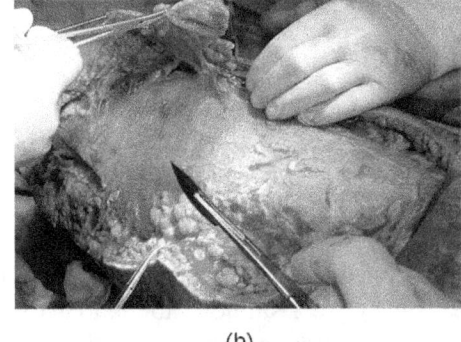

(a)   (b)

Figure 21.15 (a) The left leg of this patient shows the clinical features of necrotizing fasciitis. (b) The same patient's leg is surgically debrided to remove the infection. (credit a, b: modification of work by Piotr Smuszkiewicz, Iwona Trojanowska, and Hanna Tomczak)

CHECK YOUR UNDERSTANDING

- How do staphylococcal infections differ in general presentation from streptococcal infections?

Clinical Focus

Part 2

Observing that Sam's wound is purulent, the doctor tells him that he probably has a bacterial infection. She takes a sample from the lesion to send for laboratory analysis, but because it is Friday, she does not expect to receive the results until the following Monday. In the meantime, she prescribes an over-the-counter topical antibiotic ointment. She tells Sam to keep the wound clean and apply a new bandage with the ointment at least twice per day.

- How would the lab technician determine if the infection is staphylococcal or streptococcal? Suggest several specific methods.
- What tests might the lab perform to determine the best course of antibiotic treatment?

Jump to the next Clinical Focus box. Go back to the previous Clinical Focus box.

Pseudomonas Infections of the Skin

Another important skin pathogen is *Pseudomonas aeruginosa*, a gram-negative, oxidase-positive, aerobic bacillus that is commonly found in water and soil as well as on human skin. *P. aeruginosa* is a common cause of opportunistic infections of wounds and burns. It can also cause hot tub rash, a condition characterized by

---

[7] Nuwayhid, Z.B., Aronoff, D.M., and Mulla, Z.D.. "Blunt Trauma as a Risk Factor for Group A Streptococcal Necrotizing Fasciitis." *Annals of Epidemiology* (2007) 17:878–881.

folliculitis that frequently afflicts users of pools and hot tubs (recall the Clinical Focus case in Microbial Biochemistry). *P. aeruginosa* is also the cause of **otitis externa** (swimmer's ear), an infection of the ear canal that causes itching, redness, and discomfort, and can progress to fever, pain, and swelling (Figure 21.16).

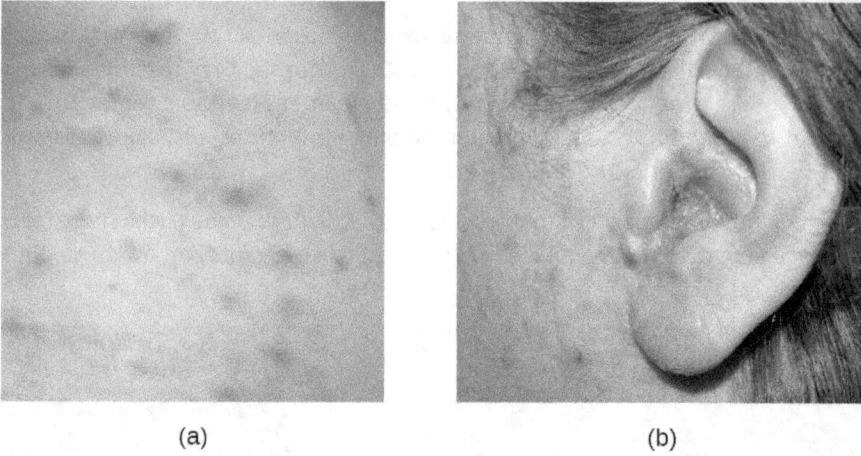

Figure 21.16 (a) Hot tub folliculitis presents as an itchy red rash. It is typically caused by *P. aeruginosa*, a bacterium that thrives in wet, warm environments such as hot tubs. (b) Otitis externa (swimmer's ear) may also be caused by *P. aeruginosa* or other bacteria commonly found in water. Inflammation of the outer ear and ear canal can lead to painful swelling. (credit b: modification of work by Klaus D. Peter)

Wounds infected with *P. aeruginosa* have a distinctive odor resembling grape soda or fresh corn tortillas. This odor is caused by the 2-aminoacetophenone that is used by *P. aeruginosa* in quorum sensing and contributes to its pathogenicity. Wounds infected with certain strains of *P. aeruginosa* also produce a blue-green pus due to the pigments **pyocyanin** and **pyoverdin**, which also contribute to its virulence. Pyocyanin and pyoverdin are siderophores that help *P. aeruginosa* survive in low-iron environments by enhancing iron uptake. *P. aeruginosa* also produces several other virulence factors, including phospholipase C (a hemolysin capable of breaking down red blood cells), exoenzyme S (involved in adherence to epithelial cells), and exotoxin A (capable of causing tissue necrosis). Other virulence factors include a slime that allows the bacterium to avoid being phagocytized, fimbriae for adherence, and proteases that cause tissue damage. *P. aeruginosa* can be detected through the use of cetrimide agar, which is selective for *Pseudomonas* species (Figure 21.17).

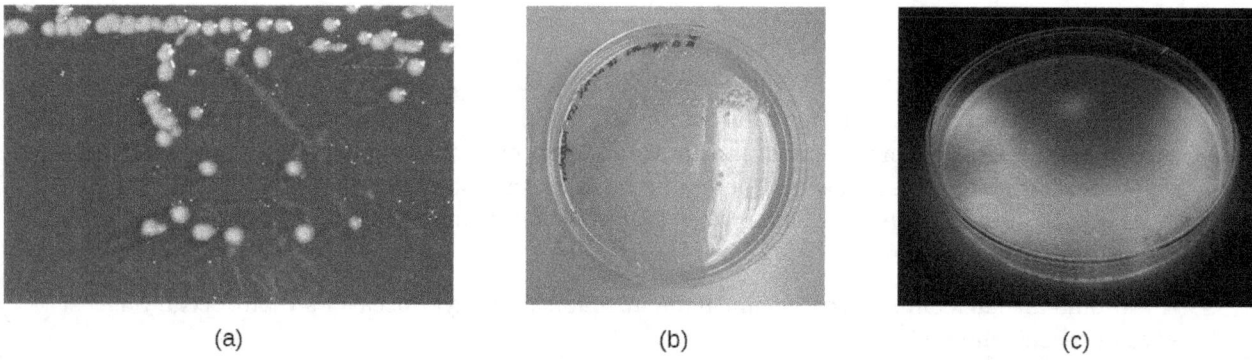

Figure 21.17 (a) These *P. aeruginosa* colonies are growing on xylose lysine sodium deoxycholate (XLD) agar. (b) *Pseudomonas* spp. can produce a variety of blue-green pigments. (c) *Pseudomonas* spp. may produce fluorescein, which fluoresces green under ultraviolet light under the right conditions. (credit a: modification of work by Centers for Disease Control and Prevention)

Pseudomonas spp. tend to be resistant to most antibiotics. They often produce β-lactamases, may have mutations affecting porins (small cell wall channels) that affect antibiotic uptake, and may pump some antibiotics out of the cell, contributing to this resistance. Polymyxin B and gentamicin are effective, as are some fluoroquinolones. Otitis externa is typically treated with ear drops containing acetic acid, antibacterials, and/or steroids to reduce inflammation; ear drops may also include antifungals because fungi can sometimes cause or contribute to otitis externa. Wound infections caused by *Pseudomonas* spp. may be treated with topical antibiofilm agents that disrupt the formation of biofilms.

CHECK YOUR UNDERSTANDING

- Name at least two types of skin infections commonly caused by *Pseudomonas* spp.

Acne

One of the most ubiquitous skin conditions is **acne**. Acne afflicts nearly 80% of teenagers and young adults, but it can be found in individuals of all ages. Higher incidence among adolescents is due to hormonal changes that can result in overproduction of sebum.

Acne occurs when hair follicles become clogged by shed skin cells and sebum, causing non-inflammatory lesions called comedones. Comedones (singular "comedo") can take the form of whitehead and blackhead pimples. Whiteheads are covered by skin, whereas blackhead pimples are not; the black color occurs when lipids in the clogged follicle become exposed to the air and oxidize (Figure 21.18).

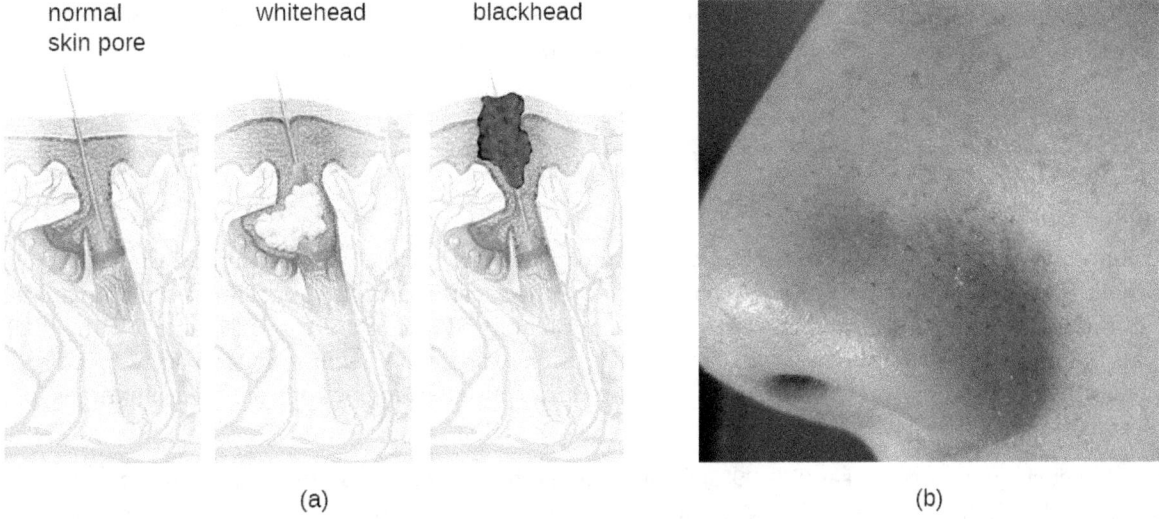

Figure 21.18 (a) Acne is characterized by whitehead and blackhead comedones that result from clogged hair follicles. (b) Blackheads, visible as black spots on the skin, have a dark appearance due to the oxidation of lipids in sebum via exposure to the air. (credit a: modification of work by Bruce Blaus)

Often comedones lead to infection by *Propionibacterium acnes*, a gram-positive, non-spore-forming, aerotolerant anaerobic bacillus found on skin that consumes components of sebum. *P. acnes* secretes enzymes that damage the hair follicle, causing inflammatory lesions that may include papules, pustules, nodules, or pseudocysts, depending on their size and severity.

Treatment of acne depends on the severity of the case. There are multiple ways to grade acne severity, but three levels are usually considered based on the number of comedones, the number of inflammatory lesions, and the types of lesions. Mild acne is treated with topical agents that may include salicylic acid (which helps to remove old skin cells) or retinoids (which have multiple mechanisms, including the reduction of inflammation). Moderate acne may be treated with antibiotics (erythromycin, clindamycin), acne creams (e.g., benzoyl peroxide), and hormones. Severe acne may require treatment using strong medications such as isotretinoin (a retinoid that reduces oil buildup, among other effects, but that also has serious side effects such as photosensitivity). Other treatments, such as phototherapy and laser therapy to kill bacteria and possibly reduce oil production, are also sometimes used.

CHECK YOUR UNDERSTANDING

- What is the role of *Propionibacterium acnes* in causing acne?

> **Clinical Focus**
>
> **Resolution**
>
> Sam uses the topical antibiotic over the weekend to treat his wound, but he does not see any improvement. On Monday, the doctor calls to inform him that the results from his laboratory tests are in. The tests show evidence of both *Staphylococcus* and *Streptococcus* in his wound. The bacterial species were confirmed using several tests. A passive agglutination test confirmed the presence of *S. aureus*. In this type of test, latex beads with antibodies cause agglutination when *S. aureus* is present. *Streptococcus pyogenes* was confirmed in the wound based on bacitracin (0.04 units) susceptibility as well as latex agglutination tests specific for *S. pyogenes*.
>
> Because many strains of *S. aureus* are resistant to antibiotics, the doctor had also requested an antimicrobial susceptibility test (AST) at the same time the specimen was submitted for identification. The results of the AST indicated no drug resistance for the *Streptococcus* spp.; the *Staphylococcus* spp. showed resistance to several common antibiotics, but were susceptible to cefoxitin and oxacillin. Once Sam began to use these new antibiotics, the infection resolved within a week and the lesion healed.
>
> *Go back to the previous Clinical Focus box.*

Anthrax

The zoonotic disease **anthrax** is caused by *Bacillus anthracis*, a gram-positive, endospore-forming, facultative anaerobe. Anthrax mainly affects animals such as sheep, goats, cattle, and deer, but can be found in humans as well. Sometimes called wool sorter's disease, it is often transmitted to humans through contact with infected animals or animal products, such as wool or hides. However, exposure to *B. anthracis* can occur by other means, as the endospores are widespread in soils and can survive for long periods of time, sometimes for hundreds of years.

The vast majority of anthrax cases (95–99%) occur when anthrax endospores enter the body through abrasions of the skin.[8] This form of the disease is called cutaneous anthrax. It is characterized by the formation of a nodule on the skin; the cells within the nodule die, forming a black **eschar**, a mass of dead skin tissue (Figure 21.19). The localized infection can eventually lead to bacteremia and septicemia. If untreated, cutaneous anthrax can cause death in 20% of patients.[9] Once in the skin tissues, *B. anthracis* endospores germinate and produce a capsule, which prevents the bacteria from being phagocytized, and two binary exotoxins that cause edema and tissue damage. The first of the two exotoxins consists of a combination of protective antigen (PA) and an enzymatic lethal factor (LF), forming lethal toxin (LeTX). The second consists of protective antigen (PA) and an edema factor (EF), forming edema toxin (EdTX).

---

8 Shadomy, S.V., Traxler, R.M., and Marston, C.K. "Infectious Diseases Related to Travel: Anthrax" 2015. Centers for Disease Control and Prevention. http://wwwnc.cdc.gov/travel/yellowbook/2016/infectious-diseases-related-to-travel/anthrax. Accessed Sept 14, 2016.

9 US FDA. "Anthrax." 2015. http://www.fda.gov/BiologicsBloodVaccines/Vaccines/ucm061751.htm. Accessed Sept 14, 2016.

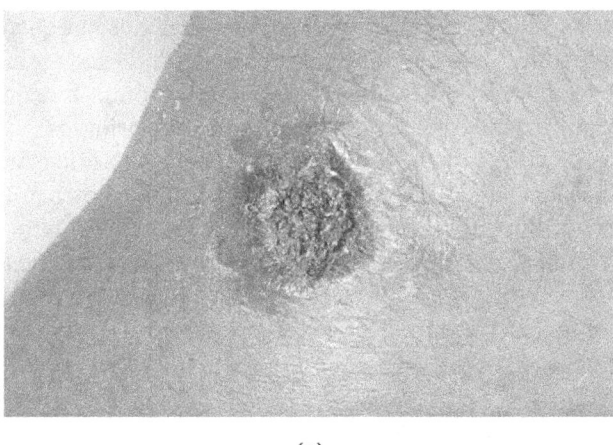 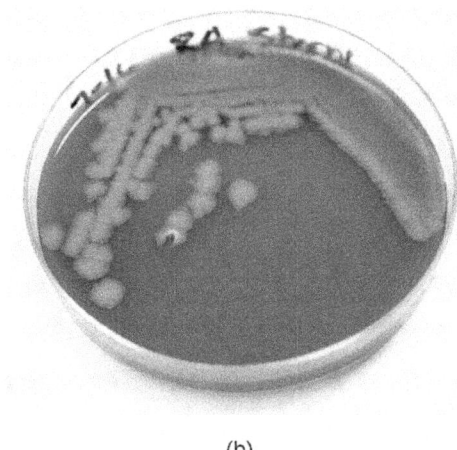

(a) (b)

Figure 21.19 (a) Cutaneous anthrax is an infection of the skin by *B. anthracis*, which produces tissue-damaging exotoxins. Dead tissues accumulating in this nodule have produced a small black eschar. (b) Colonies of *B. anthracis* grown on sheep's blood agar. (credit a, b: modification of work by Centers for Disease Control and Prevention)

Less commonly, anthrax infections can be initiated through other portals of entry such as the digestive tract (gastrointestinal anthrax) or respiratory tract (pulmonary anthrax or inhalation anthrax). Typically, cases of noncutaneous anthrax are more difficult to treat than the cutaneous form. The mortality rate for gastrointestinal anthrax can be up to 40%, even with treatment. Inhalation anthrax, which occurs when anthrax spores are inhaled, initially causes influenza-like symptoms, but mortality rates are approximately 45% in treated individuals and 85% in those not treated. A relatively new form of the disease, injection anthrax, has been reported in Europe in intravenous drug users; it occurs when drugs are contaminated with *B. anthracis*. Patients with injection anthrax show signs and symptoms of severe soft tissue infection that differ clinically from cutaneous anthrax. This often delays diagnosis and treatment, and leads to a high mortality rate.[10]

B. anthracis colonies on blood agar have a rough texture and serrated edges that eventually form an undulating band (Figure 21.19). Broad spectrum antibiotics such as penicillin, erythromycin, and tetracycline are often effective treatments.

Unfortunately, *B. anthracis* has been used as a biological weapon and remains on the United Nations' list of potential agents of bioterrorism.[11] Over a period of several months in 2001, a number of letters were mailed to members of the news media and the United States Congress. As a result, 11 individuals developed cutaneous anthrax and another 11 developed inhalation anthrax. Those infected included recipients of the letters, postal workers, and two other individuals. Five of those infected with pulmonary anthrax died. The anthrax spores had been carefully prepared to aerosolize, showing that the perpetrator had a high level of expertise in microbiology.[12]

---

10 Berger, T., Kassirer, M., and Aran, A.A.. "Injectional Anthrax—New Presentation of an Old Disease." *Euro Surveillance* 19 (2014) 32. http://www.ncbi.nlm.nih.gov/pubmed/25139073. Accessed Sept 14, 2016.

11 United Nations Office at Geneva. "What Are Biological and Toxin Weapons?" http://www.unog.ch/80256EE600585943/%28httpPages%29/29B727532FECBE96C12571860035A6DB?. Accessed Sept 14, 2016.

12 Federal Bureau of Investigation. "Famous Cases and Criminals: Amerithrax or Anthrax Investigation." https://www.fbi.gov/history/famous-cases/amerithrax-or-anthrax-investigation. Accessed Sept 14, 2016.

A vaccine is available to protect individuals from anthrax. However, unlike most routine vaccines, the current anthrax vaccine is unique in both its formulation and the protocols dictating who receives it.[13] The vaccine is administered through five intramuscular injections over a period of 18 months, followed by annual boosters. The US Food and Drug Administration (FDA) has only approved administration of the vaccine prior to exposure for at-risk adults, such as individuals who work with anthrax in a laboratory, some individuals who handle animals or animal products (e.g., some veterinarians), and some members of the United States military. The vaccine protects against cutaneous and inhalation anthrax using cell-free filtrates of microaerophilic cultures of an avirulent, nonencapsulated strain of *B. anthracis*.[14] The FDA has not approved the vaccine for routine use *after* exposure to anthrax, but if there were ever an anthrax emergency in the United States, patients could be given anthrax vaccine after exposure to help prevent disease.

✓ CHECK YOUR UNDERSTANDING

- What is the characteristic feature of a cutaneous anthrax infection?

Disease Profile

Bacterial Infections of the Skin

Bacterial infections of the skin can cause a wide range of symptoms and syndromes, ranging from the superficial and relatively harmless to the severe and even fatal. Most bacterial skin infections can be diagnosed by culturing the bacteria and treated with antibiotics. Antimicrobial susceptibility testing is also often necessary because many strains of bacteria have developed antibiotic resistance. Figure 21.20 summarizes the characteristics of some common bacterial skin infections.

---

13 Centers for Disease Control and Prevention. "Anthrax: Medical Care: Prevention: Antibiotics." http://www.cdc.gov/anthrax/medical-care/prevention.html. Accessed Sept 14, 2016.

14 Emergent Biosolutions. AVA (BioThrax) vaccine package insert (Draft). Nov 2015. http://www.fda.gov/downloads/biologicsbloodvaccines/bloodbloodproducts/approvedproducts/licensedproductsblas/ucm074923.pdf.

Bacterial Infections of the Skin				
Disease	Pathogen	Signs and Symptoms	Transmission	Antimicrobial Drugs
Acne	*Propionibacterium acnes*	Comedones (whiteheads, blackheads); papules, pustules, nodules, or pseudocysts	Not transmissible; clogged pores become infected by normal skin microbiota (*P. acnes*)	Erythromycin, clindamycin
Anthrax (cutaneous)	*Bacillus anthracis*	Eschar at site of infection; may lead to septicemia and can be fatal	Entry of *B. anthracis* endospores through cut or abrasion	Penicillin, erythromycin, or tetracycline
Cellulitis	*Streptococcus pyogenes*	Localized inflammation of dermis and hypodermis; skin red, warm, and painful to the touch	Entry of *S. pyogenes* through cut or abrasion	Oral or intravenous antibiotics (e.g., penicillin)
Erysipelas	*S. pyogenes*	Inflamed, swollen patch of skin, often on face; may be suppurative	Entry of *S. pyogenes* through cut or abrasion	Oral or intravenous antibiotics (e.g., penicillin)
Erythema nodosum	*S. pyogenes*	Small red nodules, often on shins	Associated with other streptococcal infection	None or anti-inflammatory drugs for severe cases
Impetigo	*Staphylococcus aureus*, *S. pyogenes*	Vesicles, pustules, and sometimes bullae around nose and mouth	Highly contagious, especially via contact	Topical or oral antibiotics
Necrotizing fasciitis	*S. pyogenes*, *Klebsiella*, *Clostridium*, others	Infection of fascia and rapidly spreading tissue death; can lead to septic shock and death	Entry of bacteria through cut or abrasion	Intravenous broad-spectrum antibiotics
Otitis externa	*Pseudomonas aeruginosa*	Itching, redness, discomfort of ear canal, progressing to fever, pain, swelling	*P. aeruginosa* enters ear canal via pool or other water	Acidic ear drops with antibiotics, antifungals, steroids
Staphylococcal scalded skin syndrome (SSSS)	*S. aureus*	Erythema and severe peeling of skin	Infection of skin and mucous membranes, especially in children	Intravenous antibiotics, fluid therapy
Wound infections	*P. aeruginosa*, others	Formation of biofilm in or on wound	Exposure of wound to microbes in environment; poor wound hygiene	Polymyxin B, gentamicin, fluoroquinolones, topical anti-biofilm agents

Figure 21.20

Bacterial Conjunctivitis

Like the skin, the surface of the eye comes in contact with the outside world and is somewhat prone to infection by bacteria in the environment. Bacterial conjunctivitis (pinkeye) is a condition characterized by inflammation of the conjunctiva, often accompanied by a discharge of sticky fluid (described as acute purulent conjunctivitis) (Figure 21.21). Conjunctivitis can affect one eye or both, and it usually does not affect vision permanently. Bacterial conjunctivitis is most commonly caused by *Haemophilus influenzae*, but can also be caused by other species such as *Moraxella catarrhalis*, *S. pneumoniae*, and *S. aureus*. The causative agent may be identified using bacterial cultures, Gram stain, and diagnostic biochemical, antigenic, or nucleic acid profile tests of the isolated pathogen. Bacterial conjunctivitis is very contagious, being transmitted via secretions from infected individuals, but it is also self-limiting. Bacterial conjunctivitis usually resolves in a few days, but topical antibiotics are sometimes prescribed. Because this condition is so contagious, medical attention is recommended whenever it is suspected. Individuals who use contact lenses should discontinue their use when conjunctivitis is suspected. Certain symptoms, such as blurred vision, eye pain, and light sensitivity, can be associated with serious conditions and require medical attention.

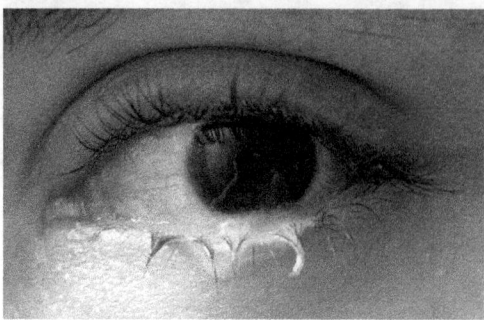

Figure 21.21 Acute, purulent, bacterial conjunctivitis causes swelling and redness in the conjunctiva, the membrane lining the whites of the eyes and the inner eyelids. It is often accompanied by a yellow, green, or white discharge, which can dry and become encrusted on the eyelashes. (credit: "Tanalai"/Wikimedia Commons)

Neonatal Conjunctivitis

Newborns whose mothers have certain sexually transmitted infections are at risk of contracting **ophthalmia neonatorum** or **inclusion conjunctivitis**, which are two forms of neonatal conjunctivitis contracted through exposure to pathogens during passage through the birth canal. Gonococcal ophthalmia neonatorum is caused by *Neisseria gonorrhoeae*, the bacterium that causes the STD gonorrhea (Figure 21.22). Inclusion (chlamydial) conjunctivitis is caused by *Chlamydia trachomatis*, the anaerobic, obligate, intracellular parasite that causes the STD chlamydia.

To prevent gonoccocal ophthalmia neonatorum, silver nitrate ointments were once routinely applied to all infants' eyes shortly after birth; however, it is now more common to apply antibacterial creams or drops, such as erythromycin. Most hospitals are required by law to provide this preventative treatment to all infants, because conjunctivitis caused by *N. gonorrhoeae*, *C. trachomatis,* or other bacteria acquired during a vaginal delivery can have serious complications. If untreated, the infection can spread to the cornea, resulting in ulceration or perforation that can cause vision loss or even permanent blindness. As such, neonatal conjunctivitis is treated aggressively with oral or intravenous antibiotics to stop the spread of the infection. Causative agents of inclusion conjunctivitis may be identified using bacterial cultures, Gram stain, and diagnostic biochemical, antigenic, or nucleic acid profile tests.

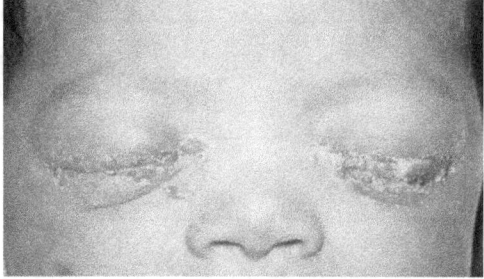

Figure 21.22 A newborn suffering from gonoccocal opthalmia neonatorum. Left untreated, purulent discharge can scar the cornea, causing loss of vision or permanent blindness. (credit: Centers for Disease Control and Prevention)

CHECK YOUR UNDERSTANDING

- Compare and contrast bacterial conjunctivitis with neonatal conjunctivitis.

Trachoma

Trachoma, or granular conjunctivitis, is a common cause of preventable blindness that is rare in the United States but widespread in developing countries, especially in Africa and Asia. The condition is caused by the same species that causes neonatal inclusion conjunctivitis in infants, *Chlamydia trachomatis*. *C. trachomatis* can be transmitted easily through fomites such as contaminated towels, bed linens, and clothing and also by direct contact with infected individuals. *C. trachomatis* can also be spread by flies that transfer infected mucous containing *C. trachomatis* from one human to another.

Infection by *C. trachomatis* causes chronic conjunctivitis, which leads to the formation of necrotic follicles and scarring in the upper eyelid. The scars turn the eyelashes inward (a condition known as trichiasis) and mechanical abrasion of the cornea leads to blindness (Figure 21.23). Antibiotics such as azithromycin are effective in treating trachoma, and outcomes are good when the disease is treated promptly. In areas where this disease is common, large public health efforts are focused on reducing transmission by teaching people how to avoid the risks of the infection.

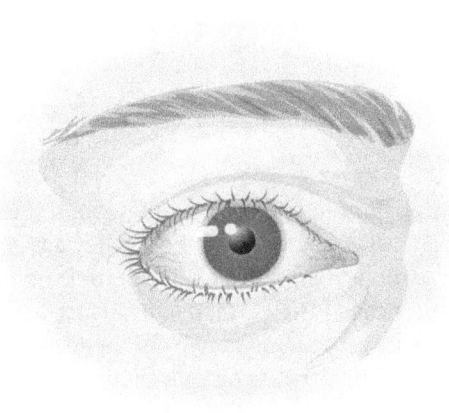
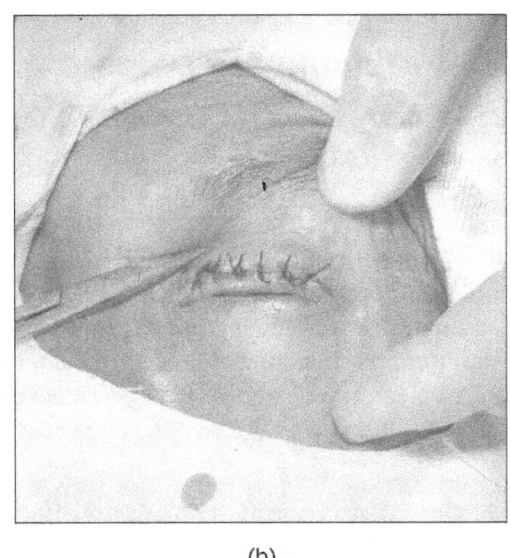

(a)   (b)

Figure 21.23 (a) If trachoma is not treated early with antibiotics, scarring on the eyelid can lead to trichiasis, a condition in which the eyelashes turn inward. (b) Trichiasis leads to blindness if not corrected by surgery, as shown here. (credit b: modification of work by Otis Historical Archives National Museum of Health & Medicine)

✓ CHECK YOUR UNDERSTANDING

- Why is trachoma rare in the United States?

 MICRO CONNECTIONS

SAFE Eradication of Trachoma

Though uncommon in the United States and other developed nations, trachoma is the leading cause of preventable blindness worldwide, with more than 4 million people at immediate risk of blindness from trichiasis. The vast majority of those affected by trachoma live in Africa and the Middle East in isolated rural or desert communities with limited access to clean water and sanitation. These conditions provide an environment conducive to the growth and spread of *Chlamydia trachomatis*, the bacterium that causes trachoma, via wastewater and eye-seeking flies.

In response to this crisis, recent years have seen major public health efforts aimed at treating and preventing trachoma. The Alliance for Global Elimination of Trachoma by 2020 (GET 2020), coordinated by the World Health Organization (WHO), promotes an initiative dubbed "SAFE," which stands for "Surgery, Antibiotics, Facial cleanliness, and Environmental improvement." The Carter Center, a charitable, nongovernment organization led by former US President Jimmy Carter, has partnered with the WHO to promote the SAFE initiative in six of the most critically impacted nations in Africa. Through its Trachoma Control Program, the Carter Center trains and equips local surgeons to correct trichiasis and distributes antibiotics to treat trachoma. The program also promotes better personal hygiene through health education and improves sanitation by funding the construction of household latrines. This reduces the prevalence of open sewage, which provides breeding grounds for the flies that spread trachoma.

Bacterial Keratitis

Keratitis can have many causes, but bacterial keratitis is most frequently caused by *Staphylococcus epidermidis* and/or *Pseudomonas aeruginosa*. Contact lens users are particularly at risk for such an infection because *S. epidermidis* and *P. aeruginosa* both adhere well to the surface of the lenses. Risk of infection can be greatly reduced by proper care of contact lenses and avoiding wearing lenses overnight. Because the infection can quickly lead to blindness, prompt and aggressive treatment with antibiotics is important. The causative agent may be identified using bacterial cultures, Gram stain, and diagnostic biochemical, antigenic, or nucleic acid profile tests of the isolated pathogen.

✓ CHECK YOUR UNDERSTANDING

- Why are contact lens wearers at greater risk for developing keratitis?

Biofilms and Infections of the Skin and Eyes

When treating bacterial infections of the skin and eyes, it is important to consider that few such infections can be attributed to a single pathogen. While biofilms may develop in other parts of the body, they are especially relevant to skin infections (such as those caused by *S. aureus* or *P. aeruginosa*) because of their prevalence in chronic skin wounds. Biofilms develop when bacteria (and sometimes fungi) attach to a surface and produce extracellular polymeric substances (EPS) in which cells of multiple organisms may be embedded. When a biofilm develops on a wound, it may interfere with the natural healing process as well as diagnosis and treatment.

Because biofilms vary in composition and are difficult to replicate in the lab, they are still not thoroughly understood. The extracellular matrix of a biofilm consists of polymers such as polysaccharides, extracellular DNA, proteins, and lipids, but the exact makeup varies. The organisms living within the extracellular matrix may include familiar pathogens as well as other bacteria that do not grow well in cultures (such as numerous obligate anaerobes). This presents challenges when culturing samples from infections that involve a biofilm. Because only some species grow *in vitro*, the culture may contain only a subset of the bacterial species involved in the infection.

Biofilms confer many advantages to the resident bacteria. For example, biofilms can facilitate attachment to surfaces on or in the host organism (such as wounds), inhibit phagocytosis, prevent the invasion of neutrophils, and sequester host antibodies. Additionally, biofilms can provide a level of antibiotic resistance not found in the isolated cells and colonies that are typical of laboratory cultures. The extracellular matrix provides a physical barrier to antibiotics, shielding the target cells from exposure. Moreover, cells within a biofilm may differentiate to create subpopulations of dormant cells called persister cells. Nutrient limitations deep within a biofilm add another level of resistance, as stress responses can slow metabolism and increase drug resistance.

Disease Profile

Bacterial Infections of the Eyes

A number of bacteria are able to cause infection when introduced to the mucosa of the eye. In general, bacterial eye infections can lead to inflammation, irritation, and discharge, but they vary in severity. Some are typically short-lived, and others can become chronic and lead to permanent eye damage. Prevention requires limiting exposure to contagious pathogens. When infections do occur, prompt treatment with antibiotics can often limit or prevent permanent damage. Figure 21.24 summarizes the characteristics of some common bacterial infections of the eyes.

Bacterial Infections of the Eyes				
Disease	Pathogen	Signs and Symptoms	Transmission	Antimicrobial Drugs
Acute bacterial conjunctivitis	*Haemophilus influenzae*	Inflammation of conjunctiva with purulent discharge	Exposure to secretions from infected individuals	Broad-spectrum topical antibiotics
Bacterial keratitis	*Staphylococcus epidermidis*, *Pseudomonas aeruginosa*	Redness and irritation of eye, blurred vision, sensitivity to light; progressive corneal scarring, which can lead to blindness	Exposure to pathogens on contaminated contact lenses	Antibiotic eye drops (e.g., with fluoroquinolones)
Neonatal conjunctivitis	*Chlamydia trachomatis*, *Neisseria gonorrhoeae*	Inflammation of conjunctiva, purulent discharge, scarring and perforation of cornea; may lead to blindness	Neonate exposed to pathogens in birth canal of mother with chlamydia or gonorrhea	Erythromycin
Trachoma (granular conjunctivitis)	*C. trachomatis*	Chronic conjunctivitis, trichiasis, scarring, blindness	Contact with infected individuals or contaminated fomites; transmission by eye-seeking flies	Azithromycin

Figure 21.24

21.3 Viral Infections of the Skin and Eyes

Learning Objectives

By the end of this section, you will be able to:
- Identify the most common viruses associated with infections of the skin and eyes
- Compare the major characteristics of specific viral diseases affecting the skin and eyes

Until recently, it was thought that the normal microbiota of the body consisted primarily of bacteria and some fungi. However, in addition to bacteria, the skin is colonized by viruses, and recent studies suggest that Papillomaviridae, Polyomaviridae and Circoviridae also contribute to the normal skin microbiota. However, some viruses associated with skin are pathogenic, and these viruses can cause diseases with a wide variety of presentations.

Numerous types of viral infections cause rashes or lesions on the skin; however, in many cases these skin conditions result from infections that originate in other body systems. In this chapter, we will limit the discussion to viral skin infections that use the skin as a portal of entry. Later chapters will discuss viral infections such as chickenpox, measles, and rubella—diseases that cause skin rashes but invade the body through portals of entry other than the skin.

Papillomas

Papillomas (warts) are the expression of common skin infections by human papillomavirus (HPV) and are transmitted by direct contact. There are many types of HPV, and they lead to a variety of different presentations, such as common warts, plantar warts, flat warts, and filiform warts. HPV can also cause sexually-transmitted genital warts, which will be discussed in Urogenital System Infections. Vaccination is available for some strains of HPV.

Common warts tend to develop on fingers, the backs of hands, and around nails in areas with broken skin. In contrast, plantar warts (also called foot warts) develop on the sole of the foot and can grow inwards, causing pain and pressure during walking. Flat warts can develop anywhere on the body, are often numerous, and are relatively smooth and small compared with other wart types. Filiform warts are long, threadlike warts that grow quickly.

In some cases, the immune system may be strong enough to prevent warts from forming or to eradicate established warts. However, treatment of established warts is typically required. There are many available treatments for warts, and their effectiveness varies. Common warts can be frozen off with liquid nitrogen. Topical applications of salicylic acid may also be effective. Other options are electrosurgery (burning), curettage (cutting), excision, painting with cantharidin (which causes the wart to die so it can more easily be removed), laser treatments, treatment with bleomycin, chemical peels, and immunotherapy (Figure 21.25).

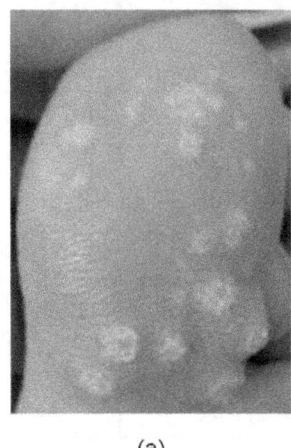
(a)
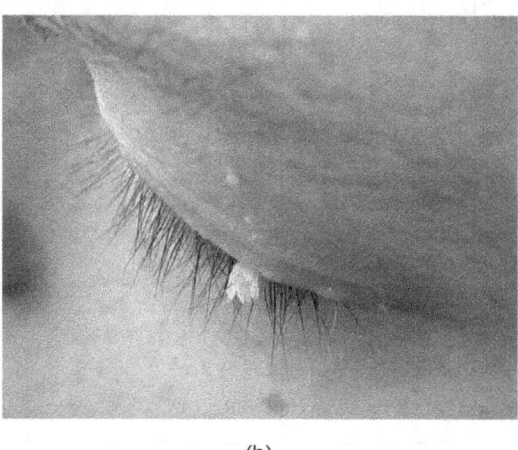
(b)

Figure 21.25　Warts can vary in shape and in location. (a) Multiple plantar warts have grown on this toe. (b) A filiform wart has grown on this eyelid.

Oral Herpes

Another common skin virus is herpes simplex virus (HSV). HSV has historically been divided into two types, HSV-1 and HSV-2. HSV-1 is typically transmitted by direct oral contact between individuals, and is usually associated with **oral herpes**. HSV-2 is usually transmitted sexually and is typically associated with genital herpes. However, both HSV-1 and HSV-2 are capable of infecting any mucous membrane, and the incidence of genital HSV-1 and oral HSV-2 infections has been increasing in recent years. In this chapter, we will limit our discussion to infections caused by HSV-1; HSV-2 and genital herpes will be discussed in Urogenital System Infections.

Infection by HSV-1 commonly manifests as cold sores or fever blisters, usually on or around the lips (Figure 21.26). HSV-1 is highly contagious, with some studies suggesting that up to 65% of the US population is infected; however, many infected individuals are asymptomatic.[15] Moreover, the virus can be latent for long periods, residing in the trigeminal nerve ganglia between recurring bouts of symptoms. Recurrence can be triggered by stress or environmental conditions (systemic or affecting the skin). When lesions are present, they may blister, break open, and crust. The virus can be spread through direct contact, even when a patient is asymptomatic.

While the lips, mouth, and face are the most common sites for HSV-1 infections, lesions can spread to other areas of the body. Wrestlers and other athletes involved in contact sports may develop lesions on the neck, shoulders, and trunk. This condition is often called herpes gladiatorum. Herpes lesions that develop on the fingers are often called herpetic whitlow.

HSV-1 infections are commonly diagnosed from their appearance, although laboratory testing can confirm the diagnosis. There is no cure, but antiviral medications such as acyclovir, penciclovir, famciclovir, and

---

15　Wald, A., and Corey, L. "Persistence in the Population: Epidemiology, Transmission." In: A. Arvin, G. Campadelli-Fiume, E. Mocarski et al. *Human Herpesviruses: Biology, Therapy, and Immunoprophylaxis*. Cambridge: Cambridge University Press, 2007. http://www.ncbi.nlm.nih.gov/books/NBK47447/. Accessed Sept 14, 2016.

valacyclovir are used to reduce symptoms and risk of transmission. Topical medications, such as creams with n-docosanol and penciclovir, can also be used to reduce symptoms such as itching, burning, and tingling.

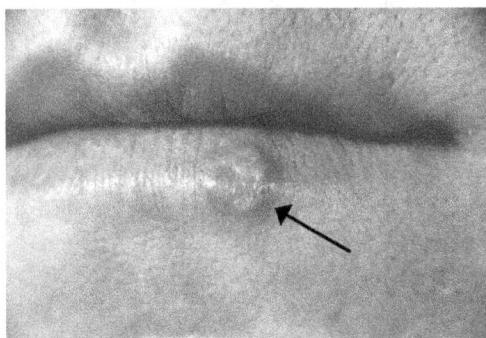

Figure 21.26 This cold sore was caused by HSV-1. (credit: Centers for Disease Control and Prevention)

CHECK YOUR UNDERSTANDING

- What are the most common sites for the appearance of herpetic lesions?

Roseola and Fifth Disease

The viral diseases **roseola** and **fifth disease** are somewhat similar in terms of their presentation, but they are caused by different viruses. Roseola, sometimes called roseola infantum or exanthem subitum ("sudden rash"), is a mild viral infection usually caused by human herpesvirus-6 (HHV-6) and occasionally by HHV-7. It is spread via direct contact with the saliva or respiratory secretions of an infected individual, often through droplet aerosols. Roseola is very common in children, with symptoms including a runny nose, a sore throat, and a cough, along with (or followed by) a high fever (39.4 °C). About three to five days after the fever subsides, a rash may begin to appear on the chest and abdomen. The rash, which does not cause discomfort, initially forms characteristic macules that are flat or papules that are firm and slightly raised; some macules or papules may be surrounded by a white ring. The rash may eventually spread to the neck and arms, and sometimes continues to spread to the face and legs. The diagnosis is generally made based upon observation of the symptoms. However, it is possible to perform serological tests to confirm the diagnosis. While treatment may be recommended to control the fever, the disease usually resolves without treatment within a week after the fever develops. For individuals at particular risk, such as those who are immunocompromised, the antiviral medication ganciclovir may be used.

Fifth disease (also known as erythema infectiosum) is another common, highly contagious illness that causes a distinct rash that is critical to diagnosis. Fifth disease is caused by parvovirus B19, and is transmitted by contact with respiratory secretions from an infected individual. Infection is more common in children than adults. While approximately 20% of individuals will be asymptomatic during infection,[16] others will exhibit cold-like symptoms (headache, fever, and upset stomach) during the early stages when the illness is most infectious. Several days later, a distinct red facial rash appears, often called "slapped cheek" rash (Figure 21.27). Within a few days, a second rash may appear on the arms, legs, chest, back, or buttocks. The rash may come and go for several weeks, but usually disappears within seven to twenty-one days, gradually becoming lacy in appearance as it recedes.

In children, the disease usually resolves on its own without medical treatment beyond symptom relief as needed. Adults may experience different and possibly more serious symptoms. Many adults with fifth disease do not develop any rash, but may experience joint pain and swelling that lasts several weeks or months. Immunocompromised individuals can develop severe anemia and may need blood transfusions or immune globulin injections. While the rash is the most important component of diagnosis (especially in children), the symptoms of fifth disease are not always consistent. Serological testing can be conducted for confirmation.

---

16 Centers for Disease Control and Prevention. "Fifth Disease." http://www.cdc.gov/parvovirusb19/fifth-disease.html. Accessed Sept 14, 2016.

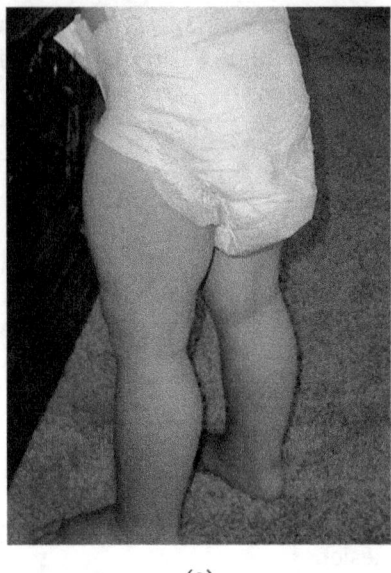
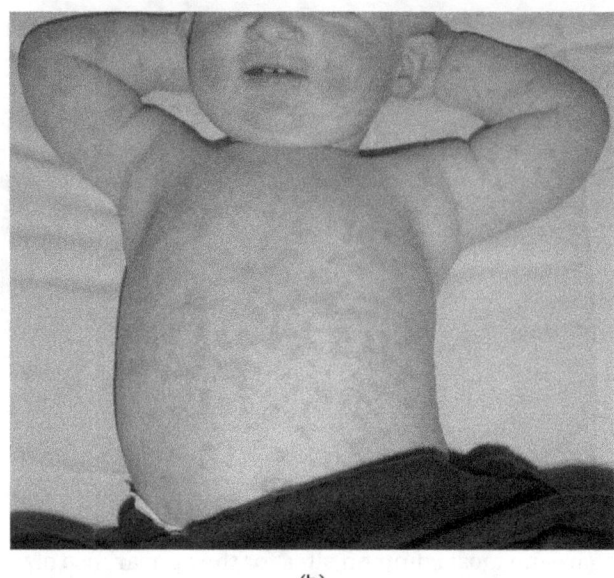

Figure 21.27 (a) Roseola, a mild viral infection common in young children, generally begins with symptoms similar to a cold, followed by a pink, patchy rash that starts on the trunk and spreads outward. (b) Fifth disease exhibits similar symptoms in children, except for the distinctive "slapped cheek" rash that originates on the face.

CHECK YOUR UNDERSTANDING

- Identify at least one similarity and one difference between roseola and fifth disease.

Viral Conjunctivitis

Like bacterial conjunctivitis viral infections of the eye can cause inflammation of the conjunctiva and discharge from the eye. However, **viral conjunctivitis** tends to produce a discharge that is more watery than the thick discharge associated with bacterial conjunctivitis. The infection is contagious and can easily spread from one eye to the other or to other individuals through contact with eye discharge.

Viral conjunctivitis is commonly associated with colds caused by adenoviruses; however, other viruses can also cause conjunctivitis. If the causative agent is uncertain, eye discharge can be tested to aid in diagnosis. Antibiotic treatment of viral conjunctivitis is ineffective, and symptoms usually resolve without treatment within a week or two.

Herpes Keratitis

Herpes infections caused by HSV-1 can sometimes spread to the eye from other areas of the body, which may result in keratoconjunctivitis. This condition, generally called **herpes keratitis** or herpetic keratitis, affects the conjunctiva and cornea, causing irritation, excess tears, and sensitivity to light. Deep lesions in the cornea may eventually form, leading to blindness. Because keratitis can have numerous causes, laboratory testing is necessary to confirm the diagnosis when HSV-1 is suspected; once confirmed, antiviral medications may be prescribed.

Disease Profile

Viral Infections of the Skin and Eyes

A number of viruses can cause infections via direct contact with skin and eyes, causing signs and symptoms ranging from rashes and lesions to warts and conjunctivitis. All of these viral diseases are contagious, and while some are more common in children (fifth disease and roseola), others are prevalent in people of all ages (oral herpes, viral conjunctivitis, papillomas). In general, the best means of prevention is avoiding contact with infected individuals. Treatment may require antiviral medications; however, several of these conditions are mild and typically resolve without treatment. Figure 21.28 summarizes the characteristics of some common viral infections of the skin and eyes.

\multicolumn{5}{c}{Viral Infections of the Skin and Eyes}
Disease	Pathogen	Signs and Symptoms	Transmission	Antimicrobial Drugs
Fifth disease	Parvovirus B19	May have initial cold-like symptoms; "slapped cheek" rash	Highly contagious via respiratory secretions of infected individuals	None
Herpes keratitis	Herpes simplex virus 1 (HSV-1)	Inflammation of conjunctiva and cornea; irritation, excess tears, sensitivity to light; lesions in cornea leading to blindness	Direct eye contact with discharge from herpes lesions elsewhere in the body or from another infected individual	Acyclovir, ganciclovir, famiclovir, valacyclovir
Oral herpes	Herpes simplex virus 1 (HSV-1)	May cause initial systemic symptoms; cold sores	Highly contagious via direct contact with infected individuals	Acyclovir, penciclovir, famiclovir, valacyclovir
Papillomas	Human papillomavirus (HPV)	Common warts, plantar warts, flat warts, filiform warts, and others	Contact with infected individuals	Topical salicylic acid, cantharidin
Roseola (roseola infantum, exanthem subitum)	Human herpesvirus 6 (HHV-6), human herpesvirus 7 (HHV-7)	Initial cold-like symptoms with high fever, followed by a macular or papular rash three to five days later	Spread by viral and respiratory secretions of infected individuals	Typically none; ganciclovir for immunocompromised patients
Viral conjunctivitis	Adenoviruses and others	Inflammation of the conjunctiva; watery, nonpurulent discharge	Associated with common cold; contagious via contact with eye discharge	None

Figure 21.28

21.4 Mycoses of the Skin

Learning Objectives

By the end of this section, you will be able to:
- Identify the most common fungal pathogens associated with cutaneous and subcutaneous mycoses
- Compare the major characteristics of specific fungal diseases affecting the skin

Many fungal infections of the skin involve fungi that are found in the normal skin microbiota. Some of these fungi can cause infection when they gain entry through a wound; others mainly cause opportunistic infections in immunocompromised patients. Other fungal pathogens primarily cause infection in unusually moist environments that promote fungal growth; for example, sweaty shoes, communal showers, and locker rooms provide excellent breeding grounds that promote the growth and transmission of fungal pathogens.

Fungal infections, also called mycoses, can be divided into classes based on their invasiveness. Mycoses that cause superficial infections of the epidermis, hair, and nails, are called **cutaneous mycoses**. Mycoses that penetrate the epidermis and the dermis to infect deeper tissues are called **subcutaneous mycoses**. Mycoses that spread throughout the body are called **systemic mycoses**.

Tineas

A group of cutaneous mycoses called **tineas** are caused by **dermatophytes**, fungal molds that require keratin, a protein found in skin, hair, and nails, for growth. There are three genera of dermatophytes, all of which can cause cutaneous mycoses: *Trichophyton*, *Epidermophyton*, and *Microsporum*. Tineas on most areas of the body are generally called **ringworm**, but tineas in specific locations may have distinctive names and symptoms (see Table 21.3 and Figure 21.29). Keep in mind that these names—even though they are Latinized—refer to locations on the body, not causative organisms. Tineas can be caused by different dermatophytes in most areas of the body.

Some Common Tineas and Location on the Body

Tinea corporis (ringworm)	Body
Tinea capitis (ringworm)	Scalp
Tinea pedis (athlete's foot)	Feet
Tinea barbae (barber's itch)	Beard
Tinea cruris (jock itch)	Groin
Tinea unguium (onychomycosis)	Toenails, fingernails

Table 21.3

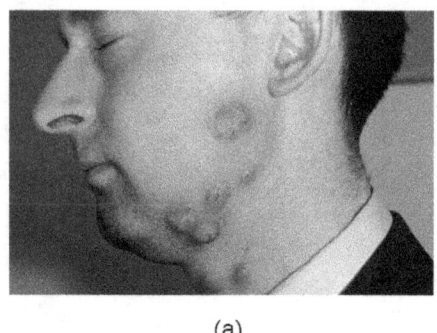
(a)

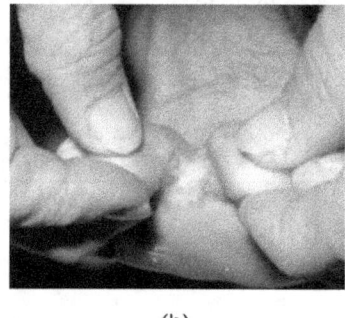
(b)

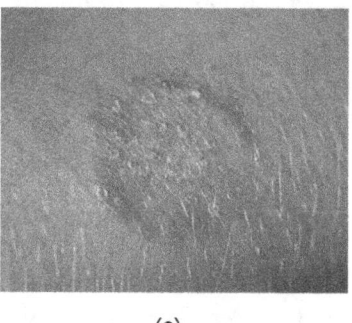
(c)

Figure 21.29 Tineas are superficial cutaneous mycoses and are common. (a) Tinea barbae (barber's itch) occurs on the lower face. (b) Tinea pedis (athlete's foot) occurs on the feet, causing itching, burning, and dry, cracked skin between the toes. (c) A close-up view of tinea corporis (ringworm) caused by *Trichophyton mentagrophytes*. (credit a, c: modification of work by Centers for Disease Control and Prevention; credit b: modification of work by Al Hasan M, Fitzgerald SM, Saoudian M, Krishnaswamy G)

Dermatophytes are commonly found in the environment and in soils and are frequently transferred to the skin via contact with other humans and animals. Fungal spores can also spread on hair. Many dermatophytes grow well in moist, dark environments. For example, **tinea pedis** (athlete's foot) commonly spreads in public showers, and the causative fungi grow well in the dark, moist confines of sweaty shoes and socks. Likewise, **tinea cruris** (jock itch) often spreads in communal living environments and thrives in warm, moist undergarments.

Tineas on the body (**tinea corporis**) often produce lesions that grow radially and heal towards the center. This causes the formation of a red ring, leading to the misleading name of ringworm recall the Clinical Focus case in The Eukaryotes of Microbiology.

Several approaches may be used to diagnose tineas. A Wood's lamp (also called a black lamp) with a wavelength of 365 nm is often used. When directed on a tinea, the ultraviolet light emitted from the Wood's lamp causes the fungal elements (spores and hyphae) to fluoresce. Direct microscopic evaluation of specimens from skin scrapings, hair, or nails can also be used to detect fungi. Generally, these specimens are prepared in a wet mount using a potassium hydroxide solution (10%–20% aqueous KOH), which dissolves the keratin in hair, nails, and skin cells to allow for visualization of the hyphae and fungal spores. The specimens may be grown on Sabouraud dextrose CC (chloramphenicol/cyclohexamide), a selective agar that supports dermatophyte growth while inhibiting the growth of bacteria and saprophytic fungi (Figure 21.30). Macroscopic colony morphology is often used to initially identify the genus of the dermatophyte; identification can be further confirmed by visualizing the microscopic morphology using either a slide culture or a sticky tape prep stained with lactophenol cotton blue.

Various antifungal treatments can be effective against tineas. Allylamine ointments that include terbinafine are commonly used; miconazole and clotrimazole are also available for topical treatment, and griseofulvin is used orally.

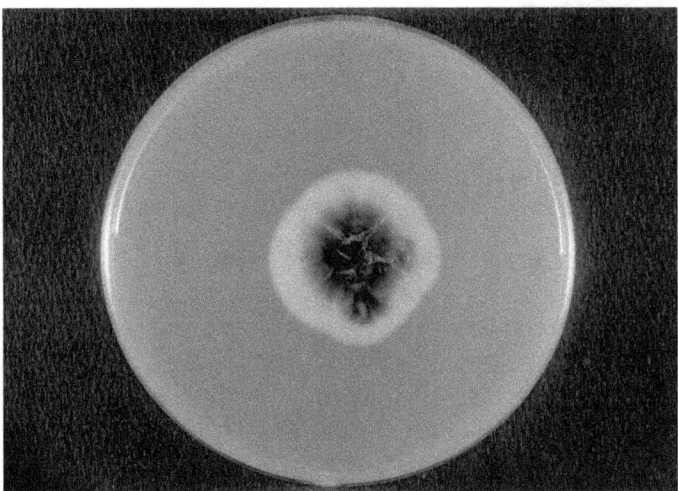

Figure 21.30 To diagnose tineas, the dermatophytes may be grown on a Sabouraud dextrose CC agar plate. This culture contains a strain of *Trichophyton rubrum*, one of the most common causes of tineas on various parts of the body. (credit: Centers for Disease Control and Prevention)

CHECK YOUR UNDERSTANDING

- Why are tineas, caused by fungal molds, often called ringworm?

Cutaneous Aspergillosis

Another cause of cutaneous mycoses is *Aspergillus*, a genus consisting of molds of many different species, some of which cause a condition called aspergillosis. Primary cutaneous aspergillosis, in which the infection begins in the skin, is rare but does occur. More common is secondary cutaneous aspergillosis, in which the infection begins in the respiratory system and disseminates systemically. Both primary and secondary cutaneous aspergillosis result in distinctive eschars that form at the site or sites of infection (Figure 21.31). Pulmonary aspergillosis will be discussed more thoroughly in Respiratory Mycoses).

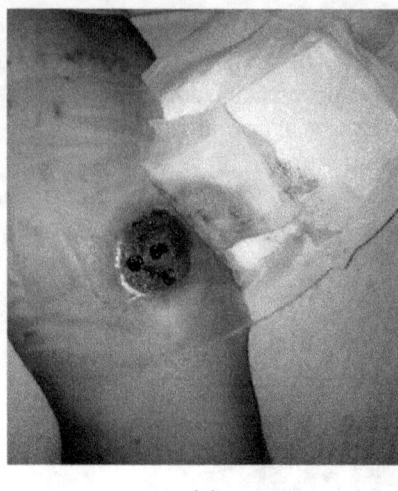 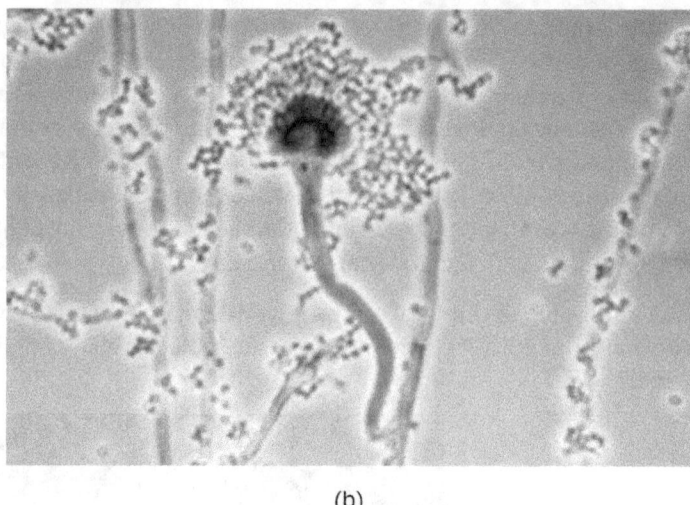

(a) (b)

Figure 21.31 (a) Eschar on a patient with secondary cutaneous aspergillosis. (b) Micrograph showing a conidiophore of *Aspergillus*. (credit a: modification of work by Santiago M, Martinez JH, Palermo C, Figueroa C, Torres O, Trinidad R, Gonzalez E, Miranda Mde L, Garcia M, Villamarzo G; credit b: modification of work by U.S. Department of Health and Human Services)

Primary cutaneous aspergillosis usually occurs at the site of an injury and is most often caused by *Aspergillus fumigatus* or *Aspergillus flavus*. It is usually reported in patients who have had an injury while working in an agricultural or outdoor environment. However, opportunistic infections can also occur in health-care settings, often at the site of intravenous catheters, venipuncture wounds, or in association with burns, surgical wounds, or occlusive dressing. After candidiasis, aspergillosis is the second most common hospital-acquired fungal infection and often occurs in immunocompromised patients, who are more vulnerable to opportunistic infections.

Cutaneous aspergillosis is diagnosed using patient history, culturing, histopathology using a skin biopsy. Treatment involves the use of antifungal medications such as voriconazole (preferred for invasive aspergillosis), itraconazole, and amphotericin B if itraconazole is not effective. For immunosuppressed individuals or burn patients, medication may be used and surgical or immunotherapy treatments may be needed.

CHECK YOUR UNDERSTANDING

- Identify the sources of infection for primary and secondary cutaneous aspergillosis.

Candidiasis of the Skin and Nails

Candida albicans and other yeasts in the genus *Candida* can cause skin infections referred to as cutaneous candidiasis. *Candida* spp. are sometimes responsible for **intertrigo**, a general term for a rash that occurs in a skin fold, or other localized rashes on the skin. *Candida* can also infect the nails, causing them to become yellow and harden (Figure 21.32).

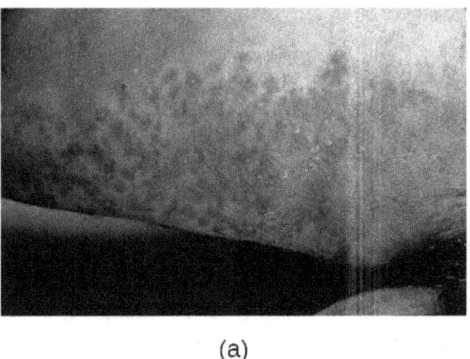 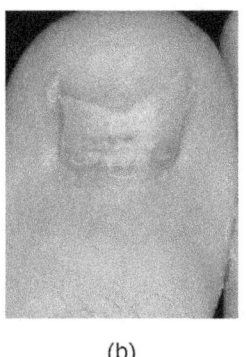 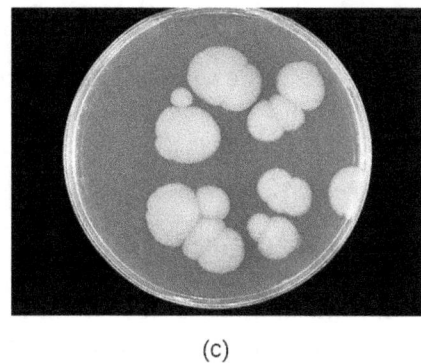

(a) (b) (c)

Figure 21.32 (a) This red, itchy rash is the result of cutaneous candidiasis, an opportunistic infection of the skin caused by the yeast *Candida albicans*. (b) Fungal infections of the nail (tinea unguium) can be caused by dermatophytes or *Candida* spp. The nail becomes yellow, brittle, and prone to breaking. This condition is relatively common among adults. (c) *C. albicans* growing on Sabouraud dextrose agar. (credit a: modification of work by U.S. Department of Veterans Affairs; credit c: modification of work by Centers for Disease Control and Prevention)

Candidiasis of the skin and nails is diagnosed through clinical observation and through culture, Gram stain, and KOH wet mounts. Susceptibility testing for anti-fungal agents can also be done. Cutaneous candidiasis can be treated with topical or systemic azole antifungal medications. Because candidiasis can become invasive, patients suffering from HIV/AIDS, cancer, or other conditions that compromise the immune system may benefit from preventive treatment. Azoles, such as clotrimazole, econazole, fluconazole, ketoconazole, and miconazole; nystatin; terbinafine; and naftifine may be used for treatment. Long-term treatment with medications such as itraconazole or ketoconazole may be used for chronic infections. Repeat infections often occur, but this risk can be reduced by carefully following treatment recommendations, avoiding excessive moisture, maintaining good health, practicing good hygiene, and having appropriate clothing (including footwear).

Candida also causes infections in other parts of the body besides the skin. These include vaginal yeast infections (see Fungal Infections of the Reproductive System) and oral thrush (see Microbial Diseases of the Mouth and Oral Cavity).

✓ CHECK YOUR UNDERSTANDING

- What are the signs and symptoms of candidiasis of the skin and nails?

Sporotrichosis

Whereas cutaneous mycoses are superficial, subcutaneous mycoses can spread from the skin to deeper tissues. In temperate regions, the most common subcutaneous mycosis is a condition called **sporotrichosis**, caused by the fungus *Sporothrix schenkii* and commonly known as rose gardener's disease or rose thorn disease (recall Case in Point: Every Rose Has Its Thorn). Sporotrichosis is often contracted after working with soil, plants, or timber, as the fungus can gain entry through a small wound such as a thorn-prick or splinter. Sporotrichosis can generally be avoided by wearing gloves and protective clothing while gardening and promptly cleaning and disinfecting any wounds sustained during outdoor activities.

Sporothrix infections initially present as small ulcers in the skin, but the fungus can spread to the lymphatic system and sometimes beyond. When the infection spreads, nodules appear, become necrotic, and may ulcerate. As more lymph nodes become affected, abscesses and ulceration may develop over a larger area (often on one arm or hand). In severe cases, the infection may spread more widely throughout the body, although this is relatively uncommon.

Sporothrix infection can be diagnosed based upon histologic examination of the affected tissue. Its macroscopic morphology can be observed by culturing the mold on potato dextrose agar, and its microscopic morphology can be observed by staining a slide culture with lactophenol cotton blue. Treatment with itraconazole is generally recommended.

CHECK YOUR UNDERSTANDING

- Describe the progression of a *Sporothrix schenkii* infection.

Disease Profile

Mycoses of the Skin

Cutaneous mycoses are typically opportunistic, only able to cause infection when the skin barrier is breached through a wound. Tineas are the exception, as the dermatophytes responsible for tineas are able to grow on skin, hair, and nails, especially in moist conditions. Most mycoses of the skin can be avoided through good hygiene and proper wound care. Treatment requires antifungal medications. Figure 21.33 summarizes the characteristics of some common fungal infections of the skin.

Mycoses of the Skin				
Disease	Pathogen	Signs and Symptoms	Transmission	Antimicrobial Drugs
Aspergillosis (cutaneous)	*Aspergillus fumigatus*, *Aspergillus flavus*	Distinctive eschars at site(s) of infection	Entry via wound (primary cutaneous aspergillosis) or via the respiratory system (secondary cutaneous aspergillosis); commonly a hospital-acquired infection	Itraconazole, voriconazole, amphotericin B
Candidiasis (cutaneous)	*Candida albicans*	Intertrigo, localized rash, yellowing of nails	Overgrowth of normal skin microbiota, especially in moist, dark areas	Azoles
Sporotrichosis (rose gardener's disease)	*Sporothrix schenkii*	Subcutaneous ulcers and abscesses; may spread to a large area, e.g., hand or arm	Entry via thorn prick or other wound	Itraconazole
Tineas	*Trichophyton* spp., *Epidermophyton* spp., *Microsporum* spp.	Itchy, ring-like lesions (ringworm) at sites of infection	Contact with dermatophytic fungi, especially in warm, moist environments conducive to fungal growth	Terbinafine, miconazole, clotrimazole, griseofulvin

Figure 21.33

21.5 Protozoan and Helminthic Infections of the Skin and Eyes

Learning Objectives

By the end of this section, you will be able to:
- Identify two parasites that commonly cause infections of the skin and eyes
- Identify the major characteristics of specific parasitic diseases affecting the skin and eyes

Many parasitic protozoans and helminths use the skin or eyes as a portal of entry. Some may physically burrow into the skin or the mucosa of the eye; others breach the skin barrier by means of an insect bite. Still others take advantage of a wound to bypass the skin barrier and enter the body, much like other opportunistic pathogens. Although many parasites enter the body through the skin, in this chapter we will limit our discussion to those for which the skin or eyes are the primary site of infection. Parasites that enter through the skin but travel to a different site of infection will be covered in other chapters. In addition, we will limit our discussion to microscopic parasitic infections of the skin and eyes. Macroscopic parasites such as lice, scabies, mites, and ticks are beyond the scope of this text.

Acanthamoeba Infections

Acanthamoeba is a genus of free-living protozoan amoebae that are common in soils and unchlorinated bodies of fresh water. (This is one reason why some swimming pools are treated with chlorine.) The genus contains a few parasitic species, some of which can cause infections of the eyes, skin, and nervous system. Such infections can sometimes travel and affect other body systems. Skin infections may manifest as abscesses, ulcers, and nodules. When acanthamoebae infect the eye, causing inflammation of the cornea, the condition is called **Acanthamoeba keratitis**. Figure 21.34 illustrates the *Acanthamoeba* life cycle and various modes of infection.

While *Acanthamoeba* keratitis is initially mild, it can lead to severe corneal damage, vision impairment, or even blindness if left untreated. Similar to eye infections involving *P. aeruginosa*, *Acanthamoeba* poses a much greater risk to wearers of contact lenses because the amoeba can thrive in the space between contact lenses and the cornea. Prevention through proper contact lens care is important. Lenses should always be properly disinfected prior to use, and should never be worn while swimming or using a hot tub.

Acanthamoeba can also enter the body through other pathways, including skin wounds and the respiratory tract. It usually does not cause disease except in immunocompromised individuals; however, in rare cases, the infection can spread to the nervous system, resulting in a usually fatal condition called granulomatous amoebic encephalitis (GAE) (see Fungal and Parasitic Diseases of the Nervous System). Disseminated infections, lesions, and *Acanthamoeba* keratitis can be diagnosed by observing symptoms and examining patient samples under the microscope to view the parasite. Skin biopsies may be used.

Acanthamoeba keratitis is difficult to treat, and prompt treatment is necessary to prevent the condition from progressing. The condition generally requires three to four weeks of intensive treatment to resolve. Common treatments include topical antiseptics (e.g., polyhexamethylene biguanide, chlorhexidine, or both), sometimes with painkillers or corticosteroids (although the latter are controversial because they suppress the immune system, which can worsen the infection). Azoles are sometimes prescribed as well. Advanced cases of keratitis may require a corneal transplant to prevent blindness.

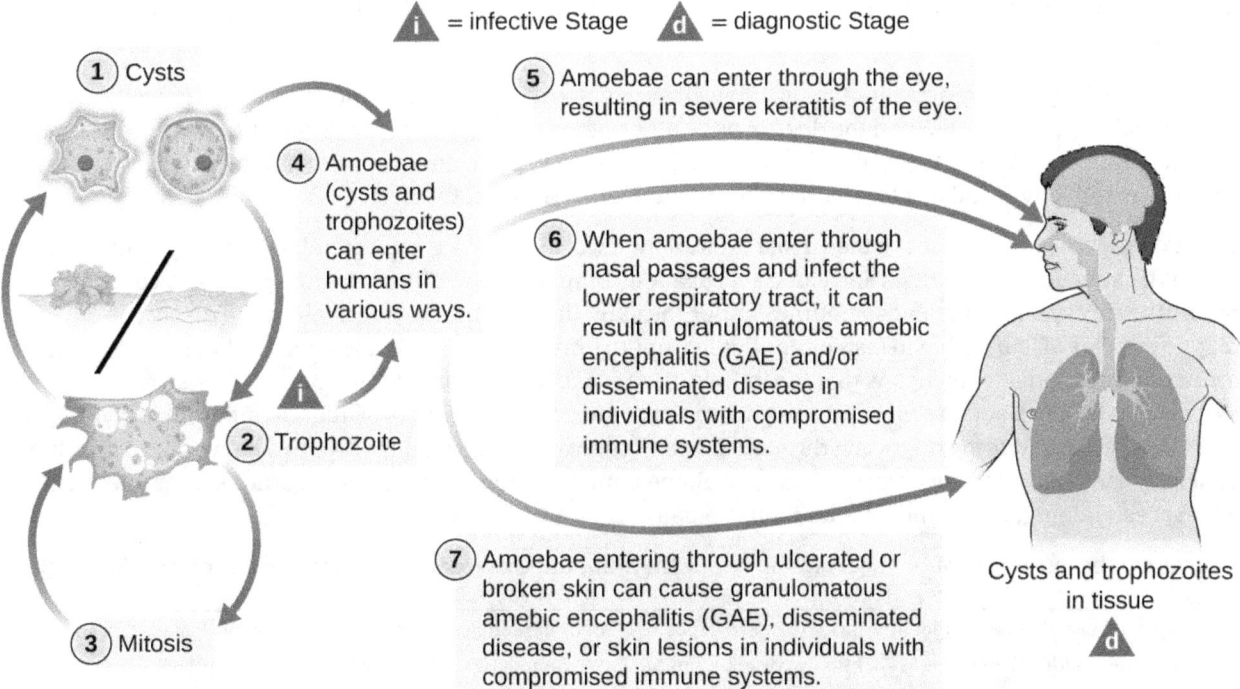

Figure 21.34 *Acanthamoeba* spp. are waterborne parasites very common in unchlorinated aqueous environments. As shown in this life cycle, *Acanthamoeba* cysts and trophozoites are both capable of entering the body through various routes, causing infections of the eye,

skin, and central nervous system. (credit: modification of work by Centers for Disease Control and Prevention)

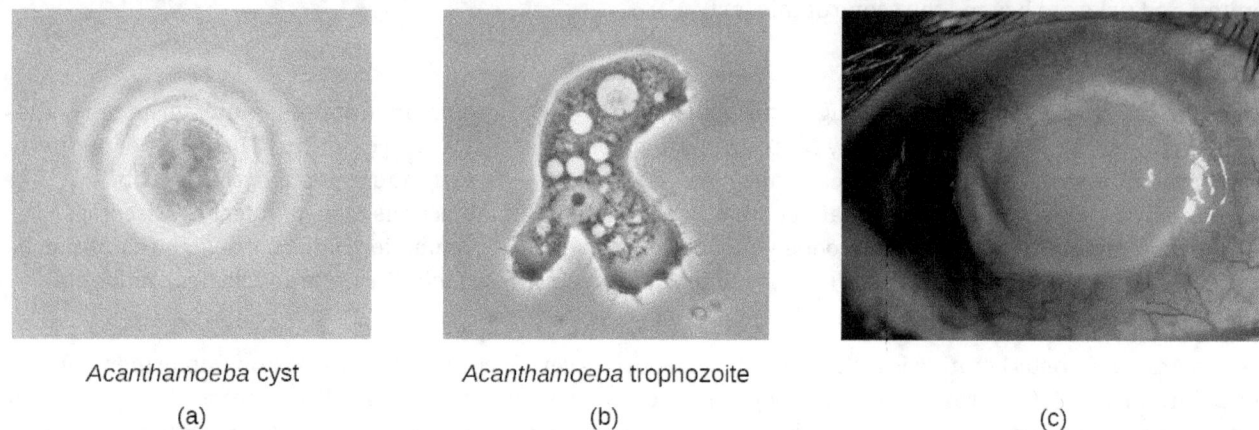

Figure 21.35 (a) An *Acanthamoeba* cyst. (b) An *Acanthamoeba* trophozoite (c) The eye of a patient with *Acanthamoeba* keratitis. The fluorescent color, which is due to sodium fluorescein application, highlights significant damage to the cornea and vascularization of the surrounding conjunctiva. (credit a: modification of work by Centers for Disease Control and Prevention; credit b, c: modification of work by Jacob Lorenzo-Morales, Naveed A Kahn and Julia Walochnik)

CHECK YOUR UNDERSTANDING

- How are *Acanthamoeba* infections acquired?

Loiasis

The helminth *Loa loa*, also known as the African eye worm, is a nematode that can cause **loiasis**, a disease endemic to West and Central Africa (Figure 21.36). The disease does not occur outside that region except when carried by travelers. There is evidence that individual genetic differences affect susceptibility to developing loiasis after infection by the *Loa loa* worm. Even in areas in which *Loa loa* worms are common, the disease is generally found in less than 30% of the population.[17] It has been suggested that travelers who spend time in the region may be somewhat more susceptible to developing symptoms than the native population, and the presentation of infection may differ.[18]

The parasite is spread by deerflies (genus *Chrysops*), which can ingest the larvae from an infected human via a blood meal (Figure 21.36). When the deerfly bites other humans, it deposits the larvae into their bloodstreams. After about five months in the human body, some larvae develop into adult worms, which can grow to several centimeters in length and live for years in the subcutaneous tissue of the host.

The name "eye worm" alludes to the visible migration of worms across the conjunctiva of the eye. Adult worms live in the subcutaneous tissues and can travel at about 1 cm per hour. They can often be observed when migrating through the eye, and sometimes under the skin; in fact, this is generally how the disease is diagnosed. It is also possible to test for antibodies, but the presence of antibodies does not necessarily indicate a current infection; it only means that the individual was exposed at some time. Some patients are asymptomatic, but in others the migrating worms can cause fever and areas of allergic inflammation known as Calabar swellings. Worms migrating through the conjunctiva can cause temporary eye pain and itching, but generally there is no lasting damage to the eye. Some patients experience a range of other symptoms, such as widespread itching, hives, and joint and muscle pain.

Worms can be surgically removed from the eye or the skin, but this treatment only relieves discomfort; it does

---

17 Garcia, A.. et al. "Genetic Epidemiology of Host Predisposition Microfilaraemia in Human Loiasis." *Tropical Medicine and International Health* 4 (1999) 8:565–74. http://www.ncbi.nlm.nih.gov/pubmed/10499080. Accessed Sept 14, 2016.

18 Spinello, A., et al. "Imported *Loa Ioa* Filariasis: Three Cases and a Review of Cases Reported in Non-Endemic Countries in the Past 25 Years." *International Journal of Infectious Disease* 16 (2012) 9: e649–e662. DOI: http://dx.doi.org/10.1016/j.ijid.2012.05.1023.

not cure the infection, which involves many worms. The preferred treatment is diethylcarbamazine, but this medication produces severe side effects in some individuals, such as brain inflammation and possible death in patients with heavy infections. Albendazole is also sometimes used if diethylcarbamazine is not appropriate or not successful. If left untreated for many years, loiasis can damage the kidneys, heart, and lungs, though these symptoms are rare.

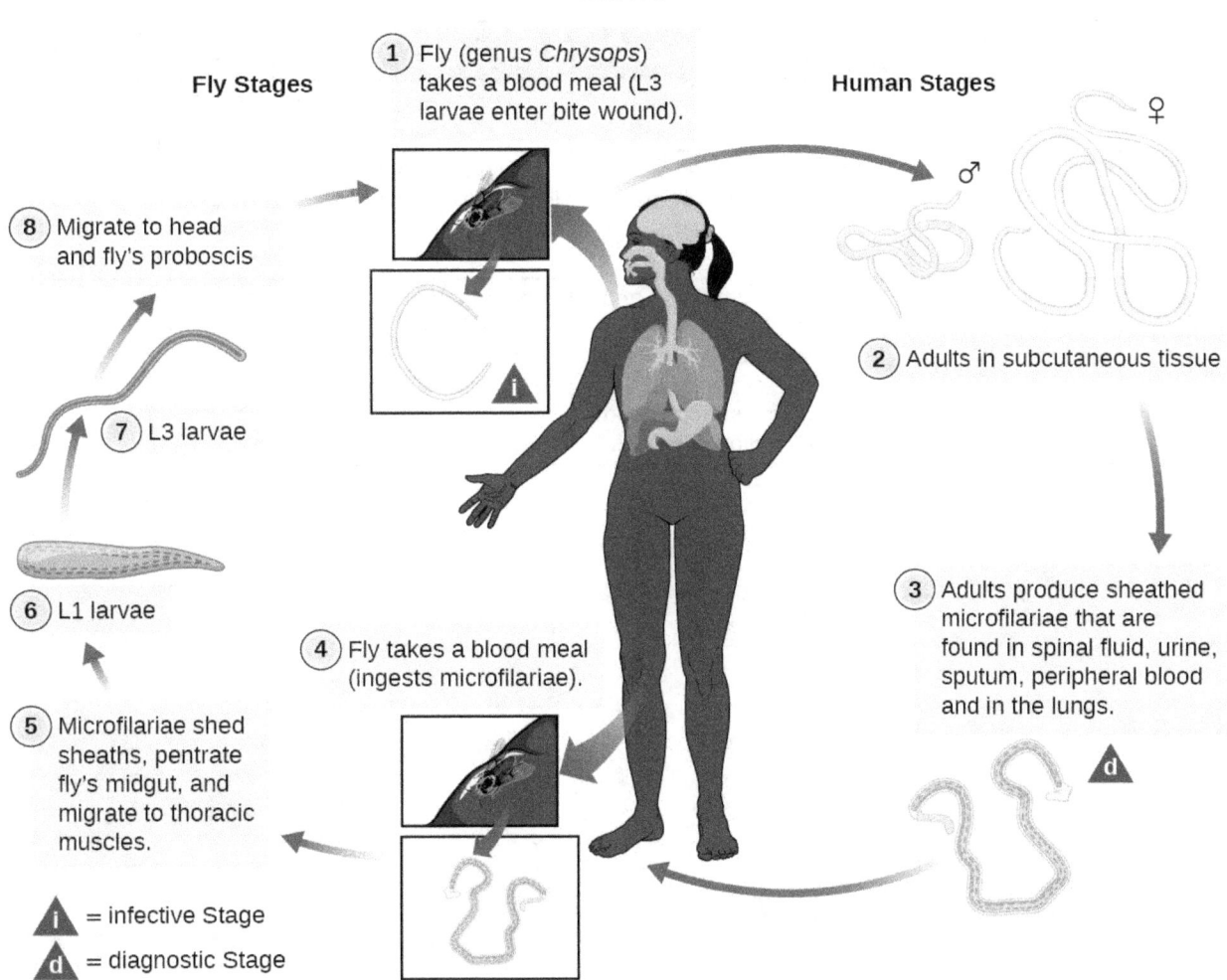

Figure 21.36 This *Loa loa* worm, measuring about 55 mm long, was extracted from the conjunctiva of a patient with loiasis. The *Loa loa* has a complex life cycle. Biting deerflies native to the rain forests of Central and West Africa transmit the larvae between humans. (credit a: modification of work by Eballe AO, Epée E, Koki G, Owono D, Mvogo CE, Bella AL; credit b: modification of work by NIAID; credit c: modification of work by Centers for Disease Control and Prevention)

CHECK YOUR UNDERSTANDING

- Describe the most common way to diagnose loiasis.

LINK TO LEARNING

See a video (https://openstax.org/l/22microfilvid) of a live *Loa loa* microfilaria under the microscope.

> **Disease Profile**
>
> ## Parasitic Skin and Eye Infections
>
> The protozoan *Acanthamoeba* and the helminth *Loa loa* are two parasites capable of causing infections of the skin and eyes. Figure 21.37 summarizes the characteristics of some common fungal infections of the skin.
>
> | Parasitic Skin and Eye Infections | | | | |
> |---|---|---|---|---|
> | Disease | Pathogen | Signs and Symptoms | Transmission | Antimicrobial Drugs |
> | *Acanthamoeba* keratitis | *Acanthamoeba* | Inflammation and damage to cornea; vision impairment or blindness | Exposure to pathogens in contaminated water or on contact lenses | Polyhexamethylene biguanide, chlorhexidine, azoles |
> | Loiasis | *Loa loa* | Recurring fever and localized Calabar swelling, itching, and skin or eye pain during subcutaneous migration of worms | Larvae transmitted between humans by deerfly vector | Diethylcarbamazine, albendazole |
>
> Figure 21.37

SUMMARY

21.1 Anatomy and Normal Microbiota of the Skin and Eyes

- Human skin consists of two main layers, the **epidermis** and **dermis**, which are situated on top of the **hypodermis**, a layer of connective tissue.
- The skin is an effective physical barrier against microbial invasion.
- The skin's relatively dry environment and normal microbiota discourage colonization by transient microbes.
- The skin's normal microbiota varies from one region of the body to another.
- The **conjunctiva** of the eye is a frequent site for microbial infection, but deeper eye infections are less common; multiple types of conjunctivitis exist.

21.2 Bacterial Infections of the Skin and Eyes

- *Staphylococcus* and *Streptococcus* cause many different types of skin infections, many of which occur when bacteria breach the skin barrier through a cut or wound.
- *S. aureus* are frequently associated with purulent skin infections that manifest as **folliculitis**, **furuncles**, or **carbuncles**. *S. aureus* is also a leading cause of staphylococcal scalded skin syndrome (SSSS).
- *S. aureus* is generally drug resistant and current MRSA strains are resistant to a wide range of antibiotics.
- Community-acquired and hospital-acquired staphyloccocal infections are an ongoing problem because many people are asymptomatic carriers.
- **Group A streptococci (GAS)**, *S. pyogenes*, is often responsible for cases of **cellulitis**, **erysipelas**, and **erythema nosodum**. GAS are also one of many possible causes of **necrotizing fasciitis**.
- *P. aeruginosa* is often responsible for infections of the skin and eyes, including wound and burn infections, **hot tub rash**, **otitis externa**, and bacterial **keratitis**.
- **Acne** is a common skin condition that can become more inflammatory when *Propionibacterium acnes* infects hair follicles and pores clogged with dead skin cells and sebum.
- Cutaneous **anthrax** occurs when *Bacillus anthracis* breaches the skin barrier. The infection results in a localized black **eschar** on skin. Anthrax can be fatal if *B. anthracis* spreads to the bloodstream.
- Common bacterial **conjunctivitis** is often caused by *Haemophilus influenzae* and usually resolves on its own in a few days. More serious forms of conjunctivitis include gonococcal **ophthalmia neonatorum**, **inclusion conjunctivitis** (chlamydial), and **trachoma**, all of which can lead to blindness if untreated.
- **Keratitis** is frequently caused by *Staphylococcus epidermidis* and/or *Pseudomonas aeruginosa*, especially among contact lens users, and can lead to blindness.
- Biofilms complicate the treatment of wound and eye infections because pathogens living in biofilms can be difficult to treat and eliminate.

21.3 Viral Infections of the Skin and Eyes

- **Papillomas** (warts) are caused by human papillomaviruses.
- **Herpes simplex virus** (especially HSV-1) mainly causes **oral herpes**, but lesions can appear on other areas of the skin and mucous membranes.
- **Roseola** and **fifth disease** are common viral illnesses that cause skin rashes; roseola is caused by HHV-6 and HHV-7 while fifth disease is caused by parvovirus 19.
- **Viral conjunctivitis** is often caused by adenoviruses and may be associated with the common cold. **Herpes keratitis** is caused by herpesviruses that spread to the eye.

21.4 Mycoses of the Skin

- **Mycoses** can be **cutaneous**, **subcutaneous**, or **systemic**.
- Common cutaneous mycoses include **tineas** caused by **dermatophytes** of the genera *Trichophyton*, *Epidermophyton*, and *Microsporum*. **Tinea corporis** is called **ringworm**. Tineas on other parts of the body have names associated with the affected body part.
- **Aspergillosis** is a fungal disease caused by molds of the genus *Aspergillus*. Primary cutaneous aspergillosis enters through a break in the skin, such as the site of an injury or a surgical wound; it is a common hospital-

acquired infection. In secondary cutaneous aspergillosis, the fungus enters via the respiratory system and disseminates systemically, manifesting in lesions on the skin.
- The most common subcutaneous mycosis is **sporotrichosis** (rose gardener's disease), caused by *Sporothrix schenkii*.
- Yeasts of the genus *Candida* can cause opportunistic infections of the skin called **candidiasis**, producing **intertrigo**, localized rashes, or yellowing of the nails.

21.5 Protozoan and Helminthic Infections of the Skin and Eyes

- The protozoan *Acanthamoeba* and the helminth *Loa loa* are two parasites that can breach the skin barrier, causing infections of the skin and eyes.
- **Acanthamoeba keratitis** is a parasitic infection of the eye that often results from improper disinfection of contact lenses or swimming while wearing contact lenses.
- **Loiasis**, or eye worm, is a disease endemic to Africa that is caused by parasitic worms that infect the subcutaneous tissue of the skin and eyes. It is transmitted by deerfly vectors.

REVIEW QUESTIONS

Multiple Choice

1. _____ glands produce a lipid-rich substance that contains proteins and minerals and protects the skin.
    a. Sweat
    b. Mammary
    c. Sebaceous
    d. Endocrine

2. Which layer of skin contains living cells, is vascularized, and lies directly above the hypodermis?
    a. the stratum corneum
    b. the dermis
    c. the epidermis
    d. the conjunctiva

3. *Staphylococcus aureus* is most often associated with being
    a. coagulase-positive.
    b. coagulase-negative.
    c. catalase-negative.
    d. gram-negative

4. M protein is produced by
    a. *Pseudomonas aeruginosa*
    b. *Staphylococcus aureus*
    c. *Propionibacterium acnes*
    d. *Streptococcus pyogenes*

5. _____ is a major cause of preventable blindness that can be reduced through improved sanitation.
    a. Ophthalmia neonatorum
    b. Keratitis
    c. Trachoma
    d. Cutaneous anthrax

6. Which species is frequently associated with nosocomial infections transmitted via medical devices inserted into the body?
    a. *Staphylococcus epidermidis*
    b. *Streptococcus pyogenes*
    c. *Proproniobacterium acnes*
    d. *Bacillus anthracis*

7. Warts are caused by
    a. human papillomavirus.
    b. herpes simplex virus.
    c. adenoviruses.
    d. parvovirus B19.

8. Which of these viruses can spread to the eye to cause a form of keratitis?
    a. human papillomavirus
    b. herpes simplex virus 1
    c. parvovirus 19
    d. circoviruses

9. Cold sores are associated with:
    a. human papillomavirus
    b. roseola
    c. herpes simplex viruses
    d. human herpesvirus 6

10. Which disease is usually self-limiting but is most commonly treated with ganciclovir if medical treatment is needed?
    a. roseola
    b. oral herpes
    c. papillomas
    d. viral conjunctivitis

11. Adenoviruses can cause:
    a. viral conjunctivitis
    b. herpetic conjunctivitis
    c. papillomas
    d. oral herpes

12. _____ is a superficial fungal infection found on the head.
    a. Tinea cruris
    b. Tinea capitis
    c. Tinea pedis
    d. Tinea corporis

13. For what purpose would a health-care professional use a Wood's lamp for a suspected case of ringworm?
    a. to prevent the rash from spreading
    b. to kill the fungus
    c. to visualize the fungus
    d. to examine the fungus microscopically

14. Sabouraud dextrose agar CC is selective for:
    a. all fungi
    b. non-saprophytic fungi
    c. bacteria
    d. viruses

15. The first-line recommended treatment for sporotrichosis is:
    a. itraconazole
    b. clindamycin
    c. amphotericin
    d. nystatin

16. Which of the following is most likely to cause an *Acanthamoeba* infection?
    a. swimming in a lake while wearing contact lenses
    b. being bitten by deerflies in Central Africa
    c. living environments in a college dormitory with communal showers
    d. participating in a contact sport such as wrestling

17. The parasitic *Loa loa* worm can cause great pain when it:
    a. moves through the bloodstream
    b. exits through the skin of the foot
    c. travels through the conjunctiva
    d. enters the digestive tract

18. A patient tests positive for *Loa loa* antibodies. What does this test indicate?
    a. The individual was exposed to *Loa loa* at some point.
    b. The individual is currently suffering from loiasis.
    c. The individual has never been exposed to *Loa loa*.
    d. The individual is immunosuppressed.

19. _____ is commonly treated with a combination of chlorhexidine and polyhexamethylene biguanide.
    a. *Acanthamoeba* keratitis
    b. Sporotrichosis
    c. Candidiasis
    d. Loiasis

Fill in the Blank

20. The _____ is the outermost layer of the epidermis.

21. The mucous membrane that covers the surface of the eyeball and inner eyelid is called the _____.

22. A purulent wound produces _____.

23. Human herpesvirus 6 is the causative agent of _____.

24. The most common subcutaneous mycosis in temperate regions is _____.

25. Eye worm is another name for _____.

26. The _____ is the part of the eye that is damaged due to *Acanthamoeba* keratitis.

Short Answer

27. What is the role of keratin in the skin?

28. What are two ways in which tears help to prevent microbial colonization?

29. Which label indicates a sweat gland?

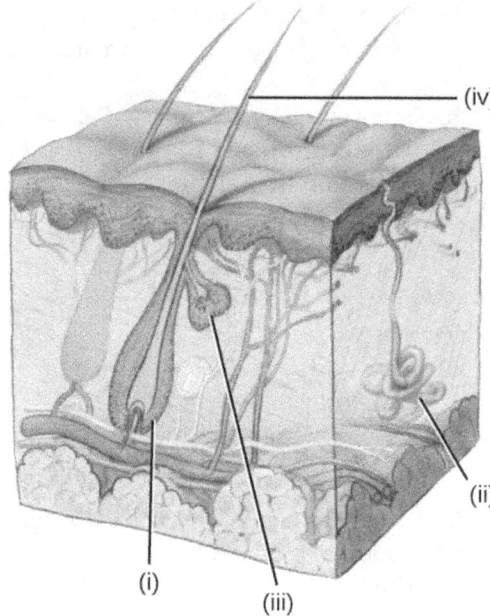

Figure 21.38 (credit: modification of work by National Cancer Institute)

30. How are leukocidins associated with pus production?
31. What is a good first test to distinguish streptococcal infections from staphylococcal infections?
32. Compare and contrast bacterial and viral conjunctivitis.
33. What yeasts commonly cause opportunistic infections?

Critical Thinking

34. Explain why it is important to understand the normal microbiota of the skin.
35. Besides the presence or absence of ulceration, how do acute ulcerative and nonulcerative blepharitis differ?
36. What steps might you recommend to a patient for reducing the risk of developing a fungal infection of the toenails?
37. Why might a traveler to a region with *Loa loa* worm have a greater risk of serious infection compared with people who live in the region?
38. What preventative actions might you recommend to a patient traveling to a region where loiasis is endemic?

CHAPTER 22
Respiratory System Infections

Figure 22.1 Aerosols produced by sneezing, coughing, or even just speaking are an important mechanism for respiratory pathogen transmission. Simple actions, like covering your mouth when coughing or sneezing, can reduce the spread of these microbes. (credit: modification of work by Centers for Disease Control and Prevention)

Chapter Outline

22.1 Anatomy and Normal Microbiota of the Respiratory Tract

22.2 Bacterial Infections of the Respiratory Tract

22.3 Viral Infections of the Respiratory Tract

22.4 Respiratory Mycoses

INTRODUCTION The respiratory tract is one of the main portals of entry into the human body for microbial pathogens. On average, a human takes about 20,000 breaths each day. This roughly corresponds to 10,000 liters, or 10 cubic meters, of air. Suspended within this volume of air are millions of microbes of terrestrial, animal, and human origin—including many potential pathogens. A few of these pathogens will cause relatively mild infections like sore throats and colds. Others, however, are less benign. According to the World Health Organization, respiratory tract infections such as tuberculosis, influenza, and pneumonia were responsible for more than 4 million deaths worldwide in 2012.[1]

At one time, it was thought that antimicrobial drugs and preventive vaccines might hold respiratory infections in check in the developed world, but recent developments suggest otherwise. The rise of multiple-antibiotic resistance in organisms like *Mycobacterium tuberculosis* has rendered many of our modern drugs ineffective.

---

1 World Health Organization. "The Top Ten Causes of Death." May 2014. http://www.who.int/mediacentre/factsheets/fs310/en/

In addition, there has been a recent resurgence in diseases like whooping cough and measles, once-common childhood illnesses made rare by effective vaccines. Despite advances in medicine and public health programs, it is likely that respiratory pathogens will remain formidable adversaries for the foreseeable future.

22.1 Anatomy and Normal Microbiota of the Respiratory Tract

Learning Objectives

By the end of this section, you will be able to:
- Describe the major anatomical features of the upper and lower respiratory tract
- Describe the normal microbiota of the upper and lower respiratory tracts
- Explain how microorganisms overcome defenses of upper and lower respiratory-tract membranes to cause infection
- Explain how microbes and the respiratory system interact and modify each other in healthy individuals and during an infection

---

Clinical Focus

Part 1

John, a 65-year-old man with asthma and type 2 diabetes, works as a sales associate at a local home improvement store. Recently, he began to feel quite ill and made an appointment with his family physician. At the clinic, John reported experiencing headache, chest pain, coughing, and shortness of breath. Over the past day, he had also experienced some nausea and diarrhea. A nurse took his temperature and found that he was running a fever of 40 °C (104 °F).

John suggested that he must have a case of influenza (flu), and regretted that he had put off getting his flu vaccine this year. After listening to John's breathing through a stethoscope, the physician ordered a chest radiography and collected blood, urine, and sputum samples.

- Based on this information, what factors may have contributed to John's illness?

Jump to the next Clinical Focus box.

The primary function of the respiratory tract is to exchange gases (oxygen and carbon dioxide) for metabolism. However, inhalation and exhalation (particularly when forceful) can also serve as a vehicle of transmission for pathogens between individuals.

Anatomy of the Upper Respiratory System

The respiratory system can be conceptually divided into upper and lower regions at the point of the **epiglottis**, the structure that seals off the lower respiratory system from the **pharynx** during swallowing (Figure 22.2). The upper respiratory system is in direct contact with the external environment. The nares (or nostrils) are the external openings of the nose that lead back into the **nasal cavity**, a large air-filled space behind the nares. These anatomical sites constitute the primary opening and first section of the respiratory tract, respectively. The nasal cavity is lined with hairs that trap large particles, like dust and pollen, and prevent their access to deeper tissues. The nasal cavity is also lined with a mucous membrane and Bowman's glands that produce mucus to help trap particles and microorganisms for removal. The nasal cavity is connected to several other air-filled spaces. The sinuses, a set of four, paired small cavities in the skull, communicate with the nasal cavity through a series of small openings. The **nasopharynx** is part of the upper throat extending from the posterior nasal cavity. The nasopharynx carries air inhaled through the nose. The middle ear is connected to the nasopharynx through the **eustachian tube**. The middle ear is separated from the outer ear by the **tympanic membrane**, or ear drum. And finally, the lacrimal glands drain to the nasal cavity through the **nasolacrimal ducts** (tear ducts). The open connections between these sites allow microorganisms to move from the nasal cavity to the sinuses, middle ears (and back), and down into the lower respiratory tract from the nasopharynx.

The oral cavity is a secondary opening for the respiratory tract. The oral and nasal cavities connect through the fauces to the pharynx, or throat. The pharynx can be divided into three regions: the nasopharynx, the **oropharynx**, and the **laryngopharynx**. Air inhaled through the mouth does not pass through the nasopharynx; it proceeds first through the oropharynx and then through the laryngopharynx. The **palatine tonsils**, which consist of lymphoid tissue, are located within the oropharynx. The laryngopharynx, the last portion of the pharynx, connects to the **larynx**, which contains the vocal fold (Figure 22.2).

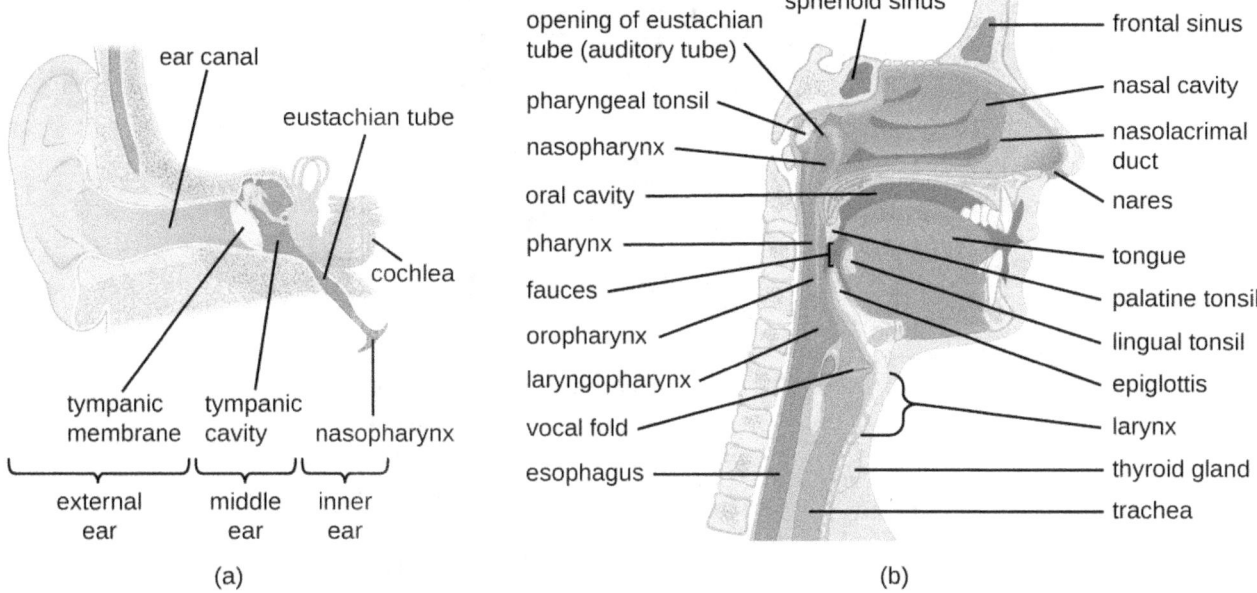

Figure 22.2 (a) The ear is connected to the upper respiratory tract by the eustachian tube, which opens to the nasopharynx. (b) The structures of the upper respiratory tract.

✓ CHECK YOUR UNDERSTANDING

- Identify the sequence of anatomical structures through which microbes would pass on their way from the nares to the larynx.
- What two anatomical points do the eustachian tubes connect?

Anatomy of the Lower Respiratory System

The lower respiratory system begins below the epiglottis in the larynx or voice box (Figure 22.3). The **trachea**, or windpipe, is a cartilaginous tube extending from the larynx that provides an unobstructed path for air to reach the lungs. The trachea bifurcates into the left and right **bronchi** as it reaches the lungs. These paths branch repeatedly to form smaller and more extensive networks of tubes, the **bronchioles**. The terminal bronchioles formed in this tree-like network end in cul-de-sacs called the **alveoli**. These structures are surrounded by capillary networks and are the site of gas exchange in the respiratory system. Human lungs contain on the order of 400,000,000 alveoli. The outer surface of the lungs is protected with a double-layered pleural membrane. This structure protects the lungs and provides lubrication to permit the lungs to move easily during respiration.

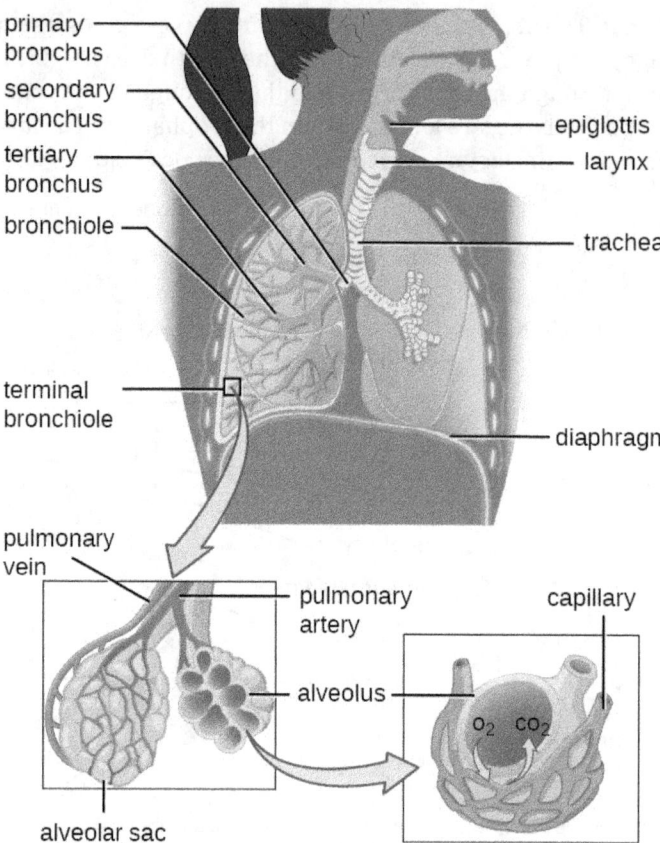

Figure 22.3 The structures of the lower respiratory tract are identified in this illustration. (credit: modification of work by National Cancer Institute)

Defenses of the Respiratory System

The inner lining of the respiratory system consists of mucous membranes (Figure 22.4) and is protected by multiple immune defenses. The goblet cells within the respiratory epithelium secrete a layer of sticky mucus. The viscosity and acidity of this secretion inhibits microbial attachment to the underlying cells. In addition, the respiratory tract contains ciliated epithelial cells. The beating cilia dislodge and propel the mucus, and any trapped microbes, upward to the epiglottis, where they will be swallowed. Elimination of microbes in this manner is referred to as the mucociliary escalator effect and is an important mechanism that prevents inhaled microorganisms from migrating further into the lower respiratory tract.

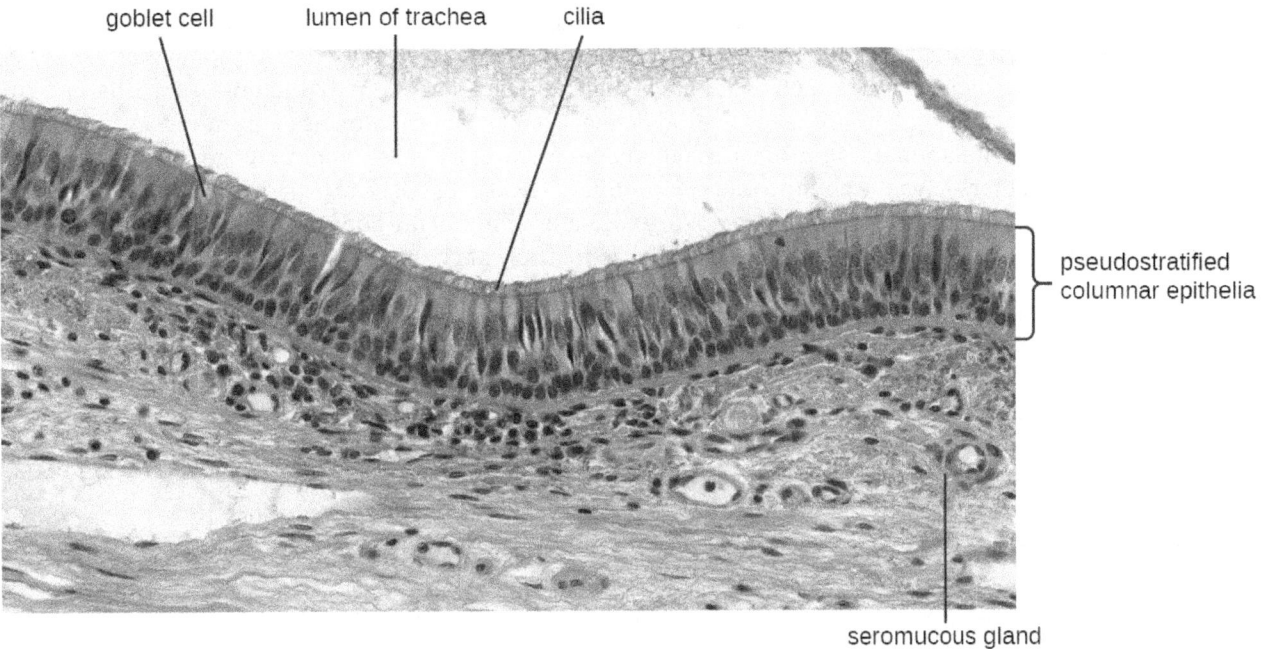

Figure 22.4 This micrograph shows the structure of the mucous membrane of the respiratory tract. (credit: modification of micrograph provided by the Regents of University of Michigan Medical School © 2012)

The upper respiratory system is under constant surveillance by mucosa-associated lymphoid tissue (MALT), including the adenoids and tonsils. Other mucosal defenses include secreted antibodies (IgA), lysozyme, surfactant, and antimicrobial peptides called defensins. Meanwhile, the lower respiratory tract is protected by alveolar macrophages. These phagocytes efficiently kill any microbes that manage to evade the other defenses. The combined action of these factors renders the lower respiratory tract nearly devoid of colonized microbes.

CHECK YOUR UNDERSTANDING

- Identify the sequence of anatomical structures through which microbes would pass on their way from the larynx to the alveoli.
- Name some defenses of the respiratory system that protect against microbial infection.

Normal Microbiota of the Respiratory System

The upper respiratory tract contains an abundant and diverse microbiota. The nasal passages and sinuses are primarily colonized by members of the Firmicutes, Actinobacteria, and Proteobacteria. The most common bacteria identified include *Staphylococcus epidermidis*, viridans group streptococci (VGS), *Corynebacterium* spp. (diphtheroids), *Propionibacterium* spp., and *Haemophilus* spp. The oropharynx includes many of the same isolates as the nose and sinuses, with the addition of variable numbers of bacteria like species of *Prevotella*, *Fusobacterium*, *Moraxella*, and *Eikenella*, as well as some *Candida* fungal isolates. In addition, many healthy humans asymptomatically carry potential pathogens in the upper respiratory tract. As much as 20% of the population carry *Staphylococcus aureus* in their nostrils.[2] The pharynx, too, can be colonized with pathogenic strains of *Streptococcus*, *Haemophilus*, and *Neisseria*.

The lower respiratory tract, by contrast, is scantily populated with microbes. Of the organisms identified in the lower respiratory tract, species of *Pseudomonas*, *Streptococcus*, *Prevotella*, *Fusobacterium*, and *Veillonella* are the most common. It is not clear at this time if these small populations of bacteria constitute a normal microbiota or if they are transients.

---

[2] J. Kluytmans et al. "Nasal Carriage of *Staphylococcus aureus*: Epidemiology, Underlying Mechanisms, and Associated Risks." *Clinical Microbiology Reviews* 10 no. 3 (1997):505–520.

Many members of the respiratory system's normal microbiota are opportunistic pathogens. To proliferate and cause host damage, they first must overcome the immune defenses of respiratory tissues. Many mucosal pathogens produce virulence factors such as adhesins that mediate attachment to host epithelial cells, or polysaccharide capsules that allow microbes to evade phagocytosis. The endotoxins of gram-negative bacteria can stimulate a strong inflammatory response that damages respiratory cells. Other pathogens produce exotoxins, and still others have the ability to survive within the host cells. Once an infection of the respiratory tract is established, it tends to impair the mucociliary escalator, limiting the body's ability to expel the invading microbes, thus making it easier for pathogens to multiply and spread.

Vaccines have been developed for many of the most serious bacterial and viral pathogens. Several of the most important respiratory pathogens and their vaccines, if available, are summarized in Table 22.1. Components of these vaccines will be explained later in the chapter.

Some Important Respiratory Diseases and Vaccines

Disease	Pathogen	Available Vaccine(s)[3]
Chickenpox/shingles	Varicella-zoster virus	Varicella (chickenpox) vaccine, herpes zoster (shingles) vaccine
Common cold	Rhinovirus	None
Diphtheria	*Corynebacterium diphtheriae*	DtaP, Tdap, DT, Td, DTP
Epiglottitis, otitis media	*Haemophilus influenzae*	Hib
Influenza	Influenza viruses	Inactivated, FluMist
Measles	Measles virus	MMR
Pertussis	*Bordetella pertussis*	DTaP, Tdap
Pneumonia	*Streptococcus pneumoniae*	Pneumococcal conjugate vaccine (PCV13), pneumococcal polysaccharide vaccine (PPSV23)
Rubella (German measles)	Rubella virus	MMR
Severe acute respiratory syndrome (SARS)	SARS-associated coronavirus (SARS-CoV)	None
Tuberculosis	*Mycobacterium tuberculosis*	BCG

Table 22.1

---

[3] Full names of vaccines listed in table: *Haemophilus influenzae* type B (Hib); Diphtheria, tetanus, and acellular pertussis (DtaP); tetanus, diphtheria, and acellular pertussis (Tdap); diphtheria and tetanus (DT); tetanus and diphtheria (Td); diphtheria, pertussis, and tetanus (DTP); Bacillus Calmette-Guérin; Measles, mumps, rubella (MMR)

CHECK YOUR UNDERSTANDING

- What are some pathogenic bacteria that are part of the normal microbiota of the respiratory tract?
- What virulence factors are used by pathogens to overcome the immune protection of the respiratory tract?

Signs and Symptoms of Respiratory Infection

Microbial diseases of the respiratory system typically result in an acute inflammatory response. These infections can be grouped by the location affected and have names ending in "itis", which literally means *inflammation of.* For instance, **rhinitis** is an inflammation of the nasal cavities, often characteristic of the common cold. Rhinitis may also be associated with hay fever allergies or other irritants. Inflammation of the sinuses is called **sinusitis** inflammation of the ear is called **otitis**. Otitis media is an inflammation of the middle ear. A variety of microbes can cause **pharyngitis**, commonly known as a sore throat. An inflammation of the larynx is called **laryngitis**. The resulting inflammation may interfere with vocal cord function, causing voice loss. When tonsils are inflamed, it is called **tonsillitis**. Chronic cases of tonsillitis may be treated surgically with tonsillectomy. More rarely, the epiglottis can be infected, a condition called **epiglottitis**. In the lower respiratory system, the inflammation of the bronchial tubes results in **bronchitis**. Most serious of all is **pneumonia**, in which the alveoli in the lungs are infected and become inflamed. Pus and edema accumulate and fill the alveoli with fluids (called consolidations). This reduces the lungs' ability to exchange gases and often results in a productive cough expelling phlegm and mucus. Cases of pneumonia can range from mild to life-threatening, and remain an important cause of mortality in the very young and very old.

CHECK YOUR UNDERSTANDING

- Describe the typical symptoms of rhinitis, sinusitis, pharyngitis, and laryngitis.

Case in Point

Smoking-Associated Pneumonia

Camila is a 22-year-old student who has been a chronic smoker for 5 years. Recently, she developed a persistent cough that has not responded to over-the-counter treatments. Her doctor ordered a chest radiograph to investigate. The radiological results were consistent with pneumonia. In addition, *Streptococcus pneumoniae* was isolated from Camila's sputum.

Smokers are at a greater risk of developing pneumonia than the general population. Several components of tobacco smoke have been demonstrated to impair the lungs' immune defenses. These effects include disrupting the function of the ciliated epithelial cells, inhibiting phagocytosis, and blocking the action of antimicrobial peptides. Together, these lead to a dysfunction of the mucociliary escalator effect. The organisms trapped in the mucus are therefore able to colonize the lungs and cause infections rather than being expelled or swallowed.

22.2 Bacterial Infections of the Respiratory Tract

Learning Objectives

By the end of this section, you will be able to:
- Identify the most common bacteria that can cause infections of the upper and lower respiratory tract
- Compare the major characteristics of specific bacterial diseases of the respiratory tract

The respiratory tract can be infected by a variety of bacteria, both gram positive and gram negative. Although the diseases that they cause may range from mild to severe, in most cases, the microbes remain localized within the respiratory system. Fortunately, most of these infections also respond well to antibiotic therapy.

Streptococcal Infections

A common upper respiratory infection, **streptococcal pharyngitis** (**strep throat**) is caused by *Streptococcus pyogenes*. This gram-positive bacterium appears as chains of cocci, as seen in Figure 22.5. Rebecca Lancefield serologically classified streptococci in the 1930s using carbohydrate antigens from the bacterial cell walls. *S. pyogenes* is the sole member of the Lancefield group A streptococci and is often referred to as GAS, or group A strep.

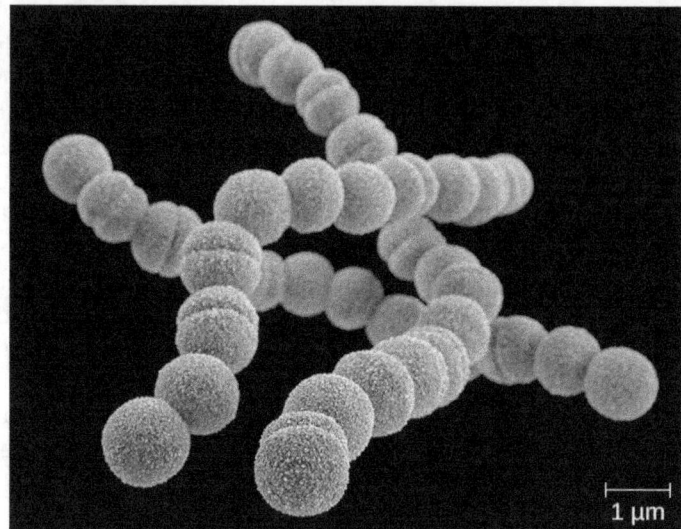

Figure 22.5 This scanning electron micrograph of *Streptococcus pyogenes* shows the characteristic cellular phenotype resembling chains of cocci. (credit: modification of work by U.S. Centers for Disease Control and Prevention - Medical Illustrator)

Similar to streptococcal infections of the skin, the mucosal membranes of the pharynx are damaged by the release of a variety of exoenzymes and exotoxins by this extracellular pathogen. Many strains of *S. pyogenes* can degrade connective tissues by using hyaluronidase, collagenase and streptokinase. Streptokinase activates plasmin, which leads to degradation of fibrin and, in turn, dissolution of blood clots, which assists in the spread of the pathogen. Released toxins include streptolysins that can destroy red and white blood cells. The classic signs of streptococcal pharyngitis are a fever higher than 38 °C (100.4 °F); intense pharyngeal pain; erythema associated with pharyngeal inflammation; and swollen, dark-red palatine tonsils, often dotted with patches of pus; and petechiae (microcapillary hemorrhages) on the soft or hard palate (roof of the mouth) (Figure 22.6). The submandibular lymph nodes beneath the angle of the jaw are also often swollen during strep throat.

Some strains of group A streptococci produce **erythrogenic toxin**. This exotoxin is encoded by a temperate bacteriophage (bacterial virus) and is an example of phage conversion (see The Viral Life Cycle). The toxin attacks the plasma membranes of capillary endothelial cells and leads to **scarlet fever** (or scarlatina), a disseminated fine red rash on the skin, and strawberry tongue, a red rash on the tongue (Figure 22.6). Severe cases may even lead to streptococcal toxic shock syndrome (STSS), which results from massive superantigen production that leads to septic shock and death.

S. pyogenes can be easily spread by direct contact or droplet transmission through coughing and sneezing. The disease can be diagnosed quickly using a rapid enzyme immunoassay for the group A antigen. However, due to a significant rate of false-negative results (up to 30%[4]), culture identification is still the gold standard to confirm pharyngitis due to *S. pyogenes*. *S. pyogenes* can be identified as a catalase-negative, beta hemolytic bacterium that is susceptible to 0.04 units of bacitracin. Antibiotic resistance is limited for this bacterium, so most β-lactams remain effective; oral amoxicillin and intramuscular penicillin G are those most commonly prescribed.

---

4 WL Lean et al. "Rapid Diagnostic Tests for Group A Streptococcal Pharyngitis: A Meta-Analysis." *Pediatrics* 134, no. 4 (2014):771–781.

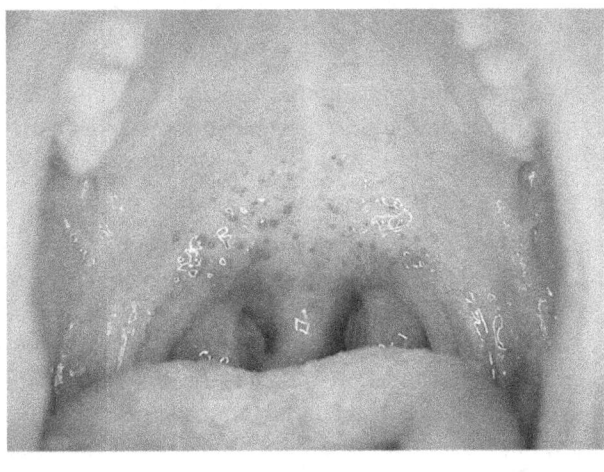
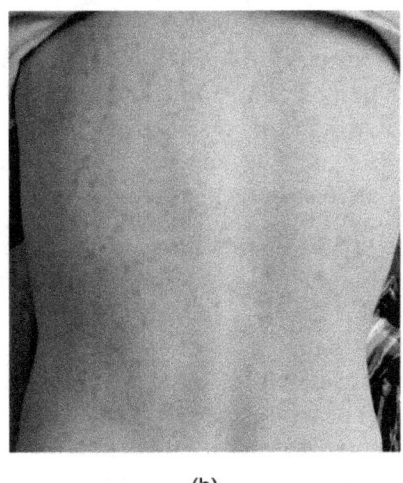

(a)   (b)

Figure 22.6 Streptococcal infections of the respiratory tract may cause localized pharyngitis or systemic signs and symptoms. (a) The characteristic appearance of strep throat: bright red arches of inflammation with the presence of dark-red spots (petechiae). (b) Scarlet fever presents as a rash on the skin. (credit a: modification of work by Centers for Disease Control and Prevention; credit b: modification of work by Alicia Williams)

Sequelae of *S. pyogenes* Infections

One reason strep throat infections are aggressively treated with antibiotics is because they can lead to serious **sequelae**, later clinical consequences of a primary infection. It is estimated that 1%–3% of untreated *S. pyogenes* infections can be followed by nonsuppurative (without the production of pus) sequelae that develop 1–3 weeks after the acute infection has resolved. Two such sequelae are **acute rheumatic fever** and **acute glomerulonephritis**.

Acute rheumatic fever can follow pharyngitis caused by specific rheumatogenic strains of *S. pyogenes* (strains 1, 3, 5, 6, and 18). Although the exact mechanism responsible for this sequela remains unclear, molecular mimicry between the M protein of rheumatogenic strains of *S. pyogenes* and heart tissue is thought to initiate the autoimmune attack. The most serious and lethal clinical manifestation of rheumatic fever is damage to and inflammation of the heart (carditis). Acute glomerulonephritis also results from an immune response to streptococcal antigens following pharyngitis and cutaneous infections. Acute glomerulonephritis develops within 6–10 days after pharyngitis, but can take up to 21 days after a cutaneous infection. Similar to acute rheumatic fever, there are strong associations between specific nephritogenic strains of *S. pyogenes* and acute glomerulonephritis, and evidence suggests a role for antigen mimicry and autoimmunity. However, the primary mechanism of acute glomerulonephritis appears to be the formation of immune complexes between *S. pyogenes* antigens and antibodies, and their deposition between endothelial cells of the glomeruli of kidney. Inflammatory response against the immune complexes leads to damage and inflammation of the glomeruli (glomerulonephritis).

CHECK YOUR UNDERSTANDING

- What are the symptoms of strep throat?
- What is erythrogenic toxin and what effect does it have?
- What are the causes of rheumatic fever and acute glomerulonephritis?

Acute Otitis Media

An infection of the middle ear is called **acute otitis media** (**AOM**), but often it is simply referred to as an earache. The condition is most common between ages 3 months and 3 years. In the United States, AOM is the second-leading cause of visits to pediatricians by children younger than age 5 years, and it is the leading indication for antibiotic prescription.[5]

---

5   G. Worrall. "Acute Otitis Media." *Canadian Family Physician* 53 no. 12 (2007):2147–2148.

AOM is characterized by the formation and accumulation of pus in the middle ear. Unable to drain, the pus builds up, resulting in moderate to severe bulging of the tympanic membrane and otalgia (ear pain). Inflammation resulting from the infection leads to swelling of the eustachian tubes, and may also lead to fever, nausea, vomiting, and diarrhea, particularly in infants. Infants and toddlers who cannot yet speak may exhibit nonverbal signs suggesting AOM, such as holding, tugging, or rubbing of the ear, as well as uncharacteristic crying or distress in response to the pain.

AOM can be caused by a variety of bacteria. Among neonates, *S. pneumoniae* is the most common cause of AOM, but *Escherichia coli*, *Enterococcus* spp., and group B *Streptococcus* species can also be involved. In older infants and children younger than 14 years old, the most common bacterial causes are *S. pneumoniae*, *Haemophilus influenzae*, or *Moraxella catarrhalis*. Among *S. pneumoniae* infections, encapsulated strains are frequent causes of AOM. By contrast, the strains of *H. influenzae* and *M. cattarhalis* that are responsible for AOM do not possess a capsule. Rather than direct tissue damage by these pathogens, bacterial components such as lipopolysaccharide (LPS) in gram-negative pathogens induce an inflammatory response that causes swelling, pus, and tissue damage within the middle ear (Figure 22.7).

Any blockage of the eustachian tubes, with or without infection, can cause fluid to become trapped and accumulate in the middle ear. This is referred to as **otitis media with effusion (OME)**. The accumulated fluid offers an excellent reservoir for microbial growth and, consequently, secondary bacterial infections often ensue. This can lead to recurring and chronic earaches, which are especially common in young children. The higher incidence in children can be attributed to many factors. Children have more upper respiratory infections, in general, and their eustachian tubes are also shorter and drain at a shallower angle. Young children also tend to spend more time lying down than adults, which facilitates drainage from the nasopharynx through the eustachian tube and into the middle ear. Bottle feeding while lying down enhances this risk because the sucking action on the bottle causes negative pressure to build up within the eustachian tube, promoting the movement of fluid and bacteria from the nasopharynx.

Diagnosis is typically made based on clinical signs and symptoms, without laboratory testing to determine the specific causative agent. Antibiotics are frequently prescribed for the treatment of AOM. High-dose amoxicillin is the first-line drug, but with increasing resistance concerns, macrolides and cephalosporins may also be used. The pneumococcal conjugate vaccine (PCV13) contains serotypes that are important causes of AOM, and vaccination has been shown to decrease the incidence of AOM. Vaccination against influenza has also been shown to decrease the risk for AOM, likely because viral infections like influenza predispose patients to secondary infections with *S. pneumoniae*. Although there is a conjugate vaccine available for the invasive serotype B of *H. influenzae*, this vaccine does not impact the incidence of *H. influenzae* AOM. Because unencapsulated strains of *H. influenzae* and *M. catarrhalis* are involved in AOM, vaccines against bacterial cellular factors other than capsules will need to be developed.

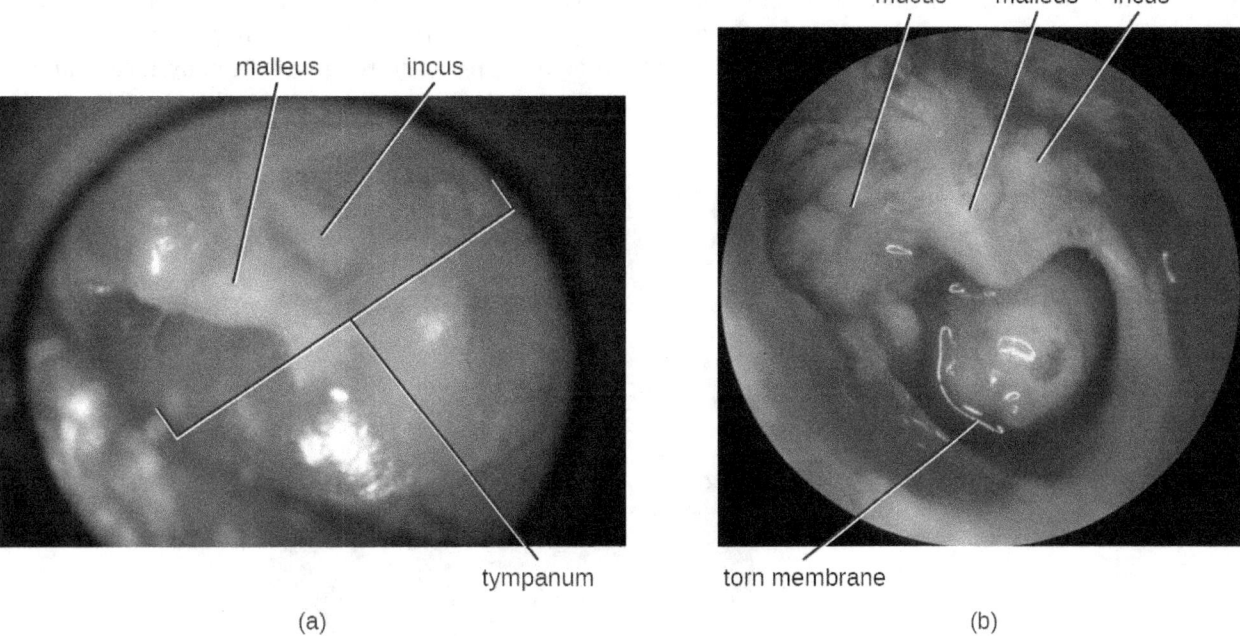

Figure 22.7 (a) A healthy tympanic membrane; the middle ear bones can be seen behind the membrane. (b) An ear with chronic inflammation that has resulted in a torn membrane, erosion of the inner ear bones, and mucus buildup. (credit a: modification of work by "DrER.tv"/YouTube; credit b: modification of work by Li Mg, Hotez PJ, Vrabec JT, Donovan DT)

Bacterial Rhinosinusitis

The microbial community of the nasopharynx is extremely diverse and harbors many opportunistic pathogens, so it is perhaps not surprising that infections leading to rhinitis and sinusitis have many possible causes. These conditions often occur as secondary infections after a viral infection, which effectively compromises the immune defenses and allows the opportunistic bacteria to establish themselves. Bacterial sinusitis involves infection and inflammation within the paranasal sinuses. Because bacterial sinusitis rarely occurs without rhinitis, the preferred term is rhinosinusitis. The most common causes of bacterial rhinosinusitis are similar to those for AOM, including *S. pneumoniae*, *H. influenzae*, and *M. catarrhalis*.

✓ CHECK YOUR UNDERSTANDING

- What are the usual causative agents of acute otitis media?
- What factors facilitate acute otitis media with effusion in young children?
- What factor often triggers bacterial rhinosinusitis?

Diphtheria

The causative agent of **diphtheria**, *Corynebacterium diphtheriae*, is a club-shaped, gram-positive rod that belongs to the phylum Actinobacteria. Diphtheroids are common members of the normal nasopharyngeal microbiota. However, some strains of *C. diphtheriae* become pathogenic because of the presence of a temperate bacteriophage-encoded protein—the diphtheria toxin. Diphtheria is typically a respiratory infection of the oropharynx but can also cause impetigo-like lesions on the skin. Although the disease can affect people of all ages, it tends to be most severe in those younger than 5 years or older than 40 years. Like strep throat, diphtheria is commonly transmitted in the droplets and aerosols produced by coughing. After colonizing the throat, the bacterium remains in the oral cavity and begins producing the diphtheria toxin. This protein is an A-B toxin that blocks host-cell protein synthesis by inactivating elongation factor (EF)-2 (see Virulence Factors of Bacterial and Viral Pathogens). The toxin's action leads to the death of the host cells and an inflammatory response. An accumulation of grayish exudate consisting of dead host cells, pus, red blood cells, fibrin, and infectious bacteria results in the formation of a **pseudomembrane**. The pseudomembrane can cover mucous membranes of the nasal cavity, tonsils, pharynx, and larynx (Figure 22.8). This is a classic sign of diphtheria.

As the disease progresses, the pseudomembrane can enlarge to obstruct the fauces of the pharynx or trachea and can lead to suffocation and death. Sometimes, **intubation**, the placement of a breathing tube in the trachea, is required in advanced infections. If the diphtheria toxin spreads throughout the body, it can damage other tissues as well. This can include myocarditis (heart damage) and nerve damage that may impair breathing.

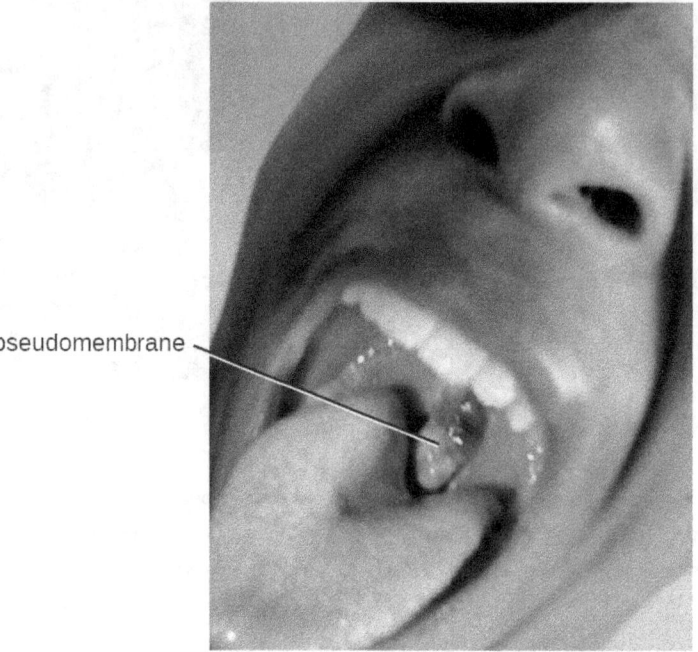

Figure 22.8 The pseudomembrane in a patient with diphtheria presents as a leathery gray patch consisting of dead cells, pus, fibrin, red blood cells, and infectious microbes. (credit: modification of work by Putnong N, Agustin G, Pasubillo M, Miyagi K, Dimaano EM)

The presumptive diagnosis of diphtheria is primarily based on the clinical symptoms (i.e., the pseudomembrane) and vaccination history, and is typically confirmed by identifying bacterial cultures obtained from throat swabs. The diphtheria toxin itself can be directly detected in vitro using polymerase chain reaction (PCR)-based, direct detection systems for the diphtheria *tox* gene, and immunological techniques like radial immunodiffusion or Elek's immunodiffusion test.

Broad-spectrum antibiotics like penicillin and erythromycin tend to effectively control *C. diphtheriae* infections. Regrettably, they have no effect against preformed toxins. If toxin production has already occurred in the patient, antitoxins (preformed antibodies against the toxin) are administered. Although this is effective in neutralizing the toxin, the antitoxins may lead to serum sickness because they are produced in horses (see Hypersensitivities).

Widespread vaccination efforts have reduced the occurrence of diphtheria worldwide. There are currently four combination toxoid vaccines available that provide protection against diphtheria and other diseases: DTaP, Tdap, DT, and Td. In all cases, the letters "d," "t," and "p" stand for diphtheria, tetanus, and pertussis, respectively; the "a" stands for acellular. If capitalized, the letters indicate a full-strength dose; lowercase letters indicate reduced dosages. According to current recommendations, children should receive five doses of the DTaP vaccine in their youth and a Td booster every 10 years. Children with adverse reactions to the pertussis vaccine may be given the DT vaccine in place of the DTaP.

✓ CHECK YOUR UNDERSTANDING

- What effect does diphtheria toxin have?
- What is the pseudomembrane composed of?

Bacterial Pneumonia

Pneumonia is a general term for infections of the lungs that lead to inflammation and accumulation of fluids and white blood cells in the alveoli. Pneumonia can be caused by bacteria, viruses, fungi, and other organisms, although the vast majority of pneumonias are bacterial in origin. Bacterial pneumonia is a prevalent, potentially serious infection; it caused more 50,000 deaths in the United States in 2014.[6] As the alveoli fill with fluids and white blood cells (consolidation), air exchange becomes impaired and patients experience respiratory distress (Figure 22.9). In addition, pneumonia can lead to pleurisy, an infection of the pleural membrane surrounding the lungs, which can make breathing very painful. Although many different bacteria can cause pneumonia under the right circumstances, three bacterial species cause most clinical cases: *Streptococcus pneumoniae*, *H. influenzae*, and *Mycoplasma pneumoniae*. In addition to these, we will also examine some of the less common causes of pneumonia.

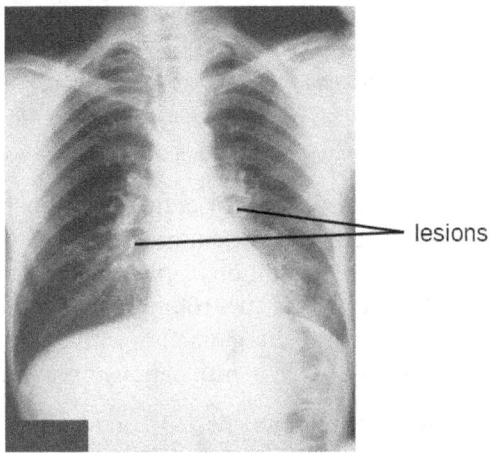

Figure 22.9 A chest radiograph of a patient with pneumonia shows the consolidations (lesions) present as opaque patches. (credit: modification of work by Centers for Disease Control and Prevention)

Pneumococcal Pneumonia

The most common cause of community-acquired bacterial pneumonia is *Streptococcus pneumoniae*. This gram-positive, alpha hemolytic streptococcus is commonly found as part of the normal microbiota of the human respiratory tract. The cells tend to be somewhat lancet-shaped and typically appear as pairs (Figure 22.10). The pneumococci initially colonize the bronchioles of the lungs. Eventually, the infection spreads to the alveoli, where the microbe's polysaccharide capsule interferes with phagocytic clearance. Other virulence factors include autolysins like Lyt A, which degrade the microbial cell wall, resulting in cell lysis and the release of cytoplasmic virulence factors. One of these factors, pneumolysin O, is important in disease progression; this pore-forming protein damages host cells, promotes bacterial adherence, and enhances pro-inflammatory cytokine production. The resulting inflammatory response causes the alveoli to fill with exudate rich in neutrophils and red blood cells. As a consequence, infected individuals develop a productive cough with bloody sputum.

---

6 KD Kochanek et al. "Deaths: Final Data for 2014." *National Vital Statistics Reports* 65 no 4 (2016).

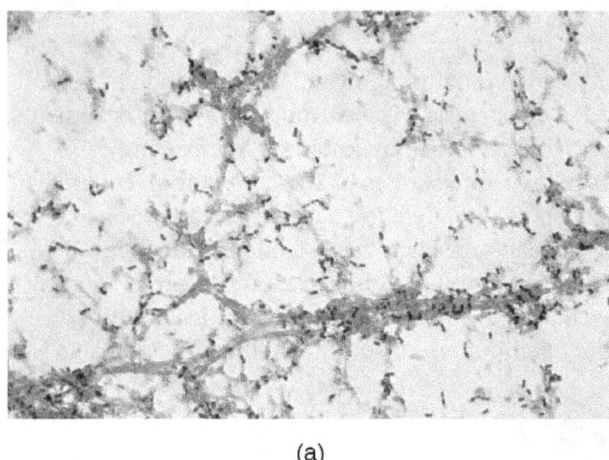 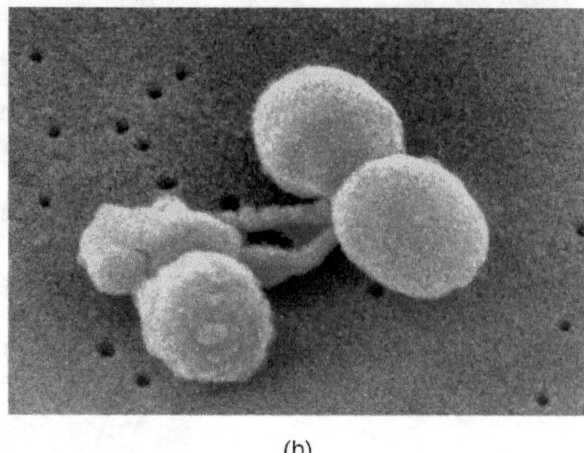

Figure 22.10 (a) This micrograph of *Streptococcus pneumoniae* grown from a blood culture shows the characteristic lancet-shaped diplococcal morphology. (b) A colorized scanning electron micrograph of *S. pneumoniae*. (credit a: modification of work by Centers for Disease Control and Prevention; credit b: modification of work by Janice Carr, Centers for Disease Control and Prevention)

Pneumococci can be presumptively identified by their distinctive gram-positive, lancet-shaped cell morphology and diplococcal arrangement. In blood agar cultures, the organism demonstrates alpha hemolytic colonies that are autolytic after 24 to 48 hours. In addition, *S. pneumoniae* is extremely sensitive to optochin and colonies are rapidly destroyed by the addition of 10% solution of sodium deoxycholate. All clinical pneumococcal isolates are serotyped using the quellung reaction with typing antisera produced by the CDC. Positive quellung reactions are considered definitive identification of pneumococci.

Antibiotics remain the mainstay treatment for pneumococci. β-Lactams like penicillin are the first-line drugs, but resistance to β-lactams is a growing problem. When β-lactam resistance is a concern, macrolides and fluoroquinolones may be prescribed. However, *S. pneumoniae* resistance to macrolides and fluoroquinolones is increasing as well, limiting the therapeutic options for some infections. There are currently two pneumococcal vaccines available: pneumococcal conjugate vaccine (PCV13) and pneumococcal polysaccharide vaccine (PPSV23). These are generally given to the most vulnerable populations of individuals: children younger than 2 years and adults older than 65 years.

Haemophilus Pneumonia

Encapsulated strains of *Haemophilus influenzae* are known for causing meningitis, but nonencapsulated strains are important causes of pneumonia. This small, gram-negative coccobacillus is found in the pharynx of the majority of healthy children; however, *Haemophilus* pneumonia is primarily seen in the elderly. Like other pathogens that cause pneumonia, *H. influenzae* is spread by droplets and aerosols produced by coughing. A fastidious organism, *H. influenzae* will only grow on media with available factor X (hemin) and factor V (NAD), like chocolate agar (Figure 22.11). Serotyping must be performed to confirm identity of *H. influenzae* isolates.

Infections of the alveoli by *H. influenzae* result in inflammation and accumulation of fluids. Increasing resistance to β-lactams, macrolides, and tetracyclines presents challenges for the treatment of *Haemophilus* pneumonia. Resistance to the fluoroquinolones is rare among isolates of *H. influenzae* but has been observed. As discussed for AOM, a vaccine directed against nonencapsulated *H. influenzae*, if developed, would provide protection against pneumonia caused by this pathogen.

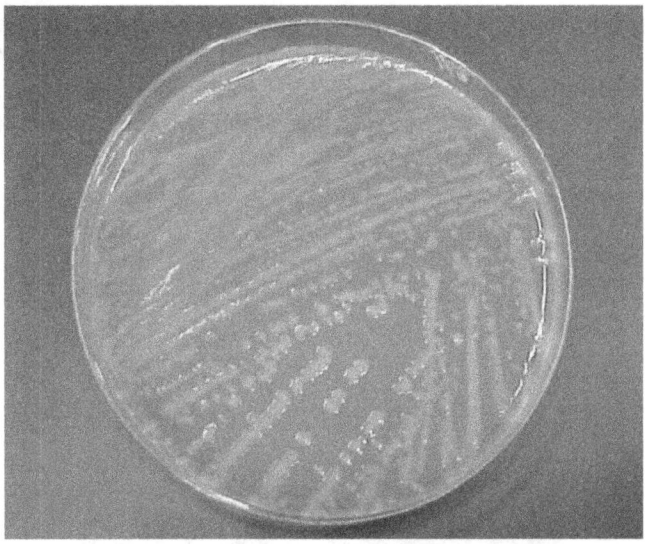

Figure 22.11 Culture of *Haemophilus influenzae* on a chocolate agar plate. (credit: modification of work by Centers for Disease Control and Prevention)

> ### Case in Point
>
> **Why Me?**
> Tracy is a 6-year old who developed a serious cough that would not seem to go away. After 2 weeks, her parents became concerned and took her to the pediatrician, who suspected a case of bacterial pneumonia. Tests confirmed that the cause was *Haemophilus influenzae*. Fortunately, Tracy responded well to antibiotic treatment and eventually made a full recovery.
>
> Because there had been several other cases of bacterial pneumonia at Tracy's elementary school, local health officials urged parents to have their children screened. Of the children who were screened, it was discovered that greater than 50% carried *H. influenzae* in their nasal cavities, yet all but two of them were asymptomatic.
>
> Why is it that some individuals become seriously ill from bacterial infections that seem to have little or no effect on others? The pathogenicity of an organism—its ability to cause host damage—is not solely a property of the microorganism. Rather, it is the product of a complex relationship between the microbe's virulence factors and the immune defenses of the individual. Preexisting conditions and environmental factors such as exposure to secondhand smoke can make some individuals more susceptible to infection by producing conditions favorable to microbial growth or compromising the immune system. In addition, individuals may have genetically determined immune factors that protect them—or not—from particular strains of pathogens. The interactions between these host factors and the pathogenicity factors produced by the microorganism ultimately determine the outcome of the infection. A clearer understanding of these interactions may allow for better identification of at-risk individuals and prophylactic interventions in the future.

Mycoplasma Pneumonia (Walking Pneumonia)

Primary atypical pneumonia is caused by *Mycoplasma pneumoniae*. This bacterium is not part of the respiratory tract's normal microbiota and can cause epidemic disease outbreaks. Also known as walking pneumonia, **mycoplasma pneumonia** infections are common in crowded environments like college campuses and military bases. It is spread by aerosols formed when coughing or sneezing. The disease is often mild, with a low fever and persistent cough. These bacteria, which do not have cell walls, use a specialized attachment organelle to bind to ciliated cells. In the process, epithelial cells are damaged and the proper function of the cilia is hindered (Figure 22.12).

Mycoplasma grow very slowly when cultured. Therefore, penicillin and thallium acetate are added to agar to prevent the overgrowth by faster-growing potential contaminants. Since *M. pneumoniae* does not have a cell wall, it is resistant to these substances. Without a cell wall, the microbial cells appear pleomorphic. *M. pneumoniae* infections tend to be self-limiting but may also respond well to macrolide antibiotic therapy. β-lactams, which target cell wall synthesis, are not indicated for treatment of infections with this pathogen.

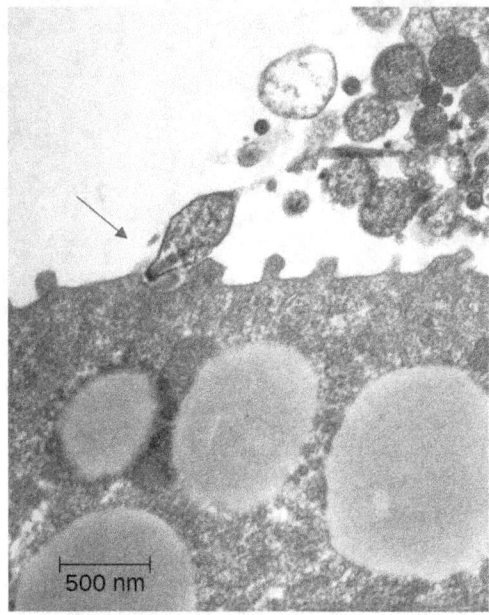

Figure 22.12 The micrograph shows *Mycoplasma pneumoniae* using their specialized receptors to attach to epithelial cells in the trachea of an infected hamster. (credit: modification of work by American Society for Microbiology)

Chlamydial Pneumonias and Psittacosis

Chlamydial pneumonia can be caused by three different species of bacteria: *Chlamydophila pneumoniae* (formerly known as *Chlamydia pneumoniae*), *Chlamydophila psittaci* (formerly known as *Chlamydia psittaci*), and *Chlamydia trachomatis*. All three are obligate intracellular pathogens and cause mild to severe pneumonia and bronchitis. Of the three, *Chlamydophila pneumoniae* is the most common and is transmitted via respiratory droplets or aerosols. *C. psittaci* causes **psittacosis**, a zoonotic disease that primarily affects domesticated birds such as parakeets, turkeys, and ducks, but can be transmitted from birds to humans. Psittacosis is a relatively rare infection and is typically found in people who work with birds. *Chlamydia trachomatis,* the causative agent of the sexually transmitted disease chlamydia, can cause pneumonia in infants when the infection is passed from mother to baby during birth.

Diagnosis of chlamydia by culturing tends to be difficult and slow. Because they are intracellular pathogens, they require multiple passages through tissue culture. Recently, a variety of PCR- and serologically based tests have been developed to enable easier identification of these pathogens. Tetracycline and macrolide antibiotics are typically prescribed for treatment.

Health Care-Associated Pneumonia

A variety of opportunistic bacteria that do not typically cause respiratory disease in healthy individuals are common causes of health care-associated pneumonia. These include *Klebsiella pneumoniae*, *Staphylococcus aureus*, and proteobacteria such as species of *Escherichia*, *Proteus*, and *Serratia*. Patients at risk include the elderly, those who have other preexisting lung conditions, and those who are immunocompromised. In addition, patients receiving supportive therapies such as intubation, antibiotics, and immunomodulatory drugs may also be at risk because these interventions disrupt the mucociliary escalator and other pulmonary defenses. Invasive medical devices such as catheters, medical implants, and ventilators can also introduce opportunistic pneumonia-causing pathogens into the body.[7]

Pneumonia caused by *K. pneumoniae* is characterized by lung necrosis and "currant jelly sputum," so named

---

7 SM Koenig et al. "Ventilator-Associated Pneumonia: Diagnosis, Treatment, and Prevention." *Clinical Microbiology Reviews* 19 no.

because it consists of clumps of blood, mucus, and debris from the thick polysaccharide capsule produced by the bacterium. *K. pneumoniae* is often multidrug resistant. Aminoglycoside and cephalosporin are often prescribed but are not always effective. *Klebsiella* pneumonia is frequently fatal even when treated.

Pseudomonas Pneumonia

Pseudomonas aeruginosa is another opportunistic pathogen that can cause serious cases of bacterial pneumonia in patients with cystic fibrosis (CF) and hospitalized patients assisted with artificial ventilators. This bacterium is extremely antibiotic resistant and can produce a variety of exotoxins. Ventilator-associated pneumonia with *P. aeruginosa* is caused by contaminated equipment that causes the pathogen to be aspirated into the lungs. In patients with CF, a genetic defect in the cystic fibrosis transmembrane receptor (CFTR) leads to the accumulation of excess dried mucus in the lungs. This decreases the effectiveness of the defensins and inhibits the mucociliary escalator. *P. aeruginosa* is known to infect more than half of all patients with CF. It adapts to the conditions in the patient's lungs and begins to produce alginate, a viscous exopolysaccharide that inhibits the mucociliary escalator. Lung damage from the chronic inflammatory response that ensues is the leading cause of mortality in patients with CF.[8]

⊘ CHECK YOUR UNDERSTANDING

- What three pathogens are responsible for the most prevalent types of bacterial pneumonia?
- Which cause of pneumonia is most likely to affect young people?
- In what contexts does *Pseudomonas aeruginosa* cause pneumonia?

> **Clinical Focus**
>
> **Part 2**
>
> John's chest radiograph revealed an extensive consolidation in the right lung, and his sputum cultures revealed the presence of a gram-negative rod. His physician prescribed a course of the antibiotic clarithromycin. He also ordered the rapid influenza diagnostic tests (RIDTs) for type A and B influenza to rule out a possible underlying viral infection. Despite antibiotic therapy, John's condition continued to deteriorate, so he was admitted to the hospital.
>
> - What are some possible causes of pneumonia that would not have responded to the prescribed antibiotic?
>
> Jump to the next Clinical Focus box. Go back to the previous Clinical Focus box.

Tuberculosis

Tuberculosis (TB) is one of the deadliest infectious diseases in human history. Although **tuberculosis** infection rates in the United States are extremely low, the CDC estimates that about one-third of the world's population is infected with *Mycobacterium tuberculosis*, the causal organism of TB, with 9.6 million new TB cases and 1.5 million deaths worldwide in 2014.[9]

M. tuberculosis is an acid-fast, high G + C, gram-positive, nonspore-forming rod. Its cell wall is rich in waxy mycolic acids, which make the cells impervious to polar molecules. It also causes these organisms to grow slowly. *M. tuberculosis* causes a chronic granulomatous disease that can infect any area of the body, although it is typically associated with the lungs. *M. tuberculosis* is spread by inhalation of respiratory droplets or aerosols from an infected person. The infectious dose of *M. tuberculosis* is only 10 cells.[10]

---

4 (2006):637–657.

8   R. Sordé et al. "Management of Refractory *Pseudomonas aeruginosa* Infection in Cystic Fibrosis." *Infection and Drug Resistance* 4 (2011):31–41.

9   Centers for Disease Control and Prevention. "Tuberculosis (TB). Data and Statistics." http://www.cdc.gov/tb/statistics/default.htm

10   D. Saini et al. "Ultra-Low Dose of *Mycobacterium tuberculosis* Aerosol Creates Partial Infection in Mice." *Tuberculosis* 92 no. 2

After inhalation, the bacteria enter the alveoli (Figure 22.13). The cells are phagocytized by macrophages but can survive and multiply within these phagocytes because of the protection by the waxy mycolic acid in their cell walls. If not eliminated by macrophages, the infection can progress, causing an inflammatory response and an accumulation of neutrophils and macrophages in the area. Several weeks or months may pass before an immunological response is mounted by T cells and B cells. Eventually, the lesions in the alveoli become walled off, forming small round lesions called **tubercles**. Bacteria continue to be released into the center of the tubercles and the chronic immune response results in tissue damage and induction of apoptosis (programmed host-cell death) in a process called liquefaction. This creates a caseous center, or air pocket, where the aerobic *M. tuberculosis* can grow and multiply. Tubercles may eventually rupture and bacterial cells can invade pulmonary capillaries; from there, bacteria can spread through the bloodstream to other organs, a condition known as **miliary tuberculosis**. The rupture of tubercles also facilitates transmission of the bacteria to other individuals via droplet aerosols that exit the body in coughs. Because these droplets can be very small and stay aloft for a long time, special precautions are necessary when caring for patients with TB, such as the use of face masks and negative-pressure ventilation and filtering systems.

Eventually, most lesions heal to form calcified **Ghon complexes**. These structures are visible on chest radiographs and are a useful diagnostic feature. But even after the disease has apparently ended, viable bacteria remain sequestered in these locations. Release of these organisms at a later time can produce **reactivation tuberculosis** (or secondary TB). This is mainly observed in people with alcoholism, the elderly, or in otherwise immunocompromised individuals (Figure 22.13).

(2012):160–165.

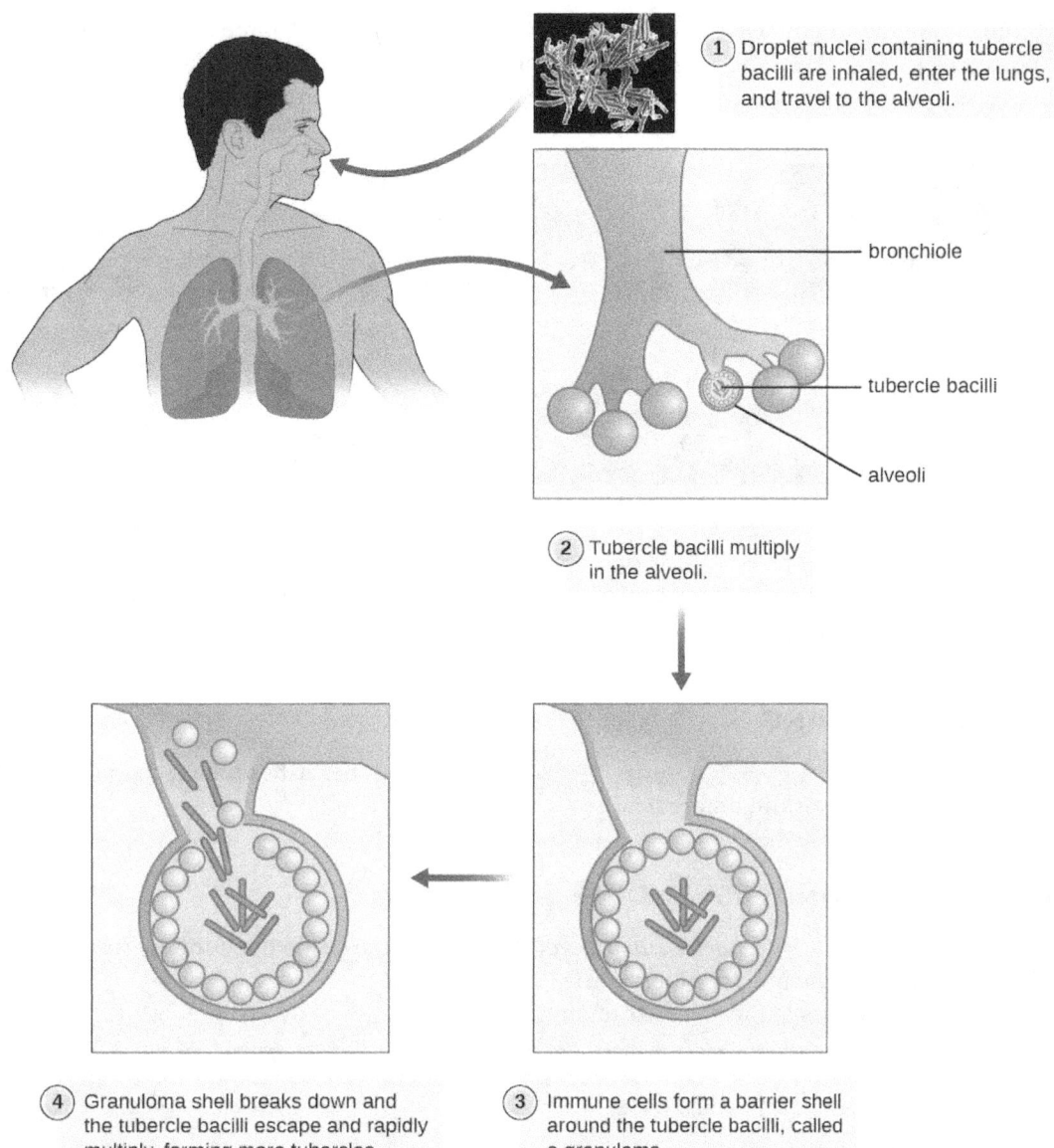

Figure 22.13 In the infectious cycle of tuberculosis, the immune response of most infected individuals (approximately 90%) results in the formation of tubercles in which the infection is walled off.[11] The remainder will suffer progressive primary tuberculosis. The sequestered bacteria may be reactivated to form secondary tuberculosis in immunocompromised patients at a later time. (credit: modification of work by Centers for Disease Control and Prevention)

Because TB is a chronic disease, chemotherapeutic treatments often continue for months or years. Multidrug resistant (MDR-TB) and extensively drug-resistant (XDR-TB) strains of *M. tuberculosis* are a growing clinical concern. These strains can arise due to misuse or mismanagement of antibiotic therapies. Therefore, it is imperative that proper multidrug protocols are used to treat these infections. Common antibiotics included in these mixtures are isoniazid, rifampin, ethambutol, and pyrazinamide.

A TB vaccine is available that is based on the so-called bacillus Calmette-Guérin (BCG) strain of *M. bovis* commonly found in cattle. In the United States, the BCG vaccine is only given to health-care workers and members of the military who are at risk of exposure to active cases of TB. It is used more broadly worldwide. Many individuals born in other countries have been vaccinated with BCG strain. BCG is used in many countries with a high prevalence of TB, to prevent childhood tuberculous meningitis and miliary disease.

---

11 G. Kaplan et al. "*Mycobacterium tuberculosis* Growth at the Cavity Surface: A Microenvironment with Failed Immunity." *Infection and Immunity* 71 no.12 (2003):7099–7108.

The Mantoux tuberculin skin test (Figure 22.14) is regularly used in the United States to screen for potential TB exposure (see Hypersensitivities). However, prior vaccinations with the BCG vaccine can cause false-positive results. Chest radiographs to detect Ghon complex formation are required, therefore, to confirm exposure.

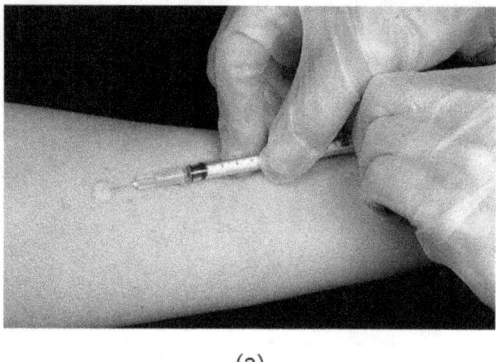 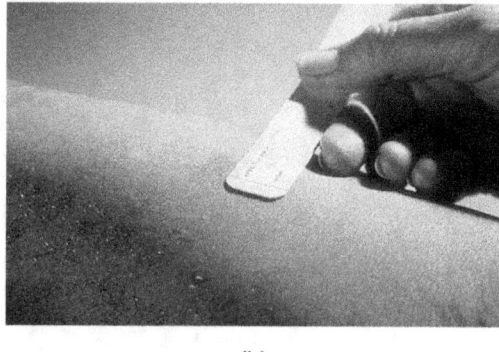

(a) (b)

Figure 22.14 (a) The Mantoux skin test for tuberculosis involves injecting the subject with tuberculin protein derivative. The injection should initially produce a raised wheal. (b) The test should be read in 48–72 hours. A positive result is indicated by redness, swelling, or hardness; the size of the responding region is measured to determine the final result. (credit a, b: modification of work by Centers for Disease Control and Prevention)

LINK TO LEARNING

This short animation (https://openstax.org/l/22mycotublegpnean) describes the mechanisms of infection associated with Mycobacterium tuberculosis.

CHECK YOUR UNDERSTANDING

- What characteristic of *Mycobacterium tuberculosis* allows it to evade the immune response?
- What happens to cause miliary tuberculosis?
- Explain the limitations of the Mantoux tuberculin skin test.

Pertussis (Whooping Cough)

The causative agent of **pertussis**, commonly called **whooping cough**, is *Bordetella pertussis*, a gram-negative coccobacillus. The disease is characterized by mucus accumulation in the lungs that leads to a long period of severe coughing. Sometimes, following a bout of coughing, a sound resembling a "whoop" is produced as air is inhaled through the inflamed and restricted airway—hence the name whooping cough. Although adults can be infected, the symptoms of this disease are most pronounced in infants and children. Pertussis is highly communicable through droplet transmission, so the uncontrollable coughing produced is an efficient means of transmitting the disease in a susceptible population.

Following inhalation, *B. pertussis* specifically attaches to epithelial cells using an adhesin, filamentous hemagglutinin. The bacteria then grow at the site of infection and cause disease symptoms through the production of exotoxins. One of the main virulence factors of this organism is an A-B exotoxin called the **pertussis toxin (PT)**. When PT enters the host cells, it increases the cyclic adenosine monophosphate (cAMP) levels and disrupts cellular signaling. PT is known to enhance inflammatory responses involving histamine and serotonin. In addition to PT, *B. pertussis* produces a tracheal cytotoxin that damages ciliated epithelial cells and results in accumulation of mucus in the lungs. The mucus can support the colonization and growth of other microbes and, as a consequence, secondary infections are common. Together, the effects of these factors produce the cough that characterizes this infection.

A pertussis infection can be divided into three distinct stages. The initial infection, termed the **catarrhal stage**, is relatively mild and unremarkable. The signs and symptoms may include nasal congestion, a runny nose, sneezing, and a low-grade fever. This, however, is the stage in which *B. pertussis* is most infectious. In the

paroxysmal stage, mucus accumulation leads to uncontrollable coughing spasms that can last for several minutes and frequently induce vomiting. The paroxysmal stage can last for several weeks. A long **convalescence stage** follows the paroxysmal stage, during which time patients experience a chronic cough that can last for up to several months. In fact, the disease is sometimes called the 100-day cough.

In infants, coughing can be forceful enough to cause fractures to the ribs, and prolonged infections can lead to death. The CDC reported 20 pertussis-related deaths in 2012,[12] but that number had declined to five by 2015.[13]

During the first 2 weeks of infection, laboratory diagnosis is best performed by culturing the organism directly from a nasopharyngeal (NP) specimen collected from the posterior nasopharynx. The NP specimen is streaked onto Bordet-Gengou medium. The specimens must be transported to the laboratory as quickly as possible, even if transport media are used. Transport times of longer than 24 hours reduce the viability of *B. pertussis* significantly.

Within the first month of infection, *B. pertussis* can be diagnosed using PCR techniques. During the later stages of infection, pertussis-specific antibodies can be immunologically detected using an enzyme-linked immunosorbent assay (ELISA).

Pertussis is generally a self-limiting disease. Antibiotic therapy with erythromycin or tetracycline is only effective at the very earliest stages of disease. Antibiotics given later in the infection, and prophylactically to uninfected individuals, reduce the rate of transmission. Active vaccination is a better approach to control this disease. The DPT vaccine was once in common use in the United States. In that vaccine, the P component consisted of killed whole-cell *B. pertussis* preparations. Because of some adverse effects, that preparation has now been superseded by the DTaP and Tdap vaccines. In both of these new vaccines, the "aP" component is a pertussis toxoid.

Widespread vaccination has greatly reduced the number of reported cases and prevented large epidemics of pertussis. Recently, however, pertussis has begun to reemerge as a childhood disease in some states because of declining vaccination rates and an increasing population of susceptible children.

LINK TO LEARNING

This web page (https://openstax.org/l/22pertussaudio) contains an audio clip of the distinctive "whooping" sound associated with pertussis in infants.

This interactive map (https://openstax.org/l/22intmapprevacc) shows outbreaks of vaccine preventable diseases, including pertussis, around the world.

CHECK YOUR UNDERSTANDING

- What accounts for the mucus production in a pertussis infection?
- What are the signs and symptoms associated with the three stages of pertussis?
- Why is pertussis becoming more common in the United States?

Legionnaires Disease

An atypical pneumonia called **Legionnaires disease** (also known as legionellosis) is caused by an aerobic gram-negative bacillus, *Legionella pneumophila*. This bacterium infects free-living amoebae that inhabit moist environments, and infections typically occur from human-made reservoirs such as air-conditioning cooling towers, humidifiers, misting systems, and fountains. Aerosols from these reservoirs can lead to infections of susceptible individuals, especially those suffering from chronic heart or lung disease or other

---

12 Centers for Disease Control and Prevention. "2012 Final Pertussis Surveillance Report." 2015. http://www.cdc.gov/pertussis/downloads/pertuss-surv-report-2012.pdf. Accessed July 6, 2016.

13 Centers for Disease Control and Prevention. "2015 Provisional Pertussis Surveillance Report." 2016. http://www.cdc.gov/pertussis/downloads/pertuss-surv-report-2015-provisional.pdf. Accessed July 6, 2016.

conditions that weaken the immune system.

When *L. pneumophila* bacteria enter the alveoli, they are phagocytized by resident macrophages. However, *L. pneumophila* uses a secretion system to insert proteins in the endosomal membrane of the macrophage; these proteins prevent lysosomal fusion, allowing *L. pneumophila* to continue to proliferate within the phagosome. The resulting respiratory disease can range from mild to severe pneumonia, depending on the status of the host's immune defenses. Although this disease primarily affects the lungs, it can also cause fever, nausea, vomiting, confusion, and other neurological effects.

Diagnosis of Legionnaires disease is somewhat complicated. *L. pneumophila* is a fastidious bacterium and is difficult to culture. In addition, since the bacterial cells are not efficiently stained with the Gram stain, other staining techniques, such as the Warthin-Starry silver-precipitate procedure, must be used to visualize this pathogen. A rapid diagnostic test has been developed that detects the presence of *Legionella* antigen in a patient's urine; results take less than 1 hour, and the test has high selectivity and specificity (greater than 90%). Unfortunately, the test only works for one serotype of *L. pneumophila* (type 1, the serotype responsible for most infections). Consequently, isolation and identification of *L. pneumophila* from sputum remains the defining test for diagnosis.

Once diagnosed, Legionnaire disease can be effectively treated with fluoroquinolone and macrolide antibiotics. However, the disease is sometimes fatal; about 10% of patients die of complications.[14] There is currently no vaccine available.

✓ CHECK YOUR UNDERSTANDING

- Why is Legionnaires disease associated with air-conditioning systems?
- How does *Legionella pneumophila* circumvent the immune system?

Q Fever

The zoonotic disease **Q fever** is caused by a rickettsia, *Coxiella burnetii*. The primary reservoirs for this bacterium are domesticated livestock such as cattle, sheep, and goats. The bacterium may be transmitted by ticks or through exposure to the urine, feces, milk, or amniotic fluid of an infected animal. In humans, the primary route of infection is through inhalation of contaminated farmyard aerosols. It is, therefore, largely an occupational disease of farmers. Humans are acutely sensitive to *C. burnetii*—the infective dose is estimated to be just a few cells.[15] In addition, the organism is hardy and can survive in a dry environment for an extended time. Symptoms associated with acute Q fever include high fever, headache, coughing, pneumonia, and general malaise. In a small number of patients (less than 5%[16]), the condition may become chronic, often leading to endocarditis, which may be fatal.

Diagnosing rickettsial infection by cultivation in the laboratory is both difficult and hazardous because of the easy aerosolization of the bacteria, so PCR and ELISA are commonly used. Doxycycline is the first-line drug to treat acute Q fever. In chronic Q fever, doxycycline is often paired with hydroxychloroquine.

Disease Profile

Bacterial Diseases of the Respiratory Tract

Numerous pathogens can cause infections of the respiratory tract. Many of these infections produce similar signs and symptoms, but appropriate treatment depends on accurate diagnosis through laboratory testing. The tables in Figure 22.15 and Figure 22.16 summarize the most important bacterial respiratory

---

14 Centers for Disease Control and Prevention. "*Legionella* (Legionnaires' Disease and Pontiac Fever: Diagnosis, Treatment, and Complications)." http://www.cdc.gov/legionella/about/diagnosis.html. Accessed Sept 14, 2016.

15 WD Tigertt et al. "Airborne Q Fever." *Bacteriological Reviews* 25 no. 3 (1961):285–293.

16 Centers for Disease Control and Prevention. "Q fever. Symptoms, Diagnosis, and Treatment." 2013. http://www.cdc.gov/qfever/symptoms/index.html. Accessed July 6, 2016.

infections, with the latter focusing specifically on forms of bacterial pneumonia.

| \multicolumn{7}{c|}{**Bacterial Infections of the Respiratory Tract**} |
Disease	Pathogen	Signs and Symptoms	Transmission	Diagnostic Tests	Antimicrobial Drugs	Vaccine
Acute otitis media (AOM)	*Haemophilus influenzae, Streptococcus pneumoniae, Moraxella catarrhalis,* others	Earache, possible effusion; may cause fever, nausea, vomiting, diarrhea	Often a secondary infection; bacteria from respiratory tract become trapped in eustachian tube, cause infection	None	Cephalosporins, fluoroquinolones	None
Diphtheria	*Corynebacterium diphtheriae*	Pseudomembrane on throat, possibly leading to suffocation and death	Inhalation of respiratory droplets or aerosols from infected person	Identification of bacteria in throat swabs; PCR to detect diphtheria toxin in vitro	Erythromycin, penicillin, antitoxin produced in horses	DtaP, Tdap, DT, Td, DTP
Legionnaires disease	*Legionella pneumophila*	Cough, fever, muscle aches, headaches, nausea, vomiting, confusion; sometimes fatal	Inhalation of aerosols from contaminated water reservoirs	Isolation, using Warthin-Starry procedure, of bacteria in sputum	Fluoroquinolones, macrolides	None
Pertussis (whooping cough)	*Bordetella pertussis*	Severe coughing with "whoop" sound; chronic cough lasting several months; can be fatal in infants	Inhalation of respiratory droplets from infected person	Direct culture of throat swab, PCR, ELISA	Macrolides	DTaP, Tdap
Q fever	*Coxiella burnetii*	High fever, coughing, pneumonia, malaise; in chronic cases, potentially fatal endocarditis	Inhalation of aerosols of urine, feces, milk, or amniotic fluid of infected cattle, sheep, goats	PCR, ELISA	Doxycycline, hydroxychloroquine	None
Streptococcal pharyngitis, scarlet fever	*Streptococcus pyogenes*	Fever, sore throat, inflammation of pharynx and tonsils, petechiae, swollen lymph nodes; skin rash (scarlet fever), strawberry tongue	Direct contact, inhalation of respiratory droplets or aerosols from infected person	Direct culture of throat swab, rapid enzyme immunoassay	β-lactams	None
Tuberculosis	*Mycobacterium tuberculosis*	Formation of tubercles in lungs; rupture of tubercles, leading to chronic, bloody cough; healed tubercles (Ghon complexes) visible in radiographs; can be fatal	Inhalation of respiratory droplets or aerosols from infected person	Mantoux tuberculin skin test with chest radiograph to identify Ghon complexes	Isoniazid, rifampin, ethambutol, pyrazinamide	BCG

Figure 22.15

Bacterial Causes of Pneumonia

Disease	Pathogen	Signs and Symptoms	Transmission	Diagnostic Tests	Antimicrobial Drugs	Vaccine
Chlamydial pneumonia	*Chlamydophila pneumoniae*, *C. psittaci*, *Chlamydia trachomatis*	Bronchitis; mild to severe respiratory distress	Inhalation of respiratory droplets or aerosols from infected person (*C. pneumoniae*); exposure to infected bird (*C. psittaci*); exposure in the birth canal (*Chlamydia trachomatis*)	Tissue culture, PCR	Tetracycline, macrolides	None
Haemophilus pneumonia	*Haemophilus influenzae*	Cough, fever or low body temperature, chills, chest pain, headache, fatigue	Inhalation of respiratory droplets or aerosols from infected person or asymptomatic carrier	Culture on chocolate agar, serotyping of blood or cerebrospinal fluid samples	Cephalosporins, fluoroquinolones	Hib
Klebsiella pneumonia	*Klebsiella pneumoniae*, others	Lung necrosis, "currant jelly" sputum; often fatal	Health care associated; bacteria introduced via contaminated ventilators, intubation, or other medical equipment	Culture, PCR	Multidrug resistant; antibiotic susceptibility testing necessary	None
Mycoplasma pneumonia (walking pneumonia)	*Mycoplasma pneumoniae*	Low fever, persistent cough	Inhalation of respiratory droplets or aerosols from infected person	Culture with penicillin, thallium acetate	Macrolides	None
Pneumococcal pneumonia	*Streptococcus pneumoniae*	Productive cough, bloody sputum, fever, chills, chest pain, respiratory distress	Direct contact with respiratory secretions	Gram stain, blood agar culture with optichin and sodium deoxycholate, quellung reaction	β-lactams, macrolides, fluoroquinolones	Pneumococcal conjugate vaccine (PCV13), pneumococcal polysaccharide vaccine (PPSV23)
Pseudomonas pneumonia	*Pseudomonas aeruginosa*	Viscous fluid and chronic inflammation of lungs; often fatal	Health care associated; bacteria introduced via contaminated ventilators; also frequently affects patients with cystic fibrosis	Culture from sputum or other body fluid	Multidrug resistant; antibiotic susceptibility testing necessary	None

Figure 22.16

22.3 Viral Infections of the Respiratory Tract

Learning Objectives

By the end of this section, you will be able to:
- Identify the most common viruses that can cause infections of the upper and lower respiratory tract
- Compare the major characteristics of specific viral diseases of the respiratory tract

Viruses are the most frequent cause of respiratory tract infections. Unlike the bacterial pathogens, we have few effective therapies to combat viral respiratory infections. Fortunately, many of these diseases are mild and self-limiting. A few respiratory infections manifest their primary symptoms at other locations in the body.

The Common Cold

The **common cold** is a generic term for a variety of mild viral infections of the nasal cavity. More than 200 different viruses are known to cause the common cold. The most common groups of cold viruses include rhinoviruses, coronaviruses, and adenoviruses. These infections are widely disseminated in the human population and are transmitted through direct contact and droplet transmission. Coughing and sneezing efficiently produce infectious aerosols, and rhinoviruses are known to persist on environmental surfaces for up to a week.[17]

Viral contact with the nasal mucosa or eyes can lead to infection. Rhinoviruses tend to replicate best between 33 °C (91.4 °F) and 35 °C (95 °F), somewhat below normal body temperature (37 °C [98.6 °F]). As a consequence, they tend to infect the cooler tissues of the nasal cavities. Colds are marked by an irritation of the mucosa that leads to an inflammatory response. This produces common signs and symptoms such as nasal excess nasal secretions (runny nose), congestion, sore throat, coughing, and sneezing. The absence of high fever is typically used to differentiate common colds from other viral infections, like influenza. Some colds may progress to cause otitis media, pharyngitis, or laryngitis, and patients may also experience headaches and body aches. The disease, however, is self-limiting and typically resolves within 1–2 weeks.

There are no effective antiviral treatments for the common cold and antibacterial drugs should not be prescribed unless secondary bacterial infections have been established. Many of the viruses that cause colds are related, so immunity develops throughout life. Given the number of viruses that cause colds, however, individuals are never likely to develop immunity to all causes of the common cold.

⊘ CHECK YOUR UNDERSTANDING

- How are colds transmitted?
- What is responsible for the symptoms of a cold?

Clinical Focus

Part 3

Since antibiotic treatment had proven ineffective, John's doctor suspects that a viral or fungal pathogen may be the culprit behind John's case of pneumonia. Another possibility is that John could have an antibiotic-resistant bacterial infection that will require a different antibiotic or combination of antibiotics to clear.

The RIDT tests both came back negative for type A and type B influenza. However, the diagnostic laboratory identified the sputum isolate as *Legionella pneumophila*. The doctor ordered tests of John's urine and, on the second day after his admission, results of an enzyme immunoassay (EIA) were positive for the *Legionella* antigen. John's doctor added levofloxacin to his antibiotic therapy and continued to monitor him. The doctor also began to ask John where he had been over the past 10 to 14 days.

---

17 AG L'Huillier et al. "Survival of Rhinoviruses on Human Fingers." *Clinical Microbiology and Infection* 21, no. 4 (2015):381–385.

- Do negative RIDT results absolutely rule out influenza virus as the etiologic agent? Why or why not?
- What is John's prognosis?

Jump to the next Clinical Focus box. Go back to the previous Clinical Focus box.

Influenza

Commonly known as the flu, **influenza** is a common viral disease caused by an orthomyxovirus that primarily affects the upper respiratory tract but can also extend into the lower respiratory tract. Influenza is pervasive worldwide and causes 3,000–50,000 deaths each year in the United States. The annual mortality rate can vary greatly depending on the virulence of the strain(s) responsible for seasonal epidemics.[18]

Influenza infections are most typically characterized by fever, chills, and body aches. This is followed by symptoms similar to the common cold that may last a week or more. Table 22.2 compares the signs and symptoms of influenza and the common cold.

Comparing the Common Cold and Influenza

Sign/Symptom	Common Cold	Influenza
Fever	Low (37.2 °C [99 °F])	High (39 °C [102.2 °F])
Headache	Common	Common
Aches and pains	Mild	Severe
Fatigue	Slight	Severe
Nasal congestion	Common	Rare
Sneezing	Common	Rare

Table 22.2

In general, influenza is self-limiting. However, serious cases can lead to pneumonia and other complications that can be fatal. Such cases are more common in the very young and the elderly; however, certain strains of influenza virus (like the 1918–1919 variant discussed later in this chapter) are more lethal to young adults than to the very young or old. Strains that affect young adults are believed to involve a cytokine storm—a positive feedback loop that forms between cytokine production and leukocytes. This cytokine storm produces an acute inflammatory response that leads to rapid fluid accumulation in the lungs, culminating in pulmonary failure. In such cases, the ability to mount a vigorous immune response is actually detrimental to the patient. The very young and very old are less susceptible to this effect because their immune systems are less robust.

A complication of influenza that occurs primarily in children and teenagers is **Reye syndrome**. This sequela causes swelling in the liver and brain, and may progress to neurological damage, coma, or death. Reye syndrome may follow other viral infections, like chickenpox, and has been associated with the use of aspirin. For this reason, the CDC and other agencies recommend that aspirin and products containing aspirin never be used to treat viral illnesses in children younger than age 19 years.[19]

---

18 Centers for Disease Control and Prevention. "Estimating Seasonal Influenza-Associated Deaths in the United States: CDC Study Confirms Variability of Flu." 2016. http://www.cdc.gov/flu/about/disease/us_flu-related_deaths.htm. Accessed July 6, 2016.

19 ED Belay et al. "Reye's Syndrome in the United States From 1981 Through 1997." *New England Journal of Medicine* 340 no. 18 (1999):1377–1382.

The influenza virus is primarily transmitted by direct contact and inhalation of aerosols. The RNA genome of this virus exists as seven or eight segments, each coated with ribonucleoprotein and encoding one or two specific viral proteins. The influenza virus is surrounded by a lipid membrane envelope, and two of the main antigens of the influenza virus are the spike proteins hemagglutinin (H) and neuraminidase (N), as shown in Figure 22.17. These spike proteins play important roles in the viral infectious cycle.

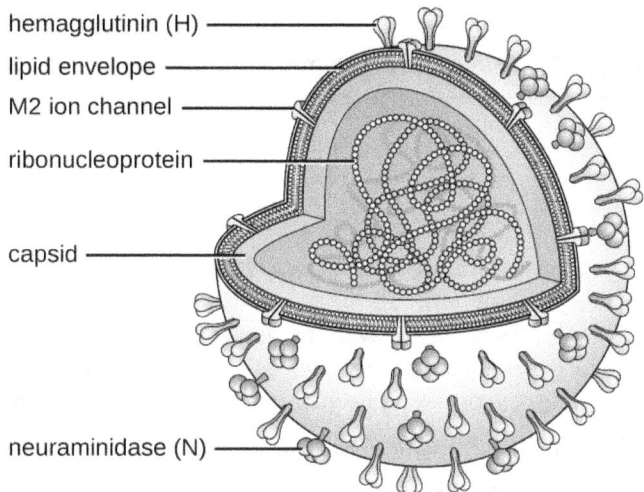

Figure 22.17 The illustration shows the structure of an influenza virus. The viral envelope is studded with copies of the proteins neuraminidase and hemagglutinin, and surrounds the individual seven or eight RNA genome segments. (credit: modification of work by Dan Higgins, Centers for Disease Control and Prevention)

Following inhalation, the influenza virus uses the hemagglutinin protein to bind to sialic acid receptors on host respiratory epithelial cells. This facilitates endocytosis of the viral particle. Once inside the host cell, the negative strand viral RNA is replicated by the viral RNA polymerase to form mRNA, which is translated by the host to produce viral proteins. Additional viral RNA molecules are transcribed to produce viral genomic RNA, which assemble with viral proteins to form mature virions. Release of the virions from the host cell is facilitated by viral neuraminidase, which cleaves sialic-acid receptors to allow progeny viruses to make a clean exit when budding from an infected cell.

There are three genetically related influenza viruses, called A, B, and C. The influenza A viruses have different subtypes based on the structure of their hemagglutinin and neuraminidase proteins. There are currently 18 known subtypes of hemagglutinin and 11 known subtypes of neuraminidase. Influenza viruses are serologically characterized by the type of H and N proteins that they possess. Of the nearly 200 different combinations of H and N, only a few, such as the H1N1 strain, are associated with human disease. The influenza viruses A, B, and C make up three of the five major groups of orthomyxoviruses. The differences between the three types of influenza are summarized in Table 22.3. The most virulent group is the influenza A viruses, which cause seasonal pandemics of influenza each year. Influenza A virus can infect a variety of animals, including pigs, horses, pigs, and even whales and dolphins. Influenza B virus is less virulent and is sometimes associated with epidemic outbreaks. Influenza C virus generally produces the mildest disease symptoms and is rarely connected with epidemics. Neither influenza B virus nor influenza C virus has significant animal reservoirs.

The Three Major Groups of Influenza Viruses

	Influenza A virus	Influenza B virus	Influenza C virus
Severity	Severe	Moderate	Mild
Animal reservoir	Yes	No	No

The Three Major Groups of Influenza Viruses

	Influenza A virus	Influenza B virus	Influenza C virus
Genome segments	8	8	7
Population spread	Epidemic and pandemic	Epidemic	Sporadic
Antigenic variation	Shift/drift	Drift	Drift

Table 22.3

Influenza virus infections elicit a strong immune response, particularly to the hemagglutinin protein, which would protect the individual if they encountered the same virus. Unfortunately, the antigenic properties of the virus change relatively rapidly, so new strains are evolving that immune systems previously challenged by influenza virus cannot recognize. When an influenza virus gains a new hemagglutinin or neuraminidase type, it is able to evade the host's immune response and be successfully transmitted, often leading to an epidemic.

There are two mechanisms by which these evolutionary changes may occur. The mechanisms of antigen drift and antigenic shift for influenza virus have been described in Virulence Factors of Bacterial and Viral Pathogens. Of these two genetic processes, it is viruses produced by antigenic shift that have the potential to be extremely virulent because individuals previously infected by other strains are unlikely to produce any protective immune response against these novel variants.

The most lethal influenza pandemic in recorded history occurred from 1918 through 1919. Near the end of World War I, an antigenic shift involving the recombination of avian and human viruses is thought to have produced a new H1N1 virus. This strain rapidly spread worldwide and is commonly claimed to have killed as many as 40 million to 50 million people—more than double the number killed in the war. Although referred to as the Spanish flu, this disease is thought to have originated in the United States. Regardless of its source, the conditions of World War I greatly contributed to the spread of this disease. Crowding, poor sanitation, and rapid mobilization of large numbers of personnel and animals facilitated the dissemination of the new virus once it appeared.

Several of the most important influenza pandemics of modern times have been associated with antigenic shifts. A few of these are summarized in Table 22.4.

Historical Influenza Outbreaks[20][21][22]

Years	Common Name	Serotype	Estimated Number of Deaths
1918–1919	Spanish flu	H1N1	20,000,000–40,000,000
1957–1958	Asian flu	N2N2	1,000,000–2,000,000
1968–1969	Hong Kong flu	H3N2	1,000,000–3,000,000
2009–2010	Swine flu	H1N1/09	152,000–575,000

Table 22.4

---

20 CE Mills et al. "Transmissibility of 1918 Pandemic Influenza." *Nature* 432, no. 7019 (2004):904–906.

21 E. Tognotti. "Influenza Pandemics: A Historical Retrospect." *Journal of Infection in Developing Countries* 3, no. 5 (2009):331–334.

22 FS Dawood et al. "Estimated Global Mortality Associated with the First 12 Months of 2009 Pandemic Influenza A H1N1 Virus

Laboratory diagnosis of influenza is typically performed using a variety of RIDTs. These tests are inoculated by point-of-care personnel and give results within 15–20 minutes. Unfortunately, these tests have variable sensitivity and commonly yield false-negative results. Other tests include hemagglutination of erythrocytes (due to hemagglutinin action) or complement fixation. Patient serum antibodies against influenza viruses can also be detected in blood samples. Because influenza is self-limiting disease, diagnosis through these more time-consuming and expensive methods is not typically used.

Three drugs that inhibit influenza neuraminidase activity are available: inhaled zanamivir, oral oseltamivir, and intravenous peramivir. If taken at the onset of symptoms, these drugs can shorten the course of the disease. These drugs are thought to impair the ability of the virus to efficiently exit infected host cells. A more effective means of controlling influenza outbreaks, though, is vaccination. Every year, new influenza vaccines are developed to be effective against the strains expected to be predominant. This is determined in February by a review of the dominant strains around the world from a network of reporting sites; their reports are used to generate a recommendation for the vaccine combination for the following winter in the northern hemisphere. In September, a similar recommendation is made for the winter in the southern hemisphere.[23] These recommendations are used by vaccine manufacturers to formulate each year's vaccine. In most cases, three or four viruses are selected—the two most prevalent influenza A strains and one or two influenza B strains. The chosen strains are typically cultivated in eggs and used to produce either an inactivated or a live attenuated vaccine (e.g., FluMist). For individuals 18 years or older with an allergy to egg products, a recombinant egg-free trivalent vaccine is available. Most of the influenza vaccines over the past decade have had an effectiveness of about 50%.[24]

Case in Point

Flu Pandemic

During the spring of 2013, a new strain of H7N9 influenza was reported in China. A total of 132 people were infected. Of those infected, 44 (33%) died. A genetic analysis of the virus suggested that this strain arose from the reassortment of three different influenza viruses: a domestic duck H7N3 virus, a wild bird H7N9 virus, and a domestic poultry H9N2 virus. The virus was detected in the Chinese domestic bird flocks and contact with this reservoir is thought to have been the primary source of infection. This strain of influenza was not able to spread from person to person. Therefore, the disease did not become a global problem. This case does, though, illustrate the potential threat that influenza still represents. If a strain like the H7N9 virus were to undergo another antigenic shift, it could become more communicable in the human population. With a mortality rate of 33%, such a pandemic would be disastrous. For this reason, organizations like the World Health Organization and the Centers for Disease Control and Prevention keep all known influenza outbreaks under constant surveillance.

⊘ CHECK YOUR UNDERSTANDING

- Compare the severity of the three types of influenza viruses.
- Why must new influenza vaccines be developed each year?

Viral Pneumonia

Viruses cause fewer cases of pneumonia than bacteria; however, several viruses can lead to pneumonia in children and the elderly. The most common sources of viral pneumonia are adenoviruses, influenza viruses, parainfluenza viruses, and respiratory syncytial viruses. The signs and symptoms produced by these viruses

---

Circulation: A Modelling Study." *The Lancet Infectious Diseases* 12, no. 9 (2012):687–695.

23 World Health Organization. "WHO Report on Global Surveillance of Epidemic-Prone Infectious Diseases." 2000. http://www.who.int/csr/resources/publications/surveillance/Influenza.pdf. Accessed July 6, 2016.

24 Centers of Disease Control and Prevention. "Vaccine Effectiveness - How Well Does the Flu Vaccine Work?" 2016. http://www.cdc.gov/flu/about/qa/vaccineeffect.htm. Accessed July 6, 2016.

can range from mild cold-like symptoms to severe cases of pneumonia, depending on the virulence of the virus strain and the strength of the host defenses of the infected individual. Occasionally, infections can result in otitis media.

Respiratory syncytial virus (RSV) infections are fairly common in infants; most people have been infected by the age of 2 years. During infection, a viral surface protein causes host cells to fuse and form multinucleated giant cells called **syncytia**. There are no specific antiviral therapies or vaccines available for viral pneumonia. In adults, these infections are self-limiting, resemble the common cold, and tend to resolve uneventfully within 1 or 2 weeks. Infections in infants, however, can be life-threatening. RSV is highly contagious and can be spread through respiratory droplets from coughing and sneezing. RSV can also survive for a long time on environmental surfaces and, thus, be transmitted indirectly via fomites.

✓ CHECK YOUR UNDERSTANDING

- Who is most likely to contract viral pneumonia?
- What is the recommended treatment for viral pneumonia?

SARS and MERS

Severe acute respiratory syndrome (**SARS**) and Middle East respiratory syndrome (**MERS**) are two acute respiratory infections caused by coronaviruses. In both cases, these are thought to be zoonotic infections. Bats and civet cats are thought to have been the reservoirs for SARS; camels seem to be the reservoir for MERS.

SARS originated in southern China in the winter of 2002 and rapidly spread to 37 countries. Within about 1 year, more than 8,000 people experienced influenza-like symptoms and nearly 800 people died. The rapid spread and severity of these infections caused grave concern at the time. However, the outbreak was controlled in 2003 and no further cases of SARS have been recorded since 2004.[25] Signs and symptoms of SARS include high fever, headache, body aches, and cough, and most patients will develop pneumonia.

MERS was first reported in Saudi Arabia in 2013. Although some infected individuals will be asymptomatic or have mild cold-like symptoms, most will develop a high fever, aches, cough and a severe respiratory infection that can progress to pneumonia. As of 2015, over 1,300 people in 27 countries have been infected. About 500 people have died. There are no specific treatments for either MERS or SARS. In addition, no vaccines are currently available. Several recombinant vaccines, however, are being developed.

✓ CHECK YOUR UNDERSTANDING

- What is the cause of SARS?
- What are the signs and symptoms of MERS?

Viral Respiratory Diseases Causing Skin Rashes

Measles, rubella (German measles), and chickenpox are three important viral diseases often associated with skin rashes. However, their symptoms are systemic, and because their portal of entry is the respiratory tract, they can be considered respiratory infections.

Measles (Rubeola)

The measles virus (MeV) causes the highly contagious disease **measles**, also known as rubeola, which is a major cause of childhood mortality worldwide. Although vaccination efforts have greatly reduced the incidence of measles in much of the world, epidemics are still common in unvaccinated populations in certain countries.[26]

---

25 Y. Huang. "The SARS Epidemic and Its Aftermath in China: A Political Perspective." In *Learning from SARS: Preparing for the Next Disease Outbreak*. Edited by S. Knobler et al. Washington, DC: National Academies Press; 2004. Available at: http://www.ncbi.nlm.nih.gov/books/NBK92479/

26 Centers for Disease Control and Prevention. "Global Health - Measles, Rubella, and CRS, Eliminating Measles, Rubella &

The measles virus is a single-stranded, negative-strand RNA virus and, like the influenza virus, it possesses an envelope with spikes of embedded hemagglutinin. The infection is spread by direct contact with infectious secretions or inhalation of airborne droplets spread by breathing, coughing, or sneezing. Measles is initially characterized by a high fever, conjunctivitis, and a sore throat. The virus then moves systemically through the bloodstream and causes a characteristic rash. The measles rash initially forms on the face and later spreads to the extremities. The red, raised macular rash will eventually become confluent and can last for several days. At the same time, extremely high fevers (higher than 40.6 °C [105 °F]) can occur. Another diagnostic sign of measles infections is **Koplik's spots**, white spots that form on the inner lining of inflamed cheek tissues (Figure 22.18).

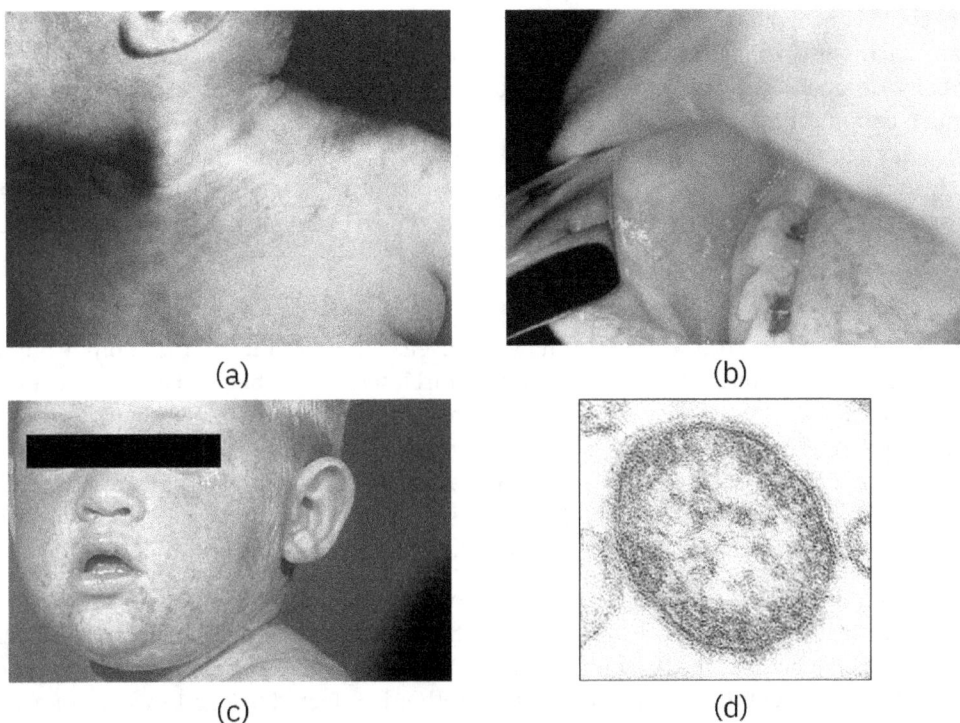

Figure 22.18   (a and b) Measles typically presents as a raised macular rash that begins on the face and spreads to the extremities. (c) Koplik's spots on the oral mucosa are also characteristic of measles. (d) A thin-section transmission electron micrograph of a measles virion. (credit a: a. Betty G. Partin /CDC.; b, c: modification of work by Centers for Disease Control and Prevention)

Although measles is usually self-limiting, it can lead to pneumonia, encephalitis, and death. In addition, the inhibition of immune system cells by the measles virus predisposes patients to secondary infections. In severe infections with highly virulent strains, measles fatality rates can be as high as 10% to 15%. There were more than 145,000 measles deaths (mostly young children) worldwide in 2013.[27]

The preliminary diagnosis of measles is typically based on the appearance of the rash and Koplik's spots. Hemagglutination inhibition tests and serological tests may be used to confirm measles infections in low-prevalence settings.

There are no effective treatments for measles. Vaccination is widespread in developed countries as part of the measles, mumps, and rubella (MMR) vaccine. As a result, there are typically fewer than 200 cases of measles in the United States annually.[28] When it is seen, it is often associated with children who have not been vaccinated.

Congenital Rubella Syndrome (CRS) Worldwide." 2015. http://www.cdc.gov/globalhealth/measles/. Accessed July 7, 2016.

27   World Health Organization. "Measles Factsheet." 2016. http://www.who.int/mediacentre/factsheets/fs286/en/. Accessed July 7, 2016.

28   Centers for Disease Control and Prevention. "Measles Cases and Outbreaks." 2016. http://www.cdc.gov/measles/cases-outbreaks.html. Accessed July 7, 2016.

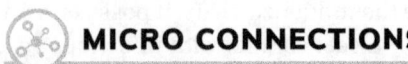

MICRO CONNECTIONS

Preventable Measles Outbreaks

In December 2014, a measles epidemic began at Disneyland in southern California. Within just 4 months, this outbreak affected 134 people in 24 states.[29] Characterization of the virus suggests that an unidentified infected individual brought the disease to the United States from the Philippines, where a similar virus had sickened more than 58,000 people and killed 110.[30] Measles is highly communicable, and its spread at Disneyland may have been facilitated by the low vaccination rate in some communities in California.[31]

Several factors could conceivably lead to a strong comeback of measles in the U.S. Measles is still an epidemic disease in many locations worldwide. Air travel enables infected individuals to rapidly translocate these infections globally. Compounding this problem, low vaccination rates in some local areas in the United States (such as in Amish communities) provide populations of susceptible hosts for the virus to establish itself. Finally, measles has been a low-prevalence infection in the U.S. for some time. As a consequence, physicians are not as likely to recognize the initial symptoms and make accurate diagnoses. Until vaccination rates become high enough to ensure herd immunity, measles is likely to be an ongoing problem in the United States.

Rubella (German Measles)

Rubella, or the German measles, is a relatively mild viral disease that produces a rash somewhat like that caused by the measles, even though the two diseases are unrelated. The rubella virus is an enveloped RNA virus that can be found in the respiratory tract. It is transmitted from person to person in aerosols produced by coughing or sneezing. Nearly half of all infected people remain asymptomatic. However, the virus is shed and spread by asymptomatic carriers. Like rubeola, **rubella** begins with a facial rash that spreads to the extremities (Figure 22.19). However, the rash is less intense, shorter lived (2–3 days), not associated with Koplik's spots, and the resulting fever is lower (101 °F [38.3 °C]).

Congenital rubella syndrome is the most severe clinical complication of the German measles. This occurs if a woman is infected with rubella during pregnancy. The rubella virus is **teratogenic**, meaning it can cause developmental defects if it crosses the placenta during pregnancy. There is a very high incidence of stillbirth, spontaneous abortion, or congenital birth defects if the mother is infected before 11 weeks of pregnancy and 35% if she is infected between weeks 13–16; after this time the incidence is low.[32] For this reason, prenatal screening for rubella is commonly practiced in the United States. Postnatal infections are usually self-limiting and rarely cause severe complications.

Like measles, the preliminary diagnosis of rubella is based on the patient's history, vaccination records, and the appearance of the rash. The diagnosis can be confirmed by hemagglutinin inhibition assays and a variety of other immunological techniques. There are no antiviral therapies for rubella, but an effective vaccine (MMR) is widely available. Vaccination efforts have essentially eliminated rubella in the United States; fewer than a dozen cases are reported in a typical year.

---

29 Ibid.

30 World Health Organization. "Measles-Rubella Bulletin." Manila, Philippines; Expanded Programme on Immunization Regional Office for the Western Pacific World Health Organization; 9 no. 1 (2015). http://www.wpro.who.int/immunization/documents/mrbulletinvol9issue1.pdf

31 M. Bloch et al. "Vaccination Rates for Every Kindergartener in California." *The New York Times* February 6, 2015. http://www.nytimes.com/interactive/2015/02/06/us/california-measles-vaccines-map.html?_r=1. Accessed July 7, 2016.

32 E. Miller et al. "Consequences of Confirmed Maternal Rubella at Successive Stages of Pregnancy." *The Lancet* 320, no. 8302 (1982):781–784.

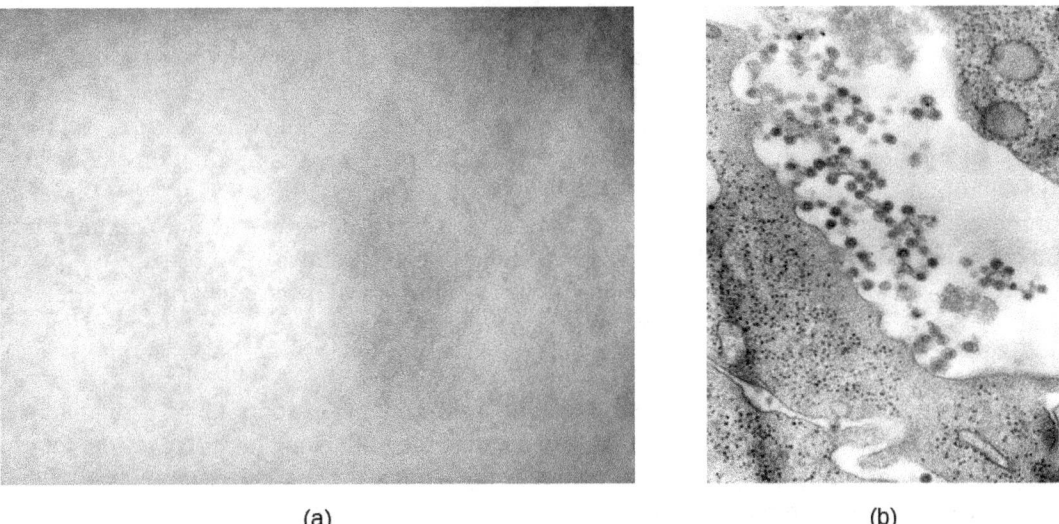

Figure 22.19 (a) This photograph shows the appearance of the German measles (rubella) rash. Note that this is less intense than the rash of measles and the lesions are not confluent. (b) This transmission electron micrograph shows rubella virus virions just budding from a host cell. (credit a, b: modification of work by Centers for Disease Control and Prevention)

Chickenpox and Shingles

Chickenpox, also known as varicella, was once a common viral childhood disease. The causative agent of **chickenpox**, the varicella-zoster virus, is a member of the herpesvirus family. In children, the disease is mild and self-limiting, and is easily transmitted by direct contact or inhalation of material from the skin lesions. In adults, however, chickenpox infections can be much more severe and can lead to pneumonia and birth defects in the case of infected pregnant women. Reye syndrome, mentioned earlier in this chapter, is also a serious complication associated with chickenpox, generally in children.

Once infected, most individuals acquire a lifetime immunity to future chickenpox outbreaks. For this reason, parents once held "chickenpox parties" for their children. At these events, uninfected children were intentionally exposed to an infected individual so they would contract the disease earlier in life, when the incidence of complications is very low, rather than risk a more severe infection later.

After the initial viral exposure, chickenpox has an incubation period of about 2 weeks. The initial infection of the respiratory tract leads to viremia and eventually produces fever and chills. A pustular rash then develops on the face, progresses to the trunk, and then the extremities, although most form on the trunk (Figure 22.20). Eventually, the lesions burst and form a crusty scab. Individuals with chickenpox are infectious from about 2 days before the outbreak of the rash until all the lesions have scabbed over.

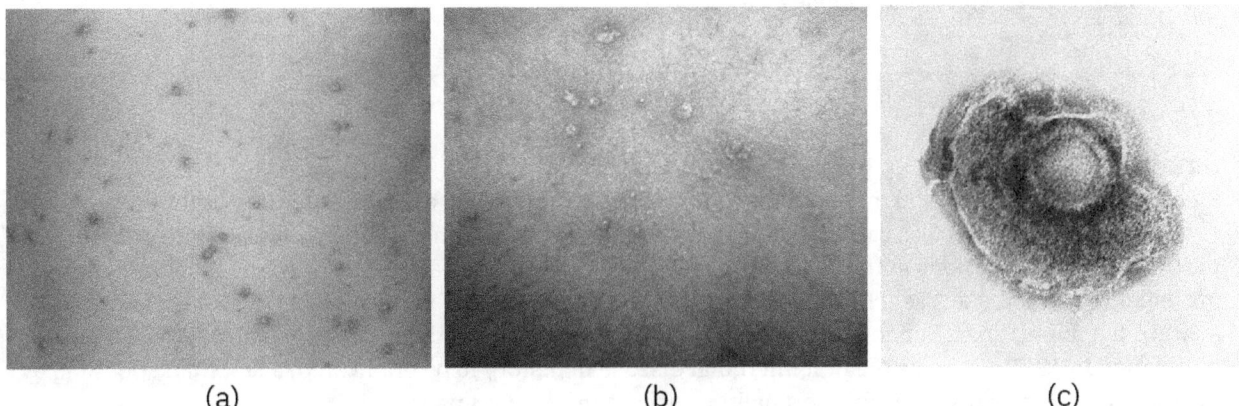

Figure 22.20 (a) and (b) The characteristic appearance of the pustular chickenpox rash is concentrated on the trunk region. (c) This transmission electron micrograph shows a viroid of human herpesvirus 3, the virus that causes chickenpox in children and shingles when it is reactivated in adults. Credit b: John Noble/CDC. c: modification of work by Centers for Disease Control and Prevention)

Like other herpesviruses, the varicella-zoster virus can become dormant in nerve cells. While the pustular vesicles are developing, the virus moves along sensory nerves to the dorsal ganglia in the spinal cord. Once there, the varicella-zoster virus can remain latent for decades. These dormant viruses may be reactivated later in life by a variety of stimuli, including stress, aging, and immunosuppression. Once reactivated, the virus moves along sensory nerves to the skin of the face or trunk. This results in the production of the painful lesions in a condition known as **shingles** (Figure 22.21). These symptoms generally last for 2–6 weeks, and may recur more than once. Postherpetic neuralgia, pain signals sent from damaged nerves long after the other symptoms have subsided, is also possible. In addition, the virus can spread to other organs in immunocompromised individuals. A person with shingles lesions can transmit the virus to a nonimmune contact, and the newly infected individual would develop chickenpox as the primary infection. Shingles cannot be transmitted from one person to another.

The primary diagnosis of chickenpox in children is mainly based on the presentation of a pustular rash of the trunk. Serological and PCR-based tests are available to confirm the initial diagnosis. Treatment for chickenpox infections in children is usually not required. In patients with shingles, acyclovir treatment can often reduce the severity and length of symptoms, and diminish the risk of postherpetic neuralgia. An effective vaccine is now available for chickenpox. A vaccine is also available for adults older than 60 years who were infected with chickenpox in their youth. This vaccine reduces the likelihood of a shingles outbreak by boosting the immune defenses that are keeping the latent infection in check and preventing reactivation.

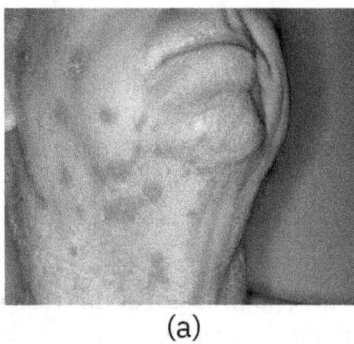
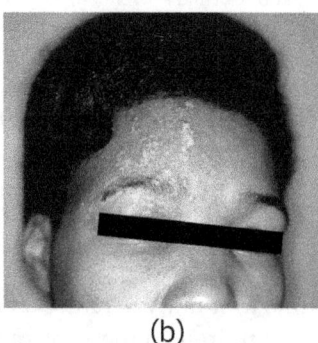
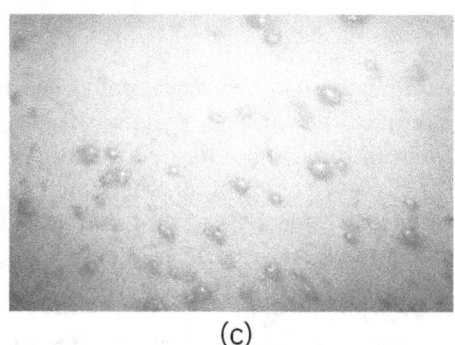

(a) (b) (c)

Figure 22.21 (a and b) An individual suffering from shingles. (c) The rash is formed because of the reactivation of a varicella-zoster infection that was initially contracted in childhood. (credit a: modification of work by National Institute of Allergy and Infectious Diseases (NIAID); credit b: Robert E. Sumpter/CDC.; credit c: modification of work by Centers for Disease Control and Prevention)

CHECK YOUR UNDERSTANDING

- Why does measles often lead to secondary infections?
- What signs or symptoms would distinguish rubella and measles?
- Why can chickenpox lead to shingles later in life?

Eye on Ethics

Smallpox Stockpiles

Smallpox has probably killed more humans than any other infectious disease, with the possible exception of tuberculosis. This disease, caused by the variola major virus, is transmitted by inhalation of viral particles shed from lesions in the throat. The smallpox virus spreads systemically in the bloodstream and produces a pustular skin rash. Historical epidemics of smallpox had fatality rates of 50% or greater in susceptible populations. Concerted worldwide vaccination efforts eradicated smallpox from the general population in 1977. This was the first microbial disease in history to be eradicated, a feat made possible by the fact that the only reservoir for the smallpox virus is infected humans.

Although the virus is no longer present in the wild, laboratory samples of the virus still exist in the United States and Russia.[33] The question is, why do these samples still exist? Some claim that these stocks should be maintained for research purposes. Should the smallpox virus ever reappear, they say, we would need access to such stocks for development of vaccines and treatments. Concerns about a re-emergence of the virus are not totally unfounded. Although there are no living reservoirs of the virus, there is always the possibility that smallpox could re-emerge from mummified human bodies or human remains preserved in permafrost. It is also possible that there are as-yet undiscovered samples of the virus in other locations around the world. An example of such "lost" samples was discovered in a drawer in a Food and Drug Administration lab in Maryland.[34] If an outbreak from such a source were to occur, it could lead to uncontrolled epidemics, since the population is largely unvaccinated now.

Critics of this argument, including many research scientists and the World Health Organization, claim that there is no longer any rational argument for keeping the samples. They view the "re-emergence scenarios" as a thinly veiled pretense for harboring biological weapons. These scenarios, they say, are less probable than an intentional reintroduction of the virus from militarized stocks by humans. Furthermore, they point out that if we needed to research smallpox in the future, we could rebuild the virus from its DNA sequence.

What do you think? Are there legitimate arguments for maintaining stockpiles of smallpox, or should all forms of this deadly disease be eradicated?

Disease Profile

Viral Infections of the Respiratory Tract

Many viruses are capable of entering and causing disease in the respiratory system, and a number are able to spread beyond the respiratory system to cause systemic infections. Most of these infections are highly contagious and, with a few exceptions, antimicrobial drugs are not effective for treatment. Although some of these infections are self-limiting, others can have serious or fatal complications. Effective vaccines have been developed for several of these diseases, as summarized in Figure 22.22.

---

[33] Centers for Disease Control and Prevention. "CDC Media Statement on Newly Discovered Smallpox Specimens." July 8, 2014. http://www.cdc.gov/media/releases/2014/s0708-nih.html. Accessed on July 7, 2016.

[34] Ibid.

Viral Infections of the Respiratory Tract

Disease	Pathogen	Signs and Symptoms	Transmission	Vaccine
Chickenpox (varicella)	Varicella-zoster virus	In children, fever, chills, pustular rash of lesions that burst and form crusty scabs; in adults, more severe symptoms and complications (e.g., pneumonia)	Highly contagious via contact with aerosols, particles, or droplets from infected individual's blisters or respiratory secretions	Varicella (chickenpox) vaccine
Common cold	Rhinoviruses, adenoviruses, coronaviruses, others	Runny nose, congestion, sore throat, sneezing, headaches and muscle aches; may lead to otitis media, pharyngitis, laryngitis	Highly contagious via contact with respiratory secretions or inhalation of droplets or aerosols	None
Influenza	Influenza viruses A, B, C	Fever, chills, headaches, body aches, fatigue; may lead to pneumonia or complications such as Reye syndrome. Highly virulent strains may cause lethal complications	Highly contagious between humans via contact with respiratory secretions or inhalation of droplets or aerosols. Influenza A virus can be transmitted from animal reservoirs.	Vaccines developed yearly against most prevalent strains
Measles	Measles virus (MeV)	High fever, conjunctivitis, sore throat, macular rash becoming confluent, Koplik's spots on oral mucosa; in severe cases, can lead to fatal pneumonia or encephalitis, especially in children	Highly contagious via contact with respiratory secretions, skin rash, or eye secretions of infected individual	MMR
MERS	Middle East respiratory syndrome coronavirus (MERS-CoV)	Fever, cough, shortness of breath; in some cases, complications such as pneumonia and kidney failure; can be fatal	Contact with respiratory secretions or inhalation of droplets or aerosols	None
Rubella (German measles)	Rubella virus	Facial rash spreading to extremities, followed by low-grade fever, headache, conjunctivitis, cough, runny nose, swollen lymph nodes; congenital rubella may cause birth defects, miscarriage, or stillbirth	Contagious via inhalation of droplets or aerosols from infected person or asymptomatic carrier; transplacental infection from mother to fetus	MMR
SARS	SARS-associated coronavirus (SARS-CoV)	High fever, headache, body aches, dry cough, pneumonia; can be fatal	Contact with respiratory secretions or inhalation of droplets or aerosols	None
Shingles	Varicella-zoster virus	Painful lesions on face or trunk lasting several weeks; may cause postherpetic neuralgia (chronic pain) or spread to organs in severe cases	Nontransmissible; occurs when dormant virus is reactivated, generally many years after initial chicken-pox infection	Herpes zoster (shingles) vaccine
Viral pneumonia	Adenoviruses, influenza viruses, parainfluenza viruses, respiratory syncytial viruses, others	From mild cold-like symptoms to severe pneumonia; in infants, RSV infections may be life-threatening	Highly contagious via contact with respiratory secretions or inhalation of droplets or aerosols	None

Figure 22.22

22.4 Respiratory Mycoses

Learning Objectives

By the end of this section, you will be able to:
- Identify the most common fungi that can cause infections of the respiratory tract
- Compare the major characteristics of specific fungal diseases of the respiratory tract

Fungal pathogens are ubiquitous in the environment. Serological studies have demonstrated that most people have been exposed to fungal respiratory pathogens during their lives. Yet symptomatic infections by these microbes are rare in healthy individuals. This demonstrates the efficacy of the defenses of our respiratory system. In this section, we will examine some of the fungi that can cause respiratory infections.

Histoplasmosis

Histoplasmosis is a fungal disease of the respiratory system and most commonly occurs in the Mississippi Valley of the United States and in parts of Central and South America, Africa, Asia, and Australia. The causative agent, *Histoplasma capsulatum*, is a dimorphic fungus. This microbe grows as a filamentous mold in the environment but occurs as a budding yeast during human infections. The primary reservoir for this pathogen is soil, particularly in locations rich in bat or bird feces.

Histoplasmosis is acquired by inhaling microconidial spores in the air; this disease is not transmitted from human to human. The incidence of **histoplasmosis** exposure is high in endemic areas, with 60%–90% of the population having anti-*Histoplasma* antibodies, depending on location;[35] however, relatively few individuals exposed to the fungus actually experience symptoms. Those most likely to be affected are the very young, the elderly, and immunocompromised people.

In many ways, the course of this disease is similar to that of tuberculosis. Following inhalation, the spores enter the lungs and are phagocytized by alveolar macrophages. The fungal cells then survive and multiply within these phagocytes (see Figure 5.26). Focal infections cause the formation of granulomatous lesions, which can lead to calcifications that resemble the Ghon complexes of tuberculosis, even in asymptomatic cases. Also like tuberculosis, histoplasmosis can become chronic and reactivation can occur, along with dissemination to other areas of the body (e.g., the liver or spleen).

Signs and symptoms of pulmonary histoplasmosis include fever, headache, and weakness with some chest discomfort. The initial diagnosis is often based on chest radiographs and cultures grown on fungal selective media like Sabouraud's dextrose agar. Direct fluorescence antibody staining and Giemsa staining can also be used to detect this pathogen. In addition, serological tests including a complement fixation assay and histoplasmin sensitivity can be used to confirm the diagnosis. In most cases, these infections are self-limiting and antifungal therapy is not required. However, in disseminated disease, the antifungal agents amphotericin B and ketoconazole are effective; itraconazole may be effective in immunocompromised patients, in whom the disease can be more serious.

CHECK YOUR UNDERSTANDING

- In what environments is one more likely to be infected with histoplasmosis?
- Identify at least two similarities between histoplasmosis and tuberculosis.

Coccidioidomycosis

Infection by the dimorphic fungus *Coccidioides immitis* causes **coccidioidomycosis**. Because the microbe is endemic to the San Joaquin Valley of California, the disease is sometimes referred to as Valley fever. A related species that causes similar infections is found in semi-arid and arid regions of the southwestern United States, Mexico, and Central and South America.[36]

Like histoplasmosis, coccidioidomycosis is acquired by inhaling fungal spores—in this case, arthrospores formed by hyphal fragmentation. Once in the body, the fungus differentiates into spherules that are filled with endospores. Most *C. immitis* infections are asymptomatic and self-limiting. However, the infection can be very serious for immunocompromised patients. The endospores may be transported in the blood, disseminating the infection and leading to the formation of granulomatous lesions on the face and nose (Figure 22.23). In severe cases, other major organs can become infected, leading to serious complications such as fatal meningitis.

---

35 NE Manos et al. "Geographic Variation in the Prevalence of Histoplasmin Sensitivity." *Dis Chest* 29, no. 6 (1956):649–668.
36 DR Hospenthal. "Coccioidomycosis." Medscape. 2015. http://emedicine.medscape.com/article/215978-overview. Accessed July 7, 2016.

Coccidioidomycosis can be diagnosed by culturing clinical samples. *C. immitis* readily grows on laboratory fungal media, such as Sabouraud's dextrose agar, at 35 °C (95 °F). Culturing the fungus, however, is rather dangerous. *C. immitis* is one of the most infectious fungal pathogens known and is capable of causing laboratory-acquired infections. Indeed, until 2012, this organism was considered a "select agent" of bioterrorism and classified as a BSL-3 microbe. Serological tests for antibody production are more often used for diagnosis. Although mild cases generally do not require intervention, disseminated infections can be treated with intravenous antifungal drugs like amphotericin B.

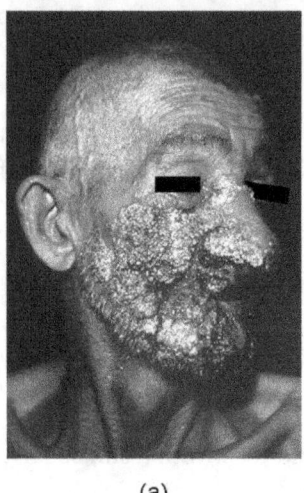
(a)

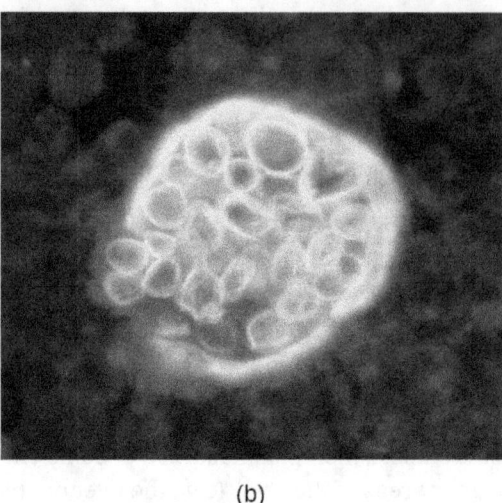
(b)

Figure 22.23 (a) This patient has extensive facial lesions due to a disseminated *Coccidioides* infection. (b) This fluorescent micrograph depicts a spherule of *C. immitis* containing endospores. (credit a, b: modification of work by Centers for Disease Control and Prevention)

Clinical Focus

Resolution

John's negative RIDT tests do not rule out influenza, since false-negative results are common, but the *Legionella* infection still must be treated with antibiotic therapy and is the more serious condition. John's prognosis is good, provided the physician can find an antibiotic therapy to which the infection responds.

While John was undergoing treatment, three of the employees from the home improvement store also reported to the clinic with very similar symptoms. All three were older than 55 years and had *Legionella* antigen in their urine; *L. pneumophila* was also isolated from their sputum. A team from the health department was sent to the home improvement store to identify a probable source for these infections. Their investigation revealed that about 3 weeks earlier, the store's air conditioning system, which was located where the employees ate lunch, had been undergoing maintenance. *L. pneumophila* was isolated from the cooling coils of the air conditioning system and intracellular *L. pneumophila* was observed in amoebae in samples of condensed water from the cooling coils as well (Figure 22.24). The amoebae provide protection for the *Legionella* bacteria and are known to enhance their pathogenicity.[37]

In the wake of the infections, the store ordered a comprehensive cleaning of the air conditioning system and implemented a regular maintenance program to prevent the growth of biofilms within the cooling tower. They also reviewed practices at their other facilities.

After a month of rest at home, John recovered from his infection enough to return to work, as did the other three employees of the store. However, John experienced lethargy and joint pain for more than a year after his treatment.

---

[37] HY Lau and NJ Ashbolt. "The Role of Biofilms and Protozoa in *Legionella* Pathogenesis: Implications for Drinking Water." *Journal of Applied Microbiology* 107 no. 2 (2009):368–378.

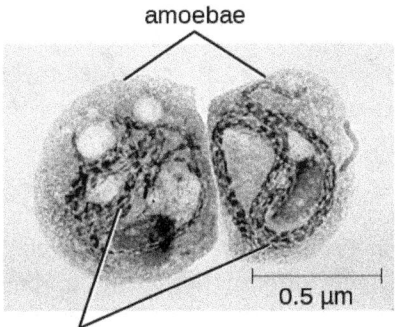

Figure 22.24 *Legionella pneumophila* (red intracellular rods) infecting amoebae from a contaminated water sample. (credit: modification of work by Centers for Disease Control and Prevention)

Go back to the previous Clinical Focus box.

Blastomycosis

Blastomycosis is a rare disease caused by another dimorphic fungus, *Blastomyces dermatitidis*. Like *Histoplasma* and *Coccidioides*, *Blastomyces* uses the soil as a reservoir, and fungal spores can be inhaled from disturbed soil. The pulmonary form of **blastomycosis** generally causes mild flu-like symptoms and is self-limiting. It can, however, become disseminated in immunocompromised people, leading to chronic cutaneous disease with subcutaneous lesions on the face and hands (Figure 22.25). These skin lesions eventually become crusty and discolored and can result in deforming scars. Systemic blastomycosis is rare, but if left untreated, it is always fatal.

Preliminary diagnosis of pulmonary blastomycosis can be made by observing the characteristic budding yeast forms in sputum samples. Commercially available urine antigen tests are now also available. Additional confirmatory tests include serological assays such as immunodiffusion tests or EIA. Most cases of blastomycosis respond well to amphotericin B or ketoconazole treatments.

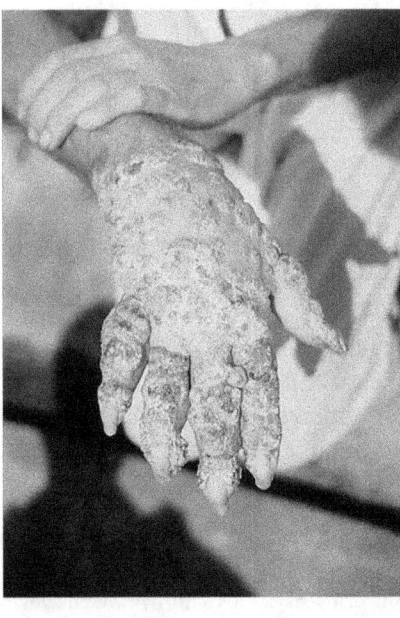
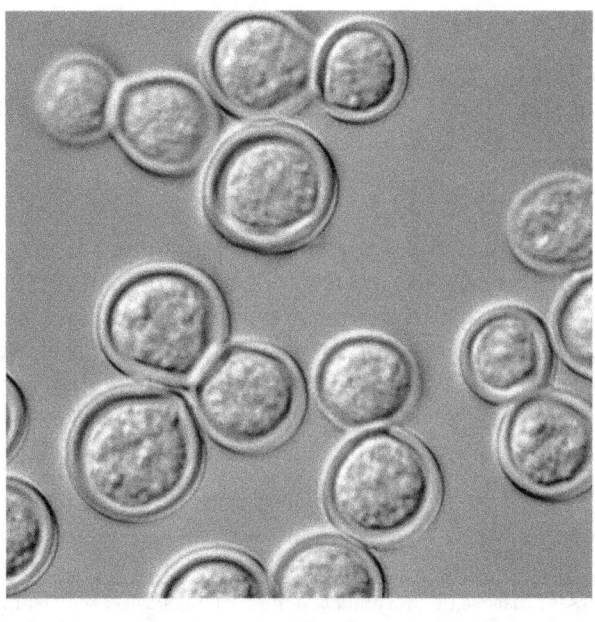

(a) (b)

Figure 22.25 (a) These skin lesions are the result of disseminated cutaneous blastomycosis. (b) A differential interference contrast micrograph of *B. dermatitidis* yeast cultured on blood agar. (credit a: modification of work by Centers for Disease Control and Prevention)

🔗 LINK TO LEARNING

Watch this profile (https://openstax.org/l/22blastlunginf) of a blastomycosis lung infection.

Mucormycosis

A variety of fungi in the order Mucorales cause **mucormycosis**, a rare fungal disease. These include bread molds, like *Rhizopus* and *Mucor*; the most commonly associated species is *Rhizopus arrhizus (oryzae)* (see Figure 5.28). These fungi can colonize many different tissues in immunocompromised patients, but often infect the skin, sinuses, or the lungs.

Although most people are regularly exposed to the causative agents of mucormycosis, infections in healthy individuals are rare. Exposure to spores from the environment typically occurs through inhalation, but the spores can also infect the skin through a wound or the gastrointestinal tract if ingested. Respiratory mucormycosis primarily affects immunocompromised individuals, such as patients with cancer or those who have had a transplant.[38]

After the spores are inhaled, the fungi grow by extending hyphae into the host's tissues. Infections can occur in both the upper and lower respiratory tracts. Rhinocerebral mucormycosis is an infection of the sinuses and brain; symptoms include headache, fever, facial swelling, congestion, and tissue necrosis causing black lesions in the oral cavity. Pulmonary mucormycosis is an infection of the lungs; symptoms include fever, cough, chest pain, and shortness of breath. In severe cases, infections may become disseminated and involve the central nervous system, leading to coma and death.[39]

Diagnosing mucormycosis can be challenging. Currently, there are no serological or PCR-based tests available to identify these infections. Tissue biopsy specimens must be examined for the presence of the fungal pathogens. The causative agents, however, are often difficult to distinguish from other filamentous fungi. Infections are typically treated by the intravenous administration of amphotericin B, and superficial infections are removed by surgical debridement. Since the patients are often immunocompromised, viral and bacterial secondary infections commonly develop. Mortality rates vary depending on the site of the infection, the causative fungus, and other factors, but a recent study found an overall mortality rate of 54%.[40]

✓ CHECK YOUR UNDERSTANDING

- Compare the modes of transmission for coccidioidomycosis, blastomycosis, and mucormycosis.
- In general, which are more serious: the pulmonary or disseminated forms of these infections?

Aspergillosis

Aspergillus is a common filamentous fungus found in soils and organic debris. Nearly everyone has been exposed to this mold, yet very few people become sick. In immunocompromised patients, however, *Aspergillus* may become established and cause **aspergillosis**. Inhalation of spores can lead to asthma-like allergic reactions. The symptoms commonly include shortness of breath, wheezing, coughing, runny nose, and headaches. Fungal balls, or aspergilloma, can form when hyphal colonies collect in the lungs (Figure 22.26). The fungal hyphae can invade the host tissues, leading to pulmonary hemorrhage and a bloody cough. In severe cases, the disease may progress to a disseminated form that is often fatal. Death most often results from pneumonia or brain hemorrhages.

Laboratory diagnosis typically requires chest radiographs and a microscopic examination of tissue and

---

38 Centers for Disease Control and Prevention. "Fungal Diseases. Definition of Mucormycosis." 2015 http://www.cdc.gov/fungal/diseases/mucormycosis/definition.html. Accessed July 7, 2016.

39 Centers for Disease Control and Prevention. "Fungal Diseases. Symptoms of Mucormycosis." 2015 http://www.cdc.gov/fungal/diseases/mucormycosis/symptoms.html. Accessed July 7, 2016.

40 MM Roden et al. "Epidemiology and Outcome of Zygomycosis: A Review of 929 Reported Cases." *Clinical Infectious Diseases* 41 no. 5 (2005):634–653.

respiratory fluid samples. Serological tests are available to identify *Aspergillus* antigens. In addition, a skin test can be performed to determine if the patient has been exposed to the fungus. This test is similar to the Mantoux tuberculin skin test used for tuberculosis. Aspergillosis is treated with intravenous antifungal agents, including itraconazole and voriconazole. Allergic symptoms can be managed with corticosteroids because these drugs suppress the immune system and reduce inflammation. However, in disseminated infections, corticosteroids must be discontinued to allow a protective immune response to occur.

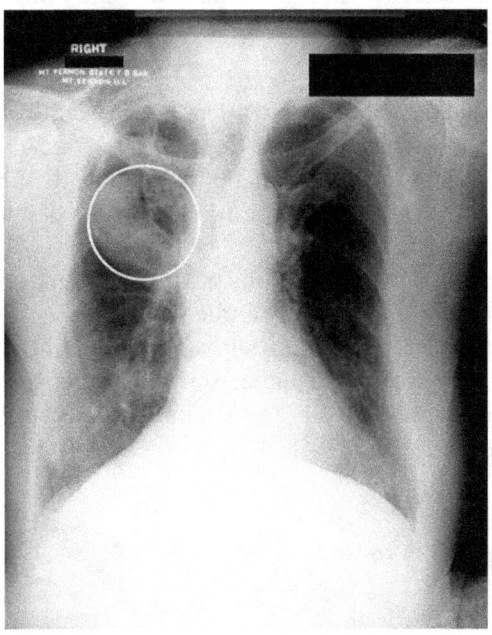

Figure 22.26   A fungal ball can be observed in the upper lobe of the right lung in this chest radiograph of a patient with aspergilloma. (credit: modification of work by Centers for Disease Control and Prevention)

Pneumocystis Pneumonia

A type of pneumonia called **Pneumocystis pneumonia** (PCP) is caused by *Pneumocystis jirovecii*. Once thought to be a protozoan, this organism was formerly named *P. carinii* but it has been reclassified as a fungus and renamed based on biochemical and genetic analyses. *Pneumocystis* is a leading cause of pneumonia in patients with acquired immunodeficiency syndrome (AIDS) and can be seen in other compromised patients and premature infants. Respiratory infection leads to fever, cough, and shortness of breath. Diagnosis of these infections can be difficult. The organism is typically identified by microscopic examination of tissue and fluid samples from the lungs (Figure 22.27). A PCR-based test is available to detect *P. jirovecii* in asymptomatic patients with AIDS. The best treatment for these infections is the combination drug trimethoprim-sulfamethoxazole (TMP/SMZ). These sulfa drugs often have adverse effects, but the benefits outweigh these risks. Left untreated, PCP infections are often fatal.

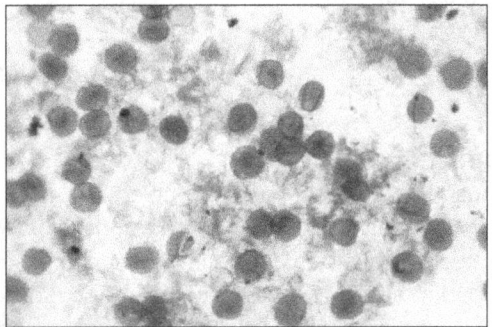

Figure 22.27   A light micrograph of a smear containing *Pneumocystis jirovecii* (dark purple cells) obtained from human lung tissue and stained with toluidine blue. (credit: Centers for Disease Control and Prevention)

Cryptococcosis

Infection by the encapsulated yeast *Cryptococcus neoformans* causes **cryptococcosis**. This fungus is ubiquitous in the soil and can be isolated from bird feces. Immunocompromised people are infected by inhaling basidiospores found in aerosols. The thick polysaccharide capsule surrounding these microbes enables them to avoid clearance by the alveolar macrophage. Initial symptoms of infection include fever, fatigue, and a dry cough. In immunocompromised patients, pulmonary infections often disseminate to the brain. The resulting meningitis produces headaches, sensitivity to light, and confusion. Left untreated, such infections are often fatal.

Cryptococcus infections are often diagnosed based on microscopic examination of lung tissues or cerebrospinal fluids. India ink preparations (Figure 22.28) can be used to visualize the extensive capsules that surround the yeast cells. Serological tests are also available to confirm the diagnosis. Amphotericin B, in combination with flucytosine, is typically used for the initial treatment of pulmonary infections. Amphotericin B is a broad-spectrum antifungal drug that targets fungal cell membranes. It can also adversely impact host cells and produce side effects. For this reason, clinicians must carefully balance the risks and benefits of treatments in these patients. Because it is difficult to eradicate cryptococcal infections, patients usually need to take fluconazole for up to 6 months after treatment with amphotericin B and flucytosine to clear the fungus. Cryptococcal infections are more common in immunocompromised people, such as those with AIDS. These patients typically require life-long suppressive therapy to control this fungal infection.

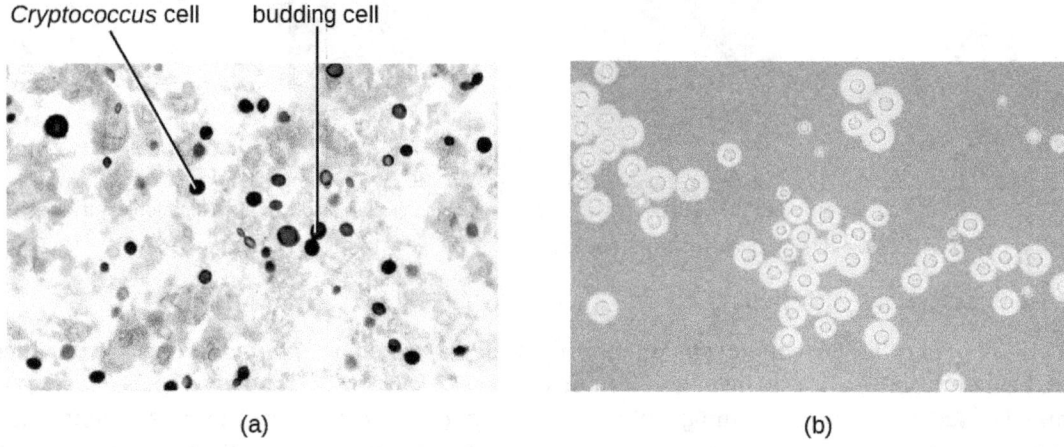

Figure 22.28 (a) The micrograph shows stained budding *Cryptococcus* yeast cells from the lungs of a patient with AIDS. (b) The large capsule of *Cryptococcus neoformans* is visible in this negative stain micrograph. (credit a, b: modification of work by Centers for Disease Control and Prevention)

✓ CHECK YOUR UNDERSTANDING

- What populations are most at risk for developing *Pneumocystis* pneumonia or cryptococcosis?
- Why are these infections fatal if left untreated?

Disease Profile

Fungal Diseases of the Respiratory Tract

Most respiratory mycoses are caused by fungi that inhabit the environment. Such infections are generally transmitted via inhalation of fungal spores and cannot be transmitted between humans. In addition, healthy people are generally not susceptible to infection even when exposed; the fungi are only virulent enough to establish infection in patients with HIV, AIDS, or another condition that compromises the immune defenses. Figure 22.29 summarizes the features of important respiratory mycoses.

| Fungal Infections of the Respiratory Tract |||||
Disease	Pathogen	Signs and Symptoms	Diagnostic Tests	Antimicrobial Drugs
Aspergillosis	*Aspergillus fumigatus*	Shortness of breath, wheezing, coughing, runny nose, headaches; formation of aspergillomas causing severe pneumonia and pulmonary or brain hemorrhages; can be fatal	Chest radiograph, skin test, microscopic observation of sputum samples	Itraconazole, voriconazole
Blastomycosis	*Blastomyces dermatitidis*	Fever, chills, cough, headache, fatigue, chest pain, body aches; in disseminated infections, chronic, crusted lesions on face and hands with permanent scarring; can be fatal	Microscopic observation of sputum samples; urine antigen test; EIA	Amphotericin B, ketoconazole
Coccidioidomycosis (Valley fever)	*Coccidioides immitis*	Granulomatous lesions on face and nose; may spread to organs or brain, causing fatal meningitis	Culture (in BSL-3 lab only), serological antibody tests	Amphotericin B
Cryptococcosis	*Cryptococcus neoformans*	Fever, cough, shortness of breath; can cause fatal meningitis if disseminated to brain	Microscopic examination of lung tissue or cerebrospinal fluid	Amphotericin B, fluconazole, flucytosine
Histoplasmosis	*Histoplasma capsulatum*	Fever, headache, weakness, chest pain, lesions on lungs	Chest radiograph, culture, direct fluorescence antibody staining, complement fixation assay, histoplasmin sensitivity test	Amphotericin B, ketoconazole, itraconazole
Mucormycosis	*Rhizopus arrhizus*, other *Rhizopus* spp., *Mucor* spp.	Headache, fever, facial swelling, congestion, black lesions in oral cavity, cough, chest pain, shortness of breath; often fatal	Microscopic examination of tissue biopsy specimens	Amphotericin B
Pneumocystis pneumonia (PCP)	*Pneumocystis jirovecii*	Fever, cough, shortness of breath; can be fatal if untreated	Microscopic examination of lung tissue and fluid, PCR	Trimethoprim-sulfamethoxazole

Figure 22.29

SUMMARY

22.1 Anatomy and Normal Microbiota of the Respiratory Tract

- The respiratory tract is divided into upper and lower regions at the **epiglottis**.
- Air enters the upper respiratory tract through the **nasal cavity** and mouth, which both lead to the **pharynx**. The lower respiratory tract extends from the **larynx** into the **trachea** before branching into the **bronchi**, which divide further to form the **bronchioles**, which terminate in **alveoli**, where gas exchange occurs.
- The upper respiratory tract is colonized by an extensive and diverse normal microbiota, many of which are potential pathogens. Few microbial inhabitants have been found in the lower respiratory tract, and these may be transients.
- Members of the normal microbiota may cause opportunistic infections, using a variety of strategies to overcome the innate nonspecific defenses (including the mucociliary escalator) and adaptive specific defenses of the respiratory system.
- Effective vaccines are available for many common respiratory pathogens, both bacterial and viral.
- Most respiratory infections result in inflammation of the infected tissues; these conditions are given names ending in -itis, such as **rhinitis**, **sinusitis**, **otitis**, **pharyngitis**, and **bronchitis**.

22.2 Bacterial Infections of the Respiratory Tract

- A wide variety of bacteria can cause respiratory diseases; most are treatable with antibiotics or preventable with vaccines.
- *Streptococcus pyogenes* causes **strep throat**, an infection of the pharynx that also causes high fever and can lead to **scarlet fever**, **acute rheumatic fever**, and **acute glomerulonephritis**.
- **Acute otitis media** is an infection of the middle ear that may be caused by several bacteria, including *Streptococcus pneumoniae*, *Haemophilus influenzae*, and *Moraxella catarrhalis*. The infection can block the eustachian tubes, leading to **otitis media with effusion**.
- **Diphtheria**, caused by *Corynebacterium diphtheriae*, is now a rare disease because of widespread vaccination. The bacteria produce exotoxins that kill cells in the pharynx, leading to the formation of a **pseudomembrane**; and damage other parts of the body.
- **Bacterial pneumonia** results from infections that cause inflammation and fluid accumulation in the alveoli. It is most commonly caused by *S. pneumoniae* or *H. influenzae*. The former is commonly multidrug resistant.
- *Mycoplasma* **pneumonia** results from infection by *Mycoplasma pneumoniae*; it can spread quickly, but the disease is mild and self-limiting.
- **Chlamydial pneumonia** can be caused by three pathogens that are obligate intracellular parasites. *Chlamydophila pneumoniae* is typically transmitted from an infected person, whereas *C. psittaci* is typically transmitted from an infected bird. *Chlamydia trachomatis*, may cause pneumonia in infants.
- Several other bacteria can cause pneumonia in immunocompromised individuals and those with cystic fibrosis.
- **Tuberculosis** is caused by *Mycobacterium tuberculosis*. Infection leads to the production of protective **tubercles** in the alveoli and calcified **Ghon complexes** that can harbor the bacteria for a long time. Antibiotic-resistant forms are common and treatment is typically long term.
- **Pertussis** is caused by *Bordetella pertussis*. Mucus accumulation in the lungs leads to prolonged severe coughing episodes (whooping cough) that facilitate transmission. Despite an available vaccine, outbreaks are still common.
- **Legionnaires disease** is caused by infection from environmental reservoirs of the *Legionella pneumophila* bacterium. The bacterium is endocytic within macrophages and infection can lead to pneumonia, particularly among immunocompromised individuals.
- **Q fever** is caused by *Coxiella burnetii*, whose primary hosts are domesticated mammals (zoonotic disease). It causes pneumonia primarily in farm workers and can lead to serious complications, such as endocarditis.

22.3 Viral Infections of the Respiratory Tract

- Viruses cause respiratory tract infections more frequently than bacteria, and most viral infections lead to mild symptoms.
- The **common cold** can be caused by more than

- 200 viruses, typically rhinoviruses, coronaviruses, and adenoviruses, transmitted by direct contact, aerosols, or environmental surfaces.
- Due to its ability to rapidly mutate through **antigenic drift** and **antigenic shift**, **influenza** remains an important threat to human health. Two new influenza vaccines are developed annually.
- Several viral infections, including **respiratory syncytial virus** infections, which frequently occur in the very young, can begin with mild symptoms before progressing to viral pneumonia.
- **SARS** and **MERS** are acute respiratory infections caused by coronaviruses, and both appear to originate in animals. SARS has not been seen in the human population since 2004 but had a high mortality rate during its outbreak. MERS also has a high mortality rate and continues to appear in human populations.
- **Measles**, **rubella**, and **chickenpox** are highly contagious, systemic infections that gain entry through the respiratory system and cause rashes and fevers. Vaccines are available for all three. Measles is the most severe of the three and is responsible for significant mortality around the world. Chickenpox typically causes mild infections in children but the virus can reactivate to cause painful cases of **shingles** later in life.

22.4 Respiratory Mycoses

- Fungal pathogens rarely cause respiratory disease in healthy individuals, but inhalation of fungal spores can cause severe pneumonia and systemic infections in immunocompromised patients.
- Antifungal drugs like amphotericin B can control most fungal respiratory infections.
- **Histoplasmosis** is caused by a mold that grows in soil rich in bird or bat droppings. Few exposed individuals become sick, but vulnerable individuals are susceptible. The yeast-like infectious cells grow inside phagocytes.
- **Coccidioidomycosis** is also acquired from soil and, in some individuals, will cause lesions on the face. Extreme cases may infect other organs, causing death.
- **Blastomycosis**, a rare disease caused by a soil fungus, typically produces a mild lung infection but can become disseminated in the immunocompromised. Systemic cases are fatal if untreated.
- **Mucormycosis** is a rare disease, caused by fungi of the order Mucorales. It primarily affects immunocompromised people. Infection involves growth of the hyphae into infected tissues and can lead to death in some cases.
- **Aspergillosis**, caused by the common soil fungus *Aspergillus*, infects immunocompromised people. Hyphal balls may impede lung function and hyphal growth into tissues can cause damage. Disseminated forms can lead to death.
- ***Pneumocystis*** **pneumonia** is caused by the fungus *P. jirovecii*. The disease is found in patients with AIDS and other immunocompromised individuals. Sulfa drug treatments have side effects, but untreated cases may be fatal.
- **Cryptococcosis** is caused by *Cryptococcus neoformans*. Lung infections may move to the brain, causing meningitis, which can be fatal.

REVIEW QUESTIONS
Multiple Choice

1. Which of the following is not directly connected to the nasopharynx?
   A. middle ear
   B. oropharynx
   C. lacrimal glands
   D. nasal cavity

2. What type of cells produce the mucus for the mucous membranes?
   A. goblet cells
   B. macrophages
   C. phagocytes
   D. ciliated epithelial cells

3. Which of these correctly orders the structures through which air passes during inhalation?
   A. pharynx → trachea → larynx → bronchi
   B. pharynx → larynx → trachea → bronchi
   C. larynx → pharynx → bronchi → trachea
   D. larynx → pharynx → trachea → bronchi

4. The _____ separates the upper and lower respiratory tract.
   A. bronchi
   B. larynx
   C. epiglottis
   D. palatine tonsil

5. Which microbial virulence factor is most important for attachment to host respiratory tissues?
   A. adhesins
   B. lipopolysaccharide
   C. hyaluronidase
   D. capsules

6. Which of the following does not involve a bacterial exotoxin?
   A. diphtheria
   B. whooping cough
   C. scarlet fever
   D. Q fever

7. What disease is caused by *Coxiella burnetii*?
   A. Q fever
   B. tuberculosis
   C. diphtheria
   D. walking pneumonia

8. In which stage of pertussis is the characteristic whooping sound made?
   A. convalescence
   B. catarrhal
   C. paroxysmal
   D. prodromal

9. What is the causative agent of Q fever?
   A. *Coxiella burnetii*
   B. *Chlamydophila psittaci*
   C. *Mycoplasma pneumoniae*
   D. *Streptococcus pyogenes*

10. Which of these microbes causes "walking pneumonia"?
    A. *Klebsiella pneumoniae*
    B. *Streptococcus pneumoniae*
    C. *Mycoplasma pneumoniae*
    D. *Chlamydophila pneumoniae*

11. Which of the following viruses is not commonly associated with the common cold?
    A. coronavirus
    B. adenovirus
    C. rhinovirus
    D. varicella-zoster virus

12. Which of the following viral diseases has been eliminated from the general population worldwide?
    A. smallpox
    B. measles
    C. German measles
    D. influenza

13. What term refers to multinucleated cells that form when many host cells fuse together during infections?
    A. Ghon elements
    B. Reye syndrome
    C. Koplik's spots
    D. syncytia

14. Which of the following diseases is not associated with coronavirus infections?
    A. Middle East respiratory syndrome
    B. German measles
    C. the common cold
    D. severe acute respiratory syndrome

15. Which of these viruses is responsible for causing shingles?
    A. rubella virus
    B. measles virus
    C. varicella-zoster virus
    D. variola major virus

16. Which of these infections is also referred to as Valley fever?
    A. histoplasmosis
    B. coccidioidomycosis
    C. blastomycosis
    D. aspergillosis

17. Which of the following is not caused by a dimorphic fungus?
    A. histoplasmosis
    B. coccidioidomycosis
    C. blastomycosis
    D. aspergillosis

18. Which of the following is caused by infections by bread molds?
    A. mucormycosis
    B. coccidioidomycosis
    C. cryptococcosis
    D. *Pneumocystis* pneumonia

19. In the United States, most histoplasmosis cases occur
    A. in the Pacific northwest.
    B. in the desert southwest.
    C. in the Mississippi river valley.
    D. in Colorado river valley.

20. Which of the following infections can be diagnosed using a skin test similar to the tuberculin test?
    A. histoplasmosis
    B. cryptococcosis
    C. blastomycosis
    D. aspergillosis

Fill in the Blank

21. Unattached microbes are moved from the lungs to the epiglottis by the _____ effect.
22. Many bacterial pathogens produce _____ to evade phagocytosis.
23. The main type of antibody in the mucous membrane defenses is _____.
24. _____ results from an inflammation of the "voice box."
25. _____ phagocytize potential pathogens in the lower lung.
26. Calcified lesions called _____ form in the lungs of patients with TB.
27. An inflammation of the middle ear is called _____.
28. The _____ is used to serologically identify *Streptococcus pneumoniae* isolates.
29. _____ is a zoonotic infection that can be contracted by people who handle birds.
30. The main virulence factor involved in scarlet fever is the _____.
31. The _____ virus is responsible for causing German measles.
32. A(n) _____ is an uncontrolled positive feedback loop between cytokines and leucocytes.
33. In cases of shingles, the antiviral drug _____ may be prescribed.
34. The slow accumulation of genetic changes to an influenza virus over time is referred to as _____.
35. The _____ vaccine is effective in controlling both measles and rubella.
36. In coccidioidomycosis, _____ containing many endospores form in the lungs.
37. In cryptococcosis, the main fungal virulence factor is the _____, which helps the pathogen avoid phagocytosis.
38. In some mycoses, fungal balls called _____ form in the lungs
39. Most US cases of coccidioidomycosis occur in _____.
40. Coccidioidomycosis may develop when *Coccidioides immitis* _____ are inhaled.

Short Answer

41. Explain why the lower respiratory tract is essentially sterile.
42. Explain why pneumonia is often a life-threatening disease.
43. Name three bacteria that commonly cause pneumonia. Which is the most common cause?
44. How does smoking make an individual more susceptible to infections?
45. How does the diphtheria pathogen form a pseudomembrane?
46. Since we all have experienced many colds in our lifetime, why are we not resistant to future infections?
47. Which pulmonary fungal infection is most likely to be confused with tuberculosis? How can we discriminate between these two types of infection?
48. Compare and contrast aspergillosis and mucormycosis.

Critical Thinking

49. Name each of the structures of the respiratory tract shown, and state whether each has a relatively large or small normal microbiota.

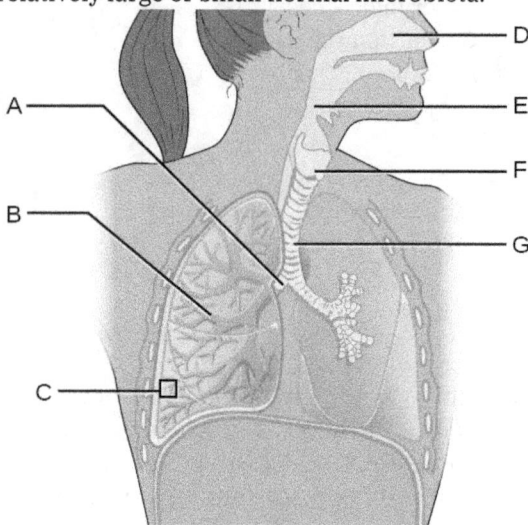

Figure 22.30 (credit: modification of work by National Cancer Institute)

50. Cystic fibrosis causes, among other things, excess mucus to be formed in the lungs. The mucus is very dry and caked, unlike the moist, more-fluid mucus of normal lungs. What effect do you think that has on the lung's defenses?
51. Why do you think smokers are more likely to suffer from respiratory tract infections?
52. Why might β-lactam antibiotics be ineffective against *Mycoplasma pneumoniae* infections?
53. Why is proper antibiotic therapy especially important for patients with tuberculosis?
54. What role does the common cold have in the rise of antibiotic-resistant strains of bacteria in the United States?
55. Why is it highly unlikely that influenza A virus will ever be eradicated, like the smallpox virus?
56. Why are fungal pulmonary infections rarely transmissible from person to person?

CHAPTER 23
Urogenital System Infections

Figure 23.1 Many pathogens that cause infections of the urogenital system can be detected in urine samples (left). The top sample in the culture (right) was prepared from the urine of a patient with a urinary tract infection. (credit b: modification of work by Nathan Reading)

Chapter Outline

23.1 Anatomy and Normal Microbiota of the Urogenital Tract

23.2 Bacterial Infections of the Urinary System

23.3 Bacterial Infections of the Reproductive System

23.4 Viral Infections of the Reproductive System

23.5 Fungal Infections of the Reproductive System

23.6 Protozoan Infections of the Urogenital System

INTRODUCTION The urogenital system is a combination of the urinary tract and reproductive system. Because both systems are open to the external environment, they are prone to infections. Some infections are introduced from outside, whereas others result from imbalances in the microbiota of the urogenital tract.

Urinary tract infections (UTIs) are one the most common bacterial infections worldwide, affecting over 100 million people each year. During 2007 in the United States, doctor office visits for UTIs exceeded 10 million, and an additional 2–3 million emergency department visits were attributed to UTIs. Sexually transmitted infections (STIs) also primarily affect the urogenital system and are an important cause of patient morbidity. The Centers for Disease Control and Prevention (CDC) estimates that there are approximately 20 million new cases of reportable STIs annually in the United States, half of which occur in people aged 15–24 years old. When STIs spread to the reproductive organs, they can be associated with severe morbidity and loss of fertility.

Because males and females have different urogenital anatomy, urogenital infections may affect males and females differently. In this chapter, we will discuss the various microbes that cause urogenital disease and the factors that contribute to their pathogenicity.

23.1 Anatomy and Normal Microbiota of the Urogenital Tract

Learning Objectives

By the end of this section, you will be able to:
- Compare the anatomy, function, and normal microbiota associated with the male and female urogenital systems
- Explain how microorganisms, in general, overcome the defenses of the urogenital system to cause infection
- Name, describe, and differentiate between general signs and symptoms associated with infections of the urogenital tract

---

Clinical Focus

Part 1

Nadia is a newly married 26-year-old graduate student in economics. Recently she has been experiencing an unusual vaginal discharge, as well as some itching and discomfort. Since she is due for her annual physical exam, she makes an appointment with her doctor hoping that her symptoms can be quickly treated. However, she worries that she may have some sort of sexually transmitted infection (STI). Although she is now in a monogamous relationship, she is not fully certain of her spouse's sexual history and she is reluctant to ask him about it.

At her checkup, Nadia describes her symptoms to her primary care physician and, somewhat awkwardly, explains why she thinks she might have an STI. Nadia's doctor reassures her that she regularly sees patients with similar concerns and encourages her to be fully transparent about her symptoms because some STIs can have serious complications if left untreated. After some further questioning, the doctor takes samples of Nadia's blood, urine, and vaginal discharge to be sent to the lab for testing.

- What are some possible causes of Nadia's symptoms?
- Why does the doctor take so many different samples?

Jump to the next Clinical Focus box.

The urinary system filters blood, excretes wastes, and maintains an appropriate electrolyte and water balance. The reproductive system is responsible for the production of gametes and participates in conception and, in females, development of offspring. Due to their proximity and overlap, these systems are often studied together and referred to as the urogenital system (or genitourinary system).

Anatomy of the Urinary Tract

The basic structures of the urinary tract are common in males and females. However, there are unique locations for these structures in females and males, and there is a significant amount of overlap between the urinary and genital structures in males. Figure 23.2 illustrates the urinary anatomy common to females and males.

The **kidneys** carry out the urinary system's primary functions of filtering the blood and maintaining water and electrolyte balance. The kidneys are composed of millions of filtration units called nephrons. Each nephron is in intimate contact with blood through a specialized capillary bed called the **glomerulus** (plural *glomeruli*). Fluids, electrolytes, and molecules from the blood pass from the glomerulus into the nephron, creating the filtrate that becomes urine (Figure 23.3). Urine that collects in each kidney empties through a **ureter** and drains to the **urinary bladder**, which stores urine. Urine is released from the bladder to the **urethra**, which transports it to be excreted from the body through the **urinary meatus**, the opening of the urethra.

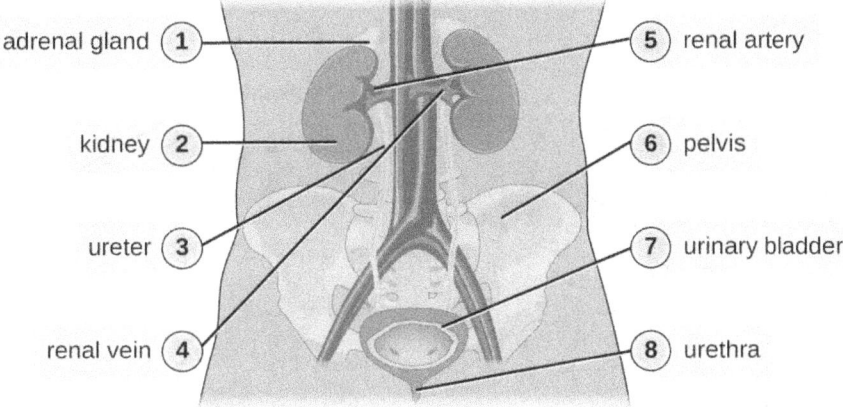

Figure 23.2 These structures of the human urinary system are present in both males and females.

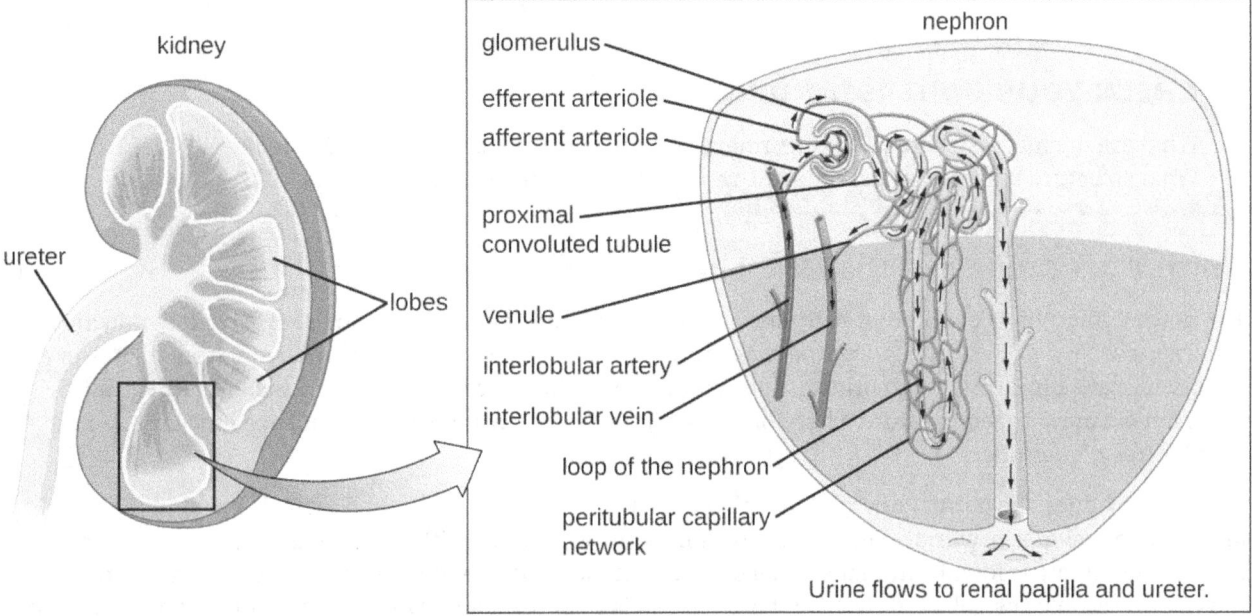

Figure 23.3 The kidney contains several lobes, each of which contains millions of nephrons. The nephron is the functional unit of the kidney, filtering the blood and removing water and dissolved compounds. The filtrate first enters the glomerulus and then enters the proximal convoluted tubule. As it passes through the tubule, the filtrate is further modified by osmosis and active transport until it reaches the larger ducts as urine.

Anatomy of the Reproductive System

The male reproductive system (Figure 23.4) is located in close proximity to the urinary system, and the urethra is part of both systems. The **testes** are responsible for the production of sperm. The **epididymis** is a coiled tube that collects sperm from the testes and passes it on to the vas deferens. The epididymis is also the site of sperm maturation after they leave the testes. The **seminal vesicles** and **prostate** are accessory glands that produce fluid that supports sperm. During ejaculation, the **vas deferens** releases this mixture of fluid and sperm, called semen, into the urethra, which extends to the end of the **penis**.

The female reproductive system is located near the urinary system (Figure 23.4). The external genitalia (**vulva**) in females open to the **vagina**, a muscular passageway that connects to the cervix. The **cervix** is the lower part of the **uterus** (the organ where a fertilized egg will implant and develop). The cervix is a common site of infection, especially for viruses that may lead to cervical cancer. The uterus leads to the fallopian tubes and eventually to the ovaries. Ovaries are the site of ova (egg) production, as well as the site of estrogen and progesterone production that are involved in maturation and maintenance of reproductive organs, preparation of the uterus for pregnancy, and regulation of the menstrual cycle.

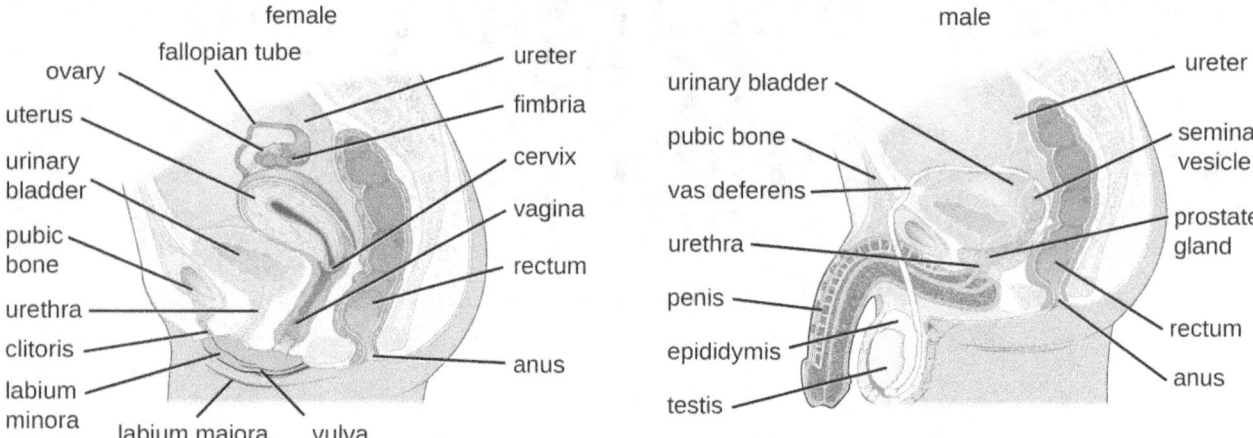

Figure 23.4 The female reproductive system is located in close proximity to the urinary system. In males, the urethra is shared by the reproductive and urinary systems.

CHECK YOUR UNDERSTANDING

- What are the major structures of the urinary system, starting where urine is formed?
- What structure in males is shared by the reproductive and the urinary systems?

Normal Microbiota of the Urogenital System

The normal microbiota of different body sites provides an important nonspecific defense against infectious diseases (see Physical Defenses), and the urogenital tract is no exception. In both men and women, however, the kidneys are sterile. Although urine does contain some antibacterial components, bacteria will grow in urine left out at room temperature. Therefore, it is primarily the flushing action that keeps the ureters and bladder free of microbes.

Below the bladder, the normal microbiota of the male urogenital system is found primarily within the distal urethra and includes bacterial species that are commonly associated with the skin microbiota. In women, the normal microbiota is found within the distal one third of the urethra and the vagina. The normal microbiota of the vagina becomes established shortly after birth and is a complex and dynamic population of bacteria that fluctuates in response to environmental changes. Members of the vaginal microbiota play an important role in the nonspecific defense against vaginal infections and sexually transmitted infections by occupying cellular binding sites and competing for nutrients. In addition, the production of lactic acid by members of the microbiota provides an acidic environment within the vagina that also serves as a defense against infections. For the majority of women, the lactic-acid–producing bacteria in the vagina are dominated by a variety of species of *Lactobacillus*. For women who lack sufficient lactobacilli in their vagina, lactic acid production comes primarily from other species of bacteria such as *Leptotrichia* spp., *Megasphaera* spp., and *Atopobium vaginae*. *Lactobacillus* spp. use glycogen from vaginal epithelial cells for metabolism and production of lactic acid. This process is tightly regulated by the hormone estrogen. Increased levels of estrogen correlate with increased levels of vaginal glycogen, increased production of lactic acid, and a lower vaginal pH. Therefore, decreases in estrogen during the menstrual cycle and with menopause are associated with decreased levels of vaginal glycogen and lactic acid, and a higher pH. In addition to producing lactic acid, *Lactobacillus* spp. also contribute to the defenses against infectious disease through their production of hydrogen peroxide and bacteriocins (antibacterial peptides).

CHECK YOUR UNDERSTANDING

- What factors affect the microbiota of the female reproductive tract?

General Signs and Symptoms of Urogenital Infections

Infections of the urinary tract most commonly cause inflammation of the bladder (**cystitis**) or of the urethra (**urethritis**). Urethritis can be associated with cystitis, but can also be caused by sexually transmitted infections. Symptoms of urethritis in men include burning sensation while urinating, discharge from the penis, and blood in the semen or the urine. In women, urethritis is associated with painful and frequent urination, vaginal discharge, fever, chills, and abdominal pain. The symptoms of cystitis are similar to those of urethritis. When urethritis is caused by a sexually transmitted pathogen, additional symptoms involving the genitalia can occur. These can include painful vesicles (blisters), warts, and ulcers. Ureteritis, a rare infection of the ureter, can also occur with cystitis. These infections can be acute or chronic.

Pyelonephritis and **glomerulonephritis** are infections of the kidney that are potentially serious. Pyelonephritis is an infection of one or both of the kidneys and may develop from a lower urinary tract infection; the upper urinary tract, including the ureters, is often affected. Signs and symptoms of pyelonephritis include fever, chills, nausea, vomiting, lower back pain, and frequent painful urination. Pyelonephritis usually only becomes chronic in individuals who have malformations in or damage to the kidneys.

Glomerulonephritis is an inflammation of the glomeruli of the nephrons. Symptoms include excessive protein and blood in urine, increased blood pressure, and fluid retention leading to edema of face, hands, and feet. Glomerulonephritis may be an acute infection or it can become chronic.

Infections occurring within the reproductive structures of males include epididymitis, orchitis, and prostatitis. Bacterial infections may cause inflammation of the epididymis, called **epididymitis**. This inflammation causes pain in the scrotum, testicles, and groin; swelling, redness, and warm skin in these areas may also be observed. Inflammation of the testicle, called **orchitis**, is usually caused by a bacterial infection spreading from the epididymis, but it can also be a complication of mumps, a viral disease. The symptoms are similar to those of epididymitis, and it is not uncommon for them both to occur together, in which case the condition is called epididymo-orchitis. Inflammation of the prostate gland, called **prostatitis**, can result from a bacterial infection. The signs and symptoms of prostatitis include fever, chills, and pain in the bladder, testicles, and penis. Patients may also experience burning during urination, difficulty emptying the bladder, and painful ejaculation.

Because of its proximity to the exterior, the vagina is a common site for infections in women. The general term for any inflammation of the vagina is **vaginitis**. Vaginitis often develops as a result of an overgrowth of bacteria or fungi that normally reside in the vaginal microbiota, although it can also result from infections by transient pathogens. Bacterial infections of the vagina are called bacterial **vaginosis**, whereas fungal infections (typically involving *Candida* spp.) are called **yeast infections**. Dynamic changes affecting the normal microbiota, acid production, and pH variations can be involved in the initiation of the microbial overgrowth and the development of vaginitis. Although some individuals may have no symptoms, vaginosis and vaginitis can be associated with discharge, odor, itching, and burning.

Pelvic inflammatory disease (PID) is an infection of the female reproductive organs including the uterus, cervix, fallopian tubes, and ovaries. The two most common pathogens are the sexually transmitted bacterial pathogens *Neisseria gonorrhoeae* and *Chlamydia trachomatis*. Inflammation of the fallopian tubes, called **salpingitis**, is the most serious form of PID. Symptoms of PID can vary between women and include pain in the lower abdomen, vaginal discharge, fever, chills, nausea, diarrhea, vomiting, and painful urination.

✓ CHECK YOUR UNDERSTANDING
- What conditions can result from infections affecting the urinary system?
- What are some common causes of vaginitis in women?

General Causes and Modes of Transmission of Urogenital Infections

Hormonal changes, particularly shifts in estrogen in women due to pregnancy or menopause, can increase susceptibility to urogenital infections. As discussed earlier, estrogen plays an important role in regulating the availability of glycogen and subsequent production of lactic acid by *Lactobacillus* species. Low levels of estrogen are associated with an increased vaginal pH and an increased risk of bacterial vaginosis and yeast infections. Estrogen also plays a role in maintaining the elasticity, strength, and thickness of the vaginal wall, and keeps the vaginal wall lubricated, reducing dryness. Low levels of estrogen are associated with thinning of the vaginal wall. This thinning increases the risk of tears and abrasions, which compromise the protective barrier and increase susceptibility to pathogens.

Another common cause of urogenital infections in females is fecal contamination that occurs because of the close proximity of the anus and the urethra. *Escherichia coli*, an important member of the digestive tract microbiota, is the most common cause of urinary tract infections (urethritis and cystitis) in women; it generally causes infection when it is introduced to the urethra in fecal matter. Good hygiene can reduce the risk of urinary tract infections by this route. In men, urinary tract infections are more commonly associated with other conditions, such as an enlarged prostate, kidney stones, or placement of a urinary catheter. All of these conditions impair the normal emptying of the bladder, which serves to flush out microbes capable of causing infection.

Infections that are transmitted between individuals through sexual contact are called sexually transmitted infections (STIs) or sexually transmitted diseases (STDs). (The CDC prefers the term STD, but WHO prefers STI,[1] which encompasses infections that result in disease as well as those that are subclinical or asymptomatic.) STIs often affect the external genitalia and skin, where microbes are easily transferred through physical contact. Lymph nodes in the genital region may also become swollen as a result of infection. However, many STIs have systemic effects as well, causing symptoms that range from mild (e.g., general malaise) to severe (e.g., liver damage or serious immunosuppression).

CHECK YOUR UNDERSTANDING

- What role does *Lactobacillus* play in the health of the female reproductive system?
- Why do urinary tract infections have different causes in males and females?

23.2 Bacterial Infections of the Urinary System

Learning Objectives

By the end of this section, you will be able to:
- Identify the most common bacterial pathogens that can cause urinary tract infections
- Compare the major characteristics of specific bacterial diseases affecting the urinary tract

Urinary tract infections (UTIs) include infections of the urethra, bladder, and kidneys, and are common causes of urethritis, cystitis, pyelonephritis, and glomerulonephritis. Bacteria are the most common causes of UTIs, especially in the urethra and bladder.

Cystitis

Cystitis is most often caused by a bacterial infection of the bladder, but it can also occur as a reaction to certain treatments or irritants such as radiation treatment, hygiene sprays, or spermicides. Common symptoms of cystitis include **dysuria** (urination accompanied by burning, discomfort, or pain), **pyuria** (pus in the urine), **hematuria** (blood in the urine), and bladder pain.

In women, bladder infections are more common because the urethra is short and located in close proximity to the anus, which can result in infections of the urinary tract by fecal bacteria. Bladder infections are also more common in the elderly because the bladder may not empty fully, causing urine to pool; the elderly may also have weaker immune systems that make them more vulnerable to infection. Conditions such as prostatitis in men or kidney stones in both men and women can impact proper drainage of urine and increase risk of bladder infections. Catheterization can also increase the risk of bladder infection (see Case in Point: Cystitis in

---

1 World Health Organization. "Guidelines for the Management of Sexually Transmitted Infections." World Health Organization, 2003. http://www.who.int/hiv/pub/sti/en/STIGuidelines2003.pdf.

the Elderly).

Gram-negative bacteria such as *Escherichia coli* (most commonly), *Proteus vulgaris*, *Pseudomonas aeruginosa*, and *Klebsiella pneumoniae* cause most bladder infections. Gram-positive pathogens associated with cystitis include the coagulase-negative *Staphylococcus saprophyticus*, *Enterococcus faecalis*, and *Streptococcus agalactiae*. Routine manual urinalysis using a urine dipstick or test strip can be used for rapid screening of infection. These test strips (Figure 23.5) are either held in a urine stream or dipped in a sample of urine to test for the presence of nitrites, leukocyte esterase, protein, or blood that can indicate an active bacterial infection. The presence of nitrite may indicate the presence of *E. coli* or *K. pneumonia*; these bacteria produce nitrate reductase, which converts nitrate to nitrite. The leukocyte esterase (LE) test detects the presence of neutrophils as an indication of active infection.

Low specificity, sensitivity, or both, associated with these rapid screening tests require that care be taken in interpretation of results and in their use in diagnosis of urinary tract infections. Therefore, positive LE or nitrite results are followed by a urine culture to confirm a bladder infection. Urine culture is generally accomplished using blood agar and MacConkey agar, and it is important to culture a clean catch of urine to minimize contamination with normal microbiota of the penis and vagina. A clean catch of urine is accomplished by first washing the labia and urethral opening of female patients or the penis of male patients. The patient then releases a small amount of urine into the toilet bowl before stopping the flow of urine. Finally, the patient resumes urination, this time filling the container used to collect the specimen.

Bacterial cystitis is commonly treated with fluoroquinolones, nitrofurantoin, cephalosporins, or a combination of trimethoprim and sulfamethoxazole. Pain medications may provide relief for patients with dysuria. Treatment is more difficult in elderly patients, who experience a higher rate of complications such as sepsis and kidney infections.

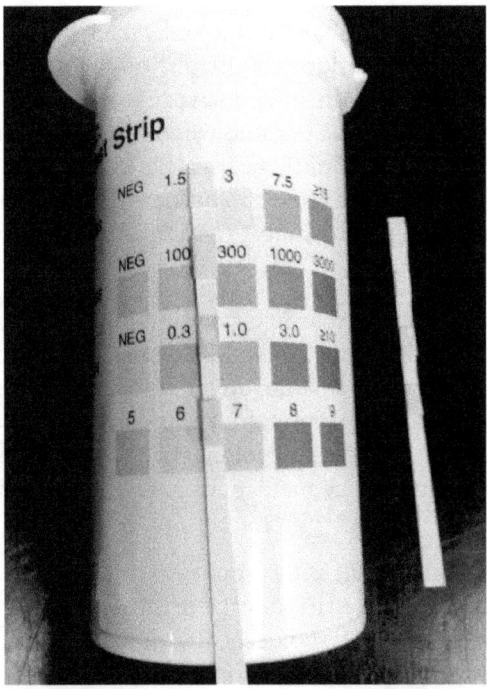

Figure 23.5 A urine dipstick is compared against a color key to determine levels of various chemicals, proteins, or cells in the urine. Abnormal levels may indicate an infection. (credit: modification of work by Suzanne Wakim)

> **Case in Point**
>
> **Cystitis in the Elderly**
> Robert, an 81-year-old widower with early onset Alzheimer's, was recently moved to a nursing home because he was having difficulty living on his own. Within a few weeks of his arrival, he developed a fever

and began to experience pain associated with urination. He also began having episodes of confusion and delirium. The doctor assigned to examine Robert read his file and noticed that Robert was treated for prostatitis several years earlier. When he asked Robert how often he had been urinating, Robert explained that he had been trying not to drink too much so that he didn't have to walk to the restroom.

All of this evidence suggests that Robert likely has a urinary tract infection. Robert's age means that his immune system has probably begun to weaken, and his previous prostate condition may be making it difficult for him to empty his bladder. In addition, Robert's avoidance of fluids has led to dehydration and infrequent urination, which may have allowed an infection to establish itself in his urinary tract. The fever and dysuria are common signs of a UTI in patients of all ages, and UTIs in elderly patients are often accompanied by a notable decline in mental function.

Physical challenges often discourage elderly individuals from urinating as frequently as they would otherwise. In addition, neurological conditions that disproportionately affect the elderly (e.g., Alzheimer's and Parkinson's disease) may also reduce their ability to empty their bladders. Robert's doctor noted that he was having difficulty navigating his new home and recommended that he be given more assistance and that his fluid intake be monitored. The doctor also took a urine sample and ordered a laboratory culture to confirm the identity of the causative agent.

- Why is it important to identify the causative agent in a UTI?
- Should the doctor prescribe a broad-spectrum or narrow-spectrum antibiotic to treat Robert's UTI? Why?

Kidney Infections (Pyelonephritis and Glomerulonephritis)

Pyelonephritis, an inflammation of the kidney, can be caused by bacteria that have spread from other parts of the urinary tract (such as the bladder). In addition, pyelonephritis can develop from bacteria that travel through the bloodstream to the kidney. When the infection spreads from the lower urinary tract, the causative agents are typically fecal bacteria such as *E. coli*. Common signs and symptoms include back pain (due to the location of the kidneys), fever, and nausea or vomiting. Gross hematuria (visible blood in the urine) occurs in 30–40% of women but is rare in men.[2] The infection can become serious, potentially leading to bacteremia and systemic effects that can become life-threatening. Scarring of the kidney can occur and persist after the infection has cleared, which may lead to dysfunction.

Diagnosis of pyelonephritis is made using microscopic examination of urine, culture of urine, testing for leukocyte esterase and nitrite levels, and examination of the urine for blood or protein. It is also important to use blood cultures to evaluate the spread of the pathogen into the bloodstream. Imaging of the kidneys may be performed in high-risk patients with diabetes or immunosuppression, the elderly, patients with previous renal damage, or to rule out an obstruction in the kidney. Pyelonephritis can be treated with either oral or intravenous antibiotics, including penicillins, cephalosporins, vancomycin, fluoroquinolones, carbapenems, and aminoglycosides.

Glomerulonephritis occurs when the glomeruli of the nephrons are damaged from inflammation. Whereas pyelonephritis is usually acute, glomerulonephritis may be acute or chronic. The most well-characterized mechanism of glomerulonephritis is the post-streptococcal sequelae associated with *Streptococcus pyogenes* throat and skin infections. Although *S. pyogenes* does not directly infect the glomeruli of the kidney, immune complexes that form in blood between *S. pyogenes* antigens and antibodies lodge in the capillary endothelial cell junctions of the glomeruli and trigger a damaging inflammatory response. Glomerulonephritis can also occur in patients with bacterial endocarditis (infection and inflammation of heart tissue); however, it is currently unknown whether glomerulonephritis associated with endocarditis is also immune-mediated.

Leptospirosis

Leptospira are generally harmless spirochetes that are commonly found in the soil. However, some pathogenic species can cause an infection called **leptospirosis** in the kidneys and other organs (Figure 23.6).

---

2 Tibor Fulop. "Acute Pyelonephritis" *Medscape*, 2015. http://emedicine.medscape.com/article/245559-overview.

Leptospirosis can produce fever, headache, chills, vomiting, diarrhea, and rash with severe muscular pain. If the disease continues to progress, infection of the kidney, meninges, or liver may occur and may lead to organ failure or meningitis. When the kidney and liver become seriously infected, it is called **Weil's disease**. Pulmonary hemorrhagic syndrome can also develop in the lungs, and jaundice may occur.

Leptospira spp. are found widely in animals such as dogs, horses, cattle, pigs, and rodents, and are excreted in their urine. Humans generally become infected by coming in contact with contaminated soil or water, often while swimming or during flooding; infection can also occur through contact with body fluids containing the bacteria. The bacteria may enter the body through mucous membranes, skin injuries, or by ingestion. The mechanism of pathogenicity is not well understood.

Leptospirosis is extremely rare in the United States, although it is endemic in Hawaii; 50% of all cases in the United States come from Hawaii.[3] It is more common in tropical than in temperate climates, and individuals who work with animals or animal products are most at risk. The bacteria can also be cultivated in specialized media, with growth observed in broth in a few days to four weeks; however, diagnosis of leptospirosis is generally made using faster methods, such as detection of antibodies to *Leptospira* spp. in patient samples using serologic testing. Polymerase chain reaction (PCR), enzyme-linked immunosorbent assay (ELISA), slide agglutination, and indirect immunofluorescence tests may all be used for diagnosis. Treatment for leptospirosis involves broad-spectrum antibiotics such as penicillin and doxycycline. For more serious cases of leptospirosis, antibiotics may be given intravenously.

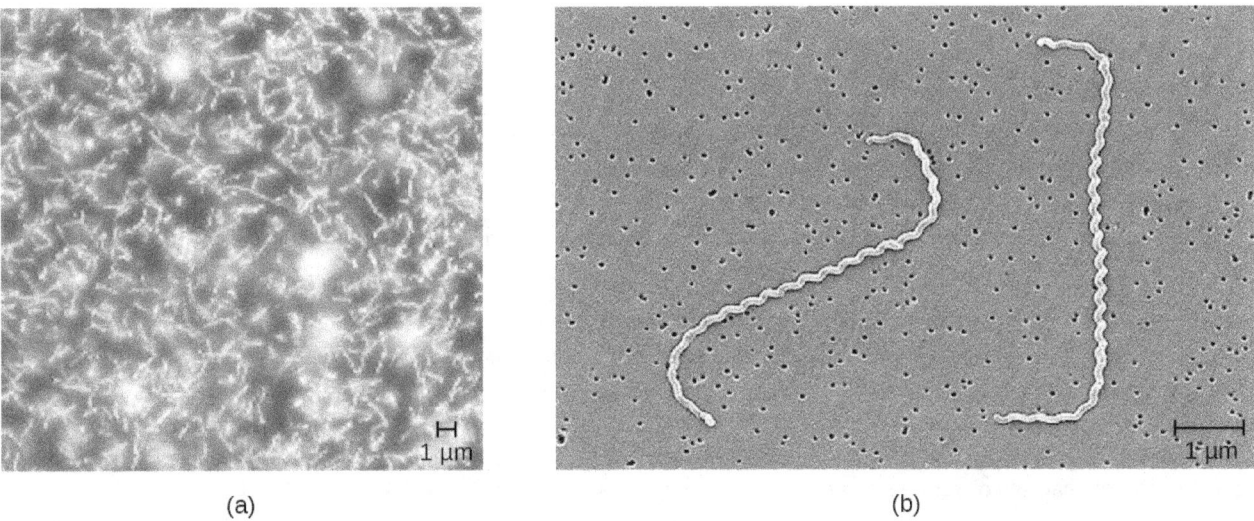

Figure 23.6 (a) Dark field view of *Leptospira* sp. (b) A scanning electron micrograph of *Leptospira interrogans*, a pathogenic species, shows the distinctive spirochete morphology of this genus. (credit b: modification of work by Janice Carr, Centers for Disease Control and Prevention)

CHECK YOUR UNDERSTANDING

- What is the most common cause of a kidney infection?
- What are the most common symptoms of a kidney infection?

Nongonococcal Urethritis (NGU)

There are two main categories of bacterial urethritis: gonorrheal and nongonococcal. Gonorrheal urethritis is caused by *Neisseria gonorrhoeae* and is associated with gonorrhea, a common STI. This cause of urethritis will be discussed in Bacterial Infections of the Reproductive System. The term **nongonococcal urethritis (NGU)** refers to inflammation of the urethra that is unrelated to *N. gonorrhoeae*. In women, NGU is often asymptomatic. In men, NGU is typically a mild disease, but can lead to purulent discharge and dysuria. Because the symptoms are often mild or nonexistent, most infected individuals do not know that they are

---

3 Centers for Disease Control and Prevention. "Leptospirosis." 2015. http://www.cdc.gov/leptospirosis/health_care_workers.

infected, yet they are carriers of the disease. Asymptomatic patients also have no reason to seek treatment, and although not common, untreated NGU can spread to the reproductive organs, causing pelvic inflammatory disease and salpingitis in women and epididymitis and prostatitis in men. Important bacterial pathogens that cause nongonococcal urethritis include *Chlamydia trachomatis, Mycoplasma genitalium, Ureaplasma urealyticum*, and *Mycoplasma hominis*.

C. trachomatis is a difficult-to-stain, gram-negative bacterium with an ovoid shape. An intracellular pathogen, *C. trachomatis* causes the most frequently reported STI in the United States, chlamydia. Although most persons infected with *C. trachomatis* are asymptomatic, some patients can present with NGU. *C. trachomatis* can also cause non-urogenital infections such as the ocular disease trachoma (see Bacterial Infections of the Skin and Eyes). The life cycle of *C. trachomatis* is illustrated in Figure 4.12.

C. trachomatis has multiple possible virulence factors that are currently being studied to evaluate their roles in causing disease. These include polymorphic outer-membrane autotransporter proteins, stress response proteins, and type III secretion effectors. The type III secretion effectors have been identified in gram-negative pathogens, including *C. trachomatis*. This virulence factor is an assembly of more than 20 proteins that form what is called an injectisome for the transfer of other effector proteins that target the infected host cells. The outer-membrane autotransporter proteins are also an effective mechanism of delivering virulence factors involved in colonization, disease progression, and immune system evasion.

Other species associated with NGU include *Mycoplasma genitalium, Ureaplasma urealyticum*, and *Mycoplasma hominis*. These bacteria are commonly found in the normal microbiota of healthy individuals, who may acquire them during birth or through sexual contact, but they can sometimes cause infections leading to urethritis (in males and females) or vaginitis and cervicitis (in females).

M. genitalium is a more common cause of urethritis in most settings than *N. gonorrhoeae*, although it is less common than *C. trachomatis*. It is responsible for approximately 30% of recurrent or persistent infections, 20–25% of nonchlamydial NGU cases, and 15%–20% of NGU cases. *M. genitalium* attaches to epithelial cells and has substantial antigenic variation that helps it evade host immune responses. It has lipid-associated membrane proteins that are involved in causing inflammation.

Several possible virulence factors have been implicated in the pathogenesis of *U. urealyticum* (Figure 23.7). These include the ureaplasma proteins phospholipase A, phospholipase C, multiple banded antigen (MBA), urease, and immunoglobulin α protease. The phospholipases are virulence factors that damage the cytoplasmic membrane of target cells. The immunoglobulin α protease is an important defense against antibodies. It can generate hydrogen peroxide, which may adversely affect host cell membranes through the production of reactive oxygen species.

Treatments differ for gonorrheal and nongonococcal urethritis. However, *N. gonorrhoeae* and *C. trachomatis* are often simultaneously present, which is an important consideration for treatment. NGU is most commonly treated using tetracyclines (such as doxycycline) and azithromycin; erythromycin is an alternative option. Tetracyclines and fluoroquinolones are most commonly used to treat *U. urealyticum*, but resistance to tetracyclines is becoming an increasing problem.[4] While tetracyclines have been the treatment of choice for *M. hominis*, increasing resistance means that other options must be used. Clindamycin and fluoroquinolones are alternatives. *M. genitalium* is generally susceptible to doxycycline, azithromycin, and moxifloxacin. Like other mycoplasma, *M. genitalium* does not have a cell wall and therefore β-lactams (including penicillins and cephalosporins) are not effective treatments.

---

4 Ken B Waites. "Ureaplasma Infection Medication." *Medscape*, 2015. http://emedicine.medscape.com/article/231470-medication.

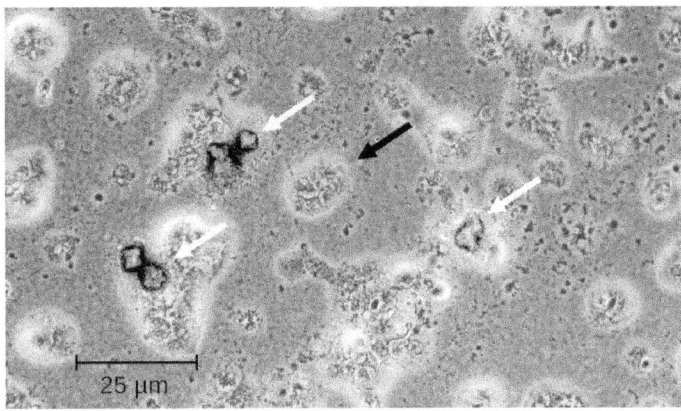

Figure 23.7 *Ureaplasma urealyticum* microcolonies (white arrows) on agar surface after anaerobic incubation, visualized using phase contrast microscopy (800×). The black arrow indicates cellular debris. (credit: modification of work by American Society for Microbiology)

CHECK YOUR UNDERSTANDING

- What are the three most common causes of urethritis?
- What three members of the normal microbiota can cause urethritis?

Disease Profile

Bacterial Infections of the Urinary Tract

Urinary tract infections can cause inflammation of the urethra (urethritis), bladder (cystitis), and kidneys (pyelonephritis), and can sometimes spread to other body systems through the bloodstream. Figure 23.8 captures the most important features of various types of UTIs.

Bacterial Infections of the Urinary Tract

Disease	Pathogen	Signs and Symptoms	Transmission	Diagnostic Tests	Antimicrobial Drugs
Cystitis	*Escherichia coli, Enterococcus faecalis, Streptococcus agalactiae, Klebsiella pneumoniae, Staphylococcus saprophyticus,* others	Dysuria, pyuria, hematuria, and bladder pain; most common in females due to the shorter urethra and abundant normal vaginal microbiota	Nontransmisible; opportunistic infections occur when fecal bacteria are introduced to urinary tract or when normal urination or immune function is impaired	Urine dipstick, urine culture for confirmation	Fluoroquinolones, nitrofurantoin, cephalosprins, trimethoprim, sulfamethoxazole
Leptospirosis	*Leptospira* spp.	Fever, headache, chills, vomiting, diarrhea, rash, muscular pain; in disseminated infections, may cause jaundice, pulmonary hemorrhaging, meningitis	From animals to humans via contact with urine or body fluids	PCR, ELISA, slide agglutination, indirect immunofluorescence	Doxycycline, amoxicillin, ampicillin, erythromycin, penicillin
Nongonococcal urethritis (NGU)	*Chlamydia trachomatis, Mycoplasma genitalium, Mycoplasma hominis, Ureaplasma urealyticum*	Mild or asymptomatic; may cause purulent discharge and dysuria	Transmitted sexually or from mother to neonate during birth	Urethral swabs and urine culture, PCR, NAAT	Azithromycin, doxycycline, erythromycin, fluoroquinolones
Pyelonephritis, glomerulonephritis	*E. coli, Proteus* spp., *Klebsiella* spp., *Streptococcus pyogenes,* others	Back pain, fever, nausea, vomiting, blood in urine; possible scarring of the kidneys and impaired kidney function; severe infections may lead to sepsis and death	Nontransmissible; infection spreads to kidneys from urinary tract or through bloodstream	Urinalysis, urine culture, radioimaging of kidneys	Penicillins, cephalosprins, fluoroquinolones, aminoglycosides, others

Figure 23.8

23.3 Bacterial Infections of the Reproductive System

Learning Objectives

By the end of this section, you will be able to:
- Identify the most common bacterial pathogens that can cause infections of the reproductive system
- Compare the major characteristics of specific bacterial diseases affecting the reproductive system

In addition to infections of the urinary tract, bacteria commonly infect the reproductive tract. As with the urinary tract, parts of the reproductive system closest to the external environment are the most likely sites of infection. Often, the same microbes are capable of causing urinary tract and reproductive tract infections.

Bacterial Vaginitis and Vaginosis

Inflammation of the vagina is called vaginitis, often caused by a bacterial infection. It is also possible to have an imbalance in the normal vaginal microbiota without inflammation called **bacterial vaginosis (BV)**. Vaginosis may be asymptomatic or may cause mild symptoms such as a thin, white-to-yellow, homogeneous vaginal discharge, burning, odor, and itching. The major causative agent is *Gardnerella vaginalis*, a gram-variable to gram-negative pleomorphic bacterium. Other causative agents include anaerobic species such as members of the genera *Bacteroides* and *Fusobacterium*. Additionally, ureaplasma and mycoplasma may be involved. The disease is usually self-limiting, although antibiotic treatment is recommended if symptoms develop.

G. vaginalis appears to be more virulent than other vaginal bacterial species potentially associated with BV. Like *Lactobacillus* spp., *G. vaginalis* is part of the normal vaginal microbiota, but when the population of *Lactobacillus* spp. decreases and the vaginal pH increases, *G. vaginalis* flourishes, causing vaginosis by attaching to vaginal epithelial cells and forming a thick protective biofilm. *G. vaginalis* also produces a cytotoxin called vaginolysin that lyses vaginal epithelial cells and red blood cells.

Since *G. vaginalis* can also be isolated from healthy women, the "gold standard" for the diagnosis of BV is direct examination of vaginal secretions and not the culture of *G. vaginalis*. Diagnosis of bacterial vaginosis from vaginal secretions can be accurately made in three ways. The first is to use a DNA probe. The second method is to assay for sialidase activity (sialidase is an enzyme produced by *G. vaginalis* and other bacteria associated with vaginosis, including *Bacteroides* spp., *Prevotella* spp., and *Mobiluncus* spp.). The third method is to assess gram-stained vaginal smears for microscopic morphology and relative numbers and types of bacteria, squamous epithelial cells, and leukocytes. By examining slides prepared from vaginal swabs, it is possible to distinguish lactobacilli (long, gram-positive rods) from other gram-negative species responsible for BV. A shift in predominance from gram-positive bacilli to gram-negative coccobacilli can indicate BV. Additionally, the slide may contain so-called clue cells, which are epithelial cells that appear to have a granular or stippled appearance due to bacterial cells attached to their surface (Figure 23.9). Presumptive diagnosis of bacterial vaginosis can involve an assessment of clinical symptoms and evaluation of vaginal fluids using Amsel's diagnostic criteria which include 3 out of 4 of the following characteristics:

1. white to yellow discharge;
2. a fishy odor, most noticeable when 10% KOH is added;
3. pH greater than 4.5;
4. the presence of clue cells.

Treatment is often unnecessary because the infection often clears on its own. However, in some cases, antibiotics such as topical or oral clindamycin or metronidazole may be prescribed. Alternative treatments include oral tinidazole or clindamycin ovules (vaginal suppositories).

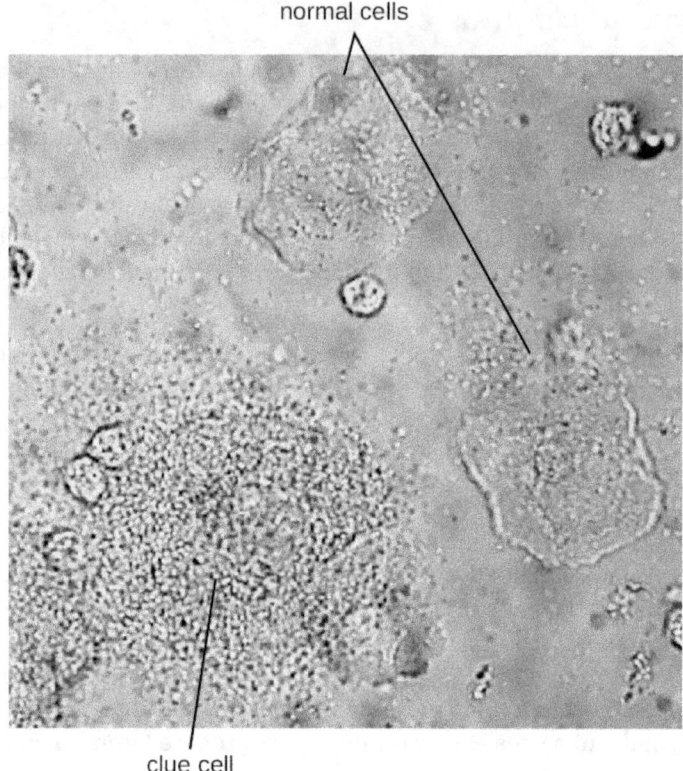

Figure 23.9 In this vaginal smear, the cell at the lower left is a clue cell with a unique appearance caused by the presence of bacteria on the cell. The cell on the right is a normal cell.

CHECK YOUR UNDERSTANDING

- Explain the difference between vaginosis and vaginitis.
- What organisms are responsible for vaginosis and what organisms typically hold it at bay?

Clinical Focus

Part 2

There is no catch-all test for STIs, so several tests, in addition to a physical exam, are necessary to diagnose an infection. Nadia tries to relax in the exam room while she waits for the doctor to return, but she is nervous about the results.

When the doctor finally returns, she has some unexpected news: Nadia is pregnant. Surprised and excited, Nadia wants to know if the pregnancy explains her unusual symptoms. The doctor explains that the irritation that Nadia is experiencing is vaginitis, which can be caused by several types of microorganisms. One possibility is bacterial vaginosis, which develops when there is an imbalance in the bacteria in the vagina, as often occurs during pregnancy. Vaginosis can increase the risk of preterm birth and low birth weight, and a few studies have also shown that it can cause second-trimester miscarriage; however, the condition can be treated. To check for it, the doctor has asked the lab to perform a Gram stain on Nadia's sample.

- What result would you expect from the Gram stain if Nadia has bacterial vaginosis?
- What is the relationship between pregnancy, estrogen levels, and development of bacterial vaginosis?

Jump to the next Clinical Focus box. Go back to the previous Clinical Focus box.

Gonorrhea

Also known as the clap, **gonorrhea** is a common sexually transmitted disease of the reproductive system that is especially prevalent in individuals between the ages of 15 and 24. It is caused by *Neisseria gonorrhoeae*, often called gonococcus or GC, which have fimbriae that allow the cells to attach to epithelial cells. It also has a type of lipopolysaccharide endotoxin called lipooligosaccharide as part of the outer membrane structure that enhances its pathogenicity. In addition to causing urethritis, *N. gonorrhoeae* can infect other body tissues such as the skin, meninges, pharynx, and conjunctiva.

Many infected individuals (both men and women) are asymptomatic carriers of gonorrhea. When symptoms do occur, they manifest differently in males and females. Males may develop pain and burning during urination and discharge from the penis that may be yellow, green, or white (Figure 23.10). Less commonly, the testicles may become swollen or tender. Over time, these symptoms can increase and spread. In some cases, chronic infection develops. The disease can also develop in the rectum, causing symptoms such as discharge, soreness, bleeding, itching, and pain (especially in association with bowel movements).

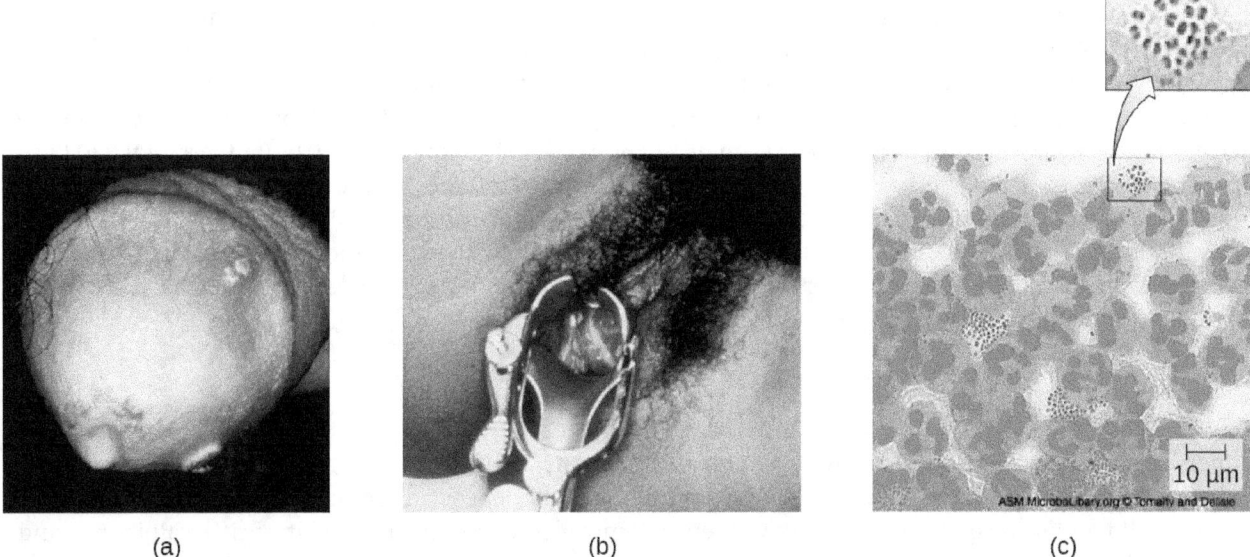

Figure 23.10 (a) Clinical photograph of gonococcal discharge from penis. The lesions on the skin could indicate co-infection with another STI. (b) Purulent discharge originating from the cervix and accumulating in the vagina of a patient with gonorrhea. (c) A micrograph of urethral discharge shows gram-negative diplococci (paired cells) both inside and outside the leukocytes (large cells with lobed nuclei). These results could be used to diagnose gonorrhea in a male patient, but female vaginal samples may contain other *Neisseria* spp. even if the patient is not infected with *N. gonorrhoeae*. (credit a, b: modification of work by Centers for Disease Control and Prevention; credit c: modification of work by American Society for Microbiology)

Women may develop pelvic pain, discharge from the vagina, intermenstrual bleeding (i.e., bleeding not associated with normal menstruation), and pain or irritation associated with urination. As with men, the infection can become chronic. In women, however, chronic infection can cause increases in menstrual flow. Rectal infection can also occur, with the symptoms previously described for men. Infections that spread to the endometrium and fallopian tubes can cause pelvic inflammatory disease (PID), characterized by pain in the lower abdominal region, dysuria, vaginal discharge, and fever. PID can also lead to infertility through scarring and blockage of the fallopian tubes (salpingitis); it may also increase the risk of a life-threatening ectopic pregnancy, which occurs when a fertilized egg begins developing somewhere other than the uterus (e.g., in the fallopian tube or ovary).

When a gonorrhea infection disseminates throughout the body, serious complications can develop. The infection may spread through the blood (bacteremia) and affect organs throughout the body, including the heart (gonorrheal endocarditis), joints (gonorrheal arthritis), and meninges encasing the brain (meningitis).

Urethritis caused by *N. gonorrhoeae* can be difficult to treat due to antibiotic resistance (see Micro Connections). Some strains have developed resistance to the fluoroquinolones, so cephalosporins are often a first choice for treatment. Because co-infection with *C. trachomatis* is common, the CDC recommends treating with a combination regimen of ceftriaxone and azithromycin. Treatment of sexual partners is also recommended to avoid reinfection and spread of infection to others.[5]

CHECK YOUR UNDERSTANDING

- What are some of the serious consequences of a gonorrhea infection?
- What organism commonly coinfects with *N. gonorrhoeae*?

MICRO CONNECTIONS

Antibiotic Resistance in *Neisseria*

Antibiotic resistance in many pathogens is steadily increasing, causing serious concern throughout the public health community. Increased resistance has been especially notable in some species, such as *Neisseria gonorrhoeae*. The CDC monitors the spread of antibiotic resistance in *N. gonorrhoeae*, which it classifies as an urgent threat, and makes recommendations for treatment. So far, *N. gonorrhoeae* has shown resistance to cefixime (a cephalosporin), ceftriaxone (another cephalosporin), azithromycin, and tetracycline. Resistance to tetracycline is the most common, and was seen in 188,600 cases of gonorrhea in 2011 (out of a total 820,000 cases). In 2011, some 246,000 cases of gonorrhea involved strains of *N. gonorrhoeae* that were resistant to at least one antibiotic.[6] These resistance genes are spread by plasmids, and a single bacterium may be resistant to multiple antibiotics. The CDC currently recommends treatment with two medications, ceftriaxone and azithromycin, to attempt to slow the spread of resistance. If resistance to cephalosporins increases, it will be extremely difficult to control the spread of *N. gonorrhoeae*.

Chlamydia

Chlamydia trachomatis is the causative agent of the STI **chlamydia** (Figure 23.11). While many *Chlamydia* infections are asymptomatic, chlamydia is a major cause of nongonococcal urethritis (NGU) and may also cause epididymitis and orchitis in men. In women, chlamydia infections can cause urethritis, salpingitis, and PID. In addition, chlamydial infections may be associated with an increased risk of cervical cancer.

Because chlamydia is widespread, often asymptomatic, and has the potential to cause substantial complications, routine screening is recommended for sexually active women who are under age 25, at high risk (i.e., not in a monogamous relationship), or beginning prenatal care.

Certain serovars of *C. trachomatis* can cause an infection of the lymphatic system in the groin known as **lymphogranuloma venereum**. This condition is commonly found in tropical regions and can also co-occur in conjunction with human immunodeficiency virus (HIV) infection. After the microbes invade the lymphatic system, buboes (large lymph nodes, see Figure 23.11) form and can burst, releasing pus through the skin. The male genitals can become greatly enlarged and in women the rectum may become narrow.

Urogenital infections caused by *C. trachomatis* can be treated using azithromycin or doxycycline (the recommended regimen from the CDC). Erythromycin, levofloxacin, and ofloxacin are alternatives.

---

5 Centers for Disease Control and Prevention. "2015 Sexually Transmitted Diseases Treatment Guidelines: Gonococcal Infections," 2015. http://www.cdc.gov/std/tg2015/gonorrhea.htm.

6 Centers for Disease Control and Prevention. "Antibiotic Resistance Threats in the United States, 2013," 2013. http://www.cdc.gov/drugresistance/pdf/ar-threats-2013-508.pdf.

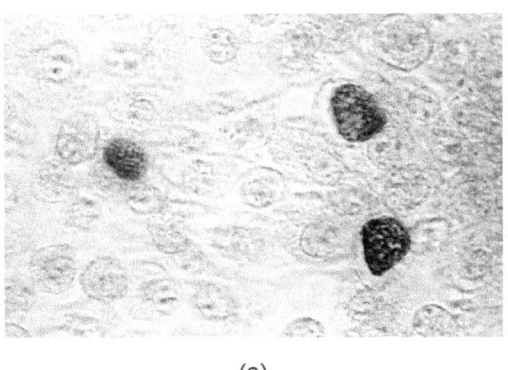 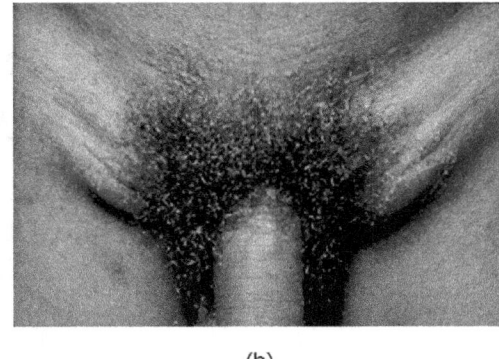

(a) (b)

Figure 23.11 (a) *Chlamydia trachomatis* inclusion bodies within McCoy cell monolayers. Inclusion bodies are distinguished by their brown color. (b) Lymphogranuloma venereum infection can cause swollen lymph nodes in the groin called buboes. (credit a: modification of work by Centers for Disease Control and Prevention; credit b: modification of work by Herbert L. Fred and Hendrik A. van Dijk)

CHECK YOUR UNDERSTANDING

- Compare the signs and symptoms of chlamydia infection in men and women.

Syphilis

Syphilis is spread through direct physical (generally sexual) contact, and is caused by the gram-negative spirochete *Treponema pallidum*. *T. pallidum* has a relatively simple genome and lacks lipopolysaccharide endotoxin characteristic of gram-negative bacteria. However, it does contain lipoproteins that trigger an immune response in the host, causing tissue damage that may enhance the pathogen's ability to disseminate while evading the host immune system.

After entering the body, *T. pallidum* moves rapidly into the bloodstream and other tissues. If not treated effectively, syphilis progresses through three distinct stages: primary, secondary, and tertiary. Primary syphilis appears as a single lesion on the cervix, penis, or anus within 10 to 90 days of transmission. Such lesions contain many *T. pallidum* cells and are highly infectious. The lesion, called a **hard chancre**, is initially hard and painless, but it soon develops into an ulcerated sore (Figure 23.12). Localized lymph node swelling may occur as well. In some cases, these symptoms may be relatively mild, and the lesion may heal on its own within two to six weeks. Because the lesions are painless and often occur in hidden locations (e.g., the cervix or anus), infected individuals sometimes do not notice them.

The secondary stage generally develops once the primary chancre has healed or begun to heal. Secondary syphilis is characterized by a rash that affects the skin and mucous membranes of the mouth, vagina, or anus. The rash often begins on the palms or the soles of the feet and spreads to the trunk and the limbs (Figure 23.12). The rash may take many forms, such as macular or papular. On mucous membranes, it may manifest as mucus patches or white, wartlike lesions called condylomata lata. The rash may be accompanied by malaise, fever, and swelling of lymph nodes. Individuals are highly contagious in the secondary stage, which lasts two to six weeks and is recurrent in about 25% of cases.

After the secondary phase, syphilis can enter a latent phase, in which there are no symptoms but microbial levels remain high. Blood tests can still detect the disease during latency. The latent phase can persist for years.

Tertiary syphilis, which may occur 10 to 20 years after infection, produces the most severe symptoms and can be fatal. Granulomatous lesions called **gummas** may develop in a variety of locations, including mucous membranes, bones, and internal organs (Figure 23.12). Gummas can be large and destructive, potentially causing massive tissue damage. The most deadly lesions are those of the cardiovascular system (cardiovascular syphilis) and the central nervous system (neurosyphilis). Cardiovascular syphilis can result in a fatal aortic aneurysm (rupture of the aorta) or coronary stenosis (a blockage of the coronary artery). Damage to the central nervous system can cause dementia, personality changes, seizures, general paralysis, speech impairment, loss of vision and hearing, and loss of bowel and bladder control.

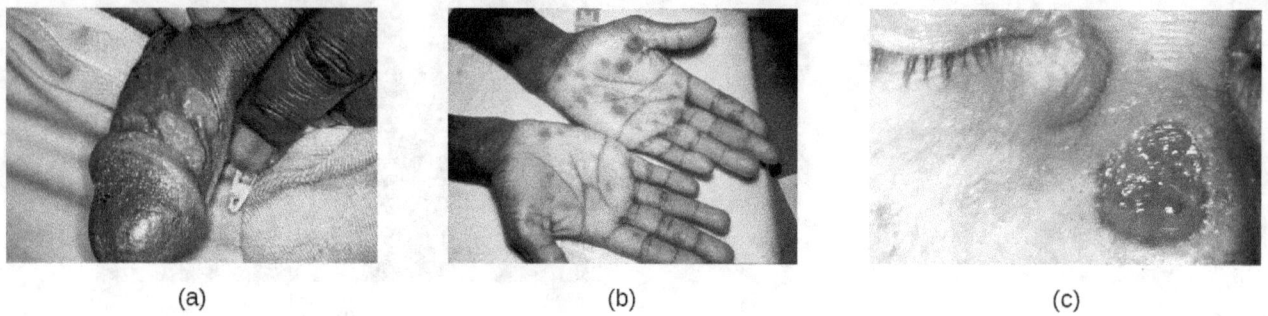

Figure 23.12 (a) This ulcerated sore is a hard chancre caused by syphilis. (b) This individual has a secondary syphilis rash on the hands. (c) Tertiary syphilis produces lesions called gummas, such as this one located on the nose. (credit a, b, c: modification of work by Centers for Disease Control and Prevention)

The recommended methods for diagnosing early syphilis are darkfield or brightfield (silver stain) microscopy of tissue or exudate from lesions to detect *T. pallidum* (Figure 23.13). If these methods are not available, two types of serologic tests (treponemal and nontreponemal) can be used for a presumptive diagnosis once the spirochete has spread in the body. **Nontreponemal serologic tests** include the Venereal Disease Research Laboratory (VDRL) and rapid plasma reagin (RPR) tests. These are similar screening tests that detect nonspecific antibodies (those for lipid antigens produced during infection) rather than those produced against the spirochete. **Treponemal serologic tests** measure antibodies directed against *T. pallidum* antigens using particle agglutination (*T. pallidum* passive particle agglutination or TP-PA), immunofluorescence (the fluorescent *T. pallidum* antibody absorption or FTA-ABS), various enzyme reactions (enzyme immunoassays or EIAs) and chemiluminescence immunoassays (CIA). Confirmatory testing, rather than screening, must be done using treponemal rather than nontreponemal tests because only the former tests for antibodies to spirochete antigens. Both treponemal and nontreponemal tests should be used (as opposed to just one) since both tests have limitations than can result in false positives or false negatives.

Neurosyphilis cannot be diagnosed using a single test. With or without clinical signs, it is generally necessary to assess a variety of factors, including reactive serologic test results, cerebrospinal fluid cell count abnormalities, cerebrospinal fluid protein abnormalities, or reactive VDRL-CSF (the VDRL test of cerebrospinal fluid). The VDRL-CSF is highly specific, but not sufficiently sensitive for conclusive diagnosis.

The recommended treatment for syphilis is parenteral penicillin G (especially long-acting benzathine penicillin, although the exact choice depends on the stage of disease). Other options include tetracycline and doxycycline.

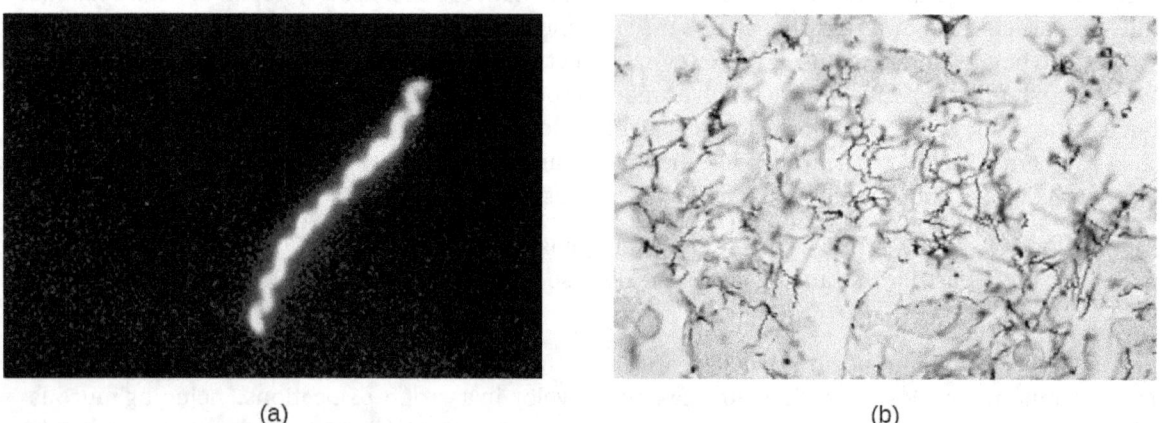

Figure 23.13 (a) Darkfield micrograph of *Treponema pallidum*. (b) Silver stain micrograph of the same species. (credit a, b: modification of work by Centers for Disease Control and Prevention)

Congenital Syphilis

Congenital syphilis is passed by mother to fetus when untreated primary or secondary syphilis is present. In many cases, infection may lead to miscarriage or stillbirth. Children born with congenital syphilis show symptoms of secondary syphilis and may develop mucus patches that deform the nose. In infants, gummas can cause significant tissue damage to organs and teeth. Many other complications may develop, such as osteochondritis, anemia, blindness, bone deformations, neurosyphilis, and cardiovascular lesions. Because congenital syphilis poses such a risk to the fetus, expectant mothers are screened for syphilis infection during the first trimester of pregnancy as part of the TORCH panel of prenatal tests.

CHECK YOUR UNDERSTANDING

- What aspect of tertiary syphilis can lead to death?
- How do treponemal serologic tests detect an infection?

Chancroid

The sexually transmitted infection **chancroid** is caused by the gram-negative rod *Haemophilus ducreyi*. It is characterized by **soft chancres** (Figure 23.14) on the genitals or other areas associated with sexual contact, such as the mouth and anus. Unlike the hard chancres associated with syphilis, soft chancres develop into painful, open sores that may bleed or produce fluid that is highly contagious. In addition to causing chancres, the bacteria can invade the lymph nodes, potentially leading to pus discharge through the skin from lymph nodes in the groin. Like other genital lesions, soft chancres are of particular concern because they compromise the protective barriers of the skin or mucous membranes, making individuals more susceptible to HIV and other sexually transmitted diseases.

Several virulence factors have been associated with *H. ducreyi*, including lipooligosaccharides, protective outer membrane proteins, antiphagocytic proteins, secretory proteins, and collagen-specific adhesin NcaA. The collagen-specific adhesion NcaA plays an important role in initial cellular attachment and colonization. Outer membrane proteins DsrA and DltA have been shown to provide protection from serum-mediated killing by antibodies and complement.

H. ducreyi is difficult to culture; thus, diagnosis is generally based on clinical observation of genital ulcers and tests that rule out other diseases with similar ulcers, such as syphilis and genital herpes. PCR tests for *H. ducreyi* have been developed in some laboratories, but as of 2015 none had been cleared by the US Food and Drug Administration (FDA).[7] Recommended treatments for chancroid include antibiotics such as azithromycin, ciprofloxacin, erythromycin and ceftriaxone. Resistance to ciprofloxacin and erythromycin has been reported.[8]

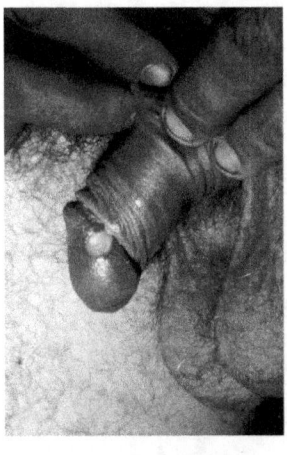
(a)

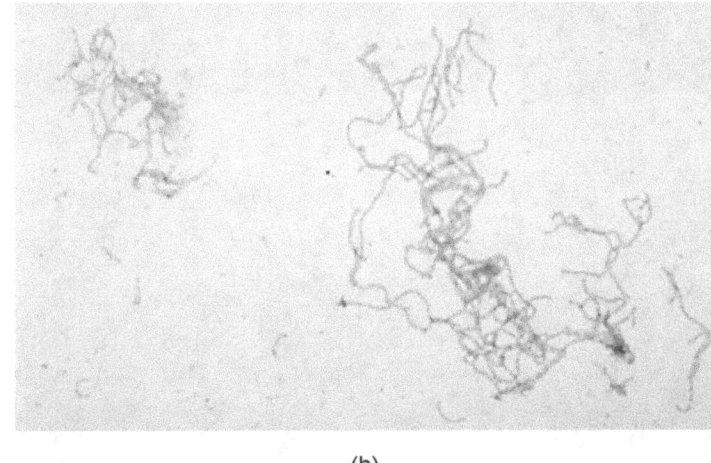
(b)

Figure 23.14 (a) A soft chancre on the penis of a man with chancroid. (b) Chancroid is caused by the gram-negative bacterium

---

7 Centers for Disease Control and Prevention. "2015 Sexually Transmitted Diseases Treatment Guidelines: Chancroid," 2015. http://www.cdc.gov/std/tg2015/chancroid.htm.

8 Ibid.

Haemophilus ducreyi, seen here in a gram-stained culture of rabbit blood. (credit a, b: modification of work by Centers for Disease Control and Prevention)

✓ CHECK YOUR UNDERSTANDING

- What is the key difference between chancroid lesions and those associated with syphilis?
- Why is it difficult to definitively diagnose chancroid?

Disease Profile

Bacterial Reproductive Tract Infections

Many bacterial infections affecting the reproductive system are transmitted through sexual contact, but some can be transmitted by other means. In the United States, gonorrhea and chlamydia are common illnesses with incidences of about 350,000 and 1.44 million, respectively, in 2014. Syphilis is a rarer disease with an incidence of 20,000 in 2014. Chancroid is exceedingly rare in the United States with only six cases in 2014 and a median of 10 cases per year for the years 2010–2014.[9] Figure 23.15 summarizes bacterial infections of the reproductive tract.

---

9 Centers for Disease Control and Prevention. "2014 Sexually Transmitted Disease Surveillance," 2015. http://www.cdc.gov/std/stats14/default.htm.

Bacterial Infections of the Reproductive Tract					
Disease	Pathogen	Signs and Symptoms	Transmission	Diagnostic Tests	Antimicrobial Drugs
Bacterial vaginosis (BV)	*Gardnerella vaginalis*, *Bacteroides* spp., *Fusobacterium* spp., others	Often asymptomatic; vaginal discharge, burning, odor, or itching	Opportunistic infection caused by imbalance of normal vaginal microbiota	Vaginal smear	Clindamycin, metronidazole, tinidazole
Chancroid	*Haemophilus ducreyi*	Soft, painful chancres on genitals, mouth, or anus; swollen lymph nodes; pus discharge	Sexual contact or contact with open lesions or discharge	Observation of clinical symptoms and negative tests for syphilis and herpes	Azithromycin, ceftriaxone, erythromycin, ciprofloxacin
Chlamydia	*Chlamydia trachomatis*	Often asymptomatic; in men, urethritis, epididymitis, orchitis; in women, urethritis, vaginal discharge or bleeding, pelvic inflammatory disease, salpingitis, increased risk of cervical cancer	Sexual contact or from mother to neonate during birth	NAAT, urine sample, vaginal swab, culture	Azithromycin, doxycycline, erythromycin, ofloxacin, or levofloxacin
Gonorrhea	*Neisseria gonorrhoeae*	Urethritis, dysuria, penile or vaginal discharge, rectal pain and bleeding; in females, pelvic pain, intermenstrual bleeding, pelvic inflammatory disease, salpingitis, increased risk of infertility or ectopic pregnancy; in disseminated infections, arthritis, endocarditis, meningitis	Sexual contact	Urine sample or culture, NAAT, PCR, ELISA	Ceftriaxone, azithromycin
Syphilis	*Treponema pallidum*	Primary: hard chancre; Secondary: rash, cutaneous lesions, condylomata, malaise, fever, swollen lymph nodes; Tertiary: gummas, cardiovascular syphilis, neurosyphilis, possibly fatal	Sexual contact or from mother to neonate during birth	Darkfield or brightfield silver stain examination of lesion tissue or exudate, treponemal and non-treponemal serological testing, VDRL-CSF for neurosyphilis, prenatal TORCH panel	Penicillin G, tetracycline, doxycycline

Figure 23.15

23.4 Viral Infections of the Reproductive System

Learning Objectives

By the end of this section, you will be able to:
- Identify the most common viruses that cause infections of the reproductive system
- Compare the major characteristics of specific viral diseases affecting the reproductive system

Several viruses can cause serious problems for the human reproductive system. Most of these viral infections are incurable, increasing the risk of persistent sexual transmission. In addition, such viral infections are very common in the United States. For example, human papillomavirus (HPV) is the most common STI in the country, with an estimated prevalence of 79.1 million infections in 2008; herpes simplex virus 2 (HSV-2) is the next most prevalent STI at 24.1 million infections.[10] In this section, we will examine these and other major viral infections of the reproductive system.

Genital Herpes

Genital herpes is a common condition caused by the herpes simplex virus (Figure 23.16), an enveloped, double-stranded DNA virus that is classified into two distinct types. Herpes simplex virus has several virulence factors, including infected cell protein (ICP) 34.5, which helps in replication and inhibits the maturation of dendritic cells as a mechanism of avoiding elimination by the immune system. In addition, surface glycoproteins on the viral envelope promote the coating of herpes simplex virus with antibodies and complement factors, allowing the virus to appear as "self" and prevent immune system activation and elimination.

There are two herpes simplex virus types. While herpes simplex virus type 1 (HSV-1) is generally associated with oral lesions like cold sores or fever blisters (see Viral Infections of the Skin and Eyes), **herpes simplex virus type 2 (HSV-2)** is usually associated with genital herpes. However, both viruses can infect either location as well as other parts of the body. Oral-genital contact can spread either virus from the mouth to the genital region or vice versa.

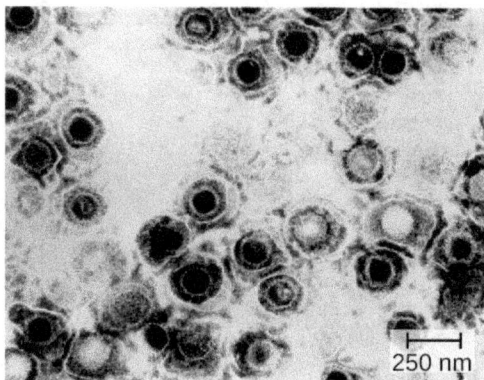

Figure 23.16 Virions of the herpes simplex virus are shown here in this transmission electron micrograph. (credit: modification of work by Centers for Disease Control and Prevention)

Many infected individuals do not develop symptoms, and thus do not realize that they carry the virus. However, in some infected individuals, fever, chills, malaise, swollen lymph nodes, and pain precede the development of fluid-filled vesicles that may be irritating and uncomfortable. When these vesicles burst, they release infectious fluid and allow transmission of HSV. In addition, open herpes lesions can increase the risk of spreading or acquiring HIV.

In men, the herpes lesions typically develop on the penis and may be accompanied by a watery discharge. In women, the vesicles develop most commonly on the vulva, but may also develop on the vagina or cervix (Figure 23.17). The symptoms are typically mild, although the lesions may be irritating or accompanied by urinary discomfort. Use of condoms may not always be an effective means of preventing transmission of genital herpes since the lesions can occur on areas other than the genitals.

---

10 Catherine Lindsey Satterwhite, Elizabeth Torrone, Elissa Meites, Eileen F. Dunne, Reena Mahajan, M. Cheryl Bañez Ocfemia, John Su, Fujie Xu, and Hillard Weinstock. "Sexually Transmitted Infections Among US Women and Men: Prevalence and Incidence Estimates, 2008." *Sexually Transmitted Diseases* 40, no. 3 (2013): 187–193.

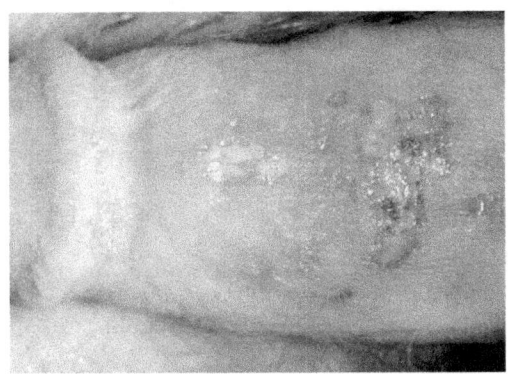 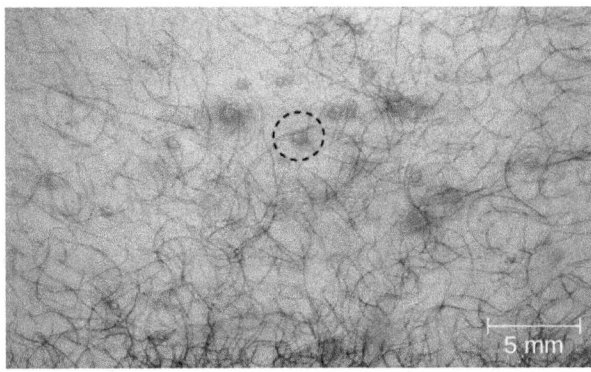

Figure 23.17 Genital herpes is typically characterized by lesions on the genitals (left), but lesions can also appear elsewhere on the skin or mucous membranes (right). The lesions can be large and painful or small and easily overlooked. (credit b: modification of work by Schiffer JT, Swan D, Al Sallaq R, Magaret A, Johnston C, Mark KE, Selke S, Ocbamichael N, Kuntz S, Zhu J, Robinson B, Huang ML, Jerome KR, Wald A, and Corey)

Herpes simplex viruses can cause recurrent infections because the virus can become latent and then be reactivated. This occurs more commonly with HSV-2 than with HSV-1.[11] The virus moves down peripheral nerves, typically sensory neurons, to ganglia in the spine (either the trigeminal ganglion or the lumbar-sacral ganglia) and becomes latent. Reactivation can later occur, causing the formation of new vesicles. HSV-2 most effectively reactivates from the lumbar-sacral ganglia. Not everyone infected with HSV-2 experiences reactivations, which are typically associated with stressful conditions, and the frequency of reactivation varies throughout life and among individuals. Between outbreaks or when there are no obvious vesicles, the virus can still be transmitted.

Virologic and serologic techniques are used for diagnosis. The virus may be cultured from lesions. The immunostaining methods that are used to detect virus from cultures generally require less expertise than methods based on cytopathic effect (CPE), as well as being a less expensive option. However, PCR or other DNA amplification methods may be preferred because they provide the most rapid results without waiting for culture amplification. PCR is also best for detecting systemic infections. Serologic techniques are also useful in some circumstances, such as when symptoms persist but PCR testing is negative.

While there is no cure or vaccine for HSV-2 infections, antiviral medications are available that manage the infection by keeping the virus in its dormant or latent phase, reducing signs and symptoms. If the medication is discontinued, then the condition returns to its original severity. The recommended medications, which may be taken at the start of an outbreak or daily as a method of prophylaxis, are acyclovir, famciclovir, and valacyclovir.

Neonatal Herpes

Herpes infections in newborns, referred to as **neonatal herpes**, are generally transmitted from the mother to the neonate during childbirth, when the child is exposed to pathogens in the birth canal. Infections can occur regardless of whether lesions are present in the birth canal. In most cases, the infection of the newborn is limited to skin, mucous membranes, and eyes, and outcomes are good. However, sometimes the virus becomes disseminated and spreads to the central nervous system, resulting in motor function deficits or death.

In some cases, infections can occur before birth when the virus crosses the placenta. This can cause serious complications in fetal development and may result in spontaneous abortion or severe disabilities if the fetus survives. The condition is most serious when the mother is infected with HSV for the first time during pregnancy. Thus, expectant mothers are screened for HSV infection during the first trimester of pregnancy as part of the TORCH panel of prenatal tests (see How Pathogens Cause Disease). Systemic acyclovir treatment is recommended to treat newborns with neonatal herpes.

---

11 Centers for Disease Control and Prevention. "2015 Sexually Transmitted Disease Treatment Guidelines: Genital Herpes," 2015. http://www.cdc.gov/std/tg2015/herpes.htm.

CHECK YOUR UNDERSTANDING

- Why are latent herpes virus infections still of clinical concern?
- How is neonatal herpes contracted?

Human Papillomas

Warts of all types are caused by a variety of strains of **human papillomavirus (HPV)** (see Viral Infections of the Skin and Eyes). Condylomata acuminata, more commonly called **genital warts** or venereal warts (Figure 23.18), are an extremely prevalent STI caused by certain strains of HPV. Condylomata are irregular, soft, pink growths that are found on external genitalia or the anus.

HPV is a small, non-enveloped virus with a circular double-stranded DNA genome. Researchers have identified over 200 different strains (called types) of HPV, with approximately 40 causing STIs. While some types of HPV cause genital warts, HPV infection is often asymptomatic and self-limiting. However, genital HPV infection often co-occurs with other STIs like syphilis or gonorrhea. Additionally, some forms of HPV (not the same ones associated with genital warts) are associated with cervical cancers. At least 14 oncogenic (cancer-causing) HPV types are known to have a causal association with cervical cancers. Examples of oncogenic HPV are types 16 and 18, which are associated with 70% of cervical cancers.[12] Oncogenic HPV types can also cause oropharyngeal cancer, anal cancer, vaginal cancer, vulvar cancer, and penile cancer. Most of these cancers are caused by HPV type 16. HPV virulence factors include proteins (E6 and E7) that are capable of inactivating tumor suppressor proteins, leading to uncontrolled cell division and the development of cancer.

HPV cannot be cultured, so molecular tests are the primary method used to detect HPV. While routine HPV screening is not recommended for men, it is included in guidelines for women. An initial screening for HPV at age 30, conducted at the same time as a Pap test, is recommended. If the tests are negative, then further HPV testing is recommended every five years. More frequent testing may be needed in some cases. The protocols used to collect, transport, and store samples vary based on both the type of HPV testing and the purpose of the testing. This should be determined in individual cases in consultation with the laboratory that will perform the testing.

Because HPV testing is often conducted concurrently with Pap testing, the most common approach uses a single sample collection within one vial for both. This approach uses liquid-based cytology (LBC). The samples are then used for Pap smear cytology as well as HPV testing and genotyping. HPV can be recognized in Pap smears by the presence of cells called koilocytes (called koilocytosis or koilocytotic atypia). Koilocytes have a hyperchromatic atypical nucleus that stains darkly and a high ratio of nuclear material to cytoplasm. There is a distinct clear appearance around the nucleus called a perinuclear halo (Figure 23.19).

---

12 Lauren Thaxton and Alan G. Waxman. "Cervical Cancer Prevention: Immunization and Screening 2015." *Medical Clinics of North America* 99, no. 3 (2015): 469–477.

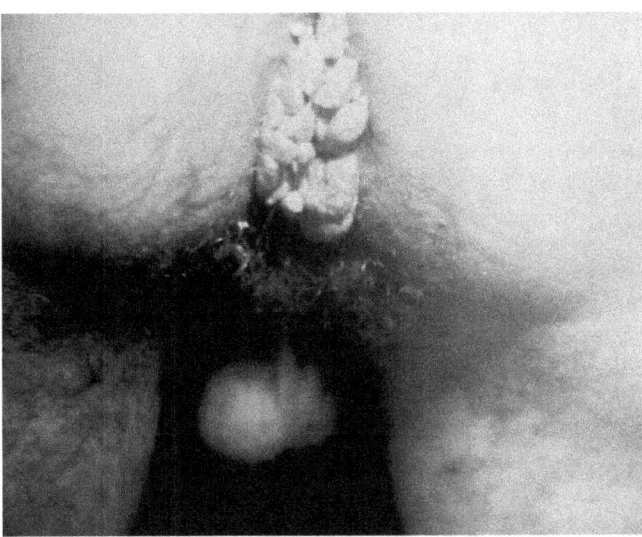
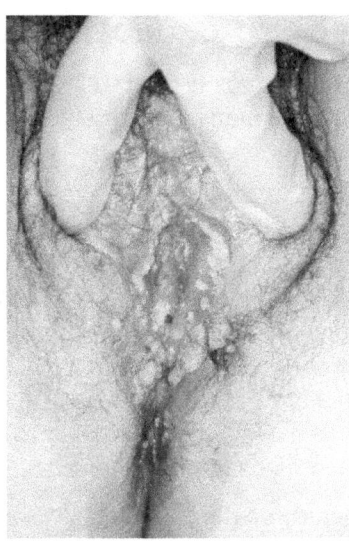

Figure 23.18 Genital warts may occur around the anus (left) or genitalia (right). (credit left, right: modification of work by Centers for Disease Control and Prevention)

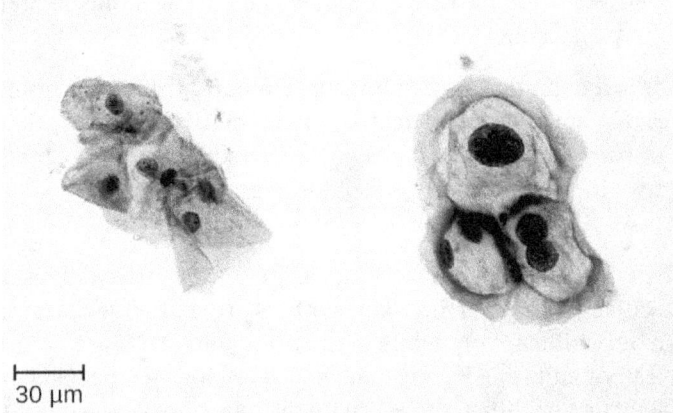

Figure 23.19 In this image, the cervical cells on the left are normal and those on the right show enlarged nuclei and hyperchromasia (darkly stained nuclei) typical of HPV-infected koilocytes. (credit: modification of work by Ed Uthman)

Most HPV infections resolve spontaneously; however, various therapies are used to treat and remove warts. Topical medications such as imiquimod (which stimulates the production of interferon), podofilox, or sinecatechins, may be effective. Warts can also be removed using cryotherapy or surgery, but these approaches are less effective for genital warts than for other types of warts. Electrocauterization and carbon dioxide laser therapy are also used for wart removal.

Regular Pap testing can detect abnormal cells that might progress to cervical cancer, followed by biopsy and appropriate treatment. Vaccines for some of the high risk HPV types are now available. Gardasil vaccine includes types 6, 11, 16 and 18 (types 6 and 11 are associated with 90% of genital wart infections and types 16 and 18 are associated with 70% of cervical cancers). Gardasil 9 vaccinates against the previous four types and an additional five high-risk types (31, 33, 45, 52, and 58). Cervarix vaccine includes just HPV types 16 and 18. Vaccination is the most effective way to prevent infection with oncogenic HPV, but it is important to note that not all oncogenic HPV types are covered by the available vaccines. It is recommended for both boys and girls prior to sexual activity (usually between the ages of nine and fifteen).

LINK TO LEARNING

Watch a video (https://openstax.org/l/22HPVpercep) of how perceptions of HPV affect vaccination rates.

CHECK YOUR UNDERSTANDING

- What is diagnostic of an HPV infection in a Pap smear?
- What is the motivation for HPV vaccination?

MICRO CONNECTIONS

Secret STIs

Few people who have an STI (or think they may have one) are eager to share that information publicly. In fact, many patients are even uncomfortable discussing the symptoms privately with their doctors. Unfortunately, the social stigma associated with STIs makes it harder for infected individuals to seek the treatment they need and creates the false perception that STIs are rare. In reality, STIs are quite common, but it is difficult to determine exactly *how* common.

A recent study on the effects of HPV vaccination found a baseline HPV prevalence of 26.8% for women between the ages of 14 and 59. Among women aged 20–24, the prevalence was 44.8%; in other words, almost half of the women in this age bracket had a current infection.[13] According to the CDC, HSV-2 infection was estimated to have a prevalence of 15.5% in younger individuals (14–49 years of age) in 2007–2010, down from 20.3% in the same age group in 1988–1994. However, the CDC estimates that 87.4% of infected individuals in this age group have not been diagnosed by a physician.[14]

Another complicating factor is that many STIs can be asymptomatic or have long periods of latency. For example, the CDC estimates that among women ages 14–49 in the United States, about 2.3 million (3.1%) are infected with the sexually transmitted protozoan *Trichomonas* (see Protozoan Infections of the Urogenital System); however, in a study of infected women, 85% of those diagnosed with the infection were asymptomatic.[15]

Even when patients are treated for symptomatic STIs, it can be difficult to obtain accurate data on the number of cases. Whereas STIs like chlamydia, gonorrhea, and syphilis are notifiable diseases—meaning each diagnosis must be reported by healthcare providers to the CDC—other STIs are not notifiable (e.g., genital herpes, genital warts, and trichomoniasis). Between the social taboos, the inconsistency of symptoms, and the lack of mandatory reporting, it can be difficult to estimate the true prevalence of STIs—but it is safe to say they are much more prevalent than most people think.

Disease Profile

Viral Reproductive Tract Infections

Figure 23.20 summarizes the most important features of viral diseases affecting the human reproductive tract.

---

13 Eileen F. Dunne, Elizabeth R. Unger, Maya Sternberg, Geraldine McQuillan, David C. Swan, Sonya S. Patel, and Lauri E. Markowitz. "Prevalence of HPV Infection Among Females in the United States." *Journal of the American Medical Association* 297, no. 8 (2007): 813–819.

14 Centers for Disease Control and Prevention. "Genital Herpes - CDC Fact Sheet," 2015. http://www.cdc.gov/std/herpes/stdfact-herpes-detailed.htm.

15 Centers for Disease Control and Prevention. "Trichomoniasis Statistics," 2015. http://www.cdc.gov/std/trichomonas/stats.htm.

Viral Infections of the Reproductive Tract					
Disease	Pathogen	Signs and Symptoms	Transmission	Diagnostic Tests	Antimicrobial Drugs/Vaccines
Cervical cancer	HPV types 16, 18, and others	Development of cancer in cervix (or elsewhere)	Direct contact, including sexual	Pap smear	Gardasil vaccine, Cervarix vaccine
Genital herpes	Herpes simplex virus (HSV-1 or HSV-2)	Recurring outbreaks of skin vesicles on genitalia and elsewhere; asymptomatic in many individuals	Sexual contact or contact with open lesions	Viral culture, PCR, ELISA	Acyclovir, famciclovir, valacyclovir
Human papillomas	Human papillomavirus (HPV) (various strains)	Genital warts or warts in other areas	Direct contact, including sexual	Pap smear	Imiquimod, podofilox, sinecatechins
Neonatal herpes	Herpes simplex virus (HSV-1 or HSV-2)	Vesicles on the skin, mucous membranes, eyes; in disseminated infections, motor impairment and possible death of fetus or newborn	Exposure to pathogens in the birth canal; transplacental infection in some cases	Viral culture or PCR	Acyclovir

Figure 23.20

23.5 Fungal Infections of the Reproductive System

Learning Objectives

By the end of this section, you will be able to:
- Summarize the important characteristics of vaginal candidiasis

Only one major fungal pathogen affects the urogenital system. *Candida* is a genus of fungi capable of existing in a yeast form or as a multicellular fungus. *Candida* spp. are commonly found in the normal, healthy microbiota of the skin, gastrointestinal tract, respiratory system, and female urogenital tract (Figure 23.21). They can be pathogenic due to their ability to adhere to and invade host cells, form biofilms, secrete hydrolases (e.g., proteases, phospholipases, and lipases) that assist in their spread through tissues, and change their phenotypes to protect themselves from the immune system. However, they typically only cause disease in the female reproductive tract under conditions that compromise the host's defenses. While there are at least 20 *Candida* species of clinical importance, *C. albicans* is the species most commonly responsible for fungal vaginitis.

As discussed earlier, lactobacilli in the vagina inhibit the growth of other organisms, including bacteria and *Candida*, but disruptions can allow *Candida* to increase in numbers. Typical disruptions include antibiotic therapy, illness (especially diabetes), pregnancy, and the presence of transient microbes. Immunosuppression can also play a role, and the severe immunosuppression associated with HIV infection often allows *Candida* to thrive. This can cause genital or vaginal **candidiasis**, a condition characterized by vaginitis and commonly known as a yeast infection. When a yeast infection develops, inflammation occurs along with symptoms of pruritus (itching), a thick white or yellow discharge, and odor.

Other forms of candidiasis include cutaneous candidiasis (see Mycoses of the Skin) and oral thrush (see Microbial Diseases of the Mouth and Oral Cavity). Although *Candida* spp. are found in the normal microbiota, *Candida* spp. may also be transmitted between individuals. Sexual contact is a common mode of transmission, although candidiasis is not considered an STI.

Diagnosis of vaginal candidiasis can be made using microscopic evaluation of vaginal secretions to determine whether there is an excess of *Candida*. Culturing approaches are less useful because *Candida* is part of the normal microbiota and will regularly appear. It is also easy to contaminate samples with *Candida* because it is so common, so care must be taken to handle clinical material appropriately. Samples can be refrigerated if there is a delay in handling. *Candida* is a dimorphic fungus, so it does not only exist in a yeast form; cultivation can be used to identify chlamydospores and pseudohyphae, which develop from germ tubes (Figure 23.22). The presence of the germ tube can be used in a diagnostic test in which cultured yeast cells are combined with rabbit serum and observed after a few hours for the presence of germ tubes. Molecular tests are also available if needed. The Affirm VPII Microbial Identification Test, for instance, tests simultaneously for the vaginal microbes *C. albicans*, *G. vaginalis* (see Bacterial Infections of the Urinary System), and *Trichomonas vaginalis* (see Protozoan Infections of the Urogenital System).

Topical antifungal medications for vaginal candidiasis include butoconazole, miconazole, clotrimazole, tioconazole, and nystatin. Oral treatment with fluconazole can be used. There are often no clear precipitating factors for infection, so prevention is difficult.

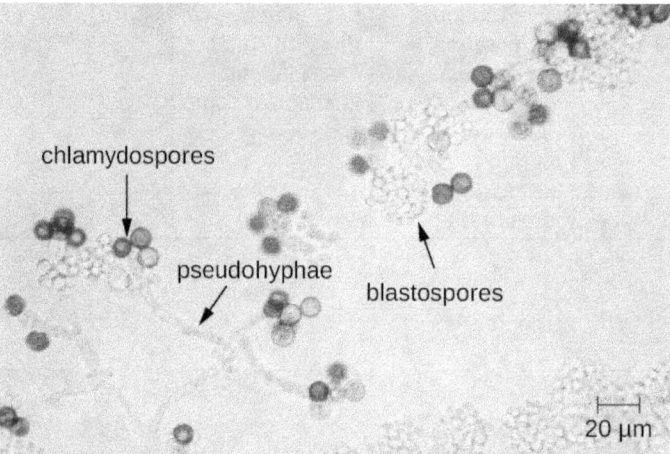

Figure 23.21 *Candida* blastospores (asexual spores that result from budding) and chlamydospores (resting spores produced through asexual reproduction) are visible in this micrograph. (credit: modification of work by Centers for Disease Control and Prevention)

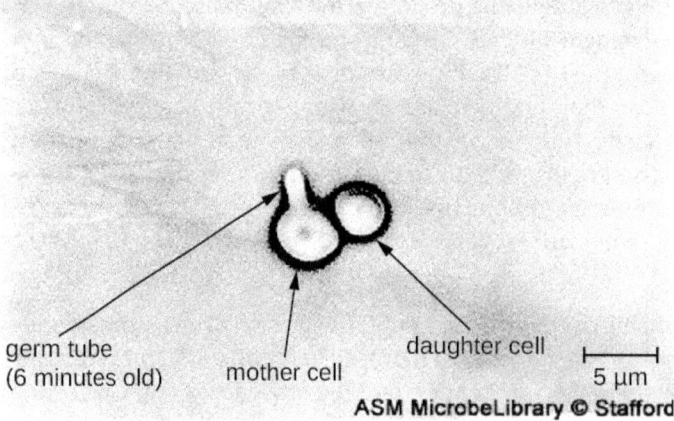

Figure 23.22 *Candida* can produce germ tubes, like the one in this micrograph, that develop into hyphae. (credit: modification of work by American Society for Microbiology)

CHECK YOUR UNDERSTANDING

- What factors can lead to candidiasis?
- How is candidiasis typically diagnosed?

Clinical Focus

Part 3

The Gram stain of Nadia's vaginal smear showed that the concentration of lactobacilli relative to other species in Nadia's vaginal sample was abnormally low. However, there were no clue cells visible, which suggests that the infection is not bacterial vaginosis. But a wet-mount slide showed an overgrowth of yeast cells, suggesting that the problem is candidiasis, or a yeast infection (Figure 23.23). This, Nadia's doctor assures her, is good news. Candidiasis is common during pregnancy and easily treatable.

- Knowing that the problem is candidiasis, what treatments might the doctor suggest?

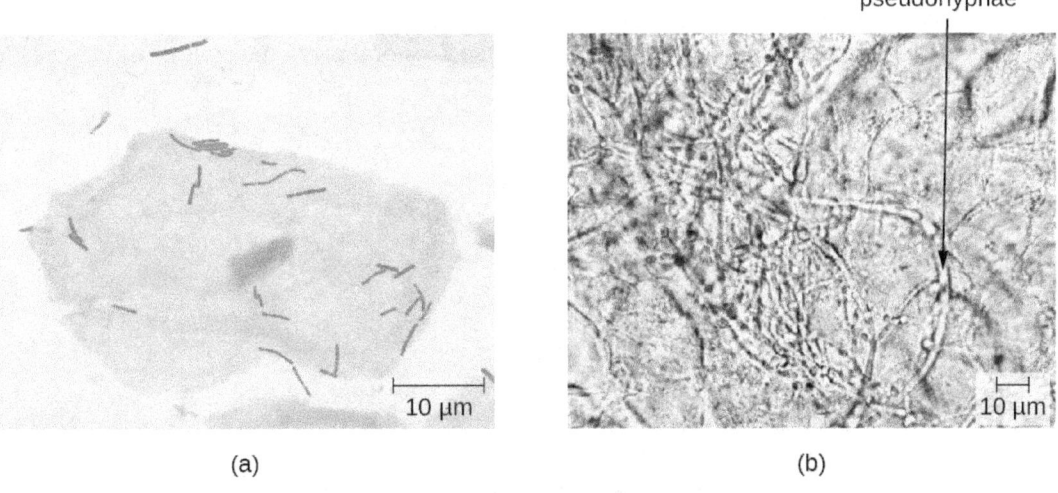

Figure 23.23 (a) Lactobacilli are visible as gram-positive rods on and around this squamous epithelial cell. (b) This wet mount prepared with KOH shows *Candida albicans* pseudohyphae and squamous epithelial cells in a vaginal sample from a patient with candidiasis. (credit a: modification of work by Centers for Disease Control and Prevention; credit b: modification of work by Mikael Häggström)

Jump to the next Clinical Focus box. Go back to the previous Clinical Focus box.

23.6 Protozoan Infections of the Urogenital System

Learning Objectives

By the end of this section, you will be able to:
- Identify the most common protozoan pathogen that causes infections of the reproductive system
- Summarize the important characteristics of trichomoniasis

Only one major protozoan species causes infections in the urogenital system. **Trichomoniasis**, or "trich," is the most common nonviral STI and is caused by a flagellated protozoan *Trichomonas vaginalis*. *T. vaginalis* has an undulating membrane and, generally, an amoeboid shape when attached to cells in the vagina. In culture, it has an oval shape.

T. vaginalis is commonly found in the normal microbiota of the vagina. As with other vaginal pathogens, it can cause vaginitis when there is disruption to the normal microbiota. It is found only as a trophozoite and does not form cysts. *T. vaginalis* can adhere to cells using adhesins such as lipoglycans; it also has other cell-surface virulence factors, including tetraspanins that are involved in cell adhesion, motility, and tissue invasion. In addition, *T. vaginalis* is capable of phagocytosing other microbes of the normal microbiota, contributing to the development of an imbalance that is favorable to infection.

Both men and women can develop trichomoniasis. Men are generally asymptomatic, and although women are more likely to develop symptoms, they are often asymptomatic as well. When symptoms do occur, they are characteristic of urethritis. Men experience itching, irritation, discharge from the penis, and burning after urination or ejaculation. Women experience dysuria; itching, burning, redness, and soreness of the genitalia; and vaginal discharge. The infection may also spread to the cervix. Infection increases the risk of transmitting or acquiring HIV and is associated with pregnancy complications such as preterm birth.

Microscopic evaluation of wet mounts is an inexpensive and convenient method of diagnosis, but the sensitivity of this method is low (Figure 23.24). Nucleic acid amplification testing (NAAT) is preferred due to its high sensitivity. Using wet mounts and then NAAT for those who initially test negative is one option to improve sensitivity. Samples may be obtained for NAAT using urine, vaginal, or endocervical specimens for women and with urine and urethral swabs for men. It is also possible to use other methods such as the OSOM *Trichomonas* Rapid Test (an immunochromatographic test that detects antigen) and a DNA probe test for multiple species associated with vaginitis (the Affirm VPII Microbial Identification Test discussed in section 23.5).[16] *T. vaginalis* is sometimes detected on a Pap test, but this is not considered diagnostic due to high rates of false positives and negatives. The recommended treatment for trichomoniasis is oral metronidazole or tinidazole. Sexual partners should be treated as well.

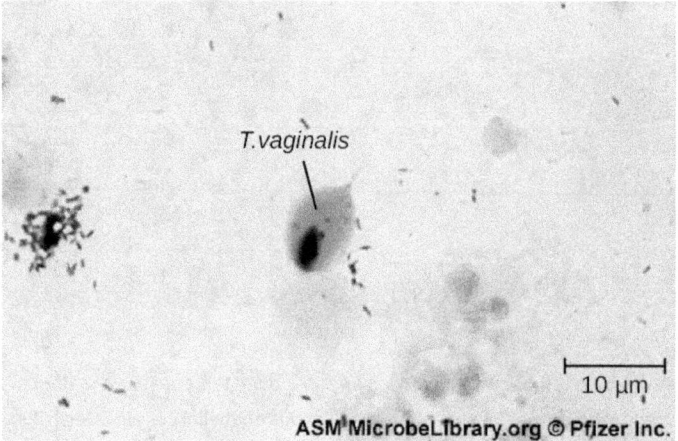

Figure 23.24 *Trichomonas vaginalis* is visible in this Gram stained specimen. (credit: modification of work by American Society for Microbiology)

CHECK YOUR UNDERSTANDING

- What are the symptoms of trichomoniasis?

Eye on Ethics

STIs and Privacy

For many STIs, it is common to contact and treat sexual partners of the patient. This is especially important when a new illness has appeared, as when HIV became more prevalent in the 1980s. But to contact sexual partners, it is necessary to obtain their personal information from the patient. This raises difficult questions. In some cases, providing the information may be embarrassing or difficult for the patient, even though withholding such information could put their sexual partner(s) at risk.

---

16 Association of Public Health Laboratories. "Advances in Laboratory Detection of *Trichomonas vaginalis*," 2013. http://www.aphl.org/AboutAPHL/publications/Documents/ID_2013August_Advances-in-Laboratory-Detection-of-Trichomonas-vaginalis.pdf.

Legal considerations further complicate such situations. The Health Insurance Portability and Accountability Act (HIPPA), passed into law in 1996, sets the standards for the protection of patient information. It requires businesses that use health information, such as insurance companies and healthcare providers, to maintain strict confidentiality of patient records. Contacting a patient's sexual partners may therefore violate the patient's privacy rights if the patient's diagnosis is revealed as a result.

From an ethical standpoint, which is more important: the patient's privacy rights or the sexual partner's right to know that they may be at risk of a sexually transmitted disease? Does the answer depend on the severity of the disease or are the rules universal? Suppose the physician knows the identity of the sexual partner but the patient does not want that individual to be contacted. Would it be a violation of HIPPA rules to contact the individual without the patient's consent?

Questions related to patient privacy become even more complicated when dealing with patients who are minors. Adolescents may be reluctant to discuss their sexual behavior or health with a health professional, especially if they believe that healthcare professionals will tell their parents. This leaves many teens at risk of having an untreated infection or of lacking the information to protect themselves and their partners. On the other hand, parents may feel that they have a right to know what is going on with their child. How should physicians handle this? Should parents always be told even if the adolescent wants confidentiality? Does this affect how the physician should handle notifying a sexual partner?

Clinical Focus

Resolution

Vaginal candidiasis is generally treated using topical antifungal medications such as butoconazole, miconazole, clotrimazole, ticonozole, nystatin, or oral fluconazole. However, it is important to be careful in selecting a treatment for use during pregnancy. Nadia's doctor recommended treatment with topical clotrimazole. This drug is classified as a category B drug by the FDA for use in pregnancy, and there appears to be no evidence of harm, at least in the second or third trimesters of pregnancy. Based on Nadia's particular situation, her doctor thought that it was suitable for very short-term use even though she was still in the first trimester. After a seven-day course of treatment, Nadia's yeast infection cleared. She continued with a normal pregnancy and delivered a healthy baby eight months later.

Higher levels of hormones during pregnancy can shift the typical microbiota composition and balance in the vagina, leading to high rates of infections such as candidiasis or vaginosis. Topical treatment has an 80–90% success rate, with only a small number of cases resulting in recurrent or persistent infections. Longer term or intermittent treatment is usually effective in these cases.

Go back to the _previous_ Clinical Focus box.

Disease Profile

Fungal and Protozoan Reproductive Tract Infections

Figure 23.25 summarizes the most important features of candidiasis and trichomoniasis.

Fungal and Protozoan Infections of the Reproductive Tract					
Disease	Pathogen	Signs and Symptoms	Transmission	Diagnostic Tests	Antimicrobial Drugs
Trichomoniasis	Trichomonas vaginalis	Urethritis, vaginal or penile discharge; redness or soreness of female genitalia	Sexual contact	Wet mounts, NAAT of urine or vaginal samples; OSOM Trichomonas Rapid Test, Affirm VPII Microbial Identification Test	Metronidazole, tinidazole
Vaginal candidiasis (yeast infection)	Candida spp., especially C. albicans	Dysuria; vaginal burning, itching, discharge	Transmissible by sexual contact, but typically only causes opportunistic infections after immunosuppresion or disruption of vaginal microbiota	Culture, Affirm VPII Microbial Identification Test	Fluconazole, miconazole, clotrimazole, tioconazole, nystatin

Figure 23.25

LINK TO LEARNING

Take an online quiz (https://openstax.org/l/22quizstireview) for a review of sexually transmitted infections.

SUMMARY

23.1 Anatomy and Normal Microbiota of the Urogenital Tract

- The urinary system is responsible for filtering the blood, excreting wastes, and helping to regulate electrolyte and water balance.
- The urinary system includes the **kidneys, ureters, urinary bladder**, and **urethra**; the bladder and urethra are the most common sites of infection.
- Common sites of infection in the male reproductive system include the urethra, as well as the testes, **prostate** and **epididymis**.
- The most commons sites of infection in the female reproductive system are the **vulva, vagina**, **cervix**, and **fallopian tubes**.
- Infections of the urogenital tract can occur through colonization from the external environment, alterations in microbiota due to hormonal or other physiological and environmental changes, fecal contamination, and sexual transmission (STIs).

23.2 Bacterial Infections of the Urinary System

- Bacterial **cystitis** is commonly caused by fecal bacteria such as *E. coli*.
- Pyelonephritis is a serious kidney infection that is often caused by bacteria that travel from infections elsewhere in the urinary tract and may cause systemic complications.
- **Leptospirosis** is a bacterial infection of the kidney that can be transmitted by exposure to infected animal urine, especially in contaminated water. It is more common in tropical than in temperate climates.
- **Nongonococcal urethritis (NGU)** is commonly caused by *C. trachomatis, M. genitalium, Ureaplasma urealyticum*, and *M. hominis*.
- Diagnosis and treatment for bacterial urinary tract infections varies. Urinalysis (e.g., for leukocyte esterase levels, nitrite levels, microscopic evaluation, and culture of urine) is an important component in most cases. Broad-spectrum antibiotics are typically used.

23.3 Bacterial Infections of the Reproductive System

- **Bacterial vaginosis** is caused by an imbalance in the vaginal microbiota, with a decrease in lactobacilli and an increase in vaginal pH. *G. vaginalis* is the most common cause of bacterial vaginosis, which is associated with vaginal discharge, odor, burning, and itching.
- **Gonorrhea** is caused by *N. gonorrhoeae*, which can cause infection of the reproductive and urinary tracts and is associated with symptoms of urethritis. If left untreated, it can progress to epididymitis, salpingitis, and pelvic inflammatory disease and enter the bloodstream to infect other sites in the body.
- **Chlamydia** is the most commonly reported STI and is caused by *C. trachomatis*. Most infections are asymptomatic, and infections that are not treated can spread to involve the epididymis of men and cause salpingitis and pelvic inflammatory disease in women.
- **Syphilis** is caused by *T. pallidum* and has three stages, primary, secondary, and tertiary. Primary syphilis is associated with a painless hard chancre lesion on genitalia. Secondary syphilis is associated with skin and mucous membrane lesions. Tertiary syphilis is the most serious and life-threatening, and can involve serious nervous system damage.
- **Chancroid** is an infection of the reproductive tract caused by *H. ducreyi* that results in the development of characteristic **soft chancres**.

23.4 Viral Infections of the Reproductive System

- **Genital herpes** is usually caused by **HSV-2** (although HSV-1 can also be responsible) and may cause the development of infectious, potentially recurrent vesicles
- **Neonatal herpes** can occur in babies born to infected mothers and can cause symptoms that range from relatively mild (more common) to severe.
- **Human papillomaviruses** are the most common sexually transmitted viruses and include strains that cause **genital warts** as well as strains that cause **cervical cancer**.

23.5 Fungal Infections of the Reproductive System

- *Candida* spp. are typically present in the normal microbiota in the body, including the skin, respiratory tract, gastrointestinal tract, and female urogenital system.
- Disruptions in the normal vaginal microbiota can lead to an overgrowth of *Candida*, causing vaginal **candidiasis**.

- Vaginal candidiasis can be treated with topical or oral fungicides. Prevention is difficult.

23.6 Protozoan Infections of the Urogenital System

- **Trichomoniasis** is a common STI caused by *Trichomonas vaginalis*.
- *T. vaginalis* is common at low levels in the normal microbiota.
- Trichomoniasis is often asymptomatic. When symptoms develop, trichomoniasis causes urinary discomfort, irritation, itching, burning, discharge from the penis (in men), and vaginal discharge (in women).
- Trichomoniasis is treated with the antiflagellate drugs tinidazole and metronidazole.

REVIEW QUESTIONS
Multiple Choice

1. When it first leaves the kidney, urine flows through
    a. the urinary bladder.
    b. the urethra.
    c. the ureter.
    d. the glomeruli.

2. What part of the male urogenital tract is shared by the urinary and reproductive systems?
    a. the prostate gland
    b. the seminal vesicles
    c. the vas deferens
    d. the urethra

3. Which species is not associated with NGU?
    a. *Neisseria gonorrhoeae*
    b. *Mycoplasma hominis*
    c. *Chlamydia trachomatis*
    d. *Mycoplasma genitalium*

4. A strain of bacteria associated with a bladder infection shows gram-negative rods. What species is most likely to be the causative agent?
    a. *Mycoplasma hominis*
    b. *Escherichia coli*
    c. *Neisseria gonorrhoeae*
    d. *Chlamydia trachomatis*

5. Treponemal and non-treponemal serological testing can be used to test for
    a. vaginosis.
    b. chlamydia.
    c. syphilis.
    d. gonorrhea.

6. Lymphogranuloma venereum is caused by serovars of
    a. *Neisseria gonorrhoeae*.
    b. *Chlamydia trachomatis*.
    c. *Treponema pallidum*.
    d. *Haemophilus ducreyi*.

7. The latent stage of syphilis, which may last for years, can occur between
    a. the secondary and tertiary stages.
    b. the primary and secondary stages.
    c. initial infection and the primary stage.
    d. any of the three stages.

8. Based on its shape, which microbe is this?

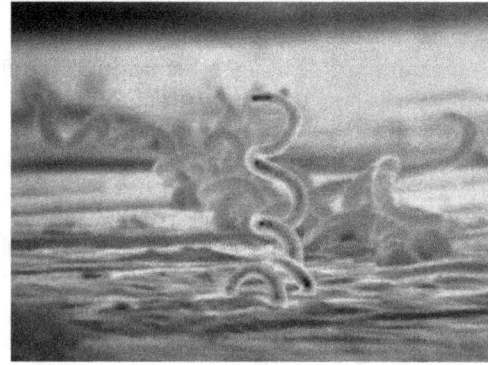

Figure 23.26 (credit: modification of work by Centers for Disease Control and Prevention)

    a. *Neisseria gonorrhoeae*
    b. *Chlamydia trachomatis*
    c. *Treponema pallidum*
    d. *Haemophilus ducreyi*

9. Genital herpes is most commonly caused by
    a. herpes simplex virus 1.
    b. varicella-zoster virus.
    c. herpes simplex virus 2.
    d. cytomegalovirus.

10. Koilocytes are characteristic of
    A. cells infected with human papillomavirus
    B. cells infected with herpes simplex virus 2
    C. cells infected with all forms of herpesviruses
    D. cervical cancer cells

11. Which oral medication is recommended as an initial topical treatment for genital yeast infections?
    a. penicillin
    b. acyclovir
    c. fluconazole
    d. miconazole

12. What is the only common infection of the reproductive tract caused by a protozoan?
    a. gonorrhea
    b. chlamydia
    c. trichomoniasis
    d. candidiasis

13. Which test is preferred for detecting *T. vaginalis* because of its high sensitivity?
    a. NAAT
    b. wet mounts
    c. Pap tests
    d. all of the above are equally good

Fill in the Blank

14. The genus of bacteria found in the vagina that is important in maintaining a healthy environment, including an acidic pH, is _____.
15. Pyelonephritis is a potentially severe infection of the _____.
16. Soft chancres on the genitals are characteristic of the sexually transmitted disease known as _____.
17. Condylomata are _____.
18. The most common *Candida* species associated with yeast infections is _____.
19. Trichomoniasis is caused by _____.

Short Answer

20. When the microbial balance of the vagina is disrupted, leading to overgrowth of resident bacteria without necessarily causing inflammation, the condition is called _____.
21. Explain the difference between a sexually transmitted infection and a sexually transmitted disease.
22. In the figure shown here, where would cystitis occur?

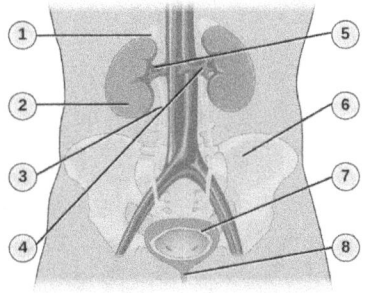

23. What is pyuria?
24. Compare gonococcal and nongonoccocal urethritis with respect to their symptoms and the pathogens that cause each disease.
25. Is it true that human papillomaviruses can always be detected by the presence of genital warts?
26. How is neonatal herpes transmitted?
27. Name three organisms (a bacterium, a fungus, and a protozoan) that are associated with vaginitis.

Critical Thinking

28. Epidemiological data show that the use of antibiotics is often followed by cases of vaginosis or vaginitis in women. Can you explain this finding?
29. What are some factors that would increase an individual's risk of contracting leptospirosis?
30. Chlamydia is often asymptomatic. Why might it be important for an individual to know if he or she were infected?

31. Why does the CDC recommend a two-drug treatment regimen to cover both *C. trachomatis* and *N. gonorrhoeae* if testing to distinguish between the two is not available? Additionally, how does the two-drug treatment regimen address antibiotic resistance?

32. Recently, studies have shown a reduction in the prevalence of some strains of HPV in younger women. What might be the reason for this?

CHAPTER 24
Digestive System Infections

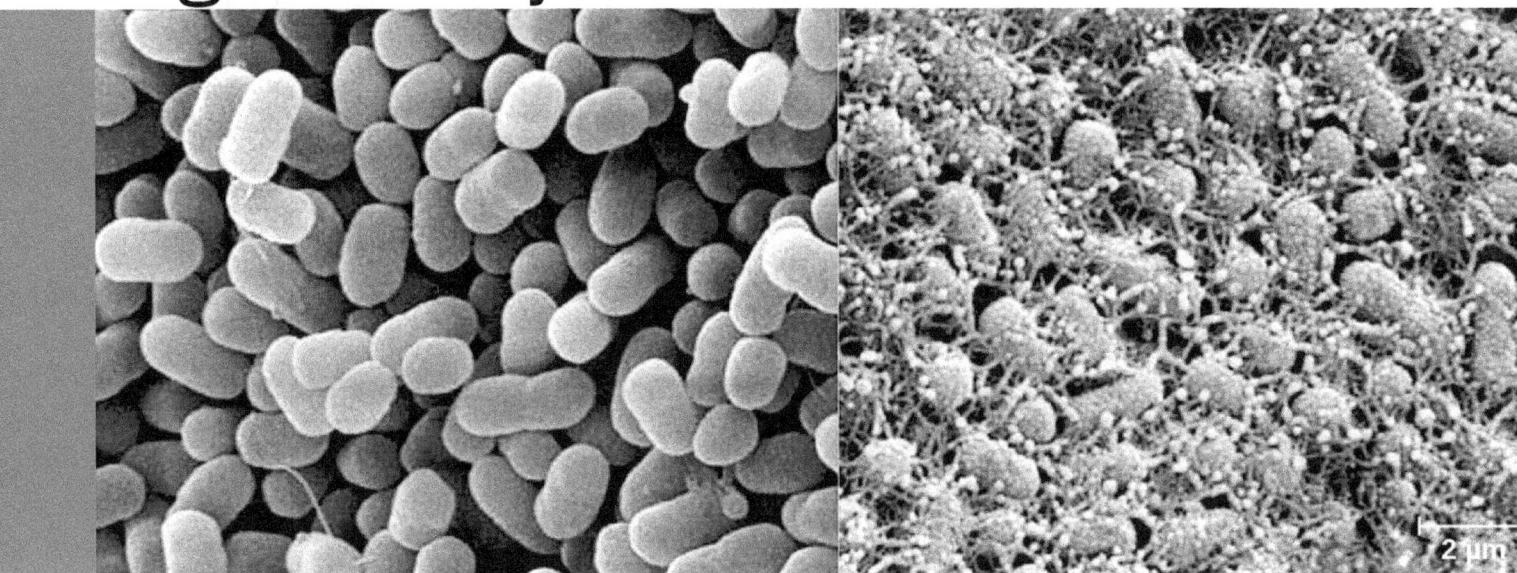

Figure 24.1 *E. coli* O157:H7 causes serious foodborne illness. Curli fibers (adhesive surface fibers that are part of the extracellular matrix) help these bacteria adhere to surfaces and form biofilms. Pictured are two groups of cells, curli non-producing cells (left) and curli producing cells (right). (credit left, right: modification of work by USDA)

Chapter Outline

24.1 Anatomy and Normal Microbiota of the Digestive System

24.2 Microbial Diseases of the Mouth and Oral Cavity

24.3 Bacterial Infections of the Gastrointestinal Tract

24.4 Viral Infections of the Gastrointestinal Tract

24.5 Protozoan Infections of the Gastrointestinal Tract

24.6 Helminthic Infections of the Gastrointestinal Tract

INTRODUCTION Gastrointestinal (GI) diseases are so common that, unfortunately, most people have had first-hand experience with the unpleasant symptoms, such as diarrhea, vomiting, and abdominal discomfort. The causes of gastrointestinal illness can vary widely, but such diseases can be grouped into two categories: those caused by infection (the growth of a pathogen in the GI tract) or intoxication (the presence of a microbial toxin in the GI tract).

Foodborne pathogens like *Escherichia coli* O157:H7 are among the most common sources of gastrointestinal disease. Contaminated food and water have always posed a health risk for humans, but in today's global economy, outbreaks can occur on a much larger scale. *E. coli* O157:H7 is a potentially deadly strain of *E. coli* with a history of contaminating meat and produce that are not properly processed. The source of an *E. coli* O157:H7 outbreak can be difficult to trace, especially if the contaminated food is processed in a foreign country. Once the source is identified, authorities may issue recalls of the contaminated food products, but by then there are typically numerous cases of food poisoning, some of them fatal.

24.1 Anatomy and Normal Microbiota of the Digestive System

Learning Objectives

By the end of this section, you will be able to:
- Describe the major anatomical features of the human digestive system
- Describe the normal microbiota of various regions in the human digestive system
- Explain how microorganisms overcome the defenses of the digestive tract to cause infection or intoxication
- Describe general signs and symptoms associated with infections of the digestive system

---

Clinical Focus

Part 1

After a morning of playing outside, four-year-old Carli ran inside for lunch. After taking a bite of her fried egg, she pushed it away and whined, "It's too slimy, Mommy. I don't want any more." But her mother, in no mood for games, curtly replied that if she wanted to go back outside she had better finish her lunch. Reluctantly, Carli complied, trying hard not to gag as she choked down the runny egg.

That night, Carli woke up feeling nauseated. She cried for her parents and then began to vomit. Her parents tried to comfort her, but she continued to vomit all night and began to have diarrhea and run a fever. By the morning, her parents were very worried. They rushed her to the emergency room.

- What could have caused Carli's signs and symptoms?

Jump to the next Clinical Focus box.

The human digestive system, or the gastrointestinal (GI) tract, begins with the mouth and ends with the anus. The parts of the mouth include the teeth, the gums, the tongue, the oral vestibule (the space between the gums, lips, and teeth), and the oral cavity proper (the space behind the teeth and gums). Other parts of the GI tract are the pharynx, esophagus, stomach, small intestine, large intestine, rectum, and anus (Figure 24.2). Accessory digestive organs include the salivary glands, liver, gallbladder, spleen, and pancreas.

The digestive system contains normal microbiota, including archaea, bacteria, fungi, protists, and even viruses. Because this microbiota is important for normal functioning of the digestive system, alterations to the microbiota by antibiotics or diet can be harmful. Additionally, the introduction of pathogens to the GI tract can cause infections and diseases. In this section, we will review the microbiota found in a healthy digestive tract and the general signs and symptoms associated with oral and GI infections.

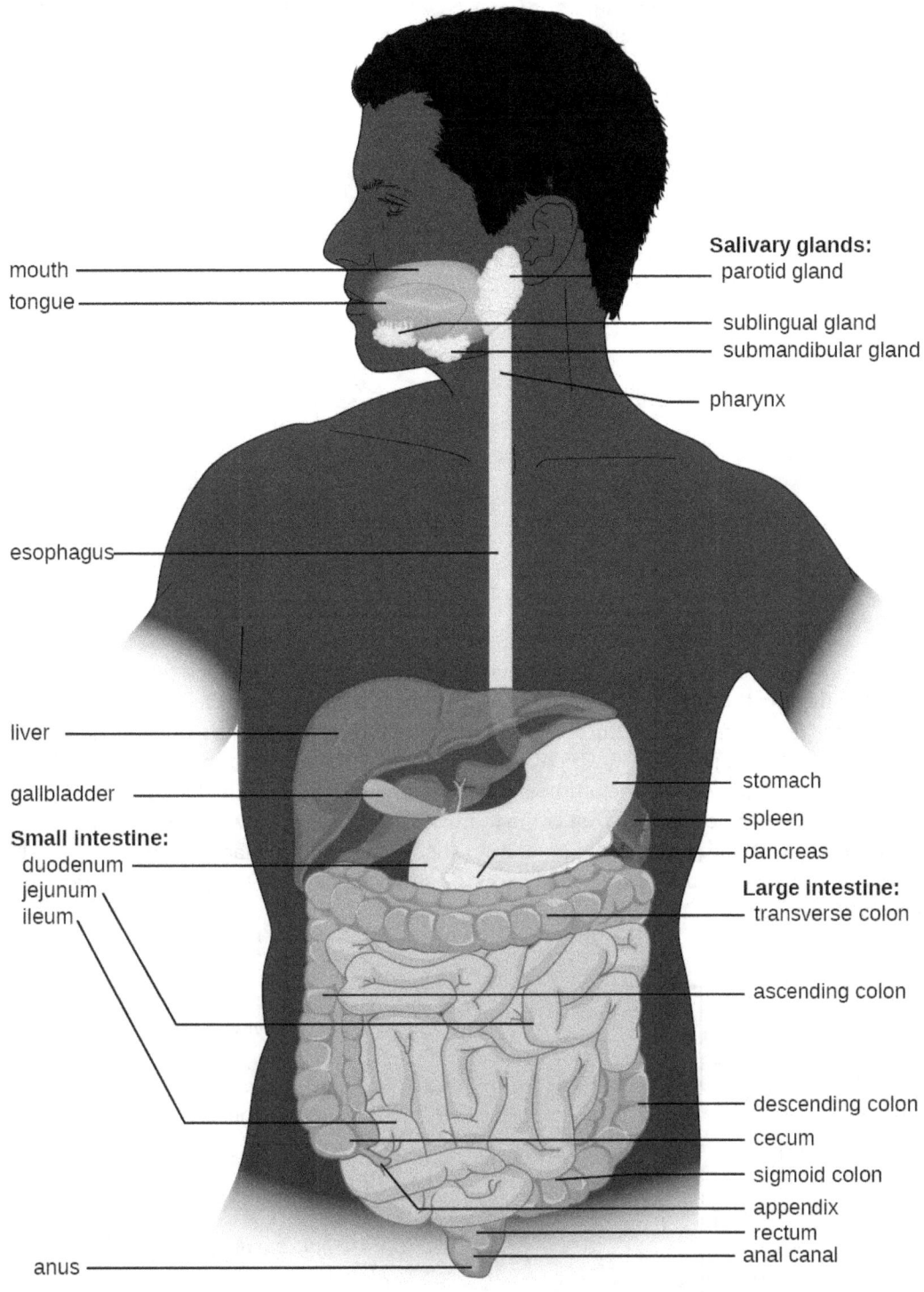

Figure 24.2 The digestive system, or the gastrointestinal tract, includes all of the organs associated with the digestion of food.

Anatomy and Normal Microbiota of the Oral Cavity

Food enters the digestive tract through the mouth, where mechanical digestion (by chewing) and chemical digestion (by enzymes in saliva) begin. Within the mouth are the tongue, teeth, and salivary glands, including the parotid, sublingual, and submandibular glands (Figure 24.3). The salivary glands produce saliva, which lubricates food and contains digestive enzymes.

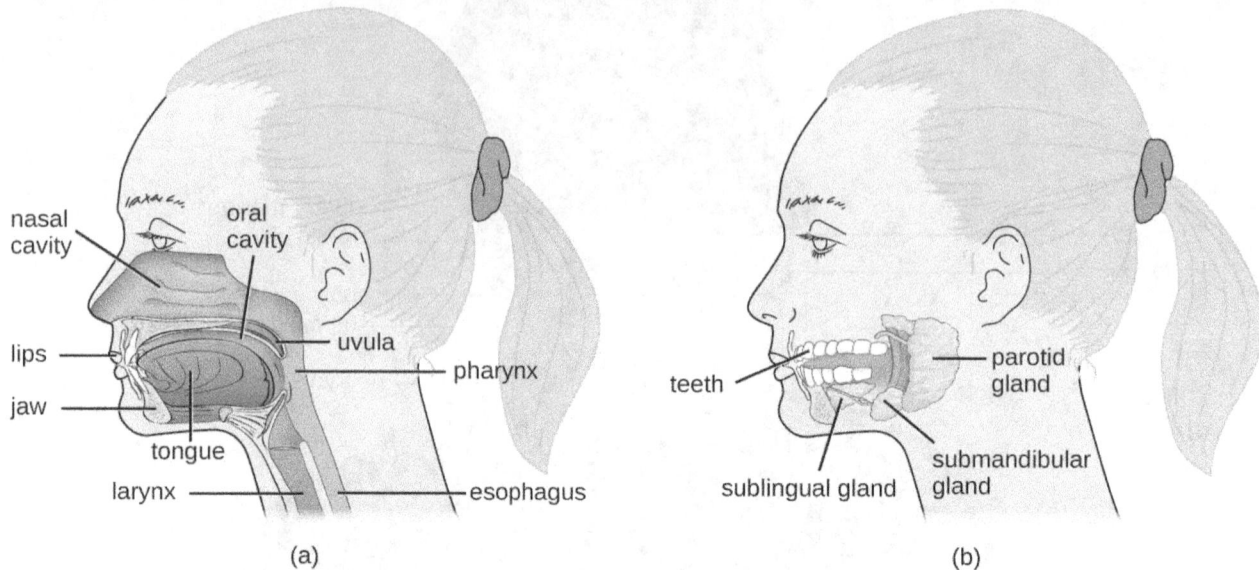

Figure 24.3 (a) When food enters the mouth, digestion begins. (b) Salivary glands are accessory digestive organs. (credit: modification of work by National Cancer Institute)

The structure of a tooth (Figure 24.4) begins with the visible outer surface, called the crown, which has to be extremely hard to withstand the force of biting and chewing. The crown is covered with enamel, which is the hardest material in the body. Underneath the crown, a layer of relatively hard dentin extends into the root of the tooth around the innermost pulp cavity, which includes the pulp chamber at the top of the tooth and pulp canal, or root canal, located in the root. The pulp that fills the pulp cavity is rich in blood vessels, lymphatic vessels, connective tissue, and nerves. The root of the tooth and some of the crown are covered with cementum, which works with the periodontal ligament to anchor the tooth in place in the jaw bone. The soft tissues surrounding the teeth and bones are called gums, or gingiva. The gingival space or gingival crevice is located between the gums and teeth.

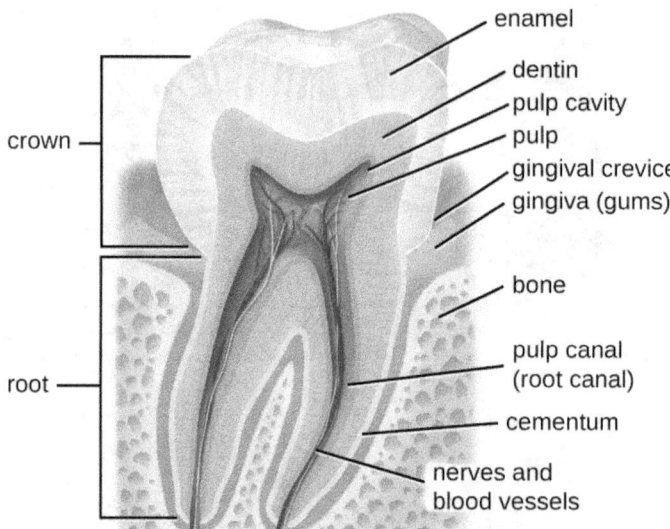

Figure 24.4 The tooth has a visible crown with an outer layer of enamel, a layer of dentin, and an inner pulp. The root, hidden by the gums, contains the pulp canal (root canal). (credit: modification of work by Bruce Blaus)

Microbes such as bacteria and archaea are abundant in the mouth and coat all of the surfaces of the oral cavity. However, different structures, such as the teeth or cheeks, host unique communities of both aerobic and anaerobic microbes. Some factors appear to work against making the mouth hospitable to certain microbes. For example, chewing allows microbes to mix better with saliva so they can be swallowed or spit out more easily. Saliva also contains enzymes, including lysozyme, which can damage microbial cells. Recall that lysozyme is part of the first line of defense in the innate immune system and cleaves the β-(1,4) glycosidic linkages between N-acetylglucosamine (NAG) and N-acetylmuramic acid (NAM) in bacterial peptidoglycan (see Chemical Defenses). Additionally, fluids containing immunoglobulins and phagocytic cells are produced in the gingival spaces. Despite all of these chemical and mechanical activities, the mouth supports a large microbial community.

CHECK YOUR UNDERSTANDING

- What factors make the mouth inhospitable for certain microbes?

Anatomy and Normal Microbiota of the GI Tract

As food leaves the oral cavity, it travels through the pharynx, or the back of the throat, and moves into the esophagus, which carries the food from the pharynx to the stomach without adding any additional digestive enzymes. The stomach produces mucus to protect its lining, as well as digestive enzymes and acid to break down food. Partially digested food then leaves the stomach through the pyloric sphincter, reaching the first part of the small intestine called the duodenum. Pancreatic juice, which includes enzymes and bicarbonate ions, is released into the small intestine to neutralize the acidic material from the stomach and to assist in digestion. Bile, produced by the liver but stored in the gallbladder, is also released into the small intestine to emulsify fats so that they can travel in the watery environment of the small intestine. Digestion continues in the small intestine, where the majority of nutrients contained in the food are absorbed. Simple columnar epithelial cells called enterocytes line the lumen surface of the small intestinal folds called villi. Each enterocyte has smaller microvilli (cytoplasmic membrane extensions) on the cellular apical surface that increase the surface area to allow more absorption of nutrients to occur (Figure 24.5).

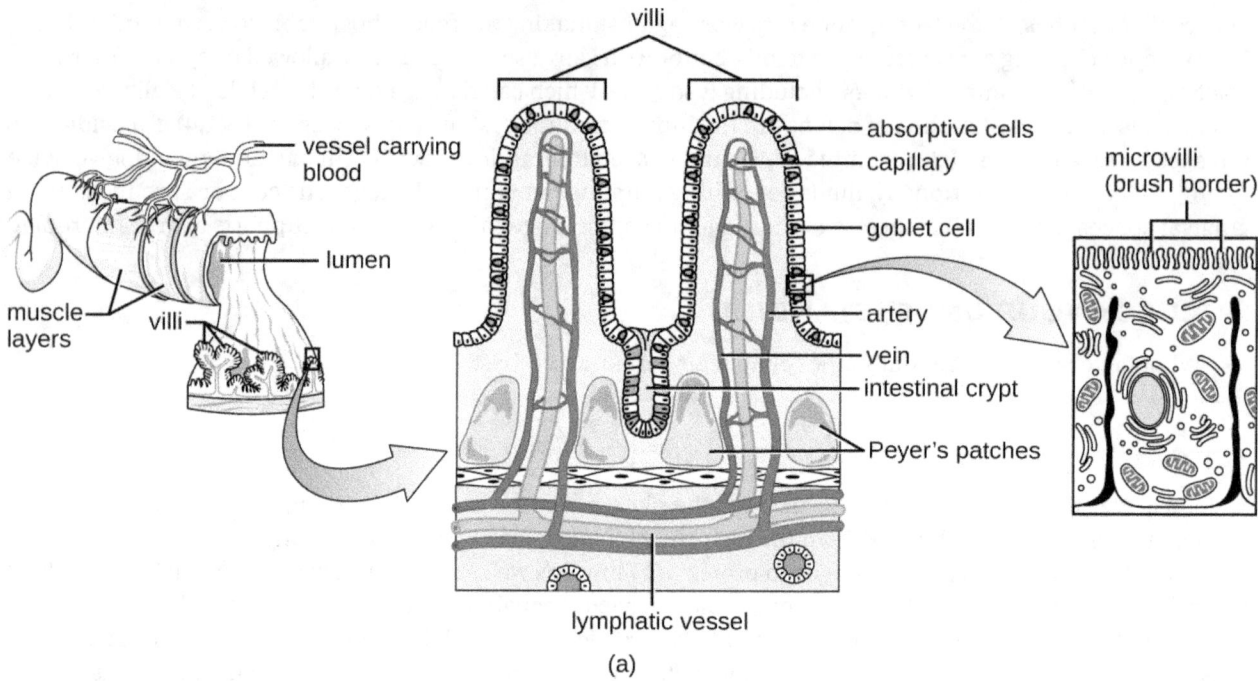

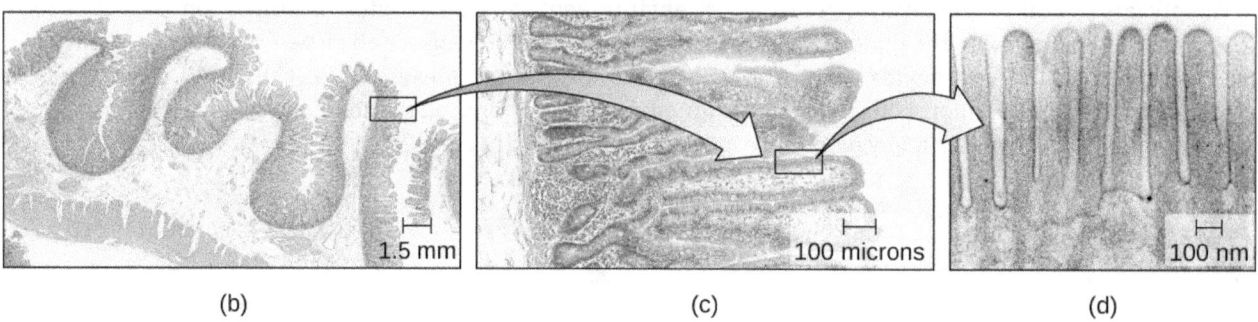

Figure 24.5 (a) The structure of the wall of the small intestine allows for the majority of nutrient absorption in the body. (b) Villi are folds in the surface of the small intestine. Microvilli are cytoplasmic extensions on individual cells that increase the surface area for absorption. (c) A light micrograph shows the shape of the villi. (d) An electron micrograph shows the shape of the microvilli. (credit b, c, d: Modification of micrographs provided by the Regents of University of Michigan Medical School © 2012)

Digested food leaves the small intestine and moves into the large intestine, or colon, where there is a more diverse microbiota. Near this junction, there is a small pouch in the large intestine called the cecum, which attaches to the appendix. Further digestion occurs throughout the colon and water is reabsorbed, then waste is excreted through the rectum, the last section of the colon, and out of the body through the anus (Figure 24.2).

The environment of most of the GI tract is harsh, which serves two purposes: digestion and immunity. The stomach is an extremely acidic environment (pH 1.5–3.5) due to the gastric juices that break down food and kill many ingested microbes; this helps prevent infection from pathogens. The environment in the small intestine is less harsh and is able to support microbial communities. Microorganisms present in the small intestine can include lactobacilli, diptherioids and the fungus *Candida*. On the other hand, the large intestine (colon) contains a diverse and abundant microbiota that is important for normal function. These microbes include *Bacteriodetes* (especially the genera *Bacteroides* and *Prevotella*) and *Firmicutes* (especially members of the genus *Clostridium*). Methanogenic archaea and some fungi are also present, among many other species of bacteria. These microbes all aid in digestion and contribute to the production of feces, the waste excreted from the digestive tract, and flatus, the gas produced by microbial fermentation of undigested food. They can also produce valuable nutrients. For example, lactic acid bacteria such as bifidobacteria can synthesize vitamins, such as vitamin B12, folate, and riboflavin, that humans cannot synthesize themselves. *E. coli* found in the intestine can also break down food and help the body produce vitamin K, which is important for blood coagulation.

The GI tract has several other methods of reducing the risk of infection by pathogens. Small aggregates of underlying lymphoid tissue in the ileum, called **Peyer's patches** (Figure 24.5), detect pathogens in the intestines via microfold (M) cells, which transfer antigens from the lumen of the intestine to the lymphocytes on Peyer's patches to induce an immune response. The Peyer's patches then secrete IgA and other pathogen-specific antibodies into the intestinal lumen to help keep intestinal microbes at safe levels. Goblet cells, which are modified simple columnar epithelial cells, also line the GI tract (Figure 24.6). Goblet cells secrete a gel-forming mucin, which is the major component of mucus. The production of a protective layer of mucus helps reduce the risk of pathogens reaching deeper tissues.

The constant movement of materials through the gastrointestinal tract also helps to move transient pathogens out of the body. In fact, feces are composed of approximately 25% microbes, 25% sloughed epithelial cells, 25% mucus, and 25% digested or undigested food. Finally, the normal microbiota provides an additional barrier to infection via a variety of mechanisms. For example, these organisms outcompete potential pathogens for space and nutrients within the intestine. This is known as competitive exclusion. Members of the microbiota may also secrete protein toxins known as bacteriocins that are able to bind to specific receptors on the surface of susceptible bacteria.

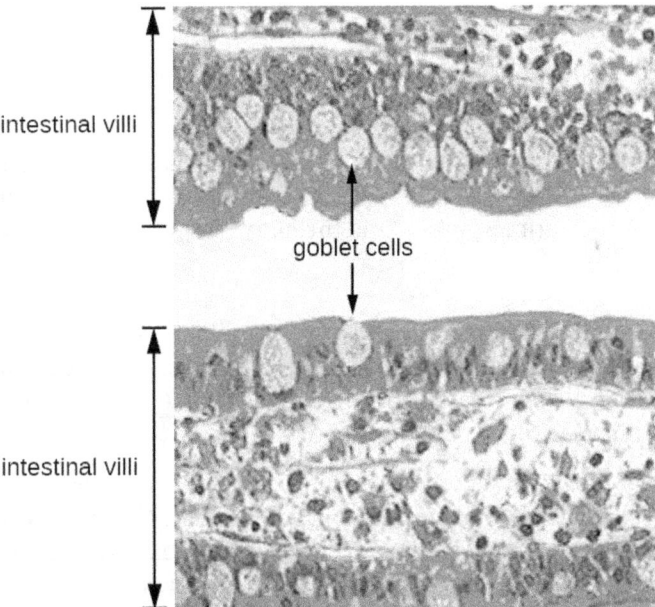

Figure 24.6 A magnified image of intestinal villi in the GI tract shows goblet cells. These cells are important in producing a protective layer of mucus.

✓ CHECK YOUR UNDERSTANDING

- Compare and contrast the microbiota of the small and large intestines.

General Signs and Symptoms of Oral and GI Disease

Despite numerous defense mechanisms that protect against infection, all parts of the digestive tract can become sites of infection or intoxication. The term food poisoning is sometimes used as a catch-all for GI infections and intoxications, but not all forms of GI disease originate with foodborne pathogens or toxins.

In the mouth, fermentation by anaerobic microbes produces acids that damage the teeth and gums. This can lead to tooth decay, cavities, and **periodontal disease**, a condition characterized by chronic inflammation and erosion of the gums. Additionally, some pathogens can cause infections of the mucosa, glands, and other structures in the mouth, resulting in inflammation, sores, cankers, and other lesions. An open sore in the mouth or GI tract is typically called an **ulcer**.

Infections and intoxications of the lower GI tract often produce symptoms such as nausea, vomiting, diarrhea, aches, and fever. In some cases, vomiting and diarrhea may cause severe dehydration and other complications that can become serious or fatal. Various clinical terms are used to describe gastrointestinal symptoms. For example, **gastritis** is an inflammation of the stomach lining that results in swelling and **enteritis** refers to inflammation of the intestinal mucosa. When the inflammation involves both the stomach lining and the intestinal lining, the condition is called **gastroenteritis**. Inflammation of the liver is called **hepatitis**. Inflammation of the colon, called **colitis**, commonly occurs in cases of food intoxication. Because an inflamed colon does not reabsorb water as effectively as it normally does, stools become watery, causing diarrhea. Damage to the epithelial cells of the colon can also cause bleeding and excess mucus to appear in watery stools, a condition called **dysentery**.

✓ CHECK YOUR UNDERSTANDING

- List possible causes and signs and symptoms of food poisoning.

24.2 Microbial Diseases of the Mouth and Oral Cavity

Learning Objectives

By the end of this section, you will be able to:
- Explain the role of microbial activity in diseases of the mouth and oral cavity
- Compare the major characteristics of specific oral diseases and infections

Despite the presence of saliva and the mechanical forces of chewing and eating, some microbes thrive in the mouth. These microbes can cause damage to the teeth and can cause infections that have the potential to spread beyond the mouth and sometimes throughout the body.

Dental Caries

Cavities of the teeth, known clinically as **dental caries**, are microbial lesions that cause damage to the teeth. Over time, the lesion can grow through the outer enamel layer to infect the underlying dentin or even the innermost pulp. If dental caries are not treated, the infection can become an abscess that spreads to the deeper tissues of the teeth, near the roots, or to the bloodstream.

Tooth decay results from the metabolic activity of microbes that live on the teeth. A layer of proteins and carbohydrates forms when clean teeth come into contact with saliva. Microbes are attracted to this food source and form a biofilm called plaque. The most important cariogenic species in these biofilms is *Streptococcus mutans*. When sucrose, a disaccharide sugar from food, is broken down by bacteria in the mouth, glucose and fructose are produced. The glucose is used to make dextran, which is part of the extracellular matrix of the biofilm. Fructose is fermented, producing organic acids such as lactic acid. These acids dissolve the minerals of the tooth, including enamel, even though it is the hardest material in the body. The acids work even more quickly on exposed dentin (Figure 24.7). Over time, the plaque biofilm can become thick and eventually calcify. When a heavy plaque deposit becomes hardened in this way, it is called **tartar** or **dental calculus** (Figure 24.8). These substantial plaque biofilms can include a variety of bacterial species, including *Streptococcus* and *Actinomyces* species.

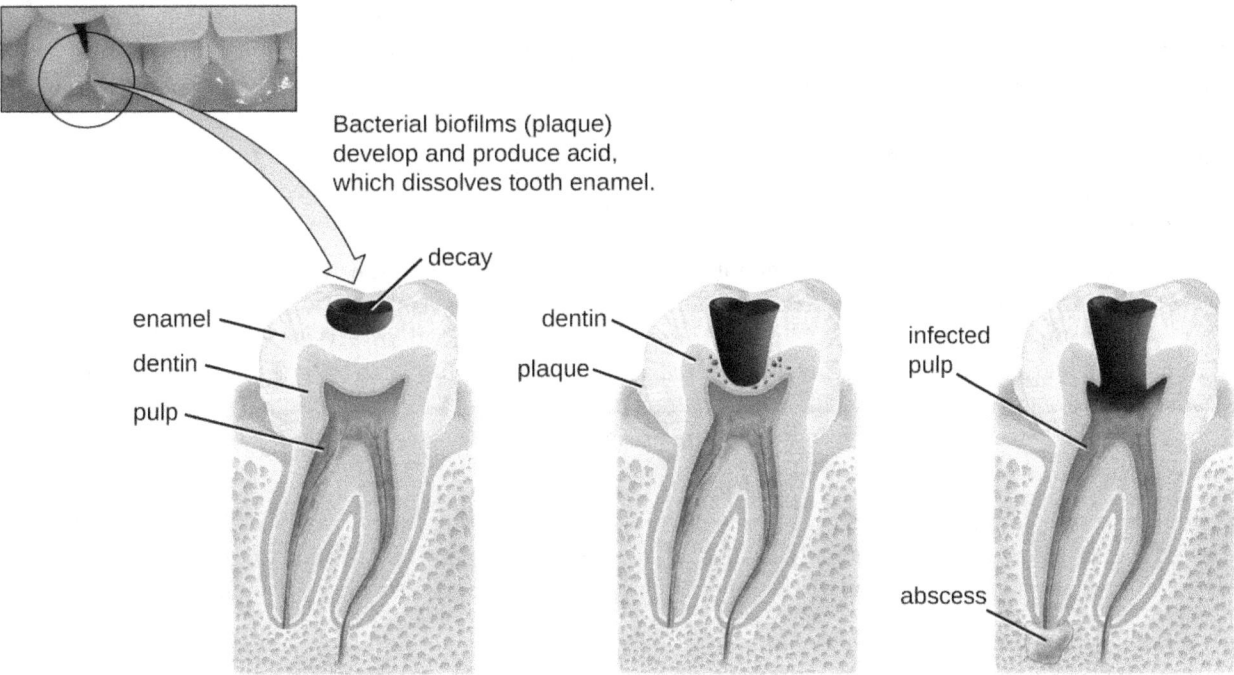

Figure 24.7 Tooth decay occurs in stages. When bacterial biofilms (plaque) develop on teeth, the acids produced gradually dissolve the enamel, followed by the dentin. Eventually, if left untreated, the lesion may reach the pulp and cause an abscess. (credit: modification of work by "BruceBlaus"/Wikimedia Commons)

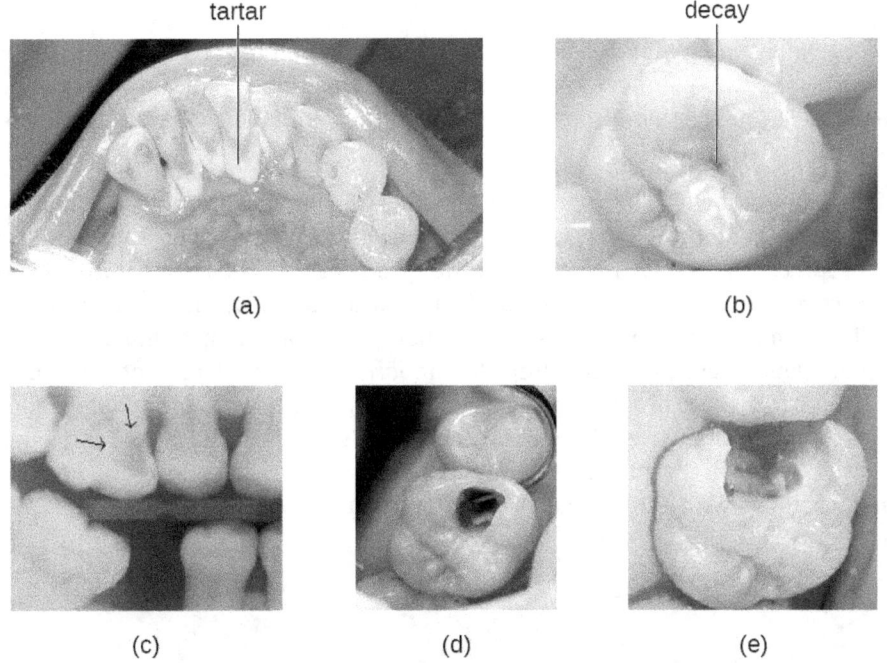

Figure 24.8 (a) Tartar (dental calculus) is visible at the bases of these teeth. The darker deposits higher on the crowns are staining. (b) This tooth shows only a small amount of visible decay. (c) An X-ray of the same tooth shows that there is a dark area representing more decay inside the tooth. (d) Removal of a portion of the crown reveals the area of damage. (e) All of the cavity must be removed before filling. (credit: modification of work by "DRosenbach"/Wikimedia Commons)

Some tooth decay is visible from the outside, but it is not always possible to see all decay or the extent of the decay. X-ray imaging is used to produce radiographs that can be studied to look for deeper decay and damage to the root or bone (Figure 24.8). If not detected, the decay can reach the pulp or even spread to the bloodstream. Painful abscesses can develop.

To prevent tooth decay, prophylactic treatment and good hygiene are important. Regular tooth brushing and flossing physically removes microbes and combats microbial growth and biofilm formation. Toothpaste contains fluoride, which becomes incorporated into the hydroxyapatite of tooth enamel, protecting it against acidity caused by fermentation of mouth microbiota. Fluoride is also bacteriostatic, thus slowing enamel degradation. Antiseptic mouthwashes commonly contain plant-derived phenolics like thymol and eucalyptol and/or heavy metals like zinc chloride (see Using Chemicals to Control Microorganisms). Phenolics tend to be stable and persistent on surfaces, and they act through denaturing proteins and disrupting membranes.

Regular dental cleanings allow for the detection of decay at early stages and the removal of tartar. They may also help to draw attention to other concerns, such as damage to the enamel from acidic drinks. Reducing sugar consumption may help prevent damage that results from the microbial fermentation of sugars. Additionally, sugarless candies or gum with sugar alcohols (such as xylitol) can reduce the production of acids because these are fermented to nonacidic compounds (although excess consumption may lead to gastrointestinal distress). Fluoride treatment or ingesting fluoridated water strengthens the minerals in teeth and reduces the incidence of dental caries.

If caries develop, prompt treatment prevents worsening. Smaller areas of decay can be drilled to remove affected tissue and then filled. If the pulp is affected, then a root canal may be needed to completely remove the infected tissues to avoid continued spread of the infection, which could lead to painful abscesses.

CHECK YOUR UNDERSTANDING

- Name some ways that microbes contribute to tooth decay.
- What is the most important cariogenic species of bacteria?

Periodontal Disease

In addition to damage to the teeth themselves, the surrounding structures can be affected by microbes. Periodontal disease is the result of infections that lead to inflammation and tissue damage in the structures surrounding the teeth. The progression from mild to severe periodontal disease is generally reversible and preventable with good oral hygiene.

Inflammation of the gums that can lead to irritation and bleeding is called **gingivitis**. When plaque accumulates on the teeth, bacteria colonize the gingival space. As this space becomes increasingly blocked, the environment becomes anaerobic. This allows a wide variety of microbes to colonize, including *Porphyromonas*, *Streptococcus*, and *Actinomyces*. The bacterial products, which include lipopolysaccharide (LPS), proteases, lipoteichoic acids, and others, cause inflammation and gum damage (Figure 24.9). It is possible that methanogenic archaeans (including *Methanobrevibacter oralis* and other *Methanobrevibacter* species) also contribute to disease progression as some species have been identified in patients with periodontal disease, but this has proven difficult to study.[1,2,3] Gingivitis is diagnosed by visual inspection, including measuring pockets in the gums, and X-rays, and is usually treated using good dental hygiene and professional dental cleaning, with antibiotics reserved for severe cases.

---

1   Hans-Peter Horz and Georg Conrads. "Methanogenic *Archaea* and Oral Infections—Ways to Unravel the Black Box." *Journal of Oral Microbiology* 3(2011). doi: 10.3402/jom.v3i0.5940.

2   Hiroshi Maeda, Kimito Hirai, Junji Mineshiba, Tadashi Yamamoto, Susumu Kokeguchi, and Shogo Takashiba. "Medical Microbiological Approach to Archaea in Oral Infectious Diseases." *Japanese Dental Science Review* 49: 2, p. 72–78.

3   Paul W. Lepp, Mary M. Brinig, Cleber C. Ouverney, Katherine Palm, Gary C. Armitage, and David A. Relman. "Methanogenic *Archaea* and Human Periodontal Disease." *Proceedings of the National Academy of Sciences of the United States of America* 101 (2003): 16, pp. 6176–6181. doi: 10.1073/pnas.0308766101.

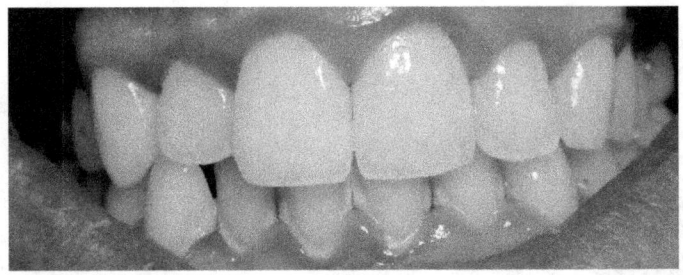

Figure 24.9 Redness and irritation of the gums are evidence of gingivitis.

Over time, chronic gingivitis can develop into the more serious condition of **periodontitis** (Figure 24.10). When this happens, the gums recede and expose parts of the tooth below the crown. This newly exposed area is relatively unprotected, so bacteria can grow on it and spread underneath the enamel of the crown and cause cavities. Bacteria in the gingival space can also erode the cementum, which helps to hold the teeth in place. If not treated, erosion of cementum can lead to the movement or loss of teeth. The bones of the jaw can even erode if the infection spreads. This condition can be associated with bleeding and halitosis (bad breath). Cleaning and appropriate dental hygiene may be sufficient to treat periodontitis. However, in cases of severe periodontitis, an antibiotic may be given. Antibiotics may be given in pill form or applied directly to the gum (local treatment). Antibiotics given can include tetracycline, doxycycline, macrolides or β-lactams. Because periodontitis can be caused by a mix of microbes, a combination of antibiotics may be given.

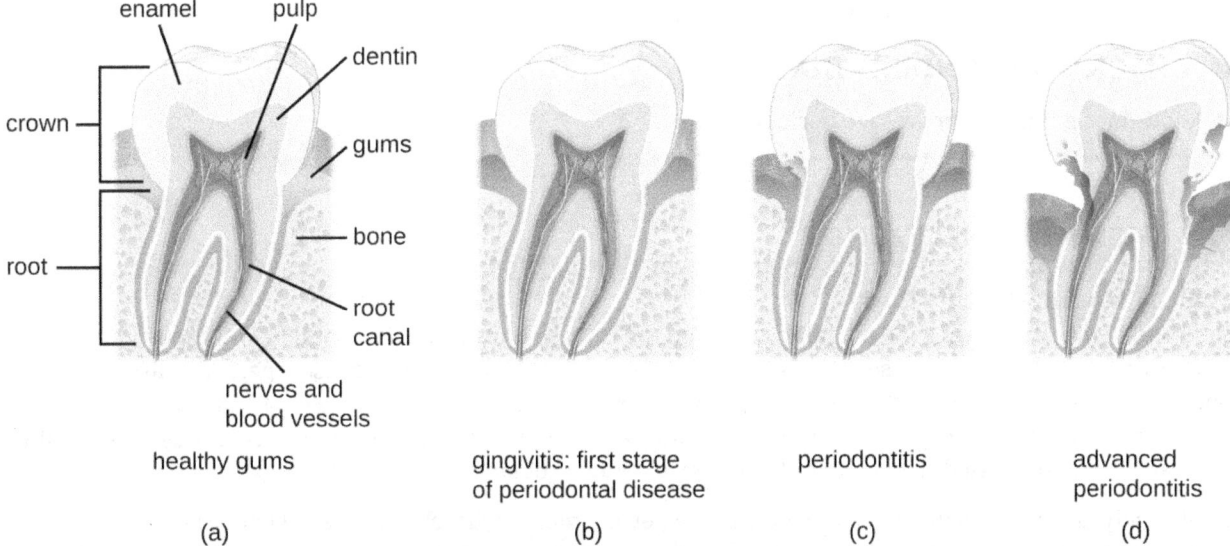

Figure 24.10 (a) Healthy gums hold the teeth firmly and do not bleed. (b) Gingivitis is the first stage of periodontal disease. Microbial infection causes gums to become inflamed and irritated, with occasional bleeding. (c) In periodontitis, gums recede and expose parts of the tooth normally covered. (d) In advanced periodontitis, the infection spreads to ligaments and bone tissue supporting the teeth. Tooth loss may occur, or teeth may need to be surgically removed. (credit: modification of work by "BruceBlaus"/Wikimedia Commons)

Trench Mouth

When certain bacteria, such as *Prevotella intermedia*, *Fusobacterium* species, and *Treponema vicentii*, are involved and periodontal disease progresses, **acute necrotizing ulcerative gingivitis** or **trench mouth**, also called Vincent's disease, can develop. This is severe periodontitis characterized by erosion of the gums, ulcers, substantial pain with chewing, and halitosis (Figure 24.11) that can be diagnosed by visual examination and X-rays. In countries with good medical and dental care, it is most common in individuals with weakened immune systems, such as patients with AIDS. In addition to cleaning and pain medication, patients may be prescribed antibiotics such as amoxicillin, amoxicillin clavulanate, clindamycin, or doxycycline.

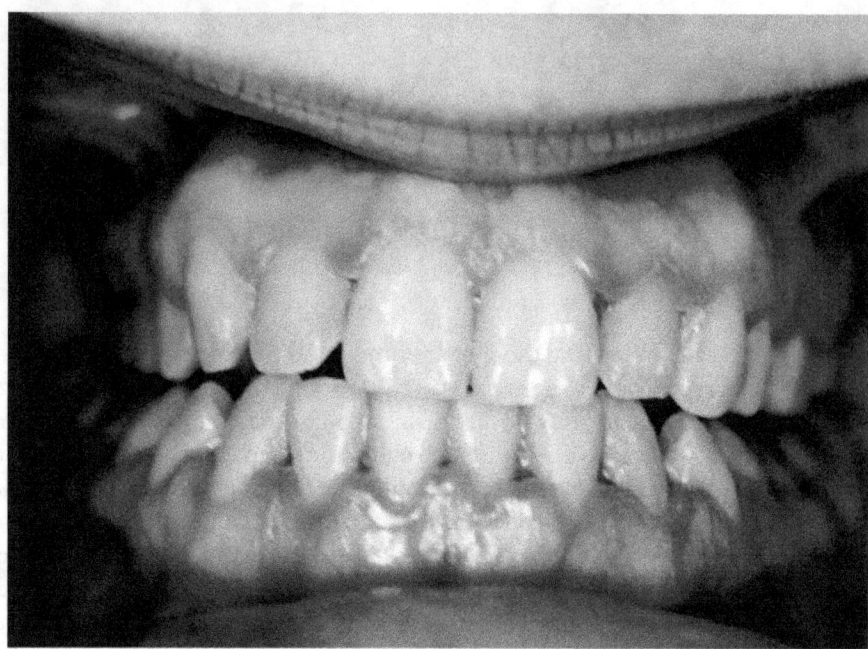

Figure 24.11 These inflamed, eroded gums are an example of a mild case of acute necrotizing ulcerative gingivitis, also known as trench mouth. (credit: modification of work by Centers for Disease Control and Prevention)

CHECK YOUR UNDERSTANDING

- How does gingivitis progress to periodontitis?

MICRO CONNECTIONS

Healthy Mouth, Healthy Body

Good oral health promotes good overall health, and the reverse is also true. Poor oral health can lead to difficulty eating, which can cause malnutrition. Painful or loose teeth can also cause a person to avoid certain foods or eat less. Malnutrition due to dental problems is of greatest concern for the elderly, for whom it can worsen other health conditions and contribute to mortality. Individuals who have serious illnesses, especially AIDS, are also at increased risk of malnutrition from dental problems.

Additionally, poor oral health can contribute to the development of disease. Increased bacterial growth in the mouth can cause inflammation and infection in other parts of the body. For example, *Streptococcus* in the mouth, the main contributor to biofilms on teeth, tartar, and dental caries, can spread throughout the body when there is damage to the tissues inside the mouth, as can happen during dental work. *S. mutans* produces a surface adhesin known as P1, which binds to salivary agglutinin on the surface of the tooth. P1 can also bind to extracellular matrix proteins including fibronectin and collagen. When *Streptococcus* enters the bloodstream as a result of tooth brushing or dental cleaning, it causes inflammation that can lead to the accumulation of plaque in the arteries and contribute to the development of atherosclerosis, a condition associated with cardiovascular disease, heart attack, and stroke. In some cases, bacteria that spread through the blood vessels can lodge in the heart and cause endocarditis (an example of a focal infection).

Oral Infections

As noted earlier, normal oral microbiota can cause dental and periodontal infections. However, there are number of other infections that can manifest in the oral cavity when other microbes are present.

Herpetic Gingivostomatitis

As described in Viral Infections of the Skin and Eyes, infections by herpes simplex virus type 1 (HSV-1) frequently manifest as oral herpes, also called acute herpes labialis and characterized by cold sores on the lips, mouth, or gums. HSV-1 can also cause acute **herpetic gingivostomatitis**, a condition that results in ulcers of the mucous membranes inside the mouth (Figure 24.12). Herpetic gingivostomatitis is normally self-limiting except in immunocompromised patients. Like oral herpes, the infection is generally diagnosed through clinical examination, but cultures or biopsies may be obtained if other signs or symptoms suggest the possibility of a different causative agent. If treatment is needed, mouthwashes or antiviral medications such as acyclovir, famciclovir, or valacyclovir may be used.

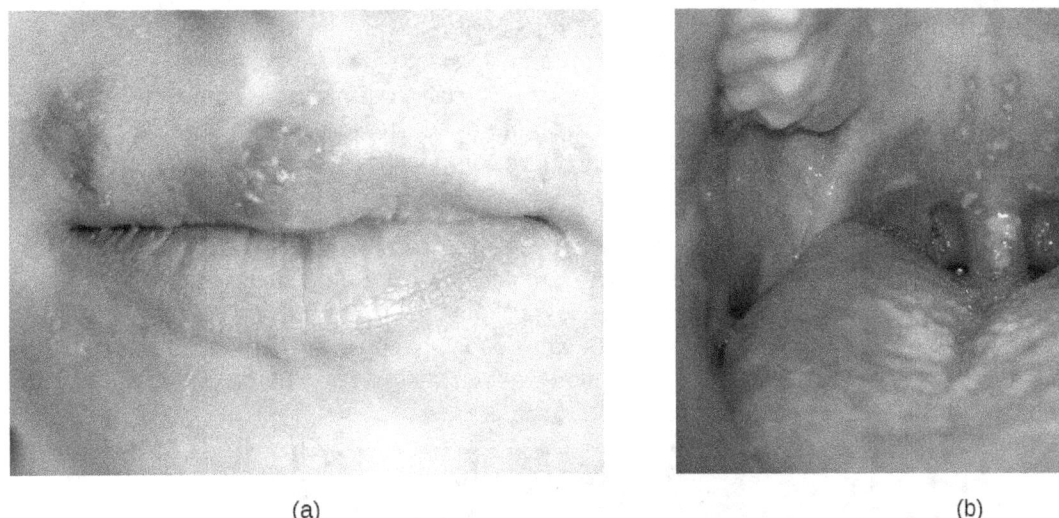

Figure 24.12 (a) This cold sore is caused by infection with herpes simplex virus type 1 (HSV-1). (b) HSV-1 can also cause acute herpetic gingivostomatitis. (credit b: modification of work by Klaus D. Peter)

Oral Thrush

The yeast *Candida* is part of the normal human microbiota, but overgrowths, especially of *Candida albicans*, can lead to infections in several parts of the body. When *Candida* infection develops in the oral cavity, it is called **oral thrush**. Oral thrush is most common in infants because they do not yet have well developed immune systems and have not acquired the robust normal microbiota that keeps *Candida* in check in adults. Oral thrush is also common in immunodeficient patients and is a common infection in patients with AIDS.

Oral thrush is characterized by the appearance of white patches and pseudomembranes in the mouth (Figure 24.13) and can be associated with bleeding. The infection may be treated topically with nystatin or clotrimazole oral suspensions, although systemic treatment is sometimes needed. In serious cases, systemic azoles such as fluconazole or itraconazole (for strains resistant to fluconazole), may be used. Amphotericin B can also be used if the infection is severe or if the *Candida* species is azole-resistant.

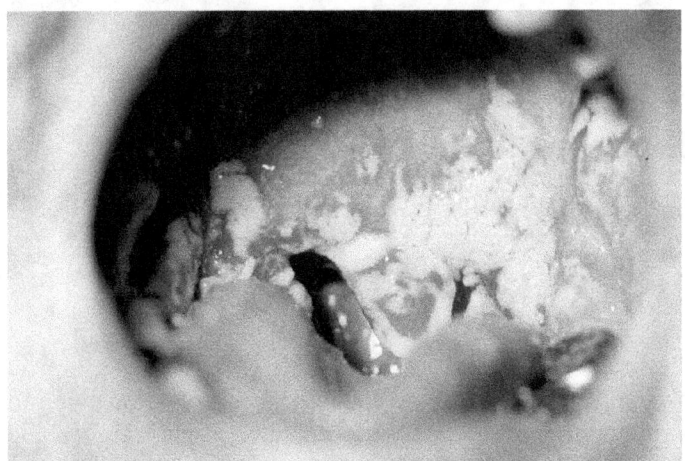

Figure 24.13 Overgrowth of *Candida* in the mouth is called thrush. It often appears as white patches. (credit: modification of work by Centers for Disease Control and Prevention)

Mumps

The viral disease **mumps** is an infection of the parotid glands, the largest of the three pairs of salivary glands (Figure 24.3). The causative agent is mumps virus (MuV), a paramyxovirus with an envelope that has hemagglutinin and neuraminidase spikes. A fusion protein located on the surface of the envelope helps to fuse the viral envelope to the host cell plasma membrane.

Mumps virus is transmitted through respiratory droplets or through contact with contaminated saliva, making it quite contagious so that it can lead easily to epidemics. It causes fever, muscle pain, headache, pain with chewing, loss of appetite, fatigue, and weakness. There is swelling of the salivary glands and associated pain (Figure 24.14). The virus can enter the bloodstream (viremia), allowing it to spread to the organs and the central nervous system. The infection ranges from subclinical cases to cases with serious complications, such as encephalitis, meningitis, and deafness. Inflammation of the pancreas, testes, ovaries, and breasts may also occur and cause permanent damage to those organs; despite these complications, a mumps infection rarely cause sterility.

Mumps can be recognized based on clinical signs and symptoms, and a diagnosis can be confirmed with laboratory testing. The virus can be identified using culture or molecular techniques such as RT-PCR. Serologic tests are also available, especially enzyme immunoassays that detect antibodies. There is no specific treatment for mumps, so supportive therapies are used. The most effective way to avoid infection is through vaccination. Although mumps used to be a common childhood disease, it is now rare in the United States due to vaccination with the measles, mumps, and rubella (MMR) vaccine.

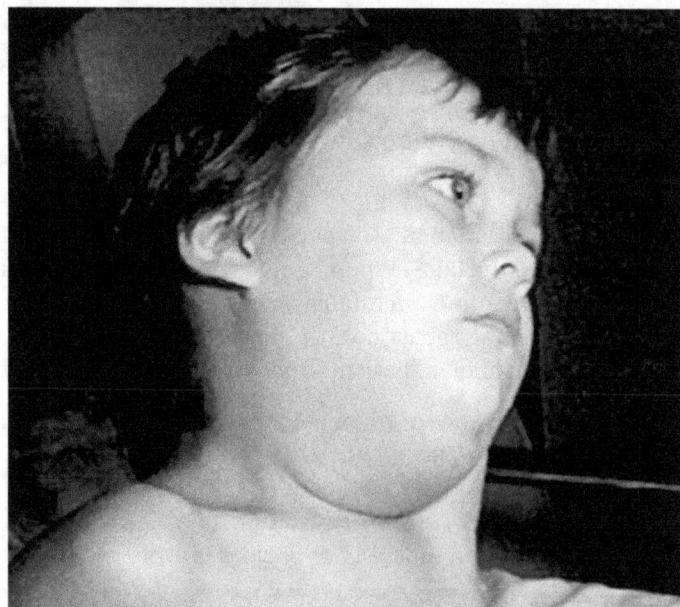

Figure 24.14 This child shows the characteristic parotid swelling associated with mumps. (credit: modification of work by Centers for Disease Control and Prevention)

✓ CHECK YOUR UNDERSTANDING

- Compare and contrast the signs and symptoms of herpetic gingivostomatitis, oral thrush, and mumps.

Disease Profile

Oral Infections

Infections of the mouth and oral cavity can be caused by a variety of pathogens, including bacteria, viruses, and fungi. Many of these infections only affect the mouth, but some can spread and become systemic infections. Figure 24.15 summarizes the main characteristics of common oral infections.

Oral Infections					
Disease	Pathogen	Signs and Symptoms	Transmission	Diagnostic Tests	Antimicrobial Drugs
Dental caries	*Streptococcus mutans*	Discoloration, softening, cavities in teeth	Non-transmissible; caused by bacteria of the normal oral microbiota	Visual examinations, X-rays	Oral antiseptics (e.g., Listerine)
Gingivitis and periodontitis	*Porphyromonas, Streptococcus, Actinomyces*	Inflammation and erosion of gums, bleeding, halitosis; erosion of cementum, leading to tooth loss in advanced infections	Non-transmissible; caused by bacteria of the normal oral microbiota	Visual examination, X-rays, measuring pockets in gums	Tetracycline, doxycycline, macrolides or beta-lactams. Mixture of antibiotics may be given.
Herpetic gingivostomatitis	Herpes simplex virus type 1 (HSV-1)	Lesions in mucous membranes of mouth	Contact with saliva or lesions of an infected person	Culture or biopsy	Acyclovir, famcyclovir, valacyclovir
Mumps	Mumps virus (a paramyxovirus)	Swelling of parotid glands, fever, headache, muscle pain, weakness, fatigue, loss of appetite, pain while chewing; in serious cases, encephalitis, meningitis, and inflammation of testes, ovaries, and breasts	Contact with saliva or respiratory droplets of an infected person	Virus culture or serologic tests for antibodies, enzyme immunoassay, RT-PCR	None for treatment; MMR vaccine for prevention
Oral thrush	*Candida albicans*, other *Candida* spp.	White patches and pseudomembranes in mouth, may cause bleeding	Non-transmissible; caused by overgrowth of *Candida* spp. in the normal oral microbiota; primarily affects infants and the immuno-compromised	Microscopic analysis of oral samples	Clotrimazole, nystatin, fluconazole, or itraconazole; amphotericin B in severe cases
Trench mouth (acute necrotizing ulcerative gingivitis)	*Prevotella intermedia Fusobacterium* species, *Treponema vincentii*, others	Erosion of gums, ulcers, substantial pain with chewing, halitosis	Non-transmissible; caused by members of the normal oral microbiota	Visual examinations, X-rays	Amoxicillin, amoxicillin clavulanate, clindamycin, or doxycycline

Figure 24.15

24.3 Bacterial Infections of the Gastrointestinal Tract

Learning Objectives

By the end of this section, you will be able to:
- Identify the most common bacteria that can cause infections of the GI tract
- Compare the major characteristics of specific bacterial diseases affecting the GI tract

A wide range of gastrointestinal diseases are caused by bacterial contamination of food. Recall that **foodborne disease** can arise from either infection or intoxication. In both cases, bacterial toxins are typically responsible for producing disease signs and symptoms. The distinction lies in where the toxins are produced. In an infection, the microbial agent is ingested, colonizes the gut, and then produces toxins that damage host cells. In an intoxication, bacteria produce toxins in the food before it is ingested. In either case, the toxins cause damage to the cells lining the gastrointestinal tract, typically the colon. This leads to the common signs and symptoms of diarrhea or watery stool and abdominal cramps, or the more severe dysentery. Symptoms of foodborne diseases also often include nausea and vomiting, which are mechanisms the body uses to expel the toxic materials.

Most bacterial gastrointestinal illness is short-lived and self-limiting; however, loss of fluids due to severe diarrheal illness can lead to dehydration that can, in some cases, be fatal without proper treatment. Oral rehydration therapy with electrolyte solutions is an essential aspect of treatment for most patients with GI disease, especially in children and infants.

Staphylococcal Food Poisoning

Staphylococcal food poisoning is one form of food intoxication. When *Staphylococcus aureus* grows in food, it may produce enterotoxins that, when ingested, can cause symptoms such as nausea, diarrhea, cramping, and vomiting within one to six hours. In some severe cases, it may cause headache, dehydration, and changes in blood pressure and heart rate. Signs and symptoms resolve within 24 to 48 hours. *S. aureus* is often associated with a variety of raw or undercooked and cooked foods including meat (e.g., canned meat, ham, and sausages) and dairy products (e.g., cheeses, milk, and butter). It is also commonly found on hands and can be transmitted to prepared foods through poor hygiene, including poor handwashing and the use of contaminated food preparation surfaces, such as cutting boards. The greatest risk is for food left at a temperature below 60 °C (140 °F), which allows the bacteria to grow. Cooked foods should generally be reheated to at least 60 °C (140 °F) for safety and most raw meats should be cooked to even higher internal temperatures (Figure 24.16).

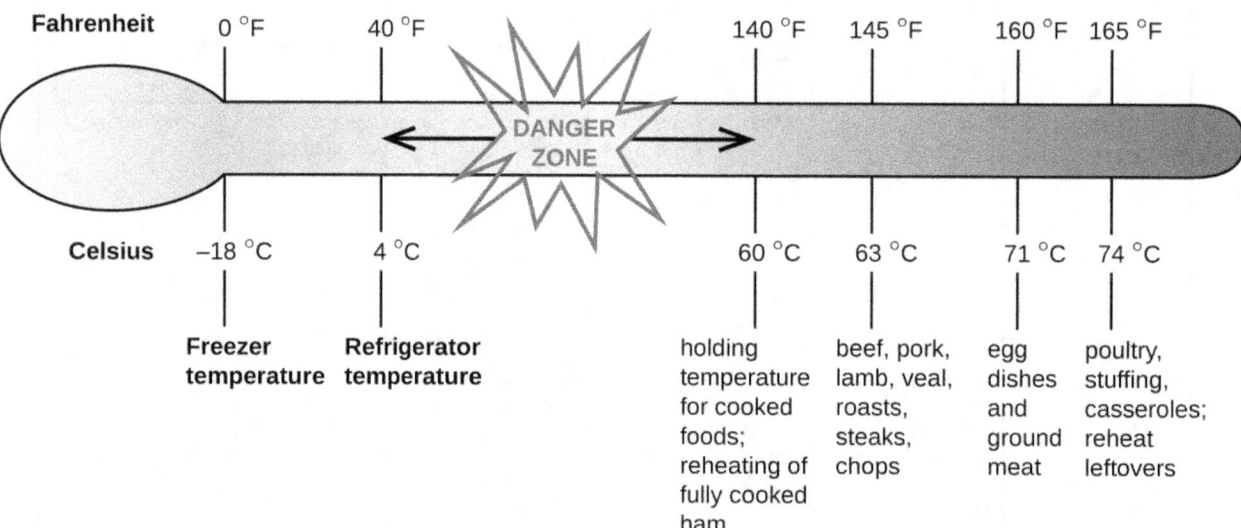

Figure 24.16 This figure indicates safe internal temperatures associated with the refrigeration, cooking, and reheating of different foods. Temperatures above refrigeration and below the minimum cooking temperature may allow for microbial growth, increasing the likelihood of foodborne disease. (credit: modification of work by USDA)

There are at least 21 *Staphylococcal* enterotoxins and *Staphylococcal* enterotoxin-like toxins that can cause food intoxication. The enterotoxins are proteins that are resistant to low pH, allowing them to pass through the stomach. They are heat stable and are not destroyed by boiling at 100 °C. Even though the bacterium itself may be killed, the enterotoxins alone can cause vomiting and diarrhea, although the mechanisms are not fully understood. At least some of the symptoms may be caused by the enterotoxin functioning as a superantigen and provoking a strong immune response by activating T cell proliferation.

The rapid onset of signs and symptoms helps to diagnose this foodborne illness. Because the bacterium does not need to be present for the toxin to cause symptoms, diagnosis is confirmed by identifying the toxin in a food sample or in biological specimens (feces or vomitus) from the patient. Serological techniques, including

ELISA, can also be used to identify the toxin in food samples.

The condition generally resolves relatively quickly, within 24 hours, without treatment. In some cases, supportive treatment in a hospital may be needed.

CHECK YOUR UNDERSTANDING

- How can *S. aureus* cause food intoxication?

Shigellosis (Bacillary Dysentery)

When gastrointestinal illness is associated with the rod-shaped, gram-negative bacterium *Shigella*, it is called **bacillary dysentery**, or **shigellosis**. Infections can be caused by *S. dysenteriae*, *S. flexneri*, *S. boydii*, and/or *S. sonnei* that colonize the GI tract. Shigellosis can be spread from hand to mouth or through contaminated food and water. Most commonly, it is transmitted through the fecal-oral route.

Shigella bacteria invade intestinal epithelial cells. When taken into a phagosome, they can escape and then live within the cytoplasm of the cell or move to adjacent cells. As the organisms multiply, the epithelium and structures with M cells of the Peyer's patches in the intestine may become ulcerated and cause loss of fluid. Stomach cramps, fever, and watery diarrhea that may also contain pus, mucus, and/or blood often develop. More severe cases may result in ulceration of the mucosa, dehydration, and rectal bleeding. Additionally, patients may later develop hemolytic uremic syndrome (HUS), a serious condition in which damaged blood cells build up in the kidneys and may cause kidney failure, or reactive arthritis, a condition in which arthritis develops in multiple joints following infection. Patients may also develop chronic post-infection irritable bowel syndrome (IBS).

S. dysenteriae type 1 is able to produce Shiga toxin, which targets the endothelial cells of small blood vessels in the small and large intestine by binding to a glycosphingolipid. Once inside the endothelial cells, the toxin targets the large ribosomal subunit, thus affecting protein synthesis of these cells. Hemorrhaging and lesions in the colon can result. The toxin can target the kidney's glomerulus, the blood vessels where filtration of blood in the kidney begins, thus resulting in HUS.

Stool samples, which should be processed promptly, are analyzed using serological or molecular techniques. One common method is to perform immunoassays for *S. dysenteriae*. (Other methods that can be used to identify *Shigella* include API test strips, Enterotube systems, or PCR testing. The presence of white blood cells and blood in fecal samples occurs in about 70% of patients[4] (Figure 24.17). Severe cases may require antibiotics such as ciprofloxacin and azithromycin, but these must be carefully prescribed because resistance is increasingly common.

---

4 Jaya Sureshbabu. "Shigella Infection Workup." *Medscape*. Updated Jun 28, 2016. http://emedicine.medscape.com/article/968773-workup.

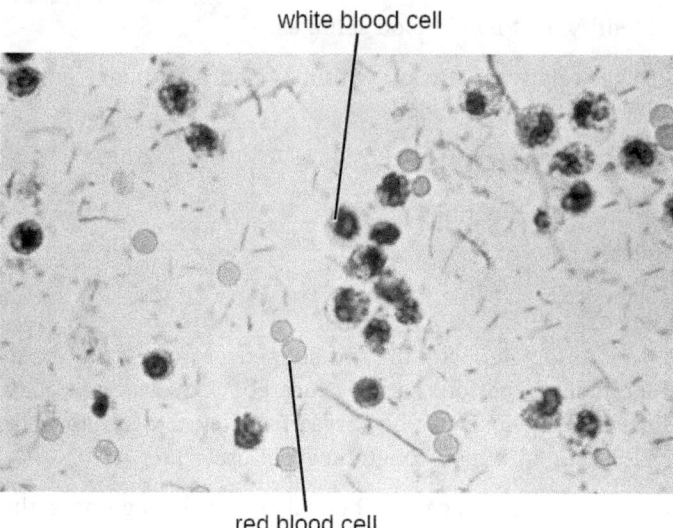

Figure 24.17 Red and white blood cells can be seen in this micrograph of a stool sample from a patient with shigellosis.

CHECK YOUR UNDERSTANDING

- Compare and contrast *Shigella* infections and intoxications.

Salmonellosis

Salmonella gastroenteritis, also called **salmonellosis**, is caused by the rod-shaped, gram-negative bacterium *Salmonella*. Two species, *S. enterica* and *S. bongori*, cause disease in humans, but *S. enterica* is the most common. The most common serotypes of *S. enterica* are Enteritidis and Typhi. We will discuss typhoid fever caused by serotypes Typhi and Paratyphi A separately. Here, we will focus on salmonellosis caused by other serotypes.

Salmonella is a part of the normal intestinal microbiota of many individuals. However, salmonellosis is caused by exogenous agents, and infection can occur depending on the serotype, size of the inoculum, and overall health of the host. Infection is caused by ingestion of contaminated food, handling of eggshells, or exposure to certain animals. *Salmonella* is part of poultry's microbiota, so exposure to raw eggs and raw poultry can increase the risk of infection. Handwashing and cooking foods thoroughly greatly reduce the risk of transmission. *Salmonella* bacteria can survive freezing for extended periods but cannot survive high temperatures.

Once the bacteria are ingested, they multiply within the intestines and penetrate the epithelial mucosal cells via M cells where they continue to grow (Figure 24.18). They trigger inflammatory processes and the hypersecretion of fluids. Once inside the body, they can persist inside the phagosomes of macrophages. *Salmonella* can cross the epithelial cell membrane and enter the bloodstream and lymphatic system. Some strains of *Salmonella* also produce an enterotoxin that can cause an intoxication.

Infected individuals develop fever, nausea, abdominal cramps, vomiting, headache, and diarrhea. These signs and symptoms generally last a few days to a week. According to the Centers for Disease Control and Prevention (CDC), there are 1,000,000 cases annually, with 380 deaths each year.[5] However, because the disease is usually self-limiting, many cases are not reported to doctors and the overall incidence may be underreported. Diagnosis involves culture followed by serotyping and DNA fingerprinting if needed. Positive results are reported to the CDC. When an unusual serotype is detected, samples are sent to the CDC for further analysis. Serotyping is important for determining treatment. Oral rehydration therapy is commonly used. Antibiotics are only recommended for serious cases. When antibiotics are needed, as in immunocompromised patients, fluoroquinolones, third-generation cephalosporins, and ampicillin are recommended. Antibiotic resistance is a serious concern.

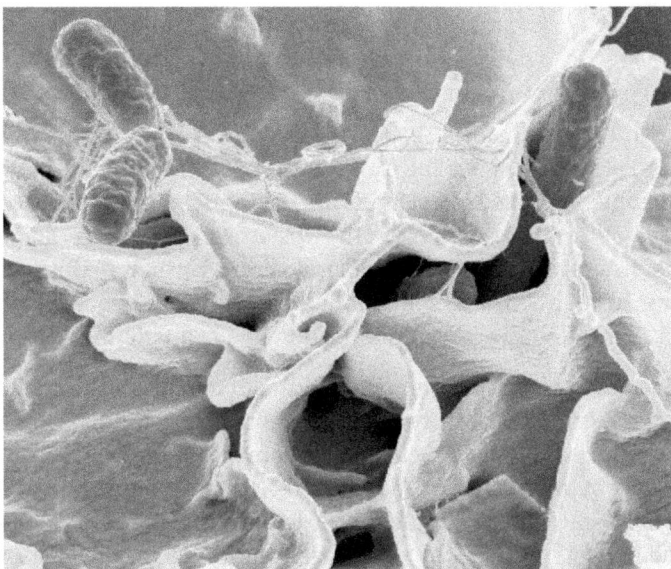

Figure 24.18 Salmonella entering an intestinal epithelial cell by reorganizing the host cell's cytoskeleton via the trigger mechanism. (credit: modification of work by National Institutes for Health)

Typhoid Fever

Certain serotypes of *S. enterica*, primarily serotype Typhi (*S. typhi*) but also Paratyphi, cause a more severe type of salmonellosis called **typhoid fever**. This serious illness, which has an untreated mortality rate of 10%, causes high fever, body aches, headache, nausea, lethargy, and a possible rash.

Some individuals carry *S. typhi* without presenting signs or symptoms (known as asymptomatic carriers) and continually shed them through their feces. These carriers often have the bacteria in the gallbladder or intestinal epithelium. Individuals consuming food or water contaminated with these feces can become infected.

S. typhi penetrate the intestinal mucosa, grow within the macrophages, and are transported through the body, most notably to the liver and gallbladder. Eventually, the macrophages lyse, releasing *S. typhi* into the bloodstream and lymphatic system. Mortality can result from ulceration and perforation of the intestine. A wide range of complications, such as pneumonia and jaundice, can occur with disseminated disease.

S. typhi have *Salmonella* pathogenicity islands (SPIs) that contain the genes for many of their virulence factors. Two examples of important typhoid toxins are the Vi antigen, which encodes for capsule production, and chimeric A2B5 toxin, which causes many of the signs and symptoms of the acute phase of typhoid fever.

Clinical examination and culture are used to make the diagnosis. The bacteria can be cultured from feces, urine, blood, or bone marrow. Serology, including ELISA, is used to identify the most pathogenic strains, but confirmation with DNA testing or culture is needed. A PCR test can also be used, but is not widely available.

The recommended antibiotic treatment involves fluoroquinolones, ceftriaxone, and azithromycin. Individuals must be extremely careful to avoid infecting others during treatment. Typhoid fever can be prevented through vaccination for individuals traveling to parts of the world where it is common.

CHECK YOUR UNDERSTANDING

- Why is serotyping particularly important in *Salmonella* infections and typhoid fever?

---

5 Centers for Disease Control and Prevention. *Salmonella*. Updated August 25, 2016. https://www.cdc.gov/salmonella.

> ### Eye on Ethics
>
> **Typhoid Mary**
>
> Mary Mallon was an Irish immigrant who worked as a cook in New York in the early 20th century. Over seven years, from 1900 to 1907, Mallon worked for a number of different households, unknowingly spreading illness to the people who lived in each one. In 1906, one family hired George Soper, an expert in typhoid fever epidemics, to determine the cause of the illnesses in their household. Eventually, Soper tracked Mallon down and directly linked 22 cases of typhoid fever to her. He discovered that Mallon was a carrier for typhoid but was immune to it herself. Although active carriers had been recognized before, this was the first time that an asymptomatic carrier of infection had been identified.
>
> Because she herself had never been ill, Mallon found it difficult to believe she could be the source of the illness. She fled from Soper and the authorities because she did not want to be quarantined or forced to give up her profession, which was relatively well paid for someone with her background. However, Mallon was eventually caught and kept in an isolation facility in the Bronx, where she remained until 1910, when the New York health department released her under the condition that she never again work with food. Unfortunately, Mallon did not comply, and she soon began working as a cook again. After new cases began to appear that resulted in the death of two individuals, the authorities tracked her down again and returned her to isolation, where she remained for 23 more years until her death in 1938. Epidemiologists were able to trace 51 cases of typhoid fever and three deaths directly to Mallon, who is unflatteringly remembered as "Typhoid Mary."
>
> The Typhoid Mary case has direct correlations in the health-care industry. Consider Kaci Hickox, an American nurse who treated Ebola patients in West Africa during the 2014 epidemic. After returning to the United States, Hickox was quarantined against her will for three days and later found not to have Ebola. Hickox vehemently opposed the quarantine. In an editorial published in the British newspaper *The Guardian*,[6] Hickox argued that quarantining asymptomatic health-care workers who had not tested positive for a disease would not only prevent such individuals from practicing their profession, but discourage others from volunteering to work in disease-ridden areas where health-care workers are desperately needed.
>
> What is the responsibility of an individual like Mary Mallon to change her behavior to protect others? What happens when an individual believes that she is not a risk, but others believe that she is? How would you react if you were in Mallon's shoes and were placed in a quarantine you did not believe was necessary, at the expense of your own freedom and possibly your career? Would it matter if you were definitely infected or not?
>
>

E. coli Infections

The gram-negative rod *Escherichia coli* is a common member of the normal microbiota of the colon. Although the vast majority of *E. coli* strains are helpful commensal bacteria, some can be pathogenic and may cause dangerous diarrheal disease. The pathogenic strains have additional virulence factors such as type 1 fimbriae that promote colonization of the colon or may produce toxins (see Virulence Factors of Bacterial and Viral Pathogens). These virulence factors are acquired through horizontal gene transfer.

Extraintestinal disease can result if the bacteria spread from the gastrointestinal tract. Although these bacteria can be spread from person to person, they are often acquired through contaminated food or water. There are six recognized pathogenic groups of *E. coli*, but we will focus here on the four that are most commonly transmitted through food and water.

Enterotoxigenic *E. coli* (ETEC), also known as **traveler's diarrhea**, causes diarrheal illness and is common in

---

6 Kaci Hickox. "Stop Calling Me the 'Ebola Nurse.'" *The Guardian*. November 17, 2014. http://www.theguardian.com/commentisfree/2014/nov/17/stop-calling-me-ebola-nurse-kaci-hickox.

less developed countries. In Mexico, ETEC infection is called Montezuma's Revenge. Following ingestion of contaminated food or water, infected individuals develop a watery diarrhea, abdominal cramps, **malaise** (a feeling of being unwell), and a low fever. ETEC produces a heat-stable enterotoxin similar to cholera toxin, and adhesins called colonization factors that help the bacteria to attach to the intestinal wall. Some strains of ETEC also produce heat-labile toxins. The disease is usually relatively mild and self-limiting. Diagnosis involves culturing and PCR. If needed, antibiotic treatment with fluoroquinolones, doxycycline, rifaximin, and trimethoprim-sulfamethoxazole (TMP/SMZ) may shorten infection duration. However, antibiotic resistance is a problem.

Enteroinvasive *E. coli* (EIEC) is very similar to shigellosis, including its pathogenesis of intracellular invasion into intestinal epithelial tissue. This bacterium carries a large plasmid that is involved in epithelial cell penetration. The illness is usually self-limiting, with symptoms including watery diarrhea, chills, cramps, malaise, fever, and dysentery. Culturing and PCR testing can be used for diagnosis. Antibiotic treatment is not recommended, so supportive therapy is used if needed.

Enteropathogenic *E. coli* (EPEC) can cause a potentially fatal diarrhea, especially in infants and those in less developed countries. Fever, vomiting, and diarrhea can lead to severe dehydration. These *E. coli* inject a protein (Tir) that attaches to the surface of the intestinal epithelial cells and triggers rearrangement of host cell actin from microvilli to pedestals. Tir also happens to be the receptor for Intimin, a surface protein produced by EPEC, thereby allowing *E. coli* to "sit" on the pedestal. The genes necessary for this pedestal formation are encoded on the locus for enterocyte effacement (LEE) pathogenicity island. As with ETEC, diagnosis involves culturing and PCR. Treatment is similar to that for ETEC.

The most dangerous strains are **enterohemorrhagic *E. coli* (EHEC)**, which are the strains capable of causing epidemics. In particular, the strain O157:H7 has been responsible for several recent outbreaks. Recall that the O and H refer to surface antigens that contribute to pathogenicity and trigger a host immune response ("O" refers to the O-side chain of the lipopolysaccharide and the "H" refers to the flagella). Similar to EPEC, EHEC also forms pedestals. EHEC also produces a Shiga-like toxin. Because the genome of this bacterium has been sequenced, it is known that the Shiga toxin genes were most likely acquired through transduction (horizontal gene transfer). The Shiga toxin genes originated from *Shigella dysenteriae*. Prophage from a bacteriophage that previously infected *Shigella* integrated into the chromosome of *E. coli*. The Shiga-like toxin is often called verotoxin.

EHEC can cause disease ranging from relatively mild to life-threatening. Symptoms include bloody diarrhea with severe cramping, but no fever. Although it is often self-limiting, it can lead to hemorrhagic colitis and profuse bleeding. One possible complication is HUS. Diagnosis involves culture, often using MacConkey with sorbitol agar to differentiate between *E. coli* O157:H7, which does not ferment sorbitol, and other less virulent strains of *E. coli* that can ferment sorbitol.

Serological typing or PCR testing also can be used, as well as genetic testing for Shiga toxin. To distinguish EPEC from EHEC, because they both form pedestals on intestinal epithelial cells, it is necessary to test for genes encoding for both the Shiga-like toxin and for the LEE. Both EPEC and EHEC have LEE, but EPEC lacks the gene for Shiga toxin. Antibiotic therapy is not recommended and may worsen HUS because of the toxins released when the bacteria are killed, so supportive therapies must be used. Table 24.1 summarizes the characteristics of the four most common pathogenic groups.

Some Pathogenic Groups of *E. coli*

Group	Virulence Factors and Genes	Signs and Symptoms	Diagnostic Tests	Treatment
Enterotoxigenic *E. coli* (ETEC)	Heat stable enterotoxin similar to cholera toxin	Relatively mild, watery diarrhea	Culturing, PCR	Self-limiting; if needed, fluoroquinolones, doxycycline, rifaximin, TMP/SMZ; antibiotic resistance is a problem
Enteroinvasive *E. coli* (EIEC)	*Inv* (invasive plasmid) genes	Relatively mild, watery diarrhea; dysentery or inflammatory colitis may occur	Culturing, PCR; testing for *inv* gene; additional assays to distinguish from *Shigella*	Supportive therapy only; antibiotics not recommended
Enteropathogenic *E. coli* (EPEC)	Locus of enterocyte effacement (LEE) pathogenicity island	Severe fever, vomiting, nonbloody diarrhea, dehydration; potentially fatal	Culturing, PCR; detection of LEE lacking Shiga-like toxin genes	Self-limiting; if needed, fluoroquinolones, doxycycline, rifaximin (TMP/SMZ); antibiotic resistance is a problem
Enterohemorrhagic *E. coli* (EHEC)	Verotoxin	May be mild or very severe; bloody diarrhea; may result in HUS	Culturing; plate on MacConkey agar with sorbitol agar as it does not ferment sorbitol; PCR detection of LEE containing Shiga-like toxin genes	Antibiotics are not recommended due to the risk of HUS

Table 24.1

✓ CHECK YOUR UNDERSTANDING

- Compare and contrast the virulence factors and signs and symptoms of infections with the four main *E. coli* groups.

Cholera and Other Vibrios

The gastrointestinal disease **cholera** is a serious infection often associated with poor sanitation, especially following natural disasters, because it is spread through contaminated water and food that has not been heated to temperatures high enough to kill the bacteria. It is caused by *Vibrio cholerae* serotype O1, a gram-negative, flagellated bacterium in the shape of a curved rod (vibrio). According to the CDC, cholera causes an estimated 3 to 5 million cases and 100,000 deaths each year.[7]

Because *V. cholerae* is killed by stomach acid, relatively large doses are needed for a few microbial cells to survive to reach the intestines and cause infection. The motile cells travel through the mucous layer of the intestines, where they attach to epithelial cells and release cholera enterotoxin. The toxin is an A-B toxin with activity through adenylate cyclase (see Virulence Factors of Bacterial and Viral Pathogens). Within the intestinal cell, cyclic AMP (cAMP) levels increase, which activates a chloride channel and results in the release of ions into the intestinal lumen. This increase in osmotic pressure in the lumen leads to water also entering the lumen. As the water and electrolytes leave the body, it causes rapid dehydration and electrolyte imbalance. Diarrhea is so profuse that it is often called "rice water stool," and patients are placed on cots with a hole in them to monitor the fluid loss (Figure 24.19).

Cholera is diagnosed by taking a stool sample and culturing for *Vibrio*. The bacteria are oxidase positive and show non-lactose fermentation on MacConkey agar. Gram-negative lactose fermenters will produce red colonies while non-fermenters will produce white/colorless colonies. Gram-positive bacteria will not grow on MacConkey. Lactose fermentation is commonly used for pathogen identification because the normal microbiota generally ferments lactose while pathogens do not. *V. cholerae* may also be cultured on thiosulfate citrate bile salts sucrose (TCBS) agar, a selective and differential media for *Vibrio* spp., which produce a distinct yellow colony.

Cholera may be self-limiting and treatment involves rehydration and electrolyte replenishment. Although antibiotics are not typically needed, they can be used for severe or disseminated disease. Tetracyclines are recommended, but doxycycline, erythromycin, orfloxacin, ciprofloxacin, and TMP/SMZ may be used. Recent evidence suggests that azithromycin is also a good first-line antibiotic. Good sanitation—including appropriate sewage treatment, clean supplies for cooking, and purified drinking water—is important to prevent infection (Figure 24.19)

(a)
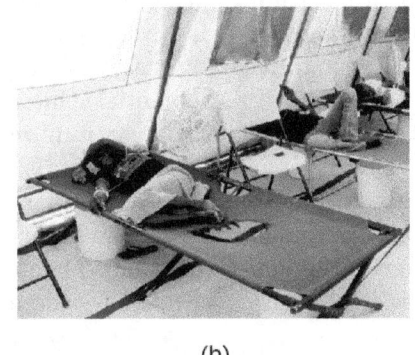
(b)
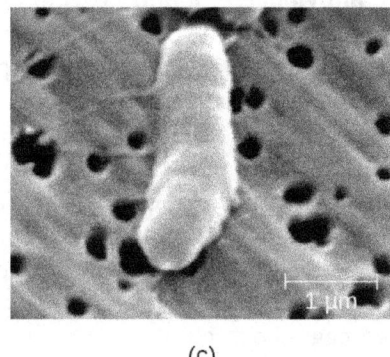
(c)

Figure 24.19 (a) Outbreaks of cholera often occur in areas with poor sanitation or after natural disasters that compromise sanitation infrastructure. (b) At a cholera treatment center in Haiti, patients are receiving intravenous fluids to combat the dehydrating effects of this disease. They often lie on a cot with a hole in it and a bucket underneath to allow for monitoring of fluid loss. (c) This scanning electron micrograph shows *Vibrio cholera*. (credit a, b: modification of work by Centers for Disease Control and Prevention; credit c: modification of work by Janice Carr, Centers for Disease Control and Prevention)

V. cholera is not the only *Vibrio* species that can cause disease. *V. parahemolyticus* is associated with consumption of contaminated seafood and causes gastrointestinal illness with signs and symptoms such as watery diarrhea, nausea, fever, chills, and abdominal cramps. The bacteria produce a heat-stable hemolysin, leading to dysentery and possible disseminated disease. It also sometimes causes wound infections. *V. parahemolyticus* is diagnosed using cultures from blood, stool, or a wound. As with *V. cholera*, selective medium (especially TCBS agar) works well. Tetracycline and ciprofloxacin can be used to treat severe cases, but antibiotics generally are not needed.

Vibrio vulnificus is found in warm seawater and, unlike *V. cholerae*, is not associated with poor sanitary conditions. The bacteria can be found in raw seafood, and ingestion causes gastrointestinal illness. It can also

---

[7] Centers for Disease Control and Prevention. *Cholera—Vibrio cholerae Infection.* Updated November 6, 2014. http://www.cdc.gov/cholera/general. Accessed Sept 14, 2016.

be acquired by individuals with open skin wounds who are exposed to water with high concentrations of the pathogen. In some cases, the infection spreads to the bloodstream and causes septicemia. Skin infection can lead to edema, ecchymosis (discoloration of skin due to bleeding), and abscesses. Patients with underlying disease have a high fatality rate of about 50%. It is of particular concern for individuals with chronic liver disease or who are otherwise immunodeficient because a healthy immune system can often prevent infection from developing. *V. vulnificus* is diagnosed by culturing for the pathogen from stool samples, blood samples, or skin abscesses. Adult patients are treated with doxycycline combined with a third generation cephalosporin or with fluoroquinolones, and children are treated with TMP/SMZ.

Two other vibrios, *Aeromonas hydrophila* and *Plesiomonas shigelloides*, are also associated with marine environments and raw seafood; they can also cause gastroenteritis. Like *V. vulnificus*, *A. hydrophila* is more often associated with infections in wounds, generally those acquired in water. In some cases, it can also cause septicemia. Other species of *Aeromonas* can cause illness. *P. shigelloides* is sometimes associated with more serious systemic infections if ingested in contaminated food or water. Culture can be used to diagnose *A. hydrophila* and *P. shigelloides* infections, for which antibiotic therapy is generally not needed. When necessary, tetracycline and ciprofloxacin, among other antibiotics, may be used for treatment of *A. hydrophila*, and fluoroquinolones and trimethoprim are the effective treatments for *P. shigelloides*.

✓ CHECK YOUR UNDERSTANDING

- How does *V. cholera* infection cause rapid dehydration?

Campylobacter jejuni Gastroenteritis

Campylobacter is a genus of gram-negative, spiral or curved bacteria. They may have one or two flagella. **Campylobacter jejuni gastroenteritis**, a form of campylobacteriosis, is a widespread illness that is caused by *Campylobacter jejuni*. The primary route of transmission is through poultry that becomes contaminated during slaughter. Handling of the raw chicken in turn contaminates cooking surfaces, utensils, and other foods. Unpasteurized milk or contaminated water are also potential vehicles of transmission. In most cases, the illness is self-limiting and includes fever, diarrhea, cramps, vomiting, and sometimes dysentery. More serious signs and symptoms, such as bacteremia, meningitis, pancreatitis, cholecystitis, and hepatitis, sometimes occur. It has also been associated with autoimmune conditions such as Guillain-Barré syndrome, a neurological disease that occurs after some infections and results in temporary paralysis. HUS following infection can also occur. The virulence in many strains is the result of hemolysin production and the presence of *Campylobacter* cytolethal distending toxin (CDT), a powerful deoxyribonuclease (DNase) that irreversibly damages host cell DNA.

Diagnosis involves culture under special conditions, such as elevated temperature, low oxygen tension, and often medium supplemented with antimicrobial agents. These bacteria should be cultured on selective medium (such as Campy CV, charcoal selective medium, or cefaperazone charcoal deoxycholate agar) and incubated under microaerophilic conditions for at least 72 hours at 42 °C. Antibiotic treatment is not usually needed, but erythromycin or ciprofloxacin may be used.

Peptic Ulcers

The gram-negative bacterium *Helicobacter pylori* is able to tolerate the acidic environment of the human stomach and has been shown to be a major cause of **peptic ulcers**, which are ulcers of the stomach or duodenum. The bacterium is also associated with increased risk of stomach cancer (Figure 24.20). According to the CDC, approximately two-thirds of the population is infected with *H. pylori,* but less than 20% have a risk of developing ulcers or stomach cancer. *H. pylori* is found in approximately 80% of stomach ulcers and in over 90% of duodenal ulcers.[8]

---

8 Centers for Disease Control and Prevention. "*Helicobacter pylori*: Fact Sheet for Health Care Providers." Updated July 1998. http://www.cdc.gov/ulcer/files/hpfacts.pdf.

H. pylori colonizes epithelial cells in the stomach using pili for adhesion. These bacteria produce urease, which stimulates an immune response and creates ammonia that neutralizes stomach acids to provide a more hospitable microenvironment. The infection damages the cells of the stomach lining, including those that normally produce the protective mucus that serves as a barrier between the tissue and stomach acid. As a result, inflammation (gastritis) occurs and ulcers may slowly develop. Ulcer formation can also be caused by toxin activity. It has been reported that 50% of clinical isolates of *H. pylori* have detectable levels of exotoxin activity *in vitro*.[9] This toxin, VacA, induces vacuole formation in host cells. VacA has no primary sequence homology with other bacterial toxins, and in a mouse model, there is a correlation between the presence of the toxin gene, the activity of the toxin, and gastric epithelial tissue damage.

Signs and symptoms include nausea, lack of appetite, bloating, burping, and weight loss. Bleeding ulcers may produce dark stools. If no treatment is provided, the ulcers can become deeper, more tissues can be involved, and stomach perforation can occur. Because perforation allows digestive enzymes and acid to leak into the body, it is a very serious condition.

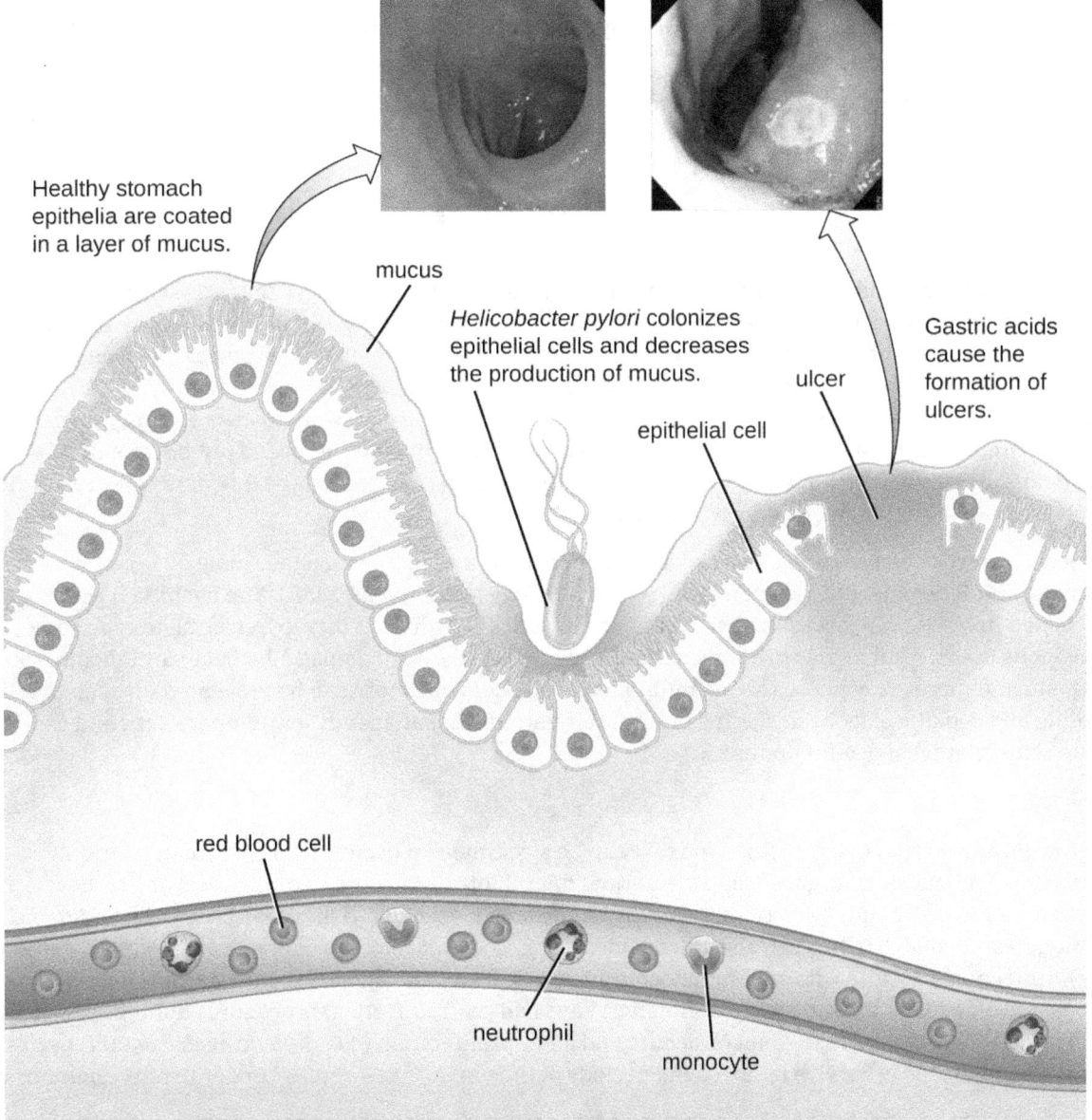

Figure 24.20 *Helicobacter* infection decreases mucus production and causes peptic ulcers. (credit top left photo: modification of work by

---

9 T. L. Cover. "The Vacuolating Cytotoxin of *Helicobacter pylori*." Molecular Microbiology 20 (1996) 2: pp. 241–246. http://www.ncbi.nlm.nih.gov/pubmed/8733223.

"Santhosh Thomas"/YouTube; credit top right photo: modification of work by Moriya M, Uehara A, Okumura T, Miyamoto M, and Kohgo Y)

To diagnose *H. pylori* infection, multiple methods are available. In a breath test, the patient swallows radiolabeled urea. If *H. pylori* is present, the bacteria will produce urease to break down the urea. This reaction produces radiolabeled carbon dioxide that can be detected in the patient's breath. Blood testing can also be used to detect antibodies to *H. pylori*. The bacteria themselves can be detected using either a stool test or a stomach wall biopsy.

Antibiotics can be used to treat the infection. However, unique to *H. pylori*, the recommendation from the US Food and Drug Administration is to use a triple therapy. The current protocols are 10 days of treatment with omeprazole, amoxicillin, and clarithromycin (OAC); 14 days of treatment with bismuth subsalicylate, metronidazole, and tetracycline (BMT); or 10 or 14 days of treatment with lansoprazole, amoxicillin, and clarithromycin (LAC). Omeprazole, bismuth subsalicylate, and lansoprazole are not antibiotics but are instead used to decrease acid levels because *H. pylori* prefers acidic environments.

Although treatment is often valuable, there are also risks to *H. pylori* eradication. Infection with *H. pylori* may actually protect against some cancers, such as esophageal adenocarcinoma and gastroesophageal reflux disease.[10][11]

CHECK YOUR UNDERSTANDING

- How does *H. pylori* cause peptic ulcers?

Clostridium perfringens Gastroenteritis

***Clostridium perfringens* gastroenteritis** is a generally mild foodborne disease that is associated with undercooked meats and other foods. *C. perfringens* is a gram-positive, rod-shaped, endospore-forming anaerobic bacterium that is tolerant of high and low temperatures. At high temperatures, the bacteria can form endospores that will germinate rapidly in foods or within the intestine. Food poisoning by type A strains is common. This strain always produces an enterotoxin, sometimes also present in other strains, that causes the clinical symptoms of cramps and diarrhea. A more severe form of the illness, called pig-bel or enteritis necroticans, causes hemorrhaging, pain, vomiting, and bloating. Gangrene of the intestines may result. This form has a high mortality rate but is rare in the United States.

Diagnosis involves detecting the *C. perfringens* toxin in stool samples using either molecular biology techniques (PCR detection of the toxin gene) or immunology techniques (ELISA). The bacteria itself may also be detected in foods or in fecal samples. Treatment includes rehydration therapy, electrolyte replacement, and intravenous fluids. Antibiotics are not recommended because they can damage the balance of the microbiota in the gut, and there are concerns about antibiotic resistance. The illness can be prevented through proper handling and cooking of foods, including prompt refrigeration at sufficiently low temperatures and cooking food to a sufficiently high temperature.

Clostridium difficile

Clostridium difficile is a gram-positive rod that can be a commensal bacterium as part of the normal microbiota of healthy individuals. When the normal microbiota is disrupted by long-term antibiotic use, it can allow the overgrowth of this bacterium, resulting in **antibiotic-associated diarrhea** caused by *C. difficile*. Antibiotic-associated diarrhea can also be considered a nosocomial disease. Patients at the greatest risk of *C. difficile* infection are those who are immunocompromised, have been in health-care settings for extended periods, are older, have recently taken antibiotics, have had gastrointestinal procedures done, or use proton pump inhibitors, which reduce stomach acidity and allow proliferation of *C. difficile*. Because this species can form endospores, it can survive for extended periods of time in the environment under harsh conditions and is

---

10 Martin J. Blaser. "Disappearing Microbiota: *Helicobacter pylori* Protection against Esophageal Adenocarcinoma." *Cancer Prevention Research* 1 (2008) 5: pp. 308–311. http://cancerpreventionresearch.aacrjournals.org/content/1/5/308.full.pdf+html.

11 Ivan F. N. Hung and Benjamin C. Y. Wong. "Assessing the Risks and Benefits of Treating *Helicobacter pylori* Infection." *Therapeutic Advances in Gastroenterology* 2 (2009) 3: pp, 141–147. doi: 10.1177/1756283X08100279.

a considerable concern in health-care settings.

This bacterium produces two toxins, *Clostridium difficile* toxin A (TcdA) and *Clostridium difficile* toxin B (TcdB). These toxins inactivate small GTP-binding proteins, resulting in actin condensation and cell rounding, followed by cell death. Infections begin with focal necrosis, then ulceration with exudate, and can progress to **pseudomembranous colitis**, which involves inflammation of the colon and the development of a pseudomembrane of fibrin containing dead epithelial cells and leukocytes (Figure 24.21). Watery diarrhea, dehydration, fever, loss of appetite, and abdominal pain can result. Perforation of the colon can occur, leading to septicemia, shock, and death. *C. difficile* is also associated with necrotizing enterocolitis in premature babies and neutropenic enterocolitis associated with cancer therapies.

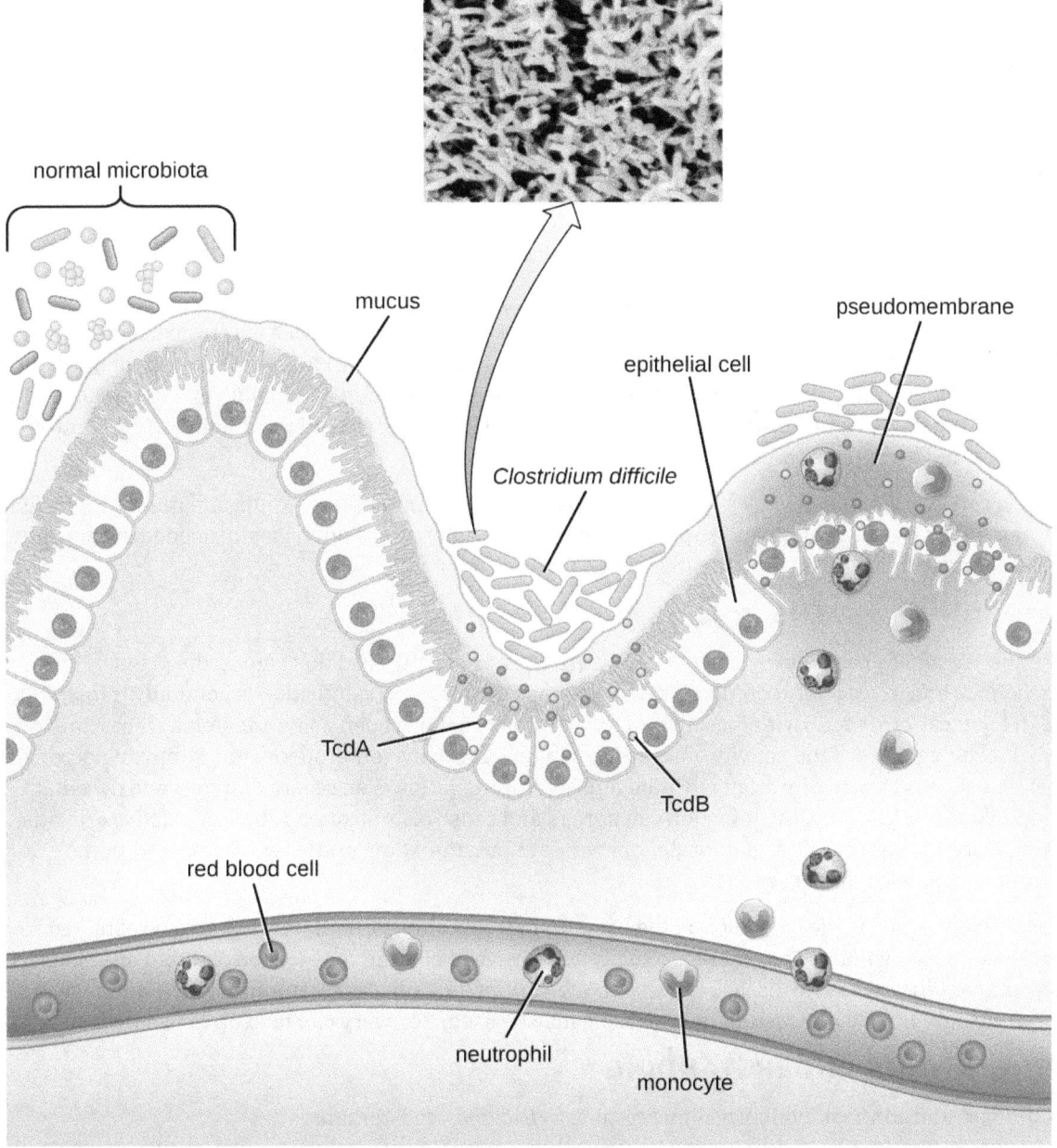

Figure 24.21 *Clostridium difficile* is able to colonize the mucous membrane of the colon when the normal microbiota is disrupted. The toxins TcdA and TcdB trigger an immune response, with neutrophils and monocytes migrating from the bloodstream to the site of infection. Over time, inflammation and dead cells contribute to the development of a pseudomembrane. (credit micrograph: modification of work by Janice Carr, Centers for Disease Control and Prevention)

Diagnosis is made by considering the patient history (such as exposure to antibiotics), clinical presentation, imaging, endoscopy, lab tests, and other available data. Detecting the toxin in stool samples is used to confirm diagnosis. Although culture is preferred, it is rarely practical in clinical practice because the bacterium is an obligate anaerobe. Nucleic acid amplification tests, including PCR, are considered preferable to ELISA testing for molecular analysis.

The first step of conventional treatment is to stop antibiotic use, and then to provide supportive therapy with electrolyte replacement and fluids. Metronidazole is the preferred treatment if the *C. difficile* diagnosis has been confirmed. Vancomycin can also be used, but it should be reserved for patients for whom metronidazole was ineffective or who meet other criteria (e.g., under 10 years of age, pregnant, or allergic to metronidazole).

A newer approach to treatment, known as a fecal transplant, focuses on restoring the microbiota of the gut in order to combat the infection. In this procedure, a healthy individual donates a stool sample, which is mixed with saline and transplanted to the recipient via colonoscopy, endoscopy, sigmoidoscopy, or enema. It has been reported that this procedure has greater than 90% success in resolving *C. difficile* infections.[12]

CHECK YOUR UNDERSTANDING

- How does antibiotic use lead to *C. difficile* infections?

Foodborne Illness Due to *Bacillus cereus*

Bacillus cereus, commonly found in soil, is a gram-positive endospore-forming bacterium that can sometimes cause foodborne illness. *B. cereus* endospores can survive cooking and produce enterotoxins in food after it has been heated; illnesses often occur after eating rice and other prepared foods left at room temperature for too long. The signs and symptoms appear within a few hours of ingestion and include nausea, pain, and abdominal cramps. *B. cereus* produces two toxins: one causing diarrhea, and the other causing vomiting. More severe signs and symptoms can sometimes develop.

Diagnosis can be accomplished by isolating bacteria from stool samples or vomitus and uneaten infected food. Treatment involves rehydration and supportive therapy. Antibiotics are not typically needed, as the illness is usually relatively mild and is due to toxin activity.

Foodborne Illness Due to *Yersinia*

The genus *Yersinia* is best known for *Yersinia pestis*, a gram-negative rod that causes the plague. However, *Y. enterocolitica* and *Y. pseudotuberculosis* can cause gastroenteritis. The infection is generally transmitted through the fecal-oral route, with ingestion of food or water that has been contaminated by feces. Intoxication can also result because of the activity of its endotoxin and exotoxins (enterotoxin and cytotoxin necrotizing factor). The illness is normally relatively mild and self-limiting. However, severe diarrhea and dysentery can develop in infants. In adults, the infection can spread and cause complications such as reactive arthritis, thyroid disorders, endocarditis, glomerulonephritis, eye inflammation, and/or erythema nodosum. Bacteremia may develop in rare cases.

Diagnosis is generally made by detecting the bacteria in stool samples. Samples may also be obtained from other tissues or body fluids. Treatment is usually supportive, including rehydration, without antibiotics. If bacteremia or other systemic disease is present, then antibiotics such as fluoroquinolones, aminoglycosides, doxycycline, and trimethoprim-sulfamethoxazole may be used. Recovery can take up to two weeks.

CHECK YOUR UNDERSTANDING

- Compare and contrast foodborne illnesses due to *B. cereus* and *Yersinia*.

---

12 Faith Rohlke and Neil Stollman. "Fecal Microbiota Transplantation in Relapsing *Clostridium difficile* Infection," *Therapeutic Advances in Gastroenterology* 5 (2012) 6: 403–420. doi: 10.1177/1756283X12453637.

> **Disease Profile**
>
> **Bacterial Infections of the Gastrointestinal Tract**
> Bacterial infections of the gastrointestinal tract generally occur when bacteria or bacterial toxins are ingested in contaminated food or water. Toxins and other virulence factors can produce gastrointestinal inflammation and general symptoms such as diarrhea and vomiting. Bacterial GI infections can vary widely in terms of severity and treatment. Some can be treated with antibiotics, but in other cases antibiotics may be ineffective in combating toxins or even counterproductive if they compromise the GI microbiota. Figure 24.22 and Figure 24.23 the key features of common bacterial GI infections.

Bacterial Infections of the GI Tract

Disease	Pathogen	Signs and Symptoms	Transmission	Diagnostic Tests	Antimicrobial Drugs
Bacillus cereus infection	Bacillus cereus	Nausea, pain, abdominal cramps, diarrhea, or vomiting	Ingestion of contaminated rice or meat, even after cooking	Testing stool sample, vomitus, or uneaten food for presence of bacteria	None
Campylobacter jejuni gastroenteritis	Campylobacter jejuni	Fever, diarrhea, cramps, vomiting, and sometimes dysentery; sometimes more severe organ or autoimmune effects	Ingestion of unpasteurized milk, undercooked chicken, or contaminated water	Culture on selective medium with elevated temperature and low oxygen concentration	Generally none; erythromycin or ciprofloxacin if necessary
Cholera	Vibrio cholerae	Severe diarrhea and fluid loss, potentially leading to shock, renal failure, and death	Ingestion of contaminated water or food	Culture on selective medium (TCBS agar); distinguished as oxidase positive with fermentative metabolisms	Generally none; tetracyclines, azithromycin, others if necessary
Clostridium difficile infection	Clostridium difficile	Pseudomembranous colitis, watery diarrhea, fever, abdominal pain, loss of appetite, dehydration; in severe cases, perforation of the colon, septicemia, shock, and death	Overgrowth of C. difficile in the normal microbiota due to antibiotic use; hospital-acquired infections in immunocompromised patients	Detection of toxin in stool, nucleic acid amplification tests (e.g., PCR)	Discontinuation of previous antibiotic treatment; metronidazole or vancomycin
Clostridium perfringens gastroenteritis	Clostridium perfringens (especially type A)	Mild cramps and diarrhea in most cases; in rare cases, hemorrhaging, vomiting, intestinal gangrene, and death	Ingestion of undercooked meats containing C. perfringens endospores	Detection of toxin or bacteria in stool or uneaten food	None
E. coli infection	ETEC, EPEC, EIEC, EHEC	Watery diarrhea, dysentery, cramps, malaise, fever, chills, dehydration; in EHEC, possible severe complications such as hematolytic uremic syndrome	Ingestion of contaminated food or water	Tissue culture, immunochemical assays, PCR, gene probes	Not recommended for EIEC and EHEC; fluoroquinolones, doxycycline, rifaximin, and TMP/SMZ possible for ETEC and EPEC

Figure 24.22

Bacterial Infections of the GI Tract (continued)					
Disease	Pathogen	Signs and Symptoms	Transmission	Diagnostic Tests	Antimicrobial Drugs
Peptic ulcers	Helicobacter pylori	Nausea, bloating, burping, lack of appetite, weight loss, perforation of stomach, blood in stools	Normal flora, can also be acquired via saliva; fecal-oral route via contaminated food and water	Breath test, detection of antibodies in blood, detection of bacteria in stool sample or stomach biopsy	Amoxicillin, clarithromycin metronidazole, tetracycline, lansoprazole; antacids may also be given in combination with antibiotics
Salmonellosis	Salmonella enterica, serotype Enteritides	Fever, nausea, vomiting, abdominal cramps, headache, diarrhea; can be fatal in infants	Ingestion of contaminated food, handling of eggshells or contaminated animals	Culturing, serotyping and DNA fingerprinting	Not generally recommended; fluoroquinolones, ampicillin, others for immunocompromised patients
Shigella dysentery	Shigella dysenteriae, S. flexneri, S. boydii, and S. sonnei	Abdominal cramps, fever, diarrhea, dysentery; possible complications: reactive arthritis and hemolytic uremic syndrome	Fecal-oral route via contaminated food and water	Testing of stool samples for presence of blood and leukocytes; culturing, PCR, immunoassay for S. dysenteriae	Ciprofloxacin, azithromycin
Staphylococcal food poisoning	Staphylococcus aureus	Rapid-onset nausea, diarrhea, vomiting lasting 24–48 hours; possible dehydration and change in blood pressure and heart rate	Ingestion of raw or undercooked meat or dairy products contaminated with staphylococcal enterotoxins	ELISA to detect enterotoxins in uneaten food, stool, or vomitus	None
Typhoid fever	S. entrica, subtypes Typhi or Paratyphi	Aches, headaches, nausea, lethargy, diarrhea or constipation, possible rash; lethal perforation of intestine can occur	Fecal-oral route; may be spread by asymptomatic carriers	Culture of blood, stool, or bone marrow, serologic tests; PCR tests when available	Fluoroquinolones, ceftriaxone, azithromycin; preventive vaccine available
Yersinia infection	Yersinia enterocolitica, Y. pseudotuberculosis	Generally mild diarrhea and abdominal cramps; in some cases, bacteremia can occur, leading to severe complications	Fecal-oral route, typically via contaminated food or water	Testing stool samples, tissues, body fluids	Generally none; fluoroquinolones, aminoglycosides, others for systemic infections

Figure 24.23

Clinical Focus

Part 2

At the hospital, Carli's doctor began to think about possible causes of her severe gastrointestinal distress.

One possibility was food poisoning, but no one else in her family was sick. The doctor asked about what Carli had eaten the previous day; her mother mentioned that she'd had eggs for lunch, and that they may have been a little undercooked. The doctor took a sample of Carli's stool and sent it for laboratory testing as part of her workup. She suspected that Carli could have a case of bacterial or viral gastroenteritis, but she needed to know the cause in order to prescribe an appropriate treatment.

In the laboratory, technicians microscopically identified gram-negative bacilli in Carli's stool sample. They also established a pure culture of the bacteria and analyzed it for antigens. This testing showed that the causative agent was *Salmonella*.

- What should the doctor do now to treat Carli?

Jump to the next Clinical Focus box. Go back to the previous Clinical Focus box.

24.4 Viral Infections of the Gastrointestinal Tract

Learning Objectives

By the end of this section, you will be able to:
- Identify the most common viruses that can cause infections of the GI tract
- Compare the major characteristics of specific viral diseases affecting the GI tract and liver

In the developing world, acute viral gastroenteritis is devastating and a leading cause of death for children.[13] Worldwide, diarrhea is the second leading cause of mortality for children under age five, and 70% of childhood gastroenteritis is viral.[14] As discussed, there are a number of bacteria responsible for diarrhea, but viruses can also cause diarrhea. *E. coli* and rotavirus are the most common causative agents in the developing world. In this section, we will discuss rotaviruses and other, less common viruses that can also cause gastrointestinal illnesses.

Gastroenteritis Caused by Rotaviruses

Rotaviruses are double-stranded RNA viruses in the family Reoviridae. They are responsible for common diarrheal illness, although prevention through vaccination is becoming more common. The virus is primarily spread by the fecal-oral route (Figure 24.24).

---

13 Caleb K. King, Roger Glass, Joseph S. Bresee, Christopher Duggan. "Managing Acute Gastroenteritis Among Children: Oral Rehydration, Maintenance, and Nutritional Therapy." *MMWR* 52 (2003) RR16: pp. 1–16. http://www.cdc.gov/mmwr/preview/mmwrhtml/rr5216a1.htm.

14 Elizabeth Jane Elliott. "Acute Gastroenteritis in Children." *British Medical Journal* 334 (2007) 7583: 35–40, doi: 10.1136/bmj.39036.406169.80; S. Ramani and G. Kang. "Viruses Causing Diarrhoea in the Developing World." *Current Opinions in Infectious Diseases* 22 (2009) 5: pp. 477–482. doi: 10.1097/QCO.0b013e328330662f; Michael Vincent F Tablang. "Viral Gastroenteritis." *Medscape*. http://emedicine.medscape.com/article/176515-overview.

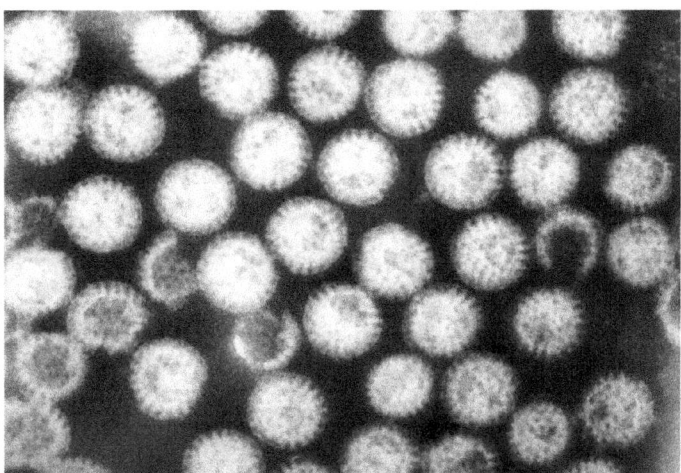

Figure 24.24 Rotaviruses in a fecal sample are visualized using electron microscopy. (credit: Dr. Graham Beards)

These viruses are widespread in children, especially in day-care centers. The CDC estimates that 95% of children in the United States have had at least one rotavirus infection by the time they reach age five.[15] Due to the memory of the body's immune system, adults who come into contact with rotavirus will not contract the infection or, if they do, are asymptomatic. The elderly, however, are vulnerable to rotavirus infection due to weakening of the immune system with age, so infections can spread through nursing homes and similar facilities. In these cases, the infection may be transmitted from a family member who may have subclinical or clinical disease. The virus can also be transmitted from contaminated surfaces, on which it can survive for some time.

Infected individuals exhibit fever, vomiting, and diarrhea. The virus can survive in the stomach following a meal, but is normally found in the small intestines, particularly the epithelial cells on the villi. Infection can cause food intolerance, especially with respect to lactose. The illness generally appears after an incubation period of about two days and lasts for approximately one week (three to eight days). Without supportive treatment, the illness can cause severe fluid loss, dehydration, and even death. Even with milder illness, repeated infections can potentially lead to malnutrition, especially in developing countries, where rotavirus infection is common due to poor sanitation and lack of access to clean drinking water. Patients (especially children) who are malnourished after an episode of diarrhea are more susceptible to future diarrheal illness, increasing their risk of death from rotavirus infection.

The most common clinical tool for diagnosis is enzyme immunoassay, which detects the virus from fecal samples. Latex agglutination assays are also used. Additionally, the virus can be detected using electron microscopy and RT-PCR.

Treatment is supportive with oral rehydration therapy. Preventive vaccination is also available. In the United States, rotavirus vaccines are part of the standard vaccine schedule and administration follows the guidelines of the World Health Organization (WHO). The WHO recommends that all infants worldwide receive the rotavirus vaccine, the first dose between six and 15 weeks of age and the second before 32 weeks.[16]

Gastroenteritis Caused by Noroviruses

Noroviruses, commonly identified as Norwalk viruses, are caliciviruses. Several strains can cause gastroenteritis. There are millions of cases a year, predominately in infants, young children, and the elderly. These viruses are easily transmitted and highly contagious. They are known for causing widespread infections in groups of people in confined spaces, such as on cruise ships. The viruses can be transmitted through direct contact, through touching contaminated surfaces, and through contaminated food. Because the virus is not

---

15 Centers for Disease Control and Prevention. "Rotavirus," *The Pink Book*. Updated September 8, 2015. http://www.cdc.gov/vaccines/pubs/pinkbook/rota.html.

16 World Health Organization. "Rotavirus." *Immunization, Vaccines, and Biologicals*. Updated April 21, 2010. http://www.who.int/immunization/topics/rotavirus/en/.

killed by disinfectants used at standard concentrations for killing bacteria, the risk of transmission remains high, even after cleaning.

The signs and symptoms of norovirus infection are similar to those for rotavirus, with watery diarrhea, mild cramps, and fever. Additionally, these viruses sometimes cause projectile vomiting. The illness is usually relatively mild, develops 12 to 48 hours after exposure, and clears within a couple of days without treatment. However, dehydration may occur.

Norovirus can be detected using PCR or enzyme immunoassay (EIA) testing. RT-qPCR is the preferred approach as EIA is insufficiently sensitive. If EIA is used for rapid testing, diagnosis should be confirmed using PCR. No medications are available, but the illness is usually self-limiting. Rehydration therapy and electrolyte replacement may be used. Good hygiene, hand washing, and careful food preparation reduce the risk of infection.

Gastroenteritis Caused by Astroviruses

Astroviruses are single-stranded RNA viruses (family Astroviridae) that can cause severe gastroenteritis, especially in infants and children. Signs and symptoms include diarrhea, nausea, vomiting, fever, abdominal pain, headache, and malaise. The viruses are transmitted through the fecal-oral route (contaminated food or water). For diagnosis, stool samples are analyzed. Testing may involve enzyme immunoassays and immune electron microscopy. Treatment involves supportive rehydration and electrolyte replacement if needed.

✓ CHECK YOUR UNDERSTANDING

- Why are rotaviruses, noroviruses, and astroviruses more common in children?

Disease Profile

Viral Infections of the Gastrointestinal Tract

A number of viruses can cause gastroenteritis, characterized by inflammation of the GI tract and other signs and symptoms with a range of severities. As with bacterial GI infections, some cases can be relatively mild and self-limiting, while others can become serious and require intensive treatment. Antimicrobial drugs are generally not used to treat viral gastroenteritis; generally, these illnesses can be treated effectively with rehydration therapy to replace fluids lost in bouts of diarrhea and vomiting. Because most viral causes of gastroenteritis are quite contagious, the best preventive measures involve avoiding and/or isolating infected individuals and limiting transmission through good hygiene and sanitation.

| Viral Causes of Gastroenteritis ||||||
Disease	Pathogen	Signs and Symptoms	Transmission	Diagnostic Tests	Vaccine
Astrovirus gastroenteritis	Astroviruses	Fever, headache, abdominal pain, malaise, diarrhea, vomiting	Fecal-oral route, contaminated food or water	Enzyme immunoassays, immune electron microscopy	None
Norovirus gastroenteritis	Noroviruses	Fever, diarrhea, projectile vomiting, dehydration; generally self-limiting within two days	Highly contagious via direct contact or contact with contaminated food or fomites	Rapid enzyme immunoassay confirmed with RT-qPCR	None
Rotavirus gastroenteritis	Rotaviruses	Fever, diarrhea, vomiting, severe dehydration; recurring infections can lead to malnutrition and death	Fecal-oral route; children and elderly most susceptible	Enzyme immunoassay of stool sample, latex agglutination assays, RT-PCR	Preventive vaccine recommended for infants

Figure 24.25

Hepatitis

Hepatitis is a general term meaning inflammation of the liver, which can have a variety of causes. In some cases, the cause is viral infection. There are five main hepatitis viruses that are clinically significant: hepatitisviruses A (HAV), B (HBV), C (HCV), D, (HDV) and E (HEV) (Figure 24.26). Note that other viruses, such as Epstein-Barr virus (EBV), yellow fever, and cytomegalovirus (CMV) can also cause hepatitis and are discussed in Viral Infections of the Circulatory and Lymphatic Systems.

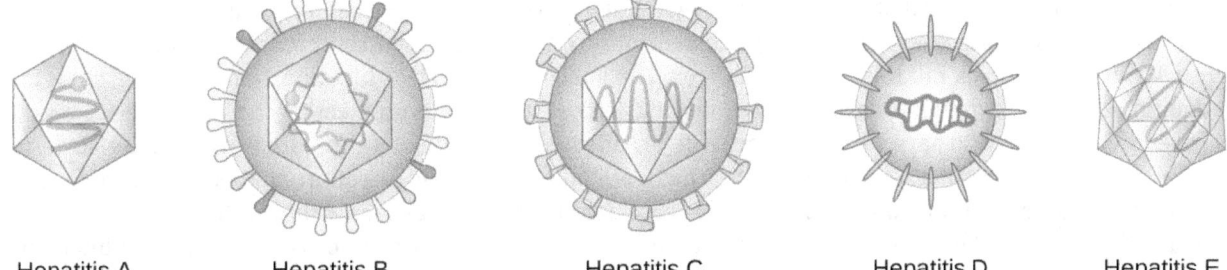

Hepatitis A   Hepatitis B   Hepatitis C   Hepatitis D   Hepatitis E

Figure 24.26 Five main types of viruses cause hepatitis. HAV is a non-enveloped ssRNA(+) virus and is a member of the picornavirus family (Baltimore Group IV). HBV is a dsDNA enveloped virus, replicates using reverse transcriptase, and is a member of the hepadnavirus family (Baltimore Group VII). HCV is an enveloped ssRNA(+) virus and is a member of the flavivirus family (Baltimore Group IV). HDV is an enveloped ssRNA(−) that is circular (Baltimore Group V). This virus can only propagate in the presence of HBV. HEV is a non-enveloped ssRNA(+) virus and a member of the hepeviridae family (Baltimore Group IV).

Although the five hepatitis viruses differ, they can cause some similar signs and symptoms because they all have an affinity for hepatocytes (liver cells). HAV and HEV can be contracted through ingestion while HBV, HCV, and HDV are transmitted by parenteral contact. It is possible for individuals to become long term or chronic carriers of hepatitis viruses.

The virus enters the blood (viremia), spreading to the spleen, the kidneys, and the liver. During viral replication, the virus infects hepatocytes. The inflammation is caused by the hepatocytes replicating and releasing more hepatitis virus. Signs and symptoms include malaise, anorexia, loss of appetite, dark urine, pain in the upper right quadrant of the abdomen, vomiting, nausea, diarrhea, joint pain, and gray stool. Additionally, when the liver is diseased or injured, it is unable to break down hemoglobin effectively, and bilirubin can build up in the body, giving the skin and mucous membranes a yellowish color, a condition called **jaundice** (Figure 24.27). In severe cases, death from liver necrosis may occur.

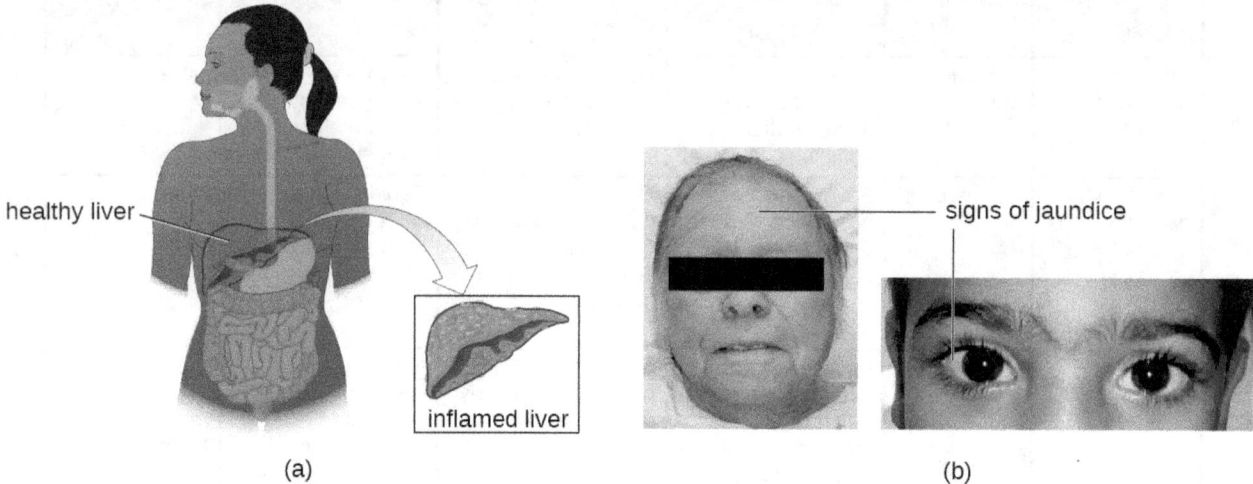

Figure 24.27 (a) Hepatitis is inflammation of the liver resulting from a variety of root causes. It can cause jaundice. (b) Jaundice is characterized by yellowing of the skin, mucous membranes, and sclera of the eyes. (credit b left: modification of work by James Heilman, MD; credit b right: modification of work by "Sab3el3eish"/Wikimedia Commons)

Despite having many similarities, each of the hepatitis viruses has its own unique characteristics. HAV is generally transmitted through the fecal-oral route, close personal contact, or exposure to contaminated water or food. Hepatitis A can develop after an incubation period of 15 to 50 days (the mean is 30). It is normally mild or even asymptomatic and is usually self-limiting within weeks to months. A more severe form, fulminant hepatitis, rarely occurs but has a high fatality rate of 70–80%. Vaccination is available and is recommended especially for children (between ages one and two), those traveling to countries with higher risk, those with liver disease and certain other conditions, and drug users.

Although HBV is associated with similar signs and symptoms, transmission and outcomes differ. This virus has a mean incubation period of 120 days and is generally associated with exposure to infectious blood or body fluids such as semen or saliva. Exposure can occur through skin puncture, across the placenta, or through mucosal contact, but it is not spread through casual contact such as hugging, hand holding, sneezing, or coughing, or even through breastfeeding or kissing. Risk of infection is greatest for those who use intravenous drugs or who have sexual contact with an infected individual. Health-care workers are also at risk from needle sticks and other injuries when treating infected patients. The infection can become chronic and may progress to cirrhosis or liver failure. It is also associated with liver cancer. Chronic infections are associated with the highest mortality rates and are more common in infants. Approximately 90% of infected infants become chronic carriers, compared with only 6–10% of infected adults.[17] Vaccination is available and is recommended for children as part of the standard vaccination schedule (one dose at birth and the second by 18 months of age) and for adults at greater risk (e.g., those with certain diseases, intravenous drug users, and those who have sex with multiple partners). Health-care agencies are required to offer the HBV vaccine to all workers who have occupational exposure to blood and/or other infectious materials.

HCV is often undiagnosed and therefore may be more widespread than is documented. It has a mean incubation period of 45 days and is transmitted through contact with infected blood. Although some cases are asymptomatic and/or resolve spontaneously, 75%–85% of infected individuals become chronic carriers. Nearly all cases result from parenteral transmission often associated with IV drug use or transfusions. The risk is greatest for individuals with past or current history of intravenous drug use or who have had sexual contact with infected individuals. It has also been spread through contaminated blood products and can even be transmitted through contaminated personal products such as toothbrushes and razors. New medications have recently been developed that show great effectiveness in treating HCV and that are tailored to the specific

---

17 Centers for Disease Control and Prevention. "The ABCs of Hepatitis." Updated 2016. http://www.cdc.gov/hepatitis/resources/professionals/pdfs/abctable.pdf.

genotype causing the infection.

HDV is uncommon in the United States and only occurs in individuals who are already infected with HBV, which it requires for replication. Therefore, vaccination against HBV is also protective against HDV infection. HDV is transmitted through contact with infected blood.

HEV infections are also rare in the United States but many individuals have a positive antibody titer for HEV. The virus is most commonly spread by the fecal-oral route through food and/or water contamination, or person-to-person contact, depending on the genotype of the virus, which varies by location. There are four genotypes that differ somewhat in their mode of transmission, distribution, and other factors (for example, two are zoonotic and two are not, and only one causes chronic infection). Genotypes three and four are only transmitted through food, while genotypes one and two are also transmitted through water and fecal-oral routes. Genotype one is the only type transmitted person-to-person and is the most common cause of HEV outbreaks. Consumption of undercooked meat, especially deer or pork, and shellfish can lead to infection. Genotypes three and four are zoonoses, so they can be transmitted from infected animals that are consumed. Pregnant women are at particular risk. This disease is usually self-limiting within two weeks and does not appear to cause chronic infection.

General laboratory testing for hepatitis begins with blood testing to examine liver function (Figure 24.28). When the liver is not functioning normally, the blood will contain elevated levels of alkaline phosphatase, alanine aminotransferase (ALT), aspartate aminotransferase (AST), direct bilirubin, total bilirubin, serum albumin, serum total protein, and calculated globulin, albumin/globulin (A/G) ratio. Some of these are included in a complete metabolic panel (CMP), which may first suggest a possible liver problem and indicate the need for more comprehensive testing. A hepatitis virus serological test panel can be used to detect antibodies for hepatitis viruses A, B, C, and sometimes D. Additionally, other immunological and genomic tests are available.

Specific treatments other than supportive therapy, rest, and fluids are often not available for hepatitis virus infection, except for HCV, which is often self-limited. Immunoglobulins can be used prophylactically following possible exposure. Medications are also used, including interferon alpha 2b and antivirals (e.g., lamivudine, entecavir, adefovir, and telbivudine) for chronic infections. Hepatitis C can be treated with interferon (as monotherapy or combined with other treatments), protease inhibitors, and other antivirals (e.g., the polymerase inhibitor sofosbuvir). Combination treatments are commonly used. Antiviral and immunosuppressive medications may be used for chronic cases of HEV. In severe cases, liver transplants may be necessary. Additionally, vaccines are available to prevent infection with HAV and HBV. The HAV vaccine is also protective against HEV. The HBV vaccine is also protective against HDV. There is no vaccine against HCV.

LINK TO LEARNING

Learn more information about hepatitis virus (https://www.cdc.gov/hepatitis/resources/professionals/pdfs/abctable.pdf) infections.

CHECK YOUR UNDERSTANDING

- Why do the five different hepatitis viruses all cause similar signs and symptoms?

 MICRO CONNECTIONS

Preventing HBV Transmission in Health-Care Settings

Hepatitis B was once a leading on-the-job hazard for health-care workers. Many health-care workers over the years have become infected, some developing cirrhosis and liver cancer. In 1982, the CDC recommended that health-care workers be vaccinated against HBV, and rates of infection have declined since then. Even though vaccination is now common, it is not always effective and not all individuals are vaccinated. Therefore, there is still a small risk for infection, especially for health-care workers working with individuals who have chronic infections, such as drug addicts, and for those with higher risk of needle sticks, such as phlebotomists. Dentists are also at risk.

Health-care workers need to take appropriate precautions to prevent infection by HBV and other illnesses. Blood is the greatest risk, but other body fluids can also transmit infection. Damaged skin, as occurs with eczema or psoriasis, can also allow transmission. Avoiding contact with body fluids, especially blood, by wearing gloves and face protection and using disposable syringes and needles reduce the risk of infection. Washing exposed skin with soap and water is recommended. Antiseptics may also be used, but may not help. Post-exposure treatment, including treatment with hepatitis B immunoglobulin (HBIG) and vaccination, may be used in the event of exposure to the virus from an infected patient. Detailed protocols are available for managing these situations. The virus can remain infective for up to seven days when on surfaces, even if no blood or other fluids are visible, so it is important to consider the best choices for disinfecting and sterilizing equipment that could potentially transmit the virus. The CDC recommends a solution of 10% bleach to disinfect surfaces.[18] Finally, testing blood products is important to reduce the risk of transmission during transfusions and similar procedures.

> Disease Profile
>
> **Viral Hepatitis**
> Hepatitis involves inflammation of the liver that typically manifests with signs and symptoms such as jaundice, nausea, vomiting, joint pain, gray stool, and loss of appetite. However, the severity and duration of the disease can vary greatly depending on the causative agent. Some infections may be completely asymptomatic, whereas others may be life threatening. The five different viruses capable of causing hepatitis are compared in Figure 24.28. For the sake of comparison, this table presents only the unique aspects of each form of viral hepatitis, not the commonalities.

---

18 Centers for Disease Control and Prevention. "Hepatitis B FAQs for Health Professionals." Updated August 4, 2016. http://www.cdc.gov/hepatitis/HBV/HBVfaq.htm.

Viral Forms of Hepatitis					
Disease	Pathogen	Signs and Symptoms	Transmission	Antimicrobial Drugs	Vaccine
Hepatitis A	Hepatitisvirus A (HAV)	Usually asymptomatic or mild and self-limiting within one to two weeks to a few months, sometimes longer but not chronic; in rare cases leads to serious or fatal fulminant hepatitis	Contaminated food, water, objects, and person to person	None	Vaccine recommended for one year olds and high-risk adults
Hepatitis B	Hepatitisvirus B (HBV)	Similar to Hepatitis A, but may progress to cirrhosis and liver failure; associated with liver cancer	Contact with infected body fluids (blood, semen, saliva), e.g., via IV drug use, sexual transmission, health-care workers treating infected patients	Interferon, entecavir, tenofovir, lamivudine, adefovir	Vaccine recommended for infants and high-risk adults
Hepatitis C	Hepatitisvirus C (HCV)	Often asymptomatic, with 75%–85% chronic carriers; may progress to cirrhosis and liver failure; associated with liver cancer	Contact with infected body fluids, e.g., via IV drug use, transfusions, sexual transmission	Depends on genotype and on whether cirrhosis is present; interferons, new treatment such as simeprevir plus sofosbuvir, ombitasvir/paritaprevir/ritonavir and dasabuvir	None available
Hepatitis D	Hepatitisvirus D (HDV)	Similar to hepatitis B; usually self-limiting within one to two weeks but can become chronic or fulminant in rare cases	Contact with infected blood; infections can only occur in patients already infected with hepatitis B	None	Hepatitis B vaccine protects against HDV
Hepatitis E	Hepatitisvirus E (HEV)	Generally asymptomatic or mild and self-limiting; typically does not cause chronic disease	Fecal-oral route, often in contaminated water or undercooked meat; most common in developing countries	Supportive treatment; usually self-limiting, but some strains can become chronic; antiviral and immunosuppressive possible for chronic cases	Vaccine available in China only

Figure 24.28

24.5 Protozoan Infections of the Gastrointestinal Tract

Learning Objectives

By the end of this section, you will be able to:
- Identify the most common protozoans that can cause infections of the GI tract
- Compare the major characteristics of specific protozoan diseases affecting the GI tract

Like other microbes, protozoa are abundant in natural microbiota but can also be associated with significant illness. Gastrointestinal diseases caused by protozoa are generally associated with exposure to contaminated food and water, meaning that those without access to good sanitation are at greatest risk. Even in developed countries, infections can occur and these microbes have sometimes caused significant outbreaks from contamination of public water supplies.

Giardiasis

Also called backpacker's diarrhea or beaver fever, **giardiasis** is a common disease in the United States caused by the flagellated protist *Giardia lamblia*, also known as *Giardia intestinalis* or *Giardia duodenalis* (Figure 1.16). To establish infection, *G. lamblia* uses a large adhesive disk to attach to the intestinal mucosa. The disk is comprised of microtubules. During adhesion, the flagella of *G. lamblia* move in a manner that draws fluid out from under the disk, resulting in an area of lower pressure that promotes its adhesion to the intestinal epithelial cells. Due to its attachment, *Giardia* also blocks absorption of nutrients, including fats.

Transmission occurs through contaminated food or water or directly from person to person. Children in day-care centers are at risk due to their tendency to put items into their mouths that may be contaminated. Large outbreaks may occur if a public water supply becomes contaminated. *Giardia* have a resistant cyst stage in their life cycle that is able to survive cold temperatures and the chlorination treatment typically used for drinking water in municipal reservoirs. As a result, municipal water must be filtered to trap and remove these cysts. Once consumed by the host, *Giardia* develops into the active tropozoite.

Infected individuals may be asymptomatic or have gastrointestinal signs and symptoms, sometimes accompanied by weight loss. Common symptoms, which appear one to three weeks after exposure, include diarrhea, nausea, stomach cramps, gas, greasy stool (because fat absorption is being blocked), and possible dehydration. The parasite remains in the colon and does not cause systemic infection. Signs and symptoms generally clear within two to six weeks. Chronic infections may develop and are often resistant to treatment. These are associated with weight loss, episodic diarrhea, and malabsorption syndrome due to the blocked nutrient absorption.

Diagnosis may be made using observation under the microscope. A stool ova and parasite (O&P) exam involves direct examination of a stool sample for the presence of cysts and trophozoites; it can be used to distinguish common parasitic intestinal infections. ELISA and other immunoassay tests, including commercial direct fluorescence antibody kits, are also used. The most common treatments use metronidazole as the first-line choice, followed by tinidazole. If the infection becomes chronic, the parasites may become resistant to medications.

Cryptosporidiosis

Another protozoan intestinal illness is **cryptosporidiosis**, which is usually caused by *Cryptosporidium parvum* or *C. hominis*. (Figure 24.29) These pathogens are commonly found in animals and can be spread in feces from mice, birds, and farm animals. Contaminated water and food are most commonly responsible for transmission. The protozoan can also be transmitted through human contact with infected animals or their feces.

In the United States, outbreaks of cryptosporidiosis generally occur through contamination of the public water supply or contaminated water at water parks, swimming pools, and day-care centers. The risk is greatest in areas with poor sanitation, making the disease more common in developing countries.

Signs and symptoms include watery diarrhea, nausea, vomiting, cramps, fever, dehydration, and weight loss. The illness is generally self-limiting within a month. However, immunocompromised patients, such as those with HIV/AIDS, are at particular risk of severe illness or death.

Diagnosis involves direct examination of stool samples, often over multiple days. As with giardiasis, a stool O&P exam may be helpful. Acid fast staining is often used. Enzyme immunoassays and molecular analysis (PCR) are available.

The first line of treatment is typically oral rehydration therapy. Medications are sometimes used to treat the diarrhea. The broad-range anti-parasitic drug nitazoxanide can be used to treat cryptosporidiosis. Other anti-parasitic drugs that can be used include azithromycin and paromomycin.

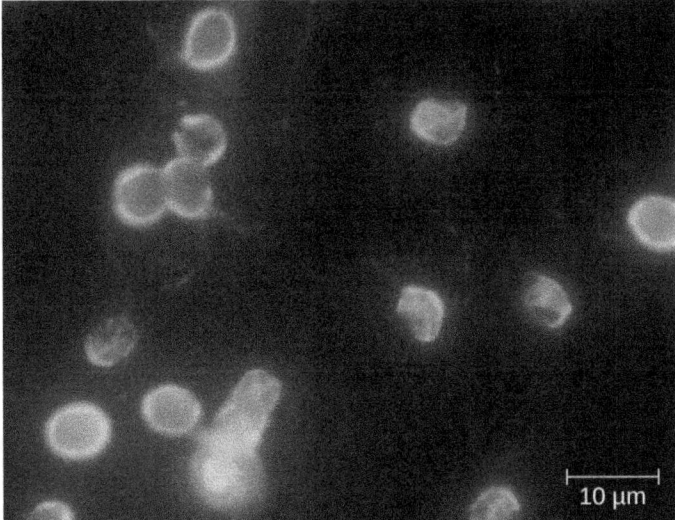

Figure 24.29 Immunofluorescent staining allows for visualization of *Cryptosporidium* spp. (credit: modification of work by EPA/H.D.A. Lindquist)

Amoebiasis (Amebiasis)

The protozoan parasite *Entamoeba histolytica* causes **amoebiasis**, which is known as **amoebic dysentery** in severe cases. *E. histolytica* is generally transmitted through water or food that has fecal contamination. The disease is most widespread in the developing world and is one of the leading causes of mortality from parasitic disease worldwide. Disease can be caused by as few as 10 cysts being transmitted.

Signs and symptoms range from nonexistent to mild diarrhea to severe amoebic dysentery. Severe infection causes the abdomen to become distended and may be associated with fever. The parasite may live in the colon without causing signs or symptoms or may invade the mucosa to cause colitis. In some cases, the disease spreads to the spleen, brain, genitourinary tract, or lungs. In particular, it may spread to the liver and cause an abscess. When a liver abscess develops, fever, nausea, liver tenderness, weight loss, and pain in the right abdominal quadrant may occur. Chronic infection may occur and is associated with intermittent diarrhea, mucus, pain, flatulence, and weight loss.

Direct examination of fecal specimens may be used for diagnosis. As with cryptosporidiosis, samples are often examined on multiple days. A stool O&P exam of fecal or biopsy specimens may be helpful. Immunoassay, serology, biopsy, molecular, and antibody detection tests are available. Enzyme immunoassay may not distinguish current from past illness. Magnetic resonance imaging (MRI) can be used to detect any liver abscesses. The first line of treatment is metronidazole or tinidazole, followed by diloxanide furoate, iodoquinol, or paromomycin to eliminate the cysts that remain.

Cyclosporiasis

The intestinal disease **cyclosporiasis** is caused by the protozoan *Cyclospora cayetanensis*. It is endemic to tropical and subtropical regions and therefore uncommon in the United States, although there have been outbreaks associated with contaminated produce imported from regions where the protozoan is more common.

This protist is transmitted through contaminated food and water and reaches the lining of the small intestine, where it causes infection. Signs and symptoms begin within seven to ten days after ingestion. Based on limited data, it appears to be seasonal in ways that differ regionally and that are poorly understood.[19]

Some individuals do not develop signs or symptoms. Those who do may exhibit explosive and watery diarrhea,

---

19 Centers for Disease Control and Prevention. "Cyclosporiasis FAQs for Health Professionals." Updated June 13, 2014. http://www.cdc.gov/parasites/cyclosporiasis/health_professionals/hp-faqs.html.

fever, nausea, vomiting, cramps, loss of appetite, fatigue, and bloating. These symptoms may last for months without treatment. Trimethoprim-sulfamethoxazole is the recommended treatment.

Microscopic examination is used for diagnosis. A stool O&P examination may be helpful. The oocysts have a distinctive blue halo when viewed using ultraviolet fluorescence microscopy (Figure 24.30).

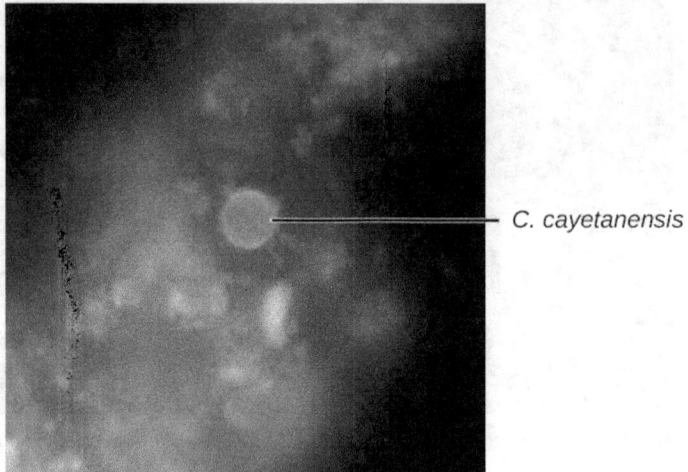

Figure 24.30 *Cyclospora cayetanensis* are autofluorescent under ultraviolet light. (credit: modification of work by Centers for Disease Control and Prevention)

CHECK YOUR UNDERSTANDING

- Which protozoan GI infections are common in the United States?

Disease Profile

Protozoan Gastrointestinal Infections

Protozoan GI infections are generally transmitted through contaminated food or water, triggering diarrhea and vomiting that can lead to dehydration. Rehydration therapy is an important aspect of treatment, but most protozoan GI infections can also be treated with drugs that target protozoans.

Protozoan Infections of the GI Tract					
Disease	Pathogen	Signs and Symptoms	Transmission	Diagnostic Tests	Antimicrobial Drugs
Amoebiasis (amoebic dysentery)	*Entamoeba histolytica*	From mild diarrhea to severe dysentery and colitis; may cause abscess on the liver	Fecal-oral route; ingestion of cysts from fecally contaminated water, food, or hands	Stool O&P exam, enzyme immunoassay	Metronidazole, tinidazole, diloxanide furoate, iodoquinol, paromomycin
Cryptosporidiosis	*Cryptosporidium parvum*, *Cryptosporidium hominis*	Watery diarrhea, nausea, vomiting, cramps, fever, dehydration, and weight loss	Contact with feces of infected mice, birds, farm animals; ingestion of contaminated food or water; exposure to contaminated water while swimming or bathing	Stool O&P exam, enzyme immunoassay, PCR	Nitazoxanide, azithromycin, and paromomycin
Cyclosporiasis	*Cyclospora cayetanensis*	Explosive diarrhea, fever, nausea, vomiting, cramps, loss of appetite, fatigue, bloating	Ingestion of contaminated food or water	Stool O&P exam using ultraviolet fluorescence microscopy	Trimethoprim-sulfmethoxazole
Giardiasis	*Giardia lamblia*	Diarrhea, nausea, stomach cramps, gas, greasy stool, dehydration if severe; sometimes malabsorption syndrome	Contact with infected individual or contaminated fomites; ingestion of contaminated food or water	Stool O&P exam; ELISA, direct fluorescence antibody assays	Metronidazole, tinidazole

Figure 24.31

24.6 Helminthic Infections of the Gastrointestinal Tract

Learning Objectives

By the end of this section, you will be able to:
- Identify the most common helminths that cause infections of the GI tract
- Compare the major characteristics of specific helminthic diseases affecting GI tract

Helminths are widespread intestinal parasites. These parasites can be divided into three common groups: round-bodied worms also described as nematodes, flat-bodied worms that are segmented (also described as cestodes), and flat-bodied worms that are non-segmented (also described as trematodes). The nematodes include roundworms, pinworms, hookworms, and whipworms. Cestodes include beef, pork, and fish tapeworms. Trematodes are collectively called flukes and more uniquely identified with the body site where the adult flukes are located. Although infection can have serious consequences, many of these parasites are so well adapted to the human host that there is little obvious disease.

Ascariasis

Infections caused by the large nematode roundworm *Ascaris lumbricoides*, a soil-transmitted helminth, are called **ascariasis**. Over 800 million to 1 billion people are estimated to be infected worldwide.[20] Infections are most common in warmer climates and at warmer times of year. At present, infections are uncommon in the United States. The eggs of the worms are transmitted through contaminated food and water. This may happen if food is grown in contaminated soil, including when manure is used as fertilizer.

When an individual consumes embryonated eggs (those with a developing embryo), the eggs travel to the intestine and the larvae are able to hatch. *Ascaris* is able to produce proteases that allow for penetration and degradation of host tissue. The juvenile worms can then enter the circulatory system and migrate to the lungs where they enter the alveoli (air sacs). From here they crawl to the pharynx and then follow the gut lumen to return to the small intestine, where they mature into adult roundworms. Females in the host will produce and release eggs that leave the host via feces. In some cases, the worms can block ducts such as those of the pancreas or gallbladder.

The infection is commonly asymptomatic. When signs and symptoms are present, they include shortness of breath, cough, nausea, diarrhea, blood in the stool, abdominal pain, weight loss, and fatigue. The roundworms may be visible in the stool. In severe cases, children with substantial infections may experience intestinal blockage.

The eggs can be identified by microscopic examination of the stool (Figure 24.32). In some cases, the worms themselves may be identified if coughed up or excreted in stool. They can also sometimes be identified by X-rays, ultrasounds, or MRIs.

Ascariasis is self-limiting, but can last one to two years because the worms can inhibit the body's inflammatory response through glycan gimmickry (see Virulence Factors of Eukaryotic Pathogens). The first line of treatment is mebendazole or albendazole. In some severe cases, surgery may be required.

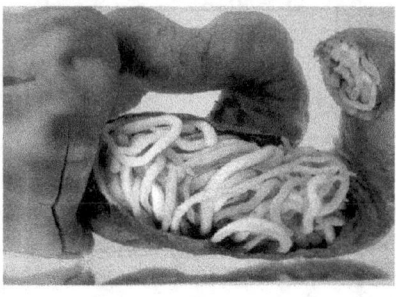

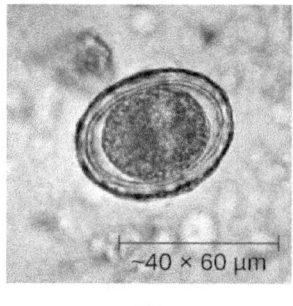

(a) (b) (c)

Figure 24.32 (a) Adult *Ascaris lumbricoides* roundworms can cause intestinal blockage. (b) This mass of *A. lumbricoides* worms was excreted by a child. (c) A micrograph of a fertilized egg of *A. lumbricoides*. Fertilized eggs can be distinguished from unfertilized eggs because they are round rather than elongated and have a thicker cell wall. (credit a: modification of work by South African Medical Research Council; credit b: modification of work by James Gathany, Centers for Disease Control and Prevention; credit c: modification of work by Centers for Disease Control and Prevention)

✓ CHECK YOUR UNDERSTANDING

- Describe the route by which *A. lumbricoides* reaches the host's intestines as an adult worm.

Hookworm

Two species of nematode worms are associated with **hookworm infection**. Both species are found in the Americas, Africa, and Asia. *Necator americanus* is found predominantly in the United States and Australia. Another species, *Ancylostoma doudenale*, is found in southern Europe, North Africa, the Middle East, and Asia.

---

20 Centers for Disease Control and Prevention. "Parasites–Ascariasis." Updated May 24, 2016. http://www.cdc.gov/parasites/ascariasis/index.html.

The eggs of these species develop into larvae in soil contaminated by dog or cat feces. These larvae can penetrate the skin. After traveling through the venous circulation, they reach the lungs. When they are coughed up, they are then swallowed and can enter the intestine and develop into mature adults. At this stage, they attach to the wall of the intestine, where they feed on blood and can potentially cause anemia. Signs and symptoms include cough, an itchy rash, loss of appetite, abdominal pain, and diarrhea. In children, hookworms can affect physical and cognitive growth.

Some hookworm species, such as *Ancylostoma braziliense* that is commonly found in animals such as cats and dogs, can penetrate human skin and migrate, causing cutaneous larva migrans, a skin disease caused by the larvae of hookworms. As they move across the skin, in the subcutaneous tissue, pruritic tracks appear (Figure 24.33).

The infection is diagnosed using microscopic examination of the stool, allowing for observation of eggs in the feces. Medications such as albendazole, mebendazole, and pyrantel pamoate are used as needed to treat systemic infection. In addition to systemic medication for symptoms associated with cutaneous larva migrans, topical thiabendazole is applied to the affected areas.

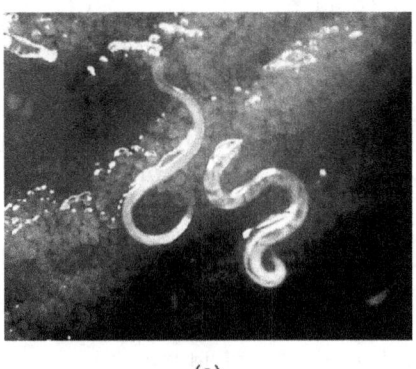
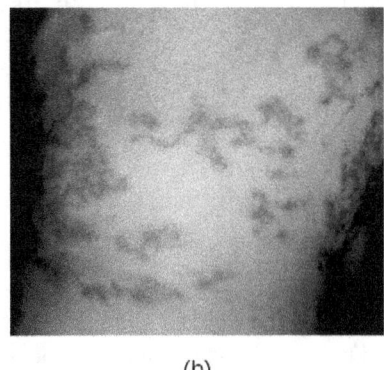
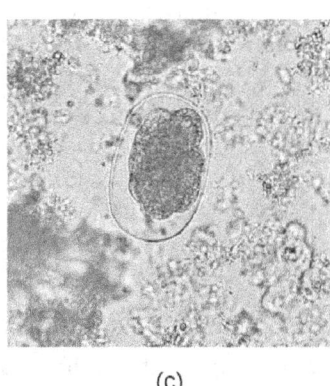

(a)   (b)   (c)

Figure 24.33 (a) This animal hookworm, *Ancylostoma caninum*, is attached to the intestinal wall. (b) The tracks of hookworms are visible in this individual with cutaneous larva migrans. (c) This micrograph shows the microscopic egg of a hookworm. (credit a, c: modification of work by Centers for Disease Control and Prevention)

Strongyloidiasis

Strongyloidiasis is generally caused by *Strongyloides stercoralis*, a soil-transmitted helminth with both free-living and parasitic forms. In the parasitic form, the larvae of these nematodes generally penetrate the body through the skin, especially through bare feet, although transmission through organ transplantation or at facilities like day-care centers can also occur. When excreted in the stool, larvae can become free-living adults rather than developing into the parasitic form. These free-living worms reproduce, laying eggs that hatch into larvae that can develop into the parasitic form. In the parasitic life cycle, infective larvae enter the skin, generally through the feet. The larvae reach the circulatory system, which allows them to travel to the alveolar spaces of the lungs. They are transported to the pharynx where, like many other helminths, the infected patient coughs them up and swallows them again so that they return to the intestine. Once they reach the intestine, females live in the epithelium and produce eggs that develop asexually, unlike the free-living forms, which use sexual reproduction. The larvae may be excreted in the stool or can reinfect the host by entering the tissue of the intestines and skin around the anus, which can lead to chronic infections.

The condition is generally asymptomatic, although severe symptoms can develop after treatment with corticosteroids for asthma or chronic obstructive pulmonary disease, or following other forms of immunosuppression. When the immune system is suppressed, the rate of autoinfection increases, and huge amounts of larvae migrate to organs throughout the body.

Signs and symptoms are generally nonspecific. The condition can cause a rash at the site of skin entry, cough (dry or with blood), fever, nausea, difficulty breathing, bloating, pain, heartburn, and, rarely, arthritis, or cardiac or kidney complications. Disseminated strongyloidiasis or hyperinfection is a life-threatening form of the disease that can occur, usually following immunosuppression such as that caused by glucocorticoid treatment (most commonly), with other immunosuppressive medications, with HIV infection, or with malnutrition.

As with other helminths, direct examination of the stool is important in diagnosis. Ideally, this should be continued over seven days. Serological testing, including antigen testing, is also available. These can be limited by cross-reactions with other similar parasites and by the inability to distinguish current from resolved infection. Ivermectin is the preferred treatment, with albendazole as a secondary option.

CHECK YOUR UNDERSTANDING

- How does an acute infection of *S. stercoralis* become chronic?

Pinworms (Enterobiasis)

Enterobius vermicularis, commonly called pinworms, are tiny (2–13 mm) nematodes that cause **enterobiasis**. Of all helminthic infections, enterobiasis is the most common in the United States, affecting as many as one-third of American children.[21] Although the signs and symptoms are generally mild, patients may experience abdominal pain and insomnia from itching of the perianal region, which frequently occurs at night when worms leave the anus to lay eggs. The itching contributes to transmission, as the disease is transmitted through the fecal-oral route. When an infected individual scratches the anal area, eggs may get under the fingernails and later be deposited near the individual's mouth, causing reinfection, or on fomites, where they can be transferred to new hosts. After being ingested, the larvae hatch within the small intestine and then take up residence in the colon and develop into adults. From the colon, the female adult exits the body at night to lay eggs (Figure 24.34).

Infection is diagnosed in any of three ways. First, because the worms emerge at night to lay eggs, it is possible to inspect the perianal region for worms while an individual is asleep. An alternative is to use transparent tape to remove eggs from the area around the anus first thing in the morning for three days to yield eggs for microscopic examination. Finally, it may be possible to detect eggs through examination of samples from under the fingernails, where eggs may lodge due to scratching. Once diagnosis has been made, mebendazole, albendazole, and pyrantel pamoate are effective for treatment.

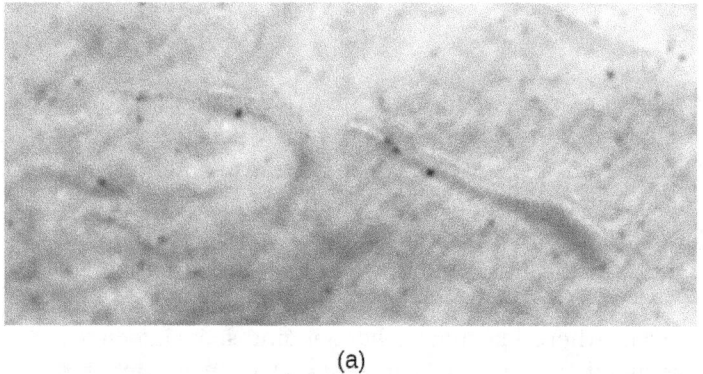
(a)

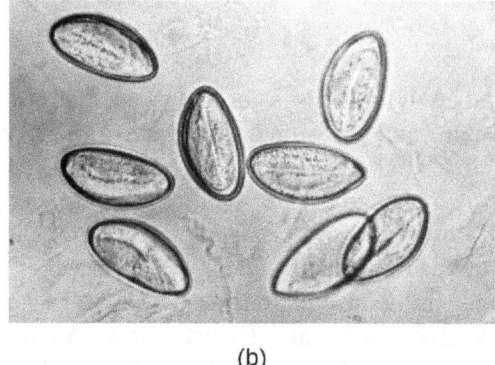
(b)

Figure 24.34 (a) *E. vermicularis* are tiny nematodes commonly called pinworms. (b) This micrograph shows pinworm eggs.

Trichuriasis

The nematode whipworm *Trichuris trichiura* is a parasite that is transmitted by ingestion from soil-contaminated hands or food and causes **trichuriasis**. Infection is most common in warm environments, especially when there is poor sanitation and greater risk of fecal contamination of soil, or when food is grown in soil using manure as a fertilizer. The signs and symptoms may be minimal or nonexistent. When a substantial infection develops, signs and symptoms include painful, frequent diarrhea that may contain mucus and blood. It is possible for the infection to cause rectal prolapse, a condition in which a portion of the

---

21 "Roundworms." *University of Maryland Medical Center Medical Reference Guide.* Last reviewed December 9, 2014. https://umm.edu/health/medical/altmed/condition/roundworms.

rectum becomes detached from the inside of the body and protrudes from the anus (Figure 24.35). Severely infected children may experience reduced growth and their cognitive development may be affected.

When fertilized eggs are ingested, they travel to the intestine and the larvae emerge, taking up residence in the walls of the colon and cecum. They attach themselves with part of their bodies embedded in the mucosa. The larvae mature and live in the cecum and ascending colon. After 60 to 70 days, females begin to lay 3000 to 20,000 eggs per day.

Diagnosis involves examination of the feces for the presence of eggs. It may be necessary to use concentration techniques and to collect specimens on multiple days. Following diagnosis, the infection may be treated with mebendazole, albendazole, or ivermectin.

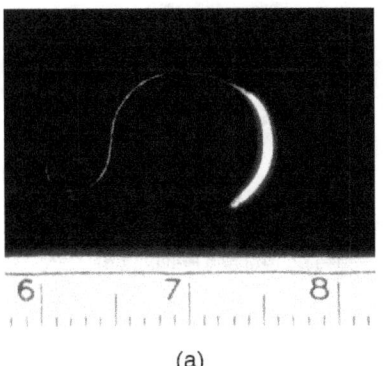
(a)

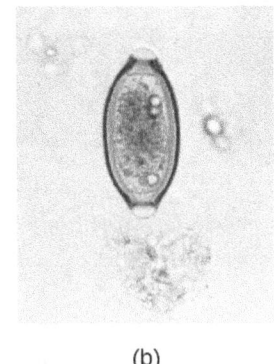
(b)

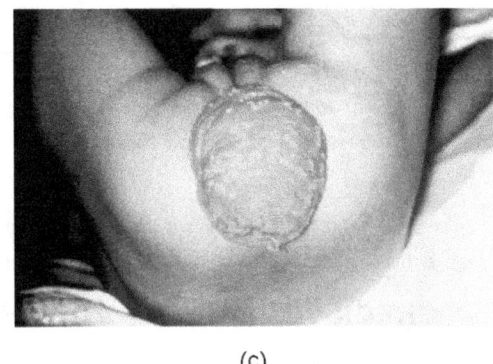
(c)

Figure 24.35 (a) This adult female *Trichuris* whipworm is a soil-transmitted parasite. (b) *Trichuris trichiura* eggs are ingested and travel to the intestines where the larvae emerge and take up residence. (c) Rectal prolapse is a condition that can result from whipworm infections. It occurs when the rectum loses its attachment to the internal body structure and protrudes from the anus. (credit a, b, c: modification of work by Centers for Disease Control and Prevention)

Trichinosis

Trichinosis (trichenellosis) develops following consumption of food that contains *Trichinella spiralis* (most commonly) or other *Trichinella* species. These microscopic nematode worms are most commonly transmitted in meat, especially pork, that has not been cooked thoroughly. *T. spiralis* larvae in meat emerge from cysts when exposed to acid and pepsin in the stomach. They develop into mature adults within the large intestine. The larvae produced in the large intestine are able to migrate into the muscles mechanically via the stylet of the parasite, forming cysts. Muscle proteins are reduced in abundance or undetectable in cells that contain *Trichinella* (nurse cells). Animals that ingest the cysts from other animals can later develop infection (Figure 24.36).

Although infection may be asymptomatic, symptomatic infections begin within a day or two of consuming the nematodes. Abdominal symptoms arise first and can include diarrhea, constipation, and abdominal pain. Other possible symptoms include headache, light sensitivity, muscle pain, fever, cough, chills, and conjunctivitis. More severe symptoms affecting motor coordination, breathing, and the heart sometimes occur. It may take months for the symptoms to resolve, and the condition is occasionally fatal. Mild cases may be mistaken for influenza or similar conditions.

Infection is diagnosed using clinical history, muscle biopsy to look for larvae, and serological testing, including immunoassays. Enzyme immunoassay is the most common test. It is difficult to effectively treat larvae that have formed cysts in the muscle, although medications may help. It is best to begin treatment as soon as possible because medications such as mebendazole and albendazole are effective in killing only the adult worms in the intestine. Steroids may be used to reduce inflammation if larvae are in the muscles.

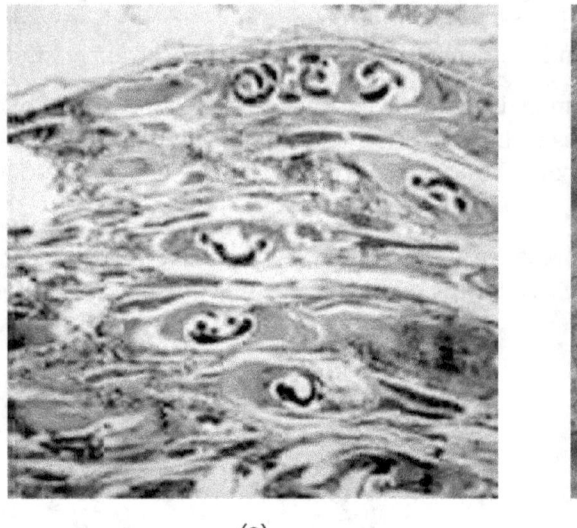
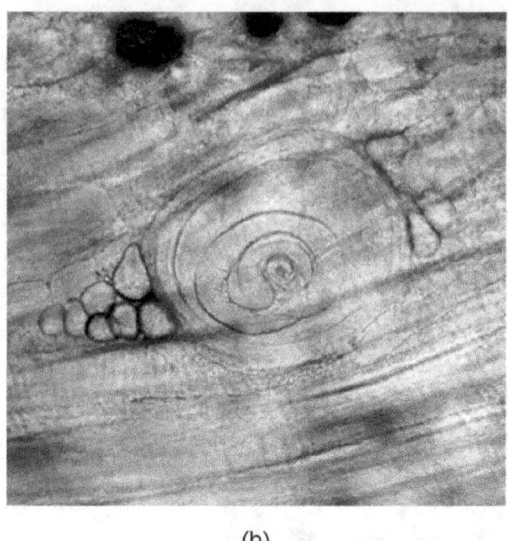

Figure 24.36 (a) This image shows larvae of *T. spiralis* within muscle. (b) In meat, the larvae have a characteristic coiled appearance, as seen in this partially digested larva in bear meat. (credit a, b: modification of work by Centers for Disease Control and Prevention)

CHECK YOUR UNDERSTANDING

- Compare and contrast the transmissions of pinworms and whipworms.

Tapeworms (Taeniasis)

Taeniasis is a tapeworm infection, generally caused by pork (*Taenia solium*), beef (*Taenia saginata*), and Asian (*Taenia asiatica*) tapeworms found in undercooked meat. Consumption of raw or undercooked fish, including contaminated sushi, can also result in infection from the fish tapeworm (*Diphyllobothrium latum*). Tapeworms are flatworms (cestodes) with multiple body segments and a head called a scolex that attaches to the intestinal wall. Tapeworms can become quite large, reaching 4 to 8 meters long (Figure 24.37). Figure 5.23 illustrates the life cycle of a tapeworm.

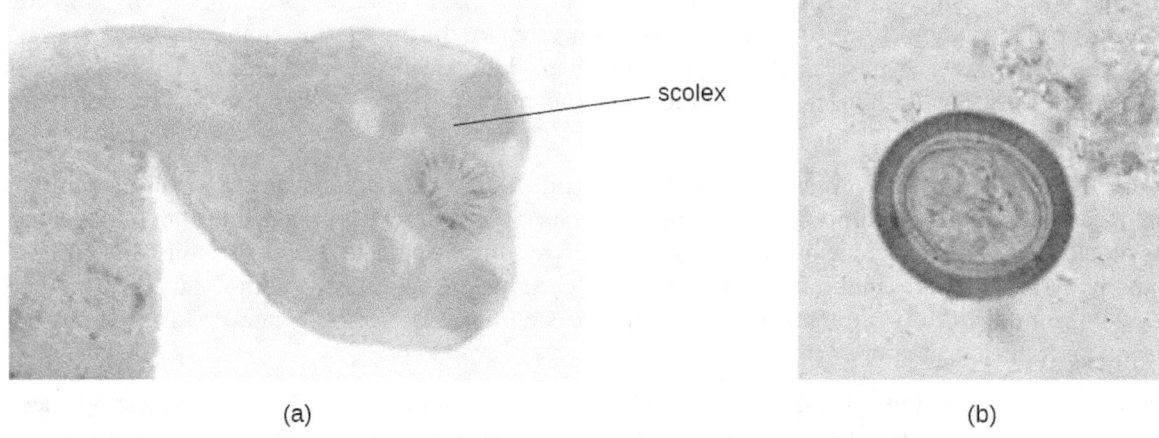

Figure 24.37 (a) An adult tapeworm uses the scolex to attach to the intestinal wall. (b) The egg of a pork tapeworm (*Taenia solium*) is visible in this micrograph. (credit a, b: modification of work by Centers for Disease Control and Prevention)

Tapeworms attached to the intestinal wall produce eggs that are excreted in feces. After ingestion by animals, the eggs hatch and the larvae emerge. They may take up residence in the intestine, but can sometimes move to other tissues, especially muscle or brain tissue. When *T. solium* larvae form cysts in tissue, the condition is called cysticercosis. This occurs through ingestion of eggs via the fecal-oral route, not through consumption of undercooked meat. It can develop in the muscles, eye (ophthalmic cysticercosis), or brain (neurocysticercosis).

Infections may be asymptomatic or they may cause mild gastrointestinal symptoms such as epigastric discomfort, nausea, diarrhea, flatulence, or hunger pains. It is also common to find visible tapeworm segments passed in the stool. In cases of cysticercosis, symptoms differ depending upon where the cysts become established. Neurocysticercosis can have severe, life-threatening consequences and is associated with headaches and seizures because of the presence of the tapeworm larvae encysted in the brain. Cysts in muscles may be asymptomatic, or they may be painful.

To diagnose these conditions, microscopic analysis of stool samples from three separate days is generally recommended. Eggs or body segments, called proglottids, may be visible in these samples. Molecular methods have been developed but are not yet widely available. Imaging, such as CT and MRI, may be used to detect cysts. Praziquantel or niclosamide are used for treatment.

 MICRO CONNECTIONS

What's in Your Sushi Roll?

As foods that contain raw fish, such as sushi and sashimi, continue to increase in popularity throughout the world, so does the risk of parasitic infections carried by raw or undercooked fish. *Diphyllobothrium* species, known as fish tapeworms, is one of the main culprits. Evidence suggests that undercooked salmon caused an increase in *Diphyllobothrium* infections in British Columbia in the 1970s and early 1980s. In the years since, the number of reported cases in the United States and Canada has been low, but it is likely that cases are underreported because the causative agent is not easily recognized.[22]

Another illness transmitted in undercooked fish is herring worm disease, or anisakiasis, in which nematodes attach to the epithelium of the esophagus, stomach, or small intestine. Cases have increased around the world as raw fish consumption has increased.[23]

Although the message may be unpopular with sushi lovers, fish should be frozen or cooked before eating. The extremely low and high temperatures associated with freezing and cooking kill worms and larvae contained in the meat, thereby preventing infection. Ingesting fresh, raw sushi may make for a delightful meal, but it also entails some risk.

Hydatid Disease

Another cestode, *Echinococcus granulosus*, causes a serious infection known as **hydatid disease (cystic echinococcosis)**. *E. granulosus* is found in dogs (the definitive host), as well as several intermediate hosts (sheep, pigs, goats, cattle). The cestodes are transmitted through eggs in the feces from infected animals, which can be an occupational hazard for individuals who work in agriculture.

Once ingested, *E. granulosus* eggs hatch in the small intestine and release the larvae. The larvae invade the intestinal wall to gain access to the circulatory system. They form hydatid cysts in internal organs, especially in the lungs and liver, that grow slowly and are often undetected until they become large. If the cysts burst, a severe allergic reaction (anaphylaxis) may occur.

Cysts present in the liver can cause enlargement of the liver, nausea, vomiting, right epigastric pain, pain in the right upper quadrant, and possible allergic signs and symptoms. Cysts in the lungs can lead to alveolar disease. Abdominal pain, weight loss, pain, and malaise may occur, and inflammatory processes develop.

E. granulosus can be detected through imaging (ultrasonography, CT, MRI) that shows the cysts. Serologic tests, including ELISA and indirect hemagglutinin tests, are used. Cystic disease is most effectively treated with surgery to remove cysts, but other treatments are also available, including chemotherapy with anti-helminthic drugs (albendazole or mebendazole).

---

22 Nancy Craig. "Fish Tapeworm and Sushi." *Canadian Family Physician* 58 (2012) 6: pp. 654–658. http://www.ncbi.nlm.nih.gov/pmc/articles/PMC3374688/.

23 Centers for Disease Control and Prevention. "Anisakiasis FAQs." Updated November 12, 2012. http://www.cdc.gov/parasites/anisakiasis/faqs.html.

CHECK YOUR UNDERSTANDING

- Describe the risks of the cysts associated with taeniasis and hydatid disease.

Flukes

Flukes are flatworms that have a leaflike appearance. They are a type of trematode worm, and multiple species are associated with disease in humans. The most common are liver flukes and intestinal flukes (Figure 24.38).

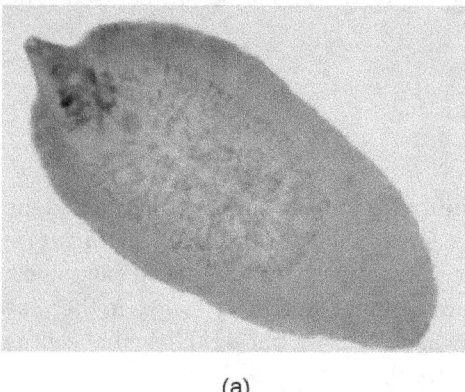 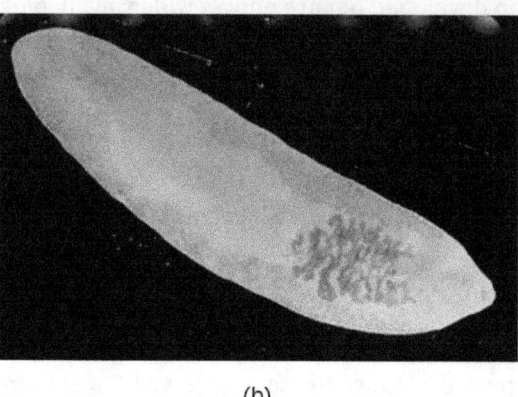

(a) (b)

Figure 24.38 (a) A liver fluke infects the bile ducts. (b) An intestinal fluke infects the intestines. (credit a: modification of work by Shafiei R, Sarkari B, Sadjjadi SM, Mowlavi GR, and Moshfe A; credit b: modification of work by Georgia Division of Public Health)

Liver Flukes

The **liver flukes** are several species of trematodes that cause disease by interfering with the bile duct. Fascioliasis is caused by *Fasciola hepatica* and *Fasciola gigantica* in contaminated raw or undercooked aquatic plants (e.g., watercress). In *Fasciola* infection, adult flukes develop in the bile duct and release eggs into the feces. Clonochiasis is caused by *Clonorchis sinensis* in contaminated freshwater fish. Other flukes, such as *Opisthorchis viverrini* (found in fish) and *Opisthorchis felineus* (found in freshwater snails), also cause infections. Liver flukes spend part of their life cycle in freshwater snails, which serve as an intermediate host. Humans are typically infected after eating aquatic plants contaminated by the infective larvae after they have left the snail. Once they reach the human intestine, they migrate back to the bile duct, where they mature. The life cycle is similar for the other infectious liver flukes, (see Figure 5.22).

When *Fasciola* flukes cause acute infection, signs and symptoms include nausea, vomiting, abdominal pain, rash, fever, malaise, and breathing difficulties. If the infection becomes chronic, with adult flukes living in the bile duct, then cholangitis, cirrhosis, pancreatitis, cholecystitis, and gallstones may develop. Symptoms are similar for infections by other liver flukes. Cholangiocarcinoma can occur from *C. sinensis* infection. The *Opisthorchis* species can also be associated with cancer development.

Diagnosis is accomplished using patient history and examination of samples from feces or other samples (such as vomitus). Because the eggs may appear similar, immunoassay techniques are available that can help distinguish species. The preferred treatment for fascioliasis is triclabendazole. *C. sinensis* and *Opisthorchis* spp. infections are treated with praziquantel or albendazole.

Intestinal Flukes

The **intestinal flukes** are trematodes that develop in the intestines. Many, such as *Fasciolopsis buski*, which causes fasciolopsiasis, are closely related to liver flukes. Intestinal flukes are ingested from contaminated aquatic plants that have not been properly cooked. When the cysts are consumed, the larvae emerge in the duodenum and develop into adults while attached to the intestinal epithelium. The eggs are released in stool.

Intestinal fluke infection is often asymptomatic, but some cases may involve mild diarrhea and abdominal pain. More severe symptoms such as vomiting, nausea, allergic reactions, and anemia can sometimes occur, and high parasite loads may sometimes lead to intestinal obstructions.

Diagnosis is the same as with liver flukes: examination of feces or other samples and immunoassay.

Praziquantel is used to treat infections caused by intestinal flukes.

✓ CHECK YOUR UNDERSTANDING

- How are flukes transmitted?

Disease Profile

Helminthic Gastrointestinal Infections
Numerous helminths are capable of colonizing the GI tract. Many such infections are asymptomatic, but others may cause signs and symptoms ranging from mild GI stress to severe systemic infection. Helminths have complex and unique life cycles that dictate their specific modes of transmission. Most helminthic infections can be treated with medications.

Common Helminthic Infections of the GI Tract

Disease	Causative Agent(s)	Mode of Transmission	Laboratory Tests	Symptoms	Treatments
Ascariasis	*Ascaris lumbricoides*	Eggs in fecally contaminated food or water	Microscopic examination of the stool, imaging	Shortness of breath, cough, nausea, diarrhea, blood in stool, abdominal pain, weight loss, fatigue	Self-limiting within 1 to 2 years; albendazole and mebendazole if needed
Hookworm	*Necator americanus*, *Ancyclostoma doudenale*	Larvae in soil contaminated by dog or cat feces penetrate skin	Microscopic examination of stool (may require a concentration procedure)	Cough, itchy rash, loss of appetite, abdominal pain, diarrhea; in children, may affect physical and cognitive growth	Albendazole and mebendazole; pyrantel pamoatemay if needed
Strongyloidiasis	*Strongyloides stercoralis*	Soil-dwelling larvae penetrate the skin, usually bare feet	Microscopic examination of stool over several days (ideally at least 7); some serologic testing available	Often asymptomatic; cough (sometimes bloody), skin rash, abdominal pain, and diarrhea; in immunosuppressed patients, may become disseminated, causing serious and potentially fatal complications	Ivermectin (preferred), albendazole
Enterobiasis (pinworm)	*Enterobius vermicularis*	Fecal–oral route	Observation of eggs or worms from anal area; examination of samples under fingernails	Itching around the anus, abdominal pain, insomnia, irritation of female genital tract	Mebendazole, albendazole, pyrantel pamoate
Trichiuriasis (whipworm)	*Trichuris trichiura*	Fecal contamination or fertilization in soil	Microscopic examination of stool	Abdominal pain, anemia, diarrhea that may be bloody	Albendazole, mebendazole, ivermectin if needed
Trichinosis	*Trichinella spiralis*	Eating raw or undercooked pork or other meat of infected animal	Clinical history, muscle biopsy, serological testing, enzyme immunoassay	Diarrhea, constipation, abdominal pain, headache, cough, chills, light sensitivity, muscle pain, fever, conjunctivitis; in severe cases may affect motor coordination, breathing, heart function	Albendazole, mebendazole if needed
Taeniasis and cysticercosis	*Taenia solium*, *T. saginata*, *T. asiatica*, *Diphyllobothrium latum*	Eating raw or undercooked beef or pork from infected animal	Observation of worm segments or microscopic eggs in stool samples	Asymptomatic or mild GI distress; cysts in muscle, eye, or brain (cysticercosis); brain cysts can cause headaches, seizures, or death	Praziquantel or niclosamide
Cystic echinococcosis (hydatid disease)	*Echinococcus granulosus* (cystic)	Exposure to eggs in feces of infected dogs or livestock	Imaging; serological testing including ELISA and indirect hemagglutinin test	Cysts in lungs, liver, and other organs causing nausea, GI distress, and weight loss; severe anaphylaxis or death if cysts burst	Surgical removal or aspiration of cysts or chemotherapy with albendazole or mebenazole
Liver fluke infections	*Fasciola hepatica*, *F. gigantica*, *Clonorchis sinensis*, *Opisthorchis viverrini*, *O. felineus*	Eating raw or undercooked aquatic plants (*Fasciola* spp.) or freshwater fish (*Clonorchis* spp.) contaminated with eggs or cysts	Microscopic examination of eggs in stool or other samples; immunoassays	Fever, malaise, anemia, abdominal symptoms, transaminitis; cholangitis, cirrhosis, pancreatitis, cholecystitis, gall stones in chronic phase	Triclabendazole (preferred) for *Fasciola* spp.; praziquantel and albendazole for *C. sinensis* and *Opisthorchis* spp.
Fasciolopiasis (intestinal fluke)	*Fasciola buski*	Eating raw or undercooked aquatic plants containing cysts	Microscopic examination of eggs in stool or other samples; immunoassays	Diarrhea, abdominal pain; in severe cases, vomiting, nausea, intestinal obstruction, anemia, allergic reactions	Praziquantel

Figure 24.39

Helminthic Infections of the GI Tract (continued)					
Disease	Pathogen	Signs and Symptoms	Transmission	Diagnostic Tests	Antimicrobial Drugs
Strongyloidiasis	*Strongyloides stercoralis*	Often asymptomatic; cough (sometimes bloody), skin rash, abdominal pain, diarrhea; in immunosuppressed patients, may become disseminated, causing serious and potentially fatal complications	Soil-dwelling larvae penetrate the skin, usually bare feet	Microscopic observation of larvae in stool; serological testing for antigens	Ivermectin, albendazole
Tapeworms (taeniasis)	*Taenia solium, T. saginata, T. asiatica, Diphyllobothrium latum*	Asymptomatic or mild GI distress; cysts in muscle, eye, or brain (cysticercosis); brain cysts can cause headaches, seizures, or death	Ingestion of raw or undercooked pork or beef from infected animal	Observation of worm segments or microscopic eggs in stool; CT or MRI to detect cysts	Praziquantel, niclosamide
Trichinosis	*Trichinella spiralis*, other *Trichinella* spp.	Diarrhea, constipation, abdominal pain, headache, cough, chills, light sensitivity, muscle pain, fever, conjunctivitis; in severe cases may affect motor coordination, breathing, heart function	Ingestion of raw or undercooked pork or other meat of infected animal	Observation of cysts in muscle biopsy, enzyme immunoassay	Albendazole, mebendazole
Whipworm (trichuriasis)	*Trichuris trichiura*	Abdominal pain, anemia, diarrhea (possibly bloody), rectal prolapse	Ingestion of eggs in fecally contaminated food	Microscopic observation of eggs in stool	Albendazole, mebendazole, ivermectin

Figure 24.40

Clinical Focus

Resolution

Carli's doctor explained that she had bacterial gastroenteritis caused by *Salmonella* bacteria. The source of these bacteria was likely the undercooked egg. Had the egg been fully cooked, the high temperature would have been sufficient to kill any *Salmonella* in or on the egg. In this case, enough bacteria survived to cause an infection once the egg was eaten.

Carli's signs and symptoms continued to worsen. Her fever became higher, her vomiting and diarrhea continued, and she began to become dehydrated. She felt thirsty all the time and had continual abdominal cramps. Carli's doctor treated her with intravenous fluids to help with her dehydration, but did not prescribe antibiotics. Carli's parents were confused because they thought a bacterial infection should always be treated with antibiotics.

The doctor explained that the worst medical problem for Carli was dehydration. Except in the most vulnerable and sick patients, such as those with HIV/AIDS, antibiotics do not reduce recovery time or improve outcomes in *Salmonella* infections. In fact, antibiotics can actually delay the natural excretion of bacteria from the body. Rehydration therapy replenishes lost fluids, diminishing the effects of dehydration and improving the patient's condition while the infection resolves.

After two days of rehydration therapy, Carli's signs and symptoms began to fade. She was still somewhat thirsty, but the amount of urine she passed became larger and the color lighter. She stopped vomiting. Her fever was gone, and so was the diarrhea. At that point, stool analysis found very few *Salmonella* bacteria. In one week, Carli was discharged as fully recovered.

Go back to the previous Clinical Focus box.

SUMMARY

24.1 Anatomy and Normal Microbiota of the Digestive System

- The digestive tract, consisting of the oral cavity, pharynx, esophagus, stomach, small intestine, and large intestine, has a normal microbiota that is important for health.
- The constant movement of materials through the gastrointestinal canal, the protective layer of mucus, the normal microbiota, and the harsh chemical environment in the stomach and small intestine help to prevent colonization by pathogens.
- Infections or microbial toxins in the oral cavity can cause **tooth decay**, **periodontal disease**, and various types of **ulcers**.
- Infections and intoxications of the gastrointestinal tract can cause general symptoms such as nausea, vomiting, diarrhea, and fever. Localized inflammation of the GI tract can result in **gastritis**, **enteritis**, **gastroenteritis**, **hepatitis**, or **colitis**, and damage to epithelial cells of the colon can lead to **dysentery**.
- **Foodborne illness** refers to infections or intoxications that originate with pathogens or toxins ingested in contaminated food or water.

24.2 Microbial Diseases of the Mouth and Oral Cavity

- **Dental caries**, **tartar**, and **gingivitis** are caused by overgrowth of oral bacteria, usually *Streptococcus* and *Actinomyces* species, as a result of insufficient dental hygiene.
- Gingivitis can worsen, allowing *Porphyromonas*, *Streptococcus*, and *Actinomyces* species to spread and cause **periodontitis**. When *Prevotella intermedia*, *Fusobacterium* species, and *Treponema vicentii* are involved, it can lead to **acute necrotizing ulcerative gingivitis**.
- The herpes simplex virus type 1 can cause lesions of the mouth and throat called **herpetic gingivostomatitis**.
- Other infections of the mouth include **oral thrush**, a fungal infection caused by overgrowth of *Candida* yeast, and **mumps**, a viral infection of the salivary glands caused by the mumps virus, a paramyxovirus.

24.3 Bacterial Infections of the Gastrointestinal Tract

- Major causes of gastrointestinal illness include *Salmonella* spp., *Staphylococcus* spp., *Helicobacter pylori*, *Clostridium perfringens*, *Clostridium difficile*, *Bacillus cereus*, and *Yersinia* bacteria.
- *C. difficile* is an important cause of hospital acquired infection.
- *Vibrio cholerae* causes **cholera**, which can be a severe diarrheal illness.
- Different strains of *E. coli*, including **ETEC**, **EPEC**, **EIEC**, and **EHEC**, cause different illnesses with varying degrees of severity.
- *H. pylori* is associated with **peptic ulcers**.
- *Salmonella enterica* serotypes can cause **typhoid fever**, a more severe illness than **salmonellosis**.
- Rehydration and other supportive therapies are often used as general treatments.
- Careful antibiotic use is required to reduce the risk of causing *C. difficile* infections and when treating antibiotic-resistant infections.

24.4 Viral Infections of the Gastrointestinal Tract

- Common viral causes of gastroenteritis include rotaviruses, noroviruses, and astroviruses.
- Hepatitis may be caused by several unrelated viruses: hepatitis viruses A, B, C, D, and E.
- The hepatitis viruses differ in their modes of transmission, treatment, and potential for chronic infection.

24.5 Protozoan Infections of the Gastrointestinal Tract

- **Giardiasis**, **cryptosporidiosis**, **amoebiasis**, and **cyclosporiasis** are intestinal infections caused by protozoans.
- Protozoan intestinal infections are commonly transmitted through contaminated food and water.
- Treatment varies depending on the causative agent, so proper diagnosis is important.
- Microscopic examination of stool or biopsy specimens is often used in diagnosis, in combination with other approaches.

24.6 Helminthic Infections of the Gastrointestinal Tract

- Helminths often cause intestinal infections after transmission to humans through exposure to contaminated soil, water, or food. Signs and symptoms are often mild, but severe

- complications may develop in some cases.
- *Ascaris lumbricoides* eggs are transmitted through contaminated food or water and hatch in the intestine. Juvenile larvae travel to the lungs and then to the pharynx, where they are swallowed and returned to the intestines to mature. These nematode roundworms cause **ascariasis**.
- *Necator americanus* and *Ancylostoma doudenale* cause **hookworm infection** when larvae penetrate the skin from soil contaminated by dog or cat feces. They travel to the lungs and are then swallowed to mature in the intestines.
- *Strongyloides stercoralis* are transmitted from soil through the skin to the lungs and then to the intestine where they cause **strongyloidiasis**.
- *Enterobius vermicularis* are nematode pinworms transmitted by the fecal-oral route. After ingestion, they travel to the colon where they cause **enterobiasis**.
- *Trichuris trichiura* can be transmitted through soil or fecal contamination and cause **trichuriasis**. After ingestion, the eggs travel to the intestine where the larvae emerge and mature, attaching to the walls of the colon and cecum.
- *Trichinella* spp. is transmitted through undercooked meat. Larvae in the meat emerge from cysts and mature in the large intestine. They can migrate to the muscles and form new cysts, causing **trichinosis**.
- *Taenia* spp. and *Diphyllobothrium latum* are tapeworms transmitted through undercooked food or the fecal-oral route. *Taenia* infections cause **taeniasis**. Tapeworms use their scolex to attach to the intestinal wall. Larvae may also move to muscle or brain tissue.
- *Echinococcus granulosus* is a cestode transmitted through eggs in the feces of infected animals, especially dogs. After ingestion, eggs hatch in the small intestine, and the larvae invade the intestinal wall and travel through the circulatory system to form dangerous cysts in internal organs, causing **hydatid disease**.
- Flukes are transmitted through aquatic plants or fish. **Liver flukes** cause disease by interfering with the bile duct. **Intestinal flukes** develop in the intestines, where they attach to the intestinal epithelium.

REVIEW QUESTIONS
Multiple Choice

1. Which of the following is NOT a way the normal microbiota of the intestine helps to prevent infection?
    a. It produces acids that lower the pH of the stomach.
    b. It speeds up the process by which microbes are flushed from the digestive tract.
    c. It consumes food and occupies space, outcompeting potential pathogens.
    d. It generates large quantities of oxygen that kill anaerobic pathogens.

2. What types of microbes live in the intestines?
    a. Diverse species of bacteria, archaea, and fungi, especially *Bacteroides* and *Firmicutes* bacteria
    b. A narrow range of bacteria, especially *Firmicutes*
    c. A narrow range of bacteria and fungi, especially *Bacteroides*
    d. Archaea and fungi only

3. What pathogen is the most important contributor to biofilms in plaque?
    a. *Staphylococcus aureus*
    b. *Streptococcus mutans*
    c. *Escherichia coli*
    d. *Clostridium difficile*

4. What type of organism causes thrush?
    a. a bacterium
    b. a virus
    c. a fungus
    d. a protozoan

5. In mumps, what glands swell to produce the disease's characteristic appearance?
    a. the sublingual glands
    b. the gastric glands
    c. the parotid glands
    d. the submandibular glands

6. Which of the following is true of HSV-1?
   a. It causes oral thrush in immunocompromised patients.
   b. Infection is generally self-limiting.
   c. It is a bacterium.
   d. It is usually treated with amoxicillin.

7. Which type of *E. coli* infection can be severe with life-threatening consequences such as hemolytic uremic syndrome?
   a. ETEC
   b. EPEC
   c. EHEC
   d. EIEC

8. Which species of *Shigella* has a type that produces Shiga toxin?
   a. *S. boydii*
   b. *S. flexneri*
   c. *S. dysenteriae*
   d. *S. sonnei*

9. Which type of bacterium produces an A-B toxin?
   a. *Salmonella*
   b. *Vibrio cholera*
   c. ETEC
   d. *Shigella dysenteriae*

10. Which form of hepatitisvirus can only infect an individual who is already infected with another hepatitisvirus?
    a. HDV
    b. HAV
    c. HBV
    d. HEV

11. Which cause of viral gastroenteritis commonly causes projectile vomiting?
    a. hepatitisvirus
    b. Astroviruses
    c. Rotavirus
    d. Noroviruses

12. Which protozoan is associated with the ability to cause severe dysentery?
    a. *Giardia lamblia*
    b. *Cryptosporidium hominis*
    c. *Cyclospora cayetanesis*
    d. *Entamoeba histolytica*

13. Which protozoan has a unique appearance, with a blue halo, when viewed using ultraviolet fluorescence microscopy?
    a. *Giardia lamblia*
    b. *Cryptosporidium hominis*
    c. *Cyclospora cayetanesis*
    d. *Entamoeba histolytica*

14. The micrograph shows protozoans attached to the intestinal wall of a gerbil. Based on what you know about protozoan intestinal parasites, what is it?

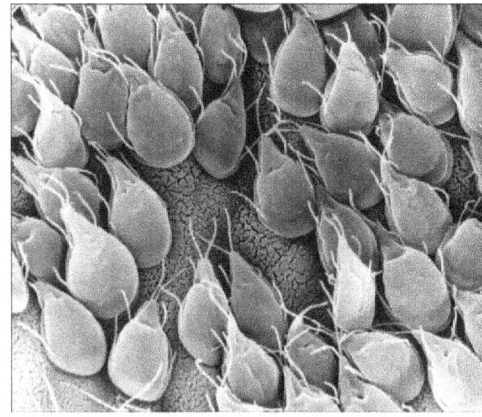

Figure 24.41 (credit: Dr. Stan Erlandsen, Centers for Disease Control and Prevention)

   a. *Giardia lamblia*
   b. *Cryptosporidium hominis*
   c. *Cyclospora cayetanesis*
   d. *Entamoeba histolytica*

15. What is another name for *Trichuris trichiura*?
    a. pinworm
    b. whipworm
    c. hookworm
    d. ascariasis

16. Which type of helminth infection can be diagnosed using tape?
    a. pinworm
    b. whipworm
    c. hookworm
    d. tapeworm

Fill in the Blank

17. The part of the gastrointestinal tract with the largest natural microbiota is the _____.
18. When plaque becomes heavy and hardened, it is called dental calculus or _____.
19. Antibiotic associated pseudomembranous colitis is caused by _____.
20. Jaundice results from a buildup of _____.

21. Chronic _____ infections cause the unique sign of disease of greasy stool and are often resistant to treatment.
22. Liver flukes are often found in the _____ duct.

Short Answer

23. How does the diarrhea caused by dysentery differ from other types of diarrhea?
24. Why do sugary foods promote dental caries?
25. Which forms of viral hepatitis are transmitted through the fecal-oral route?
26. What is an O&P exam?
27. Why does the coughing up of worms play an important part in the life cycle of some helminths, such as the roundworm *Ascaris lumbricoides*?

Critical Thinking

28. Why does use of antibiotics and/or proton pump inhibitors contribute to the development of *C. difficile* infections?
29. Why did scientists initially think it was unlikely that a bacterium caused peptic ulcers?
30. Does it makes a difference in treatment to know if a particular illness is caused by a bacterium (an infection) or a toxin (an intoxication)?
31. Based on what you know about HBV, what are some ways that its transmission could be reduced in a health-care setting?
32. Cases of strongyloidiasis are often more severe in patients who are using corticosteroids to treat another disorder. Explain why this might occur.

CHAPTER 25
Circulatory and Lymphatic System Infections

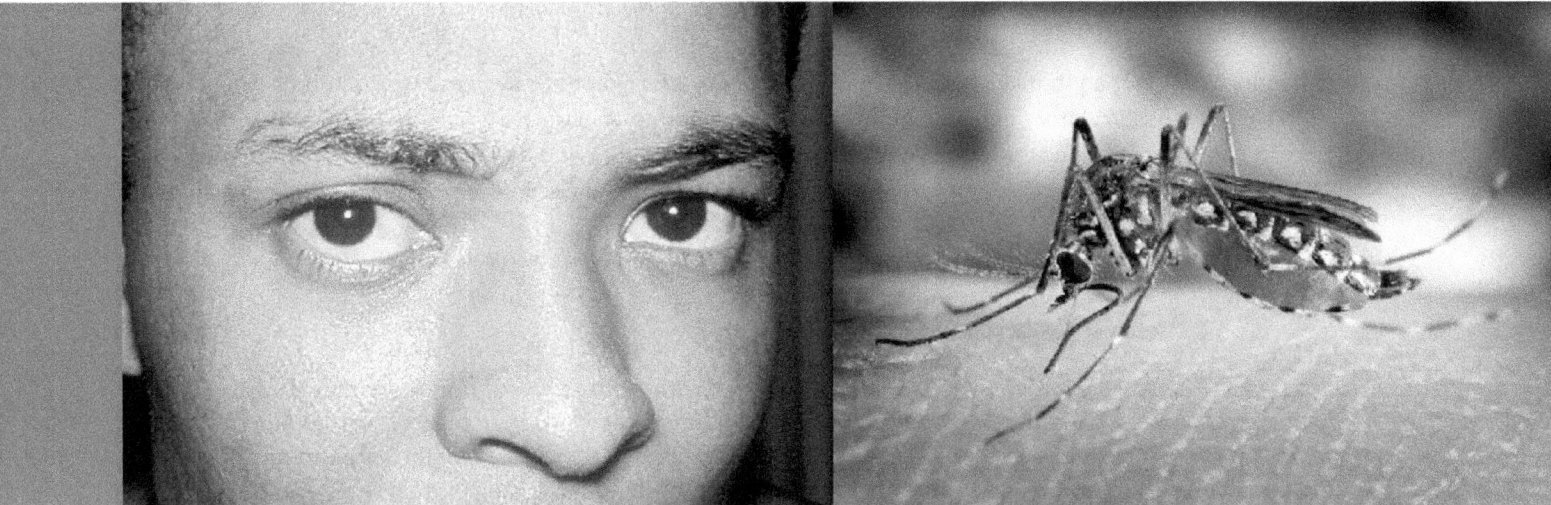

Figure 25.1 Yellow fever is a viral hemorrhagic disease that can cause liver damage, resulting in jaundice (left) as well as serious and sometimes fatal complications. The virus that causes yellow fever is transmitted through the bite of a biological vector, the *Aedes aegypti* mosquito (right). (credit left: modification of work by Centers for Disease Control and Prevention; credit right: modification of work by James Gathany, Centers for Disease Control and Prevention)

Chapter Outline

25.1 Anatomy of the Circulatory and Lymphatic Systems

25.2 Bacterial Infections of the Circulatory and Lymphatic Systems

25.3 Viral Infections of the Circulatory and Lymphatic Systems

25.4 Parasitic Infections of the Circulatory and Lymphatic Systems

INTRODUCTION Yellow fever was once common in the southeastern US, with annual outbreaks of more than 25,000 infections in New Orleans in the mid-1800s.[1] In the early 20th century, efforts to eradicate the virus that causes yellow fever were successful thanks to vaccination programs and effective control (mainly through the insecticide dichlorodiphenyltrichloroethane [DDT]) of *Aedes aegypti*, the mosquito that serves as a vector. Today, the virus has been largely eradicated in North America.

Elsewhere, efforts to contain yellow fever have been less successful. Despite mass vaccination campaigns in some regions, the risk for yellow fever epidemics is rising in dense urban cities in Africa and South America.[2] In an increasingly globalized society, yellow fever could easily make a comeback in North America, where *A. aegypti* is still present. If these mosquitoes were exposed to infected individuals, new outbreaks would be possible.

---

1 Centers for Disease Control and Prevention. "The History of Yellow Fever." http://www.cdc.gov/travel-training/local/HistoryEpidemiologyandVaccination/page27568.html

Like yellow fever, many of the circulatory and lymphatic diseases discussed in this chapter are emerging or re-emerging worldwide. Despite medical advances, diseases like malaria, Ebola, and others could become endemic in the US given the right circumstances.

25.1 Anatomy of the Circulatory and Lymphatic Systems

Learning Objectives

By the end of this section, you will be able to:
- Describe the major anatomical features of the circulatory and lymphatic systems
- Explain why the circulatory and lymphatic systems lack normal microbiota
- Explain how microorganisms overcome defenses of the circulatory and lymphatic systems to cause infection
- Describe general signs and symptoms of disease associated with infections of the circulatory and lymphatic systems

---

Clinical Focus

Part 1

Barbara is a 43-year-old patient who has been diagnosed with metastatic inflammatory breast cancer. To facilitate her ongoing chemotherapy, her physician implanted a port attached to a central venous catheter. At a recent checkup, she reported feeling restless and complained that the site of the catheter had become uncomfortable. After removing the dressing, the physician observed that the surgical site appeared red and was warm to the touch, suggesting a localized infection. Barbara's was also running a fever of 38.2 °C (100.8 °F). Her physician treated the affected area with a topical antiseptic and applied a fresh dressing. She also prescribed a course of the antibiotic oxacillin.

- Based on this information, what factors likely contributed to Barbara's condition?
- What is the most likely source of the microbes involved?

Jump to the next Clinical Focus box.

The circulatory and lymphatic systems are networks of vessels and a pump that transport blood and lymph, respectively, throughout the body. When these systems are infected with a microorganism, the network of vessels can facilitate the rapid dissemination of the microorganism to other regions of the body, sometimes with serious results. In this section, we will examine some of the key anatomical features of the circulatory and lymphatic systems, as well as general signs and symptoms of infection.

The Circulatory System

The circulatory (or cardiovascular) system is a closed network of organs and vessels that moves blood around the body (Figure 25.2). The primary purposes of the circulatory system are to deliver nutrients, immune factors, and oxygen to tissues and to carry away waste products for elimination. The heart is a four-chambered pump that propels the blood throughout the body. Deoxygenated blood enters the right atrium through the superior vena cava and the inferior vena cava after returning from the body. The blood next passes through the tricuspid valve to enter the right ventricle. When the heart contracts, the blood from the right ventricle is pumped through the pulmonary arteries to the lungs. There, the blood is oxygenated at the alveoli and returns to the heart through the pulmonary veins. The oxygenated blood is received at the left atrium and proceeds through the mitral valve to the left ventricle. When the heart contracts, the oxygenated blood is pumped throughout the body via a series of thick-walled vessels called arteries. The first and largest **artery** is called the aorta. The arteries sequentially branch and decrease in size (and are called arterioles) until they end in a network of smaller vessels called capillaries. The **capillary** beds are located in the interstitial spaces within tissues and release nutrients, immune factors, and oxygen to those tissues. The capillaries connect to a series of vessels called venules, which increase in size to form the **vein**s. The veins join together into larger vessels as they transfer blood back to the heart. The largest veins, the superior and inferior vena cava, return the blood to

---

2 C.L. Gardner, K.D. Ryman. "Yellow Fever: A Reemerging Threat." *Clinical Laboratory Medicine* 30 no. 1 (2010):237–260.

the right atrium.

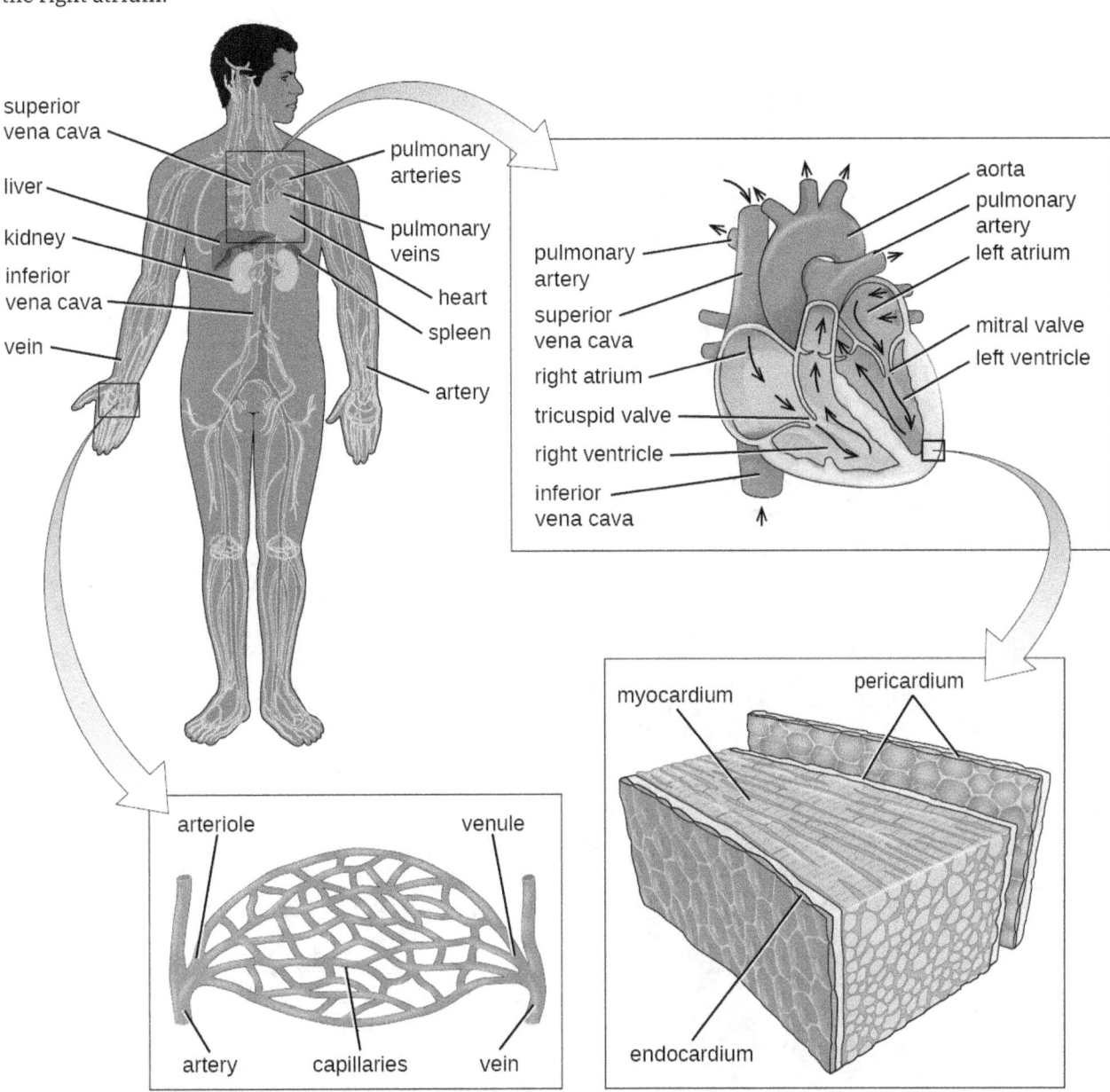

Figure 25.2 The major components of the human circulatory system include the heart, arteries, veins, and capillaries. This network delivers blood to the body's organs and tissues. (credit top left: modification of work by Mariana Ruiz Villareal; credit bottom right: modification of work by Bruce Blaus)

Other organs play important roles in the circulatory system as well. The kidneys filter the blood, removing waste products and eliminating them in the urine. The liver also filters the blood and removes damaged or defective red blood cells. The spleen filters and stores blood, removes damaged red blood cells, and is a reservoir for immune factors. All of these filtering structures serve as sites for entrapment of microorganisms and help maintain an environment free of microorganisms in the blood.

The Lymphatic System

The lymphatic system is also a network of vessels that run throughout the body (Figure 25.3). However, these vessels do not form a full circulating system and are not pressurized by the heart. Rather, the lymphatic system is an open system with the fluid moving in one direction from the extremities toward two drainage points into veins just above the heart. Lymphatic fluids move more slowly than blood because they are not pressurized.

Small lymph capillaries interact with blood capillaries in the interstitial spaces in tissues. Fluids from the tissues enter the lymph capillaries and are drained away (Figure 25.4). These fluids, termed lymph, also contain large numbers of white blood cells.

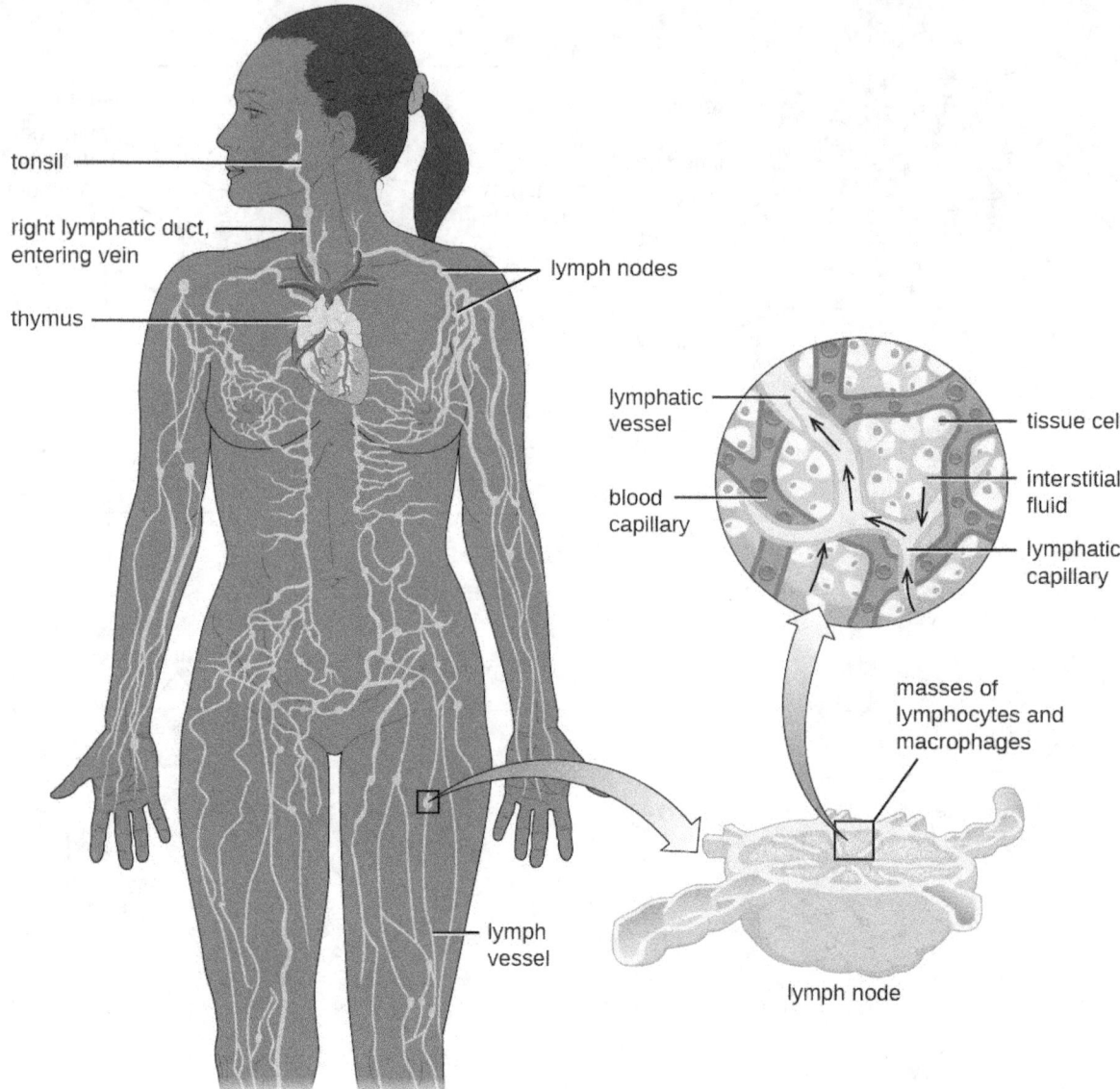

Figure 25.3   The essential components of the human lymphatic system drain fluid away from tissues.

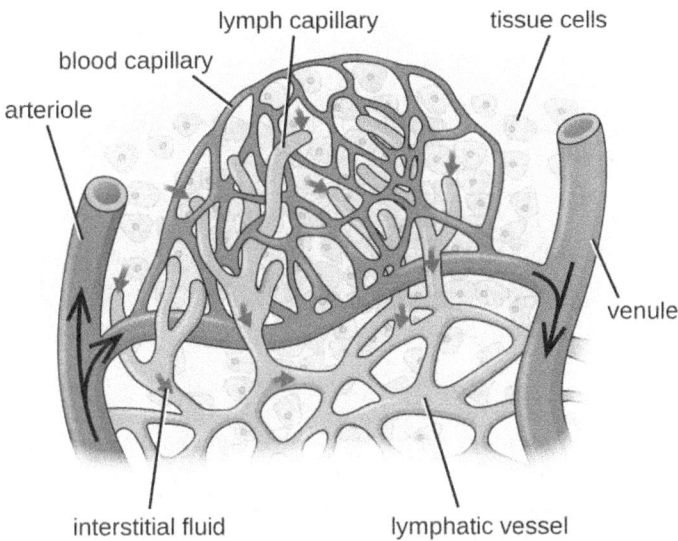

Figure 25.4 Blood enters the capillaries from an arteriole (red) and leaves through venules (blue). Interstitial fluids may drain into the lymph capillaries (green) and proceed to lymph nodes. (credit: modification of work by National Cancer Institute, National Institutes of Health)

The lymphatic system contains two types of lymphoid tissues. The **primary lymphoid tissue** includes bone marrow and the thymus. Bone marrow contains the hematopoietic stem cells (HSC) that differentiate and mature into the various types of blood cells and lymphocytes (see Figure 17.12). The **secondary lymphoid tissues** include the spleen, lymph nodes, and several areas of diffuse lymphoid tissues underlying epithelial membranes. The **spleen**, an encapsulated structure, filters blood and captures pathogens and antigens that pass into it (Figure 25.5). The spleen contains specialized macrophages and dendritic cells that are crucial for antigen presentation, a mechanism critical for activation of T lymphocytes and B lymphocytes (see Major Histocompatibility Complexes and Antigen-Presenting Cells). Lymph nodes are bean-shaped organs situated throughout the body. These structures contain areas called germinal centers that are rich in B and T lymphocytes. The **lymph nodes** also contain macrophages and dendritic cells for antigen presentation. Lymph from nearby tissues enters the lymph node through afferent lymphatic vessels and encounters these lymphocytes as it passes through; the lymph exits the lymph node through the efferent lymphatic vessels (Figure 25.5).

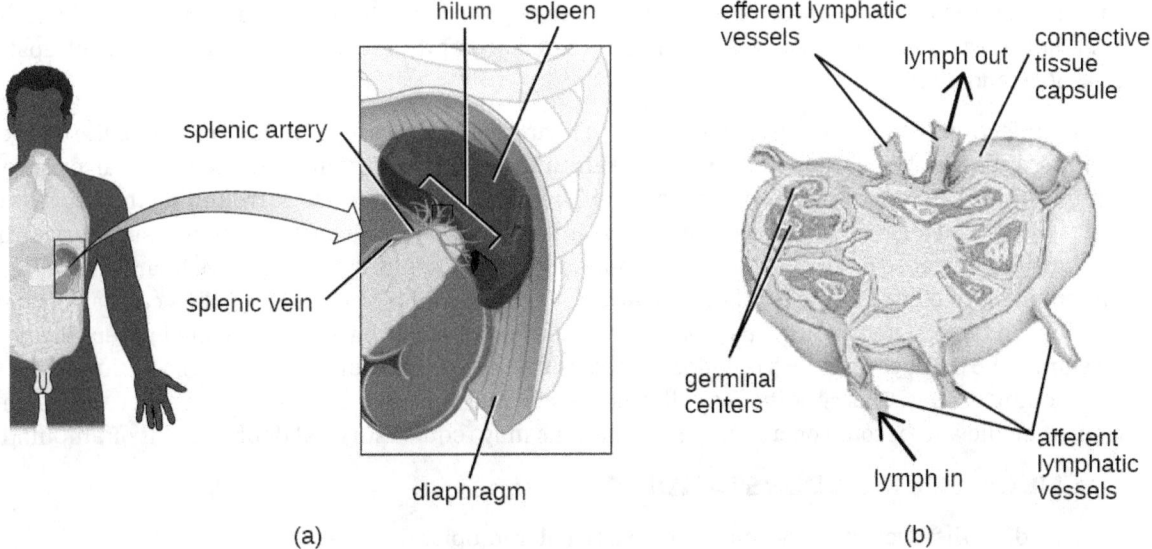

Figure 25.5 (a) The spleen is a lymphatic organ located in the upper left quadrant of the abdomen near the stomach and left kidney. It contains numerous phagocytes and lymphocytes that combat and prevent circulatory infections by killing and removing pathogens from the blood. (b) Lymph nodes are masses of lymphatic tissue located along the larger lymph vessels. They contain numerous lymphocytes that

kill and remove pathogens from lymphatic fluid that drains from surrounding tissues.

🔗 LINK TO LEARNING

The lymphatic system filters fluids that have accumulated in tissues before they are returned to the blood. A brief overview of this process is provided at this (https://openstax.org/l/22lymphatic) website.

✓ CHECK YOUR UNDERSTANDING

- What is the main function of the lymphatic system?

Infections of the Circulatory System

Under normal circumstances, the circulatory system and the blood should be sterile; the circulatory system has no normal microbiota. Because the system is closed, there are no easy portals of entry into the circulatory system for microbes. Those that are able to breach the body's physical barriers and enter the bloodstream encounter a host of circulating immune defenses, such as antibodies, complement proteins, phagocytes, and other immune cells. Microbes often gain access to the circulatory system through a break in the skin (e.g., wounds, needles, intravenous catheters, insect bites) or spread to the circulatory system from infections in other body sites. For example, microorganisms causing pneumonia or renal infection may enter the local circulation of the lung or kidney and spread from there throughout the circulatory network.

If microbes in the bloodstream are not quickly eliminated, they can spread rapidly throughout the body, leading to serious, even life-threatening infections. Various terms are used to describe conditions involving microbes in the circulatory system. The term **bacteremia** refers to bacteria in the blood. If bacteria are reproducing in the blood as they spread, this condition is called **septicemia**. The presence of viruses in the blood is called **viremia**. Microbial toxins can also be spread through the circulatory system, causing a condition termed **toxemia**.

Microbes and microbial toxins in the blood can trigger an inflammatory response so severe that the inflammation damages host tissues and organs more than the infection itself. This counterproductive immune response is called **systemic inflammatory response syndrome (SIRS)**, and it can lead to the life-threatening condition known as **sepsis**. Sepsis is characterized by the production of excess cytokines that leads to classic signs of inflammation such as fever, vasodilation, and edema (see Inflammation and Fever). In a patient with sepsis, the inflammatory response becomes dysregulated and disproportionate to the threat of infection. Critical organs such as the heart, lungs, liver, and kidneys become dysfunctional, resulting in increased heart and respiratory rates, and disorientation. If not treated promptly and effectively, patients with sepsis can go into shock and die.

Certain infections can cause inflammation in the heart and blood vessels. Inflammation of the endocardium, the inner lining of the heart, is called **endocarditis** and can result in damage to the heart valves severe enough to require surgical replacement. Inflammation of the pericardium, the sac surrounding the heart, is called **pericarditis**. The term **myocarditis** refers to the inflammation of the heart's muscle tissue. Pericarditis and myocarditis can cause fluid to accumulate around the heart, resulting in congestive heart failure. Inflammation of blood vessels is called **vasculitis**. Although somewhat rare, vasculitis can cause blood vessels to become damaged and rupture; as blood is released, small red or purple spots called **petechiae** appear on the skin. If the damage of tissues or blood vessels is severe, it can result in reduced blood flow to the surrounding tissues. This condition is called **ischemia**, and it can be very serious. In severe cases, the affected tissues can die and become necrotic; these situations may require surgical debridement or amputation.

✓ CHECK YOUR UNDERSTANDING

- Why does the circulatory system have no normal microbiota?
- Explain why the presence of microbes in the circulatory system can lead to serious consequences.

Infections of the Lymphatic System

Like the circulatory system, the lymphatic system does not have a normal microbiota, and the large numbers of immune cells typically eliminate transient microbes before they can establish an infection. Only microbes with an array of virulence factors are able to overcome these defenses and establish infection in the lymphatic system. However, when a localized infection begins to spread, the lymphatic system is often the first place the invading microbes can be detected.

Infections in the lymphatic system also trigger an inflammatory response. Inflammation of lymphatic vessels, called **lymphangitis**, can produce visible red streaks under the skin. Inflammation in the lymph nodes can cause them to swell. A swollen lymph node is referred to as a **bubo**, and the condition is referred to as **lymphadenitis**.

25.2 Bacterial Infections of the Circulatory and Lymphatic Systems

Learning Objectives

By the end of this section, you will be able to:
- Identify and compare bacteria that most commonly cause infections of the circulatory and lymphatic systems
- Compare the major characteristics of specific bacterial diseases affecting the circulatory and lymphatic systems

Bacteria can enter the circulatory and lymphatic systems through acute infections or breaches of the skin barrier or mucosa. Breaches may occur through fairly common occurrences, such as insect bites or small wounds. Even the act of tooth brushing, which can cause small ruptures in the gums, may introduce bacteria into the circulatory system. In most cases, the bacteremia that results from such common exposures is transient and remains below the threshold of detection. In severe cases, bacteremia can lead to septicemia with dangerous complications such as toxemia, sepsis, and septic shock. In these situations, it is often the immune response to the infection that results in the clinical signs and symptoms rather than the microbes themselves.

Bacterial Sepsis, Septic and Toxic Shock

At low concentrations, pro-inflammatory cytokines such as interleukin 1 (IL-1) and tumor necrosis factor-α (TNF-α) play important roles in the host's immune defenses. When they circulate systemically in larger amounts, however, the resulting immune response can be life threatening. IL-1 induces vasodilation (widening of blood vessels) and reduces the tight junctions between vascular endothelial cells, leading to widespread edema. As fluids move out of circulation into tissues, blood pressure begins to drop. If left unchecked, the blood pressure can fall below the level necessary to maintain proper kidney and respiratory functions, a condition known as **septic shock**. In addition, the excessive release of cytokines during the inflammatory response can lead to the formation of blood clots. The loss of blood pressure and occurrence of blood clots can result in multiple organ failure and death.

Bacteria are the most common pathogens associated with the development of sepsis, and septic shock.[3] The most common infection associated with sepsis is bacterial pneumonia (see Bacterial Infections of the Respiratory Tract), accounting for about half of all cases, followed by intra-abdominal infections (Bacterial Infections of the Gastrointestinal Tract) and urinary tract infections (Bacterial Infections of the Urinary System).[4] Infections associated with superficial wounds, animal bites, and indwelling catheters may also lead to sepsis and septic shock.

These initially minor, localized infections can be caused by a wide range of different bacteria, including *Staphylococcus*, *Streptococcus*, *Pseudomonas*, *Pasteurella*, *Acinetobacter*, and members of the Enterobacteriaceae. However, if left untreated, infections by these gram-positive and gram-negative pathogens

---

3 S.P. LaRosa. "Sepsis." 2010. http://www.clevelandclinicmeded.com/medicalpubs/diseasemanagement/infectious-disease/sepsis/.

4 D.C. Angus, T. Van der Poll. "Severe Sepsis and Septic Shock." *New England Journal of Medicine* 369, no. 9 (2013):840–851.

can potentially progress to sepsis, shock, and death.

Toxic Shock Syndrome and Streptococcal Toxic Shock-Like Syndrome

Toxemia associated with infections caused by *Staphylococcus aureus* can cause staphylococcal **toxic shock syndrome (TSS)**. Some strains of *S. aureus* produce a superantigen called toxic shock syndrome toxin-1 (TSST-1). TSS may occur as a complication of other localized or systemic infections such as pneumonia, osteomyelitis, sinusitis, and skin wounds (surgical, traumatic, or burns). Those at highest risk for staphylococcal TSS are women with preexisting *S. aureus* colonization of the vagina who leave tampons, contraceptive sponges, diaphragms, or other devices in the vagina for longer than the recommended time.

Staphylococcal TSS is characterized by sudden onset of vomiting, diarrhea, myalgia, body temperature higher than 38.9 °C (102.0 °F), and rapid-onset hypotension with a systolic blood pressure less than 90 mm Hg for adults; a diffuse erythematous rash that leads to peeling and shedding skin 1 to 2 weeks after onset; and additional involvement of three or more organ systems.[5] The mortality rate associated with staphylococcal TSS is less than 3% of cases.

Diagnosis of staphylococcal TSS is based on clinical signs, symptoms, serologic tests to confirm bacterial species, and the detection of toxin production from staphylococcal isolates. Cultures of skin and blood are often negative; less than 5% are positive in cases of staphylococcal TSS. Treatment for staphylococcal TSS includes decontamination, debridement, vasopressors to elevate blood pressure, and antibiotic therapy with clindamycin plus vancomycin or daptomycin pending susceptibility results.

A syndrome with signs and symptoms similar to staphylococcal TSS can be caused by *Streptococcus pyogenes*. This condition, called **streptococcal toxic shock-like syndrome (STSS)**, is characterized by more severe pathophysiology than staphylococcal TSS,[6] with about 50% of patients developing *S. pyogenes* bacteremia and necrotizing fasciitis. In contrast to staphylococcal TSS, STSS is more likely to cause acute respiratory distress syndrome (ARDS), a rapidly progressive disease characterized by fluid accumulation in the lungs that inhibits breathing and causes hypoxemia (low oxygen levels in the blood). STSS is associated with a higher mortality rate (20%–60%), even with aggressive therapy. STSS usually develops in patients with a streptococcal soft-tissue infection such as bacterial cellulitis, necrotizing fasciitis, pyomyositis (pus formation in muscle caused by infection), a recent influenza A infection, or chickenpox.

✓ CHECK YOUR UNDERSTANDING

- How can large amounts of pro-inflammatory cytokines lead to septic shock?

Clinical Focus

Part 2

Despite oxacillin therapy, Barbara's condition continued to worsen over the next several days. Her fever increased to 40.1 °C (104.2 °F) and she began to experience chills, rapid breathing, and confusion. Her doctor suspected bacteremia by a drug-resistant bacterium and admitted Barbara to the hospital. Cultures of the surgical site and blood revealed *Staphylococcus aureus*. Antibiotic susceptibility testing confirmed that the isolate was methicillin-resistant *S. aureus* (MRSA). In response, Barbara's doctor changed her antibiotic therapy to vancomycin and arranged to have the port and venous catheter removed.

- Why did Barbara's infection not respond to oxacillin therapy?

---

5 Centers for Disease Control and Prevention. "Toxic Shock Syndrome (Other Than Streptococcal) (TSS) 2011 Case Definition." https://wwwn.cdc.gov/nndss/conditions/toxic-shock-syndrome-other-than-streptococcal/case-definition/2011/. Accessed July 25, 2016.

6 Centers for Disease Control and Prevention. "Streptococcal Toxic Shock Syndrome (STSS) (*Streptococcus pyogenes*) 2010 Case Definition." https://wwwn.cdc.gov/nndss/conditions/streptococcal-toxic-shock-syndrome/case-definition/2010/. Accessed July 25, 2016.

- Why did the physician have the port and catheter removed?
- Based on the signs and symptoms described, what are some possible diagnoses for Barbara's condition?

Jump to the next Clinical Focus feature box. Go back to the previous Clinical Focus feature box.

Puerperal Sepsis

A type of sepsis called **puerperal sepsis**, also known as puerperal infection, puerperal fever, or childbed fever, is a nosocomial infection associated with the period of puerperium—the time following childbirth during which the mother's reproductive system returns to a nonpregnant state. Such infections may originate in the genital tract, breast, urinary tract, or a surgical wound. Initially the infection may be limited to the uterus or other local site of infection, but it can quickly spread, resulting in peritonitis, septicemia, and death. Before the 19th century work of Ignaz Semmelweis and the widespread acceptance of germ theory (see Modern Foundations of Cell Theory), puerperal sepsis was a major cause of mortality among new mothers in the first few days following childbirth.

Puerperal sepsis is often associated with *Streptococcus pyogenes*, but numerous other bacteria can also be responsible. Examples include gram-positive bacterial (e.g. *Streptococcus* spp., *Staphylococcus* spp., and *Enterococcus* spp.), gram-negative bacteria (e.g. *Chlamydia* spp., *Escherichia coli*, *Klebsiella* spp., and *Proteus* spp.), as well as anaerobes such as *Peptostreptococcus* spp., *Bacteroides* spp., and *Clostridium* spp. In cases caused by *S. pyogenes*, the bacteria attach to host tissues using M protein and produce a carbohydrate capsule to avoid phagocytosis. *S. pyogenes* also produces a variety of exotoxins, like streptococcal pyrogenic exotoxins A and B, that are associated with virulence and may function as superantigens.

Diagnosis of puerperal fever is based on the timing and extent of fever and isolation, and identification of the etiologic agent in blood, wound, or urine specimens. Because there are numerous possible causes, antimicrobial susceptibility testing must be used to determine the best antibiotic for treatment. Nosocomial incidence of puerperal fever can be greatly reduced through the use of antiseptics during delivery and strict adherence to handwashing protocols by doctors, midwives, and nurses.

Infectious Arthritis

Also called **septic arthritis**, **infectious arthritis** can be either an acute or a chronic condition. Infectious arthritis is characterized by inflammation of joint tissues and is most often caused by bacterial pathogens. Most cases of acute infectious arthritis are secondary to bacteremia, with a rapid onset of moderate to severe joint pain and swelling that limits the motion of the affected joint. In adults and young children, the infective pathogen is most often introduced directly through injury, such as a wound or a surgical site, and brought to the joint through the circulatory system. Acute infections may also occur after joint replacement surgery. Acute infectious arthritis often occurs in patients with an immune system impaired by other viral and bacterial infections. *S. aureus* is the most common cause of acute septic arthritis in the general population of adults and young children. *Neisseria gonorrhoeae* is an important cause of acute infectious arthritis in sexually active individuals.

Chronic infectious arthritis is responsible for 5% of all infectious arthritis cases and is more likely to occur in patients with other illnesses or conditions. Patients at risk include those who have an HIV infection, a bacterial or fungal infection, prosthetic joints, rheumatoid arthritis (RA), or who are undergoing immunosuppressive chemotherapy. Onset is often in a single joint; there may be little or no pain, aching pain that may be mild, gradual swelling, mild warmth, and minimal or no redness of the joint area.

Diagnosis of infectious arthritis requires the aspiration of a small quantity of synovial fluid from the afflicted joint. Direct microscopic evaluation, culture, antimicrobial susceptibility testing, and polymerase chain reaction (PCR) analyses of the synovial fluid are used to identify the potential pathogen. Typical treatment includes administration of appropriate antimicrobial drugs based on antimicrobial susceptibility testing. For nondrug-resistant bacterial strains, β-lactams such as oxacillin and cefazolin are often prescribed for staphylococcal infections. Third-generation cephalosporins (e.g., ceftriaxone) are used for increasingly prevalent β-lactam-resistant *Neisseria* infections. Infections by *Mycobacterium* spp. or fungi are treated with appropriate long-term antimicrobial therapy. Even with treatment, the prognosis is often poor for those infected. About 40% of patients with nongonnococcal infectious arthritis will suffer permanent joint damage and mortality rates range from 5% to 20%.[7] Mortality rates are higher among the elderly.[8]

Osteomyelitis

Osteomyelitis is an inflammation of bone tissues most commonly caused by infection. These infections can either be acute or chronic and can involve a variety of different bacteria. The most common causative agent of **osteomyelitis** is *S. aureus*. However, *M. tuberculosis, Pseudomonas aeruginosa, Streptococcus pyogenes, S. agalactiae*, species in the Enterobacteriaceae, and other microorganisms can also cause osteomyelitis, depending on which bones are involved. In adults, bacteria usually gain direct access to the bone tissues through trauma or a surgical procedure involving prosthetic joints. In children, the bacteria are often introduced from the bloodstream, possibly spreading from focal infections. The long bones, such as the femur, are more commonly affected in children because of the more extensive vascularization of bones in the young.[9]

The signs and symptoms of osteomyelitis include fever, localized pain, swelling due to edema, and ulcers in soft tissues near the site of infection. The resulting inflammation can lead to tissue damage and bone loss. In addition, the infection may spread to joints, resulting in infectious arthritis, or disseminate into the blood, resulting in sepsis and thrombosis (formation of blood clots). Like septic arthritis, osteomyelitis is usually diagnosed using a combination of radiography, imaging, and identification of bacteria from blood cultures, or from bone cultures if blood cultures are negative. Parenteral antibiotic therapy is typically used to treat osteomyelitis. Because of the number of different possible etiologic agents, however, a variety of drugs might be used. Broad-spectrum antibacterial drugs such as nafcillin, oxacillin, or cephalosporin are typically prescribed for acute osteomyelitis, and ampicillin and piperacillin/tazobactam for chronic osteomyelitis. In cases of antibiotic resistance, vancomycin treatment is sometimes required to control the infection. In serious cases, surgery to remove the site of infection may be required. Other forms of treatment include hyperbaric oxygen therapy (see Using Physical Methods to Control Microorganisms) and implantation of antibiotic beads or pumps.

✓ CHECK YOUR UNDERSTANDING

- What bacterium the most common cause of both septic arthritis and osteomyelitis?

---

7 M.E. Shirtliff, Mader JT. "Acute Septic Arthritis." *Clinical Microbiology Reviews* 15 no. 4 (2002):527–544.

8 J.R. Maneiro et al. "Predictors of Treatment Failure and Mortality in Native Septic Arthritis." *Clinical Rheumatology* 34, no. 11 (2015):1961–1967.

9 M. Vazquez. "Osteomyelitis in Children." *Current Opinion in Pediatrics* 14, no. 1 (2002):112–115.

Rheumatic Fever

Infections with *S. pyogenes* have a variety of manifestations and complications generally called sequelae. As mentioned, the bacterium can cause suppurative infections like puerperal fever. However, this microbe can also cause nonsuppurative sequelae in the form of acute **rheumatic fever** (ARF), which can lead to rheumatic heart disease, thus impacting the circulatory system. Rheumatic fever occurs primarily in children a minimum of 2–3 weeks after an episode of untreated or inadequately treated pharyngitis (see Bacterial Infections of the Respiratory Tract). At one time, rheumatic fever was a major killer of children in the US; today, however, it is rare in the US because of early diagnosis and treatment of streptococcal pharyngitis with antibiotics. In parts of the world where diagnosis and treatment are not readily available, acute rheumatic fever and rheumatic heart disease are still major causes of mortality in children.[10]

Rheumatic fever is characterized by a variety of diagnostic signs and symptoms caused by nonsuppurative, immune-mediated damage resulting from a cross-reaction between patient antibodies to bacterial surface proteins and similar proteins found on cardiac, neuronal, and synovial tissues. Damage to the nervous tissue or joints, which leads to joint pain and swelling, is reversible. However, damage to heart valves can be irreversible and is worsened by repeated episodes of acute rheumatic fever, particularly during the first 3–5 years after the first rheumatic fever attack. The inflammation of the heart valves caused by cross-reacting antibodies leads to scarring and stiffness of the valve leaflets. This, in turn, produces a characteristic heart murmur. Patients who have previously developed rheumatic fever and who subsequently develop recurrent pharyngitis due to *S. pyogenes* are at high risk for a recurrent attacks of rheumatic fever.

The American Heart Association recommends[11] a treatment regimen consisting of benzathine benzylpenicillin every 3 or 4 weeks, depending on the patient's risk for reinfection. Additional prophylactic antibiotic treatment may be recommended depending on the age of the patient and risk for reinfection.

Bacterial Endocarditis and Pericarditis

The endocardium is a tissue layer that lines the muscles and valves of the heart. This tissue can become infected by a variety of bacteria, including gram-positive cocci such as *Staphylococcus aureus*, viridans streptococci, and *Enterococcus faecalis*, and the gram-negative so-called HACEK bacilli: *Haemophilus* spp., *Actinobacillus actinomycetemcomitans*, *Cardiobacterium hominis*, *Eikenella corrodens*, and *Kingella kingae*. The resulting inflammation is called endocarditis, which can be described as either acute or subacute. Causative agents typically enter the bloodstream during accidental or intentional breaches in the normal barrier defenses (e.g., dental procedures, body piercings, catheterization, wounds). Individuals with preexisting heart damage, prosthetic valves and other cardiac devices, and those with a history of rheumatic fever have a higher risk for endocarditis. This disease can rapidly destroy the heart valves and, if untreated, lead to death in just a few days.

In **subacute bacterial endocarditis**, heart valve damage occurs slowly over a period of months. During this time, blood clots form in the heart, and these protect the bacteria from phagocytes. These patches of tissue-associated bacteria are called vegetations. The resulting damage to the heart, in part resulting from the immune response causing fibrosis of heart valves, can necessitate heart valve replacement (Figure 25.6). Outward signs of subacute endocarditis may include a fever.

Diagnosis of infective endocarditis is determined using the combination of blood cultures, echocardiogram, and clinical symptoms. In both acute and subacute endocarditis, treatment typically involves relatively high doses of intravenous antibiotics as determined by antimicrobial susceptibility testing. Acute endocarditis is

---

10 A. Beaudoin et al. "Acute Rheumatic Fever and Rheumatic Heart Disease Among Children—American Samoa, 2011–2012." *Morbidity and Mortality Weekly Report* 64 no. 20 (2015):555–558.

11 M.A. Gerber et al. "Prevention of Rheumatic Fever and Diagnosis and Treatment of Acute Streptococcal Pharyngitis: A Scientific Statement From the American Heart Association Rheumatic Fever, Endocarditis, and Kawasaki Disease Committee of the Council on Cardiovascular Disease in the Young, the Interdisciplinary Council on Functional Genomics and Translational Biology, and the Interdisciplinary Council on Quality of Care and Outcomes Research: Endorsed by the American Academy of Pediatrics." *Circulation* 119, no. 11 (2009):1541–1551.

often treated with a combination of ampicillin, nafcillin, and gentamicin for synergistic coverage of *Staphylococcus* spp. and *Streptococcus* spp. Prosthetic-valve endocarditis is often treated with a combination of vancomycin, rifampin, and gentamicin. Rifampin is necessary to treat individuals with infection of prosthetic valves or other foreign bodies because rifampin can penetrate the biofilm of most of the pathogens that infect these devices.

Staphylcoccus spp. and *Streptococcus* spp. can also infect and cause inflammation in the tissues surrounding the heart, a condition called acute pericarditis. Pericarditis is marked by chest pain, difficulty breathing, and a dry cough. In most cases, pericarditis is self-limiting and clinical intervention is not necessary. Diagnosis is made with the aid of a chest radiograph, electrocardiogram, echocardiogram, aspirate of pericardial fluid, or biopsy of pericardium. Antibacterial medications may be prescribed for infections associated with pericarditis; however, pericarditis can also be caused other pathogens, including viruses (e.g., echovirus, influenza virus), fungi (e.g., *Histoplasma* spp., *Coccidioides* spp.), and eukaryotic parasites (e.g., *Toxoplasma* spp.).

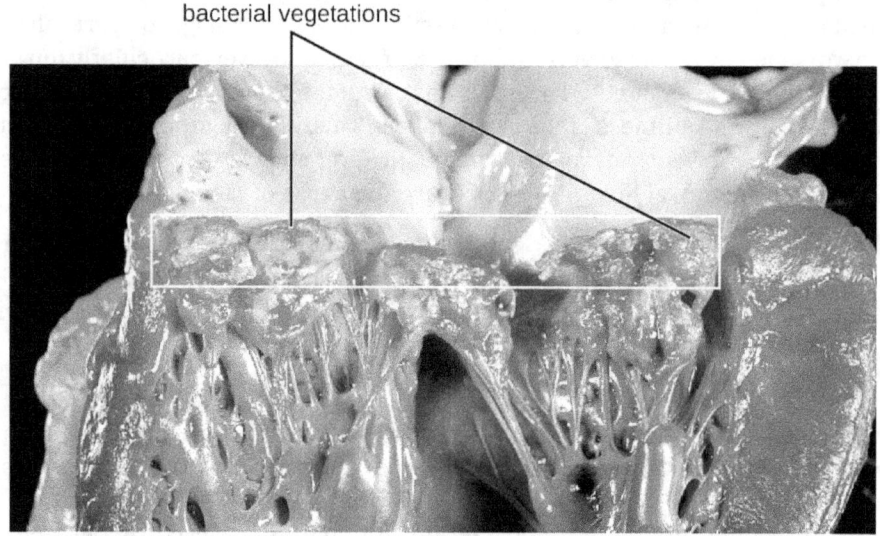

Figure 25.6 The heart of an individual who had subacute bacterial endocarditis of the mitral valve. Bacterial vegetations are visible on the valve tissues. (credit: modification of work by Centers for Disease Control and Prevention)

CHECK YOUR UNDERSTANDING

- Compare acute and subacute bacterial endocarditis.

Gas Gangrene

Traumatic injuries or certain medical conditions, such as diabetes, can cause damage to blood vessels that interrupts blood flow to a region of the body. When blood flow is interrupted, tissues begin to die, creating an anaerobic environment in which anaerobic bacteria can thrive. This condition is called ischemia. Endospores of the anaerobic bacterium *Clostridium perfringens* (along with a number of other *Clostridium* spp. from the gut) can readily germinate in ischemic tissues and colonize the anaerobic tissues.

The resulting infection, called **gas gangrene**, is characterized by rapidly spreading myonecrosis (death of muscle tissue). The patient experiences a sudden onset of excruciating pain at the infection site and the rapid development of a foul-smelling wound containing gas bubbles and a thin, yellowish discharge tinged with a small amount of blood. As the infection progresses, edema and cutaneous blisters containing bluish-purple fluid form. The infected tissue becomes liquefied and begins sloughing off. The margin between necrotic and healthy tissue often advances several inches per hour even with antibiotic therapy. Septic shock and organ failure frequently accompany gas gangrene; when patients develop sepsis, the mortality rate is greater than 50%.

α-Toxin and theta (θ) toxin are the major virulence factors of *C. perfringens* implicated in gas gangrene. α-

Toxin is a lipase responsible for breaking down cell membranes; it also causes the formation of thrombi (blood clots) in blood vessels, contributing to the spread of ischemia. θ-Toxin forms pores in the patient's cell membranes, causing cell lysis. The gas associated with gas gangrene is produced by *Clostridium*'s fermentation of butyric acid, which produces hydrogen and carbon dioxide that are released as the bacteria multiply, forming pockets of gas in tissues (Figure 25.7).

Gas gangrene is initially diagnosed based on the presence of the clinical signs and symptoms described earlier in this section. Diagnosis can be confirmed through Gram stain and anaerobic cultivation of wound exudate (drainage) and tissue samples on blood agar. Treatment typically involves surgical debridement of any necrotic tissue; advanced cases may require amputation. Surgeons may also use vacuum-assisted closure (VAC), a surgical technique in which vacuum-assisted drainage is used to remove blood or serous fluid from a wound or surgical site to speed recovery. The most common antibiotic treatments include penicillin G and clindamycin. Some cases are also treated with hyperbaric oxygen therapy because *Clostridium* spp. are incapable of surviving in oxygen-rich environments.

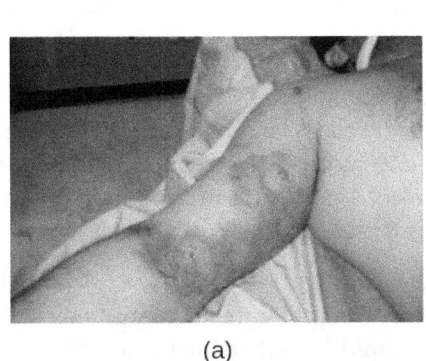
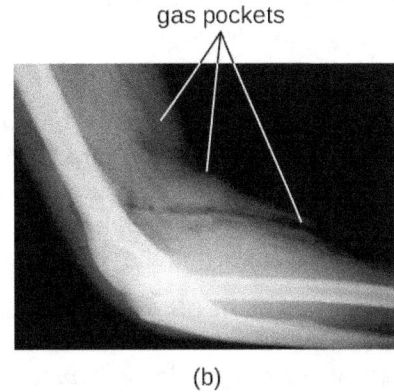

Figure 25.7 (a) In this image of a patient with gas gangrene, note the bluish-purple discoloration around the bicep and the irregular margin of the discolored tissue indicating the spread of infection. (b) A radiograph of the arm shows a darkening in the tissue, which indicates the presence of gas. (credit a, b: modification of work by Aggelidakis J, Lasithiotakis K, Topalidou A, Koutroumpas J, Kouvidis G, and Katonis P)

Tularemia

Infection with the gram-negative bacterium *Francisella tularensis* causes **tularemia** (or rabbit fever), a zoonotic infection in humans. *F. tularensis* is a facultative intracellular parasite that primarily causes illness in rabbits, although a wide variety of domesticated animals are also susceptible to infection. Humans can be infected through ingestion of contaminated meat or, more typically, handling of infected animal tissues (e.g., skinning an infected rabbit). Tularemia can also be transmitted by the bites of infected arthropods, including the dog tick (*Dermacentor variabilis*), the lone star tick (*Amblyomma americanum*), the wood tick (*Dermacentor andersoni*), and deer flies (*Chrysops* spp.). Although the disease is not directly communicable between humans, exposure to aerosols of *F. tularensis* can result in life-threatening infections. *F. tularensis* is highly contagious, with an infectious dose of as few as 10 bacterial cells. In addition, pulmonary infections have a 30%–60% fatality rate if untreated.[12] For these reasons, *F. tularensis* is currently classified and must be handled as a biosafety level-3 (BSL-3) organism and as a potential biological warfare agent.

Following introduction through a break in the skin, the bacteria initially move to the lymph nodes, where they are ingested by phagocytes. After escaping from the phagosome, the bacteria grow and multiply intracellularly in the cytoplasm of phagocytes. They can later become disseminated through the blood to other organs such as the liver, lungs, and spleen, where they produce masses of tissue called granulomas (Figure 25.8). After an incubation period of about 3 days, skin lesions develop at the site of infection. Other signs and symptoms include fever, chills, headache, and swollen and painful lymph nodes.

---

12 World Health Organization. "WHO Guidelines on Tulareamia." 2007. http://www.cdc.gov/tularemia/resources/whotularemiamanual.pdf. Accessed July 26, 2016.

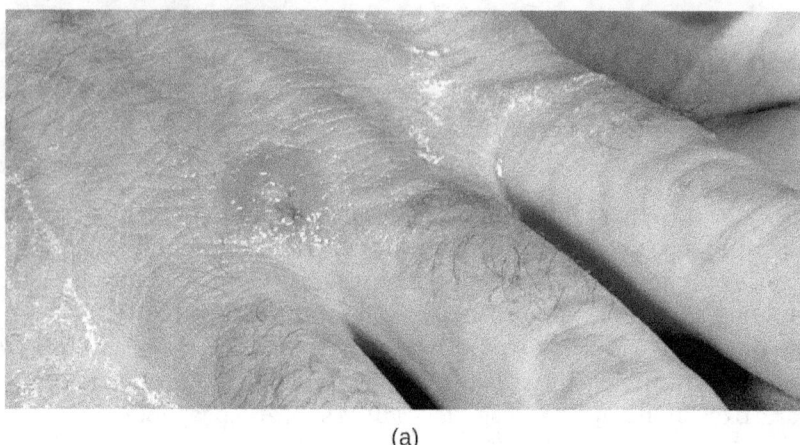
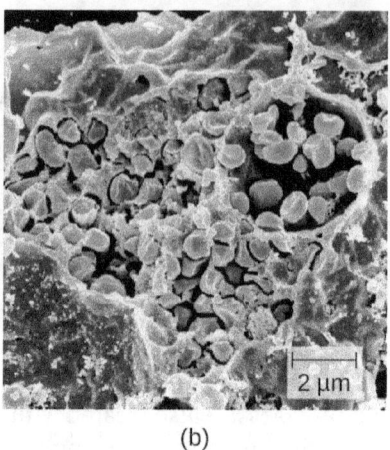

Figure 25.8 (a) A skin lesion appears at the site of infection on the hand of an individual infected with *Francisella tularensis*. (b) A scanning electron micrograph shows the coccobacilli cells (blue) of *F. tularensis*. (credit a: modification of work by Centers for Disease Control and Prevention; credit b: modification of work by NIAID)

A direct diagnosis of tularemia is challenging because it is so contagious. Once a presumptive diagnosis of tularemia is made, special handling is required to collect and process patients' specimens to prevent the infection of health-care workers. Specimens suspected of containing *F. tularensis* can only be handled by BSL-2 or BSL-3 laboratories registered with the Federal Select Agent Program, and individuals handling the specimen must wear protective equipment and use a class II biological safety cabinet.

Tularemia is relatively rare in the US, and its signs and symptoms are similar to a variety of other infections that may need to be ruled out before a diagnosis can be made. Direct fluorescent-antibody (DFA) microscopic examination using antibodies specific for *F. tularensis* can rapidly confirm the presence of this pathogen. Culturing this microbe is difficult because of its requirement for the amino acid cysteine, which must be supplied as an extra nutrient in culturing media. Serological tests are available to detect an immune response against the bacterial pathogen. In patients with suspected infection, acute- and convalescent-phase serum samples are required to confirm an active infection. PCR-based tests can also be used for clinical identification of direct specimens from body fluids or tissues as well as cultured specimens. In most cases, diagnosis is based on clinical findings and likely incidents of exposure to the bacterium. The antibiotics streptomycin, gentamycin, doxycycline, and ciprofloxacin are effective in treating tularemia.

Brucellosis

Species in the genus *Brucella* are gram-negative facultative intracellular pathogens that appear as coccobacilli. Several species cause zoonotic infections in animals and humans, four of which have significant human pathogenicity: *B. abortus* from cattle and buffalo, *B. canis* from dogs, *B. suis* from swine, and *B. melitensis* from goats, sheep, and camels. Infections by these pathogens are called brucellosis, also known as undulant fever, "Mediterranean fever," or "Malta fever." Vaccination of animals has made brucellosis a rare disease in the US, but it is still common in the Mediterranean, south and central Asia, Central and South America, and the Caribbean. Human infections are primarily associated with the ingestion of meat or unpasteurized dairy products from infected animals. Infection can also occur through inhalation of bacteria in aerosols when handling animal products, or through direct contact with skin wounds. In the US, most cases of brucellosis are found in individuals with extensive exposure to potentially infected animals (e.g., slaughterhouse workers, veterinarians).

Two important virulence factors produced by *Brucella* spp. are urease, which allows ingested bacteria to avoid destruction by stomach acid, and lipopolysaccharide (LPS), which allows the bacteria to survive within phagocytes. After gaining entry to tissues, the bacteria are phagocytized by host neutrophils and macrophages. The bacteria then escape from the phagosome and grow within the cytoplasm of the cell. Bacteria phagocytized by macrophages are disseminated throughout the body. This results in the formation of granulomas within many body sites, including bone, liver, spleen, lung, genitourinary tract, brain, heart, eye, and skin. Acute infections can result in undulant (relapsing) fever, but untreated infections develop into chronic disease that usually manifests as acute febrile illness (fever of 40–41 °C [104–105.8 °F]) with recurring flu-like signs and symptoms.

Brucella is only reliably found in the blood during the acute fever stage; it is difficult to diagnose by cultivation. In addition, *Brucella* is considered a BSL-3 pathogen and is hazardous to handle in the clinical laboratory without protective clothing and at least a class II biological safety cabinet. Agglutination tests are most often used for serodiagnosis. In addition, enzyme-linked immunosorbent assays (ELISAs) are available to determine exposure to the organism. The antibiotics doxycycline or ciprofloxacin are typically prescribed in combination with rifampin; gentamicin, streptomycin, and trimethoprim-sulfamethoxazole (TMP-SMZ) are also effective against *Brucella* infections and can be used if needed.

✓ CHECK YOUR UNDERSTANDING

- Compare the pathogenesis of tularemia and brucellosis.

Cat-Scratch Disease

The zoonosis **cat-scratch disease (CSD)** (or cat-scratch fever) is a bacterial infection that can be introduced to the lymph nodes when a human is bitten or scratched by a cat. It is caused by the facultative intracellular gram-negative bacterium *Bartonella henselae*. Cats can become infected from flea feces containing *B. henselae* that they ingest while grooming. Humans become infected when flea feces or cat saliva (from claws or licking) containing *B. henselae* are introduced at the site of a bite or scratch. Once introduced into a wound, *B. henselae* infects red blood cells.

B. henselae invasion of red blood cells is facilitated by adhesins associated with outer membrane proteins and a secretion system that mediates transport of virulence factors into the host cell. Evidence of infection is indicated if a small nodule with pus forms in the location of the scratch 1 to 3 weeks after the initial injury. The bacteria then migrate to the nearest lymph nodes, where they cause swelling and pain. Signs and symptoms may also include fever, chills, and fatigue. Most infections are mild and tend to be self-limiting. However, immunocompromised patients may develop bacillary angiomatosis (BA), characterized by the proliferation of blood vessels, resulting in the formation of tumor-like masses in the skin and internal organs; or bacillary peliosis (BP), characterized by multiple cyst-like, blood-filled cavities in the liver and spleen. Most cases of CSD can be prevented by keeping cats free of fleas and promptly cleaning a cat scratch with soap and warm water.

The diagnosis of CSD is difficult because the bacterium does not grow readily in the laboratory. When necessary, immunofluorescence, serological tests, PCR, and gene sequencing can be performed to identify the bacterial species. Given the limited nature of these infections, antibiotics are not normally prescribed. For immunocompromised patients, rifampin, azithromycin, ciprofloxacin, gentamicin (intramuscularly), or TMP-SMZ are generally the most effective options.

Rat-Bite Fever

The zoonotic infection **rat-bite fever** can be caused by two different gram-negative bacteria: *Streptobacillus moniliformis*, which is more common in North America, and *Spirillum minor*, which is more common in Asia. Because of modern sanitation efforts, rat bites are rare in the US. However, contact with fomites, food, or water contaminated by rat feces or body fluids can also cause infections. Signs and symptoms of rat-bite fever include fever, vomiting, myalgia (muscle pain), arthralgia (joint pain), and a maculopapular rash on the hands and feet. An ulcer may also form at the site of a bite, along with some swelling of nearby lymph nodes. In most cases, the infection is self-limiting. Little is known about the virulence factors that contribute to these signs and symptoms of disease.

Cell culture, MALDI-TOF mass spectrometry, PCR, or ELISA can be used in the identification of *Streptobacillus moniliformis*. The diagnosis *Spirillum minor* may be confirmed by direct microscopic observation of the pathogens in blood using Giemsa or Wright stains, or darkfield microscopy. Serological tests can be used to detect a host immune response to the pathogens after about 10 days. The most commonly used antibiotics to treat these infections are penicillin or doxycycline.

Plague

The gram-negative bacillus *Yersinia pestis* causes the zoonotic infection **plague**. This bacterium causes acute febrile disease in animals, usually rodents or other small mammals, and humans. The disease is associated with a high mortality rate if left untreated. Historically, *Y. pestis* has been responsible for several devastating pandemics, resulting in millions of deaths (see Micro Connections: The History of the Plague). There are three forms of plague: **bubonic plague** (the most common form, accounting for about 80% of cases), **pneumonic plague**, and **septicemic plague**. These forms are differentiated by the mode of transmission and the initial site of infection. Figure 25.9 illustrates these various modes of transmission and infection between animals and humans.

In bubonic plague, *Y. pestis* is transferred by the bite of infected fleas. Since most flea bites occur on the legs and ankles, *Y. pestis* is often introduced into the tissues and blood circulation in the lower extremities. After a 2- to 6-day incubation period, patients experience an abrupt onset fever (39.5–41 °C [103.1–105.8 °F]), headache, hypotension, and chills. The pathogen localizes in lymph nodes, where it causes inflammation, swelling, and hemorrhaging that results in purple buboes (Figure 25.10). Buboes often form in lymph nodes of the groin first because these are the nodes associated with the lower limbs; eventually, through circulation in the blood and lymph, lymph nodes throughout the body become infected and form buboes. The average mortality rate for bubonic plague is about 55% if untreated and about 10% with antibiotic treatment.

Septicemic plague occurs when *Y. pestis* is directly introduced into the bloodstream through a cut or wound and circulates through the body. The incubation period for septicemic plague is 1 to 3 days, after which patients develop fever, chills, extreme weakness, abdominal pain, and shock. Disseminated intravascular coagulation (DIC) can also occur, resulting in the formation of thrombi that obstruct blood vessels and promote ischemia and necrosis in surrounding tissues (Figure 25.10). Necrosis occurs most commonly in extremities such as fingers and toes, which become blackened. Septicemic plague can quickly lead to death, with a mortality rate near 100% when it is untreated. Even with antibiotic treatment, the mortality rate is about 50%.

Pneumonic plague occurs when *Y. pestis* causes an infection of the lungs. This can occur through inhalation of aerosolized droplets from an infected individual or when the infection spreads to the lungs from elsewhere in the body in patients with bubonic or septicemic plague. After an incubation period of 1 to 3 days, signs and symptoms include fever, headache, weakness, and a rapidly developing pneumonia with shortness of breath, chest pain, and cough producing bloody or watery mucus. The pneumonia may result in rapid respiratory failure and shock. Pneumonic plague is the only form of plague that can be spread from person to person by infectious aerosol droplet. If untreated, the mortality rate is near 100%; with antibiotic treatment, the mortality rate is about 50%.

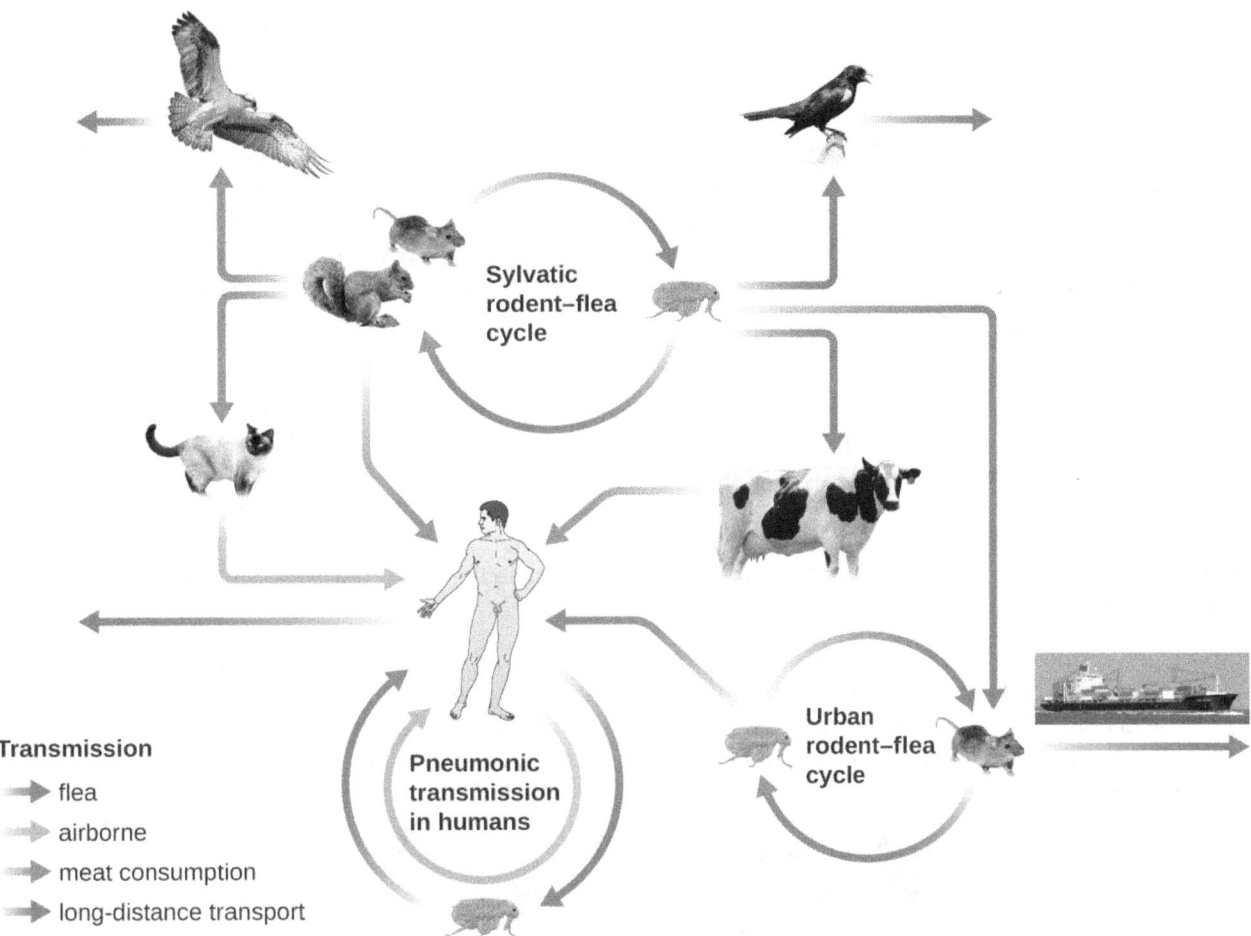

Transmission
→ flea
→ airborne
→ meat consumption
→ long-distance transport

Figure 25.9 *Yersinia pestis*, the causative agent of plague, has numerous modes of transmission. The modes are divided into two ecological classes: urban and sylvatic (i.e., forest or rural). The urban cycle primarily involves transmission from infected urban mammals (rats) to humans by flea vectors (brown arrows). The disease may travel between urban centers (purple arrow) if infected rats find their way onto ships or trains. The sylvatic cycle involves mammals more common in nonurban environments. Sylvatic birds and mammals (including humans) may become infected after eating infected mammals (pink arrows) or by flea vectors. Pneumonic transmission occurs between humans or between humans and infected animals through the inhalation of *Y. pestis* in aerosols. (credit "diagram": modification of work by Stenseth NC, Atshabar BB, Begon M, Belmain SR, Bertherat E, Carniel E, Gage KL, Leirs H, and Rahalison L; credit "cat": modification of work by "KaCey97078"/Flickr)

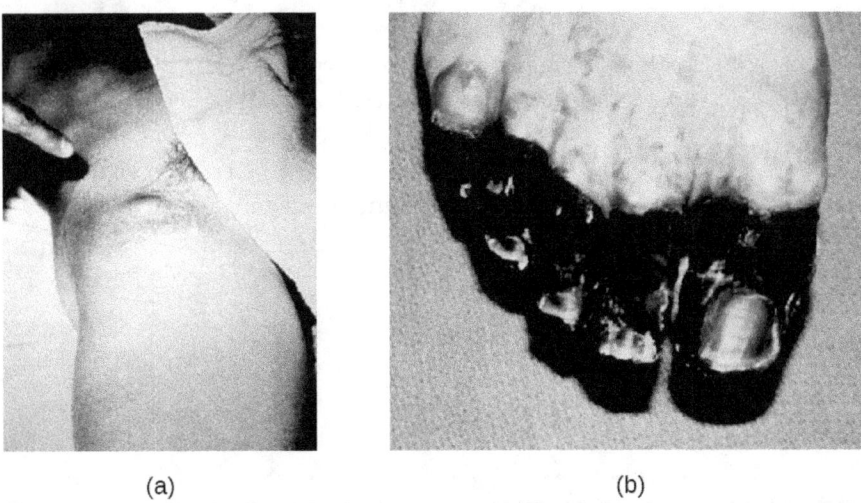

(a) (b)

Figure 25.10 (a) *Yersinia pestis* infection can cause inflamed and swollen lymph nodes (buboes), like these in the groin of an infected patient. (b) Septicemic plague caused necrotic toes in this patient. Vascular damage at the extremities causes ischemia and tissue death. (credit a: modification of work by American Society for Microbiology; credit b: modification of work by Centers for Disease Control and Prevention)

The high mortality rate for the plague is, in part, a consequence of it being unusually well equipped with virulence factors. To date, there are at least 15 different major virulence factors that have been identified from *Y. pestis* and, of these, eight are involved with adherence to host cells. In addition, the F1 component of the *Y. pestis* capsule is a virulence factor that allows the bacterium to avoid phagocytosis. F1 is produced in large quantities during mammalian infection and is the most immunogenic component.[13] Successful use of virulence factors allows the bacilli to disseminate from the area of the bite to regional lymph nodes and eventually the entire blood and lymphatic systems.

Culturing and direct microscopic examination of a sample of fluid from a bubo, blood, or sputum is the best way to identify *Y. pestis* and confirm a presumptive diagnosis of plague. Specimens may be stained using either a Gram, Giemsa, Wright, or Wayson's staining technique (Figure 25.11). The bacteria show a characteristic bipolar staining pattern, resembling safety pins, that facilitates presumptive identification. Direct fluorescent antibody tests (rapid test of outer-membrane antigens) and serological tests like ELISA can be used to confirm the diagnosis. The confirmatory method for identifying *Y. pestis* isolates in the US is bacteriophage lysis.

Prompt antibiotic therapy can resolve most cases of bubonic plague, but septicemic and pneumonic plague are more difficult to treat because of their shorter incubation stages. Survival often depends on an early and accurate diagnosis and an appropriate choice of antibiotic therapy. In the US, the most common antibiotics used to treat patients with plague are gentamicin, fluoroquinolones, streptomycin, levofloxacin, ciprofloxacin, and doxycycline.

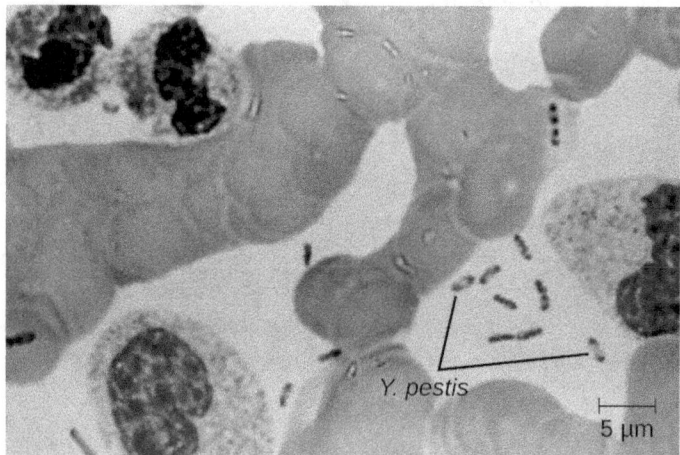

Figure 25.11 This Wright's stain of a blood sample from a patient with plague shows the characteristic "safety pin" appearance of *Yersinia pestis*. (credit: modification of work by Centers for Disease Control and Prevention)

✓ CHECK YOUR UNDERSTANDING

- Compare bubonic plague, septicemic plague, and pneumonic plague.

---

13 MOH Key Laboratory of Systems Biology of Pathogens. "Virulence Factors of Pathogenic Bacteria, *Yersinia*." http://www.mgc.ac.cn/cgi-bin/VFs/genus.cgi?Genus=Yersinia. Accessed September 9, 2016.

MICRO CONNECTIONS

The History of the Plague

The first recorded pandemic of plague, the Justinian plague, occurred in the sixth century CE. It is thought to have originated in central Africa and spread to the Mediterranean through trade routes. At its peak, more than 5,000 people died per day in Constantinople alone. Ultimately, one-third of that city's population succumbed to plague.[14] The impact of this outbreak probably contributed to the later fall of Emperor Justinian.

The second major pandemic, dubbed the Black Death, occurred during the 14th century. This time, the infections are thought to have originated somewhere in Asia before being transported to Europe by trade, soldiers, and war refugees. This outbreak killed an estimated one-quarter of the population of Europe (25 million, primarily in major cities). In addition, at least another 25 million are thought to have been killed in Asia and Africa.[15] This second pandemic, associated with strain *Yersinia pestis* biovar Medievalis, cycled for another 300 years in Europe and Great Britain, and was called the Great Plague in the 1660s.

The most recent pandemic occurred in the 1890s with *Yersinia pestis* biovar Orientalis. This outbreak originated in the Yunnan province of China and spread worldwide through trade. It is at this time that plague made its way to the US. The etiologic agent of plague was discovered by Alexandre Yersin (1863–1943) during this outbreak as well. The overall number of deaths was lower than in prior outbreaks, perhaps because of improved sanitation and medical support.[16] Most of the deaths attributed to this final pandemic occurred in India.

LINK TO LEARNING

Visit this link (https://openstax.org/l/22blackdeath) to see a video describing how similar the genome of the Black Death bacterium is to today's strains of bubonic plague.

Zoonotic Febrile Diseases

A wide variety of zoonotic febrile diseases (diseases that cause fever) are caused by pathogenic bacteria that require arthropod vectors. These pathogens are either obligate intracellular species of *Anaplasma, Bartonella, Ehrlichia, Orientia,* and *Rickettsia*, or spirochetes in the genus *Borrelia*. Isolation and identification of pathogens in this group are best performed in BSL-3 laboratories because of the low infective dose associated with the diseases.

Anaplasmosis

The zoonotic tickborne disease **human granulocytic anaplasmosis (HGA)** is caused by the obligate intracellular pathogen *Anaplasma phagocytophilum*. HGA is endemic primarily in the central and northeastern US and in countries in Europe and Asia.

HGA is usually a mild febrile disease that causes flu-like symptoms in immunocompetent patients; however, symptoms are severe enough to require hospitalization in at least 50% of infections and, of those patients, less than 1% will die of HGA.[17] Small mammals such as white-footed mice, chipmunks, and voles have been identified as reservoirs of *A. phagocytophilum*, which is transmitted by the bite of an *Ixodes* tick. Five major virulence factors[18] have been reported in *Anaplasma*; three are adherence factors and two are factors that allow the pathogen to avoid the human immune response. Diagnostic approaches include locating intracellular microcolonies of *Anaplasma* through microscopic examination of neutrophils or eosinophils stained with Giemsa or Wright stain, PCR for detection of *A. phagocytophilum*, and serological tests to detect antibody titers against the pathogens. The primary antibiotic used for treatment is doxycycline.

---

14. Rosen, William. Justinian's Flea: Plague, Empire, and the Birth of Europe. Viking Adult; pg 3; ISBN 978-0-670-03855-8.
15. Benedictow, Ole J. 2004. The Black Death 1346-1353: The Complete History. Woodbridge: Boydell Press.
16. Centers for Disease Control and Prevention. "Plague: History." http://www.cdc.gov/plague/history/. Accessed September 15, 2016.

Ehrlichiosis

Human monocytotropic ehrlichiosis (HME) is a zoonotic tickborne disease caused by the BSL-2, obligate intracellular pathogen *Ehrlichia chaffeensis*. Currently, the geographic distribution of HME is primarily the eastern half of the US, with a few cases reported in the West, which corresponds with the known geographic distribution of the primary vector, the lone star tick (*Amblyomma americanum*). Symptoms of HME are similar to the flu-like symptoms observed in anaplasmosis, but a rash is more common, with 60% of children and less than 30% of adults developing petechial, macula, and maculopapular rashes.[19] Virulence factors allow *E. chaffeensis* to adhere to and infect monocytes, forming intracellular microcolonies in monocytes that are diagnostic for the HME. Diagnosis of HME can be confirmed with PCR and serologic tests. The first-line treatment for adults and children of all ages with HME is doxycycline.

Epidemic Typhus

The disease **epidemic typhus** is caused by *Rickettsia prowazekii* and is transmitted by body lice, *Pediculus humanus*. Flying squirrels are animal reservoirs of *R. prowazekii* in North America and can also be sources of lice capable of transmitting the pathogen. Epidemic typhus is characterized by a high fever and body aches that last for about 2 weeks. A rash develops on the abdomen and chest and radiates to the extremities. Severe cases can result in death from shock or damage to heart and brain tissues. Infected humans are an important reservoir for this bacterium because *R. prowazekii* is the only *Rickettsia* that can establish a chronic carrier state in humans.

Epidemic typhus has played an important role in human history, causing large outbreaks with high mortality rates during times of war or adversity. During World War I, epidemic typhus killed more than 3 million people on the Eastern front.[20] With the advent of effective insecticides and improved personal hygiene, epidemic typhus is now quite rare in the US. In the developing world, however, epidemics can lead to mortality rates of up to 40% in the absence of treatment.[21] In recent years, most outbreaks have taken place in Burundi, Ethiopia, and Rwanda. For example, an outbreak in Burundi refugee camps in 1997 resulted in 45,000 illnesses in a population of about 760,000 people.[22]

A rapid diagnosis is difficult because of the similarity of the primary symptoms with those of many other diseases. Molecular and immunohistochemical diagnostic tests are the most useful methods for establishing a diagnosis during the acute stage of illness when therapeutic decisions are critical. PCR to detect distinctive genes from *R. prowazekii* can be used to confirm the diagnosis of epidemic typhus, along with immunofluorescent staining of tissue biopsy specimens. Serology is usually used to identify rickettsial infections. However, adequate antibody titers take up to 10 days to develop. Antibiotic therapy is typically begun before the diagnosis is complete. The most common drugs used to treat patients with epidemic typhus are doxycycline or chloramphenicol.

Murine (Endemic) Typhus

Murine typhus (also known as endemic typhus) is caused by *Rickettsia typhi* and is transmitted by the bite of

---

17   J.S. Bakken et al. "Diagnosis and Management of Tickborne Rickettsial Diseases: Rocky Mountain Spotted Fever, Ehrlichioses, and Anaplasmosis–United States. A Practical Guide for Physicians and Other Health Care and Public Health Professionals." *MMWR Recommendations and Reports* 55 no. RR04 (2006):1–27.

18   MOH Key Laboratory of Systems Biology of Pathogens, "Virulence Factors of Pathogenic Bacteria, Anaplasma" 2016. http://www.mgc.ac.cn/cgi-bin/VFs/jsif/main.cgi. Accessed July, 26, 2016.

19   Centers for Disease Control and Prevention. "Ehrlichiosis, Symptoms, Diagnosis, and Treatment." 2016. https://www.cdc.gov/ehrlichiosis/symptoms/index.html. Accessed July 29, 2016.

20   Drali, R., Brouqui, P. and Raoult, D. "Typhus in World War I." *Microbiology Today* 41 (2014) 2:58–61.

21   Centers for Disease Control and Prevention. *CDC Health Information for International Travel 2014: The Yellow Book*. Oxford University Press, 2013. http://wwwnc.cdc.gov/travel/yellowbook/2016/infectious-diseases-related-to-travel/rickettsial-spotted-typhus-fevers-related-infections-anaplasmosis-ehrlichiosis. Accessed July 26, 2016.

22   World Health Organization. "Typhus." 1997. http://www.who.int/mediacentre/factsheets/fs162/en/. Accessed July 26, 2016.

the rat flea, *Xenopsylla cheopis*, with infected rats as the main reservoir. Clinical signs and symptoms of **murine typhus** include a rash and chills accompanied by headache and fever that last about 12 days. Some patients also exhibit a cough and pneumonia-like symptoms. Severe illness can develop in immunocompromised patients, with seizures, coma, and renal and respiratory failure.

Clinical diagnosis of murine typhus can be confirmed from a biopsy specimen from the rash. Diagnostic tests include indirect immunofluorescent antibody (IFA) staining, PCR for *R. typhi*, and acute and convalescent serologic testing. Primary treatment is doxycycline, with chloramphenicol as the second choice.

Rocky Mountain Spotted Fever

The disease **Rocky Mountain spotted fever** (RMSF) is caused by *Rickettsia rickettsii* and is transmitted by the bite of a hard-bodied tick such as the American dog tick (*Dermacentor variabilis*), Rocky Mountain wood tick (*D. andersoni*), or brown dog tick (*Rhipicephalus sanguineus*).

This disease is endemic in North and South America and its incidence is coincident with the arthropod vector range. Despite its name, most cases in the US do not occur in the Rocky Mountain region but in the Southeast; North Carolina, Oklahoma, Arkansas, Tennessee, and Missouri account for greater than 60% of all cases.[23] The map in Figure 25.12 shows the distribution of prevalence in the US in 2010.

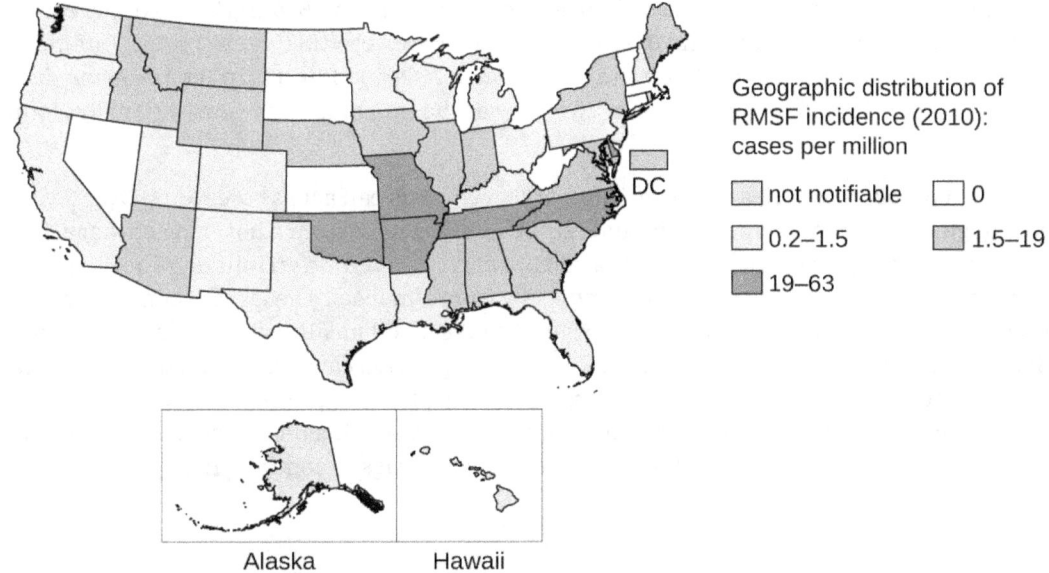

Figure 25.12 In the US, Rocky Mountain spotted fever is most prevalent in the southeastern states. (credit: modification of work by Centers for Disease Control and Prevention)

Signs and symptoms of RMSF include a high fever, headache, body aches, nausea, and vomiting. A petechial rash (similar in appearance to measles) begins on the hands and wrists, and spreads to the trunk, face, and extremities (Figure 25.13). If untreated, RMSF is a serious illness that can be fatal in the first 8 days even in otherwise healthy patients. Ideally, treatment should begin before petechiae develop, because this is a sign of progression to severe disease; however, the rash usually does not appear until day 6 or later after onset of symptoms and only occurs in 35%–60% of patients with the infection. Increased vascular permeability associated with petechiae formation can result in fatality rates of 3% or greater, even in the presence of clinical support. Most deaths are due to hypotension and cardiac arrest or from ischemia following blood coagulation.

Diagnosis can be challenging because the disease mimics several other diseases that are more prevalent. The diagnosis of RMSF is made based on symptoms, fluorescent antibody staining of a biopsy specimen from the rash, PCR for *Rickettsia rickettsii*, and acute and convalescent serologic testing. Primary treatment is doxycycline, with chloramphenicol as the second choice.

---

23 Centers for Disease Control and Prevention. "Rocky Mountain Spotted Fever (RMSF): Statistics and Epidemiology." http://www.cdc.gov/rmsf/stats/index.html. Accessed Sept 16, 2016.

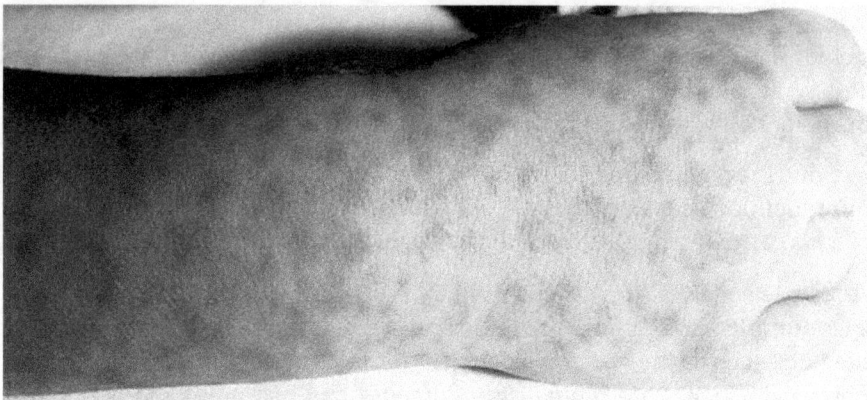

Figure 25.13 Rocky Mountain spotted fever causes a petechial rash. Unlike epidemic or murine typhus, the rash begins at the hands and wrists and then spreads to the trunk. (credit: modification of work by Centers for Disease Control and Prevention)

Lyme Disease

Lyme disease is caused by the spirochete *Borrelia burgdorferi* that is transmitted by the bite of a hard-bodied, black-legged *Ixodes* tick. *I. scapularis* is the biological vector transmitting *B. burgdorferi* in the eastern and north-central US and *I. pacificus* transmits *B. burgdorferi* in the western US (Figure 25.15). Different species of *Ixodes* ticks are responsible for *B. burgdorferi* transmission in Asia and Europe. In the US, Lyme disease is the most commonly reported vectorborne illness. In 2014, it was the fifth most common Nationally Notifiable disease.[24]

Ixodes ticks have complex life cycles and deer, mice, and even birds can act as reservoirs. Over 2 years, the ticks pass through four developmental stages and require a blood meal from a host at each stage. In the spring, tick eggs hatch into six-legged larvae. These larvae do not carry *B. burgdorferi* initially. They may acquire the spirochete when they take their first blood meal (typically from a mouse). The larvae then overwinter and molt into eight-legged nymphs in the following spring. Nymphs take blood meals primarily from small rodents, but may also feed on humans, burrowing into the skin. The feeding period can last several days to a week, and it typically takes 24 hours for an infected nymph to transmit enough *B. burgdorferi* to cause infection in a human host. Nymphs ultimately mature into male and female adult ticks, which tend to feed on larger animals like deer or, occasionally, humans. The adults then mate and produce eggs to continue the cycle (Figure 25.14).

---

[24] Centers for Disease Control and Prevention. "Lyme Disease. Data and Statistics." 2015. http://www.cdc.gov/lyme/stats/index.html. Accessed July 26, 2016.

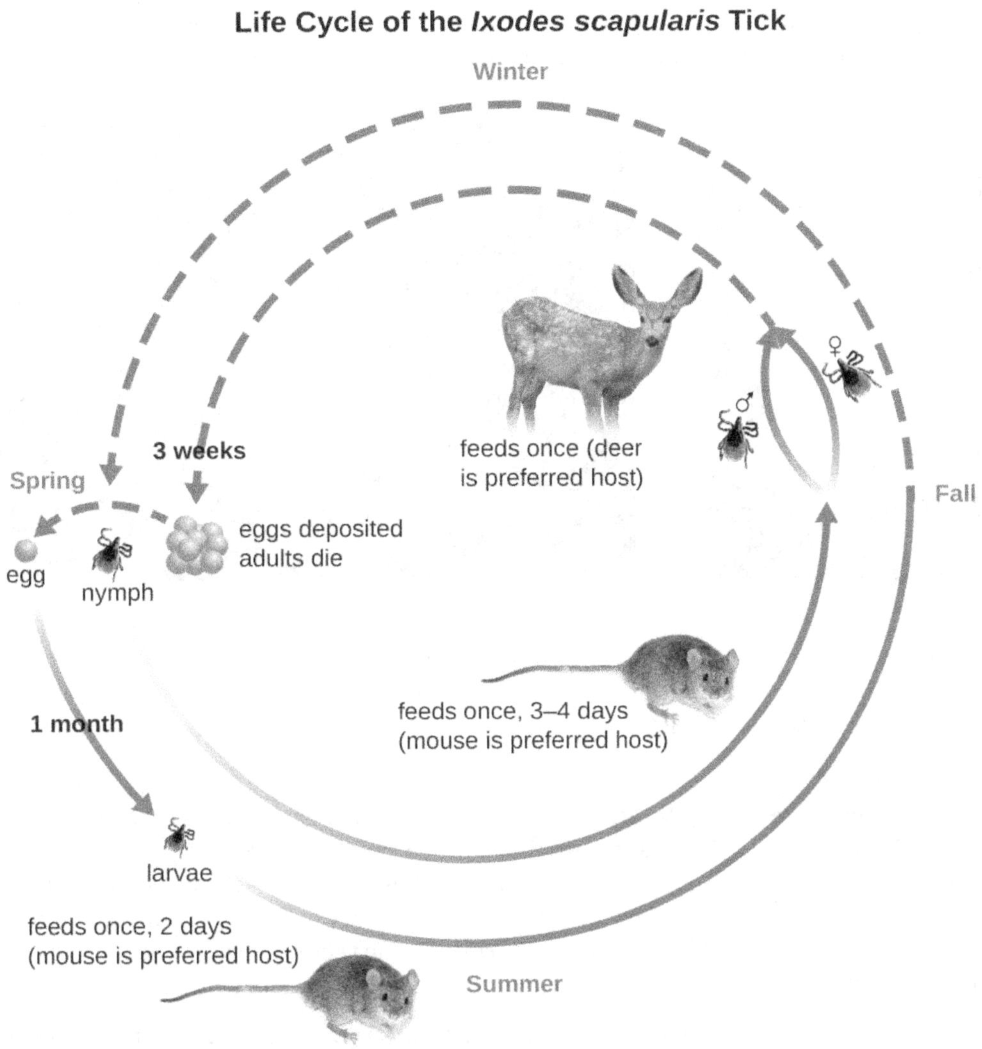

Life Cycle of the *Ixodes scapularis* Tick

Figure 25.14 This image shows the 2-year life cycle of the black-legged tick, the biological vector of Lyme disease. (credit "mouse": modification of work by George Shuklin)

The symptoms of Lyme disease follow three stages: early localized, early disseminated, and late stage. During the early-localized stage, approximately 70%–80%[25] of cases may be characterized by a bull's-eye rash, called erythema migrans, at the site of the initial tick bite. The rash forms 3 to 30 days after the tick bite (7 days is the average) and may also be warm to the touch (Figure 25.15).[26] This diagnostic sign is often overlooked if the tick bite occurs on the scalp or another less visible location. Other early symptoms include flu-like symptoms such as malaise, headache, fever, and muscle stiffness. If the patient goes untreated, the second early-disseminated stage of the disease occurs days to weeks later. The symptoms at this stage may include severe headache, neck stiffness, facial paralysis, arthritis, and carditis. The late-stage manifestations of the disease may occur years after exposure. Chronic inflammation causes damage that can eventually cause severe arthritis, meningitis, encephalitis, and altered mental states. The disease may be fatal if untreated.

A presumptive diagnosis of Lyme disease can be made based solely on the presence of a bull's-eye rash at the site of infection, if it is present, in addition to other associated symptoms (Figure 25.15). In addition, indirect immunofluorescent antibody (IFA) labeling can be used to visualize bacteria from blood or skin biopsy specimens. Serological tests like ELISA can also be used to detect serum antibodies produced in response to

---

25 Centers for Disease Control and Prevention. "Signs and Symptoms of Untreated Lyme Disease." 2015. http://www.cdc.gov/lyme/signs_symptoms/index.html. Accessed July 27, 2016.

26 Centers for Disease Control and Prevention. "Ticks. Symptoms of Tickborne Illness." 2015. http://www.cdc.gov/ticks/symptoms.html. Accessed July 27, 2016.

infection. During the early stage of infection (about 30 days), antibacterial drugs such as amoxicillin and doxycycline are effective. In the later stages, penicillin G, chloramphenicol, or ceftriaxone can be given intravenously.

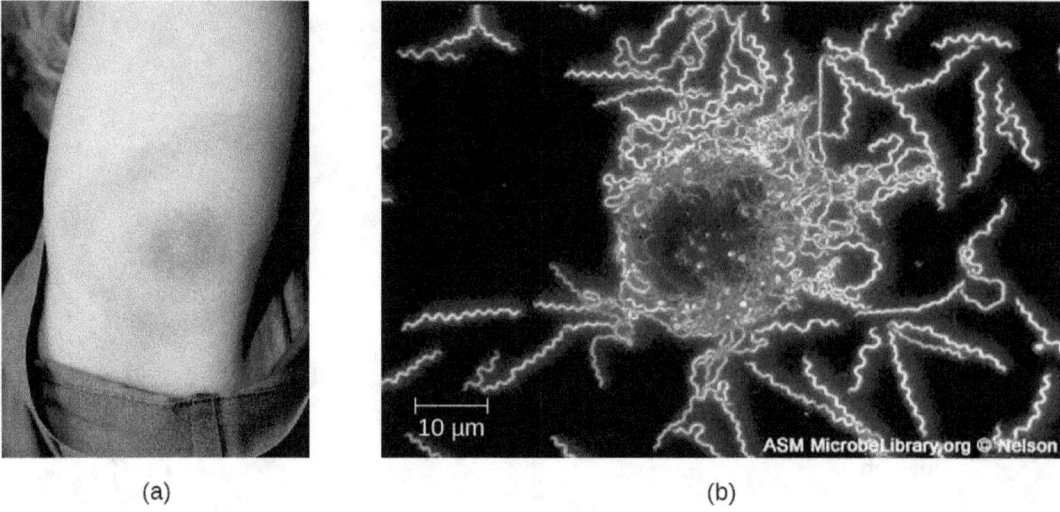

Figure 25.15 (a) A characteristic bull's eye rash of Lyme disease forms at the site of a tick bite. (b) A darkfield micrograph shows *Borrelia burgdorferi*, the causative agent of Lyme disease. (credit a: modification of work by Centers for Disease Control and Prevention; credit b: modification of work by American Society for Microbiology)

Relapsing Fever

Borrelia spp. also can cause **relapsing fever**. Two of the most common species are *B. recurrentis*, which causes epidemics of louseborne relapsing fever, and *B. hermsii*, which causes tickborne relapsing fevers. These *Borrelia* species are transmitted by the body louse *Pediculus humanus* and the soft-bodied tick *Ornithodoros hermsi*, respectively. Lice acquire the spirochetes from human reservoirs, whereas ticks acquire them from rodent reservoirs. Spirochetes infect humans when *Borrelia* in the vector's saliva or excreta enter the skin rapidly as the vector bites.

In both louse- and tickborne relapsing fevers, bacteremia usually occurs after the initial exposure, leading to a sudden high fever (39–43 °C [102.2–109.4 °F]) typically accompanied by headache and muscle aches. After about 3 days, these symptoms typically subside, only to return again after about a week. After another 3 days, the symptoms subside again but return a week later, and this cycle may repeat several times unless it is disrupted by antibiotic treatment. Immune evasion through bacterial antigenic variation is responsible for the cyclical nature of the symptoms in these diseases.

The diagnosis of relapsing fever can be made by observation of spirochetes in blood, using darkfield microscopy (Figure 25.16). For louseborne relapsing fever, doxycycline or erythromycin are the first-line antibiotics. For tickborne relapsing fever, tetracycline or erythromycin are the first-line antibiotics.

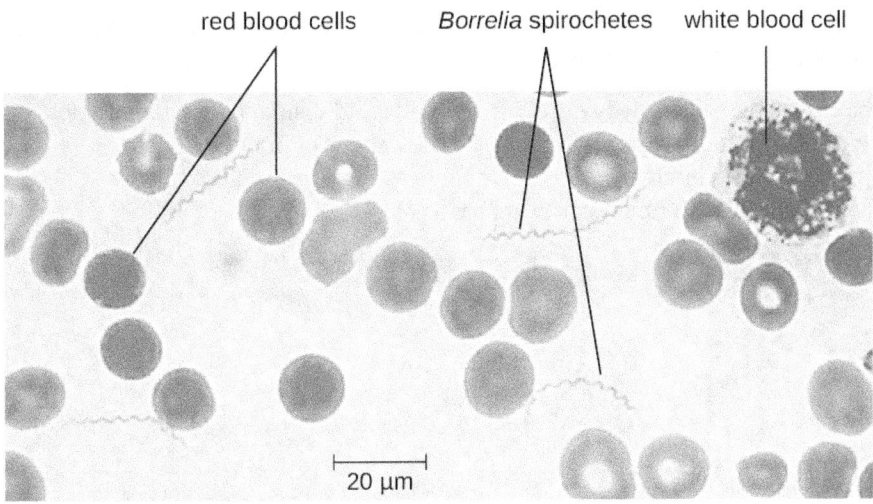

Figure 25.16 A peripheral blood smear from a patient with tickborne relapsing fever. *Borrelia* appears as thin spirochetes among the larger red blood cells. (credit: modification of work by Centers for Disease Control and Prevention)

Trench Fever

The louseborne disease **trench fever** was first characterized as a specific disease during World War I, when approximately 1 million soldiers were infected. Today, it is primarily limited to areas of the developing world where poor sanitation and hygiene lead to infestations of lice (e.g., overpopulated urban areas and refugee camps). Trench fever is caused by the gram-negative bacterium *Bartonella quintana*, which is transmitted when feces from infected body lice, *Pediculus humanus* var *corporis*, are rubbed into the louse bite, abraded skin, or the conjunctiva. The symptoms typically follow a 5-day course marked by a high fever, body aches, conjunctivitis, ocular pain, severe headaches, and severe bone pain in the shins, neck, and back. Diagnosis can be made using blood cultures; serological tests like ELISA can be used to detect antibody titers to the pathogen and PCR can also be used. The first-line antibiotics are doxycycline, macrolide antibiotics, and ceftriaxone.

✓ CHECK YOUR UNDERSTANDING

- What is the vector associated with epidemic typhus?
- Describe the life cycle of the deer tick and how it spreads Lyme disease.

 MICRO CONNECTIONS

Tick Tips

Many of the diseases covered in this chapter involve arthropod vectors. Of these, ticks are probably the most commonly encountered in the US. Adult ticks have eight legs and two body segments, the cephalothorax and the head (Figure 25.17). They typically range from 2 mm to 4 mm in length, and feed on the blood of the host by attaching themselves to the skin.

Unattached ticks should be removed and eliminated as soon as they are discovered. When removing a tick that has already attached itself, keep the following guidelines in mind to reduce the chances of exposure to pathogens:

- Use blunt tweezers to gently pull near the site of attachment until the tick releases its hold on the skin.
- Avoid crushing the tick's body and do not handle the tick with bare fingers. This could release bacterial pathogens and actually increase your exposure. The tick can be killed by drowning in water or alcohol, or frozen if it may be needed later for identification and analysis.
- Disinfect the area thoroughly by swabbing with an antiseptic such as isopropanol.
- Monitor the site of the bite for rashes or other signs of infection.

Many ill-advised home remedies for tick removal have become popular in recent years, propagated by social media and pseudojournalism. Health professionals should discourage patients from resorting to any of the following methods, which are NOT recommended:

- using chemicals (e.g., petroleum jelly or fingernail polish) to dislodge an attached tick, because it can cause the tick to release fluid, which can increase the chance of infection
- using hot objects (matches or cigarette butts) to dislodge an attached tick
- squeezing the tick's body with fingers or tweezers

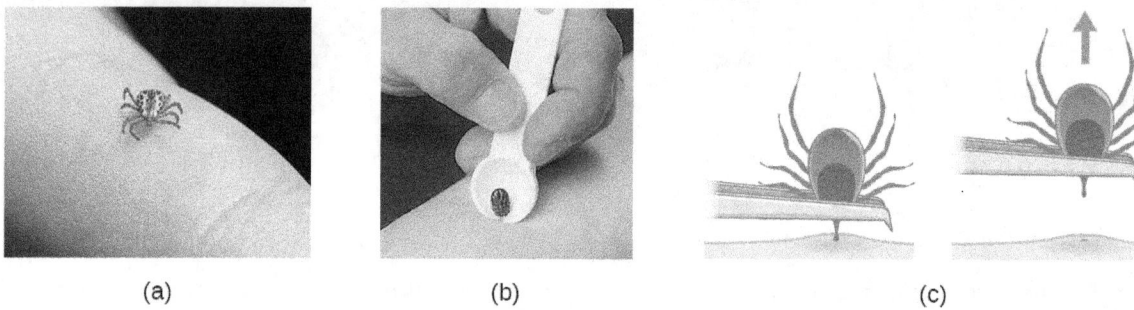

Figure 25.17 (a) This black-legged tick, also known as the deer tick, has not yet attached to the skin. (b) A notched tick extractor can be used for removal. (c) To remove an attached tick with fine-tipped tweezers, pull gently on the mouth parts until the tick releases its hold on the skin. Avoid squeezing the tick's body, because this could release pathogens and thus increase the risk of contracting Lyme disease. (credit a: modification of work by Jerry Kirkhart; credit c: modification of work by Centers for Disease Control and Prevention)

Disease Profile

Bacterial Infections of the Circulatory and Lymphatic Systems

Although the circulatory system is a closed system, bacteria can enter the bloodstream through several routes. Wounds, animal bites, or other breaks in the skin and mucous membranes can result in the rapid dissemination of bacterial pathogens throughout the body. Localized infections may also spread to the bloodstream, causing serious and often fatal systemic infections. Figure 25.18 and Figure 25.19 summarize the major characteristics of bacterial infections of the circulatory and lymphatic systems.

Bacterial Infections of the Circulatory and Lymphatic Systems

Disease	Pathogen	Signs and Symptoms	Transmission	Diagnostic Tests	Antimicrobial Drugs
Anaplasmosis (HGA)	*Anaplasma phagocytophilum*	Fever, flu-like symptoms	From small-mammal reservoirs via tick vector	Blood smear, PCR	Doxycycline
Brucellosis	*Brucella melitensis, B. abortus, B. canis, B. suis*	Granuloma, undulating fever, chronic flu-like symptoms	Direct contact with infected livestock or animals	Agglutination tests, ELISA	Doxycycline, rifampin
Cat-scratch disease	*Bartonella henselae*	Lymph-node swelling and pain, fever, chills, fatigue	Bite or scratch from domestic cats	Immunofluorescence, serological tests, PCR	None for immunocompetent patients
Ehrlichiosis (HME)	*Ehrlichia chaffeensis*	Flu-like symptoms, rash	Lone star tick vector	Serologic tests, PCR	Doxycycline
Endocarditis/pericarditis	S*Staphylococcus* spp., *Streptococcus* spp., *Enterococcus* spp., HACEK bacilli	Chest pain, difficulty breathing, dry cough, fever; potentially fatal damage to heart valves	Pathogens introduced to bloodstream via contaminated catheters, dental procedures, piercings, or wounds	Echocardiogram, blood culture	Ampicillin, nafcillin, gentamicin, others; based on susceptibility testing
Epidemic typhus	*Rickettsia prowazekii*	High fever, body aches, rash; potentially fatal damage to heart and brain	From rodent reservoir via body louse vector	PCR, immunofluorescence	Doxycycline, chloramphenicol
Gas gangrene	*Clostridium perfringens*, other *Clostridium* spp.	Rapidly spreading myonecrosis, edema, yellowish and then purple discharge from wound, pockets of gas in tissues, septic shock and death	Germination of endospores in ischemic tissues, typically due to injury or chronic disease (e.g., diabetes)	Wound culture	Penicillin G, clindamycin, metronidazole
Infectious arthritis (septic arthritis)	*Staphylococcus aureus, Neisseria gonorrhoeae*	Joint pain and swelling, limited range of motion	Infection spreads to joint via circulatory system from wound or surgical site	Synovial fluid culture	Oxacillin, cefazolin, cephtriaxone
Lyme disease	*Borrelia burgdorferi*	Early localized: bull's eye rash, malaise, headache, fever, muscle stiffness; early disseminated: stiff neck, facial paralysis, arthritis, carditis; late-stage: arthritis, meningitis, possibly fatal	From deer, rodent, bird reservoirs via tick vector	IFA, serology, and ELISA	Amoxicillin, doxycycline, penicillin G, chloramphenicol, ceftriaxone
Murine (endemic) typhus	*Rickettsia typhi*	Low-grade fever, rash, headache, cough	From rodents or between humans via rat flea vector	Biopsy, IFA, PCR	Doxycycline, chloramphenicol
Osteomyelitis	*Staphylococcus aureus, Streptococcus pyogenes*, others	Inflammation of bone tissue, leading to fever, localized pain, edema, ulcers, bone loss	Pathogens introduced through trauma, prosthetic joint replacement, or from other infected body site via bloodstream	Radiograph of affected bone, culture of bone biopsy specimen	Cephalosporin, penicillins, others

Figure 25.18

Bacterial Infections of the Circulatory and Lymphatic Systems (continued)

Disease	Pathogen	Signs and Symptoms	Transmission	Diagnostic Tests	Antimicrobial Drugs
Plague	*Yersinia pestis*	Bubonic: buboes, fever, internal hemorrhaging; septicemic: fever, abdominal pain, shock, DIC, necrosis in extremities; pneumonic: acute pneumonia, respiratory failure, shock. All forms have high mortality rates.	Transmitted from mammal reservoirs via flea vectors or consumption of infected animal; transmission of pneumonic plague between humans via respiratory aerosols	Culture of bacteria from lymph, blood, or sputum samples; DFA, ELISA	Gentamycin, fluoroquinolones, others
Puerperal sepsis	*Streptococcus pyogenes*, many others	Rapid-onset fever, shock, and death	Pathogens introduced during or immediately following childbirth	Wound, urine, or blood culture	As determined by susceptibility testing
Rat-bite fever	*Streptobacillus moniliformis, Spirillum minor*	Fever, muscle and joint pain, rash, ulcer	Bite from infected rat or exposure to rat feces or body fluids in contaminated food or water	Observation of the organism from samples and antibody tests	Penicillin
Relapsing fever	*Borrelia recurrentis, B. hermsii*, other *Borrelia* spp.	Recurring fever, headache, muscle aches	From rodent or human reservoir via body louse or tick vector	Darkfield microscopy	Doxycycline, tetracycline, erythromycin
Rheumatic fever	*Streptococcus pyogenes*	Joint pain and swelling, inflammation and scarring of heart valves, heart murmur	Sequela of streptococcal pharyngitis	Serology, electrocardiogram, echocardiogram	Benzathine benzylpenicillin
Rocky Mountain spotted fever	*Rickettsia rickettsii*	High fever, headache, body aches, nausea and vomiting, petechial rash; potentially fatal hypotension and ischemia due to blood coagulation	From rodent reservoir via tick vectors	Biopsy, serology, PCR	Doxycycline, chloramphenicol
Toxic shock syndrome (TSS)	*Staphylococcus aureus*	Sudden high fever, vomiting, diarrhea, hypotension, death	Pathogens from localized infection spread to bloodstream; pathogens introduced on tampons or other intravaginal products	Serology, toxin identification from isolates	Clindamycin, vancomycin
Toxic shock-like syndrome (STSS)	*Streptococcus pyogenes*	Sudden high fever, vomiting, diarrhea, acute respiratory distress syndrome (ARDS), hypoxemia, necrotizing fasciitis, death	Sequela of streptococcal skin or soft-tissue infection	Serology, blood culture, urinalysis	Penicillin, cephalosporin
Trench fever	*Bartonella quintana*	High fever, conjunctivitis, ocular pain, headaches, severe pain in bones of shins, neck, and back	Between humans via body louse vector	Blood culture, ELISA, PCR	Doxycycline, macrolide antibiotics, ceftriaxone
Tularemia (rabbit fever)	*Francisella tularensis*	Skin lesions, fever, chills, headache, buboes	Eating or handling infected rabbit; transmission from infected animal via tick or fly vector; aerosol transmission (in laboratory or as bioweapon)	DFA	Streptomycin, gentamycin, others

Figure 25.19

25.3 Viral Infections of the Circulatory and Lymphatic Systems

Learning Objectives

By the end of this section, you will be able to:
- Identify common viral pathogens that cause infections of the circulatory and lymphatic systems
- Compare the major characteristics of specific viral diseases affecting the circulatory and lymphatic systems

Viral pathogens of the circulatory system vary tremendously both in their virulence and distribution worldwide. Some of these pathogens are practically global in their distribution. Fortunately, the most ubiquitous viruses tend to produce the mildest forms of disease. In the majority of cases, those infected remain asymptomatic. On the other hand, other viruses are associated with life-threatening diseases that have impacted human history.

Infectious Mononucleosis and Burkitt Lymphoma

Human herpesvirus 4, also known as Epstein-Barr virus (EBV), has been associated with a variety of human diseases, such as mononucleosis and Burkitt lymphoma. Exposure to the human herpesvirus 4 (HHV-4) is widespread and nearly all people have been exposed at some time in their childhood, as evidenced by serological tests on populations. The virus primarily resides within B lymphocytes and, like all herpes viruses, can remain dormant in a latent state for a long time.

When uninfected young adults are exposed to EBV, they may experience **infectious mononucleosis**. The virus is mainly spread through contact with body fluids (e.g., saliva, blood, and semen). The main symptoms include pharyngitis, fever, fatigue, and lymph node swelling. Abdominal pain may also occur as a result of spleen and liver enlargement in the second or third week of infection. The disease typically is self-limiting after about a month. The main symptom, extreme fatigue, can continue for several months, however. Complications in immunocompetent patients are rare but can include jaundice, anemia, and possible rupture of the spleen caused by enlargement.

In patients with malaria or HIV, Epstein-Barr virus can lead to a fast-growing malignant cancer known as **Burkitt lymphoma** (Figure 25.20). This condition is a form of non-Hodgkin lymphoma that produces solid tumors chiefly consisting of aberrant B cells. Burkitt lymphoma is more common in Africa, where prevalence of HIV and malaria is high, and it more frequently afflicts children. Repeated episodes of viremia caused by reactivation of the virus are common in immunocompromised individuals. In some patients with AIDS, EBV may induce the formation of malignant B-cell lymphomas or oral hairy leukoplakia. Immunodeficiency-associated Burkitt lymphoma primarily occurs in patients with HIV. HIV infection, similar to malaria, leads to polyclonal B-cell activation and permits poorly controlled proliferation of EBV$^+$ B cells, leading to the formation of lymphomas.

Infectious mononucleosis is typically diagnosed based on the initial clinical symptoms and a test for antibodies to EBV-associated antigens. Because the disease is self-limiting, antiviral treatments are rare for mononucleosis. Cases of Burkitt lymphoma are diagnosed from a biopsy specimen from a lymph node or tissue from a suspected tumor. Staging of the cancer includes computed tomography (CT) scans of the chest, abdomen, pelvis, and cytologic and histologic evaluation of biopsy specimens. Because the tumors grow so rapidly, staging studies must be expedited and treatment must be initiated promptly. An intensive alternating regimen of cyclophosphamide, vincristine, doxorubicin, methotrexate, ifosfamide, etoposide, and cytarabine (CODOX-M/IVAC) plus rituximab results in a cure rate greater than 90% for children and adults.

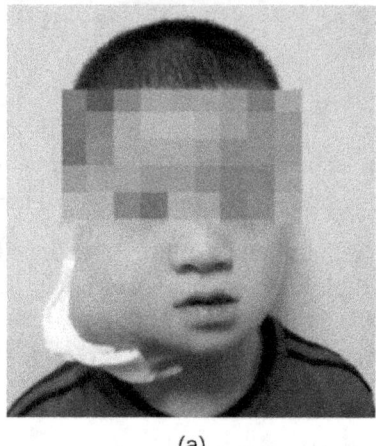
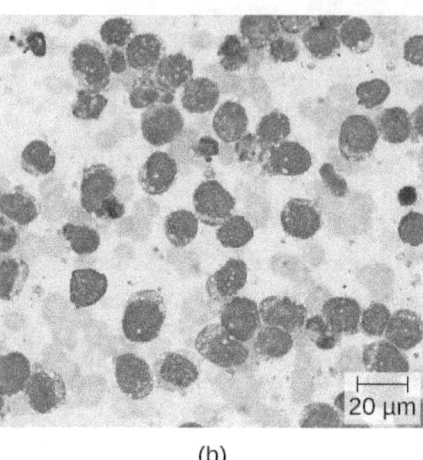

Figure 25.20 (a) Burkitt lymphoma can cause large tumors. (b) Characteristic irregularly shaped abnormal lymphocytes (large purple cells) with vacuoles (white spots) from a fine-needle aspirate of a tumor from a patient with Burkitt lymphoma. (credit a: modification of work by Bi CF, Tang Y, Zhang WY, Zhao S, Wang XQ, Yang QP, Li GD, and Liu WP; credit b: modification of work by Ed Uthman)

Cytomegalovirus Infections

Also known as cytomegalovirus (CMV), human herpesvirus 5 (HHV-5) is a virus with high infection rates in the human population. It is currently estimated that 50% of people in the US have been infected by the time they reach adulthood.[27] CMV is the major cause of non-Epstein-Barr infectious mononucleosis in the general human population. It is also an important pathogen in immunocompromised hosts, including patients with AIDS, neonates, and transplant recipients. However, the vast majority of CMV infections are asymptomatic. In adults, if symptoms do occur, they typically include fever, fatigue, swollen glands, and pharyngitis.

CMV can be transmitted between individuals through contact with body fluids such as saliva or urine. Common modes of transmission include sexual contact, nursing, blood transfusions, and organ transplants. In addition, pregnant women with active infections frequently pass this virus to their fetus, resulting in congenital CMV infections, which occur in approximately one in every 150 infants in US.[28] Infants can also be infected during passage through the birth canal or through breast milk and saliva from the mother.

Perinatal infections tend to be milder but can occasionally cause lung, spleen, or liver damage. Serious symptoms in newborns include growth retardation, jaundice, deafness, blindness, and mental retardation if the virus crosses the placenta during the embryonic state when the body systems are developing in utero. However, a majority (approximately 80%) of infected infants will never have symptoms or experience long-term problems.[29] Diagnosis of CMV infection during pregnancy is usually achieved by serology; CMV is the "C" in prenatal TORCH screening.

Many patients receiving blood transfusions and nearly all those receiving kidney transplants ultimately become infected with CMV. Approximately 60% of transplant recipients will have CMV infection and more than 20% will develop symptomatic disease.[30] These infections may result from CMV-contaminated tissues but also may be a consequence of immunosuppression required for transplantation causing reactivation of prior CMV infections. The resulting viremia can lead to fever and leukopenia, a decrease in the number of white blood cells in the bloodstream. Serious consequences may include liver damage, transplant rejection, and death. For similar reasons, many patients with AIDS develop active CMV infections that can manifest as encephalitis or progressive retinitis leading to blindness.[31]

---

27 Centers for Disease Control and Prevention. "Cytomegalovirus (CMV) and Congenital CMV Infection: About CMV." 2016. http://www.cdc.gov/cmv/transmission.html. Accessed July 28, 2016.

28 Centers for Disease Control and Prevention. "Cytomegalovirus (CMV) and Congenital CMV Infection: Babies Born with CMV (Congenital CMV Infection)." 2016. http://www.cdc.gov/cmv/congenital-infection.html. Accessed July 28, 2016.

29 ibid.

30 E. Cordero et al. "Cytomegalovirus Disease in Kidney Transplant Recipients: Incidence, Clinical Profile, and Risk Factors."

Diagnosis of a localized CMV infection can be achieved through direct microscopic evaluation of tissue specimens stained with routine stains (e.g., Wright-Giemsa, hematoxylin and eosin, Papanicolaou) and immunohistochemical stains. Cells infected by CMV produce characteristic inclusions with an "owl's eye" appearance; this sign is less sensitive than molecular methods like PCR but more predictive of localized disease (Figure 25.21). For more severe CMV infection, tests such as enzyme immunoassay (EIA), indirect immunofluorescence antibody (IFA) tests, and PCR, which are based on detection of CMV antigen or DNA, have a higher sensitivity and can determine viral load. Cultivation of the virus from saliva or urine is still the method for detecting CMV in newborn babies up to 3 weeks old. Ganciclovir, valganciclovir, foscarnet, and cidofovir are the first-line antiviral drugs for serious CMV infections.

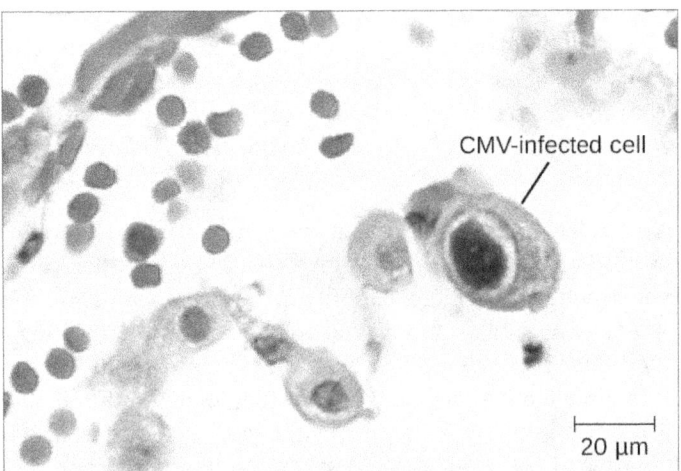

Figure 25.21 Cells infected with CMV become enlarged and have a characteristic "owl's eye" nucleus. This micrograph shows kidney cells from a patient with CMV. (credit: modification of work by Centers for Disease Control and Prevention)

✓ CHECK YOUR UNDERSTANDING

- Compare the diseases caused by HHV-4 and HHV-5.

Arthropod-Borne Viral Diseases

There are a number of arthropod-borne viruses, or **arboviruses**, that can cause human disease. Among these are several important hemorrhagic fevers transmitted by mosquitoes. We will discuss three that pose serious threats: yellow fever, chikungunya fever, and dengue fever.

Yellow Fever

Yellow fever was once common in the US and caused several serious outbreaks between 1700 and 1900.[32] Through vector control efforts, however, this disease has been eliminated in the US. Currently, **yellow fever** occurs primarily in tropical and subtropical areas in South America and Africa. It is caused by the yellow fever virus of the genus *Flavivirus* (named for the Latin word *flavus* meaning *yellow*), which is transmitted to humans by mosquito vectors. Sylvatic yellow fever occurs in tropical jungle regions of Africa and Central and South America, where the virus can be transmitted from infected monkeys to humans by the mosquitoes *Aedes africanus* or *Haemagogus* spp. In urban areas, the *Aedes aegypti* mosquito is mostly responsible for transmitting the virus between humans.

Transplantation Proceedings 44 no. 3 (2012):694–700.

31 L.M. Mofenson et al. "Guidelines for the Prevention and Treatment of Opportunistic Infections Among HIV-Exposed and HIV-Infected Children: Recommendations From CDC, the National Institutes of Health, the HIV Medicine Association of the Infectious Diseases Society of America, the Pediatric Infectious Diseases Society, and the American Academy of Pediatrics." *MMWR Recommendations and Reports* 58 no. RR-11 (2009):1–166.

32 Centers for Disease Control and Prevention. "History Timeline Transcript." http://www.cdc.gov/travel-training/local/HistoryEpidemiologyandVaccination/HistoryTimelineTranscript.pdf. Accessed July 28, 2016.

Most individuals infected with yellow fever virus have no illness or only mild disease. Onset of milder symptoms is sudden, with dizziness, fever of 39–40 °C (102–104 °F), chills, headache, and myalgias. As symptoms worsen, the face becomes flushed, and nausea, vomiting, constipation, severe fatigue, restlessness, and irritability are common. Mild disease may resolve after 1 to 3 days. However, approximately 15% of cases progress to develop moderate to severe yellow fever disease.[33]

In moderate or severe disease, the fever falls suddenly 2 to 5 days after onset, but recurs several hours or days later. Symptoms of jaundice, petechial rash, mucosal hemorrhages, oliguria (scant urine), epigastric tenderness with bloody vomit, confusion, and apathy also often occur for approximately 7 days of moderate to severe disease. After more than a week, patients may have a rapid recovery and no sequelae.

In its most severe form, called malignant yellow fever, symptoms include delirium, bleeding, seizures, shock, coma, and multiple organ failure; in some cases, death occurs. Patients with malignant yellow fever also become severely immunocompromised, and even those in recovery may become susceptible to bacterial superinfections and pneumonia. Of the 15% of patients who develop moderate or severe disease, up to half may die.

Diagnosis of yellow fever is often based on clinical signs and symptoms and, if applicable, the patient's travel history, but infection can be confirmed by culture, serologic tests, and PCR. There are no effective treatments for patients with yellow fever. Whenever possible, patients with yellow fever should be hospitalized for close observation and given supportive care. Prevention is the best method of controlling yellow fever. Use of mosquito netting, window screens, insect repellents, and insecticides are all effective methods of reducing exposure to mosquito vectors. An effective vaccine is also available, but in the US, it is only administered to those traveling to areas with endemic yellow fever. In West Africa, the World Health Organization (WHO) launched a Yellow Fever Initiative in 2006 and, since that time, significant progress has been made in combating yellow fever. More than 105 million people have been vaccinated, and no outbreaks of yellow fever were reported in West Africa in 2015.

 MICRO CONNECTIONS

Yellow Fever: Altering the Course of History

Yellow fever originated in Africa and is still most prevalent there today. This disease is thought to have been translocated to the Americas by the slave trade in the 16th century.[34] Since that time, yellow fever has been associated with many severe outbreaks, some of which had important impacts upon historic events.

Yellow fever virus was once an important cause of disease in the US. In the summer of 1793, there was a serious outbreak in Philadelphia (then the US capitol). It is estimated that 5,000 people (10% of the city's population) died. All of the government officials, including George Washington, fled the city in the face of this epidemic. The disease only abated when autumn frosts killed the mosquito vector population.

In 1802, Napoleon Bonaparte sent an army of 40,000 to Hispaniola to suppress a slave revolution. This was seen by many as a part of a plan to use the Louisiana Territory as a granary as he reestablished France as a global power. Yellow fever, however, decimated his army and they were forced to withdraw. Abandoning his aspirations in the New World, Napoleon sold the Louisiana Territory to the US for $15 million in 1803.

The most famous historic event associated with yellow fever is probably the construction of the Panama Canal. The French began work on the canal in the early 1880s. However, engineering problems, malaria, and yellow fever forced them to abandon the project. The US took over the task in 1904 and opened the canal a decade later. During those 10 years, yellow fever was a constant adversary. In the first few years of work, greater than 80% of the American workers in Panama were hospitalized with yellow fever. It was the work of Carlos Finlay

---

33 Centers for Disease Control and Prevention. "Yellow Fever, Symptoms and Treatment." 2015 http://www.cdc.gov/yellowfever/symptoms/index.html. Accessed July 28, 2016.

34 J.T. Cathey, J.S. Marr. "Yellow fever, Asia and the East African Slave Trade." *Transactions of the Royal Society of Tropical Medicine and Hygiene* 108, no. 5 (2014):252–257.

and Walter Reed that turned the tide. Taken together, their work demonstrated that the disease was transmitted by mosquitoes. Vector control measures succeeded in reducing both yellow fever and malaria rates and contributed to the ultimate success of the project.

Dengue Fever

The disease **dengue fever**, also known as breakbone fever, is caused by four serotypes of dengue virus called dengue 1–4. These are *Flavivirus* species that are transmitted to humans by *A. aegypti* or *A. albopictus* mosquitoes. The disease is distributed worldwide but is predominantly located in tropical regions. The WHO estimates that 50 million to 100 million infections occur yearly, including 500,000 dengue hemorrhagic fever (DHF) cases and 22,000 deaths, most among children.[35] Dengue fever is primarily a self-limiting disease characterized by abrupt onset of high fever up to 40 °C (104 °F), intense headaches, rash, slight nose or gum bleeding, and extreme muscle, joint, and bone pain, causing patients to feel as if their bones are breaking, which is the reason this disease is also referred to as breakbone fever. As the body temperature returns to normal, in some patients, signs of dengue hemorrhagic fever may develop that include drowsiness, irritability, severe abdominal pain, severe nose or gum bleeding, persistent vomiting, vomiting blood, and black tarry stools, as the disease progresses to DHF or dengue shock syndrome (DSS). Patients who develop DHF experience circulatory system failure caused by increased blood vessel permeability. Patients with dengue fever can also develop DSS from vascular collapse because of the severe drop in blood pressure. Patients who develop DHF or DSS are at greater risk for death without prompt appropriate supportive treatment. About 30% of patients with severe hemorrhagic disease with poor supportive treatment die, but mortality can be less than 1% with experienced support.[36]

Diagnostic tests for dengue fever include serologic testing, ELISA, and reverse transcriptase-polymerase chain reaction (RT-PCR) of blood. There are no specific treatments for dengue fever, nor is there a vaccine. Instead, supportive clinical care is provided to treat the symptoms of the disease. The best way to limit the impact of this viral pathogen is vector control.

Chikungunya Fever

The arboviral disease **chikungunya fever** is caused by chikungunya virus (CHIKV), which is transmitted to humans by *A. aegypti* and *A. albopictus* mosquitoes. Until 2013, the disease had not been reported outside of Africa, Asia, and a few European countries; however, CHIKV has now spread to mosquito populations in North and South America. Chikungunya fever is characterized by high fever, joint pain, rash, and blisters, with joint pain persisting for several months. These infections are typically self-limiting and rarely fatal.

The diagnostic approach for chikungunya fever is similar to that for dengue fever. Viruses can be cultured directly from patient serum during early infections. IFA, EIA, ELISA, PCR, and RT-PCR are available to detect CHIKV antigens and patient antibody response to the infection. There are no specific treatments for this disease except to manage symptoms with fluids, analgesics, and bed rest. As with most arboviruses, the best strategy for combating the disease is vector control.

LINK TO LEARNING

Use this interactive map (https://openstax.org/l/22denguemap) to explore the global distribution of dengue.

CHECK YOUR UNDERSTANDING

- Name three arboviral diseases and explain why they are so named.
- What is the best method for controlling outbreaks of arboviral diseases?

---

35 Centers for Disease Control and Prevention. "Dengue, Epidemiology." 2014. http://www.cdc.gov/dengue/epidemiology/index.html. Accessed July 28, 2016.

36 C.R. Pringle "Dengue." MSD Manual: Consumer Version. https://www.msdmanuals.com/home/infections/viral-infections/dengue. 2016. Accessed Sept 15, 2016.

Ebola Virus Disease

The Ebola virus disease (EVD) is a highly contagious disease caused by species of *Ebolavirus*, a BSL-4 filovirus (Figure 25.22). Transmission to humans occurs through direct contact with body fluids (e.g., blood, saliva, sweat, urine, feces, or vomit), and indirect contact by contaminated fomites. Infected patients can easily transmit Ebola virus to others if appropriate containment and use of personal protective equipment is not available or used. Handling and working with patients with EVD is extremely hazardous to the general population and health-care workers. In almost every EVD outbreak there have been Ebola infections among health-care workers. This ease of Ebola virus transmission was recently demonstrated in the Ebola epidemic in Guinea, Liberia, and Sierra Leone in 2014, in which more than 28,000 people in 10 countries were infected and more than 11,000 died.[37]

After infection, the initial symptoms of Ebola are unremarkable: fever, severe headache, myalgia, cough, chest pain, and pharyngitis. As the disease progresses, patients experience abdominal pain, diarrhea, and vomiting. Hemorrhaging begins after about 3 days, with bleeding occurring in the gastrointestinal tract, skin, and many other sites. This often leads to delirium, stupor, and coma, accompanied by shock, multiple organ failure, and death. The mortality rates of EVD often range from 50% to 90%.

The initial diagnosis of Ebola is difficult because the early symptoms are so similar to those of many other illnesses. It is possible to directly detect the virus from patient samples within a few days after symptoms begin, using antigen-capture ELISA, immunoglobulin M (IgM) ELISA, PCR, and virus isolation. There are currently no effective, approved treatments for Ebola other than supportive care and proper isolation techniques to contain its spread.

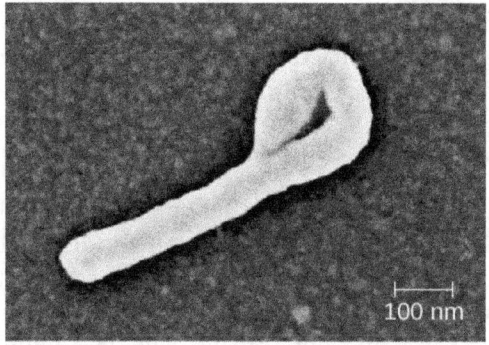

Figure 25.22 An Ebola virus particle viewed with electron microscopy. These filamentous viruses often exhibit looped or hooked ends. (credit: modification of work by Centers for Disease Control and Prevention)

CHECK YOUR UNDERSTANDING

- How is Ebola transmitted?

Hantavirus

The genus *Hantavirus* consists of at least four serogroups with nine viruses causing two major clinical (sometimes overlapping) syndromes: **hantavirus pulmonary syndrome** (HPS) in North America and **hemorrhagic fever with renal syndrome** (HFRS) in other continents. Hantaviruses are found throughout the world in wild rodents that shed the virus in their urine and feces. Transmission occurs between rodents and to humans through inhalation of aerosols of the rodent urine and feces. Hantaviruses associated with outbreaks in the US and Canada are transmitted by the deer mouse, white-footed mouse, or cotton rat.

---

37 HealthMap. "2014 Ebola Outbreaks." http://www.healthmap.org/ebola/#timeline. Accessed July 28, 2016.

HPS begins as a nonspecific flu-like illness with headache, fever, myalgia, nausea, vomiting, diarrhea, and abdominal pain. Patients rapidly develop pulmonary edema and hypotension resulting in pneumonia, shock, and death, with a mortality rate of up to 50%.[38] This virus can also cause HFRS, which has not been reported in the US. The initial symptoms of this condition include high fever, headache, chills, nausea, inflammation or redness of the eyes, or a rash. Later symptoms are hemorrhaging, hypotension, kidney failure, shock, and death. The mortality rate of HFRS can be as high as 15%.[39]

ELISA, Western blot, rapid immunoblot strip assay (RIBA), and RT-PCR detect host antibodies or viral proteins produced during infection. Immunohistological staining may also be used to detect the presence of viral antigens. There are no clinical treatments other than general supportive care available for HPS infections. Patients with HFRS can be treated with ribavirin.[40]

CHECK YOUR UNDERSTANDING

- Compare the two Hantavirus diseases discussed in this section.

Human Immunodeficiency Virus

Human T-lymphotropic viruses (HTLV), also called human immunodeficiency viruses (HIV) are retroviruses that are the causative agent of acquired immune deficiency syndrome (AIDS). There are two main variants of **human immunodeficiency virus (HIV)**. HIV-1 (Figure 25.23) occurs in human populations worldwide, whereas HIV-2 is concentrated in West Africa. Currently, the most affected region in the world is sub-Saharan Africa, with an estimated 25.6 million people living with HIV in 2015.[41] Sub-Saharan Africa also accounts for two-thirds of the global total of new HIV infections (Figure 25.24).[42]

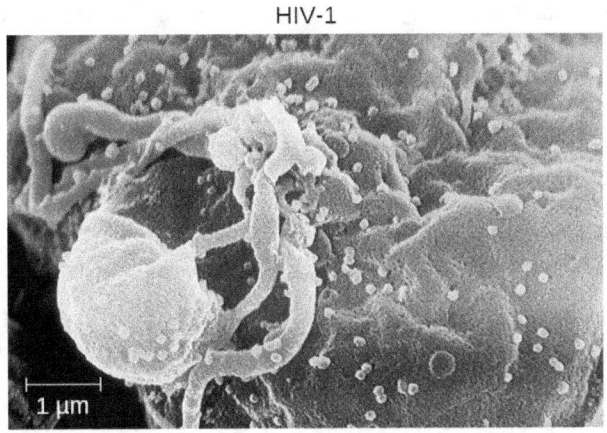

Figure 25.23 This micrograph shows HIV particles (green) budding from a lymphocyte (top right). (credit: modification of work by Centers for Disease Control and Prevention)

---

38 World Health Organization. "Hantavirus Diseases." 2016. http://www.who.int/ith/diseases/hantavirus/en/. Accessed July 28, 2016.

39 ibid.

40 Centers for Disease Control and Prevention. "Hantavirus: Treatment." 2012. http://www.cdc.gov/hantavirus/technical/hps/treatment.html. Accessed July 28, 2016.

41 World Health Organization. "HIV/AIDS: Fact Sheet." 2016.http://www.who.int/mediacentre/factsheets/fs360/en/. Accessed July 28, 2016.

42 ibid.

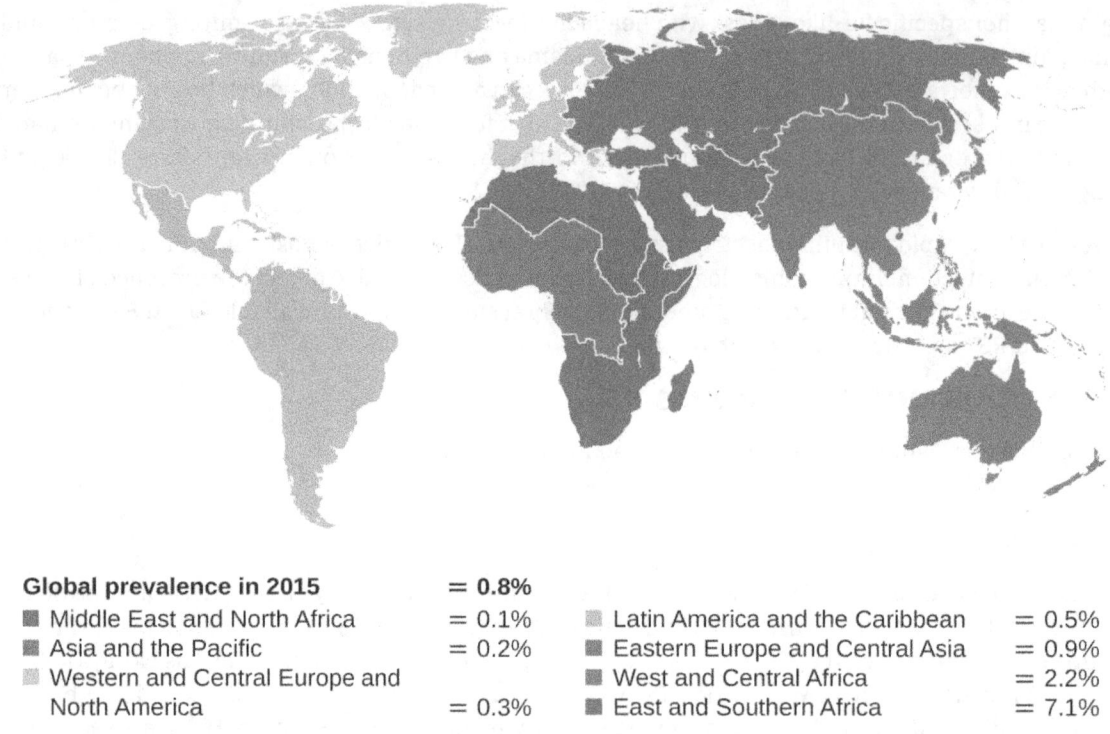

Figure 25.24 This map shows the prevalence of HIV worldwide in 2015 among adults ages 15–49 years.

HIV is spread through direct contact with body fluids. Casual contact and insect vectors are not sufficient for disease transmission; common modes of transmission include sexual contact and sharing of needles by intravenous (IV) drug users. It generally takes many years before the effects of an HIV infection are detected. HIV infections are not dormant during this period: virions are continually produced, and the immune system continually attempts to clear the viral infection, while the virus persistently infects additional CD4 T cells. Over time, the CD4 T-cell population is devastated, ultimately leading to AIDS.

When people are infected with HIV, their disease progresses through three stages based on CD4 T-cell counts and the presence of clinical symptoms (Figure 25.25).

- **Stage 1: Acute HIV infection.** Two to 4 weeks after infection with HIV, patients may experience a flu-like illness, which can last for a few weeks. Patients with acute HIV infection have more than 500 cells/μL CD4 T cells and a large amount of virus in their blood. Patients are very contagious during this stage. To confirm acute infection, either a fourth-generation antibody-antigen test or a nucleic acid test (NAT) must be performed.
- **Stage 2: Clinical latency.** During this period, HIV enters a period of dormancy. Patients have between 200 and 499 cells/μL CD4 T cells; HIV is still active but reproduces at low levels, and patients may not experience any symptoms of illness. For patients who are not taking medicine to treat HIV, this period can last a decade or longer. For patients receiving antiretroviral therapy, the stage may last for several decades, and those with low levels of the virus in their blood are much less likely to transmit HIV than those who are not virally suppressed. Near the end of the latent stage, the patient's viral load starts to increase and the CD4 T-cell count begins to decrease, leading to the development of symptoms and increased susceptibility to opportunistic infections.
- **Stage 3: Acquired immunodeficiency syndrome (AIDS).** Patients are diagnosed with AIDS when their CD4 T-cell count drops below 200 cells/μL or when they develop certain opportunistic illnesses. During this stage, the immune system becomes severely damaged by HIV. Common symptoms of AIDS include chills, fever, sweats, swollen lymph glands, weakness, and weight loss; in addition, patients often develop rare cancers such as Kaposi's sarcoma and opportunistic infections such as *Pneumocystis* pneumonia, tuberculosis, cryptosporidiosis, and toxoplasmosis. This is a fatal progression that, in the terminal stages, includes wasting syndrome and dementia complex. Patients with AIDS have a high viral load and are highly infectious; they typically survive about 3 years without treatment.

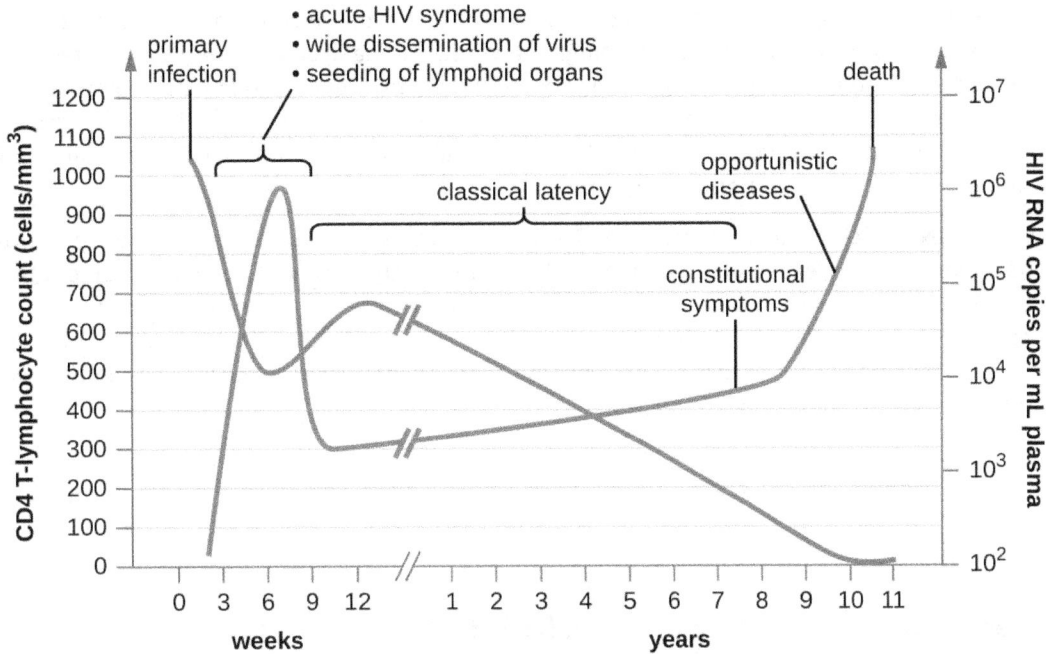

Figure 25.25 This graph shows the clinical progression of CD4 T cells (blue line), clinical symptoms, and viral RNA (red line) during an HIV infection. (credit: modification of work by Kogan M, and Rappaport J)

The initial diagnosis of HIV is performed using a serological test for antibody production against the pathogen. Positive test results are confirmed by Western blot or PCR tests. It can take weeks or months for the body to produce antibodies in response to an infection. There are fourth-generation tests that detect HIV antibodies and HIV antigens that are present even before the body begins producing antibodies. Nucleic acid tests (NATs) are a third type of test that is relatively expensive and uncommon; NAT can detect HIV in blood and determine the viral load.

As a consequence of provirus formation, it is currently not possible to eliminate HIV from an infected patient's body. Elimination by specific antibodies is ineffective because the virus mutates rapidly—a result of the error-prone reverse transcriptase and the inability to correct errors. Antiviral treatments, however, can greatly extend life expectancy. To combat the problem of drug resistance, combinations of antiretroviral drugs called antiretroviral therapy (ART), sometimes called highly active ART or combined ART, are used. There are several different targets for antiviral drug action (and a growing list of drugs for each of these targets). One class of drugs inhibits HIV entry; other classes inhibit reverse transcriptase by blocking viral RNA-dependent and DNA-dependent DNA polymerase activity; and still others inhibit one of the three HIV enzymes needed to replicate inside human cells.

✓ CHECK YOUR UNDERSTANDING

- Why is it not yet possible to cure HIV infections?

Eye on Ethics

HIV, AIDS, and Education

When the first outbreaks of AIDS in the US occurred in the early 1980s, very little was known about the disease or its origins. Erroneously, the disease quickly became stigmatized as one associated with what became identified as at-risk behaviors such as sexual promiscuity, homosexuality, and IV drug use, even though mounting evidence indicated the disease was also contracted through transfusion of blood and blood products or by fetuses of infected mothers. In the mid-1980s, scientists elucidated the identity of the virus, its mode of transmission, and mechanisms of pathogenesis. Campaigns were undertaken to educate the public about how HIV spreads to stem infection rates and encourage behavioral changes that reduced the risk for infection. Approaches to this campaign, however, emphasized very different strategies. Some groups favored educational programs that emphasized sexual abstinence, monogamy, heterosexuality, and "just say no to drugs." Other groups placed an emphasis on "safe sex" in sex education programs and advocated social services programs that passed out free condoms to anyone, including sexually active minors, and provided needle exchange programs for IV drug users.

These are clear examples of the intersection between disease and cultural values. As a future health professional, what is your responsibility in terms of educating patients about behaviors that put them at risk for HIV or other diseases while possibly setting your own personal opinions aside? You will no doubt encounter patients whose cultural and moral values differ from your own. Is it ethical for you to promote your own moral agenda to your patients? How can you advocate for practical disease prevention while still respecting the personal views of your patients?

Disease Profile

Viral Diseases of the Circulatory and Lymphatic Systems

Many viruses are able to cause systemic, difficult-to-treat infections because of their ability to replicate within the host. Some of the more common viruses that affect the circulatory system are summarized in Figure 25.26.

Viral Diseases of the Circulatory and Lymphatic Systems					
Disease	Pathogen	Signs and Symptoms	Transmission	Diagnostic Tests	Antimicrobial Drugs
AIDS/HIV infection	Human immunodeficiency virus (HIV)	Flu-like symptoms during acute stage, followed by long period of clinical latency; final stage (AIDS) includes fever, weight loss, wasting syndrome, dementia, and opportunistic secondary infections leading to death	Contact with body fluids (e.g., sexual contact, use of contaminated needles)	Serological tests for antibodies and/or HIV antigens; nucleic acid test (NAT) for presence of virus	Antiretroviral therapy (ART) using various combinations of drugs
Burkitt lymphoma	Epstein-Barr virus (human herpesvirus-4 [HHV-4])	Rapid formation of malignant B-cell tumors, oral hairy leukoplakia; fatal if not promptly treated	Contact with body fluids (e.g., saliva, blood, semen); primarily affects patients immunocompromised by HIV or malaria	CT scans, tumor biopsy	Intensive alternating chemotherapy regimen
Chikungunya fever	Chikungunya virus	Fever, rash, joint pain	Transmitted between humans by Aedes aegypti and A. albopictus vectors	Viral culture, IFA, EIA, ELISA, PCR, RT-PCR	None
Cytomegalovirus infection	Cytomegalovirus (HHV-5)	Usually asymptomatic but may cause non-Epstein-Barr mononucleosis in adults; may cause developmental issues in developing fetus; in transplant recipients, may cause fever, transplant rejection, death	Contact with body fluids, blood transfusions, organ transplants; infected mothers can transmit virus to fetus transplacentally or to newborn in breastmilk, saliva	Histology, culture, EIA, IFA, PCR	Ganciclovir, valganciclovir, foscarnet, cidofovir
Dengue fever (breakbone fever)	Dengue fever viruses 1-4	Fever, headache, extreme bone and joint pain, abdominal pain, vomiting, hemorrhaging; can be fatal	Transmitted between humans by A. aegypti and A. albopictus vectors	Serologic testing, ELISA, and PCR	None
Ebola virus disease (EVD)	Ebola virus	Fever, headache, joint pain, diarrhea, vomiting, hemorrhaging in gastrointestinal tract, organ failure; often fatal	Contact with body fluids (e.g., blood, saliva, sweat, urine, feces, vomit); highly contagious	ELISA, IgM ELISA, PCR, virus isolation	None
Hantavirus pulmonary syndrome (HPS)	Hantavirus	Initial flu-like symptoms followed by pulmonary edema and hypotension leading to pneumonia and shock; can be fatal	Inhalation of dried feces, urine from infected mouse or rat	ELISA, Western blot, RIBA, RT-PCR	None
Hemorrhagic fever with renal syndrome	Hantavirus	Fever, headache, nausea, rash, or eye inflammation, followed by hemorrhaging and kidney failure; can be fatal	Inhalation of dried feces, urine from infected mouse or rat	ELISA, Western blot, RIBA, RT-PCR	None
Infectious mononucleosis	Epstein-Barr virus (HHV-4), cytomegalovirus (HHV-5)	Pharyngitis, fever, extreme fatigue; swelling of lymph nodes, spleen, and liver	Contact with body fluids (e.g., saliva, blood, semen)	Tests for antibodies to various EBV-associated antigens	None
Yellow fever	Yellow fever virus	Dizziness, fever, chills, headache, myalgia, nausea, vomiting, constipation, fatigue; moderate to severe cases may include jaundice, rash, mucosal hemorrhaging, seizures, shock, and death	From monkeys to humans or between humans via Aedes or Haemagogus mosquito vectors	Culture, serology, PCR	None for treatment; preventive vaccine available

Figure 25.26

25.4 Parasitic Infections of the Circulatory and Lymphatic Systems

Learning Objectives

By the end of this section, you will be able to:
- Identify common parasites that cause infections of the circulatory and lymphatic systems
- Compare the major characteristics of specific parasitic diseases affecting the circulatory and lymphatic systems

Some protozoa and parasitic flukes are also capable of causing infections of the human circulatory system. Although these infections are rare in the US, they continue to cause widespread suffering in the developing world today. Fungal infections of the circulatory system are very rare. Therefore, they are not discussed in this chapter.

Malaria

Despite more than a century of intense research and clinical advancements, **malaria** remains one of the most important infectious diseases in the world today. Its widespread distribution places more than half of the world's population in jeopardy. In 2015, the WHO estimated there were about 214 million cases of malaria worldwide, resulting in about 438,000 deaths; about 88% of cases and 91% of deaths occurred in Africa.[43] Although malaria is not currently a major threat in the US, the possibility of its reintroduction is a concern. Malaria is caused by several protozoan parasites in the genus *Plasmodium*: *P. falciparum*, *P. knowlesi*, *P. malariae*, *P. ovale*, and *P. vivax*. *Plasmodium* primarily infect red blood cells and are transmitted through the bite of *Anopheles* mosquitoes.

Currently, *P. falciparum* is the most common and most lethal cause of malaria, often called falciparum malaria. Falciparum malaria is widespread in highly populated regions of Africa and Asia, putting many people at risk for the most severe form of the disease.

The classic signs and symptoms of malaria are cycles of extreme fever and chills. The sudden, violent symptoms of malaria start with malaise, abrupt chills, and fever (39–41° C [102.2–105.8 °F]), rapid and faint pulse, polyuria, headache, myalgia, nausea, and vomiting. After 2 to 6 hours of these symptoms, the fever falls, and profuse sweating occurs for 2 to 3 hours, followed by extreme fatigue. These symptoms are a result of *Plasmodium* emerging from red blood cells synchronously, leading to simultaneous rupture of a large number of red blood cells, resulting in damage to the spleen, liver, lymph nodes, and bone marrow. The organ damage resulting from hemolysis causes patients to develop sludge blood (i.e., blood in which the red blood cells agglutinate into clumps) that can lead to lack of oxygen, necrosis of blood vessels, organ failure, and death.

In established infections, malarial cycles of fever and chills typically occur every 2 days in the disease described as tertian malaria, which is caused by *P. vivax* and *P. ovale*. The cycles occur every 3 days in the disease described as quartan malaria, which is caused by *P. malariae*. These intervals may vary among cases.

Plasmodium has a complex life cycle that includes several developmental stages alternately produced in mosquitoes and humans (Figure 25.27). When an infected mosquito takes a blood meal, sporozoites in the mosquito salivary gland are injected into the host's blood. These parasites circulate to the liver, where they develop into schizonts. The schizonts then undergo schizogony, resulting in the release of many merozoites at once. The merozoites move to the bloodstream and infect red blood cells. Inside red blood cells, merozoites develop into trophozoites that produce more merozoites. The synchronous release of merozoites from red blood cells in the evening leads to the symptoms of malaria.

In addition, some trophozoites alternatively develop into male and female gametocytes. The gametocytes are taken up when the mosquito takes a blood meal from an infected individual. Sexual sporogony occurs in the gut of the mosquito. The gametocytes fuse to form zygotes in the insect gut. The zygotes become motile and elongate into an ookinete. This form penetrates the midgut wall and develops into an oocyst. Finally, the oocyst

---

43 World Health Organization. "World Malaria Report 2015: Summary." 2015. http://www.who.int/malaria/publications/world-malaria-report-2015/report/en/. Accessed July 28, 2016.

releases new sporozoites that migrate to the mosquito salivary glands to complete the life cycle.

Diagnosis of malaria is by microscopic observation of developmental forms of *Plasmodium* in blood smears and rapid EIA assays that detect *Plasmodium* antigens or enzymes (Figure 25.28). Drugs such as chloroquine, atovaquone, artemether, and lumefantrine may be prescribed for both acute and prophylactic therapy, although some *Plasmodium* spp. have shown resistance to antimalarial drugs. Use of insecticides and insecticide-treated bed nets can limit the spread of malaria. Despite efforts to develop a vaccine for malaria, none is currently available.

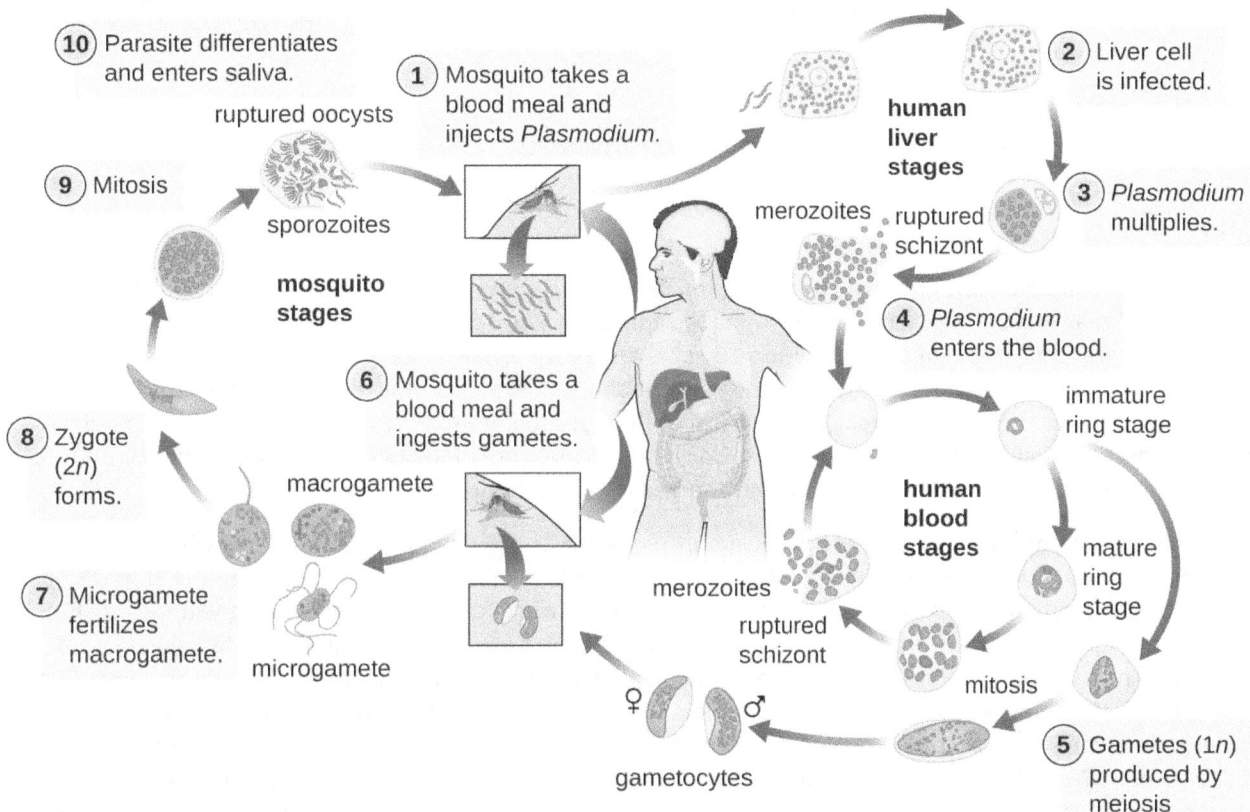

Figure 25.27 The life cycle of *Plasmodium*. (credit: modification of work by Centers for Disease Control and Prevention)

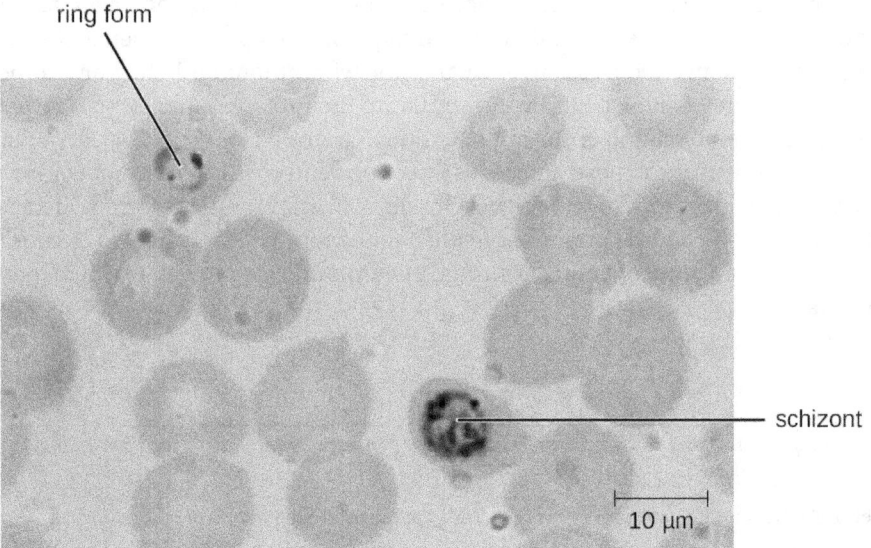

Figure 25.28 A blood smear (human blood stage) shows an early trophozoite in a delicate ring form (upper left) and an early stage schizont form (center) of *Plasmodium falciparum* from a patient with malaria. (credit: modification of work by Centers for Disease Control

and Prevention)

🔗 LINK TO LEARNING

Visit this site (https://openstax.org/l/22plasmodium) to learn how the parasite *Plasmodium* infects red blood cells.

The Nothing But Nets campaign, an initiative of the United Nations Foundation, has partnered with the Bill and Melinda Gates Foundation to make mosquito bed nets available in developing countries in Africa. Visit their website (https://openstax.org/l/22mosquitonet) to learn more about their efforts to prevent malaria.

✓ CHECK YOUR UNDERSTANDING

- Why is malaria one of the most important infectious diseases?

Toxoplasmosis

The disease **toxoplasmosis** is caused by the protozoan *Toxoplasma gondii*. *T. gondii* is found in a wide variety of birds and mammals,[44] and human infections are common. The Centers for Disease Control and Prevention (CDC) estimates that 22.5% of the population 12 years and older has been infected with *T. gondii*; but immunocompetent individuals are typically asymptomatic, however.[45] Domestic cats are the only known definitive hosts for the sexual stages of *T. gondii* and, thus, are the main reservoirs of infection. Infected cats shed *T. gondii* oocysts in their feces, and these oocysts typically spread to humans through contact with fecal matter on cats' bodies, in litter boxes, or in garden beds where outdoor cats defecate.

T. gondii has a complex life cycle that involves multiple hosts. The *T. gondii* life cycle begins when unsporulated oocysts are shed in the cat's feces. These oocysts take 1–5 days to sporulate in the environment and become infective. Intermediate hosts in nature include birds and rodents, which become infected after ingesting soil, water, or plant material contaminated with the infective oocysts. Once ingested, the oocysts transform into tachyzoites that localize in the bird or rodent neural and muscle tissue, where they develop into tissue cysts. Cats may become infected after consuming birds and rodents harboring tissue cysts. Cats and other animals may also become infected directly by ingestion of sporulated oocysts in the environment. Interestingly, *Toxoplasma* infection appears to be able to modify the host's behavior. Mice infected by *Toxoplasma* lose their fear of cat pheromones. As a result, they become easier prey for cats, facilitating the transmission of the parasite to the cat definitive host[46] (Figure 25.29).

Toxoplasma infections in humans are extremely common, but most infected people are asymptomatic or have subclinical symptoms. Some studies suggest that the parasite may be able to influence the personality and psychomotor performance of infected humans, similar to the way it modifies behavior in other mammals.[47] When symptoms do occur, they tend to be mild and similar to those of mononucleosis. However, asymptomatic toxoplasmosis can become problematic in certain situations. Cysts can lodge in a variety of human tissues and lie dormant for years. Reactivation of these quiescent infections can occur in immunocompromised patients following transplantation, cancer therapy, or the development of an immune disorder such as AIDS. In patients with AIDS who have toxoplasmosis, the immune system cannot combat the growth of *T. gondii* in body tissues; as a result, these cysts can cause encephalitis, retinitis, pneumonitis, cognitive disorders, and seizures that can eventually be fatal.

---

44 A.M. Tenter et al.. "*Toxoplasma gondii*: From Animals to Humans." *International Journal for Parasitology* 30 no. 12-13 (2000):1217–1258.

45 Centers for Disease Control and Prevention. "Parasites - Toxoplasmosis (Toxoplasma Infection). Epidemiology & Risk Factors." 2015 http://www.cdc.gov/parasites/toxoplasmosis/epi.html. Accessed July 28, 2016.

46 J. Flegr. "Effects of *Toxoplasma* on Human Behavior." *Schizophrenia Bulletin* 33, no. 3 (2007):757–760.

47 Ibid

Toxoplasmosis can also pose a risk during pregnancy because tachyzoites can cross the placenta and cause serious infections in the developing fetus. The extent of fetal damage resulting from toxoplasmosis depends on the severity of maternal disease, the damage to the placenta, the gestational age of the fetus when infected, and the virulence of the organism. Congenital toxoplasmosis often leads to fetal loss or premature birth and can result in damage to the central nervous system, manifesting as mental retardation, deafness, or blindness. Consequently, pregnant women are advised by the CDC to take particular care in preparing meat, gardening, and caring for pet cats.[48] Diagnosis of toxoplasmosis infection during pregnancy is usually achieved by serology including TORCH testing (the "T" in TORCH stands for toxoplasmosis). Diagnosis of congenital infections can also be achieved by detecting *T. gondii* DNA in amniotic fluid, using molecular methods such as PCR.

In adults, diagnosis of toxoplasmosis can include observation of tissue cysts in tissue specimens. Tissue cysts may be observed in Giemsa- or Wright-stained biopsy specimens, and CT, magnetic resonance imaging, and lumbar puncture can also be used to confirm infection (Figure 25.30).

Preventing infection is the best first-line defense against toxoplasmosis. Preventive measures include washing hands thoroughly after handling raw meat, soil, or cat litter, and avoiding consumption of vegetables possibly contaminated with cat feces. All meat should be cooked to an internal temperature of 73.9–76.7 °C (165–170 °F).

Most immunocompetent patients do not require clinical intervention for *Toxoplasma* infections. However, neonates, pregnant women, and immunocompromised patients can be treated with pyrimethamine and sulfadiazine—except during the first trimester of pregnancy, because these drugs can cause birth defects. Spiramycin has been used safely to reduce transmission in pregnant women with primary infection during the first trimester because it does not cross the placenta.

---

48 Centers for Disease Control and Prevention. "Parasites - Toxoplasmosis (Toxoplasma infection). Toxoplasmosis Frequently Asked Questions (FAQs)." 2013. http://www.cdc.gov/parasites/toxoplasmosis/gen_info/faqs.html. Accessed July 28, 2016.

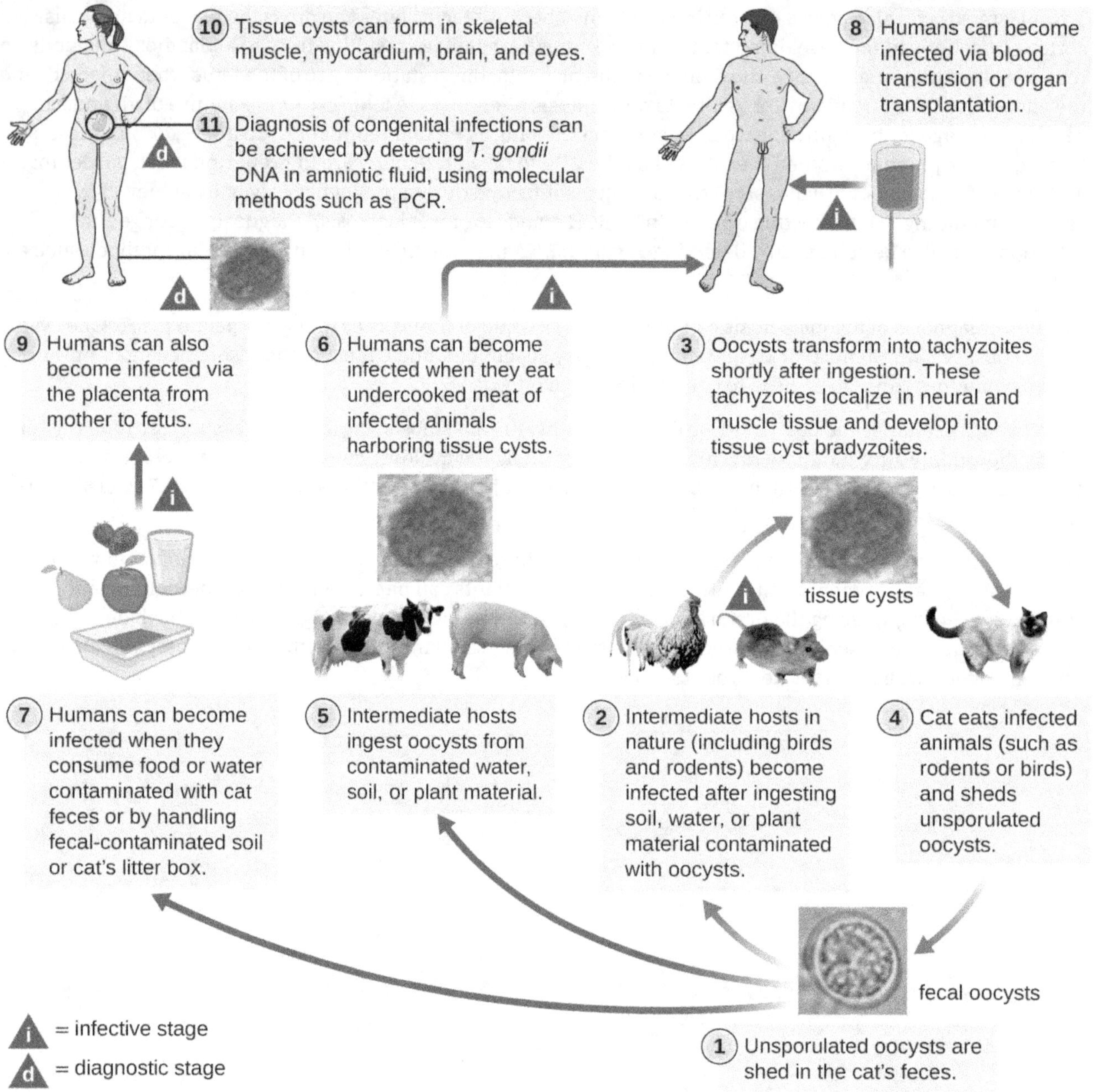

Figure 25.29 The infectious cycle of *Toxoplasma gondii*. (credit: "diagram": modification of work by Centers for Disease Control and Prevention; credit "cat": modification of work by "KaCey97078"/Flickr)

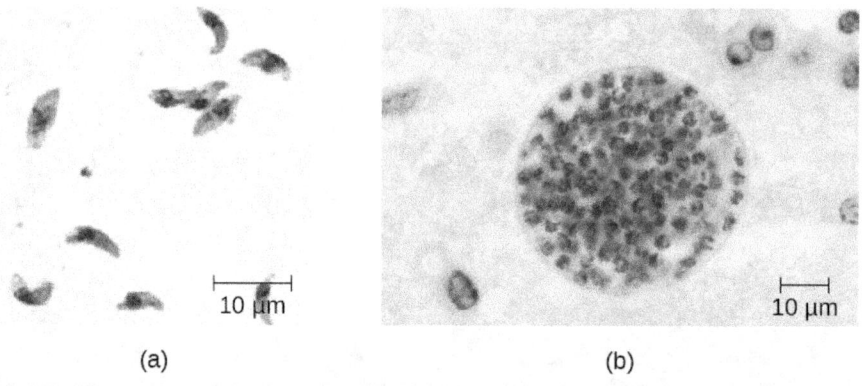

Figure 25.30 (a) Giemsa-stained *Toxoplasma gondii* tachyzoites from a smear of peritoneal fluid obtained from a mouse inoculated with *T. gondii*. Tachyzoites are typically crescent shaped with a prominent, centrally placed nucleus. (b) Microscopic cyst containing *T. gondii*

from mouse brain tissue. Thousands of resting parasites (stained red) are contained in a thin parasite cyst wall. (credit a: modification of work by Centers for Disease Control and Prevention; credit b: modification of work by USDA)

✓ CHECK YOUR UNDERSTANDING

- How does *T. gondii* infect humans?

Babesiosis

Babesiosis is a rare zoonotic infectious disease caused by *Babesia* spp. These parasitic protozoans infect various wild and domestic animals and can be transmitted to humans by black-legged *Ixodes* ticks. In humans, *Babesia* infect red blood cells and replicate inside the cell until it ruptures. The *Babesia* released from the ruptured red blood cell continue the growth cycle by invading other red blood cells. Patients may be asymptomatic, but those who do have symptoms often initially experience malaise, fatigue, chills, fever, headache, myalgia, and arthralgia. In rare cases, particularly in asplenic (absence of the spleen) patients, the elderly, and patients with AIDS, **babesiosis** may resemble falciparum malaria, with high fever, hemolytic anemia, hemoglobinuria (hemoglobin or blood in urine), jaundice, and renal failure, and the infection can be fatal. Previously acquired asymptomatic Babesia infection may become symptomatic if a splenectomy is performed.

Diagnosis is based mainly on the microscopic observation of parasites in blood smears (Figure 25.31). Serologic and antibody detection by IFA can also be performed and PCR-based tests are available. Many people do not require clinical intervention for Babesia infections, however, serious infections can be cleared with a combination of atovaquone and azithromycin or a combination of clindamycin and quinine.

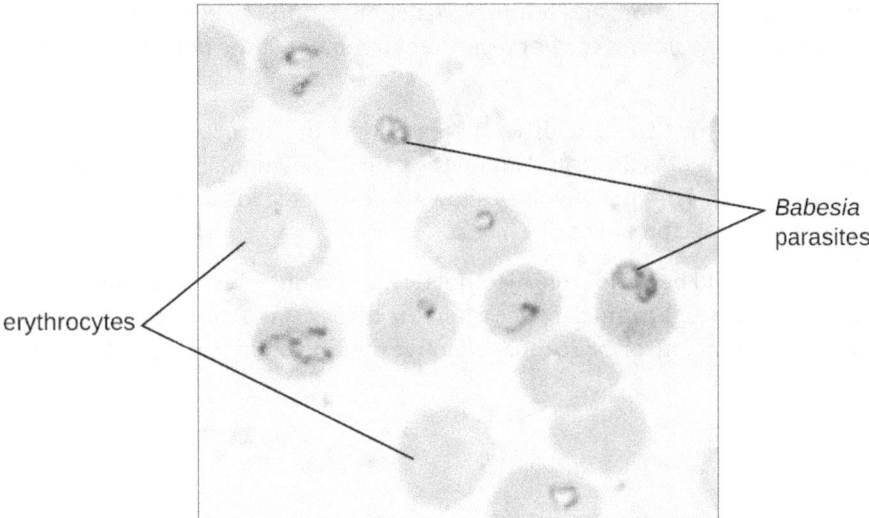

Figure 25.31 In this blood smear from a patient with babesiosis, *Babesia* parasites can be observed in the red blood cells. (credit: modification of work by Centers for Disease Control and Prevention)

Chagas Disease

Also called American trypanosomiasis, Chagas disease is a zoonosis classified as a neglected tropical disease (NTD). It is caused by the flagellated protozoan *Trypanosoma cruzi* and is most commonly transmitted to animals and people through the feces of triatomine bugs. The triatomine bug is nicknamed the kissing bug because it frequently bites humans on the face or around the eyes; the insect often defecates near the bite and the infected fecal matter may be rubbed into the bite wound by the bitten individual (Figure 25.32). The bite itself is painless and, initially, many people show no signs of the disease. Alternative modes of transmission include contaminated blood transfusions, organ transplants from infected donors, and congenital transmission from mother to fetus.

Chagas disease is endemic throughout much of Mexico, Central America, and South America, where, according to WHO, an estimated 6 million to 7 million people are infected.[49] Currently, Chagas disease is not endemic in the US, even though triatomine bugs are found in the southern half of the country.

Triatomine bugs typically are active at night, when they take blood meals by biting the faces and lips of people or animals as they sleep and often defecate near the site of the bite. Infection occurs when the host rubs the feces into their eyes, mouth, the bite wound, or another break in the skin. The protozoan then enters the blood and invades tissues of the heart and central nervous system, as well as macrophages and monocytes. Nonhuman reservoirs of *T. cruzi* parasites include wild animals and domesticated animals such as dogs and cats, which also act as reservoirs of the pathogen.[50]

There are three phases of Chagas disease: acute, intermediate, and chronic. These phases can be either asymptomatic or life-threatening depending on the immunocompetence status of the patient.

In acute phase disease, symptoms include fever, headache, myalgia, rash, vomiting, diarrhea, and enlarged spleen, liver, and lymph nodes. In addition, a localized nodule called a chagoma may form at the portal of entry, and swelling of the eyelids or the side of the face, called Romaña's sign, may occur near the bite wound. Symptoms of the acute phase may resolve spontaneously, but if untreated, the infection can persist in tissues, causing irreversible damage to the heart or brain. In rare cases, young children may die of myocarditis or meningoencephalitis during the acute phase of Chagas disease.

Following the acute phase is a prolonged intermediate phase during which few or no parasites are found in the blood and most people are asymptomatic. Many patients will remain asymptomatic for life; however, decades after exposure, an estimated 20%–30% of infected people will develop chronic disease that can be debilitating and sometimes life threatening. In the chronic phase, patients may develop painful swelling of the colon, leading to severe twisting, constipation, and bowel obstruction; painful swelling of the esophagus, leading to dysphagia and malnutrition; and flaccid cardiomegaly (enlargement of the heart), which can lead to heart failure and sudden death.

Diagnosis can be confirmed through several different tests, including direct microscopic observation of trypanosomes in the blood, IFA, EIAs, PCR, and culturing in artificial media. In endemic regions, xenodiagnoses may be used; this method involves allowing uninfected kissing bugs to feed on the patient and then examining their feces for the presence of *T. cruzi*.

The medications nifurtimox and benznidazole are effective treatments during the acute phase of Chagas disease. The efficacy of these drugs is much lower when the disease is in the chronic phase. Avoiding exposure to the pathogen through vector control is the most effective method of limiting this disease.

---

49 World Health Organization. "Chagas disease (American trypanosomiasis). Fact Sheet." 2016. http://www.who.int/mediacentre/factsheets/fs340/en/. Accessed July 29, 2016.

50 C.E. Reisenman et al. "Infection of Kissing Bugs With *Trypanosoma cruzi*, Tucson, Arizona, USA." *Emerging Infectious Diseases* 16 no. 3 (2010):400–405.

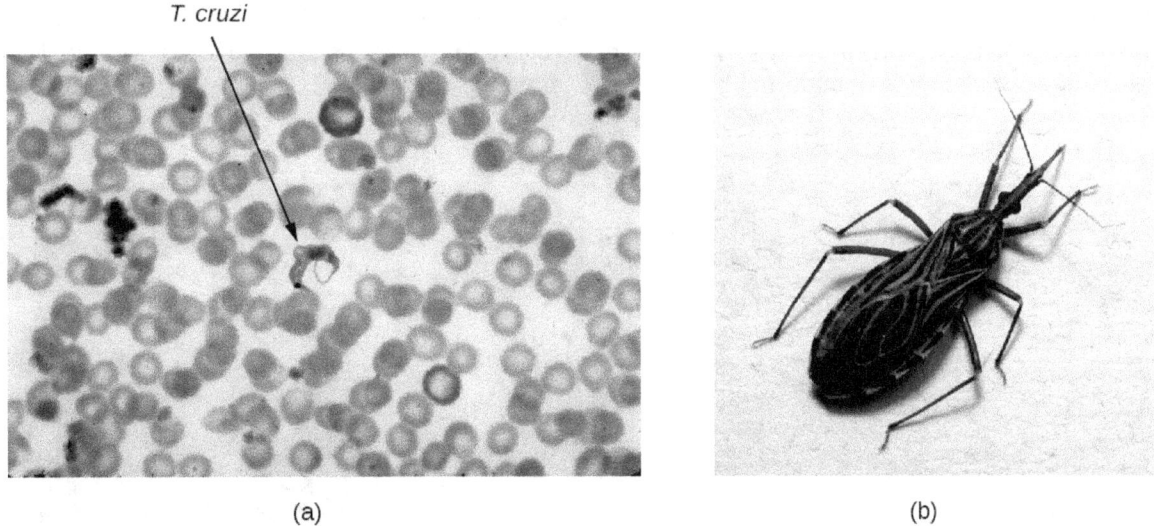

Figure 25.32 (a) *Trypanosoma cruzi* protozoan in a blood smear from a patient with Chagas disease. (b) The triatomine bug (also known as the kissing bug or assassin bug) is the vector of Chagas disease. (credit a: modification of work by Centers for Disease Control and Prevention; credit b: modification of work by Erwin Huebner)

CHECK YOUR UNDERSTANDING

- How do kissing bugs infect humans with *Trypanosoma cruzi*?

Leishmaniasis

Although it is classified as an NTD, **leishmaniasis** is relatively widespread in tropical and subtropical regions, affecting people in more than 90 countries. It is caused by approximately 20 different species of *Leishmania*, protozoan parasites that are transmitted by sand fly vectors such as *Phlebotomus* spp. and *Lutzomyia* spp. Dogs, cats, sheep, horses, cattle rodents, and humans can all serve as reservoirs.

The *Leishmania* protozoan is phagocytosed by macrophages but uses virulence factors to avoid destruction within the phagolysosome. The virulence factors inhibit the phagolysosome enzymes that would otherwise destroy the parasite. The parasite reproduces within the macrophage, lyses it, and the progeny infect new macrophages (see Micro Connections: When Phagocytosis Fails).

The three major clinical forms of leishmaniasis are cutaneous (oriental sore, Delhi boil, Aleppo boil), visceral (kala-azar, Dumdum fever), and mucosal (espundia). The most common form of disease is cutaneous leishmaniasis, which is characterized by the formation of sores at the site of the insect bite that may start out as papules or nodules before becoming large ulcers (Figure 25.33).

It may take visceral leishmaniasis months and sometimes years to develop, leading to enlargement of the lymph nodes, liver, spleen, and bone marrow. The damage to these body sites triggers fever, weight loss, and swelling of the spleen and liver. It also causes a decrease in the number of red blood cells (anemia), white blood cells (leukopenia), and platelets (thrombocytopenia), causing the patient to become immunocompromised and more susceptible to fatal infections of the lungs and gastrointestinal tract.

The mucosal form of leishmaniasis is one of the less common forms of the disease. It causes a lesion similar to the cutaneous form but mucosal leishmaniasis is associated with mucous membranes of the mouth, nares, or pharynx, and can be destructive and disfiguring. Mucosal leishmaniasis occurs less frequently when the original cutaneous (skin) infection is promptly treated.

Definitive diagnosis of leishmaniasis is made by visualizing organisms in Giemsa-stained smears, by isolating *Leishmania* protozoans in cultures, or by PCR-based assays of aspirates from infected tissues. Specific DNA probes or analysis of cultured parasites can help to distinguish *Leishmania* species that are causing simple cutaneous leishmaniasis from those capable of causing mucosal leishmaniasis.

Cutaneous leishmaniasis is usually not treated. The lesions will resolve after weeks (or several months), but may result in scarring. Recurrence rates are low for this disease. More serious infections can be treated with stibogluconate (antimony gluconate), amphotericin B, and miltefosine.

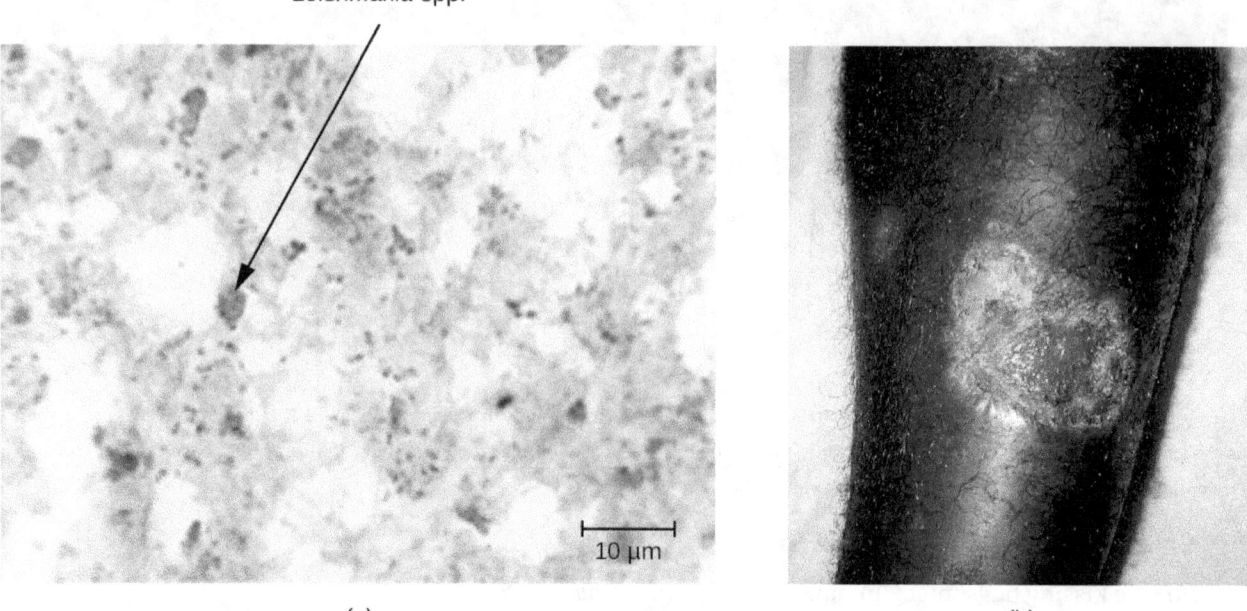

Figure 25.33 (a) A micrograph of a tissue sample from a patient with localized cutaneous leishmaniasis. Parasitic *Leishmania mexicana* (black arrow) are visible in and around the host cells. (b) Large skin ulcers are associated with cutaneous leishmaniasis. (credit a: modification of work by Fernández-Figueroa EA, Rangel-Escareño C, Espinosa-Mateos V, Carrillo-Sánchez K, Salaiza-Suazo N, Carrada-Figueroa G, March-Mifsut S, and Becker I; credit b: modification of work by Jean Fortunet)

✓ CHECK YOUR UNDERSTANDING

- Compare the mucosal and cutaneous forms of leishmaniasis.

Schistosomiasis

Schistosomiasis (bilharzia) is an NTD caused by blood flukes in the genus *Schistosoma* that are native to the Caribbean, South America, Middle East, Asia, and Africa. Most human **schistosomiasis** cases are caused by *Schistosoma mansoni*, *S. haematobium*, or *S. japonicum*. *Schistosoma* are the only trematodes that invade through the skin; all other trematodes infect by ingestion. WHO estimates that at least 258 million people required preventive treatment for schistosomiasis in 2014.[51]

Infected human hosts shed *Schistosoma* eggs in urine and feces, which can contaminate freshwater habitats of snails that serve as intermediate hosts. The eggs hatch in the water, releasing miracidia, an intermediate growth stage of the *Schistosoma* that infect the snails. The miracidia mature and multiply inside the snails, transforming into cercariae that leave the snail and enter the water, where they can penetrate the skin of swimmers and bathers. The cercariae migrate through human tissue and enter the bloodstream, where they mature into adult male and female worms that mate and release fertilized eggs. The eggs travel through the bloodstream and penetrate various body sites, including the bladder or intestine, from which they are excreted in urine or stool to start the life cycle over again (Figure 5.22).

A few days after infection, patients may develop a rash or itchy skin associated with the site of cercariae penetration. Within 1–2 months of infection, symptoms may develop, including fever, chills, cough, and myalgia, as eggs that are not excreted circulate through the body. After years of infection, the eggs become

---

51 World Health Organization. "Schistosomiasis. Fact Sheet." 2016. http://www.who.int/mediacentre/factsheets/fs115/en/. Accessed July 29, 2016.

lodged in tissues and trigger inflammation and scarring that can damage the liver, central nervous system, intestine, spleen, lungs, and bladder. This may cause abdominal pain, enlargement of the liver, blood in the urine or stool, and problems passing urine. Increased risk for bladder cancer is also associated with chronic *Schistosoma* infection. In addition, children who are repeatedly infected can develop malnutrition, anemia, and learning difficulties.

Diagnosis of schistosomiasis is made by the microscopic observation of eggs in feces or urine, intestine or bladder tissue specimens, or serologic tests. The drug praziquantel is effective for the treatment of all schistosome infections. Improving wastewater management and educating at-risk populations to limit exposure to contaminated water can help control the spread of the disease.

Cercarial Dermatitis

The cercaria of some species of *Schistosoma* can only transform into adult worms and complete their life cycle in animal hosts such as migratory birds and mammals. The cercaria of these worms are still capable of penetrating human skin, but they are unable to establish a productive infection in human tissue. Still, the presence of the cercaria in small blood vessels triggers an immune response, resulting in itchy raised bumps called **cercarial dermatitis** (also known as swimmer's itch or clam digger's itch). Although it is uncomfortable, cercarial dermatitis is typically self-limiting and rarely serious. Antihistamines and antipruritics can be used to limit inflammation and itching, respectively.

✓ CHECK YOUR UNDERSTANDING

- How do schistosome infections in humans occur?

Disease Profile

Common Eukaryotic Pathogens of the Human Circulatory System

Protozoan and helminthic infections are prevalent in the developing world. A few of the more important parasitic infections are summarized in Figure 25.34.

Parasitic Diseases of the Circulatory and Lymphatic Systems

Disease	Pathogen	Signs and Symptoms	Transmission	Diagnostic Tests	Antimicrobial Drugs
Protozoa					
Babesiosis	*Babesia* spp.	Malaise, chills, fever, headache, myalgia, arthralgia	From animals to humans via *Ixodes* tick vectors	Blood smear, serology, IFA, and PCR	Atovaquone and azithromycin or clindamycin and quinine
Chagas disease	*Trypanosoma cruzi*	Fever, headache, body aches, swollen lymph nodes; potentially fatal	Between humans or from animal reservoirs via triatomine (kissing bug) vector	Blood smear, IFA, EIA, PCR, xenodiagnosis	Nifurtimox, benznidazole
Leishmaniasis	*Leishmania* spp.	Ulcer; enlargement of the lymph nodes, liver, spleen, and other organs	Between humans or from animal reservoirs via sand fly (*Phlebotomus* spp., *Lutzomyia* spp.) vectors	Blood smear, culture, PCR, DNA probe, biopsy	Stibogluconate, amphotericin B.
Malaria	*Plasmodium vivax*, *P. malariae*, *P. falciparum*, *P. ovale*, *P. knowlesi*	Extreme fever, chills, myalgia, nausea, and vomiting, possibly leading to organ failure and death	Between humans via *Anopheles* mosquito vectors	Blood smear, EIA	Chloroquine, atovaquone, artemether, and lumefantrine
Toxoplasmosis	*Toxoplasma gondii*	Tissue cysts; in pregnant women, birth defects or miscarriage	Contact with feces of infected cat; eating contaminated vegetables or undercooked meat of infected animal	Serological tests, direct detection of pathogen in tissue sections	Sulfadiazine, pyrimethamine, spiramycin
Helminths					
Schistosomiasis	*Schistosoma* spp.	Rash, fever, chills, myalgia; chronic inflammation and scarring of liver, spleen, and other organs where cysts develop	Snail hosts release cercaria into freshwater; cercaria burrow into skin of swimmers and bathers	Eggs in stool or urine, tissue biopsy, serological testing	Praziquantel

Figure 25.34

Clinical Focus

Resolution

Despite continued antibiotic treatment and the removal of the venous catheter, Barbara's condition further declined. She began to show signs of shock and her blood pressure dropped to 77/50 mmHg. Anti-inflammatory drugs and drotrecogin-α were administered to combat sepsis. However, by the seventh day of hospitalization, Barbara experienced hepatic and renal failure and died.

Staphylococcus aureus most likely formed a biofilm on the surface of Barbara's catheter. From there, the bacteria were chronically shed into her circulation and produced the initial clinical symptoms. The chemotherapeutic therapies failed in large part because of the drug-resistant MRSA isolate. Virulence factors like leukocidin and hemolysins also interfered with her immune response. Barbara's ultimate decline may have been a consequence of the production of enterotoxins and toxic shock syndrome toxin (TSST), which can initiate toxic shock.

Venous catheters are common life-saving interventions for many patients requiring long-term administration of medication or fluids. However, they are also common sites of bloodstream infections. The World Health Organization estimates that there are up to 80,000 catheter-related bloodstream infections each year in the US, resulting in about 20,000 deaths.[52]

Go back to the previous Clinical Focus box.

---

[52] World Health Organization. "Patient Safety, Preventing Bloodstream Infections From Central Line Venous Catheters." 2016. http://www.who.int/patientsafety/implementation/bsi/en/. Accessed July 29, 2016.

SUMMARY

25.1 Anatomy of the Circulatory and Lymphatic Systems

- The **circulatory system** moves blood throughout the body and has no normal microbiota.
- The **lymphatic system** moves fluids from the interstitial spaces of tissues toward the circulatory system and filters the lymph. It also has no normal microbiota.
- The circulatory and lymphatic systems are home to many components of the host immune defenses.
- Infections of the circulatory system may occur after a break in the skin barrier or they may enter the bloodstream at the site of a localized infection. Pathogens or toxins in the bloodstream can spread rapidly throughout the body and can provoke systemic and sometimes fatal inflammatory responses such as **SIRS**, **sepsis**, and **endocarditis**.
- Infections of the lymphatic system can cause **lymphangitis** and **lymphadenitis**.

25.2 Bacterial Infections of the Circulatory and Lymphatic Systems

- Bacterial infections of the circulatory system are almost universally serious. Left untreated, most have high mortality rates.
- Bacterial pathogens usually require a breach in the immune defenses to colonize the circulatory system. Most often, this involves a wound or the bite of an arthropod vector, but it can also occur in hospital settings and result in nosocomial infections.
- **Sepsis** from both gram-negative and gram-positive bacteria, **puerperal fever, rheumatic fever, endocarditis, gas gangrene, osteomyelitis,** and **toxic shock syndrome** are typically a result of injury or introduction of bacteria by medical or surgical intervention.
- **Tularemia, brucellosis, cat-scratch fever, rat-bite fever,** and **bubonic plague** are zoonotic diseases transmitted by biological vectors
- **Ehrlichiosis, anaplasmosis, endemic** and **murine typhus, Rocky Mountain spotted fever, Lyme disease, relapsing fever,** and **trench fever** are transmitted by arthropod vectors.
- Because their symptoms are so similar to those of other diseases, many bacterial infections of the circulatory system are difficult to diagnose.
- Standard antibiotic therapies are effective for the treatment of most bacterial infections of the circulatory system, unless the bacterium is resistant, in which case synergistic treatment may be required.
- The systemic immune response to a bacteremia, which involves the release of excessive amounts of cytokines, can sometimes be more damaging to the host than the infection itself.

25.3 Viral Infections of the Circulatory and Lymphatic Systems

- Human herpesviruses such **Epstein-Barr virus (HHV-4)** and **cytomegalovirus** (HHV-5) are widely distributed. The former is associated with infectious mononucleosis and Burkitt lymphoma, and the latter can cause serious congenital infections as well as serious disease in immunocompromised adults.
- Arboviral diseases such as **yellow fever, dengue fever,** and **chikungunya fever** are characterized by high fevers and vascular damage that can often be fatal. **Ebola virus disease** is a highly contagious and often fatal infection spread through contact with bodily fluids.
- Although there is a vaccine available for yellow fever, treatments for patients with yellow fever, dengue, chikungunya fever, and Ebola virus disease are limited to supportive therapies.
- Patients infected with **human immunodeficiency virus (HIV)** progress through three stages of disease, culminating in **AIDS. Antiretroviral therapy (ART)** uses various combinations of drugs to suppress viral loads, extending the period of latency and reducing the likelihood of transmission.
- Vector control and animal reservoir control remain the best defenses against most viruses that cause diseases of the circulatory system.

25.4 Parasitic Infections of the Circulatory and Lymphatic Systems

- **Malaria** is a protozoan parasite that remains an important cause of death primarily in the tropics. Several species in the genus *Plasmodium* are responsible for malaria and all are transmitted by *Anopheles* mosquitoes. *Plasmodium* infects and destroys human red blood cells, leading to organ damage, anemia, blood vessel necrosis, and death. Malaria can be treated with various antimalarial drugs and prevented through vector control.

- **Toxoplasmosis** is a widespread protozoal infection that can cause serious infections in the immunocompromised and in developing fetuses. Domestic cats are the definitive host.
- Babesiosis is a generally asymptomatic infection of red blood cells that can causes malaria-like symptoms in elderly, immunocompromised, or asplenic patients.
- **Chagas disease** is a tropical disease transmitted by triatomine bugs. The trypanosome infects heart, neural tissues, monocytes, and phagocytes, often remaining latent for many years before causing serious and sometimes fatal damage to the digestive system and heart.
- **Leishmaniasis** is caused by the protozoan *Leishmania* and is transmitted by sand flies. Symptoms are generally mild, but serious cases may cause organ damage, anemia, and loss of immune competence.
- **Schistosomiasis** is caused by a fluke transmitted by snails. The fluke moves throughout the body in the blood stream and chronically infects various tissues, leading to organ damage.

REVIEW QUESTIONS

Multiple Choice

1. Which term refers to an inflammation of the blood vessels?
   A. lymphangitis
   B. endocarditis
   C. pericarditis
   D. vasculitis

2. Which of the following is located in the interstitial spaces within tissues and releases nutrients, immune factors, and oxygen to those tissues?
   A. lymphatics
   B. arterioles
   C. capillaries
   D. veins

3. Which of these conditions results in the formation of a bubo?
   A. lymphangitis
   B. lymphadenitis
   C. ischemia
   D. vasculitis

4. Which of the following is where are most microbes filtered out of the fluids that accumulate in the body tissues?
   A. spleen
   B. lymph nodes
   C. pericardium
   D. blood capillaries

5. Which of the following diseases is caused by a spirochete?
   A. tularemia
   B. relapsing fever
   C. rheumatic fever
   D. Rocky Mountain spotted fever

6. Which of the following diseases is transmitted by body lice?
   A. tularemia
   B. bubonic plague
   C. murine typhus
   D. epidemic typhus

7. What disease is most associated with *Clostridium perfringens*?
   A. endocarditis
   B. osteomyelitis
   C. gas gangrene
   D. rat bite fever

8. Which bacterial pathogen causes plague?
   A. *Yersinia pestis*
   B. *Bacillus moniliformis*
   C. *Bartonella quintana*
   D. *Rickettsia rickettsii*

9. Which of the following viruses is most widespread in the human population?
   A. human immunodeficiency virus
   B. Ebola virus
   C. Epstein-Barr virus
   D. hantavirus

10. Which of these viruses is spread through mouse urine or feces?
    A. Epstein-Barr
    B. hantavirus
    C. human immunodeficiency virus
    D. cytomegalovirus

11. A patient at a clinic has tested positive for HIV. Her blood contained 700/μL CD4 T cells and she does not have any apparent illness. Her infection is in which stage?
    A. 1
    B. 2
    C. 3

12. Which of the following diseases is caused by a helminth?
    A. leishmaniasis
    B. malaria
    C. Chagas disease
    D. schistosomiasis

13. Which of these is the most common form of leishmaniasis?
    A. cutaneous
    B. mucosal
    C. visceral
    D. intestinal

14. Which of the following is a causative agent of malaria?
    A. *Trypanosoma cruzi*
    B. *Toxoplasma gondii*
    C. *Plasmodium falciparum*
    D. *Schistosoma mansoni*

15. Which of the following diseases does not involve an arthropod vector?
    A. schistosomiasis
    B. malaria
    C. Chagas disease
    D. babesiosis

Fill in the Blank

16. Vasculitis can cause blood to leak from damaged vessels, forming purple spots called _____.

17. The lymph reenters the vascular circulation at _____.

18. Lyme disease is characterized by a(n) _____ that forms at the site of infection.

19. _____ refers to a loss of blood pressure resulting from a system-wide infection.

20. _____ is a cancer that forms in patients with HHV-4 and malaria coinfections.

21. _____ are transmitted by vectors such as ticks or mosquitoes.

22. Infectious mononucleosis is caused by _____ infections.

23. The _____ mosquito is the biological vector for malaria.

24. The kissing bug is the biological vector for _____.

25. Cercarial dermatitis is also known as _____.

Short Answer

26. How do lymph nodes help to maintain a microbial-free circulatory and lymphatic system?

27. What are the three forms of plague and how are they contracted?

28. Compare epidemic and murine typhus.

29. Describe the progression of an HIV infection over time with regard to the number of circulating viruses, host antibodies, and CD4 T cells.

30. Describe the general types of diagnostic tests used to diagnose patients infected with HIV.

31. Identify the general categories of drugs used in ART used to treat patients infected with HIV.

32. Describe main cause of *Plasmodium falciparum* infection symptoms.

33. Why should pregnant women avoid cleaning their cat's litter box or do so with protective gloves?

Critical Thinking

34. What term refers to the red streaks seen on this patient's skin? What is likely causing this condition?

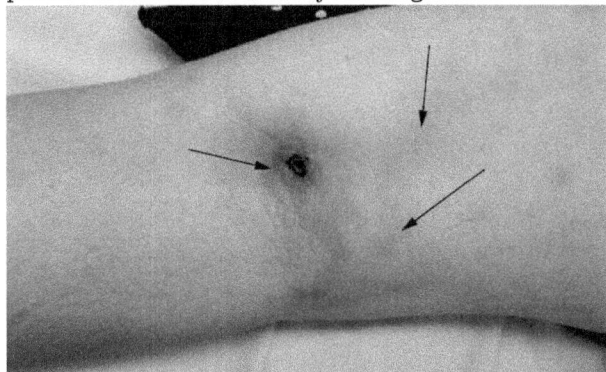

Figure 25.35 (credit: modification of work by Centers for Disease Control and Prevention)

35. Why would septicemia be considered a more serious condition than bacteremia?
36. Why are most vascular pathogens poorly communicable from person to person?
37. How have human behaviors contributed to the spread or control of arthropod-borne vascular diseases?
38. Which is a bigger threat to the US population, Ebola or yellow fever? Why?
39. What measures can be taken to reduce the likelihood of malaria reemerging in the US?

CHAPTER 26
Nervous System Infections

Figure 26.1 This dog is exhibiting the restlessness and aggression associated with rabies, a neurological disease that frequently affects mammals and can be transmitted to humans. (credit: modification of work by the Centers for Disease Control and Prevention)

Chapter Outline

26.1 Anatomy of the Nervous System

26.2 Bacterial Diseases of the Nervous System

26.3 Acellular Diseases of the Nervous System

26.4 Fungal and Parasitic Diseases of the Nervous System

INTRODUCTION Few diseases inspire the kind of fear that rabies does. The name is derived from the Latin word for "madness" or "fury," most likely because animals infected with rabies may behave with uncharacteristic rage and aggression. And while the thought of being attacked by a rabid animal is terrifying enough, the disease itself is even more frightful. Once symptoms appear, the disease is almost always fatal, even when treated.

Rabies is an example of a neurological disease caused by an acellular pathogen. The rabies virus enters nervous tissue shortly after transmission and makes its way to the central nervous system, where its presence leads to changes in behavior and motor function. Well-known symptoms associated with rabid animals include foaming at the mouth, hydrophobia (fear of water), and unusually aggressive behavior. Rabies claims tens of thousands of human lives worldwide, mainly in Africa and Asia. Most human cases result from dog bites, although many mammal species can become infected and transmit the disease. Human infection rates are low in the United States and many other countries as a result of control measures in animal populations.

However, rabies is not the only disease with serious or fatal neurological effects. In this chapter, we examine the important microbial diseases of the nervous system.

26.1 Anatomy of the Nervous System

Learning Objectives

By the end of this section, you will be able to:
- Describe the major anatomical features of the nervous system
- Explain why there is no normal microbiota of the nervous system
- Explain how microorganisms overcome defenses of the nervous system to cause infection
- Identify and describe general symptoms associated with various infections of the nervous system

---

Clinical Focus

Part 1

David is a 35-year-old carpenter from New Jersey. A year ago, he was diagnosed with Crohn's disease, a chronic inflammatory bowel disease that has no known cause. He has been taking a prescription corticosteroid to manage the condition, and the drug has been highly effective in keeping his symptoms at bay. However, David recently fell ill and decided to visit his primary care physician. His symptoms included a fever, a persistent cough, and shortness of breath. His physician ordered a chest X-ray, which revealed consolidation of the right lung. The doctor prescribed a course of levofloxacin and told David to come back in a week if he did not feel better.

- What type of drug is levofloxacin?
- What type of microbes would this drug be effective against?
- What type of infection is consistent with David's symptoms?

Jump to the next Clinical Focus box.

The human nervous system can be divided into two interacting subsystems: the **peripheral nervous system (PNS)** and the **central nervous system (CNS)**. The CNS consists of the brain and spinal cord. The peripheral nervous system is an extensive network of nerves connecting the CNS to the muscles and sensory structures. The relationship of these systems is illustrated in Figure 26.2.

The Central Nervous System

The brain is the most complex and sensitive organ in the body. It is responsible for all functions of the body, including serving as the coordinating center for all sensations, mobility, emotions, and intellect. Protection for the brain is provided by the bones of the skull, which in turn are covered by the scalp, as shown in Figure 26.3. The scalp is composed of an outer layer of skin, which is loosely attached to the aponeurosis, a flat, broad tendon layer that anchors the superficial layers of the skin. The periosteum, below the aponeurosis, firmly encases the bones of the skull and provides protection, nutrition to the bone, and the capacity for bone repair. Below the boney layer of the skull are three layers of membranes called **meninges** that surround the brain. The relative positions of these meninges are shown in Figure 26.3. The meningeal layer closest to the bones of the skull is called the **dura mater** (literally meaning *tough mother*). Below the dura mater lies the **arachnoid mater** (literally *spider-like mother*). The innermost meningeal layer is a delicate membrane called the **pia mater** (literally *tender mother*). Unlike the other meningeal layers, the pia mater firmly adheres to the convoluted surface of the brain. Between the arachnoid mater and pia mater is the subarachnoid space. The subarachnoid space within this region is filled with **cerebrospinal fluid (CSF)**. This watery fluid is produced by cells of the choroid plexus—areas in each ventricle of the brain that consist of cuboidal epithelial cells surrounding dense capillary beds. The CSF serves to deliver nutrients and remove waste from neural tissues.

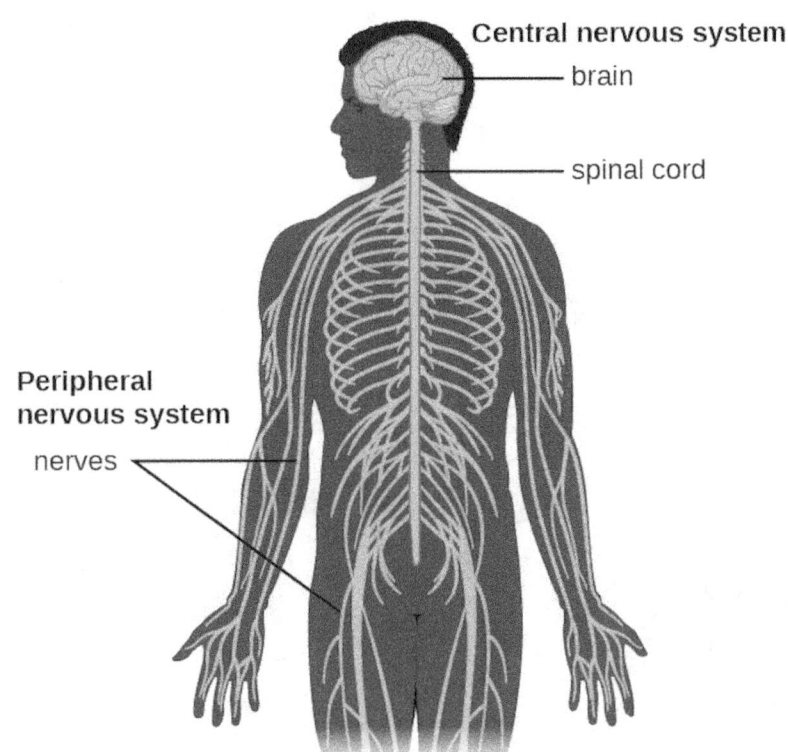

Figure 26.2 The essential components of the human nervous system are shown in this illustration. The central nervous system (CNS) consists of the brain and spinal cord. It connects to the peripheral nervous system (PNS), a network of nerves that extends throughout the body.

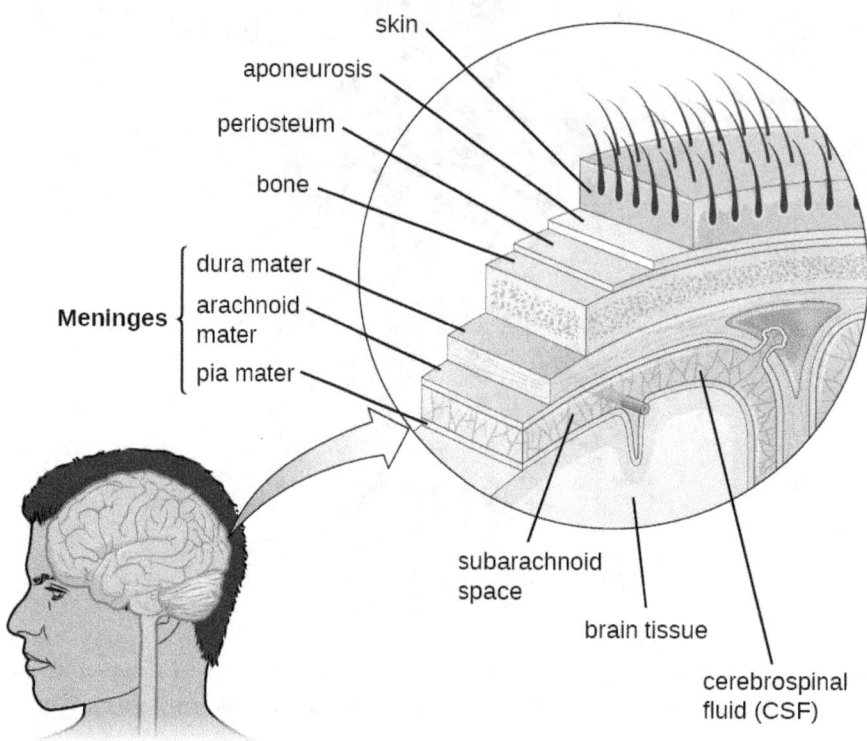

Figure 26.3 The layers of tissue surrounding the human brain include three meningeal membranes: the dura mater, arachnoid mater, and pia mater. (credit: modification of work by National Institutes of Health)

The Blood-Brain Barrier

The tissues of the CNS have extra protection in that they are not exposed to blood or the immune system in the same way as other tissues. The blood vessels that supply the brain with nutrients and other chemical substances lie on top of the pia mater. The capillaries associated with these blood vessels in the brain are less permeable than those in other locations in the body. The capillary endothelial cells form tight junctions that control the transfer of blood components to the brain. In addition, cranial capillaries have far fewer fenestra (pore-like structures that are sealed by a membrane) and pinocytotic vesicles than other capillaries. As a result, materials in the circulatory system have a very limited ability to interact with the CNS directly. This phenomenon is referred to as the blood-brain barrier.

The blood-brain barrier protects the cerebrospinal fluid from contamination, and can be quite effective at excluding potential microbial pathogens. As a consequence of these defenses, there is no normal microbiota in the cerebrospinal fluid. The blood-brain barrier also inhibits the movement of many drugs into the brain, particularly compounds that are not lipid soluble. This has profound ramifications for treatments involving infections of the CNS, because it is difficult for drugs to cross the blood-brain barrier to interact with pathogens that cause infections.

The spinal cord also has protective structures similar to those surrounding the brain. Within the bones of the vertebrae are meninges of dura mater (sometimes called the dural sheath), arachnoid mater, pia mater, and a blood-spinal cord barrier that controls the transfer of blood components from blood vessels associated with the spinal cord.

To cause an infection in the CNS, pathogens must successfully breach the blood-brain barrier or blood-spinal cord barrier. Various pathogens employ different virulence factors and mechanisms to achieve this, but they can generally be grouped into four categories: intercellular (also called paracellular), transcellular, leukocyte facilitated, and nonhematogenous. Intercellular entry involves the use of microbial virulence factors, toxins, or inflammation-mediated processes to pass between the cells of the blood-brain barrier. In transcellular entry, the pathogen passes through the cells of the blood-brain barrier using virulence factors that allow it to adhere to and trigger uptake by vacuole- or receptor-mediated mechanisms. Leukocyte-facilitated entry is a Trojan-horse mechanism that occurs when a pathogen infects peripheral blood leukocytes to directly enter the CNS. Nonhematogenous entry allows pathogens to enter the brain without encountering the blood-brain barrier; it occurs when pathogens travel along either the olfactory or trigeminal cranial nerves that lead directly into the CNS.

LINK TO LEARNING

View this video (https://www.openstax.org/l/22bldbrbarr) about the blood-brain barrier

CHECK YOUR UNDERSTANDING

- What is the primary function of the blood-brain barrier?

The Peripheral Nervous System

The PNS is formed of the nerves that connect organs, limbs, and other anatomic structures of the body to the brain and spinal cord. Unlike the brain and spinal cord, the PNS is not protected by bone, meninges, or a blood barrier, and, as a consequence, the nerves of the PNS are much more susceptible to injury and infection. Microbial damage to peripheral nerves can lead to tingling or numbness known as **neuropathy**. These symptoms can also be produced by trauma and noninfectious causes such as drugs or chronic diseases like diabetes.

The Cells of the Nervous System

Tissues of the PNS and CNS are formed of cells called **glial cells** (neuroglial cells) and **neurons** (nerve cells). Glial cells assist in the organization of neurons, provide a scaffold for some aspects of neuronal function, and aid in recovery from neural injury.

Neurons are specialized cells found throughout the nervous system that transmit signals through the nervous system using electrochemical processes. The basic structure of a neuron is shown in Figure 26.4. The cell body (or **soma**) is the metabolic center of the neuron and contains the nucleus and most of the cell's organelles. The many finely branched extensions from the soma are called **dendrites**. The soma also produces an elongated extension, called the **axon**, which is responsible for the transmission of electrochemical signals through elaborate ion transport processes. Axons of some types of neurons can extend up to one meter in length in the human body. To facilitate electrochemical signal transmission, some neurons have a **myelin sheath** surrounding the axon. Myelin, formed from the cell membranes of glial cells like the Schwann cells in the PNS and oligodendrocytes in the CNS, surrounds and insulates the axon, significantly increasing the speed of electrochemical signal transmission along the axon. The end of an axon forms numerous branches that end in bulbs called synaptic terminals. Neurons form junctions with other cells, such as another neuron, with which they exchange signals. The junctions, which are actually gaps between neurons, are referred to as **synapses**. At each synapse, there is a presynaptic neuron and a postsynaptic neuron (or other cell). The synaptic terminals of the axon of the presynaptic terminal form the synapse with the dendrites, soma, or sometimes the axon of the postsynaptic neuron, or a part of another type of cell such as a muscle cell. The synaptic terminals contain vesicles filled with chemicals called **neurotransmitters**. When the electrochemical signal moving down the axon reaches the synapse, the vesicles fuse with the membrane, and neurotransmitters are released, which diffuse across the synapse and bind to receptors on the membrane of the postsynaptic cell, potentially initiating a response in that cell. That response in the postsynaptic cell might include further propagation of an electrochemical signal to transmit information or contraction of a muscle fiber.

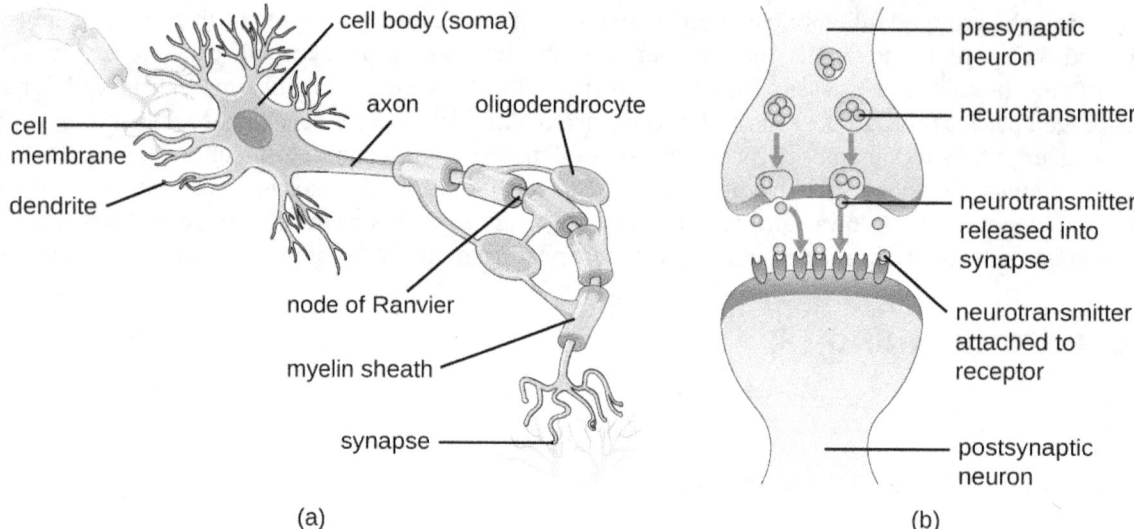

Figure 26.4 (a) A myelinated neuron is associated with oligodendrocytes. Oligodendrocytes are a type of glial cell that forms the myelin sheath in the CNS that insulates the axon so that electrochemical nerve impulses are transferred more efficiently. (b) A synapse consists of the axonal end of the presynaptic neuron (top) that releases neurotransmitters that cross the synaptic space (or cleft) and bind to receptors on dendrites of the postsynaptic neuron (bottom).

CHECK YOUR UNDERSTANDING

- What cells are associated with neurons, and what is their function?
- What is the structure and function of a synapse?

Meningitis and Encephalitis

Although the skull provides the brain with an excellent defense, it can also become problematic during infections. Any swelling of the brain or meninges that results from inflammation can cause intracranial pressure, leading to severe damage of the brain tissues, which have limited space to expand within the inflexible bones of the skull. The term **meningitis** is used to describe an inflammation of the meninges. Typical symptoms can include severe headache, fever, photophobia (increased sensitivity to light), stiff neck, convulsions, and confusion. An inflammation of brain tissue is called **encephalitis**, and patients exhibit signs and symptoms similar to those of meningitis in addition to lethargy, seizures, and personality changes. When inflammation affects both the meninges and the brain tissue, the condition is called **meningoencephalitis**. All three forms of inflammation are serious and can lead to blindness, deafness, coma, and death.

Meningitis and encephalitis can be caused by many different types of microbial pathogens. However, these conditions can also arise from noninfectious causes such as head trauma, some cancers, and certain drugs that trigger inflammation. To determine whether the inflammation is caused by a pathogen, a lumbar puncture is performed to obtain a sample of CSF. If the CSF contains increased levels of white blood cells and abnormal glucose and protein levels, this indicates that the inflammation is a response to an infectioninflinin.

CHECK YOUR UNDERSTANDING

- What are the two types of inflammation that can impact the CNS?
- Why do both forms of inflammation have such serious consequences?

MICRO CONNECTIONS

Guillain-Barré Syndrome

Guillain-Barré syndrome (GBS) is a rare condition that can be preceded by a viral or bacterial infection that results in an autoimmune reaction against myelinated nerve cells. The destruction of the myelin sheath around these neurons results in a loss of sensation and function. The first symptoms of this condition are tingling and weakness in the affected tissues. The symptoms intensify over a period of several weeks and can culminate in complete paralysis. Severe cases can be life-threatening. Infections by several different microbial pathogens, including *Campylobacter jejuni* (the most common risk factor), cytomegalovirus, Epstein-Barr virus, varicella-zoster virus, *Mycoplasma pneumoniae*,[1] and Zika virus[2] have been identified as triggers for GBS. Anti-myelin antibodies from patients with GBS have been demonstrated to also recognize *C. jejuni*. It is possible that cross-reactive antibodies, antibodies that react with similar antigenic sites on different proteins, might be formed during an infection and may lead to this autoimmune response.

GBS is solely identified by the appearance of clinical symptoms. There are no other diagnostic tests available. Fortunately, most cases spontaneously resolve within a few months with few permanent effects, as there is no available vaccine. GBS can be treated by plasmapheresis. In this procedure, the patient's plasma is filtered from their blood, removing autoantibodies.

26.2 Bacterial Diseases of the Nervous System

Learning Objectives

By the end of this section, you will be able to:
- Identify the most common bacteria that can cause infections of the nervous system
- Compare the major characteristics of specific bacterial diseases affecting the nervous system

Bacterial infections that affect the nervous system are serious and can be life-threatening. Fortunately, there are only a few bacterial species commonly associated with neurological infections.

Bacterial Meningitis

Bacterial meningitis is one of the most serious forms of meningitis. Bacteria that cause meningitis often gain access to the CNS through the bloodstream after trauma or as a result of the action of bacterial toxins. Bacteria may also spread from structures in the upper respiratory tract, such as the oropharynx, nasopharynx, sinuses, and middle ear. Patients with head wounds or cochlear implants (an electronic device placed in the inner ear) are also at risk for developing meningitis.

Many of the bacteria that can cause meningitis are commonly found in healthy people. The most common causes of non-neonatal bacterial meningitis are *Neisseria meningitidis*, *Streptococcus pneumoniae*, and *Haemophilus influenzae*. All three of these bacterial pathogens are spread from person to person by respiratory secretions. Each can colonize and cross through the mucous membranes of the oropharynx and nasopharynx, and enter the blood. Once in the blood, these pathogens can disseminate throughout the body and are capable of both establishing an infection and triggering inflammation in any body site, including the meninges (Figure 26.5). Without appropriate systemic antibacterial therapy, the case-fatality rate can be as high as 70%, and 20% of those survivors may be left with irreversible nerve damage or tissue destruction, resulting in hearing loss, neurologic disability, or loss of a limb. Mortality rates are much lower (as low as 15%) in populations where appropriate therapeutic drugs and preventive vaccines are available.[3]

---

1 Yuki, Nobuhiro and Hans-Peter Hartung, "Guillain–Barré Syndrome," *New England Journal of Medicine* 366, no. 24 (2012): 2294-304.

2 Cao-Lormeau, Van-Mai, Alexandre Blake, Sandrine Mons, Stéphane Lastère, Claudine Roche, Jessica Vanhomwegen, Timothée Dub et al., "Guillain-Barré Syndrome Outbreak Associated with Zika Virus Infection in French Polynesia: A Case-Control Study," *The Lancet* 387, no. 10027 (2016): 1531-9.

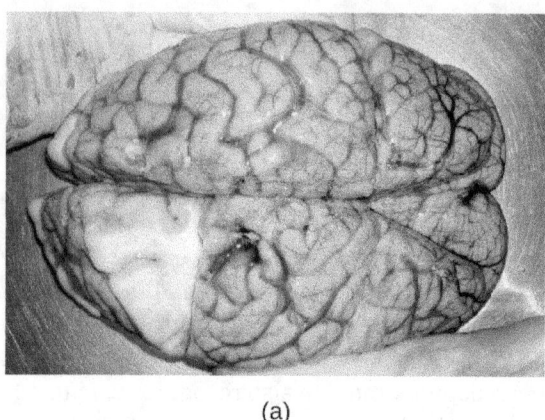
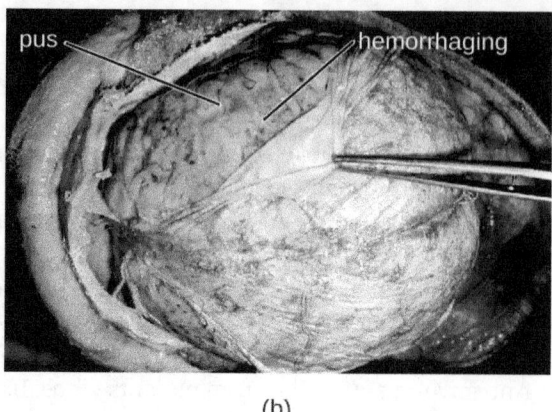

Figure 26.5 (a) A normal human brain removed during an autopsy. (b) The brain of a patient who died from bacterial meningitis. Note the pus under the dura mater (being retracted by the forceps) and the red hemorrhagic foci on the meninges. (credit b: modification of work by the Centers for Disease Control and Prevention)

A variety of other bacteria, including *Listeria monocytogenes* and *Escherichia coli*, are also capable of causing meningitis. These bacteria cause infections of the arachnoid mater and CSF after spreading through the circulation in blood or by spreading from an infection of the sinuses or nasopharynx. *Streptococcus agalactiae*, commonly found in the microbiota of the vagina and gastrointestinal tract, can also cause bacterial meningitis in newborns after transmission from the mother either before or during birth.

The profound inflammation caused by these microbes can result in early symptoms that include severe headache, fever, confusion, nausea, vomiting, photophobia, and stiff neck. Systemic inflammatory responses associated with some types of bacterial meningitis can lead to hemorrhaging and purpuric lesions on skin, followed by even more severe conditions that include shock, convulsions, coma, and death—in some cases, in the span of just a few hours.

Diagnosis of bacterial meningitis is best confirmed by analysis of CSF obtained by a lumbar puncture. Abnormal levels of polymorphonuclear neutrophils (PMNs) (> 10 PMNs/mm^3), glucose (< 45 mg/dL), and protein (> 45 mg/dL) in the CSF are suggestive of bacterial meningitis.[4] Characteristics of specific forms of bacterial meningitis are detailed in the subsections that follow.

Meningococcal Meningitis

Meningococcal meningitis is a serious infection caused by the gram-negative coccus *N. meningitidis*. In some cases, death can occur within a few hours of the onset of symptoms. Nonfatal cases can result in irreversible nerve damage, resulting in hearing loss and brain damage, or amputation of extremities because of tissue necrosis.

Meningococcal meningitis can infect people of any age, but its prevalence is highest among infants, adolescents, and young adults.[5] Meningococcal meningitis was once the most common cause of meningitis epidemics in human populations. This is still the case in a swath of sub-Saharan Africa known as the meningitis belt, but meningococcal meningitis epidemics have become rare in most other regions, thanks to meningococcal vaccines. However, outbreaks can still occur in communities, schools, colleges, prisons, and other populations where people are in close direct contact.

N. meningitidis has a high affinity for mucosal membranes in the oropharynx and nasopharynx. Contact with respiratory secretions containing *N. meningitidis* is an effective mode of transmission. The pathogenicity of *N.*

---

3 Thigpen, Michael C., Cynthia G. Whitney, Nancy E. Messonnier, Elizabeth R. Zell, Ruth Lynfield, James L. Hadler, Lee H. Harrison et al., "Bacterial Meningitis in the United States, 1998–2007," *New England Journal of Medicine* 364, no. 21 (2011): 2016-25.

4 Popovic, T., et al. World Health Organization, "Laboratory Manual for the Diagnosis of Meningitis Caused by *Neisseria meningitidis, Streptococcus pneumoniae,* and *Haemophilus influenza,*" 1999.

5 US Centers for Disease Control and Prevention, "Meningococcal Disease," August 5, 2015. Accessed June 28, 2015. http://www.cdc.gov/meningococcal/surveillance/index.html.

meningitidis is enhanced by virulence factors that contribute to the rapid progression of the disease. These include lipooligosaccharide (LOS) endotoxin, type IV pili for attachment to host tissues, and polysaccharide capsules that help the cells avoid phagocytosis and complement-mediated killing. Additional virulence factors include IgA protease (which breaks down IgA antibodies), the invasion factors Opa, Opc, and porin (which facilitate transcellular entry through the blood-brain barrier), iron-uptake factors (which strip heme units from hemoglobin in host cells and use them for growth), and stress proteins that protect bacteria from reactive oxygen molecules.

A unique sign of meningococcal meningitis is the formation of a petechial rash on the skin or mucous membranes, characterized by tiny, red, flat, hemorrhagic lesions. This rash, which appears soon after disease onset, is a response to LOS endotoxin and adherence virulence factors that disrupt the endothelial cells of capillaries and small veins in the skin. The blood vessel disruption triggers the formation of tiny blood clots, causing blood to leak into the surrounding tissue. As the infection progresses, the levels of virulence factors increase, and the hemorrhagic lesions can increase in size as blood continues to leak into tissues. Lesions larger than 1.0 cm usually occur in patients developing shock, as virulence factors cause increased hemorrhage and clot formation. Sepsis, as a result of systemic damage from meningococcal virulence factors, can lead to rapid multiple organ failure, shock, disseminated intravascular coagulation, and death.

Because meningococcoal meningitis progresses so rapidly, a greater variety of clinical specimens are required for the timely detection of *N. meningitidis*. Required specimens can include blood, CSF, naso- and oropharyngeal swabs, urethral and endocervical swabs, petechial aspirates, and biopsies. Safety protocols for handling and transport of specimens suspected of containing *N. meningitidis* should always be followed, since cases of fatal meningococcal disease have occurred in healthcare workers exposed to droplets or aerosols from patient specimens. Prompt presumptive diagnosis of meningococcal meningitis can occur when CSF is directly evaluated by Gram stain, revealing extra- and intracellular gram-negative diplococci with a distinctive coffee-bean microscopic morphology associated with PMNs (Figure 26.6). Identification can also be made directly from CSF using latex agglutination and immunochromatographic rapid diagnostic tests specific for *N. meningitidis*. Species identification can also be performed using DNA sequence-based typing schemes for hypervariable outer membrane proteins of *N. meningitidis*, which has replaced sero(sub)typing.

Meningococcal infections can be treated with antibiotic therapy, and third-generation cephalosporins are most often employed. However, because outcomes can be negative even with treatment, preventive vaccination is the best form of treatment. In 2010, countries in Africa's meningitis belt began using a new serogroup A meningococcal conjugate vaccine. This program has dramatically reduced the number of cases of meningococcal meningitis by conferring individual and herd immunity.

Twelve different capsular serotypes of *N. meningitidis* are known to exist. Serotypes A, B, C, W, X, and Y are the most prevalent worldwide. The CDC recommends that children between 11–12 years of age be vaccinated with a single dose of a quadrivalent vaccine that protects against serotypes A, C, W, and Y, with a booster at age 16.[6] An additional booster or injections of serogroup B meningococcal vaccine may be given to individuals in high-risk settings (such as epidemic outbreaks on college campuses).

---

6 US Centers for Disease Control and Prevention, "Recommended Immunization Schedule for Persons Aged 0 Through 18 Years, United States, 2016," February 1, 2016. Accessed on June 28, 2016. http://www.cdc.gov/vaccines/schedules/hcp/imz/child-adolescent.html.

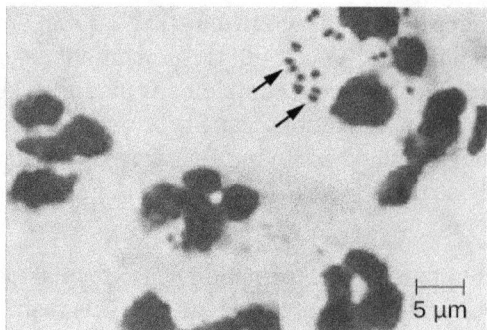

Figure 26.6 *N. meningitidis* (arrows) associated with neutrophils (the larger stained cells) in a gram-stained CSF sample. (credit: modification of work by the Centers for Disease Control and Prevention)

 MICRO CONNECTIONS

Meningitis on Campus

College students living in dorms or communal housing are at increased risk for contracting epidemic meningitis. From 2011 to 2015, there have been at least nine meningococcal outbreaks on college campuses in the United States. These incidents involved a total of 43 students (of whom four died).[7] In spite of rapid diagnosis and aggressive antimicrobial treatment, several of the survivors suffered from amputations or serious neurological problems.

Prophylactic vaccination of first-year college students living in dorms is recommended by the CDC, and insurance companies now cover meningococcal vaccination for students in college dorms. Some colleges have mandated vaccination with meningococcal conjugate vaccine for certain students entering college (Figure 26.7).

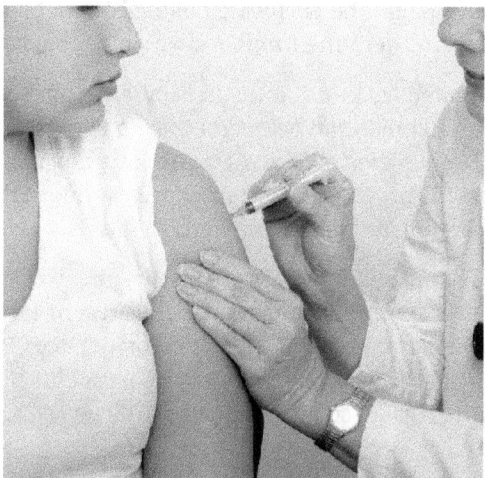

Figure 26.7 To prevent campus outbreaks, some colleges now require students to be vaccinated against meningogoccal meningitis. (credit: modification of work by James Gathany, Centers for Disease Control and Prevention)

Pneumococcal Meningitis

Pneumococcal meningitis is caused by the encapsulated gram-positive bacterium *S. pneumoniae* (pneumococcus, also called strep pneumo). This organism is commonly found in the microbiota of the pharynx of 30–70% of young children, depending on the sampling method, while *S. pneumoniae* can be found in fewer than 5% of healthy adults. Although it is often present without disease symptoms, this microbe can cross the blood-brain barrier in susceptible individuals. In some cases, it may also result in septicemia. Since

---

7 National Meningitis Association, "Serogroup B Meningococcal Disease Outbreaks on U.S. College Campuses," 2016. Accessed June 28, 2016. http://www.nmaus.org/disease-prevention-information/serogroup-b-meningococcal-disease/outbreaks/.

the introduction of the Hib vaccine, *S. pneumoniae* has become the leading cause of meningitis in humans aged 2 months through adulthood.

S. pneumoniae can be identified in CSF samples using gram-stained specimens, latex agglutination, and immunochromatographic RDT specific for *S. pneumoniae*. In gram-stained samples, *S. pneumoniae* appears as gram-positive, lancet-shaped diplococci (Figure 26.8). Identification of *S. pneumoniae* can also be achieved using cultures of CSF and blood, and at least 93 distinct serotypes can be identified based on the quellung reaction to unique capsular polysaccharides. PCR and RT-PCR assays are also available to confirm identification.

Major virulence factors produced by *S. pneumoniae* include PI-1 pilin for adherence to host cells (pneumococcal adherence) and virulence factor B (PavB) for attachment to cells of the respiratory tract; choline-binding proteins (cbpA) that bind to epithelial cells and interfere with immune factors IgA and C3; and the cytoplasmic bacterial toxin pneumolysin that triggers an inflammatory response.

With the emergence of drug-resistant strains of *S. pneumoniae*, pneumococcal meningitis is typically treated with broad-spectrum antibiotics, such as levofloxacin, cefotaxime, penicillin, or other β-lactam antibiotics. The two available pneumococcal vaccines are described in Bacterial Infections of the Respiratory Tract.

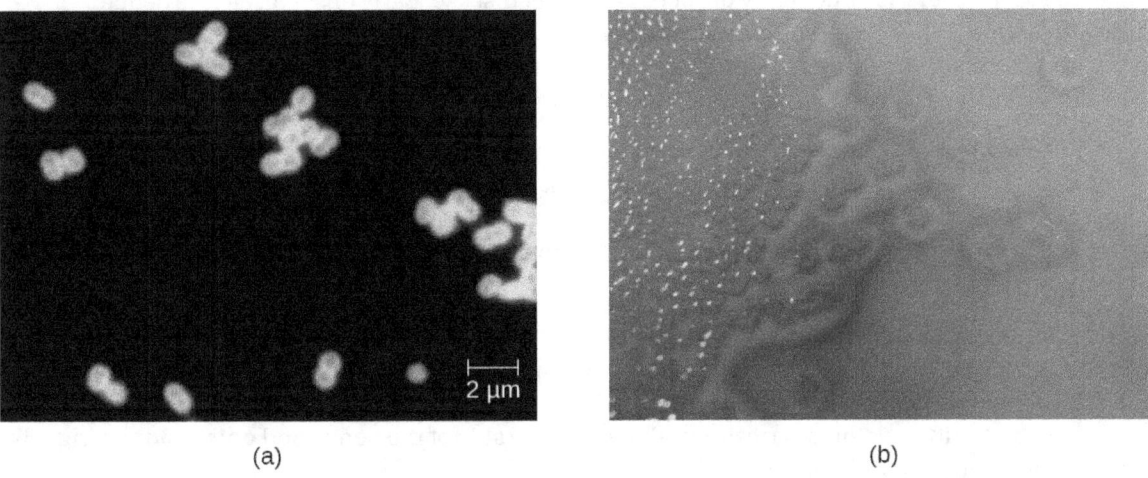

Figure 26.8 (a) Digitally colorized fluorescent antibody stained micrograph of *Streptococcus pneumoniae* in CSF. (b) *S. pneumoniae* growing on blood agar. (credit a: modification of work by the Centers for Disease Control and Prevention; credit b: modification of work by Nathan Reading)

Haemophilus influenzae Type b

Meningitis due to *H. influenzae* serotype b (Hib), an encapsulated pleomorphic gram-negative coccobacilli, is now uncommon in most countries, because of the use of the effective Hib vaccine. Without the use of the Hib vaccine, *H. influenzae* can be the primary cause of meningitis in children 2 months thru 5 years of age. *H. influenzae* can be found in the throats of healthy individuals, including infants and young children. By five years of age, most children have developed immunity to this microbe. Infants older than 2 months of age, however, do not produce a sufficient protective antibody response and are susceptible to serious disease. The intracranial pressure caused by this infection leads to a 5% mortality rate and 20% incidence of deafness or brain damage in survivors.[8]

H. influenzae produces at least 16 different virulence factors, including LOS, which triggers inflammation, and *Haemophilus* adhesion and penetration factor (Hap), which aids in attachment and invasion into respiratory epithelial cells. The bacterium also has a polysaccharide capsule that helps it avoid phagocytosis, as well as factors such as IgA1 protease and P2 protein that allow it to evade antibodies secreted from mucous membranes. In addition, factors such as hemoglobin-binding protein (Hgp) and transferrin-binding protein (Tbp) acquire iron from hemoglobin and transferrin, respectively, for bacterial growth.

---

8 United States Department of Health and Human Services, "Hib (Haemophilus Influenzae Type B)," Accessed June 28, 2016. http://www.vaccines.gov/diseases/hib/#.

Preliminary diagnosis of *H. influenzae* infections can be made by direct PCR and a smear of CSF. Stained smears will reveal intracellular and extracellular PMNs with small, pleomorphic, gram-negative coccobacilli or filamentous forms that are characteristic of *H. influenzae*. Initial confirmation of this genus can be based on its fastidious growth on chocolate agar. Identification is confirmed with requirements for exogenous biochemical growth cofactors NAD and heme (by MALDI-TOF), latex agglutination, and RT-PCR.

Meningitis caused by *H. influenzae* is usually treated with doxycycline, fluoroquinolones, second- and third-generation cephalosporins, and carbapenems. The best means of preventing *H. influenza* infection is with the use of the Hib polysaccharide conjugate vaccine. It is recommended that all children receive this vaccine at 2, 4, and 6 months of age, with a final booster dose at 12 to 15 months of age.[9]

Neonatal Meningitis

S. agalactiae, Group B streptococcus (GBS), is an encapsulated gram-positive bacterium that is the most common cause of **neonatal meningitis**, a term that refers to meningitis occurring in babies up to 3 months of age.[10] *S. agalactiae* can also cause meningitis in people of all ages and can be found in the urogenital and gastrointestinal microbiota of about 10–30% of humans.

Neonatal infection occurs as either early onset or late-onset disease. Early onset disease is defined as occurring in infants up to 7 days old. The infant initially becomes infected by *S. agalactiae* during childbirth, when the bacteria may be transferred from the mother's vagina. Incidence of early onset neonatal meningitis can be greatly reduced by giving intravenous antibiotics to the mother during labor.

Late-onset neonatal meningitis occurs in infants between 1 week and 3 months of age. Infants born to mothers with *S. agalactiae* in the urogenital tract have a higher risk of late-onset menigitis, but late-onset infections can be transmitted from sources other than the mother; often, the source of infection is unknown. Infants who are born prematurely (before 37 weeks of pregnancy) or to mothers who develop a fever also have a greater risk of contracting late-onset neonatal meningitis.

Signs and symptoms of early onset disease include temperature instability, apnea (cessation of breathing), bradycardia (slow heart rate), hypotension, difficulty feeding, irritability, and limpness. When asleep, the baby may be difficult to wake up. Symptoms of late-onset disease are more likely to include seizures, bulging fontanel (soft spot), stiff neck, hemiparesis (weakness on one side of the body), and **opisthotonos** (rigid body with arched back and head thrown backward).

S. agalactiae produces at least 12 virulence factors that include FbsA that attaches to host cell surface proteins, PI-1 pili that promotes the invasion of human endothelial cells, a polysaccharide capsule that prevents the activation of the alternative complement pathway and inhibits phagocytosis, and the toxin CAMP factor, which forms pores in host cell membranes and binds to IgG and IgM antibodies.

Diagnosis of neonatal meningitis is often, but not uniformly, confirmed by positive results from cultures of CSF or blood. Tests include routine culture, antigen detection by enzyme immunoassay, serotyping of different capsule types, PCR, and RT-PCR. It is typically treated with β-lactam antibiotics such as intravenous penicillin or ampicillin plus gentamicin. Even with treatment, roughly 10% mortality is seen in infected neonates.[11]

✓ CHECK YOUR UNDERSTANDING

- Which groups are most vulnerable to each of the bacterial meningitis diseases?

---

9 US Centers for Disease Control and Prevention, "Meningococcal Disease, Disease Trends," 2015. Accessed September 13, 2016. http://www.cdc.gov/meningococcal/surveillance/index.html.

10 Thigpen, Michael C., Cynthia G. Whitney, Nancy E. Messonnier, Elizabeth R. Zell, Ruth Lynfield, James L. Hadler, Lee H. Harrison et al., "Bacterial Meningitis in the United States, 1998–2007," *New England Journal of Medicine* 364, no. 21 (2011): 2016-25.

11 Thigpen, Michael C., Cynthia G. Whitney, Nancy E. Messonnier, Elizabeth R. Zell, Ruth Lynfield, James L. Hadler, Lee H. Harrison et al., "Bacterial Meningitis in the United States, 1998–2007," *New England Journal of Medicine* 364, no. 21 (2011): 2016-25; Heath, Paul T., Gail Balfour, Abbie M. Weisner, Androulla Efstratiou, Theresa L. Lamagni, Helen Tighe, Liam AF O'Connell et al., "Group B Streptococcal Disease in UK and Irish Infants Younger than 90 Days," *The Lancet* 363, no. 9405 (2004): 292-4.

- For which of the bacterial meningitis diseases are there vaccines presently available?
- Which organism can cause epidemic meningitis?

Clostridium-Associated Diseases

Species in the genus *Clostridium* are gram-positive, endospore-forming rods that are obligate anaerobes. Endospores of *Clostridium* spp. are widespread in nature, commonly found in soil, water, feces, sewage, and marine sediments. *Clostridium* spp. produce more types of protein exotoxins than any other bacterial genus, including two exotoxins with protease activity that are the most potent known biological toxins: botulinum neurotoxin (BoNT) and tetanus neurotoxin (TeNT). These two toxins have lethal doses of 0.2–10 ng per kg body weight.

BoNT can be produced by unique strains of *C. butyricum*, and *C. baratii*; however, it is primarily associated with *C. botulinum* and the condition of botulism. TeNT, which causes tetanus, is only produced by *C. tetani*. These powerful neural exotoxins are the primary virulence factors for these pathogens. The mode of action for these toxins was described in Virulence Factors of Bacterial and Viral Pathogens and illustrated in Figure 15.16.

Diagnosis of tetanus or botulism typically involves bioassays that detect the presence of BoNT and TeNT in fecal specimens, blood (serum), or suspect foods. In addition, both *C. botulinum* and *C. tetani* can be isolated and cultured using commercially available media for anaerobes. ELISA and RT-PCR tests are also available.

Tetanus

Tetanus is a noncommunicable disease characterized by uncontrollable muscle spasms (contractions) caused by the action of TeNT. It generally occurs when *C. tetani* infects a wound and produces TeNT, which rapidly binds to neural tissue, resulting in an intoxication (poisoning) of neurons. Depending on the site and extent of infection, cases of tetanus can be described as localized, cephalic, or generalized. Generalized tetanus that occurs in a newborn is called neonatal tetanus.

Localized tetanus occurs when TeNT only affects the muscle groups close to the injury site. There is no CNS involvement, and the symptoms are usually mild, with localized muscle spasms caused by a dysfunction in the surrounding neurons. Individuals with partial immunity—especially previously vaccinated individuals who neglect to get the recommended booster shots—are most likely to develop localized tetanus as a result of *C. tetani* infecting a puncture wound.

Cephalic tetanus is a rare, localized form of tetanus generally associated with wounds on the head or face. In rare cases, it has occurred in cases of otitis media (middle ear infection). Cephalic tetanus often results in patients seeing double images, because of the spasms affecting the muscles that control eye movement.

Both localized and cephalic tetanus may progress to generalized tetanus—a much more serious condition—if TeNT is able to spread further into body tissues. In generalized tetanus, TeNT enters neurons of the PNS. From there, TeNT travels from the site of the wound, usually on an extremity of the body, retrograde (back up) to inhibitory neurons in the CNS. There, it prevents the release of gamma aminobutyric acid (GABA), the neurotransmitter responsible for muscle relaxation. The resulting muscle spasms often first occur in the jaw muscles, leading to the characteristic symptom of lockjaw (inability to open the mouth). As the toxin progressively continues to block neurotransmitter release, other muscles become involved, resulting in uncontrollable, sudden muscle spasms that are powerful enough to cause tendons to rupture and bones to fracture. Spasms in the muscles in the neck, back, and legs may cause the body to form a rigid, stiff arch, a posture called opisthotonos (Figure 26.9). Spasms in the larynx, diaphragm, and muscles of the chest restrict the patient's ability to swallow and breathe, eventually leading to death by asphyxiation (insufficient supply of oxygen).

Neonatal tetanus typically occurs when the stump of the umbilical cord is contaminated with spores of *C. tetani* after delivery. Although this condition is rare in the United States, neonatal tetanus is a major cause of infant mortality in countries that lack maternal immunization for tetanus and where birth often occurs in unsanitary conditions. At the end of the first week of life, infected infants become irritable, feed poorly, and develop rigidity with spasms. Neonatal tetanus has a very poor prognosis with a mortality rate of 70%–100%.[12]

Treatment for patients with tetanus includes assisted breathing through the use of a ventilator, wound debridement, fluid balance, and antibiotic therapy with metronidazole or penicillin to halt the growth of *C. tetani*. In addition, patients are treated with TeNT antitoxin, preferably in the form of human immunoglobulin to neutralize nonfixed toxin and benzodiazepines to enhance the effect of GABA for muscle relaxation and anxiety.

A tetanus toxoid (TT) vaccine is available for protection and prevention of tetanus. It is the T component of vaccines such as DTaP, Tdap, and Td. The CDC recommends children receive doses of the DTaP vaccine at 2, 4, 6, and 15–18 months of age and another at 4–6 years of age. One dose of Td is recommended for adolescents and adults as a TT booster every 10 years.[13]

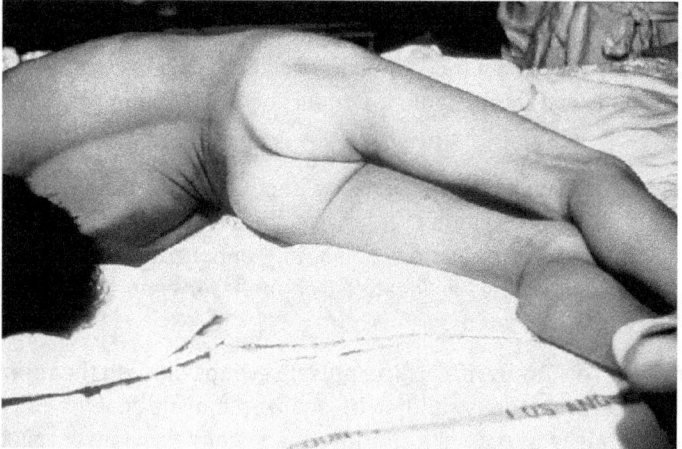

Figure 26.9 A tetanus patient exhibiting the rigid body posture known as opisthotonos. (credit: Centers for Disease Control and Prevention)

Botulism

Botulism is a rare but frequently fatal illness caused by intoxication by BoNT. It can occur either as the result of an infection by *C. botulinum*, in which case the bacteria produce BoNT *in vivo*, or as the result of a direct introduction of BoNT into tissues.

Infection and production of BoNT *in vivo* can result in wound botulism, infant botulism, and adult intestinal toxemia. Wound botulism typically occurs when *C. botulinum* is introduced directly into a wound after a traumatic injury, deep puncture wound, or injection site. Infant botulism, which occurs in infants younger than 1 year of age, and adult intestinal toxemia, which occurs in immunocompromised adults, results from ingesting *C. botulinum* endospores in food. The endospores germinate in the body, resulting in the production of BoNT in the intestinal tract.

Intoxications occur when BoNT is produced outside the body and then introduced directly into the body through food (foodborne botulism), air (inhalation botulism), or a clinical procedure (iatrogenic botulism). Foodborne botulism, the most common of these forms, occurs when BoNT is produced in contaminated food and then ingested along with the food (recall Case in Point: A Streak of Bad Potluck). Inhalation botulism is rare

---

12 UNFPA, UNICEF WHO, "Maternal and Neonatal Tetanus Elimination by 2005," 2000. http://www.unicef.org/immunization/files/MNTE_strategy_paper.pdf.

13 US Centers for Disease Control and Prevention, "Tetanus Vaccination," 2013. Accessed June 29, 2016. http://www.cdc.gov/tetanus/vaccination.html.

because BoNT is unstable as an aerosol and does not occur in nature; however, it can be produced in the laboratory and was used (unsuccessfully) as a bioweapon by terrorists in Japan in the 1990s. A few cases of accidental inhalation botulism have also occurred. Iatrogenic botulism is also rare; it is associated with injections of BoNT used for cosmetic purposes (see Micro Connections: Medicinal Uses of Botulinum Toxin).

When BoNT enters the bloodstream in the gastrointestinal tract, wound, or lungs, it is transferred to the neuromuscular junctions of motor neurons where it binds irreversibly to presynaptic membranes and prevents the release of acetylcholine from the presynaptic terminal of motor neurons into the neuromuscular junction. The consequence of preventing acetylcholine release is the loss of muscle activity, leading to muscle relaxation and eventually paralysis.

If BoNT is absorbed through the gastrointestinal tract, early symptoms of botulism include blurred vision, drooping eyelids, difficulty swallowing, abdominal cramps, nausea, vomiting, constipation, or possibly diarrhea. This is followed by progressive flaccid paralysis, a gradual weakening and loss of control over the muscles. A patient's experience can be particularly terrifying, because hearing remains normal, consciousness is not lost, and he or she is fully aware of the progression of his or her condition. In infants, notable signs of botulism include weak cry, decreased ability to suckle, and hypotonia (limpness of head or body). Eventually, botulism ends in death from respiratory failure caused by the progressive paralysis of the muscles of the upper airway, diaphragm, and chest.

Botulism is treated with an antitoxin specific for BoNT. If administered in time, the antitoxin stops the progression of paralysis but does not reverse it. Once the antitoxin has been administered, the patient will slowly regain neurological function, but this may take several weeks or months, depending on the severity of the case. During recovery, patients generally must remain hospitalized and receive breathing assistance through a ventilator.

✓ CHECK YOUR UNDERSTANDING

- How frequently should the tetanus vaccination be updated in adults?
- What are the most common causes of botulism?
- Why is botulism not treated with an antibiotic?

 MICRO CONNECTIONS

Medicinal Uses of Botulinum Toxin

Although it is the most toxic biological material known to man, botulinum toxin is often intentionally injected into people to treat other conditions. Type A botulinum toxin is used cosmetically to reduce wrinkles. The injection of minute quantities of this toxin into the face causes the relaxation of facial muscles, thereby giving the skin a smoother appearance. Eyelid twitching and crossed eyes can also be treated with botulinum toxin injections. Other uses of this toxin include the treatment of hyperhidrosis (excessive sweating). In fact, botulinum toxin can be used to moderate the effects of several other apparently nonmicrobial diseases involving inappropriate nerve function. Such diseases include cerebral palsy, multiple sclerosis, and Parkinson's disease. Each of these diseases is characterized by a loss of control over muscle contractions; treatment with botulinum toxin serves to relax contracted muscles.

Listeriosis

Listeria monocytogenes is a nonencapsulated, nonsporulating, gram-positive rod and a foodborne pathogen that causes **listeriosis**. At-risk groups include pregnant women, neonates, the elderly, and the immunocompromised (recall the Clinical Focus case studies in Microbial Growth and Microbial Mechanisms of Pathogenicity). Listeriosis leads to meningitis in about 20% of cases, particularly neonates and patients over the age of 60. The CDC identifies listeriosis as the third leading cause of death due to foodborne illness, with overall mortality rates reaching 16%.[14] In pregnant women, listeriosis can cause also cause spontaneous abortion in pregnant women because of the pathogen's unique ability to cross the placenta.

L. monocytogenes is generally introduced into food items by contamination with soil or animal manure used as fertilizer. Foods commonly associated with listeriosis include fresh fruits and vegetables, frozen vegetables, processed meats, soft cheeses, and raw milk.[15] Unlike most other foodborne pathogens, *Listeria* is able to grow at temperatures between 0 °C and 50 °C, and can therefore continue to grow, even in refrigerated foods.

Ingestion of contaminated food leads initially to infection of the gastrointestinal tract. However, *L. monocytogenes* produces several unique virulence factors that allow it to cross the intestinal barrier and spread to other body systems. Surface proteins called internalins (InlA and InlB) help *L. monocytogenes* invade nonphagocytic cells and tissues, penetrating the intestinal wall and becoming disseminating through the circulatory and lymphatic systems. Internalins also enable *L. monocytogenes* to breach other important barriers, including the blood-brain barrier and the placenta. Within tissues, *L. monocytogenes* uses other proteins called listeriolysin O and ActA to facilitate intercellular movement, allowing the infection to spread from cell to cell (Figure 26.10).

L. monocytogenes is usually identified by cultivation of samples from a normally sterile site (e.g., blood or CSF). Recovery of viable organisms can be enhanced using cold enrichment by incubating samples in a broth at 4 °C for a week or more. Distinguishing types and subtypes of *L. monocytogenes*—an important step for diagnosis and epidemiology—is typically done using pulsed-field gel electrophoresis. Identification can also be achieved using chemiluminescence DNA probe assays and MALDI-TOF.

Treatment for listeriosis involves antibiotic therapy, most commonly with ampicillin and gentamicin. There is no vaccine available.

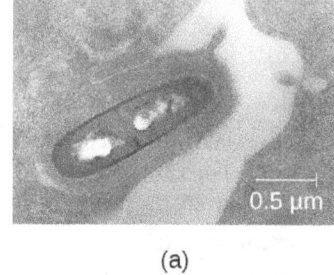
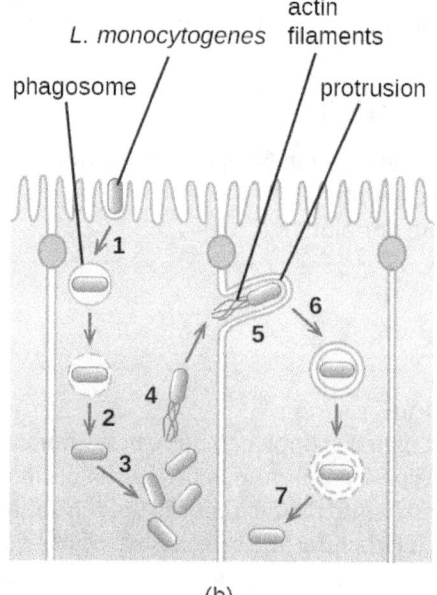

(a) (b)

Figure 26.10 (a) An electron micrograph of *Listeria monocytogenes* infecting a host cell. (b) *Listeria* is able to use host cell components to cause infection. For example, phagocytosis allows it to enter host cells, and the host's cytoskeleton provides the materials to help the pathogen move to other cells. (credit a: modification of work by the Centers for Disease Control and Prevention; credit b: modification of work by Keith Ireton)

✓ CHECK YOUR UNDERSTANDING

- How does *Listeria* enter the nervous system?

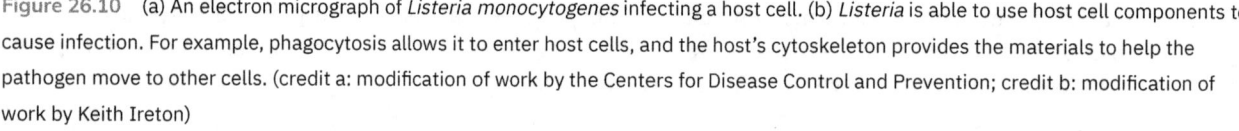

14 Scallan, Elaine, Robert M. Hoekstra, Frederick J. Angulo, Robert V. Tauxe, Marc-Alain Widdowson, Sharon L. Roy, Jeffery L. Jones, and Patricia M. Griffin, "Foodborne Illness Acquired in the United States—Major Pathogens," *Emerging Infectious Diseases* 17, no. 1 (2011): 7-15.

15 US Centers for Disease Control and Prevention, "*Listeria* Outbreaks," 2016. Accessed June 29, 2016. https://www.cdc.gov/listeria/outbreaks/index.html.

Hansen's Disease (Leprosy)

Hansen's disease (also known as **leprosy**) is caused by a long, thin, filamentous rod-shaped bacterium *Mycobacterium leprae*, an obligate intracellular pathogen. *M. leprae* is classified as gram-positive bacteria, but it is best visualized microscopically with an acid-fast stain and is generally referred to as an acid-fast bacterium. Hansen's disease affects the PNS, leading to permanent damage and loss of appendages or other body parts.

Hansen's disease is communicable but not highly contagious; approximately 95% of the human population cannot be easily infected because they have a natural immunity to *M. leprae*. Person-to-person transmission occurs by inhalation into nasal mucosa or prolonged and repeated contact with infected skin. Armadillos, one of only five mammals susceptible to Hansen's disease, have also been implicated in transmission of some cases.[16]

In the human body, *M. leprae* grows best at the cooler temperatures found in peripheral tissues like the nose, toes, fingers, and ears. Some of the virulence factors that contribute to *M. leprae*'s pathogenicity are located on the capsule and cell wall of the bacterium. These virulence factors enable it to bind to and invade Schwann cells, resulting in progressive demyelination that gradually destroys neurons of the PNS. The loss of neuronal function leads to hypoesthesia (numbness) in infected lesions. *M. leprae* is readily phagocytized by macrophages but is able to survive within macrophages in part by neutralizing reactive oxygen species produced in the oxidative burst of the phagolysosome. Like *L. monocytogenes*, *M. leprae* also can move directly between macrophages to avoid clearance by immune factors.

The extent of the disease is related to the immune response of the patient. Initial symptoms may not appear for as long as 2 to 5 years after infection. These often begin with small, blanched, numb areas of the skin. In most individuals, these will resolve spontaneously, but some cases may progress to a more serious form of the disease. Tuberculoid (paucibacillary) Hansen's disease is marked by the presence of relatively few (three or less) flat, blanched skin lesions with small nodules at the edges and few bacteria present in the lesion. Although these lesions can persist for years or decades, the bacteria are held in check by an effective immune response including cell-mediated cytotoxicity. Individuals who are unable to contain the infection may later develop lepromatous (multibacillary) Hansen's disease. This is a progressive form of the disease characterized by nodules filled with acid-fast bacilli and macrophages. Impaired function of infected Schwann cells leads to peripheral nerve damage, resulting in sensory loss that leads to ulcers, deformities, and fractures. Damage to the ulnar nerve (in the wrist) by *M. leprae* is one of the most common causes of crippling of the hand. In some cases, chronic tissue damage can ultimately lead to loss of fingers or toes. When mucosal tissues are also involved, disfiguring lesions of the nose and face can also occur (Figure 26.11).

Hansen's disease is diagnosed on the basis of clinical signs and symptoms of the disease, and confirmed by the presence of acid-fast bacilli on skin smears or in skin biopsy specimens (Figure 26.11). *M. leprae* does not grow *in vitro* on any known laboratory media, but it can be identified by culturing *in vivo* in the footpads of laboratory mice or armadillos. Where needed, PCR and genotyping of *M. leprae* DNA in infected human tissue may be performed for diagnosis and epidemiology.

Hansen's disease responds well to treatment and, if diagnosed and treated early, does not cause disability. In the United States, most patients with Hansen's disease are treated in ambulatory care clinics in major cities by the National Hansen's Disease program, the only institution in the United States exclusively devoted to Hansen's disease. Since 1995, WHO has made multidrug therapy for Hansen's disease available free of charge to all patients worldwide. As a result, global prevalence of Hansen's disease has declined from about 5.2 million cases in 1985 to roughly 176,000 in 2014.[17] Multidrug therapy consists of dapsone and rifampicin for all patients and a third drug, clofazimin, for patients with multibacillary disease.

Currently, there is no universally accepted vaccine for Hansen's disease. India and Brazil use a tuberculosis

---

16 Sharma, Rahul, Pushpendra Singh, W. J. Loughry, J. Mitchell Lockhart, W. Barry Inman, Malcolm S. Duthie, Maria T. Pena et al., "Zoonotic Leprosy in the Southeastern United States," *Emerging Infectious Diseases* 21, no. 12 (2015): 2127-34.

17 World Health Organization, "Leprosy Fact Sheet," 2016. Accessed September 13, 2016. http://www.who.int/mediacentre/factsheets/fs101/en/.

vaccine against Hansen's disease because both diseases are caused by species of *Mycobacterium*. The effectiveness of this method is questionable, however, since it appears that the vaccine works in some populations but not in others.

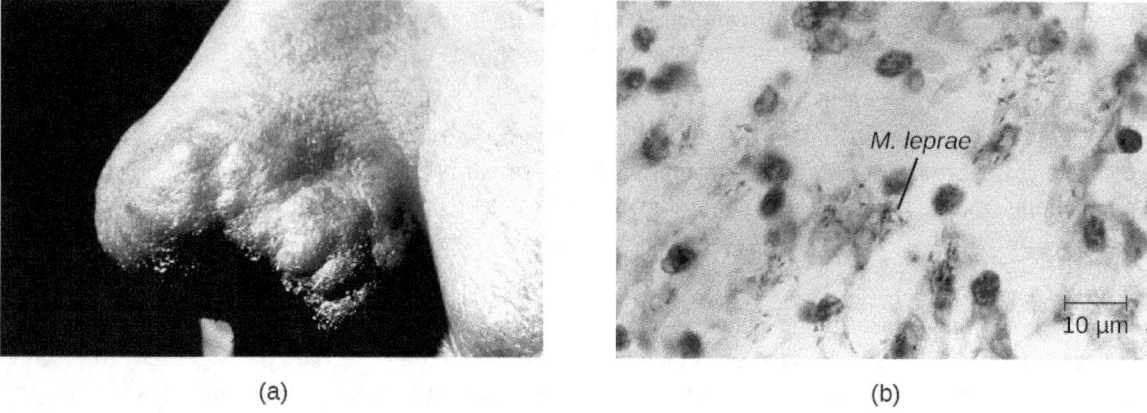

Figure 26.11 (a) The nose of a patient with Hansen's disease. Note the lepromatous/multibacillary lesions around the nostril. (b) Hansen's disease is caused by *Mycobacterium leprae*, a gram-positive bacillus. (credit a, b: modifications of work by the Centers for Disease Control and Prevention)

✓ CHECK YOUR UNDERSTANDING

- What prevents the progression from tuberculoid to lepromatus leprosy?
- Why does Hansen's disease typically affect the nerves of the extremities?

Eye on Ethics

Leper Colonies

Disfiguring, deadly diseases like leprosy have historically been stigmatized in many cultures. Before leprosy was understood, victims were often isolated in leper colonies, a practice mentioned frequently in ancient texts, including the Bible. But leper colonies are not just an artifact of the ancient world. In Hawaii, a leper colony established in the late nineteenth century persisted until the mid-twentieth century, its residents forced to live in deplorable conditions.[18] Although leprosy is a communicable disease, it is not considered contagious (easily communicable), and it certainly does not pose enough of a threat to justify the permanent isolation of its victims. Today, we reserve the practices of isolation and quarantine to patients with more dangerous diseases, such as Ebola or multiple-drug-resistant bacteria like *Mycobacterium tuberculosis* and *Staphylococcus aureus*. The ethical argument for this practice is that isolating infected patients is necessary to prevent the transmission and spread of highly contagious diseases—even when it goes against the wishes of the patient.

Of course, it is much easier to justify the practice of temporary, clinical quarantining than permanent social segregation, as occurred in leper colonies. In the 1980s, there were calls by some groups to establish camps for people infected with AIDS. Although this idea was never actually implemented, it begs the question—where do we draw the line? Are permanent isolation camps or colonies ever medically or socially justifiable? Suppose there were an outbreak of a fatal, contagious disease for which there is no treatment. Would it be justifiable to impose social isolation on those afflicted with the disease? How would we balance the rights of the infected with the risk they pose to others? To what extent should society expect individuals to put their own health at risk for the sake of treating others humanely?

---

18 National Park Service, "A Brief History of Kalaupapa," Accessed February 2, 2016. http://www.nps.gov/kala/learn/historyculture/a-brief-history-of-kalaupapa.htm.

Disease Profile

Bacterial Infections of the Nervous System

Despite the formidable defenses protecting the nervous system, a number of bacterial pathogens are known to cause serious infections of the CNS or PNS. Unfortunately, these infections are often serious and life threatening. Figure 26.12 summarizes some important infections of the nervous system.

Bacterial Infections of the Nervous System					
Disease	Pathogen	Signs and Symptoms	Transmission	Antimicrobial Drugs	Vaccine
Botulism	*Clostridium botulinum*	Blurred vision, drooping eyelids, difficulty swallowing and breathing, nausea, vomiting, often fatal	Ingestion of preformed toxin in food, ingestion of endospores in food by infants or immuno-compromised adults, bacterium introduced via wound or injection	Antitoxin; penicillin (for wound botulism)	None
Hansen's disease (leprosy)	*Mycobacterium leprae*	Hypopigmented skin, skin lesions, and nodules, loss of peripheral nerve function, loss of fingers, toes, and extremities	Inhalation, possible transmissible from armadillos to humans	Dapsone, rifampin, clofazimin	None
Haemophilus influenzae type b meningitis	*Haemophilus influenzae*	Nausea, vomiting, photophobia, stiff neck, confusion	Direct contact, inhalation of aerosols	Doxycycline, fluoroquinolones, second- and third-generation cephalosporins, and carbapenems	Hib vaccine
Listeriosis	*Listeria monocytogenes*	Initial flu-like symptoms, sepsis and potentially fatal meningitis in susceptible individuals, miscarriage in pregnant women	Bacterium ingested with contaminated food or water	Ampicillin, gentamicin	None
Meningococcal meningitis	*Neisseria meningitidis*	Nausea, vomiting, photophobia, stiff neck, confusion; often fatal	Direct contact	Cephalosporins or penicillins	Meningococcal conjugate
Neonatal meningitis	*Streptococcus agalactiae*	Temperature instability, apnea, bradycardia, hypotension, feeding difficulty, irritability, limpness, seizures, bulging fontanel, stiff neck, opisthotonos, hemiparesis, often fatal	Direct contact in birth canal	Ampicillin plus gentamicin, cefotaxime, or both	None
Pneumococcal meningitis	*Streptococcus pneumoniae*	Nausea, vomiting, photophobia, stiff neck, confusion, often fatal	Direct contact, aerosols	Cephalosporins, penicillin	Pneumococcal vaccines
Tetanus	*Clostridium tetani*	Progressive spasmatic paralysis starting with the jaw, often fatal	Bacterium introduced in puncture wound	Penicillin, antitoxin	DTaP, Tdap

Figure 26.12

26.3 Acellular Diseases of the Nervous System

Learning Objectives

By the end of this section, you will be able to:
- Identify the most common acellular pathogens that can cause infections of the nervous system
- Compare the major characteristics of specific viral diseases affecting the nervous system

A number of different viruses and subviral particles can cause diseases that affect the nervous system. Viral diseases tend to be more common than bacterial infections of the nervous system today. Fortunately, viral infections are generally milder than their bacterial counterparts and often spontaneously resolve. Some of the more important acellular pathogens of the nervous system are described in this section.

Viral Meningitis

Although it is much more common than bacterial meningitis, viral meningitis is typically less severe. Many different viruses can lead to meningitis as a sequela of the primary infection, including those that cause herpes, influenza, measles, and mumps. Most cases of viral meningitis spontaneously resolve, but severe cases do occur.

Arboviral Encephalitis

Several types of insect-borne viruses can cause encephalitis. Collectively, these viruses are referred to as arboviruses (because they are arthropod-borne), and the diseases they cause are described as **arboviral encephalitis**. Most arboviruses are endemic to specific geographical regions. Arborviral encephalitis diseases found in the United States include eastern equine encephalitis (EEE), western equine encephalitis (WEE), St. Louis encephalitis, and West Nile encephalitis (WNE). Expansion of arboviruses beyond their endemic regions sometimes occurs, generally as a result of environmental changes that are favorable to the virus or its vector. Increased travel of infected humans, animals, or vectors has also allowed arboviruses to spread into new regions.

In most cases, arboviral infections are asymptomatic or lead to a mild disease. However, when symptoms do occur, they include high fever, chills, headaches, vomiting, diarrhea, and restlessness. In elderly patients, severe arboviral encephalitis can rapidly lead to convulsions, coma, and death.

Mosquitoes are the most common biological vectors for arboviruses, which tend to be enveloped ssRNA viruses. Thus, prevention of arboviral infections is best achieved by avoiding mosquitoes—using insect repellent, wearing long pants and sleeves, sleeping in well-screened rooms, using bed nets, etc.

Diagnosis of arboviral encephalitis is based on clinical symptoms and serologic testing of serum or CSF. There are no antiviral drugs to treat any of these arboviral diseases, so treatment consists of supportive care and management of symptoms.

Eastern equine encephalitis (EEE) is caused by eastern equine encephalitis virus (EEEV), which can cause severe disease in horses and humans. Birds are reservoirs for EEEV with accidental transmission to horses and humans by *Aedes, Coquillettidia,* and *Culex* species of mosquitoes. Neither horses nor humans serve as reservoirs. EEE is most common in US Gulf Coast and Atlantic states. EEE is one of the more severe mosquito-transmitted diseases in the United States, but fortunately, it is a very rare disease in the United States (Figure 26.13).[19][20]

---

[19] US Centers for Disease Control and Prevention, "Eastern Equine Encephalitis Virus Disease Cases and Deaths Reported to CDC by Year and Clinical Presentation, 2004–2013," 2014. http://www.cdc.gov/EasternEquineEncephalitis/resources/EEEV-Cases-by-Year_2004-2013.pdf.

[20] US Centers for Disease Control and Prevention, "Eastern Equine Encephalitis, Symptoms & Treatment, 2016," Accessed June 29, 2016. https://www.cdc.gov/easternequineencephalitis/tech/symptoms.html.

Western equine encephalitis (WEE) is caused by western equine encephalitis virus (WEEV). WEEV is usually transmitted to horses and humans by the *Culex tarsalis* mosquitoes and, in the past decade, has caused very few cases of encephalitis in humans in the United States. In humans, WEE symptoms are less severe than EEE and include fever, chills, and vomiting, with a mortality rate of 3–4%. Like EEEV, birds are the natural reservoir for WEEV. Periodically, for indeterminate reasons, epidemics in human cases have occurred in North America in the past. The largest on record was in 1941, with more than 3400 cases.[21]

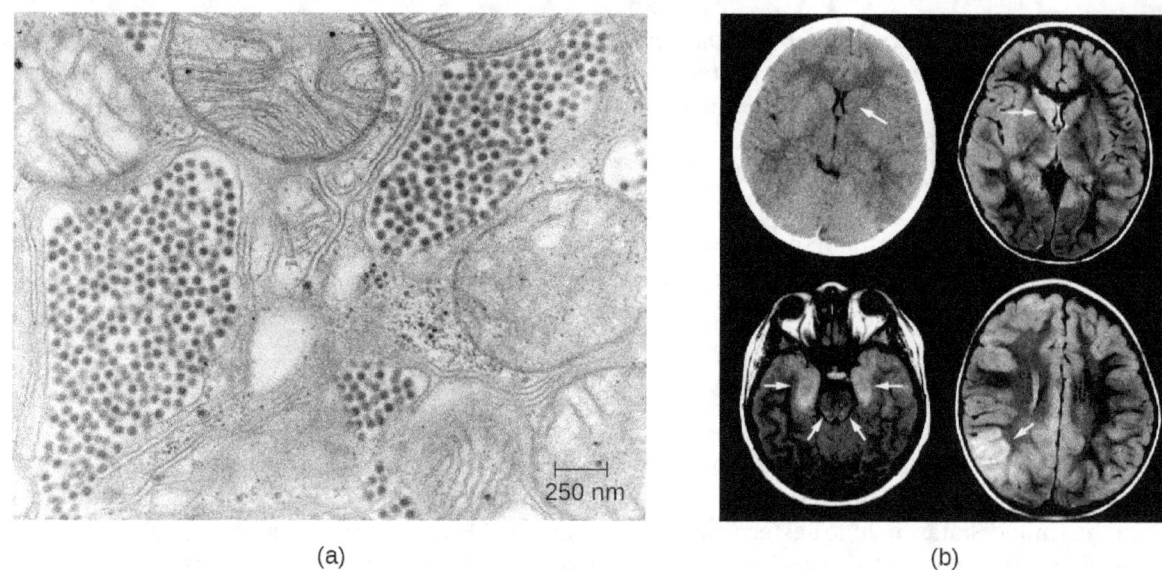

(a)   (b)

Figure 26.13 (a) A false color TEM of a mosquito salivary gland cell shows an infection of the eastern equine encephalitis virus (red). (b) CT (left) and MRI (right) scans of the brains of children with eastern equine encephalitis infections, showing abnormalities (arrows) resulting from the infection. (credit a, b: modifications of work by the Centers for Disease Control and Prevention)

St. Louis encephalitis (SLE), caused by St. Louis encephalitis virus (SLEV), is a rare form of encephalitis with symptoms occurring in fewer than 1% of infected patients. The natural reservoirs for SLEV are birds. SLEV is most often found in the Ohio-Mississippi River basin of the central United States and was named after a severe outbreak in Missouri in 1934. The worst outbreak of St. Louis encephalitis occurred in 1975, with over 2000 cases reported.[22] Humans become infected when bitten by *C. tarsalis*, *C. quinquefasciatus*, or *C. pipiens* mosquitoes carrying SLEV. Most patients are asymptomatic, but in a small number of individuals, symptoms range from mild flu-like syndromes to fatal encephalitis. The overall mortality rate for symptomatic patients is 5–15%.[23]

Japanese encephalitis, caused by Japanese encephalitis virus (JEV), is the leading cause of vaccine-preventable encephalitis in humans and is endemic to some of the most populous countries in the world, including China, India, Japan, and all of Southeast Asia. JEV is transmitted to humans by *Culex* mosquitoes, usually the species *C. tritaeniorhynchus*. The biological reservoirs for JEV include pigs and wading birds. Most patients with JEV infections are asymptomatic, with symptoms occurring in fewer than 1% of infected individuals. However, about 25% of those who do develop encephalitis die, and among those who recover, 30–50% have psychiatric, neurologic, or cognitive impairment.[24] Fortunately, there is an effective vaccine that can prevent infection with JEV. The CDC recommends this vaccine for travelers who expect to spend more than one month in endemic areas.

---

21 US Centers for Disease Control and Prevention, "Western Equine Encephalitis—United States and Canada, 1987," *Morbidity and Mortality Weekly Report* 36, no. 39 (1987): 655.

22 US Centers for Disease Control and Prevention, "Saint Louis encephalitis, Epidemiology & Geographic Distribution," Accessed June 30, 2016. http://www.cdc.gov/sle/technical/epi.html.

23 US Centers for Disease Control and Prevention, "Saint Louis encephalitis, Symptoms and Treatment," Accessed June 30, 2016. http://www.cdc.gov/sle/technical/symptoms.html.

24 US Centers for Disease Control and Prevention, "Japanese Encephalitis, Symptoms and Treatment," Accessed June 30, 2016.

As the name suggests, West Nile virus (WNV) and its associated disease, **West Nile encephalitis (WNE)**, did not originate in North America. Until 1999, it was endemic in the Middle East, Africa, and Asia; however, the first US cases were identified in New York in 1999, and by 2004, the virus had spread across the entire continental United States. Over 35,000 cases, including 1400 deaths, were confirmed in the five-year period between 1999 and 2004. WNV infection remains reportable to the CDC.

WNV is transmitted to humans by *Culex* mosquitoes from its natural reservoir, infected birds, with 70–80% of infected patients experiencing no symptoms. Most symptomatic cases involve only mild, flu-like symptoms, but fewer than 1% of infected people develop severe and sometimes fatal encephalitis or meningitis. The mortality rate in WNV patients who develop neurological disease is about 10%. More information about West Nile virus can be found in Modes of Disease Transmission.

LINK TO LEARNING

This interactive map (https://www.openstax.org/l/22arboviralUS) identifies cases of several arboviral diseases in humans and reservoir species by state and year for the United States.

CHECK YOUR UNDERSTANDING

- Why is it unlikely that arboviral encephalitis viruses will be eradicated in the future?
- Which is the most common form of viral encephalitis in the United States?

Clinical Focus

Part 2

Levofloxacin is a quinolone antibiotic that is often prescribed to treat bacterial infections of the respiratory tract, including pneumonia and bronchitis. But after taking the medication for a week, David returned to his physician sicker than before. He claimed that the antibiotic had no effect on his earlier symptoms. In addition, he now was experiencing headaches, a stiff neck, and difficulty focusing at work. He also showed the doctor a rash that had developed on his arms over the past week. His doctor, more concerned now, began to ask about David's activities over the past two weeks.

David explained that he had been recently working on a project to disassemble an old barn. His doctor collected sputum samples and scrapings from David's rash for cultures. A spinal tap was also performed to examine David's CSF. Microscopic examination of his CSF revealed encapsulated yeast cells. Based on this result, the doctor prescribed a new antimicrobial therapy using amphotericin B and flucytosine.

- Why was the original treatment ineffective?
- Why is the presence of a capsule clinically important?

Jump to the previous Clinical Focus box. Jump to the next Clinical Focus box.

Zika Virus Infection

Zika virus infection is an emerging arboviral disease associated with human illness in Africa, Southeast Asia, and South and Central America; however, its range is expanding as a result of the widespread range of its mosquito vector. The first cases originating in the United States were reported in 2016. The Zika virus was initially described in 1947 from monkeys in the Zika Forest of Uganda through a network that monitored yellow fever. It was not considered a serious human pathogen until the first large-scale outbreaks occurred in Micronesia in 2007;[25] however, the virus has gained notoriety over the past decade, as it has emerged as a cause of symptoms similar to other arboviral infections that include fever, skin rashes, conjunctivitis, muscle and joint pain, malaise, and headache. Mosquitoes of the *Aedes* genus are the primary vectors, although the virus can also be transmitted sexually, from mother to baby during pregnancy, or through a blood transfusion.

---

http://www.cdc.gov/japaneseencephalitis/symptoms/index.html.

Most Zika virus infections result in mild symptoms such as fever, a slight rash, or conjunctivitis. However, infections in pregnant women can adversely affect the developing fetus. Reports in 2015 indicate fetal infections can result in brain damage, including a serious birth defect called microcephaly, in which the infant is born with an abnormally small head (Figure 26.14).[26]

Diagnosis of Zika is primarily based on clinical symptoms. However, the FDA recently authorized the use of a Zika virus RNA assay, Trioplex RT-PCR, and Zika MAC-ELISA to test patient blood and urine to confirm Zika virus disease. There are currently no antiviral treatments or vaccines for Zika virus, and treatment is limited to supportive care.

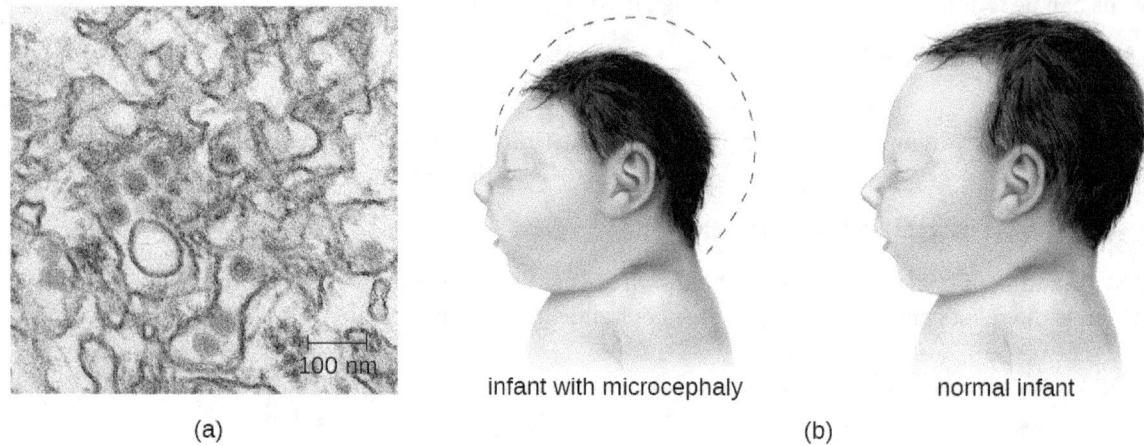

Figure 26.14 (a) This colorized electron micrograph shows Zika virus particles (red). (b) Women infected by the Zika virus during pregnancy may give birth to children with microcephaly, a deformity characterized by an abnormally small head and brain. (credit a, b: modifications of work by the Centers for Disease Control and Prevention)

CHECK YOUR UNDERSTANDING

- What are the signs and symptoms of Zika virus infection in adults?
- Why is Zika virus infection considered a serious public health threat?

Rabies

Rabies is a deadly zoonotic disease that has been known since antiquity. The disease is caused by rabies virus (RV), a member of the family Rhabdoviridae, and is primarily transmitted through the bite of an infected mammal. Rhabdoviridae are enveloped RNA viruses that have a distinctive bullet shape (Figure 26.15); they were first studied by Louis Pasteur, who obtained rabies virus from rabid dogs and cultivated the virus in rabbits. He successfully prepared a rabies vaccine using dried nerve tissues from infected animals. This vaccine was used to first treat an infected human in 1885.

The most common reservoirs in the United States are wild animals such as raccoons (30.2% of all animal cases during 2014), bats (29.1%), skunks (26.3%), and foxes (4.1%); collectively, these animals were responsible for a total of 92.6% of animal rabies cases in the United States in 2014. The remaining 7.4% of cases that year were in domesticated animals such as dogs, cats, horses, mules, sheep, goats, and llamas.[27] While there are typically only one or two human cases per year in the United States, rabies still causes tens of thousands of human deaths per year worldwide, primarily in Asia and Africa.

---

25 Sikka, Veronica, Vijay Kumar Chattu, Raaj K. Popli, Sagar C. Galwankar, Dhanashree Kelkar, Stanley G. Sawicki, Stanislaw P. Stawicki, and Thomas J. Papadimos, "The Emergence of Zika Virus as a Global Health Security Threat: A Review and a Consensus Statement of the INDUSEM Joint Working Group (JWG)," *Journal of Global Infectious Diseases* 8, no. 1 (2016): 3.

26 Mlakar, Jernej, Misa Korva, Nataša Tul, Mara Popović, Mateja Poljšak-Prijatelj, Jerica Mraz, Marko Kolenc et al., "Zika Virus Associated with Microcephaly," *New England Journal of Medicine* 374, no. 10 (2016): 951-8.

The low incidence of rabies in the United States is primarily a result of the widespread vaccination of dogs and cats. An oral vaccine is also used to protect wild animals, such as raccoons and foxes, from infection. Oral vaccine programs tend to focus on geographic areas where rabies is endemic.[28] The oral vaccine is usually delivered in a package of bait that is dropped by airplane, although baiting in urban areas is done by hand to maximize safety.[29] Many countries require a quarantine or proof of rabies vaccination for domestic pets being brought into the country. These procedures are especially strict in island nations where rabies is not yet present, such as Australia.

The incubation period for rabies can be lengthy, ranging from several weeks or months to over a year. As the virus replicates, it moves from the site of the bite into motor and sensory axons of peripheral nerves and spreads from nerve to nerve using a process called retrograde transport, eventually making its way to the CNS through the spinal ganglia. Once rabies virus reaches the brain, the infection leads to encephalitis caused by the disruption of normal neurotransmitter function, resulting in the symptoms associated with rabies. The virions act in the synaptic spaces as competitors with a variety of neurotransmitters for acetylcholine, GABA, and glycine receptors. Thus, the action of rabies virus is neurotoxic rather than cytotoxic. After the rabies virus infects the brain, it can continue to spread through other neuronal pathways, traveling out of the CNS to tissues such as the salivary glands, where the virus can be released. As a result, as the disease progresses the virus can be found in many other tissues, including the salivary glands, taste buds, nasal cavity, and tears.

The early symptoms of rabies include discomfort at the site of the bite, fever, and headache. Once the virus reaches the brain and later symptoms appear, the disease is always fatal. Terminal rabies cases can end in one of two ways: either furious or paralytic rabies. Individuals with furious rabies become very agitated and hyperactive. Hydrophobia (a fear of water) is common in patients with furious rabies, which is caused by muscular spasms in the throat when swallowing or thinking about water. Excess salivation and a desire to bite can lead to foaming of the mouth. These behaviors serve to enhance the likelihood of viral transmission, although contact with infected secretions like saliva or tears alone is sufficient for infection. The disease culminates after just a few days with terror and confusion, followed by cardiovascular and respiratory arrest. In contrast, individuals with paralytic rabies generally follow a longer course of disease. The muscles at the site of infection become paralyzed. Over a period of time, the paralysis slowly spreads throughout the body. This paralytic form of disease culminates in coma and death.

Before present-day diagnostic methods were available, rabies diagnosis was made using a clinical case history and histopathological examination of biopsy or autopsy tissues, looking for the presence of Negri bodies. We now know these histologic changes *cannot* be used to confirm a rabies diagnosis. There are no tests that can detect rabies virus in humans at the time of the bite or shortly thereafter. Once the virus has begun to replicate (but before clinical symptoms occur), the virus can be detected using an immunofluorescence test on cutaneous nerves found at the base of hair follicles. Saliva can also be tested for viral genetic material by reverse transcription followed by polymerase chain reaction (RT-PCR). Even when these tests are performed, most suspected infections are treated as positive in the absence of contravening evidence. It is better that patients undergo unnecessary therapy because of a false-positive result, rather than die as the result of a false-negative result.

Human rabies infections are treated by immunization with multiple doses of an attenuated vaccine to develop active immunity in the patient (see the Clinical Focus feature in the chapter on Acellular Pathogens). Vaccination of an already-infected individual has the potential to work because of the slow progress of the disease, which allows time for the patient's immune system to develop antibodies against the virus. Patients may also be treated with human rabies immune globulin (antibodies to the rabies virus) to encourage passive

---

27 US Centers for Disease Control and Prevention, "Rabies, Wild Animals," 2016. Accessed September 13, 2016. http://www.cdc.gov/rabies/location/usa/surveillance/wild_animals.html.

28 Slate, Dennis, Charles E. Rupprecht, Jane A. Rooney, Dennis Donovan, Donald H. Lein, and Richard B. Chipman, "Status of Oral Rabies Vaccination in Wild Carnivores in the United States," *Virus Research* 111, no. 1 (2005): 68-76.

29 Finnegan, Christopher J., Sharon M. Brookes, Nicholas Johnson, Jemma Smith, Karen L. Mansfield, Victoria L. Keene, Lorraine M. McElhinney, and Anthony R. Fooks, "Rabies in North America and Europe," *Journal of the Royal Society of Medicine* 95, no. 1 (2002): 9-13. http://www.ncbi.nlm.nih.gov/pmc/articles/PMC1279140/.

immunity. These antibodies will neutralize any free viral particles. Although the rabies infection progresses slowly in peripheral tissues, patients are not normally able to mount a protective immune response on their own.

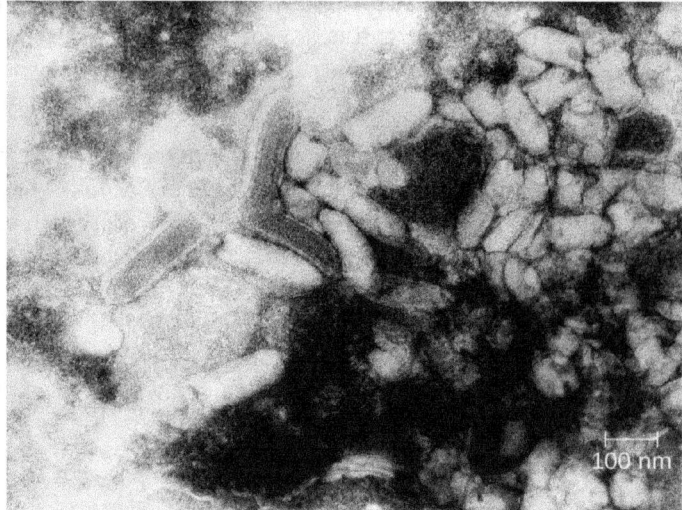

Figure 26.15 Virions of the rabies virus have a characteristic bullet-like shape. (credit: modification of work by the Centers for Disease Control and Prevention)

✓ CHECK YOUR UNDERSTANDING

- How does the bite from an infected animal transmit rabies?
- What is the goal of wildlife vaccination programs for rabies?
- How is rabies treated in a human?

Poliomyelitis

Poliomyelitis (polio), caused by poliovirus, is a primarily intestinal disease that, in a small percentage of cases, proceeds to the nervous system, causing paralysis and, potentially, death. Poliovirus is highly contagious, with transmission occurring by the fecal-oral route or by aerosol or droplet transmission. Approximately 72% of all poliovirus infections are asymptomatic; another 25% result only in mild intestinal disease, producing nausea, fever, and headache.[30] However, even in the absence of symptoms, patients infected with the virus can shed it in feces and oral secretions, potentially transmitting the virus to others. In about one case in every 200, the poliovirus affects cells in the CNS.[31]

After it enters through the mouth, initial replication of poliovirus occurs at the site of implantation in the pharynx and gastrointestinal tract. As the infection progresses, poliovirus is usually present in the throat and in the stool before the onset of symptoms. One week after the onset of symptoms, there is less poliovirus in the throat, but for several weeks, poliovirus continues to be excreted in the stool. Poliovirus invades local lymphoid tissue, enters the bloodstream, and then may infect cells of the CNS. Replication of poliovirus in motor neurons of the anterior horn cells in the spinal cord, brain stem, or motor cortex results in cell destruction and leads to flaccid paralysis. In severe cases, this can involve the respiratory system, leading to death. Patients with impaired respiratory function are treated using positive-pressure ventilation systems. In the past, patients were sometimes confined to Emerson respirators, also known as iron lungs (Figure 26.16).

Direct detection of the poliovirus from the throat or feces can be achieved using reverse transcriptase PCR (RT-PCR) or genomic sequencing to identify the genotype of the poliovirus infecting the patient. Serological tests

---

30 US Centers for Disease Control and Prevention, "Global Health – Polio," 2014. Accessed June 30, 2016. http://www.cdc.gov/polio/about/index.htm.
31 US Centers for Disease Control and Prevention, "Global Health – Polio," 2014. Accessed June 30, 2016. http://www.cdc.gov/polio/about/index.htm.

can be used to determine whether the patient has been previously vaccinated. There are no therapeutic measures for polio; treatment is limited to various supportive measures. These include pain relievers, rest, heat therapy to ease muscle spasms, physical therapy and corrective braces if necessary to help with walking, and mechanical ventilation to assist with breathing if necessary.

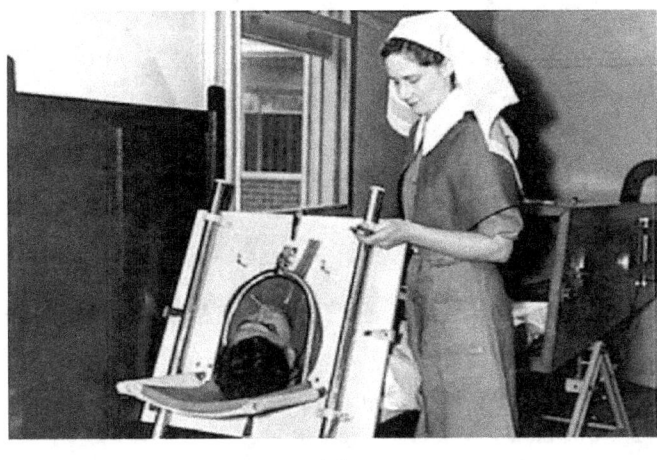 

(a)   (b)

Figure 26.16 (a) An Emerson respiratory (or iron lung) that was used to help some polio victims to breathe. (b) Polio can also result in impaired motor function. (credit b: modification of work by the Centers for Disease Control and Prevention)

Two different vaccines were introduced in the 1950s that have led to the dramatic decrease in polio worldwide (Figure 26.17). The Salk vaccine is an inactivated polio virus that was first introduced in 1955. This vaccine is delivered by intramuscular injection. The Sabin vaccine is an oral polio vaccine that contains an attenuated virus; it was licensed for use in 1962. There are three serotypes of poliovirus that cause disease in humans; both the Salk and the Sabin vaccines are effective against all three.

Attenuated viruses from the Sabin vaccine are shed in the feces of immunized individuals and thus have the potential to infect nonimmunized individuals. By the late 1990s, the few polio cases originating in the United States could be traced back to the Sabin vaccine. In these cases, mutations of the attenuated virus following vaccination likely allowed the microbe to revert to a virulent form. For this reason, the United States switched exclusively to the Salk vaccine in 2000. Because the Salk vaccine contains an inactivated virus, there is no risk of transmission to others (see Vaccines). Currently four doses of the vaccine are recommended for children: at 2, 4, and 6–18 months of age, and at 4–6 years of age.

In 1988, WHO launched the Global Polio Eradication Initiative with the goal of eradicating polio worldwide through immunization. That goal is now close to being realized. Polio is now endemic in only a few countries, including Afghanistan, Pakistan, and Nigeria, where vaccination efforts have been disrupted by military conflict or political instability.

Figure 26.17 (a) Polio is caused by the poliovirus. (b) Two American virologists developed the first polio vaccines: Albert Sabin (left) and Jonas Salk (right). (credit a: modification of work by the Centers for Disease Control and Prevention)

MICRO CONNECTIONS

The Terror of Polio

In the years after World War II, the United States and the Soviet Union entered a period known as the Cold War. Although there was no armed conflict, the two super powers were diplomatically and economically isolated from each other, as represented by the so-called Iron Curtain between the Soviet Union and the rest of the world. After 1950, migration or travel outside of the Soviet Union was exceedingly difficult, and it was equally difficult for foreigners to enter the Soviet Union. The United States also placed strict limits on Soviets entering the country. During the Eisenhower administration, only 20 graduate students from the Soviet Union were allowed to come to study in the United States per year.

Yet even the Iron Curtain was no match for polio. The Salk vaccine became widely available in the West in 1955, and by the time the Sabin vaccine was ready for clinical trials, most of the susceptible population in the United States and Canada had already been vaccinated against polio. Sabin needed to look elsewhere for study participants. At the height of the Cold War, Mikhail Chumakov was allowed to come to the United States to study Sabin's work. Likewise, Sabin, an American microbiologist, was allowed to travel to the Soviet Union to begin clinical trials. Chumakov organized Soviet-based production and managed the experimental trials to test the new vaccine in the Soviet Union. By 1959, over ten million Soviet children had been safely treated with Sabin's vaccine.

As a result of a global vaccination campaign with the Sabin vaccine, the overall incidence of polio has dropped dramatically. Today, polio has been nearly eliminated around the world and is only rarely seen in the United States. Perhaps one day soon, polio will become the third microbial disease to be eradicated from the general population [small pox and rinderpest (the cause of cattle plague) being the first two].

CHECK YOUR UNDERSTANDING

- How is poliovirus transmitted?
- Compare the pros and cons of each of the two polio vaccines.

Transmissible Spongiform Encephalopathies

Acellular infectious agents called prions are responsible for a group of related diseases known as transmissible spongiform encephalopathies (TSEs) that occurs in humans and other animals (see Viroids, Virusoids, and Prions). All TSEs are degenerative, fatal neurological diseases that occur when brain tissue becomes infected by prions. These diseases have a slow onset; symptoms may not become apparent until after an incubation period of years and perhaps decades, but death usually occurs within months to a few years after the first symptoms appear.

TSEs in animals include **scrapie**, a disease in sheep that has been known since the 1700s, and **chronic wasting disease**, a disease of deer and elk in the United States and Canada. **Mad cow disease** is seen in cattle and can be transmitted to humans through the consumption of infected nerve tissues. Human prion diseases include **Creutzfeldt-Jakob disease** and **kuru**, a rare disease endemic to Papua New Guinea.

Prions are infectious proteinaceous particles that are not viruses and do not contain nucleic acid. They are typically transmitted by exposure to and ingestion of infected nervous system tissues, tissue transplants, blood transfusions, or contaminated fomites. Prion proteins are normally found in a healthy brain tissue in a form called PrP^C. However, if this protein is misfolded into a denatured form (PrP^{Sc}), it can cause disease. Although the exact function of PrP^C is not currently understood, the protein folds into mostly alpha helices and binds copper. The rogue protein, on the other hand, folds predominantly into beta-pleated sheets and is resistant to proteolysis. In addition, PrP^{Sc} can induce PrP^C to become misfolded and produce more rogue protein (Figure 26.18).

As PrP^{Sc} accumulates, it aggregates and forms fibrils within nerve cells. These protein complexes ultimately cause the cells to die. As a consequence, brain tissues of infected individuals form masses of neurofibrillary tangles and amyloid plaques that give the brain a spongy appearance, which is why these diseases are called spongiform encephalopathy (Figure 6.26). Damage to brain tissue results in a variety of neurological symptoms. Most commonly, affected individuals suffer from memory loss, personality changes, blurred vision, uncoordinated movements, and insomnia. These symptoms gradually worsen over time and culminate in coma and death.

The gold standard for diagnosing TSE is the histological examination of brain biopsies for the presence of characteristic amyloid plaques, vacuoles, and prion proteins. Great care must be taken by clinicians when handling suspected prion-infected materials to avoid becoming infected themselves. Other tissue assays search for the presence of the 14-3-3 protein, a marker for prion diseases like Creutzfeldt-Jakob disease. New assays, like RT-QuIC (real-time quaking-induced conversion), offer new hope to effectively detect the abnormal prion proteins in tissues earlier in the course of infection. Prion diseases cannot be cured. However, some medications may help slow their progress. Medical support is focused on keeping patients as comfortable as possible despite progressive and debilitating symptoms.

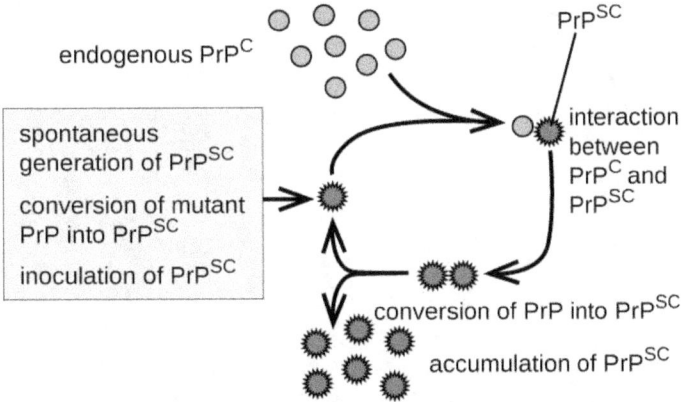

Figure 26.18 The replicative cycle of misfolded prion proteins.

🔗 LINK TO LEARNING

Because prion-contaminated materials are potential sources of infection for clinical scientists and physicians, both the World Health Organization (https://www.openstax.org/l/22WHOprion) and CDC (https://www.openstax.org/l/22CDCprion) provide information to inform, educate and minimize the risk of infections due to prions.

CHECK YOUR UNDERSTANDING

- Do prions reproduce in the conventional sense?
- What is the connection between prions and the removal of animal byproducts from the food of farm animals?

Disease Profile

Acellular Infections of the Nervous System

Serious consequences are the common thread among these neurological diseases. Several cause debilitating paralysis, and some, such as Creutzfeldt-Jakob disease and rabies, are always or nearly always fatal. Since few drugs are available to combat these infections, vector control and vaccination are critical for prevention and containment. Figure 26.19 summarizes some important viral and prion infections of the nervous system.

Acellular Infections of the Nervous System

Disease	Pathogen	Signs and Symptoms	Transmission	Diagnostic Tests	Antimicrobial Drugs	Vaccine
Arboviral encephalitis (eastern equine, western equine, St. Louis, West Nile, Japanese)	EEEV, WEEV, SLEV, WNV, JEV	In mild cases, fever, chills, headaches, and restlessness; in serious cases, encephalitis leading to convulsions, coma, and death	From bird reservoirs to humans (and horses) by mosquito vectors of various species	Serologic testing of serum or CSF	None	Human vaccine available for JEV only; no vaccines available for other arboviruses
Creutzfeldt-Jacob Disease and other TSEs	Prions	Memory loss, confusion, blurred vision, uncoordinated movement, insomnia, coma, death	Exposure to infected nerve tissue via consumption or transplant, inherited	Tissue biopsy	None	None
Poliomyelitis	Poliovirus	Asymptomatic or mild nausea, fever, headache in most cases; in neurological infections, flaccid paralysis and potentially fatal respiratory paralysis	Fecal-oral route or contact with droplets or aerosols	Culture of poliovirus, PCR	None	Attenuated vaccine (Sabin), killed vaccine (Salk)
Rabies	Rabies virus (RV)	Fever, headaches, hyperactivity, hydrophobia, excessive salivation, terrors, confusion, spreading paralysis, coma, always fatal if not promptly treated	From bite of infected mammal	Viral antigen in tissue, antibodies to virus	Attenuated vaccine, rabies immunoglobulin	Attenuated vaccine
Viral meningitis	HSV-1, HSV-2, varicella zoster virus, mumps virus, influenza virus, measles virus	Nausea, vomiting, photophobia, stiff neck, confusion, symptoms generally resolve within 7–10 days	Sequela of primary viral infection	Testing of oral, fecal, blood, or CSF samples	Varies depending on cause	Varies depending on cause
Zika virus infection	Zika virus	Fever, rash, conjunctivitis; in pregnant women, can cause fetal brain damage and microcephaly	Between humans by *Aedes* spp. mosquito vectors, also may be transmitted sexually or via blood transfusion	Zika virus RNA assay, Trioplex RT-PCR, Zika MAC-ELISA test	None	None

Figure 26.19

26.4 Fungal and Parasitic Diseases of the Nervous System

Learning Objectives

By the end of this section, you will be able to:
- Identify the most common fungi that can cause infections of the nervous system
- Compare the major characteristics of specific fungal diseases affecting the nervous system

Fungal infections of the nervous system, called **neuromycoses**, are rare in healthy individuals. However, neuromycoses can be devastating in immunocompromised or elderly patients. Several eukaryotic parasites are also capable of infecting the nervous system of human hosts. Although relatively uncommon, these infections can also be life-threatening in immunocompromised individuals. In this section, we will first discuss neuromycoses, followed by parasitic infections of the nervous system.

Cryptococcocal Meningitis

Cryptococcus neoformans is a fungal pathogen that can cause meningitis. This yeast is commonly found in soils and is particularly associated with pigeon droppings. It has a thick capsule that serves as an important virulence factor, inhibiting clearance by phagocytosis. Most *C. neoformans* cases result in subclinical respiratory infections that, in healthy individuals, generally resolve spontaneously with no long-term consequences (see Respiratory Mycoses). In immunocompromised patients or those with other underlying illnesses, the infection can progress to cause meningitis and granuloma formation in brain tissues. *Cryptococcus* antigens can also serve to inhibit cell-mediated immunity and delayed-type hypersensitivity.

Cryptococcus can be easily cultured in the laboratory and identified based on its extensive capsule (Figure 26.20). *C. neoformans* is frequently cultured from urine samples of patients with disseminated infections.

Prolonged treatment with antifungal drugs is required to treat cryptococcal infections. Combined therapy is required with amphotericin B plus flucytosine for at least 10 weeks. Many antifungal drugs have difficulty crossing the blood-brain barrier and have strong side effects that necessitate low doses; these factors contribute to the lengthy time of treatment. Patients with AIDS are particularly susceptible to *Cryptococcus* infections because of their compromised immune state. AIDS patients with cryptococcosis can also be treated with antifungal drugs, but they often have relapses; lifelong doses of fluconazole may be necessary to prevent reinfection.

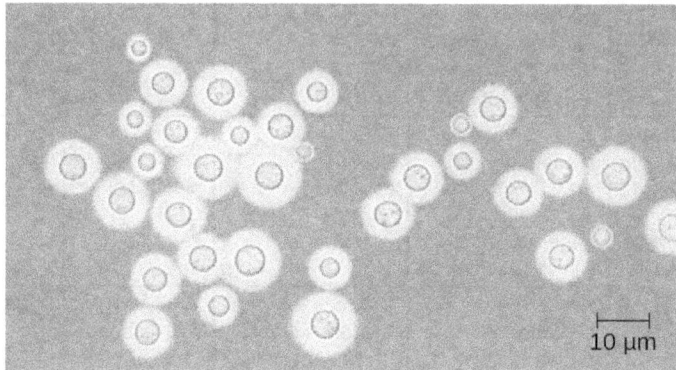

Figure 26.20 An India ink-negative stain of *C. neoformans* showing the thick capsules around the spherical yeast cells. (credit: modification of work by Centers for Disease Control and Prevention)

✓ CHECK YOUR UNDERSTANDING

- Why are neuromycoses infections rare in the general population?
- How is a cryptococcal infection acquired?

Disease Profile

Neuromycoses

Neuromycoses typically occur only in immunocompromised individuals and usually only invade the nervous system after first infecting a different body system. As such, many diseases that sometimes affect the nervous system have already been discussed in previous chapters. Figure 26.21 presents some of the most common fungal infections associated with neurological disease. This table includes only the neurological aspects associated with these diseases; it does not include characteristics associated with other body systems.

Neuromycoses					
Disease	Pathogen	Signs and Symptoms	Transmission	Diagnostic Tests	Antimicrobial Drugs
Aspergillosis	*Aspergillus fumigatus*	Meningitis, brain abscesses	Dissemination from respiratory infection	CSF, routine culture	Amphotericin B, voriconazole
Candidiasis	*Candida albicans*	Meningitis	Oropharynx or urogenital	CSF, routine culture	Amphotericin B, flucytosine
Coccidioido-mycosis (Valley fever)	*Coccidioides immitis*	Meningitis (in about 1% of infections)	Dissemination from respiratory infection	CSF, routine culture	Amphotericin B, azoles
Cryptococcosis	*Cryptococcus neoformans*	Meningitis, granuloma formation in brain	Inhalation	Negative stain of CSF, routine culture	Amphotericin B, flucytosine
Histoplasmosis	*Histoplasma capsulatum*	Meningitis, granulomas in the brain	Dissemination from respiratory infection	CSF, routine culture	Amphotericin B, itraconazole
Mucormycosis	*Rhizopus arrhizus*	Brain abscess	Nasopharynx	CSF, routine culture	Amphotericin B, azoles

Figure 26.21

Clinical Focus

Resolution

David's new prescription for two antifungal drugs, amphotericin B and flucytosine, proved effective, and his condition began to improve. Culture results from David's sputum, skin, and CSF samples confirmed a fungal infection. All were positive for *C. neoformans*. Serological tests of his tissues were also positive for the *C. neoformans* capsular polysaccharide antigen.

Since *C. neoformans* is known to occur in bird droppings, it is likely that David had been exposed to the fungus while working on the barn. Despite this exposure, David's doctor explained to him that immunocompetent people rarely contract cryptococcal meningitis and that his immune system had likely been compromised by the anti-inflammatory medication he was taking to treat his Crohn's disease. However, to rule out other possible causes of immunodeficiency, David's doctor recommended that he be tested for HIV.

After David tested negative for HIV, his doctor took him off the corticosteroid he was using to manage his Crohn's disease, replacing it with a different class of drug. After several weeks of antifungal treatments, David managed a full recovery.

Jump to the previous Clinical Focus box.

Amoebic Meningitis

Primary amoebic meningoencephalitis (PAM) is caused by *Naegleria fowleri*. This amoeboflagellate is commonly found free-living in soils and water. It can exist in one of three forms—the infective amoebic trophozoite form, a motile flagellate form, and a resting cyst form. PAM is a rare disease that has been associated with young and otherwise healthy individuals. Individuals are typically infected by the amoeba while swimming in warm bodies of freshwater such as rivers, lakes, and hot springs. The pathogenic trophozoite infects the brain by initially entering through nasal passages to the sinuses; it then moves down olfactory nerve fibers to penetrate the submucosal nervous plexus, invades the cribriform plate, and reaches the subarachnoid space. The subarachnoid space is highly vascularized and is a route of dissemination of trophozoites to other areas of the CNS, including the brain (Figure 26.22). Inflammation and destruction of gray matter leads to severe headaches and fever. Within days, confusion and convulsions occur and quickly progress to seizures, coma, and death. The progression can be very rapid, and the disease is often not diagnosed until autopsy.

N. fowleri infections can be confirmed by direct observation of CSF; the amoebae can often be seen moving while viewing a fresh CSF wet mount through a microscope. Flagellated forms can occasionally also be found in CSF. The amoebae can be stained with several stains for identification, including Giemsa-Wright or a modified trichrome stain. Detection of antigens with indirect immunofluorescence, or genetic analysis with PCR, can be used to confirm an initial diagnosis. *N. fowleri* infections are nearly always fatal; only 3 of 138 patients with PAM in the United States have survived.[32] A new experimental drug called miltefosine shows some promise for treating these infections. This drug is a phosphotidylcholine derivative that is thought to inhibit membrane function in *N. fowleri*, triggering apoptosis and disturbance of lipid-dependent cell signaling pathways.[33] When administered early in infection and coupled with therapeutic hypothermia (lowering the body's core temperature to reduce the cerebral edema associated with infection), this drug has been successfully used to treat primary amoebic encephalitis.

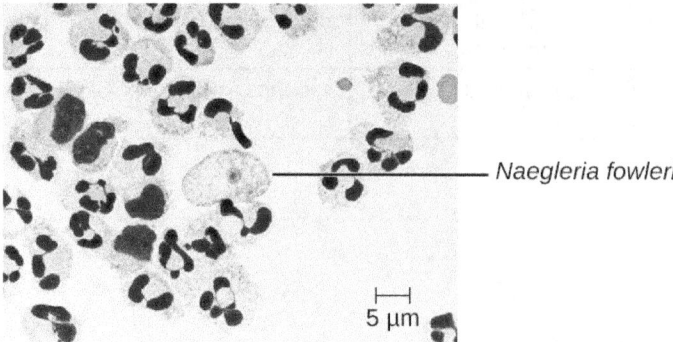

Figure 26.22   Free-living amoeba in human brain tissue from a patient suffering from PAM. (credit: modification of work by the Centers for Disease Control and Prevention)

Granulomatous Amoebic Encephalitis

Acanthamoeba and *Balamuthia* species are free-living amoebae found in many bodies of fresh water. Human infections by these amoebae are rare. However, they can cause amoebic keratitis in contact lens wearers (see Protozoan and Helminthic Infections of the Eyes), disseminated infections in immunocompromised patients, and **granulomatous amoebic encephalitis (GAE)** in severe cases. Compared to PAM, GAE tend to be subacute infections. The microbe is thought to enter through either the nasal sinuses or breaks in the skin. It is disseminated hematogenously and can invade the CNS. There, the infections lead to inflammation, formation of lesions, and development of typical neurological symptoms of encephalitis (Figure 26.23). GAE is nearly always fatal.

---

32   US Centers for Disease Control and Prevention, "*Naegleria fowleri*—Primary Amoebic Meningoencephalitis (PAM)—Amebic Encephalitis," 2016. Accessed June 30, 2016. http://www.cdc.gov/parasites/naegleria/treatment.html.

33   Dorlo, Thomas PC, Manica Balasegaram, Jos H. Beijnen, and Peter J. de Vries, "Miltefosine: A Review of Its Pharmacology and Therapeutic Efficacy in the Treatment of Leishmaniasis," *Journal of Antimicrobial Chemotherapy* 67, no. 11 (2012): 2576-97.

GAE is often not diagnosed until late in the infection. Lesions caused by the infection can be detected using CT or MRI. The live amoebae can be directly detected in CSF or tissue biopsies. Serological tests are available but generally are not necessary to make a correct diagnosis, since the presence of the organism in CSF is definitive. Some antifungal drugs, like fluconazole, have been used to treat acanthamoebal infections. In addition, a combination of miltefosine and voriconazole (an inhibitor of ergosterol biosynthesis) has recently been used to successfully treat GAE. Even with treatment, however, the mortality rate for patients with these infections is high.

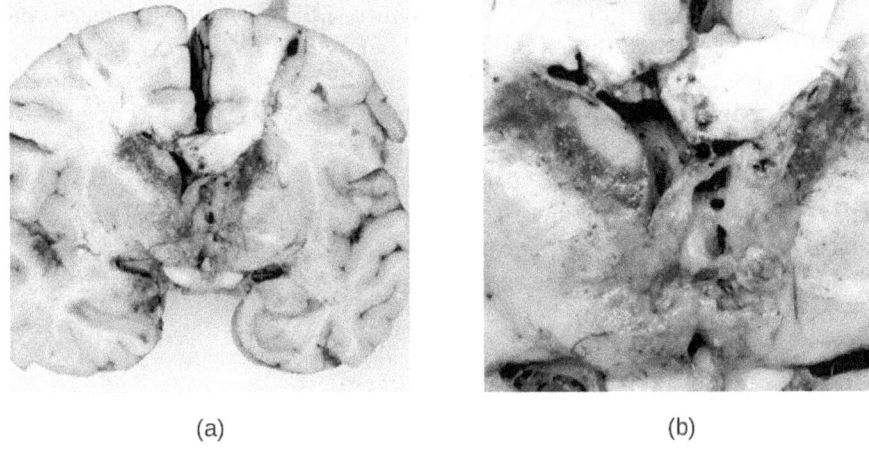

(a) (b)

Figure 26.23 (a) Brain tissue from a patient who died of granulomatous amebic encephalitis (GAE) caused by *Balamuthia mandrillaris*. (b) A close-up of the necrosis in the center of the brain section. (credit a, b: modifications of work by the Centers for Disease Control and Prevention)

CHECK YOUR UNDERSTANDING

- How is granulomatous amoebic encephalitis diagnosed?

Human African Trypanosomiasis

Human African trypanosomiasis (also known as **African sleeping sickness**) is a serious disease endemic to two distinct regions in sub-Saharan Africa. It is caused by the insect-borne hemoflagellate *Trypanosoma brucei*. The subspecies *Trypanosoma brucei rhodesiense* causes **East African trypanosomiasis** (EAT), and another subspecies, *Trypanosoma brucei gambiense* causes **West African trypanosomiasis** (WAT). A few hundred cases of EAT are currently reported each year.[34] WAT is more commonly reported and tends to be a more chronic disease. Around 7000 to 10,000 new cases of WAT are identified each year.[35]

T. brucei is primarily transmitted to humans by the bite of the tsetse fly (*Glossina* spp.). Soon after the bite of a tsetse fly, a chancre forms at the site of infection. The flagellates then spread, moving into the circulatory system (Figure 26.24). These systemic infections result in an undulating fever, during which symptoms persist for two or three days with remissions of about a week between bouts. As the disease enters its final phase, the pathogens move from the lymphatics into the CNS. Neurological symptoms include daytime sleepiness, insomnia, and mental deterioration. In EAT, the disease runs its course over a span of weeks to months. In contrast, WAT often occurs over a span of months to years.

Although a strong immune response is mounted against the trypanosome, it is not sufficient to eliminate the pathogen. Through antigenic variation, *Trypanosoma* can change their surface proteins into over 100 serological types. This variation leads to the undulating form of the initial disease. The initial septicemia

---

34 US Centers for Disease Control and Prevention, "Parasites – African Trypanosomiasis (also known as Sleeping Sickness), East African Trypanosomiasis FAQs," 2012. Accessed June 30, 2016. http://www.cdc.gov/parasites/sleepingsickness/gen_info/faqs-east.html.

35 US Centers for Disease Control and Prevention, "Parasites – African Trypanosomiasis (also known as Sleeping Sickness), Epidemiology & Risk Factors," 2012. Accessed June 30, 2016. http://www.cdc.gov/parasites/sleepingsickness/epi.html.

caused by the infection leads to high fevers. As the immune system responds to the infection, the number of organisms decrease, and the clinical symptoms abate. However, a subpopulation of the pathogen then alters its surface coat antigens by antigenic variation and evades the immune response. These flagellates rapidly proliferate and cause another bout of disease. If untreated, these infections are usually fatal.

Clinical symptoms can be used to recognize the early signs of African trypanosomiasis. These include the formation of a chancre at the site of infection and **Winterbottom's sign**. Winterbottom's sign refers to the enlargement of lymph nodes on the back of the neck—often indicative of cerebral infections. *Trypanosoma* can be directly observed in stained samples including blood, lymph, CSF, and skin biopsies of chancres from patients. Antibodies against the parasite are found in most patients with acute or chronic disease. Serologic testing is generally not used for diagnosis, however, since the microscopic detection of the parasite is sufficient. Early diagnosis is important for treatment. Before the nervous system is involved, drugs like pentamidine (an inhibitor of nuclear metabolism) and suramin (mechanism unclear) can be used. These drugs have fewer side effects than the drugs needed to treat the second stage of the disease. Once the sleeping sickness phase has begun, harsher drugs including melarsoprol (an arsenic derivative) and eflornithine can be effective. Following successful treatment, patients still need to have follow-up examinations of their CSF for two years to detect possible relapses of the disease. The most effective means of preventing these diseases is to control the insect vector populations.

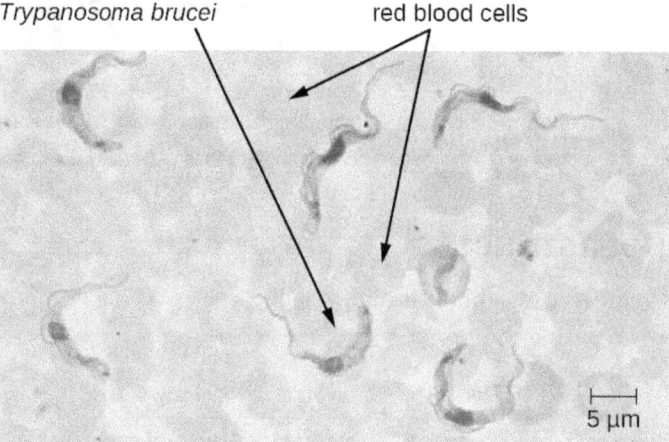

Figure 26.24 *Trypanosoma brucei*, the causative agent of African sleeping sickness, in a human blood smear. (credit: modification of work by the Centers for Disease Control and Prevention)

✓ CHECK YOUR UNDERSTANDING

- What is the symptom of a systemic *Trypanosoma* infection?
- What are the symptoms of a neurological *Trypanosoma* infection?
- Why are trypanosome infections so difficult to eradicate?

Neurotoxoplasmosis

Toxoplasma gondii is an ubiquitous intracellular parasite that can cause neonatal infections. Cats are the definitive host, and humans can become infected after eating infected meat or, more commonly, by ingesting oocysts shed in the feces of cats (see Parasitic Infections of the Circulatory and Lymphatic Systems). *T. gondii* enters the circulatory system by passing between the endothelial cells of blood vessels.[36] Most cases of toxoplasmosis are asymptomatic. However, in immunocompromised patients, **neurotoxoplasmosis** caused by *T. gondii* infections are one of the most common causes of brain abscesses.[37] The organism is able to cross the blood-brain barrier by infecting the endothelial cells of capillaries in the brain. The parasite reproduces within these cells, a step that appears to be necessary for entry to the brain, and then causes the endothelial cell to lyse, releasing the progeny into brain tissues. This mechanism is quite different than the method it uses to enter the bloodstream in the first place.[38]

The brain lesions associated with neurotoxoplasmosis can be detected radiographically using MRI or CAT scans (Figure 26.25). Diagnosis can be confirmed by direct observation of the organism in CSF. RT-PCR assays can also be used to detect *T. gondii* through genetic markers.

Treatment of neurotoxoplasmosis caused by *T. gondii* infections requires six weeks of multi-drug therapy with pyrimethamine, sulfadiazine, and folinic acid. Long-term maintenance doses are often required to prevent recurrence.

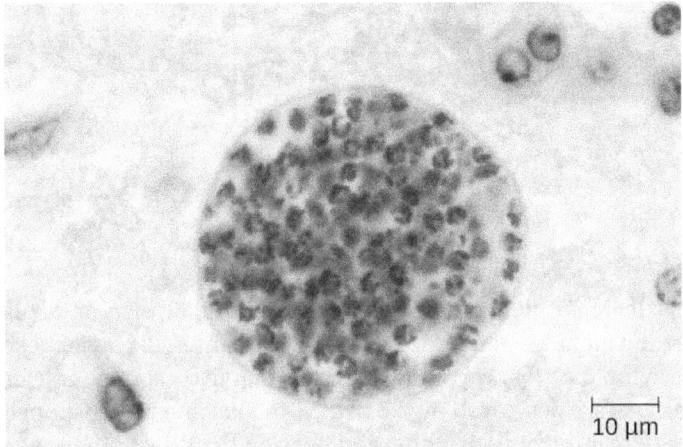

Figure 26.25   This *Toxoplasma gondii* cyst, observed in mouse brain tissue, contains thousands of inactive parasites. (credit: modification of work by USDA)

✓ CHECK YOUR UNDERSTANDING

- Under what conditions is *Toxoplasma* infection serious?
- How does *Toxoplasma* circumvent the blood-brain barrier?

Neurocysticercosis

Cysticercosis is a parasitic infection caused by the larval form of the pork tapeworm, *Taenia solium*. When the larvae invade the brain and spinal cord, the condition is referred to as **neurocysticercosis**. This condition affects millions of people worldwide and is the leading cause of adult onset epilepsy in the developing world.[39]

The life cycle of *T. solium* is discussed in Helminthic Infections of the Gastrointestinal Tract. Following ingestion, the eggs hatch in the intestine to form larvae called **cysticerci**. Adult tapeworms form in the small intestine and produce eggs that are shed in the feces. These eggs can infect other individuals through fecal contamination of food or other surfaces. Eggs can also hatch within the intestine of the original patient and lead to an ongoing autoinfection. The cystercerci, can migrate to the blood and invade many tissues in the body, including the CNS.

Neurocysticercosis is usually diagnosed through noninvasive techniques. Epidemiological information can be used as an initial screen; cysticercosis is endemic in Central and South America, Africa, and Asia. Radiological imaging (MRI and CT scans) is the primary method used to diagnose neurocysticercosis; imaging can be used to detect the one- to two-centimeter cysts that form around the parasites (Figure 26.26). Elevated levels of

---

36 Carruthers, Vern B., and Yasuhiro Suzuki, "Effects of *Toxoplasma gondii* Infection on the Brain," *Schizophrenia Bulletin* 33, no. 3 (2007): 745-51.

37 Uppal, Gulshan, "CNS Toxoplasmosis in HIV," 2015. Accessed June 30, 2016. http://emedicine.medscape.com/article/1167298-overview#a3.

38 Konradt, Christoph, Norikiyo Ueno, David A. Christian, Jonathan H. Delong, Gretchen Harms Pritchard, Jasmin Herz, David J. Bzik et al., "Endothelial Cells Are a Replicative Niche for Entry of *Toxoplasma gondii* to the Central Nervous System," *Nature Microbiology* 1 (2016): 16001.

39 DeGiorgio, Christopher M., Marco T. Medina, Reyna Durón, Chi Zee, and Susan Pietsch Escueta, "Neurocysticercosis," *Epilepsy Currents* 4, no. 3 (2004): 107-11.

eosinophils in the blood can also indicate a parasitic infection. EIA and ELISA are also used to detect antigens associated with the pathogen.

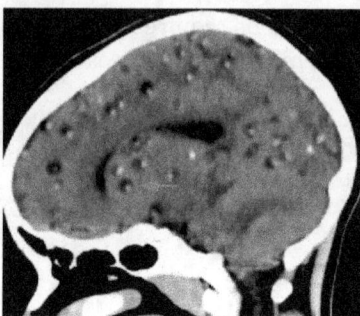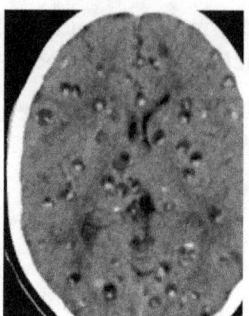

Figure 26.26 Brain CT scans of sagittal (left) and axial (right) sections of a brain with neurocysticercosis. Numerous cysts are visible in both images, as indicated by the arrows. (credit: modification of work by Segamwenge IL, Kioko NP)

The treatment for neurocysticercosis depends on the location, number, size, and stage of cysticerci present. Antihelminthic chemotherapy includes albendazole and praziquantel. Because these drugs kill viable cysts, they may acutely increase symptoms by provoking an inflammatory response caused by the release of *Taenia* cysticerci antigens, as the cysts are destroyed by the drugs. To alleviate this response, corticosteroids that cross the blood-brain barrier (e.g., dexamethasone) can be used to mitigate these effects. Surgical intervention may be required to remove intraventricular cysts.

Disease Profile

Parasitic Diseases of the Nervous System

Parasites that successfully invade the nervous system can cause a wide range of neurological signs and symptoms. Often, they inflict lesions that can be visualized through radiologic imaging. A number of these infections are fatal, but some can be treated (with varying levels of success) by antimicrobial drugs (Figure 26.27).

Parasitic Diseases of the Nervous System					
Disease	Pathogen	Signs and Symptoms	Transmission	Diagnostic Tests	Antimicrobial Drugs
Granulomatous amoebic encephalitis (GAE)	*Acanthamoeba* spp., *Balamuthia mandrillaris*	Inflammation, lesions in CNS, almost always fatal	Freshwater ameobae invade CNS via breaks in skin or sinuses	CT scan, MRI, CSF	Fluconazole, miltefosine, voriconazole
Human African trypanosomiasis	*Trypanosoma brucei gambiense*, *T. brucei rhodesiense*	Chancre, Winterbottom's sign, undulating fever, lethargy, insomnia, usually fatal if untreated	Protozoan transmitted via bite of tsetse fly	Blood smear	Pentamidine and suramine (initial phase); melarsoprol and eflornithine (final phase)
Neurocysticercosis	*Taenia solium*	Brain cysts, epilepsy	Ingestion of tapeworm eggs in fecally contaminated food or surfaces	CT scan, MRI	Albendazole, praziquantel, dexamethasone
Neurotoxoplasmosis	*Toxoplasma gondii*	Brain abscesses, chronic encephalitis	Protozoan transmitted via contact with oocytes in cat feces	CT scan, MRI, CSF	Pyrimethamine, sulfadiazine, folinic acid
Primary amoebic meningoencephalitis (PAM)	*Naegleria fowleri*	Headache, seizures, coma, almost always fatal	Freshwater ameobae invade brain via nasal passages	CSF, IFA, PCR	Miltefosine (experimental)

Figure 26.27

CHECK YOUR UNDERSTANDING

- What neurological condition is associated with neurocysticercosis?
- How is neurocysticercosis diagnosed?

SUMMARY

26.1 Anatomy of the Nervous System

- The nervous system consists of two subsystems: the **central nervous system** and **peripheral nervous system**.
- The skull and three **meninges** (the **dura mater**, **arachnoid mater**, and **pia mater**) protect the brain.
- Tissues of the PNS and CNS are formed of cells called **glial cells** and **neurons**.
- Since the **blood-brain barrier** excludes most microbes, there is no normal microbiota in the CNS.
- Some pathogens have specific virulence factors that allow them to breach the blood-brain barrier. Inflammation of the brain or **meninges** caused by infection is called **encephalitis** or **meningitis**, respectively. These conditions can lead to blindness, deafness, coma, and death.

26.2 Bacterial Diseases of the Nervous System

- **Bacterial meningitis** can be caused by several species of encapsulated bacteria, including *Haemophilus influenzae*, *Neisseria meningitidis*, *Streptococcus pneumoniae*, and *Streptococcus agalactiae* (group B streptococci). *H. influenzae* affects primarily young children and neonates, *N. meningitidis* is the only communicable pathogen and mostly affects children and young adults, *S. pneumoniae* affects mostly young children, and *S. agalactiae* affects newborns during or shortly after birth.
- Symptoms of bacterial meningitis include fever, neck stiffness, headache, confusion, convulsions, coma, and death.
- Diagnosis of bacterial meningitis is made through observations and culture of organisms in CSF. Bacterial meningitis is treated with antibiotics. *H. influenzae* and *N. meningitidis* have vaccines available.
- *Clostridium* species cause neurological diseases, including **botulism** and **tetanus**, by producing potent neurotoxins that interfere with neurotransmitter release. The PNS is typically affected. Treatment of *Clostridium* infection is effective only through early diagnosis with administration of antibiotics to control the infection and antitoxins to neutralize the endotoxin before they enter cells.
- *Listeria monocytogenes* is a foodborne pathogen that can infect the CNS, causing meningitis. The infection can be spread through the placenta to a fetus. Diagnosis is through culture of blood or CSF. Treatment is with antibiotics and there is no vaccine.
- **Hansen's disease** (**leprosy**) is caused by the intracellular parasite *Mycobacterium leprae*. Infections cause demylenation of neurons, resulting in decreased sensation in peripheral appendages and body sites. Treatment is with multi-drug antibiotic therapy, and there is no universally recognized vaccine.

26.3 Acellular Diseases of the Nervous System

- **Viral meningitis** is more common and generally less severe than bacterial menigitis. It can result from secondary sequelae of many viruses or be caused by infections of arboviruses.
- Various types of **arboviral encephalitis** are concentrated in particular geographic locations throughout the world. These mosquito-borne viral infections of the nervous system are typically mild, but they can be life-threatening in some cases.
- **Zika virus** is an emerging arboviral infection with generally mild symptoms in most individuals, but infections of pregnant women can cause the birth defect microcephaly.
- **Polio** is typically a mild intestinal infection but can be damaging or fatal if it progresses to a neurological disease.
- **Rabies** is nearly always fatal when untreated and remains a significant problem worldwide.
- **Transmissible spongiform encephalopathies** such as **Creutzfeldt-Jakob disease** and **kuru** are caused by prions. These diseases are untreatable and ultimately fatal. Similar prion diseases are found in animals.

26.4 Fungal and Parasitic Diseases of the Nervous System

- **Neuromycoses** are uncommon in immunocompetent people, but immunocompromised individuals with fungal infections have high mortality rates. Treatment of neuromycoses require prolonged therapy with antifungal drugs at low doses to avoid side effects and overcome the effect of the blood-brain barrier.
- Some protist infections of the nervous systems

are fatal if not treated, including **primary amoebic meningitis, granulomatous amoebic encephalitis, human African trypanosomiasis,** and **neurotoxoplasmosis.**
- The various forms of ameobic encephalitis caused by the different amoebic infections are typically fatal even with treatment, but they are rare.

- **African trypanosomiasis** is a serious but treatable disease endemic to two distinct regions in sub-Saharan Africa caused by the insect-borne hemoflagellate *Trypanosoma brucei.*
- **Neurocysticercosis** is treated using antihelminthic drugs or surgery to remove the large cysts from the CNS.

REVIEW QUESTIONS
Multiple Choice

1. What is the outermost membrane surrounding the brain called?
   a. pia mater
   b. arachnoid mater
   c. dura mater
   d. alma mater

2. What term refers to an inflammation of brain tissues?
   a. encephalitis
   b. meningitis
   c. sinusitis
   d. meningoencephalitis

3. Nerve cells form long projections called _____.
   a. soma
   b. axons
   c. dendrites
   d. synapses

4. Chemicals called _____ are stored in neurons and released when the cell is stimulated by a signal.
   a. toxins
   b. cytokines
   c. chemokines
   d. neurotransmitters

5. The central nervous system is made up of
   a. sensory organs and muscles.
   b. the brain and muscles.
   c. the sensory organs and spinal cord.
   d. the brain and spinal column.

6. Which of the following organisms causes epidemic meningitis cases at college campuses?
   a. *Haemophilus influenzae* type b
   b. *Neisseria meningitidis*
   c. *Streptococcus pneumoniae*
   d. *Listeria monocytogenes*

7. Which of the following is the most common cause of neonatal meningitis?
   a. *Haemophilus influenzae b*
   b. *Streptococcus agalactiae*
   c. *Neisseria meningitidis*
   d. *Streptococcus pneumoniae*

8. What sign/symptom would NOT be associated with infant botulism?
   a. difficulty suckling
   b. limp body
   c. stiff neck
   d. weak cry

9. Which of the following can NOT be prevented with a vaccine?
   a. tetanus
   b. pneumococcal meningitis
   c. meningococcal meningitis
   d. listeriosis

10. How is leprosy primarily transmitted from person to person?
    a. contaminated toilet seats
    b. shaking hands
    c. blowing nose
    d. sexual intercourse

11. Which of these diseases can be prevented with a vaccine for humans?
    a. eastern equine encephalitis
    b. western equine encephalitis
    c. West Nile encephalitis
    d. Japanese encephalitis

12. Which of these diseases does NOT require the introduction of foreign nucleic acid?
    a. kuru
    b. polio
    c. rabies
    d. St. Louis encephalitis

13. Which of these is true of the Sabin but NOT the Salk polio vaccine?
    a. requires four injections
    b. currently administered in the United States
    c. mimics the normal route of infection
    d. is an inactivated vaccine

14. Which of the following animals is NOT a typical reservoir for the spread of rabies?
    a. dog
    b. bat
    c. skunk
    d. chicken

15. Which of these diseases results in meningitis caused by an encapsulated yeast?
    a. cryptococcosis
    b. histoplasmosis
    c. candidiasis
    d. coccidiomycosis

16. What kind of stain is most commonly used to visualize the capsule of cryptococcus?
    a. Gram stain
    b. simple stain
    c. negative stain
    d. fluorescent stain

17. Which of the following is the causative agent of East African trypanosomiasis?
    a. *Trypanosoma cruzi*
    b. *Trypanosoma vivax*
    c. *Trypanosoma brucei rhodanese*
    d. *Trypanosoma brucei gambiense*

18. Which of the following is the causative agent of primary amoebic meningoencephalitis?
    a. *Naegleria fowleri*
    b. *Entameba histolyticum*
    c. *Amoeba proteus*
    d. *Acanthamoeba polyphaga*

19. What is the biological vector for African sleeping sickness?
    a. mosquito
    b. tsetse fly
    c. deer tick
    d. sand fly

20. How do humans usually contract neurocysticercosis?
    a. the bite of an infected arthropod
    b. exposure to contaminated cat feces
    c. swimming in contaminated water
    d. ingestion of undercooked pork

21. Which of these is the most important cause of adult onset epilepsy?
    a. neurocysticercosis
    b. neurotoxoplasmosis
    c. primary amoebic meningoencephalitis
    d. African trypanosomiasis

Matching

22. Match each strategy for microbial invasion of the CNS with its description.

___intercellular entry

A. pathogen gains entry by infecting peripheral white blood cells

___transcellular entry

B. pathogen bypasses the blood-brain barrier by travel along the olfactory or trigeminal cranial nerves

___leukocyte-facilitated entry

C. pathogen passes through the cells of the blood-brain barrier

___nonhematogenous entry

D. pathogen passes between the cells of the blood-brain barrier

Fill in the Blank

23. The cell body of a neuron is called the _____.
24. A signal is transmitted down the _____ of a nerve cell.
25. The _____ is filled with cerebrospinal fluid.
26. The _____ _____ prevents access of microbes in the blood from gaining access to the central nervous system.
27. The _____ are a set of membranes that cover and protect the brain.
28. The form of meningitis that can cause epidemics is caused by the pathogen _____.
29. The symptoms of tetanus are caused by the neurotoxin _____.
30. _____ is another name for leprosy.
31. Botulism prevents the release of the neurotransmitter _____.
32. _____ is a neurological disease that can be prevented with the DTaP vaccine.
33. Tetanus patients exhibit _____ when muscle spasms causes them to arch their backs.
34. The rogue form of the prion protein is called _____.
35. _____ are the most common reservoir for the rabies virus worldwide.
36. _____ was the scientist who developed the inactivated polio vaccine.
37. _____ is a prion disease of deer and elk.
38. The rogue form of prion protein exists primarily in the _____ conformation.
39. The _____ is the main virulence factor of *Cryptococcus neoformans*.
40. The drug of choice for fungal infections of the nervous system is _____.

41. The larval forms of a tapeworm are known as _____.
42. _____ sign appears as swollen lymph nodes at the back of the neck in early African trypanosomiasis.
43. _____ African trypanosomiasis causes a chronic form of sleeping sickness.
44. The definitive host for *Toxoplasma gondii* is _____.
45. Trypanosomes can evade the immune response through _____ variation.

Short Answer

46. Briefly describe the defenses of the brain against trauma and infection.
47. Describe how the blood-brain barrier is formed.
48. Identify the type of cell shown, as well as the following structures: axon, dendrite, myelin sheath, soma, and synapse.

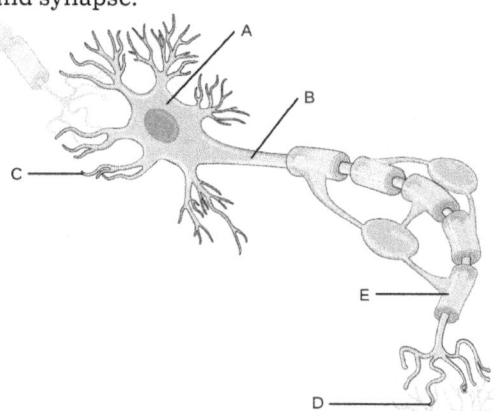

49. A physician suspects the lesion and pustule pictured here are indicative of tuberculoid leprosy. If the diagnosis is correct, what microorganism would be found in a skin biopsy?

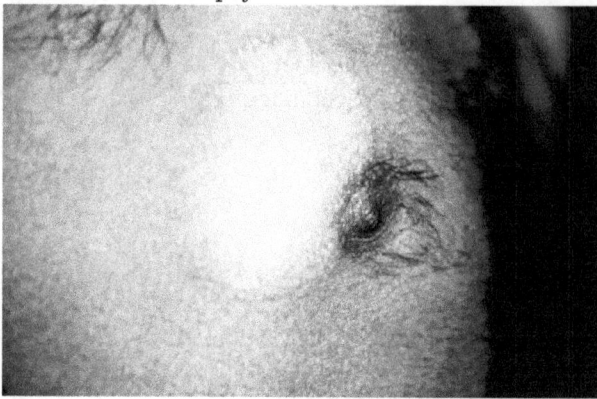

Figure 26.28 (credit: Centers for Disease Control and Prevention)

50. Explain how a person could contract variant Creutzfeldt-Jakob disease by consuming products from a cow with bovine spongiform encephalopathy (mad cow disease).
51. Why do nervous system infections by fungi require such long treatment times?
52. Briefly describe how humans are infected by *Naegleria fowleri*.
53. Briefly describe how humans can develop neurocysticercosis.

Critical Thinking

54. What important function does the blood-brain barrier serve? How might this barrier be problematic at times?
55. Explain how tetanospasmin functions to cause disease.
56. The most common causes of bacterial meningitis can be the result of infection by three very different bacteria. Which bacteria are they and how are these microbes similar to each other?
57. Explain how infant botulism is different than foodborne botulism.
58. If the Sabin vaccine is being used to eliminate polio worldwide, explain why a country with a near zero infection rate would opt to use the Salk vaccine but not the Sabin vaccine?

59. The graph shown tracks the body temperature of a patient infected with *Trypanosoma brucei*. How would you describe this pattern, and why does it occur?

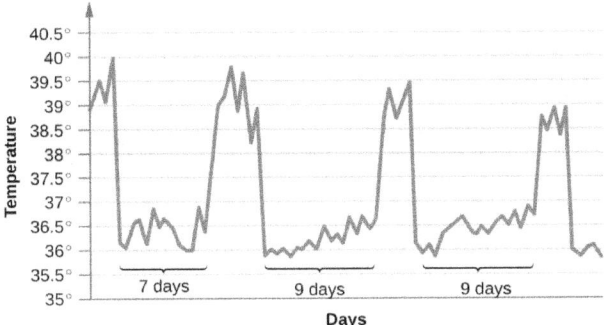

Figure 26.29 (credit: modification of work by Wellcome Images)

60. Fungal meningoencephalitis is often the ultimate cause of death for AIDS patients. What factors make these infections more problematic than those of bacterial origin?

61. Compare East African trypanosomiasis with West African trypanosomiasis.

APPENDIX A

Fundamentals of Physics and Chemistry Important to Microbiology

Like all other matter, the matter that comprises microorganisms is governed by the laws of chemistry and physics. The chemical and physical properties of microbial pathogens—both cellular and acellular—dictate their habitat, control their metabolic processes, and determine how they interact with the human body. This appendix provides a review of some of the fundamental principles of chemistry and physics that are essential to an understanding of microbiology. Many of the chapters in this text—especially Microbial Biochemistry and Microbial Metabolism—assume that the reader already has an understanding of the concepts reviewed here.

Atomic Structure

Life is made up of matter. Matter occupies space and has mass. All matter is composed of **atoms**. All atoms contain **protons**, **electrons**, and **neutrons** (Figure A1). The only exception is hydrogen (H), which is made of one proton and one electron. A proton is a positively charged particle that resides in the nucleus (the core of the atom) of an atom and has a mass of 1 atomic mass unit (amu) and a charge of +1. An electron is a negatively charged particle that travels in the space around the nucleus. Electrons are distributed in different energy levels called electron shells. Electrons have a negligible mass and a charge of –1. Neutrons, like protons, reside in the nucleus of an atom. They have a mass of 1 amu and no charge (neutral). The positive (proton) and negative (electron) charges balance each other in a neutral atom, which has a net zero charge. Because protons and neutrons each have a mass of 1 amu, the mass of an atom is equal to the number of protons and neutrons of that atom. The number of electrons does not factor into the overall mass because electron mass is so small.

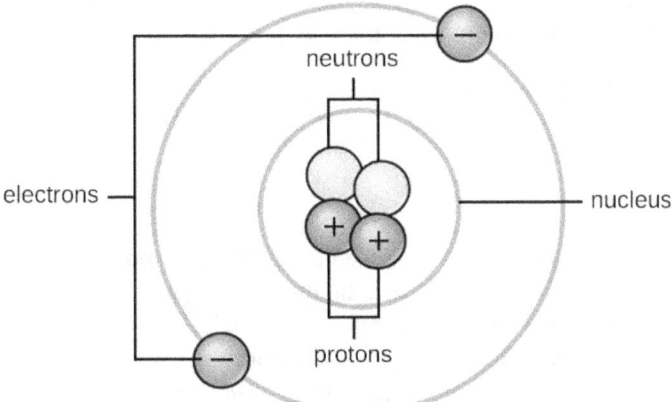

Figure A1 Atoms are made up of protons and neutrons located within the nucleus and electrons surrounding the nucleus.

Chemical Elements

All matter is composed of atoms of **elements**. Elements have unique physical and chemical properties and are substances that cannot easily be transformed either physically or chemically into other substances. Each element has been given a name, usually derived from Latin or English. The elements also have one- or two-letter symbols representing the name; for example, sodium (Na), gold (Au), and silver (Ag) have abbreviations derived from their original Latin names *natrium*, *aurum*, and *argentum*, respectively. Examples with English abbreviations are carbon (C), hydrogen (H), oxygen (O), and nitrogen (N). A total of 118 different elements (92 of which occur naturally) have been identified and organized into the periodic table of elements. Of the naturally occurring elements, fewer than 30 are found in organisms on Earth, and four of those (C, H, O, and N) make up approximately 96% of the mass of an organism.[1]

Each unique element is identified by the number of protons in its atomic nucleus. In addition to protons, each element's atomic nucleus contains an equal or greater number of neutrons (with the exception of hydrogen, which has only one proton). The total number of protons per element is described as the **atomic number**, and the combined mass of protons and neutrons is called the atomic mass or **mass number**. Therefore, it is possible to determine the number of neutrons by subtracting the atomic number from the mass number.

Isotopes are different forms of the same element that have the same number of protons, but a different number of neutrons. Many elements have several isotopes with one or two commonly occurring isotopes in nature. For example, carbon-12 (^{12}C), the most common isotope of carbon (98.6% of all C found on Earth),[2] contains six protons and six neutrons. Therefore, it has a mass number of 12 (6 protons + 6 neutrons) and an atomic number of 6.

There are two additional types of isotopes in nature: heavy isotopes, and radioisotopes. Heavy isotopes have one or more extra neutrons while still maintaining a stable atomic nucleus. An example of a heavy isotope is carbon-13 (^{13}C) (1.1% of all carbon).[3] ^{13}C has a mass number of 13 (6 protons + 7 neutrons). Since the atomic number of ^{13}C is 6, it is still the element carbon; however, it has more mass than the more common form of the element, ^{12}C, because of the extra neutron in the nucleus. Carbon-14 (^{14}C) (0.0001% of all carbon)[4] is an example of a radioisotope. ^{14}C has a mass number of 14 (6 protons + 8 neutrons); however, the extra neutrons in ^{14}C result in an unstable nucleus. This instability leads to the process of radioactive decay. Radioactive decay involves the loss of one or more neutrons and the release of energy in the form of gamma rays, alpha particles, or beta particles (depending on the isotope).

Heavy isotopes and radioisotopes of carbon and other elements have proven to be useful in research, industry, and medicine.

Chemical Bonds

There are three types of chemical bonds that are important when describing the interaction of atoms both within and between molecules in microbiology: (1) covalent bonds, which can be either polar or non-polar, (2) ionic bonds, and (3) hydrogen bonds. There are other types of interactions such as *London* dispersion forces and *van der Waals* forces that could also be discussed when describing the physical and chemical properties of the intermolecular interactions of atoms, but we will not include descriptions of these forces here.

Chemical bonding is determined by the outermost shell of electrons, called the valence electrons (VE), of an atom. The number of VE is important when determining the number and type of chemical bonds an atom will form.

Covalent Bonds

The strongest chemical bond between two or more atoms is a **covalent bond**. These bonds form when an electron is shared between two atoms, and these are the most common form of chemical bond in living organisms. Covalent bonds form between the atoms of elements that make up the biological molecules in our cells. An example of a simple molecule formed with covalent bonds is water, H_2O, with one VE per H atom and 6 VE per O atom. Because of the VE configuration, each H atom is able to accept one additional VE and each O atom is able to accept two additional VE. When sharing electrons, the hydrogen and oxygen atoms that combine to form water molecules become bonded together by covalent bonds (Figure A2). The electron from the hydrogen atom divides its time between the outer electron shell of the hydrogen atom and the outermost electron shell of the oxygen atom. To completely fill the outer shell of an oxygen atom, two electrons from two hydrogen atoms are needed, hence the subscript "2" indicating two atoms of H in a molecule of H_2O. This sharing is a lower energy state for all of the atoms involved than if they existed without their outer shells filled.

---

1 Schrijver, Karel, and Iris Schrijver. *Living with the Stars: How the Human Body Is Connected to the Life Cycles of the Earth, the Planets, and the Stars*. Oxford University Press, USA, 2015.

2 National Oceanic and Atmospheric Administration, "Stable and Radiocarbon Isotopes of Carbon Dioxide." Web page. Accessed Feb 19, 2016 [http://www.esrl.noaa.gov/gmd/outreach/isotopes/chemistry.html]

3 ibid.

4 ibid.

There are two types of covalent bonds: polar and nonpolar. **Nonpolar covalent** bonds form between two atoms of the same or different elements that share the electrons equally (Figure A2). In a **polar covalent bond**, the electrons shared by the atoms spend more time closer to one nucleus than to the other nucleus. Because of the unequal distribution of electrons between the different nuclei, a slightly positive (δ+) or slightly negative (δ−) charge develops. Water is an example of a molecule formed with **polar covalent bonds** (Figure A2).

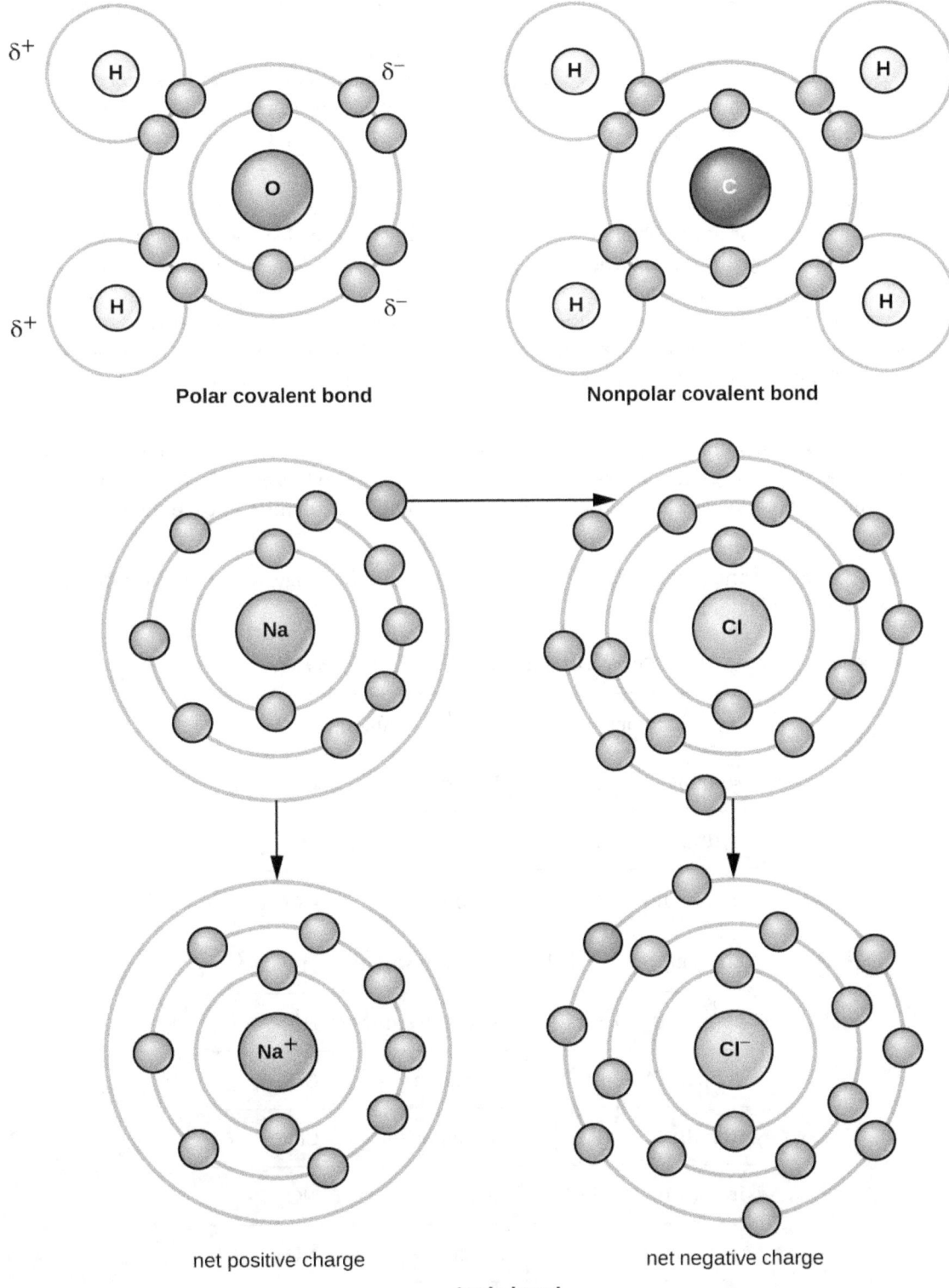

Figure A2 The water molecule (top left) depicts a polar bond with a slightly positive charge on the hydrogen atoms and a slightly negative charge on the oxygen. Methane (top right) is an example of a nonpolar covalent bond. Sodium chloride (bottom) is a substance formed from ionic bonds between sodium and chlorine.

Ions and Ionic Bonds

When an atom does not contain equal numbers of protons and electrons, it is called an **ion**. Because the number of electrons does not equal the number of protons, each ion has a net charge. Positive ions are formed by losing electrons and are called **cations**. Negative ions are formed by gaining electrons and are called **anions**.

For example, a sodium atom has only has one electron in its outermost shell. It takes less energy for the sodium atom to donate that one electron than it does to accept seven more electrons, which it would need to fill its outer shell. If the sodium atom loses an electron, it now has 11 protons and only 10 electrons, leaving it with an overall charge of +1. It is now called a sodium ion (Na^+).

A chlorine atom has seven electrons in its outer shell. Again, it is more energy efficient for the chlorine atom to gain one electron than to lose seven. Therefore, it will more likely gain an electron to form an ion with 17 protons and 18 electrons, giving it a net negative (−1) charge. It is now called a chloride ion (Cl^-). This movement of electrons from one atom to another is referred to as electron transfer. Because positive and negative charges attract, these ions stay together and form an **ionic bond**, or a bond between ions. When Na^+ and Cl^- ions combine to produce NaCl, an electron from a sodium atom stays with the other seven from the chlorine atom, and the sodium and chloride ions attract each other in a lattice of ions with a net zero charge (Figure A2).

Polyatomic ions consist of multiple atoms joined by covalent bonds; but unlike a molecule, a polyatomic ion has a positive or negative charge. It behaves as a cation or anion and can therefore form ionic bonds with other ions to form ionic compounds. The atoms in a polyatomic ion may be from the same element or different elements.

Table A1 lists some cations and anions that commonly occur in microbiology. Note that this table includes monoatomic as well as polyatomic ions.

Some Common Ions in Microbiology

Cations		Anions	
sodium	Na^+	chloride	Cl^-
hydrogen	H^+	bicarbonate	HCO_3^-
potassium	K^+	carbonate	CO_3^{2-}
ammonium	NH_4^+	hydrogen sulfate	$H_2SO_4^{2-}$
copper (I)	Cu^+	hydrogen sulfide	HS^-
copper (II)	Cu^{2+}	hydroxide	OH^-
iron (II)	Fe^{2+}	hypochlorite	ClO^-
iron (III)	Fe^{3+}	nitrite	NO_2^-
		nitrate	NO_3^-
		peroxide	O_2^{2-}

Some Common Ions in Microbiology

Cations		Anions	
		phosphate	PO_4^{3-}
		pyrophosphate	$P_2O_7^{4-}$
		sulfite	SO_3^{2-}
		thiosulfate	$S_2O_3^{2-}$

Table A1

Molecular Formula, Molecular Mass, and the Mole

For molecules formed by covalent bonds, the molecular formula represents the number and types of elemental atoms that compose the molecule. As an example, consider a molecule of glucose, which has the molecular formula $C_6H_{12}O_6$. This molecular formula indicates that a single molecule of glucose is formed from six carbon atoms, twelve hydrogen atoms, and six oxygen atoms.

The **molecular mass** of a molecule can be calculated using the molecular formula and the atomic mass of each element in the molecule. The number of each type of atom is multiplied by the atomic mass; then the products are added to get the molecular mass. For example the molecular mass of glucose, $C_6H_{12}O_6$ (Figure A3), is calculated as:

$$\text{mass of carbon} = 12 \tfrac{\text{amu}}{\text{atom}} \times 6 \text{ atoms} = 72 \text{ amu}$$
$$\text{mass of hydrogen} = 1 \tfrac{\text{amu}}{\text{atom}} \times 12 \text{ atoms} = 12 \text{ amu}$$
$$\text{mass of oxygen} = 16 \tfrac{\text{amu}}{\text{atom}} \times 6 \text{ atoms} = 96 \text{ amu}$$
$$\text{molecular mass of glucose} = 72 \text{ amu} + 12 \text{ amu} + 96 \text{ amu} = 180 \text{ amu}$$

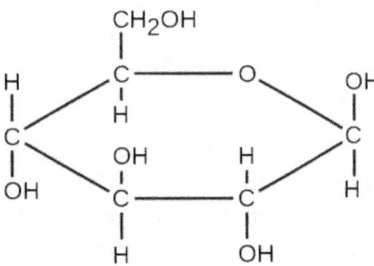

Figure A3 The molecular structure of glucose showing the numbers of carbon, oxygen, and hydrogen atoms. Glucose has a molecular mass of 180 amu.

The number of entities composing a mole has been experimentally determined to be 6.022×10^{23}, a fundamental constant named **Avogadro's number** (NA) or the Avogadro constant. This constant is properly reported with an explicit unit of "per mole."

Energy

Thermodynamics refers to the study of energy and energy transfer involving physical matter.

Matter participating in a particular case of energy transfer is called a system, and everything outside of that matter is called the surroundings. There are two types of systems: open and closed. In an **open system**, energy can be exchanged with its surroundings. A **closed system** cannot exchange energy with its surroundings. Biological organisms are open systems. Energy is exchanged between them and their surroundings as they use energy from the sun to perform photosynthesis or consume energy-storing molecules and release energy to the environment by doing work and releasing heat. Like all things in the physical world, energy is subject to physical laws. In general, energy is defined as the ability to do work, or to create some kind of change. Energy exists in different forms. For example, electrical energy, light energy, and heat energy are all different types of energy. The **first law of thermodynamics,** often referred to as the law of conservation of energy, states that the total amount of energy in the universe is constant and conserved. Energy exists in many different forms. According to the first law of thermodynamics, energy may be transferred from place to place or transformed into different forms, but it cannot be created or destroyed.

The challenge for all living organisms is to obtain energy from their surroundings in forms that they can transfer or transform into usable energy to do work. Microorganisms have evolved to meet this challenge. Chemical energy stored within organic molecules such as sugars and fats is transferred and transformed through a series of cellular chemical reactions into energy within molecules of ATP. Energy in ATP molecules is easily accessible to do work. Examples of the types of work that cells need to do include building complex molecules, transporting materials, powering the motion of cilia or flagella, and contracting muscle fibers to create movement.

A microorganism's primary tasks of obtaining, transforming, and using energy to do work may seem simple. However, the **second law of thermodynamics** explains why these tasks are more difficult than they appear. All energy transfers and transformations are never completely efficient. In every energy transfer, some amount of energy is lost in a form that is unusable. In most cases, this form is **heat energy**. Thermodynamically, heat energy is defined as the energy transferred from one system to another that is not work. For example, some energy is lost as heat energy during cellular metabolic reactions.

The more energy that is lost by a system to its surroundings, the less ordered and more random the system is. Scientists refer to the measure of randomness or disorder within a system as **entropy**. High entropy means high disorder and low energy. Molecules and chemical reactions have varying entropy as well. For example, entropy increases as molecules at a high concentration in one place diffuse and spread out. The second law of thermodynamics says that energy will always be lost as heat in energy transfers or transformations. Microorganisms are highly ordered, requiring constant energy input to be maintained in a state of low entropy.

Chemical Reactions

Chemical reactions occur when two or more atoms bond together to form molecules or when bonded atoms are broken apart. The substances used in a chemical reaction are called the **reactants** (usually found on the left side of a chemical equation), and the substances produced by the reaction are known as the **products** (usually found on the right side of a chemical equation). An arrow is typically drawn between the reactants and products to indicate the direction of the chemical reaction; this direction is not always a "one-way street."

An example of a simple chemical reaction is the breaking down of hydrogen peroxide molecules, each of which consists of two hydrogen atoms bonded to two oxygen atoms (H_2O_2). The reactant hydrogen peroxide is broken down into water, containing one oxygen atom bound to two hydrogen atoms (H_2O), and oxygen, which consists of two bonded oxygen atoms (O_2). In the equation below, the reaction includes two hydrogen peroxide molecules and two water molecules. This is an example of a balanced chemical equation, wherein the number of atoms of each element is the same on each side of the equation. According to the law of conservation of matter, the number of atoms before and after a chemical reaction should be equal, such that no atoms are, under normal circumstances, created or destroyed.

$$2H_2O_2 \text{ (hydrogen peroxide)} \longrightarrow 2H_2O \text{ (water)} + O_2 \text{ (oxygen)}$$

Some chemical reactions, such as the one shown above, can proceed in one direction until the reactants are all used up. Equations that describe these reactions contain a unidirectional arrow and are irreversible. **Reversible reactions** are those that can go in either direction. In reversible reactions, reactants are turned into products, but when the concentration of product rises above a certain threshold (characteristic of the particular reaction), some of these products will be converted back into reactants; at this point, the designations of products and reactants are reversed. The changes in concentration continue until a certain relative balance in concentration between reactants and products occurs—a state called **chemical equilibrium**. At this point, both the forward and reverse reactions continue to occur, but they do so at the same rate, so the concentrations of reactants and products do not change. These situations of reversible reactions are often denoted by a chemical equation with a double-headed arrow pointing towards both the reactants and products. For example, when carbon dioxide dissolves in water, it can do so as a gas dissolved in water *or* by reacting with water to produce carbonic acid. In the cells of some microorganisms, the rate of carbonic acid production is accelerated by the enzyme carbonic anhydrase, as indicated in the following equation:

$$CO_2 + H_2O \xrightleftharpoons{\text{carbonic anhydrase}} H_2CO_3 \rightleftharpoons H^+ + HCO_3^-$$

Properties of Water and Solutions

The hydrogen and oxygen atoms within water molecules form polar covalent bonds. There is no overall charge to a water molecule, but there is one ∂^+ on each hydrogen atom and two ∂^- on the oxygen atom. Each water molecule attracts other water molecules because of the positive and negative charges in the different parts of the molecule (Figure A4). Water also attracts other polar molecules (such as sugars), forming hydrogen bonds. When a substance readily forms hydrogen bonds with water, it can dissolve in water and is referred to as **hydrophilic** ("water-loving"). Hydrogen bonds are not readily formed with nonpolar substances like oils and fats. These nonpolar compounds are **hydrophobic** ("water-fearing") and will orient away from and avoid water.

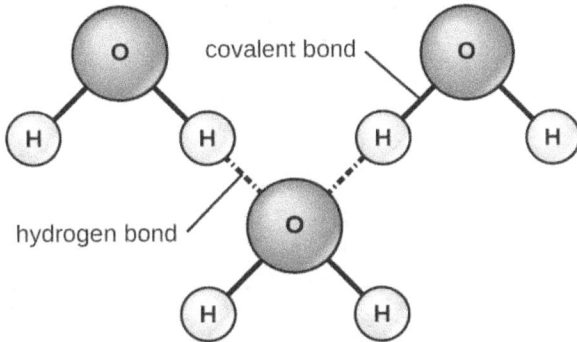

Figure A4 Hydrogen bonds form between slightly positive ($\partial+$) and slightly negative ($\partial-$) charges of polar covalent molecules such as water.

The hydrogen bonds in water allow it to absorb and release heat energy more slowly than many other substances. This means that water moderates temperature changes within organisms and in their environments. As energy input continues, the balance between hydrogen-bond formation and breaking swings toward fewer hydrogen bonds: more bonds are broken than are formed. This process results in the release of individual water molecules at the surface of the liquid (such as a body of water, the leaves of a plant, or the skin of an organism) in a process called **evaporation**.

Conversely, as molecular motion decreases and temperatures drop, less energy is present to break the hydrogen bonds between water molecules. These bonds remain intact and begin to form a rigid, lattice-like structure (e.g., ice). When frozen, ice is less dense (the molecules are farther apart) than liquid water. This means that ice floats on the surface of a body of water. In lakes, ponds, and oceans, ice will form on the surface of the water, creating an insulating barrier to protect the animal and plant life beneath from freezing in the water. If this did not happen, plants and animals living in water would freeze in a block of ice and could not move freely, making life in cold temperatures difficult or impossible.

Because water is polar, with slight positive and negative charges, ionic compounds and polar molecules can readily dissolve in it. Water is, therefore, what is referred to as a solvent—a substance capable of dissolving another substance. The charged particles will form hydrogen bonds with a surrounding layer of water molecules. This is referred to as a **sphere of hydration** and serves to keep the ions separated or dispersed in the water (Figure A5). These spheres of hydration are also referred to as hydration shells. The polarity of the water molecule makes it an effective solvent and is important in its many roles in living systems.

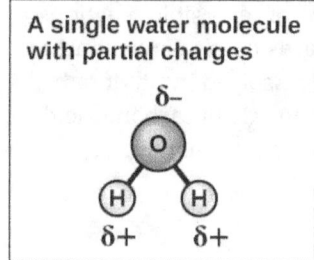
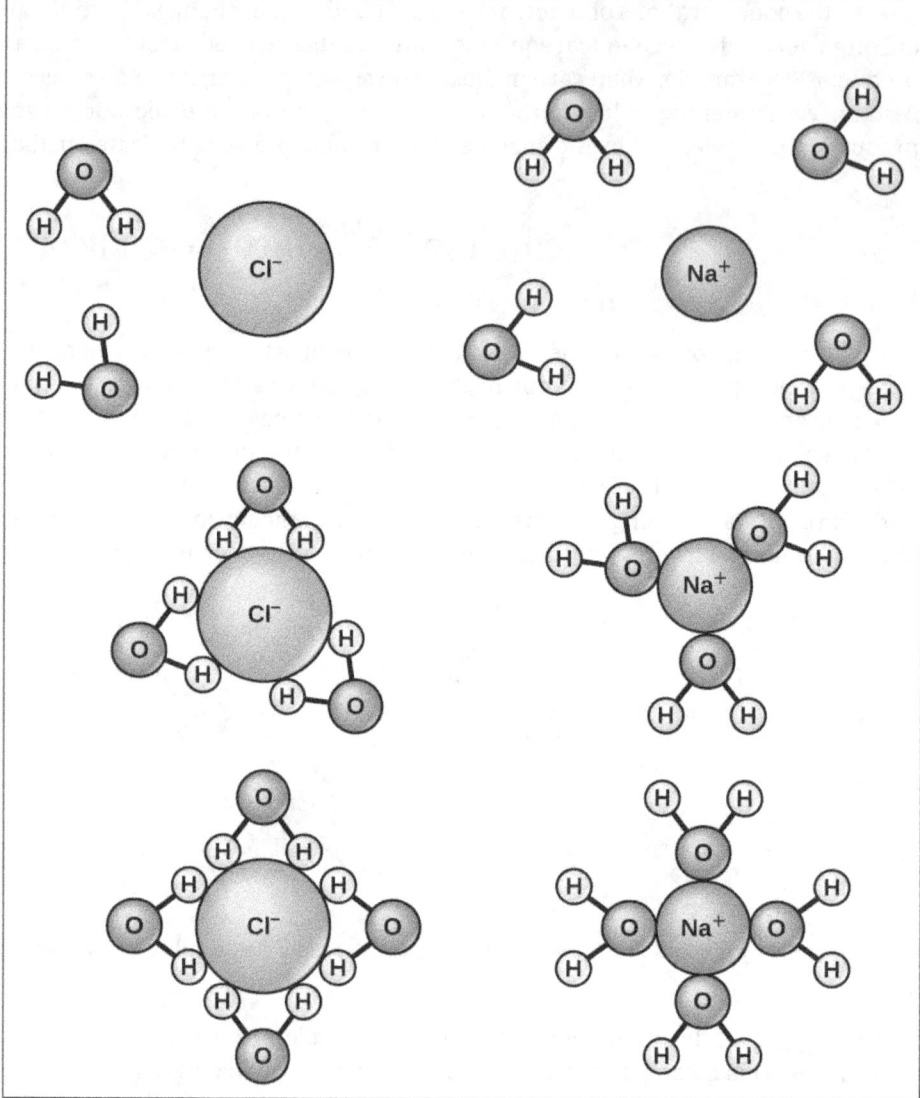

Figure A5 When table salt (NaCl) is mixed in water, spheres of hydration form around the ions.

The ability of insects to float on and skate across pond water results from the property of **cohesion**. In cohesion, water molecules are attracted to each other (because of hydrogen bonding), keeping the molecules together at the liquid-air (gas) interface. Cohesion gives rise to surface tension, the capacity of a substance to withstand rupture when placed under tension or stress.

These cohesive forces are also related to water's property of **adhesion**, or the attraction between water molecules and other molecules. This is observed when water "climbs" up a straw placed in a glass of water. You will notice that the water appears to be higher on the sides of the straw than in the middle. This is because the water molecules are attracted to the straw and therefore adhere to it.

Cohesion and adhesion are also factors in bacterial colonies and biofilm formation. Cohesion keeps the colony intact (helps it "stick" to a surface), while adhesion keeps the cells adhered to each other. Cohesive and adhesive forces are important for sustaining life. For example, because of these forces, water in natural surroundings provides the conditions necessary to allow bacterial and archaeal cells to adhere and accumulate on surfaces.

Acids and Bases

The **pH** of a solution is a measure of hydrogen ion (H^+) and hydroxide ion (OH^-) concentrations and is described as **acidity** or **alkalinity,** respectively. Acidity and alkalinity (also referred to as basicity) can be measured and calculated. pH can be simply represented by the mathematic equation, $pH = -\log_{10}[H^+]$. On the left side of the equation, the "p" means "the negative logarithm of " and the H represents the $[H^+]$. On the right side of the equation, $[H^+]$ is the concentration of H^+ in moles/L. What is not represented in this simple equation is the contribution of the OH^-, which also participates in acidity or alkalinity. Calculation of pH results in a number range of 0 to 14 called the pH scale (Figure A6). A pH value between 0 and 6.9 indicates an acid. It is also referred to as a low pH, due to a high $[H^+]$ and low $[OH^-]$ concentration. A pH value between 7.1 and 14 indicates an alkali or base. It is also referred to as a high pH, due to a low $[H^+]$ and high $[OH^-]$ concentration. A pH of 7 is described as a neutral pH and occurs when $[H^+]$ equals $[OH^-]$.

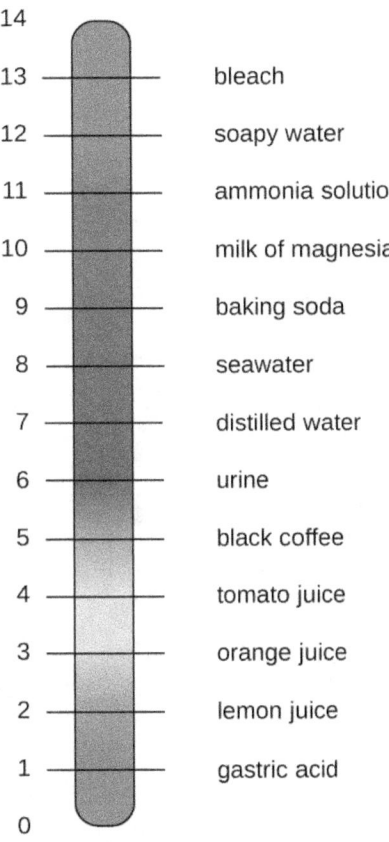

Figure A6 The pH scale measures the concentration of hydrogen ions $[H^+]$ and $[OH^-]$ in a substance. (credit: modification of work by Edward Stevens)

A change of one unit on the pH scale represents a change in the $[H^+]$ by a factor of 10, a change in two units represents a change in the $[H^+]$ by a factor of 100. Thus, small changes in pH represent large changes in $[H^+]$.

APPENDIX B

Mathematical Basics

Squares and Other Powers

An exponent, or a power, is mathematical shorthand for repeated multiplications. For example, the exponent "2" means to multiply the base for that exponent by itself (in the example here, the base is "5"):

$$5^2 = 5 \times 5 = 25$$

The exponent is "2" and the base is the number "5." This expression (multiplying a number by itself) is also called a square. Any number raised to the power of 2 is being squared. Any number raised to the power of 3 is being cubed:

$$5^3 = 5 \times 5 \times 5 = 125$$

A number raised to the fourth power is equal to that number multiplied by itself four times, and so on for higher powers. In general:

$$n^x = n \times n^{x-1}$$

Calculating Percents

A percent is a way of expressing a fractional amount of something using a whole divided into 100 parts. A percent is a ratio whose denominator is 100. We use the percent symbol, %, to show percent. Thus, 25% means a ratio of $\frac{25}{100}$, 3% means a ratio of $\frac{3}{100}$, and 100 % percent means $\frac{100}{100}$, or a whole.

Converting Percents

A percent can be converted to a fraction by writing the value of the percent as a fraction with a denominator of 100 and simplifying the fraction if possible.

$$25\% = \frac{25}{100} = \frac{1}{4}$$

A percent can be converted to a decimal by writing the value of the percent as a fraction with a denominator of 100 and dividing the numerator by the denominator.

$$10\% = \frac{10}{100} = 0.10$$

To convert a decimal to a percent, write the decimal as a fraction. If the denominator of the fraction is not 100, convert it to a fraction with a denominator of 100, and then write the fraction as a percent.

$$0.833 = \frac{833}{1000} = \frac{83.3}{100} = 83.3\%$$

To convert a fraction to a percent, first convert the fraction to a decimal, and then convert the decimal to a percent.

$$\frac{3}{4} = 0.75 = \frac{75}{100} = 75\%$$

Suppose a researcher finds that 15 out of 23 students in a class are carriers of *Neisseria meningitides*. What percentage of students are carriers? To find this value, first express the numbers as a fraction.

$$\frac{\text{carriers}}{\text{total students}} = \frac{15}{23}$$

Then divide the numerator by the denominator.

$$\frac{15}{23} = 15 \div 23 \approx 0.65$$

Finally, to convert a decimal to a percent, multiply by 100.

$$0.65 \times 100 = 65\%$$

The percent of students who are carriers is 65%.

You might also get data on occurrence and non-occurrence; for example, in a sample of students, 9 tested positive for *Toxoplasma* antibodies, while 28 tested negative. What is the percentage of seropositive students? The first step is to determine the "whole," of which the positive students are a part. To do this, sum the positive and negative tests.

$$\text{positive} + \text{negative} = 9 + 28 = 37$$

The whole sample consisted of 37 students. The fraction of positives is:

$$\frac{\text{positive}}{\text{total students}} = \frac{9}{37}$$

To find the percent of students who are carriers, divide the numerator by the denominator and multiply by 100.

$$\frac{9}{37} = 9 \div 37 \approx 0.24$$
$$0.24 \times 100 = 24\%$$

The percent of positive students is about 24%.

Another way to think about calculating a percent is to set up equivalent fractions, one of which is a fraction with 100 as the denominator, and cross-multiply. The previous example would be expressed as:

$$\frac{9}{37} = \frac{x}{100}$$

Now, cross multiply and solve for the unknown:

$9 \times 100 = 37x$	
$\frac{9 \times 100}{37} = x$	Divide both sides by 37
$\frac{900}{37} = x$	Multiply
$24 \approx x$	Divide

The answer, rounded, is the same.

Multiplying and Dividing by Tens

In many fields, especially in the sciences, it is common to multiply decimals by powers of 10. Let's see what happens when we multiply 1.9436 by some powers of 10.

$$1.9436\,(10) = 19.436$$
$$1.9436\,(100) = 194.36$$
$$1.9436\,(1000) = 1943.6$$

The number of places that the decimal point moves is the same as the number of zeros in the power of ten. Table B1 summarizes the results.

Multiply by	Zeros	Decimal point moves . . .
10	1	1 place to the right
100	2	2 places to the right

Multiply by	Zeros	Decimal point moves . . .
1,000	3	3 places to the right
10,000	4	4 places to the right

Table B1

We can use this pattern as a shortcut to multiply by powers of ten instead of multiplying using the vertical format. We can count the zeros in the power of 10 and then move the decimal point that same number of places to the right.

So, for example, to multiply 45.86 by 100, move the decimal point 2 places to the right.

$$45.86 \times 100 = 4586.$$

Sometimes when we need to move the decimal point, there are not enough decimal places. In that case, we use zeros as placeholders. For example, let's multiply 2.4 by 100. We need to move the decimal point 2 places to the right. Since there is only one digit to the right of the decimal point, we must write a 0 in the hundredths place.

$$2.4 \times 100 = 240.$$

When dividing by powers of 10, simply take the opposite approach and move the decimal to the left by the number of zeros in the power of ten.

Let's see what happens when we divide 1.9436 by some powers of 10.

$$1.9436 \div 10 = 0.19436$$
$$1.9436 \div 100 = 0.019436$$
$$1.9436 \div 1000 = 0.0019436$$

If there are insufficient digits to move the decimal, add zeroes to create places.

Scientific Notation

Scientific notation is used to express very large and very small numbers as a product of two numbers. The first number of the product, the digit term, is usually a number not less than 1 and not greater than 10. The second number of the product, the exponential term, is written as 10 with an exponent. Some examples of scientific notation are given in Table B2.

Standard Notation	Scientific Notation
1000	1×10^3
100	1×10^2
10	1×10^1
1	1×10^0
0.1	1×10^{-1}
0.01	1×10^{-2}

Table B2

Scientific notation is particularly useful notation for very large and very small numbers, such as 1,230,000,000 = 1.23×10^9, and 0.00000000036 = 3.6×10^{-10}.

Expressing Numbers in Scientific Notation

Converting any number to scientific notation is straightforward. Count the number of places needed to move the decimal next to the left-most non-zero digit: that is, to make the number between 1 and 10. Then multiply that number by 10 raised to the number of places you moved the decimal. The exponent is positive if you moved the decimal to the left and negative if you moved the decimal to the right. So

$$2386 = 2.386 \times 1000 = 2.386 \times 10^3$$

and

$$0.123 = 1.23 \times 0.1 = 1.23 \times 10^{-1}$$

The power (exponent) of 10 is equal to the number of places the decimal is shifted.

Logarithms

The common logarithm (log) of a number is the power to which 10 must be raised to equal that number. For example, the common logarithm of 100 is 2, because 10 must be raised to the second power to equal 100. Additional examples are in Table B3.

Number	Exponential Form	Common Logarithm
1000	10^3	3
10	10^1	1
1	10^0	0
0.1	10^{-1}	−1
0.001	10^{-3}	−3

Table B3

To find the common logarithm of most numbers, you will need to use the LOG button on a calculator.

Rounding and Significant Digits

In reporting numerical data obtained via measurements, we use only as many significant figures as the accuracy of the measurement warrants. For example, suppose a microbiologist using an automated cell counter determines that there are 525,341 bacterial cells in a one-liter sample of river water. However, she records the concentration as 525,000 cells per liter and uses this rounded number to estimate the number of cells that would likely be found in 10 liters of river water. In this instance, the last three digits of the measured quantity are not considered *significant*. They are rounded to account for variations in the number of cells that would likely occur if more samples were measured.

The importance of significant figures lies in their application to fundamental computation. In addition and subtraction, the sum or difference should contain as many digits to the right of the decimal as that in the *least* certain (indicated by underscoring in the following example) of the numbers used in the computation.

Suppose a microbiologist wishes to calculate the total mass of two samples of agar.

$$\begin{array}{r} 4.38\underline{3} \text{ g} \\ \underline{3.002\underline{1} \text{ g}} \\ 7.38\underline{5} \text{ g} \end{array}$$

The least certain of the two masses has three decimal places, so the sum must have three decimal places.

In multiplication and division, the product or quotient should contain no more digits than than in the factor containing the *least* number of significant figures. Suppose the microbiologist would like to calculate how much of a reagent would be present in 6.6 mL if the concentration is 0.638 g/mL.

$$0.63\underline{8}\ \frac{g}{mL} \times 6.\underline{6}\ mL = 4.1\ g$$

Again, the answer has only one decimal place because this is the accuracy of the least accurate number in the calculation.

When rounding numbers, increase the retained digit by 1 if it is followed by a number larger than 5 ("round up"). Do not change the retained digit if the digits that follow are less than 5 ("round down"). If the retained digit is followed by 5, round up if the retained digit is odd, or round down if it is even (after rounding, the retained digit will thus always be even).

Generation Time

It is possible to write an equation to calculate the cell numbers at any time if the number of starting cells and doubling time are known, as long as the cells are dividing at a constant rate. We define N_0 as the starting number of bacteria, the number at time $t = 0$. N_i is the number of bacteria at time $t = i$, an arbitrary time in the future. Finally we will set j equal to the number of generations, or the number of times the cell population doubles during the time interval. Then we have,

$$N_i = N_0 \times 2^j$$

This equation is an expression of growth by binary fission.

In our example, $N_0 = 4$, the number of generations, j, is equal to 3 after 90 minutes because the generation time is 30 minutes. The number of cells can be estimated from the following equation:

$$\begin{aligned} N_i &= N_0 \times 2^j \\ N_{90} &= 4 \times 2^3 \\ N_{90} &= 4 \times 8 = 32 \end{aligned}$$

The number of cells after 90 minutes is 32.

Most Probable Number

The table in Figure B1 contains values used to calculate the most probable number example given in How Microbes Grow.

Most Probable Number Table

Number of tubes giving a positive reaction for a 5-tube set			MPN (per 100 ml)	95% Confidence Limits	
10 ml	1 ml	0.1 ml		Low	High
0	0	0	<2	<1	7
0	1	0	2	<1	7
0	2	0	4	<1	11
1	0	0	2	<1	7
1	0	1	4	<1	11
1	1	0	4	<1	11
1	1	1	6	<1	15
2	0	0	5	<1	13
2	0	1	7	1	17
2	1	0	7	1	17
2	1	1	9	2	21
2	2	0	9	2	21
2	3	0	12	3	28
3	0	0	8	1	19
3	0	1	11	2	25
3	1	0	11	2	25
3	1	1	14	4	34
3	2	0	14	4	34
3	2	1	17	5	46
3	3	0	17	5	46
4	0	0	13	3	31
4	0	1	17	5	46
4	1	0	17	5	46
4	1	1	21	7	63
4	1	2	26	9	78
4	2	0	22	7	67
4	2	1	26	9	78
4	3	0	27	9	80
4	3	1	33	11	93
4	4	0	34	12	93
5	0	0	23	7	70
5	0	1	31	11	89
5	0	2	43	15	110
5	1	0	33	11	93
5	1	1	46	16	120
5	1	2	63	21	150
5	2	0	49	17	130
5	2	1	70	23	170
5	2	2	94	28	220
5	3	0	79	25	190
5	3	1	110	31	250
5	3	2	140	37	340
5	3	3	180	44	500

Figure B1

APPENDIX C

Metabolic Pathways

Glycolysis

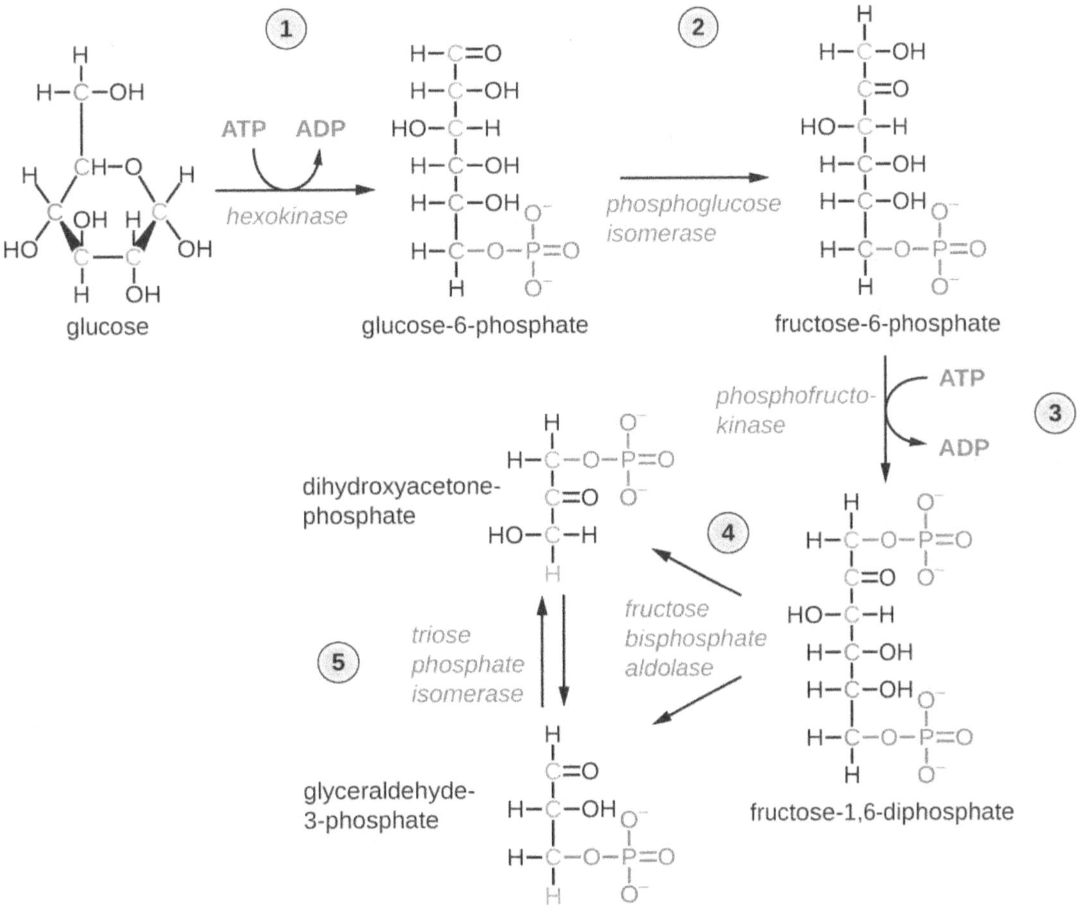

Figure C1 The first half of glycolysis uses two ATP molecules in the phosphorylation of glucose, which is then split into two three-carbon molecules.

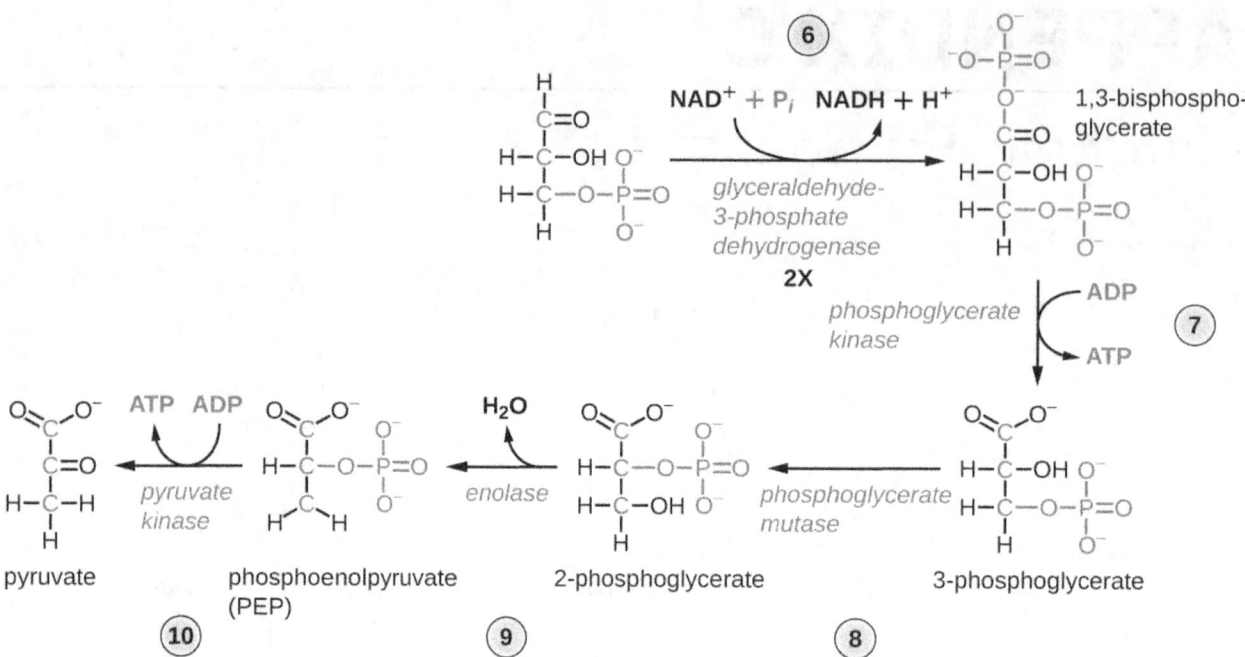

Figure C2 The second half of glycolysis involves phosphorylation without ATP investment (step 6) and produces two NADH and four ATP molecules per glucose.

Entner–Doudoroff Pathway

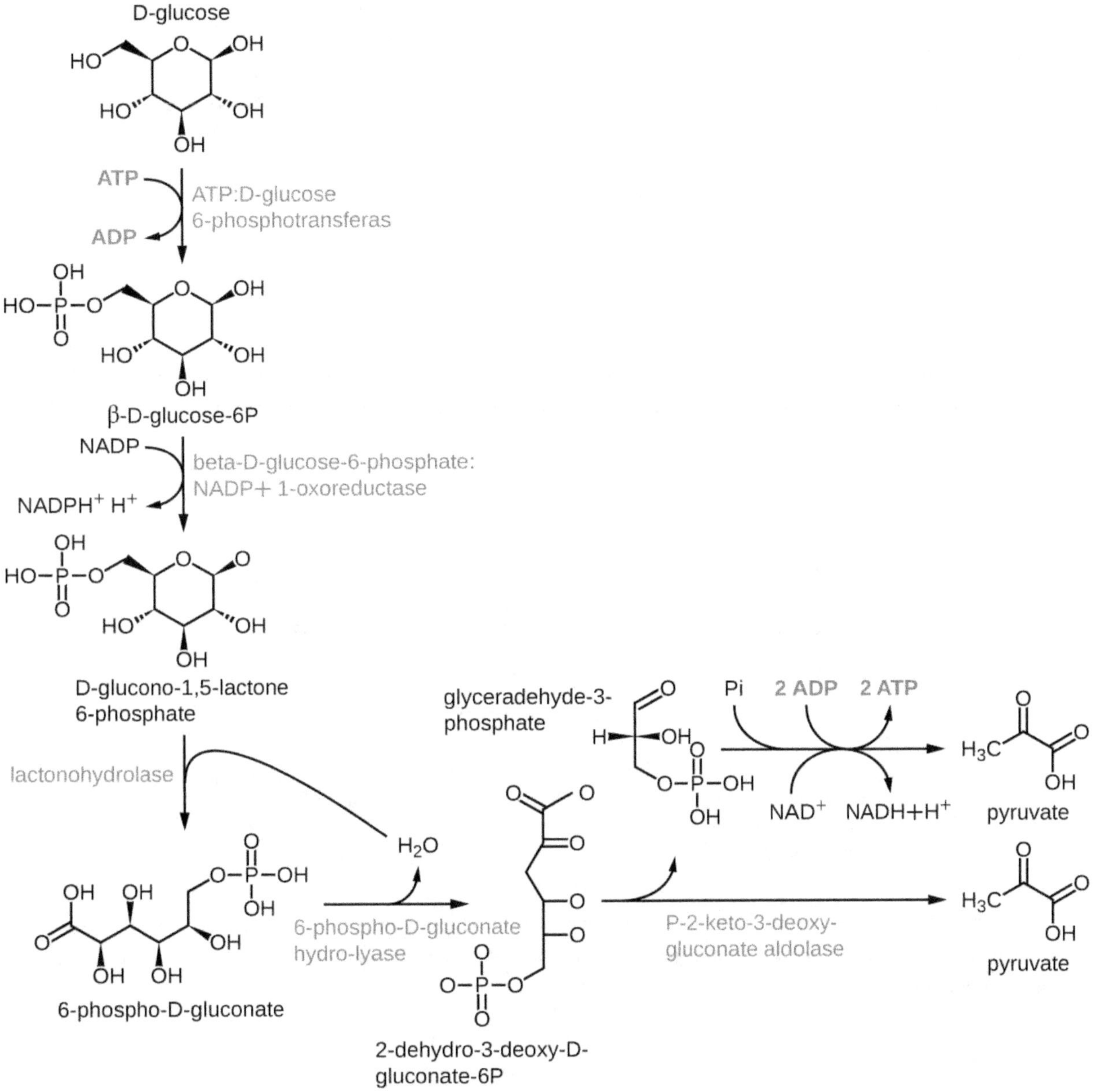

Figure C3 The Entner–Doudoroff Pathway is a metabolic pathway that converts glucose to ethanol and nets one ATP.

The Pentose-Phosphate Pathway

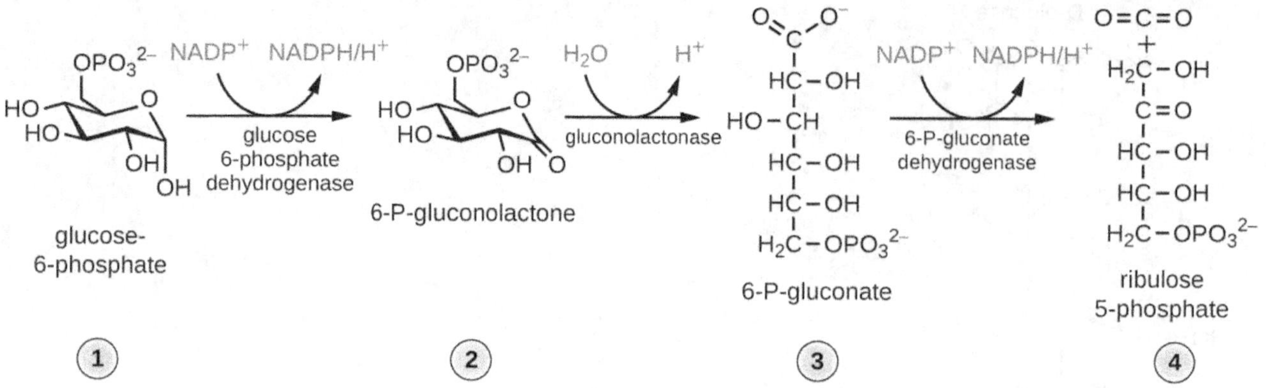

Figure C4 The pentose phosphate pathway, also called the phosphogluconate pathway and the hexose monophosphate shunt, is a metabolic pathway parallel to glycolysis that generates NADPH and five-carbon sugars as well as ribose 5-phosphate, a precursor for the synthesis of nucleotides from glucose.

TCA Cycle

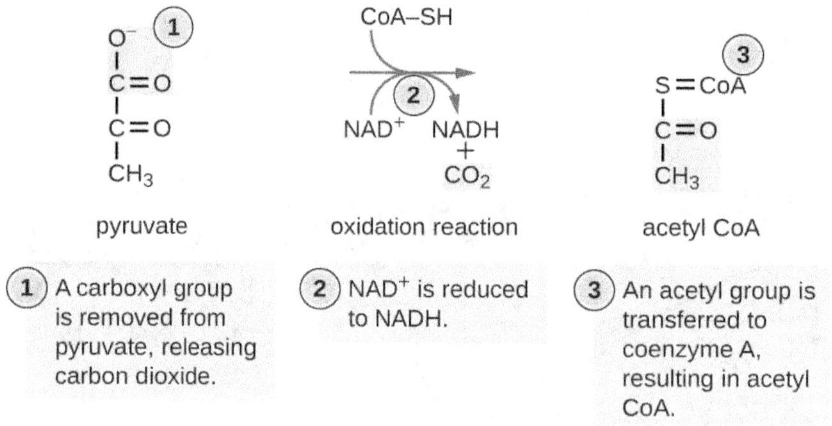

1. A carboxyl group is removed from pyruvate, releasing carbon dioxide.
2. NAD^+ is reduced to NADH.
3. An acetyl group is transferred to coenzyme A, resulting in acetyl CoA.

Figure C5 In this transition reaction, a multi-enzyme complex converts pyruvate into one acetyl (2C) group plus one carbon dioxide (CO_2). The acetyl group is attached to a Coenzyme A carrier that transports the acetyl group to the site of the Krebs cycle. In the process, one molecule of NADH is formed.

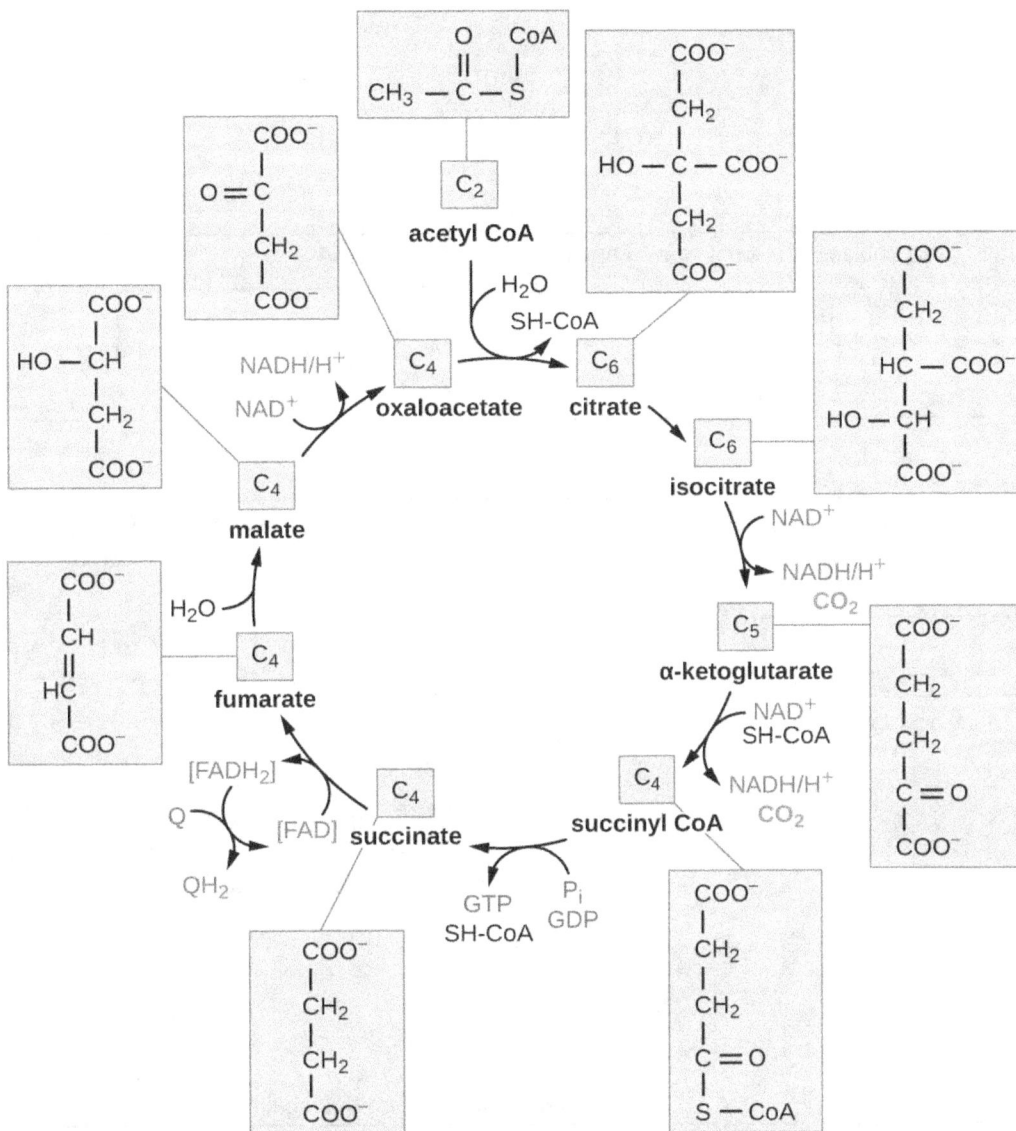

Figure C6 In the citric acid cycle, the acetyl group from acetyl CoA is attached to a four-carbon oxaloacetate molecule to form a six-carbon citrate molecule. Through a series of steps, citrate is oxidized, releasing two carbon dioxide molecules for each acetyl group fed into the cycle. In the process, three NADH, one FADH2, and one ATP or GTP (depending on the cell type) is produced by substrate-level phosphorylation. Because the final product of the citric acid cycle is also the first reactant, the cycle runs continuously in the presence of sufficient reactants. (credit: modification of work by "Yikrazuul"/Wikimedia Commons)

Beta Oxidation

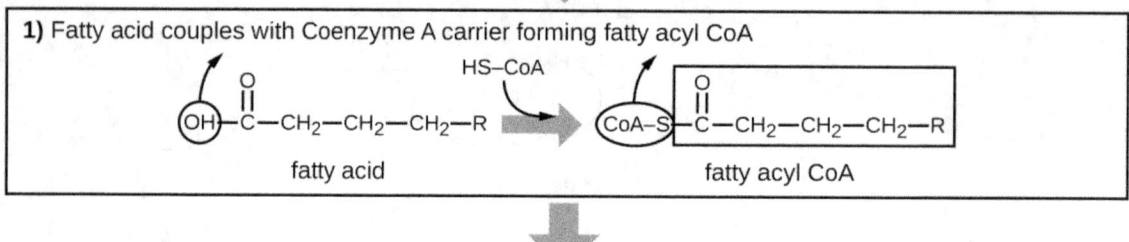

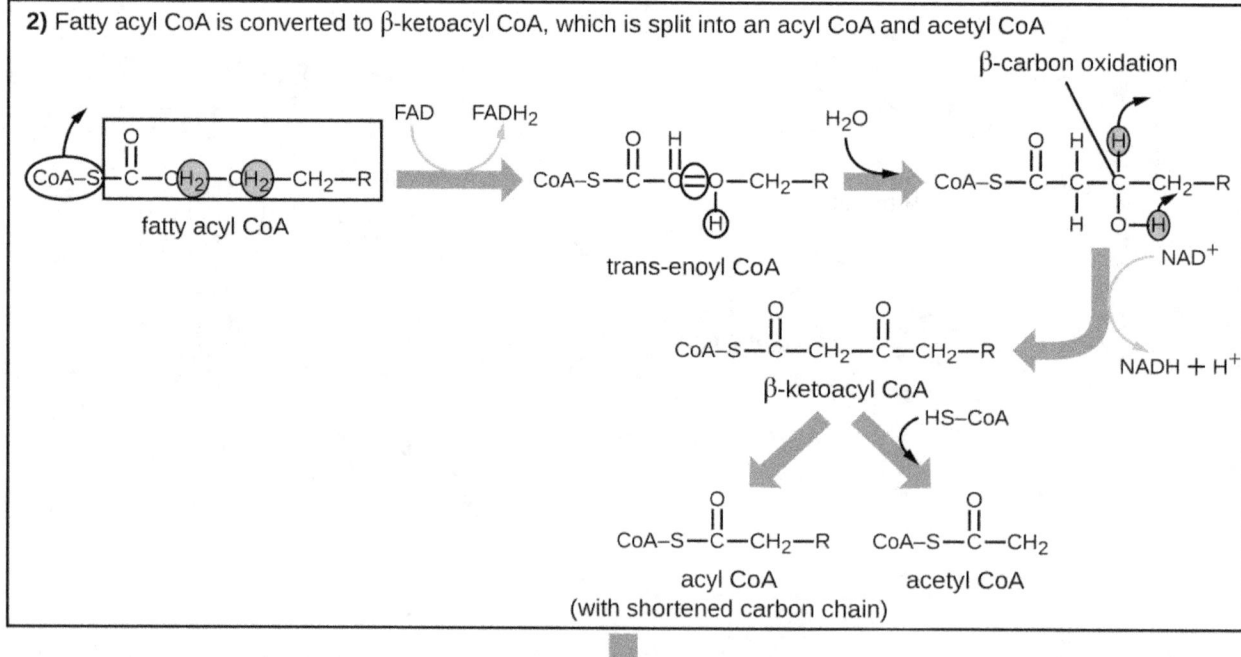

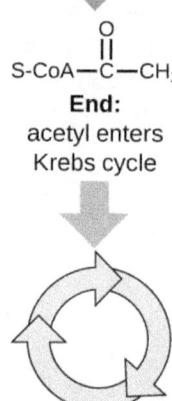

Figure C7 During fatty acid oxidation, triglycerides can be broken down into 2C acetyl groups that can enter the Krebs cycle and be used as a source of energy when glucose levels are low.

Electron Transport Chain and Oxidative Phosphorylation

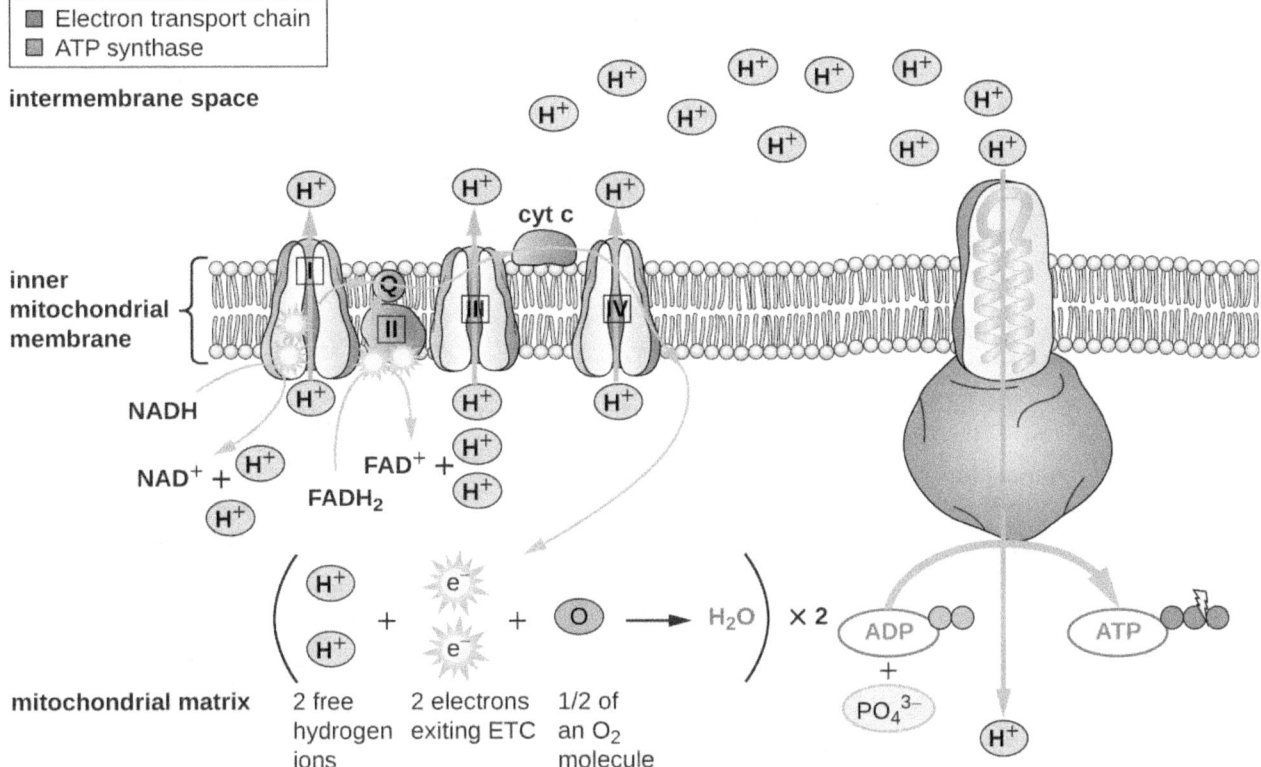

Figure C8 The electron transport chain is a series of electron carriers and ion pumps that are used to pump H⁺ ions across a membrane. H⁺ then flow back through the membrane by way of ATP synthase, which catalyzes the formation of ATP. The location of the electron transport chain is the inner mitochondrial matrix in eukaryotic cells and cytoplasmic membrane in prokaryotic cells.

Calvin-Benson Cycle

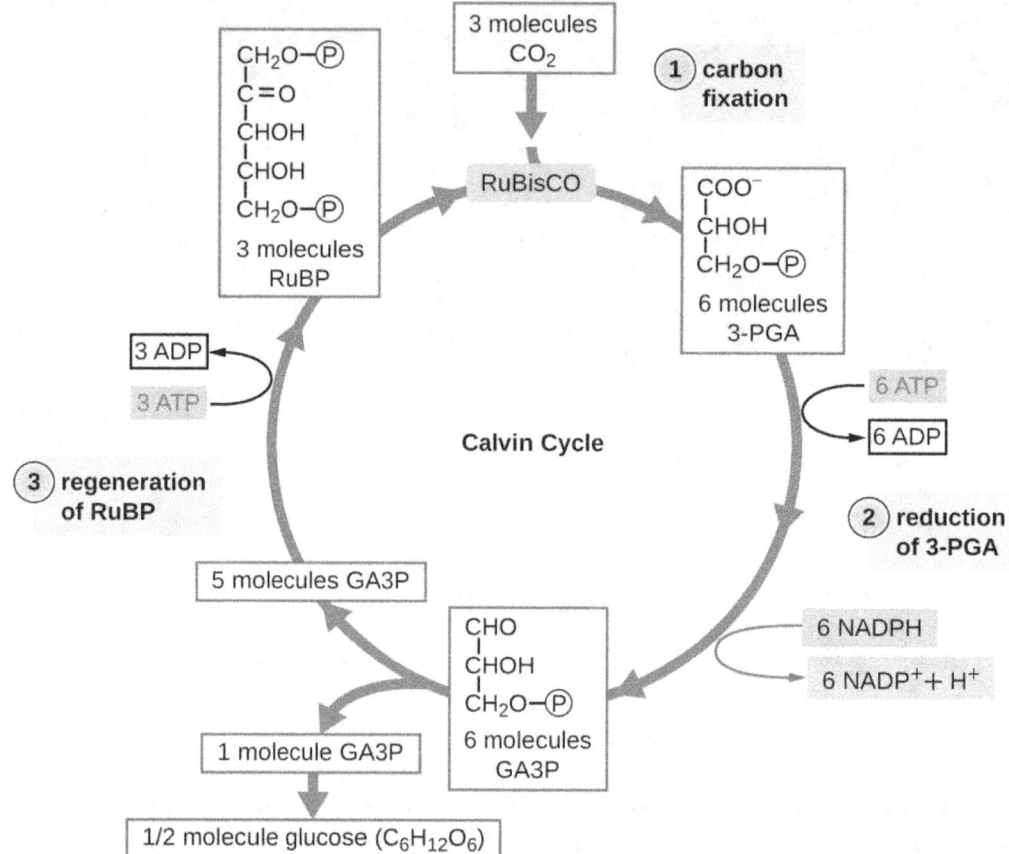

Figure C9 The Calvin-Benson cycle has three stages. In stage 1, the enzyme RuBisCO incorporates carbon dioxide into an organic molecule, 3-PGA. In stage 2, the organic molecule is reduced using electrons supplied by NADPH. In stage 3, RuBP, the molecule that starts the cycle, is regenerated so that the cycle can continue. Only one carbon dioxide molecule is incorporated at a time, so the cycle must be completed three times to produce a single three-carbon GA3P molecule, and six times to produce a six-carbon glucose molecule.

APPENDIX D

Taxonomy of Clinically Relevant Microorganisms

Bacterial Pathogens

The following tables list the species, and some higher groups, of pathogenic Eubacteria mentioned in the text. The classification of Bacteria, one of the three domains of life, is in constant flux as relationships become clearer through sampling of genetic sequences. Many groups at all taxonomic levels still have an undetermined relationship with other members of the phylogenetic tree of Bacteria. *Bergey's Manual of Systematics of Archaea and Bacteria* maintains a published list and descriptions of prokaryotic species. The tables here follow the taxonomic organization in the *Bergey's Manual* Taxonomic Outline.[1]

We have divided the species into tables corresponding to different bacterial phyla. The taxonomic rank of kingdom is not used in prokaryote taxonomy, so the phyla are the subgrouping below domain. Note that many bacterial phyla not represented by these tables. The species and genera are listed only under the class within each phylum. The names given to bacteria are regulated by the International Code of Nomenclature of Bacteria as maintained by the International Committee on Systematics or Prokaryotes.

Phylum Actinobacteria

Class	Genus	Species	Related Diseases
Actinobacteria	*Corynebacterium*	*diphtheriae*	Diphtheria
	Gardnerella	*vaginalis*	Bacterial vaginosis
	Micrococcus		Opportunistic infections
	Mycobacterium	*bovis*	Tuberculosis, primarily in cattle
	Mycobacterium	*leprae*	Hansen's disease
	Mycobacterium	*tuberculosis*	Tuberculosis
	Propionibacterium	*acnes*	Acne, blepharitis, endophthalmitis

Table D1

Phylum Bacteroidetes

Class	Genus	Species	Related Diseases
Bacteroidia	*Porphyromonas*		Periodontal disease

---

1 Bergey's Manual Trust. *Bergey's Manual of Systematics of Archaea and Bacteria, Taxonomic Outline.* 2012. http://www.bergeys.org/outlines.html

Phylum Bacteroidetes

Class	Genus	Species	Related Diseases
	Prevotella	intermedia	Periodontal disease

Table D2

Phylum Chlamydiae

Class	Genus	Species	Related Diseases
Chlamydiae	Chlamydia	psittaci	Psittacosis
	Chlamydia	trachomatis	Sexually transmitted chlamydia

Table D3

Phylum Firmicutes

Class	Genus	Species	Related Diseases
Bacilli	Bacillus	anthracis	Anthrax
	Bacillus	cereus	Diarrheal and emetic food poisoning
	Listeria	monocytogenes	Listeriosis
	Enterococcus	faecalis	Endocarditis, septicemia, urinary tract infections, meningitis
	Staphylococcus	aureus	Skin infections, sinusitis, food poisoning
	Staphylococcus	epidermidis	Nosocomial and opportunistic infections
	Staphylococcus	hominis	Opportunistic infections
	Staphylococcus	saprophyticus	Urinary tract infections
	Streptococcus	agalactiae	Postpartum infection, neonatal sepsis
	Streptococcus	mutans	Tooth decay
	Streptococcus	pneumoniae	Pneumonia, many other infections
	Streptococcus	pyogenes	Pharyngitis, scarlet fever, impetigo, necrotizing fasciittis
Clostridia	Clostridium	botulinum	Botulinum poisoning
	Clostridium	difficile	Colitis

Phylum Firmicutes

Class	Genus	Species	Related Diseases
	Clostridium	*perfringens*	Food poisoning, gas gangrene
	Clostridium	*tetani*	Tetanus

Table D4

Phylum Fusobacteria

Class	Genus	Species	Related Diseases
Fusobacteriia	*Fusobacterium*		Periodontal disease, Lemierre syndrome, skin ulcers
Fusobacteriia	*Streptobacillus*	*moniliformis*	Rat-bite fever

Table D5

Phylum Proteobacteria

Class	Genus	Species	Related Diseases
Alphaproteobacteria	*Anaplasma*	*phagocytophilum*	Human granulocytic anaplasmosis
Alphaproteobacteria	*Bartonella*	*henselae*	Peliosis hepatitis, bacillary angiomatosis, endocarditis, bacteremia
Alphaproteobacteria	*Bartonella*	*quintana*	Trench fever
Alphaproteobacteria	*Brucella*	*melitensis*	Ovine brucellosis
Alphaproteobacteria	*Ehrlichia*	*chaffeensis*	Human monocytic ehrlichiosis
Alphaproteobacteria	*Rickettsia*	*prowazekii*	Epidemic typhus
Alphaproteobacteria	*Rickettsia*	*rickettsii*	Rocky Mountain spotted fever
Alphaproteobacteria	*Rickettsia*	*typhi*	Murine typhus
Betaproteobacteria	*Bordetella*	*pertussis*	Pertussis
Betaproteobacteria	*Eikenella*		Bite-injury infections

Phylum Proteobacteria

Class	Genus	Species	Related Diseases
	Neisseria	gonorrhoeae	Gonorrhea
	Neisseria	meningitidis	Meningitis
	Spirillum	minus (alt. minor)	Sodoku (rat-bite fever)
Epsilonproteobacteria	Campylobacter	jejuni	Gastroenteritis, Guillain-Barré syndrome
	Helicobacter	pylori	Gastric ulcers
Gammaproteobacteria	Aeromonas	hydrophila	Dysenteric gastroenteritis
	Coxiella	burnetii	Q fever
	Enterobacter		Urinary and respiratory infections
	Escherichia	coli Strains: shiga toxin-producing (STEC) (e.g., O157:H7) also called enterohemorrhagic E. coli (EHEC) or verocytotoxin-producing E. coli (VTEC)	Foodborne diarrhea outbreaks, hemorrhagic colitis, hemolytic-uremic syndrome
	Escherichia	coli Strain: enterotoxigenic E. coli (ETEC)	Traveler's diarrhea
	Escherichia	coli Strain: enteropathogenic E. coli (EPEC)	Diarrhea, especially in young children
	Escherichia	coli Strain: enteroaggregative E. coli (EAEC)	Diarrheal disease in children and travelers
	Escherichia	coli Strain: diffusely adherent E. coli (DAEC)	Diarrheal disease of children
	Escherichia	coli Strain: enteroinvasive E. coli (EPEC)	Bacillary dysentery, cells invade intestinal epithelial cells

Phylum Proteobacteria

Class	Genus	Species	Related Diseases
	Francisella	tularensis	Tularemia
	Haemophilus	ducreyi	Chancroid
	Haemophilus	influenzae	Bacteremia, pneumonia, meningitis
	Klebsiella	pneumoniae	Pneumonia, nosocomial infections
	Legionella	pneumophila	Legionnaire's disease
	Moraxella	catarrhalis	Otitis media, bronchitis, sinusitis, laryngitis, pneumonia
	Pasteurella		Pasteurellosis
	Plesiomonas	shigelloides	Gastroenteritis
	Proteus		Opportunistic urinary tract infections
	Pseudomonas	aeruginosa	Opportunistic, nosocomial pneumonia and sepsis
	Salmonella	bongori	Salmonellosis
	Salmonella	enterica	Salmonellosis
	Serratia		Pneumonia, urinary tract infections
	Shigella	boydii	Dysentery
	Shigella	dysenteriae	Dysentery
	Shigella	flexneri	Dysentery
	Shigella	sonnei	Dysentery
	Vibrio	cholerae	Cholera

Phylum Proteobacteria

Class	Genus	Species	Related Diseases
	Vibrio	*parahemolyticus*	Seafood gastroenteritis
	Vibrio	*vulnificus*	Seafood gastroenteritis, necrotizing wound infections, septicemia
	Yersinia	*enterocolitica*	Yersiniosis
	Yersinia	*pestis*	Plague
	Yersinia	*pseudotuberculosis*	Far East scarlet-like fever

Table D6

Phylum Spirochaetes

Class	Genus	Species	Related Diseases
Spirochaetia	*Borrelia*	*burgdorferi*	Lyme disease
	Borrelia	*hermsii*	Tick-borne relapsing fever
	Borrelia	*recurrentis*	Louse-borne relapsing fever
	Leptospira	*interrogans*	Leptospirosis
	Treponema	*pallidum*	Syphilis, bejel, pinta, yaws

Table D7

Phylum Tenericutes

Class	Genus	Species	Related Diseases
Mollicutes	*Mycoplasma*	*genitalium*	Urethritis, cervicitis
	Mycoplasma	*hominis*	Pelvic inflammatory disease, bacterial vaginosis
	Mycoplasma	*pneumoniae*	*Mycoplasma* pneumonia
	Ureaplasma	*urealyticum*	Urethritis, fetal infections

Table D8

Viral Pathogens

There are several classification systems for viruses. The International Committee on Taxonomy of Viruses (ICTV) is the international scientific body responsible for the rules of viral classification. The ICTV system used here groups viruses based on genetic similarity and presumed monophyly. The viral classification system is separate from the classification system for cellular organisms. The ICTV system groups viruses within seven orders, which contain related families. There is, presently, a large number of unassigned families with unknown affinities to the seven orders. Three of these orders infect only Eubacteria, Archaea, or plants and do not appear in this table. Some families may be divided into subfamilies. There are also many unassigned genera. Like all taxonomies, viral taxonomy is in constant flux. The latest complete species list and classification can be obtained on the ICTV website.[2]

---

[2] International Committee on Taxonomy of Viruses. "ICTV Master Species List." http://talk.ictvonline.org/files/ictv_documents/m/msl/default.aspx

Viral Pathogens

Order	Family	Sub-family	Genus	Species	Related diseases
Herpesvirales	Herpesviridae	Betaherpesvirinae	Human cytomegalovirus group	Human herpesvirus 5	Cytomegalovirus hepatitis and other infections in immunocompromised people
		Gammaherpesvirinae	Lymphocryptovirus	Human herpesvirus 4 (HHV-4; Epstein-Barr virus)	Infectious mononucleosis
		Alphaherpesvirinae	Simplexvirus	Human herpesvirus 1, human herpesvirus 2	Herpes simplex virus 1, herpes simplex virus 2
			Varicellovirus	Human herpesvirus 3	Chicken pox, shingles
Mononegavirales	Filoviridae		Ebolavirus	Zaire ebolavirus (EBOV)	Ebola
			Marburgvirus	Marburg marburgvirus (MARV)	Marburg virus disease
	Rhabdoviridae		Lyssavirus	Rabies virus	Rabies
	Paramyxoviridae	Pneumovirinae	Pneumovirus	Human respiratory syncytial virus	Lower respiratory tract infection
		Paramyxovirinae	Morbillivirus	Measles virus	Measles (rubeola)
Nidovirales	Coronaviridae	Coronavirinae	Coronavirus		Common cold, pneumonia, SARS
Picornavirales	Picornaviridae		Hepatovirus	Hepatitis A virus	Hepatitis A
			Enterovirus	Enterovirus C	Polio
				Rhinovirus A	Common cold
				Rhinovirus B	Common cold
				Rhinovirus C	Common cold
Unassigned	Adenovirus		Mastadenovirus		Respiratory and other infections
	Arenaviridae		Mammarenavirus	Lassa mammarenavirus	Lassa fever
	Astroviridae				Gastroenteritis
	Bunyaviridae		Hantavirus	Several species	Hantavirus hemorrhagic fever with renal syndrome (HFRS), hantavirus pulmonary syndrome (HPS)
			Nairovirus	Crimean-Congo hemorrhagic fever virus (CCHF)	Crimean-Congo hemorrhagic fever
	Caliciviridae		Norovirus	Norwalk virus	Gastroenteritis

Figure D1

Viral Pathogens (continued)					
Order	Family	Sub-family	Genus	Species	Related diseases
Unassigned	Flaviviridae		Flavivirus	Dengue virus	Dengue fever
				Yellow fever virus	Yellow fever
			Hepacivirus	Hepatitis C virus	Hepatitis C
	Hepadnaviridae		Orthohepadna-virus	Hepatitis B virus	Hepatitis B
	Hepeviridae		Orthohepevirus	Hepatitis E virus	Hepatitis E
	Orthomyxoviridae		Influenzavirus A	Influenza A virus	Pandemic flu
			Influenzavirus B	Influenza B virus	Flu
			Influenzavirus C	Influenza C virus	Flu
	Papillomaviridae		Alphapapilloma-virus	Human papillomavirus	Skin warts
	Parvoviridae	Parvovirinae	Erythroparvovirus	Human parvovirus B19	Fifth disease (erythema infectosum)
	Poxviridae	Chordopoxvirinae	Orthopoxvirus	Variola virus	Variola major, Variola minor (smallpox)
				Vaccinia virus	Cowpox
	Reoviridae	Sedoreovirinae	Rotavirus	Eight species	Gastroenteritis
	Retroviridae	Orthoretrovirinae	Lentivirus	Human immuno-deficiency virus	AIDS
	Togaviridae		Alphavirus	Chikungunya virus (CHIKV)	Chikungunya
			Rubivirus	Rubella virus	Rubella (German measles)
	Unassigned		Deltavirus	Hepatitis D virus	Hepatitis D

Figure D2

Fungal Pathogens

The Fungi are one of the kingdoms of the domain Eukarya. Fungi are most closely related to the animals and a few other small groups and more distantly related to the plants and other groups that formerly were categorized as protist. At present, the Fungi are divided into seven phyla (or divisions, a hold over from when fungi were studied with plants), but there are uncertainties about some relationships.[3] Many groups of fungi, particularly those that were formerly classified in the phylum Zygomycota, which was not monophyletic, have uncertain relationships to the other fungi. The one species listed in this table that falls into this category is *Rhizopus arrhizus*. Fungal names are governed by the International Code of Nomenclature for Algae, Fungi, and Plants,[4] but the International Commission on the Taxonomy of Fungi (ICTF) also promotes taxonomic work on fungi. One activity of the ICTF is publicizing name changes for medically and otherwise important fungal species. Many species that formerly had two names (one for the sexual form and one for the asexual form) are now being brought together under one name.

---

3 D. S. Hibbett et al. "A Higher-level Phylogenetic Classification of the Fungi." *Mycological Research* 111 no. 5 (2007):509–547.

4 J. McNeill et al. *International Code of Nomenclature for Algae, Fungi, and Plants (Melbourne Code)*. Oberreifenerg, Germany. Koeltz Scientific Books; 2012. http://www.iapt-taxon.org/nomen/main.php?

Fungal Pathogens

Division	Genus	Species	Related Diseases
Ascomycota	Aspergillus	flavus	Opportunistic aspergillosis
	Aspergillus	fumigatus	Opportunistic aspergillosis
	Blastomyces	dermatitidis	Blastomycosis
	Candida	albicans	Thrush (candidiasis)
	Coccidioides	immitis	Valley fever (coccidioidomycosis)
	Epidermophyton		Tinea corporis (ringworm), tinea cruris (jock itch), tinea pedis (althlete's foot), tinea unguium (onychomycosis)
	Histoplasma	capsulatum	Histoplasmosis
	Microsporum		Tinea capitis (ringworm), tinea corpus (ringworm), other dermatophytoses
	Pneumocystis	jirovecii	Opportunistic pneumonia
	Sporothrix	schenckii	Sporotrichosis (rose-handler's disease)
	Trichophyton	mentagrophytes var. interdigitale	Tinea barbae (barber's itch), dermatophytoses
	Trichophyton	rubrum	Tinea corporis (ringworm), tinea cruris (jock itch), tinea pedis (althlete's foot), tinea unguium (onychomycosis)
Basidiomycota	Cryptococcus	neoformans	Opportunistic cryptococcosis, fungal meningitis, encephalitis
	Malassezia		Dandruff, tinea versicolor
uncertain	Rhizopus	arrhizus	Mucormycosis

Table D9

Protozoan Pathogens

The relationships among the organisms (and thus their taxonomy) previously grouped under the name Protists are better understood than they were two or three decades ago, but this is still a work in progress. In 2005, the Eukarya were divided into six supergroups.[5] The latest high-level classification combined two of the previous supergroups to produce a system comprising five supergroups.[6] This classification was developed for the Society of Protozoologists, but it is not the only suggested approach. One of the five supergroups includes the animals, fungi, and some smaller protist groups. Another contains green plants and three algal groups. The other three supergroups (listed in the three tables below) contain the other protists, many of them which cause disease. In addition, there is a large number of protist groups whose relationships are not understood. In the three supergroups represented here we have indicated the phyla to which the listed pathogens belong.

Supergroup Amoebozoa

Phylum	Genus	Species	Related Diseases
Amoebozoa	*Acanthamoeba*		Granulomatous amoebic encephalitis, acanthamoebic keratitis
	Entamoeba	*histolytica*	Enterobiasis

Table D10

Supergroup SAR (Stramenopiles, Alveolata, Rhizaria)

Phylum	Genus	Species	Related Diseases
Apicomplexa	*Babesia*		Babesiosis
	Cryptosporidium	*hominis*	Cryptosporidiosis
	Cryptosporidium	*parvum*	Cryptosporidiosis
	Cyclospora	*cayetanensis*	Gastroenteritis
	Plasmodium	*falciparum*	Malaria
	Plasmodium	*malariae*	"Benign" or "quartan" (3-day recurrent fever) malaria
	Plasmodium	*ovale*	"Tertian" (2-day recurrent fever) malaria
	Plasmodium	*vivax*	"Benign" "tertian" (2-day recurrent fever) malaria
	Plasmodium	*knowlesi*	Primate malaria capable of zoonosis, quotidian fever
	Toxoplasma	*gondii*	Toxoplasmosis

Table D11

---

5 S.M. Adl et al. "The New Higher Level Classification of Eukaryotes with Emphasis on the Taxonomy of Protists." *Journal of Eukaryotic Microbiology* 52 no. 5 (2005):399–451.

6 S.M. Adl et al. "The Revised Classification of Eukaryotes." *Journal of Eukaryotic Microbiology* 59 no. 5 (2012):429–514.

Supergroup Excavata

Phylum	Genus	Species	Related Diseases
Metamonada	Giardia	lamblia	Giardiasis
	Trichomonas	vaginalis	Trichomoniasis
Euglenozoa	Leishmania	braziliensis	Leishmaniasis
	Leishmania	donovani	Leishmaniasis
	Leishmania	tropica	Cutaneous leishmaniasis
	Trypanosoma	brucei	African sleeping sickness (African trypanosomiasis)
	Trypanosoma	cruzi	Chagas disease
Percolozoa	Naegleria	fowleri	Primary amoebic meningoencephalitis (naegleriasis)

Table D12

Parasitic Helminths

The taxonomy of parasitic worms, all of which belong to the kingdom Animalia still contains many uncertainties. The pathogenic species are found in two phyla: the Nematoda, or roundworms, and the Platyhelminthes, or flat worms. The Nematoda is tentatively divided into two classes[7], one of which, Chromadorea, probably contains unrelated groups. The parasitic flatworms are contained within three classes of flatworm, of which two are important to humans, the trematodes and the cestodes.

Phylum Nematoda

Class	Genus	Species	Related Diseases
Chromadorea	Ancylostoma	caninum	Dog hookworm infection
	Ancylostoma	duodenale	Old World hookworm infection
	Ascaris	lumbricoides	Ascariasis
	Enterobius	vermicularis	Enterobiasis (pin worm)
	Loa	loa	Loa loa filariasis (eye worm)
	Necator	americanus	Necatoriasis (New World hookworm infection)
	Strongyloides	stercoralis	Strongyloidiasis
Enoplea	Trichinella	spiralis	Trichinosis

---

7 National Center for Biotechnology Information. "Taxonomy Browser: Nematoda." http://www.ncbi.nlm.nih.gov/Taxonomy/Browser/wwwtax.cgi?id=6231

Phylum Nematoda

Class	Genus	Species	Related Diseases
	Trichuris	*trichiura*	Trichuriasis (whip worm infection)

Table D13

Phylum Platyhelminthes

Class	Genus	Species	Related Diseases
Trematoda	*Clonorchis*	*sinensis*	Chinese liver fluke
	Fasciolopsis	*buski*	Fasciolopsiasis
	Fasciola	*gigantica*	Fascioliasis
	Fasciola	*hepatica*	Fascioliasis
	Opisthorchis	*felineus*	Opisthorchiasis
	Opisthorchis	*viverrini*	Opisthorchiasis
	Schistosoma	*haematobium*	Urinary schistosomiasis
	Schistosoma	*japonicum*	Schistosomiasis
	Schistosoma	*mansoni*	Intestinal schistosomiasis
Cestoda	*Diphyllobothrium*	*latum*	Diphyllobothriosis
	Echinococcus	*granulosus*	Hydatid cysts (cystic echinococcosis)
	Echinococcus	*multilocularis*	Echinococcosis
	Taenia	*asiatica*	Intestinal taeniasis
	Taenia	*saginata*	Intestinal taeniasis
	Taenia	*solium*	Intestinal taeniasis, cysticercosis

Table D14

To download a PDF edition of this book with the Appendix E Glossary, Answer Key and Index

Please contact us at:

info@openstaxtextbooks.com